LARGE IGNEOUS PROVINCES

Large Igneous Provinces (LIPs) are intraplate magmatic events, involving volumes of mainly mafic magma upwards of 100 000 km^3, and often above 1 million km^3. Throughout Earth's history, such mega-volcanic events have occurred in both continental and oceanic environments, and are typically characterized by a short duration magmatic pulse or pulses (less than 1–5 million years). LIPs are key processes in shaping our planet over geological time, being linked to continental break-up, global environmental catastrophes including mass extinction events, regional uplift, and a variety of ore deposit types.

In this up-to-date, fascinating book, leading expert Richard Ernst explores all aspects of LIPs, beginning with a helpful introduction to their definition and essential characteristics. Topics covered include continental and oceanic LIPs (both their volcanic components and their plumbing systems); their origins, structures, and geochemistry; geological and environmental effects; association with silicic, carbonatite, and kimberlite magmatism; and analogs of LIPs in the Archean, and on other planets. The book concludes with an assessment of the influence of LIPs on natural resources such as mineral deposits, petroleum, and aquifers.

This is a vital, one-stop resource for researchers and graduate students in a wide range of disciplines, including tectonics, igneous petrology, geochemistry, geophysics, Earth history, and planetary geology. It will also be of importance to mining industry professionals.

RICHARD E. ERNST is Scientist-in-Residence in the Department of Earth Science at Carleton University, Ontario, Canada. After receiving his PhD in 1989, he spent 14 years conducting contracted research in Canada and many global regions, particularly Siberia, mainly for the Geological Survey of Canada. In 2003 he started his consulting firm Ernst Geosciences and also became co-leader of the Large Igneous Provinces (LIPs) Commission of IAVCEI (International Association of Volcanology and Chemistry of the Earth's Interior). Dr. Ernst has worked extensively to further research in the field: in 2009, he co-launched a consortium of industry sponsors funding use of the LIP record for research into reconstruction of supercontinents back into deep-time. Dr. Ernst is the author or co-author of more than 100 refereed publications, focused on all aspects of the terrestrial LIP record, and including planetary analogs.

D1211249

LARGE IGNEOUS PROVINCES

RICHARD E. ERNST

Department of Earth Sciences, Carleton University &
Ernst Geosciences

CAMBRIDGE
UNIVERSITY PRESS

CAMBRIDGE
UNIVERSITY PRESS

University Printing House, Cambridge CB2 8BS, United Kingdom

One Liberty Plaza, 20th Floor, New York, NY 10006, USA

477 Williamstown Road, Port Melbourne, VIC 3207, Australia

4843/24, 2nd Floor, Ansari Road, Daryaganj, Delhi - 110002, India

79 Anson Road, #06-04/06, Singapore 079906

Cambridge University Press is part of the University of Cambridge.

It furthers the University's mission by disseminating knowledge in the pursuit of
education, learning and research at the highest international levels of excellence.

www.cambridge.org
Information on this title: www.cambridge.org/9781108446686

© Richard E. Ernst 2014

First published 2014
First paperback edition 2017

A catalogue record for this publication is available from the British Library

Library of Congress Cataloging in Publication data
Ernst, Richard E., author.
Large igneous provinces / Richard E. Ernst, Department of Earth Sciences,
Carleton University & Ernst Geosciences.
pages cm
Summary: "Large Igneous Provinces (LIPs) are intraplate magmatic events,
involving volumes of mainly mafic magma upwards of 100 000 km^3, and often
above 1 million km^3. They are linked to continental breakup, global environmental
catastrophes, regional uplift and a variety of ore deposit types. In this up-to-date,
fascinating book, leading expert Richard Ernst explores all aspects of LIPs,
beginning by introducing their definition and essential characteristics. Topics
covered include continental and oceanic LIPs; their origins, structures, and
geochemistry; geological and environmental effects; association with silicic,
carbonatite and kimberlite magmatism; and analogues of LIPs in the
Archean, and on other planets. The book concludes with an assessment of LIPs'
influence on natural resources such as mineral deposits, petroleum and aquifers. This
is a one-stop resource for researchers and graduate students in a wide range of
disciplines, including tectonics, igneous petrology, geochemistry, geophysics, Earth
history, and planetary geology, and for mining industry
professionals"– Provided by publisher.
ISBN 978-0-521-87177-8 (Hardback)
1. Igneous rocks. 2. Magmatism. 3. Earth (Planet)–Mantle. 4. Geodynamics.
I. Title.
QE461.E76 2014
551.21–dc23 2014006194

ISBN 978-0-521-87177-8 Hardback
ISBN 978-1-108-44668-6 Paperback

Contents

Acknowledgments

I am pleased to be able to offer this overview of the growing field of Large Igneous Provinces, with the delightful acronym LIP, to the diverse groups in the Earth Sciences community that directly or indirectly intersect with this fascinating class of mega-magmatic events.

While I am solely responsible for the content of this book I owe a huge direct debt of gratitude toward many colleagues who have contributed in varied ways to the completion of this book. Comments on drafts of various chapters were provided by: Keith Bell (all chapters), Steve Bergman (Chapter 16), Scott Bryan (Chapter 8), Simon Jowitt (Chapters 1–4, 8, 10, 16), Jim Head (Chapter 7), the late Rob Kerrich (Chapters 6, 10, 16), Julian Pearce (Chapter 10), Dave Peck (Chapter 16), Paul Wignall (Chapter 14). Reviews by Peter Lightfoot and Doreen Ames, and comments by guest editors Maurice Colpron, and John Thompson on a "LIPs and Metallogeny" paper (Ernst and Jowitt, 2013) for a special Society of Economic Geologists helped also improve the similar-themed Chapter 16, which was being prepared at the same time. In addition, comments by Nicole Januszczak, Stephan Kurszlaukis, and Sebastian Tappe of De Beers on kimberlite aspects of the same Ernst and Jowitt paper were incorporated into Chapter 9.

In addition, I acknowledge the pioneers of the field, Mike Coffin and the late John Mahoney who coined the term Large Igneous Province (LIP) in the early 1990s. There are so many other colleagues over the years with whom discussions have shaped my understanding of LIPs; to produce a complete list would be impossible, but I have been privileged to co-author publications with many of these colleagues. A special group are those colleagues/friends from Russia, China, Morocco, and Southern Africa, with whom exciting collaborations have shed light on the LIP record of these regions. Let me also identify here a few colleagues with whom my association has been particularly extensive over a long period of time: Wouter Bleeker, Ken Buchan, Claire Samson, Simon Jowitt, Eric Grosfils, Sergei Pisarevsky, Don Schissel, Ulf Söderlund, and mentors: Jelle de Boer, Bob Baragar, Keith Bell, Henry Halls, Jim Head, and Franco Pirajno.

Fourth year and graduate students of my Large Igneous Province course given at the University of Ottawa in the 2010 winter term (GEO4301E), and at Carleton University in the 2012 fall term (ERTH4504/5202) provided superb feedback on all parts of the draft book. I have also been learning from my graduate students as they progress in their LIP research topics.

In most cases figures have been modified and combined from original published diagrams to fit the specific purposes of this book. Digital versions of originals were provided for parts of about 60% of the figures and this contribution by my colleagues (acknowledged in the figure captions) was a major boost to the drafting process. My student drafting team of Lindsay Coffin, Carley Crann, Jamie Cutts, and Nicole Williamson is gratefully acknowledged for their enormous effort to produce clear diagrams with a consistent look and style, and all in black and white in order to keep the book price reasonable. Jamie Cutts also obtained most of the diagram permissions. Aline Vachon and Philippe Gagnon of the Geological Survey of Canada Library found many key publications for me. John Morgan helped with assembling an initial reference list. To many at Cambridge University Press, especially Laura Clark, Susan Francis, Beata Mako, Zoë Pruce, and the copy-editor, Zoë Lewin. I have appreciated your mixture of pressure and patience during the protracted birth of this book. Finally, let me thank my parents, Richard and Mary, for launching me into geology in my youth with family rockhounding trips.

This is publication 33 of the International Government-Industrial-Academic Program "Reconstruction of Supercontinents Back to 2.7 Ga Using the Large Igneous Province (LIP) Record, with Implications for Mineral Deposit Targeting, Hydrocarbon Resource Exploration, and Earth System Evolution" (www.supercontinent.org). This program has been sponsored by industry: Anglo American/De Beers, Gold Fields, Minerals and Metals Group (MMG), Norwest Rotors–Archon Minerals (Stewart Blusson), Shell and Vale and also received support from the Geological Survey of Canada and from Canadian grant NSERC CRD grant [CRDP] 419503–11 – and all this support is gratefully acknowledged.

Enjoy your journey through the pages ahead.

1

Introduction, definition, and general characteristics

1.1 Introduction

Earth history is punctuated by numerous periods of magmatic activity during which especially large volumes of mainly mafic magma were emplaced in a short duration pulse or pulses and are not linked to normal plate-boundary processes. These Large Igneous Provinces (LIPs) can occur in either continental or oceanic settings (or a mixture of the two). Magma volumes can range from < 0.1 Mkm^3 to 80 Mkm^3 and event duration ranges from a single pulse as short as 0.5 Ma to multiple pulses extending over tens of millions of years. Huge silicic provinces can be associated with LIP events as can carbonatites and kimberlites. LIPs are notable for their association with global or regional environmental changes including extinction events, regional topographic changes including domal uplift, breakup or attempted breakup of continents. They have an association with a wide range of ore deposit types, and implications for the oil/gas industry and aquifer flow. Various origins are considered but evidence favors mantle-plume involvement for many LIPs. In this book I provide an overview of all aspects of LIPs starting with the history of the term and a review of definitions.

1.1.1 History of the term

The term "Large Igneous Province" was initially proposed by Coffin and Eldholm (1991, 1992, 1993a, 1993b, 1994) to identify a variety of mafic igneous provinces with areal extents > 0.1 Mkm^2 that represented "massive crustal emplacements of predominantly mafic (Mg- and Fe-rich) extrusive and intrusive rock, and originated via processes other than 'normal' seafloor spreading." The initial database upon which the term LIP was defined, relied almost exclusively on the relatively well-preserved Mesozoic and Cenozoic record of continental flood-basalt provinces, volcanic passive margins, and similar large-volume oceanic magmatism of intraplate origin such as oceanic plateaus, submarine ridges, seamount groups, and ocean-basin flood basalts (Coffin and Eldholm, 1994, 2005). These types of provinces were divided into those that represented massive transient basaltic volcanism occurring over a few

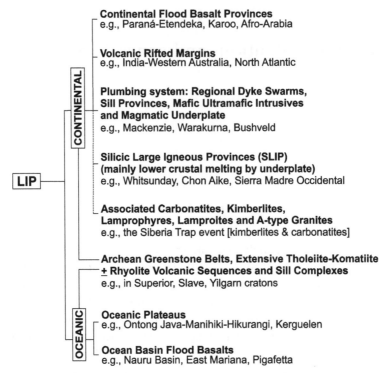

Figure 1.1 Classification of LIPs, based on the initial work of Coffin and Eldholm (1994), but modified to incorporate recent advances in the recognition of ancient LIPs and SLIPs (Bryan and Ernst, 2008). Examples of each type of LIP are given. Modified after Bryan and Ernst (2008) and Ernst and Bell (2010).With kind permission from Springer Science and Business Media.

million years (e.g. continental flood-basalt provinces, volcanic rifted margins, oceanic plateaus, and ocean basin flood basalts) and those representing persistent basaltic volcanism lasting tens to hundreds of million years (e.g. seamount groups, submarine ridges, and areas of anomalous seafloor spreading; Coffin and Eldholm, 2001). The formative work of Coffin and Eldholm, under the umbrella of the term LIP, aided in uniting previously separate research areas. This was particularly significant for linking the then newly recognized oceanic record of anomalous intra-plate magmatism with that of the longer-understood continental flood-basalt record.

Since this first categorization of LIPs by Coffin and Eldholm, substantial progress has been made in extending the LIP record back to the Paleozoic, Proterozoic, and Archean (Ernst and Buchan, 1997, 2001, 2003; Tomlinson and Condie, 2001; Arndt *et al.*, 2001; Isley and Abbott, 2002; Fig. 1.1). For many ancient LIPs, where much or all of the volcanic components of the LIP have been lost to erosion, definition has been based on the observed areal extent and inferred volume of intrusive rock (e.g. giant continental dyke swarms, sills, and layered intrusions) that form the exposed plumbing system to LIPs. Also, subsequent usage of the LIP term (e.g. Bryan and Ernst, 2008) focused on the transient LIPs of Coffin and Eldholm (1994), which allows the LIP definition to be reserved for those magmatic events with particularly significant regional to global effects. This approach excludes seamount chains and submarine ridges from the definition of LIPs (see discussion in Chapter 2).

Finally, it has also been recognized that LIPs can be directly linked with massive crustal emplacements of predominantly silicic (> 65 wt% SiO_2) extrusive and intrusive rocks (Chapter 8), with carbonatites, and in some cases with kimberlites (Chapter 9).

1.1.2 LIP definition

The increasing realization that LIPs and their associations are more varied in character, age, and composition than first defined prompted others (e.g. Sheth, 2007; Bryan and Ernst, 2008) to revise and broaden the original definition of LIP. Some alternative definitions are included in Table 1.1.

In this book I use the definition:

A LIP is a mainly mafic (+ultramafic) magmatic province with areal extent > 0.1 Mkm^2 and igneous volume > 0.1 Mkm^3, that has intraplate characteristics, and is emplaced in a short duration pulse or multiple pulses (less than 1–5 Ma) with a maximum duration of $< c.$ 50 Ma. Silicic magmatism (including that of LIP scale, termed Silicic LIPs (SLIPs)) and also carbonatites and kimberlites may be associated.

This definition follows closely that of Bryan and Ernst (2008) (see Table 1.1) with the modification that LIPs sensu stricto are considered to be of mafic (–ultramafic) composition. The related magmatic provinces of dominantly silicic composition are considered a separate but related class of magmatism called SLIPs.

It is important to also recognize LIP "fragments/remnants" (Ernst, 2007a). These are magmatic units which are small (sub-LIP scale), but are thought to belong to a LIP event and to have been reduced in size by erosion or tectonic fragmentation. Such units have essential characteristics which indicate membership in a larger-size event (see Section 2.5). For instance, dolerite dykes (regardless of their preserved length or scale of associated swarm) with an average width of > 10 m are likely to belong to a LIP.

Note, that in this book I will preferentially use the term dolerite rather than the essentially synonymous diabase, except where diabase is part of a formal name.

1.1.3 Importance of LIPs

LIPs are important for testing plume and non-plume models for the generation of LIPs (Chapter 15), as precise time markers for stratigraphic correlations,

Table 1.1 *Some definitions of Large Igneous Provinces*

- "Massive crustal emplacements of predominantly mafic (Mg- and Fe-rich) extrusive and intrusive rock, and originated via processes other than 'normal' seafloor spreading" (Coffin and Eldholm, 1994).
- LIPs "record periods when the outward transfer of material and energy from the Earth's interior operated in a *significantly different mode* than at present" (Mahoney and Coffin, 1997b, p. ix).
- Mainly mafic magma production of at least one million km^3 in less than 1 million years (K. Burke, pers. comm. 2004).
- Such events also constitute a "Pulse of the Earth" (Bleeker, 2004) in the sense of being major magmatic events that are tightly linked to many other aspects of Earth geological history.
- LIP should be used in its broadest sense to designate any igneous provinces with outcrop areas $\geq$ 50,000 km^2, regardless of composition, tectonic setting, or emplacement mechanism. Large volcanic provinces would be distinguished from large plutonic provinces and each would be further subdivided on the basis of composition (Sheth, 2007).
- "The products of transient large-scale magmatic processes 'rooted' in the Earth's mantle that are not predicted by plate tectonics" (Halls *et al.*, 2008).
- "Magmatic provinces with areal extents >0.1 Mkm^2, igneous volumes > 0.1 Mkm^3 and maximum lifespans of ~50 Myr that have intraplate tectonic settings or geochemical affinities, and are characterised by igneous pulse(s) of short duration (~1–5 Myr), during which a large proportion (> 75%) of the total igneous volume has been emplaced" (Bryan and Ernst, 2008).
- See also discussion in Cañón-Tapia (2010).
- A LIP is a mainly mafic (+ultramafic) magmatic province with areal extent > 0.1 Mkm^2 and igneous volume > 0.1 Mkm^3, that has intraplate characteristics, and is emplaced in a short duration pulse or multiple pulses (less than 1–5 Ma) with a maximum duration of < *c.* 50 Ma. Silicic magmatism (including that of LIP scale, termed Silicic LIPs (SLIPs)) and also carbonatites and kimberlites may be associated.

---

as an aid in paleocontinental reconstruction (Chapter 11), as the hosts of major Ni–Cu–platinum-group-element deposits and other metal commodities (Chapter 16). They are also useful as a potential targeting tool in exploration for diamondiferous kimberlites (Chapters 9 and 15), for rare-earth deposits in associated carbonatites (Chapters 9 and 16), and also for hydrocarbon (oil and gas) and water (aquifer) resources (Chapter 16). In addition, they may be helpful in studying climatic effects (Chapter 14) and regional uplift (Chapter 11). Emplacement of a LIP and the onset of breakup (Chapter 12) may also change the local and global plate stress framework and be associated with the initiation of arcs elsewhere in the world (Chapter 13).

1.2 Overview of LIP style through time

Below is provided an overview of the styles of LIPs through time. More detail on each is provided in later chapters.

1.2.1 Mesozoic–Cenozoic LIPs

Most research on LIPs has focused on the dramatic flood basalts (Fig. 1.2) which characterize Mesozoic–Cenozoic events, both continental flood basalts (Chapter 3) and the flood basalts within ocean basins (oceanic plateaus and ocean-basin flood basalts) (Chapter 4). These parts of the LIP record are generally well preserved and have been critical to development of many key concepts for LIPs, most importantly their large size and short duration (or short-duration pulses).

The short duration of emplacement is evident from the main stage of flood-basalt magmatism, which consists of monotonous sequences up to several kilometers thick of large tabular flow units and that commonly lack any significant interlayered sediments; high-precision dating confirms emplacement of the flood basalts in times of 1–5 Ma (Courtillot and Renne, 2003; Jerram and Widdowson, 2005).

Continental flood basalts are dominately mafic (tholeiitic) with minor ultramafic (picritic) and transitional-alkaline components lower in the sequence; silicic magmatism becomes increasingly more significant in the higher levels of the sequence (Chapter 3). Bimodal magmatism is associated with LIP-related rifted margins. Oceanic plateaus and ocean basin flood basalts have variable mantle sources, and continental flood basalts represent the same sublithospheric mantle sources with the added aspect of interaction with lithospheric mantle and crust (e.g. Hofmann, 1997; Condie, 2003: Hawkesworth and Scherstén, 2007; Kerr and Mahoney, 2007).

As mentioned above, the flood basalts of Mesozoic and Cenozoic age (Fig. 1.2; Chapters 3 and 4) are typically the best preserved and best studied. In contrast, the pre-Mesozoic record is more deeply eroded and therefore LIPs of Paleozoic and Proterozoic age are typically recognized by flood-basalt remnants and exposed plumbing systems represented by giant dyke swarms, sill provinces, and layered intrusions (Fig. 1.3; Chapter 5). Oceanic LIPs are incompletely preserved during ocean closure and can occur as obducted deformed sequences in orogenic belts. In the Archean the most promising LIP candidates are greenstone belts containing tholeiite–komatiite sequences (Fig. 1.4; Chapter 6).

1.2.2 Paleozoic–Proterozoic LIPs

Pre-Mesozoic LIPs are more greatly affected by erosion, which largely removes their flood basalts and exposes their plumbing systems (Fig. 1.3; Chapter 5). Therefore, continental LIPs of Paleozoic and Proterozoic age typically consist of giant dyke swarms (defined as those > 300 km long), sill provinces, large layered intrusions, and

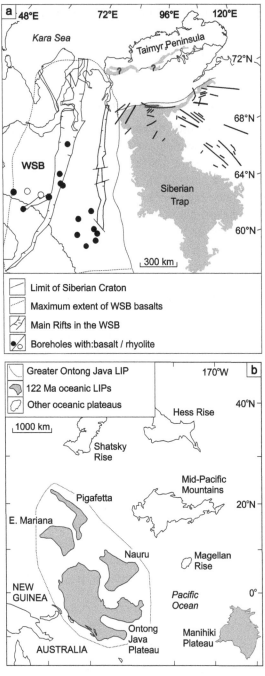

Figure 1.2 Examples of young Mesozoic–Cenozoic LIPs (a) 250 Ma Siberian Trap (b) Greater Ontong Java Plateau and other oceanic plateaus, Manihiki Plateau (122 Ma), Shatsky Rise (147 Ma), Magellan Rise (145 Ma), Mid-Pacific Mountains (*c.* 130–80 Ma), and Hess Rise (*c.* 110 and *c.* 100 Ma). WSB = West Siberian Basin. Modified from Ernst *et al.* (2005). With permission from Elsevier.

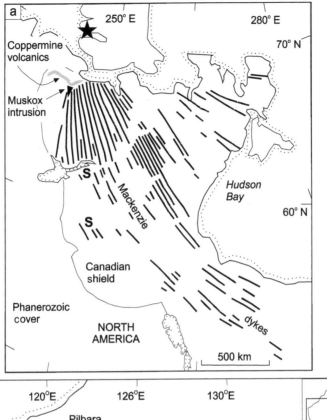

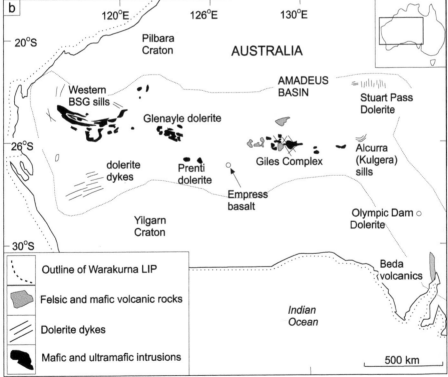

Figure 1.3 Examples of Proterozoic LIPs where erosion has exposed the 'plumbing' system of dykes, sills, and layered intrusion: (a) 1270 Ma Mackenzie event consisting of Mackenzie radiating dyke swarm, Muskox layered intrusion, and Coppermine volcanics, and widely distributed sill provinces ('S') and (b) 1070 Ma Warakurna LIP of central Australia. Modified from Ernst *et al.* (2005). With permission from Elsevier.

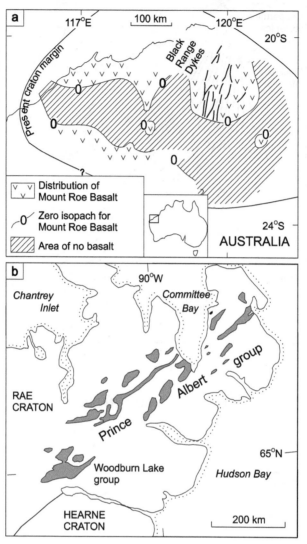

Figure 1.4 Examples of Archean LIPs (a) Part of Fortescue group of Pilbara craton, specifically showing the distribution of lowermost 2780 Ma Mount Roe sequence and associated Black Range feeder dykes (see also Section 6.2.4). (b) 2730–2700 Ma Prince Albert, Woodburn Lake, and Mary River groups in the Rae craton of northern Canada. Modified from Ernst *et al.* (2005). Original of part (a) after Thorne and Trendall (2001). With permission from Elsevier.

remnants of flood basalts (Ernst and Buchan, 1997, 2001a). Like their Mesozoic–Cenozoic flood-basalt equivalents, this class of intrusive-dominated LIPs has large areal extents and volumes, exhibits short-duration pulses, and has "intraplate" character, consistent with definition as a LIP (Coffin and Eldholm, 1994, 2001, 2005; Ernst *et al.*, 2005; Bryan and Ernst, 2008; Bryan and Ferrari, 2013).

1.2.3 Archean LIPs

Erosional remnants of typical Archean flood-basalt provinces include the Fortescue sequence of the Pilbara craton in Australia (Fig. 1.4a) and the Ventersdorp sequence of the Kaapvaal craton in southern Africa (Chapter 6). However, most Archean volcanic rocks occur as deformed and fault-fragmented packages termed greenstone belts. One class of greenstone belts contains mafic to silicic igneous rocks with calc-alkaline geochemical signatures and is interpreted to be arc-related. The other major class of greenstone belts consists of tholeiite–komatiite sequences, which are the best candidates for being remnants of Archean LIPs (e.g. the Prince Albert and related greenstone belts of the Rae craton, North America; Fig. 1.4b). In terms of setting, these Archean LIPs include accreted oceanic plateaus and also those emplaced in a continental platform setting. More details on the Archean LIPs are provided in Chapter 6.

1.3 LIPs on other planets

Intraplate magmatism, including events of LIP scale, are present on other planets. Comparison of large-volume intraplate magmatism on other planets with those on Earth provides insights to the LIP record on Earth. In Chapter 7 I review the LIP analog record of Mars, Venus, the Moon, Mercury, and Io.

1.4 Global LIP barcode record of Earth

The terrestrial global LIP record can be expressed as a barcode diagram (Fig. 1.5). From the present day to about 180 Ma the record consists of both continental LIPs and oceanic LIPs and has a combined rate of about 1 per 10 Ma (Coffin and Eldholm, 2001; Ernst *et al.*, 2005). From about 180 Ma to 2600 Ma the frequency of LIP production (mainly continental LIPs) is relatively constant back to 2.6 Ga, occurring with an average frequency of about 1 per 20 Ma (Chapter 11). Since most oceanic plateaus do not survive subduction and are difficult to recognize in pre-Mesozoic orogenic belts, the pre-Mesozoic record underestimates the LIP production. If one assumes that the frequency of oceanic LIP record observed in the Mesozoic–Cenozoic continues back in time, then average LIP production (including both continental and oceanic LIPs) back to the Archean may be closer to one event every 10 Ma. However, as noted by N. Dobretsov (personal communication, 2007), and discussed in Section 11.8, multiple independent LIPs can occur at the same time (plume clusters) and skew the average LIP production rate to smaller values. By his estimate the average continental LIP production is closer to 1 per 30 Ma, and so the combined oceanic and continental production would be closer to 1 per 15 Ma. A more detailed analysis of the LIP record through time is provided in Chapter 11, including the recognition of variations correlated to the supercontinent cycle.

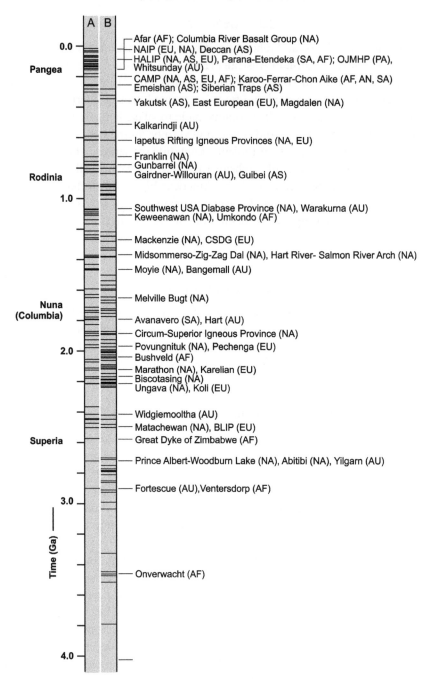

Figure 1.5 Global barcode for Earth. Column A contains events which satisfied the full criteria of Bryan and Ernst (2008) and herein and which are therefore LIPs sensu stricto. The Whitsunday and Chon Aike SLIPs are also shown. Column B contains events which are

1.5 Origin of LIPs

A variety of models have been proposed to explain the origin of LIPs (more details are provided in Chapter 15). Included among these are mantle plumes emanating from the core–mantle boundary (e.g. Richards *et al.*, 1989; Campbell and Griffiths, 1990; Campbell, 2005, 2007; Dobretsov, 2005). Other proposed models include impact-induced decompression melting (e.g. Jones *et al.*, 2002; Ingle and Coffin, 2004); lithospheric delamination (Elkins-Tanton and Hager, 2000; Elkins-Tanton, 2005, 2007; Hales *et al.*, 2005); decompression melting during rifting (White and McKenzie, 1989) or following mantle heating beneath supercontinents (Coltice *et al.*, 2007); edge-driven convection (King and Anderson, 1998); melting of fertile mantle without excess heat (Anderson, 2005); or shallow-melting anomalies generated by plate-tectonic-related processes ("Plate" model of Foulger, 2007); stress-induced lithospheric fracturing and rapid drainage of a relatively slowly accumulated sublithospheric basaltic magma reservoir (Silver *et al.*, 2006); back-arc rifting (e.g. Smith, 1992; Rivers and Corrigan, 2000); or overriding of a spreading ridge by a continent (Gower and Krogh, 2002). As shown in Chapter 15, there is strong evidence in favour of a dominant role for mantle plumes for many LIPs with contributions from these other mechanisms (especially rift-related decompression melting) causing an additional pulse or pulses of LIP activity.

1.6 Global distribution of LIPs

This introductory chapter ends with a location map of the main magmatic events being currently considered as LIPs, many of which will individually be discussed in different sections of the book (Fig. 1.6). This series of time slices shows both events that are definitely LIPs and others that are more speculatively recognized as LIPs. More details on the characteristics of each are presented in Table 1.2. In Fig. 1.6 the overall distribution of each LIP is outlined by one or more polygons that enclose units including volcanics, dykes, sills, and layered intrusions.

interpreted to be "fragments/remnants" of LIPs (Ernst, 2007a; Caption for Figure 1.5 (*cont.*) Section 2.5). Events are labeled at the starting age of the main pulse and, for multi-pulse LIPs, the arrow is placed at the oldest pulse. Only selected LIPs are labeled. Associated supercontinents are listed along the left side. LIP abbreviations are: BLIP = Baltic Large Igneous Province, CAMP = Central Atlantic Magmatic Province, CSDG = Central Scandinavian Dolerite Group, HALIP = High Arctic Large Igneous Province, NAIP = North Atlantic Igneous Province, and OJHMP = Ontong Java–Hikurangi–Manihiki Plateau event. Locations for LIP events are abbreviated as follows: NA = North America, SA = South America, EU = Europe, AF = Africa, AS = Asia, AU = Australia, AN = Antarctica, PA = Pacific Ocean. From Bryan and Ernst (2008). With permission from Elsevier.

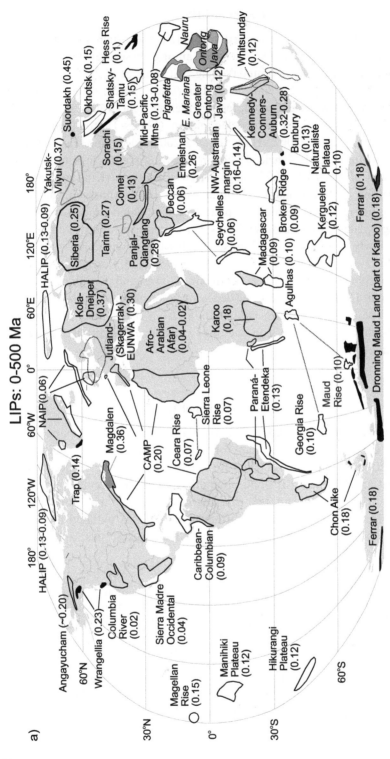

Figure 1.6 Global maps showing the schematic distribution of LIPs and SLIPs through time. (a) 0–275 Ma, (b) 275–550 Ma, (c) 550–1000 Ma, (d) 1000–1500 Ma, (e) 1500–2000 Ma, (f) 2000–2200 Ma, (g) 2200–2600 Ma, (h) 2600– (Archean). Abbreviations: CAMP = Central Atlantic Magmatic Province; HALIP = High Arctic Large Igneous Province; NAIP = North Atlantic Igneous Province; OJP = Ontong Java plateau. See Table 1.2 for more information on each LIP or SLIP including key references. Maps are in Robinson projection.

12

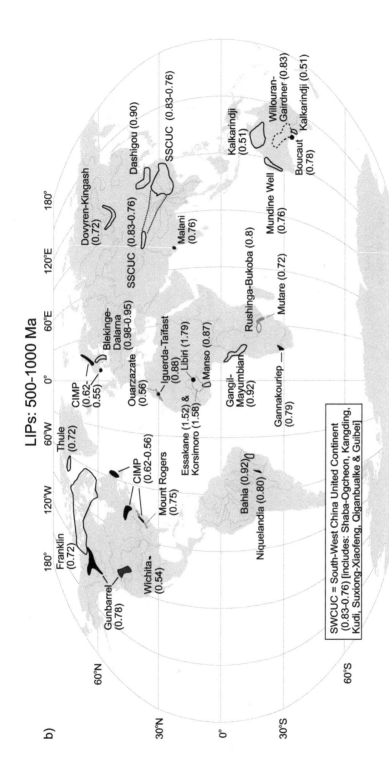

LIPs: 500-1000 Ma

Figure 1.6 (cont.)

SWCUC = South-West China United Continent (0.83-0.76) [includes: Shaba-Ogcheon, Kangding, Kudi, Suxiong-Xiaofeng, Qiganbualke & Guibei]

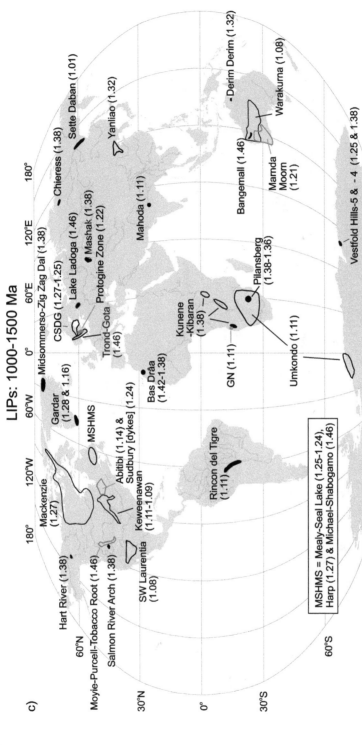

LIPs: 1000–1500 Ma

c)

Hart River (1.38)
Moyie-Purcell-Tobacco Root (1.46)
Salmon River Arch (1.38)
SW Laurentia (1.08)
Keweenawan (1.11-1.09)
Mackenzie (1.27)
MSHMS
Abitibi (1.14) & Sudbury [dykes] (1.24)
Gardar (1.28 & 1.16)
Midsommerso-Zig Zag Dal (1.38)
CSDG (1.27-1.25)
Lake Ladoga (1.46)
Trond-Gota (1.46)
Mashak (1.38)
Protogine Zone (1.22)
Chieress (1.38)
Sette Daban (1.01)
Yanliao (1.32)
Mahoda (1.11)
Derim Derim (1.32)
Warakurna (1.08)
Bangemall (1.46)
Mamda Moom (1.21)
Pilansberg (1.38-1.36)
Kunene-Kibaran (1.38)
GN (1.11)
Umkondo (1.11)
Bas Drâa (1.42-1.38)
Rincon del Tigre (1.11)
Vestfold Hills-5 & -4 (1.25 & 1.38)

MSHMS = Mealy-Seal Lake (1.25-1.24), Harp (1.27) & Michael-Shabogamo (1.46)

180°   120°W   60°W   0°   60°E   120°E   180°

60°N   30°N   0°   30°S   60°S

Figure 1.6 (*cont.*)

14

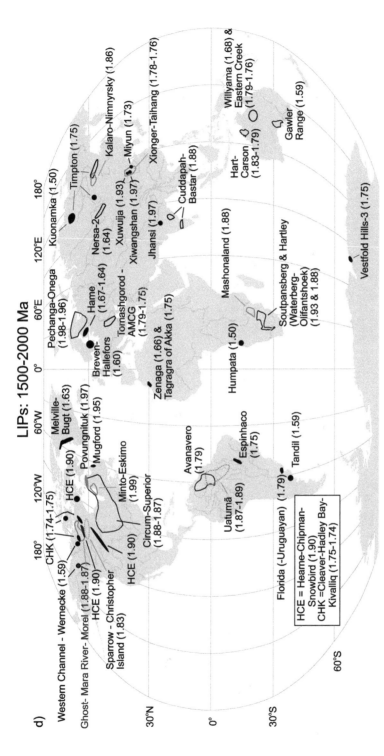

Figure 1.6 (*cont.*)

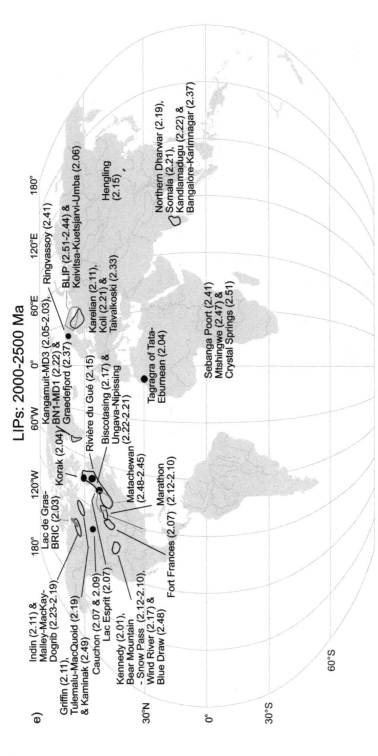

Figure 1.6 *(cont.)*

16

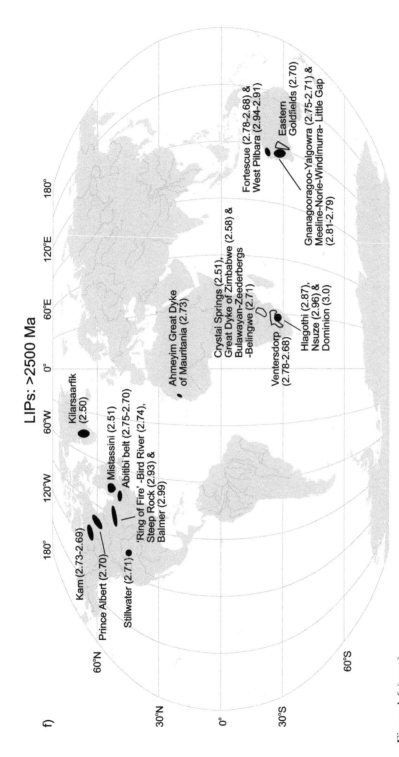

LIPs: >2500 Ma

f)

Kam (2.73–2.69)

Prince Albert (2.70)

Stillwater (2.71)

Kilarsaarfik (2.50)

Mistassini (2.51)

Abitibi belt (2.75–2.70)

'Ring of Fire' -Bird River (2.74), Steep Rock (2.93) & Balmer (2.99)

Ahmeyim Great Dyke of Mauritania (2.73)

Crystal Springs (2.51), Great Dyke of Zimbabwe (2.58) & Bulawayan-Zeederbergs -Belingwe (2.71)

Ventersdorp (2.78–2.68)

Hlagothi (2.87), Nsuze (2.96) & Dominion (3.0)

Fortescue (2.78–2.68) & West Pilbara (2.94–2.91)

Eastern Goldfields (2.70)

Gnanagooragoo-Yalgowra (2.75–2.71) & Meeline-Norie-Windimurra- Little Gap (2.81–2.79)

180°   120°W   60°W   0°   60°E   120°E   180°

60°N   30°N   0°   30°S   60°S

Figure 1.6 (cont.)

17

Table 1.2 *LIPs and interpreted LIP "fragments" and Silicic LIPs (SLIPs). See Fig. 1.6 for locations.*

No.	Age (Ga)	LIP Name [linked events, S = section, N = note at bottom]	Type	Location(s)	Area (A) Volume (V) (Mkm^2 or Mkm^3)	Figure (F) Section (S) Table (T)	Selected reference(s)
1	0.02	**Columbia River**	C	North America	A = 0.24	S(3.2.2); F(3.2, 5.1, 13.4); T(5.2, 8.1, 12.1, 14.1, 14.2)	Reidel *et al.*, 2013
2	0.03	**Afro-Arabian (Afar)**	C	Arabian Penn. & Africa	A = 2.00	S(3.2.3, S 9.2.2); F(3.3, 5.2, 9.3, 12.1); T(8.1, 9.1, 11.1, 12.1, 14.1, 14.2)	Avni *et al.*, 2012; White and McKenzie, 1989
3	0.03	**Sierra Madre Occidental**	S	Southwestern USA, Mexico	A= 0.40 V = 0.39	S(8.3.2); F(8.3); T(8.2, 16.5)	Bryan and Ferrari, 2013
4	0.06	**North Atlantic Igneous Province (NAIP)**	C	Greenland / Northern Canada & Europe (UK)	A = 1.30	S(3.2.4); F(3.4, 5.1, 5.24, 12.3); T(8.1, 11.1, 16.4)	Jerram *et al.*, 2009a, b; Storey *et al.*, 2007; White and McKenzie, 1989
5	0.07	**Deccan**	C	India	A = 1.80	S(3.2.5, 9.2.2); F(3.5, 5.1, 9.4, 12.2); T(5.2, 8.1, 9.1, 9.2, 11.1, 12.1, 14.1, 14.2, 14.4, 14.4, 16.5)	Sheth and Vanderkluysen, 2014; Hooper *et al.*, 2010
6	0.07	**Sierra Leone (-Ceara)**	O	Atlantic Ocean	A = 0.9 V = 2.5		Eldholm and Coffin, 2000; Ernst and Buchan, 2001

#	Age	Name	O/C	Location	A/V	Codes	Reference
7	0.09	**Caribbean-Colombian**	O/C	Central America, South America	A = 1.10 V = 4.50	S(4.2.3); T(14.1, 14.2, 16.1, 16.5)	Kerr, 2014
8	0.09	**Madagascar**	C	Africa	A = 1.60 V = 4.50	F(9.14); T(5.2, 9.2, 11.1, 14.2)	Storey *et al.*, 1995; Ernst and Buchan, 2001b
9	0.1	**Naturaliste**	O	Indian Ocean	V = 1.2		Eldholm and Coffin, 2000; Ernst and Buchan, 2001b
10	0.1	**Hess**	O	Pacific Ocean	A = 0.8 V = 9.1		Eldholm and Coffin, 2000; Ernst and Buchan, 2001b
11	0.13–0.09	**HALIP**	C	Circum-Arctic	A = 0.5 (Canadian portion)	F(5.1, 9.15); T(9.2)	Jowitt *et al.*, 2014
12	0.12	**Kerguelen–(Broken Ridge)** [12, 19, 20: S4.2.2]	O	Indian Ocean	V = 9.10	F(4.4); S(4.2.2); T(11.1)	Wallace *et al.*, 2002
13	0.12–0.10	**Southeast African (Agulhas, Maud, Georgia)**	O/C	Africa	A = 3.50	S(4.4.2); F(4.7)	Gohl *et al.*, 2011
14	0.12–(0.09)	**Whitsunday**	S	Australia (east)	V = 2.20	S(8.3.3); F(8.4); T(8.2)	Pirajno and Hoatson, 2012
15	0.12	**Greater Ontong Java (Ontong Java + Nauru basin, etc.)** [15, 16, 17: S4.4.2]	O	Pacific Ocean	A = 4.27 V = 58.0	S(4.2.1, 4.4.1); F(1.2, 3.4, 4.6, 5.8); T(14.1, 14.2, 16.5)	Ingle and Coffin, 2004

Table 1.2 (*cont.*)

No.	Age (Ga)	LIP Name [linked events, S = section, N = note at bottom]	Type	Location(s)	Area (A) Volume (V) (Mkm² or Mkm³)	Figure (F) Section (S) Table (T)	Selected reference(s)
16	0.12	**Manihiki** [15, 16, 17: S4.4.2]	O	Pacific Ocean	A = 0.80 V = 8.80	S(4.4.2); F(4.6); T(14.1)	Taylor, 2006
17	0.12	**Hikurangi** [15, 16, 17: S4.4.2]	O	Pacific Ocean	A = 2.70 V = 0.70	S(4.4.2); F(4.6); T(14.1)	Taylor, 2006
18	0.13	**Paraná-Etendeka**	C	South America & Africa	A = 2.00	S(3.2.6, 9.2.2); F(3.6, 9.5, 9.14); T(8.1, 9.1, 9.2, 12.1, 14.1, 14.4)	Thiede and Vasconcelos, 2010
19	0.13	Comei [12, 19, 20: S4.2.2]	C	Greater India	A = 0.04	S(4.2.2); F(4.4); T(11.1)	Zhu et al., 2009
20	0.13	Bunbury [12, 19, 20: S4.2.2]	C	Australia		S(4.2.2); F(4.4); T(11.1)	Zhu et al., 2009
21	0.14	Trap	C	Greenland (southwest)	A = 0.02		Ernst and Buchan, 2001b
22	0.16–0.14	**NW Australian margin** (Greater Exmouth, Gascoyne, Wallaby, Scott, Browse, Argo)	O	Australia (northwest margin)	A = 0.16		Rohrman, 2013; Symonds et al., 1998; Pirajno and Hoatson, 2012
23	0.13–0.08	**Mid-Pacific Mountains**	O	Pacific Ocean	A = 1.1		Ernst and Buchan, 2001b
24	0.14	**Shatsky-Tamu**	O	Pacific Ocean	A = 0.20 V = 2.50	T(14,4)	Sager et al., 2013; Heydolph et al., 2014
25	0.15	**Magellan**	O	Pacific Ocean	V = 0.5 V = 1.8		Eldholm and Coffin, 2000; Ernst and Buchan, 2001b

26	0.15	Sorachi–Okhotsk	O	Japan / Sakhalin/ Okhotsk, Russia	V = 0.6		Ernst and Buchan, 2001b; Bogdanov and Dobretsov, 2002
27	0.18	**Karoo, Ferrar**, [27, 28: S3.2.7]	C/S	Africa, South America, Antarctica	V = 5.00	S(3.2.7, 5.3.2); F(5.2, 5.1, 5.9, 5.10, 9.14); T(5.2, 5.3, 8.1, 9.2, 11.1, 14.1, 14.2, 14.4, 16.3)	Svensen et al., 2012; Neumann et al., 2011; Storey et al., 2013
28	0.18–0.15	**Chon Aike** [27, 28: S3.2.7]	S	South America, Antarctica	V = 0.23	S(8.3.4); T(8.2, 16.5)	Pankhurst et al., 1998, 2000
29	0.20	**CAMP**	C	North America, South America, Africa, Europe (Iberia)	A = 10.0	S(12.2.7); F(5.1, 12.5); T(5.2, 5.3, 11.1, 12.1, 14.1, 14.2, 16.1, 16.5)	Bertrand et al., 2014; Marzoli et al., 2011
30	0.23	**Wrangellia**	O	Western North America	V = 1.00	S(4.5.1); T(12.1, 14.1, 16.2, 16.3, 16.5)	Greene et al., 2010
31	0.20	Angayucham	O/C	Alaska	A = 0.15		Pallister et al., 1989; Ernst and Buchan, 2001b
32	0.25	**Siberian Trap**	C	Asia	A = 7.00 V = 4.00	S(3.2.8, 9.2.2); F(1.2, 3.8, 4.8, 5.1, 5.7, 5.17, 9.6, 9.15, 16.7); T(5.2, 5.3, 8.1, 9.1, 9.2, 11.2, 12.1, 14.1, 14.4, 16.1, 16.3, 16.4, 16.6)	Ivanov et al., 2013; Sobolev et al., 2011

Table 1.2 (cont.)

No.	Age (Ga)	LIP Name [linked events, S = section, N = note at bottom]	Type	Location(s)	Area (A) Volume (V) (Mkm² or Mkm³)	Figure (F) Section (S) Table (T)	Selected reference(s)
33	0.26	**Emeishan**	C	Asia	A = 0.25	S(12.2.10); F(5.25, 12.7); T(11.2, 12.1, 14.4, 16.1, 16.6)	Shellnutt et al., 2014; He et al., 2013; Ali et al., 2010
34	0.28	**Tarim**	C	Asia (central)	A = 0.25	F(16.7); T(14.1, 14.4, 16.5, 16.6)	Xu et al., 2014b; Zhang et al., 2013
35	0.28	Qiangtang-Panjal	C	Asia (Tibet)	A = 0.04		Zhai et al., 2013; Ernst and Buchan, 2001b
36	0.30	**Jutland**, Oslo, Skaggerak, EUNWA (European – NW Africa)	C	Europe	A = 0.15	F(5.1); T(14.1)	Kirstein et al., 2004; Ernst and Buchan, 2001b; see also Fig. 3.1
37	0.32–0.28	**Kennedy–Connors–Auburn**	S	Australia (east)	V = 0.50	T(8.2)	Pirajno and Hoatson, 2012
37a	0.33	Tianshan [34, 37a: N1]	C	China	A = 1.70		Xia et al., 2012
38	0.38	**Yakutsk-Vilyui**	C	Siberia	A = 0.80	S(11.4; F(5.1, 11.7, 9.15); T(9.2, 12.1, 14.1, 14.2, 14.4, 16.1)	Kiselev et al., 2012
39	0.38	**Kola-Dnieper**	C	Baltica	A = 3.00	S(9.2.2); F(9.7); T(9.1, 9.2, 14.1, 14.2, 14.4, 16.1)	Ernst and Bell, 2010
40	0.38–0.36	Magdalen (Maritimes) basin	C/S	Laurentia (east)			Dessureau et al., 2007; Murphy et al., 1999
41	0.44	Suordakh	C	Siberia			Khudoley et al., 2013

22

42	0.51	**Kalkarindji**	C	Australia	A = 2.10	S(3.5.1); F(3.14); T(12.1, 14.1, 14.2)	Pirajno and Hoatson, 2012
43	0.54	**Wichita**	C	Laurentia (south)	A = 0.04 V = 0.25		Hanson et al., 2013
44	0.56	**CIMP (Sept Iles, Catoctin, Volyn)** [44, 45]	C	Laurentia (east) & Baltica		S(9.2.2); F(9.8); T(9.1, 9.2,11.1)	Ernst and Bell, 2010
45	0.56	**CIMP (Ouarzazate)** [44, 45]	C	West Africa craton			Ernst and Bell, 2010
46	0.59	**CIMP (Grenville)**	C	Laurentia (east) & Baltica	A = 0.14	S(9.2.2); F(5.2, 9.8); T(9.1, 9.2, 11.1)	Ernst and Bell, 2010
47	0.62	**CIMP (Long Range, Baltoscandian, Egersund)**	C	Laurentia (east) & Baltica	A = 0.10 Laurentia only)	S(9.2.2); F(9.8); T(11.1)	Ernst and Bell, 2010
48	0.72	**Franklin-(Thule)** [48, 49: N2]	C	Laurentia (north)	A = 2.25	S(5.2.1, 12.2.13); F(5.1); T(5.3, 12.1)	Ernst and Bleeker, 2010
49	0.72	Dovyren-Kingash [48, 49: N2]	C	Siberia (south)	A = 0.13		Ernst et al., 2012; Ariskin et al., 2013; Polyakov et al., 2013
50	0.72	Mutare	C	Kalahari craton	A = 0.05		Ernst et al., 2008
51	0.76	**Mundine Well** [51, 52: S8.4.1]	C	Australia craton (northwest)	A = 0.18	T(11.3)	Ernst et al., 2008

Table 1.2 (*cont.*)

No.	Age (Ga)	LIP Name [linked events, S = section, N = note at bottom]	Type	Location(s)	Area (A) Volume (V) (Mkm^2 or Mkm^3)	Figure (F) Section (S) Table (T)	Selected reference(s)
52	0.76	Malani [51, 52: S8.4.1]	S	Greater India	A = 0.02	S(8.4.1); F(8.5); T(8.2, 11.3)	Meert et al., 2013; Gregory et al., 2008; Ernst et al., 2008
53	0.76	Shaba & Ogcheon	C/S	South China block	A = 0.33	S(8.4.2); F(8.6); T(11.3)	Li et al., 2003; Ernst et al., 2008
54	0.76	Mount Rogers (Southeast Laurentia)	C	Laurentia (east)	A = 0.06	T(11.3)	McClellan et al., 2012; McClellan & Gazel, in press.
55	0.78	**Gunbarrel** [55-58: S8.4.2]	C	Laurentia (west)	A = 0.49	F(5.1); T(11.3)	Ernst et al., 2008
56	0.78	Kangding [55-58: S8.4.2]	C	South China craton	A = 1.16	S(8.4.2); F(8.6); T(11.3)	Li et al., 2003; Ernst et al., 2008
57	0.78	Kudi [55-58: S8.4.2]	C	Tarim craton		T(11.3)	Ernst et al., 2008
58	0.78	Boucaut [55-58: S8.4.2]	C	Australia (south)			Ernst et al., 2008
59	0.79	Gannakouriep	C	Kalahari craton (southwest)	A = 0.03	F(5.1, 5.2)	Rioux et al., 2010
60	0.80	Rushinga group (Madondi belt)	C	Kalahari (north)			Ernst et al., 2008
61	0.80	Suxiong-Xiaofeng	C/S	South China craton		S(8.4.2); F(8.6)	Ernst et al., 2008; Li Z-X et al., 2013

24

No.	Age	Name	Type	Craton	A	Codes	References
62	0.80	Niquelandia (includes other intrusions)	C	Sao Francisco craton	A = 0.03		Correia et al., 2012
63	0.83	Qiganbulake (Kuruketage)	C	Tarim craton		T(11.3)	Ernst et al., 2008
64	0.83	**Gairdner–Willouran** [64, 65: S8.4.2]	C	Australia craton	A = 0.63	F(5.2); T(11.3)	Ernst et al., 2008
65	0.83	**Guibei** [64, 65: S8.4.2]	C/S	South China craton	A = 1.34	S(8.4.2); F(8.6); T(8.2, 11.3, 16.3, 16.1, 16.2)	Ernst et al., 2008; Li et al., 2013
66	0.88	Iguerda-Taïfast	C	West Africa craton			Kouyaté et al., 2013; Youbi et al., 2013
67	0.92	**Gangil-Mayumbian** [67, 68: N3]	C	Congo craton	A = 0.34		Ernst et al., 2008
68	0.92	Bahia [67, 68: N3]	C	Sao Francisco craton	A = 0.03		Evans et al., 2010, in press
69	0.93–0.90	**Dashigou**	C	North China craton	A = 0.65		Peng et al., 2011
70	0.98–0.95	*Blekinge-Dalarna*	C	Baltica	A = 0.09		Ernst et al., 2008
71	1.05	Sette Daban	C	Siberia (east)			Rainbird et al., 1998
72	1.08	**Warakurna**	C	Australia craton	A = 1.55	F(1.3); T(8,1, 11.3, 16.2, 16.5)	Pirajno and Hoatson, 2012
73	1.09	**SW Laurentia** (SW USA Diabase Province)	C/S	Laurentia	A = 1.13	T(11.3)	Bright et al., 2014
74	1.11	**Umkondo**	C	Kalahari craton	A = 2.08		de Kock et al., 2014; Bullen et al., 2012
75	1.11	GN	C	Congo craton			Ernst et al., 2013

Table 1.2 (*cont.*)

No.	Age (Ga)	LIP Name [linked events, S = section, N = note at bottom]	Type	Location(s)	Area (A) Volume (V) (Mkm² or Mkm³)	Figure (F) Section (S) Table (T)	Selected reference(s)
76	1.11	**Rincon del Tigre-Huanchaca**	C	Amazonia	500 km across		Hamilton *et al.*, 2012; Teixeira *et al.*, in press
77	1.11	Mahoba	C	Greater India			Pradhan *et al.*, 2012
78	1.12–1.09	**Keweenawan** [78, 79: S5.2.1]	C	Laurentia	A = 0.41	S(3.5.2; 9.2.2); F(3.15, 5.25, 9.9); T(9.1, 9.2, 11.2, 11.3, 16.1, 16.2, 16.3, 16.4, 16.5)	Miller and Nicholson, 2013; Heaman *et al.*, 2007; Ernst *et al.*, 2008
79	1.14	Abitibi [78, 79: S5.2.1]	C	Laurentia	A = 0.26	F(5.2)	Ernst *et al.*, 2008
80	1.17	Late Gardar	C	Laurentia			Ernst *et al.*, 2008
81	1.21	**Marnda Moorn**	C	Australia craton	A = 0.59		Pirajno and Hoatson, 2012
82	1.22	*Protogine Zone*	C	Baltica	A = 0.02		Ernst *et al.*, 2008
83	1.24	**Sudbury** [dyke]	C	Laurentia	A = 0.12	T(11.3)	Shellnutt and MacRae, 2012; Ernst *et al.*, 2008
84	1.25	*Mealy-Seal Lake*	C	Laurentia	A = 0.02	T(11.3)	Ernst *et al.*, 2008
85	1.25	Vestfold Hills-5	C	East Antarctica	A = 0.01		Ernst *et al.*, 2008
86	1.27	**Mackenzie**	C	Laurentia	A = 2.7	S(12.2.14); F(1.3, 5.1, 5.2, 5.22, 13.5); T(5.3, 11.3, 12.1, 16.1, 16.3, 16.5)	Baragar *et al.*, 1996; Ernst *et al.*, 2008

#	Age	Name	C/S	Craton	A	T	Reference
87	1.27	Harp	C	Laurentia	A = 0.06		Ernst et al., 2008
88	1.27, 1.26, & 1.25	Central Scandinavian Dolerite Group (CSDG)	C	Baltica	A = 0.14	T(11.3)	Ernst et al., 2008
89	1.32	Derim–Derim (Roper)	C	North Australia craton	A = 0.01		Pirajno and Hoatson, 2012
90	1.32	Yanliao	C/S	North China craton	A = 0.08		Zhang et al., 2012b
91	1.38	Hart River–Salmon Arch	C	Laurentia	A = 0.02	T(11.3)	Ernst et al., 2008
92	1.38	Midsommerso–Zig Zag Dal	C	Laurentia	A = 0.06	T(11.3)	Upton et al., 2005; Ernst et al., 2008
93	1.38	Vestfold Hills-4	C	East Antarctica	A = 0.01	T(11.3)	Ernst et al., 2008
94	1.38	Chieress	C	Siberia craton		T(11.3)	Ernst et al., 2008
95	1.38–1.42	Bas Drâa	C	West Africa craton			El Bahat et al., 2013; Söderlund et al., 2013
96	1.38	Kunene-Kibaran	C	Congo craton	A = 0.1	T(11.3)	Maier et al., 2013; Tack et al., 2010; Ernst et al., 2013
97	1.38	Pilanesberg	C	Kalahari craton	A = 0.16	T(11.3)	Ernst et al., 2008
98	1.38	Mashak	C	Baltica	A = 0.50	T(11.3)	Puchkov et al., 2013
99	1.46	Tuna–Lake Ladoga	C	Baltica	A = 0.13	T(11.3)	Ernst et al., 2008
100	1.46	West Bangemall-Edmund	C	North Australia craton	A = 0.03	T(11.3)	Pirajno and Hoatson, 2012

Table 1.2 (cont.)

No.	Age (Ga)	LIP Name [linked events, S = section, N = note at bottom]	Type	Location(s)	Area (A) Volume (V) (Mkm² or Mkm³)	Figure (F) Section (S) Table (T)	Selected reference(s)
101	1.47	**Moyie**	C	Laurentia	A = 0.12	T(11.3)	Ernst et al., 2008
101a	1.46–1.43	Michael–Shabogamo	C	Laurentia (eastern)	A = 0.05		Ernst and Buchan, 2001b; Gower et al., 2002
102	1.50	Kuonamka	C	Siberia craton	A = 0.06		Ernst et al., 2014a; Ernst et al., 2008
103	1.50	Humpata	C	Angolan block			Ernst et al., 2013
104	1.52	**Essakane**	C	southern West African craton	A = 0.38		Baratoux et al., 2014
105	1.58	**Korsimoro**	C	southern West African craton	A = 0.20		Baratoux et al., 2014
106	1.59	**Gawler Range** [95, 96: S16.3.2]	C/S	South Australian craton	A = 0.131	S(8.4.3); F(8.7); T(8.2. 16.1, 16.5)	Pirajno and Hoatson, 2012
107	1.59	Western Channel [95, 96: S16.3.2]	C	Laurentia			Ernst et al., 2008
108	1.59	*Tandil*	C	Rio de la Plata craton			Teixeira et al., 2013
109	1.60	Breven-Hallefors	C	Baltica	A = 0.059		Ernst et al., 2008
110	1.63	Taishan	C	North China craton			Xiang et al., 2012
111	1.63	**Melville-Bugt**	C	Laurentia (Greenland)	A = 0.22		Halls et al., 2011

No.	Age	Name	Type	Craton	Notes	A	Reference
112	1.64	'Nersa'	C	Siberia craton (south)			Metelkin et al., 2011
113	1.67–1.64	Hame	C	Baltica		A = 0.04	Ernst and Buchan, 2001b
114	1.66	Zenaga	C	West African craton			Kouyaté et al., 2013; Youbi et al., 2013
115	1.69–1.67	Willyama	C	South Australian craton			Pirajno and Hoatson, 2012
116	1.79–1.76	Eastern Creek	C	North Australian craton			Pirajno and Hoatson, 2012
117	1.73	Miyun	C	North China craton			Peng et al., 2012
118	1.75	Timpton	C	Siberia craton		A = 0.06	Gladkochub et al., 2010
119	1.75	Espinhaco	C	Sao Francisco craton			Ernst and Buchan, 2001b
120	1.75	Vestfold Hills-3	C	East Antarctica			Ernst and Buchan, 2001b
121	1.75	Cleaver-Hadley Bay-Kivalliq (**Pitz–Nueltin**)	C/S	Rae craton		A = 0.30	Ernst and Bleeker, 2010
122	1.75	Tagragra of Akka		West African craton		A = 0.01	Youbi et al., 2013
123	1.78	**Xiong'er–Taihang**	C	North China craton	S(8.4.4); F(8.8); T(8.2)	A = 0.27	Peng, 2010
124	1.79	**Avanavero**	C	Amazonian craton		A = 0.30	Reis et al., 2013

Table 1.2 (cont.)

No.	Age (Ga)	LIP Name [linked events, S = section, N = note at bottom]	Type	Location(s)	Area (A) Volume (V) (Mkm² or Mkm³)	Figure (F) Section (S) Table (T)	Selected reference(s)
125	1.79	Florida (Uruguayan)	C	Rio de la Plata craton	A = 0.02	S(5.2.3); F(5.5)	Teixeira et al., 2013
126	1.79–1.75	Tomashgorod–AMCG	C/S	Baltica (Volgo-Sarmatia)			Bogdanova et al., 2013
127	1.79	Libiri	C	S West African craton			Ernst et al., 2014b
128	1.83–1.79	**Hart-Carson**	C	Northern Australian craton	A = 0.11		Pirajno and Hoatson, 2012
129	1.83	**Sparrow–Christopher Island**	C	Rae craton	A = 0.10 (dykes only)		Ernst and Bleeker, 2010
130	1.87–1.85	**Kalaro–Nimnyrsky**		Siberia craton	A = 0.10		Gladkochub et al., 2010
131	1.88	Circum-Superior–Pan Superior	C	Superior craton	A = 0.60	S(9.2.2); F(9.10); T(9.1, 11.2, 16.1, 16.2, 16.3, 16.5)	Ernst and Bell, 2010; Minifie et al., 2013
132	1.88	**Ghost–Mara River–Morel**	C	Slave craton	A = 0.06		Buchan et al., 2010
133	1.88	**Mashonaland**	C	Zimbabwe craton	A = 0.16		Soderlund et al., 2010

No.	Age	Name	Type	Craton	A	T	Reference
134	1.88	Soutpansberg	C	Kaapvaal craton			Hanson et al., 2004b
135	1.89–1.87	**Uatumã**	S	Amazonian craton	A = 1.50		Klein et al., 2012
136	1.89	Cuddapah-Bastar	C	Dharwar-Bastar cratons	A = 0.03		French et al., 2008; Ernst and Srivastava, 2008; Belica et al., 2014
137	1.90	Hearne-Chipman-Snowbird	C	Slave and Rae cratons			Ernst and Bleeker, 2010
138	1.92	Hartley (Waterberg-Olifantshoek)	C	Kaapvaal craton			Hanson et al., 2004a
139	1.93	Xuwujia	C	North China craton	A = 0.01		Peng et al., 2005
140	1.95	Mugford	C	North Atlantic (Nain) craton			Ernst and Bleeker, 2010
141	1.97–1.96	**Povungnituk**	C	Superior craton			Ernst and Bleeker, 2010
142	1.98–1.96	**Pechenga-Onega**	C	Karelia-Kola craton	A = 0.60	T(16.2, 16.3)	Ernst and Bleeker, 2010
143	1.97	Xiwangshan	C	North China craton			Peng et al., 2005
144	1.98	Jhansi	C	Bundelkhand craton	A = 0.02		Pradhan et al., 2012
145	2.00	**Minto-Eskimo (Purtuniq ophiolite)**	C/O	Superior craton			Ernst and Bleeker, 2010
146	2.01	Kennedy	C	Wyoming craton			Ernst and Bleeker, 2010
147	2.03	**Lac des Gras–Booth River**	C	Slave craton	A = 0.04		Buchan et al., 2010

Table 1.2 (*cont.*)

No.	Age (Ga)	LIP Name [linked events, S = section, N = note at bottom]	Type	Location(s)	Area (A) Volume (V) (Mkm² or Mkm³)	Figure (F) Section (S) Table (T)	Selected reference(s)
148	2.05–2.03	**Kangamuit-MD3**	C	North Atlantic craton (Greenland)	A = 0.04	F(5.2)	Nilsson *et al.*, 2013
149	2.04	Tagragra of Tata–Eburnean	C/S	West African craton			Kouyaté *et al.*, 2013; Youbi *et al.*, 2013
150	2.04	Korak	C	Superior craton			Ernst and Bleeker, 2010
151	2.06	**Bushveld** (includes other intrusions)	C	Kaapvaal craton	A = 0.25	S(5.4.2, 9.2.2); F(5.19, 9.11); T(8.1, 16.1, 16.2, 16.4, 16.5)	Rajesh *et al.*, 2013
152	2.06	Keivitsa–Kuetsjärvi–Umba-	C	Karelian craton			Martin *et al.*, 2013; Malehmir *et al.*, 2014
153	2.07	**Fort Frances** (–Cauchon–Lac Esprit)	C	Superior craton	A = 0.08	F(11.12)	Ernst and Bleeker, 2010
154	2.12–2.10	**Marathon** [154, 158: N4]	C	Superior craton	A = 0.06	F(11.12)	Ernst and Bleeker, 2010
155	2.10	Indin	C	Slave craton	A = 0.03		Buchan *et al.*, 2010
156	2.10	**Karelia**	C	Karelian craton	A = 0.21		Vuollo and Huhma, 2005
157	2.11	Griffin	C	Hearne craton	A = 0.08		Ernst and Bleeker, 2010

No.	Age	Name	Type	Craton	A	Notes	Reference
158	2.11–2.10	Bear Mountain-Snowy Pass [154, 148: N4]	C	Wyoming craton			Ernst and Bleeker, 2010
159	2.15	Riviere du Gue	C	Superior craton			Ernst and Bleeker, 2010
160	2.15	Hengling	C	North China craton			Peng et al., 2005
161	2.17	Wind River	C	Wyoming craton			Ernst and Bleeker, 2010
162	2.17	**Biscotasing**	C	Superior craton	A = 0.35		Ernst and Bleeker, 2010
163	2.18	Dandeli	C	Dharwar craton	A = 0.06		French et al., 2010
164	2.19	Southwest Slave Magmatic Province (Dogrib)	C	Slave craton	A = 0.01		Buchan et al., 2010
165	2.19	Tulemalu–(MacQuoid)	C	Rae craton	A = 0.02		Ernst and Bleeker, 2010
166	2.20	Ongeluk–Hekpoort	C	Kaapvaal craton			Ernst and Buchan, 2001
167	2.21	Turee Creek–Cheela Springs	C	Pilbara craton			Müller et al., 2005; Martin and Morris, 2010
168	2.21	**Ungava–Nipissing**	C	Superior craton	A = 0.50	S(5.3.3); F(5.1, 5.11); T(5.3)	Buchan et al., 1998; Ernst and Bleeker, 2010
169	2.23–2.21	MacKay-Malley	C	Slave craton	A = 0.04		Buchan et al., 2010; Ernst and Bleeker, 2010

Table 1.2 (*cont.*)

No.	Age (Ga)	LIP Name [linked events, S = section, N = note at bottom]	Type	Location(s)	Area (A) Volume (V) (Mkm² or Mkm³)	Figure (F) Section (S) Table (T)	Selected reference(s)
170	2.21	Koli	C	Karelian craton			Vuollo and Huhma, 2005
171	2.21	Somala	C	Dharwar craton	A = 0.05		French *et al.*, 2010
172	2.22	BN1	C	North Atlantic craton (Greenland)	A = 0.05		Ernst and Bleeker, 2010
173	2.24	Vestfold Hills-2	C	East Antarctica			Ernst and Bleeker, 2010
174	2.33	Taivalkoski		Karelia			Salminen *et al.*, 2014; Vuollo and Huhma, 2005
175	2.37	**Bangalore-Karimnagar**	C	Dharwar craton	A = 0.14		Kumar *et al.*, 2012
176	2.38–2.41	Graedefjord–Scourie	C	North Atlantic craton (Greenland)			Nilsson *et al.*, 2013; Davies and Heaman, 2014; Hughes *et al.*, 2014
177	2.41	Sebanga Poort	C	Zimbabwe craton			Soderlund *et al.*, 2010
178	2.41	Ringvassoy	C	Karelia-Kola craton (northwest)			Kullerud *et al.*, 2006

34

No.	Age	Name	Type	Craton	A	Notation	Reference
179	2.41	Du Chef	C	Superior craton			Ernst and Bleeker, 2010
180	2.42	**Widgiemooltha**	C	Yilgarn craton	A = 0.64	S(5.4.2); F(5.2, 5.21)	Smirnov et al., 2013
181	2.48–2.45	**Matachewan** [181, 182: N5]	C	Superior craton	A = 0.36	S(5.2.3); F(5.1, 5.2, 5.5, 11.8, 11.12); T(8.1, 11.2, 16.2, 16.3)	Ernst and Bleeker, 2010
182	2.51–2.45	**Baltic LIP (BLIP) East Scandinavian Paleoproterozoic LIP [ESCLIP]** [181, 182, 187: N5]	C	Karelian craton	A = 0.50	S(5.4.3); F(5.23); T(11.2, 16.4)	Ernst and Buchan, 2001
183	2.45	Woongarra–Weeli Wolli	C	Pilbara craton		T(8.1, 16.5)	Pirajno and Hoatson, 2012
184	2.47	Mtshingwe	C	Zimbabwe craton			Soderlund et al., 2010
185	2.50	Kilarsaarfik	C	North Atlantic craton			Nilsson et al., 2013
186	2.50	**Kaminak**	C	Hearne craton	A = 0.18		Sandeman et al., 2013
187	2.51	**Mistassini** [182, 187: N5]	C	Superior craton	A = 0.10	F(5.1, 11.12); T(11.2)	Ernst and Bleeker, 2010
188	2.51	Crystal Springs	C	Zimbabwe craton			Soderlund et al., 2010
189	2.58	**Great Dyke of Zimbabwe**	C	Zimbabwe craton	A = 0.06	S(5.4.2); F(5.20)	Soderlund et al., 2010
190	2.70	**Eastern Goldfields**	C/O	Yilgarn craton	A = 0.10	S(6.3.4); F(6.11); T(6.1, 8.1, 16.2, 16.3, 16.5)	Barnes et al., 2012; Said et al., 2012

Table 1.2 (*cont.*)

No.	Age (Ga)	LIP Name [linked events, S = section, N = note at bottom]	Type	Location(s)	Area (A) Volume (V) (Mkm² or Mkm³)	Figure (F) Section (S) Table (T)	Selected reference(s)
191	2.73–2.70	Kam	C	Slave craton		S(6.3.3); F(6.9); T(6.1)	Bleeker and Hall, 2007
192	2.70	Bulawayan–Zeederbergs-Belingwe	C	Zimbabwe craton		T(6.1, 16.2, 16.3)	Prendergast, 2004; Ernst and Buchan, 2001b
193	2.71	Stillwater	C	Wyoming craton		T(16.1, 16.4)	Wall et al., 2012; Ernst and Buchan, 2001b
194	2.75–2.70	Abitibi belt	C/O	Superior craton		S(6.3.2); F(6.8); T(6.2, 16.3)	Ayer et al., 2002; Sproule et al., 2002
195	2.74	Ring of Fire–Bird River	C	North Caribou craton		S(16.2.2); T(16.1, 16.5)	Mungall et al., 2010; Ernst and Jowitt, 2013
196	2.74	Ahmeyim Great Dyke of Mauritania	C	Reguibat shield (West African craton)			Tait et al., 2013
197	2.70–2.65	Allanridge (Ventersdorp-3)	C	Kaapvaal craton		S(6.2.3); F(6.3, 6.4)	Olsson et al., 2012; van der Westhuizen et al., 2006
198	2.73	Maddina (Fortestcue-3)	C	Pilbara craton		S(6.2.4); F(1.4, 6.5, 6.4)	

36

199	2.70	Prince Albert Group	C	Rae craton	F(1.4); T(6.1)	MacHattie et al., 2008
200	2.72	Platberg (Ventersdorp-2) [200, 201: S6.2.5]	C	Kaapvaal craton	S(6.2.3); F(6.3, 6.4)	van der Westhuizen et al., 2006
201	2.72	Kylena (Fortescue-2) [200, 201: S6.2.5]	C	Pilbara craton	S(6.2.4); F(1.4, 6.5, 6.4)	Pirajno and Hoatson, 2012
202	2.75–2.71	Gnanagooragoo-Yalgowra	C	Youanmi terrane (Yilgarn craton)	S(6.3.4); F(6.10)	Ivanic et al., 2010
203	2.78	Derdepoort-Gaberone (Ventersdorp-1) [203, 204: S6.2.5]	C	Kaapvaal craton	F(6.3, 6.4)	Wingate et al., 1998; Pirajno & Hoatson, 2012
204	2.78	Mount Roe–Black Range (Fortescue-1) [203, 204: S6.2.5]	C	Pilbara craton	S(6.2.4); F(1.4, 6.5, 6.4)	Pirajno and Hoatson, 2012
205	2.81–2.79	Meeline-Norie-Windimurra and Little Gap	C	Youanmi terrane (Yilgarn craton)	S(6.3.4); F(6.10)	Ivanic et al., 2010
206	2.87	Hlagothi	C	Kaapvaal craton		Gumsley et al., 2013
207	2.94–2.91	West Pilbara		Pilbara craton		Ernst and Buchan, 2001b; Pirajno and Hoatson, 2012

Table 1.2 (*cont.*)

No.	Age (Ga)	LIP Name [linked events, S = section, N = note at bottom]	Type	Location(s)	Area (A) Volume (V) (Mkm² or Mkm³)	Figure (F) Section (S) Table (T)	Selected reference(s)
208	2.93	Steep Rock		N. Caribou terrane		T(6.1)	Tomlinson and Condie, 2001; Ernst and Buchan, 2001b
209	2.96	Nsuze	C	Kaapvaal craton		S(6.2.2)	Klaussen et al., 2010
210	2.99	Balmer		N. Caribou terrane		T(6.1)	Tomlinson and Condie, 2001; Ernst and Buchan, 2001b
211	3.00	Dominion	C	Kaapvaal craton		S(6.2.1); F(6.2); T(8.1)	Eriksson et al., 2002; Marsh, 2006

Notes: "LIP" column: bold = LIPs; not bold = interpreted LIP fragments; *italicized* = very speculative LIPs. LIP size, mainly from Ernst and Buchan, 2001b, 2004 and Ernst et al., 2008, and Bryan, 2007. See also Bryan and Ferrari, 2013. Type: C = continental, O = oceanic and S = silicic. Figure/Section/Table: = figure, section or table in this book where the given LIP is discussed. N1 = events linked in Xia et al., 2012; Tianshan event not shown on figure but is in a similar location to the Tarim event; N2 = events linked in Ernst et al., 2013; N3 = events linked in Ernst et al., 2008; N4 & N5 = events linked in Ernst and Bleeker, 2010. Area/Volume are minimum estimates and very approximate; see discussions in text (Sections 2.2.1, 2.2.2, and 2.5). In some cases no estimate was provided because the inhomogeneous distribution precluded a simple calculation. The Archean LIP record is less well understood and is likely underrepresented in this table.

1.7 Summary

This chapter has provided an overview of LIPs, starting with the history of the term, the evolving definition, and an overview of LIP types and their key characteristics (Fig. 1.1). I introduce flood basalts (both continental and oceanic), the plumbing system (of dyke swarms, sill provinces, and layered intrusions), Archean types, and also planetary analogs. Also discussed are associated silicic magmatism (including major silicic provinces called Silicic LIPs, or SLIPs), and the link with carbonatites and kimberlites. Also addressed were associated topographic changes (e.g. domal uplift), rifting/breakup, and implications for ore deposits, hydrocarbons (oil and gas), and aquifers. Table 1.2 and Figure 1.6 provide a summary of the LIP record as currently known. With this overview of the topics, it is time to now launch into the rest of the book with detailed treatments of each of these topics.

2

Essential criteria: distinguishing LIP from non-LIP events

2.1 Introduction

Here I look in some more detail into what constitutes a Large Igneous Province (LIP), and how to distinguish these events from other classes of intraplate magmatism. The importance of this discussion is to define the term LIP since it has been used quite broadly (and variably) in the literature since its inception in the early 1990s.

I initially discuss, in some detail, the specific attributes (area, volume, duration, and intraplate setting) that can be used to describe LIPs, and which form the basis for the definition presented in Chapter 1. As noted in Chapter 1, a number of studies have already attempted to quantify LIPs in terms of type, age, areal extent, and volume (e.g. Coffin and Eldholm, 1994; Ernst and Buchan, 2001; Courtillot and Renne, 2003; Sheth, 2007; Cañón-Tapia, 2010). The review by Bryan and Ernst (2008) forms the basis for the discussion below, and extends the earlier studies listed above. Volume, duration or pulsed character of magmatism are critical to determining magma emplacement rates and the thermal and material budgets of LIP events, from which models for the origin of LIPs can be better evaluated (Chapter 15).

2.2 Essential attributes of LIPs

2.2.1 Volume

The volume of igneous rocks formed during a magmatic event is a critical attribute. LIPs are characterized by the emplacement of tremendous volumes of magma on to the surface and throughout the crustal profile and presumably also in the lithospheric mantle (Table 2.1). However, total volumes can be difficult to constrain for the various components of a LIP, especially as volcanic components are only intact for the youngest LIPs, whereas older LIPs are dominated by intrusive components that have had their volcanic portions removed by erosion. Regional geophysics can reveal some parts of the plumbing system but seismic studies are the only way to get at the distribution of underplated components at the base of the crust.

Table 2.1 *Classification of igneous provinces by size (after Bleeker and Ernst, 2006). To be classed as a LIP a province must also satisfy the additional criteria discussed in the text: a short duration (or short-duration pulses), an intraplate setting, and/or intraplate geochemistry.*

Name	Size
Giant (=LIP)	$>10^7$ km^3 (>10 Mkm^3)
Major (=LIP)	10^6–10^7 km^3 (1–10 Mkm^3)
Substantial (=LIP)	10^5–10^6 km^3 (0.1–1 Mkm^3)
Moderate	10^3–10^5 km^3 (0.001–0.1 Mkm^3)
Small	$\leq 10^3$ km^3 (≤ 0.001 Mkm^3)

The extrusive and intrusive components and underplate of a LIP are fundamentally interrelated, as exemplified by the feeding of flood basalts by giant continental dyke swarms. For these reasons, I follow Bryan and Ernst (2008) in rejecting the approach of Sheth (2007) in subdividing LIPs into primarily volcanic or plutonic types.

Preserved thicknesses of extrusive rocks for many Mesozoic–Cenozoic LIPs range from *c*. 500 m to > 3 km (e.g. Bryan *et al.*, 2002; Jerram and Widdowson, 2005; Bryan, 2007), and the thickness of any individual section is typically ≥ 1 km. Many LIPs thus have eruptive and/or subvolcanic intrusive volumes well in excess of 1 Mkm^3 (Courtillot and Renne, 2003). LIP-related dyke swarms have average dyke thicknesses of 10–30 m, can have extents greater than 1000 km, and have an overall radiating pattern. LIP-related sills are tens to hundreds of meters in thickness, and associated layered intrusions have volumes up to 40 000 km^3.

I follow both the original definition of Coffin and Eldholm (1994) and the revised definition of Bryan and Ernst (2008) in concluding that LIPs should have a minimum extrusive/subvolcanic intrusive volume exceeding 0.1 Mkm^3. Also, for the purpose of comparing LIP-volume estimates, it is important to note whether these estimates include extrusive and subvolcanic (upper crustal) intrusive volumes and/or middle and lower crustal components revealed by geophysical methods (see also Courtillot and Renne, 2003; Bleeker and Ernst, 2006).

2.2.2 Area

Given the complexities of determining volume for LIPs, a simpler measure is areal extent and for most events there is probably a broad proportionality between LIP volume and original areal extent although oceanic plateaus appear to have a distinctly higher volume to surface area than for continental LIPs (Fig. 2.1). A reasonable reconnaissance approach is to draw a line around all units of a LIP and calculate the enclosed areal extent and convert to a volume using an assumed "thickness" of 1 km. For instance, the Ontong Java Plateau (Section 4.2.1)

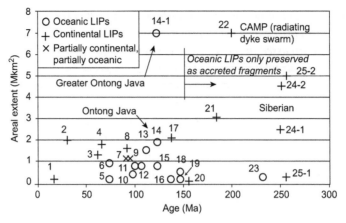

Figure 2.1 Comparison of age and estimated original size of LIPs since 260 Ma. (1) Columbia River, (2) Afar (Ethiopian and Yemen Traps), (3) North Atlantic Igneous Province (NAIP), (4) Deccan, (5) Maud Rise, (6) Sierra Leone Rise, (7) CCIP (Colombia-Caribbean Igneous Province), (8) Madagascar, (9) HALIP (High Arctic Large Igneous Province; includes Alpha Ridge), (10) Wallaby, (11) Hess Rise, (12) Hikurangi, (13) Kerguelen, (14) Ontong Java, (14-1) Greater Ontong Java, (15) Manihiki, (16) Gascoyne Margin, (17) Paraná–Etendeka, (18) Magellan Rise, (19) Shatsky, (20) Argo basin, (21) Karoo-Ferrar, (22) CAMP (Central Atlantic Magmatic Province), (23) Wrangellia, (24–1, 24-2) Siberian, and (25-1, 25-2) Emeishan. From Ernst *et al.* (2007a). With kind permission from Springer Science & Business Media.

encompasses *c.* 2 Mkm^2, which is approximately one-third of the conterminous United States or equivalent to the area of western Europe. The smallest recognized LIP, the Columbia River flood-basalt province was originally estimated to cover *c.* 0.16 Mkm^2 (Section 3.2; Coffin and Eldholm, 1994;) but has been subsequently revised upward to > 0.2 Mkm^2 (Camp *et al.*, 2003). Studies on the areal extents of the exposed plumbing systems and intrusive provinces of LIPs are also consistent with this *c.* 0.1 Mkm^2 minimum estimate (e.g. Yale and Carpenter, 1998; Marzoli *et al.*, 1999; Ernst *et al.*, 2005). For example, major regional continental dyke swarms (so-called giant dyke swarms) are > 300 km in length (Ernst and Buchan, 1997) and typically have areal extents that also exceed 0.1 Mkm^2. The largest, the 1270 Ma Mackenzie swarm of northern Canada, exceeds 2.7 Mkm^3 (Fahrig, 1987). In practice, many LIPs are much larger than 0.1 Mkm^2. A review of several classic Mesozoic–Cenozoic LIPs reveals areal extents of *c.* 1 Mkm^2 (see summary in Courtillot and Renne, 2003), and many oceanic plateaus also cover areas > 1 Mkm^2 (Kerr, 2005, 2014), and Silicic LIPs (SLIPs) also have dimensions well in excess of 0.1 Mkm^2 (Bryan *et al.*, 2002; Bryan and Ernst, 2008). I conclude that while many LIPs exceed 1 Mkm^2, it remains useful to keep the cutoff at the original value of 0.1 Mkm^2 to avoid losing some long-recognized LIPs, such as the 17 Ma Columbia River LIP of North America.

2.2.3 Duration of magmatism

A defining characteristic of LIPs is that large volumes of magma are emplaced over a geologically short and finite period and in a focused area. This allows us to distinguish LIPs from normal plate-boundary processes that can generate LIP-scale magmas (i.e. mid-ocean ridges and subduction zones) if considered over a long duration and areal extent. For example, as pointed out by Sheth (2007), the 50 000-km-long worldwide network of mid-ocean ridges, with an average half-spreading rate of 5 cm per year, creates 5 Mkm^2 of oceanic lithosphere of c. 7 km thickness in just 1 million years, although it should be noted that magma production rates for LIPs have been estimated at c. 10 to > 100% greater than mid-ocean-ridge emplacement rates.

Determining the timing and duration of LIP events is strongly dependent on the quality of the age data, sample quality and degree of weathering/alteration, the resolution of the dating techniques, and data availability, and the U–Pb method has proved to be particulary useful (e.g. Heaman and LeCheminant, 1993; Courtillot and Renne, 2003; Söderlund *et al.*, 2013). Not all LIPs have been studied to the same level of detail, and age data in terms of reliability, technique, and quantity are extremely variable. LIPs appear to have maximum possible duration of up to c. 50 Ma (Ernst and Buchan, 2001a; Fig. 2.2). However, many LIPs span less than 10–15 Ma and in many cases are emplaced in only a few Ma or less (Hofmann *et al.*, 2000; Courtillot and Renne, 2003; Jerram and Widdowson, 2005; Blackburn *et al.*, 2013). This applies both to the flood-basalt component (Chapter 3; e.g. Hofmann *et al.*, 2000) and the dyke-swarm components (Section 5.2; e.g. Heaman and LeCheminant, 1989, 1991). Layered intrusions such as the 9-km-thick Paleoproterozoic Bushveld Complex (part of the Bushveld LIP; Section 5.4) were also emplaced rapidly; Cawthorn and Walraven (1998) (see also Cawthorn and Webb, 2013) estimated emplacement of the Bushveld Complex in only c. 75 Ka. Oceanic plateaus can also have been rapidly emplaced in only 3 Ma (Section 4.2) but dating is really only available for the top few hundred meters of these plateaus sampled at Ocean Drilling Program (ODP) drill sites. For some such as the Kerguelen the ODP results indicate a minimum 25 Ma span of volcanic activity, but this is likely to represent a number of magmatic pulses and hiatuses, rather than continuous magmatism (Tejada *et al.*, 2002).

2.2.4 Pulsed nature of magmatism

Systematic dating combined with detailed stratigraphic and volcanological studies have revealed a complex igneous history for LIPs. Geochronologic studies have shown that LIPs with > 20 Ma age span were emplaced in multiple shorter duration pulses of c. 1–5 Ma rather than as a continuous longer-lasting magmatic event (e.g. Tolan *et al.*, 1989; Saunders *et al.*, 1997; Courtillot and Renne, 2003; Jerram and Widdowson, 2005; Storey *et al.*, 2007a; Ernst *et al.*, 2008). In comparison, evidence

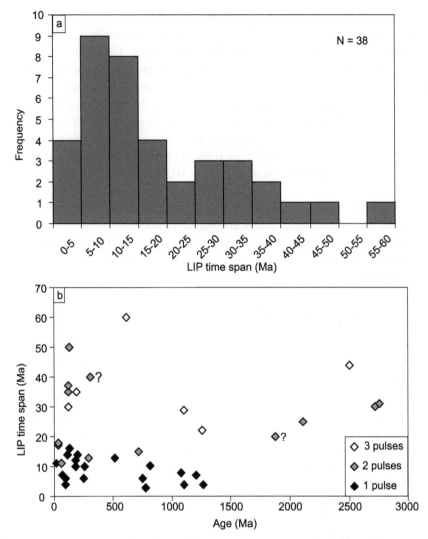

Figure 2.2 (a) Histogram showing the duration of LIP events (including both continental and oceanic). Many have an overall time span of < 15 Ma; (b) Plot illustrating three key variables of LIP events: event age, event duration, and number of pulses. Only those LIP events that have sufficient precise age data (i.e. U–Pb age dates for the Paleozoic and Precambrian LIP examples) are plotted. From Figure 2 in Bryan and Ernst (2008). With permission from Elsevier.

for a single short pulse and rapid emplacement of a large volume of magma is well expressed in the Columbia River flood-basalt province (Section 3.2), the youngest and smallest LIP where, although basaltic eruptions occurred between 17 and 6 Ma, > 90% of the total volume (*c.* 0.234 Mkm³) was erupted between 16.6 and 15.3 Ma (Tolan *et al.*, 1989; Camp *et al.*, 2003; Hooper *et al.*, 2007; Reidel *et al.*, 2013).

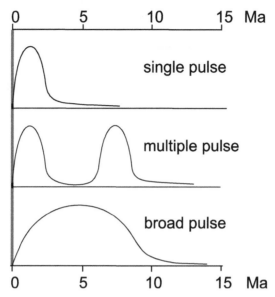

Figure 2.3. Schematic models of size vs. duration for different types of LIP events: a single event of short duration followed by a minor activity over a longer time; multiple pulses of LIPs: and a LIP with broad age range. Modified from Prokoph *et al.* (2004a).

Two distinct pulses in some continental LIPs (Fig. 2.3) may correspond to pre- and syn-rift magmatic events. This is exemplified by the North Atlantic Igneous Province (NAIP), where the first pulse of 62 to 58 Ma magmatism corresponded to the emplacement of terrestrial continental flood-basalt sequences, whereas the bulk of the volcanic sequences along the continental shelves that form the so-called "seaward-dipping reflector series" (SDRS) were emplaced during a second syn-rift pulse at 56–52 Ma (e.g. Saunders *et al.*, 1997; Jerram and Widdowson, 2005; Storey *et al.*, 2007a). Zones of high-velocity lower crust are also thought to be produced during continental-breakup-associated rifting (Menzies *et al.*, 2002a). The hiatus between distinct magmatic pulses is variable, but can be a few to tens of millions of years. The relative extrusive volumes of pulses can vary, and the volume of the second pulse may exceed the first (Campbell, 1998; Courtillot *et al.*, 1999; Storey *et al.*, 2007a). Nevertheless, the igneous volumes emplaced during such pulses represent a substantial proportion (> 75%) of the total volume of a LIP (Bryan and Ernst, 2008).

In detail, comparative studies have indicated that several continental flood-basalt provinces formed during at least three eruptive phases that vary in duration from 0.1 to > 5 Ma: (1) an initial phase of relatively low-volume transitional-alkaline basaltic eruptions; (2) the main phase of flood volcanism where the bulk of the volcanic stratigraphy is emplaced rapidly during repeated large-volume eruptions of tholeiitic basalt magmas (in some cases associated with silicic magma); and (3) a waning and more protracted phase of volcanism where the volume of eruptions

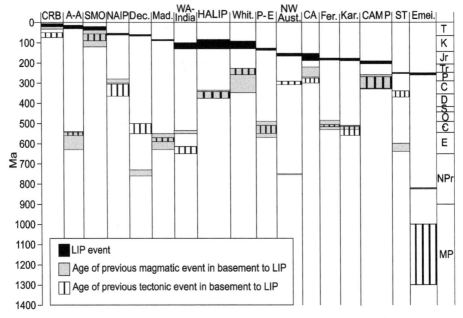

Figure 2.4 Comparing LIP and SLIP ages with the ages of the prior magmatic and tectonic events in the underlying basement. Abbreviations: CAMP = Central Atlantic Magmatic Province, CRB = Columbia River Basalt, DEC = Deccan, Emei = Emeishan, HALIP = High Arctic Large Igneous Province, CA = Chon Aike (SLIP), KAR = Karoo, FER = Ferrar, Mad. = Madagascar, NAIP = North Atlantic Igneous Province, NW Aust. = northwest Australia, P-E = Paraná-Etendeka, SMO = Sierra Madre Occidental, ST = Siberian Trap, WA = western Australia, Whit. = Whitsunday. From Figure 3 in Bryan and Ernst (2008). With permission from Elsevier.

rapidly decreases and may become more widely distributed, or focused when rifting is occurring (Bryan *et al.*, 2002; Jerram and Widdowson, 2005).

Systematic dating combined with stratigraphic studies is lacking in the majority of SLIPs (see Chapter 8), meaning that it is unknown whether these events have well-defined, pulsed magmatic characters. However, the 188–153 Ma Chon Aike province of South America–Antarctica (Section 8.3.3), has been divided on the basis of U–Pb zircon and ^{40}Ar–^{39}Ar dating into three main pulses of silicic volcanic activity each of 5–10 Ma duration, associated with the emplacement of *c*. 0.05 to > 0.1 Mkm^3 of rhyolite magma (Pankhurst *et al.*, 1998, 2000; Fig. 3.7). For the Whitsunday SLIP (Section 8.3.2), although age data indicate a main period of activity from *c*. 120–105 Ma, pulses in volcanism occurred at *c*. 118–113 Ma and *c*. 110–105 Ma (Ewart *et al.*, 1992; Fig. 3 of Bryan *et al.*, 1997). In addition, recent work has shown that the bulk of the rhyolite ignimbrites of the Sierra Madre Occcidental (SMO) SLIP of Mexico was emplaced in two pulses each of *c*. 4 Ma duration (*c*. 38 and 20 Ma; Section 8.3.1). The *c*. 1–3 Ma age ranges for many exposed ignimbrite sections > 1 km thick also emphasize a rapid emplacement (Ferrari *et al.*, 2002, 2007; Swanson *et al.*, 2006).

In summary, LIPs are finite, dominantly mafic igneous events of no longer than *c.* 50 Ma duration (Fig. 2.4), and are characterized by a magmatic pulse or pulses of ≤ 5 Ma duration in which a large proportion (> 75%) of the total igneous volume of the LIP was emplaced. Associated SLIPs have similar characteristics (Chapter 8).

2.2.5 *Intraplate tectonic setting*

An integral part of the original definition of Coffin and Eldholm (1992, 1993a, 1994) was that LIPs formed by processes not characteristic of modern plate boundaries (i.e. mid-ocean ridges and subduction zones). This independence was also emphasized by Halls *et al.* (2008) who considered LIPs to be "the products of large-scale transient magmatic processes rooted in the Earth's mantle that are not predicted by plate tectonic theory" (Table 1.1). LIP events, therefore, show the hallmarks of intraplate magmatism, albeit being extremely voluminous (and areally extensive). The following discussion is based on that in Bryan and Ernst (2008).

The term "intraplate" has been widely used in both a tectonic and petrological or geochemical sense. Intraplate settings, synonymous with plate interiors, and with areas of incipient rifting, are characterized by tectonic stability. In terms of magmatism, the term intraplate generally implies that magma generation is so remote from existing plate boundaries that it cannot be related to energy releases and tectonic processes at mid-ocean ridges, subduction zones, and, less commonly, "leaky" transform boundaries (Johnson and Taylor, 1989; Neuendorf *et al.*, 2005).

Most LIP events have an intraplate tectonic setting. Mesozoic–Cenozoic plate reconstructions demonstrate that many continental LIPs were emplaced initially into the interiors of tectonic plates and on or along the current edges of Archean cratons in areas of incipient rifting. This is also true for Proterozoic LIPs whose dyke swarms radiate into plate interiors from areas that became continental margins after associated rifting and breakup (e.g. Chapter 11). Furthermore, these dyke swarms are typically linked to associated breakup events that resulted in the formation of new continental margins (e.g. Courtillot *et al.*, 1999; Ernst and Bleeker, 2010; Chapter 11), but at the time of initial LIP emplacement (the initial pulse) the setting was intraplate.

As noted above, LIPs can have a second pulse of magmatism that can be linked to the breakup and onset of ocean opening (e.g. NAIP: Section 3.2.4 and Kerguelen Bunbury–Comei LIP; Section 4.2.2). Nevertheless, it is important to recognize that, in such cases, the initial pulse of the LIP occurred in an intraplate setting that existed prior to rifting and ocean opening. Also, not all LIPs are accompanied by complete rifting and the formation of ocean basins. Some may be associated with failed rifting; this is exemplified by the Emeishan, Siberian Trap, and Columbia River flood-basalt provinces, which provide the most prominent recent examples, with the Keweenawan LIP providing an older example of this type of setting (see discussion in Section 11.6).

In addition to being emplaced remotely from then-active plate boundaries, most continental LIPs were emplaced into stable continental regions with a long history (commonly hundreds of millions of years) of no prior magmatism or contractional deformation (Fig. 2.4). Consequently, this abrupt voluminous emplacement of magmas in stable continental regions is what makes LIP events so distinctive and anomalous, and underlines the intraplate character of these events.

Oceanic plateaus and ocean-basin flood basalts also appear to have been emplaced in intraoceanic (intraplate) settings, with many of these LIPs remaining undeformed except at their edges where they have subsequently entered subduction zones (Mann and Taira, 2004; Petterson, 2004). The Shatsky Rise appears anomalous in that it was sited along an active ridge triple junction and emplaced onto a region of oceanic crust affected by multiple spreading ridge jumps (Nakanishi *et al.*, 1999; Sager, 2005) but, importantly, it is now mid-plate following ridge migrations.

The same relationship between breakup and timing of magmatism observed in multi-pulsed continental LIPs is also present in many oceanic LIPs. While the initial pulse may be intraplate, subsequent pulses of an oceanic LIP can be associated with new spreading centers, such as the reconstructed Ontong Java–Manihiki–Hikurangi oceanic plateaus (Section 4.4.2; Taylor, 2006); Kerguelen–Broken Ridge plateaus (Section 4.2.2; Frey *et al.*, 2003); Agulhas–Maud Rise plateaus (Section 4.2.2; Jokat *et al.*, 2004; Gohl *et al.*, 2011). Where the initiation stage of such LIPs predates the development of a new mid-ocean ridge system, or where it post-dates the cessation of seafloor spreading (e.g. ocean-basin flood basalts), these LIP events can therefore be classed as "intraplate" in a tectonic sense.

Intraplate is less applicable in a tectonic sense to characterize those LIP events that initiate close to active plate boundaries; as exemplified by the Columbia River and Shatsky Rise LIPs. An intraplate setting may also be less evident for LIP events occurring in tectonically active regions such as mobile belts between cratons or along continental margins experiencing ocean closure ahead of continent–continent collision (e.g. *c.* 1880 Ma Circum-Superior LIP; Heaman *et al.*, 2009; Ernst and Bell, 2010; Minifie *et al.*, 2013).

An intraplate tectonic setting is particularly problematic for the Cenozoic LIPs of North America that were emplaced along a continental margin with a history of subduction-related magmatism and deformation immediately prior to and after the LIP events. Specifically, the Columbia River flood-basalt province (Section 3.3.2) and Sierra Madre Occidental SLIP (Section 8.3.1) geographically overlap with prior subduction-related magmatism. Given this regional setting, it has been argued that the Columbia River LIP may result from back-arc spreading (e.g. Carlson and Hart, 1988; Smith, 1992).

However, LIP events at plate-margin settings are distinctive from spatially associated and synchronous plate-margin-related magmatism in terms of their extent (both in terms of total area and extent inboard from the plate margin), volume, rapidity of eruption and melt production rates, association with extension, and composition. For the Columbia River LIP, the largest volume and earliest emplaced flood-basalt formations have geochemical similarities with oceanic island tholeiites and other intraplate magma compositions and, thus, contrast strongly with neighboring subduction-related magmatism (e.g. Hooper, 1997; Hooper *et al.*, 2007). In these cases, the intraplate characteristics of the LIP event are based more on petrological and geochemical grounds (and other features such as melt production rates, if well constrained), where the LIP has compositions distinctly different from those formed at mid-ocean ridges and subduction zones. However, for continental LIPs, the frequent addition of lithospheric geochemical signatures to flood basalts and rhyolites makes recognition of an intraplate signature difficult (Chapter 10). This is particularly true for SLIPs such as the Sierra Madre Occidental, where both rhyolites and basalts have geochemical signatures that are transitional between within-plate and convergent margin fields on trace-element discrimination diagrams (Bryan, 2007).

Finally, even LIPs that are spatially associated with plate boundaries can have an "intraplate setting" in the following sense. If it is accepted that LIPs represent magmatism derived from an ascending deep plume (see Chapter 15) then it is reasonable that the ascending plume will have no preference for a particular upper-mantle tectonic setting (see Section 14.3 in Arndt *et al.*, 2008). Thus, some percentage of deep-sourced mantle plumes will ascend into an existing upper-mantle plate-boundary setting despite having no tectonic link with such a setting (see Figure 13.5). Another plausible scenario is that an ascending plume will rise beneath a craton with a thick lithospheric root and will slide sideways and upward along the root of the craton towards lithospheric thinspots in the interior or along an edge (or edges) of the craton (e.g Thompson and Gibson, 1991; Sleep, 2003; Begg *et al.*, 2010; Bright *et al.*, 2014), a scenario that can be potentially applied to the Circum-Superior LIP (e.g. Ernst and Bell, 2010).

Important indications of an intraplate setting, then, are the petrological, geochemical, and isotopic characteristics (discussed in detail in Chapter 10), which can be used to discriminate between LIP-related within-plate tholeiites and plate-margin magmatism (mid-ocean ridge basalt (MORB) and subduction-related basalts). The volume and rate of magma generation in LIP events are also distinctive, and, combined with the geochemical characteristics of LIPs have led many workers to relate these features to hot-mantle upwellings with magmas sourced from the asthenospheric mantle or plume, subcontinental lithospheric mantle, and the depleted asthenospheric mantle (Carlson, 1991; Turner and Hawkesworth, 1995; Hofmann, 1997; Condie, 2003; Ewart *et al.*, 2004a). Crustal contamination and source heterogeneity are major causes of intra- and inter-LIP variation that in some cases can lead to apparent subduction-related signatures in flood basalts and rhyolites (Section 10.5.4). Nevertheless, the least contaminated basalts and picrites in many continental LIPs have isotope and geochemical compositions similar to oceanic island basalts (i.e. the high-Ti basaltic magma suites) that reinforce the intraplate geochemical characteristics of LIP magmatism.

To summarize, the intraplate criterion in the LIP definition after Bryan and Ernst (2008) includes any of the following:

(1) magmatism remote from any concurrently active plate boundaries;
(2) magmatism occurring in stable crustal regions with long histories of no prior magmatism or contractional deformation;
(3) magmatism occurring in plate interiors undergoing extension;

(4) magmatism initiated in an intraplate setting (as defined in points 1 and 2), but where subsequent pulses of the LIP may occur in association with newly forming ocean-ridge spreading systems;

(5) compositional characteristics ("intraplate" or "within-plate") that are distinct from plate-boundary magmatism;

(6) magmatism linked to a mantle plume originating from the deep mantle and which randomly ascends into upper-mantle tectonic settings.

2.3 Associated magmatism (silicic, carbonatitic, and kimberlitic)

As noted in Chapter 8 silicic magmatism is associated with most LIPs, and provinces of dominantly silicic magmatism of corresponding scale and duration and intraplate setting to (mafic–ultramafic) LIPs have also been identified; the latter are inferred to be generated by LIP underplating and voluminous melting of lower crustal material. Such silicic magmatism is termed a Silicic LIP (SLIP) and is discussed in more detail in Chapter 8. Two additional types of magmatism are also typically associated with LIPs, carbonatites and kimberlites, both of which are discussed in detail in Chapter 9.

2.4 Types of non-LIP magmatism

As summarized in Chapter 1, LIPs can include continental flood basalts, volcanic rifted margins, oceanic plateaus, ocean-basin flood basalts, giant dyke swarms, sill provinces, mafic–ultramafic-intrusions, and tholeiite–komatiite greenstone belts generally of Archean age. What is not included as LIPs is standard plate-boundary magmatism (cf. Sheth, 2007), i.e. magmatism related to mid-ocean-ridge basalts, orogenic-related batholiths, and transform faults. These types of magmatism fail on the intraplate criterion, as well as on the volume criterion, in the case of the transform-fault-related magmatism. Less clear a priori is the status of back-arc-related magmatism which is extension-related but also plate-boundary-related (see further discussion in Section 15.2.5).

A number of the intraplate igneous province types originally linked with LIPs by Coffin and Eldholm (1994) do not fit the revised LIP definition, i.e. submarine ridges, seamount groups, and anomalous seafloor-spreading crust (Table 2.2). These provinces are typically spatially associated, providing a spatial–temporal connection between a LIP and an active hotspot (e.g. Paraná–Etendeka flood-basalt province–Walvis Ridge-Tristan da Cunha (Fig. 2.5)). Submarine ridges and seamounts are commonly included in LIP studies because they can provide: (1) evidence of the geochemical characteristics of the underlying asthenospheric hotspot or plume; and (2) a reference point on geochemical–isotopic variation diagrams for LIP compositions (Saunders *et al.*, 1997). However, these submarine volcanic structures are post-LIP features that formed in some cases tens to hundreds of million years after the main LIP pulse, and in some cases over much longer intervals (tens to > 100 Ma).

Table 2.2 *Types of non-LIP magmatism*

Unit type	Volume	Duration	Pulse	Intraplate	Link with LIP
Submarine ridges	Yes	?	No		
Seamount groups	No	No	No	Yes	**Plume tail**
Mid-ocean-ridge basalts	Yes	No	No	Yes	No
Orogenic batholiths	Yes	?	No	No	No
Large area, small volume magmatic provinces (e.g. Eastern Australia; Fig. 2.5)	No	No?	?	Yes	No

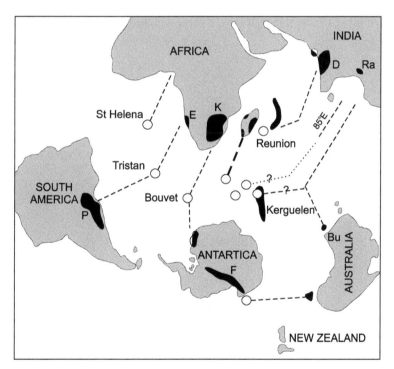

Figure 2.5 Link between LIPs, hotspot and seamount chains on background of distribution of continents. From Figure 4 in Storey (1995) based on original diagram from Duncan and Richards (1991). Digital version of original figure kindly provided by B. Storey. Generalized LIPs are in solid black. P= Paraná, E = Etendeka, K = Karoo, F = Ferrar, D = Deccan, Ra = Rajmahal and Bu = Bunbury (both part of Kerguelen–Bunbury–Comei LIP). Open circles are present day hotspots and dashed lines represent seamount chain connections between LIPs and present day hotspots through subsequent ocean opening. Reprinted by permission from Macmillan Publishers Ltd: Nature.

These submarine ridges and seamounts thus represent the "trails" and "tails" of plumes.

Seamounts and seamount groups are localized topographic features on the seafloor, and have a very different volcanic expression to other LIPs. Seamounts are centralized constructive (largely submarine) volcanoes, in contrast to the very extensive plateaus of the continental flood basalt, oceanic plateau, and ocean-basin flood-basalt provinces that are built up by repeated eruption and emplacement of large-volume tabular and extensive basaltic lavas and sills (Chapter 3). Although seamount groups may be areally extensive, they have cumulative erupted volumes that are significantly less than the LIP criteria outlined above, and/or may have been emplaced over a much longer duration (> 50 Ma). For example, the Canary Islands are the result of persistent volcanism over the last 80 Ma, which has only produced a cumulative volume of *c*. 0.12 Mkm^3 (Schmincke, 1982; Coello *et al.*, 1992; Geldmacher *et al.*, 2001), a value that is significantly less than typical LIP emplacement rates. So these are not LIPs as they fail on the large-volume, short-duration of emplacement criteria.

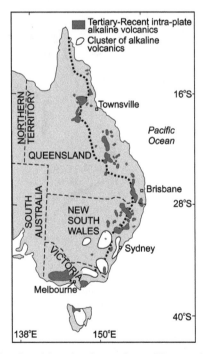

Figure 2.6 Tertiary intraplate basaltic volcanic province of Eastern Australia. This is an example of an igneous province which meets the area criteria, but fails the volume criteria for a LIP distribution of alkaline volcanic fields in eastern Australia. From Pirajno (2007), which was based on Sutherland (1998).

In addition, there are widespread but low-volume mafic magmatic events that are definitely intraplate, but are not LIPs. For example, the Tertiary intraplate basalt volcanic province of eastern Australia forms a belt of > 4400 km along the eastern highlands adjacent to the rifted margin (Johnson, 1989) (Fig. 2.6). Although the province has an areal extent of > 0.1 Mkm^2, is dominantly basaltic with intraplate compositions, and is characterized by long lava flows typical of those in continental flood-basalt provinces (Stephenson *et al.*, 1998), the province has an extrusive volume of only *c.* 20 000 km^3 that was emplaced over the last 80 Ma, indicating that the eruptive volume, melt production rates, and duration of this event contrast markedly with LIP magmatism.

2.5 LIP fragments/remnants

LIPs are characteristically associated with continental breakup or attempted continental breakup (Chapter 11), meaning that it is unsurprising that pre-Mesozoic LIPs can be intensely fragmented by continental breakup such that portions of a single LIP end up as remnants on different crustal blocks. A LIP can also be "fragmented" by erosion, which can remove semi-continuous volcanic flood-basalt to expose the elements of the magmatic plumbing system (dykes, sills, and layered intrusions) as widely scattered fragments/remnants (Ernst, 2007a; Bryan and Ernst, 2008).

In many cases, units that are dated and grouped to define a distinct magmatic event will be of insufficient volume to qualify as a LIP, even if they actually represent a fragment/remnant of a LIP. Therefore, it is important to ask whether there are characteristics of individual magmatic units that can identify them as LIP "fragments."

Many attempts have been made to identify plume-related LIPs using geochemistry in the context of identifying intraplate composition (e.g. Ernst and Buchan, 2003). But, as pointed out by Hawkesworth and Schersten (2007), magmatism on a range of scales (including sub-LIP scale) may also exhibit intraplate chemistry. Abbott and Isley (2002) have proposed that "superplume"-related layered intrusions (such as the Bushveld intrusion) should have high platinum-group-element (PGE) and Cr contents. While the PGE content of LIP mafic units can be relatively high compared to other settings (e.g. Crockett, 2002) the controls on PGE content and links to PGE levels in mantle source areas can be complicated (e.g. Rehkämper *et al.*, 1997). Coffin and Eldholm (2001) further speculated that high Cr contents in ophiolite complexes may be explained by "ophiolitic chromite originating within oceanic LIPs." On the other hand, Rollinson and Adetunji (2013) have indicated that podiform chromitites (in many ophiolites) are not spreading ridge derived, but rather must be subduction related. More research is required to test any such geochemical discriminants (e.g. high PGEs, high Cr) between LIP and non-LIP intraplate magmatism, and this topic is discussed further in Chapter 15.

A key "field" criterion for LIP-related dykes is average dyke width. Extensive research on LIP-related Canadian shield dykes suggests that swarms with average

Table 2.3 *LIP proxies*

Unit	Proxy criteria	Reference	Comment
Some ophiolites	Have characteristics of oceanic pleateau	Coffin and Eldholm, 2001; Dilek and Ernst, 2008	
Carbonatites	Coeval with nearby LIP	Ernst and Bell, 2010 (Chapter 9)	The carbonatite–LIP link is very robust
Some kimberlites	Coeval with nearby LIP	Chalapathi Rao *et al.*, 2011 (Chapter 9)	Some are associated with LIPs while others have an age progression related to a plume track (e.g. Heaman and Kjarsgaard, 2000), and for others the relationship is not clear (e.g. Schissel and Smail, 2001)
Basalt pile	> 500 m thickness of flat-lying basalts, and the {near} absence of interflow sediments		Additional criteria are needed to distinguish a lava pile that is an erosional remnant of a LIP from that only associated with a local rift
Dolerite dykes	> 10 m average width	Ernst *et al.*, 1995	
Dolerite dykes	> 70 m maximum width	Abbott and Isley, 2002	
Sill	Probably based on average thickness		Criteria need to be developed
Layered intrusions	High PGE and Cr content; absence of primary amphibole	Abbott and Isley, 2002	
Silicic magmatism	A-type chemistry	Bryan and Ernst, 2008	Those derived from melting of lower crustal rocks will have composition depending on the protolith melted (Chapter 8)
Massif anorthosites; AMCG (anorthosite–mangerite–charnockite–granite) suites	?		Uncertain link between anorthosites, AMCGs and LIPs (Chapter 9)

dyke widths of >10 m are not formed in a subduction or spreading-ridge setting, nor in association with an individual volcanic center, and are therefore candidates for LIP membership (Ernst *et al.*, 1995). Abbott and Isley (2002) suggest that swarms belonging to a "superplume" event (a LIP with an areal extent of at least 0.41 Mkm^2) should have a maximum dyke width of > 70 m. Similar physical criteria also need to be developed to determine whether an individual sill is LIP-related and what characteristics of a lava pile would be indicative of membership in a broader magmatic province (of LIP scale). Table 2.3 shows a number of additional provisional criteria for identifying LIP fragments/remnants.

2.6 Summary

This chapter identifies and discusses the essential components of mafic–ultramafic LIP magmatism: their large overall volumes (and the useful proxy measure of large areal extent), their overall short duration of emplacement (or dominance of short-duration magmatic pulses), and their intraplate setting. These same criteria are also applied to the separate class of associated SLIPs (Chapter 8). The genetic link between LIPs and carbonatites and LIPs and some kimberlites has also been introduced here, but is detailed in Chapter 9.

3

Continental flood basalts and volcanic rifted margins

3.1 Introduction

Most research on Large Igneous Provinces (LIPs) has focused on the dramatic flood basalts that characterize Mesozoic–Cenozoic events, both continental flood basalts (CFBs) and, to a lesser extent, flood basalts within existing ocean basins (oceanic plateaus and ocean-basin flood basalts). This record is generally well preserved and has been critical to the development of many key concepts for LIPs, most importantly their impressive size and overall short duration (or a short duration for each pulse of a multipulse event). This chapter focuses on the continental flood basalts and volcanic rifted margins and Chapter 4 focuses on the oceanic flood basalts including both oceanic plateaus and ocean-basin flood basalts.

3.2 Continental flood basalts (CFBs)

3.2.1 General characteristics

Continental-flood-basalt magmatism and the associated development of volcanic rifted margins form monotonous packages of large tabular flow units that are up to several kilometers thick; these units commonly lack any significant interlayed sediments (Macdougall, 1988; Coffin and Eldholm, 1994; Mahoney and Coffin, 1997; Jerram and Widdowson, 2005; White et al., 2009). High-precision dating confirms that these sequences are typically emplaced over 1–5 Ma or less (e.g. Courtillot and Renne, 2003; Bryan and Ernst, 2008; Bryan and Ferrari, 2013). CFBs are dominantly mafic (and tholeiitic) with minor ultramafic flows (picritic) and transitional-alkaline flows typically lower in the sequence, and associated silicic volcanism can become more significant higher in the sequence (see Chapter 8).

Classic examples of CFBs are the Deccan (Traps), Columbia River, Siberian Trap, Karoo, North Atlantic Igneous Province (NAIP), and Paraná–Etendeka LIPs (Fig. 3.1). (Note, in this book I use Deccan instead of the Deccan Traps as per recent literature. However, I will continue to refer to Siberian Trap, to distinguish it from other widespread older "Siberian" LIPs that are now being recognized

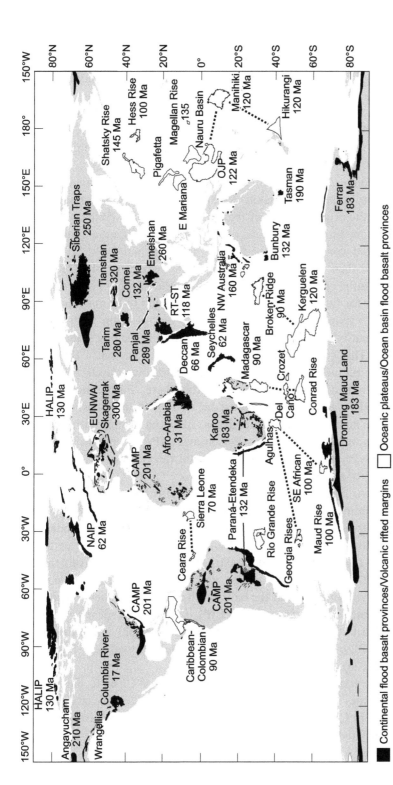

Figure 3.1 Global distribution of LIPs following assembly of Pangea *c.* 320 Ma. Annotated ages denote the onset of the main phase or first pulse of magmatism to the LIP event; note that some LIPs may have precursor magmatism at lower intensity up to 10 Ma prior. Dotted tie lines connect oceanic LIPs subsequently rifted apart by seafloor spreading. The inferred extent of some of the oldest LIP events is shown by a dashed line, as many remain poorly mapped and studied. Abbreviations: CAMP = Central Atlantic Magmatic Province; EUNWA = European, northwest Africa, part of Skaggerak–Jutland (Jutland) event; HALIP = High Arctic Large Igneous Province; NAIP = North Atlantic Igneous Province; OJP = Ontong Java plateau; RT-ST = Rajmahal Traps–Sylhet Traps (part of the Kerguelen–Bunbury–Comei LIP). Associated SLIPs are shown in Figure 8.1. Modified from Figure 1 in Bryan and Ferrari (2013) after the original version in Bryan and Ernst (2008).

57

(e.g. at *c.* 370 Ma and at *c.* 1750 Ma; Table 1.2).) In the 1980s it became clear that plate tectonics allowed reconstruction of LIPs that had been fragmented during continental breakup (e.g. Richards *et al.*, 1989). For example, the Paraná CFB of South America is rejoined with the Etendeka CFB of Namibia, Africa, after closure of the South Atlantic Ocean (Fig. 3.1; cf. Fig. 2.6). Also, the British Tertiary Igneous Province is matched with the coeval magmatism in Greenland after closure of the North Atlantic Ocean, with these two areas of magmatism together defining the NAIP.

CFBs range in size from *c.* 220 500 Mkm^3 for the Columbia River Basalt Group (Camp *et al.*, 2003) up to more than 4 Mkm^3 for the Siberian Trap LIP (Ivanov, 2007), and 5 Mkm^3 for the original extent of the Karoo–Ferrar LIP (White, 1997). All of these are minimum estimates because of the difficulties of estimating the intrusive and underplated component (Chapter 5) of continental LIPs. If these components are included then these sizes increase up to 8.6 Mkm^3 for the Deccan LIP (Eldholm and Coffin, 2000) or 10 Mkm^3 for the Karoo–Ferrar LIP (White, 1997). Estimates for the volume of oceanic plateaus tend to be much higher than for continental LIPs (typically > 1 Mkm^3 and up to tens of Mkm^3; e.g. Ernst and Buchan, 2001), primarily due to the higher degree of partial melting associated with these LIP events. This greater melting follows from the fact that a thermal anomaly (e.g. a mantle plume) can typically ascend to shallower levels beneath oceanic lithosphere than beneath the continental lithosphere. Furthermore, the volume of continental LIPs may be underestimated due to the difficulty of estimating the contributions of the intrusive component (dykes, sills, and layered intrusions) into continental crust, as well as underplating. For oceanic LIPs the volume is much easier to estimate simply by noting the excess in crustal thickness over a normal oceanic crustal thickness of about 7 km.

The pre-volcanic environment for these Mesozoic–Cenozoic CFBs can vary from shallow marine continental (e.g. the NAIP), fluvial continental (e.g. Yemen–Ethiopia parts of the Afro-Arabian LIP), to eolian continental (e.g. the Etendeka portion of the Paraná-Etendeka LIP) settings. Flood volcanism can also generate thick CFB sequences (e.g. 7 km, the Greenland part of the NAIP) or relatively thin sequences (e.g. 1.5 km, Deccan LIP). In addition, volcanism can be represented by predominantly basaltic volcanic rocks at the base and by associated silicic volcanic rocks at the top (e.g. Yemen of the Afro-Arabian LIP), or associated silicic volcanic rocks may be found distributed throughout the volcanic stratigraphy (e.g. Ethiopia of the Afro-Arabian LIP), or silicilic rocks may be essentially absent (e.g. Deccan LIP) (Menzies *et al.*, 2002a).

Here, I provide an overview of some prominent examples of CFBs to illustrate their range in characteristics including composition. Geochemical/isotopic character-istics are also presented in Chapter 10, which is devoted to LIP geochemistry of all types of LIPs and associated silicic, carbonatitic, and kimberlitic magmatism in an integrated treatment (Fig. 1.1).

3.2.2 Columbia River LIP (c. 17 Ma)

General characteristics

The smallest and youngest of the classic CFBs is the Columbia River (e.g. Swanson *et al.*, 1979; Reidel and Hooper, 1989; White and McKenzie, 1989; Ernst and Buchan, 2001 (ch. 12); Camp and Hanan, 2008; Barry *et al.*, 2010; Coble and Mahood, 2012; Camp *et al.*, 2013; Reidel *et al.*, 2013). This event began with eruption of the Steens basalts and associated dykes; this was followed by a shift in activity to the north, with the eruption of the voluminous Imnaha and Grande Ronde basalts from the Chief Joseph dyke swarm, and the eruption of the partly contemporaneous Picture Gorge basalts from the Monument dyke swarm (Fig. 3.2). Activity waned with the emplacement of Wanapum and Saddle Mountain basalts. Significant dykes and scattered volcanism are also associated with the Nevada Rift that extends to the south (e.g. Ernst and Buchan, 2001c ; Glen and Ponce, 2002). The main phase of this LIP lasted for a very short interval (16.6 to 15.0 Ma) and generated a lava volume of *c.* 220 500 km^3, with most of the magmatism occurring within a very short duration (Tolan *et al.*, 1989). This magmatism occurred in a back-arc regional tectonic setting but the anomalously high volume and short duration of magmatism has been linked to a mantle plume and potentially may also have involved plume-triggered delamination (e.g. Camp and Hanan, 2008), although non-plume origins have also been argued for this LIP (see discussions in Chapter 15). The associated dykes have an overall fanning pattern that converges towards a plume-center position near the McDermitt (silicilic) caldera. Activity continued with a linear progression of caldera ages younging to the present along the Snake River Plain, a feature that is thought to mark the path of the plume tail to the present position underlying the Yellowstone caldera.

Age

The emplacement of the whole province occurred between 17 and 6 Ma, with the voluminous Grande Ronde basalt lavas estimated to have erupted between *c.* 16.0 and *c.* 15.0 Ma (e.g. Hooper *et al.*, 2007). Recent Ar–Ar dating suggests that the duration of the main pulse of Grande Ronde basalt emplacement was even shorter (15.99 $\pm$ 0.20 to 15.57 $\pm$ 0.15 Ma) suggesting a total duration of approximately 420 000 years (Barry *et al.*, 2010). Furthermore, a review of all the available ^{40}Ar–^{39}Ar data for the Steens, Imnaha, and Grande Ronde basalts suggests that these petrochemically distinct groups may have, at times, been erupted simultaneously (Barry *et al.*, 2010).

Composition

The Steens, Imnaha, and Picture Gorge basalts are olivine tholeiites, with subordinate alkali olivine basalts exposed in the upper part of the Steens stratigraphy. Such rocks are noted in flood-basalt provinces worldwide (Camp and Hanan,

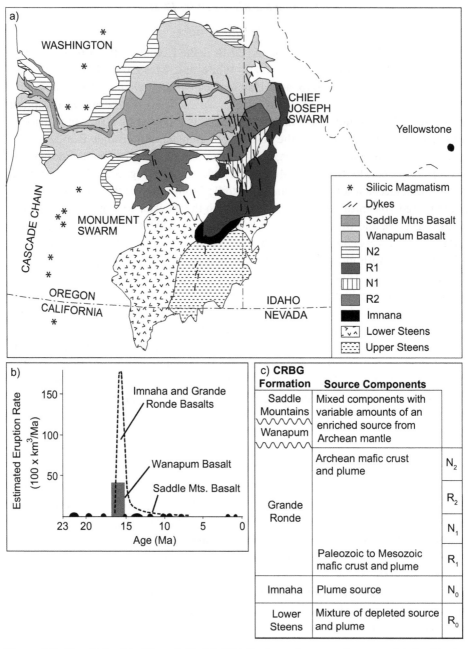

Figure 3.2. The Columbia River LIP. (a) Distribution of magmatism associated with the Columbia River LIP; asterisks locate similar-age volcanoes of the Cascades chain caused by subduction to the west; (b) age distribution; and (c) stratigraphy of the main Columbia River Basalt Group (CRBG). Modified after Camp and Hanan (2008). R1, N1, R2, and N2 refer to magnetic polarity of volcanics. Original of part (b) is after Hooper *et al.* (2002, 2007). Digital version of original figures kindly provided by V. Camp.

2008). In marked contrast, the Grande Ronde basalts and feeder dykes are composed of high-Fe/Mg, high-silica (52–58 wt%) basaltic andesites. These high-silica rocks are unusual, and perhaps unique, when compared to the rock types that dominate the bulk composition of all other flood-basalt provinces (Camp and Hanan, 2008)

3.2.3 Afro-Arabian LIP (Afar LIP) (mainly c. 30 Ma)

General characteristics and age

The Afro-Arabian (Afar) LIP (Fig. 3.3) was emplaced in the southern Arabian Peninsula, and northeastern Africa, and is associated with a broad regional uplift event (e.g. Camp and Roobol, 1992; Sengör, 2001; Avni *et al.*, 2012; Section 12.2.2), triple-junction rifting and the onset of separation of Africa and Arabia and attempted breakup along the East-African rift system (e.g. Mohr, 1983; Mohr and Zanettin, 1988; White and McKenzie, 1989; Coleman, 1993; Pik *et al.*, 1998, 1999; Courtillot *et al.*, 1999; Ukstins *et al.*, 2002; Kieffer *et al.*, 2004; Arndt and Menzies, 2005; Rogers, 2006; Corti *et al.*, 2009). The total volume of extrusive and hypabyssal rocks has been estimated by Mohr (1983) and Mohr and Zanettin (1988) to be about 0.35 Mkm^3. (In the literature both Afro-Arabian and Afar have been used to label this LIP. I will use Afro-Arabian to better encompass the significant portion of this LIP on the Arabian Peninsula.)

The magmatism associated with this LIP event occurred in three main stages, at 30–25 Ma, 24–19 Ma, and less than 5 Ma. The oldest (and main) stage of activity is marked by Yemeni and Ethiopian flood volcanism, generating a single CFB province that was separated by the rift opening of the Gulf of Aden. Although most of the Ethiopian and Yemini volcanic plateaus erupted in a short period between 31 and 30 Ma (Hofmann *et al.*, 1997), the upper parts of the flood volcanic pile have ages to around 25 Ma, suggesting a relatively protracted period of flood volcanism (Ukstins Peate *et al.*, 2002). This style of magmatism was high-flux, fissure-fed, mafic flood volcanism (29–31 Ma) that produced 1–2 km of mafic eruptives, followed by low-flux, bimodal mafic–silicic volcanism from stratovolcanoes with associated caldera-forming (Ultra-Plinian) eruptions (26–29 Ma), producing a sequence of *c.* 1-km-thick mafic–silicic eruptives (Arndt and Menzies, 2005). The Ethiopian flood basalts and associated silicic pyroclastics cover an area of about 600 000 km^2, and the thickness of the sequence produced during this event is highly variable but reaches 2 km in some regions. About 20 volcanic fields (locally called "harrats") in the Arabian Peninsula may also belong to this main magmatic pulse.

A younger 26–19 Ma tholeiitic stage of magmatism formed the major, coast-parallel, Red Sea dykes, with lesser magmatism in the Ethiopian plateau and Yemeni regions, and the development of a number of associated plutons. The youngest 5–0 Ma magmatism occured within the western portion of the Ethiopian sequence

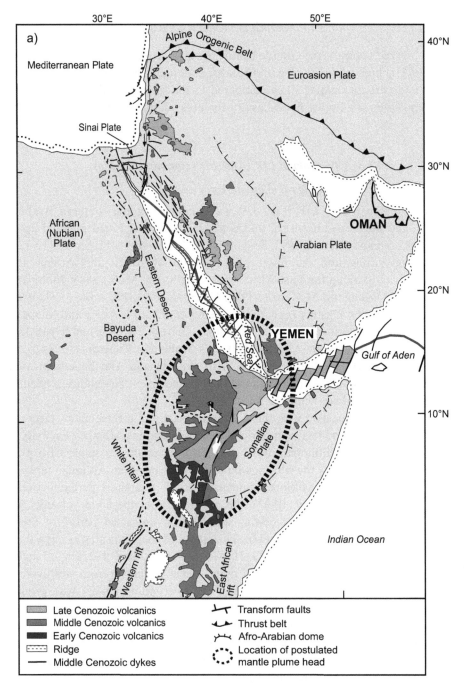

Figure 3.3 The Afro-Arabian (Afar) LIP. (a) Map showing the distribution of volcanic rocks in association with the Afar plume. Note the oval shape of the Afar dome. (b) Compilation of K–Ar (dark rectangles) and Ar–Ar (light diamonds) ages of magmatic rocks from Arabia, Sinai, and Israel, and major magmatic and tectonic events. Abbreviations: CB, cover basalt formation; Gr., Group; CFV, continental flood volcanics; HAS, Harrat Ash Shaam, including Jordan, Syria, and the Golan Heights; RTS, regional truncation surface. Modified after Figure 10 in Avni *et al.* (2012) with additional dykes along the Red Sea added after Ernst and Buchan (1997). Digital version of original figures kindly provided by Y. Avni.

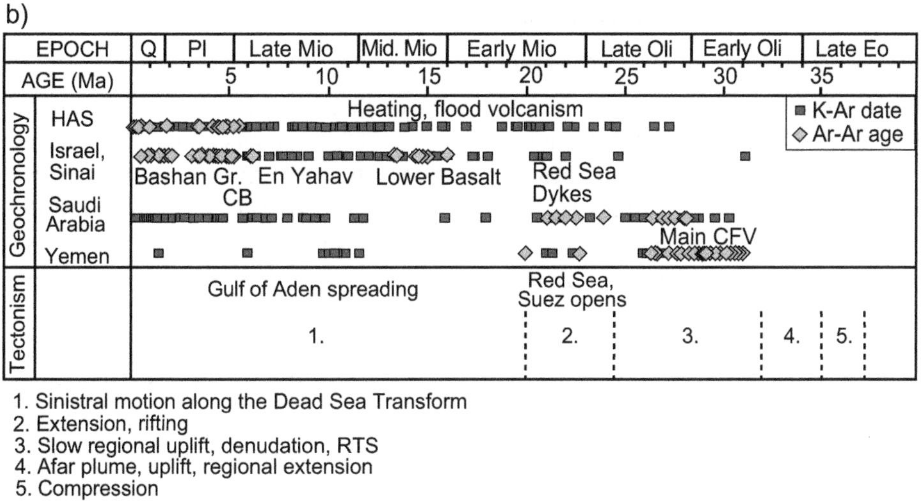

b)

EPOCH	Q	Pl	Late Mio	Mid. Mio	Early Mio	Late Oli	Early Oli	Late Eo
AGE (Ma)		5	10	15	20	25	30	35

Geochronology

HAS — Heating, flood volcanism — ■ K-Ar date / ◇ Ar-Ar age

Israel, Sinai — Bashan Gr. En Yahav Lower Basalt Red Sea Dykes / CB

Saudi Arabia — Main CFV

Yemen

Tectonism

Gulf of Aden spreading | Red Sea, Suez opens

1. | 2. | 3. | 4. | 5.

1. Sinistral motion along the Dead Sea Transform
2. Extension, rifting
3. Slow regional uplift, denudation, RTS
4. Afar plume, uplift, regional extension
5. Compression

Figure 3.3 (*cont.*)

and in the Yemen sequence, generating alkali basalts. Syn-rift volcanic activity continues to the present on the uplifted rift shoulders of the Red Sea/Gulf of Aden and on ocean islands in the Red Sea and Gulf of Aden (e.g. Deniel *et al.*, 1994; Arndt and Menzies, 2005).

In addition, uplift and magmatism as early as 45 Ma is associated with the formation of the Kenyan Dome about 1000 km to the south, with a second mantle plume inferred for this older magmatism (Rogers, 2006). A most remarkable alkaline-carbonatite province ranging in age from 45 to 0 Ma is also associated with the East African rift system (Ernst and Bell, 2010; Section 9.2.2).

Composition

The mineralogical and chemical composition of the flood basalts is relatively uniform. Most have tholeiitic to transitional compositions (Mohr, 1983; Mohr and Zanettin, 1988; Pik *et al.*, 1998). Inter-layered with the flood basalts are alkali picrites, and, particularly at upper stratigraphic levels, lavas and pyroclastic rocks of rhyolitic or, less commonly, trachytic compositions (Ukstins Peate *et al.*, 2002, 2005). A suite of low-Ti basalts characterized by relatively flat rare-earth element (REE) patterns and low concentrations of Ti and incompatible trace elements is largely restricted to the northwestern part of the province. Alkali basalts with higher concentrations of incompatible elements and more fractionated REE patterns – the so-called "high-Ti" basalts (HT1 and HT2) – are found to the south and east. The HT2 basalts are slightly more magnesian than the HT1 basalts (Arndt and Menzies, 2005).

3.2.4 North Atlantic Igneous Province (NAIP) (c. 60 Ma)

General characteristics

The NAIP has a total volume estimated at more than 6.6 Mkm^3, and has several regional components (e.g. White and McKenzie, 1989; Saunders *et al.*, 1997; Storey *et al.*, 2007a; Meyer *et al.*, 2007; Peate *et al.*, 2008) (Fig. 3.4). It consists of extrusive rocks, intrusive centers, and associated dykes of the British Tertiary province in the British Isles. Volcanic rocks (mainly flood basalts) and associated intrusions and dykes also extend for 2000 km along the east coast of Greenland, in the adjacent ocean, on the Voring Plateau (off Norway), on the Rockall Bank, and with minor accumulations in the area around Disko Island area in west Greenland and Baffin Island in Canada. Dyke swarms of the British Tertiary province have a dominant northwest trend, although some sets locally radiate from intrusive centers (Speight *et al.*, 1982; MacDonald *et al.*, 1988; Ernst and Buchan, 1997; Fig. 5.23). Numerous dyke swarms are also present along and are oriented parallel to the coast of east Greenland (e.g. Klausen and Larsen, 2002), and a minor dyke set in west Greenland is associated with the Disko Island volcanic activity (Nielsen, 1987; Buchan and Ernst, 2004).

Age

The NAIP magmatism, including central east Greenland, occurred principally during two episodes (*c*. 62–57 and 56–54 Ma: e.g. Saunders *et al.*, 1997; Tegner *et al.*, 1998; Storey *et al.*, 2007a) that are thought to have been related to the initial arrival of plume-head material, and the initial plate breakup/opening of the North Atlantic Ocean, respectively (Peate *et al.*, 2008).

Composition

The main activity within the NAIP consists of tholeiitic basaltic magmatism although alkali basalts are also common. Differentiated rocks are commonly found in continental areas and along rifted margins within the NAIP. There is good evidence that the lithosphere has strongly influenced the composition of the erupted rocks with the majority of contamination associated with assimilation of continental crust material (Saunders *et al.*, 1997). The earliest phase of magmatic activity (*c*. 62–57 Ma) is characterized by highly contaminated magmas that show a temporal change in assimilant type from amphibolite to granulite. The voluminous breakup phase of magmatism (*c*. 56–54 Ma) saw a significant decrease in the extent of assimilation because of the decreasing availability of assimilant material in the mature feeder systems, and many samples have Sr–Nd–Pb isotope compositions that overlap with those of asthenospheric melts (as represented by recent Icelandic basalts and North Atlantic mid-ocean-ridge basalt (MORB)) (Peate *et al.*, 2008).

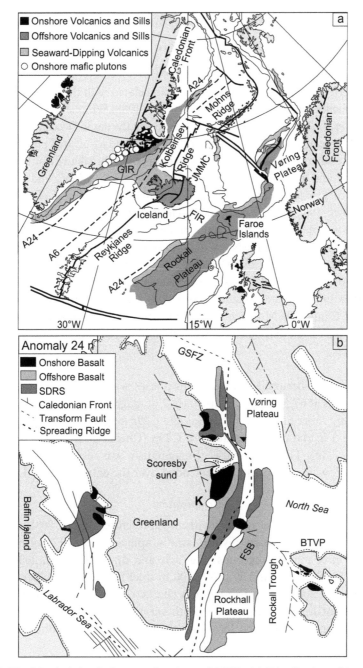

Figure 3.4 The North Atlantic Igneous Provinces (NAIP). (a) Distribution of igneous rocks related to the NAIP (black shading is subaerial lavas and intrusions; light-grey shading is offshore, subaerially erupted lavas; dark grey shading is offshore, seaward-dipping lavas). JMMC = Jan Mayen microcontinent: JMFZ = Jan Mayen fracture zone; GIR = Greenland-Iceland Ridge; FIR = Faroe-Iceland Ridge. With permission from Elsevier. (b) Pre-drift reconstruction of the North Atlantic region, at 61 Ma, and the distribution of earliest NAIP volcanic activity. Bold dashed line indicates position of magnetic chron 24r mapped along the margins of Greenland and Europe. BTVP = British Tertiary Volcanic Province, K = Kangerlussuaq Fjord (near plume center location of White and McKenzie, 1989), FSB = Faeroe Shettand Basin, GSFZ = Greenland-Senja fracture zone, SDRS = seaward-dipping reflector sequence. From Figure 1a Tegner *et al.* (2008) (a) and from Figure 2 in Larsen and Saunders (1998) (b). Digital version of original figure for (a) kindly provided by C. Tegner.

3.2.5 Deccan LIP (c. 65 Ma)

General characteristics

The Deccan LIP in India (Fig. 3.5) is spatially associated with the separation of India from the Seychelles microcontinent (e.g. Mahoney, 1988; White and McKenzie, 1989, 1995; Hooper *et al.*, 2010). (As noted above I am using the name Deccan rather than the older label of Deccan Traps.) The volume of the lavas and associated plumbing system of dykes, sills, and layered intrusions prior to continental breakaway of the Seychelles microcontinent and to erosion, probably far exceeded 1 Mkm^3, and was erupted in a million years or less, straddling the iridum-anomaly-defined K–T boundary (e.-g. Courtillot *et al.*, 1996; Chenet *et al.*, 2008). The large volume and short duration of the event is suggestive of a plume origin and a preferred location for the plume center is at the locus of the intersection of the Narmada–Tapti rift and the west coast, representing a triple-junction scenario (Burke and Dewey, 1973; Courtillot *et al.*, 1999), although non-plume origins for the Deccan LIP have been argued by some (e.g. Seth, 2005). A number of carbonatites and associated alkalic rocks are associated with the Deccan LIP (Ernst and Bell, 2010; Chapter 9). Most broadly, alkalic volcanism both preceded and followed the relatively short-lived outpouring of the Deccan flood basalts (Basu *et al.*, 1993).

Age

Deccan magmatism occurred in three main phases with the initial relatively small phase-1 eruptions at 67.5 Ma, the main phase-2 of magmatism with *c.* 80% eruptions over a relatively short time interval during the C29r magnetic polarity interval, and the last phase-3 in the early Danian base C29n (Chenet *et al.*, 2007; Keller *et al.*, 2011a, b; Fig. 3.5b). The initial pulse of magmatism was restricted to the northern part of the LIP, while the main pulse extended over the entirety of the LIP.

Composition

The main pulse of the Deccan magmatism, representing about 80% of the total present-day volume, has been divided stratigraphically into 12 formations and grouped into three major subgroups (Fig. 3.5c and d): Kalsubai (*c.* 2000 m thick), Lonavala (*c.* 525 m thick) and Wai (*c.* 1100 m thick) going from older to younger (e.g. Chenet *et al.*, 2008). Each formation is characterized by accumulations of large composite subhorizontal tholeiitic flows with varying thicknesses, and chemical and mineralogical composition. The locations of the stratigraphic columns shown in Fig. 3.5c are given in Fig. 3.5a.

3.2.6 Paraná–Etendeka LIP (c. 135)

General characteristics

The extensive Paraná flood basalts in central coastal South America and the minor Etendeka remnant in Namibia once formed a single LIP that was associated with the

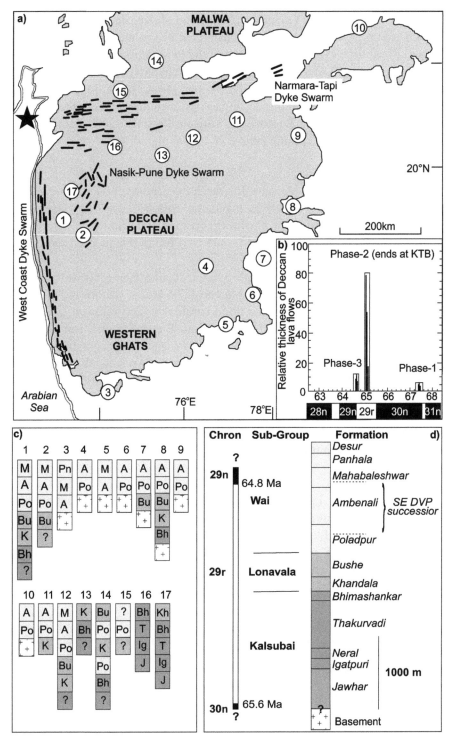

Figure 3.5 The Deccan LIP. (a) General distribution of magmatism (modified from Ju *et al.* (2003), with permission from Elsevier, with location of chemotype sections superimposed from

opening of the South Atlantic Ocean (e.g. White and McKenzie, 1989; Storey, 1995) (Fig. 3.6). The Paraná–Etendeka province has a preserved volume in excess of 1 Mkm^3 (White and McKenzie, 1989; Peate, 1997) and the Paraná basalts consist of various volcanic packages (Fig. 3.6) with a significant silicic (rhyolite) component that is also present in the Etendeka portion of this LIP. Major dyke swarms are exposed in areas of deeper erosion, along with numerous intrusions, including carbonatites, alkaline intrusions, and kimberlites (e.g. Le Roex and Lanyon, 1998; Bell, 2001; Ernst and Bell, 2010; D. Schissel, personal communication, 2013; Chapter 9).

Age

Stratigraphic data and ^{40}Ar–^{39}Ar ages for the Early Cretaceous Paraná–Etendeka flood basalts indicate that the main magmatic episode lasted for several Ma (129–134 Ma) and was linked to the northward opening of the South Atlantic Ocean, but with some precursor magmatism (138–135 Ma) found inland from the oceanic rift (Peate, 1997).

Other Ar–Ar dating suggests that small-fraction alkaline melt generation occurred beneath the region in two phases: at 145 Ma and 127.5 Ma, i.e. before and at the end of the 139–127.5 Ma Paraná–Etendeka flood-basalt eruptions (Gibson *et al.*, 2006; Fig. 3.6b). However, the age duration of the Paraná–Etendeka volcanism has been a matter of debate with some earlier Ar–Ar studies suggesting magmatism between 140 and 129 Ma, contrasting with a more recent Ar–Ar study by Thiede and Vasconcelos (2010), that indicated a short duration (< 1 Ma) magmatic event at 134.6 ± 0.6 Ma. The geochronology framework is based on Ar–Ar dating, and experience in other provinces (e.g. Karoo; Section 3.2.7) has shown that the age distribution is typically tightened when only U–Pb dating is considered. A suite of U–Pb ages is required to confirm the duration of the Paraná–Etendeka LIP.

Composition

The lava pile within the Paraná–Etendeka LIP is dominated by tholeiitic basalts (> 90%), but significant quantities of rhyolites are found along the Brazilian continental margin and in the Etendeka (Peate, 1997). Six magma types (Fig. 3.6a) have been distinguished using major and trace element abundances and ratios. The regional distribution of distinct high-Ti–Y (Urubici, Pitanga, Paranapanema, Ribeira) and low-Ti–Y (Gramado, Esmeralda) magmas within both lavas and

Caption for Figure 3.5 (*cont.*) Widdowson *et al.* (2000), by permission of Oxford University Press; see also Jay and Widdowson, 2008). Star locates inferred plume center. (b) Age distribution of three main phases (pulses) (Keller *et al.*, 2011a, b). (c) Chemotype sections: J, Ig, T, Bh, Kh, Bu, Po, A, M and Pn are Jawhar, Igatpuri, Thakurvadi, Bhimashankar, Khandala, Bushe, Poladpur, Ambenali, Mahabaleshwar, and Panhala formations, respectively (after Fig. 1 Widdowson *et al.*, 2000, by permission of Oxford University Press). (d) Ages of various chemotypes of main pulses of magma (after Jay and Widdowson, 2008). For distribution of associated carbonatites and kimberlites, see Figure 9.4. Digital versions of original figures kindly provided by W. Ju and M. Widdowson. n and r refer to normal and reversed magnetic polarity, respectively.

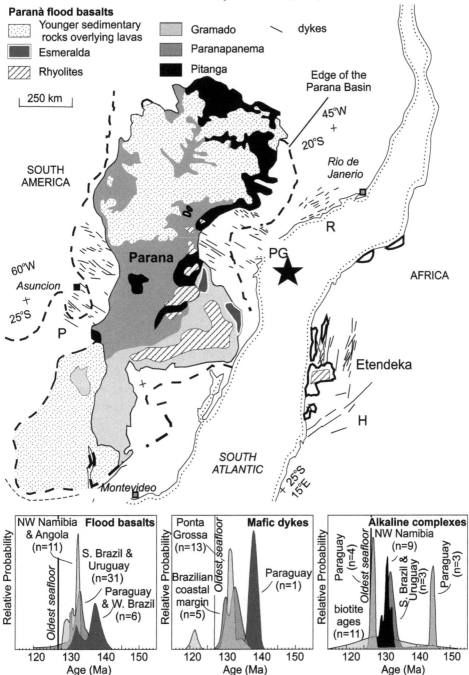

Figure 3.6 The Paraná–Etendeka LIP. Reconstructed Paraná–Etendeka LIP after closure of the South Atlantic. The Paraná portion is divided into chemostratigraphic units (from Figure 1 in Thiede and Vasconcelos (2010) based on original distributions shown in Stewart *et al.* (1996)). The distribution of associated carbonatites and kimberlites is shown in Figure 9.5. The rhyolite component of volcanics in Etendeka is shown after Stewart *et al.* (1996). The dykes of the Hentis Bay–Outjo (H) dyke swarm are added after Wiegand *et al.* (2011). Other dyke swarms are the Ponto Grosso (PG), Rio de Janeiro (R) and Paraguay (P). In geochronology diagrams, n = number of dates (by Ar–Ar method, except for K–Ar ages of Paraguayan alkaline complexes). Geochronology diagrams redrafted from Gibson *et al.* (2006).

associated dyke swarms implies that magma generation involved different mantle sources. Samples with low MgO concentrations (3–7%) are also present and provide evidence of extensive fractional crystallization, and upper crustal assimilation was also important in the evolution of the Gramado magmas. However, uncontaminated Paraná basalts have major and trace element and isotope characteristics that appear to require mantle sources distinct from typical oceanic basalts. The minor, late-stage Esmeralda magma type is an exception, as these magmas contain components derived from the incompatible element-depleted asthenosphere.

3.2.7 Karoo–Ferrar–LIP and Chon Aike SLIP (c. 180 Ma)

The initial breakup of Gondwana was broadly contemporaneous with the formation of three large coeval Middle Jurassic LIPs (Fig. 3.7):

(1) Karoo province (estimated to be 2.5 Mkm^3) in southern Africa and its extension into the Dronning Maud Land sector of East Antarctica (Cox, 1988);
(2) Ferrar province (estimated to be 0.5 Mkm^3) along the Transantarctic Mountains and its continuation into Tasmania, Australia, and New Zealand; and
(3) Chon Aike silicic province (estimated to be about 1.7 Mkm^3 in Patagonia).

Precise dating confirms that at least part of all three provinces was emplaced at *c.* 183–179 Ma, with magmatism within the Chon Aike silicic province continuing to younger ages. This section focuses on the Karoo and Ferrar events, and more information on the Chon Aike silicic province is provided in Section 8.3.3.

General characteristics

Karoo Remnants of the Karoo basalts are found across a large area of southern Africa, with smaller outcrops in Antarctica (Dronning Maud Land), and sequences associated with the breakup of Africa and Antarctica are present on the offshore continental margins of East Antarctica (e.g. Eales *et al.*, 1984; White and McKenzie, 1989; Cox, 1992; Storey, 1995; Hargraves *et al.*, 1997; Jourdan *et al.*, 2005; Neumann *et al.*, 2011; Fig. 3.7). Huge volumes of basalt (> 2 Mkm^3) were emplaced within a very short period of the Early Jurassic (183–179 Ma), most probably developing continuous basaltic cover across much of southern Africa, covering an area of more than 1 Mkm^2. However, much of this province has been eroded and the present outcrop covers about 140 000 km^2, with much of the magmatism in the central Karoo present within a large sill province, and there are associated dyke swarms (discussed in Chapter 5).

Ferrar The Ferrar CFB province extends from southern Australia (including Tasmania), through the Transantarctic Mountains and into Queen Maud Land in Antarctica, a distance of more than 4000 km. It includes units such as the

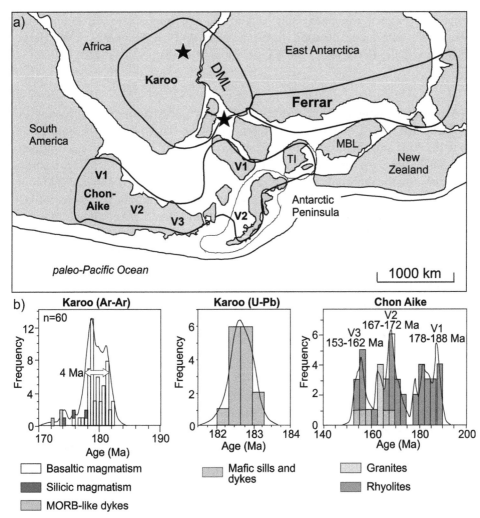

Figure 3.7 Generalized distribution of the Karoo and Ferrar flood-basalt provinces, and their relationship to the silicic Chon Aike igneous province in a Gondwana reconstruction showing the active proto-Pacific margin. As noted in Bryan *et al.* (2002) there is a general migration of the volcanic phases of Chon Aike (V1 (188–178 Ma), V2 (172–162 Ma) and V3 (157–153 Ma)) away from the locus of basaltic flood volcanism. DML is Dronning Maud Land, MBL = Marie Byrd Land, TI= Thurston Island. From Figure 5 in Bryan *et al.* (2002) and originally modified from Pankhurst *et al.* (1998, 2000). With permission from Elsevier. (b) Geochronology: Karoo Ar–Ar is from Figure 7 in Jourdan *et al.* (2007a). Karoo U–Pb dating is from Figure 4 in Svensen *et al.* (2012). Chon Aike ages are from Figure 9 in Pankhurst *et al.* (2000). See also Cúneo *et al.* (2013). With permission from Elsevier.

Kirkpatrick basalts, associated sills, dykes and the impressive Dufek–Forrestal layered mafic–ultramafic intrusion (Encarnación *et al.*, 1996; Elliot *et al.*, 1999; Hergt and Brauns, 2001; Ferris *et al.*, 2003; Leat, 2008; Muirhead *et al.*, 2012).

Ages

Ar–Ar dating suggests multiple distinct brief major pulses during the formation of the Karoo LIP: the Lesotho and southern Botswana lava piles (each centered at about 182 Ma) and the Okavango dyke swarm (centered at 179 Ma) (e.g. Jourdan *et al.*, 2007b; Fig. 3.7). The pulses have a shorter duration of 3 and 4.5 Ma, compared to the duration for the whole province of *c*. 10 Ma, between *c*. 182 and *c*. 172 Ma on the basis of the Ar–Ar dating. However, a shorter duration is indicated by recent U–Pb zircon and baddeleyite dating of fourteen Karoo sill and dyke samples that are separated by as much as 1100 km across the half-million-square-kilometer Karoo basin. The ages range from 183.0 ± 0.5 to 182.3 ± 0.6 Ma, and probability modeling indicates emplacement took place within an interval of about 0.47 Ma (Svensen *et al.*, 2012). The age duration of the Ferrar sills is thought to be similarly short with magmatism focused around 183 Ma, corresponding to the Pliensbachian–Toarcian boundary (Burgess *et al.*, 2011). The associated Chon Aike silicic magmatism occurred during three major magmatic pulses between 183 Ma and 157 Ma (see Chapter 8).

Composition

Karoo The tholeiitic basalts and dolerites that constitute the overwhelming majority of Karoo rocks, have been classified into low- and high-Ti subgroups, on the basis of their TiO_2, P_2O_5, and incompatible-trace-element contents, with the groups subdivided at a TiO_2 concentration of 2.0–2.5 wt% (e.g. Jourdan *et al.*, 2007a), yielding a strong geographic provinciality (Section 10.5.3). Rhyolites capping the basaltic lava pile in Lebombo and Mwenezi have a wide range of isotopic values from mantle-like to extreme crustal signatures (Harris and Erlank, 1992), and picrites and nephelenites in the Mwenezi area are indicative of the involvement of SCLM (subcontinental lithospheric mantle) and other sources. The MORB-like Rooi Rand dyke swarm emplaced along the southern part of the Lebombo monocline is interpreted as the final stage of Karoo magmatism, occurring just prior to the onset of ocean floor spreading (Duncan *et al.*, 1990).

Ferrar The composition of the Ferrar LIP is remarkably uniform over its 4000-km distribution; this vast area is remarkable and includes basalts and also dykes and sills (Brewer *et al.*, 1992; Elliot and Fleming, 2000; Hergt and Brauns, 2001). However, Elliot *et al.* (1999) noted that the youngest preserved flows of the Kirkpatrick basalts have a distinctive geochemical character that can be traced over a distance of 1600 km and are consistent with derivation from a single batch of magma. Overall, the Ferrar LIP magmas share important chemical features, including (i) high Si and low Fe, Ti, Na, and P concentrations at a given Mg, and (ii) incompatible trace-element patterns and isotopic ratios (Sr, Nd, and Pb) that are remarkably similar to the continental crust.

The distinct geochemical differences between Karoo- and Ferrar-type magmas favor derivation from distinct source regions in the mantle as well as different subsequent histories of interaction with the lithosphere and crust (e.g. Elliot and Fleming, 2000). The source area(s) for the Karoo may be the Nuanetsi and Zambesi plume centers, which were located on the basis of converging rifts and radiating dyke swarms (Fig. 5.10; Burke and Dewey, 1973; Ernst *et al.*, 1995a). The mantle source for the Ferrar may be marked by the Weddell Sea triple-rift junction (Fig. 5.10; e.g. Storey *et al.*, 2001 and references therein).

3.2.8 Siberian Trap LIP (aka North Asian; Uralo-Siberian) (250 Ma)

General characteristics

One of the largest LIP events in the world is the Siberian Trap event (Fig. 3.8), which is widespread in the northwestern Siberian craton, the Taimyr Peninsula to the north, to the west in the West Siberian basin and still further west as inliers within the Ural Mountains region, and to the south in the Kuznetsk basin and in the Central Asian fold belt (e.g. Gurevitch *et al.*, 1995; Fedorenko *et al.*, 1996; Dobretsov, 1997; Nikishin *et al.*, 2002; Ivanov, 2007; Reichow *et al.*, 2009; Ivanov *et al.*, 2013). The total volume has been estimated at 4 Mkm^3 (e.g. Ivanov, 2007) although previous estimates were as high as 16 Mkm^3 (Dobretsov, 2005), with a main pulse of magmatism at 251–248 Ma (e.g. Renne and Basu, 1991; Kamo *et al.*, 2003; Reichow *et al.*, 2009); with a potentially more minor younger pulse at 230–225 Ma (Ivanov, 2007). Flood basalts in the uplifted Anabar region have been removed by erosion revealing a major north–west-trending regional dyke swarm that with feeder systems in the Noril'sk region define a radiating dyke swarm (Ernst and Buchan, 1997; Fig. 3.8). The Noril'sk region also hosts a number of world-class magmatic sulfide deposits that were generated during the Siberian Trap LIP and are famously rich in Cu, Ni, and platinum-group elements (PGEs), especially Pd (e.g. Naldrett, 1999; Chapter 16). The location of the plume center for the Siberian Trap LIP is usually considered to be within the northern margin of the LIP, along the intersection of the north–south-trending West Siberian failed rift with the east–west-trending Khatanga rift (Schissel and Smail, 2001). However, an alternative location is located some 300 km further east along the Khatanga rift as inferred using associated radiating and circumferential dyke swarms (Ernst and Buchan, 1997a, 2001). As discussed in Chapter 11 (e.g. Fig. 11.10; see also Fig. 16.4), the center of triple-junction rifting need not be coincident with the focus of a radiating dyke swarm.

Age

The Siberian Trap LIP event benefits from both abundant Ar–Ar dating and U–Pb dating. ^{40}Ar–^{39}Ar ages from the Noril'sk, Tunguska, Taimyr, Kuznetsk, and Vorkuta areas, combined with previously published data, demonstrate that 250 Ma volcanic activity in Siberia encompasses an area of over 5 million km^2 (Reichow *et al.*, 2009) and a summary of ages is provided in Fig. 3.8c. A younger pulse(s) at *c.* 230–245 Ma is also suggested by Ar–Ar ages for various mafic units (Ivanov, 2007) and U–Pb ages of mafic rocks from the Noril'sk, Maimecha-Kotui, and Taimyr areas mostly group tightly at 248–251 Ma (Campbell *et al.*, 1992; Kamo *et al.*, 1996; Kamo *et al.*, 2003; Kuzmichev and Pease, 2007). U–Pb dating on silicic rocks in the Taimyr and Kuznetsk basin areas exhibits a broader range from *c.* 250–225 Ma

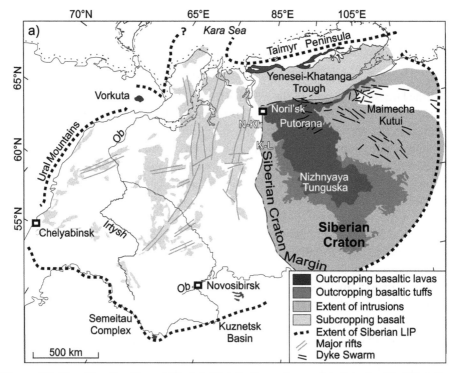

Figure 3.8 The Siberian Trap LIP. (a) Distribution of associated units. Modified from Figure 2 in Saunders and Reichow (2009) with addition of dykes from Ernst and Buchan (1997a). Permission granted by Springer. (b) Distribution of chemotypes from Figure 2 in Sharma (1997). Regions are identified in (a). (c) Chronology after Figure 21 in Ivanov *et al.* (2013). Faunal turnover from the Permian to Middle Triassic is compared to timing of the Emeishan and Siberian Trap flood-basalt volcanism. Details in Ivanov *et al.* (2013). With permission from Elsevier. Digital version of original figures for parts (a) and (c) kindly provided by M. Reichow and A. Ivanov, respectively.

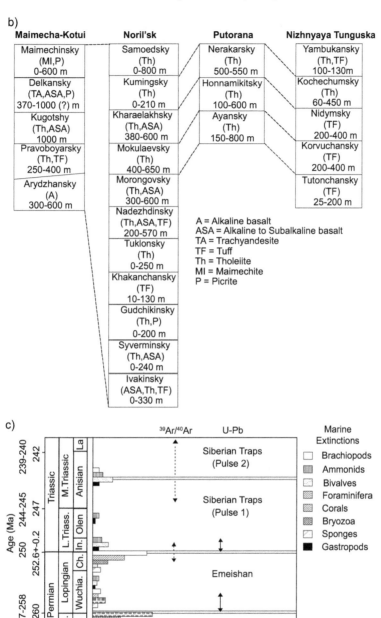

Figure 3.8 (cont.)

(Vladimirov *et al.*, 2001; Vernikovsky *et al.*, 2003). An additional younger mafic pulse at *c.* 240 Ma is suggested by Ivanov *et al.* (2013). The Siberian Trap LIP occurred close to the Permian–Triassic boundary and has been causally linked with the mass extinction event that occurred at this time; this is shown in Figure 11.10 and is further discussed in Chapter 14. Important new high-precision U–Pb geochronological constraints that support this LIP–mass extinction correlation are in progress (Section 14.3.8, Burgess *et al.*, 2014a; Burgess and Bowring, 2014; S. Burgess, pers. comm. 2014).

Composition

The Siberian Trap LIP can be divided broadly into four regions of fundamentally different volcanic sequences: Putorana, Noril'sk, Maimecha-Kotui, and Nizhnaya Tunguska. A number of discrete formations are identified in each, which can be correlated between the different regions (Fig. 3.8b). The great majority of the Putorana rocks are relatively homogeneous, aphyric, and polphyric tholeiitic basalts. The volcanism in the Noril'sk region is marked by varied rock types from picritic through tholeiitic to subalkalic basalts and basaltic andesites, in addition to cogenetic and comagmatic intrusions that are significant hosts for Ni–Cu–(PGE) mineralization (see Chapter 16).

The volcanic rocks of the Maimecha-Kotui area are quite evolved as evidenced by a wide variety of rock types that include picrite, tholeiitic basalt, alkaline-olivine basalt, trachybasalt, trachyandesite, basanite, olivine nephelinite and maimechite, with associated carbonatites and kimberlites (Egorov, 1970; Vasiliev and Zolotukhin, 1995; Arndt *et al.*, 1995). In comparison, the Nizhnaya Tunguska region is remarkable as this area is dominated by basaltic tuffs. Figure 3.8b provides an interpreted correlation between the volcanic–stratigraphic sequences present in these four regions (Sharma, 1997).

3.3 Volcanic rifted margins

CFBs are typically linked to breakup (Chapter 11) which in turn is commonly associated with the development of volcanic rifted margins that mark the transition zone between rifted continental and new oceanic crust (e.g. Menzies *et al.*, 2002a; Geoffroy, 2005). This transitional area is characterized by the formation of high-velocity lower crust and seaward-dipping reflectors, and it has the overall structure shown in Figure 3.9.

Volcanic passive margins mark continental breakup over a hotter mantle than is associated with non-volcanic passive margins, meaning the former are probably subject to local thermotectonic conditions, including small-scale convection. This leads to marked differences in the temporal and spatial relationships between tectonics, magmatism, uplift, and erosion between volcanic and non-volcanic passive margins (Table 1 in Menzies *et al.*, 2002a).

The general pattern of formation of volcanic rifted margins is characterized as follows using seismology (e.g. Menzies *et al.*, 2002a): The pre-rift to syn-rift transition is associated with (a) extended continental crust, (b) formation of a subaerial inner seaward-dipping reflector series (SDRS), (c) the development of a high-velocity

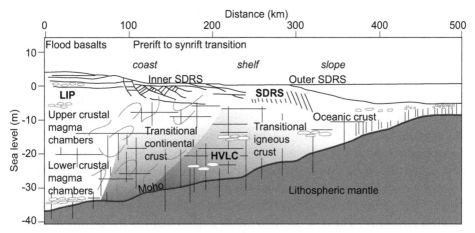

Figure 3.9 Schematic volcanic rifted margin based on data from Ethiopia–Yemen (part of the Afro–Arabian LIP) and the Atlantic margins (part of the CAMP LIP). HVLC = high-velocity lower crust, SDRS = seaward-dipping reflector series. From Figure 2 in Menzies *et al.* (2002).

lower crust (HVLC) in the transition from continental crust to oceanic domains, and (d) formation of submarine outer seaward-dipping reflector series (SDRS) (e.g. Mutter *et al.*, 1982; White *et al.*, 1987; Eldholm and Grue, 1994; Planke *et al.*, 2000). Extrusive magmatism includes pre-rift (flood basalts) and syn-rift magmatism (inner subaerial SDRS), and syn-rift and/or post-rift magmatism (outer submarine SDRS). The seismically defined layering in the SDRS is represented by both sub-aerial and submarine volcanic rocks and variable amounts of sedimentary detritus shed from the volcanic rifted margin during uplift and tectonic denudation of the kilometer-scale rift mountains. The HVLC (with a seismic velocity of ~ 7.4 km/s) forms within the continent–ocean transition, represents magmatic underplating, and can reach considerable thicknesses (10–15 km).

Eventually, stretching and heating of the magmatically modified continental lithosphere will lead to effective rupture and the commencement of seafloor spreading. This initial oceanic crust may be thicker than normal as it is associated with a hotter asthenosphere caused by the presence of a plume and/or steep gradients at the lithosphere–asthenosphere boundary (e.g. Boutilier and Keen, 1999). In addition, the interval between the first expression of volcanic-rifted-margin formation on the pre-rift continental margin and the formation of true ocean floor can be a few million years or longer.

3.4 Thematic issues related to CFBs

Having introduced a selection of the classic CFBs and profiled the most significant characteristics of each, the next section provides an overview of some general aspects of the process of CFB emplacement.

3.4.1 Flow characteristics

This section focuses on the mechanisms involved in the feeding and emplacement of flood basalts. CFB lavas are predominantly emplaced as inflated compound pahoehoe flow fields via protracted episodic eruptions (Fig. 3.10a, b; the two versions of the model illustrate complementary aspects). Such flows have the following stages (Self *et al.*, 1997): (a) flow arrival as a small, slow-moving, lobe of molten lava held inside a stretchable, chilled viscoelastic skin with brittle crust on top; (b) continued injection of lava into the lobe results in inflation (lifting of the upper crust) and new breakouts; (c) after stagnation, diapirs of vesicular residuum form vertical cylinders and horizontal sheets within the crystallizing lava core; (d) the emplacement history of the flow is preserved in vesicle distribution and jointing pattern of frozen lavas.

Figure 3.11 shows the history of emplacement of the Roza compound flow of the Columbia River LIP. The flow inflation model implies that the Roza and other CFB flows were emplaced over an extended period of time, with individual Roza flows emplaced over periods between 5 and 50 months and that the Roza flow field was constructed over a period of 6 to 14 years (Self *et al.*, 1997). In general, individual flow units can be emplaced for great distances, up to several hundred kilometers (e.g. Self *et al.*, 1997), or perhaps even up to 1000 km in the case of the Deccan LIP (Self *et al.*, 2008). Similar observations of lava characteristics in other Columbia River and Deccan LIP flows by Self *et al.* (1997) suggest that this emplacement style is typical of many, if not most, CFB flows.

Another important point illustrated by geological mapping is that the different Roza flows start from different segments along the Roza fissure and fissure segments (Fig. 3.11b); this is a very clear illustration of the role of fissure-fed eruptions, considered to be the dominant mode of flow emplacement in flood basalts. Eruption from shield volcanoes is also important in flood basalts, but is much rarer, with Arndt and Menzies (2005) describing examples of shield volcanoes belonging to the Ethiopian portion of the Afro-Arabian LIP. Although shield volcanics may be rare in LIPs, they are a key component of terrestrial hotspots (e.g. Hawaii), which can be the post-LIP stage of a mantle plume, and are an important type of LIP-scale planetary magmatism (Chapter 7).

3.4.2 Pyroclastic units

Although the majority of CFB extrusives are emplaced as basalt flows, some flood-basalt packages also have a significant pyroclastic component, which can be either (or both) mafic and silicic in composition. In the lower parts of volcanic piles, the composition of the pyroclastics is usually mafic, while in the upper parts of the volcanic pile silicic pyroclastics can be important. Mafic volcaniclastic deposits (MVDs) are present in several LIPs, with occurrences and characteristics reviewed by Ross *et al.* (2005). Most dramatically, the Siberian Trap LIP contains a considerable

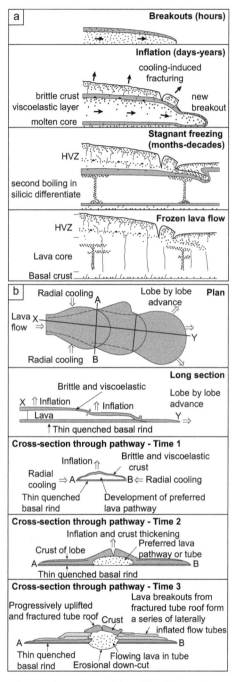

Figure 3.10 Models for the emplacement of basaltic flows. (a) Mode of eruption via the mechanism of a generic inflating pahoehoe sheet flow redrafted from Self *et al.* (1997). The vertical scale varies from 1–5 m for Hawaiian flow to 5–50 m for the Columbia River Basalt flows. (b) Endogenous growth model for propagation and development of basaltic flow lobes. Redrafted from Figure 9.2 in Arndt *et al.* (2008) after the original version of Hill (2001). Reprinted with the permission of Cambridge University Press.

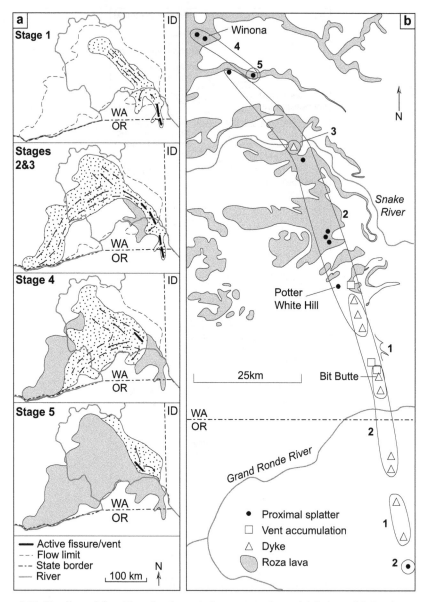

Figure 3.11 Feeding of flood basalts through fissure eruptions. (a) Sequential development of the Roza compound flow field fed from a dyke (Columbia River flood basalt). Lava travels from the vent (above the dyke) under an insulating crust in a preferred pathway (dashed lines). This flux of lava is used to both inflate the sheet flow and to feed new breakouts. Stages 1–5 are the five major flows that form the Roza flow field; the active fissure segment at each stage is shown by a thick bar. From Figure 14 in Self *et al.* (1997). (b) Sketch map of features along the fissure segments thought to be related to each Roza lava flow (1–5) (see part a). WA = Washington; OR = Oregon; ID = Idaho. Redrafted from Figure 11 in Self *et al.* (1997).

amount of mafic pyroclastics that outcrop over a wide area of the Nizhnaya Tunguska region (e.g. Ivanov, 2007; Fig. 3.8b).

Another important example of LIP-related mafic pyroclastics is within the Ferrar province, where the preserved thickness of lavas varies from 380 m to 780 m, all of which are underlain by MVDs ranging in thickness from 10 to > 400 m (Hanson and Elliot, 1996; Elliot and Hanson, 2001; White and McClintock, 2001; McClintock and White, 2005; Ross *et al.*, 2005). Most of the explosive activity within the Ferrar LIP was the result of violent magma–groundwater interaction, and the temporal evolution from explosive activity to eruption of lava flows is *not* thought to be due to a difference in magma chemistry.

Silicic pyroclastics are also associated with many LIPs, and can be distributed far from the site of eruption. For instance, dating and trace-element geochemistry of ash from drill core in the Indian Ocean (2700 km away) has allowed correlation of individual ash layers with specific Afro-Arabian LIP silicic units (Ukstins Peate *et al.*, 2003).

3.4.3 Main phases of flood-basalt volcanism

Three phases of flood-basalt volcanism are recognized in young (Mesozoic–Cenozoic) CFBs (Jerram and Widdowson, 2005) and some clear facies patterns have been defined for these (e.g. Fig. 3.12). The inception of flood-basalt volcanism is associated with the eruption of relatively low-volume transitional-alkaline eruptions onto exposed lithologies, including sediments and, in some cases, water, and the distribution of this initial volcanism is strongly controlled by pre-existing topography. The main phase of volcanism is typically characterized by repeated episodes of large-volume tholeiitic flows that predominantly generate large tabular flows and flow fields from a number of spatially restricted eruption sites and fissures. These tabular flows build a thick lava flow stratigraphy in a relatively short period of time (*c.* 1–5 Ma). The fact that the overall duration of flood volcanism can last 5–10 Ma, means that this main phase accounts for less than half the overall eruptive time in each specific case. The waning phase of flood volcanism is associated with a rapid decrease in the volume of eruptions and the development of more widely distributed centers of eruption. These late-stage eruptions are commonly associated with increasing magma silica contents and highly explosive eruptive products.

3.4.4 Facies types in continental LIPs

Volcanology and facies architecture of flood basalts

CFBs do not typically have a simple layer-cake lava flow stratigraphy (Jerram, 2002; Jerram and Widdowson, 2005), but instead exhibit variations in both vertical and horizontal stacking (e.g. Figs. 3.12 and 3.13). The architecture of CFBs and associated

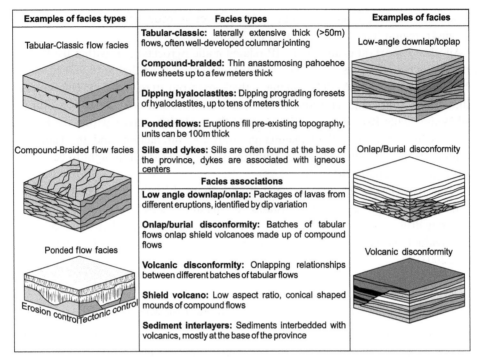

Examples of facies types	Facies types	Examples of facies
Tabular-Classic flow facies	**Tabular-classic:** laterally extensive thick (>50m) flows, often well-developed columnar jointing	Low-angle downlap/toplap
	Compound-braided: Thin anastomosing pahoehoe flow sheets up to a few meters thick	
	Dipping hyaloclastites: Dipping prograding foresets of hyaloclastites, up to tens of meters thick	
	Ponded flows: Eruptions fill pre-existing topography, units can be 100m thick	
Compound-Braided flow facies	**Sills and dykes:** Sills are often found at the base of the province, dykes are associated with igneous centers	Onlap/Burial disconformity
	Facies associations	
	Low angle downlap/onlap: Packages of lavas from different eruptions, identified by dip variation	
	Onlap/burial disconformity: Batches of tabular flows onlap shield volcanoes made up of compound flows	
Ponded flow facies	**Volcanic disconformity:** Onlapping relationships between different batches of tabular flows	Volcanic disconformity
	Shield volcano: Low aspect ratio, conical shaped mounds of compound flows	
Erosion control Tectonic control	**Sediment interlayers:** Sediments interbedded with volcanics, mostly at the base of the province	

Figure 3.12 Summary of key facies types and facies associations in flood basalts. From Jerram (2002) and Jerram and Widdowson (2005). With permission from Elsevier. See also Jerram *et al.* (2010). Digital version of original figure kindly provided by D. Jerram.

volcanic rifted margins is recorded by facies types and facies associations (Fig. 3.12). Facies types, such as tabular-classic flows, braided-compound flows, or hyaloclastites, represent building blocks of the volcanic stratigraphy. Tabular-classic facies architecture indicates that the flows were erupted as single continuous flows from fissure systems. Compound-braided flows suggest a different eruptive style where flows were not erupted continuously but over longer time intervals from separate shield volcanoes. Facies associations, such as downlaps, onlaps, and disconformities, relate how the volcanic facies are stacked together. Many of the facies associations occur on an intermediate to large basin-wide scale and may only be revealed by detailed field work, photogrammetry, and the construction of three-dimensional geological models.

There are several ways these facies variations may develop: through eruption into different environments, changes in eruption flux, and/or proximal to distal variations. Different scales of observation within single provinces also lead to the identification of different hierarchies of heterogeneity (Jerram and Widdowson, 2005; Jerram *et al.*, 2010). Volcanic centers can also shift laterally over time (e.g. over the course of a flood-basalt event) (Jerram and Widdowson, 2005), a process which must reflect changes in the plumbing system or "wandering" of the underlying heat source (e.g. plume) (Section 12.5.3).

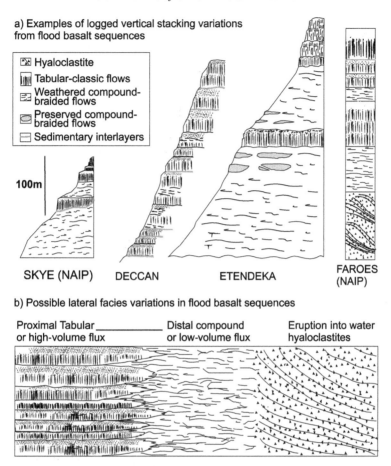

Figure 3.13 (a) Logged sections through flood-basalt provinces showing vertical heterogeneities of different facies. (b) Schematic conceptual diagram looking at possible lateral facies variations. From Nelson *et al.* (2009a and b). Digital version of original figure kindly provided by D. Jerram.

There is also a typical correlation between magma composition and lava facies type (Jerram, 2002). For example, picritic lavas in the Etendeka portion of the Paraná-Etendeka LIP ($\sim$ 45 wt% SiO_2) occur as low-volume compound pahoehoe sheets, whereas basaltic andesites in the same LIP form classic higher volume c. 60-m-thick tabular lavas, and silicic flows form similar-scale tabular-classic tabular units (50–150 m thick).

3.5 Remnants of CFBs in the older LIP record

As one looks back in time, the volcanic rock component of CFBs should be increasingly removed by erosion thus exposing the LIP plumbing system of sills, dykes, and layered intrusions. The details of this plumbing system are presented in Chapter 5, and herein I review some characteristics of the flood basalts left as erosional remnants, using, as examples, the 510 Ma Kalkarindji LIP of Australia and the 1115–1085 Ma Keweenawan magmatism within the Midcontinent region of central North America.

3.5.1 The Kalkarindji LIP (c. 510 Ma)

General characteristics

Despite being one of the largest LIPs on Earth, the Kalkarindji LIP, Australia, is poorly known (e.g. Bultitude, 1976; Hanley and Wingate, 2000; Ernst *et al.*, 2003; Glass and Phillips, 2006; Evins *et al.*, 2009; Jourdan and Evins, 2010; Fig. 3.14). This LIP comprises scattered basalt suites extending over a known area of ≥ 2.1 Mkm^2 across the Northern Territory and Western Australia and also possibly into the southernmost part of South Australia (equivalent to a total area of ≥ 3 Mkm^2) (Jourdan and Evins, 2010). The province includes flows, intrusions, volcanic tuffs, and dykes and is subdivided into a series of units, including the subaerial Antrim

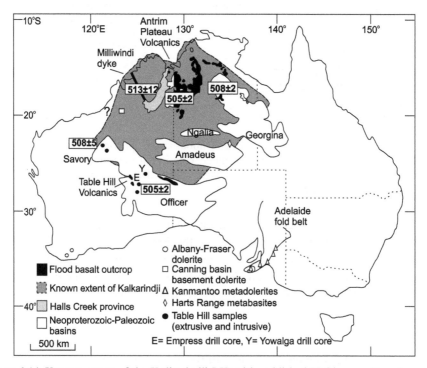

Figure 3.14 Known extent of the Kalkarindji LIP with published U–Pb ages. It perhaps also extends into the Adelaide fold belt region. Modified from Figure 1 in Evins *et al.* (2009). With permission from Elsevier.

Plateau volcanics, and the correlated Nutwood Downs, Helen Springs, Peaker Piker, and Colless volcanics in the east, and the Table Hill and other volcanics in the south. (Note these Antrim basalts of the Kalkarindji LIP are distinct from the same-named Antrim basalts of the British Tertiary Igneous Province portion of the NAIP). This vast Kalkarindji Cambrian igneous event is Australia's oldest and largest Phanerozoic LIP. Aeromagnetic maps across northern Australia reveal the presence of extensive basalt sub-crops under Paleozoic and Mesozoic cover, with flows up to 200 km long, filling paleo-valley and stratigraphic lows (Evins *et al.*, 2009). The Kalkarindji LIP thus extends across a vast area (Fig. 3.14), with maximum thickness attained in the northwestern part of the province within the Antrim Plateau volcanics (*c.* 1500 m). The original pre-erosion volume of the province is difficult to estimate, although it is likely to have exceeded 1.5 Mkm^3 (based on an areal extent of > 3 Mkm^2 and an average thickness of *c.* 0.5 km).

The Antrim Plateau basalts portion of the Kalkarindji LIP typically comprise *c.* 20–60-m (up to 200-m)-thick lava flows of mostly aphanitic massive basalt with vesicular or brecciated flow tops, and less common plagioclase–phyric porphyritic basalt. Glomeroporphyritic and typically columnar-jointed and hydrothermally altered, basalts occur locally in the west. Individual basalt flows may be intercalated with thin beds of aeolian sandstone, conglomerate, siltstone, and stromatolitic chert.

Age

Stratigraphic constraints suggest the Kalkarindji volcanics are at least Middle Cambrian in age (> 500 Ma) consistent with geochronological constraints: $^{40}Ar-^{39}Ar$ ages obtained from the Helen Springs volcanics (508 ± 2 Ma) and Antrim Plateau volcanics (505 ± 2 Ma) and Table Hill volcanics (505 ± 3 Ma) (Glass and Phillips, 2006; Evins *et al.*, 2009) are indistinguishable from SHRIMP U–Pb zircon ages obtained from the Kalkarindji province dolerites including the Milliwindi dolerite dyke (513 ± 12 Ma, Hanley and Wingate, 2000; 508 ± 10 Ma, Macdonald *et al.*, 2005). New U-Pb and Ar–Ar ages confirm the 5ll Ma age (Jourdan *et al.*, 2014).

Composition

Samples from both the Table Hill volcanics and the northern part of the province are low-Ti tholeiites with MgO content ranging from 3 to 9 wt%, and all samples are highly enriched in incompatible elements compared to primitive mantle (Evins *et al.*, 2009). This, coupled with the presence of a negative Nb anomaly, suggests crustal contamination at an early stage in magma evolution or, alternatively, a significant contribution from the subcontinental lithospheric mantle (see Section 10.5.4). Subtle differences in some incompatible-element ratios between the Table Hill volcanics and the northern Kalkarindji LIP indicate the presence of variations that are most likely related to heterogeneity of the assimilated crustal components or the mantle source(s) for these magmas.

3.5.2 Keweenawan LIP (1115–1085 Ma)

General characteristics

The Midcontinent rift system and its associated Keweenawan magmatism is one of the best-preserved examples of a plume-related Precambrian intracontinental rift (Fig. 3.15) (e.g. Burke and Dewey, 1973; Wold and Hinze, 1982; Green *et al.*, 1987; Nicholson and Shirey, 1990; Hutchinson *et al.*, 1990; Cannon *et al.*, 1992; Ojakangas *et al.*, 2001; Vervoort *et al.*, 2007; Heaman *et al.*, 2007; Merino *et al.*, 2013; Miller and Nicholson, 2013). Gravity and magnetic surveys indicate that the rift extends for more than 2000 km along strike in an arcuate pattern with a potential "third arm" marked by the Nipigon Embayment. The central basin in Lake Superior is up to 100 km wide and is filled with volcanics that may be up to 25 km thick, yielding a volume of perhaps 1.3 Mkm^3 (Hutchinson *et al.*, 1990). Given the likely erosion of significant volumes of volcanism outside the rift and the contribution from unexposed dykes and sills and magmatic underplating, the total volume of this LIP must be considerably greater.

Age

Current geochronological data splits the Keweenawan LIP magmatism into four pulses (Heaman *et al.*, 2007; Miller and Nicholson, 2013): a precursor pulse at *c.* 1140 Ma consisting of diabase dykes and lamprophyres (extending outside the rift to the northeast); a second pulse of mafic, ultramafic, and alkaline intrusions in the Lake Nipigon region at 1115–1110 Ma was followed by a major pulse of volcanism along the main rift axis at 1108–1105 Ma; a third stage at 1100–1094 Ma marked by another major pulse of magmatism in the rift along with some intrusions, including the Duluth Complex (1100–1094 Ma); and a final stage of limited magmatism that occurred after 1090 Ma.

Composition

The Keweenawan volcanic sequence within the Midcontinent rift is composed of predominately tholeiitic to subalkaline flood basalts, but also includes intermediate and silicic flows and fluvial interflow sedimentary rocks. As summarized (Miller, 2007; Miller and Nicholson, 2013), all lavas barring a few of the basal flows were erupted subaerially and the majority have the sheet-like form that is typical of flood basalts. Several notable exceptions are the intermediate to felsic composite volcanoes represented by the older Kallander Creek volcanics of the Powder Mill Volcanic Group, the younger Porcupine Mountain Formation in the Wisconsin–Michigan border area, and the Michipicoten Island Formation in eastern Lake Superior.

Geochemical and isotopic data indicate that most basalts were derived from a primitive, high-Al olivine tholeiitic primary magma that was generated from a common, chondritic to mildly enriched mantle source. The different

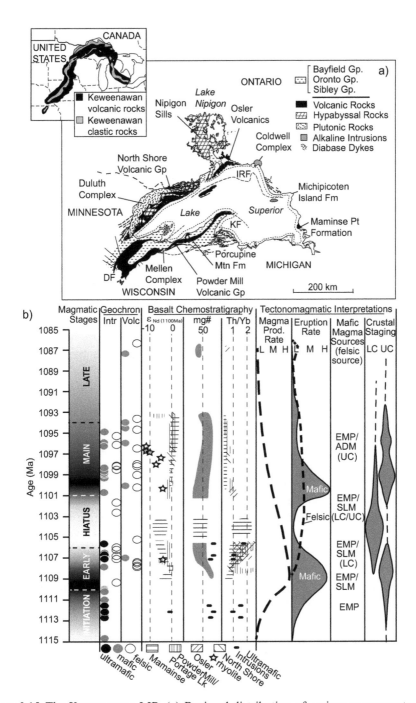

Figure 3.15 The Keweenawan LIP. (a) Regional distribution of various components of the Keweenawan LIP. For related carbonatites and kimberlites see also Figure 9.9. For distribution of precursor dykes, 1140 Ma Abitibi swarm see Figure 5.2i. (b) Age correlation chart. Abbreviation for magma sources is as follows: EMP = enriched mantle plume, SLM = subcontinental lithospheric mantle, ADM = asthenospheric depleted mantle; LC = Paleoproterozoic lower crust, and UC = Archean upper crust. After Miller and Nicholson (2013). Digital versions of original figures kindly provided by J. Miller.

pulses of magmatism within the Keweenawan LIP are associated with a range of different compositions representing involvement of contributions from plume mantle, subcontinental lithospheric mantle, asthenospheric-depleted mantle, and lower and upper crustal material (Fig. 3.15b; Miller and Nicholson, 2013).

3.5.3 Archean flood basalts

Archean CFB-dominated volcanism includes the *c.* 3.0 Ga Dominion Group lavas (at the base of the Witwatersrand Supergroup in South Africa) and two *c.* 2.7 Ga successions: the Ventersdorp Supergroup in South Africa and the Fortescue Group in Western Australia. The Ventersdorp Supergroup (an 8-km-thick succession of dominantly subaerially erupted tholeiitic basalts, komatiites, andesites, and pyroclastics) was emplaced onto the granitic gneiss basement of the Kaapvaal craton and overlying Witwatersrand Supergroup units, whereas the Fortescue was emplaced on the Pilbara craton (e.g. Eriksson *et al.*, 2002). These Archean-age flood basalts are discussed in more detail in Chapter 6 and are compared with types of Archean greenstone belts that are also considered as Archean LIPs.

3.5.4 Missing rifted-margin (passive-margin) record in older rocks

The analysis of the modern record shows that > 50% and perhaps as much as 90% of current passive continental margins are "volcanic" and can potentially be associated with a LIP (Skogseid, 2001; Menzies *et al.*, 2002a). In contrast, non-volcanic rifted volcanic margins (Wilson *et al.*, 2001; Russell and Whitmarsh, 2003) represent the LIP-free transition to normal-thickness oceanic crust. A similar proportion of continental margins throughout the geological record are also likely to be LIP-related, although the surface expression of such volcanic passive margins are likely obscured by erosion and younger tectonics. However, as discussed in Chapters 5 and 11, the radiating dolerite dyke swarms associated with LIPs can provide a robust record for the location of past volcanic rifted margins.

As noted by Coffin and Eldholm (2001) volcanic passive margins would appear to be relatively strong candidates for obduction and preservation, as the crust is commonly a mixture of mafic and more silicic components and therefore has a lower overall density than oceanic plateaus or oceanic-basin flood basalts. However, volcanic margins are highly likely to be uplifted and eroded during a continent–continent collision. Criteria need to be developed to distinguish ophiolites having a volcanic passive-margin or oceanic-plateau origin (and linked to LIPs) from those with subduction-related origins (e.g. Coffin and Eldholm, 2001; Dilek and Ernst, 2008).

3.6 Summary

CFBs and volcanic rifted margins represent the classic expression of LIPs emplaced into a continental setting. CFBs consist of volcanic sequences that are typically up to several kilometers thick and are dominated by fissure-fed compound flows. Isolated shield volcanoes may also be present in some LIPs, and those LIPs of Mesozoic and Cenozoic age share many features with the CFB remnants present in older Proterozoic and even in Archean LIPs.

4

Oceanic LIPs: oceanic plateaus and ocean-basin flood basalts and their remnants through time

4.1 Introduction

This chapter addresses Large Igneous Provinces (LIPs) in an oceanic setting, a record best expressed in Mesozoic–Cenozoic time with oceanic flood basalts emplaced onto existing oceanic crust (Fig. 4.1). Remnants of older oceanic LIPs can also be preserved during ocean closure by accretion into orogenic belts.

Two types of oceanic LIPs were identified by Coffin and Eldholm (1994, 2001) (Fig. 4.2), namely oceanic plateaus and ocean-basin flood basalts. Strictly speaking, oceanic plateaus (e.g. Ontong Java, Kerguelen) form in the deep-ocean basins as broad, more or less flat-topped plateaus, wheras the ocean-basin flood basalts (e.g. Nauru) are extensive submarine lava flows accumulating at abyssal depths in the ocean basins. The difference is largely a question of the scale of magmatism; while ocean-basin flood basalts reflect enhanced thicknesses of crust of a few kilometers, oceanic plateaus correspond to increased crustal thicknesses often of typically more than 10 km. In current usage both types are usefully grouped as oceanic plateaus, since ocean-basin flood basalts are typically genetically linked to oceanic plateaus (e.g. Kerr, 2014; Ingle and Coffin, 2004).

4.1.1 Comparison of crustal structure

Figure 4.2 shows the broad similarities in crustal structure between oceanic LIPs (oceanic plateaus and ocean-basin flood basalts) and continental LIPs, each with extrusive surface (flood basalts), underplate (lower-crustal body), and intrusive components (dykes, sills, and layered intrusions). The differences in composition and thickness of the crust between oceanic and continental LIPs have consequences for the geochemical characteristics. Since most oceanic LIPs erupt through relatively thin ($\sim$ 6–7 km) mafic and ultramafic crust, they are unlikely to be chemically modified by crustal contamination as much as continental LIPs, which traverse thicker crust of intermediate to silicic composition. Therefore, oceanic LIPs are potentially more useful than their continental flood-basalt counterparts for deciphering mantle processes and the sources involved in LIP formation. A comparison of the geochemistry of oceanic LIPs with other types of LIPs is provided in Chapter 10.

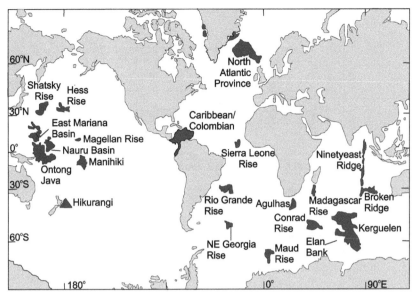

Figure 4.1 Map showing all major oceanic plateaus formed within the last 150 Ma. From Kerr (2014). With permission from Elsevier. Digital version of original figure kindly provided by A.C. Kerr.

4.2 Oceanic plateaus

Oceanic plateaus (e.g. Ontong Java, Kerguelen) form in the deep-ocean basins as broad, more or less flat-topped features lying 2000 m or more above the surrounding seafloor (e.g. Coffin and Eldholm, 1994, 2001). Commonly isolated from major continents, their crust is significantly thicker than adjacent regions of the oceanic crust (up to 38 km thick vs. 7 km thick) and their age may or may not be similar to that of the surrounding seafloor. As illustrated with specific cases below they can form in intraplate settings or even on spreading ridges and the flux of magma from the mantle may or may not be sufficient to build an oceanic plateau above sea level.

4.2.1 Ontong Java LIP (120 Ma)

General characteristics

The submarine Ontong Java Plateau (OJP) is the world's largest LIP (Coffin and Eldholm, 1994; Neal *et al.*, 1997; Fitton *et al.*, 2004). It covers an area of about 2.0 Mkm^2 (comparable in size with Western Europe), and OJP-related volcanism extends over a considerably larger area into the adjacent Nauru, East Mariana, and possibly Lyra and Pigafetta basins and also including the obducted sections of this LIP on the Solomon Islands (Figs. 4.1 and 4.3). With a maximum thickness of crust of 30–35 km (e.g. Gladczenko *et al.*, 1997; Richardson *et al.*, 2000; Miura *et al.*, 2004), the volume of igneous rock forming the plateau is approximately 44 Mkm^3 and, if the amount

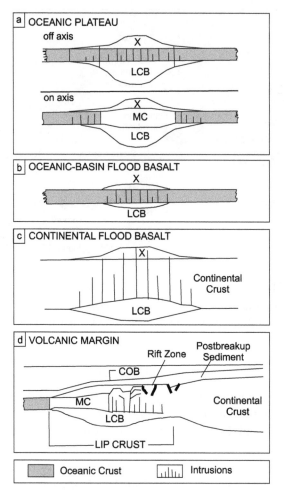

Figure 4.2 Comparison of schematic crustal structure of oceanic LIPs – (a) oceanic plateaus and (b) ocean-basin flood basalts – with (c) continental flood basalts and (d) volcanic rifted margins. LIP crustal components are extrusive cover (X), middle crust (MC), and lower-crustal body (LCB). Normal oceanic crust is grey. COB: continent–ocean boundary. Redrafted from Figure 2 in Coffin and Eldholm (2001).

filling the adjacent basins is included (see Section 4.4.1), the total amount could be as high as 60 Mkm³ (e.g. Coffin and Eldholm, 1994; Ingle and Coffin, 2004). Most of this volume of magma was erupted under deep water. However, the identification of a thick succession of volcaniclastic rocks at site 1184 in the eastern salient of the OJP shows that at least part of the plateau was erupted in a subaerial environment.

Seismic-tomography has identified a rheologically strong but seismically slow upper-mantle root extending to about 300 km depth beneath the OJP (e.g. Richardson *et al.*, 2000; Klosko *et al.*, 2001), interpreted to represent a compositional rather than thermal anomaly in the underlying mantle.

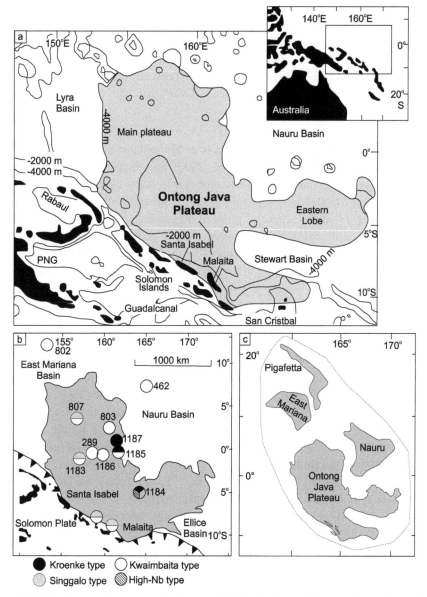

Figure 4.3 The Ontong Java oceanic plateau (a) Distribution of Ontong Java plateau (grey area) and main divisions and regional geography, including bathymetry. Modified after Figure 1a in Sano and Yamashita (2004). (b) Distribution of drillholes and their chemostratigraphy is modified after Figure 9 in Fitton and Godard (2004) with the subduction zone location added after Sikora and Bergen (2004). (c) Link with Nauru basin, East Mariana and Pigafetta ocean-basin flood basalts into "Greater Ontong Java." Redrafted from Ingle and Coffin (2004). With permission from Elsevier.

Collision of the OJP with the older Solomon arc has resulted in uplift of the OJP's southern margin to create on-land exposures of basaltic basement in the Solomon islands (Fig. 4.3), notably in Malaita, Santa Isabel, and San Cristobal (e.g. Petterson *et al.*, 2004). In addition to these exposures, the basaltic basement on the OJP and surrounding Nauru and East Mariana basins has been sampled at Deep Sea Drilling Project (DSDP) and Ocean Drilling Program (ODP) drill sites (Fig. 4.3).

Age

The OJP formed rapidly at around 120 Ma (e.g. Mahoney *et al.*, 1993; Tejada *et al.*, 2002; Chambers *et al.*, 2002; Parkinson *et al.*, 2002), and the peak magma production rate may have exceeded that of the entire global mid-ocean ridge system at the time (e.g. Tarduno *et al.*, 1991; Mahoney *et al.*, 1993; Coffin and Eldholm, 1994). Previous Ar–Ar dating suggested that a second pulse of magmatism occurred at *c.* 90 Ma (see Neal *et al.*, 1997), although some subsequent Ar–Ar geochronology (e.g. Chambers *et al.*, 2002; Tejada *et al.*, 2002) has discounted this younger pulse and suggests that all of the OJP magmas were emplaced in a single pulse at *c.* 120 Ma (e.g. Fitton *et al.*, 2004).

Composition

Three chemotypes of basalt are present within the OJP: the Kwaimbaita, Kroenke, and Singgalo groups (Fitton and Godard, 2004). All exhibit similar flat patterns on chondrite-normalized rare-earth element (REE) diagrams; more details are provided in Chapter 10. The Kwaimbaita chemotype is the most abundant type of basalt on the plateau, and is found at all but one of the OJP drill sites; it therefore represents the dominant type of magma within the OJP. Kroenke type basalts are more magnesian (8.5–11.0 wt% MgO) and have lower concentrations of incompatible trace elements than Kwaimbaita type basalts. These features, combined with radiogenic isotope ratios that are indistinguishable from Kwaimbaita-type basalts, suggest that Kroenke-type basalts represent the parental magmas of the Kwaimbaita type (Fitton *et al.*, 2004; Tejada *et al.*, 2004). The Singgalo type is volumetrically minor; it overlies the Kwaimbaita type and has a similar range in MgO contents (6–8 wt%), but is more enriched in incompatible trace elements than the other two types.

4.2.2 Kerguelen–Bunbury–Comei LIP (130–100 Ma)

General characteristics and age

The Kerguelen LIP is the second largest oceanic plateau on Earth (Fig. 4.4; e.g. Coffin and Eldholm, 1994; Frey *et al.*, 2000; Wallace *et al.*, 2002). It covers an area of *c.* 2.3 Mkm^2, has a volume of 25 Mkm^3 (Coffin *et al.*, 2002) and formed in multiple pulses over a protracted time period of about 20 Ma (*c.* 120–100 Ma). Despite its scale it probably represents the second stage of a multi-pulse LIP that began before

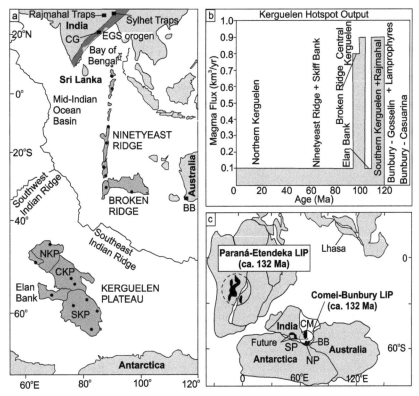

Figure 4.4 The Kerguelen–Bunbury–Comei LIP. (a) The Kergeuelen pulses of the LIP consisting of the Kerguelen plateau – north (NKP), central (CKP), southern (SKP), and Elan Bank portions, Ninetyeast Ridge, Rajmahal and Sylhet Traps. BB = Bunbury and CG = Chilka granulites in Eastern Ghats Shillong orogen (EGS). By permission of Oxford University Press. (b) Age distribution of magmatism (modified after Coffin *et al.*, 2002). By permission of Oxford University Press. (c) Generalized plate-tectonic reconstruction at *c.* 132 Ma of Gondwana (Zhu *et al.*, 2009), showing proximity of the initial pulse of the LIP (represented by the Bunbury basalts and Comei basalts). In part c, CM = remnant Comei LIP, SP = Shillong Plateau, NP = Naturaliste Plateau, and BB = Bunbury basalts.

continental breakup with the eruption of the Bunbury basalts in western Australia at *c.* 132–123 Ma (Coffin *et al.*, 2002) and the recently recognized and correlated Comei mafic intrusions of Tibet that are adjacent to western Australia in a Gondwana reconstruction (Zhu *et al.*, 2009). As the Indian Ocean opened, this initial Bunbury-Comei magmatic pulse was followed by the Rajmahal and Sylhet volcanism in eastern India at *c.* 118 Ma (Kent *et al.*, 1997; Ghatak and Basu, 2011) and the initial construction of the southern Kerguelen plateau at *c.* 119 Ma (Duncan, 2002). Another pulse of this multi-pulse LIP occurred at *c.* 100 Ma and was concentrated at Broken Ridge and the central Kerguelen plateau area (Fig. 4.4). The remaining plateaus (North Kerguelen and Ninetyeast Ridge) and the Kerguelen archipelago were constructed over the following 85 Ma before spreading began on the Southeast

Indian Ridge, dividing the Kerguelen plateaus from the Broken Ridge. Strictly speaking, the last 85 Ma of magmatism is not LIP-related given its protracted and lower-volume magmatic production. At peak output (corresponding to a LIP pulse sensu stricto), Kerguelen eruption volumes may have been 0.9 km^3/a or more (Coffin *et al.*, 2002), and the Kerguelen, Heard, and McDonald islands are the most recent products of the Kerguelen hotspot, which remains presently active.

Composition

The Kerguelen oceanic-plateau pulses of the Kerguelen–Bunbury–Comei LIP have the broadest age range and most diverse composition of the known oceanic plateaus (Weis *et al.*, 1993; Mahoney *et al.*, 1995; Frey *et al.*, 1996, 2000; Kent *et al.*, 1997; Wallace *et al.*, 2002; Neal *et al.*, 2002; Coffin *et al.*, 2002; Ingle *et al.*, 2004; Zhu *et al.*, 2008). The central Kerguelen plateau is tholeiitic, with comparable composition (major and trace elements) to the Ontong Java and Caribbean–Colombian oceanic plateaus. The younger North Kerguelen plateau and Kerguelen archipelago are more alkalic and are more large-ion lithophile-element (LILE)-enriched. Many early-stage magmatic products within the Kerguelen oceanic plateau are derived from magmas that appear to have traversed and assimilated continental lithosphere during their ascent. Continental lithosphere is now known to underlie significant regions of the plateau, particularly at Elan Bank and within the Southern Kerguelen plateau (Fig. 4.4) (e.g. Frey *et al.*, 2000). All of the < 40 Ma basalts on the Kerguelen Archipelago and the North Kerguelen plateau, together with the Ninetyeast Ridge, erupted in an entirely oceanic setting and show no sign of a continental influence. In addition to the more voluminous tholeiites, the *c.* 118 Ma Rajmahal–Sylhet volcanic province is also characterized by contemporaneous alkaline volcanism, including lamproite and kimberlite intrusions and also the formation of alkaline carbonatite complexes (Ghatak and Basu, 2011).

4.2.3 Caribbean–Colombian LIP (94–89 Ma)

General characteristics and age

The *c.* 90 Ma Caribbean–Colombian Oceanic Plateau (CCOP) is exposed around the margins of the Caribbean and along the northwestern continental margin of South America (e.g. Burke, 1988; Kerr and Tarney, 2005; Kerr, 2014) (Fig. 4.5). It also underlies the bulk of the Caribbean plate where it has been sampled by DSDP drill holes. Strictly speaking, it does not form a plateau, and was originally classified as an ocean-basin flood basalt by Coffin and Eldholm (1994, 2001). However, the absence of a preserved plateau (despite its 15 km total thickness) is probably due to this LIP being dismembered/disturbed as it was tectonically forced into the gap between North and South America (A. Kerr, personal communication, 2011). A speculative model suggests that this LIP is a juxtaposition of two different

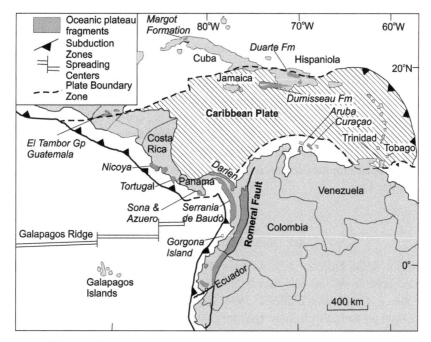

Figure 4.5 The Caribbean–Colombian LIP: main accreted outcrops of the Caribbean–Colombian oceanic plateau along with the distribution on the Caribbean plate shown by hatchuring. From Kerr (2014). With permission from Elsevier. Digital version of original figure kindly provided by A.C. Kerr.

LIPs, both at *c.* 90 Ma, with one linked with the Caribbean plateau and the other of a similar age linked to the Gorgona plateau, and these are each linked to separate present-day hotspots, Galapagos (Fig. 4.5) and Sala y Gomez (26.5° S, 105.4° W), respectively (Kerr and Tarney, 2005).

The preserved, water-covered area of the CCOP is about 0.6 Mkm^2 with a volume estimated at 4.5 Mkm^3 (Eldholm and Coffin, 2000). However, because a significant portion of the plateau appears to have accreted onto the western continental margin of Colombia and Ecuador (and some may have subcreted), the oceanic plateau may originally have been more than twice this size. Some sedimentary exposures associated with the plateau in the Western Cordillera of Colombia contain fragments of carbonized tree trunks and coral; taken together, these occurrences suggest that deposition occurred in shallow-marine, or even subaerial, environments during the Late Cretaceous (Kerr, 2014).

Composition

Most of the exposures of the plateau around the Caribbean and in northwestern South America consist of fairly homogeneous tholeiitic basalts (e.g. Kerr, 2014) with flat, chondrite-normalized REE patterns, and containing 6–10 wt% MgO (more details are found in Chapter 10). The accreted plateau material in Colombia, Ecuador, Costa Rica, Jamaica, and Hispaniola consists of fault-bounded slices of basaltic, and occasionally picritic, lavas and sills along with layered and isotropic gabbros and ultramafic rocks. Oceanic-plateau-derived high-MgO basalts (> 12 wt%), picrites, and komatiites are also found. On the island of Curaçao, 70 km off the coast of Venezuela, a 5-km-thick sequence of pillowed picrites and basalts intercalated with hyaloclastites and intrusive sheets is exposed. The small island of Gorgona located about 50 km off the western coast of Colombia, is the site of the youngest known komatiites (MgO-rich lava flows: > 15 wt%), which possess platy and blade-shaped (spinifex-textured) olivines (cf. Archean komatiites in greenstone belts; Chapter 6).

4.3 Ocean-basin flood basalts

Ocean-basin flood basalts, such as those within the Nauru basin, are extensive submarine lava flows and sills that lie above and post-date normal oceanic crustal basement (e.g. Coffin and Eldholm, 1994, 2001). Varying considerably in thickness (10 to > 1000 m) and morphology, they form at abyssal depths in the ocean basins. In some cases (as discussed later) their proximity and age-match to major oceanic plateaus suggests they are linked to and indeed fed from these plateaus.

4.3.1 Nauru basin LIP (120 Ma)

The Nauru basin in the western Pacific (Figs. 4.1 and 4.3) was initiated by seafloor spreading in late Jurassic and early Cretaceous time (as estimated from observed magnetic lineations) and was subsequently affected by Early Cretaceous flood volcanism, which roughly coincides with emplacement of the nearby *c.* 122 Ma OJP (Mochizuki *et al.*, 2005). Multi-channel seismic reflection and refraction data indicate that the Nauru igneous Complex is thickest (*c.* 5500 m) in the center of the basin. Volcaniclastic sediments in the Nauru basin have composition similar to rocks of the OJP, supporting a link with the OJP (Castillo, 2004). The total volume of the Early Cretaceous Nauru igneous Complex is 3.3 Mkm^3 (Mochizuki *et al.*, 2005).

4.4 Reconstruction of links between oceanic LIPs

As in the continental LIP record (Chapters 3 and 11), distinct oceanic LIPs of the same age can sometimes be genetically linked. Here, I consider two types of linkages: (1) the matching of oceanic plateaus with nearby ocean-basin flood basalts, and (2) reconstruction of oceanic LIPs that have been fragmented by rifting of oceanic plates (Fig. 4.6).

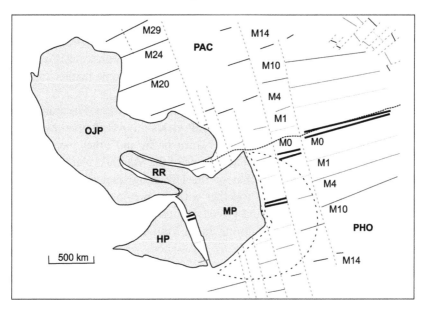

Figure 4.6 Reconstruction of the Ontong Java–Manihiki–Hikurangi plateau at M0 time (*c.* 125 Ma), just before its breakup at the end of Pacific (PAC)–Phoenix (PHO) spreading. The coarse dashed grey line depicts the possible former plateau east of Manihiki Plateau (MP) rifted away (and suggested by Kerr and Mahoney (2007) to potentially be the Gorgona plateau of Kerr and Tarney (2005) (Section 4.2.3). OJP = Ontong Java Plateau, HP = Hikurangi Plateau and RR = Robbie Ridge. Redrafted from Taylor (2006). With permission from Elsevier.

4.4.1 "Greater" Ontong Java (120 Ma)

The ocean-basin flood basalts (Nauru, East Mariana, and possibly Pigafetta) in the vicinity of the OJP, have ages similar to the OJP. This suggested to Ingle and Coffin (2004) that these oceanic flood basalts are linked to the OJP, and the term "Greater Ontong Java" LIP was applied to this broader region of magmatism (Fig. 4.3). The mechanism for feeding these Nauru basin flood basalts from the presumed locus of mantle-plume activity centered on the OJP has not been identified, but must have involved lateral flow of lava from oceanic plateaus or via laterally emplaced dykes, sills, or sublithospheric channeling (see discussion of plumbing system mechanisms in Chapter 5).

4.4.2 Reconstruction of rifted oceanic plateaus

Ontong Java–Manihiki–Hikurangi reconstructed plateau (120 Ma)

Seafloor fabric data indicate that the Ontong Java, Manihiki, and Hikurangi plateaus (Fig. 4.3) originally formed as a single oceanic plateau (Taylor, 2006) (Fig. 4.6). These data support previous interpretations that the Osbourn Trough is a relict spreading center that separated the Manihiki and Hikurangi plateaus, although this model requires a revised interpretation of the tectonic model for the Ellice basin to represent spreading between the Manihiki and Ontong Java plateaus. Geochemical and geochronological data from Manihiki plateau provide some further support for a link with the OJP (e.g. Kerr and Mahoney, 2007).

Combining the 0.8 Mkm^2 Manihiki plateau with the 0.7 Mkm^2 Hikurangi plateau and with the 1.9 Mkm^2 areal extent of the OJP yields an overall areal extent of 3.4 Mkm^2, which would be still larger if the Nauru basin and other members of the "Greater Ontong Java" are included. The volume of the Manihiki plateau is variously estimated at 8.8–13.6 Mkm^3, and that of the Hikurangi plateau at > 2.7 Mkm^3 (Coffin and Eldholm, 1994). Taken together with the previously mentioned volume estimates for the OJP (44 Mkm^3) and Greater Ontong Java (60 Mkm^3), the overall original volume of the combined Ontong Java, Manihiki, and Hikurangi plateaus would have been at least 60–76 Mkm^3.

Agulhas plateau and its links with the Northeast Georgia Rise and the Maud Rise (Southeast African LIP, 100–94 Ma)

The Agulhas plateau is an oceanic plateau south of South Africa in the southwestern Indian Ocean (Figs. 4.1 and 4.7). It rises up to 2500 m above the surrounding seafloor, covers an area of more than 0.3 Mkm^2, and hence represents a major bathymetric high in this region. The Agulhas plateau is separated from the South African continent by the up to 4700-m-deep Agulhas Passage. The volume of the Agulhas plateau has been estimated at 4 Mkm^3. It has lava-flow structures, crustal thicknesses of up to 25 km, and high seismic velocities of 7.0–7.6 km/s for the lower part of the crust, as well as an extrusive cover, an intruded middle part, and a lower-crustal body (Uenzelmann-Neben *et al.*, 1999; Parsiegla *et al.*, 2008; Gohl *et al.*, 2011). It has been further suggested that the Agulhas plateau was formed as part of a larger LIP comprising the Agulhas plateau, the Northeast Georgia Rise, and the Maud Rise, and collectively termed the Southeast African LIP (Gohl *et al.*, 2011; Fig. 4.7).

4.5 Accreted oceanic plateaus

Ocean closure can cause oceanic plateaus to be accreted into orogenic belts and thus they may be preserved. Here I present one of the best studied examples; the Wrangellia accreted oceanic plateau of northwestern North America.

4.5.1 Wrangellia accreted oceanic plateau (230 Ma)

General characteristics

The Wrangellia LIP (also known as the Nikolai LIP) is distributed along the > 2500-km length of the Wrangellia accreted terrane along the western North American

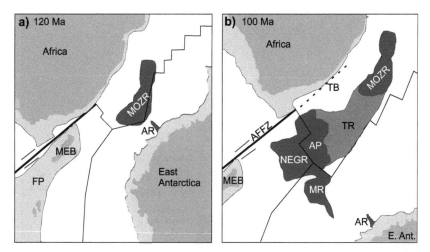

Figure 4.7 120–100 Ma Southeast African LIP. Reconstruction model for formation of the southeast African LIP between (a) 120 and (b) 100 Ma. Mozambique Ridge (MOZR), Agulhas plateau (AP), Northeast Georgia Rise (NEGR), Maud Rise (MR), and the northern Astrid Ridge (AR) are fully developed oceanic-plateau LIPs, while Transkei Rise (TR) is a partially developed LIP province. AFFZ = Agulhas–Falkland Fracture Zone, TB = Transkei basin, FP = Falkland plateau, MEB = Maurice Ewing Bank, E. Ant = East Antarctica. From Figure 5 in Gohl *et al.* (2011). Digital version of original figure kindly provided by K. Gohl.

continental margin (Fig. 4.8) (e.g. Richards *et al.*, 1991; Lassiter *et al.*, 1995; Hulbert, 1997; Schmidt and Rogers, 2007; Greene *et al.*, 2009, 2010). Prior to accretion the Wrangellia flood basalts were originally erupted onto older Paleozoic arc volcanic and marine sedimentary sequences in both shallow and deep marine settings.

Much of the original stratigraphic thickness of the Wrangellia plateau is intact and is defined as the Karmutsen Formation on Vancouver and Queen Charlotte Islands (Haida Gwaii), and as the Nikolai Formation in southwest Yukon and south-central Alaska (Fig. 4.8). On Vancouver Island, the volcanic stratigraphy is a three-part succession of submarine, volcaniclastic, and subaerial flows approximately 6 km thick. In Alaska and Yukon, the volcanic stratigraphy (*c.* 3.5 km) is predominantly massive subaerial flows with a small proportion of submarine flows along the base. Smaller units in southeast Alaska may be correlative with the Wrangellia flood basalts. Throughout areas of Wrangellia, the flood basalt stratigraphy is bounded by Middle to Late Triassic marine sediments.

While Barker *et al.* (1989) interpreted a back-arc rift setting for the Karmutsen Formation on Vancouver Island, the Wrangellia LIP is more typically interpreted as an oceanic plateau produced by a mantle plume head (e.g. Richards *et al.*, 1991). Paleontological and paleomagnetic studies indicate that the Wrangellia flood

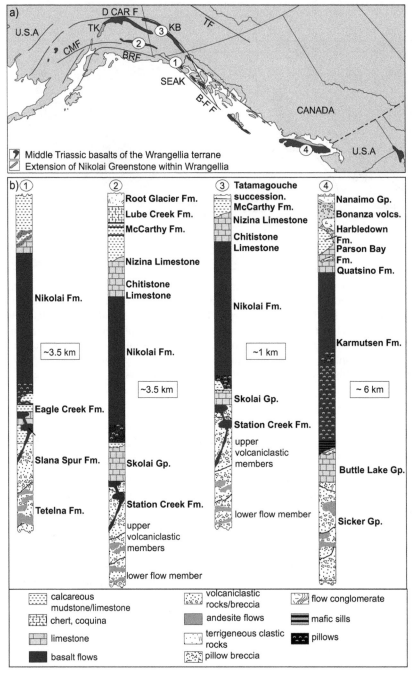

Figure 4.8 The Wrangellia accreted oceanic plateau. (a) Distribution of accreted fragments in Cordillera of western North America. (b) Stratigraphy of the accreted section of the Wrangellia oceanic plateau in Alaska, Yukon, and British Columbia. Total estimated stratigraphic thicknesses for the Nikolai and Karmutsen formations are outlined with boxes. The distribution in (a) is modified after Schmidt and Rogers (2007). The stratigraphy in (b) is simplified from Figure 3 in Greene *et al.* (2010). Digital version of original figure used for (b) was kindly provided by A. Greene.

basalts probably erupted as an oceanic plateau in the eastern Panthalassic Ocean in equatorial latitudes (Jones *et al.*, 1977; Katvala and Henderson, 2002). The subsequent accretion of the Wrangellia oceanic plateau to western North America was a major tectonic event and represents a significant addition of oceanic mantle-derived material to western North America (Condie, 2001; Greene *et al.*, 2009).

Age

Formation of the Wrangellia LIP was a relatively short-lived event based on pale-ontological grounds and on age dating. As summarized in Schmidt and Rogers (2007) and Greene *et al.* (2010), U–Pb ages include: 227.3 ± 2.6, 226.8 ± 0.5 Ma, and 228.4 ± 2.5 from gabbroic rocks on southern Vancouver Island, 232.2 ± 1.0 Ma on the Maple Creek gabbroic feeders to the basalts in the Nikolai Formation in the Yukon Territory; Ar^{40}–Ar^{39} ages include 228.3 ± 1.1 Ma and 230.4 ± 1.3 Ma from mafic–ultramafic intrusions in the central Alaska Range, and 231.1 ± 11.0 Ma from a gabbro in the northern Talkeetna Mountains. These radiometric ages, all from intrusive rocks interpreted to be co-magmatic with the flood basalts, fall within the late Ladinian (late Middle Triassic) and suggest that the entire Wrangellia LIP event lasted from *c.* 230–225 Ma, possibly as few as 2 million years.

Composition

The vast majority of the flood basalts are LREE-enriched, high-Ti basalts. In Alaska and Yukon, low-Ti basalts form the lower parts of the volcanic stratigraphy, and the remainder of the volcanic stratigraphy is comprised of high-Ti basalts (e.g. Greene *et al.*, 2009). The low-Ti basalts have compositions that indicate melting and/or interaction with subduction-modified, lithospheric mantle material, whereas the high-Ti basalts were derived from a relatively uniform (plume-type) Pacific mantle source. On Vancouver Island, the volcanic stratigraphy is dominated by tholeiitic basalts, with minor volumes of picritic pillowed basalts erupted late in the submarine phase of volcanism. Volcanological differences in the stratigraphy are primarily related to the eruption environment (deep-water to shallow-water to subaerial; Greene *et al.*, 2009, 2010).

4.6 Pre-Mesozoic oceanic LIPs

4.6.1 Missing oceanic LIP record

Oceanic plateaus are abundant in the present ocean basins (back to 180 Ma) where there are at least a dozen oceanic plateaus (and ocean-basin flood basalts) that are thought to be derived from plumes (Fig. 4.9; e.g. Coffin and Eldholm, 2001; Ernst and Buchan, 2001b; Arndt and Weis, 2002). As discussed in Chapter 6, in

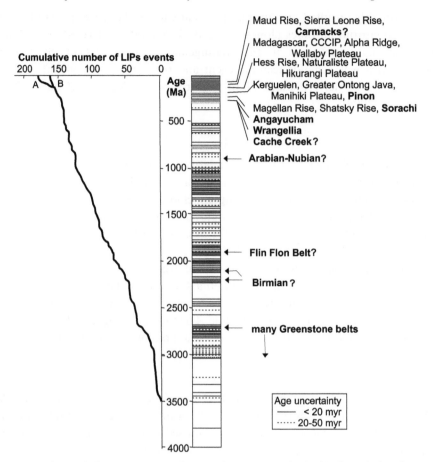

Figure 4.9 The "Missing" pre-Mesozoic oceanic LIP record. (Left) Cumulative frequency diagram of LIP events (after Ernst, 2007a). The average rate of preserved LIP production (on cumulative frequency diagram) from 2500 to about 200 Ma is about 1 per 20 Ma, a rate which continues between 200 Ma and the present for curve B (plotting only the continental record of LIPs). Curve A (which plots all the available LIPs during that interval) is steeper because it includes the full oceanic LIP record that is mostly lost during ocean closure, and therefore is greatly diminished and obscured in the pre-Mesozoic record. (Right) The LIP record through time; only Oceanic LIPs are labeled. Those that are accreted in orogenic belts during oceanic closure are in bold; most of these apart from Wrangellia are speculative. From Figure 2 in Dilek and Ernst (2008). With permission from Elsevier.

the Archean there are at last 14 greenstone belts that are interpreted to represent oceanic plateaus (Tomlinson and Condie, 2001; Arndt *et al.*, 2001; see also listing in Ernst and Buchan, 2001b). The rapid drop in the number of identified oceanic LIPs older than 180 Ma is simply due to preservation potential; i.e. the absence of oceanic crust older than about 180 Ma, and the poor preservation potential of oceanic LIPs during ocean closure and difficulty of recognizing fragments of accreted oceanic plateaus in orogenic belts.

Because the average rate of continental LIP production throughout the entire Proterozoic and Phanerozoic period (including within the last 200 Ma) was approximately constant with an average of about 1 per 20 Ma (Ernst and Buchan, 2002), it can be inferred that the oceanic LIP production rate observed between 180 and the present, 1 per 20 Ma, was also broadly constant back to 2500 Ma (Fig. 4.9). If so, it is surmised that there must be more than 100 oceanic LIP events in Proterozoic–Paleozoic time, mostly currently unrecognized (Fig. 4.9). See also discussion of LIP production through time in Section 11.8.

Strategies for recovering this "lost" LIP oceanic record include (1) recognizing which ophiolites and other accreted volcanic/plutonic packages are of oceanic LIP origin (e.g. Coffin and Eldholm, 2001), and (2) tracing giant dyke swarms (up to 2500 km in length) that are predicted to radiate from "lost" oceanic LIP centers onto formerly adjacent continental landmasses (Ernst and Buchan, 1997a). Point 1 is addressed below and point 2 in Section 5.2.7.

Because of their great crustal thickness, and increased buoyancy (compared to normal oceanic crust) oceanic plateaus are relatively unsubductable, particularly those thicker plateaus and those that reach subduction zones relatively soon after formation (e.g. Cloos, 1993; Abbott and Mooney, 1995; Arrial and Billen, 2013). This means that they can "dock" to the upper-plate margins of subduction zones and be at least partially preserved in the geological record as accreted terranes (Kerr, 2014). For example, the Caribbean plateau of the Caribbean-Colombian LIP largely resisted subduction because it collided with an arc at the entrance to the proto-Caribbean basin only a few million years after the plateau formed at *c*. 90 Ma (Burke, 1988; Kerr *et al.*, 1999) (Fig. 4.10). The OJP, most of which formed at *c*. 122 Ma, collided with the Solomon subduction zone *c*. 100 Ma later (Fig. 4.3) (e.g. Petterson, 2004). The lower crust at the edge of the plateau was subducted (Mann and Taira, 2004), but overall the plateau resisted subduction because of a combination of very thick crust, an anomalously small amount of post-emplacement subsidence, and a > 300-km-thick bouyant lithospheric mantle "root" (e.g. Section 4.2.1).

4.6.2 Criteria for recognizing oceanic plateaus in orogenic belts

Table 4.1 summarizes criteria from Kerr *et al.* (2000) that can be used to discriminate between oceanic plateau sequences and other types of volcanic rocks, in volcanic arcs, mid-ocean ridges, marginal (or back-arc) basins, ocean-islands, continental flood-basalt provinces, and seaward-dipping reflector sequences. Basically, oceanic plateaus should have the following: (a) the presence of

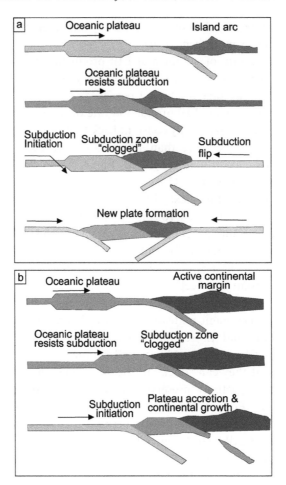

Figure 4.10 Accretion of oceanic plateaus: idealized cross sections illustrating the likely effects of the collision of an oceanic plateau with (a) an island arc and (b) an active continental margin. From Kerr (2014). With permission from Elsevier. Digital version of original figure kindly provided by A.C. Kerr.

high-MgO lavas (picrites and komatiites); (b) chemically homogeneous basalts with relatively flat chondrite-normalized REE patterns and low (primitive mantle-normalized) La/Nb ratios (<1); (c) pillowed lavas, a low abundance of volcaniclastic deposits; (d) lack of sheeted dyke complexes; and (e) a relatively thick (*c*. 5 km) extrusive section. Some plateaus show a predominance of subaerial eruptions, with associated flow morphologies. See also Safonova (2009) for a further discussion of criteria for recognizing oceanic intraplate magmatism that has accreted into orogenic belts.

Table 4.1 *Geochemical and geological characteristics of volcanic sequences from different tectonic settings (from Kerr et al., 2000)*

Tectonic setting	High-MgO lavas (> 14%)	La_{pmn}/Nb_{pmn}	$(REE)_{cn}$ pattern	Pillowed lavas	Tephra layers	Subaerial eruption
Oceanic plateau	Yes	≤ 1	Predominantly flat	May be common (e.g. OJP) or absent (e.g. Kerguelen)	Very few	Occasionally
Mid-ocean ridge	Rare	≤ 1	LREE-depleted	Common	Very few	No
Marginal basin	Rare	≤ 1	Predominantly flat	Common	Very common	No
Oceanic island	Rare	≤ 1	Predominantly LREE-enriched	Present	Very few	Frequently
Volcanic rifted margin	Yes	Contain sequences with ≤ 1 and >> 1	LREE-enriched	Not all lavas are pillowed	Very common	Common
Arc (continental and oceanic)	Rare	>> 1	LREE-enriched	Not all lavas are pillowed	Very common	Frequently
Continental flood basalts	Yes	Mostly >> 1; < 10% of flows ≤ 1	Flat to LREE-enriched	Usually absent	Occasional	Always

pmn; primitive-mantle normalized; cn, chondrite normalized.

Table 4.2 *Intraplate basalts of the Paleo-Asian and Paleo-Pacific oceans and their hosting accretionary complexes (ACs) From Safonova (2009).*

Age of intraplate magmatism	Hosting oceanic plate	Intraplate basalts and accretionary complexes
I. Paleo-Asian Ocean (640–340 Ma)		
Late Neoproterozoic–Early Cambrian	Paleo-Asian Ocean (maximal opening)	Dzhidot and Urgol paleoseamounts, Dzhida AC
Late Neoproterozoic, 600±25 Ma;	Paleo-Asian Ocean	Kurai paleoseamount, Kurai AC
Early Cambrian;		Katun paleoseamount, Katun AC
Late Cambrian–Early Ordovician		Fragments of seamounts in the Zasur'ia series; Charysh-Terekta AC
Late Devonian–Early Carboniferous	Paleo-Asian Ocean (closure)	Fragments of seamounts in the Karabaev and Verrochar Formations; Chara AC
II. Paleo-Pacific Ocean (320–140 Ma)		
Carboniferous	Farallon plate (FP)	Akiyoshi-Sawadani seamount chain; FP Akiyoshi and Khabarovsk (?) ACs
Permian	Farallon plate; Izanagi plate	Maizuru plateau (FP); Maizuru AC
		Akasaka–Kuzuu seamount chain; Mino–Tamba and Samarka ACs
Late Jurassic	Izanagi plate	Mikabu plateau and seamounts; Southern Chichibu and Taukha ACs

4.6.3 Orogenic belts with accreted oceanic LIPs

Figure 4.11 shows regions of ocean closure within the Central Asian orogenic belt since 600 Ma, an area that is expected to host numerous oceanic plateaus and also oceanic islands (plume-tail magmatism). The history of intraplate oceanic magmatism in these regions is exemplified by the distribution of accreted oceanic LIPs and ocean islands in the Altai–Sayan region of southern Siberia (Fig. 4.11, Table 4.2); full details are provided in Safonova (2009).

4.7 Summary

Oceanic LIPs consist of oceanic plateaus and ocean-basin flood basalts. Oceanic plateaus sensu stricto represent dramatic magmatic accumulations into normal

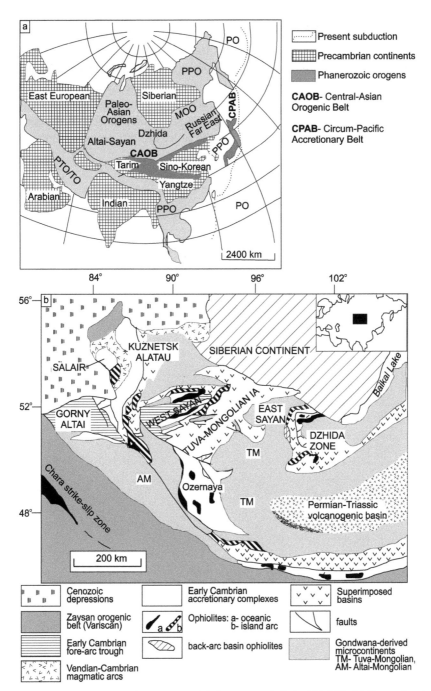

Figure 4.11 Oceanic plateaus in the Central Asian orogenic belt (CAOB). (a) Tectonic scheme of Asia showing main orogens formed in oceans over the past 600 Ma: PPO = Paleo-Pacific, PTO = Paleo-Tethys, TO = Tethys, PO = modern Pacific, MOO = Mongolian–Okhotsk orogenic belt (redrafted from Safonova, 2009). Each of these orogens is a potential host for fragments of oceanic plateaus and oceanic islands. With permission from Elsevier. (b) Early Paleozoic accretion belts and Gondwana-derived Precambrian microcontinents of Altai–Sayan at the southern margin of the Siberian continent (redrafted from Safonova, 2009). With permission from Elsevier. "Oceanic" ophiolites are those derived from oceanic plateaus and oceanic islands.

oceanic crust. Oceanic plateaus have both flood basalt and intrusive portions and presumably also have the equivalent of underplate components. Ocean-basin flood basalts are present in smaller volumes but in many cases are linked to oceanic plateaus and may represent a portion of oceanic-plateau magmatism that flowed into and filled a nearby basin. Oceanic plateaus are poorly preserved during ocean closure and so there are many (perhaps on the order of 100) oceanic LIPs yet to be identified in the pre-Mesozoic ophiolite record.

5

Plumbing system of LIPs

5.1 Introduction

This chapter considers the plumbing system for Large Igneous Provinces (LIPs). In many Proterozoic LIPs, the flood basalts have been mostly eroded revealing the underlying dolerite dykes, sills, and larger differentiated intrusions of layered mafic to ultramafic magmatism. This plumbing system (which also includes magmatic underplating) transports magma into the lithosphere from underlying mantle sources and distributes it throughout the lithospheric mantle and crust and onto the surface as lava flows. Dolerite sills are particularly prominent in sedimentary successions of the upper crust, and regional dyke swarms are common in basement rocks. Layered mafic–ultramafic intrusions commonly form at the boundary between basement and supracrustal sequence. Associated intrusive (and extrusive) components include silicic magmatism mainly derived from melting of the lower crust (Chapter 8) and also carbonatites and kimberlites (Chapter 9).

5.1.1 Definition of dykes vs. sills vs. sheets

I first address a nomenclature issue with respect to bodies having a sheet-like geometry. The traditional definition of sills vs. dykes relates to their orientation with respect to primary layering in the host rock. For instance, Neuendorf *et al.* (2011) define a dyke as "a tabular igneous intrusion that cuts across the bedding or foliation of the country rock," and a sill as "a tabular igneous intrusion that parallels the bedding or foliation of the sedimentary or metamorphic country rock, respectively." However, it is also possible to define dykes and sills in terms of the orientation *at the time of emplacement* regardless of the relationship with bedding/foliation (cf. Hall, 1996). According to Ernst and Buchan (2004), a dyke (dike) "is a tabular igneous body that was sub-vertical at the time of emplacement. A dyke swarm is a set of coeval dykes which typically display a linear, radiating or arcuate geometry" and sill could be correspondingly defined as an originally sub-horizontal tabular body. Given improvements in geochronology, the age of sheet-like intrusions with respect to that of their host rocks and with respect to any deformation can be generally determined and thus the primary orientation (vertical, horizontal, or inclined) can be much more easily determined than previously. The advantage of considering the primary orientation of sheet-like bodies is that it can be related to the regional stress pattern, which is particularly important for dyke swarms (aligned in the direction of maximum compression). In this book I use this nomenclature based on the orientation of magmatic sheets *at the time of emplacement*: dykes are tabular (sheet-like) bodies intruded sub-vertically, sills are tabular bodies intruded sub-horizontally, and sheets are tabular bodies of any orientation but are typically applied to bodies of intermediate dip.

5.2 Dolerite dyke swarms

The most dramatic component of the plumbing system of LIPs is the giant dolerite dyke swarms that are very prominent in most basement terrains and can be traced over areas of up to several Mkm^2.

5.2.1 Giant dolerite dyke swarms

These swarms are at least 300 km long (by definition; e.g. Ernst *et al.*, 1995b) but can be more than 2000 km long, have linear or radiating geometry and consist of

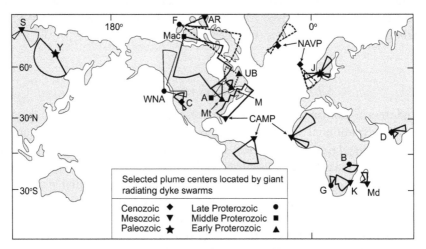

Figure 5.1 Radiating dyke swarms and their mantle plume centers. Selected examples: Columbia River (C), North Atlantic Volcanic Province (NAVP), Deccan (D), Madagascar (Md), Alpha Ridge (AR), Karoo–Ferrar (K), Central Atlantic Magmatic Province (CAMP), Jutland (J), Yakutsk (Y), Gannakouriep (G), Franklin (F), Western North America (WNA), Bukoban (B), Abitibi (A), Mackenzie (Mac), Ungava Bay (UB), Matachewan (Mt), Mistassini (M), and Siberian Trap (S). Modified from Figure 4 in Ernst *et al.* (2001a). See also Ernst and Buchan (1997a, 2001c). Some speculative radiating swarms are presented in Goldberg (2010).

individual dykes with average widths 10 to 40 m (Parker *et al.*, 1990; Ernst *et al.*, 1995a). These are distinct from the dykes associated with volcanic edifices and with mid-ocean-ridge spreading centers and ophiolites, which have widths typically in the range of 0.5 to 2 m with maximum sizes up to a few meters (Ernst *et al.*, 1995b).

Individual dykes in giant dyke swarms have been traced for distances up to 1000 km (Eyal and Eyal, 1987). Moreover, the magnetic fabric evidence for long-distance lateral flow from the central area in the Mackenzie radiating swarm (Section 5.2.4; Ernst and Baragar, 1992; Baragar *et al.*, 1996) suggests that individual dykes can have a length greater than 2000 km. The maximum known widths of dolerite dykes in some swarms are (Figs. 5.1 and 5.2): 150 m for the 2500–2450 Ma Matachewan swarm/LIP, 220 m for the 1140 Ma Abitibi swarm, precursor swarm to the Keweenawan LIP (Ernst and Bell, 1992), 100 m for the 1238 Ma Sudbury swarm/LIP, 100 m for the 1267 Ma Mackenzie swarm/LIP, 200 m for the 590 Ma Grenville swarm, part of the Central Iapetus magmatic province (CIMP), 600 m for the "Great Dyke of Carolina" associated with the Eastern North America portion of the Central Atlantic Magmatic Province (CAMP) event (e.g. Nomade *et al.*, 2007) and 200 m for the Yakutsk swarm of the Yakutsk–Vilyui LIP (Shpount and Oleinikov, 1987).

Giant swarms can be subdivided into six types based on their primary geometry (Fig. 5.3). Dykes of types I, II, and III show convergence and are thought to radiate from the point above the center of a mantle plume and/or triple-junction rifting (e.g. May, 1971; Burke and Dewey, 1973; Fahrig, 1987). Therefore, all of these three types are essentially giant radiating swarms, although type III can also be thought of as rift parallel swarms associated with the arms of a rift triple junction (Section 11.4). Some rift margins contain two sets of coast parallel swarms, a pre-rifting set which is part of a radiating swarm and a syn-rifting swarm. Type I and the subswarms of type II swarms may mark overlying rifts that have been removed by erosion. Types IV and V are linear, but in some cases they may also represent distal (far from focus) portions of giant radiating dyke swarms. Specifically, far from the center, a radiating swarm may sweep into a more linear pattern as a result of regional stress fields at the time of emplacement (see Section 5.2.2). In addition to these well-constrained types, there is an additional, more speculative, type VI, arcuate swarm, which partially or completely circumscribes a magmatic center.

Dykes of type-I swarms form a continuous fan, whereas dykes of type II swarms occur in fanning subswarms separated by dyke-poor regions. The most dramatic example of a type I swarm is the Mackenzie swarm/LIP, Canada (Fig. 5.2c), which fans continuously over an angle of 100°. The most dramatic example of a type II swarm is the Matachewan swarm/LIP, Canada, with dykes concentrated in three prominent subswarms with an overall fan-angle of 60° (Fig. 5.2h), and, in this case, each of the subswarms may reflect the deep exposure level of a rift arm of a triple junction (Chapter 11). An example of a type III swarm is the *c.* 20 Ma dyke swarm associated with the Red Sea rift of the Afro-Arabian LIP (Fig. 5.2b).

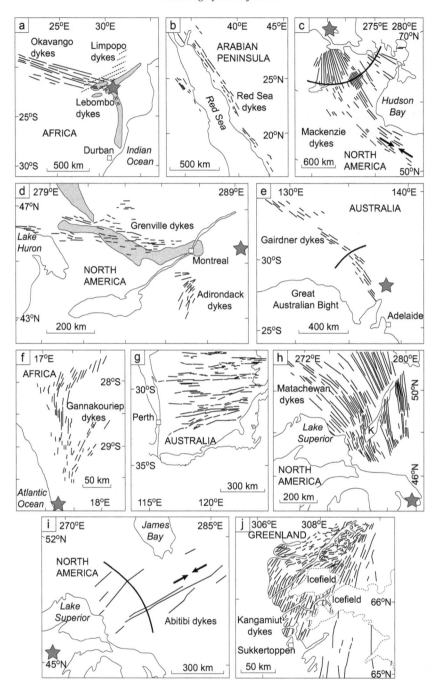

Figure 5.2 Giant dyke swarms. (a) *c.* 180 Ma Karoo dykes; (b) 23 Ma Red Sea dykes of the Afro-Arabian LIP; (c) 1270 Ma Mackenzie dykes; (d) 590 Ma Grenville and in-part related Adirondack dykes of the CIMP; (e) 825 Ma Gairdner dykes of Gairdner–Willouran LIP; (f) 790 Ma Gannakouriep dykes; (g) 2400 Ma Widgiemooltha dykes; (h) 2490–2450 Ma

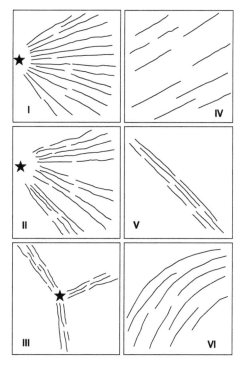

Figure 5.3 Types of regional-scale dyke swarms. Six characteristic geometries of giant radiating dyke swarms: I, continuous fanning pattern; II, fanning pattern divided into separate subswarms; III, subswarms of subparallel dykes that radiate from a common point; IV, subparallel dykes over a broad area; V, subparallel dykes in a narrow zone; VI, arcuate pattern. Stars locate probable mantle plume centers defined by convergence of radiating patterns of dykes. While types IV and V could be subsets of types I–III, they can also have distinct origins as discussed in the text. From Ernst and Buchan (2001c).

Subparallel swarms can either be broadly distributed (type IV) or restricted to a relatively narrow zone (type V) and may represent linear portions of radiating systems (Fig. 5.4). An example of type IV swarm is the 2415 Ma Widgiemooltha swarm/LIP of Australia (Fig. 5.2g). Examples of type V swarms include the 590 Ma Grenville swarm of North America, part of the CIMP event (Fig. 5.2d), and the 825 Ma Gairdner swarm (Gairdner–Willouran LIP) of Australia (Fig. 5.2e).

Caption for Figure 5.2 (*cont.*) Matachewan dykes of the Matachewan (-East Bull Lake) LIP; (i) 1140 Ma Abitibi dykes, precursor to the Keweenawan LIP; (j) Kangâmiut dykes of Kangâmiut–MD3 radiating swarm. Stars locate plume centers based on convergence of the radiating pattern. Circular arcs in (c), (e), and (i) locate change in the trend of dykes, linked to proximity to the plume center. Arrows define the axis of regional stress field at the time of the dyke emplacement. Rift zones and volcanic sequences contemporaneous with dyke emplacement are shaded grey. In (h), K= Kapuskasing Structural Zone, an area of *c*. 1900 Ma thrusting and uplift. Modified from Ernst *et al.* (1995a).

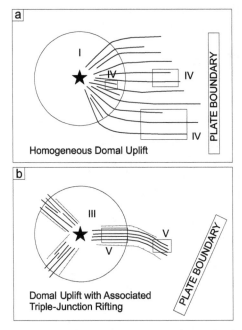

Figure 5.4 Giant dyke swarms and their relationships with mantle plume centers (marked by stars) under two scenarios: (a) homogeneous domal uplift and (b) domal uplift with associated triple-junction rifting. Solid lines are dykes, and lines with hachures are rift boundaries. Uplift region is outlined by a circle. Parts of each swarm are labelled with their geometric types using the terminology of Figure 5.3. Distal plate boundaries that may control dyke orientations beyond the plume-related uplift are illustrated. From Ernst and Buchan (2001c).

The additional type VI (circumferential) swarms are certainly present on the scale of individual volcanoes (e.g. Galapagos; Chadwick and Howard, 1991), but it is not known whether they occur on the giant swarm scale. A speculative example has the arcuate Kochikha swarm circumscribing a proposed plume center for the Siberian Trap event (see Figure 6 of Ernst and Buchan, 2001c). Another possible example is discussed by Denyszyn *et al.* (2009a, b) in connection with the 725 Ma Franklin LIP of northern Canada although his suggestion for the location of the plume center of the Franklin LIP is not consistent with that shown in Ernst and Buchan (1997a). Chapter 9 discusses evidence for type VI (circumferential) swarms on Venus and Mars. See also the 1.37 Ga Lake Victoria (Africa) circumferential swarm (Mäkitie *et al.*, 2014).

5.2.2 *Mapping regional paleostress fields*

The distribution and origin of present-day regional stress fields is well established. From mapping of thousands of stress indicators worldwide, Zoback (1992) demonstrated that the present-day regional stress fields are consistent throughout large intraplate areas, and are linked to pressure from current plate-boundary stresses (particularly controlled by the orientation of spreading ridges). It is reasonable to assume that estimates of paleostress directions can similarly reveal the orientations of ancient plate boundaries (spreading ridges) (Fig. 5.4; Ernst and Buchan, 1999, 2001c; Hou *et al.*, 2010).

Dyke swarms can potentially be used to map such paleostress trends (e.g. Ernst *et al.*, 1995a; Hou *et al.*, 2010). Examples are the 1270 Ma Mackenzie and 825 Ma Gairdner swarms (Fig. 5.2) which swing in trend beyond about 1000 km from the inferred plume center, i.e. at a distance when regional stresses begin to dominate over central radial stresses caused by plume uplift. Such primary dyke-swarm patterns must be distinguished from the effect of secondary deformation (next section).

5.2.3 Secondary deformation of swarms

Some geometric patterns in dyke swarms are due to later deformation. For example, a sigmoidal flexure in the overall radiating pattern of the Matachewan swarm/LIP (southern Superior craton) is due to deformation along the intervening Kapuskasing Structural Zone (Fig. 5.5a; Evans and Halls, 2010 and references therein). As another example, dextral dragging of the eastern end of the 1790 Ma Uruguay (Florida) swarm/LIP of the Rio de la Plata craton occurs along the Sarandí del Yí megashear (Fig. 5.5b; Teixeira *et al.*, 2012). Other examples include the apparent en echelon offset pattern in the 2190 Ma Dogrib dykes of the Slave craton, which is due to systematic sinistral offset along north–south faults (Ernst *et al.*, 1995a). In addition, the 1240 Ma Sudbury swarm of the southwest Superior craton beomes progressively deformed upon entering the Grenville province (Bethune, 1993).

5.2.4 Characteristics of regional dyke swarms

Research studies over the past couple of decades have burst various long-standing misunderstandings about regional dyke swarms, and have resulted in important insights such as the following:

Trend matters Crosscutting trends of regional-scale dykes in a single region are commonly assumed all to have the same age. However, from detailed studies, particularly in the Superior craton of Canada, it is now recognized that each dyke swarm (distinguished on the basis of age) typically has a consistent linear trend or regional radiating pattern (e.g. Halls, 1982; Fahrig, 1987; Halls and Fahrig, 1987; Buchan and Ernst, 2004; Buchan *et al.*, 2010). Therefore, regional dyke sets which have crosscutting trends will almost certainly have different ages and belong to different swarms. Overall the "take-home" message is that trend matters. In addition, there are also situations where a single dyke trend can include more than one age of dyke, i.e. more than one swarm. Another observation is that regional dyke swarm

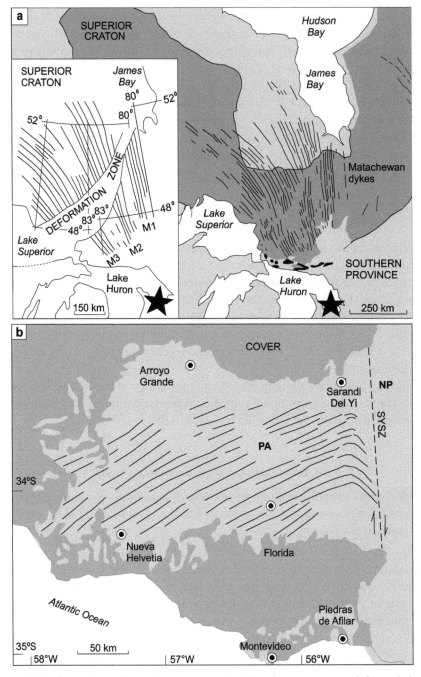

Figure 5.5 Deformation of a dyke swarm. (a) Matachewan swarm deformed by the Kapuskasing Structural Zone "Deformation Zone." From Ernst and Buchan (2001c). See also West and Ernst (1991) and Bates and Halls (1991). (b) Dextral dragging of the eastern end of the 1790 Ma Uruguay (Florida) swarm of the Rio de la Plata craton due to dextral strike-slip movement along the Sarandí del Yí shear zone (SYSZ). From Teixeira *et al.* (2013). With permission from Elsevier. Digital version of original figure for (b) kindly provided by W. Teixeira.

trends are not significantly influenced by planes of weakness in the host rock (except at the local scale), but instead are controlled by regional stresses. Such stresses include those associated with domal uplift producing radiating dyke patterns and those associated with plate-boundary stresses that cause swarms to swing into regional linear trends. The simple geometry of swarms is most true for dykes emplaced into basement rocks. However, the LIP-related dykes cutting the shallow supracrustal package and volcanics of the LIP can have multiple trends (e.g. Muirhead *et al.*, 2012) as shown for the Beacon Group of Antarctica that also hosts sills and volcanics of the Ferrar portion of the Karoo-Ferrar LIP. Note that secondary deformation can distort primary patterns (as discussed above, Section 5.2.3).

Dominant emplacement as Mode 1 cracks (i.e. normal to sigma 3) Regional dyke sets that intersect at an acute angle have often been inferred to be coeval and to be emplaced along a conjugate shear set (e.g. the Kangâmiut dykes of Greenland; Escher *et al.*, 1975; Hanmer *et al.*, 1997). However, in every proposed case, subsequent study has revealed that the two "conjugate" trends of dykes have different ages and so are unrelated (e.g. Ernst and Bleeker, 2010). Regional dyke swarms are typically emplaced along principal stress directions (i.e. parallel to the maximum compressive stress, sigma 1, and normal to the minimum compressive stress, sigma 3) (e.g. Pollard, 1987) and are not emplaced along shear stress directions, except perhaps locally. Again, note that secondary deformation can distort primary patterns (Section 5.2.3).

Horizontal emplacement can be important It had been typically assumed that dykes were fed vertically from underlying source areas. However, starting in the 1980s (e.g. Halls, 1982), it became increasingly clear, based on magnetic fabric studies (e.g. Ernst and Baragar, 1992; Hastie *et al.*, 2014), as well as modeling (e.g. Lister and Kerr, 1991) textural indicators of flow direction (e.g. Rickwood, 1990) and comparison with seismically monitored real-time lateral dyke injection in Hawaii and Iceland (e.g. Rubin, 1995) and analog studies on Venus and Mars (Sections 7.2.4 and 7.3.6), that dyke swarms can also be emplaced laterally for long distances through the crust into cratonic interiors (i.e. up to > 2000 km) (e.g. Halls, 1982; Fahrig, 1987; Lister and Kerr, 1991; Ernst and Baragar, 1992; Baragar *et al.*, 1996; Ernst *et al.*, 2001; Wilson and Head, 2002). Based particularly on magnetic fabric and geochemical study of the 1270 Ma Mackenzie swarm (Ernst and Baragar, 1992; and Baragar *et al.* 1996) (Fig. 5.1 and 5.2c) it was inferred that vertical emplacement occurs near the focus of a radiating swarm (above a mantle plume head) and lateral flow at all greater distances from the focus (e.g. Baragar *et al.*, 1996). Furthermore it was proposed that the lateral flow originates from shallow-level source chambers located in the focal region (e.g. Baragar *et al.*, 1996; Ernst *et al.*, 2005).

Each dyke is a unique event Each dyke in a swarm represents a distinct emplacement event. This is evident from studies that have shown that pairs of long dykes (e.g. 200

km long) that were only separated by a narrow gap (e.g. 10–20 km) could each exhibit a consistent chemistry and paleomagnetic direction along their complete length, and yet be distinct in composition and paleomagnetic direction from that of its neighbouring dyke (e.g. Halls, 1986; Buchan *et al.*, 1993; see Section 10.6.2).

Radiating regional dyke swarms are the norm The Superior craton of Canada has major Paleoproterozoic dyke swarms that radiate into the cratonic interior from points along its margin (Fig. 11.12). This observation is consistent with these swarms being part of plume-generated LIPs marking progressive breakout (or attempted breakout) of the Superior craton from a larger Archean continent. As intraplate magmatic units from around the world continue to be precisely dated, such radiating dyke patterns are also being recognized in other cratons such as the Slave and Indian cratons (Ernst and Srivastava, 2008; French *et al.*, 2010; Ernst and Bleeker, 2010). It is predicted that every major crustal block through time will similarly host radiating dyke swarms (and their associated LIPs) focused along the margins of the crustal block, which locate mantle plumes and help sort out the breakup history of Precambrian supercontinents (Chapter 11).

5.2.5 Differences between basement dykes and dyke–sill complexes in supracrustal sequence

It has been noted that the simple geometry of major regional dyke swarms observed mainly in basement terranes may not apply within shallow level supracrustal sequences where shallow sills and dykes of variable trend and short length are observed (Muirhead *et al.*, 2012; J. Muirhead, personal communication, 2013). The Ferrar magmatism (Ferrar sills and Kirkpatrick volcanics) in Antarctica provide an excellent example (Fig. 5.6). The following is extracted from Muirhead *et al.* (2012).

Field observations from South Victoria Land, Antarctica, demonstrate that magma is transported through a complex pattern of interconnected sills and inclined sheets. These sills and sheets represent the upper-crustal (top-4-km) plumbing system of the 183 Ma Ferrar magmatism of the Karoo-Ferrar LIP and are interpreted to have supplied magma for the Mawson Formation pyroclastic rocks in various parts of South Victoria Land and the Kirkpatrick flood-basalt lavas. The relative paucity of related dykes in the supracrustal sequence associated with Ferrar sills is apparent also in other sill provinces (e.g. Karoo sills of the Karoo portion of the Karoo-Ferrar LIP, Section 5.3.2, and Nipissing sills of the Nipissing–Ungava LIP, Section 5.3.3).

5.2.6 Fracture zones facilitating magma transport

It is often described that magma can ascend along a fracture zone, fault or lithospheric break (cf. Begg *et al.*, 2009, 2010). A prominent example is associated with the Siberian Traps LIP. Heavy gray lines in Figure 5.7 are feeder zones for flows

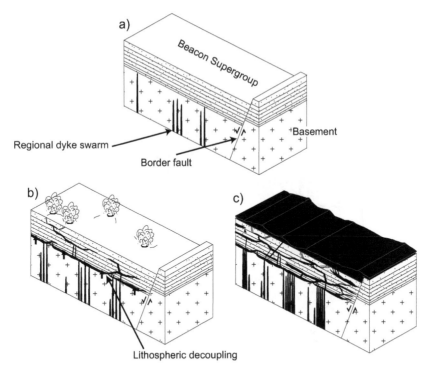

Figure 5.6 LIP plumbing system at shallow levels versus basement using the evolution of the Ferrar magmatism (part of the Karoo-Ferrar LIP) plumbing system of South Victoria Land, Antarctica as an example. (a) Giant dyke swarm propagates both laterally and vertically in basement (cf. Figs. 5.1–5.4). (b) The giant dyke swarm intersects the Beacon Supergroup and upper basement levels, and magma begins to flow laterally as sills and the aerially extensive Ferrar sills detach the shallow (upper-4-km) plumbing system from the basement. (c) The final expression of the Ferrar magmatism: a network of extensive sills connected by moderately dipping sheets and isolated dykes, which overlie a giant dyke swarm in the basement. Modified from Figure 14 in Muirhead et al. (2012). Digital version of original figure kindly provided by J. Muirhead.

of the Noril'sk area (Fedorenko et al., 1996). Specifically, they correspond to the locations of zones of maximum thickness of the lavas, which also correlate with the locations of fault zones of the Kharaelakh fault system. Another example of magma ascent along faults is related to the Bushveld LIP of southern Africa. The Bushveld intrusion was potentially fed by magmas ascending along the east–west-trending Thabazimbi–Murchison lineament/fault (e.g. Good and de Wit, 1997; Kinnaird, 2005). To the extent that such faults are (locally or extensively) filled with magma, they could also be termed dykes. On the other hand, magma may pass through such fractures, but if the fault is not extensional, and has a significant trans-tensional or compressional component, then the magma may be completely expelled after passage through the fault.

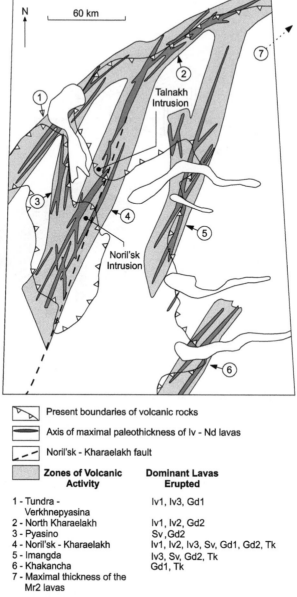

Figure 5.7 Concentrations of volcanic rocks along faults that are broadly associated with the Noril'sk–Kharaelakh lithospheric scale fault. Example from the Siberian Trap LIP event in the Noril'sk area. It can be inferred that these individual "faults" represent dykes feeding volcanics by a fissure flow style (cf. Figure 3.11). Other potential examples of magma emplacement along major faults are discussed in the text (Section 5.2.6). Redrafted from Federenko *et al.* (1996). Reprinted by permission of the publisher (Taylor & Francis Ltd, http://www.tandf.co.uk/journals). Original from Lightfoot *et al.* (1994). © Queen's Printer for Ontario, 1994. Reproduced with permission.

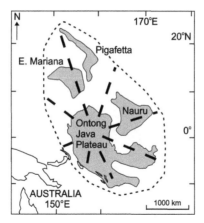

Figure 5.8 Speculation about a radiating dyke swarm associated with the Greater Ontong Java LIP.

5.2.7 Oceanic radiating swarms

Apart from the (sheeted) dyke complexes associated with the spreading ridges, the distribution of dykes cutting oceanic basement is unknown. It is suspected that regional dyke swarms associated with oceanic plateaus would have the same geometries described above for continental LIPs; i.e. radiating swarms (Fig. 5.8). This would also be predicted from a comparison with Venus and Mars (Chapter 7) where radiating swarms extend from volcanic edifices (LIP analogs) into adjacent volcanic plains consisting of basalts (like terrestrial ocean floor). The surface expression of dykes on Venus and Mars is as grabens (Fig. 7.7). Furthermore, if the speculated radiating swarms associated with oceanic plateaus exist they might potentially reach adjacent continental areas and appear as dyke swarms cutting continental crust (e.g. Ernst *et al.*, 1995b), an idea that remains to be tested with continued geochronology of the many undated swarms around the world.

5.3 Dolerite sill provinces

Next is considered another major component of the plumbing system of LIPs, the often vast dolerite sill provinces. These can range in size from a few tens of square kilometers associated with local volcanic centers to great provinces underlying thousands of square kilometers (e.g. the Karoo sills of southern Africa, Fig. 5.9). The thickness of individual sills also varies from less than a meter to hundreds of meters thick and, at the upper size, they verge on being sill-like differentiated intrusions (see Section 5.4.1). Most are emplaced into sedimentary basins, but seismic images also reveal sills in the underlying basement (e.g. Mandler and Clowes, 1997; Welford and Clowes, 2004).

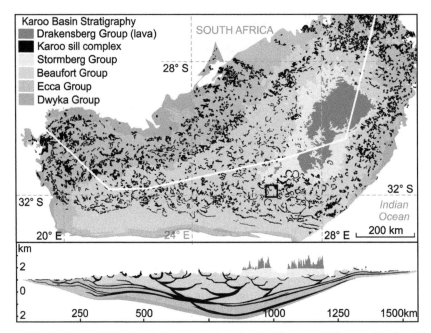

Figure 5.9 The Karoo LIP. (Top) Map showing the distribution of Karoo sills (black) and lavas (Drankenberg group) in the Karoo sedimentary basin of southern Africa. The gray line represents the regional cross section in the bottom diagram. (Bottom) Cross section showing the interpreted pattern of sill emplacement in the Karoo basin. Modified from Svensen *et al.* (2012). With permission from Elsevier. Digital version of original figure kindly provided by H. Svensen. The associated Karoo dykes are shown in Figure 5.10.

5.3.1 Basement sills

Examples of sills emplaced in basement regions include the Siljan Ring area in Sweden, where bright seismic reflectors have been drilled and identified as dolerite sills with thicknesses from a few meters to as much as 60 m (Juhlin, 1990), and the Bagdad reflector sequence in Arizona, where the reflection character of the signals has been modeled as diabase sheets (Litak and Hauser, 1992) and has been potentially linked to the southwest USA diabase province (event 89 in Ernst and Buchan (2001b) also now called the SouthWestern Laurentia LIP (SWLLIP; Bright *et al.*, 2014)). The Winagami reflectors represent sills at 3.5 to 18.5 km depth underlying 120 000 km^2 of the Western Canada basin of western Canada with an age constrained between *c.* 1890–1760 Ma (Ross and Eaton, 1997; event 149 in Ernst and Buchan, 2001b; see also Welford and Clowes, 2004; Welford *et al.*, 2007), which are potentially correlated with the 1740 Ma Cleaver dyke swarm of the Slave craton region, part of a regional LIP in northern Canada (Ernst and Bleeker, 2010). Several hundred kilometers to the east is the 160-km-long sequence of bright reflections that comprise the Wollaston Lake reflector, and are potentially related to those sills exposed at the surface in the overlying Athabasca basin and considered to belong to the 1270 Ma Mackenzie LIP (Mandler and Clowes, 1997).

5.3.2 Karoo sills (180 Ma), South Africa

The Karoo portion of the Karoo-Ferrar LIP (Section 3.2.7) includes a classic sill province in the sediments of the Karoo basin of southern Africa. As shown in Figure 5.9 the sills are widespead in southern Africa and have been the source of many classic studies on sills which reveal the complexity of this sill province. Some of the most detailed recent work has been on the Golden Valley sills (see box in Fig. 5.9). Karoo sills are linked to the feeding of the overlying Drakensburg lavas. The distribution of the associated Karoo-aged dykes (Fig. 5.10) points to the complexity of the plumbing system.

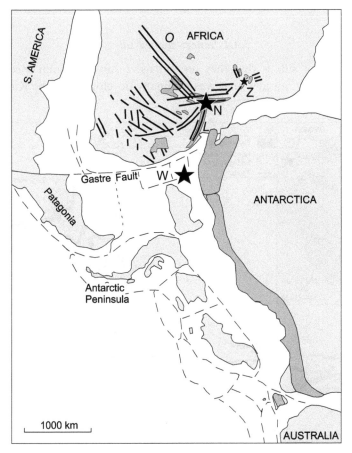

Figure 5.10 Dyke swarms associated with Karoo LIP. O = Okavango swarm, L = Lebombo swarm and monocline, N = Nuanetsi, Z = Zambesi plume center, and W = Weddell triple junction. Modified from Ernst and Buchan (1997a). Weddell Sea triple junction added from Elliot and Fleming (2000).

5.3.3 Nipissing sills (2215 Ma), Canada

Nipissing sills occur over an area of about 35 000 km^2 in the sedimentary units of the Paleoproterozoic Huronian Supergroup along the southern margin of the Superior craton (Fig. 5.11). The form of the Nipissing intrusions includes sills, undulating sheets, and somewhat irregular, dyke-like bodies. In the northern half of the outcrop area a few annular-shaped outcrop patterns are suggestive of cone sheets or saucer-shaped sills (e.g. Palmer *et al.*, 2007). Because of the age equivalence of the Senneterre dykes and the Nipissing sills, Buchan *et al.* (1998) suggested that the Senneterre dykes acted as feeders for the Nipissing sills. More broadly, the Senneterre dykes are part of the Ungava radiating dyke swarm focused to the northeast in the Ungava Bay area of northeastern Canada (Fig. 5.11), and marking the plume center inferred to be responsible for the entire Ungava–Nipissing LIP (Ernst and Buchan, 1997a; Buchan *et al.*, 1998).

5.3.4 Saucer-shaped sills

The classic view of sills is that they are extensive sub-horizontal sheets with an undulatory geometry, an idea that was originally developed for the Karoo sills of

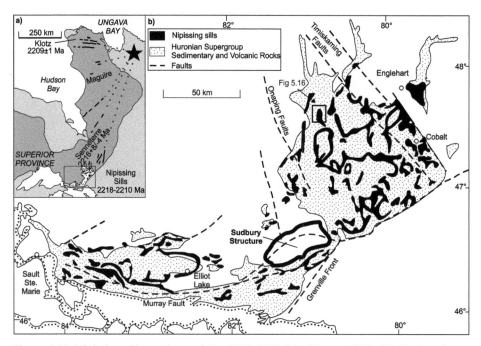

Figure 5.11 Nipissing sill province of the 2220–2210 Ma Ungava LIP. (a) Regional map showing radiating dyke swarms of Ungava LIP (modified from Buchan *et al.* (1998), with updates from Buchan *et al.* (2007)). (b) Distribution of Nipissing sills and host Huronian sediments after Lightfoot *et al.* (1993a).

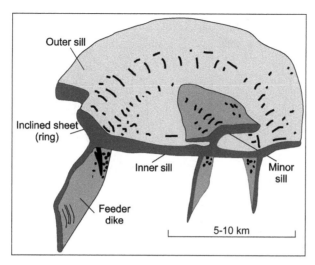

Figure 5.12 Saucer-shaped mafic sills. Typical geometry of mafic sills in sedimentary basins. Modified from Chevallier and Woodford (1999).

Africa (e.g. Du Toit, 1920; Scholtz, 1936). Bradley (1965) and Meyboom and Wallace (1978) offered an explanation in terms of variations in a "compensation surface" where magma pressure would be equal to the lithostatic pressure. Gretener (1969) suggested that sills were generated by horizontal compression.

A revolution in our understanding of sill-province geometry derives from more recent detailed mapping, seismic imaging, magnetic fabric studies and experimental modeling (Chevallier and Woodford, 1999; Davies et al., 2002; Smallwood and Maresh, 2002; Thomson and Hutton, 2004; Planke et al., 2005; Hansen and Cartwright, 2006; Thomson and Schofield, 2008; Polteau et al., 2008a; Galland et al., 2009; Miles and Cartwright, 2010; Schofield et al., 2010). Three-dimensional (3D) seismic data and detailed field mapping have shown that sills in sedimentary basins are generally saucer-shaped, and circular or elliptical in plan view with an inner flat dish surrounded by an inward-dipping arcuate rim and an outer sill (Fig. 5.12). The 3D seismic data also show that individual sills appear to be constructed from a series of lobes (analogous to those seen in lava flows) that branch out from a central source or feeder. The lobes may be constructed from smaller structures (e.g. Fig. 5.13).

Another point is that the diameter of the inner sill (flat portion) increases with increasing emplacement depth (e.g. Malthe-Sørenssen et al., 2004; Polteau et al., 2008b; Galland et al., 2009). There is a correlation factor of 4–5 between the depth of emplacement and the inner sill diameter. This relationship explains the greater lateral extent of sills emplaced in deeper crust, particularly in the basement. Another aspect is that emplacement of saucer-shaped sills is correlated with local domal uplift of the overlying crust (Polteau et al., 2008b; Galland et al., 2009).

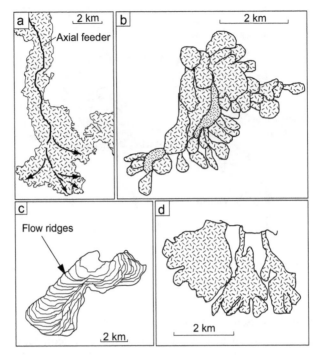

Figure 5.13 Patterns of shallow-level sill emplacement. (a) Shallow-level sill (emplaced 200–400 m below contemporaneous sea bed). Note the lava-like form of the intrusion and ragged outer edge of the magma lobes. (b) Occurrence of magma lobes with secondary and tertiary breakout structures within a shallow-level sill (emplaced 150–550 m below contemporaneous sea bed). (c) Shallow-level sill (emplaced under 400-m depth), displaying a ridged flow morphology attributed to fluidization of host sediments. (d) Magma lobes developed in a shallow-level intrusion. From Figure 7 in Schofield *et al.* (2012) after original diagrams by Miles and Cartwright (2010), Hansen and Cartwright (2006), Trude (2004), and Thomson and Hutton (2004), for parts (a) to (d), respectively. Digital version of the original figure kindly provided by N. Schofield.

5.3.5 Feeding and emplacement of sills

There are some important questions regarding the emplacement of saucer-shaped sills:

(1) How is the magma emplaced within the sill and between the inner and outer sills (Figs. 5.14, 5.15)?
(2) How are the saucer sills fed? Are they fed from underlying dykes or underlying pipe intrusions? Or are they fed from the side by other sills or by dykes (Fig. 5.14)?

Flow pattern of magma within saucer-shaped sills

Several models propose that saucer-shaped sills are fed from the outer sills or from the inclined sheets followed by buoyancy-controlled, downward flow of the magma

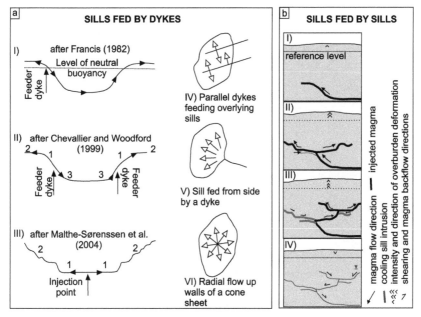

Figure 5.14 Feeding of sills by dykes and other sills. (a) (I, II, and III) Schematic cross section view of feeding of saucer-shaped sills. From Figure 3 in Polteau *et al.* (2008b). With permission from Elsevier. (IV, V, and VI) Schematic plan view of feeding of sills: IV, dyke feeding an overlying sill yielding bidirectional flow in the overlying sill; V, a fanning pattern of flow in the sill due to being fed from the side by an adjacent dyke; VI, pattern due to feeding from an underlying centrally located point source producing a cone sheet. From Figure 2 in Palmer *et al.* (2007). (b) Schematic representation of a sill complex emplacement (sills feeding sills) divided into four stages. From Figure 8 in Polteau *et al.* (2008a). With permission from Elsevier.

to the inner sills (Fig. 5.14) (Bradley, 1965; Francis, 1982; Chevallier and Woodford, 1999). However, recent anisotropy of magnetic susceptibility (AMS) (Polteau *et al.*, 2008a) and seismic (e.g. Thomson and Hutton, 2004; Hansen and Cartwright, 2006; Thomson, 2007) data suggest that the direction of magma flow is outward and upward in saucer-shaped sill complexes and that the outer parts of the saucers are fed from the inner sills. In particular, AMS data from the Golden Valley sill of the Karoo system suggests that saucer-shaped sills are emplaced radially and that the overburden influenced the magnetic fabric (Polteau *et al.*, 2008a). Support for such a model derives from the observation of magmatic fingers radially disposed trending away from the inner sill (Fig. 5.15). Specifically, as observed by Schofield *et al.* (2010), the transgressive rims of the Golden Valley sill are composed of a series of lobes constructed from coalesced magma fingers, similar to those originally described by Pollard *et al.* (1975). However, in their AMS study of the Nipissing sills (Fig. 5.11), Palmer *et al.* (2007) identified consistent north–northwest to south–southeast flow trajectories on the east and west sides of a saucer-shaped sill that had

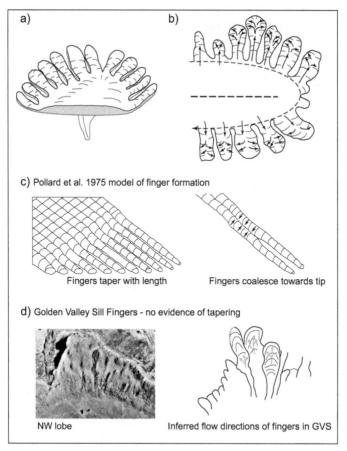

a)   b)

c) Pollard et al. 1975 model of finger formation

Fingers taper with length   Fingers coalesce towards tip

d) Golden Valley Sill Fingers - no evidence of tapering

NW lobe   Inferred flow directions of fingers in GVS

Figure 5.15 Importance of finger-style feeding of sills. (a, b) Schematic illustrating the possible ways in which fingers inflate. (c, d) Previous mechanism proposed by Pollard *et al.* (1975) in which magma fingers zip up from the rear (base), which appears incompatible with the observations from the Golden Valley sill of the Karoo sills in (d), where magma fingers show a bulbous nature appearing to have coalesced in a down-flow direction. Location of GVS shown by box in Figure 5.9. From Schofield (2009).

been previously interpreted as a cone sheet. In this Nipissing sill case, and in contrast to the Karoo example above, the AMS pattern was inconsistent with outward radial injection and therefore was not supportive of cone-sheet origin (Fig. 5.16); instead, it was more consistent with feeding from the side (north or south side) of the Nipissing sills.

Feeding of saucer-shaped sills

Geochemistry and paleomagnetism are useful for sorting out links between different saucer-shaped sills. Incompatible-trace-element ratios are not significantly modified during transport in the upper crust and therefore geochemistry of each magma batch

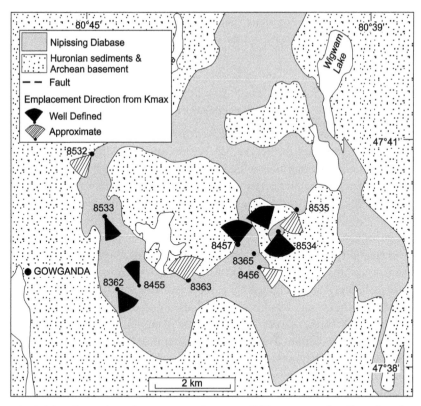

Figure 5.16 Directions of maximum magnetic susceptibility representing flow directions in a Nipissing sill. Data used for these three possible models for emplacement: a cone sheet, saucer-shaped sill, or an undulating sheet. Solid fans represent sites with well-defined AMS patterns, whereas fans with diagonal ruling are based on sites with less-stringent criteria. From Palmer *et al.* (2007). There is a consistent north–northwest trend of flow axes on the east and west sides of the circular-appearing portion of the sill, a pattern that is NOT consistent with the cone-sheet interpretation. See Fig. 5.11 for location.

will have a distinct "fingerprint" that can be used to track its path. Similarly, the emplacement of dolerites is relatively rapid and so the emplacement and cooling of a single magma batch is short with respect to secular variation of the Earth's magnetic field and so each magma batch can potentially have a distinct paleomagnetic direction that can be used to trace the "downstream" distribution of each magma batch.

Geochemical data from the 183 Ma Golden Valley sill Complex were used to compare the different sill "saucers" and two potential feeder dykes to see which are similar and therefore genetically linked (Galerne *et al.*, 2008). In this example stacked sills had slight but distinct differences in composition, indicating that they represented different magma batches fed by separate plumbing systems (see Fig. 10.23).

Paleomagnetic study of the Nipissing sills reveals three distinct paleomagnetic directions, which indicate that at least this many distinct magma pulses are present (Palmer *et al.*, 2007).

5.3.6 *Vent complexes and sills*

Detailed seismic studies of the 62–55 Ma North Atlantic Igneous Province (NAIP) and complementary studies in the 183 Ma Karoo and 252 Ma Siberian Trap LIPs reveal thousands of hydrothermal vent complexes (HVCs) (Fig. 5.17). Up to 5–10 km across at the paleosurface, these vents connect to underlying dolerite sills at paleo-depths of up to 8 km (Jamtveit *et al.*, 2004; Planke *et al.*, 2005; Svensen *et al.*, 2006; Svensen *et al.*, 2007, 2009; Neumann *et al.*, 2011). HVCs originate from explosive release of gases generated when thick sills (> 50 m) are emplaced into volatile-rich, but low-permeability, sedimentary strata. They are phreatomagmatic in origin. Their architecture, economic potential for forming iron oxide–copper–gold (IOCG)-type deposits (see Chapter 16), and effects on climate (see Chapter 14) strongly depend on the types of host rocks (e.g. black shales at Karoo and evaporites at Siberian Trap LIPs) and their fluid (brine) saturation at the time of emplacement. About 250 HVCs associated with the Siberian Trap LIP are mineralized having magnetite in the matrix. Some (e.g. Korshunovskoe and Rudnogorskoe) are being mined for Fe (Svensen *et al.*, 2009).

These observations from the Phanerozoic LIP record suggest that HVCs should also be an essential component of sill provinces associated with Proterozoic LIPs, with a potential for causing major climatic shifts and ore deposits (Sections 14.6.6 and 16.3.3), particularly if the host sediments include substantial evaporites.

As one example, the 725 Ma Franklin LIP covers 1.1 Mkm^2 in northern Canada (Fig. 5.1; Buchan and Ernst, 2004); in the Minto Inlier of Victoria Island this event includes volcanics, sills, and breccia pipes (Jefferson *et al.*, 1994; Rainbird, 1998; Bedard *et al.*, 2012). The breccia pipes appear identical to HVCs and, furthermore, the presence of evaporites in the host sediments of the Shaler Supergroup suggests, based on the Siberian Traps example, the potential for associated climatic effects and ore deposits (Sections 14.6.6 and 16.3.3).

5.3.7 *Sills emplaced subglacially*

Another type of magmatic emplacement is that emplaced subglacially. As noted by Smellie (2008) basaltic volcanic sequences erupted subglacially are observed to be of two major types, corresponding to eruptions under "thick" and "thin" ice, respectively. The latter are termed the Mount Pinafore type; the former, the Dalsheidi type, represents products of eruption under much thicker ice (probably > 1000 m). Eruptions that form the Dalsheidi type of sequence commence with the injection

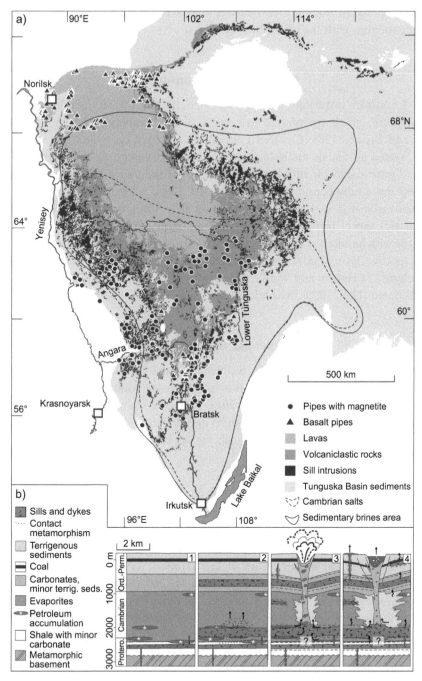

Figure 5.17 Hydrothermal vent complexes (HVCs) associated with the Siberian Trap LIP. (a) Geological map of the Tunguska basin in eastern Siberia, Russia. Note the high abundance of phreatomagmatic pipes with magnetite south of latitude 64°, and the numerous basalt-filled pipes north of 68°. Modified after Figures 1 and 6 in Svensen *et al.* (2009). With permission from Elsevier. Digital version of original figure modified for (a) kindly provided by A. Polozov.

and inflation of a sill along the ice–bedrock interface. See Section 7.2.4 for discussion of sill interactions with the cryosphere on Mars.

5.4 Differentiated intrusions

Herein I consider intrusions that are large enough to exhibit significant internal differentiation, which are inferred to be part of the plumbing system of a LIP (e.g. Wager and Brown, 1968; Cawthorn, 1996). Generalized cross sections for several classic layered intrusions are shown in Figure. 5.18, illustrating their cumulus patterns and general division into lower U (ultramafic) and upper T (tholeiite) portions, an aspect more fully considered in Chapter 10. There is abundant literature on the processes associated with this differentiation (e.g. Campbell, 1996), and I don't consider this aspect further here. Instead I am focusing on the geometry of these intrusions with an interest in how they relate to other parts of the LIP system. From this perspective, layered intrusions are classified as sill-like or dyke-like following the nomenclature of Hatton and von Gruenewaldt (1990). In addition, funnel intrusions are considered (Hall, 1996), which in certain cases may also be linked to LIPs.

5.4.1 Sill-like layered intrusions

Many large intrusions, such as the Stillwater or the Bushveld Complexes have a horizontal sheet-like geometry (Hatton and Von Gruenewaldt, 1990; Ashwal, 1993). For instance, the 2710 Ma (Wall *et al.*, 2012) Stillwater Complex is about 6 km thick and covers an areas of > 4400 km^2 of which only 194 km^2 is exposed. It can be traced down dip and therefore likely has a much greater extent. The Bushveld Complex is much larger, as outlined below.

The Bushveld LIP (c. 2060 Ma)

The 2055–2060 Ma Bushveld Complex of South Africa includes the Earth's largest, layered igneous intrusion, which is regarded as the intrusive equivalent of a flood-basalt province, given its extensive volume of 0.38 Mkm^3 and short duration of emplacement of a few million years or less (Hatton, 1995; Eales and Cawthorn, 1996; Cawthorn and Walraven, 1998; Kinnaird, 2005; Wall *et al.*, 2012). More specifically, the Bushveld Complex consists of an older Rooiberg basalt–rhyolite lava group, younger Lebowa Granites, and Rashoop Granophyre suites, as well as the world's largest (*c.* 65 000 km^2; 500-km-across, 7–9-km-thick) layered intrusion (the Rustenburg Layered Suite), which is in the form of a sub-horizontal soup-dish-shaped intrusion (sill-like layered intrusion). U–Pb baddeleyite ages produced by Scoates *et al.* (2012) indicate the lower ultramafic portion below the Merensky reef (a well-known platinum-rich horizon) are *c.* 2060 Ma while those from above are *c.* 2055 Ma, indicating that this intrusion represents the intrusion of at least two

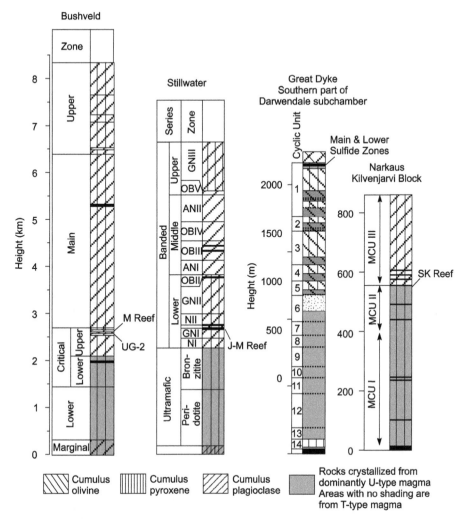

Figure 5.18 Typical stratigraphic profiles for the 2060 Ma Bushveld, 2710 Ma Stillwater, 2575 Ma Great Dyke of Zimbabwe, and the 2450 Ma Narkaus block intrusions. All are LIP-related including the latter, which belongs to the Portimo intrusion of the BLIP (Baltica LIP). The gray shading indicates portions of the profile for which the magma was dominantly U-type. The proportion of U-type shown for the Great Dyke is exaggerated, because much of the profile composed of T-type rocks has been eroded. The proportion of U-type for Portimo is also exaggerated because of the choice of the Narkhaus block for the illustration. Note that in all of the profiles, the principal PGE reef lies close to this changeover between U- and T-type magma. From Figure 9 in Naldrett (2010a). For the Stillwater OB = olivine-bearing, AN = anorthosite, GN = gabbronorite, N = norite. For the Portimo intrusion MCU = mega cycle unit.

pulses. This new geochronology is in apparent conflict with the estimate that the entire intrusion was filled in 75 Ka based on mineralogy and textural parameters (Cawthorn and Walraven, 1998).

The Bushveld intrusion is part of more widely distributed coeval magmatism in the Kaapvaal craton collectively representing the Bushveld LIP (Fig. 5.19), which includes the Molopo Farms layered intrusion, the Okwa basement complex, and smaller intrusive bodies of the Bushveld high-Ti suite near the Vredefort Structure and associated carbonatites (Shiel and Phalborwa) (e.g. Reichardt, 1994; Kinnaird, 2005; Mapeo *et al.*, 2006; Ernst and Bell, 2010; Walker *et al.*, 2010; Prendergast, 2012; Rajesh *et al.*, 2013).

5.4.2 *Dyke-like layered intrusions*

Dyke-like layered intrusions are vertical sheet-like bodies that are much thicker than ordinary regional dolerite dykes. They often have a Y-shaped cross section and are

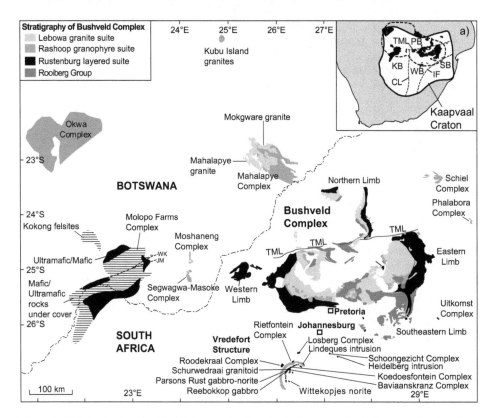

Figure 5.19 The Bushveld LIP. (a) Generalized geological map of the different units that form part of the Bushveld LIP. Only the presently exposed (and preserved) remnants of the different units are shown; the original extent was far wider.

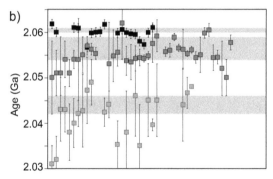

Figure 5.19 (*cont.*) (b) Available ages of individual units other than the Bushveld Complex proper and Rooiberg Group volcanic suite are indicated. Shading shows inferred timing of the three pulses. The inset shows the location of Bushveld large igneous province with respect to the Kaapvaal craton. The dotted line in the inset surrounds the region of seismically slow cratonic mantle at 150 km depth resulting from the Bushveld magmatism. The inset shows the different blocks (Kimberley block (KB), Witwatersrand block (WB), Pietersburg block (PB) and Swaziland block (SB)) of the Kaapvaal craton and also shows the known suture zones (Colesberg lineament (CL), Thabazimbi–Murchinson lineament (TML), and Inyoka fault (IF)) between the blocks. The shear zones (shown by dotted lines; east–northeast-trending Jwaneng–Makopong (JM) and northeast-trending Werda–Kgare (WK) shear zones; also known as the Dikgomodikae Lineament. Modified after Figure 1 of Rajesh *et al.* (2013). With permission from Elsevier. Digital version of original figures modified for both parts kindly provided by H. Rajesh.

also sometimes termed "funnel dykes" (e.g. Hall, 1996). Commonly, they also narrow down with depth to a dyke known as the keel or feeder dyke (e.g. Podmore and Wilson, 1987). In some instances, the dyke-like intrusions can be part of an associated dolerite dyke swarm (e.g. the Muskox intrusion and associated Mackenzie swarm/LIP; e.g. Baragar *et al.*, 1996).

As summarized in Ernst and Buchan (1997c), dyke-like layered intrusions include: the 550-km-long 2575 Ma Great Dyke in Zimbabwe, which is up to 11 km wide (Fig. 5.20; Podmore and Wilson, 1987; Wilson, 1996); the Jimberlana dyke, which is up to 2.5 km wide and belongs to the 2410 Ma Widgiemooltha LIP in western Australia (Fig. 5.21; McClay and Campbell, 1976); the Muskox feeder dyke (picrites and tholeiites), which is up to 0.5 km wide and belongs to the 1270 Ma Mackenzie LIP (Fig. 5.22; Irvine, 1980; Francis, 1994); and the *c.* 1160 Ma giant Tugtutoq dykes of southern Greenland, which are between 0.5 and 0.8 km wide (Upton and Thomas, 1980; Upton *et al.*, 1996). Most of these large dyke-like layered intrusions have a Y-shaped geometry and thin rapidly with depth. In particular, gravity modeling over the Great Dyke of Zimbabwe has revealed a Y-shaped geometry in which surface widths of up to 11 km typically contract to less than 1 km within less than 3 km of the surface (Podmore and Wilson, 1987). A similar

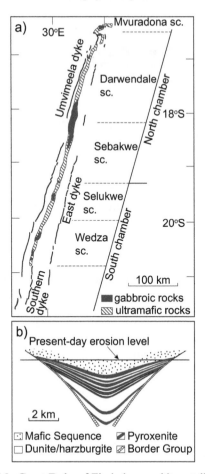

Figure 5.20 (a) The 2575 Ma Great Dyke of Zimbabwe and its satellite dykes. The Great Dyke continues to the south as a much thinner set of en echelon dyke segments, and is flanked to the east and west by the East and Umvimeela dykes. sc. = sub-chamber (b) Cross section of the Great Dyke at the Darwendale subchamber. From Figure 5 in Ernst and Buchan (1997b).

trough-shaped geometry is inferred for the Muskox layered intrusion (which narrows from > 11 km to the feeder dyke width of 500 m), and also for the Jimberlana intrusion.

Great Dyke of Zimbabwe LIP (2575 Ma)

The 2575 Ma Great Dyke of Zimbabwe extends for 550 km and varies between 1 and 11 km wide (Fig. 5.20) Based on distinctive layered styles, the Great Dyke has been divided into two distinct magma chambers, the north and south chambers (e.g. Wilson, 1996). These have been further subdivided into the Darwendale, Sebakwe, Selukwe, and Wedza subchambers, with perhaps a fifth subchamber, the

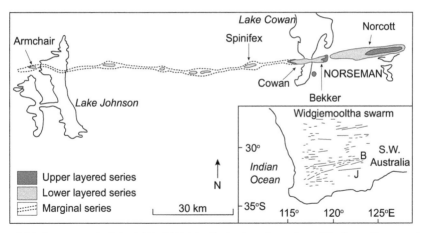

Figure 5.21 Eastern end of the 2410–2418 Ma Jimberlana intrusions. The inset diagram shows Widgiemooltha dyke swarm and the location of two giant dykes that belong to the swarm, the Binneringie (B) and the Jimberlana (J) intrusions. From Figure 6 in Ernst and Buchan (1997b).

Mvuradona, located at the northern end of the Great Dyke. The four main subchambers increase in preserved volume toward the north (Wilson, 1996). Based on modelling of gravity profiles, the cross-section geometry is Y-shaped (Podmore and Wilson, 1987) and a feeder dyke underlies most of the Great Dyke with varying widths up to about a kilometer. An en-echelon dyke continues the Great Dyke trend beyond its southern end and may represent the feeder to an additional subchamber of the Great Dyke, now entirely eroded. Two narrower satellite dykes, the Umvimeela and the East Main dykes, parallel the length of the Great Dyke (Fig. 5.20).

Jimberlana and Binneringie intrusions of the 2410 Ma Widgiemooltha LIP

The two widest dykes of the *c.* 2410 Ma Widgiemooltha dyke swarm of western Australia (e.g. Smirnov *et al.*, 2013), the 180-km-long Jimberlana and the 320-km-long Binneringie intrusions, locally widen upward into layered intrusions up to 2.5 and 3.2 km wide, respectively (Fig. 5.21). The seven local layered complexes distributed along the Jimberlana dyke each have a canoe-shaped cross section (Campbell, 1987; Hatton and von Gruenewaldt, 1990). The spacing between complexes is generally 20–30 km and the two easternmost bodies are much larger than the others. Unlike the complexes of the Jimberlana dyke, in which the layering is inward dipping, the widest portions of the Binneringie dyke show consistent vertical layering (McCall and Peers, 1971). Both the Binneringie and Jimerlana dykes are generally much wider toward their eastern ends, which could indicate proximity to a source plume or variation in depth of exposure, or it could be correlated with a change in host rock type (the latter is suggested by McCall and Peers, 1971).

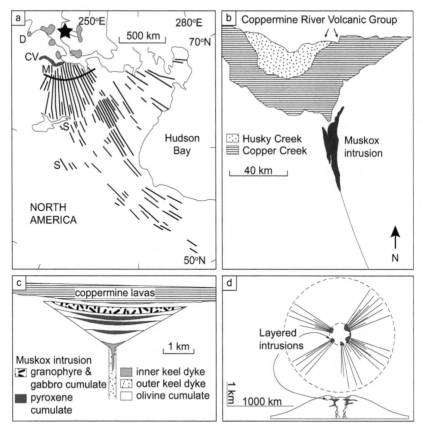

Figure 5.22 1270 Ma Mackenzie LIP: (a) Mackenzie event showing the Mackenzie radiating dyke swarm, the Coppermine River volcanics (CV), the Muskox layered intrusion (M1), and the ring-of-gravity anomalies ("x" pattern) about the plume center identified as the focal point ("star") of the Mackenzie swarm. D is Darnley Bay gravity anomaly, which may be related to the 0.72 Ga Franklin LIP (Fig. 5.1) rather than Mackenzie LIP. S marks the location of sills. The arc marks the boundary between vertical and horizontal flow regimes in the Mackenzie swarm (Ernst and Baragar, 1992). (b) The Muskox intrusion and its Keel dyke (extending south of the intrusion). (c) Muskox intrusion cross section (after Francis, 1994). (d) An interpretation of the ring-of-gravity anomalies as mafic–ultramafic bodies (black areas) emplaced during apical graben collapse of the plume-uplifted crust and each spawning a subswarm. The apical graben model is after Baragar *et al.* (1996), although the suggested association with circumferential dykes is from Ernst and Buchan (1998). See the discussion of circumferential dyke swarms on Mars and Venus in Chapter 7. From Figure 4 in Ernst and Buchan (1997a).

5.4.3 Links between dyke-like and sill-like layered intrusions

Koillismaa intrusion (c. 2440 Ma)

Another important elongate, layered intrusion is the integrated dyke–sill complex of the Syöte–Näränkävaara belt of Finland and Russia (Fig. 5.23), termed the Koillismaa layered intrusion (Alapieti, 1982; Smolkin, 1997). Five separate massifs

Table 5.1 *Components of LIP plumbing systems: dykes, sills, layered intrusions, remnant volcanic rocks, magma chambers (layered intrusions, zoned intrusions), deep crustal chambers, and crustal underplating*

Component of system	Chapter with information
Remnant flood basalts	Chapter 3
Giant dyke swarms	This chapter
Sills	This chapter
Differentiated (usually layered) mafic-ultramafic intrusions	This chapter
Magmatic underplating	This chapter and Chapter 3
Associated silicic magmatism	Chapter 8
Associated carbonatites and kimberlites	Chapter 9

are grouped in two complexes located about 50 km apart, which are linked by the gravity trace of an unexposed feeder dyke which varies in trend from northwest to west (Alapieti, 1982). The feeder dyke reaches to within only 1 or 2 km of the present surface. At its top it is several kilometers wide, but is thought to taper down gradually with increasing depth. The two complexes are fed from this feeder dyke. The Näränkävaara intrusion at the eastern end is a wide, elongated (trough-shaped) dyke-like layered intrusion. In contrast, the western end of the feeder dyke is connected with a fault-disrupted, sill-like layered intrusion 30 or 40 km wide and a few kilometers thick. The Koillismaa layered intrusions belong to the *c.* 2450–2500 Ma Baltic LIP (BLIP), which can be linked with coeval Matachewan and Mistassini LIPs of eastern and southeastern Superior craton in a paleocontinental reconstruction of Superia (e.g. Ernst and Bleeker, 2010).

Emplacement at basement-supracrustal interface

Many of the dyke-like intrusions discussed in the previous section are emplaced near the basement–sedimentary cover interface and may themselves have spawned major sill-like layered intrusions along that interface. For instance, on the basis of geochemical modeling, some units within the Great Dyke of Zimbabwe originally extended beyond the intrusions, probably as sills (Podmore and Wilson, 1987). The sill complex spawned from the western end of the Koillismaa feeder dyke (Fig. 5.23) is a particularly clear example of emplacement of sills along the basement–supracrustal interface. The Bushveld intrusion was emplaced at shallow levels beneath its own volcanic carapace.

Synformally layered dykes

There are a number of examples of synformally layered dykes that may represent deep exposure levels of trough-shaped intrusions. Two of the more prominent

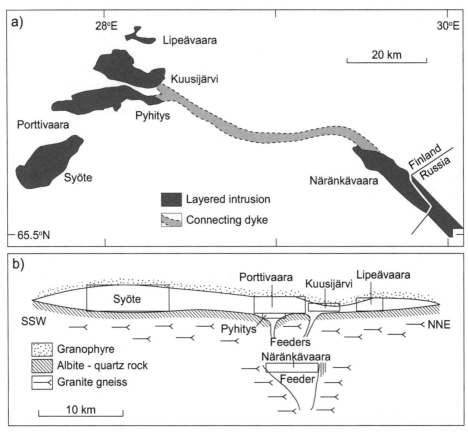

Figure 5.23 The 2.44 Ga Koillismaa complex of northern Finland. (a) The distribution of the fragmented pieces of a single sill-like intrusion and the connecting dyke-like layered intrusion at depth. (b) Schematic diagram showing the primary cross-section geometry of the complex before it was segmented by erosion. From Figure 7 in Ernst and Buchan (1997c) modified after original diagrams in Alapieti (1982) and Smolkin (1997).

examples are the giant Gardar dykes of southwestern Greenland, and the Great Abitibi dyke of the Superior province (Fig. 5.2i). The *c.* 1.17 Ga giant gabbro and syenite dykes of the Gardar province of southwestern Greenland include the 500-m-wide Giant Older Dyke Complex and the up to 800-m-wide Giant Younger Dyke Complex (Upton *et al.*, 1996; Buchan *et al.*, 2000). They exhibit prominent synformal layering in the axial zone of the dyke below a presumed sill-like upper portion now eroded (Upton *et al.*, 1996). The giant dykes are likely to be related to a parallel swarm of narrower diabase dykes of similar age (Piper, 1995). The 1140 Ma Great Abitibi dyke of the Superior province, Canadian shield, is 700 km long and varies in width from 100 to 250 m wide (Ernst and Bell, 1992; Fig. 5.2i). At its widest it is strongly differentiated, showing a dramatic enrichment in incompatible elements and a crude synformal layering reminiscent of funnel dykes. This dyke, along with several other major dykes, define the Abitibi dyke swarm, which converges slightly to the southwest (Ernst and Buchan, 1993) and is likely related to a precursor stage of the Keweenawan LIP of the Midcontinent rift system (Fig. 3.25; Heaman *et al.*, 2007).

5.4.4 Funnel-shaped differentiated intrusions

An important class of layered intrusions are those with roughly equant shape in plan view, whose cross-section geometry is variously described as funnel-, cone-, or bowl-shaped (Petraske *et al.*, 1978; Loney and Himmelberg, 1983; Hall, 1996). Other terms that can apply are plug, stock, lopolith, or laccolith depending on the details of the geometry (e.g. Corry, 1988; Bunger and Cruden, 2011). Examples include the Kiglapait intrusion of the Nain Plutonic suite of eastern Canada (e.g. Morse, 1979), and the Skaergaard intrusion of the east Greenland portion of the NAIP (e.g. Wager and Brown, 1968; McBirney, 1996). The intrusions such as Rum and Skye of the British Tertiary Igneous Province of the NAIP are also examples (Fig. 5.24). Although feeder zones are rarely observed, the inference is that the bodies are fed from pipe-like feeders located beneath the deepest part of the funnel (Loney and Himmelberg, 1983; Hall, 1996). However, it is also possible that in some cases they represent localized areas of upflow along a dyke (Section 5.6.6).

Some funnel intrusions, such as Skaergaard and other intrusions along the east Greenland margin, are linked to a LIP, in that case the link is with the NAIP. However, other funnel intrusions, e.g. the Alaskan-type intrusions (concentrically zoned mafic–ultramafic complexes) are linked to a subduction setting (e.g. Loney and Himmelberg, 1983; Eyuboglu *et al.*, 2011).

5.5 Magmatic underplating

Here we discuss an intrusive component of the LIP that is very much out of sight and difficult to constrain. As originally proposed by Cox (1980, 1993), ponding and polybaric fractionation of primary picritic magmas can occur at the base of the crust and may be an important process in the generation of continental flood basalts. This hypothesis has received supporting evidence from geophysical, geological, and geochemical/petrologic observations (see the summary in Xu and He, 2007; see also Artemieva and Thybo, 2013). Seismic data consistently indicate the existence of HVLC (high-velocity lower crust) at the base of the crust associated with LIPs and commonly also on volcanic rifted margins (see Fig. 3.9). The thickness of the HVLC ranges from a few kilometers to more than 20 km, and their seismic velocities are intermediate between those of the mantle (> 8 km/s) and those typical of the lower

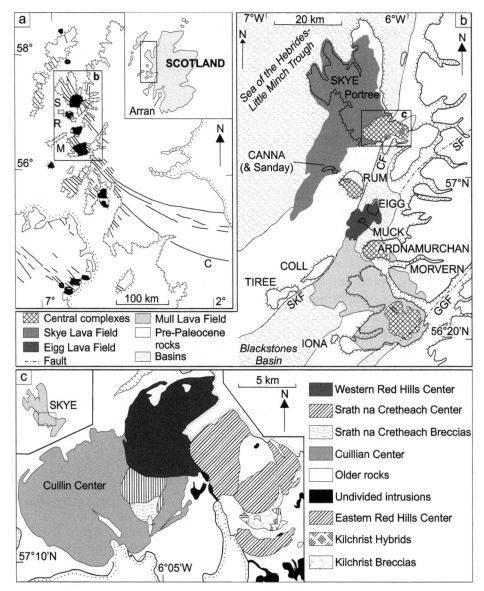

Figure 5.24 British Tertiary Igneous Province (BTIP) of the NAIP. (a) Distribution of regional dykes and location of intrusions (black): S, Skye: R, Rum (Rhum); and M, Mull. From Figure 11 in Ernst and Buchan (1997b) and based on Speight *et al.* (1982) and MacDonald *et al.* (1998). Reproduced with permission. (b) More detailed geology of intrusions, from Figure 1 in Brown *et al.* (2009). Reproduced with permission. (c) Detailed map of the Skye intrusion (modified from Figure 2 in Brown *et al.* (2009)).

crust (< 7 km/s). Examples of seismic sections showing the underplate component are provided for the Keweenawan and Emeishan LIPs (Fig. 5.25). Additional evidence for the presence of magmatic underplating is given by uplift history. Many basins in volcanic rifted margins of the North Atlantic underwent permanent uplift of several hundred meters, which was likely caused by magmatic underplating associated with the NAIP (Saunders *et al.*, 2007; see also Tiley *et al.*, 2004). Underplating is suspected also beneath the Rajmahal basaltic volcanic rocks of the Kerguelen Bunbury–Comei LIP (Singh *et al.*, 2004).

5.5.1 Xenoliths from the lower crust

Another way to study magmatic underplating is through xenoliths of lower crust. For example, mafic xenoliths from young Phanerozoic kimberlites of the Lac de Gras area of the Slave province of northern Canada contain zircon (and rutile), which yield both Archean and Proterozoic ages. The Proterozoic ages are similar to the post-cratonization 1.27 Ga Mackenzie and 2.23–2.21 Ga Malley and MacKay dyke swarm/LIP events, which are prominent in the same region (Davis, 1997). These coeval xenoliths are inferred to be derived from magmatic underplating associated with these LIP events or represent isotopic resetting caused by the thermal and fluid input from underplating (Ernst and Buchan, 2003). Another example is from the Kirkland Lake area of the Abitibi belt of the Superior province of Canada. A kimberlite pipe yielded xenoliths with Archean and 2.4–2.5 Ga zircon ages. The latter are similar to the regionally distributed Matachewan dyke swarm/LIP event (Section 5.2.3; Moser and Heaman, 1997).

5.6 Relations between different components of the LIP plumbing system

Here I show possible relationships between different components of the plumbing system of LIPs. By components I mean the different types of magmatic units that belong to a LIP, namely, volcanic rocks, dykes, sills, layered intrusions, and underplating; or those units that are associated with a LIP, namely, silicic magmatism, carbonatites, and kimberlites. First, I consider possible patterns for lithospheric entry points for magma from the underlying asthenospheric mantle; then, the links between the various plumbing-system components.

5.6.1 Lithospheric entry points

It is important to consider how the magma generated in the underlying asthenosphere initially enters the lithosphere. Figure 5.26 shows a variety of possible patterns (see discussion in Ernst *et al.*, 2005), such as: (a) from widespread sources above the mantle plume head (cf. White and McKenzie, 1989); (b) centrally

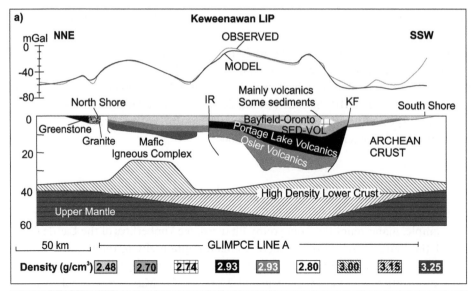

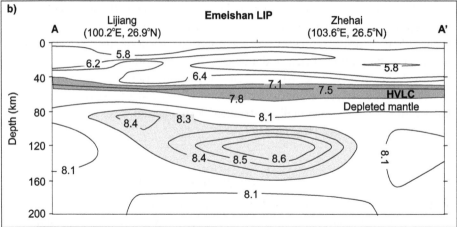

Figure 5.25 Crustal underplating associated with the 1115–1085 Ma Keweenawan and the 260 Ma Emeishan LIPs. (a) Velocity structure beneath the Keweenawan (Midcontinent rift system) of Lake Superior region of North America showing the underplate associated with the Midcontinent rift system. From Miller and Nicholson (2013), modified after original by Thomas and Teskey (1994). Digital version kindly provided by J. Miller. (b) The GLIMPCE seismic tomographic velocity structure of the crust and upper mantle beneath the Emeishan LIP of the west Yangtze craton. HVLC = high velocity lower crust. From Figure 3 in Xu and He (2007). As noted by Xu and He (2007) there is also a high-velocity lens-shaped body (8.3–8.6 km/s) in the upper mantle at a depth of 110–160 km. This body is *c.* 200 km long in an east–west direction and about 50 km thick, and it is surrounded by normal-velocity mantle (*c.* 8.1 km/s).

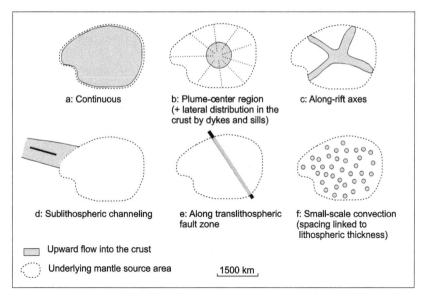

Figure 5.26 Patterns by which magma enters the lithosphere from below. (a) Widespread and closely spaced areas of magma access. (b) Access above a plume head and subsequent sideways dispersement by lateral dyke injection. (C) Magma emplaced upward along rift axes. (d) Lateral magma transport via sublithospheric channeling related to the topography at the base of the lithosphere. (e) Lithospheric access along translithospheric fault zones. (f) Ingress controlled by small-scale convection with the spacing related to the thickness of the lithosphere. The contribution of each of these styles of injection may vary from one LIP event to another.

located conduits near the plume center from which magma is distributed vertically and laterally throughout the crust as dykes and sills (Ernst *et al.*, 2005); (c) vertical access along zones of triple-junction rifting; (d) initial concentration at the plume center, followed by distribution along sublithospheric channels (Oyarzun *et al.*, 1997; Ebinger and Sleep, 1998; Duggen *et al.*, 2009); (e) vertical access along translithospheric fracture zones (e.g. Begg *et al.*, 2009, 2010) or (f) widespread points of ingress from a diapirically destabilized asthenospheric thermal boundary layer (Geoffroy *et al.*, 2007).

5.6.2 Sublithospheric channeling

Buoyant mantle (e.g. that associated with an arriving plume) can move sideways and upward along the base of the lithospheric root, and can also be channeled toward areas of thinner lithosphere ("thinspots" in the terminology of Thompson and Gibson (1991)) where partial melting can occur. An example is associated with the Afro-Arabian LIP and plume in which sublithospheric channeling from the plume-center region explains distal local concentrations of similar-age magmatism in central and western Africa (Ebinger and Sleep, 1998). Duggen *et al.* (2009) argued for sublithospheric flow of a mantle plume from beneath the Canary Islands to the Mediterranean.

5.6.3 Package of sills or sill-like layered intrusions

While funnel intrusions such as the Skaergaard intrusion of east Greenland, part of the NAIP LIP, are often emplaced as a single pulse (e.g. McBirney, 1996), other larger intrusions are typically filled by multiple pulses. For instance, two important complexes of the Midcontinent rift system (Keweenawan event), the Duluth and Mellen intrusions, are composed of tholeiitic mafic layered intrusions containing troctolitic, gabbroic, and anorthositic cumulates and some granophyric bodies. These tholeiitic complexes were emplaced as multiple sheet-like intrusions into the basal part of the comagmatic volcanic pile (Miller *et al.*, 2002).

Whether a layered intrusion or sill complex is produced depends on the time gap between the initial pulses. If the time gap between pulses is short, then there is insufficient time for crystalization of each pulse, and the emplacement of each pulse along the earlier emplaced sill would result in a magma chamber (e.g. Cawthorn, 2012). However, if each sill pulse has time to cool before the next sill is emplaced then the result is a sill complex.

5.6.4 Lateral dyke emplacement and distal feeding of sills and lava flows

The model of lateral dyke emplacement for regional swarms (Sections 5.2.1 and 5.2.4) has significant implications for the distribution of flood basalts and sills. Flows and sills can be emplaced at great distances from the plume center (Tables 5.2 and 5.3) via lateral flow in giant dyke swarms (Ernst and Buchan, 1997a, 1997b, 2001a; Fig. 5.27).

Specifically, lava flows can be formed when the upper edge of a dyke intersects with the Earth's surface. This can happen in a pre-existing or concurrently developing topographic low (i.e. basin); for example, in the 200 Ma dykes of the CAMP-swarm-fed lava flows located in the Newark rift basins of the eastern USA (e.g. Hill, 1991). As another example, the dyke swarms that fed the Columbia River lavas were emplaced into the pre-existing Pasco basin (Hooper, 1997). A distinctive type of lava flow within the 183 Ma Ferrar event of the Karoo-Ferrar LIP is found at three sites distributed over a distance of > 1600 km and is inferred by Elliot *et al.* (1999) to have originated from a single reservoir and to have been dispersed by a system of sills or possibly underlying dykes.

Laterally propagating dykes that intersect a sedimentary basin continue into the basin as sills (Table 5.3). A prominent example discussed earlier (Section 5.3.3) is the case of the Senneterre dykes of the Ungava radiating swarm, injected laterally from

Table 5.2 *The distance of flood basalts from their plume center (after Ernst and Buchan, 2001b)*

Flood-basalt unit	Associated dyke swarm	Distance from plume center	Plume/LIP event: n: name l: plume-center location a: age
ENA rift basin flows (North America)	ENA subswarm of CAMP radiating swarm	1300–1700 & 2200 km	n: CAMP l: 31° N, 78° W a: 0.20 Ga
Maranhão, Anari & Tapirapuã flows (South America)	CAMP radiating swarm	c.1500–2200 km	n: CAMP l: 10° N, 55° W a: 0.20 Ga
Siberian Trap flood basalts (E. Russia)	Ebekhaya and Maymecha subswarms of Siberian Trap radiating swarm	Up to 1500 km	n: Siberian Trap l: 72.5° N, 96° E a: 0.25 Ga
Scarab Peak lavas of Kirkpatrick Basalt (Antarctica)	Ferrar dykes?	2500–4000 km	n: Karoo–Ferrar l: 72° S, 20° W a: 0.18 Ga
Madagascar flows (Madagascar)	Madagascar radiating swarm	0–1000 km	n: Madagascar l: 24.4° S, 47.3° E a: 0.088 Ga
Deccan flows (India)	Deccan radiating swarm	0–1000	n: Deccan l: 21° N, 74° E a: 0.065 Ga
Columbia River flows (W. North America)	Columbia River dykes	c. 400 km	n: Columbia River LIP (Yellowstone plume) l: 42° N, 118° W a: 0.015 Ga

149

Table 5.3 *Distances of sill provinces from their plume center (after Ernst and Buchan, 2001b)*

Flat-lying sheets	Associated dyke swarm	Distance from plume center	Plume/LIP event n: name l: plume-center location a: age
Nipissing sills (S. Canada)	Senneterre (subswarm of Ungava Bay radiating swarm)	c. 1500 km	n: Ungava Bay l: 58° N, 66° W a: 2.21 Ga
Christie Bay sills (N. Canada)	Mackenzie radiating swarm	c. 1000 km	n: Mackenzie l: 71° N, 116° W a: 1.27 Ga
Coronation sills (N. Canada)	Franklin radiating swarm	c. 800 km	n: Franklin l: 75° N, 120° W a: 0.72 Ga
Minto Inlier sills (N. Canada)	Franklin radiating swarm	c. 400 km	n: Franklin l: 75° N, 120° W a: 0.72 Ga
Siberian Trap sills (E. Russia)	Ebekhaya and Maymecha subswarms of Siberia radiating swarm	up to 2000 km	n: Siberian Trap l: 72.5° N, 96° E a: 0.25 Ga
Mali sills (W. Africa)	CAMP radiating swarm	c. 500 km	n: CAMP l: 18° N, 20° W a: 0.20 Ga
Taoudenni sills (W. Africa)	CAMP radiating swarm	c. 1200 km	n: CAMP l: 18° N, 20° W a: 0.20 Ga
Amazon basin sills (E. South America)	CAMP radiating swarm	c. 1500 km	n: CAMP l: 10° N, 55° W a: 0.20 Ga
Ferrar sills (Antarctica)	Ferrar dykes?	100 s to 3000 km	n: Karoo–Ferrar l: 72° S, 20° W a: 0.18 Ga

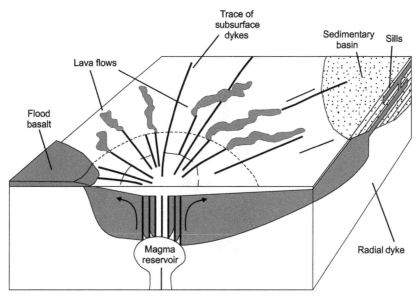

Figure 5.27 Role of lateral emplacement of dykes in feeding distal sills and volcanics.

the Ungava plume center for *c.* 1300 km before entering the Huronian sedimentary basin and producing the Nipissing sill province.

5.6.5 Funnel intrusions can spawn dykes

In parts of the NAIP, individual funnel intrusions have spawned local fanning swarms: e.g. the Imilik Complex of eastern Greenland (Klausen and Larsen, 2002) and the Rhum, Mull, and other intrusions of the British Tertiary Igneous Province (Speight *et al.*, 1982; MacDonald *et al.*, 1988). It is also envisioned that subswarms of the major radiating swarms are spawned by magma chambers located in the vicinity of the plume focus (Baragar *et al.*, 1996).

Progress is being made on developing the theoretical basis for feeding dykes from sills and, specifically, injection of lateral dykes from oblate reservoirs (e.g. Grosfils (2007; Grosfils *et al.*, 2013; McGovern *et al.*, 2014).

5.6.6 Feeding of funnel intrusions along dykes

A model is presented in Ernst and Buchan (1997b) for the emplacement of funnel intrusions as localized upwellings along feeder dykes. Figure 5.28 shows a series of cinder cones emplaced along underlying dykes. Figure 5.29 shows various models for how an underlying dyke can produce funnel intrusions and even wider dyke-like layered intrusions.

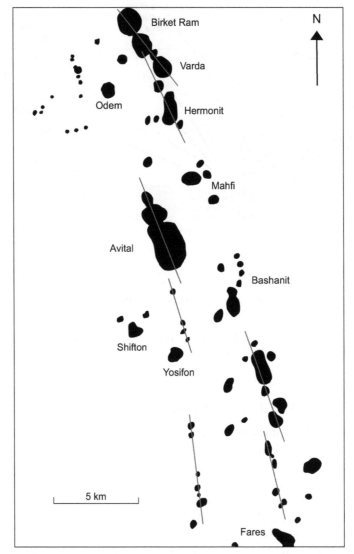

Figure 5.28 Distribution of cinder cones associated with the Golan volcanic plateau of Israel and Syria. The bodies are interpreted to be aligned along subsurface dykes. From Figure 9 in Ernst and Buchan (1997b).

5.7 Summary

This chapter has reviewed the evidence for components of the plumbing system of LIPs. The important regional dyke swarms that can extend over areas of up to several million square kilometers usually have regional linear or radiating geometry. The latter are associated with plume centers and breakup (or attempted breakup) of

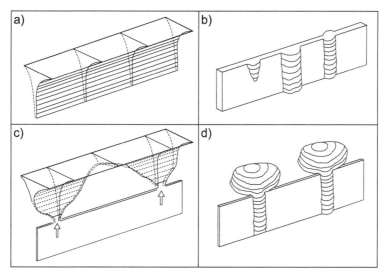

Figure 5.29 Layered intrusions fed by dykes: (a) dyke-like (funnel-dyke) intrusion emanating from a continuous dyke; (b) feeders coalescing above a dyke-like intrusion; (c) dyke-feeding plugs; and (d) funnel-shaped intrusions and pipes issued from a dyke. From Ernst and Buchan (1997b).

cratons (Chapter 11). Dolerite sill provinces can also be spread over vast areas, and the individual sills have characteristic saucer-shaped geometry. In addition, layered intrusions are a key part of the plumbing system. Interactions between different parts of the plumbing system have been discussed. Magmatic underplating at the base of the crust represents a volumetrically important, but poorly understood, portion of the plumbing system.

6

Archean LIPs

6.1 Introduction

As earlier chapters have shown, Large Igneous Provinces (LIPs) are prominent throughout the Mesozoic and Cenozoic (as flood basalts), and in the Paleozoic and Proterozoic (typically eroded to expose the plumbing system of LIPs). The extension of the LIP record back into the Archean is the subject of this chapter. Erosional remnants of typical Archean flood basalt provinces include the Fortescue sequence of the Pilbara craton in Australia and the Ventersdorp sequence of the Kaapvaal craton in southern Africa. However, most Archean basic volcanic rocks occur as deformed and fault-fragmented packages that have been weakly to strongly metamorphosed and are termed greenstone belts. One class of greenstone belts contains mafic to silicic igneous rocks with calc-alkaline geochemical signatures and has been interpreted to correspond to remnants of volcanic arcs. However, the other major class of greenstone belts consists of tholeiite sequences that commonly contain komatiites, and these are the best candidates for being remnants of Archean LIPs (e.g. Arndt *et al.*, 2001; Ernst and Buchan, 2001b; Prendergast, 2004; Ernst *et al.*, 2005; Barnes *et al.*, 2012; Kerr, 2014). Both types of Archean LIPs are discussed below: Archean flood basalts and Archean greenstone belts containing komatiites.

6.2 Archean flood basalts

Given the early stabilization of the Kaapvaal and Pilbara cratons, there is potential in these regions for the preservation of post-cratonization cycles of LIPs associated with the various sedimentary cover sequences (Fig. 6.1). The Dominion Group of the Kaapvaal craton is the oldest cover sequence, and it is unconformably overlain by the Witwatersrand/Pongola and Ventersdorp supergroups. The more easterly occurring Pongola Supergroup correlates with the coeval Witwatersrand Supergroup.

As shown in Fig. 6.1a there are major mafic (and ultramafic) magmatic packages associated with each of these sedimentary-cover sequences. For the Kaapvaal craton, I discuss the following: the *c*. 3.0 Ga Dominion Group lavas (at the base of the Witwatersrand), the slightly younger *c*. 2.95 Ga Nsuze volcanics (in the lower part of

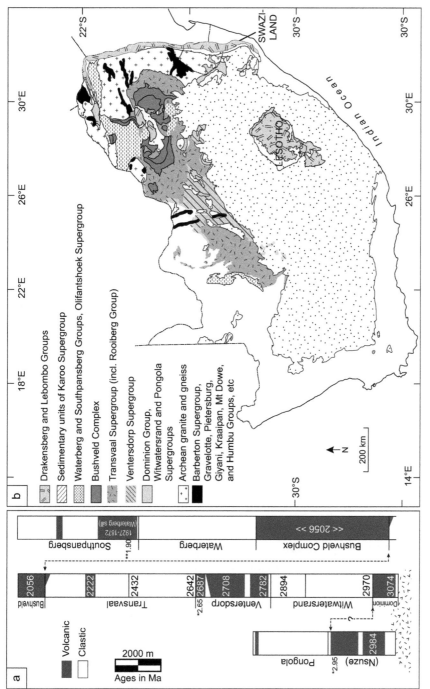

Figure 6.1 Proterozoic and Archean LIPs of the Kaapvaal craton (Dominion, Ventersdorp, Bushveld, Waterberg, Karoo) in the context of associated sedimentary sequences. (a) Cumulative Mesoarchean–Paleoproterozoic stratigraphy on the Kaapvaal craton, compiled by combining maximum-thickness estimates and ages (in Ma). From Figure 2 in Klausen *et al.* (2010). With permission from Elsevier. (b) Map and legend are modified from frontispiece of Johnson *et al.* (2006).

the Pongola Supergroup), and *c.* 2.78–2.7 Ga multi-pulsed events in the Ventersdorp Supergroup. From the Pilbara craton I consider the multi-pulsed 2.8–2.7 Ga Fortescue Group magmatism.

6.2.1 The Dominion Group LIP (c. 3.0 Ga)

Volcano-sedimentary strata of the Dominion Group (maximum preserved thickness 2250 m) non-conformably overlie the gneissic and granite–greenstone basement of the Kaapvaal craton and represent the initial phase of volcanic activity and sedimentation within the Witwatersrand basin (e.g. Eriksson *et al.*, 2002; Marsh, 2006). As reviewed by Marsh (2006), the volcanic rocks of the Dominion Group constitute a bimodal suite, consisting of a mafic–intermediate suite (Rhenosterhoek Formation), and overlying silicic suite (Syferfontein Formation). Rocks of the mafic–intermediate suite have been referred to as "andesite" because of their high-silica content, but their least differentiated samples also have high magnesium, and these have been interpreted as a strongly differentiated tholeiite suite (Fig. 6.2; Marsh, 2006).

The interpretation of the tectonic setting is varied, ranging from arc volcanism to a bimodal tholeiitic basalt–rhyolite association; the latter is characteristic of

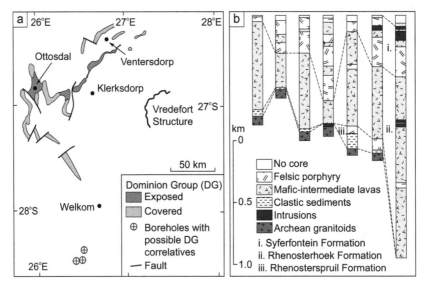

Figure 6.2 The *c.* 3.0 Ga Dominion LIP of the Kaapvaal craton. (a) Map showing the location of known surface and subsurface exposures of the Dominion Group. (b) Lithological and stratigraphic variations in the Dominion Group encountered in closely spaced borehole cores (locations given in (a)). The variation in thickness of the Syferfontein Formation is, in part, due to erosion but the variation in thickness of the lower units is primary. Redrafted from Figures 1 and 2 in Marsh (2006). See Figure 6.1 for location of Dominion Group.

deposition in a failed continental-rift-basin tectonic setting (Marsh, 2006). Dominion Group magmatism is generally considered to be the erosional remnant of one of the earliest flood basalts emplaced on cratonized crust.

The currently accepted age for the Dominion Group is 3074 ± 6 Ma (Armstrong *et al.*, 1991), which is considerably older than ages obtained to date for the Nsuze Group volcanics (see below). However, strong tectonic and lithological similarities exist between these two major volcanic occurrences and on this basis the linkage of the Dominion Group and Nsuze Group into a single LIP is plausible (Wilson and Hofmann, 2013).

6.2.2 The Nsuze LIP (c. 2.95 Ga)

The Pongola Supergroup (Fig. 6.1) is characterized by a lower, volcanogenic *c.* 2.95 Ga Nsuze Group which is up to 8 km thick and overlain by the clastic Mozaan Group which is up to 5 km thick (Goodwin, 1996; Wilson and Grant, 2006; Gold, 2006). In addition, the mafic–ultramafic rocks of the Usushwana intrusive suite may also be related (J. Olsson, personal communication, 2011). As far as dating of the Pongola Supergroup is concerned, Hegner *et al.* (1994) obtained the most precise age on a basaltic andesite in the Mhlangatsha area in Swaziland of 2985 ± 1 Ma. The most recent U–Pb isotopic age determinations of the Pongola volcanic rocks (Mukasa *et al.*, 2013) using secondary ion mass spectrometry (SIMS) on zircons give a progression of ages from the base of the Nsuze Group in the Hartland area of 2980 ± 10 Ma, to a rhyolite in the mid-sequence of 2968 ± 6 Ma, and an age in the upper part of the Mozaan Group (Tobolsk volcanics) of 2954 ± 9 Ma.

The importance of this Nsuze magmatic event as an Archean LIP has recently increased due to the recognition that a major regional dolerite dyke swarm can be linked to it, on the basis of geochronology, geochemistry, and paleomagnetism (Olsson *et al.*, 2010, 2011; Klausen *et al.*, 2010; Lubnina *et al.*, 2010). The 250-km-wide Badplaas–Barberton swarm of southeast-trending dolerite dykes along with associated rift-basin structures (Burke *et al.*, 1985) sweep in from the southeastern corner of the Kaapvaal craton (Hunter and Halls, 1992; Uken and Watkeys, 1997; Klausen *et al.*, 2010). Dykes in the southeast-trending swarm exhibit a dominant northwest direction of bifurcation, indicating emplacement from the southeast (Hunter and Halls, 1992). These northwest-trending dykes and adjacent geochemically similar west-trending dykes may be parts of a radiating dyke swarm focused on a mantle plume center, located near a hypothetical passive margin along the southeast side of the Kaapvaal craton (Klausen *et al.*, 2010).

6.2.3 The Ventersdorp LIP(s) (2.78–2.7 Ga)

The Ventersdorp Supergroup of the Kaapvaal craton (Figs. 6.3 and 6.4) is host to a multi-pulsed sequence of Archean continental flood basalts (CFBs; e.g. White, 1997;

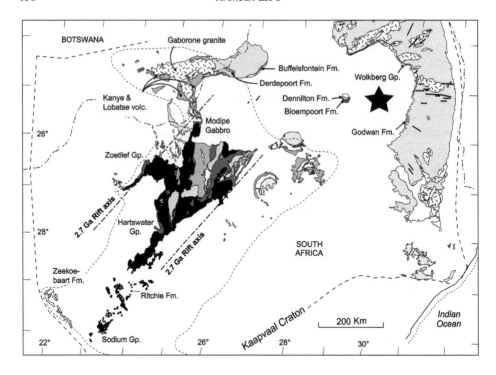

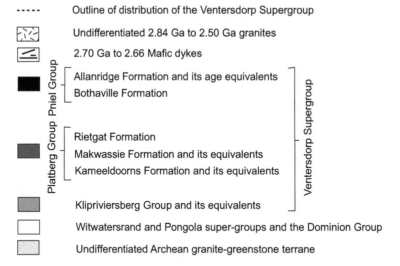

Figure 6.3 The 2780–2660 Ma Ventersdorp LIP(s) of southern Africa. Multiple magmatic pulses are present that have been collectively grouped as the Ventersdorp Supergroup, but which individually may represent separate LIPs (see discussion in text). Ages are given in Figure 6.4. Modified after unpublished digital diagram kindly provided by M. de Kock with addition of fanning dyke swarm after Olsson *et al.* (2010, 2011). With permission from Nature Publishing Group. Focal point of swarm (star) interpreted to mark the plume center for the *c.* 2680 Ma stage (Allanridge). Note the overall northeast-trending rift structure.

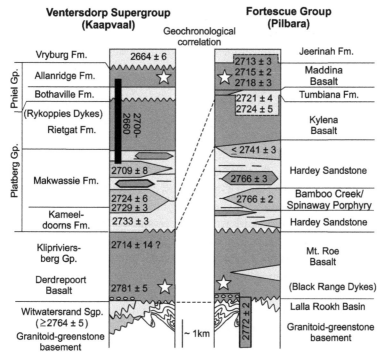

Figure 6.4 Age correlations of the Ventersdorp Supergroup of the Kaapvaal craton of South Africa and the Fortescue Group of the Pilbara craton, northwestern Australia. Stars indicate paleomagnetic studies (Wingate, 1998; Strik *et al.*, 2003; de Kock *et al.*, 2009) to test the reconstruction of Kaapvaal and Pilabara cratons at this time (in supercraton Vaalbara). Modified from Figure 1 in de Kock *et al.* (2009). With permission from Elsevier.

Eriksson *et al.*, 2002; van der Westhuizen *et al.*, 1991, 2006). The Ventersdorp Supergroup comprises a 5100-m-thick succession of subaerial volcanic rocks including up to 2900 m of sedimentary units identified in borehole sections. The volcanism occupies an area greater than 0.3 Mkm^2 within the central part of the Kaapvaal craton, and has a minimum eruptive volume of 0.66 Mkm^3, which falls within the definition of a LIP (> 0.1 Mkm^2 and > 0.1 Mkm^3; Bryan and Ernst, 2008) (Chapters 1 and 2).

The Ventersdorp Supergroup is typically divided into three magmatic groups: the basal Klipriviersberg, overlying Platberg (2709 ± 4 Ma (U–Pb)), and the uppermost Pniel Group, which include the Allanridge volcanics. In addition, the Derdrepoort volcanics were dated by Wingate (1998) at 2782 ± 5 Ma (U–Pb), who argues that these basalts represent the most basal rocks of the Ventersdorp Supergroup and that the earlier-obtained 2714 ± 8 Ma (U–Pb) age for the Klipriviersberg should be regarded as a minimum age and, as discussed by de Kock *et al.* (2009), could be as old as 2764 ± 5 Ma (U–Pb). Questions regarding the 2714 Ma age of the Klipriviersberg lavas are reinforced by the presence of an angular unconformity between the two apparently coeval units (Klipriviersberg and Platberg), characterized by well-developed paleosols (van der Westhuizen *et al.*, 1991; White, 1997; Eriksson *et al.*, 2002). Furthermore, this earliest 2780 Ma pulse is augmented by the similar age for the Gaborone Granite Complex and Kanye volcanic formation which have been dated at 2781 ± 2 Ma (Grobler and Walraven, 1993). It is also possible that the Modipe Gabbro is the same age (Wingate, 1998). These various 2.78 Ga components – volcanics, granites, etc. – may together constitute either a precursor or early stage of the Ventersdorp LIP (Wingate, 1998) or an entirely independent LIP (Section 15.6).

The bulk of the Klipriviersberg Group consists of 1500–2000 m of lavas, with basal komatiitic rocks passing up into tholeiitic basalts with slight calc-alkaline affinities (due to assimilation and fractional crystallization (AFC); e.g. Kerrich, personal communication, 2011; Eriksson *et al.*, 2002; van der Westhuizen *et al.*, 2006). Klipriviersberg lavas are typical flood basalts; and rare sedimentary interbeds and a lack of paleosols support rapid and continuous eruptive activity. Flows are predominately subaerial and amygdaloidal.

The Platberg Group consists of clastic and chemical sediments and also a bimodal assemblage of mafic and felsic lavas that were deposited during graben development (van der Westhuizen *et al.*, 2006).

The overlying 750-m-thick Allanridge volcanic rocks (Winter, 1976; van der Westhuizen *et al.*, 1991, 2006) represent the final flood-basaltic event in the Ventersdorp Supergroup. Locally, at the base of this formation, komatiitic lavas occur but most of the sequence consists of amygdaloidal and porphyritic basaltic andesites. The Allanridge volcanic rocks are undated, but have recently been linked to a *c.* 2.65 Ga radiating Rykoppies dyke swarm on the basis of age constraints, and geochemical and paleomagnetic similarity (Olsson *et al.*, 2010; Klausen *et al.*, 2010; Lubnina *et al.*, 2010). Upper Buffelsfontein Group lavas associated with so-called "protobasinal fills" (discrete sedimentary basin-fills along the northern and eastern exposed base of the Transvaal Supergroup) (Fig. 6.1), have ages of 2657–2659 Ma and 2664 ± 0.7 Ma (e.g. Eriksson *et al.*, 2002, and references therein) and can also be part of the Allanridge event.

The rapidly emplaced large volumes of Ventersdorp Supergroup flood basalts (mainly Klipriviersberg Group lavas; McCarthy *et al.*, 1990) appear to have initially erupted from a northeast-trending feeder dyke swarm (McCarthy *et al.*, 1990; Klausen *et al.*, 2010), and the younger Allanridge Formation lavas are linked to a radiating east-to-southeast-trending Rykoppies feeder dyke swarm at *c.* 2.65 Ga (Klausen *et al.*, 2010; Olsson *et al.*, 2010, 2011) (Fig. 6.3). In addition, a northeast–southwest rift structure associated with the bulk of the Ventersdorp volcanics (Stanistreet and McCarthy, 1991) trends toward the Rykoppies plume center; this geometry of a failed rift aligned toward a plume center would be suggestive of a triple-junction rift system (cf. Burke and Dewey, 1973). As discovered by Olsson *et al.* (2011), the locus of the

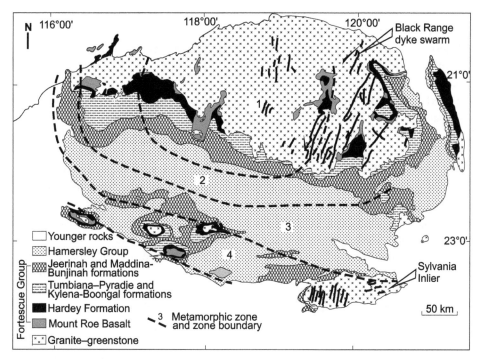

Figure 6.5 The 2780–2710 Ma Fortescue LIP(s) of the Pilbara craton. Simplified geological map of the Hamersley basin and Fortescue Group of the Pilbara craton of northwestern Australia. The distribution of the various Fortescue magmatic pulses are shown and each may represent a separate LIP (see discussion in text). Superimposed are burial metamorphic zones (1 = prehnite–pumpellyite, 2 = prehnite–pumpellyite–epidote, 3 = prehnite–pumpellyite–epidote–actinolite, 4 = actinolite). From Pirajno and Hoatson (2012) after Thorne and Trendall (2001). With permission from Elsevier. Dykes (Black Range swarm and those of Sylvania Inlier) are added from Wingate (1999). Digital version of original figure (not including the dykes) kindly provided by F. Pirajno.

Allanridge radiating swarm was the locus of the eastern lobe of the 2.06 Ga Bushveld intrusion of the Bushveld LIP.

6.2.4 The Fortescue LIP(s) (2.78–2.7 Ga)

The Fortescue Group is a thick succession of ultramafic, mafic, intermediate, and felsic volcanic and associated volcaniclastic and sedimentary rocks, which unconformably overlies the granite–greenstone rocks of the Archean Pilbara craton, in the northwest of Western Australia (e.g. Thorne and Trendall, 2001; Eriksson *et al.*, 2002; Blake *et al.*, 2004; Figs. 6.4 and 6.5). The outcrops of the Fortescue Group cover an area of about 40 000 km^2.

The lowermost magmatic unit, the 2775–2663 Ma Mount Roe volcanic rocks, consists dominantly of subaerial tholeiitic basaltic flows, massive to amygdaloidal, with lesser hyaloclastite and pillow lavas. Also present are silicic and mafic lapilli tuffs at or near the base of the Mount Roe Formation. The Black Range dykes are coeval with the Mount Roe basalts and are probably its feeders (Wingate, 1999). The overlying Hardey Formation contains silicic volcanism; it is up to 3 km thick and is associated with a coeval granitic plutonic complex. Overlying units Kylena, Tumbiana, and Maddina Formations in the north (with equivalent units Boongal, Pyradie, and Bunjinah Formations in the south; Thorne and Trendall, 2001), are from 1.5 to 3 km thick and consist mostly of tholeiitic basaltic, high-Mg basalt, and komatiite rocks. The units in the north form as stacked subaerial flows, and those in the south are submarine lavas. Age constraints are provided in Fig. 6.4.

The apparent lateral continuity of the mafic–ultramafic lavas, the general lack of interbedded sedimentary rocks within each volcanic package, and the massive character of the lava units have been collectively interpreted to indicate the rapid eruption of large volumes of magma analogous to the Mesozoic–Cenozoic CFBs. However, there are differences between Archean and Phanerozoic CFBs. As noted by Flament *et al.* (2011) many of the interpreted Archean CFBs (and platform-type greenstone belts; see below) are subaqueous or episodically returned to sea level in contrast to Phanerozoic CFBs, which are more typically subaerial. Flament *et al.* (2011) offer an explanation for this pattern in suggesting that the gravity-driven lower-crustal flow enabled by a higher Archean geotherm may have contributed to maintaining Archean CFBs close to sea level.

6.2.5 Archean reconstructions: Ventersdorp and Fortescue LIPs

It has long been postulated that the Kaapvaal and Pilbara cratons were joined in the late Archean into a supercraton named 'Vaalbara' (Cheney *et al.*, 1988; Nelson *et al.*, 1992; Cheney, 1996; Wingate, 1998; Zegers *et al.*, 1998; Bleeker, 2003; de Kock *et al.*, 2009). The Vaalbara reconstruction is based on the age correlation between magmatic pulses of the Fortescue (Pilbara craton) and Ventersdorp (Kaapvaal craton) *c.* 2.7–2.8 Ga flood basalts (Fig. 6.4); paleomagnetism on these units provides additional constraints on the reconstruction (e.g. de Kock *et al.*, 2009).

6.3 Archean greenstone belts of the tholeiite–komatiite association

As demonstrated in the previous section, Archean CFBs exist as far back as 3.0 Ga, in areas of already-cratonized crust. Here I discuss another population of probable Archean LIPs emplaced on non-cratonized crust that is subsequently deformed and cratonized (cf. discussion Wyman and Kerrich (2009) regarding the late timing for attachment of continental lithospheric mantle keels/mantle roots marking the timing of cratonization).

Greenstone belts have been comprehensively reviewed (de Wit and Ashwal, 1997; de Wit, 1998) and also assessed from a mantle plume/LIP perspective (e.g. Isley and Abbot, 1999; Tomlinson and Condie, 2001; Condie, 2001; Arndt *et al.*, 2001, 2008; Ernst and Buchan, 2001b; Ernst *et al.*, 2005; Bryan and Ernst, 2008; Wyman and Kerrich, 2009, 2010; Barnes *et al.*, 2012).

6.3.1 Tholeiitic–komatiite greenstone belts as LIPs

Based on 51 Archean greenstones belts for which lithologic proportions were available, 65% were found to have arc affinities and 35% non-arc affinities (Condie, 2001, p. 198). Those with non-arc greenstone belts have tholeiite $\pm$ komatiite sequences (Fig. 6.6) and, as discussed below, the proposal has been made that these are examples of Archean LIPs (e.g. Ernst and Buchan, 2001b; Ernst *et al.*, 2005). The proportion of komatiites in a greenstone belt is variable, ranging up to perhaps 30% (Fig. 6.7). (Note that komatiites are much rarer in the post-Archean record, with the youngest, on Gorgona island, associated with *c.* 90 Ma Caribbean–Colombia LIP (Fig. 4.5) (e.g. Arndt *et al.*, 2008)). I evaluate the interpretation of these tholeiitic–komatiite Archean greenstone belts as LIPs with respect to the criteria discussed in Chapter 2 (after Bryan and Ernst, 2008): intraplate setting/characteristics, large area/volume, and short duration or short-duration pulses.

Intraplate setting

As originally proposed by Thurston and Chivers (1990) and discussed subsequently (e.g. Abbott and Mooney, 1995; Tomlinson and Condie, 2001; Condie, 2001; Chapter 14 in Arndt *et al.*, 2008), there are two main Archean greenstone associations not linked to subduction and therefore potentially linked to both LIPs and plumes.

(1) The platform association (platform type) comprises tonalite conglomerates, quartz arenites, carbonates and BIF (banded ironstone formation), overlain by komatiites and basalts and overlying tonalitic basement. The tholeiitic-komatiite volcanic rocks can show geochemical evidence for crustal contamination and contemporaneous emplacement with significant volumes of silicic igneous rocks, and were emplaced onto submerged continental platforms, submerged highly extended continental crust, or into intracontinental rift basins (Table 6.1). These settings have many similarities to CFBs, and may represent plume-generated magmas erupted through continental crust, but an important difference is that in most cases the magmatism is subaqueous rather than subaerial as mentioned above (e.g. Arndt, 1999; Flament *et al.*, 2011). (These greenstone belts of the platform association also differ from younger flood basalts in that the crust had not yet cratonized.)

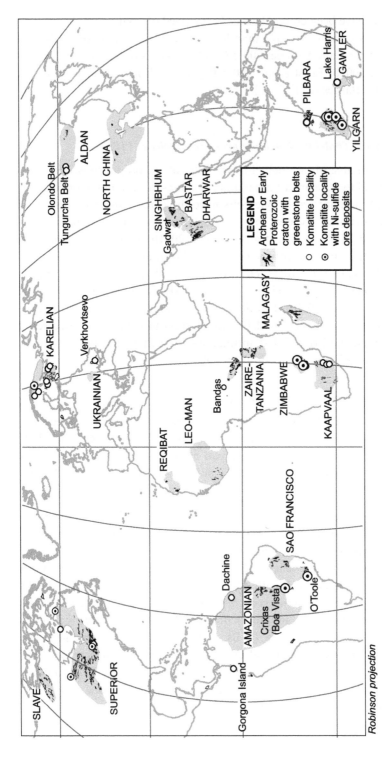

Figure 6.6 Global distribution of Archean greenstone belts. Those that include komatiites are interpreted to be part of LIPs. Modified from Figure 1.1 in Arndt *et al.* (2008). Digital version of original figure kindly provided by S. J. Barnes.

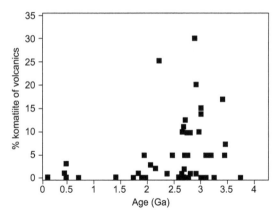

Figure 6.7 Percent komatiite versus age for greenstone belts. Note that many greenstone belts contain no komatiites. Redrafted from Figure 2 in de Wit and Ashwal (1997).

(2) The mafic-plains association (mafic-plains type) (Table 6.2) comprises predominantly pillowed basalts with variable amounts of komatiite, and small amounts of chert and BIF, and greenstone belts with these affinities are interpreted to be oceanic plateaus that were accreted during ocean closure (e.g. Abbott and Mooney, 1995; Tomlinson and Condie, 2001; Condie, 2001; Kerr, 2014).

Some examples of the mafic-plains type of greenstone belts are also associated with arc magmatism (e.g. the Abitibi greenstone belt) and have been interpreted to represent intermixed oceanic plateaus and arcs caused by plume ascent into a region of subduction and ocean closure (e.g. Arndt *et al.*, 2008) or oceanic plateaus captured by migrating oceanic arcs, as for the Ontong–Java plateau and Solomon arc (e.g. Wyman and Kerrich, 2009, 2010).

Plume origin of komatiites

As can be seen in Tables 6.3 and 6.4 (see also Arndt *et al.*, 2008), komatiites are moderate- to high-degree partial melts from particularly hot and from moderate to deep sources. These characteristics are best explained in the context of a mantle plume and, in particular, of melting in the hot central conduit of the plume (Chapter 15 and e.g. Campbell *et al.*, 1989, Campbell, 2001; Condie, 2001; Sproule *et al.*, 2002; Arndt *et al.*, 2003, 2008). Although some (e.g. Parman and Grove, 2005, p. 3) have suggested an origin by melting of wet mantle under subduction conditions, this process arguably occurs only rarely (e.g. Arndt *et al.*, 1998, 2008), and komatiites are compositionally distinct from high-MgO arc picrites (Polat and Kerrich, 2006). So the presence of komatiites indicates a major thermal mantle anomaly with a significant production of high- to moderate-degree partial melts in

Table 6.1 *Examples of Archean greenstone belts interpreted as platform-type (erupted through non-cratonized continental crust). From Tomlinson and Condie (2001); event no. links after Ernst and Buchan (2001b).*

Greenstone belt (age)	LIP association	Location
Steep Rock greenstone belt (c. 2930 Ma)	Steep Rock LIP (event no. 252)	Central Wabigoon terrane, Superior craton (Canada)
Lumby Lake greenstone belt (< 2963–2898 Ma)	Steep Rock LIP (event no. 252)	Central Wabigoon terrane, Superior craton (Canada)
D'Alton Lake–Toronto Lake belt (c. 2920 Ma)	Steep Rock LIP (event no. 252)	Central Wabigoon terrane, Superior craton (Canada)
North Caribou Lake greenstone belt (<2980–2932 Ma)	Steep Rock LIP (event no. 252)	North Caribou terrane, Superior craton (Canada)
Balmer assemblage, Red Lake greenstone belt (2992–2964 Ma)	Balmer LIP (event no. 256)	North Caribou terrane, Superior craton (Canada)
Mtshingwe Group, Belingwe greenstone belt (c. 2900 Ma)	Bulawayan LIP (event no. 259)	Zimbabwe
Ngezi Group, Belingwe greenstone belt (2962 Ma)	Bulawayan LIP (event no. 259)	Zimbabwe
Kambalda, Norseman–Wiluna belt (c. 2.7 Ga)	Eastern Goldfields (Section 6.3.4)	Eastern Goldfields terrane, Yilgarn craton (Australia)
Forrestania greenstone belt (c. 2.9 Ga)	?Forrestania-Lake Johnston LIP (event no. 250	Southern Cross terrane (Australia)
Kam group, Yellowknife greenstone belt (2.7 Ga)	Kam LIP (Section 6.3.3)	Slave craton (Canada)
North Star basalts (c. 3450 Ma)	?	Pilbara craton (Australia)
Kolar schist belt (c. 2.9 Ga)	?	Dharwar craton (India)
Prince Albert Group and Woodburn Lake group (2.7 Ga)	Prince Albert LIP (Section 6.3; event no. 220)	Rae craton (Canada)

the hottest part of the plume, and which have access to the crust (via plumbing-system pathways, and sufficiently thin lithosphere). (The rarity of komatiites in the post-Archean record must reflect secular cooling of the deep-mantle-source areas from where the plumes originate.) Under the model, in which komatiites are produced from the hottest portion of the plume, there would be complementary melt production of basalts from the cooler annulus of the plume head which had entrained some ambient asthenosphere (e.g. Campbell *et al.*, 1989; Tomlinson *et al.*, 1999; Said *et al.*, 2012).

Table 6.2 *Examples of Archean greenstone belts interpreted to represent mafic-plains-type, i.e., accreted oceanic, plateaus. From Tomlinson and Condie (2001); event no. links after Ernst and Buchan (2001b).*

Greenstone belt (age)	LIP association	Location
Heaven Lake greenstone belt (2954 Ma)	? Steep Rock LIP (event 252)	Central Wabigoon terrane, Superior craton (Canada)
Southern Onaman–Tashota terrane (> 2740 Ma)	? Southern Onaman-Tashota LIP (event no. 229)	Eastern Wabigoon terrane, Superior craton (Canada)
Vizien greenstone belt (2786 Ma)	? Vizean LIP (event no. 236)	Minto block, Superior craton (Canada)
Abitibi greenstone belt (2750–2700 Ma)	Abitibi belt LIP(s) (Section 6.3.2)	Superior craton (Canada)
Wawa greenstone belt (2.7 Ga)	? part of Abitibi belt LIP(s)	Wawa terrane, Superior craton (Canada)
Pickle Crow assemblage, Pickle Lake greenstone belt (2.86 Ga)	? Pickle Crow LIP (event no. 244)	Uchi terrane, Superior craton (Canada)
Lower Onverwacht Group, Barberton greenstone belt (3472 Ma)	? Lower Onverwacht LIP (event no. 273)	South Africa
Kostomuksha belt (> 2.8 Ga)	? Kostomucksha LIP (event no. 241)	Karelian Domain, northwest Baltic shield (Russia)
Fiskenaesset (> 2.85 Ga?)		Southeest Greenland
Tungurcha greenstone belt (> 2.8 Ga)	? Tungurcha LIP (event no. 266)	Western Aldan shield (Russia)

Inferred large size

The question of whether tholeiitic–komatiitic greenstone belts are of LIP scale can be addressed in two ways: first, indirectly on the basis of the event size implied by the presence of komatiites, and on the basis of some rare examples in which the large scale is preserved despite the deformation and faulting that typically fragment greenstone belts.

While deformation and faulting generally prevent the tracing of Archean tholeiite–komatiite greenstone belts over large distances, there are some cases in which LIP scales are preserved. An example is the Rae craton of northern Canada where the *c.* 2700 Ma Prince Albert, Woodburn Lake, and Mary River groups define a linear belt (Fig. 1.4b), which extends for a distance of 1500 km (> 0.2 Mkm^2), and may be linked with a mantle plume and/or a late Archean breakup margin (e.g. MacHattie *et al.*, 2004; MacHattie, 2008). Another example of an extensively

Table 6.3 *Main types of komatiites and conditions of their formation. After Sproule et al., 2002 (see also Table 6.4).*

Geochemical type (Type area)	Al-depleted (Munro)	Al-depleted (Barberton)	Ti-enriched (Finland)	Ti-depleted (Shining Tree)	
Al_2O_3/TiO_2	15–25	< 15	< 15	25–35	
$(Gd/Yb)_{MN}$	c. 1.0	1.2–2.8	> 1.2	0.6–0.8	
Source	Garnet peridotite	Garnet peridotite	Garnet peridotite	Garnet peridotite	Stage 1: Garnet peridotite Stage 2: Al-rich peridotite
Degree of partial melting	30–50%	20–40%	> 20%	Single-stage dynamic melting 30–50%	Two stage 20–40%
Source residuum	Dunite + Harzburgite	Garnet peridotite	Dunite + Harzburgite	Harzburgite	Harzburgite
Separation depth and region	Plume head at 2–8 GPa	Plume tail at 6–9 GPa	Plume head at 2–8 GPa	Plume head at 2–8 GPa	Plume tail at 6–9 GPa

MN = primitive mantle normalized (after McDonough and Sun, 1995).

Table 6.4 *Conditions of formation of komatiites and related rocks. From Figure 13.1 in Arndt* et al. *(2008).*

Type	Conditions
Barberton-type komatiite	Batch melting of fertile peridotite at very high pressure (> 9 GPa)
Munro-type komatiite	Critical melting of depleted peridotite at high pressure (*c.* 5–7 GPa)
Gorgona-type komatiite	Critical melting of depleted peridotite at moderate pressure (*c.* 3–4 GPa)
Gorgona picrite	Advanced critical melting of depleted peridotite at moderate pressure (*c.* 3–4 GPa)
Karasjok-type komatiite	Critical melting of depleted peridotite
Karasjok picrite	Partial melting of enriched peridotite
Commondale komatiite	Extreme critical melting of depleted peridotite
Komatiitic basalts	Fractional crystallization of parental komatiite, or crystallization plus crustal contamination, or independent melting of the mantle source

preserved Archean LIP is the *c.* 2.7 Ga Bulawayan Supergroup, which contains 4–6-km-thick mafic–ultramafic–silicic volcanic sequences that extend for *c.* 0.25 Mkm^2 across the Zimbabwe craton (Prendergast, 2004). Other extensive (> 800-km strike length) submarine volcanic sequences containing tholeiites and komatiites occur in the Yilgarn craton (Norseman–Wiluna belt) and Canadian Superior province (Abitibi belt). Event volume is even more difficult to assess given the deformation in most greenstone belts. The full scale of tholeiitic–komatiitic greenstone belts, and how many are of LIP scale, is likely to only become clearer when robust late-Archean reconstructions are achieved (Bleeker, 2003). Consider Superia reconstructions in which the southeastern Superior craton (Abitibi belt), Karelian (Baltica), Hearne, and possibly Zimbabwe cratons are joined, thus potentially linking all their *c.* 2.70–2.75 Ga tholeiitic–komatiitic greenstone belts into a single LIP (e.g. Bleeker and Ernst, 2006; Söderlund *et al.*, 2010).

Duration and pulses

A characteristic of many Archean greenstone belts is the presence of multiple pulses over a protracted period of time. For example, the Abitibi belt has pulses of tholeiitic–komatiitic magmatism at 2750–2735 Ma, 2725–2720 Ma, 2718–2710 Ma, and 2710–2703 Ma (Ayer *et al.*, 2002). The Fortescue flood basalt of the Pilbara craton has pulses at *c.* 2770, 2720, and 2690 Ma (Thorne and Trendall, 2001; Eriksson *et al.*, 2002; Blake *et al.*, 2004). The presence of multiple pulses and their apparent short duration has strong similarities to younger LIP records (Chapter 3; see also Section 15.6). However, the meaning of the protracted duration (50 My or more) for each of these multi-pulse Archean LIPs remains unclear, as it does for comparable examples in the younger record (e.g. the Matachewan event, which spans from 2490–2450 Ma and consists of multiple pulses, 2475, 2459, 2445 Ma; Chapter 5). The possible causes of multiple pulses over a long duration in some LIPs are discussed in Section 15.6; there, it is noted that some examples of multiple pulses spanning tens of million years may represent the contribution of pulses for more than one independent LIP whose plume centers could be widely separated.

Below I present the Abitibi greenstone belt as an illustration of accreted oceanic plateaus (mafic-plains type) and of the Kam Group in the Slave craton as an illustration of emplacement on non-cratonized continental crust (the platform type).

6.3.2 Abitibi belt LIP(s) (2.75–2.7 Ga)

The example of the Abitibi belt (Fig. 6.8) of the Superior craton of Canada is useful in illustrating the complexities of oceanic plateaus interacting with arcs (e.g. Ayer *et al.*, 2002; Sproule *et al.*, 2002; Wyman and Kerrich, 2009, 2010). Many greenstone belts represent a tectonic collage of separate assemblages, and the Abitibi greenstone belt is no exception. Currently, the Abitibi belt can be subdivided on the basis of geochronological and lithological criteria into nine assemblages. The seven oldest of these are predominantly volcanic, and were deposited in almost continuous volcanism from *c.* 2.75 to 2.70 Ga (Ayer *et al.*, 2002; Sproule *et al.*, 2002). Two of the volcanic-dominated assemblages were dominantly calc-alkaline and are interpreted to be arc-related, whereas five were dominantly mafic–ultramafic and interpreted to be linked to mantle plumes, and, in the context of this chapter, are considered as LIP-related.

An important question is the relationship between the five different plume/LIP pulses and useful insights are available from the geochemical study of the komatiites belonging to each of the LIP/plume-related assemblages. Based on lithogeochemical analysis of over 2000 komatiite samples from throughout the southern Abitibi greenstone belt, Sproule *et al.* (2002) characterized each of the assemblages and concluded that overall the komatiite data demonstrate a decreasing influence of garnet in the source regions. This pattern through time can be interpreted in various ways, including progressively shallower depths of melt extraction associated with the rising plume, or a decrease in the amount of magma derived from deeper levels in the plume relative to that derived from shallower levels. The major- and trace-element data of Sproule *et al.* (2002) indicate greater amounts of crustal contamination in the younger komatiites, a feature which is consistent with initial emplacement on non-continental crust, but with subsequent autochthonous stratigraphic evolution, with later assemblages partly being built on the earlier assemblages.

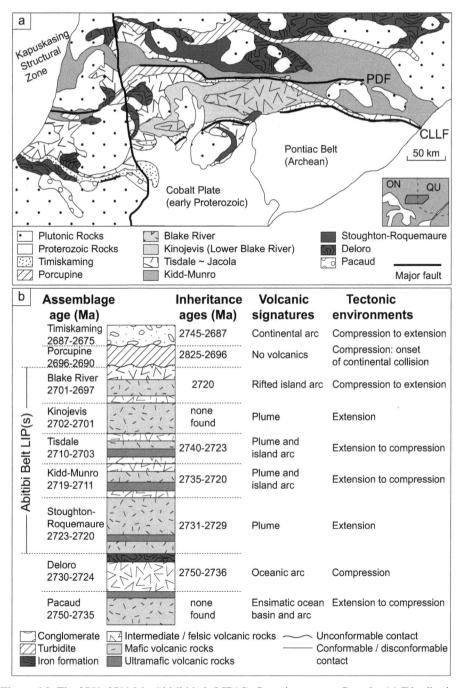

Figure 6.8 The 2750–2700 Ma Abitibi belt LIP(s?), Superior craton, Canada. (a) Distribution of magmatic and sedimentary sequences in the Abitibi subprovince in the Superior province craton. K = Kapuskasing Structural Zone. From Wyman and Kerrich (2009) which was modified after Ayer *et al.* (2002) and Sproule *et al.* (2002). With permission from Elsevier. (b) Idealized stratigraphic column for the southern Abitibi greenstone belt, from Figure 8 in Ayer *et al.* (2002). With permission from Elsevier. The units Stoughton–Roquemaure, Kidd–Munro, Tisdale, Kinojevis, and perhaps Blake River are plume-related and each is provisionally considered to represent LIP magmatism. Given the age range 2730 to 2720 Ma, they may represent multiple pulses of a single LIP event. The Pacaud magmatism may represent a separate LIP or the earliest pulse of the Abitibi belt LIP. PDF = Porcupine-Destor Fault. CLLF = Cadillac Larder Lake Fault.

In terms of setting, the geochemical data favor mantle-plume ascent in the vicinity of an active arc, so that there would have been much more direct interaction between these two geodynamic regimes, arc and plume (e.g. Wyman, 1999; Wyman and Kerrich, 2009). Then, during subsequent ocean closure, there would have been some imbrication of the oceanic plateaus and arcs (e.g. Saunders *et al.*, 1996; Polat and Kerrich, 1999; Hollings *et al.*, 1999; Wyman and Kerrich, 2009, 2010). This interaction would be consistent with the general statement by Arndt *et al.* (2008, p. 407) that "a mantle plume [rising from the deep mantle] does not know what it will meet at the surface," so plume ascent in the vicinity of an active arc will happen randomly some percentage of the time.

6.3.3 Kam Group LIP (c. 2.7 Ga), Slave craton

Neoarchean tholeiitic–komatiitic greenstone belts are scattered across the Slave craton, northern Canada, and are interpreted as remnants of a single LIP placed on continental crust and subsequently deformed (Fig. 6.9). In the portion of the Slave craton near Yellowknife, this basalt-dominated, volcanic sequence is known as the Kam Group (Helmstaedt and Padgham, 1986; Bleeker *et al.*, 1999b; Cousens, 2000; Bleeker and Hall, 2007). Possible correlative basalt successions are known across the basement domain from east to west (Fig. 6.9), and at least as far north as the Acasta area (Fig. 6.9) (Bleeker and Hall, 2007).

This basalt sequence typically consists of several hundred meters to several kilometers of pillowed and massive flows, intercalated with thin felsic volcaniclastic horizons and komatiite flows. Well-dated components of this basalt-dominated sequence yield ages from 2.72 to 2.70 Ga (Bleeker and Hall, 2007). The presence of feeder dykes, one dated at 2738 Ma (after J. Ketchum, cited in Bleeker and Hall (2007)) indicates that the basaltic volcanism started even earlier.

If the broad regional correlation of these basalts is valid (Fig. 6.9), the magnitude of volcanism (areal distribution >100 000 km^2, typical thickness 1–6 km) approaches LIP proportions (Chapters 1 and 2). This widespread basalt-dominated volcanism can be linked with protracted rifting of the basement complexes, and can plausibly be caused by a mantle plume. After the *c.* 2.7 Ga basaltic (LIP) volcanism most areas in the Slave craton show a transition to calc-alkaline greenstone belts (2.69–2.66 Ga) related to subduction beneath the eastern margin of the craton (e.g. Bleeker and Hall, 2007).

6.3.4 Yilgarn craton LIPs (3.0–2.7 Ga)

The Yilgarn craton, Australia, is also host to a number of greenstone packages of postulated LIP affinities and a number of these represent tholeiitic–komatiitic associations, which have been linked to LIPs (Ernst and Buchan, 2001b; Ivanic *et al.*, 2010) ranging in age from 3 to 2.7 Ga (Fig. 6.10).

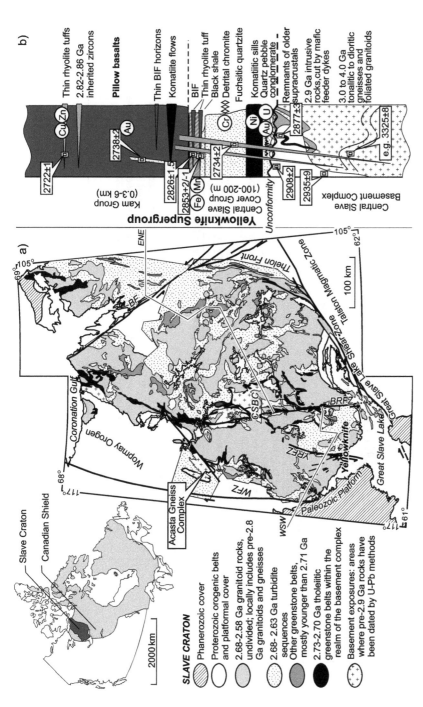

Figure 6.9 The 2730–2700 Ma Kam Group LIP, Slave craton, northwestern Canada. (a) Simplified map of the Slave craton showing the distribution of c. 2700 Ma greenstone belts including those linked to the Kam LIP and also younger (c. 2680 Ma) greenstones which are arc related (and do not represent a LIP). (b) Generalized stratigraphic column of the Central Slave Cover Group, and the autochthonous cover of the Central Slave Basement Complex. (c) Structural stratigraphic cross section across the Slave craton. The Kam LIP and correlated volcanics in the Eastern Domain are part of the cover sequence and are deformed and preserved only as folded remnants. A younger arc-related sequence of greenstone belts and associated intrusions are also present. Modified after Bleeker and Hall (2007), which is based on earlier versions in Bleeker et al. (1999a,b) and Bleeker (2002). Digital version of original figures kindly provided by W. Bleeker.

173

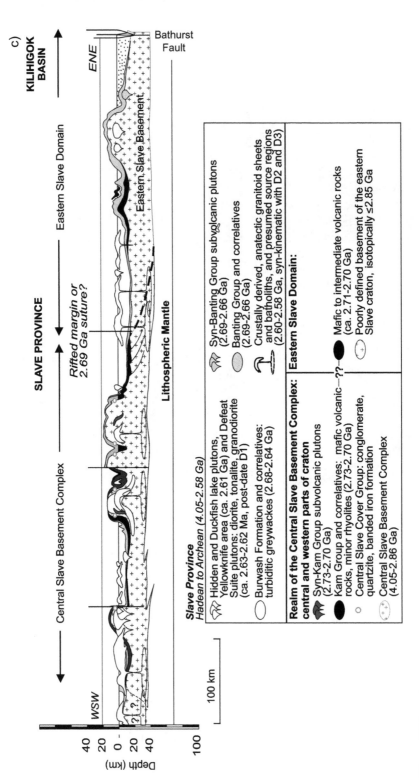

Figure 6.9 (*cont.*)

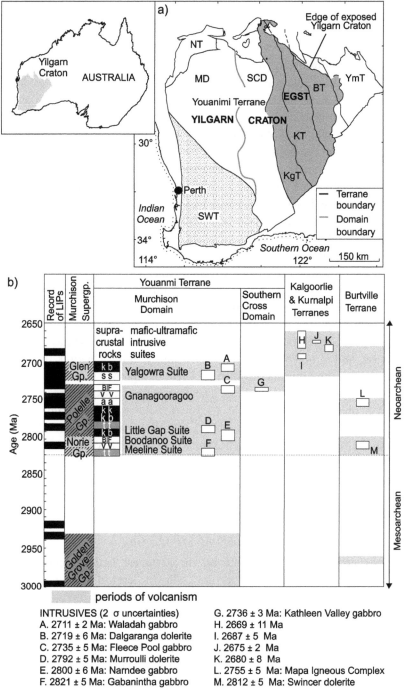

Figure 6.10 Archean LIPs of the Yilgarn craton. (a) Yilgarn craton and its terranes. (b) Mafic intrusions and associated volcanic rocks of the Youanmi terrane compared with those in the Kalgoorlie, Kurnalpi and Burtville terranes. On the left side, this record is compared with the global mafic magmatism LIP "bar code" (Ernst *et al.*, 2005). Legend for supracrustal rocks: s s = clastic sedimentary rocks, BIF = banded iron formation layers in clastic rocks, v v = silicic volcaniclastics and volcanics, a a = andesitic volcanics, t t = tholeiitic basalts, k b = komatiitic basalts, k k = komatiite and komatiitic basalts. Regional diagram part (a) is modified from Figure 1 in Pawley *et al.* (2012). Digital version of the original figure for the regional diagram kindly provided by M.J. Pawley. Combined from Figures 2 and 9 in Ivanic *et al.* (2010). Reprinted by permission of Australian Journal of Earth Sciences (Taylor & Francis Ltd, http://www.tandf.co.uk/journals).

Mafic–ultramafic intrusions

Ivanic *et al.* (2010) considered the mafic–ultramafic intrusions in this area from the perspective of LIPs. They concluded these mafic–ultramafic intrusive suites are additional evidence (complementary to the volcanic record) for a protracted LIP history. Mafic–ultramafic rocks in structurally dismembered layered intrusions comprise approximately 40% by volume of greenstone belts in the Murchison Domain of the Youanmi terrane. As shown in Figure 6.10, the mafic–ultramafic intrusions can be divided into five suites: (i) the *c.* 2810 Ma Meeline suite, which includes the large Windimurra Igneous Complex; (ii) the 2800 ± 6 Ma Boodanoo suite, which includes the Narndee Igneous Complex; (iii) the 2792 ± 5 Ma Little Gap suite; (iv) the *c.* 2750 Ma Gnanagooragoo Igneous Complex; and (v) the 2735–2710 Ma Yalgowra suite of layered gabbroic sills. These intrusions are typically layered, tabular bodies of gabbroic rock with ultramafic basal units which, in places, are more than 6 km thick and up to 2500 km^2 in areal extent (Ivanic *et al.*, 2010).

The suites are anhydrous except for the Boodanoo suite, which contains a large volume of hornblende gabbro, and therefore each, except for the Boodanoo suite, can be linked to LIPs (the presence of hydrous minerals is more consistent with an arc setting for the Boodanoo suite). These intrusions also host significant vanadium mineralization, and minor Ni–Cu–(PGE) mineralization (cf. Chapter 16). All suites are demonstrably contemporaneous with packages of high-Mg tholeiitic lavas and/or silicic volcanic rocks in greenstone belts. The distribution, ages, and compositions of these suites of mafic–ultramafic rocks (except for the Boodanoo suite which has hornblende gabbros) are most consistent with genesis as part of the plumbing system for at least two LIPs. The *c.* 2810 Meeline suite (including the Windimurra Complex) and the 2792 Ma Little Gap suite could together represent one LIP, and the 2750 Gnanagooragoo and 2735–2710 Ma Yalgowra suites could represent a separate multi-pulse LIP (with the latter linked to the Eastern Goldfields LIP).

Eastern Goldfields LIP (2.7 Ga)

Detailed geochemistry has been done on the youngest LIP associated with the Eastern Goldfields superterrane of the Yilgarn craton, western Australia (Fig. 6.11). This superterrane comprises elongated, narrow to arcuate greenstone belts, composed of deformed and metamorphosed volcanic and sedimentary rocks, intruded by numerous granitoids and high-level intermediate to silicic porphyries. The superterrane is divided into three main north–northwest-trending main terranes: Kalgoorlie, Kurnalpi, and Burtville (Figs. 6.10 and 6.11). Like other Archean LIP analogs the magmatism is associated with several pulses, in this case spanning an age from 2710 to 2670 Ma.

As noted by Barnes *et al.* (2012), the most widespread group of the Eastern Goldfields LIP, the low-Th basalt, is remarkably homogeneous across terranes and domains. This unit in particular is interpreted as the Archean analog of

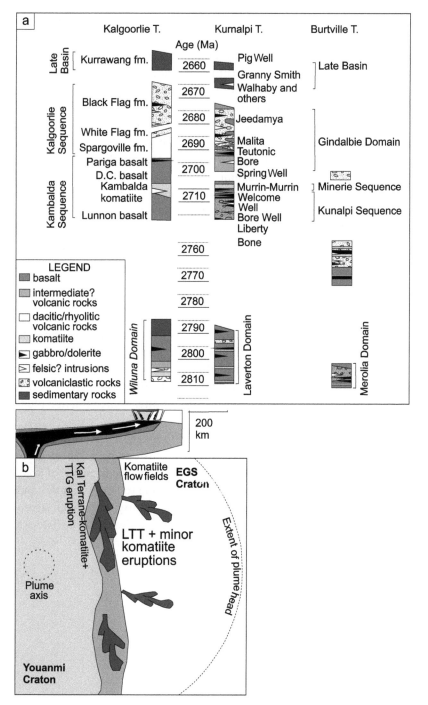

Figure 6.11 The 2700 Ma Eastern Goldfields LIP of the Eastern Goldfields terrane, Yilgarn craton, Australia. (a) Simplified stratigraphic columns for the various terranes of the Eastern Goldfields superterrane. See locations in Figure 6.10a. From Figure 2 in Barnes *et al.* (2012). (b) Interpreted plume center and pattern of magma emplacement into the Eastern Goldfields terrane. Upper image of (b) shows cross-section view of plume tail flowing beneath Youanmi lithospheric root to shallow level at the cratonic margin; then, low-pressure melting produces voluminous komatiite. The lower image is a plan view showing the interpreted location of the plume head, and the main flux of komatiite magmatism along the craton margin in the Kalgoorlie terrane. Note eruption of LTT basalt and less voluminous komatiite further east from the plume head. From Figure 13 in Barnes *et al.* (2012). Reprinted by permission of Australian Journal of Earth Sciences (Taylor & Francis Ltd, http://www.tandf.co.uk/journals). Digital version of original figures kindly provided by S.J. Barnes.

plume-head-related LIP basalts, showing a close geochemical match to flood basalts associated with late-stage continental rifting. An additional unit, the high-Th siliceous basalt group displays a distinctive geochemical signature, which is evidently unique to Archean greenstone terranes. Derivation by contamination during fractionation of komatiites, probably deep in the crust, followed by midcrustal homogenization in magma chambers, is the most likely hypothesis.

Furthermore, as illustrated in Fig. 6.11, it is proposed by Barnes *et al.* (2012) that the plume arrived beneath the Youanmi terrane (of the Yilgarn craton) and was deflected by the thick lithospheric root of the Youanmi terrane to the thinner eastern cratonic margin where it induced continental rifting and basaltic and komatiitic magmatism across the Eastern Goldfields superterrane.

6.4 Summary

The recognition of LIPs extends back into the Archean. There are flood basalts (Dominion/Nsuze and Ventersdorp of the Kaapvaal craton, and Fortescue of the Pilbara craton) that have been long recognized as the equivalents of more modern CFBs. Given the nearly 100 Ma duration of activity in the case of both the Ventersdorp and Fortescue LIPS, it is suggested that each should be better considered as consisting of more than one LIP rather than a single protracted LIP. Furthermore, the widespread presence of komatiites testifies to a hotter source (mantle plume in a higher temperature Archean mantle).

In addition, there are many deformed volcanic sequences that are developed on less-stable (uncratonized) cratons and constitute greenstone belts. Those greenstone belts of tholeiitic–komatiite affinity are considered LIPs. The presence of komatiites is the distinctive feature of these Archean LIPs, testifying to hotter mantle conditions at that time. These Archean LIPs represent fragments of originally much larger LIPs, and their original scale and distribution will only become clear once the late-Archean global reconstruction history has been sorted out.

7

Planetary LIPs

7.1 Introduction

Other planetary bodies in the Solar System also have intraplate magmatism on the scale of terrestrial Large Igneous Provinces (LIPs). These planetary LIP analogs can provide valuable insights into both the origin and evolution of terrestrial LIPs. Specifically:

(1) The terrestrial planets, Mars, Venus, Mercury, and the Moon are one-plate bodies, and lack plate tectonics (except possibly for Mars in its earliest history). Therefore, all the observed magmatism on these bodies is of an intraplate type. Characterization of magmatic events on these other planetary bodies can be compared with Earth, and helps us understand intraplate magmatism (and especially that of LIP scale) on Earth as distinct from plate-boundary-related magmatism.
(2) The terrestrial bodies (with the exception of Venus and Earth) preserve a rich surface geologic record of much of early Solar System history, and therefore potentially provide a more complete record of intraplate magmatism through time and under different geodynamic regimes from that available on Earth (Fig. 7.1).
(3) Io, a moon of Jupiter, exhibits the highest-temperature volcanism in the Solar System and the distribution and characteristics of this magmatism might have implications for Archean komatiite volcanism on Earth (Fig. 7.1).

Herein I provide a brief review of intraplate magmatism on Mars, Venus, the Moon, Mercury, and Io, followed by a discussion of some key implications for understanding LIPs on Earth. A classic overall source of information on planetary LIPs is Head and Coffin (1997).

7.2 Mars

7.2.1 Introduction

Mars is about half the size of the Earth, and its surface area is only slightly less than the total area of the Earth's continents (Figs. 7.2 and 7.3). It has low atmospheric pressure, with a water content about 1/1000 of Earth's, but there is evidence for

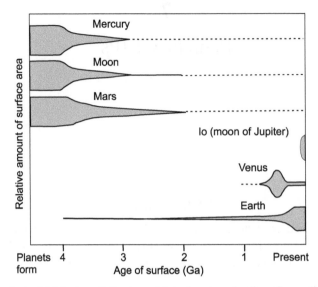

Figure 7.1 Geological histories of the terrestrial planetary bodies. Generalized plot of the approximate percentage of presently exposed surface area that formed at different times in the history of the planet. Modified after Figure 1 in Head and Coffin (1997) to include Io, an inner moon of Jupiter.

a past denser atmosphere that allowed major water flow and perhaps even a northern ocean. The planet exhibits a crustal dichotomy being divided between a smooth northern hemisphere (northern lowlands) and heavily cratered southern highlands that are on average about 5 km higher. This northern lowlands may have developed through early plate tectonics, but through most of its history Mars has been a single-plate planet (Figs. 7.2–7.4). The oldest period of Mars' history is the Noachian, beginning at the origin of Mars and extending to *c.* 3.9 Ga, followed by the Hesperian, extending from *c.* 3.9–3.0 Ga, and the Amazonian, from *c.* 3.0 Ga up to the present.

Volcanic rocks cover about 60% of the surface of Mars and volcanic features include lava flows up to 1300 km long, edifices ranging up to hundreds of kilometers in width and tens of kilometers in height, elongate and circular vent-like features, sinuous rilles representing lava thermal erosion, and mantling blankets interpreted to be pyroclastic in origin (e.g. Tanaka *et al.*, 1992; Wilson and Head, 1994; Head and Coffin, 1997; Fuller and Head, 2003; Keszthelyi *et al.*, 2006).

Volcanism (mainly basaltic) was dominant in the Early and Middle Noachian, presumably related to initial crustal formation processes, then decreased in the Late Noachian, increased again in the Early Hesperian, and then decreased to the present (Tanaka *et al.*, 1992). In terms of spatial distribution, the volcanism was globally extensive in the Noachian and again in the Early Hesperian, but became localized in the Amazonian, with activity focused in the Tharsis and Elysium regions (Fig. 7.2).

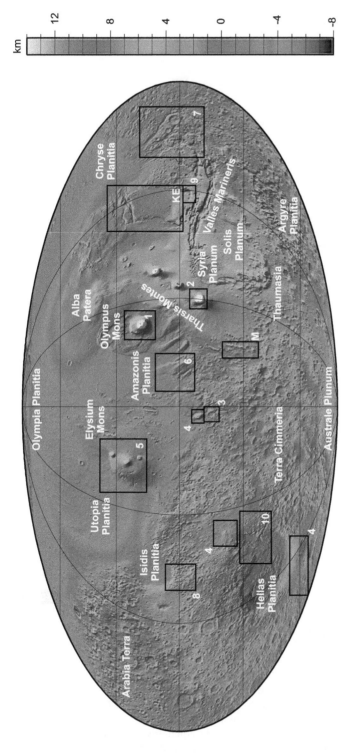

Figure 7.2 Topographic map of Mars derived from Mars Orbiter Laser Altimeter (MOLA) data with major features located. Modified from Solomon et al. (2005). Superimposed boxes and numbers locate areas discussed in Fig. 7.5. M = Mangala Valles Region, KE = Kasei Valles/ Echus Chasma Region, 1 = Olympus Mons and its vicinity, 2 = Arsia Mons, 3 = Gusev Crater Region, 4 = Highland Paterae, 5 = Elysium Region, 6 = Medusae Fossae Formation, 7 = Iani Chaos/Tiu/Ares Valles, 8 = Libya Montes Valley Network (western part), 9 = Juventae Chasma (sulphate mountain), 10 = Dao/Niger Valles. Mollweide equal-area projection with the central meridian at 180° longitude. Digital version of topographic base map kindly provided by S. Solomon. Modified from Figure 1b in Solomon et al. (2005), reprinted with permission from AAAS.

a)

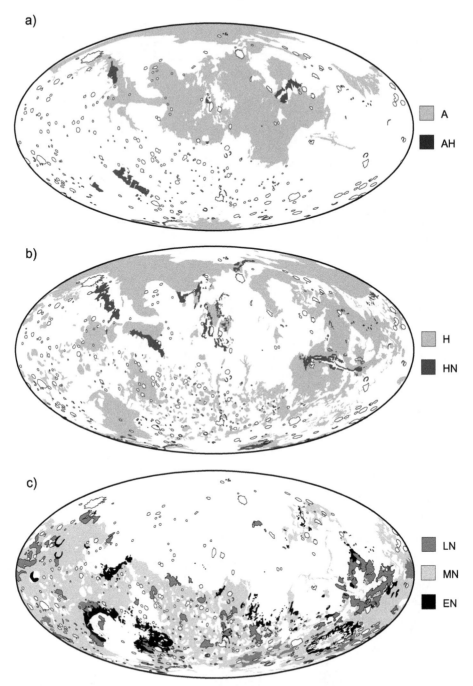

b)

c)

Figure 7.3 Geological map of Mars grouped into time intervals. A = Amazonian, AH = Amazonian–Hesperian, H = Hesperian, HN = Hesperian-Noachian, LN = Late Noachian, MN = Middle Noachian, EN = Early Noachian. Not shown on any of the diagrams are impact craters and their ejecta deposits. Mollweide equal-area projection with a central meridian at 180° longitude. Redrafted from Solomon *et al*. (2005) and separated into three parts on the basis of age in order to allow the geology to be shown in black and white format. Modified from Figure 1b in Solomon *et al*. (2005), reprinted with permission from AAAS.

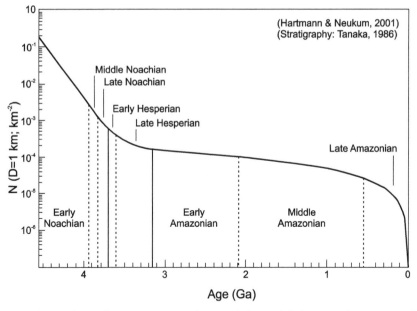

Figure 7.4 Chronology of Mars. From Neukum and the HRSC Co-Investigator Team (2008) based on Hartmann & Neukum (2001) and Tanaka (1986).

In this section, I consider several aspects of Mars in the context of LIPs: (1) the work of Neukum *et al.* (2010), which recognized episodisity in the intraplate magmatic record of Mars; (2) the magmatism associated with two major and long-lived centers, Tharsis and Elysium, which produced clusters of LIP events; (3) the widespread pulse of increased volcanism in the Early Hesperian; and (4) aspects of the LIP plumbing system (dykes, sills, and layered intrusions).

7.2.2 *Global history of magmatic and fluvial activity on Mars*

Twelve prominent regions on Mars were analyzed in detail by Neukum *et al.* (2010) on the basis of detailed geologic mapping and age determination through crater-counting techniques, all using various types of high-resolution imagery. Their data (combined with previous data) showed that volcanic and fluvial/glacial geologic activity spans Martian history from > 4 Ga ago until the present day, with main pulses at *c.* 3.8–3.3 Ga, 2.0–1.8 Ga, 1.6 to 1.2 Ga, 0.8–0.3 Ga, 0.2 Ga, and 0.1 Ga, along with a possible weaker phase around *c.* 2.5–2.2 Ga ago. In between these episodes, there was relative quiescence of volcanic and/or fluvial/glacial activity. Furthermore, some of these episodes of magmatic activity coincide with some of the age groups of the Martian meteorites that have reached Earth (*c.* 1.3 Ga, *c.* 0.6–0.3, and 0.17 Ga) (Neukum *et al.*, 2010).

It is notable that the volcanic and fluvial activity coincide for many intervals and such a correlation is consistent with the idea of a link between magmatic and fluvial activity. Pulses of Martian magmatism are thought to have been the cause of dramatic melting of the permafrost leading to major fluvial activity. However, the correlation between magmatic and fluvial/glacial activity is weaker for the heavily cratered highlands of Mars, in which fluvial activity was apparently more controlled by climate than initiated by internal magmatic processes (Neukum *et al.*, 2010).

7.2.3 Long-lived superplume centers: Tharsis and Elysium regions

In spite of Mars being only about one-half the size of Earth, it has several volcanoes that surpass the scale of the largest terrestrial volcanoes. The most massive volcanoes are located on huge regional uplifts in the Tharsis and Elysium regions of Mars, and the former region has been described as a long-lived superplume (Baker *et al.*, 2007; Dohm *et al.*, 2007). Michalski and Bleacher (2013) propose significant magmatic activity also in the Arabia Terra region (Fig. 7.2).

Tharsis region

The Tharsis region is a dome about 5000 km across that includes the Tharsis Montes, three shield volcanoes that are 14 to 18 km high (Arsia Mons, Pavonis Mons, and Ascraeus Mons), but also older volcanoes and magmatic centers of more subdued height. The volcanoes are surrounded by extensive flood-lava plateaus and plains. Beyond the dome's northwest edge is Olympus Mons, the largest of the Tharsis shield volcanoes. It is approximately 25 kilometers high, 550 kilometers in diameter, and is rimmed by a 6-km-high scarp. Geomorphology and stratigraphy suggest that the earliest volcanic activity in the Tharsis region began at the Syria Planum magma center during the Early Hesperian and lasted until the Late Hesperian (Figs. 7.2–7.5). However, the bulk of the Tharsis volcanism occurred later, during Late Hesperian to Late Amazonian (< 0.2 Ga). Some Tharsis magmatic centers and plateaus may have begun forming in the late Noachian (e.g. Dohm *et al.*, 2007).

The volcanoes in the Tharsis region are many times larger than those on Earth and Martian flows are longer, up to 1300 km long (e.g. Fuller and Head, 2003). Assuming basaltic compositions, this is probably due to larger eruption rates and to lower gravity, but is also explained by the absence of plate tectonics and the presence of a stable lithospheric shell. On Mars, a rising mantle plume will remain fixed beneath the same lithosphere spot throughout its lifetime, both through the plume-head stage, where voluminous magmatism is produced in a short period of time, and through the protracted plume-tail stage, during which much lower rates of magmatism are produced over a period of perhaps 200 Ma (if the maximum observed age of terrestrial hotspots is applicable). So the magmatism produced by both plume head and plume tail will contribute to the single volcanic edifice.

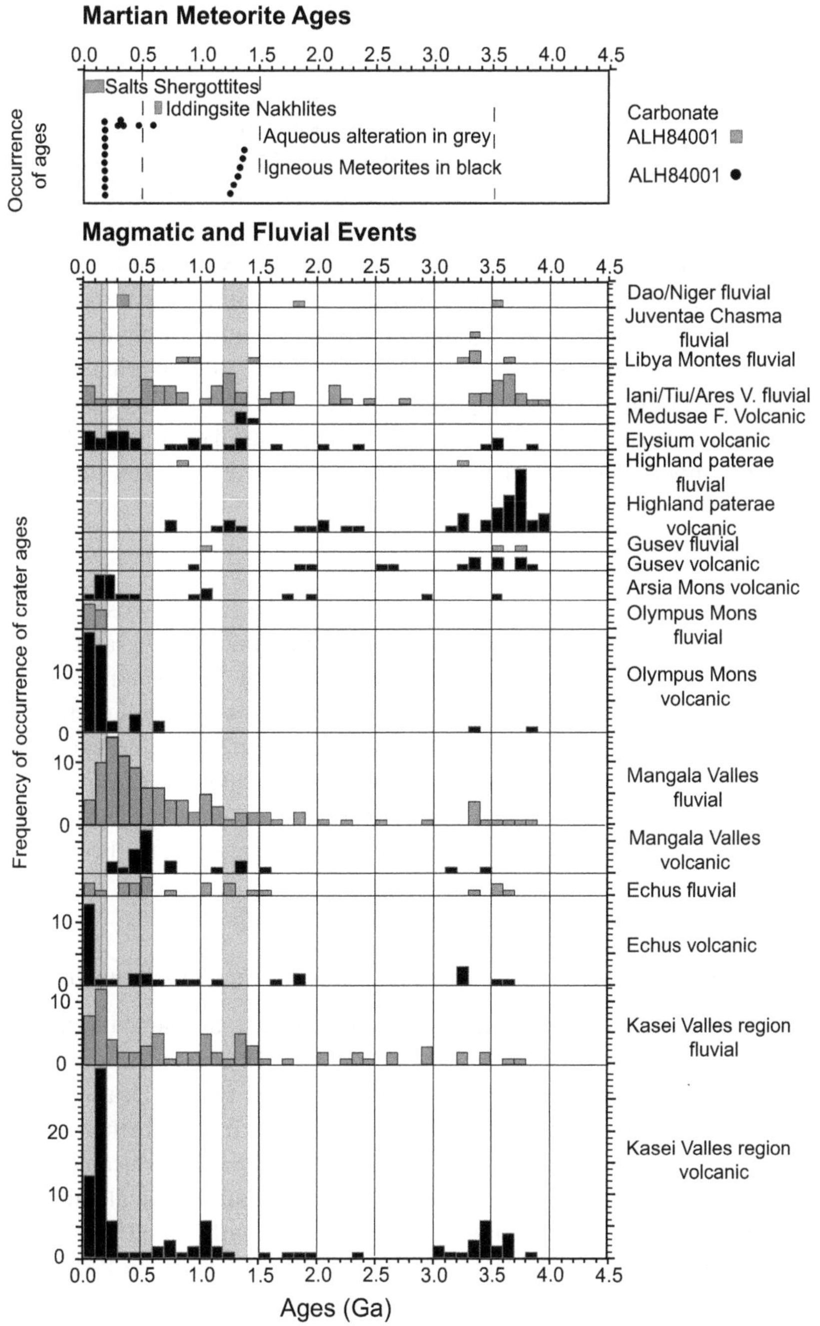

Figure 7.5 Ages of magmatic and fluvial events on Mars based on ages extracted from measurements on HRSC (High/Super Resolution Stereo Colour Imager), MOC (Mars Orbiter Camera), and THEMIS (Thermal Emission Imaging System) imagery in twelve regions of Mars. For comparison, gray-tone vertical columns show radiometric age ranges for Martian meteorites. Both magmatic and fluvial events are of interest from a magmatic perspective as many of the fluvial events are linked to melting of permafrost by magmatism. Locations shown in Fig. 7.2. See discussion in Neukum *et al.* (2010). Redrafted from Figure 9 in Neukum *et al.* (2010). With permission from Elsevier.

Although the overall duration of magmatic activity in the Tharsis region is a billion years or more, the building of individual volcanoes (e.g. Olympus Mons, Arsia Mons, Ascraeus Mons, and Pavonis Mons) could occur in a much shorter period of time. In this sense each volcanic edifice (e.g. Olympus Mons, Arsia Mons, Ascraeus Mons, or Pavonis Mons) can be nominally considered to represent a single LIP (voluminous magmatism produced in a short period of time) followed by a lower-rate production associated with the post-LIP hotspot phase. While the location of the Tharsis center has been constant for much of Mars' history it has been recently proposed that the locus of global magmatism was in the southern highlands prior to *c*. 3.8 Ga. Specifically, Hynek *et al.* (2011) proposed that an early plume migration (about 3.8 Ga) from the southern highlands of Mars led to the development of the Tharsis bulge in its current location along the Martian crustal dichotomy. The path is recognized by smooth volcanic plains embaying ancient massifs and infilling large impact craters.

Elysium region

The other main area of central volcanism on Mars is the Elysium Rise (centered at 25° N, 215° W), which is contemporaneous with volcanic activity at the Tharsis Montes. The Elysium Rise is a broad dome that is 1700 by 2400 km in size, and has three volcanoes, Hecates Tholus, Elysium Mons, and Albor Tholus, each of which can also be considered to represent separate LIP pulses (following the same logic as discussed above for individual volcanoes of the Tharsis region).

There are also giant radiating and circumferential dyke swarms (Table 7.1) associated with the volcanoes in both the Tharsis and Elysium regions, further supporting the notion of underlying mantle plumes (Chapter 5).

7.2.4 Hesperian Ridged Plains: a global Large Igneous Province (GLIP)

The geological record of Mars indicates a major increase in magmatic flux in the Early Hesperian period (Fig. 7.2–7.5; Tanaka *et al.*, 1992). The most prominent unit emplaced during that period is labeled Hesperian Ridged Plains (Head *et al.*, 2002, 2006; Head, 2006). This unit may underlie younger units in the northern lowlands, and the Utopia, Isidis, and Hellas basins as well as be linked with volcanic centers in several regions including Hadriaca Patera, Tyrrhena Patera, Apollinaris Patera, Syrtis Major, and Alba Patera (Fig. 7.2). Also, the southeast (Thaumasia) and eastern (Hesperia Planum) parts of Tharsis are formed of Hesperian Ridged Plains, as are other portions on the flanks and surrounding Tharsis Hesperian Ridged Plains. Volcanism may also be underlying the younger Amazonian-aged lavas higher on the Tharsis rise. As estimated by Head (2006) and Head *et al.* (2006) the percentage of Mars resurfaced during the Early Hesperian is about 30%, for a total area of about 43.5 Mkm^2. Using average-thickness estimates, the volume of the Hesperian Ridged Plains volcanism is estimated to be of the order of 33 Mkm^3.

Table 7.1 *Giant dyke swarms on Mars. Ages: l = lower, m = middle, u = upper, N = Noachian, H = Hesperian, A = Amazonian. Dyke trends given with respect to the magmatic center (e.g. SW means trending to the southwest from the magmatic center). Modified after Table 1 in Ernst et al. (2001) and Mège and Masson (1996a, b). See also Anderson et al. (2001). Locations for those in the Tharsis region are shown in Fig. 7.7. Some terms are defined at the bottom of the table.*

	Dyke swarms	Length (× width), or diameter (km)	Geometry (trend)	Associated magmatic center	Age
	THARSIS REGION				
1	Thaumasia fossae	800 km	Radial: weakly fanning S-trending	Thaumasia "A"	lH
2	Valles Marineris fossae and catenae	> 3000 km	Radial: ESE-trending subswarm	Syria Planum ("B")	lH
3	Parts of Sirenum and Memnonia fossae	3000 km (of which 1500 km buried by younger lavas)	Radial: SW–WSW-fanning subswarm	Syria Planum ("B")	lH
4	Most of Tempe and Mareotis fossae and Tractus catenae	> 3000 km	Radial: NE-trending subswarm	Pavonis Mons in Tharsis Montes ("C")	uH-A
5	Parts of Icaria, Sirenum, and Memnonia fossae	> 3000 km	Radial: SSW–W-trending fanning subswarm	Pavonis Mons in Tharsis Montes ("C")	uH-A
6	Alba and part of Tantalus fossae	3000 km	Radial: N–NE-trending subswarm	Alba Patera ("D")	uH-A
7	Catenae within Ceraunius fossae		N–S-trending swarm	Tharsis Montes OR Alba Patera ("C" or "D")	
8	Olympus Mons swarm	?	?	Olympus Mons	uH-A
9	Arsia Mons swarm	500 km	Circumferential	Arsia Mons	uA
10	Pavonis Mons swarm	500 km	Circumferential	Pavonis Mons	uA
11	Ascraeus Mons swarm	500 km	Circumferential	Ascraeus Mons	uA
12	Noctis Labyrinthus	1000 km	Circumferential	Syria Planum	lH
13, 14	Valles Marineris	over 600 km (× 300 km)	Two preferential orientations, NE and NW		N

187

Table 7.1 (cont.)

	Dyke swarms	Length (× width), or diameter (km)	Geometry (trend)	Associated magmatic center	Age
	ELYSIUM REGION				
15	Cerberus fossae	1500 km	Radial: ESE–E-trending subswarm	Elysium Rise	IA or later
16	Elysium fossae	1200 km	Radial: NW-trending subswarm	Elysium Rise	
17	Elysium–Utopia ridges	400 km (× 250 km)	Radial: NW-trending subswarm	Elysium Rise	IA
18	Stygis fossae, Zephyrus fossae, Elysium chasma, and Hyblaeus chasma	600 km	Circumferential	Elysium Mons	IA or alter
	SYRTIS MAJOR REGION	1300 km			
19	NE-trending set of graben	1200 km	Radial: NE-trending subswarm	Syrtis Major	N or after
20	NNW-trending set of graben	1300 km	Radial: NNW-trending subswarm	Syrtis Major	N or after
21	WNW–NW-trending set of graben	1200 km (× 300 km)	Linear: WNW–NW-trending subswarm	North of Syrtis Major	N or after
22	NNE-trending graben	1800 km (× 400 km)	Linear: NNE-trending subswarm	East of Syrtis Major	H or after
	TERRA TYRRHENA REGION				
23	Huygens–Helles	600–700 km	Linear: NW-trending subswarm		IH

Note: Catena (catenae) = pit crater chain; chasma (chasmata) = rift; fossa (fossae) = long narrow trough, graben; mons (montes) = mountain; planum = plateau or plain.

Most of the Hesperian Ridged Plains unit is not characterized by obvious individual sources, and this is one reason it has been thought to have been emplaced in a flood-basalt mode from widely distributed sources. Dykes of the Huygens–Hellas swarm and other swarms of Early Hesperian age (Table 7.1; see next section) could be its feeders (Head *et al.*, 2006).

The huge scale of this Early Hesperian magmatic event and its relatively short duration (≤ 100 Ma) qualify it as a possible LIP and, given its wide distribution, it could be interpreted as a global LIP, possibly linked with mantle overturn (Head *et al.*, 2006; Stein and Hofmann, 1994).

7.2.5 Magmatic plumbing system on Mars: mafic dyke swarms, sills, and magma chambers

Voluminous (intraplate) volcanism on Mars must be fed via a plumbing system of dyke swarms, sills, and layered intrusions, and herein I review evidence for this.

Radiating and circumferential dyke swarms

Numerous swarms of extensional tectonic structures are observed to be either radial or circumferential about the individual volcanoes of the central Tharsis area and, to a smaller extent, the Elysium Rise (e.g. Mège and Masson, 1996a, b; Anderson *et al.*, 2001; Ernst *et al.*, 2001; Montési, 2001; Wilson and Head, 2002; Fig. 7.6a; Table 7.1). These extensional structures are narrow grabens with a maximum width of 5 km and maximum length of 3000 km and are termed fossae in the planetary literature. Some of the radiating and circumferential graben sets are associated with pit craters chains (termed catenae), ovoid and linear troughs, shallow ovoid flat-floored depressions, and spatter cones, all thought to be of volcanic origin and to reflect underlying dykes (Davis *et al.*, 1995; Mège and Masson, 1996a; Liu and Wilson, 1998; Ferrill *et al.*, 2011; cf. Davey *et al.*, 2013). Figure 7.7 shows an interpretation of graben–fissure systems in terms of underlying dyke swarms.

Additional orbital images have revealed positive linear features that may represent outcropping dykes (Mège, 1999; Wilson and Mouginis-Mark, 1999; Ernst *et al.*, 2001; Head *et al.*, 2006; Pedersen *et al.*, 2010). For instance, Head *et al.* (2006) documented the presence of two low, narrow, and broadly arcuate ridges that extend for 600–700 km in a northwest trend across western Terra Tyrrhena (Fig. 7.6b). These Huygens–Hellas dykes crosscut a Noachian terrane and are closely associated with the Early Hesperian Ridged Plains for which they were probably feeders. Also, hundreds of narrow, linear ridge segments are found on the northwest side of the Elysium Rise and represent a northwest-trending dyke swarm that is about 400 km long and 250 km wide and these may be feeders to Early Amazonian flows and flood-plain deposits (Pedersen *et al.*, 2010).

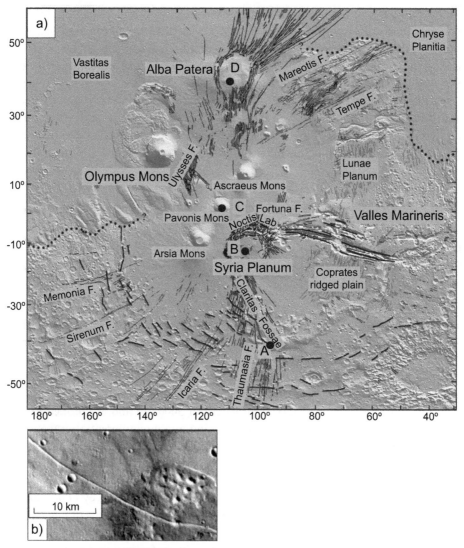

Figure 7.6 Dyke swarms on Mars. (a) Dyke swarm map of Tharsis volcanic province (dotted line = Martian Dichotomy boundary). Dykes are shown as thin black lines, and wrinkle ridges (see also Fig. 13.2) are shown as thick grey lines, and thick black lines, the latter belonging to the South Syria Planum ridge belt. (A)–(D) Convergence centers of dyke swarms identified by Mège and Masson (1996a): (A) Hypothetical early center at Thaumasia Planum; (B) Syria Planum magma center; (C) subsequent center in the Tharsis Montes area; (D) Alba Patera center. 1° latitude = 59 km, F = Fossae). Most of these dykes are present as narrow extensive graben (termed fossae in the planetary literature), that are interpreted to overlie dykes (Fig. 7.7) (after Ernst *et al.*, 2001; Wilson and Head, 2002). Modified with permission from the Annual Review of Earth and Planetary Sciences, volume 29 © 2001 by Annual Reviews, http://www.annualreviews.org. (b) In other regions linear ridges represent dykes exposed by erosion. An example is shown from *c.* 7° S, 63° E, northeast of Huygens crater on the northern rim of the Hellas basin. From Head *et al.* (2006).

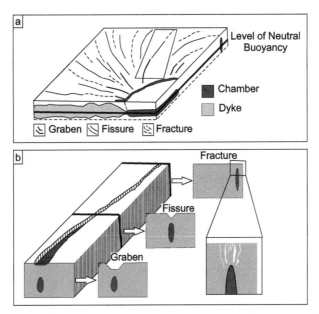

Figure 7.7 Giant dyke swarms and their relationship to graben, fissure, and fracture systems which are common on Mars and Venus. (a) Lateral emplacement of a radiating swarm; (b) graben, fissure, and fracture association with a dyke of this swarm. Part (b) is located on part (a) by the rectangle drawn on the upper surface. Redrafted from Figure 5 in Grosfils and Head (1994b). With kind permission from Springer Science and Business Media.

Most of the swarms in Table 7.1 are identified as narrow, linear graben interpreted to overlie dykes, but the Olympus Mons dyke swarms, Huygens–Hellas, and the Elysium–Utopia swarms are associated with positive features that have been interpreted as outcropping dykes exposed by erosion.

Sills

Sills are undoubtedly present on Mars, but the evidence is only indirect. Various disrupted surface features are thought to be the result of water/steam produced as sill magma interacts with ice in the cryosphere (e.g. Schultz and Glicken, 1979; Squyres *et al.*, 1987; Tanaka *et al.*, 2002; Wilson and Head, 2002; Wilson and Mouginis-Mark, 2003; Leask *et al.*, 2006).

Magma chambers: mafic–ultramafic intrusions, layered intrusions

It is commonly inferred that radiating dyke swarms are fed from a magma chamber located in the focal region. Each of the radiating systems in Table 7.1 can be used to locate a magma chamber which is usually beneath the volcanic edifice (Mège and Masson, 1996a). Specifically, in the case of Tharsis region, four magma chamber(s) are identified, at locations A to D, with inferred age order oldest to youngest (Fig. 7.6a). Magma chambers A and B underlie the Thaumasia and Syria Planum regions, respectively. C and D correspond with the younger volcanoes, Pavonis Mons and Alba Patera, respectively.

Possible evidence for an exhumed mafic–ultramafic layered intrusion in the Columbia Hills area has been discussed in Francis (2011), who suggests that the stratification observed by the Spirit Mars Rover reflects magmatic layering in a large mafic–ultramafic intrusion consisting of harzburgites, olivine norite, and gabbronorites. If this interpretation is correct, the Columbia Hills would represent a Noachian-age layered intrusion that was exhumed by rebound following the meteorite impact event that formed Gusev crater in the Hesperian. Other evidence for intrusions on Mars include domal structures interpreted as volcanic cryptodomes (e.g. Farrand *et al.*, 2011) or as domal uplift above intrusive bodies (Michaut *et al.*, 2013).

7.3 Venus

7.3.1 Introduction

Venus is very similar to the Earth in size and density, but is dramatically different in a number of characteristics including having hot (475 °C) temperatures at the surface, and a dense 90-atm CO_2-rich atmosphere (representing global warming to the extreme), and also the absence of water, lack of significant surface erosion, and absence of plate tectonics. However, it has a vigorous history of intraplate magmatism; this record is reviewed below especially those aspects of LIP scale.

The Magellan mission in the early 1990s building on earlier planetary missions to Venus, produced detailed radar images of nearly the entire surface of Venus to a resolution of about 100 m, and revealed that the surface consists of volcanic plains, remnants of "basement" (termed tesserae), rift systems, individual volcanoes, and circular features known as coronae, with nearly all regions thought to be basaltic in composition; only minor silicic magmatism is recognized. While the first-order magmatic and tectonic structures have been cataloged on Venus, and their general characteristics determined, their age distribution is only known in the broadest terms (e.g. Head *et al.*, 1992; Hansen *et al.*, 1997; Crumpler and Aubele, 2000; Ernst *et al.*, 2007). Crater counting is of limited value for detailed dating studies, given the inhibited cratering resulting from Venus' thick atmosphere (cf. McKinnon *et al.*, 1997). However, it is clear that the majority of magmatism is concentrated at young times (mean surface age based on crater counting of about 750 Ma) indicating much more extensive volcanic resurfacing at younger times than observed on the other planets (e.g. Basilevsky and Head, 2002; Fig. 7.1). The current Venus Express mission has revealed regions of high thermal emissivity interpreted to reflect young, unweathered volcanics indicating that Venus could be actively resurfacing (Smrekar *et al.*, 2010a).

7.3.2 Magmatic record of Venus

Venus preserves a vigorous history of intraplate (mainly basaltic) magmatism (Fig. 7.8) consisting of (1) *individual volcanoes*, with those of 100–1000 km in diameter interpreted to have a plume origin; (2) *annular structures termed coronae* (*and arachnoids*) with diameters averaging 300 km, but ranging up to 2400 km, and having a diapir or plume origin; (3) *radiating graben–fissure systems* (*including novae*) caused by underlying laterally propagating dykes and extending up to 2000 km (or perhaps up to 6000 km; see below) away from volcanoes or coronae at their foci; (4) *lava flow fields* of scale comparable to terrestrial flood basalts; (5) *regions of small shield volcanoes*, often a few hundred kilometers across; and (6) *canali*, narrow sinuous lava channels of lengths up to more than 7000 km. All these tectonomagmatic features are superimposed on earlier plains' volcanism and deformed terranes (crustal plateaus and tesserae) (e.g. Head *et al.*, 1992; Grosfils and Head, 1994; Crumpler *et al.*, 1997; Baker *et al.*, 1997; Herrick, 1999; Crumpler and Aubele, 2000; Stofan *et al.*, 2001; Ernst and Desynoyers, 2004; Hansen, 2007; Ivanov and Head, 2013; Head, 2014).

The lack of water and the absence of plate tectonics explain some differences in intraplate magmatic style from that found on Earth, but many useful cross-comparisons can be made with terrestrial LIPs and hotspots. Triple-junction rifts (chasmata) are associated with plume-generated volcanic rises especially in the BAT (Beta–Atla–Themis) region (e.g. Hamilton and Stofan, 1996; Martin *et al.*, 2007; Smrekar *et al.*, 2010b).

Some tectonomagmatic features may be much larger than previously realized. Artemis, a 2400-km-diameter corona is linked by Hansen and Olive (2010) with a > 5000-km-diameter outer trough, a 12 000-km-diameter radiating graben–fissure system (dyke swarm), and a 13 000-km-diameter concentric wrinkle ridge system. Other large wrinkle ridge systems (2600–8000 km in diameter) circumscribing volcanic rises may belong to the same class of huge plume-related tectonomagmatic features.

Fundamental questions remain (e.g. Hansen and Young, 2007; Bjonnes *et al.*, 2012; Ivanov and Head, 2013; Head, 2014). What is the detailed age distribution on the planet? Crater counting provides a global average age estimate of about 750 Ma, but crater density is insufficient for regional chronology. Is there a strict lithological age progression following a major planetary resurfacing event, wherein the most/ least deformed units are always the oldest/youngest? Alternatively, are available constraints satisfied by a progressive time-transgressive resurfacing? Finally, has lithospheric thickness increased toward younger times?

7.3.3 Volcanoes

The Crumpler and Aubele (2000) catalog lists 168 large volcanoes with diameters > 100 km (Figs. 7.9 and 7.10), and 289 intermediate volcanoes with diameters ranging from 20 to 100 km. Small (< 20-km-diameter) volcanoes are also abundant

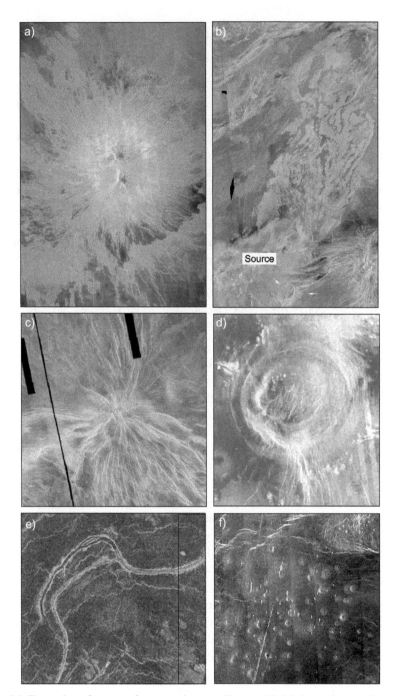

Figure 7.8 Examples of types of magmatism on Venus. (a) Major volcano, Sapas Mons volcano, is 400 km across and rises 1.5 km above surrounding terrain. (b) Major flow field, Mylitta Flutus is 1000 km long by 460 km wide. (c) Nova (radiating graben–fissure system) in Themis Regio (250 km across), part of giant radiating graben–fissure system (Ernst *et al.*, 2003). (d) Corona, circular features, common on Venus, but not recognized on Earth. Fatua corona is 400 km in diameter. (e) Canali, long sinuous lava channels. This example is from south of Ishtar Terra. Width of image 50 km. (f) Shield fields (width of image 120 km). Images from http://volcano.oregonstate.edu/book/export/html/989 except for the Corona in (d), which is after Squyres *et al.* (1992).

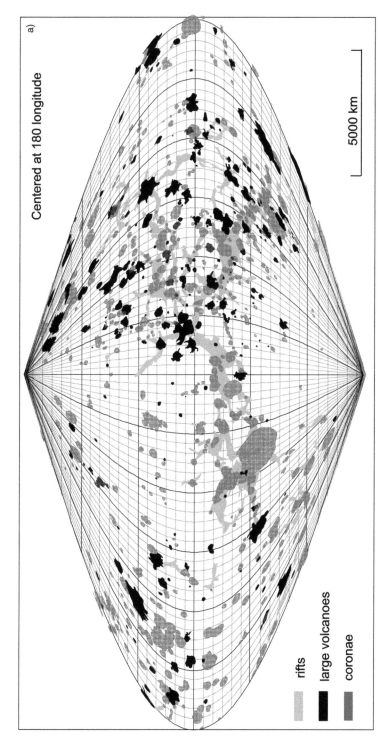

Figure 7.9 Maps of Venus. (a) Sinusoidal projection showing distribution of large volcanos, rifts, and corona. Feature locations originally obtained from global geomorphologic map of Price and Suppe (1995). Note that the large volcanos are located atop most geoid highs (not shown, but present in original Herrick (1999) diagram). Rifts connect the geoid highs and the corona generally lie along rifts. From Figure 2 in Herrick (1999). Central meridian is 180°. Digital version of original figure kindly provided by R. Herrick. (b) Molliwede projection of Venus showing average model surface age provinces, exposures of ribbon tessera terrain, volcanic rises, crustal plateaus, and selected geographic regions. Top inset illustrates the stages of impact crater modification with time. Central meridian is 90° E. After Figure 2 in Bjonnes *et al.* (2012). With permission from the American Astronomical Society. Digital version of original figure kindly provided by V. Hansen. Note different projections and central meridians for (a) and (b).

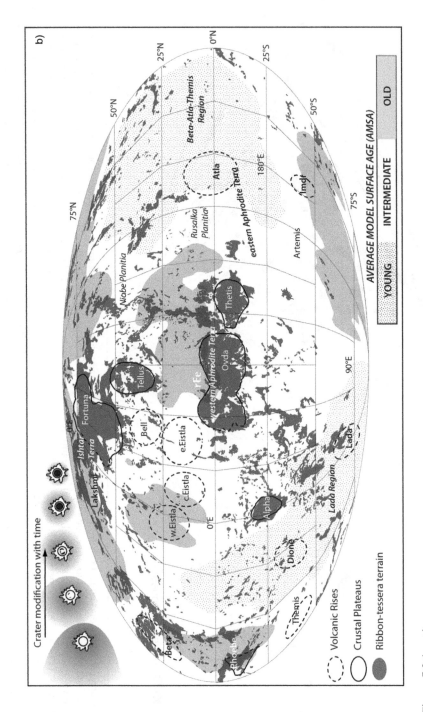

Figure 7.9 (*cont.*)

196

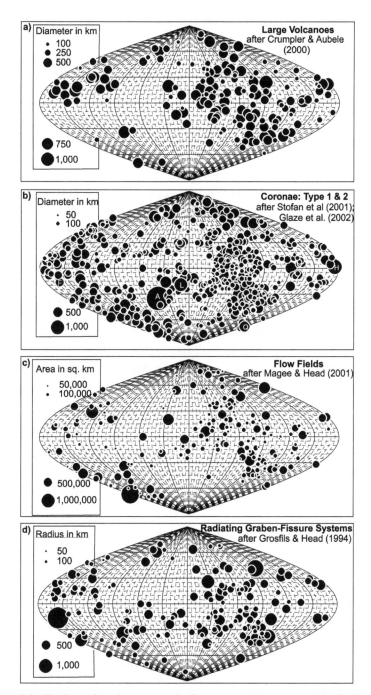

Figure 7.10 Distribution of major magmatic features on Venus. (a) Distribution of large volcanoes. Extracted from catalog of Crumpler and Aubele (2000). (b) Distribution of coronae on Venus, after Stofan *et al.* (2001) as modified in Glaze *et al.* (2002), A = Artemis, F = Fatua, H = Heng-O, L = Latona, and Q = Quetzalpetlatl. (c) Flow fields after Magee and Head (2001). (d) Radiating graben–fissure systems (dyke swarms). Displayed in a sinusoidal projection with a central meridian of 180°. Symbol size is greater than the actual feature size; this was done to allow a greater dynamic range in symbol size. From Figure 4 in Ernst *et al.* (2007). With kind permission from Springer Science and Business Media.

on Venus and are grouped into "shield fields." Large volcanoes are defined as eruptive centers with diameters > 100 km, extending up to 1000 km (Crumpler *et al.*, 1997, p. 703) and can be correlated with plumes (e.g. Head *et al.*, 1992; Stofan *et al.*, 1995; Herrick, 1999; Crumpler and Aubele, 2000). Most large volcanoes on Venus are relatively low in relief, averaging approximately 1.5 km in height, and are considered to represent basaltic shield volcanoes. They have a wide distribution on Venus, but show a preferred association with broad topographic rises and at the junctions of extensive chasmata (deep linear valleys with steep sides generally interpreted as young rifts) and belts of fractures (interpreted as older rifts; Basilevsky and Head, 2000; or alternately interpreted as dense graben–fissure systems, e.g. Ernst *et al.*, 2003). Crumpler and Aubele (2000) list 60 volcanoes that are > 500 km in diameter. With corresponding volumes > 0.1 Mkm^3, they are of LIP scale.

Venus is also a one-plate planet and its large volcanoes (like those on Mars) are thought (by analogy with terrestrial hotspots) to be a product of a mantle plume arriving beneath stationary lithosphere (stagnant lithospheric lid) and producing initial plume-head magmatism followed by plume-tail magmatism, all erupted at the same spot. So the volcanoes are predicted to have a LIP portion (large volume emplaced in a short duration) followed by younger volcanic sequence (plume-tail stage) emplaced over a much longer period. However, the relative magnitude of the plume-head stage vs. plume-tail stage magmatism is unknown for the large volcanoes. On Earth, plume-tail magmatism occurs as oceanic hotspot tracks, but is rarely observed on continents. So it is possible that only limited plume-tail magmatism can penetrate the Venusian lithosphere and that the bulk of the volcanoes are built during a plume-head stage.

Even more dramatic than the individual volcanoes are the volcanic rises (see below) in which groups of large volcanoes and corona can be perched on geoid and topographic highs and associated with triple-junction rifting (e.g. Basilevsky and Head, 2007).

7.3.4 Coronae

An important set of circular tectonomagmatic structures on Venus are termed coronae (Figs. 7.8–7.12). They are defined by the presence of a raised annulus of circumferential fractures or ridges, a peripheral moat or trough, and an interior region that can be either topographically positive, neutral, or negative (Stofan *et al.*, 1992, 1997; Janes *et al.*, 1992; Squyres *et al.*, 1992; Crumpler and Aubele, 2000). An additional class of coronae has the raised annular topographic signature, but largely lack the fractured annuli of classic coronae (Stofan *et al.*, 2001, 2003). Overall the coronae range in diameter from 60 km to over 2400 km, but average about 300 km in diameter, and although they appear to lack terrestrial analogs they are probably related to plumes or diapirs (Janes *et al.*, 1992; Herrick, 1999; Crumpler and Aubele, 2000; Stofan *et al.*, 2001; Johnson and Richards, 2003; Hoogenboom *et al.*, 2005; Jurdy and Stoddard, 2007). A comprehensive compilation (Stofan *et al.*, 2001, as slightly modified in Glaze *et al.*, 2002) identifies 513 coronae (Fig. 7.10).

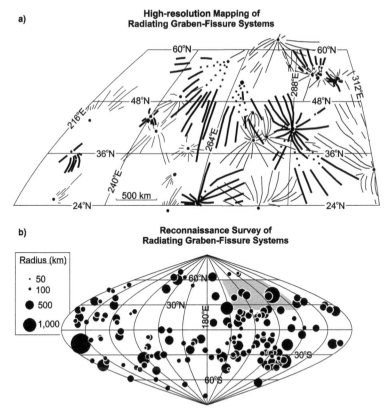

Figure 7.11 (a) Radiating graben–fissure systems from full-resolution mapping in an area of about 28 million km. Map is from Figure 12 in Ernst and Desnoyers (2004). (b) Radiating systems from the reconnaissance scale mapping of Grosfils and Head (1994a). With permission from Elsevier. Area of part (a) shown by shading. Both parts are displayed in a sinusoidal projection with a central meridian of 294 and 180°, respectively, for (a) and (b).

Coronae occur at volcanic rises, as isolated features in the plains, and, most commonly, along chasmata (rift) systems (e.g. Stefanick and Jurdy, 1996; Hansen *et al.*, 1997; Crumpler and Aubele, 2000). The association with volcanic rises may reflect thermal diapirs resulting from the breakup of a deep-sourced thermal plume (see Section 7.3.7) (Stofan *et al.*, 1995), whereas coronae along chasmata may represent compositional diapirs (Hansen *et al.*, 1997). (See further discussion of thermal and compositional plumes in Chapter 15 on LIP origins).

Many coronae are associated with chasmata in the BAT region (Fig. 7.12), specifically along the Parga and Hecate chasmata. There are 46 coronae along the 8000-km-long Hecate Chasma (Hamilton and Stofan, 1996) and at least 27 coronae along the 10 000-km-long Parga Chasma (Hamilton and Stofan, 1996; Martin *et al.*, 2007; Smrekar *et al.*, 2010b). There is also a significant distribution of coronae along the rifts in the Aphrodite Terra region (Stefanick and Jurdy, 1996), and along the

BAT REGION

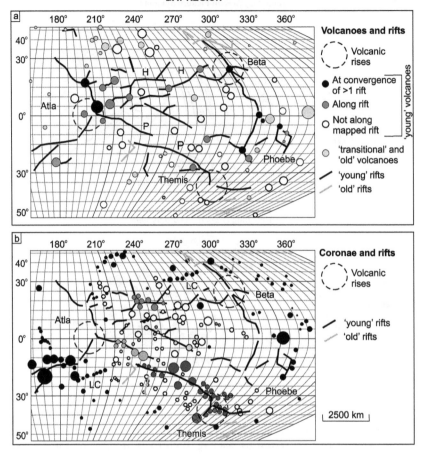

Figure 7.12 Magmatism of the BAT (Beta–Atla–Themis) region. (a) Distribution of volcanoes and rifts in the BAT region. (b) Schematic distribution of coronae and rifts extracted from Herrick (1999) and originally from Price (1995) and Price and Suppe (1995). Regional rises are located with large circles. Separation of rifts into "young" and "old," and separation of volcanoes into "young" and "transitional/old" is based on Plate 1 in Basilevsky and Head (2000). Symbol size is approximately the minimum dimension of the feature. Sinusoidal projection with a central meridian of 180°. Thin latitude and longitude lines are spaced at 5° intervals. Thick lines are at 30° intervals. LC = lines of coronae not associated with a mapped rift, and may represent cryptic rifts. Hecate Chasma connects Atla and Beta, and Parga Chasma connects Atla and Themis. From Figure 3 in Ernst and Desnoyers (2004). With permission from Elsevier.

Alpha–Lada and Derceto–Quetzalpetlatl extensional belts in northern Lada Terra (Baer *et al.*, 1994). Other linear distributions of coronae are not localized along a mapped chasma (Fig. 7.12). As noted by Hansen *et al.* (1997), older chasmata will lose their thermal buoyancy and sink, and may be progressively flooded by volcanism. Therefore, linear distributions of coronae may be a clue to the identification of older rift systems in which the rift has been obscured by younger flooding events (Ernst and Desynoyers, 2004; Ernst *et al.*, 2007).

In general, only a portion of coronae can be classed as LIPs, i.e. those subtypes with significant associated volcanism and also those with associated giant radiating graben–fissure (dyke) systems (e.g. Ernst *et al.*, 2003; Hansen and Olive, 2010).

7.3.5 Volcanic flow fields

On Venus, large flow fields represent direct analogs of terrestrial flood basalts (Head and Coffin, 1997). A global survey of Venus (Magee and Head, 2001) recognizes 208 volcanic flow fields that exceed 50 000 km^2 in size (Figs. 7.9 and 7.10). These "large" flow fields average 220 000 km^2, but range up to 1.6 Mkm^2, constituting 11% of the plains areas which cover 80% of Venus. Of these, there are 140 with a size > 0.1 Mkm^2, which on their own represent LIPs. There is a sub-population of 81 classified as exceptionally large or "great" flow fields on the basis of having a maximum flow length greater than 500 km (Magee and Head, 2001). As summarized by Magee and Head (2001), the flow fields most typically derive from coronae (37%), large volcanoes (25%), and fractures (graben–fissure systems) within rifts and fracture belts (20%). Approximately 36% are associated with (and presumably fed by) radiating fissure–graben systems.

The majority (up to 74%) of all large flow fields are located within zones of extension (major rifts and fracture belts), and largely post-date the onset of extensional deformation. Many have also been deformed by subsequent fracturing along their associated rifts (Magee and Head, 2001). This association suggests that the rift-generated decompression–melting model of White and McKenzie (1989) is applicable (Section 15.2.2).

7.3.6 Radiating graben–fissure systems

The presence of giant radiating graben–fissure systems that superficially resemble the geometry of radiating dyke swarms on Earth was recognized early during the Magellan mission (e.g. Head *et al.*, 1992; McKenzie *et al.*, 1992; Parfit and Head, 1992; Grosfils and Head, 1994a, b). (These are also equivalent to the radiating graben–fissure systems discussed above for Mars in Section 7.2.4.) Using the Magellan data at reconnaissance scale (225 m/pixel), an initial global reconnaissance map was produced for Venus (Grosfils and Head, 1994a) with 163 radiating systems of which 118 were confidently interpreted to be underlain by dykes (Figs. 7.7 and 7.11); for the other, an origin by extension alone (linked to uplift) could not be ruled out. Subsequent systematic mapping using full-resolution images (75 m/pixel) is revealing six times as many radiating graben–fissure systems (e.g. Ernst *et al.*, 2003; Grosfils *et al.*, 2011; Studd *et al.*, 2011; see also Galgana *et al.*, 2013; McGovern *et al.*, 2014). The more detailed studies are also revealing an increased size of individual systems, including systems up to 2500 km in radius (Ernst *et al.*, 2003). Recently, a huge radiating system (6000 km in radius) centered on Artemis corona was proposed (Hansen and Olive, 2010). While there was initially debate about the origin of graben–fissure systems (and purely tectonic central uplift models were considered; see discussion in Grosfils and Head (1994a)), most graben–fissure systems are now recognized to be underlain by dykes, as was also concluded for the similar features on Mars (see Section 7.2.4).

Some of the evidence in support of a dyke-swarm origin are the following: (a) the systematic transition from grabens to fissures to fractures as a function of distance from the swarm focus, consistent with a gradual decrease in dyke width or depth; (b) the presence of grabens that transition into pit chains; (c) the alignment of small shield volcanoes along the lineaments; and (d) lava flows that clearly emanate from the fractures, particularly in the distal portions of the system away from any central topography.

The radiating graben–fissure systems are commonly centered on known volcanoes and coronae, but in other cases, locate previously unknown magmatic centers (Grosfils and Head, 1994a, b; Ernst *et al.*, 2003; Studd *et al.*, 2011). Radiating systems greater that 300 km in radius are, by analogy with terrestrial dyke swarms, considered to be LIP-related (see Chapter 5).

In addition, linear graben–fissure systems may be associated with rifts trending toward magmatic centers (e.g. Krassilnikov and Head, 2003). There are also graben–fissure systems with a circumferential pattern and these are typically interpreted to be associated with the annulae of coronae (e.g. Ernst *et al.*, 2003). In many cases these linear and circumferential systems are also thought to be underlain by dykes and therefore to represent part of the magmatic budget of their volcanoes and coronae (cf. related story on Mars, Section 7.2.4).

7.3.7 Volcanic rises – LIP clusters

Volcanic rises range in diameter from 1000 to 2500 km, and in height from about 1 to 2.5 km (Senske *et al.*, 1992; Stofan *et al.*, 1995; Hansen *et al.*, 1997; Smrekar *et al.*, 1997; Hansen, 2007; Basilevsky and Head, 2007). These rises are interpreted as the surface manifestation of mantle upwellings, and represent the clearest expression of terrestrial-style, deep-sourced mantle plumes on Venus. Volcanic rises include three morphologically distinct types, rift-, volcano-, and coronae-dominated rises (e.g. Senske *et al.*, 1992), each possibly caused by differences in lithospheric properties, or representing different stages in a common process, or both (e.g. Hansen, 2007). The rises comprise a topographic rise, a geoid high, and, commonly, triple-junction rifting, and can be associated with LIP-scale magmatism in the form of volcanoes and coronae.

Particularly significant is the Beta–Atla–Themis (BAT) region (Fig. 7.12) where several volcanic rises are each associated with triple-junction rifting; in this area an additional 10 large volcanoes occur along or at the termination of single rifts, and an additional 13 large volcanoes are not associated with rifts and may be older. Evidence for their associated rifts is interpreted to have been obscured by younger flows.

In the broad Eistla region (Fig. 7.9), there are four regional upwellings (topographic rises), western Eistla, central Eistla, eastern Eistla, and Bell Regio. Each appears to be represented by a clustering of volcanoes and coronae, along with an associated geoid high. However, unlike in the BAT region, these rises are not associated with rifts, apart from one major rift (Guor Linea) associated with western Eistla.

7.3.8 Artemis

New geologic mapping of Hansen and Olive (2010) has concluded that Artemis, a unique 2400-km-diameter feature on Venus, originally classed as a corona, can be linked with additional elements that make it a much larger tectonomagmatic feature than previously thought (Fig. 7.9b). These additional elements include a wide outer trough (> 5000 km diameter), a radial dyke swarm (12 000 km diameter), and a concentric wrinkle ridge suite (13 000 km diameter). Hansen and Olive (2010) suggest that Artemis represents the magmatic signature of a deep mantle plume acting on relatively thin lithosphere. The cumulative volume of magmatism in the associated dykes would certainly be of LIP scale. Wrinkle ridge suites occur with large diameters circumferential to the volcanic centers at Lada (8000 km), Themis (4200 km), Bell (3200 km), and western (3700 km), central (2600 km), and eastern Eistla (3200 km) (Bilotti and Suppe, 1999). Could these centers also have associated huge radiating graben–fissure systems as appear to be associated with Artemis?

7.3.9 Beta–Atla–Themis (BAT) region – as a plume cluster

Magmatic activity is concentrated in the BAT region (Figs. 7.9 and 7.12) corresponding to about 92 Mkm^2 or about 20% of the surface of Venus (Head and Coffin, 1997). This area of regionally abundant volcanoes largely postdates the earliest plains emplacement and corresponds to the region of three major rifted volcanic rises, Beta Regio, Atla Regio, and Themis Regio, all of which are presumed to mark young mantle plumes. The chasmata, which radiate from (in triple-junction style; Burke and Dewey, 1973; Şengör and Natal'in, 2001) and connect the plume centers of the BAT region, are also young (Basilevsky and Head, 2000). Furthermore, coronae in the BAT region have largely uncompensated gravity anomalies confirming that this region is geologically young (Smrekar *et al.*, 2010).

It is notable that the BAT region is marked by clustering of young coronae (representing transient plume heads; "thermals" in the terminology of Jellinek *et al.*, 2002; see also Dombard *et al.*, 2007) and also persistent plume conduits marked by large volcanoes (Johnson and Richards, 2003). On Earth, such a grouping of related plumes would be termed a plume cluster or "superplume event" (Condie, 2001; Ernst and Buchan, 2002; Schubert *et al.*, 2004; cf. also Maruyama, 1994), and could be analogous to those LIP clusters on Earth linked to the long-lived, deep-mantle, geoid highs in the South Pacific and under Africa (see Chapters 12 and 15).

7.3.10 Plains volcanism

The volcanic plains on Venus potentially represent a global LIP (Head and Coffin, 1997). Volcanic plains units cover 80% of the surface area of Venus or *c*. 368 Mkm^2. Assuming an average plains thickness of 2.5 km, the volume would be about 920 Mkm^3 (Head and Coffin, 1997). The magmatism responsible for flooding the plains would be equivalent to about 21 plumes of the scale of the largest LIP on Earth (Ontong Java, 44 Mkm^3; Coffin and Eldholm, 2001), or 184 of the largest continental LIPs (estimated to be about 5 Mkm^3 in volume; see Table 1 in Ernst and Buchan (2001b)). Critical to assessment as a LIP is the duration of magmatism (Chapter 2).

According to Price *et al.* (1996), the duration of this plains volcanism may be between 400 and 200 Ma (values based on a mean plains age of 300 Ma). However, using the revised mean-age estimate of 750 Ma with uncertainties (McKinnon *et al.*, 1997), an age of duration from 1000 to 500 Ma or even 2000 to the present would be possible (e.g. Basilevsky and Head, 2002; Ernst and Desnoyers, 2004).

However, if the plains were resurfaced instead by a thin volcanic "veil" derived from widespread volcanism from small (1–15 km diameter) volcanoes (Hansen and Bleamaster, 2002) then the volume of resurfacing would be greatly reduced, and a catastrophic plains resurfacing event would not be required; therefore, the plains magmatism would not necessarily represent LIP-scale magmatism.

This fundamental debate persists regarding the duration and magnitude of magmatism on Venus: is there a strict lithological age progression (following a major planetary resurfacing event) where the most deformed units are the oldest (e.g. Basilevsky and Head, 1994, 2002; Ivanov and Head, 2001; Ivanov and Head, 2013; Head, 2014), or can the available constraints be satisfied by a progressive time-transgressive resurfacing (e.g. Guest and Stofan, 1999; Stofan *et al.*, 2005; Hansen and Young, 2007; Hansen and López, 2010; Bjonnes *et al.*, 2012)? Further mapping will be required to determine the relative age relationships and estimates of the thickness of volcanic plains to determine the existence or not of a global LIP event(s) associated with the plains.

7.3.11 Tesserae/crustal plateaus

The oldest crust on Venus includes regions of a distinctive unit marked by faulting and deformation termed tesserae (e.g. Ivanov and Head, 1996; Gilmore *et al.*, 1998; Hansen *et al.*, 1999, 2000). Tesserae occur as large clusters (e.g. Aphrodite Terra) and arc-like segments that may extend for thousands of kilometers (Ivanov and Head, 1996). About 8% of Venus is currently covered by tesserae, but Ivanov and Head

(1996) conclude that tesserae represent a basement unit distributed over at least 55% of the surface of Venus, which is largely covered by younger volcanic units. Hansen *et al.* (1999) disagree and consider that tesserae are more spatially restricted and represent local thickening of a globally thin and ancient Venusian lithosphere. In any case, the largest areas of tesserae are associated with crustal plateaus, which represent steep-sided, flat-topped quasi-circular regions 1600–2500 km in diameter (e.g. Hansen *et al.*, 1999). Seven crustal plateaus are identified: Fortuna Tesserae, Tellus Tesserae, Alpha Regio, eastern Ovda Regio, western Ovda Regio, Phoebe Regio, and Thetis Regio. In addition, some of the arc-like segments of tesserae observed elsewhere may represent the outer boundary of crustal plateaus that experienced central collapse and flooding (Tewksbury, 2003).

Crustal plateaus are distinct from volcanic rises by having small gravity anomalies, low gravity-to-topography ratios, and shallow apparent depths of compensation. Two general end-member models have been developed for plateau formation: "hotspot" or mantle upwelling (e.g. Phillips and Hansen, 1994, 1998) and "coldspot," or mantle downwelling (Bindschadler *et al.*, 1992; Bindschadler, 1995; Ivanov and Head, 1996). Hansen (2007) suggested that crustal plateaus may represent the surface scum of huge lava ponds formed by massive partial melting in the mantle due to large bolides impacting ancient thin lithosphere.

7.4 Mercury

Mercury is about one-third the diameter of Earth, but has a similar density. It has a surface (dominated by volcanic plains, impact craters, and major contractional fault scarps), a global magnetic field, and an interior dominated by an iron core having a radius that is at least three-quarters of the radius of the planet (e.g. Kiefer and Murray, 1987; Spudis and Guest, 1997; Head and Coffin, 1997; Head *et al.*, 2007, 2008, 2009a, b; Wilson and Head, 2008; Watters *et al.*, 2009). It was suggested that major volcanic provinces were emplaced (Spudis and Guest, 1988), but the available images were insufficient to characterize such provinces, or to allow assessment in a LIP context.

However, with the recent MESSENGER spacecraft, the nature of the smooth plains areas and other aspects of the volcanism of Mercury could begin to be sorted out. Earlier Mariner 10 data revealed heavily cratered plains units and also the presence of younger smooth plains units. These widespread plains deposits, occurring as relatively smooth surfaces between craters (intercrater plains), and as apparently ponded material (smooth plains and Caloris floor plains) – see Figs. 7.13 and 7.14 – were proposed by most investigators to be basaltic lavas, but their interpretation as impact ejecta also remained possible.

New data from the current MESSENGER mission support the interpretation of significant volcanic activity in the early history of Mercury (Wilson and Head, 2008; Head *et al.*, 2008, 2009a, b; Head *et al.*, 2011; Head and Wilson, 2012). Volcanic vents

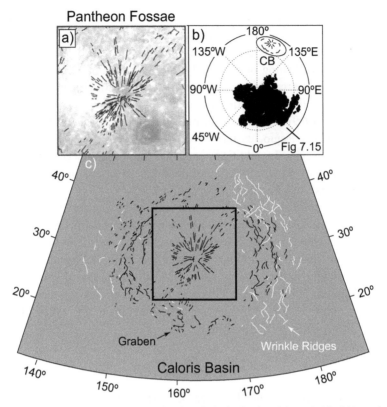

Figure 7.13 Map of tectonic features in the Caloris basin. (a) and (c) 560 grabens and 96 wrinkle ridges digitized from MESSENGER and Mariner 10 image mosaics. The center of the basin is dominated by the radial graben system (Pantheon fossae), which extends outward to a band of concentric (circumferential) and radially oriented graben arrayed near the margin of the Caloris basin. Concentric and radially oriented wrinkle ridges near the basin margin coincide with the band of polygonally patterned graben. Modified from Figure 6 in Watters *et al.* (2009). Box in part (c) locates part (a). With permission from Elsevier. (b) Inset map showing location of Caloris basin and flood basalt of Fig. 7.14. Digital version of original figure for (c) kindly provided by T. Watters.

and a shield-like structure were found. Also, mineralogical interpretations indicate that the smooth plains were produced by volcanic activity of a composition different from that of the crater and basin ejecta and deposits. Furthermore, dating based on impact-crater size–frequency distributions showed that many smooth plains units are younger than their surrounding craters and basins. All these new results from the MESSENGER mission are confirming the important role of volcanism in the early history of Mercury. In addition, a radiating graben–fissure system, Pantheon fossae, was discovered in the center of the Caloris basin (Fig. 7.13; Watters *et al.*, 2009; Head *et al.*, 2009b) and was interpreted to be similar to radial dyke swarms on Venus, Mars, and Earth (see Sections 5.2, 7.2.4, 7.3.6). Pantheon fossae is also circumscribed

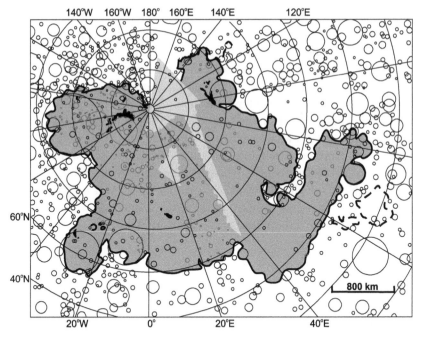

Figure 7.14 Potential flood-basalt province in the northern high latitudes of Mercury. From Figure 1 in Head *et al.* (2011). Evidence for lava flooding includes craters that are flooded (those shown are "ghost" craters). Light-gray areas denote gaps in MDIS (Mercury Dual Imaging System) coverage. See location in Fig. 7.13b. Reprinted with permission from AAAS.

by circumferential graben and wrinkle ridges, again with similarities to magmatic systems on Mars and Venus (cf. Mège and Ernst, 2001).

In addition, it has been recently discovered from MESSENGER observations that a large contiguous region of smooth plains covers much of Mercury's high northern latitudes and occupies more than 6% of the planet's surface area; it is interpreted as a flood-basalt province with a composition (based on X-ray spectrometric data) intermediate between that of basalts and komatiites (Fig. 7.14; Head *et al.*, 2011).

7.5 The Moon

7.5.1 Introduction

The Moon is about one-quarter the diameter of the Earth and is of lower density, has not retained an atmosphere, has no history of plate tectonics, and presently has a very thick lithosphere. Most of its geological surface activity took place in the first half of the Solar System history (Wilhelms, 1987). As summarized by Head (2004) and discussed in Head and Coffin (1997), the most important magmatic elements are the individual maria themselves, with extensive flow fronts, some stretching for distances of over 1200 km (Schaber, 1973). Other magmatic components include dykes, sills, and sinuous rilles. Dykes are recognized where they reach the surface as narrow grabens (Head and Wilson, 1993), and are also required for the transport of magma from deep sources. Sill emplacement is recognized in the form of floor-fractured craters, impact craters whose floors have been fractured, uplifted, and commonly partially flooded by lava flows (Schultz, 1976). Sinuous rilles are meandering channels occurring primarily in the lunar maria. They range in widths up to about 3 km and in length from a few kilometers to more than 300 km, and are explained as high effusion-rate eruptions involving thermal erosion of the substrate (Head and Coffin, 1997, and references therein). These are similar to the canali seen on Venus (Section 7.3.2). No large shield volcanoes, such as those seen on the Earth (e.g. Hawaii), Mars (e.g. Olympus Mons; Fig. 7.6), or Venus (e.g. Sapas Mons; Fig. 7.8), are observed on the Moon (Head and Wilson, 1991).

7.5.2 Mare

Lunar maria are inferred to be dominantly basaltic in composition and cover about 17% of the surface, mainly on the near side of the Moon (Fig. 7.15) (e.g. Head and Coffin, 1997; Head, 2004; Shearer *et al.*, 2006; Whitten *et al.*, 2011; Hiesinger *et al.*, 2011). Lunar maria cover a total area of *c*. 6.3 Mkm^2, have an estimated volume of 10 Mkm^3. The average lunar global volcanic flux was low, about 10^{-2} km^3/a, even during peak periods of mare emplacement (in the Imbrian Period, 3.8–3.2 Ga). However, isolated mare basalt ponds in the highlands have magma volumes in the range 10–9000 km^3; values similar to those of terrestrial basaltic flood eruption units (Yingst and Head, 1997). A similar estimate comes from sinuous rilles, which can produce flows of extremely high volumes, in the range 300–1200 km^3 (Head and Coffin, 1997). Individual lava flows are extensive, such as the flows extending hundreds of kilometers into Mare Imbrium (Schaber, 1973).

7.5.3 Timing of magmatism

The chronology of lunar volcanism is based on radiometric ages determined from *Apollo* and *Luna* landing-site samples and regional stratigraphic relationships and used to calibrate the crater-counting method for relative age dating. The lunar age distributions were largely defined prior to the end of the *Apollo* program in the 1970s. However, there has been a recent reappraisal of the timing of mare magmatism on the Moon (Hiesinger *et al.*, 2011, and earlier papers). Hiesinger *et al.* (2011) used ages derived from crater size–frequency distribution measurements to reassess ages for exposed mare basalt units on the lunar nearside hemisphere. Using this method their analysis shows the following.

(1) Lunar volcanism was active for almost 3 Ga, starting at *c*. 3.9–4.0 Ga and ceasing at *c*. 1.2 Ga.

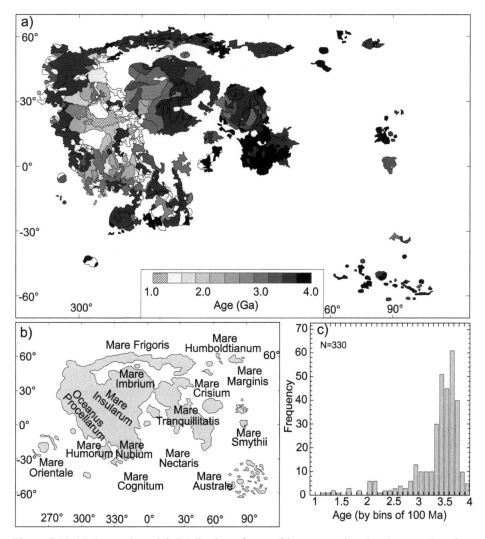

Figure 7.15 (a) Age and spatial distribution of ages of lunar mare basalts. Ages are based on crater counts on spectrally and morphologically defined mare units (Hiesinger *et al.* (2011) and earlier publications). (b) Labelling of mare. (c) Histogram of ages. Ages are in Ga; bin size is 100 Ma. Map coverage is ~ 90° W to 120° E, ~ 75° S to 75° N. From Hiesinger *et al.* (2011). See also Whitten and Head (2011).

(2) Most basalts erupted during the late Imbrian Period at *c.* 3.6–3.8 Ga.

(3) Basalts of possible Copernican age (younger than 3.3 Ga) have been found only in limited areas in Oceanus Procellarum.

(4) Episodically, there might have been intervals with increased volcanic activity, for example, at *c.* 2.0–2.2 Ga with some peaks in mare production also at 1.4, 1.7, 2.1, 3.1, and 3.6 Ga.

(5) Thickness estimates for 58 basalt units based on crater size–frequency measurements revealed average thicknesses of 30–60 m.

(6) Volume estimates in combination with ages of individual basalt units indicate a steep decline in erupted basalt volumes since *c.* 3.6 Ga.

These results underscore the predominance of older mare basalt ages in the eastern and southern nearside of the Moon, and also in patches of maria peripheral to the larger maria. This constrasts with the younger basalt ages on the western nearside of the Moon, i.e. in Oceanus Procellarum.

7.5.4 Additional concentrations of magmatism on the Moon

As summarized by Head and Coffin (1997) several areas show unusual concentrations of volcanic features, which may be the surface manifestation of hotspots and could be possible analogs for terrestrial LIPs. Two of the most significant of these are the Marius Hills area (35 000 km^2) (latitude: 14° N, longitude: 56° W), which displays 20 sinuous rilles and over 100 domes and cones, and the Aristarchus Plateau/Rima Prinz region (40 000 km^2), which is dominated by 36 sinuous rilles (latitude: 26° N, longitude: 46° W).

7.5.5 Model for emplacement of basaltic mare

Basaltic mare are superimposed on the ancient, globally continuous, and thick low-density anorthositic highland crust derived primarily from global-scale melting associated with planetary accretion. As summarized by Head and Coffin (1997), the low-density highlands crust provided a density barrier to basaltic melts ascending from the mantle. This deep zone of neutral buoyancy for rising magma could only be overcome by over-pressurization events in which magma was transported via dykes to the surface.

7.5.6 Evidence that mare volcanism postdated impact-basin formation

Given proposed models for the role of bolide impacts in generating LIPs on Earth (see Section 15.2.6) it is useful to consider the relationship between impacts and flood volcanism (mare) on the Moon. Early theories (e.g. Alt *et al.*, 1988) suggested a causal relationship between lunar impacts, basin formation, and basaltic mare filling. However, the results of the *Apollo* and *Luna* exploration programs, and models of basin formation and evolution (see review in Head and Coffin, 1997), showed a time gap in many cases between impact basin formation and mare basalt filling. For instance, the extensive lava flows of Mare Imbrium (Fig. 7.15) were emplaced at least a billion years after the formation of the impact basin. Although the Orientale impact basin formed at *c.* 3.68 Ga its 200 000 km^3 of mare magmatism was emplaced in several pulses between 3.47–1.66 Ga (Whitten and Head, 2011). However, the possibility remains that larger impacts earlier in lunar history, such as the South Pole–Aitken basin, could have generated mare because of the greater scale of the impacts and the thinner lithosphere characterizing the Moon during this earliest period of lunar history.

7.6 Io: satellite of Jupiter

Io is the innermost of the four Galilean moons of Jupiter with a diameter of 3642 kilometers, and is the fourth largest moon in the Solar System. Unlike most satellites in the outer Solar System (which have a thick coating of ice), Io is primarily composed of silicate rock surrounding a molten iron or iron sulfide core. Io has one of the most geologically active surfaces in the Solar System, with over 400 active volcanoes (Fig. 7.16b; e.g. Lopes-Gautier, 2000; Geissler, 2003; Davies, 2007). This extreme geologic activity is the result of tidal heating from friction generated within Io's interior by Jupiter's varying gravitational pull. Several volcanoes on Io produce plumes of sulfur and sulfur dioxide that reach as high as 500 km.

The volcanoes found on Io are hotter than those on other planets, many with optically measured temperatures in excess of 1500 K and the corresponding magma types are inferred to be ultramafic lavas or superheated basalt. The possibility of ultramafic magmatism represents an analogy with Archean Earth where greenstone with komatiites (ultramafic melts) are common (Chapter 6). Io has a number of types of volcanic feature classes: paterae (caldera-like features), shields, flows and mountains, lava lakes, all of intraplate origin (no plate tectonics on Io), which are useful for comparison with intraplate magmatism on Earth. Some notable magmatic features on Io include an active persistent lava lake at Pele and an impressive flow field at Prometheus. Also, Loki Patera is associated with a quiescent periodically over-turning lava sea covering 20 000 km^2 (Davies, 2007). Figure 7.16a shows the range of types of mainly silicate (basalt and komatiite) volcanism and minor sulfur volcanism on Io. Silicate magma is inferred to originate from deep sources, to rise along faults, and be emplaced in superheated, short-lived, gas-rich, explosive events. The effusive eruption of silicate magma from shallow reservoirs mobilizes surface volatiles to form dust-rich plumes. The lithosphere, at least 30 km thick, overlies a partially molten upper mantle of silicate composition with a melt fraction of perhaps 20% or more.

7.7 Summary

Intraplate magmatism of LIP scale is common on the Moon, Mars, and Venus, and potentially on Mercury, and their presence, characteristics, and geologic and temporal settings offer a complementary perspective for interpreting LIPs on Earth. On

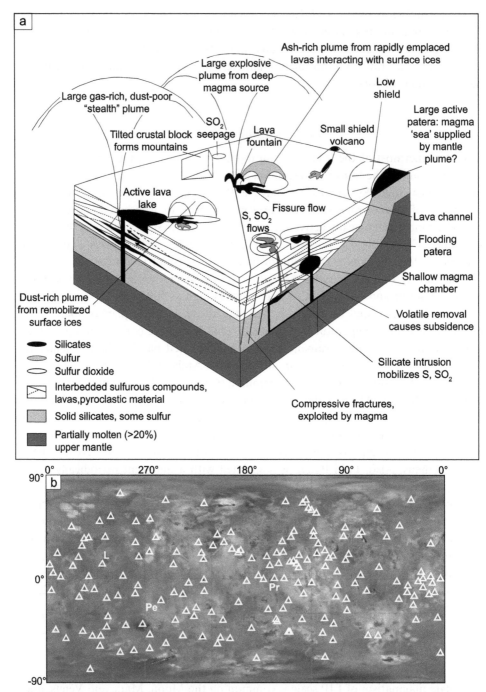

Figure 7.16 Magmatism on Io (a) Styles of intraplate magmatism on Io, a moon of Jupiter. From Figure 18.1 in Davies (2007). Copyright © 2007 A.G. Davies. Reprinted with the permission of Cambridge University Press. (b) Global distribution of hotspots (simplified from Williams *et al.* (2011) using base from http://gallery.usgs.gov/images/03_15_2012/xcs1VIh77P_03_15_2012/ large/Io_figure_press_2xglobal_300dpi.jpg). Those locations referred to in the text are identified: L = Loki, Pe = Pele, Pr = Prometheus.

Mars, massive shield volcanoes have formed on the stable lithosphere particularly in the Tharsis region. Although the evolution of the Tharsis region extended over a 1 Ga or more, each volcano may represent a separate LIP event formed by a plume arriving beneath the stationary lithosphere. In addition, widespread plains magmatism of early Hesperian age, constitutes a potential global LIP on Mars.

Individual volcanoes on Venus, especially those associated with volcanic rises, may represent separate LIPs (as on Mars), generated by plume arrival beneath the stationary lithosphere. The nature of the plains volcanism on Venus is a source of controversy. Is it a product of rapid and massive planet-wide volcanic resurfacing centered about 750 Ma ago, or can plains magmatism be explained by progressive resurfacing to form the plains over a longer period of time?

New data from the current MESSENGER mission to Mercury is revealing the details of plains magmatism, some of which is clearly analogous to terrestrial flood basalts, and has also revealed a giant radiating graben (dyke?) system. On the Moon, large shield volcanoes are unknown. The relatively low-density, thick anorthositic crust creates a density trap for ascending basaltic magmatism, which is thought to collect in reservoirs at the base of the crust. Reservoir over-pressurization is required to cause dykes to propagate to the surface, which feed the surface mare. Individual flows are of flood-basalt scale, but mare form over much longer periods of time than terrestrial LIPs. Some of the hotspots on Io (a moon of Jupiter) produce high-temperature komatiitic magmatism.

On Mars, there is a concentration of magmatic activity in the Tharsis and Elysium regions. On Venus there is a concentration of younger activity in the BAT region and the central Eistla region. These are potential analogs of the regional concentrations of terrestrial LIPs in two regions, the Pacific and Africa, which are associated with long-lived deep-mantle anomalies and associated geoid highs (Section 15.2.1).

Are there times of global resurfacing (global LIPs) perhaps related to mantle overturn (Stein and Hofmann, 1994)? As discussed above, on Mars, this could have occurred in the Early Hesperian and on Venus this may have occurred at *c.* 750 Ma. These can be compared with the terrestrial record where there are some periods, *c.* 1880 Ma and 2700 Ma (Sections 11.8.2 and 11.8.3) during which LIP magmatism occurred on many blocks, and may have been essentially global.

Earth is unique in having plate tectonics, and so all the magmatism that I reviewed above on the other planetary bodies is, by definition, intraplate. For Mars, plate tectonics may have occurred only in its earliest history and, for Venus, the extensive rifts of the BAT region represent attempted, but failed, plate breakup. Therefore, the planetary bodies reviewed in this chapter have provided an opportunity for evaluating intraplate magmatism (including that of LIP scale) without the superimposed effects of plate tectonics. The implications of this planetary record for models of LIP formation will be discussed in Section 15.7.

8

Silicic LIPs

8.1 Introduction

Silicic magmatism producing granites and rhyolites occurs in a variety of tectonic settings: subduction (oceanic and continental arcs particularly when undergoing extension), intraplate, continental rifts, MORB (mid-ocean ridge basalt), hotspot-modified mid-ocean ridges (e.g. Iceland), and post-collisional. Of these, intraplate, continental rift, and extending subduction settings are the most important in generating large volumes of silicic magma.

Here I focus on silicic magmatism in intraplate and continental rift systems potentially associated with LIPs, where two types have been recognized: (1) silicic volcanics and intrusions that are ubiquitously present in small to moderate proportions in all continental LIPs, and (2) huge, dominantly silicic provinces that have a non-subduction origin and which are termed Silicic Large Igneous Provinces (SLIPs) (Fig. 8.1).

Before proceeding further, a point about terminology needs attention. In the literature, the second class has been referred to as SLIPs, notably by S. Bryan who, along with R. Pankhurst and others, has done significant pioneering work on this style of magmatism (e.g. Pankhurst and Rapela, 1995; Bryan et al., 2000, 2002, 2010; Bryan, 2007; Bryan and Ernst, 2008; Pankhurst et al., 1998, 2000, 2010, 2011; Bryan and Ferrari, 2013). However, in the broader geologic literature, granite and rhyolitic magmatism has also been described as "felsic." There is some debate regarding which usage is more correct for LIP-scale high-silica events (e.g. Pankhurst et al., 2011). The relevant definitions in the *Glossary of Geology*, 5th edition (Neuendorf et al., 2011), say felsic = "applied to an igneous rock having abundant light-colored minerals in its mode; also, applied to those minerals (quartz, feldspars, feldspathoids, muscovite) as a group"; silicic = "said of a silica-rich igneous rock or magma. Although there is no firm agreement among petrologists, the amount of silica is usually said to constitute at least 65 percent or two-thirds of the rock. In addition to the combined silica in feldspars, silicic rocks generally contain free silica in the form of quartz. Granite and rhyolite are typical silicic rocks." By these definitions, felsic rock types can include low-silica rocks which have a high proportion of feldspars (e.g. anorthosites, syenites) or can contain feldspathoids (e.g. nepheline syenites). Neither anorthosites nor

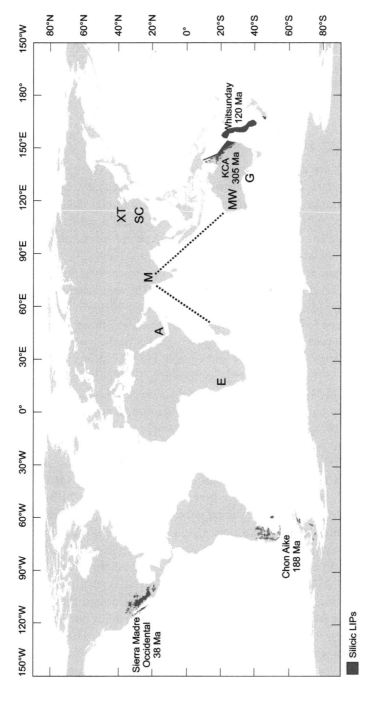

Figure 8.1 Global distribution of SLIPs (Silicic LIPs) following assembly of Pangea *c.* 320 Ma. SLIPS or LIPs with significant silicic component: A = Afro-Arabian, E = Etendeka, G = Gawler, KCA = Kennedy-Connors-Auburn, M = Malani, SC = South China, X-T = Xiong'er-Taihang. Modified from Bryan (2007). Digital version of original figure was kindly provided by S. Bryan.

feldspathoid-bearing rocks are part of the high-silica units of interest herein. Volumetrically, the SLIPs are overwhelmingly dominated by igneous rocks with SiO_2 contents > 65 wt% (e.g. Bryan, 2007). So, strictly speaking, in the context of LIP magmatism, silicic is the more appropriate label for the high silica suites being discussed in this chapter, and I follow the usage and precedent of Bryan and refer to SLIPs rather than Felsic LIPs (FLIPs).

8.2 Silicic magmatism associated with LIPs

Silicic igneous rocks are associated with all continental LIPs (see reviews in Bryan *et al.*, 2002; Bryan, 2007; Bryan and Ernst, 2008; Pankhurst *et al.*, 2011; Table 8.1) from the oldest Precambrian to the youngest Cenozoic examples. Silicic magmatism can be well preserved in the Mesozoic–Cenozoic continental flood-basalt provinces, and these LIPs, silicic volcanic and volcaniclastic rocks can form substantial parts of the eruptive stratigraphy (Bryan *et al.*, 2002; White *et al.*, 2009).

The volume of individual silicic eruptive units can reach 9000 km^3 (dense-rock equivalent; Bryan *et al.*, 2010, and references therein). The areal extent of silicic units can reach > 0.1 Mkm^2 (e.g. Milner *et al.*, 1995; Bryan *et al.*, 2010). In the older record, one of the largest is the Rooiberg silicic unit that overlies the Rustenberg layered suite of the 2.06 Ga Bushveld intrusion; although it remains unclear whether this is the product of a single eruption (e.g. Twist and French, 1983; Vantongeren *et al.*, 2010); this extrusive sheet, of rhyolitic to dacitic composition, is 3 km thick over an area of 0.3 Mkm^3. In the 250-Ma Siberian Trap event (Section 3.2.8), synchronous syenite–granite intrusions and bimodal volcanic sequences are present around the periphery of the Siberian Trap LIP, specifically in the Taimyr (to the north), and the Altai–Sayan and Transbaikalia regions to the south and southeast sides of the Siberian craton (Dobretsov and Vernikovsky, 2001; Dobretsov, 2005; Bryan and Ernst, 2008). In contrast, the Deccan LIP (Section 3.2.5) has by comparison a very small volume of rhyolite preserved (*c.* 500 km^3) (S. Bryan, personal communication, 2013).

As shown in the examples from Yemen (Afro-Arabian LIP) (Section 3.2.3) and Etendeka (Paraná–Etendeka LIP, Section 3.2.6) (Fig. 8.2) the silicic component can be concentrated toward the upper part of the stratigraphic pile, and as a consequence is preferentially lost to erosion in comparison to the mafic component. However, in the NAIP (Section 3.2.4) some of the oldest units are rhyolites (Bryan *et al.*, 2002 and references therein). On the other hand, deeper erosion through the volcanic pile exposes the intrusive silicic component ("granitic rocks"). Silicic magmatism that is associated with LIPs generally consists of high temperature (≤ 1100 °C) rhyolites/ignimbrites and equivalent A-type granites, which are linked to differentiation (fractionation ± assimilation) of mafic magma or partial melting of anhydrous lower crust as a response to underplating by high-temperature mafic magmas (e.g. Huppert and Sparks, 1988; Harris *et al.*, 1990; Bryan *et al.*, 2002; Pankhurst *et al.*, 2011).

Table 8.1 *Examples of LIPs with associated silicic magmatism*

Silicic unit	LIP (Location)	Age (Ma)	References
Silicic volcanism, coeval with LIP and associated with subsequent hotspot trail of Snake River plain	Columbia River	17–0 Ma	McCurry *et al.* (2008); Coble and Mahood (2012)
Welded ignimbrite	Afro-Arabian (Yemen–Ethiopian; Afar–East African)	30 Ma	Ukstins Peate *et al.* (2005); Ayalew and Gibson (2009)
Rhyolites and trachytes	Deccan	65 Ma	Lightfoot *et al.* (1987)
Ignimbrite, and also granites associated with the British Tertiary intrusive (e.g. granites on Arran, the Red Cuillin of Skye)	NAIP (North Atlantic Igneous Province)	62–55 Ma	Meyer *et al.* (2009)
Rhyolites and felsic igneous complexes	Parana–Etendeka	135 Ma	Milner *et al.* (1995); Garland *et al.* (1995); Kirstein *et al.* (2001); Ewart *et al.* (2004a,b)
Rheoignimbrite	Karoo	180 Ma	Harris and Erlank (1992); Turner and Rushmer (2009); Miller and Harris (2007).
Granites in margins of the Siberian craton	Siberian Trap	250 Ma	Dobretsov and Vernikovsky (2001); Dobretsov (2005)
Intrusions and volcanic in the Giles complex	Warakurna	1070 Ma	Wingate *et al.* (2004); Evins *et al.* (2010); Smithies *et al.* (2013)
Silicic lavas and rheoignimbrites	Keweenawan	c. 1100 Ma	Green and Fritz (1993)
Rooiberg felsites, Nebo granite	Bushveld	2060 Ma	Twist and French (1983); Vantongeren *et al.* (2010)
Woongara felsic sheet	Woongara	2450 Ma	Trendall (1995); Barley *et al.* (1997)
Copper Cliff rhyolites and Murray and Creighton, granite	Matachewan	2450–2490 Ma	Event no. 206 in Ernst and Buchan (2001b); Smith *et al.* (1999)
Associated granitoids	Kambalda	2700 Ma	Campbell and Hill (1988)
Hardey Formation	Fortescue	2700 Ma	Thorne and Trendall (2001)
Syferfontein Formation	Dominion	c. 3000 Ma	Marsh (2006)

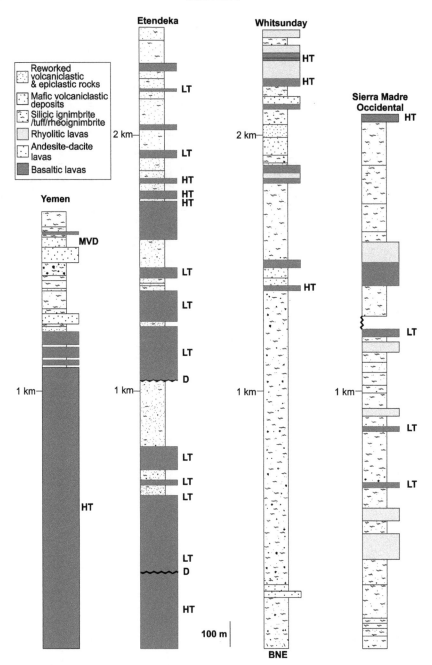

Figure 8.2 Generalized composite stratigraphic sections comparing two continental flood basalt provinces (Afro-Arabian and Etendeka) with two SLIPs (Whitsunday and Sierra Madre Occidental). The intrusive components (dykes, sills, and other intrusions) are not shown. HT and LT refer to high-Ti and low-Ti compositional types. D = disconformity, MVD = Mafic volcaniclastic deposit, BNE = base not exposed. Major silicic units are labeled in italic font. From Figure 5 in Bryan (2007). Digital version of original figure kindly provided by S. Bryan.

8.3 Silicic LIPs (SLIPs)

8.3.1 Introduction

In addition to representing a subordinate igneous component in LIPs (as discussed above), additional suites of intraplate magmatism have been recognized in which silicic magmatism has dominated. With respect to LIP criteria (Chapters 1 and 2), these silicic events are of large volume (> 0.1 Mkm^3), consist of short-duration pulses, and have an intraplate origin or setting (Bryan *et al.*, 2002; Bryan, 2007; Bryan and Ernst, 2008). These are termed Silicic LIPs or SLIPs, and have subordinate proportions of basalt and/or intrusive equivalent.

Throughout this book I am following traditional usage in considering LIPs to be mainly mafic events (with minor ultramafic and silicic components). In contrast, when I am referring to dominantly silicic provinces, I will refer to SLIPs, and consider SLIPs to be affiliated or associated with LIPs. This is slight modification to the view in Bryan and Ernst (2008) in which SLIPs were effectively considered a subclass of LIPs. However, I think that it remains important to retain the traditional focus of LIPs on the dominant mafic component to reinforce the direct link with mantle melting. In contrast, SLIPs have an indirect link with LIPs and mantle sources in that they are mainly linked to lower crustal anatexis caused by LIP underplating (see below). In using SLIPs as an affiliated class to LIPs (rather than as a subclass of LIPs), I am effectively grouping them with carbonatites (and kimberlites) which can be similarly viewed as commonly affiliated/associated with LIPs (Chapter 9).

To better understand the composition of SLIPs and their link with LIPs, I briefly review the classification and origin of large-scale silicic magmatism. Silicic rocks have been classified in a number of schemes, and one of the most popular has been the "alphabetic system." In this system, granites are divided into different geochemical types (mainly A, I, and S) that largely reflect the composition/nature of the crustal source materials involved in partial melting (Pearce, 1996b; Chappell and White, 2001; Frost *et al.*, 2001; Bonin, 2007; Clemens *et al.*, 2011; Brown, 2013; Bryan and Ferrari, 2013). To a rough approximation, I-type (igneous-type) granites (metaluminous to weakly peraluminous, relatively sodic and with a wide range of silica content, 56–77 wt% SiO_2) can be derived from hydrous melting of mafic protoliths, typically arc volcanics or greywackes, and are metaluminous. S-type (sedimentary-type) granites (strongly peraluminous, relatively potassic and restricted to higher silicic compositions, 64–77 wt% SiO_2) are derived from melting of sedimentary rocks, particularly hydrous shales and alumina-rich sediments. A-type (anorogenic-type) granites (anhydrous, relatively potassic, with high FeO/FeO + MgO), and high-Zr and other high-field-strength elements) are linked to lower crustal melting under extremely dry conditions and are typically peralkaline. A-types are associated with continental rifts, but can also be associated with LIPs (as discussed above), and can also be derived by differentiation of mafic magmas. The other main types, I and S, are typically linked with subduction-related settings, but, in fact granites and rhyolites of both I and S type can also be associated with SLIPs. In a LIP setting, I and S types can be derived by partial melting of suitable lower crust by underplating associated with a LIP.

The number of confirmed SLIPs is small at present (Table 8.2) but is expected to grow along with the growing identification of LIPs in the older record owing to improving geochronological coverage. The realization that major silicic magmatism can be associated with LIP/mantle-plume events has significance for the detrital zircon studies that are typically inferred to record only orogenic events (e.g. Condie and Aster, 2010). As noted in Bryan and Ferrari (2013) zircon generally only appears as a new crystallizing phase in silicic magmas (around $c.>70$ wt% SiO_2). In contrast, supra-subduction-zone magmatism is dominantly basaltic andesite to andesite–dacite at modern oceanic and continental arcs, respectively; the consequence is that these magma compositions are zircon-undersaturated and will not crystallize new zircon. Therefore, as noted by Bryan and Ferrari (2013), large-volume silicic magmatism that will have a measurable effect on the detrital zircon age record occurs in intraplate continental regions, as well as along continental margins or island arcs undergoing rifting. So it must be considered that the peaks in zircon ages are perhaps more likely to reflect extensional events, and should not be so closely tied to periods of supercontinent assembly (cf. Condie and Aster, 2010; Iizuka *et al.*, 2010). As a consequence, it should be considered that the long-recognized $c.$ 2.7 and $c.$ 1.9 Ga peaks in detrital zircon populations may at least partly reflect LIP-associated silicic magmatism (Bryan and Ferrari, 2013).

As summarized in Bryan and Ernst (2008), the Mesozoic–Cenozoic examples of SLIPs are the best preserved, and their key unifying characteristics are summarized as follows:

(1) Areal extents > 0.1 Mkm^2 (many are > 0.5 Mkm^2), which is the minimum areal extent for LIPs defined by Coffin and Eldholm (1994) and Eldholm and Coffin (2000), and affirmed in Bryan and Ernst (2008); extrusive volumes are > 0.25 Mkm^3 (up to $c.$ 3 Mkm^3).
(2) The provinces comprise $> 80\%$ by volume of dacite–rhyolite, with transitional calc-alkaline I-type (Chappell and White, 1992, 2001) to A-type granites (e.g. Whalen *et al.*, 1987; Turner *et al.*, 1992; Trumball *et al.*, 2004; Bonin, 2007). S-type granites are also present, for example, as part of the Guibei LIP (Table 8.2).
(3) Rhyolitic ignimbrite is the dominant volcanic lithology.
(4) The duration of igneous activity is up to 40 Ma, during which a large proportion of the magma volume was erupted during shorter intervals or pulses (3–10 Ma). The relatively long life span of some SLIPs is likely due to the presence of multiple magmatic pulses, some of which may be pre-rift and syn-rift (e.g. Ferrari *et al.*, 2007; Bryan *et al.*, 2014; Bryan and Ferrari, 2013).

Table 8.2 *Examples of SLIPs*

Province	Age	Volume, area, dimensions	Related mafic magmatism	References
Whitsunday (eastern Australia)	132–95 Ma	> 2.2 Mkm^3 > 2500 × 200 km	Ontong Java oceanic plateau	Bryan *et al.* (2012); Section 8.3
Kennedy–Connors–Auburn (northeastern Australia)	320–280 Ma	> 0.5 Mkm^3 > 190 × 300 km		Bain & Draper (1997); Bryan *et al.* (2003)
Sierra Madre Occidental (Mexico)	c.38–20 Ma	> 0.39 Mkm^3 > 2000 × 2500 km	Basin and Range mafic magmatism	Bryan and Ferrari (2013); Section 8.3.2
Chon Aike (South America–Antarctica)	188–153 Ma	> 0.23 Mkm^3 > 3000 × 1000 km	Karoo (Africa) and Ferrar (Antarctica LIPs)	Pankhurst *et al.* (1998, 2000); Section 8.3.4
Malani (India) including related magmatism in Seychelles and Madagascar	750 Ma	> 0.1 Mkm^2	Dolerite sills and dykes in possibly formerly connected blocks	Sharma (2005); Bhushan (2000); Section 8.4.1
Guibei	825 Ma	1.34 Mkm^2	Guibei	Li *et al.* (2013); Section 8.4.2
Gawler Range felsic volcanic and Hiltaba suite granites	1590 Ma	0.11 Mkm^3	Western Channel diabase of northwestern Laurentia (formerly attached)	Wade *et al.* (2012); Pankhurst *et al.* (2010); Section 8.4.3
Xiong'er	1780 Ma	> 0.1 Mkm^3	Associated radiating dolerite dyke swarm	Peng (2010); Section 8.4.4

(5) In terms of crustal setting, SLIPs are exclusively continental and sited along paleo- and active continental margins as they are produced by large-scale crustal anatexis of such fusible lower crust, caused by a thermal pulse of non-subduction origin. Some examples were clearly emplaced into intraplate settings, far removed from active plate boundaries (e.g. Whitsunday; see Bryan *et al.*, 1997, 2012) whereas others have been emplaced in closer proximity to a subducting plate margin (e.g. Sierra Madre Occidental; see Ferrari *et al.*, 2007; Bryan *et al.*, 2014), where igneous activity was extensive (up to 1000 km) inboard of the plate margin.

Considering Figure 8.2, which compares the stratigraphy of some SLIPs to mafic–ultramafic LIPs, rhyolitic ignimbrite forms the overwhelming proportion of volcanic rock in all the SLIPs, typically representing > 80% of the total stratigraphic thickness. Although rhyolite ignimbrites dominate, the upper portions are bimodal (where basalt and rhyolite are interbedded). This is similar to the continental flood basalts, which are dominated by basalts, but where the upper portions are also bimodal.

A selection of SLIPs are compiled in Table 8.2, and several are discussed in detail below and located in Fig. 8.1. The Sierra Madre Occidental of Mexico is a representative example of general SLIP architecture, being an extensive, relatively flat-lying ignimbrite plateau covering an enormous area (> 0.5 Mkm^2) to *c.* 1-km thickness (e.g. Ferrari *et al.*, 2007; Bryan and Ernst, 2008). The Whitsunday igneous province is the largest of the world's SLIPs with an eruptive output of > 2.5 Mkm^3 (Bryan *et al.*, 2012). Older examples occur as continental caldera systems and major batholiths (e.g. Bryan and Ernst, 2008), such as the 320–280 Ma Kennedy–Connors–Auburn province of northeast Australia (Table. 8.2).

Extensive silicic and bimodal dyke swarms are exposed in the more deeply eroded provinces (e.g. Whitsunday, Kennedy–Connors–Auburn; Ewart *et al.*, 1992; Stephenson, 1990), or become exposed during tectonic exhumation associated with syn-rift phases (e.g. Sierra Madre Occidental; Aguirre Díaz and Labarthe-Hernández, 2003; Ramos Rosique, 2012). The 750 Ma Malani event of India, the 1590 Ma Gawler Range event of South Australia, and the 1780 Ma Xiong'er event of the North China craton, illustrate the tracking of SLIPs into the Precambrian record (Table 8.2).

8.3.2 Sierra Madre Occidental (40–20 Ma), Mexico

General characteristics

The Tertiary-aged Sierra Madre Occidental magmatism of Mexico at 396 000 km^3 represents the largest SLIP in North America (Fig. 8.3; Table 8.1; McDowell and Keizer, 1977; McDowell and Clabaugh, 1979; Ward, 1995; Bryan *et al.*, 2002; Ferrari *et al.*, 2007; Bryan *et al.*, 2008; Cather *et al.*, 2009). Magmatism is distributed over an area 200–500 km wide and extends for more than 2000 km principally within western Mexico, but is contiguous with silicic volcanism through the Basin and Range

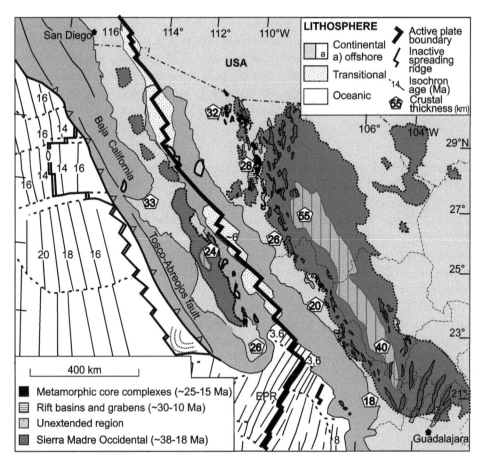

Figure 8.3 38 Ma Sierra Madre Occidental (SMO) SLIP of northwestern Mexico showing (1) the preserved extents of the Oligocene–early Miocene silicic-dominant volcanic activity of SMO; (2) extents of the dominantly bimodal early Miocene pulse that coincided with the wide development of grabens and rift basins, and a restricted belt of metamorphic core complexes. Lithospheric variation across the region is also shown, including unextended and extended continental regions, and transitional to new oceanic crust formed by the propagating spreading center in the Gulf of California. Abbreviations: EPR, East Pacific Rise. Modified from Bryan and Ferrari (2013). Digital version of original figure kindly provided by S. Bryan.

province of western USA to the north (Lipman *et al.*, 1972; Gans *et al.*, 1989; Best and Christiansen, 1991) and also with the ignimbrite province of the Sierra Madre Sur, south of the Trans Mexican Volcanic Belt (Morán-Zenteno *et al.*, 2007).

Age constraints

The rhyolitic ignimbrite that comprises the Sierra Madre Occidental was mostly erupted, in general between *c.* 38 and 18 Ma, with two main pulses or "flare-ups" of ignimbrite activity at *c.* 34–28 Ma and *c.* 24–18 Ma (Ferrari *et al.*, 2002, 2007; Bryan *et al.*, 2008; McDowell and McIntosh, 2012). The younger Early Miocene pulse is clearly syn-rift, being related to the opening up of numerous grabens across a 500-km width of the province, while the Oligocene pulse is thought to represent the main volumetric outpouring with *c.* 300 000 km^3 of rhyolite being erupted (Ferrari *et al.*, 2002, 2007; Bryan *et al.*, 2008, 2014).

Composition

Ignimbrite compositions are dominantly in the range 65–75 wt% SiO_2. For the Sierra Madre Occidental SLIP, a broad silicic peak exists for both the Oligocene and Early Miocene pulses but two compositional groupings can be recognized of dacitic to low-silica rhyolite and high-silica rhyolite suites, with the latter being dominant and occurring as lavas/domes and ignimbrites (Bryan *et al.*, 2014). The syn-rift Early Miocene pulse was distinctly bimodal with a paucity of volcanic compositions between 64 and 68 wt% SiO_2. The intermediate to silicic composition volcanic rocks generally have medium- to high-K calc-alkaline compositions (Bryan *et al.*, 2008; Bryan, 2007), but zircon inheritance data indicate that much of this subduction-related chemical affinity has been derived from the crustal materials involved in their petrogenesis (Bryan *et al.*, 2008, 2014). Associated basalts with within-plate, asthenospheric compositions began erupting in the Oligocene (*c.* 30 Ma, Southern Cordillera basaltic andesites of Cameron *et al.* (1989)) as well as during the syn-rift Early Miocene pulse (e.g. Bryan *et al.*, 2014).

8.3.3 Whitsunday (c. 120 Ma), Australia

General characteristics

The Early Cretaceous Whitsunday LIP is the world's largest SLIP. It is a within-plate, silicic-dominated ignimbrite–granite belt, > 2500 km long and *c.* 300 km wide, with an eruptive output (> 2.5 Mkm^3) roughly coincident with the present eastern Australian plate margin (Fig. 8.4; Bryan *et al.*, 1997, 2000; Bryan, 2007; Bryan *et al.*, 2012). It consists of the following igneous, volcano–sedimentary, and tectonic elements: (1) the Whitsunday Volcanic Province of northeast Australia; (2) scattered igneous intrusions and volcanic rocks along the southeast Australian margin; (3) the sedimentary fill in the Great Australian and Otway–Gippsland–Bass sedimentary basin systems of northeast and southeast Australia, respectively; and (4) submerged volcanic and rift-fill sequences on marginal continental plateaus and troughs. The Whitsunday Volcanic Province is the preserved section of a linear SLIP that formed at least 1000 km inboard of the eastern Australian continental margin in the Early Cretaceous, and thus resembles the *c.* 183 Ma Ferrar LIP (part of the broader Karoo–Ferrar LIP) along strike to the south in terms of province geometry and relative position to the continental margin (Bryan *et al.*, 2012; Cook *et al.*, 2013).

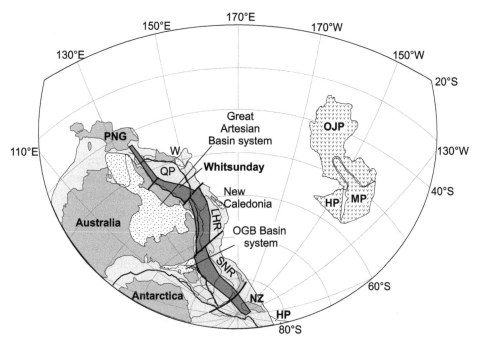

Figure 8.4 120 Ma Whitsunday LIP and regional setting. Plate reconstruction map showing the extent and relative location of the Whitsunday SLIP to the coeval Ontong Java oceanic plateau (part of the Ontang Java–Hikurangi–Manihiki OJHMP reconstructed oceanic plateau; Taylor, 2006; see Figs. 4.3 and 4.6). Both the Whitsunday SLIP and OJHMP formed rapidly at ∼ 120 Ma. Regional-scale crustal lineaments, which have partitioned deformation and influenced subsequent rifting along the eastern Gondwanaland margin are also shown (see also Bryan *et al.*, 2012). Recent studies have recognized a Whitsunday-aged and related detrital volcanic zircon contribution to Cretaceous sedimentary rocks in the Woodlark Rift (W) (Zirakparvar *et al.*, 2013) and New Caledonia regions (Adams *et al.*, 2009) indicating volumetrically significant resedimentation of pyroclastic material also occurred along the eastern margin of the LIP. Other abbreviations: HP = Hikurangi Plateau, LHR = Lord Howe Rise, NZ = New Zealand, OJP = Ontong Java Plateau, OGB = Otway-Gippsland Basin, PNG = Papua New Guinea, SNR = southern Norfolk Ridge. Modified from Bryan and Ernst (2008). With permission from Elsevier. Digital version of original figure kindly provided by S. Bryan.

SLIP magmatism (*c*. 35 Ma duration) and continental rifting (15–30 Ma duration) were protracted before complete rupture of the continent and onset of seafloor spreading at *c*. 84 Ma off southeastern Australia (Bryan *et al.*, 2012).

The Whitsunday Volcanic Province is volumetrically dominated by welded dacitic, rhyolitic, and relatively lithic-rich ignimbrite (Bryan *et al.*, 2000). Most exposures present monotonous successions of stacked, welded ignimbrite units (Fig. 8.4) up to 1 km thick, particularly where deposited into tectonic depressions, or occurring as intracaldera fills. Exhumed caldera sections reveal multiple caldera-forming ignimbrite eruptions characterized by coarse lithic-lag breccias containing clasts up to 6 m in diameter (Ewart *et al.*, 1992; Cook *et al.*, 2013) that typically cap the ignimbrites and record collapse episodes to calderas *c*. 10–20 km wide. Intrusions are a significant part of the Whitsunday Volcanic Province. As emphasized by Ewart *et al.* (1992), numerous dyke swarms (with composition ranging from dolerite to rhyolite) are present across the province, and these are consistent with the rifting environment of the volcanism.

Age constraints

Relatively limited K–Ar and Rb–Sr isotopic dating has established an age range of 132 to 95 Ma, but with a main period of igneous activity between 120 and 105 Ma (Ewart *et al.*, 1992; Bryan *et al.*, 1997). This is a striking age match with the Ontong Java oceanic plateau (Section 4.2.1) located more than 3000 km away at the time (Fig. 8.4), but a geodynamic connection has yet to be established. The age span of the Whitsunday SLIP also overlaps with LIP events along the west Australian margin from 130–120 Ma (Kerguelen–Bunbury–Comei and Paraná–Etendeka LIPs; Sections 4.2.2. and 3.2.6, respectively). The majority of granites onshore in the Whitsunday SLIP have U–Pb and K–Ar ages of 130–120 Ma (Bryan *et al.*, 2012).

Detrital studies indicate that a significant volume of Whitsunday magmatism was eroded and is present in the adjacent sedimentary basins of eastern Australia. Detrital zircon geochronology study (Bryan *et al.*, 2012) of the coeval (latest Albian–Cenomanian) volcanogenic Winton Formation in the Eromanga basin (Great Australian superbasin) has recognized a dominant volcanic and euhedral detrital zircon component of *c*. 105–95 Ma. These ages and their apparent general absence from the Whitsunday Volcanic Province but widespread occurrence in age spectra of Mesozoic and Cenozoic basins in eastern Australia (Sircombe, 1999; Cross *et al.*, 2010), combined with the exhumation history of the Queensland margin (Bryan *et al.*, 2012), imply that a substantial missing section to the Whitsunday Volcanic Province is preserved in adjacent sedimentary basins (Cook *et al.*, 2013).

Composition

Most of the rhyolitic ignimbrites contain the phenocryst assemblage of plagioclase and Fe–Ti oxides with uncommon quartz and K-feldspar, and the ferromagnesian phases dominated by pyroxene. Biotite and/or hornblende are less common such that the silicic volcanics are predominantly pyroxene rhyolites. The granites are mostly of peraluminous I type and dominantly hornblende–biotite granites with accessory allanite, Fe–Ti oxides, and titanite (Ewart *et al.*, 1992); however, some granites are of A type and syenitic.

Chemically, the igneous suite ranges continuously from basalt to high-silica rhyolite, with calc-alkaline to high-K affinities; a relatively small percentage of mafic rocks classify as alkalic (Ewart *et al.*, 1992; Bryan 2007). In detail, however, this range of silica content is defined by dyke compositions; lavas are bimodal (basalt–andesite and rhyolite to high-silica rhyolite), and ignimbrites are dominantly rhyolitic (Bryan *et al.*, 2012). Consequently, like the Sierra Madre Occidental SLIP (Section 8.3.2), volcanic compositions are bimodal, but volumetrically silicic-dominant. Rhyolites and basalts have geochemical signatures transitional between within-plate and convergent margin fields on trace-element discrimination diagrams (Bryan, 2007; Chapter 10). The basaltic rocks tend to show more compositional variation in their minor and trace-element contents, reflecting low-Ti (≤ 1 wt%) and high-Ti (>2 wt%) basaltic magma types and a range of nepheline to quartz normative basalts (Ewart *et al.*, 1992; Bryan, 2007; Murray *et al.*, 2014). The range of compositions has been generated by two-component magma mixing and fractional crystallization superimposed to produce the high-silica rhyolites (Ewart *et al.*, 1992). The two magma components are a volumetrically dominant partial melt of relatively young (Paleozoic–Mesozoic), non-radiogenic, calc-alkaline igneous to volcaniclastic crust and a within-plate tholeiitic basalt of enriched-MORB (E-MORB) affinity, similar to the Cenozoic intraplate basalts of eastern Australia (Section 2.4; Ewart *et al.*, 1992; Bryan, 2007).

8.3.4 Chon Aike (c. 180–150 Ma), Antarctica and South America

Mesozoic volcanic rocks of Patagonia (South America) and West Antarctica are volumetrically dominated by rhyolitic ignimbrites, which compositionally form a bimodal association with minor mafic and intermediate lavas (Fig. 3.7) and three distinct pulses are present: 188–178 Ma, 172–162 Ma, 157–153 Ma, V1, V2, and V3 pulses, respectively (Pankhurst *et al.*, 1998; Pankhurst *et al.*, 2000; Bryan *et al.*, 2002). The first pulse is concentrated in eastern Patagonia (Marifil and Chon Aike Formations) and matches in age with the Karoo–Ferrar LIP of southern Africa and Antarctica (Fig. 3.7; e.g. Pankhurst *et al.*, 1988). However, the two younger pulses are concentrated in the Andean Cordillera further west (El Quemado, Ibañez, and Tobífera Formations) and are not matched with any LIP pulses of the Karoo–Ferrar event. The silicic products of the first pulse (*c.* 180 Ma) are lower-crustal melts that have incorporated upper-crustal material, and are linked to crustal heating and magmatic underplating related to emplacement of the Karoo–Ferrar LIP (Pankhurst *et al.*, 2000; Riley *et al.*, 2001; Fig. 3.7). The geochemistry of the younger two pulses of ignimbrites is more characteristic of destructive plate margins, consistent with their Andean setting but the presence of inherited zircon still points to involvement of a crustal source.

8.4 Precambrian SLIPs

I now turn to provide a brief description of some older examples of SLIPs whose volcanic component has variably been lost to erosion, and which can be recognized by their widespread distribution of the intrusive component of granitic rocks. These are identified with less certainty because, as noted by Bryan (2007), SLIPs typically show calc-alkaline affinities that resemble modern destructive plate-margin volcanic rocks, and back-arc extensional environments can also produce significant volumes of rhyolite. This composition is ambiguous when interpreting the tectonic setting of magmatism and, as a consequence, the tectonic setting of SLIPs can be wrongly interpreted as subduction- or orogeny-related. As emphasized by Bryan and Ernst (2008) other features must be observed before a potentially voluminous silicic igneous province can be considered a SLIP. Examples of Precambrian SLIPs are compiled in Table 8.2.

It should be noted that some of the examples below have a significant mafic component and their proper classification as SLIPs (as dominantly silicic) or LIPs (and therefore dominantly mafic) is somewhat ambiguous pending further studies to sort out their full extent and characteristics.

8.4.1 Malani rhyolite province of Greater India (750 Ma)

General characteristics

The *c*. 750 Ma Malani igneous province occupies an area of about 50 000 km^2 in the northwestern Indian shield (Fig. 8.5; e.g. Bhushan, 2000; Sharma, 2005; Gregory *et al.*, 2009; Meert *et al.*, 2013). Magmatism consists of initial basaltic eruptions followed by voluminous silicic lava flows, and finally with granitic plutonism and terminal dyke-swarm (both silicic and mafic) activity. Related bimodal magmatism is also present in the Seychelles and Madagascar (Tucker *et al.*, 1999, 2001; Ashwal *et al.*, 2002). The scale of the Malani event of Greater India (India, plus Seychelles and Madagascar) is estimated to be greater than 100 000 km^2 (Ashwal *et al.*, 2002; Gregory *et al.*, 2007, 2009). Details on the age and composition for units of the Greater Malani event are provided below as well as speculations regarding a possible link with the Mundine Well dyke swarm of western Australia.

Age constraints

Approximate ages including Rb–Sr suggested a *c*. 745 ± 10 Ma timing for the Malani bimodal volcanic province (e.g. Bhushan, 2000). This timing is confirmed by U–Pb ages ranging between 771 ± 2 and 751 ± 3 Ma, which were obtained for Malani rhyolites (Tucker *et al.*, personal communication in Torsvik *et al.*, 2001a).

Northwest–north-northwest-trending dolerite dykes in the Seychelles have a U–Pb zircon age of 750 ± 3 Ma (Torsvik *et al.*, 2001b), and widespread granitoids in the Seychelles have U–Pb zircon ages between 748 and 764 Ma (Stephens *et al.*, 1997; Tucker *et al.*, 2001). (Incidentally, the Seychelles also have some younger, *c*. 63 Ma, alkaline granite magmatism that is presumably linked to the Deccan; LIP; Section 3.2.5; Torsvik *et al.*, 2001b.)

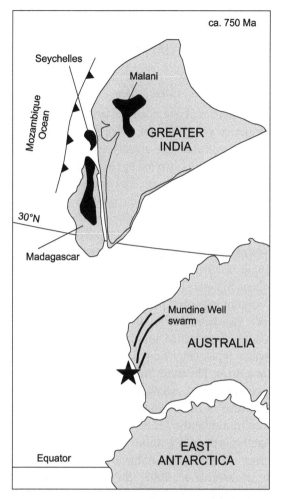

Figure 8.5 755 Ma Malani SLIP and related units in the Seychelles and Madagascar of Greater India and relationship with coeval Mundine Well radiating dyke swarm of western Australia. Diagram is modified from Ashwal *et al.* (2002) to be consistent with the paleomagnetic data from Gregory *et al.* (2009) that show India north of 30° N. The distribution of the Mundine Well swarm and the location of the mantle plume center (star) is after Ernst and Buchan (2001b). As noted in Meert *et al.* (2013) recent work suggests that the northern Australia and southwestern Australian blocks were still adjusting their positions within Australia at this time.

The magmatic province continues into northeastern Madagascar, where there are *c.* 750 Ma (*c.* 715–754 Ma from U–Pb zircon dating) intermediate to mafic metavolcanoic rocks belonging to the Daraina Complex (Ashwal *et al.*, 2002). Additional granitic and gabbroic intrusive rocks of broadly similar age (*c.* 720–820 Ma) are present more widely in Madagascar (Ashwal *et al.*, 2002).

More precise U–Pb ages need to be obtained not only in Madagascar but also in the Seychelles and India to confirm the duration of this magmatic event. Do the ages cluster closely about *c*. 750 Ma or is the longer age range (*c*. 770–750 Ma or even longer) correct?

Composition

The Malani bimodal magmatism of the northwestern Indian shield includes volcanic, plutonic, and dyke components (e.g. Bushan, 2000; Ashwal *et al.*, 2002; Sharma, 2005; Meert *et al.*, 2013). Initial volcanics are mostly basalt with occasional andesite or trachybasalts. These are subsequently covered by the voluminous outpouring of peralkaline and peraluminous rhyolite, basalt, dacite, and trachyte flows. Volcanism ended with an outburst of ash-flow deposits. Eruptions were predominantly subaerial, but aqueous conditions existed locally, as is indicated by conglomerate and arkose beds underlying the lavas, and also the local presence of pillowed basalts and the development of sedimentary features in the lower volcanics. Peralkaline and peraluminous flows sometimes alternate with each other. Trachytic flows are also associated with the peralkaline rhyolites.

The second component consists of widespread silicic plutons, which compositionally comprise anorogenic peraluminous and peralkaline granites. The silicic magmatism can be grouped into at least two petrographically and geochemically distinct groups: the dominantly peraluminous (hornblende granite) Jalore type and the more alkaline (aegirine granite) Siwana types (Eby and Kochhar, 1990; Bhushan, 2000; Ashwal *et al.*, 2002; Meert *et al.*, 2013). The third Malani phase involved the intrusion of mafic and silicic dykes of variable trend, which are considered to be associated with the development of several zones of north–south rifting (Bhushan, 2000; Sharma, 2005).

The Neoproterozoic (dominantly 752 ± 4 Ma) granitoids of the Seychelles are undeformed and unmetamorphosed granodiorites and monzogranites with metaluminous, I-type chemistry (Baker, 1963; Ashwal *et al.*, 2002). Peralkaline granitoids are apparently absent (Ashwal *et al.*, 2002). Subsolvus and lesser hypersolvus granitoids are crosscut by coeval dolerite dykes dominantly of olivine tholeiite composition. Mixing between granitoid and dolerite magmatism generated intermediate rocks. Two groups of granitoids (Mahé and Praslin groups) are thought to be derived from a juvenile mantle-derived component with variable amounts of an ancient, possibly Archean, silicic source constituent or contaminant (Ashwal *et al.*, 2002). The Mahé and Praslin types in the Seychelles are correlated with the Jalore and Siwana types of the Malani province in India, respectively (Ashwal *et al.*, 2002).

Finally, the silicic magmatism of the Daraina Complex of Madagascar closely resembles both the Mahé group granitoids (Seychelles) and the Jalore type granitic and rhyolitic rocks (India) (Ashwal *et al.*, 2002).

Link with the 755 Ma Mundine Well LIP of western Australia

Potentially related magmatism (the 755 Ma Mundine Well mafic dyke swarm) is also present in western Australia (Fig. 8.5), but paleomagnetic data indicate a latitudinal separation of nearly 25° from that in the Seychelles and India (Indian plate) (Gregory *et al.*, 2008; see also Meert *et al.*, 2013). However, a connection cannot be ruled out as LIP events can extend over several thousand kilometers (particularly in their dyke-swarm component; Chapter 5). A potential link with the 755 Ma Shaba bimodal pulse of magmatism in South China is discussed in the next section.

8.4.2 South China events (825–755 Ma)

General characteristics and age constraints

The *c.* 1.0 Ga Sibao orogeny marks the collisional boundary between the Yangtze craton and the Cathaysian block that formed the South China craton. Subsequently, the South China craton was affected by an impressive prolonged period of bimodal magmatism between 825 and 755 Ma, which has been divided into multiple pulses (Ernst *et al.*, 2008; Li *et al.*, 2008a): the 825 Ma Guibei, the 800 Ma Suxiong–Xiaofeng, the 780 Ma Kangding, and the 755 Ma Shaba pulses. There is also evidence for an earlier pulse at *c.* 880–850 Ma (that includes the Shenwu dolerite dykes of northern Zhejiang province, South China; Li *et al.* (2008b). (See also Li *et al.*, 2003a and Li *et al.*, 2013).

There has been debate regarding setting of this protracted magmatism given its significant silicic component including both I- and S-type granites. An orogenic origin (island-arc or post-collision collapse) has been suggested (e.g. Zhou *et al.*, 2006; Zheng *et al.*, 2007; Wang *et al.*, 2012a, b; see also the discussion in Deng *et al.*, 2013).

However, a mantle-plume/LIP model for this protracted multi-pulse episode of bimodal magmatism is supported on the basis of several lines of evidence (Li *et al.*, 1999, 2003b, 2005, 2008a; Ernst *et al.*, 2008; Lu *et al.*, 2008; Deng *et al.*, 2013; Li *et al.*, 2013): syn-magmatic doming, the bimodal and intraplate nature of the magmatism, and the associations with continental rifting (e.g. Li *et al.*, 1999; Li *et al.*, 2003b; Li *et al.*, 2013). More recent work by Wang *et al.* (2007, 2008a) demonstrates that both 826 ± 3 Ma Yiyang komatiitic basalts and the 811 ± 12 Ma upper Bikou basalts have mantle potential temperatures over 1500 °C (Li *et al.*, 2013), which are nearly 200 °C in excess of the ambient mantle.

Comparison with the LIP record of other crustal blocks provides support for the "Missing-Link" configuration for the Rodinia supercontinent in which the South China block is placed between western North America and Australia, thereby providing a link to both the 780 Ma Gunbarrel LIP of western Laurentia and the 825 Ma Gairdner–Willouran LIP of southern and central Australia (Li *et al.*, 2008; Fig. 8.6).

It is also significant that the same pulses of 825–760 Ma bimodal magmatism observed in the South China craton are also found in the Tarim craton (Li *et al.*, 2008a; Ernst *et al.*, 2008; Dan *et al.*, 2014). This barcode matching is consistent with more recent research that connects the Qaidam and Qilian blocks with the Tarim block to the west and the South China block to the east into a single terrane, termed the South-West China United Continent (SWCUC; Song *et al.*, 2012). In support of

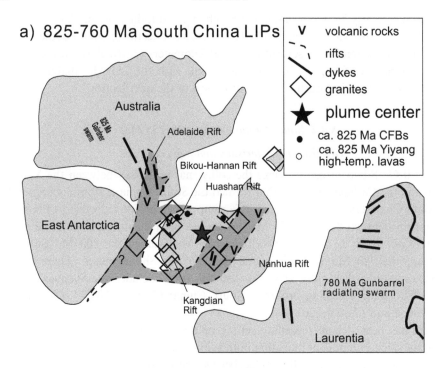

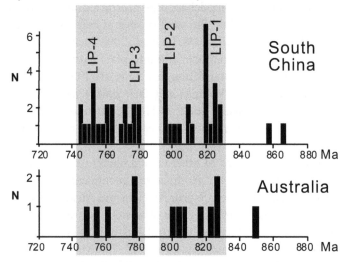

Figure 8.6 The 825–760 Ma multi-pulse South China LIP(s) that have a significant granitoid component. (a) Map showing distribution of mafic dykes, rifts, granitoids and inferred mantle plume center. (b) Age distribution of four pulses discussed in the text. N = number of U–Pb age determinations. Modified after Li *et al.* (2008a, 2013), and Li (2011).

this model 825 Ma bimodal magmatism is not only present in the South China and Tarim blocks but also in the intervening Qaidam, Qilian, and Alashan blocks (Li *et al.*, 2005a; Ernst *et al.*, 2008; Lu *et al.*, 2008; Pirajno *et al.*, 2009), which collectively constitute a single LIP (Song *et al.*, 2012; S.G. Song, personal communication, 2013; Xu and Song, 2014). Rifting occurs along the Kangdian and Nanhua rifts (Fig. 8.6); however, this is associated with the 810–800 Ma and 790–760 Ma pulses. It is notable that the earliest pulse, at *c.* 825 Ma, precedes rifting.

Composition

The composition of the mafic–ultramafic magmatism is discussed first, and then that of the associated silicic magmatism. As summarized by Li *et al.* (2013) the 860–750 Ma basaltic rocks are predominantly of a subalkaline series and a subordinate amount belong to the alkaline series. The subalkaline basalts are dominantly of tholeiitic nature; calc-alkaline basalts are rare. Fe-poor, Si-enriched, and Nb–Ta-depleted characteristics of some basaltic rocks reflect contributions from the subcontinental lithospheric mantle (SCLM). Wang *et al.* (2009) interpret that the SCLM was metasomatized by subduction-induced melts/fluids during the prior 1.0–0.9-Ga Sibao orogeny. The continental intraplate geochemical signatures (such as oceanic island basalt (OIB)-type), high-mantle potential temperatures, and recycled components (e.g. eclogite in the source) suggest the presence of a mantle plume beneath the Neoproterozoic South China block. There is an age progression in composition and setting: the early phases of basaltic rocks (825–810 Ma) were most likely formed by melting within the metasomatized SCLM heated by the rising mantle plume. The subsequent continental rifting at *c.* 810–800 Ma and 790–760 Ma allowed adiabatic decompression and partial melting of an upwelling mantle plume at a relatively shallow depth to form the widespread syn-rifting basalts.

825–820 Ma pre-rifting granitoids of the Guibei LIP pulses are widespread in the Yangtze craton, South China. As shown by Li *et al.* (2003a), mineralogical, petrographic, and geochemical characteristics indicate that there are two types of peraluminous, S-type granitoids: muscovite-bearing leucogranites (MPG) and cordierite-bearing granodiorites (CPG); and two types of I-type granitoids: K-rich calc-alkaline granitoids (KCG) and tonalite–trondhjemite–granodiorite (TTG). Sm–Nd isotopic data suggest that all were generated by partial melting of various crustal rocks without appreciable involvement of new mantle-derived magmas, i.e. pelitic and psammitic sources for MPG and CPG, and tonalitic to granodioritic and amphibolite sources for KCG and TTG, respectively.

In addition, it has been more recently discovered that some A-type granites are also present and are associated with the 780 Ma pulse. Based on the compositions and geochronological and Hf isotope data of Neoproterozoic and older igneous rocks in the region, Wang *et al.* (2010) suggested that the A-type northern Daolin-shan granites were most probably produced by high-temperature melting of slightly earlier tholeiitic rocks underplated in the lower crust as a result of the *c.* 780 Ma mantle plume (i.e. the *c.* 780 Ma Kangding LIP pulse).

8.4.3 Gawler Range volcanics (1590 Ma), Australia

General characteristics

The Gawler Range volcanics (GRV) of the Gawler craton are dominated by silicic (> 60% SiO_2) magmas and have only a subordinate mafic component (Fig. 8.7; Blissett *et al.*, 1993; Allen *et al.*, 2008; McPhie *et al.*, 2011; Pirajno and Hoatson, 2012; Wade *et al.*, 2012). The Benagerie Volcanic Suite (BVS) of the Curnamona craton has been recently dated and geochemically matched with the GRV (Wade *et al.*, 2012). A conservative estimate for the total preserved volume of the combined GRV–BVS is in the order of 0.11 Mkm^3. At least three individual silicic units are present, each representing 1000–3000 km^3. U–Pb zircon geochronology indicates emplacement over only a few million years at 1590 Ma. Associated Hiltaba suite granites are roughly coeval (extending to about 20 Ma younger). Both the volcanics (rhyolites) and the granites are compositionally A-type. This LIP is also host to the huge Olympic Dam IOCG (iron oxide–copper–gold) ore deposits (Chapter 16; Fig. 8.7). Furthermore, northwest Laurentia contains 1590 Ma sills, and the Wernecke Breccia, which may represent a portion of the Gawler Range LIP that rifted away (Hamilton and Buchan, 2010; Fig. 14 in Crawford *et al.*, 2010).

Age constraints

Isotope-dilution thermal-ionization mass-spectrometric (ID-TIMS) dating of the Waganny Dacite, one of the lowermost units of the lower GRV yielded an upper intercept date of 1591 ± 3 Ma, interpreted as the age of crystallization (Fanning *et al.*, 1988). Similarly, U–Pb ID-TIMS dating of the Yardea Dacite, the uppermost unit of the upper GRV in the Lake Everard region, yielded a crystallization age of 1592 ± 3 Ma (Fanning *et al.*, 1988), revealing that the entire GRV erupted within only a few million years.

Composition

The following summary is after Wade *et al.* (2012); see also Blissett *et al.* (1993) and Allen *et al.* (2008). The GRV are divided into lower and upper sequences. The lower GRV comprises bimodal mafic and silicic volcanic rocks, ranging in composition from basalt, andesite, dacite, rhyodacite, to rhyolite. The lower GRV forms at least three discrete volcanic centers and fractionation trends indicate that silicic units in these centers are direct fractionates of the mafic units. Assimilation and fractional crystallization (AFC) and crustal contamination processes also modified the geochemical signature of the silicic volcanic rocks (Fricke, 2005). The upper GRV consists of extensive, flat-lying, relatively undeformed silicic (rhyolite–rhyodacite–dacite)

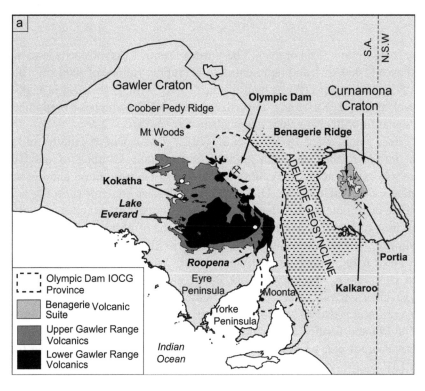

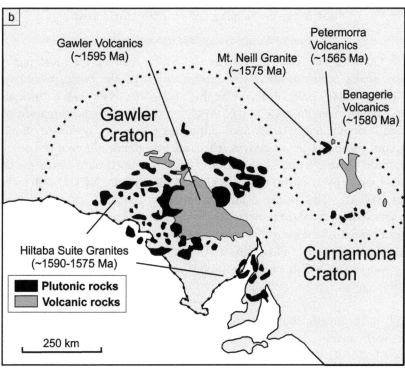

Figure 8.7 Distribution of 1590 Ma Gawler Range LIP consisting of bimodal volcanism, Hitalba granites, and Olympic Dam IOCG deposits. These suites are associated with the Gawler and Curnamona cratons, which are separated by the younger *c.* 825 Ma Gairdner–Willouran LIP and rifting event, that separated the two cratons along the Adelaide "Geosyncline." From Wade *et al.* (2012). With permission from Elsevier. Digital version of original figure kindly provided by C. Wade.

volcanic rocks up to *c*. 1.5 km thick. The Yardea Dacite is the most extensive unit, is exposed over 12 000 km^2, and represents a total erupted volume of 3000 km^3 (Blissett *et al.*, 1993). The upper GRV is high in silica (> 65% SiO_2) enriched in high-field-strength elements (HFSEs) and rare-earth elements (REEs), and has a predominantly crustal signature with εNd values ranging between −5.8 and −2.2.

As further summarized by Wade *et al.* (2012) the associated Hiltaba suite is a bimodal intrusive suite, although granites predominate. Granites of the Hiltaba suite are widespread across the central and southern Gawler craton and are mostly fractionated, enriched in HFSE, U, Th, and K, with silica contents generally > 70%. Nd isotopic data indicate that the more evolved Hiltaba suite granites are linked to Archean host rocks. Also related are mafic intrusions of hornblende-bearing quartz monzodiorite, quartz monzonite, and granodiorite from the northeastern Gawler craton. These mafic units have SiO_2 contents of < 65%, and are characterized by elevated TiO_2, Fe_2O_3, MgO, P_2O_5, CaO, Ba, Sr, and Zr. Finally, a subordinate suite of S-type intrusive rocks, the Munjeela suite, known chiefly from the western Gawler craton are also linked to the Hiltaba suite. The Munjeela suite intruded at *c*. 1585 Ma and formed by partial melting of a metasedimentary protolith, and the thermal input can be linked to the plume inferred to be responsible for the Gawler LIP.

8.4.4 Xiong'er–Taihang LIP (c. 1780 Ma), China

General characteristics and age constraints

The 1.78 Ga Xiong'er Volcanic Province (XVP) is one of the most important magmatic events occurring after the amalgamation of the North China craton (NCC) (Peng *et al.*, 2008; Peng, 2010; Fig. 8.8). The XVP has a thickness of 3–7 km, extends over an area > 0.06 Mkm^2, along the southern margin of the NCC, and is dominated by thick and continuous lava flows, with rare, thin, sedimentary and volcaniclastic interlayers. There are also minor pillowed lavas.

In addition, there is a coeval dyke swarm extending northward across the craton (Taihang–Lvliang dyke swarm or North China Dyke Swarm, NCDS) with an extent of about 1000 km. This swarm exhibits a radiating geometry after correction for block rotations in parts of the pattern based on paleomagnetic data (Peng, 2010). The convergence point of the swarm is coincident with the locus of Xiong'er volcanics. The radiating swarm was followed by a younger north-northwest–south-southeast-trending swarm with a distinct composition (the Beitai swarm, 1765 Ma).

Composition

The XVP is chemically tholeiitic and varies from basalt to andesite, dacite and rhyolite, with andesitic compositions being dominant; thus, the XVP does not resemble a bimodal association. It consists of two volcanic cycles, both varying from mafic–intermediate to more silicic compositions.

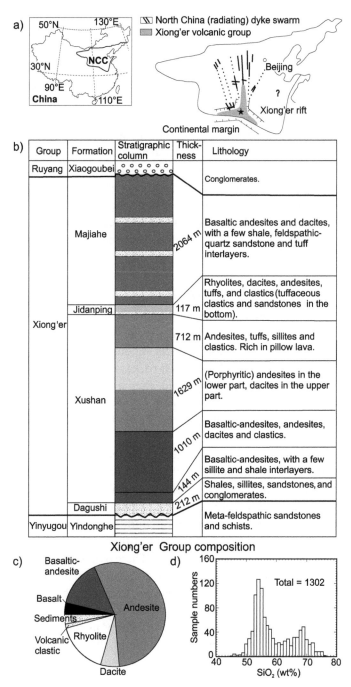

Figure 8.8 The 1790 Ma Xiong'er–Taihang LIP of North China. (a) Location of North China and the general distribution of the radiating swarm, triple-junction rifting, and inferred plume center. (b) Stratigraphy of the Xiong'er Group and its composition. (c) Pie chart showing the dominance of andesitic compositions in the volcanic rocks. (d) SiO₂ (wt%) histogram of the volcanic-bearing formations (the Xushan, Jidanping, and Majiahe Formation) in the Xiong'er Group. Modified from Peng *et al.* (2008) and Peng (2010); and Zhao *et al.* (2002). Digital version of original figures kindly provided by P. Peng.

Dyke compositions vary from basalt to andesite and dacite, but are dominantly mafic, and comprise two series of magmatism. Previous studies revealed that the NCDS recorded assimilation and fractional crystallization of an enriched mantle (EM) I-type magma source, with a minor DM contribution in the younger magmas. Both syn-collisional and intracontinental anorogenic environments have been proposed. Spatial and petrogenic correlations suggest a cogenetic relationship between the NCDS and XVP and, considered together, they define a LIP of 0.1 M km^2 in area and 0.1 M km^3 in volume, which is also notable for its continuous compositional range from mafic to felsic (with no gap at intermediate compositions). The petrology is explained by a common magma source that undergoes a silica-poor and iron-enriched fractionation trend at depth followed by a silica-rich and iron-poor fractionation trend in shallow-level magma conduits (dykes) and surface lavas.

Origin

One hypothesis for the origin of the Xiong'er volcanics represents synorogenic Andean style collision along the southern margin of the craton (e.g. He *et al.*, 2009). The hypothesis is mainly based on the subduction-influenced geochemistry (e.g. depletion in HFSE) of North China XVP rocks, and Andean-style calc-alkaline volcanism. Nevertheless, the chemistry could alternatively be interpreted as being affected by assimilation of the continental crust, or inherited from a fertilized mantle region. Also, widespread albitization occurs in the XVP, and may also have affected some of the TLS dykes.

The intraplate LIP context was confirmed with the recognition of the coeval North China radiating dyke swarm extending northward across the craton. The presence of the radiating dyke swarm is diagnostic of plume origin for the event (see the discussion in Chapters 5 and 15). Furthermore, the systematic northward branching of the dykes indicates a magma flow direction from south to north (Peng, 2010). This is consistent with a plume center at the focal point of the swarm, which coincides with the locus of XVP volcanics and rifting (Fig. 8.8). These dykes have compositions that vary from basalt to andesite and dacite, but are dominantly mafic. In addition to the age correlation there are petrogenetic correlations (including both being tholeiitic) suggesting a cogenetic relationship between the NCDS and XVP; considered together they define a LIP of > 0.1 Mkm^2 in area and > 0.1 Mkm^3 in volume, which is notable for its continuous compositional range from mafic to silicic (with no gap at intermediate compositions).

8.5 Speculative SLIPs

Numerous other examples have potential to be SLIPs but require more study to confirm their size duration and intraplate character. Examples include the *c.* 300 Ma

Angara–Vitim batholith of Transbaikalia (just east and southeast of Lake Baikal), which covers an area of 150 000 km² (e.g. Yarmolyuk *et al.*, 1997; Tsygankov *et al.*, 2007, 2010; Litvinovsky *et al.*, 2011); the coeval suites of granites and picritic intrusions of age *c.* 500 Ma associated with the central Asian fold belt in Mongolia (e.g. Izokh *et al.*, 2010); the Mid-Proterozoic Eastern (*c.* 1500–1420 Ma) and Western (1400–1340 Ma) granite–rhyolite provinces of North America (Kay *et al.*, 1989). The granite "blooms" also need to be considered; these are widespread crust-derived granite (sensu stricto) plutons, that were emplaced in relatively short (10–30 Ma) pulses that followed, by 5–20 Ma, regional tectonism in numerous Archean granite–greenstone belts (e.g. Percival and Pysklywec, 2007). These post-tectonic granites manifest widespread crustal melting that is not subduction-derived; but neither are they easily explained by lithospheric removal (delamination) or the arrival of a mantle plume, given their association with cratonization and the formation of stable lithospheric roots.

8.6 Discussion

Herein I discuss some additional thematic aspects related to SLIPs, related to the nature of the plumbing system, the importance of fusible lower crust as a source, and other aspects concerning the origin and recognition of SLIPs.

8.6.1 Eruptive sources

In Chapter 5 I reviewed the dolerite dyke and sill plumbing system for LIP provinces and noted the importance of fissures (dykes) as sources and, to a lesser extent, shield volcanoes. Here I discuss some aspects of the plumbing system for SLIPs and specifically consider the dominant role of calderas and possibly important role of silicic dyke swarms, based on the work of Bryan (e.g. Bryan, 2007; Bonin, 2007; Fig. 8.9).

Caldera-type complexes are the main source for the extensive ignimbrite fields of SLIPs (e.g. Bryan, 2007), but are part of a larger multiple-vent volcanic system that includes numerous extra-caldera (and intra-caldera) effusive vents for basaltic and silicic magmas (e.g. Bryan *et al.*, 2000, 2002; Bryan, 2007). Caldera dimensions range between 10 and 30 km in diameter, where calderas with 30 to 40 km and possibly up to 100 km diameters have been inferred for the Chon Aike province (Aragón *et al.*, 1996; Riley *et al.*, 2001). Many cauldrons are well exposed in the more deeply dissected Permo-Carboniferous Kennedy–Connors–Auburn SLIP of northeastern Australia, with individual cauldrons ranging from 10 to 40 km in diameter (Bain and Draper, 1997; Bryan *et al.*, 2003). Along the Snake River plain of the Columbia River LIP, a line of calderas likely mark the > 500-km track of the tail of the Yellowstone plume between 17 Ma and the present. In more deeply eroded SLIPs the location of caldera are likely given by the location of high-level silicic intrusions.

With greater erosion the deeper plumbing system is exposed. This is illustrated in Fig. 8.9 for A-type silicic magmatism. As the depth of erosion increases, the intrusive

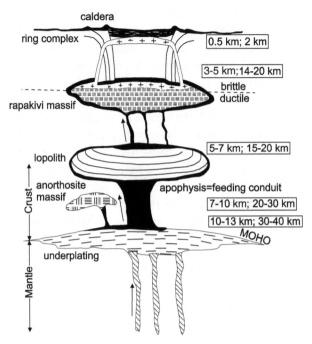

Figure 8.9 Plumbing system of A-type granite magmatism. The depth values are only approximate: Surface level: caldera volcano, feeder radial and ring dykes, and domes. Subvolcanic level: ring complex, cone sheets, and dyke swarms. Intermediate level: mixed cumulate–liquid rapakivi granite batholiths, and quartz–alkali feldspar cumulates (AMCG complex). Deep plutonic level: mafic to intermediate layered lopoliths from which silicic residual liquids can escape, and associated massif anorthosite complexes. Underplate zone (at the 30–40-km-deep mantle–crust boundary): ultramafic layered sheets. Partial melting source area zones are located at greater depths within the upper mantle. From Figure 8 in Bonin (2007). With permission from Elsevier.

style transitions through rapakivi granites, AMCG suites, mafic–intermediate intrusions and anorthosites.

In addition, ignimbrite-forming eruptions also occur from extensive volcano-tectonic fissures producing welded pyroclastic dykes (Bryan, 2007). Dykes of both silicic and mafic composition occur in the Whitsunday (Bryan *et al.*, 2000), Sierra Madre Occidental (Aguirre Díaz and Labarthe Hernández, 2003), Chon Aike (Pankhurst *et al.*, 1998), and the 320–280 Ma Kennedy–Connors–Auburn (Bryan *et al.*, 2003) SLIPs.

8.6.2 Distinguishing SLIPs from orogenic silicic rocks

As noted above (Section 8.3.1) SLIPs can be generated in a wide variety of settings but their geochemical characteristics can often be misinterpreted in terms of the tectonic setting in which the magmas were emplaced. The chemistry of granitic magmas strongly reflects the composition of their crustal sources, and thus provides an unreliable guide to the tectonic setting at the time of SLIP magmatism. So, while there may be A-type magmas (indicative of an intraplate setting), there may also be significant volumes of S- or I-type silicic rocks. Therefore, the composition of silicic magma is not a diagnostic regarding the setting. Other characteristics that are useful to distinguish SLIPs from potentially significant volumes of silicic igneous rocks produced in other tectonic settings are the regional extent, eruptive volumes, whether there is a significant presence of A-type silicic magma, and whether the associated mafic magmatism has intraplate character.

This latter point is significant because it identifies an asthenospheric mantle source/driver for magmatism, both in terms of heat and material additions to the LIP system, and indicates mantle processes and source regions unrelated directly to any active subduction. In the SLIP examples, this asthenospheric mantle component is similar to those found in LIPs (e.g. Cameron *et al.*, 1989) or younger hotspot-related volcanism (Ewart *et al.*, 1992; Bryan *et al.*, 1997, 2012). It has thus been important in recognizing LIPs such as the Columbia River and SLIPs such as the Sierra Madre Occidental as being unrelated to active subduction along the nearby plate margin, despite their apparent back-arc geographic setting. Further discussion of the geochemical character of silicic magmatism associated with LIPs is presented in Section 10.10.3.

8.6.3 Link of SLIPs with rifting and breakup

Like their LIP counterparts, many SLIPs are associated with rifting and continental rupture (Fig. 8.10). SLIPs had initially been grouped into two extensional settings: rifting breakup (e.g. Whitsunday) and back-arc rifting (Sierra Madre Occidental) (Bryan *et al.*, 2002; Pankhurst *et al.*, 2011). But recent studies are now establishing the intraplate character of the Sierra Madre Occidental SLIP, and that it was the pre- to syn-rift event to the opening of the Gulf of California (Bryan *et al.*, 2014;

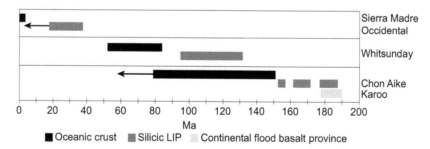

Figure 8.10 Formation of SLIPs on volcanic rifted margins and the timing of formation of oceanic crust. As with the continental flood-basalt provinces, the SLIPs precede breakup and formation of oceanic crust. Modified from Figure 2 in Bryan (2007).

Bryan and Ferrari, 2013; Ferrari *et al.*, 2013). Consequently, SLIPS are now considered to be strictly intraplate and rift-related events that, like some LIPs (Columbia River, Ferrar part of Karoo–Ferrar), may occur in proximity to actively subducting plate boundaries, and their geographic setting is in the back-arc (see also discussion on back-arc setting in Section 15.2.5).

8.6.4 Role of fusible lower crust

The large-volume silicic component has compositions that are predominately dacite–rhyolite often near the hydrous granite minimum. It originates because the heat flux from the basaltic magma results in partial melting of the lower to mid crust, unless that crust has previously been melted and is already refractory (e.g. Ramos Rosique, 2012; Bryan and Ferrari, 2013; Bryan *et al.*, 2014). In some cases, the isotopic character of the SLIP matches that of some associated mafic magmatism, suggesting remelting of the mafic underplate. A key to generating the large volumes of silicic magma preserved in SLIPs is the prior formation of hydrous, highly fusible lower crust (Bryan *et al.*, 2002). Specifically, partial melting (*c.* 20%–25%) of this fusible crust will generate melts of intermediate to silicic composition. Therefore, according to Bryan *et al.* (2002), prior subduction event(s) are crucial to the development of a hydrous, fusible lower crust (i.e. mafic–intermediate, transitional to a high-K calc-alkaline meta-igneous/sedimentary source). So the basement to the Whitsunday (e.g. Cook *et al.*, 2013), Chon Aike (Riley *et al.*, 2001), and, in large part, the Sierra Madre Occidental (Ferrari *et al.*, 2007; Bryan *et al.*, 2008) SLIPs comprises Paleozoic–Mesozoic igneous and sedimentary rocks accreted and/or deposited along the continental margin. These crustal materials have been generated by subduction up to hundreds of million years prior to the emplacement of the SLIP, and the subduction-generated igneous rocks may reside as an underplate or as widespread intrusions in the lower crust. The combined geochemical and isotopic data therefore indicate the importance of relatively young, non-radiogenic calc-alkaline crust in SLIP petrogenesis.

Some SLIPs occur in a back-arc setting, but this link may be more opportunistic than genetic. Specifically, SLIPs that have a linear distribution (up to 2500 km) can be associated with the location of paleo- or more recent subduction zones. These represent zones of fusible lower crust, which will preferentially melt in the presence of a broader thermal input from a plume. Thus, the linear distribution and apparent back-arc setting may simply reflect the distribution of the underlying fusible lower crust of the associated paleosubduction zone.

8.6.5 Silicic-magma density barriers

Magmatic systems that produce large volumes of relatively low-density, silicic magma tend to inhibit the ascent of higher-density, mafic magma, and as such the surface expression of such a system is biased towards the silicic igneous compositions (e.g. Pankhurst *et al.*, 1998; Christiansen *et al.*, 2007). Given the regional scale of magmatism and the volumes of crust-derived silicic partial melts, extensive lower to midcrustal density barriers would be expected in SLIP events, preventing the general rise and surface expression of mafic volcanism. If these silicic magmas were of suitably low viscosity, this trapping of mafic magma and eruption of silicic compositions would be exaggerated. However, although locally the system may be deficient in a mafic component, in some cases a broader look reveals a regional link with mafic magmatism (LIP or potential LIP fragment; see next subsection). In a SLIP, mafic volcanism also becomes more common over time, with basaltic lavas occurring in the younger parts of the eruptive stratigraphy (Fig. 8.1) or observed to cap and overlie thick rhyolitic ignimbrite successions (e.g. Cameron *et al.*, 1989). This increased surface expression of mafic volcanism in SLIPs is also tied to the active rift environment where crust-penetrating extensional faults begin acting as conduits for mafic magmas to reach the surface and the volcanic expression subsequently becomes more bimodal in character (e.g. Bryan *et al.*, 2000, 2008, 2012, 2014; Ferrari *et al.*, 2013).

8.6.6 Correlations between SLIPs and LIPs

In many cases, SLIPs can be spatially and temporally related to LIPs, particularly given the more regional perspective becoming available from robust paleocontinental reconstructions (Table 8.2). For example, the Chon Aike SLIP is spatially and temporally linked with the Karoo–Ferrar LIP (Pankhurst *et al.*, 1998, 2000). The Malani SLIP (India) is coeval with mafic magmatism in the Seychelles, South China, Korea (Ogcheon), and Australia (Mundine Well) (Ernst *et al.*, 2008). The 1590 Ma Gawler Range SLIP of south-central Australia is linked with mafic sills in the formerly adjacent northwestern Laurentia (Hamilton and Buchan, 2010). The largest SLIP (Whitsunday) and the largest known oceanic plateau/LIP (Ontong–Java–Manihiki–Hikurangi plateau Complex; Taylor, 2006) were emplaced at the same time, *c.* 120 Ma, with the latter occurring offshore from the fragmenting continent (Fig. 8.4), although the geodynamic link is not apparent.

Such linkages between SLIPs and LIPs offer the opportunity to understand the broader architecture of LIPs and to consider controls on the distribution of mafic and silicic portions. Factors of importance include the distribution of the hot plume center and the distribution of fusible lower crust (associated with a prior subduction history along the continental margins).

8.7 Summary

This chapter reviews the robust association of silicic magmatism with LIPs. Two aspects were considered, a LIP sensu stricto (i.e. mafic dominated) with a subordinate amount of associated silicic magmatism and, dominantly, Silicic LIPs (SLIPs).

Silicic magmatism in a LIP: The presence of silicic magmatism, often of major volumes (up to *c*. 50 000 km^3) is common in mafic–ultramafic LIPs through time. This silicic magmatism is typically of a high-temperature type (> 1000 °C), either high-temperature rhyolites or A-type granites, and is explained by high degrees of differentiation (fractionation ± assimilation) of basalts, or by melting of anhydrous and refractory lower crust caused by the magmatic underplating and consequent crustal heating of the LIP.

SLIPs: Some major provinces of predominant silicic magmatism are present in non-subduction settings, e.g. the Whitsunday, Chon Aike, Gawler SLIPs, etc. (Table 8.2). These typically are associated with melting of the hydrous lower crust to produce LIP-scale volumes and areal extents of dominantly calc-alkaline I-type and lesser S-type silicic magmas, although most SLIPs also have a significant proportion of silicic magmas showing A-type/within-plate chemical affinities.

A crucial point is that SLIPs may identify "hidden" mafic magmatism. In other words, they can be indirect evidence for mafic–ultramafic magmatism of LIP scale at depth (underplate) that caused lower-crustal anatexis and ascent of silicic magmas (in which cases the mafic–ultramafic component was stalled in the lower crust). "Hidden" mafic magmatism at depth is particularly likely in cases in which there is a coeval but spatially separated mafic LIP component (e.g. the Chon Aike SLIP vs. the Karoo–Ferrar mafic LIP). It is possible that some SLIPs may simply represent a thermal pulse (e.g. caused by a plume, or other mechanism, Chapter 15) that *would have* produced LIP magmatism but for the efficient occurrence of anatexis of lower crust.

9

Links with carbonatites, kimberlites, and lamprophyres/lamproites

9.1 Introduction

There appears to be a close association between Large Igneous Provinces (LIPs) and carbonatites, and in some cases with kimberlites, and lamprophyres and lamproites. In this chapter these links are explored and some of the implications are considered. This chapter sets the stage for inclusion of these small-volume magma types in a broader consideration of the origin of LIPs in Chapter 15.

9.2 Carbonatites and LIPs

Many provinces of carbonatites (Fig. 9.1) are linked both spatially and temporally with LIPs (Bell, 2001; Ernst and Bell, 2010), and key examples are summarized in Table 9.1. The frequency of this LIP–carbonatite association suggests that LIPs and carbonatites might be considered as different evolutionary "pathways" in a single magmatic process/system. This aspect is further explored in Section 15.4.3, but in this chapter we demonstrate the correlation and much of the discussion below is based on Ernst and Bell (2010).

9.2.1 Carbonatites (and associated alkaline intrusions)

Although carbonatites are volumetrically insignificant, they are found on all continents and range in age from 3 billion years to the present. Carbonatites are made up of at least 50% carbonate minerals, and may or may not be associated with alkaline silicate rocks (e.g. Bell, 1989, 2001; Bell and Keller, 1995; Le Maitre, 2002; Mitchell, 2005; Woolley and Church, 2005; Woolley and Kjarsgaard, 2008a, b). With distinctive trace-element chemistries, many carbonatites contain unusual accessory minerals, and some are associated with economic mineral deposits (Woolley and Kjarsgaard, 2008a, b; Chakmouradian and Zaitsev, 2012; see also Section 16.2.3). The main physical property that separates carbonatitic melts from other melts is viscosity. Because these melts are ionic liquids, there is little or no polymerization and hence these magmas have very low viscosities (Treiman, 1989). Carbonatitic melts can

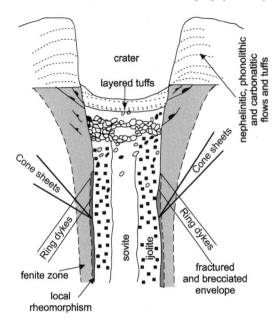

Figure 9.1 Structure of a carbonatite. From Figure 2 in Bell (2005).

therefore sample large mantle volumes (McKenzie, 1985), and estimates of ascent rates based on fluid-flow calculations suggest that carbonatitic melts can migrate rapidly to the surface at speeds of 20 to 65 m/s (Genge *et al.*, 1995) thus inhibiting contamination with surrounding wall rocks.

Once considered rare, carbonatites are now widely recognized, with 527 occurrences reported (Woolley and Kjarsgaard, 2008a, b). Carbonatites are found on all continents but only three oceanic island occurrences have so far been found: Kerguelen and the Cape Verde and the Canary Islands (Kogarko, 1993; Bell, 2001; Moine *et al.*, 2004; de Agnacio *et al.*, 2006).

Although diverse origins have been proposed for carbonate-rich magmas, including slab break-off, and lithospheric delamination or lithospheric melting (e.g. Woolley, 1989; Bailey, 1993), there is increasing evidence that many carbonatites (and associated alkaline rocks including nepheline syenites) are linked with rifts (e.g. Burke and Dewey, 1973; Bailey, 1974, 1977, 1992; Burke *et al.*, 2003), and with LIPs and mantle plumes (e.g. Gerlach *et al.*, 1988; Simonetti *et al.*, 1995; Bell, 2001; Ernst and Bell, 2010).

9.2.2 Examples of carbonatites associated with LIPs

Herein I review and focus on the link with LIPs (Fig. 9.2). I first discuss examples of the association of particular LIPs and carbonatites, and then I discuss general characteristics and implications of that association.

Table 9.1 *Selected LIPs and their associated carbonatite complexes*

LIP name (ages, to nearest 5 Ma)	Associated carbonatite complexes
Afro-Arabian (45–0 Ma); Section 3.2.3	UGANDA: Bukusu; Napak; Tororo, Toror; KENYA: Buru Hill probably coeval with Tinderet; Homa Mountain; Kisingiri and Rangwa, North and South Ruri, Nyamaji, Shombole, Tinderet and Londiani, Wasaki Peninsula; TANZANIA: Basotu; Hanang and Balangida, Kerimasi, Kwahera, Monduli-Arusha tuff cones, Mosonik, Oldoinyo Lengai, Sadiman
Deccan (65 Ma); Section 3.2.5	Amba Dongar, Danta-Langera-Mahabar, Mundwara, Sarnu-Dandali (Barmer)
Paraná–Etendeka (135 Ma); Section 3.2.6	SOUTH AMERICA: Anitápolis, Barra do Itapirapuã, Chiriguelo, Ipanema, Itanhaem, Itapirapuã, Jacupiranga, Juquiá; AFRICA: Messum
Siberian Trap (250 Ma); Section 3.2.8	MAIMECHA-KOTUI: Bor-Uryakh, Guli, Kugda, Magan, Odikhincha; ANABAR PROVINCE: Orto-Yrigakhskoe, Tundrovoye and Nomottookhskoe; SOUTHWEST SIBERIA (CHADOBETSKAYA REGION) Chadobetskaya
Kola–Dnieper (*c.* 380–360 Ma); Section 9.2.2	Afrikanda, Kovdor, Vuoriyarvi, Seblyavr, Salmagorskii, Kontozerskii, Turiy Peninsula, Kandaguba, Khibina, Ozernaya Varaka, Sokli, Sallanlatvi
CIMP (Central Iapetus Magmatic Province) (615–555 Ma); Section 9.2.2	EASTERN LAURENTIA: Aillik Bay, Arvida (Chicoutimi), Baie-Des-Moutons (Mutton Bay), Brent, Callander Bay, Manitou Islands, St-Honoré, Torngat Mountains (Abloviak Fjord); GREENLAND: Quigussaq (Umanak), Sarfartoq; BALTICA (NORWAY AND SWEDEN): Avike Bay, Alnö, Fen, Lillebukt, Seiland, Sørøy
Keweenawan (1114–1085 Ma); Section 3.5.2	Big Beaver House, Firesand River, Nemegosenda, Lackner Lake, Seabrook Lake, Schryburt Lake, Sullivan Island, Prairie Lake, Valentine
Circum-Superior (*c.* 1880–1870 Ma); Section 9.2.2	KAPUSKASING STRUCTURAL ZONE: Argor, Borden, Cargill, Goldray, Spanish River; LABRADOR TROUGH: Lac Castignon; NORTHWEST SUPERIOR CRATON: Carb Lake
Bushveld (2060–2055 Ma); Sections 5.4.1 and 9.2.2	Phalaborwa, Shiel

Note: modified after Ernst and Bell (2010) and including only those carbonatites confidently linked to the LIP. Details on all these carbonatites are provided in the global database of Woolley and Kjarsgaard (2008). LIP locations are shown in Figure 1.6.

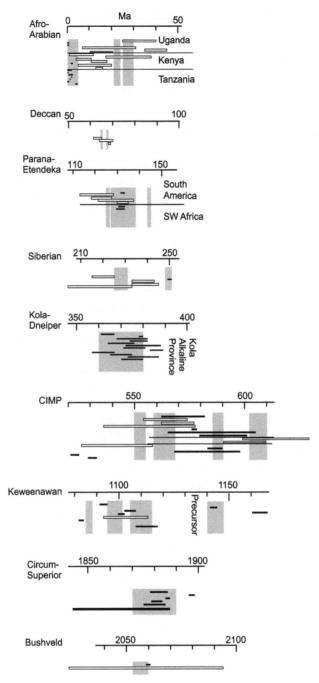

Figure 9.2 Summary of age distribution of carbonatites with respect to main episodes of LIP events. Gray rectangles indicate main episodes of LIP magmatism. The range is based on given ± 2 sigma uncertainties, or given ranges. In cases in which an uncertainty is not provided, a value of ± 10 Ma is arbitrarily assigned, which is high but appropriately conservative. U–Pb, Ar–Ar, and Rb–Sr ages are considered more reliable and are presented as solid black rectangles. K–Ar and fission track ages are left as open rectangles, except for the less than 10 Ma ages of the Afro-Arabian LIP, which are shown as solid black rectangles. Modified from Ernst and Bell (2010). With kind permission from Springer Science and Business Media.

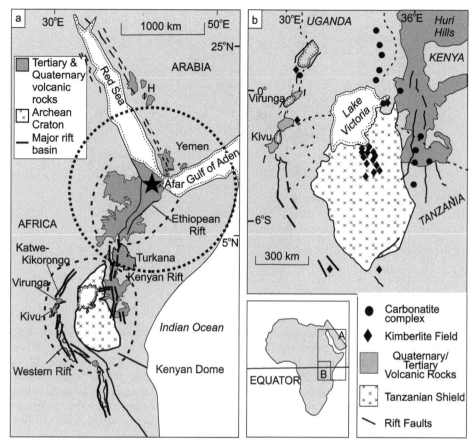

Figure 9.3 Carbonatites and kimberlites of the Afro-Arabian LIP: (a) for the full region; (b) portion of the region focused on Kenya dome/Tanzanian craton. See inset for locations of (a) and (b). Star and circle locate inferred center and 1000 km radius of underlying mantle plume. Carbonatites distribution from Figure 3 in Ernst and Bell (2010). With kind permission from Springer Science and Business Media. Kimberlite distribution, courtesy of D. Schissel.

Carbonatites associated with the Afro-Arabian LIP (45–0 Ma)

The Afro-Arabian LIP, also known as the Afar LIP (Section 3.2.3), is the youngest major LIP and was emplaced in the southern Arabian Peninsula and northeastern Africa in several pulses between 45 and 0 Ma. One of the most remarkable alkaline-carbonatite provinces is associated with the East African rift system, which is part of the LIP (Fig. 9.3). The carbonatites also range in age from 45 to 0 Ma, and the youngest is the only currently active carbonatite volcano, Oldoinyo Lengai (see Bell and Keller, 1995). The age distribution of carbonatites shows a weak correlation with the timing of the main pulses of LIP magmatism.

Spatially, the carbonatites of the East African rift system are associated with the Kenyan Dome and are distal from the main locus of LIP magmatism further north in

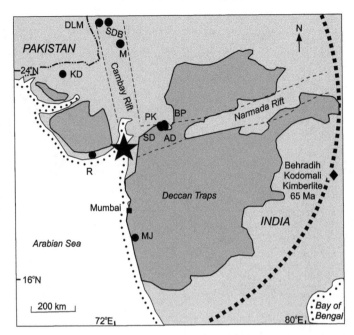

Figure 9.4 Carbonatites and kimberlites of the Deccan LIP. Carbonatite complexes are shown as solid dots. Abbreviations for carbonatite names are as follows: AD = Amba Dongar, BP = Bakhatgarh–Phulmahal, DLM = Danta–Langera–Mahabar, KD = Kala Doongar, M = Mundwara, MJ = Murud Janjira, PK = Panwad–Kawant, R = Rajula, SDB = Sarnu–Dandali (Barmer), SD = Siriwasan–Dughda. The star and circle locate inferred center and 1000-km radius of the underlying mantle plume. Location of *c.* 65 Ma kimberlites in Bastar craton is from Lehmann *et al.* (2010) and Chalapathi Rao and Lehmann (2011). Modified from Figure 4 in Ernst and Bell (2010). With kind permission from Springer Science and Business Media.

Ethiopia and the southern Yemen regions. The carbonatites are spatially linked to the thickened lithosphere of the Tanzanian craton. It has been proposed that the lack of extensive volumes of flood basalts in those parts of East Africa rich in carbonatites is because the plume head may not have reached a sufficiently shallow level in these areas, because of the presence of the thick lithospheric root to produce voluminous basaltic liquids.

Carbonatites associated with the Deccan LIP (65 Ma)

The Deccan LIP in India (Section 3.2.5; Fig. 9.4) is one of the largest continental flood-basalt provinces on Earth, and is spatially associated with the separation of India from the Seychelles microcontinent. The volume of the lavas and associated plumbing system of dykes, sills, and layered intrusions exceeded a million cubic kilometers. It consists of three pulses, at 67.5 Ma, 65 Ma, and 64.5 Ma, with the middle pulse contributing over 80% of the volcanism. A number of carbonatites and associated alkalic rocks are associated with the Deccan LIP (e.g. Viladkar, 1981; Simonetti *et al.*, 1995, 1998; Ray and Pande, 1999).

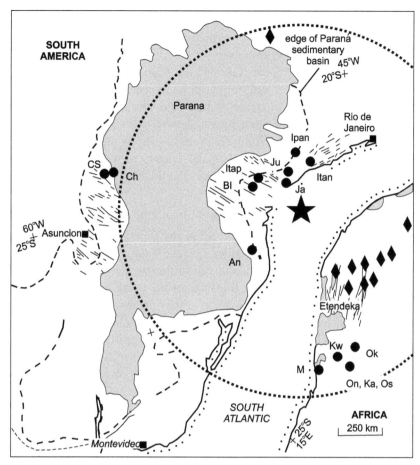

Figure 9.5 Carbonatites and kimberlites associated with Paraná–Etendeka LIP. Abbreviations for carbonatite names are as follows: SOUTH AMERICA: An = Anitapolis, BI = Barra do Itapirapua, CS = Cerro Sarambi, Ch = Chiriguelo, Ipan = Ipanema, Itan = Itanhaem, Itap = Itapirapua, Ju = Juquia, Ja = Jacupiranga; AFRICA: Ka = Kalkfeld, Kw = Kwaggaspan, M = Messum, Ok = Okorusu, On = Ondurakorume, Os = Osongombo. Kimberlite distribution, courtesy of D. Schissel. The star and circle locate the inferred center and 1000-km radius of the underlying mantle plume. Modified from Figure 5 in Ernst and Bell (2010). With kind permission from Springer Science and Business Media.

Carbonatites associated with the Paraná–Etendeka LIP (c. 135 Ma)

The extensive Paraná flood basalts in central coastal South America and the minor Etendeka remnant in Namibia once formed a single magmatic province that was related to the Tristan da Cunha plume and associated with the opening of the South Atlantic Ocean (Section 3.2.6). The emplacement of the Paraná–Etendeka LIP was accompanied by the emplacement of carbonatites and alkaline complexes (Fig. 9.5; Pirjano, 1994; Le Roex and Lanyon, 1998; Bell, 2001). Dating by Gibson *et al.* (2006)

suggests emplacement of the carbonatites and associated alkaline complexes in two phases: at 145 Ma and 127.5 Ma, i.e. before and at the end of the 139–127.5 Ma Paraná–Etendeka flood basalt eruptions.

Le Roex and Lanyon (1998) have evaluated the Tristan plume and its relationship to carbonatitic (and associated lamprophyric) magmatism in Namibia, southwest Africa. On the basis of isotope and trace-element data, in conjunction with the spatial and temporal position of the Tristan plume, a model was proposed in which the carbonatites (and associated lamprophyres) were formed by melting of metasomatic vein material resulting from incursion of alkalic fluids or melts into the subcontinental lithosphere from the upwelling Tristan plume at the time of continental breakup.

Carbonatites associated with the Siberian Trap LIP (250 Ma)

One of the largest LIP events in the world is the Siberian Trap event, which is widespread in the northwestern Siberian craton, the Taimyr Peninsula to the north, to the west in the West Siberian basin and still further west as inliers along the Ural Mountains, and in the central Asian fold belt to the south (Section 3.2.8). Flood basalts, mainly in the Maimecha–Kotui area, of the northern margin of the Siberian platform are also cut by kimberlites, carbonatites, and ultramafic silicate rocks that belong to the Maimecha–Kotui province (Fig. 9.6; Egorov, 1970; Vasiliev and Zolotukhin, 1995; Arndt *et al.*, 1995). Early Triassic sills, dykes, explosive pipes, and igneous complexes, including carbonatites, were emplaced between 250 and 220 Ma. Arndt *et al.* (1995) proposed that since the meimechitic suite lies at the fringe of the flood-basalt province then the source of heat for the meimechites (including the carbonatites) was the Siberian plume itself.

Carbonatites of the Kola Alkaline Province and association with the Kola–Dnieper LIP (380–360 Ma)

Erosional remnants of an originally widespread *c.* 380–360 Ma LIP are present in Baltica (also called the East European craton). This event was originally introduced as the East European craton event (no. 45 in Ernst and Buchan (2001b)), but herein I am labeling it the Kola–Dnieper LIP, based on the names of two key components of the event (Kola Alkaline Province and Dnieper–Donets aulacogen) located on far sides of Baltica.

Significant basalt volcanism is present in the Dnieper–Donets aulacogen, which consists of two pulses of late Frasnian and late Famennian age (Stephenson *et al.*, 2006), and elsewhere in Baltica where it is also associated with rifts (Nikishin *et al.*, 1996). There are also coeval dyke swarms of general north–south trend along the eastern margin of Baltica (Puchkov, 2012). This north–south swarm paralleling the Urals trends toward the postulated triple junction (a possible plume center) located at the mouth of the Dnieper–Donets aulacogen (Fig. 11 in Stephenson *et al.*, 2006; Fig. 9.7).

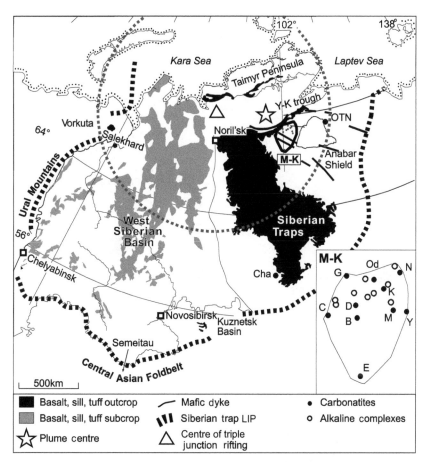

Figure 9.6 Carbonatites and alkaline complexes of the Siberian Trap LIP. Abbreviations for carbonatite names are as follows: MAIMECHA–KOTUI (M–K): B = Bor–Uryakh, C = Changit and satellites, D = Dalbykha, E = Essei, G = Guli, K = Kugda, M = Magan, N = Nemakit and satellite plugs, Od = Odikhincha, Y = Yraas (Iraas); WEST ANABAR: OTN = Orto–Yrigakhskoe, Tundrovoye, and Nomottookhskoe; southwest SIBERIA: Cha = Chadobetskaya. The star and circle locate the inferred center and 1000-km radius of the underlying mantle plume. Kimberlites associated with the Siberian Trap LIP are shown in Figure 9.15. From Figure 6 in Ernst and Bell (2010). With kind permission from Springer Science and Business Media.

The Kola Alkaline Province (Fig. 9.7) in the northern part of Baltica consists of about 14 carbonatites and related alkaline intrusions that are also *c*. 380 Ma in age (Table 9.1). For reviews see Bell and Rukhlov (2004) and Bulakh *et al.* (2004). Many of these complexes are clearly aligned along faults, implying that they are taking advantage of the fault as zones of weakness. These complexes have been linked with an underlying plume (Marty *et al.*, 1998; Tolstikhin *et al.*, 2002; Bell and Rukhlov, 2004). Much more work is required to explore the linkages between the Kola

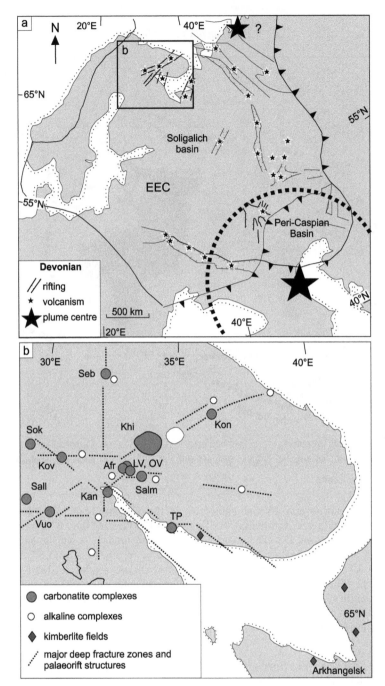

Figure 9.7 Carbonates, alkaline complexes and kimberlites associated with Kola–Dnieper LIP. (a) The East European craton (EEC, also termed Baltica) with the location of the *c*. 380 Ma volcanic and associated rifts. (b) The Kola Peninsula with associated carbonatites complexes (solid circles) and alkaline complexes (open circles). Abbreviations for carbonatite names are as follows: Afr = Afrikanda, Kan = Kandaguba, Khi = Khibiny, Kon = Kontozerskii, Kov = Kovdor, LV = Lesnaya Varaka, OV = Ozernaya Varaka, Sall = Sallanlatvi, Salm = Salmagorskii, Seb = Seblyavr, Sok = Sokli, TP = Turiy Peninsula, Vuo = Vuoriyarvi. The star and circle locate the inferred center and 1000-km radius of the underlying mantle plume. Modified after Figure 7 in Ernst and Bell (2010). With kind permission from Springer Science and Business Media. Location of associated kimberlite field at Arkhangelsk is after Mahotkin *et al.* (2000).

Alkaline Province and the rest of the coeval basalts elsewhere in the Kola–Dnieper LIP and to evaluate whether a single plume could be involved. A second plume center to the northeast from the Kola Alkaline Province may exist (Ernst and Puchkov, unpublished data).

Carbonatites associated with the Central Iapetus Magmatic Province (CIMP; 615–555 Ma)

The breakup to form the Iapetus Ocean is associated with a multi-pulse LIP event termed the CIMP (Central Iapetus Magmatic Province), and this event is best studied along the Laurentian margin (Fig. 9.8; Puffer, 2002; Ernst and Buchan, 2001b, 2004a; McCausland *et al.*, 2007; Ernst and Bell, 2010). Main pulses are at *c.* 615 Ma, 590 Ma, 565 Ma, and 550 Ma, and each is potentially linked with the progressive breakup of the eastern Laurentian margin from Baltica and other blocks. Much of the following summary is from Ernst and Bell (2010) and detailed referencing is available therein.

The earliest pulse, *c.* 615–610 Ma, consists of the extensive Long Range dykes of Labrador, Canada; but magmatism of this age is also present in Baltica (i.e. Sarek and possibly the Ottfjället dykes), which lie along the Baltoscandian failed rift margin along the Caledonides, and east–west Egersund dykes (Fig. 9.8). Alkaline magmatism in Greenland is about 600 Ma, but age uncertainties are too large to assign this magmatism to either this 615 Ma pulse or the subsequent *c.* 590 Ma pulse. The second pulse at 590 Ma consists most prominently of the Grenville–Rideau fanning dolerite dyke swarm of eastern Laurentia. In Baltica, some of the carbonatites with this age (i.e. Fen and Alnö) have been linked to Iapetus breakup and, on the basis of their most precise ages, can be grouped with this 590 Ma pulse.

The third, *c.* 565 Ma, pulse is very significant in eastern Laurentia consisting of the huge Sept Isle layered intrusion, and extensive Cactoctin volcanics (and their feeder dykes), and numerous carbonatite intrusions and ultramafic lamprophyres of eastern Laurentia along the St. Lawrence rift, and also along the Labrador Sea rift (notably, Aillik Bay, Arvida, Baie Des Moutons, and Manitou Islands). Finally, the *c.* 550 Ma pulse (e.g. Tibbit Hill basalts, Skinner Cove basalts, Mt. St.-Anselme volcanic along the Laurentian margin) is linked with the onset of major ocean crust formation, and this pulse may be present in southwest Baltica as the Volyn and related flood basalts.

Some carbonatites (along with associated carbonate-rich ultramafic lamprophyres) are associated with coeval rifting along the St. Lawrence Valley and the Labrador rift systems of eastern Canada (Fig. 9.8; Tappe *et al.*, 2008). Some generalizations can be made about the spatial distribution of different age grouping of carbonatites. Among more precisely dated carbonatites (Figs. 9.2 and 9.8) those of *c.* 560–570 Ma are concentrated in eastern Laurentia, spatially associated with the locus of the *c.* 563 Ma Sept Isle layered intrusion. The better-dated carbonatites

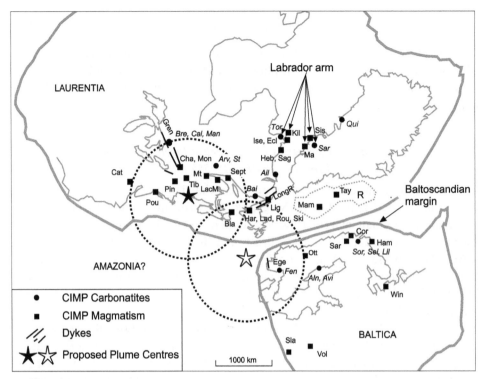

Figure 9.8 Distribution of magmatism associated with the CIMP (Central Iapetus Magmatic Province) along the eastern Laurentian margin, along the Labrador Sea (in both Greenland and adjacent Labrador), and in Baltica (after Ernst and Bell, 2010). The filled stars locate the approximate foci for the fanning 590 Ma dolerite dyke swarm (Buchan and Ernst, 2004). The open star at the triple point of the potential breakup between Laurentia, Baltica, and an unknown third block (possibly Amazonia) locates the potential plume center applicable to the broader age range of the event. Units are labeled as follows (carbonatites are in italic font): EASTERN LAURENTIA: *Arv* = *Arvida* carbonatite complex; *Bai* = *Baie Des Moutons* carbonatite complex; Bla = Blair River metagabbro dykes; *Bre* = *Brent* carbonatite complex; *Cal* = *Callander Bay* carbonatite complex; Cat = Catoctin Formation metavolcanics and dykes; Cha = Chatham–Grenville stock; Gren = Grenville–Rideau fanning dyke swarm; Har = Hare Hill granite; LacM = Lac Matapedia volcanics; *Man* = *Manitou Islands* carbonatite complex; Lad = Lady Slipper pluton; Lig = Lighthouse Cove volcanics; LongR = Long Range dykes; Mon = Mont Rigaud stock; Mt = Mt. St.-Anselme volcanics; Pin = Pinney Hollow Formation metafelsite; Pou = Pound Ridge granite gneiss and Yonkers gneiss; Rou = Round Pond granite; Sept = Sept Iles layered intrusion; Ski = Skinner Cove volcanics; *St* = *St.-Honoré* carbonatite complex; Tib = Tibbit Hill Formation metavolcanics. LAURENTIA (LABRADOR AREA – LABRADOR AND WEST GREENLAND PORTIONS): *Ail* = *Aillik Bay* carbonatite complex; Ecl= Eclipse Harbour ultramafic lamprophyre; Heb = Hebron ultramafic lamprophyre; Ise = Iselin Harbour ultramafic lamprophyre; Kil = Killinek Island ultramafic lamprophyre; Ma = Maniitsoq ultramafic complex; *Qui* = *Quigussaq* carbonatite complex; Sag = Saglek ultramafic lamprophyre; *Sar* = *Sarfartoq* carbonatite complex. Sis = Sisimiut ultramafic lamprophyre; *Tor* = *Torngat Mountains* carbonatite complex. ROCKALL AND NORTH SCOTLAND (R): Mam = Mam sill; Tay = Tayvallich volcanics. BALTICA: *Aln* = *Alnö* carbonatite complex; *Avi* = *Avike Bay* carbonatite complex; Cor = Corrovare Nappe dyke swarm; *Ege* = *Egersund swarm; *Fen* = *Fen* carbonatite complex; Ham = Hamningberg dyke; *Lil* = *Lillebukt* carbonatite complex; Ott = Ottfjället dolerite dyke swarm; Sar = Sarek swarm; *Sei* = *Seiland* carbonatite complex; *Sor* = *Sørøy* carbonatite complex; Sla = Slawatycze Formation volcanics, correlated with the Volyn volcanics; Vol = Volyn basalts; Win = Winter Coast volcanics, White Sea region. From Figure 8 in Ernst and Bell (2010). With kind permission from Springer Science and Business Media. Location of Labrador region noted in Figure 9.18.

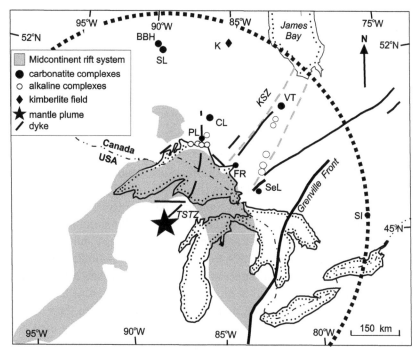

Figure 9.9 Carbonatites, alkaline complexes and kimberlites associated with Keweenawan LIP. Carbonatite complexes are labeled as follows: BBH = Big Beaver House, SL = Schryburt Lake, SI = Sullivan Island, SeL = Seabrook Lake, VT = Valentine Township, FR = Firesand River, PL = Prairie Lake, TT = Teetzel Township, CL = Chipman Lake. The location of the *c.* 1100 Ma Kyle Lake (K) kimberlite cluster is after Kjarsgaard (2007). KSZ = Kapuskasing Structural Zone. TSTZ = Trans-Superior tectonic zone. Modified from Figure 9 in Ernst and Bell (2010). With kind permission from Springer Science and Business Media.

(e.g. Fen and Alnö) of Baltica seem to coincide with the 590 Ma pulse, which is otherwise prominent in eastern Laurentia (e.g. the Grenville–Rideau radiating dyke swarm), and which has a well-defined plume center at the focus of the radiating pattern.

Carbonatites associated with the Keweenawan LIP (1115–1085 Ma)

One of the most dramatic flood-basalt events in North America is arguably the Keweenawan magmatism of the Midcontinent rift system (e.g. Nicholson *et al.*, 1997; Vervoort *et al.*, 2007; Heaman *et al.*, 2007; Miller and Nicholson, 2013; Section 3.5.2; Fig. 9.9). This LIP comprises at least 2 Mkm^3 of volcanic rocks and possibly an equal volume of intrusive rocks, and is associated with the attempted breakup of North America at *c.* 1100 Ma. The main concentration of magmatism is along the Midcontinent rift system in the Lake Superior region.

A number of carbonatites of *c.* 1100 Ma age are coeval with the Keweenawan LIP (Fig. 9.9; Sage, 1991; Heaman and Machado, 1992; Heaman *et al.*, 2007; Rukhlov and Bell, 2010). The greatest concentration are associated with the KSZ (Kapuskasing Structural Zone), along which differential rotation of the east and west halves of the Superior craton occurred in the Late Paleoproterozoic (e.g. Buchan *et al.*, 2007; Evans and Halls, 2010). Other carbonatites are located further to the north in the Superior craton, but still within the 1000-km radius for a Keweenawan plume centered in Lake Superior. Those carbonatites that are precisely dated have ages that match with the pre- *c*. 1100 Ma pulses of Keweenawan magmatism (Fig. 9.2).

Carbonatites associated with the Circum-Superior LIP (1880–1870 Ma)

A belt of *c*. 1880–1870 Ma mafic–ultramafic rocks surrounds much of the Superior craton for a distance of more than 3400 km (Fig. 9.10), the "Circum-Superior belt" of Baragar and Scoates (1987), and includes economic Ni deposits in both the Thompson and Cape Smith regions (e.g. Eckstrand and Hulbert, 2007). Remarkably extensive, and with relatively tight current age constraints (e.g. event 8 in Ernst and Buchan, 2004a; Heaman *et al.*, 2009; Hamilton *et al.*, 2009; Ernst and Bell, 2010; Minifie *et al.*, 2013), this Circum-Superior (also termed Pan-Superior in Ernst and Buchan, 2010) magmatism may represent the erosional remnant of an originally more widespread LIP.

In addition to this 1880–1870 Ma mafic–ultramafic magmatism there are coeval carbonatite complexes in the interior of the craton, most notably along the KSZ that bisects the craton (Rukhlov and Bell, 2010; Ernst and Bell, 2010). These carbonatite complexes along the KSZ include Cargill, Goldray, Borden, Spanish River, and probably Argor (Table 9.1). In addition, the Lac Castignon Complex is located along the Labrador Trough, and the Carb Lake Complex is located in northwestern Superior craton (Fig. 9.10).

Carbonatites associated with the Bushveld LIP (2060 Ma)

The 2060 Ma Bushveld Complex of South Africa includes the Earth's largest layered igneous intrusion, which is regarded as the intrusive equivalent of a flood-basalt province, given its extensive volume of 0.38 Mkm^3, short duration of emplacement, and the presence of a coeval smaller intrusion elsewhere in the Kaapvaal craton (e.g. Rajesh *et al.*, 2013). The Bushveld LIP (Section 5.4.1) is associated with two carbonatites, the Phalaborwa and Shiel complexes (Fig. 9.11), which are located along the northeastern margin of the Kaapvaal craton. The Phalaborwa carbonatite complex is particularly interesting as a significant ore deposit for Cu and Au (Section 16.2.3).

Carbonatites without an obvious LIP association

There are also some well-dated carbonatites without an obvious LIP association. For instance, in South America to the north of the Paraná portion of the *c*. 130 Ma Paraná–Etendeka LIP, there is a younger 80 Ma pulse of carbonatites have and there is no known LIP event of this age in the region. These *c*. 80 Ma carbonatites have been linked with the passage of the Trinidad hotspot (e.g. Toyoda *et al.*, 1994; Bell, 2001).

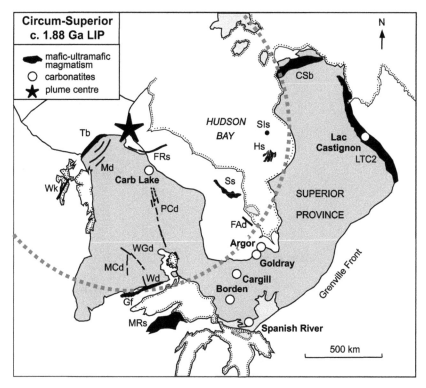

Figure 9.10 Distribution of magmatic elements of the *c.* 1880–1870 Ma Circum-Superior (also called Pan-Superior) LIP, CSb = Cape Smith belt with the Chukotat Formation, LTC2 = Labrador Trough Cycle 2, SIs = Sleeper Island sills, Hs = Haig sills of Belcher Islands, FAd = Fort Albany dykes, Ss = Sutton Inlier with sills. FRs = Fox River sill, PCd = Pickle Crow dykes, Tb = Thompson belt, Md = Molson dykes, Wk = Winnipegosis komatiites, MCd = Mine Center dyke, WGd = Wabigoon dyke, Wd = Wantelto Lake dykes, Gf = Gunflint Formation with volcanics. MRs = Marquette Range Supergroup with volcanics and sills. Carbonatite complexes (solid black circles) are labeled. The star and dotted circle locate the inferred center and 1000-km radius of the underlying mantle plume *c.* 1880 Ma ago. Modified from Figure 10 in Ernst and Bell (2010). With kind permission from Springer Science and Business Media. The Borden, Cargill, Goldray, and probably Argor carbonatite complexes are located along the Kapuskasing Structural Zone (see discussion in text).

So although not all carbonatites are associated with LIPs, in some cases, they may be associated with plume tracks.

9.2.3 Discussion

Associated alkali silicates suites

Carbonatites have been recently divided into magmatic and carbohydrothermal types (Woolley and Kjarsgaard, 2008a, b; cf. carbothermal of Mitchell, 2005). The former represent high-temperature magmatic crystallization and the latter represent

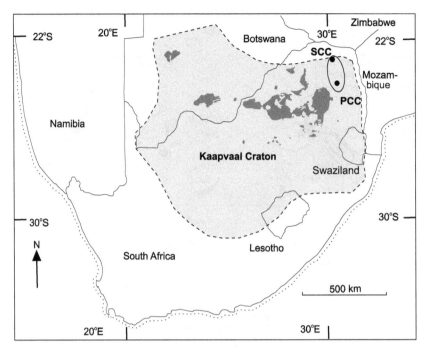

Figure 9.11 Carbonatites associated with the Bushveld LIP. BC = Bushveld Complex, UC= Uitkomst Complex, SCC = Shiel carbonatite complex, PCC = Phalaborwa carbonatite complex. For detailed labeling of units of Bushveld LIP see Fig. 5.19a. From Figure 11 in Ernst and Bell (2010). With kind permission from Springer Science and Business Media.

lower-temperature carbohydrothermal carbonatite precipitation – those that precipated at subsolidus temperatures from a mixed CO_2–H_2O fluid that can be either CO_2-rich (i.e. carbothermal) or H_2O-rich (i.e. hydrothermal) (Woolley and Kjarsgaard, 2008b). Magmatic carbonatites, the focus here, are considered to be derived in two principal ways, either directly in the mantle or via differentiation from a silicate magma at shallower depths (Bell, 1989; Woolley and Kjarsgaard, 2008b). The association between carbonatites and alkaline rocks has long been recognized, and has been recently divided into a number of types (Fig. 9.12); in decreasing order of abundance: (1) nephelinite–ijolite, (2) phonolite–feldspathoidal syenite, (3) trachyte–syenite, (4) melilitite–melilitolite, (5) lamprophyre, (6) kimberlite, and (7) basanite–alkali gabbro.

Identifying translithospheric faults and rifts

Many carbonatites are associated with rifts and also with translithospheric breaks, and several examples are discussed.

A rift association is clear for the Deccan carbonatites which are spatially associated with the Narmada–Tapti–Son and other rifts. Carbonatites of the Afro-Arabian LIP are associated with the East African rift valley system. In Canada, a number of

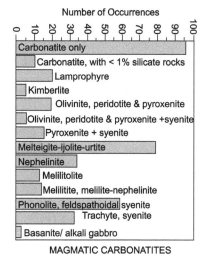

Figure 9.12 Carbonatite–silicate rock associations. From Woolley and Kjarsgaard (2008b).

1107 Ma Keweenawan and 1880 Ma Circum-Superior LIPs are emplaced along the KSZ, a significant break across the narrow "waist" of the Superior craton (e.g. Percival and West, 1994; Evans and Halls, 2010). It thus appears possible that carbonatites can be used to establish the sites of cryptic lithospheric disruption. Within complicated terranes, the identification of deformed alkaline rocks and carbonatites (DARCs) may mark ancient rift zones associated with subsequent ocean closure and orogenic deformation (Burke *et al.*, 2003).

Carbonatites are located at a range of distances with respect to their plume centers. Carbonatites belonging to the Siberian Trap (Fig. 9.6), and Circum-Superior (Fig. 9.10) LIPs are potentially located peripheral to the plume edge. In contrast, carbonatites of the Deccan (Fig. 9.4) and Paraná–Etendeka (Fig. 9.5) LIPs are mainly proximal to the plume center. The situation for the Kola–Dnieper LIP (Fig. 9.7), CIMP (Fig. 9.8), and Bushveld LIPs (Fig. 9.11), are unclear since the location of their plume center or centers is not well-constrained.

Timing of carbonatites with respect to LIP magmatism

The plume model of Bell and Rukhlov (2004) (Fig. 15.3) suggests that carbonatites should form from small-volume melts early in the magmatic history of a LIP whether at the periphery of the plume or above the plume before flood-basalt generation. Limited precise ages provide provisional support to this timing (Fig. 9.2).

For instance, the precisely dated Phalaborwa carbonatite (2059 Ma) is slightly older than the rest of the Bushveld units (which cluster between 2058 and 2055 Ma) (Table 9.1, Fig. 9.2). Similarly, some Keweenawan carbonatites are dated at 1115–1100 Ma, which represents the early and main stages of the overall

1115–1085 Ma magmatic history of the LIP (Heaman *et al.*, 2007). However, in the case of the 615–550 Ma CIMP event (Fig. 9.8), many of the associated carbonatites cluster with the 590 and 565 Ma pulses, and so far none are definitively linked with either the initial 615 Ma pulse or youngest 550 Ma (ocean-opening pulse). The East African rift system carbonatites (Afro-Arabian LIP) appear to be somewhat unique in that they have been emplaced throughout the entire LIP/plume event from 45 Ma to the present.

After more carbonatites are precisely dated it will be possible to better assess their timing with respect to magmatic pulses of associated LIP events, and allow robust generalizations regarding the timing of carbonatites within a LIP event.

9.3 Kimberlites and LIPs

9.3.1 Introduction

Kimberlites (Fig. 9.13) have been described as volatile-rich (dominantly CO_2), potassic ultramafic rocks possessing a distinctive inequigranular texture that results from the presence of both macrocrystic and groundmass phenocrystic olivine (Chalapathi Rao and Lehmann, 2011; see also Mitchell, 1995, 2006, 2008; Woolley *et al.*, 1996; Le Maître, 2002; Neuendorf *et al.*, 2011). Other minerals comprising the macrocryst/megacryst assemblage include magnesian ilmenite, pyrope, diopside (subcalcic), phlogopite, enstatite and Ti-poor chromite. Late-stage laths of phlogopite are common. Other minerals in the groundmass can include: monticellite, phlogopite, perovskite, spinel, apatite, carbonate, serpentine and diopside. Kimberlites occur as volcanic diatremes and as dykes and sills (Fig. 9.13), and are petrographically complex because of the presence of abundant xenocrysts, megacrysts, and xenoliths. They are rapidly and explosively emplaced with passage through the lithosphere as dykes, but final ascent to the surface is via explosive diatremes with ascent velocities as high as 30 to 50 m/s (e.g. Mitchell, 1995; Haggerty, 1999a, b; Schissel and Smail, 2001; Sparks *et al.*, 2007; Wilson and Head, 2007; Russell *et al.*, 2012). Russell *et al.* (2012) offered a link between kimberlites and carbonatites by suggesting that kimberlites represent carbonatite-like parental melts that were modified as they migrated through the mantle lithosphere by assimilating mantle minerals, especially orthopyroxene, causing these magmas to become more silicic, and leading to a coincident drop in carbon dioxide solubility.

Kimberlites are mainly located in Archean cratons and the Paleoproterozoic mobile belts that surround cratonic areas. They have also been interpreted to be the deep-sourced equivalent of basalts trapped beneath lithospheric roots (e.g. Campbell, 2001; Arndt, 2003). They can be classified into Group I (basaltic) and Group II (micaceous) types (e.g. Mitchell, 1995; Schissel and Smail, 2001; Arndt, 2003; Becker and le Roex, 2006; Gurney *et al.*, 2010), the latter termed "orangeites" by Mitchell (1995). Important evidence on the origin of kimberlites derives from their

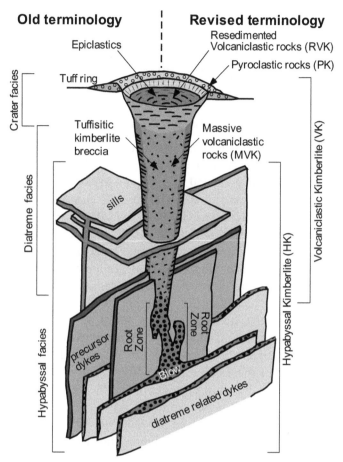

Figure 9.13 Kimberlite model and its nomenclature. PK = pyroclastic kimberlite, RVK = resedimented volcaniclastic kimberlite, MVK = massive volcaniclastic kimberlite, HK = hypabyssal kimberlite. From Figure 10 in Kjarsgaard (2007).

diamond endowments; most researchers consider that diamonds are picked up and transported during magma ascent as xenocrysts, and inclusions within diamonds indicate that the host kimberlite (and lamproite) magmas originated at depths below 180 km, perhaps from plumes that ultimately started from the core–mantle boundary (e.g. Torsvik *et al.*, 2010). Although the majority of diamonds are thought to be sourced from the deep lithospheric roots of ancient cratons, some diamonds contain inclusions of majoritic garnet (and its exsolution products) and Na pyroxene–enstatite solid solutions, which are evidence for an origin at transition-zone depths (410–660 km), whereas other inclusions in diamonds are indicative of a lower mantle origin; yet other diamonds contain Fe, FeC, and SiC inclusions that may have originated in the Earth's core (Haggerty, 1999a, b; Kerrich *et al.*, 2005; Stachel and

Harris, 2008; Harte, 2011). The deep origin of these constituents is most compatible with a mantle-plume origin for kimberlites, although non-plume origins for kimberlites have also been proposed (e.g. Mitchell, 1995; Helmstaedt and Gurney, 1997; Currie and Beaumont, 2011; Tappe *et al.*, 2013).

9.3.2 Global age distribution

On a global scale, kimberlite magmatism seems to be episodic (Haggerty, 1999b). Major kimberlite pulses occurred at about *c.* 1 Ga (Africa, Brazil, Australia, Siberia, India, and Greenland), *c.* 450 to 500 Ma (Arkhangel on the White Sea, China, Canada, South Africa, and Zimbabwe), 370 and 410 Ma (Siberia and United States), *c.* 200 Ma (Botswana, Canada, Swaziland, and Tanzania), 80 to 120 Ma (south, central, and west Africa, Brazil, Canada, India, Siberia, and the United States), and *c.* 50 Ma (Canada and Tanzania), with minor fields in Ellendale in northwest Australia at 22 Ma (Haggerty, 1994). In a more recent review (based on many additional ages from North America), Heaman *et al.* (2003) identified, in both North America and Yakutia (Siberia), three distinct short-duration (*c.* 10 Ma) periods of kimberlite magmatism at 48–60, 95–105, and 150–160 Ma. Cenozoic/Mesozoic kimberlite magmatism in southern Africa is dominated by a continuum of activity between 70–95 and 105–120 Ma with additional, less-prolific periods of magmatism in the Eocene (50–53 Ma), Jurassic (150–160 Ma), and Triassic (*c.* 235 Ma). Several discrete episodes of pre-Mesozoic kimberlite magmatism variably occur in North America, southern Africa, and Yakutia (Siberia) at 590–615, 520–540, 435–450, 400–410, and 345–360 Ma. Heaman *et al.* (2003) also noted a globally quiet time (in terms of kimberlite magmatism) from 360 to 250 Ma.

Kimberlite corridors

From their detailed compilation of kimberlite ages in North America, Heaman *et al.* (2004) proposed continental scale "corridors" of kimberlite magmatism each emplaced over a relatively short interval of 10–20 Ma. The most dramatic is the central Cretaceous (103–94 Ma) corridor extending for more than 4000 km from Somerset Island in northern Canada through the Fort à la Corne field in Saskatchewan to the kimberlites in central USA. The interpretation of such "corridors" remains speculative.

Link with superplume and supercontinent breakup events

Some of the more widespread kimberlite events have been associated with superplume events linked to the breakup of supercontinents (e.g. Haggerty, 1999a, b; Heaman *et al.*, 2003; Kerrich *et al.*, 2005), although others may be triggered by returning mantle flow in areas proximal to convergent plate margins that may not be definitively linked with LIPs (Tappe *et al.*, 2013) or low-angle subduction (Currie and Beaumont, 2011). The links between LIPs and kimberlites are exemplified by 120–80 Ma kimberlites in North America, India, Siberia, Brazil, and Africa that are broadly linked to the Pacific Cretaceous superplume and associated dispersal of Gondwana, and 590 to 550 Ma kimberlites and carbonatites in Greenland located along the periphery of the CIMP LIP events (Tappe *et al.*, 2011, 2012).

Torsvik *et al.* (2010) used plate reconstructions and tomographic images to indicate that the majority of Phanerozoic kimberlites, similar to Phanerozoic LIPs, are associated with the edges of the largest heterogeneities in the deepest mantle; these areas have been stable for at least 200 Ma and possibly for 540 Ma, with the authors concluding that both LIPs and kimberlites are derived from plumes. Additional insights are possible from more focus on kimberlite clusters and their potential linkage with specific LIPs. Studies of the links between LIPs and kimberlites have recently received a boost by the discovery of kimberlites associated with the Deccan LIP (Chalapathi Rao and Lehmann, 2011) and with the Tarim LIP (Zhang *et al.*, 2013). As discussed below, there are numerous kimberlite clusters that are spatially and temporally associated with known LIPs, as well as kimberlites that exhibit a distinct age progression over large distances and are therefore thought to be associated with a mantle plume (hotspot) tail. However, for many other kimberlite fields, their setting is not yet clear.

9.3.3 Examples of kimberlites linked with LIPs

Specific LIP–kimberlite associations are summarized in Table 9.2 and discussed below:

Kimberlites associated with the Deccan LIP (65 Ma)

A cluster of Group II kimberlites in central India have been dated by Ar–Ar and U–Pb perovskite methods as about 65 Ma, and therefore are coeval with the Deccan LIP (Lehmann *et al.*, 2010; Chalapathi Rao *et al.*, 2011). This kimberlite cluster is located in the Bastar craton, nearly 1000 km from the inferred plume center for the Deccan LIP (Fig. 9.4). Based on Nd isotopic studies of the Behradih pipe it is inferred that the Deccan plume only contributed heat but not substantial melt to the kimberlite (Chalapathi Rao *et al.*, 2011).

Kimberlites in Southern Africa and links with LIPs (c. 90, 130, and 180 Ma)

There are an impressive number of kimberlite clusters in southern Africa (e.g. Schissel and Smail, 2001). Some of these can be linked on the basis of age and

Table 9.2 *Selected LIPs and their associated kimberlites*

LIP name (ages, to nearest 5 Ma)	Associated kimberlites	Distance of kimberlites from linked plume center	Reference for link with the LIP
Deccan (c. 65 Ma); Section 3.2.5	Kimberlites (Group II) of Mainpur kimberlite field, Bastar craton (65–62 Ma)	1000 km	Lehmann et al. (2010); Chalapathi Rao et al. (2011)
Madagascar (c. 90 Ma); Section 9.3.3	90 Ma kimberlites in southern Africa (Schissel and Smail, 2001)	500–1300 km	Ernst and Jowitt (2013)
HALIP (c. 130–90 Ma)	Cretaceous kimberlites on north slope of the Olenok uplift, Siberia (Kravchinsky et al., 2002)	>500 km	Kravchinsky et al. (2002)
Paraná–Etendeka (c. 135 Ma); Section 3.2.6	Kimberlites (Group II) of Kaapvaal craton; 125–120 Ma (Schissel and Smail, 2001)	1600 km	Schissel and Smail (2001); Chalapathi and Lehmann (2011)
Karoo–Ferrar (c. 180 Ma); Section 5.3.2	Dullstrom–Elandskloof Kimberlites (South Africa)	300 km	Schissel and Smail (2001)
Siberian Trap (250 Ma); Section 3.2.8	Kharamai kimberlite field (245–228 Ma)	400–600 km	Chalapathi and Lehmann (2011); Kiselev et al. (2012)
Yatutsk–Vilyui (c. 380–360 Ma); Section 11.4	Daldyn–Alakit, Nakyn, & Mirnyi kimberlite fields	600–800 km	Kiselev et al. (2012)
Kola–Dnieper (c. 380–360–Ma); Section 9.2.2	Arkhangelsk field (c. 360 Ma; Beard et al., 2000)	? (plume center uncertain)	Mahotkin et al. (2000)
CIMP (Central Iapetus Magmatic Province) (615–555 Ma); Section 9.2.2	Eocambrian/Cambrian Labrador Sea kimberlite province	? (plume center uncertain)	Heaman et al. (2003, 2004); Tappe et al. (2006, 2007, 2008)
Keweenawan (1115–1085 Ma); Section 3.5.2	Kimberlite from Lake Superior and James Bay Lowlands (c. 1172–1035 Ma; some ages very approximate)	? (plume center uncertain)	Heaman et al. (2004, 2007); Chalapathi and Lehmann (2011)

Note: LIP locations are shown in Figure 1.6.

266

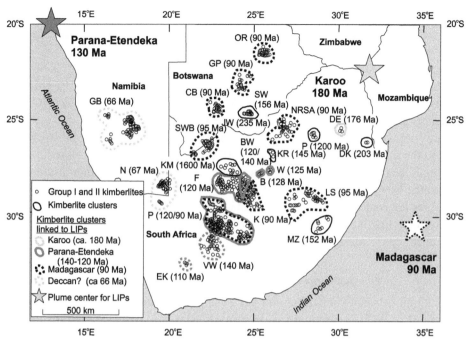

Figure 9.14 Examples of the association of kimberlites and LIPs in southern Africa. Stars locate interpreted mantle plume centers that can be linked to kimberlite clusters. Names and distribution of kimberlite clusters after Schissel and Smail (2001). B = Bosnof (128 Ma, Group II); BW = Barkley West (120/140 Ma, Group I & II); CB = Central Botswana (90 Ma, Group I); DE = Dullstrom-Elandskloof (176 Ma, Group II); DK = Dokolwaya (203 Ma, Group II); EK = Eende Kuil (110 Ma, Group II); F = Finsch (120 Ma, Group II); GB= Gibeon Eastern Namibia (66 Ma, Group I); K = Kimberley (90 Ma, Group I & II); KM = Kuruman (1600 Ma); KR = Krononstad (145 Ma, Group II); LS = Letseng (95 Ma); MZ = Mzongwea (152 Ma, Group I); N = Namaqualand (67 Ma, Group I); P = Premier (1200 Ma); PR = Prieska (120/90 Ma, Group I & II); OR = Orapa (90 Ma, Group I); GP = Gope (90 Ma, Group I); JW = Jwaneng (235 Ma, Group I); SW = Swartruggens (156 Ma, Group I); NRSA = North Republic of South Africa (90 Ma, Group I & II); SWB = southwest Botswana (95 Ma, Group I); VW = Victoria West (140 Ma, Group I); W = Winburg (125 Ma, Group II). Links to *c.* 180 Ma Karoo, *c.* 130 Ma Paraná–Etendeka, *c.* 90 Ma Madagascar, and possibly 65 Ma Deccan LIPs are made. 110, 140, 150, 235, 1290, and 1400 Ma ages are not associated with any specific LIPs or plume. However, some of these may be alternatively linked with plume tails (Schissel and Smail, 2001). From Figure 7 in Ernst and Jowitt (2013).

proximity with known plume-related LIPs of the region, the c. 90 Ma Madagascar, 183 Ma Karoo–Ferrar, and *c.* 135 Ma Paraná-Etendeka LIPs (Fig. 9.14).

Several clusters of *c.* 90 Ma kimberlites are recognized (Fig. 9.14). This is an exact age match with the Madagascar LIP (Storey *et al.*, 1997), which was attached to southeast Africa at this time (Storey, 1995; Torsvik *et al.*, 2000) inviting suggestion of a connection between this 90 Ma LIP and kimberlites. The 90 Ma kimberlite fields range from 500 km to 1300 km away from the Madagascar plume center (Table 9.2).

The 176 Ma Dullstrom–Elandskloof kimberlites at 175–165 Ma (Fig. 9.14) are probably linked to the 183 Ma Karoo event (part of the Karoo–Ferrar LIP) which is a widespread event in southern Africa (Sections 3.2.7 and 5.3.2); these 176 Ma kimberlites are located about 300 km from the Karoo plume center.

More speculatively, the 128–120 Ma kimberlites (Fig. 9.14) should probably be linked to the *c.* 135 Ma Paraná–Etendeka LIP (Fig. 9.5), whose plume center at that time should be further north than the location of Tristan. The maximum distance of these kimberlites from the plume center would be about 1600 km. Some additional kimberlites inferred to belong to the Paraná–Etendeka LIP are also proximal to the plume center (Fig. 9.5) and others are further away, in nearby Congo (D. Schissel, personal communication, 2013).

Siberian kimberlites at c. 360, 245–228, and 135 Ma

A strong link between kimberlites and LIPs has been established for the kimberlites of Siberia (Fig. 9.15). As summarized by Kravchinsky *et al.* (2002) the kimberlites of the Siberian platform were emplaced during three main epochs: (a) in the Devonian–Early Carboniferous (in the Vilyui (Viluy) and Markha basins, and the Aikhal region), (b) in the Triassic (in the Mir, Aikhal, and Olenok regions), and (c) in the Cretaceous (northern slopes of the Olenok uplift). The Devonian–Early Carboniferous (*c.* 370 Ma) kimberlites are associated with the Yakutsk–Vilyui LIP (Kiselev *et al.*, 2012), the Late Permian–Early Triassic (*c.* 250 Ma) kimberlites are associated with the Siberian Trap LIP (e.g. Kiselev *et al.*, 2012 and references therein), and the Cretaceous (135 Ma; Kravchinsky *et al.*, 2002; Rosen *et al.*, 2007) kimberlites can potentially be linked to the High Arctic LIP (HALIP) event of the Arctic (Fig. 9.15; e.g. Maher *et al.*, 2001; Buchan and Ernst, 2006; Jowitt *et al.*, 2014) based on age matching and spatial proximity.

The *c.* 360 Ma kimberlites associated with the Yakutsk–Vilyui LIP are diamondiferous, while the *c.* 250 Ma kimberlites associated with the Siberian Trap LIP are not (e.g. Kiselev *et al.*, 2012). The influence of the thermal pulse of the 370 Ma LIP/plume may have a control on destroying lithospheric diamond potential for the later *c.* 250 Ma kimberlites of the Siberian craton and elsewhere, as is discussed in more detail in Section 16.2.4.

Kimberlites associated with the Ma Keweenawan LIP (1115–1085 Ma)

As noted by Heaman *et al.* (2004) there are a number of Mesoproterozoic kimberlites and ultramafic lamprophyres that were emplaced between Lake Superior and Hudson Bay, Canada during the period 1035–1172 Ma and which can be linked with the Keweenawan LIP and associated carbonatites (Fig. 9.9). Notable among these is the Kyle Lake cluster (Heaman *et al.*, 2004; Kjarsgaard, 2007).

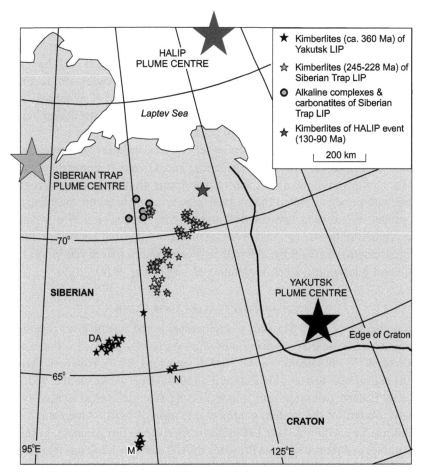

Figure 9.15 Kimberlites of Siberia and links with Yakutsk, Siberian Trap, and HALIP LIPs. Small black stars identify kimberlite fields that are linked with Yakutsk LIP (DA = Daldyn–Alakit, N = Nakyn, M = Mirnyi fields), and small gray stars identify kimberlites belonging to the Siberian Trap LIP, based on Kiselev *et al.* (2012). Kravchinsky *et al.* (2002) mentions "Cretaceous" kimberlites in the Olenok uplift region, which speculatively could belong to the HALIP (High Arctic LIP) event. The shown location of the HALIP plume center is only approximate, and depends on the appropriate (and as yet unknown) *c.* 130 Ma reconstruction of the Arctic. From Figure 7 in Ernst and Jowitt (2013).

9.3.4 Kimberlites indirectly linked to LIPs

Other kimberlite fields have a more indirect link with LIPs.

Kimberlites as tracks of plume tails

In a plume model, an initial plume-head burst of dominantly mafic magmatism (causing a LIP) should be followed by protracted plume-tail magmatism. In the oceans, the plume-head magmatism is marked by oceanic plateaus and the plume-tail magmatism is marked by seamount (hotspot) trails (Section 2.4). On the continents, plume-head magmatism is marked by continental flood basalts, but with few exceptions plume-tail magmatism cannot breach thick, continental lithosphere. However, it is possible that the record of plume tails under continental areas can be revealed, not by the basalts, but by the kimberlites that show an age progression. Kimberlites easily transit the lithosphere (e.g. Russell *et al.*, 2012) and if generated from a sublithospheric plume they should easily penetrate and traverse the lithospheric root of a moving plate and produce a hotspot trail.

Examples summarized by Heaman *et al.* (2003) include Mesozoic–Cenozoic kimberlites in Brazil, eastern North America, and Group II kimberlites in South Africa. Each of these follows a relatively narrow linear unidirectional age progression consistent with plate movement that is associated with the breakup and rifting of the supercontinent Gondwana between 180 and 50 Ma. The eastern North American example is particularly strong; precise dating of perovskite from kimberlites identifies an age progression over 2000 km in length in North America from 200 Ma to 100 Ma (Heaman and Kjarsgaard, 2000; Heaman *et al.*, 2004; Fig. 9.16).

Eocene kimberlites in the Slave craton

There is a distinct *c.* 74–48 Ma field of kimberlites in the Lac de Gras region of the Slave craton and those that are diamondiferous are tightly grouped within a narrow time span of 56–53 Ma (Heaman *et al.*, 2003; Kjarsgaard, 2007). However, there is no LIP in the region with this age. Heaman *et al.*, (2004) suggest a link with a widespread Cordilleran Eocene extension and major Eocene bimodal volcanic activity event recognized inboard of the western continental margin of North America.

An indirect link with a distant LIP is also conceivable. For example, in this case, the distant North Atlantic Igneous Province (NAIP; Section 3.2.4) has two pulses, an initial 62 Ma pulse and a second pulse at 55 Ma which is linked with the onset of rifting in the North Atlantic. Could the Slave craton kimberlites and the LIP-associated rifting in the North Atlantic be indirectly linked? The onset of rifting at 55 Ma in the North Atlantic caused a burst of compression and transpression in the global plate framework. Could such a pulse of movement along favorably oriented fractures in the Slave craton cause release of kimberlitic magma that had pooled at the base of the lithosphere, allowing it to migrate upwards?

Regardless of the link with plume heads (LIPs) or plume trails, there is abundant evidence that the control on kimberlite emplacement is related to major crustal structures (e.g. Heaman *et al.*, 2004; Kerrich *et al.*, 2005; Kjarsgaard, 2007; Jelsma *et al.*, 2009).

9.3.5 Discussion of the links between kimberlites and LIPs

Link with dolerite dyke swarms

There is evidence that individual kimberlites are preferentially emplaced along certain regional dolerite dyke swarms. In the Slave craton there has been a preferential

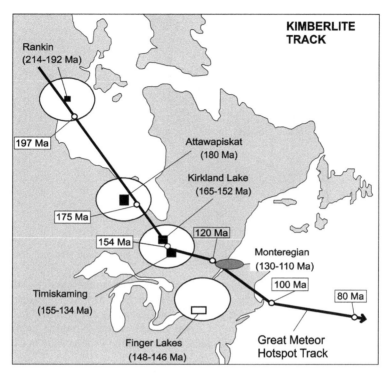

Figure 9.16 Kimberlite hotspot track. Kimberlite age progression across North America interpreted as a hotspot track that continues through the Monteregian intrusions and into the Atlantic as the Great Meteor hotspot track (Heaman and Kjarsgaard, 2000). Small open circles represent the position of the North American plate based on reconstructions. Modified after Figure 4 in Heaman and Kjasgaard (2000). With permission from Elsevier.

spatial link between Tertiary kimberlites and the 2025 Ma Lac de Gras dykes/LIP (Wilkinson *et al.*, 2001; Stubley, 2004, Nowicki *et al.*, 2004). In addition, the densest cluster of pipes in the Lac de Gras area occurs at the intersection of the major "305" dyke system with a concentration of Mackenzie dykes/LIP (Nowicki *et al.*, 2004). In the Attawapiskat region of James Bay, in northern Ontario, Canada, there are two associations noted by Stott (2003): first, Early Jurassic pipes, near the Attawapiskat River in the James Bay Lowlands, mainly form a linear northwestward trend close to a Matachewan (*c.* 2446 Ma) diabase dyke (part of the Matachewan LIP). Second, over 90 km farther west, the *c.* 20-km-wide bundle of dykes belonging to the 2121 Ma Marathon swarm (part of the Marathon LIP) are trending northwards in the vicinity of the *c.* 1100 Ma Kyle (Keweenawan LIP related) kimberlite pipes. In each case, individual pipes lie close to aeromagnetic traces of the dykes.

This kimberlite pipe–dolerite dyke relationship, if confirmed, might be explained by a model similar to that shown in Figure 9.17. Dolerite dykes can be blade-like in shape and should have a lower (and upper) edge, since they are commonly propagated laterally in the crust at a level of neutral buoyancy away from plume-center regions

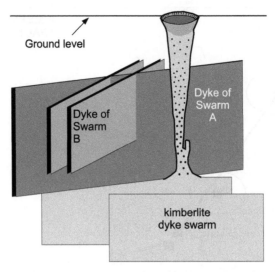

Figure 9.17 Model of interaction of dolerite dyke swarms with kimberlite dykes to explain spatial association of kimberlites along some dyke swarms. Dyke swarm "A" extends deeper than dyke swarm "B". So rising kimberlite dykes will intersect dykes of swarm "A" first and are more likely to ascend along the margins of dykes of this swarm. Note the inferred dyke geometry as laterally propagating blades of limited vertical extent (see discussion in Chapter 5).

(e.g. Fahrig, 1987; Section 5.2.4). So dykes with a lower edge that extends to greater depth than the lower edge of other dykes (of other swarms) are more likely to be the first to be intersected by (and capture) kimberlite magmas that are rising along kimberlite dykes. Hypothesize that once a kimberlite magma had intersected the bottom edge of a dolerite dyke it would preferentially continue to ascend along the side of that dolerite dyke, taking advantage of the dyke margin as a zone of weakness.

Location with respect to mantle plume heads

As illustrated in Table 9.2, kimberlites are typically located distal from the plume head of the associated LIP. The explanation for such a pattern is that the plume head rises and moves upward and along lithospheric surfaces to reach "thinspots," areas of this lithosphere which are at such shallow levels that extensive flood-basalt-scale partial melting occurs. However, associated kimberlites would only form from plume material trapped beneath thick lithospheric roots, typically in the cratonic interior away from the plume center.

9.4 Lamprophyres, lamproites, and LIPs

The classification of "exotic" alkaline rocks including lamprophyres (e.g. Rock, 1991) and lamproites (e.g. Mitchell and Bergman, 1991) is complicated and still under debate (Woolley *et al.*, 1996; Le Maitre, 2002; Tappe *et al.*, 2005). According to the

Glossary of Geology, 5th edition (Neuendorf *et al.*, 2011), *lamprophyres* are a group of porphyritic igneous rocks in which mafic minerals form the phenocrysts; feldspars, if present, are restricted to the groundmass. By a second definition, lamprophyres are a group of dark-colored, porphyritic, hypabyssal igneous rocks characterized by panidiomorphic texture, a high percentage of mafic minerals (especially biotite, hornblende, and pyroxene), which form the phenocrysts, and a fine-grained groundmass with the same mafic minerals in addition to feldspars and/or feldspathoids. Ultramafic, calc-alkaline, and alkaline varieties of lamprophyre have been distinguished (Rock, 1991; cf. Le Maitre, 2002). Rock (1991) distinguishes calc-alkaline, alkaline, and ultramafic lamprophyres. The calc-alkaline lamprophyres most likely originate in a convergent setting. However, the other two types, alkaline and ultramafic lamprophyres, are potentially LIP-related (see examples below).

Similarly, according to the *Glossary of Geology* (Neuendorf *et al.*, 2011), *lamproite* is a group name for hypabyssal or extrusive rocks that are rich in potassium and magnesium. A major survey of lamproites is provided in Mitchell and Bergman (1991) and more recently lamproites have been studied in detail in the vicinity of eastern Canada and formerly connected Greenland (Tappe *et al.*, 2006, 2007, 2008; see also Mirnejad and Bell, 2006).

These alkaline magma types (lamprophyres and lamproites) seem to be related to small-volume asthenospheric melts that can interact with the lithosphere. The composition and temperature of asthenospheric magma, the presence, or not, of a plume, the depth of interaction between the asthenosphere and the lithosphere, and the type of prior metasomatism in the lithosphere – these and other factors control the type of alkaline rock that is produced.

Some pathways in this process are illustrated through study of alkaline rocks from Labrador, Canada, and formerly attached Greenland (Tappe *et al.*, 2007, 2008; Fig. 9.18). Mesoproterozoic (*c.* 1375 Ma) lamproite melting occurred at the base of long-term enriched SCLM (subcontinental lithospheric mantle), whereas *c.* 610–550 Ma aillikites formed during extensive interaction between this ancient SCLM and the upwelling asthenosphere. In contrast, the Cretaceous nephelinite suite was produced within a shallower SCLM region that had experienced carbonate metasomatism during the Neoproterozoic aillikite/carbonatite magmatism.

9.4.1 Examples of lamproites and lamprophyres associated with LIPs

A number of lamproites and lamprophyre occurrences can be linked with LIPs. Some of these events also have associated carbonatites and kimberlites.

With reference to Table 9.3, the Leucite Hills lamproite intrusions are spatially linked to the Columbia River LIP (Fig. 3.2; Mirnejad and Bell, 2006) and it is interpreted that the LIP metasomatically enriches the lithosphere, which can melt to produce associated lamproites. Lamprophyres of the Paraná–Etendeka LIP (Fig. 9.5) in part precede and postdate LIP magmatism (e.g. Gibson *et al.*, 2006).

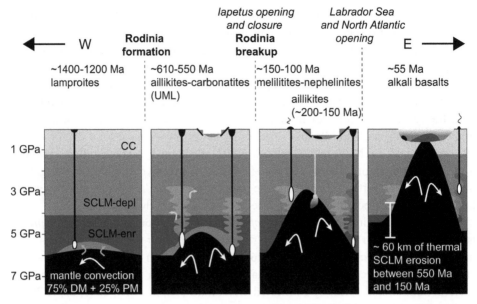

Figure 9.18 Model for the generation of various alkaline magma types in the Labrador region of eastern Canada. For location see Figure 9.8. Events at *c.* 550 Ma are associated with the CIMP LIP (Ernst and Bell, 2010). Those at 130 Ma can be linked with the 130 Ma Trap dyke swarm of southwest Greenland (Buchan and Ernst, 2004) and may represent an early pulse associated with the eventual opening of the Labrador Sea with Greenland. SCLM is subcontinental lithospheric mantle. Redrafted from Figure 9 in Tappe *et al.* (2007). With permission from Elsevier.

Ultramafic lamprophyre dykes associated with the Ferrar portion of the Karoo–Ferrar LIP are characterized by high Ti, Cr, Ni, Nb/La, La_N/Yb_N, and Mg# values, and are the most primitive rocks of the Ferrar event (Riley *et al.*, 2003). Ultramafic lamprophyres of Labrador and central west Greenland are of *c.* 606–570 Ma age and are linked to the opening of the Iapetus Ocean (Heaman *et al.*, 2003, 2004; Tappe *et al.*, 2006, 2008) and therefore to the associated CIMP (Fig. 9.8; Ernst and Bell, 2010). The Keweenawan LIP (Fig. 9.9) has a protracted history with main magmatic pulses between 1115 and 1085 Ma, but there is also a precursor fanning dolerite dyke swarm (Abitibi dyke swarm; Ernst and Buchan, 1993), which is associated with a similarly fanning distribution of *c.* 1144 Ma lamprophyres (Queen *et al.*, 1996). A globally important, 1385 Ma, LIP event has a node in northern Greenland (e.g. Upton *et al.*, 2005; Ernst *et al.*, 2008), and in a Nuna supercontinent reconstruction (e.g. Evans and Mitchell, 2011) other nodes of 1380 Ma LIP magmatism (e.g. Siberia and Baltica) may be juxtaposed adjacent to northeast Greenland. The *c.* 1375 Ma lamproites of Labrador, Canada, and from formerly attached Greenland (Tappe *et al.*, 2006, 2007), are speculatively linked to this 1385 Ma LIP node in northeast Greenland about 1500 km away.

Table 9.3 *Selected LIPs and their associated lamprophyres/lamproites*

LIP name (ages, to nearest 5 Ma)	Associated lamprophyres and lamproites (with precise U–Pb ages)	References for link with the LIP
Columbia River Basalt Group (15–0 Ma); Section 3.2.2	Leucite Hills lamproites (1.6 Ma) (Mitchell and Bergman, 1991)	Mirnejad and Bell (2006)
Rajmahal (120 Ma) (linked to Kerguelen–Bunbury–Comei); Section 4.2.2	Damador valley aillikites/lamproites (117 Ma)	Chalapathi Rao and Lehmann (2011 and references therein)
Paraná–Etendeka (130 Ma); Section 3.2.6	Lamprophyres of Paraná–Etendeka LIP (110–125 and 145–165 Ma)	Le Roex and Lanyon (1998); Gibson *et al.* (2006)
Ferrar (180 Ma), part of Karoo–Ferrar	Lamproites/orangeites of Dronning Maud Land (159 Ma)	Luttinen *et al.* (2002); Riley *et al.* (2003); Romu *et al.* (2008)
CIMP (Central Iapetus Magmatic Province) (615–555 Ma); Section 9.2.2	Ultramafic lamprophyres (Labrador)	Tappe *et al.* (2006, 2007, 2008)
Keweenawan (1115–1085); Section 3.5.2	Precursor diabase dykes and lamprophyre dykes (1144 Ma) (Queen *et al.*, 1996)	Queen *et al.* (1996); Heaman *et al.* (2007)
Midsommerso–Zig-Zag Dal (1385 Ma)	Lamproites, northern Labrador, Canada (Tappe *et al.*, 2007, 2008)	Herein
Sparrow–Uranium City (1830 Ma)	Christopher Island Formation lamproites; Dubawnt minette dyke swarm (www.largeigneousprovinces.org/10dec; Cousens *et al.*, 2001, 2004; Peterson *et al.*, 2002)	Ernst and Bleeker (2010)

In the Western Churchill province of Canada (centered in the Baker Lake region), at 1830 ± 20 Ma, there are approximately 5000 km^3 of potassic to ultrapotassic subaerial lamprophyric lavas (phlogopite–clinopyroxene minettes) (Christopher Island Formation) along with a feeder dyke swarm (Dubawnt minettes) erupted over an area of 200 000 km^2 (Peterson *et al.*, 2010). At the same time, granodioritic laccoliths (the Hudson suite) were emplaced at midcrustal levels in a > 850-km-wide, > 2000-km-long swath. This magmatism is coeval with the Sparrow dolerite dyke swarm and associated Uranium City dykes further west in the Rae craton (Ernst and Bleeker, 2010). The scale of the Sparrow dyke swarm (400 × 250 km) is sufficient to be labeled as a LIP, and the age similarity suggests the potassic alkaline suites of the Baker Lake basin to the east must be part of the same event.

Meimechites may form the magnesian end-member of a suite of continental ultrapotassic rocks that also includes lamproites and some lamprophyres (Arndt, 2003; Chapter 10). Meimechite is a rare type of ultramafic, potassium-rich volcanic rock found in northern Siberia, in the Maimechi–Kotui region along with associated carbonatites and alkaline complexes, which are all part of the Siberian Trap LIP (Section 3.2.8).

9.5 Summary

Carbonatites have a robust link with LIPs, and are commonly emplaced along associated rifts. Many kimberlite fields are similarly linked to LIPs and plume heads, but typically are distal from the plume center and are located in regions of thick lithosphere. Other kimberlite fields show a regional age progression, which can be linked to the passage of plumes under the continents, and thus represent the equivalent of hotspot trails in oceanic areas (cf. Chapter 2). More exotic alkaline magma types, the lamprophyres and lamproites, are also linked to sublithospheric mafic melts, possibly interacting with metasomatized lithosphere. Some, and perhaps many lamprophyres and lamproites can be linked with LIPs. In such cases, the magmatic underplate associated with the LIP can be a source of magma and heat that interacts with metasomatized lithosphere to produce lamprophyres and lamproites.

With this chapter I have completed the survey of magmatic types associated with LIPs, and in the next chapters I will consider more thematic aspects, starting with the geochemistry of LIPs.

10

Geochemistry of LIPs

10.1 Introduction

This chapter provides an overview of the geochemical characteristics of different components of Large Igneous Provinces (LIPs): continental LIPs (continental flood basalts, CFBs), oceanic LIPs (oceanic plateaus and ocean-basin flood basalts), LIP plumbing-system components (dolerite dykes and sills and mafic–ultramafic layered intrusions), Archean LIPs, and also related units: silicic LIPs, carbonatites and alkaline rocks, selected kimberlites, and lamprophyres/lamproites. The literature is enormous and I focus on selected geochemical aspects particularly relevant to addressing a LIP as a system of interacting components and magma pathways. Each component (extrusive, intrusive, and related silicic and alkaline units) has its own story to tell about the overall magmatic system, and collectively these components provide critical constraints on the mantle source (sublithospheric, asthenospheric, and/or plume), and the compositional modification through interaction with lithospheric mantle/crust during transport via dykes and sills (with pauses in magma chambers at various depths) toward final emplacement in the crust as underplate intrusives and extrusives. First, I consider potential mantle sources for LIPs.

10.2 Mantle sources and processes for producing LIPs

10.2.1 Mantle reservoirs

Based on the record from ocean-island basalts (OIBs), the mantle source for LIPs must include contributions from one or more of the five main mantle reservoirs: DMM (depleted MORB (mid-ocean-ridge basalt) mantle) – the shallow source of spreading ridge basalts, HIMU (so-called because it has a high ^{238}U–^{204}Pb ratio, or μ value), EM1 (enriched mantle 1), EM2 (enriched mantle 2), and FOZO (focal zone) (e.g. Zindler and Hart, 1986, Hart, 1988; Hart *et al.*, 1992; Farley *et al.*, 1992; Hauri *et al.*, 1994; Hofmann, 1997; Hilton *et al.*, 1999; van Keken *et al.*, 2002; Stracke *et al.*, 2005; Fig. 10.1).

DMM has low $^{87}Sr/^{86}Sr$, low $^{206}Pb/^{204}Pb$, and high $^{143}Nd/^{144}Nd$ ratios, and represents the long-term depleted mantle after *c.* 3 Ga of MORB extraction. This

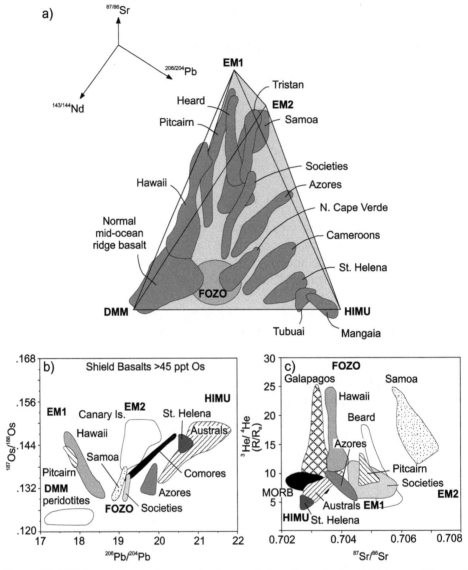

Figure 10.1 Main geochemical reservoirs in mantle based on the isotopic compositions of OIBs. (a) Mantle tetrahedron from Hofmann (2003), with permission from Elsevier, and originally after Hart *et al*. (1992), reprinted with permission from AAAS. (b) and (c) These show similar story based on Os and He isotopes from van Keken *et al*. (2002). Reproduced with permission of *Annual Review of Earth and Planetary Sciences* Volume 30 by Annual reviews, http://www.annualreviews.org. Most of the individual arrays appear to radiate from a common depleted region, FOZO, which is thought to represent the composition of the deep mantle. R/R_a = measured He ratio relative to the measured atmospheric ratio, which is 1.39×10^{-6}.

reservoir is typically considered to be located in the upper mantle. The other four reservoirs are characteristic of the sources of OIBs, and arguably are located in the deeper sublithospheric mantle. EM1 has high ^{87}Sr/^{86}Sr, low ^{206}Pb/^{204}Pb, and low ^{143}Nd/^{144}Nd ratios. The other enriched component, EM2, is characterized by high ^{87}Sr/^{86}Sr (> EM1), moderate ^{206}Pb/^{204}Pb, and low ^{143}Nd/^{144}Nd ratios. HIMU represents strong enrichment in ^{206}Pb/^{204}Pb, ^{207}Pb/^{204}Pb, ^{208}Pb/^{204}Pb and depletion in ^{87}Sr/^{86}Sr. An additional depleted reservoir (distinct from MORB) is suggested by the convergence of ocean island trends in isotopic space (Fig. 10.1). This reservoir, FOZO, is interpreted to represent a common OIB component located in the lower mantle (Hofmann, 2003; Bennett, 2003; Stracke *et al.*, 2005; Jackson *et al.*, 2007). A more recent interpretation considers FOZO (slightly modified in position from the original definition) to be a ubiquitous component in the source of MORB as well as OIB produced by the continuous recycling and aging of unmodified, oceanic crust (Stracke *et al.*, 2005). This depleted component also has high ^{176}Hf/^{177}Hf (Kempton *et al.*, 2000) and high ^{187}Os/^{188}Os isotopic ratios (Shirey and Walker, 1998; van Keken *et al.*, 2002). High ^{3}He/^{4}He ratios are observed at Hawaii, Iceland, Bouvet, Galapagos, Easter, Juan Fernandez, Pitcairn, Samoa, Reunion, and Heard islands, and they increase toward the FOZO component (Hofmann, 1997; van Keken *et al.*, 2002). According to Stracke *et al.* (2005), FOZO and DMM represent a continuum of compositions, but FOZO can be distinguished from DMM by much higher ^{3}He/^{4}He isotope values and slightly lower ^{143}Nd/^{144}Nd ratios. The influence of these mantle reservoirs can also be monitored using particular trace-element diagrams (e.g. Fig. 10.2c).

Origin of mantle reservoirs

EM1, EM2, and HIMU components can be derived from subducted oceanic lithosphere that has been carried to depth and incorporated into (or is already part of) rising plumes (e.g. Hofmann, 1997, 2003; Condie, 2001, 2003). Specifically, the HIMU component may represent the oceanic crust from which preferential loss of Pb during subduction allowed the U+Th/Pb ratio to increase with aging to produce high ^{206}Pb/^{204}Pb and ^{208}Pb/^{204}Pb ratios. EM1 had proposed links to old oceanic mantle lithosphere (plus or minus sediments) or metasomatized lower mantle and recycled plume heads, and EM2 can represent subducted oceanic sediments, essentially reflecting a continental crust signature. In short, the conventional view is that EM2 represents a continental crust component, EM1 represents mantle lithosphere, and HIMU represents oceanic crust.

An alternative view has been proposed by Collerson (2010), who suggested complementary formation of HIMU (solid restite) and EM2 (melt) from CO_2-fluxed melting in the lower mantle. Therefore, according to Collerson, HIMU does not reflect involvement of hydrothermally altered oceanic crust, and EM2 does not require entrainment of continent-derived sediment introduced during subduction of oceanic lithosphere.

The age of formation of these isotopic reservoirs is important in interpreting the isotopic data from older LIPs. The HIMU, EM1, and EM2 reservoirs (corresponding to the OIB sources) became significant in the Paleoproterozoic and probably date

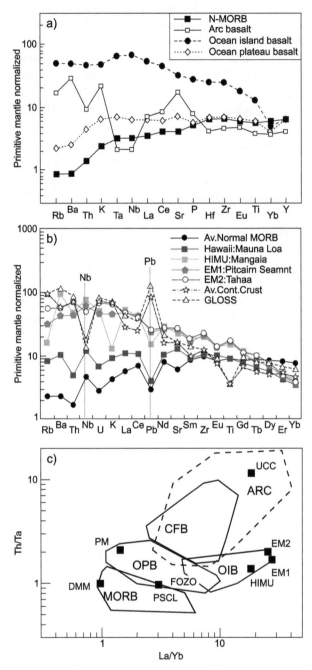

Figure 10.2 Geochemistry diagrams for basalts of different settings. Primitive mantle normalized multi-element diagrams ("spidergrams"): (a) for basalts from various tectonic settings (Figure 5.1 in Condie (2001); see also Figure 1 in Tomlinson and Condie (2001), copyright © 2001 Cambridge University Press; and (b) for representative samples of HIMU

from that time. This is reasonable in terms of models in which these reservoirs are associated with subducted slabs that accumulated in the deep mantle (Campbell, 1998). However, FOZO may represent a still older depleted component that developed during the Archean (Campbell, 1998). Hf isotopic values also suggest an Archean age for the depleted component (Kempton *et al.*, 2000). More recently Jackson and Carlson (2011) suggested that such a depleted reservoir has been in existence since nearly 4.5 Ga, and has been tapped in the NAIP (North Atlantic Igneous Province), Ontong Java, Deccan, Kerguelen, and Siberian Trap LIPs.

10.2.2 Partial melting

LIPs are the products of partial melting of these mantle sources, but the range of melt compositions reflects additional considerations, such as the depth and degree of melting, presence of fluids (H_2O and CO_2) in the source areas and subsequent interaction with metasomatized regions of lithospheric mantle (including possible partial melting of this lithospheric mantle), and contamination by interaction with the crust.

LIP-related tholeiitic magmas are the products of relatively high degrees of partial melting (commonly 20–30%) (Fig. 4 in Herzberg and Gazel, 2009; Figs. 10.3, 15.15) of a near-anhydrous peridotitic source, and melting of entrained eclogite streaks may also be important (Cordery *et al.*, 1997). Alkali basalts and other members of the alkaline magma series are produced by lower degrees of melting (Fig. 10.3), commonly of a source that is geochemically enriched or contains a relatively high proportion of basaltic/eclogitic material. The source (low-percentage melting) of alkali magmas may contain high levels of CO_2 that play an important role in the generation of silica-undersaturated members of the magma series. Calc-alkaline affinities characterize melts associated with subduction and melting in the mantle wedge above the descending slab. However, as discussed later (Section 10.5.4) similar subduction signatures in basalts can be produced by interaction with lithosphere that has been previously metasomatized by low-degree partial melts of the asthenosphere and/or subduction-derived fluids.

As summarized by Arndt (2003), a similar pattern exists for ultramafic magmas. Differing conditions produce a spectrum of ultramafic magma types. Komatiites (a key part of Archean LIPs; Chapter 6) form by high degrees of melting, some at great depths, of an essentially anhydrous source, although there is some controversy on this point in that hydrous melting has also been argued (e.g. see discussion in Section 15.2.5; Arndt *et al.*, 2008; Gurenko and Kamenetsky, 2011). Kimberlites

(Mangaia, Austral Islands), EM1 (Pitcairn), Caption for Figure 10.2 (*cont.*) EM2 (Tahaa, Society Islands); average Mauna Loa (Hawaii) tholeiite, average continental crust, average subducting sediment (GLOSS), and MORB. From Hofmann (2003). With permission from Elsevier. (c) "Condie" diagram Th–Ta versus La–Yb diagram showing fields for main basalt types and locations of reservoirs. OPB = ocean-plateau basalt and ARC = arc-related. FOZO is only approximately located. Diagram modified from Condie (2003).

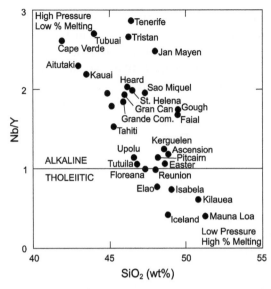

Figure 10.3 OIBs and depth of origin. Plot of average SiO_2 versus Nb/Y in OIBs. Highly alkaline magmas are products of low percentages of melting (hence high Nb/Y) at great depths (low SiO_2) below old, cold, and thick lithosphere. Tholeiites form under opposite conditions. Abbreviation: Gran Can = Gran Canaria, Canaries. From Figure 3 in Greenough *et al.* (2005).

(associated with many LIPs; Chapter 9) are magnesian, incompatible-element-, and volatile-enriched liquids resulting from low-degree melting, also at great depth, of sources rich in incompatible elements and $CO_2 + H_2O$. They become further enriched through interaction with overlying asthenospheric or lithospheric mantle. An origin by lithospheric modification of primary carbonatite magmas has been advocated by Russell *et al.* (2012).

10.2.3 Interaction with lithospheric mantle

Mantle lithosphere can have an influence in two different ways (e.g. Garfunkel, 2008). CFBs can form by melting of hydrated (metasomatized) lower subcontinental mantle lithosphere resulting from heating by plumes, while the plumes themselves do not melt. Paraná low-Ti CFBs of the Paraná–Etendeka LIP are the type example (Gallagher and Hawkesworth, 1992; Turner and Hawkesworth, 1995; Hawkesworth *et al.*, 1999). The other scenario is one in which the lithospheric contribution derives from contamination of plume-derived melts on their ascent through the lithosphere (e.g. Menzies, 1992; Arndt and Christensen, 1992; Arndt *et al.*, 1993).

Lithospheric involvement can also potentially be recognized using radiogenic isotopes. In terms of the Nd–Sr isotope diagram, the influence of the lithosphere is to move the composition to the lower right quadrant (lower εNd and higher

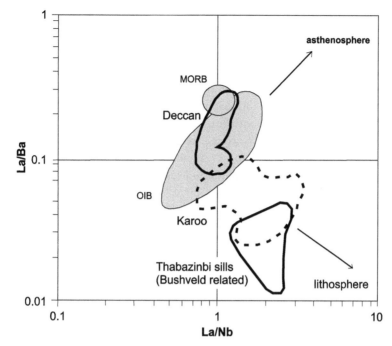

Figure 10.4 La/Ba vs. La/Nb plot for discriminating asthenospheric and lithospheric contributions. The data shown are from the Deccan and Karoo low-Ti basalts and sills and Bushveld LIPs. From Figure 8 in Jourdan *et al.* (2009) and Figure 14 in Jourdan *et al.* (2007a). Bushveld data are from Rajesh *et al.* (2013). See also Saunders *et al.* (1992).

^{87}Sr/^{86}Sr), nominally toward EM1, which is viewed as caused by lithospheric mantle. However, there can be ambiguity as to whether this represents a component derived from earlier subduction and entrained in the convecting mantle or whether it is acquired from interaction with *in situ* lithospheric mantle.

Carefully chosen trace-element diagrams can be used to distinguish an asthenospheric from a lithospheric mantle source. For instance, a La/Ba–La/Nb plot (Fig. 10.4) can be useful for CFB rocks (e.g. Saunders *et al.*, 1992). As summarized by Jourdan *et al.* (2009), positive correlations between La/Nb and La/Ba reflect OIB and/or asthenospheric mantle source(s), whereas negative correlations are diagnostic of a strong lithospheric contribution, wherein the lithosphere has a negative Nb anomaly as a result of metasomatism during a prior subduction event in which Nb was preferentially retained in the slab. These ratios (La/Nb and La/Ba) are negligibly modified by petrogenetic processes such as partial melting or fractional crystallization and thus should represent mantle-source signature(s). One drawback to this diagram is the susceptibility of Ba to alteration and another is the possible sensitivity to crustal assimilation (e.g. assimilation fractional crystallization, AFC). An example of the successful use of this diagram, Fig. 10.4, shows the substantial influence of the lithosphere on the Bushveld and Karoo LIPs, but a negligible influence of the lithosphere on the Deccan LIP.

The classic example of melting metasomatized Archean continental lithospheric mantle is the 240 000 km^2 1.8 Ga Christopher Island Formation within the Churchill province of northern Laurentia, the influence of which is seen in subsequent magmatic events passing through this lithosphere (Cousens *et al.*, 2001).

As a final point, lithospheric contributions can be highly spatially and temporally variable. They will depend on the original composition of lithospheric mantle and the subsequent history of interaction with plumes and subducting slabs, each of which will have imparted metasomatic changes that will vary significantly on a regional scale.

10.2.4 Crustal contamination

Compared with uncontaminated mantle-derived magmas, most contaminated magmas are enriched in highly incompatible elements (e.g. Th, LREE, U, alkali metals), and exhibit negative Ta, Nb, P, and Ti anomalies. These are similar to many of the same changes caused by interaction with metasomatized lithospheric mantle, i.e. increases in Ba, Rb, K, and an associated increase in Sr isotopic and a decrease in Nd isotopic ratios. Fortunately, comparison against SiO$_2$ and MgO for a suite of samples provides a method to distinguish crustal and lithospheric mantle influences (e.g. Hollings *et al.*, 2012). Continental crust has an average andesitic composition (e.g. Rudnick and Gao, 2003), which is distinctly higher in silica than the ultramafic composition of lithospheric mantle. Therefore, bulk crustal contamination will cause an increase in SiO$_2$ and a decrease in MgO. In contrast, contamination by lithospheric mantle only affects these elements (SiO$_2$ and MgO) in a minor way. So plots of contamination-sensitive variables (involving, for instance, Ba and Nd isotopes) against SiO$_2$ or MgO can reveal whether the changes are due to interaction with the crust or lithospheric mantle.

As one example, rocks of the Keweenawan (Midcontinent) LIP on the north shore of Lake Superior, southern Canada, have a negative εNd_T, which could be the result of lithospheric or crustal contamination. The lack of a correlation with other indices of fractionation (e.g. SiO$_2$) supports an interaction with subcontinental lithospheric mantle, either through assimilation or melting of the mantle, rather than crustal contamination during emplacement (Hollings *et al.*, 2012).

The specific effects of crustal contamination will depend on the process (e.g. AFC; RAFC, recharge assimilation fractional crystallization; ATA, assimilation during turbulent flow; and MASH, melting–assimilation–storage–homogenization; each having their own specific trajectories on the trace-element diagrams (e.g. Pearce, 2008).

Some additional ideas related to monitoring the crustal interaction of basaltic magmatism (of LIPs) can be extracted from a broader look at the trace-element chemistry of basalts. As summarized by Hofmann (1997), the ratios of Ba/Rb, K/U (except in HIMU magmas), Nb/Ta, Zr/Hf, Y/Ho, Ti/Sm, Sn/Sm, and P/Nd are approximately constant for oceanic basalts (both OIBs and MORBs) and continental crust. Nb/Ta, Zr/Hf, and Y/Ho ratios also approximately match the chondritic ratios of primitive mantle (Hofmann, 1997). Other elemental ratios, Nb/U, Ta/U, and Ce/Pb, are useful in studying interaction with the continental crust; these ratios are similar in MORBs and OIBs, but much lower values are found in continental crust and island-arc volcanics (and primitive mantle) (Hofmann, 1997). For example, the Nb/U ratio for OIBs and MORBs is about 47, whereas that for chondrites and primitive mantle is about 32 and that for continental crust is about 10. Similarly, the Nb/Th ratio of primitive mantle is 8 whereas the upper crust has a ratio of 1.1. So selective use of these ratios can yield diagrams that effectively allow monitoring of crustal influence on mafic magmatism. The ratios Ba/La, Rb/La, and Th/La are generally greater than (or equal to) the respective source areas (Hofmann, 1997). Thus, if a source has experienced an episode of prior melt extraction, these ratios will be lower than primitive mantle values. Ratios involving Ba, K, Rb, and U can be less useful because of the mobility of these elements during alteration events, whereas Nb/Th is robust (e.g. Kerrich and Xie, 2002; Said *et al.*, 2012; Jowitt and Ernst, 2013).

It is also possible to use Os isotopes to distinguish crustal from lithospheric contamination (e.g. Figure 5.16 in Condie, 2001; Richardson and Shirey, 2008; Day, 2013). Specifically, during mantle melting Os is highly compatible whereas Re is moderately incompatible. So continental crust will have a high Re/Os ratio and hence as the crust ages the $^{187}Os/^{188}Os$ ratio (expressed as γOs) will increase and become large. In contrast, γOs in the lithospheric mantle will remain small. So interaction with the continental crust will impart a high γOs value but interaction with the lithospheric mantle will impart a low γOs value (e.g. Fig. 10.5).

10.2.5 No unique geochemical fingerprint for LIPs

An important part of the LIP definition is an intraplate setting (see the detailed discussion in Chapter 2). However, as shown below, this notion does not imply a unique chemistry for LIPs as a group.

Recognition of intraplate geochemistry is easily defined on the basis of various classification diagrams. Starting in the 1970s, a number of diagrams, mainly based on a combination of incompatible and compatible elements, especially those that are also generally immobile during hydrothermal alteration and metamorphism, were developed to distinguish intra- (within-) plate basaltic magmatism from MORB and arc basalts and monitor the effects of alteration and crustal contamination (e.g. Pearce and Cann, 1973; Winchester and Floyd, 1977; Shervais, 1982; Thompson *et al.*, 1983; Meschede, 1986; Kerrich and Wyman, 1997; see reviews in Rollinson, 1993; Pearce, 1996a, 2008). In fact, many LIPs do plot in the intraplate field on these diagrams (e.g. Hawkesworth

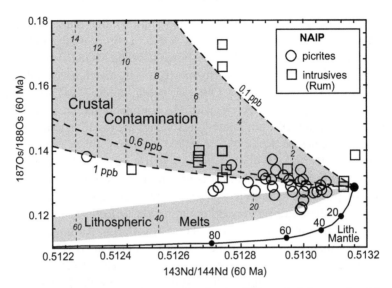

Figure 10.5 Modification by crustal contamination in CFBs and their associated intrusions. Os/Nd isotope data for the *c.* 60 Ma NAIP picrites (circles) from West Greenland and Baffin Island are shown versus intrusive rocks from the Rum intrusion, Scotland (squares). Three types of interactions are shown: contamination by the crust, contamination by lithospheric mantle melts, and contamination by interaction with unmelted mantle lithospere. The numbers on each: 0–14 for crust, 20, 40, and 60 for lithospheric melting; and 20, 40, 60, and 80 for lithosphere, all represent percentages of contamination. Values of 0.1, 0.6, and 1 ppb represent differing concentrations of Os in the parent melt for purposes of modeling. From Figure 7 in Day (2013). With permission from Elsevier.

and Scherstén, 2007). However, as noted by Hawkesworth and Scherstén (2007), other smaller non-LIP-scale intraplate events also have intraplate chemistry and so an intraplate geochemical signature is not a sufficient criterion to identify magmatism as belonging to a LIP. Furthermore, not all LIPs exhibit intraplate geochemistry; as noted in Section 10.5.4, many continental LIPs exhibit a subduction signature (e.g. a negative Nb–Ta anomaly) and, as discussed later, this subduction signature is interpreted to derive from subduction-metasomatized mantle lithosphere, and/or continental crust and not to represent a subduction setting at the time of LIP emplacement. As shown later (Section 10.3) there is a great range in the isotopic compositions of LIPs and so isotopes do not provide a defining characteristic for LIPs either.

   These facts lead to some ambiguity in the recognition of LIPs from the perspective of geochemistry alone. The other key criteria, large magma volume and short duration of magmatic pulses, become essential characteristics when the geochemisty is ambiguous. However, while there is no diagnostic chemistry for LIPs as a group, individual LIPs can have one or more distinctive magma compositions (particularly based on the trace-element patterns) and compositional trends that represent a fingerprint for that particular LIP and reveal the petrogenetic processes responsible for the observed compositions.

10.2.6 Monitoring of processes by trace-element patterns

An insightful summary of the geochemical constraints on oceanic basalts (including oceanic LIPs, and with implications for CFBs) was provided by Pearce (2008) and much of the following is drawn from this publication. In a multi-element diagram, generally normalized to an average MORB composition, elements are typically ordered in terms of increasing incompatibility from right to left (Fig. 10.6). In Th/Yb versus Nb/Yb coordinate space, at a first order, the within-plate basalt MORB–OIB array is distinguished from convergent-margin magmas. At a second order, WPB are separated into MORBs, enriched (E) MORBs, and alkaline OIBs, whereas convergent-margin magmas are divided into tholeiitic intraoceanic, calc-alkaline transitional, and shoshonitic continental margin series. Three sections of the diagram have specific information about magma generation/modification processes: Ti–Yb for monitoring depth of melting, Nb–Ti for monitoring source area and degree of melting, and La–Nb for monitoring crustal contamination (Fig. 10.6). La–Sm (over-lapping with the Nb–Ti range) can be interpreted as follows. Lower $(La/Sm)_{PM}$ (PM = normalization to primitive mantle) ratios can be caused by three different pro-cesses, namely crustal contamination or increased partial melting, both of which can increase La/Sm ratios in a melt, or melting of a depleted source, which would lower La/Sm ratios (e.g. Khudoley *et al.*, 2013).

These concepts, related to the interpretation of the slopes of different parts of the multi-element diagrams, can lead to some useful diagrams (Figs. 10.7 and 10.8). A key one is Th vs. Nb (i.e. Th/Yb–Nb/Yb) chosen to highlight the crustal contamination and also enriched sources; the second is Ti vs. Nb (i.e. Ti/Yb–Nb/Yb) chosen to highlight the melting depth. Each is discussed below. For full details see Pearce (2008).

Th/Yb–Nb/Yb diagram

The Th/Yb–Nb/Yb diagram (Fig. 10.7) exhibits an array along which nearly all oceanic basalts lie. Furthermore, the position along this array represents the range from MORBs (normal (N), enriched (E), and plume (P) types) to OIBs. Subduction (e.g. assimiliation of arc-related material) pulls the composition above the MORB–OIB array, as does nearly all crustal contamination (although some granulites represent an exception). As summarized by Pearce (2008), oceanic basalts (intraplate islands, plume–distal ocean ridges, and oceanic plateaus) predominantly plot within the MORB–OIB array (shaded) while volcanic-arc basalts plot above the array. Crustally contaminated basalts and alkalic basalts containing a large recycled crustal component mainly plot above the MORB–OIB array, or on a vector at a steep angle to the array, reflecting greater addition of Th than Nb. Although these diagrams are typically used for basalts repre-senting melt compositions, they could also be useful for many cumulate rocks, as they use ratios of immobile incompatible elements that would therefore be transparent to cumulate effects. Care is needed, however, as for example with cumulation of oxides which can be Nb-rich (J. Pearce, written communication, 2013).

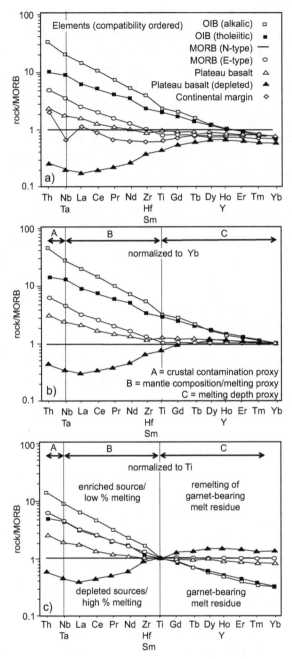

Figure 10.6 MORB-normalized geochemical patterns for types of oceanic basalts. (a) Patterns with no additional normalization. (b) The oceanic basalt patterns normalized also to Yb = 1 (which forms the basis of the discriminant diagrams) to reduce the effect of fractional crystallization. (c) The oceanic basalt patterns normalized also to $TiO_2 = 1$, which highlights the difference between the right-hand side of the pattern (garnet-dependent) and the left-hand side (partial melting and source-composition-dependent). From Pearce (2008). With permission from Elsevier.

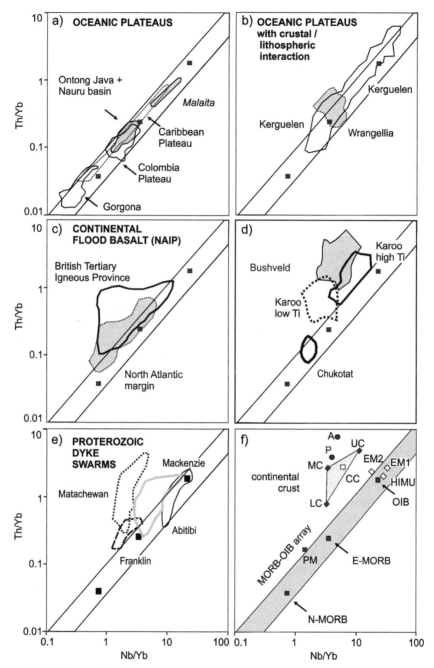

Figure 10.7 Th/Yb–Nb/Yb diagram of Pearce (2008) for LIPs of different settings. (a)–(c) Extracted from diagrams in Pearce (2008). (d) From Rajesh *et al.* (2013). (e) Proterozoic dyke swarms of Canada; data from Jowitt and Ernst (2013), with the plot prepared by S. Jowitt. (f) Petrogenetic modeling of the Ti/Yb–Nb/Yb diagram (from Figure 8 in Pearce (2008)).

Figure 10.7 provides a useful overview of the chemistry of LIPs. The oceanic LIPs plot within the MORB–OIB array indicating no interaction with lithosphere or crust, which is understandable given that oceanic LIPs are not interacting with continental crust. The only exceptions are magmatism of Kerguelen and Wrangellia LIPs, which plot near OIBs and also outside the array, indicating some involvement of the lithosphere or crust. Again, this makes sense in that Kerguelen is considered to overlie a crustal fragment and Wrangellia is considered to have been built on an island arc (Chapter 4). As exemplified by the British Tertiary province and North Atlantic margin (parts of the NAIP), continental LIPs can exhibit significant involvement of the lithosphere and continental crust. The same is true for the Bushveld LIP, and Karoo low- and high-Ti magmatism, which partly plot above the OIB–MORB array, indicating significant interaction with mantle lithosphere and crust. For the 1880 Ma Chukotat magmatism (of the Circum-Superior LIP), sills and volcanics are within the array, and therefore interaction with the mantle lithosphere or crust can be ruled out (Fig. 10.7d). However, various regional Proterozoic dyke swarms (representing the plumbing system of LIPs) of Canada (Fig. 10.7e) do show the influence of the lithosphere/crust. Only the Abitibi dyke swarm (precursor to the Keweenawan LIP) shows no interaction with the crust or lithosphere outside the OIB–MORB array. Another observation from Fig. 10.7 is that while geochemistry can provide a potential fingerprint for an event, there is no unique geochemical composition held by all LIPs that would be diagnostic of a LIP vs. other types of magmatism.

The Ti/Yb–Nb/Yb diagram

Another useful diagram is Ti/Yb–Nb/Yb (Fig. 10.8), which can decipher the depth of melting and hence is sensitive to variations in mantle temperature and lithospheric thickness (Pearce, 2008). As summarized by Pearce (2008), OIBs have higher Ti/Yb ratios than MORBs, reflecting deeper melting that results from a combination of a thicker lithospheric "lid" and hotter mantle temperatures. Furthermore, within OIBs, alkalic basalts have the higher Nb/Yb and Ti/Yb ratios, reflecting a low degree of deep melting. In contrast, MORBs, N-MORBs plot to the left of the array (defined as having Nb/Yb < 1.45, the C1 chondrite value) and E-MORBs to the right of the array (defined as having Nb/Yb > 1.45). In areas of plume–ridge interaction (near-ridge plumes, and ocean–continent transitions at volcanic-rifted margins), basalts follow a diagonal trend from the tholeiitic OIBs to the MORB–OIB array. Finally,

Caption for Figure 10.7 (*cont.*) Oceanic basalts (intraplate islands, plume–distal ocean ridges, and oceanic plateaux) predominantly plot within the MORB–OIB array (shaded) while volcanic-arc basalts plot above the array. Crustally contaminated basalts and alkalic basalts containing a large recycled crustal component mainly plot above the array. N-MORBs, E-MORBs, OIBs, and primordial mantle (PM) are from Sun and McDonough (1989); average lower crust (LC), upper crust (UC), total continental crust (CC), felsic Phanerozoic (P) and Archean (A) crust are from Rudnick and Fountain (1995). Digital version of original diagram kindly provided by J. Pearce.

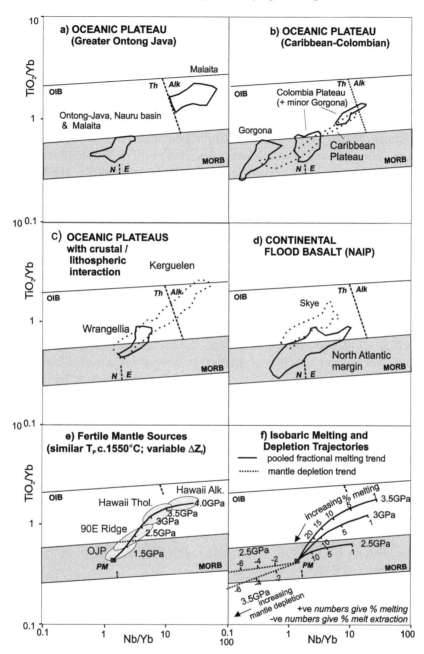

Figure 10.8 Ti/Yb–Nb/Yb OIBs have higher Ti/Yb ratios than MORBs reflecting deeper melting, which results from a combination of a thicker lithospheric cap and hotter mantle temperatures. Within OIBs, alkalic basalts have the higher Nb/Yb and Ti/Yb ratios. Within MORBs, N-MORBs lie to the left of the array and E-MORBs to the right of the array. Diagram from Pearce (2008), with permission from Elsevier, and data from Pearce (2008) generalized as fields, with permission from Elsevier. Digital version of original diagram kindly provided by J. Pearce.

oceanic plateaus typically form clusters near the center of the MORB array, which is indicative of a high degree of melting in which residual garnet plays a minor role.

10.3 Overview of geochemical variations among LIPs

A robust geochemistry dataset (Zhang *et al.*, 2008) illustrates some of the overall compositional characteristics of CFBs and their comparison with the compositions of oceanic plateaus and magmatism associated with LIP plumbing systems (Table 10.1; Figs. 10.9–10.12). Zhang *et al.* (2008) selected ten LIPs of which seven are CFBs, two oceanic LIPs (Ontong Java and Kerguelen), and one plumbing-system-dominated type (i.e. Bushveld), with the intent of identifying compositional criteria linked to fertility for Ni–Cu–platinum-group-element (PGE) mineralization (Chapter 16).

Overall, LIPs, especially continental LIPs, exhibit significant variations in major and trace elements and radiogenic isotopic signatures, well beyond the ranges shown by OIBs. This may be attributed not only to the heterogeneity of their mantle sources, but also to the complexity of subsequent differentiation (AFC) processes involved in magma evolution of continental LIPs. For example, the REE patterns for the LIPs (Fig. 10.10) vary dramatically. Also, the εNd vs. Mg# diagram (Fig. 10.11) shows data for several continental LIPs that have a large range suggestive of significant AFC processes.

Isotopic data from LIPs have a range of values extending from depleted mantle to the enriched quadrants (overlapping with MORB and OIB components), and suggestive of involvement of enriched mantle components but also suggestive of crustal or lithospheric contamination (Fig. 10.12). Zhang *et al.* (2008) suggest that some LIPs, i.e. Bushveld, Siberian Trap, Keweenawan, Emeishan, and Karoo, those that they view as fertile in terms of Ni–PGE potential (Chapter 16), display trends of Sr–Nd–Pb isotopic variation intermediate between the depleted plume and an EM1-type mantle. In contrast, they view other LIPs, i.e. Deccan, Kerguelen, Ontong Java, Paraná (part of Paraná–Etendeka), Ferrar (part of Karoo-Ferrar), which they view as barren in terms of Ni–PGE potential (Chapter 16), to have positions between depleted plume and EM2.

10.4 Geochemistry of oceanic LIPs

More detail on the geochemistry of oceanic plateaus is provided here. Oceanic LIPs are potentially simpler than CFB systems owing to their oceanic setting. While they have the same mantle sources that produce OIBs and MORBs, they do not have the superimposed and complicating effect of continental and subcontinental lithospheric influences – except for Kerguelen and Wrangellia which appear oceanic but actually have a continental affiliation (see discussion below). Some key publications on their

Table 10.1 *Compositional ranges of some key geochemical parameters of LIPs*

Province	Bushveld (Section 5.4.1)	Keweenawan (Section 3.5.2)	Siberian Trap (Section 3.2.8)	Emeishan (Section 12.2.10)	Karoo (Section 3.2.7)	Ferrar (Section 3.2.7)	Paraná, part of Paraná–Etendeka (Section 3.2.6)	Ontong Java (Section 4.2.1)	Kerguelen, part of Kerguelen–Bunbury–Comei (Section 4.2.2)	Deccan (Section 3.2.5)
MgO, wt%	18.0–3.5	16.8–3.8	39.8–2.9	30.1–3.6	21.5–4.0	10.7–3.1	20.0–3.2	11.2–5.4	10.2–3.0	15.3–3.1
Al_2O_3, wt%	17.5–8.4	17.9–8.3	21.1–1.7	18.1–5.6	17.2–7.5	21.9–11.4	18.3–9.2	17.1–13.4	21.0–13.2	18.8–10.0
Mg#	0.82–0.44	0.76–0.36	0.85–0.305	0.85–0.34	0.80–0.31	0.73–0.31	0.79–0.33	0.65–0.32	0.69–0.38	0.73–0.32
CaO/Al_2O_3	0.81–0.50	1.69–0.27	2.95–0.28	2.12–0.25	1.22–0.44	0.87–0.49	0.89–0.46	0.90–0.64	0.81–0.24	1.31–0.24
K_2O/TiO_2	4.15–0.07	4.2–0.04	2.67–0.015	1.73–0.007	11.7–0.034	2.55–0.32	2.54–0.10	1.06–0.019	1.68–0.08	1.89–0.04
Ni, ppm	720–36	876–18	2175–6	1406–4	1140–8	167–7	893–15	227–12	220–15	573–8
Ba, ppm	639–79	1160–46	1870–19	1700–5.2	1525–33	1400–97	798–38	133–4.7	636–40	1545–29
Nd, ppm	38.2–3.4	71.1–4.6	113–2.5	74–1.4	97.4–6.9	41.3–7.7	73–5.0	14.1–4.3	80–3.6	49–5.9
Yb, ppm	3.3–0.72	8.1–0.8	6.2–0.4	4.0–0.7	4.9–1.2	6.2–1.4	5.2–1.1	3.8–1.6	5.7–1.4	4.7–0.76
$(La/Sm)_n$	4.7–1.9	4.2–1.2	6.0–1.1	5.2–0.34	3.2–0.49	5.9–1.6	3.3–0.92	2.5–0.79	5.11–0.86	7.3–0.86
$(La/Yb)_n$	14.1–3.0	24.2–1.5	63–1.7	30.1–0.24	33.3–1.4	5.5–2.1	9.8–1.1	1.67–0.78	15.2–0.65	16.6–1.5
Ba/Th	2000–36	680–16	850–21	840–2.7	4200–71	568–8.8	396–20	493–24	424–32	434–15
Rb/Ba	0.34–0.005	0.30–0.005	0.32–0.005	0.35–0.002	0.19–0.006	0.41–0.007	0.87–0.007	1.42–0.004	0.49–0.004	0.35–0.002
ε_{Sr} (i)	16–90	–13–41	–24–67	–11–48	–16–86	11–123	–20–178	–15 to –1.3	–12–77	–8.4–220
ε_{Nd} (i)	–4.6–9.8	3.3–9.0	7.9–12.3	7.8–7.9	9.1–15.9	–1.3–7.0	9.0–12.1	7.5–3.7	5.5–8.2	7.4–19.9
Os, ppb	0.47–0.042	1.9–0.021	11.7–0.014	7.0–0.033	2.6–0.14	0.018–0.004	2–<0.3	No data	0.36–0.00036	0.14–0.0028
Re, ppb	0.83–0.050	0.71–0.075	0.63–0.026	0.86–0.042	1.26–0.055	0.83–0.22	0.78–0.22	No data	0.73–0.035	1.14–0.32
Re/Os	9.2–1.2	34–0.058	6.1–0.025	4.4–0.010	2.52–0.054	208–12	>2.6–0.22	No data	204–0.48	297–2.7
Pd/Ir	500–22	110–48	276–4.4	806–4.7	500–3.8	No data	92 (n = 20)*	64–6.2	19–4.4	415–7.9

Note: *, average value of Crocket (2002). $(La/Sm)_n$ and $(La/Yb)_n$ are chondrite-normalized ratios (Sun and McDonough, 1989). From Table 2 of Zhang *et al.* (2008), with some LIP names modified to match usage in this book.

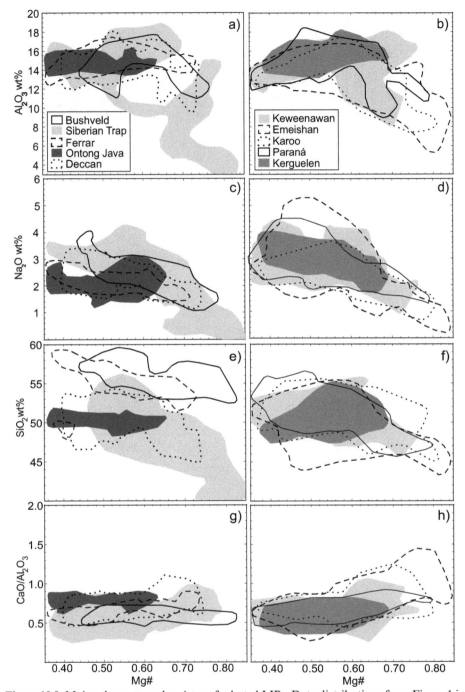

Figure 10.9 Major-element geochemistry of selected LIPs. Data distributions from Figure 1 in Zhang *et al.* (2008) generalized into fields. With permission from Elsevier. Digital version of original figure kindly provided by W. Griffin and M. Zhang.

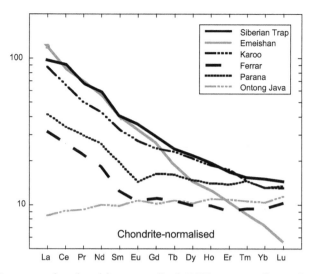

Figure 10.10 Representative-chondrite normalized REE patterns for various LIPs. From Zhang *et al.* (2008). With permission from Elsevier. Digital version of original figure kindly provided by W. Griffin and M. Zhang.

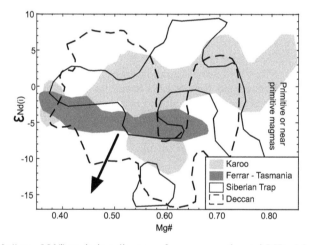

Figure 10.11 Mg# vs. εNd(i) variation diagrams for some continental LIPs. Mg# = Mg/(Mg + Fe^{2+}) (atomic ratio) calculated assuming $Fe_2O_3/FeO = 0.2$. The arrow shows the likely trend derived by crustal contamination. Fields generalized after data distributions from Zhang *et al.* (2008). With permission from Elsevier. Digital version of original figure kindly provided by W. Griffin and M. Zhang.

chemistry are from Kerr *et al.* (2000), Arndt and Weis (2002), Kerr (2003), Kerr and Mahoney (2007), Pearce (2008), Greene *et al.* (2009), Safonova (2009), and Kerr (2014). Geochemically, most lavas and intrusions of oceanic plateaus can be subdivided into three main compositional groups: basalts with restricted chemistry

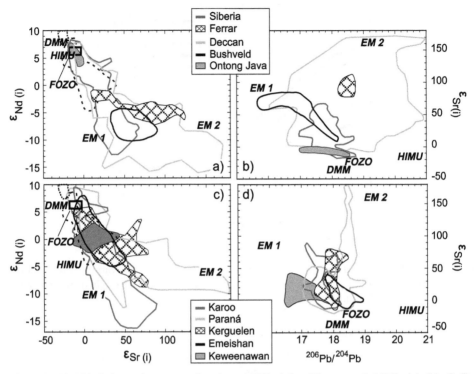

Figure 10.12 Nd–Sr isotopic patterns in selected LIPs (after Zhang *et al.* 2008). (a)–(b) εSr(i) vs. εNd(i) and (c)–(d) $^{206}Pb/^{204}$ vs. εSr(i) variation diagrams. The high-Mg basalts from the Siberian Trap, Karoo, Paraná, and Deccan LIPs have Mg# > 0.65 and/or MgO > 9 wt%; the rest are labeled as low-Mg basalts. The dotted line shows the distribution of MORB and the dashed line locates OIB; from Hofmann (1997). Outlines generalized from data distribution in Figure 5 in Zhang *et al.* (2008). With permission from Elsevier. Digital version of original figure kindly provided by W. Griffin and M. Zhang.

including a distinctive flat REE pattern, crustally contaminated basalts, and high-Mg suites (Kerr and Mahoney, 2007, and references therein; Figs. 10.13 and 10.14). These are discussed below, in turn, in Sections 10.4.1–10.4.3.

10.4.1 Oceanic LIPs with no lithospheric interaction

The first and main group comprises basalts and dolerites that display a restricted range of major-element, trace-element and radiogenic-isotope composition, and have chondrite-normalized REE patterns that are flat to slightly LREE-enriched (Figs. 10.13 and 10.14). Concentrations of the REEs and other incompatible elements tend to be similar to those in MORBs, specifically Nb/Th is generally > 8, whereas isotope ratios of Nd, Pb, Hf, and Sr are broadly ocean-island-like (Kerr, 2014 and references therein).

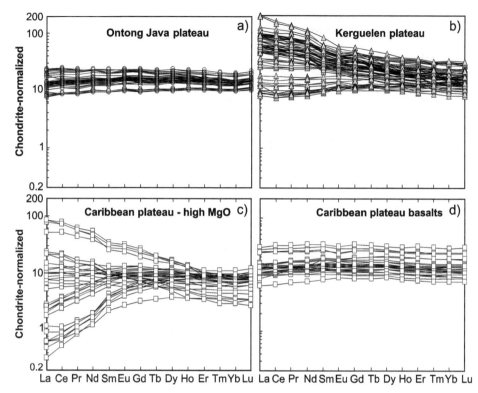

Figure 10.13 REE geochemistry of main Cretaceous oceanic plateaus; chondrite-normalized (Sun and McDonough, 1989). From Figure 8 in Kerr (2014). With permission from Elsevier. Digital version of original figure kindly provided by A. Kerr.

Most Ontong Java samples are of this type, as are many of the lavas of the Caribbean and Kerguelen oceanic plateaus (Kerr, 2014). The remarkable homogeneity of Ontong Java plateau basalts is also seen in their major- and trace-element composition (Fitton and Godard, 2004). Incompatible-element abundances in the primary Ontong Java plateau magma can be modeled by around 30% melting of a peridotitic primitive mantle source from which about 1% by mass of average continental crust had previously been extracted (Fitton and Godard, 2004).

10.4.2 Oceanic LIPs that interacted with the lithosphere

The second type of oceanic LIP chemistry exhibits evidence of interaction with lithosphere (Figs. 10.13 and 10.14). For instance, the Kerguelen LIP shows evidence for crustal contamination and the Wrangellia LIP shows evidence for interaction with a subduction-modified lithosphere. Contamination by continental crust or lithospheric mantle is evidenced by La/Nb $\gg 1$, initial $^{87}Sr/^{86}Sr > 0.705$, and

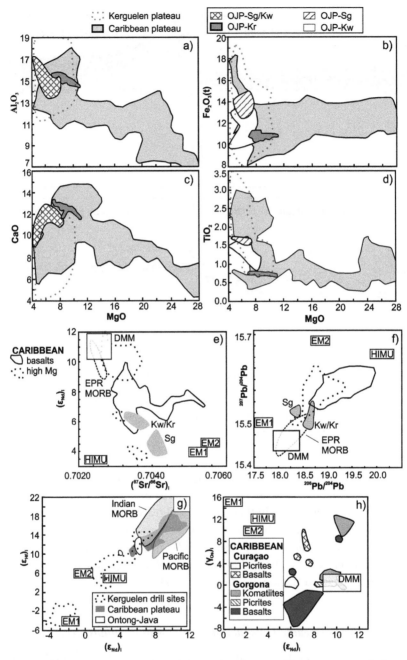

Figure 10.14 Selected geochemical and isotopic plots to illustrate general features of Cretaceous oceanic plateaus. All lavas with > 18 wt% MgO contain accumulated olivine. The Ontong Java plateau (OJP) is presented as a single field except where the Kwaimbaita (Kw), Singgalo (Sg), and Kroenke (Kr) type differ markedly in composition. (h) Data from various parts of the Caribbean–Colombian LIP. The marked shift to lower εHf–εNd values for Kerguelen samples reflects contamination with continental crust (see text for more discussion). Mantle end-member compositions are from Zindler and Hart (1986). Data fields generalized from data distribution in Kerr (2014). With permission from Elsevier.

variably negative ε_{Nd} (e.g. Frey *et al.*, 2002; Ingle *et al.*, 2002a, b; Neal *et al.*, 2002). In the case of Kerguelen, the crustal isotopic signature is evidence for a rifted continental fragment under the Kerguelen LIP (Section 4.2.2) inherited from dispersal of Pangea. The other example is provided by the lavas of the Wrangellia LIP (Section 4.5.1), which comprise both high-Ti lavas and low-Ti lavas. The high- and low-Ti lavas have differing source characteristics; e.g. the low-Ti lavas are high-field-strength-element (HFSE)-depleted, low-titanium basalts from magmas generated by the impingement of a plume head on the base of the arc lithosphere, whereas the high-Ti lavas are plume-related and not affected by lithosphere (Greene *et al.*, 2009).

10.4.3 High-Mg type of oceanic LIPs

The third compositional group of oceanic LIPs (e.g. Kerr and Mahoney, 2007; Kerr, 2014) comprises picrites and komatiites with high MgO contents (> 12 wt%). As a group, such rocks have a wide range of isotope ratios and incompatible trace-element concentrations and ratios (including both incompatible-element-depleted and -enriched types). Such high-MgO rocks are found principally in the Caribbean plateau and are suspected to present in the deeper unexposed portions of other oceanic plateaus such as Ontong Java.

10.5 Geochemistry of continental flood basalts (CFBs)

10.5.1 General geochemical characteristics of CFBs

A voluminous literature exists on the geochemical characteristics of CFBs (a selection of key papers/volumes include: Pearce and Cann, 1973; Baragar, 1977; Cox, 1980; Basaltic Volcanism Study Project, 1981; Thompson *et al.*, 1983; Macdougall, 1988; White and McKenzie, 1989, 1995; Mahoney and Coffin, 1997; and an excellent summary and dataset are presented in Zhang *et al.*, 2008)). Almost all continental LIPs are compositionally and volumetrically basic (< 54 wt% SiO_2), and usually viewed as comprising relatively homogeneous successions where large volumes of phenocryst-poor tholeiitic basalt lavas are the main rock type. However, in addition there are typically minor amounts of alkali basalts, basaltic andesites, picritic basalts, and high-Al basalts. The majority of CFB magmas have compositions (e.g. low MgO and Ni content and low Mg#) consistent with substantial fractionation from primary picritic melts (e.g. Cox, 1980; Thompson *et al.*, 1983; Garfunkel, 2008).

At the scale of individual eruptive units, flood-basalt lavas generally show remarkable chemical and mineralogical homogeneity, even across many hundreds of kilometers (Hooper, 1997) and perhaps even 1000 km (the largest known length of a single flow; Self *et al.*, 2008). At the regional scale, there is also compositional homogeneity in terms of major-element composition, but significant variations can be present in the trace elements.

10.5.2 Chemostratigraphy of volcanic piles

Significant variations in the trace elements can be used to subdivide the basalts into distinct chemostratigraphic packages, each with their unique petrogenetic history. Detailed geochemical studies undertaken over the last 40 years have recognized many (in some cases >10) different magma types within an individual continental LIP (e.g. Saunders *et al.*, 1997; Marsh *et al.*, 2001; Bryan and Ernst, 2008).

An example of such chemostratigraphy is provided from the Siberian Trap LIP at Noril'sk (Fig. 10.15). The Lower Sequence consists of sub-alkalic basalts, tholeiites, and picritic basalts (stratigraphically upwards, these are the Ivakinsky, Syverminsky, and Gudchichinsky Formations). The overlying Upper Sequence consists of picritic basalts and tholeiites interbedded with tuffs (stratigraphically upward, these are the Khakanchansky, Tuklonsky, Nadezhdinsky, Morongovsky, Mokulaevsky, and Kharayelakhsky Formations). Each formation has distinctive trace-element composition. For instance, the picritic Tuklonsky Formation lavas and the tholeiitic lavas of the Upper Sequence have compositions that are characteristic of magmas strongly influenced by lithospheric contamination, whereas the high-Ti and Nb/La, low La/Sm and radiogenic Nd-isotope signatures of the Lower Sequence are more comparable to deeper asthenospheric/mantle-plume-generated lavas with similarities to OIBs (Lightfoot *et al.*, 1993b). Incidentally, the Nadezhinsky Formation is particularly important in that it is genetically linked to the Noril'sk intrusions that host rich Ni–Cu–(PGE) ores (Section 16.2.1). The chalcophile-element depletion in the Nadezhinsky Formation is complementary to the chalcophile enrichment (in sulfides) in the Noril'sk intrusions and indicates that the Nadezhinsky basalts and the ore-bearing Noril'sk intrusions shared a common staging chamber (Lightfoot and Keays, 2005).

10.5.3 Low- and high-Ti magma types

One of the long-recognized aspects of CFBs is their division into low-Ti and high-Ti magma types, a distinction that basically reflects incompatible-element levels (e.g. Cox *et al.*, 1967; Marsh *et al.*, 2001; Jourdan *et al.*, 2007a; Bryan and Ernst, 2008). Furthermore, there is a typically marked provinciality in the distribution of the low- and high-Ti suites (e.g. Paraná–Etendeka, Karoo, and Emeishan) (Figs. 10.16 and 10.17). The low-Ti character to the tholeiitic basaltic magma types has commonly been interpreted to reflect crustal contamination, either with the subcontinental lithospheric mantle and/or with the continental crust (e.g. Carlson, 1991; Peate, 1997; Ewart *et al.*, 1998, 2004a), or mantle melting conditions such as potentially higher degrees of partial melting of the upper mantle or melting at shallower depths (Arndt *et al.*, 1993; Xu *et al.*, 2004a). In contrast, the high-Ti mafic suites commonly show greater geochemical and isotopic similarity to OIBs (i.e. intraplate geochemistry) and are interpreted to comprise a significant and relatively uncontaminated

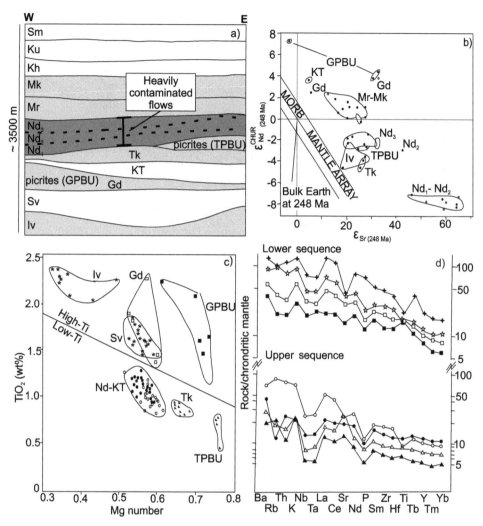

Figure 10.15 (a)–(d) Chemostratigraphy of the Siberian Trap LIP at Noril'sk. Stratigraphy projected onto a west–east section. Abbreviations: Sm = Samoedskyh, Ku = Kumginsky, Kh = Kharaelakhsky, Mk = Mokulaevsky, Mr = Morongovsky, Nd_1, Nd_2, Nd_3 = Nadezhdinsky tholeiites, TPBU = Tuklonsky picrite, Tk = Tuklonsky tholeiite, KT = Khakhanchansky Tuff, GPBU = Gudchikhinsky picrite, Gd = Gudchikhinsky tholeiite, Sv = Syverminsky tholeiite, Iv = Ivakinsky sub-alkalic basalt. The Nadezhda horizons are the most contaminated and are depleted in chalcophile elements; the corresponding enrichment is interpreted to be locate in the mineralized Noril'sk area intrusions. Parts (b), (c), and (d) are modified from Lightfoot *et al.* (1994). Queen's Printer for Ontario, 1994. Reproduced with permission. Part (a) is modified from Lightfoot and Hawkesworth (1997).

asthenospheric mantle or plume component (e.g. Arndt *et al.*, 1993; Zhao *et al.*, 1994; Ewart *et al.*, 1998, 2004a).

Another look at the meaning of this division into high- and low-Ti basalts is provided by Puffer (2001) who compiled average geochemical data for a number of

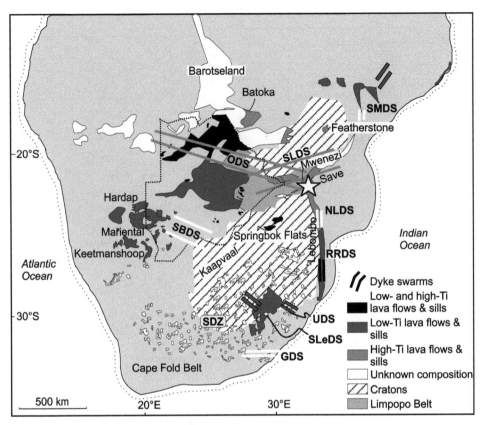

Figure 10.16 Distribution of high-TiO₂ and low-TiO₂ magma types in the Karoo LIP. Sketch map of southern Africa showing the distribution of the Karoo magmatism and related dyke swarms. The distributions of low-Ti (TiO₂ < 2 wt%), high-Ti (TiO₂ > 2 wt%), and unknown compositions are indicated. ODS, Okavango dyke swarm; SLDS, Save-Limpopo dyke swarm; SBDS, South Botswana dyke swarm; SleDS, South Lesotho dyke swarm; UDS, Underberg dyke swarm; SMDS, South Malawi dyke swarm; RRDS, Rooi Rand dyke swarm; NLDS, north Lebombo dyke swarm; GDS, Gap dyke swarm. The dotted line corresponds to the Botswana border. The sill dense zone (SDZ on map) is schematic, as sills occur as quasi-continuous outcrops in the South Africa Karoo sedimentary basin (see also Fig. 5.10). Dyke color: black, gray, or white corresponds to low-Ti, high-Ti, or unknown Ti type, respectively. The star is added to locate the inferred plume center. Modified from Figure 1 in Jourdan *et al*. (2007a).

CFBs and revealed that there are two contrasting distributions of HFSEs plotted on silicate Earth-normalized multi-element diagrams (Fig. 10.18). Many CFBs plot close to a standard OIB line and are considered the product of magma from a mantle plume head mixed to varying degrees with contributions from other sources. The compositional diversity of the OIB-like basalts (high-Ti) is consistent with the interaction of plume heads with diverse mantle layers followed by plume-tail

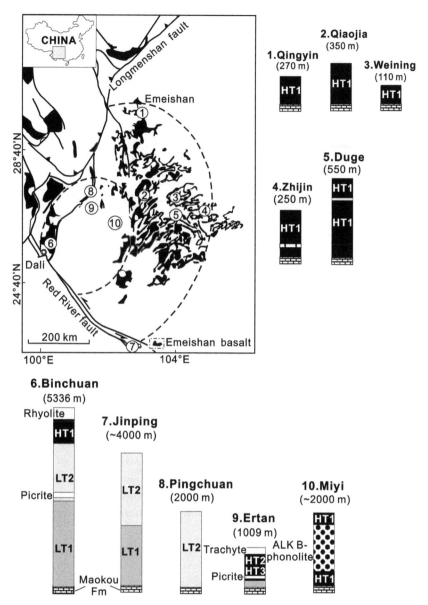

Figure 10.17 Distribution of high-TiO_2 and low-TiO_2 magma types in the Emeishan LIP. HT = high-Ti basalt; LT = low-Ti basalt; ALK = alkaline series. From Figure 1 in Xu *et al.* (2004b).

magmatism along hotspot tracks. On the other hand, at least three CFBs, including those of the 200 Ma CAMP, the Siberian Trap LIP, and the Lesotho basalt province (of the Karoo LIP) of South Africa, plot close to a standard arc-basalt line (low-Ti) including a negative Nb–Ta anomaly, which could be interpreted as evidence of a

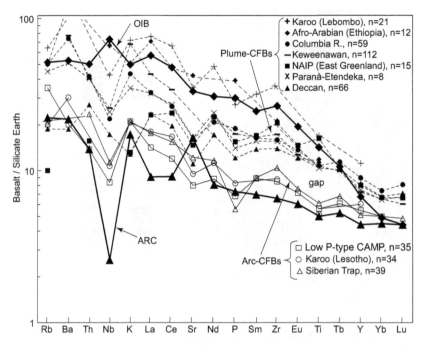

Figure 10.18 Comparison of high-TiO$_2$ and low-TiO$_2$ groups on a multi-element diagram. Note the HFSE (high field strength element) gap separating the arc-related CFB cluster (Arc-CFBs) from a separate cluster interpreted to be plume-related (Plume-CFBs, dashed lines). Standard OIB after Sun and McDonough (1989); standard arc (ARC). Basalt compositions normalized to silicate Earth of McDonough and Sun (1995). Modified from Figure 1 in Puffer (2001).

subduction setting at the time of LIP emplacment. However, some of the high-Ti basalts (representing an OIB source) also exhibit a negative Nb anomaly, leading to the suggestion that this subduction-like signature in the basalts is not reflective of a subduction setting, but indicates interaction with lithospheric mantle modified by an earlier subduction event.

10.5.4 Apparent arc signatures

To consider this point further, the trace-element geochemistry of many continental LIPs is characterized by negative anomalies in Nb, Ta, Ti, and P, and this is also true of some oceanic LIPs (e.g. Wrangellia, Section 4.5). This has also been noted by Neumann *et al.* (2011) in their study of the Karoo LIP. They note that among the geochemical features common to many CFBs are negative Nb–Ta anomalies and enriched Nd/Sr isotope ratios (e.g. Gallagher and Hawkesworth, 1992; Puffer, 2001). Proposed models for these subduction-like signatures in LIPs include (points 1, and 5–8 modified after Neumann *et al.* (2011):

(1) partial melting of heterogeneous subcontinental lithospheric mantle (SCLM; e.g. Gallagher and Hawkesworth, 1992; Jourdan *et al.*, 2007a, 2009);
(2) partial melting of variably depleted and hydrated mantle, inferred to be within the lithosphere (Turner and Hawkesworth,1995);
(3) partial melting of lithospheric mantle that has previously been modified by subduction-related fluids (e.g. Sandeman and Ryan, 2008);
(4) partial melting of the lithospheric mantle that had been modified by subduction processes during major crust-formation events in the region (Zhao and McCulloch, 1993);
(5) contamination during passage through the SCLM of magmas derived from sublithospheric plume or MORB sources, or mixing between sublithospheric and lithospheric mantle melts (e.g. Marsh and Eales, 1984; Arndt and Christensen, 1992; Riley *et al.*, 2005; Jourdan *et al.*, 2007a; Heinonen and Luttinen, 2010);
(6) assimilation of sediments (Elburg and Goldberg, 2000);
(7) polybaric fractional crystallization (Marsh and Eales, 1984);
(8) contamination by sedimentary country rocks combined with fractional crystallization in addition to processes at depth (e.g. Riley *et al.*, 2005, 2006; Jourdan *et al.*, 2007a);
(9) contamination of mantle melts by crustal material in deep–mid-level crustal "staging chambers" (e.g. Nelson *et al.*, 1990; Ihlenfeld and Keays, 2011; Ciborowski *et al.*, 2013; Jowitt and Ernst, 2013);
(10) crustal contribution derived from within an ancient region of the mantle lithosphere, i.e. from recycled sediment rather than from the overlying continental crust (Lightfoot *et al.*, 1993b); or
(11) sudden reactivation of dormant arc or backarc sources trapped under continental-plate sutures (Puffer, 2001).

What these mechanisms have in common is interaction with the lithosphere and/or crust, so a plausible conclusion is that a Nb–Ta anomaly does not require a subduction setting at the time of LIP magmatism. It could simply imply interaction with a lithosphere that has previously (perhaps many hundreds of million years previously) been modified by subduction processes in an adjacent subduction zone.

10.6 Geochemistry of the LIP plumbing system

The plumbing system for LIPs (discussed in Chapter 5) consists of mafic–ultramafic dyke swarms, sill provinces, and layered intrusions. Density will act as a filter so that denser ultramafic magmas will preferentially remain in the plumbing system and would be less likely to reach the surface as volcanic rocks. The same diagrams used for volcanic rocks can be used for the geochemical study of dyke swarms; however, care must be taken to ensure that magmatic compositions have been obtained rather than the composition of cumulate rocks.

10.6.1 Mafic–ultramafic dyke swarms

LIP-related dyke swarms are mainly dolerites/gabbros (dominantly composed of clinopyroxene and plagioclase with minor olivine and orthopyroxene) of mainly tholeiitic (but also alkaline) chemistry and having high-Ti, low-Ti, and Fe-rich subtypes. Also present can be norites and picrites. In the Archean, komatiite dykes would also be expected. Tarney and Weaver (1987) identify four types of Scourie dykes (in the Lewisian part of the North Atlantic craton) on petrological and geochemical grounds: bronzite–picrites, norites, olivine–gabbros, and quartz dolerites. (See also Hughes *et al.*, 2014, and Davis and Heaman, 2014.) Picrites and picrodolerites are common among Phanerozoic LIP-related intrusions of western Mongolia (Izokh *et al.*, 2011). Among some of the suites of dykes in Karelia there are noritic, gabbronoritic, low-Ti tholeiitic, Fe-tholeiitic, and orthopyroxene–plagioclase–phyric types (Vuollo and Huhma, 2005). However, to reiterate, most regional dyke swarms are dolerites/gabbros.

Prior to the first international dyke swarm conference in 1985 (Halls and Fahrig, 1987), dolerite dykes were not generally a focus of geochemistry studies, at least not in comparison to the level of geochemical study then focused on sill provinces and layered intrusions. However, subsequently, dyke swarms have become much better understood geochemically. Some of the key publications that address the differentiation patterns within individual dykes, geochemical characterization, and modeling of source characteristics include the following: Tarney and Weaver (1987), Condie *et al.* (1987), Gibson *et al.* (1987), Ernst and Bell (1992), Baragar *et al.* (1996), Condie (1997), Mayborn and Lesher (2004); Jowitt and Ernst (2013). Different dolerite dyke swarms typically have similar major-element characteristics (Fig. 10.19) that overlap with the typical compositions of flood basalts (Section 10.5). However, like flood basalts, the trace-element patterns for particular swarms can be distinctive and can allow comparison with the specific chemostratigraphic groupings in flood basalts to identify links between the plumbing system (dykes) and the flood basalts (e.g. Baragar *et al.*, 1996).

10.6.2 Along-dyke compositional consistency

One remarkable aspect of dyke-swarm chemistry is the evidence for along-dyke compositional consistency. LIP-related swarms tend to consist of dykes with average widths of >10 m and long lateral extents (up to > 2000 km; Chapter 5). For swarms in which the individual dykes are widely spaced it has proved possible to obtain and analyze material from long distances along individual dykes. It is shown that a dyke will have consistent ratios of incompatible elements along strike for many hundreds of kilometers, and which can be distinct from that in an adjacent dyke (e.g. Halls, 1986; Buchan *et al.*, 2007). The inference is that each dyke represents a single magmatic pulse, probably injected laterally from a magmatic chamber along strike. After a period of time, a second dyke could be injected from this chamber. During the time gap between the two dyke pulses, the magma in the chamber will have evolved and so the second dyke pulse will have a composition distinct from the first pulse.

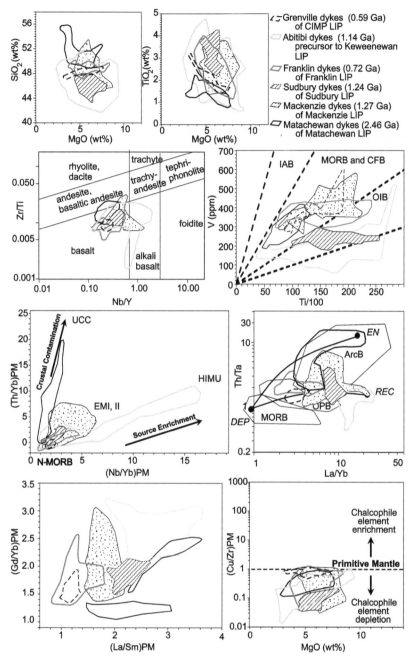

Figure 10.19 Geochemical variations in dyke swarms of the Candian shield showing how different swarms, all broadly basaltic, can have different trace-element patterns. These patterns echo the patterns in CFBs that can be grouped based on different processes. Partly from Figure 11 in Jowitt and Ernst (2013), with permission from Elsevier, and some additional diagrams were specially produced by S. Jowitt (written communication, 2013).

This is illustrated for dykes of the 2070 Ma Lac Esprit swarm in the eastern Superior craton (Fig. 10.20). Note that I am referring to ratios of incompatible elements rather than values of these elements. Of course the values of an incompatible element will typically increase along strike of individual dykes (due to in situ fractionation during lateral emplacement), but in general the ratio of incompatible-element ratios will remain constant along individual dykes.

Additional support for this model comes from complementary paleomagnetic studies (Fig. 10.20) that show that each individual dyke will have a paleomagnetic direction which is consistent along its strike length, but which can be different from the paleomagnetic direction that characterizes an adjacent dyke. Such paleomagnetic differences are explained by a time gap between the emplacement of each dyke, consistent with secular variation on the scale of at least hundreds of years.

10.6.3 Pearce element ratio (PER) diagrams

As discussed for the 1270 Ma Mackenzie LIP (Baragar *et al.*, 1996) Pearce element ratio (PER) diagrams are helpful in analyzing fractionation patterns of dyke and sill swarms in that they are sensitive to changing relationships between constituents in an evolving magma relative to a conserved element (i.e. an element that is not removed from the magma during fractionation; Fig. 10.21). The theory was developed by Pearce (1968, 1987) partly to avoid the closure problem in summing analyses to 100%. The technique and appropriate use of PER analysis was further developed in a series of publications (Ernst *et al.*, 1988; Stanley and Russell, 1989a, b; Russell and Stanley, 1990; Ernst and Bell, 1992; Nicholls and Gordon, 1994). These papers show that the concerns of Rollinson and Roberts (1986) and Rollinson (1993) regarding spurious correlations can be obviated with appropriate choice of the denominator element.

An example of the use of the PER method for the Mackenzie swarm is shown in Fig. 10.21, which is designed to distinguish between compositional trends caused by fractionation of olivine and orthopyroxene. Plots can be developed to distinguish between fractionation/accumulation of other combinations of minerals and collectively provide a basis for calculating the proportions of the extracted minerals responsible for a given geochemical trend (Stanley and Russell, 1989a, b). The data in Fig. 10.21 are from a set of dykes in the Mackenzie swarm 800 km from the plume center (Baragar *et al.*, 1996). They are typical of the main group of Mackenzie dyke data in demonstrating that olivine is not a significant fractionating mineral in this part of the Mackenzie system. According to additional PER plots (see Baragar *et al.* (1996) for details), the fractionating minerals are plagioclase, clinopyroxene, and orthopyroxene, but not olivine, and the fractionating assemblage for dykes at each distance from the plume center (500, 600, 800, 1000, and 2100 km) are determined using similar plots and the results summarized are in Figure 10.21. (In interpreting these data it is important to appreciate the evidence for lateral injection of Mackenzie dykes from magma chambers at the north end of the swarm (near the plume center;

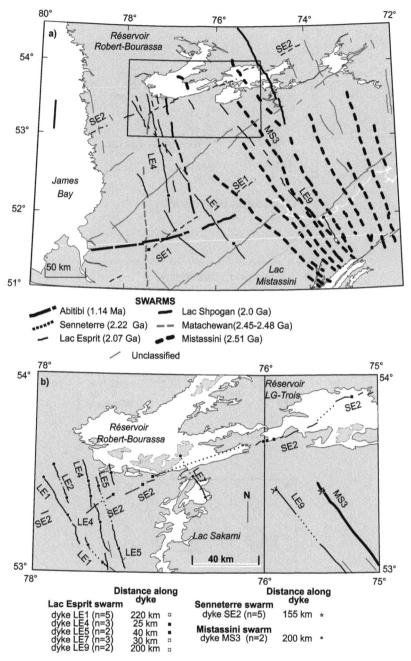

Figure 10.20 (a)–(e) Demonstration of along-dyke compositional consistency in dolerite dykes. Data from the southern Superior craton, Canada, used as an example. Consistency in paleomagnetic data (e) are also observed along individual dykes, indicating that individual dykes represent a distinct magma pulse and that there is a time gap between the emplacement of different dykes between which paleomagnetic secular variation occurs. Modified from Figures 2, 4, 5, 12, and 13 in Buchan et al. (2007).

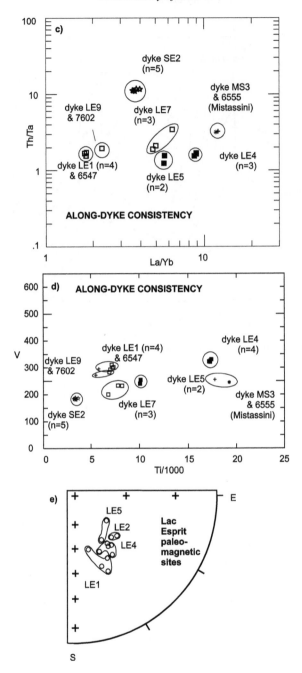

Figure 10.20 (*cont.*)

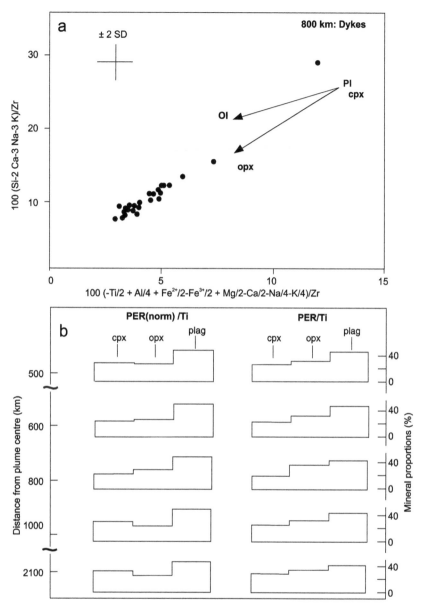

Figure 10.21 Pearce element ratio (PER) diagrams. (a) PER diagram designed to test relative control on fractionation by olivine (ol) vs. orthopyroxene (opx), and which is also designed to be insensitive to removal or addition of plagioclase (pl) or clinopyroxene (cpx). Elemental abundances are in molar proportions and Zr is the conserved element. Data from Mackenzie dykes located at a distance of 800 km from the plume center; the data show clear control of the fractionation by orthopyroxene (opx) and not by olivine (ol). (b) Other PER diagrams using CIPW normative mineral abundances (PER norm) and using multi-element numerators (PER), each ratioed with Ti as the conserved element, are used to distinguish the relative proportions of clinopyroxene, orthopyroxene, and plagioclase fractionated. Modified from Baragar *et al.* (1996). By permission of Oxford University Press. For full details on the method see Stanley and Russell (1989a, b), Ernst *et al.* (1988), Ernst and Bell (1992).

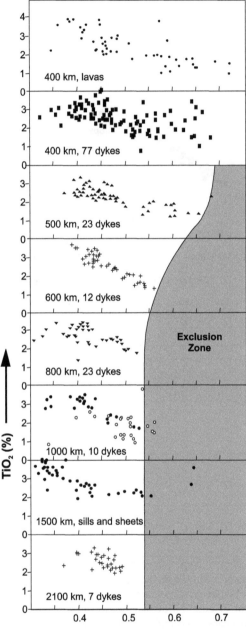

Figure 10.22 Comparison between along-swarm composition and compositions upward in coeval flood-basalt sequence. Along-swarm compositional variation and links with flood-basalt composition for the 1.27 Ga Mackenzie LIP. The shaded area is an "exclusion zone" marking the diminishing range of fractionation within the dyke magmas with increasing distance from the focus of the swarm. Modified from original diagram to remove the olivine dolerite data from 1500 km, which subsequent dating showed to at least in part belong to a younger 1108 Ma event (French *et al.*, 2008). Modified from Baragar *et al.* (1996). By permission of Oxford University Press.

Ernst and Baragar, 1992; Baragar *et al.*, 1996)). It is evident from this figure that the mode of fractionation at all sites was surprisingly constant. Augite, orthopyroxene, and plagioclase are fractionating in proportions that show only minor variation from site to site. Using Zr instead of Ti as the conserved element produces essentially the same results. Two principal points emerge from the PER analysis: (1) olivine fractionation is not a major factor in compositional variation along the main sampled trend of the Mackenzie swarm (from 400 km to 2100 km away from the plume center); (2) the mode of fractionation represented by dyke rocks at all points along this swarm is generally similar. If the parent magma of these dykes emerged from a Muskox-type of magma chamber, it must have evolved beyond the point at which olivine was on the liquidus, or else been contaminated by continental crust as shown by Francis (1994) for parts of the Muskox intrusion.

10.6.4 Links between dykes of the Mackenzie swarm and Coppermine lavas

Baragar *et al.* (1996) showed regional compositional variation in the Mackenzie dyke swarm, tying compositional variation in the dykes to compositional variation in the associated Coppermine volcanic (Fig. 10.22). They further emphasized that the range of composition decreased radially outward in the swarm so that the furthest traveled dykes showed only the upper range in composition of Coppermine lavas; these observations are most consistent with a model in which the later magma pulses flowing outward in dykes reached to greater distances away from the plume center than did the earlier magma pulses. These data were interpreted by Baragar *et al.* (1996) to reflect increasing uplift of the central (plume-center) region and increased hydraulic head during the period over which dyke emplacement occurred.

10.6.5 Mafic–ultramafic sills

As noted in Chapter 5, many LIPs have impressive dolerite sill provinces. As with the dyke swarms (as discussed above) these sill provinces also show a compositional range from basaltic to more Mg-rich compositions. Few dolerite sill provinces have been studied in as much detail as the Karoo LIP (e.g. Jourdan *et al.*, 2007a, 2009; Galerne *et al.*, 2008; Neumann *et al.*, 2011). On a regional scale the sills can be divided compositionally (see the high-Ti vs. low-Ti distribution of Figs. 10.16 and 10.17).

In addition, on a more local scale, more minor variations in trace elements (and isotope composition) can be used to fingerprint particular magma batches and trace their progression through an area. Such data can reveal surprising facts about the connections between sills. For instance, data from the Golden Valley Sill Complex (of the Karoo LIP) show that the various nearby sills in fact belong to distinct magma batches. Furthermore, as shown in Figure 10.23, the compositions of the

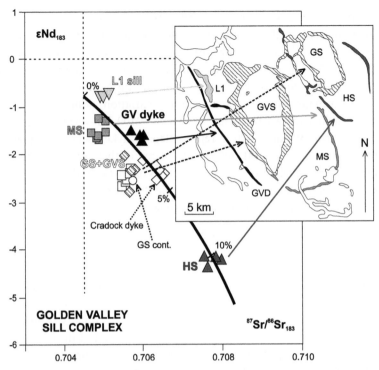

Figure 10.23 Variations in initial Sr–Nd isotope compositions for Karoo dykes and sills in the Golden Valley area. The original grouping of pulses is based on the geochemistry of pulses. Such data can allow tracking of the different magma pulses through the plumbing system. GVS = Golden Valley sill; GS = Glen sill; HS = Harmony sill; MS = Morning Sun sill; L1 = L1 sill; GV dyke = Golden Valley dyke. The *c.* 100-km-long, up to 30-m-wide Cradock dyke is located 70 km west of the Golden Valley Sill Complex. Sills that were not part of this study are shown outlined with a white interior. Tick marks on the model curve are 5% increments of crustal contamination. From Neumann *et al.* (2011). By permission of Oxford University Press. See also Galerne *et al.* (2008) for characterizing GV sills and dykes using geochemistry.

various sills lie along an evolution line reflecting increasing crustal contamination, presumably in the magma chamber (or chambers) from which the various magma pulses have been expelled.

For the Karoo province as a whole, Neumann *et al.* (2011) provide an illustration of the complexity of the plumbing system that allows magma to evolve in staging chambers at various depths and the role of host rocks at each depth (Fig. 10.24). Neumann *et al.* (2011) propose that the primary melts were derived from an asthenospheric source mantle and had acquired a weak subduction signature (relative depletion in Nb–Ta, mildly enriched Sr–Nd isotopic ratios) through interaction with metasomatized lithospheric mantle. In the deep crust, and assuming the magmas underwent further AFC processes, the melt has to assimilate about 10% of granulites

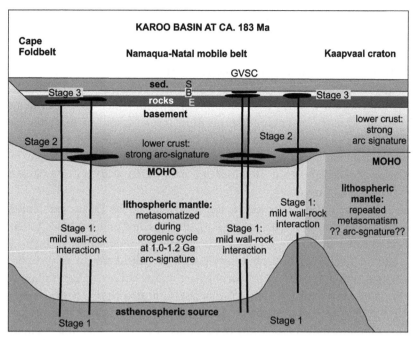

Figure 10.24 Schematic presentation of the ascent through the lithosphere of the melts that gave rise to the dolerites in the GVSC (Golden Valley Sill Complex) and drill cores, and the processes believed to have taken place at different depths and in different areas. The figure is not to scale and the possibility of significant lateral magma transport as dykes, sills, or via sublithospheric channeling (Chapter 5) is not considered. Shading variations between the intrusions in the deep crust indicate different degrees of contamination and fractional crystallization (Stage 2); vertical shading variations in the shallow sills indicate a new stage of fractional crystallization locally combined with contamination that reflects the local sedimentary country-rocks (Stage 3). E = Ecca Group; B = Beaufort Group; S = Stormberg Group. From Figure 18 in Neumann *et al.* (2011). By permission of Oxford University Press.

with strong arc-type geochemical signatures. During and/or after intrusion into the sedimentary rocks of the Karoo basin, the magmas underwent a second stage of fractional crystallization (50–60%) and local contamination by sedimentary host rocks. It is likely that other LIPs have similar complicated evolutionary histories to that proposed for the Karoo sills.

10.6.6 Mafic–ultramafic (M-UM) layered intrusions

A most dramatic portion of the plumbing system of LIPs are the major layered mafic–ultramafic intrusions ranging in size up to the Bushveld LIP, which is up to 9 km thick and covers an area of 65 000 km². The main characteristics of mafic-ultramafic intrusions were covered in Chapter 5 and here I discuss some geochemical aspects. Mafic–ultramafic layered intrusions have long been a focus of study and there is a voluminous literature on them (e.g. Wager and Brown, 1968; Cawthorn, 1996 and papers therein). Much of the research has been focused on the processes that yield the complicated compositional (modal, geochemical) layering on a variety of scales. Complexities include the difficulty of identifying the primary magmatic composition (because chilled margins are not well preserved for such thick, slow-cooling bodies) and the likelihood of multiple injections of magma into the chamber (most do not represent a single pulse). Also, magma expelled from the chamber can be injected as dykes or sills and can reach the surface as flows.

Three important layered intrusions, the Bushveld, Great Dyke, and Stillwater, each of which can be interpreted as part of a LIP (e.g. Ernst and Buchan, 2001b), can be considered to have crystallized from two magma types: an unusual, high- MgO, and -Cr, and relatively high SiO_2, but low-Al_2O_3 type (ultramafic or U-type; olivine-saturated) that was emplaced at an early stage; and a later, normal tholeiitic-type magma (T-type), also called A-type (plagioclase-saturated) (e.g. Irvine *et al.*, 1983; Campbell and Turner, 1986; Naldrett, 2010a). The U-type can be interpreted as a PGE-rich, komatiitic magma (possibly the product of two-stage mantle melting) that has interacted to varying degrees with the crust, becoming SiO_2-enriched in this way (Naldrett, 2010a).

In support of the multiple pulse interpretation, precise U–Pb dating has revealed a small age difference between the ultramafic and more mafic portions of both the Bushveld (Scoates *et al.*, 2012) and the Stillwater Complexes (Wall *et al.*, 2012). In the case of the Bushveld Complex, the age difference of about 5 Ma corresponded to 2060 Ma for the Lower Zone and Critical Zone vs. 2055 Ma for the rest of the Bushveld Complex above the Merensky reef. In the case of the Stillwater Complex, the age difference was perhaps only a couple of million years, corresponding to 2711 Ma for the lower ultramafic portion (feldspathic orthopyroxenite at the top of the Bronzitite zone) vs. 2709 Ma in the overlying Banded Series (see cross sections in Fig. 5.18).

10.6.7 Role of staging chambers

Figure 10.25 illustrates some different pathways for feeding surface lavas by distinct magma conduits. The data from west Greenland is consistent with switching of magma sources from more picritic magma arriving more directly from the mantle alternating with more differentiated magmas from staging chambers in the middle/upper crust. The Noril'sk model has some differentiation happening in the conduit system before, during, and after passage through the magma chambers. In this case, a

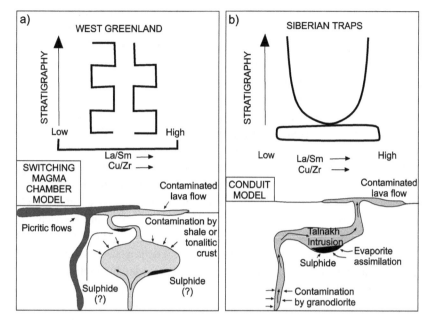

Figure 10.25 Sketch sections illustrating differences in magma plumbing systems and resulting differences in Cu/Zr and La/Sm ratios with height in the lava pile based on data in Greenland (NAIP) and Noril'sk. From Figure 14 in Lightfoot and Hawkesworth (1997). Copyright © 1997, John Wiley and Sons.

major sulfide-accumulation event in the magma chamber corresponds to major depletion in chalcophile elements in the corresponding surface lava fed from the chamber (Nadezhdinsky series). Additional discussion of staging chambers is presented in Ihlenfeld and Keays (2011).

10.7 Geochemistry of Archean LIPs

10.7.1 Overview

As presented in Chapter 6, Archean LIPs correspond to basalts-komatiites emplaced on a stable continental craton (essentially equivalent to CFBs), and also those greenstone belts with associated tholeiitic–komatiitic sequences. As discussed earlier (Section 6.3.1) these greenstone belts of LIP affinity are of two types: (1) those emplaced onto continental crust yet to be cratonized followed by subsequent deformation ("platform" type) and (2) those that were inferred to have been emplaced onto oceanic crust ("mafic plains" type). The latter are considered equivalent to oceanic plateaus. Here are some general comments regarding the geochemistry of Archean analogs of LIPs (modified after R. Kerrich, personal communication, 2011).

(1) Tholeiitic basalts of the "mafic plain" type are closely comparable to basalts of Phanerozoic ocean plateaus such as Ontong Java and Caribbean–Colombian (e.g. Chapter 4; Fig. 10.13). Given the flat heavy REE (HREE) patterns in such cases, one may assume that, with reasonable temperature estimates, plumes melted under an Archean ocean lithosphere of less than *c.* 90 km thick, i.e at a depth above the garnet stability field; see Fig. 15.12.

(2) Many greenstone tholeiite–komatiite associations appear to have erupted on rifted continental lithosphere, and represent the "platform" type. Plumes were likely "steered" towards thinner craton margins (Kerrich *et al.*, 2005). Proximity to rifted continental margins is indicated by basalt–komatiite suites having both uncontaminated and lithosphere-contaminated members. In all cases, a thinned lithosphere is revealed by flat HREE patterns with no garnet signature (Manikyamba *et al.*, 2008; Said and Kerrich, 2010; Manikyamba and Kerrich, 2011).

(3) For the dominantly intracratonic Fortesque and Ventersdorp lavas there is evidence for prevalent contamination by the lithosphere. There is also evidence for plume melting locally beneath a thick craton lithosphere provided by fractionated HREEs, which is consistent with residual garnet at > 90 km, as for many Phanerozoic CFBs. (Note, given that the HREEs are also temperature dependent, one cannot make a precise depth interpretation (J. Pearce, personal communication, 2013)).

10.7.2 Komatiites

The presence of komatiites is a diagnostic feature of higher-temperature plumes in the Archean. Several types of Archean komatiites can be distinguished geochemically (Tables 6.3 and 6.4) (Xie *et al.*, 1993; Xie and Kerrich, 1994; Fan and Kerrich; 1997; Kerrich and Wyman, 1997; Condie, 2001; Sproule *et al.*, 2002; Arndt, 2003; Arndt *et al.*, 2008; Pearce, 2008). The different depths of melting for each of the main types are shown in Figure 10.26, and the different distributions of majorite garnet fractionation and crustal contamination are shown in Figure 10.27.

Barberton type

The first type, Al-depleted Barberton-type komatiites, have low Al_2O_3/TiO_2 and Ti/Zr ratios with fractionated REE patterns (e.g. Condie, 2001; Arndt *et al.*, 2008). Some Al-depleted komatiites exhibit high $(La/Yb)_N$ (LREE-enriched, N = ratio normalized to primitive mantle) slopes along with negative Zr and Hf anomalies consistent with majorite garnet remaining in the restite during partial melting at depths between 300 and 600 km (Xie *et al.*, 1993; Agee, 1993; Condie, 2001, p. 202).

Munro type

Munro-type Al-undepleted komatiites have near-chondritic Al_2O_3/TiO_2 and Ti/Zr ratios and relatively flat HREE patterns (Arndt *et al.*, 2008).

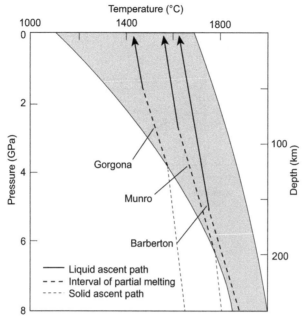

Figure 10.26 Pressure–temperature diagram summarizing the paths taken by anomalously hot mantle sources that melted to form the three main types of komatiite. From Figure 13.6 in Arndt *et al.* (2008).

Karasjok (Boston) type

The ferro-picrites and ferro-komatiites are analogous to the ferro-picrites that are present in some Phanerozoic flood basalts (Gibson, 2002). These are distinctively Fe–Ti-rich and have been termed the "Karasjok-type komatiites" (Arndt *et al.*, 2008). The ferro-picrites are also observed as thin layers at the base of LIP sequences of Archean age (e.g. the Onverwacht Group, South Africa, and in the Superior province, Canada) and of Phanerozoic age (the Siberian Trap LIP, the Paraná–Etendeka LIP, and the NAIP) (Gibson, 2002). These ferro-picrites are interpreted to represent melting of plume asthenosphere mantle with streaks of eclogite, which represent paleosubduction zone material entrained in the plume head.

Other types

Some komatiites have assimilated continental crust, producing siliceous high magnesium basalts (SHMBs), which are characterized by enrichment of incompatible trace elements, negative Nb–Ta anomalies, high $^{87}Sr/^{86}Sr$ ratios, and low $^{143}Nd/^{144}Nd$ ratios (Sun *et al.*, 1989; Arndt *et al.*, 2008). Caution is needed to distinguish these rocks from magmas formed in subduction settings as all these characteristics are also consistent with the interaction of komatiitic liquids with metasomatized continental mantle lithosphere (Said and Kerrich, 2010).

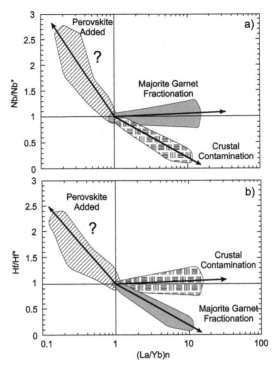

Figure 10.27 Recognition of processes in komatiites from key geochemical ratios. The Nb/
Nb* and Hf/Hf* ratios are very useful when plotted against (La/Yb)n ratios to disiguish three
major evolution paths recorded in komatiites: majorite garnet fractionation, perovskite
addition, and crustal contamination. Majorite garnet signature reflects magma segregation
at depths of 425 to 540 km. Speculative perovskite addition trend reflects melting of perovskite
at depths below the 660-km seismic discontinuity. Nb and Hf are analyzed values, and Nb*
and Hf* are interpolated values from adjacent elements on a primitive-normalized multi-
element plot. From Figure 7.8 in Condie (2001). Copyright © 2001 Cambridge University
Press.

10.7.3 Basalts of Archean LIPs

Geochemical patterns are shown for selected Archean basalts (LIP-related) in
Figure 10.28. Figs. 10.28a–d show a representative set of Archean greenstones of
proposed continental provenance (platform type) plotted on the Th–Nb proxy
diagram. All the data plot above the MORB–OIB array and/or have steep vectors
oblique to the array, demonstrating extensive magma–crust interaction. Figs. 10.28e
and f show two Archean greenstone suites of proposed subduction provenance, showing that they plot in the volcanic-arc field with clear displacement from the MORB–OIB
array. The examples in Fig.10.28 do not include basalts of the mafic plains type (i.e.
emplaced onto oceanic crust and lacking interaction with continental crust).

Figure 10.29 shows representative MORB and Ti-normalized patterns for the four
main types of Archean oceanic basalt. The first three basaltic types (Barberton,

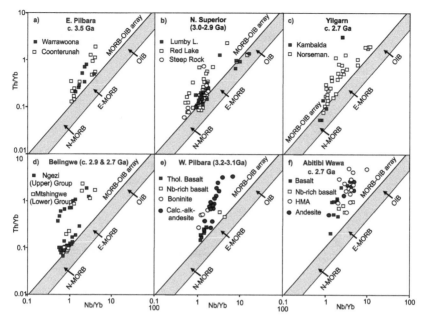

Figure 10.28 Archean greenstone suites (a)–(d) Non-subduction (possibly LIP-type) Archean greenstones exhibiting crustal interaction on the Th–Nb proxy diagram. (e)–(f). Archean greenstone suites of proposed subduction origin. The MORB–OIB array is labeled; HMA = high-Mg andesite. From Figure 10 in Pearce (2008). With permission from Elsevier.

Munro, and Boston) can be likened to the corresponding komatiite types (Tables 6.3 and 6.4) and the fourth, Isua, exhibits similarities to the crustally contaminated Commondale type of komatiite (J. Pearce, personal communication, 2013) on the basis that depletion in garnet and then remelting gives low Ti/Yb, their distinctive feature.

An additional type of high-Mg series rocks is important for the LIP record of the Archean and also for the Paleoproterozoic. This is the boninite type, which can have both orogenic and anorogenic types; the latter can be linked to mantle plumes and LIPs. This class of units are discussed in the next section.

10.8 Geochemistry of anorogenic boninite series rocks

This section is focused on units that can be termed boninites or boninitic- like and which have high-magnesium (MgO > 8%), high-silica (SiO_2 > 52%), and low-titanium (TiO_2 < 0.5%) contents (Table 10.2). They represent a class of high-MgO rocks which can be distinguished from other mafic and ultramafic magmatic types as follows:

According to the IUGS classification scheme (Fig. 10.30) (Le Bas, 2000; Le Maitre, 2002), igneous rocks with MgO >12%, SiO_2 between 30 and 52%, and $Na_2O + K_2O$ < 3% are classified as boninite, picrite, or komatiite on the basis of differences in these oxides and TiO_2: boninites have MgO > 8%, SiO_2 > 52%, and

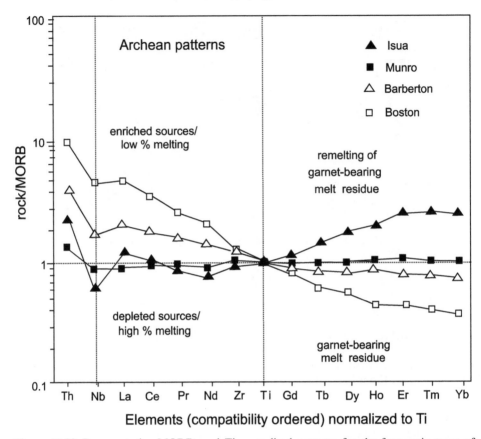

Figure 10.29 Representative MORB- and Ti-normalized patterns for the four main types of Archean oceanic basalt: Al-depleted komatiite (ADK) type (Barberton), Al-undepleted komatiite (AUK) type (Munro), high-Ti, low-Al type (Boston), and low-Ti type (Isua). The first three patterns can be likened to corresponding komatiite types (Tables 6.3 and 6.4), and the fourth, Isua, has similarities to the crustally contaminated Commondale types (J. Pearce, personal communication, 2013). From Figure 12 in Pearce (2008). With permission from Elsevier.

$TiO_2 < 0.5\%$, whereas basalts have $MgO > 12\%$, SiO_2 between 30 and 52%, and $Na_2O + K_2O < 3\%$ without any consideration of TiO_2. On the other hand, komatiite/meimechites have $MgO > 18\%$, SiO_2 between 30 and 52%, and $Na_2O + K_2O < 2\%$; if TiO_2 is less than 1%, it is regarded as komatiite, and if TiO_2 is greater than 1%, it is meimechite. As shown in Pearce and Robinson (2010) most boninites constitute a boninitic series extending from boninites to bronzite andesites, and to dacites and rhyolites (Pearce and Robinson, 2010). On a diagram of MgO vs. SiO_2 (Fig. 10.31), they are further distinguished from tholeiitic series rocks.

Boninite series rocks are typically inferred to have a subduction setting. However, broadly similar compositions can be obtained through interaction with the lithosphere, as discussed below (e.g. Sun *et al.*, 1989), and therefore it is useful to consider

Table 10.2 *Comparison of geochemical compositional ranges of different high-Si, high-Mg mafic rocks*

Rock types → Chemical composition ↓	1 Phanerozoic boninites	2 Archean Mallina basin boninites (Whitney type)	3 Paleoproterozoic high-Mg norites	4 Paleoproterozoic norites (Labrador)	5 Archean second-stage melt (Whundo type)	6 Archean SHMB	7 High-Mg andesites	8 Bastar boninites (High-Ca type)	9 Bastar norites (Low-Ca type)
SiO_2	52.40–61.30	52.00–54.00	47.00–56.00	48.00–54.00	47.30–53.20	51.00–57.00	51.68–53.18	52.33–54.12	52.63–55.10
TiO_2	0.07–0.50	0.24–0.28	0.30–1.00	0.22–0.53	0.30–0.50	0.40–1.00	0.93–0.94	0.26–0.39	0.35–0.55
Al_2O_3	6.10–15.00	16.60–17.60	6.80–16.10	10.00–12.00	16.00–17.00	9.90–13.00	13.68–14.36	8.97–16.55	8.55–13.20
$Fe_2O_3^{total}$	7.10–11.10	8.40–9.70	8.60–13.00	9.40–12.30	9.77–10.69	9.50–11.40	8.36–9.70	7.51–10.73	9.08–12.01
MgO	4.50–21.70	7.80–8.60	5.70–21.60	16.00–19.00	8.90–9.90	9.50–16.50	8.64–10.19	8.31–13.10	9.14–18.35
Mg#	42–76	62–67	56–80	75–78	61–66	69–75	–	66–74	60–81
Zr	8–55	33–41	37–128	37–96	11–53	41–74	114–118	35–50	43–79
Nb	<0.5–2.0	1.1–1.5	1.0–8.0	0.3–3.0	–	2.8–3.5	6	0.0–1.0	3.0–5.0
Sc	29–53	35–42	22–40	21–30	–	28–43	26	31–45	28–32
V	131–343	157–174	120–312	132–150	–	147–208	–	146–187	150–184
Yb	0.30–2.40	1.09–1.67	–	–	0.60–1.60	1.00–2.80	1.88	0.80–1.10	0.98–1.70
Al_2O_3/TiO_2	27–133	62–73	13–34	18–52	35–58	20–30	14.71–15.28	28.03–63.65	21.38–29.57
Ti/Zr	22–153	40–44	26–87	33–43	–	44–85	48–49	44.54–51.84	37.19–53.43
Ti/V	3–15	9–10	9–23	9–20	–	13–24	–	10.26–13.59	12.20–19.12
Ti/Sc	11–84	34–45	70–211	48–120	37–60	73–125	214–217	42.62–75.42	65.56–103.90

Mg# = 100 × $MgO/(MgO + FeO^{total})$. After Srivastava (2006).

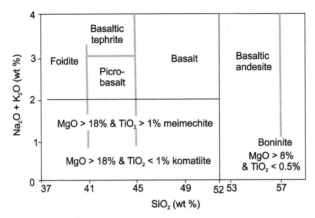

Figure 10.30 Geochemical differences between various types of mafic and ultramafic rocks. From Le Bas (2000). By permission of Oxford University Press.

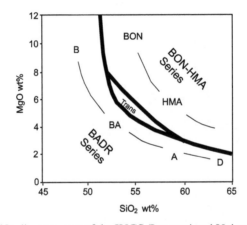

Figure 10.31 MgO–SiO$_2$ diagram, part of the IUGS (International Union of Geological Sciences) protocol for distinguishing the Boninite–HMA volcanic rock series (HMA = high-Mg andesite) from the BADR series (B = basalt; A = andesite; D = dacite; R = rhyolite). "Trans" marks the transition zone between the two fields. For rocks belonging to LIPs this will separate anorogenic boninite from the more dominant tholeiitic series. Modified from Figure 4 in Pearce and Robinson (2010). With permission from International Association for Gondwana Research.

both orogenic and anorogenic boninites. Anorogenic types of boninites may be particularly important in the Archean and in the Paleoproterozoic.

10.8.1 Archean anorogenic boninites

Smithies *et al.* (2004) distinguish two types of Archean boninitic-like rocks: Whundo-type and Whitney-type. Whundo-type rocks are most like orogenic boninites in terms of their composition and association with tholeiitic to calc-alkaline mafic to inter-mediate volcanics. Small compositional differences compared to modern boninites, including higher Al_2O_3 and HREEs, probably reflect secular changes in mantle temperatures and a more garnet-rich residual source. Whundo-type rocks are known from 3.1 and 2.8 Ga assemblages and are the most likely contenders for true Archean analogs of modern subduction-related (orogenic) boninites.

Whitney-type rocks occur throughout the Archean, as far back as *c.* 3.8 Ga, and are closely associated with ultramafic magmatism, including komatiites, in an affili-ation unlike that of modern subduction zones. They are characterized by very high Al_2O_3 and HREE concentrations, and their extremely depleted compositions require a source which at some stage was more garnet-rich than the source for either modern boninites or Whundo-type second-stage melts. Low La/Yb$_N$ and La/Gd$_N$ ratios compared to Whundo-type rocks and modern boninites either reflect very weak subduction-related metasomatism of the mantle source or very limited crustal assimi-lation by a refractory-mantle-derived melt. Regardless, the petrogenesis of the Whitney-type rocks appears either directly or indirectly related to the juxtaposition of convergent margin and plume magmatism (e.g. Wyman and Kerrich, 2012). In the sense that I have been considering (i.e. a plume link), the Whitney-type rocks are provisionally considered to be LIP-related.

10.8.2 Paleoproterozoic anorogenic boninites

Many mafic–ultramafic intrusions of late Archean to early Paleoproterozoic age and linked to LIPs are of the anorogenic boninite type. Examples of these high-MgO, high-SiO$_2$ units include the lower units (first pulses) of some layered intrusions and dykes. Examples are the Great Dyke (Zimbabwe), the Stillwater Complex (USA), and the Bushveld Complex (South Africa) as well as a number of dyke swarms in Greenland, Wyoming, and elsewhere. These anorogenic boninites, also termed BN (boninite–noritic) rocks, or "high-Mg norites," and their extrusive volcanic equiva-lent, SHMBs, are reported from Neoarchean to Paleoproterozoic and emplaced mostly in intracratonic rift or plume settings (Hall and Hughes, 1990, 1993; Smithies, 2002; Smithies *et al.*, 2004; Srivastava, 2006, 2008; Barnes *et al.*, 2012; Srivastava and Ernst, 2013) and are linked to LIPs (e.g. Hall and Hughes, 1993; Cadman *et al.*, 1997; Ernst and Buchan, 2001b). The following section is mostly based on Srivastava and Ernst (2013).

10.8.3 Siliceous high-Mg rocks

Apart from the general classification of units as belonging to the boninite series (on the basis of high MgO, high SiO$_2$, and low TiO$_2$), there are additional geochemical arguments, which partially distinguish the various types from each other. The REE

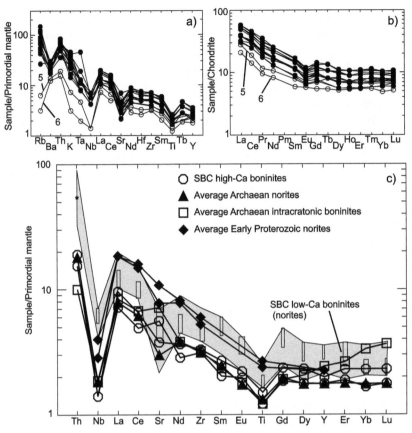

Figure 10.32 (a)–(c) Comparison of primordial-mantle-normalized multi-element spidergrams of average values of the boninite–norite (anorogenic boninite) suites. (a) and (b) Data from the Bastar craton (open circles are high-Ca boninites, and filled circles are low-Ca boninites. (c) Comparison of boninite–norite (anorogenic boninite) suites, Baster craton, with units from elsewhere. Primordial mantle values are from McDonough *et al.* (1992). Note all intracratonic norites (anorogenic boninites) are characterized by $(Gd/Yb)_{PN} > 1$ whereas (orogenic) boninites have $(Gd/Yb)_{PN} < 1$, where PN is primitive mantle normalized. SBC is southern Bastar craton. Parts (a) and (b) are from Figure 7 and part (c) is from Figure 8 in Srivastava (2006).

and spidergram patterns of BN rocks are very similar (Figs. 10.32 and 10.33). Boninites are also distinguished by $(Gd/Yb)_{PN} < 1$.

These various classes of boninitic rocks can be weakly distinguished on various trace-element classification diagrams. On a TiO_2 vs. Zr diagram (Fig. 10.33), subduction-related boninites define a field at low-TiO_2 contents, MORBs have higher abundances of both Zr and TiO_2, and Precambrian anorogenic boninitic dykes fall in a field between MORBs and subduction-related boninites. High-Mg andesite has higher values of TiO_2 and Zr and a lower Al_2O_3/TiO_2 ratio than true boninite. The Al_2O_3/CaO ratio is also used to classify high-Mg mafic igneous rocks into high-Ca

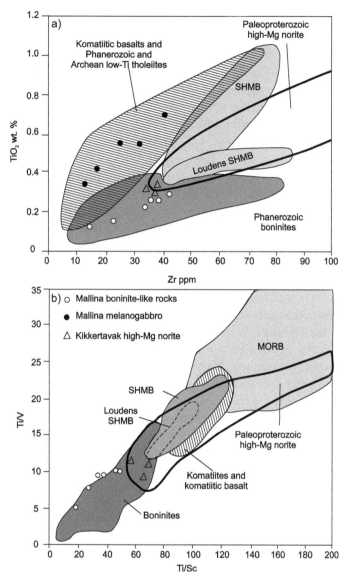

Figure 10.33 (a) and (b) Compositional differences between the various high-Mg suites. SHMB = siliceous high-magnesian basalt. Loudens and Mallina are Archean units from the Pilbara craton. Kikkertavak = 2.2 Ga dyke swarm from Nain craton. From Smithies (2002). With permission from Elsevier.

and low-Ca boninites (Crawford, 1989); high-Ca boninites show typical subduction boninitic geochemical characteristics, whereas the low-Ca variety show geochemical characteristics similar to anorogenic boninites (BN magmatism/high-Mg norites). It is believed that some SHMBs may also be a volcanic equivalent of BN rocks (Sun *et al.*, 1989; Hall and Hughes, 1990). Most SHMBs are Precambrian in age. However, recently a Permian SHMB was also reported from the Xinjiang, northwest China (Gao and Zhou, 2013). Various constraints on the origin of SHMBs have been proposed (Sun *et al.*, 1989) and different SHMBs may have different origins. Models for their origin include: (1) they are the Archean equivalent of modern boninites and form in an Archean subduction setting; (2) they are derived from metasomatized refractory mantle sources (Sun and Nesbit, 1978); and (3) they are derived from a parental komatiite magma through AFC processes (e.g. Arndt and Jenner, 1986; Arndt *et al.*, 1987; Sun *et al.*, 1989; Barnes *et al.*, 2012).

For subduction-related boninites, the refractory mantle wedge melts in response to the input of subduction-derived fluids. However, the anorogenic boninites (Table 10.2) are considered to originate in an intracratonic rift or hotspot (plume) setting and be derived from a previously metasomatized refractory mantle (Hall and Hughes, 1987, 1990; Kuehner, 1989; Srivastava, 2006, 2008).

It is noted that the BN suites differ significantly in composition from Archean komatiites and no genetic association has been established between them (Hall and Hughes, 1987, 1990). Archean komatiites are derived from a high degree of partial melting of a peridotite source (e.g. Cattell and Taylor, 1990; Bickle, 1990; Arndt *et al.*, 2008), whereas the BN suite of rocks is derived from a depleted refractory mantle, which is enhanced by metasomatism through subduction during the assembly at *c.* 2.7 Ga of many Archean cratons (e.g. Hall and Hughes, 1990, Srivastava, 2006).

The boninitic units are inferred to originate as follows; a mantle source was depleted by earlier basalt extraction (first melting), but the mantle was subsequently metasomatized and variably enriched in incompatible trace elements before second melting under anomalously high mantle temperatures and/or low pressures. In the case of anorogenic boninites, the initial source enrichment may be related to subduction but second melting occurs significantly later, in a non-subduction setting (e.g. Cadman *et al.*, 1997). This would be distinct from orogenic boninites whose source enrichment and associated melting are linked in the same convergent-margin environment (e.g. Poidevin, 1994; Piercey *et al.*, 2001; Smithies, 2002; Srivastava, 2008).

10.9 Geochemistry of carbonatites, kimberlites, and lamprophyres/lamproites

Carbonatites, and some kimberlites, lamproites, and lamprophyres can be linked with LIPs and therefore should share some compositional characteristics.

10.9.1 Carbonatites and associated alkaline complexes

The link with LIPs is strongest for carbonatites and the following is adapted from Bell and Simonetti (2010) and Ernst and Bell (2010).

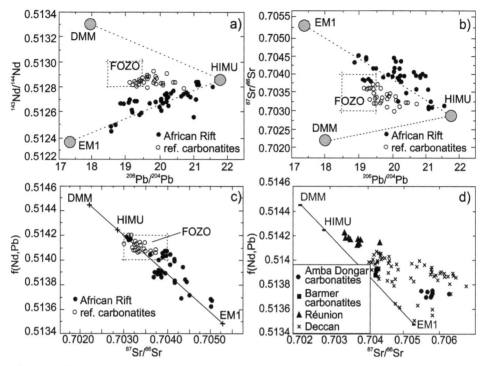

Figure 10.34 Isotopic values for carbonatites. (a)–(c) Data from the carbonatites in the East African rift and also from so-called reference carbonatites, based on data from seven carbonatite localities in Africa, Eurasia, North America, and New Zealand. (d) Data from Deccan carbonatites. (a)–(c) From Bell and Tilton (2001). By permission of Oxford University Press. (d) From Figure 14 in Bell (2001). By permission of Oxford University Press.

(1) HIMU, EM1, and FOZO, mantle components found in OIBs, are also found in many young (< 200 Ma) carbonatites (e.g. Tilton and Bell, 1994; Simonetti *et al.*, 1995; 1998; Bell and Simonetti, 1996; Bell and Tilton, 2002).

(2) Noble-gas data from some carbonatites indicate derivation from a relatively primitive mantle (Sasada *et al.*, 1997; Marty *et al.*, 1998; Dauphas and Marty, 1999; Tolstikhan *et al.*, 2002; Bell and Rukhlov, 2004).

(3) There is little evidence for the involvement of DMM in the source for carbonatites, ruling out any major involvement of the upper mantle asthenosphere source of MORBs in the source regions of carbonatites.

(4) Some carbonatites are marked by an unradiogenic Hf isotope composition, suggesting an origin from an enriched, ancient deep mantle reservoir (Bizzarro *et al.*, 2002), e.g. FOZO.

As shown in Fig. 10.34, carbonatite data can exhibit remarkably simple patterns. The carbonatites from the East African rift define a mixture with varying proportions between just two end-members HIMU and EM1 (Bell and Tilton, 2001), even at single volcanic centers within the rift (e.g. Oldoinyo Lengai; Bell and Dawson, 1995.).

In addition, Bell and Tilton (2001) have defined a second group of so-called Reference Carbonatites, based on data from seven carbonatite localities in Africa, Eurasia, North America, and New Zealand (Fig. 10.34a–c). The isotope ratios from these reference samples are characterized by mixture patterns between HIMU and FOZO. Many oceanic island groups tend to converge on FOZO (Fig. 10.1) and FOZO is also considered to characterize some plumes (e.g. Hauri *et al.*, 1994). These carbonatite data suggest that there is a widespread mantle reservoir with reasonably uniform isotopic composition that can be considered as further confirmation of the validity of FOZO as an approximate mantle end-member under both oceans and continents. Figure 10.34d shows that the isotopic data from the Deccan flood basalts do not lie on the EMM–EM1–HIMU mantle plane but must have an additional component relative to EM1 and HIMU, which is probably continental crust (K. Bell, personal communication, 2013).

10.9.2 Kimberlites

As shown in Section 9.3.3, many LIPs have associated kimberlites and therefore the composition of such kimberlites should have implications for the overall magmatic system. Geochemical differences between kimberlites and other high-Mg types are illustrated in Fig. 10.35. As summarized by Arndt (2003), kimberlites occur as volcanic diatremes and as dykes and sills, and have complex

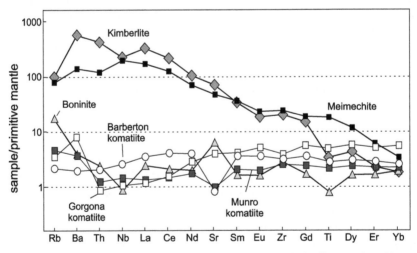

Figure 10.35 Trace-element comparison of high-Mg suites. Kimberlites, meimechites, three types of komatiites (Barberton, Munro, and Gorgona), and boninites. Trace-element concentrations are normalized to the primitive mantle of Hofmann (1988). Note the distinctive Nb, Sr, and Ti anomalies and negatively fractionated HREEs of boninites and the absence of these anomalies in komatiites. From part of Figure 3 in Arndt (2003). Data sources in Arndt (2003).

petrography because of the presence of abundant xenocrysts, megacrysts, and xenoliths (e.g. Mitchell, 1986, 1995; Section 9.3.1). Their compositions are characterized by high MgO contents and low SiO_2 and Al_2O_3 contents, extremely high concentrations of incompatible trace elements, and by high H_2O and CO_2 contents (as hydrous minerals and carbonates).

Kimberlites result from low-degree melting, at great depth, of sources rich in incompatible elements and $CO_2 + H_2O$. They become further enriched through interaction with the lithospheric mantle. Kimberlites have been subdivided into Group I (basaltic) and Group II (micaceous). Their mineralogy can be used to distinguish between the two types. Group I are relatively rich in ilmenite and contain a characteristic suite of low-Cr megacrysts and high-T sheared xenoliths (e.g. Arndt, 2003). Group II have abundant mica, and have higher concentrations of incompatible elements, and their isotopic compositions match those of an old, geochemically enriched source (high $^{87}Sr/^{86}Sr$, low $^{143}Nd/^{144}Nd$ ratios). In contrast, Group I kimberlites have the isotopic features of a more depleted source like a mantle plume or convecting upper mantle.

10.9.3 Lamprophyres and lamproites

The association between lamprophyres and lamproites (see definitions and discussion in Section 9.4) and LIPs is more speculative and indirect. The influence of a metasomatized lithospheric mantle is important in their development and so lamprophyre and lamproite compositions may help map out the subcontinental lithospheric mantle underlying associated LIP events. For example, according to Riley *et al.* (2003), ultramafic lamprophyre (UML) dykes from the Ferrar magmatic province, Antarctica (part of the Karoo–Ferrar LIP) are characterized by high Ti, Cr, Ni, Nb/La, La/Yb$_N$, and Mg# values, and are the most primitive rocks of the Ferrar event. The dykes have initial (183 Ma) $^{87}Sr/^{86}Sr$ ratios of 0.7044–0.7055, εNd of 4.6–4.8, $^{208}Pb/^{204}Pb$ of 39.6–40.3, and $^{187}Os/^{188}Os$ of 0.120–0.146, which contrast markedly with even the most primitive rocks of the Karoo–Ferrar LIP. The trace-element and isotope characteristics have similarities to OIBs and the highly radiogenic character of $^{208}Pb/^{204}Pb$ and $^{206}Pb/^{204}Pb$ bears closest resemblance to the Bouvet OIB, which has been suggested as the plume responsible for the Ferrar event. The ultramafic lamprophyres are believed to be the result of melting of enriched Bouvet mantle-plume material and represent one of the mantle compositional end-members for the Karoo–Ferrar LIP.

The Leucite Hills lamproites, which are linked to the Columbia River LIP, have the following characteristics as summarized by Mirnejad and Bell (2006). They exhibit negative Nb, Ta, and Ti anomalies in mantle-normalized trace-element diagrams and have low time-integrated U/Pb, Rb/Sr, and Sm/Nd ratios. These results indicate an ancient metasomatic enrichment (> 1.0 Ga) of the mantle source associated with the subduction of carbonate-bearing sediments. Other chemical characteristics of the Leucite Hills lamproites, especially their high K_2O and volatile contents, are attributed to more recent metasomatism (< 100 Ma) involving influx from upwelling mantle during back-arc extension or plume activity, linked to the Columbia River LIP.

The proposed relationships between silicate and carbonatite melts, and silicate-melt differentiation, are illustrated schematically in Fig. 9.18 (see also Woolley and Kjarsgaard, 2008b). Note that carbonatite originating in the mantle can pass directly to the surface, and need not be involved in differentiation at a higher level. Deeper-level metasomatic events are considered a necessary precursor to carbonatite generation and probably to many alkaline silicate rocks.

10.10 Geochemistry of associated silicic magmatism

10.10.1 Silicic magmatism associated with the Bushveld LIP

Extensive silicic magmatism of both intrusive and extrusive type is associated with the Bushveld LIP (Section 5.4.1; Fig. 10.36). The silicic magmatism comprises the Lebowa suite (including the Nebo granite), granophyres from the Rashoop suite, granites and nepheline syenites from the Moshaneng Complex, and silicic and basaltic units from the Rooiberg volcanics. The silicic rocks associated with these groups have geochemical characteristics similar to ferroan (A-type) granitoids. They are likely formed by the differentiation and incorporation of more continental crust material by the earlier parent mafic rocks, with increased amounts of crustal con-tamination and increased differentiation.

10.10.2 Silicic magmatism in LIPs

Silicic magmatism occurs throughout the geologic record in association with LIPs (dominantly mafic) or as voluminous Silicic LIPs (SLIPs; Chapter 8). The former is discussed here and the latter in the next section. As summarized by Pankhurst *et al.* (2011), rhyolites observed within LIPs are likely to be primarily the product of efficient fractional crystallization from mafic mantle sources. They bear a suite of diagnostic characteristics: high temperature (900–1100 °C), low water content, high F (in thousands of parts per million), high SiO_2/Al_2O_3 ratios (*c.* 5–7), high K_2O/Na_2O ratios (> 1.5), and high Ga/Al_2O_3 ratios (> 1.5). They are erupted as rheoignimbrites or lavas with large lateral extents and are erupted rapidly along with the (usually) volumetrically dominant mafic LIP material. They represent some of the largest single events to produce silicic crustal material (e.g. Bryan et al., 2010). As noted by Bryan and Ferrari (2013), these flood rhyolites in CFB provinces have been interpreted by some as due to crustal melting, including remelting of the basaltic igneous underplate (e.g. Garland *et al.*, 1995; Ewart *et al.*, 2004b; Miller and Harris, 2007).

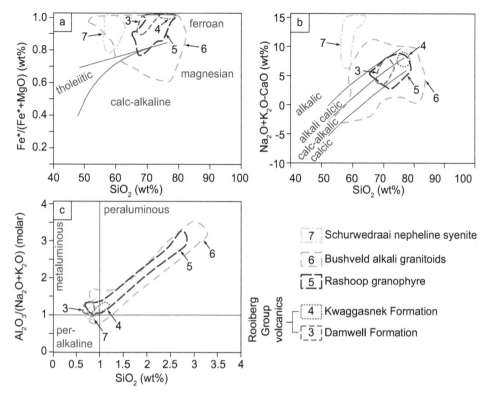

Figure 10.36 Silicic magmatism associated with the Bushveld LIP. (a)–(c) Representative major-element-based diagrams illustrating the general geochemical characteristics of the different silicic rocks from the three pulses of the Bushveld LIP. Digital version kindly provided by H. Rajesh, which he modified from the original version in Rajesh *et al.* (2013). Permission provided by Elsevier.

In terms of a connection with AMCG (anorthosite–mangerite–charnockite granite) and rapakivi suites (Fig. 8.9), Bonin (2007) notes that mafic melts can pond at the Moho density filter, fractionating into gabbro–anorthosite complexes at deeper crustal levels with A-type granites as highly fractionated derivatives that undergo AFC processes with crust en route to shallow emplacement. In such cases Bonin (2007) notes that the magma can have the signature of a depleted mantle, the low-velocity zone at the base of the continental lithospheric mantle, and also the lower crust.

10.10.3 Silicic LIPs (SLIPs)

In contrast to most LIPs, the SLIPs are compositionally and volumetrically dominated by silicic (> 65 wt% SiO_2) igneous compositions, but commonly have a

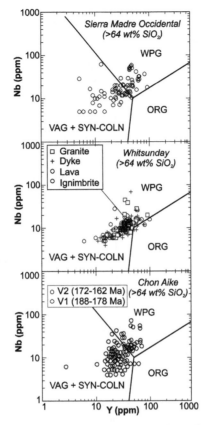

Figure 10.37 Composition of selected Silicic LIPs on the Nb–Y discrimination plots of Pearce *et al.* (1984). Data from the Whitsunday, Sierra Madre Occidental, and Chon Aike SLIPs show the characteristic transitional signatures from within-plate (A-type) to convergent margin fields (I- and S-types). WPG = Within-plate granites; ORG = ocean-ridge granites; VAG + SYN-COLN = volcanic arc and syncollisional granites. Two pulses, V1 (188–178 Ma) and V2 (172–162 Ma) of the Chon Aike data are shown (Fig. 3.7; Section 8.3.4). From Figure 7 in Bryan *et al.* (2007).

spectrum of extrusive and intrusive compositions from basalt to high-silica rhyolite (e.g. Ewart *et al.*, 1992; Bryan *et al.*, 2000; Riley *et al.*, 2001; Bryan, 2007; Ferrari *et al.*, 2007; Chapter 8).

The geochemistry of three important SLIPs is shown in Figs. 10.37 and 10.38. Note that the Whisunday, Sierra Madre Occidental, and Chon Aike SLIPs show the characteristic transitional signatures from within-plate (corresponding to A-type granites) to convergent-margin fields (I- and S-types). In Figure 10.38, some SLIPS have compositions similar to I-type granites while others plot closer to A-type granites. Rhyolites within (mafic) LIPs consistently plot closer to A-type granites than I-type granites.

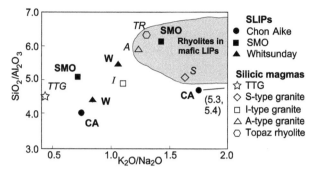

Figure 10.38 Major-element discrimination diagram for typical rhyolites within mafic LIPs and SLIPs compared with well-constrained silicic magma types. Note that within SLIPs some compositions are similar to I-type granites while others plot closer to A-type granites. Rhyolites within mafic LIPs consistently plot closer to A-type granites than I-type granites. TTG is Tonalite-Trondhjemite-Granodiorite, which are dominant in Archean granite-greenstone belts, and are linked to subduction. Topaz rhyolites (TR) are potentially the extrusive equivalents of rapakivi granites (e.g. Christiansen *et al.*, 2007). SMO = Sierra Madre Occidental. Modified from Pankhurst *et al.* (2011). With permission from Elsevier.

As noted in Chapter 8, these compositions are consistent with SLIPs being mainly derived from partial melting of the lower crust due to heat provided by the arrival of voluminous LIP mafic–ultramafic magmatism at the base of the crust (e.g. Pankhurst and Rapela, 1995; Riley *et al.*, 2001; Bryan *et al.*, 2002; Bryan, 2007; Bryan and Ernst, 2008). SLIPs, at least in the Phanerozoic, are restricted to continental margins where fertile, hydrous metasomatism was caused by earlier subduction. As a general rule it seems that pre-hydration of the lithosphere by an earlier subduction episode appears to be important in developing SLIPs. In areas where thinned lithosphere lacks such refertilized crust, lower crustal melting by a LIP underplate would not occur and instead the mainly mafic magmas would transit the crust to produce a LIP (i.e. mainly mafic).

10.11 Geochemistry of the LIP as a system

This section integrates the discussion of previous sections in order to consider how the chemistry of the different units of a LIP can be used in a synergistic way to infer characteristics of the source and plumbing-system pathways. These are just some initial thoughts and represent an area of future research.

10.11.1 Relationships between different parts of the LIP system

Here I summarize a few of the interrelationships that are now recognized between different components of the LIP system. It was shown using geochemistry (and paleomagnetism) that each individual dyke typically represents injection of a distinct and separate magma pulse (Section 10.6.2). Similarly, geochemistry from the Karoo sill province showed that the pathways for distinct magma batches can be tracked using geochemistry (Section 10.6.5). The composition of particular volcanic packages can be linked to particular subswarms of dykes (e.g. for the Mackenzie swarm; Section 10.6.4) and to individual subvolcanic intrusions (Siberian Trap LIP; Section 3.2.8). It was also pointed out that mafic magmatism can be transported laterally in the crust as dykes for more than 1000 km (Section 5.2.4) and also for long distances via sills; and furthermore, that plume material can slide along the base of the lithosphere (sublithospheric channeling; Section 5.6.2). Owing to their higher density, ultramafic magmatism is more likely to be present in the intrusive component of the LIPs than in the extrusive component. Maximum MgO content in LIPs decreases through time (toward the present), with LIPs containing komatiites being mainly restricted to the Archean. Anorogenic boninites are common in intrusives of late-Archean and early-Paleoproterozoic time. More detailed geochemical studies across all units of a LIP can be used to distinguish various geochemical types and track the movement of magma batches corresponding to each geochemical type spatially through the LIP system. A step in this direction are the studies that have distinguished the distribution of high- and low-Ti sills of the Karoo LIP (Section 10.5.3).

10.11.2 Carbonatites and kimberlites

The recognition that carbonatites are strongly linked with LIPs (Section 9.2; Ernst and Bell, 2010) indicates that carbonatites are also being generated (directly or indirectly) by the same plume source. Carbonatites seem to preserve their primary composition better than mafic–ultramafic magmatism, which is typically quite modified on passage through the continental lithosphere. Therefore carbonatites may better preserve the record of the isotopic components in the plume than the LIP magmatism. The dominance of OIB reservoirs associated with carbonatites is strong confirmation for the involvement of OIB mantle in the generation of those LIPs that are associated with carbonatites. The association of some kimberlites with LIPs may potentially be used in a similar synergistic way, to infer about mantle sources for LIPs based on the geochemical and isotopic characteristics of associated kimberlites.

10.11.3 Location of mantle-source areas

It is important to recognize that some units associated with a LIP are strictly emplaced vertically. This is certainly true for carbonatites and kimberlites, so their source areas lie directly underneath at some depth. Kimberlites indicate the presence of a thick lithosphere. Carbonatites can be emplaced under a range of thicknesses of lithosphere. Similarly, silicic magmatism (given its buoyancy) should also essentially rise vertically and so the lower crust that is involved in melting should

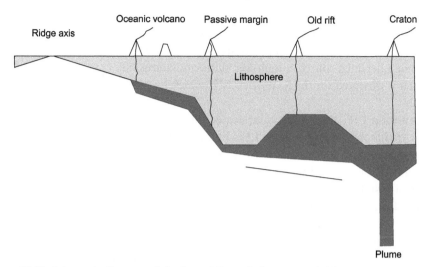

Figure 10.39 Schematic diagram of the lateral flow of plume material beneath a slow-moving continent. The plume material ponds beneath locally thin lithosphere of an old rift. It arrives beneath a region of thick lithosphere and moves upwards across a passive margin. It then flows toward the ridge axis. From Figure 2 in Sleep (2006). With permission from Elsevier.

directly underlie the silicic magmatism. Furthermore, the distribution of silicic magmatism can be used to locate underlying fusible lower crust (possibly linked with a prior subduction event).

However, for mafic magma the situation can be rather different, given the possibility of lateral flow in giant, radiating dyke swarms for distances of hundreds of kilometers to > 2000 km (Section 5.6.4). Therefore, the source of the dykes (and any volcanic and sills they might feed) are not necessarily located in the mantle immediately beneath. Instead, the appropriate source mantle could be > 2000 km away toward the plume center, marked by the focus of a radiating dyke swarm. So, laterally emplaced dykes can be spatially juxtaposed with vertically emplaced kimberlites. Dolerite sills may also be laterally emplaced for great distances. Sublithospheric channeling (Section 5.6.2) can result in magmatism displaced a long distance from its source plume – perhaps > 1000 km (dependent on the size of the crustal block and topographies within its mantle root (Fig. 10.39). Picrites and high-Mg rocks originate early and are preferentially concentrated above the plume-center region.

10.12 Summary

LIPs represent large-volume, short-duration pulses of intraplate magmatism, but there is no unique geochemical signature that is diagnostic of all LIPs. While many LIPs do have an intraplate signature, others have a subduction-like character interpreted to be derived from interaction with wall rocks during passage of magma through the lithosphere mantle or through contamination by overlying crust.

Oceanic LIPs have three compositional types: low-Mg, with a flat REE pattern indicating a depleted source, a high-Mg suite, and a crustally contaminated pattern. The latter is observed for the Kerguelen LIP, which has evidence of continental crust at depth.

Continental LIPs have more complicated patterns. Presumably, the same mantle sources are involved as for oceanic LIPs but, in addition, interactions with continental lithosphere (both mantle and crust) can significantly modify the primary compositions of LIP magmas. These impart some additional variability in composition, which can result in a number of compositionally distinct magma batches that can be tracked through the plumbing system and linked to surface lavas. Dyke swarms are mainly of dolerite composition and exhibit along-dyke geochemical consistency, indicating that each dyke represents a separate magmatic pulse. In a sill province, the geochemical fingerprint of particular magma batches can be traced through the region. Results can be surprising in which nested overlying sills may not belong to the same magmatic batch. Associated carbonatites exhibit compositions that strongly correspond to the major mantle reservoirs. Kimberlites can be viewed as products of the mantle plume ascending through thick lithospheric roots. Associated silicic magmatism has a variety of compositions reflecting melting of fusible lower crust (resulting in I-, S-, or, most typically, A-type melts). The composition depends on the lower crustal rocks being melted.

An important goal for future geochemical studies of LIPs is to integrate the chemistry from the different magmatic components of a LIP to determine the number of distinct compositions present, the petrogenetic history of each, and the pathways of magmatic batches through the LIP plumbing system from mantle source areas to final emplacement in, on, or at the base of the crust.

11

LIPs, rifting, and the supercontinent cycle

11.1 Introduction

This chapter addresses in more detail the links between Large Igneous Provinces (LIPs) and rifting and continental breakup. The Mesozoic/Cenozoic record is used to demonstrate the remarkable correlation between LIPs, continental breakup, and formation of new ocean basins. That association includes the characteristic "triple-junction" style of rifting and the formation of aulacogens. There is frequently a small timing gap between LIP pulses and breakup, and this chapter also addresses why some major LIPs did not result in successful continental breakup. Rifting and breakup also affect oceanic plateaus. In more deeply eroded terranes, the plumbing-system record of LIPs can be used to identify the timing and location of Proterozoic breakup margins of crustal blocks. This leads to a natural role for LIPs in sorting out paleocontinental reconstructions, and allows consideration of the role of LIPs in understanding the supercontinent cycle.

11.2 LIPs, continental breakup, and formation of ocean basins

The association of LIPs with continental breakup and formation of new ocean basins is the rule rather than the exception (e.g. White and McKenzie, 1989; Courtillot *et al.*, 1999). For example, opening of the Red Sea and Gulf of Aden, Arabian Sea, South Atlantic, Central Atlantic, North Atlantic, South Atlantic, and southwest Indian ocean basins can all be linked with emplacement of new continental LIPs (Table 11.1). However, there are notable exceptions in which major LIPs are not associated with ocean opening (e.g. the Siberian Trap LIP) and the reasons are discussed in Section 11.6.

The progressive breakup of the Gondwana portion of the Pangea supercontinent is associated with a sequence of discrete LIP events (Table 11.1, Fig. 11.1; see also Chapter 3). The youngest example is the Afro-Arabian LIP with a main pulse at 30 Ma, which is associated with the early stages of opening of a new ocean between the Arabian Peninsula (+Asia) and Africa. The Deccan LIP is associated with the separation of the Seychelles microcontinent and the opening of the Arabian Sea

Table 11.1 *LIPs associated with post-Paleozoic continental breakup and ocean opening*

LIP name (age)	Blocks separated	Associated ocean opening (age)
Afro-Arabian (31 Ma and younger pulses); Section 3.2.3	Arabian Peninsula–Africa	Gulf of Aden, Red Sea (21–25 Ma)
Deccan (*c.* 65 Ma); Section 3.2.5	India–Seychelles microcontinent	Arabian Sea, Indian Ocean
NAIP (62–58 and 55 Ma); Section 3.2.4	Northern Europe–Greenland (North America)	North Atlantic Ocean (55 Ma)
Madagascar (93 Ma); Section 11.2	Madagascar–southern Africa, Madagascar–India	Mozambique Channel (Indian Ocean) Arabian Sea, Indian Ocean
Paraná–Etendeka (mainly 135 Ma); Section 3.2.6	South America–Africa	South Atlantic Ocean (*c.* 122 Ma)
Kerguelen–Comei–Bunbury (130–117 Ma); Section 4.2.2	India–Antarctica	Bay of Bengal, Indian Ocean (130–117 Ma)
Karoo–Ferrar (183–177 Ma); Section 3.2.7	Africa–Antarctica	Indian Ocean
CAMP (200 Ma); Section 12.2.7	North America, Africa, South America	Central Atlantic Ocean
CIMP (615, 590, and 560 Ma pulses); Fig. 9.8	Laurentia, Baltica, Amazonia (?)	Iapetus Ocean

portion of the Indian Ocean. The 90 Ma Madagascar LIP is associated with the separation of Madagascar from both southern Africa and India linked to opening of the Indian Ocean (Mozambique Channel and Arabian Sea portions, respectively). The *c.* 135 Ma Paraná–Etendeka LIP is associated with the opening of the South Atlantic Ocean between Africa and South America. The *c.* 118 Ma Kerguelen oceanic plateau is associated with the India–Antarctica–Australia breakup, but was emplaced after initiation of the Indian Ocean (Bay of Bengal portion) had already begun. However, the 130 Ma initiation of this LIP is associated with initial ocean opening and is preserved in southwestern Australia as the Bunbury basalts and in the Himalayas as the Comei basalts and I refer to the entire event as the Kerguelen-Bunbury-Comei LIP (Section 4.2.2). The 183 Ma Karoo-Ferrar LIP is associated with separation of Antarctica from Australia and the opening of the Indian Ocean.

Three LIP events are linked to the opening of the full length of the Atlantic Ocean (Fig. 11.2). In addition to the already mentioned Paraná–Etendeka LIP, which is associated with *c.* 120 Ma opening of the South Atlantic Ocean between Africa and South America, the opening of the Central Atlantic Ocean (between North America

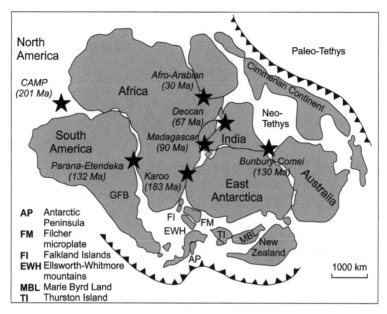

Figure 11.1 Gondwanaland reconstruction and location of mantle plume centers associated with LIPs. Modified from Fig. 1 in Storey and Kyle (1997) to which the location of mantle plume centers has been added. The CAMP plume center is associated with the separation of North America and the opening of the Central Atlantic.

and Africa) starting at *c.* 195 Ma can be linked to the 201 Ma Central Atlantic Magmatic Province (CAMP), and the *c.* 55 Ma opening of the North Atlantic Ocean between Europe and Greenland is linked to the North Atlantic Igneous Province (NAIP).

Having established the LIP–continental breakup link with a number of examples, the next step is to consider some of the characteristics of the LIP–rift association.

11.3 Active/passive rifting and rift classification

The association of LIPs and breakup suggests a component of active rifting. As discussed in many papers (e.g. White and McKenzie, 1989, 1995; Ruppell, 1995; Courtillot *et al.*, 1999; Şengör and Natal'in, 2001; Bialas *et al.*, 2010) there are two types of rifting, active and passive (Fig. 11.3). During passive rifting the crust and lithosphere undergo extension as a result of plate-boundary forces and this process includes the following stages (e.g. Courtillot *et al.*, 1999): (1) incipient rifting occurs; (2) far field (plate-boundary) stresses thin the crust and lithospheric mantle; and (3) the hot asthenospheric mantle passively enters the thinned area. By contrast, in active rifting: (1) a rising mantle-plume forces doming of the crust; (2) this is followed

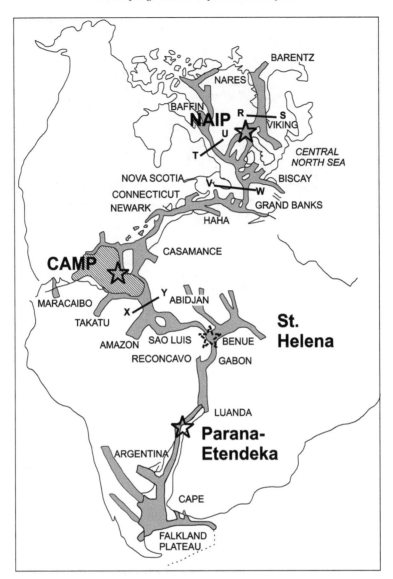

Figure 11.2: Rifting associated with the opening of the Atlantic Ocean and relationship to plumes and LIPs. The complex distribution of rifting is associated with the initiation of the breakup. The locations of plume centers for the NAIP (62–55 Ma), Paraná–Etendeka (*c.* 135 Ma), and CAMP (200 Ma) LIPs are shown. The St. Helena plume center (also shown) is not known to have an associated LIP. The timing of rifts are indicated as follows: 210–170 Ma ages for rifts between lines V–W and X–Y (corresponding to the CAMP LIP). Rifts south of line X–Y formed between 145–125 Ma (corresponding to the Paraná–Etendeka LIP). Rifts north of line T–U formed at about 80 Ma, and rifts north of R–S formed at about 80–60 Ma (roughly corresponding to the NAIP LIP). The collective rift system is termed a taphrogen according to the terminology of Şengör (1995) and Şengör and Natal'in (2001). Background rift diagram from Figure 2.7B in Şengör (1995). Permission granted by Wiley.

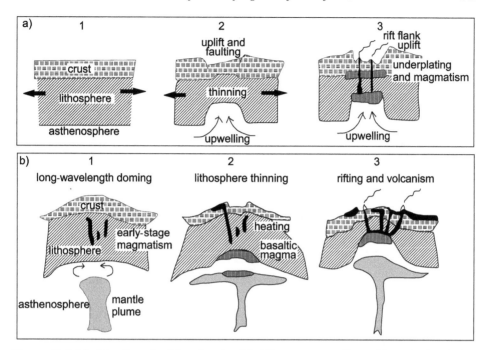

Figure 11.3 Models of passive (a) and active (b) rifting processes. From Figure 4.3 in Pirajno (2000), which was originally after Ruppell (1995). Permission granted by Springer. Digital version of original figure kindly provided by F. Pirajno.

by thinning of the crust and lithosphere resulting in adiabatic melting and under-plating; (3) rifting occurs at the crest of the domed crust and LIP magmatism is associated. The main difference is that the lithosphere is thinned by plate-boundary forces under passive rifting, and by a mantle plume under active rifting.

It is generally concluded that, in most situations, rifting that forms a new ocean basin is probably a combination of both mechanisms. Specifically, an active component (a plume and resulting LIP) is a prerequisite for the breakup to form a major oceanic basin. But rifting must be allowed by plate-boundary forces and is influenced by pre-existing heterogeneities in the lithospheric structure. This is especially clear in the case of the breakup to open the Central Atlantic Ocean (see Section 11.6).

The topic of rifting has been considered in great detail by Şengör (1995) and Şengör and Natal'in (2001). These authors use several systems to classify rifts. The dynamical system classifies rifts as related to passive or active extension (as discussed above). The geometric classification includes solitary rifts, rift stars, and other more-complicated patterns (geometries) of rifting. Kinematic classification is on the basis of plate-tectonic setting and includes intra-plate rifts as well as those associated with divergent, conservative, and convergent plate boundaries (Fig. 11.4). These authors

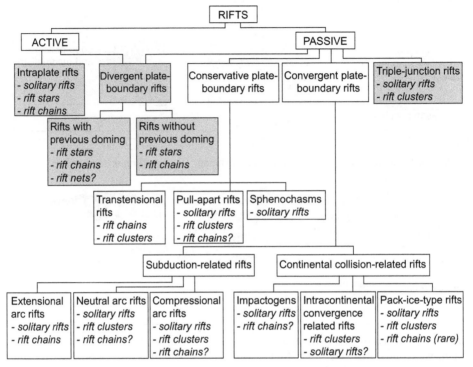

Figure 11.4 Classifications of rifts as follows: dynamic classification ("ACTIVE" or "PASSIVE"), geometric classification (italicized text), kinematic classification (other text). Shaded in grey are those types that can be linked to a LIP. From Şengör (1995), permission granted by Wiley, and Şengör and Natal'in (2001).

also identify rift systems (representing linked systems of rifts and grabens) which they term taphrogens by analogy with the name of orogens used for collections of structures associated with ocean closure events.

Şengör and Natal'in (2001) consider all those rifts associated with uplift to be of plume origin, and they cataloged 181 such rifts (or 32% of the total). This would suggest that at least 30% of rifts are plume-related, and given the strong link between plumes and LIPs (see discussion in Chapter 15) this would suggest that about 30% of rifts back through time are potentially LIP-related.

Other aspects associated with the rift-to-drift transition are summarized in Figure 11.5 (after Armitage and Allen, 2010), in which rifts are characterized on the basis of degree of stretching and rate of extensional strain. Note that intracratonic basins (Section 12.6.2) are distinguished by a low stretching factor and low strain rate. Failed rifts (aulacogens) are the product of greater stretching and strain rate, and the onset of ocean opening has the largest values of each.

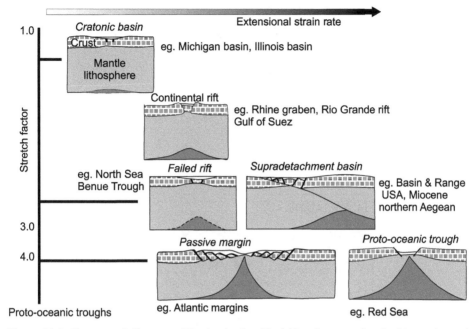

Figure 11.5 Conceptual diagram of basins in the rift–drift suite, associated with continental extension. Cratonic basins are viewed as basins whose primary mechanism for subsidence is low strain-rate stretching, leading to low final stretching factors. From Figure 1 in Armitage and Allen (2010).

11.4 Triple-junction rifting

The most important model for linking LIPs with breakup is the classic "triple-junction" model of Burke and Dewey (1973). In the terminology of Şengör (1995) and Şengör and Natal'in (2001) (Fig. 11.4), triple-junction rifts can also be termed rift stars. Such rifts consist of three (or more) rift arms that develop in association with breakup. Typically, one arm fails to progress to breakup and is termed a "failed rift" or aulacogen (Burke, 1977; Şengör, 1995). The full Wilson cycle that begins with ocean opening ultimately leads back to ocean closure marked by an orogen developing along two of the rift arms. This process can be illustrated using the dyke-swarm distributions after Fahrig (1987; Fig. 11.6), which shows the importance of the dyke-swarm component of LIPs in identifying the location of failed arms, particularly in older, more deeply eroded cases (as discussed below).

A classic example of triple-junction rifting is associated with the Afro-Arabian LIP. Two main arms led to opening of the Gulf of Aden and the Red Sea, while the third arm of the East African rift system has attempted but failed to open (Fig. 3.3). In the case of the Deccan LIP (Fig. 3.5), the failed arm continues to the east at the Narmada–Tapti rift. In the case of the Paraná–Etendeka LIP (Fig. 3.6), the Ponta

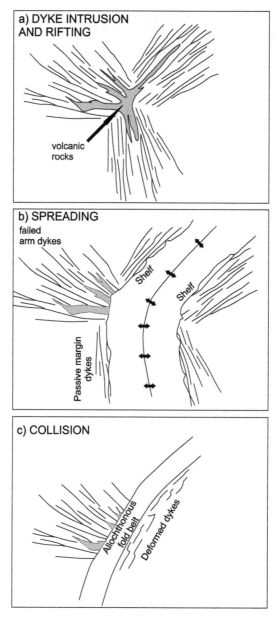

Figure 11.6 (a)–(c) Three-stage plate-tectonic cycle in the development of mafic continental dyke swarms. From Figure 11B in Ernst and Buchan (1997a), which is modified after original version Figure 1 in Fahrig (1987).

Grossa dyke swarm may mark the failed rift (Renne *et al.*, 1996; Courtillot *et al.*, 1999). A third arm for the Karoo event is along the Okavango (Botswana) swarm (Fig. 5.10). Additionally, the long linear extent of the Ferrar magmatism in Antarctica may suggest it also represents an additional failed rift of a combined

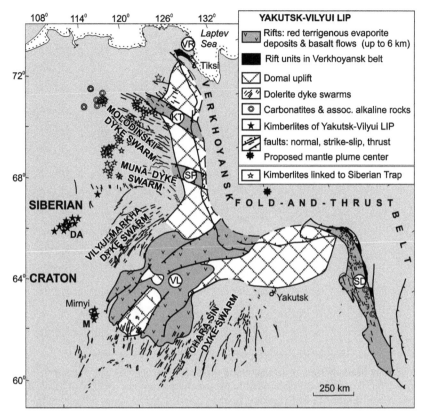

Figure 11.7 Triple-junction rifting associated with Yakutsk plume and LIP. Rifts (with volcanism) are: Vilyui (VL), Sobopol' (SP), Kyutungda (KT), Verkhoyansk (VR), and Sette–Daban (SD). Dyke swarms are: Sette Daban (SD), Chara–Sin (CS), Vilyui–Markha (VM), Muna (Mu), and Molodinski (Mo). Associated kimberlites fields are Mirni (M), Daldyn (DA), and Nakyn (N). Modified after Kiselev *et al.* (2012). With permission from Elsevier.

Karoo–Ferrar LIP (Courtillot *et al.*, 1999). One of the most remarkable triple-junction rifting systems is that of the *c.* 370 Ma Yakutsk–Vilyui event of eastern Siberia (Fig. 11.7).

The 370 Ma Yakutsk–Vilyui LIP of eastern Siberia exhibits a dramatic triple-junction rifting with five arms (Fig. 11.7; Shpount and Oleinikov, 1987; Kiselev *et al.*, 2012). The north–south arms (Tiksi and Sette–Daban rifts) led to the successful breakup on the eastern margin of the Siberian craton, and the other arms (the Kyutungda, Sobopol', and Vilyui rifts) represent aulacogens that pinch out into the cratonic interior. The Vilyui rift is slightly more complicated, consisting of a pair of grabens (Vilyui–Markha and Chara–Sin) separated by a horst. Associated magmatism consists of flood-basalt volcanism along the rifts, and a giant radiating dyke swarm paralleling the rifts and extending beyond them. Associated kimberlites and carbonatites are also present.

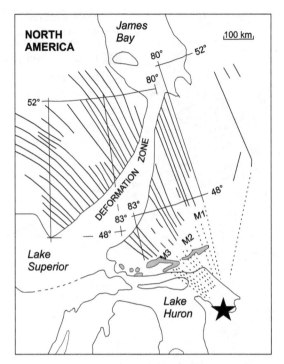

Figure 11.8 The Matachewan swarm showing three subswarms that may represent failed rifts (aulacogens) that have been deeply eroded to expose their basement rocks with associated dyke subswarms. From Figure 11 in Ernst and Buchan (2001c). Star locates mantle plume center.

Another way that dykes may reveal the pattern of triple-junction rifting is exhibited by the 2480–2450 Ma Matachewan swarm/LIP (Fig. 11.8) in which an overall radiating swarm is grouped into subswarms (labeled M1, M2, and M3) (originally recognized by West and Ernst, 1991). It is speculated that each of these three subswarms represents deep erosion of rifts, very similar to that observed in the Yakutsk–Vilyui triple-junction rifting, particularly in the pair of rift arms (Vilyui–Marcha and Chara–Sinsk) extending to the southwest.

11.4.1 Apical graben

An additional type of extensional structure (Fig. 11.9) is an "apical" graben a few hundred kilometers across (morphologically similar to a large caldera collapse structure). This has been tentatively identified above the plume-center region of the 1270 Ma Mackenzie LIP based on gravity data (Baragar *et al.*, 1996). It is suggested that the group of gravity anomalies circumscribing the plume center mark layered intrusions, which are distributed along the rim of such an apical graben. It is further

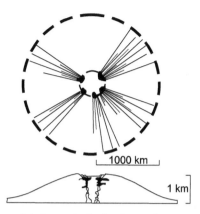

Figure 11.9 Apical graben model shown in both plan and cross-section views. The extent of the domal uplift is indicated by the dashed line. The radiating lines are dyke swarms, and the black areas toward the center are mafic–ultramafic intrusions interpreted from their gravity anomalies. They lie along a circular region, interpreted as an apical graben. From Baragar *et al.* (1996). By permission of Oxford University Press.

proposed that subswarms of dykes emanate from and are emplaced laterally away from these layered intrusions. Whether this model has application to the plume-center regions of other LIPs is currently unknown.

11.4.2 Triple-junction rifting vs. LIP distribution

One method of estimating the location of the plume center is based on the overall distribution of associated flood-basalt magmatism. A circle (of typically 1000-km radius) can be drawn around the magmatism and the centre of this circle can approximate the plume center (e.g. White and McKenzie, 1989; Courtillot *et al.*, 1999). However, the location of subsequent rifting/breakup is not necessarily symmetrically disposed around the plume center (Fig. 11.10). For instance, the small Yemen, British, Seychelles, and Etendeka rifted-away remnants of their respective LIPs (Afro-Arabian, NAIP, Deccan, and Paraná–Etendeka) are clear evidence of this. As a generalization, the largest portion of the LIP is preserved on the side associated with the failed rift (Courtillot *et al.*, 1999). For instance, in the case of the Deccan LIP, failed rifts may have been produced in the Cambay and Narmada grabens on the Indian side (Fig. 11.10), which then retained the largest portion of the LIP, as the smaller Seychelles LIP segment was rifted away on the other side of the newly established spreading ridge. A similar story applies to the Paraná segment of the Paraná–Etendeka LIP, with the possibility for a failed rift along the Ponta Grossa dyke swarm and arch, and the tiny Etendeka fragment of the flood basalts that rifted away on the other side of the new South Atlantic Ocean basin (Fig. 3.6).

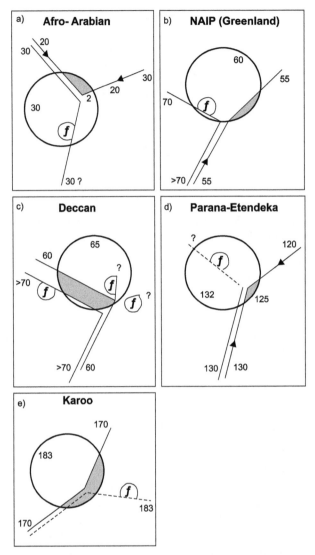

Figure 11.10 (a)–(e) Schematic plan-view outline of the flood-basalt distribution for Cenozoic and Mesozoic LIPs. Lines indicate locations of rifts. Circle locates inferred plume center with 1000 km radius. Ages of LIP event and rifting event are indicated: "f" indicates a failed rift and the arrows indicates direction of rift propagation. The gray part indicates the minor part of the flood basalts rifted away after successful ocean-basin opening. Modified from Figure 10 in Courtillot *et al.* (1999). With permission from Elsevier.

The case of the Karoo event (part of the broader Karoo–Ferrar LIP) (Figs. 3.7 and 5.10) is more complicated. The dominant volume of Karoo volcanics is preserved on the African side, consistent with the presence of the Okavango failed rift, and only a minor fragment is preserved on the Antarctic side. However, this interpretation is more complicated for a couple of reasons. The Karoo volcanic rocks represent just part of an overall LIP (Figs. 3.7, 5.9, and 5.10); the distribution of magma in the plumbing systems (dykes, sills, layered intrusions) should also be taken into account in considering whether rifting has split the distribution of Karoo magmatism symmetrically or not. In fact, given the great extent of Karoo sills and dykes in southern Africa, the asymmetry is even more pronounced with respect to the smaller Karoo component on the Antarctic side. Another complexity is that despite its difference in geochemistry, the Ferrar magmatism of Antarctica is identical in age and should be considered as part of the overall distribution of a single Karoo–Ferrar LIP prior to breakup.

An additional point is that there may be ridge jumps associated with the LIP magmatism (Courtillot *et al.*, 1999). For instance, in the case of the NAIP, shortly after the flood basalts were emplaced, the Labrador rift died and a new successful plate boundary was established as the North Atlantic Ocean began to open. As another example, prior to the Deccan LIP, rifting had occurred south of the Seychelles but was apparently starved and jumped north shortly after the Deccan magmatism began.

Finally, an interesting observation from the NAIP, Paraná–Etendeka, and Afar-Arabian LIPs is that the propagation of rift arms typically occurs toward the plume center, rather than away, suggesting a role for passive rifting (Fig. 11.10; Courtillot *et al.*, 1999).

11.5 Timing of LIPs and rifting

Here the timing between LIP emplacement and rifting is considered.

11.5.1 Multiple-pulsed LIPs and relation to rifting

The well-preserved Mesozoic–Cenozoic record reveals that many LIPs exhibit two magmatic pulses (and, in some cases, more; Section 15.6) and that the latest pulse can often be associated with rift-related decompression and the onset of ocean opening (White and McKenzie, 1989; Campbell, 1998; Courtillot *et al.*, 1999; Ernst *et al.*, 2005). For instance, the NAIP has two pulses, 62 Ma and 55 Ma, and it is the latter pulse which is associated with breakup (e.g. Courtillot *et al.*, 1999). Usually the volume of magmatism drops dramatically after the onset of rifting, and may continue at a much lower rate as a hotspot track. However, in some cases post-breakup magmatism can also be of LIP scale. For instance, the *c*. 118 Ma Kergeulen oceanic plateau represents a major pulse emplaced after initial opening of the Indian Ocean, which is associated with the initial Bunbury–Comei LIP pulse (Section 4.2.2).

11.5.2 Time lag between LIP pulse and onset of rifting

The magmatic pulse associated with the onset of rifting is associated with seaward dipping reflectors (SDRs; Chapter 3) and the lag between the SDR pulse and the

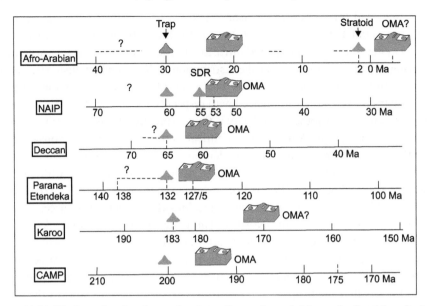

Figure 11.11 Timing of volcanism related to Cenozoic and Mesozoic flood-basalt (trap) events. Time is in Ma. The main episode of flood-basalt emplacement is plotted with a dark volcano symbol. Occurrence of SDRs is labeled. Dashed lines (sometimes with question marks) indicate other minor volcanism. Appearance of the first oceanic magnetic anomalies (OMAs) is also indicated. Modified from Figure 9 in Courtillot *et al.* (1999), with permission from Elsevier. Timing for CAMP is modified on the basis of Sahabi *et al.* (2004); see also Labails *et al.* (2010).

onset of ocean opening is variable (Fig. 11.11). In the Afro-Arabian case it is about 4 Ma and in the NAIP case it is about 2 Ma. For other examples the time gap between youngest LIP emplacement and formation of oceanic crust ranges from about 5 Ma (Deccan and Paraná), to 13 Ma (Karoo–Ferrar). For the CAMP event the lag is variably estimated at 10–25 Ma (Benson, 2003) or perhaps only 5–10 Ma (Sahabi *et al.*, 2004; Labails *et al.*, 2010).

Factors that may influence the time lag between the flood-basalt event, rifting, SDR formation, and oceanic crust production include the plume characteristics, the lithospheric thickness, the velocity of plate over the plume, lateral variations in lithospheric and crustal structure and the stress field (Courtillot *et al.*, 1999).

11.6 LIPs that did not result in continental breakup

As summarized in Table 11.2 there are examples in the young (Mesozoic–Cenozoic) record where a LIP did not progress to breakup. The most prominent is the 250 Ma Siberian Trap event which, although arguably the largest continental LIP (variously estimated at 4–16 Mkm^3; Ivanov, 2007), is not associated with breakup. There was attempted breakup in the West Siberian basin (Fig. 3.8; Saunders *et al.*, 2005), but it may

Table 11.2 *LIPs that failed to lead to ocean opening. See discussion in the text.*

LIP name	Attempted breakup (rift zone)	Orogen preventing breakup
Siberian Trap (250 Ma); Section 3.2.8	Asia–Europe (West Siberian basin)	Orogens maintaining Pangea
Emeishan (260 Ma); Section 12.2.10		Orogens maintaining Pangea
Keweenawan (1115–1085 Ma); Section 3.5.2	Canada–USA (Midcontinent rift system)	Grenville orogen
Circum-Superior (c. 1880 Ma); Section 9.2.2	Northwestern and northern margin of the Superior craton	Plume centered on cratonic margin at time of ocean closure
Matachewan, Mistassini and Baltic LIPs (2500–2450 Ma); Fig. 11.12	Superior, Karelia and Kola cratons	?

have been aborted by a concurrent regional compressional environment associated with the final assembly of the supercontinent Pangea. A second example is the Keweenawan LIP (of the Midcontinent rift system), in the Great Lakes region of North America (Fig. 3.15). Despite an impressive start to rifting, breakup was thwarted by a coeval pulse of compression during the formation of the nearby similar-aged Grenville orogen. In support of this interpretation there are inversion structures in the Keweenawan rift that are linked to Grenville compression (e.g. Manson and Halls, 1997).

These observations lead to a hypothesis that the reason that a LIP would fail to cause breakup is if the plate-boundary stresses at the time indicated an overall regional compressional environment, and did not favor breakup. Another possible example is the 2.5–2.45 Ga Matachewan and Mistassini LIPs of the southeastern Superior craton that are linked to the similar age Baltica LIP of the Karelian (and Kola) craton (Fig. 11.12; Bleeker and Ernst, 2006; Ernst and Bleeker, 2010; Kulikov *et al.*, 2010). Given the scale of this combined magmatism in Superior and Karelian/Kola cratons, breakup could be expected. However, it is clear that the Superior and Karelia/Kola cratons remained attached for another few hundred million years based on the presence of two younger LIPs (at 2215 and 2100–2070 Ma) that are also shared by both crustal blocks (Superior and Karelia cratons) (Bleeker and Ernst, 2006; Ernst and Bleeker, 2010). The intervening Paleoproterozoic Huronian basin of the Superior craton and the corresponding Jatulian sedimentary sequence of Karelia must therefore represent a failed rift basin. The failure of an ocean to form despite the huge 2.5–2.45 Ga LIPs would therefore suggest that the regional plate-tectonic stress regime at this earliest Paleoproterozoic time was unfavorable to breakup.

An additional reason that a LIP might not be associated with breakup is if that LIP was emplaced on the edge of an already existing continental margin. A good

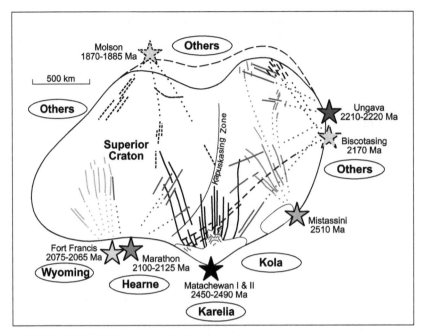

Figure 11.12 Radiating dyke swarms which converge to the margin of the Superior craton and are associated with Paleoproterozoic breakup of a supercontinent or supercraton of which the Superior craton was a core piece. Modified from Bleeker and Ernst (2006) and Ernst and Bleeker (2010).

example is the 1880–1870 Ma Circum-Superior LIP, which circumscribes the northwestern, northern, northeastern, and southwestern sides of the Superior craton, and which can be linked to a mantle plume located on the northwestern margin of the craton (Figs. 11.12 and 9.10).

In some cases, LIPs/plumes can provide the critical additional extension to lead to continental breakup. An example is given from eastern North America. The *c.* 230 Ma Newark rift basins are distributed along nearly the entire coast of eastern North America (Fig. 12.5; Chapter 12), indicating that conditions were favorable to rifting, but not for breakup. However, once the CAMP LIP/plume arrived at 200 Ma, breakup began only a few Ma later, at 195–190 Ma (Sahabi *et al.*, 2004; Labails *et al.*, 2010). It can be inferred that while conditions were favourable for extension at 230 Ma, it required the extra extensional forces (e.g. domal uplift) provide by the plume arrival to cause successful breakup.

11.7 LIPs and continental breakup

11.7.1 Volcanic rifted margins vs. non-volcanic rifted margins

In order to further assess the link between LIPs, breakup, and forming of new oceans, I consider the distribution of volcanic and non-volcanic rifted margins

(VRMs and non-VRMs). As discussed in Chapter 3, VRMs can be recognized by SDRs, and also by significant thicknesses (up to 15 km) of juvenile high-velocity lower crust (HVLC) located seaward from the continental rifted margin. VRMs can be linked to plume-generated LIP magmatism (e.g. Menzies *et al.*, 2002b). Plume-associated VRMs may postdate the main pulse of flood-basalt magmatism, but predate ocean opening and formation of true oceanic mid-ocean-ridge basalt (MORB) crust. Prominent examples of VRMs are along most of the borders of the Atlantic Ocean with VRMs associated with the 201 Ma CAMP, 135 Ma Paraná–Etendeka, or the 62–55 Ma NAIP LIP/plume events. Examples of non-VRMs include southeast China, Iberia, and the Newfoundland basin/Labrador Sea (Wilson *et al.*, 2001; Russell and Whitmarsh, 2003).

The analysis of the modern record shows that > 50% and perhaps as much as 90% of current passive continental margins are "volcanic" and can potentially be associated with a LIP (Skogseid, 2001; Menzies *et al.*, 2002a). In contrast, non-VRMs represent the LIP-free transition to an oceanic crust of normal thickness.

The predominance of VRMs over non-VRMs provides additional confirmation of the importance of LIPs to the continental breakup process. This makes sense in the context of the huge areal extent of LIPs, which commonly have a diameter of 2000 km or more (e.g. White and McKenzie, 1989), and so each LIP can potentially generate 2000 km of VRMs on each side of the newly opening ocean.

11.7.2 Essential role of LIPs in continental breakup

The next question to consider is whether the breakup and formation of new ocean basins can occur without the input of LIPs. This is important in understanding how strongly to infer the link between LIPs and formation of new oceans in assessing the breakup history of pre-Pangea supercontinents.

In the previous section it was noted that many Phanerozoic LIPs are associated with continental breakup and that a majority (at least 50%) of the current oceans have volcanic margins, in most cases probably linked with a LIP. Let's assess the LIP–new-ocean link from another perspective. One can consider each major present-day ocean basin in turn, and try to identify the LIP(s) that were responsible for its formation. For example, the opening of the Atlantic Ocean formed as a consequence of three LIPs: the Paraná–Etendeka LIP in the South Atlantic (at 135 Ma), the CAMP event in the central Atlantic (at 200 Ma), and the NAIP in the North Atlantic (62–55 Ma) (Fig. 11.2). The onset of the High Arctic LIP (HALIP) event at 130 Ma matches the interpreted Hauterivian stage of opening of the Canada basin (Grantz and May, 1983). However, the role of the HALIP (and its interpreted extensions along the Alpha Mendeleev ridge of the central Arctic) in the opening history of the Arctic basin is still being sorted out (e.g. Døssing *et al.*, 2013). The Indian Ocean basin formed as a consequence of Gondwana breakup, which was associated with LIPs at 183 Ma (Karoo–Ferrar), 130–117 Ma (Keguelen–Bunbury–Comei), and

67–65 Ma (Deccan). The Pacific Ocean is a long-lived ocean basin with a complex plate-tectonic history. Its initial opening (as the proto-Pacific Ocean) is thought to have been associated with Neoproterozoic rifting of the supercontinent Rodinia; the associated LIPs were emplaced at *c.* 820 and 780 Ma (e.g. Section 8.4.2; Li *et al.*, 2008a).

Looking back in time, and based on the examples from the more recent record, it is hypothesized that each major paleo-ocean will have been associated with a LIP that just preceded and contributed to its opening. So it should be possible to compare the LIP record from different blocks to identify conjugate margins of past oceans. For example (Table 11.1), the 615–555 Ma Central Iapetus Magmatic Province (CIMP) magmatism (representing several LIP pulses) (Fig. 9.8) is present on both eastern Laurentia and Baltica consistent with these being conjugate margins of the Iapetus Ocean (e.g. Bingen *et al.*, 1998; McCausland *et al.*, 2007; Ernst and Bell, 2010; Fig. 9.8).

11.7.3 Links between LIPs and ancient breakup margins

There is increasing evidence that radiating dyke swarms typically converge to points along a cratonic margin (Fig. 11.12). These focal points are inferred to be the location of plume centers and to identify a pulse of rifting along that margin potentially representing breakup (or attempted breakup) out of a larger crustal block.

11.7.4 Oceanic LIPs and rifting

Oceanic LIPs can also be affected by rifting. As illustrated in Fig. 3.1 and discussed in Chapter 4, there are several oceanic plateaus that have been fragmented by rifting. The most prominent is associated with the Ontong Java, Manihiki, and Hikurangi plateaus. Seafloor fabric data indicate that these three currently separate plateaus originally formed as a single oceanic plateau in the vicinity of a spreading ridge and were subsequently fragmented by rifting along that ridge (Section 4.4.2; Fig. 4.6; Taylor, 2006). Similarly it has been suggested that the Agulhas plateau (south of South Africa) was formed as part of a larger LIP that includes the Northeast Georgia Rise and the Maud Rise, which then suffered breakup; the original pre-breakup LIP has been termed the Southeast African LIP (e.g. Section 4.4.2; Fig. 4.7; Parsiegla *et al.*, 2008).

11.8 LIPs and supercontinents

Here I assess the distribution of LIPs through time and any periodicities/cyclicities in that record and whether there is a link with the supercontinent cycle. Then in Section 11.9 a methodology is presented for using the LIP record to fast-track progress on constraining Precambrian supercontinent reconstructions.

11.8.1 Supercontinents through time

As shown in Fig. 11.13, Earth's history is dominated by periods of supercontinent formation (e.g. Bleeker, 2003; Rogers and Santosh, 2004; Li *et al.*, 2008a; Meert, 2012; Zhang *et al.*, 2012a; Evans, 2013). The most recent 0.3–0.25 Ga Pangea (–Gondwana) supercontinent is best understood. The previous cycles include the *c.* 1.0–0.8 Ga Rodinia supercontinent, 1.8–1.4 Ga Columbia (Nuna) supercontinent, and latest Archean–earliest Proterozoic supercontinent (Kenorland). The latter is also alternatively conceived as multiple supercratons (Superia, Sclavia, and Vaalbara) (Bleeker, 2003).

The reconstruction history of Pangea (and its precursor step, Gondwana) are well known, based on a variety of approaches including seafloor spreading, paleomagnetism, and faunal and floral correlations. However, these tools, apart from paleomagnetism, are not available for Pre-Pangea (Precambrian) reconstructions. The nature and basic configuration of the prior pre-Pangea supercontinents are hypothesized mainly from the distribution of major orogenic belts indicating assembly of supercontinents at particular times. Paleomagnetism has also been providing some clues to the configurations of these supercontinents/supercratons, but has been hampered by the lack of precisely dated units for paleomagnetic studies (see discussion in Ernst *et al.*, 2013b). Currently, the best versions of the reconstruction of Rodinia are provided by the summary of the IGCP 440 project (Li *et al.*, 2008a). Recent reviews of Columbia (Nuna) reconstructions are provided by Meert (2012) and Evans (2013), and new versions are provided by Zhang *et al.* (2012a), Pesonen *et al.* (2012), and Pisarevsky *et al.* (2014). The nature of the late Archean/earliest Proterozoic reconstructions are robust for only some few pieces of the "puzzle" (Bleeker and Ernst, 2006; Söderlund *et al.*, 2010; French and Heaman, 2010; Nilsson *et al.*, 2013; Evans, 2013).

Although the broad timing of these pre-Pangea (Precambrian) supercontinents/supercratons is well constrained, the exact configurations of these supercontinents remains largely speculative. As discussed in the next sections, LIPs represent a robust tool for helping sort out the configurations of pre-Pangea supercontinents.

11.8.2 LIP distribution through time

Average LIP rate

A preliminary analysis of the LIP record (Ernst and Buchan, 2001b, 2002) indicated that the rate of LIP emplacement was semi continuous (Fig. 4.9; but also see Fig. 11.13). Estimates of average LIP frequency based on this record indicate an average frequency of one LIP/plume every 20 million years since the Archean (Ernst and Buchan, 2002), and for the more completely preserved oceanic plus continental LIP record of the past 250 Ma, an estimate of one per 10 Ma (Coffin and Eldholm, 2001).

There is a caveat to this calculation. As noted by N.L. Dobretsov (personal communication, 2007), the average rate of LIP emplacement can be different from

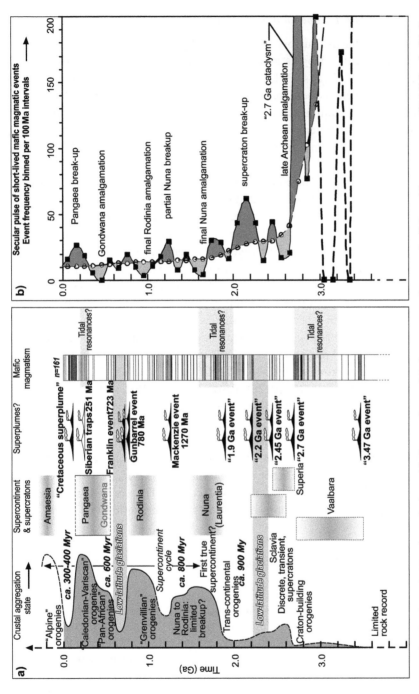

Figure 11.13 Distribution of LIP's through time. (a) Comparison with the supercontinent cycle. Selected key events are highlighted. (b) Squares with black fill plot data that have been normalized by available continental crust. Circles with white fill locate smoothed distribution of events (binned on a 100 ma scale). Note the remarkable correlation between peaks in LIP production and supercontinent breakup. Note the dramatic ("cataclysmic") scale of the 2.7 Ga greenstone-belt-type LIPs. (a) From Bleeker (2003), with permission from Elsevier, and (b) from Ernst and Bleeker (2010). Since this diagram was originally prepared, c. 1880 Ma events have been recognized in more regions of the world (Minifie et al., 2013) and so the peak should be larger at this time. Also, events at 1380 Ma have been found on more blocks (e.g. Ernst et al., 2008; Puchkov et al., 2013; El Bahat et al., 2013; Soderlund et al., 2013) and also should have enhanced significance on this diagram. See discussion in Section 11.8.3.

the characteristic interval between emplacement of LIPs. Given the existence of coeval LIPs that are widely separated and interpreted to be unrelated (Ernst and Buchan, 2002; see Section 11.8.1) then the average rate of LIP emplacement (based on the number of LIPs divided by the time span) will be a smaller number than that based on the number of distinct LIP age groupings per time span. For instance, in such a calculation all the 93 Ma LIPs (Caribbean–Colombian, Madagascar, second pulse of HALIP, and possible second pulse of Ontong Java) would count as one event. So the real average interval between continental LIP events could be closer to 30 Ma and for combined continental and oceanic LIP events is probably closer to 15 Ma. The precise numbers can only be determined once our understanding of LIPs and of global reconstructions (Section 11.8.1) has become more robust.

LIPs and the supercontinent cycle

A different calculation of the LIP record through time Figure 11.13 emphasizes the variations in LIP frequency, and reveals a link between the supercontinent cycle and the LIP record. The data (number of LIPs per 100 Ma) has been normalized to the area of available continental crust. There is a significant increase in the frequency of LIP events associated with times of supercontinent breakup, and there is a decrease in LIPs during the supercontinent assembly. Specifically, in Figure 11.13, a peak in LIP production is observed at *c.* 200 Ma, which corresponds to the timing of Gondwanaland breakup. Another peak at 800–600 Ma, links to Rodinia breakup. The peak at *c.* 1300 Ma links to a major late stage of Columbia (Nuna) breakup. Finally the peak at *c.* 2200 Ma links to the breakup of a late Archean supercontinent (or supercratons). When the data are weighted against available crust, there is an overall secular decrease in the average rate of LIP production since the beginning of the Proterozoic (Fig. 11.13).

11.8.3 Cyclicity in the LIP record – a statistical analysis

Another approach to assessing the link with the supercontinent cycle is provided by a statistical analysis of the data. There have been a number of attempts based on the available LIP database. Rampino and Caldeira (1993) used a moving window analysis/Gaussian filtering/Fourier analysis. Isley and Abbott (2002) used spectral analysis on their database (Isley and Abbott, 1999). Yale and Carpenter (1998) used slopes on cumulative percentage diagrams weighted by inferred LIP volumes based on the areal extent of dyke swarms from Ernst *et al.* (1996). The most sophisticated analysis by Prokoph *et al.* (2004) combines wavelet, spectral, and cross-spectral techniques to analyze the LIP data for 154 events over the last 3.5 Ga (Fig. 11.14). The Prokoph *et al.* (2004) analysis was based on an update of the first detailed global LIP database (Ernst and Buchan, 2001b). An extension of this approach to look at

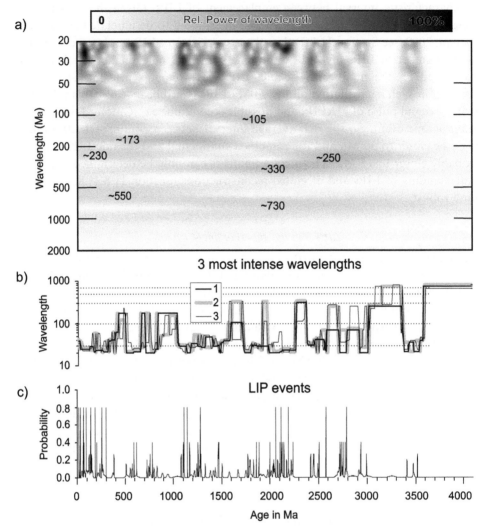

Figure 11.14 Wavelet analysis of "AB50" LIP data (154 events) in Prokoph *et al.* (2004a); data from Ernst and Buchan (2001b) database. (a) Wavelet scalogram, with continuous gray shading. (b) Three most significant periodicities. (c) Gaussian distribution of data. The three most intense wavelengths are determined as follows: after the peak value is identified, the second most intensive wavelength must be more than 5% shorter than the first, and the third must be at least another 5% shorter than the second. This approach respects bandwidth uncertainties. Peaks at 730–550 Ma may reflect a supercontinent cycle. Meaning of shorter 330, 250, 170, and 105 Ma cyclicities are unclear. From Figure 3 in Prokoph *et al.* (2004a).

the co-variation with environmental proxies after Prokoph *et al.* (2013) is presented in Section 14.4.3.

Prokoph *et al.* (2004) have shown that the average frequency of LIPs is relatively constant and supports an overall model of semi-continuous temporal emplacement of LIPs. However, several weak cycles were observed in the analysis. Those of longest duration are a *c.* 170 Ma cycle from 1600 Ma to the present and in the late Archean, a *c.* 330 Ma cycle from about 3000 to 1000 Ma, and a cycle decreasing from 900 to 1000 Ma over the interval 2000 Ma to the present. Additional cycles of shorter duration include a *c.* 250 Ma cycle in the late Archean, a *c.* 230 Ma cycle in the Phanerozoic, a *c.* 105 Ma cycle in the early Proterozoic, and several < 60 Ma cycles occurring in scattered intervals. The uncertainties in the cycle patterns preclude a simple correlation with previously interpreted forcing functions (as discussed above): a *c.* 30 Ma cometary impact, a 270 Ma galactic year, a *c.* 600 Ma supercontinent cycle, and a *c.* 800 Ma resonance between tidal and free oscillations of the core. Future time-series analysis on improved versions of the LIP database (and appropriate subsets) will be required to test the robustness of the observed cycles and to identify underlying forcing functions.

11.8.4 State of LIP record

Future work on cyclicities will require an improved LIP database to ensure that no events are missing and that the full extent of each event is known so that volumes can be determined. An essential aspect will be working towards robust paleocontinental reconstructions that allow LIPs fragments to be correctly linked via reconstructions, and to allow recognition of widely separated coeval LIPs which are independent. These points are elaborated on below.

Coeval LIPs on separate blocks

The reconstruction of Pangea is well established, and so it is readily apparent which LIPs on different blocks belong together and which had an independent origin (Ernst and Buchan, 2002). For instance, it is clear the Deccan LIP (*c.* 67–65 Ma) is independent of the NAIP (*c.* 62–55 Ma), which is about 10 000 km away at the time of emplacement. On the other hand, the Phanerozoic record shows numerous examples of LIPs on different crustal blocks which can be linked. For instance, it has long been known that the Paraná magmatism of South America and the Etendeka magmatism of western Africa are part of the same *c.* 135 Ma LIP when the South Atlantic Ocean is closed (Section 3.2.6). Until more robust Precambrian supercontinent reconstructions are produced, it will remain uncertain which occurrences of a given LIP age located on different crustal blocks should be grouped into a single LIP and count as one in estimates of cyclicity, and which should be counted as separate LIPs. Some examples of multiple LIP "nodes" of the same age are compiled in Table 11.3.

An additional complexity is illustrated by widely separated LIPs which nevertheless could be connected by movement of a plume along the base of the lithosphere (via sublithospheric channeling; Section 5.6.2). For instance, *c.* 1380 Ma LIP magmatism is

Table 11.3 *Possible multi-node LIPs during the interval 1600–700 Ma (details in Ernst et al., 2008)*

Age	Node 1	Node 2	Node 3
1460–1430 Ma	Western Laurentia (Moyie), +? Australia (Edmund)	Eastern Laurentia: (Michael–Shabagamo)	? Baltica (Tuna–Trond Göta and Lake Ladoga)
1380 Ma	Western Laurentia (associated with rifts +? Antarctica (Vestfold Hills, and Kalahari (Pilanesberg)	Northern Greenland (Midsommersø/Zig-Zag Dal + ? Siberia (Chieress) + ? southeast Baltica (Mashak)	? Africa, Congo craton (Kunene–Zebra River and related intrusions)
1270 Ma	Northern Laurentia (Mackenzie)	Eastern Laurentia (1280–1235 Ma)–Baltica (Central Scandinavian Dolerite Group)	
1115–1070 Ma	Central Laurentia (Keweenawan)	Southwestern Laurentia (Southwest USA diabase province) + central-western Australia (Warakurna LIP)	Kalahari craton (Umkondo)
825 Ma	Australia (Gairdner–Willouran), South China (Guibei)	? Tarim (Qiganbulake)	
780 Ma	Western Laurentia (Gunbarrel), South China (Kangding)	? Tarim (Kudi)	
755 Ma	Western Australia (Mundine Well) + Greater India (Malani–Seychelles)	? South China (Shaba), South Korea (Ogcheon), southern Siberia (Sharyzhalgai)	? Southeastern Laurentia

found both in western Laurentia and northeastern Laurentia (northern Greenland), and these represent separate "nodes" of 1380 Ma activity (Table 11.3; Ernst *et al.*, 2008). They have been interpreted as representing entirely independent LIP/plume events (Ernst *et al.*, 2008). However, it is also possible that a single 1380 Ma plume arising underneath the Laurentia craton could spread beneath the craton (Section 5.6.2) and end up rising into lithospheric "thinspots" along the northeastern and western sides of the craton forming the two nodes. See Rogers *et al.* (2013) and Bright *et al.* (2014) for a similar story for the multiple LIP nodes of ages, *c.* 1460 and *c.* 1100 Ma, respectively, also associated with the Laurentian craton).

Robustness of the current database

The pace of discovery of totally new LIPs has decreased (compare databases in 2001 (Ernst and Buchan, 2001b), in 2008 (Ernst *et al.*, 2008), in 2013 (Ernst *et al.*, 2013b), and the current database (Table 1.2, Fig. 1.6)), suggesting that the current database is reasonably robust. However, it should be acknowledged that many LIP events in Table 1.2 are recognized by only a few dated units (and strictly speaking should be considered as LIP fragments, per critieria in Section 2.5). Therefore many of the events in Table 1.2 require much more work to confirm that they are of LIP scale. At the same time it is also important to note that major gaps identified in compilations of the LIP record have been filled in (Table 1.2; Fig. 1.6). For instance, the major gap between about 2400 and 2200 Ma (Ernst and Buchan, 2001b) has contracted with the discovery of the Bangalore-Karimnagar LIP in India at 2370 Ma (Kumar *et al.*, 2012) and a more speculative LIP at 2330 Ma in Baltica (Vuollo and Huhma, 2005; Salminen *et al.*, 2014; see Table 1.2).

Calculating LIP volumes

In any analysis of the LIP record through time it would be ideal to weight the record by LIP volumes. However, this is currently difficult to achieve for three reasons. First, the recognized size of LIP events on individual crustal blocks is continually increasing due to ongoing geochronology that identifies additional units belonging to LIPs. This is true even for young events. For instance, the size of the Siberian Trap LIP has nearly doubled in the past decade with the confirmation of coeval magmatism under younger cover in the West Siberian basin and also in the Urals and Central Asia (Reichow *et al.*, 2002, 2009; Section 3.2.8). Second, until robust paleocontinental reconstructions are achieved it will be uncertain which coeval units on different blocks actually belong to the same LIP and therefore whether their volumes should be combined into a single LIP. Third, criteria need to be developed for determining and integrating volume estimates for all the different magmatic components of LIPs (volcanic rocks, mafic sills and dykes, mafic–ultramafic intrusions, and magmatic underplate). In the meanwhile, the areal extent of the event (embracing the areal extent of all related units volcanics, dykes, sills, etc.) can continue to be used as a proxy for the volume (Ernst and Buchan, 2001b; Prokoph *et al.*, 2013).

11.9 Using LIPs to reconstruct supercontinents

Given the strong link between LIPs and breakup or attempted breakup (Section 11.7) and the secular changes in the LIP record correlated to the supercontinent cycle (Section 11.8.3; Fig. 11.13), it is natural to consider use of the LIP record for constraining the breakup of Precambrian supercontinents. A method for utilizing the LIP record for reconstructions is outlined below (Section 11.9.2), but first I review some of the difficulties in sorting out the pre-Pangea supercontinent record.

11.9.1 Problems with traditional approaches to continental reconstruction

As addressed in Bleeker and Ernst (2006), prior to 250 Ma, the paleogeographic record of Earth's continental crust becomes increasingly speculative, although there is optimism that this problem may be tractable, in principle, back to *c.* 2.6 Ga, the age of "cratonization" of a considerable fraction of continental crust existing today (e.g. Bleeker, 2003).

In the pre-Pangea world, there is no preserved record of ocean-floor spreading (with magnetic striping) to help guide paleogeographic reconstructions. Therefore, reconstruction of Precambrian supercontinents must rely on matching details in continental geology from one craton to the next (Fig. 11.15). Many such details can be (1) fuzzy (e.g. ages of granitoid belts and metamorphism), (2) variable or diachronous along strike (e.g. orogenic belts and their structural trends, ages of structural events), or (3) susceptible to modification as a function of depth of erosion (e.g. the outlines of sedimentary basin and the "piercing points" they provide). As a consequence, reconstructions based on such data can have large uncertainties (Bleeker and Ernst, 2006; Ernst and Bleeker, 2010).

11.9.2 Using the LIP record for continental reconstruction

The LIP record can play a useful role in constraining Precambrian reconstructions.

Paleomagnetism

Many of the key units used for paleomagnetic study are LIP-related dykes and sills, because dolerite dykes and sills are particularly well-behaved paleomagnetically. Considerable progress is being made in producing and comparing the key paleomagnetic poles of units on different blocks in order to constrain the relative positions of these blocks (e.g. Buchan *et al.*, 2001; Wingate *et al.*, 2002; Meert, 2002; Pisarevsky *et al.*, 2003 Personen *et al.*, 2012, 2014; Evans, 2013; Buchan, 2014; Pisarevsky *et al.*, 2014).

The LIP barcode – piercing-point method for continental reconstruction

More recently, a two-part approach has been proposed (Bleeker and Ernst, 2006; Ernst *et al.*, 2013b; Fig. 11.16). The first part of the technique is comparison of the ages of LIPs present on different blocks. Specifically a "barcode" presentation of those ages is constructed and compared. Those pairs of blocks with a number of LIP age matches over a given interval were probably nearest neighbors during that time

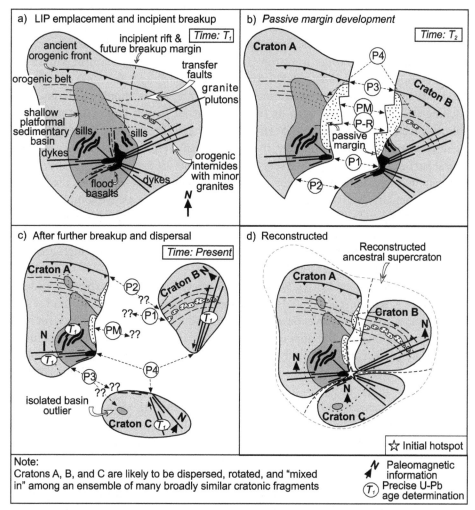

Figure 11.15 Craton reconstruction and key role of LIPs and their dyke swarms. (a) A hypothetical supercraton with various geological elements, just prior to breakup. A LIP with flood basalts and associated dykes and sills is emplaced along the incipient rift. (b) Breakup of the supercraton has spawned two cratons (A and B). P–R = promontories and re-entrants along the rifted margins; PM, conjugate passive margins; P1, piercing points and reconstruction of the LIP; P2, piercing points provided by older sedimentary basins; P3, piercing points provided by an ancient orogenic front or fold-thrust belt; and P4, the non-precise piercing points provided by orogenic internides. (c) The more general case where further breakup has occurred (craton C) and craton margins have been abraded, modified, and differentially uplifted. Craton B was strongly uplifted and its sedimentary cover has been eroded. Piercing point P3, if still recognizable as such, has strongly shifted, and an exhumed granitoid belt is unmatched in craton A. Craton C was also uplifted, erasing piercing point P2. Dykes related to the LIP, however, remain on all three cratons and precise age dating yields a critical clue that they might be part of a single event. Primary paleomagnetic data may yield additional geometrical clues (north (N) arrows), if not paleolatitudes. (d) Reconstruction of the original supercraton, based only on the precise piercing points and other information derived from the dyke swarms. From Bleeker and Ernst (2006) and Ernst and Bleeker (2010).

interval. The second step is to use the geometry of dyke swarms (linear, radiating) as precise piercing points to orient the crustal blocks which had been previously determined to be nearest neighbors. Specifically, crustal blocks can be oriented with respect to each other in order to restore an overall radiating dyke-swarm pattern or to align components of a linear dyke-swarm pattern.

Reconstructing ancient supercontinents could thus be as simple as dating and matching the magmatic LIP records (barcodes) of all (*c.* 50) large cratonic pieces applicable back through time to the end of the Archean (Bleeker and Ernst, 2006). In this respect, giant dyke swarms are of particular interest because (1) they are an integral part of LIPs; (2) they have very large footprints (300–3000 km); (3) they were emplaced in short time pulses that can now be dated precisely using the U–Pb method particularly on baddeleyite; (4) they are relatively insensitive to uplift owing to their steep dips; (5) they project deep into cratonic hinterlands; (6) they contain rich geometrical and paleostress information; (7) they provide superior piercing points; and finally, (8) they provide the target rocks of choice for high-quality, precisely dated, paleomagnetic poles ("key poles").

As noted in Bleeker and Ernst (2006) the improvements in the ability to precisely date mafic units (e.g. Krogh *et al.*, 1987; Heaman and LeCheminant, 1993; Söderlund *et al.*, 2013), commonly with a precision of ±2 Ma or better, has paved the way for the increasing use of integrated mapping, high-precision age dating, and paleomagnetism of short-lived LIP events and their dyke swarms (e.g. Halls, 1982; Fahrig, 1987; Buchan *et al.*, 2000; Wingate and Giddings, 2000; Ernst *et al.*, 2005; Vuollo and Huhma, 2005; Buchan 2014).

A systematic application of this methodology using the LIP record for paleocontinental reconstruction is underway (Ernst *et al.*, 2013b). A broad-based group of scientists (led by R. Ernst and W. Bleeker), established in 2009 an industry–academia–government collaborative project, "Reconstruction of Supercontinents back to 2.7 Ga using the Large Igneous Province Record: With Implications For Mineral Deposit Targeting, Hydrocarbon Resource Exploration, and Earth System Evolution" (www. supercontinent.org), for U–Pb dating of mafic–ultramafic units (particularly dolerite dykes and sills) around the world, and especially in poorly understood regions. The goal of this multi-year project is to fast-track LIP barcoding of all major crustal blocks to enable robust reconstructions to be determined, an effort that is involving the broader research community.

Strategies

There are various criteria which can be employed in using LIPs to constrain continental reconstructions (modified after Ernst *et al.*, 2008; see also Bleeker and Ernst, 2006; Ernst and Srivastava, 2008):

(1) **Key event comparisons:** Identify particularly extensive and distinctive LIP events that can be matched between crustal blocks, and thereby identify that these blocks were nearest neighbors (see also point 4).

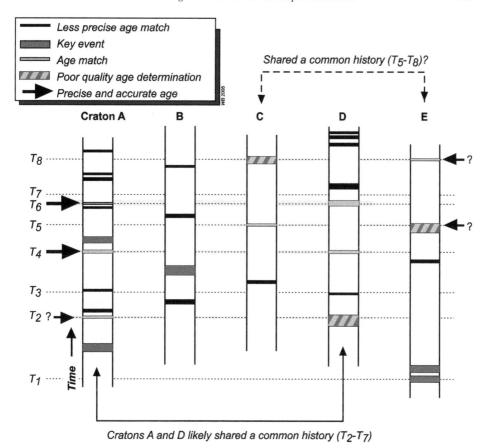

Figure 11.16 Illustration of the methodology of using the barcode matching step of barcode-piercing point method. Mutiple barcode matches are required to confirm that two cratons were nearest neighbors. See discussion in text. From Bleeker and Ernst (2006).

(2) **Paleomagnetic studies:** Compare paleomagnetic directions of coeval LIP units on different blocks to determine their relative paleolatitude and paleo-orientation.

(3) **Geometry of dyke swarms:** Reconstruct crustal blocks to restore the primary geometry of giant dyke swarms (belonging to LIPs): restoring either a primary radiating or linear pattern.

(4) **Geochemical fingerprinting:** Compare the trace-element geochemistry "fingerprints" of LIPs on different blocks to identify matches that would suggest the blocks (and their associated LIPs) were formerly adjacent.

(5) **Location of plume center:** Use LIPs to locate mantle-plume centers (e.g. at the focus of giant radiating dyke swarms). If plume centers of the same age can be identified on the margins of more than one crustal block, then the proper reconstruction of those blocks would have the plume-center locations overlapping.

(6) **Comparison of barcode of LIP ages between crustal blocks:** Summarize the LIP record for a crustal block as a barcode (e.g. Fig. 11.16) and the barcodes of

different blocks can be compared to determine which blocks have matching barcodes and were therefore probably nearest neighbors during the interval of matching. The more barcode matches between a pair of crustal blocks, the more likely the crustal blocks were connected over the time period spanned by the matching LIP ages.

(7) **Determination of breakup timing on a particular cratonic margin:** Date the LIPs associated with a cratonic margin. Cratonic margins with a similar timing of breakup were conceivably previously conjugate to each other.

(8) **Timing of accretion (assembly) of currently adjacent crustal blocks:** Note the timing at which barcodes on different blocks start to match. Prior to their accretion, crustal blocks will have independent LIP histories, but after assembly/accretion, the LIP barcodes should match. In this way, the timing of accretion can be approximated from the LIP barcode record.

Recognition of coeval but independent LIPs

Some LIPs are present on multiple blocks that are unlikely to have been nearest neighbors, suggesting that these are coeval but independent LIPs. In the younger record, there are many examples of such coeval but independent LIPs (plume clusters of Ernst and Buchan (2002); "superplume" of Larson (1991a, b); "nodes" of Ernst *et al.* (2008)). For example, at *c.* 135–130 Ma LIPs are present in four different regions that are far enough apart as to be clearly unrelated: (1) the Paraná–Etendeka LIP of the South America–Africa breakup, (2) the Trap dykes/LIP of southern Greenland, the (3) Bunbury basalts of southwestern Australia and the Comei basalts of Tibet, both precursors of the Kerguelen LIP magmatism (Sections 11.5.1; and 4.2.2), and (4) the initial pulse of the HALIP.

A second example is the essentially coeval 65 Ma Deccan and 62–55 Ma North Atlantic LIPs of India and Europe–Greenland, respectively, which are about 10 000 km apart and clearly unrelated.

Similar examples are observed for older times. For instance, 1380 Ma magmatism occurs in northern Greenland, and also in two places in western Canada about 1000 km apart (Ernst *et al.*, 2008). These two distinct nodes of 1380 Ma magmatism in Laurentia (Greenland and western Canada) are so widely separated (about 5000 km), that they likely represent independent events. However, there is also the possibility (as mentioned above) that they could be linked to a single plume rising beneath Laurentia and spreading out along the base of the lithosphere to rise in distal thinspots along different margins of Laurentia (e.g. the northeast and western margins; cf. Bright *et al.*, 2014).

Such LIP multi-node events are now cataloged at 65–62, 90, 120, 133, 1115–1070, 1270, 1380, and 1460 Ma, and possibly at 755, 780, and 825 Ma (Ernst and Buchan, 2002; Ernst *et al.*, 2007). This observation that coeval, but widely separated (and possibly independent), LIP events have occurred mutiple times over the last 1.5 Ga, has implications for the LIP barcoding method (Fig. 11.16). For the barcode matching method, a single matching age is therefore not sufficient; mutiple barcode matches are required for confidence that two blocks were nearest neighbors.

11.10 Summary

This chapter summarizes the link between LIPs and rifting/breakup, and in particular the link with triple-junction rifting. Some LIPs are linked to breakup and others not. Both continental rifting and oceanic rifting can fragment LIPs. The link between the LIP record and the supercontinent cycle is explored. The LIP record is being continually improved through U–Pb dating programs, most notably through a consortium industry–academia–government project (www.supercontinent.org). The record of known LIPs is increasing although it is thought that most events have now been recognized. The frontier goal is to map out the full distribution of each LIP in order to be able to provide a robust estimate of the volume of magmatism and the timing and distribution of any multiple pulses. Robust paleocontinental reconstructions are key to the correct linking of coeval LIP fragments on different crustal blocks. A methodology based on the LIP record is allowing fast progress toward robust supercontinent reconstructions. The method involves matching the LIP barcode record on different crustal blocks to identify those blocks which were nearest neighbors and then using the simple geometry of dyke swarms as a criterion for precisely reconstructing these nearest neighbors so that the primary radiating or linear dyke swarm geometry is restored.

12

LIPs and topographic changes

12.1 Introduction

There are several types of topographic changes typically associated with Large Igneous Provinces (LIPs). Broad-scale domal uplift can occur prior to the onset of magmatism. Progressive and sudden changes in topography can occur during the period of LIP emplacement (especially coeval with different pulses of magmatism). There is also associated uplift of rift shoulders. Finally, post-LIP topography must be considered. LIP uplift can remain permanent (due to underplating) or, in some cases, there is a potential transition to a broad sedimentary (intracratonic) basin. Also, the mechanisms proposed for the topographic variations are assessed. I start with reviewing the topographic changes associated with a number of specific LIPs.

12.2 Domal uplift associated with LIPs

A sedimentary record of domal uplift has been recognized for magmatic events ranging from 2770 to 60 Ma (e.g. Table 12.1; Rainbird and Ernst, 2001; Saunders *et al.*, 2007). Selected examples are summarized in Table 12.1 and discussed below.

12.2.1 Columbia River LIP (17–0 Ma)

Much of the western Cordillera of the United States is elevated more than 1 km above sea level (Parsons *et al.*, 1994). This reflects broad domal uplift that can be linked to the influence of a starting mantle plume. In addition, the younger hotspot track along the Snake River plain is marked by an associated topographic "wake," reflecting the buoyancy (hotspot-related) that continued during the plume-tail stage of the Columbia River LIP (Section 3.2.2) (e.g. Fig. 10 in Saunders *et al.*, 2007; see also Pierce and Morgan, 1992).

12.2.2 Afro-Arabian LIP (45–0 Ma)

The classic Afro-Arabian (Afar) LIP consists of flood basalts distributed over a wide area, along with associated sills, dykes, and carbonatite/alkaline complexes

Table 12.1 *Examples of LIPs with associated domal uplift. LIP ages quoted to the nearest 5 Ma.*

LIP name (age)	Criteria (references)
Columbia River (15 Ma); Section 12.2.1	Regional topographic uplift (Parsons *et al.*, 1994)
Afro-Arabian (*c.* 30 Ma); Section 12.2.2	Domal uplift (Şengör, 2001; Saunders *et al.*, 2007)
Deccan (65 Ma); Section 12.2.3	Drainage patterns (Cox, 1989; Widdowson and Cox, 1996)
	Post-eruption uplift monitored by laterite surface (Widdowson, 1997)
Paraná–Etendeka (*c.* 135 Ma); Section 12.2.5	Drainage patterns (Cox, 1989; Moore and Blenkinsop, 2002)
Rattray Formation (165–155 Ma); Section 12.2.6	Sedimentary patterns (Underhill and Partington, 1994; Rainbird and Ernst, 2001)
CAMP (200 Ma); Section 12.2.7	Missing sediments in basins close to plume center (Hill, 1991; Rainbird and Ernst, 2001)
Wrangellia (230 Ma); Section 12.2.8	Lithology changes (Richards *et al.*, 1991; Rainbird and Ernst, 2001)
Siberian Trap (250 Ma); Section 12.2.9	No uplift on craton (Czamanske *et al.*, 1988), but Permian strata missing from West Siberian basin (Saunders *et al.*, 2007)
Emeishan (260 Ma); Section 12.2.10	Controversial – see discussion in text
Yakutsk–Vilyui (*c.* 360 Ma); Section 12.2.11	Regional unconformity pattern (Kiselev *et al.*, 2011)
Kalkarindji (510 Ma); Section 12.2.12	Incised canyons (e.g. Williams and Gostin, 2000)
Franklin (725–715 Ma); Section 12.2.13	Stratigraphic evidence (Rainbird and Ernst, 2001)
Mackenzie (1270 Ma); Section 12.2.14	Karst topography in central region (Rainbird and Ernst, 2001)
	Interpreted from dyke swarm characteristics (Baragar *et al.*, 1996)

(Sections 3.2.3 and 9.2.2). There is also a link with triple-junction rifting (Yemen, Red Sea, and East African rift system arms) (Section 11.4), and broad domal uplift centered on the inferred plume center (Fig. 12.1; Şengör, 2001; Avni *et al.*, 2012). The Afar dome has a radius of almost 1000 km and began rising after the early Eocene, and probably reached an elevation of more than 1 km in the Oligocene time (Şengör, 2001; Menzies *et al.*, 2001; Avni *et al.*, 2012). Afterward, both basalt extrusion and rifting began.

12.2.3 Deccan LIP (c. 65 Ma)

The Deccan LIP (Section 3.2.5) is associated with dramatic evidence for regional topographic uplift. Domal uplift is marked by a radial drainage pattern away from

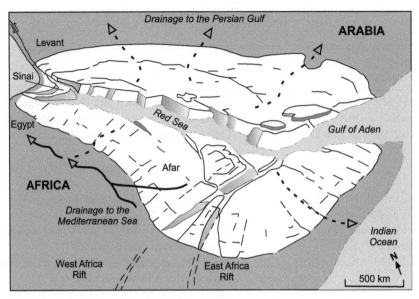

Figure 12.1 Afro-Arabian domal uplift of Cloos (1953) as redrawn and modified by Avni *et al.* (2012). Digital version of original figure kindly provided by Y. Avni.

the proposed plume center along with the superimposed effects of flow eastward from the rift flank uplift of the Western Ghats and flow along the east–west Narmada Tapti rift valley (Fig. 12.2; Cox, 1989). Widdowson (1997) tracked post-eruptive uplift through two laterite-covered paleosurfaces. Cox (1989) suggested that the pattern of uplift was preserved by crustal thickening caused by basaltic underplating.

12.2.4 North Atlantic Igneous Province (NAIP) (c. 60 Ma)

As summarized by Saunders *et al.* (2007) the North Atlantic Igneous Province (NAIP) provides a clear example of how the sedimentological response to regional transient surface uplift depends critically on paleogeography. Because the NAIP was emplaced across a variety of basement blocks and sedimentary basins of varying age and original water depths, the effect of surface uplift was recorded by a continuous sedimentary succession in some areas and by an erosional unconformity in others (Fig. 12.3). More than half of the total transient regional kilometer-scale uplift grew in much less than 1 Ma immediately before the *c.* 55 Ma Paleocene–Eocene Thermal Maximum global climatic change (Section 14.3.1). In detail, the NAIP was not associated with a single domal swell. Instead, surface uplift occurred in discrete events: Phase 1 magmatism (62–58 Ma) was associated with minor uplift localized around igneous centers, and Phase 2 magmatism (*c.* 55 Ma) was associated with widespread kilometer-scale uplift. The role of magmatic underplating and timing of Paleogene

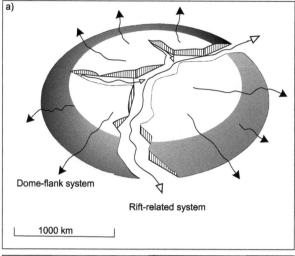

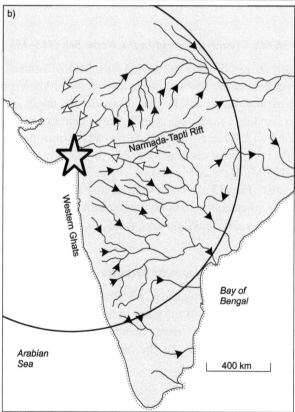

Figure 12.2 Evidence of domal uplift recorded in river flow directions. (a) Distinction between centripetal drainage of the dome flank system (solid arrowheads) and the radial drainage of the rift-related system (open arrowheads). From Figure 1 in Moore and Blenkinsop (2002).

uplift and denudation is explored by Tiley *et al.* (2004). Some of this uplift is also recognized through provenance studies of the Irish Sea which reveal that a radial system of drainage developed, partly recognized in a southeastern tilt of Britain (Cope, 1994; Sleep, 2009). Part of this radial drainage system, including the Thames, persists in the British Isles even though the center of the uplift was subsequently beveled by erosion into a shallow sea, the Irish Sea.

12.2.5 *Paraná–Etendeka LIP (c. 135 Ma)*

The Early Cretaceous river systems of southern Africa have a distribution and trend consistent with uplift associated with the Paraná–Etendeka LIP (Moore and Blenkinsop, 2002; cf. Cox, 1989). Additional influences on drainage cited by Moore and Blenkinsop (2002) include structural controls, the role of erosionally exhumed surfaces and processes of river capture; these can explain the imperfections in the radiating pattern.

12.2.6 *Middle Cimmerian event in the North Sea (165–155 Ma)*

A regional thermally driven doming associated with Middle Jurassic magmatism is present in the North Sea basin (Hallam and Sellwood, 1976; Underhill and Partington, 1994; Rainbird and Ernst, 2001). The pre-volcanic uplift is marked by an unconformity (the "middle Cimmerian event") which defines an elliptical elevation anomaly *c.* 1500 km by 1250 km, with a concentric pattern of isopachs corresponding to maximum uplift and erosion centered over North Sea rifting. The unconformity underlies the 165–155 Ma Rattray Formation volcanic rocks, which are confined to the Moray Firth area (Fig. 12.4). The extent of the volcanics is not known and therefore it is unclear whether the basaltic magmatism is of LIP size. However, the size and extent of uplift are of mantle-plume scale.

12.2.7 *Central Atlantic Magmatic Province (CAMP) LIP (200 Ma)*

The Central Atlantic Magmatic Province (CAMP) is linked to the arrival of a mantle plume located by the convergence of the giant radiating dyke swarm (Fig. 12.5). The effects of thermal uplift due to the arriving mantle plume are recorded by a regional change in the age of sedimentation in the Newark rift basins of eastern North

Caption for Figure 12.2 (*cont.*) (b) Example: drainage pattern of peninsular India with the postulated Deccan plume (circle) superimposed (redrafted from Cox, 1989). Reprinted by permission from Macmillan Publishers Ltd: Nature. Black arrows highlight drainage away from the domal uplift. White arrows depict the eastward drainage pattern in a rift that cuts the domal uplift. There is also drainage eastward away from the Western Ghats marking rift-flank uplift along the western coast of India. Star locates inferred mantle plume center.

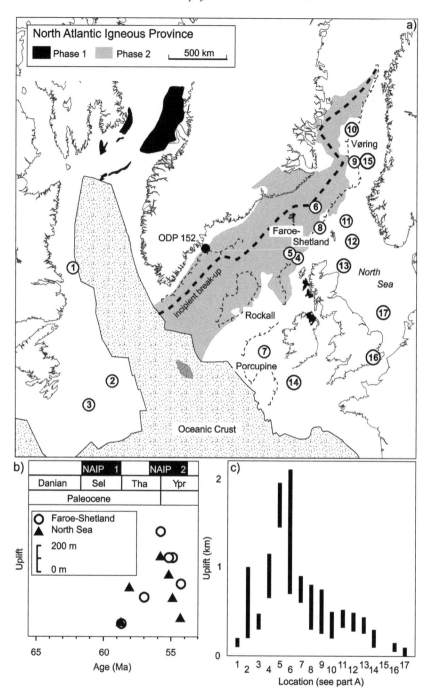

Figure 12.3 Uplift pattern associated with the NAIP in reconstructed North Atlantic. Numbered circles mark locations of dynamic support estimates in (c). (a) From Figure 7 in Saunders *et al.* (2007) and (b) and (c) are from Figures 8E and F in Saunders *et al.* (2007), with permission from Elsevier. Danian, Selandian (Sel), Thanetian (Tha), and Ypresian (Ypr) are stratigraphic stages.

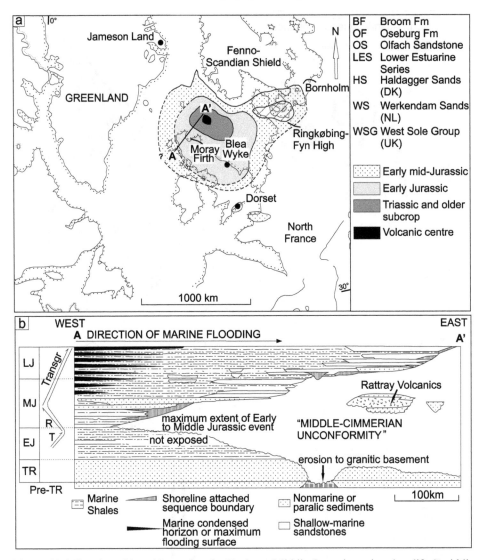

Figure 12.4 Stratigraphic evidence for the Early to Middle Jurassic regional uplift ("middle Cimmerian event") in response to a mantle plume head, and associated volcanics (Rattray volcanics). T = transgression and R = regression. It is uncertain whether these volcanics could be LIP-related. The cross section in (b) is labeled as A–A' in (a). From Underhill and Partington (1994).

America (Fig. 12.5; Hill, 1991; Rainbird and Ernst, 2001). Basins within about 1000 km of the plume center ceased sedimentation between 225 to 220 Ma, whereas sedimentation in the northern basins continued into the Early Jurassic (200–190 Ma). Based on theoretical arguments of Griffiths and Campbell (1991), an arriving plume begins to cause uplift 20–30 Ma prior to flood volcanism (Section 12.3.1). Therefore,

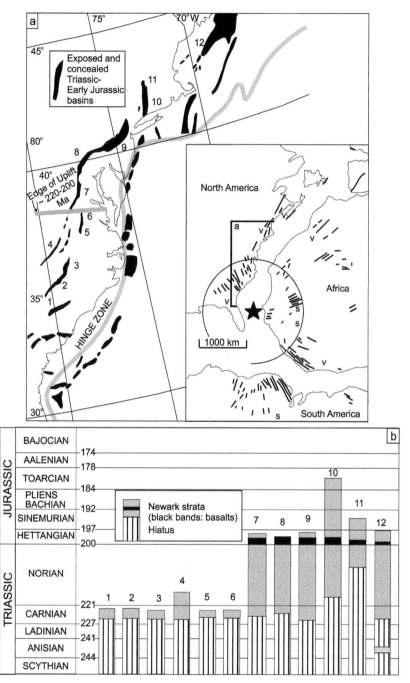

Figure 12.5 Uplift associated with the CAMP as recorded in the Newark sedimentary basins. (a) General distribution of CAMP dykes. (b) Distribution of Late Triassic–Early Jurassic basins of eastern North America, both exposed and those blanketed by younger cover. Basin numbers keyed to chart in (c). (c) Age distribution of Newark Supergroup strata in selected basins. From Figure 4 in Rainbird and Ernst (2001) modified after original version in Hill (1991).

the cessation of sedimentation in the southern basins may be linked to uplift generated by the arriving plume and suggests that the radius of that uplift was about 1000 km.

12.2.8 Wrangellia LIP (230 Ma)

The 232 Ma Wrangellia LIP represents an oceanic plateau that was accreted onto western North America (Section 4.5.1). There is an abrupt but conformable super-position of terrestrial and locally pillowed flood basalts above radiolarian cherts and marine mudstones containing deep-water bivalves (Richards *et al.*, 1991; Rainbird and Ernst, 2001; Campbell, 2001). Several lines of evidence inferred maximum water depths of *c.* 500 m at the end of the Permian (middle Ladinian), lack of intervening shallow-water deposits, and volcanism beginning only about 1 Ma after sedimenta-tion ceased; all of which imply an extremely rapid rate of uplift (*c.* 5 mm/year) prior to flood volcanism. This is inferred to indicate the arrival of a mantle plume. After accumulation of about 5 km of flood basalts there was a return to subsidence.

12.2.9 Siberian Trap LIP (250 Ma)

On the Siberian craton where the sedimentary record is most complete there is little evidence for either pre-volcanic or syn-volcanic uplift and erosion, and in fact there may even have been slight subsidence in some locations during these periods prior to eruption of the Siberian Trap LIP (Czamanske *et al.*, 1988). See Section 3.2.8 for background on this LIP. However, the sedimentary sequences on the thick, stable craton do not provide the full picture. As pointed out by Saunders *et al.* (2007) the absence of Permian strata beneath large tracts of the West Siberian basin suggest that significant pre-volcanic uplift occurred there, although the magnitude and detailed timing are not yet constrained.

12.2.10 Emeishan LIP (260 Ma)

The Emeishan LIP has been inferred to exhibit classic domal uplift due to the arrival of a mantle plume (He *et al.*, 2003, 2006; Xu *et al.*, 2004a; Saunders *et al.*, 2007). Surface uplift prior to basaltic eruption has been interpreted from the erosion pattern of the Maokou limestones that underlie the basalts (Figs. 12.6 and 12.7). Detailed biostratigraphy established for limestones shows that uplift occurred less than 2.5 Ma before the eruption of the basalts.

However, there is current debate on the interpretation of pre-volcanic uplift based particularly on evidence for the widespread presence of mafic hydromagmatic deposits (MHDs) produced by explosive magma–water interaction. These MHDs are widespread in the region originally considered to be the locus of domal uplift (e.g. Ukstins Peate and Bryan, 2008, 2009a, b; He *et al.*, 2009; Ali *et al.*, 2010; Sun *et al.*,

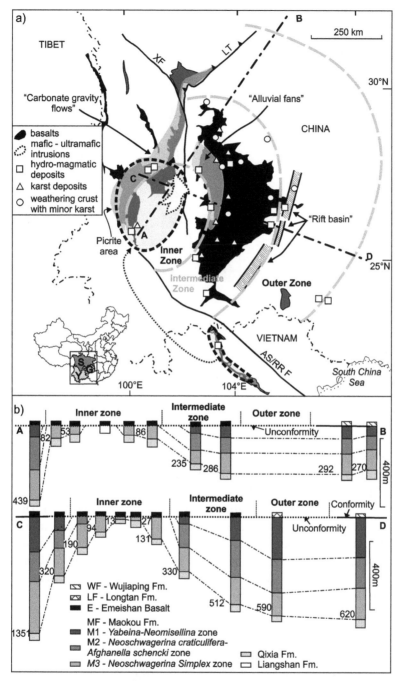

Figure 12.6 The Emeishan LIP of South China. (a) Regional map. A portion of central region (above the inferred plume center inside the dashed circle) has been fault offset into northern Vietnam along the AS/RRF (Ailao Shan/Red River fault). The detailed distribution of LIP

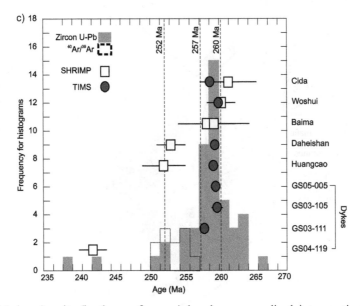

Figure 12.6 (*cont.*) units (in these references) has been generalized into continuous areas. (b) Uplift profiles. (c) U–Pb geochronology demonstrating the short duration for the Emeishan LIP. Zircon U–Pb and ^{40}Ar–^{39}Ar dates are from the literature. Comparison of the zircon U–Pb SHRIMP and ID–TIMS results for selected rocks. Note that TIMS ages give a tight grouping despite the large range in SHRIMP ages. (a) and (b) From various sources: He *et al.* (2009); Ali *et al.* (2010), and Sun *et al.* (2010). (c) Modified from Figure 5 in Shellnut *et al.* (2012). With permission from International Association for Gondwana Research. Digital version of original figure kindly provided by S. Denyszyn.

2010). The most recent contribution to this debate considers the additional evidence for rifting within the area of uplift, which has downdropped some areas below sea level, allowing the formation of MHDs in these areas (e.g. He *et al.*, 2011). The domal uplift story is affirmed by He *et al.* (2011) in a five-stage model (Fig. 12.7). This model incorporates the data based on the missing Maokou strata, paleokarst morphology, erosional features, diagnostic rocks on the surface such as basal conglomerates, kaolins, bauxites, and ferruginous duricrust. However, it is likely that the debate on the scale and nature of domal uplift associated with the Emeishan LIP will continue (Shellnutt, 2014).

12.2.11 Yakutsk–Vilyui LIP (c. 370 Ma)

The 370 Ma Yakutsk–Vilyui LIP of eastern Siberia (Fig. 11.7) consists of significant volcanic accumulations, a giant radiating dyke swarm, and associated kimberlites and carbonatites (Fig. 9.15), and also dramatic triple-junction rifting (Fig. 11.7) along with impressive topographic changes (e.g. Kiselev *et al.*, 2012). Most significantly, in

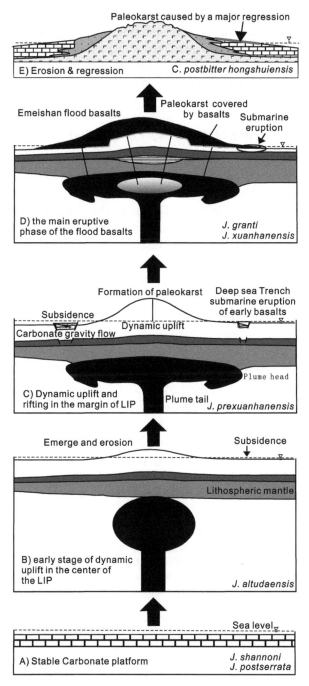

Figure 12.7 Schematic cross section through the Emeishan LIP from west to east illustrating the surface responses to the Emeishan mantle plume. Conodont zones on the right of the diagram (in italics) representing time are modified from Sun *et al.* (2010). (a) Stable,

early Frasnian time, prior to LIP emplacement, there was widespread domal uplift of about 500 m marked by an erosional unconformity between the Middle and Upper Devonian sequences. This uplift is preserved along rift arms. There may also have been an earlier period (before the domal uplift) in the Middle Devonian in which widespread basins formed containing terrigenous deposits (thickness 20–120 m).

12.2.12 Kalkarindji LIP (510 Ma)

Regional uplift exceeding 1 km is preserved across an area more than 1000 km wide in late Neoproterozoic strata in South Australia (Williams and Gostin, 2000). Spectacular paleocanyons, which locally attain depths exceeding 1 km were cut into shallow-marine strata of the Wilpena Group in the Adelaide fold belt. Similar canyons, as deep as 700 m, have been identified in the subsurface in correlative strata of the eastern Officer basin. To explain these features Williams and Gostin (2000) proposed thermal uplift of the crust beneath the Gawler craton by a mantle plume that preceded eruption of the widespread flood-basalt event now termed the Kalkarindji LIP (Section 3.5.1), which is widespread across northern, central, and southern Australia.

12.2.13 Franklin LIP (c. 725–715 Ma)

The Franklin LIP of northern Canada consists of volcanics, sills, and most prominently a giant radiating dyke swarm that extends across Canada from west of the Slave craton into formerly attached Greenland (Fig. 5.1; Fahrig, 1987; Heaman *et al.*, 1992; Ernst and Buchan, 1997a; Buchan and Ernst, 2004; Denyszyn *et al.*, 2009a, b; Buchan and Ernst, 2013). Based on the radiating pattern the plume center was located northwest of Banks Island (Ernst and Buchan, 1997a; Buchan *et al.*, 2010, but cf. Denyszyn *et al.* (2009a) for a different inferred location for the plume center). An

Caption for Figure 12.7 (*cont.*) homogeneous carbonate platform in the middle Permian. (b) The carbonate platform collapsed as a result of the rising plume head. Domal uplift (above sea level) took place only in the center of the LIP and subsidence occurred on the margins. (c) With further ascent, the plume head collided with the base of lithosphere and spread out, forming a flattened head 800–1000 km in diameter. Domal uplift occurred and a paleokarst formed on the Maokou Formation. At the same time, rifting occurred at the edge of the flattened plume head, resulting in gravity flow deposits. Some basalts were erupted in a submarine setting in these rifts. (d) The main phase of Emeishan flood-basalt volcanism began with subaerial eruption (covering the paleokarst), with continuing eruption of submarine basalts at the margins of the uplift. The thickness of the flood basalts decreases from the inner zone to the outer zone. (e) Crustal uplift continued until the Late Triassic, and the Chuandian (old land) was eroded continually, and eroded clasts were deposited in the intermediate–outer zones. Redrafted from He *et al.* (2011) and modified from an earlier version in He *et al.* (2006). With permission from Elsevier.

erosional remnant of the Natkusiak Formation basalts remains on northwestern Victoria Island in the Minto Inlier. Here the basalts have a preserved thickness of *c*. 1 km and form the capping unit of a *c*. 4-km-thick succession of platform marine and terrestrial sedimentary rocks known as the Shaler Supergroup. Strata immediately beneath the flood basalts display stratigraphic and sedimentalogical features suggesting that differential crustal uplift just preceded eruption. The uplift is postulated to have been a consequence of thermal doming by a mantle plume (Rainbird, 1993; Rainbird and Ernst, 2001). The specific relationships in the Minto Inlier suggest that uplift was progressively greater toward the northeast (along a southwestern–northeastern profile). The preserved cross section in the Minto Inlier might be intersecting the outer edge of a much larger uplift, centered several hundred kilometers farther north, near the inferred plume center near Banks Island.

12.2.14 Mackenzie LIP (1270 Ma)

The late Mesoproterozoic (1267 Ma) Mackenzie magmatic event comprises flood basalts (of which the Coppermine sequence is the largest erosional remnant), the Muskox layered igneous intrusion, scattered sill provinces, and a huge mafic dyke swarm that radiates across the Canadian shield with a focus on western Victoria Island marking the plume center (Fig. 13.5; Fahrig, 1987; LeCheminant and Heaman, 1989; Baragar *et al.*, 1996).

Regional domal uplift is inferred by the regional change in the trends of the radiating dyke swarm. Within about 1000 km of the inferred plume center the swarm has a clear radiating pattern, but at further distances away the dykes swing into a common linear northwest–southeast trend, and this is inferred to mark the transition from a stress pattern controlled by domal uplift, to the dominant influence of a regional linear stress field generated by a distant plate boundary (e.g. Fig. 5.2).

Another line of evidence in support of uplift associated with the Mackenzie LIP is the presence of karstic unconformity (in the Greenhorn Lake Formation) in the vicinity of the Muskox intrusion and also about 250 km to the east (in the Parry Bay Formation; Rainbird and Ernst, 2001). The latter are marked by spectacular cave deposits. The uplift marked by formation of karst topography is inferred to have occurred prior to eruption of Coppermine and other basalts of the Mackenzie LIP.

12.2.15 Rift-flank uplift

Many rifted continental margins associated with LIPs are bordered by eroded mountain ranges reflecting rift-flank (rift-shoulder) uplift (e.g. Braun and Beaumont, 1989; Menzies *et al.*, 2002a). Examples are in the Paraná–Etendeka, Afro-Arabian, NAIP, and Deccan LIPs (Chapter 3). In many volcanic rifted margins marine sediments are now up to several kilometers above present sea level. Rift-flank uplift is also seen in river drainage patterns, such as for the Western Ghats mountains of the Deccan LIP where young rivers flow eastward from this rift-flank uplift on the west coast of India (Fig. 12.2; Cox, 1989). As noted by Sengör (1995) rift formation is almost always followed by uplift of rift flanks. This is explained by isostasy and elastic bending of the faulted plate and to extension-related pressure drop and consequent adiabatic partial melting and volumetric expansion in the underlying mantle. Sengör (1995) also notes that after the rift becomes inactive, uplift caused by the partial melting mechanism commonly reverses and leads to subsidence, to produce young Atlantic-type rift continental margins.

12.2.16 Oceanic plateaus

As a final point I consider the oceanic plateaus, which represent topographic highs in the ocean basins as a combination of both magmatic accumulation, dynamic support from the underlying thermal anomaly, and a change in mantle density associated with development of an underlying lithospheric root. An interesting problem is related to the largest oceanic plateau, Ontong Java (Section 4.2.1). As noted by Ingle and Coffin (2004) arguments from isostasy would suggest that emplacement of this 35-km-thick oceanic plateau onto pre-existing uplifted *c.* 7-km-thick oceanic crust should have resulted in major, widespread subaerial volcanism. Yet Ontong Java basalts seem to be mostly erupted below sea level. This has led to consideration of non-plume models for Ontong Java (e.g. bolide impact; Ingle and Coffin, 2004); further discussion is found in Chapter 15.

12.3 Mechanisms that control uplift and variation through time

The discussion above illustrates that topographic uplift has occurred in association with many LIPs. Below are considered possible mechanisms to explain the observed domal uplift. Potential controls on topography during LIP events include the following: mantle plumes, lithospheric delamination, decompression melting associated with rifting, and possibly plate-boundary models. In this chapter, I consider each of these mechanisms strictly from the point of view of their effect on topography. These potential mechanisms are visited again in Chapter 15 in the broader context on the origin of LIPs.

12.3.1 Mantle-plume origin

From a theoretical perspective, the arrival of mantle plumes at the base of the lithosphere can form domal uplift within a time span of several million years (Figs. 12.8 and 12.9). The dimensions of the uplift depend on the size of the plume head and thus on the depth from which the plume originates. For a plume rising from the deep mantle, the lithosphere is uplifted by one or two kilometers over an area comparable to the size of the flattened plume head, i.e. about 2000–2500-km diameter (e.g. White and McKenzie, 1989; Campbell and Griffiths, 1990; Sleep, 1990, 2006; Hill, 1991; Monnereau *et al.*, 1993; Farnetani and Richards, 1994; Şengör, 2001; Campbell, 2001; Rainbird and Ernst, 2001; Campbell and Kerr, 2007; Davies, 2011).

For plumes rising from a shallower level, such as the 660-km discontinuity, the plume-head size (after flattening against the lithosphere), and hence the uplift area, should be smaller, about 500 km in diameter. In either case, uplift is a combination of initial purely dynamical support, followed by erosion of the lithosphere, magmatic underplating, and thermal expansion caused by longer-term heating of the lithosphere. According to modeling, the uplift begins when the plume is still substantially below the lithosphere, and the maximum uplift occurs about 5 Ma before plume arrival at the base of the lithosphere (Fig. 12.9) and before major melting to produce flood basalts.

Complexities may occur. Leng and Zhong (2010) suggested that a plume head temporarily ponding below the 660-km phase-change boundary causes significant subsidence (up to 400 m) at the Earth's surface over an extended period before the eruption. This may explain pre-existing sedimentary basins in the vicinity of some LIPs, such as the Pasco basin predating the Columbia River LIP (Section 5.6.4), and the subsidence prior to the Yakutsk–Vilyui event (Section 12.2.11). Leng and Zhong (2010) also modeled that loading from erupted basalts causes surface subsidence at the periphery of the eruption area.

12.3.2 Thermochemical plumes

The dynamics of plumes with purely thermal origin has been well established (as discussed above). However, buoyancy can be affected by addition of other components into the plume such as eclogite from recycled oceanic crust (e.g. Cordery *et al.*, 1997; Campbell, 1998; Leitch and Davies, 2001; Sobolev *et al.*, 2011), or by components such as hydrogen and methane potentially entering the lower mantle from the core (e.g. Dobretsov *et al.*, 2008). Compositional changes to the plume can lead to an interplay between the thermal and compositional buoyancy forces (e.g. Lin and van Keken, 2006a–c). For example, the inclusion of large amounts (e.g. 15%) of dense recycled oceanic crust into a rising plume head results in extensive melting and enhanced effectiveness of erosion of thick cratonic lithosphere, but predicts no regional uplift and requires no lithospheric extension (Sobolev *et al.*, 2011; see also Cordery *et al.*, 1997; Campbell, 1998; Leitch and Davies, 2001).

12.3.3 Lithospheric delamination

Delamination of a dense lithosphere fragment can produce magmatism and uplift (e.g. Schott and Schmeling, 1998; Elkins-Tanton and Hager, 2000; Şengör, 2001;

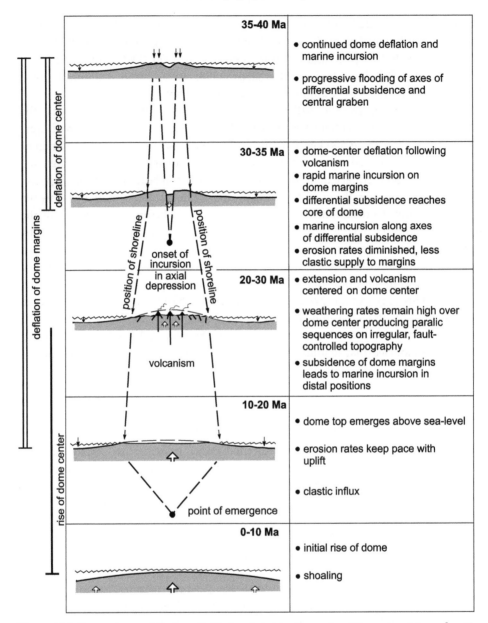

Figure 12.8 Succession and timing of effects related to thermal uplift produced by a mantle plume beneath a marine basin. This example is from the Jurassic North Sea basin (Underhill and Partington, 1994) (Section 12.3.7). From Figure 8 in Rainbird and Ernst (2001).

Elkins-Tanton, 2005, 2007). In combination with an upwelling thermal anomaly (Figs. 12.10 and 12.11) delamination has been linked to the production of the Siberian Trap LIP, but the magnitude of associated uplift would be modeled as minor, matching with the observed absence of pre-volcanic uplift in the portion of

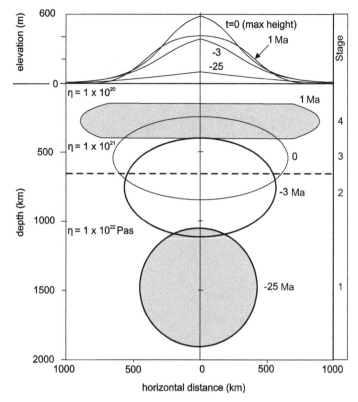

Figure 12.9 Uplift at surface related to the rise and flattening of a mantle plume. Stages can be identified by number. Stages 1 to 4 correspond to −25 Ma, −3 Ma, 0 Ma, and +1 Ma, respectively. Uplift reaches a maximum at stage 3, then subsides and spreads (stage 4) as the plume head flattens and spreads. Modified from Figure 12 in Griffiths and Campbell (1991). η = viscosity of the mantle in units of pascal seconds (Pa s).

this LIP which is located on the Siberian craton (Elkins-Tanton, 2007). Note, however, the evidence for associated uplift in the western part of this LIP, in the West Siberian basin (Section 12.2.9). The role of delamination on its own (in the absence of a thermal anomaly) is to cause an initial slight depression due to traction on the crust during downward movement of delaminated lithosphere (Fig. 12.10). Beyond a certain point traction will end and the crust can slightly rebound to its original level. In the presence of a thermal anomaly (e.g. plume) the initial surface sag will still occur but there will also be a subsequent uplift with magnitude depending on the thermal anomaly (Fig. 12.11).

12.3.4 Decompression melting related to extension

As shown in the classic White and McKenzie (1989) paper, extension associated with rifting results in superimposed subsidence caused by lithospheric thinning along with

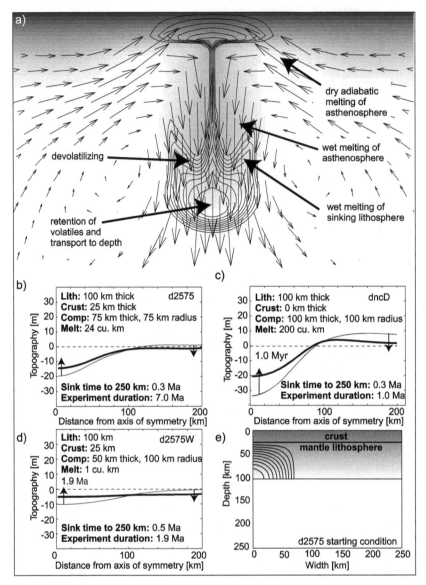

Figure 12.10 Models for surface topographic evolution during delamination by a Rayleigh–Taylor instability. As the lithospheric fragment sinks, the asthenosphere moves into the resulting lithospheric domal hole and may melt adiabatically. If the instability is hydrous, the delaminated fragment may dewater as it sinks and heats up, resulting in melting of the adjacent mantle or of itself. Representative models are shown in parts b–d, under various conditions of **Lith** (lithospheric thickness measured from surface down to the adiabatic mantle): **Crust** (crustal thickness); **Comp** (height/radius of added lithosphere measured from adiabatic mantle upward); **Melt** (volume of melt produced during instability fall); **Sink time** (time required for main delaminating mass to sink below the bottom of the model box at 250 km); and **Experiment duration**. In parts b–d, the thin line indicates the topography with the density anomaly in place but before significant growth of the instability has occurred. The dashed line indicates surface prior to density injection. Arrows show the sense

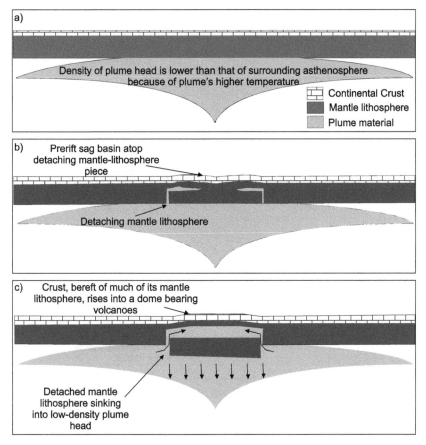

Figure 12.11 Role of delamination in association with the arrival of a mantle plume. From Figure 8 in Şengör (2001).

uplift due to the addition of melt to the crust. Further uplift results from reduced density in the abnormally hot asthenospheric mantle and also results from reduced density in the residual mantle from which melt has been extracted. In the absence of anomalous thermal temperatures (i.e. ambient temperature of 1280 °C) the overall

of time evolution of the topography; the bold line is the final topographic profile at the end of the numerical simulation. Part (e) shows example of starting conditions for the modeling (specifically for part b) and a two-dimensional cross section through the lithosphere and mantle, with an axis of symmetry at the left side. Temperature increases with depth as shown in shades of gray, from 20 °C at the surface to the mantle potential temperature (in white). Added material with density and temperature contrast is shown in contour lines of 10% of Δρ each, with the maximum addition at the axis of symmetry (at the lower left side) on the interface between the asthenospheric mantle and lithosphere. The maximum addition in each model corresponds to 1 or 5% density contrast with the mantle. Modified from Elkins-Tanton (2007). Copyright 2007 by the American Geophysical Union. Digital version of original figure kindly provided by L. Elkins-Tanton.

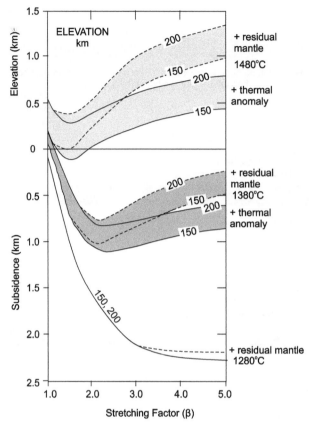

Figure 12.12 Rifting of the lithosphere and resulting topographic changes. The beta-factor defined as the extended width divided by the unextended width. Curves are calculated for an initial lithosphere thickness of 118 km and for the ambient temperatures given. Numbers refer to the depth of compensation (in km) and span the depth range of anomalous mantle associated with a mantle plume. The regions bound by solid lines (and labeled "+thermal anomaly") include the buoyancy due to the hot asthenospheric mantle. The regions bound by dashed lines (and labeled "+residual mantle") additionally include the buoyancy due to reduced density in the residual mantle from which melt has been extracted. Redrafted from Figure 4.14 in Condie (2001); modified from original Figure 7 in White and McKenzie (1989). Copyright © 2001 Cambridge University Press.

topographic change is of slight subsidence (Fig. 12.12). With an anomalous positive temperature anomaly of 100 °C above ambient mantle (i.e. temperature of 1380 °C), there is basically no subsidence or uplift. With a greater positive thermal anomaly uplift occurs. For example an excess temperature of 200 °C above ambient (i.e. temperature of 1480 °C), results in significant uplift. Increasing the amount of stretching (i.e. increasing "beta factor") increases the effect in each case to a factor of about three, at which point breakup occurs and a new ocean forms.

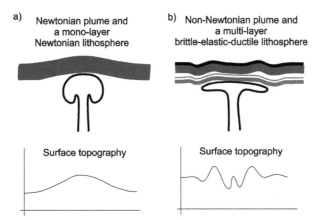

Figure 12.13 Sketches of plume–lithosphere interactions. (a) Plume arrival beneath a single-layer viscous lithosphere, resulting in a single, long-wavelength topographic signature. (b) Plume arrival beneath a region with more complicated lithospheric rheology resulting in a more heterogeneous pattern of local uplift and subsidence. From Figure 5 in Foulger (2007) after original in Burov and Guillou-Forttier (2005).

12.3.5 Plate-boundary-related convection models

As an alternative to plume models, plate-boundary mechanisms have been offered to explain LIP production (e.g. Foulger, 2007) and such models should have uplift implications (e.g. Saunders *et al.*, 2007). In one model (edge convection; see Section 15.2.5; e.g. King and Anderson, 1995) convection cells are generated between lithosphere of different thicknesses, and would be predicted to generate linear zones of uplift. For a more complete discussion see Saunders *et al.* (2007).

12.3.6 Complexities due to lithospheric heterogeneity

In reality, uplift above a plume may be more complex than the simple models described above (e.g. Fig. 12.13). Complications include the possible presence of elongated uplift related to "hot lines" (e.g. Al-Kindi *et al.*, 2003), the variation in topography caused as the plume flattens and expands beneath the lithosphere (Fig. 12.14; Griffiths and Campbell, 1991), the different size of the swell depending on the depth at which the plume head stalls (shallower beneath oceanic than under continental lithosphere). Since the capacity for broad domal uplift depends on sufficient strength of the lithosphere, then in areas with rheological layering of the lithosphere, several wavelengths of surface topographic undulations would be predicted leading to both local uplift and subsidence (Fig. 12.13b; Burov and Guillou-Frottier, 2005a; Foulger, 2007). This would be particularly true for areas with weak lower crust, but Saunders *et al.* (2007) note that in many continental regions there is evidence in support of strong lower crust and weak underlying asthenosphere, and hence more consistent with a broad domal uplift above a plume.

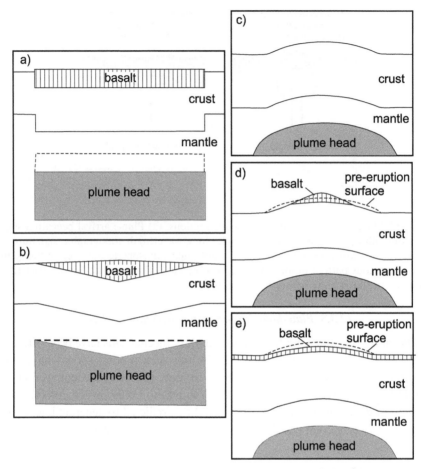

Figure 12.14 (a)–(e) Various controls on the patterns of subsidence and uplift associated with the eruption of a flood basalt. See discussion in the text. From Figures 5 and 6 in Campbell (2001).

12.3.7 Comparison of proposed mechanisms to observed uplift

Many of these examples in Section 12.2 indicate early uplift prior to the onset of LIP magmatism. This pre-LIP timing for uplift is particularly well constrained for the CAMP, Yakutsk–Vilyui, Emeishan LIPs. The evidence for pre-LIP uplift is consistent with the plume uplift model, and is in distinct opposition to models like the edge-convection or delamination model (Sections 15.2.5 and 15.2.3), in which uplift must occur after breakup.

The onset of rifting (leading to a second LIP pulse, per the White and McKenzie (1989) model, produces an immediate decrease in topography leading to the formation of volcanic rifted margins. This is the origin of some of the topographic complexities related to the NAIP LIP (Section 12.2.4).

The topographic effects of delamination are relatively minor, and so this mechanism is not a significant factor in the main topographic variations of LIPs. However, in one special case it may be important. A magmatic underplate can dramatically increase in density during conversion to eclogite, and if so can potentially delaminate. However, if delaminataion doesn't occur then the increase in density can be a potential mechanism for the formation of intracratonic basins (Section 12.6.2).

This discussion leads to a thematic aspect to be more fully developed below: the pattern of associated topographic changes may vary with time and, broadly, there may be uplift that is transient and uplift that is permanent (e.g. Saunders *et al.*, 2007). The domal uplift associated with a plume may be transient in that once the thermal anomaly associated with the plume has decayed then so will the uplift. A key control on whether early uplift is preserved relates to the scale of underplating (Section 5.5), which preserves the uplift patterns even after the thermal pulse associated with a mantle plume fades. There are additional causes of variation in uplift over time and those aspects are considered below in three time intervals: early uplift prior to LIP magmatism, changes in uplift during the period of LIP magmatism, and topographic changes happening post-LIP emplacement.

12.4 Regional domal uplift prior to LIP magmatism

As already shown in examples described above a number of LIPs have early domal uplift, which is most consistent with a rising mantle plume approaching the base of the lithosphere. The scale of uplift can be up to several kilometers, consistent with models of Campbell (Fig. 12.9). An important observation is that since this uplift is preserved it means that the original transient uplift was made permanent by underplating, even after the thermal anomaly that caused the original uplift had decayed away.

12.5 Uplift changes during LIP magmatism

Here, I consider the uplift variations that potentially occur during the period of LIP emplacement and are mainly related to processes of magma redistribution within the lithosphere and implications for isostatic and topographic readjustments (Campbell, 2001). These are in addition to transient topographic changes associated with flattening of the buoyant plume head against the lithosphere during the period of LIP magmatism (Fig. 12.9).

12.5.1 Vertical redistribution of magma

Magma redistribution can occur both vertically and horizontally above the plume, and can lead to topographic changes (Fig. 12.14; Campbell, 2001), with some specific scenarios highlighted below.

As magma is drained from the mantle and emplaced at higher levels in the crust, the bottom of the crust must sink (Fig. 12.14a, b). If it is assumed that magma is moved vertically from the mantle to the overlying crust (with no net lateral movement of melt), then isostatic balance is maintained. However, the height of the upper surface will increase by 10% owing to the density difference between the basalts and the mantle. So, as noted by Campbell (2001), eruption of a 2-km-thick sequence of flood basalt will cause the lower surface of the volcanic pile to subside by 1800 m but the upper surface is raised by 200 m, relative to the pre-eruption level.

12.5.2 Lateral redistribution of magma

If there is also lateral movement of magma away from the underlying source area, then isostatic adjustments will occur. Basaltic lavas can flow significant distances (up to 1000 km; Section 10.5.1), and sills and particularly radiating dyke swarms can potentially transport magma laterally in the crust for distances of more than 2000 km (e.g. Section 5.2.1). As estimated by Fahrig (1987) the volume of magma in the giant radiating Mackenzie swarm (Fig. 5.22) is 80 000 km^3 and so this represents a significant amount of magma potentially removed from the plume-center region.

If the area covered by the basaltic magmatism is not the same as the area of melt extraction in the mantle, then isostatic adjustment will be required in both the region above the mantle source area, which has lost magma (the surface will drop; Fig. 12.14e) and outside the mantle source area, which has gained magma (the surface will rise). It is also possible to envision a scenario in which basaltic lavas were concentrated in a smaller area than the area drained in the underlying mantle, and in such a case extra topographic uplift will occur (Fig. 12.14d). More details are provided in Campbell (2001).

12.5.3 Systematic shifts in the locus of magmatism

In some young LIPs there can be a spatial migration in the locus of volcanism. For instance, as shown in Fig. 12.15, lava successions in the Deccan LIP shift southward and this is linked to the Indian plate moving rapidly northward over the Deccan plume (Saunders *et al.*, 2007). Similarly, the Paraná lava pile consists of an overlapping sequence of units dipping to the north, which suggests northward migration of the magma source (e.g. Peate, 1997). A changing locus of magmatism can correspond to transient topographic changes.

12.5.4 Gravity-induced sliding off a domal uplift

A further process is required by observations from the Afar uplift associated with the Afro-Arabian LIP (Fig. 12.1). Modeling of the inferred uplift yields a conundrum in terms of the relation between uplift and formation of triple-junction rifts (Şengör, 2001). No amount of doming within reasonable limits is sufficient to create the east

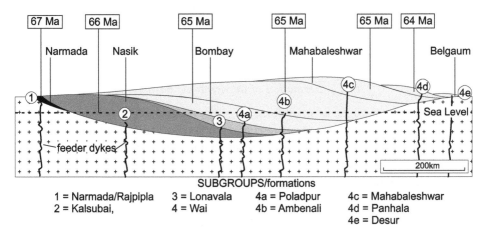

Figure 12.15 Shifting of locus of Deccan magmatism to the south due to northward movement of India over the Reunion plume. North–south sketch section showing structure and stratigraphy of Deccan lava sequences along the Western Ghats escarpment. Vertical exaggeration ×25. From Figure 6 in Saunders *et al.* (2007). With permission from Elsevier.

African rift troughs by the extension resulting from the areal increase associated with Afar doming. Models of simple triple-junction rifting (e.g. Cloos, 1953; Burke and Whiteman, 1973; Burke and Dewey, 1973) do not work, and an alternative explanation is needed. A simple model that would work is that of gravity-induced sliding off the topographic uplift (Şengör, 2001; Section 13.2.5).

A similar model was also inferred for the 1270 Ma Mackenzie LIP based on unrealistic uplift inferred from the magnitude of extension calculated from dyke-intrusion density (Section 13.2.4; Baragar *et al.*, 1996). Gravity-induced sliding off a domal uplift introduces an additional element into topographic variations. The model also predicts associated compressional features on the flanks and periphery of the sliding crustal panels, which are further discussed in Section 13.2.

12.6 Post-LIP topographic changes

12.6.1 Permanent uplift due to underplating

Continued maintenance of uplift in post-LIP time results from balancing isostatic compensation for (1) erosional unroofing of the uplifted crust and/or (2) thickening at the base of the lithosphere by magmatic underplating (Cox, 1989; Widdowson, 1997; Rainbird and Ernst, 2001; Xu and He, 2007). Many basins in North Atlantic volcanic rifted margins underwent permanent uplift of several hundred meters, which was likely caused by magmatic underplating (Saunders *et al.*, 2007). For example, Brodie and White (1994) showed that many of the extensional sedimentary basins in the vicinity of the British Isles underwent permanent exhumation during the Tertiary.

After having studied drainage patterns of the Deccan, Paraná–Etendeka, and Karoo continental LIPs, Cox (1989) suggested topographic doming associated with plume activity can be preserved even after *c.* 200 Ma, and crustal thickening by magmatic underplating is the most likely cause for the persistence of such features.

12.6.2 Formation of intracratonic basins

Post-LIP subsidence can be ascribed to decay of the thermal anomaly associated with cooling of the plume, deflation due to magma extraction, and/or rifting and thermal cooling of thinned crust. In addition, this subsidence process can be enhanced by conversion of crustal mafic underplate to denser eclogite. Intracratonic basins (broad basins in the interior of continents) can be caused by eclogitization of the lower crust, and resulting isostatic readjustment (e.g. Baird *et al.*, 1995). For instance, the Michigan basin has been linked with the conversion of basalts of the Keweenawan LIP (Section 3.5.2) to eclogite (Haxby *et al.*, 1976). However, the link is not straightforward as basin formation (and presumably eclogite conversion) in the Michigan case occurred at *c.* 460 Ma, over 600 Ma after the *c.* 1100 Ma Keweenawan LIP. Also, not all intracratonic basins are associated with prior LIP events (Fig. 12.16). A model for the Williston basin envisions a residual portion of crustal root that developed during the prior (*c.* 1.8) Hudsonian orogeny being subsequently eclogitized more than 1 billion years later (Baird *et al.*, 1995). Baird *et al.* (1995) also postulate a link between the similar timing of Cambrian–Mid-Ordovician intracratonic basins around the world (Michigan, Williston, Hudson Bay, Illinois, Baltic, Moscow, and Paraná) with the effect of thermal blanketing beneath the Rodinia supercontinent causing eclogitization of deeper parts of basaltic lower crust. They note that the affected basaltic crust may have been produced much earlier and at different times for each basin.

Another potential example of eclogitization on basin formation is the location of the 2060 Ma Bushveld Complex at the site of the earlier 2.7 Ga Ventersdorp mantle-plume center identified on the basis of a radiating swarm (Olsson *et al.*, 2011). The density increase caused by eclogite conversion of the underplate associated with the Ventersdorp LIP (Section 6.2.3) is postulated to have triggered subsidence to form the Transvaal basin. Eventual delamination of this dense lithosphere led to upflow of hot mantle, which partially melted to produce the Bushveld LIP according to the Olsson *et al.* (2011) model.

As an alternative model for the formation of intracratonic basins, Kaminski and Jaupart (2000) suggest that the density anomaly is produced by localized thinning of otherwise thick lithosphere ($\geq$ 170 km) by plume penetration that results in the emplacement of compositionally denser mantle into the lithosphere. As the plume cools, density contrast becomes important and produces the necessary load, which drives permanent flexure and basin formation. Kaminski and Jaupart (2000) interpreted this to mean that the buoyancy of a plume was a controlling factor regarding the effectiveness of plume penetration of the lithosphere. One might further note that

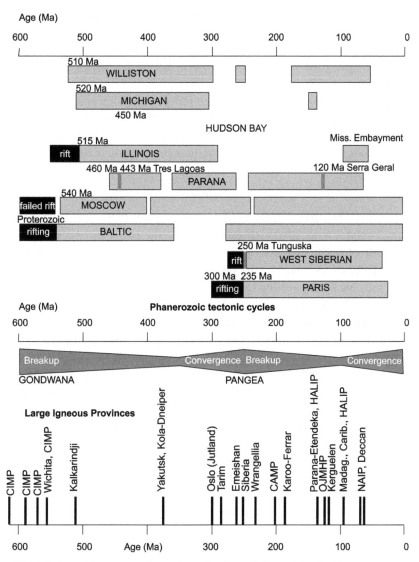

Figure 12.16 Selected intracratonic basins, showing timing of basin-fill megasequences in relation to the two great tectonic cycles of the Phanerozoic. Some intracratonic basin megasequences are preceded by rifting (black), and some have important magmatic episodes (medium gray bands). Modified from Figure 30.2 in Allen and Armitage (2012) with LIP barcode record added at the bottom. Copyright © 2012, John Wiley and Sons.

plume penetration into shallow lithosphere inevitably leads to melting to produce LIPs (Chapter 10), so the model of Kaminski and Jaupart (2000) should be expanded to consider production of LIPs and thereby associated production of an underplate component which can be subsequently converted to eclogite.

More work is required to consider the intracratonic-basin record through time and assess which basins can be spatially linked with prior LIP events and also to consider important LIP events that are not linked with subsequent intracratonic basins. Did eclogitization fail to occur in these circumstances, or has the eclogite layer delaminated in such cases?

A recent treatment by Armitage and Allen (2010) and Allen and Armitage (2012) inferred that a model of low-strain-rate extension followed by cooling of the underlying lithosphere explains the long-term subsidence history of intracratonic basins. Their model does not ascribe a role to eclogitization or to associated LIPs.

12.7 Summary

I have considered the effects of topographic changes potentially associated with LIPs. Early domal uplift is associated with many LIPs. In the examples reviewed in Section 12.2 the effect of uplift on patterns of sedimentation are similar and predictable (Fig. 12.8). Differences relate to the degree of preservation and the initial depositional and tectonic settings of the basins that were affected by the uplift. Two of the main effects of initial thermal uplift are localized shoaling (shallowing) and thinning of overlying strata over the uplifted area. These effects can result in zero depositional thickness of affected formations at the center of uplift. With continued uplift, the zero-thickness edge moves outward (offlap), and the core of the uplift is exposed to subaerial weathering and erosion, producing an unconformity, with potentially associated karst paleotopography and deeply incised canyons. Fluvial sandstones that directly underlie the flood basalts may record radial paleocurrent patterns.

These topographic effects typically start up to tens of millions of years before magmatism and may be preserved for hundreds of millions of years afterward providing that underplating has preserved the uplift against the later decay of the thermal anomaly associated with uplift. Rift flank uplift is associated with zones of triple-junction rifting.

There are also potential major topographic changes during the period of LIP magmatism associated with how the magma is redistributed during the overall magmatic event. The simple transfer of magma from source area to surface has a topographic implication as does the presence of any coeval rifting in the central region. In addition, lateral transport of magma away from the central region via laterally emplaced dykes, or transport through sills, or via sublithospheric transport of magma, all cause adjustments in the topography of the central region. In addition, the multiple pulses present in many LIPs can be linked to complicated further adjustments in the topography. Also, the onset of rifting will cause an immediate and local transition to subaqueous magmatism.

The final stage considered is the post-LIP stage. The amount of underplating that has occurred has an effect on whether the transient uplift associated with a plume is preserved. A poorly understood aspect is whether there is any link between LIPs and the formation of intracratonic basins.

13

LIPs and links with contractional structures

13.1 Introduction

An important component of the definition of Large Igneous Provinces (LIPs) is their origin via extensional processes, and earlier chapters (especially Chapters 1, 2, and 11) have outlined the links with an intraplate setting, continental rifting, and breakup. In this chapter I focus on a complementary theme, direct and indirect links with contractional structures in compressional settings. These include the following: (1) LIPs that are associated with domal uplift (due to plume arrival), which can be linked with gravity-induced sliding off the uplift to produce contractional structures on the flanks of the uplift and around its periphery; (2) LIPs that are spatially associated with orogenic belts, which may represent back-arc magmatism or plume arrival into a convergent setting; and (3) LIPs with an indirect relationship with compressional structures. Since LIPs are typically associated with breakup and extension and ocean opening, then through the plate-tectonic circuit, the timing of LIP emplacement should correlate with coeval contractional pulses expressed as deformation, arc initiation, and ophiolite obduction, elsewhere on Earth. Each of these topics is addressed in turn.

13.2 Contractional structures associated with domal uplift

One of the key contractional structures observed on terrestrial planets is "wrinkle ridges" (Fig. 13.1; Chapter 7). Although various models of wrinkle-ridge origin have been published (e.g. Watters, 1988, 2004; Mège and Ernst, 2001; Okubo and Schultz, 2004; Schultz et al., 2010), there is basic agreement that wrinkle-ridge formation involved folding and thrusting, more or less at the same time as flood-basalt out-pourings. Thrusts have been interpreted to be either emerging or blind, and can be associated with coeval strike-slip faults. No general agreement has been found on the deep-crustal structure of planetary wrinkle ridges, which have been interpreted as either thin-skinned features or thick-skinned features. As discussed below, wrinkle ridges and other contractional features are associated with plume-generated LIPs on Mars, Venus, and Earth.

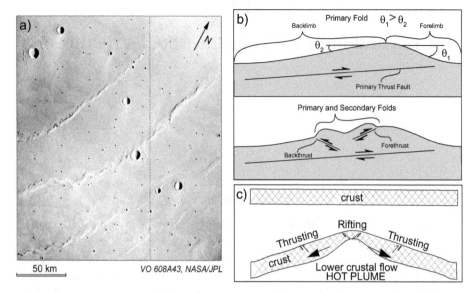

Figure 13.1 Wrinkle ridges. (a) Close-up of wrinkle ridges in the Coprates ridged plain (located south of Valles Marineris, Mars in Fig. 13.2). (b) General structural relationships in wrinkle ridges. The thrust is shallowest below the steeper-dipping forelimb and dips toward and below the shallower-dipping backlimb. These geometries allow the fold topography to reveal the dip direction and relative fault length of the primary thrust (upper part) and to the geometry of secondary forethrusts or backthrusts (lower part). After Fig. 1 in Okubo and Schultz (2004). (c) Model for crustal spreading associated with Beta Regio and other plume-generated volcanic centers on Venus showing postulated formation of central graben and peripheral thrusts (after McKenzie, 1994). Modified from Mège and Ernst (2001).

13.2.1 *Contractional structures on Mars*

A general analysis of contractional structures (including wrinkle ridges) associated with plume-related LIPs was presented by Mège and Ernst (2001), and was prompted by work on the Tharsis volcanic province of Mars (Mège, 2001). The 5000-km-wide Tharsis plume province is associated with both circumscribing wrinkle ridges (on the eastern flank of the uplift) and also by a set of peripheral thrust belts along the southern portion of the uplift (Figs. 7.6 and 13.2). These can be linked with different evolutionary stages of the Tharsis plume province corresponding to different timings and locations of individual plume centers within an overall super-plume event (see Chapter 7). Specifically, the wrinkle ridges along the eastern side of the uplift (Figs. 7.6 and 13.2) are circumferential to Tharsis plume center C, and the ridge belt along the southern margin is circumscribing the Syria Planum plume center (center B). In each of these cases, these contractional structures are oriented perpendicular to the geometry of contemporaneous radiating dyke swarms associated with these plume centers (Chapter 7), and the contractional features can be related to gravitational sliding off the domal uplift (Mège and Ernst, 2001; Fig. 13.1c).

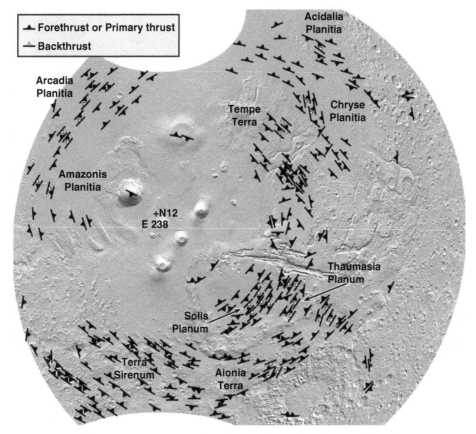

Figure 13.2 Wrinkle ridges circumscribing the Tharsis region of Mars. After Fig. 7 in Okubo and Schultz (2004).

13.2.2 Contractional structures on Venus

Venus is host to a variety of contractional features which can be related to mantle plumes (details in Mège and Ernst, 2001): (1) wrinkle ridges that are concentric about LIP-scale volcanic rises (see below), (2) radial, parallel, or anastomosing contractional ridges associated with crustal plateaus (dominated by tesserae terrane), and (3) both circumferential extensional and contractional features associated with the annular zones which characterize coronae. Here I focus on point (1) as these ridges have the clearest LIP connection. Details on points (2) and (3) are provided in Mège and Ernst (2001).

Wrinkle ridges are widespread on Venus and have a strong association with topography (e.g. Bilotti and Suppe, 1999; McGill *et al.*, 2010) and in many cases circumscribe young plume-generated uplifts (called "volcanic rises") (Fig. 13.3), but wrinkle ridges are also associated with the volcanic plains. Subsidence due to both topographic loading and sinking mantle currents may have played a role in

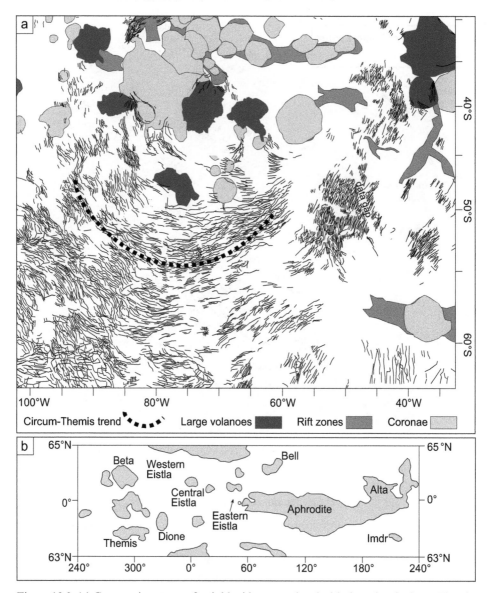

Figure 13.3 (a) Concentric pattern of wrinkle ridges associated with the volcanic rise at Themis Regio, Venus, modified after Bilotti and Suppe (1999). (b) Location map for volcanic rises modified after Stofan *et al.* (1992). From Figure 5 in Mège and Ernst (2001).

wrinkle-ridge formation on Venus. One dramatic model proposed that global production of wrinkle ridges was caused by density changes (expansion) in the uppermost crust as the atmosphere rapidly warmed to the current $> 460\ ^{\circ}\text{C}$ surface temperatures through the major global pulse of magmatic activity (Solomon *et al.*, 1999).

As discussed in Section 7.3, volcanic rises represent topographic and geoid highs with associated individual volcanoes up to 1000 km across, and are located at the convergence of triple-junction rifts. Volcanic rises on Venus therefore represent classic plume-related uplifts. The voluminous magmatism at volcanic rises is partly due to an apparent absence of movement of Venus's single lithospheric shell, such that the plume-head and plume-tail magmatism remain coincident. Volcanic rises display wrinkle ridges that are similar in morphology, size, and structure to Martian wrinkle ridges. Like the latter, they are periodically spaced and some are associated with conjugate strike-slip faults (Watters, 1993). At volcanic rises such as Themis Regio, western Eistla, and Bell (Fig. 13.3) wrinkle ridges are concentric about the volcanic centers (e.g. Bilotti and Suppe, 1999) and linked to crustal sliding off the uplift (Fig. 13.1c).

13.2.3 Columbia River LIP (17 Ma)

The Yakima folds (Fig. 13.4) are a series of regularly spaced (*c.* 20-km-apart), thrust-faulted anticlines in the western Columbia Plateau of the Columbia River LIP of the northwestern United States (Mège and Ernst, 2001). The folds are evidence of contractional deformation throughout the period of flood-basalt activity. The peak of Yakima ridge growth corresponds to the onset of the Columbia River LIP volcanic activity, and the ridge-growth rate was closely dependent on the rate of magma supply. Seismic, borehole, and magnetotelluric data from the Columbia Plateau have shown that the Yakima ridges are thin-skinned and the bulk of the Yakima folds are orientated east–west. Mège and Ernst (2001) suggested that the Yakima folds may be a set of wrinkle ridges which are oriented circumferentially to the interpreted plume center for the Columbia River LIP at 17 Ma (Fig. 13.4). Alternative explanations for the Yakima ridges are reviewed by Mège and Ernst (2001), and relate to the influence of the subducting Juan de Fuca plate to the west.

13.2.4 Mackenzie LIP (1270 Ma) and gravity sliding

A major LIP with inferred domal uplift is the 1270 Ma Mackenzie LIP of northern Canada (Fig. 13.5). The amount of uplift in the Mackenzie swarm central region can be calculated from the overall extension implied by emplacement of the radiating dyke swarm. Such measurements lead to unreasonably high estimates of uplift (Baragar *et al.*, 1996; see also discussion in Mège and Ernst (2001) and Şengör (2001)). Calculations were performed along two arcs, each spanning 30° at the 400-km and 900-km distances from the plume center (Fig. 13.5). Aggregate dyke thickness (number of dykes times average thickness of 30 m) were calculated for the arcs and the implied uplift was calculated using a "trap-door" model yielding absurd uplift estimates ranging from 17.5 to 124.0 km. An alternative calculation involved

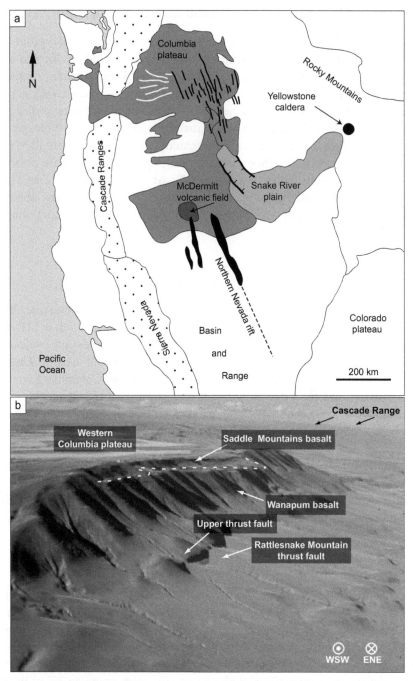

Figure 13.4 (a) Yakima fold belt, possible wrinkle ridge associated with the Columbia River LIP. The McDermitt volcanic field marks the location of the Columbia LIP/plume arrival. Dark gray, McDermitt volcanic field; medium gray, Columbia River Basalt Group (to the north) and other tholeiitic basalts of similar age (to the south); light gray, Snake River Plain volcanism; black, Northern Nevada rift; dotted pattern, coastal ranges; dark lines, dyke trends; white lines, Yakima ridge trends; lines with hatchures, western Snake River Plain graben. From Figure 11 in Mège and Ernst (2001). (b) Rattlesnake Mountain, Columbia Plateau, Columbia River LIP. Photograph of S.P. Reidel; from Figure 10 in Mège and Ernst (2001).

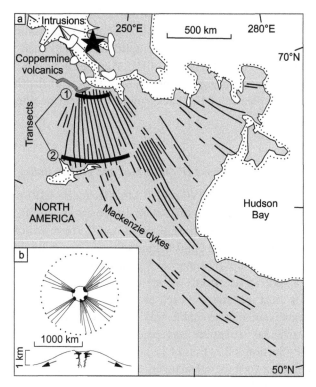

Figure 13.5 Gravity-induced spreading off topographic uplifts. (a) The 1270 Ma Mackenzie dyke swarm of the Canadian shield. The focal point of the swarm (star) is interpreted to locate the plume center and is surrounded by a ring of magma chambers identified by gravity anomalies (outlined areas). An additional gravity anomaly marks the Darnley Bay intrusion, which may be younger (westernmost anomaly). The dark gray pattern locates coeval Coppermine volcanic rocks. Traverses at 400 and 900 km from plume center (bold arcs) are used to calculate crustal dilation due to dyke injection. The magnitude of crustal dilation as indicated by emplacement of the Mackenzie dyke swarm (determined at 400- and 900-km distances) requires unrealistic domal uplift and, therefore, implies crustal spreading off the uplift. See discussion in text. (b) Cross section of inferred sliding off domal uplift; for scale, see Fig. 11.9. Redrafted from Baragar *et al.* (1996). By permission of Oxford University Press.

modeling the uplift with a spherical cap instead of a trap-door model (Şengör, 2001) and yielded uplift estimates ranging from 3.7 to 54 km (Mège and Ernst, 2001; Şengör, 2001).

To avoid the unreasonable uplift required by the dyke emplacement data, it was concluded that during uplift the crust must have slid outward off the domal uplift, which also led to the formation of a central apical graben consistent with the observed pattern of gravity data (Figs. 13.5 and 13.1c; Baragar *et al.*, 1996). The maximum amount of outward spreading was estimated at 6.8 km, which would also lead to downdrop of about 20 m for a 200-km-radius central graben (Mège and Ernst, 2001). This outward spreading should be accommodated on the flanks or along the periphery of the domal uplift. Although such wrinkle ridges have not yet been observed in the region of the Mackenzie swarm, they would be difficult to recognize given the significant (many kilometers) depth of erosion in this area, and the complexity of the background basement geology of the Slave craton.

13.2.5 Afro-Arabian LIP (45–0 Ma) and gravity sliding

This example and the previous one (the 1270 Ma Mackenzie LIP) provide indirect evidence for gravitational spreading off a domal topographic uplift. Şengör (2001) investigated the structural consequences of domal uplift above various African and Australian hotspots by estimating uplift magnitudes and assuming different doming mechanisms (cone-shaped uplift, spherical cap-shaped uplifts, and spherical cap-shaped uplift on a spherical earth) to calculate the extension due to doming alone. The latter model is most realistic, but each of the types of calculations indicate that the amount of extension due to uplift was insufficient to explain the observed rift extension. Particular attention was paid to the uplift associated with the Afro-Arabian LIP (Section 3.2.3, Fig. 12.1).

The inconsistency between the calculated extension due to uplift and that observed can be explained by the effect of gravitational stresses generated by the topographic potential of the dome. The influence of gravity on a density-layered outer shell is complex, but Şengör (2001) considers the simple end-member case of gravity-induced sliding off the topographic high to produce the observed extension at the Afro-Arabian LIP.

13.2.6 Discussion

Given the association of many LIPs with plume-generated central uplift (Chapter 11), then crustal spreading off the domal uplift with associated formation of an apical graben, wrinkle ridges on the uplift flank and a contractional peripheral belt could be a common occurrence above plumes associated with giant radiating dyke systems. Şengör (2001) notes the absence of peripheral zones of thrusting around the Afro-Arabian LIP and other young hotspot centers in Africa and Australia and prefers accommodation of crustal spreading off plumes along normal plate boundaries. While this may be a correct interpretation, in this case, the identification of small-scale thrust faulting may be cryptic when the topographic expression has been lost to erosion. The presence of crustal spreading on non-plate-tectonic planets (Venus, Mars, Mercury) confirms that wrinkle ridges should also be possible in terrestrial examples. As support I reiterate the interpretation above in Section 13.2.4 that potential wrinkle ridges are associated with the young Columbia River LIP.

13.3 LIPs in a back-arc setting

As further discussed in Section 15.2.5 a number of LIPs are emplaced in a back-arc setting. Some are interpreted to be generated by back-arc extension (e.g. Rivers and Corrigan, 2000), but others can reflect the arrival of mantle plumes into a compressional setting (e.g. Fig. 13.6). Regarding the latter, a proportion of plumes (generating LIPs) randomly sourced and rising from the lower mantle would be expected to ascend into upper-mantle regions with convergent plate tectonics (cf. Section 14.3 in Arndt *et al.*, 2008). Several examples are discussed.

13.3.1 Back-arc setting to the evolving southeastern margin of Laurentia

A continental magmatic arc is inferred to have existed on the southeastern margin of Laurentia from Labrador, Canada to Texas, USA, from *c*. 1500–1230 Ma, with part of the arc subsequently being incorporated into the 1190–990 Ma collisional Grenville orogen (e.g. Rivers and Corrigan, 2000). Several intraplate magmatic events are present along this long-lived accretionary arc, most notably, the *c*. 1460 Ma Michael Gabbro/ Shabogamo Gabbro LIP of western Labrador and eastern Quebec, which Gower *et al.* (1990) interpreted to be a single magmatic province with a mapped strike length in excess of 750 km and a width of over 100 km. This and other Mesoproterozoic magmatic suites paralleling the Grenville Province are considered by Rivers and Corrigan (2000) to have been emplaced during pulses of extension in a back-arc setting with respect to a long-lived zone of convergence along that margin of Laurentia.

13.3.2 Keweenawan LIP and a back-arc setting

The 1115–1085 Ma Keweenawan LIP (Section 3.5.2) of the Midcontinent rift system is also emplaced into a back-arc setting with respect to the Grenville orogen located on the Laurentian margin and marking collision of Laurentia with other crustal blocks, postulated to be Amazonia and Baltica. In earlier models Keweenawan magmatism was thought to have been generated by compression related to the Grenville compression (see discussion in Van Schmus and Hinze, 1985). A recent microplate model for the Midcontinent rift system suggests that it is part of an evolving regional plate boundary system (Merino *et al.*, 2013).

However, it is generally viewed that active rifting associated with a mantle plume was responsible for the Keweenawan magmatism, a conclusion based on the total volume of magma, and short pulses of emplacement, and also based on geochemistry (e.g. Nicholson and Shirey, 1990; Miller and Nicholson, 2013). So the Keweenawan

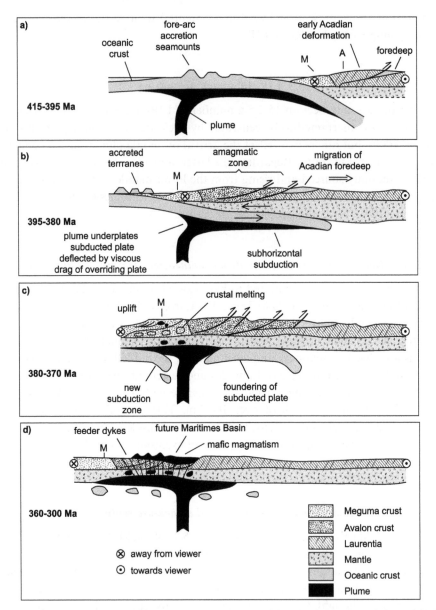

Figure 13.6 (a)–(d) Plume-modified plate-tectonic model for the formation of the *c.* 360 Ma Maritimes flood basalt event in the Canadian Appalachians. See discussion in the text. From Figure 2 in Murphy *et al.* (1999). M and A = Meguma and Avalon terranes, respectively. The "away from viewer" and "towards viewer" symbols indicate that the upper crust was undergoing dextral shear at the time of emplacement. Digital version of original figure kindly provided by B. Murphy.

LIP is an example of a mantle plume arising randomly into a region that is experiencing regional compression in a back-arc setting with respect to the Grenville orogen. As the extension related to dynamic uplift of the Keweenawan plume waned, the Grenville convergence reasserted itself and led to an inversion of structures in the Keweenawan rift (Manson and Halls, 1997).

13.3.3 Columbia River LIP and a back-arc setting (17 Ma)

The 17 Ma Columbia River LIP is emplaced in a back-arc setting with respect to subduction of the Juan de Fuca plate under the western United States, and some early models suggested a back-arc origin for the magmatism (e.g. Carlson and Hart, 1988). However, as with the Keweenawan example, various lines of evidence (Section 3.2.2) indicate plume involvement. Furthermore, more recent seismic tomography supports a model of interaction between the ascending plume and the descending slab of the Juan de Fuca plate with eventual breaching of the slab and continued ascent of the plume, ultimately causing the outpouring of Columbia River LIP magmatism (e.g. Obrebski *et al.*, 2010).

13.3.4 Maritimes event (c. 380–360 Ma), eastern Canada

Tholeiitic magmatism (up to 1.5 km thick) underlies the Maritimes basin of eastern Canada (Dessureau *et al.*, 2000), and was emplaced at the same time as convergent tectonics was occurring in the Canadian Appalachians. The model of Murphy *et al.* (1999) suggests the arrival of a mantle plume that is initially prevented from reaching shallow levels by an intervening subducting plate (Fig. 13.6). When the plume finally breaches the subducting plate, the result is the emplacement of the Maritimes event (*c.* 380–370 Ma with magmatism continuing to *c.* 300 Ma). The time sequence of activity is shown in Fig. 13.6 and presented in detail in Murphy *et al.* (1999). As the plume eventually dies, cooling of the thermal anomaly associated with the plume leads to subsidence and formation of the Maritimes basin.

13.4 Convergent zones (orogenic/deformation belts) linked to distal LIPs through the plate-tectonic circuit

Normally, the formation of magmatic arcs and LIPs are viewed as entirely separate tectonic phenomena, with arcs resulting from convergent plate tectonics and LIPs being a product of intraplate processes (e.g. the arrival of a mantle plume). However, as noted in Chapter 11, LIPs can initiate plate breakup and, because of the spherical nature of the Earth, extensional tectonic activity in one region can be linked with convergent tectonic processes along plate boundaries in other regions. Evidence for the onset of convergence includes the initiation of arcs, formation of ophiolites (obduction of oceanic crust), and initiation or reactivation of orogenic activity. Examples of such potential linkages between LIPs and distal collisional processes are discussed below.

13.4.1 Distal deformation induced by LIPs

The initiation of a new ocean basin associated with LIP arrival can cause reactivation of faults and folding on distal plate boundaries (e.g. Fig. 13.7). As discussed in Section 16.6, such deformation zones are potentially associated with various classes of ore deposits, including gold.

NAIP and the Eureken orogeny in Arctic Canada and northern Greenland

An example of such a feature is given by Harrison *et al.* (1999) who noted that the outpouring of flood basalts in East and West Greenland and belonging to the North Atlantic Igneous Province (NAIP) coincided with the Eureken orogeny in the Arctic Islands. This suggests that the opening of the North Atlantic and westward driving of Greenland by the Iceland mantle plume, which is responsible for the NAIP, was a significant, if not the primary, driving force for the Eureken orogeny. Furthermore, Harrison *et al.* (1999) more speculatively suggest that a plume-head push acting on the Labrador and Baffin margin of North America, beginning at about 61 Ma, may also partly account for the simultaneous far-field development of the Beaufort foldbelt and other Laramide thrust belts of the North American Cordillera.

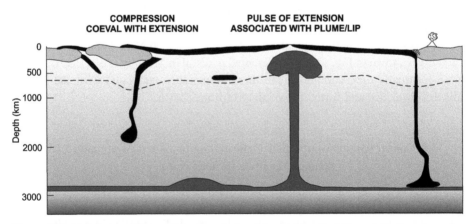

Figure 13.7 Link between locus of plume arrival and formation of a LIP, and distal compression via the plate-tectonic circuit. In other words, a plume from the deep mantle arriving beneath the plate can cause a LIP and a pulse of plate extension which can lead to obduction of oceanic crust at a distal converging plate boundary and associated compression or transpression. Modified from Figure 3c in Vaughan and Scarrow (2003). With permission from Elsevier.

Regional intraplate compression

Plate-boundary stresses can be transmitted far into cratonic interiors (e.g. Zoback, 1992; Pilger, 2003). Determining paleostress distributions in cratonic areas can help us to understand the plate-tectonic setting of the time. It was mentioned in Section 5.2.2 that radiating swarms typically swing into a regional stress direction outside the influence of domal uplift associated with a plume, and thus can reveal the orientation of past intracratonic stress configurations. For instance, the 1270 Ma Mackenzie dykes swing into a stress direction that aligns with the convergence direction associated with the Grenville orogen on the southeastern margin of Laurentia (Section 5.2.2).

13.4.2 Initiation of arcs

As pointed out by Pearce (Pearce, 2007; J. Pearce, personal communication, 2007), the initiation of arcs can be caused by two mechanisms: (1) continent-collision, which causes the ending of one arc setting and requires the initiation of others elsewhere on the globe; and (2) the onset of major rifting (typically initiated by a LIP), which causes the start of subduction somewhere else on the globe.

CAMP (200 Ma) as an example

An example is the 200 Ma Central Atlantic Magmatic Province (CAMP), which is associated with the onset of opening of the Central Atlantic Ocean basin at *c*. 190–175 Ma (Section 11.5). Andean volcanism of South America begins at *c*. 170 Ma. So one could conclude that emplacement of the CAMP at 200 Ma and subsequent opening of the Atlantic Ocean area are correlated with the birth of this Andean arc magmatism. Also, as noted by Pearce (2007), SSZ (supra subduction zone) ophiolites representing the birth of arcs off North America and Eurasia give ages of 170–160 Ma. This example also illustrates the role of ophiolites as a proxy for identifying arc initiation, a topic more extensively developed in the next section.

13.4.3 LIPs and ophiolite obduction

Here I address the use of ophiolites to monitor pulses of plate-boundary compression potentially correlated with LIPs. On the basis of the type of oceanic crust involved, a range in ophiolite types is offered by Dilek and Furnes (2011). They divide ophiolites into subduction-related and subduction-unrelated types. Subduction-unrelated ophiolites include continental-margin, mid-ocean-ridge, and plume (oceanic-plateau) types. Subduction-related types include suprasubduction-zone, volcanic-arc, and accretionary types. In the context of this book, the most interesting example of the first type would be ophiolites that are generated from oceanic plateaus. As an example, the Piñón ophiolite in Ecuador, South America has a timing that matches the formation of the Ontong Java LIP and can be interpreted to represent a portion of the Ontong Java LIP that was carried away by rifting and partially accreted when it reached the South American subduction zone (event 22 in Ernst and Buchan (2001b); see also Reynaud *et al.* (1999)).

The subduction-setting types of ophiolites are obviously not directly related to LIPs, but the timing of their obduction (and becoming an ophiolite) could be relevant (as discussed above). Obducted fragments of ocean lithosphere can be emplaced (obducted) when the plate-setting switches from extensional or strike-slip to compressional. Therefore, obduction of ophiolites is sensitive to the global plate-tectonic circuit and can therefore be a sensitive monitor of the LIP effect that is being investigated here. In considering the timing between LIPs and compression forces elsewhere in the global plate-tectonic circuit, note that there should be a time lag, because of the typically variable time gap (5–25 Ma) between the onset of LIP magmatism and the onset of rifting (Section 11.5.2).

Ophiolite summary

Two compilations of ophiolite distribution through time are considered (Figs. 13.8 and 13.9). Vaughan and Scarrow (2003) identify seven ophiolite pulses back to 500 Ma and, as illustrated in Fig. 13.8, these can be strongly correlated in time with LIPs, allowing for the variable lag time of 5–20 Ma between LIP emplacement, onset of ocean opening, and ophiolite emplacement.

The more recent comprehensive survey (Fig. 13.9; Dilek and Ernst, 2008; Dilek and Furnes, 2011) shows a more continuous distribution of ophiolites. In this compilation, the times of enhanced ophiolite genesis and emplacement in Earth history appear to coincide with the timing of major collisional events during the assembly of supercontinents (basin collapse and closure), dismantling of these super-continents via continental rifting, and widespread development of LIPs (Dilek and Ernst, 2008; Dilek and Furnes, 2011).

In comparison, between the two ophiolite compilations (Figs. 13.8 and 13.9), only three of the pulses in Fig. 13.8 are distinctly present in Fig. 13.9: 70 Ma, 90–100 Ma, and 170 Ma. The other pulses (275–290 Ma, 375–400 Ma, 440–450 Ma, 475–490 Ma) of Fig. 13.8, while present in Fig. 13.9, do not stand out. More rigorous statistical comparison of the LIP and ophiolite timing record is required.

Ophiolite pulses (180–140 Ma): initial opening of Gondwana

The most discrete ophiolite pulse during 180–140 Ma coincides with the formation and emplacement of the Tethyan, Caribbean, and some of the circum-Pacific (western Pacific and North American Cordillera) ophiolites. In the Tethyan system this timing marks the collapse of restricted basins between various Gondwana-derived subcontinents prior to the terminal closure of oceans and major continental collisions. Key LIPs during this time that can be correlated with these contractional events are those associated with the opening of the Central Atlantic ocean basin (200 Ma CAMP) and of the Indian ocean basin (182–177 Ma Karoo LIP).

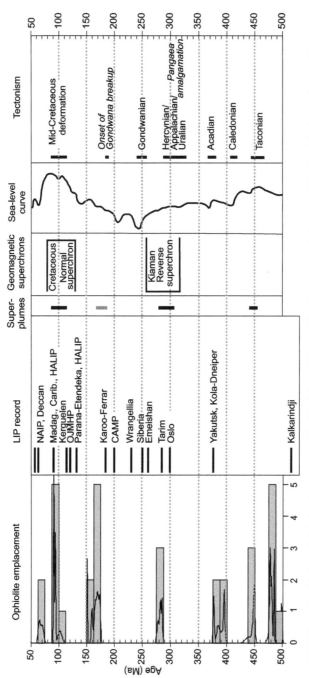

Figure 13.8 Plot of ophiolite obduction frequency. Modified from Figure 2 in Vaughan and Scarrow (2003) with the addition of LIP events in a middle left panel. Digital version of original figure kindly provided by A.P.M. Vaughan. With permission from Elsevier.

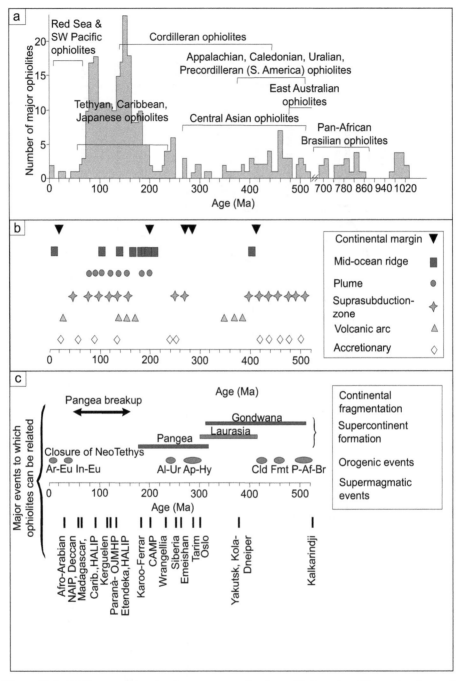

Figure 13.9 (a) Histogram showing the occurrence of major ophiolite pulses, (b) associated tectonic environments for the ophiolites, and (c) the life spans of supercontinents and major collisional–orogenic events that led to their assembly, and formation of LIPs (including giant dyke swarms) through time. Abbreviations for orogenic events (from youngest to oldest): Ar–Eu = Arabia–Eurasia collision; In–Eu = India–Eurasia collision; Al–Ur = Altaid–Uralian orogenies of central Asia; Ap–Hy = Appalachian–Hercynian orogenies; Cld = Caledonian orogeny; Fmt = Famatinian orogeny; P–Af–Br = Pan-African–Brasiliano orogenies. Phanerozoic portion redrafted from Figure 2 in Dilek and Furnes (2011) and Neoproterozoic portion redrafted from Figure 1 in Dilek and Ernst (2008) with updated distribution of LIP events. Dilek and Furnes, 2011: Courtesy of the Geological Society of America; Dilek and Ernst (2008): With permission from Elsevier.

Ophiolite pulses (100–90 Ma)

The second important ophiolite pulse during the Late Cretaceous follows the Mid-Cretaceous "superplume event" (Larson, 1991a), which consists of 120–115 Ma oceanic plateaus (e.g. Ontong Java, Kerguelen, etc.), and is coeval with 90 Ma oceanic and continental LIPs (i.e. possible second pulse of the Ontong Java LIP, Madagascar and Caribbean–Colombian LIPs, and second pulse of the High Arctic LIP (HALIP)) (e.g. Ernst and Buchan, 2002). This ophiolite pulse coincides with the breakup of Pangea, and the closure of Neo-Tethyan seaways. Some of the Alpine–Apennine ophiolites that formed during the breakup of Pangea represent rift-related mafic–ultramafic assemblages (i.e. exhumed subcontinental mantle fragments) and/or remnants of embryonic ocean floor. The enhanced LIP formation and ophiolite generation in the Late Cretaceous seem to be linked in space and time with increased seafloor spreading rates, extensive oceanic-plateau formation, and widespread compression at convergent margins (Larson, 1991a; Vaughan, 1995; Dalziel *et al.*, 2000; Dilek, 2003; Vaughan and Scarrow, 2003).

Paleozoic and Proterozoic ophiolite pulses

Figure 13.9 shows ophiolite pulses extending into the Neoproterozoic. The peak at *c.* 700 Ma could correspond to the 725 Ma Franklin LIP (northern Laurentia) that may have been responsible for the opening of an ocean between Siberia and northern Laurentia (e.g. Ernst and Bleeker, 2010). The 780 Ma pulse could be associated with 780 Ma Gunbarrel LIP of western Laurentia, which could be associated with the opening of the proto-Pacific Ocean (Park *et al.*, 1995; Harlan *et al.*, 2003; Li *et al.*, 2008a; Ernst *et al.*, 2008) with South China or Australia as the conjugate margin. Similarly, the 820 Ma ophiolite pulse correlates with the Gairdner–Willouran LIP of southern Australia (Wingate *et al.*, 1998) and the related magmatism in South China (Section 8.4.2; e.g. Li *et al.*, 2008a) which also would reflect proto-Pacific Ocean opening. The 1000 Ma ophiolite pulse can correspond to the poorly known Sette Daban event of the southwest Siberian craton (Ernst *et al.*, 2008).

13.4.4 Correlation with orogenies

A complementary look at speculative links between extension (marked by LIPs) and pulses of convergence is afforded by the comparison of LIPs and orogenies (Fig. 13.9c), which tracks a particular timing of compressional pulses; the closure of ocean basins. This differs from Section 13.4.2 where I considered the speculative link between LIPs and the initiation of arc magmatism.

The Grenville orogen of eastern Laurentia is a key event in the amalgamation of the Rodinia supercontinent, and in terms of timing there is an approximate match with the 1110 Ma Umkondo LIP of the Kalahari craton and elsewhere. The collision of East and West Gondwana and the construction of Pannotia (*c.* 700 and 600 Ma) can potentially be correlated with the 725 Ma Franklin event of northern Canada and southern Siberia (e.g. Ernst and Bleeker, 2010). The pan-African–Brasiliano orogenies (520–500 Ma) are time-correlated with the *c*. 510 Ma Kalkarindji LIP of Australia. The Caledonian–Famatinian orogenies (460–440 Ma) may be associated with a *c*. 450 Ma magmatic event, Suordakh, newly identified in eastern Siberia (Khudoley *et al.*, 2013). The Appalachian–Hercynian orogenies (*c*. 300–270 Ma) have a time match with the *c*. 290 Ma Tarim LIP associated with the opening of the Tethys. The Altaid–Uralian orogenies in central Asia (*c*. 240 Ma) correlate with opening of the West Siberian basin associated with the Siberian Trap event. Finally, the sequential collisions of India (In–Eu) and Arabia (Ar–Eu) with Eurasia during the Neogene are part of the current assembly of a new supercontinent, and time-correlated LIPs include the Deccan (65 Ma), NAIP (62 Ma), and Afar (30 Ma). However, to rigorously assess whether these correlations between LIPs and orogenies have any tectonic significance requires more rigorous analysis and robust reconstructions for the relevant time periods.

13.4.5 Slab avalanche events

An additional link between LIPs and orogenic belts can also be considered. Major orogenic events supply cold subducted slabs into the lower mantle, and this has been hypothesized to be linked to subsequent large upwellings and thermal anomalies that spawn mantle plumes, which then produce extensive continental and oceanic LIPs (e.g. Maruyama, 1994; Moores *et al.*, 2000; Nikishin, 2002; Condie, 2004; Dilek and Ernst, 2008). Under this hypothetical scenario, bursts of orogenic activity would precede LIPs.

This view of orogenic pulses preceding LIPs, is different from the above models, which have coeval timing (with a modest 5–25 Ma lag) between LIPs and collisional events. If both types of correlations (an orogenic pulse can both precede or be coeval with LIP emplacement) are applicable, then fully characterizing links between LIPs and orogenesis requires a particularly careful analysis.

13.4.6 Link with superplume events

The story can be carried further, with a consideration of so-called superplume events (Section 15.2.1) and their potential link with orogenic/collisional events. Vaughan and Livermore (2005) note several links between the timing of episodes of Mesozoic terrane accretion along the Pacific margin of Gondwana with LIPs and superplume events. For example, they note that Late Triassic–Early Jurassic deformation appears to be concentrated in the period 202–197 Ma and was therefore coeval with emplacement of the CAMP, with the onset of Pangea breakup, a period of extended normal magnetic polarity, and with a major mass extinction event, all of which are possible expressions of a "superplume event." As another example they cite a Mid-Cretaceous deformation which occurred in two brief periods, the first from approximately 116 Ma to 110 Ma in the west paleo-Pacific Ocean and the second from roughly 105 Ma to 99 Ma in the east paleo-Pacific Ocean. Related deformation was also possibly present in northeast Siberia. These two pulses of deformation were also coeval with eruption of major oceanic plateaus, core-complex formation, and rifting of New Zealand from Gondwana, the Cretaceous normal polarity epoch, and a major radiation of flowering plants and several animal groups, and all linked with the Mid-Cretaceous superplume event (an event earlier identified by Larson (1991a, b, 1997). In the model of Vaughan and Livermore (2005) a simple unifying mechanism is presented suggesting that a large continental or oceanic plate, when impacted by a superplume (with associated emplacement of LIPs), will tend to be associated with gravitational spreading away from a broad, thermally generated topographic high and with a resulting short-lived pulse of plate-margin deformation and terrane accretion.

13.5 Summary

Based on the planetary record it is suggested that domal uplift above a plume must be associated with modest sliding of crustal panels off the uplift. This is concluded for both planetary (Mars and Venus) examples and also terrestrial examples (e.g. the Afar and Mackenzie LIPs). In the case of the planetary examples deformation structures (e.g. wrinkle ridges and small thrust belts) occur on the flank and around the periphery of the uplift. This suggests a similar story for Earth and possible wrinkle ridges are identified for the Columbia River LIP. However, Şengör (2001) considers the sliding off the Afar dome of the Afro-Arabian LIP to be accommodated along plate-boundaries, rather than along the flanks or periphery of the uplift.

The time correlation of LIPs and contractional events (orogenic activity, initiation of arcs, ophiolite emplacement, orogenic activity, etc.) is extremely complicated, highly speculative, and requires a more systematic analysis to confirm which apparent temporal matches actually reflect a real geodynamic link. A more complete study will require analysis of the global plate-stress configuration, which should be possible for the Mesozoic and Cenozoic. However, detailed consideration of links between LIPs and contractional events in the Proterozoic and the older part of the Paleozoic record will need to await development of a framework of robust continental reconstructions for those time periods. Some potential economic implications of the link between orogenic activity and LIPs are considered in Section 16.6.

14

LIPs and environmental changes and catastrophes

14.1 Introduction

The distribution of life through the Phanerozoic is highly variable, primarily as a result of environmental changes. The most dramatic and sudden changes are associated with extinction events; these define many of the boundaries in the biostratigraphic time scale. Less extreme environmental changes are also recorded by excursions in the isotopic composition of seawater and in the distribution of anoxia events, and by sea-level changes. In this chapter I provide a snapshot of this fast changing field and introduce the environmental impact of LIPs on climate change, and as catalysts for faunal and floral collapse and extinction events.

14.2 Link between LIPs and global extinction events

The temporal link between LIPs and extinction events appears robust (Figs. 14.1–14.3; Table 14.1). Many major, and some minor, LIP events occur within several million years or less of global extinctions (e.g. Stothers, 1993; Wignall, 2001; Courtillot and Renne, 2003). For example, the Deccan (*c.* 65 Ma), Central Atlantic Magmatic Province (CAMP; 201 Ma), and Siberian Trap (252 Ma) LIPs match in age to the Cretaceous–Tertiary, Triassic–Jurassic, and Permian–Triassic boundary extinctions, respectively. Only the major Ordovician–Silurian boundary extinction has not been correlated with a LIP event, although recent research has identified a *c.* 440 Ma dolerite Suordakh event in eastern Siberia (Khudoley *et al.*, 2013), but the scale of this event remains unknown pending dating of additional mafic units in the region.

Although a number of LIP–extinction correlations have been identified, the eruption volume of the LIP appears to be unrelated to the intensity of the extinction event (Fig. 14.4). Most dramatically, the largest LIP event, the reconstructed *c.* 120 Ma Ontong Java–Manihiki–Hikurangi oceanic plateau at *c.* 80 Mkm^3, is not associated with an extinction event, but is associated with an anoxia event, the Aptian-aged Selli event (Section 14.6.9). The absence of a strong relationship between LIP size and magnitude of extinction event underlines the complexity of this relationship. Other parameters besides volume presumably have a very important role in determining the

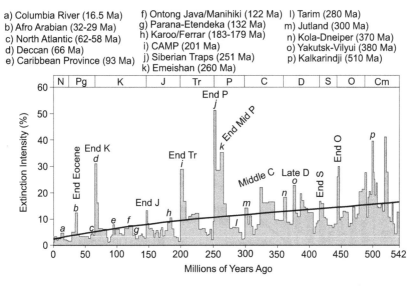

Figure 14.1. Correlation of LIP events with extinction events. This figure shows the genus extinction intensity, i.e. the fraction of genera that are present in each interval of time but do not exist in the following interval. The data are from Rohde and Muller (2005), and are based on Sepkoski (2002). The curve is based on marine genera with the LIP record superimposed. Modified from Figure A2 in supplementary files of Rohde and Muller (2005) to include links with the LIP record.

environmental impact of a LIP, and these factors are considered in subsequent sections of this chapter. But first I profile the environmental consequences of selected LIPs.

14.3 LIPs and their environmental consequences

This section considers in detail the links between specific LIPs and environmental catastrophes; these are presented in age order, from youngest to oldest, and illustrate a range of specific environmental effects, which are addressed more fully in subsequent sections (e.g. Section 14.4).

14.3.1 NAIP (62–55 Ma) and Paleocene–Eocene Thermal Maximum (PETM)

The marine sedimentary record provides evidence of a combined sudden warming (superimposed on a much broader thermal high), negative $\delta^{13}C$ excursion, and carbonate dissolution event that occurred at about 55 Ma (Fig. 14.5), termed the Paleocene–Eocene Thermal Maximum (PETM) (e.g. Kennett and Stott, 1991; Zachos *et al.*, 2003; Tripati and Elderfield, 2004). Earth's surface temperature rapidly increased by 5–10 °C in a geologically short time (2000–15 000 years), and generated a warm climate that prevailed for about 200 000 years (e.g. Kennett and Stott, 1991).

Table 14.1 *Selected LIPs and potentially related biostratigraphic stage boundaries.*
Boundary names and ages after Ogg et al. *(2008).*

LIP	Age	Extinction
Columbia River	16.5–14.5 Ma	Burdigalian–Langhian (15.97 Ma)
Afro-Arabian	32–29 Ma	Rupelian–Chattian (28.4 Ma)
NAIP	62–58 and 55 Ma	Danian–Selandian (61.1 Ma) and Thanetian–Ypresian (55.8 Ma)
Deccan	67 Ma	Maastrichtian–Danian (Cretaceous–Paleogene) (65.5 Ma)
Caribbean–Madagascar and others	93 Ma	Cenomanian–Turonian (93.6 Ma) and/or Turonian-Coniacian (*c.* 88.6 Ma)
OJMHP (reconstructed Ontong Java–Manihiki Plateau–Hikurangi plateaus)	122 Ma	–
Paraná–Etendeka, HALIP	135 Ma	Valanginian–Hauterivian (*c.* 133.9 Ma) and/or Hauterivian–Barremian (130.0 Ma)
Karoo–Ferrar	184 Ma and *c.* 179 Ma	Pliensbachian–Toarcian (183.0 Ma) and Toarcian–Aalenian Lower–Middle Jurassic (175.6 Ma)
CAMP	201 Ma	Rhaetian–Hettangian Triassic-Jurassic (199.6 Ma)
Wrangellia	232 Ma	Ladinian–Carnian (Middle–Upper Triassic) (*c.* 228.7 Ma) or Carnian–Norian (216.5 Ma)
Siberian Trap	251 Ma	Changhsingian–Induan (Permian–Triassic) (251.0 Ma)
Emeishan	260 Ma	Capitanian–Wuchiapingian (Guadalupian–Lopingian) (260.4 Ma) and perhaps Wuchiapingian–Changhsingian (253.8 Ma)
Tarim	*c.* 285 Ma	Sakmarian–Artinskian (284.4 Ma) or Artinskian–Kungurian (275.6 Ma)
Jutland	*c.* 300 Ma	Gzhelian–Asselian (Carboniferous–Permian) (299.0 Ma)
Kola–Dnieper and Yakutsk–Vilyui	*c.* 390–360 Ma	Eifelian–Givetian (391.8 Ma) or Givetian–Frasnian (385.3 Ma) or Frasnian–Famennian (374.5 Ma) or Famennian–Tournaisian (Devonian–Carboniferous) (359.2 Ma)
Kalkarindji	*c.* 510 Ma	Series 2–Series 3, Stage 4–5 (*c.* 510 Ma)

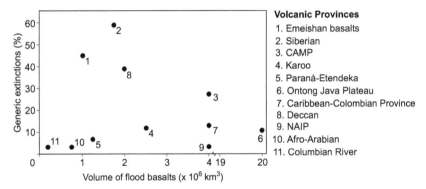

Figure 14.4 Comparison of generic extinction percentages, from Sepkoski (2002), with estimated original volumes of coeval igneous provinces. From Figure 1 in Wignall (2001). With permission from Elsevier.

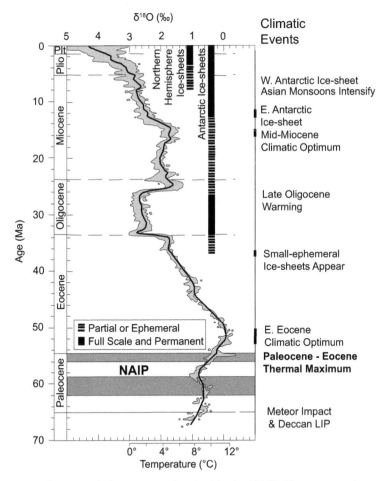

Figure 14.5 Environmental changes associated with the NAIP. Temperature increase of the PETM. Redrafted from Figure 2 in Zachos *et al.* (2001) with timing of two pulses of NAIP added. With permission from AAAS.

This extinction event also affected the majority of plankton and many tropical invertebrates, especially reef-dwellers, and also significantly affected land plants (e.g. Cowen, 1994; Kaiho and Lamolda, 1999).

Two causal mechanisms have been proposed for this major K–T extinction event: meteorite (bolide) impact and the Deccan LIP, and both mechanisms are discussed in turn below).

Alvarez *et al.* (1980) first reported anomalous enrichment of iridium in a centimeter-thick layer of white clay that outcrops in K–T boundary section in Italy and used this as evidence that the impact of a large asteroid caused the K–T extinction event. This Ir anomaly has been subsequently identified in many K–T boundary sections throughout the world (e.g. Alvarez *et al.*, 1990), and the bolide-impact hypothesis is supported by other lines of evidence, most importantly the discovery of the coeval Chicxulub impact crater in the Yucatan (e.g. Hildebrand *et al.*, 1991).

The Deccan LIP of India (Section 3.2.5) is also nearly simultaneous with the K–T extinction and has been linked with the extinction and associated environmental shifts (e.g. Courtillot and Renne, 2003; Keller *et al.*, 2008; Keller, 2012). The Deccan LIP could have produced SO_2 at a rate of several gigatons per annum, yielding an effect that is comparable in magnitude to that released by the Chicxulub impact (Self *et al.*, 2006; Chenet *et al.*, 2008). The most likely scenario is that the coincidence of a major impact (Chicxulub) and LIP (Deccan) represented a one–two "punch" of climate change (e.g. Saunders *et al.*, 2005).

14.3.3 *Madagascar, Caribbean, and other 93 Ma LIPs and an oceanic anoxic event (OAE2)*

The 93 Ma Cenomanian–Turonian boundary is associated with the major but short ($c. < 500$ ka) Bonarelli oceanic anoxic event 2 (OAE2) marked by the development of widespread black shales, drops in sea level and sea-surface temperatures, and associated marine extinctions (Fig. 14.6; e.g. Arthur *et al.*, 1987; Snow *et al.*, 2005; Bardet *et al.*, 2008; Jarvis *et al.*, 2011). During the Cenomanian–Turonian oceanic anoxic event ~ 0.12–0.36×10^3 Gt sulfur was removed by pyrite burial leading to a $\delta^{34}S_{SO4}$ increase of 5‰ to 7‰ across the Bonarelli black-shale horizon (Ohkouchi *et al.*, 1999). In addition, the onset of the OAE2, Bonarelli event, was marked by an abrupt shift in Os isotopic ratios in organic-rich sediments interpreted to indicate magmatic input (Turgeon and Creaser, 2008). Calculated initial $^{187}Os/^{188}Os$ ratios reflecting contemporaneous seawater Os isotopic values drop abruptly from 0.7–1.2 to unradiogenic values of ~ 0.14–0.15 at the onset of OAE2. This magmatic input at the start of the anoxic event is likely due to the numerous LIPs emplaced globally at this time (93 Ma), namely the Caribbean and Madagascar LIPs, a younger pulse of the High Arctic LIP (HALIP), and a potential second pulse of the Ontong Java LIP (Kerr, 1998, 2005). In detail, trace element enrichments (marking LIP influence) occur in the middle of the associated positive carbon isotopic excursion revealing complexities in the relationship between LIPs and anoxia events (Eldrett *et al.*, 2014).

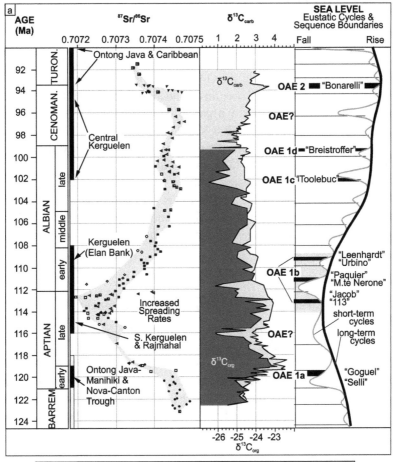

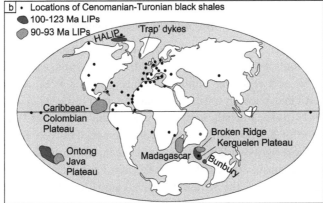

Figure 14.6 Anoxia events. (a) Distribution of anoxia events in comparison with other proxies for climate. (b) Cenomanian–Turonian plate-tectonic reconstruction showing the location of *c.* 90–93 Ma and 100–123 Ma LIPs and Cenomanian–Turonian boundary black shale deposits. Part (a) from Leckie *et al.* (2002), copyright 2002 by the American Geophysical Union, Wiley, and (b) from Figure 1 in Kerr (2005), reprinted with permission from Mineralogical Society of America. Digital version of original figures used for (a) and (b) kindly provided by R.M. Leckie and A.C. Kerr, respectively.

14.3.4 Ontong Java (120 Ma) LIP and the Aptian (Selli) anoxia event

The onset of the early Aptian (Selli) oceanic anoxic event 1a (OAE1a) at *c.* 120 Ma (Fig. 14.6) coincided with a major perturbation of the carbon cycle, which is reflected in the sedimentary carbon isotope record (Kuhnt *et al.*, 2011). The main positive carbon isotope shift at the onset of the OAE1a was previously regarded as continuous but actually occurred stepwise over an extended period of > 300 Ka. In addition, the carbon isotope record reveals a negative, low-amplitude $\delta^{13}C$ excursion that preceded the OAE1a and lasted > 100 Ka. This suggests that enhanced volcanic CO_2 emission and/or pulsed methane dissociation over a prolonged time span were instrumental in triggering the OAE1a (Kuhnt *et al.*, 2011).

The largest LIP, the reconstructed Ontong Java–Manihiki–Hikurangi oceanic plateau, occurs in the middle of the Aptian and is not associated with any extinction event. However, it matches in time with this major anoxia event, the Selli anoxia event, and this reconstructed mega-LIP may well have been the main cause of this anoxia event (e.g. Kerr, 2005).

14.3.5 Paraná–Etendeka LIP (135 Ma) and the Valanginian Weissert OAE

The precise timing of the Paraná–Etendeka magmatism (Section 3.2.6) has been a matter of debate, although the main pulse probably occurred at *c.* 132 Ma, with potential evidence for a precursor *c.* 140 Ma pulse. However, recent Ar–Ar dating by Thiede and Vasconcelos (2010) suggests that the older pulse does not exist and that all the magmatism was concentrated at 134.6 ± 0.6 Ma, the timing of which coincides with a minor anoxia event associated with the Hauterivian–Valanginian biostratigraphic boundary (e.g. Erba *et al.*, 2004; Fig. 14.7).

14.3.6 Karoo–Ferrar LIP (183 Ma) and the Pliensbachian–Toarcian boundary

The Pliensbachian–Toarcian boundary is marked on the basis of the second-order global extinction event. Two sharp $\delta^{13}C_{org}$ negative excursions, each with a magnitude of *c.* −2.5 ‰ and reaching minimum values of −28.5 ‰, are recorded in sediments of latest Pliensbachian to earliest Toarcian age (Littler *et al.*, 2010). Precise U–Pb dating of 14 dyke and sill samples by Svensen *et al.* (2012) confirms the very short duration of Karoo emplacement, yielding ages between 183.0 ± 0.5 and 182.3 ± 0.6 Ma. These ages overlap the timing of the initiation of the Toarcian OAE (182.16 ± 0.6 Ma) (Mazzini *et al.*, 2010), and the rapid emplacement of magma into the Karoo basin strengthens the hypothesis of a causal link between the resulting contact metamorphism (via hydrothermal vent complexes; Sections 5.3.6 and 14.6.6) and Toarcian carbon-cycle perturbation. Similar precise U–Pb dating by Burgess *et al.* (2011) of Ferrar sills also affirms a link between the Karoo and Ferrar magmatism, and the Pliensbachian–Toarcian boundary event.

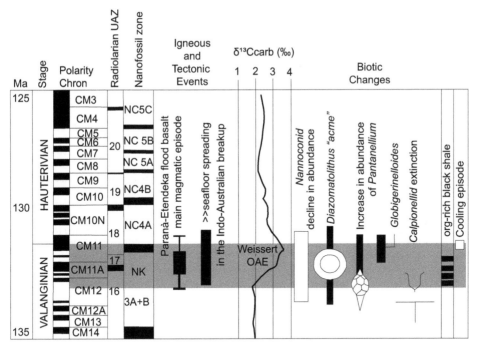

Figure 14.7 Environmental changes in association with the Paraná–Etendeka LIP. Synthesis of igneous-tectonic events, biotic changes, occurrence of organic C-rich black shales, and paleotemperature variations during the Valanginian Weissert OAE. UAZ = Unitary Association Zone. From Elba *et al.* (2004) where additional details can be found.

A second pulse for the Karoo portion of the Karoo–Ferrar LIP has also been proposed on the basis of a 179 ± 1 Ma age obtained by Ar–Ar dating of the Okavango dyke swarm (Jourdan *et al.*, 2009), and may correlate with the younger of the sharp $\delta^{13}C_{org}$ negative excursions described above, although this younger age needs to be confirmed by U–Pb dating.

14.3.7 CAMP LIP (201 Ma) and Rhaetian (Tr)–Hettangian (J) boundary

High-precision U–Pb dating of volcanic ash layers shows that the Triassic–Jurassic boundary and end-Triassic biological crisis correlate with the onset of terrestrial flood volcanism in the CAMP to within 150 Ka (Schoene *et al.*, 2010, and references therein). The precise correlation between the CAMP event and this Triassic–Jurassic boundary is noted in Figure 14.8 based on the high-precision dating of Blackburn *et al.* (2013). It is hypothesized that CAMP volcanism may have destabilized methane hydrates or accessed large carbon reservoirs by erupting through organic-rich sediments, resulting in a massive input of light carbon into the oceans and atmosphere and creating a < 200 Ka negative $\delta^{13}C$ spike (Pálfy *et al.*, 2001; Retallack, 2001; Beerling and Berner, 2002; Schoene *et al.*, 2010).

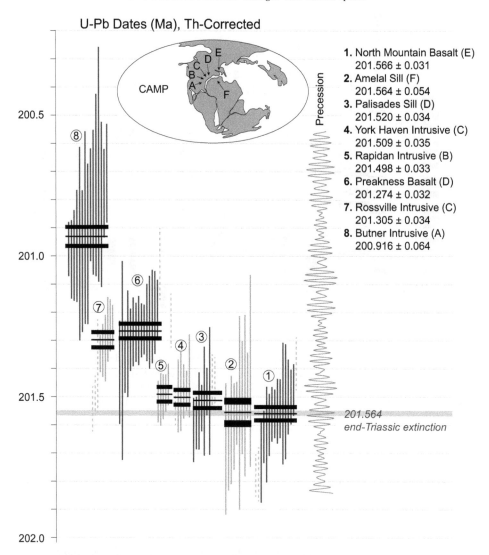

Figure 14.8 Precise U–Pb ages for the CAMP LIP and link with the end-Triassic extinction event and link with astronomical cycles ("precession"). The U–Pb ages are ^{238}U–^{206}Pb weighted mean dates with ±2σ analytical uncertainties corrected for initial ^{230}Th disequilibrium. Modified from Blackburn *et al.* (2013). Digital versions of original figures kindly provided by T.J. Blackburn. Reprinted with permission from AAAS.

14.3.8 Siberian Trap LIP (251 Ma) and Permian–Triassic extinction event

The end-Permian crisis is the biggest known marine and terrestrial extinction event, which eliminated up to 96% of marine species – more than any other event in Earth history (Erwin, 1994; Wignall, 2001; Benton *et al.*, 2004; Erwin, 2006; Xu *et al.*, 2007;

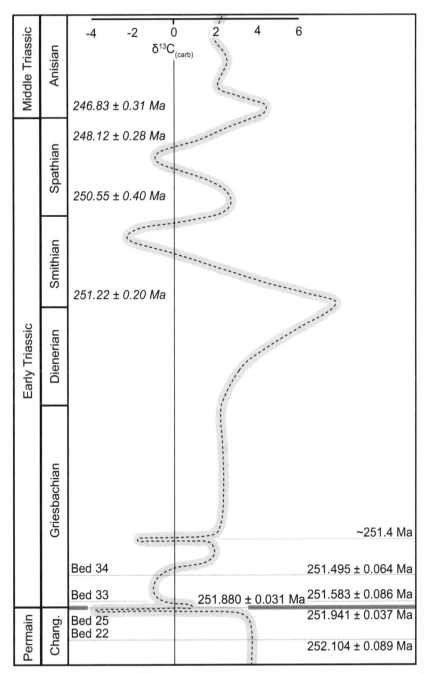

Figure 14.9 Carbon isotope (δ^{13}C) variation through the Permian–Triassic (P–Tr) section at Meishan, China and U-Pb geochronological constraints. Note that the extinction timing is now constrained between 251.941 ± 0.037 and 251.880 ± 0.031 Ma, an interval of 60 ± 48 ka (for details see Burgess *et al.*, 2014b; ages younger than 251.3 are from Ovtcharova *et al.*, 2006). New high-precision U/Pb geochronology on more than 25 samples from lavas, sills, and pyroclastic rocks indicates Siberian Trap magmatism is coincident with this mass extinction interval (Burgess et al., 2014a; Burgess and Bowring, 2014; S. Burgess, personal communication, 2014). Figure slightly modified from Figure 3 in Burgess *et al.* (2014b, corrected version). Digital version of original figure kindly provided by S. Burgess. Permission granted by Springer.

Reichow et al., 2009; Saunders and Reichow, 2009; Fig. 14.9). In addition, abrupt ecosystem changes at the Permian–Triassic boundary have been documented (see the summaries in Xu et al., 2007; Black et al., 2012, 2014), such as marine anoxia (Wignall and Twitchett, 1996), an increase in atmospheric CO_2 content (Berner, 2002), global warming (Huey and Ward, 2005), carbon isotopic excursions (Payne et al., 2004), and acid rains (Liang, 2002; Maruoka et al., 2003). The main stage of the extinction occurred during a short time span of about 60 000 years which is contemporaneous with the Siberian Trap LIP volcanism (Section 3.2.8; Kamo et al., 2003; Mundil et al., 2004; Burgess et al., 2014a, b; Burgess and Bowring, 2014; S. Burgess, personal communication, 2014).

Most hypotheses for the cause of the end-Permian extinction are linked to the Siberian Trap LIP, although a bolide-impact model has also been proposed (Kaiho et al., 2001). With respect to the Siberian Trap LIP, specific kill mechanisms could include volcanic degassing of CO_2 (e.g. Wignall, 2001; Reichow et al., 2002; Grard et al., 2005; Xie et al., 2007), associated mercury release (Sanei et al., 2012), possible release of methane from gas hydrates destabilized by a warming climate (Wignall, 2001; Berner, 2002), and contact metamorphism (by emplacement of sills) of coal and other carbonaceous sediments in the Tunguska basin in eastern Siberia that generated carbon gases and possibly halocarbons (Svensen et al., 2009 and references therein). In terms of the latter, basin-scale gas production potential estimates show that metamorphism of organic matter and petroleum could have generated $> 100\,000\,Gt\,CO_2$, which was released to the end-Permian atmosphere partly through hundreds of hydrothermal vent complexes belonging to the Siberian Trap LIP (Sections 5.3.6 and 14.6.6; Svensen et al., 2009).

Grasby et al. (2011) reported that a substantial amount of char (coal fly ash) was deposited in Permian aged rocks in the Canadian High Arctic immediately before the mass extinction. The geochemistry and petrology of this material suggested that it was derived from the global dispersion of char generated by the combustion of Siberian coal and organic-rich sediments by flood basalts.

14.3.9 Kalkarindji LIP (510 Ma) and the early Cambrian extinction

A mass extinction near the end of the Early Cambrian affected reef fauna worldwide (Zhuravlev and Wood, 1996; Hallam, 2005). This event is associated with the anoxic Sinsk event in the mid-Botomian (Zhuravlev and Wood, 1996), an event that is marked by a major positive sulfur isotope ($\delta^{34}S\%$) excursion (Hough et al., 2006; Fig. 14.10) and the Toyonian aged sea-level fall, termed the Hawke Bay regression (e.g. Palmer and James, 1980; Evins et al., 2009).

These events broadly coincide with the still poorly documented Kalkarindji LIP that extends across parts of central Australia within the Northern Territory and Western Australia and covers a currently known area of $\geq 2.1\ Mkm^2$, with recent research suggesting this event may have streched as far as the southernmost part of South Australia, over a total area of $\geq 3\ Mkm^2$ (Section 3.5.1; Evins et al., 2009; Jourdan

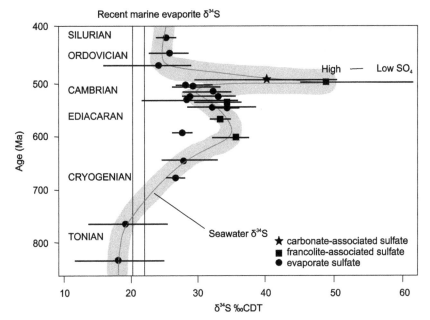

Figure 14.10 Secular evolution of marine sedimentary sulfate $\delta^{34}S$ from 800 to 400 Ma using evaporite sulfate, carbonate-associated sulfate, and francolite-associated sulfate. For each datum, the horizontal lines represent data range, internal circles show mean values, while age uncertainty is given by vertical lines. Phosphorite $\delta^{34}S$ data suggest a major shift to uniquely high seawater $\delta^{34}S$ during the Cambrian period. Modified from Hough *et al.* (2006). © 2013 The Authors–Wiley.

and Evins, 2010). In addition to its great size, the environmental impact of this event may have been enhanced by the fact that Kalkarindji basalts rest on and were presumably emplaced through thick, sulfate-rich evaporite layers and carbonate rocks of the Precambrian-age basins in central Australia (Jourdan and Evins, 2010). Metamorphism of evaporite and carbonate units by magmatism of the Kalkarindji LIP event may have released significant amounts of CO_2, SO_2, and halocarbons in a similar fashion (HVCs) as during the Siberian Trap LIP event (Svensen *et al.*, 2009). In addition, pyroclastic eruptions during the event, as evidenced by the presence of tuff layers within the LIP, may have also increased the amount of volatiles that were injected into the stratosphere.

14.3.10 Bushveld LIP (2058–2053 Ma) and Lomagundi–Jatuli isotopic event

The last example considered here occurred during the Precambrian, and indicates that environmental catastrophes also likely accompanied and were genetically associated with Precambrian LIP events. Melezhik *et al.* (2007) state that the Paleoproterozoic sedimentary carbonate Lomagundi–Jatuli positive $\delta^{13}C$ excursion represents an event whose magnitude and duration is unique in Earth history (Fig. 14.11). A *c.* 1300-m-

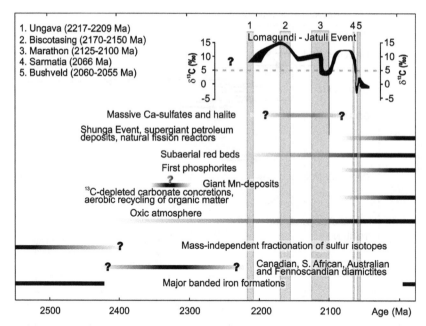

Figure 14.11 Major global paleoenvironmental and tectonic events during the early Paleoproterozoic. Modified from Figure 1.1 in Melezhik *et al.* (2013) to include LIP record. Digital version of original figure kindly provided by V.A. Melezhik. Permission granted by Springer.

thick sedimentary–volcanic succession of the Pechenga greenstone belt in northeastern Fennoscandia records the decline of this isotopic excursion and a return to more normal $\delta^{13}C$ values. Zircons from these sedimentary rocks within the declining part of this isotopic excursion have yielded ^{207}Pb–^{206}Pb dates around 2058 ± 2 Ma and provide the first maximum age constraint on the termination of the Lomagundi–Jatuli event.

The 2058 Ma age for the end of the Lomagundi–Jatuli event exactly corresponds with significant mafic magmatism in Karelia (as noted by Melezhik *et al.*, 2007) and also with the major Bushveld LIP of the Kaapvaal craton (Section 5.4.1). Since many LIP events cause negative carbon isotope excursions (see Section 14.5) it is reasonable to consider that the similar-age Bushveld and Fennoscandia magmatic events at *c.* 2058 Ma may have caused a negative shift in carbon isotopes that resulted in the termination of the high carbon isotope values associated with the Lomagundi–Jatuli event.

14.4 Effects of LIPs on climate

14.4.1 An array of environmental effects

LIP events are associated with a wide array of environmental impacts that can be monitored by proxies, including changes in the isotopic composition of seawater

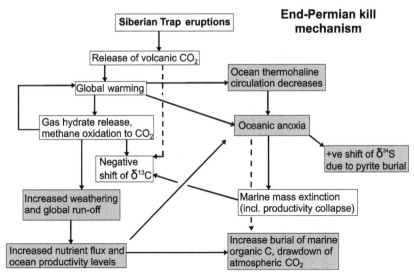

Figure 14.12 Web of events caused by the eruption of the Siberian Trap LIP which could serve as a model for the effect of continental LIPs on the environment. These indicate the effects linked to the end-Permian mass-extinction event associated with the Siberian Trap LIP. Boxes with gray fill indicate effects which are not part of the end-Triassic extinction (associated with the CAMP LIP). Version revised from Wignall (2007) and P. Wignall (personal communication, 2012).

preserved in sediments (Sr, O, C, S, N, and Os), the distribution of black shales (marking anoxia events), and sea-level changes recorded in the sedimentary record. As summarized in Figs. 14.12 and 14.13 the web of possible environmental effects is complex. The potential effects of continental LIP events are illustrated using the Siberian Trap LIP as an example (Fig 14.12). The effects associated with the CAMP event are slightly different (Fig. 14.12) illustrating the point that each LIP should be considered for its own environmental consequences (P. Wignall, personal communication, 2012), the majority of which are controlled by the differences in the hosting terrane (see Section 14.6.7). The environmental effects of oceanic plateau events can be different to those associated with continental LIPs (Fig. 14.13); for instance, seawater cover can have a "buffering" effect on compositional changes (Coffin and Eldholm, 1994; Racki, 2010).

An additional environmental effect is due to Silicic LIPs (SLIPs). As noted by Cather *et al.* (2009), Phanerozoic cool-climate episodes were coeval with major explosive SLIP volcanism (e.g. Sierra Madre Occidental SLIP), suggesting a link between them. Geochronological and biogeochemical data suggest that iron fertilization by great volumes of silicic volcanic ash was an effective climatic forcing mechanism that helped to establish "icehouse" conditions (Section 14.7). This is supported by a model where the availability of large amounts of micronutrients such as iron could have stimulated plankton to produce large quantities of organic matter and use up atmospheric carbon dioxide through photosynthesis.

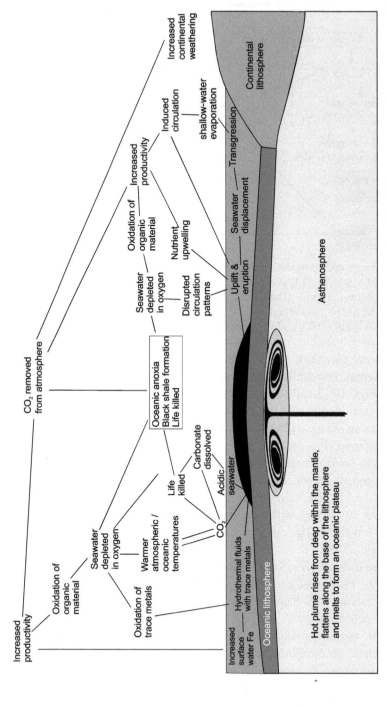

Figure 14.13 Physical and chemical environmental effects of oceanic-plateau formation. Redrafted from Figure 3 in Kerr (2005). Digital version of figure kindly provided by A.C. Kerr. Reprinted with permission from the Mineralogical Society of America.

14.4.2 Proxies for measuring the environmental impact of LIPs

I provide a review of the effects that LIPs can have on global seawater composition, as monitored through isotopic fluctuations (Sr, O, C, S, N, and Os isotopes) in the marine sedimentary record (e.g. Veizer *et al.*, 1999; Taylor and Lasaga, 1999; Prokoph and Veizer, 1999; Prokoph *et al.*, 2008; Hannisdal and Peters, 2011; Kidder and Worsley, 2012) (Figs. 14.14 and 14.15; Table 14.2). Oceanic plateaus release Sr into the ocean through hydrothermal brine exchange between seafloor basalts and the ocean, and continental LIPs release Sr into the ocean through weathering and transport by rivers. A LIP event, whether oceanic or continental, should shift the $^{87}Sr/^{86}Sr$ ratio to lower, more mantle-like values, i.e. produce a negative Sr isotopic excursion. Major troughs in the $^{87}Sr/^{86}Sr$ isotopic curves at 180 and 270 Ma are matched by major LIP events, but the troughs at 160, 330, 390, and 450 Ma do not correspond with known LIPs (Fig. 14.15). Furthermore, many known LIPs do not correspond with troughs, although the 200 Ma LIP may correspond with an inflection in the data.

The oxygen isotope record ($^{18}O/^{16}O$) is primarily a measure of paleotemperature and will change as a result of increased sea-surface temperature associated with LIP events (lower $\delta^{18}O$, i.e. decreases in the $^{18}O/^{16}O$ isotopic ratio) owing to greenhouse-gas release and global warming. $\delta^{18}O$ values are also a good proxy for the growth and decrease of continental ice sheets (e.g. Miller *et al.*, 2005). Positive temperature excursions (recorded by decreases in $\delta^{18}O$ values) have been reported following some major LIP events: 55 Ma (second pulse of) NAIP, 183 Ma Karoo–Ferrar, 93 Ma Caribbean-Colombian, 201 Ma CAMP, 252 Ma Siberian Trap, 258 Ma Emeishan, *c.* 380–360 Ma Kola-Dnieper, *c.* 380-360 Ma Yakutsk-Vilyui, and *c.* 510 Ma Kalkarindji LIPs (Table 14.2; Figs. 14.14 and 14.15; Stanley, 2010; Kidder and Worsley, 2010, 2012; Hannisdal and Peters, 2011).

Positive $\delta^{13}C$ excursions (i.e. increases in $^{13}C/^{12}C$ isotopic ratios) are indicative of increased burial of C (anoxic events), whereas negative excursions indicate greenhouse atmospheric conditions (greenhouse forcing caused by major carbon release, e.g. from the LIP itself and by methane hydrate release). Both positive and negative changes are assigned to the *c.* 120 Ma Ontong Java, *c.* 380–360 Ma Yakutsk–Vilyui, 93 Ma Madagascar, and *c.* 93 Ma Caribbean–Colombian LIPs. Negative excursions are associated with the 55 Ma pulse of the NAIP, 67 Ma Deccan, 183 Ma Karoo–Ferrar, 201 Ma CAMP, 252 Ma Siberian Trap, 258 Ma Emeishan, and *c.* 510 Ma Kalkarindji LIPs (Table 14.2; Figs. 14.14 and 14.15; Stanley, 2010; Kidder and Worsley, 2010, 2012; Hannisdal and Peters, 2011).

Negative shifts in $\delta^{34}S$ sulfur isotope ratios (i.e. a decrease in $^{34}S/^{32}S$ isotopic ratios) are indicative of release and injection of isotopically light sulfur to the ocean–atmosphere system. S isotopic compositions in the oceans are changed by the abundance of oxygen, which in turn correlates to the prevailing climate, as warm oceans can dissolve less oxygen. In addition, the fact that ocean anoxia favors bacterial sulfate reduction means that higher pyrite burial rates can correlate with

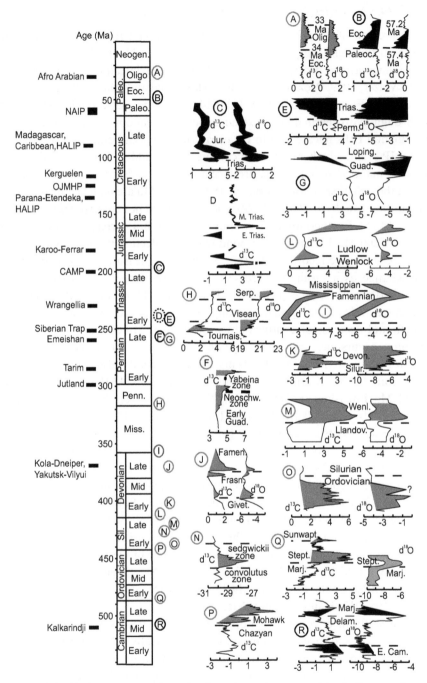

Figure 14.14 Stable isotope excursions that have been documented in shallow-marine strata and are associated with mass extinctions and comparison with the timing of LIP events. Eighteen intervals (A–R) contain a total of 26 such $\delta^{13}C$ excursions. Corresponding to these, and trending in the same direction, are 19 $\delta^{18}O$ excursions, which are displayed in the plots to the right of those depicting $\delta^{13}C$. Temporal positions of excursions are indicated by encircled

an increase in marine sulfate $\delta^{34}S$ values (Kampschulte and Strauss, 2004; Hough *et al.*, 2006). Positive spikes in the seawater $\delta^{34}S$ are linked to the 93 Ma Madagascar, *c.* 93 Ma Caribbean–Colombian, 183 Ma Karoo–Ferrar, 252 Ma Siberian Trap, *c.* 380–360 Ma Yakutsk–Vilyui, and *c.* 510 Ma Kalkarindji, and both positive and negative changes in $\delta^{34}S$ are linked to the CAMP (Table 14.2; Fig. 14.15; Kidder and Worsley, 2010, 2012; Hannisdal and Peters, 2011; Kampschulte and Strauss, 2004; Hough *et al.*, 2006).

Low $\delta\ ^{15}N$ is linked to increased anoxia in the oceans that can lead to a nitrogen fixation gap (see Kidder and Worsley, 2010, 2012). As shown in Table 14.2, this effect can be correlated with number of LIPs. The association between anoxia, euxinia (anoxic conditions in the presence of hydrogen sulfide (H_2S)), major environmental changes, and LIPs is summarized in Table 14.2 (Kidder and Worsley, 2010, 2012).

The improving marine sedimentary Os isotope record has the potential to provide an additional monitor for contributions from mantle sources (Peucker-Ehrenbrink and Ravizza, 2000; Turgeon and Creaser, 2008). For instance, the marine Os isotope data of Turgeon and Creaser (2008) documents an abrupt shift at the beginning of the Cenomanian–Turonian OAE2, which is interpreted to mark a magmatic Os isotope signal associated with the 93 Ma LIPs (Caribbean, Madagascar, etc.; Section 14.3.3). Furthermore, this Os signal appears slightly before the OAE2 event, suggesting a time lag of up to *c.* 23 Ka between magmatism and the onset of significant organic-carbon burial.

The eustatic sea-level curve shows a broad 400 Ma cyclicity (Worsley *et al.*, 1986) with second-order peaks at 530, 480, 270, 220, 160, and 100 Ma, and minor peaks at 420 and 60 Ma that only broadly correlate with some LIP events (Fig. 14.15). The additional complexities in the curve reflect other factors, in addition to mantle-plume arrival, that contribute to sea-level uplift, and these factors can potentially be separated on the basis of duration. For example, glacial events operate on time scales of up to tens of thousands of years, whereas continent breakup and collisional events operate on a time scale of 1 to 100 Ma and plume/LIP events operate on a similar scale of 1 to < 50 Ma.

It is reasonable to expect that major LIP events (or clusters of simultaneous LIPs on different blocks) would have a greater effect on seawater composition, sea level,

Caption for Figure 14.14 (*cont.*) letters on the left. Gray fill indicates association with global cooling and black fill with global warming; absence of fill indicates the absence of published evidence of associated climate change. Horizontal scales represent magnitudes of $\delta^{13}C$ and $\delta^{18}O$ excursions in parts per thousand. Light $\delta^{13}C$ in N is for organic carbon rather than carbonates, and heavy $\delta^{18}O$ in H is for conodonts rather than bulk or skeletal carbonate. Vertical axes represent relative stratigraphic positions of samples and are neither precisely linear with respect to time nor scaled the same for all graphs. Modified from Figure 1 in Stanley (2010) where additional details can be found. Corresponding LIP events have been added to the left side of the diagram. Digital version of original figure kindly provided by S.M. Stanley.

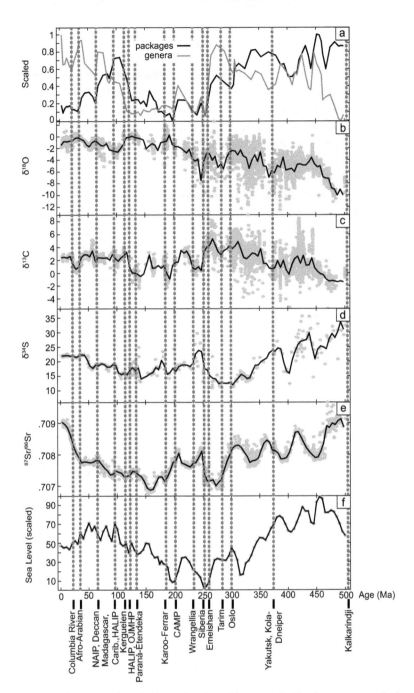

Figure 14.15 Environmental variables related to Phanerozoic Earth system evolution and compared with LIP record. (a) Total number of marine genera (gray), based on North American fossil occurrences, and total number of marine sedimentary packages (black) in North America. (b to e) Isotope ratios from marine carbonates. $\delta^{18}O$ and $\delta^{13}C$ records are the low-latitude subsets. Solid black lines are averages for the time bins in (A), which are used for time-series analyses. (f) Global estimates of continental flooding in 5-Ma bin averages resampled (black line) at the bin midpoints of 80 time intervals (median duration, 5.3 Ma) from the Late Cambrian through the Pliocene. From Figure 1 in Hannisdal and Peters (2011) with the LIP record added at the bottom. Digital black-and-white version of original figure kindly provided by B. Hannisdal. Reprinted with permission from AAAS.

Table 14.2 *HEATT episodes (and two non HEATTS), LIPs, and selected hothouse-related effects (from Kidder and Worsley, 2012, with some modifications to the LIP information)**

HEATT and extinction age (Ma)	LIP and age of peak eruption (Ma)	Approximate extinction intensity (%)	Transgression (Third order)	Warming	Anoxia	Euxinia (sub-photic)	Euxinia (photic zone)	δ¹³C	δ³⁴S (sulfate)	δ¹⁵N < 2‰
Mid-Miocene Climate Optimum no HEATT	Columbian River (15.6–16.0)	5	Y	Y	Y					
No HEATT	(Afro-Arabian) (29–31)	5								
Paleocene–Eocene Thermal Maximum (55)	NAIP (58–62)	5	Y	Y	Lim	Y	Y	N		Y
End-Cretaceous (65)	Deccan (66)	30	Y	Y	Lim	Y	Y	N		Y
Cenomanian–Turonian OAE2 (93)	Madagascar & Caribbean–Colombian (93)	10	Y	Y	Y	Y	Y	P/N	P	Y
Early Aptian OAE1a	Ontong Java I (119–125)	5	Y	Y	Y	Y	Y	P/N		Y
Toarcian–Pliensbachian (183)	Karoo-Ferrar (179–183)	10	Y	Y	Y	Y	Y	N	P	Y
End-Triassic (200)	CAMP (200)	30	Y	Y	Y	Y	Y	N	P/N	Y
End-Permian (251)	Siberian Trap (249–251)	55	Y	Y	Y	Y	Y	N	P	Y
Hangenberg (359)	Kola–Dnieper (c. 370)	20	Y	Y	Y	Y	Y	P		
Frasnian–Famennian (374)	Yakutsk–Vilyui (c. 370)	25	Y	Y	Y	Y	Y	P/N	P	Y
Late Ordovician (444)		30	Y	Y	Y	Y	Y	N		Y
SPICE (499)		40	Y	Y	Y	Y		P	P	
Botomian (c. 520)	Kalkarindji (c. 510)	40	Y	Y	Y			N	P	
Ediacaran (542)		?	Y		Y			N		Y

* SPICE event is late Cambrian (Elrick *et al.*, 2011). Y = yes, P = positive, N = negative, P/N = positive and negative, and Lim = limited.

and other aspects of the marine sedimentary record than LIP magmatism associated with a single plume. The timing of potential LIP clusters (Section 15.2.1) is compared with these other data types in Figures 14.14 and 14.15. There is no obvious correlation between the timing of potential LIP clusters (e.g. at 93 Ma and at *c.* 380–360 Ma) and the Sr, O, and C isotopic curves, suggesting that other controls on isotopic composition, such as supercontinent breakup, production of new buoyant crust, accretion of blocks into continents, and glacial events (e.g. the "Snowball Earth"), complicate the recognition of the effect of clusters of LIPs.

14.4.3 Cross-wavelet analysis

Prokoph *et al.* (2013) compared the LIP record to other environmental proxies using a combination of cross-wavelet and other time-series analysis methods to quantify potential linkages between LIP-emplacement periodicity, geochemical changes, and the Phanerozoic marine genera record. Prokoph *et al.* (2013) concluded, based on the analysis illustrated in Figs. 14.16 and 14.17, and other results, that there is a mantle-plume cyclicity represented in the LIP volume (*V*) data which follows the formula: $V = -(350 - 770) \times 10^3 \, km^3 \sin(2\pi t/170 \, Ma) + (300 - 650) \times 10^3 \, km^3 \sin(2\pi t/64.5 \, Ma + 2.3)$ for *t* = time in Ma. (The difference terms "350 – 700" and "300 – 650" reflect the range in model results.) The analysis shows a shift from a 64.5 Ma to a weaker *c.* 28–35 Ma LIP cyclicity during the Jurassic contributes, together with probable changes in the marine sulfur cycle, to reduce ocean anoxia, and a general stabilization of ocean chemistry and increasing marine biodiversity throughout the last *c.* 135 Ma. The LIP cycle pattern is coherent with marine-biodiversity fluctuations corresponding to a reduction of marine biodiversity of *c.* 120 genera/Ma at *c.* $600 \times 10^3 \, km^3$ LIP eruption volume. The 62–65 Ma LIP cycle pattern as well as excursions in $\delta^{34}S_{sulfate}$ data and marine genera reductions suggest that there should be a LIP event at ~450–440 Ma. No such LIP-scale event has been discovered yet, although scattered occurrences of Suordakh dolerite sills of this age in eastern Siberia (e.g. Khudoley *et al.*, 2013) are a possible clue to the existence of such an event.

Research by Prokoph *et al.* (2013) reveals a periodic pattern of LIP emplacement in comparison with marine isotope records, and quantifies the effects on ocean chemistry and marine biodiversity over the last 520 Ma based on compiled LIP, stable isotope, and marine genera records at a data resolution of *c.* 1 Ma. Time-series analysis using wavelet and cross-wavelet transforms not only shows that the *c.* 140 Ma and *c.* 65 Ma cycles are significant in LIP timing, ocean chemistry, and marine biodiversity records throughout the Phanerozoic, but also highlights that a strong *c.* 32 Ma cyclicity in all related records occurs simultaneously at *c.* 135 Ma. The link between LIPs and biodiversity at *c.* 65 Ma periodicity is particularly strong when correlating the volume of the LIPs with the marine genera record. The strong link

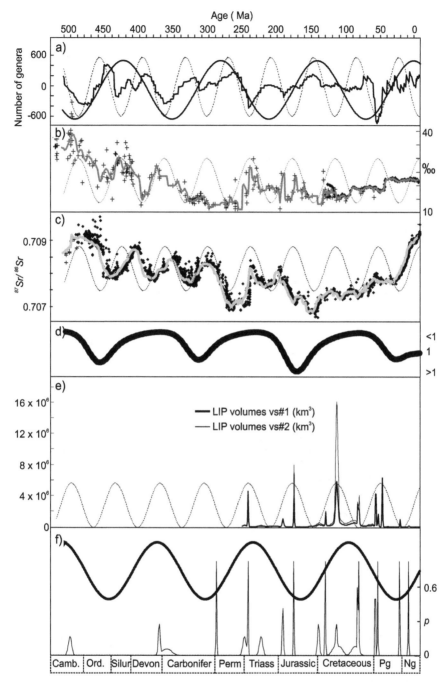

Figure 14.16 Geological records of last 520 Ma with best-fit 140–170 Ma and 65 Ma sine waves. (a) Third-order polygonal of detrended well-dated marine genera (Rhode and Muller, 2005); (b) diamonds: $\delta^{34}S_{barite}$ (Paytan *et al.*, 2004), crosses: $\delta^{34}S$ of structurally substituted sulfate (SSS) (Kampschulte and Strauss, 2004), line: Gaussian filtered time series; (c) $^{87}Sr/^{86}Sr$ data: diamonds: low-Mg fossil shell data (Prokoph *et al.*, 2008), crosses: whole-rock samples (Shields and Veizer, 2002); (d) simplified cosmic-ray flux (CRF) ratio model (Shaviv and Veizer, 2003); (e) LIP volume (Ernst and Buchan, 2001b; Berner, 2002) times probability of occurrence (Prokoph *et al.*, 2004a); (f) probability (of occurrence) of Gaussian-filtered precisely dated LIPs, mostly flood and oceanic plateau basalts (Prokoph *et al.*, 2004b), bottom: time scale (Gradstein *et al.*, 2005). From Figure 2 in Prokoph *et al.* (2013). Digital version kindly provided by A. Prokoph.

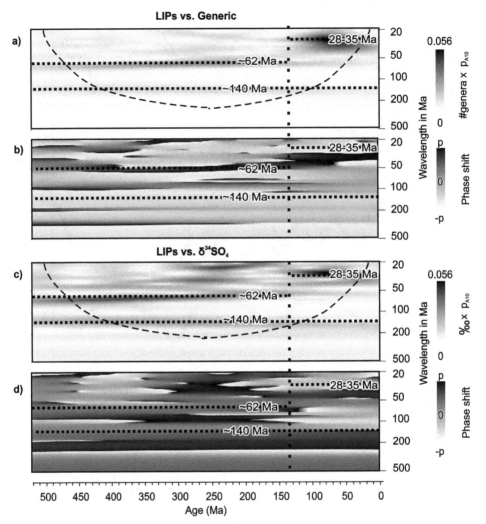

Figure 14.17 Cross-wavelet analysis of LIP occurrences. (a) Cross-wavelet scalogram of LIP_{A10} times # marine genera. (b) Phase-shifts between frequencies of LIP_{A10} and # marine genera. (c) Cross-wavelet scalogram of LIP_{A10} times $\delta^{34}S_{sulfate}$. (d) Phase-shifts between frequencies of LIP_{A10} and $\delta^{34}S_{sulfate}$. Stripped curves separate frequency time space of edge effect of < 20% (above line) from > 20% (below line), horizontal dotted lines mark cycle bands, vertical dotted line marks the onset of *c.* 32 Ma cyclicity in LIP and fossil records at *c.* 135 Ma, gray code for cross-wavelet coefficient = uncorrected cross-amplitude and phase-shifts on right. From Figure 6 in Prokoph *et al.* (2013). Digital version kindly provided by A. Prokoph.

between oceanic $\delta^{34}S_{sulfate}$ and $^{87}Sr/^{86}Sr$ cycles and LIPs also suggests the timing for several Paleozoic LIPs which have not yet been discovered (Prokoph *et al.*, 2013).

14.5 Problem of different time scales

There is clearly a broad link in time between LIP events and extinction events and anoxia events, and also with seawater isotopic compositions. But in order to infer a causal relationship the time matches must be exact. So for instance, if a LIP event entirely postdates the onset of extinction then the link will be less likely.

Nearly all LIP events have been isotopically dated, frequently to a resolution of a few million years or less; the majority of this dating has been undertaken using the U–Pb system, although some LIPs, particularly of younger age, have been well dated by the Ar–Ar method. However, much of the seawater isotope data and Phanerozoic sea-level, impact, and extinction data are assigned ages based on interpolation from dated horizons. Unfortunately, as demonstrated in Table 14.3, there can be distinct differences between the ages assigned to stage boundaries (Harland *et al.*, 1990; Gradstein and Ogg, 1996; Okulitch, 2002; Gradstein *et al.*, 2005; Ogg *et al.*, 2008; www.stratigraphy.org, the official website of the International Commission on Stratigraphy (ICS)). However, these concerns are becoming less significant given the recent advances in comparing biostratigraphic and isotopic dating (see http://www. earth-time.org/intro.html; Zalasiewicz *et al.*, 2012), and it is probable that the most recent versions of the biostratigraphic time scale are converging to the isotopically correct ages for most stage boundaries.

14.6 Environmental effects of LIPs

14.6.1 Meteorite impact versus LIP

There has been a heated debate for more than two decades over whether global extinctions are caused by meteorite impacts or LIP events. Those supporting the impact connection cite the close correspondence in age between the large Chicxulub impact, associated with a large iridium anomaly, and the K–T extinction event (e.g. Alvarez *et al.*, 1980; Hildebrand *et al.*, 1991). Other correlations between large impacts and extinctions are less clear because of the uncertainty in the ages of most large impacts or, in some cases, because of uncertainty in the age of the extinction events (e.g. Section 14.5; Fig. 5 in Ernst and Buchan, 2003). In some cases, it has been proposed that multiple impacts are required for an observable effect, and that the extinction event itself lags the impacts by several million years (e.g. McGhee, 2001). It is also proposed that global extinctions may be caused by more than one mechanism, and that even individual extinction events may have multiple causes. For instance, White and Saunders (2005) suggested that the combined "one–two punch" of a bolide impact and a LIP may be required for the largest extinctions.

Table 14.3 *Time-scale comparison (in Ma) modified after Table 2 in Ernst and Buchan (2003)*

Period	Harland et al. (1990) (H)	Difference between H & G1	Gradstein and Ogg (1996) (G1)	Difference between G1 & O	Okulitch (2002) (O)	Difference between O and G2	Gradstein et al. (2005); Ogg et al. (2008) (G2)
Neogene							
	23.3	0.5	23.8	0.9	22.9	0.13	23.03
Paleogene							
	65.0	0.0	65.0	0.0	65.0	0.5	65.5
Cretaceous							
	145.6	3.6	142.0	2.8	144.8	0.7	145.5
Jurassic							
	208.0	2.3	205.7	5.7	200.0	0.4	199.6
Triassic							
	245.0	3.2	248.2	4.8	253.0	2.0	251.0
Permian							
	290.0	0.0	290.0	10	300.0	1.0	299.0
Carboniferous							
	362.5	8.5	354.0	6.0	360.0	0.8	359.2
Devonian							
	408.5	8.5	417.0	1.0	418.0	2.0	416.0
Silurian							
	439.0	4.0	443.0	0.0	443.0	0.7	443.7
Ordovician							
	510.0	15	495.0	6.0	489.0	0.7	488.3
Cambrian							
	570.0	25	545.0	1.0	544.0	2.0	542.0

A mechanism proposed for linking the effects of impacts and LIPs in time (ensuring a one–two punch) is through the genetic model of impact-triggered LIPs (Rampino and Stothers, 1988; Isley and Abbott, 2002; Jones *et al.*, 2002, 2005; Hagstrum, 2005). However, the ability of an impact to cause flood-basalt magmatism has been questioned by the analyses of Loper and McCartney (1990) and Ivanov and Melosh (2003). See the additional discussion in Section 15.2.6.

As noted in Section 14.3, the link between LIPs and extinction events has become remarkably robust. At the same time the only bolide impact with a robust link to an extinction is the Chicxulub impact in Yucatan, Mexico, linked with the Deccan LIP in a one–two "punch" to cause the end Cretaceous extinction

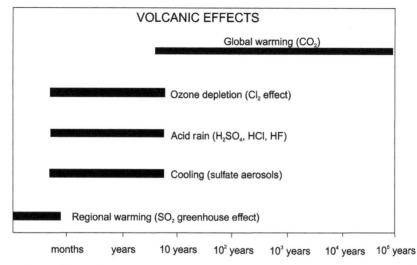

Figure 14.18 Effects of volcanic gases and the intervals over which they operate. Apart from CO_2, most gases are rapidly removed from the atmosphere. Redrafted from Figure 1 in Wignall (2001). With permission from Elsevier.

14.6.2 Stratospheric injection by volcanoes

Individual volcanic eruptions can dramatically affect the climate system on a range of time scales from days to years (e.g. Robock, 2000; Wignall, 2001). An important factor is the release altitude, since gases and particles that are released into the stratosphere as aerosols will have a greater and longer lasting effect on climate (1–3 years) than material reaching only the tropospheric level (1–3 weeks). Also, aerosol clouds produce cooling at the surface but heating in the stratosphere (Fig. 14.18; Robock, 2000; Robock and Oppenheimer, 2003; see also Self *et al.*, 2006).

14.6.3 Greenhouse-gas warming

The geological record shows that abrupt increases in the atmospheric concentration of greenhouse gases have occurred many times during the Phanerozoic (e.g. Jenkyns, 1988; Hesselbo *et al.*, 2000; Kemp *et al.*, 2005; McElwain *et al.*, 2005). The release of several thousand gigatons of isotopically light carbon gases has been proposed to be the cause of warm periods at the Permian–Triassic boundary (*c.* 251 million years ago; Siberian Trap LIP), the Toarcian stage of the Early Jurassic (*c.* 183 Ma; Karoo–Ferrar LIP), and in the initial Eocene (*c.* 55.5 Ma; second pulse of NAIP LIP). A thermal pulse caused by the greenhouse effect could also destabilize methane hydrate (clathrate) mega-reservoirs causing massive release of the greenhouse gas methane into oceans and the atmosphere, representing a strong positive feedback (e.g. Wignall, 2001; Jahren, 2002), and creating a short-duration negative $\delta^{13}C$ spike (e.g. 200 ka) (Pálfy *et al.*, 2001; Retallack, 2001; Beerling and Berner, 2002; Schoene *et al.*, 2010).

14.6.4 Weathering and CO_2 drawdown

Global warming due to greenhouse gas build-up in the atmosphere (particularly CO_2) is transient (Fig. 14.18), and weathering can be an efficient method of CO_2 drawdown, and can even be a cause of global cooling by removal of these greenhouse gases (Dessert *et al.*, 2003; Goddéris *et al.*, 2003). Goddéris *et al.* (2003) discuss the following sequence of events: coeval with the beginning of the LIP event, atmospheric CO_2 first rises due to degassing of hot basaltic lavas. Subsequent global warming is then slowly counteracted by the increasing consumption of atmospheric CO_2 due to continental silicate weathering (including the weathering of flood-basalt material). Note that weathering of continental silicates is enhanced under a warmer and (assumed) wetter climate, and basaltic volcanic rocks weather about five to ten times faster than granitic rocks (Dessert *et al.*, 2003). This model indicates that weathering of LIP events can cause significant CO_2 drawdown and temperature decreases, as well as a positive shift in $\delta^{13}C$ values.

14.6.5 Link with glaciations

The notion of CO_2 drawdown due to weathering of LIPs leads to the interesting idea of whether erosion of flood basalts can initiate glaciations. Based on weathering laws for basaltic lithologies (Dessert *et al.*, 2003) and on climatic model results, weathering of a 6 Mkm^2 basaltic province located within the equatorial region (where weathering and consumption of CO_2 are optimal) could be sufficient to trigger a snowball glaciation (Goddéris *et al.*, 2003). Goddéris *et al.* (2003) noted that the Sturtian "snowball" glaciation is broadly associated with magmatic events (LIPs) associated with the breakup of the supercontinent Rodinia. Specifically, improved dating of the Sturtian glaciation at 730–705 Ma (Macdonald *et al.*, 2010; Young, 2013; see also Evans and Raub, 2011) demonstrates a link with the 725–715 Ma Franklin LIP of northern Canada (Macdonald *et al.*, 2010) and coeval Dovyren LIP in formerly attached southern Siberia (Ernst and Bleeker, 2010; Ernst *et al.*, 2012).

As another example, Taylor and Lasaga (1999) calculate the CO_2 drawdown rates for the weathering of the Columbia River LIP basalts and suggest that, on a several-million-year time scale, the formation of LIPs represents a net sink for atmospheric CO_2. Furthermore, the removal of CO_2 via the rapid dissolution of the Columbia River basalts represents an explanation for the glacial period believed to have followed its formation.

Figure 14.19 Schematic cross section of a sedimentary basin illustrating the release of carbon gas caused by emplacement of sills and dykes and contact metamorphism of organic-rich host rocks. The resulting pressure buildup leads to hydrofracturing and the formation of hydrothermal vent complexes, transporting greenhouse gases from the aureoles to the atmosphere. From Svensen and Planke (2008). Digital version of original figure kindly provided by S. Planke.

One important qualification on this CO_2 drawdown model is whether the basalt flows are quickly buried beneath younger rocks. If so, then the basalts are not available for weathering and their important contribution to CO_2 drawdown is lost (e.g. P. Wignall, personal communication, 2011). For instance, a substantial portion of the Siberian Trap LIP is under younger rocks of the West Siberia basin and much of the NAIP is beneath shelf sediments.

14.6.6 Vent complexes

As discussed in Chapter 5 (Section 5.3.6), hydrothermal vent complexes (HVCs) are an essential component of LIPs. They originate from explosive release of gases generated when thick sills (> 50 m) are emplaced into volatile-rich but low-permeability sedimentary strata (Fig. 14.19). When the host rocks have favorable lithologies for melting they can potentially contribute significant volumes of gases which can contribute to the extinction-causing effect of LIPs.

The most detailed calculation of the climatic effect of such HVCs is based on those associated with the Siberian Trap LIP. The end-Permian crisis can be attributed to the effect of Siberian Trap sills on host evaporites (producing halocarbons) and organic-rich deposits (producing greenhouse gases like CH_4 and CO_2), releasing gases which are then transported to the surface and into the atmosphere via the HVCs. Basin-scale gas-production-potential estimates show that metamorphism of organic matter and petroleum could have generated $>$ 100 000 Gt CO_2 (e.g. Svensen *et al.*, 2009). The discovery of the HVCs in the Siberian Trap LIP (as well as in other LIPs) shows that LIPs contribute gases to the atmosphere not only from their volcanic component but also in a very significant way from the intrusive (sill) component and the interaction between these sills and their host rocks.

14.6.7 Importance of the host terrane

As illustrated above, the host terrane of a LIP exerts significant controls on the environmental consequences of the LIP. Contact metamorphism around intrusions within dolomite, evaporite, coal, or organic-rich shale host rocks can generate large quantities of greenhouse and toxic gases (CO_2, CH_4, SO_2) (Fig. 14.20; Ganino and Arndt, 2009). The pressure of these gases may build up and be released explosively to the surface through HVCs, or may also be trapped if the overlying rocks are not permeable. Trapping of these gases could potentially lead to the formation of hydrocarbon resources (see Section 16.7).

14.6.8 Sea-level changes

The arrival of a mantle plume (causing a LIP) beneath oceanic lithosphere should result in a rise in eustatic sea level because of isostatic uplift and displacement of water by the oceanic LIP itself (e.g. Kerr, 1998; Lithgow-Bertelloni and Silver, 1998; Condie *et al.*, 2001; Miller *et al.*, 2005). Emplacement of oceanic plateaus produces moderately rapid sea-level rises (60 m/Ma), but then sea level slowly falls as a result of the decay of the thermal anomaly (10 m/Ma) (Miller *et al.*, 2005). For example, emplacement of the Ontong Java oceanic plateau would have caused an estimated sea-level rise of at least 10 m (Coffin and Eldholm, 1994). However, controls on sea-level changes are complex and occur over a range of time scales (e.g. Miller *et al.*, 2005; Hannisdal and Peters, 2011; Figs. 14.15 and 14.21). One potential method of distinguishing which eustatic sea-level rises are linked to mantle plumes may be provided by the iron formation record. Isley and Abbott (1999) and Abbott and Isley (2001) have suggested that iron formations can be linked to the presence of oceanic plateaus (as a source of iron) and sea-level rise (increasing the extent of

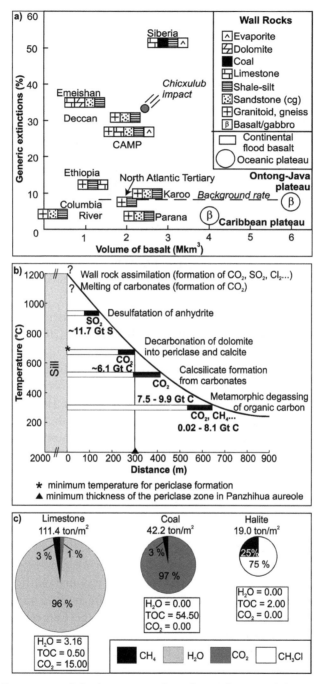

Figure 14.20 Importance of the host terrane for release of gases. (a) Percentage of generic extinctions versus volume of erupted basalt from major LIPs. Those with high values intrude sedimentary rocks that released abundant greenhouse or toxic gases. (b) Effect of dolerite sill on gas release from various types of host rocks. Panzhihua is a major mafic intrusion of the Emeishan LIP. (c) Gas release from different lithologies (TOC = total organic carbon). Parts (a) and (b) from Ganino and Arndt (2009); see also comment by Racki (2010), and reply by Ganino and Arndt (2010). Part (c) from Aarnes *et al.* (2011); copyright 2011 by the American Geophysical Union, Wiley. Digital version of original figures for (a) and (b) kindly provided by C. Ganino.

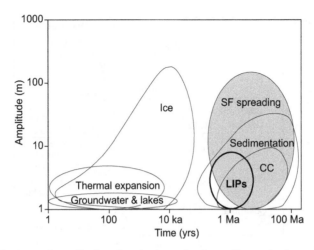

Figure 14.21 Timing and amplitudes of geologic mechanisms of eustatic change. SF = seafloor. CC = continental collision. Modified after Fig. 1 in Miller *et al.* (2004) to add a "ballpark" distribution for LIPs; a more rigorous analysis is needed. With permission from AAAS.

shallow-marine shelves available for deposition of iron). Examples of correlations between LIPs and iron formations are noted in Bekker *et al.* (2010, 2014) and Ernst and Jowitt (2013); see also Section 16.3.5.

Locally, if the occurrence of a LIP is linked to a plume then the arrival of that plume beneath the continental lithosphere will contribute to relative sea-level fall owing to domal uplift above the plume (Chapters 11 and 15). However, on a global scale, continental LIPs only contribute to sea-level rise if the LIP is linked with continental breakup and the formation of extensive new buoyant oceanic crust and lithosphere, which causes widespread transient flooding of shallow platform environments and sedimentary transgression onto the continents.

14.6.9 Anoxia events

Periods of oceanic environmental crisis through time are identified by the development of black shales that are indicative of low, or oxygen-absent, deep-ocean conditions (e.g. Kerr, 1998, 2005). The Cretaceous was marked by a number of global OAEs, black-shale deposition, and an increase in $\delta^{13}C$ values correlated with oceanic-plateau formation, particularly around the Cenomanian–Turonian boundary (93.5 Ma) and during the Aptian (124–112 Ma) (Sections 14.3.3 and 14.3.4; Fig. 14.6). In addition, as noted by Kerr (2005), important Kimmeridgian to Tithonian (155–146 Ma) oil source rocks correlate with the formation of the Sorachi plateau in the western Pacific (Kimura *et al.*, 1994) and perhaps also the Shatsky–Tamu LIP. Furthermore, the formation of Toarcian (187–178 Ma) black shales corresponds with the eruption of the Karoo–Ferrar LIP (Section 3.2.7).

This link can be extended back to the Precambrian, with the observation by Condie *et al.* (2001) that significant black-shale formation events occur at *c*. 1.9 and 2.7 Ga and that these correlate with the formation of LIPs and times of a warmer paleoclimate. Thus, there seems to be a temporal association, throughout a significant proportion of geological history, between periods of global oceanic environmental crises, black-shale formation, and oceanic-plateau formation (Kerr, 2005).

14.6.10 Links between geomagnetic reversals and the LIP record

It has been proposed that the changes in the circulation pattern within the Earth's core can be linked with superplume events. One possible way of testing this link is by comparing the ages of plume-derived LIPs with the ages of magnetic superchrons (e.g. Condie, 2001). Larson and Olson (1991) pointed out that the *c*. 120–80 Ma Pacific superplume event correlates with the Cretaceous normal magnetic polarity superchron. They proposed that plumes leaving from the base of the mantle cool the adjacent outer core and thus increase core circulation. This in turn stabilizes the circulation against instabilities that cause reversals of the magnetic field. A prior comparison with the reversal frequency of Johnson *et al.* (1995) did not reveal a clear correlation (Ernst and Buchan, 2003), although a comparison with the more detailed reversal frequency diagram of Didenko (2011) does reveal plausible correlations (Fig. 14.22; Table 14.4).

14.7 HEATT (haline euxinic acidic thermal transgression) model

A framework for understanding climatic variations is provided by the recent work of Kidder and Worsley (2010, 2012) who proposed that the Phanerozoic consists of three mutually exclusive climate states: Icehouse, Greenhouse, and Hothouse (Figs. 14.23 and 14.24; Tables 14.2 and 14.5). The Greenhouse is the default state, representing *c*. 70% of the Phanerozoic era and is characterized by a weak thermal mode, with little or no continental sea ice, and weakly oxic cool-water sinking near the poles. In addition, Greenhouse oceans are characterized by nitrogen limitation where ammonium may be the dominant bioavailable form of nitrogen. Continent–continent collision and resulting uplift (mountain-building) caused intense chemical weathering that draws down atmospheric CO_2 and results in the Icehouse state that comprises 20–25% of the Phanerozoic (Section 14.6.4). With reference to Table 14.2 and Fig. 14.3, of the 23 most conspicuous mass extinctions of the Phanerozoic (including the late Ediacaran), at least 16 occurred during transitions from a Greenhouse climate to a Hothouse climate.

How does a LIP contribute to a change to a Hothouse climate? An important insight by Kidder and Worsley (2010, 2012) is that LIPs can trigger what they term

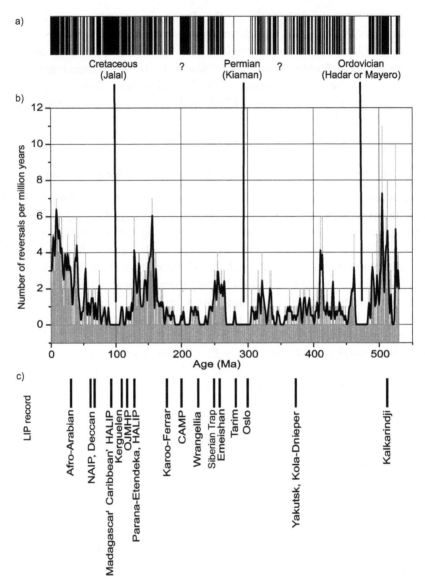

Figure 14.22 Magnetic-polarity reversal frequency and comparison with the LIP record. (a) Phanerozoic geomagnetic polarity scale of Pechersky *et al.* (2010). Black and white colors correspond to normal and reverse polarity intervals, respectively. The scale above is for Cenozoic. (b) The gray histogram is initial time series of reversal frequency calculated according to the Phanerozoic geomagnetic polarity scale of Pechersky *et al.* (2010) and the black line is its spline interpolation. (c) LIP record. Modified from Figure 1 in Didenko (2011) with addition of the LIP record. Digital version of original figure kindly provided by A. Didenko. With permission from Elsevier.

Table 14.4 Time correlation of superchrons, quasiperiodic $^{87}Sr/^{86}Sr$ variations, and events of mantle magmatism through the Phanerozoic. From Didenko (2011).

Superchron	Onsets and ends of superchrons, according to Molostovsky et al. (2007), Ma		Minimums of reversal frequency and their spacing*, Ma	Minimums of $\Delta^{87}Sr/^{86}Sr$ and their spacing*, Ma	Time delay between reversal frequency and $\Delta87Sr/^{86}Sr$ minimums, Ma	LIPs, modified after Ernst and Buchan (2001b)
Jalal	123	83	92.5 108*	54.5 107*	38	Deccan (66 Ma), Tien Shan (68 Ma), Sierra Leone (73 Ma)
?	214	186	200.5 96*	161.5 103	39	Paraná–Etendeka (134 Ma), Shatsky Rise (145 Ma), Karoo–Ferrar (183 Ma), Gondwana breakup, Atlantic opening
Kiaman	315	258	296.5 66*	264.5 78*	32	Siberian Trap (251 Ma), Emeishan (258 Ma), Cache Creek (265), Tarim (280 Ma)
?	375	347	362.5 102*	342.5 110*	20	Yakutsk–Vilyui (c. 370 Ma), East European / Kola–Dnieper (c. 370 Ma)
Hadar	494	468	464.5	452.5	12	Maritimes (470 Ma), opening of Ural and Mongolia–Okhotsk paleo-oceans

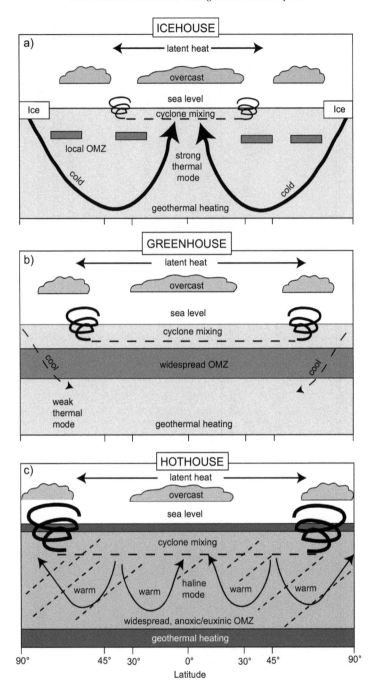

Figure 14.23 Schematic overview of Icehouse versus Greenhouse versus Hothouse conditions, and the role of LIPs as a cause for the transition to Hothouse conditions (Kidder and Worsley, 2010, 2012). OMZ = oxygen minimum zone. From Figure 2 in Kidder and Worsley (2012). Digital version of original figure kindly provided by D. Kidder.

15

Assessing the origin of LIPs

15.1 Introduction

The debate on the origin of LIPs has intensified in recent years and there have been many challenges to the interpretation that mantle plumes are a primary control for the formation of LIPs (Fig. 15.1; pro-plume: e.g. Campbell, 2005; Saunders, 2005; Ernst *et al.*, 2005; Campbell and Kerr, 2007; plume alternative: e.g. Foulger *et al.*, 2005; Foulger and Jurdy, 2007; Foulger, 2007, 2010, 2012; Anderson, 2013; Smith, 2013). Here I discuss and characterize a range of models for the origin of intraplate magmatism and assess which of those could apply to LIPs (Table 15.1). The discussion builds on the background information on LIPs from previous chapters.

15.2 Overview of models for the origin of LIPs

15.2.1 Plumes and LIPs

Most LIPs have been linked to the arrival at the base of the lithosphere of a mantle plume that originated in the deep mantle (Figs. 15.2 and 15.3; Richards *et al.*, 1989; Campbell and Griffiths, 1990; Maruyama, 1994; Ernst and Buchan, 2001b, 2003; Courtillot *et al.*, 2003). In these models (e.g. the plume-head hypothesis of Richards *et al.* (1989), and Campbell and Griffiths (1990)), LIPs are produced by melting within the plume as it reaches the base of the lithosphere, and a number of different types of plumes have been proposed; these are discussed below.

Plumes, plume clusters, and superplumes

In his review, Courtillot *et al.* (2003) distinguished three types of plumes (Fig. 15.1, left half): individual plumes originating from the deep mantle, superplumes in the deep mantle, and plumes originating from the transition zone between the upper and lower mantle. Others have offered a similar set of models (e.g. Cserepes and Yuen, 2000; Maruyama *et al.*, 2007).

There are several criteria (adapted from Courtillot *et al.*, 2003; Kerr *et al.*, 2000; Ernst and Buchan, 2001a, b, 2003) that aid in identifying LIPs with deep-mantle

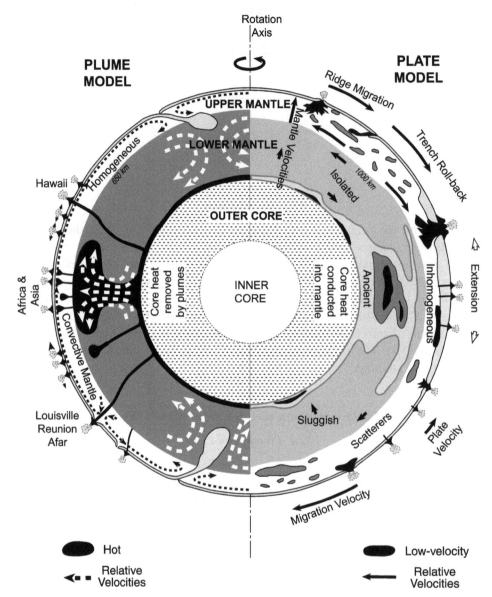

Figure 15.1 A schematic cross section of the Earth showing the plume model (on the left modified after Courtillot *et al.*, 2003) and the plate model (to the right). Not shown on the right side are the effects of decompression melting associated with rifting (e.g. White and McKenzie (1989)) or delamination (Elkins-Tanton, 2007). From Figure 1 in Anderson (2005).

origins, including the following: evidence from mantle tomography for thermal anomalies extending into the deep mantle, a link with an age-progressive hotspot trail, links to uplift (marked in the young record by a large buoyancy flux), compositional arguments (such as the presence of high-Mg rocks, and high ^{3}He/^{4}He ratios),

Table 15.1 *Models for production of intraplate magmatism that have been applied to LIPs*

Model for LIP origin	Key tests	Key reference(s) (cited example)
Plumes originating from the deep mantle	Seismic tomography, prior domal uplift, radiating dyke swarm	Richards *et al.*, 1989; Campbell and Griffiths, 1990; Campbell, 2005; Dobretsov, 2005; Campbell, 2007; Courtillot *et al.*, 2003 (proposed for most LIPs)
Lithospheric delamination	Seismic and electromagnetic (EM) evidence for missing lithosphere; the LIP precedes associated subsidence	Elkins-Tanton and Hager, 2005; Elkins-Tanton, 2007; Hales *et al.*, 2005; Lustrino, 2005 (Siberian Trap)
Plume + lithospheric delamination	Seismic and EM evidence for missing lithosphere; evidence for elevated temperature of LIP source area	Şengör, 2001 (Afro-Arabian)
Decompression melting during rifting (with plume)	Evidence of magma generation coeval with rifting. Composition indicative of shallow melting	White and McKenzie, 1989 (Afro-Arabian, Deccan, NAIP, Paraná–Etendeka, Karoo–Ferrar, CAMP)
Decompression melting during rifting (without plume)	?	van Wijk *et al.*, 2001 (NAIP)
Sublithospheric convection driven by surface cooling that brings up dense fertile mantle without a thermal anomaly	?	Korenaga, 2004 (NAIP)
Decompression melting following internal mantle heating beneath supercontinents	Compare robust supercontinent reconstructions (and pattern of blanketing) against LIP distribution	Gurnis, 1988; Doblas *et al.*, 2002; Coltice *et al.*, 2007, 2009
Impact-induced decompression melting	Evidence for bolide impact associated with LIP	Jones *et al.*, 2002; Ingle and Coffin, 2004 (Ontong Java LIP)
Edge-driven convection	Step change in lithospheric thickness prior to LIP	King and Anderson, 1998; McHone *et al.*, 2005 (CAMP)

Table 15.1 (*cont.*)

Model for LIP origin	Key tests	Key reference(s) (cited example)
Lowering of solidus above transition zone by release of water from subducting slabs	Trace-element evidence for hydrous minerals in LIP source	Ivanov, 2007; Ivanov *et al.*, 2013 (Siberian Trap) Parman and Grove, 2005 (Archean LIPs)
Shallow melting anomalies generated by plate-tectonic-related processes	Seismic tomography	"Plate" model of Foulger, 2007, 2010
Stress-induced lithospheric fracturing and drainage of a relatively slowly accumulated sublithospheric basaltic magma reservoir	Seismic evidence for LIP-scale reservoir of melt in lithosphere	Silver *et al.*, 2006; cf. Cañón-Tapia, 2010
Hot sheet/line model	Seismic tomography	Al-Kindi *et al.*, 2003 (NAIP)
Back-arc rifting	LIP located in back-arc setting; see discussion in Section 15.2.5	e.g. Smith, 1992; Rivers and Corrigan, 2000 (LIPs in eastern margin of Laurentia in Mesoproterozoic)
Plume in a back-arc setting	LIP located in back-arc setting; evidence of plume	Murphy *et al.*, 1999 (Maritimes basin LIP)
Slab window	?	Li *et al.*, 2012 (intrusions associated with Tarim LIP)

Note: In the text it is indicated some of these models on their own are only capable of producing intraplate magmatism of sub-LIP scale. Also note that the LIP examples are those cited by the given author, but in many cases alternative origins for those LIPs have been proposed by others.

and the presence of a giant radiating dyke swarm. These criteria are discussed in more detail below.

LIPs magmas with a deep origin are inferred to originate from the seismically determined "D" layer at the base of the mantle, from heat transferred from the core, and plume ascent controlled by thermal buoyancy (e.g. Campbell and Griffiths, 2014). During ascent these thermal plumes develop a bulbous head, entraining host mantle along the way and spreading out to a diameter of about 2000 km upon arrival at the base of the lithosphere (Fig. 15.2; e.g. Campbell, 2001 and earlier papers).

The notion of plumes originating from the middle mantle has also been proposed (e.g. Wilson and Patterson, 2001), although in these cases plumes that are spawned from

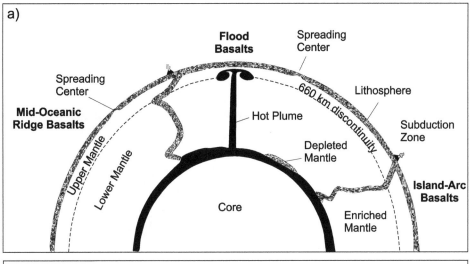

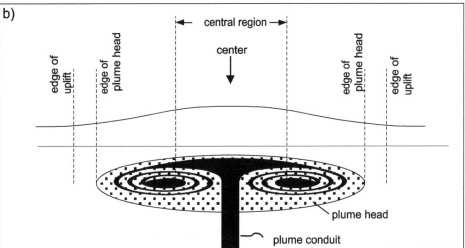

Figure 15.2 Mantle plumes of deep origin and LIPs. (a) Campbell model for the relationship between individual deep mantle plumes, subduction zones, and spreading centers. (b) Schematic diagram showing the location of the geographic elements of a plume. Note that uplift is highly exaggerated. From Figure 1 in Ernst and Buchan (2003) and (a) originally after Figure 2 in Campbell (2001).

the middle mantle may ultimately be associated with plume material ascending from the deep mantle and "ponding" at the mid-mantle boundary (Leng and Zhong, 2010).

Plume-cluster type "superplumes"

Clusters of plumes that arise from the deep mantle and have a particularly significant effect are called superplumes. These superplumes are exemplified in the

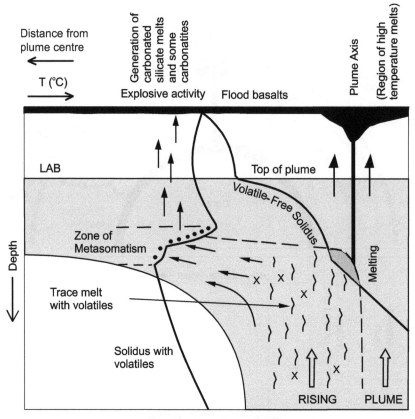

Figure 15.3 Carbonatites, LIPs, and mantle plumes. Sketch through a plume head showing the relationship between carbonated melts and LIPs. Shown are the volatile-free and volatile-rich solidii. Position of plume marked by the lightly shaded area. LAB = lithosphere–asthenosphere boundary. Volatile-rich melts migrate laterally and intersect the volatile-rich mantle solidus. Explosive activity and metasomatism (circles) at depths of about 75 km are accompanied by upward migration of small-volume, low-degree partial melts. The heterogeneous plume head contains source material from deeper parts of the mantle (X), as well as trace melts with volatiles, and possibly lithosphere at higher levels. From Ernst and Bell (2010), modified after Bell and Rukhlov (2004) and the original Wyllie (1988).

model of Larson (1991a, b) who postulated such a 120–80 Ma superplume is associated with expanded seafloor spreading and the emplacement of numerous oceanic plateaus including Ontong Java and Kerguelen. Another proposed super-plume was inferred by Vaughan and Storey (2007) between 230 and 185 Ma; this superplume was recognized as consisting of LIPs, kimberlites, and to be associated with plate reorganization and ophiolite obduction, and also to be accompanied by high reversal-rate frequency of the geomagnetic field, marine anoxia, deposition of carbon-rich sediments, major mass extinctions, and global sea-level changes.

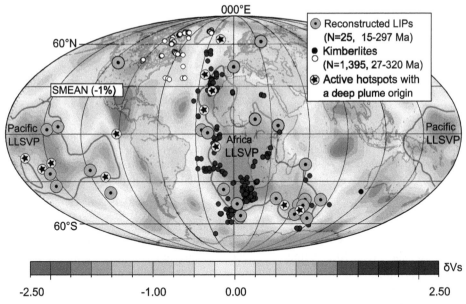

Figure 15.4 Reconstructed LIPs and kimberlites for the past 320 Ma with respect to shear-wave anomalies at the base of the mantle. LLSVP = large low-shearwave-velocity province. The 1% slow contour, approximating to the plume generation zones (PGZ), is shown as a thick line. Present-day continents are shown as a background, to illustrate the distribution of hotspots classified as being of deep-plume origin and present-day shear-wave velocity anomalies (percentage δVs), but bear no geographical relationship to reconstructed kimberlites or LIPs. Black symbols indicate Kimberlites associated with the LLSVP. From Figure 1 in Torsvik *et al.* (2010). Digital version of original figure kindly provided by T. Torsvik. Reprinted by permission from Macmillan Publishers Ltd: Nature.

The group of Asian LIPs/plumes: Tarim (270 Ma), Emeishan (260 Ma), and Siberian Trap (250 Ma) can also be grouped as the Permian Central Asian super-plume (e.g. Borisenko *et al.*, 2006). Also, Li *et al.* (2003b, 2008a) postulated that the 825, 780, and 755 Ma LIPs were connected with a superplume associated with the breakup of Rodinia.

LLSVP-type "superplumes"

These plume clusters are distinct from another use of the term superplume, which is associated with major and persistent deep-mantle anomalies in two broad areas underneath Africa and the Pacific (large low-shear-wave-velocity provinces (LLSVPs); Fig. 15.4; e.g. Cserepes and Yuen, 2000; Maruyama *et al.*, 2007; Yuen *et al.*, 2007; Li and Zhong, 2009). These deep-mantle "superplumes" are associated with geoid highs and were considered the locus of the source for hotspots (Crough and Jurdy, 1980). More recent models link present-day hotspots and reconstructed LIPs to the periphery of these deep-mantle anomalies (e.g. Burke *et al.*, 2008; Torsvik

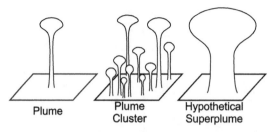

Figure 15.5 Plumes, plume clusters, and superplume events (modified after Schubert *et al.*, 2004). The heads of hypothetical superplumes would have a scale of several thousand kilometers to 10 000 km, while the heads of individual plumes would range from 100 to 2000 km. From Figure 1 in Ernst *et al.* (2007). With kind permission from Springer Science and Business Media.

et al., 2010). This source region has been proposed to correspond to a graveyard for subducting slabs (e.g. Maruyama *et al.*, 2007) or alternatively may be linked with a region of primordial mantle (e.g. Kellogg *et al.*, 1999; Collerson *et al.*, 2010). It is inferred that these types of superplumes are long-lived features, at least extending back to 200 Ma in the case of the African superplume and perhaps to 750 Ma in the case of the Pacific superplume (Maruyama *et al.*, 2007). The association between these deep-mantle superplumes (and the cluster of plumes and LIPs spawned by them) with the breakup of supercontinents may have occurred repeatedly throughout Earth history (e.g. Maruyama *et al.*, 2007; Li and Zhong, 2009).

These different uses of the term superplume overlap. In order to reduce the confusion I will refer to the spatial groupings of plumes with a *c*. 100 Ma or less age range, and having a particular significant geodynamic effect (e.g. supercontinent breakup) as "plume clusters," and follow Torsvik *et al.* (2010) in referring to the deep-mantle type of "superplumes" as LLSVPs.

Plume/superplumes and the supercontinent cycle

Some of the proposed plume cluster ("superplume") events are linked to the breakup of supercontinents. The breakup of the Pangea/Gondwana supercontinent is associated with both the proposed 230–185 Ma superplume of Vaughan and Storey (2007), and a subsequent group of plumes/LIPs spanning 185–65 Ma is associated with the progressive breakup of the Gondwana portion of the supercontinent (specific events at 183, 145, 135, 90, and 65 Ma; Storey, 1995). This Gondwana plume cluster is generally associated with the African superplume (Fig. 15.4). The breakup of the earlier Rodinia supercontinent in stages by plumes at 825, 780, and 755 Ma (Li *et al.*, 2003b, 2008; Li and Zhong, 2009) can also be considered a plume cluster (Fig. 15.5) and is linked with the deep-mantle superplume that has persisted since that time and is recognized as the present-day Pacific superplume (Fig. 15.4; Maruyama *et al.*, 2007).

In other cases the link between a plume cluster and the two known deep-mantle superplumes (LLSVPs) is not obvious. This is exemplified by the potential Permian Central Asian "superplume" consisting of the 260 Ma Emeishan and 250 Ma Siberian Trap LIPs (Dobretsov, 1997, 2003; Dobretsov and Vernikovsky, 2001) and also should include the 290–270 Ma Tarim LIP (cf. Borisenko *et al.*, 2006). Similarly, the formation of the Iapetus Ocean is marked by magmatic events from 615 to 550 Ma, termed CIMP, which may represent three plumes/LIPs (events 52–54 in Table 1 of Ernst and Buchan (2001b)). It is currently unclear whether these plume clusters (290–250 Ma and 615–550 Ma) are linked with either of the persistent deep-mantle anomalies under Africa or the Pacific.

Thermochemical plumes

The initial modeling of plume ascent was based on thermal buoyancy (e.g. Griffiths and Campbell, 1990, 1991; Campbell, 2001), and led to the classic plume-tail, plume-head model (Fig. 15.2). However, buoyancy can also be generated by compositional differences and can also lead to the ascent of thermochemical plumes (e.g. Cordery *et al.*, 1997; Tackley, 1998; Schubert *et al.*, 2001; McNamara and Zhong, 2004; Schott and Yuen, 2004; Lin and van Keken, 2005, 2006a, 2006b, 2006c; Farnetani and Samuel, 2005; Samuel and Bercovici, 2006; Zhong, 2006; Dobretsov *et al.*, 2008). In many models the role of eclogite (derived from subducted ocean crust) is important; for instance, Sobolev *et al.* (2011) present petrological evidence for a large amount (15 wt%) of dense recycled oceanic crust within the head of the plume (e.g. the Siberian Trap LIP). These authors present a thermomechanical model that predicts no pre-magmatic uplift and requires no lithospheric extension. Other effects that are relevant include the post-perovskite phase transition in the deep mantle (discovered in 2004; e.g. Nakagawa and Tackley, 2005) and the possible incorporation of an outer-core fluid phase (H_2 + CH_4 + hydrides of K, Na + SiO_2) that oxidizes to H_2O + CO_2 once it enters the mantle (e.g. Dobretsov *et al.*, 2008). The latter would increase the buoyancy of the plume as it is starting its ascent from the lower mantle. One aspect that can distinguish thermochemical plumes from purely thermal plumes is the potential absence of a plume head in the former (Fig. 15.6).

Seismic evidence for plumes

Montelli *et al.* (2006) produced a comprehensive evaluation of P- and S-seismic-wave identification of deep-mantle plumes below a large number of known hotspots. Deep-mantle plumes are present beneath the Ascension Islands, Azores, Canary Islands, Cape Verde, Cook Islands, Crozet Islands, Easter Island, Kerguelen Islands, Hawaii, Samoa, and Tahiti, and the Afar, Atlantic Ridge, Bouvet (Shona), Cocos/Keeling, Louisville, and Reunion plumes originate at least below the upper mantle if not from much deeper. Many of these examples represent the plume tails that succeed known LIPs, e.g. Afar with the Afro-Arabian LIP, Reunion with the Deccan

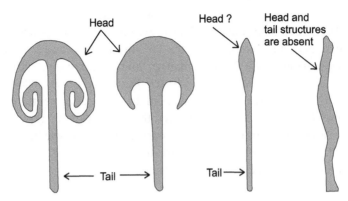

Figure 15.6 Experimentally derived morphology of plumes based on different models reported in Dobretsov *et al.* (2008). With permission from Elsevier.

LIP, etc. The seismically imaged plumes are wide in the lower mantle, with radii of at least 300–400 km, and do not show the typical head–tail shape associated with the classic model of a plume. Such broad, headless plumes suggest that the viscosity in the mantle is only weakly dependent upon temperature, that compositional buoyancy variations act to slow the rise of the plume (Montelli *et al.*, 2006). However, once the plume reaches the lithosphere it must spread out over a wide area beneath the continental lithosphere (Chapter 5); the theoretical estimates of a radius of *c.* 1000 km (e.g. Campbell, 2007) seem reasonable.

15.2.2 Decompression melting associated with rifting above plumes

The classic paper by White and McKenzie (1989) linked numerous flood basalts to decompression melting at the onset of extension (Fig. 15.7: e.g. the Afro-Arabian, Deccan, NAIP, Karoo–Ferrar, and CAMP LIPs). However, their models also required a thermal anomaly 50–200 °C above ambient mantle temperatures. More recently, van Wijk *et al.* (2001) considered the influence of confining high beta-factor rifting to a narrow zone (175 km) and concluded that these conditions combined with normal-temperature mantle (1333 °C) could produce significant volumes of magma (900–1700 km^3 per km of margin length) consistent with the magma production observed along the rifted margins of the NAIP. This modeling suggests that a mantle plume is not always a prerequisite to the formation of rifted volcanic margins, although this model does not address the consistent evidence for an initial pulse of pre-rifting LIP magmatism (Section 11.5).

15.2.3 Delamination

Another important process in the generation of magmas is delamination whereby material at the base of the mantle lithosphere can become denser than adjacent

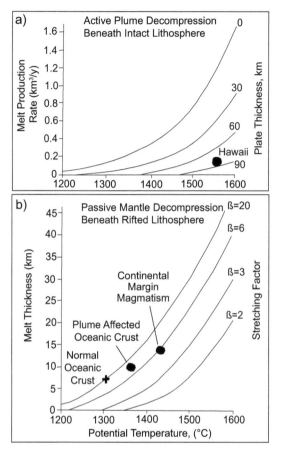

Figure 15.7 (a) Rates of melt production within a mantle plume lying beneath an intact lithosphere for a range of mantle temperatures in the core of the plume and a range of lithospheric thicknesses. (b) Thicknesses of melt generated by decompression of the asthenospheric mantle upwelling passively beneath a lithosphere thinned by different stretching factors (from an initial thickness of 120 km) over a range of potential temperatures. Redrafted from Figure 5 in White and McKenzie (1995). See also discussion of passive rifting in Section 11.3.

asthenospheric mantle through melt injection, conversion to eclogitic phase assemblages, or thickening and cooling of a lithospheric root (e.g. Elkins-Tanton, 2005; Lustrino, 2005). As the lithosphere detaches and sinks, there is a corresponding flow upward of asthenospheric material into the lithospheric hole from which the delaminating material has fallen. This asthenospheric material can melt adiabatically during descent (Fig. 12.10). In addition, the delaminated material may dehydrate as it sinks and heats, producing a fluid that can trigger melting and production of hydrous magmas (e.g. Elkins-Tanton, 2005). Delamination can occur over a range of scales, from localized events under mountain ranges or above subduction zones to larger regions in cratonic interiors (e.g. Conrad and Molnar, 1997; Percival and Pysklywec, 2007), and a wide range of mantle melt volumes can be produced, from tens or hundreds of cubic kilometers to thousands or even, in rare and catastrophic cases, on flood-basalt scales. However, the latter requires an associated thermal anomaly; Şengör (2001) presented a model of delamination above a mantle plume, which allows the plume to reach the base of the crust leading to voluminous partial melting of the plume (Fig. 12.11).

15.2.4 Thermal blanketing model

A separate idea for the origin of LIPs is formation through the thermal blanketing of continents, as part of the normal cycle of supercontinent formation. In this model, the supercontinent sows the seeds of its own destruction through thermal-blanketing effects (Gurnis, 1988; Coltice *et al.*, 2007, 2009). By this model the greatest amount of thermal blanketing should be in the center of the superconti-nent which should, therefore, be the locus of major plumes that subsequently break apart the supercontinent (Fig. 11.1). This seems to be broadly true in considering the distribution of plume centers for LIPs associated with the breakup of Gondwanaland (Fig. 11.1). However, the Siberian Trap plume center (see Fig. 15.10) is toward the edge of the supercontinent. The thermal blanketing model can only be part of the explanation for the origin of LIPs. The viability of this thermal blanketing model can also be better assessed once pre-Pangea supercontinents are better constrained so that the distribution of plume centers can be compared with the area of maximal thermal blanketing that could then be predicted for Precambrian supercontinents.

15.2.5 Plate models

There has been considerable debate regarding the role of mantle plumes in the initiation and genesis of LIP events; alternatives to the plume model include mag-matism associated with upper-mantle plate and plate boundary processes (e.g. Anderson, 2001; Foulger, 2007, 2010; Foulger *et al.*, 2005; Foulger and Jurdy, 2007; Foulger, 2012; Smith, 2013; Anderson, 2013). Specifically, Foulger (2010) suggests that propagating cracks, internal plate deformation, membrane tectonics, self-perpetuating volcanic chains, recycled subducted slabs, and continent breaking can all cause LIP-scale magmatism. However, the plate mechanisms discussed by Foulger will not be sufficient on their own to produce LIPs. (See also the Great Plume Debate – http://www.geolsoc.org.uk/plumesdebate.) The volumes of magma that are produced by most plate boundary mechanisms alone are too small and the magmatism is not produced at the right time. For instance, it is shown that in many

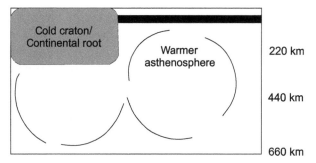

Figure 15.8 Edge-driven convection based on the King and Anderson (1998) model. Redrafted from Figure 1.19 in Foulger (2010).

cases magma generation occurs *after* domal uplift (Section 12.2) and prior to the onset of major rifting (Section 11.5), in contrast to the predictions of the plate models.

The discussion below presents various additional "plate" models and discusses their limitations with respect to producing LIP magmatism. I conclude that these mechanisms are certainly sources of additional magmatism that can augment LIPs, and perhaps can explain some differences between LIPs, but are not the primary cause of LIPs.

Edge-driven convection

The edge-driven convection model involves material exchange from a thick lithosphere (typically beneath Archean cratons) to adjacent thinner regions of the lithosphere, generating circulation patterns that bring hotter material from depth beneath the area of thicker lithosphere to shallower levels beneath the thinned lithosphere where partial melting occurs (Anderson, 1998; King and Anderson, 1998). The essential feature for this model is the presence of a step-change in lithospheric thickness prior to LIP generation (Fig. 15.8). However, in many cases, and as is discussed in Section 11.5, the initial pulse of LIP magmatism typically precedes breakup, suggesting that this mechanism cannot be generally applied to LIPs.

Fracture-controlled magmatism

Seamount chains are generally thought to be linked to age-progressive magmatism associated with plate movement over an underlying hotspot (e.g. Morgan, 1971; Richards *et al.*, 1989). However, in some cases the age progression is not well defined and on this basis a mechanism of decompression melting linked to ocean-floor cracking has been proposed (Foulger, 2007, 2010; cf. Clouard and Bonneville, 2005). Although this process might explain some small-scale hot-spot magmatism, it can only produce small volumes of melt, and scaling up to LIP size would require a significant thermal anomaly (e.g. a plume).

One interesting variation on this theme is presented by Silver *et al.* (2006) (see also Cañón-Tapia, 2010) who proposed that intracratonic flood basalts do not represent a melting event, but rather a short-duration drainage event that taps a previously created sublithospheric reservoir of molten basalt formed over a longer time scale. This model requires storage of large volumes of melt in the lithosphere for a long period; however, the feasibility of this mechanism is not clear as even small volumes of melt do not stay at their place of generation but are efficiently removed by buoyancy-driven draining from their residue (e.g. Ahern and Turcotte, 1979; McKenzie, 1984).

Role of subduction

Subduction can carry fluids into the mantle in a process that can enhance later intraplate melting; specifically, subducting slabs can transport significant amounts of water into the mantle transition zone (e.g. Ivanov, 2007; Maruyama *et al.*, 2007). The subsequent release of water from this transition zone lowers the solidus of the overlying mantle and potentially leads to voluminous melting. A similar model has been suggested for the CAMP event by McHone *et al.* (2005). I note that a plume model has also been applied to the CAMP on the basis of the giant radiating dyke swarm (Section 15.4.1); and in such cases the apparent arc geochemical signature can be explained by plume interaction with lithosphere that was modified during a prior subduction event (Section 10.5.4).

A related discussion is whether komatiites (characteristic of Archean LIPs; Section 6.3) are the result of hydrous melting. Parman and Grove (2005) suggested a subduction model for the genesis of komatiites. Vesicles, indicating a significant volatile component (although perhaps acquired during assimilation of sediments) have been observed in some komatiites (e.g. Beresford *et al.*, 2000). A detailed assessment of the controls on komatiite generation by Arndt *et al.* (1998, 2008) suggests that most komatiites were not significantly hydrous and were not generated in subduction-zone settings.

Another consideration is the fact that arc-type and LIP/plume-type greenstone belts are spatially juxtaposed in many Archean greenstone belts (Sections 6.3.1 and 6.3.2). However, this may not have any genetic significance but may simply reflect the fact that during ocean-closure arcs and oceanic plateaus are jammed together. This is exemplified by the present-day effect of the Ontong Java plateau jamming the Solomon and New Ireland arcs (MacInnes *et al.*, 1999; Kerrich *et al.*, 2005).

Back-arc setting

A mechanism for producing LIPs in a back-arc setting is outlined in Figure 15.9. However, it remains to be demonstrated whether such a mechanism can produce LIP-scale magmatism. There are in fact a number of examples where a LIP could be interpreted to lie in a back-arc setting, including the 17 Ma Columbia River basalt

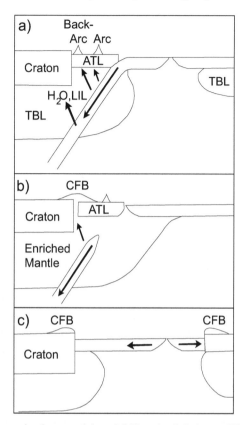

Figure 15.9 Model for a back-arc origin of LIPs. A slightly modified version of a model proposed by Anderson (1995). (a) The mantle wedge fluxed and enriched during subduction produces typical arc and back-arc magma. (b) When the tectonic setting changes from compression to extension the continental LIP forms as a result of plate pull-apart of the accreted terrane layer (ATL) and lithospheric delamination. (c) Resumption of spreading at the mid-ocean ridge. TBL is thermal boundary layer. CFB is continental flood basalt. Redrafted from Figure 5 of Puffer (2003).

province (Section 3.2.2), which is within a back arc to the subducting Juan de Fuca plate (Carlson and Hart, 1988; Smith, 1992; cf. Figure D1 in Hooper *et al.*, 2007). In addition, the 183 Ma Karoo–Ferrar LIP is conspicuously close to the southern Gondwana subduction zone (Fig. 11.1) and the 252 Ma Siberian Trap LIP is linked by Ivanov (2007) to a subduction zone now marked by the Mongolia–Okhotsk suture zone (Fig. 15.10). Similarly, a back-arc setting has been suggested for the 260 Ma Emeishan LIP (Fig. 15.10; Zhu *et al.*, 2005), and the *c.* 1110 Ma Keweenawan LIP formed in a back-arc setting with respect to ocean closure during the Grenville orogeny (Section 3.5.2). The 1880–1870 Ma Circum-Superior LIP magmatism that circumscribes the Superior craton (Section 9.2.2) was emplaced during a period of

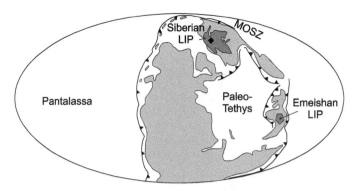

Figure 15.10 Paleotectonic map of Pangea in the late Permian (*c.* 255 Ma). The diamond marks the location of Noril'sk. MOSZ is the Mongolia–Okhotsk suture zone. Modified after Figure 5 in Ivanov (2007).

oceanic closure on the northeast, north, northwest and south sides of the Superior craton (cf. Heaman *et al.*, 2009).

Although these various LIPs described above may have occurred in back-arc locations, it is likely that these LIPs are not solely related to back-arc processes (cf. Pearce and Stern, 2006). As noted in other chapters (see also Section 14.3 in Arndt *et al.* (2008)) plumes arising from the deep mantle are not "aware" of the upper-mantle setting and can therefore ascend into all upper-mantle environments, including back arcs, some percentage of the time.

In addition, there is also the question of what constitutes a back-arc setting. How close to the active margin is close enough to count as a back arc? For instance, the Midcontinent rift system of North America and associated 1115–1085 Ma Keweenawan magmatism are located more than 500 km away from the Grenville orogen (Section 3.5.2). In addition, the Mongolia–Okhotsk subduction zone (Fig. 15.10) is 800 km away from the southeastern end of the Siberian Trap LIP and 2500 km away from Noril'sk (Ivanov, 2007). In these and the other examples, the LIP is many hundreds of kilometers to >1000 km away from the paleosubduction zone, complicating the interpretation of a "back-arc" setting for these events.

15.2.6 Bolide impact

Given observations from other planetary bodies, the Earth must have experienced major impacts throughout its history, particularly during the early Archean period that corresponds to the period of heavy bombardment noted on the Moon (ending about 3.8 Ga). Numerous impacts have also been recognized in the younger geological record of the Earth. This argument has been taken further by some researchers who have proposed that LIPs can be caused by impact, either directly (Glikson, 1999, 2005; Abbott and Isley, 2002; Jones *et al.*, 2005) or antipodally

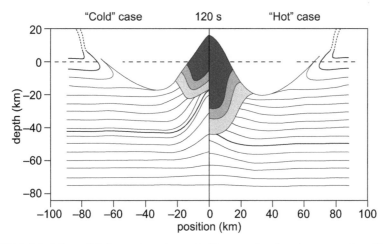

Figure 15.11 Modeling of deformation, thermal effects, and melt volumes produced by bolide impact. Displacement of initially horizontal layers of Lagrangian tracer particles in cold (left panel) and hot targets (right panel) at the intermediate stage of a transient crater collapse (120 seconds after impact). Gray shading shows 0–50%, 50–100%, and 100% of melting of rocks at this moment. Solid lines show the position of layers buried initially at base of melt zone (*c*. 42 and 55 km). Redrafted from Figure 2 in Ivanov and Melosh (2003).

(i.e. due to impact on the opposite side of the planet) (Boslough *et al.*, 1996; Hagstrum, 2005). However, the only major magmatic event conclusively linked to an impact on Earth is that of the 1850 Ma Sudbury impact (southern Superior craton, Canada). This impact generated a bulk crustal melt rather than mantle melt (Mungall *et al.*, 2004) and the resulting magma volume is on a sub-LIP scale. An impact origin has also been proposed for the largest LIP, the Ontong Java oceanic plateau (Ingle and Coffin, 2004); however, several distinct problems exist with this impact interpretation, including an overall composition that is not consistent with melting of upper mantle, effectively ruling out an impact origin for this LIP (e.g. Tejada *et al.*, 2004; Korenaga, 2005; Fitton *et al.*, 2005).

Numerical simulations by Ivanov and Melosh (2003; Fig. 15.11) reveal that an asteroid with a diameter of 20 km striking at 15 km/s can create a 250–300-km-diameter crater and about *c*. 10 000 km^3 of impact melt. However, this crater would collapse almost flat and the pressure field associated with the impact would return to the initial lithostatic pressure. This indicates that even an impact this large cannot raise mantle material above the peridotite solidus by decompression. As noted by Ivanov and Melosh (2003), to really trigger a volcanic eruption, a large impact must strike a young hotspot, such as a newly arrived plume head, that is already on the verge of producing LIP magmatism.

So, an impact into continental crust can potentially produce crustal impact melts in an area already of elevated temperature due to recent orogeny, and the type (and only) example is the Sudbury Complex in Ontario Canada; e.g. Mungall

Table 15.2 *Effect of impact into hot oceanic crust. Values are melt volumes in km³. From Jones* et al. *(2005).*

Model	Ivanov and Melosh (2003a, b)	Jones *et al.* (2005)
Model 20-km projectile		
Cold oceanic crust	3×10^4	-
Hot oceanic crust (20 Ma), as used in Ivanov and Melosh (2003b)	6×10^4	1.2×10^5
Hot oceanic crust (20 Ma)	-	5.0×10^5
Hotter oceanic crust (10 Ma)	-	8.0×10^5
Model 30-km projectile		
Hotter oceanic crust (10 Ma)	-	2.5×10^6

et al., 2004; Naldrett, 2004; Ames and Farrow, 2007). However, even in such cases, the temperature of the underlying mantle will not be raised enough to cause mantle melting. Jones *et al.* (2005) confirm that continental LIPs cannot be generated by bolide impact, but suggested that impact into young oceanic crust may be able to melt sufficient mantle to produce LIP-scale magmatism (Table 15.2). However, there are no known terrestrial examples of this type of impact-related magmatism, and, as noted above, the proposed impact origin for the Ontong Java oceanic plateau also now seems untenable.

Additional constraints on the impact model are possible by comparing the timing of the Deccan LIP, the Chicxulub bolide impact, and the K–T boundary iridium anomaly. Bhandari *et al.* (1995) and Courtillot *et al.* (2000) located and confirmed the existence of the K–T iridium anomaly in interflow sediments in the Kutch region of India, demonstrating that Deccan volcanism and the Chicxulub bolide impact coincided in time but, importantly, that there was no causal connection between the two as the volcanics are underneath (i.e. were earlier than) as well as above (younger than) the iridium anomaly (Chenet *et al.*, 2008). Additional negative evidence regarding the role of impacts in generating LIPs comes from other planets (Section 15.7).

15.3 Geochemistry and origin of LIPs

This section discusses the geochemistry of the mantle-derived mafic–ultramafic component of LIPs; the fact that the silicic components of LIPs are mainly (but not always) derived from melting of the lower crust means that such silicic material is not as useful for assessing the origin of LIPs.

At a basic level, the mafic–ultramafic components of LIPs are the product of anomalously high rates of mantle melt production. Three scenarios, namely elevated water content, a drop in pressure (decompression), and an increase in mantle

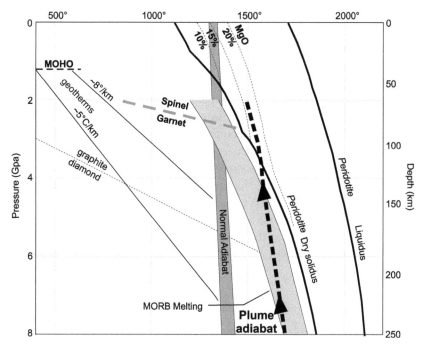

Figure 15.12 Geotherms in the lithosphere and stability melting conditions of peridotite and of eclogite. Modified after Figures 1 and 2 in Garfunkel (2008). The values 5–8 °C/km represent a range in lithospheric geotherms. Plume adiabat is from Ivanov *et al.* (2013). With permission from Elsevier.

temperature, can lead to high melt production rates in the mantle. One additional proposed model for LIP formation (Section 15.2.5) is the slow accumulation of magma over a longer period and its fast release. Here, I consider the geochemical evidence from the mafic–ultramafic components of LIPs (cf. Table 15.1, which is a summary of possible mechanisms) and compare this evidence to normal mantle geotherms (Fig. 15.12). This comparison enables the determination of how these geotherms need to be perturbed to produce LIP-scale melting of both mantle peridotite and entrained eclogite (former oceanic crust).

The addition or presence of water in the mantle lowers the mantle solidus and can potentially lead to significant amounts of melt production (Campbell, 2001; Ernst *et al.*, 2005). However, water (and/or CO_2) contents are low in the upper mantle, particularly in intraplate settings away from subduction zones, indicating that water-induced melting is unlikely to be a general cause of LIP magmatism. Decompression melting of the mantle, as occurs during mid-ocean ridge magmatism, does produce large volumes of melt but at relatively low degrees of partial melting compared to the majority of LIPs and at melt rates that are not anomalously high because, regardless of the rate of spreading, they produce normal oceanic crust with a thickness of 7 ± 1 km (White *et al.*, 1992; Fig. 15.7). By definition, "normal" oceanic crust does not represent a LIP (e.g. Chapters 1 and 2; Bryan and Ernst, 2008).

Although the extension and associated decompression of typical upper-mantle material may contribute to high melt production rates, these mechanisms in themselves are insufficient to form LIPs (e.g. Campbell, 2001; Campbell and Kerr, 2007; Garfunkel, 2008). Any successful hypothesis for the production of LIPs must include melting of anomalously hot mantle. One alternative is that the upper mantle may have a considerable range in ambient temperatures (e.g. Anderson, 1995; Foulger, 2010), which would make plumes unnecessary for the formation of LIPs. However, computer models of mantle convection consistently show that temperature variations within the upper mantle are small compared to the variations associated with plumes and subduction zones (see Davies, 1999). Finally, it is noted that fertile zones within the mantle will produce more melt than refractory zones, and variations in mantle fertility may contribute to variations in melt production between LIPs. For instance, entrained eclogite streaks will tend to melt first (e.g. Campbell, 1998; Gibson, 2002; Garfunkel, 2008). Here, I look at the geochemical evidence that constrains the origin of LIPs, building on the discussion of LIP geochemistry presented in Chapter 10.

15.3.1 Diagnostic characteristics of plume involvement

This section considers aspects of the geochemistry that are diagnostic of (or consistent with) plume involvement in LIP generation.

Ocean-island basalt (OIB) composition

Magmas with ocean-island basalt (OIB) compositions are considered to be the products of plume magmatism, and as such, OIB compositions are considered to be a characteristic plume signature. As discussed in Section 10.2.1, the isotopic components in OIBs (e.g. enriched mantle 1 (EM1), EM2, and HIMU) are distinct from those of upper-mantle mid-ocean-ridge basalts (MORBs). In addition, depleted component focal zone (FOZO) is also an important component of oceanic LIPs. The presence of OIB components (and also FOZO) in LIPs is evidence of involvement of sources from beneath the upper mantle MORB source. One important complication to the recognition of an OIB or FOZO signature is contamination of LIP magmas during transit of mantle lithosphere and crust (Sections 10.2.3 and 10.2.4).

High temperature

In order to have a consistent reference frame for describing mantle temperatures geochemists talk in terms of potential temperature (the temperature the solid material would have if transported adiabatically to the surface; McKenzie and Bickle, 1988). The potential temperature of olivine-bearing basalts can be determined using methodologies that are reviewed (along with associated uncertainties) in Herzberg (2011), and potential temperatures for a number of LIPs and OIBs were calculated by Herzberg and Gazel (2009) (see also Coogan *et al.*, 2014; Fig. 15.13). It is evident

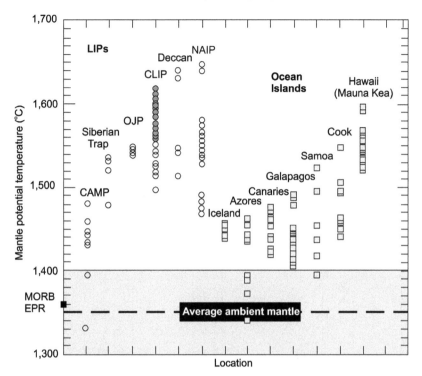

Figure 15.13 Mantle potential temperatures inferred for lavas from some LIPs and ocean islands. CLIP is the Caribbean–Colombian LIP. OJP is the Ontong Java LIP. NAIP is the North Atlantic Igneous Province. EPR is the East Pacific Rise. From Figure 2 in Herzberg and Gazel (2009). Reprinted by permission from Macmillan Publishers Ltd: Nature.

that LIPs typically have potential temperatures in excess of that of the ambient mantle, with some LIPs (e.g. the CAMP) having minimal temperature excesses (<100 °C), whereas others (e.g. the Caribbean, Deccan, and North Atlantic LIPs) have temperatures that are in excess of the ambient mantle by up to 250 °C.

With respect to the mantle geotherm diagram (Fig. 15.12), it is clear that temperatures in excess of the ambient mantle will produce significant melts as plume material rises adiabatically. A review by Garfunkel (2008) favored temperatures ⩾ 300 °C above the ambient mantle adiabat, leading to upper estimates of potential temperatures of *c.* 1750 °C and 1670 °C if melting is to begin above the peridotite solidus at depths of 180 and 150 km (below Proterozoic lithosphere; Fig. 15.12).

Depth range of melting

The classic analysis (Fig. 15.14) by White and McKenzie (1995) calculates the degree of partial melting as a function of depth using rare-earth element (REE) data. The depth of onset of melting is given by the slope of the heavy REE and is indicative

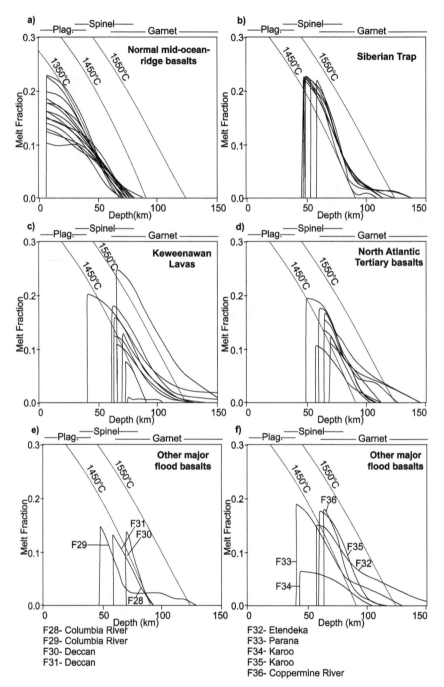

Figure 15.14 Inferred melt distributions from REE inversions of basalts from various provinces. Coppermine River volcanoes are part of the Mackenzie LIP. Modified from Figure 16 in White and McKenzie (1995).

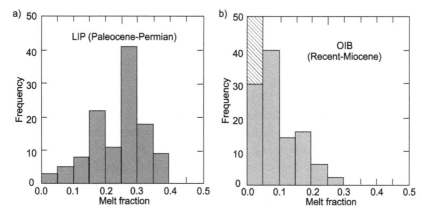

Figure 15.15 Melt fractions inferred for lavas for some LIPs (a) and ocean islands (b). Melt fractions have been computed using PRIMELT2, and refer to the total melt fraction with respect to source mass for accumulated fractional melting of fertile peridotite. In (b), solid bars indicate primary magma solutions from ocean islands. The hatched region indicates an abundance of OIB melted from volatile-enriched sources at very low melt fractions; these are generally more abundant than volatile-deficient lavas, and cannot be modeled with PRIMELT2). Frequency is the number of primary magma solutions. From Figure 4 in Herzberg and Gazel (2009). Reprinted by permission from Macmillan Publishers Ltd: Nature.

of melting within the garnet stability field. Initial partial melting can start at depths of 120 km, considerably greater than the 70 km depth for onset of melting beneath spreading ridges, and again supports a plume origin for LIPs. The shallowest depths of melting vary from about 30 to 70 km, providing evidence for the importance of lithospheric thinning (White and McKenzie, 1995) or ascent into lithospheric thinspots (cf. Thompson and Gibson, 1991).

Degree of melting

The melt fraction of OIBs and LIPs can be calculated using the methodology described in Herzberg and Gazel (2009) (Figure 15.15). Although OIB signatures are generally linked to a plume (or plume tail), production by non-plume processes is also possible (e.g. Herzberg and Gazel, 2009), for instance, by volatile-induced melting of ambient mantle and transport through lithospheric fractures (e.g. Hirano *et al.*, 2006). However, the high degrees of partial melting for some LIPs (up to 40% in Fig. 15.15) require a thermal anomaly (i.e. a plume; Herzberg and Gazel, 2009).

High-Mg melts

In this section, I address high-Mg magmas (komatiites and meimechites with MgO > 18% and picrites with MgO = 12–18%) (Le Bas, 2000), which are an important component of LIPs, and are considered diagnostic of plume-derived magmatism (e.g. Campbell, 1998, 2001; Arndt *et al.*, 2001; Tomlinson and Condie, 2001; Condie,

2001; Isley and Abbott, 2002). Komatiites comprise at least 5% of the volcanic pile in about 60% of Archean greenstone belts (Fig. 6.7; de Wit and Ashwal, 1997). Picrites are present within continental LIPs (Deccan, NAIP, Paraná–Etendeka, Karoo–Ferrar), oceanic plateaus (e.g. Gorgona Island of the Colombian–Caribbean event), and oceanic islands (e.g. Hawaii, Iceland), and are also associated with Archean komatiites (Campbell, 2001; Gibson, 2002).

High-Mg magmas, such as komatiites and picrites, are generated during high-degree partial melting that is indicative of elevated temperatures. Although there is some debate about their origin (e.g. Burke, 1997; de Wit, 1998; Smith and Lewis, 1999; Parman and Grove, 2005), high-Mg magmas can generally be linked with a thermal mantle plume. They are interpreted to be generated from the hottest portion of a plume immediately above the plume conduit (e.g. Campbell *et al.*, 1989). The generation of high-Mg anhydrous melts requires mantle potential temperatures that are about 200 °C–300 °C in excess of those of ambient upper mantle. For the modern mantle, Campbell and Griffiths (1992) have proposed that an MgO threshold of 14 wt% be used for identification of plume-related magmas, whereas an 18 wt% threshold is given for the hotter Archean mantle; Gibson (2002) applied a similar Archean threshold but used 12 wt% for the Phanerozoic and Proterozoic.

A non-plume setting can be argued only for certain classes of high-Mg rocks. Some picrites are produced by the melting of water-rich mantle, but such arc picrites are distinguished by their low Ni, Nb, Ta, and Ti, and high K, Rb, Ba, and Pb content (Campbell, 2001). As discussed above, the proposal that some Archean komatiites were produced by melting wet mantle in a subduction setting (e.g. Grove *et al.*, 1997; Parman and Grove, 2005), has been addressed and refuted by Arndt *et al.* (1998, 2008).

High-Mg melts can be produced by high-degree partial melting of peridotite at shallower levels or lower-degree partial melting at greater depths, and, as pressure increases, melting of peridotite produces increasingly MgO-rich magmas (Fig. 15.12; Garfunkel, 2008). Therefore, an important consequence of plume melting of peridotite at sublithospheric depths is that the primary magmas are picritic to komatiitic, i.e. significantly richer in MgO than the basaltic magmas that comprise the bulk of the observable high-level portions of continental LIPs. Thus, if the melting of peridotite is the dominant source of LIP magmas, then it must be inferred that the primary magmas are generally significantly modified (by fractionation of olivine) as they ascend through the lithospheric mantle (e.g. Wooden *et al.*, 1993; Herzberg and O'Hara, 1998; Garfunkel, 2008). On the other hand, if the source of these magmas contains significant amounts of eclogite, as noted by Condie (2001, p. 137), large volume LIP basalts can be generated without requiring extensive fractional crystallization of picrites (see Section 15.3.2). However, peridotite-derived melts seem to predominate in the generation of continental flood basalts, although eclogite-derived magmas also contribute for some flood-basalt events (Garfunkel, 2008).

Deep-mantle origin

As noted above, the inversion of REE data for basalts from LIPs (Fig. 15.14) reveal the range of depth of partial melting (i.e. garnet vs. spinel stability fields) during the formation of these LIP magmas; the depth of melting can also be assessed by examining high-Mg magmas. At pressures above 3 GPa (about 90 km depth), the melting of mantle material produces picrites and komatiites, and variations in normative enstatite and Al_2O_3 compositions reflect plume potential temperatures and the depth of melting (Herzberg and O'Hara, 1998). In particular, Barberton-type komatiites represent moderately high-degree melts from a particularly hot and deep source, whereas Munro-type komatiites are very high-degree melts of slightly cooler sources (e.g. Arndt *et al.*, 2008; Tables 6.3 and 6.4).

Two minerals that are of particular importance in assessing the depth of origin in komatiites are majorite garnet (left in the restite during melting at depths greater than about 350–450 km), and the high-pressure Mg-silicate perovskite, which becomes a major liquidus phase at 24 GPa (about 720 km). As summarized by Condie (2001) melting of perovskite may be recognized by positive Nb–Ta and Zr–Hf anomalies and negative La–Yb$_N$; leaving residual perovskite in the source area will generate the opposite signature. In comparison, negative Nb–Ta anomalies combined with positive La–Yb$_N$ values and an absence of Zr–Hf anomalies is indicative of crustal contamination (Figure 7.8 in Condie (2001)).

High $^3He/^4He$ anomalies have also been discussed as a possible indicator of a plume origin for LIP magmas, primarily as this provides evidence of plumes accessing deep regions of the mantle that contain primordial ^3He (e.g. Courtillot *et al.*, 2003), supporting the model of Collerson *et al.* (2010). Complexities with this interpretation are discussed in Meibom *et al.* (2005) and White (2010).

Evidence of core involvement

The presence of core-derived components within LIP melts is one possible geochemical tracer of deep-seated mantle plumes. The marked compositional differences between the core and the mantle mean that it should be possible to identify material that contains a contribution from the core and hence may be inferred to be derived from the core–mantle boundary area. The core contains high concentrations of Os and W with distinctive isotope ratios (Hawkesworth and Schersten, 2007); however, White (2010) indicates that although the Os isotope composition of material from Hawaii is indicative of the incorporation of material derived from Earth's core within the Hawaiian plume (e.g. Brandon *et al.*, 1999), this observation does not apply to other OIBs and has also not been confirmed by W isotopes. Ryabchikov (2003) suggested that high Ni concentrations within mantle-derived magmas could be evidence for material transfer from the Earth's core. In addition, Day (2013) also considers the evidence for core contribution in platinum group element (PGE) data and notes that although Pt–Os isotope systematics can potentially trace core–mantle interaction, other reservoirs in the crust and mantle may have highly siderophile element (HSE) concentrations and long-term Pt–Os isotopic anomalies sufficient to generate the ^{186}Os–^{188}Os anomalies observed in hotspot lavas.

15.3.2 Superimposed geochemical characteristics

As shown above, two important geochemical aspects are key to the identification of plumes: evidence for anomalously high potential temperature and evidence for deep origin. Uncontaminated plume-generated basaltic rocks should have flat REE patterns or OIB-enriched patterns and lack negative Nb, Ta, and Ti anomalies (Chapter 10). The presence of high-MgO magmas (picrites and komatiites) indicating high-temperature melts is considered diagnostic of plumes. Similarly, low-CaO/ Al_2O_3 and high-Fe values indicate a deep-mantle origin. High $^{3}He/^{4}He$ ratios > 10 R/R_A (measured ratio relative to the measured atmospheric ratio which is 1.39×10^{-6}) and high Os isotopic ratios are possibly significant as they can imply a deep primordial source, but may also reflect recycled oceanic crust.

The evidence for plume involvement (and melting of peridotite) as discussed above can be augmented or overprinted by other processes which are discussed below.

Role of eclogite

As mentioned above, the presence of eclogite in plume source regions can have a dramatic effect on melting (Cordery *et al.*, 1997; Garfunkel, 2008; Sobolev *et al.*, 2011). This component is derived from descending oceanic slabs whose basaltic mineralogy changes to eclogite in the mantle with a corresponding increase in density. Eclogite melts more easily than peridotite (Figure 15.12); however, since it is denser than ambient peridotite it can comprise a maximum of about 15%–25% of composite aggregates and remain buoyant (e.g. Garfunkel, 2008). In addition, as initially pointed by Campbell (1998), there are complexities that can arise during melting of such a composite of peridotite and eclogite patches (Garfunkel, 2008). The silica-saturated eclogite-derived melts will react with the enclosing peridotite, produce dispersed streaks of a hybrid "refertilized" orthopyroxene-rich and Fe-rich peridotite. With further ascent and decompression the hybrid rocks will begin to melt and produce Fe-rich picritic melts.

Interaction with the lithosphere

The final point to consider is interaction with the lithosphere. Because of high temperatures, the MgO-rich primary magmas can be considerably modified during ascent, and because of their high heat content they are predicted to interact with and become contaminated by components derived from the lithosphere.

In addition to being a contaminant, it has also been proposed that lithosphere can represent an even more significant contribution to LIPs, through its dehydration partial melting (e.g. Gallagher and Hawkesworth, 1992). Based on xenolith studies, the amount of water and K_2O content is generally low in the lithosphere (Garfunkel, 2008). So for fusible lithosphere to exist requires metasomatism that refertilizes the lithosphere mantle, allowing it to melt more easily in the presence of an underlying heat source (e.g. a plume). For instance, the CAMP event shows a lithospheric signature but does not show an OIB signature. So it represents a good candidate for lithospheric melting (Callegaro *et al.*, 2013).

Superimposed "subduction" characteristics

As noted in Section 10.5.4, a number of LIPs exhibit some geochemical characteristics that are consistent with a subduction origin. A negative Nb–Ta anomaly is produced in subduction-related basalts because Nb and Ta remain in the descending slab during transfer of fluids to the overlying mantle wedge, which then partially melts. However, the apparent subduction signature associated with many major LIPs (e.g. Karoo) is linked to lithosphere that was metasomatised in a prior subduction event (Section 10.5.4).

The related topic of anorogenic boninitic (also termed boninitic–noritic) type magmas which appear common for LIPs of late Archean to earliest Proterozoic age is also addressed in Section 10.8. While similar to boninites generated in a suprasubduction-zone setting, those associated with LIPs appear most likely to represent asthenospheric melts with a component generated from melting of the refertilized depleted lithosphere (whose depletion happened in an earlier unrelated subduction event). A component of crustal contamination can also be involved.

15.4 Assessing the origin of LIPs from geometric arguments

This section addresses the critical types of information available for LIPs from a geometric perspective and considers this evidence against the models discussed in the previous section. The idea is to identify particular critical geometric measures and points within a LIP that pertain to its origin.

15.4.1 Locating a magmatic center or center(s) for the LIP

Many LIPs have a unique central region (marking a plume center) that can be recognized using a variety of criteria including the convergence of a radiating dyke swarm, or of triple-junction rifting (rift stars), marked by a concentration of high-Mg magma, and/or marked by the center of domal uplift. Each of these aspects is addressed in more detail below.

Center of magmatic distribution

A simple approach is to circumscribe the magmatism and take the center of the distribution to overlie the plume center. This is essentially the technique used by White and McKenzie (1989) for Mesozoic/Cenozoic LIPs. However, as noted earlier (Section 5.6.4), magma may be transported by dykes or sills within the crust laterally for great distances (dykes can travel > 2000 km laterally). Also, plume material can be transported laterally via sublithospheric channeling for vast distances also away from the plume center (Section 5.6.2). If this is not occurring equally in all directions, then the sub-circular distribution may be skewed, making the circumscribing method less reliable. Furthermore, the original geographic distribution can be much less clear in older LIPs, which have been deeply eroded, fragmented by continental breakup and deformed in subsequent continent–continent collisions.

Giant radiating mafic dyke swarms

A more definite approach for locating plume centers is to use giant radiating dyke swarms whose focal point marks the plume center (Sections 5.2.1, 5.2.4, 7.2.4, 7.3.6, and 11.7.3). Giant radiating dyke swarms on Earth have been linked to mantle plumes because of the following characteristics: (1) a radiating pattern, which is suggestive of a centrally located magma source; (2) evidence for rapid emplacement within a few million years of all parts of a giant radiating swarm (LeCheminant and Heaman, 1989); (3) the presence of coeval volcanics and plutonic rocks in the focal region (Fahrig, 1987), which may represent the remnants of a LIP emplaced above a mantle plume (e.g. Coffin and Eldholm, 1994); (4) associated uplift of the focal region in response to the arrival of the mantle plume (see under high-Mg in Section 15.3.1; LeCheminant and Heaman, 1989; Rainbird, 1993; Rainbird and Ernst, 2001; Saunders *et al.*, 2007); (5) outward-directed lateral magma flow in radiating swarms except near the focal region, which contains the source magma chambers (e.g. Gibson *et al.*, 1987; Ernst and Baragar, 1992; Baragar *et al.*, 1996; Hastie *et al.*, 2014).

Non-plume views regarding the origin of radiating swarms have been proposed (Şengör, 2001); as a spreading center attempts to propagate inland, it may branch out following the older fabric within the continent and give rise to a fanning dyke swarm (see Figure 18 in Şengör (2001)). Some arguments against Şengör's model are that radiating swarms on Earth and other planets can fan more than 90 degrees (Chapter 5), and typically precede ocean opening so there is no spreading center from which to propagate.

McHone *et al.* (2005) also offered alternative models for radiating swarms. (1) For the CAMP event, which is typically cited as one of the best examples of a radiating swarm (Figure 12.5), McHone *et al.* (2005) instead interpreted it as actually being composed of sets of linear swarms reflecting discrete stages of ocean opening that, by virtue of the pattern and distribution of opening stages, mimicked a radiating swarm. (2) They also noted that plate boundary stresses (e.g. continental collision) can also cause radial stress patterns that could be associated with dykes.

Against McHone's first model is the robust evidence for lateral magma emplacement from the swarm focal region (e.g. for the Mackenzie swarm, Baragar *et al.*, 1996; for swarms of the Karoo event, Hastie *et al.*, 2014) and evidence for lateral magma flow given by graben–fissure systems overlying dykes on Venus and Mars (Chapter 9). Against McHone's collision model is the typical association of giant radiating dyke swarms with extensional events (e.g. CAMP and the opening of the Atlantic Ocean; Chapter 11).

Triple-junction rifting

As reviewed in Chapter 11, numerous plumes have associated triple-junction rifting, and these have also been termed rift stars (Burke and Dewey, 1973; Şengör and Natal'in, 2001). In the case of the Siberian Trap LIP, the focal point based on triple-junction rifting (intersection of east–west-trending Khatanga trough and north–south West Siberian trough; Schissel and Smail, 2001), is about 300 km to the west from the focal point determined by the radiating swarm (Fig. 16.4; Ernst and Buchan, 2001c). This indicates that the center of uplift (given by the focus of a radiating dyke swarm) can be offset for the locus of triple-junction rifting, that is probably explained by preferential breaking along preexisting zones of weakness. This offset of plume center and rifting for the Siberian Trap LIP is consistent with the observation that the locus of rifting is not necessarily through the center of the LIP distribution (Section 11.4.2; e.g. Courtillot *et al.*, 1999).

Locus of ultramafic magmatism

As noted in Section 15.3 (high-Mg melts) the presence of picrites and komatiites is a characteristic of plume-related LIPs. They are generated within the hot central portion of the plume head and also in the plume tail (Campbell, 1998), and thus this high-Mg magmatism might be expected to be concentrated in the central region (Fig. 15.2) of an underlying plume. For instance, picrites of the 183 Ma Karoo LIP have been found near the Nuanetsi plume center (Fig. 5.10; Eales *et al.*, 1984). The komatiites on Gorgona Island west of Colombia (associated with the 90 Ma Caribbean–Colombian LIP (Section 4.2.3) may broadly mark the plume center associated with the initiation of the Galapagos hotspot. Picrites of the Emeishan LIP of China are concentrated in the interpreted plume-center region (Fig. 12.6). Similarly, the Song La komatiites in northwest Vietnam, which belong to the Emeishan LIP (Walker, 2002), overlie the plume-center region after correcting for Cenozoic major sinistral movement along the major Red River fault (Fig. 12.6; cf. Izokh *et al.*, 2005).

Domal uplift

It has been argued by Şengör (2001) that domal uplift is a diagnostic criterion for the presence of a plume. A number of examples were offered (Section 12.2) which show evidence of domal uplift, including the 200 Ma CAMP (North American portion; Fig. 12.5) and the 1270 Ma Mackenzie LIP of northern Canada. The uplift story for the Siberian Trap event was more equivocal (Saunders *et al.*, 2005) with uplift potentially concentrated in the West Siberian basin, i.e. off craton. Also, in this case, geochemical evidence supports a significant eclogite component in the Siberian Trap LIP source that reduces the magnitude of domal uplift (Sobolev *et al.*, 2011). Paleocurrent patterns and incised canyons will radiate away from the uplifted plume center (Cox 1989; Williams and Gostin, 2000; Rainbird and Ernst, 2001). Another classic example of plume-related domal uplift has been the 260 Ma Emeishan of South China (He *et al.*, 2006, 2009). However, recognition of phreatomagmatic-style volcanism in the lower Emeishan (Ukstins Peate *et al.*, 2008, 2009a, b) required modification of this simple domal uplift model. A modified plume-uplift model was subsequently offered by He *et al.* (2011) (Fig. 12.7).

Offset of plume conduit and LIP center

The location where the plume conduit feeds the plume head will be directly below the center of the plume. An exception is where the plume head has slid along a lithospheric root upward into a lithospheric thinspot (Thompson and Gibson, 1991; Fig. 10.39).

It should also be noted that the upper end of the plume conduit, where it feeds the plume head, does not necessarily overlie the source region for the plume in the lower mantle because seismic tomography indicates that plume conduits are often inclined (e.g. Zhao, 2001; Shen *et al.*, 2002). The seismic images of Montelli *et al.* (2006) illustrate the effect of the mantle "wind" (Section 15.2.1).

Shifting of the center of magmatism

The depocenter of magmatism for the Deccan and Paraná–Etendeka LIP, systematically shifted with time (Section 12.5.3). In the case of the Deccan LIP (Fig. 12.15), its southward shifting was linked to northward movement of India, toward collision with Asia, with respect to the underlying mantle and its plume.

15.4.2 *Location of translithospheric zones*

As shown by the lithospheric architecture mapping of Begg *et al.* (2009), translithospheric fault zones can be important in controlling magma ascent and associated ore deposits of a variety of commodity types. For instance, the Kharaelakh fault controls the Ni–Cu–(PGE) mineralized magmatism of the Noril'sk portion of the Siberian Trap LIP (e.g. Schissel and Smail, 2001). The locus of magmatism ascending along such faults is a guide to the distribution of the underlying mantle source (e.g. plume) – see, e.g., Fig. 16.5.

15.4.3 Locating the extent and outer edge of the underlying mantle anomaly
(e.g. plume)

Another geometric element of importance is the extent of the underlying mantle-source area that produces the LIP magmatism. For a plume this would be the extent of the area of the underlying plume head after arriving and flattening along the base of the lithosphere. Because the flattening and spreading of the plume beneath the lithosphere is a dynamic process, the size of the plume head, and hence the location of its outer edge (Fig. 15.3), will vary with time. With this caveat, I outline several strategies to mapping the extent of the underlying plume head.

Distribution of basaltic magmatic component

The distribution of flood basalts, dyke swarms, and sill provinces of a LIP and the associated silicic, carbonatitic, and kimberlitic magmatism are a guide to the distribution of the underlying source areas.

White and McKenzie (1989) looked at the main distribution of volcanics associated with the Afro-Arabian, Deccan, NAIP, Paraná–Etendeka, and Karoo–Ferrar LIPs and concluded that most of the magmatism in each case fits within a 1000-km-radius circle, a plume size favored from theoretical considerations (see earlier discussions). This approach assumes that magma generation occurred throughout the top of the plume head. However, if magma generation was restricted to the central region of the plume head (Section 5.6.1), then the estimate of the size of plume head based on distribution of magmatic products could be an underestimate. Alternatively, because magma can be transported laterally for long distances via dykes (e.g. Ernst and Baragar, 1992; Ernst and Buchan, 2001c; Section 5.6.4) the areal extent reached by lateral dyke injection (and capable of producing surface lavas; Figs. 5.1 and 5.2) may overestimate the plume-head size. Sublithospheric channeling of plume material away from the plume-head region can skew the apparent plume-head size in one direction (Ebinger and Sleep, 1998; Wilson, 1997).

Edge of domal uplift

The outer edge of a domal uplift associated with a plume-generated LIP reflects the outer edge of the underlying flattened plume (Fig. 15.2). This interpretation assumes that the lithosphere is thin enough to allow doming and that lithospheric thickness is homogeneous to allow a simple domal uplift pattern (see Section 12.3.6). I mention two approaches to measuring that outer boundary of domal uplift: giant radiating dyke swarms and patterns of sedimentation.

Far from its focal point a giant radiating dyke swarm gives way to a subparallel pattern (Section 5.2.2). This transition is related to the increasing influence of the regional stress field compared to that of the stress field associated with domal uplift above the plume, and has been inferred to mark the edge of uplift and the approximate edge of the flattened plume head (Ernst and Buchan, 1997a, 2003). The transition in the Mackenzie swarm occurs at a radius of about 1000 km (Fig. 5.2; Section 5.2.2).

The extent of uplift can be marked by an interruption in the sedimentation pattern (Chapter 12). An example is cited regarding the Newark basins along the east coast of North America (Section 12.2.7). Newark basins in the south, located within 1000 km of the plume center for the CAMP event, ceased to accumulate sediments at *c.* 220 Ma. However, Newark basins from farther north along the coast, up to about 3000 km away from the plume center near Florida, do not show an interruption in sedimentation. These observations suggest uplift associated with the CAMP event extended for a distance of 1000 km along the future Atlantic margin of North America from the plume center position near present-day Florida.

Using the various techniques described above, the maximum plume-head size has been estimated for two of the largest continental LIPs, the 1270 Ma Mackenzie and 200 Ma CAMP events, as well as two of the largest oceanic plateaus, Ontong Java and Manihiki (Ernst and Buchan, 2002). Each is consistent with an underlying flattened plume-head radius of about 1000 km, a value considered normal for a plume rising from the deep mantle (Campbell, 2001). Note that if Ontong Java is combined with Manihiki and Hikurangi plateaus into a single LIP as proposed by Taylor (2006) its areal extent (and inferred size of underlying flattened plume head) becomes somewhat larger (approaching nearly 1500 km in radius).

Location of associated carbonatites and kimberlites

As discussed earlier, mafic magma associated with LIPs can be injected laterally (sideways) via dykes for distances of up to 2500 km away from their plume center, and similar lateral transport of mantle material can occur due to sublithospheric channeling. So the location of mafic flows and intrusions in the crust can be significantly offset from their underlying source areas (plume center). However, carbonatites and kimberlites are injected vertically. So carbonatites and kimberlites that are coeval with and linked to a LIP event (Ernst and Bell, 2010; Chapter 9) are a reliable guide to the location of the underlying mantle source area. For instance, carbonatites coeval with the Circum-Superior LIP are found in the Kapuskasing Structural Zone, which is distal from mafic–ultramafic magmatism along the northwestern to northern to northeastern margins of the Superior craton (Fig. 9.10), and indicate a much wider extent of the underlying plume than previously realized (Ernst *et al.*, 2008). Kimberlites associated with the Deccan LIP are found at the far east end of the Indian subcontinent under the Bastar craton, confirming the presence of the Deccan plume further east than previously recognized (Fig. 9.4; Chalapathi Rao *et al.*, 2011).

Deep crustal xenoliths

The regional extent of a LIP underplate at the base of the lithosphere can be linked to the distribution of deep crustal xenoliths and, in turn, linked to magmatic or thermal pulses from the LIP. For example, mafic xenoliths from young Phanerozoic kimberlites of the Lac de Gras area of the Slave province of northern Canada contain zircon (and rutile), which yields both Archean and Proterozoic U–Pb ages (Davis, 1997; Davis *et al.*, 2003). The Proterozoic ages match known regional LIP events (dominated by dyke swarms) and the xenoliths presumably represent sampling of underplate components related to these LIP events. Specifically, the xenoliths of age *c.* 1.27 Ga link to the Mackenzie LIP, the plume center of which is located at about 800 km to the north (Fig. 13.5). The *c.* 2.23–2.21 Ga xenoliths could correlate with the Malley and MacKay dyke swarms, the plume centers of which are more speculatively located on the east side of the Slave craton, approximately 300 km away (Ernst and Bleeker, 2010). Another example is from the Kirkland Lake area of the Abitibi belt of the Superior province of Canada. A kimberlite pipe yielded xenoliths with Archean and 2.4–2.5 Ga zircon ages. The latter are similar to the regionally distributed Matachewan LIP (Moser and Heaman, 1997) the plume center of which is located about 250 km away at the southern margin of the Superior craton (Fig. 11.12; e.g. Fahrig, 1987; Ernst and Bleeker, 2010).

These metamorphic zircon (and rutile) ages from the Slave and southern Superior provinces of Canada are interpreted as evidence for thermal pulses due to intrusion of plume-generated magmas along the crust–mantle interface, and provide an estimate of the spatial extent of underplating due to an underlying plume head.

Xenoliths from the subcratonic lithospheric mantle

Xenoliths from the lithospheric mantle provide an even more direct estimate of the distribution of the underlying plume, which has been accreted to, injected into, or thermally perturbed the lithosphere. The temperature, pressure, and composition of the subcontinental lithosphere mantle (SCLM) can be mapped using xenolith "probes" found in kimberlites (e.g. Griffin *et al.*, 1999; Wyman and Kerrich, 2002; Heaman and Pearson, 2010) such as the PGE and Re–Os isotopic analyses of peridotite xenoliths in kimberlites (e.g. Pearson, 1999; Pearson *et al.*, 2007). Although such studies are still relatively new, exciting results have already been obtained from the Slave craton of the northern Canadian shield (Fig. 15.16). Pearson *et al.* (2007) showed a statistical analysis of these data, combined with other samples of the upper mantle, which show that depletion ages are not evenly distributed but cluster in distinct periods, around 1.2, 1.9, and 2.7 billion years. These events can be approximately associated with major LIP events: the *c.* 1.2 Ga depletion ages can be correlated with the 1270 Ma Mackenzie LIP (Fig. 5.2; e.g. Baragar *et al.*, 1996). The *c.* 1.9 Ga depletion age can be linked with the widespread 1880 Ma Mara-Ghost event of the Slave craton (e.g. Buchan *et al.*, 2010). The 2.7 Ga depletion event can be linked with the 2.7 Ga Kam LIP of the Slave craton (Section 6.3.3), event 227 in Ernst and Buchan (2001). However, a more rigorous comparison is required to test the role for LIPs vs. subduction-related effects with respect to these depletion ages.

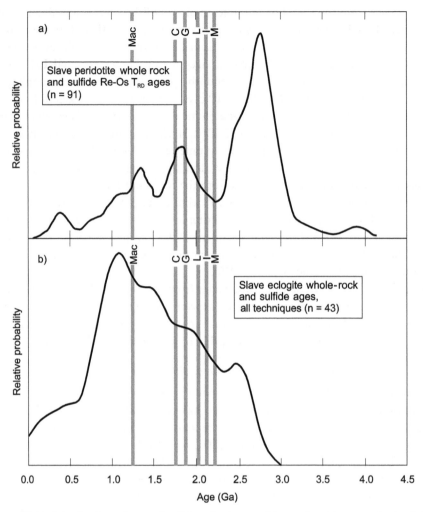

Figure 15.16 Distribution of ages for lithospheric xenoliths and xenocrysts obtained from kimberlites in the Slave craton. Superimposed LIP events are Mac = 1.27 Ga Mackenzie, C = 1.74 Ga Cleaver, G = 1.88 Ga Ghost, L = 2.03 Ga Lac des Gras, I = 2.10 Ga Indin, M = 2.21–2.23 Ga MacKay and Malley swarms. From Figure 2 in Heaman and Pearson (2010).

Subcretion of fossil plume heads to the lithosphere

Slow seismic anomalies are located in the mantle lithosphere immediately beneath the Paraná portion of the Paraná–Etendeka LIP (Van Decar *et al.*, 1995), the Deccan LIP (Kennett and Widiyantoro, 1999), and Ontong Java LIP (Richardson *et al.*, 2000; Klosko *et al.*, 2001). The first two of these are preserved beneath continental lithosphere, whereas the third is wholly beneath oceanic lithosphere. In each case, the slow seismic anomaly extends down to great depths: 500 km (Paraná), 250 km and possibly 500 km (Deccan), and 300 km (Ontong Java). Anomaly widths are 300 km (Paraná), 800 km (Deccan), and 1200 km (Ontong Java). These seismic anomalies can be interpreted as fossil mantle plume heads, which represent a combination of a residual thermal anomaly and a restite root remaining after a partial melt was extracted over this depth range to produce the overlying flood-basalt province. Given that thermal anomalies associated with a plume head will persist for only 500–1000 Ma (Griffiths and Campbell, 1990; Campbell, 2001), the potential recognition of Proterozoic and Archean fossil plume heads will require the existence of a persistent compositional anomaly.

It has also been proposed that plume heads which have been accreted to the lithosphere and are now traveling with their host plate will continue to influence subsequent magmatism (Condie, 2001, pp. 185–86). For example, the source of basalts in the Arabian–Nubian shield for the past 900 Ma may have been a fossil plume head that was accreted to the lithosphere during the formation of the shield (Stein and Hofmann, 1992; Condie, 2001, p. 186). More generally, using geochemical criteria, Condie (2001, p. 186–90) suggests that a minimum of one-third of the lower crust dating from post-Archean time comprises mafic rocks from plume sources, either as accreted oceanic plateaus or as underplated mafic magmatism.

15.5 Assessing LIP origin from associated carbonatites and kimberlites

Another approach to assessing LIP origins is to consider their links with carbonatites and kimberlites (Chapter 9). There are two aspects of this relationship that are relevant. First, many LIPs have associated carbonatites and/or kimberlites (Chapter 9). Second, as discussed below, there is considerable evidence for a plume origin for carbonatites and kimberlites. Assuming a plume origin for carbonatites and kimberlites, LIPs that have associated carbonatites and kimberlites can be considered plume-related by virtue of that association.

15.5.1 Carbonatites

The plume origin for carbonatites is supported by the similarity of isotopic signatures between OIBs and many carbonatites, the primitive nature of the noble-gas data from some carbonatites, the association of carbonatites plus alkaline rocks with known plumes, and the temporal and spatial relationships of carbonatites, lamprophyres, and kimberlites to some LIPs (Section 9.2; Bell, 2001, 2002; Ernst and Bell, 2010). Specifically, data from most young carbonatites closely plot along the mantle plane defined by three end-members, HIMU, EM1, and DMM (depleted MORB mantle; Fig. 10.1). On the basis of Pb, Sr, and Nd isotopic compositions most young carbonatites (< 200 Ma) can be broadly divided into two distinct groups, one representing mixtures between HIMU and EM1 (e.g. the East African rift carbonatites), and another group of carbonatites involving HIMU and FOZO.

Bell and Rukhlov (2004) present a model (Fig. 15.3) for carbonatites based on Wyllie (1988), in which, because of their preference for low degrees of partial melting, carbonatites are produced at the outer, cooler edges of plumes. However, for the same reasons (formation in the cooler edges of a plume) it is also possible to expect that carbonatites could form above the ascending plume, but ahead of major LIP melting (cf. Ernst and Bell, 2010).

15.5.2 Kimberlites

The relationship with kimberlites and LIPs is also robust (Section 9.3; Torsvik *et al.*, 2010; Fig. 15.5). There is a remarkable correspondence between reconstructed LIPs and kimberlites for the past 320 Ma with respect to shear-wave anomalies at the base of the mantle, which supports a deep origin for both kimberlites and mantle plumes. As shown in Fig. 15.4, 80% of all reconstructed kimberlite locations (black dots) of the past 320 Ma erupted near or over the sub-African LLSVP. The most "anomalous" kimberlites (17%) are from Canada (white dots) (Torsvik *et al.*, 2010).

15.6 Origin of multiple pulses

One of the notable characteristics of many LIPs is the presence of multiple pulses (Ernst and Buchan, 2002). While each pulse can be of very short (< 1 Ma) duration, the pulses can be distributed over a longer time period up to tens of millions of years. So in consideration of the origin of LIPs one must also address the origin of multiple pulses in LIPs. Some possible explanations are considered below.

The initial pulse is often inferred to have a plume origin (Fig. 15.12), but the second pulse can be linked to decompression melting associated with the onset of rifting (Campbell, 1998). An example is the NAIP where the initial 62–58 Ma pulse can be plume-related, while the 55 Ma pulse marks the onset of rifting and the formation of a volcanic rifted margin.

In other cases, even more pulses are observed. While these could represent multiple pulses of a single LIP, they can also represent the juxtaposition of separate LIPs. As an illustration of this possibility consider the magmatism which can be linked to the breakup history of India within Gondwana. Southwestern India contains dyke swarms of K–Ar age 144 ± 6 Ma, 105 ± 2 Ma, 81 ± 3 Ma, and 61 ± 9 Ma (Radhakrishna *et al.*, 1990). This dataset could appear to be evidence for a single multi-pulse LIP with ages 144, 105, 81, and 61 Ma. However (and keeping in mind that these are K–Ar ages with true uncertainties greater than the analytical uncertainties), these dyke swarms can be linked to three different known plume/LIP/ breakup events affecting India: *c.* 144 (and possibly *c.* 105) Ma dykes with the separation of Antarctica at about 128–116 Ma (Kerguelen–Comei–Bunbury LIP), the *c.* 81 Ma dykes with separation of Madagascar after mantle-plume arrival at *c.* 90 Ma (Madagascar LIP), and the *c.* 61 Ma age with the separation of the Seychelles–Mascarene microcontinent in association with Deccan LIP magmatism at 65 Ma (Storey, 1995; Ernst and Buchan, 2001c; Jerram and Widdowson, 2005).

As another example consider the Laurentian Iapetus-margin-related CIMP LIP pulses at 615, 590, and 563 Ma (Puffer, 2002; Ernst and Buchan, 2004a; Shumlyansky *et al.*, 2007). It is possible that these multiple pulses represent sequential breakup of other blocks from Laurentia: (1) Baltica from Laurentia (615 Ma pulse), (2) Amazonia from Laurentia at *c.* 570 Ma and linked to the 590 Ma pulse, and (3) the Dashwoods terrane from Laurentia at *c.* 550–540 Ma, and linked to the 563 Ma pulse (Bingen *et al.*, 1998; Waldron and van Staal, 2001; Cawood *et al.*, 2001; Ernst and Buchan, 2004a).

Multiple pulses are also characteristic of tholeiite–komatiite packages in many Archean greenstone belts (Section 6.3). For example, the Abitibi belt has pulses of tholeiitic–komatiitic magmatism at 2750–2735 Ma, 2725–2720 Ma, 2718–2710 Ma, and 2710–2703 Ma (e.g. Ayer *et al.*, 2002). The Fortescue flood basalt of the Pilbara craton has pulses at *c.* 2770, 2720, and 2690 Ma (Thorne and Trendall, 2001; Eriksson *et al.*, 2002; Blake *et al.*, 2004). For instance, a model for these Archean examples might be an underlying plume intermittently breaking through a shallow subducting slab (see discussion in Wyman and Kerrich (2010) for the Abitibi belt). A Phanerozoic example of such an interaction between a rising plume and subducting slab causing magmatic pulses is discussed in Section 13.3.4 (Murphy *et al.*, 1999).

There are several additional mechanisms that can have been invoked to explain multiple LIP pulses. For example, a portion of a plume can become stalled at the mid-mantle boundary and its renewed ascent can lead to a second pulse (Bercovici and Mahoney, 1994). The interaction between thermal and compositional parts of a plume can result in multiple pulses (Lin and van Kekan, 2005). Other variations on these themes are discussed in Montelli *et al.* (2006).

15.7 Evidence from planetary studies

Two aspects regarding the origin of LIPs are important from the perspective of planetary geology. As emphasized in Chapter 7, magmatism on Mars, Venus, Mercury, and the Moon is intraplate with the notable absence of plate tectonics, except perhaps in the earliest history of Mars. On Venus, there is unambiguous plume-type magmatism especially in the BAT region with plume centers at the locus of triple-junction rifting and associated geoid highs reflecting dynamic support from an underlying plume (Figs. 7.9 and 7.12). Plume-style magmatism can also be invoked for Mars in the Tharsis and Elysium regions (Chapter 7).

The planetary record is useful in further testing the idea of a link between bolide impact and LIPs. On Venus, the locations of impacts are distinct from the distribution of individual volcanoes and plains (flood-type) volcanism. Specifically, the locations of the approximately 1000 bolide impacts are clearly morphologically distinct from the distribution of individual volcanoes and plains units and so a bolide origin is not applied to tectonomagmatic features on Venus. The alternative interpretation is offered by Hamilton (2007) who suggests that widespread coronae and other circular tectonomagmatic features on Venus are caused by bolide impact. However, this is a minority view and most researchers recognize distinct morphologies of impact craters and endogenic volcanism on Venus and do not consider coronae to be impact related (e.g. Jurdy and Stoddard, 2007). However, Hansen (2007) suggests that bolide impact is responsible for crustal plateaus. Specifically, she envisions that crustal plateaus may represent the surface scum of huge lava ponds formed by massive partial melting in the mantle due to large bolide impact into an ancient thin lithosphere.

Another useful test for the link between impact and magmatism comes from studies of the lunar maria (Section 7.5). In most mare basins, the vast majority of the exposed volcanic plains were emplaced over an extended period of several hundred million years following the impact event (e.g. Basaltic Volcanism Study Project, 1981; Hiesinger *et al.*, 2011). So, on the Moon, mare volcanism (flood-basalt analogue) fills ancient impact basins, but at younger times long after the original impact, and is therefore not directly related to the impact.

Another aspect potentially of interest from planetary geology is the idea of global LIP events. On Venus, the notion of a global resurfacing event is used to explain the constancy of surface ages as given by impact-crater distribution, and on Mars, a pulse of magmatism in the Hesperian was widespread (Section 7.2.4). Such dramatic intraplate magmatic events might be an example of the foundering of vertical density instabilities leading to mantle overturn events (e.g. Head, 2006). A terrestrial analog could be the widespread magmatism at the end of the Archean (Fig. 11.13), which may have similarities to the Stein and Hofmann (1994) model of MOMO (mantle overturn, major orogeny) episodes.

15.8 Review of critical parameters related to LIP origin

Various origins have been proposed for LIPs. A mantle-plume origin is most compelling for LIPs associated with broad domal uplift, a radiating dyke swarm, and a central region of high-Mg magmatism marking a hot central axis of upwelling mantle. Short-duration LIPs (< 5 Ma) are typically associated with both plume arrival but also lithospheric delamination models. Compositional anomalies (e.g. eclogite in the source) can allow generation of LIPs without significant uplift. Longer-duration LIPs (> 25 Ma) and especially those that have a linear/elongate spatial distribution (e.g. the Ferrar portion of the Karoo–Ferrar LIP) suggest a back-arc or a rift setting, but the magma volumes instead suggest plume involvement. Multiple-pulse LIPs can indicate initial plume arrival followed by rifting event(s). Key parameters for classifying LIPs in terms of setting and origin are: the total magma volume (or, more easily estimated, the areal extent), the age distribution (single vs. multiple pulses, continuous), overall geographic distribution of the LIP (elongate or circular), maximum MgO content, the geometry of associated dyke swarms (radiating or linear), link with rifting/breakup, and link with regional uplift.

Much of the debate on the cause of LIPs has taken place with respect to Mesozoic–Cenozoic LIPs. For this record there is potential to use the architecture of the flood-basalt pile, e.g. subtle systematic sideways shifts in the position of maximum volcanism through the duration of the LIP event, in order to locate the underlying mantle source and any movement of that source over time (Section 12.5.3; e.g. Jerram and Widdowson, 2005).

The plumbing system exposed in most Proterozoic LIPs also has great potential to help discriminate between possible causes of a LIP. For instance, if a large 1 Mkm^2 province can be shown to be fed from a single localized region (say 500 km in diameter, as shown, for instance, by the lateral magma flow patterns in the 1270 Ma Mackenzie radiating dyke swarm), this would favor a plume or potentially delamination origin. However, as noted earlier in this chapter, delamination will not produce LIP-scale magmatism unless it is accompanied by a thermal anomaly, e.g. a mantle plume. If the interpreted source region is elongate or linear, then some control by a plate or other lithospheric boundary would seem likely, e.g. a translithospheric fracture zone. Increased attention to characterizing the plumbing system (dykes, sills, and layered intrusions) of a LIP will undoubtedly contribute to our understanding of the plume or non-plume cause of that LIP.

The outer boundary of a plume head circumscribes the main flood-basalt distribution and approximately coincide with the edge of domal uplift that causes shoaling (shallowing) and offlap in regional sedimentation. In the older record, where the flood-basalt signature is lost, the identification of giant dyke swarms and the presence of high-Mg rocks (picrites and komatiites) is key to locating plume centers. Furthermore, small-volume melts, carbonatites, and kimberlites may also be indicative of plumes, and their location is a direct indication of underlying magmatic sources (flattened mantle plume head) of that age.

Although complicated, the application of geochemical criteria can confirm a plume origin (e.g. elevated temperatures and depths of melting in the asthenospheric source areas), and confirm the amount and nature of lithospheric (mantle and crust) involvement, identify the component mantle sources, and track changes in source characteristics through time. Dynamic domal uplift above the plume and associated rifting and peripheral compressional structures can also be diagnostic. Fossil plume heads that are subcreted to the base of the lithosphere may be recognized through seismic studies and geochemical studies of xenolith populations in kimberlites.

15.9 Summary

Typically, LIPs have been linked to the ascent of a plume from the deep mantle and its arrival and partial melting upon reaching the lithosphere (e.g. Morgan, 1971, 1983; Richards *et al.*, 1989; Griffiths and Campbell, 1989, 1990; Coffin and Eldholm, 1994). The melting effect can be enhanced and a second LIP pulse produced by decompression melting associated with the onset of rifting (White and McKenzie, 1989), by the association with delamination (Şengör, 2001), and also with the presence of an eclogite component (recycled subducted basalts) into the plume (e.g. Cordery *et al.*, 1997; Campbell, 1998; Sobolev *et al.*, 2011).

Plume-alternative mechanisms have been proposed for Mesozoic–Cenozoic examples. While the Siberian Trap LIP is typically linked with a deep-mantle-plume origin (e.g. Sharma, 1997; Saunders *et al.*, 2005; Sobolev *et al.*, 2011), non-plume origins have also been offered: such as delamination (Elkins-Tanton and Hager, 2000; Elkins-Tanton, 2005), subduction, which introduces water into the source area allowing for enhanced melting (Ivanov, 2007), bolide impact (Jones *et al.*, 2002), or enhanced mantle convection at the edge of the Siberian craton (e.g. King and Anderson, 1998; Puffer, 2001).

Continued study to refine the time/space distribution of LIPs will provide more constraints on the nature of plume involvement and the additional roles of delamination and rifting, and also the potential role of fluid-rich sources. Such studies will lead to a more comprehensive understanding of the controls on LIP magmatism spatially throughout a LIP event, and any secular changes in LIP origin through time.

16

LIPs and implications for mineral, hydrocarbon, and water resources

16.1 Introduction

Recent years have witnessed a dramatic increase in our understanding of the economic implications of LIPs for mineral, hydrocarbon, and even groundwater resources (Fig. 16.1). Below I consider the main ore-deposit commodity types associated with LIPs, then discuss the relevance of LIPs to the petroleum industry, and, finally, the influence of LIP plumbing systems on groundwater circulation and aquifer distributions.

16.1.1 Links with ore deposits

LIPs are associated with some of the largest ore deposits on Earth (Table 16.1). LIP-related resources can be broadly split into four different categories, according to the genetic relationship between LIPs and the resources in question (see review in Ernst and Jowitt, 2013):

(1) LIPs as a primary host for mineral deposits (Section 16.2). This is exemplified by LIP-related orthomagmatic Ni–Cu–platinum group element (PGE) sulfide, Fe–Ti–V oxide, and Cr deposit formation, where mineral deposits are formed as a direct consequence of mafic–ultramafic magmatism during a LIP event. The links between LIPs and carbonatites and some kimberlites also mean that the commodities associated with these rocks, namely the rare earth elements (REEs), Nb, and Ta (carbonatites), and diamonds (kimberlites), can also be directly linked with LIP events.

(2) LIPs can contribute to ore formation in hydrothermal systems (Section 16.3). The links between LIPs and hydrothermal deposits can be explained by three differing models that are not necessarily mutually exclusive: (a) LIPs can provide a source of energy for circulating hydrothermal systems, e.g. development of hydrothermal iron oxide–copper–gold (IOCG), volcanogenic massive sulfide (VMS), and other mineralizing systems; (b) LIPs can also be a source of metals and ligands for post-magmatic circulating hydrothermal systems via

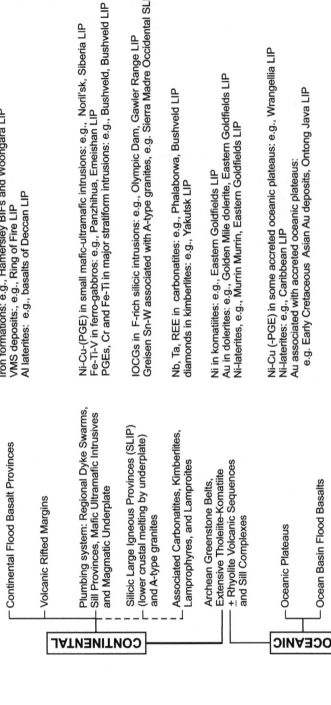

Figure 16.1 Types of LIPs and associated ore deposit types. LIP classification modified after Bryan and Ernst (2008) and Ernst and Bell (2010). Ore-deposit types discussed in the text. From Figure 1 in Ernst and Jowitt (2013). With permission from the Society of Economic Geologists.

Table 16.1 *Selected major ore deposits and their associated LIPs*

LIP (for locations see Figure 1.6)	Commodity	Region/Host tectonic terrane	Age (Ma)	Type/Location/Example
Caribbean–Colombian	Ni	Caribbean	93	3: in Caribbean, northern South America
CAMP	Al	Africa and South America	200	3: in Guinea, west Africa
Siberian Trap	Ni–Cu–(PGE)	Siberia, Ural Mountains, Central Asian fold belt	251	1: Noril'sk
	Mo–W, Sn–W, Hg, Au–Hg			2: in Kuznetsk basin
Emeishan	Fe–Ti–V	China and Vietnam	258	1: Panzhihua and related gabbroic intrusions
	Ni–Cu–(PGE)			1: ultramafic intrusions
Yakutsk	Diamonds	Siberian craton (eastern margin)	c. 360	1: kimberlites
Guibei	Ni–Cu–(PGE)	Alashan block (part of combined South China– Qaidam–Qilian–Tarim craton,	825	1: Jinchuan
Keweenawan	Cu–Ni–(PGE)	Midcontinent rift system	1114–1085	1: Duluth
Mackenzie	Cu–Ni–(PGE)	Laurentia	1267	1: Muskox intrusion
Gawler Range (SLIP)	Cu–Au–Ag–U–REE	Gawler Block, Australia	1590	1, 2: Olympic Dam
Circum-Superior	Ni–Cu–(PGE)	Circum-Superior craton, North America	1885–1865	1: Thompson, Raglan
Bushveld	Ni–Cu–(PGE)	Kaapvaal craton, southern Africa	2060	1: Bushveld
	Cu			1: Phalaborwa carbonatite
	Au			2: Witwatersrand
Stillwater	Cr	Beartooth Mountains, Wyoming craton	2710	1: Various, e.g., Chrome Mountain
Ring of Fire (–Bird River)	Ni–Cu–(PGE)	Caribou block, Superior craton	c. 2730 Ma	1: Eagle's Nest
	Cr			1: Blackbird
	V–Ti–Fe			1: Thunderbird
	VMS			2: McFaulds

Note: the 1850 Ma Sudbury event is not listed despite its rich Ni–Cu metal endowment because its size is sub-LIP scale, and because it has a bolide impact origin. Types refer to the list outlined in Section 16.1, where 1 = magmatic, 2 = hydrothermal, and 3 = weathering processes (e.g. laterite formation). LIP locations are shown in Figure 1.6.

hydrothermal alteration of rocks formed during LIP events. Again, some IOCG, VMS, and potentially Au deposits are examples of these links; (c) in some cases, LIP units act as structural/impermeable barriers or as reactive precipitation mechanisms during hydrothermal fluid flow, such as in Archean LIPs associated with the formation of orogenic Au deposits (Section 16.3.1).

(3) Tropical weathering of mafic–ultramafic LIP units to form economically import-ant Ni–Co laterites and Al bauxites, and weathering of associated carbonatites to yield Nb, Ta, and REE laterites (Section 16.4).

(4) Indirect links between LIPs and ore deposits. Here two aspects are considered: (a) the systematic use of LIPs as a tool for generating robust pre-Pangea reconstructions that allow the tracing of known ore deposits from one crustal block into "greenfield" areas on a formerly adjacent crustal block (Section 16.5); and (b) the nature of the plate-tectonic cycle means that a pulse of rifting and breakup (characteristically associated with LIP emplacement) should be correl-ated with corresponding pulses of transpression and compression (and associated mineralization, e.g. orogenic Au) on favorably oriented plate boundaries elsewhere in the world (Section 16.6).

Below (in Sections 16.2–16.6) I detail each of these categories, assess the strength of the LIP–commodity association in each case, and consider how a LIP context has predictive value for mineral exploration.

16.1.2 Link with petroleum resources

The hydrocarbon link (with oil and gas) is based on three aspects (each of which is described in more detail below; Section 16.7):

(1) LIP events, particularly those of oceanic plateaus, can cause anoxic conditions and produce black shales, which are an important source rock for hydrocarbons.

(2) The regional thermal input from LIPs can cause hydrocarbon composition to change from more complex hydrocarbons to "mature" simpler types desired by the oil and gas industry, and finally to over-mature types of less use to the industry (beyond the oil and gas windows).

(3) Dolerite sills of LIP events can act as structural traps for oil and gas.

16.1.3 Link with water resources

Dolerite dykes and sills provinces that belong to LIP events can be widespread in sedimentary basins (Section 5.3), and they can have a major effect on aquifer transport by being barriers to water flow (Section 16.8). In addition, some dolerites with extensive jointing (columnar jointing) and fracturing can also serve as an aquifer.

16.2 Magmatic ore deposits linked to LIPs

Mafic–ultramafic intrusions of LIPs are highly prospective for Ni–Cu–(PGE), Fe–Ti–V, and Cr ore. In fact, with the exception of Sudbury ores (sub-LIP scale and linked to bolide impact), the majority of major Ni, Pd, and Pt resources are directly related to LIP events. There is also a spatial and temporal association between LIPs and diamondiferous kimberlites and between LIPs and carbonatites that host Nb–Ta–REE phosphate, iron ore, lime, Cu, Zr, Th, U, fluorine/fluorite, and vermiculite resources. Below, I detail links between these different types of ores and LIPs and also discuss how the LIP context can provide insights supportive of exploration targeting. The following is mainly drawn from Ernst and Jowitt (2013).

16.2.1 Types of Ni–Cu–(PGE) deposits and links with LIPs

The most important LIP-related magmatic sulfide deposits are the Ni–Cu–(PGE) deposits, typically subdivided based upon their economic importance into mainly Ni–Cu- vs. PGE-dominated types, although some, like Duluth (Keweenawan LIP) and Noril'sk (Siberian Trap LIP), plot in between in terms of both Ni-Cu and PGEs being important (e.g. Naldrett, 2004; Eckstrand and Hulbert, 2007). Both Ni–Cu- and PGE-dominated types of deposits form through similar processes, whereby immiscible sulfide liquids that are preferentially enriched in chalcophile elements, such as Ni–Cu, and the PGEs are segregated from mafic or ultramafic silicate magmas that became S-saturated.

Various processes have been invoked to explain how magmas become S-saturated, including crustal contamination and the assimilation of crustal sulfides (e.g. Lightfoot and Keays, 2005; Wilson and Chunnett, 2006; Keays and Lightfoot, 2010), magma mixing (e.g. Naldrett and von Grunewaldt, 1989), and fractionation (Andersen *et al.*, 1998); the exact process involved varies from deposit to deposit.

The association between magmatic Ni–Cu–(PGE) mineralization and flood-basalt events is well established (e.g. Naldrett, 1997, 2010a, b; Pirajno, 2000; Schissel and Smail, 2001; Borisenko *et al.*, 2006; Eckstrand and Hulbert, 2007; Tables 16.2 and 16.3). An example is the world-class magmatic Ni–Cu–(PGE) sulfide mineralization at Noril'sk–Talnakh, associated with the end-Permian Siberian Trap LIP (Section 3.2.8; e.g. Naldrett *et al.*, 1992; Hawkesworth *et al.*, 1995; Lightfoot and Keays, 2005; Arndt *et al.*, 2008). Other examples include Ni–Cu–(PGE) mineralization associated with the 260 Ma Emeishan LIP (Section 12.2.10; e.g. Izokh *et al.*, 2005; Borisenko *et al.*, 2006; Pirajno *et al.*, 2009) and mineralization within the Duluth Complex and other intrusions that formed as part of the 1115–1085 Ma Keweenawan LIP event (Section 3.5.2; e.g. Miller and Ripley, 1996; Gál *et al.*, 2011). Furthermore, given the breadth of the LIP family beyond continental and oceanic flood basalts, as summarized in Figs. 1.1 and 16.1, then the range of magmatic

Table 16.2 *Selected Ni–Cu–(PGE) deposits and their associated LIPs*

Ore-bearing intrusion (location)	Associated LIP (age Ma)	Key references for ore deposits
Wellgreen (Yukon, Canada)	Wrangellia (232 Ma)	Hulbert (2002); Schmidt and Rogers (2007)
Noril'sk–Talnakh (Siberia)	Siberian Trap (250 Ma)	e.g. Lightfoot and Keays (2005);
Jinchuan (Alashan terrane, China)	Gubei (825 Ma)	e.g. Pirajno *et al.* (2009)
Giles Complex (central Australia)	Warakurna (1075 Ma)	Evins *et al.* (2010)
Duluth, Eagle, Current Lake (Great Lakes region, central North America)	Keweenawan (1115–1085 Ma)	Ding *et al.* (2010); Peterson and Peck (2010)
Thompson (Manitoba, Canada); Raglan-Katinniq (northern Quebec, Canada)	Circum-Superior (1880 Ma)	Eckstrand and Hulbert (2007); Layton-Matthews *et al.* (2007); Heaman *et al.* (2009)
Pechenga (Karelia and Kola peninsula, northern Europe)	Pechanga–Onega (1970 Ma)	Alapieti and Lahtinen (2002); Dedeev *et al.* (2002)
Bushveld (southern Africa)	Bushveld (2060 Ma)	Barnes and Maier (2002); Cawthorn (2010)
East Bull Lake intrusive suite (southern Ontario, Canada)	Matachewan (2480–2450 Ma)	Peck *et al.* (2001); James *et al.* (2002a,b)
Kambalda and Agnew (Western Australia)	Eastern Goldfields (2700 Ma)	Barnes *et al.* (2012)
Shangani, Trojan and Hunter's Road (Zimbabwe)	Bulawayan–Belingwe (2700 Ma)	Prendergast (2004)

Note: LIP locations are shown in Figure 1.6.

Ni–Cu–(PGE) deposits that can be linked to LIPs is greatly expanded. For example, dyke swarm dominated LIPs can represent eroded flood basalts and have the potential to host additional "Noril'sk"-type Ni–Cu–(PGE) deposits.

Ni–Cu deposit types and links with LIPs

The connections between the full LIP family (Fig. 16.1) and Ni–Cu–(PGE) deposits can be looked at using the classification system of Naldrett (2004, 2010a, b) who grouped Ni–Cu–(PGE) deposits into seven distinct types (Fig. 16.2; Table 16.3). Four of the classes (NC-1, NC-2, NC-3, and NC-5; Naldrett, 2010a, b) are all robustly linked to LIPs: Komatiitic NC-1-type magmatic sulfide deposits are formed from high-Mg ultramafic magmas associated with Archean komatiite–tholeiitic greenstone belts (e.g. Kambalda and Abitibi belts of Western Australia and Canada, respectively) that are considered as Archean LIPs (Chapter 6; e.g. Ernst and Buchan,

Table 16.3 *Types of Ni–Cu–(PGE) magmatic sulfide deposits and their associations with LIPs*

Class	Related magmatism	Camps and deposit (name and location)	Tectonic setting of magmatism	Associated LIP
NC-1	Komatiite	Wiluna–Norseman greenstone belt (Kambalda, Mt. Keith, Perseverance and others) (Australia)	Greenstone belts (rift?)	Belong to Archean LIPs in each area
		Abitibi belt (Canada)		
		Zimbabwe (Zimbabwe)		
		Thompson (Canada)	Rifted continental margin	Circum-Superior (1.88–1.87 Ga)
		Raglan (Canada)		
NC-2	Flood basalt	Noril'sk (Siberia)	Rift (triple junction)	Siberian Trap (0.25 Ga)
		Duluth (Midcontinent North America)	Rifted continental margin	Keweenawan (1.12–1.09 Ga)
		Muskox (northern Canada)	Rifted continental margin	Mackenzie (1.27 Ga)
		Insizwa (South Africa)	Rifted continental margin	Karoo (0.18 Ga)
		Wrangellia (western Canada and Alaska)	Rifted island arc	Wrangellia (0.23 Ga)
NC-3	Ferropicrite	Pechenga (Scandinavia)	Rifted continental margin	Pechenga–Onega (1.97 Ga)
NC-4	Troctolite–anorthosite–granite	Voisey's Bay (eastern Canada)	Rift	?
NC-5	Miscellaneous picrite–tholeiite	Montcalm (Ontario Canada)	Greenstone belts (rift?)	–
		Jinchuan (China)	Rifted continental margin	Gubei (0.83 Ga)
		Niquelandia (Brazil)	Continental rift	?
		Moxie (NE USA)	Orogenic (compressive)	–
		Aberdeenshire Gabbros (UK)		–
		Rona (Scandinavia)		–
		Acoje (Indonesia)	Ophiolite belt (Oceanic)	–
NC-6	Impact melt	1.85 Ga Sudbury (Canada)	Meteorite impact	Target rocks include: East Bull Lake intrusive suite of Matachewan LIP (2.49–2.45 Ga) and Nipissing sills of Ungava–Nipissing (2.21 Ga)
NC-7	Ural–Alaskan	Duke Island (Alaska)	Convergent margin	–
		Turnagain Arm (Alaska)		–
		Salt Chuck (Alaska)		–
		Quetico (?) (Canada)	Possible forearc setting	–

Note: adapted from Naldrett (2010a) with information added on LIP associations. Locations of LIPs are shown in Figure 1.6. "?" in the last column indicates a LIP association is suspected but not yet identified. "–" indicates a LIP association is considered unlikely.

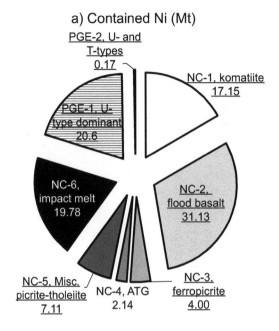

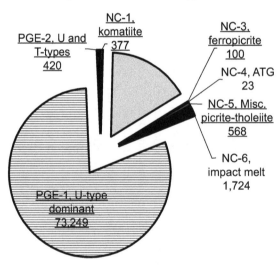

Figure 16.2 Distribution of Ni (a) and Pt + Pd (b) resources within Ni- and PGE-dominated magmatic sulfide systems; underlining indicates deposit classes where the majority, if not all, of the deposits have definite links to LIP events, and plain text indicates classes with no, or an uncertain, LIP link. ATG = Anorthosite–troctolite–granite. Numbers in (a) are in mega-tonnes (Mt), and in (b) are in tonnes (t). From Figure 3 in Ernst and Jowitt (2013) and adapted from Naldrett (2010a). With permission from the Society of Economic Geologists.

2001b; Prendergast, 2004; Ernst *et al.*, 2005; Barnes *et al.*, 2012). An additional komatiitic basalt-related example is the *c.* 1880 Ma Circum-Superior event of the Superior craton that has also been interpreted as a LIP (Fig 9.10; e.g. Heaman *et al.*, 2009; Ernst and Bell, 2010; Minifie *et al.*, 2013). Flood-basalt-related high-Mg ultramafic NC-2 type deposits are associated with the feeders to flood basalts and are exemplified by magmatic sulfide deposits at Noril'sk–Talnakh (belonging to the 250 Ma Siberian Trap LIP; Section 3.2.8), Duluth, Eagle, and Tamarack (all associated with the 1115–1085 Ma Keweenawan LIP; Section 3.5.2), and within the Muskox intrusion (1270 Ma Mackenzie LIP; Fig. 13.5). The NC-3 class is associated with feeders for overlying ferropicritic lavas, as is the case for the Pechenga deposits in far northwestern Russia which belong to the regionally extensive *c.* 1970 Ma Pechenga–Onega LIP. This LIP includes volcanics and sills in the Lake Onega area, the economically important intrusions in the Pechenga-type locality within the Kola Peninsula, and into southern Karelia as a major north-northwest-trending dyke swarm (e.g. event 160 in Ernst and Buchan, 2001b; Vuollo and Huhma, 2005). The type (and only economic) example of the NC-4 troctolite–anorthosite–granite-related class of magmatic sulfide mineralization is the Voisey's Bay deposit in Canada. This deposit is associated with the Nain plutonic suite, which, although an important magmatic event, is not currently classified as a LIP as it is too small (15–25000km^2) and this event may have spanned too large a time range (1.34–1.29 Ga, e.g. Li *et al.*, 2000). Also, see "Fe–Ti–V deposits" subsection of Section 16.2.2 for a discussion of anorthosite–mangerite–charnockite–rapakivi granite (AMCG) magmatism, including an example with associated Ni mineralization.

The NC-5 miscellaneous picrite–tholeiite class of deposits is associated with high-Mg basalts, as exemplified by the Jinchuan deposit of the Alashan block, north China. Although the Jinchuan intrusion (and associated Ni–Cu–(PGE) mineralization) has been linked with the adjacent North China craton (Li and Ripley, 2011), more recent research has linked the host Alashan block with the adjacent Qaidam and Qilian blocks that connect the Tarim block to the west and the South China block to the east into a single terrane, termed the South-West China United Continent (SWCUC; Song *et al.*, 2012). Furthermore, the South China, Tarim, and intervening portions all share 825 Ma intraplate magmatism that can be considered to represent a single LIP (Li *et al.*, 2005b; Ernst, 2007; Ernst *et al.*, 2008; Lu *et al.*, 2008; Pirajno *et al.*, 2009; Song *et al.*, 2013; S.G. Song, personal communication, 2013).

The two remaining classes, NC-6 (meteorite impact related) and NC-7 (Ural-Alaskan type) are not related to LIPs, at least not directly. In particular, the Sudbury event, in Ontario, Canada (Naldrett, 2004; Ames and Farrow, 2007), is the type example of the production of significant volumes of mafic magma (although on a sub-LIP scale) and associated Ni–Cu–(PGE) ore deposits by crustal melting resulting from a meteorite impact. However, it is interesting that the metallogeny of the Sudbury event may be inherited from the impact target rocks that would likely have included Ni–Cu-rich intrusions of the East Bull Lake intrusive suite that are associated with the 2470–2450 Ma Matachewan LIP (Section 5.2.3; Figs. 5.5 and 11.8), with an interpreted mantle-plume center nearby (Ernst and Bleeker, 2010), and the Nipissing sills of the 2215 Ma Ungava–Nipissing LIP (Section 5.3.3). The final NC-7 Ural-Alaskan class contains ore-bearing intrusions formed in arc settings that are not LIP-associated; the members of this class generally contain subeconomic nickel but can have modest PGE abundances (see PGE-4 below). In summary, this strong link between classes NC-1, NC-2, NC-3, and NC-5 and LIP events affirms the importance of a LIP context for Ni–Cu-dominated magmatic deposits. In fact, as summarized in Fig. 16.2, the global magmatic sulfide-hosted Ni and Ni–Cu resources are dominated by LIP-related mineral deposits (Fig. 16.2; Mudd *et al.*, 2013).

Oceanic Ni–Cu–(PGE) deposits

The majority of known Ni–Cu–(PGE) ore deposits are associated with continental LIPs; the only significant mineralization associated with oceanic LIPs occurs within the Wrangellia accreted-oceanic-plateau LIP of Alaska and Canada, as exemplified by the Wellgreen deposit in the Yukon Territory of Canada (e.g. Marcantonio *et al.*, 1994; Schmidt and Rogers, 2007). The Wrangellia LIP was built on an earlier island arc, suggesting that only oceanic plateaus that have interacted with continental crust/lithosphere are likely to host coeval magmatic sulfide mineralization.

PGE-dominated magmatic sulfide deposits

PGE-dominated magmatic sulfide deposits are commonly hosted by layered mafic–ultramafic intrusions (e.g. Naldrett, 2004, 2010a; Eckstrand and Hulbert, 2007), with Naldrett (2004) splitting these deposits into six classes (Table 16.4; Fig. 16.3). Naldrett (2004, 2010a) suggested that the majority of PGE-dominated magmatic sulfide deposits are associated with two distinct magma types, an early siliceous high-Mg basalt magma (U-type) and a later more tholeiitic-type magma (T-type). U-type magmas may be similar to modern boninites (e.g. Hamlyn and Keays, 1986), although it is more likely that they represent primary mantle melts that have been crustally contaminated (e.g. Eales and Costin, 2012) or that have interacted with subduction-modified regions of the continental lithospheric mantle. See also discussion of boninite-like magmas in Section 10.8.

The PGE-1 class of magmatic sulfide deposits consists of intrusions formed from both U and T types of magma, but in systems that are dominated by U-type magmas, and with PGE mineralization concentrated in areas of mixing between the two magma types (Naldrett, 2004). Deposits of this class host PGE mineralization within reef-type or stratiform deposits associated with chromitite and sulfide-rich layers in large, well-layered mafic–ultramafic intrusions (Naldrett, 2004; Eckstrand and Hulbert, 2007). Examples of reef-type mineralization include the Merensky reef and UG-2 chromitite layers within the Bushveld Complex of South Africa, the J-M reef of the Stillwater complex, Montana, USA and the Main Sulfide Zone of the Great Dyke, Zimbabwe (Naldrett, 2004).

Table 16.4 *Types of PGE-dominated magmatic sulfide deposits and their associations with LIPs*

Class	Related magmatism	Intrusive complexes	Deposits	Tectonic setting of magmatism	Associated LIP
PGE-1	U-type magmatism with minor tholeiites	Bushveld	Merensky, UG-2 reefs, Platreef, Dunite pipes	Intracratonic	Bushveld (2.05 Ga)
		Stillwater	J-M reef	Not known	Stillwater (2.71 Ga)
		Great Dyke	Main sulfide zone	Intracratonic rift	?
		Lac des Iles	Robie zone	Not known	?
PGE-2	U-type and tholeiitic	Munni–Munni	Porphyritic websterite layer	Intracratonic?	West Pilbara e (2.94–2.91 Ga)
		Penikat	SJ reef	Rifted continental margin	BLIP (2.5-2.44 Ga)
		Portimo	SK and RK reefs, marginal ore		
PGE-3	Tholeiitic	East Bull Lake	Marginal ore		Matachewan (2.49-2.45 Ga)
		River Valley	Marginal ore		NAIP (0.06 Ga)
		Skaergaard	Platinova reef		
		Cap Edvard Holm	Willow Ridge reef		
		Sonju Lake	Sonju Lake reef	Rift (triple junction)	Keweenawan (1.11-1.09 Ga)
		Coldwell	Marathon, Bermuda		
PGE-4	Calc-alkaline mafic	Longwoods	Longwoods area	Orogenic (island arc)	?
		Volkovsky	Volkovsky, Baron		?
PGE-5	Ural–Alaskan alkaline mafic–ultramafic	Urals platinum belt	Soloviev Hills, Urals placers		?
		Koryakia region	Seynav–Galmoznav		?
		Kondyor	Kondyor	Cratonic	?
PGE-6	Alkaline mafic–ultramafic	Guli	Ingarinda	Rift	Siberian Trap (0.25 Ga)

Note: Adapted from Naldrett (2004) with information on associated LIPs and the composition of the Lac des Iles deposit added; LIP locations are shown in Figure 1.6.

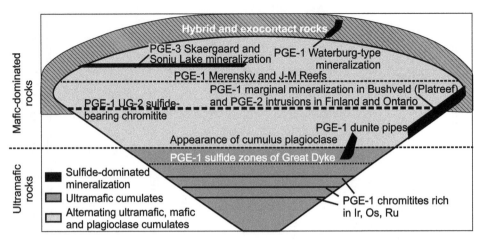

Figure 16.3 Schematic model of a LIP-related layered intrusion, showing the relative position and petrological affinities (e.g. chromite vs. sulfide-dominated; ultramafic vs. mafic; reef vs. contact styles of mineralization) of the differing types of LIP-related PGE-dominated magmatic sulfide deposits discussed in the text; adapted from Hoatson *et al.* (2006) and Naldrett (2010a). Note that a single layered intrusion is unlikely to host all of these styles of mineralization, and that PGE deposits with differing magmatic affinities can occur in similar positions within an intrusive system, as exemplified by the similar positions of PGE-1 contact-type mineralization of the Platreef section of the Bushveld Complex, part of the Bushveld LIP, and PGE-2 contact-type mineralization hosted by 2500–2450 Ma intrusions of the BLIP (Baltic LIP) of Finland and Russia and the 2490–2450 Ma Matachewan LIP of Ontario and Quebec in Canada. From Figure 4 in Ernst and Jowitt (2013). With permission from the Society of Economic Geologists.

All of these classic layered intrusions have associated coeval magmatism in the region and are considered to belong to LIP events. Specifically, the 2060 Ma Bushveld intrusion is part (albeit the dominant part by volume) of a magmatic LIP event that includes minor mafic intrusions in the Vredefort area, the subsurface Molopo Farm Complex of Botswana, silicic magmatism in the Okwa inlier (northwestern Kaapvaal craton), and associated carbonatitic magmatism, including the mineralized Phalaborwa carbonatite (Sections 5.4.1 and 9.2.2; e.g. Rajesh *et al.*, 2013). The 2710 Ma Stillwater Complex is plausibly linked to widespread mafic dykes of similar age (2710–2680 Ma) that are distributed across the Wyoming craton (K. Chamberlain, personal communication, 2013). In addition, the approximately 550-km-long 2575 Ma Great Dyke of Zimbabwe is flanked along its entire length by two nearby coeval dolerite dykes (e.g. Wilson *et al.*, 1987; Wilson, 1996). Also given that the Great Dyke is essentially truncated at the north and south edges of the Zimbabwe craton then its continuity into other formerly adjacent blocks seems likely.

The PGE-1 class also includes magmatic breccia-type deposits within stock-like or layered bodies, such as the Lac des Iles deposit of Ontario, Canada, which currently have no demonstrable links with LIP events (e.g. Hinchey *et al.*, 2005; D. Peck, personal communication, 2011), although it should be noted that the resources within this breccia type of deposits are minute compared to the PGE-1-type deposits with demonstrable LIP links.

The PGE-2 class is associated with intrusions formed from both U- and T-types of magma, but where the T-type of magma is dominant. This class includes mineralization hosted by the Penikat, Portimo, and Fedorova intrusions that all belong to the well-known 2500 to 2450 Ma Baltic LIP (BLIP) of Finland and Russia (e.g. event 207 of Ernst and Buchan, 2001b; Sharkov *et al.*, 2005; Bayanova *et al.*, 2009; Kulikov *et al.*, 2010).

The PGE-3 class is associated with T-type, tholeiite-only intrusions, including the East Bull Lake and River Valley intrusions of the 2490 to 2450 Ma Matachewan LIP (Section 5.2.3; Fahrig, 1987; Ernst and Bleeker, 2010), mineralization within the Sonju Lake intrusion of the Duluth Complex and the Coldwell intrusion in Ontario, both related to the Keweenawan LIP event (Section 3.5.2), as well as the Platinova reef of the Skaergaard intrusion of the North Atlantic Igneous Province (NAIP; Section 3.2.4). These examples again demonstrate a close link between this class of deposits and LIP events.

Calc-alkaline mafic–ultramafic intrusions host PGE-4, PGE-5, and PGE-6 class mineralization, as exemplified by the Volkovsky deposit and the Baron prospect of the Urals platinum belt (Naldrett, 2004). Although the majority of these are not LIP-related by virtue of a calc-alkaline setting, one example, the Guli deposit, is associated with the final stages of the Siberian Trap LIP (Kamo *et al.*, 2003).

The first class of PGE deposit (PGE-1) dominates the world's PGE production and resources (Fig. 16.3), with the Bushveld Complex alone representing some 70.9% of global PGE resources (Mudd, 2012). It is also worth noting that the second-largest single PGE resource (hosted by the Siberian Trap LIP) comprises the NC-2-type Ni-dominated Noril'sk–Talnakh deposits; these Pd-dominated deposits host *c.* 12% of current global PGE resources (Mudd, 2012). PGE-4 deposits (not LIP-related) are generally subeconomic, with the exception of placer prospects associated with the PGE-4 class of deposit found within the Urals, Kamchatka, and Colombia (e.g. Weiser, 2002). These data indicate that the PGE-1 and -2 ore classes, as well as some PGE-3 deposits, are dominantly LIP-related. In short, the global producers of PGE are dominated by LIP-related magmas (Table 16.4).

Implications for exploration of LIP context for Ni–Cu–(PGE) deposits

Here I consider some for the implications from a LIP context for exploration for Ni–Cu–(PGE) deposits.

Finding the small "sweet spot" for deposits in a huge LIP An outstanding question is whether LIP-related Ni–Cu–(PGE) deposits have a preferential location or setting within a given LIP. It is apparent that Ni–Cu-mineralized bodies represent a very small target in an otherwise huge magmatic event (Fig. 16.4). To illustrate the scale of the problem, the Siberian Trap LIP covers an area of approximately 7 Mkm^2 with a volume of 4 km^3 (Ivanov, 2007). Yet despite this vast extent, the only operating Ni–Cu–(PGE) mines are located in the *c.* 30-km-long × 20-km-wide Noril'sk and Talnakh ore junctions, representing a very small fraction (*c.* 0.01%) of the total area of the Siberian Trap LIP.

There are two exploration implications from this realization: (1) given the scale of the LIP event, if one small "sweet spot" is found, perhaps there are others remaining to be discovered; and (2) strategies are required for vectoring toward the small prospective regions (sweet spots) within such extensive events.

Proximity to plume center As discussed in Chapter 15, for a variety of reasons I favor a plume origin for a majority of major LIP events, and that context is assumed in the discussion below. Figure 16.4 illustrates that a first-order control on the location of Ni–Cu–(PGE) mineralization is proximity to a mantle-plume center, as has been previously noted for the Sibarian Trap LIP (e.g. Lightfoot *et al.*, 1993b). The Siberian Trap, Mackenzie, and CAMP LIPs host the respective mineralized Noril'sk–Talnakh, Muskox, and Freetown intrusions that are each located within a few hundred kilometers of their respective plume centers. Magmatic sulfide deposits within the Thompson Ni belt of Canada belong to the *c.* 1880 Ma Circum-Superior LIP and are located proximal to the plume center for this event, which is defined on the basis of a radiating dyke swarm (Fig. 9.10).

The other economically important part of the Circum-Superior LIP, the Raglan and related deposits of the Cape Smith belt are more distal from the proposed plume center (Fig. 9.10). However, given a 10–20° relative rotation between the eastern and western halves of the Superior craton (e.g. Evans and Halls, 2010), and if further geochronology reveals that it occurred post-1880 Ma, then the Raglan ores would originally have been within about 600 km of the plume center (about 500 km closer than at present) (e.g. Ernst and Bell, 2010).

More broadly it seems that large layered intrusions (potential hosts for PGE-1 type ores, as discussed above) are typically located near their plume centers. For instance, the Bushveld intrusion is thought to overlie its plume center (e.g. Hatton, 1995; Olsson *et al.*, 2011). Large gravity anomalies (inferred to mark layered intrusions) partially circumscribe the well-defined Mackenzie plume center at a distance of about 300 km (Baragar *et al.*, 1996). Also as noted above, the large Freetown layered intrusion is located within a hundred kilometers of the CAMP plume center (Fig. 16.4).

The spatial association of major magmatic sulfide mineralization and proximity to the plume-center is related to the greater volumes of magma flow above a mantle plume head (e.g. Naldrett *et al.*, 1992). These greater volumes of flow lead to a more dynamic environment and a higher R factor (i.e. increased interaction between magmatic sulfides and silicate magma), leading to higher tenor sulfides (Campbell and Naldrett, 1979). (Tenor equals the concentration of metals within sulfides in a

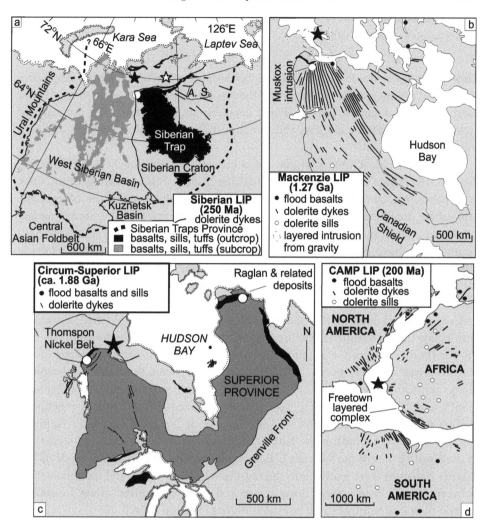

Figure 16.4 Locating a small Ni–Cu–PGE-mineralized region (the "sweet spot") marked by white circles in an areally extensive LIP. Four examples are shown: (a) the 250 Ma Siberian Trap LIP with its rich Noril'sk–Talnakh deposits; (b) the 1270 Ma Mackenzie event with its associated Muskox intrusion; (c) the 1880 Ma Circum-Superior LIP with its associated ore-rich areas of the Thompson belt, and the Raglan and related deposits of the Cape Smith belt; (d) the 200 Ma CAMP LIP with the associated Freetown intrusion of Sierra Leone. A.S. = Anabar shield. The stars locate inferred mantle plumes. For the Siberian LIP the solid star is at the convergence of north–south rifting (of the East Siberian basin) and east–west rifting (of the Khatanga trough) (Schissel and Smail, 2001). The non-filled star marks the plume center inferred from a radiating dyke swarm (Ernst and Buchan, 1997a). From Figure 5 in Ernst and Jowitt (2013). With permission from the Society of Economic Geologists.

given rock/sample.) (Note, R-factor = the number of volumes of magma interacting with a given volume of sulfide.)

Link to thinned cratonic margins and translithospheric fractures While I emphasize the link with plume center regions, two other aspects are considered important in the localization of magmatic Ni–Cu–(PGE) deposits: an association with thinned and metasomatized cratonic margins and the importance of magma transport up along translithospheric faults.

Craton and paleocraton margins are a critical location for Ni–Cu–(PGE) ore deposits (Kerrich, 2005; Begg *et al.*, 2010) as plumes cannot melt beneath thick lithosphere (e.g. at 250 km depth) but as the plume slides along the base of the lithosphere it can start melting when it reaches thinner margins of cratons; melting can begin upon reaching depths of ~ 100 km. Cratonic margins are also important because they have been typically previously metasomatized in earlier subduction events. As a consequence craton margin regions are also fertile metasomatic sources for boninitic-like intrusions, which are an important magma type for intrusions that host Ni–Cu–(PGE)s (see discussion below and also in Section 10.8).

In continental regions, the margins of Archean cratons are often associated with major translithospheric fractures; and these may facilitate the transport of such Ni–Cu–(PGE) prospective magmas through the lithosphere (e.g. Naldrett *et al.*, 1992; Naldrett, 1999; Begg *et al.*, 2010). Magma transported through the lithosphere along translithospheric fractures, can interact with subcontinental lithospheric mantle and crustal material (e.g. with sulfur- and silica-rich sediments), potentially leading to sulfide concentration and segregation of Ni and PGEs (e.g. Naldrett, 1997). The classic example is the Kharaelakh fault which is the locus of magmatism of the Noril'sk ores of the Siberian LIP. I would note that the dominant mode of LIP magma transport from mantle into the crust is via dyke swarms and the relationship between magma transported along dykes and that transported along translithospheric fractures requires some investigation (e.g. Section 5.2.6).

Importance of conduits and chonoliths for Ni–Cu-dominated deposits The feeder systems into or out of a layered intrusion are the loci of many Ni–Cu–(PGE) ore deposits. In particular, as noted by Beresford and Hronsky (2012), mafic Ni–Cu sulfide deposits are hosted in a diversity of types of igneous conduits including small concentrically zoned pipe-like intrusions called chonoliths, along blows in dykes, or within breccia pipes within layered intrusions. Notably, the Noril'sk ores of the Siberian Trap LIP reside in chonoliths.

The importance of conduits can be explained by the concept of the R factor (Campbell and Naldrett, 1979). The higher the ratio, the greater the volume of magma that can contribute to the enriching of the sulfides in metal content.

Conduits are important because of their ability to channel the flow of large amounts of hot magma and, in response to changes in flow rate and direction,

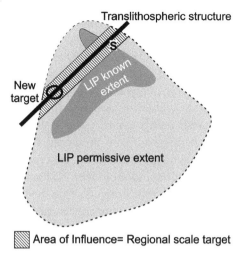

Figure 16.5 The potential target regions guided by the intersection of translithospheric faults with the distribution of a LIP. S = known sulfide ore deposit. Unpublished diagram of S. Beresford, kindly provided by him.

high-tenor, magmatic sulfides can be concentrated in structurally favorable locations (D. Peck, written communication, 2011).

Geochemical assessment of LIPs for Ni–Cu–(PGE) deposit potential An important question that arises is whether there are geochemical characteristics that can be observed in the non-mineralized portions of the broader LIP event which can allow assessment of whether a given LIP event has potential for Ni–Cu–(PGE) ore deposits.

There are three main factors that affect the Ni–Cu–(PGE) prospectivity of a LIP (e.g. Ernst and Jowitt, 2013; Jowitt and Ernst, 2013; Jowitt *et al.*, 2014): (1) magma fertility for these metals; (2) opportunities for crustal contamination and chalcophile element segregation through sulfur saturation; and (3) magma (plumbing system) pathways through the crust, including vertical and lateral transport of magma and corresponding links between metal-depleted and metal-enriched parts of the system.

Additional effects that are important include: (1) depth and high degree of partial melting in the asthenosphere source to "melt out" PGE phases (Naldrett, 2010a); and (2) earlier subduction-related metasomatic alteration of SCLM to generate the source of norites and boninitic-like magmas that are characteristic of many late Archean, Paleoproterozoic Ni–Cu–(PGE) ore deposits (R. Kerrich, personal communication, 2012; see also Section 10.8).

Assessing fertility One approach to assessing magma fertility was provided by Zhang *et al.* (2008) who compiled geochemical data from ten major LIPs, five of which are considered barren: the Deccan, Kerguelen, Ontong Java, Paraná (part of the Paraná–Etendeka LIP), Ferrar, and five of which are prospective: Karoo,

Emeishan, Siberian Trap, Midcontinent (Keweenawan), and Bushveld LIPs, to determine possible geochemical signatures favorable for Ni–Cu–(PGE) mineralization. (Note that Karoo and Ferrar LIP magmatism have identical 183 Ma ages (Sections 3.2.7 and 5.3.2), and are inferred to belong to the same LIP; but the distinct compositions of Karoo and Ferrar magmas indicate a distinct petrogenetic history for each, which also explains their differing potential for Ni–Cu–(PGE)s.)

Zhang *et al.* (2008) focused mainly on the source characteristics of magmas, concluding that LIP parental magmas were generated from deep-seated, mantle plumes but that there are diagnostic differences between those associated with known Ni–Cu–(PGE) mineralization and those with no known mineralization. LIPs associated with Ni–Cu–(PGE) mineralization contain high proportions of high-MgO and -Ni primitive melts, which have low concentrations of Al_2O_3 and Na_2O, are enriched in incompatible elements, and have isotopic signatures that vary between depleted-plume and EM1-type mantle compositions. This suggests that the interaction between plume magmas and ancient cratonic lithospheric mantle may significantly contribute to the formation of Ni–Cu–(PGE) magmatic sulfide deposits (Zhang *et al.*, 2008). In contrast, barren LIPs have fewer high-MgO magmas, and have isotopic compositions that vary between plume and EM2-type mantle, suggesting involvement of deep recycled material in the mantle-source region for these LIPs or crustal contamination but, in any case, have little interaction with old lithospheric mantle (Zhang *et al.*, 2008).

Chalcophile-element depletion A second important factor in assessing Ni–Cu–(PGE) potential relates to chalcophile-element segregation from primary fertile chalcophile-element-undepleted magmas (e.g. Naldrett, 2004; Zhang *et al.*, 2008; Jowitt and Ernst, 2013). Chalcophile-element depletion can be caused by crustal contamination together with the assimilation of crustal sulfur and the segregation of immiscible magmatic sulfides. The presence of magmas that have undergone chalcophile-element segregation is a clue to the presence of chalcophile enrichment elsewhere in the magmatic system. For example, the Nadezhdinsky volcanic formation of the Siberian Trap LIP exhibits chalcophile-element depletion which is related to chalcophile-element enrichment in the ores associated with the Noril'sk intrusions (Fig. 10.15; e.g. Lightfoot and Keays, 2005). In contrast, even though the Deccan LIP event included significant volumes of fertile magma, there was insignificant assimilation of crustal sulfur via crustal contamination. Therefore, Deccan magmas did not reach sulfur saturation, and did not develop Ni–Cu–(PGE) sulfide mineralization (Keays and Lightfoot, 2010). So the Deccan LIP is an example of a fertile yet barren LIP event.

Exploring "upstream" in the plumbing system The third aspect of assessing Ni–Cu–(PGE) potential relates to the characteristics of the magmatic plumbing system and geometric relationship between chalcophile-depleted and -enriched parts of the system. Specifically, if the geochemistry reveals that a given LIP is fertile in terms of Ni–Cu–(PGE) potential and has experienced local chalcophile-element depletion then it can be inferred that there is metal enrichment elsewhere in the system and strategies for vectoring toward such areas (potential ore deposits) are necessary.

The vectoring approach is to first identify units within the LIP that are depleted in chalcophile elements; these should then be prospected "upstream" in the plumbing system for corresponding sites of metal enrichments (i.e. potential ore deposits). Several types of such "downstream–upstream" relationships are considered: One type, incorporating linkages between "downstream" chalcophile-depleted and contaminated sections of a magmatic plumbing system with "upstream" S saturation and deposition of magmatic sulfides, has been developed for komatiitic systems (e.g. Lesher *et al.*, 2001).

Another type links volcanic units with their "upstream" magma source in a staging chamber at depth. In this case metal depletion in the volcanic rocks can be linked to "upstream" areas of magmatic sulfide deposition in metal-enriched magma chambers. This vectoring approach is exemplified by the Siberian Trap LIP where the voluminous and highly chalcophile-element-depleted Nadezhdinsky Formation volcanic rocks have been linked to a large "upstream" magma chamber that also fed the Noril'sk and Talnakh ore-bearing intrusions (e.g. Lightfoot and Keays, 2005).

Another more speculative type of vectoring method is related to dyke swarms, where lateral emplacement of radiating dyke swarms has been linked to magma chambers in the focal region of the swarm. This approach can be exemplified with reference to the Mackenzie LIP (Section 5.2.4). Here, different subswarms of the overall Mackenzie radiating dyke swarm were spawned and laterally injected from magma chambers in the plume focal region (Fig. 5.22). It is postulated that identification of a chalcophile-depleted and crustally contaminated subswarm of Mackenzie dykes implies that the original metal endowment of these magmas was lost "upstream," in a layered intrusion somewhere along strike in the "upstream direction," i.e. toward the Mackenzie plume center (e.g. Ernst and Jowitt, 2013).

Summary LIPs can be used in a two-part strategy for Ni–Cu–(PGE) exploration (e.g. Jowitt and Ernst, 2013): (1) geochemistry can be used to assess the overall prospectivity of LIPs, the fertility of magmas and extent of chalcophile element depletion/enrichment; and (2) an understanding of LIP plumbing systems can allow vectoring "upstream" toward regions or identification of magmatic subpulses within a given LIP that have greater ore deposit potential. I want to emphasize that this geochemical approach allows assessment of non-mineralized portions of a LIP, to predict whether and where mineralized units are likely present elsewhere in the LIP system.

16.2.2 Magmatic oxide deposits

LIPs can also host major Fe–Ti–V and Cr magmatic oxide deposits, as exemplified by the Bushveld and Emeishan LIPs (Table 16.1). Here I discuss these and other examples of the links between LIP events and the formation of significant magmatic oxide mineral deposits. This section is also mainly based on Ernst and Jowitt (2013).

Fe-Ti-V deposits

The Main and Upper Series of the Bushveld Complex contain a number of Ti–V-bearing magnetite reefs that form 30.1 and 9.7% of the world's V and ilmenite (Ti) resources (Laznicka, 2010); the Main Magnetite Layer, located near the base of the Upper Series, hosts the world's largest resource of vanadium (Kruger, 2005). The Emeishan LIP of China and Vietnam hosts Fe–Ti–V ores in mafic–ultramafic intrusions (e.g. Izokh *et al.*, 2005; Pirajno *et al.*, 2009), the most prominent of which are the Hongge, Panzhizhua, Taihe, and Baima intrusions.

The magmas that formed the Emeishan LIP are split into high- and low-Ti series, with the high-Ti series magmas associated with Fe–Ti–V oxide mineralization and the low-Ti series magmas associated with Ni–Cu–(PGE) sulfide mineralization. The low-Ti series magmas underwent crustal contamination that led to sulfur saturation, whereas the high-Ti magmas underwent long-lived fractionation that caused eventual Fe oxide saturation (Hou *et al.*, 2011). The fractionation of the latter formed the highly evolved ferrobasaltic or ferropicritic LIP magmas that generated the world-class Fe–Ti–V ore bodies of the Emeishan LIP (e.g. Zhou *et al.*, 2005). Similarly, fractionation and Fe oxide saturation, in addition to potential pressure changes during these processes, are thought to be responsible for the formation of the Bushveld Complex Fe–Ti–V reefs (e.g. Kruger, 2005; Cawthorn and Ashwal, 2009).

An important class of anorogenic magmas, AMCG (anorthosite–mangerite–charnockite–rapakivi granite), and associated anorogenic granites, are also known to host Fe–Ti–V deposits (e.g. Hébert *et al.*, 2005) but have uncertain relationships to LIPs and mantle plumes (Ashwal, 1993; McLelland *et al.*, 2010). Current models suggest that AMCG suites of magmas can form during orogenic thickening of the lithosphere and subsequent delamination (e.g. McLelland *et al.*, 2010). However, an interesting example that illustrates a potential LIP and plume connection is the *c.* 1800–1750 Ma Korosten and Korsun–Novomirgorod AMCG complexes of the Ukrainian shield of Sarmatia (Duchesne *et al.*, 2006; Bogdanova *et al.*, 2013); here, an early (1790 Ma) pulse of AMCG magmatism is coeval with Ni-sulfide-bearing dolerites/gabbros (Amelin *et al.*, 1994; Shumlyanskyy *et al.*, 2012; Bogdanova *et al.*, 2013). Furthermore, the 1790 Ma mafic magmatism is potentially widespread in Sarmatia and may be linked to the > 300 000 km^2 Avanavero LIP within the Amazonian craton and dykes in the West African craton (e.g. Ernst *et al.*, 2013a, 2014b) if the South America–Baltica (SAMBA) continental reconstruction of Johansson (2009, 2014) is correct. This indicates that the AMCG magmatism within the Ukrainian shield is potentially part of a LIP event potentially extending over an area greater than 1 Mkm^2.

The AMCG type of Fe–Ti–V ores tends to be massive, rich in apatite and monazite, and may be REE-enriched, as exemplified by AMCG complexes within the Ukrainian shield portion of Sarmatia described above (Bogdanova *et al.*, 2013). In comparison, layered, intrusion-hosted Fe–Ti–V ores (e.g. Bushveld) are generally meter-thick layers of magnetite-ilmenite or titanomagnetite that have higher grades than typical AMCG-type ores (Cawthorn and Ashwal, 2009; D. Peck, personal communication, 2011).

Chromium (Cr) deposits

There are two major sources of chromite, namely podiform chromitites within ophiolites (e.g. Melcher *et al.*, 1997) that are not LIP-related, and stratiform chromitites within layered complexes that are generally LIP-related. These stratiform chromitites consist of chromite-rich layers within large layered intrusions.

The world's largest Cr resources are located within the Critical zone of the Bushveld Complex; these deposits contain around 75% of the world's known Cr resources (e.g. Laznicka, 2010). The other main sources of Cr are hosted by the 2570 Ma Great Dyke of Zimbabwe (Selukwe deposit), the 2710 Ma Stillwater Complex, and the *c.* 2450 Ma Kemi intrusion of Karelia. As noted elsewhere (Section 5.4.2), the Bushveld, Great Dyke of Zimbabwe and Stillwater intrusions are each inferred to belong to a LIP. The Kemi intrusion is also part of a LIP, the BLIP (also known as the East Scandinavian LIP or Sumian LIP; Sharkov *et al.*, 2005; Bayonova *et al.*, 2009; Kulikov *et al.*, 2010, and references therein).

It should be emphasized that not all LIP-related layered intrusions contain significant chromite. For instance, Vogel *et al.* (1998) divided the 2.5 to 2.45 Ga Fennoscandian mafic intrusions of the BLIP on the basis of age, chromium content, and incompatible-element compositions, and noted that Cr-poor intrusions are dominated by gabbronorite and leucogabbronorite, whereas Cr-rich intrusions contain much higher proportions of ultramafic to mafic rocks (almost 1:1) along with the presence of chromitite layers.

The *c.* 2730 Ma Ring of Fire Complex, a potential Archean LIP within the McFauld's greenstone belt of northwestern Ontario (Canada) on the eastern edge of the older Caribou block (Houlé *et al.*, 2012; Laarman *et al.*, 2012; Metsaranta and Houlé, 2012) also hosts world-class Cr mineralization. There are two reasons for linking this complex to a LIP event. Firstly, the Ring of Fire Complex represents an Archean komatiite–tholeiite association that is considered indicative of a LIP (Section 6.3; cf. Ernst *et al.*, 2005; Bryan and Ernst, 2008). Secondly, the Ring of Fire Complex occurs on the same microcontinent (Caribou terrane) as the contemporaneous Ni–Cu sulfide and Cr-mineralized Bird River sill, suggesting that these two complexes may form part of a wider LIP-scale event (e.g. Ernst and Jowitt, 2013). The Ring of Fire Complex hosts a number of world-class Cr deposits, namely the Blackbird, Big Daddy, Black Thor, and Black Label deposits, which are located within the same complex as Ni–Cu–(PGE) magmatic sulfide deposits (Eagle's Nest, Eagle Two, AT12) and a large Ti–V deposit (e.g. Mungall et al., 2010; Metsaranta and Houlé, 2012). These deposits are also associated with contemporaneous VMS mineralization (Section 16.3.4).

All of the magmatic mineral deposits within the Ring of Fire Complex formed from stratigraphically differentiated komatiitic magmas in distinct metallogenic associations, namely lowermost zones where magmatic sulfides accumulated at the base of dyke-like structures, intermediate-level chromitites within sill-like feeder zones, and residual magmas stripped of Ni–Cu–(PGE) and Cr that fractionated to produce Ti–V mineralization within layered intrusions (Mungall *et al.*, 2010). The Ring of Fire Complex provides a classic example of a single suite of potentially LIP-associated magmas that underwent differing igneous evolutionary paths to produce numerous different types of mineral deposits.

Implications for the relationship between LIPs and Fe–Ti–V and Cr deposits

The presence within the Ring of Fire event of differing types of magmatic Ni–Cu–(PGE) sulfide and Ti–V and Cr oxide mineralization offers the possibility that an improved understanding of the sourcing and evolution of magmas within a LIP context can enable vectoring toward prospective areas for both sulfide (Ni–Cu–(PGE)) and oxide (Fe–Ti–V and Cr) mineralization, and also the delineation of prospective or entirely unprospective LIP events (e.g. Jowitt and Ernst, 2013; Ernst and Jowitt, 2013).

16.2.3 Carbonatite-related deposits

The links between LIPs and carbonatites (and associated alkaline complexes) have been explored in detail in Chapter 9 (see also Ernst and Bell, 2010; Ernst and Jowitt, 2013). The strong link between carbonatites and rifting that has been previously documented (e.g. Burke *et al.*, 2003) is consistent with a LIP affinity for carbonatites, as the majority of LIP events are also rift-related (e.g. Courtillot *et al.*, 1999; Ernst and Bleeker, 2010). Carbonatites are genetically linked to a number of important mineral deposit types and a wide variety of commodities, including the REEs, Nb, F, P, Fe, Th, U, Cu, Zr, Ta, Au, Ag, the PGEs, and industrial minerals such as vermiculite and lime (Woolley and Kjarsgaard, 2008a, b). For instance, the 2.06 Ga Phalaborwa carbonatite complex of South Africa, which belongs to the Bushveld LIP (Section 9.2.2), is rich in Cu sulfides. Interestingly, it also has some affinities with IOCG deposits; see Section 16.3.2 (e.g. Corriveau, 2007; Ernst and Bell, 2010). In addition, the Bayan Obo carbonatite REE–Nb–Fe deposit of the North China craton, currently the world's most important source of the REEs, has an uncertain age but was provisionally linked by Peng *et al.* (2011) to the 920–900 Ma Dashigou LIP, and by Zhang *et al.* (2012b) to the 1320 Ma Yanliao LIP, both located in the North China craton.

Implications of a LIP context for carbonatites

The strong link between LIPs and carbonatites is compelling (Ernst and Bell, 2010), but the nature of that link is at an early stage of understanding. For instance, it is not yet known whether carbonatite geochemistry can be used to fingerprint a family of carbonatites and demonstrate a link with a particular LIP. Another question is whether carbonatites associated with particular LIPs are more likely to be more REE- or Nb–Ta-rich, and whether this can be predicted from studying the geochemistry of the associated LIP. Another area of speculation is whether the (1) setting of a carbonatite: e.g. along a rift, near the plume center, or distal from the plume center, (2) magma source depth/lithospheric thickness, or (3) emplacement timing with respect to the plume/LIP history (early or late), have any predictive value in terms of the endowment of economically important elements/minerals in carbonatites.

16.2.4 Kimberlites and LIPs (diamondiferous vs. barren)

LIPs and diamond potential of kimberlites

In Section 9.3 it was shown that many LIPs have associated kimberlites (see also Ernst and Jowitt, 2013). A key question, addressed in this section, is whether a LIP context can be used to assess the diamond potential of kimberlites. Whether or not a kimberlite is diamondiferous is controlled by the nature of the lithosphere that the kimberlite magma passes through. This is because diamonds within these magmas are present as xenocrysts and were extracted from the lithospheric mantle root of ancient cratons within the diamond stability field (at depths of >200 km). Thus, whether a kimberlite is diamondiferous is dependent on the thermal structure and composition of the subcontinental lithospheric root; in order for this root to contain diamonds it must be deep, cool, and contain carbon, therefore providing the necessary pressure, but not at temperatures so high as to move outside of the diamond stability field (Fig. 15.12; e.g. Gurney *et al.*, 2005; Kerrich *et al.*, 2005). The deep cool roots needed for this are characteristic of ancient Archean cratons (e.g. Artemieva, 2011) and, therefore, ancient roots are the classic key target areas for diamondiferous kimberlite exploration (e.g. Clifford, 1966).

Helmstaedt and Gurney (1995) introduced the concept of processes that can be friendly or unfriendly (destructive) to cratonic roots. The diamond potential of a lithospheric root can be potentially destroyed by a thermal event (e.g. impingement of a mantle plume) that can heat up the lithosphere, or by delamination, that effectively drops potentially diamondiferous regions of the lithospheric mantle into the deeper asthenosphere. An excellent example of an "unfriendly" impact is the effect of the 1270 Ma Mackenzie plume on the root of the northern Slave craton (Helmstaedt and Gurney, 1995; Gurney *et al.*, 2010), where seismic tomography suggests destruction of the cratonic root out to a distance of about 700 km south from the Mackenzie plume center, but still to the north of the diamondiferous kimberlites of the Lac de Gras area. As noted in Ernst and Jowitt (2013), the "unfriendly" effects of a mantle plume can also include propagating a thermal pulse into the lithospheric root potentially destroying any diamonds by raising the temperature out of the diamond stability field.

It is observed that the first generation of successively spatially overlapping generations of kimberlites is normally the most diamond-prospective, potentially because the finite resource of diamonds within the root of a craton is more likely to be tapped by early-formed kimberlitic magmas (Gurney *et al.*, 2010). As an example, in South Africa *c.* 200–110 Ma Mesozoic Group II kimberlites of South Africa are more consistently diamondiferous than later *c.* 100–85 Ma megacryst-bearing Group I kimberlites (Gurney *et al.*, 2010). This effect can be explained by Figure 9.14 showing potential craton-unfriendly events in the form of the 180 Ma Karoo LIP/plume and the 130 Ma Paraná–Etendeka LIP/plume that may have affected the diamond potential for the younger *c.* 90 Ma kimberlites located in the area broadly between the Karoo and Paraná-Etendeka plume centers. A similar relationship is evident in the Churchill kimberlite province, near Rankin Inlet, Nunavut, Canada, where highly diamondiferous 234 Ma kimberlite dykes are succeeded by weakly diamondiferous to barren kimberlite pipes that formed between 228 and 170 Ma (Gurney *et al.*, 2010), although in this case an "unfriendly" plume event has not as yet been identified.

Another clear example is from Siberia (Fig. 9.15) where a temporal control on the diamond potential of kimberlites is evident; diamondiferous *c.* 370 Ma kimberlites are linked with the Yakutsk–Vilyui LIP, but younger (*c.* 245−210 Ma) barren kimberlites are linked with a pulse(s) of the Siberian Trap LIP (e.g. Kiselev *et al.*, 2012; Ivanov *et al.*, 2013). This observation suggests that the "unfriendly" event was the earlier 370 Ma Yakutsk–Vilyui LIP. This recognition that the relationship between LIP events and kimberlites can enhance or reduce kimberlite diamond prospectivity can potentially be used as an exploration tool as follows.

The location of plume centers can be determined using radiating dyke swarms associated with LIPs (Sections 5.2.4 and 11.7.3), and these plume centers can be circumscribed with circles marking the predicted extent of plume influence in terms of destroying diamond potential in the lithospheric roots. The extent of each circle represents a "diamond exclusion zone" that is predicted to host barren post-LIP kimberlites, whereas kimberlites older than the LIP may be diamondiferous.

However, there are important qualifications in this approach: (1) the area of influence of a plume center on the lithosphere has not yet been modeled in the context of its influence on diamonds within lithospheric roots, meaning that the appropriate radius for the circles used within this exploration technique has not yet been determined; (2) a lag time is expected for any plume-related thermal pulse to propagate into the adjacent lithospheric root and affect any diamonds, meaning that the diamond potential would be destroyed closer to the plume center earlier, as this diamond unfriendly thermal pulse propagates outward to greater distances with time; (3) finally, there is the possibility that diamonds in the lithospheric root reform and the diamond potential of the root is restored after the thermal anomaly associated with the plume has decayed back to the ambient levels stage. Bedini *et al.* (2004) estimated a slow cooling rate of 40° to 105 °C/Ga for the South African lithosphere based on garnet Sm–Nd ages. If typical, this would suggest a > 1 Ga lag for any potential return of diamond potential.

16.3 Hydrothermal ore deposits linked to LIPs

LIP events can be linked to the formation of hydrothermal and secondary mineral deposits in a number of ways (Tables 16.5 and 16.6); here I provide an overview of the links between LIP events and these types of mineral deposits. This section is also based on Ernst and Jowitt (2013).

16.3.1 High-temperature hydrothermal deposits

Burial and metamorphism of LIP lithologies can enable the release of base metals (prominantly Cu, Zn, Ni, Pb) and precious (noble) metals (Au, Ag, PGE) that can be incorporated into ore-forming systems, with the large thermal anomalies generated during LIP emplacement also potentially driving large-scale fluid circulation and the potential formation of several differing types of hydrothermal ore deposits in adjacent host rocks (e.g. Pirajno, 2000; Ernst and Jowitt, 2013). Lower-temperature, generally sediment-related systems can also be affected by thermal pulses from LIP events, causing or enhancing mineral deposit formation.

Native Cu deposits in flood basalts and overlying sediments

A class of hydrothermal ore deposits that is clearly linked with LIPs is the native Cu type that is common within flood-basalt sequences; this style of mineralization is formed during low-grade burial metamorphism when Cu and Ag are mobilized from the basalts by metamorphic fluids that migrate up along permeable strata or along structures (e.g. Brown, 2007). These elements are then deposited as native metals or high-Cu/S-ratio sulfides in oxidized traps such as open fractures, and amygdules, and are also deposited interstitially to conglomerates and breccias (Schmidt and Rogers, 2007; Laznicka, 2010). The same mechanism that produces these basalt-hosted sulfide occurrences can also produce Cu–Ag mineralization within sedimentary rocks overlying the flood basalts, forming sedimentary copper or sediment-hosted Cu $\pm$ Ag $\pm$ Co deposits (Schmidt and Rogers, 2007; Laznicka, 2010). The best-known example of flood-basalt-hosted Cu mineralization is that associated with the 1115–1085 Ma Keweenawan LIP of the Midcontinent region of North America, with over 5 Mt of Cu and by-product Ag produced before 1960 from flood basalts of the Keweenaw peninsula of northern Michigan (Nicholson *et al.*, 1992). These deposits continue to be productive, with some hosted by the Middle Keweenawan Portage Lake volcanic sequence, but the most significant deposit is the White Pine deposit that is hosted in shales overlying the Keweenawan flood basalts (e.g. Brown, 2007). Native copper is also hosted by the Coppermine basalts of the

Table 16.5 *Examples of LIP-related hydrothermal and secondary mineral deposits*

Deposit type	Deposit	Associated LIPs	Age (Ma)	Association	Notes
VMS	McFaulds VMS deposits	Ring of Fire (Bird River)	c. 2735?	LIPs as energy plus potential source rocks	
Orogenic Au	Various, e.g. Golden Mile Dolerite, Western Australia	Archean LIPs	Archean	LIPs as chemical/structural traps for fluids	
	Au deposits within North China, Yangtze (South China), and Siberian craton margins	Ontong Java	c.120–90	LIPs as far-field tectonic effects	
Ni–Co laterite	Wingellina Several deposits, e.g. Murrin Murrin	Warakurna Eastern Goldfields	c. 1075 2715–2670	Weathering of LIP rocks	
	Cerro Matoso deposit in northwest Colombia	Caribbean–Colombian	90		Uncertain link – may be related to weathering of ophiolitic rocks
Bauxite	Central and western Indian deposits	Deccan	67–60		
	Guinea bauxite district	CAMP	c. 200		c.9.1 Gt of bauxite
Banded iron formations	Hamersley BIFs	Woongarra–Weeli Wolli	2450	Changes to global ocean chemistry caused by major LIP magmatic events, leading to contemporaneous global deposition of BIFs	
	Superior granular iron formations	Circum-Superior	c. 1880		

Basalt-hosted Cu	Various deposits, e.g. White Pine Coppermine basalts Eastern Tianshan Nikolai greenstones	Keweenawan Mackenzie Tarim Wrangellia	1115–1085 1270 290–275 230–225	Native Cu hosted by LIP-related flood basalts; deposits form during low-grade metamorphism and hydrothermal mobilization of metals	5 Mt of Cu, by-product Au
IOCGs	Olympic IOCG district, e.g. Olympic Dam, Prominent Hill	Gawler Range	1598–1583	LIPs as energy plus potential source rocks	
Granite-related and greisen Sn–W	Various deposits associated with Sierra Madre Occidental	Sierra Madre Occidental	38–20	Exsolution of mineralizing magmatic fluids from LIP-related felsic magmas plus LIPs as potential energy sources	
	Mineralization associated with granites of Bushveld Complex	Bushveld	2060	Exsolution of mineralizing magmatic fluids from LIP-related felsic magmas plus LIPs as potential energy sources	
Epithermal Au–Ag	Various deposits associated with Sierra Madre Occidental	Sierra Madre Occidental	38–20	Exsolution of mineralizing magmatic fluids from LIP-related felsic magmas plus LIPs as potential energy sources	
	Deposits of the Chon Aike province in Patagonia	Chon Aike	188–153	Exsolution of mineralizing magmatic fluids from LIP-related felsic magmas plus LIPs as potential energy sources	

Note: The ages refer to the LIP formation age; mineral deposit formation may in some cases significantly post-date this magmatism. LIP locations are shown in Figure 1.6.

Table 16.6 *Permian–Triassic mineralization linked to Siberian Trap, Emeishan, and Tarim LIPs of Asia*

Mineralization	Siberian Trap LIP				Emeishan LIP	Tarim LIP		
	Siberian craton, Taimyr	KTFZ, Kuznetsk basin	Altai, eastern Kazakhstan	Orkhon–Selenga basin	Southeast China, Vietnam	Eastern Tien Shan, southern Mongolia	Western Tien Shan	Kalba–Narym area
Mo–W Greisen		214–213						
Sn–W (Ta–Nb)		220–218 236–233	266					231–225
Hg, Au–Hg	(T)	< 238	234–231	T	(219)		236–231, 271	
Porphyry Cu–Mo	229–223	240–232, 274–273		225–200 234–232	230	233–225, 275–270, 296	294–290	
Mo–W (Cu–Au) Greisen		240–232	242–237	238	234	248–244		
Ni–Co–As	<248	255–252	258–250	256	250	248–244, 285–282	290	
Cu–Ni–Pt	250–248		(250–240)					

Note: Adapted from Borisenko *et al.* (2006). LIP locations are shown in Figure 1.6. All ages in million years (Ma).

1270 Ma Mackenzie LIP (Jones *et al.*, 1992) and basalts of the eastern Tienshan area of China that may be linked with the *c.* 280 Ma Tarim LIP (e.g. Yuan *et al.*, 2008; Zhang *et al.*, 2008). The Nikolai greenstone and associated intrusions of the 230 Ma Wrangellia accreted oceanic LIP and their metamorphosed equivalents also have potential to host strata-bound disseminated Cu deposits, specifically in the Denali prospect (Schmidt and Rogers, 2007). Basaltic sequences in the Russian Far East also host Cu deposits (Nokleberg *et al.*, 2005), although the LIP affinity of these basalts remains undetermined.

Gold and oceanic plateaus

There is also a speculative link between oceanic plateaus and orogenic gold; the following is based on Ernst and Jowitt (2013). Bierlein and Pisarevsky (2008) suggest that the formation of world-class orogenic Au mineralization of the Sierra Nevada in California and the Jiaodong peninsula of China is related to the remobilization of Au from oceanic-plateau material and the formation of substantial orogenic gold deposits during the late stages of accretionary orogens (Kerrich and Fyfe, 1981; Kerrich and Wyman, 1990; Kerrich *et al.*, 2005). This model postulates metamorphic devolatilization following subduction and partial accretion of an oceanic plateau to a continental margin. Such devolatilization would cause the release of gold from the oceanic plateau into the overlying crust (Bierlein and Pisarevsky, 2008). Devolatilization and liberation of gold and siderophile metals from gold- and siderophile-enriched plume-related plateau material would increase gold concentrations within hydrothermal ore-forming fluids, essentially fertilizing the orogenic gold systems associated with this devolatilization (Bierlein and Pisarevsky, 2008).

This model involving sourcing of gold from the oceanic plateau itself is supported by the ease of mobilization of Au within mafic rocks, especially if those mafic rocks are sulfide-bearing (e.g. Jowitt *et al.*, 2012). An alternative model for the role of oceanic-plateau accretion in the formation of orogenic Au deposits involves the stalling of subduction–accretion complexes as thicker mantle lithosphere (associated with oceanic plateaus) gets jammed within the subduction zone, as is currently occurring with the Ontong Java plateau (Section 4.2.1; MacInnes *et al.*, 1999; Kerrich *et al.*, 2000). This jamming of the subduction zone and cessation of accretionary complex formation causes both subcreted and hydrated mantle-wedge material that was cooled during subduction to heat up and become dehydrated. Metamorphic heating and associated dehydration is caused by the migration of geotherms up the jammed subduction–accretionary complex and leads to remobilization and liberation of gold within this complex (R. Kerrich, personal communication, 2012). See full discussion in Ernst and Jowitt (2013).

Remobilization of gold by a LIP event

Further evidence for a relationship between LIP events and gold mineralization is provided by the Witwatersrand basin of southern Africa, the largest gold province in the world. Rasmussen *et al.* (2007) provided *in situ* U–Pb SHRIMP data on monazite and xenotime ages of *c.* 2.06–2.03 Ga and *c.* 2.14–2.12 Ga for the timing of gold deposition and/or remobilization in the northwestern-central parts and southern parts of the Witwatersrand basin, respectively. The former is linked with the *c.* 2.06 Ga Bushveld event and suggests that emplacement of the Bushveld Complex may have produced a widespread thermal and fluid pulse that remobilized the existing gold in those portions of the Witwatersrand basin within about 100 km of the Bushveld Complex. The older 2.14–2.12 Ga ages reflect a separate thermal pulse of unknown origin more proximal to the southern side of the Witwatersrand basin (perhaps another as yet unidentified LIP event affecting the southern Kaapvaal craton). It should also be noted that other studies have suggested that the initial mineralization within the Witwatersrand basin occurred earlier than these dates. For example, Schaefer *et al.* (2010) provided an Re–Os date of 2.3–2.2 Ga for mineralization within the Vaal reefs, meaning that it is most likely that the 2.06–2.03 Ga (Bushveld-related) and *c.* 2.14–2.12 Ga thermal/fluid pulses merely modified the world-class Au mineralization already present within the Witwatersrand basin.

Orogenic gold

Orogenic gold deposits are generated at midcrustal (4–16 km) levels in tectonic settings that include areas adjacent to terrane boundaries within transpressional subduction–accretion complexes in orogenic belts, or in areas that are inboard of these settings during mineralizing events associated with either the delamination of mantle lithosphere material or by impingement of a mantle plume on an area (Kerrich *et al.*, 2000).

For example, differentiated dolerite sills, such as the Golden Mile dolerite in the Yilgarn craton (2680 Ma; Rasmussen *et al.*, 2009), are the single most important host for orogenic Au; approximately 50% of gold deposits and 75% of the Au within the craton are hosted by these sills (Hergt *et al.*, 2000). These sills are related to low-Th basalts within the Eastern Goldfields superterrane; these basalts and associated sills define an Archean plume-head-related LIP (e.g. Barnes *et al.*, 2012; Section 6.3.4). The differentiated nature of these sills means that they contained significant concentrations of Fe, meaning they were ideal traps for Au (e.g. Phillips and Groves, 1983; Goldfarb *et al.*, 2005). This in turn means that the Fe-rich root zones of the differentiated sections of LIPs, in general, may also be ideal traps for Au precipitating from fluids within orogenic Au systems. See further discussion in Ernst and Jowitt (2013).

16.3.2 Iron oxide–copper–gold (IOCG) deposits

A number of iron oxide–copper–gold (IOCG) and associated deposits have been linked with silicic or silicic parts of LIPs, including the world-class Olympic Dam IOCG deposit in South Australia (Tables 16.1 and 16.5; Fig. 16.6); this deposit is a

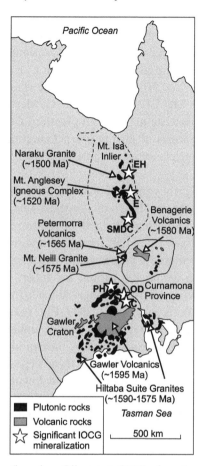

Figure 16.6 Reconstruction (based on Giles *et al.*, 2004) of north- and south-central Australia showing the location and timing of definitively (e.g. Gawler Range) (and potentially LIP-related) A-type magmatism in eastern Australia; adapted from Betts *et al.* (2007). Ernst and Jowitt (2013) suggest that the IOCG deposits shown within this figure either formed from hydrothermal systems driven by the LIP or that the metals and/or ligands (e.g. F) within the hydrothermal fluids which formed these deposits were sourced from LIP-related igneous rocks. Note the south-to-north progression of magmatism, with initial formation of the Gawler craton SLIP before subsequent events in the Curnamona province and around Mount Isa; the dashed line indicates the geophysical extent of the Mount Isa Inlier. The stars indicate the location of significant IOCG mineralization: C = Carrapateena, E = Eloise, EH = Ernest Henry, OD = Olympic Dam, PH = Prominent Hill, SMDC = deposits of Selwyn–Mount Dore corridor. From Figure 8 in Ernst and Jowitt (2013). With permission from the Society of Economic Geologists.

supergiant iron oxide Cu–U–Au–Ag ore deposit that represents the world's largest uranium deposit and fourth largest copper deposit, containing some 9.075 Gt of ore at grades of 0.87% Cu, 0.32 g/t Au, 1.5 g/t Ag, and 0.027% U_3O_8, with a production to date of *c.* 3 Mt of Cu metal (Mudd *et al.*, 2013). The Olympic Dam deposit is the most important example of the IOCG class of mineral deposit (e.g. Corriveau, 2007; Groves *et al.*, 2010; McPhie *et al.*, 2011). At Olympic Dam, the immediate host to the ore is hydrothermal breccia within the granitic and volcanic rocks of the Mesoproterozoic (1590 Ma) Gawler craton bimodal LIP (e.g. Pirajno and Bagas, 2008; McPhie *et al.*, 2011; Pirajno and Hoatson, 2012; Wade *et al.*, 2012).

Furthermore, this 1590 Ma LIP event in the Gawler craton has been linked reconstructed with coeval dolerite magmatism (Western Channel Diabase) and the hydrothermal Wernecke breccias of northwestern Laurentia on the basis of paleomagnetism and U–Pb geochronology (Thorkelson *et al.*, 2001; Hamilton and Buchan, 2010).

Contemporaneous or slightly younger IOCG deposits also occur within the Eastern Succession of the Mount Isa Inlier of Australia (e.g. Cloncurry belt, Ernest Henry; Betts *et al.*, 2009). IOCG deposits in all of these areas have been linked to a *c*. 1500-km continental hotspot track that formed plume-related volcanic rocks and granites in the Gawler craton and Curnamona province, and also intrusions of the Williams–Naraku batholith within the Mount Isa Inlier, suggesting a widespread *c*. 1600 Ma LIP event associated with a younger hotspot trail, or events in a plume-modified orogenic setting (Betts *et al.*, 2007, 2009; Fig. 16.6).

High-fluorine-content magmas may have a positive influence on the formation of IOCG and associated deposits, as exemplified by the causal link between the high-F Gawler craton LIP and the high-F Olympic Dam deposit (Agangi *et al.*, 2010; McPhie *et al.*, 2011). See full discussion in Ernst and Jowitt (2013).

Other events, such as the Sierra Madre Occidental SLIP and the Warakurna and Paraná–Etendeka LIPs are also associated with the formation of significant volumes of F-enriched silicic rocks (e.g. Frindt *et al.*, 2004; Pankhurst *et al.*, 2011; Ernst and Jowitt, 2013). These rocks and magmas provide an ideal source for F-rich fluids that could either be directly derived from the SLIP or LIP magmas or could be formed during postmagmatic hydrothermal alteration of F-rich silicic magmatic rocks.

This connection between high-F silicic magmas of LIPs and SLIPs and increased mobility and mobilization of elements that are often found concentrated within IOCG deposits (e.g. REEs, U; Agangi *et al.*, 2010) suggests that such F-rich LIP/-SLIP-derived magmas may have an important role in the formation of some IOCG deposits. This also suggests that intrusions that formed from high-F magmas, such as those of the Sierra Madre Occidental LIP in Mexico, silicic parts of the Warakurna LIP of central-west Australia, and the topaz-bearing Spitzkoppe granites of the Paraná–Etendeka LIP (e.g. Pankhurst *et al.*, 2011), may be prospective targets for IOCG and associated deposits. In contrast, F-poor A-type silicic magmas, such as those of the Snake River plain hotspot-trail magmatism (subsequent to the Columbia River LIP) in the United States (e.g. Ellis, 2009), may be unprospective for IOCG mineralization, as shown by the paucity of mineral occurrences/prospects in the Snake River plain area. More research is needed into these speculative links between high-F magmatism and IOCG formation.

Further evidence of possible links between IOCG deposits and LIP events is provided by the LIP–carbonatite links described above; carbonatites have been described as magmatic end-members of the IOCG family, as exemplified by the LIP-related Phalaborwa carbonatite (Corriveau, 2007; Groves *et al.*, 2010).

16.3.3 Mineralized vent complexes

Detailed seismic studies of the 62–55 Ma North Atlantic Igneous Province (NAIP) and complementary studies in the 183 Ma Karoo LIP (part of the Karoo-Ferrar LIP) and 250 Ma Siberian Trap LIP have revealed many hundreds of hydrothermal vent complexes (Fig. 5.17). These vents have widths of up to 5–10 km at the paleosurface and connect to underlying dolerite sills at paleodepths of up to 8 km (see Sections 5.3.6 and 14.6.6).

Such vent complexes are distinct from zones of brecciation associated with both modern and ancient VMS systems, where hydrothermal fluids react at depth with basaltic rocks, before being vented at the seafloor (e.g. Franklin *et al.*, 2005), and they are distinct from brecciation associated with the formation of peperites where basaltic magmas interact with overlying wet sediments (e.g. Skilling *et al.*, 2002).

As discussed in Sections 5.3.6 and 14.6.6, hydrothermal vent complexes (HVCs) originate from the explosive release of gases generated during the emplacement of thick sills (> 50 m) into volatile-rich but low-permeability sedimentary strata, meaning that the hydrothermal vent complexes are phreatomagmatic in origin. Their architecture and economic potential is dependent on the host rocks they were emplaced into (black shales at Karoo and evaporites, marls, carbonates, shales, and coals in Siberia) and the level of fluid (brine) saturation in the host rocks during emplacement. About 250 HVCs associated with the Siberian Trap LIP are mineralized and have magnetite matrixes (Svensen *et al.*, 2009) and some are being mined for Fe, e.g. the Korshunovskoe and Rudnogorskoe mines.

These HVCs have also been proposed to be analogous to IOCG deposits (Ernst *et al.*, 2009), although the sourcing of the metals within these HVCs is unclear, and any enrichment in copper and gold remains both understudied and unconfirmed.

In any case, the presence of HVCs associated with Phanerozoic LIPs suggests that they should also be an essential component of sill provinces associated with Proterozoic LIPs and have potential for hosting magnetite and potentially IOCG mineral deposits (Ernst *et al.*, 2009).

16.3.4 Other high-temperature hydrothermal deposits

There are a number of other examples of hydrothermal deposit types that are spatially and temporally associated with LIPs and several of particular significance are highlighted next.

Hydrothermal deposits associated with the Ring of Fire Complex

The potential LIP represented by the Ring of Fire Complex of northwestern Ontario is associated with magmatic deposits (as already discussed, Section 16.2.2), but significant VMS deposits are also present (Table 16.5; Mungall *et al.*, 2010; Metsaranta and Houlé, 2012). Silicic volcanic rocks that host the McFaulds VMS deposits yielded an age of 2737 ± 7 Ma (U–Pb), which is within error of a ferrogabbro within the Ring of Fire Complex that has been dated to 2734.5 ± 1.0 Ma (U–Pb). This contemporaneous magmatism and VMS deposit formation suggests that the energy and heat required to form the hydrothermal systems may have been supplied by the intrusion of the Ring of Fire Complex (e.g. Mungall *et al.*, 2010; Metsaranta and Houlé, 2012), which, as discussed above (Section 16.2.2), is potentially LIP-related.

Hydrothermal deposits associated with Siberian Trap, Emeishan, and Tarim LIPs of Asia

Borisenko *et al.* (2006) identified a variety of magmatic and hydrothermal ore-deposit types associated with the three major Phanerozoic LIPs in central Asia: the 250 Ma Siberian Trap LIP (Section 3.2.8), the 260 Ma Emeishan LIP (Section 12.2.10), and the 290–275 Ma Tarim LIP (Fig. 16.7; Table 16.6; see also Pavlova and Borisenko, 2009; Pirajno *et al.*, 2009). In addition to magmatic Ni–Cu–(PGE) deposits in all three areas, and magmatic Fe–Ti–V deposits in the Emeishan LIP, there is also a range of hydrothermal deposits associated with each (Table 16.6).

The Siberian Trap LIP event is associated with a wide range of mineral deposits, including apatite–magnetite, REEs, Cu–Ni, and Au mineralization within the Maimecha–Kotui alkaline ultramafic intrusions, Ni–Co–Fe and Sb–Ag–Hg–Au mineralization in the Noril'sk, Taimyr, and North Urals areas and *c.* 249–245 Ma porphyry Cu–Mo mineralization in the Taimyr region (Pirajno *et al.*, 2009). Other porphyry Cu ± Mo ± Au ± W deposits are located in the Kuznetsk basin, Kolyvan–Tomsk fold zone, and Salair areas on the southern and southwestern areas of the Siberian craton (Pirajno *et al.*, 2009). These porphyry-type deposits are unusual in that they did not form in subduction or immediate postsubduction settings; instead, they formed in within-plate rift or orogenic zones at the periphery of the Siberian Trap LIP. Borisenko *et al.* (2006) suggested that the spatial relationship between plume-related magmatism and within-plate rifting was a key ingredient in the formation of these porphyry deposits in an atypical setting. Sediment and dolerite dyke-hosted Hg and Au–Sb–Hg mineralization is also present in the Kolyvan–Tomsk fold zone, and similar mineralization is hosted by Siberian Trap flood basalts in the Kuznetsk basin and the Salair area, implying a mineralizing event at 238 Ma or later (Pirajno *et al.*, 2009). Mineralization contemporaneous with the formation of the Siberian Trap LIP has also been reported from eastern Kazakhstan (alkaline-related Hg–Au and granodiorite–granite–leucogranite-related Mo–W; Borisenko *et al.*, 2006), the southeastern Altay and southwestern Mongolia (Hg, Cu–Hg–Ba, Ag–Sb,

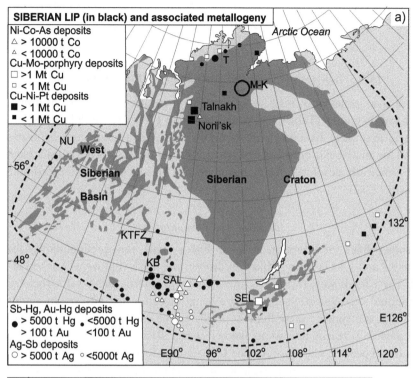

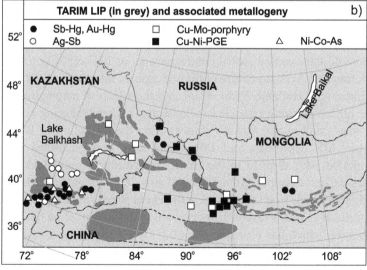

Figure 16.7 Range of ore deposits associated with (a) the 250 Ma Siberian LIP and (b) the *c.* 270 Ma Tarim LIP of central Asia. The distribution for each LIP includes both surface and subsurface occurrences; details in Pirajno *et al.* (2009). Abbreviations in (a): KTZF = Kolyvan–Tomsk fold zone; KB = Kuznetsk basin; SAL = Salair; SEL = Selenga; M–K = Maimecha–Kotui; NU = North Urals; T = Taimyr. From Figure 9 in Ernst and Jowitt (2013), and modified from Figures 5, 6, and 8 in Pirajno *et al.* (2009). With permission from the Society of Economic Geologists.

Ni–Co–As, and Cu–Co–As; Borisenko *et al.*, 2006), and also from northern Mongolia (Ni–Cu–(PGE); Pirajno *et al.*, 2009).

16.3.5 Low-temperature and sedimentary LIP-related deposits

LIPs are also genetically related to lower-temperature and sediment-dominated hydrothermal systems (Tables 16.5 and 16.6); here I discuss the links between LIPs and the formation of banded iron formations (BIFs) and manganese deposits. The potential for an association with U-type mineralization is also being considered but is not yet sufficiently developed for inclusion herein. This section is also based on Ernst and Jowitt (2013).

Banded iron formations

Sedimentary BIFs are common in Earth history and were predominantly formed between 3.8 and 1.9 Ga from reduced hydrothermal fluids that were enriched in Fe^{3+} and Si. These fluids were formed by interaction with submarine basaltic rocks within hydrothermal convection systems before being incorporated into ocean circulation systems that transported these fluids to shallower basins and shelf regions where Fe precipitated (e.g. recent reviews in Bekker *et al.*, 2010, 2014). A series of temporal links between oceanic plateaus (oceanic LIP) and the formation of BIFs was noted by Isley and Abbott (1999), with these links further explored by Abbott and Isley (2001) and discussed in Kerrich *et al.* (2005) and Bekker *et al.* (2010, 2014). The reducing atmosphere prevalent at the time of BIF development (3.8–1.9 Ga) meant that major LIP magmatic events at these times were more likely to have a global effect on ocean chemistry and could have caused global contemporaneous BIF deposition. In comparison, the increasingly oxidized nature of the atmosphere after the *c.* 1.85 Ga Great Oxygenation Event (e.g. Holland, 2006) meant that the effects of magmatism, either LIP-related or otherwise, were restricted to local areas where reducing conditions were retained (e.g. small anoxic basins), thus limiting the formation of BIFs. Age matches between BIF deposition and LIP events have been identified at 2600, 2500–2450, and 1880 Ma, all prior to the Great Oxygenation Event (Bekker *et al.*, 2010, 2014). The earliest example of the linking of BIF development and a LIP is the formation of Superior-type iron formations in the Hamersley province of the Pilbara craton of Australia at *c.* 2600 Ma, contemporaneous with the formation of the Great Dyke of Zimbabwe, which has a precise age of 2575 Ma (e.g. Olsson *et al.*, 2010) although the distance between the Pilbara and Zimbabwe cratons at this time is uncertain (e.g. Smirnov *et al.*, 2013; Evans, 2013). Even more extensive development of Superior-type BIFs at 2500–2450 Ma in the Hamersley and Transvaal basins (Krapež *et al.*, 2003) is also linked with global 2500–2450 Ma plume and LIP events (e.g. Heaman, 1997; Ernst and Buchan, 2001b), with approximately coeval iron formations also located in the Quadrilátero Ferrífero region of Brazil, the Krivoy

Rog area in the Ukraine, and the Kursk Magnetic Anomaly region in Russia (Bekker *et al.*, 2010). There may be multiple distinct 2500–2450 Ma LIP events that occurred at this time, depending on paleocontinental reconstructions; however, it is notable that the Hamersley BIF with a pulse at 2450 Ma is proximal to the Woongarra–Weeli Wolli LIP (Barley *et al.*, 1997), and both this BIF and LIP events are associated with the southern margin of the Pilbara craton. In addition, the widespread development of granular iron formations (GIFs) at the margins of the Superior craton in North America at 1880 Ma was contemporaneous with the Circum-Superior craton LIP (Fig. 9.10; Heaman *et al.*, 2009; Ernst and Bell, 2010; Minifie *et al.*, 2013). Finally, the formation of the Rapitan BIFs at *c.* 715 Ma is contemporaneous with the Franklin LIP (Bekker *et al.*, 2010) that was associated with the rifting and breakup of Rodinia (e.g. Ernst *et al.*, 2008; Li *et al.*, 2008a).

Manganese deposits

LIPs are potentially associated with world-class manganese deposits, as exemplified by the Kalahari manganese field of the northern Cape province of South Africa (Tsikos *et al.*, 2006; Ernst and Jowitt, 2013). These ores are hosted by the Hotazel Formation within the 2.65–2.05 Ga uppermost Paleoproterozoic Transvaal Supergroup and consist of three laminated ore horizons interbedded with BIFs. The Hotazel Formation is gently folded and underwent low-grade metamorphism and supergene alteration related to exposure and weathering (along a basal unconformity) that caused formation of the earliest Mn ores shortly after *c.* 2.22 Ga (Evans *et al.*, 2002). Further hydrothermal and/or contact metamorphic upgrading of both Mn and Fe horizons occurred at *c.* 1900 Ma, forming high-grade Mn ores (Evans *et al.*, 2002); this supergene upgrading may be related to either the 1930–1920 Ma Hartley or *c.* 1880 Ma Mashonaland LIP event.

16.3.6 Implications of a LIP context for hydrothermal ore deposits

Although LIP-related hydrothermal deposits can be split into high- and low-temperature categories, it is possible that individual LIPs can be associated with the synchronous formation of both high- and low-temperature mineral deposits, as suggested by the separately identified genetic link between LIP events and the deposition of BIFs, and between LIP events and the formation of VMS deposits. Bekker *et al.* (2010) outlined an underlying mechanism that links all three, with mantle plumes and associated LIP magmatism delivering the energy to drive submarine hydrothermal cells associated with the formation of VMS deposits. These VMS deposits contain Cu, Zn, Pb, and other base metals that were trapped within sulfides at the volcanic seafloor–water interface, whereas the Fe and Si that were released with hydrothermal fluids were entrained within a buoyant hydrothermal plume that spread to continental and oceanic-plateau shelf areas (Bekker *et al.*,

2010, 2014). In this case, Fe was oxidized and precipitated out with Si, forming lower-temperature Fe deposits as a distal counterpart to the cogenetic and contemporaneous VMS deposition; this relationship between LIP emplacement and the formation of high-temperature VMS and low-temperature BIF deposits illustrates the potential for near- and far-field metallogenic effects of LIP events, and demonstrates how LIPs can be related to synchronous yet geochemically diverse and geographically isolated hydrothermal mineral deposits (Bekker *et al.*, 2010).

It could be further speculated that the relative timing and spatial distribution of high- and low-temperature hydrothermal deposits may relate to the original spatial distribution of a LIP event, especially its plumbing system, and gradations in intensity of magmatism as a function of distance from a plume center (Ernst and Jowitt, 2013).

16.4 LIPs and laterite deposits (secondary enrichment by weathering)

Weathering of rocks formed during LIP events can form a range of differing laterites (e.g. Table 16.1) that are exploited for a range of differing commodities, including iron, aluminum, nickel, gold, phosphorus, and/or niobium, depending on the source rock (e.g. Freyssinet *et al.*, 2005; Retallack, 2010). Here, I focus on the relationship between LIP events and the development of Ni–Co laterites and bauxites (exploited for Al), with the latter being derived from the weathering of basalts and dolerites (Bárdossy and Aleva, 1990; Laznicka, 2010) and the former being derived from the weathering of ultramafic rocks (e.g. Golightly, 1981; Gaudin *et al.*, 2005; Lewis *et al.*, 2006). It is also important to note that LIP-related carbonatites, as well as being associated with primary mineralization (Section 16.2.3), can also weather to yield niobium and phosphorous laterites (Freyssinet *et al.*, 2005). Each of these source-rock types (basalts, dolerites, ultramafic rocks, and carbonatites) can be formed during LIP events, with the broad scale of LIPs making them favorable targets for the identification of economic laterites.

16.4.1 Ni–Co laterites

Although the majority of the historic Ni production has come from Ni–Cu–(PGE) sulfide ores, the majority of identified Ni resources are hosted by Ni–(Co) laterites (e.g. Mudd, 2010). Nickel laterites form by the deep weathering of Mg-rich ultramafic rocks, such as komatiites and ophiolites, in tropical environments, meaning that the ultramafic units of LIPs are ideal for the generation of Ni-laterites, providing the weathering conditions are appropriate. A number of Australian Ni laterites have formed from weathering of rocks formed during LIP events, including thick olivine cumulates within Archean LIP-associated komatiitic and ultramafic systems in the Yilgarn craton of Western Australia, as seen at Murrin Murrin, Bulong, and Cawse

(Elias, 2006; Arndt *et al.*, 2008; Laznicka, 2010), and the Wingellina laterite of central Australia that formed over ultramafics associated with the 2075 Ma Warakurna LIP (Ernst *et al.*, 2008; Pirajno and Bagas, 2008). In addition, the Caribbean region hosts widespread Ni laterite deposits (e.g. Lewis *et al.*, 2006), including the Cerro Matoso Ni laterite deposit in northwest Colombia. Although the 90 Ma Caribbean–Colombian LIP is a major geodynamic element in the region, it is currently unclear which ultramafic bodies are linked to this oceanic-plateau LIP, and which are linked to coeval arc-related ophiolites (e.g. Lewis *et al.*, 2006).

16.4.2 Bauxites

Weathering of basalt and dolerite under tropical conditions may lead to the development of bauxites, the main source of Al (Bárdossy and Aleva, 1990; Laznicka, 2010; Retallack, 2010). About 19% of bauxite deposits worldwide are formed on basalt flows, the majority of which are LIP-related, with another 17% associated with dominantly LIP-related dolerite sills and dykes. For example, bauxites of central and western India formed from weathering of basalts and dolerites associated with the Deccan LIP (Bárdossy and Aleva, 1990). In addition, the supergiant Guinea bauxite district of West Africa hosts 9.1 Gt of bauxite derived from the weathering of sills and dykes that formed during the 200 Ma CAMP (Laznicka, 2010).

16.4.3 LIP controls on laterite and bauxite formation

Retallack (2010) documented the fact that peak times of laterite and bauxite ore formation coincided with elevated global temperatures, amounts of precipitation, and atmospheric carbon dioxide levels, and also with oceanic anoxia events (Section 14.6.9), exceptional fossil preservation, and mass extinction events (Section 14.2). These key conditions are reflected by the association of laterite and bauxite resources with particular stratigraphic horizons formed during greenhouse crises, indicating the usefulness of a stratigraphic approach to exploration and exploitation of these laterite resources.

As discussed in Chapter 14, LIPs have a particularly robust link with major climate-change events, particularly mass extinctions (e.g. Courtillot and Renne, 2003; Kidder and Worsley, 2010), indicating a potentially strong association in general between the timing of LIP events and peak laterite and bauxite formation. This indicates that not only are the rocks formed during LIP events ideal protoliths for weathering to form laterite and bauxite resources, but also that these voluminous magmatic events may have a causative association with climatic changes that increased the chances of the ideal chemical weathering conditions conducive to the formation of laterites and bauxites.

16.5 LIPs and supercontinent reconstruction (indirect links between LIPs and metallogeny)

Next, I consider two ways in which LIPs can be useful in exploration in a plate-tectonic context, and which represent an indirect link between LIPs and metallogeny. See also discussion in Ernst and Jowitt (2013).

16.5.1 Tracing of metallogenic belts

Major metallogenic belts are often truncated at the margins of cratonic blocks, as shown by the truncation of the Abitibi greenstone belt, with its impressive endowment of Au (4470 t Au in production and reserves; Dubé and Gosselin, 2007), at the eastern margin of the Superior craton (Fig. 11.12), indicating that this highly prospective belt must continue into a formerly neighboring crustal block (Bleeker, 2003; Bleeker and Ernst, 2006; Ernst, 2007). As discussed in Chapter 11, the pre-Pangea reconstruction framework is poorly constrained, but significant reconstruction progress is being achieved through use of the LIP barcode and dyke-swarm piercing-point methods (Section 11.9.2).

Producing reconstructions has implications for tracing of metallogenic belts. For instance, 2500 to 2100 Ma LIPs are used to reconstruct the Karelia–Kola, Hearne, and Wyoming cratons adjacent to the southern Superior craton (Fig. 11.12). This reconstruction has the following metallogenic implications. Intrusions such as the PGE-mineralized PGE-2-type Penikat, Portimo, and Fedorova intrusions of the BLIP are linked with the PGE-mineralized PGE-3-type East Bull Lake and River Valley intrusions of the Matachewan (–East Bull Lake) LIP of the Superior craton (Section 5.2.3; Figs. 5.5 and 11.8). This reconstruction would also link the *c.* 2490 Ma Cr-rich Monchegorsk intrusion (belonging to BLIP) in Finland to the 2505 Ma Mistassini LIP of the southeastern Superior craton and suggests that this poorly explored region of the Superior craton should be considered highly prospective for both Cr and PGE mineralization. In addition, the world-class Cr mineralization within the 2.44 Ga Baltic LIP-related Kemi intrusion of Baltica (belonging to BLIP) should be linked with the Matachewan (-East Bull Lake) LIP/plume event (but not the older Mistassini LIP/plume center) and indicates that the Matachewan LIP units of southern Superior craton should also be considered prospective for Cr mineralization.

The same LIP barcode/piercing-point approach (Section 11.9.2) is used by Söderlund *et al.* (2010) to speculate that the Zimbabwe craton may have been formerly adjacent to the eastern side of the Superior craton near the eastern end of the Abitibi belt (Fig. 11.12). This juxtaposition would correlate the 2.75–2.70 Ga LIP-related greenstone belts of the Abitibi belt (Section 6.3.2; e.g. Ernst and Buchan, 2001b; Sproule *et al.*, 2002) with similar age and economically important greenstone belts of the Zimbabwe craton (e.g. Prendergast, 2004). See further discussion in Ernst and Jowitt (2013).

16.6 Emplacement of LIPs, extensional pulses, and distal compression/transpression

There is also an even more indirect application of LIPs to the understanding of Au mineralization (Ernst and Jowitt, 2013). Specifically, orogenic gold can be linked with pulses of transpression or collision along plate boundaries (e.g. Goldfarb *et al.*, 2005) and it has been well established that LIPs are associated with the breakup (or attempted but failed breakup) of continents (Section 11.2). Under either scenario (successful or failed breakup), the arrival of a LIP can therefore be linked with a pulse of extension which in the closed circuit of a global plate-tectonic framework should be linked with a corresponding pulse of compression or transpression along favorably oriented distal plate boundaries. This proposed linkage between LIP events (as a proxy for the timing of breakup events) and far-field tectonic changes, including transpressive/compressive pulses, suggests the possibility that the LIP record could globally correlate with the timing of major pulses of orogenic Au mineralization (Ernst and Jowitt, 2013).

This type of link is exemplified by Goldfarb *et al.* (2007), who noted that *c.* 125 Ma orogenic gold deposits of the north China, Yangtze (South China), and Siberian craton margins, as well as in young terranes in California, may relate to emplacement of the 122 Ma Ontong Java plateau and related LIPs in the southern Pacific basin, due to the dramatic tectonic consequences resulting from such a massive addition of magmatism to the Pacific plate.

Future research will test this model through a detailed examination of the LIP record in combination with periods of formation of major orogenic Au camps with reference to plate reconstructions, in order to test the correlations between the timing of LIP events and orogenic Au metallogenesis. The comparison will be easiest for the Mesozoic–Cenozoic record where the reconstructions are well established. Precambrian tests of any correlation between LIPs and orogenic gold will have to await robust Precambrian reconstructions.

16.7 Implications for the oil industry

Three aspects of LIPs that are of particular significance to the oil industry are addressed: (1) oceanic plateaus as a cause of anoxia events that produce black shales, which are source rocks for oil, (2) the maturation or over-maturation of regional hydrocarbons by LIPs, and (3) associated dolerite sills (belonging to LIP events) acting as cap rocks or otherwise contributing to structural traps for hydrocarbons.

16.7.1 Oceanic plateaus and anoxia events and black shales

Ninety percent of source rocks for hydrocarbons are organic-rich shales; the remainder are carbonates (e.g. Jahn *et al.*, 2008). Those black-shale horizons represent

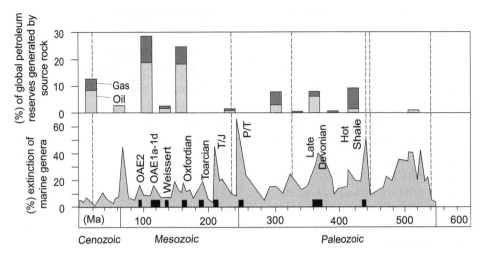

Figure 16.8 Link between anoxia events and petroleum source rocks. From Figure 1 in Takashima *et al.* (2006).

anoxic events, and in particular, those at the oceanic anoxic event 2 (OAE2) at 94 Ma and OAE1 at 120 Ma (Section 14.6.8) are key targets for the oil industry. It has been well demonstrated (e.g. Kerr, 1998, 2005) that the eruption of oceanic LIPs at *c*. 90 Ma is linked to widespread Cenomanian–Turonian boundary (94 Ma) black-shale deposits (Fig. 16.8). Similarly one could argue a link between *c*. 123–110 Ma oceanic LIPs and widespread Aptian (124–112 Ma) black shales. This association can be tracked further back in time. Important Kimmeridgian–Tithonian (155–146 Ma) oil source rocks correlate with the formation of the Sorachi plateau and potentially the Shatsky and Magellan rises in the western Pacific. Finally, the Toarcian (187–178 Ma) black shales correlate with the eruption of the Karoo–Ferrar and Weddell Sea LIPs (Kerr, 1998).

Most shale source rocks are Phanerozoic in age. However, some are Neoproterozoic and even Mesoproterozoic (Ghori *et al.*, 2009). An impressive example is the extensive valuable hydrocarbons of eastern Siberia whose source rocks are Riphean shales. The Velkerri Formation of approximately 1.43 Ga in northern Australia is one of the organically richest Proterozoic successions in Australia and has become a target for oil exploration (Warren *et al.*, 1998). The hydrocarbon-bearing Nunesuch shale is associated with the Midcontinent rift system and its associated Keweenawan LIP (Ghori *et al.*, 2009). The oldest preserved oilfield is found in a thick succession of 2.0 Ga siliciclastic and volcanic rocks along a rifted passive margin of Kenorland in northwestern Russia. The estimated carbon reserve is 250 Gt, which is present as shungite and carbonaceous shales (Melezhik *et al.*, 2004; R. Kerrich, written communication, 2012). In Laurentia, a metamorphosed oilfield is inferred to have been present in a 2.2–1.9 Ga sedimentary rock sequence from the Great Lakes region of

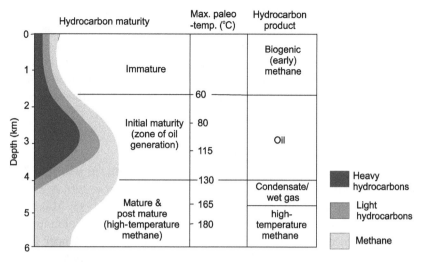

Figure 16.9 Thermal maturation of hydrocarbons. Presence of individual intrusions and the regional thermal anomaly associated with a LIP and the concentration of this thermal pulse near an interpreted plume center are all factors which can cause maturation of hydrocarbons. Redrafted from Figure 3.4 in Jahn *et al.* (2008). With permission from Elsevier.

the Superior craton passive margin (Mancuso *et al.*, 1989; R. Kerrich, written communication).

16.7.2 *Thermal pulse from LIPs affecting maturation of hydrocarbons*

The thermal input from LIPs can raise temperatures over a wide area and can affect the thermal maturation of hydrocarbons, representing a spectrum of decreasing complexity from organic matter through oil to gas (Fig. 16.9). The spatial juxtaposition of LIPs and hydrocarbons is quite common since hydrocarbon-bearing sedimentary basins and LIPs are both commonly linked with successful continental breakup and the development of passive margins as the new ocean grows. The thermal maturation of hydrocarbons should increase broadly along continental margins toward the vicinity of the plume-center region, however this remains to be systematically tested.

The Caribbean–Colombian LIP (90 Ma)

The accretion of still-warm oceanic plateaus to continental margins where black shales have formed can profoundly influence the thermal maturation of these oil source rocks. Residual heat from the formation of plateaus (from both crust and mantle) is likely to influence the thermal maturation of proximal black shales for up to 20 Ma after the cessation of volcanism (Kerr, 2010).

The best example is the Caribbean oceanic plateau, which caused oceanic anoxia and the formation of black shales. The subsequent collision with the coast of South America is likely to have influenced the thermal maturation of the La Luna Formation. Cenomanian–Turonian black shales are found worldwide but the La Luna Formation is one of the most prolific petroleum source rocks in the world (Bralower and Lorente, 2003) and this may be the result of its interaction with a warm oceanic plateau (A. Kerr, written communication, 2010).

The CAMP (200 Ma)

A clear example of LIP-associated oil maturation is associated with the 200 Ma CAMP event in South America and along the margin of northwestern Africa (Mello *et al.*, 1994; Wilson and Guiraud, 1998; Logan and Duddy, 1998; Makhous and Galushkin, 2003). Intrusion of 200 Ma sills in the Upper Amazon Solimões basin of South America generated high levels of thermal maturity within Upper Devonian petroleum source rocks (Mello *et al.*, 1994; Wilson and Guiraud, 1998).

According to Logan and Duddy (1998) apatite- and zircon-fission track analysis on rocks in the Ahmet and Reggane basins (central Algeria) showed clear evidence for a major heating event at *c.* 200 Ma. They noted a pre-Hercynian phase, in which chiefly liquid hydrocarbons were expelled, and a later phase, associated with a "heat-spike" at *c.* 200 Ma, in which significant quantities of dry gas were generated and expelled. This heat spike matches the timing of the CAMP LIP.

16.7.3 Associated dolerite sills acting as cap rocks

Hydrocarbons migrate through rocks from their source areas and can be trapped and accumulated in structural and lithological traps. One type of trap is associated with dolerite sills associated with LIPs. A prominent example of this effect is illustrated in the North Sea and Rockall Trough where dolerite sills belonging to the NAIP act as structural traps for oil (e.g. Planke *et al.*, 2005; Archer *et al.*, 2005).

16.7.4 Gas accumulation from heating of favorable host rocks by sills

As noted in Chapter 14, significant quantities of greenhouse gases can be released from volatile-rich sediments due to heating by sills associated with LIPs. These gases can be vented to the surface through hydrothermal vent complexes (Sections 5.3.6 and 14.6.6) and are considered a significant mechanism by which LIPs cause extinction-scale environmental changes. However, there is also the possibility that some portion of the generated gases will remain trapped in the sediments. Under such a scenario and given a carbon-rich source rock, potential natural-gas deposits could conceivably be generated.

16.8 LIPs and water resources

The intrusive components of LIPs, sills, and dykes, have a major influence on groundwater circulation patterns (e.g. Chevallier and Woodford, 1999; Chevallier *et al.*, 2001; Cook, 2003). There are two modes in which dykes and sills can affect groundwater flow. Dykes can disrupt and segregate groundwater flow in the space between dykes. Also, sills can potentially trap groundwater beneath them and form aquifers. Second, fractured dolerite dykes and sills can be effective aquifers on their own. Examples of both modes are discussed below.

As summarized in Cook (2003), massive and unweathered dykes can form barriers to lateral groundwater movement. In the Witwatersrand Goldfields area of South Africa, a series of north–south-trending dykes between 6 and 60 m in thickness (of unknown age) that intruded volcanic and sedimentary rocks have created a series of isolated aquifer compartments. In southwestern Australia, dolerite dykes impede the lateral flow of groundwater, forcing it to the surface where evaporation can cause salinization (Engel *et al.*, 1989). Pumping tests in Botswana indicate that dykes that are thicker than 10 m serve as groundwater barriers, but those of smaller width are permeable as they develop cooling joints and fractures (Bromley *et al.*, 1994).

Some dolerite sills are very thick and extend over large areas. Due to their low permeability (except when fractured; see below) sills may support perched water bodies (Cook, 2003).

Fractured dykes may form good aquifers. In South Africa, boreholes are often sited in highly fractured dykes for good water supplies. One of the dykes in the Palaghat Gap region of the Western Ghats of India, which is traced for a strike length of about 14 km, is highly fractured. The discharges from some of the wells drilled into this dyke are between 240 and 840 m^3 per day (Singhal and Gupta, 1999). Due to thermal effects, dykes can also cause fracturing of adjacent rock. For instance, the boreholes adjacent to dolerite dykes intruding the fractured sandstone/mudstone of the Karoo aquifer, South Africa, yield greater flow than elsewhere in the basin (Sami, 1996; Cook, 2003). Basaltic flows can serve as aquifers for water resources. For instance, lava flows of the Columbia River LIP serve as an important aquifer source in the Columbia basin and in the Columbia River gorge of the Pacific northwest of the USA (e.g. Burns *et al.*, 2012; C. Kingsbury, personal communication, 2013).

16.9 Summary

LIPs represent significant reservoirs of energy and metals that can either drive or be incorporated into a variety of differing metallogenic systems and also have significant controls on the maturity of hydrocarbon source rocks, the formation of oil and gas reservoirs, and the development of important aquifer systems. The relationships between LIPs and these differing systems (ore deposit, hydrocarbon, and water) can be split into five distinct, although partially overlapping, types. This is modification of the four-part classification at the beginning of this chapter (and also in Ernst and Jowitt, 2013) in which type 2 has been split into two parts in order to better reflect the integration of hydrocarbon and aquifer stories into this framework (S. Jowitt, 2013, personal communication).

(1) LIPs are the primary source of commodities within mineral deposits (e.g. ortho-magmatic Ni–Cu–(PGE) sulfides), and can be linked to carbonatite-associated REE–Nb–Ta deposits, and diamondiferous kimberlites.

(2) LIPs provide the energy to drive hydrothermal systems, act as source rocks for hydrothermal ore deposits (e.g. VMS), are associated with the formation of IOCG deposits linked with silicic components of LIPs, or entirely Silicic LIPs (SLIPs), or drive hydrocarbon source rocks to maturation or over-maturation.

(3) LIP rocks, particularly sills and dykes, act as barriers to fluid flow and/or as reaction zones, leading to mineralization (e.g. orogenic Au); these barriers can also control aquifer formation and may act as structural traps for oil and gas resources.

(4) Weathering of LIP rocks concentrates elements such as Ni–Co and Al within laterites or bauxites, with additional surficial interactions that causally link oceanic-plateau formation with the anoxic events that form black shales, key source rocks for oil and gas.

(5) Indirect links are proposed between LIPs and ore deposits, such as far-field tectonic changes caused by LIP events during the plate-tectonic cycle (e.g. orogenic Au).

For each classification I have discussed specific ways where the identification of links between LIP events and the formation of these various deposits or resources can be used to enhance exploration strategies.

In addition to these documented links between LIPs and a variety of resource types, LIP events are a key tool in reconstructing Precambrian supercontinents and enabling the tracing of metallogenic belts between presently separated, but formerly contiguous, crustal blocks. This final chapter has shown how our understanding of LIPs, and the processes that affect LIP magmas and rocks, have direct consequences for resource exploration and economic geology.

References

Aarnes, I, Fristad, K., Planke, S., & Svensen, H. (2011). The impact of host-rock composition on devolatilization of sedimentary rocks during contact metamorphism around mafic sheet intrusions. *Geochemistry, Geophysics, Geosystems* 12, G10019.

Abbott, D.H. & Isley, A.E. (2001). Oceanic upwelling and mantle-plume activity: paleomagnetic tests of ideas on the source of the Fe in early Precambrian iron formations. In Ernst, R.E. & Buchan, K.E. (eds.), *Mantle Plumes: Their Identification through Time*. GSA Special Publication 352, pp. 323–339.

Abbott, D.H. & Isley, A.E. (2002). Extraterrestrial influences on mantle plume activity. *Earth and Planetary Science Letters*, 205: 53–42.

Abbott, D. & Mooney, W. (1995). The structural and geochemical evolution of the continental crust: support for the oceanic plateau model of continental growth. *Reviews of Geophysics*, 33 (S1): 231–242.

Adams, C.J., Cluzel, D., & Griffin, W.L. (2009). Detrital-zircon ages and geochemistry of sedimentary rocks in basement Mesozoic terranes and their cover rocks in New Caledonia, and provenances at the Eastern Gondwanaland margin. *Australian Journal of Earth Sciences*, 56: 1023–1047.

Agangi, A., Kamenetsky, V.S., & McPhie, J. (2010). The role of fluorine in the concentration and transport of lithophile trace elements in felsic magmas: insights from the Gawler Range Volcanics, South Australia. *Chemical Geology*, 273: 314–325.

Agashev, A.M., Pokhilenko, N.P., Tolstov, A.V., *et al.* (2004). New age data on kimberlites from the Yakutian diamondiferous province. *Doklady Earth Sciences*, 399: 1142–1145.

Agee, C.B. (1993). Petrology of the mantle transition zone. *Annual Review of Earth and Planetary Sciences*, 21: 19–41.

Aguirre-Díaz, G. & Labarthe-Hernández, G. (2003). Fissure ignimbrites: fissure-source origin for voluminous ignimbrites of the Sierra Madre Occidental and its relationship with basin and range faulting. *Geology*, 31: 773–776.

Ahern, J.L. & Turcotte, D. (1979). Magma migration beneath an ocean ridge. *Earth and Planetary Science Letters*, 45: 115–122.

Alapieti, T.T. (1982). The Koillismaa layered igneous complex, Finland – its structure, mineralogy and geochemistry, with special emphasis on the distribution of chromium. *Bulletin of the Geological Survey Finland*, 319.

Alapieti, T.T. & Lahtinen, J.J. (2002). Platinum-group element mineralization in layered intrusions of northern Finland and the Kola Peninsula, Russia. In Cabri, L.J. (ed.), *The Geology, Geochemistry, Mineralogy and Mineral Beneficiation of Platinum-Group Elements*. Canadian Institute of Mining, Metallurgy and Petroleum, Special Volume, 54, pp. 507–546.

Ali, J.R., Fitton, J.G., & Herzberg, C. (2010). Emeishan large igneous province (SW China) and the mantle-plume up-doming hypothesis. *Journal of the Geological Society, London*, 167: 953–959.

Al-Kindi, S., White, N., Sinha, M., England, R., & Tiley, R. (2003). Crustal trace of a hot convective sheet. *Geology*, 31: 207–210.

Allen, P.A. & Armitage, J.J. (2012). Cratonic basins. In Busby, C. & Azor, A. (eds.), *Tectonics of Sedimentary Basins: Recent Advances*, 1st edition. Oxford: Blackwell, pp. 602–620.

Allen, S.R., McPhie, J., Ferris, G., & Simpson, C. (2008). Evolution and architecture of a large felsic igneous province in western Laurentia: the 1.6 Ga Gawler Range volcanics, South Australia. *Journal of Volcanology and Geothermal Research*, 172: 132–147.

Alt, D., Sears, J.M., & Hyndman, D.W. (1988). Terrestrial maria: the origins of large basaltic plateaus, hotspot tracks and spreading ridges. *Journal of Geology*, 96: 647–662.

Alvarez, L.W., Alvarez, W., Asaro, F., & Michel, H.V. (1980). Extraterrestrial cause for the Cretaceous-Tertiary extinction: experimental results and theoretical interpretation. *Science*, 208: 1095–1108.

Amelin, Yu.V., Heaman, L.M., Verchogliad, V.M., & Skobelev, V.M. (1994). Contributions to geochronological constraints on the emplacement history of an anorthosite–rapakivi granite suite: U–Pb zircon and baddeleyite study of the Korosten complex of the Ukraine. *Contributions to Mineralogy and Petrology*, 116: 411–419.

Ames, D.E. & Farrow, C.E.G. (2007). Metallogeny of the Sudbury mining camp, Ontario. In Goodfellow, W.D. (ed.), *Mineral Deposits of Canada: A Synthesis of Major Deposit-Types, District Metallogeny, the Evolution of Geological Provinces, and Exploration Methods*. St John's, NL: Geological Association of Canada, Mineral Deposits Division, Special Publication 5, pp. 329–350.

Andersen, J.C.Ø., Rasmussen, H., Nielsen, T.F.D., & Rønsbo, J.G. (1998). The triple group and the Platinova gold and palladium reefs in the Skaergaard intrusion: stratigraphic and petrographic relations. *Economic Geology*, 93: 488–509.

Andersen, J.C.Ø., Power, M.R., & Momme, P. (2002). Platinum-group elements in the Palaeogene North Atlantic Igneous Province. In Cabri, L.J. (ed.), *The Geology, Geochemistry, Mineralogy and Mineral Beneficiation of Platinum-Group Elements*. Canadian Institute of Mining, Metallurgy and Petroleum, Special Volume, 54, pp. 637–667.

Anderson, D.L. (1995). Lithosphere, asthenosphere, and perisphere. *Reviews of Geophysics*, 33: 125–149.

Anderson, D.L. (1998). The EDGES of the mantle. In Gurnis, M., Wysession, M.E., Knittle, E., & Buffett, B.A. (eds.), *The Core–Mantle Boundary Region*. Washington, DC: American Geophysical Union, Geodynamics Series, Volume 28, pp. 255–271.

Anderson, D.L. (2001). Top-down tectonics. *Science*, 293: 2016–2018.

Anderson, D.L. (2005). Large igneous provinces, delamination, and fertile mantle. *Elements*, 1: 271–275.

Anderson, D.L. (2013). The persistent mantle plume myth. *Australian Journal of Earth Sciences*, 60: 657–673.

Anderson, R.C., Dohm, J.M., Golombek, M.P., *et al.* (2001). Primary centers and secondary concentrations of tectonic activity through time in the western hemisphere of Mars. *Journal of Geophysical Research*, 106 (E9): 20563–20585, doi: 10.1029/2000JE001278.

Aragón, E., Rodriguez, A. M. I., & Benialgo, A. (1996). A caldera field at the Marifil Formation: new volcanogenic interpretation, North Patagonian massif, Argentina. *Journal of South American Earth Sciences*, 9: 321–328.

Archer, S.G., Bergman, S.C., Iliffe, J., Murphy, C.M., & Thornton, M. (2005). Palaeogene igneous rocks reveal new insights into the geodynamic evolution and petroleum potential of the Rockall Trough, NE Atlantic margin. *Basin Research*, 17: 171–201.

Ariskin, A.A., Kostitsyn, Yu.A., *et al.* (2013). Geochronology of the Dovyren Intrusive Complex, Northwestern Baikal Area, Russia, in the Neoproterozoic. *Geochemistry International*, 51: 859–875.

Armitage, J.J. & Allen, P.A. (2010). Cratonic basins and the long-term subsidence history of continental interiors. *Journal of the Geological Society, London*, 167: 61–70.

Armstrong, R.A., Compston, W., Retief, E.A., William, L.S., & Welke, H.J. (1991). Zircon ion microprobe studied bearing on the age and evolution of the Witwatersrand triad. *Precambrian Research*, 53: 243–266.

Arndt, N.T. (1999). Why was flood volcanism on submerged continental platforms so common in the Precambrian? *Precambrian Research*, 97: 155–164.

Arndt, N.T. (2003). Komatiites, kimberlites, and boninites. *Journal of Geophysical Research*, 108, (B6).

Arndt, N.T. (2012). Book review: *Plates vs. Plumes: a Geological Controversy* by Gillian R. Foulger. *Lithos*, 128–131: 148–149.

Arndt, N.T. & Christensen, U. (1992). Role of lithospheric mantle in continental volcanism: thermal and geochemical constraints. *Journal of Geophysical Research*, 97: 10,967–10,981.

Arndt, N.T. & Jenner, G.A. (1986). Crustally contaminated komatiites and basalts from Kambalda, western Australia. *Chemical Geology*, 56: 229–255.

Arndt, N.T. & Menzies, M.A. (2005). The Ethiopian Large Igneous Province. *LIP of the Month*, January 2005. See: www.largeigneousprovinces.org/LOM.html.

Arndt, N.T. & Weis, D. (2002). Oceanic plateaus as windows to the Earth's interior: an ODP success story. *JOIDES Journal, Special Issue*, 28: 79–84.

Arndt, N.T., Brögmann, G.E., Lenhert, K., Chappel, B.W., & Chauvel, C. (1987). Geochemistry, petrogenesis and tectonic environment of Circum-Superior Belt basalts, Canada. In Pharaoh, T.C., Beckinsdale, R.D., & Rickard, D. (eds.), *Geochemistry and Mineralization of Proterozoic Volcanic Suites*. Boulder, CO: Geological Society, Special Publication 33, pp. 133–146.

Arndt, N.T., Czamanske, G.K., Wooden, J.L., & Fedorenko, V.A. (1993). Mantle and crustal contributions to continental flood volcanism. *Tectonophysics*, 223: 39–52.

Arndt, N., Lehnert, K., & Vasilyev, Y. (1995). Meimechites: highly magnesian lithosphere-contaminated alkaline magmas from deep subcontinental mantle. *Lithos*, 34: 41–59.

Arndt, N.T., Ginibre, C., Chauvel, C., Albarède, F., & Cheadle, M. (1998). Were komatiites wet? *Geology*, 26: 739–742.

Arndt, N.T., Bruzak, G., & Reischmann, T. (2001). The oldest continental and oceanic plateaus: geochemistry of basalts and komatiites of the Pilbara craton, Australia. In Ernst, R.E. & Buchan, K.E. (eds.), *Mantle Plumes: Their Identification through Time*. GSA Special Publication 352, pp. 359–387.

Arndt, N.T., Czamanske, G.K., Walker, R.J., Chauvel, C., & Fedorenko, V.A. (2003). Geochemistry and origin of the intrusive hosts of the Noril'sk-Talnakh Cu–Ni–PGE sulfide deposits. *Economic Geology*, 98: 495–516.

Arndt, N., Lesher, C.M., & Barnes, S. (2008). *Komatiites*. Cambridge: Cambridge University Press.

Arrial, P.A. & Billen, M.I. (2013). Influence of geometry and eclogitization on oceanic plateau subduction. *Earth and Planetary Science Letters*, 363: 34–43.

Artemieva, I.M. (2011). *The Lithosphere: An Interdisciplinary Approach*. Cambridge: Cambridge University Press.

Artemieva, I.M. & Thybo, H. (2013). EUNAseis: A seismic model for Moho and crustal structure in Europe, Greenland, and the North Atlantic region. *Tectonophysics*, 609: 97–153.

Arthur, M.A., Schlanger, S.O., & Jenkyns, H.C. (1987). The Cenomanian–Turonian oceanic anoxic event; II: palaeoceanographic controls on organic-matter production and preservation. In Brooks, J. & Fleet, A.J. (eds.), *Marine Petroleum Source Rocks*. London: Geological Society of London Special Publication 26, pp. 401–420.

Ashwal, L.D. (1993). *Anorthosites*. New York: Springer-Verlag.

Ashwal, L.D., Demaiffe, D., & Torsvik, T.H. (2002). Petrogenesis of Neoproterozoic granitoids and related rocks from the Seychelles: evidence for the case of an Andean-type arc origin. *Journal of Petrology*, 43: 45–83.

Avni, Y., Segev, A., & Ginat, H. (2012). Oligocene regional denudation of the northern Afar dome: pre- and syn-breakup stages of the Afro-Arabian plate. *Geological Society of America Bulletin*, 124: 1871–1897.

Ayalew, D. & Gibson, S.A. (2009). Head-to-tail transition of the Afar mantle plume: geochemical evidence from a Miocene bimodal basalt-rhyolite succession in the Ethiopian Large Igneous Province. *Lithos*, 112: 461–476.

Ayer, J., Amelin, Y., Corfu, F., *et al.* (2002). Evolution of the southern Abitibi greenstone belt based on U–Pb geochronology: autochthonous volcanic construction followed by plutonism, regional deformation and sedimentation. *Precambrian Research*, 115: 63–95.

Baer, G., Schubert, G., Bindschadler, D.L., & Stofan, E.R. (1994). Spatial and temporal relations between coronae and extensional belts, northern Lada Terra, Venus. *Journal of Geophysical Research*, 99: 8355–8369.

Bailey, D.K. (1974). Continental rifting and alkaline magmatism. In Sorensen, H. (ed.), *The Alkaline Rocks*. New York: Wiley, pp. 148–159.

Bailey, D.K. (1977). Lithospheric control of continental rift magmatism. *Geological Society of London Journal*, 133: 103–106.

Bailey, D.K. (1992). Episodic alkaline activity across Africa: implications for the causes of continental break-up. In Storey, B.C., Alabaster, T., & Pankhurst, R.J. (eds.), *Magmatism and the Causes of Continental Break-Up*. London: Geological Society of London, Special Publication 68: 91–98.

Bailey, D.K. (1993). Petrogenetic implications of the timing of alkaline, carbonatite, and kimberlite igneous activity in Africa. *South African Journal of Geology*, 96: 67–74.

Bain, J.H.C. & Draper, J.J., eds. (1997). North Queensland Geology. Australian Geological Survey Organisation (AGSO), Queensland Geology, volume 9.

Baird, D.J., Knapp, J.H., Steer, D.N., Brown, L.D., & Nelson, K.D. (1995). Upper-mantle reflectivity beneath the Williston basin, phase-change Moho, and the origin of intracratonic basins. *Geology*, 23: 431–434.

Baker, B.H. (1963). Geology and mineral resources of the Seychelles archipelago. *Geological Survey of Kenya Memoir*, 3.

Baker, V.R., Komatsu, G., Gulick, V.C., & Parker, T.J. (1997). Channels and valleys. In Bougher, S.W., Hunten, D.M., & Phillips, R.J. (eds.), *Venus II*. Tucson, AZ: University of Arizona Press, pp. 757–793.

Baker, V.R., Maruyama, S., & Dohm, J.M. (2007). Tharsis superplume and the geological evolution of Early Mars. In Yuen, D.A., Maruyama, S., Karato, S-i., & Windley, B.F. (eds.), *Superplumes: Beyond Plate Tectonics*. Berlin: Springer-Verlag, pp. 507–522.

Baragar, W.R.A. (1977). Volcanism of the stable crust. In Baragar, W.R.A., Coleman, L.C., & Hall, J.M. (eds.), *Volcanic Regimes in Canada*, Geological Association of Canada Special Publication 16, pp. 377–405.

Baragar, W.R.A., Ernst, R.E., Hulbert, L., & Peterson, T. (1996). Longitudinal petrochemical variation in the Mackenzie dyke swarm, northwestern Canadian Shield. *Journal of Petrology*, 37: 317–359.

Baratoux, L., Söderlund, U., Ernst, R.E., *et al.* (2014). New U-Pb age and geochemical constraints on the doleritic dyke swarms from the Leo-Man craton. 25th Colloquium of African Geology (CAG25), 11–14 August 2014. Dar Es Salaam,Tanzania.

Barbarin, B. (1999). A review of the relationships between granitoid types, their origins and their geodynamic environments. *Lithos*, 46: 605–626.

Bardet, N., Houssaye, A., Rage, J.-C., & Pereda Suberbiola, X. (2008). The Cenomanian–Turonian (late Cretaceous) radiation of marine squamates (Reptilia): the role of the Mediterranean Tethys. *Bulletin de la Société Géologique de France*, 179 (6): 605–622.

Bárdossy, G. & Aleva, G.J.J. (1990). *Lateritic Bauxites*. Amsterdam: Elsevier.

Barker, F.A., Sutherland Brown, J.R., Budahn, J.R., & Plafker, G. (1989). Backarc with frontal-arc component origin of Triassic Karmutsen basalt, British Columbia, Canada. *Chemical Geology*, 75 (1–2): 81–102.

Barley, M.E., Pickard, A.L., & Sylvester, P.J. (1997). Emplacement of a large igneous province as a possible cause of banded iron formation 2.45 billion years ago. *Nature*, 385: 55–58.

Barnes, S.-J. & Maier, W.D. (2002). Platinum-group element distributions in the Rustenburg Layered Suite of the Bushveld Complex, South Africa. In Cabri, L.J. (ed.), *The Geology, Geochemistry, Mineralogy and Mineral Beneficiation of Platinum-Group Elements*. Canadian Institute of Mining, Metallurgy and Petroleum, Special Volume, 54, pp. 431–458.

Barnes, S.-J., Van Kranendonk, M.J., & Sonntag, I. (2012). Geochemistry and tectonic setting of basalts from the Eastern Goldfields Superterrane. *Australian Journal of Earth Sciences*, 59: 707–735.

Barry, T.L., Self, S., Kelley, S.P., *et al.* (2010). New $^{40}Ar/^{39}Ar$ dating of the Grande Ronde lavas, Columbia River Basalts, USA: implications for duration of flood basalt eruption episodes. *Lithos*, 118: 213–222.

Basaltic Volcanism Study Project (1981). *Basaltic Volcanism on the Terrestrial Planets*. New York: Pergamon Press.

Basilevsky, A.T. & Head, J. (1994). Global stratigraphy of Venus: analysis of a random sample of thirty-six rest areas. *Earth Moon and Planets*, 66, 285–336.

Basilevsky, A.T. & Head, J.W. (2000). Rifts and large volcanoes on Venus: global assessment of their age relations with regional plains. *Journal of Geophysical Research*, 105: 24583–24611.

Basilevsky, A.T. & Head, J.W. (2002). Venus: timing and rates of geologic activity. *Geology*, 30: 1015–1018.

Basilevsky, A.T. & Head, J.W. (2007). Beta Regio, Venus: evidence for uplift, rifting, and volcanism due to a mantle plume. *Icarus*, 192: 167–186.

Basu, A.R., Renne, P., DasGupta, D.K., Teichmann, F., & Poreda, R. (1993). Early and late alkali igneous pulses and a high-[3]He plume origin for the Deccan flood basalts. *Science*, 261: 902–906.

Bates, M.P. & Halls, H.C. (1991). Broad-scale deformation of the central Superior Province revealed by paleomagnetism of the 2.45 Ga Matachewan dyke swarm. *Canadian Journal of Earth Sciences*, 28: 1780–1796.

Bayanova, T., Ludden, J., & Mitrofanov, F. (2009). Timing and duration of Palaeoproterozoic events producing ore-bearing layered intrusions of the Baltic Shield: metallogenic, petrological and geodynamic implications. London: Geological Society, Special Publication 323, pp. 165–198.

Beard, A.D., Downes, H., Hegner, E., & Sablukov, S.M. (2000). Geochemistry and mineralogy of kimberlites from the Arkhangelsk Region, NW Russia: evidence for transitional kimberlite magma types. *Lithos*, 51: 47–73.

Becker, M. & le Roex, A.P. (2006). Geochemistry of South African on- and off-craton, Group I and Group II kimberlites: petrogenesis and source region evolution. *Journal of Petrology*, 47: 673–703.

Bedini, R.-M., Blichert-Toft, J., Boyet, M., & Albarède, F. (2004). Isotopic constraints on the cooling of the continental lithosphere. *Earth and Planetary Science Letters*, 223: 99–111.

Beerling, D.J. & Berner, R.A. (2002). Biogeochemical constraints on the Triassic–Jurassic boundary carbon cycle event. *Global Biogeochemical Cycles*, 16: 1036.

Begg, G.C., Griffin, W.L., Natapov, L.M., *et al.* (2009). The lithospheric architecture of Africa: seismic tomography, mantle petrology, and tectonic evolution. *Geosphere*, 5: 23–50.

Begg, G.C., Hronsky, J.A.M., Arndt, N.T., *et al.* (2010). Lithospheric, cratonic, and geodynamic setting of Ni–Cu–PGE sulfide deposits. *Economic Geology*, 105: 1057–1070.

Bekker, A., Slack, J.F., Planavsky, N., *et al.* (2010). Iron formation: the sedimentary product of a complex interplay among mantle, tectonic, oceanic and biospheric processes. *Economic Geology*, 105: 467–508.

Bekker, A., Planavsky, N.J., Krapež, B., *et al.* (2014). Iron formations: their origins and implications for ancient seawater chemistry. In Mackenzie, F.T. (volume ed.), *Sediments, Diagenesis and Sedimentary Rocks (Treatise of Geochemistry)*. Amsterdam: Elsevier, pp. 561–628.

Belica, M.E., Piispa, E.J., Meert, J.G., *et al.* (2014). Paleoroterozoic mafic dyke swarms from the Dharwar craton; paleomagnetic poles for India from 2.37 to 1.88 Ga and rethinking the Columbia supercontinent. *Precambrian Research*, 244: 100–122.

Bell, K. (ed.) (1989). *Carbonatites: Genesis and Evolution*. London: Unwin Hyman.

Bell, K. (2001). Carbonatites: relationship to mantle–plume activity. In Ernst, R.E. & Buchan, K.L. (eds.), *Mantle Plumes: Their Identification through Time.* Boulder, CO: Geological Society of America, Special Publication 352, pp. 267–290.

Bell, K. (2002). Carbonatites and related alkaline rocks, lamprophyres, and kimberlites – indicators of mantle-plume activity. In Extended Abstracts and Photos of Superplume Workshop, Tokyo, January 28–31. *Electronic Geosciences*, 7 (1): 1–2.

Bell, K. (2005). Carbonatites. In Selley, R.C., Cocks, L.R.M., & Plimer, I.R. (eds.), *Encyclopedia of Geology*, Amsterdam: Elsevier, pp. 217–233.

Bell, K. & Dawson, J.B. (1995). Nd and Sr isotope systematics of the active carbonatite volcano, Oldoinyo Lengai. In Bell, K. & Keller, J. (eds.), *Carbonatite Volcanism: IAVCEI Proceeding in Volcanology 4.* Berlin: Springer-Verlag, pp. 100–112.

Bell, K. & Keller, J., eds. (1995). *Carbonatite Volcanism.* Berlin: Springer-Verlag.

Bell, K. & Rukhlov, A.S. (2004). Carbonatites from the Kola Alkaline Province: origin, evolution and source characteristics. In Zaitsev, A. & Wall, F. (eds.), *Phoscorites and Carbonatites from Mantle to Mine: the Key Example of the Kola Alkaline Province.* Mineralogical Society, London, Book Series, Volume 10, pp. 433–468.

Bell, K. & Simonetti, A. (1996). Carbonatite magmatism and plume activity: implications from the Nd, Pb, and Sr isotope systematics of Oldoinyo Lengai. *Journal of Petrology*, 37: 1321–1339.

Bell, K. & Simonetti, A. (2010). Source of parental melts to carbonatites–critical isotopic constraints. *Mineralogy and Petrology*, 98: 77–89.

Bell, K. & Tilton, G.R. (2001). Nd, Pb and Sr isotopic compositions of East African carbonatites: evidence for mantle mixing and plume inhomogeneity. *Journal of Petrology*, 42: 1927–1945.

Bell, K. & Tilton, G.R. (2002). Probing the mantle: the story from carbonatites. *EOS, Transactions of the American Geophysical Union*, 83: 273–277.

Bell, K., Kjarsgaard, B.A., & Simonetti, A. (1998). Carbonatites—Into the twenty-first century. *Journal of Petrology*, 39: 1839–1845.

Bennett, V.C. (2003). Compositional evolution of the mantle. In Carlson, R.W. (ed.), *The Mantle and Core.* New York: Elsevier, Treatise on Geochemistry, volume 2, pp. 493–519.

Benson, R.N. (2003). Age estimates of the seaward-dipping volcanic wedge, earliest oceanic crust, and earliest drift-stage sediments along the North American Atlantic continental margin. *AGU Geophysical Monograph*, 136: 61–75.

Benton, M.J., Tverdokhlebov, V.P., & Surkov, M.V. (2004). Ecosystem remodeling among vertebrates at the Permian-Triassic boundary in Russia. *Nature*, 432: 97–100.

Bercovici, D. & Mahoney, J. (1994). Double flood basalts and plume head separation at 660-kilometer discontinuity. *Science*, 266: 1367–1369.

Beresford, S. & Hronsky, J. (2012). The hunt for Oktyabrysky: understanding the camp to deposit scale footprint of mafic Ni–Cu sulfide deposits. *Proceedings of the 34th International Geological Congress Unearthing our Past and Future – Resourcing Tomorrow*, 5–10 August 2012, Brisbane, Australia.

Beresford, S.W., Cas, R.A.F., Lambert, D.D., & Stone, W.E. (2000). Vesicles in thick komatiite lava flows, Kambalda, Western Australia. *Journal of the Geological Society, London*, 157: 11–14.

Berner, R.A. (2002). Examination of hypotheses for the Permo–Triassic boundary extinction by carbon cycle modeling. *Proceedings of the National Academy of Science*, 99: 4172–4177.

Bertrand, H., Fornari, M., Marzoli, A., García-Duarte, R., & Sempere, T. (2014). The Central Atlantic Magmatic Province extends into Bolivia, *Lithos*, 188: 33–43.

Best, M.G. & Christiansen, E.H. (1991). Limited extension during peak Tertiary volcanism, Great Basin of Nevada and Utah. *Journal of Geophysical Research*, 96(B8): 13509–13528.

Bethune, K.M. (1993). Evolution of the Grenville Front in the Tyson Lake area, southwest of Sudbury, Ontario with emphasis on the tectonic significance of the Sudbury diabase dykes. Ph.D. Thesis, Department of Geological Sciences, Queen's University, Canada, p. 263.

Betts, P.G., Giles, D., Schaefer, B.F., & Mark, G. (2007). 1600–1500 Hotspot track in eastern Australia: implications for Mesoproterozoic continental reconstructions. *Terra Nova*, 19: 496–501.

Betts, P.G., Giles, D., Foden, J., *et al.* (2009). Mesoproterozoic plume-modified orogenesis in eastern Precambrian Australia. *Tectonics*, 28, TC3006.

Bhandari, N., Shukla, P.N., Ghevariya, Z.G., & Sundaram, S.M. (1995). Impact did not trigger Deccan volcanism: evidence from Anjar K/T boundary intertrappean sediments. *Geophysical Research Letters*, 22: 433–436.

Bhushan, S.K. (2000). Malani rhyolites: a review. *Gondwana Research*, 3: 65–77.

Bialas, R.W., Buck, W.R., & Qin, R., (2010). How much magma is required to rift a continent? *Earth and Planetary Science Letters*, 292: 68–78.

Bickle, M.J. (1990). Mantle evolution. In Hall, R.P. & Hughes, D.J. (eds.), *Early Precambrian Basic Magmatism*. Glasgow: Blackie, pp. 111–135.

Bierlein, F.P. & Pisarevsky, S. (2008). Plume-related oceanic plateaus as a potential source of gold mineralization. *Economic Geology*, 103: 425–430.

Bilotti, F. & Suppe, J. (1999). The global distribution of wrinkle ridges on Venus. *Icarus*, 139: 137–157.

Bindschadler, D.L. (1995). Magellan – a new view of Venus geology and geophysics. *Reviews of Geophysics*, 33: 459–467.

Bindschadler, D.L., Schubert, G., & Kaula, W.M. (1992). Coldspots and hotspots: global tectonics and mantle dynamic of Venus. *Journal of Geophysical Research*, 97: 13,495–13,532.

Bingen, B., Demaiffe, D., & van Breemen, O. (1998). The 616 Ma old Egersund basaltic dike swarm, SW Norway, and Late Neoprotorozoic opening of the Iapetus Ocean. *Journal of Geology*, 106: 565–574.

Bizzarro, M., Simonetti, A., Stevenson, R.K., & David, J. (2002). Hf isotope evidence for a hidden mantle reservoir. *Geology*, 30: 771–774.

Bjonnes, E.E., Hansen, V.L., James, B., & Swenson, J.B. (2012). Equilibrium resurfacing of Venus: results from new Monte Carlo modeling and implications for Venus surface histories. *Icarus*, 217: 451–461.

Black, B.A., Elkins-Tanton, L.T., Rowe, M.C., & Ukstins Peate, I. (2012), Magnitude and consequences of volatile release from the Siberian Traps. *Earth and Planetary Science Letters*, 317–318: 363–373.

Black, B.A., Lamarque, J.-F., Shields, C., Elkins-Tanton, L.T., & Kiehl, J.T. (2014). Acid rain and ozone depletion from pulsed Siberian Traps magmatism. *Geology*, 42: 67–70.

Blackburn, T.J., Olsen, P.E., Bowring, S.A., *et al.* (2013). Zircon U–Pb geochronology links the end-Triassic extinction with the Central Atlantic Magmatic Province. *Science*, 340: 941–945.

Blake, T.S., Buick, R., Brown, S.J.A., & Barley, M.E. (2004). Geochronology of a Late Archaean flood basalt province in the Pilbara Craton, Australia: constraints on basin evolution, volcanic and sedimentary accumulation, and continental drift rates. *Precambrian Research*, 133: 143–173.

Bleeker, W. (2002). Archaean tectonics: a review, with illustrations from the Slave craton. In Fowler, C.M.R., Ebinger, C.J., & Hawkesworth, C.J. (eds.), *The Early Earth: Physical, Chemical and Biological Development*. London: Geological Society of London Special Publication 199, pp. 151–181.

Bleeker, W. (2003). The late Archean record: a puzzle in ca. 35 pieces. *Lithos*, 71: 99–134.

Bleeker, W. (2004). Taking the pulse of planet Earth: a proposal for a new multidisciplinary flagship project in Canadian solid Earth sciences. *Geoscience Canada*, 31: 179–190.

Bleeker, W. & Ernst, R. (2006). Short-lived mantle generated magmatic events and their dyke swarms: the key unlocking Earth's paleogeographic record back to 2.6 Ga. In Hanski, E., Mertanen, S., Rämö, T., Vuollo, J. (eds.), *Dyke Swarms – Time Markers of Crustal Evolution*. London: Taylor and Francis/Balkema, pp. 3–26.

Bleeker, W. & Hall, B. (2007). The Slave craton: geology and metallogenic evolution. In Goodfellow, W.D. (ed.), *Mineral Deposits of Canada: A Synthesis of Major Deposit-Types, District Metallogeny, The Evolution of Geological Provinces, and Exploration Methods*. Geological Association of Canada, Mineral Deposits Division, Special Publication, Number 5, pp. 849–879.

Bleeker, W., Ketchum, J.W.F., Jackson, V.A., & Villeneuve, M.E. (1999a). The Central Slave Basement Complex, Part I: its structural topology and autochthonous cover. *Canadian Journal of Earth Sciences*, 36: 1083–1109.

Bleeker, W., Parrish, R.R., & Sager-Kinsman, A. (1999b). High-precision U–Pb geochronology of the Late Archean Kidd Creek deposit and Kidd Volcanic Complex. *Society of Economic Geologists, Economic Geology, Monograph* 10: 43–70.

Blissett, A.H., Creaser, R.A., Daly, S.J., Flint, R.B., & Parker, A.J. (1993). Gawler Range volcanics. In Drexel, J.F., Preiss, W.V., & Parker, A.J. (eds.), *The Geology of South Australia. Vol. 1, The Precambrian*. Adelaide: Geological Survey of South Australia, Bulletin 54, pp. 107–131.

Bogdanov, N.A. & Dobretsov, N.L. (2002). The Okhotsk volcanic oceanic plateau. *Russian Geology and Geophysics*, 43: 87–99.

Bogdanova, S.V., Gintov, O.B., Kurlovich, D.M., *et al.* (2013). Late Palaeoproterozoic mafic dyking in the Ukrainian shield of Volgo-Sarmatia caused by rotation during the assembly of supercontinent Columbia (Nuna). *Lithos*, 174: 196–216.

Bonin, B. (2007). A-type granites and related rocks: evolution of a concept, problems and prospects. *Lithos*, 97: 1–29.

Borisenko, A.S., Sotnikov, V.I., Izokh, A.E., Polyakov, G.V., & Obolensky, A.A. (2006). Permo–Triassic mineralization in Asia and its relation to plume magmatism. *Russian Geology and Geophysics*, 47: 166–182.

Boslough, M.B., Chael, E.P., Trucano, T.G., Crawford, D.A., & Campbell, D.L. (1996). Axial focusing of impact energy in the Earth's interior: a possible link to flood basalts and hotspots. In Ryder, G., Fastovsky, D., & Gartner, S. (eds.),

The Cretaceous-Tertiary Event and Other Catastrophes in Earth History. Boulder, CO: Geological Society of America, Special Paper 307, pp. 541–550.

Boutilier, R.R., & Keen, C.E. (1999). Small-scale convection and divergent plate boundaries. *Journal of Geophysical Research*, 104 (B4): 7389–7403.

Bradley, J. (1965). Occurrence and origin of ring-shaped dolerite outcrops in the Eastern Cape Province and Western Transkei. *Transactions of the Royal Society of New Zealand (Geology)*, 3: 27–55.

Bralower, T.J. & Lorente, M.A. (2003). Paleogeography and stratigraphy of the La Luna Formation and related cretaceous anoxic depositional systems. *Palaios*, 18: 301–304.

Bralower, T.J., Fullagar, P.D., Paull, C.K., Dwyer, G.S., & Leckie, R.M. (1997). Middle Cretaceous strontium isotope stratigraphy of deep sea sections. *Geological Society of America Bulletin*, 109: 1421–1442.

Brandon, A.D., Norman, M.D., Walker, R.J., & Morgan, J.W. (1999). ^{186}Os–^{187}Os systematics of Hawaiian picrites. *Earth and Planetary Science Letters*, 174: 25–42.

Braun, J. & Beaumont, C. (1989). A physical explanation of the relation between flank uplifts and the breakup unconformity at rifted continental margins. *Geology*, 17: 760–764.

Brewer, T.S., Hergt, J.M., Hawkesworth, C.J., Rex, D., & Storey, B.C. (1992). Coats Land dolerites and the generation of Antarctic continental flood basalts. In Storey B., Alabaster, T., & Pankhurst, R. (eds.), *Magmatism and the Causes of Continental Break-up*. London: Geological Society of London, Special Publication 68, pp. 185–208.

Bright, R.M., Amato, J.M., Denyszyn, S.W., & Ernst, R.E. (2014). U–Pb geochronology of 1.1 Ga diabase in the southwestern United States: testing models for the origin of a post-Grenville Large Igneous Province. *Lithosphere*, 6, 135–156.

Brodie, J. & White, N. (1994). Sedimentary basin inversion due to igneous underplating: Northwest European continental shelf. *Geology*, 22: 147–150.

Bromley, J., Mannström, B., Nisca, D., & Jamtlid, A. (1994). Airborne geophysics: application to a ground-water study in Botswana. *Ground Water*, 32: 79–90.

Brown, A.C. (2007). Genesis of native copper lodes in the Keweenaw district, Northern Michigan: a hybrid evolved meteoric and metamorphogenic model. *Economic Geology*, 101: 1437–1444.

Brown, D.J., Holohan, E.P., & Bell, B.R. (2009). Sedimentary and volcano-tectonic processes in the British Paleocene Igneous Province: a review. *Geological Magazine*, 146: 326–352.

Brown, M. (2013). Granite: from genesis to emplacement. *Geological Society of America Bulletin*, 125, 1079–1113.

Bryan, S. (2007). Silicic large igneous provinces. *Episodes*, 30: 20–31.

Bryan, S. & Ernst, R.E. (2008). Revised definition of Large Igneous Provinces (LIPs). *Earth-Science Reviews*, 86: 175–202.

Bryan, S.E. & Ferrari, L. (2013). Large Igneous Provinces and Silicic Large Igneous Provinces: progress in our understanding over the last 25 years. *Geological Society of America Bulletin*, 125: 1053–1078.

Bryan, S.E., Constantine, A.E., Stephens, C.J., *et al.* (1997). Early Cretaceous volcano–sedimentary successions along the eastern Australian continental margin: implications for the break-up of eastern Gondwana. *Earth and Planetary Science Letters*, 153: 85–102.

Bryan, S.E., Ewart, A., Stephens, C.J., Parianos, J., & Downes, P.J. (2000). The Whitsunday Volcanic Province, Central Queensland, Australia: lithological and stratigraphic investigations of a silicic dominated large igneous province. *Journal of Volcanology and Geothermal Research*, 99: 55–78.

Bryan, S.E., Riley, T.R., Jerram, D.A., Leat, P.T., & Stephens, C.J. (2002). Silicic volcanism: an under-valued component of large igneous provinces and volcanic rifted margins. In Menzies, M.A., Klemperer, S.L., Ebinger, C.J., & Baker, J. (eds.), *Magmatic Rifted Margins*. Boulder, CO: Geological Society of America, Special Paper 362, pp. 97–118.

Bryan, S.E., Holcombe, R.J., & Fielding, C.R. (2003). Reply to: "The Yarrol terrane of the northern New England Fold Belt: Forearc or backarc?" Discussion by Murray, C.G., Blake, P.R., Hutton, L.J., Withnall, I.W., Hayward, M.A., Simpson, G.A., & Fordham, B.G. *Australian Journal of Earth Sciences*, 50: 271–293.

Bryan, S.E., Ferrari, L., Reiners, P.W., *et al.* (2008). New insights into large volume rhyolite generation in the Mid-Tertiary Sierra Madre Occidental Province, Mexico, revealed by U–Pb geochronology. *Journal of Petrology*, 49: 47–77.

Bryan, S., Ukstins Peate, I., Self, S., *et al.* (2010). The largest volcanic eruptions on Earth. *Earth-Science Reviews*, 102: 207–229.

Bryan, S.E., Cook, A.G., Allen, C.M., *et al.* (2012). Early–Mid Cretaceous tectonic evolution of eastern Gondwana: from silicic LIP magmatism to continental rupture. *Episodes*, 35(1): 142–152.

Bryan, S.E., Orozco-Esquivel, T., Ferrari, L., & López-Martínez, M. (2014). Pulling apart the mid to late Cenozoic magmatic record of the Gulf of California: is there a comondú arc? In Gómez-Tuena, A., Straub, S.M., & Zellmer, G.F. (eds.), *Orogenic Andesites and Crustal Growth*. London: Geological Society, Special Publication 385, pp. 389–407.

Buchan, K.L. (2014) Reprint of "Key paleomagnetic poles and their use in Proterozoic continent and supercontinent reconstructions: A review". *Precambrian Research*, 244: 5–22.

Buchan, K.L. & Ernst, R.E. (2004). Dyke swarms and related units in Canada and adjacent regions. *Geological Survey of Canada Map 2022A* (scale 1:5,000,000) and accompanying booklet.

Buchan, K.L. & Ernst, R.E. (2006). Giant dyke swarms and the reconstruction of the Canadian Arctic islands, Greenland, Svalbard and Franz Josef Land. In Hanski, E., Mertanen, S., Rämö, T., & Vuollo, J. (eds.), *Dyke Swarms: Time Markers of Crustal Evolution*. Amsterdam: Taylor & Francis/Balkema, pp. 27–48.

Buchan, K.L. & Ernst, R.E. (2013). Diabase dyke swarms of Nunavut, Northwest Territories, and Yukon, Canada. *Geological Survey of Canada, Open File 7464*.

Buchan, K.L., Mortensen, J.K., & Card, K.D. (1993). Northeast-trending Early Proterozoic dykes of southern Superior Province: multiple episodes of emplacement recognized from integrated paleomagnetism and U–Pb geochronology. *Canadian Journal of Earth Sciences*, 30: 1286–1296.

Buchan, K.L., Mortensen, J.K., Card, K.D., & Percival, J.A. (1998). Paleomagnetism and U–Pb geochronology of diabase dyke swarms of Minto Block, Superior Province, Quebec, Canada. *Canadian Journal of Earth Sciences*, 35: 1054–1069.

Buchan, K.L., Mertanen, S., Park, R.G., *et al.* (2000). Comparing the drift of Laurentia and Baltica in the Proterozoic: the importance of key paleomagnetic poles. *Tectonophysics*, 319: 167–198.

Buchan, K.L., Ernst, R.E., Hamilton, M.A., *et al.* (2001). Rodinia: the evidence from integrated paleomagnetism and U–Pb geochronology. *Precambrian Research*, 110: 9–32.

Buchan, K.L., Goutier, J., Hamilton, M.A., Ernst, R.E., & Matthews, W.A. (2007). Paleomagnetism, U–Pb geochronology and geochemistry of Lac Esprit and other dyke swarms, James Bay area, Quebec, and implications for Paleoproterozoic deformation of the Superior Province. *Canadian Journal of Earth Sciences*, 44: 643–664.

Buchan, K.L., Ernst, R.E., Bleeker, W., *et al.* (2010). Proterozoic magmatic events of the Slave craton, Wopmay Orogen and environs. *Geological Survey of Canada Open File*, 5985.

Bulakh, A.G., Ivanikov, V.V., & Orlova, M.P. (2004). Overview of carbonatite phoscorite complexes of the Kola Alkaline Province in the context of a Scandinavian North Atlantic Alkaline Province. In Zaitsev, A.N. & Wall, F. (eds.), *Phoscorites and Carbonatites from Mantle to Mine: the Key Example of the Kola Alkaline Province*. Mineralogical Society, London, Book Series, 10, pp. 1–43.

Bullen, D.S., Hall, R.P., & Hanson, R.E. (2012). Geochemistry and petrogenesis of mafic sills in the 1.1 Ga Umkondo large igneous province, southern Africa. *Lithos*,142–143: 116–129.

Bultitude, R.J. (1976). Flood basalts of probable Early Cambrian age in northern Australia. In Johnson, R.W. (ed.), *Volcanism in Australia*. New York: Elsevier Science, pp. 1–20.

Bunger, A.P. & Cruden, A.R. (2011). Modeling the growth of laccoliths and large mafic sills: the role of magma body forces. *Journal of Geophysical Research*, 116, B02203, doi:10.1029/2010JB007648.

Bunger, A.P., Menand, T., Cruden, A., Zhang, X., & Halls, H. (2013). Analytical predictions for a natural spacing within dyke swarms. *Earth and Planetary Science Letters*, 375: 270–279.

Burgess, S.D. & Bowring, S.A. (2014). Linking Earth's biggest volcanic eruption with its most severe mass extinction. Manuscript submitted for publication.

Burgess, S.D., Fleming, T.H., Elliot, D.H., & Bowring, S.A. (2011). High-precision U–Pb zircon geochronology of the Ferrar large igneous province: can sill emplacement be linked with the Pliensbachian–Toarcian extinction event? *Geological Society of America Abstracts with Programs*, 43, (5): 372.

Burgess, S., Bowring, S., Pavlov, V.E. & Veselovsky, R.V. (2014a). High-precision temporal constraints on intrusive magmatism of the Siberian Traps. *Geophysical Research Abstracts*, vol. 16, EGU2014-14342-1. EGU General Assembly, 2014.

Burgess, S.D., Bowring, S., & Shen, S.-Z. (2014b). High-precision timeline for Earth's most severe extinction. *Proceedings of the National Academy of Sciences*, 111: 3316–3321 [correction 2014, 111: 5050].

Burke, K. (1977). Aulacogens and continental breakup. *Annual Review of Earth and Planetary Sciences*, 5: 371–396.

Burke, K. (1988). Tectonic evolution of the Caribbean. *Annual Review of Earth and Planetary Sciences*, 16: 201–230.

Burke, K. (1997). Foreword. In de Wit, M.J. & Ashwal, L.D. (eds.), *Greenstone Belts*. Oxford: Clarendon, pp. v–vii.

Burke, K. & Dewey, J. (1973). Plume-generated triple junctions: key indicators in applying plate tectonics to old rocks. *Journal of Geology*, 81: 406–433.

Burke, K. & Whiteman, A.J. (1973). Uplift, rifting, break-up of Africa, In Tarling, D.H. & Runcorn, S.K. (eds.), *Implications of Continental Drift to the Earth Sciences*. London: Academic Press, pp. 735–755.

Burke, K., Kidd, W.S.F., & Kusky, T.M. (1985). The Pongola structure of southern Africa: the World's oldest rift? *Journal of Geodynamics*, 2: 35–49.

Burke, K., Ashwal, L.D., & Webb, S.J. (2003). New way to map old sutures using deformed alkaline rocks and carbonatites. *Geology*, 31: 391–394.

Burke, K., Steinberger, B., Torsvik, T.H., & Smethurst, M.A. (2008). Plume generation zones at the margins of Large Low Shear Velocity Provinces on the core–mantle boundary. *Earth and Planetary Science Letters*, 265: 49–60.

Burns, E.R., Morgan, D.S., Lee, K.K., Haynes, J.V., & Conlon, T.D. (2012). Evaluation of long-term water-level declines in basalt aquifers near Mosier, Oregon. US Geological Survey Scientific Investigations Report 2012–5002.

Burov, E. & Guillou-Frottier, L. (2005). The plume head-continental lithosphere interaction using a tectonically realistic formulation for the lithosphere. *Geophysical Journal International*, 161: 469–490.

Cadman, A.C., Tarney, J., & Hamilton, M.A. (1997). Petrogenetic relationships between Palaeoproterozoic tholeiitic dykes and associated high-Mg noritic dykes, Labrador, Canada. *Precambrian Research*, 82: 63–84.

Callegaro, S., Marzoli, A., Bertrand, H., *et al.* (2013). Upper and lower crust recycling in source of CAMP basaltic dykes from SE North America. *Earth and Planetary Science Letters*, 376: 186–199.

Cameron, K.L., Nimz, G.J., Kuentz, D., Niemeyer, S., & Gunn, S. (1989). Southern Cordilleran basaltic andesite suite, southern Chihuahua, Mexico: a link between Tertiary continental arc and flood basalt magmatism in North America. *Journal of Geophysical Research*, 94: 7817–7840.

Camp, V.E. & Hanan, B.B. (2008). A plume-triggered delamination origin for the Columbia River Basalt Group. *Geosphere*, 4: 480–495.

Camp, V.E. & Roobol, M.J. (1992). Upwelling asthenosphere beneath western Arabia and its regional implications. *Journal of Geophysical Research*, 15 (B11): 255–271.

Camp, V.E. & Ross, M.E. (2004). Mantle dynamics and genesis of mafic magmatsim in the intermontane Pacific Northwest. *Journal of Geophysical Research*, 109, (B8), B8204, doi 10.1029/2003JB2838.

Camp, V.E., Ross, M.E., & Hanson, W.E. (2003). Genesis of flood basalts and Basin and Range volcanic rocks from Steens Mountain to the Malheur River Gorge, Oregon. *Geological Society of America Bulletin*, 115: 105–128.

Campbell, I.H. (1987). Distribution of orthocumulate textures in the Jimberlana intrusion. *Journal of Geology*, 95: 35–54.

Campbell, I.H. (1996). Fluid dynamic processes in basaltic magma chambers. In Cawthorn, R.G. (ed.), *Layered Intrusions*. Amsterdam: Elsevier, pp. 45–76.

Campbell, I.H. (1998). The mantle's chemical structure: insights from the melting products of mantle plumes. In Jackson, I.N.S. (ed.), *The Earth's Mantle: Composition, Structure and Evolution*. New York: Cambridge University Press, pp. 259–310.

Campbell, I.H. (2001). Identification of ancient mantle plumes. In Ernst, R.E., Buchan, K.L. (eds.), *Mantle Plumes: Their Identification through Time*. Boulder, CO: Geological Society of America, Special Paper 352, pp. 5–21.

Campbell, I.H. (2005). Large igneous provinces and the mantle plume hypothesis. *Elements*, 1: 265–269.

Campbell, I.H. (2007). Testing the plume theory. *Chemical Geology*, 241: 153–176.

Campbell, I.H. & Griffiths, R.W. (1990). Implications of mantle plume structure for the evolution of flood basalts. *Earth and Planetary Science Letters*, 99: 79–93.

Campbell, I.H. & Griffiths, R.W. (1992). The changing nature of mantle hotspots through time: implications for the geochemical evolution of the mantle. *Journal of Geology*, 92: 497–523.

Campbell, I.H. & Griffiths, R.W. (2014). Did the formation of D'' cause the Archaean–Proterozoic transition? *Earth and Planetary Science Letters*, 388: 1–8.

Campbell, I.H. & Hill, R.I. (1988). A two-stage model for the formation of the granite-greenstone terrains of the Kalgoorlie-Norseman area, Western Australia. *Earth and Planetary Science Letters*, 90: 11–25.

Campbell, I.H. & Kerr, A.C. (2007). The great plume debate: testing the plume theory. *Chemical Geology*, 241: 149–152.

Campbell, I.H. & Naldrett, A.J. (1979). The influence of silicate: sulfide ratio on the geochemistry of the magmatic sulfides. *Economic Geology*, 74: 1503–1505.

Campbell, I.H. & Turner, J.S. (1986). The role of convection in the formation of platinum and chromitite deposits in layered intrusions. In Scarfe, C.M. (ed.), *Short Course in Silicate Melts*. Mineralogical Association of Canada, volume 12, pp. 236–278.

Campbell, I.H., Griffiths, R.W., & Hill, R.I. (1989). Melting in an Archaean mantle plume: heads it's basalts, tails it's komatiites. *Nature*, 339: 697–699.

Campbell, I.H., Czamanske, G.K., Fedorenko, V.A., Hill, R.I., & Stepanov, V. (1992). Synchronisn of the Siberian Traps and the Permian–Triassic boundary. *Science*, 258: 1760–1763.

Campbell, I.H., Compston, D.M., Richards, J.P., Johnson, J.P., & Kent, A.J.R. (1998). Review of the application of isotopic studies to the genesis of Cu–Au mineralisation at Olympic Dam and Au mineralisation at Porgera, the Tennant creek district and Yilgarn Craton. *Australian Journal of Earth Sciences*, 45: 201–218.

Cannon, W.F. (1992). The Midcontinent rift in the Lake Superior region with emphasis on its geodynamic evolution. *Tectonophysics*, 213: 41–48.

Cañón-Tapia, E. (2010). Origin of Large Igneous Provinces: the importance of a definition. In Cañón-Tapia, E. & Szakács, A. (eds.), *What Is a Volcano?* Boulder, CO: Geological Society of America Special Paper 470, pp. 77–101.

Carlson, R.W. & Hart, W.K. (1988). Flood basalt volcanism in the northwestern U.S. In Macdougall, J.D. (ed.), *Continental Flood Basalts*. Dordrecht: Kluwer Academic Publishers, pp. 35–61.

Carlson, R.W. (1991). Physical and chemical evidence on the cause and source characteristics of flood basalt volcanism. *Australian Journal of Earth Sciences*, 38: 525–544.

Castillo, P.R. (2004). Geochemistry of Cretaceous volcaniclastic sediments in the Nauru and East Mariana basins provides insights into the mantle sources of giant oceanic plateaus. In Fitton, J.G., Mahoney, J.J., Wallace, P.J., & Saunders, A.D. (eds.), *Origin and Evolution of the Ontong Java Plateau*. London: Geological Society of London, Special Publication 229, pp. 353–368.

Cather, S.M., Dunbar, N.W., McDowell, F.W., McIntosh, W.C., & Scholle, P.A. (2009). Climate forcing by iron fertilization from repeated ignimbrite eruptions: the icehouse-silicic large igneous province (SLIP) hypothesis. *Geosphere*, 5: 315–324.

Cattell, A.C. & Taylor, R.N. (1990). Archean basic magma. In Hall, R.P. and Hughes, D.J. (eds.), *Early Precambrian Basic Magmatism*. Glasgow: Blackie, pp. 11–39.

Cawood, P.A., McCausland, P.J.A., & Dunning, G.R. (2001). Opening Iapetus: constraints from the Laurentian margin in Newfoundland. *Geological Society of America Bulletin*, 113: 443–453.

Cawthorn, R.G., ed. (1996). *Layered Intrusions. Developments in Petrology 15*. Amsterdam: Elsevier.

Cawthorn, R.G., 2010. The Platinum group element deposits of the Bushveld Complex. *South Africa Platinum Metals Review*, 54 (4): 205–215.

Cawthorn, R.G. (2012). Multiple sills or a layered intrusion? Time to decide. *South African Journal of Geology*, 115: 283–290.

Cawthorn, R.G. & Ashwal, L.D. (2009). Origin of anorthosite and magnetitite layers in the Bushveld Complex, constrained by major element compositions of plagioclase. *Journal of Petrology*, 50: 1607–1637.

Cawthorn, R.G. & Walraven, F. (1998). Emplacement and crystallization time for the Bushveld Complex. *Journal of Petrology*, 39: 1669–1687.

Cawthorn, R.G. & Webb, S.J. (2013). Cooling of the Bushveld Complex, South Africa: implications for paleomagnetic reversals. *Geology*, 41: 687–690.

Chadwick, W.W., Jr. & Howard, K.A. (1991). The pattern of circumferential and radial eruptive fissures on the volcanoes of Fernandina and Isabela Islands, Galapagos. *Bulletin of Volcanology*, 53: 259–275.

Chakmouradian, A.R. & Zaitsev, A.N. (2012). Rare earth mineralization in igneous rocks: sources and processes. *Elements*, 8: 347–353.

Chalapathi Rao, N.V. & Lehmann, B. (2011). Kimberlites, flood basalts and mantle plumes: new insights from the Deccan Large Igneous Province. *Earth-Science Reviews*, 107: 315–324.

Chalapathi Rao, N.V. Lehmann, B., Mainkar, D., & Belyatsky, B. (2011). Petrogenesis of the end-Cretaceous diamondiferous Behradih orangeite pipe: implication for mantle plume–lithosphere interaction in the Bastar craton, Central India. *Contributions to Mineralogy and Petrology*, 161: 721–742.

Chambers, L.M., Pringle, M.S., & Fitton, J.G. (2002). Age and duration of magmatism on the Ontong Java Plateau: ^{40}Ar–^{39}Ar results from ODP Leg 192. Abstract V71B-1271. *EOS Transactions of the American Geophysical Union*, 83: F47.

Chappell, B.W. & White, A.J.R. (1992). I- and S-type granites in the Lachlan Fold Belt. *Transactions of the Royal Society of Edinburgh: Earth Sciences*, 83: 1–26.

Chappell, B.W. & White, A.J.R. (2001). Two contrasting granite types: 25 years later. *Australian Journal of Earth Sciences*, 48: 489–499.

Chenet, A.L., Quidelleur, X., Fluteau, F., Courtillot, V., & Bajpai, S. (2007). ^{40}K–^{39}Ar dating of the main Deccan large igneous province: further evidence of KTB age and short duration. *Earth and Planetary Science Letters*, 263: 1–15.

Chenet, A.-L., Fluteau, F., Courtillot, V., Gérard, M., & Subbarao, K.V. (2008). Determination of rapid Deccan eruptions across the Cretaceous–Tertiary boundary using paleomagnetic secular variation: results from a 1200-m-thick section in the Mahabaleshwar escarpment. *Journal of Geophysical Research*, 113: B04101.

Cheney, E.S. (1996). Sequence stratigraphy and plate tectonic significance of the Transvaal succession of southern Africa and its equivalent in Western Australia. *Precambrian Research*, 79: 3–24.

Cheney, E.S., Roering, C., & Steller, E. (1988). Vaalbara. *Extended Abstracts, Geocongress '88, Geological Society of South Africa, Durban*, 85–88.

Chevallier, L. & Woodford, A. (1999). Morpho-tectonics and mechanism of emplacement of the dolerite rings and sills of the western Karoo, South Africa. *South African Journal of Geology*, 102: 43–54.

Chevallier, L., Goedhart, M., & Woodford, A. (2001). The influence of dolerite sill and ring complexes on the occurrence of groundwater in the Karoo fractured aquifers: a morpho-tectonic approach. Water Resources Commission South Africa, Research Report, 937.

Christiansen, E.H., Haapala, I., & Hart, G.L. (2007). Are Cenozoic topaz rhyolites the erupted equivalents of Proterozoic rapakivi granites? Examples from the Western United States and Finland. *Lithos*, 97: 219–246.

Ciborowski, T.J.R., Kerr, A.C., McDonald, I., Ernst, R.E., & Minife, M.J. (2013). The geochemistry and petrogenesis of the Blue Draw Metagabbro. *Lithos*, 174: 271–290.

Clemens, J.D., Stevens, G., & Farina, F. (2011). The enigmatic sources of I-type granites: the peritectic connexion. *Lithos*, 126: 174–181.

Clifford, T.N. (1966). Tectono-metallogenetic units and metallogenic provinces of Africa. *Earth and Planetary Science Letters*, 1: 421–434.

Cloos, H. (1953). *Conversation with the Earth*. New York: Knopf.

Cloos, M. (1993). Lithospheric buoyancy and collisional orogenesis: subduction of oceanic plateaus, continental margins, island arcs, spreading ridges and seamounts. *Geological Society of America Bulletin*, 105: 715–737.

Clouard, V. & Bonneville, A. (2005). Ages of seamounts, islands, and plateaus on the Pacific plate. In Foulger, G.R., Natland, J.H., Presnall, D.C., & Anderson, D.L. (eds.), *Plates, Plumes, and Paradigms*. Boulder, CO: Geological Society of America, Special Paper 388, pp. 71–90.

Coble, M.A. & Mahood, M.A. (2012). Initial impingement of the Yellowstone plume located by widespread silicic volcanism contemporaneous with Columbia River flood basalts. *Geology*, 4: 655–658.

Coello, J., Cantagrel, J.M., Hernán, F., *et al.* (1992). Evolution of the eastern volcanic ridge of the Canary Islands based on new K–Ar data. *Journal of Volcanology and Geothermal Research*, 53: 251–274.

Coffin, M.F., & Eldholm, O., eds. (1991). *Large Igneous Provinces: JOI/ USSAC Workshop Report*. Austin, TX: The University of Texas at Austin Institute for Geophysics, Cannon Technical Report.

Coffin, M.F. & Eldholm, O. (1992). Volcanism and continental break-up: A global compilation of large igneous provinces. In Storey, B.C., Alabaster, T., & Pankhurst, R.J., (eds.), *Magmatism and the Causes of Continental Break-up*. London, Geological Society, Special Publication 68, pp. 17–30.

Coffin, M.F. & Eldholm, O. (1993a). Scratching the surface: estimating dimensions of large igneous provinces. *Geology*, 21: 515–518.

Coffin, M.F. & Eldholm, O. (1993b). Large igneous provinces. *Scientific American*, 269: 42–49.

Coffin, M.F. & Eldholm, O. (1994). Large igneous provinces: crustal structure, dimensions, and external consequences. *Review of Geophysics*, 32: 1–36.

Coffin, M.F. & Eldholm, O. (2001). Large igneous provinces: progenitors of some ophiolites? In Ernst, R.E. & Buchan, K.L. (eds.), *Mantle Plumes: Their Identification through Time*. Boulder, CO: Geological Society of America Special Paper 352, pp. 59–70.

Coffin, M.F. & Eldholm, O. (2005). Large igneous provinces. In Selley, R.C., Cocks, R., & Plimer, I.R. (eds.), *Encyclopedia of Geology*. Elsevier: Oxford, pp. 315–323.

Coffin, M.F., Pringle, M.S., Duncan, R.A., *et al.* (2002). Kerguelen hotspot magma output since 130 Ma. *Journal of Petrology*, 43: 1121–1139.

Coleman, R.C. (1993). *Geological Evolution of the Red Sea*. Oxford: Oxford University Press.

Collerson, K.D., Williams, Q., Ewart, A.E., & Murphy, D.T. (2010). Origin of HIMU and EM-1 domains sampled by ocean island basalts, kimberlites and carbonatites: the role of CO_2-fluxed lower mantle melting in thermochemical upwellings. *Physics of the Earth and Planetary Interiors*, 181: 112–131.

Coltice, N., Phillips, B.R., Bertrand, H., Ricard, Y., & Rey, P. (2007). Global warming of the mantle at the origin of flood basalts over supercontinents. *Geology*, 35: 391–394.

Coltice, N., Bertrand, H., Rey, P., *et al.* (2009). Global warming of the mantle beneath continents back to the Archean. *Gondwana Research*, 15: 254–266.

Condie, K.C. (1997). Sources of Proterozoic mafic dyke swarms: constraints from Th/Ta and La/Yb ratios. *Precambrian Research*. 81: 3–14.

Condie, K.C. (2001). *Mantle Plumes and Their Record in Earth History*. Cambridge: Cambridge University Press.

Condie, K.C. (2003). Incompatible element ratios in oceanic basalts and komatiites: tracking deep mantle sources and continental growth rates with time. *Geochemistry, Geophysics, Geosystems*, 4: 1005.

Condie, K.C. (2004). Supercontinents and superplume events: distinguishing signals in the geologic record. *Physics of the Earth and Planetary Interiors*, 146: 319–332.

Condie, K.C. & Aster, R.C. (2010). Episodic zircon age spectra of orogenic granitoids: the supercontinent connection and continental growth. *Precambrian Research*, 180: 227–236.

Condie, K.C., Bobrow, D.J., & Card, K.D. (1987). Geochemistry of Precambrian mafic dykes from the southern Superior province of the Canadian Shield. In Halls, H.C. & Fahrig, W.F. (eds.), *Mafic Dyke Swarms*. St John's, NL: Geological Association of Canada, Special Paper 34, pp. 95–108.

Condie, K.C., DesMarais, D.J., & Abbott, D. (2001). Precambrian superplumes and supercontinents: a record in black shales, carbon isotopes, and paleoclimates? *Precambrian Research*, 106: 239–260.

Conrad, C.P. & Molnar, P. (1997). The growth of Rayleigh–Taylor-type instabilities in the lithosphere for various rheological and density structures. *Geophysics Journal International*, 129: 95–112.

Coogan, L.A., Saunders, A.D., & Wilson, R.N. (2014). Aluminum-in-olivine thermometry of primitive basalts: evidence of an anomalously hot mantle source for large igneous provinces. *Chemical Geology*, 368: 1–10.

Cook, A.G., Bryan, S.E., & Draper, J.J. (2013). Post-orogenic Mesozoic basins and magmatism. In Jell, P. (ed.), *Geology of Queensland. Brisbane*. Geological Survey of Queensland, (in press).

Cook, P.G. (2003). *A Guide to Regional Groundwater Flow in Fractured Rock Aquifers*. Clayton South, Victoria: CSIRO.

Cope, J.C.W. (1994). A latest Cretaceous hotspot and the southeasterly tilt of Britain. *Journal of the Geological Society*, 151: 905–908.

Cordery, M. J., Davies, G. F., & Campbell, I. H. (1997). Genesis of flood basalts from eclogite bearing mantle plumes. *Journal of Geophysical Research Solid Earth*, 102 (B9): 20179–20197.

Correia, C.T., Sinigoi, S., Girardi, V.A.V., *et al.* (2012). The growth of large mafic intrusions: comparing Niquelândia and Ivrea igneous complexes. *Lithos*, 155: 167–182.

Corriveau, L. (2007). Iron oxide copper–gold deposits: a Canadian perspective. In Goodfellow, W.D. (ed.), *Mineral Deposits of Canada: A Synthesis of Major Deposit-Types, District Metallogeny, the Evolution of Geological Provinces, and Exploration Methods*. St John's, NL: Geological Association of Canada, Mineral Deposits Division, Special Publication 5, pp. 307–328.

Corry, C.E. (1988). *Laccoliths: Mechanics of Emplacement and Growth*. Boulder, CO: Geological Society of America, Special Paper 220.

Corti, G. (2009). Continental rift evolution: from rift initiation to incipient break-up in the Main Ethiopian Rift, East Africa. *Earth-Science Reviews*, 96: 1–53.

Courtillot, V. & Renne, P.R. (2003). On the ages of flood basalt events. *Comptes Rendus Géoscience*, 335: 113–140.

Courtillot, V., Jaupart, C., Manighetti, I., Tapponier, P., & Besse, J. (1999). On causal links between flood basalts and continental breakup. *Earth and Planetary Science Letters*, 166: 177–195.

Courtillot, V., Gallet, Y., Rocchia, R., *et al.* (2000). Cosmic markers, $^{40}Ar/^{39}Ar$ dating and paleomagnetism of the KT section in the Anjar Area of the Deccan large igneous province. *Earth and Planetary Science Letters*, 182: 137–156.

Courtillot, V., Davaille, A., Besse, J., & Stock, J. (2003). Three distinct types of hotspots in the Earth's mantle. *Earth and Planetary Science Letters*, 205: 295–308.

Cousens, B.L. (2000). Geochemistry of the Archean Kam Group, Yellowknife greenstone belt, Slave Province, Canada. *Journal of Geology*, 108: 181–197.

Cousens, B.L., Aspler, L.B., Chiarenzelli, J.R., *et al.* (2001). Enriched Archean lithospheric mantle beneath western Churchill Province tapped during Paleoproterozoic orogenesis. *Geology*, 29: 827–830.

Cousens, B.L., Chairenzelli, J.R., & Aspler, L.B. (2004). An unusual Paleoproterozoic magmatic event, the ultrapotassic Christopher Island Formation, Baker Lake Group, Nunavut, Canada: Archean mantle metasomatism and Paleoproterozoic mantle reactivation. In Eriksson, P.G., Altermann, W., Nelson, D.R., Mueller, W.U., & Catuneanu, O. (eds.), *The Precambrian Earth: Tempos and Events*. Amsterdam: Elsevier Science.

Cowen, R. (1994). *History of Life*, 2nd edition. Cambridge, MA: Blackwell Scientific Publications.

Cox, K.G. (1980). A model for flood basalt vulcanism. *Journal of Petrology*, 21: 629–650.

Cox, K.G. (1989). The role of mantle plumes in the development of continental drainage patterns. *Nature*, 342: 873–877.

Cox, K.G. (1992). Karoo igneous activity, and the early stages of the break-up of Gondwanaland. In Storey, B.C., Alabaster, T., Pankhurst, R.J. (eds.), *Magmatism and the Causes of Continental Break-up*. London: Geological Society of London, Special Publication 68, pp. 37–148.

Cox, K.G. (1993). Continental magmatic underplating. *Philosophical Transactions of the Royal Society London*, A342: 155–166.

Cox, K.G., MacDonald, R., & Hornung, G. (1967). Geochemical and petrological provinces in the Karoo basalts of southern Africa. *American Mineralogist*, 52: 1451–1474.

Crawford, B.L., Betts, P.G., & Aillères, L. (2010). An aeromagnetic approach to revealing buried basement structures and their role in the Proterozoic evolution of the Wernecke inlier, Yukon Territory, Canada. *Tectonophysics*, 490: 28–46.

Crawford, J. (1989). *Boninites*. London: Unwin Hyman.

Crocket, J.H. (2002). Platinum-group element geochemistry of mafic and ultramafic rocks. In Cabri, L.J. (ed.), *The Geology, Geochemistry, Mineralogy, and Mineral Beneficiation of Platinum-Group Elements*. Canadian Institute of Mining, Metallurgy and Petroleum, Special Volume, 54, pp. 177–210.

Cross, A., Jaireth, S., Hore, S., Michaelsen, B., & Schofield, A. (2010). SHRIMP U–Pb detrital zircon results, Lake Frome region, South Australia. *Geoscience Australia, Record* 2009/46.

Crough, S.T. & Jurdy, D.M. (1980). Subducted lithosphere, hot spots and the geoid. *Earth and Planetary Science Letters*, 48: 15–22.

Crumpler, L.S. & Aubele, J.C. (2000). Volcanism on Venus. In Sigurdsson, H. (ed.), *Encyclopedia of Volcanoes*. San Diego, CA: Academic Press, pp. 727–769.

Crumpler, L.S., Aubele, J.C., Senske, D.A., *et al.* (1997). Volcanoes and centers of volcanism on Venus. In Bougher, S.W., Hunten, D.M., & Phillips, R.J. (eds.), *Venus II: Geology, Geophysics, Atmosphere, and Solar Wind Environment*. Tucson, AZ: University of Arizona Press, pp. 697–756.

Cserepes, L. & Yuen, D.A. (2000). On the possibility of a second kind of mantle plume. *Earth and Planetary Science Letters*, 183: 61–71.

Cúneo, R., Ramezani, J., & Scasso, R. (2013). High-precision U–Pb geochronology and a new chronostratigraphy for the Cañadón Asfalto Basin, Chubut, central Patagonia: implications for terrestrial faunal and floral evolution in Jurassic. *Gondwana Research*, 24: 1267–1275.

Currie, C.A. & Beaumont, C. (2011). Are diamond-bearing Cretaceous kimberlites related to low-angle subduction beneath western North America? *Earth and Planetary Science Letters*, 303: 59–70.

Czamanske, G.K., Gurevitch, A.B., Fedorenko, V., & Simonov, O. (1998). Demise of the Siberian plume: paleogeographic and paleotectonic reconstruction from the prevolcanic and volcanic record, north-central Siberia. *International Geology Reviews*, 40: 95–115.

Dalziel, I.W.D., Lawver, L.A., & Murphy, J.B. (2000). Plumes, orogenesis, and supercontinent fragmentation. *Earth and Planetary Science Letters*, 178: 1–11.

Dan, W., Li, X.-H., Wang, Q., Tang, G-J., & Liu, Y. (2014). An Early Permian (*ca.* 280 Ma) silicic igneous province in the Alxa Block, NW China: a magmatic flare-up triggered by a mantle-plume? *Lithos*. doi: 10.1016/j. lithos.2014.01.018.

Davey, S.C., Ernst, R.E., Samson, C., & Grosfils, E.B. (2013). Hierarchical clustering of pit crater chains on Venus. *Canadian Journal of Earth Sciences*, 50: 109–126.

Davies, A.G. (2007). *Volcanism on Io: A Comparison with Earth*. Cambridge: Cambridge University Press.

Davies, G.F. (1999). *Dynamic Earth: Plates, Plumes and Mantle Convection*. Cambridge: Cambridge University Press.

Davies, G.F. (2011). *Mantle Convection for Geologists.* Cambridge: Cambridge University Press.

Davies, J.H.F.L. & Heaman, L.M. (2014). New U-Pb baddeleyite and zircon ages for the Scourie dyke swarm: a long-lived large igneous province with implications for the Paleoproterozoic evolution of NW Scotland. *Precambrian Research* (in press).

Davis, A.S., Gunn, S.H., Bohrson, W.A., Gray, L.B., & Hein, J.R. (1995). Chemically diverse, sporadic volcanism at seamounts offshore southern and Baja California. *Geological Society of America Bulletin*, 107: 554–570.

Davis, W.J. (1997). U–Pb zircon and rutile ages from granulite xenoliths in the Slave Province: evidence for mafic magmatism in the lower crust coincident with Proterozoic dike swarms. *Geology*, 25: 343–346.

Davis, W.J., Canil, D., MacKenzie, J.M., & Carbno, G.B. (2003). Petrology and U–Pb geochronology of lower crustal xenoliths and the development of a craton, Slave Province, Canada. *Lithos*, 71: 541–573.

Day, M.D. (2013). Hotspot volcanism and highly siderophile elements. *Chemical Geology*, 341: 50–74.

de Agnacio, C., Munoz, M., Sagredo, J., Fernandez-Santin, S., & Johansson, A. (2006). Isotope geochemistry and FOZO mantle component of the alkaline-carbonatitic association of Fuerteventura, Canary Islands, Spain. *Chemical Geology*, 232: 99–113.

de Kock, M.O., Evans, D.A.D., & Beukes, N.J. (2009). Validating the existence of Vaalbara in the Neoarchean. *Precambrian Research*, 174: 145–154.

de Kock, M.O., Ernst, R.E., Söderlund, U., *et al.* (2014). Dykes of the 1.11 Ga Umkondo LIP, Southern Africa: clues to a complex plumbing system. *Precambrian Research* (in press).

de Wit, M.J. (1998). On Archean granites, greenstones, cratons and tectonics: does the evidence demand a verdict? *Precambrian Research*, 91: 181–226.

de Wit, M.J. & Ashwal, L.D. (1997). Preface: convergence toward divergent models of greenstone belts. In de Wit, M.J. & Ashwal, L.D. (eds.), *Greenstone Belts*. Oxford: Clarendon Press, pp. ix–xvii.

Dedeev, A.V., Khashkovskaya, T.N., & Galkin, A.S. (2002). PGE mineralization of the Monchegorsk layered mafic–ultramafic intrusion of the Kola Peninsula. In Cabri, L.J. (ed.), *The Geology, Geochemistry, Mineralogy and Mineral Beneficiation of Platinum-Group Elements*. Canadian Institute of Mining, Metallurgy and Petroleum, Special Volume, 54, pp. 569–577.

Deng, Q., Wang, J., Wang, Z.-J., *et al.* (2013). Continental flood basalts of the Huashan Group, northern margin of the Yangtze block – implications for the breakup of Rodinia. *International Geology Review*, 55: 1865–1884.

Deniel, C., Vidal, P., Coulon, C., Vellutini, P. J., & Piguet, P. (1994). Temporal evolution of mantle sources through continental rifting: the volcanism of Djibouti (Afar). *Journal of Geophysical Research*, 99: 2853–2869.

Denyszyn, S.W., Davis, D.W., & Halls, H.C. (2009a). Paleomagnetism and U-Pb geochronology of the Clarence Head dykes, Arctic Canada: orthogonal emplacement of mafic dykes in a large igneous province. *Canadian Journal of Earth Sciences*, 46: 155–167.

Denyszyn, S.W., Halls, H.C., Davis, D.W., & Evans, D.A.D. (2009b). Paleomagnetism and U-Pb geochronology of Franklin dykes in High Arctic Canada and Greenland: a revised age and paleomagnetic pole constraining block rotations in the Nares Strait region. *Canadian Journal of Earth Sciences*, 46: 689–705.

Dessert, C., Dupre, B., Gaillardet, J., François, L.M., & Allègre, C.J. (2003). Basalt weathering laws and the impact of basalt weathering on the global carbon cycle. *Chemical Geology*, 202: 257–273.

Dessureau, G., Piper, D.J.W., & Pe-Piper, G. (2007). Geochemical evolution of earliest Carboniferous continental tholeiitic basalts along a crustal-scale shear zone, southwestern Maritimes basin, eastern Canada. *Lithos*, 50: 27–50.

Didenko, A.N. (2011). Possible causes of quasiperiodic variations in geomagnetic reversal frequency and $^{87}Sr/^{86}Sr$ ratios in marine carbonates through the Phanerozoic. *Russian Geology and Geophysics*, 52: 1530–1538.

Dilek, Y. (2003). *Ophiolites, plumes and orogeny*. In Dilek, Y. & Robinson, P.T. (eds.), *Ophiolites in Earth History*. London: Geological Society, Special Publication 218, pp. 9–19.

Dilek, Y. & Ernst, R. (2008). Links between ophiolites and Large Igneous Provinces (LIPs) in Earth history: Introduction. *Lithos*, 100: 1–13.

Dilek, Y. & Furnes, H. (2011). Ophiolite genesis and global tectonics: geochemical and tectonic fingerprinting of ancient oceanic lithosphere. *Geological Society of America Bulletin*, 123: 387–411.

Ding, X., Li, C., Ripley, E.M., Rossell, D., & Kamo, S. (2010). The Eagle and East Eagle sulfide ore-bearing mafic–ultramafic intrusions in the Midcontinent rift system, upper Michigan: geochronology and petrologic evolution. *Geochemistry, Geophysics, Geosystems*, 11, Q03003.

Doblas, M., Lopez-Ruiz, J., Cebria, J.M., Youbi, N., & Degroote, E. (2002). Mantle insulation beneath the West African craton during the Precambrian–Cambrian transition. *Geology*, 30: 839–842.

Dobretsov, N.L. (1997). Permo-Triassic magmatism and sedimentation in Eurasia as expression of a mantle superplume. *Doklady Earth Sciences*, 354: 497–500.

Dobretsov, N.L. (2003). Mantle plumes and their role in the formation of anorogenic granitoids. *Russian Geology and Geophysics*, 44: 1199–1218.

Dobretsov, N.L. (2005). 250 Ma Large Igneous Provinces of Asia: Siberian and Emeishan traps (Plateau Basalts) and associated granitoids. *Russian Geology and Geophysics*, 46: 847–868.

Dobretsov, N.L. & Vernikovsky, V.A. (2001). Mantle plumes and their geological manifestations. *International Geology Review*, 43: 771–788.

Dobretsov, N.L., Kirdyashkin, A.A., Kirdyashkin, A.G., Vernikovsky, V.A., & Gladkov, I.N. (2008). Modelling of thermochemcial plumes and implications for the origin of the Siberian traps. *Lithos*, 100: 66–92.

Dohm, J.M., Baker, V.R., Maruyama, S., & Anderson, R.C. (2007). Traits and evolution of the Tharsis Superplume, Mars. In Yuen, D.A., Maruyama, S., Karato, S.-i., & Windley, B.F. (eds.), *Superplumes: Beyond Plate Tectonics*. Dordrecht: Springer-Verlag, pp. 523–536.

Dombard, A.J., Johnson, C.L., Richards, M.A., & Solomon, S.C. (2007). A magmatic loading model for coronae on Venus. *Journal of Geophysical Research (Planets)*, 112, doi:10.1029/2006JE002731, 2007.

Døssing, A., Jackson, H.R., Matzka, J., *et al.* (2013). On the origin of the Amerasia basin and the High Arctic large igneous province: results of new aeromagnetic data: *Earth and Planetary Science Letters*, 363: 219–230.

Du Toit, A.L. (1920). The Karoo dolerites – a study in hypabyssal intrusion. *Transaction of the Geological Society of South Africa*, 23: 1–42.

Dubé, B. & Gosselin, P. (2007). Greenstone-hosted quartz-carbonate vein deposits. In Goodfellow, W.D. (ed.), *Mineral Deposits of Canada: A Synthesis of Major*

Deposit-Types, District Metallogeny, the Evolution of Geological Provinces, and Exploration Methods. St John's, NL: Geological Association of Canada, Mineral Deposits Division, Special Publication 5, pp. 49–73.

Duchesne, J.C., Shumlyanskyy, L., & Charlier, B. (2006). The Fedorivka layered intrusion (Korosten Pluton, Ukraine): an example of highly differentiated ferrobasaltic evolution. *Lithos*, 89 (3–4): 353–377.

Duggen, S., Hoernle, K.A., Hauff, F., *et al.* (2009). Flow of Canary mantle plume material through a subcontinental lithospheric corridor beneath Africa to the Mediterranean. *Geology*, 37: 283–286.

Duncan, A.R., Armstrong, R.A., Erlank, A.J., Marsh, J.S., & Watkins, R.T. (1990). MORB-related dolerites associated with the final phases of Karoo flood basalt volcanism in southern Africa. In Parker, A.J., Rickwood, P.C., & Tucker, D.H. (eds.), *Mafic Dykes and Emplacement Mechanisms*. Rotterdam: Balkema, pp. 119–129.

Duncan, R.A. (2002). A timeframe for construction of the Kerguelen Plateau and Broken Ridge. *Journal of Petrology*, 43: 1109–1119.

Duncan, R.A. & Richards, M.A. (1991). Hotspots, mantle plumes, flood basalts and true polar wander. *Reviews of Geophysics*, 29: 31–50.

Eales, H.V. & Cawthorn, R.G. (1996). The Bushveld Complex. In Cawthorn, R.G. (ed.), *Layered Intrusions*. Amsterdam: Elsevier, pp. 181–229.

Eales, H.V. & Costin, G. (2012). Crustally contaminated komatiite: primary source of the chromitites and Marginal, Lower, and Critical zone magmas in a staging chamber beneath the Bushveld Complex. *Economic Geology*, 107: 645–665.

Eales H.V., Marsh, J.S., & Cox, K.G. (1984). The Karoo igneous province: an introduction. *Geological Society of South Africa Special Publication*, 13: 1–26.

Ebinger, C.J. & Sleep, N.H. (1998). Cenozoic magmatism throughout East Africa resulting from impact of a single plume. *Nature*, 395: 788–791.

Eby, G.N. & Kochhar, N. (1990). Geochemistry and petrogenesis of the Malani Igneous Suite, northern India. *Journal of the Geological Society of India*, 36: 109–130.

Eckstrand, O.R. & Hulbert, L.J. (2007). Magmatic nickel–copper–platinum group elements deposits. In Goodfellow, W.D. (ed.), *Mineral Deposits of Canada: A Synthesis of Major Deposit-types, District Metallogeny, the Evolution of Geological Provinces, and Exploration Methods*. St. John's, NL: Geological Association of Canada, Mineral Deposits Division, Special Publication 5, pp. 205–222.

Egorov, L.S. (1970). Carbonatites and ultrabasic-alkaline rocks of the Maimecha-Kotui region. N. Siberia. *Lithos*, 3: 341–359.

El Bahat, A., Ikenne, M., Söderlund, U., *et al.* (2013). U–Pb ages of dolerite dykes in the Bas Drâa Inlier of the Anti-Atlas of Morocco: newly identified 1380 Ma event in the West African Craton. *Lithos*, 174: 85–98.

Elburg, M. & Goldberg, A. (2000). Age and geochemistry of Karoo dolerite dykes from northeast Botswana. *Journal of African Earth Sciences*, 31: 539–554.

Eldholm, O. & Coffin, M.F. (2000). Large igneous provinces and plate tectonics. *AGU Geophysical Monograph*, 121: 309–326.

Eldholm, O. & Grue, K. (1994). North Atlantic volcanic margins: dimensions and production rates. *Journal of Geophysical Research*, 99 (B2): 2955–2968.

Eldrett, J.S., Minisini, D., & Bergman, S.C. (2014). Decoupling of the carbon cycle during Ocean Anoxic Event 2. *Geology*, in press doi:10.1130/G35520.1.

Elias, M. (2006). Lateritic nickel mineralization of the Yilgarn craton. *Society of Economic Geologists Special Publication*, 13: 195–210.

Elkins-Tanton, L.T. (2005). Continental magmatism caused by lithospheric delamination. In Foulger, G.R., Natland, J.H., Presnall, D.C., & Anderson, D.L. (eds.), *Plates, Plumes, and Paradigms*. Boulder, CO: Geological Society of America, Special Paper 388, pp. 449–461.

Elkins-Tanton, L.T. (2007). Continental magmatism, volatile recycling, and a heterogeneous mantle caused by lithospheric gravitational instabilities. *Journal of Geophysical Research*, 112: B03405.

Elkins-Tanton, L.T. & Hager, B.H. (2000). Melt intrusion as a trigger for lithospheric foundering and the eruption of the Siberian flood basalt. *Geophysical Research Letters*, 27: 3937–3940.

Elkins-Tanton, L.T. & Hager, B.H. (2005). Giant meteoroid impacts can cause volcanism. *Earth and Planetary Science Letters*, 239: 219–232.

Elliot, D.H. & Fleming, T.H. (2000). Weddell triple junction: the principal focus of Ferrar and Karoo magmatism during the initial breakup of Gondwana. *Geology*, 28: 539–542.

Elliot, D.H. & Hanson, R.E. (2001). Origin of widespread, exceptionally thick basaltic phreatomagmatic tuff breccia in the Middle Jurassic Prebble and Mawson Formations, Antarctica. *Journal of Volcanology and Geothermal Research*, 111: 183–201.

Elliot, D.H., Fleming, T.H., Kyle, P.R., & Foland, K.A. (1999). Long-distance transport of magmas in the Jurassic Ferrar large igneous province, Antarctica. *Earth and Planetary Science Letters*, 167: 89–104.

Ellis, B.S. (2009). Rhyolitic explosive eruptions of the central Snake River Plain Idaho: investigations of the Lower Cassia Mountains succession and surrounding areas. Unpublished Ph.D. thesis, University of Leicester.

Elrick, M., Rieboldt, S., Saltzman, M., & McKay, R.M. (2011). Oxygen-isotope trends and seawater temperature changes across the Late Cambrian Steptoean positive carbon-isotope excursion (SPICE event). *Geology*, 39: 987–990.

Encarnación, J., Fleming, T.H., Elliot, D.H., & Eales, H.V. (1996). Synchronous emplacement of Ferrar and Karoo dolerites and the early break-up of Gondwana. *Geology*, 24: 535–538.

Engel, R., McFarlane, D.J., & Street, G.J. (1989). Using geophysics to define recharge and discharge areas associated with saline seeps in south-western Australia. In Sharma, M.L. (ed.), *Groundwater Recharge*. Rotterdam: A.A. Balkema, pp. 25–39.

Erba, E., Bartolini, A., & Larson, R.L. (2004). Valanginian Weissert oceanic anoxic event. *Geology*, 32: 149–152.

Eriksson, P.G., Condie, K.C., van der Westhuizen, W., *et al.* (2002). Late Archaean superplume events: a Kaapvaal–Pilbara perspective. *Journal of Geodynamics*, 34: 207–247.

Ernst, R.E. (1994). Mapping the magma flow pattern in the Sudbury dyke swarm in Ontario using magnetic fabric analysis. *Geological Survey of Canada Current Research*, 1994-E: 183–192.

Ernst, R.E. (2007a). Mafic–ultramafic Large Igneous Provinces (LIPs): importance of the pre-Mesozoic record. *Episodes*, 30: 107–113.

Ernst, R.E. (2007b). Large Igneous Provinces (LIPs) in Canada through time and their metallogenic potential. In Goodfellow, W.D. (ed.), *Mineral Deposits of Canada: A Synthesis of Major Deposits Types, District Metallogeny, the Evolution of Geological Provinces and Exploration Methods*. Geological Association of Canada, Mineral Deposits Division, Special Publication 5, pp. 929–937.

Ernst, R.E. (2008). Carbonatites and Large Igneous Province (LIPs). *Geochimica et Cosmochimica Acta*, 72, 12S): A246.

Ernst, R.E. & Baragar, W.R.A. (1992). Evidence from magnetic fabric for the flow pattern of magma in the Mackenzie giant radiating dyke swarm. *Nature*, 356: 511–513.

Ernst, R.E. & Bell, K. (1992). Petrology of the Great Abitibi Dyke, Superior Province, Canada. *Journal of Petrology*, 33: 423–469.

Ernst, R.E. & Bell, K. (2010). Large Igneous Provinces (LIPs) and carbonatites. *Mineralogy and Petrology*, 98: 55–76.

Ernst, R.E. & Bleeker, W. (2010). Large igneous provinces (LIPs), giant dyke swarms, and mantle plumes: significance for breakup events within Canada and adjacent regions from 2.5 Ga to the Present. *Canadian Journal of Earth Sciences*, 47: 695–739.

Ernst, R.E. & Buchan, K.L. (1993). Paleomagnetism of the Abitibi dyke swarm, southern Superior Province, and implications for the Logan Loop. *Canadian Journal of Earth Sciences*, 30: 1886–1897.

Ernst, R.E. & Buchan, K.L. (1997a). Giant radiating dyke swarms: their use in identifying pre-Mesozoic large igneous provinces and mantle plumes. *AGU Geophysical Monograph*, 100: 297–333.

Ernst, R.E. & Buchan, K.L. (1997b). Layered mafic intrusions: a model for their feeder systems and relationship with giant dyke swarms and mantle plume centers. *South African Journal of Geology*, 100: 319–334.

Ernst, R.E. & Buchan, K.L. (1998). Arcuate dyke swarms associated with mantle plumes on Earth: implications for Venusian coronae. *Lunar and Planetary Science Conference*, XXIX, Abstract 1021 (CD-ROM).

Ernst, R.E. & Buchan, K.L. (1999). Paleo-stress patterns from giant dyke swarms. Extended Abstract. *Lunar and Planetary Science Conference*, XXX, Abstract 1737 (CD-ROM).

Ernst, R.E. & Buchan, K.L., eds. (2001a). *Mantle Plumes: Their Identification through Time*. Boulder, CO: Geological Society of America, Special Paper 352.

Ernst, R.E. & Buchan, K.L. (2001b). Large mafic magmatic events through time and links to mantle-plume heads. In Ernst, R.E. & Buchan, K.L. (eds.), *Mantle Plumes: Their Identification through Time*. Boulder, CO: Geological Society of America, Special Paper 352, pp. 483–575.

Ernst, R.E. & Buchan, K.L. (2001c). The use of mafic dike swarms in identifying and locating mantle plumes. In Ernst, R.E. & Buchan, K.L. (eds.), *Mantle Plumes: Their Identification through Time*. Boulder, CO: Geological Society of America, Special Paper 352, pp. 247–265.

Ernst, R.E. & Buchan, K.L. (2002). Maximum size and distribution in time and space of mantle plumes: evidence from large igneous provinces. *Journal of Geodynamics*, 34: 309–342. [Erratum, *Journal of Geodynamics*, 34: 711–714.]

Ernst, R.E. & Buchan, K.L. (2003). Recognizing mantle plumes in the geological record. *Annual Reviews Earth and Planetary Science*, 31: 469–523.

Ernst, R.E. & Buchan, K.L. (2004a). Igneous rock associations in Canada 3: Large Igneous Provinces (LIPs) in Canada and adjacent regions: 3 Ga to Present. *Geoscience Canada*, 31 (3): 103–126.

Ernst, R.E. & Buchan, K.L. (2004b). Dyke (dike) swarm. In Goudie, A.S. (ed.), *Encyclopedia of Geomorphology*. London: Routledge, p. 305.

Ernst, R.E. & Desnoyers, D.W. (2004). Lessons from Venus for understanding mantle plumes on Earth. *Special Issue of Physics of the Earth and Planetary Interiors*, 146: 195–229. [Corrigendum, 2005, 149: 371].

Ernst, R.E. & Jowitt, S.M. (2013). Large Igneous Provinces (LIPs) and Metallogeny. *Society of Economic Geologists Special Publication*, 17: 17–51.

Ernst, R.E. & Srivastava, R.K. (2008). India's place in the Proterozoic world: constraints from the large igneous provinces (LIP) record. In Srivastava, R.K., Sivaji, Ch., & Chalapathi Rao, N.V. (eds.), *Indian Dykes: Geochemistry, Geophysics, and Geochronology*. New Delhi, India: Narosa Publishing House Pvt. Ltd, pp. 41–56.

Ernst, R.E., Fowler, A.D., & Pearce, T.H. (1988). Modelling of igneous fractionation and other processes using Pearce diagrams. *Contributions to Mineralogy and Petrology*, 100: 12–18.

Ernst, R.E., Buchan, K.L., & Palmer, H.C. (1995a). Giant dyke swarms: characteristics, distribution and geotectonic applications. In Baer, G. & Heimann, A. (eds.), *Physics and Chemistry of Dykes*. Rotterdam: Balkema, pp. 3–21.

Ernst, R.E., Head, J.W., Parfitt, E., Grosfils, E.B., & Wilson, L. (1995b). Giant radiating dyke swarms on Earth and Venus. *Earth Science Reviews*, 39: 1–58.

Ernst, R.E., Buchan, K.L., West, T.D., & Palmer, H.C. (1996). Diabase (dolerite) dyke swarms of the world, first edition. *Geological Survey of Canada Open-File 3241*, includes map, scale 1:35 000 000 at Equator.

Ernst, R.E., Grosfils, E.B., & Mège, D. (2001). Giant dike swarms: Earth, Venus and Mars. *Annual Reviews Earth and Planetary Science*, 29: 489–534.

Ernst, R.E., Desnoyers, D.W., Head, J.W., & Grosfils, E.B. (2003). Graben–fissure systems in Guinevere Planitia and Beta Regio (264–312º E, 24–60º N), Venus, and implications for regional stratigraphy and mantle plumes/diapirs. *Icarus*, 164: 282–316.

Ernst, R.E., Buchan, K.L., & Prokoph, A. (2004). Large igneous province record through time. In Eriksson, P.G., Altermann, W., Nelson, D.R., Mueller, W.U. & Catuneanu, O. (eds.), *The Precambrian Earth: Tempos and Events. Developments in Precambrian Geology Series*, 12, Elsevier, pp. 173–180.

Ernst, R.E., Buchan, K.L., & Campbell, I.H. (2005). Frontiers in large igneous province research. *Lithos*, 79: 271–297.

Ernst, R.E., Buchan, K.L., & Desnoyers, D.W. (2007). Plumes and plume clusters on Earth and Venus: evidence from large igneous provinces (LIPs). In Yuen, D.A., Maruyama, S., Karato, S.-i., & Windley, B.F. (eds.), *Superplumes: Beyond Plate Tectonics*. Berlin: Springer-Verlag, pp. 537–562.

Ernst, R.E., Wingate, M.T.D., Buchan, K.L., & Li, Z.X. (2008). Global record of 1600–700 Ma Large Igneous Provinces (LIPs): implications for the reconstruction of the proposed Nuna (Columbia) and Rodinia supercontinents. *Precambrian Research*, 160: 159–178.

Ernst, R.E., Bleeker, W., Svensen, H., Planke, S., & Polozov, A.G. (2009). Vent complexes above dolerite sills in Phanerozoic LIPs: implications for Proterozoic LIPs and IOCG deposits. American Geological Union–Geological Assocation of Canada–Mineralogical Association of Canada Annual Meeting, Toronto, Canada, 24–27 May.

Ernst, R., Srivastava, R., Bleeker, W., & Hamilton, M. (2010). Precambrian Large Igneous Provinces (LIPs) and their dyke swarms: new insights from high-precision geochronology integrated with paleomagnetism and geochemistry. *Precambrian Research*, 183: vii–xi.

Ernst, R.E., Hamilton, M.A., & Söderlund, U. (2012). A proposed 725 Ma Dovyren–Kingash LIP of southern Siberia, and possible reconstruction link with the 725–715 Ma Franklin LIP of northern Laurentia. Abstract volume 35, Geological Association of Canada (GAC)—Mineralogical Association of Canada (MAC) Joint Annual Meeting, Geoscience at the Edge, May 27–29, 2012, St. John's, Newfoundland and Labrador, Canada.

Ernst, R.E., Pereira, E., Hamilton, M.A., *et al.* (2013a). Mesoproterozoic intraplate magmatic "barcode" record of the Angola portion of the Congo craton: newly dated magmatic events at 1500 and 1110 Ma and implications for Nuna supercontinent reconstructions. *Precambrian Research*, 230: 103–118.

Ernst, R.E., Bleeker, W., Söderlund, U., & Kerr, A.C. (2013b). Large Igneous Provinces and supercontinents: Toward completing the plate tectonic revolution. *Lithos*, 174: 1–14.

Ernst, R.E., Bleeker, W., Söderlund, U., & Kerr, A.C., eds. (2013c). Large Igneous Provinces (LIPs) and supercontinents. *Lithos Special Issue*, 174.

Ernst, R.E., Hamilton, M.A., Kamo, S.L., *et al.* (2014a). The 1498–1503 Ma Kuonamka LIP of northern Siberia; new precise U-Pb baddeleyite dating. Abstract for Annual GAC-MAC meeting, 21–24 May 2014.

Ernst, R.E., Söderlund, U., Baratoux, L., *et al.* (2014b). 1790 Ma dyke swarm in southwest Niger (West African craton): part of a regional LIP that extends into formerly attached Amazonia and Sarmatia. 25th Colloquium of African Geology (CAG25), 11–14 August 2014. Dar Es Salaam,Tanzania

Erwin, D.H. (1994). The Permo-Triassic extinction. *Nature*, 367: 231–236.

Erwin, D.H. (2006). *Extinction. How Life on Earth Nearly Ended 250 Million Years Ago.* Princeton, NJ: Princeton University Press.

Escher, A., Escher, J.C., & Watterson, J. (1975). The reorientation of the Kangamiut dyke swarm, West Greenland. *Canadian Journal of Earth Sciences*, 12: 158–173.

Evans, D.A.D. (2013). Reconstructing pre-Pangean supercontinents. *Geological Society of America Bulletin*, 125: 1735–1751.

Evans, D.A.D. & Halls, H.C. (2010). Restoring Proterozoic deformation within the Superior craton. *Precambrian Research*, 183: 474–489.

Evans, D.A.D. & Mitchell, R.N. (2011). Assembly and breakup of the core of Paleoproterozoic–Mesoproterozoic supercontinent Nuna. *Geology*, 39: 443–446.

Evans, D.A.D. & Raub, T.D. (2011). Neoproterozoic glacial palaeolatitudes: a global update. In Arnaud, E., Halverson, G. P., & Shields-Zhou, G. (eds.), *The Geological Record of Neoproterozoic Glaciations*. Geological Society, London, Memoirs, 36, pp. 93–112.

Evans, D.A.D., Beukes, N.J., & Kirschvink, J.L. (2002). Paleomagnetism of a lateritic paleoweathering horizon and overlying Paleoproterozoic red beds from South Africa: implications for the Kaapvaal apparent polar wander path and a confirmation of atmospheric oxygen enrichment. *Journal of Geophysical Research*, 107: doi: 10.1029/2001JB000432.

Evans, D.A.D., Heaman, L.M., Trindade, R.I.F., *et al.* (2010). Precise U–Pb baddeleyite ages from Neoproterozoic mafic dykes in Bahia, Brazil, and their paleomagnetic/paleogeographic implications. AGU Brazil Abstract, GP31E-07 American Geophysical Union, Joint Assembly, Meeting of the Americas, Iguassu Falls, August 2010.

Evans, D., Trindade, R., Catelani, E., *et al.* (in press). Return to Rodinia? Moderate to high paleolatitude of the Sao Francisco /Congo craton at 920 Ma. In Li, Z-X.,

Evans, D.A.D., & Murphy, J.B. (eds.), *Supercontinent Cycles Through Earth History*. London, Geological Society of London, Special Publication, in press.

Evins, L.Z., Jourdan, F., & Phillips, D. (2009). The Cambrian Kalkarindji Large Igneous Province: extent and characteristics based on new ^{40}Ar/^{39}Ar and geochemical data. *Lithos*, 110: 294–304.

Evins, P.M., Smithies, R.H., Howard, H.M., *et al.* (2010). Devil in the detail: the 1150–1000 Ma magmatic and structural evolution of the Ngaanyatjarra Rift, west Musgrave Province, Central Australia. *Precambrian Research*, 183: 572–588.

Ewart, A., Schön, R.W., & Chappell, B.W. (1992). The Cretaceous volcanic–plutonic province of the central Queensland (Australia) coast: a rift related "calcalkaline" province. *Transactions of the Royal Society of Edinburgh: Earth Sciences*, 83: 327–345.

Ewart, A., Milner, S.C., Armstrong, R.A., & Duncan, A.R. (1998). Etendeka volcanism of the Goboboseb Mountains and Messum Igneous Complex, Namibia. Part I: geochemical evidence of Early Cretaceous Tristan plume melts and the role of crustal contamination in the Paraná–Etendeka CFB. *Journal of Petrology*, 39: 191–225.

Ewart, A., Marsh, J.S., Milner, S.C., *et al.* (2004a). Petrology and geochemistry of early cretaceous bimodal continental flood volcanism of the NW Etendeka, Namibia: Part 1. Introduction, mafic lavas and reevaluation of mantle source components. *Journal of Petrology*, 45: 59–105.

Ewart, A., Marsh, J.S., Milner, S.C., *et al.* (2004b). Petrology and geochemistry of early cretaceous bimodal continental flood volcanism of the NW Etendeka, Namibia: Part 2. Characteristics and petrogenesis of the high-Ti Latite and high-Ti and low-Ti voluminous quartz Latite Eruptives. *Journal of Petrology*, 45: 107–138.

Eyal, Y. & Eyal, M. (1987). Mafic dyke swarms in the Arabian-Nubian Shield. *Israel Journal of Earth Sciences*, 36: 195–211.

Eyuboglu, Y., Santosh, M., Bektas, O., Chung, S.-L., (2011). Late Triassic subduction-related ultramafic–mafic magmatism in the Amasya region (eastern Pontides, N. Turkey): implications for the ophiolite conundrum in Eastern Mediterranean. *Journal of Asian Earth Sciences*, 42: 234–257.

Fahrig, W.F. (1987). The tectonic setting of continental mafic dyke swarms: failed arm and early passive margin. In Halls, H.C. & Fahrig, W.F. (eds.), *Mafic Dyke Swarms*. St John's, NL: Geological Association of Canada, Special Publication 34, pp. 331–348.

Fan, J. & Kerrich, R. (1997). Geochemical characteristics of aluminum depleted and undepleted komatiites and HREE-enriched low-Ti tholeiites, western Abitibi greenstone belt: a heterogeneous mantle plume-convergent margin environment. *Geochimica et Cosmochimica Acta*, 61: 4723–4744.

Fanning, C.M., Flint, R.B., Parker, A.J., Ludwig, K.R., & Blissett, A.H. (1988). Refined Proterozoic evolution of the Gawler Craton, South Australia, through U–Pb zircon geochronology. *Precambrian Research*, 40–41: 363–386.

Farley, K.A., Natland, J.H., & Craig, H. (1992). Binary mixing of enriched and undegassed (primitive?) mantle components (He, Sr, Nd and Pb) in Samoan lavas. *Earth and Planetary Science Letters*, 111: 183–199.

Farnetani, C.G. & Hofmann, A.W. (2009). Dynamics and internal structure of a lower mantle plume conduit. *Earth and Planetary Science Letters*, 282: 314–322.

Farnetani, C.G. & Richards, M.A. (1994). Numerical investigations of the mantle plume initiation model for flood basalt events. *Journal of Geophysical Research*, 99 (B7): 13813–13833.

Farnetani, C.G. & Samuel, H. (2005). Beyond the thermal plume paradigm. *Geophysical Research Letters*, 32: L07311.

Farrand, W.H., Lane, M.D., Edwards, B.R., & Yingst, R.A. (2011). Spectral evidence of volcanic cryptodomes on the northern plains of Mars. *Icarus*, 211: 139–156.

Fedorenko, V.A., Lightfoot, P.C., Naldrett, A.J., et al. (1996). Petrogenesis of flood-basalt sequence at Noril'sk, north central Siberia. *International Geology Review*, 38: 99–135.

Ferrari, L., López-Martínez, M., & Rosas-Elguera, J. (2002). Ignimbrite flare up and deformation in the southern Sierra Madre Occidental, western Mexico: implications for the late subduction history of the Farallon plate. *Tectonics*, 21: 17–1/24.

Ferrari, L., Valencia-Moreno, M., & Bryan, S. (2007). Magmatism and tectonics of the Sierra Madre Occidental and its relation with the evolution of western margin of North America. In Alaniz-Álvarez, S.A., & Nieto-Samaniego, Á.F. (eds.), *Geology of México: Celebrating the Centenary of the Geological Society of México.* Boulder, CO: Geological Society of America, Special Paper 442, pp. 1–39.

Ferrari, L., López-Martínez, M., Orozco, M.T., et al. (2013). Miocene rifting and syn-extensional magmatism in the southwestern Sierra Madre Occidental (Sinaloa and Nayarit states): the early history of the Gulf of California. *Geosphere*, 9: 1161–1200.

Ferrill, D.A., Wyrick, D.Y., & Smart, K.J. (2011). Coseismic dilational fault and extension-fracture related pit chain formation in Iceland: analogue for pit chains on Mars. *Lithosphere*, 3: 133–142.

Ferris, J.K., Storey, B.C., Vaughan, A.P.M., Kyle, P.R., & Jones, P.C. (2003). The Dufek and Forrestal intrusions, Antarctica: A centre for Ferrar large igneous province dike emplacement? *Geophysical Research Letters*, 30: 81.1–81.4.

Fitton, J.G. & Godard, M. (2004). Origin and evolution of magmas on the Ontong Java Plateau. In Fitton, J.G., Mahoney, J.J., Wallace, P.J., & Saunders, A.D. (eds.), *Origin and Evolution of the Ontong Java Plateau.* London: Geological Society, Special Publication 229, pp. 151–178.

Fitton, J.G., Mahoney, J.J., Wallace, P.J., & Saunders, A.D., eds. (2004). *Origin and Evolution of the Ontong Java Plateau.* London: Geological Society of London, Special Publication 229, pp. 1–8.

Fitton, J.G., Mahoney, J.J., Wallace, P.J., & Saunders, A.D. (2005). The Ontong Java plateau. *LIP of the Month*, April 2005. See: http://www.largeigneousprovinces.org/05apr.

Flament, N., Rey, P.F., Coltice, N., Dromart, G., & Olivier, N. (2011). Lower crustal flow kept Archean continental flood basalts at sea level. *Geology*, 39: 1159–1162.

Foulger, G.R. (2007). The "plate" model for the genesis of melting anomalies. In Foulger, G.R. & Jurdy, D.M. (eds.), *Plates, Plumes, and Planetary Processes.* Boulder, CO: Geological Society of America Special Paper 430, pp. 1–28.

Foulger, G.R. (2010). *Plates vs. Plumes: A Geological Controversy.* Chichester: Wiley-Blackwell.

Foulger, G.R. (2012). Are "hot spots" hot spots? *Journal of Geodynamics*, 58: 1–28.

Foulger, G.R. & Jurdy, D.M., eds. (2007). *Plates, Plumes, and Planetary Processes.* Boulder, CO: Geological Society of America, Special Paper 430.

Foulger, G.R., Natland, J.H., Presnall, D.C., & Anderson, D.L., eds. (2005). *Plates, Plumes, and Paradigms.* Boulder, CO: Geological Society of America, Special Paper 388.

Francis, D. (1994). Chemical interaction between picritic magmas and upper crust along the margins of the Muskox intrusion, Northwest Territories. *Geological Survey of Canada Paper*, 92–12.

Francis, D. (2011). Columbia Hills – an exhumed layered igneous intrusion on Mars? *Earth and Planetary Science Letters*, 310: 59–64.

Francis, E.H. (1982). Emplacement mechanism of late Carboniferous tholeiite sills in northern Britain. *Geological Society of London Journal*, 139: 1–20.

Franklin, J.M., Gibson, H.L., Jonasson, I.R., & Galley, A.G. (2005). Volcanogenic massive sulfide deposits. *Economic Geology, 100th Anniversary Volume*: 523–560.

French, J.E. & Heaman, L.M. (2010). Precise U–Pb dating of Paleoproterozoic mafic dyke swarms of the Dharwar craton, India: implications for the existence of the Neoarchean supercraton, Sclavia. *Precambrian Research*, 183: 416–441.

French, J.E., Heaman, L.M., Chacko, T., & Srivastava, R.K. (2008). 1891–1883 Ma Southern Bastar–Cuddapah mafic igneous events, India: a newly recognized large igneous province. *Precambrian Research*, 160: 308–322.

Frey, F.A., McNaughton, N.J., Nelson, D.R., de Laeter, J.R., & Duncan, R.A. (1996). Petrogenesis of the Bunbury Basalt, Western Australia: interaction between the Kerguelen plume and Gondwana lithosphere? *Earth and Planetary Science Letters*, 144: 163–183.

Frey, F.A., Coffin, M.F., Wallace, P.J., *et al.* (2000). Origin and evolution of a submarine large igneous province: the Kerguelen Plateau and Broken Ridge, southern Indian Ocean. *Earth and Planetary Science Letters*, 176: 73–89.

Frey, F.A., Weiss, D., Borisova, A.Y., & Xu, G. (2002). Involvement of continental crust in formation of the Kerguelen plateau: new perspectives from ODP Leg 120 Sites. *Journal of Petrology*, 43: 1207–1239.

Frey, F.A., Coffin, M.F., Wallace, P.J., & Weis, D. (2003). Leg 183 synthesis: Kerguelen Plateau–Broken Ridge — a large igneous province. In Frey, F.A., Coffin, M.F., Wallace, P.J., & Quilty, P.G. (eds.), *Proceedings of the ODP. Scientific Results*, vol. 183, pp. 1–48.

Freyssinet, P., Butt, C.R.M., Morris, R.C., & Piantone, P. (2005). Ore-forming processes related to laterite weathering. *Economic Geology, 100th Anniversary Volume*: 681–721.

Frindt, S., Trumbull, R.B., & Romer, R.L. (2004). Petrogenesis of the Gross Spitzkoppe topaz granite, central western Namibia: a geochemical and Nd–Sr–Pb isotope study. *Chemical Geology*, 206: 43–71.

Frost, B.R., Barnes, C.G., Collins, W.J., *et al.* (2001). A geochemical classification for granitic rocks. *Journal of Petrology*, 42: 2033–2048.

Fuller, E.R. & Head, J.W. (2003). Olympus Mons, Mars: detection of extensive preaureole volcanism and implications for initial mantle plume behavior. *Geology*, 31: 175–178.

Gál, B., Molnár, F., & Peterson, D.M. (2011). Cu–Ni–PGE mineralization in the South Filson Creek area, south Kawishiwi intrusion, Duluth Complex: mineralization styles and magmatic and hydrothermal processes. *Economic Geology*, 106: 481–509.

Galerne, C.Y., Neumann, E.-R., & Planke, S. (2008). Emplacement mechanisms of sill complexes: information from the geochemical architecture of the Golden Valley Sill Complex, South Africa. *Journal of Volcanology and Geothermal Research*, 177: 425–440.

Galgana, G.A., Grosfils, E.B., & McGovern, P.J. (2013). Radial dike formation on Venus: insights from models of uplift, flexure, and magmatism. *Icarus*, 225: 538–547.

Gallagher, K. & Hawkesworth, C. (1992). Dehydration melting and the generation of continental flood basalts. *Nature*, 358: 57–59.

Galland, O., Planke, S., Neumann, E.-R., & Malthe-Sørenssen, A. (2009). Experimental modelling of shallow magma emplacement: application to saucer-shaped intrusions. *Earth and Planetary Science Letters*, 277: 373–383.

Ganino, C. & Arndt, N.T. (2009). Climate changes caused by degassing of sediments during the emplacement of large igneous provinces. *Geology*, 37: 323–326.

Ganino, C. & Arndt, N.T. (2010). Climate changes caused by degassing of sediments during the emplacement of large igneous provinces: reply. *Geology*, 38: e211.

Gans, P.B., Mahood, G.A., & Schermer, E. (1989). Synextensional magmatism in the Basin and Range Province; a case study from the eastern Great Basin. Boulder, CO: Geological Society of America, Special Paper 233, pp. 1–53.

Gao, J.-F. & Zhou, M.-F. (2013). Generation and evolution of siliceous high magnesium basaltic magmas in the formation of the Permian Huangshandong intrusion (Xinjiang, NW China). *Lithos*, 162–163: 128–139.

Garfunkel, Z. (2008). Formation of continental flood volcanism – the perspective of setting of melting. *Lithos*, 100: 49–65.

Garland, F., Hawkesworth, C.J., & Mantovani, M.S.M. (1995). Description and petrogenesis of the Parana rhyolites, southern Brazil. *Journal of Petrology*, 36: 1193–1227.

Gaudin, A., Decarreau, A., Noack, Y., & Grauby, O. (2005). Clay mineralogy of the nickel laterite ore developed from serpentinized peridotites at Murrin Murrin, Western Australia. *Australian Journal of Earth Sciences*, 52: 231–241.

Geissler, P.E. (2003). Volcanic activity on Io during the Galileo Era. *Annual Review of Earth and Planetary Sciences*, 31: 175–211.

Geldmacher, J., Hoernle, K., van den Bogaard, P., Zanki, G., & Garbe- Schonberg, D. (2001). Earlier history of the ≥70-Ma-old Canary hotspot based on the temporal and geochemical evolution of the Selvagen Archipelago and neighbouring seamounts in the eastern North Atlantic. *Journal of Volcanology and Geothermal Research*, 111: 55–87.

Genge, M.J., Price, G.D., & Jones, A.P. (1995). Molecular dynamics simulation of $CaCO_3$ melts to mantle pressures and temperatures: implications for carbonatite magmas. *Earth and Planetary Science Letters*, 131: 225–238.

Geoffroy, L. (2005). Volcanic passive margins. *Comptes Rendus Geoscience*, 337: 1395–1408.

Geoffroy, L., Aubourg, C., Callot, J.-P., & Barrat, J.-A. (2007). Mechanisms of crustal growth in large igneous provinces: the North Atlantic province as a case study. In Foulger, G.R. & Jurdy, D.M. (eds.), *Plates, Plumes, and Planetary Processes*. Boulder, CO: Geological Society of America, Special Paper 430, pp. 747–774.

George, R., Rogers, N., & Kelley, S. (1998). Earliest magmatism in Ethiopia: evidence for two mantle plumes in one flood basalt province. *Geology*, 26: 923–926.

Gerlach, D.C., Cliff, R.A., Davies, G.R., Norry, M., & Hodgson, N. (1988). Magma sources of the Cape Verdes archipelago: isotopic and trace element constraints. *Geochimica et Cosmochimica Acta*, 52: 2979–2992.

Ghatak, A. & Basu, A.R. (2011). Vestiges of the Kerguelen plume in the Sylhet traps, northeastern India. *Earth and Planetary Science Letters*, 308: 52–64.

Ghori, K.A.R., Craig, J., Thusu, B., Lüning, S., & Geiger, M. (2009). Global Infracambrian petroleum systems: a review. In Craig, J., Thurow, J., Thusu, B., Whitham, A., & Aubutarruma, Y. (eds.), *Global Neoproterozoic Petroleum Systems: A Review*. London: Geological Society of London, Special Publication 326, pp. 109–136.

Gibson, I.L., Sinah, M.N., & Fahrig, W.F. (1987). The geochemistry of the Mackenzie dike swarm, Canada. In Halls, H.C. & Fahrig, W.F. (eds.), *Mafic Dyke Swarms*. St John's, NL: Geological Association of Canada, Special Publication 34, pp. 109–121.

Gibson, S.A. (2002). Major element heterogeneity in Archean to recent mantle plume starting heads. *Earth and Planetary Science Letters*, 195: 59–74.

Gibson, S.A., Thompson, R.N., & Day, J.A. (2006). Timescales and mechanisms of plume–lithosphere interactions: $^{40}Ar/^{39}Ar$ geochronology and geochemistry of alkaline igneous rocks from the Paraná–Etendeka large igneous province. *Earth and Planetary Science Letters*, 251: 1–17.

Gilmore, M.S., Collins, G.C., Ivanov, M.A., Marinangeli, L., & Head, J.W. (1998). Style and sequence of extensional structures in tessera terrain, Venus. *Journal of Geophysical Research*, 103: 16813–16840.

Gladczenko, T.P., Coffin, M.F., & Eldholm, O. (1997). Crustal structure of the Ontong Java Plateau: modelling of new gravity and existing seismic data. *Journal of Geophysical Research*, 102: 22711–22729.

Gladkochub, D.P., Pisarevsky, S.A., Donskaya, T.V., *et al.* (2010). Proterozoic mafic magmatism in Siberian craton: an overview and implications for paleocontinental reconstruction. *Precambrian Research*, 183: 660–668.

Glass, L.M. & Phillips, D. (2006). The Kalkarindji continental flood basalt province: a new Cambrian large igneous province in Australia with possible links to faunal extinctions. *Geology*, 34: 461–464.

Glaze, L.S., Stofan, E.R., Smrekar, S.E., & Baloga, S.M. (2002). Insights into corona formation through statistical analyses. *Journal of Geophysical Research*, 107: 5135.

Gleeson, S.A., Herrington, R.J., Durango, J., Velásquez, C.A., & Koll, G. (2004). The Mineralogy and Geochemistry of the Cerro Matoso S.A. Ni Laterite Deposit, Montelíbano, Colombia. *Economic Geology*, 99: 1197–1213.

Glen, J.M.G. & Ponce, D.A. (2002). Large-scale fractures related to inception of Yellowstone hot spot. *Geology*, 30: 647–650.

Glikson, A.Y. (1999). Oceanic mega-impacts and crustal evolution. *Geology*, 27: 387–390.

Glikson, A. (2005). Asteroid/comet impact clusters, flood basalts and mass extinctions: significance of isotopic age overlaps. *Earth and Planetary Science Letters*, 236: 933–937.

Goddéris, Y., Donnadieu, Y., Nédélec, A., *et al.* (2003). The Sturtian "snowball" glaciation: fire and ice. *Earth and Planetary Science Letters*, 211: 1–12.

Gohl, K., Uenzelmann-Neben, G., & Grobys, N. (2011). Growth and dispersal of a Southeast African Large Igneous Province. *South African Journal of Geology*, 114: 379–386.

Gold, D.J.C. (2006). The Pongola Supergroup. In Johnson, M.R., Anhaeusser, C.R., & Thomas, R.J. (eds.), *The Geology of South Africa*. Johannesburg/Pretoria: Geological Society of South Africa/Council of Geoscience, pp. 135–147.

Goldberg, A.S. (2010). Dyke swarms as indicators of major extensional events in the 1.9–1.2 Ga Columbia supercontinent. *Journal of Geodynamics*, 50: 176–190.

Goldfarb, R.J., Groves, D.I., & Gardoll, S. (2001). Orogenic gold and geologic time: a global synthesis. *Ore Geology Reviews*, 18: 1–75.

Goldfarb, R.J., Baker, T., Dubé, B., Groves, D.I., Hart, C.R., & Gosselin, P. (2005). Distribution, character, and genesis of gold deposits in metamorphic terranes. *Economic Geology, 100th Anniversary Volume*: 407–450.

Goldfarb, R.J., Hart, C., Davies, G., & Groves, D. (2007). East Asian gold, deciphering the anomaly of Phanerozoic gold in Precambrian cratons, *Economic Geology*, 102: 345.

Golightly, J.P. (1981). Nickeliferous laterite deposits. *Economic Geology, 75th Anniversary Volume*: 710–735.

Good, N. & De Wit, M.J. (1997). The Thabazimbi-Murchison Lineament of the Kaapvaal Craton, South Africa: 2700 Ma of episodic deformation. *Journal of the Geological Society*, 154: 93–97.

Goodwin, A.M. (1996). *Principles of Precambrian Geology*. New York: Academic Press.

Gower, C.F. & Krogh, T.E. (2002). A U–Pb geochronological review of the Proterozoic history of the eastern Grenville Province. *Canadian Journal of Earth Sciences*, 39: 795–829.

Gower, C.F., Rivers, T., & Brewer, T.S. (1990). Middle Proterozoic mafic magmatism in Labrador, eastern Canada. In Gower, C.F., Rivers, T., & Ryan, A.B. (eds.), *Mid-Proterozoic Laurentia-Baltica*. St John's, NL: Geological Association of Canada, Special Paper 38, pp. 485–506.

Gradstein, F.M. & Ogg, J.G. (1996). A Phanerozoic time scale. *Episodes [IUGS]* 19: 1–2 (comes with detachable insert).

Gradstein, F.M., Ogg, J.G., Smith, A.G., *et al.* (2005). *A Geologic Time Scale 2004*. Cambridge: Cambridge University Press.

Grantz, A. & May, S.P. (1983). Rifting history and structural development of the continental margin north of Alaska: In Watkins, J.S. & Drake, C.L. (eds.), *Studies in Continental Margin Geology*. Tulsa, OK: American Association of Petroleum Geologists, Memoir 34, pp. 77–100.

Grard, A., Francois, L.M., Dessert, C., Dupre, B., & Godderis, Y. (2005). Basaltic volcanism and mass extinction at the Permo–Triassic boundary: environmental impact and modeling of the global carbon cycle. *Earth and Planetary Science Letters*, 234: 207–221.

Grasby, S.E., Sanei, H., & Beauchamp, B. (2011). Catastrophic dispersion of coal fly ash into oceans during the latest Permian extinction. *Nature Geoscience*, 4: 104–107.

Green, J.C. & Fritz, III, T.J. (1993). Extensive felsic lavas and rheoignimbrites in the Keweenawan Midcontinent Rift plateau volcanics, Minnesota; petrographic and field recognition. *Journal of Volcanology and Geothermal Research*, 54: 177–196.

Green, J.C., Bornhorst, T.J., Chandler, V.W., *et al.* (1987). Keweenawan dykes of the Lake Superior region: evidence for evolution of the Middle Proterozoic Midcontinent rift of North America. In Halls, H.C. & Fahrig, W.F. (eds.), *Mafic Dyke Swarms*. Geological Association of Canada, Special Paper 34, pp. 289–302.

Greene, A.R., Scoates, J.S., Weis, D., Nixon, G.T., & Kieffer, B. (2009). Melting history and magmatic evolution of basalts and picrites from the accreted Wrangellia oceanic plateau, Vancouver Island, Canada. *Journal of Petrology* 50: 467–505.

Greene, A.R., Scoates, J.S., Weis, D., Katvala, E.C., Israel, S., & Nixon, G.T. (2010). The architecture of oceanic plateaus revealed by the volcanic stratigraphy of the accreted Wrangellia oceanic plateau. *Geosphere*, 6 (1): 47–73.

Greenough, J.D., Dostal, J., & Mallory-Greenough, L.M. (2005). Igneous rock associations 5. Oceanic island volcanism II: mantle processes. *Geoscience Canada*, 32: 77–90.

Greff-Lefftz, M. & Legros, H. (1999). Core rotational dynamics and geological events. *Science*, 286: 1707–1709.

Gregory, L.C., Meert, J.G., Bingen, B., Pandit, M.K., & Torsvik, T.H. (2007). Paleomagnetism and geochronology of the Malani Igneous Suite, Northwest India: implications for the configuration of Rodinia and the assembly of Gondwana. *LIP of the Month*, July 2007. See http://www.largeigneousprovinces.org/07jul.

Gregory, L.C., Meert, J.G., Bingen, B., Pandit, M.K., & Torsvik, T.H. (2009). Paleomagnetism and geochronology of the Malani igneous suite, NW India: implications for the configuration of Rodinia and the assembly of Gondwana. *Precambrian Research*, 170: 13–26.

Gretener, P.E. (1969). On the mechanics of sills. *Canadian Journal of Earth Sciences*, 6: 1415–1419.

Griffin, W.L., Doyle, B.J., Ryan, C.G., *et al.* (1999). Layered mantle lithosphere in the Lac de Gras area, Slave craton: composition, structure, and origin. *Journal of Petrology*, 40: 705–727.

Griffiths, R.W. & Campbell, I.H. (1990). Stirring and structure in mantle starting plumes. *Earth and Planetary Science Letters*, 99: 66–78.

Griffiths, R.W. & Campbell, I.H. (1991). Interaction of mantle plume heads with the Earth's surface and onset of small-scale convection. *Journal of Geophysical Research*, 96 (B11): 18295–18310.

Grobler, D.F. & Walraven, F. (1993). Geochronology of Gaborone Granite Complex extensions in the area north of Mafikeng, South Africa. *Chemical Geology*, 105: 319–337.

Grosfils, E.B. (2007). Magma reservoir failure on the terrestrial planets: assessing the importance of gravitational loading in simple elastic models. *Journal of Volcanology and Geothermal Research*, 166: 47–75.

Grosfils, E.B. & Head, J.W. (1994a). The global distribution of giant radiating dike swarms on Venus: implications for the global stress state. *Geophysics Research Letters*, 21: 701–704.

Grosfils, E.B. & Head, J.W. (1994b). Emplacement of a radiating dike swarm in western Vinmara Planatia, Venus: interpretation of the regional stress field orientation and subsurface magmatic configuration. *Earth, Moon and Planets*, 66: 153–171.

Grosfils, E.B., Long, S.M., Venechuk, E.M., *et al.* (2011). Geologic map of the Ganiki Planitia quadrangle (V-14), Venus. *US Geological Survey Scientific Investigations Map 3121* (scale 1:5000,000) 1 sheet, includes pamphlet. See http://pubs.usgs.gov/sim/3121/.

Grosfils, E.B., McGovern, P.J., Gregg, P.M., *et al.* (2014). Elastic models of magma reservoir mechanics: a key tool for investigating planetary volcanism. In Platz, T., Massironi, M., Byrne, P.K., & Hiesinger, H. (eds.), *Volcanism and Tectonism across the Inner Solar System*. London: Geological Society, Special Publication 401, doi:10.1144/SP401.2.

Grove, T.L., de Wit, M.J., & Dann, J. (1997). Komatiites from the Komati type section. Barberton, South Africa. In de Wit, M.J. & Ashwal, L.D. (eds.), *Greenstone Belts*. Oxford: Clarendon, pp. 422–437.

Groves, D.I., Bierlein, F.P., Meinert, L.D., & Hitzman, M.W. (2010). Iron oxide copper–gold (IOCG) deposits through Earth history: implications for origin, lithospheric setting, and distinction from other epigenetic iron oxide deposits. *Economic Geology*, 105: 641–654.

Guest, J.E. & Stofan, E.R. (1999). A new view of the stratigraphic history of Venus. *Icarus*, 139: 55–66.

Gumsley, A.P., de Kock, M.O., Rajesh, H.M., *et al.* (2013). The Hlagothi Complex: the identification of fragments from a Mesoarchaean Large Igneous Province on the Kaapvaal Craton. *Lithos*, 174: 333–348.

Gurenko, A.A. & Kamenetsky, V.S. (2011). Boron isotopic composition of olivine-hosted melt inclusions from Gorgona komatiites, Colombia: new evidence supporting wet komatiite origin. *Earth and Planetary Science Letters*, 312: 201–212.

Gurevitch, E., Westphal, M., Daragan-Suchov, J., *et al.* (1995). Paleomagnetism and magnetostratigraphy of the traps from Western Taimyr (northern Siberia) and the Permo-Triassic crisis. *Earth and Planetary Science Letters*, 136: 461–473.

Gurney, J.J., Helmstaedt, H.H., Richardson, S.H., & Shirey, S.B. (2010). Diamonds through time. *Economic Geology*, 105: 689–712.

Gurney, J.J., Helmstaedt, H.H., le Roex, A.P., *et al.* (2005). Diamonds, crustal distributions and formation processes in time and space and an integrated deposit model. *Economic Geology, 100th Anniversary Volume*: 143–177.

Gurnis, M. (1988). Large-scale mantle convection and the aggregation and dispersal of supercontinents. *Nature*, 332: 695–699.

Haggerty, S.E. (1994). Superkimberlites: a geodynamic diamond window to the Earth's core. *Earth and Planetary Science Letters*, 12: 57–69.

Haggerty, S.E. (1999a). Diamond formation and kimberlite-clan magmatism in cratonic settings. In Fei, Y.-W., Bertka, C.M., & Mysen, B.O. (eds.), *Mantle Petrology: Field Observation and High Pressure Experimentation: A Tribute to Francis R. (Joe) Boyd.* University Park, PA: The Geochemical Society, Special Publication 6, pp. 105–123.

Haggerty, S.E. (1999b). A diamond trilogy: superplumes, supercontinents, and supernovae. *Science*, 285: 851–860.

Hagstrum, J.T. (2005). Antipodal hotspots and bipolar catastrophes: were oceanic large-body impacts the cause? *Earth and Planetary Science Letters*, 236: 13–27.

Hales, T.C., Abt, D.L., Humphreys, E.D., & Roering, J.J. (2005). A lithospheric instability origin for Columbia River flood basalts and Wallowa Mountains uplift in northeast Oregon. *Nature*, 438: 842–845.

Hall, A. (1996). *Igneous Petrology*, 2nd edition. Harlow: Longman Science & Technology.

Hall, R.P. & Hughes, D.J. (1987). Norite dykes of southern Greenland, early Proterozoic boninitic magmatism. *Contributions to Mineralogy and Petrology*, 97: 169–182.

Hall, R.P. & Hughes, D.J. (1990). Noritic magmatism. In Hall, R.P. & Hughes, D.J. (eds.), *Early Precambrian Basic Magmatism*. Glasgow: Blackie, pp. 83–110.

Hall, R.P. & Hughes, D.J. (1993). Early Precambrian crustal development: changing styles of mafic magmatism. *Journal of the Geological Society*, 150: 625–635.

Hallam, A. (2005). *Catastrophes and Lesser Calamities*. Oxford: Oxford University Press.

Hallam, A. & Sellwood, B.W. (1976). Middle Mesozoic sedimentation in relation to tectonics in the British area. *Journal of Geology*, 84: 302–321.

Halls, H.C. (1982). The importance and potential of mafic dyke swarms in studies of geodynamic process. *Geoscience Canada*, 9: 145–154.

Halls, H.C. (1986). Paleomagnetism, structure and longitudinal correlation of Middle Precambrian dykes from northwestern Ontario and Minnesota. *Canadian Journal of Earth Sciences*, 23: 142–157.

Halls, H.C. & Fahrig, W.F. (eds.), (1987). *Mafic Dyke Swarms. St John's, NL.* Geological Association of Canada, Special Publication 34.

Halls, H.C., Burns, K.G., Bullock, S.J., & Batterham, P.M. (1987). Mafic dyke swarms of Tanzania interpreted from aeromagnetic data. In Halls, H.C. & Fahrig, W.F. (eds.), *Mafic Dyke Swarms. St John's, NL.* Geological Association of Canada, Special Publication 34, pp. 173–186.

Halls, H.C., Davis, D.W., Stott, G.M., Ernst, R.E., & Hamilton, M.A. (2008). The Paleoproterozoic Marathon Large Igneous Province: new evidence for a 2.1 Ga long-lived mantle plume event along the southern margin of the North American Superior Province. *Precambrian Research*, 162: 327–353.

Halls, H.C., Hamilton, M.A., & Denyszyn, S.W. (2011). The Melville Bugt dyke swarm of Greenland: a connection to the 1.5–1.6 Ga Fennoscandian Rapakivi Granite Province. In Srivasatava, R.K. (ed.), *Keys for Geodynamic Interpretation*. Berlin: Springer-Verlag, pp. 509–535.

Hamilton, M.A. & Buchan, K.L. (2010). U–Pb geochronology of the Western Channel diabase, northwestern Laurentia: implications for a large 1.59 Ga magmatic province, Laurentia's APWP and paleocontinental reconstructions of Laurentia, Baltica and Gawler craton of southern Australia. *Precambrian Research*, 183: 463–473.

Hamilton, M.A., Buchan, K.L., Ernst, R.E., & Stott, G.M. (2009). Widespread and short-lived 1870 Ma mafic magmatism along the northern Superior craton margin. Abstract, Joint Assembly (AGU, CGU, GS, GAC, IAH-CNC, MAC, MSA, SEG), 24–27 May 2009, Toronto, Canada.

Hamilton, M.A., Sadowski, G.R., Teixeira, W., Ernst, R.E., & Ruiz, A.S. (2012). Precise, matching U–Pb ages for the Rincon del Tigre mafic layered intrusion and Huanchaca gabbro sill, Bolivia: evidence for a late Mesoproterozoic LIP in SW Amazonia? GAC MAC Joint Annual Meeting, St. John's, Newfoundland. *Geoscience at the Edge*, vol. 35.

Hamilton, V.E. & Stofan, E.R. (1996). The geomorphology and evolution of Hecate Chasma, Venus. *Icarus*, 121: 171–194.

Hamilton, W.B. (2007). An alternative Venus. In Foulger, G.R. & Jurdy, D.M. (eds.), *Plates, Plumes and Planetary Processes*. Boulder, CO: Geological Society of America, Special Paper 430, pp. 879–911.

Hamlyn, P.R. & Keays, R.R. (1986). Sulfur saturation and second-stage melts: application to the Bushveld platinum metal deposits. *Economic Geology*, 81: 1431–1445.

Handke, M.J., Tucker, R.D., & Ashwal, L.D. (1999). Neoproterozoic continental arc magmatism in west-central Madagascar. *Geology*, 27: 351–354.

Hanley, L.M. & Wingate, M.T.D. (2000). SHRIMP zircon age for an Early Cambrian dolerite dyke: an intrusive phase of the Antrim Plateau volcanics of northern Australia. *Australian Journal of Earth Sciences*, 47: 1029–1040.

Hanmer, S., Mengel, F., Connelly, J., & van Gool, J. (1997). Significance of crustal scale shear zones and synkinematic mafic dikes in the Nagssugtoqidian orogeny, SW Greenland. A reexamination. *Journal of Structural Geology*, 19: 59–75.

Hannisdal, B. & Peters, S.E. (2011). Phanerozoic Earth system evolution and marine biodiversity. *Science*, 334: 1121–1124.

Hansen, D.M., & Cartwright, J.A. (2006). Saucer shaped sill with lobate morphology revealed by 3-D seismic data: implications for resolving a shallow-level sill emplacement mechanism. *Geological Society of London Journal*, 163: 509–523.

Hansen, V.L. (2007). LIPs on Venus. *Chemical Geology*, 241: 354–374.

Hansen, V.L. & Bleamaster, L.F., III. (2002). Distributed point source volcanism: a mechanism for "regional plains" resurfacing, Venus. *Lunar and Planetary Science Conference*, XIII, Abstract 1061 (CD-ROM).

Hansen, V.L. & López, I. (2010). Venus records a rich early history. *Geology*, 38: 311–314.

Hansen, V.L. & Olive, A. (2010). Artemis, Venus: the largest tectonomagmatic feature in the Solar System? *Geology*, 38: 467–470.

Hansen, V.L. & Young, D.A. (2007). Venus's evolution: a synthesis. In Cloos, M., Carlson, W.D., Gilbert, M.C., Liou, J.G., & Sorensen, S.S. (eds.), *Convergent Margin Terranes and Associated Regions: A Tribute to W.G. Ernst*. Boulder, CO: Geological Society of America, Special Paper 419, pp. 255–273.

Hansen, V.L., Willis, J.J., & Banerdt, W.B. (1997). Tectonic overview and synthesis. In Bougher, S.W., Hunten, D.M., & Phillips, R.J. (eds.), *Venus II*. Tucson: University of Arizona Press, pp. 797–844.

Hansen, V.L., Banks, B.K., & Ghent, R.R. (1999). Tessera terrain and crustal plateaus, Venus. *Geology*, 27: 1071–1074.

Hansen, V.L., Phillips, R.J., Willis, J.J., & Ghent, R.R. (2000). Structures in tessera terrain, Venus: issues and answers. *Journal of Geophysical Research*, 105: 4135–4152.

Hanson, R.E. & Elliot, D.H. (1996). Rift-related Jurassic basaltic phreatomagmatic volcanism in the central Transantarctic Mountains: precursory stage to flood-basalt effusion. *Bulletin of Volcanology*, 58: 327–347.

Hanson, R.E., Crowley, J.L., Bowring, S.A., *et al.* (2004a). Coeval large-scale magmatism in the Kalahari and Laurentian cratons during Rodinia assembly. *Science*, 304: 1126–1129.

Hanson, R.E., Gose, W.A., Crowley, J.L., *et al.* (2004b). Paleoproterozoic intraplate magmatism and basin development on the Kaapvaal craton: age, paleomagnetism and geochemistry of ~1.93 to ~1.87 Ga post-Waterberg dolerites. *South African Journal of Geology*, 107: 233–254.

Hanson, R.E., Puckett Jr., R.E., Keller, G.R., *et al.* (2013). Intraplate magmatism related to opening of the southern Iapetus Ocean: Cambrian Wichita igneous province in the Southern Oklahoma rift zone. *Lithos*, 174: 57–70.

Hargraves, R.B., Rehacek, J., & Hooper, P.R. (1997). Palaeomagnetism of the Karoo igneous rocks in Southern Africa. *South African Journal of Geology*, 100: 195–212.

Harlan, S.S., Heaman, L., LeCheminant, A.N., & Premo, W.R. (2003). Gunbarrel mafic magmatic event: a key 780 Ma time marker for Rodinia plate reconstructions. *Geology*, 31: 1053–1056.

Harland, W.B., Armstrong, R.L., Cox, A.V., *et al.* (1990). *A Geological Time Scale 1989*. Cambridge: Cambridge University Press.

Harris, C. & Erlank, A.J. (1992). The production of large-volume low-δ^{18}O rhyolites during the rifting of Africa and Antarctica: the Lebombo Monocline, southern Africa. *Geochimica et Cosmochimica Acta*, 56: 3561–3570.

Harris, C., Whittingham, A.M., Milner, S.C., & Armstrong, R.A. (1990). Oxygen isotope geochemistry of the silicic volcanic rocks of the Etendeka–Parana Province: source constraints. *Geology*, 18: 1119–1121.

Harrison, J.C., Mayr, U., NcNeil, D.H., *et al.* (1999). Correlation of Cenozoic sequences of the Canadian Arctic region and Greenland: implications for the tectonic history of northern North America. *Bulletin of Canadian Petroleum Geology*, 47: 223–254.

Hart, S.R. (1988). Heterogeneous mantle domains: signatures, genesis and mixing chronologies. *Earth and Planetary Science Letters*, 90: 273–296.

Hart, S.R., Hauri, E.H., Oschmann, L.A., & Whitehead, J.A. (1992). Mantle plumes and entrainment: istopic evidence. *Science*, 256: 517–520.

Harte, B. (2011). Diamond window into the lower mantle. *Science*, 334: 51–52.

Hartmann, W. K. and Neukum, G. (2001). Cratering chronology and the evolution of Mars. *Space Science Reviews*, 96: 165–194.

Hastie, W.W., Watkeys, M.K., & Aubourg, C. (2014). Magma flow in dyke swarms of the Karoo LIP: implications for the mantle-plume hypothesis. *Gondwana Research*, 25: 736–755.

Hatton, C.J. (1995). Mantle plume origin for the Bushveld and Ventersdorp magmatic provinces. *Journal of African Earth Sciences*, 21: 571–577.

Hatton, C.J. & von Gruenewaldt, G. (1990). Early Precambrian layered intrusions. In Hall, R.P. & Hughes, D.J. (eds.), *Early Precambrian Basic Magmatism*. Glasgow: Blackie, pp. 56–82.

Hauri, E.H., Whitehead, J.A., & Hart, S.R. (1994). Fluid dynamic and geochemical aspects of entrainment in mantle plumes. *Journal of Geophysical Research*, 99: 275–300.

Hawkesworth, C. & Schersténn, A. (2007). Mantle plumes and geochemistry. *Chemical Geology*, 241: 319–331.

Hawkesworth, C.J., Lightfoot, P.C., Fedorenko, V.A., *et al.* (1995). Magma differentiation and mineralization in the Siberian continental basalts. *Lithos*, 34: 61–88.

Hawkesworth, C., Kelley, S., Turner, S., Le Roex, A., & Storey, B. (1999). Mantle processes during Gondwana break-up and dispersal. *Journal of African Earth Sciences*, 28: 239–261.

Haxby, W.F., Turcotte, D.L., & Bird, J.M. (1976). Thermal and mechanical evolution of the Michigan Basin. *Tectonophysics*, 36: 57–75.

He, B., Xu, Y.-G., Chung, S.-L., Xiao, L., & Wang, Y.M. (2003). Sedimentary evidence for a rapid, kilometer scale crustal doming prior to the eruption of the Emeishan flood basalts. *Earth and Planetary Science Letters*, 213: 391–405.

He, B., Xu, Y.-G., Wang, Y.M., & Luo, Z.Y. (2006). Sedimentation and lithosfacies paleogeography in SW China before and after the Emeishan flood volcanism: new insights into surface response to mantle plume activity. *Journal of Geology*, 114: 117–132.

He, B., Xu, Y.-G., & Campbell, I. (2009). Pre-eruptive uplift in the Emeishan? *Nature Geosciences*, 2: 530–531.

He, B., Xu, Y.-G., Guan, J.-P., & Zhong, Y.-T. (2011). Reply to comment on "Paleokarst on the top of the Maokou Formation: further evidence for domal crustal uplift prior to the Emeishan flood volcanism" by Bin He, Yi-Gang Xu, Jun-Peng Guan & Yu-Ting Zhong. Lithos 119 1–9, 2010. *Lithos*, 125: 1009–1011.

Head, J.W. (2004). Lunar mare basalt volcanism: formation of Large Igneous Provinces (LIPs) on a one-plate planet. *LIP of the Month*, April 2004. See http://www.largeigneousprovinces.org/04apr.

Head, J.W. (2006). Magmatic and volcanic phases in planetary evolution as global Large Igneous Provinces (GLIPS): the Hesperian ridged plains of Mars. *LIP of the Month*, July 2006. See www.largeigneousprovinces.org/06jul.

Head, J.W. (2014). The geologic evolution of Venus: insights into Earth history. *Geology*, 42: 95–96.

Head, J.W. & Coffin, M.F. (1997). Large Igneous Provinces: a planetary perspective. *AGU Geophysical Monograph*, 100: 411–438.

Head, J.W. & Wilson, L. (1991). Absence of large shield volcanoes and calderas on the Moon: consequence of magma transport phenomena? *Geophysical Research Letters*, 18: 2121–2124.

Head, J.W. & Wilson, L. (1993). Lunar graben formation due to near-surface deformation accompanying dike emplacement. *Planetary and Space Science*, 41: 719–727.

Head, J.W. & Wilson, L. (2012). A flood lava large igneous province on Mercury. *LIP of the Month*, March 2012. See http://www.largeigneousprovinces.org/12mar.

Head, J.W., Crumpler, L.S., Aubele, J.C., Guest, J.E., & Saunders, R.S. (1992). Venus volcanism: classification of volcanic features and structures, associations, and global distribution from Magellan data. *Journal of Geophysical Research*, 97: 13153–13197.

Head, J.W., Kreslavsky, M.A., & Pratt, S. (2002). Northern lowlands of Mars: evidence for widespread volcanic flooding and tectonic deformation in the Hesperian Period. *Journal of Geophysical Research*, 107 (E1).

Head, J.W., Wilson, L., Dickson, J.G., & Neukum, G. (2006). The Huygens–Hellas giant dike system on Mars: implications for Late Noachian–Early Hesperian volcanic resurfacing and climatic evolution. *Geology*, 34: 285–288.

Head, J.W., Chapman, C.R., Domingue, D.L., *et al.* (2007). The geology of Mercury: the view prior to MESSENGER mission. *Space Science Reviews*, 131: 41–84.

Head, J.W., Murchie, S.L., Prockter, L.M., *et al.* (2008). Volcanism on Mercury: evidence from the first MESSENGER flyby. *Science*, 321: 69–72.

Head, J.W., Murchie, S.L., Prockter, L.M., *et al.* (2009a). Volcanism on Mercury: evidence from the first MESSENGER flyby for extrusive and explosive activity and the volcanic origin of plains. *Earth and Planetary Science Letters*, 285: 227–242.

Head, J.W., Murchie, S.L., Prockter, L.M., *et al.* (2009b). Evidence for intrusive activity on Mercury from the first MESSENGER flyby. *Earth and Planetary Science Letters*, 285: 251–262.

Head, J.W., Chapman, C.R., Strom, R.G., *et al.* (2011). Flood volcanism in the northern high latitudes of Mercury revealed by MESSENGER. *Science*, 333: 1853–1856.

Heaman, L.M. (1997). Global mafic magmatism at 2.45 Ga: remnants of an ancient large igneous province? *Geology*, 25: 299–302.

Heaman, L.M. & Kjarsgaard, B.A. (2000). Timing of eastern North American kimberlite magmatism: continental extension of the Great Meteor hotspot track? *Earth and Planetary Science Letters*, 178: 253–268.

Heaman, L.M. & LeCheminant, A.N. (1993). Paragenesis and U–Pb systematics of baddeleyite (ZrO_2). *Chemical Geology*, 119: 95–126.

Heaman, L.M. & Machado, N. (1992). Timing and origin of Midcontinent Rift alkaline magmatism, North America: evidence from the Coldwell Complex. *Contributions to Mineralogy and Petrology*, 110: 289–293.

Heaman, L.M. & Pearson, D.G. (2010). Nature and evolution of the Slave province subcontinental lithospheric mantle. *Canadian Journal of Earth Sciences*, 47: 369–388.

Heaman, L.M., LeCheminant, A.N., & Rainbird, R.H. (1992). Nature and timing of Franklin igneous events. Canada: implications for a Late Proterozoic mantle plume and the break-up of Laurentia. *Earth and Planetary Science Letters*, 109: 117–131.

Heaman, L.M., Kjarsgaard, B.A., & Creaser, R.A. (2003). The timing of kimberlite magmatism in North America: implications for global kimberlite genesis and diamond exploration. *Lithos*, 71: 153–184.

Heaman, L.M., Kjarsgaard, B.A., & Creaser, R.A. (2004). The temporal evolution of North American kimberlites. *Lithos*, 76: 377–397.

Heaman, L.M., Easton, R.M., Hart, T.R., Hollings, P., MacDonald, C.A., & Smyk, M.C. (2007). Further refinement to the timing of Mesoproterozoic magmatism, Lake Nipigon region, Ontario. *Canadian Journal of Earth Sciences*, 44: 1055–1086.

Heaman, L.M., Peck, D., & Toope, K. (2009). Timing and geochemistry of 1.88 Ga Molson Igneous Events, Manitoba: insights into the formation of a craton-scale magmatic and metallogenic province. *Precambrian Research*, 172: 143–162.

Hébert, C., Cadieux, A.M., & Breemen, O.V. (2005). Temporal evolution and nature of Ti–Fe–P mineralization in the anorthosite mangerite charnockite granite (AMCG) suites of the south-central Grenville province, Saguenay Lac St. Jean area, Quebec, Canada. *Canadian Journal of Earth Sciences*, 42: 1865–1880.

Hegner, E., Kröner, A., & Hunt, P. (1994). A precise U–Pb age for the Archaean Pongola Supergroup volcanics in Swaziland. *Journal of African Earth Science*, 18: 339–341.

Heinonen, J.S. & Luttinen, A.V. (2010). Mineral chemical evidence for extremely magnesian subalkaline melts from the Antarctic extension of the Karoo large igneous province. *Mineralogy and Petrology*, 99: 201–217.

Helmstaedt, H.H. & Gurney, J.J. (1995). Geotectonic controls of primary diamond deposits: implications for area selection. *Journal of Geochemical Exploration*, 53: 125–144.

Helmstaedt, H.H. & Gurney, J.J. (1997). Geodynamic controls of kimberlites: what are the roles of hotspot and plate tectonics? *Russian Geology and Geophysics*, 38: 492–508.

Helmstaedt, H. & Padgham, W.A. (1986). A new look at the stratigraphy of the Yellowknife Supergroup at Yellowknife, N.W.T.: implications for the age of gold-bearing shear zones and Archean basin evolution. *Canadian Journal of Earth Sciences*, 23: 454–475.

Hergt, J. & Brauns, C.M. (2001). On the origin of Tasmanian dolerites. *Australian Journal of Earth Sciences*, 48: 543–549.

Hergt, J., Peate, D., & Hawkesworth, C.J. (1991). The petrogenesis of Mesozoic Gondwana low-Ti flood basalts. *Earth and Planetary Science Letters*, 105: 134–148.

Hergt, J.M., Preston, P., Bright, R.M.J., & Phillips, G.N. (2000). Differentiated dolerite sills in the Yandal greenstone belt. *AIG Bulletin*, 32: 55–67.

Herrick, R.R. (1999). Small mantle upwellings are pervasive on Venus and Earth. *Geophysical Research Letters*, 26: 803–806.

Herzberg, C. (2011). Basalts as temperature probes of Earth's mantle. *Geology*, 39: 1179–1180.

Herzberg, C. & Gazel, E. (2009). Petrological evidence for secular cooling in mantle plumes. *Nature*, 458: 619–622.

Herzberg, C. & O'Hara, M.J. (1998). Phase equilibrium constraints on the origin of basalts, picrites, and komatiites. *Earth-Science Reviews*, 44: 39–79.

Herzberg, C., Asimow, P.D., Arndt, N., *et al.* (2007). Temperatures in ambient mantle and plumes: constraints from basalts, picrites, and komatiites. *Geochemistry, Geophysics, Geosystems*, 8, 2006GC001390.

Hesselbo, S.P., Gröcke, D.R., Jenkyns, H.C., *et al.* (2000). Massive dissociation of gas hydrate during a Jurassic oceanic anoxic event. *Nature*, 406: 392–395.

Heydolph, K., Murphy, D.T., Geldmacher, J., *et al.* (2014). Plume versus plate origin for the Shatsky Rise oceanic plateau (NW Pacific): insights from Nd, Pb and Hf isotopes. *Lithos*, 200–201: 49–63.

Hiesinger, H., Head, J.W., Wolf, U., Jaumann, R., & Neukum, G. (2011). Ages and stratigraphy of lunar mare basalts: a synthesis. In Ambrose, W.A. & Williams, D.A. (eds.), *Recent Advances and Current Research Issues in Lunar Stratigraphy*. Boulder, CO: Geological Society of America, Special Paper 477, pp. 1–51.

Hildebrand, A.R., Penfield, G.T., Kring, D.A., *et al.* (1991). Chicxulub crater: a possible Cretaceous/Tertiary boundary impact crater on the Yucatan Peninsula, Mexico. *Geology*, 19: 867–871.

Hill, R.I. (1991). Starting plumes and continental break-up. *Earth and Planetary Science Letters*, 104: 398–416.

Hilton, D.R., Grönvold, K., Macpherson, C.G., & Castillo, P.R. (1999). Extreme 3He/4He ratios in northwest Iceland: constraining the common component in mantle plumes. *Earth and Planetary Science Letters*, 173: 53–60.

Hinchey, J.G., Hattori, K.H., & Lavigne, M.J. (2005). Geology, petrology, and controls on PGE mineralization of the Southern Roby and Twilight zones, Lac des Iles Mine, Canada. *Economic Geology*, 100: 43–61.

Hirano, N., Takahashi, E., Yamamoto, J., *et al.* (2006). Volcanism in response to plate flexure. *Science*, 313: 1426–1428.

Hitzman, M.C. (2000). Iron oxide–Cu–Au deposits: what, where, when, and why? In Porter, T.M. (ed.), *Hydrothermal Iron Oxide Copper–Gold and Related Deposits: A Global Perspective*. Adelaide: PGC Publishing, pp. 9–25.

Hoatson, D.M., Jaireth, S., & Jaques, A.L. (2006). Nickel sulfide deposits in Australia: characteristics, resources, and potential. *Ore Geology Reviews*, 29: 177–241.

Hofmann, A.W. (1988). Chemical differentiation of the Earth: the relationship between mantle, continental crust, and oceanic crust. *Earth and Planetary Science Letters*, 90: 297–314.

Hofmann, A.W. (1997). Mantle geochemistry: the message from oceanic volcanism. *Nature*, 385: 219–229.

Hofmann, A.W. (2003). Sampling mantle heterogeneity through oceanic basalts: isotopes and trace elements. In Carlson, R.W. (ed.), *The Mantle and Core*. New York: Elsevier, Treatise on Geochemistry, volume 2, pp. 61–101.

Hofmann, C., Courtillot, V., Feraud, G., *et al.* (1997). Timing of the Ethiopian flood basalt event and implications for plume birth and global change. *Nature*, 389: 838–841.

Hofmann, C., Feraud, G., & Courtillot, V. (2000). $^{40}Ar/^{39}Ar$ dating of mineral separates and whole rocks from the Western Ghats lava pile: further constraints on duration and age of the Deccan traps. *Earth and Planetary Science Letters*, 180: 13–27.

Holland, H.D. (2006). The oxygenation of the atmosphere and oceans. *Philosophical Transactions of the Royal Society: Biological Sciences*, 361: 903–915.

Hollings, P., Wyman, D., & Kerrich, R. (1999). Komatiite-basalt-rhyolite volcanic associations in Northern Superior Province greenstone belts: significance of plume-arc interaction in the generation of the proto-continental Superior Province. *Lithos*, 46: 137–161.

Hollings, P., Smyk, M., & Cousens, B. (2012). The radiogenic isotope characteristics of dikes and sills associated with the Mesoproterozoic Midcontinent Rift near Thunder Bay, Ontario, Canada. *Precambrian Research*, 214–215: 269–279.

Hoogenboom, T., Houseman, G., & Martin, P. (2005). Elastic thickness estimates for coronae associated with chasmata on Venus. *Journal of Geophysical Research*, 110: doi: 10.1029/2004JE002394.

Hooper, P.R. (1997). The Columbia River Flood Basalt Province: current status. *AGU Geophysical Monograph*, 10: 1–27.

Hooper, P.R., Camp, V.E., Reidel, S.P., & Ross, M.E. (2007). The origin of the Columbia River flood basalt province: plume versus nonplume models. In Foulger, G.R. & Jurdy, D.M. (eds.), *Plates, Plumes, and Planetary Processes*. Boulder, CO: Geological Society of America Special Paper 430, pp. 635–668.

Hooper, P.R., Widdowson, M., & Kelley, S. (2010). Tectonic setting and timing of the final Deccan flood basalt eruptions. *Geology*, 38: 839–842.

Hou, G.T., Kusky, T.M., Wang, C., & Wang, Y.X. (2010). Mechanics of the giant radiating Mackenzie dyke swarm: a palaeostress field modeling. *Journal of Geophysical Research*, 115 (B02402): 1–14.

Hou, T., Zhang, Z., Kusky, T., Du, Y., Liu, J., & Zhao, Z. (2011). A reappraisal of the high-Ti and low-Ti classification of basalts and petrogenetic linkage between basalts and mafic-ultramafic intrusions in the Emeishan large igneous province, SW China. *Ore Geology Reviews*, 43: 133–143.

Hough, M.L., Shields, G.A., Evins, L.Z., *et al.* (2006). A major sulphur isotope event at *c.* 510 Ma: a possible anoxia–extinction–volcanism connection during the Early–Middle Cambrian transition? *Terra Nova*, 18: 257–263.

Huey, R.B. & Ward, P.D. (2005). Hypoxia, global warming, and terrestrial late Permian extinctions. *Science*, 308: 398–401.

Hughes, H.S.R., McDonald, I., Goodenough, K.M., *et al.* (2014). Enriched lithospheric mantle keel below the Scottish margin of the North Atlantic Craton: evidence from the Palaeoproterozoic Scourie dyke swarm and mantle xenoliths. *Precambrian Research* (in press).

Hulbert, L.J. (1997). Geology and metallogeny of the Kluane mafic–ultramafic belt, Yukon Territory, Canada: Eastern Wrangellia – a new Ni–Cu–PGE metallogenic terrane. *Geological Survey of Canada Bulletin*, 506.

Hulbert, L.J. (2002). Magmatic platinum group element environments in Canada: present and future exploration target areas. Geological Association of Canada Robinson Lecture 2002, MP#2 (CD-ROM).

Hunter, D.R. & Halls, H.C. (1992). A geochemical study of a Precambrian mafic dyke swarm, Eastern Transvaal, South Africa. *Journal of African Earth Sciences*, 15: 153–168.

Huppert, H.E. & Sparks, R.S.J. (1988). The generation of granitic magmas by intrusion of basalt into continental crust. *Journal of Petrology*, 29: 599–624.

Hutchinson, D.R., White, R.S., Cannon, W.F., & Schulz, K.J. (1990). Keweenawan hot spot: geophysical evidence for a 1.1 Ga mantle plume beneath the Midcontinent Rift system. *Journal of Geophysical Research*, 95(B7): 10869–10884.

Hynek, B.M., Robbins, S.J., Sramek, O., & Zhong, S.J. (2011). Geological evidence for a migrating Tharsis plume on early Mars. *Earth and Planetary Science Letters*, 310: 327–333.

Ihlenfeld, C. & Keays, R.R. (2011). Crustal contamination and PGE mineralization in the Platreef, Bushveld Complex, South Africa: evidence for multiple contamination events and transport of magmatic sulfides. *Mineralium Deposita*, 46: 813–832.

Iizuka, T., Komiya, T., Rino, S., Maruyama, S., & Hirata, T. (2010). Detrital zircon evidence for Hf isotopic evolution of granitoid crust and continental growth. *Geochimica et Cosmochimica Acta*, 74: 2450–2472.

Ingle, S. & Coffin, M.F. (2004). Impact origin for the greater Ontong Java Plateau? *Earth and Planetary Science Letters*, 218: 123–134.

Ingle, S., Weis, D., & Frey, F.A. (2002a). Indian continental crust recovered from Elan Bank, Kerguelen Plateau (ODP Leg 183, Site 1137). *Journal of Petrology*, 43: 1241–1257.

Ingle, S., Weis, D., Scoates, J.S., & Frey, F.A. (2002b). Relationship between the early Kerguelen plume and continental flood basalts of the paleo-Eastern Gondwanan margins. *Earth and Planetary Science Letters*, 197: 35–50.

Ingle, S., Scoates, J.S., Weis, D., Brügmann, G., & Kent, R.W. (2004). Origin of Cretaceous continental tholeiites in southwestern Australia and eastern India: insights from Hf and Os isotopes. *Chemical Geology*, 209: 83–106.

Irvine, T.N. (1977). Origin of chromitite and other layers in the Muskox intrusion and other stratiform intrusions: a new interpretation: *Geology*, 5: 273–277.

Irvine, T.N. (1980). Magmatic infiltration metasomatism, double-diffusive fractional crystallization, and adcumulus growth in the Muskox intrusion and other layered intrusions. In Hargraves, R.B. (ed.), *Physics of Magmatic Processes*. Princeton, NJ: Princeton University Press, pp. 325–383.

Irvine, T.N., Keith, D.W., & Todd, S.G. (1983). The J–M platinum–palladium reef of the Stillwater Complex, Montana. II. Origin by double-diffusive convective magma mixing and implications for the Bushveld Complex. *Economic Geology*, 78: 1287–1334.

Isley, A.E. & Abbott, D.H. (1999). Plume-related mafic volcanism and the deposition of banded iron formation. *Journal of Geophysical Research*, 104(B7): 15461–15477.

Isley, A.E. & Abbott, D.H. (2002). Implications of the temporal distribution of high-Mg magmas for mantle plume volcanism through time. *Journal of Geology*, 110: 141–158.

Ivanic, T.J., Wingate, M.T.D., Kirkland, C.L., van Kranendonk, M.J., & Wyche, S. (2010). Age and significance of voluminous mafic–ultramafic magmatic events in the Murchison Domain, Yilgarn Craton. *Australian Journal of Earth Sciences*, 57: 597–614.

Ivanov, A.V. (2007). Evaluation of different models for the origin of the Siberian Traps. In Foulger, G.R. & Jurdy, D.M. (eds.), *The Origins of Melting Anomalies: Plates, Plumes, and Planetary Processes*. Boulder, CO: Geological Society of America, Special Paper 430, pp. 669–691.

Ivanov, A.V., He, H.-Y., Yan, L.-K., *et al.* (2013). Siberian Traps large igneous province: evidence for two flood basalt pulses around the Permo-Triassic boundary and in the Middle Triassic, and contemporaneous granitic magmatism. *Earth Science Reviews*, 122: 58–76.

Ivanov, M.A., & Head, J.W. (1996). Tessera terrain on Venus: a survey of the global distribution, characteristics, and relation to surrounding units from Magellan data. *Journal of Geophysical Research*, 101: 14861–14908.

Ivanov, M.A. & Head, J.W. (2001). Geology of Venus: mapping of a global geotraverse at 30N latitude. *Journal of Geophysical Research*, 106 (E8): 17515–17566.

Ivanov, M.A. & Head III, J.W. (2013). The history of volcanism on Venus. *Planetary and Space Science*, 84, 66–92.

Ivanov, B.A. & Melosh, H.J. (2003). Impacts do not initiate volcanic eruptions: eruptions close to the crater. *Geology*, 31: 869–872.

Izokh, A.E., Polyakov, G.V., Hoa, T.T., Balykin, P.A., & Phuong, N.T. (2005). Permian–Triassic ultramafic–mafic magmatism of northern Vietnam and southern China as expressions of plume magmatism. *Russian Geology and Geophysics*, 46 (9): 942–951.

Izokh, A.E., Vishnevskii, A.V., Polyakov, G.V., *et al.* (2010). The Ureg Nuur Pt-bearing volcanoplutonic picrite–basalt association in the Mongolian Altay as evidence for a Cambrian–Ordovician Large Igneous Province. *Russian Geology and Geophysics*, 51: 521–533.

Izokh, A.E., Vishnevskii, A.V., Polyakov, G.V., & Shelepaev, R.A. (2011). Age of picrite and picrodolerite magmatism in western Mongolia. *Russian Geology and Geophysics*, 52: 7–23.

Jackson, M.G. & Carlson, R.W. (2011). An ancient recipe for flood-basalt genesis. *Nature*, 476: 316–319.

Jackson, M.G., Kurz, M.D., Hart, S.R., & Workman, R.K. (2007). New Samoan lavas from Ofu Island reveal a hemispherically heterogeneous high ^{3}He/^{4}He mantle. *Earth and Planetary Science Letters*, 264: 360–374.

Jahn, F., Cook, M., & Graham, M. (2008). *Hydrocarbon Exploration and Production*, 2nd edition. Amsterdam: Elsevier.

Jahren, A.H. (2002). The biogeochemical consequences of the mid-Cretaceous super-plume. *Journal of Geodynamics*, 34: 177–191.

James, R.S., Easton, R.M., Peck, D.C., & Hrominchuk, J.L. (2002a). The East Bull Lake intrusive suite: remnants of a ~ 2.48 Ga large igneous and metallogenic province in the Sudbury area of the Canadian shield. *Economic Geology*, 97: 1577–1606.

James, R.S., Jobin-Bevans, S., Easton, R.M., *et al.* (2002b). Platinum group element mineralization in Paleoproterozoic basic intrusions in central and northeastern Ontario, Canada. In Cabri, L.J. (ed.), *The Geology, Geochemistry, Mineralogy, and Mineral Beneficiation of Platinum-Group Elements*. Canadian Institute of Mining, Metallurgy and Petroleum, Special Volume 54, pp. 339–365.

Jamtveit, B., Svensen, H., Podladchikov, Y.Y., & Planke, S. (2004). Hydrothermal vent complexes associated with sill intrusions in sedimentary basins. In Breitkreuz, C. & Petford, N. (eds.), *Physical Geology of High-Level Magmatic Systems*. London: Geological Society of London, Special Publication 234, pp. 233–241.

Janes, D.M., Squyres, S.W., Bindshadler, D.L., *et al.* (1992). Geophysical models for the formation and evolution of coronae on Venus. *Journal of Geophysical Research*, 97: 16055–16067.

Jarvis, I., Lignum, J.S., Gröcke, D.R., Jenkyns, H.C., & Pearce, M.A. (2011). Black shale deposition, atmospheric CO_2 drawdown, and cooling during the Cenomanian–Turonian Oceanic Anoxic Event. *Palaeoceanography*, 26, PA3201.

Jay, A.E. & Widdowson, M. (2008). Stratigraphy, structure and volcanology of the SE Deccan continental flood basalt province: implications for eruptive extent and volumes. *Journal of the Geological Society of London* 165: 177–188.

Jefferson, C.W., Hulbert, L.J., Rainbird, R.H., *et al.* (1994). Mineral resource assessment of the Neoproterozoic Franklin igneous events of Arctic Canada: comparison with the Permo-Triassic Noril'sk Talnakh Ni–Cu–PGE deposits of Russia. *Geological Survey of Canada Open File* 2789.

Jellinek, A.M., Lenardic, A., & Manga, M. (2002). The influence of interior mantle temperature on the structure of plumes: heads for Venus, tails for the Earth. *Geophysical Research Letters*, 29(11): 1532, doi:10.1029/2001GL014624.

Jelsma, H.A., Barnett, W., Richards, S., & Lister, G. (2009). Tectonic setting of kimberlites. *Lithos*, 112: 155–165.

Jenkyns, H.C. (1988). The Early Toarcian (Jurassic) anoxic event: stratigraphic, sedimentary and geochemical evidence. *American Journal of Science*, 288: 101–151.

Jerram, D.A. (2002). Volcanology and facies architecture of flood basalts. In Menzies, M.A., Klemperer, S.L., Ebinger, C.J., & Baker, J. (eds.), *Volcanic Rifted Margins*. Boulder, CO: Geological Society of America, Special Paper 362, pp. 119–132.

Jerram, D.A. & Widdowson, M. (2005). The anatomy of continental flood basalt provinces: geological constraints on the processes and products of flood volcanism. *Lithos*, 79: 385–405.

Jerram, D.A., Single, R.T., Hobbs, R.W., & Nelson, C.E. (2009a). Understanding the offshore flood basalt sequence using onshore volcanic facies analogues: an example from the Faroe–Shetland basin. *Geological Magazine*, 146: 353–367.

Jerram, D.A., Goodenough, K.M., & Troll, V.R. (2009b). Introduction: from the British Tertiary into the future – modern perspectives on the British Palaeogene and North Atlantic Igneous provinces. *Geological Magazine*, 146: 305–308, doi:10.1017/S001675680900627X.

Jerram, D.A., Nelson, C.E., & Hobbs, R.W. (2010). How do your flood basalts grow? Understanding the facies architecture and rock properties of flood basalts. *LIP of the Month*, January 2010. See: www.largeigneousprovinces.org/10jan.

Johansson, Å. (2009). Baltica, Amazonia and the SAMBA connection: 1000 million years of neighbourhood during the Proterozoic? *Precambrian Research*, 175: 221–234.

Johansson, Å. (2014). From Rodinia to Gondwana with the "SAMBA" model: a distant view from Baltica towards Amazonia and beyond. *Precambrian Research*, 244: 226–235.

John, E.H., Wignall, P.B., Newton, R.J., & Bottrell, S.H. (2010). $\delta34SCAS$ and $\delta18OCAS$ records during the Frasnian–Famennian (Late Devonian) transition and their bearing on mass extinction models. *Chemical Geology*, 275: 221–234.

Johnson, R.W., ed. (1989). *Intraplate Volcanism in Eastern Australia and New Zealand*. Sydney: Cambridge University Press.

Johnson, C.L. & Richards, M.A. (2003). A conceptual model for the relationship between coronae and large-scale mantle dynamics on Venus. *Journal of Geophysical Research*, 108 (E6): 5058.

Johnson, H.P., Van Patten, D., Tivey, M., & Sager, W.W. (1995). Geomagnetic polarity reversal rate for the Phanerozoic. *Journal of Geophysical Research*, 22: 231–234.

Johnson, M.R., Anhaeusser, C.R., & Thomas, R.J., eds. (2006). *The Geology of South Africa*. Johannesburg/Pretoria: Geological Society of South Africa/Council of Geoscience.

Jokat, W., Ritzmann, O., Reichert, C., & Hinz, K. (2004). Deep crustal structure of the continental margin off the Explora Escarpment and in the Lazarev Sea, East Antarctica. *Marine Geophysical Researches*, 25: 283–304.

Jones, A.P., Price, G.D., Price, N.J., DeCarli, P.S., & Clegg, R.A. (2002). Impact induced melting and the development of large igneous provinces. *Earth and Planetary Science Letters*, 202: 551–561.

Jones, A.P., Wünemann, K., & Price, G.D. (2005). Modeling impact volcanism as a possible origin for the Ontong Java Plateau. In Foulger, G.R., Natland, J.H., Presnall, D.C., & Anderson, D.L. (eds.), *Plates, Plumes, and Paradigms*. Boulder, CO: Geological Society of America, Special Paper 388, pp. 711–720.

Jones, D.L., Silberling, N.J., & Hillhouse, J. (1977). Wrangellia: a displaced terrane in northwestern North America. *Canadian Journal of Earth Sciences*, 14: 2565–2577.

Jones, T.A., Jefferson, C.W., & Morrell, G.R. (1992). Assessment of mineral and energy resource potential in the Brock Inlier–Bluenose Lake area, N.W.T. *Geological Survey of Canada Open File* 2434.

Jourdan, F. & Evins, L. (2010). The Kalkarindji Large Igneous Province and the Early–Middle Cambrian Extinction. Conference Proceedings, EGU General Assembly 2010, 2–7 May, Vienna, Austria.

Jourdan, F., Féraud, G., Bertrand, H., et al. (2004). The Karoo triple junction questioned: evidence from $^{40}Ar/^{39}Ar$ Jurassic and Proterozoic ages and geochemistry of the Okavango dyke swarm (Botswana). *Earth and Planetary Science Letters*, 222: 989–1006.

Jourdan, F., Féraud, G., Bertrand, H., et al. (2005). Karoo large igneous province: brevity, origin, and relation to mass extinction questioned by new $^{40}Ar/^{39}Ar$ age data. *Geology*, 33: 745–748.

Jourdan, F., Bertrand, H., Sharer, U., et al. (2007a). Major-trace element and Sr–Nd–Hf–Pb isotope compositions of the Karoo large igneous province in Botswana–Zimbabwe. *Journal of Petrology*, 48: 1043–1077.

Jourdan, F., Féraud, G., Bertrand, H., Watkeys, M.K., & Renne, P.R. (2007b). Distinct brief major events in the Karoo large igneous province clarified by new $^{40}Ar/^{39}Ar$ ages on the Lesotho basalts. *Lithos*, 98: 195–209.

Jourdan, F., Marzoli, A., Bertrand, H., et al. (2009). $^{40}Ar/^{39}Ar$ ages of CAMP in North America: implications for the Triassic–Jurassic boundary and the 40 K decay constant bias. *Lithos*, 110: 167–180.

Jourdan, F., Hodges, K., Sell, B., et al. (2014). High-precision dating of the Kalkarindji large igneous province, Australia, and synchrony with the Early–Middle Cambrian (Stage 4-5) extinction. *Geology*, 42: 543–546.

Jowitt, S.M. & Ernst, R.E. (2013). Geochemical assessment of the metallogenic potential of Proterozoic LIPs of Canada. *Lithos*, 174: 291–307.

Jowitt, S.M., Keays, R.R., Jackson, P.G., Hoggart, C.R., & Green, A.H. (2012). Mineralogical and geochemical controls on the formation of the Woods Point dike swarm, Victoria, Australia: evidence from the Morning Star dike and implications for sourcing of Au within orogenic gold systems. *Economic Geology*, 107: 251–273.

Jowitt, S.M., Williamson, M.-C., & Ernst, R.E. (2014). Geochemistry of the 130 to 80 Ma Canadian High Arctic Large Igneous Province (HALIP) event and implications for Ni–Cu–PGE prospectivity. *Economic Geology*, 109: 281–307.

Ju, W., Hou, G.T., & Hari, K.R. (2013). Mechanics of mafic dyke swarms in the Deccan Large Igneous Province: palaeostress field modelling. *Journal of Geodynamics*, 66: 79–91.

Juhlin, C. (1990). Interpretation of the reflections in the Siljan Ring area based on results from the Gravberg-1 borehole. *Tectonophysics*, 173: 345–360.

Jurdy, D.M. & Stoddard, P.R. (2007). The coronae of Venus: impact, plume, or other origin? In Foulger, G.R. & Jurdy, D.M. (eds.), *Plates, Plumes and Planetary Processes*. Boulder, CO: Geological Society of America, Special Paper 430, pp. 859–878.

Kaiho, K. & Lamolda, M.A. (1999). Catastrophic extinction of planktonic foraminifera at the Cretaceous–Tertiary boundary evidenced by stable isotopes and foraminiferal abundance at Caravaca, Spain. *Geology*, 27: 355–358.

Kaiho, K., Kajiwara, Y., Nakano, T., *et al.* (2001). End-Permian catastrophe by a bolide impact: evidence of a gigantic release of sulfur from the mantle. *Geology*, 29: 815–818.

Kaminski, E. & Jaupart, C. (2000). Lithosphere structure beneath the Phanerozoic intracratonic basins of North America. *Earth and Planetary Science Letters*, 178: 139–149.

Kamo, S.L., Czamanske, G.K., & Krogh, T.E. (1996). A minimum U–Pb age for Siberian flood-basalt volcanism. *Geochimica et Cosmochimica Acta*, 60: 3505–3511.

Kamo, S.L., Czamanske, G.K., Amelin, Y., *et al.* (2003). Rapid eruption of Siberian flood-volcanic rocks and evidence for coincidence with the Permian–Triassic boundary and mass extinction at 251 Ma. *Earth and Planetary Science Letters*, 214: 75–91.

Kampschulte, A. & Strauss, H. (2004). The sulfur isotopic evolution of Phanerozic seawater based on the analysis of structurally substituted sulfate in carbonates. *Chemical Geology*, 204: 255–286.

Katvala, E.C. & Henderson, C.M. (2002). Conodont sequence biostratigraphy and paleogeography of the Pennsylvanian-Permian Mount Mark and Fourth Lake Formations, southern Vancouver Island. In Hills, L.V., *et al.* (eds.), *Carboniferous and Permian of the World*. Calgary, Alberta: Canadian Society of Petroleum Geologists, Memoir 19, pp. 461–478.

Kay, S.M., Ramos, V.A., Mpodozis, C., & Sruoga, P. (1989). Late Paleozoic to Jurassic silicic magmatism at the Gondwana margin: analogy to the Middle Proterozoic in North America? *Geology*, 23: 13–16.

Keays, R.R. (1995). The role of komatiitic and picritic magmatism and S-saturation in the formation of ore deposits. *Lithos*, 34: 1–18.

Keays, R.R. & Lightfoot, P.C. (2010). Crustal sulfur is required to form magmatic Ni–Cu sulfide deposits: evidence from chalcophile element signatures of Siberian and Deccan Trap basalts. *Mineralium Deposita*, 45: 241–257.

Keller, G. (2012). *The Cretaceous–Tertiary Mass Extinction, Chicxulub Impact, and Deccan Volcanism*. Berlin: Springer, pp. 759–793.

Keller, G., Adatte, T., Gardin, S., Bartolini, A., & Bajpai, S. (2008). Main Deccan volcanism phase ends near the K–T boundary: evidence from the Krishna–Godavari basin, SE India. *Earth and Planetary Science Letters*, 268: 293–311.

Keller, G., Adatte, T., Bhowmick, P.K., *et al.* (2011a). Deccan volcanism and the KT mass extinction: a new perspective on global effects of volcanism. *Search and Discovery Article*, 30156. See: http://www.searchanddiscovery.com/documents/2011/30156keller/ndx_keller.pdf.

Keller, G., Bhowmick, P.K., Upadhyay, H., *et al.* (2011b). Deccan volcanism linked to the Cretaceous–Tertiary Boundary (KTB) mass extinction: new evidence from ONGC wells in the Krishna–Godavari Basin. *Journal of the Geological Society of India*, 78: 399–428.

Kellogg, L.H., Hager, B.H., & van der Hilst, R.D. (1999). Compositional stratification in the deep mantle. *Science*, 283: 1881–1884.

Kemp, D.B., Coe, A.L., Cohen, A.S., & Schwark, L. (2005). Astronomical pacing of methane release in the Early Jurassic period. *Nature*, 437: 396–399.

Kempton, P.D., Fitton, J.G., Saunders, A.D., *et al.* (2000). The Iceland plume in space and time: a Sr–Nd–Pb–Hf study of the North Atlantic rifted margin. *Earth and Planetary Science Letters*, 177: 255–271.

Kennett, B.L.N. & Widiyantoro, S. (1999). A low seismic wavespeed anomaly beneath northwestern India: a seismic signature of the Deccan plume? *Earth and Planetary Science Letters*, 165: 145–155.

Kennett, J.P. & Stott, L.D. (1991). Abrupt deep-sea warming, palaeoceanographic changes and benthic extinctions at the end of the Palaeocene. *Nature*, 353: 225–229.

Kent, R.W., Saunders, A.D., Kempton, P.D., & Ghose, N.C. (1997). Rajmahal Basalts, eastern India: mantle sources and melt distribution at a volcanic rifted margin. *AGU Geophysical Monograph*, 100: 145–182.

Kerber, L.A., Head, J.W., Solomon, S.C., *et al.* (2009). Explosive volcanic eruptions on Mercury: eruption conditions, magma volatile content, and implications for mantle volatile abundances. *Earth and Planetary Science Letters*, 285: 263–271.

Kerr, A.C. (1998) Oceanic plateau formation: a cause of mass extinction and black shale deposition around the Cenomanian–Turonian boundary. *Journal of the Geological Society, London*, 155: 619–626.

Kerr, A.C. (2005). Oceanic LIPs: the kiss of death. *Elements*, 1: 289–292.

Kerr, A. (2010). LIP links to the formation and maturation of oil source rocks at the Cenomanian-Turonian Boundary (~ 93 Ma). Unpublished Report (5 April 2010) of the Project "Reconstruction of Supercontinents Back to 2.7 Ga Using The Large Igneous Province (LIP) Record: with Implications for Mineral Deposit Targeting, Hydrocarbon Resource Exploration, and Earth System Evolution"; see www.supercontinent.org.

Kerr, A.C. (2014). Oceanic plateaus. In Rudnick, R. (volume ed.), *The Crust* (Treatise on Geochemistry, 2nd edn.). Amsterdam: Elsevier.

Kerr, A.C. and Mahoney, J.J. (2007). Oceanic plateaus: problematic plumes, potential paradigms. *Chemical Geology*, 241: 332–353.

Kerr, A.C. & Tarney, J. (2005). Tectonic evolution of the Caribbean and northwestern South America: the case for accretion of two Late Cretaceous oceanic plateaus. *Geology*, 33: 269–272.

Kerr, A.C., Iturralde-Vinent, M.A., Saunders, A.D., Babbs, T.L., & Tarney, J. (1999). A new plate tectonic model of the Caribbean: implications from a geochemical reconnaissance of Cuban Mesozoic volcanic rocks. *Geological Society of America Bulletin*, 111: 1581–1599.

Kerr, A.C., White, R.V., & Saunders, A.D. (2000). LIP reading: recognizing oceanic plateaux in the geological record. *Journal of Petrology*, 41: 1041–1056.

Kerr, A.C., White, R.V., Thompson, P.M.E., Tarney, J., & Saunders, A.D. (2003). No oceanic plateau – no Caribbean plate? The seminal role of an oceanic plateau in Caribbean plate evolution. In Bartolini, C., Buffler, R.T., & Blickwede, J.

(eds.), *The Circum-Gulf of Mexico and the Caribbean: Hydrocarbon Habitats, Basin Formation, and Plate Tectonics*. Tulsa, OK: American Association of Petroleum Geologists, Memoir 79, pp. 126–168.

Kerrich, R. (2011). Book review: *Plates vs. Plumes: a Geological Controversy* by Gillian R. Foulger. *Economic Geology*, 30: 525–526.

Kerrich, R. & Fyfe, W.S. (1981). The gold–carbonate association: source of CO_2 and CO_2 fixation reactions in Archaean lode deposits. *Chemical Geology*, 33: 265–294.

Kerrich, R. & Said, N. (2011). Extreme positive ce-anomalies in a 2.8 Ga submarine volcanic sequence, Murchison Province: oxygenated marine bottom waters. *Chemical Geology*, 280: 232–241, /10.1016/j.chemgeo.2010.11.012.

Kerrich, R. & Wyman, D. (1990). Geodynamic setting of mesothermal gold deposits. An association with accretionary tectonic regimes. *Geology*, 18: 882–885.

Kerrich, R. & Wyman, D.A. (1997). Review of developments in trace-element fingerprinting of geodynamic settings and their implications for mineral exploration. *Australian Journal of Earth Sciences*, 44: 465–487.

Kerrich, R. & Xie, Q. (2002). Compositional recycling structure of an Archean superplume: Nb–Th–U–LREE systematics of Archean komatiites and basalts revisited. *Contributions to Mineralogy and Petrology*, 142: 476–484.

Kerrich, R., Goldfarb, R., Groves, D., & Garwin, S. (2000a). The geodynamics of world-class gold deposits: characteristics, space-time distribution, and origins. *SEG Reviews*, 13: 501–551.

Kerrich, R., Goldfarb, R., Groves, D., Garwin, S., & Jia, Y. (2000b). The characteristics, origins, and geodynamic settings of supergiant gold metallogenic provinces. *Science in China Series D: Earth Sciences*, 43: 1–68.

Kerrich, R., Goldfarb, R.J., & Richards, J.P. (2005). Metallogenic provinces in an evolving geodynamic framework. *Economic Geology, 100th Anniversary Volume*, 1097–1136.

Keszthelyi, L., Self, S., & Thordarson, T. (2006). Flood lavas on Earth, Io and Mars. *Journal of the Geological Society, London*, 163: 253–264.

Khudoley, A.K., Prokopiev, A.V., Chamberlain, K.R., *et al*. (2013). Early Paleozoic mafic magmatic events on the eastern margin of the Siberian Craton. *Lithos*, 174: 44–56.

Kidder, D.L. & Worsley, T.R. (2010). Phanerozoic Large Igneous Provinces (LIPs), HEATT (haline euxinic acidic thermal transgression) episodes, and mass extinctions. *Palaeogeography, Palaeoclimatology, Palaeoecology*, 295: 162–191.

Kidder, D.L. & Worsley, T.R. (2012). A human-induced hothouse climate? *GSA Today*, 22: 4–11.

Kiefer, W.S. & Murray, B.C. (1987). The formation of Mercury's smooth plains. *Icarus*, 72: 477–491.

Kieffer, B., Arndt, N., Lapierre, H., Bastien, F., *et al*. (2004). Flood and shield basalts from Ethiopia: magmas from the African Superswel. *Journal of Petrology*, 45: 793–834.

Kimura, G., Sakakibara, M., & Okamura, M. (1994). Plumes in central Panthalassa? Deductions from accreted oceanic fragments in Japan. *Tectonics*, 13: 905–916.

King, S.D. & Anderson, D.L. (1998). Edge-driven convection. *Earth and Planetary Science Letters*, 160: 289–296.

Kinnaird, J.A. (2005). The Bushveld Large Igneous Province. *LIP of the Month*, May 2005. See www.largeigneousprovinces.org/05may.

Kirstein, L.A., Kelley, S., Hawkesworth, C., *et al.* (2001). Protracted felsic magmatic activity associated with the opening of the South Atlantic. *Journal of the Geological Society London*, 158: 583–592.

Kirstein, L.A., Dunai, T.J., Davies, G.R., Upton, B.G.J., & Nikogosian, I.K. (2004). Helium isotope signature of lithospheric mantle xenoliths from Permo-Carboniferous magmatic province in Scotland: no evidence for a lower-mantle plume. London: Geological Society, Special Publication 223, pp. 243–258.

Kiselev, A.I., Ernst, R.E., Yarmolyuk, V.V., & Egorov, K.N. (2012). Radiating rifts and dyke swarms of the middle Paleozoic Yakutsk plume of eastern Siberian craton. *Journal of Asian Earth Sciences*, 45: 1–16.

Kjarsgaard, B.A. (2007). Kimberlite diamond deposits. In Goodfellow, W.D. (ed.), *Mineral Deposits of Canada: A Synthesis of Major Deposit-Types, District Metallogeny, the Evolution of Geological Provinces, and Exploration Methods*. St. John's, NL: Geological Association of Canada, Mineral Deposits Division, Special Publication Number 5, pp. 245–272.

Klausen, M.B. & Larsen, H.C. (2002). East Greenland coast – parallel dike swarm and its role in continental breakup. In Menzies, M.A., Klemperer, S.L., Ebinger, C.J., & Baker, J. (eds.), *Volcanic Rifted Margins*. Boulder, CO: Geological Society of America, Special Paper 362, pp. 133–158.

Klausen, M.B., Söderlund, U., Olsson, J.R., *et al.* (2010). Petrological discrimination among Precambrian dyke swarms: eastern Kaapvaal craton (South Africa). *Precambrian Research*, 183: 501–522.

Klein, E.L., Almeida, M.E., Rosa-Costa, L.T. (2012). The 1.89-1.87 Ga Uatumã Silicic Large Igneous Province, northern South America. *LIP of the Month*, November 2012. See http://www.largeigneousprovinces.org/12nov.

Klosko, E.R., Russo, R.M., Okal, E.A., & Richardson, W.P. (2001). Evidence for a rheologically strong chemical mantle root beneath the Ontong-Java Plateau. *Earth and Planetary Science Letters*, 186: 347–361.

Kogarko, L.N. (1993). Geochemical characteristics of oceanic carbonatites from the Cape Verde Islands. *South African Journal of Geology*, 96: 119–125.

Korenaga, J. (2004). Mantle mixing and continental breakup magmatism. *Earth and Planetary Science Letters*, 218: 463–473.

Korenaga, J. (2005). Why didn't the Ontong Java Plateau form subaerially? *Earth and Planetary Science Letters*, 234: 385–399.

Kouyaté, D., Söderlund, U., Youbi, N., *et al.* (2013). U–Pb baddeleyite ages of 2040 Ma, 1650 Ma and 885 Ma on dolerites in the West African Craton (Anti-Atlas inliers): possible links to break-ups of Precambrian supercontinents. *Lithos*, 174: 71–84.

Krapež, B., Barley, M.E., & Pickard, A.L. (2003). Hydrothermal and resedimented origins of the precursor sediments to banded iron formation: sedimentological evidence from the Early Palaeoproterozoic Brockman supersequence of Western Australia. *Sedimentology*, 50: 979–1011.

Krassilnikov, A.S. & Head, J.W. (2003). Novae on Venus: geology, classification and evolution. *Journal of Geophysical Research*, 108 (E9): 5108.

Kravchinsky, V.A., Konstantinov, K.M., Courtillot, V., *et al.* (2002). Palaeomagnetism of East Siberian traps and kimberlites: two new poles and palaeogeographic reconstructions at about 360 and 250 Ma. *Geophysical Journal International*, 148: 1–33.

Krogh, T.E., Corfu, F., Davis, D.W., *et al.* (1987). Precise U–Pb isotopic ages of diabase dykes and mafic to ultramafic rocks using trace amounts of baddeleyite and zircon. In Halls, H.C. & Fahrig, W.F. (eds.), *Mafic Dyke Swarms*. St John's, NL: Geological Association of Canada, Special Paper 34, pp. 147–152.

Kruger, F.J. (2005). Filling the Bushveld Complex magma chamber: lateral expansion, roof and floor interaction, magmatic unconformities, and the formation of giant chromitite, PGE and Ti–V–magnetitite deposit. *Mineralium Deposita*, 40: 451–470.

Kuehner, S.M. (1989). Petrology and geochemistry of early Proterozoic high-Mg dykes from the Vestfold Hills, Antarctica. In Crawford, A.J. (ed.), *Boninites and Related Rocks*. London: Unwin Hyman, pp. 208–231.

Kuhnt, W., Holbourn, A., & Moullade, M. (2011). Transient global cooling at the onset of early Aptian oceanic anoxic event (OAE) 1a. *Geology*, 39: 323–326.

Kulikov, V.S., Bychkova, Y.V., Kulikova, V.V., & Ernst, R.E. (2010). The Vetreny Poyas (Windy Belt) subprovince of southeastern Fennoscandia: an essential component of the *ca.* 2.5–2.4 Ga Sumian large igneous provinces. *Precambrian Research*, 183: 589–601.

Kullerud, K., Skjerlie, K.P., Corfu, F., & de la Rosa, J. (2006). The 2.40 Ringvassøy mafic dykes, West Troms Basement Complex, Norway: the concluding act of early Palaeoproterozoic continental breakup. *Precambrian Research*, 150: 183–200.

Kumar, A., Hamilton, M.A., & Halls, H.C. (2012). A Paleoproterozoic giant radiating dyke swarm in the Dharwar Craton, southern India., *Geochemistry, Geophysics, Geosystems*, 13: Q02011.

Kuzmichev, A.B. & Pease, V.L. (2007). Siberian trap magmatism on the New Siberian Islands: constraints for Arctic Mesozoic plate tectonic reconstructions. *Journal of the Geological Society of London*, 164: 959–968.

Laarman, J., Barnett, R., & Duke, N. (2012). Preliminary results on the chromite geochemistry at the Black Label, Black Thor and Big Daddy chromite deposits in the McFaulds Lake greenstone belt, Ontario. In Summary of Field Work and Other Activities 2012. Ontario: Ontario Geological Survey Open File Report 6280, pp. 44-1–46-8.

Labails, C., Olivet, J-L., Aslanian, D., & Roest, W.R. (2010). An alternative early opening scenario for the Central Atlantic Ocean. *Earth and Planetary Science Letters*, 297: 355–368.

Larsen, H.C. & Saunders, A.D. (1998). Tectonism and volcanism at the SE Greenland rifted margin: a record of plume impact and later continental rupture. In Wise, S. (ed.), *Proceedings of the Ocean Drilling Program, Scientific Results 152*. College Station, TX: Ocean Drilling Program, pp. 503–533.

Larson, R.L. (1991a). Latest pulse of Earth: evidence for a Mid-Cretaceous superplume. *Geology*, 19: 547–550.

Larson, R.L. (1991b). Geological consequences of superplumes. *Geology*, 19: 963–966.

Larson, R. (1997). Superplumes and ridge interactions between Ontong Java and Manihiki Plateaus and the Nova-Canton Trough. *Geology*, 25: 779–782.

Larson, R.L. & Olson, P. (1991). Mantle plumes control magnetic reversal frequency. *Earth and Planetary Science Letters*, 107: 437–447.

Lassiter, J.C., DePaolo, D.J., & Mahoney, J.J. (1995). Geochemistry of the Wrangellia flood basalt province: implications for the role of continental and oceanic lithosphere in flood basalt genesis. *Journal of Petrology*, 36: 983–1009.

Layton-Matthews, D., Lesher, C.M., Burnham, O.M., *et al.* (2007). Magmatic Ni–Cu–(PGE) deposits of the Thompson Nickel Belt. In Goodfellow, W.D. (ed.), *Mineral Deposits of Canada: A Synthesis of Major Deposit-types, District*

Metallogeny, the Evolution of Geological Provinces, and Exploration Methods. Geological Association of Canada, Mineral Deposits Division, Special Publication 5, pp. 409–432.

Laznicka, P. (2010). *Giant Metallic Deposits: Future Sources of Industrial Metals*, 2nd edition. Berlin: Springer-Verlag.

Le Bas, M.J. (2000). IUGS reclassification of the high-Mg and picritic rocks. *Journal of Petrology*, 41: 1467–1470.

Le Maitre, R.W. (2002). *Igneous Rocks: A Classification and Glossary of Terms: Recommendations of the International Union of Geological Sciences Subcommission on the Systematics of Igneous Rocks.* Cambridge: Cambridge University Press.

Leask, H.J., Wilson, L., & Mitchell, K.L. (2006). Formation of Aromatum Chaos, Mars: morphological development as a result of volcano–ice interactions. *Journal of Geophysical Research – Planets*, 111, (E8): (E08071), doi: 10.1029/2005JE002549.

Leat, P.T. (2008). On the long-distance transport of Ferrar magmas. In Thomson, K. and Petford, N. (eds.), *Structure and Emplacement of High-Level Magmatic Systems.* London: Geological Society of London, Special Publication 302, pp. 45–61.

LeCheminant, A.N. & Heaman, L.M. (1989). Mackenzie igneous events, Canada: middle Proterozoic hotspot magmatism associated with ocean opening. *Earth and Planetary Science Letters*, 96: 38–48.

Leckie, R.M., Bralower, T.J., & Cashman, R. (2002). Oceanic anoxic events and plankton evolution: biotic response to tectonic forcing during the mid-Cretaceous. *Paleoceanography*, 17: 1301–1329.

Lehmann, B., Burgess, R., Frei, D., *et al.* (2010). Diamondiferous kimberlites in central India synchronous with Deccan flood basalts. *Earth and Planetary Science Letters*, 290: 142–149.

Leitch, A.M. & Davies, G.F. (2001). Mantle plumes and flood basalts: enhanced melting from plume ascent and an eclogite component. *Journal of Geophysical Research*, 106: 2047–2059.

Leng, W. & Zhong, S.J. (2010). Surface subsidence caused by mantle plumes and volcanic loading in large igneous provinces. *Earth and Planetary Science Letters*, 291: 207–214.

Lesher, C.M. (2007). Ni–Cu–(PGE) deposits in the Raglan area, Cape Smith Belt, New Québec. In Goodfellow, W.D. (ed.), *Mineral Deposits of Canada: A Synthesis of Major Deposit-types, District Metallogeny, the Evolution of Geological Provinces, and Exploration Methods.* Geological Association of Canada, Mineral Deposits Division, Special Publication 5, pp. 351–386.

Lewis, J.F., Draper, G., Proenza, J.A., Espaillat, J., & Jimenez, J. (2006). Ophiolite-related ultramafic rocks (serpentinites) in the Caribbean region: a review of their occurrence, composition, origin, emplacement and Ni-laterite soil formation. *Geologica Acta*, 4: 237–263.

Li, C. & Ripley, E.M. (2011). The Giant Jinchuan Ni–Cu–(PGE) deposit: tectonic setting, magma evolution, ore genesis and exploration implications. *Reviews in Economic Geology*, 17: 163–180.

Li, C., Lightfoot, P.C., Amelin, Y., & Naldrett, A.J. (2000). Contrasting petrological and geochemical relationships in the Voisey's Bay and Mushuau intrusions, Labrador, Canada: implications for ore genesis. *Economic Geology*, 95: 771–799.

Li, C., Zhang, M.-J., Fu, P., Qian, Z.-Z., Hu, P.-Q., & Ripley, E.M. (2012). The Kalatongke magmatic Ni–Cu deposits in the Central Asian Orogenic Belt, NW China: product of slab window magmatism. *Mineralium Deposita*, 47: 51–67.

Li, W.-X., Li, X.-H., & Li, Z.X. (2005a). Neoproterozoic bimodal magmatism in the Cathaysia Block of South China and its tectonic significance. *Precambrian Research*, 136 (1): 51–66.

Li, X.-H., Li, Z.-X., Zhou, H., Liu, Y., & Kinny, P.D. (2002). U–Pb zircon geochronology, geochemistry and Nd isotopic study of Neoproterozoic bimodal volcanic rocks in the Kangdian Rift of South China: implications for the initial rifting of Rodinia. *Precambrian Research*, 113: 135–154.

Li, X.-H., Li, Z.-X., Ge, W.C., *et al.* (2003a). Neoproterozoic granitoids in South China: crustal melting above a mantle plume at *ca.* 825 Ma? *Precambrian Research*, 122: 45–83.

Li, X.-H., Su, L., Chung, S.-L., *et al.* (2005b). Formation of the Jinchuan ultramafic intrusion and the world's third largest Ni–Cu sulfide deposit: associated with the *c.* 825 Ma South China mantle plume? *Geochemistry, Geophysics, Geosystems*, 6: Q11004.

Li, X.-H., Li, W.-X., Li, Z.-X., & Liu, Y. (2008b). 850 790 Ma bimodal volcanic and intrusive rocks in northern Zhejiang, South China: a major episode of continental rift magmatism during the breakup of Rodinia. *Lithos*, 102: 341–357.

Li, Z.-X. & Zhong, S.-J. (2009). Supercontinent–superplume coupling, true polar wander and plume mobility: plate dominance in whole-mantle tectonics. *Physics of the Earth and Planetary Interiors*, 176: 143–156.

Li, Z.-X., Li, X.-H., Kinny, P.D., & Wang, J. (1999). The breakup of Rodinia: did it start with a mantle plume beneath South China? *Earth and Planetary Science Letters*, 173: 171–181.

Li, Z.-X., Li, X.-H., Kinny, P.D., *et al.* (2003b). Geochronology of Neoproterozoic syn-rift magmatism in the Yangtze Craton, South China, and correlations with other continents: evidence for a mantle superplume that broke up Rodinia. *Precambrian Research*, 122: 85–109.

Li, Z.-X., Bogdanova, S.V., Collins, A.S., *et al.* (2008a). Assembly, configuration, and break-up history of Rodinia: a synthesis. *Precambrian Research*, 160: 179–210.

Li, Z.-X., Chen, H., Li, X.-H., & Zhang, F. (2013). *A Tectonic History of South China in Nine Days*. Beijing: Science Press.

Liang, H. (2002). End-Permian catastrophic event of marine acidification by hydrated sulfuric acid: mineralogical evidence from Meishan section of south China. *Chinese Science Bulletin*, 47: 1393–1397.

Lightfoot, P.C. & Hawkesworth, C.J. (1997). Flood basalts and magmatic Ni, Cu, and PGE sulphide mineralization: comparative geochemistry of the Noril'sk (Siberian Traps) and West Greenland sequences. *AGU Geophysical Monograph*, 100: 357–380.

Lightfoot, P.C. & Keays, R.R. (2005). Siderophile and chalcophile metal variations in flood basalts from the Siberian Trap: Noril'sk region: implications for the origin of the Ni–Cu–PGE sulfide ores. *Economic Geology*, 100: 439–462.

Lightfoot, P.C., Hawkesworth, C.J., & Sethna, S.F. (1987). Petrogenesis of rhyolites and trachytes from the Deccan Trap: Sr, Nd and Pb isotope and trace element evidence. *Contributions to Mineralogy and Petrology*, 95: 44–54.

Lightfoot, P.C., De Souza, H., & Doherty, W. (1993a). Differentiation and source of the Nipissing Diabase intrusions, Ontario, Canada. *Canadian Journal of Earth Sciences*, 30: 1123–1140.

Lightfoot, P.C., Hawkesworth, C.J., Herg, J., *et al.* (1993b). Remobilisation of the continental lithosphere by a mantle plume: major-, trace-element and Sr-, Nd-, and Pb isotope evidence from picritic and tholeiitic lavas of the Noril'sk District, Siberian Trap, Russia. *Contributions to Mineralogy and Petrology*, 114: 171–188.

Lightfoot, P.C., Naldrett, A.J., Gorbachev, N.S., *et al.* (1994). Chemostratigraphy of Siberian Trap lavas, Noril'sk District, Russia: implications for the source of flood basalt magmas and their associated Ni–Cu mineralization. In Lightfoot, P.C. & Naldrett, A.J. (eds.), *Proceedings of the Sudbury-Noril'sk Symposium*, Toronto: Ontario Geological Survey, vol. 5, pp. 283–312.

Lin, S.-C. & van Keken, P.E. (2005). Multiple volcanic episodes of flood basalts caused by thermochemical mantle plumes. *Nature*, 436: 250–252.

Lin, S.-C. & van Keken, P.E. (2006a). Dynamics of thermochemical plumes: 1. Plume formation and entrainment of a dense layer. *Geochemistry, Geophysics, Geosystems*, 7: Q02006.

Lin, S.-C. & van Keken, P.E. (2006b). Dynamics of thermochemical plumes: 2. Complexity of plume structures and implications for the mapping of mantle plumes. *Geochemistry, Geophysics, and Geosystems*, 7: Q03003, doi:10.1029/2005GC001072.

Lin, S.-C. & van Keken, P.E. (2006c). Deformation, stirring and material transport in thermochemical plumes. *Geophysical Research Letters*, 33 (L20306).

Lipman, P.W., Prostka, H.J., & Christiansen, R.L. (1972). Cenozoic volcanism and plate-tectonic evolution of the western United States, I. Early and Middle Cenozoic. *Philosophical Transactions of the Royal Society of London*, A271: 217–248.

Lister, J.R. & Kerr, R.C. (1991). Fluid-mechanical models of crack propagation and their application to magma in dykes. *Journal of Geophysical Research*, 96: 10049–10077.

Litak, R.K. & Hauser, E.C. (1992). The Bagdad reflection sequence as tabular mafic intrusions: evidence from seismic modeling of mapped exposures. *Geological Society of American Bulletin*, 104: 1315–1325.

Lithgow-Bertelloni, C. & Silver, P.G. (1998). Dynamic topography, plate driving forces, and the African superswell. *Nature*, 395: 269–272.

Littler, K., Hesselbo, S.P., & Jenkyns, H.C. (2010). A carbon-isotope perturbation at the Pliensbachian–Toarcian boundary: evidence from the Lias Group, NE England. *Geological Magazine*, 147: 181–192.

Litvinovsky, B.A., Tsygankov, A.A., Jahn, B.M., Katzir, Y., & Be'eri-Shlevin, Y. (2011). Origin and evolution of overlapping calc-alkaline and alkaline magmas: the Late Palaeozoic post-collisional igneous province of Transbaikalia (Russia). *Lithos*, 1: 845–874.

Liu, S.Y. & Wilson, L. (1998). Collapse pits due to gas release from shallow dikes on Mars. *Lunar and Planetary Science Conference*, IX, Abstract 1602 (CD-ROM).

Logan, P. & Duddy, I. (1998). An investigation of the thermal history of the Ahnet and Reggane basins, central Algeria, and the consequences for hydrocarbon generation and accumulation. In MacGregor, D.S., Moody, R.T.J. & Clark-Lowes, D.D. (eds.), *Petroleum Geology of North Africa*. London: Geological Society of London, Special Publication 132, pp. 131–155.

Loney, R.A. & Himmelberg, G.R. (1983). Structure and petrology of the La Perouse gabbro intrusion, Fairweather Range, southeastern Alaska. *Journal of Petrology*, 24: 377–423.

Loper, D.E. & McCartney, K. (1990). On impacts as a cause of geomagnetic field reversals or flood basalts. In Sharpton, V.L. & Ward, P.D. (eds.), *Global Catastrophes in Earth History: An Interdisciplinary Conference on Impacts, Volcanism, and Mass Mortality*. Boulder, CO: Geological Society of America, Special Paper 247, pp. 19–25.

Lopes-Gautier, R. (2000). Volcanism on Io. In Sigurdsson, H., Houghton, B., McNutt, S.R., Rymer, H., & Stix, J. (eds.), *Encyclopedia of Volcanoes*. San Diego, CA: Academic Press, pp. 709–726.

Lu, S.-N., Li, H.-K., Zhang, C.-L., & Niu, G.-H. (2008). Geological and geochronological evidence for the Precambrian evolution of the Tarim Craton and surrounding continental fragments. *Precambrian Research*, 160: 94–107.

Lubnina, N., Ernst, R., Klausen, M., & Söderlund, U. (2010). Paleomagnetic study of NeoArchean–Paleoproterozoic dykes in the Kaapvaal Craton. *Precambrian Research*, 183: 523–552.

Lustrino, M. (2005). How the delamination and detachment of lower crust can influence basaltic magmatism. *Earth-Science Reviews*, 72: 21–38.

Luttinen, A.V., Zhang, X., & Foland, K.A. (2002). 159 Ma Kjakebeinet lamproites (Dronning Maud Land, Antarctica) and their implications for Gondwana breakup processes. *Geological Magazine*, 139: 525–539.

Macdonald, F.A., Wingate, M.T.D., & Mitchell, K. (2005). Geology and age of the Glikson impact structure, Western Australia. *Australian Journal of Earth Sciences*, 52: 641–651.

MacDonald, R., Wilson, L., Thorpe, R.S., & Martin, A. (1988). Emplacement of the Cleveland dyke: evidence from geochemistry, mineralogy and physical modelling. *Journal of Petrology*, 29: 559–583.

Macdougall, J.D., ed. (1988). *Continental Flood Basalts*. Dordrecht: Kluwer Academic Publishers.

MacHattie, T.G. (2008). Geochemistry and geochronology of the late Archean Prince Albert group (PAg). Ph.D. Thesis, University of Alberta (Edmonton), Nunavut, Canada, NR46373.

MacHattie, T., Heaman, L., Creaser, R., Skulski, T., & Sandeman, H. (2004). Remnants of an Archean Large Igneous Province: evidence from the komatiite-bearing greenstone belts of the northern Rae domain, Nunavut, Canada. *LIP of the Month*, July 2004. See www.largeigneousprovinces.org/04jul.

MacInnes, B.I.A., McBride, J.S., Evans, N.J., Lambert, D.D., & Andrew, A.S. (1999). Osmium isotope constraints on ore metal recycling in subduction zones. *Science*, 286: 512–516.

Magee, K.P. & Head, J.W. (2001). Large flow fields on Venus: implications for plumes, rift associations, and resurfacing. In Ernst, R.E. & Buchan, K.L., (eds.), *Mantle Plumes: Their Identification through Time*. Boulder, CO: Geological Society of America, Special Paper 352, pp. 81–101.

Maier, W.D., Rasmussen, B., Fletcher, I.R., *et al.* (2013). The Kunene Anorthosite Complex, Namibia, and its satellite intrusions: geochemistry, geochronology and economic potential. *Economic Geology*, 108: 953–986.

Mahoney, J.J. & Coffin, M.F. (1997). Preface. *AGU Geophysical Monograph* 100, ix–x.

Mahoney, J.J., Storey, M., Duncan, R.A., Spencer, K.J., & Pringle, M. (1993). Geochemistry and geochronology of the Ontong Java Plateau. *AGU Geophysical Monograph*, 77: 233–261.

Mahoney, J. J., Jones, W. B., Frey, F.A., *et al.* (1995). Geochemical characteristics of lavas from Broken Ridge, the Naturaliste Plateau and southernmost Kerguelen Plateau: Cretaceous plateau volcanism in the southeast Indian Ocean. *Chemical Geology*, 120: 315–345.

Mahotkin, I.L., Gibson, S.A., Thompson, R.N., Zhuravlev, D.Z., & Zherdev, P.U. (2000). Late Devonian diamondiferous kimberlite and alkaline picrite (proto-kimberlite?) magmatism in the Arkhangelsk region, NW Russia. *Journal of Petrology*, 41: 201–227.

Mäkitie, H., Data, G., Isabirye, E., *et al.* (2014). Petrology, geochronology and emplacement model of the giant 1.37 Ga arcuate Lake Victoria Dyke Swarm on the margin of a large igneous province in eastern Africa. *Journal of African Earth Sciences*, 97: 273–296.

Makhous, M. & Galushkin, Yu.I. (2003). Burial history and thermal evolution of the northern and eastern Saharan basins. *Association of American Petrologists and Geologists Bulletin*, 87: 1623–1651.

Malehmir, A., Koivisto, E., Manzi, M., *et al.* (2014). A review of reflection seismic investigations in three major metallogenic regions: the Kevitsa Ni–Cu–PGE district (Finland), Witwatersrand goldfields (South Africa), and the Bathurst Mining Camp (Canada). *Ore Geology Reviews*, 56: 423–441.

Malthe-Sørenssen, A., Planke, S., Svensen, H., & Jamtveit, B. (2004). Formation of saucer-shaped sills. In Breitkreuz, C., & Petford, N. (eds.), *Physical Geology of High-level Magmatic Systems*. London: Geological Society of London, Special Publication 234, pp. 215–227.

Mancuso, J.J., Kneller, W.A., & Quick, J.C. (1989). Precambrian vein pyrobitumen: evidence for petroleum generation and migration 2 Ga ago. *Precambrian Research*, 44: 137–146.

Mandler, H.A.F. & Clowes, R.M. (1997). Evidence for extensive tabular intrusions in the Precambrian shield of western Canada: a 160 km long sequence of bright reflections. *Geology*, 25: 271–274.

Mandler, H.A.F. & Clowes, R.M. (1998). The HSI bright reflector: further evidence for extensive magmatism in the Precambrian of western Canada. *Tectonophysics*, 288: 71–81.

Manglik, A. & Christensen, U.R. (2006). Effect of lithospheric root on decompression melting in plume–lithosphere interaction models. *Geophysical Journal International*, 164: 259–270.

Manikyamba, C. & Kerrich, R. (2011). Geochemistry of alkaline basalts and associated high-Mg basalts from the 2.7 Ga Penakacherla Terrane, Dharwar Craton, India: an Archean depleted mantle–OIB array. *Precambrian Research*, 188: 104–122.

Manikyamba, C., Kerrich, R., Khanna, T.C., Krishna, A.K., & Satyanarayanan, M. (2008). Geochemical systematics of komatiite–tholeiite and adakitic-arc basalt associations: the role of a mantle plume and convergent margin in formation of the Sandur Superterrane, Dharwar craton, India. *Lithos*, 106: 155–172.

Mann, P. & Taira, A. (2004). Global tectonic significance of the Solomon Islands and Ontong Java Plateau convergent zone. *Tectonophysics*, 389: 137–190.

Manson, M.L. & Halls, H.C. (1997). Proterozoic reactivation of the southern Superior province and its role in the evolution of the Midcontinent rift. *Canadian Journal of Earth Sciences*, 34 (4): 562–575.

Mapeo, R.B.M., Ramokate, L.V., Corju, F., Davis, D.W., & Kampunzu, A.B. (2006). The Okwa basement complex, western Botswana: U–Pb zircon geochronology and implications for Eburnean processes in southern Africa. *Journal of African Earth Sciences*, 46: 253–262.

Marcantonio, F., Riesberg, L., Zindler, A., Wyman, D., & Hulbert, L.J. (1994). An isotopic study of the Ni–Cu–PGE-rich Wellgreen intrusion of the Wrangellia terrane: evidence for hydrothermal mobilization of rhenium and osmium. *Geochimica et Cosmochimica Acta*, 58: 1007–1017.

Marsh, J.S. (2006). The Dominion Group. In Johnson, M.R., Anhaeusser, C.R., & Thomas, R.J. (eds.), *The Geology of South Africa*. Johannesburg/Pretoria: Geological Society of South Africa/Council of Geoscience, pp. 149–154.

Marsh, J.S. & Eales, H.V. (1984). Chemistry and petrogenesis of igneous rocks of the Karoo Central area, southern Africa. *Special Publication of the Geological Society of South Africa*, 13: 27–68.

Marsh, J.S., Ewart, A., Milner, S.C., Duncan, A.R., & Miller, R.McG. (2001). The Etendeka igneous province: magma types and their stratigraphic distribution with implications for the evolution of the Paraná–Etendeka flood basalt province. *Bulletin of Volcanology*, 62: 464–486.

Martin, A.P., Condon, D.J., Prave, A.R., *et al.* (2013). Dating the termination of the Palaeoproterozoic Lomagundi-Jatuli carbon isotopic event in the North Transfennoscandian Greenstone Belt. *Precambrian Research*, 224: 160–168.

Martin, D.McB. & Morris, P.A. (2010). Tectonic setting and regional implications of ca. 2.2 Ga mafic magmatism in the southern Hamersley Province, Western Australia. *Australian Journal of Earth Sciences*, 57: 911–931.

Martin, P., Stofan, E.R., Glaze, L.S., & Smrekar, S.E. (2007). Coronae of Parga Chasma, Venus. *Journal of Geophysical Research*, 112 (E4).

Marty, B., Tolstikhin, I., Kamensky, I.L., *et al.* (1998). Plume-derived rare gases in 380 Ma carbonatites from the Kola region (Russia) and the argon isotopic composition in the deep mantle. *Earth and Planetary Science Letters*, 164: 179–192.

Maruoka, T., Koeberl, C., Hancox, P.J., & Reimold, W.U. (2003). Sulfur geochemistry across a terrestrial Permian–Triassic boundary section in the Karoo Basin, South Africa. *Earth and Planetary Science Letters*, 206: 101–117.

Maruyama, S. (1994). Plume tectonics. *Journal of Geological Society of Japan*, 100: 24–49.

Maruyama, S., Yuen, D.A., & Windley, B.F. (2007). Dynamics of plumes and superplumes through time. In Yuen, D.A., Maruyama, S., Karato, S.-i., & Windley, B.F. (eds.), *Superplumes: Beyond Plate Tectonics*. Berlin: Springer-Verlag, pp. 441–502.

Marzoli, A., Renne, P.R., Piccirillo, E.M., *et al.* (1999). Extensive 200-million-year-old continental flood basalts of the Central Atlantic Magmatic Province. *Science*, 284: 616–618.

Marzoli, A., Jourdan, F., Puffer, J.H., *et al.* (2011). Timing and duration of the Central Atlantic magmatic province in the Newark and Culpeper basins, eastern U.S.A. *Lithos*, 122: 175–188.

May, P.R. (1971). Pattern of Triassic–Jurassic diabase dikes around the North Atlantic in the context of predrift position of the continents. *Geological Society of America Bulletin*, 82: 1285–1292.

Mayborn, K.R. & Lesher, C.E. (2004). Paleoproterozoic mafic dike swarms of northeast Laurentia: products of plumes or ambient mantle? *Earth and Planetary Science Letters*, 225: 305–317.

Mazzini, A., Svensen, H., Leanza, H.A., Corfu, F., & Planke, S. (2010). Early Jurassic shale chemostratigraphy and U–Pb ages from the Neuquén Basin (Argentina): implications for the Toarcian Oceanic Anoxic Event. *Earth and Planetary Science Letters*, 297: 633–645.

McBirney, A.R. (1996). The Skaergaard Intrusion. In Cawthorn, R.G. (ed.), *Layered Intrusions*. Amsterdam: Elsevier, pp. 147–180.

McCall, G.J.H. & Peers, R. (1971). Geology of the Binneringie Dyke, Western Australia. *Geologische Rundschau*, 60: 1174–1263.

McCarthy, T.S., McCallum, K., Myers, R.E., & Linton, P. (1990). Stress states along the northern margin of the Witwatersrand Basin during Klipriviersberg Group volcanism. *South African Journal of Geology*, 93: 245–260.

McCausland, P.J.A., Van der Voo, R., & Hall, C.M. (2007). Circum-Iapetus paleo-geography of the Precambrian–Cambrian transition with a new paleomagnetic constraint from Laurentia. *Precambrian Research*, 156: 125–152.

McClay, K.R. & Campbell, I.H. (1976). The structure and shape of the Jimberlana Intrusion, Western Australia, as indicated by an investigation of the Bronzite Complex. *Geological Magazine*, 113: 129–139.

McClellan, E.A., All, J., Sublett, D.M., Jessee, M., & Yonts, J. (2012). Basement-cover contact relationships between Mesoproterozoic rocks and the Neoproter-ozoic Mount Rogers Fm., southern Appalachian Blue Ridge. *Geological Society of America Abstracts with Programs*, 44: 257.

McClellan, E., and Gazel, E. (in press). The Cryogenian intra-continental rifting of Rodinia: evidence from the Laurentian margin in Eastern North America. *Lithos*, in press.

McClintock, M. & White, J.D.L. (2005). Large phreatomagmatic vent complex at Coombs Hills, Antarctica: wet, explosive initiation of flood basalt volcanism in the Ferrar-Karoo LIP. *Bulletin of Volcanology*, 68: 215–239.

McCurry, M., Hayden, K.P., Morse, L.H., & Mertzman, S. (2008). Genesis of post-hotspot, A type rhyolite of the Eastern Snake River Plain volcanic field by extreme fractional crystallization of olivine tholeiite. *Bulletin of Volcanology*, 70: 361–383.

McDonough, W.F. & Sun, S.S. (1995). The composition of the Earth. *Chemical Geology*, 120: 223–254.

McDonough, W.F., Sun, S.-S., Ringwood, A.E., Jagoutz, E., & Hofmann, A.W. (1992). K, Rb, and Cs in the earth and moon and the evolution of the Earth's mantle: *Geochimica et Cosmochimica Acta*, 56: 1001–1012.

McDowell, F. & Clabaugh, S.E. (1979). Ignimbrites of the Sierra Madre Occidental and their relation to the tectonic history of western Mexico. In Chapin, C.E. & Elston, W.E. (eds.), *Ash-Flow Tuffs*. Boulder, CO: Geological Society of America, Special Paper 180, pp. 113–124.

McDowell, F. & Keizer, R.P. (1977). Timing of mid-Tertiary volcanism in the Sierra Madre Occidental between Durango city and Mazatlán, Mexico. *Geological Society of America Bulletin*, 88: 1479–1487.

McDowell, F. & McIntosh, W.C. (2012). Timing of intense magmatic episodes in the northern and central Sierra Madre Occidental, western Mexico. *Geosphere*, 8: 1505–1526.

McElwain, J.C., Wade-Murphy, J., & Hesselbo, S.P. (2005). Changes in carbon dioxide during an oceanic anoxic event linked to intrusion into Gondwana coals. *Nature*, 435: 479–482.

McGhee, G.R. Jr. (2001). The "multiple impacts hypothesis" for mass extinction: a comparison of the Late Devonian and the Late Eocene. *Palaeogeography, Palaeoclimatology, Palaeoecology*, 176: 47–58.

McGovern, P.J., Galgana, G.A., Verner, K.R., & Herrick, R.R. (2014). New constraints on volcano-tectonic evolution of large volcanic edifices on Venus from stereo topography-derived strain estimates. *Geology*, 42: 59–62.

McHone, J.G., Anderson, D.L., Beutel, E.K., & Fialko, Y.A. (2005). Giant dikes, rifts, flood basalts, and plate tectonics: a contention of mantle models. In Foulger, G.R., Natland, J.H., Presnall, D.C., & Anderson, D.L. (eds.), *Plates, Plumes and Paradigms*. Boulder, CO: Geological Society of America, Special Paper 388, pp. 401–420.

McKenzie, D. (1984). The generation and compaction of partial melts. *Journal of Petrology*, 25: 713–765.

McKenzie, D. (1985). The extraction of magma from the crust and mantle. *Earth and Planetary Science Letters*, 74: 81–91.

McKenzie, D. (1994). The relationship between topography and gravity on Earth and Venus. *Icarus*, 112: 55–88.

McKenzie, D. & Bickle, M.J. (1988). The volume and composition of melt generated by extension of the lithosphere, *Journal of Petrology*, 29: 625–679.

McKenzie, D., McKenzie, J.M., & Saunders, R.S. (1992). Dyke emplacement on Venus and on Earth. *Journal of Geophysical Research*, 97: 15977–15990.

McKinnon, W.B., Zahnle, K.J., Ivanov, B.A., & Melosh, H.J. (1997). Cratering on Venus: models and observations. In Bougher, S.W., Hunten, D.M., & Phillips, R.J. (eds.), *Venus II: Geology, Geophysics, Atmosphere, and Solar Wind Environment*. Tucson, AZ: University of Arizona Press, pp. 969–1014.

McLelland, J.M., Selleck, B.W., Hamilton, M.A., & Bickford, M.E. (2010). Late- to post-tectonic setting of some major Proterozoic anorthosite–mangerite–charnockite–granite (AMCG) suites. *Canadian Mineralogist*, 48: 1025–1046.

McNamara, A.K. & Zhong, S. (2004). The influence of thermochemical convection on the fixity of mantle plumes. *Earth and Planetary Science Letters*, 222: 485–500.

McPhie, J., Kamenetski, V., Allen, S., *et al.* (2011). The fluorine link between a supergiant ore deposit and silicic large igneous province. *Geology*, 39: 1003–1006.

Meert, J.G. (2002). Paleomagnetic evidence for a Paleo-Mesoproterozoic supercontinent Columbia. *Gondwana Research*, 5 (1): 207–215.

Meert, J.G. (2012). What's in a name? The Columbia (Paleopangaea/Nuna) supercontinent. *Gondwana Research*, 21: 987–993.

Meert, J.G., Pandit, M.K., & Kamenov, G.D. (2013). Further geochronological and paleomagnetic constraints on Malani (and pre-Malani) magmatism in NW India. *Tectonophysics*, 608: 1254–1267.

Mège, D. (1999). Dikes on Mars: (1) What to look for? (2) A first survey of possible dikes during the Mars Global Surveyor aerobraking and science phasing orbits. Proceedings, 5th International Mars Conference, Pasadena, CA, p. 6207.

Mège, D. (2001). Uniformitarian plume tectonics: the post-Archean Earth and Mars. In Ernst, R.E. & Buchan, K.L. (eds.), *Mantle Plumes: Their Identification through Time*. Boulder, CO: Geological Society of America, Special Paper 352, pp. 141–164.

Mège, D. & Ernst, R.E. (2001). Contractional effects of mantle plumes on Earth, Mars and Venus. In Ernst, R.E. & Buchan, K.L. (eds.), *Mantle Plumes: Their Identification through Time*. Boulder, CO: Geological Society of America, Special Paper 352, pp. 103–140.

Mège, D. & Masson, P. (1996a). A plume tectonics model for the Tharsis province, Mars. *Planetary and Space Sciences*, 44: 1499–1546.

Mège, D. & Masson, P. (1996b). Stress models for Tharsis formation, Mars. *Planetary and Space Sciences*, 44: 1471–1497.

Meibom, A., Sleep, N.H., Zahnle, K., & Anderson, D.L. (2005). Models for noble gases in mantle geochemistry: some observations and alternatives. In Foulger, G.R., Natland, J.H., Presnall, D.C., & Anderson, D.L. (eds.), *Plates, Plumes and Paradigms*. Boulder, CO: Geological Society of America, Special Paper 388, pp. 347–363.

Melcher, F., Grum, W., Simon, G., Thalhammer, T.V., & Stumpel, E.F. (1997). Petrogenesis of the ophiolitic giant chromite deposits of Kempirsai, Kazakhstan: a study of solid and fluid inclusions in chromite. *Journal of Petrology*, 38: 1419–1458.

Melezhik, V.A., Filippov, M.M., & Romashkin, A.E. (2004). A giant Paleoproterozoic deposit of shungite in NW Russia: genesis and practical applications. *Ore Geology Reviews*, 24: 135–154.

Melezhik, V.A., Huhma, H., Condon, D.J., Fallick, A.E., & Whitehouse, M.J. (2007). Temporal constraints on the Paleoproterozoic Lomagundi–Jatuli carbon isotopic event. *Geology*, 35: 655–658.

Melezhik, V.A., Kump, L.R., Hanski, E.J., Fallick, A.E., & Prave, A.R. (2013). Tectonic evolution and major global Earth-surface palaeoenvironmental events in the Palaeoproterozoic. In Melezhik, V.A., Prave, A.R., Hanski, E.J., *et al.* (eds.), *Reading the Archive of Earth's Oxygenation. Volume 1: The Palaeoproterozoic of Fennoscandia as Context for the Fennoscandian Arctic Russia – Drilling Early Earth Project*. Heidelberg: Springer-Verlag, pp. 3–21.

Mello, M.R., Koutsoukos, E.A.M., Mohriak, W.U., & Bacoccoli, G. (1994). Selected petroleum systems in Brazil. In Magoon, L.B. & Dow, W.G. (eds.), *The Petroleum System: From Source to Trap*. Tulsa, OK: American Association of Petroleum Geologists, Memoir 60, pp. 499–512.

Menzies, M.A. (1992). The lower lithosphere as a major source for continental flood basalts: a re-appraisal. In Storey, B.C., Alabaster, T., & Pankhurst, R.J. (eds.), *Magmatism and the Causes of Continental Break-up*. London: Geological Society, Special Publication 68, pp. 33–41.

Menzies, M.A., Baker, J., Bosence, D., *et al.* (1992). The timing of magmatism, uplift and crustal extension: preliminary observations from Yemen. In Storey, B.C., Alabaster, T., & Pankhurst, R.J. (eds.), *Magmatism and the Causes of Continental Break-up*. London: Geological Society of London, Special Publication 68, pp. 293–304.

Menzies, M., Baker, J., & Chazot, G. (2001). Cenozoic plume evolution and flood basalts in Yemen: a key to understanding older examples. In Ernst, R.E. & Buchan, K.L. (eds.), *Mantle Plumes: Their Identification through Time*. Boulder, CO: Geological Society of America, Special Paper 352, pp. 23–36.

Menzies, M.A., Klemperer, S.L., Ebinger, C.J., & Baker, J., eds. (2002a). *Volcanic Rifted Margins*. Boulder, CO: Geological Society of America, Special Paper 362.

Menzies, M.A., Klemperer, S.L., Ebinger, C.J., & Baker, J. (2002b). Characteristics of volcanic rifted margins. In Menzies, M.A., Klemperer, S.L., Ebinger, C.J., &

Baker, J. (eds.), *Volcanic Rifted Margins*. Boulder, CO: Geological Society of America, Special Paper 362: 1–14.

Merino, M., Keller, G.R., Stein, S., & Stein, C. (2013). Variations in Mid-Continent Rift magma volumes consistent with microplate evolution. *Geophysical Research Letters*, 40: 1513–1516.

Meschede, M. (1986). A method of discriminating between different types of mid-ocean ridge basalts and continental tholeiites with the Nb–Zr–Y diagram. *Chemical Geology*, 56: 207–218.

Metelkin, D.V., Ernst, R.E., & Hamilton, M.A. (2011). A *ca.* 1640 Ma mafic magmatic event in southern Siberia, and links with northern Laurentia. Geological Society of America Annual Meeting, 9–12 October 2011, abstract no. 101-1, Minneapolis.

Metsaranta, R.T. & Houlé, M.G. (2012). Progress on the McFaulds Lake ("Ring of Fire") region data compilation and bedrock geology mapping project. In Summary of Field Work and Other Activities 2012. Ontario: *Ontario Geological Survey, Open File Report 6280*, pp. 43-1–43-12.

Meyer, R., van Wijk, J., & Gernigon, L. (2007). The North Atlantic Igneous Province: a review of models for its formation. In Foulger, G.R. & Jurdy, D.M. (eds.), *Plates, Plumes and Planetary Processes*. Boulder, CO: Geological Society of America, Special Paper 430, pp. 525–552.

Meyer, R., Nicoll, G.R., Hertogen, J., *et al.* (2009). Trace element and isotope constraints on crustal anatexis by upwelling mantle melts in the North Atlantic igneous province: an example from the Isle of Rum, NW Scotland. *Geological Magazine*, 146: 382–399.

Michalski, J.R. & Bleacher, J.E. (2013). Supervolcanoes within an ancient volcanic province in Arabia Terra, Mars. *Nature*, 502: 47–52.

Michaut, C., Baratoux, D., & Thorey, C. (2013). Magmatic intrusions and deglaciation at mid-latitude in the northern plains of Mars. *Icarus*, 225: 602–613.

Miles, A. & Cartwright, J. (2010). Hybrid flow sills: a new mode of igneous sheet intrusion. *Geology*, 38: 343–346.

Miller, J.D., Jr. (2007). The Midcontinent Rift in the Lake Superior Region: a 1.1 Ga Large Igneous Province. *LIP of the Month*, November 2007. See: www.largeigneousprovinces.org/07nov.

Miller, J.D. & Nicholson, S.W. (2013). Geology and mineral deposits of the Midcontinent Rift – an overview. In Miller, J.D. (ed.), *Field Guide to the Cu-Ni-PGE Deposits of the Lake Superior Region, Precambrian Research Center Guidebook Series 13–1*. Duluth, MN: University of Minnesota Press.

Miller, J.D., Jr. & Ripley, E.M. (1996). Layered intrusions of the Duluth Complex, Minnesota, USA. In Cawthorn, R.G. (ed.), *Layered Intrusions*. Amsterdam: Elsevier, pp. 257–301.

Miller, J.D., Green, J.C., & Severson, M.J., eds. (2002). *Geology and Mineral Potential of the Duluth Complex and Related Rocks of Northeastern Minnesota*. Minnesota: Minnesota Geological Survey, Report of Investigations 58.

Miller, K.G., Kominz, M.A., Browning, J.V., *et al.* (2005). The Phanerozoic record of global sea-level change. *Science*, 310: 1293–1298.

Milner, S.C., Duncan, A.R., Whittingham, A.M., & Ewart, A. (1995). Trans-Atlantic correlation of eruptive sequences and individual silicic volcanic units within the Parana–Etendeka igneous province. *Journal of Volcanology and Geothermal Research*, 69: 137–157.

Minifie, M.J., Kerr, A.C., Ernst, R.E., *et al.* (2013). The northern and southern sections of the western ca. 1880 Ma Circum-Superior Large Igneous Province, North America: the Pickle Crow dyke connection? *Lithos*, 174: 217–235.

Mirnejad, H. & Bell, K. (2006). Origin and source evolution of the Leucite Hills lamproites: evidence from Sr–Nd–Pb–O isotopic compositions. *Journal of Petrology*, 47: 2463–2489.

Mitchell, R.H. (1986). *Kimberlites: Mineralogy, Geochemistry, and Petrology.* New York: Plenum.

Mitchell, R.H. (1995). *Kimberlites, Orangeites and Related Rocks.* New York: Plenum Press.

Mitchell, R.H. (2005). Carbonatites and carbonatites and carbonatites. *Canadian Mineralogist*, 43: 2049–2068.

Mitchell, R.H. (2008). Petrology of hypabyssal kimberlites: relevance to primary magma compositions. *Journal of Volcanology and Geothermal Research*, 174: 1–8.

Mitchell, R.H. & Bergman, S.C. (1991). *Petrology of Lamproites.* New York: Plenum Press.

Miura, S., Suyehiro, K., Shinohara, M., *et al.* (2004). Seismological structure and implications of collision between the Ontong Java Plateau and Solomon Island Arc from ocean bottom seismometer–airgun data. *Tectonophysics*, 389: 191–220.

Mochizuki, K., Coffin, M.F., Eldholm, O., & Taira, A. (2005). Massive Early Cretaceous volcanic activity in the Nauru Basin related to emplacement of the Ontong Java Plateau. *Geochemistry, Geophysics, Geosystems*, 6: Q10003.

Mohr, P. (1983). Ethiopian flood basalt province. *Nature*, 303: 577–584.

Mohr, P. & Zanettin, B. (1988). The Ethiopian flood basalt province. In MacDougall, J.D. (ed.), *Continental Flood Basalts.* Dordrecht: Kluwer Academic Publications, pp. 63–110.

Moine, B.N., Grégoire, M., O'Reilly, S.Y., *et al.* (2004). Carbonatite melt in oceanic upper mantle beneath the Kerguelen Archipelago. *Lithos*, 75: 239–252.

Molostovskii, E.A., Pechersky, D.M., & Frolov, I.Yu. (2007). The Phanerozoic magnetostratigraphic scale and its description on the basis of the cumulative distribution function. *Izvestiya Physics of the Solid Earth (Fizika Zemli)*, 43: 811–818.

Monnereau, M., Rabinowicz, M., & Arquis, E. (1993). Mechanical erosion and reheating of the lithosphere: a numerical model for hotspot swells. *Journal of Geophysical Research*, 98: 809–823.

Montelli, R., Nolet, G., Dahlen, F.A., & Masters, G. (2006). A catalogue of deep mantle plumes: new results from finite frequency tomography. *Geochemistry, Geophysics, Geosystems*, 7: Q11007.

Montési, L.G.J. (2001). Concentric dikes on the flanks of Pavonis Mons: implications for the evolution of Martian shield volcanoes and mantle plumes. In Ernst, R.E & Buchan, K.L. (eds.), *Mantle Plumes: Their Identification through Time.* Boulder, CO: Geological Society of America, Special Paper 352, pp. 165–181.

Moore, A. & Blenkinsop, T. (2002). The role of mantle plumes in the development of continental-scale drainage patterns: the southern African example revisited. *South African Journal of Geology*, 105: 353–360.

Moores, E.M., Kellogg, L.H., & Dilek, Y. (2000). Tethyan ophiolites, mantle convection, and tectonic "historical contingency": a resolution of the "ophiolite conundrum." In Dilek, Y., Moores, E.M., Elthon, D., & Nicolas, A. (eds.), *Ophiolites and Oceanic Crust: New Insights from Field Studies and the Ocean Drilling Program.* Boulder, CO: Geological Society of America, Special Paper 349, pp. 3–12.

Morán-Zenteno, D.J., Cerca, M., & Keppie, J.D. (2007). The Cenozoic tectonic and magmatic evolution of southwestern Mexico: advances and problems of interpretation. In Alaniz-Alvarez, S.A. & Nieto-Samaniego, A.F. (eds.), *Geology of México: Celebrating the Centenary of the Geological Society of México*. Boulder, CO: Geological Society of America, Special Paper 422, pp. 71–91.

Morelli, R.M., Hartlaub, R.P., Ashton, K.E., & Andsell, K.M. (2009). Evidence for enrichment of subcontinental lithospheric mantle from Paleoproterozoic intracratonic magmas: geochemistry and U–Pb geochronology of Martin Group igneous rocks, western Rae craton, Canada. *Precambrian Research*, 175: 1–15.

Morgan, W.J. (1971). Convection plumes in the lower mantle. *Nature*, 230: 42–43.

Morgan, W.J. (1983). Hotspot tracks and the early rifting of the Atlantic. *Tectonophysics*, 94: 123–139.

Moser, D.E. & Heaman, L.M. (1997). Proterozoic zircon growth in Archean lower crustal xenoliths, southern Superior craton – a consequence of Matachewn ocean opening. *Contributions to Mineralogy and Petrology*, 128: 164–175.

Mudd, G.M. (2010). Global trends and environmental issues in nickel mining: sulfides versus laterites. *Ore Geology Reviews*, 38: 9–26.

Mudd, G.M. (2012). Key trends in the resource sustainability of platinum group elements. *Ore Geology Reviews*, 46: 106–117.

Mudd, G.M., Weng, Z., & Jowitt, S.M. (2013). A detailed assessment of global Cu resource trends and endowments. *Economic Geology*, 108: 1163–1183.

Muirhead, J.D., Airoldi, G., Rowland, J.V., & White, J.D.L. (2012). Interconnected sills and inclined sheet intrusions control shallow magma transport in the Ferrar large igneous province, Antarctica. *Geological Society of America Bulletin*, 124: 162–180.

Mukasa, S.B., Wilson, A.H., & Young, K.R. (2013). Geochronological constraints on the magmatic and tectonic development of the Pongola Supergroup (Central Region), South Africa. *Precambrian Research*, 224: 268–286.

Müller, S.G., Krapež, B., Barley, M.E., & Fletcher, I.R. (2005). Giant iron-ore deposits of the Hamersley province related to the breakup of Paleoproterozic Australia: new insights from in situ SHRIMP dating of baddeleyite from mafic intrusions. *Geology*, 33: 577–580.

Mundil, R., Ludwig, K.R., Metcalfe, I., & Renne, P.R. (2004). Age and timing of the Permian mass extinctions: U/Pb dating of closed-system zircons. *Science*, 305: 1760–1763.

Mungall, J.E., Ames, D.E., & Hanley, J. J., *et al.* (2004). Geochemical evidence from the Sudbury structure for crustal redistribution by large bolide impacts. *Nature*, 429: 546–548.

Mungall, J.E., Harvey, J.D., Balch, S.J., *et al.* (2010). Eagle's Nest: a magmatic Ni–Cu–PGE deposit in the James Bay Lowlands, Ontario, Canada. *Society of Economic Geologists Special Publication*, 15: 539–557.

Murphy, J.B., van Staal, C.R., & Keppie, J.D. (1999). Middle to late Paleozoic Acadian orogeny in the northern Appalachian: a Laramide-style plume-modified orogeny? *Geology*, 27: 653–656.

Murray, B.P., Busby, C.J., Ferrari, L., & Solari, L.A. (2014). Synvolcanic crustal extension during the mid-Cenozoic ignimbrite flare-up in the northern Sierra Madre Occidental, Mexico: evidence from the Guazapares Mining District region, western Chihuahua. *Geosphere*, 9: doi:10.1130/GES00862.1.

Mutter, J.C., Talwani, M., & Stoffa, P.L. (1982). Origin of seaward-dipping reflectors in oceanic crust off the Norwegian margin by subaerial sea-floor spreading. *Geology*, 10: 353–357.

Nakagawa, T. & Tackley, P.J. (2005). The interaction between the post-perovskite phase change and a thermo-chemical boundary layer near the core–mantle boundary. *Earth and Planetary Science Letters*, 238: 204–216.

Nakanishi, M., Sager, W.W., & Klaus, A. (1999). Magnetic lineations within Shatsky Rise, Northwest Pacific Ocean; implications for hot spot-triple junction interaction and oceanic plateau formation. *Journal of Geophysical Research*, 104: 7539–7556.

Naldrett, A.J. (1997). Key factors in the genesis of Noril'sk, Sudbury, Jinchuan, Voisey's Bay and other world-class Ni–Cu–PGE deposits: implications for exploration. *Australian Journal of Earth Sciences*, 44: 283–315.

Naldrett, A.J. (1999). World-class Ni–Cu–PGE deposits: key factors in their genesis. *Mineralium Deposita*, 34: 227–240.

Naldrett, A.J. (2004). *Magmatic Sulfide Deposits: Geology, Geochemistry and Exploration*. Heidelberg: Springer-Verlag.

Naldrett, A.J. (2010a). Secular variation of magmatic sulfide deposits and their source magmas. *Economic Geology*, 105: 669–688.

Naldrett, A.J. (2010b). From the mantle to the bank: the life of a Ni–Cu–(PGE) sulfide deposit. *South African Journal of Geology*, 113: 1–32.

Naldrett, A.J. & von Gruenewaldt, G. (1989). The association of PGE with chromite in layered intrusions and ophiolite complexes. *Economic Geology*, 84: 180–187.

Naldrett, A.J., Lightfoot, P.C., Fedorenko, V.A., Doherty, W., & Gorbachev, N.A. (1992). Geology and geochemistry of intrusions and flood basalts of the Noril'sk region, USSR, with implications for the origin of the Ni–Cu ores. *Economic Geology*, 87: 975–1004.

Neal, C.R., Mahoney, J.J., Kroenke, L.W., Duncan, R.A., & Petterson, M.G. (1997). The Ontong Java plateau. *AGU Geophysical Monograph*, 100: 183–216.

Neal, C.R., Mahoney, J., Kroenke, L.W., *et al.* (2002). Mantle sources and the highly variable role of continental lithosphere in basalt petrogenesis of the Kerguelen plateau and Broken Ridge LIP: results from ODP Leg 183. *Journal of Petrology*, 43: 1177–1205.

Nelson, C.E., Jerram, D.A., & Hobbs, R.W. (2009a). Flood basalt facies from borehole data: implications for prospectivity and volcanology in volcanic rifted margins. *Petroleum Geoscience*, 15: 313–324.

Nelson, C.E., Jerram, D.A., Single, R.T., & Hobbs, R.W. (2009b). Understanding the facies architecture of flood basalts and volcanic rifted margins and its effect on geophysical properties. *Faroe Islands Exploration Conference: Proceedings of the 2nd Conference, Annales Societatis Scientarium Faroensis, supplementum*, 48: 84–103.

Nelson, D.O., Morrison, D.A., & Phinney, W.C. (1990). Open-system evolution versus source control in basaltic magmas: Matachewan–Hearst dike swarm, Superior Province, Canada. *Canadian Journal of Earth Science*, 27: 767–783.

Nelson, D.R., Trendall, A.F., de Laeter, J.R., Grobler, N.J., & Fletcher, I.R. (1992). A comparative study of the geochemical and isotopic systematics of late Archaean flood basalts from the Pilbara and Kaapvaal cratons. *Precambrian Research*, 54: 231–257.

Neuendorf, K.K.E., Mehl, J.P., & Jackson, J.A., eds. (2005). *Glossary of Geology*. Virginia, VA: American Geological Institute.

Neuendorf, K.K.E., Mehl, J.P., Jr., & Jackson, J.A., eds. (2011). *Glossary of Geology*, 5th edition. Alexandria, VA: American Geological Institute.

Neukum, G., Basilevsky, A.T., Kneissl, T., *et al.* (2010). The geologic evolution of Mars: episodicity of resurfacing events and ages from cratering analysis of image data and correlation with radiometric ages of Martian meteorites. *Earth and Planetary Science Letters*, 294: 204–222.

Neumann, E.-R., Svensen, H., Galerne, C.Y., & Planke, S. (2011). Multistage evolution of dolerites in the Karoo Large Igneous Province, Central South Africa. *Journal of Petrology*, 52: 959–984.

Nicholls, J. & Gordon, T.M. (1994). Procedures for the calculation of axial ratios on Pearce element-ratio diagrams. *The Canadian Mineralogist*, 32: 969–977.

Nicholson, S.W. & Shirey, S.B. (1990). Midcontinent rift volcanism in the Lake Superior region: Sr, Nd, and Pb isotopic evidence for a mantle plume origin. *Journal of Geophysical Research*, 95 (B7): 10851–10868.

Nicholson, S.W., Cannon, W.F., & Schulz, K.J. (1992). Metallogeny of the Midcontinent rift system of North America. *Precambrian Research*, 58: 355–386.

Nicholson, S.W., Shirey, S.B., Schulz, K.J., & Green, J.C. (1997). Rift-wide correlation of 1.1 Ga Midcontinent rift system basalts: implications for multiple mantle sources during rift development. *Canadian Journal of Earth Sciences*, 34: 504–520.

Nikishin, A.M., Ziegler, P.A., Stephenson, R.A., *et al.* (1996). Late Precambrian to Triassic history of the East European craton: dynamics of sedimentary basin evolution. *Tectonophysics*, 268: 23–63.

Nikishin, A.M., Ziegler, P.A., Abbott, D., Brunet, M.-F., & Cloetingh, S. (2002). Permo-Triassic intraplate magmatism and rifting in Eurasia: implications for mantle plumes and mantle dynamics. *Tectonophysics*, 351: 3–39.

Nilsson, M.K.M., Klausen, M.B., Söderlund, U., & Ernst, R.E. (2013). Precise U–Pb ages and geochemistry of Paleoproterozoic mafic dykes from southern West Greenland: linking the North Atlantic and the Dharwar Cratons. *Lithos*, 174: 255–270.

Nokleberg, W.J., Bundtzen, T.K., Eremin, R.A., *et al.* (2005). Metallogenesis and tectonics of the Russian Far East, Alaska, and the Canadian Cordillera. *US Geological Survey Professional Paper*, 1697.

Nomade, S., Knight, K.B., Beutel, E., *et al.* (2007). Chronology of the Central Atlantic Magmatic Province: implications for the Central Atlantic rifting processes and the Triassic–Jurassic biotic crisis. *Palaeogeography, Palaeoclimatology, Palaeoecology*, 244: 326–344.

Nowicki, T., Crawford, B., Dyck, D., *et al.* (2004). The geology of kimberlite pipes of the Ekati property, Northwest Territories, Canada. *Lithos*, 76: 1–27.

Obrebski, M., Allen, R.M., Xue, M., & Hung, S.-H. (2010). Slab–plume interaction beneath the Pacific Northwest. *Geophysical Research Letters*, 37, L14305.

Ogg, J.G., Ogg, G., & Gradstein, F.M. (2008). *The Concise Geologic Time Scale*. Cambridge: Cambridge University Press.

Ohkouchi, N., Kawamura, K., Kajiwara, Y., *et al.* (1999). Sulfur isotope records around Livello Bonarelli (northern Apennines Italy) black shale at the Cenomanian–Turonian boundary. *Geology*, 27: 535–538.

Ojakangas, R.W., Morey, G.B., & Green, J.C. (2001). The Mesoproterozoic Midcontinent rift system, Lake Superior region, USA. *Sedimentary Geology*, 141–142: 421–442.

Okubo, C.H. & Schultz, R.A. (2004). Mechanical stratigraphy in the western equatorial region of Mars based on thrust fault-related fold topography and implications for near-surface volatile reservoirs. *Geological Society of America Bulletin*, 116: 594–605.

Okulitch, A.V. (2002). Geological Time Chart, 2002. *Geological Survey of Canada Open File 3040*.

Olsson, J.R., Söderlund, U., Klausen, M.B., & Ernst, R.E. (2010). U–Pb baddeleyite ages of major Archean dyke swarms and the Bushveld Complex, Kaapvaal Craton (South Africa): correlations to volcanic rift forming events. *Precambrian Research*, 183: 490–500.

Olsson, J.R., Söderlund, U., Hamilton, M.A., Klausen, M.B., & Helffrich, G.R. (2011). A late Archaean radiating dyke swarm as possible clue to the origin of the Bushveld Complex. *Nature Geoscience*, 4: 865–869.

Ovtcharova, M., Bucher, H., Schaltegger, U., *et al.* (2006). New Early to Middle Triassic U–Pb ages from South China: calibration with ammonoid biochronozones and implications for the timing of the Triassic biotic recovery. *Earth and Planetary Science Letters*, 243: 463–475.

Oyarzun, R., Doblas, M., López-Ruiz, J., & Cebra, J.M. (1997). Opening of the central Atlantic and asymmetric mantle upwelling phenomena: implications for long-lived magmatism in western North Africa and Europe. *Geology*, 25: 727–730.

Pálfy, J., Demény, A., Haas, J., *et al.* (2001). Carbon isotope anomaly and other geochemical changes at the Triassic–Jurassic boundary from a marine section in Hungary. *Geology*, 29: 1047–1050.

Pallister, J.S., Budahn, J.R., & Murchey, B.L. (1989). Pillow basalts of the Angayucham Terrane: Oceanic plateau and island crust accreted to the Brooks Range. *Journal of Geophysical Research*, 94: 15901–15923.

Palmer, A.R. & James, N.P. (1980). The Hawke Bay event: a circum Iapetus regression near the Lower–Middle Cambrian boundary. *Virginia Polytechnic Institute and State University Memoir*, 2: 15–18.

Palmer, H.C., Ernst, R.E., & Buchan, K.L. (2007). Magnetic fabric studies of the Nipissing sill province and Senneterre dykes, Canadian Shield, and implications for emplacement. *Canadian Journal of Earth Sciences*, 44: 507–528.

Pankhurst, R.J. & Rapela, C.R. (1995). Production of Jurassic rhyolite by anatexis of the lower crust of Patagonia. *Earth and Planetary Science Letters*, 134: 23–36.

Pankhurst, R.J., Leat, P.T., Sruoga, P., *et al.* (1998). The Chon Aike province of Patagonia and related rocks in West Antarctica: a silicic large igneous province. *Journal of Volcanology and Geothermal Research*, 81: 113–136.

Pankhurst, R.J., Riley, T.R., Fanning, C.M., & Kelley, S.P. (2000). Episodic silicic volcanism in Patagonia and the Antarctic Peninsula: chronology of magmatism associated with the break-up of Gondwana. *Journal of Petrology*, 41: 605–625.

Pankhurst, M.J., Schaefer, B.F., Betts, P.G., Phillips, N., & Hand, M. (2010). A Mesoproterozoic continental flood rhyolite province, the Gawler Ranges, Australia: the end member example of the Large Igneous Province clan. *Solid Earth Discussions*, 2: 251–274.

Pankhurst, M.J., Schaefer, B.F., & Betts, P.G. (2011). Geodynamics of rapid voluminous felsic magmatism through time. *Lithos*, 123: 92–101.

Parfitt, E.A. & Head, J.W. (1992). A survey of radial fracture systems on Venus. Abstract. *Lunar Planetary Science*, 23: 1029–1030.

Park, J.K., Buchan, K.L., & Harlan, S.S. (1995). A proposed giant radiating dyke swarm fragmented by the separation of Laurentia and Australia based on paleomagnetism of ca. 780 Ma mafic intrusions in western North America. *Earth and Planetary Science Letters*, 132: 129–139.

Parker, A.J., Rickwood, P.C., & Tucker, D.H., eds. (1990). *2nd International Dyke Conference Australia, Adelaide.* Rotterdam: Balkema, IGCP Project 257, Publication No. 23.

Parman, S.W. & Grove, T.L. (2005). Komatiites in the plume debate. In Foulger, G.R., Natland, J.H., Presnall, D.C., & Anderson, D.L. (eds.), *Plates, Plumes, and Paradigms.* Boulder, CO: Geological Society of America, Special Paper 388, pp. 249–256.

Parsiegla, N., Gohl, K., & Uenzelmann-Neben, G. (2008). The Agulhas Plateau: structure and evolution of a large igneous province and its impact on the Cretaceous environment. *Geophysical Journal International*, 174 (1): 336–350.

Parsons, T., Thompson, G.A., & Sleep, N.H. (1994). Mantle plume influence on the Neogene uplift and extension of the U.W. western Cordillera. *Geology*, 22: 83–86.

Pavlova, G.G. & Borisenko, A.S. (2009). The age of Ag–Sb deposits of central Asia and their correlation with other types of ore systems and magmatism. *Ore Geology Reviews*, 35: 164–185.

Pawley, M.J., Wingate, M.T.D., Kirkland, C.L., *et al.* (2012). Adding pieces to the puzzle: episodic crustal grown and a new terrane in the northeast Yilgarn Craton, Western Australia. *Australian Journal of Earth Sciences*, 59: 603–623.

Payne, J.L., Lehrmann, D.J., Wei, J., *et al.* (2004). Large perturbations of the carbon cycle during recovery from the end-Permian extinction. *Science*, 305: 506–509.

Paytan, A., Kastner, M., Campbell, D., & Thiemens, M.H. (2004). Seawater sulfur isotope fluctuations in the Cretaceous. *Science*, 304: 1663–1665.

Pearce, J.A. (1996a). A user's guide to basalt discrimination diagrams. In Wyman, D.A. (ed.), *Trace Element Geochemistry of Volcanic Rocks: Applications for Massive Sulphide Exploration.* Geological Association of Canada, Short Course Notes 12, pp. 79–113.

Pearce, J.A. (1996b). Sources and settings of granitic rocks. *Episodes*, 19: 120–125.

Pearce, J.A. (2007). "Life cycles" of volcanic arcs. In Bleeker, W. & Sylvester, P. (eds.), *The Pulse of the Earth & Planetary Evolution: Rates and Rhythms, Cycles and Cataclysms.* GAC NUNA Conference, Sudbury, 13–17 June 2007, Program & Abstract Volume, p. 48.

Pearce, J.A. (2008). Geochemical fingerprinting of oceanic basalts with applications to ophiolite classification and the search for Archean oceanic crust. *Lithos*, 100: 14–48.

Pearce, J.A. & Cann, J.R. (1973). Tectonic setting of basic volcanic rocks determined using trace element analyses. *Earth and Planetary Science Letters*, 19: 290–300.

Pearce, J.A. & Robinson, P.T. (2010). The Troodos ophiolitic complex probably formed in a subduction initiation, slab edge setting. *Gondwana Research*, 18: 60–81.

Pearce, J.A. & Stern, R.J. (2006). Origin of back-arc basin magmas: trace element and isotope perspectives. *AGU Geophysical Monograph*, 166: 63–86.

Pearce, J.A., Harris, N.W.B., & Tindle, A.G. (1984). Trace element determination diagrams for the tectonic interpretation of granitic rocks. *Journal of Petrology*, 25: 956–983.

Pearce, T.H. (1968). A contribution of the theory of variation diagrams. *Contributions to Mineralogy and Petrology*, 19: 142–157.

Pearce, T.H. (1987). The identification and assessment of spurious trends in Pearce-type ratio variation diagrams: a discussion of some statistical arguments. *Contributions to Mineralogy and Petrology*, 97: 529–534.

Pearson, D.G. (1999). The age of continental roots. *Lithos*, 48: 171–194.

Pearson, D.G., Parman, S.W., & Nowell, G.M. (2007). A link between large mantle melting events and continent growth seen in osmium isotopes. *Nature*, 449: 202–205.

Peate, D.W. (1997). The Paraná–Etendeka Province. *AGU Geophysical Monograph*, 100: 217–245.

Peate, D.W., Barker, A.K., Riishuus, M.S., & Andreasen, R. (2008). Temporal variations in crustal assimilation of magma suites in the East Greenland flood basalt province: tracking the evolution of magmatic plumbing systems. *Lithos*, 102: 179–197.

Pechersky, D.M., Lyubushin, A.A., & Sharonova, Z.V. (2010). Synchronicity of events in the core and on the Earth's surface: changes in the organic world and geomagnetic polarity through the Phanerozoic. *Izvestiya Physics of the Solid Earth (Fizika Zemli)*, 7: 64–74.

Peck, D.C., Keays, R.R., James, R.S., Chubb, P.T., & Reeves, S.J. (2001). Controls on the formation of contact-type platinum-group element mineralization in the East Bull Lake intrusion. *Economic Geology*, 96: 559–581.

Pedersen, G.B.M., Head, J., & Wilson, L. (2010). Formation, erosion and exposure of Early Amazonian dikes, dike swarms and possible subglacial eruptions in the Elysium Rise/Utopia Basin Region, Mars. *Earth and Planetary Science Letters*, 294: 424–439.

Peng, P. (2010). Reconstruction and interpretation of giant mafic dyke swarms: a case study of 1.78 Ga magmatism in the North China craton. In Kusky, T.M., Zhai, M.-G., & Xiao, W. (eds.), *The Evolving Continents: Understanding Processes of Continental Growth*. London: Geological Society of London, Special Publication 338, pp. 163–178.

Peng, P., Zhai, M.-G., Zhang, H.-F., & Guo, J.-H. (2005). Geochronological constraints on the Paleoproterozoic evolution of the North China craton: SHRIMP zircon ages of different types of mafic dykes. *International Geology Review*, 47: 492–508.

Peng, P., Zhai, M.-G., Ernst, R.E., *et al.* (2008). A 1.78 Ga large igneous province in the North China craton: the Xiong'er Volcanic Province and the North China dyke swarm. *Lithos*, 101: 260–280.

Peng, P., Bleeker, W., Ernst, R.E., Söderlund, U., & McNicoll, V. (2011). U–Pb baddeleyite ages, distribution and geochemistry of 925 Ma mafic dykes and 900 Ma sills in the North China craton: evidence for a Neoproterozoic mantle plume. *Lithos*, 127: 210–221.

Peng, P., Liu, F., Zhai, M.-G., & Guo, J.-H. (2012). Age of the Miyun dyke swarm: constraints on the maximum depositional age of the Changcheng System. *Chinese Science Bulletin*, 57: 105–110.

Percival, J.A. & Pysklywec, R.N. (2007). Are Archean lithospheric keels inverted? *Earth and Planetary Science Letters*, 254: 393–403.

Percival, J.A. & West, G.F. (1994). The Kapuskasing uplift: a geological and geophysical synthesis. *Canadian Journal of Earth Sciences*, 31: 1256–1286.

Pesonen, L.J., Mertanen, S., & Veikkolainen, T. (2012). Paleo-Mesoproterozoic Supercontinents: a paleomagnetic view. *Geophysica*, 48: 5–47.

Pesonen, L.J., Halls, H.C., & Mertanen, S. (2014). Precambrian supercontinents. *Precambrian Research*, 244: 1–4.

Peterson, D.M. & Peck, D.C. (2010). PGE-associated polymetallic deposits in the Mid-Continent Rift: an overview. Extended Abstract, 11th International Platinum Symposium, 21–24 June, Ontario Geological Survey, Miscellaneous Release–Data 269.

Peterson, T.D., van Breemen, O., Sandeman, H.A., & Cousens, B. (2002). Proterozoic (1.85–1.75 Ga) igneous suites of the Western Churchill Province: granitoid and ultrapotassic magmatism in a reworked Archean hinterland. *Precambrian Research*, 119: 73–100.

Peterson, T.D., Pehrsson, S., Jefferson, C., Scott, J., & Rainbird, R. (2010). The Dubawnt Supergroup. *LIP of the Month*, December 2010. See: www.largeigneousprovinces.org/10dec.

Petraske, A.K., Hodge, D.S., & Shaw, R. (1978). Mechanics of emplacement of basic intrusions. *Tectonophysics*, 46: 41–63.

Petterson, M.G. (2004). The geology of north and central Malaita, Solomon Islands: the thickest and most accessible part of the world's largest (Ontong Java) ocean plateau. In Fitton, J.G., Mahoney, J.J., Wallace, P.J., & Saunders, A.D. (eds.), *Origin and Evolution of the Ontong Java Plateau*. London: Geological Society, Special Publication 229, pp. 63–81.

Peucker-Ehrenbrink, B. & Ravizza, G. (2000). The marine osmium isotope record. *Terra Nova*, 12: 205–219.

Phillips, R.J. & Hansen, V.L. (1994). Tectonic and magmatic evolution of Venus. *Annual Review of Earth and Planetary Sciences*, 22: 597–654.

Phillips, R.J. & Hansen, V.L. (1998). Geological evolution of Venus: rises, plains, plumes and plateaus. *Science*, 279: 1492–1497.

Pierce, K.L. & Morgan, L.A. (1992). The track of the Yellowstone hotspot: volcanism, faulting and uplift. In Link, P.K., Kuntz, M.A., & Platt, L.B. (eds.), *Regional Geology of Eastern Idaho and Western Wyoming*. Boulder, CO: Geological Society of America, Memoir 179, pp. 1–53.

Piercey, S.J., Murphy, D.C., Mortensen, J.K., & Paradis, S. (2001). Boninite magmatism in a continental margin setting, Yukon–Tanana terrane, southeastern Yukon, Canada. *Geology*, 29: 731–734.

Pik, R., Deniel, C., Coulon, C., *et al.* (1998). The northwestern Ethiopian Plateau flood basalts: classification and spatial distribution of magma types. *Journal of Volcanology and Geothermal Research*, 81: 91–111.

Pik, R., Deniel, C., Coulon, C., Yirgu, G., & Marty, B. (1999). Isotopic and trace element signatures of Ethiopian basalts: evidence for plume–lithospheric interactions. *Geochimica et Cosmochimica Acta*, 63: 2263–2279.

Pilger, R.H., Jr. (2003). *Geokinematics: Prelude to Geodynamics*. Berlin: Springer-Verlag.

Piper, J.D.A. (1995). The palaeomagnetism of middle Proterozoic dyke swarms of the Gardar Province and Mesozoic dykes in SW Greenland. *Geophysical Journal International*, 120: 339–355.

Pirajno, F. (1994). Mineral resources of anorogenic alkaline complexes in Namibia. *Australian Journal of Earth Sciences*, 41: 157–168.

Pirajno, F. (2000). *Ore Deposits and Mantle Plumes*. Dordrecht: Kluwer Academic.

Pirajno, F. (2007). Alkaline intraplate volcanism in eastern Australia. *LIP of the Month*, August 2007. See: www.largeigneousprovinces.org/07aug.

Pirajno, F. & Bagas, L. (2008). A review of Australia's Proterozoic mineral systems and genetic models. *Precambrian Research*, 166: 54–80.

Pirajno, F. & Hoatson, D.M. (2012). A review of Australia's large igneous provinces and associated mineral systems: implications for mantle dynamics through geological time. *Ore Geology Reviews*, 48: 2–54.

Pirajno, F., Ernst, R.E., Borisenko, A.S., Fedoseev, G., & Naumov, E.A. (2009). Intraplate magmatism in Central Asia and China and associated metallogeny. *Ore Geology Reviews*, 35: 114–136.

Pisarevsky, S.A., Wingate, M.T.D., Powell, C. McA., Johnson, S., & Evans, D.A.D. (2003). Models of Rodinia assembly and fragmentation. In Yoshida, M., Windley, B.F., & Dasgupta, S. (eds.), *Proterozoic East Gondwana: Supercontinent Assembly and Breakup*. London: Geological Society of London, Special Publication 206, pp. 35–55.

Pisarevsky, S.A., Elming, S.-Å., Pesonen, L.J., & Li, Z.-X. (2014). Mesoproterozoic paleogeography: supercontinent and beyond. *Precambrian Research*, 244: 207–225.

Pitcairn, I.K. (2011). Background concentrations of gold in different rock types. *Applied Earth Science: Transactions of the IMM B*, 120: 31–38.

Planke, S., Symonds, P.A., Alvestad, E., & Skogseid, J. (2000). Seismic volcanostratigraphy of large-volume basaltic extrusive complexes on rifted margins. *Journal of Geophysical Research Solid Earth*, 105 (B8): 19335–19351.

Planke, S., Rasmussen, T., Rey, S.S., & Myklebust, R. (2005). Seismic characteristics and distribution of volcanic intrusions and hydrothermal vent complexes in the Vøring and Møre basins. In Doré, A.G. & Vining, B.A. (eds.), *Petroleum Geology: North-West Europe and Global Perspectives*. Proceedings of the 6th Petroleum Geology Conference (Geological Society of London), pp. 833–844.

Podmore, F. & Wilson, A.H. (1987). A reappraisal of the structure, geology and emplacement of the Great Dyke, Zimbabwe. In Halls, H.C. & Fahrig, W.F. (eds.), *Mafic Dyke Swarms*. Geological Association of Canada, Special Paper 34, pp. 317–330.

Poidevin, J.L. (1994). Boninite-like rocks from the Palaeoproterozoic greenstone belt of Bogoin, Central African Republic: geochemistry and petrogenesis. *Precambrian Research*, 68: 97–113.

Polat, A. & Kerrich, R. (1999). Formation of an Archean tectonic melange in the Schreiber-Hemlo greenstone belt, Superior Province, Canada: implications for Archean subduction-accretion processes. *Tectonics*, 18: 733–755.

Polat, A. & Kerrich, R. (2006). Reading the geochemical fingerprints of Archean hot subduction volcanic rocks: evidence for accretion and crustal recycling in a mobile tectonic regime. In Benn, K., Mareschal, J.-C., & Condie, K.C. (eds.), *Archean Geodynamics and Environments. AGU Geophysical Monograph*, 164: 184–213.

Pollard, D.D. (1987). Elementary fracture mechanics applied to the structural interpretation of dykes. In Halls, H.C., & Fahrig, W.F. (eds.), *Mafic Dyke Swarms*. Geological Association of Canada, Special Paper 34, pp. 5–24.

Pollard, D.D., Muller, O.H., & Dockstader, D.R. (1975). The form and growth of fingered sheet intrusions. *Geological Society of America Bulletin*, 86 (3): 351–363.

Polteau, S., Ferré, E.C., Planke, S., Neumann, E.-R., & Chevallier, L. (2008a). How are saucer-shaped sills emplaced? Constraints from the Golden Valley Sill, South Africa. *Journal of Geophysical Research*, 113: B12104.

Polteau, S., Mazzini, A., Galland, O., Planke, S., & Malthe-Sorenssen, A. (2008b). Saucer-shaped intrusions: occurrences, emplacement and implications. *Earth and Planetary Science Letters*, 266: 195–204.

Polyakov, G.V., Tolstykh, N.D., Mekhonoshin, A.S., *et al.* (2013). Ultramafic–mafic igneous complexes of the Precambrian East Siberian metallogenic province (southern framing of the Siberian craton): age, composition, origin, and ore potential. *Russian Geology and Geophysics*, 54: 1319–1331.

Pradhan, V.R., Meert, J.G., Pandit, M.K., Kamenov, G., & Mondal, Md.E.A. (2012). Paleomagnetic and geochronological studies of the mafic dyke swarms of Bundelkhand craton, central India: implications for the tectonic evolution and paleogeographic reconstructions. *Precambrian Research*, 198–199: 51–76.

Prendergast, M.D. (2004). The Bulawayan Supergroup: a late Archaean passive margin-related large igneous province in the Zimbabwe craton. *Journal of the Geological Society*, London, 161: 431–445.

Prendergast, M.D. (2012). The Molopo Farms Complex, southern Botswana – a reconsideration of structure, evolution, and the Bushveld connection. *South African Journal of Geology*, 115: 77–90.

Price, M. (1995). *Tectonic and Volcanic Map of Venus*. Princeton, NJ: Deptartment of Geological Sciences, Princeton University.

Price, M. & Suppe, J. (1995). Constraints on the resurfacing history of Venus from the hypsometry and distribution of volcanism, tectonism, and impact craters. *Earth, Moon and Planets*, 71: 99–145.

Price, M.H., Watson, G., Suppe, J., & Brankman, C. (1996). Dating volcanism and rifting on Venus using impact crater densities. *Journal of Geophysical Reseearch*, 101, 4657–4671.

Prokoph, A. & Veizer, J. (1999). Trends, cycles and nonstationarities in isotope signals of Phanerozoic seawater. *Chemical Geology*, 161: 225–240.

Prokoph, A., Ernst, R.E., & Buchan, K.L. (2004a). Time-series analysis of Large Igneous Provinces: 3500 Ma to present. *Journal of Geology*, 112: 1–22.

Prokoph, A., Rampino, M.R., & El Bilali, H. (2004b). Periodic components in the diversity of calcareous plankton and geological events over the past 230 Myr. *Palaeogeography, Palaeoclimatology, Palaeoecology*, 207: 105–125.

Prokoph, A., Shields, G.A., & Veizer, J. (2008). Compilation and time-series analysis of a marine carbonate $\delta^{18}O$, $\delta^{13}C$, $^{87}Sr/^{86}Sr$ and $\delta^{34}S$ database through Earth history. *Earth-Science Reviews*, 87: 113–133.

Prokoph, A., El Bilali, H., & Ernst, R. (2013). Periodicities in the emplacement of large igneous provinces through the Phanerozoic: relations to ocean chemistry and marine biodiversity evolution. *Geoscience Frontiers*, 4: 263–276.

Puchkov, V.N. (2012). Dyke swarms and associated magmatic complexes. *Geotectonics*, 1: 42–53.

Puchkov, V.N., Bogdanova, S.V., Ernst, R.E., *et al.* (2013). The *c.* 1380 Ma Mashak igneous event of the Southern Urals. *Lithos*, 174: 109–124.

Puffer, J.H. (2001). Contrasting HFSE contents of plume sourced and reactivated arc-sourced continental flood basalts. *Geology*, 29: 675–678.

Puffer, J.H. (2002). A Late Neoproterozoic Eastern Laurentian superplume: location, size, chemical composition and environmental impact. *American Journal of Science*, 302: 1–27.

Puffer, J.H. (2003). A reactivated back-arc source for CAMP Magma. *AGU Geophysical Monograph*, 136: 151–162.

Queen, M., Heaman, L.M., Hanes, J.A., Archibald, D.A., & Farrar, E. (1996). $^{40}Ar/^{39}Ar$ phlogopite and U–Pb perovskite dating of lamprophyre dykes from the eastern Lake Superior region: evidence for a 1.14 Ga magmatic precursor to Midcontinent rift volcanism. *Canadian Journal of Earth Sciences*, 33: 957–965.

Racki, G. (2010). Climate changes caused by degassing of sediments during the emplacement of large igneous provinces: comment. *Geology*, 38: e210.

Radhakrishna, T., Joseph, M., Thampi, P.K., & Mitchell, J.G. (1990). Phanerozoic mafic dyke intrusions from the high grade terrain of southwestern India: K–Ar isotope and geochemical implications. In Parker, A.J., Rickwood, P.C., Tucker, D.H. (eds.), *Mafic Dykes and Emplacement Mechanisms*. Rotterdam: Balkema, pp. 363–372.

Rainbird, R.H. (1993). The sedimentary record of mantle plume uplift preceding eruption of the Neoproterozoic Natkusiak flood basalt. *Journal of Geology*, 101: 305–318.

Rainbird, R.H. (1998). Bedrock and surficial geology, Wynniatt Bay, District of Franklin, Northwest Territories. *Geological Survey of Canada Open File 3671*.

Rainbird, R.H. & Ernst, R.E. (2001). The sedimentary record of mantle-plume uplift. In Ernst, R.E. & Buchan, K.L. (eds.), *Mantle Plumes: Their Identification through Time*. Boulder, CO: Geological Society of America, Special Paper 352, pp. 227–245.

Rainbird, R.H., Stern, R.A., Khudoley, A.K., *et al.* (1998). U–Pb geochronology of Riphean sandstone and gabbro from southeast Siberia and its bearing on the Laurentia-Siberia connection. *Earth and Planetary Science Letters*, 164: 409–420.

Rajesh, H.M., Chisonga, B.C., Shindo, K., Beukes, N.J., & Armstrong, R.A. (2013). Petrographic, geochemical and SHRIMP U–Pb titanite age characterization of the Thabazimbi mafic sills: extended time frame and a unifying petrogenetic model for the Bushveld Large Igneous Province. *Precambrian Research*, 230: 79–102.

Ramos Rosique, A. (2012). Timing and evolution of Late Oligocene to Early Miocene magmatism and epithermal mineralization in the central Bolaños graben, southern Sierra Madre Occidental, México. Ph.D. Thesis, Kingston University, UK.

Rampino, M.R. & Caldeira, K. (1993). Major episodes of geologic change: correlations, time structure and possible causes. *Earth and Planetary Science Letters*, 114: 215–227.

Rampino, M.R. & Stothers, R.B. (1988). Flood basalt volcanism during the past 250 million years. *Science*, 241: 663–668.

Rasmussen, B., Fletcher, I.R., Muhling, J.R., Mueller, A.G., & Hall, G.C. (2007). Bushveld-aged fluid flow, peak metamorphism, and gold mobilization in the Witwatersrand basin, South Africa: constraints from *in situ* SHRIMP U–Pb dating of monazite and xenotime. *Geology*, 35: 931–934.

Rasmussen, B., Mueller, A.G., & Fletcher, I.R. (2009). Zirconolite and xenotime U–Pb age constraints on the emplacement of the Golden Mile Dolerite sill and gold mineralization at the Mt Charlotte mine, Eastern Goldfields province, Yilgarn craton, Western Australia. *Contributions to Mineralogy and Petrology*, 157: 559–572.

Ray, J.S. & Pande, K. (1999) Carbonatite alkaline magmatism associated with continental flood basalts at stratigraphic boundaries: causes for mass extinctions. *Geophysical Research Letters*, 26: 1917–1920.

Rehkämper, M., Halliday, A.N., Barfod, D., Fitton, J.G., & Dawson, J.B. (1997). Platinum-Group Element abundance patterns in different mantle environments. *Science*, 278: 1595–1598.

Reichardt, F.J. (1994). The Molopo Farms Complex, Botswana: history, stratigraphy, petrography, petrochemistry, and Ni-Cu-PGE mineralisation. *Exploration and Mining Geology*, 3: 264–284.

Reichow, M.K., Saunders, A.D., White, R.V, *et al.* (2002). $^{40}Ar/^{39}Ar$ dates from the West Siberian basin: Siberian flood basalt province doubled. *Science*, 296: 1846–1849.

Reichow, M.K., Pringle, M.S., Al'Mukhamedov, A.I., *et al.* (2009) The timing and extent of the eruption of the Siberian Traps large igneous province: implications for the end-Permian environmental crisis. *Earth and Planetary Science Letters*, 277: 9–20.

Reidel, S.P. & Hooper, P.R., eds. (1989). *Volcanism and Tectonism in The Columbia River Flood-Basalt Province*. Boulder, CO: Geological Society of America, Special Paper 239.

Reidel, S.P., Camp, V.E., Ross, M.E., *et al.*, eds. (2013). *The Columbia River Flood Basalt Province*. Boulder, CO: Geological Society of America, Special Paper 497.

Reis, N.J., Teixeira, W., Hamilton, M.A., *et al.* (2013). Avanavero mafic magmatism, a late Paleoproterozoic LIP in the Guiana Shield, Amazonian Craton: U–Pb ID-TIMS baddeleyite, geochemical and paleomagnetic evidence. *Lithos*, 174: 175–195.

Renne, P.R. & Basu, A.R. (1991). Rapid eruption of the Siberian Traps flood basalts at the Permo-Triassic boundary. *Science*, 253: 176–179.

Renne, P.R., Ernesto, M., Pacca, I.G., *et al.* (1996). Age of the Ponta Grossa dike swarm (Brazil), and implications for Parana flood volcanism. *Earth and Planetary Science Letters*, 144: 199–211.

Retallack, G.J. (2001). A 300-million year record of atmospheric carbon dioxide from fossil plant cuticles. *Nature*, 411: 287–290.

Retallack, G.J. (2010). Lateritization and bauxitization events. *Economic Geology*, 105: 655–667.

Reynaud, C., Jaillard, É., Lapierre, H., Mamberti, M., & Mascle, G.H. (1999). Oceanic plateau and island arcs of southwestern Ecuador: their place in the geodynamic evolution of northwestern South America. *Tectonophysics*, 307: 235–254.

Rhode, R.A. & Muller, R.A. (2005). Cycles in fossil diversity. *Nature*, 434: 208–210.

Richards, M.A., Duncan, R.A., & Courtillot, V.E. (1989). Flood basalts and hotspot tracks – plume heads and tails. *Science*, 246: 103–107.

Richards, M.A., Jones, D.L., Duncan, R.A., & DePaolo, D.J. (1991). A mantle plume initiation model for the Wrangellia flood basalt and other oceanic plateaus. *Science*, 254: 263–267.

Richardson, S.H. & Shirey, S.B. (2008). Continental mantle signature of Bushveld magmas and coeval diamonds. *Nature*, 453: 910–913.

Richardson, W.P., Okal, E.A., & Vanderlee, S. (2000). Rayleigh-wave tomography of the Ontong-Java Plateau. *Physics of the Earth and Planetary Interiors*, 118: 29–51.

Rickwood, P.C. (1990). The anatomy of a dyke and the determination of propagation and magma flow directions. In Parker, A.J., Rickwood, P.C., & Tucker, D.H. (eds.), *Mafic Dykes and Emplacement Mechanisms*. Rotterdam: Balkema, pp. 81–100.

Riley, T.R., Leat, P.T., Pankhurst, R.J., & Harris, C. (2001). Origins of large volume rhyolitic volcanism in the Antarctic Peninsula and Patagonia by crustal melting. *Journal of Petrology*, 42: 1043–1065.

Riley, T.R., Leat, P.T., Storey, B.C., Parkinson, I.J., & Milar, I.L. (2003). Ultramafic lamprophyres of the Ferrar large igneous province: evidence for a HIMU mantle component. *Lithos*, 66: 63–76.

Riley, T.R., Leat, P.T., Curtis, M.L., Millar, I.L., & Fazel, A. (2005). Early-Middle Jurassic dolerite dykes from western Dronning Maud land (Antarctica): identifying mantle sources in the Karoo large igneous province. *Journal of Petrology*, 46: 1489–1524.

Riley, T.R., Curtis, M.L., Leat, P.T., *et al.* (2006). Overlap of Karoo and Ferrar magma types in KwaZulu Natal, South Africa. *Journal of Petrology*, 47: 541–566.

Rioux, M., Bowring, S., Dudás, F., & Hanson, R. (2010). Characterizing the U–Pb systematics of baddeleyite through chemical abrasion: application of multi-step digestion methods to baddeleyite geochronology. *Contributions to Mineralogy and Petrology*, 160: 777–801.

Rivers, T. & Corrigan, D. (2000). Convergent margin on southeastern Laurentia during the Mesoproterozoic: tectonic implications. *Canadian Journal of Earth Sciences*, 37: 359–383.

Robock, A. (2000). Volcanic eruptions and climate. *Reviews of Geophysics*, 38: 191–219.

Robock, A. & Oppenheimer, C., eds. (2003). Volcanism and the Earth's atmosphere. *AGU Geophysical Monograph*, 139.

Rock, N.M.S. (1991). *Lamprophyres*. Glasgow: Blackie.

Rogers, C., Ernst, R., Cousens, B., Gower, C., & Harlan, S. (2013). Characterization of distinct nodes of *ca*. 1460 Ma intraplate magmatism. Poster, Belt Symposium V, July 31–August 4, Salmon, ID, USA.

Rogers, J.J.W. & Santosh, M. (2004). *Continents and Supercontinents*. New York: Oxford University Press.

Rogers, N.W. (2006). Basaltic magmatism and the geodynamics of the East African Rift System. In Yirgu, G., Ebinger, C.J., & Maguire, P.K.H. (eds.), *The Afar Volcanic Province within the East African Rift System*. London: Geological Society of London, Special Publication 259, pp. 77–93.

Rohde, R.A., & Muller, R.A. (2005). Cycles in fossil diversity. *Nature*, 434: 208–210

Rohrman, M. (2013). Greater Exmouth LIP (NW Australia). *LIP of the Month*, November 2013. See http://www.largeigneousprovinces.org/13nov.

Rollinson, H. (1993). *Using Geochemical Data: Evaluation, Presentation, Interpretation*. Harlow: Longman.

Rollinson, H. & Adetunji, J. (2013). Mantle podiform chromitites do not form beneath mid-ocean ridges: a case study from the Moho transition zone of the Oman ophiolite. *Lithos*, 177: 314–327.

Rollinson, H.R. & Roberts, C.R. (1986). Ratio correlations and major element mobility in altered basalts and komatiites. *Contributions to Mineralogy and Petrology*, 93: 89–97.

Romu, K.R.I., Luttinen, A.V., & O'Brien, E. (2008). Lamproite–orangeite transition in 159 Ma dykes of Dronning Maud Land, Antarctica? Extended abstract, 9th International Kimberlite Conference, No. 91KC-A-00362.

Rosen, O.M., Manakov, V., & Gorev, N.I. (2007). Siberian superplume – study of kimberlite distribution in time and space. In *Large Igneous Provinces of Asia: Mantle Plumes and Metallogeny. Abstracts of the International Symposium 13–16 August 2007*. Novosibirsk, Russia: Novosibirsk Publishing House, pp. 58–60.

Ross, G.M. & Eaton, D.W. (1997). Winagami reflection sequence: seismic evidence for postcollisional magmatism in the Proterozoic of Western Canada. *Geology*, 25: 199–202.

Ross, P.-S., Ukstins Peate, I., McClintock, M.K., *et al.* (2005). Mafic volcaniclastic deposits in flood basalt provinces: a review. *Journal of Volcanology and Geothermal Research*, 145: 281–314.

Rubin, A.M. (1995). Propagation of magma-filled cracks. *Annual Reviews of Earth and Planetary Science*, 23: 287–336.

Rudnick, R.L. & Fountain, D.M. (1995). Nature and composition of the continental crust: a lower crustal perspective. *Reviews of Geophysics*, 33: 267–309.

Rudnick, R.L. & Gao, S. (2003). Composition of the continental crust. In Rudnick, R.L. (ed.), *The Crust*. New York: Elsevier. Treatise on Geochemistry, volume 3, pp.1–64.

Rukhlov, A.S. & Bell, K. (2010). Geochronology of carbonatites from the Canadian and Baltic Shields, and the Canadian Cordillera: clues to mantle evolution. *Mineralogy and Petrology*, 98: 11–54.

Ruppell, C. (1995). Extensional processes in continental lithosphere. *Journal of Geophysical Research*, 100: 24187–24215.

Russell, J.K., & Stanley, C.R., eds. (1990). Theory and application of Pearce Element Ratios to geochemical data analysis. *Geological Association of Canada Short Course*, vol. 8.

Russell, J.K., Porritt, L.A., Lavallée, Y., & Dingwell, D.B. (2012). Kimberlite ascent by assimilation-fuelled buoyancy. *Nature*, 481: 352–356.

Russell, S.M. & Whitmarsh, R.B. (2003). Magmatism at the west Iberia non-volcanic rifted continental margin: evidence from analyses of magnetic anomalies. *Geophysical Journal International*, 154: 706–730.

Ryabchikov, I.D. (2003). High NiO content in mantle-derived magmas as evidence for material transfer from the Earth's core. *Doklady Earth Science*, 389: 437–439.

Safonova, I. Yu. (2009). Intraplate magmatism and oceanic plate stratigraphy of the Paleo-Asian and Paleo-Pacific oceans from 600 to 140 Ma. *Ore Geology Reviews*, 35: 137–154.

Sage, R.P. (1991). Alkalic rock, carbonatite and kimberlite complexes of Ontario, Superior Province, In Thurston, P.C., Williams, H.R., Sutcliffe, R.H., & Stott, G.M. (eds.), *Geology of Ontario, Part 1*. Ontario: Geological Survey (Sudbury ON), Special Volume, Part 1, pp. 683–709.

Sager, W.W. (2005). What built Shatsky Rise, a mantle plume or ridge tectonics? In Foulger, G.R., Natland, J.H., Presnall, D.C., & Anderson, D.L. (eds.), *Plates, Plumes, and Paradigms*. Boulder, CO: Geological Society of America, Special Paper 388, pp. 721–733.

Sager, W.W., Sano, T., & Geldmacher, J. (2011). How do oceanic plateaus form? Clues from drilling at Shatsky Rise. *EOS Transactions of the American Geological Society*, 92: 37–44.

Sager, W.W., Zhang, J.C., Korenaga, J., Sano, T., *et al.* (2013). An immense shield volcano within the Shatsky Rise oceanic plateau, northwest Pacific Ocean. *Nature Geoscience*, 6: 976–981.

Sahabi, M., Aslanian, D., & Olivet, J.-L. (2004). Un nouveau point de départ pour l'histoire de l'Atlantique central (a new starting point for the history of the central Atlantic). *Comptes rendus Geoscience*, 336: 1041–1052.

Said, N. & Kerrich, R. (2010). Elemental and Nd isotope systematics of the Upper Basalt Unit, 2.7 Ga Kambalda Sequence: quantitative modeling of progressive crustal contamination of plume athenosphere. *Chemical Geology*, 273: 193–211.

Said, N., Kerrich, R., Cassidy, K., & Champion, D.C. (2012). Characteristics and geodyamic setting of the 2.7 Ga Yilgarn heterogenous plume and its interaction with contintal lithosphere: evidence from komatiitic basalt and basalt geochemistry of the Eastern Goldfields Superterrane. *Australian Journal of Earth Sciences*, 59: 737–763.

Salminen, J., Halls, H.C., Mertanen, S., *et al.* (2014). Paleomagnetic and geochronological studies on Paleoproterozoic diabase dykes of Karelia, East Finland: key for testing the Superia supercraton. *Precambrian Research*, 244: 87–99.

Sami, K. (1996). Evaluation of the variations in borehole yield from a fractured Karoo Aquifer, South Africa. *Ground Water*, 34: 114–120.

Samuel, H. & Bercovici, D. (2006). Oscillating and stagnating plumes in the Earth's lower mantle. *Earth and Planetary Science Letters*, 248: 90–105.

Sandeman, H.A. & Ryan, J.J. (2008). The Spi Lake Formation of the central Hearne domain, western Churchill Province, Canada: an axial intracratonic continental tholeiite trough above the cogenetic Kaminak dyke swarm. *Canadian Journal of Earth Sciences*, 45: 745–767.

Sandeman, H.A., Heaman, L.M., & LeCheminant, A.N. (2013). The Paleoproterozoic Kaminak dykes, Hearne craton, western Churchill Province, Nunavut, Canada: preliminary constraints on their age and petrogenesis. *Precambrian Research*, 232: 119–139.

Sanei, H., Grasby, S.E., & Beauchamp, B. (2012). Latest Permian mercury anomalies. *Geology*, 40: 63–66.

Sano, T. & Yamashita, S. (2004). Experimental petrology of basement lavas from Ocean Drilling Program Leg 192: implications for differentiation processes in Ontong Java Plateau magmas. In Fitton, J.G., Mahoney, J.J., Wallace, P.J., & Saunders, A.D. (eds.), *Origin and Evolution of the Ontong Java Plateau.* London: Geological Society of London, Special Publication 229, pp. 185–218.

Sasada, T., Hiyagon, H., Bell, K., & Ebihara, M. (1997). Mantle-derived noble gases in carbonatites. *Geochimica et Cosmochimica Acta*, 61: 4219–4228.

Saunders, A.D. (2005). Large igneous provinces: origin and environmental consequences. *Elements*, 1: 259–263.

Saunders, A.D. & Reichow, M.K. (2009). The Siberian Traps and the end-Permian mass extinction: a critical review. *Chinese Science Bulletin*, 54: 20–37.

Saunders, A.D., Storey, M., Kent, R.W., & Norry, M.J. (1992). Consequences of plume–lithosphere interactions. In Alabaster, T., Storey, B.C., & Pankhurst, R.J. (eds.), *Magmatism and the Causes of Continental Break-up.* London: Geological Society of London, Special Publication 68, pp. 41–60.

Saunders, A.D., Tarney, J., Kerr, A.C., & Kent, R.W. (1996). The formation and fate of large igneous provinces. *Lithos*, 37: 81–95.

Saunders, A.D., Fitton, J.G., Kerr, A.C., Norry, M.J., & Kent, R.W. (1997). The North Atlantic Igneous Province. *AGU Geophysical Monograph*, 100: 45–93.

Saunders, A.D., England, R.W., Reichow, M.K., & White, R.V. (2005). A mantle plume origin for the Siberian traps: uplift and extension in the West Siberian Basin, Russia. *Lithos*, 79, 407–424.

Saunders, A.D., Jones, S.M., Morgan, L.A., *et al.* (2007). Regional uplift associated with continental large igneous provinces: the role of mantle plumes and the lithosphere. *Chemical Geology*, 241: 282–318.

Schaber, G.G. (1973). Lava flows in Mare Imbrium: geologic evaluation from *Apollo* orbital photography. Houston, TX: Lunar and Planetary Institute, Lunar and Planetary Science Conference IV, pp. 73–92.

Schaefer, B.F., Pearson, D.G., Rogers, N.W., & Barnicoat, A.C. (2010). Re–Os isotope and PGE constraints on the timing and origin of gold mineralization in the Witwatersrand basin. *Mineralium Deposita*, 276: 88–94.

Schissel, D. & Smail, R. (2001). Deep-mantle plumes and ore deposits. In Ernst, R.E. & Buchan, K.L. (eds.), *Mantle Plumes: Their Identification through Time*. Boulder, CO: Geological Society of America, Special Paper 352, pp. 291–322.

Schmidt, J.M. & Rogers, R.K. (2007). Metallogeny of the Nikolai large igneous province (LIP) in southern Alaska and its influence on the mineral potential of the Talkeetna Mountains. In Ridgway, K.D., Trop, J.M., Glenn, J.M.G., & O'Neill, J.M. (eds.), *Tectonic Growth of a Collisional Continental Margin*. Boulder, CO: Geological Society of America, Special Paper 431, pp. 623–648.

Schmincke, H.-U. (1982). Volcanic and chemical evolution of the Canary Islands. In von Rad, U., Hinz, K., Sarnthein, M., Siebold, E. (eds.), *Geology of the Northwest African Continental Margin*. Berlin: Springer-Verlag, pp. 273–306.

Schoene, B., Guex, J., Bartolini, A., Schaltegger, U., & Blackburn, T.J. (2010). Correlating the end-Triassic mass extinction and flood basalt volcanism at the 100 ka level. *Geology*, 38: 387–390.

Schofield, N. (2009) Linking sill morphology to emplacement mechanisms. Ph.D. Thesis, University of Birmingham.

Schofield, N., Stevenson, C., & Reston, T. (2010). Magma fingers and host rock fluidization in the emplacement of sills. *Geology*, 38: 63–66.

Schofield, N.J., Brown, D.J., Magee, C., & Stevenson, C.T. (2012) Sill morphology and comparison of brittle and non-brittle emplacement mechanism. *Journal of the Geological Society, London*, 169: 127–141.

Schofield, N., Alsop, I., Warren, J., *et al.* (2014). Mobilizing salt: magma–salt interactions. *Geology*, doi:10.1130/G35406.1.

Scholtz, D.L. (1936). The magmatic nikeliferous ore deposits of East Griqualand and Pondoland. *Transactions of the Geological Society of South Africa*, 39, 210 pp + 37 plates.

Schott, B. & Schmeling, H. (1998). Delamination and detachment of a lithospheric root. *Tectonophysics*, 296: 225–247.

Schott, B. & Yuen, D.A. (2004). Influences of dissipation and rheology on mantle plumes coming from the D''-layer. *Physics of the Earth and Planetary Interiors*, 146: 139–145.

Schubert, G., Turcotte, D.L., & Olson, P. (2001). *Mantle Convection in the Earth and Planets*. Cambridge: Cambridge University Press.

Schubert, G., Masters, G., Olson, P., & Tackley, P. (2004). Superplumes or plume clusters? *Physics of the Earth and Planetary Interiors*, 146: 147–162.

Schultz, P.H. (1976). Floor-fractured lunar craters. *Moon*, 15: 241–273.

Schultz, P.H. & Glicken, H. (1979). Impact crater and basin control of igneous processes on Mars. *Journal of Geophysical Research*, 84: 8033–8047.

Schultz, R.A., Soliva, R., Okubo, C.H., & Mège, D. (2010). Fault populations. In Watters, T.R. & Schultz, R.A. (eds.), *Planetary Tectonics*. Cambridge: Cambridge University Press, pp. 457–510.

Scoates, J.S., Wall, C.J., Friedman, R.M., Van Tongeren, J.A., & Mathez, E.A. (2012). Age of the Bushveld Complex. Abstract, 22nd V.M. Goldschmidt Conference "Earth in Evolution," 24–29 June, Montreal, Quebec, Canada.

Self, S., Thordarson, T., & Keszthelyi, L. (1997). Emplacement of continental flood basalt lava flows. *AGU Geophysical Monograph*, 100: 381–410.

Self, S., Widdowson, M., Thordarson, T., & Jay, A.E. (2006). Volatile fluxes during flood basalt eruptions and potential effects on the global environment: a Deccan perspective. *Earth and Planetary Science Letters*, 248: 518–532.

Self, S., Jay, A.E., Widdowson, M., & Keszthelyi, L.P. (2008). Correlation of the Deccan and Rajahmundry Trap lavas: are these the longest and largest lava flows on Earth? *Journal of Volcanology and Geothermal Research*, 172: 2–19.

Şengör, A.M.C. (1995). Sedimentation and tectonics of fossil rifts. In Busby, C.J., & Ingersoll, R.-V. (eds.), *Tectonics of Sedimentary Basins*. Oxford: Blackwell, pp. 53–117.

Şengör, A.M.C. (2001). Elevation as indicator of mantle-plume activity. In Ernst, R.E. & Buchan, K.L. (eds.), *Mantle Plumes: Their Identification through Time*. Boulder, CO: Geological Society of America, Special Paper 352, pp. 183–225.

Şengör, A.M.C. & Natal'in, B.A. (2001). Rifts of the world. In Ernst, R.E. and Buchan, K.L. (eds.), *Mantle Plumes: Their Identification through Time*. Boulder, CO: Geological Society of America, Special Paper 352, pp. 389–482.

Senske, D.A., Schaber, G.G., & Stofan, E.R. (1992). Regional topographic rises on Venus: geology of western Eistla Regio and comparison to Beta Regio and Atla Regio. *Journal of Geophysical Research*, 97: 13395–13420.

Sepkoski, J. (2002). A compendium of fossil animal genera. In Jablonski, D. & Foote, M. (eds.), *Bulletin of American Paleontology no. 363*. Ithaca, NY: Paleontological Research Institution.

Sharkov, E.V., Evseeva, K.A., Krassivskaya, I.S., & Chistyakov, A.V. (2005). Magmatic systems of the early Paleoproterozoic large Baltic igneous province of siliceous high-magnesian (boninite-like) series. *Russian Geology and Geophysics*, 46: 952–965.

Sharma, K.K. (2005). Malani magmatism: an extensional lithospheric tectonic origin. In Foulger, G.R., Natland, J.H., Presnall, D.C., & Anderson, D.L. (eds.), *Plates, Plumes, and Paradigms*. Boulder, CO: Geological Society of America, Special Paper 388, pp. 463–476.

Sharma, M. (1997). Siberian Traps. *AGU Geophysical Monograph*, 100: 273–295.

Shaviv, N. & Veizer, J. (2003). Celestial driver of Phanerozoic climate. *GSA Today*, 13: 4–10.

Shearer, C.K., Hess, P.C., Wieczorek, M.A., *et al.* (2006). Thermal and magmatic evolution of the Moon. *Reviews of Mineralogy and Geochemistry*, 60: 365–518.

Sheehan, P.M. (2001). The Late Ordovician mass extinction. *Annual Review of Earth and Planetary Science*, 29: 331–364.

Shellnutt, J.G. (2014). The Emeishan large igneous province: a synthesis. *Geoscience Frontiers*, 5: 369–394.

Shellnutt, J.G. & MacRae, N.D. (2012). Petrogenesis of the Mesoproterozoic (1.23 Ga) Sudbury dyke swarm and its questionable relationship to plate separation. *International Journal of Earth Science*, 101: 3–23.

Shellnutt, J.G., Denyszyn, S.W., & Mundil, R. (2012). Precise age determination of mafic and felsic intrusive rocks from the Permian Emeishan large igneous province (SW China). *Gondwana Research*, 22: 118–126.

Shen, Y., Soloman, S.C., Bjarnason, I.T., *et al.* (2002). Seismic evidence for a tilted mantle plume and north–south mantle flow beneath Iceland. *Earth and Planetary Science Letters*, 197: 261–272.

Shen J., Algeo, T.J., Hu, Q., *et al.* (2012). Negative C-isotope excursions at the Permian-Triassic boundary linked to volcanism. *Geology*, 40: 963–966.

Shervais, J.W. (1982). Ti–V plots and the petrogenesis of modern and ophiolitic lavas. *Earth and Planetary Science Letters*, 59: 101–118.

Sheth, H. (2007). Large Igneous Provinces (LIPs): definition, recommended terminology, and a hierarchical classification. *Earth Science Reviews*, 85: 117–124.

Sheth, H.C. & Vanderkluysen, L. (2014). Flood basalts of Asia. *Journal of Asian Earth Sciences, Special Issue*, 84: 1–200.

Shields, G. & Veizer, J. (2002). Precambrian Marine Carbonate Isotopic Database: version 1.1. *Geochemistry, Geophysics, Geosystems*, 3: 1–12.

Shirey, S.B. & Walker, R.J. (1998). The Re–Os isotope system in cosmochemistry and high-temperature geochemistry. *Annual Review of Earth and Planetary Science*, 26: 423–500.

Shpount, B.R. & Oleinkikov, B.V. (1987). A comparison of mafic dyke swarms from the Siberian and Russian platforms. In Halls, H.C. & Fahrig, W.F. (eds.), *Mafic Dyke Swarms*. Geological Association of Canada, Special Paper 34, pp. 393–400.

Shumlyanskyy, L.V., Andreasson, P.-G., Buchan, K.L., & Ernst, R.E. (2007). The Volynian flood basalt province and coeval (Ediacaran) magmatism in Baltoscandia and Laurentia. *Mineralogical Journal (Ukraine)*, 29: 47–53.

Sikora, P.J. & Bergen, J.A. (2004). Lower Cretaceous planktonic foraminiferal and nannofossil biostratigraphy of Ontong Java Plateau sites from DSDP Leg 30 and ODP Leg 192. In Fitton, J.G., Mahoney J.J., Wallace, P.J., & Saunders, A.D. (eds.), *Origin and Evolution of the Ontong Java Plateau*. London: Geological Society, Special Publication 229, pp. 83–111.

Silver, P.G., Behn, M.D., Kelley, K., Schmitz, M., & Savage, B. (2006). Understanding cratonic flood basalts. *Earth and Planetary Science Letters*, 245: 190–201.

Simonetti, A., Bell, K., & Viladkar, S. (1995). Isotopic data from the Amba Dongar carbonatite complex, west-central India: evidence for an enriched mantle source. *Chemical Geology (Isotope Geoscience Section)*, 122: 185–198.

Simonetti, A., Goldstein, S.L., Schmidberger, S.S., & Viladkar, S.L. (1998). Geochemical and Nd, Pb, and Sr isotope data from Deccan alkaline complexes: inferences for mantle sources and plume–lithosphere interaction. *Journal of Petrology*, 39: 1847–1864.

Singh, A.P., Kumar, N., & Singh, B. (2004). Magmatic underplating beneath the Rajmahal Traps: gravity signature and derived 3-D configuration. *Proceedings of the Indian Academy of Sciences, Earth and Planetary Sciences*, 113: 759–769. See: www.ias.ac.in/jess/dec2004/ESB-SP16.pdf.

Singhal, B.B.S. & Gupta, R.P. (1999). *Applied Hydrogeology of Fractured Rocks*. Dordrecht: Kluwer.

Sircombe, K.N. (1999). Tracing provenance through the isotope ages of littoral and sedimentary detrital zircon, eastern Australia. *Sedimentary Geology*, 124: 47–67.

Skilling, I.P., White, J.D.L., & McPhie, J. (2002). Peperite: a review of magma–sediment mingling: *Journal of Volcanology and Geothermal Research*, 114: 1–17.

Skogseid, J. (2001). Volcanic margins: geodynamic and exploration aspects. *Marine and Petroleum Geology*, 18: 457–461.

Sleep, N.H. (1990). Hotspots and mantle plumes: some phenomenology. *Journal of Geophysical Research*, 95: 6715–6736.

Sleep, N.H. (2003). Fate of mantle plume material trapped within a lithospheric catchment with reference to Brazil. *Geochemistry, Geophysics, Geosystems*, 4: 8509.

Sleep, N.H. (2006) Mantle plumes from top to bottom. *Earth-Science Reviews*, 77: 231–271.

Sleep, N.H. (2009). Stagnant lid convection and the thermal subsidence of sedimentary basins with reference to Michigan. *Geochemistry, Geophysics, Geosystems*, 10 (12), Q12015, doi:10.1029/2009GC002881.

Smallwood, J.R. & Maresh, J. (2002). The properties, morphology and distribution of igneous sills. In Jolley, D.W. & Bell, B.R. (eds.), *The North Atlantic Igneous Province: Stratigraphy, Tectonic, Volcanic and Magmatic Processes*. London, Geological Society, Special Publication 197, pp. 271–306.

Smellie, J.L. (2008). Basaltic subglacial sheet-like sequences: evidence for two types of different implications for the inferred thickness of associated ice. *Earth-Science Reviews*, 88: 60–88.

Smirnov, A.V., Evans, D.A.D., Ernst, R.E., Söderlund, U., & Li, Z.X. (2013). Trading partners: tectonic ancestry of southern Africa and western Australia. *Precambrian Research*, 224: 11–22.

Smith, A.D. (1992). Back-arc convection model for Columbia River basalt genesis. *Tectonophysics*, 207: 269–285.

Smith, A.D. (2013). Recycling of oceanic crust and the origin of intraplate volcanism. *Australian Journal of Earth Sciences*, 60: 675–680.

Smith, A.D. & Lewis, C. (1999). The planet beyond the plume hypothesis. *Earth-Science Reviews*, 48: 135–182.

Smith, M.D., Heaman, L.M., & Breakhouse, G.P. (1999). Constraints on the timing of felsic magmatism associated with Matachewan igneous events: preliminary U–Pb results for the Creighton granite, Ontario. Abstract, GAC-MAC Joint Annual Meeting, Sudbury, volume 24, p. 119.

Smithies, R.H. (2002). Archaean boninite-like rocks in an intracratonic setting. *Earth and Planetary Science Letters*, 197: 19–34.

Smithies, R.H., Champion, D.C., & Sun, S.-S. (2004). The case of Archaean boninites. *Contributions to Mineralogy and Petrology*, 147: 705–721.

Smithies, R.H., Howard, H.M., Kirkland, C.L., *et al.* (2013). Geochemical evolution of rhyolites of the Talbot sub-basin and associated felsic units of the Warakurna supersuite. *Geological Survey of Western Australia, Report 118*.

Smolkin, V.F. (1997). The Paleoproterozoic (2.5–1.7 Ga) Midcontinent rift system of the northeastern Fennoscandian Shield. *Canadian Journal of Earth Sciences*, 34: 426–443.

Smrekar, S.E. & Stofan, E.R. (1997). Corona formation and heat loss on Venus by coupled upwelling and delamination. *Science*, 277: 1289–1294.

Smrekar, S.E., Kiefer, W.S., & Stofan, E.R. (1997). Large volcanic rises on Venus. In Bougher, S.W., Hunten, D.M., & Phillips, R.J. (eds.), *Venus II: Geology, Geophysics, Atmosphere, and Solar Wind Environment*. Tucson, AZ: University of Arizona Press, pp. 845–878.

Smrekar, S.E., Stofan, E.R., Mueller, N., *et al.* (2010a). Recent hotspot volcanism on Venus from VIRTIS emissivity data. *Science*, 328: 605–608.

Smrekar, S.E., Hoogenboom, T., Stofan, E.R., & Martin, P. (2010b). Gravity analysis of Parga and Hecate chasmata: implications for rift and corona formation. *Journal of Geophysical Resarch*, 115 (E7).

Snow, L.J., Duncan, R.A., & Bralower, T.J. (2005). Trace element abundances in the Rock Canyon Anticline, Pueblo, Colorado, marine sedimentary section and their relationship to Caribbean plateau construction and oxygen anoxic event 2. *Paleoceanography*, 20: PA3005.

Sobolev, S.V., Sobolev, A.V., Kuzmin, D.R., *et al.* (2011). Linking mantle plumes, large igneous provinces and environmental catastrophes. *Nature*, 477: 312–316.

Söderlund, U., Hofmann, A., Klausen, M.B., *et al.* (2010). Towards a complete magmatic barcode for the Zimbabwe craton: baddeleyite U–Pb dating of regional dolerite dyke swarms and sill provinces. *Precambrian Research*, 183: 388–398.

Söderlund, U., Ibanez-Mejia, M., El Bahat, A. *et al.* (2013). Reply to Comment on "U–Pb baddeleyite ages and geochemistry of dolerite dykes in the Bas-Drâa Inlier of the Anti-Atlas of Morocco: newly identified 1380 Ma event in the West African Craton" by André Michard and Dominique Gasquet. *Lithos*, 174: 101–108.

Solomon, S.C., Bullock, M.A., & Grinspoon, D.H. (1999). Climate change as a regulator of tectonics on Venus. *Science*, 286: 87–90.

Solomon, S.C., Aharonson, O., Aurnou, J.M., *et al.* (2005). New perspectives on ancient Mars. *Science*, 307: 1214–1220.

Song, S.G., Su, L., Li, X.H., Niu, Y.L., & Zhang, L.F. (2012). Grenville-age orogenesis in the Qaidam–Qilian block: the link between South China and Tarim. *Precambrian Research*, 220: 9–22.

Sparks, R.S.J., Brown, R.J., Field, M., & Gilbertson, M.A. (2007). Kimberlite ascent and eruption. *Nature*, 450: E21.

Speight, J.M., Skelhorn, R.R., Sloan, T., & Knapp, R.J. (1982). The dyke swarms of Scotland. In Sutherland, D.S. (ed.), *Igneous Rocks of the British Isles*. Chichester: John Wiley, pp. 449–459.

Sproule, R.A., Lesher, C.M., Ayer, J.A., Thurston, P.C., & Herzberg, C.T. (2002). Spatial and temporal variations in the geochemistry of komatiites and komatiitic basalts in the Abitibi greenstone belt. *Precambrian Research*, 115: 153–186.

Spudis, P.D. & Guest, J.E. (1988). Stratigraphy and geologic history of Mercury. In Vilas, F., Chapman, C.R., & Matthews, M.S. (eds.), *Mercury*. Tucson, AZ: University of Arizona Press, pp. 118–164.

Spudis, P.D. & Guest, J.E. (1997). Stratigraphy and geologic history of Mercury. In Vilas, F., Chapman, C.R., & Shapley Matthews, M. (eds.), *Mercury*. Tucson, AZ: University of Arizona Press, pp. 118–164.

Squyres, S.W., Wilhelms, D.E., & Moosman, A.C. (1987). Large-scale volcano-ground ice interaction on Mars. *Icarus*, 70: 385–408.

Squyres, S.W., Janes, D.M., Baer, G., *et al.* (1992). The morphology and evolution of coronae on Venus. *Journal of Geophysical Research*. 97: 13611–13634.

Srivastava, R.K. (2006). Geochemistry and petrogenesis of Neoarchaean high-Mg low-Ti mafic igneous rocks in an intracratonic setting, central India craton: evidence for boninite magmatism. *Geochemical Journal*, 40: 15–31.

Srivastava, R.K. (2008). Global intracratonic boninite–norite magmatism during the Neoarchean–Palaeoproterozoic: evidence from the central Indian Bastar craton. *International Geology Review*, 50: 61–74.

Srivastava, R.K. & Ernst, R.E. (2013). Global intracratonic boninitic–norite magmatism during the Neoarchean–Paleoproterozoic: revisited. *LIP of the Month*, July 2013. See http://www.largeigneousprovinces.org/13jul.

Srivastava, R.K., Ernst, R., Bleeker, W., & Hamilton, M.A., eds. (2010). Precambrian Large Igneous Provinces (LIPs) and their dyke swarms: new insights from high-precision geochronology integrated with paleomagnetism and geochemistry. *Precambrian Research Special Issue*, 183 (3).

Stachel, T. & Harris, J.W. (2008). The origin of cratonic diamonds: constraints from mineral inclusions. *Ore Geology Reviews*, 34: 5–32.

Stanistreet, I.G. & McCarthy, T.S. (1991). Changing tectono-sedimentary scenarios relevant to the development of the late Archean Witwatersrand Basin. *Journal of African Earth Sciences*, 13: 65–82.

Stanley, C.R. & Russell, J.K. (1989a). Petrologic hypothesis testing with Pearce element ratio diagrams: derivation of diagram axes. *Contributions to Mineralogy and Petrology*, 103: 78–89.

Stanley, C.R. & Russell, J.K. (1989b). PEARCE PLOT: a Turbo-PASCAL program for the analysis of rock compositions with Pearce element ratio diagrams. *Computers and Geosciences*, 15: 905–926.

Stanley, S.M. (2009). Evidence from ammonoids and conodonts for multiple Early Triassic mass extinctions. *Proceedings of the National Academy of Sciences*, 106 (36): 15264–15267.

Stanley, S.M. (2010). Relation of Phanerozoic stable isotope excursions to climate, bacterial metabolism, and major extinctions. *Proceedings of the National Academy of Sciences*, 107: 19185–19189.

Stefanick, M. & Jurdy, D.M. (1996). Venus coronae, craters and chasmata. *Journal of Geophysical Research*, 101: 4637–4643.

Stein, M. & Hofmann, A.W. (1992). Fossil plume head beneath the Arabian lithosphere? *Earth and Planetary Science Letters*, 114: 193–209.

Stein, M. & Hofmann, A.W. (1994). Mantle plumes and episodic crustal growth. *Nature*, 372: 63–68.

Stephens, W.E., Jemielita, R.A., & Davis, D. (1997). Evidence for ca. 750 Ma intraplate extensional tectonics from granite magmatism on the Seychelles: new geochronological data and implications for Rodinia reconstructions and fragmentation. *Terra Nova*, 9: 166.

Stephenson, P.J. (1990). Some aspects of dyke emplacement and characteristics in the Townsville–Ingham district, North Queensland, Australia. In Parker, A.J., Rickwood, P.C., & Tucker, D.H. (eds.), *Mafic Dykes and Emplacement Mechanisms*. Rotterdam: Balkema, pp. 421–434.

Stephenson, R.A., Yegorova, T., Brunet, M.-F., *et al.* (2006). Late Palaeozoic intra- and pericratonic basins on the East European craton and its margins. In Gee, D.G. & Stephenson, R.A. (eds.), *European Lithosphere Dynamics*. London: Geological Society of London, Memoir 32, pp. 463–479.

Stewart, K., Turner, S., Kelley, S., *et al.* (1996). 3-D, ^{40}Ar–^{39}Ar geochronology in the Paraná continental flood basalt province. *Earth and Planetary Science Letters*, 143: 95–109.

Stofan, E.R., Sharpton, V.L., Schubert, G., *et al.* (1992). Global distribution and characteristics of coronae and related features on Venus: implications for origin and relation to mantle processes. *Journal of Geophysical Research*, 97: 13347–13378.

Stofan, E.R., Smrekar, S.E., Bindschadler, D.L., & Senske, D.A. (1995). Large topographic rises on Venus: implications for mantle upwelling. *Journal of Geophysical Research*, 100: 23317–23327.

Stofan, E.R., Hamilton, V.E., Janes, D.M., & Smrekar, S.E. (1997). Coronae on Venus: morphology and origin. In Bougher, S.W., Hunten, D.M., & Phillips, R.J. (eds.), *Venus II: Geology, Geophysics, Atmosphere, and Solar Wind Environment*. Tucson, AZ: University of Arizona Press, pp. 931–965.

Stofan, E.R., Smrekar, S.E., Tapper, S.W., Guest, J.E., & Grinrod, P.M. (2001). Preliminary analysis of an expanded corona database for Venus. *Geophysical Research Letters*, 28: 4267–4270.

Stofan, E.R., Glaze, L.S., Smrekar, S.E., & Baloga, S.M. (2003). A statistical analysis of corona topography: insights into corona formation and evolution. *Lunar and Planetary Science Conference, XIV*, Abstract 1594 (CD-ROM).

Stofan, E.R., Brian, A.W., & Guest, J.E. (2005). Resurfacing styles and rates on Venus – assessment of 18 Venusian quadrangles. *Icarus*, 173, 312–321.

Storey, B.C. (1995). The role of mantle plumes in continental breakup: case histories from Gondwanaland. *Nature*, 377: 301–308.

Storey, B.C. & Kyle, P.R. (1997). An active mantle mechanism for Gondwana breakup. *South African Journal of Geology*, 100: 283–290.

Storey, B.C., Leat, P.T., & Ferris, J.K. (2001). The location of mantle plume centers during the initial stages of Gondwana break-up. In Ernst, R.E. and Buchan, K.L. (eds.), *Mantle Plumes: Their Identification through Time*. Boulder, CO: Geological Society of America, Special Paper 352, pp. 71–80.

Storey, B.C., Vaughan, A.P.M., & Riley, T.R. (2013). The links between large igneous provinces, continental break-up and environmental change: evidence reviewed from Antarctica. *Transactions of the Royal Society, Edinburgh*, 104: 17–30.

Storey, M., Mahoney, J.J., Saunders, A.D., *et al.* (1995). Timing of hot spot-related volcanism and the breakup of Madagascar and India. *Science*, 267: 852–855.

Storey, M., Mahoney, J.J., & Saunders, A.D. (1997). Cretaceous basalts in Madagascar and the transition between plume and continental lithosphere mantle sources. *AGU Geophysical Monograph*, 100: 95–122.

Storey, M., Duncan, R.A., & Tegner, C. (2007a). Timing and duration of volcanism in the North Atlantic Igneous Province: implications for geodynamics and links to the Iceland hotspot. *Chemical Geology*, 241: 264–281.

Storey, M., Duncan, R.A., & Swisher III, C.C. (2007b). Paleocene–Eocene Thermal Maximum and the opening of the Northeast Atlantic. *Science*, 316: 587–589.

Stothers, R.B. (1993). Flood basalts and extinction events, *Geophysical Research Letters*, 20: 1399–1402.

Stott, G.M. (2003). Diabase dyke swarms as structural controls for kimberlite pipes under the James Bay and Hudson Bay lowlands, Ontario. Long abstract, 8th International Kimberlite Conference, Victoria, Canada.

Stracke, A., Hofmann, A.W., & Hart, S.R. (2005). FOZO, HIMU and the rest of the mantle zoo. *Geochemistry, Geophysics, Geosystems*, 6: 1525–2027.

Strik, G., Blake, T.S., Zegers, T.E., White, S.H., & Langereis, C.G. (2003). Palaeomagnetism of flood basalt in the Pilbara Craton, Western Australia: late Archaean continental drift and the oldest known reversal of the geomagnetic field. *Journal of Geophysical Research*, 108: doi:10.1029/2003JB002475.

Stubley, M. (2004). Spatial distribution of kimberlite in the Slave Province: a geometrical approach. *Lithos*, 77: 683–694.

Studd, D., Ernst, R.E., & Samson, C. (2011). Radiating graben–fissure systems in the Ulfrun Regio area, Venus. *Icarus*, 215, 279–291.

Sun, S.-S. & McDonough, W.F. (1989). Chemical and isotopic systematics of oceanic basalts: implications for mantle composition and processes. In Saunders, A.D. & Norry, M.J. (eds.), *Magmatism in the Ocean Basins*. London: Geological Society of London, Special Publication 42, pp. 313–345.

Sun, S.-S. & Nesbitt, R.W. (1978). Petrogenesis of Archean ultrabasic and basic volcanism: evidence from rare earth elements. *Contributions to Mineralogy and Petrology*, 65: 301–325.

Sun, S.-S., Nesbitt, R.W., & McCulloch, M.T. (1989). Geochemistry and petrogenesis of Archaean and early Proterozoic siliceous high-magnesian basalts. In Crawford, A.J. (ed.), *Boninites*. London: Unwin-Hyman, pp. 149–173.

Sun, Y.-D., Lai, X.-I., Wignall, P.B., *et al.* (2010). Dating the onset and nature of the Middle Permian Emeishan large igneous province eruptions in SW China using conodont biostratigraphy and its bearing on mantle plume uplift models. *Lithos*, 119: 20–33.

Sutherland, F.L. (1998). Origin of north Queensland Cenozoic volcanism: relationships to long lava flow basaltic fields, Australia. *Journal of Geophysical Research*, 103(B11): 27347–27358.

Svensen, H. & Planke, S. (2008). The North Atlantic volcanic province (NAVP) and the Paleocene–Eocene Thermal Maximum (PETM). *LIP of the Month*, August 2008. See http://www.largeigneousprovinces.org/08aug.

Svensen, H., Planke, S., Malthe-Sørenssen, A., *et al.* (2004). Release of methane from a volcanic basin as a mechanism for the initial Eocene global warming. *Nature*, 429: 542–545.

Svensen, H., Jamtveit, B., Planke, S., & Chevallier, L. (2006). Structure and evolution of hydrothermal vent complexes in the Karoo Basin. *South African Journal of Geology*, 163: 671–682.

Svensen, H., Planke, S., Chevallier, L., *et al.* (2007). Hydrothermal venting of greenhouse gases triggering Early Jurassic global warming. *Earth and Planetary Science Letters*, 256: 554–566.

Svensen, H., Planke, S., Polozov, A.G., *et al.* (2009). Siberian gas venting and the end-Permian environmental crisis. *Earth and Planetary Science Letters*, 277: 490–500.

Svensen, H., Corfu, F., Polteau, S., Hammer, Ø., & Planke, S. (2012). Rapid magma emplacement in the Karoo Large Igneous Province. *Earth and Planetary Science Letters*, 325–326: 1–9.

Swanson, D.A., Wright, T.L., Hooper, P.R., & Bentley, R.D. (1979). Revisions in stratigraphic nomenclature of the Columbia River Basalt Group. *US Geological Survey Bulletin*, 1457-G.

Swanson, E.R., Kempter, K.A., McDowell, F.W., & McIntosh, W.C. (2006). Major ignimbrites and volcanic centers of the Copper Canyon area: a view into the core of Mexico's Sierra Madre Occidental. *Geosphere*, 2: 125.

Tack, L., Wingate, M.T.D., De Waele, B., *et al.* (2010). The 1375 Ma Kibaran event in Central Africa: prominent emplacement of bimodal magmatism under extensional regime. *Precambrian Research*, 180: 63–84.

Tackley, P.J. (1998). Three-dimensional simulations of mantle convection with a thermochemical basal boundary layer: D″? The core–mantle boundary region. *Geophysical Monograph Series*, 28: 231–253.

Tait, J., Straathof, G., Söderlund, U., *et al.* (2013). The Ahmeyim Great Dyke of Mauritania: A newly dated Archaean intrusion. *Lithos*, 174: 323–332.

Takashima, R., Nishi, H., Huber, B.T., & Leckie, R.M. (2006). Greenhouse world and the Mesozoic Ocean. *Oceanography*, 19: 82–92.

Tanaka, K.L. (1986). The stratigraphy of Mars. *Journal of Geophysical Research*, 91: 139–158.

Tanaka, K.L., Scott, D.H., & Greeley, R. (1992). Global stratigraphy. In Kieffer, H.H., Jakosky, B.M., Snyder, C.W., Matthews, M.S. (eds.), *Mars*. Tucson, AZ: University of Arizona Press, pp. 345–382.

Tanaka, K.L., Kargel, J.S., MacKinnon, D.J., Hare, T.M., & Hoffman, N. (2002). Catastrophic erosion of Hellas basin rim on Mars induced by magmatic intrusion into volatile-rich rocks. *Geophysical Research Letters*, 29 (8).

Tappe, S., Foley, S.F., Jenner, G.A., & Kjarsgaard, B.A. (2005). Integrating ultramafic lamprophyres into the IUGS classification of igneous rocks: rationale and implications. *Journal of Petrology*, 46: 1893–1900.

Tappe, S., Foley, S.F., Jenner, G.A., *et al.* (2006). Genesis of ultramafic lamprophyres and carbonatites at Aillik Bay, Labrador: a consequence of incipient lithospheric thinning beneath the North Atlantic craton. *Journal of Petrology*, 47: 1261–1315.

Tappe, S., Foley, S.F., Stracke, A., *et al.* (2007). Craton reactivation on the Labrador Sea margins: $^{40}Ar/^{39}Ar$ age and Sr–Nd–Hf–Pb isotope constraints from alkaline and carbonatite intrusives. *Earth and Planetary Science Letters*, 256: 433–454.

Tappe, S., Foley, S.F., Kjarsgaard, B.A., *et al.* (2008). Between carbonatite and lamproite – diamondiferous Torngat ultramafic lamprophyres formed by carbonate-fluxed melting of cratonic MARID-type metasomes. *Geochimica et Cosmochimica Acta*, 72: 3258–3286.

Tappe, S., Pearson, D.G., Nowell, G., *et al.* (2011). A fresh isotopic look at Greenland kimberlites: cratonic mantle lithosphere imprint on deep source signal. *Earth and Planetary Science Letters*, 305: 235–248.

Tappe, S., Steenfelt, A., & Nielsen, T.F.N. (2012). Asthenospheric source of Neoproterozoic and Mesozoic kimberlites from the North Atlantic craton, West Greenland: new high-precision U–Pb and Sr–Nd isotope data on perovskite. *Chemical Geology*, 320–321: 113–127.

Tappe, S., Pearson, D.G., Kjarsgaard, B.A., Nowell, G., & Dowall, D. (2013). Mantle transition zone input to kimberlite magmatism near a subduction zone: origin of anomalous Nd–Hf isotope systematics at Lac de Gras, Canada. *Earth and Planetary Science Letters*, 371–372: 235–251.

Tarduno, J.A., Sliter, W.V., Kroenke, L.W., *et al.* (1991). Rapid formation of the Ontong Java Plateau by Aptian mantle plume volcanism. *Science*, 254: 399–403.

Tarney, J. & Weaver, B.L. (1987). Mineralogy, petrology and geochemistry of the Scourie dykes: petrogenesis and crystallization processes in dykes intruded at depth. In Park, A.J. & Tarney, J. (eds.), *Evolution of the Lewisian and Comparable Precambrian High Grade Terrains*. London: Geological Society of London, Special Publication 27, pp. 217–233.

Taylor, A.S. & Lasaga, A.C. (1999). The role of basalt weathering in the Sr isotope budget of the oceans. *Chemical Geology*, 161: 199–214.

Taylor, B. (2006). The single largest oceanic plateau: Ontong Java–Manihiki–Hikurangi. *Earth and Planetary Science Letters*, 241: 372–380.

Tegner, C., Duncan, R.A., Bernstein, S., *et al.* (1998). Ar-40–Ar-39 geochronology of Tertiary mafic intrusions along the East Greenland rifted margin: relation to flood basalts and the Iceland hotspot track. *Earth and Planetary Science Letters*, 156: 75–88.

Tegner, C., Brooks, C.K., Duncan, R.A., Heister, L.E., & Bernstein, S. (2008). $^{40}Ar–^{39}Ar$ ages of intrusions in East Greenland: rift-to-drift transition over the Iceland hotspot. *Lithos*, 101: 480–500.

Teixeira, W., D'Agrella-Filho, M.S., Hamilton, M.A., *et al.* (2013). U–Pb (ID-TIMS) baddeleyite ages and paleomagnetism of 1.79 and 1.50 Ga tholeiitic dyke swarms, and position of the Rio de la Plata craton within the Columbia supercontinent. *Lithos*, 174: 157–174.

Teixeira, W., Hamilton, M.A., Lima, G.A., *et al.* (2014). Precise ID-TIMS U-Pb baddeleyite ages (1110-1112 Ma) for the Rincón del Tigre-Huanchaca large igneous province (LIP) of the Amazonian Craton: implications for the Rodinia supercontinent. *Precambrian Research*, in press.

Tejada, M.L.G., Mahoney, J.J., Neal, C.R., Duncan, R.A., & Petterson, M.G. (2002). Basement geochemistry and geochronology of Central Malaita, Solomon Islands, with implications for the origin and evolution of the Ontong Java Plateau. *Journal of Petrology*, 43: 449–484.

Tejada, M.L.G., Mahoney, J.J., Castillo, P.R., *et al.* (2004). Pin-pricking the elephant: evidence on the origin of the Ontong Java Plateau from Pb–Sr–Hf–Nd isotopic characteristics of ODP Leg 192 basalts. In Fitton, J.G., Mahoney, J.J., Wallace, P.J., & Saunders, A.D. (eds.), *Origin and Evolution of the Ontong Java Plateau*, London: Geological Society, Special Publication 229, pp. 133–150.

Tewksbury, C.M. (2003). Crustal plateau collapse on Venus: evidence from the Pasom-mana region. *Lunar and Planetary Science Conference*, XIV, Abstract 1291 (CD-ROM).

Thiede, D.S. & Vasconcelos, P.M. (2010). Paraná flood basalts: rapid extrusion hypothesis confirmed by new $^{40}Ar/^{39}Ar$ results. *Geology*, 38: 747–750.

Thomas, M.D. & Teskey, D.J. (1994). An interpretation of gravity anomalies over the Midcontinent rift, Lake Superior, constrained by GLIMPCE seismic and aeromagnetic data. *Canadian Journal of Earth Sciences*, 31: 682–697.

Thompson, R.N. & Gibson, S.A. (1991). Subcontinental mantle plumes, hotspots and pre-existing thinspots. *Journal of the Geological Society of London*, 148: 973–977.

Thompson, R.N., Morrison, M.A., Dickin, A.P., & Hendry, G.L. (1983). Continental flood basalts... arachnids rule OK? In Hawkesworth, C.J. & Norry, M.J. (eds.), *Continental Basalts and Mantle Xenoliths*. Nantwich: Shiva Publishing Ltd., Shiva Geology Series, 272, pp. 158–185.

Thomson, K. (2007). Determining magma flow in sills, dykes and laccoliths and their implications for sill emplacement mechanisms. *Bulletin of Volcanology*, 70: 183–201.

Thomson, K. & Hutton, D. (2004). Geometry and growth of sill complexes: insights using 3-D seismic from the North Rockall Trough. *Bulletin of Volcanology*, 66: 364–375.

Thomson, K. & Schofield, N. (2008). Lithological and structural controls on the emplacement and morphology of sills in sedimentary basins. In Thomson, K. & Petford, N. (eds.), *Structure and Emplacement of High-Level Magmatic Systems*. *Geological Society of London Special Publication*, 302: 31–44.

Thorkelson, D.J., Mortensen, J.K., Davidson, G.J., *et al.* (2001). Early Mesoproterozoic intrusive breccias in Yukon, Canada: the role of hydrothermal systems in reconstructions of North America and Australia. *Precambrian Research*, 111: 31–55.

Thorne, A.M. & Trendall, A.F. (2001). Geology of the Fortescue Group, Pilbara Craton, Western Australia. *Geological Survey of Western Australia Bulletin*, 144.

Thurston, P.C. & Chivers, K.M. (1990). Secular variation in greenstone sequence development, emphasizing Superior Province, Canada. *Precambrian Research*, 46: 21–58.

Tiley, R., White, N., & Al-Kindi, S. (2004). Linking Paleogene denudation and magmatic underplating beneath the British Isles. *Geological Magazine*, 3: 345–351.

Tilton, G.R. & Bell, K. (1994). Sr–Nd–Pb isotope relationships in Late Archean carbonatites and alkaline complexes: applications to the geochemical evolution of Archean mantle. *Geochimica et Cosmochimica Acta*, 58: 3145–3154.

Tolan, T.L., Reidel, S.P., Beeson, M.H., *et al.* (1989). *Revisions to the estimates of the areal extent and volume of the Columbia River Basalt Group*. Boulder, CO: Geological Society of America, Special Paper 239, 1–20.

Tollo, R.P., Aleinikoff, J.N., Bartholomew, M.J., & Rankin, D.W. (2004). Neoproterozoic A-type granitoids of the central and southern Appalachians: intraplate magmatism associated with episodic rifting of the Rodinian supercontinent. *Precambrian Research*, 128: 3–38.

Tolstikhin, I.N., Kamensky, I.L., Marty, B., *et al.* (2002). Rare gas isotopes and parent trace elements in ultrabasic-alkaline-carbonatite complexes. Kola

Peninsula: identification of lower mantle plume component. *Geochimica et Cosmochimica Acta*, 66: 881–901.

Tomlinson, K.Y. & Condie, K.C. (2001). Archean mantle plumes: evidence from greenstone belt geochemistry. In Ernst, R.E. & Buchan, K.L. (eds.), *Mantle Plumes: Their Identification through Time*. Boulder, CO: Geological Society of America, Special Paper 352, pp. 341–357.

Tomlinson, K.Y., Hughes, D.J., Thurston, P.C, & Hall, R.P. (1999). Plume magmatism and crustal growth at 2.9 to 3.0 Ga in the Steep Rock and Lumby Lake area, western Superior Province. *Lithos*, 146: 103–136.

Torsvik, T.H., Tucker, R.D., Ashwal, L.D., *et al.* (2000). Late Cretaceous India–Madagascar fit and timing of break-up related magmatism. *Terra Nova*, 12: 220–224.

Torsvik, T.H., Carter, L.M., Ashwal, L.D., *et al.* (2001a). Rodinia refined or obscured: palaeomagnetism of the Malani igneous suite (NW India). *Precambrian Research*, 108: 319–333.

Torsvik, T.H., Ashwal, L.D., Tucker, R.D., & Eide, E.A. (2001b). Neoproterozoic geochronology and palaeogeography of the Seychelles microcontinent: the India link. *Precambrian Research*, 110: 47–59.

Torsvik, T.H., Burke, K., Steinberger, B., Webb, S.J., & Ashwal, L.D. (2010). Diamonds sampled by plumes from the core–mantle boundary. *Nature*, 466: 352–355.

Toyoda, K., Horiuchi, H., & Tokonami, M. (1994). Dupal anomaly of Brazilian carbonatites: geochemical correlations with hotspots in the South Atlantic and implications for the mantle source. *Earth and Planetary Science Letters*, 126: 315–331.

Treiman, A.H. (1989). Carbonatite magma: properties and processes. In Bell, K. (ed.), *Carbonatites: Genesis and Evolution*. London: Unwin Hyman, pp 89–104.

Trendall, A.F. (1995). The Woongarra Rhyolite – a giant lavalike felsic sheet in the Hamersley Basin. *Geological Survey of Western Australia Report* 42.

Tripati, A.K. & Elderfield, H. (2004). Abrupt hydrographic changes in the equatorial Pacific and subtropical Atlantic from foraminiferal Mg/Ca indicate greenhouse origin for the thermal maximum at the Paleocene-Eocene Boundary. *Geochemistry, Geophysics, Geosystems*, 5: 2003GC000631.

Trude, K.J. (2004). Kinematic indicators for shallow level igneous intrusion from 3D seismic data; evidence of flow direction and feeder location. In Davies, R.J., Cartwright, J.A., Stewart, S.A., Lappin, M., & Underhill, J.R. (eds.), *3D Seismic Technology: Application to the Exploration of Sedimentary Basins*. London: Geological Society, Memoirs 29: 209–217.

Trumball, R.B., Harris, C., Frindt, S., & Wigand, M. (2004). Oxygen and neodymium isotope evidence for source diversity in Cretaceous anorogenic granites from Namibia and implications for A-type granite genesis. *Lithos*, 73: 21–40.

Tsikos, H., Beukes, N.J., Moore, J.M., & Harris, C. (2003). Deposition, diagenesis, and secondary enrichment of metals in the Paleoproterozoic Hotazel Iron Formation, Kalahari manganese field, South Africa. *Economic Geology*, 98: 1449–1462.

Tsygankov, A.A., Matukov, D.I., Berezhnaya, N.G., *et al.* (2007). Late Paleozoic granitoids of western Transbaikalia: magma sources and stages of formation. *Russian Geology and Geophysics*, 1: 156–180.

Tsygankov, A.A., Litvinovsky, B.A., Jahn, B.M., *et al.* (2010). Complex sequence of magmatic events in the Late Palaeozoic of Transbaikalia, Russia (U–Pb isotopic data). *Russian Geology and Geophysics*, 51: 901–922.

Tucker, R.D., Ashwal, L.D., Handke, M.J., *et al.* (1999). U–Pb geochronology and isotope geochemistry of the Archean and Proterozoic rocks of North-Central Madagascar. *Journal of Geology*, 107: 135–153.

Tucker, R.D., Ashwal, L.D., & Torsvik, T.H. (2001). U–Pb geochronology of Seychelles granitoids: a Neoproterozoic continental arc fragment. *Earth and Planetary Science Letters*, 187: 27–38.

Turgeon, S.C. & Creaser, R.A. (2008). Cretaceous oceanic anoxic event 2 triggered by a massive magmatic episode. *Nature*, 454: 323–326.

Turner, S. & Hawkesworth, C. (1995). The nature of the sub-continental mantle; constraints from the major-element composition of continental flood basalts. *Chemical Geology*, 120: 295–314.

Turner, S.P. & Rushmer, T. (2009). Similarities between mantle-derived A-type granites and voluminous rhyolites in continental flood basalt provinces. *Earth and Environmental Science Transactions of the Royal Society of Edinburgh*, 100: 1–10.

Turner, S.P., Foden, J.D., & Morrison, R.S. (1992). Derivation of some A-type magmas by fractionation of basaltic magma – an example from the Padthaway Ridge, South Australia. *Lithos*, 28: 151–179.

Twist, D. & French, B. (1983). Voluminous acid volcanism in the Bushveld Complex: a review of the Rooiberg Felsite. *Bulletin of Volcanology*, 46: 225–242.

Uenzelmann-Neben, G., Gohl, K., Ehrhardt, A., & Seargent, M. (1999). Agulhas Plateau, SW Indian Ocean: new evidence for excessive volcanism. *Geophysical Research Letters*, 26 (13): 1941–1944.

Uken, R. & Watkeys, M.K. (1997). An interpretation of mafic dyke swarms and their relationship with major mafic magmatic events on the Kaapvaal Craton and Limpopo Belt. *South African Journal of Geology*, 100: 341–348.

Ukstins, I.A., Renne, P.R., Wolfenden, E., *et al.* (2002). Matching conjugate volcanic rifted margins: ^{40}Ar/^{39}Ar chrono-stratigraphy of pre- and syn-rift bimodal flood volcanism in Ethiopia and Yemen. *Earth and Planetary Science Letters*, 198: 289–306.

Ukstins Peate, I. & Bryan, S.E. (2008). Re-evaluating plume-induced uplift in the Emeishan large igneous province. *Nature Geoscience*, 1: 625–629.

Ukstins Peate, I. & Bryan, S.E. (2009a). Pre-eruptive uplift in the Emeishan? *Nature Geoscience*, 2: 531– 532.

Ukstins Peate, I. & Bryan, S.E. (2009b). Reply to "Pre-eruptive uplift in the Emeishan?" by He, B., Xu, Y.-G., & Campbell, I. 2009, v. 2, pp. 530–531. *Nature Geoscience*, 2: 531–532.

Ukstins Peate, I.A., Baker, J.A., Kent, A.J.R., *et al.* (2003). Correlation of Indian Ocean tephra to individual Oligocene silicic eruptions from Afro-Arabian flood volcanism. *Earth and Planetary Science Letters*, 211: 311–327.

Ukstins Peate, I., Baker, J.A., Al-Kadasi, M., *et al.* (2005). Volcanic stratigraphy of large-volume silicic pyroclastic eruptions during Oligocene Afro-Arabian flood volcanism in Yemen. *Bulletin of Volcanology*, 68: 135–156.

Underhill, J.R. & Partington, M.A. (1994). Use of genetic sequence stratigraphy in defining and determining a regional tectonic control on the "mid-Cimmerian unconformity": implications for North Sea basin development and the global sea-level chart. In Weimer, P. & Posamentier, H. (eds.), *Siliciclastic Sequence Stratigraphy*. Tulsa, OK: American Association of Petroleum Geologists, Memoir 58, pp. 449–484.

Upton, B.G.J. & Thomas, J.E. (1980). The Tugtutôq Younger Giant Dyke Complex, South Greenland: fractional crystallisation of transitional olivine basalt magma. *Journal of Petrology*, 21: 167–198.

Upton, B.G.J., Parsons, I., Emeleus, C.H., & Hodson, M.E. (1996). Layered alkaline igneous rocks of the Gardar Province, South Greenland. In Cawthorn, R.G. (ed.), *Layered Intrusions*. Amsterdam: Elsevier, pp. 331–363.

Upton, B.G.J., Rämö, O.T., Heaman, L.M., *et al.* (2005). The Mesoproterozoic Zig-Zag Dal basalts and associated intrusions of eastern North Greenland: mantle plume–lithosphere interaction. *Contributions to Mineralogy and Petrology*, 149: 40–56.

Van Decar, J.C., James, D.E., & Assumpção, M. (1995). Seismic evidence for a fossil mantle plume beneath South America and implications for plate driving forces. *Nature*, 378: 25–31.

van der Westhuizen, W.A., De Bruiyn, H., & Meintjes, P.G. (1991). The Ventersdorp Supergroup: an overview. *Journal of African Earth Sciences*, 13: 83–105.

van der Westhuizen, W.A., de Bruiyn, H., & Meintjes, P.G. (2006). The Ventersdorp Supergroup. In Johnson, M.R., Anhaeusser, C.R., & Thomas, R.J. (eds.), *The Geology of South Africa*. Johannesburg/Pretoria: Geological Society of South Africa/Council for Geoscience, pp. 187–208.

van Keken, P.E., Hauri, E.H., & Ballentine, C.J. (2002). Mantle mixing: the generation, preservation and destruction of chemical heterogeneity. *Annual Review of Earth and Planetary Sciences*, 30: 493–525.

Van Schmus, W.R. & Hinze, W.J. (1985). The Midcontinent Rift System. *Annual Review of Earth and Planetary Sciences*, 13 (1): 345–383.

van Wijk, J.W., Huismans, R.S., Ter Voorde, M., & Cloetingh, S.A.P.L. (2001). Melt generation at volcanic continental margins: no need for a mantle plume? *Geophysical Research Letters*, 28: 3995–3998.

Vantongeren, J.A., Mathez, E.A., & Kelemen, P.B. (2010). A felsic end to Bushveld differentiation. *Journal of Petrology*, 51: 1891–1912.

Vasiliev, Yu.R. & Zolotukhin, V.V. (1995). The Maimecha–Kotui alkaline-ultramafic province of the northern Siberian platform, Russia. *Episodes*, 18: 155–164.

Vaughan, A.P.M. (1995). Circum-Pacific Mid-Cretaceous deformation and uplift: a superplume-related event? *Geology*, 23: 491–494.

Vaughan, A.P.M. & Livermore, R.A. (2005). Episodicity of Mesozoic terrane accretion along the Pacific margin of Gondwana: implications for superplume–plate interactions. In Vaughan, A.P.M., Leat, P.T., & Pankhurst, R.J. (eds.), *Terrane Processes at the Margins of Gondwana*. London: Geological Society, Special Publication 246, pp. 143–178.

Vaughan, A.P.M. & Scarrow, J.H. (2003). Ophiolite obduction pulses as a proxy indicator of superplume events? *Earth and Planetary Science Letters*, 213: 407–416.

Vaughan, A.P.M. & Storey, B.C. (2007). A new supercontinent self-destruct mechanism: evidence from the Late Triassic–Early Jurassic. *Journal of the Geological Society, London*, 164: 383–392.

Veizer, J., Ala, D., Azmy, K., *et al.* (1999). $^{87}Sr/^{86}Sr$, $\delta^{13}C$ and $\delta^{18}O$ evolution of Phanerozoic seawater. *Chemical Geology*, 161: 59–88.

Vernikovsky, V.A., Pease, V.L., Vernikovskaya, A.E., *et al.* (2003). First report of early Triassic A-type granite and syenite intrusions from Taimyr: product of the northern Eurasian superplume. *Lithos*, 66: 23–36.

Vervoort, J.D., Wirth, K., Kennedy, B., Sandland, T., & Harpp, K.S. (2007). The magmatic evolution of the Midcontinent rift: new geochronologic and geochemical evidence from felsic magmatism. *Precambrian Research*, 157: 235–268.

Viladkar, S.G. (1981). The carbonatites of Amba Dongar, Gujarat, India. *Geological Society of Finland Bulletin*, 53: 17–28.

Vladimirov, A.G., Kozlov, M.S., Shokal'skii, S.P., *et al.* (2001). Major epochs of the intrusive magmatism of Kuznetsk Alatau, Altai, and Kalba (from U–Pb isotope dates). *Geologiya u Geofizika*, 42: 1157–1178 (in Russian).

Vogel, D.C., Vuollo, J.I., Alapieti, T.T., & James, R.S. (1998). Tectonic, stratigraphic and geochemical comparison between ca. 2600–2440 Ma mafic igneous events in the Canadian and Fennoscandian Shields. *Precambrian Research*, 92: 89–116.

Vuollo, J. & Huhma, H. (2005). Paleoproterozoic mafic dikes in NE Finland. In Lehtinen, M., Nurmi, P.A., & Rämö, O.T. (eds.), *Precambrian Geology of Finland – Key to the Evolution of the Fennoscandian Shield*. Amsterdam: Elsevier, pp. 195–236.

Wade, C.E., Reid, A.J., Wingate, M.T.D., Jagodzinski, E.A., & Barovich, K. (2012). Geochemistry and geochronology of the *c.* 1585 Ma Benagerie Volcanic Suite, southern Australia: relationship to the Gawler Range volcanics and implications for the petrogenesis of a Mesoproterozoic silicic large igneous province. *Precambrian Research*, 206–207: 17–35.

Wager, L.R. & Brown, G.M. (1968). *Layered Igneous Rocks*. Edinburgh: Oliver and Boyd.

Waldron, J.W.F. & van Staal, C.R. (2001). Taconic orogeny and the accretion of the Dashwoods block: a peri-Laurentian microcontinent in the Iapetus Ocean. *Geology*, 29: 811–814.

Walker, A.S.D., Key, R.M., Pouliquen, G., *et al.* (2010). Geophysical modelling of the Molopo Farms Complex; implications for its emplacement within the ~2 Ga large igneous provinces of southern and central Africa. *South African Journal of Geology*, 113: 381–400.

Walker, R.J. (2002). Tracing the osmium isotopic compositions of large plumes through Earth history (Extended Abstracts and Photos of Superplume Workshop, Tokyo, January 28–31). *Electronic Geosciences*, 7 (1): 1–2.

Wall, C.J., Scoates, J.S., Freidman, R.M., & Meurer, W.P. (2012). Age of the Stillwater Complex. Age of the Bushveld Complex. Abstract, 22nd V.M. Goldschmidt Conference "Earth in Evolution", 24–29 June, Montreal, Quebec, Canada.

Wallace, P.J., Frey, F.A., Weis, D., & Coffin, M.F. (2002). Origin and evolution of the Kerguelen Plateau, Broken Ridge and Kerguelen Archipelago: editorial. *Journal of Petrology*, 43: 1105–1108.

Wang Q., Wyman, D.A., Li, Z.-X., *et al.* (2010). Petrology, geochronology and geochemistry of *ca.* 780 Ma A-type granites in South China: petrogenesis and implications for crustal growth during the breakup of supercontinent Rodinia. *Precambrian Research*, 178: 185–208.

Wang, W., Zhou, M.F., Yan, D.P., & Li, J.W. (2012b). Depositional age, provenance, and tectonic setting of the Neoproterozoic Sibao Group, southeastern Yangtze Block, South China. *Precambrian Research*, 192–195: 107–124.

Wang, X.C., Li, X.H., Li, W.X., & Li, Z.X. (2007). Ca. 825 Ma komatiitic basalts in South China: First evidence for > 1500 degrees C mantle melts by a Rodinian mantle plume. *Geology*, 35: 1103–1106.

Wang, X.-C., Li, X.-H., Li, W.-X., *et al.* (2008a). The Bikou basalts in the northwestern Yangtze block, South China: remnants of 820–810 Ma continental flood basalts. *Geological Society of America Bulletin*, 120: 1478–1492.

Wang, X.C., Li, X.H., Li, W.X., & Li, Z.X. (2009). Variable involvements of mantle plumes in the genesis of mid-Neoproterozoic basaltic rocks in South China: a review. *Gondwana Research*, 15: 381–394.

Wang, X.L., Shu, L.S., Xing, G.F., *et al.* (2012a). Post-orogenic extension in the eastern part of the Jiangnan orogen: evidence from ca 800–760 Ma volcanic rocks. *Precambrian Research*, 222–223: 404–423.

Wang, Y.-J., Zhao, G.-C., Cawood, P.A., Fan, W.-M., Peng, T.-P., & Sun, L.-H. (2008b). Geochemistry of Palaeoproterozoic (~1770 Ma) mafic dikes from the Trans-North China Orogen and tectonic implications. *Journal of Asian Earth Sciences*, 33: 61–77.

Ward, P.L. (1995). Subduction cycles under western North America during the Mesozoic and Cenozoic Eras. In Miller, D.M. & Busby, C. (eds.), *Jurassic Magmatism and Tectonics of the North American Cordillera*. Boulder, CO: Geological Society of America, Special Paper 299, pp. 1–46.

Warren, J.K., George, S.C., Hamilton, P.J., & Tingate, P. (1998). Proterozoic source rocks; sedimentology and organic characteristics of the Velkerri Formation, Northern Territory, Australia. *American Association of Petroleum Geologists Bulletin*, 82: 442–463.

Watters, T.R. (1988). Wrinkle ridge assemblages on the terrestrial planets. *Journal of Geophysical Research*, 93(B9): 10236–10254.

Watters, T.R. (1993). Compressional tectonism on Mars. *Journal of Geophysical Research*, 98: 17049–17060.

Watters, T.R. (2004). Elastic dislocation modeling of wrinkle ridges on Mars. *Icarus*, 171: 284–294.

Watters, T.R., Murchie, S.C., Robinson, M.S., *et al.* (2009). Emplacement and tectonic deformation of smooth plains in the Caloris basin, Mercury. *Earth and Planetary Science Letters*, 285: 309–319.

Weis, D., Frey, F.A., Leyrit, H., & Gautier, I. (1993). Kerguelen Archipelago revisited: geochemical and isotopic study of the Southeast Province lavas. *Earth and Planetary Science Letters*, 118: 101–119.

Weis, D., Frey, F.A., Schlich, R., *et al.* (2002). Trace of the Kerguelen Mantle Plume: evidence from seamounts between the Kerguelen Archipelago and Heard Island, Indian Ocean. *Geochemistry, Geophysics and Geosystems*, 3: 10.1029/2001GC000251.

Weiser, T.W. (2002). Platinum-group mineral (PGM) in placer deposits. In Cabri, LJ. (ed.), *The Geology, Geochemistry, Mineralogy and Mineral Beneficiation of Platinum-Group Elements*. Canadian Institute of Mining, Metallurgy and Petroleum, Special Volume, 54, pp. 721–756.

Welford, J.K. & Clowes, R.M. (2004). Deep 3-D seismic reflection imaging of Precambrian sills in southwestern Alberta, Canada. *Tectonophysics*, 388: 161–172.

Welford, J.K., Hearn, E.H., & Clowes, R.M. (2007). The possible role of mid-crustal igneous sheet intrusions in cratonic arch formation. *Tectonics*, 26: TC5012.

West, G.F. & Ernst, R.E. (1991). Evidence from aeromagnetics on the configuration of the Matachewan dykes and the tectonic evolution of the Kapuskasing Structural Zone. *Canadian Journal of Earth Sciences*, 28: 1797–1811.

Whalen, J.B., Currie, K.L., & Chappell, B.W. (1987). A-type granites: geochemical characteristics, discrimination and petrogenesis. *Contributions to Mineral Petrology*, 95: 407–419.

White, J.D.L. & McClintock, M.K. (2001). Immense vent complex marks flood-basalt eruption in a wet, failed rift: Coombs Hills, Antarctica. *Geology*, 29: 935–938.

White, J.D.L., Bryan, S.E., Ross, P.-S., Self, S., & Thordarson, T. (2009). Physical volcanology of continental large igneous provinces: update and review. In

Thordarson, T., Self, S., Larsen, G., Rowland, S.K., & Hoskuldsson, A. (eds.), *Studies in Volcanology: The Legacy of George Walker*. London: Geological Society, Special Publication of IAVCEI, 2, pp. 291–321.

White, R.S. (1997). Mantle plume origin for the Karoo and Ventersdorp flood basalts, South Africa. *South African Journal of Geology*, 100: 271–282.

White, R.S. & McKenzie, D. (1989). Magmatism at rift zones – the generation of volcanic continental margins and flood basalts. *Journal of Geophysical Research: Solid Earth and Planets*, 94: 7685–7729.

White, R.S. & McKenzie, D. (1995). Mantle plumes and flood basalts. *Journal of Geophysical Research*, 100B: 17543–17585.

White, R.S., Westbrook, G.K., Fowler, S.R., *et al.* (1987). Hatton Bank (northwest U.K.) continental margin structure. *Geophysical Journal of the Royal Astronomical Society*, 89: 265–272.

White, R.V. & Saunders, A.D. (2005). Volcanism, impact and mass extinctions: incredible or credible coincidences? *Lithos*, 79: 299–316.

White, W.M. (2010). Oceanic island basalts and mantle plumes: the geochemical perspective. *Annual Review of Earth and Planetary Sciences*, 38: 133–160.

Whitten, J. & Head, J. (2011). Impact basins and impact-induced volcanism: the Lunar Orientale basin as a case study on the Moon. *LIP of the Month*, January 2011. See http://www.largeigneousprovinces.org/11jan.

Whitten, J., Head, J.W., Staid, M., *et al.* (2011). Lunar mare deposits associated with the Orientale impact basin: new insights into mineralogy, history, mode of emplacement, and relation to Orientale basin evolution from Moon Mineralogy Mapper (M^3) data from Chandrayaan-1. *Journal of Geophysical Research: Planets*, 116: E00G09.

Widdowson, M. (1997). Tertiary palaeosurfaces of the SW Deccan, Western India: implications for passive margin uplift. In Widdowson, M. (ed.), *Palaeosurfaces: Recognition, Reconstruction and Palaeoenvironmental Interpretation*. London: Geological Society, Special Publication 120: 221–248.

Widdowson, M. & Cox, K.G. (1996). Uplift and erosional history of the Deccan Traps, India: Evidence from laterites and drainage patterns of the Western Ghats and Konkan Coast. *Earth and Planetary Science Letters*, 137: 57–69.

Widdowson, M., Pringle, M.S., & Fernandez, O.A. (2000). A post K–T boundary (Early Palaeocene) age for Deccan-type feeder dykes, Goa, India. *Journal of Petrology*, 41: 1177–1194.

Wiegand, M., Trumbull, R.B., Greiling, R.O., & Vietor, T. (2011). Magnetic fabric studies on mafic dykes at a volcanic rifted margin in the Henties Bay–Outjo dyke swarm, NW Namibia. In Srivastava, R.K. (ed.), *Dyke Swarms: Keys for Geodynamic Interpretation*. Berlin: Springer-Verlag, pp. 377–399.

Wignall, P.B. (2001). Large igneous provinces and mass extinctions. *Earth Sciences Review*, 53: 1–33.

Wignall, P.B. (2005). The link between Large Igneous Province eruptions and mass extinctions. *Elements*, 1: 293–297.

Wignall, P.B. (2007). The end-Permian mass extinction: how bad did it get? *Geobiology*, 5: 303–309.

Wignall, P.B. & Twitchett, R.J. (1996). Oceanic anoxia and the end Permian mass extinction. *Science*, 272: 1155–1158.

Wilhelms, D.E. (1987). The Geologic History of the Moon: US Geological Survey Professional Paper 1348.

Wilkinson, L., Kjarsgaard, B.A., LeCheminant, A.N., & Harris, J. (2001). Diabase dyke swarms in the Lac de Gras area, Northwest Territories, and their significance to kimberlite exploration: initial results. *Geological Survey of Canada, Current Research*, 2000-C8.

Williams, G.E. & Gostin, V.A. (2000). Mantle plume uplift in the sedimentary record: origin of kilometre-deep canyons within late Neoproterozoic successions, South Australia. *Journal of the Geological Society of London*, 157: 759–768.

Williams, P.J., Barton, M.D., Johnson, D.A., *et al.* (2005). Iron oxide copper–gold deposits: geology, space-time distribution and possible modes of origin. In Hedenquist, J.W., Thompson, J.F.H., Goldfarb, R.J., & Richards, J.P. (eds.), *Economic Geology, 100th Anniversary Volume*. Littleton, CO: Society of Economic Geologists, pp. 371–405.

Wilson, A.H. (1996). The Great Dyke of Zimbabwe. In Cawthorn, R.G. (ed.), *Layered Intrusions*. Amsterdam: Elsevier, pp. 365–402.

Wilson, A.H. & Grant, C.E. (2006). Physical volcanology and compositions of the basaltic lavas in the Archean Nzuse Group, White Mfolozi inlier, South Africa. In Reimold, W.U. & Gibson, R.L. (eds.), *Processes on the Early Earth*. Boulder, CO: Geological Society of America, Special Paper 405, pp. 255–290.

Wilson, A.H. & Chunnett, G. (2006). Trace element and platinum group element distributions and the genesis of the Merensky reef, Western Bushveld Complex, South Africa. *Journal of Petrology*, 47: 2369–2403.

Wilson, A.H. & Hofmann, A. (2013). The Meso-Archaean Pongola Supergroup in South Africa and Swaziland: the oldest known large igneous province on stable continental crust? *LIP of the Month*, June 2013. See: www.largeigneousprovinces. org/jun13.

Wilson, J.F., Jones, D.L., & Kramers, J.D. (1987). Mafic dyke swarms of Zimbabwe. In Halls, H.C. & Fahrig, W.F. (eds.), *Mafic Dyke Swarms*. St John's, NL: Geological Association of Canada, Special Paper 34, pp. 433–444.

Wilson, L. & Head, J.W. (1994). Mars: review and analysis of volcanic eruption theory and relationships to observed landforms. *Reviews of Geophysics*, 32: 221–263.

Wilson, L. & Head, J.W. (2002). Tharsis–radial graben systems as the surface manifestation of plume-related dike intrusion complexes: models and implications. *Journal of Geophysical Research*, 107 (E8): 5057.

Wilson, L. & Head, J.W. (2007). An integrated model of kimberlite ascent and eruption. *Nature*, 447: 53–57.

Wilson, L. & Head, J.W. (2008). Volcanism on Mercury: a new model for the history of magma ascent and eruption. *Geophysical Research Letters*, 35: L23205.

Wilson, L. & Mouginis-Mark, P.J. (1999). Widespread occurrence of dikes within the Olympus Mons aureole materials. Proceedings, 5th International Mars Conference, Pasadena, CA, abstract no. 6050.

Wilson, L. & Mouginis-Mark, P.J. (2003). Phreatomagmatic explosive activity at Hrad Vallis, Mars. *Journal of Geophysical Research: Planets*, 108 (E8): 5082.

Wilson, M. (1997). Thermal evolution of the Central Atlantic passive margins: continental break-up above a Mesozoic super-plume. *Journal of the Geological Society of London*, 154: 491–495.

Wilson, M. & Guiraud, R. (1998). Late Permian to recent magmatic activity on the African–Arabian margin of Tethys. In MacGregor, D.S., Moody, R.T.J. & Clark-Lowes, D.D. (eds.), *Petroleum Geology of North Africa*. London: Geological Society of London, Special Publication 132, pp. 231–263.

Wilson, M. & Patterson, R. (2001). Intraplate magmatism related to short-wavelength convective instabilities in the upper mantle: evidence from the Tertiary–Quaternary volcanic province of western and central Europe. In Ernst, R.E. & Buchan, K.L. (eds.), *Mantle Plumes: Their Identification through Time.* Boulder, CO: Geological Society of America, Special Paper 352, pp. 37–58.

Wilson, R.C.L., Whitmarsh, R.B., Froitzheim, N., & Taylor, B. (2001). Introduction; the land and sea approach. In Wilson, R.C.L., Whitmarsh, R.B., Taylor, B., & Froitzheim, N. (eds.), *Non-Volcanic Rifting of Continental Margins: A Comparison of Evidence from Land and Sea.* London: Geological Society of London, Special Publication 187, pp. 1–8.

Winchester, J.A. & Floyd, P.A. (1977). Geochemical discrimination of different magma series and their differentiation products using immobile elements. *Chemical Geology,* 20: 325–343.

Wingate, M.T.D. (1998). A palaeomagnetic test of the Kaapvaal–Pilbara (Vaalbara) connection at 2.78 Ga. *South African Journal of Geology,* 101: 257–274.

Wingate, M.T.D. & Giddings, J.W. (2000). Age and palaeomagnetism of the Mundine Well dyke swarm, Western Australia: implications for an Australia–Laurentia connection at 755 Ma. *Precambrian Research,* 100, 335–357.

Wingate, M.T.D, Campbell, I.H., Compston, W., & Gibson, G.M. (1998). Ion microprobe U–Pb ages for Neoproterozoic basaltic magmatism in south-central Australia and implications for the breakup of Rodinia. *Precambrian Research,* 87: 135–159.

Wingate, M.T.D., Pisarevsky, S.A., & Evans, D.A.D. (2002). Rodinia connections between Australia and Laurentia: no SWEAT, no AUSWUS? *Terra Nova,* 14: 121–128.

Wingate, M.T.D., Pirajno, F., & Morris, P.A. (2004). Warakurna large igneous province: a new Mesoproterozoic large igneous province in west-central Australia. *Geology,* 32: 105–108.

Wold, R.J. & Hinze, W.J., eds. (1982). Geology and tectonics of the Lake Superior basin. Boulder, CO: Geological Society of America, Memoir 156.

Wooden, J.L. Czamanske, G.K., Fedorenko, V.A., *et al.* (1993). Isotopic and trace-element constraints on mantle and crustal contributions to Siberian continental flood basalts, Norilsk Area, Siberia. *Geochimica et Cosmochimica Acta,* 57: 3677–3704.

Woolley, A.R. (1989). The spatial and temporal distribution of carbonatites. In Bell, K. (ed.), *Carbonatites: Genesis and Evolution.* London: Unwin Hyman, pp. 15–37.

Woolley, A.R. & Church, A.A. (2005). Extrusive carbonatites: a brief review. *Lithos,* 85: 1–14.

Woolley, A.R. & Kjarsgaard, B.A. (2008a). Carbonatite occurrences of the world: map and database. *Geological Survey of Canada, Open File 5796.*

Woolley, A.R. & Kjarsgaard, B.A. (2008b). Paragenetic types of carbonatite as indicated by the diversity and relative abundances of associated silicate rocks: evidence from a global database *Canadian Mineralogist,* 46: 741–752.

Woolley, A.R., Bergman, S.C., Edgar, A.D., *et al.* (1996). Classification of lamprophyres, lamproites, kimberlites, and the kalsilitic, melilitic, and leucitic rocks. *Canadian Mineralogist,* 34: 175–186.

Worsley, T.R., Nance, R.D., & Moody, J.B. (1986). Tectonic cycles and the history of the Earth's biogeochemical and paleoceanographic record. *Paleoceanography,* 1: 233–263.

Wyllie, P.J. (1988). Solidus curves, mantle plumes, and magma generation beneath Hawaii. *Journal of Geophysical Research*, 93: 4171–4181.

Wyman, D.A. (1999). A 2.7 Ga depleted tholeiite suite: evidence of plume–arc interaction in the Abitibi belt, Canada. *Precambrian Research*, 97: 27–42.

Wyman, D.A. & Kerrich, R. (2002). Formation of Archean continental lithospheric roots: the role of mantle plumes. *Geology*, 30: 543–546.

Wyman, D.A. & Kerrich, R. (2009). Plume and arc magmatism in the Abitibi Subprovince: implications for the origin of Archean Continental Lithospheric Mantle. *Precambrian Research*, 168: 4–22.

Wyman, D.A. & Kerrich, R. (2010). Mantle plume–volcanic arc interaction: consequences for magmatism, metallogeny, and cratonization in the Abitibi and Wawa Subprovinces, Canada. *Canadian Journal of Earth Sciences*, 47: 565–589.

Wyman, D.A. & Kerrich, R. (2012). Geochemical and isotopic characteristics of Youanmi terrane volcanism: the role of mantle plumes and subduction tectonics in the western Yilgarn Craton. *Australian Journal of Earth Sciences*, 59: 671–694.

Xia, L.-Q., Xu, X.-Y., Li, X.-M., Ma, Z.-P., & Xia, Z.-C. (2012). Reassessment of petrogenesis of Carboniferous – Early Permian rift-related volcanic rocks in the Chinese Tianshan and its neighboring areas. *Geoscience Frontiers*, 3: 445–471.

Xiang, Z.Q., Li, H.K., Lu, S.N., et al. (2012). Emplacement age of the gabbro diabase dike in the Hongmen scenic region of Mount Tai, Shandong Province, North China: addeleyite U–Pb precise dating. *Acta Petrologica Sinica*, 28: 2831–2842.

Xie, Q. & Kerrich, R. (1994). Silicate–perovskite and majorite signature komatiites from Archean Abitibi Belt: implications for early mantle differentiation and stratification. *Journal of Geophysical Research*, 99: 15799–15812.

Xie, Q., Kerrich, R., & Fan, J. (1993). HFSE/REE fractionations recorded in three komatiite–basalt sequences, Archean Abitibi greenstone belt: implications for multiple plume sources and depths. *Geochimica et Cosmochimica Acta*, 57: 4111–4118.

Xie, S.C., Pancost, R.D., Huang, J.H., et al. (2007). Changes in the global carbon cycle occurred as two episodes during the Permian–Triassic crisis. *Geology*, 35: 1083–1086.

Xu, L., Lin, Y.-T., Shen, W., et al. (2007). Platinum-group elements of the Meishan Permian–Triassic boundary section: evidence for flood basaltic volcanism. *Chemical Geology*, 246: 55–64.

Xu, X. & Song, S.-G. (2014). Potential superplume-related Neoproterozoic (850–820 Ma) LIPs in Central–Western China. *LIP of the Month*, January 2014. See http://www.largeigneousprovinces.org/14jan.

Xu, Y.-G. & He, B. (2007). Thick, high-velocity crust in the Emeishan large igneous province, southwestern China: evidence for crustal growth by magmatic underplating or intraplating. In Foulger, G.R. & Jurdy, D.M. (eds.), *Plates, Plumes, and Planetary Processes*. Boulder, CO: Geological Society of America, Special Paper 430, pp. 841–858.

Xu, Y.-G., He, B., Chung, S.-L., Menzies, M.A., & Frey, F.A. (2004a). Geologic, geochemical, and geophysical consequences of plume involvement in the Emeishan flood-basalt province. *Geology*, 32: 917–920.

Xu, Y.-G., He, B., & Chung, S.-L. (2004b). The Emeishan large igneous province: integrated geologic, geophysical and geochemical evidence for a fossil mantle plume. *LIP of the Month*, October 2004. See http://www.largeigneousprovinces. org/04oct

Xu, Y.-G., Wei, X., Luo, Z.Y., Liu, H.Q., & Cao, J. (2014b). The Early Permian Tarim Large Igneous Province: a synthesis. *Lithos*, in press.

Yale, L.B. & Carpenter, S.J. (1998). Large igneous provinces and giant dike swarms: proxies for supercontinent cyclicity and mantle convection. *Earth and Planetary Science Letters*, 163: 109–122.

Yarmolyuk, V.V., Budnikov, S.V., Kovalebnko, V.I., *et al.* (1997). Geochronology and geodynamic position of the Angara-Vitim batholith. *Petrologiya*, 5: 451–466.

Yingst, R.A. & Head, J.W. (1997). Volumes of lunar lava ponds in South Pole–Aitken and Orientale basins: implications for eruption conditions, transport mechanisms, and magma source regions. *Journal of Geophysical Research*, 102: 10909–10931.

Youbi, N., Kouyaté, D., Söderlund, U., *et al.* (2013). The 1750 Ma Magmatic Event of the West African Craton (Anti-Atlas, Morocco). *Precambrian Research*, 236: 106–123.

Young, G.M. (2013). Evolution of Earth's climatic system: evidence from ice ages, isotopes, and impacts. *GSA Today*, 23: 4–10.

Yuan, F., Zhou, T.F., & Zhang, D.Y. (2008). Genesis of the native copper mineralization in eastern Tianshan. *Geochimica et Cosmochimica Acta*, 72: A1066.

Yuen, D.A., Maruyama, S., Karato, S-i., & Windley, B.R., eds. (2007). *Superplumes: Beyond Plate Tectonics*. Berlin: Springer.

Zachos, J., Pagani, M., Sloan, L., Thomas, E., & Billups, K. (2001). Trends, rhythms, and aberrations in global climate 65 Ma to present. *Science*, 292: 686–693.

Zachos, J.C., Wara, M.W., Bohaty, S., *et al.* (2003). A transient rise in tropical sea surface temperature during the Paleocene-Eocene Thermal Maximum. *Science*, 302: 1551–1554.

Zachos, J.C., Röh, U., Schellenberg, S.A., *et al.* (2005). Rapid acidification of the ocean during the Paleocene–Eocene Thermal Maximum. *Science*, 308: 1611–1615.

Zalasiewicz, J., Cita, M.B., Hilgen, F., *et al.* (2012). Chronostratigraphy and geochronology: a proposed realignment. *GSA Today*, 23: 4–8.

Zegers, T.E., de Wit, M.J., Dann, J., & White, S.H. (1998). Vaalbara, Earth's oldest assembled continent? A combined structural, geochronological and palaeomagnetic test. *Terra Nova*, 10: 250–259.

Zentilli, M., Brooks, R.R., Helgason, J., Ryan, D.E., & Zhang, H. (1985). The distribution of gold in volcanic rocks of eastern Iceland. *Chemical Geology*, 48: 17–28.

Zhai, Q.-G., Jahn, B.-M., Su, L., *et al.* (2013). SHRIMP zircon U–Pb geochronology, geochemistry and Sr–Nd–Hf isotopic compositions of a mafic dyke swarm in the Qiangtang Terrane, northern Tibet and geodynamic implications. *Lithos*, 174: 28–43.

Zhang, D.-Y., Zhang, Z.-C., Santosh, M., *et al.* (2013). Perovskite and baddeleyite from kimberlitic intrusions in the Tarim large igneous province signal onset of an end-Carboniferous mantle plume. *Earth and Planetary Science Letters*, 361: 238–248.

Zhang, M., O'Reilly, S.Y., Wang, K.-L., Hronsky, J., & Griffin, W.L. (2008). Flood basalts and metallogeny: the lithospheric mantle connection. *Earth-Science Reviews*, 86: 145–174.

Zhang, S.-H., Zhao, Y., & Santosh, M. (2012b). Mid-Mesoproterozoic bimodal magmatic rocks in the northern North China craton: implications for magmatism related to breakup of the Columbia supercontinent. *Precambrian Research*, 222–223: 339–367.

Zhang, S.-L., Li, Z.-X., Evans, D.A.D., *et al.* (2012a). Pre-Rodinia supercontinent Nuna shaping up: a global synthesis with new paleomagnetic results from North China. *Earth and Planetary Science Letters*, 353–354: 145–155.

Zhao, D. (2001). Seismic structure and origin of hotspots and mantle plumes. *Earth and Planetary Science Letters*, 192: 251–265.

Zhao, J.-X., Malcolm, M.T., & Korsch, R.J. (1994). Characterisation of a plume-related *c.* 800 Ma magmatic event and its implications for basin formation in central-southern Australia. *Earth and Planetary Science Letters*, 121: 349–367.

Zhao, J.-X. & McCulloch, M.T. (1993). Melting of a subduction-modified continental lithospheric mantle: evidence from Late Proterozoic mafic dike swarms, in central Australia. *Geology*, 21: 463–466.

Zhao, T.-P., Zhou, M.-F., Zhai, M.-G., & Xia, B. (2002). Paleoproterozoic rift-related volcanism of the Xiong'er group, North China craton: implications for the breakup of Columbia. *International Geology Review*, 44: 336–351.

Zhong, S. (2006). Constraints on thermochemical convection of the mantle from plume heat flux, plume excess temperature and upper mantle temperature. *Journal of Geophysical Research*, 111 (B04409).

Zhou, M.-F., Robinson, P.T., Lesher, C.M., *et al.* (2005). Geochemistry, petrogenesis and metallogenesis of the Panzhihua gabbroic intrusion and associated Fe–Ti–V oxide deposits. Sichuan provinces, SW China. *Journal of Petrology*, 46: 2253–2280.

Zhou, M.-F., Ma, Y., Yan, D.P., *et al.* (2006). The Yanbian Terrane (Southern Sichuan Province, SW China): a Neoproterozoic arc assemblage in the western margin of the Yangtze Block. *Precambrian Research*, 144: 19–38.

Zhu, B.-G., Hu, Y.-G., Chang, X.-Y., Xie, J., & Zhang, Z.-W. (2005). The Emeishan large igneous province originated from magmatism of a primitive mantle plus subducted slab. *Russian Geology and Geophysics*, 46: 904–921.

Zhu, D.-C., Mo, X.X., Pan, G.T., *et al.* (2008). Petrogenesis of the earliest Early Cretaceous basalts and associated diabases from Cona area, eastern Tethyan Himalaya in south Tibet: interaction between the incubating Kerguelen plume and eastern Greater India lithosphere? *Lithos*, 100: 147–173.

Zhu, D.-C., Chung, S.-L., Mo, X.-X., *et al.* (2009). The 132 Ma Comei–Bunbury large igneous province: remnants identified in present-day southeastern Tibet and southwestern Australia, *Geology*, 37: 583–586.

Zhuravlev, A.Y. & Wood, R.A. (1996). Anoxia as the cause of the mid–Early Cambrian (Botomian) extinction event. *Geology*, 24: 311–314.

Zindler, A. & Hart, S.R. (1986). Chemical geodynamics. *Annual Review of Earth and Planetary Sciences*, 14: 493–571.

Zirakparvar, N.A., Baldwin, S.L., & Vervoort, J.D. (2013). The origin and geochemical evolution of the Woodlark Rift of Papua New Guinea. *Gondwana Research*, 23: 931–943.

Zoback, M.L. (1992). First- and second-order patterns of stress in the lithosphere: the World Stress Map project. *Journal of Geophysical Research*, 97: 11703–11728.

Index

Lightning Source UK Ltd.
Milton Keynes UK
UKHW02f1157300318
320227UK00010B/201/P

Stephen Simpson/Divine Images/Media Bakery

PART V
Adolescence

Pixel-Shot/Shutterstock.com

Teaching and Learning Aids

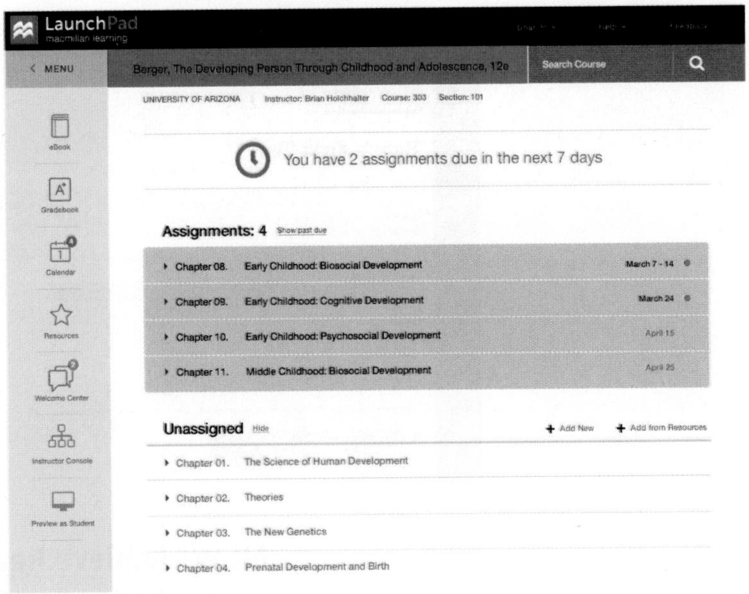

Supplements can make or break a class, as I and every other experienced instructor know. Instructors use many electronic tools that did not exist a few decades ago. The publisher's representatives are trained every year to guide students and professors in using the most effective media for their classes. I have adopted texts from many publishers; the Worth representatives are a cut above the rest. Ask them for help with media, with testing, and with content.

LaunchPad with Developing Lives, LearningCurve Quizzing, and Data Connections Activities

Built to solve key challenges in the course, LaunchPad provides everything students need to prepare for class and exams, and it gives instructors everything they need to set up a course, shape the content, craft lectures, assign homework, and monitor the learning of each student and the class as a whole.

LaunchPad (preview at launchpadworks.com) includes:

- An **interactive e-book**, which integrates videos and short self-assessments, and includes an interface that simplifies highlighting and notetaking.
- **Developing Lives**, the robust interactive experience in which students "raise" their own virtual child. This simulation integrates more than 200 videos and animations, with quizzes and questions to assign and assess.
- **Data Connections**, interactive activities that allow students to interpret data.
- **LearningCurve adaptive quizzing**, based on current research on learning and memory. It combines individualized question selection, immediate and valuable feedback, and a gamelike interface to engage students. Each LearningCurve quiz is fully integrated with other resources in LaunchPad through the Personalized Study Plan, so students can review using Worth's extensive library of videos and activities. Question analysis reports allow instructors to track the progress of individuals and the entire class.
- Worth's **Video Collection for Human Development** is an extensive archive of video clips and activities that covers the full range of the course, from classic experiments (like Ainsworth's Strange Situation and Piaget's conservation task) to illustrations of many topics. Instructors can assign these videos to students through LaunchPad or choose from 50 activities that combine videos with short-answer and multiple-choice questions. (For presentations, our videos are also available on flash drive.) The Video Assignment Tool makes it easy to assign and assess video-based activities and projects, and provides a convenient way for students to submit video coursework.
- **iClicker** offers active learning simplified, and now includes the REEF mobile app (iClicker.com). iClicker's simple, flexible tools in LaunchPad help you give students a voice and facilitate active learning in the classroom. Students can use iClicker remotes, or the REEF mobile app on their phone, tablet, or laptop to participate more meaningfully. LaunchPad includes a robust collection of iClicker questions for each chapter—readily available for use in your class.

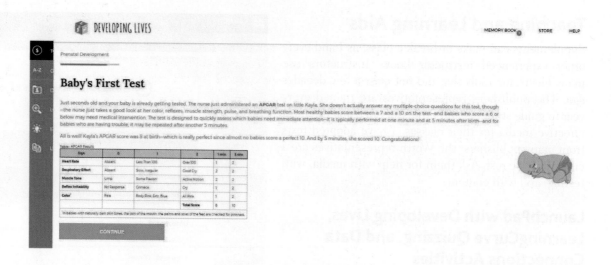

NEW! Achieve Read & Practice

Achieve Read & Practice combines LearningCurve adaptive quizzing and our mobile, accessible e-book in one easy-to-use and affordable product. Among the advantages of Achieve Read & Practice are the following:

- It is easy to get started;
- Students are better prepared: They can read and study in advance;
- Instructors can use analytics to help their students; and
- Students learn more.

Instructor's Resources

Now fully integrated with LaunchPad, and available with Achieve Read & Practice, this collection has been hailed as the richest collection of instructor's resources in developmental psychology. Included are learning objectives, topics for discussion and debate, handouts for student projects, course-planning suggestions, ideas for term projects, and a guide to videos and other online materials.

Test Bank

The test bank includes at least 100 multiple-choice and 70 fill-in-the-blank, true–false, and essay questions for every chapter. Good test questions are crucial; each has been carefully crafted. More challenging questions are included, and all questions are keyed to the textbook by topic, page number, and level of difficulty. Questions are also organized by NCLEX, NAEYC, and APA goals and Bloom's taxonomy. Rubrics for grading short-answer and essay questions are also suggested.

The **Macmillan Learning Test Bank** is an online assessment system that allows you to create and deliver tests through a secure online test center. You can:

- Create paper or online tests that you can export to your LMS using your web browser;
- Drag and drop questions to create tests; and
- Create and edit your own questions and edit publisher-created question sets.

The gradebook in LaunchPad enables instructors to: (1) record students' grades; (2) sort student records; (3) view detailed analyses of test items; (4) curve tests; (5) generate reports; and (6) weight some items more than others.

Thanks

I would like to recognize the academic reviewers who have read this book in every edition and who have provided suggestions, criticisms, references, and encouragement. They have all made this a better book. I want to thank those of you who have reviewed this edition and the prior editions, including:

Chris Alas, *Houston Community College*

Adrienne Armstrong, *Long Star College*

William Robert Aronson, *Florida International University*

T.M. Barratt, *Arizona State University*

Gina Brelsford, *Pennsylvania State University — Harrisburg*

Melissa Bright, *University of Florida*

Carol Carielli, *CUNY Kingsborough Community College*

Alda Cekrezi, *Lone Star College*

Margaret Cooley, *Eastern Michigan University*

Kristi Cordell-McNulty, *Angelo State University*

Shannon Coulter, *Moorpark College*

Barbara Crosby, *Baylor University*

Faith T. Edwards, *University of Wisconsin — Oshkosh*

Naomi Ekas, *Texas Christian University*

Michael Erickson, *Hawaii Pacific University*

Diane Kleiger Feibel, *University of Cincinnati — Blue Ash College*

Lori Neal Fernald, *The Citadel Military College*

Valerie C. Flores, *Loyola University Chicago*

Stacie Foster, *Arizona State University*

Christopher Gade, *Berkeley City College*

Erica Gelven, *Quinnipiac University*

Dan Grangaard, *Austin Community College*

Jerry Green, *Tarrant County College*

Jiansheng Guo, *California State University — East Bay*

Pinar Gurkas, *Clayton State University*

E. Allison Hagood, *Arapahoe Community College*

Tony Stepter Harris, *Virginia State University*

Raquel Henry, *Lone Star College*

Danelle Hodge, *California State University — San Bernadino*

Regina Kijewski, *Naugatuck Valley Community College*

Vernell Larkin, *Hopkinsville Community College*

Richard Marmer, *American River College*

Jerry Marshall, *Green River College*

T. Darin Matthews, *The Citadel Military College*

Elizabeth McCarroll, *Texas Women's University*

Alejandra Albarran Moses, *California State Univeristy — Los Angeles*

Cindy Sledge, *Lone Star College — CyFair*

Bernadette Stunder, *Holy Family University*

Donna Vaught, *University of North Carolina — Wilmington*

Christie Verarde, *Chabot College*

Kelly A. Warmuth, *Providence College*

Debbie Whitmer, *Shasta College*

Angela Williamson, *Tarrant County College*

Kathleen Stassen Berger

New York
August 2020

APPLICATION TO DEVELOPING LIVES PARENTING SIMULATION INTRODUCTION AND PRENATAL DEVELOPMENT

In the Introduction module of Developing Lives, you will begin to customize the developmental journey of your child with information about your personality, cognitive abilities, and demographic characteristics. Next, as you progress through the Prenatal simulation module, how you decide the following will impact the biosocial, cognitive, and psychosocial development of your baby.

	Biosocial	Cognitive	Psychosocial
	• Will you modify your behaviors and diet during pregnancy? • Will you find out the gender of your baby prior to delivery? • What kind of delivery will you and your partner plan for (in the hospital with medication, at home with a doula, etc.)?	• Are you going to talk to your baby while he or she is in the womb? • How much does your baby understand during prenatal development?	• How will your relationship with your partner change as a result of the pregnancy? • Will you begin bonding with your baby prior to birth?

The Beginnings

The science of human development includes many beginnings. Each of the first four chapters of this text forms one corner of a solid foundation for our study.

Chapter 1 introduces definitions and dimensions, explaining research strategies and methods that help us understand how people develop. The need for science, the power of culture, and insights from the life-span perspective are explained.

Without theories, our study would be only a jumble of haphazard observations. **Chapter 2** provides organizing guideposts: Five clusters of theories, each leading to hypotheses and controversies, are described.

Heredity is explained in **Chapter 3**. Genes never act alone, yet no development—anywhere in the body or brain of anyone at any time—is unaffected by epigenetics.

Chapter 4 details the growth of each developing person from a single-celled zygote to a breathing, grasping newborn. Many circumstances—from the mother's diet to the father's care, from the place of birth to the customs of the culture—affect every moment of embryonic and fetal growth.

As you see, the science and the wonder of human life begin long before the first breath. Understanding the beginnings prepares us to understand all the rest. ●●

Left: Leren Lu/DigitalVision/Getty Images
Right: GlobalStock/E+/Getty Images

The Science of Development

What Will You Know?

1. Why is science crucial for understanding how people develop?
2. Are all children the same, or does time and place make each unique?
3. Can we experiment on people to learn how they develop?
4. What questions about child development are hard to ask?

When I was 4 years old, journalists came to our house to photograph my mother and me, wearing matching dresses. I was bathed and dressed, toys were put away, my mother wore lipstick and perfume. While she was getting ready, I found a scissors and cut my hair. My mother stopped me before I could finish, but some tufts were short. She laughed, tying bows to make my hair presentable. I remember none of this, but Mom told this anecdote many times. There are photographs to prove it.

Several aspects of this event suggest another era or place, including photographers coming to homes, mother–daughter dresses, lipstick, ribbons, sharp scissors that a child can reach.

Are you puzzled? Troubled? Do you think it normal for children to misbehave and for mothers to laugh, or do you see defiance or neglect? Would you have punished me if I were your child? Why did Mom laugh?

Everyone was once a child and likely experienced dozens of incidents like this one. How powerful are genes, culture, or child rearing? Is each person's life and experience unique, or is every child like every other child? This chapter shows how to find answers.

Understanding How and Why

The **science of human development** seeks to understand how and why people—all kinds of people, everywhere, of every age—change or remain the same over time. Development is *multidirectional, multicontextual, multicultural, multidisciplinary,* and *plastic,* five terms that will be explained soon. The goal is for the 7.8 billion people on Earth, especially the 2.6 billion under age 20, to fulfill their potential (population estimates are from United Nations, Department of Economic and Social Affairs, Population Division, 2019).

First, however, we need to emphasize that developmental study is a *science.* It depends on theories, data, analysis, critical thinking, and sound methodology, just like every other science. All scientists ask questions and seek answers in order to ascertain "how and why."

This is especially useful when scientists study people: Lives depend on it.

- Should mothers give birth at home or in hospitals?
- Should babies sometimes cry themselves to sleep?

- How, when, and for what should children be punished?
- Should schools encourage independence or obedience, be optional or required, begin at age 2 or 6?
- Should education continue through high school or college, be free or expensive?

People disagree about almost everything related to child development. That is why we need science.

The Scientific Method

Science begins with curiosity. Biologists seek to understand how cells function, chemists to investigate the interactions of the elements, physicists to fathom the force of black holes, astronomers to learn what is beyond our galaxy, developmentalists to wonder about everything people do and think.

Step by Step

As you surely realize, facts may be twisted, opinions may lead people astray, and false assumptions may overwhelm data. To counter prejudices and narrow observations, scientists follow the five steps of the **scientific method** (see **Figure 1.1**):

1. *Begin with curiosity.* Ask questions: Consider theory, study research, and gather observations, all of which suggest issues to be studied.
2. *Develop a hypothesis.* Form a **hypothesis**, a prediction that can be verified. Theories are comprehensive ideas; hypotheses are specific predictions.
3. *Test the hypothesis.* Design a study and conduct research to gather **empirical evidence** (data) to test the hypothesis. When human development is the focus, pay special attention to the selection of participants, the measurement of variables, and the assumptions of the researchers. (More on this later in the chapter.)
4. *Draw conclusions.* Use evidence (Step 3) to support or refute the hypothesis (Step 2). Consider alternative interpretations, unexpected results, and possible biases.
5. *Report the results.* Share data, conclusions, and limitations. Suggest further research to explore new questions and to corroborate conclusions (back to Step 1).

Thus, developmental scientists begin with curiosity and then collect data, drawing conclusions after analyzing the evidence. The process arises not only from observations and personal experience but also from careful study of the reports (Step 5) of other scientists. Each researcher builds on the past, considers the present, and predicts the future—always using critical thinking.

science of human development The science that seeks to understand how and why people of all ages and circumstances change or remain the same over time.

scientific method A way to answer questions using empirical research and data-based conclusions.

hypothesis A specific prediction that can be tested.

empirical evidence Evidence that is based on observation, experience, or experiment; not theoretical.

FIGURE 1.1

Process, Not Proof Built into the scientific method—in questions, hypotheses, tests, and replication—is a passion for possibilities, especially unexpected ones.

1. Curiosity

2. Hypothesis

3. Test

4. Analyze data and draw conclusions

5. Report the results

Replication

replication Repeating a study, usually using different participants, perhaps of another age, SES, or culture.

Repeating a study's procedures and methods with different participants is called **replication**. Scientists study the reports of other scientists (Step 5) and build on what has gone before (back to Step 1). Sometimes they try to duplicate a study exactly, with different participants, to find out if the results apply to everyone. Often they follow up with closely related research (Stroebe & Strack, 2014). Conclusions are revised, refined, rejected, or confirmed after replication.

Obviously, the scientific method is not foolproof. Scientists may draw conclusions too hastily, misinterpret data, or ignore alternative perspectives. Very rarely (about 1 in 5,000 reports), data are falsified (Brainard, 2018). Conclusions are not generally accepted (called *robust*) until several studies confirm them. Ideally, results are replicated not only by other researchers repeating the design of the study (Step 3) but also by other studies with somewhat different designs to verify, refine, and extend the hypothesis (Larzelere et al., 2015).

Of course, no study will exactly mirror another because humans and cultures differ (De Boeck & Jeon, 2018). Perfect replication is impossible; that is why new research is inspired by what has gone before. A View from Science shows this process in more detail.

A VIEW FROM SCIENCE

Music and the Brain

An illustration of the scientific method at work comes from research on music and the mind. Does music make people smarter? That question arose from the results of one study 30 years ago, in which 36 college students who listened to Mozart for 10 minutes had higher scores on tests of spatial intelligence than other students who did not hear the music (Rauscher & Shaw, 1993).

That study did not replicate or generalize, but the popular press did not wait. Nor did some politicians. The governor of Georgia in the mid-1990s proposed a budget item of $105,000 to give every Georgia newborn a free CD of classical music. (The legislature voted it down.) Some popular manuals for parents advocated playing classical music for children to help their minds.

Although the original study was flawed in many ways, the topic intrigued scientists, who asked new questions (Step 1). Dozens of studies have since investigated the relationship between music and cognition (Perlovsky et al., 2013).

Sometimes the researchers investigate tiny details. For example, one study compared what happens in the brain when someone listens to Mozart versus listening to Bach (Verrusio et al., 2015).

Another group of researchers examined how rats react to hearing Mozart (Sampaio et al., 2017). Studying rats may seem odd; however, the reactions of rodents, dogs, and other primates might suggest hypotheses regarding people. The impact of music on learning has become a thriving area of research within psychology.

The evidence shows that the noises of hospital nurseries affect the newborn brain, and preterm infants benefit from hearing music, not machines (Lordier et al., 2019). Further, children with special needs may benefit from learning via music (Sharda et al., 2018).

Does music benefit typical children? Decades of research have found that merely listening to music does not aid cognition, but that learning to play a musical instrument may advance intelligence and academic achievement (Herholz & Zatorre, 2012; Rose et al., 2019). The hypothesis (Step 2) is that learning to play an instrument also teaches emotional control, creativity, and memory, which are all crucial cognitive skills. But tests of that hypothesis find more nuanced results.

In one study, three groups of 6-year-olds, all from similar low-income families, were compared. Each week, one group spent six to seven hours learning to play an instrument (usually the violin), one group spent three to four hours in a special sports program (soccer or swimming), and a third group had no after-school programing. After two years, the researchers found (Step 4) that children in the music group improved on several measures of brain development and cognition (Habibi et al., 2018). Surprisingly, the sports group also improved more than the children with no special lessons. The researchers suggested that:

> participation in activities other than music may in fact be associated with . . . [cognitive skills] . . . provided that the activities are socially interactive and comparably motivating and engaging.

[Habibi et al., 2018, p. 79]

This research demonstrates a basic lesson of science: Do not leap to conclusions. Any one study—even one that fails to replicate—can raise new questions. Questions, data, replication, surprises, and alternative explanations are the basics of science.

The Nature–Nurture Controversy

An easy example of the need for science concerns a great puzzle of life, the *nature–nurture debate*. **Nature** refers to the influence of the genes that people inherit. **Nurture** refers to environmental influences, beginning with the health, diet, and stress of the future person's mother at conception, and continuing lifelong, including experiences in the family, school, community, and nation.

The nature–nurture debate has many manifestations, among them *heredity–environment*, *maturation–learning*, and *sex–gender*. Under whatever name, the basic question is, "How much of any characteristic, behavior, or emotion is the result of genes, and how much is the result of experience?"

Born That Way?

Some people believe that most traits are inborn, that children are innately good ("an innocent child") or bad ("beat the devil out of them"). Others stress nurture, crediting or blaming parents, or neighborhoods, or drugs, or even additives in the food, when someone is good or bad, a hero or a villain.

Neither belief is accurate. As one group of scholars explain, human characteristics are neither born nor made (Hambrick et al., 2018). Genes and the environment *both* affect every characteristic: Nature always affects nurture, and then nurture affects nature. Any attempt to decide exactly how much of a trait is genetic and how much is environmental is bound to fail, because genetic and environmental influences continuously interact, sometimes increasing the influence of the other and sometimes decreasing it.

A further complication is that the impact of any good or bad experience might be magnified or inconsequential, depending on particular genes or past events. For example, a beating, or a beer, or a blessing might be a turning point, or it might not matter.

Every adult can remember a remark, or an incident, in their childhood that is seared in memory because it caused pain, or shame, or joy. That very remark might have been forgotten by the person who said it, or it might have been said to another child with no effect. Each aspect of nature and nurture depends on other aspects of nature and nurture in ways that, themselves, vary because of the nature and nurture of each individual.

Differential Susceptibility

This illustrates **differential susceptibility** (Ellis et al., 2011a). People vary in how sensitive they are to particular words, or drugs, or experiences, either because of the particular genes they have inherited or because of events years earlier.

Asthma is an obvious example. Because of genes, some people begin wheezing when they are near a cat, but others never do. Yet because of past experience (primarily the reactions of adults), some children are terrified at the first signs of an attack; others are nonchalant.

A more dramatic example involves dogs as well as cats (Krzych-Fałta et al., 2018). If a person lives in a rural area, fur-bearing pets reduce the rate of asthma, but in urban areas, such animals increase the incidence. That is differential susceptibility.

Developmentalists use a floral metaphor to express this idea. Some people are like *dandelions*—hardy, growing and thriving in good soil or bad, with or without ample sun and rain. Other people are like *orchids*—quite wonderful, but only under ideal growing conditions (Ellis & Boyce, 2008; Laurent, 2014). The child who takes asthma or any other illness in stride is a dandelion; the one who experiences terror may be an orchid.

Chopin's First Concert Frederick Chopin, at age 8, played his first public concert in 1818, before photographs. But this photo shows Piotr Pawlak, a contemporary prodigy playing Chopin's music in the same Polish Palace where that famous composer played as a boy. How much of talent is genetic, and how much is cultural? This is a nature–nurture question that applies to both boys, 200 years apart.

JANEK SKARZYNSKI/AFP/Getty Images

nature In development, nature refers to the traits, capacities, and limitations that each individual inherits genetically from their parents.

nurture In development, nurture includes all of the environmental influences that affect the individual after conception. This includes everything from the mother's nutrition while pregnant to the cultural influences in the nation.

differential susceptibility The idea that people vary in how sensitive they are to particular experiences. Often such differences are genetic, which makes some people affected "for better or for worse" by life events. (Also called *differential sensitivity*.)

THINK CRITICALLY: Why not assign a percent to nature and a percent to nurture so that they add up to 100 percent?

WHAT HAVE YOU LEARNED?

1. What are the five steps of the scientific method?
2. What is the difference between asking a question (Step 1) and developing a hypothesis (Step 2)?
3. Why is replication important for scientific progress?
4. What basic question is at the heart of the nature–nurture controversy?
5. When in development does nature begin to influence nurture?
6. Are you more like a dandelion or an orchid? Give an example.

Childhood and Adulthood

In the first half of the twentieth century, the focus of developmental psychology was on children. It was thought that once a person was fully grown, development stopped. But as the study of development expanded to include cognitive and emotional development, it became clear that development occurs lifelong.

Each life reflects what has happened in earlier years. This text focuses on children, but you will see many reminders that what happens before age 18 affects what happens after that. Accordingly, when we study children, we note how childhood influences adulthood. The five perspectives that arose from the study of the lifespan—multidirectional, multicontextual, multicultural, multidisciplinary, and plasticity—apply to every age of human development.

Development Is Multidirectional

The study of development highlights "how and why people change over time." At every stage, gains and losses occur. A simplistic understanding of the direction of development from birth to death—up, steady, and down—is imprecise and sometimes flat-out wrong.

Continuity and Discontinuity

Instead, multiple changes, in every direction, characterize life. Traits appear and disappear, with increases, decreases, and zigzags (see **Figure 1.2**). An earlier idea—that growth advances until about age 18, steadies, and then declines—has been soundly rejected. Do not let anyone tell you that a child was "born bad" or that someone will never change.

FIGURE 1.2
Patterns of Developmental Growth Many patterns of developmental growth have been discovered. Although linear progress seems most common, scientists now find that almost no aspect of human change follows the linear pattern exactly.

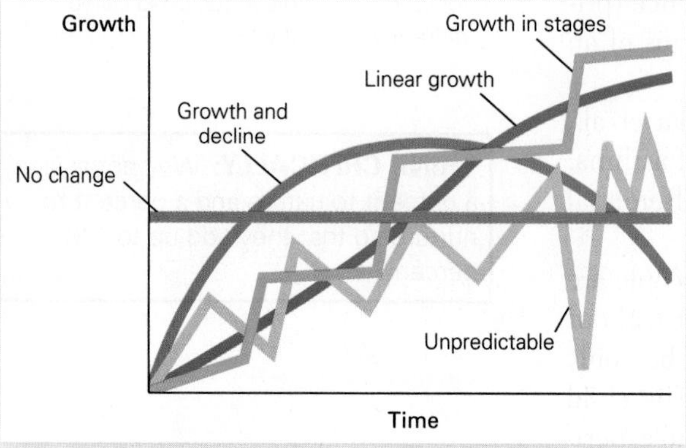

Sometimes *discontinuity* is evident. Change can occur rapidly and dramatically: Caterpillars become butterflies, shy children become social teenagers, and loners fall in love.

Sometimes *continuity* is found. Growth can be gradual: Redwoods add rings over hundreds of years; toddlers stand, take steps, walk, and run. Humans experience simple growth, radical transformation, improvement, and decline as well as stability, stages, and continuity—day to day, year to year, and generation to generation.

Critical and Sensitive Periods

The timing of losses and gains, impairments or improvements, is affected by age and maturation. Some changes are sudden and profound because of a **critical period**, which is either when something *must* occur to ensure normal development or the *only* time when an abnormality might occur.

For instance, the human embryo grows arms and legs, hands and feet, fingers and toes, each over a critical period between 28 and 54 days after conception. After that, it is too late: Unlike some insects, humans never grow replacement limbs.

We know this because of a tragedy. Between 1957 and 1961, thousands of newly pregnant women in 30 nations took *thalidomide*, an antinausea drug. This change in nurture (via the mother's bloodstream) disrupted nature (the embryo's genetic program). If an expectant woman ingested thalidomide during the critical period for limb formation, her newborn's arms or legs were malformed or absent (Moore et al., 2015). Whether all four limbs or just arms, hands, or fingers were missing depended on exactly when the drug was taken. If thalidomide was ingested only after day 54, no harm occurred.

Life has few critical periods. Often, however, a particular development occurs more easily—but not exclusively—at a certain time. That is called a **sensitive period**.

An example is learning language (Werker & Hensch, 2015). If children do not communicate in their first language between ages 1 and 3, they might do so later (these years are not critical), but their grammar is impaired (these years are sensitive). Childhood is a particularly sensitive period for learning to pronounce a second or third language with a native accent: Some brilliant immigrants master the nuances of the English language, sometimes publishing profound and riveting novels, and yet their speech still reflects their native tongue.

One review of the brain structures for second-language learning strongly recommends hearing both languages during the first year of life. The evidence suggests that adults who learn a new language never master a native accent "even after years of practice, and despite high proficiency in all other aspects of language function" (Berken et al., 2017, p. 222). The critical period for hearing and repeating nuances of accent was over.

Development Is Multicontextual

The second insight is that "human development is fundamentally contextual" (Pluess, 2015, p. 138). Among the many contexts are three obvious clusters: physical surroundings (including climate, noise, streets, trees), family structures (legal status, family size, ages of relatives), and community characteristics (urban, suburban, or rural; diverse or not).

The Social Context

Humans are intensely social creatures, so a fourth category, the social context, is especially crucial. The social context includes everyone who influences each individual, both immediately and over time, both directly and indirectly.

For example, a 20-year-old college student might be persuaded to stop by a party instead of heading to the library. The host furnishes the setting, with drinks, food, and music. The people—friends, acquaintances, strangers—provide the immediate social context, and that affects whether the student will stay till 3 A.M. or leave soon after arriving. The next morning, that social context will affect the student in class—another social context.

In the same way, we each experience several contexts each day, some by choice and some involuntarily, and they affect our later thoughts and actions. A hard day

Domov Is Home? This is Milan Kundera, world-famous author of *The Unbearable Lightness of Being*. He left his Czech *domov* (home) at age 46, becoming a French citizen but writing in Czech until age 64. He now writes in French, with a literate French native to check his work. For example, native speakers know that la maison, le domicile, and chez moi all mean home; they use the appropriate one with no problem. Now Kundera writes fluent French, and speaks it too—with an accent.

critical period A time when something *must* happen for normal development to occur.

sensitive period A time when something (such as a toxin, or nutrient, or experience) has the greatest impact on development, but recovery is possible if it occurs later.

ecological-systems approach A perspective on human development that considers all of the influences from the various contexts of development. (Later renamed *bioecological theory*.)

microsystem In Bronfenbrenner's ecological approach, the immediate social contexts that directly affect each person, such as family, peer group, work team.

exosystem In Bronfenbrenner's ecological approach, the community institutions that affect the immediate contexts, such as churches and temples, schools and colleges, hospitals and courts.

macrosystem In Bronfenbrenner's ecological approach, the overarching national or cultural policies and customs that affect the more immediate systems, such as the effect of the national economy on local hospitals (an exosystem) or on families (a microsystem).

chronosystem In Bronfenbrenner's ecological approach, the impact of historical conditions (wars, inventions, policies) on the development of people who live in that era.

mesosystem In Bronfenbrenner's ecological approach, a connection between one system and another, such as parent–teacher conferences (connecting home and school) or workplace schedules (connecting family and job).

at work might make a parent impatient with the children—and then the children might pick a fight with a friend.

In these examples, the effect is almost immediate, but much of development is affected by contexts over the long term. Childhood social play may affect adult work habits, for instance. A child's family and neighborhood contexts predispose adult psychological disorders—with differential susceptibility (Keers & Pluess, 2017).

Ecological Systems

A leading developmentalist, Urie Bronfenbrenner (1917–2005), emphasized contexts. Just as a naturalist studying an organism examines the ecology (the relationship between the organism and its environment) of a tiger, or tree, or trout, Bronfenbrenner told developmentalists to take an **ecological-systems approach** (Bronfenbrenner & Morris, 2006).

This approach recognizes three nested levels (see **Figure 1.3**).

- Most obvious is the **microsystem**—each person's immediate social contexts, such as family and peer group.
- Also important is the **exosystem** (local institutions such as school and church).
- Beyond that is the **macrosystem** (the larger setting, including cultural values, economic policies, and political processes).

Two more systems affect these three.

- One is the **chronosystem** (literally, "time system"), which is the historical context.
- The other is the **mesosystem**, consisting of the connections among the other systems.

FIGURE 1.3

The Ecological Model According to developmental researcher Urie Bronfenbrenner, each person is significantly affected by interactions among a number of overlapping systems, which provide the context of development. Microsystems—family, peer group, classroom, neighborhood, house of worship—intimately and immediately shape human development. Surrounding and supporting the microsystems are the exosystems, which include all the external networks (such as community structures and local educational, medical, employment, and communications systems) that affect the microsystems. Influencing both of these systems is the macrosystem, which includes cultural patterns, political philosophies, economic policies, and social conditions. Mesosystem refers to interactions among systems, as when parents and teachers coordinate to educate a child. Bronfenbrenner eventually added a fifth system, the chronosystem, to emphasize the importance of historical time.

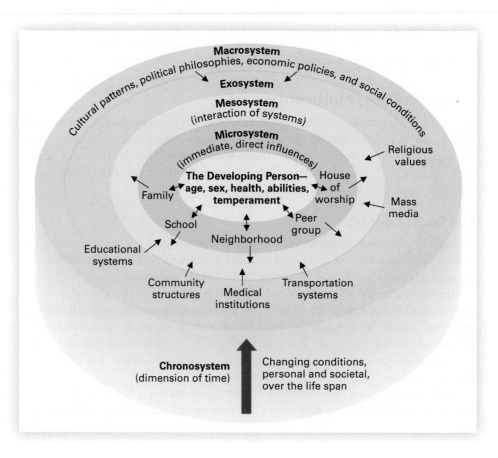

Before he died, Bronfrenbrenner renamed his theory *bioecological* to emphasize the impact of the person, including their genes and body systems. Increasingly, he and all developmentalists recognized biological systems, because signals from the stomach, lungs, heart, hormones, and neurotransmitters affect all the other systems. Recently, the *microbiome*, which is all the microbes within each person, is also recognized as a crucial part of each person (more on this in Chapter 3).

In the student-at-party example above, the student's immediate social circle is the microsystem, the college culture (is it a "party school?") is the exosystem, and the national emphasis on higher education is part of the macrosystem. The party itself is a mesosystem, in that it connects the microsystem and the exosystem.

Every gathering reflects the chronosystem. Party food and drink vary with each decade, and even the fact of a college party reflects history: Many 20-year-olds a century ago were already parents, and if they were in college, they spent evenings studying, not partying. Further, the food, drink, and air at the party affect the biosystem, with some substances changing how people feel and interact.

Bronfenbrenner's perspective remains useful, especially when considering children within families (Ferguson & Evans, 2019). For example, children who have been sexually abused are likely to be abused again, in childhood and adulthood. Why? Is there something amiss in the person, or is the fault in the culture?

Psychologists have used Bronfenbrenner's systems approach to answer this question, finding causes in all systems simultaneously (Pittenger et al., 2016). For example, an abused adult is likely to have been abused in their family as a child (macrosystem), to belong to a church that is silent about sexuality (exosystem), and to live within a culture that does not prosecute abuse (macrosystem and chronosystem). Past abuse affects brain networks (biosystem).

The Historical Context

Bronfenbrenner's chronosystem is useful in understanding social movements, including current ones such as Black Lives Matter or Extinction Rebellion or Me Too. In each case, a historical shift occurred. For example, people are now separating plastic and paper, composting scraps, and recycling bags. That illustrates historical change.

All persons born within a few years of one another are called a **cohort**, a group defined by its members' shared age. Cohorts travel through life together, at each age experiencing particular values, events, and technologies. Imagine being a child before contraception, or cell phones, or social media. A person's entire life would be affected. Although many of the elderly now have electronic devices and modern medical care, their assumptions and values reflect an earlier era, which gives rise to the dismissive saying "OK, boomer."

Even something that seems private and personal reflects history. Consider a person's name (see **Table 1.1**). If you know someone named Emma, she is probably young: Emma is the most common name for girls born between 2014 and 2018 but was not in the top 1,000 in 1990. If you know someone named Mary, she is probably old: Mary was the first or second most popular name from 1900 to 1965; now only 1 baby in 800 is named Mary.

These rankings are from the United States (U.S. Social Security Administration, 2019). Variation is evident by nation, by culture, and by region—within the United States and elsewhere. For example, Wyatt is the third most common boy's name in Alaska, but it's not in the first 100 in California.

cohort People born within the same historical period who therefore move through life together, experiencing the same events, new technologies, and cultural shifts at the same ages. For example, the effect of the Internet varies depending on what cohort a person belongs to.

TABLE 1.1

Most Popular First Names by Cohort

Girls

2018: Emma, Olivia, Ava, Isabella, Sophia

1998: Emily, Hannah, Sarah, Samantha, Ashley

1978: Jennifer, Melissa, Jessica, Amy, Heather

1958: Mary, Susan, Linda, Karen, Patricia

1938: Mary, Barbara, Patricia, Betty, Shirley

Boys

2018: Liam, Noah, William, James, Oliver

1998: Michael, Jacob, Matthew, Joshua, Christopher

1978: Michael, Jason, Christopher, David, James

1958: Michael, David, James, Robert, John

1938: Robert, James, John, William, Richard

Information from U.S. Social Security Administration, 2019.

The Socioeconomic Context

Some scholars believe that the economic context is even more important than the historical one. Each person's **socioeconomic status**, abbreviated **SES** (SES is basic to understanding human development, so the abbreviation is used often in this text), affects every stage of development. SES is not solely about income. Socioeconomic status also reflects education, occupation, and neighborhood. The combination affects development much more than money alone.

Imagine two U.S. families comprised of an infant, an unemployed mother, and a father who earns less than $20,000 a year. That family's income is markedly below the 2019 federal poverty line for a family of three ($21,330).

Suppose one family depends on the father's minimum wage job, which provides $16,965 annual income ($7.25 per hour × 45 hours a week × 52 weeks), with neither parent a high school graduate, living in a run-down neighborhood. Their SES would be low. Suppose the other family is headed by a father who is a postdoctoral student living on campus with his college graduate wife. Even though his income might be a grant of $20,000 a year, that family is middle class.

Particularly significant for our study of human development are SES differences by age and cohort (see **Figure 1.4**). In the United States, 50 years ago, the poor were most often elderly, but Medicare changed that. Now young children with young parents suffer most: Poverty in early childhood reduces academic achievement even more than poverty during adolescence (Wagmiller et al., 2015).

There are many reasons. Before age 5, children are almost totally dependent on their families and communities. Since excellent early education is pivotal for later success, and since low-SES parents have little money, limited education, and inadequate housing and thus cannot afford good preschools or provide excellent learning at home, their young children suffer.

This varies by microsystem (nation and community). Some jurisdictions provide excellent early education for all children. It also varies by family. If young children have no cousins or siblings who might need attention from

socioeconomic status (SES) A person's position in society as determined by income, occupation, education, and place of residence. (Sometimes called *social class*.)

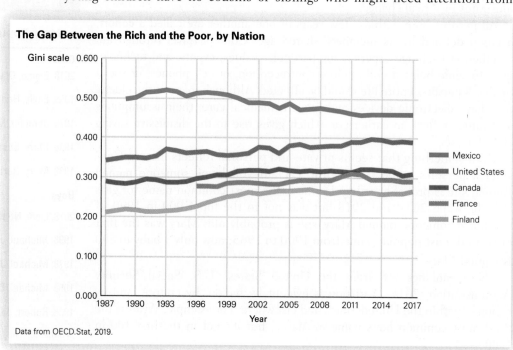

FIGURE 1.4

Top and Bottom 1 Percent The gap between rich and poor (measured by the Gini coefficient) is greatest in developing nations, because the poor are below subsistence. Worldwide, inequality is reduced recently, in that there are fewer of the most destitute families. The United States, as you see, is trending toward greater inequality. If both the United States and Mexico continue in their respective trends, Mexico will be closer to income equality than the United States by 2030.

The Gap Between the Rich and the Poor, by Nation

Mexico
United States
Canada
France
Finland

Data from OECD.Stat, 2019.

nearby grandparents and great-grandparents, they might benefit from family love and money, even though their parents struggle to pay the rent.

In this example, it is apparent that deprivation begins with income, the *capital* referred to in capitalism. However, *family capital* (the support that families provide for each other) and *social capital* (the support that comes from neighbors, religious institutions, and the larger society) create the context for development (Lin, 2017).

> **DATA CONNECTIONS: Economic Equality and Human Development** explores various nations' Gini indices, which measure economic equality in a particular society.

Development Is Multicultural

In order to learn about "all kinds of people, everywhere, at every age," it is necessary to study people of many cultures. For social scientists, **culture** is "the system of shared beliefs, conventions, norms, behaviors, expectations and symbolic representations that persist over time and prescribe social rules of conduct" (Bornstein et al., 2011, p. 30).

Social Constructions

Thus, culture is far more than food or clothes; it is a set of ideas, beliefs, and patterns. Culture is a powerful **social construction**, that is, a concept created, or *constructed*, by a society. Social constructions affect how people think and act—what they value, ignore, and punish.

Each group of people creates a culture, which means there are ethnic cultures, national cultures, family cultures, college cultures, and so on. Thus, everyone is multicultural, sometimes finding that one culture clashes with another. One of my students wrote:

> My mom was outside on the porch talking to my aunt. I decided to go outside; I guess I was being nosey. While they were talking I jumped into their conversation which was very rude. When I realized what I did it was too late. My mother slapped me in my face so hard that it took a couple of seconds to feel my face again.
>
> *[C., personal communication]*

Notice that my student reflects her family culture; she labels her own behavior "nosey" and "very rude." She later wrote that she expects children to be seen but not heard and that her own son makes her "very angry" when he interrupts.

However, her "rude" behavior may have been encouraged by the culture of her school, because she attended a public school far from her mother's native land. In the United States, many teachers want children to speak up; children's talk is welcomed. Do you think my student was nosey or, on the contrary, that her mother should not have slapped her? Your answer reflects your culture.

Deficit or Just Difference?

As with my student's mother, everyone is inclined to believe that their culture is better than others. This tendency has benefits: Generally, people who appreciate their own culture are happier, prouder, and more willing to help strangers. However, that belief becomes destructive if it reduces respect for people from other groups. Too quickly and without thought, differences are assumed to be problems (Akhtar & Jaswal, 2013).

culture A system of shared beliefs, norms, behaviors, and expectations that persist over time and prescribe social behavior and assumptions.

social construction An idea that is built on shared perceptions, not on objective reality. Many age-related terms (such as *childhood, adolescence, yuppie,* and *senior citizen*) are social constructions, connected to biological traits but strongly influenced by social assumptions.

> **CHAPTER APP 1**
>
> SDGs in Action
>
> **IOS:**
> https://tinyurl.com/u6hpjwl
>
> **ANDROID:**
> https://tinyurl.com/uo4vaqp
>
> **RELEVANT TOPIC:**
> Income inequality and social justice
>
> Organized around the United Nations' Sustainable Development Goals (a global to-do list for tackling poverty, inequality, and climate change), this app educates users about challenges in the world today and motivates them to take action.

Myrto Papadopoulos/The Washington Post/Getty Images

Difference, But Not Deficit This Syrian refugee living in a refugee camp in Greece is quite different from the aid workers who assist there. But the infant, with a pacifier in her mouth and a mother who tries to protect her, illustrates why developmentalists focus on similarities rather than on differences.

⬤⬤ **Observation Quiz** Two religions that sometimes clash are evident here. What are they? (see answer, page 31) ↑

difference-equals-deficit error The mistaken belief that a deviation from some norm is necessarily inferior to behavior or characteristics that meet the standard.

ethnic group People whose ancestors were born in the same region and who often share a language, culture, and religion.

race A group of people who are regarded by themselves or by others as distinct from other groups on the basis of physical appearance, typically skin color. Social scientists think race is a misleading concept, as biological differences are not signified by outward appearance.

Developmentalists recognize the **difference-equals-deficit error**, which is the belief that people unlike us (different) are inferior (deficit). Sadly, when humans realize that their ways of thinking and acting are not universal, they may believe that people who think or act differently are to be pitied, feared, and encouraged to change.

The difference-equals-deficit error is one reason that a careful multicultural approach is necessary. Never assume that another culture is wrong and inferior—or the opposite, right and superior. Assumptions can be harmful.

For example, one immigrant child, on her first day in a U.S. school, was teased about the food she brought for lunch. The next day, she dumped the contents of her lunchbox in the garbage soon after she arrived at school, choosing hunger over being different.

This example illustrates the problem of difference and deficit. A child's lunch of another culture might, or might not, be more nutritious than conventional peanut butter and jelly sandwiches. In this example, the children noticed only difference; the mothers thought the lunch they packed was best; the student wanted to be accepted. The difference was misjudged as a deficit, and then it harmed that girl (hungry children learn less).

Ethnic and Racial Groups

Cultural clashes fuel wars and violence when differences are seen as deficits. To prevent that, we need to understand the terms *ethnicity* and *race*. Members of an **ethnic group** almost always share ancestral heritage and often have the same national origin, religion, and language. Ethnicity is a social construction, a product of the social context, not biology.

Ethnic groups often share a culture, but not necessarily. There are "multiple intersecting and interacting dimensions" to ethnic identity (Sanchez & Vargas, 2016, p. 161). People may share ethnicity but differ culturally (e.g., people of Irish descent in Ireland, Australia, and North America), and people of one culture may come from several ethnic groups (consider British culture). [**Developmental Link:** The major discussion of ethnic identity is in Chapter 16.]

Historically, most North Americans thought that ethnicity was cultural but that race was an inborn biological characteristic that differentiated one group from another. A person could not be multiracial. Races were categorized by skin color: white, black, red, and yellow (Coon, 1962). It was not obvious that color words are a gross exaggeration of relatively minor physical differences. No one has skin that is white (like this page) or black (like these letters) or red or yellow.

Biologists now recognize **race** as a social construction. Genetic diversity is evident within groups, and genes are shared by people with quite different appearance. One team writes:

> We believe the use of biological concepts of race in human genetic research—so disputed and so mired in confusion—is problematic at best and harmful at worst. It is time for biologists to find a better way.

[*Yudell et al., 2016, p. 564*]

The fact that race is a social construction does not make it meaningless, given history. Adolescents who are proud of their racial ancestry are likely to achieve academically, resist drug addiction, and feel better about themselves (Zimmerman, 2013).

The relationship between racial pride and racial prejudice is complex. Ethnic pride, but not personal experiences with discrimination, predicts more positive attitudes about oneself and about other groups. In the United States, this was found in a study of 15- to 25-year-old Black and Hispanic youth but was not apparent among White youth—who usually did not consider themselves as belonging to a racial group (Sullivan & Ghara, 2015).

There is an interesting historical twist on that. A hundred years ago, Americans of Irish, Italian, and Greek ethnicity were *not* considered part of the White race (Gordon, 2017). This confirms, of course, that race is social construction.

Intersectionality

Intersectionality begins with the idea that we each are pushed and pulled—sometimes strongly, sometimes weakly, sometimes by ourselves, sometimes by authorities—by our gender, religion, generation, nation, age, culture, and ethnicity. Our many identities interact with and influence each other (see **Figure 1.5**). Intersectionality then recognizes that those identities can be used to divide—White women versus Black women, Asian men versus Latino men, immigrant children versus immigrant elders, and so on—instead of uniting us, as all women, or all immigrants, or, even better, as all people.

Intersectionality focuses attention on power differences between groups, bringing special attention to the needs of people who are simultaneously in several marginalized groups. They are most harmed when their intersectional identities are ignored. For example, when Crenshaw (1989) first introduced the term "intersectionality," she recognized that the courts allowed discrimination against African American women because the laws did not acknowledge that racism and sexism combined to harm them.

Intersectionality highlights discrimination in many institutions. For example, do judges give African Americans harsher prison sentences than European Americans for the same crimes? The data say yes. That is unfair, but unfairness may be deeper than that.

A careful study of all sentences meted out to convicted criminals in Pennsylvania found more age and gender disparities than ethnic ones. For the same crime, young adults were sentenced more harshly than older ones, and men more harshly than women (Steffensmeier et al., 2017). Thus, young Black men may be particularly misjudged; older White women particularly protected. Note that this was in Pennsylvania. Would this be better or worse in Minnesota or Montana, Bermuda or Brazil? We need more data to know.

More generally, like interlocking gears, systems of social categorization and group power intersect to influence everyone, every day. How would your life be different if you were of another gender, ethnicity, family background, sexual orientation, health, ability, and so on? Cultures matter for everyone.

Development Is Multidisciplinary

In order to examine each aspect of human growth, development is often considered in three domains—*biosocial*, *cognitive*, and *psychosocial*. Each domain is the focus of several academic disciplines:

- Biosocial includes biology, neuroscience, and medicine.
- Cognitive includes psychology, linguistics, and education.
- Psychosocial includes economics, sociology, and history.

Typically, each scholar follows a particular thread within one discipline and one domain, using clues and conclusions from other scientists

Mike Coppola/Getty Images Entertainment/Getty Images

Fitting In The best comedians are simultaneously outsider and insider, giving them a perspective that helps people laugh at the absurdities in their lives. Trevor Noah—son of a Xhosa South African mother and a German Swiss father—grew up within, yet outside, his native culture. For instance, he was seen as "Coloured" in his homeland but as "White" on a video, which once let him escape arrest!

intersectionality The idea that the various identities need to be combined. This is especially important in determining if discrimination occurs.

FIGURE 1.5

Identities Interacting We all are in the middle, with many identities. Our total selves are affected by them all, with variation by culture and context as to which are more salient.

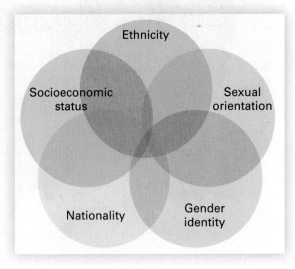

Ethnicity

Socioeconomic status

Sexual orientation

Nationality

Gender identity

Diverse Complexities

It is often repeated that "the United States is becoming more diverse," a phrase that usually refers only to ethnic diversity and not to economic and religious diversity (which are also increasing and merit attention). From a developmental perspective, two other diversities are also important—age and region, as shown below. What are the implications for schools, colleges, employment, health care, and nursing homes in the notable differences in the ages of people of various groups? And are attitudes about immigration, or segregation, or multiracial identity affected by the ethnicity of one's neighbors?

THE CHANGING ETHNIC AND RACIAL MAKEUP OF THE UNITED STATES

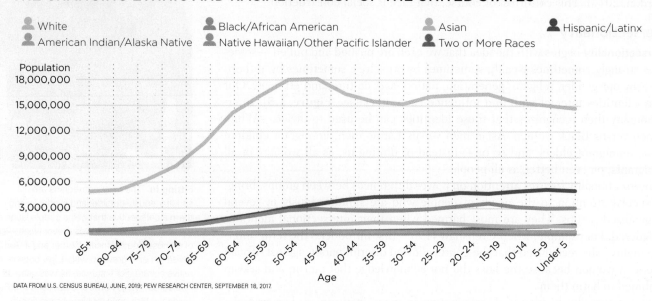

White
American Indian/Alaska Native
Black/African American
Native Hawaiian/Other Pacific Islander
Asian
Two or More Races
Hispanic/Latinx

DATA FROM U.S. CENSUS BUREAU, JUNE, 2019; PEW RESEARCH CENTER, SEPTEMBER 18, 2017.

Regional Differences in Ethnicity Across the United States

In the United States, there are both regional and age differences in ethnicity. This map shows which counties have an ethnic population greater than the national average. Counties where more than one ethnicity or race is greater than the national average are shown as multiethnic. Areas for which data are unavailable are left unshaded.

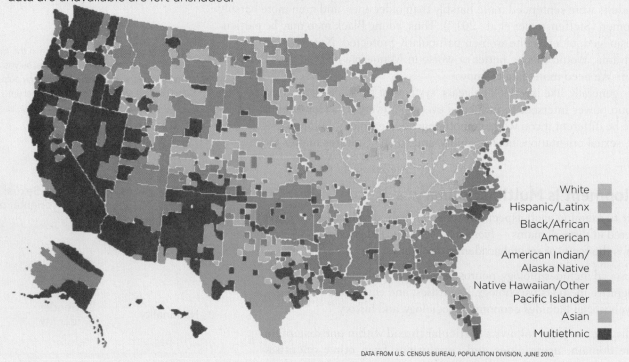

White
Hispanic/Latinx
Black/African American
American Indian/Alaska Native
Native Hawaiian/Other Pacific Islander
Asian
Multiethnic

DATA FROM U.S. CENSUS BUREAU, POPULATION DIVISION, JUNE 2010.

who have concentrated on that same thread. They try to avoid the *silo effect*, named after the silos on farms that store grain until it is needed. Those silos typically are tall, round towers, standing alone. In academia, public health, and business, the silo effect is the tendency of each discipline and organization to communicate only with others in the same silo. That reduces their scope, insight, and effectiveness (Tett, 2015).

By contrast, a comprehensive understanding of development requires insights and information from many scientists, past and present, in many disciplines. This is evident when trying to understand any one person. A child who has trouble learning in school might need family counseling, glasses, treatment for chronic illness, recognition of religious prohibitions, tutoring to overcome dyslexia, and more. No single element will solve the child's problem. As in this example, our understanding of every person, and every topic, benefits from a multidisciplinary understanding.

Criminology and Medicine Punishment of criminals and the health of newborns are central concerns of the disciplines of criminology and medicine, not psychology or political science. However, here you see 100 psychologists about to inform the U.S. Congress about the hazards of using restraints (e.g., handcuffs) on women giving birth, which is the practice in many prisons. As in this example, many human problems require a multidisciplinary approach.

Lloyd Wolf

Genetics

Avoiding silos is ideal but not always achieved. Consider genetics. When the human genome was first mapped nearly two decades ago, some people assumed that humans became whatever their genes destined them to be—heroes, killers, or ordinary people. However, multidisciplinary research quickly showed otherwise.

Yes, genes affect every aspect of behavior. But even monozygotic twins, with identical genes, differ biologically, cognitively, and socially. [**Developmental Link**: Twins are discussed in Chapter 3.] Among the many reasons each is unique are position in the womb, non-DNA influences in utero, birthweight, birth order, and then dozens of influences throughout life (Carey, 2012). Thus, to understand any person, one must consider findings from many disciplines: genes, nutrition, psychology, sociology, education, and more. Breaking out of silos can illuminate topics that mystified scientists who were stuck in their own discipline.

A dramatic example comes from autism spectrum disorder (ASD). Autism was not recognized as distinct from schizophrenia until about 1940, and then it remained in the silo of psychiatry for decades. For most of the twentieth century, psychiatrists traced psychological disorders to early mother–child relationships.

Accordingly, since many children with ASD do not look at people, or let their mothers hug them, some blamed cold, unaffectionate "refrigerator mothers," reflecting a perspective of many psychiatrists (Bettelheim, 1972). Thousands of mothers suffered from that judgment (Jack, 2014). Then a reversal came from biology, specifically genetics.

Geneticists collected data proving that autism is inborn. A discipline once thought to be irrelevant to mental disorders was now considered pivotal for autism spectrum disorder, schizophrenia, mood disorders, and so on (Kandel, 2018). However, although the genetic evidence released mothers from guilt, it convinced people that nothing could be done to help such children. Then another silo weighed in: Educators found that children "on the spectrum" can be taught language and social skills.

Soon neuroscientists found signs of autism in the brain; nutritionists suggested dietary changes; medical doctors prescribed drugs; public health workers were appalled when parents blamed vaccinations; demographers traced the increase in cases (now estimated among U.S. 8-year-old boys at about 1 in 23 [Baio et al., 2018]); and so on.

We still do not know all the causes and treatments of ASD. [**Developmental Link**: Autism spectrum disorder is discussed in Chapter 11.] But, we do know that insights from many disciplines are needed. If anyone suggests *the* cause or *the* cure for autism spectrum disorder, that is a mistaken silo approach.

Overall, multidisciplinary research broadens and deepens our knowledge of human development. People are complex, and to properly grasp all the systems—from the workings of the microbiome in the gut to the effects of climate change throughout the entire world—requires a multidisciplinary approach. Adding to this complexity: People change. That leads to the final theme of the life-span perspective, plasticity.

Development Is Plastic

plasticity The idea that abilities, personality, and other human characteristics can change over time. Plasticity is particularly evident during childhood, but even older adults are not always "set in their ways."

The term **plasticity** denotes two complementary aspects of development: Human traits can be molded (as plastic can be), yet people maintain a certain durability of identity (as plastic does). The concept of plasticity in development provides both hope and realism—hope because change is possible, and realism because development builds on what has come before.

Dynamic Systems

dynamic-systems approach A view of human development as an ongoing, ever-changing interaction between the physical, cognitive, and psychosocial influences. The crucial understanding is that development is never static but is always affected by, and affects, many systems of development.

Plasticity is basic to our contemporary understanding of human development. This is evident in the **dynamic-systems approach**. The idea is that human development is an ongoing, ever-changing interaction between the body and mind and between each person and every aspect of the environment. This includes all the systems of the ecological-systems approach.

Note the word *dynamic*: Physical contexts, emotional influences, the passage of time, each person, and every aspect of the ecosystem are always active, always in flux, always in motion.

Thus, a dynamic-systems approach to understanding the role of fathers takes into account the sex and age of the child, the role of the mother, and the cultural norms of fatherhood. The result is a complex mix of complementary effects—and this dynamic mix affects each child in diverse ways (Cabrera, 2015).

Plasticity is also apparent in social norms. Inside the Brain highlights plasticity in culture as well as in the brain.

INSIDE THE BRAIN

Thinking About Marijuana

Brains are affected by drugs, for better or worse. This is evident both in structure (parts of the brain change size because of drug exposure) and activity (connections between neurons are strengthened or weakened).

Consider marijuana. First, the brain. Fear arising from the amygdala and pleasure from the basal ganglia precede drug use, and the size and activity of both are powerfully affected by childhood. Then those parts of the brain continue to be molded, evident in how various generations of adults think about marijuana as well as whether they use the drug.

The plasticity of attitudes is dramatic. In the United States in the 1930s, marijuana was declared illegal. The

1936 movie *Reefer Madness* was shown until about 1960, with vivid images connecting marijuana with a warped brain, suicide, and insanity. Before 1960, most Americans feared and shunned "weed."

Then, that 1960s cohort of adolescents (the baby boomers) had many rebels. They listened to music from the Beatles, Bob Dylan, James Brown, and Bob Marley (all of whom sang about smoking marijuana) and rejected adult fears, values, and rules about drugs, sex, and everything else. This is charted in the annual *Monitoring the Future* report (Johnston et al., 1978–2014; Miech et al., 2015 to present).

By 1978, only 12 percent of high school students thought experimental use of marijuana was harmful, and more than half had tried the drug. That frightened the adults, whose emotional reactions to marijuana had been formed decades earlier. President Nixon declared drug abuse (especially marijuana, but not cigarettes or alcohol) "Public Enemy Number One" and Nancy Reagan (First Lady from 1981 to 1989) exhorted, "Just Say No" to drugs.

Marijuana was declared a "gateway drug," because it was thought that if someone walked through that gate they would be led to heroin use and addiction (Kandel, 2002). People were arrested and jailed for possession of even a few grams. By 1991, 80 percent of high school seniors thought there was "great risk" in regular use of marijuana, and only 21 percent of high school seniors had ever smoked it.

Obviously, attitudes continue to shift, in part because the anti-marijuana generation are now quite old (see **Figure 1.6**). Few teenagers think regular use of marijuana is "a great risk." By 2018, marijuana had become legal in Canada, Uruguay, Argentina, Ecuador, and in 30 U.S. states (often only for medical use). According to a Gallup poll, two-thirds of Americans believe smoking marijuana should be legal for adults. About half of all high school seniors have smoked marijuana.

Since we know that the brain is plastic, affected by many drugs, scientists ask "How does marijuana change the brain?" Some find that marijuana relieves pain by decreasing pain neurotransmitters, with less addiction or neurological risk than opioids and improved thinking (Gruber et al., 2018). By contrast, some find a correlation between marijuana use and "structural abnormalities in the brains of young people" (DuPont & Lieberman, 2014, p. 557).

Evidently, not only are attitudes about marijuana plastic, so are its effects on the brain, influenced by the age of the smoker, their genes, and perhaps the specifics of the marijuana (where grown, additives) (Dow-Edwards & Silva, 2017; Mandelbaum & de la Monte, 2017). Marijuana may impair brain structures that support memory and motivation—but not for everyone. Plasticity!

In all these ebbs and flows of public attitudes, scientists have no consensus about the long-term effects of marijuana because the federal government made it illegal for scientists to undertake objective research. Currently, many researchers are concerned that the political push to accept marijuana did not consider the neurological effects.

We know that the effects of the drug on the brain vary and that simply thinking about marijuana triggers neurotransmitters, causing phobias, ecstasy, and emotions between those extremes. We know that adults who smoked marijuana decades ago are a select group, so we cannot be sure their current situation is caused by their past drug use. Are current attitudes (mostly positive) more rational than those (mostly negative) of our great-grandparents? More science is needed.

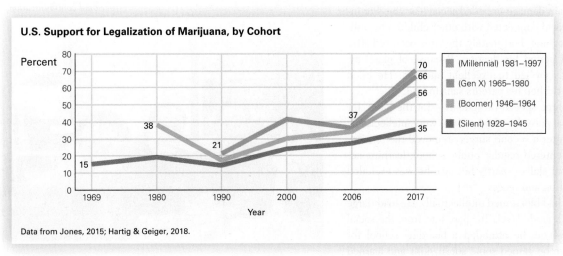

U.S. Support for Legalization of Marijuana, by Cohort

Legend:
- (Millennial) 1981–1997
- (Gen X) 1965–1980
- (Boomer) 1946–1964
- (Silent) 1928–1945

Data from Jones, 2015; Hartig & Geiger, 2018.

FIGURE 1.6

Double Trends Both cohort and generational trends are evident. Note that people of every age are becoming more accepting of marijuana, but the effect is most obvious for adults who never heard about "reefer madness."

Observation Quiz Why is the line for the 1981–1997 cohort much shorter than the line for the older cohorts? (see answer, page 31) ↑

Plasticity underlies all the other four aspects of the life-span perspective: multi-directional, multicontextual, multicultural, and multidisciplinary. With any topic, stage, or problem, the dynamic-systems approach urges consideration of all the interrelated aspects, every social and cultural factor, over days and years.

Individuals, families, and societies are moldable. In the United States 20 years ago, no one predicted that teen births, cigarette smoking, and midlife cardiovascular death would be down by half, or that rates of same-sex marriage and single parenthood would double. Each of those changes has altered development for children and adults.

Plasticity is apparent when developmentalists consider any individual. Every brain and body would have been different if that same person had been born in another culture, grown up in another era, eaten other food, witnessed particular events, with different parents, friends, teachers, and so on. My nephew David is one example.

My Brother's Son David

My sister-in-law, Dot, contracted *rubella* (also called *German measles*) early in her third pregnancy. The disease was not diagnosed until David was born, blind and dying. Immediate heart surgery saved his life. Then surgery to remove a cataract destroyed one eye.

The doctor was horrified that operating had disturbed the virus and killed the eye, so he did not remove the other cataract until David was 6. One dead eye and one thick cataract meant that young David's visual system was severely impaired. It soon became apparent that other parts of his body—thumbs, ankles, teeth, toes, spine, and brain—were also affected.

Blindness and brain damage impeded social learning. For instance, 3-year-old David connected with other children by pulling their hair and laughing. Fortunately, the virus occurred after the critical prenatal period for hearing. Because of plasticity, David's diminished sight led to excellent audition. He cried often but was quieted when his parents sang to him.

David attended three special preschools—for the blind, for children with cerebral palsy, for children who were intellectually disabled. At age 6, after a second surgery removed the remaining cataract, David entered regular public school, learning academic but not social skills—partly because he was excluded from physical education and recess.

By age 10, David had blossomed intellectually. He read (large print) at the eleventh-grade level. To spare him from the social demands of adolescence, he attended a boarding school for blind children, where he gained some social skills and learned to swim. Before age 20, he spoke a second and a third language. In young adulthood, he enrolled in college.

The interplay of developmental systems was evident. David's family and community contexts (a rural Appalachian region, where the culture is more collectivistic) helped him to become a productive and happy adult. He told me, "I try to stay in a positive mood."

David's listening skills continue to be impressive. He once told me:

> I am generally quite happy, but secretly a little happier lately, especially since November, because I have been consistently getting a pretty good vibrato when I am singing, not only by myself but also in congregational hymns in church. [*I asked, what is vibrato? David answered*] When a note bounces up and down within a quartertone either way of concert pitch, optimally between 5.5 and 8.2 times per second.

[personal communication]

My Brother's Children Michael, Bill, and David *(left to right)* are adults now, with quite different personalities, abilities, numbers of offspring (4, 2, and none), and contexts (in Massachusetts, Pennsylvania, and California). Yet despite genes, prenatal life, and contexts, I see the shared influence of Glen and Dot, my brother and sister-in-law—evident here in their similar, friendly smiles.

David works as a translator of German texts, which he enjoys because, as he says, "I like providing a service to scholars, giving them access to something they would otherwise not have."

Plasticity does not undo a person's genes, childhood, or permanent damage. The prenatal brain destruction remains, and David, age 52, still lives with his mother. When his father died in 2014, David was better than the rest of us in accepting the death (he said, "Dad is in a better place"). He comforted his mother. She applied to enter a senior citizen residence, but they would not allow David; he is too young. Instead of getting angry, Dot found another apartment for both of them near one of her other children. She laughingly said that David gives her a reason for living, because in another decade, they will be able to stay in senior housing together.

As his aunt, I have seen him repeatedly overcome handicaps. Plasticity is dramatically evident; my intellectually disabled nephew became a very intelligent adult. His case illustrates all five aspects of the life-span perspective (see **Table 1.2**).

TABLE 1.2

Five Characteristics of Development

Characteristic	Application in David's Story
Multidirectional. Change occurs in every direction, not always in a straight line. Gains and losses, predictable growth, and unexpected transformations are evident.	David's development seemed static (or even regressive, as when early surgery destroyed one eye), but then it accelerated each time he entered a new school or college.
Multidisciplinary. Numerous academic fields—especially psychology, biology, education, and sociology, but also neuroscience, economics, religion, anthropology, history, medicine, genetics, and many more—contribute insights.	Two disciplines were particularly critical: medicine (David would have died without advances in surgery on newborns) and education (special educators guided him and his parents many times).
Multicontextual. Human lives are embedded in many contexts, including historical conditions, economic constraints, and family patterns.	The high SES of David's family made it possible for him to receive daily medical and educational care. His two older brothers protected him.
Multicultural. Many cultures—not just between nations but also within them—affect how people develop.	In Appalachian regions, like the one in which David lived, the culture is more collectivistic and therefore supportive of all members of the community.
Plasticity. Every individual, and every trait within each individual, can be altered at any point in the life span. Change is ongoing, although it is neither random nor easy.	David's measured IQ changed from about 40 (severely intellectually disabled) to about 130 (far above average), and his physical disabilities became less crippling as he matured.

WHAT HAVE YOU LEARNED?

1. What aspects of development show continuity?
2. What is the difference between a critical period and a sensitive period?
3. Why is it useful to know when sensitive periods occur?
4. What did Bronfenbrenner emphasize in his ecological-systems approach?
5. How does cohort differ from age group?
6. What factors comprise a person's SES?
7. How are culture, race, and ethnicity distinct from each other?
8. What is the problem with each discipline having its own silo?
9. What does it mean to say that human development is plastic?

Using the Scientific Method

To verify or refute a hypothesis (Step 2), researchers seek the best of hundreds of research designs, choosing exactly who and what to study, how and when (Step 3), in order to gather results that will lead to valid conclusions (Step 4) that are worth publishing (Step 5). Often they use statistics to discover relationships between various aspects of the data.

To understand the science of development, it helps to understand three research designs, and three strategies to learn how people change or remain the same over time.

Observation

scientific observation A method of testing a hypothesis by unobtrusively watching and recording participants' behavior in a systematic and objective manner—in a natural setting, in a laboratory, or in searches of archival data.

Scientific observation requires researchers to record behavior systematically and objectively. Observations can occur in a naturalistic setting such as a home, or in a laboratory, where scientists observe what people do. It is also possible to analyze data collected for some other reason but observable to researchers, such as records of Twitter words, or cell phone data, or birth statistics.

Observations suggest hypotheses for further study. Especially when children are observed, a major benefit of this method is that the scientist watches what naturally occurs, without artificial conditions.

For example, in one study (Colditz et al., 2019), mothers and their 2-year-olds were video-taped in a playroom set up to encourage interaction. Only the mother–child dyad was in the room, but the tapes were viewed later by observers who recorded negative (child complains, disobeys, and so on) or positive (child is affectionate, engaged, happy) behaviors every 10 seconds.

Behaviors were carefully defined. For example, disobedience was counted if it was direct, such as responding "no" to a specific request. (If the child simply ignored a suggestion, that did not count as disobedience.)

Beyond clear definitions, note two other ways these scientists sought to make observations objective and scientific. First, raters were not in the room, so mothers or toddlers were not affected by having an observer nearby. Second, ratings were made every 10 seconds, so the observers were not influenced by their overall perceptions or by what had happened a minute before. Scientists use many such methods to make observations accurate.

The rating scale used in this study has been replicated, and is now part of a standard observation measure (Carter & Briggs-Gowan, 2006). As in previous research, toddlers displayed about three times as many positive as negative behaviors. This illustrates the benefits of observation: It is good to know that overall, positive behaviors predominate. Perhaps the "terrible twos" is a stereotype.

This particular study used observation partly because these toddlers were a special group: They all had been born two months before they were due. The observation data found that their mix of positive and negative behaviors were similar to the mix for typical 2-year-olds.

The same study examined language and motor skills, again via observation. For that, the researchers used a scale developed by Nancy Bayley, who published a set of norms for the development of infants and young children (Bayley, 1966). Researchers observed how well the toddlers could talk, walk, run, stack blocks and so on. They corrected for gestational age, which means that they subtracted the number of weeks of prematurity from chronological age in order to accurately assess developmental abilities.

Some cognitive differences between preterm and typical children were dramatic: 3 percent of the preterm 2-year-olds could not walk normally, and most were slower in language development compared to babies born on time. Was this nature or nurture? Observation does not tell us, which is why experiments are sometimes needed.

The Experiment

An **experiment** establishes causality. In the social sciences, experimenters typically impose a particular treatment or condition on a group of participants and note how they react. If they respond unlike similar people who have no special treatment, then the treatment made a difference.

Standard Terminology

In technical terms, the experimenters manipulate an **independent variable**, which is the extra treatment or special condition (also called the *experimental variable*; a *variable* is anything that can vary). They note whether this independent variable affects whatever they are studying, called the **dependent variable** (which *depends* on the independent variable).

Thus, the independent variable is the new, special treatment; any change in the dependent variable is the result. The goal is to find out what causes what, that is, whether an independent variable affects the dependent variable.

A typical experiment (as diagrammed in **Figure 1.7**) has two groups of participants. One group, the *experimental group*, experiences the particular treatment or condition (the independent variable); the other group, the *comparison group* (also called the *control group*), does not.

experiment A research method in which the scientist tries to determine the cause-and-effect relationship between two variables by manipulating one (called the *independent variable*) and then observing and recording the ensuing changes in the other (called the *dependent variable*).

independent variable In an experiment, the variable that is introduced to see what effect it has on the dependent variable. (Also called *experimental variable*.)

dependent variable In an experiment, the variable that may change as a result of whatever new condition or treatment the experimenter adds. Thus, the dependent variable *depends* on the independent variable.

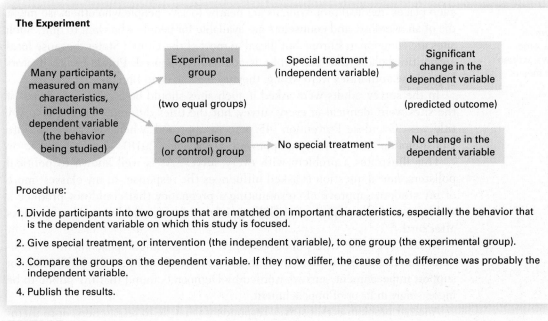

The Experiment

Many participants, measured on many characteristics, including the dependent variable (the behavior being studied)

→ Experimental group → Special treatment (independent variable) → Significant change in the dependent variable

(two equal groups) (predicted outcome)

→ Comparison (or control) group → No special treatment → No change in the dependent variable

Procedure:

1. Divide participants into two groups that are matched on important characteristics, especially the behavior that is the dependent variable on which this study is focused.

2. Give special treatment, or intervention (the independent variable), to one group (the experimental group).

3. Compare the groups on the dependent variable. If they now differ, the cause of the difference was probably the independent variable.

4. Publish the results.

FIGURE 1.7

How to Conduct an Experiment The basic sequence diagrammed here applies to all experiments. Many additional features, especially statistical measures and various ways of reducing experimenter bias, affect whether publication occurs. (Scientific journals reject reports of experiments that were not rigorous.)

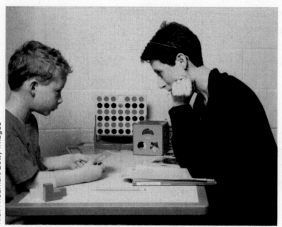

Rick Friedman/Getty Images

What Can You Learn? Scientists first establish what is, and then they try to change it. In one recent experiment, Deb Kelemen (shown here) established that few children under age 12 understand a central concept of evolution (natural selection). Then she showed an experimental group a picture book illustrating the idea. Success! The independent variable (the book) affected the dependent variable (the children's ideas), which confirmed Kelemen's hypothesis: Children can understand natural selection if instruction is tailored to their ability.

◗) Especially for Nurses In the field of medicine, why are experiments conducted to test new drugs and treatments? (see response, page 31)

survey A research method in which information is collected from a large number of people by interviews, written questionnaires, or some other means.

Experimental Design

The observation study on the previous page, of the positive and negative behaviors of the preterm 2-year-olds, also had an experimental and control group. All had received the usual specialized care for underweight infants, but the parents of the experimental group attended hospital classes regarding expectations and appropriate care, and then had lengthy phone consultations with experts after they brought their babies home.

The result: The two groups of babies did not differ in behavior (the positive and negative actions at age 2, as just described), but they did differ in cognition and motor skills (the Bayley scales). More of the intervention group were able to stack three blocks on top of each other, for instance.

You can see why a control group was needed. The crucial question was not whether the experimental group were behind the norms. [They were.] The question was whether they were as behind as another group who were also premature. [They were not.] This experiment confirmed that parents of fragile newborns need extra support, so they can support their infants.

Details of the experimental intervention, such as content and timing, may be crucial (Milgrom et al., 2019). Experiments have revealed that many well-intentioned interventions are not really helpful. That is why we need experiments, to know exactly what interventions cause positive results.

The Survey

A third research method is the **survey**, in which information is collected from a large number of people by interview, questionnaire, or some other means. This is a quick, direct way to obtain data. Surveys keep people from assuming that everyone is like people we know. The wording of a survey may be crucial.

For example, a survey described the epidemic of opioid deaths and then explained life-saving sites, where people can take drugs purchased elsewhere. At such places, medical professionals are nearby to save people who otherwise might die of an overdose, and counselors are available for people who want to quit. Such sites are common in Europe but illegal in most of the United States, because local authorities believe that the public is adamantly opposed. Public health doctors hope those authorities were wrong; they used a survey to find out.

In the survey, adults were asked if such sites should be legal. Descriptions of the sites were identical in every survey, but the titles differed. When the survey title was "Overdose Prevention" 45 percent approved; when the title was "Safe Consumption" only 29 percent approved (Davis et al., 2018).

This illustrates a problem with every survey: As is well known to political pollsters, *how* a question is asked influences the response. In my classes, more of my students approve of "terminating a pregnancy that could not produce a viable infant" than approve of "killing a fetus who could live only a few days after birth."

In late 2019, the U.S. public was surveyed regarding impeachment of President Trump. Surveys worded by Trump supporters found that the public did not support impeachment; surveys worded by Democrats found that the public to be increasingly in favor of impeachment.

Another problem is that survey respondents may lie to researchers, or to themselves. For instance, every two years since 1991, high school students in the United States have been surveyed confidentially in the Youth Risk Behavior Survey. Always a wide and diverse sample is sought: 14,956 students from public, private, and parochial schools answered the 2017 survey (MMWR, June 15, 2018).

One question asked of those 15,000 students was whether they had sexual intercourse *before* age 13. Every year, more ninth-grade boys than eleventh-grade boys say yes, yet those eleventh-graders were ninth-graders two years before (see **Figure 1.8**). Why? Do some teenage boys lie, with ninth-graders bragging? Or have eleventh-graders forgotten? Surveys do not tell us.

Meta-Analysis

As you see, every research design has strengths as well as weaknesses. No single study is conclusive. Some studies might have too few participants, or some unknown bias in the design, or some oddity in the results. Accordingly, some scholars use **meta-analysis**, which combines the results of many studies. When dozens of studies are gathered, a meta-analysis reveals significant trends.

In the best meta-analyses, the researchers begin by stating exactly how they chose studies that are the grist for their analysis. Care is taken to find all relevant studies, published and unpublished, on a particular topic. Only those that are rigorous are included. A meta-analysis "has become widely accepted as a standardized, less biased way to weigh the evidence" (de Vrieze, 2018, p. 1186).

However, even meta-analyses do not always reach clear conclusions. For instance, does media violence encourage actual violence? Some politicians blamed video games for the massing shooting at Marjory Stoneman Douglas High School in Florida. Were they right? Does shooting in video games provoke real killing? One meta-analysis says yes (Anderson et al., 2010), another no (Ferguson & Kilburn, 2010), and a third says yes but not very much (Hilgard et al., 2017).

The main reason for these differences is that, in order to avoid publication bias (journals tend to publish only dramatic results, called the *file-drawer problem*), the authors included unpublished studies. However, they disagreed about which unpublished studies to use.

For scientists, disagreements are healthy: All the foregoing descriptions of methods and analysis confirm that alternate explanations are part of science. Researchers scrutinize the details and data of published studies (Step 5) in order to interpret the results (Step 4).

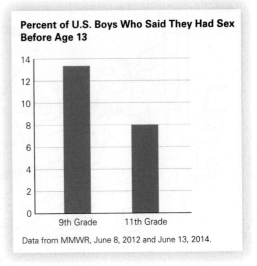

Percent of U.S. Boys Who Said They Had Sex Before Age 13

Data from MMWR, June 8, 2012 and June 13, 2014.

FIGURE 1.8

I Forgot? If these were the only data available, you might conclude that ninth-graders have suddenly become more sexually active than eleventh-graders. But we have 20 years of data—those who are ninth-graders now will answer differently by eleventh grade.

meta-analysis A technique of combining results of many studies to come to an overall conclusion. Meta-analysis is powerful, in that small samples can be added together to lead to significant conclusions, although variations from study to study sometimes make combining them impossible.

Studying Development over the Life Span

In addition to conducting observations, experiments, and surveys, developmentalists must measure how people *change or remain the same over time*, as the definition stresses. Remember that systems are dynamic, ever-changing. To capture that dynamism, developmental researchers use one of three basic research designs: *cross-sectional*, *longitudinal*, and *cross-sequential*.

Cross-Sectional Research

The quickest and least expensive way to study development over time is with **cross-sectional research**, in which groups of people of one age are compared with people of another age. You saw that with attitudes about marijuana. Younger people are more likely to approve of legalization than older people.

Cross-sectional design seems simple. However, the people being compared may differ in several ways, not just age. Attitudes about many political and social issues originate in childhood and are solidified in early adulthood. Since many contemporary older adults were taught that marijuana was a dangerous gateway drug, their opinions are still influenced by what Nancy Reagan said when they were teenagers. Their attitudes about same-sex marriage, about war, about religion, and so on are influenced by what their cohort believed decades ago.

cross-sectional research A research design that compares groups of people who differ in age but are similar in other important characteristics.

"It's a one-year timer. It gives an added sense of urgency to my research grant."

Not Long Enough For understanding the human life span, scientists wish for grants that are renewed for decades.

longitudinal research A research design in which the same individuals are followed over time, as their development is repeatedly assessed.

cross-sequential research A hybrid research design in which researchers first study several groups of people of different ages (a cross-sectional approach) and then follow those groups over the years (a longitudinal approach). (Also called *cohort-sequential research* or *time-sequential research*.)

Longitudinal Research

To help discover whether age itself rather than cohort causes a developmental change, scientists undertake **longitudinal research**. They collect data repeatedly on the same individuals over time. Longitudinal research can span a few years, or the same individuals may be studied for decades. Long-term research requires patience and dedication from many scientists, but it can pay off.

For example, a longitudinal study of 790 infants born in Baltimore to low-income parents found that only 4 percent of them had graduated from college by age 28 (Alexander et al., 2014). Why don't more low-SES young adults complete college? Many people blame high school guidance counselors, or college admission practices, or university faculty, for this result.

However, this longitudinal study began with infants. Researchers traced when children were pushed away from college or toward it. It began very early in life. Early education and friendly neighbors were more influential than high school!

Longitudinal data are cherished by developmentalists; they show when in the life span to focus on better health practices, education, family interactions. For instance, we now know that the best way to reduce lung cancer deaths is not better diagnosis, or advanced chemotherapy, but prevention of adolescent smoking. Longitudinal data show that if people do not start smoking before age 20, they are unlikely to develop lung cancer in later adulthood, because the sensitive period is over. Almost no one begins smoking in adulthood, but many who became addicted as teenagers are still damaging their lungs as adults.

Another study considered when teaching parents better ways to discipline their children is most likely to improve child behavior. It is widely assumed that earlier is better, but a meta-analysis of many studies using longitudinal research found that interventions during middle childhood were as effective as interventions earlier on (Gardner et al., 2019).

Unfortunately, technology, culture, and politics alter life experiences, so data collected on children years ago might not be relevant for today's children. For example, many recent substances and practices might be harmful or beneficial, among them *phthalates* and *bisphenol A (BPA)* (chemicals used in manufacturing plastic containers), *hydrofracking* (a process used to extract natural gas or oil from rocks), *e-waste* (from old computers and cell phones), *chlorpyrifos* (an insecticide), and *electronic cigarettes* (vaping). Some nations and states ban or regulate each of these; others do not. Data showing the long-term effects of these substances and practices are not yet possible.

A current conundrum is climate change. Some predict global civil wars, agriculture failure, disaster deaths—and that life on our planet may be snuffed out because of atmospheric warming that is happening now. Others call that alarmist, suggesting that mitigation can protect humanity (C. Murphy et al., 2018). Waiting for longitudinal data means we will not know until it is too late.

Cross-Sequential Research

Developmentalists use a third strategy, combining cross-sectional and longitudinal methods. This is **cross-sequential research** (also referred to as *cohort-sequential* or *time-sequential*). With this design, scientists begin with people of different ages (a cross-sectional approach), follow them for years (a longitudinal approach), and combine the results.

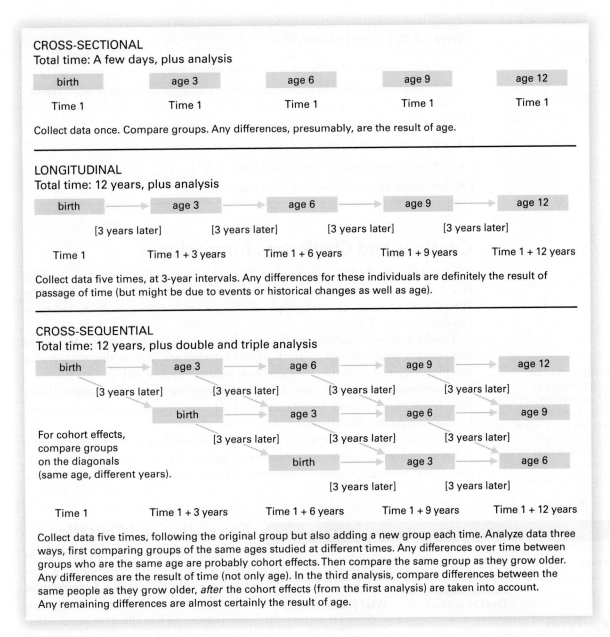

FIGURE 1.9

Which Approach Is Best? Cross-sequential research is the most time-consuming and complex, but it yields the best information. One reason that hundreds of scientists conduct research on the same topics, replicating one another's work, is to gain some advantages of cohort-sequential research without waiting for decades.

A cross-sequential design lets researchers compare findings for, say, 3-year-olds with findings for the same individuals as newborns, as well as with data for people who were 3 long ago, who are now ages 6, 9, and 12 (see **Figure 1.9**). Cross-sequential research is complicated, in recruitment and analysis, but it enables scientists to disentangle age from history.

Cross-sequential research is useful for young adults as well. For example, drug addiction (called *substance use disorder*, or *SUD*) is most common in the early 20s and decreases by the late 20s. But one cross-sequential study found that the origins of SUD are in adolescent behaviors and in genetic predispositions (McGue et al., 2014). Other research finds that heroin deaths are more common after age 30, but the best time to prevent such deaths is before age 20 (Carlson et al., 2016).

● ● **Especially for Future Researchers** What is the best method for collecting data? (see response, page 31)

Cautions and Challenges from Science

The scientific method illuminates and illustrates human development as nothing else does. Facts, consequences, and possibilities have emerged that would not be known without science—and people of all ages are healthier, happier, and more capable because of it.

Thanks to science, infectious diseases, illiteracy, and racism are much reduced compared to a century ago. Early death—from violence, war, or disease—is also less likely; science and education are likely reasons (Pinker, 2018).

Developmentalists have also discovered unexpected sources of harm. Furry baby blankets, childhood hospitalization, lead, asbestos, and vaping are all less benign than people first thought.

Thus, the benefits of science are many, in improving lives and discovering hazards. However, science can lead us all astray. We now discuss three complications: misinterpreting correlation, depending on numbers, and ignoring ethics.

correlation A number between +1.0 and –1.0 that indicates the degree of relationship between two variables, expressed in terms of the likelihood that one variable will (or will not) occur when the other variable does (or does not). A correlation indicates only that two variables are somehow related, not that one variable causes the other to occur.

TABLE 1.3

Quiz on Correlation

Two Variables	Positive, Negative, or Zero Correlation?	Why? (Third variable)
1. Ice cream sales and murder rate	_____	_____
2. Reading ability and number of baby teeth	_____	_____
3. Adult's sex assigned at birth and average number of offspring	_____	_____

For each of these three pairs of variables, indicate whether the correlation between them is positive, negative, or nonexistent. Then try to think of a third variable that might determine the direction of the correlation. The correct answers appear below.

3. Zero. Each child must begin with a sperm from a male and an ovum from a female. No third variable.

2. Negative; third variable: age

1. Positive; third variable: heat

Answers:

Correlation and Causation

Probably the most common mistake in interpreting research is confusing correlation with causation. A **correlation** exists between two variables if one variable is more (or less) likely to occur when the other does. A correlation is *positive* if both variables tend to increase together or decrease together, *negative* if one variable tends to increase while the other decreases, and *zero* if no connection is evident. (Try the quiz in **Table 1.3**.)

Expressed in numerical terms, correlations vary from +1.0 (the most positive) to –1.0 (the most negative). Correlations are almost never that extreme; a correlation of +.3 or –.3 is noteworthy; a correlation of +.8 or –.8 is astonishing.

Many correlations are unexpected. For instance: First-born children are more likely to develop asthma than are later-born children; teenage girls have higher rates of mental health problems than do teenage boys; and counties in the United States with more dentists have fewer obese residents. That last study controlled for the number of medical doctors and the poverty of the community. The authors suggest that dentists provide information about nutrition that improves health (Holzer et al., 2014).

That explanation may be wrong. Every scientist knows the mantra: *Correlation is not causation.* Just because two variables are correlated does not mean that one causes the other—even if it seems logical that it does. It proves only that the variables are connected somehow.

In the dentist correlation, the cause could be from either direction. Seeing the dentist might encourage people to eat better, or dentists may choose to practice where people are healthier. Or a third variable might cause both if, for instance, a particular community values both exercise and preventive medicine. Unless people remember that correlation is not causation, they may draw mistaken conclusions.

Quantity and Quality

A second caution concerns whether scientists should rely on data produced by **quantitative research** (from the word *quantity*). Quantitative research data are ranked or numbered, allowing easy translation across cultures. One example of quantitative research is using children's achievement scores to assess education within a school or a nation.

Since quantities can be easily summarized, compared, charted, and replicated, many scientists prefer quantitative research. Statistics require numbers. Quantitative data are easier to replicate and less open to bias.

However, when data are presented in categories and numbers, some nuances and individual distinctions are lost. Moreover, exactly how the raw data are turned into statistics may itself reflect bias. Many developmental researchers thus turn to **qualitative research** (from the word *quality*)—asking open-ended questions, reporting answers in narrative (not numerical) form.

Qualitative research reflects cultural and contextual diversity, and it may reflect the biases of the scientist. Qualitative research is harder to replicate. Both types of research are needed (Morgan, 2018).

Some studies now use both methods, which provides richer, but also more verifiable, details. For example, when a child has a psychological disorder that would benefit from professional help, less than half of all parents access such care. A review that included both quantitative studies (22 of them) and qualitative ones (24 of them) found many factors that reduced care-seeking, including cost (quantitative) and trust in providers (qualitative) (Reardon et al., 2017).

Ethics

The most important challenge for all scientists is to follow ethical standards. Each professional society involved in research of human development has a *code of ethics* (a set of moral principles). Most colleges and hospitals have an *Institutional Review Board* (IRB) that requires research to follow ethical guidelines set by the federal government.

Some research conducted before IRBs were established was clearly unethical, especially when the participants were children, members of minority groups, prisoners, or animals. Even with IRBs, serious ethical dilemmas remain, particularly when research occurs in developing nations (Leiter & Herman, 2015).

quantitative research Research that provides data that can be expressed with numbers, such as ranks or scales.

qualitative research Research that considers qualities instead of quantities. Descriptions of particular conditions and participants' expressed ideas are often part of qualitative studies.

● Especially for Future Researchers and Science Writers Do any ethical guidelines apply when an author writes about the experiences of family members, friends, or research participants? (see response, page 31)

LaunchPad
macmillan learning

VIDEO ACTIVITY: Eugenics and the "Feebleminded": A Shameful History illustrates what can happen when scientists fail to follow a code of ethics.

Risky Shot? Most vaccines undergo years of testing before they are used on people, but vaccines protecting against Ebola were not ready until the 2014 West African epidemic finally waned after 11,000 deaths. Thus, the effectiveness of Ebola vaccines is unknown. However, when deadly Ebola surfaced again in the Democratic Republic of Congo in 2018, public health doctors did not wait for longitudinal data. Here Dr. Mwamba, a representative of Congo's Expanded Program on Immunization, receives the vaccine. He hopes that it will protect him and thousands of other Congolese. We will know by 2021 if the vaccine halted a new epidemic.

AP Images/John Bompengo

Ebola and Coronavirus

Many ethical dilemmas arose in the 2014–2015 West African Ebola epidemic (Gillon, 2015; Rothstein, 2015; Sabeti & Sabeti, 2018). Those problems have reemerged in the recent Ebola crisis (Gostin et al., 2019) and the coronavirus pandemic (Wang et al., 2020). Among them:

- Should vaccines be given before their safety is demonstrated with large control and experimental groups?
- What kind of informed consent is needed to avoid both false hope and false fears?
- Should children with Ebola or coronavirus be isolated from family, even though social isolation is harmful?
- Is it fair for public health care systems to be inadequate in some countries and high-tech in others?
- When should each nation be responsible for the health of their own people, and when and how should other nations or the United Nations intervene?
- Should quarantine restrict the travel of healthy people?

Medicine tends to focus on individuals, ignoring the customs and systems that make some people in some nations more vulnerable. One observer noted:

> When people from the United States and Europe working in West Africa have developed Ebola, time and again the first thing they wanted to take was not an experimental drug. It was an airplane that would cart them home.
>
> [Cohen, 2014, p. 911]

These issues have led to some international efforts, including funding secure biocontainment laboratories in many nations, in order to quickly recognize deadly diseases (Le Duc & Yuan, 2018). Nonetheless, no developmentalist believes that the political and economic cooperation necessary for world health is adequate. This is evident not only for diseases such as Ebola, Zika, coronavirus, and polio, but also for many developmental concerns, such as early-childhood care, drug control, and public education. Multidirectional, multicontextual, multicultural, multidisciplinary, and plastic perspectives are needed.

An underlying problem is that people everywhere have strong opinions about children that they expect research to confirm. Scientists might try (sometimes without noticing it) to achieve results that confirm their national and cultural values.

Obviously, collaboration, replication, and transparency are essential ethical safeguards. Hundreds of questions regarding human development need answers, and researchers have yet to find them. That is the most important ethical mandate of all. For instance:

- Do we know enough about prenatal drugs to protect every fetus?
- Do we know enough about education to ensure that every child is ready for the future?
- Do we know enough about poverty to enable everyone to be healthy?
- Do we know enough about transgender children, or contraception, or romance to ensure that every child is loved?
- Do we know enough about single parenthood, or divorce, or same-sex marriage to ensure optimal family life?

The answer to these questions is *NO, NO, NO, NO, and NO.*

Scientists and funders tend to avoid questions that might produce unwanted answers. People have strong opinions about drugs, schools, economics, sexuality, and families (the five questions above) that may conflict with scientific findings and conclusions. Religion, politics, and customs shape scientific research, sometimes stopping investigation before it begins.

Consider research on guns, a leading cause of child death from age 2 to 20. In 1996, the U.S. Congress, in allocating funds for the Centers for Disease Control and Prevention, passed a law stating that "None of the funds made available for injury prevention and control at the Centers for Disease Control and Prevention may be used to advocate or promote gun control." The National Rifle Association (NRA) interpreted this as a reason to stop research on guns, even though guns are the most common means of suicide in the United States and the most used weapon of homicide.

Ten highly respected scientists summarize research on gun deaths:

> There is only very sparse scientific evidence [regarding] . . . which policies will be effective. . . . Even the seemingly popular view that violent crime would be reduced by laws prohibiting the purchase or possession of guns by people with mental illness was deemed to have only moderate supporting evidence.

> *[Leshner & Dzau, 2018, p. 1195]*

It is unfair to blame Congress, or to focus on guns. Indeed, there are unanswered questions about almost every aspect of human development. Culture, context, and politics distort research on many topics. That is why everyone, left, right, and center, needs science.

Awareness of bias is only the first step for the next cohort of developmental scientists. They will build on what is known, mindful of what needs to be explored, raising questions that earlier cohorts have not asked. The goal of our study remains the same, to help everyone fulfill their potential. The next 15 chapters are a beginning.

> **THINK CRITICALLY:** Can you think of an additional question that researchers should answer?

WHAT HAVE YOU LEARNED?

1. Why does correlation not prove causation?

2. What are the advantages and disadvantages of quantitative research?

3. What are the advantages and disadvantages of qualitative research?

4. What is the role of the IRB?

5. Why might a political leader avoid funding developmental research?

6. What questions about human development remain to be answered?

SUMMARY

Understanding How and Why

1. The study of human development is a science that seeks to understand how people change or remain the same over time. As a science, it begins with questions and hypotheses and then gathers empirical data. The data are analyzed, and conclusions are drawn.

2. Research is published with sufficient detail to allow other scientists to confirm or refute the conclusions. Replication is needed before a particular result is considered solid.

3. The universality of human development and the uniqueness of each individual's development are evident in both nature (the genes) and nurture (the environment); no person is quite like another. Nature and nurture always interact, and each human characteristic is affected by that interaction.

4. Differential susceptibility is evident when we study nature and nurture. Genes and experiences affect our vulnerability to developmental change, for better or worse. Some people are

impervious to conditions that affect others, as the metaphor of dandelions and orchids illustrates.

Childhood and Adulthood

5. The assumption that growth is linear has been replaced by the realization that both continuity and discontinuity are part of every life. Developmental gains and losses are apparent lifelong, because development is multidirectional.

6. Time is a crucial variable. Everyone changes with age. Critical periods are times when something *must* occur for normal development; sensitive periods are times when a particular kind of development occurs most easily.

7. Development is multicontextual and multicultural; it occurs within many contexts and cultures. Those ideas were emphasized by Urie Bronfenbrenner's ecological-systems approach. Each person is situated within larger systems of family, school, community, and culture.

8. Each cohort is influenced by the innovations and events of their historical period, and each person is affected by his or her socioeconomic status (SES), with effects of both cohort and SES varying depending on the age and circumstances of the person.

9. Culture includes beliefs and patterns; ethnicity refers to ancestral heritage. Race is a social construction, sometimes mistakenly thought to be biological. Developmentalists guard against the assumption that if something is unusual, it is not as good as the usual, called the difference-is-deficit error.

10. A multidisciplinary, dynamic-systems approach is needed because each person develops in many ways—biosocially, cognitively, and psychosocially—simultaneously, yet scientists tend to focus on a small part of a larger issue. A multidisciplinary approach has corrected many narrow ideas of the past.

11. Throughout life, human development is plastic. Brains and behaviors are molded by experiences. Cultures as well as individuals adjust to social needs.

Using the Scientific Method

12. Commonly used research methods are observation, experiments, and surveys. Each of these methods use techniques, such as having experimental and control groups, that reduce bias. Each study can provide insight, yet each is limited. Meta-analyses summarize conclusions of many studies.

13. Developmentalists study change over time, often with cross-sectional and longitudinal research. Cross-sequential research, which combines the other two methods, attempts to avoid the pitfalls of cross-sectional and longitudinal studies.

Cautions and Challenges from Science

14. A correlation is a statistic that indicates that two variables are connected, both increasing or both decreasing together, or, with a negative correlation, one increasing as the other decreases. Correlation does not prove which of two variables causes the other. Sometimes neither is the cause, because two variables correlate because of a third variable.

15. Quantitative research provides numerical data, which is best for comparing contexts and cultures via verified statistics. By contrast, more nuanced data come from qualitative research, which reports on individual lives. Both are useful.

16. Ethical behavior is crucial in all of the sciences. Results must be fairly gathered, reported, and interpreted. Participants must be informed and protected.

17. The most important ethical question is whether scientists are designing, conducting, analyzing, publishing, and applying the research that is most critically needed. The next cohort of developmental scholars will add to our scientific knowledge.

KEY TERMS

science of human development (p. 3)
scientific method (p. 3)
hypothesis (p. 3)
empirical evidence (p. 3)
replication (p. 4)
nature (p. 5)
nurture (p. 5)
differential susceptibility (p. 5)
critical period (p. 7)
sensitive period (p. 7)

ecological-systems approach (p. 8)
microsystem (p. 8)
exosystem (p. 8)
macrosystem (p. 8)
chronosystem (p. 8)
mesosystem (p. 8)
cohort (p. 9)
socioeconomic status (SES) (p. 10)
culture (p. 11)
social construction (p. 11)

difference-equals-deficit error (p. 12)
ethnic group (p. 12)
race (p. 12)
intersectionality (p. 13)
plasticity (p. 16)
dynamic-systems approach (p. 16)
scientific observation (p. 20)
experiment (p. 21)
independent variable (p. 21)
dependent variable (p. 21)

survey (p. 22)
meta-analysis (p. 23)
cross-sectional research (p. 23)
longitudinal research (p. 24)
cross-sequential research (p. 24)
correlation (p. 26)
quantitative research (p. 27)
qualitative research (p. 27)

APPLICATIONS

1. It is said that culture is pervasive but that people are unaware of it. List 30 things you did *today* that you might have done differently in another culture. Begin with how and where you woke up.

2. How would your life be different if your parents were much higher or lower in SES than they are? Consider all three domains.

3. A longitudinal case study can be insightful but is also limited in generality. Interview one of your older relatives and explain what aspects of his or her childhood are unique and what might be relevant for everyone.

Especially For ANSWERS

Response for Nurses (from p. 22) Experiments are the only way to determine cause-and-effect relationships. If we want to be sure that a new drug or treatment is safe and effective, an experiment must be conducted to establish that the drug or treatment improves health.

Response for Future Researchers (from p. 25) There is no best method for collecting data. The method used depends on many factors, such as the age of participants (infants can't complete questionnaires), the question being researched, and the time frame.

Response for Future Researchers and Science Writers (from p. 27) Yes. Anyone you write about must give consent and be fully informed about your intentions. They can be identified by name only if they give permission. For example, family members gave permission before anecdotes about them were included in this text. My nephew David read the first draft of his story (see pages 18–19) and is proud to have his experiences used to teach others.

Observation Quiz ANSWERS

Answer to Observation Quiz (from p. 12) Islam (notice the hijab) and Christianity (notice the cross).

Answer to Observation Quiz (from p. 17) Surveys rarely ask children their opinions, and the youngest cohort on this graph did not reach adulthood until about 2005.

Theories

What Will You Know?

1. What is practical about a theory?
2. Do childhood experiences affect adults?
3. Would you be a different person if you grew up in another place or century?
4. Why do we need so many theories?

I have become the head of a family of nine, the other eight of whom are my descendants. Technically one of them is not a descendant, my son-in-law Oscar, but I include him in the count, not only because he is the father of two of my grandchildren, but also because he is an integral part of the family. For instance, his humor lightens our family gatherings.

One family gathering is the springboard of this chapter, because it illustrates why theories are needed. It began when we were all together on a hot summer day. Our thoughts turned to winter. "Maybe we should all go to Iowa for Christmas," someone said.

Iowa for the holidays may be surprising for a professor who teaches in the Bronx, but Rachel recently moved to Iowa, and neither her mother (me) nor her three sisters have seen her new home. That "maybe we should" started dozens of conversations and emails, beginning with questioning the premise (why should we?) and with Oscar laughing at "another Berger trip." He knows by now that we propose a dozen versions of every gathering, with different ideas of who, when, how, and where. Even "why" is debated: Why should all nine of us gather for Christmas? Travel is more expensive then. One previous year, we opened presents together on Skype.

Months of discussion ensued. What about holiday traffic, snow, ice, money, time off work? Does the Iowa daughter want us there in December, an intense work month for her? Flying uses lots of fossil fuel, but the train takes a day and a night and goes only to Chicago; driving is more direct but less safe; planes will fly with or without us; we should reduce our carbon emissions; we should not do that if it harms our family; the money could go to better causes; family bonding is a good cause. . . . It went on for months, Oscar smiling and everyone else considering options but no one deciding. How did we choose? Theories helped; so did facts.

Finally we agreed that we should go to Iowa; then "when" became an issue. Who has days off from work, when is best to avoid crowds, how many days together is long enough but not too long?

As in this example, every human decides what to do with each minute, each day, each month of their lives, selecting among many possibilities, and thus, everyone must set priorities and clarify perspectives. With nine of us, this quickly gets complex. Without theories we would be confused and scattered, unable to agree and each stuck in their own place. The eight of us are used to the complexity of alternate views; Oscar is not. (Is that because he is an only child? That question prompts another set of theories. But not now.)

This chapter explains five comprehensive theories that guide human behavior. Each theory helped us with our Iowa dilemma, as we assessed (in order of the theories, not in order of our months-long, far-ranging discussions) family memories, past experiences, personal beliefs, cultural pressures, innate emotions. At the end of this chapter, you will learn what we did.

What Theories Do

Theories organize scattered facts and confusing observations into patterns, weaving the details into a meaningful whole. A **developmental theory** is a comprehensive statement of general principles that provides a framework for understanding how and why people change as they grow older. This is much more than a hunch or speculation: Developmental theories emerge from data, survive analysis, lead to experiments, and raise new questions.

developmental theory A group of ideas, assumptions, and generalizations that interpret and illuminate thousands of observations about human growth. A developmental theory provides a framework for explaining the patterns and problems of development.

Theory and Practice

Sometimes people think of theories as impractical. Not at all. As a leading social scientist once quipped, "Nothing is as practical as a good theory" (Lewin, 1945). He used humor to counter the idea that theory and practice are opposites. Like many other scientists, he knew not only that theories organize thoughts and experiences to help scientists grasp generalities and reach solid conclusions, but also that they aid everyone in daily life.

For example, if a child yells "I hate you," a mother might be shocked, stunned, sad, and angry. Each of the five theories in this chapter might suggest different responses, but all agree that a thoughtless, reactive "I hate you, too" is not ideal. Without theories, we would be lost with a jumble of observations and reactions, confused as to how they fit together to make a life.

Of course, theories differ; some are more comprehensive than others, and some reflect one culture more than another. In some cultures, children would never say "I hate you" to a parent: In every culture a developmental scientist would ask how old the child was, because "I hate you" means something different at age 2 than at 20.

Overall, children's needs change with age: Theories help adults respond. What should adults do when infants cry, or 3-year-olds refuse to eat their dinner, or 6-year-olds crawl into bed with their parents, or 10-year-olds multiply 7×8 correctly . . . or incorrectly? Answers reflect theories, ideally theories that have stood the tests of time and science.

It may be easier to understand the role of theory with a metaphor. Imagine trying to build a house without a design. You might have willing workers and all the raw materials: the bricks, the wood, the nails. But without tools and a plan you could not proceed. Science provides the tools; theories provide the plan; scientists are the workers who follow the plan.

Sometimes, over the years, the house needs more work—another bedroom, a new roof, an additional door. Likewise, theories are revised over time. Theories are meant to be tested and refined. Sometimes a theory leads to a hypothesis that turns out to be false, an outcome that is considered a benefit, not a liability, of theory.

"I'm going to refer to an educational theory which was first published in February and is still applicable today."

Aaron Bacall/Cartoon Collections/Cartoon Stock

The Test of Time Grand theories have endured for decades and still guide contemporary scientists.

Theories can be very useful. To be more specific:

- Theories produce *hypotheses*.
- Theories generate *discoveries*.
- Theories offer *practical guidance*.

Remember from Chapter 1 that testing a hypothesis is the third step of science, the middle of the five steps. To get to that pivot, scientists need a question and a hypothesis. That is one reason that theories are needed, to stimulate questions.

Once the question is framed as a hypothesis, the actual research begins, which then leads to analysis and conclusions to be shared with other scientists, to confirm, extend, revise, or refute the theory. For scientists, that is a practical result.

Facts and Possibilities

Sometimes people say dismissively, "that's just a theory," as if theories were disconnected from facts. In truth, facts are essential: A good theory begins with facts and discovers more of them. As one scientist explained, imagine a world without facts, "a world of ignorance where many possibilities seem equally likely . . . [with] unreliable conclusions . . . [and] shoddy evidence" (Berg, 2018, p. 379). Theories and facts work together. Facts lead to theories, and then theories lead to discovery of previously unrecognized possibilities, which need testing to discover new facts.

It is a fact, not a theory, that everywhere and for all time, part of being a person is to be "perpetually driven to look for deeper explanations of our experience, and broader and more reliable predictions about it" (Gopnik, 2001, p. 66). The perpetual need to understand is evident in history. All five of the theories in this chapter echo ideas written by ancient sages in Greece, China, India, and elsewhere.

Dozens of other theories have been formulated to explain some aspect of development. The five theories here are comprehensive, relevant at every age. None is the final word. As explained in Chapter 1, human growth is dynamic, always affected by cohort and culture.

Science begins with questions, which often spring from theory. Among the thousands of important questions are the following, each central to one of the five theories in this chapter:

1. Do early experiences—of breast-feeding or attachment or abuse—shape adult personality?
2. Does learning depend on encouragement, punishment, and/or role models?
3. Do children learn best if they figure out ideas for themselves?
4. Does culture determine parents' behavior, such as how to respond to an infant's cry?
5. Is survival an inborn instinct, underlying all personal and social decisions?

The answer to each of these questions is "yes" when examined in order by the following theories: psychoanalytic, behaviorism, cognitive, sociocultural, and evolutionary. Each question is answered "no" or "not necessarily" by several others. For every answer, more questions arise: Why or why not? When and how? SO WHAT? This last question is crucial; implications and applications affect everyone's daily life.

WHAT HAVE YOU LEARNED?

1. How are facts and theories connected?
2. What three things do theories do?
3. Why do people need theories to move forward with their lives?
4. Who develops theories—everyone or just scientists?
5. What is the relationship between theories and facts?

Grand Theories

In the first half of the twentieth century, two opposing theories—psychoanalytic and behaviorism—dominated psychology, each with extensive applications to child development. In about 1960, a third theory—cognitive—arose, and it, too, was widely applied.

These three are called "grand theories." They are explained here because they are comprehensive, enduring, and far-reaching. In developmental research, they continue to be useful. But be forewarned: None is now considered as grand as once believed.

Psychoanalytic Theory: Freud and Erikson

Inner drives, deep motives, and unconscious needs rooted in childhood—especially the first six years—are the focus of **psychoanalytic theory.** These unconscious forces are thought to influence every aspect of thinking and behavior, from the smallest details of daily life to the crucial choices of a lifetime.

Freud's Ideas

Psychoanalytic theory originated with Sigmund Freud (1856–1939), an Austrian physician who treated patients suffering from mental illness. He listened to their remembered dreams and uncensored thoughts. From that, he constructed an elaborate, multifaceted theory.

AKG/Science Source

Freud at Work In addition to being the world's first psychoanalyst, Sigmund Freud was a prolific writer. His many papers and case histories, primarily descriptions of his patients' symptoms and sexual urges, helped make the psychoanalytic perspective a dominant force for much of the twentieth century.

psychoanalytic theory Freud's theory of the stages of development, each of which emphasizes the sexual nature of the child. As its first grand theorist, Freud believed that irrational, unconscious drives and motives, often originating in childhood erotic impulses, underlie human behavior.

According to Freud, development in the first six years of life occurs in three *psychosexual* stages, each characterized by sexual interest and pleasure arising from a particular part of the body.

In infancy, the erotic body part is the mouth (the *oral stage*); in early childhood, the anus (the *anal stage*); in the preschool years, the penis (the *phallic stage*), a source of pride and fear among boys and a reason for sorrow and envy among girls. Then, after a quiet period (*latency*), the *genital stage* arrives at puberty, lasting throughout adulthood. (**Table 2.1** on page 36 describes stages in Freud's theory.)

Freud maintained that sensual satisfaction (from stimulation of the mouth, anus, or penis) is linked to major developmental stages, needs, and challenges. During the oral stage, for example, sucking provides not only nourishment for the infant but also erotic joy and attachment to the mother. Kissing between lovers is a vestige of the oral stage. Next, during the anal stage, pleasures arise from self-control, initially with toileting but later with wanting everything to be clean, neat, and regular (an "anal personality").

One of Freud's most influential ideas was that each stage includes its own struggles. Conflict occurs, for instance, when parents wean their babies (oral stage), toilet-train their toddlers (anal stage), deflect the sexual curiosity and fantasies of their 5-year-olds (phallic stage), and limit the sexual interests of adolescents (genital stage). Freud thought that the experiences surrounding these conflicts determine later personality.

Freud did not believe that any new stage occurred after puberty; rather, he believed that adult personalities and habits reflected childhood. As Freud explained it, unconscious conflicts rooted in early life were evident in adult behavior—for instance, cigarette smoking (oral) or meticulous housekeeping (anal) or driving big cars (phallic).

TABLE 2.1

Comparison of Freud's Psychosexual and Erikson's Psychosocial Stages

Approximate Age	Freud (psychosexual)	Erikson (psychosocial)
Birth to 1 year	*Oral Stage* The lips, tongue, and gums are the focus of pleasurable sensations in the baby's body, and sucking and feeding are the most stimulating activities.	*Trust vs. Mistrust* Babies either trust that others will satisfy their basic needs, including nourishment, warmth, cleanliness, and physical contact, **or** develop mistrust about the care of others.
1–3 years	*Anal Stage* The anus is the focus of pleasurable sensations in the baby's body, and toilet training is the most important activity.	*Autonomy vs. Shame and Doubt* Children either become self-sufficient in many activities, including toileting, feeding, walking, exploring, and talking, **or** feel shame, **or** doubt their own abilities.
3–6 years	*Phallic Stage* The phallus, or penis, is the most important body part, and pleasure is derived from genital stimulation. Boys are proud of their penises; girls wonder why they don't have them.	*Initiative vs. Guilt* Children either try to undertake many adultlike activities **or** internalize the limits and prohibitions set by parents. They feel either adventurous **or** guilty.
6–11 years	*Latency* Not really a stage, latency is an interlude. Sexual needs are quiet; psychic energy flows into sports, schoolwork, and friendship.	*Industry vs. Inferiority* Children busily practice and then master new skills **or** feel inferior, unable to do anything well.
Adolescence	*Genital Stage* The genitals are the focus of pleasurable sensations, and the young person seeks sexual stimulation and satisfaction in heterosexual relationships.	*Identity vs. Role Confusion* Adolescents ask themselves "Who am I?" They establish sexual, political, religious, and vocational identities **or** are confused about their roles.
Adulthood	Freud believed that the genital stage lasts throughout adulthood. He also said that the goal of a healthy life is "to love and to work."	*Intimacy vs. Isolation* Young adults seek companionship and love **or** become isolated from others, fearing rejection. *Generativity vs. Stagnation* Middle-aged adults contribute to future generations through work, creative activities, and parenthood **or** they stagnate. *Integrity vs. Despair* Older adults try to make sense of their lives, either seeing life as a meaningful whole **or** despairing at goals never reached.

Erikson's Ideas

Many of Freud's followers became famous theorists themselves—Carl Jung, Alfred Adler, and Karen Horney among them. They agreed with Freud that early-childhood experiences affect everyone, often unconsciously, but they also expanded and modified his ideas. Many prefer to call their approach *psychodynamic*, to emphasize that they do not adhere strictly to Freud's approach.

For scholars in human development, one psychodynamic theorist, Erik Erikson (1902–1994), is particularly insightful. He was born in Denmark, spent his childhood in Germany, and traveled in Italy as a young man. His interest in psychoanalytic theory was fostered by his work as a teacher in Austria in a school for children of patients who were analyzed by Freud. When Hitler's rise forced all the Freudians to flee, Erikson and his American wife, Joan, came to the United States.

Those multicultural experiences led him to develop his **psychosocial theory** of the entire life span.

Erikson described eight developmental stages, each characterized by a particular challenge, or *developmental crisis*. Although Erikson named two polarities at each crisis, he recognized a wide range of outcomes between those opposites. Typically, development at each stage leads to neither extreme but to something in between.

In the stage of *initiative versus guilt*, for example, 3- to 6-year-olds undertake activities that exceed the limits set by their parents and their culture. They leap into swimming pools, pull their pants on backward, make cakes according to their own recipes, and wander off alone.

Erikson thought that those preschool initiatives produce feelings of pride or failure, depending on adult reactions. Should adults pretend to like the cake that a preschooler made or, instead, punish that child for wasting food and messing up the kitchen? According to Erikson's theory, a child will feel guilty lifelong if adults are too critical or if social norms are too strict regarding the young child's initiatives.

As you can see from Table 2.1, Erikson's first five stages are closely related to Freud's stages. Like Freud, Erikson believed that unresolved childhood conflicts echo throughout life, causing problems in adulthood.

Erikson considered the first stage, *trust versus mistrust*, particularly crucial. For example, an adult who has difficulty establishing a secure, mutual relationship with a life partner may never have resolved that first crisis of life. If you know people who are "too trusting" or "too suspicious," Erikson might suggest that you ask about their early care.

In his emphasis on childhood, Erikson agreed with Freud. However, in two crucial aspects, Erikson was unlike his mentor.

1. Erikson's stages emphasized family and culture, not sexual urges.
2. Erikson recognized the entire life span, with three stages after adolescence.

Behaviorism: Conditioning and Learning

The comprehensive theory that dominated psychology in the United States for most of the twentieth century was **behaviorism.** This theory began in Russia, with Pavlov, who first described conditioning.

Classical Conditioning

Early in the twentieth century, Ivan Pavlov (1849–1936) did hundreds of experiments to examine the link between something that affected a living creature (such as a sight, a sound, a touch) and how that creature reacted. Technically, he was interested in how a *stimulus* effects a *response*.

While studying salivation in his laboratory, Pavlov began by studying dogs. He noticed that his research dogs drooled (response) not only at the smell of food (stimulus) but also, eventually, at signals that food was coming, even before any sight or smell of the food. This observation led to a series of famous experiments: Pavlov conditioned dogs to salivate (response) when hearing a particular noise (stimulus).

Pavlov began by sounding a tone just before presenting food. After a number of repetitions of the tone-then-food sequence, dogs began salivating at the sound

psychosocial theory Erikson's theory of the stages of development, emphasizing the interaction between the psychic needs of the individual and the surrounding social network of family and community.

Ted Streshinsky/The LIFE Images Collection/Getty Images

A Legendary Couple In his first 30 years, Erikson never fit into a particular local community, since he frequently changed nations, schools, and professions. Then he met Joan. In their first five decades of marriage, they raised a family and wrote several books. If Erikson had published his theory at age 73 (when this photograph was taken) instead of in his 40s, would he still have described life as a series of crises?

● **Especially for Teachers** Your kindergartners are talkative and always moving. They almost never sit quietly and listen to you. What would Erik Erikson recommend? (see response, page 61)

behaviorism A grand theory of human development that studies observable behavior. Behaviorism is also called *learning theory* because it describes the laws and processes by which behavior is learned.

classical conditioning The learning process in which a meaningful stimulus (such as the smell of food to a hungry animal) is connected with a neutral stimulus (such as the sound of a tone) that had no special meaning before conditioning. (Also called *respondent conditioning*.)

Hulton Deutsch/Getty Images

A Contemporary of Freud Ivan Pavlov was a physiologist who received the Nobel Prize in 1904 for his research on digestive processes. It was this line of study that led to his discovery of classical conditioning, when his research on dog saliva led to insight about learning.

● **Observation Quiz** How is Pavlov similar to Freud in appearance, and how do both look different from the other theorists pictured? (see answer, page 61) ↑

even when there was no food. This simple experiment demonstrated **classical conditioning** (also called *respondent conditioning*).

In classical conditioning, a person or animal learns to associate a neutral stimulus with a meaningful one, gradually responding to the neutral stimulus in the same way as to the meaningful one. In Pavlov's original experiment, the dog associated the tone (the neutral stimulus) with food (the meaningful stimulus) and eventually responded to the tone as if it were the food itself. The conditioned response to the tone, no longer neutral but now a conditioned stimulus, was evidence that learning had occurred.

Behaviorists see dozens of examples of classical conditioning. Infants learn to smile at their parents because they associate them with food and play; toddlers learn to fear busy streets if the noise of traffic repeatedly frightens them; students learn to enjoy—or hate—school, depending on their kindergarten experience.

This theory alerts us to many reactions linked to stimuli that once were neutral. Think of how some people react to a wasp, or a final exam, or a police car in the rearview mirror.

In my early years teaching seventh-graders, I once strode up to the desk of one boy to tell him directly to get out his notebook and start writing, because he was not responding to my general request to the entire class. He put up his arm in self-defense. That startled me. I would never hit any student, but he had been conditioned to expect a blow.

All our reactions are learned; the mere announcement about a future exam triggers sweat or chills in some students—as would not happen to a child with no exam experience. Many students find that the stress hormones triggered by seeing the exam paper makes them forget what they know—an unwelcome conditioned response.

Behaviorism in the United States

Pavlov's ideas seemed to bypass most Western European developmentalists but were welcomed in the United States, because many North Americans disputed the psychoanalytic emphasis on the unconscious. The first of three famous Americans who championed behaviorism was John B. Watson (1878–1958). He argued that if psychology was to be a true science, psychologists should examine only what they could measure, not invisible, unconscious impulses. In his words:

> Why don't we make what we can *observe* the real field of psychology? Let us limit ourselves to things that can be observed, and formulate laws concerning only those things. . . . We can observe *behavior—what the organism does or says*.
>
> *[Watson, 1924/1998, p. 6]*

According to Watson, if the focus is on behavior, it is apparent that everything is learned. He wrote:

> Give me a dozen healthy infants, well-formed, and my own specified world to bring them up in and I'll guarantee to take any one at random and train him to become any type of specialist I might select—doctor, lawyer, artist, merchant-chief, and yes, even beggar-man and thief, regardless of his talents, penchants, tendencies, abilities, vocations, and race of his ancestors.
>
> *[Watson, 1924/1998, p. 82]*

Other American psychologists agreed. They chose to study observable behavior, objectively and scientifically. For every creature at every age, behaviorists believe there are natural laws of behavior, and they seek to discover those laws to understand how simple, repeated actions become complex competencies, because stimuli in the environment affect each response.

Learning in behaviorism is far more comprehensive than the narrow definition of learning, which focuses on academics, such as learning to read or multiply. Instead, for behaviorists, everything that people think, do, and feel is learned, step by step, via conditioning.

For example, newborns *learn* to suck on a nipple; infants *learn* to smile at a caregiver; preschoolers *learn* to hold hands when crossing the street. Such learning is conditioned and can endure when no longer useful. That explains why children suck lollipops, adults smile at strangers, and I still grab my children's hands in traffic.

Operant Conditioning

The most influential North American proponent of behaviorism was B. F. Skinner (1904–1990). Skinner agreed with Watson that psychology should focus on observable behavior. He did not dispute Pavlov's classical conditioning, but, as a good scientist, he built on Pavlov's conclusions. His most famous contribution was to recognize another type of conditioning—**operant conditioning** (also called *instrumental conditioning*)—in which animals (including people) act and then something follows that action.

In other words, Skinner went beyond learning by association, in which one stimulus is paired with another stimulus (in Pavlov's experiment, the tone with the food). He focused instead on what happens *after* the response. If the consequence that follows is enjoyable, the creature (any living thing—a bird, a mouse, a child) tends to repeat the behavior; if the consequence is unpleasant, the creature does not do it again.

Consequences that increase the frequency or strength of a particular action are called *reinforcers*; the process is called **reinforcement** (Skinner, 1953). Pleasant consequences are sometimes called *rewards*, but behaviorists do not call them that because they want to avoid confusion. What some people consider a "reward" may actually be a *punishment*, an unpleasant consequence. For instance, a teacher might reward good behavior by giving the class extra recess time, but if a child hates recess, that is punishment not reinforcement.

The opposite is true as well: Something thought to be a punishment may be reinforcing. For example, parents "punish" their children by withholding dessert. But a particular child might dislike the dessert, so being deprived of it is no punishment.

Culture matters, too. Japanese parents threaten to punish their children by refusing to let them come home; American parents threaten to make the children stay home. Whether these opposite strategies are really punishments depends on the child as well as the culture (Bornstein, 2017).

The crucial question is, "What works as a reinforcement or punishment for that individual?" The answer varies by personal history as well as by age. For instance, adolescents find risk and excitement particularly reinforcing, and they consider punishments much less painful than adults do. That was one conclusion of a study of teenagers who were violent: For them, the thrill of breaking the law was reinforcing, outweighing the pain of getting caught (Shulman et al., 2017).

Consider a common practice in schools: Teachers send misbehaving children out of the classroom. Then, principals suspend the worst violators from school.

However, if a child hates the teacher, leaving class is rewarding; and if a child hates school, suspension is a reinforcement. Indeed, research on school discipline finds that some measures, including school suspension, *increase* later disobedience (Osher et al., 2010). Educators have learned that, to stop misbehavior, it is often more effective to encourage good behavior, to "catch them being good" (Polirstok, 2015, p. 932).

AP Photo

Rats, Pigeons, and People B. F. Skinner is best known for his experiments with rats and pigeons, but he also applied his knowledge to human behavior. For his daughter, he designed a glass-enclosed crib in which temperature, humidity, and perceptual stimulation could be controlled to make her time in the crib enjoyable and educational. He encouraged her first attempts to talk by smiling and responding with words, affection, or other positive reinforcement.

operant conditioning The learning process by which a particular action is followed by something desired (which makes the person or animal more likely to repeat the action) or by something unwanted (which makes the action less likely to be repeated). (Also called *instrumental conditioning*.)

reinforcement When a behavior is followed by something desired, such as food for a hungry animal or a welcoming smile for a lonely person.

LaunchPad
macmillan learning

VIDEO ACTIVITY: Modeling: Learning by Observation features the original footage of Albert Bandura's famous experiment.

social learning theory An extension of behaviorism that emphasizes the influence that other people have over a person's behavior. Even without specific reinforcement, every individual learns many things through observation and imitation of other people. (Also called *observational learning.*)

modeling The central process of social learning, by which a person observes the actions of others and then copies them.

cognitive theory A grand theory of human development that focuses on changes in how people think over time. According to this theory, our thoughts shape our attitudes, beliefs, and behaviors.

In the United States, the chance of an African American child being suspended from school is three times higher than for a European American child. The rate is also high for children designated as needing special education (Tajalli & Garba, 2014; Shah, 2011). Those statistics raise a question: Is suspension a child's punishment or a teacher's reinforcer?

The data show that children who are suspended from school are more likely than other children to be incarcerated years later. That is a correlation. It does not prove that suspension *causes* later imprisonment; it may be that children who do not obey teachers become adults who do not obey laws. But, behaviorist theory suggests that suspension does not improve child behavior, and it might make it worse (Mallett, 2016). [**Developmental Link:** Correlation and causation are discussed in Chapter 1.]

Remember, behaviorists focus on the *effect* that a consequence has on future behavior, not whether it is intended to be a reward or not. They contend that children who repeatedly misbehave have been reinforced, not punished, for misbehavior—perhaps by their parents or teachers, perhaps by their friends or themselves.

Social Learning

At first, behaviorists thought that all behavior arose from a chain of learned responses, the result of (1) the association between one stimulus and another (classical conditioning) or (2) past reinforcement (operant conditioning). Thousands of experiments inspired by learning theory have demonstrated both classical conditioning and operant conditioning in everyday life.

However, behaviorists realized people do more than respond to their personal associations, reinforcements, and punishments. They "act on the environment. They create it, preserve it, transform it, and even destroy it. . . . [in] a socially embedded interplay" (Bandura, 2006, p. 167).

That social interplay is the foundation of **social learning theory** (see **Table 2.2**), which holds that humans sometimes learn from what they see, even if they themselves have not been conditioned. As Albert Bandura, the primary proponent of this theory, explained, this learning often occurs through **modeling,** when people copy what they see others do (also called *observational learning*) (Bandura, 1986, 1997).

Modeling is not simple imitation: Some people are more likely to follow, or to be, role models than others. Indeed, humans copy only some actions, of some individuals, in some contexts. If people do the opposite of what they have seen, that could be social learning as well: Humans learn from other humans' mistakes.

Modeling is especially likely when the observer is uncertain or inexperienced and the model is admired, powerful, nurturing, or similar to the observer. That's why role models are crucial in childhood.

Social learning occurs not only for behavior and preferences (haircuts, clothing styles, slang) but also for morals. Deciding what is right is less a personal decision than one profoundly affected by other people (Bandura, 2016).

Cognitive Theory: Piaget and Information Processing

According to **cognitive theory,** each person's thoughts and expectations profoundly affect their attitudes, values,

TABLE 2.2

Three Types of Learning

Behaviorism is also called *learning theory* because it emphasizes the learning process, as shown here.

Type of Learning	Learning Process	Result
Classical conditioning	Learning occurs through association.	Neutral stimulus becomes conditioned stimulus leading to conditioned response.
Operant conditioning	Learning occurs through reinforcement and punishment.	Weak or rare responses become strong and frequent—or, with punishment, unwanted responses become extinct.
Social learning	Learning occurs through modeling what others do.	Observed behaviors become copied behaviors.

emotions, and actions. The core idea is that *how* and *what* people think is a crucial influence on human behavior.

Cognitive theory diverged from psychoanalytic theory (which emphasized hidden impulses) and behaviorism (which emphasized observed actions) to stress that thoughts are the crucial link between those impulses and actions.

Piaget's Stages of Development

Jean Piaget (1896–1980) transformed our understanding of cognition. He may have been "the greatest developmental psychologist of all time" (Haidt, 2013, p. 6). His academic training was in biology, with a focus on shellfish, which taught him to look closely at small details.

Before Piaget, most scientists believed that babies could not yet think. But Piaget used scientific observation with his own three infants. He took meticulous notes, finding infants curious and thoughtful; babies develop new concepts month by month.

Later Piaget studied hundreds of schoolchildren. From this work emerged the central thesis of cognitive theory: *How* children think changes with time and experience. According to cognitive theory, to understand the behavior of humans of any age, we need to understand their thinking.

Piaget maintained that cognitive development occurs in four age-related periods, or stages: *sensorimotor, preoperational, concrete operational,* and *formal operational* (see **Table 2.3**). Each period is characterized by particular mental processes: Infants think via their senses; preschoolers have language but not logic; school-age children have simple logic; adolescents and adults can use formal, abstract logic (Inhelder & Piaget, 1958/2013b; Piaget, 1952/2011).

At every age, Piaget found that intellectual advancement occurs because humans seek **cognitive equilibrium**—a state of mental balance. The easiest way to achieve this balance is to interpret new experiences through the lens of preexisting ideas.

Cognition is easier when the mind simplifies ideas. For instance, once children grasp the concept of "dog," they can see unfamiliar animals on the street, from

Would You Talk to This Man? Children loved talking to Jean Piaget, and he learned by listening carefully—especially to their incorrect explanations, which no one had paid much attention to before. All his life, Piaget was absorbed with studying the way children think. He called himself a "genetic epistemologist"—one who studies how children gain knowledge about the world as they grow.

cognitive equilibrium In cognitive theory, a state of mental balance in which people are not confused because they can use their existing thought processes to understand current experiences and ideas.

TABLE 2.3

Piaget's Periods of Cognitive Development

	Name of Period	Characteristics of Period	Major Gains During Period
Birth to 2 years	Sensorimotor	Infants use senses and motor abilities to understand the world. Learning is active, without reflection.	Infants learn that objects still exist when out of sight (*object permanence*) and begin to think through mental actions. (The sensorimotor period is discussed further in Chapter 6.)
2–6 years	Preoperational	Children think symbolically, with language, yet children are *egocentric*, perceiving from their own perspective.	The imagination flourishes, and language becomes a significant means of self-expression and social influence. (The preoperational period is discussed further in Chapter 9.)
6–11 years	Concrete operational	Children understand and apply logic. Thinking is limited by direct experience.	By applying logic, children grasp concepts of conservation, number, classification, and many other scientific ideas. (The concrete-operational period is discussed further in Chapter 12.)
12 years through adulthood	Formal operational	Adolescents and adults use abstract and hypothetical concepts. They can use analysis, not only emotion.	Ethics, politics, and social and moral issues become fascinating as adolescents and adults use abstract, theoretical reasoning. (The formal-operational period is discussed further in Chapter 15.)

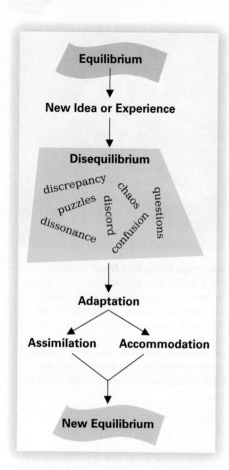

FIGURE 2.1

Challenge Me Most of us, most of the time, prefer the comfort of our conventional conclusions. According to Piaget, however, when new ideas disturb our thinking, we have an opportunity to expand our cognition with a broader and deeper understanding.

assimilation The reinterpretation of new experiences to fit into old ideas.

accommodation The restructuring of old ideas to include new experiences.

information-processing theory A perspective that compares human thinking processes, by analogy, to computer analysis of data, including sensory input, connections, stored memories, and output.

Great Danes to Chihuahuas, and say "doggie." They also expect dogs to sniff, bark, wag tails, and so on. Some children want to pet every dog they see; some fear them all—but in either case, generalities of "dogness" are evident.

Achieving cognitive equilibrium is not always easy, however. Sometimes a new experience or question is jarring or incomprehensible—such as learning that some dogs (Basenjis) do not bark. Then the individual experiences *cognitive disequilibrium*, an imbalance that creates confusion.

As **Figure 2.1** illustrates, disequilibrium advances cognition if it leads to adaptive thinking. Piaget describes two types of adaptation:

- **Assimilation:** New experiences are reinterpreted to fit, or *assimilate*, into old ideas. [A Basenji could bark if it wanted to, or Basenjis are not really dogs.]
- **Accommodation:** Old ideas are restructured to include, or *accommodate*, new experiences. [Some dogs do not bark.]

Ideally, when two people disagree, adaptation is mutual. Think of a lovers' quarrel. If both parties listen sympathetically to the other, they both accommodate. Then the quarrel strengthens their relationship, and they reach a new, better equilibrium.

As you see, accommodation requires more effort than assimilation, but it advances thought. Children—and everyone else—actively develop new concepts when the old ones fail.

In Piagetian terms, they *construct* ideas based on their experiences. He thought that knowledge is constructed (built) in the mind, an inner construction as ideas link together, not unlike the external process of construction by linking wood, metal, or whatever. For example, infants first assimilate everything using their senses—they taste and touch everything. But experience requires accommodation: They learn what should *not* be put in their mouths.

Information Processing

A newer version of cognitive theory is called **information-processing theory,** inspired by the input, programming, memory, and output of the computer. When conceptualized in that way, thinking is affected by the neurons, synapses, and neurotransmitters of the brain.

Information processing is "a framework characterizing a large number of research programs" (Miller, 2011, p. 266). Instead of interpreting *responses* by infants and children, as Piaget did, this cognitive theory focuses on the *processes* of thought—that is, when, why, and how neurons fire to activate a thought.

Information-processing theorists contend that cognition begins when *input* is picked up by one of the senses. It proceeds to brain reactions, connections, and stored memories, and it concludes with some form of *output*. For infants, output consists of moving a hand, making a sound, or staring a split second longer at one stimulus than at another. As children mature, output studied by information-processing scientists include words, hesitations, neuronal activity, and bodily reactions (heartbeat, blood pressure, hormones, and the like) (see Inside the Brain).

Information-processing theory is particularly helpful in understanding social reactions. For example, one study began by studying how parents interpreted social information before their babies were born. Later, when their 6-month-olds cried, parents reverted to their earlier assumptions: Some interpreted crying as an attempt to communicate, and others as selfish and angry. Abuse sometimes followed that interpretation (Rodriguez et al., 2019). This research advises parents to change their cognition—breathe deeply, sleep, take a walk, whatever works—when their thoughts go negative.

INSIDE THE BRAIN

Measuring Mental Activity

A hundred years ago, people thought that emotions came from the heart. That's why we still send hearts on Valentine's Day and why we speak of "broken hearts" or people who are "soft-hearted" or who have "hardened their hearts."

But now we know that everything begins inside the brain. It is foolish to dismiss a sensation with "It's all in your head." Of course it is in your head; everything is.

Until quite recently, the only way scientists assessed brains was to measure heads. Of course, measuring produced some obvious discoveries—babies with shrunken brains (microcephaly) suffered severe intellectual disability, and brains grew bigger as children matured.

Measuring also led to some obvious errors. In the nineteenth and early twentieth centuries, many scientists believed the theory that bumps on the surface of the head reflected intelligence and character, a theory known as *phrenology.*

Psychiatrists would run their hands over a person's skull to measure 27 traits, including spirituality, loyalty, and aggression. Another discredited example was suggesting that women could never be professors because their brains were too small (Swaab & Hofman, 1984).

Within the past half-century, neuroscientists developed ways to use electrodes, magnets, light, and computers to measure brain activity, not just brain size (see **Table 2.4**).

Raised areas on the head and head size (within limits) were proven irrelevant to intellectual processes. Researchers now study cognitive processes between input and output. Some results are cited later. In this feature we describe methods.

TABLE 2.4

Some Techniques Used by Neuroscientists to Understand Brain Function

EEG (electroencephalogram)

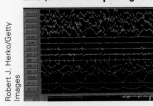

Robert J. Herko/Getty Images

The EEG measures electrical activity in the cortex. This can differentiate active brains (beta brain waves—very rapid, 12 to 30 per second) from sleeping brains (delta waves—1 to 3 per second) and brain states that are half-awake, or dreaming. Complete lack of brain waves, called flat-line, indicates brain death.

ERP (event-related potential)

Langlois Social Development Lab

The amplitude and frequency of brain electrical activity changes when a particular stimulus (called an event) occurs. First, the ERP establishes the usual patterns, and then researchers present a stimulus (such as a sound, an image, a word) that causes a blip in electrical activity. ERP indicates how quickly and extensively people react—although this method requires many repetitions to distinguish the response from the usual brain activity.

MRI (magnetic resonance imaging)

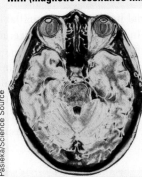

Pasieka/Science Source

The water molecules in various parts of the brain each have a magnetic current, and measuring that current allows measurement of myelin, neurons, and fluid in the brain.

fMRI (functional magnetic resonance imaging)

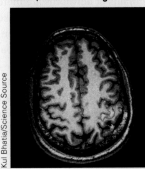

Kul Bhatia/Science Source

In advanced MRI, function is measured as more oxygen is added to the blood flow when specific neurons are activated. The presumption is that increased blood flow means that the person is using that part of the brain. fMRI has revealed that several parts of the brain are active at once—seeing something activates parts of the visual cortex, and may also activate parts of the brain far from the visual areas.

PET (positron emission tomography)

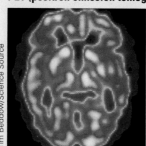

Tim Beddow/Science Source

When a specific part of the brain is active, the blood flows more rapidly in that part. If radioactive dye is injected into the bloodstream and a person lies very still within a scanner while seeing pictures or other stimuli, changes in blood flow indicate thought. PET can reveal the volume of neurotransmitters; the rise or fall of brain oxygen, glucose, amino acids; and more. PET is almost impossible to use with children (who cannot stay still).

fNIRS (functional near-infrared spectroscopy)

Pat Greenhouse/The Boston Globe via Getty Images

This method also measures changes in blood flow. But it depends on light rather than magnetic charge and can be done with children, who merely wear a special cap connected to sensors and do not need to lie still in a noisy machine (as they do for PET or fMRI). By measuring how each area of the brain absorbs light, neuroscientists infer activity of the brain (Ferrari & Quaresima, 2012).

DTI (diffusion tensor imaging)

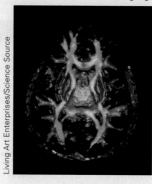

Living Art Enterprises/Science Source

DTI is another technique that builds on the MRI. It measures the flow (diffusion) of water molecules within the brain, which shows connections between one area and another. This is particularly interesting to developmentalists because life experiences affect which brain areas connect with which other ones. Thus, DTI is increasingly being used by clinicians who want to individualize treatment and monitor progress (Van Hecke et al., 2016).

For both practical and ethical reasons, it is difficult to use these techniques on large, representative samples. One of the challenges of neuroscience is to develop methods that are harmless, quick, acceptable to parents and babies, and comprehensive. A more immediate challenge is to depict the data in ways that are easy to interpret and understand.

As you see from the table, measurement and interpretation of brain activity is still difficult, but newer techniques are developing. Neuroscientists and developmentalists often disagree about the specific meaning of various results.

Nonetheless, brain imaging has revealed many surprises. For example, fNIRS finds that the brains of newborns are more active when they hear the language that their mother spoke when they were in the womb than when they hear another language (May et al., 2011). Brain scans of new mothers reveal that babies change their mothers' brains (P. Kim et al., 2016).

With adolescents, fMRI has found that a fully grown brain does not mean a fully functioning brain: The prefrontal cortex is not completely connected to the rest of the brain until about age 25, with substantial variation from one teenager to another (Foulkes & Blakemore, 2018).

Interpreting brain scans is not straightforward. Even with advanced techniques, all that scientists know for certain is which parts of the brain are functioning and active. Changes in light absorption, or magnetism, or oxygenated blood flow in the brain are miniscule from one moment to the next.

For example, the conventional lie detector is unreliable. Could brain imaging replace it? In theory, yes; in practice, no. Current technology is not ready (Rose, 2016).

Variations within and between people make it difficult to know what someone is thinking via brain scans. Once again, this confirms the need for theory: Without an idea of what to look for, or what it might mean, the millions of data points from all brain images might lead to the same trap as earlier measurements of the skull—human bias.

The latest techniques to study the brain have produced insights from neuroscience on the sequence and strength of neuronal communication. With the aid of sensitive technology, information-processing research has overturned some of Piaget's findings, as you will later read. However, the basic tenet of cognitive theory is equally true for Piaget, neuroscience, and information processing: *Ideas matter*.

Countless studies find that how children interpret a hypothetical social situation, such as whether they anticipate acceptance or rejection, affects their actual friendships; how teenagers conceptualize heaven and hell influences their sexual activity; how stressed adults feel (about their work, or income, or health) affects how they relate to children. For everyone, cognition frames situations and affects actions (see A View from Science).

A VIEW FROM SCIENCE

Walk a Mile

The folk saying "Walk a mile in my shoes," memorialized in a song that asks if "I could be you and you could be me for just an hour," reflects social perspective-taking, a cognitive accomplishment that is absent in babies and evident by adulthood. When humans understand the circumstances of each other's lives, that might lead to more caring behavior because, according to cognitive theory, thoughts guide actions.

Piaget wondered at what age are children able to understand the perspective of someone else. In one famous study, children sat in front of a sculpture of three mountains and were asked how that display would appear to someone else. Preoperational children assumed that a person sitting on the opposite side of the table would view the display exactly as they themselves did. With cognitive maturation that egocentrism disappeared: Concrete operational children could imagine the viewpoint of another (Piaget & Inhelder, 1956/2013).

Many other researchers have studied perspective-taking, building on Piaget. For example, thousands of studies focus on how to diminish bias, from Gordon Allport's work on Black–White prejudice (Allport, 1954) and continuing to "cognitive liberalization" (Hodson et al., 2018). Researchers in the past few decades have focused on *theory of mind*, which is how children understand the thoughts and emotions of other people. (e.g., Ebert, 2020; Bowman et al., 2019).

As you have read, the information-processing perspective encourages scientists to connect thinking, behavior, and brain structures not in stages but bit by bit. An international team of seven scientists did just that, with 293 participants, aged 7 to 26 (Tamnes et al., 2018).

In that study, the main test of perspective-taking was a modification of the dictator task (Keysar et al., 2000). In that task, the participants view objects in a display case that has 16 cubicles, 5 of which have backs so that the contents are not seen by someone on the other side while the other 11 are open front and back. A man on the other side (the dictator) supposedly tells the participant to move one of the objects. (See **Figure 2.2**.)

If participants are adept at theory of mind, they realize that the dictator cannot see the objects in some of the boxes. For example, three of the cubicles hold balls—large, medium, and small—but the dictator can see only the first two.

Thus, when the dictator commands "Move the small ball," the participant must decide whether to move the medium ball or the smallest ball. The correct answer would be the medium ball, because from the dictator's restricted view that ball is smallest.

Perspective-taking increases as children mature. In this study, children (up to age 11) were wrong about one-third of the time; adolescents (ages 12 to 18) were wrong about one-seventh of the time; and emerging adults were wrong about one-twentieth of the time (Tamnes et al., 2018).

This replicates many other studies: Maturation brings gradual increases in perspective-taking, although even in adulthood, sometimes people do not understand another's viewpoint. Contrary to Piaget, theory of mind increases gradually; there is no sudden transformative stage when perspective-taking appears.

The researchers in the dictator study asked the participants to answer the Strength and Difficulties questionnaire, a questionnaire often vetted and used (Goodman et al., 1998). The focus was on six questions that measured social awareness, including whether the person never, sometimes, or always was "helpful if someone is hurt, upset, or feeling ill."

Accuracy on perspective-taking in the dictator task correlated with prosocial behavior on Strength and Difficulties. Participants who were better at figuring out what the dictator could see were also better at caring for other people.

Finally—and this is the feature that makes this study an example of information processing—the scientists measured the brain, specifically the distance was between gray and white matter in various regions of the cortex.

Each of the three components above had been studied in isolation, but here the scientists combined them. They found that cognition, behavior, and brain were correlated: Those with the

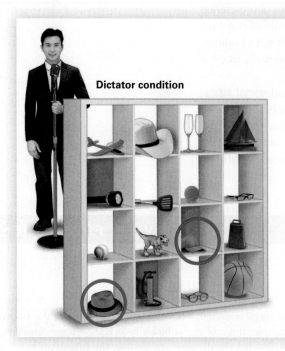
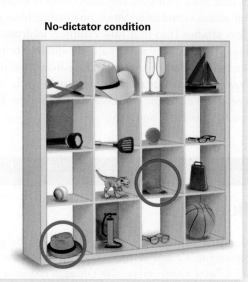

Dictator condition

No-dictator condition

FIGURE 2.2

Do You See What I See? If a man behind a set of shelves, some open and some closed, told you to move the small hat or the big ball, what would you do? Probably you would move the mid-sized objects (red) because you would realize that he could not see what you see. Young children, however, might move the wrong objects (blue).

thinner cortexes—especially in regions of the brain known for social cognition (the medial prefrontal, lateral prefrontal, and anterior cingulate)—were most likely to help other people and to understand the views of others.

This raises the possibility that, by developing perspective-taking (perhaps by having people read biographies, or engage in conversations with people unlike them), prosocial behavior might increase. Longitudinal research on the brain might verify that development.

WHAT HAVE YOU LEARNED?

1. What is the basic emphasis of psychoanalytic theory?

2. What are the similarities and differences between psychosexual and psychosocial theory?

3. How does the central focus of behaviorism differ from psychoanalytic theory?

4. When is a reward actually a punishment?

5. At what time in human development is social learning most powerful?

6. What did Piaget discover that earlier psychologists did not realize?

7. What is the emphasis of information processing?

Newer Theories

The three theories just explained are comprehensive and enduring, which is why they are called *grand*. But they all share a major limitation: They began with men in Europe a century ago. Of course, background factors limit women and non-Europeans as well, but at least newer theories reflect broader perspectives.

Developmental scientists today are as often women as men, benefiting from extensive global, historical, and multidisciplinary research.

This is evident in the two newer theories we now describe. Sociocultural theorists reflect insight from anthropologists who report on cultures in every part of the globe; evolutionary psychologists use data from archeologists who examine the bones of humans who died 200,000 years ago.

Neither has been refined by decades of later application and clarification. Nonetheless, each provides what theories do best: a practical way to organize the many thoughts, observations, and sensations that appear in our everyday life.

Sociocultural Theory: Vygotsky and Beyond

The central thesis of **sociocultural theory** is that human development results from the dynamic interaction between developing persons and their surrounding society. Culture is not something external that impinges on developing persons; it is internalized, integral to everyday attitudes and actions. This idea is so central to understanding the life span that it was first explained in the multicultural perspective discussion in Chapter 1. Now we explain it in more detail.

Teaching and Guidance

The pioneer of the sociocultural perspective was Lev Vygotsky (1896–1934). Like the three original grand theorists, he was born at the end of the nineteenth century, but unlike them, he studied Asian and European groups of many faiths, languages, and social contexts. That was possible when Russia spanned two continents in one nation, not the 14 nations it now is.

Vygotsky noted that people everywhere were taught whatever beliefs and habits were valued within their community. He studied those variations. For example, his research included how Asian farmers used tools, how illiterate people thought of abstract ideas, and how deaf children in Moscow learned in school.

He was both celebrated and marginalized during his lifetime, as the turbulent politics of Russia lurched from one government to another. After his death from tuberculosis at age 38, his work was admired and then banned. Partly because of international politics, his writing was not translated and widely read in the United States until the end of the twentieth century (e.g., Vygotsky, 2012).

Vygotsky stressed that people do not develop in isolation but rather in relationship to the culture of their community, as transmitted by the words and actions of other people. In Vygotsky's view, everyone, schooled or not, is guided by mentors in an **apprenticeship in thinking** (Vygotsky, 2012).

The word *apprentice* once had a specific meaning, often spelled out in a legal contract that detailed what a novice would learn from a master, and what that learner must do. For example, in earlier centuries, a boy wanting to repair shoes would be apprenticed to a cobbler, learning the trade while assisting his teacher.

Vygotsky believed that children become apprentices as they learn, guided by knowledgeable parents, teachers, and other people. Mentors teach children how to think within their culture by explaining ideas, asking questions, demonstrating actions.

To describe this process, Vygotsky developed the concept of **guided participation,** the method used by parents, teachers, and entire societies to teach novices expected skills, values, and habits. Tutors engage learners (*apprentices*) in joint activities, offering "mutual involvement in several widespread cultural practices with great importance for learning: narratives, routines, and play" (Rogoff, 2003, p. 285).

Guided participation is particularly useful in the medical profession. Guidance, not merely instruction, teaches new parents how to feed their fragile newborns,

sociocultural theory A newer theory that holds that development results from the dynamic interaction of each person with the surrounding social and cultural forces.

Dr. James Wertsch

Affection for Children Vygotsky lived in Russia from 1896 to 1934, when war, starvation, and revolution led to the deaths of millions. Throughout this turmoil, Vygotsky focused on learning. His love of children is suggested by this portrait: He and his daughter have their arms around each other.

apprenticeship in thinking Vygotsky's term for how cognition is stimulated and developed in people by more skilled members of society.

guided participation The process by which people learn from others who guide their experiences and explorations.

helps children develop habits that keep them healthy, and allows adolescents with special needs to monitor their own care (Pridham et al., 2018).

Active apprenticeship and sensitive guidance are central to sociocultural theory. Sociocultural theorists contend that most human beliefs are social constructions, not natural laws, and thus societies need to develop and incorporate them, ideally via mutual interaction.

For example, Vygotsky thought that children with disabilities should be educated (Vygotsky, 1994b), a belief that was not part of U.S. culture until about 1970. Then parents insisted that their children with special needs could, and should, learn. That propelled a sociocultural revolution, and now children with special needs are educated rather than excluded from school. Many other social constructions—about women's work, about same-sex relationships, about divorce—have been revised in the past half century.

Sociocultural theory stresses that customs are shaped by people, as well as vice versa. The culture provides tools, or *artifacts*, that aid a particular kind of learning. In contemporary North America, smartphones are such an artifact, teaching children patterns of thought, behavior, and skills. In past generations, televisions, paper diapers, and even rubber balls were tools that changed how children develop.

The Zone of Proximal Development

According to sociocultural theory, all learning is social, whether people are learning a manual skill, a social custom, or a language. As part of the apprenticeship of thinking, a mentor (parent, peer, or professional) finds the learner's **zone of proximal development (ZPD),** an imaginary area surrounding the learner that contains the skills, knowledge, and concepts that are close (proximal) to being grasped but not yet reached.

Vygotsky wrote that ZPD is "the distance between the actual developmental level as determined by independent problem solving and the level of potential development as determined through problem solving under adult guidance or in collaboration with more capable peers" (Vygotsky, 1980, p. 86). The ZPD pulls together all the other concepts of sociocultural theory (Eun, 2019).

Through sensitive assessment of each learner, mentors engage mentees within their zone. Together, in a "process of joint construction," new knowledge is attained. The mentor must avoid two opposite dangers: boredom and failure. Some frustration is permitted, but the learner must be actively engaged, never passive or overwhelmed (see **Figure 2.3**).

A mentor must sense whether support or freedom is needed and how peers can help (they may be the best mentors). Skilled teachers know when a person's zone of proximal development expands and shifts. The shared language of both mentor and mentee is an integral part of the sociocultural process of education. Words are the tools of thought.

Excursions into and through the zone of proximal development are everywhere. At the thousand or so science museums in the United States, exhibits are designed to guide children's scientific learning (Haden, 2010). Fifty years ago, there were no science museums: Now almost every city has one—a sociocultural shift as the culture recognizes the significance of STEM (science, technology, engineering, math).

Consider another example. North American children are expected to know how to ride a bicycle. How do they learn that? Many possibilities, but to illustrate the sociocultural understanding of apprenticeship, consider how a father might teach a daughter.

The dad begins by rolling his child along on a small bike, supporting her weight while telling her to keep her hands on the handlebars, to push the right and left pedals in rhythm, and to look straight ahead. As she becomes more comfortable

zone of proximal development (ZPD) In sociocultural theory, a metaphorical area, or "zone," surrounding a learner that includes all of the skills, knowledge, and concepts that the person is close ("proximal") to acquiring but cannot yet master without help.

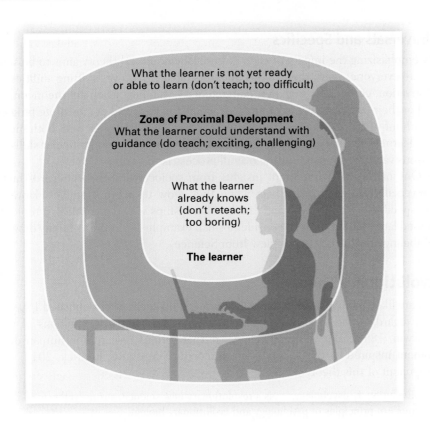

and confident, he jogs beside the bike, still holding it upright himself, praising her for steadily pedaling.

In later days or weeks, he runs beside her, lightly holding only the handlebars. When he senses that she can maintain her balance, he urges her to pedal faster while he loosens his grip. Perhaps without realizing it, she rides on her own, ideally in a wide place without cars or hard pavement. If she falls, he picks her up, reassuring her that she is not seriously hurt and that she is getting better. Someday soon she waves goodbye and bikes around the block.

Note that this is not instruction by preset rules. Sociocultural learning is interactive: No one learns bike-riding by reading and memorizing written instructions, and no good teacher merely repeats a memorized script. Guided practice in the zone of proximal development is essential: The mentor must know exactly what, when, and how support is needed.

Role models and cultural tools also teach, according to sociocultural theory. The bicycle-riding child wants to learn because she has seen other children biking, and stores sell tricycles, training wheels, and small bikes without pedals. Thus, cultural artifacts guide learning.

In another culture, everything might be different. In some nations, no females ride bikes, and no fathers teach their daughters, or even allow them outside the house without a female companion. Recognizing such cultural differences is crucial for understanding development, according to this theory. Children in every culture are taught to walk, dress, and behave in ways that their culture believes are proper.

Every day I witness the power of sociocultural learning. On crowded subways, usually someone sees my white hair and offers a seat. This is the result of apprenticeship: People respond to cultural norms rather than biology (I may be more able to stand than the person who gets up for me). Some sociocultural norms come from childhood (one young man said, in response to my thank you, "My mother would kill me if I didn't give you a seat"), and some are broadcast by loudspeaker announcements: "Offer your seat to someone who is pregnant or elderly."

Universals and Specifics

By emphasizing the impact of each culture, sociocultural theory aims to be sensitive to everyone, everywhere. Thus, mentors, attuned to ever-shifting abilities and motivation, continually urge new competence—the next level, not the moon.

For their part, learners ask questions, show interest, and demonstrate progress, which informs and inspires the mentors. When education goes well, both mentor and learner are fully engaged and productive within the zone. Particular skills and lessons vary enormously, but the overall process is the same.

One of the most important insights from sociocultural theory regards family. Universally, children thrive best when they grow up within families. However, the specifics of family type and family relationships vary a great deal, a topic discussed later. Now, however, consider a small example—children's family drawings (Rübeling et al., 2011; see A View from Science).

Evolutionary Theory

You are likely familiar with Charles Darwin and his ideas, first published 150 years ago, regarding the evolution of plants, insects, and birds over billions of years (Darwin, 1859). But you may not realize that serious research on human development inspired by this theory is quite recent (Simpson & Kenrick, 2013). As a proponent of this theory wrote:

> Evolutionary psychology . . . is a revolutionary new science, a true synthesis of modern principles of psychology and evolutionary biology.
>
> *[Buss, 2015, p. xv]*

The basic idea of evolutionary psychology is that in order to understand the emotions, impulses, and habits of humans over the life span, we must appreciate how those same emotions, impulses, and habits developed within *Homo sapiens* over the past 200,000 years.

A VIEW FROM SCIENCE

Children's Drawings

Children's drawings may reflect their emotions. In the United States and Western Europe, well-adjusted children draw their families with smiling people, holding on to each other or with their arms raised. By contrast, if a child draws a family with small people, neutral facial expressions, and arms downward and not touching, that suggests a troubled child (Fury et al., 1997).

This interpretation is increasingly challenged by drawings produced by children in Africa (Gernhardt et al., 2013), as explored in detail in a study that compared the family drawing of 32 middle-class 6-year-olds from Berlin, Germany, with the family drawings of 31 children from rural areas of Cameroon. This study allowed a triple sociocultural contrast: by nation, by livelihood, and by income (Gernhardt et al., 2016). The Cameroonian children were no less happy or less cherished, but they drew small people (see **Figure 2.4**).

The drawings of the children were rated on 20 features used to describe the attachment of children to their caregivers

(Kaplan & Main, 1986). (Attachment is considered a universal indicator of the caregiver–child relationship, and thus, measuring it is part of many studies).

The drawings of children from the two cultures were quite different. Yet other research finds that children in Cameroon are as securely attached to their families as children elsewhere. From this and many other sociocultural studies comes a clear message: Before anyone intervenes to improve the families, consider the culture (Morelli et al., 2018).

This research illustrates why we need sociocultural theory: The cultural context needs to be considered in order to understand people of any age, from newborns to centenarians. The authors "substantiated children's family drawings as an important cultural document for learning more about children's representation of their social world. However, the interpretation of drawing signs has to be derived from local cultural models of relationships" (Gernhardt et al., 2016, p. 1076).

FIGURE 2.4

Standing Firm When children draw their families, many child therapists look for signs of trouble—such as small, frowning people with hands down floating in space. But cross-cultural research shows that such depictions reflect local norms: The Cameroonian 6-year-olds were as well adjusted in their local community as the German children.

Prof. i.R. Dr. Heidi Keller & Dr. rer. nat. Ariane Gernhardt

Why We Fear Snakes More Than Cars

Evolutionary theory has intriguing explanations for many issues in human development, including 1-year-olds' attachment to their parents, pregnant women's nausea, and adult attraction to major sports contests. These may have evolved to help human survival.

evolutionary theory When used in human development, the idea that many current human emotions and impulses are a legacy from thousands of years ago.

Another example comes from phobias, which are hard to understand without considering human life in prehistoric times. You know that terror of snakes makes some people scream and sweat upon seeing one. People with that phobia may be frightened by photos of snakes, by plastic ones, or by snakes behind glass at a zoo. However, snakes currently cause less than one death in a million. That compares to a death rate a thousand times higher from motor vehicles (OECD, 2014). Why is no one terrified of automobiles?

The answer from evolutionary theory is that human fears began when snakes were common killers. Thus,

> ancient dangers such as snakes, spiders, heights, and strangers appear on lists of common phobias far more often than do evolutionarily modern dangers such as cars and guns, even though cars and guns are more dangerous to survival in the modern environment.

[*Confer et al., 2010, p. 111*]

Since our fears have not caught up to automobiles, evolutionary theory explains that our instincts will not automatically protect us. Instead, we need laws: infant seats, child-safety restraints, seat belts, red lights, speed limits, air bags, guard rails. Such measures are succeeding: The 2018 U.S. motor-vehicle death rate was 11 per 100,000, half the rate of 40 years earlier.

Other modern killers—climate change, drug addiction, obesity, pollution—also require social management, because instincts are contrary to what we know. Evolutionary theory lets us recognize the origins of destructive urges—such as the desire to eat calorie-dense cake—in order to control them (King, 2013). For instance, many adolescents find fast cars irrationally attractive rather than instinctively frightening. That is why we restrict teen drivers, set speed limits, mandate seat belts, and so on.

Why We Protect Babies

According to evolutionary theory, every species has two long-standing, biologically based drives: (1) survival and (2) reproduction. Understanding these provides insight into protective parenting, newborn deaths, infant dependency, child immaturity, puberty onset, family formation, and much more (Konner, 2010).

Here is one example. Adults see babies as cute, even though babies have little hair, no chins, stubby legs, and round stomachs—none of which is attractive in adults. The reason, evolutionary theory contends, is that adults instinctually protect and cherish infants. That was essential when survival of the species was in doubt, 200,000 years ago.

But humans do not protect every baby. Indeed, another evolutionary instinct is that all creatures seek to perpetuate their own descendants more than those who are unrelated. They might kill newborns who are not their own.

Some primates do exactly that: Chimpanzee males who take over a troop murder babies of the deposed male. This occurred among ancient humans as well. The Christian Bible chronicles several examples, including two in the story of Moses and one in the birth of Jesus. Modern humans, of course, have created laws against infanticide—a necessity to control evolutionary instincts (Hrdy, 2009).

Critics point out that people do not always act as evolutionary theory predicts: Parents sometimes abandon newborns, adults sometimes handle snakes, political leaders sometimes separate parents and children, and so on. However, evolutionary theorists counter that ancient impulses within our species need to be understood in order to protect the children of today.

Sanjay Kanoja/AFP/Getty Images

A Sacred River? This is the Ganges river at Allahabad in 2013, which the Indian government is working to clean—a monumental task. No nation is working to rid the Pacific Ocean of a much bigger garbage site. What would evolutionary theory recommend?

● **Observation Quiz** Beyond the pollution of the Ganges by humans' garbage, what characteristics of the river, visible here, contribute to the pollution? (see answer, page 61) ↑

THINK CRITICALLY: What would happen if lust were the only reason one person would mate with another?

Genetic Links

This inborn urge to protect is explained by another concept from evolutionary theory: **selective adaptation.** Biology as well as culture favors traits that promote survival of the next generation. This explains some ancient practices that we now condemn, such as letting a newborn die because it was a girl, or a twin, or a burden on the family in some other way. Biology also worked to help the healthiest children survive and grow to become parents themselves.

Some of the best qualities of people—cooperation, spirituality, and self-sacrifice—may have originated thousands of years ago when tribes and then nations became prosperous because they took care of one another (Rand & Nowak, 2016).

Selective adaptation works as follows: If one person happens to inherit a trait that makes survival more likely, the gene (or combination of genes) responsible for that trait is passed on to the next generation, because that person will live long enough to reproduce (see **Figure 2.5**). Anyone with such a fortunate genetic inheritance has a better chance than those without that gene to survive, mate, and bear many children—half of whom would inherit the genes for that desirable trait.

For example, originally almost all human babies lost the ability to digest lactose at about age 2. All older children and adults were *lactose intolerant*, unable to digest milk from cows or goats (Suchy et al., 2010). This was no problem, as non-dairy foods are nourishing, and our hunter-gatherer ancestors had no farm animals. Indeed, they had no farms until the agricultural revolution, about 12,000 years ago. At that point, cattle were domesticated and raised for their meat. In those places, "killing the fatted calf" provided a rare feast for the entire community.

As you will see in Chapter 3, genes are not always copied exactly from one generation to the next; *spontaneous mutations* occur. In those cattle-raising regions, occasionally a girl would have an aberrant but beneficial gene for the enzyme that allows digestion of cow's milk. If she drank milk intended for a calf, she could digest it and be better nourished than other girls. Her body fat would allow earlier puberty, successful pregnancy, and then ample breast milk.

For all of those reasons, her mutant gene allowing lactose tolerance would spread to half of her many descendants. Here malnourished sisters would have fewer children. Thus, the next generation would include more people who inherited that odd gene, becoming lactose tolerant like their mother. Because of the reproductive advantages, with each generation their numbers would increase. Eventually, that gene would become the new norm.

Interestingly, there are several distinct genetic versions of lactose tolerance: Apparently in each cattle-raising region, when a mutant gene allowed digestion

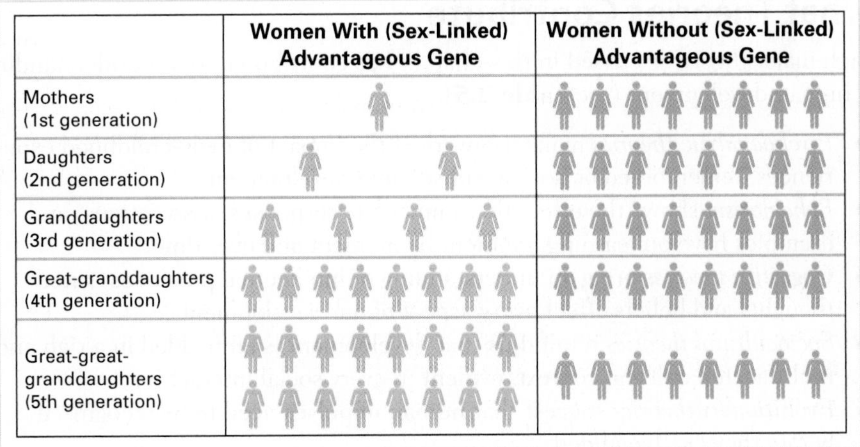

selective adaptation The process by which living creatures (including people) adjust to their environment. Genes that enhance survival and reproductive ability are selected, over the generations, to become more prevalent.

	Women With (Sex-Linked) Advantageous Gene	Women Without (Sex-Linked) Advantageous Gene
Mothers (1st generation)		
Daughters (2nd generation)		
Granddaughters (3rd generation)		
Great-granddaughters (4th generation)		
Great-great-granddaughters (5th generation)		

FIGURE 2.5

Selective Adaptation Illustrated Suppose only one of nine mothers happened to have a gene that improved survival. The average woman had only one surviving daughter, but this gene mutation might mean more births and more surviving children such that each woman who had the gene bore two girls who survived to womanhood instead of one. As you see, in 100 years, the "odd" gene becomes more common, making it a new normal.

of milk, selective adaptation increased the prevalence of that spontaneous mutation (Ranciaro et al., 2014).

For humans (unlike fast-breeding creatures, such as insects) selective adaptation takes centuries, but lactose tolerance has gradually become widespread. That is why few Scandinavians are lactose intolerant but many Africans are—but not those Africans in regions of Kenya and Tanzania where cattle are raised (Ranciaro et al., 2014).

Once it was understood that milk might make some African and Asian children sick, better ways to relieve hunger were found. Fewer children are malnourished today than decades ago, partly because nutritionists know which foods are digestible, nourishing, and tasty for whom. Evolutionary psychology has helped with that.

For groups as well as individuals, evolutionary theory notices how the interaction of genes and environment affects survival and reproduction. Genetic variations are particularly beneficial when the environment changes, which is one reason genetic diversity benefits humanity as a whole.

Evolutionary adaptation is in the human genes that allow people to adapt to many climates (Tattersall, 2017). That explains why the Inuit, an ethnic group who have survived in the Arctic for thousands of years, have genes that make them relatively short and plump. That body shape is protective against frigid temperatures. Ordinarily that would make them at high risk for heart disease, except selective adaptation has also bestowed genes that reduce heart damage (Rudkowska et al., 2013).

If a species' gene pool does not include variants that allow survival in difficult circumstances (such as exposure to a new disease or to an environmental toxin), the entire species becomes extinct, true for many nonhuman species. That has not occurred for *Homo sapiens*, because every human has some gene variants that are rare, making us all distinct from our neighbors. Evolutionary theory recognizes and praises such diversity which might mean survival.

Especially for Teachers and Counselors of Teenagers Teen pregnancy is destructive of adolescent education, family life, and sometimes even health. According to evolutionary theory, what can be done about this? (see response, page 61)

> ## WHAT HAVE YOU LEARNED?
>
> 1. Why is the sociocultural perspective particularly relevant within the United States?
> 2. How do mentors and mentees interact within the zone of proximal development?
> 3. How do the customs and manufactured items in a society affect human development?
> 4. Why are behaviors and emotions that benefited ancient humans still apparent today?
> 5. How does understanding ancient peoples protect modern humans?

What Theories Contribute

Each major theory discussed in this chapter has contributed to our understanding of human development (see **Table 2.5**):

- *Psychoanalytic theories* make us aware of the impact of early-childhood experiences, remembered or not, on subsequent development.
- *Behaviorism* shows the effect that immediate responses, associations, and examples have on learning, moment by moment and over time.
- *Cognitive theories* bring an understanding of intellectual processes, since thoughts and beliefs affect every aspect of our development.
- *Sociocultural theories* remind us that development is embedded in a rich and multifaceted cultural context, evident in every social interaction.
- *Evolutionary theories* suggest that human impulses need to be recognized before they can be guided.

TABLE 2.5

Five Perspectives on Human Development

Theory	Area of Focus	Fundamental Depiction of What People Do	Relative Emphasis on Nature or Nurture?
Psychoanalytic	Psychosexual (Freud) or psychosocial (Erikson) stages	Battle unconscious impulses and overcome major crises.	More nature (biological, sexual impulses, and parent–child bonds)
Behaviorism	Conditioning through stimulus and response	Respond to stimuli, reinforcement, and models.	More nurture (direct environment produces various behaviors)
Cognitive	Thinking, remembering, analyzing	Seek to understand experiences while forming concepts.	More nature (mental activity and motivation are key)
Sociocultural	Social control, expressed through people, language, customs	Learn the tools, skills, and values of society through apprenticeships.	More nurture (interaction of mentor and learner, within cultures)
Evolutionary	Needs and impulses that originated thousands of years ago	Develop impulses, interests, and patterns to survive and reproduce.	More nature (needs and impulses apply to all humans)

Remember that each theory is designed to be practical. This is evident with a very practical issue for many parents: how to toilet-train their children (see Opposing Perspectives).

No comprehensive view of development can ignore any of these theories, yet each has encountered severe criticism: *psychoanalytic theory* for being too subjective; *behaviorism* for being too mechanistic; *cognitive theory* for undervaluing emotions; *sociocultural theory* for neglecting individual choice; *evolutionary theory* for ignoring the power of current morals, laws, and norms.

Most developmentalists prefer an **eclectic perspective,** choosing what they consider to be the best aspects of each theory. Rather than adopt any one of these theories exclusively, they make selective use of all of them.

All theories reflect the personal background of the theorist, as do all criticisms of theories. If you find yourself drawn to one theory, or dismissive of another, ask yourself what in your own background led to your reaction.

Ideally, you can see the reasons each of these five is relevant to the study of child development. Being eclectic, not tied to any one theory, is beneficial, because everyone, scientist as well as layperson, is biased. However, even being eclectic may be limiting: Choosing the best from each theory may be too picky or the opposite, too tolerant.

For developmentalists, all of these theories merit study and respect. It is easy to dismiss any one of them, but using several perspectives opens our eyes and minds to aspects of development that we might otherwise ignore. As one overview of seven developmental theories (including those explained here) concludes,

> Because no one theory satisfactorily explains development, it is critical that developmentalists be able to draw on the content, methods, and theoretical concepts of many theories.

[Miller, 2016, p. 434]

eclectic perspective The approach taken by most developmentalists, in which they apply aspects of each of the various theories of development rather than adhering exclusively to one theory.

CHAPTER APP 2

 My Token Board

IOS:
https://tinyurl.com/yy6hmbuq

RELEVANT TOPIC:
Behaviorism and positive reinforcement for children

My Token Board is a visual reward system for children of all ages and abilities that helps motivate them to learn and complete tasks. The app's use of reinforcers is based on the principles of operant conditioning and effective with children who are on the autism spectrum.

Toilet Training—How and When?

Parents hear opposite advice about almost everything regarding infant care, including feeding, responding to cries, bathing, and exercise. Often a particular parental response springs from one of the theories explained in this chapter—no wonder advice is sometimes contradictory.

One practical example is toilet training. In the nineteenth century, many parents believed that bodily functions should be controlled as soon as possible in order to distinguish humans from lower animals. Pushed by that theory (as opposed to the theory of evolution) many parents began toilet training in the first months of life (Accardo, 2006). Then, Freud pegged the first year as the oral stage, not anal (when toilet training was supposed to occur), and Erikson stressed that infants need to develop trust.

Accordingly, many psychologists and pediatricians recommended postponing training to avoid serious personality problems later on. This was soon part of many manuals on child rearing. For example, a leading pediatrician, Barry Brazelton, wrote a popular book for parents. He advised delaying toilet training until the child was cognitively, emotionally, and biologically ready—around age 2 for daytime training and age 3 for nighttime dryness.

> As a society, we are far too concerned about pushing children to be toilet trained early. I don't even like the phrase "toilet training." It really should be toilet learning.
>
> *[Brazelton & Sparrow, 2006, p. 193]*

When the second grand theory took hold in the United States, behaviorists explained that learning results from conditioning. That led to the belief that toilet training could occur whenever the parent wished, not at a particular age.

Parents often wished for early training before disposable diapers (about 1970) because parents experienced unpleasant consequences from changing diapers for years. But now stores carry diapers designed for 4-year-olds, so parents no longer are inconvenienced by late toilet training (that is, by having to contend with dirty diapers). That is a sociocultural shift: A practical impact on parents, not a deep psychoanalytic need of babies, affected the age of toilet training.

In one application of behaviorism, children drank quantities of their favorite juice, sat on the potty with a parent nearby to keep them entertained, and then, when the inevitable occurred, the parent praised and rewarded them—a powerful reinforcement for the child and soon for the adult as well. Children were conditioned (in one day, according to some behaviorists) to head for the potty whenever the need arose (Azrin & Foxx, 1974).

Cognitive theory would consider such a concerted effort unnecessary, suggesting that parents wait until the child can understand why to urinate and defecate in the toilet. Thinking leads to behavior—which explains why older adults are mortified by urinary incontinence.

This raises the importance of modeling, as well as the sociocultural theory. In some African communities, children toilet-train themselves by following slightly older children to the surrounding trees and bushes. This is easier, of course, if toddlers wear no diapers—a practice that makes sense in some climates and ecosystems.

Finally, evolutionary theory notes that control of urination and defecation is part of every human culture, because it promotes survival by reducing the spread of pathogens. That is why dogs and cats readily learn where to eliminate waste (and instinctively scratch to bury their feces) and why humans worldwide learn bathroom hygiene. The contemporary emphasis on handwashing (a cognitive innovation) is the latest manifestation of ancient impulses.

Each of these theories would be critical of one U.S. mother who began training her baby just 33 days after birth. She noticed when her son was about to defecate, held him above the toilet, and had trained him by 6 months (Sun & Rugolotto, 2004).

- Psychoanalytic theorists would wonder what made her such an anal person, valuing cleanliness and order without considering the child's needs.
- Behaviorists would say that the mother was trained, not the son. She taught herself to be sensitive to his body; she was reinforced when she read his clues correctly.
- Cognitive theory would question the mother's thinking. For instance, did she have an irrational fear of normal body functions?
- Sociocultural theory would be aghast that the U.S. drive for personal control took such a bizarre turn.
- Evolutionary theory would expect that such early training would be a rare event unless it was adaptive to the species—and would note that every mother reading about this woman would probably consider her more whacky than wise.

What is best? Some parents are reluctant to train, and the result, according to one book, is that many children are still in diapers at age 5 (Barone, 2015). Dueling theories and diverse parental practices have led the authors of an article for pediatricians to conclude that "despite families and physicians having addressed this issue for generations, there still is no consensus regarding the best method or even a standard definition of toilet training" (Howell et al., 2010, p. 262).

One comparison study of toilet-training methods found that the behaviorist approach was best for older children with serious disabilities but that almost every method succeeded with the average young child. Many sources explain that human diversity means that there is no "right" way: "the best strategy for implementing training is still unknown" (Colaco et al., 2013, p. 49).

That may return us to the multicontextual, multicultural perspectives of Chapter 1 that explain the vast differences from one community to another. A study of parents' opinions in Belgium found that single mothers who were of low socioeconomic status waited until age 3 or so to train their children, which was too long according to those researchers (van Nunen et al., 2015). Of course, both "too soon" and "too late" are matters of opinion.

Many parents firmly believe in one approach or another. Everyone has theories, sometimes strongly held, whether they know it or not. That is one goal of this chapter, to help us recognize the theories that affect our everyday lives.

As you will see in many later chapters, theories provide a fresh look at behavior. Imagine a mother, father, teacher, coach, and grandparent discussing the problems of a particular child. Each might suggest a possible explanation that makes the others say, "I never thought of that." If they listen to each other with an open mind, together they might understand the child better and agree on a beneficial strategy.

Using five theories is like having five perceptive observers. All five are not always on target, but it is better to use theory to consider alternate possibilities than to stay in one narrow groove. A hand functions best with five fingers, although each finger is different and some fingers are more useful than others.

What finally happened with my family, Iowa, and travel?

- The children enjoyed the overnight train ride (*psychoanalytic*).
- The culture provides days off school and work, and people ask "what are you doing for the holidays?" (*sociocultural*).
- We need to protect the planet (four took the train to Chicago; four flew to Minneapolis; all eight drove many hours to avoid a multi-stop flight) (*cognitive*).
- We ignored the possibility of ice and blizzards (we have never driven in the Midwest in December, so no past conditioning) (*behaviorist*).
- Family togetherness is a priority (*evolutionary*).

What happened with all that thinking and theorizing? One happy grandmother, who was told by the other eight that our Iowa trip was a good one. But a lesson from this chapter is that perspectives differ, sometimes radically: My perception may be biased. We all need to consider alternate perspectives on whatever we think about child development!

WHAT HAVE YOU LEARNED?

1. What are the criticisms of each of the five theories?

2. Why are most developmentalists eclectic in regard to theories?

3. Why is it useful to know more than one theory to explain human behavior?

Historical Highlights of Developmental Science

As evident throughout this textbook, much more research and appreciation of the brain, social context, and the non-Western world has expanded our understanding of human development in the 21st century. This timeline lists a few highlights of the past.

200,000-50,000 BCE With their large brains, long period of child development, and extensive social and family support, early humans were able to sustain life and raise children more effectively than other primates.

c. 400 BCE In ancient Greece, ideas about children from philosophers like Plato (c. 428-348 BCE) and Aristotle (384-322 BCE) influenced further thoughts about children. Plato believed children were born with knowledge. Aristotle believed children learn from experience.

1650-1800 European philosophers like John Locke (1632-1704) and Jean Jacques Rousseau (1712-1778) debate whether children are born as "blank slates" and how much control parents should take in raising them.

1797 First European vaccination: Edward Jenner (1749-1823) publicizes smallpox inoculation, building on vaccination against smallpox in Asia, the Middle East, and Africa.

1750-1850 Beginning of Western laws regulating child labor and protecting the rights of children.

1879 First experimental psychology laboratory established in Leipzig, Germany.

1885 Sigmund Freud (1856-1939) publishes *Studies on Hysteria*, one of the first works establishing the importance of the subconscious and marking the beginning of the theories of psychoanalytic theory.

1895 Ivan Pavlov (1849-1936) begins research on dogs' salivation response.

1905 Max Weber (1864-1920), the founder of sociology, writes *The Protestant Work Ethic*, about human values and adult work.

1905 Alfred Binet's (1857-1911) intelligence test published.

1907 Maria Montessori (1870-1952) opens her first school in Rome.

1913 John B. Watson (1878-1958) publishes *Psychology As the Behaviorist Views It*.

50,000 BCE — 400 BCE — 0 — 500 — 1000 — 1500 — 1650 — 1700 — 1750

---

140 BCE In China, imperial examinations are one of the first times cognitive testing is used on young people.

500-1500 During the Middle Ages in Europe, many adults believed that children were miniature adults.

1100-1200 First universities founded in Europe. Young people pay to be educated together.

1837 First kindergarten opens in Germany, part of a movement to teach young children before they entered the primary school system.

1859 Charles Darwin (1809-1882) publishes *On the Origin of Species*, sparking debates about what is genetic and what is environmental.

1900 Compulsory schooling for children is established for most children in the United States and Europe.

1903 The term "gerontology," the branch of developmental science devoted to studying aging, first coined.

1920 Lev Vygotsky (1896-1934) develops sociocultural theory in the former Soviet Union.

1923 Jean Piaget (1896-1980) publishes *The Language and Thought of the Child*.

1933 Society for Research on Child Development, the preeminent organization for research on child development, founded.

1939 Mamie (1917-1983) and Kenneth Clark (1914-2005) receive their research grants to study race in early childhood.

NICHOLAS VEASEY/PHOTOGRAPHER'S CHOICE/GETTY IMAGES

1943 Abraham Maslow (1908–1970) publishes *A Theory of Motivation*, establishing the hierarchy of needs.

1950 Erik Erikson (1902–1994) expands on Freud's theory to include social aspects of personality development with the publication of *Childhood and Society*.

1951 John Bowlby (1907–1990) publishes *Maternal Care and Mental Health*, one of his first works on the importance of parent–child attachment.

1953 Publication of the first papers describing DNA, our genetic blueprint.

1957 Harry Harlow (1905–1981) publishes *Love in Infant Monkeys*, describing his research on attachment in rhesus monkeys.

1961 The morning sickness drug thalidomide is banned after children are born with serious birth defects, calling attention to the problem of teratogens during pregnancy.

1961 Alfred Bandura (b. 1925) conducts the Bobo Doll experiments, leading to the development of social learning theory.

1979 Urie Bronfenbrenner (1917–2005) publishes his work on ecological systems theory

1986 John Gottman (b. 1942) founded the "Love Lab" at the University of Washington to study what makes relationships work.

1987 Carolyn Rovee-Collier (1942–2014) shows that even young infants can remember in her classic mobile experiments.

1990–Present New brain imaging technology allows pinpointing of brain areas involved in everything from executive function to Alzheimer's disease.

1994 Steven Pinker (b. 1954) publishes *The Language Instinct*, focusing attention on the interaction between neuroscience and behavior.

1996 Giacomo Rizzolatti publishes his discovery of mirror neurons.

2000 Jeffrey Arnett conceptualizes emerging adulthood.

2003 Mapping of the human genome is completed.

2013 DSM-5, which emphasizes the role of context in understanding mental health problems, is published.

| 1800 | 1850 | 1900 | 1950 | 2000 |

1953 B.F. Skinner (1904–1990) conducts experiments on rats and establishes operant conditioning.

1955 Emmy Werner (b. 1929) begins her Kauai study, which focuses on the power of resilience.

1956 K. Warner Schaie's (b. 1928) Seattle Longitudinal Study of Adult Intelligence begins.

1965 Head Start, an early childhood education program, launched in the United States.

1965 Mary Ainsworth (1913–1999) starts using the "Strange Situation" to measure attachment.

1966 Diana Baumrind (1928–2018) publishes her first work on parenting styles.

1972 Beginning of the Dunedin, New Zealand, study—one of the first longitudinal studies to include genetic markers.

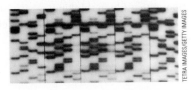

1990 Barbara Rogoff (b. 1950) publishes *Apprenticeship in Thinking*, making developmentalists more aware of the significance of culture and context. Rogoff provided new insights and appreciation of child-rearing in Latin America.

1993 Howard Gardner (b. 1943) publishes *Multiple Intelligences*, a major new understanding of the diversity of human intellectual abilities. Gardner has since revised and expanded his ideas in many ways.

2013 U.S. President Barack Obama announces his administration's Brain Research through Advancing Innovative Neurotechnologies (BRAIN) Initiative.

2017 Several U.S. cities expand public funding for early-childhood education (universal pre-K).

2020 and beyond Onward. Many more discoveries are chronicled in this book, as researchers continue to test and explore.

SUMMARY

What Theories Do

1. A theory provides general principles to guide research and explain observations. Each of the five major developmental theories—psychoanalytic, behaviorist, cognitive, sociocultural, and evolutionary—interprets human development from a distinct perspective, providing a framework for understanding human emotions, experiences, and actions.

2. Theories are neither true nor false. They begin with facts and lead to facts, but they are not themselves facts. Instead, they generate hypotheses to be tested and interpretations of the myriad human behaviors. Good theories are practical: They aid inquiry, interpretation, and daily life.

Grand Theories

3. Psychoanalytic theory emphasizes that adult actions and thoughts originate from unconscious impulses and childhood conflicts. Freud's theory is called psychosexual, because he recognized sexual urges that arise during three stages of childhood—oral, anal, and phallic—and continue, after latency, in the genital stage.

4. Erikson's psychosocial theory described eight successive stages of development, each involving a crisis to be resolved. The early stages are crucial, with lifelong effects, but unlike Freud, Erikson emphasized social more than sexual needs.

5. Behaviorists, or learning theorists, believe that scientists should study observable and measurable behavior. Behaviorism emphasizes conditioning—a lifelong learning process in which an association between one stimulus and another (classical conditioning) or the consequences of reinforcement and punishment (operant conditioning) guide behavior.

6. Social learning theory recognizes that people learn by observing others, even if they themselves have not been reinforced or punished. Children are particularly susceptible to social learning, but all humans are affected by what they notice in other people.

7. Cognitive theorists believe that thoughts and beliefs powerfully affect attitudes, actions, and perceptions, which in turn affect behavior. Piaget proposed four age-related periods of cognition, each propelled by an active search for cognitive equilibrium. Like Freud, his last stage occurred at adolescence.

8. Information processing focuses on each aspect of cognition—input, processing, and output. This perspective has benefited from technology, first from understanding computer functioning and more recently by the many ways scientists monitor the brain.

Newer Theories

9. Sociocultural theory explains human development in terms of the guidance, support, and structure provided by each social group through culture and mentoring. Vygotsky described how learning occurs through social interactions in which mentors guide learners through their zone of proximal development.

10. Sociocultural learning is also encouraged by the examples and tools that each society provides. These are social constructions, which guide everyone but also which can change.

11. Evolutionary theory contends that contemporary humans inherit genetic tendencies that have fostered survival and reproduction of the human species for tens of thousands of years. Through selective adaptation, the fears, impulses, and reactions that were useful 100,000 years ago for *Homo sapiens* continue to this day.

12. Evolutionary theory provides explanations for many human traits, from lactose intolerance to affection toward babies. Selective adaptation is the process by which genes enhance human development over thousands of years. Societies use laws and customs to protect people from some genetic impulses.

What Theories Contribute

13. Psychoanalytic, behavioral, cognitive, sociocultural, and evolutionary theories have aided our understanding of human development. However, no single theory describes the full complexity and diversity of human experience. Most developmentalists are eclectic, drawing on many theories, aware that any one perspective might be too narrow.

KEY TERMS

developmental theory (p. 33)
psychoanalytic theory (p. 35)
psychosocial theory (p. 37)
behaviorism (p. 37)
classical conditioning (p. 38)
operant conditioning (p. 39)

reinforcement (p. 39)
social learning theory (p. 40)
modeling (p. 40)
cognitive theory (p. 40)
cognitive equilibrium (p. 41)
assimilation (p. 42)

accommodation (p. 42)
information-processing theory (p. 42)
sociocultural theory (p. 47)
apprenticeship in thinking (p. 47)

guided participation (p. 47)
zone of proximal development (ZPD) (p. 48)
evolutionary theory (p. 51)
selective adaptation (p. 53)
eclectic perspective (p. 55)

APPLICATIONS

1. Developmentalists sometimes talk about "folk theories," which are developed by ordinary people. Choose three sayings in your culture, such as (from the dominant U.S. culture) "A penny saved is a penny earned" or "As the twig is bent, so grows the tree." Explain the underlying assumptions, or theory, that each saying reflects. Why might the theory be wrong?

2. Cognitive theory suggests the power of thoughts, and socio-cultural theory emphasizes the power of context. Find someone who disagrees with you about some basic issue (e.g., abortion, immigration, socialism) and listen carefully to their ideas and reasons (encourage them to explain; don't contradict). Then analyze how cognition and experience shaped their ideas *and* your own.

3. Ask three people to tell you their theories about male–female differences in mating and sexual behaviors. Which of the theories described in this chapter is closest to each explanation, and which theory is not mentioned?

Especially for ANSWERS

Response for Teachers (from p. 37) Erikson would note that the behavior of 5-year-olds is affected by their developmental stage and by their culture. Therefore, you might design your curriculum to accommodate active, noisy children.

Response for Teachers and Counselors of Teenagers (from p. 54) Evolutionary theory stresses the basic human drive for reproduction, which gives teenagers a powerful sex drive. Thus, merely informing teenagers of the difficulty of caring for a newborn (some high school sex-education programs simply give teenagers a chicken egg to nurture) is not likely to work. A better method would be to structure teenagers' lives so that pregnancy is impossible—for instance, with careful supervision or readily available contraception.

Observation Quiz ANSWERS

Answer to Observation Quiz (from p. 38) Both are balding, with white beards. Note also that none of the other theorists in this chapter have beards—a cohort difference, not an ideological one.

Answer to Observation Quiz (from p. 52) The river is slow-moving (see the boat) and shallow (see the man standing). A fast-moving, deep river is able to flush out contaminants more quickly.

The New Genetics

What Will You Know?

1. Genetically, how is each zygote unique?
2. How do twins differ from other siblings?
3. Who is likely to carry genes that they do not know they have?
4. When should people see a genetic counselor?

For 30 years I have spent Thanksgiving with my four daughters. Now adults, they fly or drive many miles to gather together; they have their own jobs and homes. One tradition is that each of us says what we are thankful for in the past year, and that typically includes each other. The family connections are strong.

It is apparent that we are closely related. Strangers say we look alike, people on the phone mistake us for one another, we laugh in the same way at the same jokes. Our similarities go far deeper. We all express strong opinions (disagreeing on specifics but sharing values); we all make our living by teaching and writing (in different places and professions); and we all have similar habits (we pick up litter; yesterday we had a robust discussion about the best strategy to encourage our neighbors to do so).

Our differences are obvious, too, at least to us. When we are together, each prepares her own coffee—caffeinated or decaf, brewed or instant, black or not, canned or bottled milk, sugar or honey or neither. One daughter was upset this year that I didn't have 2-percent milk in a carton. Without telling her, I asked our neighbor, who gave me some. She was grateful that I did that, but another daughter was troubled that I bothered that neighbor. That illustrates the general truth: We each are distinct—in cooking, eating, cleaning, sleeping, socializing . . . and so on.

In all of this, my family reflects the themes of this chapter. Genes and family background shape human lives, yet each person is unique. The rest of this chapter continues those themes—nature and nurture interwoven, from conception onward.

The Genetic Code

First, we begin with biology. All living things are composed of cells. The work of cells is done by proteins, aided by other cells. Cells manufacture proteins according to a code of instructions stored by molecules of **deoxyribonucleic acid (DNA)** at the heart of the cell. These coding DNA molecules are on a **chromosome**.

46 to 21,000 to 3 Billion

Humans have 23 pairs of chromosomes (46 in all, half from each parent), which direct the manufacture of proteins needed for life and growth (see **Figure 3.1**).

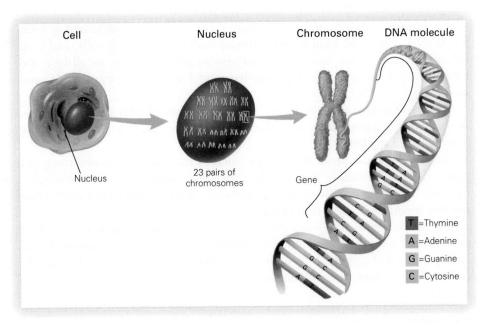

T =Thymine
A =Adenine
G =Guanine
C =Cytosine

FIGURE 3.1
How Proteins Are Made The genes on the chromosomes in the nucleus of each cell instruct the cell to manufacture the proteins needed to sustain life and development. The code for a protein is the particular combination of four bases, T-A-G-C (thymine, adenine, guanine, and cytosine).

The instructions on the 46 chromosomes are organized into genes, with every **gene** usually situated at a precise location on a particular chromosome.

When one **gamete** (a reproductive cell, sperm from a man or ovum from a woman) combines with another, those two gametes create a new cell called a **zygote**. Each gamete is formed from a particular variation of chromosomes and genes. Each man or woman produces many possible variations of their 46 chromosomes on each gamete—8 million possible combinations of 23 chromosomes (23^{23} or 8,388,608).

Zygotes have a total of about 21,000 coding genes, each directing the formation of specific proteins made from a string of 20 amino acids. The instructions on those 21,000 genes are on about 3 billion *base pairs*, which are pairs of four chemicals (adenine paired with thymine and guanine paired with cytosine).

The entire packet of instructions to make a living organism is called the **genome**. There is a genome for every species and variety of plant and animal—even for every bacterium and virus. Most human genes are also present in other creatures, yet genes define each species.

Members of the same species are similar genetically—more than 99 percent of any human's base pairs are identical to those of any other person because, although they differ in tiny details, all human chromosomes contain the same basic genes. The definition of a species is that its members can interbreed successfully. Since humans are one species, a man and a woman from opposite ends of the world can mate and produce a new human with all the genes that define a person.

deoxyribonucleic acid (DNA) The chemical composition of the molecules that contain the genes, which are the chemical instructions for cells to manufacture various proteins.

chromosome One of the 46 molecules of DNA (in 23 pairs) that virtually each cell of the human body contains and that, together, contain all the genes. Other species have more or fewer chromosomes.

gene A small section of a chromosome; the basic unit for the transmission of heredity. A gene consists of a string of chemicals that provide instructions for the cell to manufacture certain proteins.

gamete A reproductive cell. In humans it is a sperm or an ovum.

zygote The single cell formed from the union of two gametes, a sperm and an ovum.

genome The full set of genes that are the instructions to make an individual member of a certain species.

Same and Different

Decoding the genome of *Homo sapiens* was a major breakthrough in genetic research in 2003. Since then, the genomes of thousands of other species have been decoded and thousands of variations in the human genome have been discovered.

It is human nature to notice differences. But remember our shared genetic bonds. Not only do all humans have similar bodies (two eyes, hands, and feet; the same organs, blood, and bones) but also we all use words, we all love and hate, we all hope and fear, for ourselves and for descendants, friends, and strangers.

Yet none of us is exactly like anyone else, and our individuality starts with genes. Any variation in a gene, such as a differences in the precise sequence of the base pairs, is called an **allele**.

allele A variation that makes a gene different in some way from other genes for the same characteristics. Many genes never vary; others have several possible alleles.

Most alleles cause small differences (such as the shape of an eyebrow), but some are crucial. Another way to state this is that some genes are *polymorphic* (literally, "many forms"). Because each person's variations differ from every other person's variations, each of us is unique.

You can recognize at a glance that two people are not identical, even if both are the same age, gender, and ethnicity. When you search for a well-known friend among a crowd of thousands, you do not mistakenly greet someone who superficially looks like them. Tiny variations distinguish each face and each body—inside and out.

Beyond the Genes

RNA (*ribonucleic acid*, another molecule) and additional DNA surround each gene. RNA material does not code, which means RNA does not directly determine human traits. This noncoding material used to be called *junk*—but no longer.

In a process called *methylation*, the material surrounding each gene enhances, transcribes, connects, empowers, silences, and alters genetic instructions. Methylation continues throughout life, from conception until death. Obviously, genes are crucial, but even more crucial is whether and how genes are expressed. RNA regulates and transcribes genetic instructions, turning some genes and alleles on or off.

Not All Genetic Every child becomes their own person—not what their parents fantasize.

THE WORLD'S FIRST GENETICALLY ENGINEERED HUMAN HITS ADOLESCENCE

We buy you the best genes in the world—FOR THIS?

So, I got my nose pierced. So what, man.

I remember checking "genius" on the order form—AND NOW LOOK!

Roz Chast/The New Yorker Collection/The Cartoon Bank

In other words, a person can have the genetic tendency for a particular trait, disease, or behavior, but that tendency might never appear because it was never turned on. Think of a light switch: A lamp might have a new bulb and be plugged into power, but the room stays dark unless the switch is flipped.

Scientists continually discover new functions—for good or ill—of the noncoding genetic material (Iorio & Palmieri, 2019; Larsen, 2018). Applications are many—treating and controlling diseases, genetically modifying crops to resist pathogens, developing vaccines to protect people and animals.

Most scientists are hopeful that research on genes and development will help humankind. Many others are fearful. This divergence is reflected in national and international guidelines.

As one article explains, we have "fast science and sluggish policy," which is "enveloped by UN procedures that promote inclusiveness but are typically slow, lumbering, and inflexible, making it difficult to adapt to fast-moving events in the outside world" (Wynberg & Laird, 2018, p. 1).

Epigenetics

The science of the interaction between nature and nurture is explored in **epigenetics**, the study of how the environment alters genetic expression, beginning at conception and continuing lifelong. The Greek root *epi-*, meaning "around, above, below," focuses attention on the vital noncoding elements.

Events and circumstances surrounding (around, above, below) the genes determine whether genes are expressed or silenced (Ayyanathan, 2014). Contrary to what some people assume—the assumption that "genetic" means that a trait is unchangeable—we now know that nutrients, toxins, and experiences affect prenatal and postnatal development in the brain as well as in every genetic function. Plasticity is lifelong.

It might be easy to believe that biological forces, such as malnutrition, affect genes. But it is harder to recognize that social experiences, such as chronic loneliness, can change the brain (Cacioppo et al., 2014).

One new area of research is hospital care for preterm infants. Protective factors (being held and comforted by the mother, skin-to-skin) and stressful factors (painful intubations, punctures) affect the genes of the tiny person, with epigenetic changes evident years later (Provenzi et al., 2018). Thanks to research on

epigenetics The study of how environmental factors affect genes and genetic expression—enhancing, halting, shaping, or altering the expression of genes.

A CASE TO STUDY

Women Engineers

It was once thought that genes made females inferior in math, specifically that prenatal neurological connections made their brains less adept at spatial understanding. That explained why women scored lower than men on tests of math, and why no woman was allowed to study engineering.

The only way a woman could be an engineer was as a dutiful wife. For example, Elizabeth Roebling was called the "secret engineer" of the Brooklyn Bridge (completed in 1883), because the man designated as chief engineer was her husband, and he became so ill that he could not leave his bed (Dougherty, 2019; Wagner, 2017). She told people that he told her what to do (although he was actually too sick to do so).

A century later another woman, Sheri Sheppard, became a professor of engineering, as her father had done. But at first she thought that she did not belong. She remembers:

> My second year in college [at the University of Michigan], I hit my first engineering class. And at the time I was, in most of my engineering classes, the only woman. And the professor got in front of us and was lecturing and he was using words that I had no idea what he was speaking. And I'm thinking that every guy in the room knows exactly what he's talking about, I mean they were born knowing that. [Still feeling that I did not belong] I went to the next class and one of the other students . . . raised his hand right at the beginning of the class and he said to the professor, "I don't have the foggiest notion of what you're talking about." And it was just like, "they don't know it either!" And after that it became really fun because education is all about asking questions.

> *[Retelas, 2017]*

Beginning in the 1960s, millions of women insisted that nurture, not nature, kept women from advanced math, and thus

from professions that required it. Sheppard was one of the first women to study engineering; now she is a professor of engineering at Stanford, with many female students.

Many college women now major in engineering, physics, math, and chemistry (Brown & Lent, 2016). This does not erase genetic tendencies: Some people excel at math and others struggle with it, but those genes are not exclusive to people of one gender or another, or, for that matter, one family or ethnic background or another.

In the twenty-first century, many women who were hired as engineers quit. A third of them cite hostility from male co-workers as the reason. Some male co-workers were helpful, especially younger ones. However, some female engineers left because of another genetic force—the urge to become a mother.

Fortunately, a recent study of female engineers in large British companies found that social support and role models are changing (Fernando et al., 2018). One young woman engineer said:

> My boss' boss is a woman. She has a daughter. The team leader for one of the other big assets in the North Sea is a woman and she works four days a week. She's very well respected as an engineer. . . . so there is hope for me—in the beginning I didn't think there was any hope especially if I wanted to have children. But now it makes sense to develop a career here.

> *[Rosie, quoted in Fernando et al., 2018, p. 491]*

Currently, not only are more women employed as engineers, but their brains also reflect their profession: They can envision spatial relations with the best of them—and far better than their mothers and grandmothers. That is an epigenetic result.

epigenetics and preterm infants, newborn care is changing: Lights are dimmed, noises quieted, and parental touch encouraged.

More generally, even when a particular person inherits genes for a serious disease, epigenetic factors matter. As one review explains, "there are, indeed, individuals whose genetics indicate exceptionally high risk of disease, yet they never show any signs of the disorder" (Friend & Schadt, 2014, p. 970). Why? Epigenetics.

The Microbiome

microbiome All the microbes (bacteria, viruses, and so on) with all their genes in a community; here the millions of microbes of the human body.

One aspect of both nature and nurture that profoundly affects each person is the **microbiome**, which refers to all of the microbes (bacteria, viruses, fungi, archaea, yeasts) that live within every part of the body. The microbiome includes "germs," the target of disinfectants and antibiotics. Nonetheless, most microbes are helpful, enhancing life, not harming it.

Microbes have their own DNA, influencing immunity, weight, diseases, moods, and much more (Dominguez-Bello et al., 2019; Dugas et al., 2016). Particularly crucial is how the microbiome affects nutrition, since gut bacteria break down food for nourishment.

In one telling study, researchers in Malawi studied twins when one was malnourished and the other was not, even though both lived together and were fed the same foods. Did a greedy twin grab food from his brother? No! When scientists analyzed each twin's microbiome, they found crucial differences. That is why only one suffered (H. Smith et al., 2014).

Genetic Diversity

It might seem as if each child is half mother and half father, since each child has 23 chromosomes from each parent. But that is not true, because the genes and alleles of each chromosome interact with those on the other pair (chromosome 1 from the father with chromosome 1 from the mother, 2 with 2, and so on), and that interaction produces a zygote unlike either parent. Thus, each new person is a product of two parents but is not like either one.

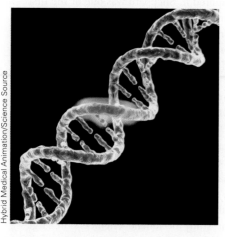

Hybrid Medical Animation/Science Source

Twelve of 3 Billion Pairs This is a computer illustration of a small segment of one gene. Even a tiny difference in one gene can cause major changes in a person's phenotype.

That diversity is added to the epigenetic diversity just described. And there is more! Mutations occur before, during, and after conception. Each zygote carries genes that "are themselves transmitted to individual cells with large apparent mistakes—somatically acquired deletions, duplications, and other mutations" (Macosko & McCarroll, 2013, p. 564). Small variations, mutations, or repetitions in the 3 billion base pairs could make a notable difference in the proteins and thus, eventually, in the person.

copy number variations Genes with various repeats or deletions of base pairs.

Attention has focused on **copy number variations**, which are genes with repeats or deletions (from one to hundreds) of base pairs. Copy number variations are at least five times as common as variations in single genes (Saitou & Gokcumen, 2020). Copy number variations correlate with heart disease, intellectual disability, mental illness, and many cancers.

Usually, all this genetic diversity helps the species, because creativity, prosperity, and survival are enhanced when one person is unlike another. Two economists suggest that there is an optimal balance between diversity and similarity: Human societies prosper when they are close to that ideal (Ashraf & Galor, 2013). That idea is controversial. Nonetheless, most scholars agree that too much genetic insularity is problematic (Spolaore & Wacziarg, 2018).

Matching Genes and Chromosomes

genotype An organism's entire genetic inheritance, or genetic potential.

The genes on the chromosomes constitute an organism's genetic inheritance, or **genotype**. The code of the original cell is duplicated again and again, in every cell.

Autosomes

In 22 of the 23 pairs of chromosomes, both members of the pair (one from each parent) are closely matched. Some of the genes have alternate alleles, but each chromosome finds its comparable chromosome, making a pair. Those 44 chromosomes are called *autosomes*, which means that they are independent (*auto-* means "self") of the sex chromosomes (the 23rd pair).

Each autosome, from number 1 to number 22, contains hundreds of genes in the same positions and sequence. If the code of a gene from one parent is exactly like the code on the same gene from the other parent, the gene pair is **homozygous** (literally, "same-zygote").

However, the match is not always letter-perfect because the mother might have a different allele of a particular gene than the father has. If a gene's code differs from that of its counterpart, the two genes still pair up, but the zygote (and, later, the person) is **heterozygous**.

Only half of a man's genes are on each sperm and only half of a woman's genes are on each ovum, so the combination creates siblings who will be, genetically, both similar and different. Thus, which particular homozygous or heterozygous genes my brother and I inherited on our autosomes from our parents is a matter of chance. It is irrelevant, genetically, that I am a sister, not a brother.

Sex Chromosomes

However, for the **23rd pair** of chromosomes, it matters which chromosome I inherited from my father. When his 23rd pair split to make gametes, half carried an X and half a Y.

This is how it happens. In males, the 23rd pair has one X-shaped chromosome and one Y-shaped chromosome. It is called **XY**. In females, the 23rd pair is composed of two X-shaped chromosomes. It is called **XX**.

The X chromosome is bigger, with about 100 more genes, but the Y has a crucial gene, called *SRY*, that directs the embryo to make male hormones and organs. Thus, sex at birth depends on which sperm penetrates the ovum—a Y sperm with the SRY gene, creating a boy (XY), or an X sperm, creating a girl (XX) (see **Figure 3.2**).

It sometimes matters whether a zygote is XY or XX, because genes on all the chromosomes might be affected, a phenomenon called *parental imprinting*. The best-known example occurs with a small deletion or duplication on chromosome 15 (Kalsner & Chamberlain, 2015).

homozygous Referring to two genes of one pair that are exactly the same in every letter of their code. Most gene pairs are homozygous.

heterozygous Referring to two genes of one pair that differ in some way. Typically one allele has only a few base pairs that differ from the other member of the pair.

23rd pair The chromosome pair that, in humans, determines sex. The other 22 pairs are autosomes, inherited equally by males and females.

XY A 23rd chromosome pair that consists of an X-shaped chromosome from the mother and a Y-shaped chromosome from the father. XY zygotes become males.

XX A 23rd chromosome pair that consists of two X-shaped chromosomes, one each from the mother and the father. XX zygotes become females.

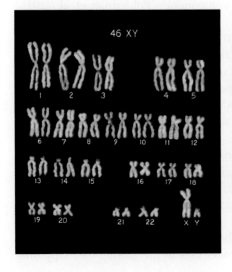
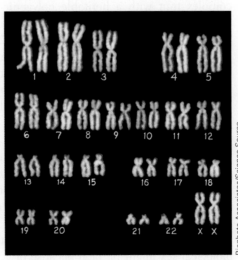

Biophoto Associates/Science Source.

Intersex Every now and then, a baby is born with "ambiguous genitals," meaning that the child's sex is not abundantly clear. When this happens, analysis of the chromosomes may reveal that the zygote was XX or XY. The karyotypes shown here indicate a typical baby boy *(left)* and girl *(right)*.

Determining a Zygote's Sex Any given couple can produce four possible combinations of sex chromosomes; two lead to female children and two, to male. In terms of the future person's sex, it does not matter which of the mother's Xs the zygote inherited. All that matters is whether the father's Y sperm or X sperm fertilized the ovum. However, for X-linked conditions it matters a great deal because typically one, but not both, of the mother's Xs carries the trait.

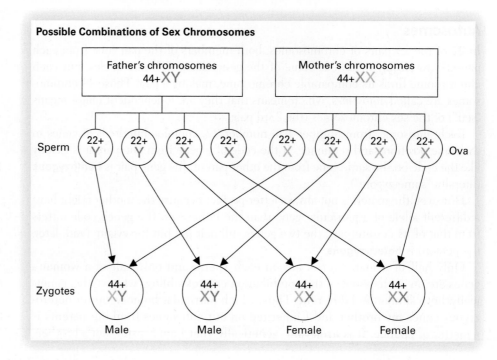

Possible Combinations of Sex Chromosomes

Father's chromosomes 44+XY · Mother's chromosomes 44+XX · Sperm: 22+Y, 22+Y, 22+X, 22+X · Ova: 22+X, 22+X, 22+X, 22+X · Zygotes: 44+XY Male, 44+XY Male, 44+XX Female, 44+XX Female

Especially for Medical Doctors Can you look at a person and then write a prescription that will personalize medicine to their particular genetic susceptibility? (see response, page 89)

If that harmful allele on chromosome 15 came from the father, the child will develop Prader-Willi syndrome and be obese, slow-moving, and stubborn. If that same allele came from the mother's chromosome 15, the child will have Angelman syndrome and be thin, hyperactive, and happy—sometimes too happy, laughing when no one else does. Other diseases and conditions are affected by imprinting, again sometimes in opposite ways (Couzin-Frankel, 2017).

More Than Sex Organs

Although biology and culture (nature and nurture) constantly interact, the word *sex* generally refers to one's assigned sex at birth, and the word *gender* refers to social and cultural constructs.

The sex chromosomes direct the development of hormones that affect the brain, skeleton, body fat, and muscles, beginning in the first weeks of prenatal development and continuing to old age. Of course, culture (gender) affects every sex difference as well. One review suggests "gender identity is a multifactorial complex trait with a heritable polygenic component" (Polderman et al., 2018, p. 95).

For example, at conception, about 120 XY zygotes are conceived for every 100 XX, probably because the sperm with fewer genes have a slight advantage in the race to the ovum. From that moment on, male life is more fragile. At birth, the male/female ratio is 105:100 in developed nations and 103:100 in the poorest ones. That ratio, no longer 120:100, reveals that male embryos die at higher rates, and that, when pregnant women are sick and malnourished, surviving newborns are more often female.

For most of history people could not learn the sex of the fetus until birth, when someone looked at the genitals and shouted "It's a ------!" Because no one knew that the 23rd pair is present at conception, millions of pregnant women ate special foods, slept on one side, or repeated certain prayers, all to control their baby's sex—which was already determined.

Now prospective parents can learn whether a fetus is male or female. Prenatally, biological sex might be fatal because of gender bias in the culture (see Opposing Perspectives).

Too Many Boys?

In past centuries, millions of newborns were killed because they were the wrong sex, a practice that would be considered murder today. Now, advances in science are enabling the same goal long before birth in various ways, such as inactivating X or Y sperm before conception.

Recently, millions of couples have used these methods to choose their newborn's sex. Should this be legal? It is against the law, in at least 36 nations. It is legal in the United States (Murray, 2014).

To some prospective parents, those 36 nations are unfair—most allow similar measures to avoid severely disabled newborns. Why is that legal but sex selection is not? There are moral reasons. But, should governments legislate morals? People disagree (Wilkinson, 2015).

One nation that recently tried to forbid prenatal sex selection is China. In about 1979, China began a "one-child" policy, urging, and sometimes forcing, couples to have only one child. That achieved the intended goal: fewer children to feed . . . or starve. Severe poverty was almost eliminated.

But the Chinese tradition is that sons care for aging parents, so couples wanted their only child to be male. Among the unanticipated results:

- Since 1980, an estimated 9 million abortions of female fetuses
- Adoption of thousands of newborn Chinese girls by Western families
- By 2010, far more unmarried young men than women

In 1993, the Chinese government forbade prenatal testing for sex selection. In 2013, China rescinded the one-child policy. Yet in 2020, there are 113 preschool boys for every 100 girls. Apparently, not every Chinese family follows the law.

The argument in favor of sex selection is freedom from government interference. Some fertility doctors and many individuals believe that each couple should be free to decide when, how many, and the sex of their progeny (Murray, 2014).

But one argument against is social harm. In China, many more young Chinese men than women die. The developmental explanation is that unmarried men take risks to attract women. They become depressed if they remain alone, risking early death

My Strength, My Daughter That's the slogan these girls in New Delhi are shouting at a demonstration against abortion of female fetuses in India. The current sex ratio of children in India suggests that this campaign has not convinced every couple.

from accidents, suicide, drug overdoses, and poor health practices (Srinivasan & Li, 2018).

This is a warning to other nations. A society with an excess of males might also have an excess of problems, since males are more likely to abuse drugs, commit crimes, kill each other, die of heart attacks, and start wars. This is true in every nation.

But wait: Chromosomes and genes do not *determine* behavior. Even traits that originate with biology are affected more by environment. Heart attacks, for instance, correlate more with diet and cigarette-smoking. Already, improved diagnosis and declines in smoking (changes in nurture) have reduced male heart attacks. In 1950, four times as many middle-aged men as women died of heart disease; by 2015, the rate was lower for both sexes, but especially for men (2:1, not 4:1) (Centers for Disease Control and Prevention, July 3, 2018).

Historically, some cultures have adjusted to battle deaths of young men (and thus too many young single women) by encouraging polygamy. Societies could change customs to adapt to an excess of males as well. But should they?

> **THINK CRITICALLY:** Might laws prohibiting prenatal sex selection be unnecessary if culture shifted?

WHAT HAVE YOU LEARNED?

1. How many chromosomes and genes do people have?

2. What is an allele?

3. What effect does the microbiome have?

4. Does it matter whether a gene pair is homozygotic or heterozygotic?

5. What determines a baby's sex?

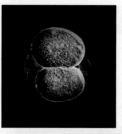

(a)

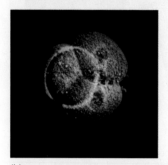

(b)

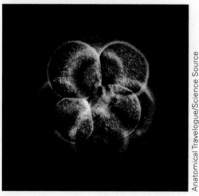

(c)

Anatomical Travelogue/Science Source

First Stages of the Germinal Period The original zygote as it divides into *(a)* two cells, *(b)* four cells, and *(c)* eight cells. Occasionally at this early stage, the cells separate completely, forming the beginning of monozygotic twins, quadruplets, or octuplets.

stem cells Cells from which any other specialized type of cell can form.

New Cells, New People

Within hours after conception, the zygote begins *duplication* and *division*. First, the 23 pairs of chromosomes (carrying all of the genes) duplicate to form two complete sets of the genome. These two sets move toward opposite sides of the zygote, and the single cell splits neatly down the middle into two cells, each containing the original genetic code.

These first two cells duplicate and divide, becoming four, which duplicate and divide, becoming eight, and so on. The name of the developing mass changes as cells multiply—the zygote (one cell) becomes a *morula*, then a *blastocyst*, then an *embryo*, then a *fetus*—and then, at birth, a *baby*. [**Developmental Link**: Prenatal growth is detailed in Chapter 4.]

Cells and Identity

Nine months after conception, a newborn has about 26 billion cells, influenced by nutrients, drugs, hormones, viruses, microbes, and so on from the pregnant woman. Adults have about 37 trillion cells, each with the same 46 chromosomes and the same thousands of genes of the original zygote. This explains why DNA testing of any body cell, even from a drop of saliva or a snip of hair, can identify "the real father," "the guilty criminal," "the long-lost brother."

Because the Y chromosome is passed down to every male descendant, and because the Y changes very little from one generation to the next, men today have the Y of their male ancestors who died thousands of years ago.

Female ancestors also live on. Each zygote has *mitochondria*, biological material that provides energy for the cell. The mitochondria come from the mother, and her mother, and her mother, and thus each person carries evidence of maternal lineage.

As with all genetic material, mitochondria sometimes predispose a person to disease, especially problems with the heart (Bonora et al., 2019). Although males are more vulnerable to heart attacks than females are, the problem may originate with their mothers!

Stem Cells

The cells that result from the early duplication and division are called **stem cells**; they can produce any other cell and thus become a complete person. After about the eight-cell stage, although duplication and division continue, a third process, *differentiation*, begins.

In differentiation, cells specialize, taking different forms and reproducing at various rates depending on where they are located. For instance, some cells become part of an eye, others part of a finger, still others part of the brain. They are no longer stem cells.

Another method, called CRISPR, has been developed to edit genes so that the stem cells have additions or deletions. CRISPR has been used to develop rice that is better adapted to various growing conditions, with the potential to feed billions of people (Endo et al., 2019). Medical researchers are excited about the possibility of using CRISPR to control the mosquitos that spread malaria, Zika, and other diseases (Liu et al., 2019).

In theory, CRISPR could be used to edit human genes as well, but that raises serious ethical questions (Boggio et al., 2020; McConnell & Blasimme, 2019). At the moment, CRISPR is forbidden for human organisms. A scientist's attempt in

China to use CRISPR in human embryos caused a storm of protest and a prison sentence (Cohen, 2018; Mills, 2019).

In Vitro Fertilization

The ethical implications of CRISPR raise the issue of **in vitro fertilization (IVF)**, which was pronounced "sacrilegious" and "against God" when first attempted in 1960. That did not stop the scientists, who finally succeeded with a live baby, Louise Brown, in 1978.

Over the past half-century, procedures have improved and IVF has become "a relatively routine way to have children" (C. Thompson, 2014, p. 361). The European Society of Human Reproduction and Embryology estimated in 2018 that 8 million IVF babies have been born, some from every nation.

The two daughters of former President Barack Obama were both conceived via IVF. So was the younger sister of Louise Brown. Both Brown daughters have had babies of their own, conceived naturally, but said they would use IVF if need be.

A study of more than 2 million ninth-graders in Sweden reported that specifics of conception made little difference in school achievement (Norrman et al., 2018). Psychologically, IVF children have also fared well. For all children, genes and family life are far more influential than details of conception.

Nonetheless, IVF differs markedly from typical conceptions, almost half of which are unintended (Finer & Zolna, 2016). When a couple opts for IVF, the woman takes hormones to increase the number of ripe ova, and several are surgically removed while the man ejaculates into a sterile container. Then a technician combines ova and sperm in a laboratory dish (*in vitro* means "in glass"), often by inserting one viable sperm into each ovum.

Zygotes that fail to duplicate properly are rejected, but several days after conception, one or more blastocysts are inserted into the uterus. Implantation succeeds about half the time, with the age of the couple an important predictor. Some older women choose IVF with an ovum donated by a younger woman in order to increase the chance of pregnancy.

Since young eggs are more viable, women in their 20s who do not yet want children may freeze their ova for IVF years later. This has prompted a debate. Some suggest that women who don't want motherhood before age 35 or so are selfish, and others contend that such women are making a responsible choice, becoming mothers when they are ready to care for a new life (Carroll & Krøløkke, 2018).

Several European nations limit the numbers of blastocysts inserted into the uterus at one time, partly because national health care pays for both IVF and infant care. The United States has no legal restrictions, although income matters. The cost varies from clinic to clinic within the United States, perhaps $20,000 for drugs, monitoring, and the procedure itself.

IVF offers many possibilities, often raising ethical and legal questions. One woman miscarried seven times and was in her early 40s when her mother, Anastassia Ontou, offered to carry her embryo and fetus. The 67-year-old woman gave birth in the Ukraine (where IVF has no legal restrictions) to her granddaughter.

Medical societies provide some oversight. For example, the California Medical Board removed the license from the physician who inserted 12 blastocysts in Nadya Suleman. She gave birth to eight surviving babies in 2009, a medical miracle but a developmental disaster.

Stuart Atkins/Shutterstock

From Miracle to Routine This is Louise Brown, whose birth was an international front-page headline when she became the first IVF baby born decades ago. Here she is holding twins who were also born via IVF, now a routine event at this clinic and thousands of others worldwide. Or, are all births, IVF or not, always a miracle? Perhaps you agree with an Indian mystic who once said that "every newborn is a sign that God has not yet given up on the world."

in vitro fertilization (IVF) Fertilization that takes place outside a woman's body (as in a glass laboratory dish). The procedure involves adding sperm to ova that have been surgically removed from the woman's ovary. If a zygote is produced, it is inserted into a woman's uterus, where it may implant and develop into a baby.

THINK CRITICALLY: When is it selfish to add another baby to the world?

sarahwolfephotography/Moment/Getty Images

Not Exactly Alike These two 4-year-old boys in South Carolina are identical twins, which means they originated from one zygote. But one was born first and heavier, and, as you see here, one appears to be more affectionate to his brother.

monozygotic (MZ) twins Twins who originate from one zygote that splits apart very early in development. (Also called *identical twins*.) Other monozygotic multiple births (such as triplets and quadruplets) can occur as well.

Another IVF miracle is that adults can have children who have no genetic connection to them because others have donated the sperm, the ova, and/or the womb. (The word *donate* is misleading, since most donors—often college students—are paid.) Some couples even travel to other nations with less restrictive laws and more women willing to be surrogate mothers (Reddy et al., 2018).

Is that international exploitation? Many aspects of fertility and infertility raise moral questions, within nations and between them.

Twins and More

Thus far, we have described conception as if one sperm and one ovum resulted in one baby. About once in 350 human conceptions, one zygote becomes two fetuses, or even four (see Visualizing Development on page 74).

Monozygotic Twins

Remember that each stem cell contains the entire genetic code. In IVF, before implantation, one stem cell can be removed and analyzed. If that cell carries a known destructive gene, or if the chromosomes are not exactly 46, then that blastocyst is not inserted into the uterus.

If genetic testing finds no problems, the remaining cell mass is inserted, where it might implant and grow normally. Removing one stem cell does not harm development, because each of the remaining stem cells has all of the instructions needed to create a person.

Twins *could* be created by separating the cells of the blastocyst before implantation, resulting in two or more identical babies if the separated cells were implanted and then grew. This is illegal with human zygotes in IVF.

However, nature within the woman's body sometimes does what doctors are forbidden to do—it splits those early cells. One separation can create **monozygotic (MZ) twins**, so called because they came from one (*mono-*) zygote (also called *identical twins*). Separations at the four- or eight-cell stage create monozygotic quadruplets or octuplets.

Because monozygotic multiples originate from one zygote, they have identical genetic instructions. Remember, however, that epigenetic influences begin as soon as conception occurs: Monozygotic twins look and act very much alike, but their prenatal environment is not identical.

For example, the particular spot in the uterus where each twin implants may allow one fetus to be better nourished than the other: Malnutrition affects the fetal brain, so it affects a twin lifelong. Experiences after birth can differ markedly, as occurred with monozygotic triplets who were adopted by three different families (as shown in the film *Three Identical Strangers*).

Monozygotic multiples are fortunate in several ways. They can donate a kidney or other organ to each other with no organ rejection. They can also befuddle their parents and teachers, who may need visible ways (such as different earrings or haircuts) to tell them apart.

Usually, monozygotic twins establish their own identities. For instance, both might inherit athletic ability, but one might choose to play basketball and the other, soccer.

As one monozygotic twin writes:

> Twins put into high relief *the* central challenge for all of us: self-definition.
> How do we each plant our stake in the ground, decide how sensitive, callous,
> ambitious, conciliatory, or cautious we want to be every day? . . . Twins

come with a built-in constant comparison, but defining oneself against one's twin is just an amped-up version of every person's life-long challenge: to individuate—to create a distinctive persona in the world.

[Pogrebin, 2010, p. 9]

That woman and her twin sister married and had a son and then a daughter within months of each other. Coincidence? Genetic? Sister pressure?

Dizygotic Twins

About once in 60 natural conceptions, **dizygotic (DZ) twins** are conceived. They begin life when two ova are fertilized by two sperm at about the same time. Usually, women release only one ovum per month, so most human newborns are singletons. However, when multiple ovulation occurs, dizygotic twins are possible.

IVF in the United States produces DZ twins about half the time. All those statistics are for natural conceptions. Rates rise with IVF (see **Figure 3.3**). Indeed, multiple pregnancy increases the risk of preterm birth, but many infertile couples think two babies is a welcome miracle.

Dizygotic twins are sometimes called *fraternal twins*, although because *fraternal* means "brotherly" (as in *fraternity*), fraternal is inaccurate. Of course, MZ twins are always the same sex (their 23rd chromosomes are either XX or XY), but for DZ twins (as with any two siblings) some are brothers, some are sisters, and some are brother and sister.

People say that twinning "skips a generation," but actually it skips fathers, not mothers. Since dizygotic twinning requires multiple ovulation, the likelihood of a woman ovulating two ova and thus conceiving twins depends on her genes from her parents. Her husband's genes are irrelevant.

> **LaunchPad**
> macmillan learning
>
> **VIDEO ACTIVITY: Identical Twins: Growing Up Apart** gives a real-life example of how genes play a significant role in people's physical, social, and cognitive development.

dizygotic (DZ) twins Twins who are formed when two separate ova are fertilized by two separate sperm at roughly the same time. (Also called *fraternal twins*.)

FIGURE 3.3

More Is Not Always Better Twinning is more common in Africa, and less common in East Asia. That has been true historically and continues to be the case in the twenty-first century, as this map shows. In medically advanced nations such as the United States, fertility drugs and IVF doubled the number of twins in the early twenty-first century, reaching a peak of 33.9 per 1,000 births in 2014. Recently, U.S. rates have fallen slightly as the challenges of low-birthweight newborns become more apparent.

One Baby or More?

Humans usually have one baby at a time, but sometimes twins are born. Most often they are from two ova fertilized by two sperm (*lower left*), resulting in dizygotic twins.

Sometimes, however, one zygote splits in two (*lower right*), resulting in monozygotic twins; if each of these zygotes splits again, the result is monozygotic quadruplets.

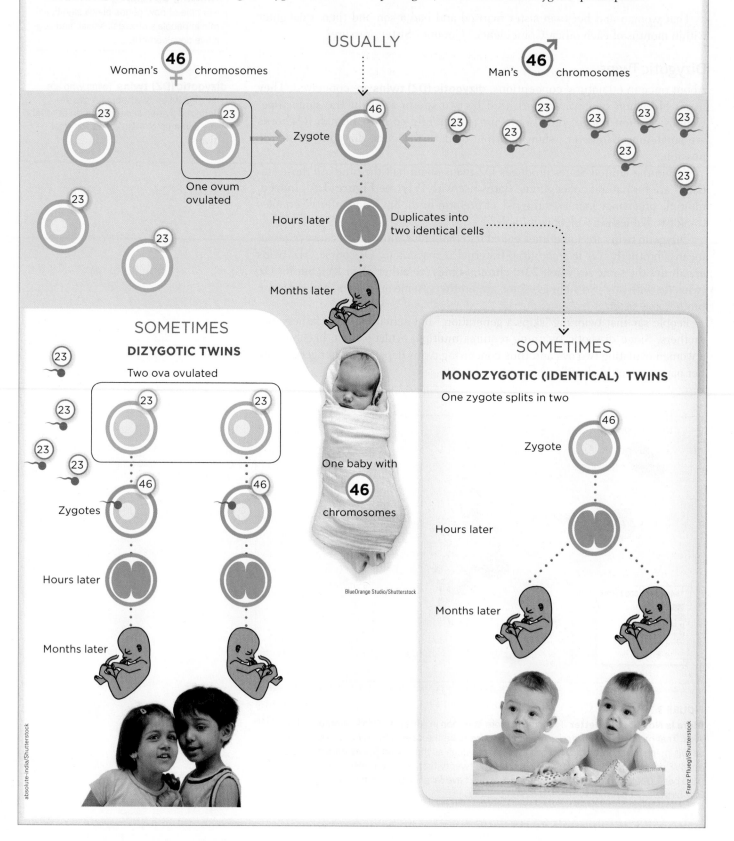

USUALLY

Woman's **46** chromosomes

Man's **46** chromosomes

23

23
One ovum
ovulated

→ Zygote 46 ←

23

23

23

23

23

23

23

23

23

Hours later — Duplicates into two identical cells

Months later

One baby with **46** chromosomes

BlueOrange Studio/Shutterstock

SOMETIMES
DIZYGOTIC TWINS

Two ova ovulated

23

23

23

23

23

23

Zygotes 46 46

Hours later

Months later

absolute-india/Shutterstock

SOMETIMES
MONOZYGOTIC (IDENTICAL) TWINS

One zygote splits in two

Zygote 46

Hours later

Months later

Franz Pfluegl/Shutterstock

However, half of a man's genes are from his mother. If he inherited her gene for multiple ovulation, there is a 50-percent chance that his genes include the multiple ovulating one. Then his daughters may have twins because their paternal grandmother did.

When dizygotic twinning occurs naturally, the incidence varies by ethnicity either because of genes or diet (Smits & Monden, 2011). Age matters, too: Older women more often double-ovulate and thus have more twins.

Like all children from the same parents, dizygotic twins have about half of their genes in common. They can differ markedly in appearance, or they can look so much alike that only genetic tests determine whether they are MZ or DZ. In the rare incidence that a woman releases two ova at once *and* has sex with two men over a short period, twins can have different fathers. Then they share only one-fourth of their genes.

WHAT HAVE YOU LEARNED?

1. How does DNA establish identity?

2. What makes a cell a "stem cell"?

3. Why is CRISPR illegal for humans?

4. What is similar and different in an IVF pregnancy and a traditional pregnancy?

5. What is the difference between monozygotic twins and dizygotic twins?

6. For whom is twinning more likely?

From Genotype to Phenotype

As already explained, when a sperm and an ovum create a zygote, the *genotype* (all the genes of the developing person) is established. This initiates several complex processes that form the **phenotype**—the person's actual appearance, behavior, and brain and body functions.

phenotype The observable characteristics of a person, including appearance, personality, intelligence, and all other traits.

Many Factors

The phenotype depends on the genotype and thousands of factors in the environment. Completely accurate prediction of the phenotype is impossible, even if the genotype is entirely known (Lehner, 2013). One reason is *differential susceptibility*, as explained in Chapter 1. Because of a seemingly minor allele, or a transient environmental influence, a particular person may be profoundly changed—or not affected at all—by experiences.

Diabetes is a notable example. Several distinct alleles put people at risk for diabetes, but weight and exercise awaken that genetic risk. Yet some people are overweight and inactive but never develop diabetes because it is not in their genotype.

The same may be true for other developmental changes over the life span. Substance abuse—cocaine, cigarettes, alcohol, and so on—may produce epigenetic changes. Once addicted, people who have not used the drug for years are still vulnerable and cannot use that drug again as an unaffected person could (Bannon et al., 2014). Their brain has changed; they are "clean" but still have a substance use disorder.

Human Genome Project An international effort to map the complete human genetic code. This effort was essentially completed in 2001, though analysis is ongoing.

Gene–Gene Interactions

Many discoveries have followed the completion of the **Human Genome Project** in 2003. Once it was apparent that few disorders or talents arose from any single gene, increased attention focused on the complex interactions between one human gene and another.

Additive Heredity

Interactions among genes and alleles are often called *additive* because their effects *add up* to influence the phenotype. The phenotype then reflects the contributions of every additive gene. Height, hair curliness, and skin color, for instance, are usually the result of additive genes. Indeed, height is probably influenced by 700 genes, each contributing a very small amount (Marouli et al., 2017).

Most Americans have ancestors of varied height, hair curliness, skin color, and so on, with background variations much more nuanced than any simple idea of race. A child's phenotype may not mirror the parents' phenotypes (although the phenotype always reflects the genotype), in part because of the interactions of their unique set of genes.

I see this in my family: Our daughter Rachel is of average height, shorter than her father and me but taller than either of our mothers. She apparently inherited some of her grandmothers' height genes via our gametes. And none of my four daughters has exactly my shape or coloring—apparent when we borrow clothes from each other and notice that a shirt that is flattering to one is ugly on another.

Genetic variations are apparent in every family, particularly among African Americans in the United States. Historically, the continent of Africa was, genetically, the most diverse (Choudhury et al., 2018). Added to that, current North Americans who identify as Black are not only the product of African diversity but also carry genes from many parts of Europe and from many tribes of indigenous Americans.

Especially for Future Parents Suppose you wanted your daughters to be short and your sons to be tall. Could you achieve that? (see response, page 89)

Dominant–Recessive Heredity

Not all genes are additive. In one nonadditive form, some alleles are **dominant** and others are **recessive**. In a heterozygotic dominant/recessive pair, the *dominant gene* is more influential. It overpowers the *recessive gene*, hidden on the genotype, not apparent in the phenotype.

A person is called a **carrier** of the recessive gene when it is on the genotype but not the phenotype. In other words, people might carry a gene in their DNA, which they will transmit to half of their gametes. Only if someone inherits a recessive gene from each parent, which means there are two recessive genes but no dominant gene for that trait, does the recessive trait emerge on the phenotype.

Most recessive genes are harmless. One example is eye color. Brown eyes are dominant. Everyone with brown eyes has at least one dominant brown-eye gene.

Blue eyes are recessive. If both parents have blue eyes, then every eye-color gene on every gamete will have one or the other of their pair of blue-eye genes. In that case, all their children will all have blue eyes. If a child has one blue-eyed parent (who always has two recessive blue-eye genes) and one brown-eyed parent, that child will usually have brown eyes.

Usually, not always. Brown-eyed parents all have one brown-eye gene (otherwise they would not have brown eyes), but they might carry a blue-eye gene. In

dominant Reflected in the phenotype. Dominant genes have more influence on traits than recessive genes.

recessive Hidden, not dominant. Recessive genes are carried in the genotype and are not evident in the phenotype, except in special circumstances.

carrier A person whose genotype includes a gene that is not expressed in the phenotype. The carried gene occurs in half of the carrier's gametes and thus is passed on to half of the carrier's children. If such a gene is inherited from both parents, the characteristic appears in the phenotype.

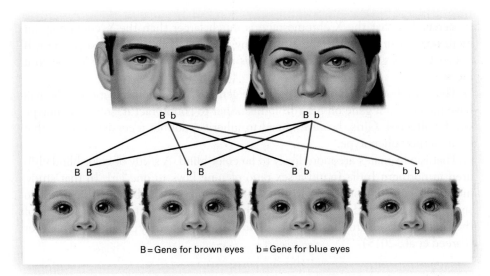

B = Gene for brown eyes   b = Gene for blue eyes

FIGURE 3.4

Changeling? No. If two brown-eyed parents both carry the blue-eye gene, they have one chance in four of having a blue-eyed child. Other recessive genes include the genes for red hair, Rh negative blood, and many genetic diseases.

⬤ **Observation Quiz** Why do these four offspring look identical except for eye color? (see answer, page 89) ↑

that case, in a blue-eyed/brown-eyed couple, every child has at least one blue-eye gene (from the blue-eyed parent), and half of the children will have a blue-eye recessive gene (from the carrier parent).

That half will have blue eyes because they have no dominant brown-eye gene. The other half will have a brown-eye dominant gene. Their eyes will be brown, but they will be carriers of the blue-eye gene, just like their brown-eyed parent. Sometimes both parents are carriers. If two brown-eyed parents both have the blue-eye recessive gene, the chances are one in four that their child will have blue eyes (see **Figure 3.4**).

This eye color example presumes that only one pair of genes determines eye color. However, as with almost every trait, eye color is polygenic, so other genes affect eye color, making eyes various shades of blue and brown, greenish or greyish.

Mother to Son

A special case of the dominant–recessive pattern occurs with genes that are **X-linked** (located on the X chromosome). If an X-linked gene is recessive—as are the genes for red–green color blindness, several allergies, a few diseases, and some learning disorders—the fact that it is on the X chromosome is critical (see **Table 3.1**).

X-linked A gene carried on the X chromosome. If a male inherits an X-linked recessive trait from his mother, he expresses that trait because the Y from his father has no counteracting gene. Females are more likely to be carriers of X-linked traits but are less likely to express them.

TABLE 3.1

The 23rd Pair and X-Linked Color Blindness

23rd Pair	Phenotype	Genotype	Next Generation
1. XX	Typical woman	Not a carrier	No color blindness
2. XY	Typical man	Typical X from mother	No color blindness
3. ⊗X	Typical woman	Carrier from father	Half her children will inherit her ⊗. The girls with her ⊗ will be carriers; the boys with her ⊗ will have color blindness.
4. X⊗	Typical woman	Carrier from mother	Half her children will inherit her ⊗. The girls with her ⊗ will be carriers; the boys with her ⊗ will have color blindness.
5. ⊗Y	Color-blind man	Inherited from mother	All his daughters will have his ⊗. None of his sons will have his ⊗. All his children will have typical vision, unless their mother also had an ⊗ for color blindness.
6. ⊗⊗	Color-blind woman (rare)	Inherited from both parents	Every child will have one ⊗ from her. Therefore, every son will have color blindness. Daughters will be only carriers, unless they also inherit an ⊗ from the father, as their mother did.

⊗ = X that carries recessive gene for color blindness

Remember that the X chromosome is much larger than the Y, containing far more genes. Consequently, the X has many genes that are unmatched on the Y. If one of those X genes is recessive, there is no dominant gene on the Y to keep it hidden.

Thus, if a boy (XY) inherits a recessive gene on his X from his mother, he will have no dominant gene on his Y from his father to counteract it, so his phenotype will be affected. Girls, however, need to inherit a double recessive to have that train on their phenotype.

That is why males are more likely to be color blind. A study of color-blind children in northern India found a sex ratio of nine boys to one girl. In that study, marriages were almost always within a small group of neighbors and relatives, so specific genes tended to stay within each group. That explains why 7 percent of the children in one group were color blind, compared to only 3 percent in another (Fareed et al., 2015).

WHAT HAVE YOU LEARNED?

1. Why do humans vary so much in skin color and height?

2. What is the difference between additive and dominant–recessive inheritance?

3. How can a blue-eyed child have brown-eyed parents?

4. Why are sons more likely to inherit recessive conditions from their mothers than their fathers?

Nature and Nurture

One goal of this chapter is for readers to grasp the complex interaction between genotype and phenotype. This is not easy. For decades, millions of scientists have struggled to understand this complexity. Each year brings advances in statistics and molecular analysis with new data to uncover various patterns, all resulting in hypotheses to be explored.

Now we examine three specific disorders: alcohol addiction, nearsightedness, and schizophrenia. As you will see, understanding the progression from genotype to phenotype has many practical implications.

Alcohol Use Disorder

At various times throughout history, people have considered the abuse of drugs to be a moral weakness, a social scourge, or a personality defect. Historically and internationally, the main focus has been on alcohol, since people everywhere discovered fermentation thousands of years ago—to the joy of many. About 12 percent of the drinkers, however, become addicted, unable to stop drinking even when it interferes with the rest of their life.

Alcohol has been declared illegal (as in the United States from 1919 to 1933) or considered sacred (as in many religious rituals). Those who cannot control their drinking have been jailed, jeered, or burned at the stake.

We now know that inherited biochemistry affects alcohol metabolism (Preedy, 2019). Punishing those with the genes does not stop addiction. There is no single "alcoholism gene," but dozens of genes and alleles make alcohol use disorder more or less likely (Edenberg et al., 2019). Alleles create an addictive pull that can be overpowering, extremely weak, or somewhere in between, or can render alcohol repulsive. Each person's biochemistry reacts to alcohol, causing sleep, nausea, aggression, joy, relaxation, forgetfulness, or tears.

Especially for Drug Counselors Is the wish for excitement likely to lead to addiction? (see response, page 89)

If metabolism allows people to "hold their liquor," they might drink too much; others (including many East Asians) sweat and become red-faced after just a few sips. This inherited "flushing" tendency not only reduces the risk of alcohol use disorder but also improves metabolism (Kuwahara et al., 2014).

Although scientists first sought the genes that affected the *biology* of addiction, they soon learned that genetic personality traits (including a quick temper, sensation seeking, and high anxiety) are as crucial as biology (Macgregor et al., 2009). Age matters, too. Teenagers feel the pressure to drink if their best friends are drinking, while adults find it easier to say "no thank you."

Social contexts matter. Fraternity parties encourage drinking; church socials in a "dry" county make it difficult to swallow anything stronger than lemonade.

Finally, both sex and gender are relevant. For biological reasons (body size, fat composition, metabolism), women become drunk on less alcohol than men. Heavy-drinking females double their risk of mortality compared to heavy-drinking males (C. Wang et al., 2014). That may be why many cultures encourage men to drink but not women (Chartier et al., 2014). Or, sexism and chauvinism may be the reason.

As you see by all this, explaining a particular habit—drinking alcohol—quickly becomes very complex. A review of many current studies from every continent emphasizes that "genetic risk is only a piece of the complex architecture of risk and protective factors" for alcohol abuse (Edenberg et al., 2019). Again, phenotype is affected by both genotype and environment.

Nearsightedness

Age, genes, and culture affect vision as well. Consider *myopia* (nearsightedness), the most common visual problem.

The effects of age are universal. Newborns focus only on things within 1 to 3 feet of their eyes; vision improves steadily until about age 10. The eyeball changes shape at puberty (increasing myopia) and again in middle age (decreasing myopia). Vision of all kinds becomes less acute in late adulthood, and serious problems increase.

Added to those developmental patterns, nearsightedness is strongly influenced by genes. Heritability is about 75 percent—which is quite high (Williams & Hammond, 2016). If one monozygotic twin is nearsighted, the other twin is virtually always nearsighted, too.

However, **heritability** indicates only how much of the variation in a particular trait, *within a particular population, in a particular context and era*, can be traced to genes. For example, the heritability of height is very high (about 95 percent) when children receive good medical care and nutrition, but it is low (about 20 percent) when children are severely malnourished. Children who are chronically underfed are short, no matter what their genes.

Thus, although nearsightedness is highly heritable, nurture may be crucial. Indeed, it is. In some African and Asian communities, heritability of nearsightedness is close to zero because of diet. Severe vitamin A deficiency blinds more than 250,000 children every year (Ehrenberg, 2016). For them, the genetic tendency to be nearsighted is irrelevant.

To prevent blindness, scientists have developed strains of local staples such as "golden rice" that are high in vitamin A, although use is limited by fears of genetically modified food (Ehrenberg, 2016). Some nations (Zambia and Cameroon) avoid genetic food modification by adding vitamin A directly to cooking oil and sugar. This must be carefully done—excessive, nonfood vitamin A causes health problems, but the risks are far less serious than blindness (Tanumihardjo et al., 2016).

What about children who consume adequate vitamin A in their food? Is their vision entirely inherited? No. Changes in the environment led one ophthalmologist

Hero Images Inc./Alamy

Welcome Home For many women in the United States, white wine is part of the celebration and joy of a house party, as shown here. Most people can drink alcohol harmlessly; there is no sign that these women are problem drinkers. However, danger lurks. Women get drunk on less alcohol than men, and females with alcoholism tend to drink privately and secretly, often at home, feeling more shame than bravado. All of that makes their addiction difficult to recognize.

heritability A statistic that indicates what percentage of the variation in a particular trait within a particular population, in a particular context and era, can be traced to genes.

Lai xinlin/AP Images

Applauding Success These eager young men are freshmen at the opening convocation of Shanghai Jiao Tong University. They have studied hard in high school, scoring high on the national college entrance exam. Now their education is heavily subsidized by the government. Although China has more college students than the United States, the proportions are far lower, since the population of China is more than four times that of the United States.

⬤ **Observation Quiz** Name three visible attributes of these young men that differ from a typical group of freshmen in North America. (see answer, page 89) ⬆

to predict "an epidemic of pathological myopia . . . in the next few decades in Asia" (Saw, quoted in Seppa, 2013a, p. 23).

That prediction is based on three decades of research. Nearsightedness increased from 26 percent to 43 percent in one decade (1980 to 1990) in the army-mandated medical exams of all 17-year-old males in Singapore (Tay et al., 1992). Similar increases are evident in China, India, and the United States (Holden, 2010).

An article in the leading British medical journal (*The Lancet*) suggests that although genes are usually to blame for severe myopia, "any genetic differences may be small" for common nearsightedness (I. Morgan et al., 2012, p. 1739). Nurture must somehow be involved. But how?

One possible culprit is homework. Fifty years ago, most Asian children were working outside; now almost all are diligent students. As their developing eyes focus on their books, those with a genetic vulnerability to myopia may lose acuity for objects far away—which is exactly what nearsightedness means.

A related culprit is too much time indoors. Data from the United States on children playing sports have led some ophthalmologists to suggest that the underlying cause is inadequate exposure to daylight (I. Morgan et al., 2012). If North American children spent more time outside, would fewer need glasses?

Between the early 1970s and the early 2000s, nearsightedness in the U.S. population increased from 25 percent to 42 percent (Vitale et al., 2009). Urbanization, screen time, and fear of strangers have kept many U.S. children indoors most of the time, unlike children of earlier generations. The result may be reduced visual flexibility.

Schizophrenia

In the United States, a 2017 survey of adults found the prevalence of mental illness to be 19 percent, with the highest rates (26 percent) among young adults (see **Figure 3.5**). The rate of depression among college students is said to be 16 percent (American College Health Association, 2017).

FIGURE 3.5

Not Just Genetic Details here may be questioned, as the diagnosis of mental illness is complex. But general trends have been found in other studies as well: Women have higher rates of major depression than men, and some mental illness (such as schizophrenia) "age out," which means they are less common among older adults.

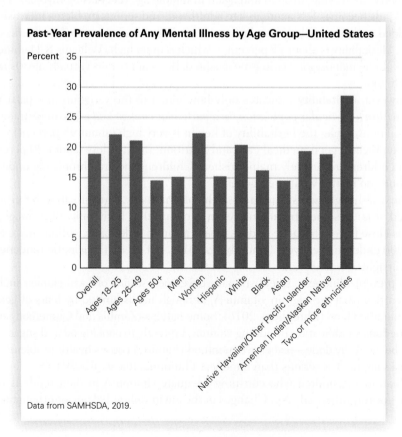

Past-Year Prevalence of Any Mental Illness by Age Group—United States

Data from SAMHSDA, 2019.

Among the common disorders with genetic origins are depression, anxiety, bipolar disorder, schizophrenia, and autism spectrum disorder (Knopik et al., 2017; Sandstrom et al., 2019). Although both genes and environment are crucial for all of these, we focus here on the relatively rare disorder of schizophrenia, because scientists have seesawed between thinking schizophrenia was caused entirely by destructive parenting, to thinking is was entirely genetic. Because of that seesaw, extensive research has focused on the origins of this particular disorder.

One study reported on the entire population of Denmark, where good medical records and decades of free public health care make research accurate (Gottesman et al., 2010). If *both* Danish parents developed schizophrenia, 27 percent of their children developed it; if one parent had it, 7 percent of their children developed it. Evidence for genes?

These same statistics can be presented in another way: Even if both parents had schizophrenia, almost three-fourths (73 percent) of their children did not. Thus, the data confirm both nature and nurture. Similarly, if one monozygotic twin develops schizophrenia, often—but not always—the other twin also develops a psychological disorder of some sort, not necessarily schizophrenia.

As for nurture, research on MZ twins found specific risk factors (Pepper et al., 2018). One is the stress of adolescence. If one MZ twin is diagnosed with schizophrenia before age 23, almost half the time (54 percent) the other twin will be diagnosed with the disorder. However, if the first diagnosis does not occur until after age 30, the incidence for the other twin is less than a third (30 percent).

Another risk factor is the family context. If one parent has schizophrenia, and one MZ twin develops it as well, chances are 63 percent that the genetic twin will also develop the disease. However, if neither parent had schizophrenia but one MZ twin does, odds are reduced by a third.

Many factors in the environment trigger schizophrenia in vulnerable people, including prenatal viruses, early maltreatment, adolescent drug use, and moving from one culture to another during early adulthood. Indeed, considering all the environmental influences led one team to write that schizophrenia was a *pseudogenetic* disease, with the inherited aspects overestimated (Torrey & Yolken, 2019).

No doubt there is a genetic risk, but also no doubt that childhood experiences matter for schizophrenia as well as for all psychological disorders (Ottesen et al., 2018; Sullivan et al., 2000). It is a mistake to blame any condition *entirely* on genes or *entirely* on childhood or *entirely* on adult experiences. That leads to many questions: How much, for whom, and what are the triggers, the interactions, the treatments, the outcomes?

Practical Applications

Since genes affect every disorder, no one should be blamed or punished for inherited problems. We all have them: They are not our fault, nor the fault of our parents.

However, knowing that genes never act alone opens the door to prevention. For instance, if alcohol use disorder is in the genes, parents can keep alcohol out of the home and explain the dangers of addiction (neither exaggerating nor ignoring), hoping their children become cognitively and socially mature before imbibing. If nearsightedness runs in the family, parents can play outdoors with their children every day. If mental illness is a problem, children need stability and reassurance.

Of course, avoiding alcohol, playing outdoors, and maintaining stability are recommended for all children, as are dozens of other behaviors, such as flossing twice a day, saying "please," sleeping 10 hours each night, eating five servings of vegetables, and promptly writing thank-you notes. It is unrealistic to expect parents to make their children do all of these; awareness of genetic risks can guide priorities.

The same principles apply lifelong. Recognizing that each of us has a genetic pull toward some unwanted behaviors or conditions is the beginning of self-acceptance as well as a path toward prevention. Adults who know that diabetes is in their genes can avoid overeating—good advice for everyone, but especially needed for some.

> ### WHAT HAVE YOU LEARNED?
>
> 1. Is alcohol dependence a genetic condition or a cultural one?
> 2. What suggests that nearsightedness is affected by nurture?
> 3. What does *heritability* mean?
> 4. If a particular genetic problem is common in a family, what should the parents do with their children?

LaunchPad
macmillan learning

VIDEO: Genetic Disorders offers an overview of various genetic disorders.

Down syndrome A condition in which a person has 47 chromosomes instead of the usual 46, with 3 rather than 2 chromosomes at the 21st site. People with Down syndrome typically have distinctive characteristics, including unusual facial features, heart abnormalities, and language difficulties. (Also called trisomy-21.)

Claudia Daut/REUTERS

Universal Happiness All young children delight in painting brightly colored pictures on a big canvas, but this scene is unusual for two reasons: Daniel has trisomy-21, and this photograph was taken at the only school in Chile where typical and special-needs children share classrooms.

Chromosomal and Genetic Problems

We now focus on conditions caused by an extra chromosome or a single destructive gene. If all notable anomalies and disorders are included, 92 percent of people do <u>not</u> develop a serious genetic condition by early adulthood—but that means 8 percent do (Chong et al., 2015).

Not Exactly 46

As you know, most humans have 46 chromosomes, created by two gametes, each with 23 chromosomes. However, the 46 chromosomes on the parents' cell do not always split exactly in half to make sperm or ova.

About half of all zygotes have more than or fewer than 46 chromosomes (Milunsky & Milunsky, 2016). Most of them fail to duplicate, divide, differentiate, and implant. They stop growing before anyone knew that conception occurred. Those who survive that initial period usually are spontaneously aborted (miscarried). Ninety-nine percent of newborns have the usual 46 chromosomes (Benn, 2016).

Trisomies

If an entire chromosome is added, that leads to a recognizable *syndrome*, a cluster of distinct characteristics that tend to occur together. Usually the cause is three chromosomes at a particular location instead of the typical two (a condition called a *trisomy*). One in 10,000 newborns has three chromosomes at the 13th site (called Patau syndrome), and 1 in 5,000 has three at the 18th (called Edwards syndrome). For those trisomies, if a fetus survives prenatal life, death usually occurs soon after birth (Acharya et al., 2017).

A much more common trisomy is at the 21st site, which occurs about once in 700 births (Parker et al., 2010). Trisomy 21 is called **Down syndrome**, named after Dr. and Mrs. Langdon Down. In 1868, they opened the first home for such children (then called "Mongolian Idiots").

No individual with Down syndrome is just like another, but this trisomy usually produces telltale characteristics—a thick tongue, round face, and slanted eyes, as well as distinctive hands, feet, and fingerprints. The hippocampus (important for memory) is usually smaller.

Many people with Down syndrome also have hearing problems, heart abnormalities, muscle weakness, and short stature. They are slow to develop intellectually, especially in language, with a notable deficit in the ability to rhyme (Næss, 2016).

By middle age, many people with Down syndrome develop Alzheimer's disease (AD), because one gene on chromosome 21 increases amyloid plaque—a sign of AD. Since everyone with Down syndrome has an extra chromosome 21, many also have too much amyloid. A blood test may indicate their risk early enough to forestall Alzheimer's disease, and thus research on adults with Down syndrome may help all adults avoid dementia (Hamlett et al., 2018).

A group of researchers asked parents about the positive and negative aspects of having a child with Down syndrome. Happily, positives outweighed negatives. One parent said:

> He not only is a positive force in my life, but in the world. When I push him around the neighborhood in his stroller he loves to wave at people, including those who are homeless. He has a big heart that does not discriminate.

[Farkas et al., 2019, p. 525]

Especially for Teachers Suppose you know that one of your students has a sibling who has Down syndrome. What special actions should you take? (see response, page 89)

Problems of the 23rd Pair

Every human has at least 44 autosomes and one X chromosome; an embryo cannot develop without those 45. However, miscounts at the 23rd pair are more common and less deadly than miscounts in the autosomes. About 1 in every 300 infants is born with only one sex chromosome (no Y) or with three or more (not just two) (Benn, 2016). Most of them reach adulthood.

Each particular combination of sex chromosomes results in a specific syndrome (see **Table 3.2**), but all affect cognition, fertility, and sexual maturation (Hong & Reiss, 2014). One problem with the 23rd pair, called Turner syndrome, occurs when only one sex chromosome is present. The result is a girl with a wide neck and low hairline. But many children with an extra sex chromosome (XXY, for instance) seem normal until adolescence or adulthood, when sexual maturation and fertility are impaired.

Gene Disorders

More common than chromosomal problems are gene disorders. Each person carries about 40 alleles that *could* cause serious diseases. The phenotype is seriously affected only when the inherited gene is dominant or when both parents carry the same allele. Even then, only one child in two (for dominant disorders), or in four (for recessive disorders), inherits the genetic problem.

TABLE 3.2

Common Abnormalities Involving the Sex Chromosomes

Chromosomal Pattern	Physical Appearance	Psychological Characteristics	Incidence*
XXY (Klinefelter Syndrome)	Males. Usual male characteristics at puberty do not develop—penis does not grow, voice does not deepen. Usually sterile. Breasts may develop.	Can have some learning disabilities, especially in language skills.	1 in 700 males
XYY (Jacob's syndrome)	Males. Typically tall.	Risk of intellectual impairment, especially in language skills.	1 in 1,000 males
XXX (Triple X syndrome)	Females. Normal appearance.	Impaired in most intellectual skills.	1 in 1,000 females
XO (only one sex chromosome) (Turner syndrome)	Females. Short, often "webbed" neck. Secondary sex characteristics (breasts, menstruation) do not develop.	Some learning disabilities, especially related to math and spatial understanding; difficulty recognizing facial expressions of emotion.	1 in 6,000 females

*Incidence is approximate at birth.
Information from Aksglaede et al., 2013; Benn, 2016; Hamerton & Evans, 2005; Powell, 2013; Stochholm et al., 2010.

Nathan Morgan/The New York Times/Redux

Who Has the Fatal Dominant Disease?
The mother, but not the children. Unless a cure is found, Amanda Kalinsky will grow weak and experience significant cognitive decline, dying before age 60. She and her husband, Bradley, wanted children without Amanda's dominant gene for a rare disorder, Gerstmann-Straussler-Scheinker disease. Accordingly, they used IVF and pre-implantation testing. Only zygotes without the dominant gene were implanted. This photo shows the happy result.

fragile X syndrome A genetic disorder in which part of the X chromosome seems to be attached to the rest of it by a very thin string of molecules. The cause is a single gene that has more than 200 repetitions of one triplet.

Dominant Disorders

Many of the 7,000 *known* single-gene disorders are dominant (Milunsky & Milunsky, 2016). Most are relatively mild. Severe dominant disorders are rare because people who have them usually die before puberty, and thus never pass that gene to another generation.

However, if a dominant disorder does not appear until adulthood it could stay in the gene pool, because adults might have several children before they know of their condition. That is the case for adult-onset Alzheimer's disease, one form of muscular dystrophy, Marfan syndrome, and Huntington's chorea, a fatal central nervous system condition caused by a copy number variation—more than 35 repetitions of a particular set of three base pairs.

Although the dominant gene for all these conditions sometimes affects children or adolescents (Milunsky & Milunsky, 2016), serious symptoms usually do not arise until midlife. The folk singer Woody Guthrie, who wrote "This Land Is Your Land," died of Huntington's, as did two of his daughters. Fortunately, his son, Arlo, inherited his talent but not his disorder.

Recessive Disorders

Recessive diseases are more numerous than dominant ones because they are passed down by carriers who are unaware of their recessive genes. There are thousands of recessive diseases; advance carrier detection is now possible for several hundred.

A few recessive conditions are X-linked, including hemophilia, Duchenne muscular dystrophy, and **fragile X syndrome**, which is caused by more than 200 repetitions on one stretch of one gene (Plomin et al., 2013). (Some repetitions are normal, but not this many.)

The deficits caused by fragile X syndrome are the most common form of *inherited* intellectual disability. (Many other forms, such as trisomy-21, begin with gametes, not genes.) Boys are more often impaired by fragile X syndrome than are girls, because they have only one X.

Common Recessive Disorders

About 1 in 10 North American adults carries an allele on their autosomes for cystic fibrosis, thalassemia, or sickle-cell disease. Why so common? Because carriers benefited from the gene, which makes the gene frequent in the population.

Consider the most studied example: sickle-cell disease. Carriers of the sickle-cell gene die less often from malaria, which is prevalent and lethal in parts of Africa. Indeed, four distinct alleles cause sickle-cell disease, each originating in a malaria-prone region.

The gene protected more people (the carriers) than it harmed. If a carrier had children with a noncarrier, half of the children would be carriers and half not, and none of the children would have the disease. If a malaria epidemic swept through that region, and that family had many children, those who were carriers would be more likely to survive to become parents themselves, so the next generation would have more carriers.

What if both parents were carriers? If they had four children, odds were that only one would inherit the double recessive and die young. The other three would not have the disease, and two of the three would be carriers, somewhat protected against an early malaria death. Consequently, each generation included more carriers, and thus the gene became widespread.

This connection between genes and local diseases is common. Almost every lethal disease is more frequent in one place than in another (Weiss & Koepsell, 2014). If a particular allele is protective, it becomes more common in that population.

About 8 percent of Americans with African ancestors have the recessive gene for sickle-cell disease—they are protected against malaria. Cystic fibrosis is more common among people with ancestors from northern Europe, because carriers may have been protected from cholera.

Additive genes can also be beneficial. Dark skin is protective against skin cancer if a person is often exposed to direct sun. Light skin allows more vitamin D to be absorbed from the sun, if a person lives where sunlight is scarce. Being relatively short is beneficial in cold climates, or when food is scarce.

Modern Europeans inherited between 1 and 4 percent of their genes from Neanderthals, who became extinct about 30,000 years ago. Neanderthal genes protect contemporary humans against some skin conditions and other diseases but may also make them vulnerable to allergies and depression—depending on which bits of Neanderthal genes they happen to inherit (Saey, 2016).

Some genetic risks are simultaneously beneficial. Anxiety, for instance, causes people to anticipate problems. Communities benefit if that genetic risk leads to better preparation for hurricanes, earthquakes, wars, and epidemics.

The Greek mythological character Cassandra envisioned wars and other devastation in the future. No one believed her; they thought she was insane. She became depressed, was shunned and eventually murdered. According to the myth, she was later considered a prophet, not a madwoman.

Even schizophrenia may have a benefit. Bruce Springsteen's father was an angry, emotionally abusive man, who probably had schizophrenia (Springsteen, 2017). Did that help his son become a creative musical genius? Indeed, many great artists and writers apparently suffered from serious mental illness, from Vincent Van Gogh to Virginia Woolf, from Anthony Bourdain to Andrew Solomon—you can think of many more. Is there any connection here?

> **DATA CONNECTIONS: Common Genetic Diseases and Conditions** describes several different types of gene disorders.

Genetic Counseling and Testing

Until recently, after the birth of a child with a severe disorder, couples blamed witches or fate, not genes or chromosomes. Those children usually died in infancy. Currently, however, most children with genetic disorders survive, which means that everyone has relatives with a disease that is partly genetic.

A simple blood test reveals some recessive diseases, including sickle cell, and the cost of analyzing a person's entire genome has plummeted, from more than a million dollars to less than a hundred. However, there are many problems with genome-wide analysis (Tam et al., 2019).

- Data can reveal risks, but not reality: Almost every condition is better or worse depending on nongenetic factors, as apparent in our discussion of nature and nurture.
- Most reported research has been on people with western European heritage: Much more needs to be learned about the other 80 percent of humankind.
- Genetic research is proceeding faster than practical care: People might learn that they are vulnerable to a particular problem, but not how to prevent or treat it.
- Most genetic oddities are "variants of unknown significance." That means that even experts do not know what to make of them, causing carriers to worry.

Another danger arises from genetic testing that people pay for privately, mailing a saliva sample to learn about their ancestry and genes. The results may be misleading. Worse is that the emotional needs of the person are not addressed.

Reassurance and Gratitude Karen is pregnant and just learned that a relative has the gene for a serious condition. Here, a genetic counselor is explaining all the possibilities. Several months after this photo was taken, she gave birth to a healthy baby!

VIDEO: Genetic Testing examines the pros and cons of knowing what diseases may eventually harm us or our offspring.

Genetic Counseling

A rapidly expanding profession, much needed, is genetic counseling. There are an estimated 7,000 professional genetic counselors in 28 nations (Ormond et al., 2018). Their job is vital but not easy.

The task is to explain that we all have genes that push us in one direction or another, and that part of life is not only accepting but also celebrating this fact in ourselves and our children. At the same time, counseling can help people understand their personal risks of having a child with a serious genetic disorder and to help all of us realize that spontaneous mutations occur, so that prediction is never certain.

Both reassurance and warning may be needed. For example, a genetic counselor may explain that serious problems are unusual. On the other hand, if there is a specific risk, counselors alert prospective parents to organizations for caregivers of children with special needs, to adoption (both nationally and internationally) risks and joy, and to treatment that is experimental but that might soon be proven (a couple might decide to postpone conception), and so on. Counselors are trained to help clients understand and explore many possibilities.

This requires up-to-date information. New genetic disorders — and treatments — are revealed almost weekly. An inherited disorder that once meant lifelong neurological impairment (e.g., phenylketonuria, or PKU) might now mean a normal life.

Accuracy, effect, and interpretation of testing varies: Sometimes a particular gene increases the risk of a problem by only a tiny amount, perhaps 0.1 percent. Counselors need to help clients understand both risks and uncertainty.

CAREER ALERT   The Genetic Counselor

An understanding of life-span development is useful for every career. As students contemplate their future work, they should consult career counselors and check the Occupational Handbook of the Department of Labor, which lists prospects, salary, and qualifications.

Beyond those basics, however, these Career Alerts raise questions and issues that arise from a developmental perspective, issues that might not be found in a standard description of the career. You will see this in this discussion of genetic counseling.

There is far greater demand for genetic counselors than there are people trained in this area, so job prospects are good. Salary is good, too: The median in 2017 was $77,500. Training requires a master's degree and then passing an exam to be certified (though this is not required in every state). That all seems simple, but the reality is much more challenging.

The first challenge is to understand and communicate complex biological and statistical material, so that clients understand the implications of whatever genes they have. This is difficult: Not only are new discoveries made every day, but every disorder is polygenic and multifactorial, and mosaicism, methylation, and the microbiome are all relevant.

One reason this is a rapidly growing career is that many people are curious about their ancestry and pay for commercial tests (such as 23 and Me) to identify where their ancestors lived. In the process, they may discover confusing implications for their health, making genetic counseling essential (Smart et al., 2017). Further, since is it now apparent that almost every disease is partly genetic, many people are concerned about their own health, the health of their family members, or the health of their prospective children, and they want answers.

This is complex regarding the genes of an adult, but it is doubly difficult when discussing a prospective child, who will inherit only half of the genes from each parent.

Facts, medical treatments, and quality of life for a future, not-yet-conceived child are difficult to explain, but genetic counselors must consider much more than that. Each adult has emotions, assumptions, and values that differ between a husband and wife, and those values are not identical to those of the counselor. Not only does each person have a particular attitude about risk, religion, and abortion but also communication is complex. People misunderstand results; counselors must draw charts, rephrase results, and repeat basic facts.

Thus, counselors must not only know facts, recent discoveries, and explain odds and consequences but also must be sensitive to complex social dynamics, respecting everyone — especially when a couple decides to terminate or continue a pregnancy when the counselor would have made the opposite choice.

Theoretical decisions often conflict with reality. If a woman knows that her embryo has trisomy-21, should she terminate the pregnancy? About two-thirds of prospective parents say no, but about two-thirds of pregnant women at high risk (e.g., over age 35) say yes, as do almost all (87 to 96 percent) women who know they are carrying a Down syndrome embryo (Choi et al., 2012).

Similarly, variation was evident when 152 pregnant women in Wisconsin learned that their embryo had trisomy-13 or trisomy-18. Slightly more than half of the women decided to abort; most of the rest decided to give birth but provide only comfort care for the newborn; and three chose full intervention to preserve life (their three babies lived for a few days but died within the first weeks) (Winn et al., 2018).

Many factors—including childhood memories, prior children, religion, opinions of others—make a difference (Choi et al., 2012). Unfortunately, no matter what the decision, outsiders sometimes tell parents they made the wrong choice—something genetic counselors never do.

Before deciding on this profession, ask yourself what you would do in the following situations, each of which has occurred:

- Parents of a child with a disease caused by a recessive gene from both parents ask whether another baby will suffer the same condition. Tests reveal that the husband does not carry that gene. Should the counselor tell the parents that their next child won't have this disease because the husband is not the father of the first child?
- A woman learns that she is at high risk for breast cancer because she carries the BRAC1 gene. She wants to have her breasts removed, but she refuses to inform her four sisters, half of whom probably carry BRAC1.
- A pregnant couple are both "little people," with genes for short stature. They want to know whether their embryo will have typical height. They plan to abort such a fetus.

- A person is tested for a genetic disease that runs in the family. The results are good (not a carrier) and bad (the person carries another serious condition). Should the counselor reveal a risk that the client did not ask about?

This fourth issue is new: Even a few years ago, the cost of testing precluded learning about unrequested results. But now *genome-wide association study* (*GWAS*) is routine, capturing the entire genome, so counselors learn about thousands of unsuspected conditions.

Even with careful counseling, people with identical genetic conditions often make opposite choices. For instance, 108 women who had one child with fragile X syndrome were told that another pregnancy would have a 50/50 chance of fragile X. Most decided to avoid pregnancy, but some (20 percent) deliberately conceived another child (Raspberry & Skinner, 2011).

In another study, pregnant women learned that their fetus had an extra sex chromosome. Half the women aborted; half did not (Suzumori et al., 2015). That highlights why this career is not for everyone. Professionals explain facts and probabilities; people decide. Can you live with that?

Who Needs Counseling?

Rational and informed understanding is elusive, not only among prospective parents but also among professionals not trained in counseling. Consider the experience of one of my students. A month before she became pregnant, Jeannette's employer required her to have a rubella vaccination. Hearing that Jeannette had had the shot, her obstetrician gave her the following prognosis:

> My baby would be born with many defects, his ears would not be normal, he would be intellectually disabled. . . . I went home and cried for hours and hours. . . . I finally went to see a genetic counselor. Everything was fine, thank the Lord, thank you, my beautiful baby is okay.
>
> *[Jeannette, personal communication]*

Jeannette may have misunderstood, but that is exactly why the doctor should have spoken more carefully. Genetic counselors are trained to make information clear. If sensitive counseling is available, then preconception, prenatal, or even prenuptial (before marriage) testing is especially useful for:

- individuals who have a parent, sibling, or child with a serious genetic condition;
- couples who have had several spontaneous abortions or stillbirths;
- couples who are infertile;
- women over age 35 and men over age 40; and
- couples from the same ethnic group, particularly if they are relatives.

The latter is especially crucial among populations who often intermarry. This is true for Greeks in Cyprus, where about one-third of the population carries the recessive gene for thalassemia. In the 1970s, one Cypriot baby in 158 was born with thalassemia, which led to repeated hospitalization and premature death.

Then Cyprus encouraged testing, before conception or at least prenatally. Some lovers decided not to marry, some married couples decided on adoption or IVF with preconception testing. Now virtually no newborns in Cyprus have thalassemia (Hvistendahl, 2013).

CHAPTER APP 3

Gene Screen

IOS:
https://tinyurl.com/yxvxbzun

RELEVANT TOPIC:
Genetic disorders, counseling, and testing

The Gene Screen app provides information about the inheritance and prevalence of recessive genetic diseases in different cultures and ethnicities.

THINK CRITICALLY: Instead of genetic counseling, should we advocate health counseling?

This chapter raised many ethical questions. There is one more, that arises throughout. Does our current emphasis on genes distracts us from public health hazards (poverty, pollution, pesticides, and so on) that harm development more than genes? Developmentalists recognize genes, but they also know that plasticity characterizes development. Context and culture matter.

As you have read many times in this chapter, genes are part of the human story, influencing every page, but they do not determine the plot or the final paragraph. The remaining chapters describe the rest of the story.

WHAT HAVE YOU LEARNED?

1. What chromosomal miscounts might result in a surviving child?
2. What is the cause and consequence of Down syndrome?
3. How common are recessive conditions?
4. Why is sickle-cell disease very common in some parts of Africa?
5. What is the role of the genetic counselor?
6. What ethical mandates are required of genetic counselors?

SUMMARY

The Genetic Code

1. Genes are the foundation for all development, first instructing the developing creature to form the body and brain, and then affecting thought, behavior, and health lifelong. Human conception occurs when two gametes (a sperm with 23 chromosomes and an ovum with 23 chromosomes) combine to form a single cell called a zygote.

2. A zygote usually has 46 chromosomes (half from each parent), which carry a total of about 21,000 genes. Genes and chromosomes from each parent match up to make the zygote, but the match is not always letter-perfect because of genetic variations called alleles, or polymorphisms.

3. Genetic variations occur in many ways, from the chromosomes of the parent to the epigenetic material surrounding the zygote and the microbiome of every body part. Spontaneous mutations, copy number variations, and much more make each person unique.

4. The most notable mismatch is in the 23rd pair of chromosomes, which is XX in females and XY in males. The sex of the embryo depends on the sperm, since only men have a Y chromosome and thus can make Y gametes.

New Cells, New People

5. The first duplications of the one-celled zygote create stem cells, each of which could become a person if it developed.

6. Monozygotic twins occur if those first stem cells split completely. Dizygotic twins occur if two ova are fertilized by two sperm at about the same time. Genetically, they have half of their genes in common, as do all full siblings.

7. In vitro fertilization (IVF) has led to millions of much-wanted babies and also to an increase in multiple births, who often are preterm and of low birthweight. Ethical concerns regarding standard IVF have quieted, but new dilemmas appear. CRISPR is the most recent example.

From Genotype to Phenotype

8. Genes interact in many ways, sometimes additively with each gene contributing to development and sometimes in a dominant–recessive pattern. If a recessive trait is X-linked, it is passed from mother to son.

Nature and Nurture

9. Genetic makeup can make a person susceptible to many conditions. Examples include substance use disorder (especially alcohol use disorder) and poor vision (especially nearsightedness). Culture and family affect both of these conditions dramatically.

10. Every adult is a carrier of harmful genes. Their expression depends on the genes of the other parent, as well as on many influences from the environment. Genetic understanding can help caregivers protect people from potentially harmful genes.

Chromosomal and Genetic Problems

11. Often a gamete has fewer or more than 23 chromosomes, which may create a zygote with 45, 47, or 48 chromosomes. Usually such zygotes do not duplicate, implant, or grow.

12. Infants may survive if they have three chromosomes at the 21st site (Down syndrome). These individuals may have fulfilling lives, although they are vulnerable to heart and lung problems, and, in midlife, to Alzheimer's disease.

13. Another possible problem is a missing or extra sex chromosome. Such people have intellectual disabilities or other problems, but they may also lead a fulfilling life.

14. Everyone is a carrier for genetic abnormalities. Usually these conditions are recessive, not apparent unless the mother and the father both carry the gene. Serious dominant disorders usually do not appear until midlife.

15. Serious recessive diseases can become common if carriers have a health advantage. This is true for sickle-cell disease, which protected carriers against malaria.

16. Genetic testing and counseling can help many couples. Testing provides information about possibilities, which are difficult for people to understand when their emotions are overwhelming. The final decision about what to do with the information rests with the client, not the counselor.

KEY TERMS

deoxyribonucleic acid (DNA) (p. 63)
chromosome (p. 63)
gene (p. 63)
gamete (p. 63)
zygote (p. 63)
genome (p. 63)
allele (p. 64)

epigenetics (p. 65)
microbiome (p. 66)
copy number variations (p. 66)
genotype (p. 66)
homozygous (p. 67)
heterozygous (p. 67)
23rd pair (p. 67)
XY (p. 67)

XX (p. 67)
stem cells (p. 70)
in vitro fertilization (IVF) (p. 71)
monozygotic (MZ) twins (p. 72)
dizygotic (DZ) twins (p. 73)
phenotype (p. 75)

Human Genome Project (p. 76)
dominant (p. 76)
recessive (p. 76)
carrier (p. 76)
X-linked (p. 77)
heritability (p. 79)
Down syndrome (p. 82)
fragile X syndrome (p. 84)

APPLICATIONS

1. Pick one of your traits and explain the influences that both nature *and* nurture have on it. For example, if you have a short temper, explain its origins in your genetics, your culture, and your childhood experiences.

2. Many adults have a preference for having a son or a daughter. Interview adults of several ages and backgrounds about their preferences. If they give the socially preferable answer ("It does not matter"), ask how they think the two sexes differ. Listen and take notes—don't debate. Analyze the implications of the responses you get.

3. Draw a genetic chart of your biological relatives, going back as many generations as you can, listing all serious illnesses and causes of death. Include ancestors who died in infancy. Do you see any genetic susceptibility? If so, how can you overcome it?

4. List a dozen people you know who need glasses (or other corrective lenses) and a dozen who do not. Are there any patterns? Is this correlation or causation?

Especially For ANSWERS

Response for Medical Doctors (from p. 68): No. Personalized medicine is the hope of many physicians, but appearance (the phenotype) does not indicate alleles, recessive genes, copy number variations, and other genetic factors that affect drug reactions. Many medical researchers seek to personalize chemotherapy for cancer, but although this is urgently needed, success is still experimental, even when the genotype is known.

Response for Future Parents (from p. 76): Possibly, but you wouldn't want to. You would have to choose one mate for your sons and another for your daughters, and you would have to use sex-selection methods. Even so, it might not work, given all the genes on your genotype. More important, the effort would be unethical, unnatural, and possibly illegal.

Response for Drug Counselors (from p. 78): Maybe. Some people who love risk become addicts; others develop a healthy lifestyle that includes adventure, new people, and exotic places. Any trait can lead in various directions. You need to be aware of the connections so that you can steer your clients toward healthy adventures.

Response for Teachers (from p. 83): As the text says, "information combats prejudice." Your first step would be to make sure you know about Down syndrome, by reading material about it. You would learn, among other things, that it is not usually inherited (your student need not worry about his or her progeny) and that some children with Down syndrome need extra medical and educational attention. This might mean you need to pay special attention to your student, whose parents might focus on the sibling.

Observation Quiz ANSWERS

Answer to Observation Quiz (from p. 77): This is a figure drawn to illustrate the recessive inheritance of blue eyes, and thus eyes are the only difference shown. If this were a real family, each child would have a distinct appearance.

Answer to Observation Quiz (from p. 80): Not nearsightedness! Rates of corrective lenses (estimated at 85 percent) might be as high among university students in the United States, but Americans

would typically have contacts. Two other visible differences: uniforms and gender. Except for in the military, no U.S. university issues uniforms and the majority of North American students are women. A fourth difference may be inferred from their attentiveness: The graduation rate of incoming college students in China is about 90 percent, compared to about 50 percent in the United States.

Prenatal Development and Birth

What Will You Know?

1. Why do most zygotes never become babies?
2. How are home births better and worse than hospital births?
3. What can a pregnant woman do to ensure a healthy newborn?
4. Why do new mothers and fathers sometimes become depressed?

90

I did not tell her, but I was worried about my eldest daughter's pregnancy. She was almost 40, and in the previous few years had a miscarriage as well as surgery to remove a fibroid tumor. If all went well, this would be her first birth. I kept my worries quiet (pregnant women have more than enough worries of their own), but I clung to every reassuring report: The fetus gained weight on schedule, measurements of the fetal neck precluded Down syndrome, blood tests confirmed a typically developing boy. My daughter was eating well, avoiding drugs, counting kicks. I was honored and grateful that she asked me to be her birth partner.

The due date was weeks after final exams, but I recruited one of my colleagues to proctor for me if the birth happened early, as about 12 percent of U.S. births do. I waited for the call telling me that labor had started. Exams were over, grades were calculated, due date had come and gone. No call. Waiting became hard.

So I drove to stay with her, two hours away, to be with her as soon as contractions began. For three days I cooked, read, took the dog to the vet. My daughter heard that walking might start labor; we hiked up a small mountain. One night a thunderstorm woke me; future mother and baby stayed asleep. The next day, another prenatal checkup. The midwife monitored the fetal heartbeat (fine), the maternal blood pressure (fine), the birth position (head down, fine), and the amniotic fluid. Not fine! Too low.

She recommended induction; we checked into the hospital. Another midwife examined her, with surprising news. Labor had started hours earlier. That's why the fluid was low. The nurse said the thunderstorm did it; something about the air.

The actual birth also held surprises, detailed later in this chapter. Spoiler alert: Mother and babe were fine. The grandmother was not.

The scientific study of human development is not only about how individuals change over time; it is about historical change (Bronfenbrenner's chronosystem), family support (microsystem), and social contexts (macro- and exosystems). All are evident in this chapter, as well as in the text above. Elective surgery to remove fibroids? Measuring the fetal neck? Midwives? Sonogram of amniotic fluid? Induction? Never in other times and places, and uncommon in most nations now.

Currently in the United States it is uncommon for the grandmother to be the birth partner, but that was not unusual in past centuries. It is apparent in this chapter, as well as in my daughter's experience, that each birth is affected by every system, reflecting unique experiences (climbing that mountain after the due date) and universal forces (that thunderstorm?). This chapter describes both.

Prenatal Development

Universally, the months before birth are the time of the most dramatic and extensive transformation of the entire life span. To make it easier to study, prenatal development is often divided into three main periods. The first two weeks are the **germinal period;** the third through the eighth week is the **embryonic period;** from then until birth is the **fetal period.** (Alternative terms are presented in **Table 4.1**.)

Germinal: The First 14 Days

You learned in Chapter 3 that the one-celled zygote duplicates, divides, multiplies, and differentiates, when those early cells take on distinct characteristics. About a week after conception, the developing mass forms two distinct parts—a shell that will become the *placenta* and a nucleus that will become the *embryo*. The first task of the shell is *implantation*, embedding into the nurturing lining of the uterus. This is far from automatic; most zygotes never implant (Kim & Kim, 2017) (see **Table 4.2**).

Embryo: From the Third Week Through the Eighth Week

The start of the third week after conception initiates the *embryonic period*, during which the mass of inner cells takes shape—not recognizably human but worthy of the name, **embryo.** (The word *embryo* is often used loosely, but here, embryo refers to day 14 to day 56.)

Day by Day

About two weeks after conception, a thin line called the *primitive streak* appears down the middle of the embryo; it forms the neural tube, which develops into the central nervous system, that is, the brain and spinal column. Soon the head

germinal period The first two weeks of prenatal development after conception, characterized by rapid cell division and the beginning of cell differentiation.

embryonic period The stage of prenatal development from approximately the third week through the eighth week after conception, during which the basic forms of all body structures, including internal organs, develop.

fetal period The stage of prenatal development from the ninth week after conception until birth, during which the fetus gains about 7 pounds (more than 3,000 grams) and organs become more mature, gradually able to function on their own.

embryo The name for a developing human organism from about the third week through the eighth week after conception.

TABLE 4.1

Timing and Terminology

Popular and professional books use various phrases to segment the stages of pregnancy. The following comments may help to clarify the phrases used.

Beginning of pregnancy: Sometimes pregnancy is said to begin at conception, which is the start of *gestational age.* However, the organism does not become an *embryo* until about two weeks later, and pregnancy does not affect the woman (and is not confirmed by blood or urine testing) until implantation, so some do not consider the woman pregnant until then. On the other hand, some obstetricians and publications count from the woman's last menstrual period (LMP), usually about 14 days *before* conception.

Length of pregnancy: Full-term pregnancies last 266 days, or 38 weeks, or 9 months. If the LMP is used as the starting time, pregnancy lasts 40 weeks, sometimes expressed as 10 lunar months. (A lunar month is 28 days long.)

Trimesters: Instead of *germinal period, embryonic period,* and *fetal period,* as used in this text, some writers divide pregnancy into three-month periods called *trimesters.* Months 1, 2, and 3 are the *first trimester;* months 4, 5, and 6, the *second trimester;* and months 7, 8, and 9, the *third trimester.*

Due date: Although a specific due date based on the LMP is calculated, only 5 percent of babies are born on that exact day. Babies born between two weeks before and one week after that date are considered *full term.* [This is recent; until 2012, three weeks before and two weeks after were considered full term.] Because of increased risks for postmature babies, labor is often induced if the baby has not arrived a week after the due date.

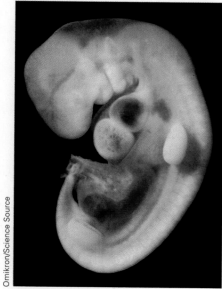

Omikron/Science Source

(a)

Petit Format/Science Source

(b)

The Embryonic Period *(a)* At 4 weeks past conception, the embryo is only about 1/8 inch (3 millimeters) long, but already the head has taken shape. *(b)* By 7 weeks, the organism is somewhat less than an inch (2 centimeters) long. Eyes, nose, the digestive system, and even the first stage of toe formation can be seen.

cephalocaudal Growth and development that occurs from the head down.

proximodistal Growth or development that occurs from the center or core in an outward direction.

fetus The name for a developing human organism from the start of the ninth week after conception until birth.

TABLE 4.2

Vulnerability During Prenatal Development

The Germinal Period

An estimated 65 percent of all zygotes do not grow or implant properly and thus do not survive the germinal period.

The Embryonic Period

About 20 percent of all embryos are aborted spontaneously. This is usually called an early *miscarriage,* a term that falsely implies something wrong with the woman. In fact the most common reason for a spontaneous abortion is a chromosomal abnormality.

The Fetal Period

About 5 percent of all fetuses are aborted spontaneously before viability at 22 weeks or are *stillborn,* defined as born dead after 22 weeks. This is much more common in poor nations.

Birth

Because of all these factors, only about 27 percent of all zygotes grow and survive to become living newborn babies.

Information from Cunningham et al., 2014; Kim & Kim, 2017.

appears, as eyes, ears, nose, and mouth start to form and a minuscule blood vessel that will become the heart begins to pulsate.

By the fifth week, buds that will become arms and legs emerge. Upper arms and then forearms, palms, and webbed fingers grow. Legs, knees, feet, and webbed toes, in that order, appear a few days later, each with the beginning of a skeleton. Then, 52 and 54 days after conception, respectively, the fingers and toes separate (Sadler, 2015).

At the end of the eighth week after conception (56 days), the embryo weighs just one-thirtieth of an ounce (1 gram) and is about 1 inch (2½ centimeters) long. It moves frequently, about 150 times per hour, but this is imperceptible to the woman. Random arm and leg movements are more frequent early in pregnancy than later on (Rakic et al., 2016).

By eight weeks post conception, the developing embryo has all of the organs and body parts of a human being, including elbows and knees. Development is **cephalocaudal** (literally, "head-to-tail") and **proximodistal** (literally, "near-to-far"): The head forms first and the extremities last. This directional pattern continues until puberty, when it reverses. (Feet first, brain last!)

The early embryo has both male (via *Wolffian ducts*) and female (via *Müllerian ducts*) potential, in a tiny intersex gonad. At the end of the embryonic period, hormonal and genetic influences typically cause one or the other to shrink, and then ovaries or testes, and a vagina or penis, grow from that omnipotent gonad (Zhao et al., 2017).

Fetus: From the Ninth Week Until Birth

The organism is called a **fetus** from the beginning of the ninth week after conception until birth. The fetal period encompasses dramatic change, from a tiny creature smaller than the final joint of your thumb to a newborn about 20 inches (51 centimeters) long.

Early growth is rapid. By three months, the fetus weighs about 3 ounces (87 grams) and is about 3 inches (7.5 centimeters) long. Those numbers—3 months, 3 ounces, 3 inches—are approximate. (Metric measures—100 days, 100 grams, 100 millimeters—are similarly imprecise, but useful as a memory guide.)

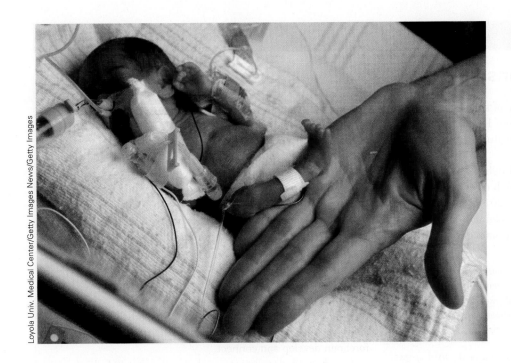

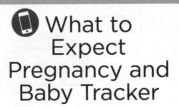

One of the Tiniest Rumaisa Rahman was born after 26 weeks and 6 days, weighing only 8.6 ounces (244 grams). Nevertheless, now age 16, she is living a full, normal life. Rumaisa gained 5 pounds (2,270 grams) in the hospital and then, six months after her birth, went home. Her twin sister, Hiba, who weighed 1.3 pounds (590 grams) at birth, had gone home two months earlier. At their one-year birthday, the twins seemed typical, with Rumaisa weighing 15 pounds (6,800 grams) and Hiba 17 pounds (7,711 grams) (Nanji, 2005).

The Middle Three Months

The 4-month-old fetus is very active, with "large body movements—whole body flexion and extension, stretching and writhing, and vigorous leg kicks that somersault the fetus through the amniotic fluid" (Adolph & Franchak, 2017). The heartbeat becomes stronger and faster when the fetus is awake and moving. Digestion and elimination develop. Fingernails, toenails, and buds for teeth form, and hair grows, including eyelashes!

Those developments inspire awe, but the crucial mid-pregnancy development is that the central nervous system becomes active, regulating heart rate, breathing, and sucking. This brain maturation allows the fetus to reach the **age of viability,** when a fetus born far too early might survive.

Every day of prenatal life within the uterus increases viability. If birth occurs before 22 weeks, death is certain because advanced technology cannot maintain life without some brain response. (Reports of survivors born before 22 weeks are unreliable, because the date of conception is unknown.)

After the age of viability, life is still fragile. Currently, if birth occurs in an advanced neonatal unit, some very preterm babies survive, and some of those reach age 2 without major impairments. Rates vary by nation, by sex (girls do better), and by medical advances.

Among the best survival rates are those in Japan and France. In Japan, about half of 22-week-old newborns born alive survive the first week of life, and about half of those survivors escape major cognitive disabilities (Kono et al., 2018). In France, when membranes are ruptured prematurely because continued pregnancy might be fatal to the woman, only 11 percent of 22-week fetuses survive to age 2 (most die at birth or soon thereafter) but 66 percent do so if they were born 25 weeks past conception (Lorthe et al., 2018).

The Final Three Months

Each day of the final three months benefits the fetal brain and body. Many aspects of prenatal life are awe-inspiring; the fact that any ordinary woman provides a far better home for a fetus than the most advanced medical technology is one of them.

age of viability The age (about 22 weeks after conception) at which a fetus might survive outside the mother's uterus if specialized medical care is available.

CHAPTER APP 4

What to Expect Pregnancy and Baby Tracker

IOS:
https://tinyurl.com/y6esrn5a

ANDROID:
https://tinyurl.com/nufnyqg

RELEVANT TOPICS:
Prenatal health and development, birth

Based on the well-known book *What to Expect When You're Expecting*, this app guides users through pregnancy day-by-day. Users receive personalized updates, tips, and articles on their baby's development, and are joined with a virtual community of other parents-to-be.

Essential Connections

In earlier decades, a newborn's chance of survival was pegged to weight. Today we know better: Some 1-pound babies live and some 3-pound babies die. Brain maturation is more crucial than body fat.

Accordingly, research on prenatal development has become "a frontier of science." However, research on the fetal brain is "exceptionally difficult" (Vasung et al., 2019). Remember the need for interdisciplinary research, explained in Chapter 1? That is essential for learning about early brain development. As one review explains:

> [C]ollaboration between at least three major expert groups is necessary: 1. biological experts (ex. neuroscientists, geneticists and developmental biologists), 2. technical experts (ex. physicists, engineers, computer scientists and mathematicians) and 3. clinical experts (ex. radiologists, neonatologists, neurologists, cardiologists and obstetricians).
>
> *[Vasung et al., 2019, p. 245]*

This collaboration has led to some verified aspects of prenatal growth. The central nervous system is the first body system to begin development. The embryonic stage starts with the primitive streak, which becomes the neural tube even before the facial features are formed and the first pulsating blood vessel appears.

Already by the third week after conception some cells specialize to become *neural progenitor cells,* which duplicate and multiply many times until some of them create neurons (brain cells). Neurons do not duplicate but may endure lifelong. Those early neurons migrate to a particular part of the brain (brainstem, cerebellum, hypothalamus, visual cortex, and so on) and specialize. For example, some neurons are dedicated to seeing faces, others to seeing red and green, others to blue and yellow, and so on.

By mid-pregnancy, in *neurogenesis,* the brain has developed billions of neurons, and folding allows the cortex to be larger and more complex than the brains of other animals (Stiles & Jernigan, 2010). [**Developmental Link:** The cortex and other brain structures are described in Chapter 5.]

Following the proximodistal (near-to-far) sequence, the six layers of the cortex are produced, with the bottom (sixth) layer first and then each new layer above the previous one so that the top, outer layer is the last to form. Similarly, the brainstem above the back of the neck, then the midbrain, and finally the forebrain develop and connect.

Synchronized connections between parts of the brain develop in the second half of pregnancy. That indicates they are working together—as in an adult who sees and smells something delicious, experiences hunger, and reaches for the food, all in a flash.

These four responses arise from different brain regions, a synchrony that begins prenatally (see **Figure 4.1**). In the later months of pregnancy, and not before, the reflexes of heartbeat, breath, and movement coordinate, giving a preterm infant a chance at survival. A relative lack of synchrony after 20 weeks predicts a newborn at risk (Thomason et al., 2017).

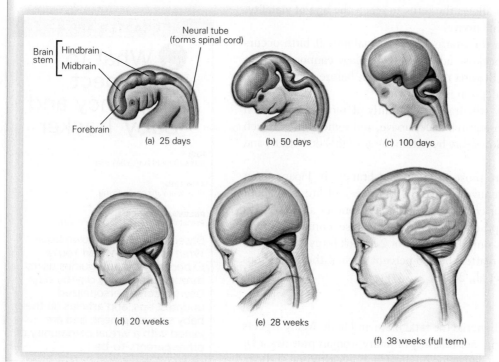

(a) 25 days

(b) 50 days

(c) 100 days

(d) 20 weeks

(e) 28 weeks

(f) 38 weeks (full term)

FIGURE 4.1

Prenatal Growth of the Brain Just 25 days after conception *(a)*, the central nervous system is already evident. The brain looks distinctly human by day 100 *(c)*. By the 28th week of gestation *(e)*, at the very time brain activity begins, the various sections of the brain are recognizable. When the fetus is full term *(f)*, all parts of the brain, including the cortex (the outer layers), are formed, folding over one another and becoming more convoluted, or wrinkled, as the number of brain cells increases.

What makes prenatal brain development a particularly complex area of study is that brain development switches from the first six months to the final three (Vasung et al., 2019). In the first part, basic structures are formed, including the limbic system (for emotions), the prefrontal cortex (for analysis), and the six layers of the cortex (for conscious awareness). Impairments in any of these structures affect the mind and behavior lifelong.

But then the environment becomes more crucial, as the brain of the future person is affected by the prenatal environment, such as maternal stress, nutrition, patterns of sleep, and exercise.

Detailed study of one crucial brain region, the *hippocampus* (the major site for memory formation), reveals an explosion of new cells during the fourth prenatal month, followed by a gradual slowdown, while other parts of the fetal body (legs, arms, etc.) increase (Ge et al., 2015).

In the final months, when somewhat fewer new neurons in the brain are formed, existing neurons begin to be affected by the environment. For example, if the mother is worried, sleeps fitfully, and/or is addicted to drugs, her newborn might have the same problems in the early weeks after birth. The fetal brain is responsive to many signals from maternal actions, and that can affect the future child's social development (Manczak & Gotlib, 2019).

This mid-gestation burst of neurogenesis and then slowdown is characteristic of the entire brain, with each area following its own timetable, prenatally and postnatally (Gilmore et al., 2018).

This is an example of the biosocial development detailed in every later biosocial chapter: Each age has a pattern of growth, and each age has particular challenges. Prenatally, the challenge is for the brain to develop as well as genes allow.

In the final three prenatal months, the neurological, respiratory, and cardiovascular systems all develop. The lungs expand and contract, and breathing muscles strengthen as the fetus swallows and spits out amniotic fluid. The valves of the heart go through a final maturation, as do the arteries and veins throughout the body; the testicles of the male fetus descend.

The various lobes and areas of the brain are also established, and pathways between one area and another are forged. For instance, sound and sight become coordinated: Newborns quickly connect voices heard during pregnancy with faces. That may be why they recognize their mother after seeing her only once or twice.

That phenomenal accomplishment (recognizing Mother) occurs within a day or two after birth. Neurological plasticity allows the fetus to recognize the voices of familiar people by the sixth month after conception (Webb et al., 2015). If a baby is born prematurely, impairments are more common in movement, intelligence, and/or vision than in hearing (Kono et al., 2018).

By full term, human brain growth is so extensive that the cortex has become wrinkled, with *gyri* and *sulci*, as the hills and valleys of the cortex are called (see **Figure 4.2**). Although some huge mammals (whales, for instance) have bigger brains than humans, no other creature needs as many folds because, relative to body size, the human brain is much larger.

FIGURE 4.2

Bigger Brains, Better Connections From 15 weeks (*far left*) to full term (*far right*), the fetal brain increases not only in size, but also in connections. The top row shows size and complexity—those folds in the cortex are necessary for all aspects of human thought. The bottom row shows the thickness of the subplate compartments, needed mid-gestation to keep one lobe from interfering with another, and then needed to disappear by full-term birth.

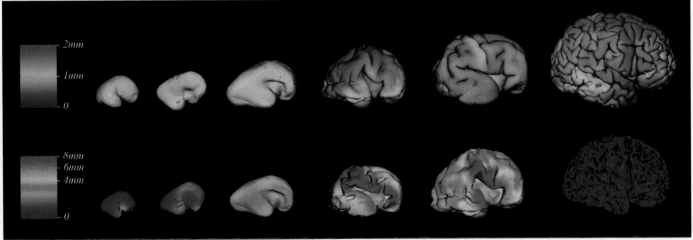

Lana Vasung

LaunchPad
macmillan learning

VIDEO: Brain Development Animation: Prenatal shows how the brain develops from just after conception until birth.

Beyond brain growth, with an estimated 86 billion neurons at birth, another process occurs in the final months of pregnancy—cell death, called *apoptosis*. Research on apoptosis has primarily centered on nonhuman creatures. It is apparent that programmed cell death is widespread in prenatal development, and the effects may be transmitted from one generation to the next (Hamada & Matthews, 2019).

Some apoptosis is easy to understand: Abnormal and immature neurons, such as those with missing or extra chromosomes, are lost. So are neurons if the woman is exposed to toxins, such as drugs or stress. Sometimes, however, seemingly normal neurons die: One estimate finds that almost half of all brain cells are gone before birth (Underwood, 2013). Why?

Perhaps death of any neuron that is not fully functioning creates space in the brain for the remaining neurons to coordinate thinking, remembering, and responding. Surviving preterm babies often have intellectual and emotional deficits, with gestational age the chief predictor (Heitzer et al., 2020). There are many hypotheses for this correlation; less time for prenatal brain development and apoptosis is one of them.

That is speculation. However, it is known that protections of the brain occur in the final months of pregnancy. One critical development is thickening of the membranes and bones covering the brain. This helps prevent "brain bleeds," a hazard of preterm birth if paper-thin blood vessels in the cortex collapse.

Another development pertains to the *fontanels*, which are areas on the top of newborn head where the bones of the skull have not yet fused. Fontanels enable the fetal head to become narrower as it moves through the vagina during birth. During the first weeks of life, the fontanels gradually close. Preterm babies have larger fontanels, making them vulnerable to brain damage (Frémondière et al., 2019).

The average fetus in the United States gains about 4½ pounds (2.1 kilograms) in the third trimester, increasing to about 7½ pounds (about 3.4 kilograms) at birth, with boys a few ounces heavier than girls. Of course, some quite healthy babies born to well-nourished women weigh more, and some weigh less. Variation depends on genes and national diet; northern European newborns tend to be heavier, East Asian to be lighter (Kiserud et al., 2018).

Reflexes (listed at the end of this chapter) are a better indication of health than weight. Unless something is amiss, most newborns are ready by the end of prenatal life to thrive at home on mother's milk—no expert help, oxygenated air, or special feeding required. For thousands of years, that is how humans survived: We would not be alive if any of our ancestors had been born before the last three months of pregnancy.

WHAT HAVE YOU LEARNED?

1. What are the three stages of prenatal development?

2. Why are the first days of life the most hazardous?

3. What parts of the embryo form first?

4. When do sex organs appear?

5. What distinguishes a fetus from a baby?

6. What is the prognosis of a baby born before 25 weeks of gestation?

7. What occurs in the final three months of pregnancy?

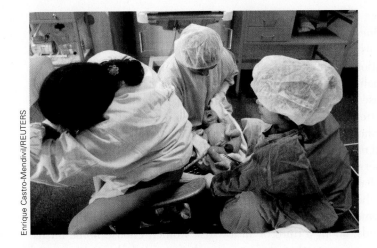

Enrique Castro-Mendivil/REUTERS

Frank Herholdt/The Image Bank/Getty Images

Choice, Culture, or Cohort? Why do it that way? Both of these women (in Peru, on the *left,* in England, on the *right*) chose methods of labor that are unusual in the United States, where birth stools (Peru) and birthing pools (England) are uncommon. However, in all three nations, most births occur in hospitals—a rare choice a century ago.

Birth

About 38 weeks (266 days) after conception, the fetal brain signals massive increases in the hormone *oxytocin* to start labor. Birth occurs after 12 hours of active labor for first births and 7 hours for subsequent ones. These are averages: Birth is still *full term* two weeks before or after the due date, and labor may take twice or half as long. Some women believe they are in active labor for days, and others say 10 minutes.

Variations in How and Where

When considering birth from a life-span perspective, the multicultural and multicontextual variations are astounding. Birth may be universal, but specifics vary tremendously.

Birth Positions

Women's birthing positions vary—sitting, squatting, lying down. Some women labor in a tub of warm water, which helps the woman relax (the fetus continues to get oxygen via the umbilical cord). Preferences and opinions on birthing positions (as on almost every other aspect of prenatal development and birth) are partly cultural and partly personal.

In general, physicians find it easier to see the head emerge if the woman lies on her back. However, women find it easier to push the fetus out if they sit up.

Home or Hospital?

Once, all family members and neighbors were nearby as the mother labored at home, sometimes with an experienced older woman (called a "granny midwife" in the United States) to help. Then, in developed nations in the middle of the twentieth century, almost all babies were born in hospitals, attended only by doctors and nurses. Now most hospitals allow one family member, usually the father, sometimes a friend or grandmother (as in my case).

An innovation in the United States and other nations is to involve a **doula,** a person trained to support the laboring woman. Doulas time contractions, use massage, provide encouragement, and do whatever else is helpful. Often doulas come to the woman's home during early labor and return after birth to encourage breast-feeding. Doulas are chosen by many women, but they have proven to be particularly helpful for immigrant, low-income, or unpartnered women who may be intimidated by doctors (Kang, 2014; Saxbe, 2017).

In many poor nations, births often occur at home. Lack of prenatal care and absence of trained birth attendants correlate with higher death rates, for both

doula A woman who helps with the birth process. Traditionally in Latin America, a doula was the only professional who attended childbirth. Now doulas are likely to arrive at the woman's home during early labor and later work alongside a hospital's staff.

They Called It "Catching" the Baby Midwife Mahala Couch shows her strong hands that "caught" thousands of newborns in the back woods of Southern Appalachia. Midwife births became illegal in about 1920, but many women still preferred home birth with Mahala over hospital birth with a doctor. Currently, midwives are trained, certified, and legal in most states and nations.

Apgar scale A quick assessment of a newborn's health, from 0 to 10. Below 5 is an emergency—a neonatal pediatrician is summoned immediately. Most babies are at 7, 8, or 9—almost never a perfect 10.

babies and mothers (Tekelab et al., 2019). However, uncomplicated home births, facilitated by a trained attendant, may be safer than births in poorly staffed and ill-equipped hospitals (Kunkel et al., 2019; Manasyan et al., 2019).

In developed nations, home births are rarely chosen. In the United States, less than 1 percent of babies are born at home, about half of them unexpected, which is hazardous because no one is nearby to rescue a newborn in distress. Planned home births are better, although sometimes illegal and not covered by insurance. Some contend that home births are risky for newborns (Wendland, 2018).

Even in advanced nations that support home births, few women choose it. In Great Britain, "based on her wishes and cultural preferences and any medical and obstetric needs she and her baby may have" the national health care system pays for home births. Nonetheless, hospital births are chosen by about 90 percent; birthing centers by 8 percent; and home births by only 2 percent (quoted in Hinton et al., 2018).

In the Netherlands, about a third of all low-risk women choose home births. National health care supports that choice, not only financially but with special ambulances called *flying storks*, which speed mother and newborn from home to a hospital if sudden complications arise. In that nation, compared to hospital births, mothers have fewer complications, and infants have better survival rates if born at home (de Jonge et al., 2015).

The Newborn's First Minutes

Newborns usually breathe and cry on their own. The first breaths of air bring oxygen to the lungs and blood, and the infant's color changes from bluish to reddish. ("Reddish" refers to blood color, visible beneath the skin, and applies to newborns of all skin tones.) Eyes open wide; tiny fingers grab; even tinier toes stretch and retract. Usually full-term babies are instantly, zestfully, ready for life.

Newborn health is often measured by the **Apgar scale,** first developed by Dr. Virginia Apgar (see Visualizing Development). When she earned her M.D. in 1933, Apgar wanted to work in a hospital but was told that only men did surgery. She became an anesthesiologist, present at many births but never the doctor in charge.

Apgar saw that "delivery room doctors focused on mothers and paid little attention to babies. Those who were small and struggling were often left to die" (Beck, 2009, p. D1). To save those young lives, Apgar developed a simple rating scale of five vital signs—color, heart rate, cry, muscle tone, and breathing. Nurses could use the scale and sound the alarm immediately if the score was 6 or lower.

Since 1950, birth attendants worldwide have used the Apgar (often using the name as an acronym: Appearance, Pulse, Grimace, Activity, and Respiration) at one minute and again at five minutes after birth, assigning each vital sign a score.

A study comparing Apgar scores in 23 nations found that the ratings of birth attendants were influenced by their culture: In some nations, 97 percent of newborns scored 9 or 10, whereas in other nations only 73 percent had such scores. Whether attendants tended to rate high or low did not correlate with quality of obstetrics. However, in every nation, an Apgar below 7 always indicated risk (Siddiqui et al., 2017).

Medical Assistance at Birth

The specifics of birth depend on the fetus, the mother, the birth attendant, the birthplace, and the culture. (See Opposing Perspectives on page 101.)

Surgery

Fifty years ago, in developed nations, hospitals banned midwives, but now many hospitals allow certified midwives to deliver babies. Midwife births have lower

The Apgar

Just moments after birth, babies are administered their very first test. The APGAR score is an assessment tool used by doctors and nurses to determine whether a newborn requires any medical intervention. It tests five specific criteria of health, and the medical professional assigns a score of 0, 1, or 2 for each category. A perfect score of 10 is rare—most babies will show some minor deficits at the 1-minute mark, and many will still lose points at the 5-minute mark.

GRIMACE RESPONSE/REFLEXES

(2) Healthy babies indicate their displeasure when their airways are suctioned—they will grimace, pull away, cough, or sneeze.

(1) Baby grimaces during suctioning.

(0) Baby shows no response to being suctioned and requires immediate medical attention.

RESPIRATION

(2) A good, strong cry indicates a normal breathing rate.

(1) Baby has a weak cry or whimper, or slow/irregular breathing.

(0) Baby is not breathing and requires immediate medical intervention.

PULSE

(2) A pulse of 100 or more beats per minute is healthy for a newborn.

(1) Baby's pulse is less than 100 beats per minute.

(0) Baby has no heartbeat and requires immediate medical attention.

APPEARANCE/COLOR

(2) Body and extremities show good color, with pink undertones indicating good circulation.

(1) Baby has some blueness in the palms and soles of the feet. Many babies exhibit some blueness at both the 1- and 5-minute marks; most warm up soon after.

(0) Baby's entire body is blue, grey, or very pale and requires immediate medical intervention.

ACTIVITY AND MUSCLE TONE

(2) Baby exhibits active motion of arms, legs, and body.

(1) Baby shows some movement of arms and legs.

(0) Baby is limp and motionless and requires immediate medical attention.

REFLEXES IN INFANTS

Never underestimate the power of a reflex. For developmentalists, newborn reflexes are mechanisms for survival, indicators of brain maturation, and vestiges of evolutionary history. For parents, they are mostly delightful and sometimes amazing.

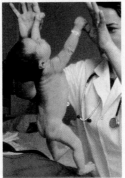

THE SUCKING REFLEX A newborn, just a few minutes old, demonstrates that he is ready to nurse by sucking on a doctor's finger.

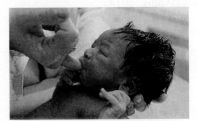

THE GRASPING REFLEX When the doctor places a finger on the palm of a healthy infant, he or she will grasp so tightly that the baby's legs can dangle.

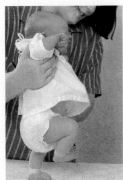

THE STEP REFLEX A 1-day-old girl steps eagerly forward on legs too tiny to support her body.

Sekundator/Shutterstock.com; Astier/BSIP/Science Source; Petit Format/Science Source; Jennie Woodcock; Reflections Photolibrary/Corbis Documentary/Getty Images

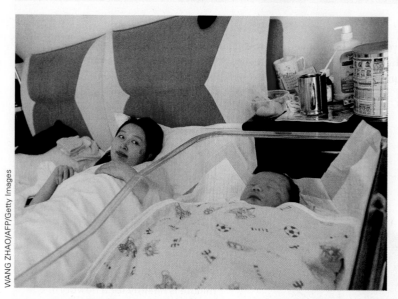

WANG ZHAO/AFP/Getty Images

Pick Up Your Baby! Probably she can't. In this maternity ward in Beijing, China, most patients are recovering from cesarean sections, making it difficult to cradle, breast-feed, or carry a newborn until the incision heals.

cesarean section (c-section) A surgical birth in which incisions through the mother's abdomen and uterus allow the fetus to be removed quickly instead of being delivered through the vagina. (Also called simply *section*.)

🌓 **Especially for Conservatives and Liberals** Do people's attitudes about medical intervention at birth reflect their attitudes about medicine at other points in their life span, in such areas as assisted reproductive technology (ART), immunization, and life support? (see response, page 117)

rates of various complications and interventions than physician births, in part because midwives emphasize breathing, massage, and social support (Bodner-Adler et al., 2017; Raipuria et al., 2018; Renfrew et al., 2014).

Nonetheless, most U.S. births are attended by physicians. They are the only ones allowed to perform certain medical measures, including **cesarean sections (c-sections),** when the fetus is removed though incision in the mother's abdomen instead of being pushed by contractions through the vagina.

C-sections were once rare, a last-ditch effort to save a new life when the mother was dying and vaginal birth was impossible. Now, with much better anesthesia and fetal monitoring, more than one birth in five worldwide (21 percent in 2015, compared to 12 percent in 2000) is a c-section (Boerma et al., 2018).

Cesareans are medically indicated for about 10 percent of births. Multiple births (twins or more), breech births (fetus is not positioned head down), prior c-section, long active labor (more than 24 hours), a narrow pelvis, a large fetus, and advanced maternal age are all conditions that suggest surgery. However, none of these *requires* a c-section.

For instance, a large study of all births (78,880) in the state of Washington focused on the relationship between age and various complications. Sixty percent of new mothers aged 50 or older delivered via c-section; 40 percent delivered vaginally (Richards et al., 2016).

Public health experts are troubled by increases in c-sections in the past decade and by international disparities. Too few (4 percent) cesareans are performed in central Africa, including about 1 percent in South Sudan, where childbirth is still a leading cause of death. Too many occur in the Caribbean and Latin America (44 percent), with the highest of all in the Dominican Republic (58 percent) (Boerma et al., 2018).

Nations with very low cesarean rates also have high numbers of childbirth deaths, but nations with high cesarean rates are not necessarily healthier. In the United States, the cesarean rate was 32 percent in 2018 (Martin et al., 2019).

Cesareans have immediate advantages for hospitals (easier to schedule, quicker, and more expensive than vaginal deliveries) and for women (advance planning, quick birth). Disadvantages appear later.

After a c-section, breast-feeding is harder and medical complications more likely. Babies born via c-section are more likely to develop asthma or become obese as children (Chu et al., 2017; Mueller et al., 2017). One reason: Vaginal deliveries provide newborns with beneficial microbiomes (Wallis, 2014).

WHAT HAVE YOU LEARNED?

1. What is the typical birth process?

2. Who was Virginia Apgar and what did she do?

3. What are the immediate and long-term results of a cesarean birth?

4. Why do cesarean rates vary internationally?

5. What are the advantages and disadvantages of a hospital birth versus a home birth?

Interventions in the Birth Process

As you just read, some nations seem to have too few cesareans and others too many. International variations are also evident in drugs to reduce pain, or to start or speed labor.

Particular concern has focused on two aspects of pain relief. First, in developed nations, many hospital births include prescription of opioids, which can become addictive. Second, more than half of all hospital births in the United States include an *epidural*, which is an injection in the spine to stop sensation in the lower body while allowing the birthing woman to be awake. Opioids and epidurals increase the rate of cesarean sections and decrease the rate of breast-feeding (Kjerulff, 2014; Mahomed et al., 2019; Mourad et al., 2019).

Another frequent intervention is the injection of Pitocin, a synthetic version of the hormone *oxytocin*, to start or speed up labor contractions. This hormone is produced naturally to start labor, promote breast-feeding, and encourage infant care. Pitocin correlates with cesarean births and other complications (Grivell et al., 2012; Mikolajczyk et al., 2016).

Someone who has never given birth might conclude that birth should be allowed to proceed without intervention. However, many laboring women and attending professionals want to ease the pain and avoid the risks. Those are reasonable goals, much more possible today than a few decades ago. In 1950, in the world, birth led to death five times more often than it does today. The difference is primarily medical intervention.

I saw this with my daughter. Remember that she was admitted to the hospital because a sonogram revealed that the amniotic fluid was low. That could harm the fetus, a problem that could not have been detected a few decades ago. To avert the problem, the midwife advised that birth should be soon, and she prescribed

Pitocin. After four hours of hard labor, my daughter requested an epidural.

The answer: "You can have an epidural, but the fetal position means that if you can't push, you might need a cesarean."

"No epidural," my daughter responded. She had listened to other women, and her birth plan said no surgery.

Yet, hearing about someone else's experience is not the best way to decide about medical care. Further, correlation is not causation. Does intervention (drugs, surgery, and so on) cause complications, or is it the other way around, with complications causing the intervention? Many doctors believe that concerns about harm to the fetus are considered "premature" when induction is medically indicated (Lønfeldt et al., 2018).

However, the data raised troubling questions. A study of 750,000 births in the United States divided hospitals into three categories—low, average, and high quality. Compared to high-quality hospitals, low-quality hospitals had five times more complications after a cesarean (20 percent versus 4 percent) and twice as many after a vaginal birth (23 versus 11 percent) (Glance et al., 2014). Is hospital quality the issue, not intervention?

Further, intervention rates vary more by doctor, region, and nation than by medical condition. Are some doctors too hasty or are some doctors are too slow to intervene?

In the United States, births are more common during weekdays than on nights and weekends because of more c-sections and Pitocin (Fischetti & Armstrong, 2017). Does that indicate better care or too much intervention? Is birth a natural process that should be left alone or a medical event that requires doctors, technology, and hospitals?

Problems and Solutions

The early days of life place the future person on the path toward health and success—or not. Problems can begin before conception, if the sperm, the ovum, or the uterus was affected by the parents' health. Indeed, the grandmother's health when the mother and father were born may affect the grandchild, although this epigenetic effect is not proven to the satisfaction of most scientists (Arshad et al., 2017).

Fortunately, healthy newborns are the norm, not the exception. However, if something is amiss, it is often part of a sequence that may become overwhelming (Rossignol et al., 2014).

Risk Analysis

Life requires risks: We routinely decide which chances to take and how to minimize harm. For example, we know the danger as well as the benefits of

cerebral palsy A disorder that results from damage to the brain's motor centers. People with cerebral palsy have difficulty with muscle control, so their speech and/or body movements are impaired.

anoxia A lack of oxygen that, if prolonged, can cause brain damage or death.

teratogen An agent or condition, including viruses, drugs, and chemicals, that can impair prenatal development and result in birth defects or even death.

behavioral teratogens Agents and conditions that can harm the prenatal brain, impairing the future child's intellectual and emotional functioning.

crossing the street, so we hold the hands of young children, teaching them how to safely cross.

That is a small illustration of risk analysis. Risks need to be taken, but they also need to be controlled. Pregnancy and birth entail many risks, but the outcome—a new baby—seems well worth it. The goal is to avoid problems and to mitigate damage of those that occur. Development is lifelong: Single events rarely cause problems, but a cascade often does.

A dramatic example is **cerebral palsy** (a condition marked by difficulties with movement. Cerebral palsy was once thought to be caused solely by birth procedures (excessive medication, slow breech birth, or misused forceps). However, it now seems that cerebral palsy begins with genetic sensitivity, prenatal insults, and maternal infection (Mann et al., 2009) and is exacerbated by insufficient oxygen to the fetal brain at birth.

This lack of oxygen is called **anoxia.** Contractions during labor may affect the fetal heart rate, indicating a moment of anoxia, but recovery occurs when the contraction is over. Similarly, anoxia often occurs for a second or two during birth, indicated by a slower fetal heart rate, with no harm done. Because their births take longer, twins and breech births are more likely to experience anoxia.

To prevent prolonged anoxia, the fetal heart rate is monitored during labor. If anoxia lasts too long, that can harm the fetus. To prevent prolonged anoxia, a cesarean may be indicated. After birth, if the Apgar indicates slow breathing or bluish color, immediate oxygen is given.

How long anoxia can continue without harming the brain depends on genes, birthweight, gestational age, drugs in the bloodstream (either taken by the mother before birth or given by the doctor during birth), and many other factors. Thus, anoxia is part of a cascade that may cause cerebral palsy. Almost every other birth complication is also the result of many factors.

Harmful Substances

Monthly, even weekly, scientists discover another **teratogen,** which is anything—drugs, viruses, pollutants, malnutrition, stress, and more—that increases the risk of prenatal abnormalities and birth complications.

But don't be like one of my students, who said that now that she knew all the things that can go wrong, she never wants a baby. As I told her, problems can be avoided, and damage can be remedied. Pregnancy is not a dangerous period to be feared; it is a natural process to be protected.

Visible and Invisible Damage

People once thought that the placenta protected the fetus against every insult. Then, about six decades ago, rubella and thalidomide (both mentioned in Chapter 1) proved otherwise.

Since then, it was apparent that teratogens caused birth defects—such as blindness from rubella or missing limbs from thalidomide—but few realized that teratogens might cause no visible harm but instead make a child hyperactive, antisocial, or intellectually disabled. Those **behavioral teratogens** are common, and, over the life span, more disabling than teratogens that do not affect the brain.

The Critical Time

Timing is crucial. Some teratogens cause damage only during a *critical period*, which may occur before a woman knows she is pregnant (see **Figure 4.3**). [**Developmental**

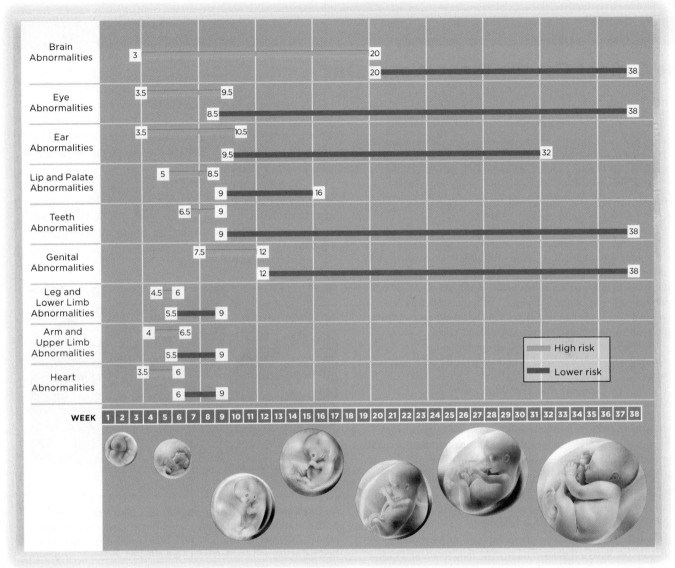

FIGURE 4.3

One More Reason to Plan a Pregnancy The embryonic period, before a woman knows she is pregnant, is the most sensitive time for causing structural birth defects. However, at no time during pregnancy is the fetus completely safe from harm. Individual differences in susceptibility to teratogens may be caused by a fetus's genetic makeup or peculiarities of the mother, including the effectiveness of her placenta or her overall health. The dose and timing of the exposure are also important.

● **Observation Quiz** What part of the embryo and fetus has the longest period of vulnerability? What has the shortest? (see answer, page 117) ↑

Link: Critical and sensitive periods are described in Chapter 1.] Consequently, women need to avoid drugs, supplement a balanced diet with folic acid and iron, update their immunizations, and gain or lose weight if needed *before* pregnancy occurs.

The first days and weeks after conception (the germinal and embryonic periods) are critical for body formation, but behavioral teratogens affect the fetus at any time. Cigarettes, for instance, are harmful when the mother smokes before or during pregnancy, but quitting by mid-pregnancy may allow the fetus to grow normally (Kvalvik et al., 2017). Similarly, a longitudinal study of 7-year-olds who were exposed to alcohol prenatally found surprising results: Although alcohol

A CASE TO STUDY

He Cannot Get the Right Words Out

Many children have difficulties in thinking and behavior that *could* be connected to teratogens. One of my students wrote:

> I was nine years old when my mother announced she was pregnant. I was the one who was most excited. . . . My mother was a heavy smoker, Colt 45 beer drinker and a strong caffeine coffee drinker.
>
> One day my mother was sitting at the dining room table smoking cigarettes one after the other. I asked "Isn't smoking bad for the baby?"
>
> She made a face and said "Yes, so what?"
>
> I said "So why are you doing it?"
>
> She said, "I don't know.". . .
>
> During this time I was in the fifth grade and we saw a film about birth defects. My biggest fear was that my mother was going to give birth to a fetal alcohol syndrome (FAS) infant. . . . My baby brother was born right on schedule. The doctors claimed a healthy newborn. . . . Once I heard healthy, I thought everything was going to be fine. I was wrong, then again I was just a child. . . .
>
> My baby brother never showed any interest in toys . . . he just cannot get the right words out of his mouth . . . he has no common sense . . .
>
> Why hurt those who cannot defend themselves?
>
> *[J., personal communication]*

As you remember from Chapter 1, one case is not proof. J. blames her mother, but genetic risks, inadequate prenatal care, and troubling postnatal experiences may be part of her brother's sorry cascade. Moreover, her mother was of low SES (itself a correlate of harm) and was poorly nourished. Boys and later-born children are more vulnerable, which may explain why J. was a good student, unlike her brother.

It is not unusual for a newborn to seem to have escaped a teratogen, yet have a damaged brain. The long reach of a seemingly harmless teratogen is evident in the *Zika virus* (ZIKV), caused by the bite of an infected mosquito.

ZIKV was not recognized until 2015, when an epidemic led to dozens of Brazilian newborns with abnormally small brains (*microcephaly*). That epidemic spread to several nations and led to research suggesting that ZIKV has infected humans for hundreds of years, undetected.

Only the most severely affected ZIKV infants have small heads. Many appear typical but have impaired vision, hearing, and emotions (Rosen, 2016; van den Pol, 2017). When pregnant monkeys were infected with Zika, brain patterns suggested schizophrenia, depression, and Alzheimer's disease (Christian et al., 2018; Waldorf et al., 2018).

Schneyder Mendoza/European Pressphoto Agency/CÃºcuta/Norte de Santander/Colombia/Newscom

No One Knows Dozens of newborns in northern Brazil alerted doctors that the Zika virus could cause microcephaly. More is now known: Zika brain damage is sometimes invisible, and newborns in North, Central, and South America are affected. However, long-term damage is still unknown. No wonder these pregnant women in a clinic in Colombia are worried, especially Sandra Ovallos (*middle*), who recently had a fever and rash.

affects body structures early in pregnancy, it is the second half of pregnancy when binge drinking is an especially potent behavioral teratogen (Niclasen et al., 2014).

Timing may be important in another way. When pregnancy occurs soon after a previous pregnancy, risk increases. For example, one study found that second-born children are twice as likely to have autism spectrum disorder if they are born within a year of the first-born child (Cheslack-Postava et al., 2011).

How Much Is Too Much?

threshold effect In prenatal development, when a teratogen is relatively harmless in small doses but becomes harmful once exposure reaches a certain level (the threshold).

A second factor that affects the harm from teratogens is the dose and/or frequency of exposure. Many teratogens have a **threshold effect;** they are virtually harmless until exposure reaches a certain level, and then they "cross the threshold" to damage the fetus.

Is there a safe dose or timing for psychoactive drugs? Research has focused on alcohol, a drug ingested by most young women in many nations. Early in pregnancy, a woman's heavy drinking can cause **fetal alcohol syndrome (FAS),** which distorts the facial features of a child (especially the eyes, ears, and upper lip). Later in pregnancy, behavior can be affected; *fetal alcohol effects (FAE)* occur, not FAS (Hoyme et al., 2016).

Currently, pregnant women are advised to avoid all alcohol, but many women in France (between 12 and 63 percent, depending on specifics of the research) do not heed that message (Dumas et al., 2014). Most of their babies seem fine. Should all women who *might* become pregnant refuse a legal substance that most men use routinely? Wise? Probably. Necessary? Maybe not.

Innate Vulnerability

Genes are a third factor that influences the effects of teratogens. When a woman carrying dizygotic twins drinks alcohol, for example, the twins' blood alcohol levels are equal, yet one twin may be more severely affected because of different alleles for the enzyme that metabolizes alcohol (Hemingway et al., 2019). Similar differential susceptibility occurs for many teratogens (McCarthy & Eberhart, 2014).

Although the links from genes to teratogens to damage are sometimes difficult to verify, two examples of genetic susceptibility are proven. First, male fetuses are more often spontaneously aborted, stillborn, or harmed by teratogens than are female fetuses. The male–female hazard rate differs from one teratogen to another (Lewis & Kestler, 2012).

Second, one maternal allele reduces folic acid, and that deficit can produce *neural-tube defects*—either *spina bifida*, in which the tail of the spine is not enclosed properly (enclosure normally occurs at about week 7), or *anencephaly*, when part of the brain is missing. Neural-tube defects are more common in certain ethnic groups (e.g., Irish, English, and Egyptian), but even those with the gene usually have healthy babies.

In one study (R. Smith et al., 2011), about half of a group of 550 mothers of a child with a neural-tube disorder (and hence genetically at risk) took folic acid supplements. The other half ate normally. The rate of newborns with neural-tube defects was 1 in 250 among the supplemented mothers and 13 in 300 in the non-supplemented ones, proof that folic acid helps. But, note that almost 96 percent of the women who were at genetic risk and did *not* take supplements had healthy babies. Also, one supplemented woman bore a second child with a neural-tube defect. Why? Was the dose too low, or did she skip taking the pills, or was some other genetic risk the problem?

> **DATA CONNECTIONS: Teratogens** examines the effects of various teratogens and the preventive measures that reduce their risk to a developing fetus. **LaunchPad.**

Prenatal Diagnosis

Early prenatal care has many benefits: Women learn what to eat, what to do, and what to avoid. Some serious conditions, syphilis and HIV among them, can be diagnosed and treated in the first prenatal months before they harm the fetus.

Tests of blood, urine, fetal heart rate, and ultrasound reassure parents, facilitating the crucial parent–child bond. It is now possible to know the sex of the fetus within the first few months. This allows parents to name the fetus, bonding with that small individual long before birth.

fetal alcohol syndrome (FAS) A cluster of birth defects, including abnormal facial characteristics, slow physical growth, and reduced intellectual ability, that may occur in the fetus of a woman who drinks alcohol while pregnant.

Smoke-Free Babies Posters such as this one have had an impact. Smoking among adults is only half of what it was 30 years ago. One-third of women smokers quit when they know they are pregnant, while the other two-thirds cut their smoking in half. Unfortunately, the heaviest smokers are least likely to quit—they need more than posters to motivate them to break the habit.

Especially For Future Doctors and Nurses How can medical professionals explain tests without causing anxiety? (see answer, page 117).

In general, early care protects fetal growth, connects women to their future child, makes birth easier, and renders parents better able to cope. When complications (such as twins, gestational diabetes, and infections) arise, early recognition increases the chance of a healthy birth.

Unfortunately, however, about 20 percent of early pregnancy tests *increase* anxiety instead of reducing it. For instance, the level of alpha-fetoprotein (AFP) may be too high or too low, or ultrasound may indicate multiple fetuses, abnormal growth, Down syndrome, or a mother's narrow pelvis. Many such warnings are **false positives;** that is, they falsely suggest a problem that does not exist. On the other hand, advice about fetal growth can alert women to avoid underweight newborns, a serious problem.

false positive The result of a laboratory test that reports something as true when in fact it is not true. This can occur for pregnancy tests, when a woman might not be pregnant even though the test says she is, or during pregnancy, when a problem is reported that actually does not exist.

Safe During Pregnancy?

As explained in Chapter 1, the scientific method is designed to be cautious. It takes years—for replication, for alternate designs, and for exploration of conflicting hypotheses—to reach sound conclusions. Scientists, governments, and the internet disagree about what is safe for pregnant women, as you will now see.

Which Drugs Are Harmless?

Many women assume that herbal medicines or over-the-counter drugs are safe. Not so. As pediatrics professor Allen Mitchell explains, "Many over-the-counter drugs were grandfathered in with no studies of their possible effects during pregnancy" (quoted in Brody, 2013, p. D5). ("Grandfathered" means that if they were legal in days past, they remain legal—no testing needed.) And, many herbal products are not classed as drugs and hence are not studied and regulated.

To learn which medications are safe, women consult the internet. However, 235 medications were listed as safe on at least one of 25 Web sites but TERSIS (a group of expert teratologists who analyze drug safety) had declared only 60 (25 percent) of them safe, and the others had insufficient evidence to confirm safety (Peters et al., 2013). On those 25 sites, some drugs declared safe on one site were on the danger list of another.

Pesticides

No biologist doubts that pesticides harm frogs, fish, and bees, but the pesticide industry insists that careful use (e.g., spraying plants, not workers) benefits people, in the form of fresh, low-cost food. That certain benefit may outweigh any possible risk.

Developmentalists, however, fear that pregnant women who breathe or ingest these toxins will bear brain-damaged babies (Heyer & Meredith, 2017). One scientist said, "Pesticides were designed to be neurotoxic. Why should we be surprised if they cause neurotoxicity?" (Lanphear, quoted in Mascarelli, 2013, p. 741).

Since 2000, the United States removed one pesticide, *chlorpyrifos*, from household use (it had been used to kill roaches and ants). It was banned from U.S. agriculture in the last month of the Obama administration but reinstated in the first year of the Trump administration—to the distress of many scientists and doctors (Lipton, 2017; Rauh, 2018).

Is that a developmental issue, an economic issue, or a political issue? Chlorpyrifos is widely used in other nations, in homes as well as farms, and is very profitable.

Analysis of umbilical cord blood finds that fetuses exposed to chlorpyrifos become children with lower IQs and more behavior problems than other children (Horton et al., 2012). However, the companies that sell chlorpyrifos argue that confounding factors need to be considered (Mascarelli, 2013).

What might those confounding factors be? For one thing, pregnant women who use roach spray are more likely to live in stressful, inner-city neighborhoods, a context that reduces children's intelligence whether pesticides are used or not. Likewise, parents who harvest sprayed crops are often migrants who move from place to place and fear deportation. Moving, and fear, disrupts children's schooling.

Could factors such as these be a third variable that explains the correlation between pregnant women exposed to pesticides and their children's education? Further, even if chlorpyrifos is a teratogen, does that outweigh the economic benefits for farmers, chemical companies, and parents who need to buy fruits and vegetables? Risk analysis is needed.

Food

Pregnant women hear conflicting advice about what to eat and how much to eat. For example, the U.S. government advises them to eat less fish, but the United Kingdom advises them to eat more fish. The reason for these opposite messages is that fish contains mercury (a teratogen) but also DHA (an omega-3 fatty acid that promotes brain development) (Lando & Lo, 2014).

Should governments protect the unborn, or should each woman do her own risk analysis to judge fish, fruits, and vegetables? Some will choose organic produce, while others will choose the cheapest products. Should pregnant women be allowed to take that risk? Should chemical companies decide what to sell, to whom?

To make all of this more difficult, stress and anxiety affect the fetus, yet pregnancy itself increases fear (Rubertsson et al., 2014). Prospective parents want clear, immediate answers about diet, habits, and circumstances. Scientists take years to find them; laws take even longer. We do know, at least, that women should not eat too little, as now explained.

Low Birthweight: Causes and Consequences

The World Health Organization defines **low birthweight (LBW)** as under 2,500 grams (5 ½ pounds). LBW babies are further grouped into **very low birthweight (VLBW),** under 1,500 grams (3 pounds, 5 ounces), and **extremely low birthweight (ELBW),** under 1,000 grams (2 pounds, 3 ounces).

About 8 percent of babies born in the United States are low birthweight, a rate similar to Brazil, Greece, and Lebanon. About 50 nations have fewer low-birthweight newborns than the United States, with the lowest rate of all in Sweden (less than 4 percent). About 100 nations are worse than the United States, with several (including Bangladesh, Nepal, and Mozambique) having rates of 20 percent or more (Blencowe et al., 2019).

It would be better for everyone—mother, father, baby, and society—if all newborns were in the womb for at least 36 weeks and weighed more than 2,500 grams (5½ pounds). (Usually, this text gives pounds before grams. But hospitals worldwide report birthweight using the metric system, so grams precede pounds and ounces here.) Being underweight at birth has lifelong consequences, including heart disease, less education, and more obesity, and since such marked variability occurs, we need to understand this better.

Too Soon or Too Small

Babies born **preterm** (two or more weeks early; no longer called *premature*) are often LBW, because fetal weight normally doubles in the last trimester of pregnancy, with 900 grams (about 2 pounds) of that gain occurring in the final three

No More Pesticides Carlos Candelario, shown here at age 9 months, was born without limbs, a birth defect that occurred when his mother (Francisca, show here) and father (Abraham) worked in the Florida fields. Since his birth in 2004, laws prohibit spraying pesticides while people pick fruit and vegetables, but developmentalists worry about the effect of the residue on developing brains.

low birthweight (LBW) A body weight at birth of less than 2,500 grams (5½ pounds).

very low birthweight (VLBW) A body weight at birth of less than 1,500 grams (3 pounds, 5 ounces).

extremely low birthweight (ELBW) A body weight at birth of less than 1,000 grams (2 pounds, 3 ounces).

preterm A birth that occurs two or more weeks before the full 38 weeks of the typical pregnancy—that is, at 36 or fewer weeks after conception.

Percentage of LBW Babies—Selected U.S. States

Data from U.S. Department of Health and Human Services, February 1, 2018.

FIGURE 4.4

Where Were You Born? Rates of low birthweight vary by nation, from about 4 to 20 percent, and, as you see here, within nations. Why? Poverty is a correlate—is it also a cause?

small for gestational age (SGA) A term for a baby whose birthweight is significantly lower than expected, given the time since conception. For example, a 5-pound (2,265-gram) newborn is considered SGA if born on time but not SGA if born two months early. (Also called *small-for-dates*.)

immigrant paradox The surprising, paradoxical fact that low-SES immigrant women tend to have fewer birth complications than native-born peers with higher incomes.

weeks. As already mentioned, every week past week 22 adds weight and maturation.

Other LBW babies have gained weight slowly throughout pregnancy and are *small-for-dates*, or **small for gestational age (SGA).** A full-term baby weighing only 2,600 grams and a 30-week-old fetus weighing only 1,000 grams are both SGA, even though the first is not technically low birth-weight. Low birthweight varies dramatically from nation to nation, and, within the United States, from county to county as both a cause and a pre-dictor of lifelong poverty (Robertson & O'Brien, 2018) (see **Figure 4.4**).

In most nations, malnutrition is the most com-mon reason for slow fetal growth. Women who begin pregnancy underweight, who eat poorly during pregnancy, or who gain less than 3 pounds (1.3 kilo-grams) per month in the last six months often have underweight infants. This problem is particularly common in Africa and South Asia, but it can occur in developed nations as well.

The second common reason, particularly in devel-oped nations, is drug use. Almost every psychoactive drug—including legal ones such as cigarettes and alcohol—reduces nutrition and birthweight.

A third reason is multiple births. Twins gain weight more slowly in pregnancy and are born, on average, three weeks early. As you remember from Chapter 3, multiple births are more common in IVF, with some hopeful parents choosing to implant sev-eral blastocysts—against medical advice.

Unfortunately, many risk factors tend to occur together. For example, undernourished mothers often live in urban neighborhoods where pollution is high—another risk factor (Erickson et al., 2016). Women in rural areas have yet another cascade of risks—distance from prenatal care, unwanted pregnancies, and exposure to pesticides (American College of Obstetricians and Gynecologists, 2011).

What About the Father?

The causes of low birthweight rightly focus on the pregnant woman. However, "Fathers' attitudes regarding the pregnancy, fathers' behaviors during the prenatal period, and the relationship between fathers and mothers . . . influence risk for adverse birth outcomes" (Misra et al., 2010, p. 99).

Indeed, everyone who affects a pregnant woman also affects a fetus. Her mother, her boss, her mother-in-law, her doctor, and especially her partner can add to her stress, or reduce it. Thus, it is not surprising that unintended pregnan-cies increase the incidence of low birthweight and birth defects, a link strongest in women of low income (Finer & Zolna, 2016). Obviously, intentions are in the mind, not in the body, and are affected by the father and the community, influenc-ing a woman's nutrition and drug use.

Evidence for this is in the **immigrant paradox.** As already mentioned, low SES correlates with low birthweight, especially in the United States (Martinson & Reichman, 2016). Many immigrants have difficulty with high school, college, and employment; hence their socioeconomic status is low. Thus, one might imagine that their babies are underweight. However, the opposite is true.

Babies born to immigrants tend to be heavier and healthier than newborns of native-born women of the same SES and ethnicity (Marks et al., 2014). This is true not only when women from Latin America are compared to Hispanic women whose families have lived in the United States for generations, but also when women born in the Caribbean, Africa, eastern Europe, and Asia are compared to U.S.-born women with similar genes.

Why? One hypothesis is that immigrant fathers and communities cherish pregnant women, keeping them drug-free and well-fed, appreciated and healthy, buffering the stress of poverty (Luecken et al., 2013). Ironically, in families of the same groups who lived in the United States for longer periods, community protection decreases and prenatal as well as postnatal development suffers (Fox et al., 2018).

Consequences of Low Birthweight

Every developmental milestone—smiling, holding a bottle, walking, talking—is delayed in low-birthweight infants, and rates of cognitive, visual, and hearing impairments increase. As toddlers, LBW children cry often, pay attention less, and disobey more (Aarnoudse-Moens et al., 2009; Stolt et al., 2014).

Problems continue, especially if birthweight was very low. Children who were extremely SGA or preterm tend to have neurological problems in middle childhood, including smaller brain volume, lower IQ, and behavioral difficulties (Clark et al., 2013; Howe et al., 2016; Hutchinson et al., 2013). Even in adulthood, risks persist, with higher rates of diabetes, obesity, heart disease, and depression (Lyall et al., 2016).

However, remember plasticity. By age 4, some ELBW infants exhibit typical brain development, especially if they had no medical complications and their mother was well educated. For some adults, early birth may no longer matter. This was true for those with good experiences, including happy marriages, even when the adult weighed less than 3 pounds at birth (R. Xu et al., 2019).

International Comparisons

As you remember from Chapter 1, scientists collect empirical data and then draw conclusions based on facts. Regarding low birthweight, the facts are clear; the conclusions are not. No less than six hypotheses might explain a puzzling fact: Low birthweight is less common in most nations than it was, with a worldwide average of 14.6 percent in 2015 compared to 17.5 percent in 2000. However, it is increasing in some nations—the United States among them. We begin with what is known.

In some northern European nations, only 4 percent of newborns weigh under 2,500 grams; in several South Asian and African nations, more than 20 percent do. Two conclusions are proven: First, fewer pregnant women are severely malnourished, so their fetuses weight more. Because babies have more body fat, they are more likely to survive the first month of life, when 36 per thousand babies died in 1990, and only 19 per thousand in 2017 (World Bank, 2018).

Second, national goals matter. In China, Cuba, and Chile low birthweight has plummeted since 2000 because prenatal care has become a national priority. That is one conclusion of a study provocatively titled *Low birth weight outcomes: Why better in Cuba than Alabama?* (Neggers & Crowe, 2013).

In other nations, the LBW rate is rising. Many of those nations are in sub-Saharan Africa, and their rise is troubling but not puzzling: Global warming, HIV/AIDS, food shortages, and civil wars are all worse problems in that part of the world than elsewhere.

Not the Fetus, the Mother! Alicia Beltran, age 28, shown here pregnant with her first child, confided at her initial prenatal visit that she had been addicted to a painkiller but was now clean (later confirmed by a lab test). She refused a prescription to keep her away from illegal drugs. But that led to the police taking her to court in handcuffs and shackles when she was 14 weeks pregnant. She was not represented nor allowed to defend herself, but a state-appointed lawyer for the fetus argued that she should be detained. After more than two months in involuntary confinement, a nonprofit lawyer got her released. More than a year later, a judge finally considered her petition that her constitutional rights had been violated but dismissed the case because the state had dropped the charges.

DARREN HAUCK/The New York Times/Redux

Watch **VIDEO: Low Birthweight in India,** which discusses the causes of LBW among babies in India.

THINK CRITICALLY: Food scarcity, drug use, and unmarried parenthood have all been suggested as reasons for the LBW rate in the United States. Which is it—or are there other factors?

● **Especially for Judges and Juries**
How much protection, if any, should the legal system provide for fetuses? Should women with alcohol use disorder who are pregnant be jailed to prevent them from drinking? What about people who enable them to drink, such as their partners, their parents, bar owners, and bartenders? (see response, page 117)

However, in other nations, an increasing rate of LBW is unexpected. The LBW rate in the United States fell throughout most of the twentieth century, reaching a low of 7.0 percent in 1990. But then it rose, dipping slightly around 2010 but increasing every year since 2012 and reaching 8.28 percent in 2018 (Martin et al., 2019). That is higher than almost every other developed nation.

Added to the puzzle is the belief that several changes in maternal ethnicity, age, and health since 1990 should have *decreased* LBW. For instance, although the rate of LBW among African Americans is higher than the national average (14 percent compared with 8 percent), and although teenagers have smaller babies than do women in their 20s, the birth rate among both groups was much lower in 2018 than in 1990.

Similarly, unintended pregnancies are less common (Finer & Zolna, 2016), and two conditions that produce heavier babies (maternal obesity and diabetes) have increased since 1990. Yet, more underweight babies are born in the United States currently than decades ago. What could the explanation be?

Is prenatal care the crucial variable? Perhaps in some nations, but not in the United States. The rates of women giving birth without prenatal care have decreased, but prenatal care as currently offered does not seem to reduce the occurrence of low birthweight (Krans & Davis, 2012).

Another hypothesis is that multiple births have increased because of fertility measures. However, LBW rates are rising for naturally conceived singletons as well as for IVF twins, and the rate of multiple births has decreased over the past five years, while the LBW rate continues to increase (Martin et al., 2019).

Perhaps the problem is nutrition. The U.S. Department of Agriculture (Coleman-Jensen et al., 2017) reported an increase in the rate of *food insecurity* (measured by skipped meals, use of food stamps, and outright hunger) between the first seven years of the twenty-first century and the next seven, from about 11 percent to about 15 percent (see **Figure 4.5**).

The group most likely to be food insecure are young mothers. Some undereat so that their children have food—unaware that they are harming a future child. Their undernourishment adds stress to their children, who become stressed in return, affecting everyone—including weight gain of a future baby (King, 2018).

FIGURE 4.5

And Recovery? As you can see, all family types were affected by the Great Recession that began in 2007—especially single fathers, who were most likely to lose their jobs and not know how to get food stamps. But why are children of single mothers hungry more often than children of single fathers and three times as often as children of married parents? The data show correlation; researchers do not agree about causes.

● **Observation Quiz** Is the gap between single mothers and single fathers increasing or decreasing? (see answer, page 117) ↑

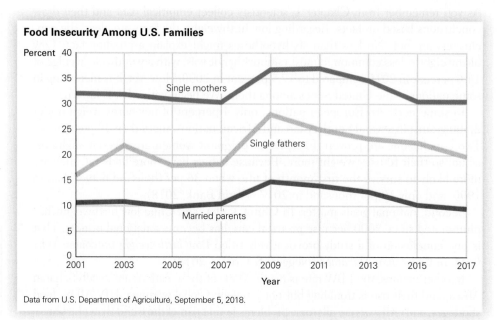

Food Insecurity Among U.S. Families

Data from U.S. Department of Agriculture, September 5, 2018.

A related possibility is lack of health care among the poorest Americans, especially young adults. Since untreated infections and chronic illness correlate with LBW, health care may be an explanation.

A fifth possible culprit is drug use, more common among young women in the United States than in most other nations (Natarajan, 2017). In the United States, the birth rate is highest among 20- to 24-year-olds, and so is drug use—in that age group, 11 percent smoke during pregnancy (Drake et al., 2018).

Looking beyond the United States, some trends are ominous. In recent years, low birthweight has decreased markedly in Asia, but smoking and drinking among young women are now increasing in those nations.

In Japan, low birthweight was slightly more than 6 percent in 2000 but almost 10 percent in 2015. Smoking and drinking are among the possible culprits, but so are low weight gain during pregnancy, increasing mercury in food, and more births after age 35 (Tamura et al., 2018).

In every hypothesis, we must distinguish correlation from causation. Since low birthweight varies from nation to nation and year to year, obviously forces beyond normal, natural variation are at work. Since LBW correlates with problems throughout life, we need to know more about causes so that we can prevent the consequences. For developing nations, the first steps are obvious—less hunger and better prenatal care. But for developed nations, more science is needed: Many hypotheses need to be explored.

WHAT HAVE YOU LEARNED?

1. How do we know that the placenta does not screen out all harmful substances?

2. What factors increase the harm from a teratogen?

3. Why is it difficult to be certain that a behavioral teratogen affected a child?

4. What is the difference between low, very low, and extremely low birthweight?

5. What are the causes and consequences of low birthweight?

6. What is puzzling about national and ethnic differences in low birthweight?

The New Family

Humans are social creatures, seeking interaction with their families and their societies. That begins before birth, and it may have a crucial impact on parents and baby after birth.

The Newborn

Before birth, the fetus already affects their mother by moving, and by hormones that affect hunger, sleep, emotions. The newborn's appearance (big hairless head, tiny feet, and so on) stirs the human heart. Both parents may be enraptured by their scraggly newborns; mothers and their partners may appreciate each other more than ever, for hormonal and practical reasons.

Newborns respond to people, even in the first hours of life (Zeifman, 2013). They listen, stare, cry, stop crying, and cuddle. In the first day or two, a professional might administer the **Brazelton Neonatal Behavioral Assessment Scale (NBAS),** which records 46 behaviors, including 20 reflexes. Watching the NBAS may help parents appreciate and understand their infant (Barlow et al., 2018).

Brazelton Neonatal Behavioral Assessment Scale (NBAS) A test that is often administered to newborns; it measures responsiveness and records 46 behaviors, including 20 reflexes.

reflex An unlearned, involuntary action or movement in response to a stimulus. A reflex occurs without conscious thought.

Technically, a **reflex** is an involuntary response to a particular stimulus. The strength of reflexes varies, depending on genes, drugs, and health. Three sets of reflexes aid survival:

- *Reflexes that maintain oxygen supply*. The *breathing reflex* begins even before the umbilical cord, with its supply of oxygen, is cut. Additional reflexes that maintain oxygen are reflexive *hiccups* and *sneezes*, as well as *thrashing* (moving the arms and legs about) to escape something that covers the face.
- *Reflexes that maintain constant body temperature*. When infants are cold, they *cry*, *shiver*, and *tuck their legs* close to their bodies. When they are hot, they try to *push away* blankets and then stay still.
- *Reflexes that manage feeding*. The *sucking reflex* causes newborns to suck whatever touches their lips—fingers, toes, blankets, and rattles, as well as nipples of various textures and shapes. In the *rooting reflex*, babies turn their mouths toward anything brushing against their cheeks—a reflexive search for a nipple—and start to suck. *Swallowing* also aids feeding, as does *crying* when the stomach is empty and *spitting up* when too much is swallowed quickly.

> **LaunchPad**
> macmillan learning
>
> **VIDEO: Newborn Reflexes** shows several infants displaying the reflexes discussed in this section.

Other reflexes promoted survival of the species in ancient times but now they signify healthy brain development. Among them:

- *Babinski reflex*. When a newborn's feet are stroked, the toes fan upward.
- *Stepping reflex*. When newborns are held upright, feet touching a flat surface, they move their legs as if to walk.
- *Swimming reflex*. When held horizontally on their stomachs, newborns stretch out their arms and legs.
- *Palmar grasping reflex*. When something touches newborns' palms, they grip it tightly.
- *Moro reflex*. When someone bangs on the table they are lying on, they fling their arms out and then bring them together on their chests, crying with wide-open eyes.

postpartum depression A new mother's feelings of inadequacy and sadness in the days and weeks after giving birth.

These 18 reflexes (in italics), and all the senses, are present at birth. If the baby tested on the Brazelton NBAS were your own, you would be proud and amazed.

New Mothers

When birth hormones decrease, between 8 and 15 percent of women experience **postpartum depression,** a sense of inadequacy and sadness (called *baby blues* in the mild version and *postpartum psychosis* in the most severe form). That 8 to 15 percent may be an underestimate: Some sources say that for low-SES and adolescent mothers the rate is 25 percent (Kozhimannil & Kim, 2014).

With postpartum depression, baby care (feeding, diapering, bathing) feels very burdensome. The newborn's cry may not compel the mother to carry and nurse her infant. Instead, the mother may have thoughts of neglect or abuse, thoughts so terrifying that she is afraid of herself. She may be overprotective, insisting that no one else care for the baby. This signifies a fearful mother, not a healthy one.

A mother who experiences postpartum depression feels terrible ("the worst time of my life," one said). She usually recovers by the time the baby is 6 months old. Unfortunately, if she does not have help in the early months, her limited interaction with the baby can affect that developing human lifelong, because those early months are crucial for brain development.

Marty Bucella/CartoonStock

"Of course I know what he wants when he cries. He wants you."

The first sign that something is amiss may be euphoria after birth. A new mother may be unable to sleep, stop talking, or eat normally. After that initial high, severe depression may set in, with long-term impact on the child. Postpartum depression may not be evident right away; anxiety and depression symptoms are stronger two or three months after birth (Kozhimannil & Kim, 2014).

Postpartum depression is not due to hormonal changes alone. From a developmental perspective, some causes (such as financial stress) predate pregnancy. Others (such as marital problems) occur during pregnancy; still others correlate with birth itself (especially if the mother is alone and expected a different birth than what actually occurred).

Finally, the baby may be disappointing. One single mother who experienced postpartum depression thought:

> My only problem in life was that I didn't have a baby. On the day I had a baby, I discovered that no, I had other problems. I hadn't any money, I was in debt, the family was fighting about the debt, it was partly my fault . . . and I started to see I wasn't such a good mother as I had thought I would be. I used to think what could be difficult? It's enough for you to love the baby and everything will be fine. This didn't happen because the baby didn't respond. I'm affectionate, I'd come and take her, hug her and the baby didn't like this. She didn't like to be hugged, she didn't like affection.

> *[O'Dougherty, 2013, p. 190]*

Successful breast-feeding mitigates maternal depression, but while most new mothers try to nurse their newborn, many quit—which increases depression. A supportive family member, friend, midwife, or lactation consultant may hasten recovery.

Fortunately, postpartum depression can be prevented, diagnosed, and treated (O'Hara & McCabe, 2013). To help with diagnosis, consider the questions of the Edinburgh Postnatal Depression scale (see **Table 4.3**).

New Fathers

Not every depressed mother reduces her baby's development. The research finds that if she manages to respond sensitively to her baby's needs, within a well-functioning, supportive family (with good emotional management, communication, and clear roles and routines) the baby may develop well (Parade et al., 2018). Fathers may be crucial in keeping the family supportive and caring for the baby.

Fathers may experience pregnancy and birth biologically, not just psychologically. Many fathers have symptoms of pregnancy and birth, including weight gain and indigestion during pregnancy and pain during labor (Leavitt, 2009).

Paternal experiences of pregnancy and birth are called **couvade,** expected in some cultures, a normal variation in many, and considered pathological in others (M. Sloan, 2009). A study in India found that couvade was common (Ganapathy, 2014). In the United States, many fathers are intensely involved during prenatal development, birth, and infancy (Brennan et al., 2007; Raeburn, 2014).

Fathers are usually the first responders when the mother experiences postpartum depression; they may be the support that the mother and baby need. But fathers are vulnerable to depression, too, with the same stresses that mothers feel (Gutierrez-Galve et al., 2015). Indeed, sometimes the father experiences more emotional problems than the mother (Bradley & Slade, 2011). Friends and relatives need to help both parents in the first weeks after birth.

Many new fathers not only feel stressed but also do not talk about their mixed emotions. In reality, both partners are affected by pregnancy and birth. Some fathers do not sleep well during pregnancy, and many fear serious problems, or even deaths,

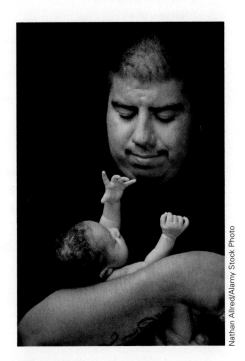

Nathan Allred/Alamy Stock Photo

Mutual Joy For thousands of years hormones and instincts have propelled fathers and babies to reach out to each other, developing lifelong connections.

couvade Symptoms of pregnancy and birth experienced by fathers.

● **Especially for Nurses in Obstetrics**
Can the father be of any practical help in the birth process? (see response, page 117)

TABLE 4.3

The Edinburgh Postnatal Depression Scale

The Edinburgh Postnatal Depression Scale asks women how they felt *in the past week*.

1. *I have been able to laugh and see the funny side of things*
   - 0-As much as I always could
   - 1-Not quite so much now
   - 2-Definitely not so much now
   - 3-Not at all

2. *I have looked forward with enjoyment to things*
   - 0- As much as I ever did
   - 1-Rather less than I used to
   - 2-Definitely less than I used to
   - 3-Hardly at all

3. *I have blamed myself unnecessarily when things went wrong*
   - 3-Yes, most of the time
   - 2-Yes, some of the time
   - 1-Not very often
   - 0-No, never

4. *I have been anxious or worried for no good reason*
   - 0-No, not at all
   - 1-Hardly ever
   - 2-Yes, sometimes
   - 3-Yes, very often

5. *I have felt scared or panicky for no very good reason.*
   - 3-Yes, quite a lot
   - 2-Yes, sometimes
   - 1-No, not much
   - 0-No, not at all

6. *Things have been getting on top of me*
   - 3-Yes, most of the time I haven't been able to cope at all
   - 2-Yes, sometimes I haven't been coping as well as usual
   - 1-No, most of the time I have coped quite well
   - 0-No, I have been coping as well as ever

7. *I have been so unhappy that I have had difficulty sleeping*
   - 3-Yes, most of the time
   - 2-Yes, sometimes
   - 1-Not very often
   - 0-No, not at all

8. *I have felt sad or miserable*
   - 3-Yes, most of the time
   - 2-Yes, quite often
   - 1-Not very often
   - 0-No, not at all

9. *I have been so unhappy that I have been crying*
   - 3-Yes, most of the time
   - 2-Yes, quite often
   - 1-Only occasionally
   - 0-No, never

10. *The thought of harming myself has occurred to me*
    - 3-Yes, quite often
    - 2- Sometimes
    - 1-Hardly ever
    - 0-Never

The total score ranges from zero to thirty. Below 9 indicates no problem; 9–12 suggests normal "baby blues"; above 12 indicates depression. High scores indicate that more intense screening is needed by a trained clinician, to discern if the new mother is truly depressed, anxious, or suicidal. (A 2 or 3 on question 10 is alarming, even if the rest are 0 or 1.)

of mothers and their babies at birth. Yet, many feel they have no right to complain. One said, "I'm always conscious that [partner]'s has it a lot worse." Another said at the birth, "I felt a bit more like a spare part . . . I just felt in the way" (quoted in Darwin et al., 2017).

Some men find relief at work, where they put the stress of fatherhood behind them or talk about their feelings with other men. An engineer said, "We probably spend half the day talking about babies and kids and that sort of thing. . . . I know that there's guys there that have had similar experiences or they know what it's like. They know how I'm feeling if I say, oh, we've had a rough night. . . . Some people have had worse experiences, so you think, what we're going through is normal" (quoted in Darwin et al., 2017).

In this discussion, we have assumed that each newborn has a mother and a father, a woman and a man who conceive a child. Biologically, that is always true: An ovum and a sperm are required to create a zygote. However, not every father is a biological father, just as not every mother is a biological mother. As one review of fathering states:

> Defining who is a father must account for the diversity of fathering that occurs. Most children have a father, whether he is currently residing with them or living separately. Some children have a single father or 2 parents who are both fathers. Children in a blended family may have both a biological nonresident father and a stepfather. Some gay men and lesbians have created families in which children have 3 or 4 adults in a parenting role, with 1 or 2 of them being fathers.

[Yogman et al., 2016, p. e2]

Whether couvade or postpartum depression occurs in nonbiological fathers is not known, but there is no doubt that children benefit from someone in a fathering role, who is involved with and committed to that child at every stage of development, from before birth through adulthood.

The crucial question regarding fathering begins with whether or not a man is responsive to the needs of family members, not with his biological connection (Brown, 2019). As you read regarding women and postpartum depression, parental emotions are not always simple and happy, evident with parents no matter what their gender or biological connection to a child.

Family Bonding

Are the first hours after birth crucial for the **parent–infant bond,** the strong, loving connection that forms when parents hold their newborn? It has been claimed that this bond develops with touch, just as sheep and goats nurture their newborns if, and only if, they immediately smell and nuzzle them (Klaus & Kennell, 1976).

However, the hypothesis that early skin-to-skin contact is *essential* for human nurturance is false (Eyer, 1992; Lamb, 1982). Substantial research on monkeys begins with *cross-fostering*, when newborns are removed from their biological mothers in the first days of life and raised by another monkey, female or male. A strong and beneficial relationship between the infant and the foster parent may develop (Suomi, 2002).

For people, bonding can begin before birth, or it may not be established until days later. It is good, but not essential, for both parents to be actively involved in pregnancy, birth, and newborn care. Encouraging parents to nurture their newborns benefits all three family members lifelong, as proven with mice, monkeys, and humans (Champagne & Curley, 2010).

The benefits of early contact are evident with **kangaroo care,** in which the newborn lies between the mother's breasts, skin-to-skin, listening to her heartbeat and feeling her body heat. A review of 124 studies confirms that kangaroo-care newborns sleep more deeply, gain weight more quickly, and spend more time alert than do infants with standard care, and they are healthier overall (Boundy et al., 2016). Father involvement, including father–infant kangaroo care, also fosters newborn's health (Feeley et al., 2013).

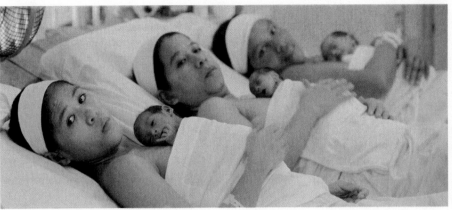

ROLEX DELA PENA/European Pressphoto Agency/MANILA/Philippines/Newscom

Better Care Kangaroo care benefits mothers, babies, and hospitals, saving space and medical costs in this ward in Manila. Kangaroo care is one reason Filipino infant mortality in 2010 was only one-fifth of what it was in 1950.

parent–infant bond The strong, loving connection that forms as parents hold, examine, and feed their newborn.

kangaroo care A form of newborn care in which mothers (and sometimes fathers) rest their babies on their naked chests, like kangaroo mothers that carry their immature newborns in a pouch on their abdomen.

Other Relatives

The birth experience may affect other family members: A birth not only changes names (a baby can make someone suddenly a brother or sister, aunt or uncle, grandmother or grandfather) but also can have a powerful effect on emotions. Time to tell more about the birth first mentioned in the beginning of this chapter.

> I held a bent right leg in place with all my strength, fighting against strong muscles to move it. A nurse strained as she held the left. The midwife commanded "push . . . push . . . push." Bethany's arm muscles bulged as she pulled a sheet tied to a mental stanchion above her. A circle of fetal skull visible, then larger, then crowning. Tissues tore, Bethany pushed once more.
>
> "Yes! Yes! Yes!" the midwife shouted. A head emerged, quickly followed by all 4,139 grams of Caleb. Apgar was a stellar 9, every other number was good.
>
> Bethany, smiling, began to nurse. Four professionals looked on, relaxed now. This is grandmother bliss. Decades of praying, studying, teaching, and mothering led me to a miracle, 6:11 A.M., my firstborn with her firstborn. Celestial music rang in my ears.
>
> The ringing grew louder, a buzzing, roaring crescendo. Bethany shimmered, overhead lights became stars, flashing bright and then dark. I was flat on the floor, looking up at four faces staring down. I had fainted.
>
> I know about birth, personally and professionally. I interpret numbers and jargon; I analyze monitors and body language; I judge doctors and nurses; I evaluate hospitals, notice stray paper on floors, hear sharp voices in corridors, see faded pictures on walls. Nothing here amiss; this hospital was excellent.

I also know Bethany: strong, healthy, drug-free. All through the night, indeed all through the nine months, I followed this pregnancy. I expected Caleb to be well-formed and Bethany to be okay. I was relieved and happy—not surprised—when my almost perfect daughter began nursing my quite perfect grandson.

Then why did I faint when all the drama was over? Indeed, why faint at all?

[Berger, 2019]

When I analyzed it, I realized that the hormones and evolutionary forces that foster bonding between parents and newborns, as well as depression in parents, affect grandmothers as well. All that aids development of everyone.

As we will see in later chapters, the relationships among family members develop over decades, not merely hours. Birth is one step of a lifelong journey.

WHAT HAVE YOU LEARNED?

1. How are newborns socially interactive?

2. What causes postpartum depression?

3. How are fathers affected by birth?

4. Why is kangaroo care beneficial?

5. When does the parent–infant bond form?

SUMMARY

Prenatal Development

1. In the first two weeks of prenatal growth (the germinal period), the single-celled zygote multiplies into more than 100 cells that will eventually form both the placenta and the embryo. About half the time, the growing organism fails to implant in the uterus, ending pregnancy.

2. The embryonic period (the third week through the eighth week after conception) begins with the primitive streak, the start of the central nervous system. The future heart begins to pulsate, and eyes, ears, nose, mouth, and brain form. By the eighth week, the first traces of all of the basic organs and features are present.

3. Early in the fetal period (ninth week until birth), male and female organs form, and hormones start to shape the brain. At 22 weeks, the brain can regulate basic body functions, and viability is possible but unlikely. Babies born before the 24th week are at high risk of death or disability.

4. In the final three months, the average fetus gains approximately 4½ pounds (2,040 grams), weighing 7½ pounds (3,400 grams) at birth. Maturation of brain, lungs, and heart ensures survival of more than 99 percent of all full-term babies.

Birth

5. Ideally, hormones (oxytocin) start labor and birth approximately 38 weeks after conception. The Apgar scale provides a quick evaluation of the newborn's health.

6. Medical assistance speeds contractions, dulls pain, and saves lives. However, many interventions, including about half of cesarean sections, have been criticized.

Problems and Solutions

7. Every birth complication, such as an unusually long and stressful labor that includes anoxia, has a cascade of causes. Long-term handicaps are not inevitable.

8. Some teratogens cause physical impairment. Others, called behavioral teratogens, harm the brain and therefore impair cognitive abilities and affect personality.

9. Whether a teratogen harms an embryo or fetus depends on timing, dose, and genes. Family members affect the pregnant woman's health.

10. Low birthweight (under 5½ pounds, or 2,500 grams) may arise from early or multiple births, placental problems, maternal illness, malnutrition, smoking, drinking, illicit drug use, and age.

11. Underweight babies experience medical difficulties and psychological problems for many years. Babies that are small for gestational age (SGA) are especially vulnerable.

The New Family

12. Newborns are primed for social interaction. The Brazelton Neonatal Behavioral Assessment Scale measures 46 newborn behaviors, 20 of which are reflexes. Some reflexes help survival; others measure brain maturation.

13. Fathers can be supportive during pregnancy as well as helpful in birth. Paternal support correlates with shorter labor and fewer complications. Some fathers become very involved with the pregnancy and birth, experiencing couvade.

14. Many women feel unhappy, incompetent, or unwell after giving birth. Postpartum depression gradually disappears with appropriate help; fathers can be crucial in baby care, or they can experience depression themselves.

15. Kangaroo care benefits all babies, but especially those who are vulnerable. The parent–infant bond depends on many factors in addition to birth practices.

KEY TERMS

germinal period (p. 91)
embryonic period (p. 91)
fetal period (p. 91)
embryo (p. 91)
cephalocaudal (p. 92)
proximodistal (p. 92)
fetus (p. 92)
age of viability (p. 93)
doula (p. 97)
Apgar scale (p. 98)

cesarean section (c-section) (p. 100)
cerebral palsy (p. 102)
anoxia (p. 102)
teratogen (p. 102)
behavioral teratogens (p. 102)
threshold effect (p. 104)
fetal alcohol syndrome (FAS) (p. 105)
false positive (p. 106)

low birthweight (LBW) (p. 107)
very low birthweight (VLBW) (p. 107)
extremely low birthweight (ELBW) (p. 107)
preterm (p. 107)
small for gestational age (SGA) (p. 108)
immigrant paradox (p. 108)

Brazelton Neonatal Behavioral Assessment Scale (NBAS) (p. 111)
reflex (p. 112)
postpartum depression (p. 112)
couvade (p. 113)
parent–infant bond (p. 115)
kangaroo care (p. 115)

APPLICATIONS

1. Go to a nearby greeting-card store and analyze the cards about pregnancy and birth. Do you see any cultural attitudes (e.g., variations depending on the sex of the newborn or of the parent)? If possible, compare those cards with cards from a store that caters to another economic or ethnic group.

2. Interview three mothers of varied backgrounds about their birth experiences. Make your interviews open-ended—let the mothers choose what to tell you, as long as they give at least a 10-minute description. Then compare and contrast the three accounts, noting especially any influences of culture, personality, circumstances, and cohort.

3. People sometimes wonder how any pregnant woman could jeopardize the health of her fetus. Consider your own health-related behavior in the past month—exercise, sleep, nutrition, drug use, medical and dental care, disease avoidance, and so on. Would you change your behavior if you were pregnant? Would it make a difference if you, your family, and your partner did not want a baby?

Especially For ANSWERS

Response for Conservatives and Liberals (from p. 100): Yes, some people are much more likely to want nature to take its course. However, personal experience often trumps political attitudes about birth and death; several of those who advocate hospital births are also in favor of spending one's final days at home.

Response for Future Doctors and Nurses (from p. 105) This is not easy, as people tend to consider the worst case more readily than the best case. Useful might be a graph that depicts the odds, or comparisons with something nonmedical. For example, chance that your birthday is tomorrow, 1:365. Chance that your fetus has Down syndrome, 1:1000.

Response for Judges and Juries (from p. 110): Some laws punish women who jeopardize the health of their fetuses, but a developmental view would consider the micro-, exo-, and macrosystems.

Response for Nurses in Obstetrics (from p. 113): Usually not, unless they are experienced, well taught, or have expert guidance. But their presence provides emotional support for the woman, which makes the birth process easier and healthier for mother and baby.

Observation Quiz ANSWERS

Answer to Observation Quiz (from p. 103): Brain. Legs.

Answer to Observation Quiz (from p. 110): Decreasing. The reason may be related to greater gender equity. Note, however, that the recession impacted fathers dramatically, as many wage-earners lost their jobs and did not immediately know how to get public or private help in feeding their families.

APPLICATION TO DEVELOPING LIVES PARENTING SIMULATION BABIES AND TODDLERS

As you progress through the Babies and Toddlers simulation module, how you decide the following will impact the biosocial, cognitive, and psychosocial development of your child.

	Biosocial	Cognitive	Psychosocial
	• Will you vaccinate your baby? • Will you breast-feed your baby? If so, for how long? • What kind of foods will you feed your baby during the first year? • How will you encourage motor skill development? • How do your baby's height and weight compare to national norms?	• What activities will you expose your baby to (music class, reading, educational videos)? • What activities will you do to promote language development? • Which of Piaget's stages of cognitive development is your child in?	• How will you soothe your baby when he or she is crying? • Can you identify your baby's temperament style? • Can you identify your baby's attachment style? • What kind of discipline will you use with your child?

PART II

The First Two Years

Adults don't change much in a year or two. They might have longer, grayer, or thinner hair; they might gain or lose weight; they might learn something new. But if you saw friends you hadn't seen for two years, you'd recognize them immediately.

Imagine caring for your sister's newborn every day for two months. You would learn everything about that baby—how to dress, when to play, what to feed, where to sleep. Toward the end of the two months, the baby would recognize you, smiling broadly, nestling comfortably in your arms, responding with happy noises when you spoke. Then imagine you had to live in another country for two years.

When you returned, your sister might ask you to pick up the toddler at the day-care center. You would need to ask the teacher which child to take, because several of them could be your sister's child. In those two years, weight quadruples, height increases by a foot, hair grows. Emotions change, too—less crying, new fear—including fear of you, now a stranger.

Two years are less than 3 percent of the average human life. However, in those 24 months, people reach half their adult height, learn to run, climb and talk in sentences, and express every emotion—not just joy and fear but perhaps jealousy and shame. Invisible growth of the brain is even more awesome; plasticity is extraordinary during infancy, enabling all these changes and more. The next three chapters describe this transformation. ●●

Left: David A Land/Blend Images/Media Bakery
Right: JBryson/iStock/Getty Images

The First Two Years: Biosocial Development

What Will You Know?

1. What part of an infant grows most in the first two years?
2. Are babies essentially blind and deaf at birth?
3. What happens if infants do not get their vaccinations?

Our first child, Bethany, was born when I was in graduate school. At 14 months, the pediatrician said she was growing well. But my husband was worried; she had not yet taken her first step. I told him that genes determine age of walking: I had read that babies in Paris are late to walk, and my grandmother was French.

To our relief, Bethany soon began to walk. A few years later, she was the fastest runner in kindergarten. Our next two children, Rachel and Elissa, were also slow to walk, and my students with Guatemalan and Ghanaian ancestors bragged about their infants who walked before a year; those from China and France were quiet. Genetic, I thought.

Fourteen years after Bethany, Sarah was born. I could finally afford a full-time caregiver, Mrs. Todd. She thought Sarah was the most advanced baby she had ever known, except for her own daughter, Gillian.

"She'll be walking by a year," Mrs. Todd told me. "Gillian walked at 10 months."

"We'll see," I graciously replied.

I underestimated Mrs. Todd. She bounced my delighted baby on her lap, day after day, and spent hours giving her "walking practice." Sarah took her first step at 12 months—late for a Todd, early for a Berger, and a humbling lesson for me.

As a scientist, I know that a single case proves nothing. My genetic explanation might be not valid, especially since Sarah shares only half her genes with Bethany and since my daughters are only one-eighth French, a fraction I had conveniently ignored.

Nonetheless, decades of research since Bethany was born confirm that caretakers influence every aspect of biosocial growth. You will soon read many examples of caregiving that enables babies to grow, move, see, and learn. Nurture is at least as important as nature.

Genes provide the scaffold, but daily circumstances shape and guide infant development. Bethany completed a marathon; her grandmother could not walk independently in her final years. Both outcomes were the result of their early experiences: Bethany admired her older cousins who won trophies in races, and Grandma had polio as a child, which weakened her legs in late adulthood.

Body Changes

In infancy, growth is so rapid and the consequences of neglect so severe that gains are closely monitored. Medical checkups, including measurement of height, weight, and head circumference, reveal whether an infant is progressing as expected—or not.

Body Size

Newborns lose several ounces in the first three days and then gain an ounce a day for months. Birthweight doubles by 4 months and triples by a year, so the average 7-pound newborn might be 21 pounds at 12 months (9,525 grams, up from 3,175 grams at birth).

Variation is substantial, depending not only on genes and nutrition but also on birthweight—small babies may double their weight in two months and quadruple by age 1, a phenomenon called *catch-up growth*. Height also increases rapidly, with variation: A typical newborn grows 10 inches (25 centimeters) by age 1, measuring about 30 inches (76 centimeters).

Physical growth then slows, but not by much. Most 24-month-olds weigh about 28 pounds (13 kilograms) and have added another 4 inches (10 centimeters) in the previous year. Typically, 2-year-olds are half their adult height and about one-fifth their adult weight (see **Figure 5.1**).

Compared to Themselves

Growth is often expressed in a **percentile**, indicating how one person compares to another. Thus, a 12-month-old's weight at the 30th percentile means that 29 percent of 12-month-old babies weigh less and 69 percent weigh more. Healthy children vary in size, so any percentile between 10 and 90 is okay, as long as the percentile is close to the previous one for that individual.

When an infant's percentile moves markedly up or down, that could signify trouble. A notable drop, say from the 50th to the 20th percentile, suggests poor nutrition. A sudden increase, perhaps from 30th to 60th signifies overfeeding, especially if their height remains at the 30th percentile.

Parents were once blamed if their infant was not gaining weight. It was thought that parents made feeding stressful, leading to *failure to thrive*. Currently, however,

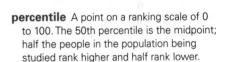

VIDEO: Physical Development in Infancy and Toddlerhood offers a quick review of the physical changes that occur during a child's first two years.

percentile A point on a ranking scale of 0 to 100. The 50th percentile is the midpoint; half the people in the population being studied rank higher and half rank lower.

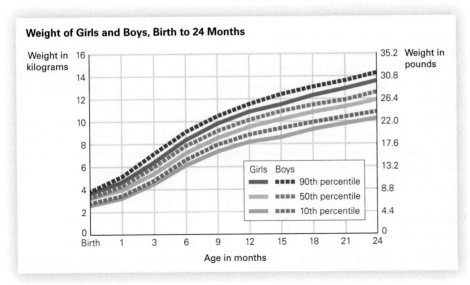

FIGURE 5.1

Averages and Individuals Norms and percentiles are useful—most 1-month-old girls who weigh 10 pounds should be at least 25 pounds by age 2. But although females weigh less than males on average lifelong, it is obvious that individuals do not always follow the norms. Do you know a 200-pound woman married to a 150-pound man?

REM (rapid eye movement) sleep
A stage of sleep characterized by flickering eyes behind closed lids, dreaming, and rapid brain waves.

Danger Here Not with the infant (although those pillows should be removed), but for the family. It is hard to maintain a happy marriage if the parents are exhausted.

co-sleeping A custom in which parents and their children (usually infants) sleep together in the same room.

bed-sharing When two or more people sleep in the same bed.

we know that many organic conditions, such as allergies, the microbiome, and liver problems, combine with nonorganic factors, in the caregiver or in infant feeding practices (Lazzara et al., 2019).

Similarly, obesity is now thought to be cultural and organic, as well as familial. Both conditions may begin during infancy; both conditions require diagnosis and intervention. Blaming the caregivers is neither accurate nor helpful.

Sleep

Throughout childhood, regular and ample sleep correlates with normal brain maturation, learning, emotional regulation, academic success, and psychological adjustment (Maski & Kothare, 2013). Sleep deprivation can cause poor health, and vice versa. As with many health habits, sleep patterns begin in the first year.

Patterns of Infant Sleep

Newborns spend most of their time sleeping, about 15 to 17 hours a day. With maturity, hours of sleep decrease rapidly and the patterns change, with longer night sleep and less daytime napping. For example, one study that included infants of many ethnic and economic backgrounds found that the average 1-month-old slept six hours during the day, and nine hours at night, waking up two to three times. By 6 months, the average was three hours of daytime sleep and ten hours at night, waking up once or twice (Ash et al., 2019).

Remember that norms are simply averages: Some infants sleep substantially more, and others less. Babies born preterm seem to be dozing most of the time, but that may be partly caused by the constant bright lights and frequent feeding of the traditional neonatal intensive care unit (NICU). When they come home, preterm babies usually adjust to a day/night schedule (Bueno & Menna-Barreto, 2016).

Newborns dream a lot: About half their sleep is **REM (rapid eye movement) sleep.** REM sleep declines over the early weeks, as does "transitional sleep," the dozing, half-awake stage. At 3 or 4 months, quiet sleep (also called *slow-wave sleep*) increases markedly.

Sleep varies not only because of biology (maturation and genes), but also because of caregiving. Sleep environment matters, since at every age, noise and light disrupt sleep. Some infants sleep where they hear others arguing, traffic noises, or television. When they are very tired, they sleep despite such disruptions, but they sleep fewer hours overall when the environment is stimulating. Pain also keeps them awake: Colicky babies sleep less, and their parents sleep less also.

Insufficient sleep becomes a problem for parents as well as for infants, because "[p]arents are rarely well-prepared for the degree of sleep disruption a newborn infant engenders, and many have unrealistic expectations about the first few postnatal months." As a result, many parents become "desperate" and institute patterns they may later regret (C. Russell et al., 2013, p. 68).

Where Should Babies Sleep?

Traditionally, most middle-class North American infants slept in cribs in their own rooms; it was feared that they would be traumatized if their parents had sex in the same room. By contrast, most infants in Asia, Africa, and Latin America slept near their parents, a practice called **co-sleeping,** and sometimes in their parents' bed, called **bed-sharing.** In those cultures, nighttime parent–child separation was considered cruel.

Adey Bryant/Cartoonstock

Asian and African mothers still worry more about separation, whereas European and North American mothers worry more about privacy. A 19-nation survey found that parents act on these fears: The extremes were 82 percent of Vietnamese babies co-sleeping compared with 6 percent in New Zealand (Mindell et al., 2010) (see **Figure 5.2**). Although co-sleeping is more common in nations with high poverty rates, culture matters, too: In Japan—one of the wealthiest nations of the world—mothers often sleep with their babies.

In the United States, bed-sharing doubled from 1993 to 2010 (6.5 percent to 13.5 percent) (Colson et al., 2013). In Canada, younger women are more likely to bed-share: A large, nationwide study found that more than 40 percent of new mothers under age 25 sleep with their infants (Gilmour et al., 2019).

The infant's feeding patterns have an effect: Bed-sharing is more common among breast-feeding mothers. A study in Sweden of preterm infants (who need to be fed every two or three hours) found that most slept with their mothers—especially if the mother had trouble getting back to sleep if she got up to feed her infant (Blomqvist et al., 2017).

The argument for co-sleeping is that the parents can quickly respond to a hungry or frightened baby. Moreover, a close physical connection at night fosters bonding. A popular book on infant care advocates "attachment parenting," advising keeping the infant nearby day and night (Sears & Sears, 2001).

Responsive attachment correlates with co-sleeping (Kim et al., 2017). However, bed-sharing doubles the risk of *sudden infant death syndrome (SIDS)*, when a baby dies unexpectedly while asleep (Vennemann et al., 2012).

For that reason, some pediatricians advise against co-sleeping. That advice may be ignored by tired mothers, especially in cultures where bed-sharing is the norm. Canadian mothers are told not to share a bed with their infant, but two-thirds of those with Asian or Caribbean heritage do so (Gilmour et al., 2019).

Consequently, many experts seek ways to safeguard the practice (Ball & Volpe, 2013). Their advice includes *never* sleeping beside a baby if the parent has been drinking, and *never* using a soft comforter, pillow, or mattress near a sleeping infant.

Developmentalists remind parents that babies learn from experience. If they become accustomed to bed-sharing, they may climb into their parents' bed long past infancy. Parents might lose sleep for years because they wanted more sleep when their babies were small. Sleeping alone may encourage independence—a trait appreciated in some cultures, abhorred in others. The entire social context matters, which explains why some studies find benefits from co-sleeping, and other studies find danger (Baddock et al., 2019).

Brain Development

From two weeks after conception to two years after birth, the brain grows more rapidly than any other organ. Brain size about 25 percent of adult weight at birth and 75 percent at age 2. Prenatal and postnatal brain growth affects later cognition (Gilles & Nelson, 2012). If teething or a stuffed-up nose temporarily slows eating, body weight is affected before brain weight, a phenomenon called **head-sparing.** That term expresses well what nature does—protect the brain.

Many other terms in neuroscience are not as self-explanatory, but they are useful to understand the brain. Accordingly, they are explained in the following.

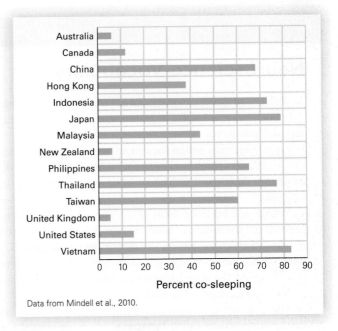

Percent co-sleeping

Data from Mindell et al., 2010.

FIGURE 5.2

Awake at Night Why the disparity between Asian and non-Asian rates of co-sleeping? It may be that Western parents use a variety of gadgets and objects—monitors, night-lights, pacifiers, cuddle cloths, sound machines—to accomplish some of what Asian parents do by having their infant next to them.

Especially for New Parents You are aware of cultural differences in sleeping practices, which raises a very practical issue: Should your newborn sleep in bed with you? (see response, page 145)

head-sparing A biological mechanism that protects the brain when malnutrition disrupts body growth. The brain is the last part of the body to be damaged by malnutrition.

INSIDE THE BRAIN

Neuroscience Vocabulary

To understand the impressive brain growth that occurs throughout childhood, it is helpful to know some basic terms of neurological development.

Communication within the *central nervous system (CNS)*—the brain and spinal cord—begins with nerve cells, called **neurons.** At birth, the human brain has about 86 billion neurons.

Within and between areas of the central nervous system, neurons are connected to other neurons by intricate networks of nerve fibers called **axons** and **dendrites** (see **Figure 5.3**). Each neuron has a single axon and numerous dendrites, which spread out like the branches of a tree. Most of the brain growth in infancy is increases in dendrites.

The axon of one neuron meets the dendrites of other neurons at intersections called **synapses,** which are critical communication links within the brain. Infancy is "characterized by overproduction of synapses followed by a period of gradual pruning" (Bernier et al., 2016, p. 1159). Synapse formation and demise is remarkably plastic, heavily dependent on experience.

Neurons communicate by *firing,* or sending electrochemical impulses through their axons to synapses to be picked up by the dendrites of other neurons. The dendrites bring the message to the cell bodies of their neurons, which, in turn, may fire, conveying messages via their axons to the dendrites of other neurons. Some firing is involuntary—such as the reflexes cited in Chapter 4. Most infant brain development requires new connections between one neuron and another, as dendrites grow (Gao et al., 2017).

Axons and dendrites do not touch at synapses. Instead, the electrical impulses in axons typically cause the release of **neurotransmitters,** which stimulate other neurons. There are about a hundred neurotransmitters.

Neurotransmitters carry information from the axon of the sending neuron, across a pathway called the *synaptic gap,* to the dendrites of the receiving neuron, a process speeded up by **myelin,** a coating on the outside of the axon. Myelin increases over childhood—lack of it is one reason infants are slow to react to something pleasurable or painful. [**Developmental link**: Myelination is discussed in Chapter 8.]

Some neurons are deep inside the brain in a region called the *hindbrain,* which controls automatic responses such as heartbeat, breathing, temperature, and arousal. Others are in the *midbrain,* in areas that affect emotions and memory. And in humans most neurons (about 70 percent) are in the *forebrain,* especially the **cortex,** the brain's six outer layers (sometimes called the *neocortex*). Most thinking, feeling, and sensing occur in the cortex (Johnson & de Haan, 2015; Kolb & Whishaw, 2015).

The forebrain has two halves and four lobes, which are general regions, each containing many parts. No important human activity is exclusively left- or right-brain, or in one lobe or another. Although each lobe and hemisphere has specialized functions, thousands of connections transmit information among the parts. The specialization of various parts is the result of various constraints and experiences, not foreordained by genes (Johnson & de Haan, 2015).

The back of the forebrain is the *occipital lobe,* where vision is located; the sides of the brain are the *temporal lobes,* for hearing; the top is the *parietal lobe,* which includes smell, touch, and spatial understanding; and the front is the *frontal lobe,* which enables people to plan, imagine, coordinate, decide, and create. Humans have a much larger frontal cortex relative to body size than any other animal.

The very front of the frontal lobe is called the **prefrontal cortex.** It is not, as once thought, "functionally silent during most of infancy" (Grossmann, 2013, p. 303), although the prefrontal cortex is very immature at birth. [**Developmental Link**: Major discussion of adolescent growth of the prefrontal cortex is in Chapter 14.]

Pleasure and pain may arise from the **limbic system,** a cluster of brain areas deep in the forebrain that is heavily involved in emotions and motivation. Two crucial parts of the limbic system are the amygdala and the hippocampus.

neuron One of billions of nerve cells in the central nervous system, especially in the brain.

axon A fiber that extends from a neuron and transmits electrochemical impulses from that neuron to the dendrites of other neurons.

dendrite A fiber that extends from a neuron and receives electrochemical impulses transmitted from other neurons via their axons.

synapse The intersection between the axon of one neuron and the dendrites of other neurons.

neurotransmitter A brain chemical that carries information from the axon of a sending neuron to the dendrites of a receiving neuron.

myelin The coating on axons that speeds transmission of signals from one neuron to another.

cortex The outer layers of the brain in humans and other mammals. Most thinking, feeling, and sensing involves the cortex.

prefrontal cortex The area of the cortex at the very front of the brain that specializes in anticipation, planning, and impulse control.

limbic system The parts of the brain that interact to produce emotions, including the amygdala, the hypothalamus, and the hippocampus. Many other parts of the brain also are involved with emotions.

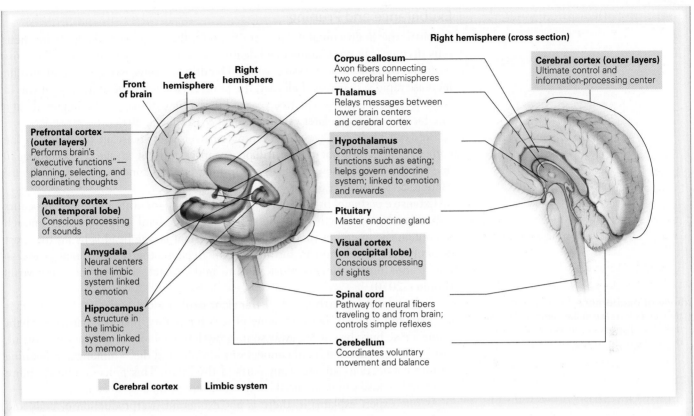

FIGURE 5.3

Connections A few of the hundreds of named parts of the brain are shown here. Although each area has particular functions, the entire brain is interconnected. The processing of emotions, for example, occurs primarily in the limbic system, where many brain areas are involved, including the amygdala, hippocampus, and hypothalamus.

The **amygdala** is a tiny structure, about the same shape and size as an almond. It registers strong emotions, both positive and negative, especially fear. The amygdala is present in infancy, but growth depends partly on early experience. Increased amygdala activity may cause terrifying nightmares or sudden terrors.

Another structure in the emotional network is the **hippocampus,** located next to the amygdala. A central processor of memory, especially memory for locations, the hippocampus responds to the amygdala by summoning memory. Some places feel comforting (perhaps a childhood room) and others evoke fear (perhaps a doctor's office), even when the experiences that originated those emotions are long gone.

Sometimes considered part of the limbic system is the **hypothalamus,** which responds to signals from the amygdala and to memories from the hippocampus by producing hormones, especially **cortisol,** a hormone that increases with stress. Another nearby brain structure, the **pituitary,** responds to the hypothalamus by sending out hormones to various body parts.

Brain research is one area of international collaboration. For example, a 5-billion-dollar, 12-year project in the United States called BRAIN (Brain Research Through Advancing Innovative Neurotechnologies) began in 2014 and is developing new tools (Huang & Luo, 2015). Given new methods and thousands of neuroscientists worldwide, the names and functions of various

parts of the brain may be described differently from one source to another.

The descriptions here are only a beginning. From a developmental perspective, what is crucial to know is that all human thoughts and actions originate in the complexity of the brain, and that understanding the brain adds insight to our effort to understand how humans live their lives. Extensive neurological plasticity is evident as all these parts of the infant brain adapt to experience (Gao et al., 2017).

amygdala A tiny brain structure that registers emotions, particularly fear and anxiety.

hippocampus A brain structure that is a central processor of memory, especially memory for locations.

hypothalamus A brain area that responds to the amygdala and the hippocampus to produce hormones that activate other parts of the brain and body.

cortisol The primary stress hormone; fluctuations in the body's cortisol level affect human emotions.

pituitary A gland in the brain that responds to a signal from the hypothalamus by producing many hormones, including those that regulate growth and that control other glands, among them the adrenal and sex glands.

Exuberance and Pruning

At birth, the brain contains far more neurons than a person needs. Some neurons disappear in programmed cell death, and some new ones develop. That loss is counterbalanced by massive gains: Dendrites, axons, synapses, and myelin increase rapidly. Because of all that, the brain is twice as large at age 1 as it was at birth (Gao et al., 2017). The first year of life is a time of massive brain growth, at a rate far more than any later year (Vannucci & Vannucci, 2019).

Between birth and age 2, an estimated fivefold increase in dendrites in the cortex occurs, with about 100 trillion synapses present at age 2. According to one expert, "40,000 new synapses are formed every second in the infant's brain" (Schore & McIntosh, 2011, p. 502).

Extensive *postnatal* brain growth is highly unusual for other mammals. It occurs in humans because birth would be impossible if the fetal head were large enough to contain the brain networks humans need. (As it is, the head is by far the most difficult part of the human birth process). Because the human brain grows so much after birth, humans must nurture and protect children for many years (Konner, 2010).

Early dendrite growth is called **transient exuberance:** *exuberant* because it is so rapid and *transient* because some of it is temporary. Just as a gardener might prune a rose bush by cutting away some growth to enable more, or more beautiful, roses to bloom, unused brain connections atrophy and die. Thinking and learning require connections among many parts of the brain. This process is made more efficient because some potential connections are pruned (Gao et al., 2017).

As one expert explains it, there is an "exuberant overproduction of cells and connections followed by a several year long sculpting of pathways by massive elimination" (Insel, 2014, p. 1727). Notice the word *sculpting*, as if a gifted artist created an intricate sculpture from raw marble or wood. Human infants are gifted artists, developing their brains to adjust to whatever family, culture, or society they are born into.

For example, to understand any sentence in this text, you need to know the letters, the words, the surrounding text, the ideas they convey, and how they relate to your other thoughts and experiences. Those connections are essential for your comprehension, which differs from other people whose infant brains developed in homes unlike yours. Thus, your brain automatically interprets these roman letters, and, for most of you, is befuddled when viewing Arabic, Cyrillic, or Chinese.

Further evidence of the benefit of cell death comes from a sad symptom of fragile X syndrome (described in Chapter 3), "a persistent failure of normal synapse pruning" (Irwin et al., 2002, p. 194). Without **pruning,** the dendrites of children with fragile X are too dense and long, making thinking difficult. Similar problems occur for children with autism spectrum disorder: Their brains are unusually large and full, making communication between neurons less efficient and some sounds and sights overwhelming (Lewis et al., 2013).

Thus, pruning is essential. As brains mature, the process of extending and eliminating dendrites is exquisitely attuned to experience, as the appropriate links in the brain are established, protected, and strengthened (Gao et al., 2017). As with the rose bush, pruning needs to be done carefully, allowing further growth.

Necessary and Possible Experiences

A scientist named William Greenough identified two experience-related aspects of brain development (Greenough et al., 1987). Understanding

transient exuberance The great but temporary increase in the number of dendrites that develop in an infant's brain during the first two years of life.

pruning When applied to brain development, the process by which unused connections in the brain atrophy and die.

Connecting The color staining on this photo makes it obvious that the two neurons (stained chartreuse) grow dendrites to connect with each other. This tangle is repeated thousands of times in every human brain. Throughout life, those fragile dendrites will grow or disappear as the person continues thinking.

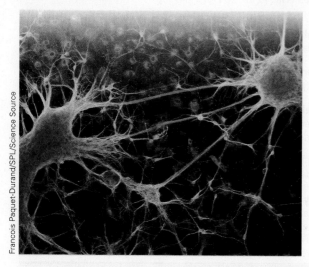

Francois Paquet-Durand/SPL/Science Source

these helps adults avoid the difference/deficit problem explained in Chapter 1, yet still provide the experiences every baby needs.

- **Experience-expectant growth.** Certain functions require basic experiences in order to develop, just as a tree requires water. Those experiences are part of almost every infant's life, and thus, almost all human brains grow as their genes direct. Brains *expect* such experiences; development suffers without them.
- **Experience-dependent growth.** Human brains are quite plastic, again as explained in Chapter 1. Particular brain connections grow as specific experiences occur. These experiences are not essential: They happen in some families and cultures but not in others.

The basic, expected experiences *must* happen for normal brain maturation to occur, and they almost always do. For example, in deserts and in the Arctic, on isolated farms and in crowded cities, almost all babies have things to see, objects to manipulate, and people to love them. Babies everywhere welcome such experiences: They look around, they grab for objects, they smile at people. As a result, babies' brains develop. Without such expected experiences, brains wither.

In contrast, dependent experiences *might* happen; because of them, one brain differs from another, even though both brains are developing normally. Experiences vary, such as the language they hear, the faces they see, the emotions their caregivers express, and, as you just read, where they sleep. *Depending* on those particulars, infant brains are structured and connected one way or another; some dendrites grow and some neurons thrive while others die (Stiles & Jernigan, 2010).

Consequently, experience-expectant events make all people similar, yet everyone is unique because each undergoes particular experience-dependent events.

The distinction between essential and variable input to the brain's networks can be made for all creatures. Some of the most persuasive research has been done with birds. All male songbirds have a brain region dedicated to listening and reproducing sounds (experience-expectant), but birds of the same species produce slightly different songs (experience-dependent) depending on where they live (Konner, 2010).

Birds inherit genes that produce the neurons they need, perhaps dedicated to learning new songs (canaries) or to finding hidden seeds (chickadees). That is experience-expectant: Songs and seeds are essential for those species. Then, depending on their ecological niche, birds *depend* on specific experiences with learning songs or finding seeds.

Indeed, human babies learn language in much the same way as birds sing (Prather et al., 2017). Unless something is seriously wrong, adults talk to babies whose brains *expect* language. But the particular language *depends* on the specific culture.

Harming the Infant Body and Brain

Thus far, we have focused on the many variations that families offer babies. Most infants develop well. Feeding and health care vary, but every family tries to ensure that their children survive in good health, and thrive within their culture.

For brain development, it does not matter whether a person learns French or Farsi, or expresses emotions dramatically or subtly (e.g., throwing themselves to

Face Lit Up, Brain Too Thanks to scientists at the University of Washington, this young boy enjoys the electroencephalogram (EEG) of his brain activity. Such research has found that babies respond to language long before they speak. Experiences of all sorts connect neurons and grow dendrites.

experience-expectant Brain functions that require certain basic common experiences (which an infant can be expected to have) in order to develop normally.

experience-dependent Brain functions that depend on particular, variable experiences and therefore may or may not develop in a particular infant.

Especially for Parents of Grown Children Suppose you realize that you seldom talked to your children until they talked to you and that you often put them in cribs and playpens. Did you limit their brain growth and their sensory capacity? (see response, page 145)

Aaron MCcoy/Photolibrary/Getty Images

shaken baby syndrome A life-threatening injury that occurs when an infant is forcefully shaken back and forth, a motion that ruptures blood vessels in the brain and breaks neural connections.

the floor or merely pursing their lips, a cultural difference). However, infant brains do not grow normally if they lack basic, expected experiences.

Necessary Stimulation

To begin with, infants need stimulation. Some adults imagine that babies need quiet, perhaps in a room painted one neutral color. That is a mistake. Babies need sights and sounds, emotional expression, and social interaction that encourages movement (arm waving, then crawling, grabbing, and walking).

Severe lack of stimulation stunts the brain. As one review explains, "enrichment and deprivation studies provide powerful evidence of . . . widespread effects of experience on the complexity and function of the developing system" of the brain (Stiles & Jernigan, 2010, p. 345).

Proof came first from rodents! Some "deprived" rats (raised alone in small, barren cages) were compared with "enriched" rats (raised in large cages with toys and other rats). At autopsy, brains of the enriched rats were larger, with more dendrites (Diamond, 1988; Greenough & Volkmar, 1973).

Subsequent research with other mammals confirms that isolation and sensory deprivation stunt development. That is now sadly evident in longitudinal studies of human orphans from Romania, described in Chapter 7.

Stress and the Brain

Some infants experience the opposite problem, too much of the wrong kind of stimulation (Bick & Nelson, 2016). If the brain produces an overabundance of *cortisol* (the stress hormone) early in life, that derails the connections from parts of the brain, causing odd responses to stress lifelong. Years later, that child or adult may be hypervigilant (always on the alert) or emotionally flat (never happy or sad).

Note that the brain is responding to fear, not directly to physical pain. Of course infants should never be hit, and occasional pain or stress—routine inoculations, temporary hunger, an unwanted bath or diaper change—is part of normal infant life. If a reassuring caregiver communicates comfort, the infant can handle the stress. However, intense and frequent stress can flood the brain with cortisol, causing damage to later cognition. For example, if infants witness a violent fight between their parents, they may suffer lasting harm, evident in brain and behavior (Mueller & Tronick, 2019).

This distinction is important for caregivers to know. All babies cry. Because the prefrontal cortex has not yet developed, telling infants to stop crying is pointless because they cannot *decide* to stop crying. Some adults yell at babies (which may terrify the baby) or even worse, shake the infant. That makes blood vessels in the brain rupture and neural connections break, causing **shaken baby syndrome,** an example of *abusive head trauma* (Christian & Block, 2009). Death is the worst consequence; lifelong intellectual impairment is the more likely one.

Not every infant who has neurological symptoms of head trauma is the victim of abuse: Legal experts worry about false accusations (Byard, 2014). Nonetheless, infants are vulnerable, so the response to a screaming, frustrating baby should be to comfort or walk away, never to shake, yell, or hit.

Lest you cannot imagine the frustration that some parents feel when their baby cries, consider what one mother in Sweden said about her colicky baby, now age 4 and much beloved.

> There were moments when, both me and my husband . . . when she was apoplectic and howling so much that I almost got this thought, 'now I'll take a pillow and put over her face just until she quietens down, until the screaming stops.'

[quoted in Landgren et al., 2012, p. 55]

Discoveries about early development have many other implications. First, since early growth is so rapid, well-baby checkups are needed often, in order to spot, and treat, any problems. Sight and hearing are springboards for growth, so sensory impairments should be remedied.

Fortunately, one characteristic of infants is called **self-righting,** an inborn drive to compensate and overcome problems. Infants with few toys develop their brains by using sticks, or empty boxes, or whatever is available. Malnourished newborns have catch-up growth, so a 5-pound newborn may gain weight faster than an 8-pound one. Plasticity is apparent from the beginning of life (Tomalski & Johnson, 2010).

self-righting The inborn drive to remedy a developmental deficit; literally, to return to sitting or standing upright after being tipped over. People of all ages have self-righting impulses, for emotional as well as physical imbalance.

WHAT HAVE YOU LEARNED?

1. What facts indicate that infants grow rapidly in the first year?

2. Why are pediatricians not troubled when an infant is consistently small, say at the 20th percentile in height and weight?

3. How do sleep patterns change from birth to 18 months?

4. What are the arguments for and against bed-sharing?

5. How can pruning increase brain potential?

6. How do experience-expectant and experience-dependent growth differ?

7. What are the effects of stress and social deprivation on early development?

8. What should caregivers remember about brain development when an infant cries?

Perceiving and Moving

Young human infants combine immature motor abilities and acute sensory awareness (Konner, 2010). That is the opposite combination for most mammals, such as kittens who are born deaf, with eyes sealed shut, yet can walk immediately. Human senses are crucial from birth on; movement skills take months and years.

The Senses

All the senses function at birth. Newborns have open eyes, sensitive ears, and responsive noses, tongues, and skin. Indeed, very young babies use all their senses to attend to everything. For instance, in the first days of life, they stare at everyone and suck almost anything in their mouths.

Sensation occurs when a sensory system detects a stimulus, as when the inner ear reverberates with sound, or the eye's retina and pupil intercept light. Thus, sensations begin when an outer organ (eye, ear, nose, tongue, or skin) meets anything that can be seen, heard, smelled, tasted, or touched.

Genetic selection over more than 200,000 years affects all the senses. Humans cannot hear what mice hear, or see what bats see, or smell what puppies smell; humans do not need to do that. However, survival requires babies to respond to people, and newborns do so with every sense (Konner, 2010; Zeifman, 2013).

sensation The response of a sensory organ (eyes, ears, skin, tongue, nose) when it detects a stimulus.

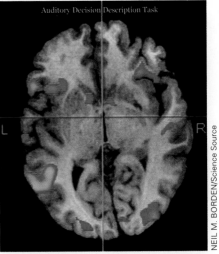

Listen, Imagine, Think, and Tap A person has just heard "banana" and "round, red fruit," and is told to tap if the two do not match. An MRI reveals that 14 areas of the brain are activated. As you see, this simple matching task requires hearing (the large region on the temporal lobe), imagined seeing (the visual cortex in the occipital lobe at the bottom), motor action (the parietal lobe), and analysis (the prefrontal cortex at the top). Imagine how much more brain activation is required for the challenges of daily life.

Hearing

The sense of hearing develops during the last trimester of pregnancy. At birth, certain sounds trigger reflexes, even without conscious perception. Sudden noises startle newborns, making them cry.

Familiar, rhythmic sounds, such as a heartbeat, are soothing: That is one reason kangaroo care reduces newborn stress, as the infant's ear rests on the mother's

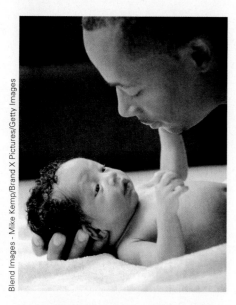

Blend Images - Mike Kemp/Brand X Pictures/Getty Images

Who's This? Newborns don't know much, but they look intensely at faces. Repeated sensations become perceptions, so in about six weeks this baby will smile at Dad, Mom, a stranger, the dog, and every other face. If this father in Utah responds like typical fathers everywhere, cognition will be apparent by 6 months: The baby will chortle with joy at seeing him but become wary of unfamiliar faces.

binocular vision The ability to focus the two eyes in a coordinated manner in order to see one image.

THINK CRITICALLY: Which is most important in the first year of life, accurate hearing or seeing?

chest. [**Developmental Link**: Kangaroo care is explained in Chapter 4.] Soon, infants turn their heads to see the source of a voice—an ability that requires instant calculation of the difference between when the sound reaches the left and right ears.

Newborn hearing is routinely checked at most hospitals in North America and Europe, since deaf infants benefit from early remediation. If they have cochlear implants by age 1, they may not be delayed in speech or comprehension—unlike for those whose deafness is remedied after age 2 (Tobey et al., 2013).

Seeing

By contrast, vision is immature at birth. Although in mid-pregnancy the eyes open and are sensitive to bright light (if the pregnant woman is sunbathing in a bikini, for instance), the fetus has nothing much to see. Consequently, newborns are legally blind; they focus only on things between 4 and 30 inches (10 and 75 centimeters) away (Bornstein et al., 2005).

Almost immediately, experience combines with maturation of the visual cortex to improve the ability to see shapes and notice details. Movement captures attention, as does contrast. For example, by 6 weeks, when babies see a person, they focus on the eyes—those colorful dots on a plain surface.

Vision improves so rapidly that researchers are hard-pressed to describe the day-by-day improvements (Dobson et al., 2009). By 2 months, infants not only stare at faces but also, with perception and cognition, smile. (Smiling can occur earlier, but not because of perception.) In many ways, the vision of the young infant is attracted to the eyes of other people, and by age 1, they interpret emotions, follow gaze, and use their own eyes to communicate (Grossman, 2017).

Because **binocular vision** (coordinating both eyes to see one image) is impossible in the womb (nothing is far enough away), many newborns seem to use their two eyes independently, momentarily appearing wall-eyed or cross-eyed. Typically, experience leads to rapid focus and binocular vision, and binocular vision emerges at about 13 weeks (Kavšek & Braun, 2016). Early screening for visual ability is as important as early screening for hearing, because good vision is crucial for learning.

Vision allows infants to develop their motor skills. For instance, unlike younger infants, by about 6 months a baby can see something and grab it.

Toddlers who are experienced crawlers and walkers scan the immediate environment to decide whether a slope is safe to cross upright or is best traversed sitting or crawling. This illustrates early coordination of the senses and motor skills (Kretch & Adolph, 2013). (This does *not* mean that toddlers can be trusted not to fall off tables or out of windows.)

Tasting and Smelling

As with vision and hearing, taste and smell rapidly adapt to the social context. Babies appreciate what their mothers eat, prenatally through amniotic fluid, then through breast milk, and finally through smells and spoonfuls of the family dinner.

The foods of a particular culture may aid survival because some natural substances are medicinal. For example, bitter foods provide some defense against malaria; hot spices help preserve food and thus work against food poisoning (Krebs, 2009). Thus, a taste for the family cuisine may save young lives.

Families pass on cultural taste preferences, despite immigration or changing historical circumstances. A feeding pattern that was protective may no longer be so. Indeed, when starvation was a threat, humans developed a taste for high-fat foods; now their descendants enjoy French fries, whipped cream, and bacon, jeopardizing their health.

Learning About a Lime As with every other typical infant, Jacqueline's curiosity leads to taste and then to a slow reaction, from puzzlement to tongue-out disgust. Jacqueline's responses demonstrate that the sense of taste is acute in infancy and that quick brain perceptions are still to come.

Adaptation also occurs for the sense of smell. When breast-feeding mothers used a chamomile balm to ease cracked nipples during the first days of their babies' lives, those babies preferred that smell almost two years later, compared with babies whose mothers used an odorless ointment (Delaunay-El Allam et al., 2010).

As babies learn to recognize each person's scent, they prefer to sleep next to their caregivers, and they nuzzle into their caregivers' chests—especially when the adults are shirtless. One way to help infants who are frightened of the bath (some love bathing, some hate it) is for the parent to join the baby in the tub. The smells of the adult's body mixed with the smell of soap, and the pleasant touch, sight, and voice of the caregiver make the warm water comforting.

Touch and Pain

The sense of touch is acute in infants. Wrapping, rubbing, massaging, and cradling are each soothing to many new babies. Even when their eyes are closed, some infants stop crying and visibly relax when held securely by their caregivers.

Gentle touch is effective to sooth an infant only a few weeks old (Tuulari et al., 2019). The infant heartbeat becomes slow and rhythmic, not rapid and erratic (as with stress), when stroked gently and rhythmically on the arm (Fairhurst et al., 2014).

Pain and temperature are not among the traditional five senses, but they are often connected to touch. Some babies cry when being changed, distressed at the sudden coldness on their skin. Some touches are unpleasant—a poke, pinch, or pat—although this varies from one baby to another.

Facial expressions, heart rate, and body movements suggest that pain can be felt, but less so than later on when the axons of the brain are better connected. Digestive difficulty (colic) and teething are said to be painful. However, this is unproven: Crying or lack of crying is an imperfect measure of pain at every stage of life.

Physiological measures, including hormones, heartbeat, and rapid brain waves, are studied to assess infant pain: The conclusions are mixed. Infant brains are immature: They have some similar responses to pain and some dissimilar ones when compared to adults (Moultrie et al., 2016).

The First Blood Test This baby will cry, but most experts believe the heel prick shown here is well worth it. The drops of blood will reveal the presence of any of several genetic diseases, including sickle-cell disease, cystic fibrosis, and phenylketonuria. Early diagnosis allows early treatment, and the cries subside quickly with a drop of sugar water or a suck of breast milk.

However, the past assumption that newborns feel no pain is replaced by the idea that pain in the newborn is not the same intensity, or for the same reasons, as adult pain, but that medical professionals nevertheless need to assess and relieve infant pain (Koress et al., 2019; Maxwell et al., 2019).

For many newborn medical procedures, from a pinprick to minor surgery, a taste of sugar right before the event is an anesthetic. Breast-feeding also relieves pain (Gad et al., 2019). There are two possible reasons for that: (1) Distraction is a way to manage pain (the brain processes one sensation at a time), and (2) breast milk

THINK CRITICALLY: What political controversy makes objective research on newborn pain difficult?

is thought to contain a mild anesthetic. Doctors hesitate to use drugs, because that may slow down breathing.

Many newborn care units in hospitals have adopted practices that make the first days of life better for babies, including allowing parents to touch their fragile infants, eliminating bright lights and noisy monitors, reducing pain and distress through careful swaddling, positioning, and so on. The result is better social and cognitive development later on (Montirosso et al., 2017).

Motor Skills

motor skill The learned abilities to move some part of the body, in actions ranging from a large leap to a flicker of the eyelid. (The word *motor* here refers to movement of muscles.)

A **motor skill** is any skilled movement of the body, from the newborn's head-lifting to the toddler's stair-climbing and, later, from the teenager's legible and sometimes artistic handwriting to the latest dance move. Every skill requires basic maturation and then depends on motivation and practice.

Motor skills begin with reflexes, explained in Chapter 4. Reflexes become skills if they are practiced and encouraged. As you saw in the chapter's beginning, Mrs. Todd set the foundation for my fourth child's walking when Sarah was only a few months old. Similarly, some very young babies can swim—if adults build on the swimming reflex by floating with them in calm, warm water.

Gross Motor Skills

gross motor skills Physical abilities involving large body movements, such as walking and jumping. (The word *gross* here means "big.")

Deliberate actions that coordinate many parts of the body, producing large movements, are called **gross motor skills.** These skills emerge directly from reflexes and proceed in a *cephalocaudal* (head-down) and *proximodistal* (center-out) direction. Infants first control their heads, lifting them up to look around. Then they control their upper bodies, their arms, and finally their legs and feet. (See At About This Time, which shows age norms for gross motor skills.)

Sitting requires muscles to steady the torso, no simple feat. By 3 months, most babies can sit propped up in a lap. By 6 months, they can usually sit unsupported, but "novice sitting and standing infants lose balance just from turning their heads

Observation Quiz Which of these skills has the greatest variation in age of acquisition? Why? (see answer, page 145) ➜

AT ABOUT THIS TIME

Age Norms (in Months) for Gross Motor Skills

	When 50% of All Babies Master the Skill	When 90% of All Babies Master the Skill
Sit unsupported	6	7.5
Stand holding on	7.4	9.4
Crawl (creep)	8	10
Stand not holding	10.8	13.4
Walk well	12.0	14.4
Walk backward	15	17
Run	18	20
Jump up	26	29

Note: As the text explains, age norms are affected by culture and cohort. The first five norms are based on babies from five continents [Brazil, Ghana, Norway, United States, Oman, and India] (World Health Organization, 2006). The next three are from a U.S.-only source [Coovadia & Wittenberg, 2004; based on Denver II (Frankenburg et al., 1992)]. Mastering skills a few weeks earlier or later does not indicate health or intelligence. Being very late, however, is a cause for concern.

or lifting their arms" (Adolph & Franchak, 2017). Babies never propped up (as in some institutions for orphaned children) sit much later, as do babies who cannot use vision to adjust their balance.

Crawling is another example of the head-down and center-out direction of skill mastery. As they gain muscle strength, infants wiggle, attempting to move forward by pushing their arms, shoulders, and upper bodies against whatever surface they are lying on. Motivation is crucial: Babies want to move forward to explore objects just out of reach.

Usually by 5 months, infants add their legs to this effort, inching forward (or backward) on their bellies. Exactly when this occurs depends partly on how much "tummy time" the infant has had to develop the muscles, and that, of course, is affected by the caregiver's culture (Zachry & Kitzmann, 2011).

Between 8 and 10 months after birth, most infants can lift their midsections and move forward—or sometimes backward first. Some babies never crawl, but they all find some way to move before they can walk (inching, bear-walking, scooting, creeping, or crawling). As soon as they are able, babies walk (falling frequently but getting up undaunted and trying again), since walking is quicker than crawling, and has another advantage—free hands (Adolph et al., 2012). That illustrates the drive that underlies every motor skill: Babies are powerfully motivated to do whatever they can as soon as they can.

Beyond motivations, the dynamic-systems perspective highlights the interaction of strength, maturation, and practice. We illustrate these three with walking.

Advancing and Advanced At 8 months, she is already an adept crawler, alternating hands and knees, intent on progress. She will probably be walking before a year.

1. *Muscle strength.* Newborns with skinny legs and 3-month-olds buoyed by water make stepping movements, but 6-month-olds on dry land do not; their legs are too chubby for their underdeveloped muscles. As they gain strength, they stand and then walk—easier for thin babies than heavy ones.
2. *Brain maturation.* The first leg movements—kicking (alternating legs at birth and then both legs together or one leg repeatedly at about 3 months)—occur without much thought. As the brain matures, deliberate and coordinated leg action becomes possible.
3. *Practice.* Unbalanced, wide-legged, short strides become a steady, smooth gait.

Once toddlers are able to walk by themselves, they practice obsessively, barefoot or not, at home or in stores, on sidewalks or streets, on lawns or in mud. This depends a great deal on caregivers providing the opportunity—holding them to walk—in the bath, after diapering, around the house, on the sidewalk.

Indeed, "Practice, not merely maturation, underlies improvements. . . . In 1 hour of free play, the average toddler takes about 2400 steps, travels the length of about 8 U.S. football fields, and falls 17 times" (Adolph & Franchak, 2017).

Fine Motor Skills

Small body movements are called **fine motor skills.** The most valued fine motor skills are finger movements, enabling humans to write, draw, type, tie, and so on. Movements of the tongue, jaw, lips, and toes are fine movements, too.

Regarding hand skills, newborns have a strong reflexive grasp but lack control. During their first 2 months, babies excitedly stare and wave their arms at objects dangling within reach. By 3 months, they can usually touch such objects, but because of limited eye–hand coordination, they cannot yet grab and hold on unless an object is placed in their hands.

By 4 months, infants sometimes grab, but their timing is off: They close their hands too early or too late. Finally, by 6 months, with a concentrated, deliberate

fine motor skills Physical abilities involving small body movements, especially of the hands and fingers, such as drawing and picking up a coin. (The word *fine* here means "small.")

VIDEO: Fine Motor Skills in Infancy and Toddlerhood shows the sequence in which babies and toddlers acquire fine motor skills.

AT ABOUT THIS TIME

Age Norms (in Months) for Fine Motor Skills

	When 50% of All Babies Master the Skill	When 90% of All Babies Master the Skill
Grasp rattle when placed in hand	3	4
Reach to hold an object	4.5	6
Thumb and finger grasp	8	10
Stack two blocks	15	21
Imitate vertical line (drawing)	30	39

Data from World Health Organization, 2006.

stare, most babies can reach, grab, and grasp almost any object that is the right size. Some can even transfer an object from one hand to the other.

Toward the end of the first year and throughout the second, finger skills improve as babies master the pincer movement (using thumb and forefinger to pick up tiny objects). They become able to feed themselves, first with hands, then fingers, then utensils (Ho, 2010). (See At About This Time.)

Cultural Variations

Practice of every motor skill advances development, not only of the skill but over-all (Leonard & Hill, 2014). When U.S. infants are grouped by ethnicity, generally African American babies are ahead of Latinx babies when it comes to walking. In turn, Latino babies are ahead of those of European descent.

Internationally, the earliest walkers are in several nations of Africa. For example, about a third of the infants in Benin walk at 10 months (Dossou et al., 2019). As evident in the opening anecdote of this chapter, variations that might seem to be genetic may instead be cultural.

Nonetheless, some infants in every culture need early intervention. World Health Organization guidelines now reflect local norms for development, because infants with disabilities need to be recognized and treated before cognitive impairment occurs (Lancaster et al., 2018).

In some cultures, babies are massaged and stretched from birth onward and are encouraged to walk as soon as possible. They do so, long before a year (Adolph & Franchak, 2017). The latest walkers (15 months) may be in rural China, where infants are bundled up against the cold (Adolph & Robinson, 2013). Some cultures discourage walking because danger lurks (venomous snakes, open fires, speeding cars), making toddlers safer since they cannot wander. By contrast, some cultures encourage running over long distances: Their children can run marathons (Adolph & Franchak, 2017).

Remember that difference is not deficit. However, slow development *relative to local norms* may indicate a problem that needs attention: Lags are much easier to remedy during infancy than later on.

If one sense or motor skill is impaired, the others are affected as well. This is true throughout childhood: Fine motor skills are aided by the ability to sit, language development depends on hearing, reading depends on vision—so careful monitoring of basic sensory and motor skills in infancy is part of good infant care.

No Stopping Him Something compels infants to roll over, sit, stand, and walk as soon as their bodies allow it. This boy will fall often, despite his balancing arms, but he will get up and try again. Soon he will run and climb. What will his cautious mother (behind him) do then?

skynesher/E+/Getty Images

Sticky Mittens

As with gross motor skills, fine motor skills are shaped by practice. In the first year, when "infants flap their arms, rotate their hands, and wiggle their fingers, and exhibit bouts of rhythmical waving, rubbing, and banging while holding objects" (Adolph & Franchak, 2017), they are practicing fine motor skills that will lead to better thinking later on.

Although infants have an inner compulsion to move their bodies and practice seeing, hearing, and all their motor skills, opportunity is crucial. This has been studied in detail in dozens of studies involving "sticky mittens" (with Velcro) that reward grabbing small toys that also have Velcro on them. With those special mittens, infants master hand skills sooner than usual, and their perceptual skills advance as well (Libertus & Needham, 2010; Libertus et al., 2016; Nascimento et al., 2019; Needham et al., 2017; Soska et al., 2010).

One follow-up of the original research suggests that the special mittens are not necessary, but that play that includes reaching for objects may be. In that study, healthy 2- to 3-month-olds were assigned to one of three groups. Over 16 days, one group had daily reaching practice with sticky mittens, another group had similar reaching practice but their mittens were not sticky, and a third control group had no special practice. The results: The two experimental groups advanced in the motor skill of reaching and the perceptual skill of focusing on objects. The control group also improved a little, but much less so (Leonard & Hill, 2014).

Another follow-up provided not only active exploration with sticky mittens, but it also compared one group who explored in silence and another group who received auditory reinforcement. The details demonstrate how carefully scientists contrast independent variables:

In the less auditory feedback condition, the table was covered with foam core that was itself covered with thin foam to reduce sounds the soft toys might make when contacting the table. In the less auditory feedback condition, the toys were silent while moving. In the more auditory feedback condition, no foam was used on the table and the toys made loud impact noises when hitting the table. In addition, in the more auditory feedback condition, bells were sewn onto the sticky mittens and a few small bells were placed inside the toys to produce rattling and jingling sounds during the movement of the toys.

[Needham et al., 2017, p. 218]

The infants who could hear as well as see the effects of their activity showed more exploration and perception. That makes sense, because some *sensory redundancy* is characteristic of infant learning. Babies use "multiple sensory modalities: infants look at an object while banging it against a table surface or they explore an object manually, orally, and visually in quick succession" (Needham et al., 2017, p. 221). They not only reach to touch their caregivers, they look, listen, smell, and (if the adult lets them) taste them!

All the research on infant reaching and brain development leads to a solid conclusion: Even early in life, infant senses and motor skills reinforce each other, as steps in an ongoing, experience-expectant, learning process. As authors of one book explain:

Achievements are not ends or goals but rather they are points in a cascade of developmental change and themselves influence future developmental outcomes. . . . different abilities, functions, or behaviors at another point in time.

[Oakes & Rakison, 2019, p. xi]

To advance the cognitive and social development of a person lifelong, early sensory and motor skills are the foundation. Encourage the baby to explore!

WHAT HAVE YOU LEARNED?

1. What particular sounds and patterns do infants pay attention to?

2. How does an infant's vision change over the first year?

3. Why is hearing more acute than vision in the early weeks?

4. Why do some babies prefer certain tastes and smells that others dislike?

5. What is known and unknown about infant pain?

6. What is universal and what is cultural in the development of gross motor skills in infancy?

7. What is the relationship between motor skills and the senses?

8. Why do caregivers vary in which motor skills they encourage?

Surviving in Good Health

Public health measures have dramatically reduced infant death. United Nations statistics show that in 1950, worldwide, one infant in six died before age 1; in 2020, the rate was about 1 in 40. Marked income differences are evident, from 1 in 200 in the most developed nations to 1 in 10 in the poorest ones (United Nations, June 17, 2019). Progress is most dramatic in large developing nations (China and India).

All told, about 2 million people are alive today who would have died if they had been born 70 years ago. Improvements are everywhere. Infant mortality has been reduced by 900 percent in Poland, Japan, Chile, China, and Finland.

Better Days Ahead

Most child deaths occur in the first month. In the twenty-first century in developed nations, 99.9 percent of 1-month-olds live to adulthood. Public health measures (clean water, nourishing food, immunization of older children who might catch a disease that could kill a newborn) deserve most of the credit.

Not only survival but life itself is better for children, because parents have fewer births and thus attend more to each one. Maternal education is pivotal. Especially in low-income nations, educated women have far fewer, but much healthier, children than women who never went to school (de la Croix, 2013).

Considering Culture

Many cultural variations are simply alternate ways to raise a healthy child, a difference, not a deficit. Sometimes, however, one mode of infant care is much better than another. To discover those, international comparisons become useful. Consider the dramatic worldwide reduction in **sudden infant death syndrome (SIDS)**.

Every year until the mid-1990s, tens of thousands of infants died of SIDS, called *crib death* in North America and *cot death* in England. Tiny infants smiled at their caregivers, waved their arms at rattles that their small fingers could not yet grasp, went to sleep, and never woke up. Scientists tested hypotheses (the cat? the quilt? natural honey? homicide? spoiled milk?) to no avail. Grief-ridden parents were sometimes falsely accused. Sudden infant death was a mystery. Finally, one major risk factor—sleeping on the stomach—was discovered, thanks to the work of one scientist, described on the next page, in A Case to Study.

Immunization

Diseases that could be deadly (including measles, chicken pox, polio, mumps, rotavirus, and whooping cough) are now rare because of **immunization,** which primes the body's immune system to resist a particular disease. Immunization (often via *vaccination*) is said to have had "a greater impact on human mortality reduction and population growth than any other public health intervention besides clean water" (Baker, 2000, p. 199).

In the first half of the twentieth century, almost every child had at least one of these diseases. Usually they recovered, and then they were immune. Indeed, some parents took their toddlers to play with a child who had an active case of chicken pox, for instance, hoping their child would catch the disease and then become immune. That protected that child later in life.

Louise Gubb/Corbis Historical/Getty Images

Well Protected Disease and early death are common in Ethiopia, where this photo was taken, but neither is likely for 2-year-old Salem. He is protected not only by the nutrition and antibodies in his mother's milk but also by the large blue net that surrounds them. Treated bed nets, like this one provided by the Carter Center and the Ethiopian Health Ministry, are often large enough for families to eat, read, and sleep in together, without fear of malaria-infected mosquitoes.

sudden infant death syndrome (SIDS) A situation in which a seemingly healthy infant, usually between 2 and 6 months old, suddenly stops breathing and dies unexpectedly while asleep.

immunization A process that stimulates the body's immune system by causing production of antibodies to defend against attack by a particular contagious disease. Creation of antibodies may be accomplished naturally (by having the disease), by injection, by drops that are swallowed, or by a nasal spray.

Scientist at Work

Susan Beal, a 35-year-old scientist with five young children, began to study SIDS deaths in South Australia. She responded to phone calls, often at 5 or 6 A.M., notifying her that another baby had died. Her husband supported her work, often becoming the sole child-care provider so she could leave home at a moment's notice.

Sometimes she was the first professional to arrive, before the police or the coroner. Initially she was embarrassed to question the grief-stricken parents. But soon she learned that parents were grateful to talk, in part because they tended to blame themselves and they needed to express that emotion to someone who was not likely to accuse them. Beal reassured them that scientists shared their bewilderment.

She was more than a sympathetic listener. She was a scientist, so she took detailed, careful notes on dozens of circumstances at each of more than 500 deaths. She found that some things did not matter (such as birth order), and some increased the risk (maternal smoking and lambskin blankets).

A breakthrough came when Beal noticed an ethnic variation: She saw far more SIDS victims in Australia among people of European descent than of Chinese descent. Genetic? Most experts thought so. But Beal's notes revealed that almost all SIDS babies died while sleeping on their stomachs, contrary to the Chinese custom of placing infants on their backs to sleep. She developed a new hypothesis: Sleeping position mattered.

To test her hypothesis, Beal convinced a large group of non-Chinese parents to put their newborns to sleep on their backs. Almost none of them died suddenly.

After several years of gathering data, she drew a surprising conclusion: Back-sleeping protected against SIDS. Her published report (Beal, 1988) caught the attention of doctors in the Netherlands, where pediatricians had told parents to put their babies to sleep on their stomachs. Two Dutch scientists (Engelberts & de Jonge, 1990) recommended back-sleeping; thousands of parents took heed. SIDS was reduced in Holland by 40 percent in one year—a stunning replication.

In the United States, Benjamin Spock's *Baby and Child Care*, first published in 1946, sold more copies than any book other than the Bible. He advised stomach-sleeping, and millions of parents followed that advice. In 1984, SIDS killed 5,245 babies.

But Beal's 1988 article, and the Dutch 1990 data, spread to America. By 1994, a "Back to Sleep" campaign cut the SIDS rate dramatically in every nation (Kinney & Thach, 2009; Mitchell, 2009). By 1996, the U.S. SIDS rate was half of what it had been. By 2015, the U.S. Centers for Disease Control and Prevention reported 1,600 SIDS deaths, even though the population of infants had increased over the past decades. Consequently, in the United States alone, about 100,000 people are alive who would be dead if they had been born before 1990 (see **Figure 5.4**).

Stomach-sleeping is a proven, replicated risk, but it is not the only one. Other risks include low birthweight, winter, being male, exposure to cigarettes, soft blankets or pillows, bed-sharing, and physical abnormalities (in the brainstem, heart, mitochondria, or microbiome) (Neary & Breckenridge, 2013; Ostfeld et al., 2010). Most SIDS victims experience several risks, a cascade of biological and social circumstances.

That does not surprise Susan Beal. She sifted through all the evidence and found the main risk—stomach-sleeping—but she continues to study other factors. She praises the courage of the hundreds of parents who talked with her hours after their baby died; the entire world praises her.

LaunchPad
macmillan learning

Interview with Susan Beal https://tinyurl.com/t5xtyjr

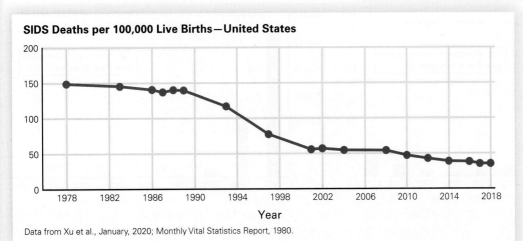

SIDS Deaths per 100,000 Live Births—United States

Data from Xu et al., January, 2020; Monthly Vital Statistics Report, 1980.

FIGURE 5.4

Alive Today As more parents learn that a baby should be on his or her "back to sleep," the SIDS rate continues to decrease. Other factors are also responsible for the decline—fewer parents smoke cigarettes in the baby's room.

Scott Eels/Redux

True Dedication This young Buddhist monk lives in a remote region of Nepal, where until recently measles was a common, fatal disease. Fortunately, a UNICEF porter carried the vaccine over mountain trails for two days so that this boy—and his whole community—could be immunized.

THINK CRITICALLY: What developmental questions are raised by the recent COVID-19 outbreak?

Success and Survival

Beginning with smallpox in the nineteenth century, doctors discovered that giving a small dose of a virus to healthy people stimulates antibodies and provides protection. Stunning successes in immunization include the following:

- Smallpox, the most lethal disease for children in the past, was eradicated worldwide as of 1980. Vaccination against smallpox is no longer needed.
- Polio, a crippling and sometimes fatal disease, has been virtually eliminated in the Americas. Only 784 cases were reported anywhere in the world in 2003. However, false rumors halted immunization in northern Nigeria. Polio reappeared, sickening 1,948 people in 2005, almost all of them in West Africa. Public health workers and community leaders rallied and Nigeria's polio rate fell again, to 6 cases in 2014. However, poverty and wars in South Asia prevented immunization there: In 2019, 116 victims were in Pakistan and Afghanistan, four times the rate in 2018. An added complication is that in nations where some children are not immunized, those who are immunized might develop a new strain of polio, as occurred with 196 children in 12 nations of Africa (Roberts, 2020).
- Measles (rubeola, not rubella) is disappearing, thanks to a vaccine developed in 1963. Prior to that time, 3 to 4 million cases occurred each year in the United States alone (Centers for Disease Control and Prevention, May 15, 2015). In 2012 in the United States, only 55 people had measles, although globally about 20 million measles cases occurred that year. (See **Figure 5.5**)

However, if a traveler brings measles back to the United States, and if some parents decide not to immunize their children, outbreaks of measles may occur. That happened in dozens of nations including the United States in 2019, where at least 1,249 people had measles (including a newborn who caught it from the mother)—the highest rate since 1994 (Patel et al., 2019).

Immunization protects not only from temporary sickness but also from complications, including deafness, blindness, sterility, and meningitis. Sometimes such damage from illness is not apparent until decades later. Having mumps in childhood, for instance, can cause sterility and doubles the risk of schizophrenia in adulthood (Dalman et al., 2008).

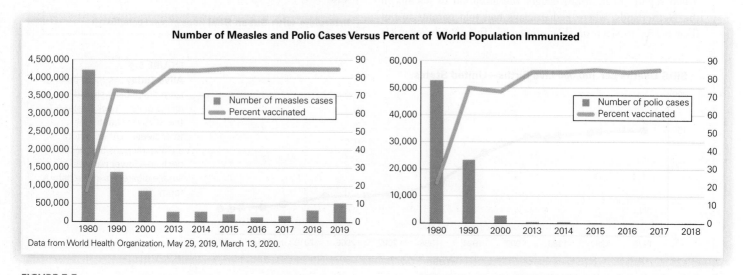

Number of Measles and Polio Cases Versus Percent of World Population Immunized

Data from World Health Organization, May 29, 2019, March 13, 2020.

FIGURE 5.5

Ask Grandma Neither polio nor measles is completely eradicated, because some parents do not realize the danger. They may never have seen the serious complications of these diseases.

CAREER ALERT   The Pediatrician and the Pediatric Nurse

Many people studying human development hope to enter the medical field as doctors, nurses, physical therapists, or in dozens of related careers. Often their ultimate goal is to become a pediatrician, a medical doctor who helps children grow strong and healthy.

From an educational perspective, that is a lofty goal indeed. To become a pediatrician, one must first become a medical doctor (M.D.) after earning a bachelor's degree. This entails entering medical school, first with two years of advanced classes and then working as an intern for two more years in many aspects of medicine.

At that point, a fledgling doctor might decide to be a pediatrician, which requires a residency in pediatrics for at least three more years. Finally, the doctor can be licensed in pediatrics. The pediatrician can then work in a hospital or other institution that treats children or begin a private practice.

If someone wants to work directly for the health of children, yet remaining a student for more than ten years after college graduation seems impossible, then the person should consider one of many other medical professions that help children, among them nurses, physician assistants, physical therapists, and pediatric psychologists. Nursing itself consists of many levels, from aide, to practical nurse (L.P.N), to registered nurse (R.N.), to nurse practitioner (who can do many of the same things as an M.D.).

For all these, the job outlook is good: The Occupational Outlook Handbook reports that the United States will need 15 percent more pediatricians in the next decade. Nurses fare even better: The projection for registered nurses is 16 percent; for nurse-midwives, 22 percent; and for nurse practitioners, 36 percent. Nurse practitioners must complete two more years of education after earning a bachelor's degree and becoming a registered nurse (RN), but the reward is higher salary, more responsibility, and job security.

How can child health be an expanding field, since the birth rate is falling? There are two reasons: First, the field of medicine is increasingly focused on prevention, and child health is the foundation for adult health. That means that well-child care is increasingly important. Second, as parents have fewer children, they have become more intensely concerned about each child.

The joy of all these professions is working with children and watching them grow into happy and successful adults. However, the children who need the most care are the ones in poor health. Often medicine can cure an illness, or at least pave a path toward a satisfying adulthood. This is now true for many common problems, such as sickle cell anemia or Down's syndrome, that once led to suffering and death. However, some children still suffer and a few die, which is devastating not only to the parents but also to medical professionals.

The other problem is that helping children requires helping caregivers. Often parents are grateful for medical advice, but sometimes they are not. Specializing in child health requires a person to also specialize in providing information and respect to whomever is responsible for the direct care of the child. That may be difficult.

For example, currently, many parents hesitate to vaccinate their children. Medical professionals know the evidence and have treated children with measles, mumps, and other childhood diseases that could have been prevented. That makes them understandably impatient with parents who choose not to protect their children. Yet parents have picketed and sued doctors who advocate vaccination. Thus, becoming a medical professional in child health requires courage as well as knowledge.

Worst of all, some parents mistreat their children, and the doctor or nurse is the first one to notice it. They are "mandated reporters"; they must report their concerns. However, as this text makes clear, reporting is only the first step toward helping a mistreated child. Medical professionals may be the best ones to support caregivers and children, making sure that the child recovers from any harm.

Thus, caring for children is a noble and joyous task, but it requires courage and strong human relations skills. As your study of human development makes clear, children do not grow in isolation: A skilled medical professional is a crucial team member, allowing every child to thrive.

Immunization also protects those who cannot be safely vaccinated, such as infants under 3 months and people with impaired immune systems (HIV-positive, aged, or undergoing chemotherapy). Fortunately, each vaccinated child stops transmission of the disease, a phenomenon called **herd immunity.** Usually, if 90 percent of the people in a community (a herd) are immunized, no one dies of that disease.

All 50 of the United States exempt vaccination requirements for children with certain medical conditions (such as undergoing chemotherapy for cancer), but 15 states allow parents to refuse vaccination because of "personal belief," and 45 states allow religious exemptions (see Visualizing Development). This horrifies public health workers, who know that the risks of the diseases—especially to babies—are far greater than the risks from immunization.

Measles, for example, can be deadly, and the "anti-vax" movement has undercut herd immunity in some nations. That makes it especially crucial for any child over 6 months who is travelling to be protected (Wong et al., 2020). The most recent outbreak was in Samoa, from October 2019 to February 2020. Vaccination rates had fallen to 31 percent for young children. After several child deaths, the government mandated vaccination, and now it is 95 percent. That was too late for some: Of the 83 deaths recorded as of January 2020, most (85 percent) were infants and children, aged 1 day to 4 years.

herd immunity The level of immunity necessary in a population (the herd) in order to stop transmission of infectious diseases. The rate is usually above 90 percent, and even higher for very infectious diseases. Newborns, and people with certain diseases (e.g., cancer patients taking immunosuppressant drugs), cannot be vaccinated; herd immunity protects them.

Immunization

Before the measles vaccine was introduced in 1963, 30 million people globally contracted measles each year. About 2 million of them died, usually because they were both malnourished and sick. (World Health Organization, April 28, 2017). Thankfully, worldwide vaccination efforts now mean that no child need die of measles.

Measles is highly infectious, so 95 percent of the population must be immunized in order for herd immunity to protect the entire community. The United States achieved that: A decade ago, measles incidence was close to zero. Experts thought it would soon be eliminated in all developed countries, so public health workers focused on the very poorest nations.

ESTIMATED MEASLES VACCINE COVERAGE – SELECTED NATIONS

- 90% or greater
- 80–89%
- 70–79%
- less than 70%

By June 2019, more than 1,000 measles cases had been reported in 28 U.S. states, the most since 1992 (measles was declared eliminated in 2000) (Centers for Disease Control and Prevention, June 17, 2019). To understand what went wrong, note that many states allow personal or religious exemptions to immunization requirements. Thus, as the U.S. map below shows, several states are not at that safe 94 percent—leaving many vulnerable, not only to discomfort but also to complications, including pneumonia, encephalitis, and even death.

Data from World Health Organization, May 29, 2019.

VACCINE EXEMPTION AND HERD IMMUNITY – UNITED STATES

MMR vaccination rate >94%

Nonmedical exemption rate above U.S. median (2%)

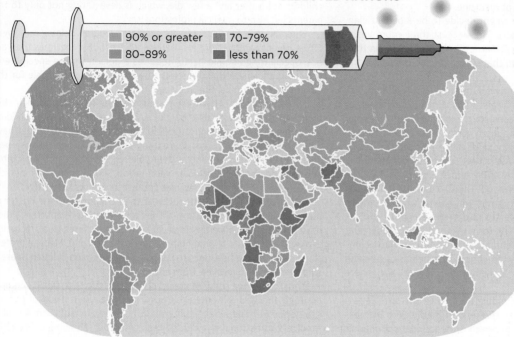

Data from Mellerson et al., 2018.

Ethicists also are concerned, because sometimes allowing individuals to make their own choices clashes with the need to protect the community. If private choices reduce herd immunity, the public is burdened with hospital costs and deaths (Giubilini & Savulescu, 2019).

Children may react to immunization by being irritable or even feverish for a day, to the distress of their parents. However, parents do not notice if their child does *not* get polio, measles, or so on. Before the varicella (chicken pox) vaccine, more than 100 people in the United States died each year from that disease, and 1 million were itchy and feverish for a week. Now, far fewer people get chicken pox, and almost no one dies of varicella.

Many parents are concerned about the potential side effects of vaccines, in part because the rare event of one person sickened by vaccination is broadcast widely. Psychologists find that a common source of irrational thinking is overestimating the frequency of a memorable case (Ariely, 2010). As a result, the rate of missed vaccinations in the United States has been rising over the past decades.

An example of the benefits of immunization comes from Connecticut, where in 2012 flu vaccination was required for all 6- to 59-month-olds in licensed day-care centers. Most children were not in day-care centers, but parents became aware of the law and many immunized their children. That winter, far fewer Connecticut children were hospitalized for flu, although rates rose everywhere else. Meanwhile, Colorado (one of the states that allowed many exemptions, with kindergartners well below herd immunity) had the highest rate of flu hospitalizations (MMWR, March 7, 2014).

> **DATA CONNECTIONS: Saving Lives: Immunization in the United States** describes the vaccines that are recommended for children and the rates at which U.S. children are immunized. **LaunchPad.**

Especially for Nurses and Pediatricians A mother refuses to have her baby immunized because she wants to prevent side effects. She wants your signature for a religious exemption, which in some jurisdictions allows the mother to refuse vaccination. What should you do? (see response, page 145)

Nutrition

As already explained, infant mortality worldwide has plummeted in recent years for several reasons: fewer sudden infant deaths, advances in prenatal and newborn care, and, as you just read, immunization. One more measure is making a huge difference: better nutrition.

Breast-Feeding

Ideally, nutrition starts with *colostrum*, a thick, high-calorie fluid secreted by the mother's breasts at birth. This benefit is not understood in some cultures, where the mother is not allowed to breast feed until her milk "comes in" two or three days after birth. (Sometimes other women nurse the newborn; sometimes herbal tea is given). Worldwide research confirmed that colostrum saves infant lives, especially if the infant is preterm (Andreas et al., 2015; Moles et al., 2015).

Compared with formula using cow's milk, human milk is sterile, more digestible, and rich in nutrients (Wambach & Riordan, 2014). Allergies and asthma are less common in children who were breast-fed, and in adulthood, their obesity, diabetes, and heart disease rates are lower.

The composition of breast milk adjusts to the age of the baby, with milk for premature babies distinct from that

Same Situation, Far Apart: Breast-Feeding Breast-feeding is universal. None of us would exist if our foremothers had not successfully breast-fed their babies for millennia. Currently, breast-feeding is practiced worldwide, but it is no longer the only way to feed infants, and each culture has particular practices.

SelectStock/Vetta/Getty Images

hadynyah/Vetta/E+/Getty Images

LaunchPad
macmillan learning

VIDEO: Nutritional Needs of Infants and Children: Breast-Feeding Promotion shows UNICEF's efforts to educate women on the benefits of breast-feeding.

for older infants. Quantity increases to meet the demand: Twins and even triplets can be exclusively breast-fed for months.

Formula is advised for medical reasons only in unusual cases, such as when the mother uses toxic drugs or is HIV-positive. Even with HIV, however, breast milk without supplementation is advised by the World Health Organization. In some nations, the infants' risk of catching HIV from their mothers is lower than the risk of dying from infections, diarrhea, or malnutrition as a result of bottle-feeding (A. Williams et al., 2016).

Doctors worldwide recommend breast-feeding with no other foods—not even juice—for the first six months of life. (**Table 5.1** lists some of the benefits of breast-feeding.) Some pediatricians suggest adding foods (rice cereal and bananas) at 4 months; others advise waiting until 6 months (Fewtrell et al., 2011).

Breast-feeding was once universal, but by the mid-twentieth century, many mothers thought formula was better. Fortunately, that has changed again. In 2016 in the United States, most (84 percent) newborns were breast-fed, as were more than half (57 percent) of all 6-month-olds and more than a third (36 percent) of all 1-year-olds (Centers for Disease Control and Prevention, 2019).

Encouragement of breast-feeding from family members, especially fathers, is crucial. In addition, ideally nurses visit new mothers weekly at home; such visits (routine in some nations, rare in others) increase the likelihood that breast-feeding will continue.

Although every expert agrees that breast milk is beneficial, given the complexity and variation of human families, mothers should not feel guilty for feeding formula. Indeed, some critics contend that breast-feeding has reached cult status, shaming those who do not do it (Jung, 2015). No single behavior, even those recommended (breast-feeding, co-sleeping, hand-washing, exercising, family planning, immunization) defines good motherhood.

TABLE 5.1

The Benefits of Breast-Feeding

For the Baby	For the Mother
Balance of nutrition (fat, protein, etc.) adjusts to age of baby	Easier bonding with baby
Breast milk has micronutrients not found in formula	Reduced risk of breast cancer and osteoporosis
Less infant illness, including allergies, ear infections, stomach upsets	Natural contraception (with exclusive breast-feeding, for several months)
Less childhood asthma	Pleasure of breast stimulation
Better childhood vision	Satisfaction of meeting infant's basic need
Less adult illness, including diabetes, cancer, heart disease	No formula to prepare; no sterilization
Protection against many childhood diseases, since breast milk contains antibodies from the mother	Easier travel with the baby
Stronger jaws, fewer cavities, advanced breathing reflexes (less SIDS)	**For the Family**
Higher IQ, less likely to drop out of school, more likely to attend college	Increased survival of other children (because of spacing of births)
Later puberty, fewer teenage pregnancies	Increased family income (because formula and medical care are expensive)
Less likely to become obese or hypertensive by age 12	Less stress on father, especially at night

Malnutrition

Protein-calorie malnutrition occurs when a person does not consume enough food to sustain normal growth. A child may suffer from **stunting,** being short for their age because chronic malnutrition kept them from growing, or **wasting,** being severely underweight for their age and height (2 or more standard deviations below average). Many nations, especially in East Asia, Latin America, and central Europe, have seen improvement in child nutrition in the past decades, with an accompanying decrease in wasting and stunting (see **Figure 5.6**).

In other nations, however, primarily in Africa, wasting has increased (Black et al., 2013). Although only 17 percent of the world's population live in Africa, a third of all the infants worldwide who suffer from malnutrition are in African nations south of the Sahara Desert (Akombi et al., 2017). Explanations include high birth rate, maternal AIDS deaths, climate change, and civil wars.

Chronically malnourished infants and children suffer in three ways:

1. Learning suffers. If malnutrition continues long enough to affect height, it also affects the brain. If hunger reduces energy and curiosity, learning suffers.
2. Diseases are more serious. About half of all childhood deaths occur because malnutrition makes a childhood disease lethal, especially from leading causes of childhood deaths—diarrhea and pneumonia—as well as from milder diseases such as measles (Imdad et al., 2011; Walker et al., 2013).
3. Some diseases result directly from malnutrition—including both *marasmus* during the first year, when body tissues waste away, and *kwashiorkor* after age 1, when growth slows down, hair becomes thin, skin becomes splotchy, and the face, legs, and abdomen swell with fluid (edema).

Prevention, more than treatment, is needed. Sadly, some children hospitalized for marasmus or kwashiorkor die even after being fed because their digestive systems were already failing (M. Smith et al., 2013). Ideally, prenatal nutrition, then breast-feeding, and then supplemental iron and vitamin A stop malnutrition before it starts. Once malnutrition is apparent, highly nutritious formula (usually fortified peanut butter) often restores weight—but not always.

Ironically, one of the factors that correlates with wasting and stunting in children from sub-Saharan African nations is breast-feeding that continues after a year, especially if the mother herself is malnourished. Thus, public health recommendations

protein-calorie malnutrition A condition in which a person does not consume sufficient food of any kind. This deprivation can result in several illnesses, severe weight loss, and even death.

stunting The failure of children to grow to a normal height for their age due to severe and chronic malnutrition.

wasting The tendency for children to be severely underweight for their age as a result of malnutrition.

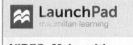

LaunchPad

VIDEO: Malnutrition and Children in Nepal shows the plight of Nepalese children who suffer from protein-energy malnutrition.

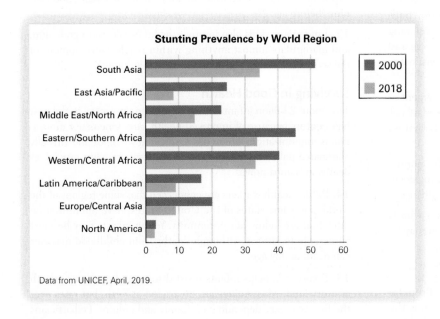

Stunting Prevalence by World Region

Data from UNICEF, April, 2019.

FIGURE 5.6

Evidence Matters Genes were thought to explain height differences among Asians and Scandinavians, until data on hunger and malnutrition proved otherwise. The result: starvation down and height up almost everywhere—especially in Asia. Despite increased world population, far fewer young children are stunted (255 million in 1970; 156 million in 2015). Evidence now finds additional problems: Civil war, climate change, and limited access to contraception have increased stunting in East and Central Africa in the past decade.

Observation Quiz Which regions have the most and least improvement since 1990? (see answer, page 145) ←

need to consider nutrition in the entire family and community: Infants survive best if breast-fed, but after 6 months, ample, digestible food is also needed (Akombi et al., 2017).

A combination of factors—genetic susceptibility, poor nutrition, infection, and abnormal bacteria in the digestive system (the microbiome)—may be fatal (M. Smith et al., 2013). Giving severely ill children an antibiotic to stop infection saves lives—but always, prevention is best (Gough et al., 2014).

WHAT HAVE YOU LEARNED?

1. Why is polio still a problem in some nations?

2. Why do doctors worry about immunization rates in the United States?

3. What are the reasons for and against breast-feeding until a child is at least 1 year old?

4. When is it advisable that a woman not breast-feed?

5. What is the relationship between malnutrition and disease?

6. Which is worse, stunting or wasting? Why?

SUMMARY

Body Changes

1. In the first two years of life, infants grow taller, gain weight, and increase in head circumference—all indicative of development. On average, birthweight doubles by 4 months, triples by 1 year, and quadruples by 2 years.

2. By age 2, the average well-nourished child weighs about 28 pounds (12.7 kilograms) and has gained more than a foot since birth, reaching about half their adult height.

3. Medical checkups in the first months of a child's life focus especially on weight, height, and head circumference because early detection of slow growth can halt later problems. Percentile changes can signify difficulties.

4. The amount of time a child sleeps decreases over the first two years. Variations in sleep patterns are normal, caused by both nature and nurture. Bed-sharing is the norm in many developing nations, although it increases the risk of SIDS. Co-sleeping and bed-sharing are increasingly common in developed nations; many breast-feeding mothers choose it.

5. Brains grow dramatically, from about 25 to about 75 percent of adult brain weight in the first two years. Complexity increases as well, with cell growth, development of dendrites, and formation of synapses.

6. Some stimulation is experience-expectant, needed for normal brain development. Both exuberant growth and pruning aid cognition, as the connections that are experience-dependent are strengthened.

7. Experience is vital for brain development. An infant who is socially isolated, overstressed, or deprived of stimulation may be impaired lifelong.

Perceiving and Moving

8. At birth, the senses already respond to stimuli. Prenatal experience makes hearing the most mature sense. Vision is the least mature sense at birth, but it improves quickly with experience. Infants use all their senses to strengthen their early social interactions.

9. The senses of smell, taste, and touch are present at birth, and they help infants respond to their social world. Pain is experienced, but infant pain is not identical to adult pain.

10. Infants gradually improve their motor skills as they begin to grow and brain maturation continues. Control of the body proceeds from the head downward (cephalocaudal) and the core to the extremities (proximodistal). Mastery of motor skills aid cognition.

11. Gross motor skills are mastered throughout infancy, depending on practice, motivation, and maturation. Major advances are sitting up (at about 6 months), walking (at about 1 year), and running (before age 2).

12. Fine motor skills also improve, as infants learn to grab, aim, and manipulate almost anything within reach. Development of the senses and motor skills are mutually reinforcing.

Surviving in Good Health

13. About 2 billion infant deaths have been prevented in the past half-century because of improved health care. One major innovation is immunization, which has eradicated smallpox and virtually eliminated polio and measles. Herd immunity remains crucial, as deaths in Samoa prove.

14. Public health workers are concerned that some regions of the world, and some states of the United States, have immunization rates that are below herd immunity. Young infants may be most vulnerable to viruses, although deaths from childhood diseases can occur at any age.

15. Breast milk helps infants resist disease and promotes growth of every kind. Most babies are breast-fed at birth, but rates over the first year vary depending on family and culture. Pediatricians now recommend breast milk as the only nourishment for the first 4–6 months.

16. Severe malnutrition stunts growth and can cause death, both directly through marasmus or kwashiorkor and indirectly

through vulnerability if a child becomes sick. Stunting and wasting are both signs of malnutrition, which is become less common worldwide except in some nations of sub-Saharan Africa.

KEY TERMS

percentile (p. 121)
REM (rapid eye movement) sleep (p. 122)
co-sleeping (p. 122)
bed-sharing (p. 122)
head-sparing (p. 123)
neuron (p. 124)
axon (p. 124)
dendrite (p. 124)
synapse (p. 124)

neurotransmitter (p. 124)
myelin (p. 124)
cortex (p. 124)
prefrontal cortex (p. 124)
limbic system (p. 124)
amygdala (p. 125)
hippocampus (p. 125)
hypothalamus (p. 125)
cortisol (p. 125)
pituitary (p. 125)

transient exuberance (p. 126)
pruning (p. 126)
experience-expectant (p. 127)
experience-dependent (p. 127)
shaken baby syndrome (p. 128)
self-righting (p. 129)
sensation (p. 129)
binocular vision (p. 130)
motor skill (p. 132)
gross motor skills (p. 132)

fine motor skills (p. 133)
sudden infant death syndrome (SIDS) (p. 136)
immunization (p. 136)
herd immunity (p. 139)
protein-calorie malnutrition (p. 143)
stunting (p. 143)
wasting (p. 143)

APPLICATIONS

1. Immunization regulations and practices vary, partly for social and political reasons. Ask at least two faculty or administrative staff members what immunizations the students at your college must have and why. If you hear "It's a law," ask why.

2. Observe three infants (whom you do not know) in public places such as a store, playground, or bus. Look closely at body size and motor skills, especially how much control each baby has over his or her legs and hands. From that, estimate the

baby's age in months, and then ask the caregiver how old the infant is.

3. *This project can be done alone, but it is more informative if several students pool responses.* Ask 3 to 10 adults whether they were bottle-fed or breast-fed and, if breast-fed, for how long. If someone does not know, or expresses embarrassment, that itself is worth noting. Do you see any correlation between adult body size and infant feeding?

Especially For ANSWERS

Response for New Parents (from p. 123): From the psychological and cultural perspectives, babies can sleep anywhere as long as the parents can hear them if they cry. The main consideration is safety: Infants should not sleep on a mattress that is too soft, nor beside an adult who is drunk or on drugs. Otherwise, families should decide for themselves.

Response for Parents of Grown Children (from p. 127): Probably not. Brain development is programmed to occur for all infants, requiring only the stimulation that virtually all families provide—warmth, reassuring touch, overheard conversation, facial expressions, movement. Extras such as baby talk, music, exercise, mobiles, and massage may be beneficial but are not essential.

Response for Nurses and Pediatricians (from p. 141): It is difficult to convince people that their method of child rearing is wrong, although you should try. In this case, listen respectfully and then describe specific instances of serious illness or death from a childhood disease. Suggest that the mother ask her grandparents whether they knew anyone who had polio, tuberculosis, or tetanus (they probably did). If you cannot convince this mother, do not despair: Vaccination of 95 percent of toddlers helps protect the other 5 percent. If the mother has genuine religious reasons, talk to her clergy adviser.

Observation Quiz ANSWERS

Answer to Observation Quiz (from p. 132): Jumping up, with a three-month age range for acquisition. The reason is that the older an infant is, the more impact both nature and nurture have.

Answer to Observation Quiz (from p. 143): Most is East Asia, primarily because China has prioritized public health. Least is western and central Africa, primarily because of civil wars. In some nations, high birth rates have dramatically increased the numbers of stunted children, even though rates in the region are lower.

The First Two Years: Cognitive Development

What Will You Know?

1. What do infants remember before they can talk?
2. Why did Piaget compare 1-year-olds to scientists?
3. When does a typical baby say a first word?

A neuroscientist learned about a Korean fortune-telling ritual for 1-year-olds:

> The unsuspecting baby is placed in front of an assortment of objects and is encouraged to pick one. . . . If the baby picks up a banana, she will never go hungry; choosing a book means she is destined for academia; a silver coin foretells wealth, or a paintbrush, creativity.
>
> I was intrigued. The very same evening I placed Olivia [her infant daughter] in front of a collection of items.
>
> A stethoscope: would she be a doctor? A stuffed dog: a vet? A plant: a Green Peace activist? A piece of pastry: a chef? And a colorful model of the brain; a neuroscientist?
>
> Olivia inspected the objects closely, took her time, and then went straight for the iPhone I had happened to leave at the corner of the table.
>
> I shouldn't have been surprised. The little girl was obsessed with this piece of machinery. She would skillfully roll herself from one side of the room to the other to grab hold of it. . . . . When she finally grabbed hold of the phone she would quickly insert it into her mouth and attempt to chew. . . . She had other bright, musical toys that she did not desire as much. The iPhone was the item she wanted because from the day she was born she had observed her parents constantly interacting with it with great interest. Although she was only a few months old and could not even say a word she was able to infer that these metal rectangles must be extremely valuable. Little Livia's fondness for iPhones tells us something important about how our brains work.

[Sharot, 2017, pp. 152–153]

This incident introduces the chapter on cognitive development during infancy for three reasons.

First, this topic is intriguing: We all wonder what thoughts, aspirations, and abilities the newest humans will have. We hope they will do well and perpetuate our own values and choices. Did you notice that the mother thought a stethoscope meant doctor (not nurse), that a plant signified Greenpeace (not farmer), and that a model of the brain was among the options (for the daughter of a neuroscientist)?

The second reason is to highlight that infants are curious and social. Of course all those Korean babies reach for something, as Olivia did—every baby would. And of course they are influenced by what they have observed in their parents.

The third is that caregiver behavior is pivotal. Without realizing it, every family and each culture molds the infant mind. This social influence on infants is obvious when babies begin to speak their native language, not any of the 6,000 other languages that infants elsewhere speak; but caregivers also affect curiosity, persistence, and logic, as you will soon learn.

This chapter describes the process of infant cognition, as well as its accomplishments. From the moment when newborns open their eyes, to the insistent grabbing, experimenting, and talking of 2-year-olds, infants are active learners.

The Eager Mind

Chapter 5 chronicled the intense human drive to use every sensory and motor ability. Newborns look and listen; toddlers run and climb. Brain growth makes sensory and motor development possible; babies obsessively use their abilities as soon as they can.

The same phenomenon occurs with cognitive development. One team suggests that infants are "scientist[s] in the crib" (Gopnik et al., 1999), a suggestion that was "frequently met with incredulity" (Halberda, 2018, p. 1215). However, "the field of developmental neuroscience has burgeoned over the last 20 years with advances in technology and methods that are well suited for measuring the human brain in vivo in infants" (Guyer et al., 2018, p. 687). Incredulity faded as evidence accumulated.

As scientists discover more about the infant brain, they are increasingly impressed by its inborn readiness to learn. One team wrote: "from early on in development, infants display perceptual biases and attentional patterns that strongly suggest a motive to acquire information" (Lucca & Wilbourn, 2018, p. 942).

Brains do much more than increase in size (Chapter 5); infant brains are preset to understand the world, using every step of the scientific method. Curiosity leads to hypothesizing, observing, experimenting, analyzing, and concluding.

Listening to Learn

Remember that newborns' hearing is acute. Infants can hear all noises—traffic on the street, clanking of dishes in the kitchen, the hum or crackle of the radiator—but they ignore most of what they hear. Thus, sensation does not usually become perception, and perception does not necessarily become cognition. However, babies listen closely to certain sounds, particularly the human voice.

Distinguishing Speech Sounds

Newborns do not understand words, of course, but they have an inborn affinity for language, probably because, for humans, most learning occurs via words.

Vast differences are audible in adult speech: Russian does not sound like a tonal language such as Chinese; English does not pronounce the *r* as French does; the cadence of German is quite different from that of Spanish. Babies need to learn from whatever language their caregivers speak, which means that every linguistic nuance must be perceived.

That is how it happens. Even in the early weeks, babies distinguish the difference between the sound of *pa* and *ba*, for example, and they hear the nuances of many other speech sounds—some insignificant in one language but crucial in another. They are called *universalists* because they hear the differences in any language (Kuhl, 2004).

By one year after birth, however, that ability to distinguish sounds in never-heard languages deteriorates, a loss that continues throughout childhood. Babies at first

"Is this the way you plan to spend your peak learning years?"

SIDNEY HARRIS/CARTOONSTOCK

Still Wrong Parents used to ignore infant cognition. Now some make the opposite mistake, assuming that infants learn via active study.

TABLE 6.1

Examples of Mispronunciations in the Bergelson and Swingley Study

Apple	*opel*
Banana	*banoona*
Milk	*mulk*
Hair	*har*
Mouth	*mith*
Nose	*nazz*

VIDEO: Event-Related Potential (ERP) Research shows a procedure in which the electrical activity of an infant's brain is recorded to see whether the brain responds differently to familiar versus unfamiliar words.

◐ Especially for Educators An infant day-care center has a new child whose parents speak a language other than the one the teachers speak. Should the teachers learn basic words in the new language, or should they expect the baby to learn the teachers' language? (see response, page 170)

attend to all linguistic sounds; by adulthood people literally cannot hear some sound differences that are crucial in languages they have never learned. That is evidence for cognitive maturation.

Babies from English-speaking families were shown pictures of 17 common objects while hearing the names of the objects (Bergelson & Swingley, 2018). Sometimes the name was deliberately mispronounced, as in the examples shown in **Table 6.1.**

Six-month-olds were tested for knowledge of these words. Performance overall was poor, but some babies already knew a few of the 17. Importantly, their understanding was equally good in three conditions: (1) their mother saying the words correctly; (2) their mother mispronouncing the words; and (3) a stranger saying the words correctly.

By 1 year of age, however, not only did they know more but also their brains had already learned correct U.S. English pronunciation. They were significantly better at understanding correct speech from strangers than mispronunciation by their own mothers.

This study shows, first, that very young babies are primed to learn language, and second, that 1-year-olds already know the accepted way to pronounce words. (English-speaking babies in Jamaica, or England, or India learn other nuances.)

Learning Two Languages

Bilingual proficiency begins in the first year of life—every young human brain can learn several languages. Ideally, parents often speak in two languages, and then their children become doubly fluent as well. The brains of bilingual 1-year-olds respond to both languages (Ramirez et al., 2017), because infant brains are primed to understand whatever speech they hear.

Infants also are attuned to the social context. Have you noticed that most bilingual adults use one language with friends and family and the other one in more formal settings? Very young infants notice that, too: They figure out which language is most important to the adults and respond preferentially to that one.

This was one conclusion from a study of 94 newborns (age 0 to 5 days) in Vancouver, Canada (Byers-Heinlein et al., 2010). For half of them, their mother spoke English and Tagalog (a language native to the Philippines); for one-third, their mothers spoke only English; and for one-sixth, their mothers spoke English and Chinese.

The infants in all three groups sucked on a pacifier connected to a recording of 10 minutes of English and 10 minutes of Tagalog. To make sure there were no subtle biases in the streams of speech, such as one language spoken with more animation, the two languages were matched for pitch, duration, and number of syllables.

As evident in the rate and intensity of their sucking (which activated the recording), babies with English-only mothers preferred English and those with bilingual mothers preferred Tagalog (Byers-Heinlein et al., 2010). They had already connected Tagalog with more animated, emotional talk—and that is what they wanted to learn first.

Looking to Learn

Developmentalists have long known that very young infants spend more time simply looking around than doing anything else. They scan their surroundings, fixate on faces, follow moving objects with their eyes. Experienced caregivers know that one way to quiet fussy 3-month-olds is to take them to see something—cars moving on the street, dogs coming to be petted, flowers in a garden, toys that move.

Gaze-Following

Until recently, however, developmentalists did not appreciate how important vision is for cognition. Very young babies choose to look at whatever is likely to advance their understanding. They wisely focus on whatever captures their caregivers' attention, via *gaze-following*, instinctively knowing that what caregivers look at might tell them something important.

For example, following adults' lead, they look at the face of someone entering the room, ignoring the ceiling, the floor, or the person's feet. Have they learned that adults look at faces because expressions are informative, or is gaze-following natural for infants?

Both. It was thought that gaze-following occurred only as a response to adults, who alert the babies to opportunities to learn. Adults say, "Javier, look, here comes Daddy," or "Sophia, here is your teddy bear." Such guides to gaze-following are part of adults' natural tendency to teach babies through *natural pedagogy* (Gergely & Csibra, 2013).

Natural pedagogy is evident whenever caregivers direct the baby's gaze, calling their name, pointing at an object, and so on. Adults try to advance infant cognition; very young babies respond by looking at whatever the adult shows.

But we now know that infants will follow an adult's gaze even without caregiver cues (Gredeback et al., 2018). If a tilt of the head and movement of the pupils indicates that something interests the caregiver, the baby follows the gaze. Thus, gaze-following arises from both nature and nurture.

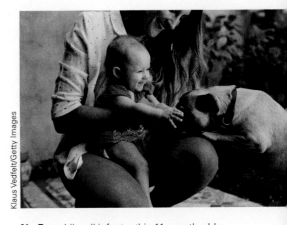

Klaus Vedfelt/Getty Images

No Fear Like all infants, this 11-month-old girl is eager to explore through sight and touch. Praise to all three—this mother for encouraging learning, this baby for reaching out, and this dog for gently licking her hand. Most dogs recognize babies, tolerating actions they would not accept from adults of any species.

Early Logic

Nature and nurture may also give babies some understanding of the laws of physics. In one study, a toy dinosaur was removed from a display where it had been next to a flower. A screen then covered the display. A moment later, the screen was lifted to reveal the dinosaur instead of the flower (Cesana-Arlotti et al., 2018).

Cameras and computers measured how long the babies looked at that unexpected event. Infants stared longer when the flower was surreptitiously replaced with the dinosaur than when the flower was still there. This indicated that they knew how things should be and were surprised when their basic understanding was wrong.

Many other events that contradict the basic laws of physics (such as a ball that is suspended in the air rather than falling, or a toy that becomes two toys) elicit the same surprise. From such research, many developmentalists believe that infants have some innate logic. Scientific reasoning may not be a "hard-won accomplishment mastered later in life," but rather an "inherent attribute of the mind" (Halberda, 2018, p. 1214).

Core Knowledge

To explain infants' cognition, scientists suggest that babies are born with an understanding of how the world works, which they call *core knowledge* (Stahl & Feigenson, 2017). Examples are that moving objects stop when they bump into a solid wall, and that adult gaze signals important information.

Core expectations not only prime learning but also alert the baby when something unexpected happens. Surprise triggers curiosity, which triggers more learning. No one thinks that babies are born with the knowledge and logic they will display in a few months or years. But core knowledge and plasticity make rapid learning possible as the brain grows (remember, it triples in size by age 2) and experiences accumulate.

Consider how this works when infants see pictures (LoBue, 2013). Infants are *not* instinctively afraid of snakes. However, when they look at pictures of flowers and snakes, they focus more on the snakes. Their brains seem to know that snakes may be important; then, during early childhood, they are taught whether or not to be afraid.

Recognizing Faces

Another example comes from attention to faces. Unless you have *prosopagnosia* (face blindness), the *fusiform face area* of your brain is astonishingly adept. This is innate. Compared to older children and adults, newborns are quicker to recognize a face that they have seen just once (Zeifman, 2013).

Because of experience-expectant brain development, every face is fascinating: Babies stare at pictures of monkey faces and photos of human ones, at drawings and toys with faces, as well as at live faces. At 6 weeks, they smile at almost anyone whose face is about 2 feet away. That fact indicates that faces are part of core knowledge.

Soon, with experience-dependent learning, babies smile more readily at familiar people, differentiate men from women, and distinguish among faces from their own ethnic group (called the *own-race effect*). This fact could be worrisome: Does it suggest that humans are naturally sexist and racist? No, as A View from Science explains. Brains are primed to pay attention to familiar faces, and most babies see faces of a particular kind.

A VIEW FROM SCIENCE

Face Recognition

The own-race effect refers to the fact that infants are better at recognizing individuals from their own ethnic group than at distinguishing individuals from other groups. Researchers test infant recognition by showing babies photos of strangers from various groups and measuring their looking time and attention, to determine if they notice differences between individuals.

This measure of looking time is used in thousands of studies that assess infant cognition. Infants pay less attention to seeing the same thing again and again (that's called *habituation*), so if a baby were to look longer at a particular picture, that suggests that the baby notices the difference between that picture and the previous one (Csibra et al., 2016).

When the experience of a baby or a child under age 12 is multiracial, the own-race effect is less evident; that is, they recognize individuality between photos of people of another race as well as of their own race. Research on children adopted from China and raised in Canada found that if a child of one ethnicity (Chinese) is raised exclusively by people of another ethnicity (European), that child recognizes differences among the people they see, not among people of their biological group (McKone et al., 2019).

The importance of experience is confirmed by two studies. The first study occurred in Malaysia, where many people of two ethnic groups (Malay and Chinese) live and where women, but not men, interact with babies. A group of Chinese infants were shown photos of people who varied by gender and ethnicity. The babies recognized individuality among Chinese women by 3 months and among Malaysian women by 8 months. However, they did not perceive individuality among Europeans or among men of any ethnicity—apparently because they had limited experience with them (Tham et al., 2019).

In another study, parents repeatedly "read" a book to their 6-month-old infants (Scott & Monesson, 2010). The book depicted six monkey faces. One-third of the parents said each monkey's name while showing the pictures; one-third said only "monkey" as they turned each page; the final one-third simply turned the pages with no verbal labeling.

At 9 months, infants in all three groups viewed pictures of six *unfamiliar* monkeys. The infants who had heard names of monkeys were better at distinguishing one new monkey from another than were the infants who saw the same picture book but did not hear each monkey's name (Scott & Monesson, 2010).

Evidently, by hearing the names, the babies realized that monkeys vary in appearance, and they learned to distinguish one monkey from another. This applies to humans' understanding of racial and national groups as well. Interacting with several *named* people of any group helps people understand that members of that group are individuals, not stereotypes (Thorup et al., 2018).

As with almost every type of infant development, experience combines with inborn brain proclivities. An intriguing study found that this innate preference is evident in all primates, not just in humans.

In that study, researchers prevented macaque monkeys from seeing faces (including those of other monkeys and of humans) for the first three months of life. Then they showed pictures, some with faces and some not. Those deprived 3-month-olds looked more attentively at photos of faces than photos of other objects, demonstrating that inborn attraction to faces. Indeed, every face—of chimpanzees, otters, humans, as well as other monkeys—was almost equally interesting.

Then, in the next few months, the monkeys were granted some experiences with other macaques. By 6 months, the researchers found that these monkeys paid more attention to the faces of their own species than of other primates (Simpson et al., 2017). That is just what human babies do: They look intently at every face-like image at first but zero in on the faces that are most important to them.

This research has important practical applications. If parents hope their children become adults who relate well with people of many ages, genders, and ethnicities, they need to expose them early on to friends, caregivers, neighbors, and so on from diverse backgrounds.

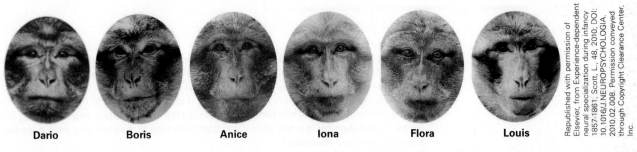

Dario   **Boris**   **Anice**   **Iona**   **Flora**   **Louis**

Republished with permission of Elsevier, from Experience-dependent neural specialization during infancy 1857-1861, Scott, L., 48, 2010; DOI: 10.1016/J.NEUROPSYCHOLOGIA. 2010.02.008. Permission conveyed through Copyright Clearance Center, Inc.

Iona Is Not Flora If you heard that Dario was not Louis or Boris, would you stare at unfamiliar monkey faces more closely in the future? For 6-month-olds, the answer is yes.

How to Learn

One other aspect of cognition merits mention—babies learn how to learn. As you remember from Chapter 4, infants are born with reflexes. Some of those reflexes fade and others build, depending on experience.

Curiosity is an inborn reflex; newborns look at and listen to everything. Then, some infants are allowed to explore and experiment (fingering their toys, reaching for people, and so on); others are not ("Don't touch!"). It matters whether caregivers encourage curiosity, respond to noises, and build on reflexes (as Mrs. Todd did in Chapter 5) . For example, if a baby utters a sound, some caregivers stop and respond, others ignore the sound, and some others tell the baby to be quiet.

By the second year of life, some toddlers are eager to explore and investigate while others are much more hesitant. One quality, sometimes called *grit* or effort, fosters learning throughout life.

In one experiment, 15-month-olds observed adults trying to get a toy from a container (Leonard et al., 2017). The adult said, "How can I get this toy out of here?" and then worked to do so. Half the infants saw the toy come out quickly, and the other half watched the adult working hard for half a minute to release the toy (see **Figure 6.1**).

Then the babies were handed another toy. The experimenter pushed a button to demonstrate that the toy played music. However, the toy was rigged to play once for the experimenter, but not after that. Not surprisingly, every baby tried to make the toy play music again, since they had just observed this could happen.

But the crucial question is how persistent the babies were. One baby quickly threw the toy to the floor in frustration and another kept trying for more than two minutes. Why the difference? Some is probably temperament, but observation mattered. If babies had seen the adult keep trying for half a minute before getting

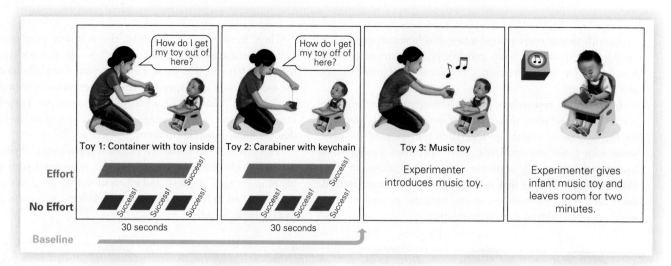

FIGURE 6.1

If At First You Don't Succeed . . . Quit? Two times, with two toys, babies watched an adult try to get toys from a container. One group saw the toys quickly released, and another group saw them released only after some effort. Then the babies were shown another toy. This third toy played music when the adult turned it on, but then (unbeknownst to the babies) the music-playing was deactivated. When babies were handed the quiet toy, how long did they try to turn on the music? (Answer is in the text.) If you don't take time to read the text, what does that suggest about your childhood?

the toy, they pushed the button an average of 22.5 times. However, if they had seen the adult succeed quickly, they pushed, on average, only 12 times (Leonard et al., 2017).

The authors conclude that adults should sometimes let children see them struggle to complete a task. The idea that hard work pays off is a learning strategy that helps throughout a person's education—and it may begin in infancy.

Theories of the Infant Mind

Two theories are relevant to this new understanding of the infant mind: information-processing theory and evolutionary theory.

As you remember from Chapter 2, *information-processing theory* originated from understanding the processing built into computers, which gather millions of bits of information and then compute a result. The central idea is that the human mind is like a computer, accumulating experiences and then establishing knowledge. Thus, information processing might lead to the hypothesis that the infant mind is programmed for cognitive development, so the myriad sights and sounds produce understanding.

The other theory that provides insight into early cognition is *evolutionary theory* (Bjorklund, 2018). The idea is that the human brain, unique among mammals, has evolved to be extraordinarily plastic so that human babies can learn everything they need within their culture. According to evolutionary theory, this occurs in two steps.

- First, infants innately attend to caregivers as well as to things that, millennia ago, were crucial for survival. That is why they listen to voices, like to be held snugly, and look at snakes more than flowers.
- Second, the diversity of human culture requires that the infant brain be amazingly plastic, allowing inborn predispositions to be shaped so people

can adapt to whatever life may bring. Over the millennia, people with flexible brains were likely to survive and procreate, advancing the genes for plasticity.

Consider one more example: the sucking reflex. For hundreds of thousands of years, the sucking/swallowing/breathing reflexes had to function well in newborns so they could survive, even though no other motor ability needed to work at birth.

You also remember that breast milk is good for the brain; babies who are breast-fed tend to have higher IQs than those who are formula-fed. That was recently proven again, this time carefully controlling for third variables, such as the mother's intelligence and education, that might cause the correlation. Babies who were breast-fed for at least a month had, by age 5, IQ scores that averaged 3 points higher than babies who were not breast-fed at all, or whose mothers gave up breast-feeding in the first few weeks (Strøm et al., 2019).

These two facts are particularly relevant for babies born very early. The sucking reflex does not begin until about 33 weeks after conception, which meant that, in prior centuries, preterm babies died. Now respirators, incubators, and so on allow viable preterm babies to live. They are tube-fed, ideally with breast milk that their mothers have expressed with a breast pump. At about 34 weeks, in the hospital, they begin to suck, either via a special bottle and nipple designed for their tiny mouths or directly from the nipples of their mothers.

NICU professionals prefer the bottle approach, since the mothers do not need to be physically present and nurses can more easily measure nutrition. Doctors thought that babies could easily switch to direct breast-feeding at home. However, that idea ignored what cognitive psychologists now know about infant learning: Sucking for nourishment is core knowledge, evident in newborns.

As evolutionary theory explains, babies refine their evolutionary impulses with experience, because plasticity is part of human nature. Tiny babies who are bottle-fed learn that this is how sucking brings milk. By contrast, those who are directly breast-fed in the hospital (for very preterm infants this means many times a day, often supplemented with tube-feeding) learn that breasts give milk. Babies who are breast-fed in the hospital are released earlier and are more often breast-fed at home (Suberi et al., 2018), which benefits their cognition.

This is an example of the infant's eager mind, because these tiny babies quickly learned to adapt an inborn ability (sucking) with experience (as evolutionary theory suggests). Sucking knowledge is plastic; it adapts.

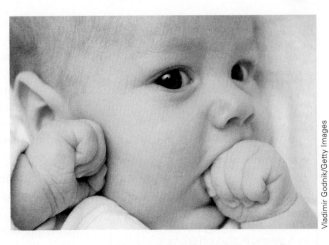

Time for Adaptation Sucking is a reflex at first, but adaptation begins as soon as an infant differentiates a pacifier from her mother's breast or realizes that her hand has grown too big to fit into her mouth. This infant's expression of concentration suggests that she is about to make that adaptation and suck just her thumb from now on.

Infant Memory

Are infant skills and cognition impressive, or are they markedly immature? Both. To further understand this, we need to look in detail at one more ability—memory, which "is crucial for the acquisition of the tremendous amount of knowledge and skills infant[s] and children acquire in the first years of life" (Vöhringer et al., 2018, p. 370).

To see the relationship between early cognition and memory, we need to appreciate that children remember what they need to remember instead of noticing

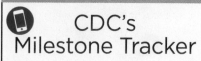

LaunchPad
macmillan learning

VIDEO: Contingency Learning in Young Infants shows Carolyn Rovee-Collier's procedure for studying instrumental learning during infancy.

their many "faults or shortcomings relative to an adult standard" (Bjorklund & Sellers, 2014, p. 142). Newborns quickly remember who their caregivers are, and they learn how their own behaviors can affect their experiences.

Of course, they remember almost nothing at first. Repeated sensations and brain maturation are required in order to recall whatever happens, as is true life-long (Bauer et al., 2010). Everyone's memory fades with time, especially if that memory was never encoded into language, never compared with similar events, or never discussed with anyone—all of which make infant memory fragile.

Forget About Infant Amnesia!

That does not mean that babies do not remember. Piaget, Freud, and other early developmentalists described *infant amnesia*, the idea that people forget everything that happened to them before age 3. However, we now know that infant brains are adapted to learn, which means that they can remember.

An insight regarding infant amnesia begins with the distinction between *implicit* and *explicit* memory. Implicit memory is not verbal; it is memory for movement, emotions, or thoughts that are not put into words. Implicit memory is evident by 3 months, begins to stabilize by 9 months, and varies from one infant to another as well as within each infant during the early months.

The reason it appears so early is partly because it comes from Implicit memory comes from the old parts of the brain—including the cerebellum and the amygdala (Vöhringer et al., 2018). Those parts mature rapidly in the first months of life, and then change relatively little, unlike the forebrain, which grows for decades. That's why implicit memory is evident so early in life.

Explicit memory takes longer to emerge, as it depends on language. It arises mainly from the cortex. Explicit memory improves dramatically throughout childhood (Hayne et al., 2015). Verbal memory, especially vocabulary, continues to increase throughout adulthood. When adults tested memory by asking questions, they were testing explicit memory, which seemed absent for the early years. Infant amnesia? Yes for some kinds of memory, especially the kind learned in school and traditionally tested by scientists. But not for many other kinds of memory.

Thus, when people say "I don't remember," they mean "I cannot recall," because something is not in explicit memory. Unconsciously and implicitly, they might remember. A person might have an irrational fear of doctors or hospitals, for instance, because of early terrifying and painful experiences that they do not recall.

Remind Me!

The most dramatic proof of very early memory comes from a series of innovative experiments in which 3-month-olds learned to move a mobile by kicking their legs (Rovee-Collier, 1987, 1990). The infants lay on their backs connected to a mobile by means of a ribbon tied to one foot.

Virtually all babies realized that kicking made the mobile move. They then kicked more vigorously and frequently, sometimes laughing at their accomplishment. So far, this is no surprise—observing self-activated movement is highly reinforcing to infants.

When infants as young as 3 months had the mobile-and-ribbon apparatus reinstalled and reconnected *one week later*, most started to kick immediately, proof that they remembered their previous experience. But when other 3-month-old infants were retested *two weeks later*, they kicked randomly. Had they forgotten? It seemed so.

But then, *two weeks after* the initial training, the lead researcher, Carolyn Rovee-Collier, allowed some infants to watch the mobile move when they were not connected to it. The next day, when a ribbon again tied their leg to the mobile, they kicked almost immediately.

Ian Boddy/SPL/Science Source

He Remembers! Infants are fascinated by moving objects within a few feet of their eyes—that's why parents buy mobiles for cribs and why Rovee-Collier tied a string to a mobile and a baby's leg to test memory. Babies not in her experiment, like this one, sometimes flail their limbs to make their cribs shake and thus make their mobiles move. Piaget's stage of "making interesting sights last" is evident to every careful observer.

● **Observation Quiz** Do you see anything here that is less than ideal? (see answer, page 171) ↑

Apparently, watching the mobile the previous day reminded them about what they had previously experienced. Other research similarly finds that reminders are powerful. If Daddy routinely plays with a 3-month-old, goes on a long trip, and the mother shows Daddy's picture and says his name on the day before his return, the baby might grin broadly when he reappears. Otherwise, he might seem to be forgotten.

At 12 months, memory improves because brains have added tens of thousands of dendrites and synapses. Babies have learned from parents and strangers, from other babies and older siblings, from picture books and family photographs, and soon from their own walking and talking (Hayne & Simcock, 2009).

Every day of their young lives, infants are processing information and storing memories. Indeed, if you saw a photo of a grandmother who cared for you every day when you were an infant and who died when you were 2, your brain would still react, even though you thought you forgot her. Information-processing research finds evidence of early memories, with visual memories particularly strong (Gao et al., 2017; Leung et al., 2016).

Evolutional theory is confirmed as well. The extraordinary plasticity of the human brain and the core knowledge of how the world might work have together astonished twenty-first-century cognitive psychologists.

Some suggest that a new theory—*evolutionary developmental psychology*—may replace the traditional approach to infant cognition, as expounded by Piaget (Bjorklund, 2018). This theory benefits from twenty-first-century neuroscience and technology that allows researchers to discover the amazing abilities of the infant brain.

WHAT HAVE YOU LEARNED?

1. What is the developmental pattern of hearing the sound of speech?

2. Why do babies look at whatever they look at?

3. What suggests that infants have an understanding of how objects move?

4. What does face recognition tell us about infant cognition?

5. What suggests that infants develop strategies for learning by watching adults?

6. Why is information-processing theory relevant for infant cognition?

7. Why is evolutionary theory relevant for infant cognition?

VIDEO: Sensorimotor Intelligence in Infancy and Toddlerhood shows how senses and motor skills fuel infant cognition.

sensorimotor intelligence Piaget's term for the way infants think—by using their senses and motor skills—during the first period of cognitive development.

Piaget's Sensorimotor Intelligence

Now we turn to Jean Piaget, the groundbreaking theorist who studied infant cognition a century ago. Of course, Piaget lacked the technological advances that undergird our current understanding of infant cognition, but his insights were revolutionary in his day. Consequently, many contemporary developmentalists consider Piaget's six stages of infant cognition a foundation on which to build.

In 1918, when Piaget earned his doctorate in biology, most scientists thought infants only ate, cried, and slept. His Ph.D. research was on shellfish, specifically how they adapt to their environment. That required meticulous observation to understand details such as how a clam interacted with sandy water, or how a snail moved along a particular surface.

When he became a father, Piaget used his scientific observation skills with his own three infants. Contrary to conventional wisdom, he detailed active learning from birth on, recording his children's cognitive development day by day.

Early reflexes, senses, and body movements are the raw materials for infant cognition, Piaget surmised. That is why he called cognition in the first two years **sensorimotor intelligence.** He subdivided this period into six stages (see **Table 6.2**). [**Developmental Link:** Piaget's theory of cognitive development over all the years of childhood is introduced in Chapter 2.]

TABLE 6.2

The Six Stages of Sensorimotor Intelligence

For an overview of the stages of sensorimotor thought, it helps to group the six stages into pairs.

Primary Circular Reactions

The first two stages involve infants' responses to their own bodies.

Stage One (birth to 1 month)	*Reflexes:* sucking, grasping, staring, listening
	Example: sucking anything that touches the lips or cheek
Stage Two (1–4 months)	*The first acquired adaptations:* accommodation and coordination of reflexes
	Examples: sucking a pacifier differently from a nipple; attempting to hold a bottle to suck it

Secondary Circular Reactions

The next two stages involve infants' responses to objects and people.

Stage Three (4–8 months)	*Making interesting sights last:* responding to people and objects
	Example: clapping hands when mother says "patty-cake"
Stage Four (8–12 months)	*New adaptation and anticipation:* becoming more deliberate and purposeful in responding to people and objects
	Example: putting mother's hands together in order to make her start playing patty-cake

Tertiary Circular Reactions

The last two stages are the most creative, first with action and then with ideas.

Stage Five (12–18 months)	*New means through active experimentation:* experimentation and creativity in the actions of the "little scientist"
	Example: putting a teddy bear in the toilet and flushing it
Stage Six (18–24 months)	*New means through mental combinations:* thinking before doing; new ways of achieving a goal without resorting to trial and error
	Example: before flushing the teddy bear again, hesitating because of the memory of the toilet overflowing and mother's anger

Stages One and Two: Primary Circular Reactions

Piaget described the interplay of sensation, perception, action, and cognition as *circular reactions*, emphasizing that, as in a circle, there is no beginning and no end. Each experience leads to the next, which loops back (see **Figure 6.2**).

In **primary circular reactions,** the circle is within the infant's body. Stage one, called the *stage of reflexes,* lasts only a month. Reflexes become deliberate; sensation leads to perception, perception leads to cognition, and then cognition leads back to sensation.

Stage two, called *first acquired adaptations* (also called *stage of first habits*), begins as the mind of the infant allows adjustment to whatever responses they elicit. Adaptation is cognitive; it includes repeating old patterns (assimilation) and developing new ones (accommodation). [**Developmental Link:** Assimilation and accommodation are explained in Chapter 2.]

Here is one example. As you remember, full-term newborns reflexively suck anything that touches their lips (stage one). They must learn to suck, swallow, and suck again without spitting up too much—a major circular reaction that takes a few days to learn.

Then, infants *adapt* their sucking reflex to bottles or breasts, pacifiers or fingers, each requiring specific types of tongue pushing. You already read about this in preterm babies who, at 33 weeks, first are able to suck (Suberi et al., 2018). This adaptation signifies that infants have begun to interpret sensations; as they *accommodate,* they are thinking—ready for stage two.

During stage two, which Piaget pegged from about 1 to 4 months of age, additional adaptation of the sucking reflex begins. Infant cognition leads babies to suck in some ways for hunger, in other ways for comfort—and not to suck fuzzy blankets or hard plastic. Once adaptation occurs, it sticks.

Adaptation is specific. For instance, 4-month-old breast-fed babies may reject milk from the nipple of a bottle if they have never experienced it. Early cognition can endure, as evolutionary theory contends. That explains why children like to suck lollipops and why some adults still like to suck things.

primary circular reactions The first of three types of feedback loops in sensorimotor intelligence, this one involving the infant's own body. The infant senses motion, sucking, noise, and other stimuli and tries to understand them.

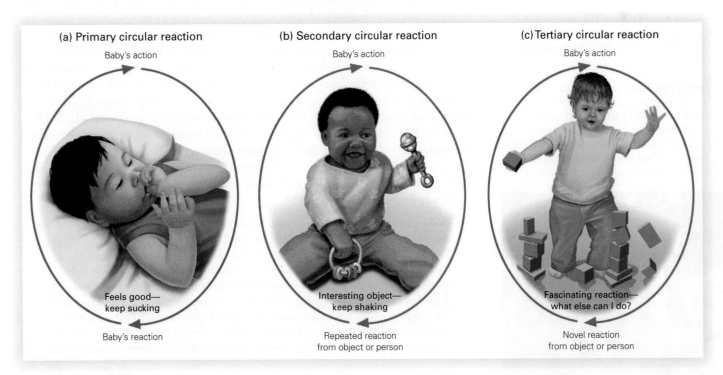

(a) Primary circular reaction
Baby's action
Feels good— keep sucking
Baby's reaction

(b) Secondary circular reaction
Baby's action
Interesting object— keep shaking
Repeated reaction from object or person

(c) Tertiary circular reaction
Baby's action
Fascinating reaction— what else can I do?
Novel reaction from object or person

FIGURE 6.2
Never Ending Circular reactions keep going because each action produces pleasure that encourages more action.

● **Especially for Parents** When should parents decide whether to feed their baby only by breast, only by bottle, or by using some combination of the two? When should they decide whether or not to let their baby use a pacifier? (see response, page 170)

secondary circular reactions The first of three types of feedback loops in sensorimotor intelligence, involving the infant and an object or another person, as with shaking a rattle or playing peek-a-boo.

object permanence The realization that objects (including people) still exist when they can no longer be seen, touched, or heard.

Family Fun Peek-a-boo makes all three happy, each for cognitive reasons. The 9-month-old is discovering object permanence, his sister (at the concrete operational stage) enjoys making her brother laugh, and their mother understands more abstract ideas—such as family bonding.

Bambu Productions/The Image Bank/Getty Images

Now, suppose 4-month-olds have discovered how to suck their thumbs and have practiced thumb sucking to their joy and satisfaction. Then, when this stage is over, suppose at 6 months the parents decide that a pacifier is better—perhaps healthier for teeth. Adaptation has already occurred; the baby might spit out the pacifier and insert the thumb.

In the same way at every age, Piaget found that older children and adults tend to stick to their early learning. This is apparent if you ever try to convince someone that something they have done all their life is wrong. Habits of thought and deed are hard to change, from infancy on.

Stages Three and Four: Secondary Circular Reactions

In stages three and four, development advances from primary to **secondary circular reactions.** These reactions extend beyond the infant's body; this circular reaction is between the baby and something else.

During stage three (4 to 8 months), infants attempt to produce exciting experiences, a stage called *making interesting sights last.* Realizing that rattles make noise, for example, they wave their arms and laugh whenever someone puts a rattle in their hand. The sight of something delightful—a favorite squeaky toy, a smiling parent—can trigger active efforts for interaction.

Next comes stage four (8 months to 1 year), called *new adaptation and anticipation* (also called the *means to the end*). Babies may ask for help (fussing, pointing, gesturing) to accomplish what they want. Thinking is more innovative because adaptation is more complex. For instance, instead of always smiling at Grandpa, an infant might first assess his mood. Stage-three babies continue an experience; stage-four babies initiate and anticipate.

Pursuing a Goal

An impressive attribute of stage four is that babies work hard to achieve their goals. A 10-month-old girl might crawl over to her mother, bringing a bar of soap to signal she loves baths, and then start to remove her clothes—finally squealing with delight when the bath water is turned on. Similarly, if a 10-month-old boy sees his father putting on a coat to leave, he might drag over his own jacket.

In both cases, infants have learned from repeated experience—Daddy may have often brought the baby along when he went out. With a combination of experience and brain maturation, babies become attuned to the goals of others, an ability that is more evident at 10 months than 8 months (Brandone et al., 2014).

According to Piaget, a major accomplishment of stage four is **object permanence**—the concept that objects or people continue to exist when they are not visible. At about 8 months—not before—infants look for toys that have fallen from the crib, rolled under a couch, or disappeared under a blanket.

As another scholar explains:

> Many parents in our typical American middle-class households have tried out Piaget's experiment in situ: Take an adorable, drooling 7-month-old baby, show her a toy she loves to play with, then cover it with a piece of cloth right in front of her eyes. What do you observe next? The baby does not know what to do to get the toy! She looks around, oblivious to the object's continuing existence under the cloth cover, and turns her attention to something else interesting in her environment. A few months later, the same baby will readily reach out and yank away the cloth cover to retrieve the highly desirable toy. This experiment has been done thousands of times and the phenomenon remains one of the most compelling in all of developmental psychology.
>
> [Xu, 2013, p. 167]

This is one of Piaget's most well-known, as well as most controversial, insights, as Opposing Perspectives explains.

Object Permanence

Piaget found that:

- Infants younger than 8 months do not search for an attractive object momentarily covered by a cloth. They do not have the concept of object permanence.
- At about 8 months, infants remove the cloth immediately after the object is covered but not if they have to wait a few seconds.
- At 18 months, infants search after a wait. However, if they have seen the object put first in one place (A) and then moved to another (B), they search in A but not B.
- By 2 years, children fully grasp object permanence, progressing through several stages, including A and B displacements (Piaget, 1954/2013a).

This research provides many practical suggestions. If young infants fuss because they see something they cannot have (keys, a cell phone, candy), caregivers can put that coveted object out of sight. Fussing stops if object permanence has not yet appeared.

By contrast, for toddlers, hiding an object is not enough. It must be securely locked up, lest the child later retrieve it, climbing onto the kitchen counter or under the bathroom sink to do so. That much is evident to every developmental scientist. Then why is this an "opposing perspective"?

The opposition comes from researchers who thought Piaget underestimated infant cognition. Piaget believed that failure to search before 8 months meant that infants had no concept of object permanence—that "out of sight" literally means "out of mind." That is where the controversy begins.

Does a baby need to be able to remove a cover from a hidden toy to demonstrate object permanence? No, according to information-processing research. The best-known example is a series of studies by Renee Baillargeon which proved that 3-month-old infants grasp object permanence, long before 8 months, when Piaget said it emerged. How did information-processing measures lead to this conclusion?

Baillargeon devised clever experiments that entailed showing infants an object, then covering it with a screen, and then removing the screen. If the object vanished behind the screen, the babies' brain waves, heart rate, or focused eyes showed surprise. This meant that they expected the object to still be present—that is, they believed an object's existence was permanent (Baillargeon & DeVos, 1991; Spelke, 1993).

Later research on object permanence has continued to question Piaget's conclusions. Many other creatures (cats, monkeys, dogs, birds) develop object permanence faster than human infants. The animal ability seems to be innate, not learned—wolves and dogs develop it, but neither is as adept as human 2-year-olds at A-not-B displacement (Fiset & Plourde, 2013). Nonhuman animal species vary in their object permanence abilities, with other primates better than dogs, for instance (Majecka & Pietraszewski, 2018).

Current research finds that early experiences combine with inherited brain dispositions, making the age of object permanence (especially A-not-B displacements) much more variable than Piaget described (MacNeill et al., 2018). The current context matters as well. When testers look at B (not A), infants are likely to guess correctly. Does that indicate gaze sensitivity, not object permanence? (Dunn & Bremner, 2019).

Most scientists agree with Baillargeon that surprise, and looking time, are evidence for early object permanence. Other scientists are less convinced (Marcovitch et al., 2016; Mareschal & Kaufman, 2012). Some stress that Piaget is essentially correct, in that object permanence is fragile until later infancy (Bremner et al., 2015).

Thus, perspectives differ on exactly how infant cognition should be measured and what gaze reveals (Dunn & Bremner, 2017). But, everyone now agrees that babies are thinking long before Piaget found that they move their hands to uncover a hidden object.

Stages Five and Six: Tertiary Circular Reactions

In their second year, infants start experimenting in thought and deed—or, rather, in the opposite sequence, deed and thought. They act first (stage five) and think later (stage six).

Tertiary circular reactions begin when 1-year-olds take independent actions to discover the properties of other people, animals, and things. Infants no longer respond only to their own bodies (primary reactions) or to other people or objects (secondary reactions). Their cognition is more like a spiral than a closed circle, increasingly creative with each discovery.

Piaget's stage five (12 to 18 months), called *new means through active experimentation*, builds on the accomplishments of stage four. Now, goal-directed and purposeful activities become more expansive.

tertiary circular reaction Piaget's description of the cognitive processes of the 1-year-old, who gathers information from experiences with the wider world and then acts on it. The response to those actions leads to further understanding, which makes this circular.

Imitation Is Lifelong As this photo illustrates, at every age, people copy what others do—often to their mutual joy. The new ability at stage six is "deferred imitation"—this boy may have seen another child lie on a tire a few days earlier.

BruesWu/Getty Images

"little scientist" Piaget's term for toddlers' insatiable curiosity and active experimentation as they engage in various actions to understand their world.

Toddlers delight in squeezing all the toothpaste out of the tube, drawing on the wall, or uncovering an anthill—activities they have never observed. Piaget referred to the stage-five toddler as a **"little scientist"** who "experiments in order to see." As you read, a scientific approach may be evident earlier, but authors who studied 12- to 19-month-olds report that "flexible and productive hypothesis testing does begin in infancy, with a vengeance" (Cesana-Arlotti et al., 2018, p. 1263).

Toddlers' preferred research method is trial and error. Their devotion to discovery is familiar to every adult scientist—and to every parent. Protection is needed. A curious toddler might swallow bleach, flush a doll down the toilet, or throw a cat out the window, all to see what happens next.

Finally, in the sixth stage (18 to 24 months), toddlers use *mental combinations*, intellectual experimentation via imagination that can supersede the active experimentation of stage five. Because they combine ideas, stage-six toddlers can pretend as well as think about the consequences of what they do, hesitating a moment before yanking the cat's tail or dropping a raw egg on the floor.

Stage-six toddlers can remember what they have seen and do it later themselves, an ability Piaget called *deferred imitation*. Newer research finds that some accomplishments that Piaget pegged for stage six—including pretending and deferred imitation—begin much earlier.

However, although he was wrong on the timing of his stages, Piaget was right to describe babies as avid and active learners who "learn so fast and so well" (Xu & Kushnir, 2013, p. 28). His main mistake was underestimating how rapidly their learning can occur.

◑ Especially for Parents One parent wants to put all breakable or dangerous objects away because the toddler is able to move around independently. The other parent says that the baby should learn not to touch certain things. Who is right? (see response, page 171)

WHAT HAVE YOU LEARNED?

1. Why did Piaget call cognition in the first two years "sensorimotor intelligence"?
2. How does stage one of sensorimotor intelligence lead to stage two?
3. In sensorimotor intelligence, what is the difference between stages three and four?
4. What is the significance of the concept of object permanence for infant cognition?
5. What does the active experimentation of the stage-five toddler suggest for parents?
6. Why did Piaget underestimate infant cognition?

Language: What Develops in the First Two Years?

Human linguistic ability by age 2 far surpasses that of full-grown adults from every other species. Very young infants listen intensely, responding as best they can. One scholar explains, "infants are acquiring much of their native language before they utter their first word" (Aslin, 2012, p. 191). How do they do it?

The Universal Sequence

The sequence of language development is the same worldwide (see At About This Time). Some children learn several languages, some only one; some learn rapidly, others slowly. But all follow the same path.

Listening and Responding

In every spoken language, adults use higher pitch, simpler words, repetition, varied speed, and exaggerated emotional tone when talking to infants. Babies respond with attention and emotion. By 7 months, they begin to recognize words that are distinctive: *Bottle*, *doggie*, and *mama*, for instance, might be differentiated, not *baby*, *Bobbie*, and *Barbie*.

Infants also like alliteration, rhymes, repetition, melody, rhythm, and varied pitch. Think of your favorite lullaby (itself an alliterative word); obviously, babies prefer sounds over content and singing over talking (Tsang et al., 2017). Early listening abilities and preferences are the result of brain function.

Babbling and Gesturing

Between 6 and 9 months, babies repeat certain syllables (*ma-ma-ma*, *da-da-da*, *ba-ba-ba*), a vocalization called **babbling** because of the way it sounds. Babbling is universal; even deaf babies babble.

babbling An infant's repetition of certain syllables, such as *ba-ba-ba*, that begins when babies are between 6 and 9 months old.

AT ABOUT THIS TIME

The Development of Spoken Language in the First Two Years

Age*	Means of Communication
Newborn	Reflexive communication—cries, movements, facial expressions.
2 months	A range of meaningful noises—cooing, fussing, crying, laughing.
3–6 months	New sounds, including squeals, growls, croons, trills, vowel sounds.
6–10 months	Babbling, including both consonant and vowel sounds repeated in syllables.
10–12 months	Comprehension of simple words; speechlike intonations; specific vocalizations that have meaning to those who know the infant well. Deaf babies express their first signs; hearing babies also use specific gestures (e.g., pointing) to communicate.
12 months	First spoken words that are recognizably part of the native language.
13–18 months	Slow growth of vocabulary, up to about 50 words.
18 months	Naming explosion—three or more words learned per day. Much variation: Some toddlers do not yet speak.*
21 months	First two-word sentence.
24 months	Multiword sentences. Half of the toddler's utterances are two or more words long.

*The ages in this table reflect norms. Many healthy, intelligent children attain each linguistic accomplishment earlier or later than indicated here. However, if a baby is late to babble, speak a word, or put words together, that may indicate a speech, hearing, or cognitive problem. A professional should determine whether something is amiss.

Who Is Babbling? Probably both the 6-month-old and the 27-year-old. During every day of infancy, mothers and babies communicate with noises, movements, and expressions.

holophrase A single word that is used to express a complete, meaningful thought.

Show Me Where Pointing is one of the earliest forms of communication, emerging at about 10 months. As you see here, pointing is useful lifelong for humans.

Caregivers usually encourage those noises, and it is wise that they do so. Babbling predicts later vocabulary, even more than the other major influence—the education of the mother (McGillion et al., 2017).

Before uttering their first word, infants notice patterns of speech, such as which sounds are commonly spoken together. A baby who often hears that something is "pretty" expects the sound of *prit* to be followed by *tee* (MacWhinney, 2015) and is startled if someone says "prit-if."

Infants also learn the relationship between mouth movements and sound. In one study, 8-month-olds watched a film of someone speaking, with the audio a fraction of a second ahead of the video. Even when the actor spoke an unknown language, babies noticed the mistiming (Pons & Lewkowicz, 2014).

Some caregivers, recognizing the power of gestures, teach "baby signs" to their 6- to 12-month-olds, who communicate with hand signs months before they move their tongues, lips, and jaws to make words. There is no evidence that baby signing accelerates talking (as had been claimed), but it may make parents more responsive, which itself is an advantage (Kirk et al., 2013). Every gesture and movement aid communication.

For deaf babies, sign language is crucial in the first year: It not only predicts later ability to communicate with signs but also advances cognitive development (Hall et al., 2017). Remember how important gaze-following was for infant learning? Deaf infants are even better at gaze-following than hearing infants, because they rely on visual signs from their caregivers (Brooks et al., 2019).

Even for hearing babies and without adult signing, gestures are a powerful means of communication (Goldin-Meadow, 2015). One early gesture is pointing and responding to pointing from someone else. The latter requires something quite sophisticated—understanding another person's perspective.

Most animals cannot interpret pointing, but before they are a year old most humans look toward wherever someone else points and already point with their tiny index fingers. Pointing is well developed by 12 months, especially when the person who is pointing also speaks (e.g., "look at that") (Daum et al., 2013).

Infants younger than a year old who are adept at pointing tend to be those who will soon begin talking. That is one reason adults need to respond to pointing as if it is intended to communicate—which it is (Bohn & Köymen, 2018).

First Words

Finally, at about a year, the average baby utters a few words, understood by caregivers if not by strangers. Those words often coincide with the age when walking begins. It may be that a certain amount of brain maturation is needed for both, or it may be that walking allows a new set of questions and makes it more important that people talk to the baby (Walle & Campos, 2014). Following the first words, spoken vocabulary increases gradually (perhaps one new word a week). Meanings are learned rapidly; babies understand about 10 times more than they can say.

Initially, the first words are merely labels for familiar things (*mama* and *dada* are common), but each early word soon becomes a **holophrase,** a single word that expresses an entire thought. That phrase is accompanied by gestures, facial expressions, and nuances of tone, loudness, and cadence. Imagine meaningful communication in "Dada," "Dada?" and "Dada!" Each is a holophrase.

The Naming Explosion

Spoken vocabulary builds rapidly once the first 50 words are mastered, with 21-month-olds typically saying twice as many words as 18-month-olds (Adamson & Bakeman, 2006). This language spurt is called the **naming explosion** because many early words are nouns, that is, names of persons, places, or things.

Before the explosion, names are already favored. Infants learn the names of each significant caregiver (often *dada, mama, nana, papa, baba, tata*), sibling, and pet. (See Visualizing Development, page 166.) Other frequently uttered words refer to favorite foods (*nana* can mean "banana" as well as "grandma") and to elimination (*pee-pee, wee-wee, poo-poo, ka-ka, doo-doo*).

Notice that all of these words have two identical syllables, a consonant followed by a vowel. Many early words follow that pattern—not just *baba* but also *bobo, bebe, bubu, bibi*. Some of the first words are only slightly more complicated—*ma-me, ama*, and so on. The meanings vary by language, but every baby says such words, and every culture assigns meaning to them. Words that are hard to say are simplified: Rabbits are "bunnies," stomachs are "tummies," and no man waits until his son or daughter can call him Father; he is Daddy or Papa instead.

> **DATA CONNECTIONS: The Development of Spoken Language in the First Two Years** reviews communication milestones for infants and toddlers. LaunchPad.

Cultural Differences

Early communication transcends culture. In one study, 102 adults listened to 40 recorded infant sounds and were asked which of five possibilities (pointing, giving, protesting, action request, food request) was the reason for each cry, grunt, or

naming explosion A sudden increase in an infant's vocabulary, especially in the number of nouns, that begins at about 18 months of age.

A CASE TO STUDY

Early Speech

As you read, sensitive caregiving is crucial for early cognition, as babies innately look at and listen to their caregivers in order to learn. For their part, caregivers are sensitive and responsive to the infant's attempts to understand words. This is evident in early language: Parents may understand what an infant is trying to say long before other people do.

Consider 13-month-old Kyle, who was advanced in language development. He knew standard words such as *mama*, but he also knew *da, ba, tam, opma*, and *daes*, which his parents knew to be, respectively, "downstairs," "bottle," "tummy," "oatmeal," and "starfish." He also had a special sound to call squirrels (Lewis et al., 1999).

When acquaintances came to visit, they were often mystified by Kyle's attempts to speak. Who would know that *daes* meant starfish, or how a person might call squirrels? Only Kyle and his very astute parents.

Even a caring grandmother might not interpret correctly. I know this personally. I was caring for my 16-month-old grandson when he said, "Mama, mama." He looked directly at me, and he didn't seem wistful.

"Mommy's not here," I told him. That didn't stop him; he repeated "mama, mama," more as a command than a complaint. I tried several things. I know that some languages use "ma" for milk. I offered some in his sippy cup. He said, "No, no."

When his father appeared, Isaac repeated "mama." Then his dad lifted him, and Isaac cuddled in his arms. I asked Oscar what "mama" means. His reply: "Pick me up."

I now understand Isaac's logic: When he saw his mother, he said "mama" and she picked him up. His parents understood and responded to his words and gestures.

Now Isaac is a proficient talker, explaining about bird families (pigeons are the parents, because they are bigger), about who should get a seat on the subway (it is Isaac, because, as he says very plaintively to other riders, "I need to sit down"; to my embarrassment, he usually gets a seat, so his words are reinforced), and about what his brother has done wrong (explained in detail to his parents, who listen sympathetically but almost never punish the older boy).

I also listen to Isaac's current chatter, repeating some of his phrases and saying "uh-huh," knowing that early adult responses continue to affect later talking. Isaac is well on his way to becoming a highly verbal adult.

What Does He See? All children stare out of windows, but only some are told the names of the various cloud formations and the landscape below. Does this matter for language learning?

whatever. Half of the sounds, and about half of the adults, were from Scotland and the other half from Uganda.

Adults in both cultures scored significantly better than chance (although no group or individual got everything right). The number correct was close to the same whether the sounds came from Scottish or Ugandan infants, or whether the adults were parents or not (Kersken et al., 2017). Human baby sounds are understood by humans no matter what language the community speaks.

However, cultures and families vary in how much child-directed speech children hear. Some parents read to their infants, teach them signs, and respond to every burp or fart as if it were an attempt to talk. Other parents are much less verbal. They use gestures and touch; they say "hush" and "no" instead of expanding vocabulary.

Traditionally, in small agricultural communities, the goal was for everyone to be "strong and silent." If adults talked too much, they might be called blabbermouths or gossips; a good worker did not waste time in conversation.

In some rural areas of the world, that notion remains. One such place is in Senegal, where mothers traditionally feared talking to their babies lest that might encourage evil spirits to take over the child (Zeitlin, 2011).

However, communication is crucial in the twenty-first-century global economy. Government, teachers, and most parents recognize this: A child's first words may be celebrated as much or more than a child's first steps. But some parents are unaware that responding to preverbal vocalizations promotes speech later on.

In one study in Senegal, professionals from the local community (fluent in Woloff, the language spoken by the people) taught mothers in some villages about infant development, including language. A year later those babies were compared to babies in similar villages where the educational intervention had not been offered.

The newly educated mothers talked more to their babies, and the babies, in turn, talked more, with 20 more utterances in five minutes than the control group (A. Weber et al., 2017). The researchers were careful not to challenge the traditional notions directly; instead they taught how early language development advanced infant cognition. The mothers applied what they learned, and the babies responded.

Putting Words Together

grammar All of the methods—word order, verb forms, and so on—that languages use to communicate meaning, apart from the words themselves.

Grammar includes all of the methods that languages use to communicate meaning. There are many ways to add letters to words and to put words together—that is grammar.

Word order, prefixes, suffixes, intonation, verb forms, pronouns and negations, prepositions and articles—all of these are aspects of grammar. Grammar can be discerned in holophrases because one word can be spoken differently depending on meaning. However, grammar becomes essential when babies combine words (Bremner & Wachs, 2010). That typically happens between 18 and 24 months.

For example, "Baby cry" and "More juice" follow grammatical word order. Children do not usually say "Juice more," and even toddlers know that "Cry baby" is not the same as "Baby cry." By age 2, children combine three words. English grammar uses subject–verb–object order; the grammar of other languages use other sequences and their toddlers do the same.

For example, English-speaking toddlers say, "Mommy read book" rather than any of the five other possible sequences of those three words. Adults might say the same three-word sentence with a few grammatical changes, "Mom reads e-books," not "e-books reads Mom." As you see, grammar changes meanings markedly.

Children's proficiency in grammar correlates with sentence length, which is why **mean length of utterance (MLU)** is used to measure a child's language progress (e.g., Miyata et al., 2013). The child who says, "The baby is crying" is more advanced than the child who says, "Baby crying" or simply, "Baby cry!"

Theories of Language Learning

Worldwide, people who are not yet 2 years old express hopes, fears, and memories—sometimes in more than one language. By adolescence, people communicate with nuanced words and gestures, some writing poems and lyrics that move thousands of their co-linguists. How is language learned so easily and so well?

Answers come from at least three schools of thought. The first theory says that infants are directly taught, the second that social impulses propel infants to communicate, and the third that infants understand language because of genetic brain structures that arose more than 100,000 years ago.

Theory One: Infants Need to Be Taught

One idea arises from behaviorism. The essential idea is that learning is acquired, step by step, through association and reinforcement.

B. F. Skinner (1957) noticed that spontaneous babbling is usually reinforced. Often, when a baby says "ma-ma-ma-ma," a grinning mother appears, repeating the sound and showering the baby with attention, praise, and perhaps food. This is especially likely if the baby is a firstborn, which may explain why later-born children, on average, have smaller vocabularies than the oldest child in a family.

Repetition strengthens associations, so infants learn language faster if parents speak to them often. Few parents know this theory, but many use behaviorist techniques by praising and responding to the toddler's simple, mispronounced speech, thus teaching language.

Behaviorists note that some 3-year-olds converse in elaborate sentences; others just barely put one simple word before another. Such variations correlate with the amount of language each child has heard.

Indeed, to some extent infants are "statistical learners" of language, deciding the meanings and boundaries of words based on how often those sounds are heard (Saffran & Kirkham, 2018). Parents of the most verbal children teach language throughout infancy—singing, explaining, listening, responding, and reading to their babies every day, giving their children a rich trove of verbal data, long before the infant utters a first spoken word (see **Figure 6.3**).

mean length of utterance (MLU) The average number of words in a typical sentence (called utterance because children may not talk in complete sentences). MLU is often used to measure language development.

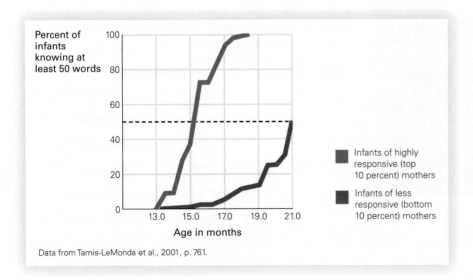

Data from Tamis-LeMonda et al., 2001, p. 761.

FIGURE 6.3

Maternal Responsiveness and Infants' Language Acquisition Learning the first 50 words is a milestone in early language acquisition, as it predicts the arrival of the naming explosion and the multiword sentence a few weeks later. Researchers found that half of the infants of highly responsive mothers (top 10 percent) reached this milestone at 15 months. The infants of less responsive mothers (bottom 10 percent) lagged significantly behind, with half of them at the 50-word level at 21 months.

Early Communication and Language

Communication Milestones: The First Two Years

These are norms. Many intelligent and healthy babies vary in the age at which they reach these milestones.

Months	Communication Milestone
0	Reflexive communication—cries, movements, facial expressions
1	Recognizes some sounds Makes several different cries and sounds Turns toward familiar sounds
3	A range of meaningful noises—cooing, fussing, crying, laughing Social smile well established Laughter begins Imitates movements Enjoys interaction with others
6	New sounds, including squeals, growls, croons, trills, vowel sounds Meaningful gestures including showing excitement (waving arms and legs) Expresses negative feelings (with face and arms) Capable of distinguishing emotion by tone of voice Responds to noises by making sounds Uses noise to express joy and unhappiness Babbles, including both consonant and vowel sounds repeated in syllables
10	Makes simple gestures, like raising arms for "pick me up" Recognizes pointing Makes a sound (not in recognizable language) to indicate a particular thing Responds to simple requests
12	More gestures, such as shaking head for "no" Babbles with inflection, intonation Names familiar people (like "mama," "dada," "nana") Uses exclamations, such as "uh-oh!" Tries to imitate words Points and responds to pointing First spoken words
18	Combines two words (like "Daddy bye-bye") Slow growth of vocabulary, up to about 50 words Language use focuses on 10–30 holophrases Uses nouns and verbs Uses movement, including running and throwing, to indicate emotion Naming explosion may begin, three or more words learned per day
24	Combines three or four words together; half the toddler's utterances are two or more words long Uses adjectives and adverbs ("blue," "big," "gentle") Sings simple songs

Information from American Academy of Pediatrics

Universal First Words

Across cultures, babies' first words are remarkably similar. The words for mother and father are recognizable in almost any language. Most children will learn to name their immediate family and caregivers between the ages of 12 and 18 months.

Language	Mother	Father
English	mama, mommy	dada. daddy
Spanish	mama	papa
French	maman, mama	papa
Italian	mamma	bebbo, papa
Latvian	mama	te-te
Syrian Arabic	mama	babe
Bantu	be-mama	taata
Swahili	mama	baba
Sanskrit	nana	tata
Hebrew	ema	abba
Korean	oma	apa

ampyang/iStock/Getty Images

Mastering Language

Children's use of language becomes more complex as they acquire more words and begin to master grammar and usage. A child's spoken words or sounds (utterances) are broken down into the smallest units of language to determine their length and complexity:

MEAN LENGTH OF UTTERANCE (MLU), ILLUSTRATED

"Baby!" = 1

"Baby + Sleep" = 2

"Baby + Sleep + ing" = 3

"Shh! + Baby + Sleep + ing" = 4

"Shh! + Baby + is + Sleep + ing" = 5

"Shh! + The + Baby + is + Sleep + ing" = 6

Theory Two: Social Impulses Foster Infant Language

The second theory arises from the sociocultural reason for language: communication. According to this perspective, infants communicate because humans are social beings, dependent on one another for survival and joy.

All human infants (and no chimpanzees) seek to master words and grammar in order to join the social world (Tomasello & Herrmann, 2010). According to this perspective, it is the social function of speech, not the words, that undergirds early language.

This theory challenges child-directed videos, CDs, and downloads named to appeal to parents (*Baby Einstein, Brainy Baby,* and *Mozart for Mommies and Daddies—Jumpstart your Newborn's I.Q.*). Since early language development is impressive, even explosive, some parents who allow infants to watch such programs believe that the rapid language learning is aided by video. Commercial apps for tablets and smartphones, such as *Shapes Game HD* and *VocabuLarry*, have joined the market.

However, developmental research finds that screen time during infancy may be harmful, because it avoids the social interaction that is essential for learning to communicate. One recent study found that toddlers could learn a word from either a book or a video but that only book-learning, not video-learning, enabled children to use the new word in another context (Strouse & Ganea, 2017). When the parents read the books, they often held their baby, spoke with clear and animated words, and allowed the child to grab pages, make noises, and point to pictures. That was an engaging social experience.

Another study focused on teaching baby signs, 18 hand gestures that refer to particular objects (Dayanim & Namy, 2015). The babies in this study were 15 months old, an age at which all babies use gestures and are poised to learn object names. The 18 signs referred to common early words, such as *baby, ball, banana, bird, cat,* and *dog.*

In this study, the toddlers were divided into four groups: (1) video only, (2) video with parent watching and reinforcing, (3) book instruction with parent reading and reinforcing, and (4) no instruction. Not surprisingly, the no-instruction group learned words (as every normal toddler does) but not signs, and the other three groups learned some signs. The two groups with parent instruction learned most, with the book-reading group remembering signs better than either video group. Why?

The crucial factor seemed to be parent interaction. When parents watch a video with their infants, they talk less than when they read a book or play with toys (Anderson & Hanson, 2016). Since adult input is essential for language learning, cognitive development is reduced by video time. Infants are most likely to understand and apply what they have learned when they learn directly from another person (Barr, 2013).

Screen time cannot "substitute for responsive, loving face-to-face relationships" (Lemish & Kolucki, 2013, p. 335). Direct social interaction is pivotal for language, according to theory two.

Theory Three: Infants Teach Themselves

A third theory holds that language learning is genetically programmed. Adults need not teach it (theory one), nor is it a byproduct of social interaction (theory two). Instead, it arises from a particular gene (FOXP2), brain maturation, and the overall human impulse to imitate.

Caught in the Middle Parents try to limit screen time, but children are beguiled and bombarded from many sides.

"Keep in mind, this all counts as screen time."

Emily Flake/The New Yorker Collection/The Cartoon Bank

language acquisition device (LAD)
Chomsky's term for a hypothesized mental structure that enables humans to learn language, including the basic aspects of grammar, vocabulary, and intonation.

Family Values Every family encourages the values and abilities that their children need to be successful adults. For this family in Ecuador, that means strong legs and lungs to climb the Andes, respecting their parents, and keeping quiet unless spoken to. A "man of few words" is admired. By contrast, many North American parents babble in response to infant babble, celebrate the first spoken word, and stop their conversation to listen to an interrupting child. If a student never talks in class, or another student blurts out irrelevant questions. Culture talks!

🔵🔵 **Observation Quiz** If this is a typical scene, what family values are evident? (see answer, page 171) ↓

Steven J. Kazlowski/Alamy

For example, English articles (*the, an, a*) signal that the next word will be the name of an object. Since babies have "an innate base" that primes them to learn, articles facilitate learning nouns (Shi, 2014, p. 9). Articles prove to be a useful clue for infants learning English but are frustrating for anyone who learns English as an adult. Adults from other linguistic communities may be highly intelligent and motivated, but their language-learning genes are past the sensitive learning time.

Our ancestors were genetically programmed to imitate for survival, but until a few millennia ago, no one needed to learn languages other than their own. Thus, human genes allow experience-dependent language learning, pruning the connections that our particular language does not need. If we want to learn another language in adulthood, our brains cannot resurrect the connections we lost in infancy.

The prime spokesman for this perspective was Noam Chomsky (1968, 1980). Although behaviorists focus on variations among children in vocabulary size, Chomsky focused on similarities in language acquisition—the evolutionary universals, not the differences.

Noting that all young children master basic grammar according to a schedule, Chomsky hypothesized that children are born with a brain structure he called a **language acquisition device (LAD),** which allows children, as their brains develop, to derive the rules of grammar quickly and effectively from the speech they hear every day.

For example, everywhere, a raised tone indicates a question, and infants prefer questions to declarative statements (Soderstrom et al., 2011). This suggests that infants are wired to talk, and caregivers universally ask them questions long before they can answer back.

According to theory three, language is experience-expectant, as the developing brain quickly and efficiently connects neurons to support whichever language the infant hears. Because of this experience-expectancy, the various languages of the world are all logical, coherent, and systematic. Then some experience-dependent learning occurs as each brain adjusts to a particular language.

The LAD works for deaf infants as well. All 6-month-olds, hearing or not, prefer to look at sign language over nonlinguistic pantomime. For hearing infants, this preference disappears by 10 months, but deaf infants begin signing at that time, which is their particular expression of the universal LAD.

All True?

A master linguist explains that "the human mind is a hybrid system," perhaps using different parts of the brain for each kind of learning (Pinker, 1999, p. 279).

The idea that every theory is partially correct may seem idealistic. However, many scientists who are working on extending and interpreting research on language acquisition have arrived at this conclusion. They contend that language learning is neither the direct product of repeated input (behaviorism) nor the result of a specific human neurological capacity (LAD). Rather, from an evolutionary perspective, "different elements of the language apparatus may have evolved in different ways," and thus, a "piecemeal and empirical" approach is needed (Marcus & Rabagliati, 2009, p. 281).

Neuroscience is the most recent method to investigate the development of language. It was once thought that language was located in two specific regions of the brain (Wernicke's area and Broca's area). But now neuroscientists are convinced that language arises from other regions as well. Some genes and regions are crucial, but hundreds of genes and many brain regions contribute to linguistic fluency.

Neuroscientists describing language development write about "connections," "networks," "circuits," and "hubs" to capture the idea that language is interrelated and complex (Dehaene-Lambertz, 2017; Pulvermüller, 2018). Even when the focus is simply on talking, one neuroscientist notes that "speech is encoded at multiple levels in different parallel pathways" (Dehaene-Lambertz, 2017, p. 52).

That neuroscientist begins her detailed description of the infant brain and language with the same amazement that traditional linguists have expressed for decades:

> For thousands of years and across numerous cultures, human infants are able to perfectly master oral or signed language in only a few years. No other machine, be it silicon or carbon based, is able to reach the same level of expertise.
>
> *[Dehaene-Lambertz, 2017, p. 48]*

Language is closely linked, in the early months, with cognition. As one review of the many pathways between learning to talk and understanding the world concludes:

> A constellation of factors that are unique to human development—infants' prolonged period of dependency, exquisite sensitivity to experience, and powerful learning strategies—collectively spark a cascade of developmental change whose ultimate result is the acquisition of language and its unparalleled interface with cognition.
>
> *[Perszyk & Waxman, 2018, p. 246]*

The words *constellation* and *cascade* signify that many brain structures and social inputs result in both language and learning.

From this we are led to an overall conclusion. Infants are amazing and active learners who advance their cognition in many ways—through understanding of people, objects, memory, and communication. Remember that before Piaget, many experts assumed that babies did not yet learn or think. How wrong they were!

WHAT HAVE YOU LEARNED?

1. What aspects of language develop in the first year?
2. When does vocabulary develop slowly and when does it develop quickly?
3. What is typical of the first words that infants speak?
4. What indicates that toddlers use some grammar?
5. According to behaviorism, how do adults teach infants to talk?
6. According to sociocultural theory, why do infants try to communicate?
7. Do people really have a language acquisition device?
8. Why do developmentalists accept several theories of language development?

SUMMARY

The Eager Mind

1. Infants learn so quickly that developmentalists now suggest that some basic understanding is programmed into the brain—no experience necessary.

2. Infants distinguish all of the possible sounds that human languages use. This ability decreases as babies become more attuned to whatever languages they hear.

3. Infants observe events and people to learn what merits their attention. Face recognition is an inborn ability that is refined by experience. They particularly attend to where adults are looking, an inborn impulse called gaze-following.

4. Information-processing theory, which looks at each step of the thinking process, helps researchers understand that every moment of experience aids cognition, enhanced by the innate programming of the infant mind.

5. Evolutionary theory is also relevant, as it emphasizes the basic cognitive capacities that humans have needed to survive. The emphasis is not only on inborn cognitive structures but also on plasticity, which allows each infant to adapt to the environment.

6. Infant memory is fragile but not completely absent. Reminder sessions help trigger memories, and young brains learn motor sequences and respond to repeated emotions (their own and those of other people) long before explicit memory, using words.

Piaget's Sensorimotor Intelligence

7. Piaget realized that very young infants are active learners who seek to understand their complex observations and experiences. The six stages of sensorimotor intelligence involve early adaptation to experience.

8. Sensorimotor intelligence begins with reflexes and ends with mental combinations. The six stages occur in pairs, with each pair characterized by a circular reaction.

9. Infants first react to their own bodies (primary), then respond to other people and things (secondary), and finally, in the stage of tertiary circular reactions, infants become more goal-oriented, creative, and experimental as "little scientists."

10. Infants gradually develop an understanding of objects. According to Piaget's classic experiments, infants understand object permanence and begin to search for hidden objects at about 8 months. Newer research finds that Piaget underestimated infant cognition.

Language: What Develops in the First Two Years?

11. Attempts to communicate are apparent in the first weeks and months, beginning with noises, facial expressions, and avid listening. Infants babble at about 6 months, understand words and gestures by 10 months, and speak their first words at about 1 year.

12. Vocabulary builds slowly until the infant knows approximately 50 words. Then the naming explosion begins. Grammar is evident in the first holophrases, and combining words together in proper sequence is further evidence that babies learn grammar as well as vocabulary.

13. Toward the end of the second year, toddlers express wishes and emotions in short sentences. Variation is evident, in part because of caregiver attention. Some babies are already bilingual.

14. Theories differ in explaining how infants learn language—whether infants must be taught or that social impulses foster language learning or that brains are genetically attuned to language as soon as the requisite maturation occurs.

15. Each theory of language learning is confirmed by research. Developmental scientists find that many parts of the brain, and many strategies for learning, result in early language accomplishments. Current research, with the benefit of advances in neuroscience, confirms that conclusion.

KEY TERMS
. . . .

sensorimotor intelligence (p. 156)

primary circular reactions (p. 157)

secondary circular reactions (p. 158)

object permanence (p. 158)

tertiary circular reactions (p. 159)

"little scientist" (p. 160)

babbling (p. 161)

holophrase (p. 162)

naming explosion (p. 163)

grammar (p. 164)

mean length of utterance (MLU) (p. 165)

language acquisition device (LAD) (p. 168)

APPLICATIONS
. . . .

1. Elicit vocalizations from an infant—babbling if the baby is under age 1, using words if the baby is older. Write down all of the baby's communication for 10 minutes. Then ask the primary caregiver to elicit vocalizations for 10 minutes, and write these down. What differences are apparent between the baby's two attempts at communication? Compare your findings with the norms described in the chapter.

2. Many educators recommend that parents read to babies every day, even before 1 year of age. What theory of language

development does this reflect and why? Ask several parents whether they did so, and why or why not.

3. Test a toddler's ability to pretend and to imitate, as Piaget would expect. Use a doll or a toy car and pretend with it, such as feeding the doll or making the car travel. Then see whether the child will do it. This experiment can be more elaborate if the child succeeds.

Especially For ANSWERS
. . . .

Response for Educators (from p. 148): Probably both. Infants love to communicate, and they seek every possible way to do so. Therefore, the teachers should try to understand the baby and the baby's parents, but they should also start teaching the baby the majority language of the school.

Response for Parents (from p. 158): Both decisions should be made within the first month, during the stage of reflexes. If parents wait until the infant is 4 months or older, they may discover that they are too late. It is difficult to introduce a bottle to a 4-month-old who has never sucked on an artificial nipple or a pacifier to a baby who has already adapted the sucking reflex to a thumb.

Response for Parents (from p. 160): It is easier and safer to babyproof the house because toddlers, being "little scientists,"

want to explore. However, it is important for both parents to encourage and guide the baby. If having untouchable items prevents a major conflict between the adults, that might be the best choice.

Observation Quiz ANSWERS

Answer to Observation Quiz (from p. 155): The mobile is a good addition—colorful and too high for the baby to reach. (Let's hope it is securely fastened and those strings are strong and tight!) But there are two things that are not what a cognitive developmentalist would recommend. First, the crib and the wall are both plain white, limiting what the baby can focus on, and second, the crib bumper is a SIDS risk.

Answer to Observation Quiz (from p. 168): Not social interaction, not talking. Instead, all quietly stare at sky and terrain; awe of nature may be a family value. Hierarchy and gender seem significant: The father is distant and above all, the mother is busy, the children are below the parents. Do only males wear hats?

The First Two Years: Psychosocial Development

What Will You Know?

1. Will a difficult newborn become a difficult child?
2. What do infants do if they are securely attached to their caregivers?
3. Is it ideal for infants to be cared for exclusively by their mothers?

Bone-tired after a day of teaching, I was grateful to find a seat on the crowded downtown train. At the next stop, more people boarded, including a mother who stood in front of me. She held a well-dressed baby, about 18 months old, in one arm, and she wrapped her other arm around a pole as she held several heavy bags. I thought of offering my seat. Too tired. But at least I could hold her bags on my lap.

"Can I help you?" I asked, offering a hand. Wordlessly she handed me . . . the baby! The toddler sat quietly and listened as I expressed admiration for her socks, pointed out the red and blue stripes, and then sang a lullaby. I could feel her body relax. Her eyes stayed on her mother.

I should not have been surprised. Mothers everywhere need help with infant care, and strangers everywhere are *allomothers*. We all attend to infant cries, bring gifts to newborns, and study, consult doctors, or volunteer to become helpers. I noticed other riders on that train watching me with that little girl. Mother, strangers, and I were doing what our culture expects. And the baby did what healthy babies do. She responded to my off-key singing, reassured by the sight of her mother.

This example opens this chapter because it illustrates infant emotions and caregiver responses. You will read about psychosocial changes over the first two years, from synchrony, to attachment, to social referencing, quite evident in this baby.

At the end of the chapter, we explore a controversy: Who should care for infants? Only mothers, or also fathers, grandmothers, day-care teachers, and strangers? Would you have handed me your baby? Families and cultures answer this question in opposite ways.

Fortunately, as this chapter explains, despite diversity of temperament and caregiving, most infants develop well if their basic emotional needs are met. This baby seemed fine.

Emotional Development

Psychosocial development during infancy can be seen as two interwoven strands—nature/nurture, or universal/particular, or experience-expectant/experience-dependent. To portray these strands with words in a book, we must pull them apart, so this chapter is a zigzag, turning from universal to particular and back, again and again.

Early Emotions

We begin with universal: In their first two years, all infants progress from reactive pain and pleasure to complex patterns of socioemotional awareness, a movement from basic instincts to learned responses (see At About This Time).

At first, comfort predominates: Newborns are content and relaxed when fed and drifting off to sleep. Discomfort is also part of daily life: Newborns cry when they are hurt or hungry, tired or frightened (as by a loud noise or a sudden loss of support).

By the second week and increasing to six weeks, some infants have bouts of uncontrollable crying, commonly referred to as *colic*, perhaps the result of immature digestion or the infant version of a migraine headache (Gelfand, 2018). Others have *reflux*, probably the result of immature swallowing. About 10 to 20 percent of babies cry "excessively," defined as more than three hours a day, more than three days a week, for more than three weeks.

Fortunately, early emotions do not necessarily predict later life. A longitudinal study of 291 infants found that, by age 2, infants with colic were no more likely to have behavioral problems than those without (G. Bell et al., 2018). As you will later read, newborn temperament is shaped by caregiver response.

Smiling and Laughing

Soon, crying decreases and additional emotions become recognizable. Colic usually subsides by 3 months. Happiness is expressed by a fleeting **social smile**, evoked by a human face at about 6 weeks (Wörmann et al., 2012).

Preterm babies smile later, because the social smile is affected by age since conception, not age since birth (White-Traut et al., 2018). The social smile is universal; all babies do it when they are old enough, evidence of the human social impulse.

Laughter builds over the first months, often in tandem with curiosity: A typical 6-month-old chortles upon discovering new things, particularly social experiences that balance familiarity and surprise, such as Daddy making a funny face. That is just what Piaget would expect, "making interesting experiences last." Very young infants prefer seeing happy faces over sad ones, even if the happy faces are not looking at them (Kim & Johnson, 2013).

Soon happiness becomes more discriminating. In one study, infants first enjoyed a video of dancing to music as it normally occurs, on the beat. Then some watched a video in which the sound track was mismatched with dancing. Eight- to twelve-month-old babies, compared to younger ones, were quite curious—but less delighted—about offbeat dancing. The researchers concluded "babies know bad dancing when they see it" (Hannon et al., 2017).

Anger and Sadness

Crying in pain and smiling in pleasure are soon joined by emotions more responsive to external experiences. Anger is notable at 6 months, usually triggered by frustration.

For example, to study infant emotions, researchers "crouched behind the child and gently restrained his or her arms for 2 min[utes] or until 20 s[econds] of hard crying ensued" (Mills-Koonce et al., 2011, p. 390). "Hard crying" was not rare: Infants hate to be strapped in, caged in, closed in, or just held in place when they want to explore.

In infancy, anger is a healthy response to frustration, unlike sadness, which also may appear in the first months (Thiam et al., 2017). Sadness indicates withdrawal instead of a bid for help, and it is accompanied by a greater increase in the body's

---

AT ABOUT THIS TIME

Developing Emotions

Birth	Distress; contentment
6 weeks	Social smile
3 months	Laughter; curiosity
4 months	Full, responsive smiles
4–8 months	Anger
9–14 months	Fear of social events (strangers, separation from caregiver)
12 months	Fear of unexpected sights and sounds
18 months	Self-awareness; pride; shame; embarrassment

As always, culture and experience influence the norms of development. This is especially true for emotional development after the first 8 months.

social smile A smile evoked by a human face, normally first evident in infants about six weeks after birth.

Developmentally Correct Both Santa's smile and Olivia's grimace are appropriate reactions for people of their age. Adults playing Santa must smile no matter what, and if Olivia smiled, that would be troubling to anyone who knows about 7-month-olds. Yet every Christmas, thousands of parents wait in line to put their infants on the laps of oddly dressed, bearded strangers.

REUTERS/Suzanne Plunkett

production of cortisol. For that reason, developmentalists are troubled if a very young baby is sad.

Fear

Fear begins with unexpected events, such as fear of falling or of loud noises, but soon it involves human interaction. Indeed, being frightened of something, from snakes to strangers, may depend on three factors: awareness of discrepancy, inborn temperament, and social context. Two kinds of social fear are typical, increasing from the middle of the first year:

- **Separation anxiety**—clinging and crying when a familiar caregiver is about to leave. Some separation anxiety is normal at age 1, may be more intense by age 2, and then usually subsides.
- **Stranger wariness**—fear of unfamiliar people, especially when they move too close, too quickly. Knowing that, in the subway incident that opened this chapter, I first distracted the baby by talking about her socks, and then sang to her. Babies everywhere respond to song.

If separation anxiety and stranger fear remain intense after age 3, impairing a child's ability to leave home, to go to school, or to play with other children, that is an emotional disorder. According to the DSM-5, separation anxiety becomes a disorder when it lingers into childhood or adolescence (American Psychiatric Association, 2013); clinicians may diagnose it in adults, as well, especially if symptoms persist for at least six months (Bögels et al., 2013).

Likewise, stranger wariness may continue, becoming social phobia or generalized anxiety (Rudaz et al., 2017). Both emotions are expected at age 1, becoming possible problems later. And both are experience-expectant and then experience-dependent, responsive to context (LoBue & Adolph, 2019). Infants at home with their mothers when a stranger comes to visit are likely to smile, not be afraid. But if a stranger yells and approaches quickly to a few inches from the babies face with an angry expression, almost any 1-year-old cries and tries to move away.

Many 1-year-olds are curious but wary of anything unexpected, from the flush of the toilet to the pop of a jack-in-the-box, from closing elevator doors to the tail-wagging approach of a dog. With repeated experience and reassurance, older infants might enjoy flushing the toilet (again and again) or calling the dog (crying if the dog does *not* come). Note the transition from instinct to learning to thought, from the amygdala to the cortex.

Toddlers' Emotions

Emotions take on new strength during toddlerhood, as both memory and mobility advance. Context is crucial for fear (LoBue & Adolph, 2019). Throughout the second year and beyond, emotions become less frequent but more focused, targeted toward infuriating or terrifying or exhilarating experiences. Both laughing and crying are louder and more discriminating.

Temper Tantrums

The new strength of emotions is apparent in temper tantrums. Toddlers are famous for fury. When something angers them, they might yell, scream, cry, hit, and throw themselves on the floor. Logic is beyond them: If adults tease or get angry, that makes it worse. Parental insistence on obedience exacerbates the tantrum (Cierpka & Cierpka, 2016).

One child said, "I don't want my feet. Take my feet off. I don't want my feet." Her mother tried logic, which didn't work, and then offered to get scissors and cut

---

separation anxiety An infant's distress when a familiar caregiver leaves; most obvious between 9 and 14 months.

stranger wariness An infant's expression of concern—a quiet stare while clinging to a familiar person, or a look of fear—when a stranger appears.

Especially for Nurses and Pediatricians Parents come to you concerned that their 1-year-old hides her face and holds onto them tightly whenever a stranger appears. What do you tell them? (see response, page 201)

off the offending feet. A new wail erupted, with a loud shriek "Nooooo!" (Katrina, quoted in Vedantam, 2011).

With temper tantrums, soon sadness comes to the fore. Then comfort—not punishment—is helpful (Green et al., 2011). Outbursts of anger are typical at ages 1 and 2, but if they persist and become destructive, that signifies trouble, in parent or child (Cierpka & Cierpka, 2016).

As with these examples, a toddler's innate reactions may evolve into moral values and psychic responses, with specifics depending on parents and experiences. For example, many children take off their clothes in public, unaware that nudity is taboo. Children are born curious and uninhibited: Shame and self-consciousness are learned.

Self and Others

Temper can be seen as an expression of selfhood, as can other common toddler emotions: pride, shame, jealousy, embarrassment, disgust, and guilt. These emotions may begin with inborn sensitivities, but they involve social awareness.

Such awareness typically emerges from family interaction, especially the relationship between caregiver and baby. For instance, in a study of infant jealousy, mothers ignored their own baby and attended to another infant. That made the babies move closer to their mothers, bidding for attention. Their brain activity also registered social emotions (Mize et al., 2014).

Positive emotions also show social awareness and learning. Many toddlers try to help a stranger who has dropped something or who is searching for a hidden object, and some express sympathy for someone who hurt themselves (Aitken et al., 2019).

Over time, children learn when and whom to help by watching adults. Some adults donate to panhandlers, others look away, and still others complain to the police. Attitudes about ethnicity, or immigration, or clothing, begin with the infant's preference for the familiar and interest in novelty, and then upbringing adds appreciation or rejection.

Recognizing the Self

In addition to social awareness, another foundation for emotional growth is **self-awareness**, the realization that one's body, mind, and activities are distinct from those of other people. Closely following the new mobility that results from walking is an emerging sense of individuality.

In a classic experiment (Lewis & Brooks, 1978), 9- to 24-month-olds looked into a mirror after a dot of rouge had been surreptitiously put on their noses. If they reacted by touching the red dot on their noses, that meant they knew the mirror showed their own faces. None of the babies younger than 12 months did that, although they sometimes smiled and touched the dot on the "other" baby in the mirror.

Between 15 and 24 months, babies become self-aware, touching their own red noses with curiosity and puzzlement. Self-recognition in the mirror/rouge test (and in photographs) usually emerges with two other advances: pretending and using first-person pronouns (*I, me, mine, myself, my*) (Lewis, 2010). Thus, "an explicit and hence reflective conception of the self is apparent at the early stage of language acquisition at around the same age that infants begin to recognize themselves in mirrors" (Rochat, 2013, p. 388).

This illustrates the interplay of infant abilities—walking, talking, social awareness, and emotional self-understanding all combine to make the 18-month-old quite unlike the 8-month-old. To repeat a now-familiar theme, timing and expression of self-awareness are affected by the social context (Ross et al., 2017).

VIDEO ACTIVITY: Self-Awareness and the Rouge Test shows the famous assessment of how and when self-awareness appears in infancy.

self-awareness A person's realization that he or she is a distinct individual whose body, mind, and actions are separate from those of other people.

My Finger, My Body, and Me Mirror self-recognition is particularly important in her case, as this 2-year-old has a twin sister. Parents may enjoy dressing twins alike and giving them rhyming names, but each baby needs to know she is an individual, not just a twin.

Some cultures value independence and others do not, and each individual within each culture may reinforce those cultural values or may resist them. If you asked a toddler to put away some toys and were told "No," would you be angry or amused? Your answer reflects your culture; babies learn to reflect that as well.

WHAT HAVE YOU LEARNED?

1. What experiences trigger happiness, anger, and fear?

2. How do emotions differ between the first and second year of life?

3. What evidence suggests that caregivers affect infant emotions?

4. What is the significance of how toddlers react to seeing themselves in a mirror?

5. How much of infant emotions is inborn, and how much depends on caregiving?

Temperament and Personality

The early smiles, curiosity, and wariness are universal, but the intensity of those emotions reflects temperament as well as the situation. The same is true for caregivers, whose personality interacts with temperament (their own as well as that of their infant) and the result is a complex cauldron, mixing nature and nurture. Accordingly, you will see that understanding temperament is an important, but complicated topic.

The Biology of Temperament

temperament Inborn differences between one person and another in emotions, activity, and self-regulation. It is measured by the person's typical responses to the environment.

Temperament is defined as the "biologically based core of individual differences in style of approach and response to the environment that is stable across time and situations" (van den Akker et al., 2010, p. 485). "Biologically based" means temperament begins with genes and prenatal determinants, the early manifestation of epigenetics.

Is that accurate? Are babies born with different temperaments? Yes! One team recorded the tone, duration, and intensity of infant cries after the first inoculation, before much experience outside the womb. Cries at this very early stage correlated with later temperament: The loudest and longest screamers become quickest to protest later on (Jong et al., 2010).

By contrast, some young infants are wonderfully sunny, smiling at everyone, fussing more than wailing when hungry or hurt. If their caregivers smile in return, they tend to become happy children (Bridgett et al., 2013). Determining the direction of this correlation, from infant-to-parent or parent-to-infant or both influenced by some third variable (such as poverty or nutrition), is difficult. But various patterns of temperament seem innate, and some stability in temperament is evident over the early months (Planalp & Goldsmith, 2019).

Temperament can make later life easier or more difficult. Even caregiver abuse and neglect seem to be weathered by some babies, with no long-lasting damage to adult personality. Of course, maltreatment is never excusable, but the effects vary, and temperament is one reason (Maltby et al., 2019).

Dimensions of Temperament

In laboratory studies of temperament, 4-month-old infants might see spinning mobiles or hear unusual sounds, and older babies might confront a clown who approaches quickly. During such experiences, some children laugh, some cry, and others are quiet.

Analyzing many studies of temperament, researchers find three distinct traits:

- Effortful control (regulating attention and emotion, self-soothing)
- Negative mood (fearful, angry, unhappy)
- Exuberance (active, social, not shy)

[Lengua et al., 2019]

Personality

Temperament is *not* the same as personality, although temperament may lead to personality differences. Generally, personality traits (e.g., honesty and humility) are heavily influenced by parents and culture, whereas temperamental traits (e.g., shyness and aggression) arise from genes. Of course, nature and nurture always interact. For both temperament and personality, family influences are powerful: Not only do parents affect their children in temperament and personality, but also children affect their parents (Lengua et al., 2019).

In general, infants with difficult temperaments are more likely than other babies to develop emotional problems, especially if their mothers had a difficult pregnancy and were depressed or anxious caregivers (Garthus-Niegel et al., 2017). In that case, the difficult baby affects the stressed parent, as well as vice versa.

Personality of the adult may be a crucial factor; some adults are more relaxed and others more stressed because of their genes and experiences. Researchers have settled on five clusters of personality in adults, known as the *Big Five*: Openness to experience, Conscientiousness, Extraversion, Agreeableness, and Neuroticism. Agreeable adults find parenthood easy; neurotic (worried, anxious) parents are stressed by difficult babies.

Temperament Over the Years

One longitudinal study analyzed temperament at least eight times, in infancy, early childhood, middle childhood, adolescence, and adulthood. The scientists designed laboratory experiments to evoke emotions appropriate for the age of the participants, collected detailed reports from mothers and later from participants themselves, and gathered observational data and physiological evidence, including brain scans (Fox et al., 2001, 2005, 2013; Jarcho et al., 2013; Shechner et al., 2018; Williams et al., 2010).

In early childhood, change was most likely for the inhibited, fearful infants and least likely for the exuberant ones (see **Figure 7.1**). Why was that? Do some parents coax frightened infants to be brave, letting exuberant babies stay happy?

When the fearful children grew up, about half were still fearful. The half who overcame their anxiety had more activation in another part of the brain (the *anterior cingulate cortex*), which signals safety (Shechner et al., 2018). Perhaps if fearful infants were quickly reassured by their caregivers, a neurological link formed between fear and comfort, so that, when anxiety rose later in life, the brain automatically counteracted it.

The researchers found unexpected gender differences. As teenagers, formerly inhibited boys had relatively high rates of drug abuse, but the opposite was found for inhibited girls (Williams et al., 2010). A likely explanation is cultural: Shy boys use drugs

FIGURE 7.1

Do Babies' Temperaments Change?
Sometimes it is possible—especially if they were fearful babies. Adults who are reassuring help children overcome fearfulness. If fearful children do not change, it is not known whether that's because their parents are not sufficiently reassuring (nurture) or because the babies themselves are temperamentally more fearful (nature).

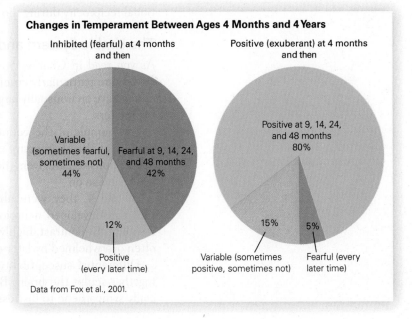

Changes in Temperament Between Ages 4 Months and 4 Years

Inhibited (fearful) at 4 months and then
- Variable (sometimes fearful, sometimes not) 44%
- Fearful at 9, 14, 24, and 48 months 42%
- Positive (every later time) 12%

Positive (exuberant) at 4 months and then
- Positive at 9, 14, 24, and 48 months 80%
- Variable (sometimes positive, sometimes not) 15%
- Fearful (every later time) 5%

Data from Fox et al., 2001.

● **Especially for Nurses** Parents come to you with their fussy 3-month-old. They have read that temperament is "fixed" before birth, and they are worried that their child will always be difficult. What do you tell them? (see response, page 201)

to mask their social anxiety, but shy girls may be more accepted as they are. Other research also finds that shyness is more stable in girls than boys (Poole et al., 2017).

Brain Variations

You read that temperament is "biologically based." That means that brain maturation is crucial for emotional development, particularly for emotions that respond to other people. Experience connects the amygdala and the prefrontal cortex (van Goozen, 2015), teaching infants to align their own feelings with those of their caregivers (Missana et al., 2014). Joy, fear, and excitement become shared, mutual experiences—as anyone who successfully makes a baby laugh knows.

Essentially, connections between innate emotional impulses from the amygdala and experience-based learning shows "dramatic age-dependent improvement," with genes, prenatal influences, and early caregiving all affecting brain growth (Gao et al., 2017). Infant experience leads to adult reactions: If you know someone who cries, laughs, or angers quickly, ask about their first two years of life.

Hormones and Temperament

All social emotions, particularly sadness and fear, affect the hormones and hence the brain. Caregiving matters. Sad and angry infants whose mothers are depressed usually become fearful toddlers and depressed children (Dix & Yan, 2014).

Abuse and unpredictable responses from caregivers are likely among the "early adverse influences [that] have lasting effects on developing neurobiological systems in the brain" (van Goozen, 2015, p. 208). Even worse is a lack of any social responses: That leads to significant brain shrinkage (Marshall, 2014; Sheridan et al., 2018). In-depth studies of the brain and emotions confirm this.

Remember that oxytocin is high in birthing mothers: It predisposes them to bond with their infants. Now consider the other half of the dyad. Infants vary in their level of oxytocin, in part because of how their parents care for them, and that affects their social responses (Feldman, 2012a).

Over the years of childhood, infants have the highest levels of oxytocin, which strongly correlates with an infant's interest in social stimuli (Nishizato et al., 2017). This suggests that hormones that begin with brain signals affect early extraversion. Oxytocin makes infants watch faces, and then people respond, an emotional reciprocity during the early months.

The Limbic System and Temperament

As you read in Chapter 5, those regions of the brain that comprise the limbic system are particularly crucial for emotions. Depending on past experience, some adults have an unusually large amygdala and small hippocampus. For this, caregiving matters.

An example of the connection between the brain and caregiving came from a study of "highly reactive" infants (i.e., intensely fearful, angry, active). If they had responsive caregivers (not hostile or neglectful), they became less fearful, less angry, and so on.

By age 4, they were able to regulate their emotions, presumably because they had developed neurological links between brain excitement and emotional response. By contrast, highly reactive toddlers with less responsive caregivers were often overwhelmed by later emotions (Ursache et al., 2013).

Differential susceptibility is apparent: Innate reactions and caregiver actions together sculpt the brain. Both are affected by culture: Some parents are especially sympathetic to distress, while others ignore crying. Caregiver responses are crucial in infancy, even though the emotional manifestations are more apparent later on (Cole & Hollenstein, 2018).

Genes and prenatal influences on the brain are evident. Some fetuses are exposed to toxic drugs; some to inherit genes, making them vulnerable to autism spectrum disorder; some newborns spend many days in the hospital. Especially for them, postnatal experiences are crucial for emotional development (Gao et al., 2017; White-Traut et al., 2018). (See **Figure 7.2**.)

The social smile, for instance, is fleeting when 2-month-olds, especially those who began life in difficult circumstances, see almost any face. As the brain develops and experience builds, those infants smile more at a familiar caregiver than a stranger, and more at responsive, smiling caregivers than at less interactive caregivers.

When neurons repeatedly fire together, the dendrites and synapses become closely connected. Experience, even at 2 months, matters.

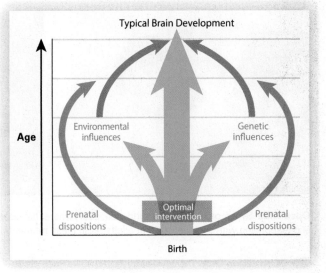

FIGURE 7.2

Seven Arrows Pointing Up This figure is intended to show the ongoing development of the brain. Prenatal, genetic, and experiential influences continue lifelong.

Mothers of Mice

Many studies of emotions begin with mice, mammals with many genetic similarities to people! In classic research, the brains of infant mice released more serotonin when their mothers licked them. That not only increased the mouselings' pleasure but also started epigenetic responses, reducing cortisol from brain and body, including the adrenal glands. The effects were lifelong; those baby mice became smarter and more social adults, with larger brains.

That research with mice has been replicated and extended to many other mammals, including humans. Neuroscientists are awed by the early "remarkable capacity for plastic changes that influence behavioural outcomes throughout the lifetime" (Kolb et al., 2017, p. 1218).

Too much fear and stress makes the brain grow more slowly. Maltreated infants develop abnormal responses to stress, anger, and other emotions, apparent in the hypothalamus, amygdala, hippocampus, and prefrontal cortex (Bernard et al., 2014; Cicchetti, 2013a).

The immune system is also impaired (Hostinar et al., 2018). Consequently, abused children typically become sickly, slow-thinking adults, with erratic emotions, because of what happened inside their brains decades earlier.

WHAT HAVE YOU LEARNED?

1. What is the difference between temperament and personality?

2. What are the three dimensions of temperament?

3. How does caregiving affect temperament?

4. What is the relationship between temperament and maltreatment?

The Development of Social Bonds

Now we return to what is universally true. Humans are, by nature, social creatures. The specifics of social interaction during infancy depend on the age of the person.

Synchrony

Early parent–infant interactions are described as **synchrony**, a mutual exchange with split-second timing. Metaphors for synchrony are often musical—a waltz,

synchrony A coordinated, rapid, and smooth exchange of responses between a caregiver and an infant.

Open Wide Synchrony is evident worldwide. Everywhere, babies watch their parents carefully, hoping for exactly what these three parents express, and responding with such delight that adults relish these moments.

Observation Quiz The universality of synchrony is evident here, not only in the babies but also in the parents, each of whom began at birth with a quite different relationship to the baby. Can you guess what those differences are? (see answer, page 201) ↑

a jazz duet—to emphasize that each partner must be attuned to the other, with moment-by-moment responses. Synchrony increases over the first year. Physiological measures, especially hearts that beat in synch, also indicate synchrony (McFarland et al., 2020).

Long before they can control their hands, infants respond excitedly to caregiver attention by waving their arms. Adults with animated expressions move close, letting that waving arm touch a face or, even better, letting a hand grab hair. This synchrony is part of early adaptation and then "making interesting events last," Piaget's stages 2 and 3 of sensorimotor development (described in Chapter 6). Thus, synchrony aids cognition.

Synchronizing adults open their eyes wide, raise their eyebrows, smack their lips, and emit nonsense sounds. Hair-grabbing might make adults bob their heads back and forth, in a playful attempt to shake off the grab, to the infant's joy. Over the early months, an adult and an infant might develop a routine of hair-grabbing in synchrony.

Another adult and infant might develop a different routine, perhaps with hand-clapping, or lip-smacking, or head-turning. Synchrony becomes a mutual dance, with both knowing the steps. Often mothers and infants engage in "social games," which are routines passed down from adult to infant in that culture. Social games soon become synchronized, with the infant anticipating and reacting to each move (Markova, 2018).

Both Partners Active

Direct observation reveals synchrony; anyone can see it when watching a caregiver play with an infant who is far too young to talk. Adults rarely smile much at newborns until that first social smile, weeks after birth. Smiling is like a switch that lights up the adult and the baby. Soon both partners synchronize smiles, eyes, noises, and movements.

Detailed research, typically with two cameras, one focused on the infant and one on the caregiver, examines the timing of every millisecond of arched eyebrows, widening eyes, pursed lips, and so on. That confirms the tight relationship between adult and infant (Messinger et al., 2010).

Physiological measures, such as heart rate and brain waves, also measure synchrony. That reveals why maternal depression leads to infant depression—the baby picks up on the mother's responses (Atzil et al., 2014).

In every interaction, infants read emotions and develop social skills. Synchrony usually begins with adults imitating infants (not vice versa), responding to barely perceptible facial and bodily motions (Beebe et al., 2016). This helps infants connect their internal state with expressions and behavior.

CHAPTER APP 7

Playground Buddy

IOS:
https://tinyurl.com/waplr56

ANDROID:
https://tinyurl.com/r9jykb7

RELEVANT TOPIC:
Psychosocial development during infancy

With a worldwide database of over 200,000, this app shows caregivers the location and street views of nearby playgrounds. Users can also send the details to others, even those who do not have the app.

The Need for Synchrony

Caregiver responsiveness to infant actions aids psychosocial and biological development, evident in heart rate, weight gain, and brain maturation. Experiments involving the **still-face technique** suggest that synchrony is experience-expectant (needed for normal brain growth) (Hari, 2017; Tronick, 1989; Tronick & Weinberg, 1997). [**Developmental Link:** Experience-expectant and experience-dependent brain function are described in Chapter 5.]

In still-face studies, at first an infant is propped in front of an adult who responds normally. Then, on cue, the adult stops all expression, staring quietly with a "still face" for a minute or two. Sometimes by 2 months, and clearly by 6 months, infants are upset when their caregiver is unresponsive. Babies frown, fuss, drool, look away, kick, cry, or suck their fingers. By 5 months, they also vocalize, as if to say, "React to me!"

One study looked in detail at 4-month-olds during and immediately after a still-face episode (Montirosso et al., 2015). The researchers found three clusters: "socially engaged" (33 percent), "disengaged" (60 percent), and "negatively engaged" (7 percent).

When the mothers were still-faced:

- The socially engaged babies remained active, looking around at other things, apparently expecting that the caregivers would soon resume connection. When the still face was over, they quickly reengaged.
- The disengaged infants became passive, looking sad. When the still face was over, they did not quickly return to normal, although they eventually did.
- The negatively engaged infants were visibly upset and angry, crying during the still face and continuing to cry when it was over.

The mothers of each type differed in how they played with their infants before and after the still face. The socially engaged mothers matched the infants' actions (bobbing heads, opening mouth, and so on), but the negatively engaged mothers almost never matched and sometimes expressed anger—not sympathy—when the baby cried (Montirosso et al., 2015). That absent synchrony is a troubling sign for future emotional and brain development.

Attachment

Attachment—the connection between one person and another, measured by how they respond to each other—is a lifelong process, beginning before birth and continuing even when an aged parent has Alzheimer's disease (Arguz Cildir et al., 2019; Lang & Fowers, 2019). Developmentalists are convinced that attachment is basic to the survival of *Homo sapiens*, with the manifestation dependent on culture and the age of the person.

Attachment continues to be studied in many nations, ages, cultures, genders, with caregivers who are not parents, children at many points in time, typical children, those with special needs, families, romances, friendships, and more (e.g., Cassidy & Shaver, 2016; Pinto & Figueiredo, 2019; Voges et al., 2019).

Thus, attachment is a universal; expression is particular. An analogy is language development, as reviewed in Chapter 6. Every human is predisposed to develop attachment, just as every infant is predisposed to develop language. However, the particular manifestations of that need, for example, which of the 6,000 languages a person speaks, or how attachment is manifest, depends on particular childhood experiences. Then that attachment style influences later life, even across generations, just as children learn a particular language and then have difficulty mastering another.

For instance, Ugandan mothers never kiss their infants, but they often massage them, unlike most Westerners. Some U.S. adults phone their mothers every

still-face technique An experimental practice in which an adult keeps their face unmoving and expressionless in face-to-face interaction with an infant.

THINK CRITICALLY: What will happen if no one plays with an infant?

attachment According to Ainsworth, "an affectional tie" that an infant forms with a caregiver—a tie that binds them together in space and endures over time.

macmillan learning

VIDEO ACTIVITY: Mother Love and the Work of Harry Harlow features classic footage of Harlow's research on attachment, showing the setup and results of his famous experiment.

Same or Different? A theme of this chapter is that babies and mothers are the same worldwide, yet dramatically different in each culture. Do you see more similarities or differences between the Huastec mother in Mexico *(left)*, the mothers waiting in a clinic in Uganda *(middle)*, and supermodel Gisele Bündchen in Boston, Massachusetts *(right)*? (Her husband is Tom Brady, star quarterback of the New England Patriots.)

day—even when the mothers are a thousand miles away—and others connect with their mothers only on holidays. Some attached family members all sit in the same room of a large house, each reading quietly. All of these signify attachment.

The concept began with research in the 1950s and 1960s by John Bowlby in England and Mary Ainsworth in Uganda and the United States on mother–infant relationships when the infants were about age 1. That is our focus in this chapter. Once infants can walk, the moment-by-moment, face-to-face synchrony is less common. Instead, attachment comes to the fore.

Two signs indicate attachment: *contact-maintaining* and *proximity-seeking*. Both take many forms in infancy. Caregivers often keep a watchful eye on their baby, initiating contact with expressions, gestures, and sounds. Before going to sleep at midnight they might tiptoe to the crib to gaze at their sleeping infant, or, in daytime, absentmindedly smooth their toddler's hair.

For their part, 1-year-olds look to their caregivers, ask to be picked up, fall asleep with their legs or arms touching them. When awake and stressed, their attachment is literal: They grab a hand, a leg, a pant leg, achieving contact and proximity.

Secure and Insecure Attachment

Attachment is classified into four types: A, B, C, and D. Infants with **secure attachment** (type B) feel comfortable and confident. The caregiver is a *base for exploration*, providing assurance and enabling discovery.

A securely attached toddler might, for example, scramble down from the caregiver's lap to play with an intriguing toy but periodically look back and vocalize (contact-maintaining), or bring the toy to the caregiver for inspection (proximity-seeking).

The caregiver's presence gives the child courage to explore; departure causes distress; return elicits positive contact (such as smiling or hugging) and then more playing. This balanced reaction—the child concerned but not overwhelmed by comings and goings—indicates security.

By contrast, insecure attachment (types A and C) is characterized by fear, anxiety, anger, or indifference. Some insecure children play independently without seeking contact; this is **insecure-avoidant attachment** (type A). The opposite reaction is **insecure-resistant/ambivalent attachment** (type C). Children with type C cling to their caregivers and are angry at being left.

Infants may be securely or insecurely attached to mothers, fathers, or other caregivers—sometimes just to one person, sometimes to several. Is secure attachment

secure attachment A relationship in which an infant obtains both comfort and confidence from the presence of his or her caregiver.

insecure-avoidant attachment A pattern of attachment in which an infant avoids connection with the caregiver, as when the infant seems not to care about the caregiver's presence, departure, or return.

insecure-resistant/ambivalent attachment A pattern of attachment in which an infant's anxiety and uncertainty are evident, as when the infant becomes very upset at separation from the caregiver and both resists and seeks contact on reunion.

innate? Are difficult children always insecurely attached? No. Every baby seeks attachment: Temperament may affect the expression, but not the need (Groh et al., 2017).

Ainsworth's original schema differentiated only types A, B, and C. Later researchers discovered a fourth category (type D), **disorganized attachment**. Type D infants may suddenly switch from hitting to kissing their mothers, from staring blankly to crying hysterically, from pinching themselves to freezing in place.

Among the general population, almost two-thirds of infants are secure (type B). About one-third are insecure, either indifferent (type A) or unduly anxious (type C), and about 5 to 10 percent are disorganized (type D).

Type D infants are especially worrisome to developmentalists, because they have no consistent strategy for social interaction, even avoidance or resistance. Without intervention to improve early attachment, type D toddlers are at risk for later psychopathology, including severe aggression and major depression (Cicchetti, 2016; Groh et al., 2012). Of course, much more information about family life is needed before type D attachment is considered evidence that a particular child needs to be in foster care (White et al., 2019).

Recently some people have begun advocating *attachment parenting*, which prioritizes the mother–infant relationship during the first three years of life far more than Ainsworth or Bowlby did (Komisar, 2017; Sears & Sears, 2001). Attachment parenting mandates that mothers should always be near their infants (co-sleeping, "wearing" the baby in a wrap or sling, breast-feeding on demand). Some experts suggest that attachment parenting is too distant from the research concept and evidence (Ennis, 2015).

Although attachment parenting may be an exaggeration, the consensus of many researchers is not only that attachment is important for infant development but also that mothers can learn some of the behaviors that increase the bond between mother and child. That is one reason that the federal government has supported more than a thousand programs of home visiting. The emphasis is on early education fostered by the mother, but guidance to increase secure attachment is sometimes included (Berlin et al., 2018).

disorganized attachment A type of attachment that is marked by an infant's inconsistent reactions to the caregiver's departure and return.

A VIEW FROM SCIENCE

Measuring Attachment

Scientists take great care to measure what they purport to measure. This is especially important when they want to measure emotions. They develop an *operational definition*, which is a measurement of an observable behavior that indicates the construct so that other scientists know what is measured and can replicate the study.

For instance, if you wanted to study love between romantic partners, what would your empirical measurement be?

- Ask each partner how much they loved each other, rated on a scale of 1 to 10?
- Ask other questions, such as whether they had, in the past year, thought of breaking up the relationship or sacrificed something for their partner?
- Video their interaction, and count how often they made eye contact, or expressed agreement, or how many inches apart they sat?

- Check details of cohabitation, marriage, physical affection, shared finances?

None of these is exactly what people call "love," and some of these are better than others as operational definitions, but all might be useful, and all could be measured.

Measuring the connection between caregiver and infant is similarly complicated. Fortunately, Mary Ainsworth (1973) developed a now-classic laboratory procedure called the *Strange Situation* to measure attachment, a measurement protocol that has been used in thousands of studies. The Strange Situation measures details on 1-year-olds' reactions to stress, with and without the mother.

The specifics of that measurement are as follows: In a well-equipped playroom, an infant is observed for eight episodes, each lasting no more than three minutes. First, the child and mother are together. Next, according to a set sequence, the mother and then a stranger come and go. Infants' responses to their mother indicate which type of attachment they have formed.

Researchers focus on the following:

Exploration of the toys. A secure toddler plays happily.

Reaction to the caregiver's departure. A secure toddler notices when the caregiver leaves and shows some sign of missing him or her. A pause in playing, a plaintive sound, a worried expression are all significant signs of attachment.

Reaction to the caregiver's return. A secure toddler welcomes the caregiver's reappearance, seeking contact, and then plays again. Typically toddlers run to their mothers for a hug, and then resume investigation of the toys.

Scientists are carefully trained in measuring attachment in 1-year-olds, via watching videos, calibrating ratings, and studying manuals. Researchers are certified only when they reach a high standard of accuracy. They learn which common behaviors signify insecurity, contrary to what untrained observers might think. For instance, clinging to the caregiver may be type C; being too friendly to the stranger suggests type A.

Many scientists who study attachment in older children and adults have developed other measures—again empirical. For instance, detailed questionnaires and interviews, calibrated to signify secure or insecure attachment, with a particular measure based on Ainsworth indicators (the *Adult Attachment Interview*, or *AAI*), are often used.

A sign of a past insecure childhood attachment is not only rejection of one's mother ("I never want to see her again"), but also sanctification of her ("she was a saint"). It is especially troubling if an adult is confused and incoherent, with few details about their awful, or perfect, childhood. This has been found in adults of every gender and cultural background, confirming that attachment is needed at every age (Haltigan et al., 2019).

The measurement of attachment via the Strange Situation has made longitudinal studies possible, assessing the later development of infants who are types A, B, C, or D. We now know that infant–caregiver attachment affects brain development and the immune system (Bernard et al., 2019), and that particular circumstances, such as domestic violence among the parents, can weaken the links between childhood attachment and adult relationships (Fearon & Roisman, 2017).

Thanks to procedures developed by Mary Ainsworth half a century ago, the importance of early caregiver–infant relationships is now recognized. Securely attached infants *are* more likely to become secure toddlers, socially competent preschoolers, high-achieving schoolchildren, partners in loving couples, capable parents, and healthy adults.

> **THINK CRITICALLY:** Is the Strange Situation a valid way to measure attachment in every culture, or is it biased toward the Western idea of the ideal mother–child relationship?

© 2016 Macmillan

Excited, Troubled, Comforted This sequence is repeated daily for 1-year-olds, which is why the same sequence is replicated to measure attachment. As you see, toys are no substitute for mother's comfort if the infant or toddler is secure, as this one seems to be. Some, however, cry inconsolably or throw toys angrily when left alone.

Orphanages in Romania

No scholar doubts that attachments should develop in the first year of life and that the lack of close caregiver–infant relationships predicts lifelong problems. Unfortunately, thousands of children born in Romania verify that conclusion.

Politics and Children Without Parents

Many nations now restrict international adoptions, in part because some children were literally snatched from their biological parents to be sent abroad. Romania halted international adoptions in 2004, although many adoptable Romanian children are still in institutions (Popescu et al., 2019).

According to government records, the number of international adoptees in the United States was 4,058 in 2018, down from 22,884 in 2004 (see **Figure 7.3**). The decrease may be influenced more by international politics than by infant needs, with Russia particularly refusing to allow any more U.S. adoptions. Ideally, every baby would be wanted, every parent would be supported, and no infant would be institutionalized. But if that ideal is not reached, scientists advocate family care over institutional care, because attachment to a dedicated caregiver is crucial for well-being (McCall, 2013).

Developmentalists of every political stripe are horrified that thousands of children of immigrants are separated from their parents at the border between Mexico and the United States (Roth et al., 2018). Attachment research confirms that children need dedicated caregivers, and that disrupting that relationship causes lifelong harm. Here again, politics, rather than what we know about attachment, interferes with good care.

International Adoptions by U.S. Citizens, All Nations

Data from U.S. Department of State, 2019.

FIGURE 7.3

Declining Need? No. More couples seek to adopt internationally, and millions of children in dozens of nations have no families. This chart does not reflect changing needs of families; it reflects increasing nationalism within and beyond the United States. Sadly, babies have become weapons in national politics.

Social Referencing

The third social connection that developmentalists look for during infancy, after synchrony and attachment, is **social referencing**. Much as a student might consult a dictionary or other reference work, social referencing means seeking emotional responses or information from other people. A reassuring glance, a string of cautionary words, a facial expression of alarm, pleasure, or dismay—those are social references.

As you read in Chapter 6, gaze-following begins in the first months of life, as part of cognition. After age 1, when infants can walk and are "little scientists," their need to consult others becomes urgent and more accurate—for emotional input, not merely cognition.

Toddlers search for clues in gazes, faces, and body position, paying close attention to emotions and intentions. They focus on their familiar caregivers, but they also use relatives, other children, and even strangers to help them assess objects and events.

From early infancy to late adolescence, children are remarkably selective, noticing that some strangers are reliable references and others are not. For instance, infants as young 14 months old were less likely to accept social referencing from a stranger whose emotions did not fit experiences (e.g., infants distrusted adults who smiled happily at opening an empty box) (Crivello & Poulin-Dubois, 2019). The infant's own emotions and desires spring from social referencing (Wellman, 2014).

Social referencing has many practical applications. Consider mealtime. Caregivers the world over pretend to taste and say "yum-yum," encouraging toddlers to eat beets, liver, or spinach. Toddlers read expressions, insisting on the foods that the adults *really* like.

If a mother enjoys eating it, and then presents some to the toddler, then they eat it—otherwise not. Some tastes (spicy, bitter, sour) are rejected by very young infants, but if they see their caregivers eat it, they learn to like it (Forestel & Mennella, 2017).

social referencing Seeking information about how to react to an unfamiliar or ambiguous object or event by observing someone else's expressions and reactions. That other person becomes a social reference.

Rotini Pasta? Look again. Every family teaches their children to relish delicacies that other people avoid. Examples are bacon (not in Arab nations), hamburgers (not in India), and, as shown here, a witchetty grub. This Australian aboriginal boy is about to swallow an insect larva.

Bill Bachman/Alamy Stock Photo

Developing Attachment

Attachment begins at birth and continues lifelong. Much depends not only on the ways in which parents and babies bond, but also on the quality and consistency of caregiving, the safety and security of the home environment, and individual and family experience. While the patterns set in infancy may echo in later life, they are not determinative.

How Many Children are Securely Attached?

The specific percentages of children who are secure and insecure vary by culture, parent responsiveness, context, and specific temperament and needs of both the child and the caregiver. Generally, about a third of all 1-year-olds seem insecure.

50–70%	10–20%	10–20%	5–10%
Secure Attachment (Type B)	Avoidant Attachment (Type A)	Ambivalent Attachment (Type C)	Disorganized Attachment (Type D)

Attachment in the Strange Situation May Influence Relationships Through the Life Span

Attachment patterns formed in infancy affect adults lifelong, but later experiences of love and rejection may change early patterns. Researchers measure attachment by examining children's behaviors in the Strange Situation where they are separated from their parent and play in a room with an unfamiliar caregiver. These early patterns can influence later adult relationships. As life goes on, people become more or less secure, avoidant, or disorganized.

Securely Attached [Type B]
In the Strange Situation, children are able to separate from caregiver but prefer caregiver to strangers.

> Later in life, they tend to have supportive relationships and positive self-concept.

Insecure-Avoidant [Type A]
In the Strange Situation, children avoid caregiver.

> Later in life, they tend to be aloof in personal relationships, loners who are lonely.

Insecure-Resistant/Ambivalent [Type C]
In the Strange Situation, children appear upset and worried when separated from caregiver; they may hit or cling.

> Later in life, their relationships may be angry, stormy, unpredictable. They have few long-term friendships.

Disorganized [Type D]
In the Strange Situation, children appear angry, confused, erratic, or fearful.

> Later in life, they can demonstrate odd behavior—including sudden emotions. They are at risk for serious psychological disorders.

The Continuum of Attachment

Avoidance and anxiety occur along a continuum. Neither genes nor cultural variations were understood when the Strange Situation was first developed (in 1965). Some contemporary researchers believe the link between childhood attachment and adult personality is less straightforward than this table suggests.

Low Avoidance

Secure   Resistant

Low Anxiety   High Anxiety

Avoidant   Disorganized

High Avoidance

Through this process, some children develop a taste for raw fish or curried goat or smelly cheese—foods that children in other cultures refuse. Similarly, toddlers use social cues to understand the difference between real and pretend eating, as well as to learn which objects, emotions, and activities are forbidden.

Fathers as Social Partners

Synchrony, attachment, and social referencing are evident with fathers as well as with mothers. Indeed, fathers tend to elicit more smiles and laughter from their infants than mothers do. They tend to play more exciting games, swinging and chasing, while mothers do more caregiving and comforting.

This gender difference should not be exaggerated: Both parents do some soothing and some exciting play. Nonetheless, fathers tend to do much more rough-and-tumble play than mothers do, and infants benefit from that (Cabrera et al., 2018). It is also true that children develop well when the roles are reversed, or when both parents are male or both parents are female. Each couple coordinates; children thrive (Shwalb et al., 2013).

Gender differences in child rearing vary more by nation, by income, by cohort, and by ideology than by natal sex or by ethnic background (Roopnarine & Yildirim, 2019). Variation is dramatic, from fathers who had nothing to do with infants to fathers who are intensely involved. The latter is common in the United States in the twenty-first century, unlike in former times (Abraham & Feldman, 2018).

Contemporary fathers worldwide tend to be more involved with their children than their own fathers were (Sriram, 2019). The effects are evident not only in infants but in fathers themselves. As one man in India said, "my child transformed me" (Kaur, 2019).

For both sexes, stress decreases parent involvement. That brings up another difference between mothers and fathers. When money is scarce and stress is high, some fathers opt out. That choice is less possible for mothers (Qin & Chang, 2013; Roopnarine & Hossain, 2013).

The reactions of both men and women to infants is also affected by their temperament, genes, and early-childhood experiences (Senese et al., 2019). This is another reason why babies of every sex need responsive caregiving: The effects will endure decades later when they have children of their own.

WHAT HAVE YOU LEARNED?

1. Why does synchrony affect early emotional development?
2. How are proximity-seeking and contact-maintaining attachment expressed by infants and caregivers?
3. How does infant behavior differ in each of the four types of attachment?
4. How might each of the four types of attachment be expressed in adulthood?
5. What has been learned from the research on Romanian orphans?
6. How is social referencing important in toddlerhood?

Theories of Infant Psychosocial Development

That infants are emotional and social creatures is one of those universal truths, recognized by everyone who studies babies. However, each of the theories discussed in Chapter 2 has a distinct perspective on this universal reality, as you will now see.

Psychoanalytic Theory

Psychoanalytic theory connects biosocial and psychosocial development. Sigmund Freud and Erik Erikson each described two distinct stages of early development, one in the first year and one beginning in the second year.

Freud: Oral and Anal Stages

According to Freud (1935/1989, 2001), the first year of life is the *oral stage*, so named because the mouth is the young infant's primary source of gratification. In the second year, with the *anal stage*, pleasure comes from the anus—particularly from the sensual satisfaction of bowel movements and, eventually, the psychological pleasure of controlling them.

Freud believed that the oral and anal stages are fraught with potential conflicts. If a mother frustrates her infant's urge to suck—weaning too early or too late, for example, or preventing the baby from sucking a thumb or a pacifier—that may later lead to an *oral fixation*. Adults with an oral fixation are stuck (fixated) at the oral stage, and therefore they eat, drink, chew, bite, or talk excessively, still seeking the mouth-related pleasures of infancy.

Similarly, if toilet training is overly strict or if it begins before maturation allows sufficient control, that causes a clash between the toddler's refusal—or inability—to comply and the wishes of the adult, who denies the infant normal anal pleasures. That may lead to an *anal personality*—an adult who seeks self-control, with a strong need for regularity and cleanliness in all aspects of life.

Erikson: Trust and Autonomy

According to Erikson, the first crisis of life is **trust versus mistrust**, when infants learn whether or not the world can be trusted to satisfy basic needs. Babies feel secure when food and comfort are provided with "consistency, continuity, and sameness of experience" (Erikson, 1993a, p. 247). If social interaction inspires trust, the child (later the adult) confidently explores the social world.

The second crisis is **autonomy versus shame and doubt**, beginning at about 18 months, when self-awareness emerges. Toddlers want autonomy (self-rule) over their own actions and bodies. Without it, they feel ashamed and doubtful. Like Freud, Erikson believed that problems in early infancy could last a lifetime, creating adults who are suspicious and pessimistic (mistrusting) or easily shamed (lacking autonomy).

Behaviorism

From the perspective of behaviorism, emotions and personality are molded as adults reinforce or punish children. Behaviorists believe that parents who respond joyously to every glimmer of a grin will have children with a sunny disposition. The opposite is also true:

> Failure to bring up a happy child, a well-adjusted child—assuming bodily health—falls squarely upon the parents' shoulders. [By the time the child is 3] parents have already determined . . . [whether the child] is to grow into a happy person, wholesome and good-natured, whether he is to be a whining, complaining neurotic, an anger-driven, vindictive, over-bearing slave driver, or one whose every move in life is definitely controlled by fear.
>
> [*Watson, 1928/1972, pp. 7, 45*]

Empathy, for instance, is an emotion that appears in direct proportion to the parents' responses (Grady & Hastings, 2018; Heyes, 2018). Shy boys, in particular,

---

● Especially for Nursing Mothers You have heard that if you wean your child too early, he or she will overeat or develop alcohol use disorder. Is it true? (see response, page 201)

trust versus mistrust Erikson's first crisis of psychosocial development. Infants learn basic trust if the world is a secure place where their basic needs (for food, comfort, attention, and so on) are met.

autonomy versus shame and doubt Erikson's second crisis of psychosocial development. Toddlers either succeed or fail in gaining a sense of self-rule over their actions and their bodies.

JOSE MIGUEL GOMEZ/REUTERS/Newscom

All Together Now Toddlers in an employees' day-care program at a flower farm in Colombia learn to use the potty on a schedule. Will this experience lead to later personality problems? Probably not.

become more outgoing if their fathers' talk with them about emotions (Grady & Hastings, 2018).

Social Learning

Behaviorists also recognize that infants' behavior reflects social learning, when infants learn from observing other people. You already saw an example, social referencing. Social learning occurs throughout life, not necessarily via direct teaching but often through observation (Shneidman & Woodward, 2016). Toddlers express emotions in various ways—from giggling to cursing—just as their parents or older siblings do.

For example, a boy might develop a hot temper if his father's outbursts seem to win his mother's respect; a girl might be coy, or passive-aggressive, if that is what she has seen at home. These examples are deliberately sexist: Gender roles, in particular, are learned, according to social learning theory.

Keeping Baby Close

Parents often unwittingly encourage certain traits in their children. Should an infant explore, or learn that danger lurks if they wander off? Should babies have many toys, or will that make them greedy? When babies cry, should their mothers pick them up, feed them, give a pacifier, ignore them? Should an infant breast-feed until age 2, switch to a bottle, or sip from a cup?

These questions highlight the distinction between **proximal parenting** (being physically close to a baby, often holding and touching) and **distal parenting** (keeping some distance—providing toys, encouraging self-feeding, talking face-to-face instead of communicating by touch). Caregivers tend to behave in proximal or distal ways very early, when infants are only 2 months old (Kärtner et al., 2010). Each pattern reinforces some behavior.

For instance, toddlers who, as infants, were often held, patted, and soothed (proximal) became toddlers who were more obedient but less likely to recognize themselves in a mirror. This finding has been replicated in Greece, Cameroon, Italy, Israel, Zambia, Scotland, and Turkey: Distal child rearing correlates with cultures that value individual independence; proximal care correlates with cultures that value collective action and family interdependence (Borke et al., 2007; Carra et al., 2013; Kärtner et al., 2011; Keller et al., 2010; Ross et al., 2017; Scharf, 2014).

Indeed, international variations of parenting practices, including not only proximal and distal parenting but also frequency of synchrony, secure attachment, and social referencing, suggest that children are taught to respond in a particular way because of how their parents treat them (Foo, 2019). That is what behaviorists believe.

Cognitive Theory

Cognitive theory holds that thoughts determine a person's perspective. Early experiences are important because beliefs, perceptions, and memories make them so, not because they are buried in the unconscious (psychoanalytic theory) or burned into the brain's patterns (behaviorism).

From this perspective, cognitive processes, including language and information, affect attachment, as children and caregivers develop a mutual understanding. Together they build (co-construct) a **working model**, which is a set of assumptions that becomes a frame of reference for later life (Posada & Waters, 2018). It is a "model" because early relationships form a prototype, or blueprint; it is "working" because it is a work in progress, not fixed or final; it is cognitive because the child's understanding and interpretation is crucial.

proximal parenting Caregiving practices that involve being physically close to the baby, with frequent holding and touching.

distal parenting Caregiving practices that involve remaining distant from the baby, providing toys, food, and face-to-face communication with minimal holding and touching.

Especially for Pediatricians A mother complains that her toddler refuses to stay in the car seat, spits out disliked foods, and almost never does what she says. How should you respond? (see response, page 201)

working model In cognitive theory, a set of assumptions that the individual uses to organize perceptions and experiences. For example, a person might assume that other people are trustworthy and be surprised by an incident in which this working model of human behavior is erroneous.

"*Which one generates the most synapses?*"

Brainy Baby Fortunately, infant brains are designed to respond to stimulation of many kinds. As long as the baby has moving objects to see (an animated caregiver is better than any mobile), the synapses proliferate.

Ideally, infants develop "a working model of the self as lovable and competent" because the parents are "emotionally available, loving, and supportive of their mastery efforts" (Harter, 2012, p. 12). However, reality does not always conform to this ideal.

A 1-year-old girl might develop a model, based on her parents' erratic actions, that people are unpredictable. She will continue to apply that model to everyone: Her childhood friendships will be insecure, and her adult relationships will be guarded.

The crucial idea, according to cognitive theory, is that an infant's early experiences themselves are less influential than the interpretation of those experiences is (Olson & Dweck, 2009). Children may misinterpret their experiences, or parents may offer inaccurate explanations, and these form ideas that affect later thinking and behavior.

In this way, working models formed in childhood echo lifelong. A hopeful message from cognitive theory is that people can rethink and reorganize their thoughts, developing new models. That mistrustful girl in the example above might marry someone who is faithful and loving, so she may gradually develop a new working model.

Evolutionary Theory

Remember that evolutionary theory stresses two needs: survival and reproduction. Human brains are extraordinarily plastic, in order to survive in a vast range of local conditions. This adjustment requires extensive experience: A human child must be nourished, protected, and taught for two decades, which is much longer than the offspring of any other species. Since survival depends on this lengthy protection, it is part of our DNA (Hrdy, 2009).

Emotions for Survival

Infant emotions are part of this evolutionary mandate. All of the reactions described in the first part of this chapter—from the hunger cry to the temper tantrum—can be seen from this perspective (Konner, 2010).

For example, newborns are extraordinarily dependent, unable to walk or talk or even sit up and feed themselves for months after birth. They must attract adult devotion—and they do. That first smile, the sound of infant laughter, and their role in synchrony are all powerfully attractive to adults—especially to parents.

Adults call their hairless, chinless, round-faced, big-stomached, small-limbed offspring "cute," "handsome," "beautiful," "adorable," yet each of these characteristics is often considered ugly in adults. Parents willingly devote hours to carrying, feeding, changing, and cleaning their infants, who never express gratitude. The love of a parent for a child is part of evolution: For hundreds of thousands of years, humans have needed that love to survive.

Why do people have children, especially now when contraception allows sex without procreation? If humans were motivated solely by money or power, no one would do so. Yet evolution has created adults who find parenting worth every sacrifice.

The same scientific advancements that make it possible to have sex without pregnancy have also compelled many infertile adults to endure pain, embarrassment, and expense to have a baby. Because of that innate desire for progeny, parents are devoted to their children, and that devoted care allows children to develop well (Narvaez et al., 2013).

Evolutionary theory holds that both caregiver and children have powerful emotions—love, jealousy, even clinginess and anger—that keep toddlers near caregivers who remain vigilant. Infants fuss at still faces, fear separation, and laugh when adults play with them—all to sustain caregiving. Emotions are our genetic legacy; we would die without them.

allocare Literally, "other-care"; the care of children by people other than the biological parents.

Allocare

Evolutionary social scientists note that if mothers were the exclusive caregivers of each child until children were adults, a given woman could rear only one or two offspring—not enough for the species to survive. Instead, before the introduction of reliable birth control, the average interval between births for humans was two to four years.

Humans birth children at relatively short intervals because of **allocare**—the care of children by *alloparents*, caregivers who are not the mother. Allocare is essential for *Homo sapiens'* survival. Compared with many other species (mother chimpanzees space births by four or five years and never let another chimp hold their babies), human mothers have evolved to let other people help with child care.

This is not only because mothers need help from, and thus prevail on, fathers and grandmothers, but also because evolution has programmed the human brain so that fathers and grandmothers want to help (Abraham & Feldman, 2018). This explains why that mother on the subway was willing to hand me her baby, and why I was happy that she did.

Allocare is universal for our species—but each community has distinct values and preferences for who should care for children and when (Konner, 2018). Indeed, cultures and theories differ in every aspect of infant care, as you have read regarding breast-feeding, co-sleeping, language, and much else. Now we consider mother care versus other care.

Same Situation, Far Apart: Safekeeping Historically, grandmothers were sometimes crucial for child survival. Now, even though medical care has reduced child mortality, grandmothers still do their part to keep children safe, as shown by these two—in the eastern United States *(top)* and Vietnam *(bottom)*.

> ### WHAT HAVE YOU LEARNED?
>
> 1. According to Freud, what might happen if a baby's oral needs are not met?
> 2. How might Erikson's crisis of "trust versus mistrust" affect later life?
> 3. How do behaviorists explain the development of emotions and personality?
> 4. What does the term *working model* mean within cognitive theory?
> 5. What is the difference between proximal parenting and distal parenting?
> 6. How does evolution explain the parent–child bond?
> 7. Why is allocare necessary for survival of the human species?

Who Should Care for Babies?

Summarizing the research on infant care is difficult, because people tend to believe that the practices of their own family or culture are best. They may be right: Each community tries to provide good care for their babies. However, that belief risks the difference-equals-deficit error.

Variations in perspectives, in norms, and in priorities lead to diverse practices. All humans seek confirmation of their opinions, and research provides it. Contradictory research may be ignored. Your opinions may be right, but that does not make others wrong.

Some cultures assume that mothers should care for young children until age 5 or so; other communities believe that early group care is better. Some cultures expect fathers to avoid infant care, with mothers as "gatekeepers" who criticize how men hold, feed, or play with babies. Mothers may discourage grandparents

Marcus McKinney/U.S. Army

Together At Last When Kristian Myrick was deployed to Afghanistan nine months ago, he was not sure he would be a father. His daughter weighed about an ounce, not yet viable. But here, Amelia and her mother (at left) welcome him back to Colorado. Father and infant will smile, laugh, and bond in the next few weeks. Only one year before this photo (in 2019), the U.S. military recognized the benefits of father–infant synchrony and attachment. Now soldiers on active duty have three weeks of paid paternal leave, a more generous policy than any other U.S. agency offers.

and others as well. By contrast, some mothers expect other relatives to help, and some nations pay professionals to provide infant and child care, just as they pay firefighters or the police.

Beyond national differences, opinions are affected by personal experience (adults who experienced nonparental care are likely to approve of it), by gender (males are more often critical of day-care centers), and by education (higher education increases support for nonparental care) (Galasso et al., 2017; Rose et al., 2018; Shpancer & Schweitzer, 2018).

In the United States

In the United States, norms do not mandate exclusive mother care. Only 20 percent of infants are cared for *exclusively* by their mothers (i.e., no other relatives or babysitters) throughout their first year (Babchishin et al., 2013). In 2018, 58 percent of the mothers of infants under age 1 were in the labor force (U.S. Bureau of Labor Statistics, April 18, 2019).

One reason why most mothers of young children work is that they need the money. The United States is the only developed nation that does not require paid parental leave. Consequently, mothers enlist family members (especially fathers and grandmothers) to care for babies. Professional infant day care, either in a center or at home, is privately paid, and thus too expensive for many families.

In the United States, after a long fight among the men and women in the U.S. Congress, laws require companies with 50 employees or more to offer *unpaid* maternity leave for 12 weeks. Some employers are better, offering more time or some pay. Paternal leave is almost never paid with one exception: As of 2018, the U.S. military allows 21 days of paid leave for new fathers who are secondary caregivers, and up to 6 weeks for mothers or fathers in the military who are primary caregivers.

Problems with Nonmaternal Care

Is there anything wrong with allocare? It depends: Some relatives provide good care, some not. Some day-care centers are excellent; some not. As you might imagine, since child care is not seen as a public right in the United States, cost is a major concern, and quality may suffer.

Beyond price, two worries are often expressed regarding babies with extensive allocare: (1) insecure attachment, and (2) emotional problems later on.

The first worry has been disproven. Secure mother–infant attachment (as measured by the Strange Situation) is as common for infants with regular father care, with professional day care, and with care from other caregivers, as with exclusive maternal care.

In retrospect, this makes sense. Even when fathers or day-care centers care for a baby 40 hours a week, mothers still spend more hours (especially nights and weekends) with their babies. Further, father–infant attachment is more likely when fathers provide major care, and infants benefit when they are securely attached to both parents (Cabrera et al., 2018).

Regarding behavioral problems, the evidence is not as clear. A mega-analysis of research on infant caregiving (Dearing & Zachrisson, 2017) reported that half of the studies of group care found that the children had more externalizing problems (such as aggression, particularly in boys) later on. However, that was not found in the other half of the studies, including one that reported better emotional adjustment when children had extensive day care.

As an earlier review explained: "This evidence now indicates that early nonparental care environments sometimes pose risks to young children and sometimes confer benefits" (Phillips et al., 2011, p. 44).

The most often cited longitudinal research comes from the Early Child Care Research Network of the National Institute of Child Health and Human Development (NICHD), which followed, from birth through adolescence, over 1,300 children born in 1991. Early day care correlated with many cognitive advances, especially in language. Children who were enrolled in high-quality preschools had higher achievement throughout elementary school and high school.

Most analyses of that group found that secure attachment was as common among infants in center care as among infants cared for at home. Like other, smaller studies, the NICHD research confirmed that the mother–child relationship was pivotal. A summary of that research found:

> higher quality of child care was linked to higher academic-cognitive skills in primary school and again at age 15. Higher hours of child care were associated with teacher reports of behavior problems in early primary school and youth reports of greater impulsivity and risk taking at age 15.

[Burchinal et al., 2014, p. 542]

Why is the research not definitive about behavior problems? Perhaps the adults, not the infants, are the deciding factor. Many studies confirm that infants with neglectful mothers benefit from day care, but if the mother is responsive and the allocare is inferior, children fare better if their mother provides care. Inferior day care usually means not enough teachers. Even if they spend most of their time engaging with children, they cannot have enough time to spend with each child to provide the quality interactions needed.

Historical Changes

Most currently reported longitudinal studies began with children who were in day care 30 years ago. Many educators and researchers believe that infant day care is of higher quality, as well as more common, than it was (Huston et al., 2015; Singer & Wong, 2019).

For example, the National Association for the Education of Young Children (NAEYC, the profession organization for educators who teach children under age 6), updated its standards for care of babies from birth to 15 months (NAEYC, 2014). Group size should be small (no more than eight infants), and the ratio of adults to babies is 1:4 or fewer (see **Table 7.2**).

TABLE 7.2

High-Quality Day Care

High-quality day care during infancy has five essential characteristics:

1. *Adequate attention to each infant*
   A small group of infants (no more than eight) needs two reliable, familiar, loving caregivers. Continuity of care is crucial.

2. *Encouragement of language and sensorimotor development*
   Infants need language — songs, conversations, and positive talk — and easily manipulated toys.

3. *Attention to health and safety*
   Cleanliness routines (e.g., handwashing), accident prevention (e.g., no small objects), and safe areas to explore are essential.

4. *Professional caregivers*
   Caregivers should have experience and degrees/certificates in early-childhood education. Turnover should be low, morale high, and enthusiasm evident.

5. *Warm and responsive caregivers*
   Providers should engage the children in active play and guide them in problem solving. Quiet, obedient children may indicate unresponsive care.

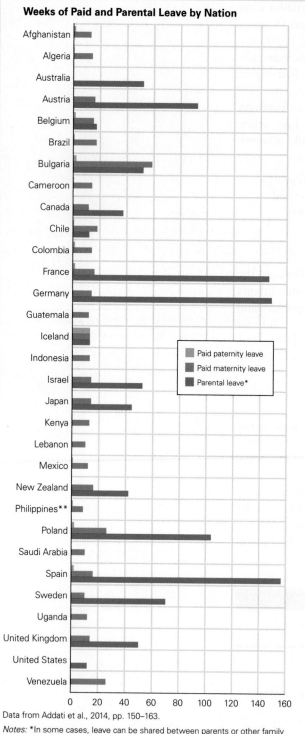

Weeks of Paid and Parental Leave by Nation

Legend:
- Paid paternity leave
- Paid maternity leave
- Parental leave*

Data from Addati et al., 2014, pp. 150–163.

Notes: *In some cases, leave can be shared between parents or other family members. Many nations have increased leave since these data were gathered.
**In the Philippines, parents must be married in order to receive paid leave.

FIGURE 7.4

A Changing World No one took maternity leave a century ago because the only jobs mothers had were unregulated ones. Now, virtually every nation has a maternity leave policy, revised every decade or so. As of 2019, only Belgium, Denmark, France, Latvia, Luxembourg, and Sweden had laws ensuring gender equality. That may be the next innovation in many nations.

Each infant is given personal attention, and the curriculum focuses on emotional and intellectual growth. For instance, teachers "engage infants in frequent face-to-face social interactions"—including talking, singing, smiling, and touching (NAEYC, 2014, p. 4). Every staff member must respect cultural differences and try to follow the mother's wishes. Maternal involvement is encouraged, including breast-feeding (bottles of pumped milk are stored for each baby).

Does that mean high-quality infant care is as good or better than mother care? People in the United States do not agree. Let us look at other nations, to see what we can learn.

Other Nations

For cultural, ideological, and economic reasons, nations differ. Publicly paid infant care is rare in most nations of South Asia, sub-Saharan Africa, and South America, where many adults believe it is harmful. Those nations also have laws that require paid maternal leave, and sometimes paid paternal leave, evidence of the belief that mothers should stay home to care for infants.

By contrast, people in some nations believe that subsidized infant care is a public right, just like fire and police protection, and thus taxes pay for child care. Government-sponsored and regulated child care is prevalent in France, Israel, China, Chile, Norway, Sweden, Denmark, Finland, Iceland, and the Netherlands, and has recently been instituted in Australia and Germany. (See **Figure 7.4**.)

In the Scandinavian nations, legislators believe that the first years are crucial for later life, so quality is high (Garvis et al., 2019), and teachers are relatively well trained and paid.

That is less true in Germany and Australia, because those nations instituted low-cost day care primarily to increase the birth rate. In both nations, many young adults were not having children, as the sacrifice seemed too great. The legislators hope that by paying for early child care, more women will give birth (Harrison et al., 2014).

France Compared to Norway

In France, subsidized care can begin at 12 weeks, but only about 10 percent of all infants receive it. The waiting list is long, parents pay part of the cost, and the adult–infant ratio is 1:5 under age 1 and 1:8 from age 1 until 3, when it increases again. That ratio reduces cost, but quality suffers. In France, "ensuring high-quality provision seems at odds with affordability and availability of places for under threes" (Fagnani, 2013, p. 92). The government emphasizes expansion so more mothers can work, rather than quality.

Norway takes the opposite course. Norwegians believe that the youngest babies are best cared for by their mothers, so they pay employed women full salary to stay home with their babies for 47 weeks after birth, and provide some paid leave for fathers as well. Beginning at age 1, high-quality, center day care is available for everyone, including in the sparsely populated rural counties. In 2016, many Norwegian 1-year-olds

(72 percent) were in center care, as were most 2-year-olds (92 percent) and almost all 3-year-olds (96 percent) (Statistics Norway, 2018).

Aggressive or Assertive?

Longitudinal results in Norway find no detrimental results of early center care. What U.S. researchers consider aggression is not aggression in Norway. Instead, Norwegians praise early experiences with other children: That makes shy children bolder (Solheim et al., 2013).

This raises the need for operational definitions, as already explained regarding *attachment*. How should "behavioral problems" be assessed? The usual measure is teacher judgment. Do teachers judge differently in Norway than in the United States? Similar questions can be raised about risk-taking. Adults consider risk-taking a negative trait, but many teenagers disagree.

A study in the Netherlands found that, if anything, children who attended day care were less likely to have emotional problems than other children (Broekhuizen et al., 2018). In that nation, day care may begin before age 1, but most children attend center care part time, two or three days a week (see **Figure 7.5**). Do children benefit from part-time preschool more than full-time preschool?

Fathers, Grandmothers, and Sisters

As you have already read, humans have evolved to allow allocare, which is help with infant care from many people in addition to the mother. Recent research in neuroscience finds that the human brain is plastic, not only in general, but also in the specifics of infant care. If a particular community encourages father–infant care, the brains of men adjust to make this a priority (Rogers & Bales, 2019).

In the nineteenth and twentieth centuries in the United States, fathers were usually designated as wage-earners, not baby minders. Families were large, and the older children cared for the younger ones. My own mother was the middle born of

Double Winner These parents brought their babies to the San Francisco legislature to advocate for paid parental leave. They won! The San Francisco Board of Supervisors voted yes, making this the first jurisdiction in the United States to mandate fully paid leave. The law went into effect in 2017—too late for both the mother and father shown here. Perhaps their next babies?

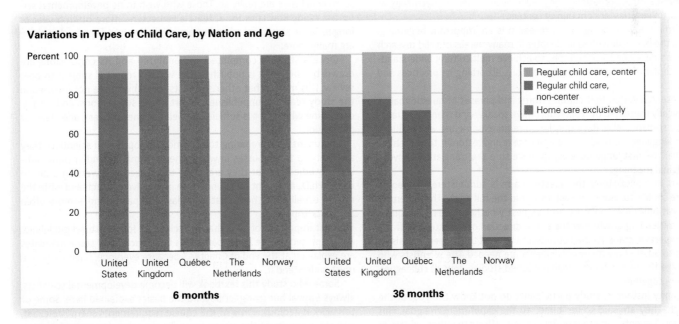

Variations in Types of Child Care, by Nation and Age

Legend:
- Regular child care, center
- Regular child care, non-center
- Home care exclusively

6 months: United States, United Kingdom, Québec, The Netherlands, Norway
36 months: United States, United Kingdom, Québec, The Netherlands, Norway

FIGURE 7.5

Who Cares for the Baby? Infants are the same everywhere, but cultures and governments differ dramatically. Does a 6-month-old need their mother more than a 3-year-old? Norway and Quebec say yes; the United States, United Kingdom, and the Netherlands say no.

Observation Quiz Which nation has the most extreme shift at age 1? (see answer, page 201) ↑

12 children. Two of her older sisters did not marry, in part because they cared for the younger ones. When I was an infant, those two maiden aunts moved in with my family, in part to help with child care as they had done when they were teenagers.

However, in the twenty-first century, families are smaller, and in 2018, 76.4 percent of the mothers whose oldest children were aged 6 to 17 were in the labor force (U.S. Bureau of Labor Statistics, April 18, 2019). Currently, fathers and grandmothers provide substantial allocare. It is unusual, but no longer rare, for a father to be the primary caregiver of an infant while the mother is employed.

Research in Europe finds that grandmothers are *more* likely to babysit in nations with extensive public care for infants (Price et al., 2018). The reason may be that, in those nations, exclusive mother care is not idealized. In nations where mothers are expected to be solely responsible for infants, if a particular woman wants or needs to have a job, her own mother is the most common caregiver (not the father or a day-care provider). Thus, in Italy for instance, fewer grandmothers babysit, but those who do put in more hours. Obviously, all of these differences are affected by culture, economics, and politics more than by universal needs of babies.

CAREER ALERT   The Developmental Scientist

The need for developmental scientists is apparent: Much more must be learned about "how and why people—all kinds of people—change over time." Would-be researchers must become scholars who know what has already been studied—earning at least a master's degree. Leaders in developmental science almost always need a Ph.D. as well.

Often scholars need further study as they work as "post-docs," researching and writing after their doctoral degree under the direction of a leading scientist at a major university. Sometimes they work in government offices such as the Centers for Disease Control or the Bureau of Labor Statistics. Sometimes they teach developmental psychology at colleges, universities, or community centers.

Learning how to design valid research is an important beginning. Beyond the basics described in Chapter 1, many details of valid research have been developed. For example, to achieve multicultural understanding, a scholar often lives in another culture, absorbing its practices and values.

In research within that culture, words and phrases of questionnaires are typically translated by a bilingual native speaker and then back-translated by someone else. Back translation requires someone who is fluent in both languages to read the translated text and restore it to English (or whatever the first language was) to make sure the translation conveys the original meaning.

Then, in a pilot study, the questionnaire is administered to people who are similar to those who will take it in the full study. They interpret the questions, raise concerns, and suggest whether a particular question is misleading or whether the entire direction of the study is off the mark. Experts in statistics are consulted to suggest the proper analysis, as well as to advise how many participants are needed. The scientists who undertake the research must read published studies, consult colleagues, and work together.

In many instances, study participants do not know the goal of the research: They are said to be "blind" to the experimenter's goal. This deception prevents conscious or unconscious efforts to either validate or undermine the study. An ethical mandate is that any deception must be explained to the participants after their involvement.

Scientists may need to be "blind," as well: They do not directly interact with the participants of the study so that they cannot inadvertently clue responses. The entire endeavor requires many people to design, implement, and interpret the research. This is where novice scientists are crucial. Depending on the particulars, novices may do most of the hard work of implementing the design.

All of this is expensive: Some of the work of the scientist is to convince other people (foundations, government, private philanthropists) to support a particular study.

Another necessity is knowledge of statistics. There are many ways to analyze numbers and verbal responses to ensure that results which seem conclusive are really so. Those who wish to be developmental scientists study statistics for at least a semester—more often a year or longer. Some statistical measures are shown in the Appendix, but there are many more.

Once a study is completed and analyzed and its conclusions are drawn, but before it is published, the written report is subject to peer review. This means that other scientists (peers) who are not involved in the study read the unpublished report, make suggestions, and, finally, determine whether it is sufficiently well designed, honest, and clear for publication.

If students are preparing to become developmental scientists, they are likely to take courses in which they summarize and critique published studies so that they are ready to be peer reviewers. In order to earn a Ph.D., a student must undertake innovative research and write the results, usually in a thesis that is at least 100 pages long—more often double or triple that.

Most important of all is that the scientists follow ethical guidelines, always protecting the participants. This is so crucial that some mandates are explained at the end of Chapter 1 because everyone, scientist or not, must understand it.

Some who study this textbook will become developmental scientists, always ethical but going far beyond the basics explained here. Some of the work is tedious, some of the course requirements seem irrelevant, and some results of the research are discouraging. However, for those who choose this career, the joy of new discovery seems well worth the effort, and the potential reward—a better life for thousands of people who benefit from the research—makes all of the work worthwhile. Go for it!

Conclusions from the Science

A strength of the scientific method is when various measures and populations lead to divergent conclusions; cumulative research rather than any one study or any personal experience makes good science. As you see, in much of the research on infant care, conclusions from one study may differ from another. However, among researchers, there is consensus on four principles:

1. Mutual attachment to one or several familiar caregivers is beneficial.
2. Quality of care, whether at home or in a day-care program, matters.
3. Babies need loving and responsive caregivers.
4. Frequent changes and instability in care are problematic.

On the last item on this list, it is now apparent that every disruption undercuts the infant's effort to understand their world. If a neighbor, a grandmother, a day-care center, and then another grandmother, each provide care on a different day of the week, or if a baby lives with the biological mother, then a foster mother, then with the biological father, then back with the biological mother, that is harmful. By age 3, children with unstable care histories are more aggressive than those with stable care (Pilarz & Hill, 2014).

As is true of many topics in child development, questions remain. But one fact is without question: Each infant needs personal responsiveness. Someone should serve as a partner in the synchrony duet, a base for secure attachment, and a social reference who encourages exploration. Then, infant emotions and experiences—cries and laughter, fears and joys—will ensure that development goes well.

 DATA CONNECTIONS: A Look at Early Child Care in the United States explores how various maternal demographics affect child-care arrangements. 📕 **LaunchPad.**

WHAT HAVE YOU LEARNED?

1. Why do people disagree about who should provide care for infants?

2. What are the advantages of maternal care for babies?

3. What are the advantages of nonmaternal care for babies?

4. What is the difference between infant care in France and Norway?

5. What is the role of fathers and grandmothers in providing infant care?

6. What distinguishes high-quality from low-quality infant day care?

7. What aspects of infant care are agreed on by everyone?

SUMMARY

Emotional Development

1. Two emotions, contentment and distress, appear in newborns. Smiles and laughter are soon evident. Between 4 and 8 months of age, anger emerges in reaction to restriction and frustration, and it becomes stronger by age 1.

2. Reflexive fear is apparent in very young infants. Fear of something specific depends on the context. Typically, some fear of strangers and of separation is strong by age 1 and continues until age 2. After that, these fears may become disorders.

3. Self-recognition (measured by the mirror/rouge test) emerges at about 18 months. Social awareness and self-awareness produce more selective and intense emotions, as well as pride, shame, affection, and explosive temper.

Temperament and Personality

4. Temperament is inborn, but the expression of temperament is influenced by the context, with evident plasticity.

The Development of Social Bonds

5. Often by 2 months, and clearly by 6 months, infants become more responsive and social, and synchrony is evident. Caregivers and infants engage in reciprocal interactions, with split-second timing.

6. Attachment is the relationship between two people who try to be close to each other (proximity-seeking and contact-maintaining). It is measured in infancy by a baby's reaction to the caregiver's presence, departure, and return in the Strange Situation.

7. Secure attachment provides encouragement for infant exploration, and its influence is lifelong. Adults are attached as well, evident not only as parents but also as romantic partners.

8. As they become more mobile and engage with their environment, infants use social referencing (looking to other people's facial expressions and body language) to detect what is safe, frightening, or fun.

9. Infants frequently use fathers as partners in synchrony, as attachment figures, and as social references, developing emotions and exploring their world. Contemporary fathers often play with their infants.

Theories of Infant Psychosocial Development

10. According to all major theories, caregivers are especially influential in the first two years. Freud stressed the mother's impact on oral and anal pleasure; Erikson emphasized trust and autonomy. Both believed that the impact of these is lifelong.

11. Behaviorists focus on learning. They note that parents teach their babies many things, including when to be fearful or joyful, and how much physical and social distance (proximal or distal parenting) is best.

12. Cognitive theory holds that infants develop working models based on their experiences. Interpretation is crucial, and that can change with maturation.

13. Evolutionary theorists recognize that both infants and caregivers have impulses and emotions that have developed over millennia to foster the survival of each new member of the human species. Attachment is one example.

Who Should Care for Babies?

14. Research confirms that every infant needs responsive caregiving, secure attachment, and cognitive stimulation. These three can occur at home or in a good day-care center. Quality matters, as does consistency of care.

15. Some people believe that infant day care benefits babies and that governments should subsidize high-quality infant care, just as governments pay professional firefighters to put out any fire. Other cultures believe the opposite—that infant care is best done by the mothers, who are solely responsible for providing it.

16. National and international research finds much variation in policies and opinions regarding infant day care. Some nations provide extensive day care for young children, and virtually all nations (except the United States) mandate paid maternal leave for employed women who have a baby.

KEY TERMS

social smile (p. 173)
separation anxiety (p. 174)
stranger wariness (p. 174)
self-awareness (p. 175)
temperament (p. 176)
synchrony (p. 179)

still-face technique (p. 181)
attachment (p. 181)
secure attachment (p. 182)
insecure-avoidant attachment (p. 182)

insecure-resistant/ambivalent attachment (p. 182)
disorganized attachment (p. 183)
social referencing (p. 187)
trust versus mistrust (p. 190)

autonomy versus shame and doubt (p. 190)
proximal parenting (p. 191)
distal parenting (p. 191)
working model (p. 191)
allocare (p. 193)

APPLICATIONS

1. One cultural factor that influences infant development is how infants are carried from place to place. Ask four mothers whose infants were born in each of the past four decades how they transported them—front or back carriers, facing out or in, strollers or carriages, in car seats or on mother's laps, and so on. Why did they choose the mode(s) they chose? What are their opinions and yours on how such cultural practices might affect infants' development?

2. Record video of synchrony for three minutes. Ideally, ask the parent of an infant under 8 months of age to play with the infant.

If no infant is available, observe a pair of lovers as they converse. Note the sequence and timing of every facial expression, sound, and gesture of both partners.

3. Contact several day-care centers to try to assess the quality of care they provide. Ask about factors such as adult/child ratio, group size, and training for caregivers of children of various ages. Is there a minimum age? Why or why not? Analyze the answers, using Table 7.2 as a guide.

Especially For ANSWERS

Response for Nurses and Pediatricians (from p. 174): Stranger wariness is normal up to about 14 months. This baby's behavior actually might indicate secure attachment.

Response for Nurses (from p. 178): It's too soon to tell. Temperament is not truly "fixed" but variable, especially in the first few months. Many "difficult" infants become happy, successful adolescents and adults, if their parents are responsive.

Response for Nursing Mothers (from p. 190): Freud thought so, but there is no experimental evidence that weaning, even when ill-timed, has such dire long-term effects.

Response for Pediatricians (from p. 191): Consider the origins of the misbehavior—probably a combination of the child's inborn temperament and the mother's distal parenting. Acceptance and consistent responses (e.g., avoiding disliked foods but always using the car seat) is more warranted than anger. Perhaps this mother is expressing hostility toward the child—a sign that intervention may be needed. Find out.

Observation Quiz ANSWERS

Answer to Observation Quiz (from p. 180) The first baby is adopted, the second was born at home, and the third parent is the father, not the mother. Synchrony is universal! Although not evident here, it is also true that each is in a different nation: United States, Ethiopia, and England.

Answer to Observation Quiz (from p. 197) Norway. Almost every mother stays home with her infant for the first year (she is paid her salary to do so), and almost every mother enrolls her 1-year-old in public day care.

APPLICATION TO DEVELOPING LIVES PARENTING SIMULATION

As you progress through the Early Childhood simulation module, how you decide the following will impact the biosocial, cognitive, and psychosocial development of your child.

	Biosocial	Cognitive	Psychosocial
	• How does your child's height and weight compare to national norms? • What foods will your child eat at this stage of development? • How much physical activity will you encourage?	• Which of Piaget's stages of cognitive development is your child in? • In what kind of school will you enroll your child? • Will your child be able to demonstrate impulse control? • How will your child compare to national averages in reading, math, and language?	• In what kind of social environment will you place your child? • How will your child react if you and your partner split up? • How will you discipline your child at this age? • How does your stress level impact your child's emotional health?

PART III

Early Childhood

From ages 2 to 6, children spend most of their waking hours discovering, creating, laughing, and imagining—all the while acquiring the skills they need. They chase each other and attempt new challenges (developing their bodies); they play with sounds, words, and ideas (developing their minds); they invent games and dramatize fantasies (learning social skills and morals).

These were once called the *preschool years* because school started in first grade. But first grade is no longer first; most children begin school long before age 6. Instead, these years are called *early childhood*; those who were once called *preschoolers* are now *young children*. By whatever name, these years are a time for extraordinary growth, impressive learning, and spontaneous play, joyful not only for young children but also for anyone who knows them. ●●

Left: LightField Studios/Shutterstock.Com
Right: baona/iStock/Getty Images

Early Childhood: Biosocial Development

What Will You Know?

1. Do young children eat too much, too little, or the right amount?
2. If children never climb trees or splash in water, do they suffer?
3. Why is injury control needed more than accident prevention?
4. Which is worse, neglect or abuse?

Many children, including my grandsons, play in the yard after school. The 4- and 5-year-olds are always in motion, running and chasing; they are not ready for the play of the older children, which includes team sports and sitting and talking.

I am struck not only by their compulsion to move and explore but also by their readiness to touch each other. They greet each other with hugs; they go down the slide together so they all bunch up at the end; they get each other's attention by grabbing an arm. Danger does not stop them: The children scramble to sit on a high ledge, not protected by a fence or safety surface.

Once my grandson came to me, crying.

"I bumped my head," he told me.

I said, "Tell me what pole you bumped it on, and I will tell that pole never to bump my grandson again."

"I bumped it on another head," he said as he ran back to play, no longer crying.

I am not surprised that two young heads bumped each other. Nature makes young children active and social. This chapter describes active growth during early childhood—in body, brain, and motor skills—and what adults do to protect it.

I hold my grandson's wrist when we cross the street, make sure he has his coat, and am ready with a Band-Aid to cover any scrape. His immaturity is evident: A bandage stops the crying. I wish every young child was as easy to comfort. As you will read in this chapter, all young children are active, but some encounter far more serious problems than another child's head.

Body Changes

In early childhood, as in infancy, the body and brain grow according to powerful epigenetic forces—biologically driven and socially guided, experience-expectant and experience-dependent.

Growth Patterns

Compare an unsteady 24-month-old with a cartwheeling 6-year-old. Physical differences are obvious. Height and weight increase in those four years by about a foot and 16 pounds, or almost 30 centimeters and 8 kilograms.

Shape changes, too: Children slim down, the lower body lengthens, fat is replaced by muscle. The average *body mass index* (or *BMI*, a ratio of weight to height) is lower at ages 5 and 6 than at any other time of life. [**Developmental Link:** Body mass index is defined in Chapter 11.]

Gone are the infant's protruding belly, round face, short limbs, and large head. The center of gravity moves from the breast to the belly, enabling cartwheels, somersaults, and many other accomplishments. The joys of dancing, gymnastics, and pumping legs on a swing become possible; changing proportions enable new achievements.

During each year of early childhood, well-nourished children grow about 3 inches (about 7½ centimeters) and gain almost 4½ pounds (2 kilograms). By age 6, the average child in a developed nation:

- is at least 3½ feet tall (more than 110 centimeters).
- weighs between 40 and 50 pounds (between 18 and 23 kilograms).
- looks lean, not chubby.
- has adultlike body proportions (legs constitute about half the total height).

Nutrition

Although they rarely starve, young children sometimes are malnourished, even in nations with abundant food. Small appetites are often satiated by unhealthy snacks, crowding out needed vitamins.

Obesity Among Young Children

Older adults often encourage overeating, protecting children against famine that no longer occurs in most places. Now, the world's most serious food problem is obesity, which eventually reduces immunity and causes disease (Rook et al., 2014).

Obesity correlates with poverty for many reasons. Families with little money or education are more likely to get less exercise, watch more TV, eat fewer vegetables, drink more sugary drinks, and buy more fast food (Cespedes et al., 2013). This is partly related to where they live, often in communities that are less safe for walking or bike-riding, with fewer stores selling fresh vegetables.

Problems endure lifelong. Children who grow up in food-insecure households learn to eat whenever food is available, becoming less attuned to hunger and satiety signals in their bodies. Consequently, as adults, they overeat when they are not hungry (S. Hill et al., 2016).

Appetite naturally decreases between ages 2 and 6, as growth slows down. Parents need to know this, neither enticing children to eat nor giving them junk food to fill them up. However, parents often underestimate their children's weight. A review of 69 studies found that half of the parents of overweight children believe their children are thinner than they actually are (Lundahl et al., 2014).

Early childhood is the best time for prevention, because overweight increases with age. In 2016, 14 percent of 2- to 5-year-olds, 18 percent of 6- to 11-year-olds, and 21 percent of 12- to 19-year-olds in the United States were obese (Fryar et al., 2018).

Surprisingly, parental recognition that a child is overweight may lead to an *increase* in that child's weight, according to research on 2,823 Australian 4- and 5-year-olds followed until mid-adolescence (Robinson & Sutin, 2017). Similar results were found in Ireland and the United States.

The reason may be, according to the authors of that Australian study, that parents add stress without changing family eating patterns. If parents criticize the child, the child may react by cycling between dieting and overeating. Childhood

Short and Chubby Limbs No Longer
Siblings in New Mexico, ages 7 and almost 1, illustrate the transformation of body shape and skills during early childhood. Head size is almost the same, but the older child's arms are twice as long, evidence of proximodistal growth.

Observation Quiz Can this toddler pedal the tricycle? (see answer, page 227) ↑

obesity is linked to childhood depression (Sutaria et al., 2019) as well as to lack of exercise and excess screen time, both of which make the problems worse.

Immigration may also increase childhood weight. Few adults realize that traditional diets in low-income nations are healthier than foods advertised in developed nations (de Hoog et al., 2014). Instead, they may urge destructive patterns. One study of immigrants from many nations reported that "universally, caregivers described a concern for whether or not their children were eating enough and a perceived need to force feed them" (Dawson-Hahn et al., 2019, p. 279).

The problem is a major export, as people in many nations are adopting Western diets, buying cars and labor-saving devices, and moving to cities where long walks outside are unusual. As a result, "childhood obesity is one of the most serious public health challenges of the 21st century. The problem is global and is steadily affecting many low- and middle-income countries, particularly in urban settings" (Sahoo et al., 2015, pp. 187–188).

There has been some good news, however. In the United States, school lunches include more fruit and less fat, and the rate of obesity in 2- to 5-year-olds has not shown a steady increase. Pediatric awareness, corporate policies, and parental action are all credited with this improvement, as is former First Lady Michelle Obama, who made child nutrition and exercise her major goal. The real credit may go to her pediatrician, who said that her daughters were gaining weight more quickly than ideal. She resolved to do something about it, first for her daughters and later, as First Lady, for all children (Obama, 2018).

Many day-care centers now promote exercise and provide healthy snacks— carrot sticks and apple slices, not cookies and chocolate milk (Sisson et al., 2016). Obesity among 2- to 6-year-olds fell slightly in about 2015. Similar trends are apparent in Germany, with rates of childhood obesity steady among older children but dropping among young children (Schienkiewitz et al., 2018).

However, after leveling off for a few years, rates of preschool obesity in the United States have risen again to 13.9 percent (Fryar et al., 2018). That bad news has a hopeful twist: Since weight gain in early childhood is fluid, parents and communities can make a difference.

 DATA CONNECTIONS: Food Insecurity Around the World and in the United States explores how rates vary by nation and by U.S. household. 📖 **LaunchPad**

Nutritional Deficiencies

Although many young children consume more than enough calories, they do not always obtain adequate iron, zinc, and calcium. For example, North American children now drink less milk than formerly. Less calcium means weaker bones later on.

Compared with the average child, young children who eat more dark-green and orange vegetables and less fried food benefit in many ways. They gain bone mass but not fat, according to a study that controlled for other factors that correlate with body fat, such as gender (girls have more), ethnicity (people of some ethnic groups are genetically thinner), and income (poor children have worse diets) (Wosje et al., 2010).

Sugar is a major problem. Many cultural customs entice children to eat sweets— in birthday cake, holiday candy, desserts, sweetened juice, soda, and so on. Children prefer tastes that are more sweet and less bitter than what adults prefer, which means that adults should not let children eat what they wish. Some mothers add sugar to increase appetite: Those children develop a lifelong sweet tooth that can harm them (Mennella & Bobowski, 2015).

In early childhood, the American Heart Association recommends no more than six teaspoons of natural or added sugars, including high-fructose corn

"IT SAYS RIGHT HERE IN THE INGREDIENTS, THIS PRODUCT CONTAINS NO YUCKY STUFF!"

Who Is Fooling Whom? He doesn't believe her, but maybe she shouldn't believe what the label says, either. For example, "low fat" might also mean high salt.

syrup. The average child consumes three times that. Too much sugar causes poor circulation—with heart attacks likely 50 years later (Vos et al., 2016).

Advertisements may mislead. For example, vitamin C is usually found in abundance in the normal diet of young children and is a cheap vitamin to add to food, so sweetened juice with 100 percent of the daily requirement of vitamin C is no bargain. Many nutrients in food have not yet been identified, much less listed on food labels.

Oral Health

The most immediate harm from sugar is cavities and decaying teeth before age 6. Thus, children should see a dentist and brush their teeth regularly during early childhood—practices that were unnecessary before widespread sugar consumption (Gibbons, 2012).

"Baby" teeth are replaced naturally from ages 6 to 10. The schedule is genetic, with girls averaging a few months ahead of boys. However, tooth brushing and dentist visits should become habitual in early childhood because poor oral health harms those permanent teeth (forming below the first teeth) and can cause jaw malformation, chewing difficulties, and speech problems.

Teeth are affected by diet and illness. Thus, a young child's teeth alert the dentist to other health problems. The process works in reverse as well: Infected teeth can affect the rest of the child's body. In adulthood, tooth infections can cause preterm births, another reason good oral health habits need to be established early (Puertas et al., 2018).

Allergies and Food

An estimated 6 to 8 percent of children are allergic to a specific food, almost always a common, healthy one: Milk, eggs, peanuts, tree nuts (such as almonds and walnuts), soy, wheat, fish, and shellfish are the usual culprits. Diagnostic standards for allergies vary (which explains the range of estimates), and treatment varies even more.

Some experts advocate total avoidance of the offending food—there are peanut-free schools, where no one is allowed to bring a peanut-butter sandwich for lunch. However, some allergists give children who are allergic to peanuts a tiny bit of peanut powder (under medical supervision), which may decrease a child's allergic reaction (Couzin-Frankel, 2018).

Indeed, exposure to peanuts can begin before birth: A study of pregnant women who ingested peanuts found that their children were less likely to be allergic (Frazier et al., 2014). Many childhood food allergies are outgrown, but ongoing allergies make a balanced diet even harder to maintain.

Other allergies increase as children grow older, depending on preventive measures, such as avoiding air pollution or having a pet (Nowak & Schaub, 2018). Diet matters. Children who eat many fast foods (which have more saturated fatty acids, trans fatty acids, sodium, carbohydrates, and sugar than home-cooked food) have higher rates of asthma, nasal congestion, watery eyes, and itchy skin.

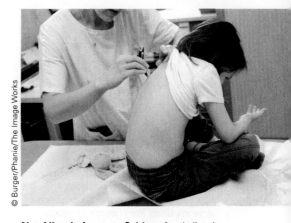

© Burger/Phanie/The Image Works

Not Allergic Anymore? Many food allergies are outgrown, so young children are more likely to have them than older ones. This skin prick will insert a tiny amount of a suspected allergen. If a red welt develops in the next half hour, the girl is still allergic. Hopefully, no reaction will occur; but if her breathing is affected, an EpiPen is within reach.

WHAT HAVE YOU LEARNED?

1. About how much does a well-nourished child grow in height and weight from age 2 to age 6?

2. Why do some adults overfeed children?

3. Why are today's children more at risk of obesity than children 50 years ago?

4. How do childhood allergies affect nutrition?

Brain Growth

By age 2, most neurons have connected to other neurons and substantial pruning has occurred. The 2-year-old's brain already weighs 75 percent of what it will weigh in adulthood; the 6-year-old's brain is 90 percent of adult weight.

Since most of the brain is already present and functioning by age 2, what remains to develop? The most important parts! Most important for people, that is.

Myelination

myelination The process by which axons become coated with myelin, a fatty substance that speeds the transmission of nerve impulses from neuron to neuron.

After infancy, the increase in brain weight is primarily in **myelination**. Sometimes called the *white matter* of the brain, *myelin* is a fatty coating on the axons that protects and speeds signals between neurons. (The *gray matter* is the neurons themselves.)

Myelin is far more than mere insulation around the axons. It smooths the connections between neurons that are far from each other. "Myelin organizes the very structure of network connectivity . . . and regulates the timing of information flow through individual circuits" (Fields, 2014, p. 266).

Connecting Left and Right

Myelination is especially evident in the major link between the left and the right halves of the brain, the **corpus callosum** (see Inside the Brain).

corpus callosum A long, thick band of nerve fibers that connects the left and right hemispheres of the brain and allows communication between them.

Left-handed people tend to have thicker, better myelinated corpus callosa than right-handed people do, perhaps because they often switch between the two sides of their bodies. When they can choose which hand to use, they prefer their left (e.g., brushing their teeth), but the culture often requires using the right hand (e.g., shaking hands). Therefore, the corpus callosum must be well-developed.

Acceptance of left-handedness is more widespread now than a century ago. More adults in Great Britain and the United States claim to be left-handed today (about 10 percent) than in 1900 (about 3 percent) (McManus et al., 2010). Developmentalists advise against forcing a left-handed child to become right-handed, since the brain is the origin of handedness.

lateralization Literally, sidedness, referring to the specialization in certain functions by each side of the brain, with one side dominant for each activity. The left side of the brain controls the right side of the body, and vice versa.

Indeed, the brain is the source of all types of **lateralization** (literally, *sidedness*). The entire human body is lateralized, apparent not only in right- or left-handedness but also in the feet, the eyes, the ears, and the brain itself.

Genes, prenatal hormones, and early experiences all affect which side does what, and then the corpus callosum puts it all together. Left-handedness (which means a stronger right brain) is an advantage in some professions, especially those involving creativity and split-second, emotional responses.

A disproportionate number of artists, musicians, and sports stars were/are left-handed, including Pele, Babe Ruth, Monica Seles, Bill Gates, Oprah Winfrey, Jimi Hendrix, Lady Gaga, and Justin Bieber. Five of the past seven presidents of the United States were/are lefties: Gerald Ford, Ronald Reagan, George H.W. Bush, Bill Clinton, and Barack Obama. Each was able to coordinate with many nations and opinions in Congress: Perhaps flexibility is a trait that correlates with a strong corpus callosum.

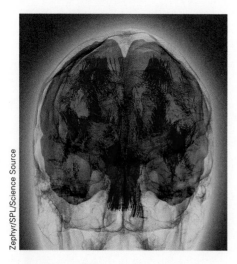

Zephyr/SPL/Science Source

Mental Coordination? This brain scan of a 38-year-old depicts areas of myelination (the various colors) within the brain. As you see, the two hemispheres are quite similar, but not identical. For most important skills and concepts, both halves of the brain are activated.

Humans and Other Animals

Although the brains and bodies of other primates are better than those of humans in some ways (they climb trees earlier and faster, for instance), and although many animals have abilities that humans lack (smell in dogs, hearing in bats), humans have intellectual capacities far beyond any other animal. Much of this is the ability to use many parts of the brain at once, an ability aided by extensive white matter.

Connected Hemispheres

The brain is divided into two halves called *hemispheres*. Those two are connected by the corpus callosum, a long, thick band of nerve fibers that myelinates and grows particularly rapidly in early childhood (Ansado et al., 2015). For that reason, compared to toddlers, young children become much better at coordinating the two sides of their brains and, hence, both sides of their bodies. They can hop, skip, and gallop at age 5; they cannot at age 2.

Both sides of the brain are usually involved in every skill, not only gross motor skills such as hopping, skipping, and galloping but also fine motor skills such as eating with utensils or buttoning one's coat. Intellectual skills also use many parts of the brain, as do social responses to other people.

As you see, the corpus callosum is crucial. With myelination, signals between the two hemispheres become quicker and clearer, enabling children to become better thinkers, to be less clumsy, and eventually to read, write, and add — all skills that require the whole brain.

The development of the corpus callosum is easy to see when comparing infants and older children. For example, no 2-year-old can hop on one foot, but most 6-year-olds can — an example of brain balancing. Many songs, dances, and games that young children love involve moving the body in some coordinated way — challenging, but fun because of that. Logic (left brain) without emotion (right brain) is a severe impairment, as is the opposite (Damasio, 2012).

Emotions also need to be balanced. In adulthood, depression is more common in people with impaired balance between the two sides of the brain (Bruder et al., 2017).

Serious disorders are caused when the corpus callosum fails to develop. That almost always results in intellectual disability. Abnormal growth of the corpus callosum is one symptom of autism spectrum disorder, as well as of dozens of other disorders (Al-Hashim et al., 2016; Travers et al., 2015; J. Wolff et al., 2015).

This imbalance is partly genetic, but it also is prenatal, as evidenced by smaller corpus callosa when the mother drank heavily during pregnancy (Biffen et al., 2018).

Astonishing studies of humans whose corpus callosa were severed to relieve severe epilepsy, as well as research on humans and other vertebrates with intact corpus callosa, reveal how the brain's hemispheres specialize.

- The left half controls the body's right side, as well as areas dedicated to logical reasoning, detailed analysis, and the basics of language.
- The brain's right half controls the body's left side and areas dedicated to emotional and creative impulses, including appreciation of music, art, and poetry.

Thus, the left half notices details, and the right half grasps the big picture. This left–right distinction has been exaggerated, especially when applied to people. No one is exclusively left-brained or right-brained, except individuals with severe brain injury in childhood, who may use half of their brain to do all of the necessary thinking. Nonetheless, we should all be grateful that many parts of our brains coordinate with other parts.

With other creatures, evolution is sometimes called "survival of the fittest." But humans have developed "a mode of living built on social cohesion, cooperation and efficient planning. It was a question of survival of the smartest" (Corballis, 2011, p. 194). Myelination from ages 3 to 6 is crucial for that, because it enables the parts of the brain to communicate with each other (Forbes & Gallo, 2017).

Maturation of the Prefrontal Cortex

The entire frontal lobe continues to develop for many years after early childhood; dendrite density and myelination continue increasing after age 20. Nonetheless, neurological control advances significantly between ages 2 and 6, evident in several ways:

- Sleep becomes more regular.
- Emotions become more nuanced and responsive.
- Temper tantrums subside.
- Uncontrollable laughter and tears are less common.

One example of the maturing brain is evident in the game Simon Says. Players are supposed to follow the leader *only* when orders are preceded by the words

Especially for Early-Childhood Teachers You know you should be patient, but frustration rises when your young charges dawdle on the walk to the playground a block away. What should you do? (see response, page 227)

Open Your Arms! But four children keep their arms closed because Simon didn't say to do so. You can almost see their prefrontal cortices (above the eyes) hard at work.

"Simon says." Thus, if leaders touch their noses and say, "Simon says touch your nose," players touch their noses; but when leaders touch their noses and say, "Touch your nose," no one is supposed to follow that example. The youngest children lose at this game because they impulsively do what they see and hear; their prefrontal cortex is too slow.

Another example is overwhelming fear. Because the amygdala is not well connected to more reflective and rational parts of the brain, many young children become suddenly terrified—even of something that exists only in imagination. Common phobias in young children are of the dark, of the ocean, of strangers. Since the emotional parts of the brain are immature, comfort and reassurance are needed, not logic. Then the fear will diminish.

Inhibition and Flexibility

Neurons have only two kinds of impulses: on–off or, in neuroscience terms, *activate–inhibit*. Each is signaled by biochemical messages from dendrites to axons to neurons. (The consequences are evident in *executive function* and *emotional regulation*, both crucial aspects of cognition that are discussed in the next two chapters.)

Impulsiveness

A balance of activation and inhibition (on/off) is necessary for thoughtful adults, who neither leap too quickly nor hesitate too long, neither lash out angrily nor freeze in fear. This is needed lifelong: One sign of cognitive loss in adulthood is a person who is too cautious or too impulsive.

The immaturity of the young brain is evident in this on/off function because many young children are notably unbalanced neurologically. They are impulsive, flitting from one activity to another. That explains why many 3-year-olds cannot stay quietly on one task, even in "circle time" in preschool, where each child is supposed to sit in place.

impulse control The ability to postpone or deny the immediate response to an idea or behavior.

Some preschool teachers tell the children to sit on their hands, or assign each child to a designated spot on the carpet, because children have difficulty not touching their neighbor. Poor **impulse control** signifies a personality disorder in adulthood but not in early childhood. Few 3-year-olds are capable of sustained attention to tasks that adults want them to attend to.

Perseveration

perseverate To stay stuck, or persevere, in one thought or action for a long time. The ability to be flexible, switching from one task to another, is beyond most young children.

The opposite reaction is also apparent. Some young children **perseverate**, which is to stick to (persevere) one thought or action. A child might play with one toy or hold one fantasy for hours. This is characteristic of children on the autism spectrum but also occurs in some typical 3-year-olds.

Young children may repeat a phrase or question again and again, or they may not be able to stop giggling once they start. That is perseveration. Crying may become uncontrollable because the child is stuck in whatever triggered the outburst.

These behaviors are linked to brain maturation. No young child is perfect at regulating attention, because immaturity of the prefrontal cortex makes controlling the limbic system almost impossible. Consequently, impulsiveness and perseveration occur.

A study of children from ages 3 to 6 found that the ability to attend to adult requests gradually increased. That correlated with academic learning and behavioral control (fewer outbursts or tears) (L. Metcalfe et al., 2013). Development continues as brain maturation (innate) and emotional regulation (learned) allow most children to pay attention and switch activities as needed, with neurological maturation related to cultural demands (Posner & Rothbart, 2017).

In childhood as well as adulthood, perseveration leads to procrastination ("Not now, I am busy"). Impulsivity and procrastination seem to be opposites, but they are closely correlated, because the same brain regions (particularly the left dorso-lateral prefrontal cortex) govern both (Liu & Feng, 2017).

Caregivers need to be patient as well as provide guidance. The comforting truth is that brains mature. Most North American teenagers are able to change tasks at the sound of the school bell, although at every age, people resist demands to stop mid-thought and attend to some other task.

Stress and the Brain

The relationship between stress and brain activity depends partly on the age of the person and partly on the degree of stress. Both too much stress and too little stress impair learning.

In an experiment, brain scans and hormone measurements were taken of 4- to 6-year-olds immediately after a fire alarm (Teoh & Lamb, 2013). As measured by their cortisol levels, some children were upset and some were not. Two weeks later, they were questioned about the event. Those with higher cortisol reactions to the alarm remembered more details. That is found in other research as well—some stress, but not too much, aids cognition (Keller et al., 2012).

However, especially with children, if an adult creates stress by asking questions sternly, demanding immediate yes-or-no answers, memories are less accurate. There are evolutionary reasons for that: People need to remember experiences that arouse emotions so that they can avoid, or adjust to, similar experiences in the future. At the same time, brains need to shut down to protect from excess stress.

Generally, a balance between arousal and reassurance is needed, again requiring coordination among many parts of the brain. For instance, if children are witnesses to a crime (a stressful experience) or experience abuse, memory is more accurate when an interviewer is warm and attentive, listening carefully but not suggesting answers (Johnson et al., 2016).

Direct maltreatment is worse, causing not only shrinkage of various regions of the brain but also decreases in white matter—and thus reduced connections between parts of the brain (Puetz et al., 2017; Rock et al., 2018). Shutdown might be best. Otherwise, child abuse might make an adult get stuck—on fear, or fantasy, or on an irrational thought, unable to feel the mixed emotions of most experiences.

Sadly, this topic leads again to the Romanian children mentioned in Chapter 7. When some adopted Romanian children saw pictures of happy, sad, frightened, or angry faces, their limbic systems were less reactive than were those of Romanian children who were never institutionalized. Their brains were also less lateralized, suggesting less efficient thinking (C. Nelson et al., 2014). Thus, institutional life, without stress reduction provided by loving caretakers, impaired their brains.

Advancing Motor Skills

Maturation and myelination allow children to move with greater speed, agility, and grace as they age. Brain growth, motivation, and guided practice undergird all motor skills. In addition, the size of the child and skeletal maturation (measured by assessing the growth of the wrist bones) affect the development of motor skills, maturing faster in girls than in boys.

Gross Motor Skills

Gross motor skills improve dramatically during early childhood. When playing, many of the youngest children fall down and bump into each other. By contrast, some 5-year-olds perform coordinated dance steps, tumbling tricks, or sports moves.

LaunchPad
macmillan learning

VIDEO ACTIVITY: The Childhood Stress–Cortisol Connection examines how high cortisol levels can negatively impact a child's overall health.

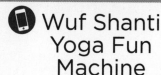
There remains much to learn, especially in adjusting to other people and situations. Thus, some 5-year-olds can kick a ball with precision but cannot be good team players on a soccer team. (Learning about people and controlling emotions are discussed in Chapters 9 and 10.)

Many North American 5-year-olds can ride a tricycle, climb a ladder, and pump a swing, as well as throw, catch, and kick a ball. A few can do these things by age 3, and some 5-year-olds can already skate, ski, dive, and ride a bike—activities that demand balanced coordination and both brain hemispheres. Elsewhere, some 5-year-olds swim in oceans or climb cliffs. All this depends on practice—some 5-year-olds can do none of these, because they have never had a chance.

Adults need to make sure that children have a safe space to play, with time, appropriate equipment, and playmates. Children learn best from peers who do whatever the child is ready to try, from catching a ball to climbing a tree. Culture and locale matter: Some children learn skateboarding, others sailing.

Urbanization worries many scientists (Acuto & Parnell, 2016; Wigginton et al., 2016). A century ago, children with varied skills played together in empty lots or fields without supervision, but now most of the world's children live in cities, with few places to throw balls, to play tag, to hide-and-seek.

Busy or violent streets not only impede development of gross motor skills but also add to the natural fears of the immature amygdala, compounded by the fears of adults. Gone are the days when parents told their children to go out and play, only to return when hunger, rain, or nightfall brought them home.

Parents may fear strangers and traffic, keeping their 3- to 5-year-olds from playing freely with peers. That distresses developmentalists. As one team, expressing the official opinion of the American Academy of Pediatrics, wrote, "Play is not frivolous; it is brain building" (Yogman et al., 2018).

Learning from Nature

Many childhood educators believe that children need space and freedom in order to develop well (Moore & Sabo-Risley, 2017). Indeed, some view nature as "another teacher having the power to enhance children's sense of wonder and capacity for learning" (Stremmel, 2012, p. 136). Balancing on branches and jumping over fences, squeezing mud and throwing pebbles, chasing birds and catching bugs—each forbidden now by some adults—educated millions of children in former cohorts.

Play is considered crucial for every aspect of child development, cognitive and social as well as physical. Accordingly, it is discussed in detail in Chapter 10. Here we need to acknowledge that motor skills—running, climbing, leaping, jumping—develop when children play freely with other children.

Some researchers analyze which environments are best for active play. A large (over 4,000 children) study of 4- to 15-year-olds in 130 U.S. communities found that children were most likely to play actively outside on dead-end streets, cul-du-sacs, or side streets where there was some visible litter (Kaczynski et al., 2018). The "counterintuitive" discovery that streets with litter were also streets with more play was explained by the researchers as a sign of fewer adults and more children.

Fine Motor Skills

Fine motor skills improve dramatically in early childhood, again with practice. This is particularly apparent for today's children in eating. Most 1-year-olds grab food in their hands, and they prefer finger foods—sandwiches, chips, carrots, cookies. Indeed, many very young children eat spaghetti, or ice cream, or mashed potatoes with their hands, despite the obvious difficulties.

Developing Motor Skills

Every child can do more with each passing year. These examples detail what one child might be expected to accomplish from ages 2 to 6. But each child is unique, and much depends on culture, practice, and maturity.

SKILLS

Jose Luis Pelaez Inc/Getty Images

AVERAGE HEIGHT IN INCHES
BOYS 45.5   GIRLS 45.0

Draw and paint recognizable images
Write simple words
Read a page of print   **6 years**
Tie shoes
Catch a small ball

Ramona Heim/Shutterstock.com

BOYS 43.0   GIRLS 42.5

Skip and gallop in rhythm
Clap, bang, sing in rhythm
Copy difficult shapes and letters   **5 years**
Climb trees, jump over things
Use a knife to cut
Wash face, comb hair

naluwan/Shutterstock.com

BOYS 40.5   GIRLS 40.0

Catch a beach ball
Use scissors
Hop on either foot
Feed self with fork   **4 years**
Dress self
Copy most letters
Pour juice without spilling
Brush teeth

Nejron Photo/Shutterstock.com

BOYS 37.5   GIRLS 37.0

Kick and throw a ball
Jump with both feet
Pedal a tricycle
Copy simple shapes   **3 years**
Walk down stairs
Climb ladders

Artem Efimov/Shutterstock.com

BOYS 34.1   GIRLS 33.5

Run without falling
Climb out of crib
Walk up stairs   **2 years**
Feed self with spoon
Draw spirals

Same Situation, Far Apart: Finger Skills
Children learn whatever motor skills their culture teaches. Some master chopsticks, with fingers to spare; others cut sausage with a knife and fork. Unlike these children in Japan *(above left)* and Germany *(above right),* some never master either, because about one-third of adults worldwide eat directly with their hands.

Gradually, however, they learn to use spoons, and then forks, and eventually knives. This is culture-specific. Asian 4-year-olds master chopsticks, unlike many 40-year-olds in North America.

Other fine motor skills are gradually developed. Young children try to mash a puzzle piece into a space, not only because their spatial visualization is immature but also because their fingers are not yet adept at rotation. That's why many puzzles for young children have wooden pieces with knobs for easier manipulation.

Sewing, knitting, drawing, buttoning, and much more take years to master. Ideally, equipment for children takes skill immaturity into account. That's why shoes are Velcro or slip-on.

Gender differences are evident; girls master fine motor skills earlier than boys. One reason is maturation: Girls are ahead of boys, evident in baby teeth eruption and word pronunciation. But another reason is practice: Girls are more often given sewing cards or dolls with clothes that require fine motor skills.

WHAT HAVE YOU LEARNED?

1. Why is myelination important for thinking and motor skills?
2. How does brain maturation affect impulsivity and perseveration?
3. When is being left-handed an asset, and when is it a liability?
4. How do stress hormones affect brain development?
5. What factors help children develop their gross motor skills?
6. What factors help children develop their fine motor skills?

Harm to Children

The goal of our study is for everyone to develop their full potential lifelong. All cultures cherish the young. Communities provide education, health care, and playgrounds; parents, grandparents, and strangers of every income, ethnicity, and nation seek to protect children while fostering their growth.

Nevertheless, more children are harmed by deliberate or accidental violence, by acts of commission or omission (action or neglect), than from disease. In the United States, almost four times as many 1- to 4-year-olds die of accidents than of cancer, which is the leading cause of disease death during these years (the numbers in one recent year were 1,267 and 325, respectively) (Kochanek et al., 2019).

Same Situation, Far Apart: Keeping Everyone Safe Preventing child accidents requires action by both adults and children. In the United States *(above left)*, adults passed laws and taught children to use seat belts—including this boy who buckles his stuffed companion. In France *(above right)*, teachers stop cars while children hold hands to cross the street—each child keeping his or her partner moving ahead.

Avoidable Injury

Worldwide, injuries that could have been prevented cause millions of premature deaths. Not until age 40 does any specific disease overtake accidents as a cause of mortality.

In some nations, malnutrition, malaria, and other infectious diseases *combined* cause more infant and child deaths than injuries do, but those nations also have high rates of child accidents. For example, southern Asia and sub-Saharan Africa have the highest rates of child motor-vehicle deaths, even though they have fewer cars per person (World Health Organization, 2015). Most children who die are pedestrians, or passengers—with no helmets—on motorcycles.

Age-Related Dangers

When accidents occur, 2- to 6-year-olds are more often seriously hurt than 6- to 10-year-olds. Why are they so vulnerable?

Immaturity of the prefrontal cortex makes young children impulsive; they plunge into danger. Unlike infants, their motor skills allow them to run, leap, scramble, and grab in a flash, before a caregiver can stop them. Their curiosity is boundless; their impulses uninhibited; their analysis limited.

Meanwhile, adults overestimate what children understand (Morrongiello, 2018). Parents teach not to run with scissors, for example, and their children can repeat that rule. They obey when the parents are nearby. Those same parents think that their children know the reason for that rule—but most do not. A child who suddenly feels angry may deliberately break a rule—unaware of the danger. Children are infamous for playing with fire, and then trying to put a fire out themselves if it gets out of hand.

Parents think they can predict what hazards will be attractive to their children, yet "young children routinely do unpredictable things that lead to injuries" (Morrongiello, 2018, p. 218). A child who has never seemed interested in, for instance, the kitchen knife rack might one day carve, cut, or stab, never having

● **Especially for Urban Planners**
Describe a neighborhood park that would benefit 2- to 5-year-olds. (see response, page 227)

injury control/harm reduction Practices that are aimed at anticipating, controlling, and preventing dangerous activities; these practices reflect the beliefs that accidents are not random and that injuries can be made less harmful if proper controls are in place.

learned that knives can be dangerous. I did exactly that when I was learning to write: One of my parents' cherished coffee tables had "KATL" carved in it.

Almost all young children do something that they know is forbidden and hide it from their parents. This could be quite harmless, such as hiding broccoli in their napkin or sweeping up shards of the glass that should have stayed on the shelf. But, it could also be dangerous: Children run away and get lost or swallow pills but never tell their parents that they feel sick.

Age-related trends are apparent. Falls are more often fatal for the youngest (under 24 months) and oldest (over 80 years) people; 1- to 4-year-olds have high rates of poisoning and drowning; motor-vehicle deaths peak from age 15 to 25.

Generally, as income falls, accident rates rise, but not for every cause. In the United States, wealthier young children drown in swimming pools six times more often than children and adults who are low SES. Usually the deadly pool is in their own backyard, a luxury few low-income children enjoy. Fire, however, is more likely to kill the youngest, poorest children, as their homes have fewer smoke alarms and sprinkler systems.

Injury Control

Instead of using the term *accident prevention*, public health experts prefer **injury control** (or **harm reduction**). Consider the implications. *Accident* implies that an injury is random, unpredictable; if anyone is at fault, it's a careless parent or an accident-prone child. Instead, *injury control* suggests that the impact of an injury can be limited, and *harm reduction*, that harm can be minimal.

If young children are allowed to play to develop their skills, minor mishaps (scratches and bruises) are bound to occur. As already explained, children benefit from play. A child with no scrapes may be overprotected.

However, playing children need protection. Serious injury is unlikely if a child falls on a safety surface instead of on concrete, if a car seat and airbags protect the body in a crash, if a bicycle helmet cracks instead of a skull, or if swallowed pills come from a tiny bottle. Reducing harm requires effort from professionals and parents, as I know from my own experience (see A Case to Study).

Less than half as many 1- to 4-year-olds in the United States were fatally injured in 2020 as in 1980, thanks to laws that limit poisons, prevent fires, and regulate cars. Control has not yet caught up with newer hazards, however.

For instance, many new homes in California, Florida, Texas, and Arizona have swimming pools: In those states, drowning is a leading cause of child death. As for poisons, children under age 5 are now less often poisoned from prescribed medicines and more often poisoned by cosmetics, recreational drugs, or personal care products (deodorant, hair colorant, etc.) (Mowry et al., 2015).

Prevention

Prevention begins long before any particular child, parent, or legislator does something foolish. Unfortunately, few people notice injuries and deaths that did *not* happen. Scientists analyze data, however, and have learned what prevention succeeds. That has reduced accidental deaths dramatically.

Levels of Prevention

Three levels of prevention apply to every health and safety issue.

primary prevention Actions that change overall background conditions to prevent some unwanted event or circumstance, such as injury, disease, or abuse.

- **Primary prevention** considers the overall conditions that affect the likelihood of harm. Laws and customs are crucial to reduce injury for people of every age.

"My Baby Swallowed Poison"

Many people think that the way to prevent injury to young children is to educate parents. However, public health research finds that laws are more effective than education, especially if parents are overwhelmed by the daily demands of child care and money management. Injury rates rise when parents have more than one small child, and not enough money.

For example, thousands of lives have been saved by infant car seats. However, before the law required it, few parents voluntarily installed car seats. Research has found that parents are more likely to use car seats if they are given them to take their newborn home from the hospital, and if an expert installs the seat and shows the parents how to use it—not simply tells them or makes them watch a video (Tessier, 2010). New laws mandating car seats and new programs at hospitals have had an effect. In 2017, in the entire United States, only 81 infant passengers died in car accidents, about one-sixth the number 15 years earlier (Insurance Institute for Highway Safety, 2019).

The research concludes that motivation and education help, but laws mandating primary prevention are more effective. I know this firsthand. Our daughter Bethany, at age 2, climbed onto the kitchen counter to find, open, and swallow most of a bottle of baby aspirin. Where was I? In the next room, nursing our second child and watching television. I did not notice what Bethany was doing until a commercial.

Bethany is alive and well today, protected by all three levels of prevention. Primary prevention included laws limiting the number of baby aspirin per container; secondary prevention included my pediatrician's written directions when Bethany was a week old to buy syrup of ipecac; tertiary prevention was my phone call to Poison Control.

I told the helpful stranger who answered the phone, "My baby swallowed poison." He calmly asked me a few questions and then advised me to give Bethany ipecac to make her throw up. I did, and she did.

That ipecac had been purchased two years before, when I was a brand-new mother and ready to follow every bit of my pediatrician's advice. If the doctor had waited until Bethany was able to climb before recommending it, I might not have followed his advice, because by then I had more confidence in my ability to prevent harm.

I still blame myself, but I am grateful for all three levels of prevention that protected my child. In some ways, my own education helped avert a tragedy. I had chosen a wise pediatrician; I knew the number for Poison Control (FYI: 1-800-222-1222). (Ipecac is no longer recommended for inducing vomiting after swallowing poison; call Poison Control for current recommendations.)

As I remember all the mistakes I made in parenting, I am grateful for every level of prevention.

- **Secondary prevention** is more targeted, averting harm in high-risk situations or for vulnerable individuals.
- **Tertiary prevention** begins after an injury has already occurred, limiting damage.

Tertiary prevention is the most visible, but primary prevention is the most effective. Much of the research has focused on sports injuries among older children and adults (Emery, 2018), but the same principles apply at every age.

An example comes from data on motor-vehicle deaths. As compared with 50 years ago, far more cars are on the road, but the rate of children killed by cars is only one-fourth of what it was (Centers for Disease Control and Prevention, 2018) (see **Figure 8.1**). How does each level of prevention contribute?

Primary prevention includes sidewalks, pedestrian overpasses, streetlights, and traffic circles. Cars have been redesigned (e.g., better headlights, windows, and brakes), and drivers' competence has improved (e.g., stronger penalties for drunk driving). Reduction of traffic via improved mass transit provides additional primary prevention.

Secondary prevention reduces danger in high-risk situations. Crossing guards and flashing lights on stopped school buses are secondary prevention, as are salt on icy roads, warning signs before blind curves, speed bumps, and walk/don't walk signals at busy intersections. Laws require safety seats and belt for child passengers, and design improvements are evident every year.

secondary prevention Actions that avert harm in a high-risk situation, such as stopping a car before it hits a pedestrian.

tertiary prevention Actions, such as immediate and effective medical treatment, that are taken after an adverse event (such as illness or injury) and that are aimed at reducing harm or preventing disability.

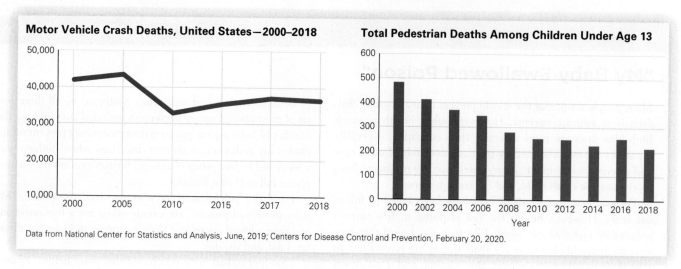

Motor Vehicle Crash Deaths, United States—2000–2018

Total Pedestrian Deaths Among Children Under Age 13

Data from National Center for Statistics and Analysis, June, 2019; Centers for Disease Control and Prevention, February 20, 2020.

FIGURE 8.1

No Matter What Statistic Motor-vehicle fatalities of pedestrians, passengers, and drivers, from cars, trucks, and motorcycles, for people of all ages, were all lower in 2018 than in 2000, a dramatic difference since the population had increased by a third and the number of cars increased as well. Proof could be shown in a dozen charts, but here is one of the most telling: deaths of child pedestrians. All three levels of prevention—in roads, cars, drivers, police, caregivers, and the children themselves—contributed to this shift.

Finally, *tertiary prevention* reduces damage after an accident. Examples include speedy ambulances, efficient emergency room procedures, effective follow-up care, and laws against hit-and-run drivers, all of which are better than decades ago. Medical personnel speak of the *golden hour*, the hour following an accident in which a victim should be treated. Of course, there is nothing magical about 60 minutes in contrast to 61 minutes, but the faster an injury victim reaches a trauma center, the better the chance of recovery (Schroeder et al., 2019).

The child death rate is lower for other reasons, as well. Air pollution has been reduced, so fewer children die of asthma. Poison control is more readily available, so fewer children die of swallowing chemicals. And many pesticides are banned from home use, so fewer children swallow them.

A most dramatic advance in harm prevention has been in accidental firearm death. In 1970, that rate was 10 per million children ages 1 to 14; in 2015, the rate was half that (National Center for Health Statistics, 2017), even though more guns had been sold.

Evidence matters. It has led to community awareness and prevention. Children are no less curious than they were, and guns are no less common. Indeed, "the civilian gun stock has roughly doubled since 1968, from one gun per every two persons to one gun per person" according to a 2012 report to the U.S. Congress (Krouse, 2012, p. 9). However, gun safety has improved: Pediatricians advise parents to hide and lock their guns, so only half as many children die of gun deaths.

The focus of this chapter is on the body, so we consider physical injury, not intellectual growth. However, pollutants in air and water, and chemicals in household products and food, may impair the brain while having no impact on the body. This is particularly true in infancy and childhood, but it continues lifelong (Babadjouni et al., 2017).

No one can prevent all harm, and government regulations are notoriously slow. The entire community needs to understand harm prevention, to protect every child. Prevention may be costly, and some efforts are not effective, which makes this topic important for every developmentalist. Unless the research is understood by everyone, thousands of children suffer. Lead is a sobering example. See A View from Science.

Solent News/Splash News/Newscom

Forget Baby Henry? Infants left in parked cars on hot days can die from the heat. Henry's father invented a disc to be placed under the baby that buzzes his cell phone if he is more than 20 feet away from the disc. He hopes all absent-minded parents will buy one.

Lead in the Environment

The need for scientists to understand the impact of various pollutants in the air, water, and food for children is particularly apparent in one sad example. Lead was recognized as a poison a century ago (Hamilton, 1914). The symptoms of *plumbism*, as lead poisoning is called, were obvious— intellectual disability, hyperactivity, and even death if the level reached 70 micrograms per deciliter of blood.

The lead industry defended the heavy metal. Correlation is not causation, they argued. Low-income children (who often had high lead levels) had lower IQs because of third variables, such as malnutrition, inadequate schools, and parents who let their children eat flaking chips of lead paint (which tastes sweet). This made sense to some developmental psychologists (Scarr, 1985) and, I confess, to me in the first edition of my textbook (Berger, 1980).

Lead remained a major ingredient in paint (it speeds drying) and in gasoline (it raises octane) for most of the twentieth century. Finally, chemical analyses of blood and teeth, with careful longitudinal and replicated research, proved that lead in dust and air (not just in eating paint chips) was indeed poisoning children (Needleman et al., 1990; Needleman & Gatsonis, 1990).

The United States banned lead in paint (in 1978) and automobile fuel (in 1996). The blood level that caused plumbism was set at 40 micrograms per deciliter, then 20, and then 10. Danger is now thought to begin at 5 micrograms, but no level has been proven to be risk-free (MMWR, April 5, 2013). Lead is especially destructive of fetal, infant, and young child brains (Hanna-Attisha et al., 2016).

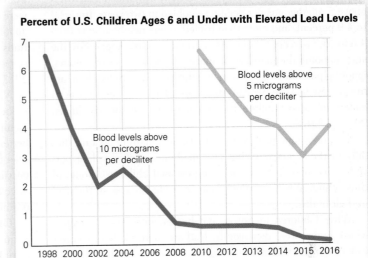

Percent of U.S. Children Ages 6 and Under with Elevated Lead Levels

Data from Child Trends Data Bank, 2015; Centers for Disease Control and Prevention, 2018.

FIGURE 8.2

Dramatic Improvement in a Decade When legislators finally accepted the research establishing the damage from lead in paint, gasoline, and water, they passed laws that helped to make it exceedingly rare for any child to die or suffer intellectual disability because of plumbism. Before 2012, 10 micrograms per deciliter of blood was thought to be completely safe; now less than 1 child in 200 tests at that level, and even 5 micrograms per deciliter alerts pediatricians and parents to find the source. These national data make the tragedy in Flint, Michigan, especially shocking.

Regulation has made a difference: The percentage of U.S. 1- to 5-year-olds with more than 5 micrograms of lead per deciliter of blood was 8.6 percent in 1999–2001, 4.1 percent in 2003–2006, 2.6 percent in 2007–2010, and less than 1 percent in 2010–2014 (Raymond & Brown, 2017) (see **Figure 8.2**).

Prevention matters. Many parents now know to wipe window ledges clean, avoid child exposure to construction dust, test drinking water, discard lead-based medicines and crockery (available in some other nations), and prevent children from eating chips of lead-based paint. A new measure is to make sure that children drink milk, since dairy products eliminate lead from the body (Kordas et al., 2018).

However, private actions alone are not sufficient to protect health. Developmentalists note many examples when blaming parents (for obesity, injury, abuse, low achievement, high blood pressure, neglect, and so on) distracts from blaming industries, laws, or the wider community.

A stark example occurred in Flint, Michigan, where in April 2014 cost-saving officials (appointed by the state to take over the city when the tax base shrunk as the auto industry left) changed the municipal drinking water from Lake Huron to

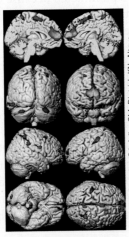

Cecil KM, Brubaker CJ, Adler CM, Dietrich KN, Altaye M, et al. (2008) Decreased Brain Volume in Adults with Childhood Lead Exposure. PLoS Med 5(5): e112. doi:10.1371/journal.pmed.0050112

Toxic Shrinkage This composite of 157 brains of adults who, as children, had high lead levels in their blood shows reduced volume. The red and yellow hot spots are all areas that are smaller than areas in a normal brain. No wonder lead-exposed children have multiple intellectual and behavioral problems.

the Flint River. That river contained chemicals from industrial waste that increased lead leaching from old pipes, contaminating tap water for drinking and for mixing infant formula.

The percent of young children in Flint with blood lead levels above 5 micrograms per deciliter doubled in two years, from 2.4 to 4.9 percent, and more than tripled in one neighborhood from 4.6 to 15.7 percent (Hanna-Attisha et al., 2016). Apparently, the state-appointed emergency manager focused on saving money, ignoring possible brain damage to children who, unlike him, are mostly low-income African Americans. This oversight has been called an "abject failure to protect public health" (Bellinger, 2016, p. 1101).

The consequences may harm these children lifelong, not only in their education but also in their activity level (hyperactivity is more common in lead-poisoned children). Some of those children will be in prison later in life because of the lead in their brains.

Why? Children in prison, not bureaucrats? Yes. Teenagers whose brains were damaged by lead are more likely to commit violent crimes than other teenagers.

At least that is what epidemiology suggests. In the United States, about 15 years after the sharp decline in blood lead levels in young children, the rates of violent crimes committed by teenagers and young adults fell sharply (Nevin, 2007).

Research in Canada, Germany, Italy, Australia, New Zealand, France, and Finland finds the same trends. Those nations that were earlier to legislate against lead had earlier crime reductions, about 20 years after the new laws. Research in many nations finds that blood lead levels predict attention deficits, school suspensions, and aggression (Amato et al., 2013; Goodlad et al., 2013; Nkomo et al., 2018).

Not surprisingly, some people think that connecting crime reduction to legislation to reduce lead is unfair, since other factors—fewer unwanted births, improved law enforcement, better education among them—have also reduced crime. But scientists in Sweden, where meticulous longitudinal research is possible, recently concluded that reduced lead levels in children directly produced a reduction of all crimes by 7 to 14 percent (Grönqvist et al., 2014).

There is now no doubt that lead, even at low levels, harms a child's brain. That makes the Flint tragedy more troubling. Developmentalists have known about the dangers of lead for decades. Why didn't the Michigan administrator know better?

---

child maltreatment Intentional harm to, or avoidable endangerment of, anyone under 18 years of age.

child abuse Deliberate action that is harmful to a child's physical, emotional, or sexual well-being.

Child Maltreatment

Accidental deaths are common worldwide, but the data reveal a related problem. Many children are harmed, not accidentally but deliberately. Indeed, in recent years, almost as many 1- to 4-year-old U.S. children have been murdered as have died of cancer. (In 2017, the numbers were 303 and 325; see **Figure 8.3**.)

In the Past 50 Years

Childhood disease deaths have plummeted because of immunization and nutrition, but childhood maltreatment death rates have not decreased. We need to understand maltreatment in order to prevent it.

Until about 1960, people thought child abuse was rare and consisted of a sudden attack by a disturbed stranger, usually a man. Today we know better, thanks to careful observation in one Boston hospital (Kempe & Kempe, 1978). A doctor saw injured children whose X-rays showed prior injuries and began to write about "battered child syndrome."

Maltreatment is neither rare nor sudden, and 90 percent of perpetrators are one or both of the child's parents—more often the mother than the father (U.S. Department of Health and Human Services, January 28, 2019). That makes it worse: Ongoing maltreatment at home, with no protector, is much more damaging than a single outside incident.

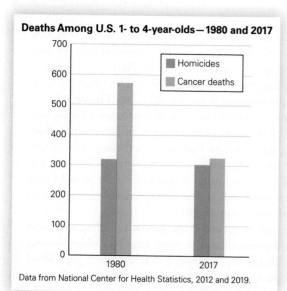

Deaths Among U.S. 1- to 4-year-olds—1980 and 2017

Data from National Center for Health Statistics, 2012 and 2019.

FIGURE 8.3

Scientists at Work Medical researchers have reduced child cancer deaths in half, primarily because new drugs have been discovered to destroy cancer cells. Why haven't scientists learned how to stop homicides?

Definitions and Statistics

Child maltreatment now refers to all intentional harm to, or avoidable endangerment of, anyone under 18 years of age. Thus, child maltreatment includes both **child abuse**, which is deliberate action that harms a child's physical, emotional, or sexual well-being, and **child neglect**, which is failure to meet essential needs, such as for food, medical care, or education.

Neglect may be worse than abuse. It is "the most common and most frequently fatal form of child maltreatment" (Proctor & Dubowitz, 2014, p. 27). About three times as many neglect cases occur in the United States as abuse cases (U.S. Department of Health and Human Services, February 1, 2018).

Data on *substantiated* maltreatment in the United States in 2018 indicate that 62 percent of cases were neglect, 11 percent physical abuse, 2 percent psychological abuse, and 7 percent sexual abuse, and 16 percent multiple forms of abuse. Ironically, neglect is often ignored by the public, who are "stuck in an overwhelming and debilitating" concept that maltreatment always causes immediate bodily harm (Kendall-Taylor et al., 2014, p. 810).

Substantiated maltreatment means that a case has been reported, investigated, and verified (see **Figure 8.4**). In 2018, 677,529 U.S. children suffered substantiated maltreatment. Every year in the United States substantiated maltreatment harms more than 1 in every 100 children, aged 2 to 5 years old.

Reported maltreatment (technically a referral) means simply that the authorities have been informed. Since 1993, the number of children referred to authorities in the United States has ranged from about 2.7 million to 4.1 million per year, with 3.6 million in 2018 (U.S. Department of Health and Human Services, January 28, 2019).

The 5-to-1 ratio of reported versus substantiated cases occurs because:

1. Each child is counted only once, so five verified reports about a single child result in one substantiated case.
2. Substantiation requires proof. Most investigations do not find unmistakable harm or a witness.
3. Many professionals are *mandated reporters*, required to report any signs of *possible* maltreatment. About two-thirds of all reports came from professionals. An investigation usually finds no harm.
4. Some reports are "screened out" as belonging to another jurisdiction, such as the military or a Native American tribe, who have their own systems, or the report is not actually about a child victim. About one-third of all reports are screened out.
5. A report may be false or deliberately misleading (though few are) (Sedlak & Ellis, 2014).

child neglect Failure to meet a child's basic physical, educational, or emotional needs.

substantiated maltreatment Harm or endangerment that has been reported, investigated, and verified.

reported maltreatment Harm or endangerment about which someone has notified the authorities.

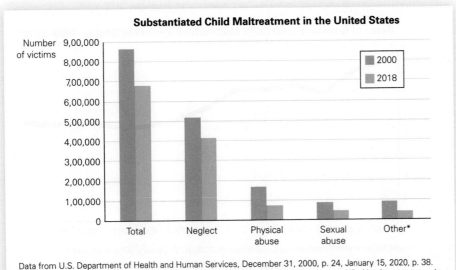

Substantiated Child Maltreatment in the United States

Data from U.S. Department of Health and Human Services, December 31, 2000, p. 24, January 15, 2020, p. 38.
*Includes emotional and medical abuse, educational neglect, and maltreatment not specified by the state records.

FIGURE 8.4

Not Good News It might seem to be good news that physical and sexual abuse are increasingly unusual. But the continued high rate of neglect is ominous. Adults can overcome memories of abuse, but neglect is likely to leave enduring traces on the brain.

Frequency of Maltreatment

How often does maltreatment occur? No one knows. Not all instances are noticed, not all that are noticed are reported, and not all reports are substantiated. Part of the problem is in distinguishing harsh discipline from abuse, and a momentary lapse from ongoing neglect. If the standard were perfect parenting all day and all night from birth to age 18, as judged by neighbors and professionals, then every child is mistreated at least once. Only severe or chronic cases are tallied.

If we rely on official U.S. statistics, positive trends are apparent. Substantiated child maltreatment increased from about 1960 to 1990 but decreased from 1990 to 2010. Other sources also report declines, particularly in sexual abuse.

Perhaps national awareness has led to more effective prevention. However, trends since 2010 suggest that rates may be increasing again (see **Figure 8.5**). There are many possible explanations. The growing gap between rich and poor families is the most plausible, since poverty correlates with neglect. However, no matter what the reason, and even if rates are declining, far too many children are mistreated.

State-by-state reports raise doubt about the data. For example, the National Administration for Children and Families reports that, between 2012 and 2016, investigations increased by 70 percent in Pennsylvania but fell by 6 percent in the neighboring state of Delaware (U.S. Department of Health and Human Services, February 1, 2018). The 2016 rate of child victims was seven times higher in Massachusetts than in Virginia, 23.3 versus 3.2 per thousand children. But no one thinks children are safer in Virginia. Maybe people in Massachusetts are quicker to report harm?

How maltreatment is defined is powerfully influenced by culture (one of my students shocked me when she asked, "When is a child too old to be beaten?"). Willingness to report also varies. The United States has become more culturally diverse, and people have become more suspicious of government but also of each other. Does that reduce reporting or increase it?

From a developmental perspective, another problem is that most maltreatment occurs early in life, before children attend school, where a teacher would be required to report it. One infant in 45 is substantiated as maltreated, as is 1 preschooler in 100 (U.S. Department of Health and Human Services, January 25, 2016). Those are substantiated cases; some of the youngest victims never reach outsiders' attention.

FIGURE 8.5

Still Far Too Many The number of substantiated cases of maltreatment of children under age 18 in the United States is too high, but there is some good news: The rate has declined significantly from its peak (15.3 per 1,000) in 1993.

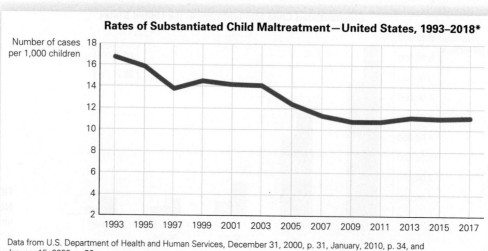

Rates of Substantiated Child Maltreatment—United States, 1993–2018*

Number of cases per 1,000 children

Data from U.S. Department of Health and Human Services, December 31, 2000, p. 31, January, 2010, p. 34, and January 15, 2020, p. 30.

*As the text explains, none of these is proof of maltreatment, but any of them requires further investigation.

An additional problem is that some children are abused often, by people who should be caregivers not abusers. A single episode of child abuse followed by parental protection and love—never blaming the child—allows children to recover. By contrast, if one family member is abusive and another family member neglects to intervene, that is likely to cause lifelong harm.

Warning Signs

Instead of relying on official statistics and mandated reporters, every reader of this book can recognize risk and prevent harm. Often the first sign is delayed development, such as slow growth, immature communication, lack of curiosity, or odd social interaction. These characteristics can be noticed in very young children. One factor that predicts maltreatment is insecure attachment: Efforts to improve parent–child relationships can also prevent serious abuse (Toth & Manly, 2019).

Table 8.1 lists signs of child maltreatment, both neglect and abuse. None of these signs *proves* maltreatment, but investigation is needed whenever they occur. The opposite is also true: Some things that many young children do (not eating much dinner, crying when they must stop playing, imagining things that are not true, throwing things in anger) are common, and usually do not signify maltreatment.

Further, community standards matter. Certain customs (such as circumcision, pierced ears, and spanking) are considered abusive among some groups but not in others; their effects vary accordingly. However, if a child is fearful, easily startled by noise, defensive and quick to attack, and confused between fantasy and reality, that suggests **posttraumatic stress disorder (PTSD)**, first identified in combat veterans, then in adults who had experienced a serious accident, natural disaster, or violent crime. Recently PTSD has been recognized in maltreated children, who suffer neurologically, emotionally, and behaviorally (E. Dunn et al., 2017).

Consequences of Maltreatment

Children suffer if they think their parents care about them less than most parents they know. If children are punished more severely, or not at all, they might feel unloved. Or if every neighbor girl has earrings, or if every boy is circumcised, a particular child might feel neglected if their parents did not follow that norm. Of course, serious abuse and gross neglect are never acceptable, but community standards matter.

Especially for Nurses While weighing a 4-year-old, you notice several bruises on the child's legs. When you ask about them, the child says nothing and the parent says that the child bumps into things. What should you do? (see response, page 227)

posttraumatic stress disorder (PTSD) An anxiety disorder that develops as a delayed reaction to having experienced or witnessed a profoundly shocking or frightening event, such as rape, severe beating, war, or natural disaster. Its symptoms may include flashbacks to the event, hyperactivity and hypervigilance, displaced anger, sleeplessness, nightmares, sudden terror or anxiety, and confusion between fantasy and reality.

TABLE 8.1
Signs of Maltreatment in Children Aged 2 to 10
Injuries that are unlikely to be accidents, such as bruises on both sides of the face or body; burns with a clear line between burned and unburned skin
Repeated injuries, especially broken bones not properly tended (visible on X-ray)
Fantasy play with dominant themes of violence or sex
Slow physical growth
Unusual appetite or lack of appetite
Ongoing physical complaints, such as stomachaches, headaches, genital pain, sleepiness
Reluctance to talk, to play, or to move, especially if development is slow
No close friendships; hostility toward others; bullying of smaller children
Hypervigilance, with quick, impulsive reactions, such as cringing, startling, or hitting
Frequent absence from school
Frequent change of address
Frequent change in caregivers
Child seems fearful, not joyful, on seeing caregiver

Loren Elliot/Reuters/Newscom

Line Up, Single File These children were separated from their parents at the border between Mexico and the United States. Here, on June 22, 2018, they are in McAllen, Texas, hoping to see their parents again soon. We do not know what happened next.

Observation Quiz Are the older children comforting the younger ones? (see answer, page 227) ↑

The long-term effects of maltreatment depend partly on the current relationship between the adult and the punishing parent. If the grown child has a good relationship with the formerly abusive parent (more common if abuse was not chronic), who expresses regret for past behavior, then recovery is likely (Schafer et al., 2014). Abused children do not necessarily become abusive parents (Widom et al., 2015a). They may avoid the mistakes of their parents, especially if friends or partners show them a better way.

In the chilling memoir *Educated*, Tara Westover (2018) describes the mistreatment she and her five siblings endured as children. With the help of many outsiders and Brigham Young University, she recovered, but three of her siblings did not.

Even for those who recover, however, the consequences of maltreatment may last for decades (Toth & Manly, 2019). Immediate impairment is obvious, when a child is bruised, broken, afraid to talk, or failing in school. More crippling effects endure lifelong, in social skills and self-esteem.

Maltreated children tend to hate themselves and distrust everyone else. Early childhood is a particularly vulnerable time (Toth & Manly, 2019). Even if maltreatment stops at age 5, emotional problems (externalizing for the boys and internalizing for the girls) linger (Godinet et al., 2014). Adult drug abuse, social isolation, and poor health results from maltreatment decades earlier.

Hate is corrosive; love is healing. A warm and enduring friendship can repair some damage, but mistreated children typically regard other people as hostile. They become less friendly, more aggressive, and more isolated than other children.

The earlier that abuse starts and the longer it continues, the worse the children's relationships are. Physically and sexually abused children are likely to be irrationally angry, and neglected children are often withdrawn. Consequently, healthy romances and close friendships are difficult in adulthood.

Finding and keeping a job is a critical aspect of adult well-being, yet adults who were maltreated suffer in this way as well. One study carefully matched 807 children who had experienced substantiated abuse with other children of the same sex, ethnicity, and family SES. About 35 years later, when maltreatment was a distant memory, those who had been mistreated were 14 percent less likely to be employed. The researchers concluded: "Abused and neglected children experience large and enduring economic consequences" (Currie & Widom, 2010, p. 111).

In that study, women had more difficulty finding and keeping a job than men. It may be that self-esteem, emotional stability, and social skills are even more important for female employees than for male ones. That study is just one of hundreds of longitudinal studies, all of which find that maltreatment affects people decades after broken bones, or skinny bodies, or medical neglect.

Preventing Harm

For accidents, child abuse, and child neglect, the ultimate goal is *primary prevention*, a social network of customs and supports for parents, neighbors, and professionals, so every child is safe. Neighborhood stability, parental education, income support, and fewer unwanted children reduce injury.

However, governments and private foundations are more likely to fund *secondary prevention* for high-risk families (Nelson & Caplan, 2014). The news media also focus on shocking examples of parental abuse or social worker neglect, which ignores the many ways in which families, communities, and professionals stop harm before it begins.

Secondary prevention includes spotting warning signs and intervening early, so a risky situation does not worsen. Many developmentalists believe that it is crucial to foster secure attachment. Relationship problems should be repaired before they become harmful (Toth & Manly, 2019).

Tertiary prevention limits harm after injury has occurred. This begins with reporting and then includes investigating. If abuse is substantiated, the most important step (tertiary prevention) is helping the caregiver provide better care. That may include treating addiction, assigning a housekeeper, locating family helpers, securing better living quarters, and helping the child recover, with special medical, psychological, or education assistance. In the process, the child must be protected, either in the same family or in a better one.

In every case, **permanency planning** is needed: This is planning that addresses how to nurture the child until adulthood (Scott et al., 2013). Uncertainty, moving, a string of temporary placements, and frequent changes in schools are all destructive, even when each one is well-intentioned.

When children are taken from their parents and entrusted to another adult, that is **foster care**. The other adult might be a relative, in which case it is called **kinship care**. Foster care sometimes is informal, as when a grandmother provides custodial care because the parents do not, or may result from Child Protective Services provided by the government.

Every year for the past decade in the United States, almost half a million children have been officially in foster care. At least another million are unofficially in kinship care, because relatives realize that the parents are unable or unwilling to provide good care.

Most foster children are from low-income, ethnic-minority families—a statistic that reveals problems in the macrosystem as well as the microsystem. In the United States, most foster children have physical, intellectual, and emotional problems that arose in their original families. Obviously, foster parents need much more than financial subsidies to provide good care for such children.

Sometimes, a child's best permanency plan is adoption by another family, who will provide care lifelong. However, adoption is difficult, for many reasons:

- Judges and biological parents are reluctant to release children for adoption.
- Most adoptive parents prefer infants, but few infants are available because some adults do not know how hard child care can be until they have tried, and failed, to provide for their children.
- Some agencies screen out families not headed by married, heterosexual couples.
- Some professionals insist that adoptive parents be of the same ethnicity and/or religion as the child.

As you have seen in this chapter, caring for the nutrition, brain development, and safety of a 2- to 6-year-old is not easy for any parent—biological, foster, or adoptive. Yet these years are vital, with consequences lasting for decades. This is as true for cognitive and psychosocial development as for biosocial, as the next two chapters explain.

Jessica Antola/Contour/Getty Images

She Recovered and Sings Maya Angelou was abused and neglected as a child, but she also was loved and protected by her brother and other family members. The result was extraordinary insight into the human condition, as she learned "why the caged bird sings."

permanency planning An effort by child-welfare authorities to find a long-term living situation that will provide stability and support for a maltreated child. A goal is to avoid repeated changes of caregiver or school, which is particularly harmful.

foster care A legal, publicly supported system in which a maltreated child is removed from the parents' custody and entrusted to another adult or family, who is reimbursed for expenses incurred in meeting the child's needs.

kinship care A form of foster care in which a relative of a maltreated child, usually a grandparent, becomes the approved caregiver.

WHAT HAVE YOU LEARNED?

1. What can be concluded from the data on rates of childhood injury?
2. How do injury deaths compare in developed and developing nations?
3. What is the difference between primary, secondary, and tertiary prevention?
4. Why have the rates of child accidental death declined?
5. Why is reported abuse five times higher than substantiated abuse?
6. Why is neglect considered worse than abuse?
7. What are the long-term consequences of childhood maltreatment?
8. When is adoption part of permanency planning?
9. Why does permanency planning rarely result in adoption?

SUMMARY

Body Changes

1. Well-nourished children gain weight and height during early childhood at a lower rate than infants do. Proportions change, allowing better body control.

2. Worldwide, an increasing number of children are obese, which puts them at risk for many health problems. Further, many children have an unbalanced diet. Sugar, particularly, is harmful, reducing appetite and harming teeth.

Brain Growth

3. The brain continues to grow in early childhood, reaching about 75 percent of its adult weight at age 2 and 90 percent by age 6. Much of the increase is in myelination, which speeds transmission of messages from one part of the brain to another.

4. Maturation of the prefrontal cortex allows more reflective, coordinated thought and memory, but this takes decades. Many young children gradually become less impulsive and less likely to perseverate.

5. Childhood trauma may create a flood of stress hormones (especially cortisol) that damage the brain and interfere with learning.

6. Gross motor skills continue to develop; clumsy 2-year-olds become 6-year-olds who move their bodies well, guided by their peers, practice, motivation, and opportunity—all varying by culture. Playing with other children develops skills that benefit children's physical, intellectual, and social development.

7. Fine motor skills are difficult to master during early childhood. Young children enjoy expressing themselves artistically, which helps them develop their body and finger control. Fortunately, self-criticism is not yet strong.

Harm to Children

8. Accidents cause more child deaths than diseases, with young children more likely to suffer a serious injury or premature death than older children. Close supervision and public safeguards can protect young children from their own eager, impulsive curiosity.

9. In the United States, various preventive measures have reduced the rate of serious injury, and medical measures have reduced disease deaths even faster. Four times as many young children die of injuries than of cancer, the leading cause of disease death in childhood.

10. Injury control occurs on many levels, including long before and immediately after each harmful incident. Primary prevention protects everyone, secondary prevention focuses on high-risk conditions and people, and tertiary prevention occurs after an injury. All three are needed.

11. Child maltreatment includes ongoing abuse and neglect, usually by a child's own parents. In 2018, about 3.6 million cases of child maltreatment were reported in the United States; about a fifth of them were substantiated, with the annual rate of maltreatment about 1 child in 100.

12. Physical abuse is the most obvious form of maltreatment, but neglect is more common and more harmful. Health, learning, and social skills are all impeded by abuse and neglect, not only during childhood but also decades later.

13. Primary prevention is needed to stop child maltreatment before it starts. Secondary prevention should begin when someone first notices a possible problem.

14. Tertiary prevention may include placement of a child in foster care, including kinship care, or in an adoptive family. Permanency planning is required because frequent changes are harmful to children.

KEY TERMS

myelination (p. 208)	injury control/harm reduction (p. 216)	child abuse (p. 220)	posttraumatic stress disorder (PTSD) (p. 223)
corpus callosum (p. 208)	primary prevention (p. 216)	child neglect (p. 221)	permanency planning (p. 225)
lateralization (p. 208)	secondary prevention (p. 217)	substantiated maltreatment (p. 221)	foster care (p. 225)
impulse control (p. 210)	tertiary prevention (p. 217)	reported maltreatment (p. 221)	kinship care (p. 225)
perseverate (p. 210)	child maltreatment (p. 220)		

APPLICATIONS

1. Keep a food diary for 24 hours, writing down what you eat, how much, when, how, and why. Did you eat at least five servings of fruits and vegetables, and very little sugar or fat? Did you get too hungry, or eat when you were not hungry? Then analyze when and why your food habits began.

2. Go to a playground or another place where young children play. Note the motor skills that the children demonstrate, including abilities and inabilities, and keep track of age and sex. What differences do you see among the children?

3. Ask several parents to describe each accidental injury of each of their children, particularly how it happened and what the consequences were. What primary, secondary, or tertiary prevention measures were in place, and what measures were missing?

4. Think back to your childhood and the friends you had at that time. Was there any maltreatment? Considering what you have learned in this chapter, why or why not?

Especially For ANSWERS

Response for Early-Childhood Teachers (from p. 209): One solution is to remind yourself that the children's brains are not yet myelinated enough to enable them to quickly walk, talk, or even button their jackets. Maturation has a major effect, as you will observe if you can schedule excursions in September and again in November. Progress, while still slow, will be a few seconds faster.

Response for Urban Planners (from p. 216): The adult idea of a park—a large, grassy, open space—is not best for young children. For them, you would design an enclosed area, small enough and with adequate seating to allow caregivers to socialize while watching their children. The playground surface would have to be protective (since young children are clumsy), with equipment that encourages motor skills. Teenagers and dogs should have their own designated areas, far from the youngest children.

Response for Nurses (from p. 223): Any suspicion of child maltreatment must be reported, and these bruises are suspicious. Someone in authority must find out what is happening so that the parent as well as the child can be helped.

Observation Quiz ANSWERS

Answer to Observation Quiz (from p. 205): No. There are no pedals! Technically this is not a tricycle; it has four wheels. The ability to coordinate both legs follows corpus callosum development in the next few years, as explained on page 208.

Answer to Observation Quiz (from p. 224) Probably not. In many detention centers, children are forbidden to touch each other except when older children must keep younger ones in line. Notice that the older boy holds the child's wrist (they are not holding hands), and the older girl is pulling the younger girl's arm back, not providing a comforting touch.

Early Childhood: Cognitive Development

What Will You Know?

1. Are young children selfish or just self-centered?
2. Do children get confused if they hear two languages?
3. Is it a mistake to let children play all day at preschool?

My grandson, when he was 3 years old, held a large rubber ball. He told me to play basketball with him.

"We can't play basketball; we don't have a hoop," I told him.

"We can imagine a hoop," he answered, throwing the ball up.

"I got it in," he said happily. "You try."

I did.

"You got it in, too," he announced, and did a little dance.

Soon I was tired, and sat down.

"I want to sit and think my thoughts," I told him.

"Get up," he urged. "You can play basketball and think your thoughts."

He was typical. Imagination came easily to him, and he aspired to the skills of older, taller people in his culture. He thought by doing, and his language was impressive, including a sentence eight words long. But he did not yet understand that my feelings differ from his, that I would rather sit than throw a ball at an imaginary basket. He did know, however, that I usually respond to his requests.

This chapter describes these characteristics of young children—imagination, active learning, language, but also their difficulty in understanding another person's perspective. I hope it also conveys the joy of understanding how young children think. When that happens, you might do what I did—get up and play.

Thinking During Early Childhood

You learned in Chapter 8 about the rapid advances in motor skills, brain development, and impulse control during ages 2 through 6, enabling the young child to do somersaults by kindergarten. Part of the reason this is possible is maturation of the body. Its changing proportions enable head-over-heels moves, because the legs are now much longer than the torso. And part of it is social influence. Few children do somersaults unless they have seen another child do one.

The same two forces allow the mental somersaults of early childhood. Soon you will read about the two major developmental theorists, Piaget and Vygotsky, who focused on young children. Piaget describes cognitive maturation, as children advance from preoperational thought to concrete operational thought. Vygotsky stressed the impact of social forces, as young children learn to talk, think, and explain.

But before describing the contributions of each of these theorists to childhood thinking, we note another aspect of thought that comes to the fore in early childhood.

Executive Function

At every age, a person's ability to think depends on what has been called *executive function*, a cognitive ability that is nascent at age 2 and that continues to improve throughout life (Diamond, 2012). It is evident, and can be measured, in early childhood (Eisenberg & Zhou, 2016; Espy et al., 2016).

Executive function has been measured at every stage of life. It protects adolescents from destructive emotional outbursts (Poon, 2018), promotes coping skills in adulthood (Nieto et al., 2019), and forestalls death in old age (Reimann et al., 2018).

Definitions

Executive function combines three essential abilities: (1) memory, (2) inhibition, and (3) flexibility. Each requires some further explanation.

The aspect of memory emphasized in executive function is *short-term*, or *working*, *memory*, which is memory for what was seen a minute ago or yesterday, or what can easily be brought to mind, not for what happened years ago. Young children who are proficient in this aspect of cognition are able to remember what they had for lunch, where they put their mittens, what they saw at the science museum.

Inhibition is the ability to control responses, to stop and think for a moment before acting or talking. Young children with this ability are able to restrain themselves from hitting or crying when someone else accidentally bumps them, and to raise their hands without blurting out an answer to a teacher's question.

Flexibility (also called *shifting*) is the ability to see things from another perspective rather than staying stuck in one idea. One example from early childhood occurs when a child wants to play with a toy that another child has. Executive function enables the child with the toy to share, and it allows the onlookers to switch to another activity or wait for a turn.

The result, as a leading expert explains, is that young children gain "core skills critical for cognitive, social, and psychological development" that allow "playing with ideas, giving a considered response rather than an impulsive one, and being able to change course or perspectives as needed, resist temptations, and stay focused" (Diamond, 2016). All this is very difficult in early childhood—and it is not easy at any point in the life span.

Developmentalists have many creative ways to measure executive function in young children. To measure memory, for instance, 3- to 5-year-olds are shown a series of barnyard animals and asked to remember them in order. For inhibition, they are asked to push a button when they see a fish but not a shark. For flexibility, they are asked to alternate stamping on a picture of a dog and one of a bone (both are presented together, again and again). Scores on all of these improve with age during early childhood (Espy, 2016).

A Sensitive Time

It is thought that early childhood is a particularly important time to develop executive function skills. Compared to older children, 2- to 6-year-olds are more open to learning, have much to learn (remember the impulsive and perseverative responses explained in Chapter 8), and are open to suggestion (Walk et al., 2018).

Many educators and parents focus on intelligence and vocabulary, since children's minds need to be engaged, and their language must be encouraged. However, for success in kindergarten and beyond, executive function is crucial.

executive function The cognitive ability to organize and prioritize the many thoughts that arise from the various parts of the brain, allowing the person to anticipate, strategize, and plan behavior.

The Joy of Rivalry Look closely at this sister and brother in Johannesburg, South Africa. Just as 1-year-olds run as soon as they are able, siblings everywhere quarrel, fight, and compromise, ideally testing their physical and intellectual skills, which, as seems to be the case here, is fun.

Grant Difford/Greatstock/Alamy

It correlates more with brain development than scores on intelligence tests (Friedman & Miyake, 2017).

When experts study executive function in preschool children (Griffin et al., 2016), they emphasize two conclusions:

1. Executive function skills are foundational. They undergird later cognitive abilities and achievements, including in reading and math.
2. Executive function skills are not inborn; they can be taught.

How might that teaching occur? Family life and early schooling are crucial. Instead of explicit, time-limited lessons, executive-function skills should permeate the entire curriculum of a preschool. Young children need to verbalize their emotions, plan their actions, sustain activity, and work with another child or two.

One suggested activity, for example, is for two children to develop a pretend scenario together. They decide on their roles and actions (advance planning) and then carry it out, responding to their playmate's actions (flexibility). They are guided and encouraged, rarely criticized or punished.

Executive function is the product of practice, at home and school. Targeted educational programs inspired by Vygotsky (e.g., Tools of the Mind) are used in preschools (Bodrava & Leong, 2018). One aspect of those programs is that children learn to think about their thinking, and that improves their ability to learn (Marulis et al., 2020).

Traditional Perspectives

Many cognitive psychologists now focus on executive function, particularly on efforts to advance it. The concept that a combination of memory, inhibition, and flexibility forms a foundation for learning is a useful one.

But this does not reduce the importance of the two traditional theorists, Piaget and Vygotsky, nor of the crucial roles of language development and early education. Indeed, they are basic to our understanding of the mind of the young child, as you will see. We begin with Jean Piaget, the "towering figure in the science of cognitive development throughout much of the 20th century" (Carey et al., 2015, p. 36).

Piaget: Preoperational Thought

preoperational intelligence Piaget's term for cognitive development between the ages of about 2 and 6; it includes language and imagination (which involve symbolic thought), but logical, operational thinking is not yet possible at this stage.

Early childhood is the time of **preoperational intelligence**, the second of Piaget's four periods of cognitive development (described in Table 2.3 on page 41). Piaget referred to early-childhood thinking as *pre*operational because children do not yet use logical operations (reasoning processes) (Inhelder & Piaget, 1964/2013a).

Symbolic Thought

symbolic thought A major accomplishment of preoperational intelligence that allows a child to think symbolically, including understanding that words can refer to things not seen and that an item, such as a flag, can symbolize something else (in this case, a country).

Preoperational children think in symbols, not just via senses and motor skills. In **symbolic thought**, an object or a word can stand for something else, including something out of sight or imagined. The rapid acquisition of vocabulary is a dramatic example of symbolic thought.

However, although vocabulary and imagination soar in early childhood, logical connections between ideas are not yet *operational*. Piaget found that young children cannot yet apply their impressive new linguistic ability to comprehend reality.

Consider how understanding the word *dog* changes over the first three levels of intelligence. During the sensorimotor level, "dog" might mean only the family dog sniffing at the child, not yet a symbol (Callaghan, 2013). By age 3, in contrast, the preoperational child can use the word *dog* as a symbol: It can refer to a remembered dog, or a plastic dog, or an imagined dog, or any new "doggie" the child sees on the street.

Nonetheless, because the child does not yet have the logic of concrete operational thought, if asked what distinguishes dogs from cats, young children mention size and coat. They cannot articulate the difference between dogs and cats, even though they do not mistake one for the other. The essence of "dogness" and "catness" is too abstract.

Animism

Symbolic thought helps explain **animism**, the belief of many young children that natural objects (such the sun and clouds) are alive and that nonhuman animals have the same characteristics as the child. Many children's stories include animals or objects that talk and listen (Aesop's fables, *Winnie-the-Pooh, Goodnight Gorilla, The Day the Crayons Quit*). Preoperational thought is symbolic and magical, not logical and realistic.

Among contemporary children, animism gradually disappears as the mind becomes more mature, by age 10 if not earlier (Kesselring & Müller, 2011).

Obstacles to Logic

Piaget described symbolic thought as characteristic of preoperational thought. He also noted four limitations that make logic difficult: *centration, focus on appearance, static reasoning,* and *irreversibility*.

Centration is the tendency to focus on one aspect of a situation to the exclusion of all others. For example, young children may insist that Daddy is a father, not a brother, because they center on the role that he fills for them. This illustrates a particular type of centration that Piaget called **egocentrism**—literally, "self-centeredness." Egocentric children contemplate the world exclusively from their personal perspective.

Egocentrism is *not* selfishness. One 3-year-old chose to buy a model car as a birthday present for his mother: His "behavior was not selfish or greedy; he carefully wrapped the present and gave it to his mother with an expression that clearly showed that he expected her to love it" (Crain, 2011, p. 133).

A second characteristic of preoperational thought is a **focus on appearance**, to the exclusion of other attributes. For instance, a girl given a short haircut might worry that she has turned into a boy. In preoperational thought, a thing is whatever it appears to be—evident in the joy young children have in wearing the hats or shoes of a grown-up, clomping noisily and unsteadily around the house.

Third, preoperational children use **static reasoning**. They believe that the world is stable, unchanging, always in the state in which they currently encounter it. Many children cannot imagine that their own parents were once children. If they are told that Grandma is their mother's mother, they still do not understand how people change with maturation. One child asked his grandmother to tell his mother not to spank him because "she has to do what her mother says."

The fourth characteristic of preoperational thought is **irreversibility**. Preoperational thinkers fail to recognize that reversing a process might restore whatever existed before. A young girl might cry because her mother put lettuce on her sandwich. She might still reject the food when the lettuce is removed because she thinks what is done cannot be undone.

Disney/Pixar/Photo 12/Alamy

All Alive Animism and egocentrism might make a 4-year-old frightened by this scene in the movie *Toy Story 3*. Very young children have no problem believing that toys (even with three eyes) are alive and have the same emotions that they themselves do.

animism The belief that natural objects and phenomena are alive, moving around, and having sensations and abilities that are human-like.

centration A characteristic of preoperational thought in which a young child focuses (centers) on one idea, excluding all others.

egocentrism Piaget's term for children's tendency to think about the world entirely from their own personal perspective.

focus on appearance A characteristic of preoperational thought in which a young child ignores all attributes that are not apparent.

static reasoning A characteristic of preoperational thought in which a young child thinks that nothing changes. Whatever is now has always been and always will be.

irreversibility A characteristic of preoperational thought in which a young child thinks that nothing can be undone. A thing cannot be restored to the way it was before a change occurred.

conservation The principle that the amount of a substance remains the same (i.e., is conserved) even when its appearance changes.

Conservation and Logic

Piaget reported many examples of the ways in which preoperational intelligence disregards logic. A famous series of experiments involved **conservation**, the notion that the amount of something remains the same (is conserved) despite changes in its appearance.

Suppose two identical glasses contain the same amount of milk, and the milk from one glass is poured into a taller, narrower glass. When children younger than 6 are asked whether one glass contains more or, alternatively, if both glasses contain the same amount, they say that the narrower glass (with the higher level) has more. (See **Figure 9.1** for other examples.)

All four characteristics of preoperational thought are evident in this failure to understand conservation. Young children focus (*center*) on what they see (*appearance*), noticing only the immediate (*static*) condition. They do not realize that pouring the milk back into the wider glass would re-create the level of a moment earlier (*irreversibility*).

Note that this reveals one of the key aspects of cognition that is central to executive function. Unless they are taught, children younger than 6 do not have the memory or flexibility to grasp conservation.

Especially for Nutritionists How can Piaget's theory help you encourage children to eat healthy foods? (see response, page 255)

Piaget's original tests of conservation required children to respond verbally to an adult's questions. When scientists make tests of logic simple and playful, young children sometimes succeed, especially when they talk to each other.

This is an important lesson for teachers of young children. To teach about quantity, they must recognize the limitations of preschoolers' thinking (McCray et al., 2018). A teacher might team up two children, ask them to pour milk from one glass to another again and again, and add a stuffed animal who asks about conservation.

Tests of Various Types of Conservation

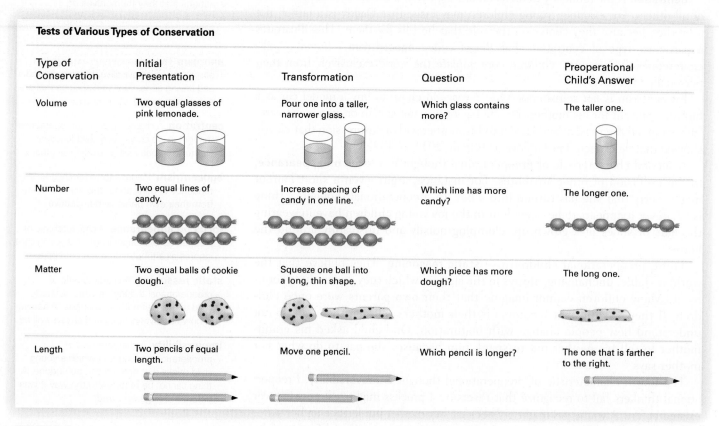

Type of Conservation	Initial Presentation	Transformation	Question	Preoperational Child's Answer
Volume	Two equal glasses of pink lemonade.	Pour one into a taller, narrower glass.	Which glass contains more?	The taller one.
Number	Two equal lines of candy.	Increase spacing of candy in one line.	Which line has more candy?	The longer one.
Matter	Two equal balls of cookie dough.	Squeeze one ball into a long, thin shape.	Which piece has more dough?	The long one.
Length	Two pencils of equal length.	Move one pencil.	Which pencil is longer?	The one that is farther to the right.

FIGURE 9.1

Conservation, Please According to Piaget, until children grasp the concept of conservation at about age 6 or 7, they cannot understand that the transformations shown here do not change the total amount of liquid, candies, cookie dough, and pencils.

Stones in the Belly

As we were reading a book about dinosaurs, my 3-year-old grandson, Caleb, told me that some dinosaurs (*sauropods*) have stones in their bellies. It helps them digest their food and then poop and pee.

"I didn't know that dinosaurs ate stones," I said.

"They don't eat them."

"Then how do they get the stones in their bellies? They must swallow them."

"They don't eat them."

"Then how do they get in their bellies?"

"They are just there."

"How did they get there?"

"They don't eat them," said Caleb. "Stones are dirty. We don't eat them."

I changed the subject, as I knew that his mother had told him not to eat pebbles, or sand, or anything else he found on the ground.

However, at dinner he asked my daughter, "Do dinosaurs eat stones?"

"Yes, they eat stones so they can grind their food," she answered.

At that, Caleb was quiet.

In all of this, preoperational cognition is evident. Caleb is bright; he can name several kinds of dinosaurs, as can many young children.

But logic eludes Caleb. He is preoperational, not operational.

It seemed obvious to me that the dinosaurs must have swallowed the stones. However, in his static thinking, Caleb said the stones "are just there." He rejected the thought that dinosaurs ate stones because he has been told that stones are too dirty to eat.

Caleb is egocentric, reasoning from his own experience, and animistic, in that he thinks other creatures think and act as he himself does. He trusts his mother, who told him never to eat stones. Static thinking prevented him for considering my authority as his mother's mother. He was skeptical that a dinosaur would do something he had been told not to do.

Nonetheless, Caleb is curious, a hallmark of preoperational thought. My question lingered rather than being completely rejected.

Should I have expected him to tell me that I was right when his mother agreed with me? No. That would have required far more understanding of reversibility and far less egocentrism than most young children can muster.

Before age 6, children indicate via eye movements or gestures that they understand conservation, although they cannot yet put their understanding into words (Goldin-Meadow & Alibali, 2013). Conservation and many other logical ideas can be grasped bit by bit, with active, guided experience. Glimmers of understanding may be apparent in children as young as age 4.

As with infants, Piaget underestimated what young children could understand. He was right about his basic idea, however: Young children are not very logical (Lane & Harris, 2014). Their cognitive limits make smart 3-year-olds sometimes foolish, as Caleb was.

Vygotsky: Social Learning

For decades, the magical, illogical, and self-centered aspects of cognition dominated research on early-childhood thought. Scientists were awed by Piaget, who demonstrated many aspects of egocentric thought in children.

Vygotsky emphasized another side of early cognition—that each person's thinking is shaped by other people. His focus on the sociocultural context contrasts with Piaget's emphasis on the individual.

Mentors

Vygotsky believed that cognitive development is embedded in the social context at every age (Vygotsky, 1987). He stressed that children are curious and observant of everything in their world, and are influenced by their social surroundings.

Young children are famous for their exasperating penchant to ask "why," "how," "where," and "why" again. They want to know how machines work, why weather

LaunchPad
macmillan learning

VIDEO ACTIVITY: Achieving Conservation focuses on the cognitive changes that enable older children to pass Piaget's conservation-of-liquid task.

Learning to Button Most shirts for 4-year-olds are wide-necked without buttons, so preschoolers can put them on themselves. But the skill of buttoning is best learned from a mentor, who knows how to increase motivation.

● **Observation Quiz** Beyond the buttoning, can you see any other signs of scaffolding? (see answer, page 255) ➔

changes, where the sky ends. They seek answers from mentors, who might be parents, teachers, older siblings, or just a stranger, who reply with their own perceptions and assumptions, which they themselves were taught in childhood. Thus, according to Vygotsky, culture shapes thought.

As you remember from Chapter 2, children learn through *guided participation*. Parents are the first guides and mentors, soon joined by many others, especially in an interactive preschool (Broström, 2017).

According to Vygotsky, mentors do the following:

- Present challenges.
- Offer assistance (without taking over).
- Add crucial information.
- Encourage motivation.

Learning from mentors indicates intelligence, according to Vygotsky: "What children can do with the assistance of others might be in some sense even more indicative of their mental development than what they can do alone" (1980, p. 85).

Scaffolding

zone of proximal development (ZPD) Vygotsky's term for the skills—cognitive as well as physical—that a person can exercise only with assistance, not yet independently.

scaffolding Temporary support that is tailored to a learner's needs and abilities and aimed at helping the learner master the next task in a given learning process.

Vygotsky believed that all individuals learn within their **zone of proximal development (ZPD)**, an intellectual arena in which new ideas and skills can be mastered. *Proximal* means "near," so the ZPD includes the ideas and skills that children are close to mastering but cannot yet demonstrate independently. Learning requires wise and willing mentors to provide **scaffolding**, or temporary support, to help children within their developmental zone (Mermelshtine, 2017).

Good mentors scaffold in many ways. They encourage children to look both ways before crossing the street (pointing out speeding trucks, cars, and buses while holding the child's hand) or let them stir the cake batter (perhaps covering the child's hand on the spoon handle, in guided participation). Scaffolding might foster emotional control as well. One study encouraged children to take deep breaths, relax, and be less egocentric (Grabell et al., 2019).

Sensitive joint engagement is crucial, because "When providing scaffolding, a teacher or a peer tutor does not make the task easier but instead makes the learner's job easier by giving the child maximum support in the beginning stages and then gradually withdrawing this support as the child's mastery of a new skill increases" (Bodrava & Leong, 2018, p. 226).

Culture matters. Educators agree that book-reading is a crucial scaffold, which adults use to transmit their understanding of what children need to learn (Hoyne & Egan, 2019). Book-reading may be time to encourage questions, or, the opposite, to teach a child to listen quietly.

A study of 27 Head Start classrooms found that children's vocabulary increased when book-reading fostered teacher–student interaction, but some teachers never asked questions and others asked one every 10 seconds. The types of question also varied: Most questions had a single answer ("did the train move?"), but some encouraged dialogue ("what might happen next?") (Hindman et al., 2019).

When reading to their children, some parents emphasize how misbehavior causes problems for the book's characters; others highlight the emotions that the characters experience (Luo et al., 2014). Although every culture teaches specific behaviors, everywhere guided participation involves conversation between children and adults, not yes/no questions.

Overimitation

Sometimes scaffolding is inadvertent, when children copy something that adults would rather the child not do. Young children curse or kick because someone showed them how.

More benignly, they imitate meaningless habits and customs, a trait called **overimitation**. Children follow mentors: They learn rituals, tool use, grammar, emotional expression. Overimitation is a "flexible and . . . highly functional phenomenon" (Hoehl et al., 2019, p. 104).

Overimitation was demonstrated in a series of experiments with 3- to 6-year-olds, 64 of them from San communities (pejoratively called "Bushmen") in South Africa and Botswana, and, for comparison, 64 from cities in Australia and 19 from aboriginal communities within Australia. Australian middle-class adults often scaffold with words and actions, but San adults rarely do. The researchers expected the urban Australian children but not the San children to follow adult demonstrations (Nielsen et al., 2014). The researchers were wrong.

In part of the study, some children watched an adult open a box, which could easily and efficiently be opened by pulling down a knob by hand. Instead, the adult waved a red stick above the box three times and used that stick to push down the knob to open the box. Then each child (one by one, so they couldn't influence each other) was given the stick and asked to open the box. No matter what their culture, they did what the adult did, waving the stick three times, not touching the knob with their hands.

Other San and Australian children did not see the demonstration. When they were given the stick and asked to open the box, they simply pulled the knob. Then they observed an adult do the stick-waving opening—and they copied those inefficient actions, even though they already knew the easy way. Children everywhere learn from others not only through explicit guidance, but also through observation.

Overimitation is universal: Young children follow what others do. They are naturally "socially motivated," which allows them to learn when adults guide them. Adults everywhere use eyes, facial expressions, and gestures to facilitate learning (Heyes, 2016).

Imitation is innate in our species. Adults enjoy transmitting knowledge, and children like imitating, as long as copying is not too difficult: They imitate adults who seem to know something, even when the adults are unaware that they are modeling (Hoehl et al., 2019; Tomasello, 2016b).

That is exactly what Vygotsky expected and explained: Children learn culture.

Tim Hall/The Image Bank/Getty Images

Count by Tens A large, attractive abacus could be a scaffold. However, in this toy store the position of the balls suggests that no mentor is nearby. Children are unlikely to grasp the number system without a motivating guide.

● **Observation Quiz** Is the girl above right-handed or left-handed? (see answer, page 255) ↑

overimitation When a person imitates an action that is not a relevant part of the behavior to be learned. Overimitation is common among 2- to 6-year-olds when they imitate adult actions that are irrelevant and inefficient.

private speech The internal dialogue that occurs when people talk to themselves (either silently or out loud).

social mediation Human interaction that expands and advances understanding, often through words that one person uses to explain something to another.

Language as a Tool

Although all of the elements of a culture guide children, Vygotsky thought language was pivotal, in two ways: private speech and social mediation.

First, talking to oneself, called **private speech**, is evident when young children talk aloud to review, decide, and explain events to themselves (and, incidentally, to anyone else within earshot) (Al-Namlah et al., 2012). Almost all young children sometimes talk to themselves; about a third sing to themselves as well (Thibodeaux et al., 2019).

With time, children become more circumspect, sometimes whispering. Many adults use private speech as they talk to themselves when alone or write down ideas.

Second, language advances thinking by facilitating social interaction (Vygotsky, 2012). This **social mediation** function of speech occurs as mentors guide mentees in their zone of proximal development, learning numbers, recalling memories, and following routines. Adults sometimes are explicit ("do it this way") but often merely encourage ("good job").

STEM Learning

A practical use of Vygotsky's theory concerns STEM (science, technology, engineering, math) education. Many adults are concerned that too few college students choose a STEM career. This is particularly true for young women and African Americans (Harris, 2019).

Developmentalists find that a person's interest in such vocations begins with learning about numbers and science (counting, shapes, fractions, molecular structure, the laws of motion) in early childhood. Spatial understanding—how one object fits with another—enhances later math skills (Verdine et al., 2017). During the preschool years, knowledge of math and physics develops month by month. Before first grade, children learn to:

- Count objects, with one number per item (called *one-to-one correspondence*).
- Remember times and ages (bedtime at 8 P.M.; a child is 4 years old).
- Understand sequence (first child wins, last child loses; play after dinner.)
- Know which numbers are greater than others (four bites are more than two).
- Understand how to make things move (toy cars, balls, game pieces).
- Appreciate temperature effects, from ice to steam.

By age 3 or 4, children are able to comprehend numbers, store memories, and recognize routines. Whether or not children actually demonstrate this comprehension depends on their families, schools, and cultures (Verdine et al., 2017).

Some 2-year-olds hear sentences such as "One, two, three, takeoff," "Here are two cookies," or "Dinner in five minutes" several times a day. They are encouraged to touch an interesting bit of moss, or are alerted to the phases of the moon outside their window, or are given large puzzle pieces that fit shapes, or hear about the relationship between pace and steepness of a hill they are climbing.

Other children never have such experiences—and they have a harder time with math in first grade, with science in the third grade, and with physics in high school. If, as Vygotsky believed, words mediate between brain potential and comprehension, STEM education begins long before formal education.

Educational software can be "a conduit for collaborative learning" (Cicconi, 2014, p. 58), as Web 2.0 programs respond to each child's abilities and needs. Several children can work together, each mentoring the others, talking aloud as the computer prompts them.

For executive function, however, interactive software needs to be carefully chosen. Video games, for instance, usually encourage rapid responses—the opposite of the planning, inhibition, and memory that are the bedrock of executive function.

Michelle Del Guercio/Science Source/Getty Images

Future Engineers in the Bronx Playing with Legos helps children learn about connecting shapes, which makes math and geometry easier to learn in school and makes STEM careers more likely. Once Legos were only marketed to boys, but no longer—there now are kits designed to appeal to girls.

Educators disapprove if a screen replaces human interaction (that is opposite to what Vygotsky advocates), but they also recognize that computers, carefully used (no more than an hour a day), might be learning tools, just as books might be (Alper, 2013; American Academy of Pediatrics, 2016).

A study that compared 226 children and their parents who used the "Bedtime Math" app on computers with 167 matched children found that the former had significantly higher math scores after a year (Berkowitz et al., 2015). The authors cited four reasons:

1. The app required parents and children to discuss math together.
2. The app had few distracting noises or animations.
3. The app helped math-phobic parents talk about numbers and shapes.
4. The app followed each story with questions that encouraged memory and reflection (that aids executive function).

Another reason might be when this interaction occurred. Children, and everyone else, learn best when they have a chance to "sleep on it."

Children's Theories

Every scholar who studies learning during early childhood notices that children are eager to understand everything. Young children do more than gain words and concepts; they develop theories to help them understand and remember—theories that arise from both brain maturation and personal experience.

Theory-Theory

In **theory-theory**, the *theory* about how children think is that they construct a *theory*. Humans always want theories (even false ones) to explain their experiences. Children ask questions, and, if they are not satisfied with the answers, they develop their own theories. For example, one child thought Grandpa died because God was lonely; another thought thunder meant God was rearranging the furniture; a third thought mountains were created to stop the world from floating away.

Children follow the same processes that scientists do: asking questions, developing hypotheses, gathering data, and drawing conclusions. As a result, "preschoolers have intuitive theories of the physical, biological, psychological, and social world" (Gopnik, 2012, p. 1623).

Of course, their cognitive methods lack the rigor of scientific experiments, but "infants and young children not only detect statistical patterns, they use those patterns to test hypotheses about people and things" (Gopnik, 2012, p. 1625). Their conclusions are not always correct: Like all good scientists, they allow new data to promote revision, although, also like all humans, they sometimes stick to their old theories instead of newer versions.

A leading developmentalist advocates answering children's questions with valid, research-based explanations, rather than simply smiling at their charming mistakes (Kelemen, 2019). Otherwise mistaken theories continue to prevail, even if later in high school and college, counter evidence is presented. That is how cultural assumptions, fake news, and prejudices, continue. Children theorize that what adults say and do is correct.

> **theory-theory** The idea that children attempt to explain everything they see and hear by constructing theories.

> **THINK CRITICALLY:** Are you aware of any misconceptions you learned in childhood that still linger in your adult thinking?

Theory of Mind

Mental processes—thoughts, emotions, beliefs, motives, and intentions—are among the most complicated and puzzling phenomena that humans encounter every day. Adults wonder why people fall in love with the particular persons they do, why they vote for the political leaders they choose, or why they make foolish decisions, from signing for a huge mortgage to buying an overripe cucumber.

theory of mind A person's theory of what other people might be thinking. In order to have a theory of mind, children must realize that other people are not necessarily thinking the same thoughts that they themselves are. That realization seldom occurs before age 4.

LaunchPad
macmillan learning

VIDEO: Theory of Mind: False-Belief Tasks demonstrates how children's theory of mind develops with age.

FIGURE 9.2

Better with Age? Could an obedient and honest 3-year-old become a disobedient and lying 5-year-old? Apparently yes, as the proportion of peekers and liars in this study more than doubled over those two years. Does maturation make children more able to think for themselves or less trustworthy?

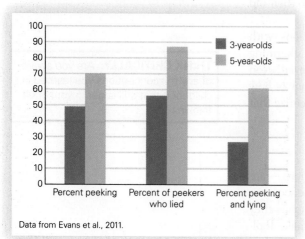

Data from Evans et al., 2011.

Children are likewise puzzled about a playmate's anger, a sibling's generosity, or an aunt's too-wet kiss.

To know what goes on in another's mind, people develop a *folk psychology*, which includes ideas about other people's thinking, called **theory of mind**. Theory of mind is an emergent ability, slow to develop but evident in most children by age 4 (Carlson et al., 2013).

Some aspects of theory of mind develop sooner than others. Longitudinal research finds that typical 2-year-olds do not know that other people think differently than they do, but 6-year-olds have a well-developed theory of mind (Wellman et al., 2011).

Part of theory of mind is understanding that someone else might have a mistaken belief. For example, a child watches a puppet named Max put a toy dog into a red box. Then Max leaves and the child sees the dog taken out of the red box and put in a blue box.

When Max returns the child is asked, "Where will Max look for the dog?" Without a theory of mind, most 3-year-olds confidently say, "In the blue box"; most 6-year-olds correctly say, "In the red box."

Theory of mind actually develops gradually, progressing from knowing that someone else might have different desires (at about age 3) to knowing that someone might hide their true feelings (about age 6). Culture, age, and experience all matter. The most notable variations, however, are neurological, not cultural: Children who are deaf or have autism spectrum disorder are remarkably slow to develop theory of mind (Peterson & Wellman, 2019).

The development of theory of mind can be seen when young children try to escape punishment by lying. Their faces often betray them: worried or shifting eyes, pursed lips, and so on. Parents sometimes say, "I know when you are lying," and, to the consternation of most 3-year-olds, parents are usually right.

In one experiment, 247 children, aged 3 to 5, were left alone at a table that had an upside-down cup covering dozens of candies (Evans et al., 2011). The children were told *not* to peek, and the experimenter left the room.

For 142 children (57 percent), curiosity overcame obedience. They peeked, spilling so many candies onto the table that they could not put them back under the cup. The examiner returned, asking how the candies got on the table. Only one-fourth of the participants (more often the younger ones) told the truth (see **Figure 9.2**).

The rest lied, and their skill at lying increased with their age. The 3-year-olds typically told hopeless lies (e.g., "The candies got out by themselves"); the 4-year-olds told unlikely lies (e.g., "Other children came in and knocked over the cup"). Some of the 5-year-olds, however, told plausible lies (e.g., "My elbow knocked over the cup accidentally").

Another study, this one of prosocial lies (saying that a disappointing gift was appreciated) found that children who were advanced in theory of mind and in executive function were also better liars, able to stick to the lie that they liked the gift (S. Williams et al., 2016). The ability to deceive advances year by year and correlates with theory of mind and executive function (Heyman et al., 2019).

Brain and Context

Evidence for crucial brain maturation comes from other research on the same 3- to 5-year-olds whose lying was studied. The children who were asked to say "day" when they saw a picture of the moon and "night" when they saw a picture of the sun needed to inhibit their automatic reaction. Their success indicated executive function, which correlated with maturation of the prefrontal cortex. That maturation depends not only on time but also on context, as Inside the Brain explains.

The Role of Experience

Why do executive control, concrete operational thought, social interaction, and even lying improve with age? There are many factors, of course, but developmentalists increasingly recognize the crucial maturation of the prefrontal cortex. With modern neuroscience, this can be traced quite precisely: A notable advance in all of these abilities occurs between ages 4 and 5, probably because the prefrontal cortex matures markedly at this point (Devine & Hughes, 2014).

Children who are slow in language development are also slow in theory of mind, a finding that makes developmentalists suggest that underlying deficits—genetic or neurological—may be crucial for both. Remember the plasticity of the brain: The early years may be particularly important for neurological control.

In studies of adults as well, many brain regions are involved in theory of mind and much depends on past history and context (Preckel et al., 2018). Developmentalists suggest that when a young child is slow in language learning, in addition to targeted work on vocabulary, articulation, and so on, therapists and teachers need to work on executive function (Nilsson & de López, 2016).

Social interactions with other children promote brain development, advancing both theory of mind and executive function. This is especially evident when those other children are siblings of about the same age (McAlister & Peterson, 2013). As one expert in theory of mind quipped, "Two older siblings are worth about a year of chronological age" (Perner, 2000, p. 383).

Indeed, many studies have found that a child's ability to develop theories correlates with neurological maturation, which also correlates with advances in executive processing—the reflective, anticipatory capacity of the mind (Baron-Cohen et al., 2013; Mar, 2011). Detailed studies find that theory of mind activates several brain regions (Koster-Hale & Saxe, 2013).

This makes sense, as theory of mind is a complex ability that humans develop in social contexts, so it is not likely to reside in just one neurological region. Brain research finds that, although each cognitive ability arises from a distinct part of the brain, experience during childhood advances neurological coordination.

Remember that experience strengthens neuronal connections. This is true not only within each neuron and the dendrites reaching another neuron but also in regions of the brain. Thus, while theory of mind promotes empathy in one brain region, and executive function promotes flexibility and memory in another, the prefrontal cortex is able to coordinate the two (Wade et al., 2018).

That is why a correlation is found between theory of mind and executive function. As a child advances in one, that child also advances in the other. Repeated coordination in the brain allows the child, for instance, to comfort a sad friend in exactly the way that sad friend is best comforted.

Even when compared to other children who were the same age, those who failed the day–night tests typically told impossible lies. Their age-mates who were higher in executive function told more plausible lies (Evans et al., 2011).

Does the crucial role of neurological maturation make culture and context irrelevant? Not at all: Nurture is always important. The reason that formal education traditionally began at about age 6 is that by the end of early childhood, maturation of the prefrontal cortex allows sustained attention. Some children are slower in this aspect of brain development (note that ADHD refers to an *attention deficit*), but for everyone experiences before age 6 advance brain development and prepare children for formal education (Blair & Raver, 2015).

Some helpful experiences before age 6 occur naturally: Children develop theory of mind in talking with adults and in playing with other children. Games that require turn-taking encourage memory and inhibitory control, two crucial components of executive control. In daily life, as brothers and sisters argue, agree, compete, and cooperate, and as older siblings fool younger ones, it dawns on 3-year-olds that not everyone thinks as they do, a thought that advances theory of mind.

By age 5, children have learned how to persuade their younger brothers and sisters to give them a toy. Meanwhile, younger siblings figure out how to gain sympathy by complaining that their older siblings have victimized them. "Who started it?" is the wrong question.

WHAT HAVE YOU LEARNED?

1. What three abilities comprise executive function?
2. What is not logical about preoperational thought?
3. What is the difference between egocentrism in a child and selfishness in an adult?
4. How does guided participation increase a child's zone of proximal development?
5. Why did Vygotsky think that talking to oneself is an aid to cognition?
6. Why does theory-theory develop?
7. What advances theory of mind?

Language Learning

Learning language is often considered the premier cognitive accomplishment of early childhood. Two-year-olds use short, telegraphic sentences ("Want cookie," "Where Daddy go?"), omitting adjectives, adverbs, and articles. By contrast, 5-year-olds seem to be able to say almost anything (see At About This Time), using every part of speech. Some preschoolers understand and speak two or three languages, an accomplishment that many adults struggle for years to achieve.

The Time for Language Learning

Brain maturation, myelination, scaffolding, and overimitation make early childhood ideal for learning language. As you remember from Chapter 1, scientists

---

AT ABOUT THIS TIME

Language in Early Childhood

Approximate Age	Characteristic or Achievement in First Language
2 years	*Vocabulary:* 100–2,000 words
	Average sentence length: 2–6 words
	Grammar: Plurals; pronouns; many nouns, verbs, adjectives
	Questions: Many "What's that?" questions
3 years	*Vocabulary:* 1,000–5,000 words
	Average sentence length: 3–8 words
	Grammar: Conjunctions, adverbs, articles
	Questions: Many "Why?" questions
4 years	*Vocabulary:* 3,000–10,000 words
	Average sentence length: 5–20 words
	Grammar: Dependent clauses, tags at sentence end ("... didn't I?" "... won't you?")
	Questions: Peak of "Why?" questions; many "How?" and "When?" questions
6 years and up	*Vocabulary:* 5,000–30,000 words
	Average sentence length: Some seem unending ("... and ... who ... and ... that ... and ...")
	Grammar: Complex, depending on what the child has heard, with some children correctly using the passive voice ("Man bitten by dog") and subjunctive ("If I were ...")
	Questions: Some about social differences (male–female, old–young, rich–poor) and many other issues

once thought that early childhood was a *critical period* for language learning—the *only* time when a first language could be mastered and the best time to learn a second or third one.

It is easy to understand why they thought so. Young children have powerful motivation and ability to sort words and sounds into meaning (theory-theory). That makes them impressive language learners. However, the critical-period hypothesis is false: A new language can be learned after age 6. A first language (even a sign language) is a scaffold for a new language, making learning possible (Mayberry & Kluender, 2018).

Even a first language can be learned after early childhood, although the only reason a child would not learn some language is extreme social isolation. Those few children who were so severely neglected, such as Genie who was rescued in 1970 at age 13 (Curtiss, 2014), never become proficient, although they can learn some language.

Although new language learning in adulthood is possible, it is not easy. Early childhood is a *sensitive period* for language learning—for rapidly mastering vocabulary, grammar, and pronunciation. Young children are language sponges; they soak up every verbal drop they encounter, some learning several languages.

One of the valuable (and sometimes frustrating) traits of young children is that they talk about many things to adults, to each other, to themselves, to their toys—unfazed by misuse, mispronunciation, ignorance, stuttering, and so on (Marazita & Merriman, 2010). Language comes easily, partly because preoperational children are not self-critical. Egocentrism has advantages; this is one of them.

LaunchPad
macmillan learning

VIDEO ACTIVITY: Language Acquisition in Young Children features video clips of a new sign language created by deaf Nicaraguan children and provides insights into how language evolves.

The Vocabulary Explosion

The average child knows about 500 words at age 2 and more than 10,000 at age 6 (Herschensohn, 2007). That's more than six new words a day.

As with many averages in development, the range is vast: The number of root words (e.g., *run* is a root word, not *running* or *runner*) that 5-year-olds know ranges from 2,000 to 6,000, with several thousand more non-root words (Biemiller, 2009). Counting a child's words is difficult, although building vocabulary is crucial (Treffers-Daller & Milton, 2013).

To understand why vocabulary is difficult to measure, consider the following: Children listened to a story about a raccoon that saw its reflection in the water. The children were asked what *reflection* means. Five of the answers:

1. "It means that your reflection is yourself. It means that there is another person that looks just like you."
2. "Means if you see yourself in stuff and you see your reflection."
3. "Is like when you look in something, like water, you can see yourself."
4. "It mean your face go in the water."
5. "That means if you the same skin as him, you blend in."

[*Hoffman et al., 2014, pp. 471–472*]

Which of the five responses indicated that the child knew what *reflection* means? None? All? Two, three, or four?

In another example, a story included "a chill ran down his spine." Children were asked what *chill* meant. One answer: "When you want to lay down and watch TV—and eat nachos" (Hoffman et al., 2014, p. 473). The child got no credit for "chill"; is that fair?

Fast-Mapping

Children develop interconnected categories for words, a kind of grid or mental map that aids speedy vocabulary acquisition. Learning a word after one exposure

Jason Lindsey/Alamy

What Is It? These two children at the Mississippi River Museum in Iowa might call this a crocodile, but really it is an alligator. Fast-mapping allows that mistake, and egocentrism might make them angry if someone tells them they chose the wrong name.

fast-mapping The speedy and sometimes imprecise way in which children learn new words by tentatively placing them in mental categories according to their perceived meaning.

is called **fast-mapping** (Woodward & Markman, 1998), because, rather than figuring out the exact definition after hearing a word used in several contexts, children hear a word once and quickly stick it into a category in their mental language grid.

Picture books offer many opportunities to advance vocabulary through scaffolding and fast-mapping. A mentor might encourage the next steps in the child's zone of proximal development, such as that tigers have stripes and leopards have spots, or, for an older child, that calico cats are almost always female and that lions with manes are always male.

This process explains children's learning of colors. Generally, 2-year-olds fast-map color names (K. Wagner et al., 2013). For instance, "blue" is used for some greens or grays. It is not that children cannot see the hues. Instead, they apply words they know to broad categories and have not yet learned the boundaries that adults use, or specifics such as chartreuse, turquoise, olive, navy. As one team of scientists explains, adults' color words are the result of slow-mapping (K. Wagner et al., 2013), which is not what young children do.

Words and the Limits of Logic

Closely related to fast-mapping is a phenomenon called *logical extension:* After learning a word, children use it to describe other objects in the same category. One child told her father that she had seen some "Dalmatian cows" on a school trip to a farm. Instead of telling her there are no such cows, he remembered the Dalmatian dog she had petted the weekend before. He realized that she saw Holstein cows, not Jersey ones.

Bilingual children who don't know a word in the language they are speaking often insert a word from the other language, code-switching in the middle of a sentence. That midsentence switch may be considered wrong, but it is actually evidence of the child's drive to communicate. By age 5, children realize who understands which language, and they avoid substitutions when speaking to a monolingual person. That illustrates theory of mind.

Some words are particularly difficult for every child, such as, in English, *who/whom, have been/had been, here/there, yesterday/tomorrow.* More than one child has awakened on Christmas morning and asked, "Is it tomorrow yet?" A child told to "stay there" or "come here" may not follow instructions because the terms are confusing. It might be better to say, "Stay there on that bench," or "Come here to hold my hand."

All languages have difficult concepts that are expressed in words; children everywhere learn them. Abstractions are particularly difficult for preoperational children; actions are easier to understand. A hole is to dig; love is hugging; hearts beat.

Acquiring Grammar

Remember from Chapter 6 that *grammar* includes structures, techniques, and rules that communicate meaning. Grammar is essential in speech, reading, and writing. A large vocabulary is useless unless a person knows how to put words together. Each language has its own grammar rules; that's one reason children speak in one-word sentences first.

Children apply rules of grammar as soon as they figure them out, using their own theories about how language works and their experience regarding when and how often various rules apply (Meltzoff & Gopnik, 2013). For example,

English-speaking children quickly learn to add an *s* to form the plural: Toddlers follow that rule, asking for two cookies or more blocks.

Soon children add an *s* to make the plural of words they have never heard before, even nonsense words. If preschoolers are shown a drawing of an abstract shape, told it is called a *wug*, and then shown two of these shapes, they say there are two *wugs* (Berko, 1958). Children comprehend grammar even before they use it.

Sometimes children apply the rules of grammar when they should not. This error is called **overregularization**. By age 4, many children overregularize that final *s*, talking about *foots*, *tooths*, and *mouses*.

This signifies knowledge, not lack of it: Many children first say words correctly (*feet, teeth, mice*), repeating what they have heard. When they are able to detect the rules of grammar, they overregularize, assuming that all constructions follow the rules (Ramscar & Dye, 2011). The child who says "I goed to the store" needs to hear "Oh, you went to the store?," not criticism.

More difficult to learn is an aspect of language called **pragmatics**—knowing which words, tones, and grammatical forms to use with whom (Siegal & Surian, 2012). In some languages, words differ when talking to someone older, or not a close friend, or when referring to maternal or paternal grandparents.

English does not make those distinctions, but pragmatics is important for early-childhood learning nonetheless. Children learn variations in vocabulary and tone depending on the context and, once theory of mind is established, on the audience.

Knowledge of pragmatics is evident when a 4-year-old pretends to be a doctor, a teacher, or a parent. Each role requires different speech. On the other hand, children often blurt out questions that embarrass their parents ("Why is that lady so fat?" or "I don't want to kiss Grandpa because his breath smells"). The pragmatics of polite speech require more social understanding than many young children possess.

Learning Two Languages

Language-minority people (those who speak a language that is not their nation's dominant one) suffer if they are not fluent in the majority language. In the United States, non-English speakers are impaired in school achievement, self-esteem, and employment.

Some of the problem is prejudice from English speakers, who think difference means deficit, but some is directly connected to language. Fluency in English erases some liabilities; fluency in another language then becomes an asset.

Early childhood is the best time to learn languages. Neuroscience finds that if adults mastered two languages before age 6, both languages are located in the same areas of the brain with no detriment to the cortex structure (Klein et al., 2014). Being bilingual then benefits the brain lifelong, because executive function is required to inhibit one language in order to speak the other (Cargnelutti et al., 2019). Indeed, the bilingual brain may provide some resistance to Alzheimer's disease in old age (Costa & Sebastián-Gallés, 2014).

By contrast, if a new language is learned in adulthood, proper use of idioms and flawless pronunciation lag behind correct grammar and vocabulary. Thus, many immigrants speak the majority language with an accent but are proficient in comprehension and reading (difference is not deficit).

From infancy on, hearing is more acute than vocalization. Almost all young children mispronounce whatever language they speak, blithely unaware of their mistakes. They comprehend more than they say, they hear better than they speak, and they learn rapidly as long as people speak to them.

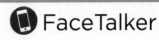
overregularization The application of rules of grammar even when exceptions occur, making the language seem more "regular" than it actually is.

pragmatics The practical use of language that includes the ability to adjust language communication according to audience and context.

Especially for Immigrant Parents
You want your children to be fluent in the language of your family's new country, even though you do not speak that language well. Should you speak to your children in your native tongue or in the new language? (see response, page 255)

Language Loss and Gains

Language-minority parents want their children to master the majority language, but they have a legitimate fear called the *language shift*, when a child becomes fluent in the school language but loses their home language. Language shift occurs whenever theory-theory leads children to conclude that their first language is inferior (Bhatia & Ritchie, 2013).

Some language-minority children in Mexico shift to Spanish; some children of Canada's First Nations shift to French; some children in the United States shift to English. In China, some children shift from Cantonese or another Chinese dialect to Mandarin. Crucial is that adults speak and value both languages, and then children will do the same.

Remember that young children are preoperational: They center on the immediate status of their language (not on future usefulness or past glory), on appearance more than substance. No wonder many children shift toward the language of the dominant culture.

Since language is integral to culture, if a child is to become fluently bilingual, everyone who speaks with the child should respect both cultures, in song, books, and daily conversation. Children learn from listening and talking, so they need to hear twice as much talk to become fluent in two languages.

The same practices make a child fluently trilingual, as some 5-year-olds are. Young children who are immersed in three languages may speak all three with no accent—except the accent of their mother, father, and friends. This was evident in one 6-year-old in the United States who spoke Korean and Farsi with his two parents and English at school, each language with whomever understood it, translating and code-switching as needed (Choi, 2019). [**Developmental Link:** Bilingual education is also discussed in Chapter 12.]

Listening, Talking, and Reading

Because understanding the printed word is crucial in the twenty-first century, a meta-analysis of about 300 studies analyzed which activities in early childhood aided reading later on. Both vocabulary and phonics (precise awareness of spoken sounds) predicted literacy (Shanahan & Lonigan, 2010). Five specific strategies and experiences were particularly effective for children of all income levels, languages, and ethnicities.

Bilingual Learners These are Chinese children learning a second language. Could this be in the United States? No, this is a class in the first Chinese Hungarian school in Budapest. There are three clues: the spacious classroom, the letters on the book, and the trees outside.

ATTILA KISBENEDEK/AFP/Getty Images

1. *Code-focused teaching.* In order for children to read, they must "break the code" from spoken to written words. One step is to connect letters and sounds (e.g., "A, alligators all around" or "B is for baby").
2. *Book-reading.* Vocabulary and print-awareness develop when adults read to children.
3. *Parent education.* When parents know how to encourage cognition (listening and talking), children become better readers. Adult vocabulary expands children's vocabulary.
4. *Language enhancement.* Within each child's zone of proximal development, mentors help children expand vocabulary. That requires teachers who know each child's zone and individualize conversation.
5. *Early-education programs.* Children learn from teachers, songs, excursions, and other children. (We discuss variations of early education next, but every study finds that preschools advance language acquisition.)

Early-Childhood Schooling

Today, virtually every nation provides some public or private early-childhood education, sometimes for a privileged few, and sometimes for every child. In some nations, as much public money is invested in early education as in primary or secondary school. In other nations, no money at all is allocated for education before age 6 (Barnett et al., 2017; Blossfeld et al., 2017).

Among developed nations, the United States invests relatively little money, and it educates comparatively few children. In 2018, 44 percent of U.S. 4-year-olds and 16 percent of 3-year-old were in publicly funded preschools (Friedman-Krauss et al., 2019). By contrast, in Northern Europe, more than 95 percent of all 3- to 5-year-olds were enrolled in government-funded schools.

The reasons for the international variations are historical, economic, and political, not developmental. The evidence regarding early cognition is no longer disputed: Young children are amazingly capable and eager to learn.

Homes and Schools

Developmental research does not translate directly into specific practices in early education, so no program can legitimately claim to follow Piaget or Vygotsky exactly (Hatch, 2012). Family practices and parent influences matter even if a young child is in a day-care center from 8 A.M. to 6 P.M., five days a week. Ideally, the teachers and parents collaborate to teach the child, but always, attachment to parents is evident.

Beyond the amazing potential of young children to learn, another robust conclusion from research on children's learning seems not yet universally understood: Quality matters.

Quality cannot be judged by the name of a program or by its sponsorship. Educational institutions for 3- to 5-year-olds are called preschools, nursery schools, day-care centers, pre-primary programs, pre-K classes, and kindergartens. Sponsors can be public (federal, state, or city), private, religious, or corporate.

Indeed, quality is notoriously difficult to judge (Votruba-Drzal & Dearing, 2017); "[B]ecause quality is hard for parents to observe, competition seems to be dominated by price" (Gambaro et al., 2014, p. 22). That is a problem, because to make a profit, programs hire fewer teachers, which means less individualized instruction. Thus, low cost indicates low quality, unless a program is subsidized by a religious, corporate, or government institution.

The converse is not true: High cost does not necessarily mean high quality. Professional assessment of quality seems inadequate (Elicker & Ruprecht, 2019; Sabol et al., 2013). However, one aspect — child–teacher interaction — correlates with learning. Effective teachers do not sit quietly. They talk, laugh, guide, and play with happy, talkative children.

Tibet, China, India, and . . . Italy? Over the past half-century, as China increased its control of Tibet, thousands of refugees fled to northern India. Tibet traditionally had no preschools, but young children adapt quickly, as in this preschool program in Ladakh, India. This Tibetan boy is working a classic Montessori board.

In order to sort through this variety, we consider the goals of preschools. Some aim to encourage each child's creative individuality (*child-centered*), some to prepare the child for formal education (*teacher-directed*).

Child-Centered Programs

Many programs are called *child-centered*, or *developmental*, because they stress each child's development and growth. The emphasize "teaching the whole child," which means individualized social and emotional growth, not academics (Hyson & Douglass, 2019). Teachers in such programs believe children need to follow their own interests rather than adult directions.

For example, they agree that "children should be allowed to select many of their own activities from a variety of learning areas that the teacher has prepared" (Lara-Cinisomo et al., 2011, p. 101). The physical space and the materials (such as dress-up clothes, art supplies, puzzles, blocks, and other toys) are arranged to allow exploration.

Most child-centered programs encourage artistic expression, including music and drama (Bassok et al., 2016). Some educators believe that young children see the world more imaginatively than older people do. According to advocates of child-centered programs, this peak of creative vision should be encouraged; children need to tell stories, draw pictures, dance, and make music.

Child-centered programs are often influenced by Piaget, who emphasized that each child discovers new ideas if given a chance, or by Vygotsky, who thought that children learn from playing, especially with other children. Neither wanted children to be rushed to adulthood.

Montessori Schools

One type of child-centered school began in the slums of Rome, Italy, in 1907, when Maria Montessori opened a nursery school (Standing, 1998). She believed that children needed structured, individualized projects to give them a sense of accomplishment. Her students completed puzzles, used sponges and water to clean tables, traced shapes, and so on.

Contemporary **Montessori schools** still emphasize individual pride and achievement, presenting many literacy-related tasks (e.g., outlining letters and looking at books) to young children. Specific materials have changed from those developed by Montessori, but the underlying philosophy is the same. Children are active learners: They do not sit quietly in groups while a teacher instructs them. That makes Montessori programs child-centered (Lillard, 2013).

Montessori schools encourage children to help other children (Lillard & Taggart, 2019). That is something Vygotsky valued as well. Children from Montessori schools seem to enjoy learning together throughout childhood, just as Vygotsky would expect.

ELIZABETH FLORES/Tribune News Service/PLEASANT PRAIRIE/WI/USA/Newscom

Reggio Emilia

Another form of early-childhood education is **Reggio Emilia**, named after the town in Italy where it began. In Reggio Emilia, children are encouraged to master skills that are not usually taught in North American schools until age 7 or so, such as writing and using tools. Although many educators admire the Reggio philosophy and practice, it is expensive to duplicate in other nations—there are few dedicated Reggio Emilia schools in the United States.

Reggio schools do not provide large-group instruction, with lessons in, say, forming letters or cutting paper. Instead, hands-on activities are chosen by each child. Drawing, cooking, and gardening are stressed. This program begins with the idea that democracy and freedom of personal expression belong in the classroom as well as in the community (McNally & Slutsky, 2017).

Learning is documented via scrapbooks, photos, and daily notes, not to measure progress but to make each child and parent proud of accomplishments. This also enhances memory, a crucial part of executive function.

Appreciation of the arts is evident. Every Reggio Emilia school originally had a studio, an artist, and space to encourage creativity (Forbes, 2012). A spacious, plant-filled playground is part of the curriculum, because "the environment is the third teacher." Big mirrors are part of the schools' décor—again, with the idea of fostering individuality and self-expression.

Waldorf Schools

A third type of child-centered program is called a **Waldorf school**, first developed by Rudolf Steiner in Germany in 1919. The emphasis again is on creativity and individuality—with no homework, no tests, no worksheets.

As much as possible, children play outdoors; appreciation of nature is basic to Waldorf education. Children of various ages learn together, because older children serve as mentors for younger ones, and the curriculum follows the interests of the child, not the age of the child.

There is a set schedule—usually circle time in the beginning and certain activities on certain days (always baking on Tuesdays, for instance)—but children are not expected to master specific knowledge at certain ages. All child-centered schools emphasize creativity; in Waldorf schools, imagination is particularly prized (Kirkham & Kidd, 2017).

Child-Centered Pride How could Rachel Koepke, a 3-year-old from a Wisconsin town called Pleasant Prairie, seem so pleased that her hands (and cuffs) are blue? The answer arises from northern Italy—Rachel attended a Reggio Emilia preschool that encourages creative expression.

Montessori schools Schools that offer early-childhood education based on the philosophy of Maria Montessori, which emphasizes careful work and tasks that each young child can do.

Reggio Emilia A program of early-childhood education that originated in the town of Reggio Emilia, Italy, and that encourages each child's creativity in a carefully designed setting.

Waldorf schools An early-childhood education program that emphasizes creativity, social understanding, and emotional growth. It originated in Germany with Rudolf Steiner and now is used in thousands of schools throughout the world.

CAREER ALERT   The Preschool Teacher

Preschool teachers are increasingly in demand, as more and more families and communities understand how much young children can learn, and more and more mothers enter the job market. Added to that is the growing realization by public leaders that social skills and self-confidence developed in early childhood continue lifelong: A child who has good early education is likely to become an adult who is a competent and compassionate member of the community.

For developmentalists, important new insights come from neurological research, which helps preschool teachers understand what, how, and when young children learn. For example, since the auditory, visual, and motor cortexes are undergoing rapid myelination, children need to coordinate both hemispheres of brain and body by running, climbing, and balancing. But the immature motor cortex is not yet ready for writing, or tying shoelaces, or sitting quietly in one place.

Research on the developing brain finds that early childhood is the best time for learning language, so the curriculum should be language-rich—talking, listening, singing, hearing stories, making rhymes, engaging in verbal play. Young children can learn to recognize and name letters just as they learn to distinguish a baseball from a soccer ball.

Fostering control of gross motor skills may be particularly important for children who are at risk for *attention-deficit/hyperactivity disorder (ADHD)*, the label given to active children who find it especially hard

to concentrate, quietly, on one activity. Such children need to exercise their bodies, which helps their brains mature (Halperin & Healey, 2011; Hillman, 2014).

That is another reason for preschool education—children are most likely to develop their brains by playing with other children. Screen time, a common activity for children who are not in preschool, does not foster the brain regulation that children need.

Moreover, preschool teachers help children learn how to cooperate with other children, a valuable life lesson that is best learned in childhood. Thus, preschool teachers can be proud that they are nurturing compassionate, prosocial adults.

The joy and satisfaction of working with young children is crucial, because at the moment, salary and working conditions are not yet what they should be. The U.S. Bureau of Labor Statistics reports that, compared to teachers overall, preschool teachers are most in demand and least well paid—the average annual salary is below $30,000 a year. There is marked variation in state-by-state certification requirements and in neighborhood-by-neighborhood salary levels.

This is changing, but students should enter this field for emotional reasons, not financial ones. If you are interested in early-childhood education, you can find more details from a professional group called the National Association for the Education of Young Children (NAEYC).

Teacher-Directed Programs

Teacher-directed preschools stress academics, often taught by one adult to the entire group. The curriculum includes learning the names of letters, numbers, shapes, and colors according to a set timetable; every child naps, snacks, and goes to the bathroom on schedule as well. Children learn to sit quietly and listen to the teacher. Praise and other reinforcements are given for good behavior, and time-outs (brief separation from activities) are imposed to punish misbehavior.

The goal of teacher-directed programs is to make all children "ready to learn" when they enter elementary school. For that reason, basic skills are stressed, including precursors to reading, writing, and arithmetic, perhaps through teachers asking questions that children answer together in unison. Behavior is also taught, as children learn to respect adults, to follow schedules, to hold hands when they go on outings, and so on.

Children practice forming letters, sounding out words, counting objects, and writing their names. If a 4-year-old learns to read, that is success. (In a child-centered program, this might arouse suspicion that there was too little time for play or socializing.)

Many teacher-directed programs were inspired by behaviorism, which emphasizes step-by-step learning and repetition, with reinforcement (praise, gold stars, prizes) for accomplishment. Another inspiration for these programs comes from information-processing research indicating that children who have not learned basic vocabulary and listening skills by kindergarten often fall behind in primary school. Many state legislatures mandate that preschoolers master specific concepts, an outcome best achieved by teacher-directed learning (Bracken & Crawford, 2010).

Intervention Programs

Several programs designed for children from low-SES families were established in the United States decades ago. Typically, they combine some elements that are child-centered and some that are teacher-directed.

Comparing Child-Centered and Teacher-Directed Preschools

Most developmentalists advocate child-centered programs (Christakis, 2016; Golinkoff & Hirsh-Pasek, 2016). They believe that from ages 3 to 6 young children learn best when they can interact in their own way with materials and ideas (Sim & Xu, 2017). On the other hand, many parents and legislators want proof that early education will improve later school achievement.

The developmental critics of teacher-directed education fear "trad[ing] emotional grounding and strong language skills known to support learning for assembly-line schooling that teaches children isolated factoids" (Hirsh-Pasek & Golinkoff, 2016, p. 1158).

As Penelope Leach wrote, "Goals come from the outside. . . . It is important that people see early learning as coming from inside children because that's what makes clear its interconnectedness with play, and therefore the inappropriateness of many 'learning goals'" (Leach, 2011, p. 17). Another developmentalist asks, "Why should we settle for unimaginative goals . . . like being able to identify triangles and squares, or recalling the names of colors and seasons" (Christakis, 2016).

Critics of child-centered programs point to data finding that children who enter kindergarten without knowing names and sounds of letters may become first-graders who cannot read (Ozernov-Palchik et al., 2017). Children are unlikely to learn literacy skills in creative play (Gellert & Elbro, 2017).

Early familiarity with numbers and shapes predicts school achievement later on. As you will soon read, Head Start programs have shifted over the past decades to be more teacher-directed, largely because national policy directives from the government have advocated that change—to the distress of many developmentalists (Walter & Lippard, 2017).

Many developmentalists resist legislative standards and academic tests for young children, arguing that social skills, emotional development, and creative play are essential but difficult to measure (Hyson & Douglass, 2019). A truly brilliant child is characterized by all the complex skills of executive function, not the easy-to-measure skills of letter recognition (Golinkoff & Hirsh-Pasek, 2016). [Children's play is discussed in Chapter 10.]

However, the divide between child-centered and teacher-directed programs goes deeper than that. It is possible that the emphasis on individual exploration is contrary to the wishes and needs of families who want their children to do as they are told, listen to adults, and master the learning tools of their culture. Indeed, child-centered education may arise from the values of middle-class, Western, European American teachers (Trawick-Smith, 2019).

Finding the right balance between formal and informal assessment, and between child-centered and teacher-directed learning, is the goal; achieving it is the challenge (see **Figure 9.3**).

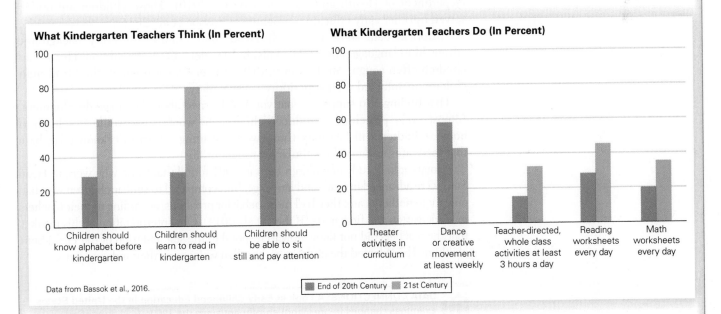

Data from Bassok et al., 2016.

FIGURE 9.3

Less Play, More Work These data come from a large survey of more than 5,000 public school teachers throughout the United States. In 1998 and 2010, kindergarten teachers were asked identical questions but gave different answers. Smaller, more recent surveys suggest that these trends continue, and they now involve preschool teachers. Some even use worksheets for 3-year-olds.

Head Start A federally funded early-childhood intervention program for low-income children of preschool age.

Head Start

In the early 1960s, millions of young children in the United States were thought to need a "head start" on their formal education to foster better health and cognition before first grade. Consequently, since 1965, the federal government has funded a massive program for 4-year-olds called **Head Start**.

The goals for Head Start have changed over the decades, from lifting families out of poverty to promoting literacy, from providing dental care and immunizations to teaching Standard English, from focusing on 4-year-olds to including 2- and 3-year-olds. In 2018, more than 9 billion dollars in federal funds were allocated to Head Start, which enrolled almost 1 million children.

Head Start is a massive program, but there are about 8 million 3- and 4-year-olds in the United States, which means that only about 12 percent of U.S. children that age attend Head Start. Many other children are in state or municipal programs, or, if they can afford it, private preschools. About twice as many 3- to 5-year-olds from high-SES families are in preschools compared to children from low-SES families.

Many U.S. states provide early education. Some do so for all 4-year-olds (Oklahoma, Florida, and Vermont), but some do so for none (Idaho, Indiana, Utah, South Dakota, and Wyoming). Curriculum, teacher training, and quality vary, with only three (Alabama, Michigan, and Rhode Island) reaching 10 quality goals (Friedman-Krauss et al., 2019).

As you read in Opposing Perspectives, although many Head Start programs were child-centered initially, they have become increasingly teacher-directed as waves of legislators have shaped them. In 2016, new requirements were put in place for Head Start, mandating that programs be open at least six hours a day (initially, most programs were half-day) and 180 days a year (that allows for 81 holidays and 104 weekend days).

Special support is required in Head Start for children who are homeless, have special needs, or are learning English. Research finds that Head Start is most beneficial for them as well as for children in poverty, or who live in rural areas (U.S. Department of Health and Human Services, 2010). Those children are underserved by other early-education programs (Crosnoe et al., 2016).

Historical data show that most Head Start children of every background advanced in language and social skills, but by elementary school non–Head Start children often caught up. However, there was one area in which the Head Start children maintained their superiority—vocabulary.

That finding also supports what you just learned about language development. Almost every preschool introduces children to words they would not learn at home. Children will fast-map those words, gaining a linguistic knowledge base that facilitates expanded vocabulary throughout life.

A longitudinal study of children born in 2001 found that those who went to Head Start were advanced in math and language, but, compared to similar children who had only their mother's care, they had more behavior problems, according to their teachers in kindergarten and first grade (R. Lee et al., 2014). One interpretation of this result is that the teachers did not know how to respond to the self-assertion of the Head Start children: They blamed the children rather than welcoming their independence.

 DATA CONNECTIONS: A Look at Early-Childhood Education in the United States examines changes in preschool enrollment since the 1960s. 🖼 LaunchPad

Long-Term Gains from Intensive Programs

This discussion of philosophies, practices, and programs may give the impression that the research on early-childhood cognition is contradictory. That is not true.

Octavio Jones/Tampa Bay Times/ZUMAPRESS/Alamy

If You're Happy and You Know It Gabby Osborne (pink shirt) has her own way of showing happiness, not the hand-clapping that Lizalia Garcia tries to teach. The curriculum of this Head Start class in Florida includes learning about emotions, contrary to the wishes of some legislators, who want proof of academics.

Specifics are debatable, but empirical evidence and longitudinal evaluation find that preschool education advances learning.

The best longitudinal evidence comes from three intensive programs that enrolled children for years—sometimes beginning with home visits in infancy, sometimes continuing in after-school programs through first grade. One program, called *Perry* (or *High/Scope*), was spearheaded in Michigan (Schweinhart & Weikart, 1997); another, called *Abecedarian*, got its start in North Carolina (Campbell et al., 2001); a third, called *Child–Parent Centers*, began in Chicago (Reynolds, 2000). They all focused on children from low-SES families.

All three programs compared experimental groups of children with matched control groups, and all reached the same conclusion: Early education has substantial long-term benefits that become most apparent when children are in the third grade or later. By age 10, children who had been enrolled in any one of these three programs scored higher on math and reading achievement tests than did other children from the same backgrounds, schools, and neighborhoods. They were less likely to be placed in classes for children with special needs, or to repeat a year of school, or to drop out of high school before graduation.

An advantage of decades of longitudinal research is that teenagers and adults who received early education can be compared with those who did not. For all three programs, early investment paid off:

- In adolescence, the children who had undergone intensive preschool education had higher aspirations, possessed a greater sense of achievement, and were less likely to have been abused.
- As young adults, they were more likely to attend college and less likely to go to jail.
- As middle-aged adults, they were more often employed, paying taxes, healthy, and not in need of government subsidies (Reynolds & Temple, 2019).

All three research projects found that providing direct cognitive training, with specific instruction in various school-readiness skills, was useful. Each child's needs and talents were considered—a circumstance made possible because the child/adult ratio was low. This combined child-centered and teacher-directed programs, with all of the teachers working together on the same goals so that children

Especially for Teachers In trying to find a preschool program, what should parents look for? (see response, page 255)

Lifetime Achievement The baby in the framed photograph escaped the grip of poverty. The woman holding the photograph proved that early education can transform children. She is Frances Campbell, who spearheaded the Abecedarian Project. The baby's accomplishments may be the more impressive of the two.

Frank Porter Graham Child Development Institute.

were not confused. The parents reinforced what the children learned. In all three, teachers deliberately involved parents, and each program included strategies to enhance the home–school connection.

These programs were expensive (ranging from $7,500 to $23,000 annually per young child in 2020 dollars). From a developmental perspective, the decreased need for special education and other social services later on made early education a "wise investment" (Duncan & Magnuson, 2013, p. 128). Additional benefits to society over the child's lifetime, including increased employment and tax revenues, as well as reduced crime, save an estimated $329,000 in later expenses per child (Schweinhart, 2019).

Ironically, the children *least* likely to be in such programs are Spanish-speaking, or from families with income slightly above poverty-level, or whose mothers are not employed. These are precisely the children for whom early education may be especially helpful, a conclusion found not only in Head Start but in other research as well. That makes early education an issue of educational justice (Morgan, 2019; Nxumalo & Adair, 2019).

Many professionals try to steer away from political issues, but this one transcends partisanship. Any cuts to preschool education not only mean less child-centered learning (which is more expensive) but also less high-quality, teacher-directed learning. In most U.S. states, full-day kindergarten programs (always locally funded) are optional and sometimes unavailable.

Scientists who know the research on early-childhood education are dismayed when the wishes of the adults (lower taxes) supersede the needs of the children. Professionals disagree about details, but they do not debate whether every young child should be offered high-quality preschool education (they should). Instead, the debate is whether the government should pay for every child, thus making education a public right like the police or fire protection, or whether the government should save money by paying only for those who cannot afford private programs.

Compared to a decade ago, much more is known about early cognition: 2- to 5-year-olds are capable of learning languages, concepts, math, theory of mind, and more. What a child learns before age 6 is pivotal for later schooling and adult life. The amazing potential of young children is also a theme of the next chapter, where we discuss other kinds of learning, such as in emotional regulation and social skills.

WHAT HAVE YOU LEARNED?

1. What do most preschools provide that most homes do not?

2. In child-centered programs, what do the teachers do?

3. What makes the Reggio Emilia program different from most other preschool programs?

4. Why are Montessori schools still functioning 100 years after the first such schools opened?

5. What are the advantages and disadvantages of teacher-directed preschools?

6. What are the goals of Head Start?

7. What are the long-term gains from intervention preschools?

Early-Childhood Schooling

Preschool can be an academic and social benefit to children. Around the world, increasing numbers of children are enrolled in early-childhood education.

Programs are described as "teacher-directed" or "child-centered," but in reality, most teachers' styles reflect a combination of both approaches. Some students benefit more from the order and structure of a teacher-directed classroom, while others work better in a more collaborative and creative environment.

TEACHER-DIRECTED APPROACH
Focused on Getting Preschoolers Ready to Learn

Direct instruction
Teacher as formal authority
Students learn by listening
Classroom is orderly and quiet
Teacher fully manages lessons
Rewards individual achievement
Encourages academics
Students learn from teacher

CHILD-CENTERED APPROACH
Focused on Individual Development and Growth

Teacher as facilitator
Teacher as delegator
Students learn actively
Classroom is designed for collaborative work
Students influence content
Rewards collaboration among students
Encourages artistic expression
Students learn from each other

Worth Publishers

DIFFERENT STUDENTS, DIFFERENT TEACHERS

There is clearly no "one right way" to teach children. Each approach has potential benefits and pitfalls. A classroom full of creative, self-motivated students can thrive when a gifted teacher acts as a competent facilitator. But students who are distracted or annoyed by noise, or who are shy or intimidated by other children, can blossom under an engaging and encouraging teacher in a more traditional environment.

Done Well

- engaging teacher
- clear, consistent assessment
- reading and math skills emphasized
- quiet, orderly classroom
- all students treated equally

- emphasizes social skills and emotion regulation
- encourages critical thinking
- builds communication skills
- fosters individual achievement
- encourages creativity and curiosity

Teacher-Directed

Child-Centered

- bored students
- passive learning
- less independent, critical thinking
- teacher may dominate

- chaotic/noisy classrooms
- students may miss important knowledge and skills
- inconclusive assessment of student progress
- some students may dominate others

Done Poorly

SUMMARY

Thinking During Early Childhood

1. An important part of developing cognition during early childhood is the emergence of executive function, or cognitive control, as children learn to regulate and control their sensory impulses in order to use their minds more effectively.

2. Three components are usually included in executive function: memory, inhibition, and flexibility. Executive function can be improved during early childhood, and that advances later learning more than a high score on an IQ test or extensive vocabulary.

3. Piaget stressed the egocentric and illogical aspects of thought during early childhood. He called this stage of thinking preoperational intelligence because young children do not yet use logical operations to think about their observations and experiences.

4. Young children, according to Piaget, sometimes focus on only one thing (centration) and see things only from their own viewpoint (egocentrism), remaining stuck on appearances and current reality. They may believe that living spirits reside in inanimate objects and that nonhuman animals have the same characteristics they themselves have, a belief called animism.

5. Vygotsky stressed the social aspects of childhood cognition, noting that children learn by participating in various experiences, guided by more knowledgeable adults or peers. Such guidance assists learning within the zone of proximal development, which encompasses the knowledge that children are close to understanding and the skills they can almost master.

6. According to Vygotsky, the best teachers use various hints, guidelines, and other tools to provide a child with a scaffold for new learning. Language is a bridge that provides social mediation between the knowledge that the child already has and the learning that society hopes to impart. For Vygotsky, words are tools for learning.

7. Children develop theories, especially to explain the purpose of life and their role in it. One theory about children's thinking is called "theory-theory"—the hypothesis that children develop theories because all humans innately seek explanations for everything they observe.

8. An example of the developing cognition of young children is theory of mind—an understanding of what others may be thinking. Theory of mind begins at around age 4, partly as a result of maturation of the brain. Culture and experiences also influence its development.

Language Learning

9. Language develops rapidly during early childhood, a sensitive period but not a critical one for language learning. Vocabulary increases dramatically, with thousands of words added between ages 2 and 6. In addition, basic grammar is mastered.

10. The child's ability to learn language is evident in fast-mapping (the quick use of new vocabulary words) and in overregularization (applying the rules of grammar even when they are not valid).

11. Many children learn to speak more than one language, gaining cognitive as well as social advantages. Early childhood is the best time to learn two languages. The benefits of bilingualism are lifelong. Pronunciation lags behind production, which lags behind comprehension.

Early-Childhood Schooling

12. Organized educational programs during early childhood advance cognitive and social skills, although specifics vary a great deal. The quality of a program cannot be judged by the name or by appearance.

13. Montessori and Reggio Emilia are two child-centered programs that began in Italy and are now offered in many nations. Waldorf, which began in Germany, is another child-centered program.

14. Behaviorist principles led to many specific practices of teacher-directed programs. Children learn to listen to teachers and become ready for kindergarten. Teacher-directed programs are preferred by many parents and legislators, and these programs are increasingly popular—to the consternation of many child developmentalists.

15. Head Start is a U.S. federal government program primarily for low-income children. Longitudinal research finds that early-childhood education reduces the risk of later problems, such as needing special education. High-quality programs increase the likelihood that a child will become a law-abiding, gainfully employed adult.

16. Many types of preschool programs are successful. It is the quality of early education that matters. The training, warmth, and continuity of early-childhood teachers benefit children in many ways.

17. Some nations provide early education for all 3- and 4-year-olds. The United States is behind on this metric, with only about half of all 4-year-olds in preschool, and far fewer 3-year-olds.

KEY TERMS

executive function (p. 229)
preoperational intelligence (p. 230)
symbolic thought (p. 230)
animism (p. 231)
centration (p. 231)
egocentrism (p. 231)

focus on appearance (p. 231)
static reasoning (p. 231)
irreversibility (p. 231)
conservation (p. 232)
zone of proximal development (ZPD) (p. 234)

scaffolding (p. 234)
overimitation (p. 235)
private speech (p. 236)
social mediation (p. 236)
theory-theory (p. 237)
theory of mind (p. 238)
fast-mapping (p. 242)

overregularization (p. 243)
pragmatics (p. 243)
Montessori schools (p. 247)
Reggio Emilia (p. 247)
Waldorf schools (p. 247)
Head Start (p. 250)

APPLICATIONS

The best way to understand thinking in early childhood is to listen to a child, as Applications 1 and 2 require. If you have no access to children, consider Application 3 or 4.

1. Replicate one of Piaget's conservation experiments. The easiest one is conservation of liquids (see Figure 9.1). Work with a child under age 5 who tells you that two identically shaped glasses contain the same amount of liquid. Then carefully pour one glass of liquid into a narrower, taller glass. Ask the child whether one glass now contains more or the glasses contain the same amount.

2. To demonstrate how rapidly language is learned, show a preschool child several objects and label one with a nonsense word that the child has never heard. (*Toma* is often used; so is *wug.*) Or choose a word that the child does not know, such as *wrench*, *spatula*, or the name of a coin from another nation. Test the child's fast-mapping.

3. Theory of mind emerges at about age 4, but many adults still have trouble understanding other people's thoughts and motives. Ask several people why someone in the news did whatever he or she did (e.g., a scandal, a crime, a heroic act). Then ask your informants how sure they are of their explanation. Compare and analyze the reasons as well as the degrees of certainty. (One person may be sure of an explanation that someone else thinks is impossible.)

4. Think about an experience in which you learned something that was initially difficult. To what extent do Vygotsky's concepts (guided participation, zone of proximal development) explain the experience? Write a detailed, step-by-step account of your learning process as Vygotsky would have described it.

Especially For ANSWERS

Response for Nutritionists (from p. 232): Take each of the four characteristics of preoperational thought into account. Because of egocentrism, having a special place and plate might assure the child that this food is exclusively theirs. Since appearance is important, the food should look tasty. Since static thinking dominates, if something healthy is added (e.g., grate carrots into the cake, add milk to the soup), do it before the food is given to the child. In the reversibility example in the text, the lettuce should be removed out of the child's sight and the "new" hamburger presented.

Response for Immigrant Parents (from p. 244): Children learn by listening, so it is important to speak with them often. Depending on how comfortable you are with the new language, you might prefer to read to your children, sing to them, and converse with them primarily in your native language and find a good preschool where they will learn the new language. The worst thing you could do is to restrict speech in either tongue.

Response for Teachers (from p. 251): Tell parents to look at the people more than the program. Parents should see the children in action and note whether the teachers show warmth and respect for each child.

Observation Quiz ANSWERS

Answer to Observation Quiz (from p. 234): The name "Jack" above the bed is an obvious scaffold. Egocentric children pay special attention to their names, and those who are as fortunate as Jack can read and spell their names before kindergarten. Jack is also learning what his culture expects of a boy, evident in his haircut, room decor, and furniture.

Answer to Observation Quiz (from p. 235): Right-handed. Her dominant hand is engaged in something more comforting than exploring the abacus.

Early Childhood: Psychosocial Development

What Will You Know?

1. Why do 2-year-olds have more sudden tempers, tears, and terrors than 6-year-olds?
2. What happens if parents let their children do whatever they want?
3. How does spanking affect children?

A t age 4, my youngest grandson considers himself a "big kid." When we wait for the bus, he begs his brother to play "the monster game." The older boy usually says no. Then the 4-year-old cries, and the big brother usually stops the wailing by pretending to be a blind monster, clomping around with arms outstretched to catch the smaller boy, who laughs as he runs ahead, never caught. Should I intervene, telling one not to cry, telling the other not to reward the tears?

In running, the 4-year-old sometimes bumps into strangers who are also waiting for the bus. Some smile, some seem annoyed, but no one expresses anger. I would not mind if they did; I wish my grandson were wiser about other people.

One stranger asked the boy his name, which he readily gave, as well as his address.

"Don't tell strangers where you live," the man said.

He repeated his address.

With strangers as well as older siblings, it is apparent that young children have much to learn. Adults—parents and, in this case, grandmothers—need to figure out when and how to guide, encourage, and discipline.

Emotional Development

As with the body growth and cognitive development, which the previous two chapters described as changing dramatically from age 2 to 6, emotions change as well.

Emotional Regulation

Controlling the expression of feelings, called **emotional regulation**, is the preeminent psychosocial task between ages 2 and 6. Emotional regulation is a lifelong endeavor, with particularly rapid development during early childhood, influenced by both peers and parents (Hu et al., 2017).

Think of what happens when you give someone a gift. If the receiver is a young child, you can probably tell whether the child liked the present, because emotions are visible (Galak et al., 2016). If the receiver is an adult, you may not be sure, because adults have learned to disguise their disappointment. People learn to regulate their emotions.

By age 6, most children can feel angry, frightened, sad, anxious, or proud without the explosive outbursts of temper, or terror, or tears of 2-year-olds. Depending on a child's training and temperament, some emotions are easier to control than others, but even temperamentally angry or fearful children learn to regulate their emotions (Moran et al., 2013; Suurland et al., 2016; Tan et al., 2013).

Among the middle class in in the United States, postponement of gratification is valued as a sign of maturity: Dieters must resist some foods today so they will be thinner next summer; students are told to study now for an exam next week, not cram at the last minute; adults are supposed to save for the future, not spend immediately. This aspect of emotional regulation is very difficult for young children (see A View from Science on page 258).

emotional regulation The ability to control when and how emotions are expressed.

Initiative Versus Guilt

Erikson's third developmental stage is **initiative versus guilt**. *Initiative* includes saying something new, beginning a project, expressing an emotion. Depending on what happens next, children feel proud or guilty. Gradually, they learn to rein in boundless pride and avoid crushing guilt.

Pride is common in early childhood. As one team expressed it:

> Compared to older children and adults, young children are the optimists of the world, believing they have greater physical abilities, better memories, are more skilled at imitating models, are smarter, know more about how things work, and rate themselves as stronger, tougher, and of higher social standing than is actually the case.

> *[Bjorklund & Ellis, 2014, p. 244]*

That *protective optimism* helps young children try new things, and thus, initiative advances learning. As Erikson predicted, their optimistic **self-concept** protects young children from guilt and shame and encourages them to learn.

Young children often brag about what they have accomplished. As long as the boast is not a lie, other young children like them for it. At about age 7, a developmental shift occurs, and children as well as adults begin to appreciate modesty more than boasting (Lockhart et al., 2018). By then, children have usually learned some emotional regulation; they can finally keep quiet about how wonderful they are.

Many young children believe that whatever they are is good. They may feel superior to children of another nationality or religion. This arises because of maturation: Cognition enables them to understand group categories, not only of ethnicity, gender, and nationality but even categories that are irrelevant.

For instance, children remember more about cartoon characters whose names begin with the same letter as theirs (Ross et al., 2011). If their parents or other adults express prejudice against another group, they may mirror those prejudices (Tagar et al., 2017).

initiative versus guilt Erikson's third psychosocial crisis, in which young children undertake new skills and activities and feel guilty when they do not succeed at them.

self-concept A person's understanding of who they are, in morality, intelligence, appearance, personality, talents, and skills.

Brain Maturation

The new initiative that Erikson described results from myelination in the limbic system, growth of the prefrontal cortex, and a longer attention span—all indicative of neurological maturation. Emotional regulation and cognitive maturation develop together, each enabling the other to advance (Bridgett et al., 2015; Frydenberg, 2017).

Already before age 2, primary emotions, such as sadness and joy, have developed, becoming more nuanced with every passing year. Now brain maturation and social awareness allow development of secondary emotions, such as pride, envy, and guilt (Frydenberg, 2017). Usually, emotional control of both primary and secondary emotions becomes more evident by age 4 or 5.

Learning Emotional Regulation Like this girl in Hong Kong, all 2-year-olds burst into tears when something upsets them—a toy breaks, a pet refuses to play, or it's time to go home. Mothers who comfort young children and help them calm down are teaching them to regulate their emotions.

Allen Brown/dbimages/Alamy

A VIEW FROM SCIENCE

Waiting for the Marshmallow

You probably have heard of the famous marshmallow test (Mischel et al., 1972; Mischel, 2014). Young children were seated in front of a marshmallow and told they could eat it immediately or wait—sometimes as long as 15 minutes—while the researcher left the room. They were promised another marshmallow if they didn't eat the first one before the adult returned.

Those who waited used various tactics—they looked away, closed their eyes, or sang to themselves. Decades later, the researchers contacted the children to see how their lives turned out. Those who delayed gobbling up one marshmallow in order to get two became more successful as teenagers, young adults, and even middle-aged adults—doing well in college, for instance, and having happy marriages.

This experiment is often replicated. The average child waits about 6 minutes; relatively few 4-year-olds sit in front of a marshmallow for 15 minutes without eating it; those who wait are likely to become more successful years later (Shoda et al., 1990).

Some cultures encourage instant gratification; others teach children to wait patiently. In a replication of the marshmallow test, children of the Nso people in Cameroon waited longer than the California children in Mischel's original experiment (Lamm et al., 2018).

In another replication, this one in the United States, children twice experienced a reliable or unreliable examiner. Specifically, the children were told they should wait for better crayons, and better stickers, but after two minutes, for half of them the examiner returned to say that no better crayons or stickers was available, and the other half got the much better crayons or stickers.

Then children were given the marshmallow test, which they could eat immediately or wait for two marshmallows, which the examiner brought after 15 minutes of waiting. Those with the reliable examiner waited, on average, 12 minutes, with 64 percent waiting the full 15 minutes. Those with the unreliable experimenter waited an average of only three minutes, with only 7 percent waiting 15 minutes (Kidd et al., 2013).

The conclusion of these experimenters is that children make rational decisions about delaying gratification. If their parents and cultures teach patience and keep their promises, children learn to control their impulses. Emotional regulation is the product of experience, not just maturation.

Jose Luis Pelaez Inc/DigitalVision/Getty Images

Both Accomplished Note the joy and pride in this father and daughter in West New York, New Jersey. Who has achieved more?

Children strengthen and develop their neuronal connections in response to the emotions of other people. The process is reciprocal and dynamic: Anger begets anger, which leads to more anger; joy begets joy, and so on.

This reciprocity is not just a matter of words and facial expression, it also directly involves the brain. For instance, researchers scanned the brains of mothers and children as they did a difficult puzzle together. When the mothers became frustrated, the children did, too—and vice versa. As the scientists explain, "mothers and children regulate or deregulate each other" (Atzaba-Poria et al., 2017, p. 551).

The practical application of shared emotionality benefits adults as well as children. If a happy young boy runs to you, you might laugh, pick him up, and swing him around; if a grinning young girl drums on the table, catch the rhythm and pound in return, smiling broadly. In both adults and children, laughter and happiness increase endorphins and lower cortisol. Emotions are infectious: Catch the good ones, and drop the bad ones.

Motivation

Motivation is the impulse that propels someone to act. It comes either from a person's own desires or from the social context.

From Within

intrinsic motivation A drive, or reason to pursue a goal, that comes from inside a person, such as the joy of reading a good book.

Intrinsic motivation arises from within, when people do something for the joy of doing it: A musician might enjoy making music even if no one else hears it; the sound is intrinsically rewarding. Intrinsic motivation is thought to advance creativity, innovation, and emotional well-being (Weinstein & DeHaan, 2014).

All of Erikson's psychosocial needs—including the young child's initiatives—are intrinsic: A child feels inwardly compelled to act. The power of that internal impulse is sometimes frustratingly evident, especially if an adult in a hurry is walking with a child. The adult's "hurry, we are late" clashes with the child's intrinsic motivation. The child may jump up to walk along a ledge, may stop to throw a snowball, or may pick up a piece of trash to explore.

When playing a game, few young children keep score; intrinsic joy is the goal, more than winning. In fact, young children often claim to have won when objective scoring would say they lost; in this case, the children may really be winners.

From the Culture

Extrinsic motivation comes from outside the person, when external praise or some other reinforcement is the goal, such as when a musician plays for applause or money. Four-year-olds might brush their teeth because they are praised and rewarded with musical toothbrushes and tasty toothpaste.

If an extrinsic reward is removed, the behavior may stop unless it has become a habit. Young children might not brush their teeth if parents don't notice and praise them. For most of us, tooth-brushing was extrinsically rewarded at first, and that continued long enough for it to become habitual. Then motivation becomes intrinsic: Tooth-brushing has become a comforting, internally motivated routine for many adults.

Spontaneous Joy

Intrinsic motivation is evident in childhood. Young children play, question, exercise, create, destroy, and explore for the sheer joy of it. That serves them well. For example, a longitudinal study found that 3-year-olds who were strong in intrinsic motivation were, two years later, advanced in early math and literacy (Mokrova et al., 2013). The probable reason: They enjoyed counting things and singing songs—when alone.

In contrast, exaggerated external praise ("Your drawing is amazingly wonderful!") undercuts motivation (Brummelman et al., 2017). If young children believe the praise, they might be afraid to try again, thinking they will not be able to do as well. If they suspect that the praise was inaccurate, they may discount the entire activity.

Imaginary Friends

Intrinsic motivation is apparent when children invent dialogues for their toys, concentrate on creating a work of art or architecture, or talk to imaginary friends. Invisible companions are rarely encouraged by adults (thus, no extrinsic motivation), but many children have them.

An international study of 3- to 8-year-olds found that about 1 child in 5 said that they had an invisible companion, with notable variation by culture: 38 percent of children in the Dominican Republic, but only 5 percent in Nepal, said they had such a friend (Wigger, 2018).

Another study, this one in the United States, found that about half the children had imaginary friends, either completely invisible ones or objects, such as stuffed animals, upon which they bestowed independent opinions, speech, and action. Many of those children also imagined entire worlds! (Taylor et al., 2020).

WHAT HAVE YOU LEARNED?

1. Why is emotional regulation an important skill that children need to develop?
2. How might protective optimism lead to new skills and competencies?
3. How are emotional regulation, emotional control, and executive function related?
4. What did Erikson think was the dominant psychic need of young children?
5. What is an example (not in the text) of intrinsic motivation?
6. What is an example (not in the text) of extrinsic motivation?

CHAPTER APP 10

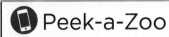 Peek-a-Zoo

IOS:
https://tinyurl.com/wdhttw6

RELEVANT TOPIC:
Emotional development in early childhood

This simple app prompts preschoolers to identify behaviors or emotions exhibited by different animals ("Who is surprised?"), helping them to learn social cues.

extrinsic motivation A drive, or reason to pursue a goal, that arises from the wish to have external rewards, perhaps by earning money or praise.

Especially for College Students Is extrinsic or intrinsic motivation more influential in your study efforts? (see response, page 281)

Play

Play is timeless and universal—apparent in every part of the world over thousands of years. Many developmentalists believe that play is children's most productive, enjoyable activity. As you will see, not everyone agrees.

The Historical Context

As Chapter 9 explained, one dispute in early education is whether child-centered creative play or teacher-directed learning is more important. To further understand that debate, we need to consider how attitudes have changed over the past 100 years.

Parten's Stages of Play

In 1932, American sociologist Mildred Parten described five stages of play, each more advanced than the previous one:

1. *Solitary*: A child plays alone, unaware of other children playing nearby.
2. *Onlooker*: A child watches other children play.
3. *Parallel*: Children play in similar ways but not together.
4. *Associative*: Children interact, sharing materials or activities, but not taking turns.
5. *Cooperative*: Children play together, creating dramas or taking turns.

Parten described play as intrinsic, with children from age 1 to 6 gradually advancing from solitary to cooperative play. This was apparent as toddlers watched older children play, and as the older children gradually learned how to play together.

Research on contemporary children finds much more age variation than Parten did, in part because children's experiences with other children vary. Those who stay home until kindergarten might be onlookers (stage 2) in kindergarten, whereas those who have been in interactive, child-centered preschools might reach cooperative play (stage 5) by age 4.

The Current Controversy

Whether play is essential or merely fun is "a controversial topic of study" (Pellegrini, 2011, p. 3; Johnson & Wu, 2019). Some educators want children to play less in order to learn reading and math; others predict emotional and academic problems if children rarely play (Golinkoff & Hirsh-Pasek, 2016).

Developmental psychologists are particularly impressed with the young child's ability to use play to express their own emotions. One source notes that play varies, with "twists and turns . . . abrupt changes in play patterns—between seriousness and frolic, earnestness and whimsy, reality and pretense, often leading to higher levels of play" (Johnson & Wu, 2019, p. 81).

Screen Time

The largest historical change is that active play time has been reduced because of *screen time*, which is when children watch television, tablets, smartphones, or computer screens. That troubles developmentalists for many reasons. Pediatricians, psychologists, and teachers all report extensive research that screen time reduces conversation, imagination, and outdoor activity (Downing et al., 2017).

A recent study in Canada of 2,441 children, followed longitudinally from before birth until age 5, found that the average 3-year-old watched more than two hours a day (Madigan et al., 2019). One result was to "disrupt interactions with caregivers"; another was to reduce cognitive and emotional development. Because this was

THINK CRITICALLY: Is "play" an entirely different experience for adults than for children?

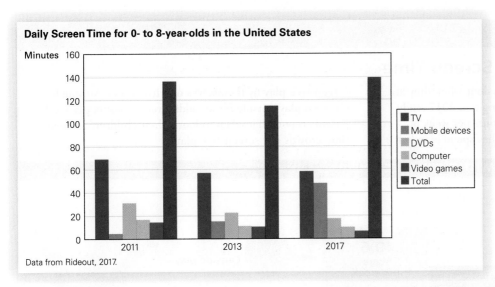

Daily Screen Time for 0- to 8-year-olds in the United States

Minutes

- TV
- Mobile devices
- DVDs
- Computer
- Video games
- Total

2011　2013　2017

Data from Rideout, 2017.

FIGURE 10.1

Learning by Playing Fifty years ago, the average child spent three hours a day in outdoor play. Screens have largely replaced that, with specifics changing over the years. In this chart "video games" referred to separate game consoles; now "mobile devices" are often used for games. Children seem safer if parents can keep an eye on them at home, but what are they learning? The long-term effects on brain and body may be dangerous.

a careful longitudinal study, the finding that screen time reduced emotional maturation was not merely correlational but also causal.

Similar results are found in many nations, with many young children using screens three hours a day, and with far-reaching consequences, in obesity, emotional immaturity, and intellectual growth (see **Figure 10.1**). Overall, the American Academy of Pediatrics (2016) recommends no more than an hour a day of any screen time for 2- to 6-year-olds. They also urge parents never to let their children watch programs or play games that include violent or sexual media, or racist and sexist stereotypes. Parent should always watch with the child, interpreting and reinforcing.

However, most young children watch more than is recommended, unsupervised, not only in the North America but also in other developed nations. If you have no children, you might wonder why any parent would allow children to look at screens more than occasionally, or, worse, give their child an iPad for their birthday. If you are a parent, you know why.

Social Play

Young children play best with *peers,* that is, people of about the same age and social status. Although infants are intrigued by other children, babies do not play together because peer play requires some social maturation (Bateson & Martin, 2013). Gradually, from age 2 to 6, most children learn how to join a peer group, manage conflict, take turns, find friends, and keep the action going (Göncü & Gaskins, 2011; Şendil & Erden, 2014).

Playmates

Children need peers. Even the most playful parent is outmatched by a child at negotiating the rules of tag, at play-fighting, at pretending to be sick, at killing dragons, and so on.

As they interact with peers, children learn emotional regulation, empathy, and cultural understanding. Specifics vary, but "play with peers is one of the most important areas in which children develop positive social skills" (Xu, 2010, p. 496).

Peers provide an audience, role models, and sometimes competition. For instance, running skills develop best when

Good Over Evil or Evil Over Good? Boys everywhere enjoy "strong man" fantasy play, as the continued popularity of Spider-Man and Superman attests. These boys follow that script. Both are Afghan refugees now in Pakistan.

REUTERS/Akhtar Soomro

More Play Time, Less Screen Time

Play is universal—all young children do it when they are with each other, if they can. According to a 2017 study, U.S. 2- to 10-year-olds average 19 hours per week of screen time, exceeding the 15 hours they spend in indoor screen-free play by themselves or with others. Although children play outside for an additional 11 hours per week, parents report that when indoors, their children's screen time crowds out screen-free play.

WHAT 2- TO 10-YEAR-OLDS DO WITH THEIR FREE TIME

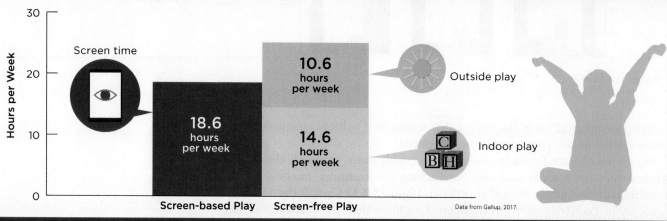

Screen time

Outside play

10.6 hours per week

18.6 hours per week

14.6 hours per week

Indoor play

Screen-based Play   Screen-free Play

Data from Gallup, 2017.

WAY TOO MUCH SCREEN TIME

Very few children have the recommended less than an hour of screen time per day. Some have much more. This is particularly evident on weekends, when they should be playing outside or interacting with their families. What did children do before 1950, without TV or computers? Talking, reading, cooking, cleaning, board games, ball games, playing music, drawing pictures, writing letters ... the list of things that some children never do could go on and on!

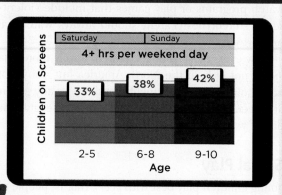

Saturday   Sunday

Children on Screens

4+ hrs per weekend day

33%   38%   42%

2-5   6-8   9-10

Age

Data from Gallup, 2017.

WHAT CAREGIVERS CAN DO TO ENCOURAGE PLAY

at Ages 1–3

Choose childcare and preschools that emphasize unstructured playtime.

Offer simple, inexpensive objects (blocks, empty containers, puzzles, etc.), not screens or fancy gadgets.

Organize caregiver-supervised playdates with peers.

Encourage make-believe play.

Sing songs and play rhythms that invite participation.

at Ages 4–6

Provide opportunities for singing and dancing.

Encourage a variety of movements in a safe environment—e.g., hopping, swinging, climbing, and somersaulting.

Blend make-believe games and reality—e.g., "playing house" and helping with chores.

Encourage school officials to offer recess and play-centered learning approaches, not just reading, memorization, and worksheets.

Information from American Academy of Pediatrics, 2018.

children chase or race each other, not when a child runs alone. Active social play—not solitary play—correlates with physical, emotional, and intellectual growth.

The importance of peer interaction has become increasingly clear, now that many children attend preschools, early-childhood centers, or other institutions of group care and learning. As you learned in Chapter 9, preschools advance language and other aspects of cognition. It is also clear that play with peers also advances social understanding.

Technology Play

Adults once feared that technology would cause children to become socially isolated. A more nuanced view considers the social setting. When a child watches a video alone, interpersonal emotional understanding is reduced. However, if children are with other children, digital play can advance development.

One detailed study of 3- to 5-year-olds in a Scotland preschool focused on digital technology such as computers, SMART boards, digital cameras, and so on. All of that was available during free play time, in the same way that puzzles, dress-up clothes, and dolls were available (Arnott, 2016).

Children used various digital tools as part of social interaction. Sometimes the digital object encouraged cooperation, as with this boy and girl, both aged 4:

> Grace begins to use the SMART board and Chris begins to use his finger to point at the screen to illustrate which selection she should choose as he verbally directs. Chris continues this process for each step of the way and Grace obeys. Then, when the game reaches a section where Grace does not need to make a selection and she needs to wait, he holds up his hand and says, "now wait."

Sometimes one child was the leader, with onlookers encouraging. For example:

> Three boys have formed a cluster around the computer. Chris is controlling the computer by using the mouse. Harvey is sitting next to him in front of the computer and Steven is hovering close by. Harvey offers encouragement to Chris, "You got 10!" he shouts with an excited tone. Steven becomes more and more excited with this activity and begins to show it by bouncing up and down while he stands next to the computer. He begins to cheer. Suddenly, Harvey turns to Steven and sharply says "Shhh—Don't do that!" Steven is immediately silent and observes quietly.

> [Arnott, 2016, p. 280]

In this study, children's technology use was almost always a social activity. Sometimes they shared, sometimes they mentored each other, sometimes they merely watched, providing encouragement, and later tried it themselves. The social context, including the physical space and the overall routines and rules, framed the interactions. It was unusual for the children to be hostile or to fight for control. (Rarely, the teachers intervened.)

Rough-and-Tumble Play

One form of social play, often unrecognized for the emotional growth it fosters, is called **rough-and-tumble**, because it looks rough and children seem to tumble over one another. The term was coined by British scientists who studied animals in East Africa (Blurton-Jones, 1976). They noticed that young monkeys often chased, attacked, rolled over in the dirt, and wrestled quite roughly without injuring one another, all while seeming to smile (showing a *play face*).

When the scientists who studied monkeys in Africa returned to London, they saw that puppies, kittens, and even their own children engaged in rough-and-tumble

LaunchPad
macmillan learning

VIDEO: The Impact of Media on Early Childhood explores how screen time can affect young children's cognition.

rough-and-tumble play Play that seems to be rough, as in play wrestling or chasing, but in which there is no intent to harm.

2016 Macmillan

Joy Supreme Pretend play in early childhood is thrilling and powerful. For this dancing 7-year-old from Park Slope, Brooklyn, pretend play overwhelms mundane realities, such as an odd scarf or awkward arm.

sociodramatic play Pretend play in which children act out various roles and themes in plots or roles that they create.

play. Children chase, wrestle, and grab each other, with established rules, facial expressions, and gestures to signify "just pretend."

Developmentalists now recognize that rough-and-tumble play happens everywhere, with every mammal species, and it has for thousands of years (Fry, 2014). It is more common among males than females, and it flourishes best in ample space with minimal supervision (Pellegrini, 2013).

Neurological benefits from such play are evident. The easiest example comes from watching rodents. Young rats try to bite the nape of another's neck. If a bite occurs, the two rats switch roles and the bitten tries to bite the other's nape. This is all playful. If rats want to hurt each other, they try to bite organs, not napes. Rat rough-and-tumble play increases rat brain development (Pellis et al., 2018). Watching puppies, kittens, and human children also shows that the participants are careful not to hurt each other.

Controlled experiments on humans, with some children allowed to play and a matched control group never playing, would be unethical, but correlations suggest that the limbic system connects more strongly with the prefrontal cortex because children engage in rough-and-tumble play. Indeed, longitudinal research on boys who played carefully but roughly with peers and parents (usually with fathers) finds that they become caring, compassionate men (Fry, 2014; Raeburn, 2014).

Sociodramatic Play

Another major type of social play is called **sociodramatic**, in which children act out various roles and plots. Through such acting, children:

- explore and rehearse social roles;
- learn to explain their ideas and persuade playmates;
- practice emotional regulation by pretending to be afraid, angry, brave, etc.; and
- develop self-concept in a nonthreatening context.

As children combine their imagination with that of their friends, they advance in theory of mind, gaining emotional regulation as they do so (Goldstein & Lerner, 2018; Kavanaugh, 2011).

WHAT HAVE YOU LEARNED?

1. What are children thought to gain from play?

2. Why do pediatricians want to limit children's screen time?

3. How can technology enhance children's learning instead of inhibiting it?

4. Why does playing with peers increase physical development and emotional regulation?

5. What do children learn from rough-and-tumble play?

6. What do children learn from sociodramatic play?

Challenges for Caregivers

Every developmentalist, and every parent, realizes that caring for a young child is difficult. Young children are energetic and curious but not wise. Some children, by temperament, are especially challenging. They, even more than compliant children, need patient parents (Ayoub et al., 2018).

Styles of Caregiving

One challenge for caregivers is choosing how strict or permissive to be. Variations in parenting styles are astonishing: Some are so strict that they seem abusive, and others are so lenient that they seem neglectful. Variations are apparent within nations, within ethnic groups, within neighborhoods, and sometimes within marriages.

Baumrind's Categories

Although thousands of researchers have traced the effects of parenting on child development, the work of one person, 60 years ago, remains influential. Diana Baumrind (1967, 1971) studied 100 preschool children, all from California, almost all middle-class European Americans.

She found that parents differed on four important dimensions:

1. *Expressions of warmth.* Some parents are warm and affectionate; others are cold and critical.
2. *Strategies for discipline.* Parents vary in how they explain, criticize, persuade, and punish.
3. *Expectations for maturity.* Parents vary in expectations for responsibility and self-control.
4. *Communication.* Some parents listen patiently; others demand silence.

On the basis of these dimensions, Baumrind identified three parenting styles: *authoritarian*, *permissive*, and *authoritative* (summarized in **Table 10.1**). A fourth style, *neglectful*, not described by Baumrind, was suggested by other researchers.

Authoritarian parenting. The authoritarian parent's word is law, not to be questioned. Misconduct brings strict punishment, usually physical. Authoritarian parents set down clear rules and hold high standards. Discussion about emotions and expressions of affection are rare. One adult raised by authoritarian parents said that "How do you feel?" had only two possible answers: "Fine" and "Tired."

authoritarian parenting An approach to child rearing that is characterized by high behavioral standards, strict punishment of misconduct, and little communication from child to parent.

Permissive parenting. Permissive parents (also called *indulgent*) make few demands. Discipline is lax, partly because expectations are low. Permissive parents are nurturing and accepting, listening to whatever their offspring say, which may include "I hate you"—a remark that authoritarian parents would not tolerate.

permissive parenting An approach to child rearing that is characterized by high nurturance and communication but little discipline, guidance, or control.

Authoritative parenting. Authoritative parents set limits, but they are flexible. They consider themselves guides, not authorities (unlike authoritarian parents) and not friends (unlike permissive parents). The goal of punishment is for the child to understand what was wrong and what should have been done differently.

authoritative parenting An approach to child rearing in which the parents set limits and enforce rules but are flexible and listen to their children.

TABLE 10.1

Characteristics of Parenting Styles Identified by Baumrind

Style	Warmth	Discipline	Expectations of Maturity	Communication	
				Parent to Child	Child to Parent
Authoritarian	Low	Strict, often physical	High	High	Low
Permissive	High	Rare	Low	Low	High
Authoritative	High	Moderate, with much discussion	Moderate	High	High

Protect Me from the Water Buffalo These two are at the Carabao Kneeling Festival. In rural Philippines, hundreds of these large but docile animals kneel on the steps of the church, part of a day of gratitude for the harvest.

⬤ **Observation Quiz** Is this father authoritarian, authoritative, or permissive? (see answer, page 281) ➜

neglectful/uninvolved parenting An approach to child rearing in which the parents seem indifferent toward their children, not knowing or caring about their children's lives.

Neglectful/uninvolved parenting. Neglectful parents are oblivious to their children's behavior; they seem not to notice. Their children do whatever they want. The child's behavior may be similar to those of the permissive parent, but the parents' attitude is quite different: Neglectful parents do not care, whereas permissive parents care very much.

Long-term effects of parenting styles have been reported in many nations. Cultural and regional differences are apparent, but everywhere authoritative parenting seems best (Pinquart & Kauser, 2018). The following trends have been found in many studies, although you will soon read that results are not as universal as the early research found.

- *Authoritarian* parents raise children who become conscientious, obedient, and quiet but not especially happy. Such children may feel guilty or depressed, internalizing their frustrations and blaming themselves when things don't go well. As adolescents, they sometimes rebel, striking out on their own. As adults, they are quick to blame and punish.
- *Permissive* parents raise children who lack self-control. Inadequate emotional regulation makes them immature and impedes friendships, so they are unhappy. They tend to continue to live at home, still dependent on their parents in adulthood.
- *Authoritative* parents raise children who are successful, articulate, happy with themselves, and generous with others. These children are usually liked by teachers and peers, especially in cultures that value individual initiative (e.g., the United States).
- *Neglectful/uninvolved* parents raise children who are immature, sad, lonely, and at risk of injury and abuse, not only in early childhood but also lifelong.

Problems with the Research

Baumrind's classification schema has been criticized, especially because she did not consider cultural norms and child temperament. How much protection, encouragement, and guidance does a particular child need? Differential susceptibility suggests that fearful children require reassurance, and impulsive ones need strong guidelines. Parents of such children may, to outsiders, seem permissive or authoritarian.

A study of parenting at age 2 and children's competence in kindergarten (including emotional regulation and friendships) found "multiple developmental pathways." The best outcomes depended on both the child and the adult (Blandon et al., 2010).

Simplistic advice—from a professional, a neighbor, or a textbook author (me) who does not know the child—may be misguided. Longitudinal, unbiased observation of parent–child interactions is needed before judging a caregiver.

Discipline

Children misbehave. Sometimes adults demand too much of an immature child, but sometimes children deliberately do exactly what they know they should not do.

Misbehavior is part of growing up, as is testing the limits of what is acceptable. However, children need guidance to keep them safe and strong. Parents must respond when the child does something forbidden, dangerous, or mean. During early childhood, most parents use several methods to discipline their children (Thompson et al., 2017). Every form of discipline has critics as well as defenders (Larzelere et al., 2017).

Physical Punishment

In the United States, young children are slapped, spanked, or beaten more often than are infants or older children, and more often than children in Canada or western Europe. Spanking is more frequent:

- in the southern United States than in New England;
- by mothers than by fathers;
- among conservative Christians than among nonreligious families;
- among African Americans than among European Americans;
- among European Americans than among Asian Americans;
- among U.S.-born Hispanics than among immigrant Hispanics; and
- in low-SES families than in high-SES families.

[*MacKenzie et al., 2011; S. Lee et al., 2015; Lee & Altschul, 2015*]

These are general trends, but do not stereotype. Contrary to these generalizations, some African American mothers living in the South never spank, and some secular, European American, high-SES fathers in New England routinely do. Local norms matter, but individual parents make their own decisions, despite what their neighbors do.

In some nations, **corporal punishment** is illegal; in other nations, it is the norm. A massive international study of low- and moderate-income nations found that 63 percent of 2- to 5-year-olds had been physically punished (slapped, spanked, hit with an object) in the past month (Deater-Deckard & Lansford, 2016).

Given a multicultural, multicontextual perspective, it is not surprising that many spanked children become fine adults, who believe they were not harmed by spanking. Nonetheless, a correlation between spanking and aggression is found in all ethnic groups, in many nations (Lansford et al., 2014; Wang & Liu, 2018).

"He's just doing that to get attention."

Harry Bliss/The New Yorker Collection/The Cartoon Bank

Pay Attention Children develop best with lots of love and attention. They shouldn't have to ask for it!

corporal punishment Disciplinary techniques that hurt the body (*corpus*) of someone, from spanking to serious harm, including death.

Smack Will the doll learn never to disobey her mother again?

Children who are *not* spanked are *more* likely to develop emotional regulation.

Although some adults believe that physical punishment will "teach a lesson," others argue that the lesson learned is that "might makes right." Children who were physically disciplined tend to become more aggressive, and more likely to use corporal punishment on others—first on their classmates, and later on their partners, and then on their children (Thompson et al., 2017).

Paddling in Schools

This dispute about corporal punishment echoes in schools. In more than 100 nations, physical punishment is illegal in any educational setting. However, each U.S. state sets its own law, and teachers may legally paddle children in 19 of them. Overall, in the United States, more than 100,000 children were corporally punished at school. A disproportionate number were African American boys (Gershoff et al., 2019; Morones, 2013).

Boys are also more likely to be harshly punished by their parents, until adolescence, when boys have more freedom and less punishment at home. However, beginning at puberty, males are at least twice as likely to be imprisoned than females, with gender differences particularly stark for violent offenses.

A study in one U.S. state that allows corporal punishment in school (Arkansas) reports that whether or not a child is physically punished depends more on the school culture than on the state or district policy (McKenzie & Ritter, 2017). The rate of discipline in Arkansas in the 2015–2016 school year was 59 per 100 students, with 5 per 100 including physical punishment.

Interpret those statistics carefully. A ratio of 59 punishments per 100 students does not mean that more than half of the students were disciplined. Instead, some students experienced more than 10 punishments (some were paddled several times), while most (especially the younger girls) were never punished. Rates were much higher in middle schools than elementary schools.

The most common infractions were "minor, non-violent," when students did not obey their teacher or follow school guidelines. The goal is for children to learn more because they are more obedient. But that may not be the result. One study found that children who were physically punished had lower grades in high school, were less likely to graduate, and were more likely to become depressed (Gershoff et al., 2019).

In Arkansas as well as nationwide, school culture is changing. Once every state allowed corporal punishment; the 19 that still allow it include Alabama, Arkansas, Arizona, Colorado, Florida, Georgia, Idaho, Indiana, Kansas, Kentucky, Louisiana, Missouri, Mississippi, North Carolina, Oklahoma, South Carolina, Tennessee, Texas, and Wyoming.

Even within those states, some school districts forbid it. However, suspensions (the school equivalent of time-out) have increased in Arkansas (McKenzie & Ritter, 2017). Longitudinally, over some of the same years (2011–2019), suspensions for the million children in the New York City public schools were cut in half (from 70,000 per year to 33,000). Data from those two states make it clear that policies regarding school discipline change politically, not based on actual misbehavior.

> **THINK CRITICALLY:** The varying rates of physical punishment in schools could be the result of prejudice, or they could be because some children misbehave more than others. Which is it?

psychological control A disciplinary technique that involves threatening to withdraw love and support, using a child's feelings of guilt and gratitude to the parents.

Alternatives to Spanking

If spanking is harmful but discipline is necessary, what is a parent to do? Some employ **psychological control**, using children's shame, guilt, and gratitude to control their behavior (Barber, 2002). But this has its own problems (Alegre, 2011).

Spare the Rod?

Opinions about spanking are influenced by past experience and cultural norms. That makes it hard for opposing perspectives to be understood by people on the other side (Ferguson, 2013). Try to suspend your own assumptions as you read this.

What might be right with spanking? Over the centuries, many parents have done it, so it has stood the test of time and has been a popular choice. Spanking is less common in the twenty-first century than in the twentieth (Taillieu et al., 2014).

However, 85 percent of U.S. adolescents who were children at the end of the twentieth century recall being slapped or spanked by their mothers (Bender et al., 2007). In low- and middle-income nations, over a third of the mothers believe that physical punishment is essential to raise a child well (Deater-Deckard & Lansford, 2016).

Those who are pro-spanking need to explain why spanking correlates with later depression, low achievement, aggression, crime, and so on. They suggest that a third variable, not spanking itself, is the reason for that connection. One possible third variable is misbehavior: Perhaps disobedient children cause spanking, not vice versa. Such children may become delinquent, depressed, and so on not *because* they were spanked but *in spite of* being spanked.

Noting problems with correlational research, one team explains, "Quite simply, parents do not need to use corrective actions when there are no problems to correct" (Larzelere & Cox, 2013, p. 284).

Further, since parents who spank their children often have less education and money than other parents, low SES may be the underlying crucial variable. Perhaps spanking is a symptom of poverty and poor parenting skills. If that is true, to reduce the low achievement, aggression, and depression, we need to increase education and income, not to ban spanking (Ferguson, 2013).

Another pro-spanking response to the troubling research is to argue that researchers need to be careful in how they define corporal punishment. If a study does not distinguish between severe corporal punishment and milder, occasional spanking, then the data will show that spanking is harmful—but that conclusion reflects the harm of severe punishment (Larzelere et al., 2017).

Now the opposing perspective.

What might be wrong with spanking? One problem is adults' emotions: Angry spankers may become abusive. Children have been seriously injured and even killed by parents who use corporal punishment.

Another problem is the child's immature cognition. Parents assume that the transgression is obvious, but children may think that the parents' anger, not the child's actions, caused spanking (Harkness et al., 2011). Most parents tell their children why they are being spanked, but when they are hit, children are less likely to listen or understand, much less learn.

Almost all of the research finds that children who are physically punished suffer overall (Grogan-Kaylo et al., 2018). Compared to children punished in other ways, they are more depressed, anti-social, and lonely. Many hate school and have few close friends. Emotional and social problems in adulthood are more common in people who were spanked as children—true for relatively mild spanking as well as for more severe spanking.

One reason for these correlations is that spanked children more often have angry, depressed, unloving parents. However, even among children of warm and loving parents, spanked children tend to be more anxious, worried they will do something to lose their parents' affection (Lansford et al., 2014).

So who is right? I know which perspective is mine. Most developmentalists believe that alternatives to spanking are better for children and a safeguard against abuse. I agree with that. The same study that found spanking common in developing nations also reported that 17 percent of the children experienced severe violence that no developmentalist would condone (Bornstein et al., 2016). That alone is reason to stop.

I recognize that there are exceptions—spanked children who become happy and successful adults. For example, one study found that conservative evangelical parents spank their children more often than other parents, but if that spanking occurred *only* in early (not middle) childhood, the children did not develop low self-esteem and increased aggression (Ellison et al., 2011).

The authors of the study suggest that, since spanking was the norm in that group, spanked children did not feel unloved. Moreover, religious leaders told the parents never to spank in anger. As a result, some children may "view mild-to-moderate corporal punishment as legitimate, appropriate, and even an indicator of parental involvement, commitment, and concern" (Ellison et al., 2011, p. 957).

Another study of conservative Christians found that many thought their faith condoned spanking. Only when they learned that the Bible opposed spanking (e.g., that "sparing the rod" refers to the guiding rod that shepherds use, which was never used to punish), and learned research on the long-term harm, did they change their minds (Perrin et al., 2017). Many then concluded that physical punishment is contrary to their belief in love and forgiveness.

A dynamic-systems, multicultural perspective reminds me that everyone is influenced by background and context. I know that I am; so is every scientist; so are you. Probably my opinions are wrong about several developmental controversies that I explain in this text. I do not think this is one of them.

time-out A disciplinary technique in which a person is separated from other people and activities for a specified time.

induction A disciplinary technique that involves explaining why a particular behavior was wrong. To be successful, explanation must be within the child's ability to understand.

⬤ **Especially for Parents** Suppose you agree that spanking is destructive, but you sometimes get so angry at your child's behavior that you hit him or her. Is your reaction appropriate? (see response, page 281)

sex differences Biological differences between males and females, in organs, hormones, and body shape.

gender differences Differences in male and female roles, behaviors, clothes, and so on that arise from society, not biology.

Consider Finland, where corporal punishment is forbidden, but psychological control is not. The higher parents scored on four measures of psychological control, the lower the children's math scores were—and this connection grew stronger over time. Moreover, the children tended to have negative emotions (depression, anger, and so on) (Aunola et al., 2013).

Another disciplinary technique is the **time-out**, in which a misbehaving child is required to sit quietly, without toys or playmates, for a short time. Time-out is not to be done in anger, or for too long; it is recommended that parents use a calm voice and that the time-out last only one to five minutes (Morawska & Sanders, 2011). Time-out is punishment if the child enjoys "time-in," when the child is engaged with parents or with peers.

Time-out is favored by many experts. They advise that time-out is part of a close parent–child relationship, a way to punish a behavior that the child knows is wrong. The message is that the child needs to stop and think, and thus indicates the parent's connection to the child, not rejection (Dadds & Tully, 2019)

Often combined with the time-out is another alternative to physical punishment and psychological control—**induction**, in which the parents discuss the infraction with their child, hoping the children themselves will realize why their behavior was wrong. Ideally, a strong and affectionate parent–child relationship allows children to express their emotions and parents to listen.

Induction takes time and patience, and, like other discipline measures, it does not always succeed. One problem is that young children confuse causes with consequences and tend to think they behaved properly, given the situation. Simple induction ("Why did he cry?") may be appropriate, but even that is hard before a child develops theory of mind. Nonetheless, induction may pay off over time. Children whose parents used induction when they were 3-year-olds became children with fewer externalizing problems in elementary school (Choe et al., 2013b).

What do parents actually do? A survey of discipline in early childhood found that most parents use more than one method (Thompson et al., 2017). In the United States, time-out is the most common punishment, and about half of the parents sometimes spank. The survey found that other methods—induction, counting, distraction, hand-smacking, removal of a toy or activity—were also used.

Specifics of parenting style and punishment seem less crucial than whether or not children know that they are loved, guided, and appreciated (Grusec et al., 2017). If a parent seems too strict, or too lenient, remember that if the children consider discipline fair, and believe they are loved and valued, they are likely to be high achievers, who are happy and proud (Pinquart & Kauser, 2018).

Becoming Boys and Girls

Another challenge for caregivers is to promote a healthy understanding of sex and gender, so children can be proud of themselves and accepting of others.

Sex and Gender

Sex differences are innate, the results of the XX or XY chromosomes and the hormones they produce. **Gender differences** are cultural, either obviously so, in clothes and hair styles, or more generally, such as differences in achievement scores on tests or an adult's participation in a group conversation.

In theory this distinction between sex and gender seems simple; in practice it is not. Culture and biology are not separate influences but are "interacting components of nature and nurture" (Eagly & Wood, 2013, p. 349). The interaction between sex and gender is such that some scholars propose that we use only one word, sex/gender, to denote both (Hyde et al., 2019).

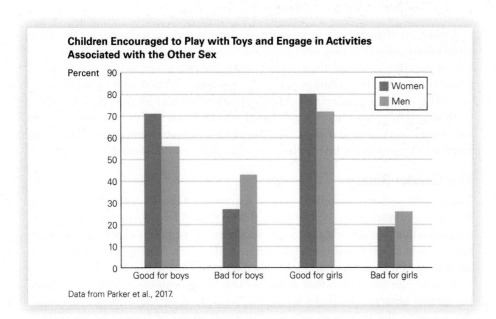

Children Encouraged to Play with Toys and Engage in Activities Associated with the Other Sex

Percent

Data from Parker et al., 2017.

FIGURE 10.2

Similarities? What is more remarkable—that most people think girls should be encouraged to play with trucks and boys encouraged to play with dolls, or that some people do not? Your answer probably depends on whether you thought gender equality has been achieved or is still far away.

Be that as it may, during early childhood most children develop distinct gender identities. By age 4, children believe that certain toys (such as dolls or trucks) and roles (Daddy, Mommy, nurse, teacher, police officer, soldier) are reserved for one sex or the other, even when their experience is otherwise. As one expert explains:

> Four year olds say that girls will always be girls and will never become boys. . . . They are often more absolute about gender than adults are. They'll tell their very own pantssuited doctor mother that girls wear dresses and women are nurses.

> *[Gopnik, 2016, p. 140]*

By age 6, children can become quite rigid. Despite their parents' and teachers' wishes, children say, "girls are stupid; boys stink, no girls [or boys] allowed."

Many parents believe they treat their sons and daughters the same and want to free their children from gender stereotypes. Few adults today would forbid a boy from wanting to push a toy stroller with a doll strapped in it, or a girl from pushing around a truck.

A 2017 survey found that most adults went one step further. They thought parents should *encourage* their children to play with toys associated with the other sex (Parker et al., 2017). Those who disagreed were mostly men who were asked about boys; 43 percent of the men thought boys should *not* be encouraged to do things usually stereotyped for girls, such as care for dolls or wear bracelets (see **Figure 10.2**).

In any case, young children may reject such encouragement. A meta-analysis of 75 studies found that girls have strong preferences for girl toys, especially dolls, and boys for boy toys, especially trucks (Davis & Hines, 2020). A child's absolutism and rejection of the other sex poses a challenge if parents want their children to be less rigid about sex roles.

Transgender Children

Sex and gender issues are particularly salient when a child is **transgender**, identifying as being a gender other than the one assigned at birth (Rahilly, 2015). It is one thing to allow young girls to play with trucks and young boys to play with dolls, but it may be harder to accept that a child has quite distinct ideas about what gender they are, when that gender is not the one on their birth certificate. One mother said:

> Since he was two, all he can say is that he wants to be a girl, or that he is a girl. He knows that he is not, but there is no way to change his mind. He is 6 now, and he still asks me everyday "Mom, can I be a girl when I grow up?"

> *[quoted in Malpas, 2011, p. 453]*

transgender Identifying oneself with a gender other than the one ascribed at birth. Thus, a transgender girl was thought to be a boy when she was born.

Mike Belleme

Not Emma In a North Carolina kindergarten, each child had an "All About Me" day in which the teacher would draw a picture of the child for all of the other children to copy. Emma was born with male sex organs but identifies as a girl. On her day, she proudly wore a light-pink shirt with a heart, pink glittery shoes, and long hair—and she came home bawling because the teacher drew this picture with her "boy name" (barely visible here). Her parents consoled her, had her edit her name and draw longer hair, with some other additions. Shouldn't children be allowed to be who they are?

gender binary The idea that there are only two (*bi-*) genders, male and female, and that they are opposites. This idea precludes intersex, gender overlap, and gender non-conformity.

phallic stage Freud's third stage of development, when the penis becomes the focus of concern and pleasure.

Oedipus complex The unconscious desire of young boys to replace their fathers and win their mothers' exclusive love.

Unlike in that family, some parents during early childhood accept their child's insistence that they are transgender. A study of 36 transgender children, whose parents accepted their transition to the other sex before age 6, compared them with their siblings and with *cisgender* children who were the same age. (The term "cisgender" refers to those whose gender identity matches their birth sex.) The researchers found that all the children had definite preferences for clothes, toys, and activities. The transgender children chose whatever conformed to their gender, just as the cisgender children did (Fast & Olson, 2018).

The Gender Binary

Transgender and intersex children raise a question: How do we define masculinity and femininity? Some people argue that male and female are on a *continuum*, which is a line that stretches from one side to the other, in this case from extremely masculine at one end to extremely feminine at the other. Most aspects of most people are not at the extremes, but somewhere in between.

This rejects the notion that male and female are opposites, a concept called the **gender binary**. (The prefix *bi-* means "two," as in *bicycle, bifurcate, bisexual*). According to the nonbinary perspective, every aspect of human brains, behavior, and bodies is somewhere along that continuum. For example, some men are shorter than the average woman, and some women are more gifted in advanced math than the average man.

That gender traits are not opposites, but instead that there is much overlap between men and women, is now understood by most adults. We expect men to sometimes cry; we cheer the U.S. women's soccer team who won the World Cup. The nonbinary perspective has also become prominent in studies of the human brain. Although male and female brains are alike on most measures, one study found a dozen neurological characteristics that differed notably between the *average* male and female brain (Joel et al., 2015).

However, that study found much overlap, with some men having neurological characteristics typical for women, and vice versa. Fewer than 10 percent of the people in this study had brains that were typical for their sex in all of the dozen traits that distinguish male and female brains (Joel et al., 2015). From studies such as this one, a team of researchers conclude that most human brains and behaviors are a male/female mosaic (Hyde et al., 2019).

The research raises an issue for developmentalists. Why are male and female distinctions recognized by most 2-year-olds, significant to most 4-year-olds, and accepted as natural by 6-year-olds? All of the major theories "devote considerable attention to gender differences. . . . the primary difference among the theories resides in the causal mechanism responsible" (Bornstein et al., 2016, pp. 10, 11). Consider the five comprehensive theories in Chapter 2: psychoanalytic, behaviorism, cognitive, sociocultural, and evolutionary.

Psychoanalytic Theory

Freud (1938/1995) called the period from about ages 3 to 6 the **phallic stage**, named after the *phallus*, the Greek word for penis. At age 3 or 4, said Freud, boys become aware of their male sexual organ. They masturbate, fear castration, and develop sexual feelings toward their mother.

These feelings make every young boy jealous of his father—so jealous, according to Freud, that he wants to replace his dad. Freud called this the **Oedipus complex**, after Oedipus, son of a king in Greek mythology. Abandoned as an infant and raised in a distant kingdom, Oedipus returned to his birthplace and, without realizing it, killed his father and married his mother. When he discovered the horror, he blinded himself.

Freud believed that this ancient story (immortalized in *Oedipus Rex,* a play written by Sophocles and first presented in Athens in 429 B.C.E.) is still presented every year somewhere in the world because it evokes unconscious wishes in everyone. Every young boy, says Freud, feels guilty about his incestuous and murderous impulses. In self-defense, young boys develop their **superego**, a powerful conscience that is quick to judge and punish.

That marks the beginning of morality, according to psychoanalytic theory. This theory contends that a small boy's fascination with superheroes, guns, kung fu, and the like arises from his unconscious impulse to kill his father. Further, an adult man's homosexuality, homophobia, or obsession with guns, pornography, prostitutes, or hell arises from problems at the phallic stage.

Freud offered several descriptions of the moral development of girls. One, called the *Electra complex,* is again named after an ancient Greek drama. Freud thought that girls also want to eliminate their same-sex parent (mother) and become intimate with the opposite-sex parent (father). For him, that explained why many 5-year-old girls dress in frills and lace and are happy to be "Daddy's girl."

Many psychologists criticize psychoanalytic theory as being unscientific. I know many adults who are gay, lesbian, or asexual, and who have happy, well-adjusted lives. However, scientists seek to reconcile theory and experience. My daughters made me reconsider. (See A Case to Study.)

> **superego** In psychoanalytic theory, the judgmental part of the personality that internalizes the moral standards of the parents.

A CASE TO STUDY

The Berger Daughters

I dressed my baby girls in blue, trying to create a unisex world for them. I wanted to free them from gender stereotypes. I failed.

My eldest, Bethany, at about 4 years old told me:

Bethany: When I grow up, I'm going to marry Daddy.
    Me: But Daddy's married to me.
Bethany: That's all right. When I grow up, you'll probably be dead.
    Me: [*Determined to stick up for myself*] Daddy's older than me, so when I'm dead, he'll probably be dead, too.
Bethany: That's OK. I'll marry him when he gets born again.

I was dumbfounded, stunned by how Freudian this sounded. Bethany saw my face fall, and pitied me:

Bethany: Don't worry, Mommy. After you get born again, you can be our baby.

A few years later, my second-born daughter Rachel told me:

Rachel: When I get married, I'm going to marry Daddy.
    Me: Daddy's already married to me.
Rachel: [*With joy at her wonderful solution*] Then we can have a double wedding!

My third daughter, Elissa, left a valentine on my husband's pillow on February 14th (see **Figure 10.3**).

Finally, when Sarah turned 5, she also said she would marry her father. I told her she couldn't, because he was married to me. Her response revealed one more hazard of screen time: "Oh, yes, a man can have two wives. I saw it on television."

FIGURE 10.3

Pillow Talk Elissa placed this artwork on my husband's pillow. My pillow, beside it, had a less colorful, less elaborate note—an afterthought. It read, "Dear Mom, I love you too."

As you remember from Chapter 1, a single example (one family with four daughters) does not prove that Freud was correct. I still think Freud was wrong on many counts, and I am proud that all four girls became college-educated, professionally employed women. But in many ways, they are following gender-specific paths. Freud's description of the phallic stage seems less bizarre than I once thought.

Behaviorism

Gender-typed behavior in play and chores (washing dishes versus fixing cars) is among the most robust findings of decades of research (Eagly & Wood, 2013). Behaviorists believe that those gender distinctions result from reinforcement, punishment, and social learning, evident in early childhood. For example, a 2-year-old boy who asks for both a train and a doll for his birthday is more likely to get the train.

Gender differentiation may be subtle, with adults unaware that they are reinforcing traditional masculine or feminine behavior. For example, parents talking to young children mention numbers and shapes more often with their sons (Pruden & Levine, 2017). Even with infants, fathers interact differently with their children, singing and talking more to their daughters but using words of achievement, such as *proud* and *win*, more with their sons (Mascaro et al., 2017).

According to social learning theory (an extension of behaviorism), people model themselves after people who are nurturing, powerful, and yet similar to themselves. For young children, those people are usually their parents.

Generally, if an employed woman is ever to leave her job to become a housewife, it is when she has a baby. Fathers tend to work longer hours—they are home less often—and mothers work fewer hours when children arrive. Since children learn gender roles from their parents, it is no surprise that they are quite sexist (Hallers-Haalboom et al., 2014).

Thus, adults are the most gender-typed of their entire lives when they are raising young children. According to social learning theory, children model themselves after what they see. They are unaware that their very existence is the reason for the father/mother divide.

Reinforcement for the gender binary extends beyond the family. The president of the Society for Research in Child Development observed, "parents, teachers, and peers . . . continue to encourage, model, and enforce traditional gender messages" (Liben, 2016, p. 24). Peers and social norms are powerful reinforcements of the binary. The boy who brings his Barbie doll to school will be punished—not physically, but with words and social exclusion—by his male classmates. As social learning increases from age 2 to 22, so does gender divergence.

Cognitive Theory

Cognitive theory offers an alternative explanation for the strong gender identity of 5-year-olds (Kohlberg et al., 1983). Remember that cognitive theorists focus on how children understand various ideas. Regarding boys and girls, they construct a **gender schema**, an understanding of male–female differences (Bem, 1981; Martin et al., 2011).

As Piaget explained, preoperational perceptions are static, not logical. Therefore, children categorize male and female as opposites. Nuances, complexities, exceptions, and gradations about gender (and about everything else) are beyond them.

During the preoperational stage, appearance is stronger than logic. One group of researchers who endorse the cognitive interpretation note that "young children pass through a stage of gender appearance rigidity; girls insist on wearing dresses, often pink and frilly, whereas boys refuse to wear anything with a hint of femininity" (Halim et al., 2014, p. 1091).

Parents who encourage unisex clothes are confronted by daughters who want a bright pink tutu and a sparkly tiara, and sons who want cowboy boots and guns. Children's gender schema overcomes the adult fight against sexism.

During these years, children are especially sensitive to language, which influences the way people think. A study of 45 of the world's languages found that,

gender schema A child's cognitive concept or general belief about male and female differences.

if the language assigns gender to asexual objects (as in Spanish, chair is *la* silla, unless it is *el* sillón, the name for a big armchair), children are more likely to see male and female as separate and distinct (DeFranza et al., 2020).

Cognitive theory explains some of the amusing mistakes that children make in their theories of gender. In one preschool, the children decided that one wash-up basin was for boys and the other for girls. That fits the cognitive schema that everything can be divided into male and female. But another cognitive mandate for 3-year-olds is to do whatever it takes to get what you want. A young girl started to use the boys' basin.

Boy: This is for the boys.
Girl: Stop it. I'm not a girl and a boy, so I'm here.
Boy: What?
Girl: I'm a boy and also a girl.
Boy: You, now, are you today a boy?
Girl: Yes.
Boy: And tomorrow what will you be?
Girl: A girl. Tomorrow I'll be a girl. Today I'll be a boy.
Boy: And after tomorrow?
Girl: I'll be a girl.

[*Ehrlich & Blum-Kulka, 2014, p. 31*]

This incident occurred in Israel, where women have been drafted into the army and have served as national leaders for decades. Probably this girl had been told that some gender restrictions are unfair to females, and she appropriated that message when she wanted to wash up. Neither she nor the boy questioned the overall binary, however, perhaps because in Hebrew every object is masculine or feminine.

Sociocultural Theory

It is evident that gender distinctions are pervasive in every culture (Starr & Zurbriggen, 2016). Some people believe this is no longer true, as women are in professions that

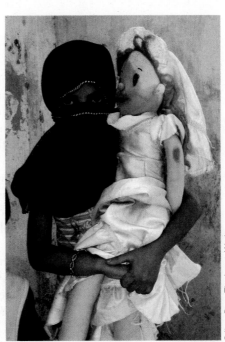

Same Situation, Far Apart: Culture Clash? He wears the orange robes of a Buddhist monk, and she wears the hijab of a Muslim girl. Although he is at a weeklong spiritual retreat led by the Dalai Lama and she is in an alley in Pakistan, both carry universal toys—a pop gun and a bride doll, identical to those found almost everywhere. Cultures differ; gender roles do not.

DIPTENDU DUTTA/AFP/GETTY IMAGES

© Ilyas Dean/The Image Works

were once exclusively male, and many fathers are active caregivers. However, in even the most gender-equal nations, women do much more child care, house cleaning, and meal preparation than do men. This is true when women are the primary wage-earners for their families (as are about 40 percent of mothers in the United States as well as worldwide) (Bornstein & Putnick, 2016).

If you think we have reached gender equality, look closely at the data on professions. Males are more often nurses than they were, but in 2020 only about 10 percent of nurses were male. That this is sociocultural is suggested by data on ethnicity: A higher percent of those male nurses were born in Africa.

In engineering, in 1976 only 2 percent of college women aspired to that profession; now it is slightly above 3 percent for women and 15 percent for college men. This gender divergence is found even when a student's mother is an engineer (Jacobs et al., 2017). Thus, culture seems to overwhelm family influence.

Everywhere, cultures socialize young girls and boys differently. For example, two 4-year-old girls might hug each other and hear "how sweet," but a boy who hugs a boy might be pushed away. Already by age 6, rough-and-tumble play is the only accepted way that boys touch each other, and girls are stopped from rough play more quickly than boys are. No wonder rough-and-tumble is more common among boys than girls.

Cultural differences are expressed by parents who do not realize that they are doing it. For instance, a massive study of 41 low- and middle-income nations found that fathers took their young boys outside more often than their girls. They also were more likely to read to, tell stories to, and count with their sons. The mothers tended to be more unisex in their activities (Bornstein & Putnick, 2016).

By age 6, children are astute "gender detectives," seeking out ways that males and females differ in their culture. They try to conform to their culture, as do all humans of every age. Mia is one example:

> On her first day of school, Mia sits at the lunch table eating a peanut butter and jelly sandwich. She notices that a few boys are eating peanut butter and jelly, but not one girl is. When her father picks her up from school, Mia runs up to him and exclaims, "Peanut butter and jelly is for boys! I want a turkey sandwich tomorrow."

[Quoted in Miller et al., 2013, p. 307]

Evolutionary Theory

Evolutionary theory holds that sexual passion is a basic human drive because all creatures must reproduce. Since conception requires an ovum and a sperm, males and females follow their evolutionary mandate by seeking to attract the other sex—walking, talking, and laughing in traditional feminine or masculine ways. That awakens sexual impulses in the other sex, assuring that the species will continue.

This evolutionary drive may explain why, already in early childhood, boys have a powerful urge to become like the men, and girls like the women. Young boys clomp around in their father's clunky shoes; young girls spray their mother's perfume all over themselves. These impulses prepare them, later on, to mate and conceive a new generation.

Evolutionary theory emphasizes the urge to survive as well as the urge to reproduce. Over millennia of human history, genes, chromosomes, and hormones dictated that young boys were more active (rough-and-tumble play) and girls more domestic (playing house). These also prepared them for adulthood, when fathers needed to defend against predators and mothers needed to care for the home and children.

Many researchers describe the women's liberation movement as a "stalled revolution" with progress evident in the public sphere (signified by employment and education) but not in the domestic one (Scarborough et al., 2019). From an evolutionary perspective, it might be that, in the intimacy of a partnership, wives want the husbands to respond as men traditionally did, even though at the workplace women fight for the same salary and power as the men.

Of course, tigers and bears no longer prowl outside in most places, but evolution takes thousands of years to reverse impulses needed 200,000 years ago. Few adolescents today follow the gendered path of teen marriage and motherhood, but almost all aspire to marry eventually, and almost all live with a sexual partner in adulthood (Sassler & Lichter, 2020). Both partners tend to fall into traditional roles, unless one or the other explicitly says otherwise. Perhaps evolution requires it.

Gender-Nonconforming The dad, not the daughter. Like many 6-year-olds, she loves wearing her frilly dress, and like many fathers he allows her to follow traditional roles, here by letting her put a tiara on his head.

What Is Best?

Each major developmental theory strives to explain the ideas that young children express and the roles they follow. No consensus has been reached. Caregivers know they should not blindly follow the norms of their culture, yet they also know that they need to guide their children regarding male–female differences and everything else.

Some recent research suggests a *gender similarities hypothesis,* the idea that our human emphasis on sex differences blinds us to the reality that the two sexes have far more in common than traditional theories recognize (Hyde, 2016). Perhaps instead of looking for sex differences, we should notice gender similarities. According to some researchers, similarities far outweigh differences in the brain, body, and behavior (Roseberry & Roos, 2016; Hutchison et al., 2019). Each child can be appreciated for themselves.

Teaching Right and Wrong

A final challenge for parents it to teach moral values. Young children are ready to learn right and wrong, an outgrowth of bonding in infancy, attachment in childhood, and now social understanding. Moral values are essential for our species; we depend on each other for protection, cooperation, and care. That is why our bodies produce hormones (oxytocin, vasopressin) that push people toward trust, love, and morality (Zak, 2012).

LaunchPad
macmillan learning

VIDEO: Interview with Lawrence Walker discusses what parents can do to encourage their children's moral development.

Empathy

With the cognitive advances of early childhood, and increased interaction with peers, these innate moral impulses are strengthened. Children develop **empathy**, an understanding of other people's feelings and concerns. Empathy depends on both experiences and brain maturation (Levy et al., 2019; Stern et al., 2019).

Empathy leads to compassion and **prosocial behavior**—"voluntary behavior meant to help another" (Padilla-Walker & Carlo, 2014, p. 6). Expressing concern, offering to share, and including a shy child in a game are examples of children's prosocial behavior. Prosocial behavior seems to result more from emotion than from intellect, more from empathy than from theory.

The link between empathy and prosocial behavior was traced longitudinally in children from 18 months to 6 years. Empathetic 2-year-olds were more likely to share, help, and play with other children in the first grade (Z. Taylor et al., 2013).

Feeling distress when someone else suffers may be a part of nature: Even infants seem distressed at the pain of another. But how and for whom distress leads to action varies by family and culture: Sacrifices are made for family members, or for peers, or not at all (Poelker & Gibbons, 2019).

empathy The ability to understand the emotions and concerns of another person, especially when they differ from one's own.

prosocial behavior Actions that are helpful and kind but that are of no obvious benefit to the person doing them.

Pinch, Poke, or Pat Antisocial and prosocial responses are actually a sign of maturation: Babies do not recognize the impact of their actions. These children have much more to learn, but they already are quite social.

Antipathy

antipathy Feelings of dislike or even hatred for another person.

Empathy appears early in life: Infants mirror the smiles of other people, and young children are happy when other children are happy. **Antipathy** develops a little later, influenced by experience and culture, as children learn to dislike some people.

Just as empathy can lead to prosocial behavior, antipathy can lead to antisocial action, such as avoidance, insults, and physical attacks. A 2-year-old might look at another child, scowl, and then kick hard without provocation.

Generally, parents and teachers teach better behavior, and children become more prosocial and less antisocial with age. A study of children's willingness to belong to a group who did antisocial things found that by age 7, children did not want to be in such a group. That was not true for 5-year-olds: They opted to stay with their group, even when others in the group were antisocial (Wilks et al., 2019).

antisocial behavior Actions that are deliberately hurtful or destructive to another person.

At every age, **antisocial behavior** indicates less empathy. That may originate in the brain. An allele or gene may have gone awry. But at least for children, lack of empathy correlates with parents who neither discuss nor respond to emotions (Richards et al., 2014; Z. Taylor et al., 2013). Antisocial parents tend to have antisocial children, a correlation that is probably both genetic and environmental, and may be particularly strong in boys (Li et al., 2017).

Aggression

Early childhood is prime time for both aggressive behavior and victimization: Almost every young child is both an aggressor and a victim at some point (Saracho, 2016). Not surprisingly, given their moral sensibilities, young children judge whether another child's actions are fair or not.

The focus at first is on effects, not motives: A child who accidentally spilled water on another's painting may be the target of that child's anger. As young children gain in social understanding, particularly theory of mind, they gradually become better at understanding intentions, and that makes them more likely to forgive an accident (Choe et al., 2013a).

The distinction between impulse and intention is critical in deciding when and how adults need to stop a child's aggression. Researchers recognize four general types of aggression, each of which is evident in early childhood (see **Table 10.2**).

TABLE 10.2

The Four Forms of Aggression

Type of Aggression	Definition	Comments
Instrumental aggression	Hurtful behavior that is aimed at gaining something (such as a toy, a place in line, or a turn on the swing) that someone else has	Apparent from age 2 to 6; involves objects more than people; quite normal; more egocentric than antisocial.
Reactive aggression	An impulsive retaliation for a hurt (intentional or accidental) that can be verbal or physical	Indicates a lack of emotional regulation, characteristic of 2-year-olds. A 5-year-old can usually stop and think before reacting.
Relational aggression	Nonphysical acts, such as insults or social rejection, aimed at harming the social connections between the victim and others	Involves a personal attack and thus is directly antisocial; can be very hurtful; more common as children become socially aware.
Bullying aggression	Unprovoked, repeated physical or verbal attack, especially on victims who are unlikely to defend themselves	In both bullies and victims, a sign of poor emotional regulation; adults should intervene before the school years. (Bullying is discussed in Chapter 13.)

Instrumental aggression is common among 2-year-olds, who often want something and try to get it. This is called *instrumental* because it is a tool, or instrument, for getting something that is desired. The harm in grabbing a toy, and hitting if someone resists, is not understood.

Because instrumental aggression occurs, **reactive aggression** also is common among young children. Almost every child reacts when hurt, whether or not the hurt was deliberate. The reaction may be aggressive—a child might punch in response to an unwelcome remark. As the prefrontal cortex matures and emotional regulation is possible, the impulse to strike back becomes controlled. Both instrumental aggression and reactive aggression are reduced with maturity (Olson et al., 2011).

Relational aggression (usually verbal) destroys self-esteem and disrupts social networks. A child might tell another, "You can't be my friend" or "You are fat," hurting another's feelings. Worse, a child might spread rumors, or tell others not to play with so-and-so.

These are examples of relational aggression, which becomes more hurtful and sometimes more common as social understanding advances. One study found that about one in every five preschool children commonly uses relational aggression (Swit & McMaugh, 2012). Before high school, almost every child has experienced some exclusion from a social group.

The fourth and most ominous type is **bullying aggression**, done to dominate. Bullying aggression occurs among young children but should be stopped before kindergarten, when it becomes more destructive. Not only does it destroy the self-esteem of victims, it also impairs the later development of the bullies, who learn behaviors that harm them lifelong.

A 3-year-old bully needs to learn the effects of their actions; a 10-year-old bully may be feared and admired; a 50-year-old bully may be hated and lonely. (An in-depth discussion of bullying appears in Chapter 13.)

Most types of aggression, including bullying, become less common from ages 2 to 6, as the brain matures, emotional regulation increases, and empathy builds.

instrumental aggression Hurtful behavior that is intended to get something that another person has.

reactive aggression An impulsive retaliation for another person's intentional or accidental hurtful action.

relational aggression Nonphysical acts, such as insults or social rejection, aimed at harming the social connection between the victim and other people.

bullying aggression Unprovoked, repeated physical or verbal attack, especially on victims who are unlikely to defend themselves.

Moreover, children understand the social context and thus use aggression selectively, which decreases victimization (Ostrov et al., 2014).

Each type of aggression is influenced by genes as well as by age (Lubke et al., 2018). Thus, some 3-year-olds are innately more aggressive than other children, and they are more likely to be antisocial children and adolescents. That should alert parents and teachers to help such children, particularly, to develop empathy and emotional regulation.

One study found that close teacher–student relationships in preschool decrease aggression and victimization in elementary school. The probable reason—children want to please the teachers, who guide them toward prosocial, not antisocial, behavior (Runions & Shaw, 2013).

WHAT HAVE YOU LEARNED?

1. What are the four main styles of parenting?

2. What are the consequences of each style of parenting?

3. Why is discipline part of being a parent?

4. What are the arguments for and against corporal punishment?

5. When is a time-out effective and when is it not?

6. What are the differences between the psychoanalytic and behaviorist theories of gender development?

7. What are the differences between the cognitive and evolutionary theories of sex-role development?

8. How might children develop empathy and antipathy as they play with one another?

9. Are prosocial and antisocial behaviors inborn or learned?

10. What are the similarities and differences of the four kinds of aggression?

SUMMARY

Emotional Development

1. Emotional regulation is crucial during early childhood. It occurs in Erikson's third developmental stage, initiative versus guilt. Children normally feel pride when they demonstrate initiative, but sometimes they feel guilt or even shame at an unsatisfactory outcome.

2. Intrinsic motivation is apparent when a child concentrates on a drawing or a conversation with an imaginary friend. It may endure when extrinsic motivation stops. Extrinsic motivation may discourage intrinsic motivation, or it may result in a habit that continues.

Play

3. All young children enjoy playing—preferably with other children of the same sex, who teach them lessons in social interaction that their parents do not.

4. Play with other children gradually changes as children mature. Peer experiences, videos, and television affect children's play as they progress from being onlookers to being cooperators.

5. Active play takes many forms, with rough-and-tumble play fostering social skills and sociodramatic play developing emotional

regulation. Boys tend to engage in the former, girls in the later, although experts disagree on whether this is nature or nurture.

Challenges for Caregivers

6. Three classic styles of parenting are authoritarian, permissive, and authoritative. Generally, children are happier and more successful when their parents express warmth and set guidelines.

7. A fourth style of parenting, neglectful/uninvolved, is always harmful. The particulars of parenting reflect the culture as well as the temperament of the child.

8. Parental punishment can have long-term consequences, with both corporal punishment and psychological control teaching lessons that few parents want their children to learn.

9. Even 2-year-olds correctly use sex-specific labels. Young children become aware of gender differences in clothes, toys, playmates, and future careers, and typically become quite strict about male–female distinction. Transgender children show strong gender preferences in early childhood.

10. Every major theory interprets children's awareness of gender differences in a particular way. Freud's emphasized attraction to the

opposite-sex parent; behaviorists stress reinforcement; cognitive theory focuses on gender schemas; sociocultural theory on social norms; and evolutionary theory on the need for species survival.

11. Young children's sense of self and social awareness become the foundation for morality, influenced by both nature and nurture. The desire to be part of a group sometimes conflicts with the desire to do the right thing.

12. Prosocial emotions lead to caring for others; antisocial behavior includes instrumental, reactive, relational, and bullying aggression.

13. Early childhood is an ideal time to teach children how to control their aggression, as well as to learn other aspects of how to navigate the social world in which they find themselves.

KEY TERMS

emotional regulation (p. 257)
initiative versus guilt (p. 257)
self-concept (p. 257)
intrinsic motivation (p. 258)
extrinsic motivation (p. 259)
rough-and-tumble play (p. 263)
sociodramatic play (p. 264)
authoritarian parenting (p. 265)

permissive parenting (p. 265)
authoritative parenting (p. 265)
neglectful/uninvolved parenting (p. 266)
corporal punishment (p. 267)
psychological control (p. 268)
time-out (p. 270)
induction (p. 270)

sex differences (p. 270)
gender differences (p. 270)
transgender (p. 271)
gender binary (p. 272)
phallic stage (p. 272)
Oedipus complex (p. 272)
superego (p. 273)
gender schema (p. 274)

empathy (p. 277)
prosocial behavior (p. 277)
antipathy (p. 278)
antisocial behavior (p. 278)
instrumental aggression (p. 279)
reactive aggression (p. 279)
relational aggression (p. 279)
bullying aggression (p. 279)

APPLICATIONS

1. Children's television programming is rife with stereotypes about ethnicity, gender, and morality. Watch an hour of children's TV, especially on a Saturday morning or through a streaming service such as Netflix, and describe the content of both the programs and the commercials. Draw conclusions about stereotyping, citing specific evidence not generalities.

2. Gender indicators often go unnoticed. Go to a public place (park, restaurant, busy street) and spend at least 10 minutes recording examples of gender differentiation, such as articles of clothing, mannerisms, interaction patterns, and activities.

Quantify what you see, such as baseball hats on eight males and two females. Or (better, but more difficult) describe four male–female conversations, indicating gender differences in length and frequency of talking, interruptions, vocabulary, and so on.

3. Ask three parents about punishment, including their preferred type, at what age, for what misdeeds, and by whom. Ask your three informants how they were punished as children and how that affected them. If your sources all agree, find a parent (or a classmate) who has a different view.

Especially For ANSWERS

Response for College Students (from p. 259): Both are important. Extrinsic motivation includes parental pressure and the need to get a good job after graduation. Intrinsic motivation includes the joy of learning, especially if you can express that learning in ways others recognize. Have you ever taken a course that was not required and was said to be difficult? That was intrinsic motivation.

Response for Parents (from p. 270): No. The worst time to spank a child is when you are angry. You might seriously hurt the child, and the child will associate anger with violence. You would do better to learn to control your anger and develop other strategies for discipline and for prevention of misbehavior.

Observation Quiz ANSWER

Answer to Observation Quiz (from p. 266): It is impossible to be certain based on one moment, but the best guess is authoritative. He seems patient and protective, providing comfort and guidance, neither forcing (authoritarian) nor letting the child do whatever he wants (permissive).

APPLICATION TO DEVELOPING LIVES PARENTING SIMULATION MIDDLE CHILDHOOD

As you progress through the Middle Childhood simulation module, how you answer the following questions will impact the biosocial, cognitive, and psychosocial development of your child.

	Biosocial	Cognitive	Psychosocial
	• How will you adjust your child's diet and activity level in middle childhood? • Will you follow the recommended immunization schedule? • Will you regulate your child's screen time?	• Which of Piaget's stages of cognitive development is your child in? • How will your child score on an intelligence test? • Will you put your child in tutoring if needed? • Will you help with your child's homework?	• Will you eat meals as a family around the table or have a different routine? • What kind of elementary school will you choose for your child? • What stage of moral development is your child in? • Will your child be popular?

Parents themselves may be overly concerned about their child's athletic success. One example from that Norwegian study:

> One father . . . had invested a lot in his daughter's talent in handball. He used to be the coach and he had been there for all her practices and games. The daughter had always been "promising" but had never really blossomed. When the development had been good, she had become injured. At the time of the interview the father was feeling lost—not able to stop his intensive support but at the same time questioning if the daughter perhaps should give up handball because of the injuries. The practice of deep involvement thus balances a very thin line between intrusion or pressure and support.
>
> [Stefansen et al., 2018, p. 169]

Most injuries of 6- to 16-year-olds involve sports. Is this because the children take risks or because the adults are not protective? Is it possible that this girl was injured in handball because she sought an acceptable way to escape her father's pressure?

A detailed study in Florida of serious athletic injuries, reported by schools, found that 46 percent of the injuries of 5- to 11-year-olds were concussions (Liller et al., 2019). This does not bode well for later brain health: Concussions before age 25 are unlikely to reveal damage at the time but are one cause of *neurocognitive disorders* (formerly called *dementia*) in late adulthood (Silver et al., 2019).

 DATA CONNECTIONS: Death at an Early Age? Almost Never! shows how middle childhood is typically the healthiest period of the entire life span. 🗞 **LaunchPad.**

The Need for Movement

It is tempting to criticize parents who are too enmeshed in their children's competitive play, or to blame schools that do not protect children's heads from concussions. However, physical activity is crucial for health and learning, and many developmentalists are more troubled when indoor activities crowd out active play. Parents used to tell their children "go out and play"; now they say, "don't leave the house."

To compensate, many parents enroll their children in after-school sports that vary by culture—tennis, karate, cricket, yoga, rugby, football, baseball, or soccer. However, the children who most need to connect their bodies and their minds—those from low-SES families or who have physical disabilities—are least likely to join Little League and the like, even when enrollment is free. The reasons are many, the consequences sad (Dearing et al., 2009).

Physical Exercise in Japan

The idea that exercise improves the brain is assumed in some cultures, while other cultures assume strong muscles indicate weak minds. Dizzy Dean, a star Major League pitcher in the 1930s, said "The Good Lord was good to me, He gave me a strong right arm, a good body, and a weak mind" (Gregory, 1993).

The opposite sentiment is evident in Japan, where people believe that physical activity promotes learning and character development (Webster & Suzuki, 2014). Many Japanese public schools have swimming pools, indoor gyms, and outdoor yards with structures for climbing, swinging, and so on. Children are allowed over an hour of recess (in several segments) over a long day, in addition to gym classes. Note that Japanese children score high on international tests of achievement.

The Japanese emphasize exercise for everyone lifelong. That may explain why they live longer, on average, than people in any other nation. Of course, longevity could result from many other factors: Correlation is not causation.

Even in Japan, however, teachers are hesitant to teach physical education to students with disabilities (Hodge et al., 2013). From what we know about the

Same Situation, Far Apart Given the contrast between the Russian children in front of their rural school *(left)* and the Japanese girls beside their urban school *(right)*, you might see the differences here. But child psychologists notice that children everywhere chase and catch, kick and throw, and as in these photos, jump rope while chanting rhymes.

brain and the body during middle childhood, all children—*especially* those who are not athletically gifted—need daily physical activity.

Motor Skills and School

Many motor skills are necessary for school achievement: Writing requires finger control; reading print requires eye control; sitting at a desk requires control of gross motor skills.

Gross Motor Skills

Some schools in the United States have cut sports, recess, and gym in order to focus on reading and math. Yet a study of all elementary schools in Illinois found that schools with low reading scores had the least time for physical activity (Kern et al., 2018).

In this example, understanding correlation provides a novel way to interpret the relationship between reading scores and recess. Remember that correlation does not indicate the direction of the connection, or whether one variable causes the other, or whether a third variable might be involved.

Some educators think that more reading instruction was needed in schools with low scores. But the correlation might occur in the opposite direction: Less physical activity might cause less learning (Kern et al., 2018). Or a third variable—perhaps less support for cluster teachers, such as coaches, counselors, and reading specialists, or the fact that schools with low scores are often low SES schools without grassy fields for recess—might underlie both restricted exercise and low reading achievement.

Unfortunately, even when school policies require exercise, schools may not follow mandates. For instance, although Alabama requires at least 30 minutes daily of physical education in primary schools, the average in one low-SES district was only 22 minutes. No school in that district had recess or after-school sports (Robinson et al., 2014). Other studies also find that as student income decreases, so does school time for physical exercise (Van Dyke et al., 2018). Children's health also declines.

Fine Motor Skills

Even at age 6, many children are frustrated if their teachers demand that they write neatly, sit still, and cut in a straight line. As you remember from Chapter 9, some adults think preschools should teach readiness for formal education; others suggest that schools should adjust to children, not vice versa.

THINK CRITICALLY: How is a person "ready" for school? Are you "ready" for your current education?

Some children seem to need to move their bodies (walking around, jiggling their feet, tapping their pencils) in order to concentrate. As you will see later in this chapter, such children may be diagnosed with *attention deficit/hyperactivity disorder* (*ADHD*). There are two opposite dangers: A typical squirmy, active child may be medicated, or an overactive child may be punished instead of treated.

Fine motor skills—like many other biological characteristics, such as bones, brains, and teeth—mature about six months earlier in girls than in boys. By contrast, boys often are ahead of girls in gross motor skills.

These gender differences may be biological, or they may result from practice: Young girls more often dress up and play with dolls (fine motor skills), while boys more often climb and kick (gross motor skills) (Saraiva et al., 2013). Either way, boys may be handicapped in traditional elementary schools. Their female classmates and teachers may criticize their immature handwriting.

Drawing and Drama

Children are imaginative and creative, developing all types of motor skills in the process. They love to express themselves, especially if their parents applaud their performances, display their artwork, and otherwise communicate approval.

The fact that their fine motor skills are immature, and thus a child's artwork lack precision, does not matter. Similarly, the emotional significance of children's theater productions—walking across the stage at the appropriate moment, making the required facial expression, gesturing in the right direction—all these require not only movement but also skill at execution and inhibition, not the impulsive actions of the younger child.

Making Music

Playing a musical instrument is a fine motor skill, in that the fingers need to move precisely. But as in all motor skills, physical movement is connected to brain function. This seems to be particularly apparent with music.

Some parents enroll their young children in music lessons, hoping they will learn to play. As a result, those children become better at listening to sounds—in speech and music alike. Neurological evidence finds that their brains reflect their new auditory abilities, a remarkable testimony to the connection between motor skills, family influences, and learning (Strait et al., 2013; Zuk et al., 2014).

Music education can also occur as part of school curriculum, with an impact on the brain. In one study, students were divided into groups: (1) music lessons, (2) visual arts, and (3) education as usual. The music and arts curricula were carefully designed for elementary schoolchildren. For example, the music curriculum included singing, clapping in rhythm, and learning to play an instrument of the student's choice.

Before and after the special curriculum, the children in all three groups were tested via several valid tasks. Some were traditional cognitive measures, and three were designed to measure the three aspects (memory, inhibition, flexibility) of executive control. [**Developmental Link:** Executive control is explained in Chapter 9.]

- For memory, children had to reproduce the configuration of dots (on a 4-by-4 matrix) they had just seen.
- For inhibition, a Go/No Go task required pushing a button to indicate whether a plane on the screen went to the left or the right, but not pushing if an X appeared after a few seconds. Those with poor inhibition pushed too soon.

Buddhism in Maine? Yes. These schoolchildren are performing a play called *Buddha Walks* on St. Patrick's Day (March 17) in 2017. There are many ways to teach children about other cultures: Drama is one of the best, as in this Lebanon, Maine, elementary school.

Shawn Patrick Ouellette/Portland Press Herald via Getty Images

- For flexibility, children were given a modified "Tower of London" task. They had to plan ahead to move colored balls, one at a time according to specific rules, from one stick to another to match a display. Several displays were presented; the children had to change tactics to match them all.

The children in the visual arts curriculum became better at drawing and specific fine motor skills, which was expected. They also became better at seeing shapes and objects and remembering what they had seen, an ability called *visual-spatial memory*. Other cognitive abilities were on par with the control group.

The effects of the music curriculum were more far-reaching. The children became better at various executive control skills, include planning ahead and inhibiting unwanted responses (Jaschke et al., 2018). The authors suggest that an exclusive focus on academic skills that exclude the arts is short-sighted. That echoes what you already read about physical exercise.

Health Problems in Middle Childhood

Some chronic health conditions, including Tourette syndrome, stuttering, and allergies, may worsen during the school years, drawing unwanted attention to the affected child. Even minor problems—wearing glasses, repeatedly coughing or blowing one's nose, or having a visible birthmark—can affect children's self-esteem. Beyond that, we now focus on two other examples of physical conditions that affect how children feel about themselves.

Childhood Obesity

childhood obesity In a child, having a BMI above the 95th percentile, according to the U.S. Centers for Disease Control's 1980 standards for children of a given age.

Childhood obesity is defined as a BMI above the 95th percentile for children of a particular age. That percentile is based on children in the United States, as measured 50 years ago. Now far more than 5 percent of children are in the obese category. For example, in 2016, 18 percent of U.S. 6- to 11-year-olds were obese (Hales et al., 2017). (See **Figure 11.2**.)

What affects children's weight? Once genes were thought to determine weight, then other biological factors (the microbiome) were suspected. Now many environmental factors are recognized as well (Albataineh et al., 2019).

Obesity rates rise if newborns are born too early, if infants are not breast-fed and are fed solid foods before 4 months, if young children have televisions in their bedrooms and drink large quantities of soda, if older children sleep too little but watch screens several hours each day, if people of any age rarely play outside.

During middle childhood, children themselves have *pester power*—the ability to get adults to do what they want (Powell et al., 2011), which often includes pestering their parents to buy calorie-dense snacks that are advertised on television or that other children eat. Parents need to say no, which is easier if they always buy and eat healthy foods, and do not bring pestering children with them as they shop for groceries.

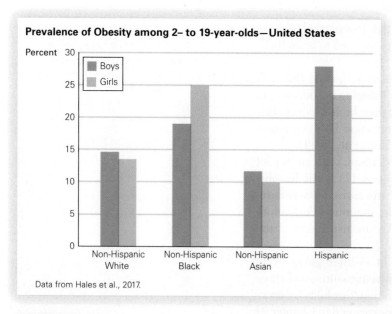

Prevalence of Obesity among 2– to 19-year-olds—United States

Data from Hales et al., 2017.

FIGURE 11.2

Ethnic or Economic? Obesity increases as income decreases. Is that obvious from this figure?

Observation Quiz Are boys more likely to be overweight than girls? (see answer, page 309) ↑

The best strategy is to be proactive, before a child is overweight. Rather than targeting parents *or* children, educating parents and their children together may improve weight and health, not just immediately but also over the long term (Yackobovitch-Gavan et al., 2018).

A dynamic-systems approach that considers individual differences, parenting practices, school lunches, fast-food restaurants, television and internet ads, and community norms is needed. Prevention must be tailored to the particular child, family, and culture (Baranowski & Taveras, 2018; Harrison et al., 2011). That makes progress slow—research has found that various interventions, in isolation, have little impact (Bleich et al., 2013).

Solutions in one culture are resisted in another. Every nation is worried about childhood obesity, but diverse practices and policies are evident (Bagchi, 2019).

For example, Mexico taxes sugar-sweetened beverages to reduce obesity, a tax that has met stiff opposition in the United States (Paarlberg et al., 2018). On the other hand, efforts to increase exercise in school have been instituted in many U.S. school districts. Given the long-term effects of childhood obesity, those who care about children must encourage every step—solutions require a concerted effort of parents, communities, and nations.

Asthma

Another childhood condition that can affect learning is **asthma**, a chronic inflammatory disorder of the airways that makes breathing difficult. Sufferers have periodic attacks, sometimes requiring a rush to the hospital emergency room, a frightening experience for children who know that asthma might kill them (although it almost never does in childhood).

If asthma continues in adulthood, which it does about half the time, it can be fatal (Banks & Andrews, 2015). But a child's most serious problem related to asthma is frequent absence from school. This impedes learning and friendships, which thrive between children who see each other every day.

In the United States, childhood asthma rates more than doubled from 1980 to 2000, increased more gradually from 2000 to 2010, and recently have decreased somewhat (probably because clean-air regulations have meant less smog) (Zahran et al., 2018).

Currently, about 1 in every 10 U.S. 5- to 11-year-olds has been diagnosed with asthma and still suffers from the condition. For more than half of them, asthma has meant missing school and having an attack in the past year, with rates somewhat higher for boys, African Americans, and children of Puerto Rican descent (Zahran et al., 2018).

Rates increase as income falls. For children whose families are under the poverty threshold, 15 percent currently have asthma, as do only 9 percent of those whose annual family income is above $100,000 (Zahran et al., 2018).

Researchers have found many causes. Some genetic alleles have been identified, as have many aspects of modern life—carpets, pollution, house pets, airtight windows, parental smoking, cockroaches, dust mites, less outdoor play. None acts in isolation. A combination of genetic sensitivity to allergies, early respiratory infections, and compromised lung functioning increases wheezing and shortness of breath (Mackenzie et al., 2014).

Some experts suggest a *hygiene hypothesis*: that "the immune system needs to tangle with microbes when we are young. . . . despite what our mothers told us, cleanliness sometimes leads to sickness" (Leslie, 2012, p. 1428).

Children may be overprotected from viruses and bacteria, especially in modern nations. In their concern about hygiene, parents prevent exposure to infections, germs, and family pets

Especially for Medical Professionals You notice that a child is overweight, but you are hesitant to say anything to the parents, who are also overweight, because you do not want to offend them. What should you do? (see response, page 309)

asthma A chronic disease of the respiratory system in which inflammation narrows the airways from the nose and mouth to the lungs, causing difficulty in breathing. Signs and symptoms include wheezing, shortness of breath, chest tightness, and coughing.

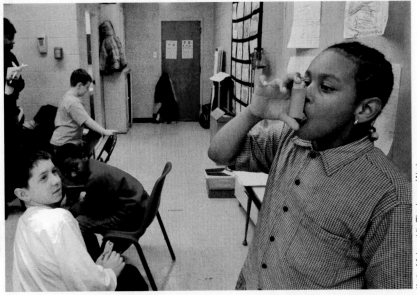

Pride and Prejudice In some city schools, asthma is so common that using an inhaler is a sign of pride, as suggested by the facial expressions of these two boys. The "prejudice" is beyond the walls of this school nurse's room, in a society that allows high rates of childhood asthma.

Kathy McLaughlin/The Image Works

Childhood Obesity Around the World

Obesity now causes more deaths worldwide than malnutrition. Reductions are possible. A multifaceted prevention effort—including parents, preschools, pediatricians, and grocery stores—has reduced obesity among U.S. 2- to 5-year-olds:

Overall, the prevalence of obesity among adolescents (20.6%) and school-aged children (18.4%) is higher than among pre-school-aged children (13.9%) (Hales et al., 2017). However, obesity rates from age 6 to 60 remain high everywhere.

Percentage of Obese 5- to 19-Year-Olds

- No data
- Less than 10%
- 10–15%
- 15–20%
- over 20%

Data from World Health Organization, September 29, 2017.

FACTORS CONTRIBUTING TO CHILDHOOD OBESITY, BY THE NUMBERS

Children's exposure to ads for unhealthy food continues to correlate with childhood obesity (e.g., Hewer, 2014), but nations differ. For instance, in stark contrast with the United States, the United Kingdom has banned television advertising of foods high in fat, sugar, and salt to children under age 16. The map above shows data from the World Health Organization; other groups' data may differ. However, the overall fact is clear: Childhood obesity is far too common.

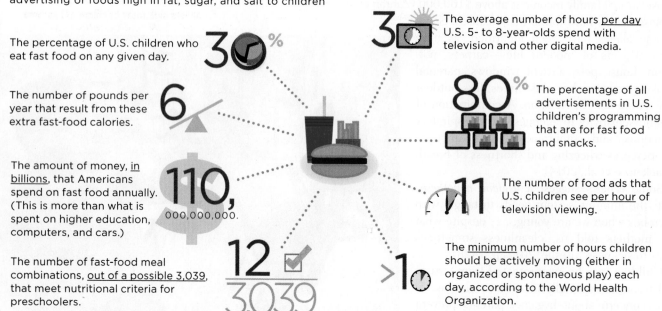

30% The percentage of U.S. children who eat fast food on any given day.

6 The number of pounds per year that result from these extra fast-food calories.

$110,000,000,000. The amount of money, <u>in billions</u>, that Americans spend on fast food annually. (This is more than what is spent on higher education, computers, and cars.)

12/3,039 The number of fast-food meal combinations, <u>out of a possible 3,039</u>, that meet nutritional criteria for preschoolers.

3 The average number of hours <u>per day</u> U.S. 5- to 8-year-olds spend with television and other digital media.

80% The percentage of all advertisements in U.S. children's programming that are for fast food and snacks.

11 The number of food ads that U.S. children see <u>per hour</u> of television viewing.

>1 The <u>minimum</u> number of hours children should be actively moving (either in organized or spontaneous play) each day, according to the World Health Organization.

Data from Council on Communications and Media, 2011, p. 202; Rideout, 2017.

that would strengthen their child's immunity. This is suggested as the reason for asthma, as for all other allergies (Liu, 2015).

This hypothesis is supported by data showing that:

- first-born children develop asthma more often than later-born ones;
- asthma and allergies are less common among farm-dwelling children; and
- children born by cesarean delivery (very sterile) have a greater incidence of asthma.

Remember the microbiome—those many bacteria within our bodies. Some microbes in the lungs affect asthma (Singanayagam et al., 2017). Accordingly, changing the microbiome—via diet, drugs, or exposure to animals—may treat asthma. However, asthma has multiple, varied causes and types; no single treatment will help everyone.

WHAT HAVE YOU LEARNED?

1. How does growth during middle childhood compare with growth earlier or later?

2. Why is middle childhood a healthy time?

3. How do motor skills and physical activity affect a child's education?

4. What are the arguments for and against giving children music lessons?

5. What are the short-term and long-term effects of childhood obesity?

6 What affects the prevalence of asthma?

Brain Development

As already mentioned many times, the most important part of the maturation process for children is in the brain, which enables the entire body to function. As you remember from Chapter 8, the child's brain develops better connections between the various parts every year. This process continues during middle childhood.

One of the distinguishing traits of humans is that the brain is not fully mature until one's mid-20s, an aspect of biological growth that enables all the cognitive and psychosocial aspects of development in middle childhood that are described in the next two chapters. Sadly, children born far too early (under 27 weeks), who escape any obvious brain malfunction (such as cerebral palsy), still have subtle signs of brain abnormalities in middle childhood that impair motor skills (Bolk et al., 2018).

Brains and Motion

Body movement improves intellectual functioning. How could this be? A review of the research suggests several possible mechanisms, including direct benefits for cerebral blood flow and neurotransmitters as well as indirect results from better moods (Singh et al., 2012). Those better moods themselves loop back to more body movement.

Many studies have found that particular regions of children's brains benefit from physical exercise (Voelcker-Rehage et al., 2018). This can be quite specific: One study found that for boys (not girls), the size of the hippocampus correlated with sports activities (Gorham et al., 2019).

While sports and aerobic exercise may directly affect brain structures, movement of every part of the body, in fine and gross motor skills, can foster learning. Children learn by doing and then express what they know by moving. This is part

LaunchPad
macmillan learning

VIDEO ACTIVITY: Brain Development: Middle Childhood depicts the changes that occur in a child's brain from age 6 to age 11.

of a concept called *embodied cognition*, the idea that thinking is connected to body movement (Pexman, 2017).

Remember that the study of music and arts education also had a control group. Their abilities improved over the two years, as you would expect, but not as much as the children in the more active learning groups. Moreover, IQ scores, which are thought to reflect brain functioning, also rose in the active groups.

In another example: The physical act of handwriting helps children learn to read (James, 2017). This has implications for students at every level: Taking a screen shot, or printing out a PowerPoint slide, is less likely to result in learning than writing out the assignment.

Paying Attention

As you remember, one specific aspect of executive control is the ability to inhibit some impulses to focus on others. Neurological advances allow children to pay special heed to the most important elements of their environment. *Selective attention*, concentrating on some stimuli while ignoring others, improves markedly at about age 7.

Selective attention is partly the result of brain maturation, but it is also greatly affected by experience, particularly active play. School-age children not only notice various stimuli (which is one form of attention), but also select appropriate responses when several possibilities conflict (Wendelken et al., 2011). In kickball, soccer, basketball, and baseball, it is crucial to attend to the ball, not to dozens of other stimuli.

For example, in baseball, young batters learn to ignore the other team's attempts to distract them; fielders start moving into position as soon as the bat connects; and pitchers adjust to the height, handedness, and past performance of the person at bat. Another physical activity that seems to foster *executive function* is karate, which requires inhibition of some reactions in order to execute others (Alesi et al., 2014).

Reaction Time

reaction time The time it takes to respond to a stimulus, either physically (with a reflexive movement such as an eyeblink) or cognitively (with a thought).

Physical play combined with brain maturation during middle childhood also improves **reaction time**, which is how long it takes to respond to a stimulus. Preschoolers are sometimes frustratingly slow in putting on their pants, eating their cereal, throwing a ball. Reaction time is reduced every year of childhood, thanks to increasing myelination. As a result, older children can react more quickly.

Skill at games is an obvious example, from scoring in a video game, to swinging at a pitch, to kicking a soccer ball toward a teammate—timing on all of these improves every year from age 6 to 11, depending partly on practice, partly on body growth, and partly on brain connections.

Measuring the Mind

In ancient times, if adults were strong and hardworking, they were respected members of the community. A child needed to be well fed, protected from injury, and tended to when sick in order to grow into a capable adult. No one was singled out if they could not think quickly, read well, or sit still. Those who had an obvious physical difference, such as being blind or deaf, received special care; no need for diagnosis.

Over the centuries, however, humans have placed more value on brain functioning. Books were printed, making reading important; money was exchanged, so calculating was needed; leaders were chosen by voters instead of kings inheriting kingdoms. This all required learning. It became evident that some people were much better at reading, at math, at analysis.

Schools were built, and some students learned more quickly than others. The slower ones struggled and quit, but adults wondered if poor students were lazy (they could be beaten or forced to stand in a corner wearing a dunce cap), or if their brains rendered learning impossible. It became important to measure intelligence.

Currently, only about 1 percent of all children are diagnosed with obvious physical differences. But in many nations, another 10 to 20 percent are thought to need special education because of something amiss in their intellectual abilities. Their needs are the focus of the last section of this chapter. But before those specifics, we need to understand how brain function is measured.

Aptitude, Achievement, and IQ

The potential to master a specific skill or to learn a certain body of knowledge is called *aptitude*. A child might have the intellectual aptitude to be a proficient reader, for instance, even though that child has never learned to read or write. Some children have the potential to become a talented athlete, artist, architect, and so on. Most aptitudes are not developed, since motivation and opportunity (or some would say, money and luck) are needed as well.

Aptitude is distinct from *achievement*, which is what is actually mastered. For children, academic achievement is measured by comparing that child with expected accomplishments at each grade. As you remember, children should be able to walk at a year, scribble with a marker at two years, catch a ball at 4 years (depending on the size and speed of the ball). These are achievements of motor skills; the same principle underlies achievement of brain skills.

Thus, a child who is at a second-grade reading level might actually be in the second grade, but it is just as likely for that child to be in the first or the third. If that second-grade reader is 9 years old, and therefore should be reading at the fourth-grade level, then something is amiss—perhaps with aptitude. That reasoning led to intelligence tests, to find out if low achievement was caused by low aptitude or something else.

People assumed that, for intelligence, one general aptitude (often referred to as **"g,"** or **general intelligence**) could be assessed by answers to a series of questions (vocabulary, memory, and so on). The number of correct answers was compared to the average for children of a particular age to compute an *IQ (intelligence quotient)*, thought to measure brain functioning.

Scores on IQ tests correlated with school achievement: Children with high IQs were able to perform above grade level, because their intellectual potential enabled them to learn quickly. On the other hand, a low IQ score indicated that a child did not have the aptitude to be a quick learner.

People with psychological disorders such as schizophrenia or obsessive-compulsive disorder tend to average lower IQ scores, because their brains do not process ideas as well as other people (Abramovitch et al., 2018). Children whose brains, for one reason or another, make them unable to do expected schoolwork need special treatment, not a dunce cap, a blow with a ruler, or other punishment.

The original IQ tests were developed by Alfred Binet in France at the beginning of the twentieth century. He sought a way to distinguish children who were unable to learn as fast as other children. Thanks to Binet, such children are less often punished or shamed. His IQ test, now called the Stanford-Binet Intelligence Scale, has been revised many times.

In the United States, David Wechsler also developed intelligence tests in the twentieth century. Because he recognized the importance of maturation, he developed different tests for young children, older children, and adults, specifically the WPPSI (Wechsler Preschool and Primary Scale of Intelligence), the WISC (Wechsler Intelligence Test for Children), and the WAIS (Wechsler Adult Intelligence Scale). Each of these tests has five indicators of verbal intelligence

g (general intelligence) The idea of *g* assumes that intelligence is one basic trait, underlying all cognitive abilities. According to this concept, people have varying levels of this general ability.

2016 Macmillan

Typical 7-Year-Old? In many ways, this boy is typical. He likes video games and school, he usually appreciates his parents, and he gets himself dressed every morning. This photo shows him using blocks to construct a design to match a picture, 1 of the 10 kinds of challenges that comprise the WISC, a widely used IQ test. His attention to the task is not unusual for children his age, but his actual performance is more like that of an older child. Because his mental age is ahead of his chronological age, his IQ is significantly above 100.

(including vocabulary, math problems, and logic) and five indicators of performance intelligence (puzzles, pictures with something missing, and so on).

Added to those standard IQ tests are dozens of other aptitude tests. Some are designed to be culture-free (such as drawing a person), some focus specifically on vocabulary, some on logic, some on spatial relationships, some on motor skills, and so on.

The ones most often used have been developed and revised over the years. Some are explicitly neurological, aimed to measure brain functioning directly, such as another Wechsler test, the WMS (Wechsler Memory Scale).

Calculating IQ

Originally, IQ tests produced a number that was literally a quotient: *Mental age* (the average age of those who answered a specific number of questions correctly) was divided by the actual age of a person taking the test. The answer from that division (the quotient) was multiplied by 100. An IQ of 100 was exactly average, because when mental age was the same as chronological age, the quotient was 1, and $1 \times 100 = 100$.

Thus, if the average 9-year-old answered, say, exactly 60 questions correctly, then everyone who got 60 questions correct—no matter what their chronological age—would have a mental age of 9. Obviously, for children whose mental age was the same as their chronological age (such as a 9-year-old who got 60 questions right), the IQ would be 100 ($9 \div 9 = 1 \times 100 = 100$), exactly average.

If a 6-year-old answered the questions as well as a typical 9-year-old, the score would be $9 \div 6 \times 100$, or 133. If a 12-year-old answered only 60 questions correctly, the IQ would be 75 ($9 \div 12 \times 100$). The current method of calculating IQ is more complex, but the basic idea is the same: *g* is calculated based on the average mental age of people of a particular chronological age. (See **Figure 11.3**.)

Plasticity and Intelligence

Probably you have already spotted the underlying problem with assumptions about mental age. Intelligence is much more plastic than people once thought. This is particularly apparent in childhood, but it also may be true in adulthood (Glenn et al., 2018; Patton et al., 2019).

Aptitude is not a fixed characteristic, present at birth, that determines how much a person can learn, or whether an individual can become an artist or architect. Young children with a low IQ sometimes become above-average or even gifted adults, like my nephew David (discussed in Chapter 1).

An added complication is the relationship between intelligence and creativity. If a child scores very high on IQ tests, that qualifies them as gifted, but how that

FIGURE 11.3

In Theory, Most People Are Average
Almost 70 percent of IQ scores fall within the "average" range. Note, however, that this is a norm-referenced test. In fact, actual IQ scores have risen in many nations; 100 is no longer exactly the midpoint. Furthermore, in practice, scores below 50 are slightly more frequent than indicated by the normal curve (shown here) because severe disability is the result not of normal distribution but of genetic and prenatal factors.

⬤⬤ **Observation Quiz** If a person's IQ is 110, what category are they in? (see answer, page 309) ➔

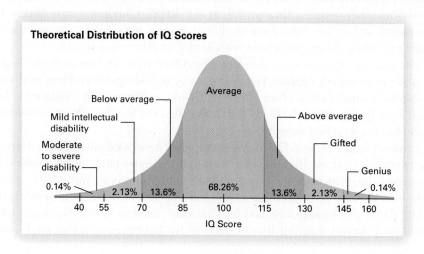

Theoretical Distribution of IQ Scores

Average
Below average
Mild intellectual disability
Above average
Moderate to severe disability
Gifted
Genius
0.14% 2.13% 13.6% 68.26% 13.6% 2.13% 0.14%
40 55 70 85 100 115 130 145 160
IQ Score

The Flynn Effect

The average IQ scores of more than 30 nations has risen substantially every decade for the past century. This phenomenon is called the **Flynn effect** (Pietschnig & Voracek, 2015; Trahan et al., 2014).

When James Flynn first suggested in the 1980s that the IQ scores had risen throughout the twentieth century, critics said he did not understand the data and was misled by biased samples. Some added that, if Flynn had realized the strong genetic inheritance of IQ, and if he recognized how genes are passed from one generation to the next, he would know that grandchildren could not be much smarter than their grandparents.

Over the past decades, those critics have grown silent, because hundreds of studies found rising scores in nation after nation. Most of those studies have been in industrialized nations of Europe, but the Flynn effect is also evident in poorer nations. For instance, a study in Kenya found that scores on the Raven's matrix (supposed to be culture-free) increased from 12.8 to 17.3 over a 14-year period (Daley et al., 2003).

Scientists no longer question *whether* the Flynn effect exists; they ask *why*. Better education? Better nutrition? Better medical care? More widespread information from newspapers, television, the internet?

The Flynn effect is more apparent for women than for men, and in southern Europe more than northern Europe, as nutrition and education improved (Weber et al., 2017). Rising average IQs are less apparent for upper-class European men. Perhaps even in the early-twentieth century, most of them had good nutrition, education, and opportunity.

Once those benefits are available to everyone, IQ will stabilize. In fact, in the most advanced nations of northern Europe, recent evidence suggests that the Flynn effect may be moving in the other direction, decreasing over time (Pietschnig & Voracek, 2015).

The fact that IQ scores vary by social conditions has changed perceptions. Most psychologists now agree that the brain is like a muscle, affected by mental exercise—which often is encouraged or discouraged by the social setting. This is proven in language and music (brains literally grow with childhood music training), and is probably true in other domains (Moreno et al., 2015; Zatorre, 2013).

Both speed and memory are crucial for *g*, and they are affected by experience, evident in the Flynn effect. Moreover, every test of intelligence is designed to measure what is considered to be intelligence within that culture. Think of someone you believe to be very smart, and then ask yourself if someone in an entirely different context (a rural village in Niger, for instance) would also think that person to be smart. When I feel proud of my intellectual ability, I ask that question. Humility needed.

---

aptitude leads to achievement depends on another trait that differs among people. Some are *divergent thinkers*, who find many solutions and even more questions for every problem, and others are *convergent thinkers*, who quickly find one, and only one, correct answer for every problem.

Flynn effect The rise in average IQ scores that has occurred over the decades in many nations.

Many Intelligences

Since scores change over time (see A View from Science), IQ tests are much less definitive than they were once thought to be, but aptitude remains a useful concept. Every human has a mix of aptitudes and abilities. Some aptitudes are nurtured, and they may become notable achievements; some aptitudes are never developed.

Aptitude is not always crucial. Some people with low aptitude for a particular achievement nonetheless accomplish their goals via extensive effort and practice. In terms of physical abilities, think of those who are relatively weak and uncoordinated who become valued members of sports teams after years of work.

An ongoing debate is whether *g* exists, and thus whether any single test can measure the complexities of the human brain. People may inherit and develop many abilities, some high and some low (e.g., Q. Zhu et al., 2010).

Two leading developmentalists (Robert Sternberg and Howard Gardner) are among those who believe that humans have **multiple intelligences**, not just one. Sternberg originally described three kinds of intelligence: *analytic, creative,* and

multiple intelligences The idea that human intelligence is composed of a varied set of abilities rather than a single, all-encompassing one.

A Gifted Child Gardner believes every person is naturally better at some of his nine intelligences, and then the social context may or may not appreciate the talent. In the twenty-first century, verbal and mathematical intelligence are usually prized far more than artistic intelligence, but Georgie Pocheptsov was drawing before he learned to speak. The reason is tragic: His father suffered and died of brain cancer when Georgie was a toddler, and his mother bought paints and canvases to help her son cope with his loss. By middle childhood (shown here), Pocheptsov was already a world-famous artist. Now a young adult, his works sell for hundreds of thousands of dollars—often donated to brain tumor research.

Especially for Teachers What are the advantages and disadvantages of using Gardner's nine intelligences to guide your classroom curriculum? (see response, page 309)

neurodiversity The idea that each person has neurological strengths and weaknesses that should be appreciated, in much the same way diverse cultures and ethnicities are welcomed. Neurodiversity seems particularly relevant for children with disorders on the autism spectrum.

practical (Sternberg, 2008, 2011). Children who are unusually creative, or very practical, are not be the best students in school, but they may flourish later on.

Gardner originally described seven intelligences: *linguistic, logical-mathematical, musical, spatial, bodily-kinesthetic* (movement), *interpersonal* (social understanding), and *intrapersonal* (self-understanding), each associated with a particular brain region (Gardner, 1983). He subsequently added an eighth (*naturalistic*: understanding nature, as in biology, zoology, or farming) and a ninth (*spiritual/existential*: thinking about life and death) (Gardner, 1999, 2006; Gardner & Moran, 2006).

Although everyone has some of all nine intelligences, Gardner believes each individual excels in particular ones. For example, someone might be gifted spatially but not linguistically (a visual artist who cannot describe her work) or might have interpersonal but not naturalistic intelligence (an astute clinical psychologist whose houseplants die).

Schools, cultures, and families dampen or expand particular intelligences. If two children are born with creative, musical aptitude, the child whose parents are musicians is more likely to develop musical intelligence than the child whose parents are tone-deaf. Gardner (2011) believes that schools often are too narrow, teaching only some aspects of intelligence and thus stunting children's learning.

Scanning the Brain

Another way to indicate aptitude is to measure the brain directly. In childhood, brain scans do not correlate with IQ scores (except for children with abnormally small brains), but they do later on (Brouwer et al., 2014). Brain scans can measure activity (reaction time, selective attention, emotional excitement) or the size of various brain areas, but they are not accurate in diagnosing cognitive disorders in childhood (Goddings & Giedd, 2014).

Neuroscientists and psychologists agree, however, on four generalities:

- *Brain development depends on experiences.* Thus, a brain scan is accurate only at the moment, not for the future.
- *Dendrites form and myelination changes throughout life.* Middle childhood is crucial, but developments before and after these years are also significant.
- *Children with disorders often have unusual brain patterns, and training may change those patterns.* Normal variation means that diagnosis based on brain patterns is imperfect.
- Each individual brain functions in a particular way, a concept called **neurodiversity**. Diverse neurological patterns are not necessarily better or worse; they are simply different, an example of the *difference is not deficit* idea explained first in Chapter 1 (Kapp et al., 2013).

WHAT HAVE YOU LEARNED?

1. What is the difference between selective attention and quick reaction time?
2. Why were intelligence tests originally developed?
3. What are the arguments for and against *g*?
4. Are aptitude and achievement distinct ideas, or are they part of the same trait?
5. How might the concept of multiple intelligences be useful in schools?

Children with Special Brains and Bodies

Developmental psychopathology links typical with atypical development, especially when the atypical results in special needs. This topic is relevant lifelong, because "[e]ach period of life, from the prenatal period through senescence, ushers in new biological and psychological challenges, strengths, and vulnerabilities" (Cicchetti, 2013b, p. 458).

Most disorders are **comorbid**, which means that more than one problem is evident in the same person. Comorbidity is now considered "the rule, rather than the exception" (Krueger & Eaton, 2015, p. 27). Turning points, opportunities, and past influences are always apparent. Many people of every age have differences that do not meet a diagnostic threshold but that nonetheless impact their lives, making other problems more likely.

At the outset, four general principles should be emphasized.

- *Abnormality is normal*, meaning that everyone has some aspects of behavior that are unusual. The opposite is also true: Everyone with a diagnosed disorder is, in many respects, like everyone else.
- *Disability changes year by year* (Clark et al., 2017). A severe childhood disorder may become insignificant, but a minor problem may become disabling. Some children with significant disabilities (e.g., blindness) become productive adults. Conversely, some conditions (e.g., conduct disorder) become more disabling.
- *Plasticity and compensation are widespread.* Many conditions, especially those that originate in the brain, seem to disappear with age and treatment (Livingston & Happe, 2017).
- *Diagnosis and treatment reflect the social context.* Each individual interacts with the surrounding settings—including family, school, community, and culture—which modify, worsen, cause, or eliminate psychopathology.

DSM-5 (the fifth edition of the *Diagnostic and Statistical Manual of Mental Disorders*, published by the American Psychiatric Association in 2013) is used as a reference here. DSM-5 is only one set of criteria—the World Health Organization has another (ICD-11); some experts use a third (RDoC) for research. Psychiatrists are discussing DSM-6, which again will redefine various disorders (Clark et al., 2017).

Because of the four principles above, it is not always obvious whether a particular child has a disorder or not. No matter what diagnostic method is used, the cutoff between what is, and what is not, a disorder requires some judgment (Clark et al., 2017).

Many Causes, Many Symptoms

To help children with special needs, it helps to know exactly what caused a problem. Perhaps a child has a chemical imbalance (a drug might correct that), or perhaps a parent or a school has turned a small vulnerability into a huge disorder (then tutoring of parent, teacher, or child is needed), or perhaps an inherited physical weakness is the problem (targeted exercise, or glasses, or a hearing aid might help). Finding the appropriate cause and treatment, however, is more complex than it appears.

One cause can have many final manifestations, a phenomenon called **multifinality** (many final forms). The opposite is also apparent: One symptom can result from several different causes, a phenomenon called **equifinality** (equal in final form). Thus, a direct line from cause to consequence cannot be drawn with certainty.

For example, in multifinality, an infant who has been flooded with stress hormones may become a child who is hypervigilant or irrationally placid, may be easily angered or quick to cry, or may not be affected. Or in equifinality, a child who

developmental psychopathology The field that uses insights into typical development to understand and remediate developmental disorders.

comorbid Refers to the presence of two or more unrelated disease conditions at the same time in the same person.

multifinality A basic principle of developmental psychopathology that holds that one cause can have many (multiple) final manifestations.

equifinality A basic principle of developmental psychopathology that holds that one symptom can have many causes.

Expressing Surprise You would be surprised as Vincent Saporita was when he was named Teacher of the Year in Huntington Beach, California. But you might not express your surprise as he did. His students are hearing impaired, so he has learned to use gestures to express emotions.

does not talk may have autism or hearing impairment, be electively mute or pathologically shy.

To illustrate the complexities in psychopathology, we home in on three conditions: *attention-deficit/hyperactivity disorder (ADHD)*, *specific learning disorders*, and *autism spectrum disorder (ASD)*. The general principles illustrated by these three apply to everyone, since we all have quirks and oddities in behavior, thoughts, and emotions.

Attention-Deficit/Hyperactivity Disorder

attention-deficit/hyperactivity disorder (ADHD) A condition characterized by a persistent pattern of inattention and/or by hyperactive or impulsive behaviors; ADHD interferes with a person's functioning or development.

Someone with **attention-deficit/hyperactivity disorder (ADHD)** is inattentive, active, and impulsive. DSM-5 says that symptoms must start before age 12 (in DSM-IV it was age 7) and must impact daily life. (DSM-IV said *impair*, not merely *impact*.)

Partly because the definition now includes older children, the rate of children diagnosed with ADHD has increased worldwide (Polanczyk et al., 2014). In 1980, about 5 percent of all U.S. 4- to 17-year-olds were diagnosed with ADHD; more recent rates are 8 percent of 4- to 11-year-olds, and 14 percent of 12- to 17-year-olds (Xu et al., 2018).

Every young child is sometimes inattentive, impulsive, and active, gradually settling down with maturation. However, those with ADHD "are so active and impulsive that they cannot sit still, are constantly fidgeting, talk when they should be listening, interrupt people all the time, can't stay on task, . . . accidentally injure themselves." All this makes them "difficult to parent or teach" (Nigg & Barkley, 2014, p. 75).

Because adults are upset by children's moods and actions, and because any physician can write a prescription to quiet a child, thousands of U.S. children may be too readily diagnosed and medicated. *But*, because many parents do not know that their child needs help, and many adults are suspicious of drugs and psychologists (Moldavsky & Sayal, 2013; Rose, 2008), thousands of children suffer needlessly. This dilemma is explored in Opposing Perspectives.

OPPOSING PERSPECTIVES

Drug Treatment for ADHD and Other Disorders

Many child psychologists believe that the public discounts the devastation and lost learning that occur when a child's serious disorder is not recognized or treated. On the other hand, many parents are suspicious of drugs and psychotherapy and avoid recommended treatment (Gordon-Hollingsworth et al., 2015).

In the United States, a non-Hispanic White child is more likely to be diagnosed with ADHD than is a non-Hispanic Black child. As for Hispanic children, their rates of diagnosed ADHD are much lower than for non-Hispanic children of any race (Xu et al., 2018). What do you make of these ethnic differences? Do they reflect genes, or prejudice, or poverty, or culture?

The question of proper diagnosis and treatment is controversial among experts as well. A leading book argues that drug companies and doctors are far too quick to push pills, making ADHD "by far, the most misdiagnosed condition in American medicine" (Schwarz, 2016, p. 2). A critical review of that book notes a failure to mention the millions of people who "have experienced life-changing, positive results" from treatment—including medication (Zametkin & Solanto, 2017, p. 9).

Most children in the United States who are diagnosed with ADHD are medicated; in England and Europe, less than half are (Polanczyk et al., 2014). In China, psychoactive medication is rarely prescribed for children: A child with ADHD symptoms is thought to need correction, not medication (Yang et al., 2013).

In many other nations, an inattentive, overactive child is more likely to be beaten than sent to a psychiatrist. Wise or cruel?

The most common drug for ADHD is Ritalin (methylphenidate), but at least 20 other psychoactive drugs are prescribed for children to treat depression, anxiety, intellectual disability, autism spectrum disorder, disruptive mood dysregulation disorder (sometimes mistaken for bipolar disorder), and many other conditions. Some parents welcome the relief that drugs may provide; others refuse to medicate their children.

Long-term benefits may include less drug abuse later on (Craig et al., 2015). It seems that children with ADHD, with or without medication, are more likely to have psychological and substance abuse problems in adulthood, but that is because of multifinality of the disorder. Those comorbid aspects should be attributed to the underlying cause, not to the medication, which seems to reduce, but not erase, the likelihood of later problems (Uchida et al., 2018).

A meta-analysis finds that medication is likely to help when it is combined with cognitive-behavioral therapy, but also that ADHD is not one simple disorder but a condition that varies from one child, one context, and one nation to another. That illustrates equifinality, and means that no particular drug or therapy, works for every child (López-Pinar et al., 2018).

Research and case study evidence supports both sides of this debate. Some research finds a correlation between medicating children and the rate of mental illness in adulthood (Moran et al., 2015). On the other hand, one expert argues that if children with ADHD are not diagnosed and treated, that increases another outcome—prison.

This outcome may be particularly likely for African American boys who are disruptive. If they are punished, not treated, for ADHD symptoms, they may enter the "school-to-prison pipeline" (Moody, 2016).

All professionals agree that finding the best drug at the right strength is difficult, in part because each child's genes and personality are unique, and in part because children's weight and metabolism change every year. Given that, why are most children who are prescribed psychoactive drugs seen only by a general practitioner who does not follow up on dose and outcome (Patel et al., 2017)? Do pharmaceutical companies mislead parents about the benefits and liabilities of ADHD drugs?

Most professionals believe that contextual interventions (instructing caregivers and schools on child management, changing the diet, increasing outdoor play, eliminating screens) should be tried before drugs. Good advice, but not easy to take if a parent or teacher is trying to manage an overactive, disruptive child every day.

Genes, culture, health care, education, religion, and stereotypes all affect ethnic and economic differences. As two experts explain, "disentangling these will be extremely valuable to improving culturally competent assessment in an increasingly diverse society" (Nigg & Barkley, 2014, p. 98). Given the emotional and practical implications of that tangle, opposing perspectives are not surprising.

In general, three problems are apparent.

- *Misdiagnosis.* If ADHD is diagnosed when another disorder is the problem, treatment might make the problem worse (Miklowitz & Cicchetti, 2010). Many psychoactive drugs alter moods, so a child with disruptive mood dysregulation disorder might be harmed by ADHD medication.
- *Drug abuse.* Although drugs may be therapeutic for true ADHD cases, some adolescents want an ADHD diagnosis in order to obtain legal amphetamines (McCabe et al., 2014). Parents or teachers may also overuse medication to quiet children.
- *Typical behavior considered pathological.* If a child's activity, impulsiveness, and curiosity are diagnosed as ADHD, that child's exuberance and self-confidence may suffer.

"Typical considered pathological" is one interpretation of data on 378,000 children in Taiwan, a Chinese nation whose rates of ADHD are increasing (M-H. Chen et al., 2016). Boys who were born in August, and hence entered kindergarten when they had just turned 5, were diagnosed with ADHD at the rate of 4.5 percent, whereas boys born in September, starting kindergarten when they were almost 6, were diagnosed at the rate of 2.8 percent.

Diagnosis for these Chinese boys typically occurred years after kindergarten, but August birthday boys were at risk throughout their school years. (See **Figure 11.4**.) The data suggest that a year of maturation would have reduced the rate of ADHD by a third.

The example in Taiwan highlights another concern. For ADHD diagnosis, one source reported that "boys outnumber girls 3-to-1 in community samples and 9-to-1 in

Observation Quiz This chart also shows medication rate. Are those August birthdays more likely to be medicated than the September birthdays? (see answer, page 309) ↓

FIGURE 11.4

One Month Is One Year In the Taiwanese school system, the cutoff for kindergarten is September 1, so some boys enter school a year later because they were born a few days later. Those who are relatively young among their classmates are less able to sit still and listen. They are almost twice as likely to be given drugs to quiet them down.

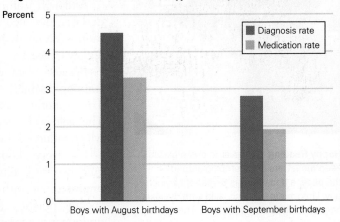

Diagnosed with Attention-Deficit/Hyperactivity Disorder (ADHD)

Data from M.-H. Chen et al., 2016.

clinical samples" (Hasson & Fine, 2012, p. 190). (Clinical samples are those children whose ADHD receives ongoing professional care.) Could typical boy activity be troubling to mothers and teachers? Could that be the reason for this male/female ratio?

Recognizing this possibility, experts recently have been diagnosing more girls with ADHD than was true a decade ago. The most recent report finds the male/female ratio close to 2:1, not 3:1. In a national survey of children's health, parents report that 14 percent of all 4- to 17-year-old boys and 6 percent of all 4- to 17-year-old girls have been diagnosed with ADHD (Xu et al., 2018).

Often a girl's main symptom is inattentiveness, not hyperactivity. Again, the question of the cut-off appears: At what point does a dreamy, distracted child merit a diagnosis? Is there sexism in this diagnosis?

Specific Learning Disorders

specific learning disorder A marked deficit in a particular area of learning that is not caused by an apparent physical disability, by an intellectual disability, or by an unusually stressful home environment.

dyslexia Unusual difficulty with reading; thought to be the result of some neurological underdevelopment.

The DSM-5 diagnosis of **specific learning disorders** now includes problems in both perception and information-processing that cause low achievement in reading, math, or writing (including spelling) (Lewandowski & Lovett, 2014). Differences in these areas undercut academic achievement, destroy self-esteem, and qualify a child for special education (according to U.S. law) or formal diagnosis (according to DSM-5).

The most commonly diagnosed learning disorder is **dyslexia**—unusual difficulty with reading. Historically, some children with dyslexia figured out themselves how to cope—as did Hans Christian Andersen and Winston Churchill. Most children with dyslexia, however, endure years of shame and low achievement, as well as punishment for not working hard enough to learn. However, we now know that the origin is in the brain, not the personality (van den Bunt et al., 2018).

Early theories hypothesized visual difficulties—for example, reversals of letters (reading *god* instead of *dog*) and mirror writing (*b* instead of *d*)—as causing dyslexia. That hypothesis was tested and found false. Instead, dyslexia more often originates with speech and hearing difficulties (Elliott & Nicolson, 2016; Swanson, 2013). Language development in the early years correlates with reading development later on.

dyscalculia Unusual difficulty with math, probably originating from a distinct part of the brain.

Another common learning disorder is **dyscalculia**, unusual difficulty with math. For example, when asked to estimate the height of a normal room, second-graders with dyscalculia might answer "200 feet," or, when shown both the 5 and 8 of hearts from a deck of playing cards and asked which is higher, schoolchildren might use their fingers to count the hearts on each card (Butterworth et al., 2011).

Although learning disorders can appear in any skill, DSM-5 recognizes only dyslexia, dyscalculia, and one more—*dysgraphia*, difficulty in writing. Few children write neatly at age 5, but practice allows most children to write easily and legibly by age 10.

Because many disorders are comorbid, it is not unusual for a child with one learning disability to also have another, or to be unusually anxious or depressed. This is another example of the interaction of domains, first described in Chapter 1. The biological aspects of brain development affect the cognitive aspects of learning, which influence the emotions.

Happy Reading Those large prism glasses keep the letters from jumping around on the page, a boon for this 8-year-old French boy. Unfortunately, each child with dyslexia needs individualized treatment: These glasses help some, but not most, children who find reading difficult.

For every child with a learning disorder, targeted help from teachers and guidance for parents is needed (Crnic et al., 2017). Remember plasticity: Skills improve, not with general practice, such as doing more homework, but with specific practice, such as sounding out letters. The current thinking is that a *multisensory* approach is needed, with hearing, vision, and motor skills all used to overcome the disability.

Autism Spectrum Disorder

Autism was once a rare disorder affecting fewer than 1 in 1,000 children, who exhibited "an extreme aloneness that, whenever possible, disregards, ignores, shuts out anything . . . from the outside" (Kanner, 1943). Children with autism were usually nonverbal and severely impaired. They were usually institutionalized, and they died young.

That is no longer true. The defining symptom is still impaired social interaction, making children with autism less adept at conversation, at social play, and at understanding emotions. Theory of mind (explained in Chapter 9) develops much later, even if general executive function skills remain (Jones et al., 2018).

However, far more children and adults have signs of autism than once was suspected. Now, in the United States, among all 3- to 17-year-olds, parents say that 1 child in every 36 has been diagnosed as having ASD (Xu et al., 2019). Parents may not be the best course for incidence data, but other sources also suggest high rates. Medical records of 8-year-olds find that 1 in 35 boys and 1 in 143 girls is on the spectrum (Maenner et al., March 27, 2020).

DSM-5 has changed the terminology from autism to **autism spectrum disorder (ASD)**. People "on the spectrum" may have a mild, moderate, or severe form. *Asperger's syndrome* (people who were highly verbal but socially inept) was once considered a separate disorder; now Asperger's is part of that spectrum.

Many scientists are searching for biological ways to detect autism early in life, perhaps with blood tests or brain scans before age 1. At the moment, behavioral signs are the best we have. ASD signs may appear in early infancy (no social smile, for example, or less gazing at faces and eyes). Two scales that measure communication attempts and play patterns are used to diagnose ASD in infants as young as 8 months (Esler et al., 2015; Kiss et al., 2017).

Early diagnosis can produce early treatment, which benefits children and their parents. Currently, some children improve by age 3; others deteriorate (Klinger et al., 2014). Indeed, some children who were diagnosed with autism in early childhood have compensated so well that they are no longer on the spectrum (Livingston & Happe, 2017).

However, diagnosis and treatment are difficult; an intervention that seems to help one child proves worthless for another. It is known, however, that biology (genes, copy number abnormalities, birth complications, prenatal injury, perhaps chemicals during fetal or infant development) is crucial; family nurture is not the cause.

Remember neurodiversity, the concept that each person's brain functions differently? That is apparent with ASD, as those with this disorder seem to have a mixture of perceptions, abilities, and deficiencies that are unlike most children. Because of their diverse abilities, adults should neither be dazzled by their talents nor saddened by their deficits. Some children with ASD have special talents in art or math. Many score above average in IQ (MMWR, March 28, 2014). Others never say a word.

The neurodiversity perspective leads to new criticisms of the many treatments for ASD. When a child is diagnosed with ASD, parental responses vary from irrational hope to deep despair, from blaming doctors and chemical additives to feeling guilty for their genes, for their behavior during pregnancy, or for the circumstances of their child's birth.

A sympathetic observer describes one child who was medicated with

Abilify, Topamax, Seroquel, Prozac, Ativan, Depakote, trazodone, Risperdal, Anafranil, Lamictal, Benadryl, melatonin, and the homeopathic remedy, Calms Forté. Every time I saw her, the meds were being adjusted again . . . [he also

VIDEO: Current Research into Autism Spectrum Disorder explores why the causes of ASD are still largely unknown.

autism spectrum disorder (ASD) A developmental disorder marked by difficulty with social communication and interaction—including difficulty seeing things from another person's point of view—and restricted, repetitive patterns of behavior, interests, or activities.

Not a Cartoon At age 3, Owen Suskind was diagnosed with autism. He stopped talking and spent hour after hour watching Disney movies. His father said his little boy "vanished," as chronicled in the Oscar-nominated documentary *Life Animated*. Now, at age 23 (shown here), Owen still loves cartoons, and he still has many symptoms of autism spectrum disorder. However, he also has learned to speak and has written a movie that reveals his understanding of himself, *The Land of the Lost Sidekicks*.

THINK CRITICALLY: Many adults are socially inept, insensitive to other people's emotions, and poor at communication—might they have been diagnosed as on the spectrum if they had been born more recently?

CHAPTER APP 11

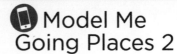

Model Me Going Places 2

IOS:
http://tinyurl.com/y6qj2j27

RELEVANT TOPIC:
Autism spectrum disorder and other childhood disorders

This app is designed for children with autism spectrum disorder and other differences, but the tutorials are useful for any child who needs help building social skills. Photos with narration demonstrate appropriate behaviors in public settings, teaching children emotion labels and characteristics, as well as cause and effect.

Kathryn Scott/The Denver Post/Getty Images

How It Should Be But Rarely Is In this well-equipped classroom in Centennial, Colorado, two teachers are attentively working with three young children, indicating that each child regularly receives individualized instruction. At this school, students with developmental disabilities learn alongside typical kids, so the earlier a child's education begins the better. Sadly, few nations have classrooms like this, and in the United States, few parents can find or afford special help for their children. Indeed, most children with special needs are not diagnosed until middle childhood.

describes] physical interventions—putting children in hyperbaric oxygen chambers, putting them in tanks with dolphins, giving them blue-green algae, or megadosing them on vitamins . . . usually neither helpful nor harmful, though they can have dangers, are certainly disorienting, and cost a lot.

[Solomon, 2012, pp. 229, 270]

Six Hypotheses

Why is autism so much more common that it once was? Two hypotheses were suggested that now are proven false. One is that unaffectionate or unavailable mothers (the so-called "refrigerator mothers") caused children to withdraw so far from social interaction as to develop autism. Before that idea was proven wrong, thousands of mothers were blamed.

The other disproven hypothesis was that infant vaccinations caused autism. Thousands of studies in many nations refute this idea (only one discredited, fraudulent study backed it), but some parents still refuse to immunize their children. The current quest of millions of health professionals is to figure out how to counter the parents' irrational fears and, thus, to prevent epidemics of measles and mumps that can harm children lifelong. The key professionals in this battle are the nurses: Because they are closest to the parents, they may be able to convince them to immunize their children to stop a "resurgence of childhood diseases" (Kubin, 2019).

Four new groups of hypotheses about increases of autism are suggested.

1. One cluster focuses on the environment, such as new chemicals in food, air, or water.
2. Another cluster considers prenatal influences: mothers who use drugs, eat foods with traces of pesticides or hormones, contract viruses such as some strains of influenza.
3. A third set of hypotheses is that ASD itself has not increased, but diagnosis has. In 2000 in the United States, education for children with ASD became publicly funded, so parents may be more willing to seek a diagnosis and doctors to provide one.
4. Finally, DSM-5 itself may be the reason. Since the definition is expanded, more children fit the category and more doctors recognize the symptoms, so children who once were overlooked are now categorized as having ASD.

Special Education

The overlap of the biosocial, cognitive, and psychosocial domains is evident to developmentalists, as is the need for parents, teachers, therapists, and researchers to work together to help each child. However, deciding whether a child should be educated differently than other children is not straightforward, nor is it closely related to individual needs. Parents, schools, and therapists often disagree.

In the United States, according to the 1975 Education of All Handicapped Children Act, all children can learn, and all must be educated in the **least restrictive environment (LRE)**. This means that children with special needs are usually educated within a regular class (a practice once called *mainstreaming*), rather than excluded from other children. Sometimes a class is an *inclusion class*, which means that children with special needs are "included" in the general classroom, with "appropriate aids and services" (ideally from a trained teacher who works with the regular teacher).

A more recent strategy is called **response to intervention (RTI)** (Al Otaiba et al., 2015; Morse, 2019; Jimerson et al., 2016). First, all children are taught

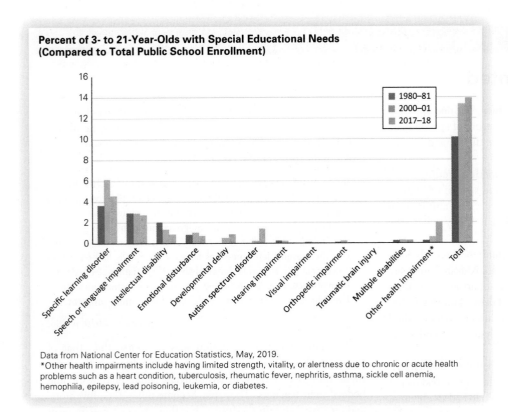

Percent of 3- to 21-Year-Olds with Special Educational Needs (Compared to Total Public School Enrollment)

Legend:
- 1980–81
- 2000–01
- 2017–18

Data from National Center for Education Statistics, May, 2019.
*Other health impairments include having limited strength, vitality, or alertness due to chronic or acute health problems such as a heart condition, tuberculosis, rheumatic fever, nephritis, asthma, sickle cell anemia, hemophilia, epilepsy, lead poisoning, leukemia, or diabetes.

FIGURE 11.5
Changing Labels Note that while fewer children have intellectual disability, more have autism. Many experts think that is a change in name, not substance.

specific skills—for instance, learning the sounds that various letters make. Then the children are tested, and those who did not master the skill receive special "intervention"—practice and individualized teaching, within the regular class.

After the first round of intervention, the children are tested again, and, if need be, intervention occurs again. According to the RTI strategy, only when children do not respond adequately to repeated, focused intervention, are they referred for special education.

At that point, the school proposes an **individual education plan (IEP)**, ideally designed for the particular child. Unfortunately, educators do not always know effective strategies, partly because research on remediation focuses on the less common problems. For example, in the United States "research funding in 2008–2009 for autistic spectrum disorder was 31 times greater than for dyslexia and 540 times greater than for dyscalculia" (Butterworth & Kovas, 2013, p. 304).

As **Figure 11.5** shows, the proportion of children designated with special needs in the United States rose from 10 percent in 1980 to 13 percent in 2016. The greatest rise was in children with learning disorders (National Center for Education Statistics, 2018).

The U.S. school system designates more children as having special needs than does any other nation: Whether or not this is a reason for national pride depends on whether one considers special education a benefit to children, to parents, or to teachers.

How many children really need special education? Some U.S. experts fear that neurodiversity, RTI, and inclusion may boomerang, limiting early and targeted help for children with special needs. If everyone is special, will that prevent help for children who desperately need it (Kauffman et al., 2017)? (See A Case to Study.)

Early Intervention

One conclusion from all of the research on special education is that diagnosis and intervention often occur too late, or not at all. The numbers of children in public schools who are designated as needing special education increase as children grow

least restrictive environment (LRE) A legal requirement that children with special needs be assigned to the most general educational context in which they can be expected to learn.

response to intervention (RTI) An educational strategy intended to help children who demonstrate below-average achievement in early grades, using special intervention.

individual education plan (IEP) A document that specifies educational goals and plans for a child with special needs.

The Gifted and Talented

Usually chapters focus on one case to study, but one apparent lesson is that children in any particular category of special need vary among themselves. Children who have the same diagnosis may have quite diverse needs.

To illustrate this point, we highlight another group of children who are sometimes thought to require special education, those who are unusually gifted and talented. Such children are so diverse that it seems best for this A Case to Study to describe several such children.

Historically, parents taught gifted children themselves or hired a special coach or tutor. For example, Mozart composed music at age 3, and Picasso created works of art at age 4. Both boys had fathers who recognized their talent. Mozart's father transcribed his earliest pieces and toured Europe with his gifted son; Picasso's father removed him from school in second grade so that he could create all day.

Although intense early education nourished their talents, neither Mozart nor Picasso had a happy adult life. Mozart had a poor understanding of math, medicine, and money. He had six children, only two of whom survived infancy, and he died in debt at age 35. Picasso regretted not learning to read or write as a child; he married at age 17 and had a total of four children by three women.

When school attendance became universal about a century ago, gifted children were allowed to skip early grades and join other children of the same mental age, not their chronological age. Many accelerated children never learned how to get along with others. As one woman remembers:

> Nine-year-old little girls are so cruel to younger girls. I was much smaller than them, of course, and would have done anything to have a friend. Although I could cope with the academic work very easily, emotionally I wasn't up to it. Maybe it was my fault and I was asking to be picked on. I was a weed at the edge of the playground.
>
> [Rachel, quoted in Freeman, 2010, p. 27]

Calling herself "a weed" suggests that she never overcame her conviction that she was less cherished than the other children. Her intellectual needs may have been met by skipping two grades, but her emotional and social needs were severely neglected.

My own father skipped three grades, graduating from high school at age 14. Because he attended a one-room school, and because he was the middle child of five, his emotional and social needs were met until he began college, where he almost failed because of his immaturity. He recovered, but some other children do not. A chilling example comes from:

> Sufiah Yusof [who] started her maths degree at Oxford [the leading University in England] in 2000, at the age of 13. She too had been dominated and taught by her father. But she ran away the day after her final exam. She was found by police but refused to go home, demanding of her father in an email: "Has it ever crossed your mind that the reason I left home was because I've finally had enough of 15 years of physical and emotional abuse?" Her father claimed she'd been abducted and brainwashed. She refuses to communicate with him. She is now a very happy, high-class, high-earning prostitute.
>
> [Freeman, 2010, p. 286]

The fate of creative children may be worse than that of intellectually gifted children. If not given an education that suits them, they joke in class, resist drudgery, ignore homework, and bedevil their teachers. They may become innovators, inventors, and creative forces in the future, but they also may use drugs or drop out of school.

They may find it hard to earn a degree or get a steady job because they are eager to try new things and feel stifled by normal life. Among the well-known creative geniuses who were less-than-stellar students were Bill Gates, Richard Branson, Steve Jobs, and hundreds of thousands of others, probably some of whom you know personally.

One such person was Charles Darwin, whose "school reports complained unendingly that he wasn't interested in studying, only shooting, riding, and beetle-collecting" (Freeman, 2010, p. 283). At the behest of his physician father, Darwin entered college to study medicine, but he dropped out.

Without a degree, he began his famous five-year trip around South America at age 22, collecting specimens and developing the theory of evolution—which disputed conventional religious dogma as only a highly creative person could do.

All these examples suggest that extraordinary children may become extraordinary adults, but that they endure problems because of their abilities. Educators are still not sure what their ideal schooling should be.

older, which is the opposite of what would occur if early intervention were successful. This is apparent in each of the disorders we have discussed.

Sometimes the current approach is called "wait to fail," when ADHD and learning disorders are not diagnosed until a child has been struggling for years without help for sensory, familial, or cultural problems. As one expert says, "We need early identification, and . . . early intervention. If you wait until third grade, kids give up" (Shaywitz, quoted in Stern, 2015, p. 1466).

A similar problem occurs with autism spectrum disorder. You read that autism appears in infancy, but children are not usually diagnosed until age 4, on average (MMWR, March 28, 2014). This is long after many parents have noticed something amiss in their child, and years after the most effective intervention can begin.

In fact, some children diagnosed with ASD before age 4 seem to overcome it later on—an outcome that seems to be related to intense social intervention in the early years (Kroncke et al., 2016). Even with early intervention, most children with ASD have some deficits in adulthood, but the fact that some children overcome social and cognitive symptoms is another argument for early intervention. Plasticity of the brain and behavior is evident.

Teaching the Gifted and Talented

Children who are unusually gifted are often thought to have special educational needs, although federal laws in the United States do not include them as a special category. Instead, each U.S. state selects and educates gifted-and-talented children in a particular way.

Should children who are unusually intelligent, talented, or creative be home-schooled, skipped, segregated, or enriched? Each of these solutions has been tried and found lacking.

Since both acceleration and intense home schooling have led to later social problems, a third education strategy has become popular, at least in the United States. (See A Case to Study.) Children who are bright, talented, and/or creative—all the same age but each with special abilities—are taught as a group in their own separate class. Ideally, such children are neither bored nor lonely; each is challenged and appreciated by classmates and teachers.

Some research supports the strategy of special education for children with exceptional music, math, or athletic gifts. Their brain structures develop in ways to support their talents (Moreno et al., 2015). Since plasticity means that children learn whatever their context teaches, perhaps some children need gifted-and-talented classes.

Such classes require unusual teachers—bright and creative, and able to individualize instruction. For example, a 7-year-old artist may need freedom, guidance, and inspiration for magnificent art but also need patient, step-by-step instruction in sounding out simple words. Similarly, 7-year-old classmates who already read at the twelfth-grade level might have immature social skills, so the teacher must find another child to befriend them and then must help both share, compromise, and take turns. The teacher must also engage the child who is advanced in reading in conversation about books that most children cannot read until college.

The argument against gifted-and-talented classes is that *every* child needs such teachers, no matter what the child's abilities or disabilities. If each school district (and sometimes each school principal) hires and assigns teachers, as occurs in the United States, then the best teachers may have the most able students, and the school districts with the most money (the most expensive homes) have the highest-paid teachers. Should it be the opposite?

High-achieving students are especially likely to have great teachers if the hidden curriculum includes *tracking*, putting children with special needs together, sorting regular classes by past achievement of the students, and allowing private or charter schools to select only certain students and expel difficult ones.

The problem is worse if the gifted students are in a separate class within the same school as the other students, or if two schools are in the same building, a "regular" school and a "special" school. Then all of the students suffer: Some feel inferior and others superior—with neither group motivated to try

Mental Age: 18. Chronological Age: 10.
Gregory Smith was a student at Randolph-Macon College in Virginia, shown here in 1999 attending physics class. At age 10, he was quite capable of understanding force and mass, and was one of the best at performing the complicated calculations required in college science. Was he ready to learn all the other things that college teaches?

Michael Williamson/The Washington Post/Getty Images

new challenges and no one learning how to work together (Herrmann et al., 2016; Van Houtte, 2016).

Every Child Special?

Mainstreaming, IEPs, and so on were developed when parents and educators saw that segregation of children with special needs led to less learning and impaired adult lives. The same may happen if gifted-and-talented children are separated from the rest.

Some nations (China, Finland, Scotland, and many others) educate all children together, assuming that all children could become high achievers if they put in the effort and are guided by effective teachers. Since every child is special, should every child have special education? But since children learn as much from peers as from adults, what does that mean for children who are notably unlike their classmates?

WHAT HAVE YOU LEARNED?

1. What are the four principles of psychopathology?

2. What is the difference between multifinality and equifinality?

3. What is the difference between ADHD and typical child behavior?

4. What are dyslexia, dyscalculia, and dysgraphia?

5. What are the similarities and differences among the symptoms of autism spectrum disorder?

6. How might the concept of neurodiversity affect treatment for special children?

7. What is the difference between mainstreaming and inclusion?

SUMMARY

A Healthy Time

1. Middle childhood is the healthiest period of the entire life span. Death and disease rates are low, and typically good health habits protect every part of the body.

2. Physical activity aids health and joy in many ways. Benefits are apparent in bodies (strength and coordination) and brains (quicker reaction time, more selective attention). However, children who most need physical activity may be least likely to have it.

3. Both gross and fine motor skills increase during middle childhood. One particular way to improve fine motor skills is through art and music.

4. Worldwide, rates of obesity and asthma are increasing. Children suffer, particularly if they miss time with friends in school and in play.

5. International research finds that family habits and national policies interact to increase or decrease the rate of obesity and asthma.

Brain Development

6. Body movement affects brain maturation through blood flow and connections between one part of the brain and another.

7. During middle childhood, children gradually become better at paying attention and at reacting quickly. These are valuable skills

in the classroom, which are developed in the team sports that also develop motor skills.

8. Intellectual aptitude is usually measured with IQ tests, with scores that can change over time. Achievement is what a person has actually learned, a product of aptitude, motivation, and opportunity.

9. Critics of IQ testing contend that intelligence is manifested in multiple ways, which makes g (general intelligence) too narrow and limited. Gardner describes nine distinct intelligences.

Children with Special Brains and Bodies

10. Developmental psychopathology uses an understanding of typical development to inform the study of unusual development. Most disorders are comorbid.

11. Four general lessons apply to psychopathology at every age: Abnormality is normal; disability changes over time; plasticity and compensation are common; the social context affects manifestation, diagnosis, and treatment. Medication should not be the first response.

12. Children with attention-deficit/hyperactivity disorder (ADHD) have potential problems in three areas: inattention, impulsiveness, and activity. DSM-5 recognizes learning disorders, specifically dyslexia (reading), dyscalculia (math), and dysgraphia (penmanship).

13. Children on the autism spectrum typically have problems with social interaction and language. ASD originates in the brain, with genetic and prenatal influences, but neurodiversity suggests caution in treatment and analysis.

14. Education of children with special needs targets needed treatment. Response to intervention is a first step, before a child is formally adjudicated as a special-needs child.

KEY TERMS

childhood obesity
(p. 290)
asthma (p. 291)
reaction time (p. 294)
g (general intelligence)
(p. 295)
Flynn effect (p. 297)

multiple intelligences (p. 297)
neurodiversity (p. 298)
developmental psychopathology
(p. 299)
comorbid (p. 299)
multifinality (p. 299)
equifinality (p. 299)

attention-deficit/hyperactivity
disorder (ADHD) (p. 300)
specific learning disorder (p. 302)
dyslexia (p. 302)
dyscalculia (p. 302)
autism spectrum disorder
(ASD) (p. 303)

least restrictive environment
(LRE) (p. 305)
response to intervention (RTI)
(p. 305)
individual education plan (IEP)
(p. 305)

APPLICATIONS

1. Compare play spaces and school design for children in different neighborhoods—ideally, urban, suburban, and rural areas. Note size, safety, and use. How might this affect children's health and learning?

2. Parents of children with special needs often consult internet sources. Pick one disorder and find 10 Web sites that describe causes and educational solutions. How valid, how accurate, and how objective is the information? What disagreements do you find? How might parents react to the information provided?

3. Should every teacher be skilled at teaching children with a wide variety of needs, or should some teachers specialize in particular kinds of learning difficulties? Ask professors in your education department. Then ask parents of children with special needs.

4. How inclusive are the elementary schools (public, charter, and private) in your community? Get data on ethnic, economic, and ability grouping. Then analyze whether this is best.

Especially For ANSWERS

Response for Medical Professionals (from p. 291): You need to speak to the parents, not accusingly (because you know that genes and culture have a major influence on body weight) but helpfully. Alert them to the potential social and health problems that their child's weight poses. Most parents are very concerned about their child's well-being and will work with you to improve the child's snacks and exercise levels.

Response for Teachers (from p. 298): The advantages are that all of the children learn more aspects of human knowledge and that many children can develop their talents. Art, music, and sports should be an integral part of education, not just a break from academics. The disadvantage is that they take time and attention away from reading and math, which might lead to less proficiency in those subjects on standardized tests and thus to criticism from parents and supervisors.

Observation Quiz ANSWERS

Answer to Observation Quiz (from p. 285): Water bottles, sun visors, and ID badges—although the latter might not be considered a healthy innovation.

Answer to Observation Quiz (from p. 290): Overall, no. But in some groups, yes. Rates of obesity among Asian American boys are almost three times higher than among Asian American girls.

Answer to Observation Quiz (from p. 296): The person is average. Anyone with a score between 85 and 115 has an average IQ.

Answer to Observation Quiz (from p. 301): Yes, not only overall but also in response to the diagnosis. When a September birthday boy is diagnosed with ADHD, he is less likely to be medicated than an August birthday boy—the opposite of what would be expected if only boys with real problems were diagnosed.

Middle Childhood: Cognitive Development

What Will You Know?

1. Does cognition improve naturally with age, or is teaching crucial to its development?
2. Why do children use slang, curse words, and bad grammar?
3. What type of school is best during middle childhood?

Middle childhood is a time of impressive intellectual capability amid stunning ignorance. Both are explained in this chapter and illustrated by Philip, a delightful 7-year-old with many cognitive skills.

Philip speaks French to his mother and English to everyone else; he can read fluently and calculate Pokémon trades; he does his schoolwork conscientiously. Everyone likes him, in part because he excels at pragmatics: He knows when to use "bathroom words" that make his peers laugh and when to use polite phrases that adults appreciate. Thus, his cognition is impressive. But not always.

Last year, Phillip's mother, Dora, needed open-heart surgery. She and her husband, Craig, told Philip that his mother would be in the hospital for a few days. They explained what this would mean for him, such as that his father would pick him up from school and several others would help with cooking and cleaning. Craig did the explaining; Dora did not want to show her fear.

Philip mirrored his parents' attitude: factual, not emotional. He had a few questions, mostly about exactly what the surgeon would do. A day later he told his parents that when he told his classmates that his mother was having heart surgery, one of them asked, "Is she going to die?" Philip reported this to show his parents how foolish his friend was; he seemed unaware of the insensitivity. His parents, wisely, exchanged wide-eyed glances but simply nodded.

Later, Craig asked Dora, "What is wrong with him? Does he have no heart?"

As you will read, Piaget described concrete operational thought, evident in 6- to 11-year-olds, and Vygotsky explained the power of the social context. Philip's thinking is typical of his age. During these years, if they are told that their parents are divorcing, children might ask, "Where will I live?" instead of expressing sympathy, surprise, or anger. Aspects of cognition that adults take for granted—empathy, emotional sensitivity, hope, and fears for the future—develop gradually.

Dora's operation went well; no repeat surgery is expected. Someday Philip might blame his 7-year-old self for his nonchalance; Craig and Dora can reassure him that he reacted as a child normally does. Later, when they are teenagers, Philip and his classmates will have more than enough "heart."

Thinking

The human mind from age 6 to 11 is ready to explore and learn from every experience, as children move from the shelter of their family to the wider world of school and community. New intellectual capacities are evident, as described somewhat differently by Piaget, Vygotsky, and information-processing theory.

Piaget on Middle Childhood

Piaget called this stage of development **concrete operational thought**, characterized by new logical abilities. *Operational* comes from the Latin verb *operare*, meaning "to work; to produce."

By calling this period *operational*, Piaget emphasized productive thinking. By using the word *concrete*, Piaget stressed that the logic of children is focused on specific experiences and observations, not on abstractions. Piaget's theory is a stage theory, expressing the discontinuity first explained in Chapter 1. Preoperational thought ends at about age 6; formal operational thought begins at about age 13; concrete operational thought is the stage between those two, from about age 7 to 12.

Think for a moment about what this means about how children learn. In middle childhood, thinking is *concrete* operational, grounded in actual experience (like the solid concrete of a cement sidewalk). That means it depends on what is visible, tangible, and real, not abstract and theoretical (as at the next stage, formal operational thought). Concrete thinking relies on personal experience. Games, excursions, classroom interactions, family circumstances—all of this is grist for concrete thought.

You read about one concrete example in Chapter 9, conservation. In middle childhood, children know that pouring a liquid into another container does not change the amount of liquid. There are other examples of concrete operational thought that Piaget described.

A Hierarchy of Categories

One logical operation is **classification**, the organization of things into groups (or *categories* or *classes*) according to some characteristic that they share. For example, *family* includes parents, siblings, and cousins. Other common classes are animals, toys, transportation, and food. Each class includes some elements and excludes others; each is part of a hierarchy.

Food, for instance, is an overarching category, with the next-lower level of the hierarchy being meat, grains, fruits, and so on. Most subclasses can be further divided: Meat includes poultry, beef, and pork, each of which can be divided again.

Adults grasp that items at the bottom of a classification hierarchy belong to every higher level: Bacon is always pork, meat, and food. They also know that each higher category includes many lower ones, but not vice versa (most food, meat, and pork are not bacon). This mental operation of moving up and down the hierarchy is beyond preoperational children.

Gradually, with personal experience (such as eating a variety of foods, meeting many relatives, and using several forms of transportation), concrete operational children can classify. Consider another category, transportation. Children begin to understand, for instance, that planes and motor vehicles are both transportation and that each can be subdivided and further subdivided. Motor vehicles can be buses, trucks, motorcycles, and cars. Then each can be subdivided again: Cars can be sedans, limousines, station wagons, and so on. They know that sedan is a kind of transportation, but not every transportation is a sedan.

Piaget devised many classification experiments. In one, he showed a child a bouquet of nine flowers—seven yellow daisies and two white roses. Then the child is asked, "Are there more daisies or more flowers?" Until about age 7, most children answer, "More daisies."

concrete operational thought Piaget's term for the ability to reason logically about direct experiences and perceptions.

Notebooks and Weeds "Write about it," said the teacher to the students on a field trip near San Diego, California. Will this boy describe the air, the plants, the trees—or will his canvas bag distract him? Concrete operational thought is not abstract: He probably won't write about Mother Earth, global warming, or God's kingdom.

● **Observation Quiz** Why are these four children sitting far apart from each other? (see answer, page 335) ↑

classification The logical principle that things can be organized into groups (or categories or classes) according to some characteristic that they have in common.

Toby Maudsley/The Image Bank/Getty Images

Numbers and Sequence Their lockers are numbered, not named, as was true in preschool. Are these children (from Stockholm, Sweden) also aware that their lockers were assigned according to how many inches tall each child is?

seriation The concept that things can be arranged in a logical series, such as the number sequence or the alphabet.

The youngest children offer no justification, but some 6-year-olds explain that "there are more yellow ones than white ones" or "because daisies are daisies, they aren't flowers" (Piaget et al., 2015). By age 8, most children can classify: "More flowers than daisies," they say.

Understanding Sequence

Another example of concrete logic is **seriation**, the knowledge that things can be arranged in a logical *series*. Seriation is crucial for using (not merely memorizing) the alphabet or the number sequence. By age 5, most children can count up to 100, but because they do not yet grasp seriation, most cannot correctly estimate where any particular two-digit number would be placed on a line that starts at 0 and ends at 100.

The ability to remember a sequence of events or actions develops gradually in middle childhood. For example, when children are shown four actions in sequence (for instance, dropping a button into a tube, moving two toys together, putting a stick through a hole, tapping a toy animal with a stick) and are asked to do them in order, most 5-year-olds are unable to do so. By age 8, children are better at this task (Loucks & Price, 2019).

This ability to follow a series may seem inconsequential, but there are dozens of practical applications. Gradually during middle childhood, children learn to get ready for school by performing a series of actions: use the toilet, eat breakfast, brush teeth, put on underwear, then pants and shirt, then socks, then shoes, and so on. This is very hard for 6-year-olds, who need to be reminded what to do next. It is easier with age—and with a list that can be checked off.

Learning Math

Math begins with familiarity with numbers. As you remember from Chapter 9, some children hear numbers often, such as 10 minutes to play, you are 4 years old, and push the elevator button for the third floor. However, to understand math, children need to grasp the logic behind the numbers, and that develops gradually during concrete operational thought.

Every logical concept helps. Concrete operational thinkers begin to understand that 15 is always 15 (conservation); that numbers from 20 to 29 are all in the 20s (classification); that 134 is less than 143 (seriation); and that because $5 \times 3 = 15$, it follows that $15 \div 5$ must equal 3 (reversibility). By age 11, children use mental categories and subcategories flexibly, inductively, and simultaneously, unlike at age 7.

This learning occurs best when hands-on math (using manipulatives such as blocks, coins, or beans) and verbal descriptions of concepts are both part of instruction (Bachman et al., 2015).

But all children need cognitive maturation before they can, for instance, perform the necessary steps of multidigit multiplication or grasp that four-fifths is greater than seven-tenths.

Vygotsky and Culture

Like Piaget, Vygotsky felt that educators should consider children's thought processes, not just the outcomes of those processes. He also believed that middle childhood was a time for much learning, with the specifics dependent on the family, school, and culture.

This is evident in the example just given, learning math. Remember that Piaget focused on maturation, with concrete operational thought enabling logic about math. Vygotsky emphasized the social context, noting that children often succeed

● **Especially for Teachers** How might Piaget's and Vygotsky's ideas help in teaching geography to a class of third-graders? (see response, page 334)

when they work out math problems together. He predicted what contemporary scholars find, that culture, preschool education, and family attitudes affect math anxiety as well as how quickly children apply math concepts (Scalise et al., 2019).

Vygotsky appreciated children's curiosity and creativity. For that reason, he believed that an educational system based on rote memorization rendered the child "helpless in the face of any sensible attempt to apply any of this acquired knowledge" (Vygotsky, 1994a, pp. 356–357).

Vygotsky would consider a repressive school particularly destructive of cognitive development, because "development depends heavily on the existing diverse social structures" (Lourenço, 2012, p. 284). Thus, the specifics of school organization and curriculum, as well as of family interactions and values, and of cultural mandates and practices, are crucial.

The Role of Instruction

Unlike Piaget, Vygotsky welcomed direct instruction. With sensitive mentoring, adults and peers provide the scaffold between potential and knowledge, engaging children in their own zone of proximal development. That would explain why children who begin school at age 4 or 5, not 6 or 7, tend to be ahead in academic achievement compared to those who enter later, an effect still apparent at age 15 (Sprietsma, 2010).

Vygotsky would expect children with more social interactions within their zone of proximal development (as would happen in any good kindergarten) to benefit intellectually. The impact of an extra year of early education will vary, because some nations have encouraged more interactive early education than others. Ideally, teachers and other children foster frequent guided participation, and thus cognition advances.

School is one arena for such guided participation, but Vygotsky noted many others. Play with peers, dinner with family, tapping on a screen, greeting neighbors—every experience, from birth on, teaches a child, according to Vygotsky.

On their own, children gradually become more logical during middle childhood, but Vygotsky emphasized the role of mentoring as changing the very structures of the mind. One summary credited Vygotsky (as well as several other twentieth-century developmentalists) with understanding

> "the brain's malleability . . . triggered and organized largely via socially enabled, emotionally driven opportunities for cognitive development. High-quality social interaction therefore presents a critical opportunity and responsibility for education"
>
> *[Immordino-Yang et al., 2019, p. 185]*

Logic about abstractions is not spontaneously understood until formal operational thought, but when children are taught with high-quality social interaction, they can master logical arguments (even counterfactual ones) by age 11. By the end of middle childhood, some children understand that *if* birds can fly, and *if* elephants are birds, *then* elephants can fly (Christoforides et al., 2016).

Vygotsky emphasized that the lessons a child learns vary by culture and school. Maturation alone is not enough.

He recognized, however, that children are limited by experience and maturation, as in comprehending philosophical issues of life and death. They tend to be quite a matter of fact, absorbing whatever their parents and culture teach rather than seeking the deeper meaning—as was true for Philip (in this chapter's opener on page 310).

ABC/Photofest

A Boy in Memphis Moziah Bridges (known as Mo Morris) created colorful bowties, which he first traded for rocks in elementary school. He then created his own company (Mo's Bows) at age 9, selling $300,000 worth of ties to major retailers by age 14. He is shown here with his mother, who encouraged his entrepreneurship.

Two or Twenty Pills a Day

As you have seen, adults often have opposite points of view on many developmental issues, from racial differences to breast-feeding, from immunization to cesarean births. Here, we present a different kind of opposing perspective, one directly related to the logic and experience of the child.

Researchers presented adults and two groups of children, some at the beginning and some at the end of middle childhood, with the following:

> Marty's ear is red and swollen and hurts a lot. Marty goes to the doctor with his badly infected ear. The doctor takes a look at Marty and gives him some medicine to solve the problem. He tells Marty to take two pills each day, one in the morning when he wakes up and one at night before he goes to bed. He should take those pills for 10 days—so that if Marty takes two pills a day for ten days that would be 20 pills in all.
>
> [Lockhart & Keil, 2018, p. 66]

The participants were asked what would happen if Marty did not take the pills as prescribed but instead:

Took one a day for 20 days.
Took two every morning and none at night for 10 days.
Took two every morning, two at night, for five days.
Took all 20 at once, immediately.

Would Marty get better (rated a 3), stay the same (2), or get worse (rated 1)? The overall results are shown in **Figure 12.1**. As you see, 20 pills at once was quite acceptable to many of the youngest children. Other data showed that only half of them thought this megadose would be harmful, as did 77 percent of the 8- to 11-year-olds and 93 percent of the adults.

This is dramatic evidence that logic by itself is not enough, nor is experience. Virtually all of these children, in their first five years of life, had taken pills prescribed over a period of days, but that experience did not teach them. At the beginning of middle childhood, some understanding began. However, simple logic (more is better) led to a dangerous perspective unless it was tempered by experience. From a developmental perspective, Piaget is inadequate without Vygotsky.

As you also see from these data, children grow wiser in middle childhood, but their concrete, literal thinking is evident. The 8- to 11-year-olds were less willing to deviate from the prescribed dose than the adults were. Most of the adults thought Marty would still get better if he took two pills every morning or only one every day, but many children disagreed. Balanced cognition, using both logic and experience, is not yet available to every 8- to 11-year-old.

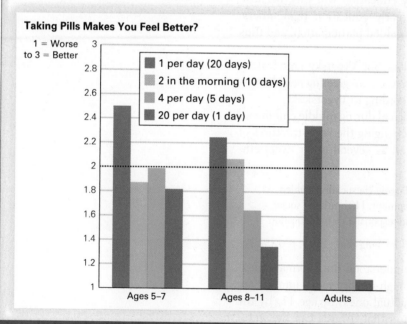

Taking Pills Makes You Feel Better?

1 = Worse to 3 = Better

- 1 per day (20 days)
- 2 in the morning (10 days)
- 4 per day (5 days)
- 20 per day (1 day)

Ages 5–7   Ages 8–11   Adults

FIGURE 12.1

Not Doctor's Orders A doctor diagnosed an infected ear and prescribed 20 pills, with instructions to take one in the morning and one in the evening for ten days. People of various ages were asked how the person would feel if they didn't follow orders. A rating of 3 means "better"; 1 means "worse." The chart shows the average responses, by age. As you see, almost every adult expected improvement if the person took two each morning, and they knew that taking 20 at once was a terrible idea. About half of the youngest children did not know that—most thought the person would still feel okay, and some thought taking 20 pills at once would be fine! Most also thought that one pill a day would be good enough. Given how children think, no wonder their accident rate is so high.

Information Processing

Contemporary educators and psychologists find both Piaget and Vygotsky insightful. International research confirms the merits of their theories (Mercer & Howe, 2012; Vince, 2018). One scholar wrote:

> [A]sked who are the two main geniuses in the field of developmental psychology, many, if not all, developmentalists would certainly point to Jean Piaget (1896–1980) and Lev Vygotsky (1896–1934).
>
> [Lourenço, 2012, p. 281]

Coordination and Capacity

Recall that emotional regulation, theory of mind, and left–right coordination emerge in early childhood. The maturing corpus callosum connects the hemispheres of the brain, enabling balance and two-handed coordination, while myelination adds speed. Maturation of the prefrontal cortex—the executive part of the brain—allows the child to plan, monitor, and evaluate.

In middle childhood, increasing maturation results in connections between the brain's various lobes and regions. Connections are crucial for the complex tasks that children must master, which require "smooth coordination of large numbers of neurons" (P. Stern, 2013, p. 577).

Certain areas of the brain, called *hubs*, are locations where massive numbers of axons meet. Hubs tend to be near the corpus callosum, and damage to them correlates with notable brain dysfunction (as in neurocognitive, autism spectrum, and other psychological disorders) (Crossley et al., 2014).

Because of hubs, brain connections forming in middle childhood are crucial. Consider learning to read. Reading is not instinctual: Our ancestors never did it, and until recent centuries, only a few scribes and scholars could interpret letters written on papyrus, carved in wood, or chiseled into stone. Consequently, the brain has no areas dedicated to reading in the way it does for talking or gesturing (Sousa, 2014).

Instead, reading uses many parts of the brain—one for sounds, another for recognizing letters, another for sequencing, another for comprehension, and more. By working together, these parts first foster listening, talking, and thinking, and then they put it all together (Lewandowski & Lovett, 2014).

Children with reading difficulties are variable as well as slow in reaction time. That makes their reading erratic (Tamm et al., 2014). Fluent reading—possible at age 8 or so—requires (1) seeing a long sequence of letters, (2) segmenting them as words, (3) differentiating words spelled alike (such as *read* and *read*, *bark* and *bark*), (4) considering context to grasp unfamiliar words, (5) recognizing oddities and ironies, and (6) understanding meaning related to the previous sentences. For fluency, reactions must be quick and automatic.

More broadly, it is now evident that the prefrontal cortex coordinates many hubs and subcortical regions of the brain "very rapidly (prior to conscious awareness)" as the brain automatically responds to stimulation (Blair, 2016, p. 3). This rapid, unconscious processing is the result of all of the neurotransmitters and neurological hubs of early and middle childhood.

The brain of a typical 6- to 11-year-old is ready to learn rapidly and well. Basic brain functions and connections are in place: Age will bring quantitative changes (increases in size and strength) but not qualitative ones (Engelhardt et al., 2019). Connections are formed that endure lifelong.

What do you remember learning during middle childhood?

Piaget described universal changes; Vygotsky noted cultural impact. However, neither grand theory describes the year-by-year details of cognition in middle childhood. Each domain of achievement may follow a particular path (Siegler, 2016). Developmentalists recognize the need for a third approach to understanding cognition.

As you read in earlier chapters, the *information-processing perspective* benefits from technology that allows much more detailed data and analysis than were possible for Piaget or Vygotsky. Like computers that process information, people accumulate large amounts of facts. They then (1) seek relevant facts (as a search engine does) for each cognitive task, (2) analyze (as software programs do), and (3) express conclusions (as the totals on a spreadsheet might). By tracing the paths and links of each of these functions, scientists better understand the learning process.

The usefulness of the information-processing approach is evident in data on children's school achievement year-by-year and even month-by-month. Absences, vacations, new schools, and even new teachers set back a child's learning because each day builds on the learning of the previous day.

Brain connections and pathways are forged from repeated experiences, allowing advances in processing. Without careful building and repetition, fragile connections between neurons break. Middle childhood is a crucial time for brain connections between one region and another, with the specifics of cultural emphasis aiding those links.

The Importance of Practice

Neither Vygotsky nor Piaget emphasized the importance of practicing intellectual skills in order to foster cognitive development. However, many more recent studies

automatization A process in which repetition of a sequence of thoughts and actions makes the sequence routine so that it no longer requires conscious thought.

of brain development note the importance of practice in reaching **automatization**, the process by which a sequence of thoughts and actions is repeated until no conscious thought is required (e.g., Y. Chen et al., 2019; Rønneberg & Torrance, 2019).

At first, almost all voluntary behaviors require careful thought. But after many repetitions, neurons fire in sequence and less thinking is needed, because the firing of one neuron sets off a chain reaction.

Consider reading again. At age 6, eyes (often aided by a guiding finger) focus intensely, painstakingly making out letters and sounding out each one. Over the years of middle childhood, this leads to the perception of syllables and then words, from emerging literacy at about age 6 to fluent reading by age 10. Eventually, the process becomes so automatic that people driving along a highway read billboards with a glance when they have no interest in doing so.

Automatization aids all academic skills. One longitudinal study of second-graders—from the beginning of the school year to the end—found that each type of academic proficiency aided each other type. Thus, learning became more automatic as automatization fostered more learning (Lai et al., 2014).

Learning a second language, reciting the multiplication tables, and writing one's name are all slow at first, but automatization makes each effortless as the school years go by. Not just academic knowledge but also habits and routines from childhood echo lifelong—and adults find them hard to break. If you have a bad habit that you can't stop, blame your brain. That's automatization.

But, also be proud of your brain's responses, evident toward the end of middle childhood. Quick reaction time and selective attention aid every social and academic skill, without taking any intellectual resources away from the next intellectual task.

For instance, being able to calculate when to utter a witty remark and when to stay quiet is something few 6-year-olds can do. By the end of middle childhood, children can (1) realize that a comment could be made and (2) decide what it could be, (3) think about the other person's possible response and, in the same split second, (4) know when something should NOT be said.

Automatic responses—saying "thank you" for a compliment, "good morning" to an acquaintance, "please" to a stranger—are learned in early childhood (often slowly, with reminders from adults) and become automatic in middle childhood.

Siegler on Math Learning

One of the leaders of the information-processing perspective is Robert Siegler, who studies day-by-day details of children's cognition in math (Siegler & Braithwaite, 2017). Siegler compares the acquisition of knowledge to waves on an ocean beach when the tide is rising. After ebb and flow, eventually a new level is reached.

Like those waves, math understanding accrues gradually, with new and better strategies for calculation tried, ignored, half-used, abandoned, and finally adopted (Siegler, 2016). The specifics are influenced by the culture, which may or may not emphasize math concepts, and the teachers, who may or may not understand the need for patience and practice. Counting itself is the product of culture: Some languages lack words for large numbers, fractions, and so on (Everett, 2017).

Using computer testing and analysis, Siegler and his colleagues have been able to pinpoint lapses in children's understanding of math concepts with precision, which was not possible a century ago for Piaget and Vygotsky. Current scholars find that many children at age 10 are quite capable of adding and subtracting whole numbers, but

VIDEO ACTIVITY: Arithmetic Strategies: The Research of Robert Siegler demonstrates how children acquire math understanding.

Following Instructions In middle childhood, children become quite capable of following adults' instructions, as these children in Tallinn, Estonia are. Their teacher told them to put out their right hand, so that Pope Francis could greet each child quickly. The teacher must not have given the most important instruction about greeting a pope: Keep your eyes open.

Abaca Press/Sipa USA via AP Images

they are woefully inadequate with fractions (Braithwaite et al., 2018). Research such as this uncovers the particular needs that children have.

Knowledge Leads to Knowledge

The more people already know, the better they can learn. Having an extensive **knowledge base**, or a broad body of knowledge in a particular subject, makes it easier to remember and understand related new information. As children gain knowledge during the school years, they become better able to judge (1) accuracy, (2) what is worth remembering, and (3) what is not important (Woolley & Ghossainy, 2013).

Past experience, current opportunity, and personal motivation all facilitate increases in the knowledge base. Motivation explains why a child may not know what facts parents or teachers prefer.

Some schoolchildren memorize words and rhythms of hit songs, calculate the worth of dozens of game cards, or recite names and statistics of basketball (or soccer, baseball, or cricket) stars. Yet they do not know whether World War I was in the nineteenth or twentieth century, or whether Pakistan is in Asia or Africa.

Concepts are learned best when linked to personal and emotional experiences. For example, children from South Asia, or those with classmates from there, learn the boundaries of Pakistan if teachers appreciate and connect with their students' heritage. On the other hand, children who are new to a nation or a neighborhood may be confused by some kinds of learning that are easy for those who have always lived there.

Control Processes

The neurological mechanisms that put memory, processing speed, and the knowledge base together are **control processes**; they regulate the analysis and flow of information within the brain. Two terms are often used to refer to cognitive control—*metacognition* (sometimes called "thinking about thinking") and *metamemory* (knowing about memory).

Control processes require the brain to organize, prioritize, and direct mental operations. For that reason, control processes are also called executive processes, and the ability to use them is called executive function (already explained in Chapter 9).

Control processes allow a person to step back from the specifics to consider more general goals and cognitive strategies, with the flexibility, memory, and inhibition that characterize executive processing. The finding, again and again, is that how children use their minds provides the foundation for learning. Maturation and experience matter.

All of these abilities develop spontaneously as the prefrontal cortex matures, but they can also be taught (de Oliveira Cardoso et al., 2018). Examples that may be familiar include spelling rules ("*i* before *e* except after *c*") and ways to remember how to turn a lightbulb ("lefty-loosey, righty-tighty").

Preschoolers ignore such rules or use them only on command; 7-year-olds begin to use them; 9-year-olds can create and master more complicated rules. Efforts to teach executive control succeed if the particular neurological maturation of the child is taken into account, which is exactly what information-processing theorists would predict (Karbach & Unger, 2014).

Theory of Mind

Closely related to control processes is the ongoing development of theory of mind (Wilson et al., 2018). As you remember, theory of mind begins in early childhood.

Carol Yepes/Getty Images

What Does She See? It depends on her knowledge base and personal experiences. Perhaps this trip to an aquarium is no more than a break from the school routine, with the teachers merely shepherding the children to keep them safe. Or, perhaps she has learned about sharks and dorsal fins, about scales and gills, about warm-blooded mammals and cold-blooded fish, so she is fascinated by the swimming creatures she watches. Or, if her personal emotions shape her perceptions, perhaps she feels sad about the fish in their watery cage or finds joy in their serenity and beauty.

knowledge base A body of knowledge in a particular area that makes it easier to master new information in that area.

control processes Mechanisms (including selective attention, metacognition, and emotional regulation) that combine memory, processing speed, and knowledge to regulate the analysis and flow of information within the information-processing system. (Also called *executive processes*.)

It continues to develop, including more nuanced beliefs and desires, in middle childhood and beyond (Wellman, 2018).

Theory of mind turns out to be pivotal for cognitive development. For example, it aids 6- to 11-year-olds in understanding the scientific process and mathematics. Both of those aspects of cognitive development are far from merely factual; they are facilitated by social awareness (Libertus et al., 2013; Peng et al., 2017; Piekny & Maehler, 2013).

WHAT HAVE YOU LEARNED?

1. What did Piaget mean when he called cognition in middle childhood *concrete operational thought*?

2. How do Vygotsky and Piaget differ in their explanations of cognitive advances in middle childhood?

3. How does information-processing theory differ from traditional theories of cognitive development?

4. According to Siegler, what is the pattern of learning math concepts?

5. How and why does the knowledge base increase in middle childhood?

6. How might control processes help a student learn?

Language

Language is crucial for cognition in middle childhood. It is the means by which children learn new concepts, and it also indicates how much children have learned. A school-age child who can explain ideas with complex sentences is a child who is thinking well. Every aspect of language—vocabulary, comprehension, communication skills, and code-switching—advances each year from age 6 to 11.

Vocabulary

Vocabulary builds during middle childhood. Because concrete operational children are logical, they can use prefixes, suffixes, compound words, phrases, and metaphors. That enables them to understand the meaning of a word they have never heard before, a decided advantage over younger children.

For example, 2-year-olds know *egg*, but 10-year-olds also know *egg salad, egg-drop soup, egghead, a good egg,* and *"last one in is a rotten egg"*—a metaphor from my childhood that a recent Google search found still relevant today. By age 10, a child who has never smelled a rotten egg nor heard that phrase can figure out the meaning.

In middle childhood, some words become pivotal for understanding the curriculum, such as *negotiate, evolve, allegation, deficit, molecule.* Consequently, vocabulary is taught in every elementary school classroom.

Adjusting Language to the Context

A crucial aspect of language that advances markedly in middle childhood is pragmatics, defined in Chapter 9. Pragmatics is evident when a child knows which words to use with teachers (never calling them a rotten egg) and informally with friends (who can be called rotten eggs or worse). For this, the social interaction foundation for cognition is apparent.

Go with the Flow This boat classroom in Bangladesh picks up students on shore and then uses solar energy to power computers linked to the internet as part of instruction. The educational context will teach skills and metaphors that peers of these students will not understand.

JONAS GRATZER/LIGHTROCKET/GETTY IMAGES

As children master pragmatics, they become more adept at making friends. Shy 6-year-olds cope far better with the social pressures of school if they use pragmatics well (Coplan & Weeks, 2009). By contrast, children with autism spectrum disorder are usually very poor at pragmatics (Klinger et al., 2014).

Mastery of pragmatics allows children to change styles of speech, or *linguistic codes*, depending on their audience. Each code includes many aspects of language—not just vocabulary but also tone, pronunciation, grammar, sentence length, idioms, and gestures.

Sometimes the switch is between *formal code* (used in academic contexts) and *informal code* (used with friends). Sometimes it is between standard (or proper) speech and dialect or vernacular (used on the street).

All children need instruction because the logic of grammar and spelling (whether *who* or *whom* is correct, or how to spell *you*) is almost impossible to deduce. Yet everyone will be judged by their ability to speak and write the formal code, so children need to learn it.

Speaking Two Languages

Code changes are obvious when children speak one language at home and another at school. Every nation includes many such children; most of the world's 6,000 languages are not school languages.

In the United States, about one school-age child in four speaks something other than English at home (see **Figure 12.2**). Many other U.S. children speak one or more of the 20 or so English dialects that reflect regional or ethnic word use, pronunciation, and grammar.

A child's comfort using the home language or regional dialect correlates with pride and motivation. At the same time, in the United States, a child's ability to use standard English correlates with school achievement.

Consequently, teachers need to respect the home language while teaching the school language (Terry et al., 2016). However, this is not as simple as it may seem, as adult ideology may conflict with what is best for the child (MacSwan, 2018). It is known, however, that emphasizing language is crucial for school achievement from early childhood on. Children can learn several codes— easily before age 5, with some help in middle childhood, and with effort after puberty.

Educators and political leaders in the United States argue about how to teach English to **English Language Learners (ELLs)**, students whose first language is not

> **THINK CRITICALLY:** Do children from some backgrounds need to become especially adept at code-switching? Does this challenge advance cognitive development?

English Language Learners (ELLs) Children in the United States whose proficiency in English is low—usually below a cutoff score on an oral or written test. Many children who speak a non-English language at home are also capable in English; they are *not* ELLs.

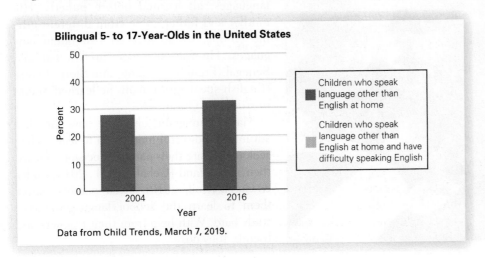

Bilingual 5- to 17-Year-Olds in the United States

Children who speak language other than English at home

Children who speak language other than English at home and have difficulty speaking English

Data from Child Trends, March 7, 2019.

FIGURE 12.2

Home and Country Do you see good news? A dramatic increase in the number of bilingual children is a benefit for the nation, but the hundreds of thousands of children who still have trouble with English suggests that more education is needed.

Standard English. About 14 percent of the children in middle childhood are designated as ELLs (McFarland et al., 2019). Note that this is significantly lower than the percent who speak another language at home. Many of them are bilingual before they reach primary school.

How should children learn the school language if they have not already done so in early childhood? One strategy is called **immersion**, in which instruction occurs entirely in the new code. The opposite strategy is to teach children in their first language initially and then to add instruction of the second as a "foreign" tongue (a strategy that is rare in the United States but common elsewhere).

Between these extremes lies **bilingual education**, with instruction in two languages, and **ESL (English as a Second Language)**, with all non-English speakers being taught English in one multilingual group, preparing them to join English-only classes. Every method for teaching a second language sometimes succeeds and sometimes fails. A major problem is that language-learning abilities change with age: The youngest children learn a new language fastest, so immersion might succeed for them.

For cognitive advances during middle childhood, the information-processing approach suggests that children should learn two languages. When bilingual individuals are asked to reason about something in their second language, they tend to be more rational and less emotional—which usually (but not always) leads to better thought (Costa et al., 2017).

Bilingual children who are fluent in both languages are able to inhibit one language while using another, advancing in cognitive control. This is obvious in language, but some say it is an asset in other aspects of executive function (Bialystok, 2018). Others dispute that (Nichols et al., 2020). The emphasis here is on proficiency: Children who are not fluent in at least one language are also impaired in cognitive skills.

Parents influence the language fluency of their children, but they are not alone. Children benefit from conversations with relatives, strangers, friends, and teachers, which is one reason going to school every day and playing with children who are proficient in other languages are recommended for language development in middle childhood.

Some bilingual children speak one language at home and learn another language elsewhere, because they spend extensive time with speakers of that second language. That is most likely in childhood; it is difficult in adulthood. One of my African students speaks five languages, all learned before puberty. His mother and father came from different ethnic groups, so he learned two home languages. He was schooled in two nations: Senegal (French-speaking) and Sierra Leone (English-speaking). Finally, he learned Arabic to study the Quran.

Many immigrant children have another advantage, powerful motivation to learn. They seek to validate their parents' decision to leave their native land (Ceballo et al., 2014; Fuller & García Coll, 2010). Their parents expect them to learn the school language and to study hard. With help from their teachers and friends, they do.

immersion A strategy in which instruction in all school subjects occurs in the second (usually the majority) language that a child is learning.

bilingual education A strategy in which school subjects are taught in both the learner's original language and the second (majority) language.

English as a Second Language (ESL) A U.S. approach to teaching English that gathers all of the non-English speakers together and provides intense instruction in English. Students' first languages are never used; the goal is to prepare students for regular classes in English.

Months or Years? ESL classes, like this one in Canada, often use pictures and gestures to foster word learning. How soon will these children be ready for instruction?

Ian Taylor/First Light/Getty Images

Poverty and Language

Every study finds that SES affects cognitive development, with poor language mastery a prominent sign and perhaps the major cause of low academic achievement in low-income children. Children from low-SES families usually have smaller vocabularies than those from higher-SES families, and they also are impaired in grammar (fewer compound sentences, dependent clauses, and conditional verbs) (Hart & Risley, 1995; Hoff, 2013). That slows down school learning in every subject.

Brain scans confirm that language proficiency is related to brain activity (Romeo et al., 2018). The development of the hippocampus (crucial for memory) is particularly affected by SES, and that may be critical for language learning (Jednoróg et al., 2012).

How can poverty affect the brain? Possibilities include inadequate prenatal care, no breakfast, lead in the bloodstream, crowded and noisy households, few books at home, teenage parents, authoritarian child rearing, inexperienced teachers, air pollution, neighborhood violence, lack of role models . . . the list could go on and on (Cree et al., 2018; Kolb & Gibb, 2015; Rowe et al., 2016). All of these conditions correlate with low SES, slower language development, and less learning. But correlation is not causation: Any one of these might reflect a third variable.

However, one factor definitely is a cause, not just a correlate: language heard early on (Neuman et al., 2018). That explains why maternal education is influential, especially if she continues her quest for learning by reading and asking questions. That also explains why children whose homes have many books accumulate, on average, three more years of education than children who live in homes with no books (Evans et al., 2010).

Of course, the books cannot merely sit on shelves. A study of 16,000 11-year-olds in England found that language proficiency was heavily influenced by how much the parents read to the child—which could begin before the first birthday and continue after the children are able to read on their own (Law et al., 2018).

Remember the plasticity of the brain. In some families, neuronal connections are strengthened, and dendrites grow to support language. This process continues in middle childhood (Perone et al., 2018).

Low income per se is not as influential as maternal talk and listening. Educated parents are more likely to take their children to museums, zoos, and libraries as well as to engage children in conversation about the interesting sights around them. Many sing to their children, not just a few simple songs but dozens of songs with varied vocabulary.

A much-cited study found that vocabulary was directly related to how much language a child heard, and found that, before age 3, children in families low in SES heard 30,000 fewer words than higher SES children (Hart & Risley, 1995). That study led to a nationwide campaign to encourage low-income parents to talk more to their children.

Many later studies also found that low-SES parents talk less to their young children. Consequently, children entered kindergarten already behind other children in their ability to learn in school. That language deficit is thought to be the crucial reason some children score lower on achievement tests, and are much more likely to drop out of high school and find only low-income employment lifelong.

Some research disputes those findings, suggesting that young children from low-SES families overhear much more language than other children. As a result,

Especially for Parents You've had an exhausting day but are setting out to buy groceries. Your 7-year-old son wants to go with you. Should you explain that you are so tired that you want to make a quick solo trip to the supermarket this time? (see response, page 334)

there is much more variation within each stratum of socioeconomic status, with some low-SES children hearing more words than some high-SES children (Sperry et al., 2019). Their eventual achievement gaps should not be blamed on their parents, but instead be blamed on the quality of their preschools and primary schools.

Some of the leading scholars in child development argue against that interpretation, contending that overheard speech is not nearly as useful to children as child-directed speech (Golinkoff et al., 2019). They argue that adults need to spend hours every day listening and talking directly to young children. This can occur no matter what the parents' SES. Then the children will develop language and do well in school.

WHAT HAVE YOU LEARNED?

1. How does language benefit from cognition in middle childhood?
2. What is the relationship between language and cognition?
3. Why would a child's linguistic code be criticized by teachers but admired by peers?
4. How does a person most readily become bilingual?
5. How does family income affect language development?

Teaching and Learning

In middle childhood, anything can be learned. Some children, by age 11, beat their elders at chess, play music that adults pay to hear, publish poems, and solve complex math problems. Others survive on the streets or fight in civil wars, skilled at firing automatic guns that are almost too big to carry.

Children worldwide learn whatever adults in their culture teach, and their brains are ready. (See the At About This Time tables for some of the universally recognized sequences of learning to read and do arithmetic.) Traditionally, children were educated at home, but now almost all of the world's 7-year-olds are in school. It is crucial that adults develop the schools and the curriculum to prepare children for adult life.

Indeed, in many developing nations, the number of students in elementary school exceeds the number of school-age children, because many older children as well as adults now seek basic education. Ghana, El Salvador, and China are among the nations with more students in primary school than the total number of children in middle childhood (UNESCO, 2014).

The Curriculum

What should children learn? Every nation seeks to teach reading, writing, and arithmetic—the classic "three R's." But beyond literacy and math, nations vary in what they expect.

For example, every nation wants children to become good citizens, but nations disagree about what good citizenship entails or how children can learn it (Cohen & Malin, 2010). Accordingly, many children simply follow their parents' example regarding everything from picking up trash to supporting a candidate for president.

AT ABOUT THIS TIME	
Math	
Age	**Norms and Expectations**
4–5 years	• Count to 20. • Understand one-to-one correspondence of objects and numbers. • Understand *more* and *less*. • Recognize and name shapes.
6 years	• Count to 100. • Understand *bigger* and *smaller*. • Add and subtract one-digit numbers.
8 years	• Add and subtract two-digit numbers. • Understand simple multiplication and division. • Understand word problems with two variables.
10 years	• Add, subtract, multiply, and divide multidigit numbers. • Understand simple fractions, percentages, area, and perimeter of shapes. • Understand word problems with three variables.
12 years	• Begin to use abstract concepts, such as formulas and algebra.

Math learning depends heavily on direct instruction and repeated practice, which means that some children advance more quickly than others. This list is only a rough guide meant to illustrate the importance of sequence.

AT ABOUT THIS TIME	
Reading	
Age	**Norms and Expectations**
4–5 years	• Understand basic book concepts. For instance, children learning English and many other languages understand that books are written from front to back, with print from left to right, and that letters make words that describe pictures. • Recognize letters—name the letters on sight. • Recognize and spell own name.
6–7 years	• Know the sounds of the consonants and vowels, including those that have two sounds (e.g., *c, g, o*). • Use sounds to figure out words. • Read simple words, such as *cat, sit, ball, jump*.
8 years	• Read simple sentences out loud, 50 words per minute, including words of two syllables. • Understand basic punctuation, consonant–vowel blends. • Comprehend what is read.
9–10 years	• Read and understand paragraphs and chapters, including advanced punctuation (e.g., the colon). • Answer comprehension questions about concepts as well as facts. • Read polysyllabic words (e.g., *vegetarian, population, multiplication*).
11–12 years	• Demonstrate rapid and fluent oral reading (more than 100 words per minute). • Vocabulary includes words that have specialized meaning in various fields. For example, in civics, *liberties, federal, parliament*, and *environment* all have special meanings. • Comprehend paragraphs about unfamiliar topics. • Sound out new words, figuring out meaning using cognates and context. • Read for pleasure.
13+ years	• Continue to build vocabulary, with greater emphasis on comprehension than on speech. Understand textbooks.

Reading is a complex mix of skills, dependent on brain maturation, education, and culture. The sequence given here is approximate; it should not be taken as a standard to measure any particular child.

Religious education also varies. The U.S. Constitution forbids overt teaching of religion in public schools (although, of course, children can pray, express beliefs, or wear religious symbols), but other nations believe religious education is part of learning. In the United States, many private schools teach religion, and parents sometimes choose them for precisely that reason.

Another variation is in how much importance is placed on each particular subject. As already mentioned in the previous chapters, the time spent on physical education, on the arts, and on other aspects of the curriculum varies from nation to nation. Most nations have national policies regarding curriculum, but within the United States variation is evident, not only state to state but also school district to school district.

The Hidden Curriculum

Differences between nations, and between schools in the United States, are stark in the **hidden curriculum**—all of the implicit values and assumptions of schools.

hidden curriculum The unofficial, unstated, or implicit patterns within a school that influence what children learn. For instance, teacher background, organization of the play space, and tracking are all part of the hidden curriculum—not formally prescribed, but instructive to the children.

Room to Learn? In the elementary school classroom in Florida *(left)*, the teacher is guiding two students who are working to discover concepts in physics—a stark contrast to the Filipino classroom *(right)* in a former storeroom. Sometimes the hidden curriculum determines the overt curriculum, as shown here.

 **Observation Quiz** How many children are in the classroom in the Philippines? (see answer, page 335) ↑

Schedules, tracking, teacher characteristics, discipline, teaching methods, sports competitions, student government, and extracurricular activities are all part of the hidden curriculum. This teaches children far beyond the formal, published curriculum that lists what is taught in each grade.

An obvious example is the physical surroundings. Some schools have spacious classrooms, wide hallways, and large, grassy playgrounds; others have cramped, poorly equipped classrooms and cement play yards. In some nations, school is held outdoors, with no chairs, desks, or books; classes are canceled when it rains. What does that tell the students?

Teacher Ethnicity

Another aspect of the hidden curriculum is who the teachers are. If their gender, ethnicity, or economic background is unlike that of their students, children may conclude that education is irrelevant to them. School organization is also significant. If the school has gifted classes, those in other classes may conclude that they are not capable of learning.

The United States is experiencing major demographic shifts. Since 2010, half of the babies born are from Hispanic, Asian, African, or Native American families, whereas more than two-thirds of adults are of European background. Because of sexual and racial discrimination, many experienced teachers are older White women.

One result is that most children never have an elementary school teacher who is a man of minority background. Of course, many older White women are excellent teachers, but schools also need more excellent male, minority teachers—not only for the minority boys. The hidden curriculum could teach that caring educators come from many backgrounds. Does it?

Teacher Expectations

Less visible, yet probably more influential, is the hidden message that comes from teacher attitudes. If a teacher expects a child to be disruptive, or unable to learn, that child is likely to confirm those expectations. Teachers who are themselves African American or Latinx have more favorable attitudes about the learning potential of students from those backgrounds (Glock & Kleen, 2019).

CHAPTER APP 12

Khan Academy

IOS:
http://tinyurl.com/y34g8v4f

ANDROID:
http://tinyurl.com/ybrxqdh3

RELEVANT TOPIC:
Schooling and academic development in middle childhood

Appropriate for students of all ages, this app contains thousands of interactive exercises, videos, and articles pertaining to arithmetic and pre-algebra, science, grammar, history, and much more. Users can take practice exercises, quizzes, and tests, receiving instant feedback and step-by-step hints. Content can be bookmarked and downloaded for access even without an internet connection.

Fortunately, teacher expectations are malleable: Learning increases and absences decrease when teachers believe all of their students are educable and they teach accordingly, encouraging every child (Sparks, 2016).

One cultural value is whether students are expected to talk or be quiet in class. In the United States, adults are encouraged to voice their opinions. Accordingly, many teachers welcome student questions, call on children who do not speak up, ask children to work in pairs so that each child talks, and grant points for participation.

As a result of their schooling, North American students learn to speak, even when they do not know the answers. Teachers say "good question" or "interesting idea" when students say something that the teacher considers wrong. Elsewhere, children are expected to be quiet.

This was dramatically apparent to me when I taught at the United Nations International School. Some of my students shouted out answers, some raised their hands, some never spoke except when I called on them. I found it hard not to favor the talkative students. When I called on one quiet student from South Asia, he immediately stood up to answer—to my surprise and the smiles of his classmates. I hope our reaction did not undercut his learning: I fear it did.

The hidden curriculum affects learning if the teachers' assumptions differ from the students' assumptions. In one study, middle-class children asked questions and requested help from their teachers more often than low-SES students did (Calarco, 2014). The researchers suggested that the low-SES students sought to avoid teacher attention, fearing it would lead to criticism. Might that have given teachers the impression that they were disinterested? Thus, the hidden curriculum might prevent students who most need encouragement from getting it.

International Testing

Every nation now wants to improve education. Longitudinal data shows that the national economy advances when school achievement rises (Hanushek & Woessmann, 2015). Better-educated children become healthier and more productive adults. That is one reason many developing nations are building more schools and colleges.

Nations seek not only education for everyone, but also more effective education. To measure that, almost 100 nations have participated in at least one massive international test of children's learning.

Achievement in science and math is tested in the **Trends in Math and Science Study (TIMSS)**. The main test of reading is the **Progress in International Reading Literacy Study (PIRLS)**. A third test is the **Programme for International Student Assessment (PISA)**, which is designed to measure the ability to apply learning to everyday issues. East Asian nations always rank high, and scores for more than a dozen nations (some in Europe, most in Asia) surpass those for the United States (see **Tables 12.1** and **12.2**).

One surprising example is that Finland's scores increased dramatically after a wholesale reform of its public education system. Reforms occurred in several waves (Sahlberg, 2011, 2015). In 1985 ability grouping was abolished, and in 1994 the curriculum began to encourage collaboration and active learning rather than competitive passive education.

Currently, in Finland, all children learn together—no tracking—and teachers are mandated to help each child. Over the past two decades, Finland has instituted strict requirements for becoming a teacher. Only the top 3 percent of Finland's high school graduates are admitted to teachers' colleges. They study for five years at the university at no charge, earning a master's degree, studying both the theory and the practice of education.

Trends in Math and Science Study (TIMSS) An international assessment of the math and science skills of fourth-graders and eighth-graders. Although the TIMSS is very useful, different countries' scores are not always comparable because sample selection, test administration, and content validity are hard to keep uniform.

Progress in International Reading Literacy Study (PIRLS) Inaugurated in 2001, a planned five-year cycle of international trend studies in the reading ability of fourth-graders.

Programme for International Student Assessment (PISA) An international test taken by 15-year-olds in 50 nations that is designed to measure problem solving and cognition in daily life.

TABLE 12.1

TIMSS Ranking and Average Scores of Math Achievement for Fourth-Graders, 2011 and 2015

	2011	2015
Singapore	606	618
Hong Kong	602	615
Korea	605	608
Chinese Taipei	591	597
Japan	585	593
N. Ireland	562	570
Russia	542	564
England	542	546
Belgium	549	546
United States	541	539
Canada (Quebec)	533	533
Finland	545	532
Netherlands	540	530
Germany	528	522
Sweden	504	519
Australia	516	517
Canada (Ontario)	518	512
Italy	508	507
New Zealand	486	491
Iran	431	431
Kuwait	342	353

Bragging or Blaming Test scores are proof for national education systems when they improve (Hong Kong, Ireland, Singapore) and as reflecting parents and the economy when they fail (Netherlands, United States). Which interpretation is better?

TABLE 12.2

PIRLS Distribution of Reading Achievement for Fourth-Graders, 2011 and 2016

	2011	2016
Hong Kong	571	569
Russian Federation	568	581
Finland	568	566
Singapore	567	576
N. Ireland	558	565
United States	556	549
Denmark	554	547
Chinese Taipei	553	559
Ireland	552	567
England	552	559
Canada	548	543
Italy	541	548
Germany	541	537
Israel	541	530
New Zealand	531	523
Australia	527	544
Poland	526	565
France	520	511
Spain	513	528
Iran	457	428

Information from Mullis et al., 2012b, 2017.

THINK CRITICALLY: Finland's success has been attributed to many factors, some mentioned here and some regarding the geography and population of the nation. What do you think is the most influential reason?

Finnish teachers are granted more autonomy within their classrooms than is typical in other nations. Since the 1990s, they have had more time and encouragement to work with colleagues (Sahlberg, 2011, 2015). They are encouraged to respond to each child's temperament as well as the child's skills. This strategy has led to achievement, particularly in math (Viljaranta et al., 2015).

As explained in Chapter 11, every educational system includes children with special educational needs. In Finland, such children are usually educated within the regular classes. Some teachers who have attained all the credentials and credits necessary for a regular teacher then gain additional credits, including classroom experience, to become a specialist for children with special needs (Takala et al., 2019).

Often teachers with such specialized training co-teach with regular teachers. However, about 3 percent of Finnish students are thought to have such pervasive special needs that they are educated in special classes (Sundqvist et al., 2019).

Problems with International Comparisons

Elaborate and extensive measures are in place to make the PIRLS, TIMSS, and PISA valid. Test items are designed to be fair and culture-free, and participating children represent the diversity (economic, ethnic, etc.) of each nation's child population. Thousands of experts work to ensure validity and reliability. Consequently, most social scientists respect the data gathered from these tests.

The tests are far from perfect, however. Creating questions that are equally valid for everyone is impossible. For example, in math, should fourth-graders be expected to understand fractions, graphs, decimals, and simple geometry? Nations introduce these concepts at different ages, and some schools stress math more than others: Should every fourth-grader be expected to divide fractions?

After such general issues are decided, items are written. The following item tested math:

> Three thousand tickets for a basketball game are numbered 1 to 3,000. People with ticket numbers ending with 112 receive a prize. Write down all the prize-winning numbers.

Only 26 percent of fourth-graders worldwide got this one right (112; 1,112; 2,112—with no additional numbers). About half of the children in East Asian nations and 36 percent of the U.S. children were correct. Those national scores are not surprising; children in Singapore, Japan, and China have been close to the top on every international test for 30 years, and the United States has been above average but not by much.

Children from North Africa did especially poorly; only 2 percent of Moroccan fourth-graders were correct. Does that suggest inferior education, or is that item biased in favor of some cultures? Is basketball, or 3,000 tickets for one game, or a random prize, rare in North Africa?

Another math item gives ingredients—4 eggs, 8 cups of flour, ½ cup of milk—and asks:

> The above ingredients are used to make a recipe for 6 people. Sam wants to make this recipe for only 3 people. Complete the table below to show what Sam needs to make the recipe for 3 people. The number of eggs he needs is shown.

The table lists 2 eggs, and the child needs to fill in amounts of flour and milk. Fourth-grade children in Ireland and England scored highest on this item (about half got it right), while those in Korea, China, and Japan scored lower (about 33 percent). The United States scored higher than East Asian nations but lower than England.

This is puzzling, since East Asians usually surpass others in math. Why not on this question? Are English and Irish children experienced with recipes that include eggs, flour, and milk, unlike Japanese children? Or are Asian children distracted by a question that assumes that a boy is cooking?

"Big deal, an A in math. That would be a D in any other country."

Eggs	2
Flour	?___
Milk	?___

Who Takes the Test?

Beyond the problem of writing items that are fair for children throughout the world is the problem of student selection. In theory, all children in school at a particular grade level take the test, but in practice the school dropout rate in some nations means that many of the lowest achieving children are not tested.

China has consistently scored far higher than the United States, but some critics say that comparison is unfair because the tested populations are not comparable (Singer & Braun, 2018). For one thing, the Chinese children who take the test are in the major cities, not rural areas. For another, some say that the American children who take the test are not motivated, since the scores are not reflected in school grades. The Chinese children, by contrast, think their personal performance reflects their nation, so they try to do well.

Related to motivation is how much a culture values high scores on achievement tests. A major controversy in the United States is whether children are overtested, with a narrow focus on high achievement. This is discussed in Chapter 15, but the underlying issue is whether or not personality traits that push toward high scores should be encouraged at home.

Gender Differences in School Performance

In addition to marked national, ethnic, and economic differences, gender differences in achievement scores are reported. The PIRLS finds fourth-grade girls ahead of boys in reading in every nation, by an average of 19 points (Mullis et al., 2017).

The 2016 female verbal advantage on the PIRLS in the United States is only 8 points, which is a difference of less than 2 percent. Several other nations are close to the U.S. norms, including France, Spain, and Hong Kong. Does this mean that those nations are more gender-equitable than the nation with the widest gender gap—Saudi Arabia, with a 65-point gap (464/399)? Maybe, maybe not.

Historically, boys were ahead of girls in math and science. However, TIMSS reported that those gender differences among fourth-graders in math have narrowed, disappeared, or reversed. In many nations, boys are still slightly ahead, with the United States showing a male advantage (7 points—less than 2 percent).

LaunchPad
macmillan learning

VIDEO ACTIVITY: Educating the Girls of the World examines the situation of girls' education around the world while stressing the importance of education for all children.

Happiness or High Grades?

Thousands of social scientists—psychologists, educators, sociologists, economists—have realized that for cognitive development from middle childhood through late adulthood, characteristics beyond IQ scores, test grades, and family SES are sometimes pivotal.

One leading proponent of this idea is Paul Tough, who wrote: "We have been focusing on the wrong skills and abilities in our children, and we have been using the wrong strategies to help nurture and teach those skills" (Tough, 2012, p. xv). Instead of focusing on test scores, Tough believes we should focus on characteristics, particularly *grit* (persistence and effort).

Many scientists agree that executive control processes with many names (grit, emotional regulation, conscientiousness, resilience, executive function, effortful control) develop over the years of middle childhood. Over the long term, these aspects of character predict achievement in high school, college, and adulthood.

Developmentalists disagree about exactly which qualities are crucial for achievement, with grit considered crucial by some but not others (Duckworth, 2016; Lam & Zhou, 2019). However, no one denies that success depends on personal traits.

This case study is actually two cases, each reflecting a culture. Does the focus on personal qualities result in blaming children instead of noting that some parents, schools, and cultures create barriers that make learning difficult? Grit could make a child feel personally at fault for low achievement, rather than help that child enjoy life.

Remember that school-age children are ready for intellectual growth (Piaget) and are responsive to mentors (Vygotsky). These universals were evident in one study that occurred in two places, 12,000 miles apart: the northeastern United States and Taiwan.

In that study, more than 200 mothers were asked to recall and then discuss with their 6- to 10-year-olds two learning-related incidents that they knew their child had experienced. In one incident, the child had a "good attitude or behavior in learning"; in the other, "not perfect" (J. Li et al., 2014).

All of the mothers were married and middle-class, and all tried to encourage their children, stressing the value of education and the importance of doing well in school. The researchers noted that the mothers differed in the attitudes they were trying to encourage in their children.

The Taiwanese mothers were far more likely to mention what the researchers called "learning virtues," such as practice, persistence, and concentration—all of which are part of grit. The American mothers were more likely to mention "positive affect," such as happiness and pride.

This distinction is evident in the following two excerpts:

First, Tim and his American mother discussed a "not perfect" incident.

Mother: I wanted to talk to you about . . . that time when you had that one math paper that . . . mostly everything was wrong and you never bring home papers like that. . . .

Tim: I just had a clumsy day.

Mother: You had a clumsy day. You sure did, but there was, when we finally figured out what it was that you were doing wrong, you were pretty happy about it . . . and then you were happy to practice it, right? . . . Why do you think that was?

Tim: I don't know, because I was frustrated, and then you sat down and went over it with me, and I figured it out right with no distraction and then I got it right.

Mother: So it made you feel good to do well?

Tim: Uh-huh.

Mother: And it's okay to get some wrong sometimes.

Tim: And I, I never got that again, didn't I?

The next excerpt occurred when Ren and his Taiwanese mother discuss a "good attitude or behavior."

Mother: Oh, why does your teacher think that you behave well?

Ren: It's that I concentrate well in class.

Mother: Is your good concentration the concentration to talk to your peer at the next desk?

Ren: I listen to teachers.

Mother: Oh, is it so only for Mr. Chang's class or is it for all classes?

Ren: Almost all classes like that. . . .

Mother: So you want to behave well because you want to get an . . . honor award. Is that so?

Ren: Yes.

Mother: Or is it also that you yourself want to behave better?

Ren: Yes. I also want to behave better myself.

[J. Li et al., 2014, p. 1218]

Both Tim and Ren are likely to be good students in their respective schools. When parents support and encourage their child's learning, the child almost always masters the basic skills required of elementary school students, and almost never does the child become crushed by life experiences. Instead, the child has sufficient strengths to overcome most challenges (Masten, 2014).

Nonetheless, do parents in Asia emphasize the hard work required for achievement, while parents in North America stress the joy of learning. Could it be, as one group of researchers contend, that U.S. children are happier but less accomplished than Asian ones (F. Ng et al., 2014)? Which is more important: high self-esteem or high grades?

However, in other nations, girls are ahead, sometimes significantly, such as 10 points in Indonesia and 20 points in Jordan. Why? Is there an anti-male bias in the schools or culture?

In middle childhood, girls in every nation have higher report card grades, including in math and science. Is that biological (girls are better able to sit still, to manipulate a pencil)? Or cultural (girls have been taught to do as they are told)? Or does the hidden curriculum favor girls (most of their elementary school teachers were women)?

The popularity of various explanations has shifted. Analysts once attributed girls' higher grades in school to their faster physical maturation. Now explanations are more often sociocultural—that parents and teachers expect girls to be good students and that schools are organized to favor female strengths. The same switch in explanations, from biology to culture, appears for male advantages in science. Is that change itself cultural?

Future Engineers After-school clubs now encourage boys to learn cooking and girls to play chess, and both sexes are active in every sport. The most recent push is for STEM (Science, Technology, Engineering, and Math) education—as in this after-school robotics club.

> 📊 **DATA CONNECTIONS: Motivation or Achievement?: A Look at Various Nations' PISA Scores** demonstrates how U.S. students compare with students in other nations.

Schooling in the United States

Some international tests indicate improvements in U.S. children's academic performance over the past decades. However, the United States has the largest disparities between income and ethnic groups. Some blame the disparity on immigration, but other nations (e.g., Canada) have more ethnic groups and immigrants than the United States, yet the achievement gap between groups in other nations is not as large.

National Standards

For decades, the United States government has sponsored the **National Assessment of Educational Progress (NAEP)**, which is a group of tests designed to measure achievement in reading, mathematics, and other subjects. The NAEP finds fewer children proficient than do state tests. For example, New York's tests reported 62 percent proficient in math, but the NAEP found only 32 percent; 51 percent were proficient in reading on New York's state tests, but only 35 percent according to NAEP (Martin, 2014).

The NAEP also finds that Hispanic American and African American fourth-graders are about 11 percent lower than their European American peers in reading and 9 percent lower in math (McFarland et al., 2019). An even wider achievement gap is evident between high- and low-SES schools.

For some statistics—high school graduation, for instance—Asian American children achieve at higher rates than European Americans. However, this contributes to the "model minority" stereotype, which obscures disadvantages for many children of Asian heritage. For instance, Asian children may suffer from parental pressure and peer jealousy (Cherng & Liu, 2017).

National Assessment of Educational Progress (NAEP) An ongoing and nationally representative measure of U.S. children's achievement in reading, mathematics, and other subjects over time; nicknamed "the Nation's Report Card."

🔵 **Especially for School Administrators** Children who wear uniforms in school tend to score higher on reading tests. Why? (see response, page 335)

FIGURE 12.3

Where'd You Go to School? Note that although home schooling is still the least-chosen option, the number of home-schooled children may be increasing. Not shown is the percentage of children attending the nearest public school, which is decreasing slightly because of charter schools and magnet schools. More detailed data indicate that the average home-schooled child is a 7-year-old European American girl living in a rural area of the South with an employed father and a stay-at-home mother.

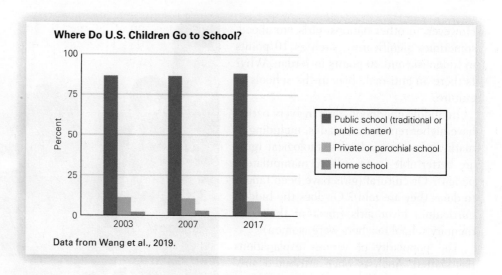

Where Do U.S. Children Go to School?

■ Public school (traditional or public charter)
■ Private or parochial school
■ Home school

Data from Wang et al., 2019.

The reason for disparities within the United States seems more economic than ethnic, because African Americans in some of the wealthier states (Massachusetts) score higher than European Americans in the poorer states (Mississippi). Indeed, on international tests, students in Massachusetts are close to the highest-scoring nations, those in East Asia (Singer & Braun, 2018).

Many suggest that the disparity in local funding for schools is at the root of the problem: High-SES children of all groups attend well-funded schools. That raises the first of several issues within U.S. education, 10 of which are mentioned now.

Ten Questions

1. Should public schools be well-supported by public funds, or should smaller class sizes, special curricula, and expensive facilities (e.g., a stage, a pool, a garden) be available only in *private schools*, paid via tuition from wealthy parents? All told, about 10 percent of students in the United States attend private schools (see **Figure 12.3**). Other nations have higher or lower rates.

2. Should parents be given *vouchers* to pay for some tuition at a private school? Each state regulates vouchers differently, but a detailed look at vouchers in Wisconsin found that most parents who used vouchers were inclined to send their children to nonpublic schools in any case, partly for religious and safety reasons (Fleming et al., 2015). Thus, vouchers subsidize schools that differ from public schools, which may allow parents to choose a school that does not follow public school policy or curriculum.

3. Should more *charter schools* open or close? Charter schools are funded and licensed by states or local districts. Thus, they are public schools but are exempt from some regulations, especially those negotiated by teacher unions (hours, class size, etc.). Most have some control over admissions and expulsions, which makes them more ethnically segregated, with fewer children with special needs (Stern et al., 2015). Quality varies, with African American boys particularly likely to leave or be expelled. However, some charters report that children who stay learn more and are more likely to go to college than their peers in regular public schools (Prothero, 2016). The most devastating criticism is that charter schools undercut the purpose of public education by being exclusive (Ladd, 2019).

4. Parents can choose to *home-school* their children, never sending them to school. In the United States, home-schooled children must learn certain subjects (reading, math, and so on), but each family decides schedules and discipline. About 2 percent of all U.S. children were home-schooled in 2003 and about

3 percent in 2007. Since then, numbers have leveled off at between 3 and 4 percent (Grady, 2017; Ray, 2013; Snyder & Dillow, 2013). Home schooling requires intense family labor, typically provided by an educated, dedicated, patient mother in a two-parent family.

The major disadvantage for home-schooled children is not academic (some have high test scores) but social: no classmates. To compensate, many parents plan activities with other home-schooling families.

5. Should public education be free of *religion* to avoid bias toward one religion or another? In the United States, thousands of parochial schools were founded when Catholics perceived Protestant bias in public schools. In the past 20 years, many Catholic schools have closed, but schools teaching other religions—Judaism, Islam, evangelical Christianity—have opened.

6. Should *the arts* be part of the curriculum? Music, drama, dance, and visual arts are essential in some places, not in others (Barton & Baguley, 2017). Many North American schools have eliminated requirements for arts education (O'Neill & Schmidt, 2017). By contrast, schools in Finland consider arts education essential (Nevanen et al., 2014).

7. Should children learn a *second language* in primary school? In Canada and in most European nations, almost every child studies two languages by age 10. In the United States, less than 5 percent of children under age 11 study a language other than English in school (Robelen, 2011).

8. Can *computers* advance education? Some enthusiasts hope that connecting schools to the internet or, even better, giving every child a laptop (as some schools do) will advance learning. The results are not dramatic, however. Sometimes computers improve achievement, particularly in math skills (Simms et al., 2019). Widespread, sustainable advances are elusive (Lim et al., 2013). Technology may be only a tool—a twenty-first-century equivalent of chalk—that depends on a creative, trained teacher to use well.

9. Are too many students in each class? Parents typically think that a smaller class size encourages more individualized education. That belief motivates many parents to choose private schools or home schooling. However, mixed evidence comes from nations where children score high on international tests. Sometimes they have large student/teacher ratios (Korea's average is 28-to-1) and sometimes small (Finland's is 14-to-1).

10. Should teachers nurture *soft skills* such as empathy, cooperation, and integrity as part of the school curriculum, even though these skills cannot be tested by multiple-choice questions? Many scholars argue that soft skills are crucial for academic success and later for employment (Reardon, 2013).

Ten Questions
1. Private schools?
2. Vouchers?
3. Charter schools?
4. Home schooling?
5. Religion?
6. The arts?
7. Second language?
8. Computers?
9. Class size?
10. Soft skills?

Who Decides?

Overall research in human development guides teachers who want to know exactly which concepts and skills are crucial foundations for mastery of reading, writing, science, math, and human relations. However, the science of child cognition is not necessarily understood by those who determine what children learn.

An underlying issue for almost any national or international school is the proper role of parents. In most nations, matters regarding public education—curriculum, funding, teacher training, and so on—are set by the central government.

Generally, when governments are responsible for education, almost all children attend the local school, whose resources and standards are similar to those of the other schools in that nation. If there are serious religious or cultural differences in the nation, public schools offer alternatives—religious or secular, in one language or another, and so on. The parents' job is to support the teachers and the child's learning.

Education in Middle Childhood

Only 20 years ago, gender differences in education around the world were stark, with far fewer girls in school than boys. Now girls have almost caught up. However, many of today's children suffer from decades of past educational inequality: Recent data find that the best predictor of childhood health is an educated mother.

WORLDWIDE PRIMARY SCHOOL ENROLLMENT, 1978–2018

This graph shows net enrollment rate, which is the ratio of enrolled school-age children to the population of children who are the same school age. Progress toward university education and gender equity is evident.

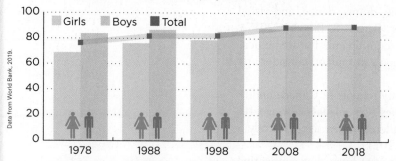

Data from World Bank, 2019.

Worldwide concerns now focus less on the existence of school and more on its quality. International tests usually find the United States is middling. Improvements are evident, but many other nations have improved even more!

WORLDWIDE, BASIC ELEMENTARY EDUCATION LEADS TO:

LESS
- Child and maternal mortality
- Transmission of HIV
- Early marriage and childbirth
- War

MORE
- Better paying jobs
- Agricultural productivity
- Use of medical care
- Voting

Information from Hanushek & Woessmann, 2007.

HOW U.S. FOURTH-GRADERS ARE DOING

Primary school enrollment is high in the United States, but not every student is learning, as these percentages from the National Assessment of Educational Progress (NAEP) show. While numbers are improving, less than half of fourth-graders are proficient in math and reading.

NAEP PROFICIENCY LEVELS FOR U.S. FOURTH-GRADERS

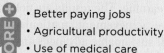

MATHEMATICS

- 40% Proficient or Better
- 40% Basic Understanding
- 20% Below Basic Level

READING

- 37% Proficient or Better
- 33% Basic Understanding
- 30% Below Basic Level

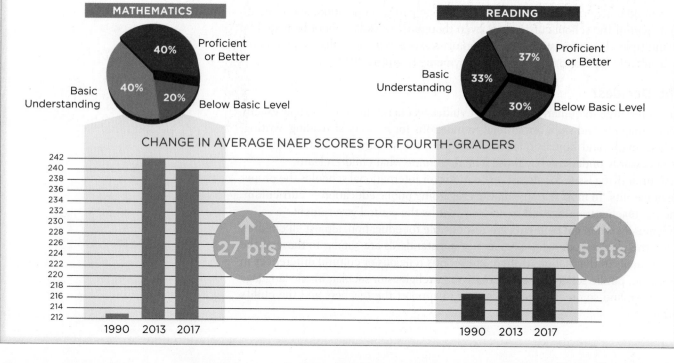

CHANGE IN AVERAGE NAEP SCORES FOR FOURTH-GRADERS

1990   2013   2017
↑ 27 pts

1990   2013   2017
↑ 5 pts

Data from National Center for Education Statistics, 2018

Data from National Center for EducationStatistics, 2018

In the United States, however, public schools are open to everyone, which means that no specific religious instruction occurs. Local districts provide most of the funds and guidelines, and parents, as voters and volunteers, are active within the school.

Although most U.S. parents send their children to the nearest public school, almost one-fifth send their children to private schools (10 percent) or charter schools (6 percent), or they educate their children at home (3 percent) (McFarland et al., 2018). Parental choices may vary for each of their children, depending on the child's characteristics, the parents' current economic status, and the political rhetoric at the time. Every option has strengths and weaknesses, both for the child and for society.

It is difficult for parents to determine the best school for their child, partly because neither the test scores of students in any of these schools, nor the moral values a particular school may espouse, correlate with the cognitive skills that developmentalists seek to foster (Finn et al., 2014).

Thus, parents may choose a school that advertises what they value, but this does not mean that the school provides the best educational experience for their child. It is not uncommon for parents to pull children from one school to enroll them in another, a phenomenon that has increased in recent years.

Statistical analysis raises questions about home schooling, vouchers, and charter schools, but continuity benefits children, so the parents' restlessness may not be wise (Finn et al., 2014; Lubienski et al., 2013). However, the data allow many interpretations.

As one review notes, "the modern day, parent-led home-based education movement . . . stirs up many a curious query, negative critique, and firm praise" (Ray, 2013, p. 261). Indeed, for all public and private schools, partisan political controversies swirl around school choice, which makes it hard to make a definitive conclusion about what school is best for a particular child (Quinn & Cheuk, 2018).

Schoolchildren's ability to be logical and teachable—now that they are no longer preoperational and egocentric—makes this a good time for primary school. They will learn whatever adults deem important. Parents, politicians, and developmental experts all agree that school is vital for development, but disagreements about teachers and curriculum—hidden or overt—abound.

Welcome Home Laura Stevens returns to her Maine elementary school after a whirlwind trip in Washington, D.C. She received the Presidential Award for Excellence in Math and Science Teaching, and $10,000. Which do you think makes her happier, the award, the hero's welcome from her students and colleagues, or the joy of teaching?

WHAT HAVE YOU LEARNED?

1. How does the hidden curriculum differ from the stated school curriculum?

2. What are the TIMSS, the PIRLS, and the PISA?

3. Which nations score highest on international tests?

4. How do boys and girls differ in school achievement?

5. How do charter schools, private schools, and home schools differ?

6. How is it decided what curriculum children should receive?

SUMMARY

Thinking

1. According to Piaget, middle childhood is the time of concrete operational thought, when egocentrism diminishes and logical thinking begins. Among the most important yet most difficult aspects are the concepts of classification and seriation.

2. By contrast, Vygotsky stressed the social context of learning, including the specific lessons of school and learning from peers, adults, and culture.

3. An information-processing approach examines each step of the thinking process, from input to output, using the computer as a model. Repeated practice is essential, as cognitive functions become automated.

4. One famous researcher who used the information-processing approach is Robert Siegler, who compared learning math to the waves lapping on a beach. Concepts are not learned suddenly.

5. Both the knowledge base and intellectual control processes are crucial aspects of the cognitive advances of middle childhood. Information-processing research has helped describe these developments.

Language

6. Language learning advances in many practical ways, including expanded vocabulary and pragmatics.

7. Most children use one code, dialect, or language with their friends and another in school. Children who are adept at code-switching, or who are fluently bilingual, have a cognitive advantage.

8. Children of low SES are usually lower in linguistic skills, primarily because they hear less language at home. Parent and teacher expectations are crucial.

Teaching and Learning

9. The hidden curriculum may be more influential on children's learning than the formal curriculum. For example, some people believe elementary schools favor girls, although internationally gender similarities seem to outweigh gender differences.

10. International assessments are useful as comparisons. Reading is assessed with the PIRLS, math and science with the TIMSS, and practical intelligence with the PISA. Culture affects answers as well as learning: East Asian scores are high, Finland has improved, and the United States is middling.

11. In the United States, each state, each district, and sometimes each school retains significant control. This makes education a controversial topic in many communities, with wide variation in curriculum, both hidden and overt. Most children attend their local public school.

12. Some parents choose charter schools, others private schools, and still others opt for home schooling. These years are prime time for basic education, but variations among children and cultures make specifics of the overt or hidden curriculum controversial in the United States.

KEY TERMS

concrete operational thought (p. 311)
classification (p. 311)
seriation (p. 312)
automatization (p. 316)
knowledge base (p. 317)
control processes (p. 317)

English Language Learners (ELLs) (p. 319)
immersion (p. 320)
bilingual education (p. 320)
ESL (English as a Second Language) (p. 320)

hidden curriculum (p. 323)
Trends in Math and Science Study (TIMSS) (p. 325)
Progress in International Reading Literacy Study (PIRLS) (p. 325)

Programme for International Student Assessment (PISA) (p. 325)
National Assessment of Educational Progress (NAEP) (p. 329)

APPLICATIONS

1. Visit a local elementary school and look for the hidden curriculum. For example, do the children line up? Why or why not, when, and how? Does gender, age, ability, or talent affect the grouping of children or the selection of staff? What is on the walls? For everything you observe, speculate about the underlying assumptions.

2. Interview a 6- to 11-year-old child to find out what that child knows *and understands* about mathematics. Relate both correct and incorrect responses to the logic of concrete operational thought and to the information-processing perspective.

3. What do you remember about how you learned to read? Compare your memories with those of two other people, one at least 10 years older and the other at least 5 years younger than you are. Can you draw any conclusions about effective reading instruction? If so, what are they? If not, why not?

4. Talk to two parents of primary school children. What do they think are the best and worst parts of their children's education? Ask specific questions and analyze the results.

Especially For ANSWERS

Response for Teachers (from p. 312): Here are two of the most obvious ways. (1) Use logic. Once children can grasp classification and class inclusion, they can understand cities within states, states within nations, and nations within continents. Organize your instruction to make logical categorization easier. (2) Make use of children's need for concrete and personal involvement. You might have the children learn first about their own location, then about

the places where relatives and friends live, and finally about places beyond their personal experience (via books, photographs, videos, and guest speakers).

Response for Parents (from p. 321): Your son would understand your explanation, but you should take him along if you can do so without losing patience. You wouldn't ignore his need for food or medicine, so don't ignore his need for learning. While shopping, you can teach vocabulary (does he know pimientos, pepperoni, polenta?), categories (root vegetables, freshwater fish), and math (which size box of cereal is cheaper?). Explain in advance that you need him to help you find items and carry them and that he can choose only one item that you wouldn't normally buy. Seven-year-olds can understand rules, and they enjoy being helpful.

Response for School Administrators (from p. 329): The relationship reflects correlation, not causation. Wearing uniforms is more common when the culture of the school emphasizes achievement and study, with strict discipline in class and a policy of expelling disruptive students.

Observation Quiz ANSWERS

Answer to Observation Quiz (from p. 311) As you know from the previous chapter, at this age children like to be near each other. That is your clue: An adult instructed them to sit apart. Their teacher is influenced by Piaget, who thought each person should discover cognitive principles on their own.

Answer to Observation Quiz (from p. 324) About 60 (6 rows, 10 in a row). Did trying to count make you realize that the children at the back cannot see or hear the teacher very well? None of them have glasses, so some of them cannot read the board.

Middle Childhood: Psychosocial Development

What Will You Know?

1. What helps children thrive in difficult family or neighborhood conditions?
2. Should parents marry, risking divorce, or not marry, risking separation?
3. What can be done to stop a bully?
4. Why would children lie to adults to protect a friend?

Ward Sutton is a professional cartoonist who won the prestigious Herblock Prize in 2018. In his acceptance speech, he thanked Kay Brown, a parent of his fifth-grade classmate. Kay noticed that Ward liked to draw, and suggested that he create a cartoon for the community newspaper. As Ward described it, "my first published editorial cartoon pulled no punches on the hard-hitting topic of . . . students leaning back in their chairs in class. Because, if you did that you might, you know, tip over. Hey, I had to start somewhere!"

Ward and Kay start this chapter because their story illustrates the nature of 10-year-olds and the importance of the community that supports them. His focus on the practical problems of fifth-graders—the tipping of chairs—is typical during middle childhood.

Later, bigger issues become salient: Ward praised his small Colorado town for upholding the freedom of speech, including in his cartoons, which were contrary to local opinions because he poked fun at national political leaders. But as a boy, Ward was much more concerned with earning the respect of his peers and teachers.

You will read more about Ward's boyhood later in this chapter. His story provides one example of the thoughts that keep some children up at night or get them out of bed each morning. Among those are friends, bullies, parents, and other adults. Middle childhood is the time when children become aware of the wider community, who might or might not channel their doodles into something meaningful. All children need caring adults like Kay to notice their talents and show them how to use them.

The Nature of the Child

As explained in the previous chapter, steady growth, brain maturation, and intellectual advances make middle childhood a time for more independence (see At About This Time). One practical result is that between ages 6 and 11, children learn to care for themselves. In contrast to younger children, they not only hold their spoon but also make their lunch, not only zip their pants but also pack their suitcases, not only walk to school but also organize games with friends. They venture outdoors alone.

AT ABOUT THIS TIME

Signs of Psychosocial Maturation Over the Years of Middle Childhood*

Children responsibly perform specific chores.

Children make decisions about a weekly allowance.

Children can tell time and have set times for various activities.

Children have homework, including some assignments over several days.

Children are punished less often than when they were younger.

Children try to conform to peers in clothes, language, and so on.

Children voice preferences about their after-school care, lessons, and activities.

Children are responsible for younger children, pets, and, in some places, work.

Children strive for independence from parents.

*Of course, culture is crucial. For example, giving a child an allowance is typical for middle-class children in developed nations since about 1960. It was rare, or completely absent, in earlier times and other places.

Industry and Inferiority

Throughout the centuries and in every culture, school-age children have been industrious. They busily master whatever skills their culture values. Their mental and physical maturation, described in the previous chapter, makes such activity possible.

With regard to his fourth psychosocial crisis, **industry versus inferiority**, Erikson noted that the child "must forget past hopes and wishes, while his exuberant imagination is tamed and harnessed to the laws of impersonal things," becoming "ready to apply himself to given skills and tasks" (Erikson, 1993a, pp. 258, 259). Simply trying new things, as in the previous stage (initiative versus guilt), is no longer sufficient. Sustained activity that leads to accomplishments that make one proud is the goal.

Think of learning to read and learning to add, both of which are painstaking and tedious. For instance, slowly sounding out "Jane has a dog" or writing "3 + 4 = 7" for the hundredth time is not exciting. Yet school-age children busily practice reading and math: They are intrinsically motivated to read a page, finish a worksheet, memorize a spelling word, color a map, and so on. Similarly, they enjoy collecting, categorizing, and counting whatever they gather—perhaps stamps, stickers, stones, or seashells. That is industry.

Overall, children judge themselves as either *industrious* or *inferior*—deciding whether they are competent or incompetent, productive or useless, winners or losers. Self-pride depends not necessarily on actual accomplishments but on how others, especially peers, view one's accomplishments. Social rejection is both a cause and a consequence of feeling inferior (Rubin et al., 2013).

industry versus inferiority The fourth of Erikson's eight psychosocial crises, during which children attempt to master many skills, developing a sense of themselves as either industrious or inferior, competent or incompetent.

Parental Reactions

Did you pause a moment ago when you read that 6- to 11-year-olds can "venture outdoors alone"? Cohort and context changes can be dramatic. Recently in the United States, many parents do not allow their children outside without an adult, even to walk to a neighbor's house, much less to go to town with money in their pocket.

Universally, in middle childhood children become capable of doing things themselves that they once could not do, but parents react in diverse ways: Some 10-year-olds

Same Situation, Far Apart: Helping at Home Sichuan, in China *(right)*, and Virginia, in the United States *(left)*, provide vastly different contexts for child development. Children everywhere help their families with household chores, as these two do, but gender expectations vary a great deal.

care for younger children, buy groceries, and make dinner while parents are away at work; some use power tools or drive tractors on the family farm; others are closely supervised if they venture outside or even turn on the kitchen stove.

Although variation is apparent, in middle childhood parents shift from providing physical care (bathing, dressing, and so on) to engaging in dialogue, discussion, and shared activities, a trend particularly apparent with boys and their fathers (Keown & Palmer, 2014).

For all children, parents gradually grant more autonomy, which helps children feel happy and capable (Yan et al., 2017). Consequently, time spent with parents decreases while time alone, and with friends, increases. One study of U.S. families found that 8-year-olds, on average, spent 95 minutes a day with their mothers, and 12-year-olds spent 70 minutes, almost half an hour less. This study found substantial variation by context and family structure (Lam et al., 2012).

Self-Concept

As children mature, they develop their *self-concept*, which is their idea about themselves, including their intelligence, personality, abilities, gender, and ethnic background. As you remember, in toddlerhood children discover that they are individuals, and in early childhood they develop a positive, global self-concept.

That rosy self-concept is modified in middle childhood. The self-concept gradually becomes more specific and logical, the result of increases in cognitive development and social awareness. Fathers, teachers, and peers become more influential, sometimes helping children feel proud of themselves but sometimes not (Verschueren, 2020).

Genes, brain maturation, and family experiences continue to affect 6- to 11-year-olds, advancing or impairing self-concept (Bick et al., 2019). One important insight, called the *adjustment-erosion model*, suggests that a child's emotional problems at the beginning of middle childhood (the aggressive or pathologically shy first-grader) affect later academic difficulties more than vice versa (Deighton et al., 2018). Notice the developmental process: Emotional regulation and social skills developed in early childhood affect school success in middle childhood.

> **THINK CRITICALLY:** When would a realistic, honest self-assessment be harmful?

Compared to Others

social comparison The tendency to assess one's abilities, achievements, social status, and other attributes by measuring them against those of other people, especially one's peers.

Crucial during middle childhood is **social comparison**—comparing oneself to others (Dweck, 2013; Lapan & Boseovski, 2017). Ideally, social comparison helps school-age children value themselves for who they are, abandoning the fantasy self-evaluation of preschoolers.

Social comparison makes the self-concept more realistic, because children incorporate comparison to peers when they judge themselves. The human tendency to think of oneself favorably is present, but it now is grounded in reality (Thomaes et al., 2017).

Some children—especially those from minority ethnic or religious groups—become newly aware of social prejudices that they need to overcome. Children also become aware of gender discrimination, with girls complaining that they are not allowed to play tougher sports and boys complaining that teachers favor the girls (C. Brown et al., 2011).

Over the years of middle childhood, children who affirm pride in their gender and ethnicity are likely to develop healthy self-esteem (Corenblum, 2014). Transgender children particularly experience discrimination. For them, parental support is crucial, but parents themselves experience stress (Hidalgo & Chen, 2019). Overall, parents who feel supported by their community are better able to affirm their own children.

Especially when the outside world seems hostile, parents and schools who teach about successful people from a variety of ethnicities, genders, or nationalities are likely to help children from those groups feel valued (Hernández et al., 2017). Much of the research regarding the impact of role models has focused on African American adolescents, but a recent review suggests that the same influences affect every group. Developing self-acceptance and pride bolsters self-confidence more than alerting children to be wary of possible prejudice (M-T. Wang et al., 2020).

Culture and Self-Esteem

Both academic and social competence are aided by realistic self-perception. That is beneficial, because unrealistically high self-esteem reduces effortful control, and without some control children are more aggressive and less conscientious than they might be.

The same consequences occur if self-esteem is too low. Obviously, the goal then is to find a middle ground. This is not easy: Children may be too self-critical or not self-critical enough. Their self-control interacts with the reactions of their parents and culture. Cultures differ on what that middle ground is.

High self-esteem is neither universally valued nor universally criticized. Many cultures expect children to be modest, not prideful. For example, Australians say that "tall poppies are cut down"; the Chinese say, "the nail that sticks up is hammered"; and the Japanese discourage social comparison aimed at making oneself feel superior.

This makes self-esteem a moral issue as well as a practical one: *Should* people believe that they are better than other people, as is typical in the United States but not in every nation? Answers vary.

One crucial component of self-concept has received considerable research attention (Dweck, 2013). As children become more self-aware, they benefit from praise for their process, not for their person: for *how* they learn, *how* they relate to others, and so on, not for static qualities such as intelligence and popularity. This encourages growth.

For example, children who fail a test are devastated if failure means they are not smart. However, process-oriented children

Learning from Each Other Middle childhood is prime time for social comparison. Swinging is done standing, or on the belly, or twisted, or head down (as shown here) if someone else does it.

Christina Kilgour/Moment Select/Getty Images

Black Panther Mythical superheroes, and the perpetual battle between good and evil, are especially attractive to boys in middle childhood but resonate with people of all ages, genders, and ethnic groups. *Black Panther* was first a comic-book hero in 1966 and then became a 2018 movie that broke records for attendance and impact. It features not only African American heroes but also an army of strong women—busting stereotypes and generating self-esteem for many children.

Pictorial Press Ltd/Alamy

Watch **VIDEO ACTIVITY: Interview with Carol Dweck** to learn about how children's mindsets affect their intellectual development.

consider failure a "learning opportunity," a time to figure out how to study the next time. They have a "growth mindset."

The self-conscious emotions (pride, shame, guilt) first evident in early childhood may also develop during middle childhood. They guide social interaction, yet they can overcome a child's self-concept, leading to psychopathology (Muris & Meesters, 2014). Especially during middle childhood (less so in adolescence), school achievement is a crucial factor in developing self-esteem, and that affects later self-concept—as someone who is inferior or not.

In addition, the onset of concrete thinking in middle childhood leads children to notice material possessions. Objects that adults find superficial (name-brand sunglasses, sock patterns) become important.

Insecure 10-year-olds might desperately want the latest jackets, smartphones, and so on. Or they may want something else that makes them seem special, such as lessons in African dance, or a brilliant light for their bicycle, or—as one of my daughters did—a bread-maker (used often for several weeks, discarded after several years).

From a life-span perspective, a developmental pattern appears. Children in many cultures tend to increase in self-esteem during middle childhood, stabilize in adolescence, and increase again in emerging adulthood. Not until very late adulthood (after age 70 or beyond) does self-esteem decrease (Orth et al., 2018). Thus, parents and grandparents are well-positioned to encourage the child's natural increase in self-esteem.

Resilience and Stress

In infancy and early childhood, children depend on their immediate families for food, learning, and life itself. In middle childhood, some children continue to benefit from supportive families, and others escape destructive families by finding their own niche in the larger world.

Surprisingly, some children seem unscathed by early experiences. They have been called "resilient" or even "invincible." Current thinking about resilience (see **Table 13.1**), with insights from dynamic-systems theory, emphasizes that no one is truly untouched by past history or current context, but some weather early storms and a few not only survive but become stronger because of them (Luthar, 2015; Masten, 2014; Rutter, 2012).

Defining Resilience

resilience The capacity to adapt well to significant adversity and to overcome serious stress.

Resilience has been defined as "a dynamic process encompassing positive adaptation within the context of significant adversity" (Luthar et al., 2000, p. 543) and "the capacity of a dynamic system to adapt successfully to disturbances that threaten system function, viability, or development" (Masten, 2014, p. 30). Note that both of these leading researchers emphasize three parts of this definition:

- Resilience is *dynamic*, not a stable trait. That means a given person may be resilient at some periods but not others, and the effects from one period reverberate as time goes on.
- Resilience is a *positive adaptation* to stress. For example, if parental rejection leads a child to a closer relationship with another adult, that is positive resilience, not passive endurance.
- Adversity must be *significant*, a threat to development.

Cumulative Stress

An important discovery is that stress accumulates over time, including minor disturbances (called "daily hassles"). A long string of hassles, day after day, takes a

TABLE 13.1	
Dominant Ideas About Resilience, 1965 to Present	
1965	All children have the same needs for healthy development.
1970	Some conditions or circumstances—such as "absent father," "teenage mother," "working mom," and "day care"—are harmful for every child.
1975	All children are *not* the same. Some children are resilient, coping easily with stressors that cause harm in other children.
1980	Nothing inevitably causes harm. Both maternal employment and preschool education, once thought to be risks, are often helpful.
1985	Factors beyond the family, both in the child (low birthweight, prenatal alcohol exposure, aggressive temperament) and in the community (poverty, violence), can be very risky for children.
1990	Risk–benefit analysis finds that some children are "invulnerable" to, or even benefit from, circumstances that destroy others.
1995	No child is invincible. Risks are always harmful—if not in education, then in emotions; if not immediately, then long term.
2000	Risk–benefit analysis involves the interplay among many biological, cognitive, and social factors, some within the child (genes, disability, temperament), the family (function as well as structure), and the community (including neighborhood, school, church, and culture).
2008	Focus on strengths, not risks. Assets in child (intelligence, personality), family (secure attachment, warmth), community (schools, after-school programs), and nation (income support, health care) must be nurtured.
2010	Strengths vary by culture and national values. Both universal ideals and local variations must be recognized and respected.
2012	Genes as well as cultural practices can be either strengths or weaknesses; differential susceptibility means that identical stressors can benefit one child and harm another.
2015	Communities are responsible for child resilience. Not every child needs help, but every community needs to encourage healthy child development.
2020	Resilience is seen more broadly as a characteristic of mothers and communities.

greater toll than an isolated major stress. Almost every child can withstand one trauma, but "the likelihood of problems increased as the number of risk factors increased" (Masten, 2014, p. 14).

The social context, especially supportive adults who do not blame the child, is crucial. A chilling example comes from the "child soldiers" in the 1991–2002 civil war in Sierra Leone (Betancourt et al., 2013). Children witnessed and often participated in murder, rape, and other traumas. When the war was over, 529 war-affected youth, then aged 10 to 17, were interviewed. Many were severely depressed, with crippling anxiety.

These war-damaged children were interviewed again two and six years later. Surprisingly, many had overcome their trauma and were functioning well. Recovery was more likely if:

- The war occurred when they were in middle childhood, not adolescence;
- at least one caregiver survived and was reunited with the child;
- their communities did not reject them, no matter which side of the civil war they had joined; and
- daily routines (school, family responsibilities) were restored.

LaunchPad
macmillan learning

VIDEO ACTIVITY: Child Soldiers and Child Peacemakers examines the state of child soldiers in the world and then explores how adolescent cognition impacts the decisions of five teenage peace activists.

Family as a Buffer

In England during World War II, many city children were sent to loving families in rural areas to escape the German bombs dropped every day. To the surprise of researchers, those children who stayed in London with their parents were more resilient, despite nights huddled in air-raid shelters, than those who were physically safe but without their parents (Freud & Burlingham, 1943).

Similar results were found in a longitudinal study of children exposed to a sudden, wide-ranging, terrifying wildfire in Australia. Almost all of the children suffered stress reactions at the time, but 20 years later the crucial factor was not proximity to the fire but whether or not it separated them from their mothers (McFarlane & Van Hooff, 2009).

Whenever war, or economic conditions, or immigration policies separate parents and children, developmentalists predict lifelong problems for the children. Longitudinal studies over the past decades have found significant emotional and social vulnerability in Holocaust survivors from World War II, refugees of African civil wars, and children in Vietnam. Recently, the same consequences have been found in immigrant children in the United States, who suffer from many health problems if their parents are not with them (Perreira & Pedroza, 2019).

For that reason, thousands of developmental scholars and dozens of professional societies have expressed their horror at the 2018 U.S. policy of separating children from their parents at the United States–Mexico border. For example, the Society of Developmental and Behavioral Pediatrics fear that "a generation of children will experience lifelong repercussions" from being forcefully separated from their parents. Their statement reads:

> Children and parents belong together. Children who are separated from their primary caregivers may experience toxic stress and a disruption of attachment that can have severe emotional, behavioral and physical implications.

[*Society for Developmental & Behavioral Pediatrics, 2018*]

Cognitive Coping

Obviously, the above examples are extreme, but the general finding is confirmed by many studies. Disasters take a toll, but resilience is possible. Factors in the child (especially problem-solving ability), in the family (consistency and care), and in the community (good schools and welcoming religious institutions) all help children recover (Masten, 2014).

The child's interpretation of events is crucial (Lagattuta, 2014). Cortisol increases in low-income children *if* they interpret events connected to their family's poverty as a personal threat and *if* the family lacks order and routines (thus increasing daily hassles) (Coe et al., 2018). If low-SES children do not feel personally to blame, and if their family is not chaotic, they may be resilient.

In general, children's interpretations of family situations (poverty, divorce, and so on) determine how they are affected. Think of people you know: Some adults from low-SES families did not feel deprived. Only later did they realize that they were poor. For them, childhood poverty is less likely to cast a shadow over adult life.

Some children consider the family they were born into a temporary hardship; they look forward to the day when they can leave childhood behind. If they also have personal strengths, such as the cognitive abilities to imagine a better life, they may shine in adulthood — evident in the United States in thousands of success stories, from Abraham Lincoln to Oprah Winfrey.

The opposite reaction is called **parentification**, when children feel responsible for the entire family, acting as caregivers of everyone, including their actual parents. Here again the child's interpretation is crucial. For instance, suppose a

---

THINK CRITICALLY: Is there any harm in having the oldest child take care of the younger ones? Why or why not?

---

parentification When a child acts more like a parent than a child. Parentification may occur if the actual parents do not act as caregivers, making a child feel responsible for the family.

child witnesses domestic abuse. This is never good, but if the child feels responsible for the abuser and the abused, recovery is less likely (Fortin et al., 2011).

One final example: Many children of immigrants exemplify parentification if they become the translators for their parents, who do not understand the language or the culture. If those children feel burdened by their role as language brokers, that increases their depression. But, if they feel they are making a positive contribution to their family well-being, they themselves benefit (Weisskirch, 2017b).

WHAT HAVE YOU LEARNED?

1. How do Erikson's stages for preschool and school-age children differ?

2. Why is social comparison particularly powerful during middle childhood?

3. Why do cultures differ in how they value pride or modesty?

4. Why and when might minor stresses be more harmful than major stresses?

5. How might a child's interpretation of events help that child cope with repeated stress?

Families During Middle Childhood

No one doubts that genes affect personality as well as ability, that peers are vital, and that schools and cultures influence what, and how much, children learn. Some have gone further, suggesting that genes, peers, and communities have so much influence that parents have little impact—unless they are grossly abusive (Harris, 1998, 2002; McLeod et al., 2007). This suggestion arose from studies about the impact of the environment on child development.

Shared and Nonshared Environments

Many studies have found that children are much less affected by *shared environment* (influences that arise from being in the same environment, such as for two siblings living in one home, raised by their parents) than by *nonshared environment* (e.g., the distinct experiences and surroundings of two people).

Even basic values and traits, or sexual orientation, which once were assumed to be heavily affected by parents, seem much more influenced by genes and nonshared environment than by growing up in a particular household (Twito & Knafo-Noam, 2020; Y. Xu et al., 2020).

Since genes and nonshared environment are more influential than shared, might parents be insignificant? That avoids "misplaced blame on parents for negative outcomes in their children . . . adding guilt to the grief parents are already feeling for their children's suffering" (Sherlock & Zietsch, 2018, p. 155). But then parents should not take credit for the accomplishments of their children.

Could it be that parents are merely caretakers, needed for basic care (food, shelter), harmful when severely neglectful or abusive, but inconsequential in what they do about household routines, prohibitions, and praise? If a child becomes a murderer or a hero, should parents be neither blamed nor credited?

That conclusion is too extreme: Parents are not the only influence, of course, but they are influential. Recent findings reassert parent power. The analysis of shared and nonshared influences was correct, but the conclusion was based on a false assumption. Siblings raised together do *not* share the same environment.

For example, if relocation, divorce, unemployment, or a new job occurs in a family, the impact on each child differs depending on that child's age, genes, resilience, and gender. Moving to another town upsets a school-age child more than an infant; divorce harms boys more than girls; poverty may hurt preschoolers the most; and so on.

Shared Environment? All three children live in the same home in Brooklyn, New York, with loving, middle-class parents. But it is not hard to imagine that family life is quite different for the 9-year-old girl than for her sister, born a year later, or their little brother, age 3.

monkeybusinessimages/iStock/Getty Images

● **Observation Quiz** Are significant gender differences evident here? (see answer, page 367) ↑

Differential susceptibility adds to the variation: One child might be more affected than another. When siblings are raised together, experiencing the same family conditions, the mix of genes, age, and gender may lead one child to become anti-social, another to be pathologically anxious, and a third to be resilient, capable, and strong. Further, not all characteristics are genetic.

For example, one study of 7-year-old twins found that the child's ability to recognize emotions (whether a particular expression or tone indicated anger or happiness, for instance) was affected by the twins' family experiences but not by genes (Schapira et al., 2019). The authors acknowledged that later on, genes may become more influential, but this was not apparent for these 7-year-olds. Another study also found parental influence on 7-year-olds, as A View from Science makes clear.

A VIEW FROM SCIENCE

"I Always Dressed One in Blue Stuff. . ."

To separate the effects of genes and environment, many researchers have studied twins. As you remember from Chapter 3, some twins are dizygotic, with only half of their genes in common, and some are monozygotic, identical in all their genes. Many scientists assumed that children growing up with the same parents would have the same nurture (shared environment).

Therefore, if dizygotic twins are less alike than monozygotic twins are, genes must be the reason. Further, if one monozygotic twin differs from their genetically identical twin, raised by their parents in the same home, those differences must arise from the nonshared environment.

Logically, everyone is influenced by three forces: genes, shared environment (same home), and nonshared environment (different schools, friends, and so on). Many people were surprised when twin research discovered that almost everything could be attributed to genes and nonshared environment, with almost nothing left over for parents.

However, that conclusion is now tempered by another finding: Twins raised in the same home may have quite different family experiences for reasons that are not genetic. A seminal study in this regard occurred with twins in England.

An expert team of scientists compared 1,000 sets of monozygotic twins reared by their biological parents (Caspi et al., 2004). Obviously, the pairs were identical in genes, sex, and age. The researchers asked the mothers to describe each twin. Descriptions ranged from very positive ("my ray of sunshine") to very negative ("I wish I never had her. . . . She's a cow, I hate her") (quoted in Caspi et al., 2004, p. 153). Some mothers noted personality differences between their twins. For example, one mother said:

> Susan can be very sweet. She loves babies . . . she can be insecure . . . she flutters and dances around. . . . There's not much between her ears. . . . She's exceptionally vain, more so than Ann. Ann loves any game involving a ball, very sporty, climbs trees, very much a tomboy. One is a serious tomboy and one's a serious girlie girl. Even when they were babies I always dressed one in blue stuff and one in pink stuff.
>
> [quoted in Caspi et al., 2004, p. 156]

Some mothers rejected one twin but not the other:

> He was in the hospital and everyone was all "poor Jeff, poor Jeff" and I started thinking, "Well, what about me? I'm the one's just had twins. I'm the one's going through this, he's a seven-week-old baby and doesn't know a thing about it" . . . I sort of detached and plowed my emotions into Mike [Jeff's twin brother].
>
> [quoted in Caspi et al., 2004, p. 156]

This mother later blamed Jeff for favoring his father: "Jeff would do anything for Don but he wouldn't for me, and no matter what I did for either of them [Don or Jeff], it wouldn't be right" (quoted in Caspi et al., 2004, p. 157). She said Mike was much more lovable.

The researchers measured personality at age 5 (assessing, among other things, antisocial behavior as reported by kindergarten teachers) and then measured each twin's personality two years later. They found that if a mother was more negative toward one of her twins, that twin *became* more antisocial, more likely to fight, steal, and hurt others at age 7 than at age 5, unlike the favored twin.

These researchers recognize that many other nonshared factors—peers, teachers, and so on—are significant. However, most developmental scientists now agree that genes, neighborhood, and parental influences are all important, and that—especially when genes or neighborhood push a child toward unhealthy development—parental intervention can be crucial (Liu & Neiderhiser, 2017).

Genes are still powerful, of course, because "a given DNA sequence operation in different environments can generate different products in different amounts at the cellular and phenotypic levels" (Waldinger & Schulz, 2018, p. 159). That expresses an underlying theme of this book, that human development is multifactorial and complex. It begins with genes (DNA), but a simple calculation of genetic and family influence is impossible.

The fact that parents sometimes treat each of a pair of monozygotic twins differently confirms that parents matter. This will surprise no one who has a brother or a sister. Children from the same family do not always experience their family in the same way.

Function and Structure

Family structure refers to the genetic and legal connections among related people. *Genetic* connections may be from parent to child, between cousins, between siblings, between grandparents and grandchildren, or more distantly, such as from great aunts, second cousins, and so on. *Legal* connections may be through marriage or adoption.

Family function is distinct from structure. It refers to how the people in a family actually care for each other. Some families function well; others are dysfunctional.

Function is more important than structure. Ideally, every family provides love and encouragement. For most people, this comes from genetic relatives, so structure and function overlap. For foster children and adopted children who share few distinct genes with their caregivers, family function is crucial (Flannery et al., 2017).

Everyone enters the world with unique genes and a particular prenatal environment, and that differential susceptibility influences how their family affects them. Beyond that, people's needs differ depending on their age: Infants need responsive caregiving, teenagers need guidance, young adults need freedom, the aged need respect. What do school-age children need?

The Needs of Children in Middle Childhood

Ideally, families that function well for children aged 6 to 11 provide five things:

1. *Physical necessities.* In middle childhood, children can eat, dress, and wash themselves, but they need food, clothing, and shelter. Ideally, their families provide these things.
2. *Learning.* These are prime years for education. Families support, encourage, and guide schooling—connecting with teachers, checking homework, and so on.
3. *Self-respect.* During these years social comparison can deflate self-esteem, so families help each child excel at something—sports, the arts, or academics.
4. *Peer relationships.* Children need friends. Families choose schools and neighborhoods with friendly children and arrange play dates, group activities, overnights, and so on.
5. *Harmony and stability.* Families provide protective, predictable routines in a home that is a safe, peaceful haven. Family conflict and chaos is destructive.

Harm from Instability

The final item on the list above may be especially significant in middle childhood: Children cherish safety and stability, not change. When children experience many changes in caregivers (e.g., mother, stepmother, aunt, father) and many changes in residence (from one neighborhood to another), they are more likely to develop emotional difficulties.

Several studies find that family instability, such as moving from place to place, and family chaos, such as no routines for sleeping, eating, or homework, increase children's internalizing and externalizing problems. Surprisingly, race might make a difference. At least one study found that harm was less apparent for African American children. The researchers suggested that grandparents and other relatives might buffer the effect of instability (Womack et al., 2019).

Nonetheless, for all children during middle childhood, having a stable network of friends and teachers is an asset, and staying in one neighborhood makes that easier. Children who are homeless suffer physiologically as well as psychologically, evident in cortisol level, blood pressure, weight, and in likelihood of hospitalization (Cutuli et al., 2017).

Also, for all children, a well-functioning family can soften the impact of change. If a child is living in a shelter for homeless families, with a mother who provides

family structure The legal and genetic relationships among relatives living in the same home. Possible structures include nuclear family, extended family, stepfamily, single-parent family, and many others.

family function The way a family works to meet the needs of its members. Children need families to provide basic material necessities, to encourage learning, to help them develop self-respect, to nurture friendships, and to foster harmony and stability.

● **Especially for Scientists** How would you determine whether or not parents treat all of their children the same? (see response, page 367)

Didn't Want to Marry This couple was happily cohabiting and strongly committed to each other but didn't wed until they learned that her health insurance would not cover them unless they were legally married. Twenty months after marriage, their son was born.

© 2016 Macmillan

stability, affection, routines, and hope, that child's school achievement may be resilient.

Another example comes from children in military families. Enlisted parents tend to have higher incomes, better health care, and more education than do civilians from the same backgrounds. But they have one major disadvantage: instability.

In general, children whose parents are in military service are no worse off than civilian children. However, for some children, parent deployment (which requires several disruptions in the child's home life) leads to higher rates of depression and aggression (Fairbank et al., 2018; Williamson et al., 2018).

Most military children learn to cope (Russo & Fallon, 2014). To help children whose parents are deployed, the U.S. military offers special programs. Caregivers of such children are encouraged to avoid changes in the child's life during the deployment: no new homes, new rules, or new schools (Lester et al., 2011). Similar transitions occur when deployed parents come home: They are welcomed, of course, but the child's life might change again—and that causes more stress.

On a broader level, children who are displaced because of storms, fire, war, and so on may suffer psychologically. They may try to comfort their parents, not telling them about their distress, but the data on health and achievement show that moving from place to place is highly stressful (Masten, 2014). All children must cope with some disruption: Some children develop good coping skills and other children do not.

Various Family Structures

A **nuclear family** is a family composed only of children and their parents (married or not). Usually the parents are the biological parents of the children, but other nuclear families are headed by adoptive parents, foster parents, stepparents, or same-sex couples, most of whom provide good care.

Rates of single parenthood vary greatly worldwide (see Visualizing Development, page 350; About a third of 6- to 11-year-olds in the United States live in a **single-parent family**. Many of them are not harmed by that, if their mother is a stable, hopeful caregiver, but obviously good caregiving is more difficult under stress.

An **extended family** includes relatives in addition to parents and children. Usually the additional persons are grandparents or sometimes uncles, aunts, or cousins of the child. The crucial distinction for official tallies is who lives under the same roof. This measures family structure, not family function.

The distinction between one-parent, two-parent, and extended families is not as simple in practice as it is on the census. Many parents of young children live near, but not with, the grandparents, who provide meals, emotional support, money, and child care, functioning as an extended family. The opposite is true as well, especially in developing nations: Some extended families share a household but create separate living quarters for each set of parents and children, making these units somewhat like nuclear families. Similarly some fathers who do not live with their children nonetheless sometimes are involved parents.

In many nations, the **polygamous family** (one husband with two or more wives) is a legal family structure. Generally in polygamous families, income per child is reduced, and education, especially for the girls, is limited—in part because girls are expected to marry young. Polygamy is rare—and illegal—in the United States. Even in nations where it is allowed—most African and many Asian nations—polygamy is less common than it was 30 years ago.

Cohort Changes

There are more single-parent households, more divorces and remarriages, and fewer children per family than in the past. Specifics vary from decade to decade

nuclear family A family that consists of a father, a mother, and their biological children under age 18.

single-parent family A family that consists of only one parent and their children.

extended family A family of relatives in addition to the nuclear family, usually three or more generations living in one household.

polygamous family A family consisting of one man, several wives, and their children.

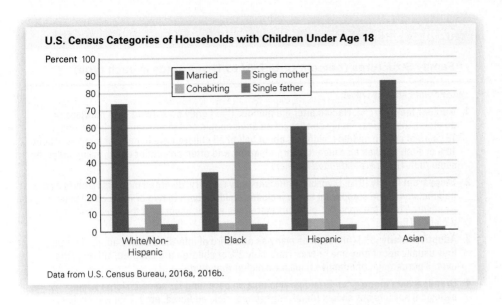

U.S. Census Categories of Households with Children Under Age 18

Data from U.S. Census Bureau, 2016a, 2016b.

FIGURE 13.1

Possible Problems As the text makes clear, structure does not determine function, but raising children is more difficult as a single parent, in part because income is lower. African American families have at least one asset, however. They are more likely to have grandparents who are actively helping with child care.

and nation to nation. That matters for children. In the United States, divorced, single-parent families were unusual in the 1960s, more common in the 1980s and 1990s, and again are less common in the twenty-first century (see **Figure 13.1**). Children were more likely to suffer from divorce when divorce was rare.

On average, however, children growing up in non-nuclear families have more emotional and academic problems than when both parents live in the same home as the child. Variations in family structures should be acknowledged, but neither exaggerated nor bemoaned. The United States has more single parents than other developed nations, yet almost two-thirds of all U.S. school-age children live with two parents (see **Table 13.2**), most often their biological parents, and some of those children have emotional or academic problems as well.

Connecting Family Structure and Function

How a family functions is more important for children than their family structure. The two are related; structure influences (but does not determine) function. Some structures increase the possibility that the five family functions mentioned earlier (physical necessities, learning, self-respect, friendship, and harmony/stability) will be fulfilled.

Two-Parent Families

On average, nuclear families function best; children living with two married parents tend to learn more in school with fewer psychological problems. Why? Does this mean that every mother and father should marry and stay married? No. Some benefits are correlates, not causes.

To understand, remember the longitudinal life-span perspective that considers the family context before the children appear. Education, earning potential, and emotional maturity increase the rate of marriage and parenthood and decrease the rate of divorce. For example, first-time mothers in the United States are usually (78 percent) married when they conceive their first child if they are highly educated, but they are usually unmarried (only 11 percent married at conception) if they are low in SES (Gibson-Davis & Rackin, 2014).

The differences between high- and low-SES individuals in marriage, child-bearing, child rearing, and divorce suggest "close to two different subsystems" of

---

TABLE 13.2

Family Structures (percent of U.S. 6- to 11-year-olds in each type)*

Two-Parent Families (70%)

1. **Nuclear family** (57%). Named after the nucleus (the tightly connected core particles of an atom), the nuclear family consists of an adult couple and their shared children under 18 years of age. In middle childhood, about half of all children live in nuclear families. About 10% of such families also include a grandparent, and often an aunt or uncle, living under the same roof. Those are *extended* families.

2. **Stepparent family** (9%). Divorced fathers usually remarry; divorced mothers remarry about half the time. If the stepparent family includes children born to two or more couples (such as children from the spouses' previous marriages and/or children of the new couple), that is a *blended* family.

3. **Adoptive family** (3%). Although as many as one-third of infertile couples adopt children, they usually adopt only one or two. Thus, only 2% of children are adopted, although the overall percentage of adoptive families is higher than that.

4. **Grandparents alone** (1%). Grandparents take on parenting for some children when biological parents are absent (dead, imprisoned, sick, addicted, etc.). That is a *skipped-generation* family.

Single-Parent Families (30%)

One-parent families are increasing, with almost half of the newborns now born to unmarried mothers, but such families average fewer children than two-parent families, and many single mothers find partners by the time their children are school age. So, in middle childhood only 31% of children have a lone parent.

1. **Single mother—never married** (14%). In 2018, 40% of all U.S. births were to unmarried mothers; but when children are school age, many such mothers have married or have entrusted their children to their parents' care. Thus, only about 14% of 6- to 11-year-olds, at any given moment, are in single-mother, never-married homes.

2. **Single mother—divorced, separated, or widowed** (12%). Although many marriages end in divorce (almost half in the United States, fewer in other nations), many divorcing couples have no children. Others remarry. Thus, only 12% of school-age children currently live with single, formerly married mothers.

3. **Single father** (3%). About 1 father in 25 has physical custody of his children and raises them without their mother or a new wife. This category increased at the start of the twenty-first century but has decreased since 2005.

4. **Grandparent alone** (1%). Sometimes a grandparent (usually the grandmother) becomes the sole caregiving adult for a child.

More Than Two Adults (10%) [Also listed as two-parent or single-parent family]

1. **Extended family** (10%). Some children live with a grandparent or other relatives, as well as with one (5%) or both (10%) of their parents. This pattern is most common with infants (20%) but occurs in middle childhood as well.

2. **Polygamous family** (0%). In some nations (not the United States), men can legally have several wives. This family structure is more favored by adults than children. Everywhere, polyandry (one woman, several husbands) is rare.

---

*Less than 1% of children under age 12 live without any caregiving adult; they are not included in this table.

The percentages in this table are estimates (based on data in U.S. Census Bureau, 2019). The category "extended family" in this table is higher than most published statistics, since some families do not tell official authorities about relatives living with them.

---

family organization in the United States today (Cherlin, 2020, p. 69). Spouses tend to have personal assets *before* they marry and become parents, and their higher income and education benefit the children that marriage might produce. The correlation between child success and married parents occurs partly because of who the spouses are, not because of fact that they marry. These two factors— selection and income—explain some of the correlation between nuclear families and child well-being.

To the surprise of some outsiders, a large study comparing male–female and same-sex couples found that the major predictor of their children's well-being was not the parents' sexual orientation but their income and stability (Cenegy et al., 2018). Similar findings come from adoptive parents, grandparents raising children, and so on. A caregiver's emotional health and the family's economic security benefit the children.

In general, married parents (of whatever gender identity and sexual orientation) are more likely to stay together than unmarried parents, and they are more likely to become wealthier and healthier than either would be alone. Further, living with one's children increases mutual bonding.

By contrast, single parenthood, especially after a bitter divorce, correlates with poor health and low income for everyone. Simply not seeing one parent very often increases internalizing and externalizing problems in children.

Contact tends to increase affection and care. Recent data come from Russia, where economic and social pressures have led many single men to drink and despair, dying years earlier than married men. The reason is thought to be that the husband/father role leads men to take better care of themselves and wives to look out for their husband's health (Ashwin & Isupova, 2014).

Shared parenting also decreases the risk of child maltreatment, because one parent is likely to protect their children if the other is abusive or neglectful. For all children, having two parents around every day makes it more likely that someone will read to them, check their homework, invite their friends over, buy them new clothes, and save for their education. Of course, having two married parents does not guarantee good care. One of my students wrote:

> My mother externalized her feelings with outbursts of rage, lashing out and breaking things, while my father internalized his feelings by withdrawing, being silent and looking the other way. One could say I was being raised by bipolar parents. Growing up, I would describe my mom as the Tasmanian devil and my father as the ostrich, with his head in the sand. . . . My mother disciplined with corporal punishment as well as with psychological control, while my father was permissive. What a pair.
>
> [C., 2013]

This student is now a single parent, having twice married, given birth, and divorced. She is one example of a general finding: The effects of childhood family function echo in adulthood, financially as well as psychologically.

Single Fathers and Stepfathers

Generally, fathers who do not live with their children become less involved with them every year. When the children reach age 18, fathers are no longer legally responsible, and divorced or unmarried fathers may no longer pay for education or other expenses.

This is a harsh reality in today's world. Emerging adults usually need substantial funds as well as emotional support in order to attend college, find jobs, and become self-sufficient adults. Stepfathers and nonresident biological fathers are less likely to provide that help, in part because each thinks the other should be responsible. There is much variation here, with the biological mother and the father's income often crucial, but generally children of separated parents are less likely to attend college than children living with both biological parents (King et al., 2020; de Leeuw & Kalmijn, 2020).

DAVID MAXWELL/The New York Times/Redux

Middle American Family This photo shows a typical breakfast in Brunswick, Ohio—Cheerios for 1-year-old Carson, pancakes that 7-year-old Carter does not finish eating, and family photos crowded on the far table.

⬤ **Observation Quiz** What do you notice about this family? (see answer, page 367) ↑

VISUALIZING DEVELOPMENT

Family Structures Around the World

Children fare best when both parents actively care for them every day. This is most likely to occur if the parents are married, although there are many exceptions. Many develop-mentalists focus on the rate of single parenthood, shown on this map. Single parents often raise children well, especially with support from their families, friends, and communities.

RATES OF SINGLE PARENTHOOD

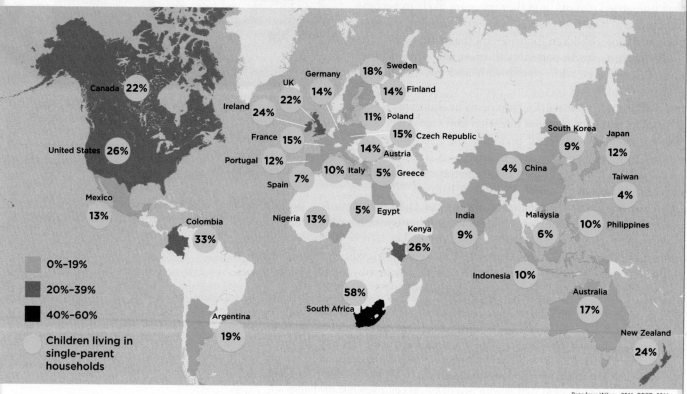

Canada **22%**
United States **26%**
Mexico **13%**
Colombia **33%**
Argentina **19%**
UK **22%**
Ireland **24%**
France **15%**
Portugal **12%**
Spain **7%**
Germany **14%**
Sweden **18%**
Finland **14%**
Poland **11%**
Czech Republic **15%**
Austria **14%**
Italy **10%**
Greece **5%**
Egypt **5%**
Nigeria **13%**
Kenya **26%**
South Africa **58%**
South Korea **9%**
Japan **12%**
Taiwan **4%**
China **4%**
India **9%**
Malaysia **6%**
Philippines **10%**
Indonesia **10%**
Australia **17%**
New Zealand **24%**

0%–19%
20%–39%
40%–60%

Children living in single-parent households

Data from Wilcox, 2011; OECD, 2011.

LIVING ARRANGEMENTS OF U.S. 0- TO 18-YEAR-OLDS

Note that, while fewer children live with their two married biological parents from birth to age 18, it is not that more children are living in stepfamilies but that more individuals have decided to raise children on their own. Another shift is evident: Single parents once were almost always mothers, but now some are single fathers.

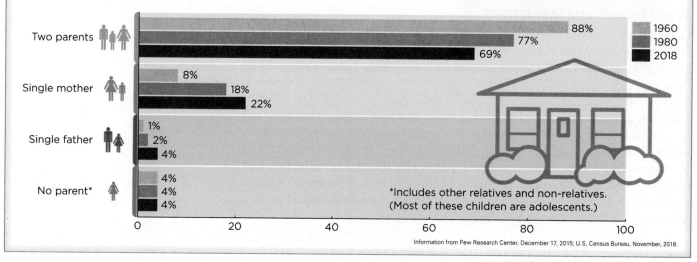

Two parents **88%** / **77%** / **69%**

1960
1980
2018

Single mother **8%** / **18%** / **22%**

Single father **1%** / **2%** / **4%**

No parent* **4%** / **4%** / **4%**

*Includes other relatives and non-relatives. (Most of these children are adolescents.)

0   20   40   60   80   100

Information from Pew Research Center, December 17, 2015; U.S. Census Bureau, November, 2018.

Fortunate Boys This single father *(left)* in Pennsylvania takes his three sons to the playground almost every day, and this nuclear family *(right)* in Mali invests time and money in their only child's education. All four boys have loving fathers. Does family function make family structure irrelevant?

Courts and social workers are increasingly recommending joint physical custody of children after a divorce. In general, when both parents are directly involved in caregiving, children of divorce are healthier, physically and emotionally, than when only one parent has custody (Baude et al., 2016; Braver & Votruba, 2018). A major reason is that the father is a more active parent, and that benefits the children.

When a father is the lone caregiver, he suffers the same problems as single mothers—too much to do and not enough money to do it. Single parents of both sexes often seek a new spouse, in part to help with parenthood. This may not work out as hoped (Booth & Dunn, 2014).

The data suggest that remarriage benefits the adults more than the children. Not only is parental support, emotional as well as financial, diminished for the child, but also the relationship suffers. Children may blame one or both biological parents for the separation, and may reject the stepparent, in part because they are loyal to their absent parent.

Not surprisingly, many stepparents find it hard to bond with a child of their new partner's former lover, especially when the child is hostile. As both a cause and a correlate, adult stepchildren are less likely than biological children to live near their parents.

Practical circumstances matter, too. A major problem for the children when parents remarry is that many parts of their lives change yet again, including such daily routines as where they sleep, what they eat, and where they go to school. Remember that stability and harmony are especially coveted in middle childhood.

One result: Children often are angry or sad, fight with friends, fail in school, rebel against family rules, and harm themselves (e.g., cutting, accidents, eating disorders, and so on). Disputes between half-siblings and stepsiblings are common, because every child is upset.

Their parents often have opposite strategies for discipline and guidance. This occurs with biological parents as well, but a solid parental alliance is particularly elusive when it includes three adults—two of whom disliked each other so much that they divorced, and a third who does not know the history of the child.

Remember that although structure affects function, structure does not determine function. Some stepparent families function well. As a review of research on stepfamilies concludes:

> The case that divorce and family instability reduce children's well-being is strong. At the same time, the magnitude of these consequences vary considerably across individuals and groups.

[Raley & Sweeney, 2020, p. 92]

Single-Parent Families

On average, the single-parent structure functions less well for children because single parents have less income, time, and stability. Most single parents fill many roles—including wage earner, daughter or son (single parents often depend on their own parents), and lover (many seek a new partner). That reduces time for emotional and academic support for children. If they are depressed (and many are), that makes it worse. Neesha, in A Case to Study, is an example.

However, many parents and communities overcome structural barriers in order to provide love and support to children. Contrary to the averages, thousands of single-parent families are wonderful, children in nuclear families are not guaranteed good care, and extended families, in particular, are sometimes protective.

Culture is always influential. In the slums of Mumbai, India, rates of psychological disorders among school-age children were *higher* in nuclear families than in extended families, presumably because grandparents, aunts, and uncles provided more care and stability in that city than two parents alone (Patil et al., 2013). The opposite is true for many families in the United States: Extended families are

---

A CASE TO STUDY

How Hard Is It to Be a Kid?

Neesha's fourth-grade teacher referred her to the school guidance team because Neesha often fell asleep in class, was late 51 days, and was absent 15 days. Testing found Neesha at the seventh-grade level in reading and writing and at the fifth-grade level in math. Since achievement was not Neesha's problem, something psychosocial must be amiss.

The counselor spoke to Neesha's mother, Tanya, a single parent who was depressed and worried about paying the rent on a tiny apartment where she had moved when Neesha's father left three years earlier. He lived with his girlfriend, now with a new baby as well. Tanya said she had no problems with Neesha, who was "more like a little mother than a kid," unlike her 15-year-old son, Tyrone, who suffered from fetal alcohol effects and whose behavior worsened when his father left.

Tyrone was recently beaten up badly as part of a gang initiation, a group he considered "like a family." He was currently in juvenile detention, after being arrested for stealing bicycle parts. Note the nonshared environment here: Although the siblings grew up together and their father left them both, 12-year-old Tyrone became rebellious whereas 7-year-old Neesha became *parentified*, "a little mother."

The school counselor also spoke with Neesha.

> Neesha volunteered that she worried a lot about things and that sometimes when she worries she has a hard time falling asleep. . . . she got in trouble for being late so many times, but it was hard to wake up. Her mom was sleeping late because she was working more nights cleaning offices. . . . Neesha said she got so far behind that she just gave up. She was also having problems with the other girls in the class, who were starting to tease her about sleeping in class and not doing her work. She said they called her names like "Sleepy"

and "Dummy." She said that at first it made her very sad, and then it made her very mad. That's when she started to hit them to make them stop.

> [Wilmshurst, 2011, pp. 152–153]

Neesha was coping with poverty, a depressed mother, an absent father, a delinquent brother, and classmate bullying. She seemed resilient—her achievement scores were impressive—but shortly after Neesha was interviewed,

> The school principal received a call from Neesha's mother, who asked that her daughter not be sent home from school because she was going to kill herself. She was holding a loaded gun in her hand and she had to do it, because she was not going to make this month's rent. She could not take it any longer, but she did not want Neesha to come home and find her dead. . . . While the guidance counselor continued to keep the mother talking, the school contacted the police, who apprehended [the] mom while she was talking on her cell phone. . . . The loaded gun was on her lap. . . . The mother was taken to the local psychiatric facility.

> [Wilmshurst, 2011, pp. 154–155]

Whether Neesha's resilience would continue depended on her ability to find support beyond her family. Perhaps the school counselor helped:

> When asked if she would like to meet with the school psychologist once in a while, just to talk about her worries, Neesha said she would like that very much. After she left the office, she turned and thanked the psychologist for working with her, and added, "You know, sometimes it's hard being a kid."

> [Wilmshurst, 2011, p. 154]

more chaotic, exactly what children do not need. But not always. One factor is immigration status. In the United States, extended families may function better for immigrant families than for the U.S. born (Areba et al., 2018).

 DATA CONNECTIONS: Family Structure in the United States and Around the World examines the many types of families.

Family Trouble

However, while it is true that structure does not determine function, two factors always correlate with dysfunction: low income and high conflict.

Wealth and Poverty

Family income affects both function and structure. Marriage rates fall in times of economic recession, and divorce increases with unemployment. Low SES correlates with many other problems, and "risk factors pile up in the lives of some children, particularly among the most disadvantaged" (Masten, 2014, p. 95).

Several scholars have developed the *family-stress model*, which holds that any risk factor (such as poverty, divorce, single parenthood, unemployment) damages a family if, and only if, it increases stress on the parents, making them less patient and responsive to the children (Masarik & Conger, 2017).

Ongoing economic hardship almost always increases family stress (Duran et al., 2019). When parents fear that they cannot provide food and shelter for their children, their worry about the future renders them tense and hostile. The parents' *reaction* to income may exacerbate or minimize stress (Evans & Kim, 2013; Lee et al., 2013; Mazza et al., 2017).

A curious correlation is evident: Children in high-income families are more likely to have developmental problems in adulthood than children of middle-SES parents (Luthar et al., 2018). Stress may be the underlying problem: Some wealthy parents are anxious about maintaining their status, and their stress makes them pressure their children to excel. Their children may have emotional problems in middle childhood and drug abuse, delinquency, and academic failure later on.

But do not conclude that children are better off poor than rich. Poverty always adds stress; high income sometimes does. The crucial question is whether economic pressures affect the parents' ability to provide children the attention and guidance they need (Roubinov & Boyce, 2017). Back to family routines: If parents set clear guidelines for sleep, eating, and home-work, children benefit. If parents are too caught up in their own financial concerns, children suffer.

The parents' stress is sometimes mitigated by national policy. In Norway, for instance, family poverty has minimal effect on children because national health care, high-quality early education, and national public schools provide a "buffering effect of the social safety net" (Bøe et al., 2018, p.1). (See **Figure 13.2**.) In the United States, parents must do that buffering themselves.

Conflict

Every researcher agrees that family conflict harms children, especially when adults fight about child rearing. Such fights are more common in stepfamilies, divorced families, and extended families, but nuclear families are not immune. Children suffer not only if they are abused, physically or

FIGURE 13.2

Families and Schools This graph shows the score gap in fourth-grade science on the 2015 TIMSS between children in schools where more than 25 percent of the children are from high-SES homes compared to children in schools where more than 25 percent are from low-SES homes. Generally, the nations with the largest gaps are also the nations with the most schools at one or the other end of the spectrum and the fewest in between. For example, 23 percent of the children in the United States attended schools where less than a fourth of the students were from high- or low-SES families, compared with 37 percent of the Japanese children.

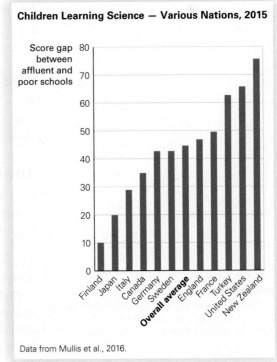

Children Learning Science — Various Nations, 2015

Data from Mullis et al., 2016.

THINK CRITICALLY: Can you describe a situation in which having a single parent would be better for a child than having two parents?

emotionally, but also if they merely witness their parents' abuse of each other or of their other children. Fights between siblings can be harmful, too (Buehler, 2020).

This correlation raises a possible hypothesis. Might genes from the parents to the children harm every child, whether or not they witness family fights? If that were so, the correlation between family conflict and child problems is caused by a third variable, specifically genes.

This hypothesis was tested in an amazing longitudinal study of 867 pairs adult twins (388 monozygotic and 479 dizygotic), who had married and had an adolescent child (Schermerhorn et al., 2011). Both parents were asked independently about marital conflict. The teenagers' problems were compared to their cousin, who was the child of their parent's twin.

Thus, this study had data from 5,202 individuals—one-third of them adult twins, one-third of them spouses of twins, and one-third of them adolescents who were genetically linked to another adolescent. If their parent was a monozygotic twin, they had one-fourth of their genes in common with their cousin; if their parent was a dizygotic twin, one-eighth of the same genes.

The researchers found that although genes had some effect, witnessing conflict itself had a powerful impact, increasing externalizing problems in the boys and internalizing problems in the girls. Quiet disagreements did little harm, but open conflict (e.g., yelling heard by the children) and divorce was harmful (Schermerhorn et al., 2011). That leads to an obvious conclusion: Parents should not fight in front of their children.

WHAT HAVE YOU LEARNED?

1. How might siblings raised together not share the same family environment?

2. What is the difference between family structure and family function?

3. Why is a harmonious, stable home particularly important during middle childhood?

4. What are the advantages for children in a nuclear family structure?

5. Why might the single-parent structure function less well than two-parent structures.

6. How are family structure and family function affected by culture?

7. Using the family-stress model, explain how family income affects family function.

The Peer Group

Peers become increasingly important in middle childhood. With their new awareness of reality (concrete operational thought), children become painfully aware of their classmates' opinions, judgments, and accomplishments.

The Culture of Children

Peer relationships, unlike adult–child relationships, involve partners who negotiate, compromise, share, and defend themselves as equals. Consequently, children learn social lessons from one another that grown-ups cannot teach (Rubin et al., 2013). Adults may follow a child's lead, but they are always much older and bigger, with their own values and experiences. They cannot substitute for a friend.

Child culture includes the customs, rules, and rituals that are passed down to younger children from slightly older ones, with no thought about the origins or implications. The child's goal is to join a culture and thus be part of the peer group. Jump-rope rhymes, insults, and superstitions are examples.

child culture The idea that each group of children has games, sayings, clothing styles, and superstitions that are not common among adults, just as every culture has distinct values, behaviors, and beliefs.

For instance, "Ring around the rosy/Pocketful of posies/Ashes, ashes/We all fall down" may have originated as children coped with the Black Death, which killed half the population of Europe in the fourteenth century. (*Rosy* may be short for *rosary*, used by Roman Catholics for prayer.) Children have passed down that rhyme for centuries, laughing together with no thought of sudden death. Rigid distinctions between boys' and girls' culture have eroded, as child culture reflects as well as forecasts changes in the overall culture (Van Rheenen, 2012).

Throughout the world, child culture may clash with adult culture. Many children reject clothes that parents buy as too loose, too tight, too long, too short, or wrong in color, style, brand, decoration, or some other aspect that adults might not notice. If their schools are multiethnic, children may choose friends from other groups, unlike their parents.

Appearance is important for child culture, but more important is independence from adults. Classmates pity those (especially boys) whose parents kiss them ("mama's boy"), tease children who please the teachers ("teacher's pet," "suck-up"), and despise those who betray children to adults ("tattletale," "grasser," "snitch," "rat"). Keeping secrets from parents and teachers is a moral mandate.

The culture of children is not always benign. For example, because communication with peers is vital, children learn whatever language or dialect the other children speak. Immigrant parents proudly note how well their children speak a second language, but all parents are distressed when their children spout their peers' curses, accents, and slang. Because they value independence, children may gravitate toward friends who defy authority, sometimes harmlessly (passing a note in class), sometimes not (shoplifting, smoking).

This is part of the nature of children, who often do what their parents do not want them to do, and it is in the nature of parents to be upset when that happens. This is easier to criticize in other cultures and centuries, as in the following example.

In 1922, the magazine *Good Housekeeping* published an article titled "Aren't you glad you are not your grandmother?" In it, a daughter quotes letters from her dead grandmother that she found in the attic. One describes an incident that occurred when that daughter's father—also long dead—was a boy and snuck out of his house to play with other boys:

> When the door was left unlocked for a moment, out he ran in his little velvet suit. We did not miss him for a while because we thought he was doing his Latin Prose, and then some wealthy ladies . . . saw him literally in the gutter, groping in the mud for a marble. . . . Horace's father was white with emotion when he heard of it. He went out, brought Horace in, gave him another whipping, and, saying that since he acted like a runaway dog he should be treated like one, he went out, bought a dog-collar and a chain, and chained Horace to the post of his little bed. He was there all the afternoon, crying so you could hear nothing else in all the house. . . . I went many times up to the hall before his door and knelt there stretching out my arms to my darling child, the tears flooding down my cheeks. But, of course, I could not open the door and go in to him, to interfere with his punishment.
>
> [Fisher, 1922, p. 8]

The author was grateful that mothers now (in 1922!) knew more than did nineteenth-century parents with their "ignorance of child-life" (Fisher, 1922, p. 15). This raises the question: What ignorance of child-life do we have today? If I knew, that would not be ignorance, but this text makes me humble; each new generation develops a child culture that may teach their elders.

No Toys Many boys in middle childhood are happiest playing outside with equipment designed for work. This wheelbarrow is perfect, especially because at any moment the pusher might tip it.

E. Hanazaki Photography/Moment/Getty Images

Friendships

Teachers sometimes separate friends, but that may be a mistake. Developmentalists find that children help each other learn both academic and social skills (Bagwell & Schmidt, 2011). The loyalty of children to their friends may work for their benefit or harm (Rubin et al., 2013).

Both aspects of friendship are expressed by these two Mexican American children.

Yolanda:

There's one friend . . . she's always been with me, in bad or good . . . She's always telling me, "Keep on going and your dreams are gonna come true."

Paul:

I think right now about going Christian, right? Just going Christian, trying to do good, you know? Stay away from drugs, everything. And every time it seems like I think about that, I think about the homeboys. And it's a trip because a lot of the homeboys are my family, too, you know?

[quoted in Nieto, 2000, pp. 220, 249]

Yolanda later went to college; Paul went to jail. This is echoed by other children. Many aspects of adult personality are influenced by the personality of childhood friends (Wrzus et al., 2016).

Indeed, quite apart from family, school, and IQ, a study found that the intelligence of a best friend in sixth grade affected intelligence at age 15 (Meldrum et al., 2018). Peers benefit children or not. As one study concludes, if low-achievers "selected similarly low-achieving students as friends, this may dampen their academic achievement over time" (Laninga-Wijnen et al., 2019, p. 347).

Friendships become more intense and intimate over the years of middle childhood, as social cognition and effortful control advance. Six-year-olds may befriend anyone of the same sex and age who will play with them. By age 10, children demand more. They choose carefully, share secrets, expect loyalty, change friends less often, are upset when they lose a friend, and find it harder to make new friends.

Older children tend to choose friends whose interests, values, and backgrounds are similar to their own. By the end of middle childhood, close friendships are almost always between children of the same sex, age, ethnicity, and socioeconomic status (Rubin et al., 2013). Both genders learn how to become good friends, with girls becoming better at sympathetic reassurance and boys becoming better at joint excitement. Everyone finds friendship increasingly satisfying over the years of childhood (Rose & Asher, 2017).

Popular and Unpopular Children

In the United States, two types of popular children and three types of unpopular children have become apparent in middle childhood (Cillessen & Marks, 2011). First, at every age, children who are friendly and cooperative are well-liked and popular. By the end of middle childhood, as status becomes important, another avenue to popularity begins: Some popular children are also aggressive (Shi & Xie, 2012).

As for the three types of unpopular children, some are *neglected*, not rejected; they are ignored, but not shunned. The other two types are actively rejected, either *aggressive-rejected*, disliked because they are antagonistic and confrontational, or *withdrawn-rejected*, disliked because they are timid and anxious. Children as young as age 6 are aware if they are rejected and decide if they should try to be more accepted or should seek other friends (Nesdale et al., 2014).

> **THINK CRITICALLY:** Do adults also choose friends who agree with them or whose background is similar to their own?

Both aggressive-rejected and withdrawn-rejected children often misinterpret social situations, lack emotional regulation, and experience mistreatment at home. Each of these problems not only cause rejection, but the rejection itself makes it worse for the child (Stenseng et al., 2015). If they do not learn when to be assertive and when to be quiet, they may become bullies or victims.

Whether a particular child is popular or not depends on cultural norms, which change over time. As you read in Chapter 7, some parts of temperament are genetic. Some people are temperamentally more aggressive, more outgoing, or more fearful than others. But culture affects whether those inborn traits are accepted, channeled, or curbed.

This is illustrated by research on shyness (X. Chen, 2019). A 1990 survey in Shanghai found that shy children were liked and respected (X. Chen et al., 1992), but 12 years later, when competition with the West became salient, shy children in the same schools were less popular (X. Chen et al., 2005).

Other research found that shyness is more accepted in rural China (Zhang & Eggum-Wilkens, 2018). In general, several aspects of the social context and culture, specifically academic achievement, friendships, and being in middle childhood (not adolescence) all make shy Chinese children more accepted by their peers (X. Chen et al., 2019; X. Chen et al., 2013; Liu et al., 2015).

Within the United States, a similar shift from acceptance to rejection of a personality trait is evident regarding aggression in males. Once that was acceptable ("Boys will be boys!"), but now teachers, both women and men themselves, teach boys to restrain their impulse to lash out. This is most apparent in the new understanding of bullies, as now explained.

Bullying

Bullying is defined as repeated, systematic attacks intended to harm those who are unable or unlikely to defend themselves. It occurs in every nation, in every community, in every kind of school (religious/secular, public/private, progressive/traditional, large/medium/small), and perhaps in every child.

Bullying is of four types:

- *Physical* (hitting, pinching, shoving, or kicking)
- *Verbal* (teasing, taunting, or name-calling)
- *Relational* (destroying peer acceptance)
- *Cyberbullying* (using electronic means to harm another)

The first three types are common in primary school and begin in preschool. Cyberbullying is more common later on and is discussed in Chapter 16.

Victims

Almost every child experiences an isolated attack or is called a derogatory name during middle childhood. Victims of bullying, however, endure shameful experiences again and again — pushed and kicked, called names, forced to do degrading things, and so on — with no defense. Victims tend to be "cautious, sensitive, quiet . . . lonely and abandoned at school. As a rule, they do not have a single good friend in their class" (Olweus, 1999, p. 15).

Even having a friend who is also a victim helps. Such friends may not be able to provide physical protection, but they can and do provide psychological defense — reassuring victims that their condition is not their fault and that the bully is mean, stupid, racist, or whatever (Schacter & Juvonen, 2018). That is crucial, because the worst harm is loss of self-respect.

LaunchPad
macmillan learning

VIDEO ACTIVITY: Bullying: Interview with Nikki Crick explores the causes and repercussions of the different types of bullying.

bullying Repeated, systematic efforts to inflict harm on other people through physical, verbal, or social attack on a weaker person.

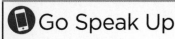
bully-victim Someone who attacks others and who is attacked as well. (Also called *provocative victims* because they do things that elicit bullying.)

Especially for Parents of an Accused Bully Another parent has told you that your child is a bully. Your child denies it and explains that the other child doesn't mind being teased. What should you do? (see response, page 367)

Although it is often thought that victims are particularly unattractive or odd, this is not necessarily the case. Victims are chosen because of their emotional vulnerability and social isolation, not their appearance. Children who are new to a school, whose background and home culture are unlike that of their peers, who are gender nonconforming, who have disabilities, or whose clothes indicate poverty, are especially vulnerable. Bullies can find something (fat or thin, glasses or unruly hair, an odd accent or unusual shoes) in almost any child to belittle.

As one boy said,

> You can get bullied because you are weak or annoying or because you are different. Kids with big ears get bullied. Dorks get bullied. You can also get bullied because you think too much of yourself and try to show off. Teacher's pet gets bullied. If you say the right answer too many times in class you can get bullied. There are lots of popular groups who bully each other and other groups, but you can get bullied within your group too. If you do not want to get bullied, you have to stay under the radar, but then you might feel sad because no one pays attention to you.

[*quoted in Guerra et al., 2011, p. 306*]

Remember the three types of unpopular children? *Neglected* children are not victimized; they are ignored, "under the radar." *Rejected* children fit into the bully network. Withdrawn-rejected children are likely victims; they are isolated, depressed, and friendless. Aggressive-rejected children may be **bully-victims** (or *provocative victims*), with neither friends nor sympathizers (Kochel et al., 2015). They suffer the most, because they strike back ineffectively, which increases the bullying.

Bullies

Unlike bully-victims, most bullies are *not* rejected. Many are proud, pleased with themselves, with friends who admire them and classmates who fear them. Some are quite popular: Bullying is a form of social dominance and authority (Pellegrini et al., 2011).

The link between bullying and popularity has long been apparent during early adolescence (Pouwels et al., 2016), but bullies are already "quite popular in middle childhood." Adults, however, have become aware that bullying is destructive. As one comprehensive summary of the research explains, bullying "is now recognized as a major and preventable public health problem" (National Academies of Sciences, Engineering, and Medicine, 2016, p. 13).

Bullying begins during the preschool years, and teachers often recognize it and try to stop it. What changes from ages 6 to 12 is that bullies become skilled at avoiding adult awareness, at picking rejected and defenseless victims, and at using nonphysical methods—which avoid adult punishment (Pouwels et al., 2017). As children become better at hiding bullying from adults, the harm to victims increases (Nelson et al., 2019).

Boys are bullies more often than girls, typically attacking smaller, weaker boys. Girl bullies usually use words to demean shyer, more soft-spoken girls. Young boys sometimes bully girls, but by puberty (about age 11), boys who bully girls are not admired, although sexual teasing is. Especially in the final years of middle childhood, boys who are thought to be gay become targets, with suicide attempts as one consequence (National Academies of Sciences, Engineering, and Medicine, 2016).

Causes and Consequences of Bullying

Bullying may begin early in life. Most toddlers try to dominate other children (and perhaps their parents) at some point. When they hit, kick, and so on, usually their parents, teachers, and peers teach them to find other ways to interact. However, if home life is chaotic, if discipline is ineffectual, if siblings are hostile, or if attachment is insecure, children do not learn how to express their frustration. Instead, vulnerable young children develop externalizing and internalizing problems, becoming bullies or victims (Turner et al., 2012).

By middle childhood, bullying is not the outburst of a frustrated child but an attempt to gain status. That makes it a social action: Bullies rarely attack victims when the two of them are alone. Instead, a bully might engage in a schoolyard fight, with onlookers who are more likely to cheer the victor than stop the fight; or a bully might utter an insult that provokes laughter in all except the target. By the end of middle childhood, bullies choose victims whom other children reject.

Siblings matter. Some brothers and sisters defend each other; children are protected if bullies fear that an older sibling will retaliate. On the other hand, if children are bullied by peers in school *and* siblings at home, they are four times more likely to develop serious psychological disorders by age 18 (Dantchev et al., 2018).

Bullies and victims risk impaired social understanding, lower school achievement, and relationship difficulties, with higher rates of mental illness in adulthood. Many victims become depressed; many bullies become increasingly cruel with higher rates of prison and death (Willoughby et al., 2014). The damage goes even further: In schools with high rates of bullying, all of the children are less likely to focus on academics and more likely to concentrate on the social dynamics of the classroom—hoping to avoid becoming the next victim.

Power to Peers Bullying is a way some children gain respect. If, instead, the school gives training and special shirts to bystanders, they can gain status by befriending victims. That seems to work in this school in Bensalem, Pennsylvania.

Can Bullying Be Stopped?

Many victimized children find ways to halt ongoing bullying—by ignoring, retaliating, defusing, or avoiding. Friendships help.

We know what does *not* work: simply increasing students' awareness of bullying, instituting zero tolerance for fighting, or putting bullies together in a therapy group or a classroom. This last measure tends to make daily life easier for teachers, but it increases aggression.

Since one cause of bullying is poor parent–child interaction, alerting parents may "create even more problems for the child, for the parents, and for their relationship" (Rubin et al., 2013, p. 267). This does not mean that parents should be kept ignorant, but it does mean that parents need help in understanding how to break the bully-victim connection (Nocentini et al., 2019).

This is a theme of an article about parents' attitudes. Many parents were naïve about bullying, some claiming that their child would not experience it because the school was small (Stives et al., 2019). Instead of recognizing that bullies gain status by bullying, some parents thought bullies needed sympathy. One said that, if his son told him he was bullied,

> I would explain to him that bullies usually are very unhappy people and have a tendency to have low self-esteem. Most bullies sometimes don't have a loving/ caring home environment. I would tell him to be kind to any bullies. Be a friend whenever possible.

[quoted in Stives et al., 2019, p. 367]

That response helps neither victim nor bully.

To decrease bullying, the entire school should be involved, with teachers, aides, children, and parents, all taught to recognize bullying and how to reduce it (Juvonen & Graham, 2014). A Spanish concept, *convivencia*, describes a culture of cooperation and positive relationships within a community. Convivencia has been applied specifically to schools. When teachers are supportive and protective of every child, and the school community encourages friendship, empathy, and cooperation among all students, bullying decreases (Zych et al., 2017).

Programs that seem good might be harmful, especially if they call attention to bullying but do nothing about it (National Academies of Sciences, Engineering, and Medicine, 2016). Longitudinal research on whole-school efforts finds variations depending on the age of the children (younger is easier), on the indicators (peer report, teacher report, absence rate, direct observation), as well as on the tactics (encouraging friendship and decreasing adult hostility is more effective than punishing overt bullying).

Bystanders are crucial: If they do not intervene—or worse, if they watch and laugh—bullying flourishes. Some children who are neither bullies nor victims feel troubled but also feel fearful and powerless. However, if they empathize with victims, feel effective (high in effortful control), and refuse to admire bullies, aggression is reduced. The best way for teachers to help victims is sometimes to encourage bystanders to befriend the victim or stop the bully (Iotti et al., 2020). Bullies ignore, avoid, or even enjoy criticism from authorities; it is harder to do so with peers.

Appreciation of human differences is not innate (remember, children seek friends who are similar to them), so adults need to encourage multicultural sensitivity. Then peers are more effective than teachers at halting bullying (Palmer & Abbott, 2018). As they mature during middle childhood, children become more socially aware, which creates a dilemma: Children become more aware of the emotional harm that bullies do to other children, but also more aware that they themselves might be harmed if they befriend a bullied child.

This explains an odd dynamic: A victimized child is less distressed if their friend is also victimized (Schacter & Juvonen, 2019). This also raises a crucial issue in middle childhood—moral development.

Children's Morality

Some moral values seem inborn. Babies prefer a puppet who is helpful to other puppets over a mean puppet, and young children believe that desired objects (cookies, stickers, candy) should be shared equally. The ideas of fairness, kindness, and equality are present in the minds of children (Rizzo & Killen, 2016; Van de Vondervoort & Hamlin, 2016).

However, the young child's idea of morality is quite limited. Middle childhood is prime time for moral development. These are

> years of eager, lively searching on the part of children . . . as they try to understand things, to figure them out, but also to weigh the rights and wrongs. . . . This is the time for growth of the moral imagination, fueled constantly by the willingness, the eagerness of children to put themselves in the shoes of others.
>
> *[Coles, 1997, p. 99]*

Many lines of research have shown that children develop their own morality, guided by peers, parents, and culture (Killen & Smetana, 2014). Children's growing interest in moral issues is guided by three forces: (1) child culture, (2) empathy, and (3) education.

Moral Rules of Child Culture

First, when child culture conflicts with adult morality, children often align themselves with peers. A child might lie to protect a friend, for instance. Friendship itself has a hostile side: Many close friends (especially girls) resist other children who want to join their play (Rubin et al., 2013). Boys are particularly likely to protect a bully if he is a friend.

THINK CRITICALLY: The text says that both former bullies and former victims suffer in adulthood. Which would you rather be, and why?

Parents Versus Peers

The fact that children place prime value on being accepted by peers sometimes results in a conflict between the moral values of society and the actions of children. A child may do something that the parents think their child would never do. This is not time to argue, or despair; it is time to teach. This is illustrated by another memory that Ward Sutton (who opened this chapter) told at his Herblock Prize acceptance speech in 2018.

> In the summer of 1974, I was seven years old.
>
> There was a kid living in my neighborhood. . . .
>
> For some reason this kid had built up animosity towards a family down the block. He started telling my friend Steve and me that this family had done all sorts of bad things, like hiding razor blades in the apples they gave out to trick-or-treaters at Halloween.
>
> Fake news for seven year olds . . . target people who had done absolutely nothing wrong.
>
> Steve became convinced of the conspiracy theory and fell in line, and then the two of them told me they were starting a club and I couldn't join unless I went along with them. . . .
>
> We snuck behind the family's house in the woods out back. We threw some rocks at the house and when there was no response we realized the family was not home. Then we escalated things, finding bricks, smashing windows, breaking in and vandalizing the home.
>
> I was a shy, introverted kid who never would have ever done anything like this on my own. But once I was swayed to join in, it was like a switch had been flipped and any sense of right or wrong was thrown by the wayside. Suddenly the unthinkable was okay. . . .
>
> Then the family came home. We ran. Police arrived. Steve was caught. I lied. To my parents. To everyone. Said I didn't know anything about it. Steve confessed, and eventually I did, too. As terrible as it all was, the worst of it was the fact that I had lied. My mother wouldn't speak to me for what felt like an eternity. . . .
>
> My father brought me back to the house to apologize to the family face-to-face. I begged him not to make me do it . . . I expected them to be angry with me, but they weren't. They were gracious, and mostly puzzled at what could have possibly possessed me to do something like this.
>
> I couldn't even explain why I had done it.
>
> As you might imagine, this episode was hugely formative for me. It awakened my moral compass and informed what I would create going forward. . . .
>
> I've . . . never spoken of it publicly until today. For the longest time, I wished I could live that day all over again, and that someone could have talked some sense into me: "Stop and think about what you're doing."

Sutton's cartoons are often ethical comments on world affairs. He says he draws them because he hopes people will smile, and then stop and think.

Three moral imperatives of child culture in middle childhood are:

- Defend your friends.
- Don't tell adults about children's misbehavior.
- Conform to peer standards of dress, talk, and behavior.

These three can explain both apparent boredom and overt defiance as well as standards of dress that mystify adults (such as jeans so loose that they fall off or so tight that they impede digestion—both styles worn by my children, who grew up in different cohorts). Given what is known about middle childhood, it is no surprise that children do not echo adult morality.

Part of child culture is that as children become more aware of their peers, they may reject other children who are outsiders as well as stay quiet about their own problems. When teachers ask, "Who threw that spitball?" or parents ask, "How did you get that bruise?" children may be mum.

Empathy

The second factor, empathy, is key. As middle childhood advances, children become more socially perceptive and more able to learn about other people (Weissberg et al., 2016). This does not always lead to increased morality as adults might define it. One example was just described: Bullies become adept at picking victims, and bystanders become better at noticing victims. However, depending on the culture of their school and home, social awareness may make them either quicker to defend or hesitant to act (Pozzoli & Gini, 2013).

Steve Russell/Toronto Star/Getty Images

Universal Morality Remarkable? Not really. By the end of middle childhood, many children are eager to express their moral convictions, especially with a friend. Chaim Ifrah and Shai Reef believe that welcoming refugees is part of being a patriotic Canadian and a devout Jewish person, so they brought a welcoming sign to the Toronto airport where Syrian refugees (mostly Muslim) will soon deplane.

The authors of a study of 7-year-olds "conclude that moral *competence* may be a universal human characteristic, but that it takes a situation with specific demand characteristics to translate this competence into actual prosocial performance" (van Ijzendoorn et al., 2010, p. 1). In other words, school-age children can think and act morally, but they do not always do so because the hidden curriculum, or adult values, or peer pressure may lead them astray.

Here, diversity in schools and neighborhoods can be helpful. Empathy is not an abstract idea as much as a recognition of the basic humanity of other people. In order to achieve that, knowing a child from another group lets children understand. Teachers and parents can help with this, not only through direct contact but, once children can read on their own, by offering books about children in other lands, centuries, and cultures.

Moral Education

Finally, cognitive development might affect moral development, at least according to Piaget (1932/2013b) and then Kohlberg (1963), who described three levels of moral reasoning and two stages at each level (see **Table 13.3**), with parallels to Piaget's stages of cognition.

preconventional moral reasoning
Kohlberg's first level of moral reasoning, emphasizing rewards and punishments.

conventional moral reasoning
Kohlberg's second level of moral reasoning, emphasizing social rules.

- **Preconventional moral reasoning** is similar to preoperational thought in that it is egocentric, with children most interested in their personal pleasure or avoiding punishment.
- **Conventional moral reasoning** parallels concrete operational thought in that it relates to current, observable practices: Children watch what their parents, teachers, and friends do and try to follow suit.

TABLE 13.3

Kohlberg's Three Levels and Six Stages of Moral Reasoning

Level I: Preconventional Moral Reasoning

The goal is to get rewards and avoid punishments; this is a self-centered level.

- *Stage one: Might makes right* (a punishment-and-obedience orientation). The most important value is to maintain the appearance of obedience to authority, avoiding punishment while still advancing self-interest. Don't get caught!
- *Stage two: Look out for number one* (an instrumental and relativist orientation). Everyone prioritizes their own needs. The reason to be nice to other people is so that they will be nice to you.

Level II: Conventional Moral Reasoning

Emphasis is placed on social rules; this is a parent- and community-centered level.

- *Stage three: Good girl and nice boy.* The goal is to please other people. Social approval is more important than any specific reward.
- *Stage four: Law and order.* Everyone must be a dutiful and law-abiding citizen, even when no police are nearby.

Level III: Postconventional Moral Reasoning

Emphasis is placed on moral principles; this level is centered on ideals.

- *Stage five: Social contract.* Obey social rules because they benefit everyone and are established by mutual agreement. If the rules become destructive or if one party doesn't live up to the agreement, the contract is no longer binding. Under some circumstances, disobeying the law is moral.
- *Stage six: Universal ethical principles.* Universal principles, not individual situations (level I) or community practices (level II), determine right and wrong. Ethical values (such as "life is sacred") are established by individual reflection and religious ideas, which may contradict egocentric (level I) or social and community (level II) values.

- **Postconventional moral reasoning** is similar to formal operational thought because it uses abstractions, going beyond what is concretely observed, willing to question "what is" in order to decide "what should be."

postconventional moral reasoning
Kohlberg's third level of moral reasoning, emphasizing moral principles.

According to Kohlberg, intellectual maturation advances moral thinking. During middle childhood, children's answers shift from being primarily preconventional to being more conventional: Concrete thought and peer experiences help children move past the first two stages (level I) to the next two (level II). Postconventional reasoning is not usually present until adolescence or adulthood, if then.

Kohlberg posed moral dilemmas to school-age boys (and eventually girls, teenagers, and adults). The most famous example of these dilemmas involves a poor man named Heinz, whose wife was dying. He could not pay for the only drug that could cure his wife, a drug that a local druggist sold for 10 times what it cost to make.

> Heinz went to everyone he knew to borrow the money, but he could only get together about half of what it cost. He told the druggist that his wife was dying and asked him to sell it cheaper or let him pay later. But the druggist said "no." The husband got desperate and broke into the man's store to steal the drug for his wife. Should the husband have done that? Why?
>
> *[Kohlberg, 1963, p. 19]*

Kohlberg's assessment of morality depends not on what a person answers but why an answer is chosen. For instance, suppose a child says that Heinz should steal the drug. That itself does not indicate the level of morality. The reason could be that Heinz needs his wife to care for him (preconventional), or that people will blame him if he lets his wife die (conventional), or that the value of a human life is greater than the law (postconventional).

Or suppose another child says Heinz should not steal. Again, the reason is crucial. If it is that he will go to jail, that is preconventional; if it is that business owners will blame him, that is conventional; if it is that no one should deprive anyone else of their livelihood, that is postconventional.

Kohlberg has been criticized for not appreciating cultural or gender differences. For example, loyalty to family overrides other values in some cultures, so some people might avoid postconventional actions that hurt their family. Also, Kohlberg's original participants were all boys, which may have led him to discount female values of nurturance and relationships (Gilligan, 1982).

Overall, Kohlberg seemed to value rational principles more than individual needs, unlike other scholars of moral development who consider emotions more influential than logic (Haidt, 2013). Regarding global warming, for instance, the facts about the world's temperature rising by a degree over a decade is less compelling for children in middle childhood than the dilemma of the stranded polar bear cub on a melting ice flow.

In general, over the years of middle childhood, children become more able to understand the difference between intentions and actions (doing harm but intending to be helpful is forgiven) and between lying deliberately and saying something that is untrue by mistake (Rizzo et al., 2019).

Maturation and Morality

As already discussed, some prosocial values are evident in infancy and early childhood. However, children have much to learn, and middle childhood is the best time for that. Maturation matters. One study measured generosity by counting how many of 10 chosen stickers 5- to 12-year-olds from five nations (United States, Canada, China, Turkey, South Africa) were willing to donate to another unknown child. Generosity increased with age: 5-year-olds gave away two and kept eight, while 12-year-olds gave away five and kept five (Cowell et al., 2017).

Beyond that, culture had an impact. Children from Toronto, Canada, were most generous, and children from Cape Town, South Africa, were least generous, a difference thought to reflect national wealth (Cowell et al., 2017). Those national differences paled when individual behavior was considered: Some children from each of the five nations kept all or almost all stickers to themselves, and some from each nation gave more than half away.

Teaching Morality

Fortunately, children enjoy thinking about and discussing moral values, and then peers help one another advance in moral behavior. Children may be more ethical than adults (once they understand moral equity, they complain when adults are not fair), and they are better at stopping a bully than adults are, because a bully is more likely to listen to other children than to adults.

Since bullies tend to be low on empathy, they need peers to teach them when their actions are not admired. During middle childhood, morality can be scaffolded just as cognitive skills are, with mentors — peers or adults — using moral dilemmas to advance moral understanding while they also advance the underlying moral skills of empathy and emotional regulation (Hinnant et al., 2013).

Usually, throughout middle childhood, moral judgment becomes more comprehensive, taking into account psychological as well as physical harm, intentions as well as consequences. For example, when 5- to 11-year-olds were presented with anecdotes that differed in whether the harm done was intended to prevent further harm or was simply mean, the younger children judged based on results, but the older children considered intention: They rated justifiable harm as less bad and unjustifiable harm as worse than the younger children did (Jambon & Smetana, 2014).

A detailed examination of the effect of peers on morality began with an update on one of Piaget's moral issues: whether punishment should seek *retribution* (hurting the transgressor) or *restitution* (restoring what was lost). Piaget found that children advance from retribution to restitution between ages 8 and 10 (Piaget, 1932/2013b), which many ethicists consider a moral advance (Claessen, 2017).

To learn how this occurs, researchers asked 133 9-year-olds:

> Late one afternoon there was a boy who was playing with a ball on his own in the garden. His dad saw him playing with it and asked him not to play with it so near the house because it might break a window. The boy didn't really listen to his dad, and carried on playing near the house. Then suddenly, the ball bounced up high and broke the window in the boy's room. His dad heard the noise and came to see what had happened. The father wonders what would be the fairest way to punish the boy. He thinks of two punishments. The first is to say: "Now, you didn't do as I asked. You will have to pay for the window to be mended, and I am going to take the money from your pocket money." The second is to say: "Now, you didn't do as I asked. As a punishment you have to go to your room and stay there for the rest of the evening." Which of these punishments do you think is the fairest?
>
> [Leman & Björnberg, 2010, p. 962]

The children were split almost equally in their answers. Then, 24 pairs were formed of children who had opposite views. Each pair was asked to discuss the issue, trying to reach agreement. (The other children did not discuss it.) Six pairs

THINK CRITICALLY: If one of your moral values differs from that of your spouse, your parents, or your community, should you still try to teach it to your children? Why or why not?

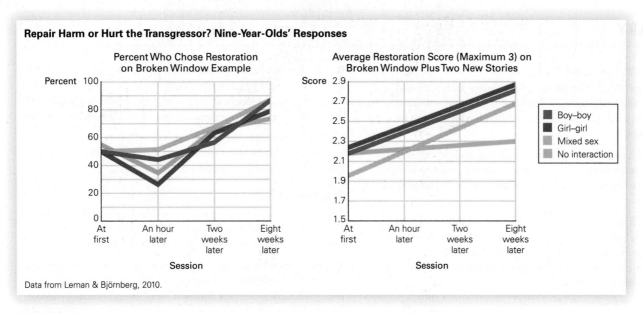

Repair Harm or Hurt the Transgressor? Nine-Year-Olds' Responses

Data from Leman & Björnberg, 2010.

FIGURE 13.3

Benefits of Time and Talking The graph on the left shows that most children, immediately after their initial punitive response, became even more likely to seek punishment rather than to repair damage. However, after some time and reflection, they affirmed the response that Piaget would consider more mature. The graph on the right indicates that children who had talked about the broken-window example moved toward restorative justice even in examples that they had not heard before, which was not true for those who had not talked about the first story.

were boy–boy, six were boy–girl with the boy favoring restitution, six were boy–girl with the girl favoring restitution, and six were girl–girl.

The conversations typically took only five minutes, and the retribution side was more often chosen. Piaget would consider a moral backslide, since more restitution than retribution advocates switched.

However, two weeks and eight weeks later all of the children were queried again, and their responses changed toward the more advanced, restitution thinking (see **Figure 13.3**). This advance occurred even for the children who merely thought about the dilemma again, but children who had discussed it with another child were particularly likely to decide that restitution was better.

The main conclusion from this study was that "conversation on a topic may stimulate a process of individual reflection that triggers developmental advances" (Leman & Björnberg, 2010, p. 969). Parents and teachers, take note: Raising moral issues and letting children talk about them may advance morality—not immediately, but soon.

WHAT HAVE YOU LEARNED?

1. How does the culture of children differ from the culture of adults?

2. What are the different kinds of popular and unpopular children?

3. What do victims and bullies have in common?

4. How might bullying be reduced?

5. What three forces affect moral development during middle childhood?

6. What are the main criticisms of Kohlberg's theory of moral development?

7. What role do adults play in the development of morality in children?

SUMMARY

The Nature of the Child

1. All theories of development acknowledge that school-age children become more independent and capable in many ways. Erikson emphasized industry, when children busily strive to master various tasks.

2. Children develop their self-concept during middle childhood, basing it on a more realistic assessment of their competence than they had in earlier years. Cultures differ in their evaluation of high self-esteem.

3. Both daily hassles and major stresses take a toll on children, with accumulated stresses more likely to impair development than any single event on its own. Resilience is aided by the child's interpretation of the situation and the availability of supportive adults, peers, and institutions.

Families During Middle Childhood

4. Families influence children in many ways, as do genes and peers. Although most siblings share a childhood home and parents, each sibling experiences different (nonshared) circumstances within the family.

5. The five functions of a supportive family are: to satisfy children's physical needs; to encourage learning; to support friendships; to protect self-respect; and to provide a safe, stable, and harmonious home.

6. The most common family structure worldwide is the nuclear family, usually with other relatives nearby and supportive. Two-parent families include adoptive, grandparent, and stepfamilies, each of which sometimes functions well for children. However, every family structure also has vulnerabilities.

7. Single-parent families have higher rates of change — for example, in where they live and who belongs to the family. Instability is particularly hard during middle childhood, as are changing routines about homework, sleep, food, and so on.

8. Income affects family function for two-parent families as well as single-parent households. Poor children are at greater risk for emotional, behavioral, and academic problems because the stresses that often accompany poverty hinder effective parenting.

9. No matter what the family SES, instability and conflict are harmful. Children suffer even when the conflict does not involve them directly but their parents or siblings fight.

The Peer Group

10. Peers teach crucial social skills during middle childhood. Each cohort of children has a culture, passed down from slightly older children. Close friends are wanted and needed.

11. Popular children may be cooperative and easy to get along with or may be competitive and aggressive. Unpopular children may be neglected, aggressive, or withdrawn, sometimes becoming victims.

12. Bullying is common among school-age children. Both bullies and victims have difficulty with social cognition; their interpretation of the normal give-and-take of childhood is impaired.

13. Bullies themselves may be admired, which makes their behavior more difficult to stop. Overall, a multifaceted, long-term, whole-school approach — with parents, teachers, and bystanders working together — seems the best way to halt bullying.

14. School-age children seek to differentiate right from wrong as moral development increases over middle childhood. Peer values, cultural standards, empathy, and education all affect their personal morality.

15. Kohlberg described three levels of moral reasoning, each related to cognitive maturity. His description has been criticized for ignoring cultural and gender differences.

16. When values conflict, children often choose loyalty to peers over adult standards of behavior. When children discuss moral issues with other children, they develop more thoughtful answers to moral questions.

KEY TERMS

industry versus inferiority (p. 337)
social comparison (p. 338)
resilience (p. 340)
parentification (p. 342)
family structure (p. 345)

family function (p. 345)
nuclear family (p. 346)
single-parent family (p. 346)
extended family (p. 346)
polygamous family (p. 346)

child culture (p. 354)
bullying (p. 357)
bully-victim (p. 358)
preconventional moral reasoning (p. 362)

conventional moral reasoning (p. 362)
postconventional moral reasoning (p. 363)

APPLICATIONS

1. Go someplace where many school-age children congregate (such as a schoolyard, a park, or a community center) and use naturalistic observation for at least half an hour. Describe what popular, average, withdrawn, and rejected children do. Note at least one potential conflict. Describe the sequence and the outcome.

2. Focusing on verbal bullying, describe at least two times when someone said something hurtful to you and two times when you said something that might have been hurtful to someone else. What are the differences between the two types of situations?

3. How would your childhood have been different if your family structure had been different, such as if you had (or had not) lived with your grandparents, if your parents had (or had not) gotten divorced, if you had (or had not) been adopted, if you had lived with one parent (or two), if your parents were both the same sex (or not)? Avoid blanket statements: Appreciate that every structure has advantages and disadvantages.

Especially For ANSWERS

Response for Scientists (from p. 345): Proof is very difficult when human interaction is the subject of investigation, since random assignment is impossible. Ideally, researchers would find identical twins being raised together and would then observe the parents' behavior over the years.

Response for Parents of an Accused Bully (from p. 358): The future is ominous if the charges are true. Your child's denial is a sign that there is a problem. (An innocent child would be worried about the misperception instead of categorically denying that any problem exists.) You might ask the teacher what the school is doing about bullying. Family counseling might help. Because bullies often have friends who egg them on, you may need to monitor your child's friendships and perhaps befriend the victim. Talk about the situation with your child. Ignoring the situation might lead to heartache later on.

Observation Quiz ANSWERS

Answer to Observation Quiz (from p. 343): Did you notice that the two males are first, and that the father carries the boy? Everyone should notice gender, ethnic, and age differences, but interpretation of such differences is not straightforward. This scene may or may not reflect male–female roles.

Answer to Observation Quiz (from p. 349): Both parents are women. The evidence shows that families with same-sex parents are similar in many ways to families with opposite-sex parents, and children in such families develop well.

APPLICATION TO DEVELOPING LIVES PARENTING SIMULATION ADOLESCENCE

As you progress through the Adolescence simulation module, how you answer the following questions will impact the biosocial, cognitive, and psychosocial development of your adolescent.

Biosocial	Cognitive	Psychosocial
• Will your child experiment with smoking, drinking, or drugs during adolescence? • How will you respond if you learn your child is experimenting with drugs? • How will you encourage your child to spend his or her free time after school (sports, part-time job)?	• Which of Piaget's stages of cognitive development is your child in? • What kind of path do you see your teenager pursuing after high school (college, military, work program)?	• How will you respond if your adolescent is struggling to fit in with peers? • How often do you think you and your teenager will have conflicts? • How social will your child be during his or her teen years? • How much privacy will you grant your teenager? • How will you respond when your teenager starts dating?

Adolescence

Acentury ago, puberty began at age 15 or so. Soon after that age, most girls married and most boys found work. It is said that *adolescence begins with biology and ends with culture*. If so, then a hundred years ago, adolescence lasted a few months.

Now, adolescence lasts for years. Puberty starts at age 10 or so, and adult responsibilities may be avoided for decades. Indeed, a few observers describe a *Peter Pan syndrome*—men who "won't grow up," too self-absorbed to love and care for anyone else (Kiley, 1983; Snow, 2015). That is unfair to men and to teenagers, but even at age 18, almost no adolescent is ready for all of the responsibilities of adulthood. If high school seniors want marriage, parenthood, and a lifelong career, they should wait at least a few years.

In the next three chapters (covering ages 11 to 18), we begin with biology (Chapter 14), consider cognition (Chapter 15), and then discuss culture (Chapter 16). Adolescence attracts the worst fears of adults, including parents, teachers, police officers, and social workers, yet many children look forward to these years and many adults wish they were young again.

Developmental researchers consider adolescence to be one interval in a long process, as patterns and events can push a teenager toward early death or toward a long, satisfying life. Understanding the possibilities and pitfalls of these years catapults us all toward happy adulthood. ●●

Adolescence: Biosocial Development

What Will You Know?

1. How can you predict when puberty will begin for a particular child?
2. Why do many teenagers ignore their nutritional needs?
3. What makes teenage sex often a problem instead of a joy?

Puberty was not easy for me. We moved two weeks before my 11th birthday. I left a suburban home with a backyard for an urban row house; I left a single-sex school and enrolled mid-year of the sixth grade, in a school with boys. In retrospect, that move served me well.

- I made friends who accepted me because they had been rejected by the popular clique. One was Mormon, one was Jewish, one was new to the school because she had just moved from Germany. They were my first introduction to the larger world; they are all still my friends, decades later.
- Because I could no longer take long walks in the woods alone, I found a public library a block away. I read dozens of books that my elders would never have chosen.
- I learned to talk to boys; 15 years later that helped me marry one of the best.

But puberty itself was stressful, and only now, as I review the biology of that process, do I realize how hard it was. I reached menarche three weeks after I arrived in the new city; I was afraid of my new male classmates and teachers; I was an awkward, pudgy girl who did not know how to sit at my small desk, to style my changing hair, or to dress my new body.

Once I wore my best taffeta dress to school; it was pale blue with black polka dots. One of the popular girls said, "You must be going somewhere special." "Yes," I lied, suddenly aware that my best dress was not appropriate school attire. Even now no one would call me a fashionable dresser, but I feel sorry for that young girl who hoped that dressing up would bring admiration.

This chapter reviews the physical aspects of puberty, and the problems many adolescents have with appearance, peers, and sexuality. Now that I understand that stress accelerates puberty, I sympathize with my 11-year-old self. Changes are difficult for children at every age, but that is particularly true in the middle of sixth grade, when the growth spurt and sexual changes appear unbidden. I hope you gain some sympathy for your younger self as well.

Puberty Begins

Puberty refers to the years of rapid physical growth and sexual maturation that end childhood, producing a person of adult size, shape, and sexuality. It all begins with a cascade of hormones that produce external growth and internal changes, including heightened emotions and sexual desires.

Sequence

The visible signs of puberty usually begin sometime between ages 9 and 15. Most biological growth ends about four years after the first signs appear, although some individuals (especially boys) add height, weight, and muscle until age 20 or so. Over the past century, the age of puberty has decreased, probably because of nutrition but perhaps because of increased sexualization of the culture (media, advertisements, clothes).

For girls, the first observable sign is usually nipple growth. Soon a few pubic hairs are visible, followed by a peak growth spurt, widening of the hips, first menstrual period (**menarche**), full pubic-hair pattern, and breast maturation. The average age of menarche in the United States is about 12 years, 4 months (Biro et al., 2013), although nutrition, genes, and stress affect the timing (Brix et al., 2019).

For boys, the usual sequence is growth of the testes, initial pubic-hair growth, growth of the penis, first ejaculation of seminal fluid (**spermarche**), appearance of facial hair, a peak growth spurt, deepening of the voice, and final pubic-hair growth (Dorn & Biro, 2011). The typical age of spermarche is 13 years, almost a year later than menarche.

Age varies markedly. The averages here are for well-nourished adolescents in the United States. Malnutrition delays the start; stress advances it; genes always matter.

Unseen Beginnings

The changes just listed are visible, but the entire process begins with an invisible event—a marked increase in hormones.

Hormone production is regulated deep within the brain, where biochemical signals from the hypothalamus signal another brain structure, the **pituitary**. The pituitary produces hormones that stimulate the **adrenal glands**, located above the kidneys at either side of the lower back.

These glands produce more hormones. Many hormones that regulate puberty follow this route, the **HPA (hypothalamus–pituitary–adrenal) axis** (see **Figure 14.1**).

Sex Hormones

Late in childhood, the pituitary activates not only the adrenal glands—the HPA axis—but also the **gonads**, or sex glands (ovaries in females; testes, or testicles,

puberty The time between the first onrush of hormones and full adult physical development. Puberty usually lasts three to five years. Many more years are required to achieve psychosocial maturity.

menarche A girl's first menstrual period, signaling that she has begun ovulation. Pregnancy is biologically possible, but ovulation and menstruation are often irregular for years after menarche.

spermarche A boy's first ejaculation of sperm. Erections can occur as early as infancy, but ejaculation signals sperm production. Spermarche may occur during sleep (in a "wet dream") or via direct stimulation.

pituitary A gland in the brain that responds to a signal from the hypothalamus by producing many hormones, including those that regulate growth and sexual maturation.

adrenal glands Two glands, located above the kidneys, that respond to the pituitary, producing hormones.

HPA (hypothalamus–pituitary–adrenal) axis A sequence of hormone production originating in the hypothalamus and moving to the pituitary and then to the adrenal glands.

gonads The paired sex glands (ovaries in females, testicles in males). The gonads produce hormones and mature gametes.

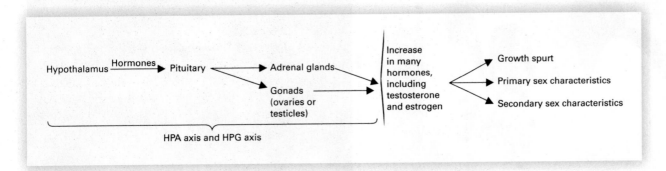

FIGURE 14.1
Biological Sequence of Puberty Puberty begins with a hormonal signal from the hypothalamus to the pituitary gland, both deep within the brain. The pituitary, in turn, sends a hormonal message through the bloodstream to the adrenal glands and the gonads to produce more hormones.

HPG (hypothalamus–pituitary–gonad) axis A sequence of hormone production originating in the hypothalamus and moving to the pituitary and then to the gonads.

estradiol A sex hormone, considered the chief estrogen. Females produce much more estradiol than males do.

testosterone A sex hormone, the best known of the androgens (male hormones); secreted in far greater amounts by males than by females.

in males), following another sequence called the **HPG (hypothalamus–pituitary–gonad) axis**.

One hormone in particular, *GnRH (gonadotropin-releasing hormone)*, causes the gonads to enlarge and dramatically increase their production of sex hormones, chiefly **estradiol** in girls and **testosterone** in boys. These hormones affect the body's shape and function, producing more hormones that regulate stress and immunity.

As you recall, hormones are body chemicals that regulate hunger, sleep, moods, stress, sexual desire, immunity, reproduction, and many other bodily functions and processes, including puberty. Throughout adolescence, hormone levels correlate with physiological changes, brain restructuring, and self-reported developments (Goddings et al., 2012; Vijayakumar et al., 2018).

Estrogens (including estradiol) are female hormones, and *androgens* (including testosterone) are male hormones, although each sex has some of both. The ovaries produce high levels of estrogens, and the testes produce dramatic increases in androgens. This "surge of hormones" affects bodies, brains, and behavior before any visible signs of puberty appear, "well before the teens" (Peper & Dahl, 2013, p. 134).

The activated gonads soon produce mature ova or sperm. Conception is possible, although peak fertility occurs four to six years later. This is crucial information for teenagers who are sexually active: Some mistakenly believe they cannot become pregnant because they had sex in the past without protection, and no pregnancy occurred. A few years later, that myth may lead to conception.

Hormonal increases may also precipitate psychopathology. Adolescence is the peak time for the emergence of many disorders, with early puberty particularly hazardous (Hamlat et al., 2019). The rush of hormones puts some vulnerable children over the edge.

Stress and genes also matter: Hormones are never the sole cause (Roberts & Lopez-Duran, 2019). Probably because of sex differences in hormones, adolescent males are almost twice as likely as females to develop schizophrenia, and females are more than twice as likely to become severely depressed.

Hormones and Relationships

If a teenager has emotional problems, no matter what their origin, relationships with other people are crucial. This is evident not only with parents and peers, but also with medical professionals and psychotherapists. When a troubled adolescent develops a good relationship with a therapist, psychopathology is limited.

Do They See Beauty? Both young women—the Mexican 15-year-old preparing for her Quinceañara and the Malaysian teen applying a rice facial mask—look wistful, even worried. They are typical of teenage girls everywhere, who do not realize how lovely they are.

Therapists who specialize in treating adolescents understand this and prioritize relationship-building.

This is good news for adults who are aware of the hazards of puberty. For example, one study found that about two-thirds of adolescents receiving psychotherapy report becoming markedly better, as compared to those who sought therapy but were put on a waiting list. Success seems about equally likely for every gender and for adolescents of all ethnic groups, although some conditions (anxiety) have a better improvement rate than others (depression) (Weisz et al., 2017).

For everyone, one psychological effect of estrogen and testosterone is new interest in sexuality. This is evident at the very beginning of puberty, when children become interested in the other gender (who used to be avoided or disparaged) or attracted to members of the same sex (again an unanticipated surprise). This onset of sexual interest can cause joy or depression, depending more on social circumstances than on hormones.

Usually the object of a young adolescent's first attraction is safely unattainable—a film star, a popular singer, a teacher—but by mid-adolescence, fantasies may settle on another young person. Sexual orientation and gender roles and identity are increasingly complex because of the combination of puberty, norms, and variations—a topic discussed in Chapter 16.

Although emotional surges, nurturant impulses, and lustful urges arise with hormones, remember that body, brain, and behavior always interact. Sexual thoughts themselves can *cause* physiological and neurological processes, not just result from them. Cortisol may rise in puberty, and that makes adolescents quick to react with passion, fury, or ecstasy (Goddings et al., 2012; Klein & Romeo, 2013). Then those emotions, in turn, increase various other hormones. Bodies, brains, and behavior all affect one another.

For example, adult reactions to a young person's emerging breasts or beards evoke adolescent thoughts and frustrations, which then raise hormone levels, propel physiological development, and trigger more emotions. Because of hormones, emotions are more likely to be expressed during adolescence (with shouts and tears), which affects everyone's reactions. Thus, the internal and external changes of puberty are reciprocal, each affecting the other.

Body Rhythms

Because of hormones, the brain of every living creature responds to environmental changes over the hours, days, and seasons. Among the known biorhythms: Children gain weight more rapidly in winter and grow taller more quickly in summer. Another example is *seasonal affective disorder (SAD)*, when people develop symptoms of depression in winter. Those are seasonal changes, but many biorhythms are on a 24-hour cycle, called the **circadian rhythm**. (*Circadian* means "about a day.") Puberty interacts with biorhythms.

For most people, daylight awakens the brain. That's why people experiencing jet lag are urged to take an early-morning walk. But at puberty, night may be more energizing, making some teens wide awake and hungry at midnight but half asleep, with no appetite or energy, all morning. Teenagers become "night owls" more than "early birds" (Gariépy et al., 2018).

In addition to circadian changes at puberty, some individuals (especially males) are naturally more alert in the evening than in the morning, a trait called *eveningness*. To some extent, this is genetic: 351 genes differ in people who are natural "larks" or "owls" (Jones et al., 2019)). Puberty plus eveningness increases risk (drugs, sex, delinquency): Many teenagers are awake when adults are asleep.

Watching late-night TV, working on a computer, or texting friends at 10 P.M. interferes with sleepiness. Any screen time in the evening correlates with later

VIDEO: The Timing of Puberty depicts the usual sequence of physical development for adolescents.

circadian rhythm A day–night cycle of biological activity that occurs approximately every 24 hours.

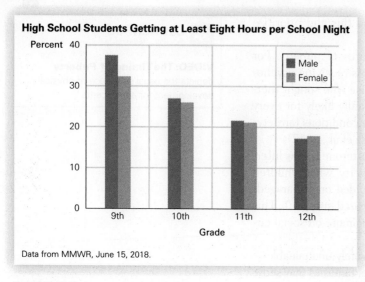

High School Students Getting at Least Eight Hours per School Night

Data from MMWR, June 15, 2018.

FIGURE 14.2

Sleepyheads Three of every four high school seniors are sleep deprived. Even if they go to sleep at midnight, as many do, they must get up before 8 A.M., as almost all do. Then they are tired all day.

 Observation Quiz As you see, the problems are worse for the girls. Why is that? (see answer, page 395) ↑

CHAPTER APP 14

📱 Sleep Cycle

IOS:
https://tinyurl.com/y36gvtum

ANDROID:
https://tinyurl.com/pgj2jyk

RELEVANT TOPIC:
Circadian rhythms and sleep

This app allows users to track their sleep patterns and get analysis that helps ensure a good night's sleep. Sleep Cycle uses a wake-up phase that ends at the desired alarm time. During this phase, the app monitors body signals to wake users softly, when they are in the lightest possible sleep state.

---

sleep, with social media and Web surfing particularly influential (Hisler et al., 2020). Probably because those activate thoughts as well as biological processes, adolescents have a harder time sleeping when they finally turn out the light.

Schools that provide each student a tablet for homework warn against bedroom use. However, the powerful adolescent urge to stay in touch with friends results in sleeping next to their smartphones—and then they are sleep deprived, nodding off in class. Further, classes are slow and boring compared to video games and human relationship dramas.

Sleep deprivation and irregular sleep schedules increase insomnia, nightmares, mood disorders (depression, conduct disorder, anxiety), and falling asleep while driving. Adolescents are particularly vulnerable to all of these, and tiredness makes it worse (see **Figure 14.2**). In addition, sleepy students do not learn as well as well-rested ones.

Oblivious to adolescent biorhythms, some parents set early curfews or stay awake until their child comes home at night. They might drag their teenager out of bed for school—the same child who, a decade earlier, was commanded to stay in bed until dawn.

Some municipalities fight adolescent biology when they implement community safety measures. In 2014, Baltimore implemented a law that required everyone under age 14 to be home by 9 P.M. and 14- to 16-year-olds to be off the streets by 10 P.M. on school nights and 11 P.M. on weekends. The assumption was that home is a safe place where teenagers will stay out of trouble.

Many high schools remain stuck in schedules set before the hazards of sleep deprivation were known. Although "the science is there; the will to change is not" (Snider, 2012, p. 25). In August 2014, the American Academy of Pediatrics concluded that high schools should not begin until 8:30 or 9 A.M., because adolescent sleep deprivation causes a cascade of intellectual, behavioral, and health problems. The doctors were distressed to report that 43 percent of high schools in the United States start *before* 8 A.M.

Some schools have instituted later start times. One study compared 29 high schools, across seven states, before and after they started school later than 8:30 A.M. Graduation rates increased from 79 to 88 percent, and average daily attendance rose from 90 to 94 percent (McKeever & Clark, 2017). Similar results were found when the city of Seattle began school at 8:45 A.M. instead of 7:50 A.M., with improvement particularly apparent in students from low-SES families (Dunster et al., 2018).

The reality that genes and family affect the circadian rhythm, and that many Western parents grant substantial freedom to their adolescents, raise a question: Is sleep deprivation more problematic for European American teenagers? If parents are lax about bedtime, computer use, and sleeping patterns, might that cause the problems just mentioned?

It seems, however, that adolescent biology is universal, changing circadian rhythms. For example, a study of Mexican American adolescents found that, although parent sleep had some influence, youths were more often awake late at night after their parents were asleep (Bai et al., 2020). Eveningness peaked at about age 16, which is also true for adolescents from other groups. The researchers reported that substance use correlated with late bedtimes, as is also found among teenagers of other groups.

Most developmentalists, pediatricians, and education researchers wonder why adult traditions are preserved while adolescent learning is ignored. There is good news here, however. As the evidence accumulates, schools are postponing their start times, and teenagers are learning more (Lo et al., 2018).

Brain Growth

A more ominous example of the disconnect between what science tells us and what adolescent brains do concerns cars, guns, sex, and drugs, all of which result in injury and even death. A chilling example comes from teenage driving (legal at age 16 in most U.S. localities). Per mile driven, teenage drivers are three times more likely to die in a motor-vehicle crash than drivers over age 20 (Insurance Institute for Highway Safety, 2018).

Sequence of Brain Changes

Many aspects of adolescent body growth are uneven. One breast, one foot, or one ear may be bigger than the other—awkward, but harmless. However, the usual sequence of brain maturation, propelled by hormones that activate the limbic system at puberty, is dangerous.

To be specific, the prefrontal cortex matures steadily, advancing gradually as time goes on. Executive function—long-term planning, postponing gratification, thinking flexibly—is better in adults than in young adolescents, because of maturation of the prefrontal cortex. The limbic system, however, is affected more by hormones (the HPG axis) than by time and thus grows dramatically in early adolescence.

Pubertal hormones target the amygdala directly (Romeo, 2013). The instinctual and emotional areas of the adolescent brain develop ahead of the reflective, analytic areas. Puberty means emotional rushes, unchecked by caution. Powerful sensations—loud music, speeding cars, strong drugs—become compelling. Adolescents brag about being wasted, smashed, out of their minds—all conditions that adults try to avoid.

Immediate impulses thwart long-term planning and reflection. My friend said to his neighbor, who had given his son a red convertible as a high school graduation gift, "Why didn't you just give him a loaded gun?" The mother of the 20-year-old who killed 20 first-graders and 7 adults in 2012 in Newtown, Connecticut, did just that. He killed her, too.

Sadly, that is not an isolated example. Guns, including those never used in hunting or target practice, are available to adolescents, who increasingly use them to kill other adolescents or themselves. About three-fourths of all suicides in the United States are by firearms, a rate higher than in other nations (MMWR, March 6, 2020). Especially in adolescence, suicides are often impulsive, which is why suicide by pills is uncommon—that takes time.

It is not that the prefrontal cortex shuts down. Actually, it continues to develop throughout adolescence and beyond (see **Figure 14.3**). Maturation doesn't stop, but the emotional hot spots of the brain zoom ahead. A study compared 886 adolescents (ages 9 to 16) and their parents (average age 44) in Hong Kong and England. All participants were asked questions to assess executive function. The adolescents were less accurate but notably quicker, another indication that the limbic system races ahead while the prefrontal cortex slowly matures (Ellefson et al., 2017).

When stress, arousal, passion, sensory bombardment, drug intoxication, or deprivation is extreme, the adolescent brain is flooded with impulses that overwhelm the cortex. Adults try to keep their thoughts straight, but adolescents may prefer such flooding. Many teenagers choose to spend a night without sleep, to eat nothing all day, to exercise in pain, to play music at deafening loudness, and to drink until they black out.

> **THINK CRITICALLY:** If a child seems to be unusually short or unusually slow in reaching puberty, would you give the child hormones? Why or why not?

FIGURE 14.3

Same People, But Not the Same Brain
These brain scans are part of a longitudinal study that repeatedly compared the proportion of gray matter from childhood through adolescence. (Gray matter refers to the cell bodies of neurons, which are less prominent with age as some neurons are unused.) Gray matter is reduced as white matter increases, in part because pruning during the teen years (the last two pairs of images here) allows intellectual connections to build. As the authors of one study that included this chart explained, teenagers may look "like an adult, but cognitively they are not there yet" (Powell, 2006, p. 865).

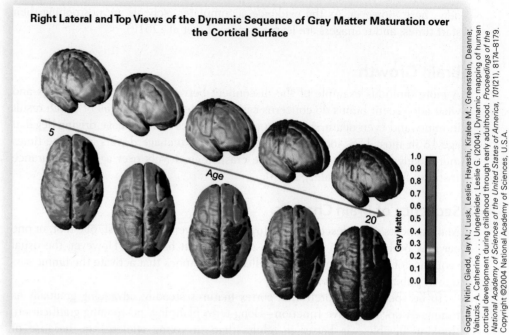

Right Lateral and Top Views of the Dynamic Sequence of Gray Matter Maturation over the Cortical Surface

Gogtay, Nitin; Giedd, Jay N.; Lusk, Leslie; Hayashi, Kiralee M.; Greenstein, Deanna; Vaituzis, A. Catherine, . . . Ungerleider, Leslie G. (2004). Dynamic mapping of human cortical development during childhood through early adulthood. *Proceedings of the National Academy of Sciences of the United States of America, 101*(21), 8174–8179. Copyright ©2004 National Academy of Sciences, U.S.A.

A common example comes from reading and sending text messages while driving. Teenagers know that this is illegal almost everywhere, but the "ping" of a text message evokes emotions that compel attention. In one survey of U.S. high school seniors who had driven a car in the past month, 39 percent had texted while driving (MMWR, June 15, 2018). This neurological disconnect is further explained in the following.

When Will Puberty Begin?

Normally, pubertal hormones begin to accelerate sometime between ages 8 and 14, and visible signs of puberty appear between ages 9 and 15. That six-year range is too great for many parents, teachers, and children, who want to know exactly when a given child will begin puberty. Fortunately, if a child's genes, gender, body fat, and stress level are known, prediction within a year or two is possible.

Genes and Gender

Genetic sex differences in hormones have a marked effect. In height, the average pubescent girl is about two years ahead of the average boy. Hormones affect sequence as well. The female height spurt occurs *before* menarche; the male increase in height occurs *after* spermarche.

Therefore, unlike height, for hormonal and sexual changes, girls are less than a year ahead of boys. Indeed, a recent study of thousands of Danish teenagers found that the boys took about four years from the beginning to the end of puberty, while the girls took five years. This means that by age 15 the two genders were about equal (Brix et al., 2019).

Thus, a short sixth-grade boy with sexual fantasies about the taller girls in his class is neither perverted nor precocious; his hormones are simply ahead of his height. It also means that, by the last years of high school, most romantic relationships are between students about the same age.

Overall, about two-thirds of age variations in onset of puberty is genetic—not only in the genes associated with the XX or XY chromosomes but also in the genes common in families and ethnic groups. If both of a child's parents were early or late to reach puberty, the child will likely be early or late as well.

INSIDE THE BRAIN

Lopsided Growth

Laurence Steinberg is a noted expert on adolescence (e.g., Steinberg, 2014, 2015). He is also a father.

> When my son, Benjamin, was 14, he and three of his friends decided to sneak out of the house where they were spending the night and visit one of their girlfriends at around two in the morning. When they arrived at the girl's house, they positioned themselves under her bedroom window, threw pebbles against her window-panes, and tried to scale the side of the house. Modern technology, unfortunately, has made it harder to play Romeo these days. The boys set off the house's burglar alarm, which activated a siren and simultaneously sent a direct notification to the local police station, which dispatched a patrol car. When the siren went off, the boys ran down the street and right smack into the police car, which was heading to the girl's home. Instead of stopping and explaining their activity, Ben and his friends scattered and ran off in different directions through the neighborhood. One of the boys was caught by the police and taken back to his home, where his parents were awakened and the boy questioned.
>
> I found out about this affair the following morning, when the girl's mother called our home to tell us what Ben had done. . . . After his near brush with the local police, Ben had returned to the house out of which he had snuck, where he slept soundly until I awakened him with an angry telephone call, telling him to gather his clothes and wait for me in front of his friend's house. On our drive home, after delivering a long lecture about what he had done and about the dangers of running from armed police in the dark when they believe they may have interrupted a burglary, I paused.
>
> "What were you thinking?" I asked.
>
> "That's the problem, Dad," Ben replied, "I wasn't."
>
> *[Steinberg, 2004, pp. 51, 52]*

Steinberg's son was right: When emotions are intense, especially when friends are nearby, cortisol floods the brain, causing the prefrontal cortex to shut down. This shutdown is not reflected in questionnaires that require teenagers to respond to paper-and-pencil questions regarding hypothetical dilemmas. On those tests, most teenagers think carefully and answer correctly.

In fact, when strong emotions are not activated, teenagers may be more logical than adults (Casey & Caudle, 2013). They remember facts learned in biology or health class about sex and drugs. They know exactly how HIV is transmitted, how pregnancy occurs, and how alcohol affects the brain. However,

the prospect of visiting a hypothetical girl from class cannot possibly carry the excitement about the possibility of surprising someone you have a crush on with a visit in the middle of the night. It is easier to put on a hypothetical condom during an act of hypothetical sex than it is to put on a real one when one is in the throes of passion. It is easier to just say no to a hypothetical beer than it is to a cold frosty one on a summer night.

[Steinberg, 2004, p. 53]

Ben reached adulthood safely. Other teenagers, with less cautious police or less diligent parents, do not. Brain immaturity makes teenagers vulnerable to social pressures and stresses, which typically bombard young people today. Emotional control, revealed by fMRI studies, is not fully developed until adulthood, because the prefrontal cortex is less connected to the limbic system (Hartley & Somerville, 2015). Thoughtful reappraisal of emotional impulses is slower in adolescence than later on (Schweizer et al., 2020; Silvers et al., 2015).

Longitudinal research finds that heightened arousal occurs in the brain's reward centers—specifically the *nucleus accumbens*, a region of the ventral striatum that is connected to the limbic system—when adolescent brains are compared to their own brains earlier or later in development (Braams et al., 2015).

When other teens are watching, adolescents thrill to take dramatic risks that produce social acclaim, risks they would not dare take alone. Interestingly, the reward regions that are highly activated when peers are watching decrease in activation when the adolescent's mother is nearby (Telzer et al., 2015).

The research on adolescent brain development confirms two insights regarding adolescent growth in general. First, physiological changes triggered by puberty are dramatic, unlike those of either childhood or adulthood. Second, the social context matters—the body and brain of humans respond not only to hormones and physical maturation but also to friends and family.

Most states in the United States now restrict drivers under age 18, such as mandating no passengers under age 20, or only family members as passengers. That saves lives, because teenagers are less likely to race a train, pass a truck, or zoom around a blind curve when no peers are watching.

On average, African Americans reach puberty about several months before European or Hispanic Americans; Chinese Americans average several months later. These are generalities: Remember that genetic differences *within* each group are greater than differences *between* groups.

The emotional significance of puberty is more gender than sex, more cultural than genetic. Nonetheless, genes exert a strong influence on the biology of puberty (Howard, 2018).

Body Fat and Chemicals

Another influence on the onset of puberty is body fat, which itself is partly genetic and partly cultural. Body fat is clearly associated with earlier puberty for girls. The effects are more variable for boys: Being overweight seems to accelerate puberty, but being obese slows it down (Reinehr & Roth, 2019). The reasons are probably hormonal: For boys, body fat may interfere with expression of male hormones.

In some nations, inadequate food delays growth of every kind. But, in developed nations, poor eating habits can result in overweight and, thus, early puberty. This is suggested by a study that found that girls who regularly drank several sugar-sweetened beverages each day were likely to experience earlier menarche (Carwile et al., 2015).

Malnutrition explains why youths reach puberty later in some parts of Africa, while their genetic relatives in North America mature much earlier. For example, in Ghana (West Africa) girls in the rural areas reach menarche more than a year later (almost age 14) than African American girls in the United States (just past 12). Even within Ghana, girls in rural areas—where malnutrition is more common—are behind those in urban areas (Ameade & Garti, 2016).

A more dramatic example arises from sixteenth-century Europe. Puberty is thought to have begun several years later than it does today, because those Europeans were often hungry. All of the data suggest that over the past three centuries, puberty has begun at younger and younger ages.

secular trend The long-term upward or downward direction of a certain set of statistical measurements, as opposed to a smaller, shorter cyclical variation. As an example, over the past two centuries, because of improved nutrition and medical care, children have tended to reach their adult height earlier and their adult height has increased.

This is an example of what is called the **secular trend**, which is earlier or greater growth as nutrition and medicine improved. Increased food availability has led to more weight gain in childhood, promoting earlier puberty and taller average height. Throughout the nineteenth and twentieth centuries, because of the secular trend, every generation reached puberty before the previous one (Brix et al., 2019; Dorn & Biro, 2011).

One curious bit of evidence of the secular trend is in the height of U.S. presidents. James Madison, the fourth president, was shortest at 5 feet, 4 inches; recent presidents have been much taller. Obama was 6 feet, 1 inch, and Trump says he is 6 feet, 3 inches (although some sources say he is not quite that tall).

The secular trend has stopped in most nations because childhood nutrition allows everyone to attain their genetic potential. Young men no longer look down

Both the Same? Yes, they are former U.S. presidents. But what a difference 150 years makes! James Madison *(left)* was the fourth president of the United States, was popular and respected, and at 5 feet, 4 inches tall weighed about 100 pounds. Barack Obama *(right)*, the 44th president, was 6 feet, 1 inch tall, and Donald Trump (#45) is said to be 6 feet, 3 inches tall. Lincoln (#16) was tallest of all—6 feet, 4 inches—which then was a reason to mock his appearance.

Photo Researchers/Science History Images/Alamy Stock Photo

JEWEL SAMAD/AFP/Getty Images

at their short fathers, or girls at their mothers, unless their parents were born in nations where hunger was common. Future presidents will not be taller than those in the recent past.

Some scientists suspect that precocious (before age 8) or delayed (after age 14) puberty may be caused by hormones in the food supply. Cattle are fed steroids to increase bulk and milk production, and hundreds of chemicals and hormones are used to produce most of the food that children consume. All of these substances *might* affect appetite, body fat, and sex hormones, with effects at puberty (Bourguignon et al., 2016).

Leptin, a hormone that is naturally produced by the human body, definitely affects the onset of puberty. Leptin is essential for appetite, energy, and puberty. However, too much leptin correlates with obesity, early puberty, and then early termination of growth. Thus, the heaviest third-grade girl may become the tallest fifth-grader and then the shortest high school graduate.

Most research on leptin has been done with mice; the effects are more complicated for humans (Bohlen et al., 2016). In fact, none of the data on the effects on humans of hormones and other chemicals, whether natural or artificial, are easy to interpret. The female body may be especially sensitive not only to leptin, but also to many other influences. Precise details require much more research, but we do know that many hormones and chemicals, both natural and artificial, affect puberty (Araki & Jensen, 2020; M. Wolff et al., 2015).

leptin A hormone that affects appetite and is believed to affect the onset of puberty. Leptin levels increase during childhood and peak at around age 12.

Especially for Parents Worried About Early Puberty Suppose your cousin's 9-year-old daughter has just had her first period, and your cousin blames hormones in the food supply for this "precocious" puberty. Should you change your young daughter's diet? (see response, page 395)

Stress

Stress hastens puberty, especially if a child's parents are sick, drug-addicted, or divorced, or if the neighborhood is violent and impoverished. One study of sexually abused girls found that they began puberty as much as a year earlier than they otherwise would have, a result attributed not only to stress but also to the hormones activated by sexual contact (Noll et al., 2017). Particularly for girls who are genetically sensitive, puberty comes early if their family interaction is stressful but late if their family is supportive.

This may explain the fact that many internationally adopted children experience early puberty, especially if their first few years of life were in an institution or an abusive home. An alternative explanation is that their age at adoption was underestimated: Puberty then seems early but actually is not (Hayes, 2013).

Developmentalists have known for decades that puberty is influenced by genes, hormones, and body fat. The effect of stress is a newer discovery, as A View from Science explains on the next page.

Too Early, Too Late

For a society's health, early puberty is problematic: It increases the rate of emotional and behavioral problems, including serious psychopathology (Dimler & Natsuaki, 2015; Hamlat et al., 2019). Early puberty is also linked to later health problems, including breast cancer, diabetes, and stroke (Day et al., 2015). Delayed puberty may also be a sign of health problems, including sickle cell anemia (Alexandre-Heymann et al., 2019).

For most adolescents, these links between puberty, stress, and health are irrelevant. Only one aspect of timing matters: their friends' schedules. No one wants to be too early or too late.

Girls

Think about the early-maturing girl. If she has visible breasts at age 10, the boys her age tease her; they are unnerved by the womanly creature in their midst. She must fit her developing body into a school chair designed for smaller children; she might hide her breasts in large T-shirts and bulky sweaters; she might refuse to undress for gym. Early-maturing girls tend to have lower self-esteem, more depression, and poorer body image than do other girls (Compian et al., 2009; Galvao et al., 2014).

Some early-maturing girls have older boyfriends, who are attracted to their womanly shape and girlish innocence. Having an older boyfriend bestows status among young adolescents, but it also promotes drug and alcohol use (Mrug et al., 2014).

A VIEW FROM SCIENCE

Stress and Puberty

Emotional stress, particularly when it has a sexual component, precipitates puberty. This is not always the case, because differential susceptibility means that some young people are more affected by family stress than others, and girls seem more affected than boys. But many lines of research agree that stress is one factor that can lead to early maturation (Ellis & Del Giudice, 2019).

For example, a large longitudinal study in England compared girls whose biological father lived at home with girls whose father was absent. Typically, in that community, when the father was absent, the mother was stressed and often another man lived with her. On average, their daughters were more depressed and reached menarche earlier (Culpin et al., 2015).

The connection between sexual stress and early puberty seems true in developing nations as well as developed ones. For example, in Peru, if a girl was physically and sexually abused, she was much more likely (odds ratio 1.56) to have her first period before age 11 than if she had not been abused (Barrios et al., 2015).

Hypothetically, the connection between stress and early puberty could be indirect. For example, perhaps children in dysfunctional families eat worse and watch TV more. That makes them overweight, which correlates with early menarche. Or, perhaps they inherit genes for early puberty from their distressed mothers, and those genes led the mothers to become pregnant too young, creating a stressful family.

Either obesity or genes *could* cause early puberty, and then stress would be a by-product, not a cause. Plausible hypothesis— but *not* correct.

Instead, longitudinal research links stress directly to early puberty (Ellis & Del Giudice, 2019). Even stress in early childhood matters. For example, one longitudinal study of 756 children found that parents who demanded respect, who often spanked, and who rarely hugged their infants and young children were, a decade later, likely to have daughters who reached puberty earlier than other girls in the same study (Belsky et al., 2007). Perhaps harsh parenting increases cortisol, which precipitates puberty.

A follow-up of the same girls at age 15, controlling for genetic differences, found that harsh treatment in childhood increased sexual problems (more sex, pregnancies, sexually transmitted infections) but *not* other risks (drugs, crime) (Belsky et al., 2010). This suggests that stress triggers earlier increases of sex hormones but not generalized rebellion. The direct impact of stress on puberty seems proven.

Why would higher cortisol accelerate puberty? The opposite effect—delayed puberty—makes more sense. Then stressed teens would still look and act childlike, which might evoke adult protection rather than lust or anger. Protection is especially needed in conflict-ridden or stressed single-parent homes, yet such homes produce earlier puberty and less parental nurturance. Is this a biological mistake? Not according to evolutionary theory:

> Maturing quickly and breeding promiscuously would enhance reproductive fitness more than would delaying development, mating cautiously, and investing heavily in parenting. The latter strategy, in contrast, would make biological sense, for virtually the same reproductive-fitness-enhancing reasons, under conditions of contextual support and nurturance.
>
> [Belsky et al., 2010, p. 121]

In other words, thousands of years ago, when harsh conditions threatened survival of the species, adolescents needed to reproduce early and often, lest the entire community become extinct. By contrast, in peaceful times, with plentiful food and loving care, puberty could occur later, allowing children to postpone maturity and instead enjoy extra years of nurturance from their biological parents and grandparents. Genes evolved to respond differently to war and peace.

Of course, this evolutionary benefit no longer applies. Today, early sexual activity and reproduction are more destructive than protective of communities. However, since the genome has been shaped over millennia, a puberty-starting allele that responds to stress will respond in the twenty-first century as it did thousands of years ago. This idea complements the idea of differential susceptibility (Harkness, 2014). Because of genetic protections, not every distressed girl experiences early puberty, but also for genetic reasons, family stress may speed up age of menarche.

Early-maturing girls enter abusive relationships more often than other girls do. Is that because their social judgment is immature?

Boys

For girls, early maturation is more harmful than helpful no matter when they were born, but for boys time and place matter. Early-maturing boys who were born around 1930 in the United States often became leaders in high school and earned more money as adults (Jones, 1965; Taga et al., 2006). Since about 1960, however, the risks associated with early male maturation have outweighed the benefits.

In the twenty-first century, early-maturing boys are more aggressive, lawbreaking, and alcohol-abusing than the average boy (Mendle et al., 2012). Although most of the research on the effects of puberty on male delinquency has been on U.S. boys, similar findings come from elsewhere, including a large study in China (Sun et al., 2016).

It is not hard to figure out why. A boy with rapidly increasing testosterone, whose body looks more like a man than a child, whose brain is more affected by emotions than logic, and who seeks approval from peers, is likely to trouble parents, schools, and the police.

Early puberty is particularly stressful if it happens suddenly: The boys most likely to become depressed are those for whom puberty was both early and quick (Mendle et al., 2010). In adolescence, depression is often masked as anger. That fuming, flailing 12-year-old may be more sad than mad.

Late puberty may also be difficult, especially for boys (Benoit et al., 2013). Slow-developing boys tend to be more anxious, depressed, and afraid of sex. Girls are less attracted to them, coaches select them for their teams, peers bully or tease them. If a 14-year-old boy still looks childish, he may react in ways (clowning, fighting, isolating) that are not healthy for him.

Ethnic Differences

The specific impact of early puberty varies by both gender and culture. For instance, one study found that, in contrast to European Americans, early-maturing African American girls were not depressed, but early-maturing African American boys were (Hamlat et al., 2014a, 2014b). Another study found that Mexican American boys thought less of themselves as pubertal changes continued, except in one domain—their relationships with girls (Harris et al., 2017).

European research found that early-maturing Swedish girls were likely to encounter problems with boys and early drug abuse, but similar Slovak girls were not, presumably because parents and social norms kept Slovak girls under tight control (Skoog & Stattin, 2014).

Social context also matters in the United States. Early-maturing Mexican American boys experienced trouble (with police and with peers) if their neighborhoods had few Mexican Americans, but not if they lived in ethnic enclaves (R. White et al., 2013). On their home turf, they were leaders, not troublemakers. They responded accordingly.

Puberty that is late by world norms, at age 14 or so, is not troubling if one's friends are late as well. However, if students in the same large high school have diverse ethnic and genetic roots, the fact that some look like tall children and others like grown adults may create tension. Contextual factors interact with biological ones. Peers, parents, and communities make off-time puberty insignificant or a major problem.

Ancient Rivals or New Friends? One of the best qualities of adolescents is that they identify more with their generation than their ethnic group, here Turkish and German. Do the expressions of these 13-year-olds convey respect or hostility? Impossible to be sure, but given that they are both about mid-puberty (face shape, height, shoulder size), and both in the same school, they may become friends.

Growth and Nutrition

Puberty entails transformation of every body part, with each change affecting all of the others. Here, we discuss biological growth and the nutrition that fuels that growth. Then we will focus on sexual maturation.

Growing Bigger and Stronger

growth spurt The relatively sudden and rapid physical growth that occurs during puberty. Each body part increases in size on a schedule: Weight usually precedes height, and growth of the limbs precedes growth of the torso.

The first set of changes is called the **growth spurt**—a sudden, uneven jump in size that turns children into adults. Growth proceeds from the extremities to the core (the opposite of the earlier proximodistal growth). Thus, fingers and toes lengthen before hands and feet, hands and feet before arms and legs, arms and legs before the torso. Because the torso is the last body part to grow, many pubescent children are temporarily big-footed, long-legged, and short-waisted.

If young teenagers complain that their jeans don't fit, they are probably correct—even if those same jeans fit when their parents bought them a month earlier. (Advance warning about rapid body growth occurs when parents first have to buy their children's shoes in the adult section.)

Sequence: Weight, Height, Muscles

As the growth spurt begins, children eat more and gain weight. Exactly when, where, and how much weight they gain depends on heredity, hormones, diet, exercise, and whether they are boys or girls. By age 17, the average girl's body has twice as much body fat as the average boy. Of course, genes and exercise influence body shape; gender and maturation are far from the only influences on body composition.

A height spurt follows the weight spurt. A year or two later, a muscle spurt occurs. Thus, the pudginess and clumsiness of early puberty are usually gone by late adolescence. Keep in mind, however, that puberty may dislodge the usual relationship between height and weight. A child may be eating too much or too little, but that may not be apparent in conventional measures of BMI (Golden et al., 2012).

In the years after puberty, all of the muscles grow. Arm muscles develop particularly in boys, doubling in strength from ages 8 to 18. Other muscles are gender-neutral. For instance, both sexes run faster with each year of adolescence, with boys not much faster than girls (unless the girls choose to slow down) (see **Figure 14.4**).

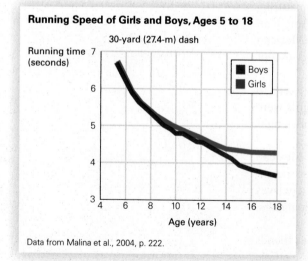

Running Speed of Girls and Boys, Ages 5 to 18

Data from Malina et al., 2004, p. 222.

FIGURE 14.4
Little Difference Both sexes develop longer and stronger legs during puberty.

Muscles are heavier than fat, so merely comparing weight and height, as BMI does, may make it seem as if a strong adolescent is overweight—but that may be inaccurate. Consider athletic activity. If a particular teenage boy is among the 35 percent who are physically active at least an hour every day, he can have a BMI above 25 (technically overweight) and still be in good physical shape. The same applies to girls: 18 percent of them are active every day (MMWR, June 15, 2018).

Organ Growth

Lungs triple in weight; consequently, adolescents breathe more deeply and slowly. The heart (another muscle) doubles in size as the heartbeat slows, decreasing the pulse rate while increasing blood pressure. Consequently, endurance improves: Some teenagers run for miles or dance for hours. Red blood cells increase, dramatically so in boys, which aids oxygen transport during intense exercise.

Both weight and height increase *before* muscles and internal organs: To protect immature muscles and organs, athletic training and weight lifting should be tailored to an adolescent's size during the previous year. Sports injuries are the most common school accidents, and they increase at puberty. One reason is that the height spurt *precedes* increases in bone mass, making young adolescents particularly vulnerable to fractures.

The other relevant factor is stress—both in competition and in personal life—that accompanies puberty. As stress increases, so does injury, with the average athlete experiencing at least one injury every year. Training should not only be adjusted to the developing body but should also include ways to decrease emotional stress via meditation, mindfulness, deep breathing, and so on (Ivarsson et al., 2017).

One organ system, the lymphoid system (which includes the tonsils and adenoids), *decreases* in size, so teenagers are less susceptible to respiratory ailments. Mild asthma, for example, may switch off at puberty.

In addition, teenagers have fewer colds and allergies than younger children. This reduction in susceptibility is aided by growth of the larynx, which also deepens the voice, dramatically so in boys but also in girls.

Another organ system, the skin, becomes oilier, sweatier, smellier, and more prone to acne—which itself is an early sign of puberty (Brix et al., 2019). Hair also changes, becoming coarser and darker. New hair grows under arms, on faces, and over genitals (pubic hair, from the same Latin root as *puberty*).

Diet Deficiencies

All of the changes of puberty depend on adequate nourishment, yet many adolescents do not eat well. Teenagers often skip breakfast, binge at midnight, guzzle down unhealthy energy drinks, and munch on salty, processed snacks. In 2017, only 14 percent of U.S. high school seniors ate the recommended three or more servings of vegetables a day (MMWR, June 15, 2018). That is even less than two years earlier (17 percent).

One reason for their eating patterns is that their hormones affect the circadian rhythm of their appetites; another reason is that their drive for independence compels them to avoid family dinners, refusing to eat what their mothers say they should.

Deficiencies of iron, calcium, zinc, and other minerals are especially problematic during adolescence. Because menstruation depletes iron, anemia is more common among adolescent girls than among any other age or sex group. This is true everywhere, especially in South Asia and sub-Saharan Africa, where teenage girls rarely eat iron-rich meat and green vegetables.

Observation Quiz Conformity among adolescents may be imposed by adults or chosen by teens. One example of each is evident here—what are they? (see answer, page 395) ↓

Next Stop: Masterpieces of the Fifteenth Century These British teens eat chips and soda before they enter the National Gallery in London. Twenty-first century fast food is causing an epidemic of diet deficiencies and disordered eating among youth in every nation.

Robert Alexander/Archive Photos/Getty Images

Reliable laboratory analysis of blood iron on a large sample of young girls in developing nations is not available, but all indications suggest that many are anemic. One study of a select group of 168 girls, ages 13 to 16, from one school in India found that two-thirds had anemia, with school grades lower among those who were iron-deficient (Tarun et al., 2016).

Another study on 18- to 23-year-old college women in Saudi Arabia found that a fourth (24 percent) were clinically anemic and another fourth (28 percent) were iron-deficient, although not technically anemic (Al-Sayes et al., 2011). These numbers are especially troubling since almost all college women in Saudi Arabia are in good health, from wealthy families, and have never been pregnant. They are among the better-nourished young women in that nation; rates of anemia are undoubtedly higher among younger, poorer girls.

Boys everywhere may also be iron-deficient, especially if they engage in physical labor or intensive sports: Muscles need iron for growth and strength. Yet, in developed as well as developing nations, adolescents spurn iron-rich foods in favor of chips, sweets, and fries.

Similarly, although the daily recommended intake of calcium for adolescents is 1,300 milligrams, the average U.S. teen consumes less than 500 milligrams a day. About half of adult bone mass is acquired from ages 10 to 20, which means that many contemporary teenagers will develop osteoporosis (fragile bones), a major cause of disability, injury, and death in late adulthood, especially for women.

One reason for calcium deficiency is that milk drinking has declined. In 1961, most North American children drank at least 24 ounces (about three-fourths of a liter) of milk each day, providing almost all (about 900 milligrams) of their daily calcium requirement. Fifty years later, only 8 percent of high school students drank that much milk, and 27 percent (more girls than boys) drank no milk at all in the previous week (MMWR, June 15, 2018).

The decline of milk drinking is one reason for the prevalent deficiency in vitamin D. Skipping breakfast and avoiding dairy products are common for adolescents of every group, particularly African Americans, affecting later health (Van Horn et al., 2011). Some are lactose-intolerant (milk is difficult for them to digest), but they could choose cheese or yogurt. Instead, many choose soda.

Choices Made

Many economists advocate a "nudge" to encourage people to make better choices, not only in nutrition but also in all other aspects of their lives (Thaler & Sunstein, 2008). Teenagers are often nudged by peers and institutions to make poor choices.

For example, fast-food establishments cluster around high schools, often with extra seating that encourages teenagers to eat and socialize (Walker et al., 2014). This is especially true for high schools with large populations of low-income students, who are most at risk for obesity. This problem seems to be getting worse, not better (Sanchez-Vaznaugh et al., 2019). Price further influences food choices, especially for adolescents: Unhealthy calories are cheaper than healthy ones.

Ethnic and gender factors are also influential. Twenty-two percent of Hispanic girls in U.S. high schools are overweight, as are 21 percent of African American girls. Rates are lower among boys and European Americans (MMWR, June 15, 2018). High schools with many Hispanic or African American students are likely to have fast-food establishments nearby (Sanchez-Vaznaugh et al., 2019).

In rural areas, new high schools are built where land is cheap, and soon a cluster of establishments that cater to students, with salty snacks, e-cigarettes, and sweetened drinks, open nearby.

Rates of obesity are falling in childhood but not in adolescence. In 2003, only three U.S. states (Kentucky, Mississippi, and Tennessee) had high-school obesity rates at

15 percent or more; in 2017, 42 states did (MMWR, June 15, 2018). In Latin America, the nutritional focus is on preventing underweight, not preventing overweight; yet overall, about one teenager in four is overweight or obese (Rivera et al., 2014).

Body Image

One reason for poor nutrition among teenagers is anxiety about **body image**—that is, the perception of how one's body looks. As one book on body image begins, each person's body "feels, conceives, imagines, represents, evaluates, loves, hates, and manipulates itself" (Cuzzolaro & Fassino, 2018, p. v). This is true lifelong, but since every part of the body changes dramatically in adolescence, the body image must change, too.

Few teenagers welcome every change in their bodies. Instead, they tend to focus on and exaggerate imperfections. They often focus on size and shape. More than half of U.S. high school girls are trying to lose weight, yet only one-sixth are actually overweight or obese (MMWR, June 15, 2018).

One problem is that almost no one has a body like those in magazines, movies, and television programs that are marketed to teenagers. Social media, which features posts of other teens at their most attractive, may lead to depression, body shame, and poor eating habits (Rodgers et al., 2020; Salomon & Brown, 2019).

Unhappiness with appearance—especially with weight for girls—is documented worldwide, including in South Korea, China, Australia, and Greece (Argyrides & Kkeli, 2015; Chen & Jackson, 2009; Kim & Kim, 2009; Murray et al., 2018). Many teenagers try to change their bodies: New diets, drugs, or intensive exercise are tried by almost every adolescent.

Eating Disorders

Dissatisfaction with body image can be dangerous, even deadly. Many teenagers, mostly girls, eat erratically or ingest drugs (especially diet pills) to lose weight; others, mostly boys, take steroids to increase muscle mass (see **Figure 14.5**). [**Developmental Link:** Teenage drug abuse is discussed in Chapter 16.]

Eating disorders are rare in childhood but increase dramatically at puberty, accompanied by distorted body image, food obsession, and depression. (See Visualizing Development, page 387.) Many adolescents switch from obsessive dieting to overeating to overexercising and back again. Although girls are most vulnerable, boys are at risk, too, especially those who aspire to be pop stars or who train to be wrestlers.

When distorted body image and excessive dieting result in severe weight loss, that indicates **anorexia nervosa**. Fewer than 1 in 100 girls develop anorexia, but those who do dramatically restrict their calorie intake and have a destructive, distorted attitude about their bodies. Their BMI may fall below 17 in cases of mild anorexia or 15 in extreme cases, but clinicians must be alert to any sudden weight loss or weight that is "less than that minimally expected" (American Psychiatric Association, 2013).

About three times as common as anorexia is **bulimia nervosa**. Sufferers overeat compulsively, consuming thousands of calories within an hour or two, and then purge through vomiting or laxatives. Most are close to typical in weight and therefore unlikely to starve. However, they risk serious health problems, including damage to their gastrointestinal system and cardiac arrest from electrolyte imbalance (Mehler, 2018).

body image A person's idea of how his or her body looks.

anorexia nervosa An eating disorder characterized by self-starvation. Affected individuals voluntarily undereat and often overexercise, depriving their vital organs of nutrition. Anorexia can be fatal.

bulimia nervosa An eating disorder characterized by binge eating and subsequent purging, usually by induced vomiting and/or use of laxatives.

FIGURE 14.5

Have You Ever...? This chart shows lifetime prevalence of eating disorders. Almost all of the adults, and many of the 17- to 18-year-olds, have recovered from an eating disorder. Thus, the .2 percent increase from age 15 to 17 suggests that only 1 adolescent is 500 developed a new eating disorder during the final years of high school. That is reassuring, but two facts from this chart are troubling. One is that almost 1 in 40 young teens (age 13–14) already had an eating disorder, and the other is that the prevalence for the oldest teens is higher than for the adults of all ages. That suggests that the rate of eating disorders is increasing in the twenty-first century.

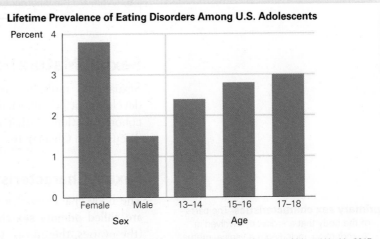

Lifetime Prevalence of Eating Disorders Among U.S. Adolescents

Data from U.S. Department of Health and Human Services, National Institute of Mental Health, 2017.

binge eating disorder Frequent episodes of uncontrollable overeating to the point that the stomach hurts. Usually the person feels shame and guilt but is unable to stop.

A disorder that is newly recognized in DSM-5 is **binge eating disorder**. Some adolescents periodically and compulsively overeat, quickly consuming large amounts of ice cream, cake, or snack food until their stomachs hurt. When bingeing becomes a disorder, overeating is typically done in private, at least weekly for several months. The sufferer does not purge (hence this is not bulimia) but feels out of control, distressed, and depressed.

Life-Span Causes and Consequences

From a life-span perspective, teenage eating disorders are not limited to adolescence, even though this is the usual age when first signs typically appear. The origins occur much earlier in family eating patterns if parents do not help their children eat sensibly—when they are hungry, without food being a punishment or a reward. Indeed, the origin could be at conception, since many studies report that anorexia is about half genetic and half environmental.

Of course, as you remember from Chapter 3, disorders that are genetic are also targets for environmental triggers and treatments. For all eating disorders, family function (not structure) is crucial (Tetzlaff & Hilbert, 2014). Peer culture is also crucial during adolescence.

Unfortunately, during the teen years, many parents are oblivious to the first signs of eating disorders. They might have given up trying to get their child to eat breakfast before school or to join the family for dinner. That may be tragic, as the treatment that seems most successful, called family-based treatment, has parents "sit with their children, requiring them to eat" (Couzin-Frankel, 2020, p. 127). The problem is that many parents feel shame and guilt regarding anorexia, and they are told to wait until the patient is ready to be cured.

Some adolescents with eating disorders die before midlife because their body systems are strained by their lack of nutrition. The most common cause of death is heart failure. Most, however, recover (Mehler, 2018).

WHAT HAVE YOU LEARNED?

1. What is the pattern of growth in adolescent bodies?
2. What complications result from the sequence of growth (weight/height/muscles)?
3. Why are many teenagers deficient in iron and calcium?
4. Why are many adolescents unhappy with their appearance?
5. What are the differences among the three eating disorders explained here?

Sexual Maturation

Sexuality is multidimensional, complicated, and variable—not unlike human development overall. Here, we consider biological changes at puberty and some cohort variations. Other aspects of adolescent sexuality and gender identity are discussed in Chapter 16.

Sexual Characteristics

primary sex characteristics The parts of the body that are directly involved in reproduction, including the vagina, uterus, ovaries, testicles, and penis.

The body characteristics that are directly involved in conception and pregnancy are called **primary sex characteristics**. During puberty, every primary sex organ (the ovaries, the uterus, the penis, and the testes) increases dramatically in size and matures in function. Reproduction becomes possible.

VISUALIZING DEVELOPMENT

Satisfied with Your Body?

Probably not, if you are a teenager. At every age, accepting who you are—not just ethnicity and gender, but also body shape, size, and strength—correlates with emotional health. During the adolescent years, when everyone's body changes dramatically, body dissatisfaction rises. As you see, this is particularly true for girls—but if the measure were satisfaction with muscles, more boys would be noted as unhappy.

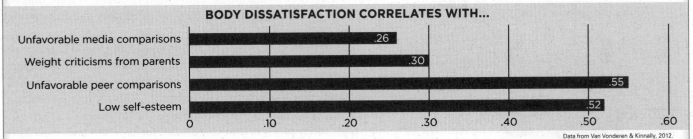

Data from Van Vonderen & Kinnally, 2012.

GENDER DIFFERENCES IN BODY DISSATISFACTION

Females of all ages tend to be dissatisfied with their bodies, but the biggest leap in dissatisfaction occurs when girls transition from early to mid-adolescence (Makinen et al., 2012).

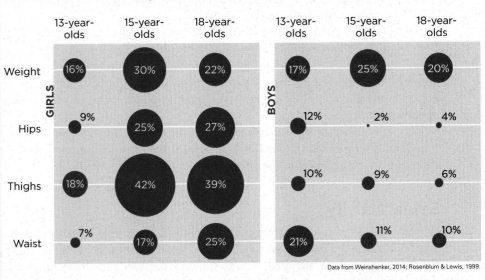

Data from Weinshenker, 2014; Rosenblum & Lewis, 1999.

SOCIAL MEDIA AND BODY DISSATISFACTION

- The more time teenage girls spend on social media, the higher their body dissatisfaction.
- 86% of teens say that social network sites hurt their body confidence.

(Information from Proud2Bme, 2012; Tiggemann & Slater, 2014)

NUTRITION AND EXERCISE

High school students are told, at home and at school, to eat their vegetables and not care about their looks. But they listen more to their peers and follow social norms.

Fortunately, some eventually learn that, no matter what their body type, good nutrition and adequate exercise make a person feel more attractive, energetic, and happy.

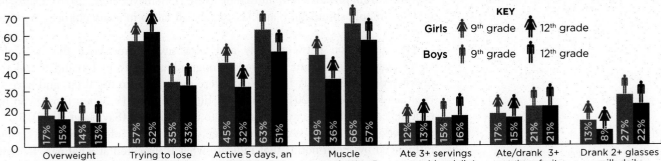

*Vegetables includes salad greens, and excludes French fries.
**Fruits include a glass of 100% fruit juice.

Data from MMWR, June 15, 2018.

When primary sex characteristics mature, another set of characteristics appear, called **secondary sex characteristics**, which are bodily features that do not directly affect reproduction (hence they are secondary) but that signify masculinity or femininity.

One secondary characteristic is body shape. Young boys and girls have similar shapes, but at puberty males widen at the shoulders and grow about 5 inches taller than females, while girls widen at the hips and develop breasts. Those female curves are often considered signs of womanhood, but neither breasts nor wide hips are required for conception; thus, they are secondary, not primary, sex characteristics.

The pattern of hair growth at the scalp line (widow's peak), the prominence of the larynx (Adam's apple), and several other anatomical features differ for men and women; all are secondary sex characteristics that few people notice. Facial and body hair increases in both sexes, affected by sex hormones as well as genes.

Girls often pluck or wax any facial hair they see and shave their legs, while boys may proudly grow sideburns, soul patches, chinstraps, moustaches, and so on—with specifics dependent on culture and cohort. Hair on the head is cut and styled to be spiky, flat, curled, long, short, or shaved. Hair is far more than a growth characteristic; it is a display of sexuality, a mark of independence.

Secondary sex characteristics are important psychologically, if not biologically. Breasts are an obvious example. Many adolescent girls buy "minimizer," "maximizer," "training," or "shaping" bras in the hope that their breasts will conform to an idealized body image. During the same years, many overweight boys are horrified to notice swelling around their nipples—a temporary result of the erratic hormones of early puberty.

The significance of breasts as a characteristic of men or women is evident in transgender individuals. Hormones and surgery are often chosen at puberty to make the breasts conform to the chosen gender (Patel et al., 2020).

Sexual Activity

Primary and secondary sex characteristics such as menarche, spermarche, hair, and body shape are not the only evidence of sex hormones. Fantasizing, flirting, hand-holding, staring, standing, sitting, walking, displaying, and touching are all done in particular ways to reflect sexuality. As already explained, hormones trigger sexual thoughts, but the culture shapes thoughts into enjoyable fantasies, shameful obsessions, frightening impulses, or actual contact.

Masturbation is common in both sexes, for instance, but culture determines attitudes, from private sin to mutual pleasure (Driemeyer et al., 2016). Caressing, oral sex, nipple stimulation, and kissing are all taboo in some cultures, expected in others.

The distinction between early and later sexual experience during adolescence may be significant. A detailed longitudinal study in Finland found that depressed and rebellious 13-year-olds were more likely to use drugs and have sex (Kaltiala-Heino et al., 2015). That had flipped by age 19, when those who had experienced intercourse were less likely to be depressed (Savioja et al., 2015).

Emotions regarding sexual experience, like the rest of puberty, are strongly influenced by social norms that indicate what is expected at what age. Recently in the United States, one study found that girls who have sex early in adolescence are likely to be depressed, but those who have sex as older adolescents tend to be

"I think I'll be more relaxed once my secondary sex characteristics kick in."

Brain Before Body Hormones affect thoughts, but visible signs reveal maturation.

secondary sex characteristics Physical traits that are not directly involved in reproduction but that indicate sexual maturity, such as a man's beard and a woman's breasts.

VIDEO: Romantic Relationships in Adolescence explores teens' attitudes and assumptions about romance and sexuality.

quite happy (Golden et al., 2016). Of course, much depends on the specific social context, whether a girl feels shamed or proud of her sexuality.

Indeed, everyone is influenced by hormones and society, biology and culture. All adolescents have sexual interests that they did not previously have (biology), and this propels teenagers in some nations to do things that teenagers in other nations would never do (culture).

Social norms regarding male–female differences are powerful. Traditionally, males were thought to have stronger sexual urges than females, which is why adolescent boys were supposed to "make the first move," from asking for a date to trying for a kiss. Then, girls were supposed to slow down the boys' advances. This was called the *double standard*, in that behaviors of boys and girls were held to different standards.

Many adolescents still expect boys and girls to approach sexual interactions differently, with boys more insistent and girls more hesitant. As one teen girl explained, "that's just how it is" (Tolman et al., 2016).

Nonetheless, many lines of research find that the double standard is less powerful than it once was, not only because of greater freedom for girls but also because many teens question the male–female binary. Interestingly, same-sex couples also find that the double standard is decreasing. A study in Italy found that lesbians experience patterns similar to those of gay males, and both are becoming more similar to straight youth (La Fauci, 2018).

In many nations including the United States, adolescent female rates of sexual activity are now almost even with male rates. For example, among high school seniors, 56 percent of the girls and 59 percent of the boys have had sexual intercourse, with most of them sexually active in the past three months. The one notable difference among high school students is in the number of partners: 8 percent of the girls and 12 percent of the boys have had four or more partners (MMWR, June 15, 2018).

Over the past two decades in the United States, every gender, ethnic, and age group is *less* sexually active than the previous cohort. Between 1993 and 2017, intercourse experience among African American high school students decreased 42 percent (to 46 percent); among European Americans, down 18 percent (to 39 percent); and among Latinos, down 27 percent (to 41 percent) (MMWR, June 15, 2018) (see **Figure 14.6**).

Everywhere Glancing, staring, and—when emotions are overwhelming—averting one's eyes are part of the universal language of love. Although the rate of intercourse among teenagers is lower than it was, passion is expressed in simple words, touches, and, as shown here, the eyes on a cold day.

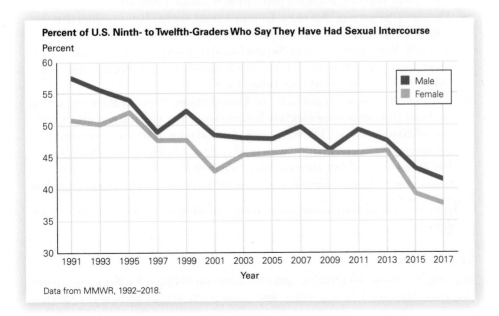

Percent of U.S. Ninth- to Twelfth-Graders Who Say They Have Had Sexual Intercourse

Data from MMWR, 1992–2018.

FIGURE 14.6

Boys and Girls Together Boys tend to be somewhat more sexually experienced than girls during the high school years, but since the Youth Risk Behavior Survey began in 1991, the overall trend has been toward equality in rates of sexual activity.

These were responses to an anonymous questionnaire. As you know from Chapter 1, some inaccuracies may have occurred, but the trends are solid because the same questions were asked over the decades. The data from the 1990s are not as reliable as current numbers, but every study over the past 20 years continues to show that more adolescents are in high school and that fewer are sexually active.

Many reasons for the trends have been suggested: sex education, fear of HIV/AIDS, awareness of the hazards of pregnancy, more female education, less intimacy. To explore these hypotheses, more research is needed.

However, it is apparent, not only in the United States but worldwide, that a universal experience (rising hormones) that produces another universal experience (growth of primary and secondary sex characteristics) is powerfully shaped by cohort, gender, and culture. The most important influence on adolescents' sexual activity is not their bodies but their close friends, who have more influence than either their parents or the norms for their gender and ethnic group (van de Bongardt et al., 2015).

 DATA CONNECTIONS: Sexual Behaviors of U.S. High School Students examines how sexually active teens really are.

Sexual Problems in Adolescence

Sexual interest and interaction are part of adolescence; healthy adult relationships are more likely to develop when adolescent impulses are not haunted by shame and fear (Tolman & McClelland, 2011). Although guidance is needed, teenagers are neither depraved nor degenerate in experiencing sexual urges. Before focusing on the hazards of adolescent sex, we should note that several "problems" are less troubling now than in earlier decades. Here are three specifics:

- *Teen births have decreased.* In the United States, the 2017 rate of births to teenage mothers (aged 15 to 19) was less than half the rate 10 years earlier, with the biggest drop among Hispanic teens (J. Martin et al., 2018). (The 2019 rate was the lowest in 50 years.) Similar declines are evident in other nations. The most dramatic results are from China, where the 2015 teen pregnancy rate was about one-tenth of the rate 50 years ago (reducing the 2015 projection of the world's population by about 1 billion).
- *The use of "protection" has risen.* Contraception, particularly condom use among adolescent boys, has increased markedly in most nations since 1990. The U.S. Youth Risk Behavior Survey found that 61 percent of sexually active high school boys used a condom during their most recent intercourse (MMWR, June 15, 2018) (see **Table 14.1**).
- *The teen abortion rate is down.* In the United States, the teen abortion rate has declined every year since abortion became legal. The rate today is about half that of 20 years earlier. Likewise, teen abortion rates as well as teen birth rates are decreasing in every nation with reliable data, whether or not that nation has liberal or restrictive abortion laws (Sedgh et al., 2015). The probable reasons are that intercourse is less frequent and contraception more prevalent.

These are positive trends, but many aspects of adolescent sexual activity remain problematic.

Sex Too Soon

Sex can, of course, be thrilling and affirming, providing a bonding experience. However, compared to a century ago, adolescent sexual activity — especially

	TABLE 14.1

Condom Use Among High School Students

Country	Sexually Active (% of total)	Used Condom at Last Intercourse (% of those sexually active)
France	20	84
England	29	83
Canada	23	78
Russia	33	75
Israel	14	72
United States	29	54

Data from Currie et al., 2009; Centers for Disease Control and Prevention, October 9, 2018.

if it results in birth—is more hazardous because four circumstances have changed:

1. Earlier puberty results in earlier fertility, and sex before age 15 correlates with depression, drug abuse, and lifelong problems (Kastbom et al., 2015).
2. If a teenager has a baby, she usually has no reliable partners to help. A century ago, teenage mothers were often married; now, in the United States, almost all are unwed.
3. Raising a child has become more complex and expensive, and family helpers are scarce. The strategy that most teenage mothers used in former times—having their mother raise the child—is less available, as most young grandmothers are employed.
4. Sexually transmitted infections (STIs) are more common and more dangerous. The rate of all STIs increased markedly over the past decade, with notable increases from 2016 to 2017 (Centers for Disease Control, 2018).

Especially for Parents Worried About Their Teenager's Risk Taking You remember the risky things you did at the same age, and you are alarmed by the possibility that your child will follow in your footsteps. What should you do? (see response, page 395)

As you read, teen births are declining, as are teen abortions. However, the U.S. rate of adolescent pregnancy is the highest of any developed nation (true among every ethnic group). If a pregnant girl is under 16 (most are not), she is more likely than older pregnant teenagers to experience complications—including spontaneous abortion, high blood pressure, stillbirth, preterm birth, and low birthweight. This is true worldwide, in wealthy as well as low-income nations (Ganchimeg et al., 2014).

There are many reasons for these hazards besides age. Poverty and lack of education correlate with teen pregnancy and with every problem just listed. Sadly, some of the problems begin with family life: Mothers who themselves had teenage births are more likely to have daughters who do the same (N. Liu et al., 2018). One reason for the frequency of complications is that younger pregnant teenagers are frequently undernourished and often postpone prenatal care.

Problems continue from one generation to the next because teenage mothers are less often the responsive mothers that newborns need, so insecure attachment is more common. Although some people imagine that the baby will benefit when an adolescent mother lives with her own mother, the opposite seems more accurate. Troubles begin in infancy, with teenagers living with their mothers less likely to breast-feed their infants than teen mothers living on their own (Pilkauskas, 2014). [**Developmental Link:** Attachment types and the importance of early attachment were discussed in Chapter 7.]

JEFF KOWALSKY/AFP/Getty Images

You, Too? Millions were shocked to learn that Larry Nassar, a physician for gymnasts training for the Olympics and at Michigan State University, sexually abused more than 150 young women. Among the victims was Kaylee Lorenz, shown here addressing Nassar in court. Nassar was convicted of multiple counts of sexual assault and sentenced to 40 to 175 years in prison, but his victims wonder why no one stopped him. The president of Michigan State University resigned in disgrace; many others are still in office.

child sexual abuse Any erotic activity that arouses an adult and excites, shames, or confuses a child, whether or not the victim protests and whether or not genital contact is involved.

Even if sexually active adolescents avoid pregnancy, early intercourse increases psychosocial problems. A study of 3,923 adult women in the United States found that those who *voluntarily* had sex before age 16 were more likely to divorce later on, whether or not they became pregnant or later married their first sexual partner. The same study found that adolescents of any age whose first sexual experience was unwanted (either "really didn't want [it]" or "had mixed feelings about [it]") were also more likely to later experience divorce (Paik, 2011, p. 477).

Sexual abuse is much more traumatic, as is now explained.

Sexual Abuse

Abuse is always devastating: It harms development lifelong. **Child sexual abuse** is defined as any sexual activity (including fondling and photographing) between a juvenile and an adult. Age 18 is the usual demarcation between adult and child (although legal age varies by state). Girls are particularly vulnerable, although boys are also at risk.

Abuse of young children gathers headlines, but young adolescents are, by far, the most frequent victims. The rate of sexual abuse increases at puberty, a particularly sensitive time because many victims are confused about their own sexual urges and identity. Virtually every adolescent problem, including unwanted motherhood, drug abuse, bulimia, and suicide, is more frequent if adolescents are sexually abused.

This is true worldwide. Although solid numbers are unknown for obvious reasons, it is apparent that millions of girls in their early teens are forced into marriage or prostitution each year. Adolescent girls are common victims of sex trafficking, not only because their youth makes them more alluring but also because their immaturity makes them more vulnerable (McClain & Garrity, 2011). Some believe they are helping their families by earning money to support them; others are literally sold by their families (Montgomery, 2015).

It is impossible to know how common child sexual abuse is, but the data suggest that ongoing forced abuse is less common than it was. We do know that fewer girls under age 16 marry older men chosen by their parents, and fewer 16-year-olds have unwanted sex with their 19-year-old boyfriends. Both of the conditions may be considered abuse, and both are less frequent. Estimates of the number of teenage girls being trafficked for sex in the United States range from 1,000 to 336,000 — hardly definitive (Miller-Perrin & Wurtele, 2017).

Trafficking is not the most common form of sexual abuse. Instead, most sexual abuse in the United States occurs at home. Typically, the victims are young adolescents who are not allowed friendships and romances. The abuser may be a biological parent, but more often he is a stepparent, older sibling, or uncle.

Young people who are sexually exploited tend to fear sex, with higher rates of virtually every developmental problem, including repeated abuse (Pittenger et al., 2018). Another developmental consequence is more frequent birth of unwanted babies, who often become mistreated themselves (Noll et al., 2019).

Our discussion of sexual abuse focuses on girls because they are the most common victims. However, teenage boys are also sometimes sexually abused, a direct attack on their fledgling identity as men. Disclosure of past abuse is particularly difficult for men, which makes gathering reliable statistics difficult (Collin-Vézina et al., 2015).

Perpetrators of abuse are often people known to the child. After puberty, although sometimes abusers are parents, coaches, religious leaders, or other authorities, often they are other teenagers. Many high school girls (15 percent) and boys (4 percent) have been kissed, touched, or forced to have sex within a dating relationship when they did not want to (MMWR, June 15, 2018). Chapter 16 discusses sex education; teenagers have much to learn.

Sexually Transmitted Infections

Unlike teen pregnancy and sexual abuse, the other major problem of teenage sex shows no signs of abating. A **sexually transmitted infection (STI)** (sometimes called a *sexually transmitted disease [STD]*) is any infection transmitted through sexual contact. Worldwide, sexually active teenagers have higher rates of the most common STIs—*gonorrhea*, *genital herpes*, and *chlamydia*—than do sexually active people of any other age group.

In the United States, half of all new STIs occur in people ages 15 to 25, even though this age group has less than one-fourth of the sexually active people (Satterwhite et al., 2013). Rates are particularly high among sexually active adolescents, ages 15 to 19. Biology provides one reason: Pubescent girls are particularly likely to catch an STI compared to fully developed women, probably because adult women have more vaginal secretions that reduce infections. Further, if symptoms appear, teens are less likely to alert their partners or seek treatment unless pain requires it, so STIs spread.

A survey of adolescents in a U.S. pediatric emergency department found that half of the teenagers (average age 15) were sexually active and 20 percent of those had an STI—although that was not why they came for medical help (Miller et al., 2015).

There are hundreds of STIs. Chlamydia is the most frequently reported one; it often begins without symptoms, yet it can cause permanent infertility.

Worse is *human papillomavirus (HPV)*, which has no immediate consequences but increases the risk of cancer in both sexes. Fortunately, in about 1990, an effective vaccine was developed that should be given before sexual activity.

However, less than half of all U.S. adolescents are fully immunized (Hirth, 2019). Among the reasons: Some state health departments do not promote it, the vaccine was originally recommended only for girls (because HPV was most closely associated with cervical cancer), and full immunization requires three doses (many teens do not see a medical professional regularly).

For those who care about infant health, *syphilis* may be the worst STI of all, because mothers transmit the virus to their newborns, who develop lifelong disabilities (and sometimes early death). The U.S. rate of *congenital syphilis* has been increasing since 2009, with the 2017 rate twice as high as the 2014 rate (Umapathi et al., 2019). Fortunately, this STI is still rare, but good, early prenatal care can prevent it, so the increase indicates that national health care does not reach the most vulnerable pregnant women.

Once again, it is apparent that a universal experience (the biology of puberty) varies markedly depending on national and family context. As we stated earlier, adolescence begins with biology and ends with culture. You will see more examples of this in the next chapter, as you learn that schools for adolescents vary a great deal in how and what they teach.

> **sexually transmitted infection (STI)** A disease spread by sexual contact; includes syphilis, gonorrhea, genital herpes, chlamydia, and HIV.

> **Especially for Health Practitioners** How might you encourage adolescents to seek treatment for STIs? (see response, page 395)

 DATA CONNECTIONS: Major Sexually Transmitted Infections: Some Basics offers more information about the causes, symptoms, and rates of various STIs.

WHAT HAVE YOU LEARNED?

1. What are examples of the difference between primary and secondary sex characteristics?

2. Why are there fewer problems caused by adolescent sexuality now than a few decades ago?

3. What is problematic regarding adolescent pregnancy?

4. What are the effects of child sexual abuse?

5. Among sexually active people, why do adolescents have more STIs than adults?

SUMMARY

Puberty Begins

1. Puberty refers to the various changes that transform a child's body into an adult one. Even before the teenage years, biochemical signals from the hypothalamus to the pituitary gland to the adrenal glands (the HPA axis) increase production of testosterone, estrogen, and various other hormones that will soon cause rapid growth and reproductive potential.

2. Some emotional reactions, such as quick mood shifts, are directly caused by hormones, as are thoughts about sex. The reactions of others to adolescents and the adolescents' own reactions to the their bodies trigger emotions, which affect hormones.

3. Hormones regulate all of the body rhythms of life, by day, by season, and by year. Changes in these rhythms in adolescence often result in sleep deprivation, partly because the natural circadian rhythm makes teenagers wide awake at night. Sleep deprivation causes numerous health and learning problems.

4. Various parts of the brain continue to mature until about age 25. The regions dedicated to emotional arousal (including the amygdala) mature before those that regulate and rationalize emotional expression (the prefrontal cortex).

5. The outward signs of puberty typically begin between ages 9 and 14. The young person's sex, genetic background, body fat, and level of stress all contribute to this variation in timing.

6. Girls generally begin and end puberty before boys do. A two-year gap is evident in height; a one-year gap is evident in sexual development. Girls from stressed families or neighborhoods are likely to reach puberty earlier.

Growth and Nutrition

7. The growth spurt is an acceleration of growth in every part of the body. Peak weight usually precedes peak height, which is then followed by peak muscle growth, a sequence that makes active adolescents particularly vulnerable to injuries. The lungs and the heart also increase in size and capacity.

8. All of the changes of puberty depend on adequate nourishment, yet adolescents do not always make healthy food choices. One reason for poor nutrition is the desire to lose (or, less often, gain) weight because of anxiety about body image.

9. The precursors of eating disorders are evident during puberty. Many adolescents eat too much of the wrong foods or too little

food overall. Deficiencies of iron, vitamin D, and calcium are common, affecting bone growth and overall development.

10. Eating disorders vary because of genes, family, and social context. Anorexia involves voluntary starvation, bulimia involves overeating and then purging, and binge eating disorder involves compulsive overeating. All these increase the risk of severe depression and medical complications.

11. Because of the sequence of brain development, many adolescents seek intense emotional experiences, unchecked by rational thought. For the same reason, adolescents are quick to react, explore, and learn. As a result, adolescents take risks, bravely or foolishly, with potential for harm as well as for good.

Sexual Maturation

12. Male–female differences in bodies and behavior become apparent at puberty. The maturation of primary sex characteristics means that by age 13 or so, after experiencing menarche or spermarche, teenagers are capable of reproducing, although peak fertility is several years later.

13. Secondary sex characteristics are not directly involved in reproduction but signify that the child is becoming a man or a woman. Body shape, breasts, voice, body hair, and numerous other features differentiate males from females. Sexual activity is influenced more by culture than by physiology.

14. In the twenty-first century, teenage sexual behavior has changed for the better in several ways. Hormones and nutrition cause sexual thoughts and behaviors at younger ages, but teen pregnancy is far less common, condom use has increased, and the average age of first intercourse has risen.

15. Among the problems that adolescents still face is intercourse before their bodies and minds are ready. Giving birth before age 16 takes a physical toll and puts the newborn at risk of physical and psychological problems.

16. Sexual abuse is more likely to occur in early adolescence than at other ages. The perpetrators are often family members or close friends. Rates of child sexual abuse are declining in the United States, but globalization has increased international sex trafficking.

17. Untreated STIs at any age can lead to infertility and even death. Rates among sexually active teenagers are rising. Immunization to prevent HPV is decreasing rates of cancer in adulthood, but most teenagers are not immunized.

KEY TERMS

puberty (p. 371)
menarche (p. 371)
spermarche (p. 371)
pituitary (p. 371)
adrenal glands (p. 371)
HPA (hypothalamus–pituitary–adrenal) axis (p. 371)

gonads (p. 371)
HPG (hypothalamus–pituitary–gonad) axis (p. 372)
estradiol (p. 372)
testosterone (p. 372)
circadian rhythm (p. 373)
secular trend (p. 378)

leptin (p. 379)
growth spurt (p. 382)
body image (p. 385)
anorexia nervosa (p. 385)
bulimia nervosa (p. 385)
binge eating disorder (p. 386)

primary sex characteristics (p. 386)
secondary sex characteristics (p. 388)
child sexual abuse (p. 392)
sexually transmitted infection (STI) (p. 393)

APPLICATIONS

1. Visit a fifth-, sixth-, or seventh-grade class. Note variations in the size and maturity of the students. Do you see any patterns related to gender, ethnicity, body fat, or self-confidence?

2. Interview two to four of your friends who are in their late teens or early 20s about their memories of menarche or spermarche, including their memories of others' reactions. Do their comments indicate that these events are, or are not, emotionally troubling for young people?

3. Talk with someone who became a teenage parent. Were there any problems with the pregnancy, the birth, or the first years of parenthood? Would the person recommend teen parenthood? What would have been different had the baby been born three years earlier or three years later?

4. Adult reactions to puberty can be reassuring or frightening. Interview two or three people about how adults prepared for, encouraged, or troubled their development. Compare that with your own experience.

Especially For ANSWERS

Response for Parents Worried About Early Puberty (from p. 379): Probably not. If she is overweight, her diet should change, but the hormone hypothesis is speculative. Genes are the main factor; she shares only one-eighth of her genes with her cousin.

Response for Parents Worried About Their Teenager's Risk Taking (from p. 391): You are right to be concerned, but you cannot keep your child locked up for the next decade or so. Since you know that some rebellion and irrationality are likely, try to minimize them by not boasting about your own youthful exploits, by reacting sternly to minor infractions to nip worse behavior in the bud, and by making allies of your child's teachers and the parents of your child's friends.

Response for Health Practitioners (from p. 393): Many adolescents are intensely concerned about privacy and fearful of adult interference. This means that your first task is to convince the teenagers that you are nonjudgmental and that everything is confidential.

Observation Quiz ANSWERS

Answer to Observation Quiz (from p. 374): Girls tend to spend more time studying, talking to friends, and getting ready in the morning. Other data show that many girls get less than seven hours of sleep per night.

Answer to Observation Quiz (from p. 383): Adults often try to control schoolchildren by making them wear uniforms. Do you see that these students all must wear blue shirts and ties—even the girls? And teens tend to buy and eat the same foods: Notice the large paper cups, all from the same store.

Adolescence: Cognitive Development

What Will You Know?

1. Why are young adolescents often egocentric?
2. Why does emotion sometimes overwhelm reason?
3. What kind of school is best for teenagers?

O n March 15, 2019, thousands of adolescents in almost 100 nations left their classrooms to gather in the streets. Their inspiration was Greta Thunberg, who began skipping school every Friday when she was 15. She stood in front of the Swedish parliament with a sign to protest adult inaction on climate change.

News media worldwide reported on the 2019 rally, often with students' own words. For example, the BBC interviewed student protesters from a dozen towns in England. In Brighton, a 13-year-old girl said:

> They're messing up our future and we're the ones who are going to have to clean it up, so I think it's important that we come and tell them about it. The school haven't let us go, they say there's consequences but it's more important than school attendance to come here and protest.

Her friend said:

> I really don't care what consequences they give us, it's more important that we fight for our future. This is the world we're going to have to live in.

A 16-year-old boy in Birmingham said:

> We are at the point that in 12 to 20 years the effects of climate change are going to be irreversible. The only way to change it is through the younger generation because the older generation don't really care.

And a 15-year-old in Stockport was torn:

> I tried to come to the last protest but my school said no. My head of year said no this time but I think it is more important to come. I think I'm going to get into trouble though.

That protest stopped public transportation in Manchester, prompting one adult to say

> it's disappointing that they've chosen to disrupt Metrolink—which, ironically, is one of the greenest and most sustainable ways to travel across Greater Manchester.

Five thousand miles away, in Missoula, Montana, students spoke again about the need for adolescents to act because the adults were oblivious. The U.S. reporters tried to put the protest into a political context (O'Brien, 2019). In response, a boy noted "a generation gap between the politicians of today and the politicians of the future."

A 15-year-old in Indiana echoed that when she said:

> This is not a Republican issue; this is not a Democrat issue. This is a human rights issue . . . I don't want to live in a world where the future is unclear. I have big plans for myself . . . People generally perceive it as us kids, we're just whining . . . Unfortunately, adults haven't done enough for us to make this future clear for me and for my peers. So we're taking it into our own hands.

[Fallahi, quoted in Van Dongen, 2019]

These quotations begin this chapter on adolescent cognition because they illustrate how adolescents think—sometimes illogically (the public transportation disruption), sometimes considering adults clueless ("they don't really care"), sometimes exaggerating their importance (the young can reverse climate change, "big plans for myself"). Yet, they are still influenced by the schools that adults created ("consequences," "get into trouble") and opinions that adults might have ("whining").

None of this is surprising. Cognition reflects generation and culture. In the United States, reporters see climate change as a political issue; in England, reporters see the young as misguided; in every nation developmental scientists consider the effects on human health. The young reject all that, instead thinking about their own lives.

This chapter attempts to avoid all these narrow judgments. Instead, we describe the facts of adolescent cognition, a mix of egocentrism and abstraction, of emotions and analysis.

We also explore the myriad structures of the schools that educate adolescents. What should the "consequences" be when adolescent thought leads to ditching school? Do the students quoted above illustrate self-centered priorities, or exaggerated self-importance, or admirable analysis?

Six months later, Greta Thunberg took a 15-day boat trip, fueled by solar and hydropower, to address adults at the United Nations. New York students were allowed to leave school to rally with her—no consequences! Reporters lauded her: She was *Time* magazine's person-of-the-year and appeared on its cover. Adults listened: wise or foolish?

Ferocious Earthlings It's hard not to admire the passion of adolescent cognition, not only on climate change, but also on drugs, religion, patriotism, sex, and many other issues. Admiration does not always mean agreement, but that's why adolescents and adults need each other.

Observation Quiz What is the meaning of the four symbols on the bottom of the "ferocious" poster? (see answer, page 422) ↑

Logic and Self

Brain maturation, additional years of schooling, moral challenges, increased independence, and intense conversations all occur between ages 11 and 18. In some ways, adolescent thought can be understood as two distinct processes, first intense focus on oneself and then moving toward rational thought.

Egocentrism

During puberty, people center on themselves, in part because body changes heighten self-consciousness. Young adolescents grapple with conflicting feelings about adults and peers, examine details of body changes, and think deeply (not realistically) about the future.

Adolescent egocentrism—when adolescents focus on themselves and on what others think of them—was first described by David Elkind (1967). He found that, egocentrically, adolescents regard themselves as much more unique, special, admired, or hated than other people consider them to be. Egocentric adolescents do not understand others' points of view.

adolescent egocentrism A characteristic of adolescent thinking that leads young people (ages 10 to 13) to focus on themselves to the exclusion of others.

For example, few girls are attracted to boys with pimples and braces, but one boy's eagerness to be seen as growing up kept him from realizing this. According to his older sister:

> Now in the 8th grade, my brother has this idea that all the girls are looking at him in school. He got his first pimple about three months ago. I told him to wash it with my face soap but he refused, saying, "Not until I go to school to show it off." He called the dentist, begging him to approve his braces now instead of waiting for a year. The perfect gifts for him have changed from action figures to a bottle of cologne, a chain, and a fitted baseball hat like the rappers wear.

> *[adapted from E., personal communication]*

Egocentrism may lead adolescents to interpret what others do as a judgment on them. A stranger's frown or a teacher's critique mean that "No one likes me," and then they deduce that "I am unlovable" or even "I can't leave the house." Positive casual reactions—a smile from a sales clerk or an extra-big hug from a younger brother—could lead to "I am great" or "Everyone loves me."

When a famous singer suddenly died, one of my students complained that everyone cared about her but would not care if he died. "I might be just as wonderful as she was, but nobody knows."

Acute self-consciousness about physical appearance may be more prevalent between ages 10 and 14 than at any other time, in part because adolescents notice changes in their body that do not exactly conform to social norms and ideals (Guzman & Nishina, 2014). Adolescents also instigate changes that they think other teenagers will admire.

For example, piercings, shaved heads, tattoos, and torn jeans—all contrary to the wishes of most parents—signify connection to youth culture. Wearing suits and ties, or dresses and pearls, would attract unwelcome attention from other youth. Notice groups of adolescents waiting in line for a midnight show or clustering near their high school: Their appearance may seem rebellious, but it conforms to teen culture.

Because adolescents are egocentric, their emotions may not be grounded in reality. A study of 1,310 Dutch and Belgian adolescents found that, for many of them, self-esteem and loneliness were closely tied to their *perception* of how others saw them, not to their actual popularity or acceptance among their peers. Gradually, after about age 15, some realized what others actually thought. Then they became less depressed (Vanhalst et al., 2013).

Rumination

Egocentrism is one reason for *rumination*, which is thinking obsessively about self-focused concerns. Some adolescents go over their problems via phone, text, conversation, social media, and private, quiet self-talk (as when they lie in bed, unable to sleep), thinking about each nuance of everything they have done, are doing, might do, and should have done. Rumination in early adolescence is likely to lead to depression later on (Krause et al., 2018).

Others act impulsively without any rumination at all, blurting out words that they regret and taking risks that they later realize were foolish. Then shame and despair can be overwhelming, again out of proportion to the actual event. Prison administrators know that rates of suicide are higher for jailed adolescent boys than for any other age or gender group (Tartaro, 2019). Impulsive action is one reason.

The Imaginary Audience

Egocentrism creates an **imaginary audience** in the minds of many adolescents. They believe that they are at center stage, with all eyes on them, and they imagine

THINK CRITICALLY: How should you judge the validity of the idea of adolescent egocentrism?

imaginary audience The other people who, in adolescents' egocentric belief, are watching and taking note of their appearance, ideas, and behavior. This belief makes many teenagers very self-conscious.

how others might react to their appearance and behavior. One of my students wrote, "If I ran out of hairspray I would refuse to go out because I wouldn't be caught dead outside with flat bangs.... I just knew in my mind that everyone would know I ran out of spray and everyone would laugh." Her mother tried to make her leave the house. She sobbed, "Never" (IP, personal reflection).

One woman who became a noted scholar remembers:

> When I was 14 and in the 8th grade, I received an award at the end-of-year school assembly. Walking across the stage, I lost my footing and stumbled in front of the entire student body. To be clear, this was not falling flat on one's face, spraining an ankle, or knocking over the school principal—it was a small misstep noticeable only to those in the audience who were paying close attention. As I rushed off the stage, my heart pounded with embarrassment and self-consciousness, and weeks of speculation about the consequence of this missed step were set into motion. There were tears and loss of sleep. Did my friends notice? Would they stop wanting to hang out with me? Would a reputation for clumsiness follow me to high school?
>
> [Somerville, 2013, p. 121]

This woman became an expert on the adolescent brain. She wrote that "adolescents are hyperaware of others' evaluations and feel they are under constant scrutiny by an imaginary audience" (Somerville, 2013, p. 124).

Fables

Egocentrism also leads naturally to a **personal fable**, the belief that one is unique, destined to have a heroic, fabled, legendary life. Some 12-year-olds plan to be star players in the NBA, or to become billionaires, or to cure cancer. The personal fable can extend to their entire generation: Some students quoted in the beginning of this chapter said not only that they understood climate change but also that adults did not care.

One fable is that they are destined to die an early, tragic death. For them, statistics about harm in midlife from STIs, junk food, vaping, or other factors are irrelevant. One of my teenage students said, "That's just a statistic," dismissing its possible application to her.

In one study, teens estimated that their chance of dying before age 20 was 1 in 5. In fact, the odds are less than 1 in 1,000. Even those at highest risk (African American males in urban areas) survive to age 20 more than 99 times in 100. Sadly, if teens think that they will die young, they are likely to risk jail, HIV, drug addiction, and so on (Haynie et al., 2014). If they know someone who did die, their fatalistic response might be "His number was up," unaware that a self-fulfilling prophecy became a nail in the coffin.

The personal fable may coexist with the **invincibility fable**, the idea that death will not occur unless it is destined. Some adolescents believe that fast driving, unprotected sex, or addictive drugs will spare them. Believing that one is invincible removes any attempt at self-control, because personal control is neither needed nor possible (Lin, 2016).

Similarly, teens post comments on Snapchat, Instagram, Facebook, and so on, and they expect others to understand, laugh, admire, or sympathize. Their imaginary audience is other teenagers, not parents, teachers, college admission officers, or future employers, who might have another interpretation (boyd, 2014).

The personal fable is evident worldwide. Boys in teen motorcycle gangs in Indonesia felt "strong and invulnerable against any possible danger while riding [a] motorcycle," which encouraged reckless driving but did not correlate with aggression, as some adults mistakenly assume it does (Saudi et al., 2018, p. 308).

Lelia Valduga/Moment/Getty Images

Oblivious? When you see a teenager with purple hair or a nose ring, or riding a bicycle and reading, do you think he or she does not imagine what others think?

personal fable An aspect of adolescent egocentrism characterized by an adolescent's belief that his or her thoughts, feelings, and experiences are unique, more wonderful, or more awful than anyone else's.

invincibility fable An adolescent's egocentric conviction that he or she cannot be overcome or even harmed by anything that might defeat a normal mortal, such as unprotected sex, drug abuse, or high-speed driving.

Indeed, many researchers assume that adolescent egocentrism is dangerous, but it might have positive effects, giving young people confidence (Hill et al., 2012).

Formal Operational Thought

formal operational thought In Piaget's theory, the fourth and final stage of cognitive development, characterized by more systematic logical thinking and by the ability to understand and systematically manipulate abstract concepts.

Now consider another aspect of adolescent thought. Piaget described a shift to **formal operational thought** as children move past concrete operational thinking and consider abstractions, including "assumptions that have no necessary relation to reality" (Piaget, 1950/2001, p. 163). Is Piaget correct? Many educators think so. They adjust the curriculum between primary and secondary school, reflecting a shift from concrete thought to formal, logical thought. Here are three examples:

- *Math*. Younger children multiply real numbers, such as $4 \times 3 \times 8$; adolescents multiply unreal numbers, such as $(2x)(3y)$ or even $(25xy^2)(-3zy^3)$.
- *Social studies*. Younger children study other cultures by considering concrete expression of daily life—drinking goat's milk or building an igloo, for instance. Adolescents consider how GNP (gross national product) and TFR (total fertility rate) affect global politics.
- *Science*. Younger students grow carrots and feed gerbils; adolescents study invisible particles and distant galaxies.

Piaget's Experiments

LaunchPad
macmillan learning

VIDEO: The Balance Scale Task shows children of various ages completing the task and gives you an opportunity to try it as well.

Piaget and his colleagues devised many tasks to assess formal operational thought (Inhelder & Piaget, 1958/2013b). They found that "in contrast to concrete operational children, formal operational adolescents imagine all possible determinants . . . [and] systematically vary the factors one by one, observe the results correctly, keep track of the results, and draw the appropriate conclusions" (P. Miller, 2016, p. 52).

One of their experiments (diagrammed in **Figure 15.1**) required balancing a scale by hooking weights onto the scale's arms. To master this task, a person must recognize the reciprocity of distance and weight.

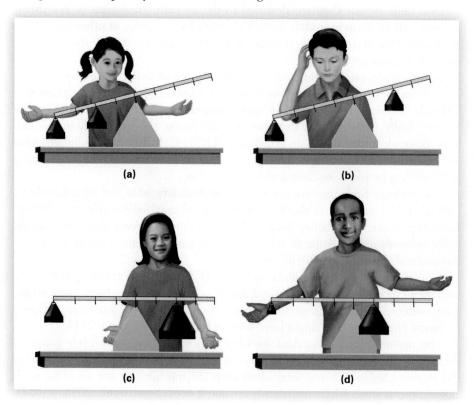

FIGURE 15.1

How to Balance a Scale Piaget's balance-scale test of formal reasoning, as it is attempted by *(a)* a 4-year-old, *(b)* a 7-year-old, *(c)* a 10-year-old, and *(d)* a 14-year-old. The key to balancing the scale is to make weight times distance from the center equal on both sides of the center; the realization of that principle requires formal operational thought.

Balancing was not understood by the 3- to 5-year-olds; when tested, they just played with the weights. By age 7, children understood the concept but balanced the scale only by putting identical weights on each arm: They didn't consider distance from the center. By age 10, children experimented, using trial and error, not logic.

By about age 13 or 14, some children hypothesized about reciprocity, realizing that a heavy weight close to the center can be counterbalanced with a light weight far from the center on the other side (Piaget & Inhelder, 1972).

Hypothetical-Deductive Reasoning

One hallmark of formal operational thought is the capacity to think of possibility, not just reality. "Here and now" is only one of many possibilities, including "there and then," "long, long ago," "not yet," and "never." As Piaget said:

> The adolescent . . . thinks beyond the present and forms theories about every-thing, delighting especially in considerations of that which is not.
>
> *[Piaget, 1950/2001, p. 163]*

Adolescents are therefore primed for **hypothetical thought**, reasoning about *if–then* propositions. Consider the following question (adapted from De Neys & Van Gelder, 2009):

> If all mammals can walk,
> And whales are mammals,
> Can whales walk?

Children answer "No!" They know that whales swim, not walk; the logic escapes them. Some adolescents answer "Yes." They understand the conditional *if,* and therefore they can use logic to interpret the phrase "if all mammals."

> *Possibility* no longer appears merely as an extension of an empirical situation or of action actually performed. Instead, it is *reality* that is now secondary to *possibility.*
>
> *[Inhelder & Piaget, 1958/2013b, p. 251; emphasis in original]*

Hypothetical thought transforms perceptions, not necessarily for the better. Adolescents might criticize everything from their mother's spaghetti (it's not *al dente*) to the Gregorian calendar (it's not the Chinese or Jewish one). They criticize what *is* because they hypothesize what might be. Added to that is a growing awareness of other families and cultures: history, anthropology, and international studies become intriguing.

In developing the capacity to think hypothetically, by age 14 or so adolescents become more capable of **deductive reasoning**, or *top-down reasoning*, which begins with an idea or premise and then uses logic to draw conclusions. In the example above, "if all mammals can walk" is a premise.

By contrast, **inductive reasoning**, or *bottom-up reasoning*, predominates during concrete operational thought. Children accumulate facts and experiences (the knowledge base) to aid their thinking. Since they know whales cannot walk, knowledge trumps logic.

In essence, a child's reasoning goes like this: "This creature waddles and quacks. Ducks waddle and quack. Therefore, this must be a duck." This is inductive: It progresses from particulars ("waddles" and "quacks") to a general conclusion ("a duck"). By contrast, deduction progresses from the general to the specific: "If it's a duck, it will waddle and quack."

hypothetical thought Reasoning that includes propositions and possibilities that may not reflect reality.

🌓 **Especially for Natural Scientists** Some ideas that were once universally accepted, such as the belief that the sun moved around Earth, have been disproved. Is it a failure of inductive or deductive reasoning that leads to false conclusions? (see response, page 422)

deductive reasoning Reasoning from a general statement, premise, or principle, through logical steps, to figure out (deduce) specifics. (Also called *top-down reasoning.*)

inductive reasoning Reasoning from one or more specific experiences or facts to reach (induce) a general conclusion. (Also called *bottom-up reasoning.*)

WHAT HAVE YOU LEARNED?

1. How does adolescent egocentrism differ from early-childhood egocentrism?
2. What perceptions arise from belief in the imaginary audience?
3. Why are the personal fable and the invincibility fable called "fables"?
4. What are the practical implications of adolescent cognition?
5. What are the advantages of using inductive rather than deductive reasoning?

Two Modes of Thinking

You just read about the sequence of thought described by Piaget. He considered intellectual development to be sequential, as children move from one cognitive stage to another (sensorimotor to preoperational to concrete to formal). Others disagree, especially when describing adolescent cognition. They suggest that thinking does not develop in sequence but in parallel, with two processes that are not tightly coordinated within the brain (Baker et al., 2015).

Imagine the adolescent as a pianist. One hand plays the high notes and the other the low notes. Sometimes both hands play together, but often the right hand plays a melody that is uncoordinated with the lower chords.

dual processing The notion that two networks exist within the human brain, one for emotional processing of stimuli and one for analytical reasoning.

intuitive thought Thought that arises from an emotion or a hunch, beyond rational explanation, and is influenced by past experiences and cultural assumptions.

analytic thought Thought that results from analysis, such as a systematic ranking of pros and cons, risks and consequences, possibilities and facts. Analytic thought depends on logic and rationality.

Dual Processing

Many scholars who describe adolescent cognition note that advanced logic (formal operational thought) is counterbalanced by advancing intuition. Thinking occurs in two ways, called **dual processing**.

The thinking described by the first half of each pair is easier and quicker, preferred in everyday life. Sometimes, however, circumstances necessitate the second mode, when deeper thought is demanded. The discrepancy between the maturation of the limbic system and the prefrontal cortex reflects this duality. So does the movement just explained in this chapter, from egocentrism to abstract logic. [**Developmental Link**: Timing differences in brain maturation are discussed in Chapter 14.]

To some extent, both modes of thinking reflect inborn temperament. Most children who are impulsive by nature learn emotional regulation in childhood, but a dual-processing perspective suggests that this regulation may break down during adolescence (Henderson et al., 2015). Even those who are naturally cautious might act impulsively—joyriding, skinny dipping, table dancing—things their parents never imagined their quiet child would do.

Dual Processing

System 1	System 2
Intuitive	Analytic
Hot	Cold
Implicit	Explicit
Creative	Factual
Gist	Specific
Experiential	Rational
Qualitative	Quantitative
Contextualized	Decontextualized

FIGURE 15.2
Two Modes Each pair describes two modes of thought. Although researchers who use each pair differ in what they emphasize, all see two contrasting ways to think.

Intuitive and Analytic Processing

Although many pairs of terms describe dual processing (see **Figure 15.2**), to eliminate confusion we use only one set here, *intuitive* and *analytic*:

- **Intuitive thought** begins with a belief, assumption, or general rule (called a *heuristic*) rather than logic. Intuition is quick and powerful; it feels "right."
- **Analytic thought** is the formal, logical, hypothetical-deductive thinking described by Piaget. It involves rational analysis of many factors whose interactions must be calculated, as in the scale-balancing problem.

To test yourself on intuitive and analytic thinking, answer the following:

1. A bat and a ball cost $1.10 in total. The bat costs $1 more than the ball. How much does the ball cost?

2. If it takes 5 minutes for 5 machines to make 5 widgets, how long would it take 100 machines to make 100 widgets?

3. In a lake, there is a patch of lily pads. Every day the patch doubles in size. If it takes 48 days for the patch to cover the entire lake, how long would it take for the patch to cover half the lake?

[From Gervais & Norenzayan, 2012, p. 494]

Answers are on page 404. As you see, the quick, intuitive responses may be wrong.

Those two processes also reflect experience. Remember *theory of mind*, which began at about age 4? It continues to develop, as children and then adolescents become more aware of what others are thinking and feeling. But distinct aspects of theory of mind are evident, as now explained.

One study assessed 11- to 25-year-olds on two measures: (1) the *Reading the Emotion in the Eyes Test*, which asks participants to looks at faces and say what emotion the person is feeling, and (2) the Social Stories Questionnaire, which asks participants to read a story and note when either of the two characters inadvertently offended the other. On both, theory of mind (ToM) advanced during adolescence, but not in lockstep.

Researchers wrote, "Advanced ToM tasks tap into distinct ToM components that may develop at different rates" (Meinhardt-Injac et al., 2020). With these two ToM tests, it was not the case that one component developed before the other: There is one person playing the piano, one mind developing. But two processes were evident: Each component developed on its own track.

Another demonstration of dual processing occurs when people are asked to report on one half of information and ignore the other half, both presented at the same time. This can be done with an emotionless task, as in the classic *Stroop test*, in which people are told to report the written color of words that spell another color. For example, they must say blue when the word RED is written in blue. This is difficult, but adolescents become better at it as they grow older.

After this "cool" [i.e., analytic] Stroop test, the same adolescents were given what is considered a "hot" task: They were supposed to report the emotion they saw on a face, ignoring the written word presented with the emotion. The word "angry" might accompany a smiling face. This was much harder than the cool task: Adolescents did worse than either children or emerging adults.

In other words, dual processing meant that unemotional processing (cool) gradually improved with age. Emotional processing (hot) created confusion and mistakes, more in adolescence that at younger ages (Aïte et al., 2018).

Age and Two Processes

When the two modes of thinking conflict, people of all ages sometimes use one and sometimes the other. To use the piano metaphor, a musician (especially a novice) sometimes uses just one hand. The first impulse of most people is to favor the melody without the balance of the chords. We are all "predictably irrational" at times (Ariely, 2010).

Increased myelination reduces reaction time, so adolescent thought and action occurs with lightning speed. They are "fast and furious" intuitive thinkers, unlike their teachers and parents, who wait to add slower, analytic thinking. No wonder "people who interact with adolescents often are frustrated by the mercurial quality of their decisions" (Hartley & Somerville, 2015, p. 112).

Observation Quiz Do you see hot or cold thinking here? (see answer, page 422) ↓

Fire Your Trebuchet! Denis Mujanovic, Anna Dim, Ahmed Kamaludeen, and Ghader Asal are all high school students participating in the Western Kentucky Physics Olympics. Here they compete with their carefully designed trebuchets, a kind of catapult related to the slingshot.

Bac Totrong/Daily News via AP Images

ANSWERS	Intuitive	Analytic
1.	10 cents	5 cents
2.	100 minutes	5 minutes
3.	24 days	47 days

Paul Klaczynski conducted dozens of studies comparing the thinking of children, young adolescents, and older adolescents (usually 9-, 12-, and 15-year-olds, respectively) (Jacobs & Klaczynski, 2005). Klaczynski found that almost every adolescent is analytical and logical on some problems but not on others, with some passing the same questions that others fail. As they grow older, adolescents sometimes gain in logic and sometimes regress, with the social context and training in statistics becoming major influences on cognition (Klaczynski & Felmban, 2014).

Interestingly, as adolescents become more aware of the impact of social structures, they also become more aware of stereotypes and more influenced by them. For some stereotypes, adolescents actually regress, with younger children better able to consider individual variations instead of overgeneralizations.

Among the stereotypes that show regression is one called *gender intensification*, such as believing that certain jobs are best filled by men and others by women. Increased gender stereotyping was temporary, evident in early adolescence until about age 15, and then reduced (Klaczynski et al., 2020).

The finding that neither age nor intelligence alone necessarily makes a person more logical has been confirmed by dozens of other studies (Kail, 2013). Being smarter as measured by an IQ test does not advance cognition as much as having more life experience, statistical knowledge, and linguistic proficiency, each of which can further analytic thought. Even though the adolescent brain is capable of logic, sometimes "social variables are better predictors . . . than cognitive abilities" (Klaczynski & Felmban, 2014, pp. 103–104).

Preferring Emotions

Why not use formal operational thought? Adolescents learn the scientific method in school, so they know the importance of empirical evidence and deductive reasoning. But they do not always think like scientists. Why?

Dozens of experiments and extensive theorizing have found some answers (Albert & Steinberg, 2011; Blakemore, 2018). Essentially, logic is more difficult than intuition: It requires questioning ideas that are comforting and familiar, and it might lead to conclusions that are rejected by friends and relatives.

Once people of any age reach an emotional conclusion (sometimes called a "gut feeling"), they resist changing their minds. Analysis might lead to an unpopular conclusion. Without deep thought, prejudice is not seen as prejudice; once people have an emotional reaction, they develop reasons to support their intuition (Haidt, 2013). This is evident in "fake news," which people believe when they suspend analysis (Pennycook & Rand, 2020).

As people gain experience in making decisions and thinking things through, they may become better at knowing when the second cognitive process is needed. For example, in contrast to younger students, when judging whether a rule is legitimate, older adolescents are more suspicious of authority and more likely to consider circumstances that might make some regulations more valid than others (Klaczynski, 2011). Ideally that reflection occurs before a quick, egocentric reaction (see A Case to Study).

What Are Your Priorities?

A developmental approach to cognition seeks to understand how people think at every age. Since adolescents are capable of logical thought, yet they seem to make foolish choices, it is helpful to know what their goals might be.

Perhaps adolescents are not irrational; instead they may rationally seek goals that are not identical to adult goals. Parents want healthy, long-lived children, so they blame faulty reasoning when adolescents risk their lives. Judges want law-abiding citizens, so they punish those who break the law. Adolescents, however, value social warmth and friendship more than old age or social conformity.

THINK CRITICALLY: When might an emotional response be better than an analytic one?

Biting the Policeman

Remember from the piano metaphor that some experience and practice may be needed before a harmony of chords and melody. Might suspicion of authority lead to impulsive and destructive action unless context is taken into account?

In midday, one of my students, herself only 18 years old, was walking in the Bronx with her younger cousin. A police officer stopped them, asking why the cousin was not in school. He patted down the boy and asked for identification. That cousin was visiting from another state; he did not have ID.

My student had done well in a college class on U.S. government. When the officer began to "stop and frisk" her cousin, she reacted emotionally, telling the cop he had no legal authority to do so. He grabbed her cousin; she bit his hand; he arrested her. After weeks in jail (Rikers Island), she was brought before a judge. Perhaps those weeks, plus a meeting with her public defender, activated her analytic mind. She had prepared a written apology; she read it out loud; the officer did not press charges.

I appeared in court on her behalf; the judge released her to me. She was shivering; the first thing I did was put a warm coat around her.

This was dual processing. In her education, my student had gained a formal understanding of the laws regarding police authority. However, her emotions overwhelmed her logic. She intuitively defended her cousin without analyzing the impact.

It is easy to conclude that more mature thought processes are wiser. The judge thought that my perspective was a needed corrective. Certainly the student should not have bitten the officer. But the entire incident, and the NYPD "stop and frisk" policy in effect at the time, shows that the authorities do not understand the adolescent mind. My student had learned a heuristic in childhood: Protect your family. In the heat of the moment, she reacted emotionally.

She is not alone in that. Do you remember something you did as a teenager that, with the wisdom of time and maturity, you would not do now?

Adolescent hormones and brains are more attuned to immediate admiration from peers than to long-term consequences (Blakemore, 2018). Instead of blaming them for foolish choices, their parents should consider that they may be responding to a value system that prioritizes independence.

An evolutionary perspective contends that all primates need to break away from their original family and find mates of their own. Adolescent cognition may push them to do so (Hartley & Somerville, 2015).

Therefore, a young person might be thrilled to ride with many friends in a speeding car, driven by an admired peer. Similarly, a 15-year-old might smoke a cigarette because peer acceptance and possible romance are more valuable than the distant risk of cancer. This is not to condone such actions, but those choices may be less foolish than they appear.

Furthermore, the systematic, analytic thought that Piaget described is slow and costly, not fast and frugal, wasting precious time when a young person seeks action. Adolescents do not want to take time to weigh alternatives and think of how they will feel in middle age. Some risks are taken impulsively, some reactions are intuitive, and that is not always bad.

Indeed, some experts suggest that the adolescent impulse to take risks, respond to peers, and explore new ideas is essential for development and beneficial for the larger society (Ernst, 2016). It may be that adolescent thinking is "adaptive and rational if one considers that a key developmental goal of this period of life is to mature into an independent adult in the context of a social world that is unstable and changing" (Blakemore, 2018, p. 116).

Societies need some people to question assumptions, and that is exactly what adolescents do. As social circumstances change, traditions need reexamination, lest old customs ossify and societies die. The generational tug between rebellion and tradition is part of human life: Neither generation should be too quick to judge the other.

CHAPTER APP 15

📱 **HappiMe for Young People**

IOS:
https://tinyurl.com/y32tsywa

RELEVANT TOPIC:
Adolescent cognition

This mindfulness app offers teens strategies for thinking positively. Teens learn to recognize and confront three distinct characters in their thought patterns and can access other tools and techniques for discouraging negative thought patterns.

WHAT HAVE YOU LEARNED?

1. When might intuition and analysis lead to contrasting conclusions?
2. What mode of thinking—intuitive or analytic—do adolescents prefer, and why?
3. How might intuitive thinking increase risk taking?
4. How can adults protect adolescents from impulsive mistakes?
5. When is intuitive thinking better than analytic thinking?

Secondary Education

What does our understanding of adolescent cognition imply about school? Educators, developmentalists, political leaders, and parents wonder exactly which curricula and school structures are best. There are dozens of options: academic or vocational, single-sex or coed, competitive or cooperative, large or small, public or private, religious or secular.

To complicate matters, adolescents are far from a homogeneous group. As a result,

> some youth thrive at school—enjoying and benefiting from most of their experiences there; others muddle along and cope as best they can with the stress and demands of the moment; and still others find school an alienating and unpleasant place to be.

> *[Eccles & Roeser, 2011, p. 225]*

A further complication arises from the research on cognition just described. Adolescents are egocentric *and* logical, intuitive *and* analytic, beset by fables *and* capable of advanced reasoning. All forms of thought advance as the brain matures during adolescence, but the connections that allow the entire brain to function as a whole are still developing.

Given all of these variations, no school structure or pedagogy is best for everyone, or even for anyone at every time. Various scientists, nations, schools, and teachers try many strategies, some based on opposite but logical hypotheses. Yet adults want to know what works and that schools are accountable, effectively instructing the teenage mind. To analyze this complexity, we present definitions, facts, and ways to measure what learning occurs.

Same Situation, Far Apart: How to Learn Although developmental psychologists find that adolescents learn best when they are actively engaged with ideas, most teenagers are easier to control when they are taking tests (*left,* Winston-Salem, North Carolina, United States) or reciting scripture (*right,* Kabul, Afghanistan).

Definitions and Facts

Secondary education—traditionally grades 7 through 12—denotes the school years after elementary or grade school (known as *primary education*) and before college or university (known as *tertiary education*). Adults are healthier and wealthier if they complete primary education, learning to read and write, and then continue on through secondary and tertiary education. This is true within nations and between them.

Partly because presidents and prime ministers recognize that educated adults advance national wealth and health, every nation is increasing the number of students in secondary schools (see Visualizing Development on the next page). Education is compulsory almost everywhere until at least age 12, and new high schools and colleges open daily in developing nations.

Each year of school advances human potential. As you have read, adolescents are capable of deep and wide-ranging thought, no longer limited by concrete experience; yet they are often egocentric and impulsive.

Quality matters: A year can propel thinking forward or can have little impact. Teachers matter, too (Hanushek et al., 2019). The most effective teachers advance children a year and a half per year, the least effective only a half.

How does anyone know enough about teacher effectiveness to decide that one teacher advances students three times the rate of another? The answer usually involves a test of student learning. As you remember from Chapter 1, scientists try to find ways to measure and evaluate outcomes in order to draw valid conclusions. In this case, many people want to know if students are learning what they should be learning in secondary school.

We want to avoid *social promotion*, promoting students simply because they have spent time in school. A high school diploma is supposed to mean that a student has mastered certain knowledge and skills, and a test might determine that.

On the other hand, as you remember from IQ testing, every test reflects the perspective of the people who write and score the tests. To explore what kind of tests are best for secondary school students, we now look at three types of tests: national "high-stakes" tests, performance assessments, and international tests.

High Stakes in the United States

Students in the United States take many more tests than they did before 2000, but fewer than they did in 2010. Many of those tests are **high-stakes tests**, so called because of the high cost of failure. In some school districts, *exit exams* determine whether or not a senior can graduate. An 18-year-old might have passed every class for the four years of high school, earning all the required credits (such as two years of a second language, three years of math, four years of English) but still not graduate because of failing a high-stakes test.

Tests also can be high stakes for teachers, who may earn extra pay or lose their jobs based on how their students score. Schools gain resources or are shuttered because of test scores. Entire school systems are rated on test scores. This is said to be one reason that widespread cheating on high-stakes tests occurred in Atlanta, Georgia, beginning in 2009 (Severson & Blinder, 2014).

High-stakes tests sometimes determine whether a child must repeat a grade, beginning at age 8. The 2016 federal education law, the Every Student Succeeds Act (ESSA), required standardized testing from the third grade on.

However, various schools interpret the results differently. For example, Florida was one of several states that mandated that third-graders who failed cannot enter fourth grade. However, in Miami, 15 percent of 28,000 third-graders failed but only half of them were actually left back (Tavassolie & Winsler, 2019).

secondary education Literally, the period after primary education (elementary or grade school) and before tertiary education (college). It usually occurs from about ages 12 to 18, although there is some variation by school and by nation.

high-stakes test An evaluation that is critical in determining success or failure. If a single test determines whether a student will graduate or be promoted, it is a high-stakes test.

How Many Adolescents Are in School?

Attendance in secondary school is a psychosocial topic as much as a cognitive one. Whether or not an adolescent is in school reflects every aspect of the social context, including national policies, family support, peer pressures, employment prospects, and other economic concerns. Rates of violence, delinquency, poverty, and births to girls younger than 17 increase as school attendance decreases.

PERCENTAGE OF ADOLESCENTS NOT ENROLLED IN SECONDARY SCHOOL

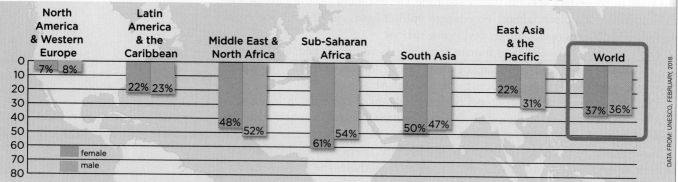

DATA FROM: UNESCO, FEBRUARY, 2018.

SELECTED SECONDARY SCHOOL GRADUATION RATES - AGE 25 AND UNDER*

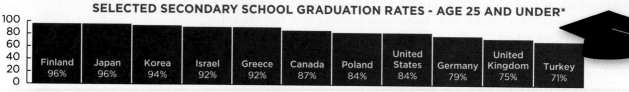

Finland	Japan	Korea	Israel	Greece	Canada	Poland	United States	Germany	United Kingdom	Turkey
96%	96%	94%	92%	92%	87%	84%	84%	79%	75%	71%

DATA FROM: OECD, 2018

*In some nations, some students do not graduate because they go directly into the job market. For example, Germany has an extensive apprenticeship program.

U.S. HIGH SCHOOL GRADUATION RATE, CLASS OF 2017

Since 2007, the dropout rate among foreign-born youth has declined much faster than for native-born youth, from 27 to 10 percent.

Males are more likely to drop out of high school than their female counterparts, a shift that has occurred since 1980. Before that, more females dropped out, usually because they were pregnant.

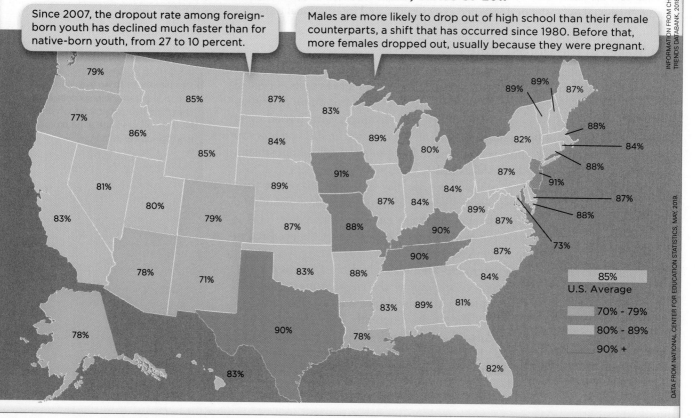

85% U.S. Average
70% - 79%
80% - 89%
90% +

INFORMATION FROM CHILD TRENDS DATABANK, 2018.

DATA FROM NATIONAL CENTER FOR EDUCATION STATISTICS, MAY, 2019.

For high school students, tests not only determine graduation, but also what college they might enter, and what credits they will earn there. Many colleges use scores on the *SAT (Scholastic Aptitude Test)* or the *ACT (American College Test)* as a crucial factor in deciding admission. Students earn college credits by scoring well on the *AP (Advance Placement)* or *IB (International Baccalaureate)* exams.

The increase in testing is evident in the AP. About 5 million AP tests were taken in 2018, with many students taking several AP tests. That is seven times as many tests as in 2007, when there were only 10 possible AP tests; now there are 38. For example, there was one test in physics and relatively few students took it; now there are three AP physics tests. Each of the 38 AP tests is aligned with a rigorous class in high school (Finn & Scanlan, 2019).

The purpose of mandatory graduation tests was to standardize and improve instruction. Ideally, a diploma earned in one city was equivalent to a diploma in another, and so teachers would try to have all their students reach a high standard.

However, many parents, teachers, and other groups complained that the test narrowed the curriculum, making teachers focus on preparing students to be good test-takers who knew facts, rather than teaching students how to work with other people, analyze problems, and imagine new solutions. In addition, tests were thought to be unfair to African Americans and to English Language Learners (Acosta et al., 2020; Dworkin & Quiroz, 2019; Koretz, 2017).

Those complaints reached lawmakers:

- In 2013, Alabama dropped its high-stakes test for graduation.
- Pennsylvania instituted such a test in that same year, but opponents postponed implementation, and in 2018 the Pennsylvania legislature voted to allow several alternate paths to graduation (including the SAT).
- In 2020, Janet Napolitano, president of the University of California, recommended that the entire California college system scrap the SAT and the ACT as admission requirements and use homegrown tests instead (Hoover, 2020).
- A 2007 law in Texas required 15 tests for graduation; in 2013, Texas law reduced that to four tests (Rich, 2013).

In 2002, more than half of all U.S. states required passing an exit exam before graduation; in 2019, only 13 did (Gewertz, April 9, 2019). However, another trend increased testing: More states (25 by 2019) require all students to take the SAT or ACT, in part to encourage them to plan for college. Overall, high school graduation rates in the United States have increased every year for the past decade (McFarland et al., 2018), reaching 85 percent of fourth-year high school students in 2019 (see **Figure 15.3**). (Graduation rates are higher when they include students who take five or six years to graduate, or who earn an equivalency diploma.) Do these trends reflect increased or decreased testing? If graduation tests are not used, how will employers, parents, or colleges know if a graduate really learned what high school is supposed to teach? Some schools (mostly private ones) use other criteria, such as portfolios or senior papers.

One group of 38 public schools in New York (36 of them in New York City), called the New York Performance Standards Consortium, requires students to

FIGURE 15.3

Mostly Good News This depicts wonderful improvements in high school graduation rates, especially among Hispanic youth, who drop out only half as often as they did 20 years ago. However, since high school graduation is increasingly necessary for lifetime success, even the rates shown here may not have kept pace with the changing needs of the economy. Future health, income, and happiness for anyone who drops out may be in jeopardy.

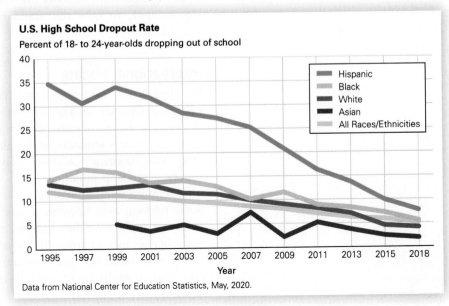

U.S. High School Dropout Rate

Percent of 18- to 24-year-olds dropping out of school

Legend: Hispanic, Black, White, Asian, All Races/Ethnicities

Data from National Center for Education Statistics, May, 2020.

| | TABLE 15.1 | | | |

Sample of New York Performance Standards Indicators for the Social Science Research Paper

	Outstanding	Good	Competent	Needs Revision
Evidence and Sources	• Supporting arguments include specific, relevant, accurate and verifiable, and highly persuasive evidence, drawn from both primary and secondary sources. • Provides specific, relevant, accurate evidence for counterargument, where appropriate. • Uses quotations and paraphrasing appropriately to sustain an argument.	• Supporting arguments include relevant, accurate and verifiable, and mostly persuasive evidence, drawn from both primary and secondary sources. • Provides relevant, accurate evidence for counterargument, where appropriate. • Uses quotations and paraphrasing appropriately to sustain an argument.	• Evidence for supporting arguments is accurate and verifiable, mostly specific and relevant, and generally persuasive drawn from secondary sources. • Use of quotations and paraphrasing is mostly evident.	• Supporting arguments may include inaccurate evidence and lack clear, persuasive, or relevant evidence. • Quotations and paraphrasing do not effectively support arguments.

Only One of Eight! The entire set of New York Performance Standards for graduation includes several disciplines (including science and literature), each with several measures. Here is just one of eight measures used for the social studies research paper. The other seven include grammar, organization, and analysis. As you see, several evaluators and five levels of success are rated. The high schools in the Consortium consider these rubrics much more demanding than traditional exit exams.

complete four tasks: an essay that analyzes literature, a research paper in the social sciences, a science project that students design and complete, and an applied demonstration of advanced math. Each of these four includes a written and oral test, judged by teachers and outsiders, according to set rubrics (see **Table 15.1**).

Initial results are very positive, with higher graduation rates and college success (Barlowe & Cook, 2016). However, whether or not this can be applied to all of the estimated 43,000 secondary schools in the United States is unknown.

The fact that high-stakes tests have increased and then decreased so dramatically over the past few decades means that change is possible. The fact that virtually every U.S. student takes many standardized achievement tests suggests that eliminating tests completely is unlikely. Even students in the Consortium schools are required, by the New York State Education Department, to take the standardized test to demonstrate proficiency in English.

International Testing

We already described several international tests, particularly the TIMSS and the PIRLS, in Chapter 12. These tests are given to fourth-, eighth-, and twelfth-grade students in dozens of nations. The results for secondary students are similar to those for younger students, with East Asian nations scoring high, Middle Eastern and South American children scoring low (with some exceptions), and North American students middling.

One more international test, the PISA (mentioned in Chapter 12), is especially relevant here because it was designed to measure high school students' ability to apply what they have learned. The PISA is taken by 15-year-olds, an age when almost all students are still in school but when many will soon stop formal education. The questions are supposed to be practical, measuring knowledge that all adults might need. Recently the emphasis has been on science and analysis.

For example, among the 2012 math questions is this one:

> Chris has just received her car driving license and wants to buy her first car. The table below shows the details of four cars she finds at a local car dealer.
> What car's engine capacity is the smallest?

       A. Alpha   B. Bolte   C. Castel   D. Dezal

Model	Alpha	Bolte	Castel	Dezal
Year	2003	2000	2001	1999
Advertised price (zeds)	4,800	4,450	4,250	3,990
Distance traveled (kilometers)	105,000	115,000	128,000	109,000
Engine capacity (liters)	1.79	1.796	1.82	1.783

For that and the other questions on the PISA, the calculations are quite simple—most 10-year-olds can do them; no calculus, calculators, or complex formulas required. However, almost half of the 15-year-olds worldwide got that question wrong. (The answer is D.) One problem is decimals: Some students do not remember how to interpret them when a practical question, not an academic one, is asked. Even in Singapore and Hong Kong, one out of five 15-year-olds got this question wrong. Another problem is that distance traveled is irrelevant, yet many students are distracted by it.

Overall, U.S. students score lower on the PISA compared to those in many other nations, including Canada, the nation most similar to the United States in ethnicity and location. Compared to peers in other nations, the 2015 results rank U.S. 15-year-olds 25th out of 72 (see **Table 15.2**).

TABLE 15.2

Average PISA Scores of 15-Year-Old Students in 2015—Selected Nations

	Mathematics	Reading	Science
Singapore	564	535	556
Japan	532	516	538
Canada	516	527	528
Netherlands	512	503	509
Germany	506	509	509
Poland	504	506	501
New Zealand	495	509	513
Russia	494	495	487
Australia	494	503	510
United Kingdom	492	498	509
Italy	490	485	481
United States	470	497	496
Chile	423	459	447
Mexico	408	423	416
Lebanon	396	347	386
Indonesia	386	397	403
Brazil	377	407	401
Dominican Republic	328	358	332

Some 2015 results were not surprising (China, Japan, Korea, and Singapore were all high), but some were unexpected (high scores for Finland, Vietnam, and Estonia). Among the lowest results were Peru, Indonesia, and the Dominican Republic. The results reflect the educational systems more than geography, since low-scoring Indonesia is close to Singapore.

International analysis finds that the following items correlate with high achievement of high school students on the PISA. The standards were first articulated a decade ago, but they continue to apply (OECD, 2010).

- Leaders, parents, and citizens value education overall, with individualized approaches to learning so that all students learn what they need.
- Standards are high and clear, so every student knows what they must do, with a "focus on the acquisition of complex, higher-order thinking skills."
- Teachers and administrators are valued, and they are given "considerable discretion . . . in determining content" and sufficient salary as well as time for collaboration.
- Learning is prioritized "across the entire system," with high-quality teachers assigned to the most challenging schools.

The PISA and international comparisons of high school students note that students who are immigrants and who are from low-income families do less well. Researchers caution that variations within nations affect results, so comparing one nation to another may be unfair. However, comparing students within nations is illuminating. Generally, girls do less well in science than boys, but this is less true in the United States than in most other nations. However, low-income students in general do especially poorly in the United States.

The scores from Finland have caught international attention, since educators in that nation do not use high-stakes tests. Nor do the students spend much time on homework or after-school education, as is true in some other high-scoring nations, notably South Korea. A Finnish expert proudly states that "schoolteachers teach in order to help their students learn, not to pass tests" (Sahlberg, 2011, p. 26).

International comparisons of 40 nations also show that students in private schools who take the PISA tend to do better than those in public schools, partly because they take the test more seriously and are less likely to guess randomly or skip questions (DeAngelis, 2019). The effort put into learning may be crucial in the United States as well, with different factors relevant depending on whether the student is in middle school or high school, as you will now see.

Middle School

Often nations provide two levels of secondary education. Traditionally, in the United States, students attended junior high (usually grades 7 and 8) and then senior high (usually grades 9 through 12). As the average age of puberty declined, **middle schools** were created for grades 5 or 6 through 8, as educators recognized that pubescent 12-year-olds are quite different human beings, cognitively and in many other ways, from 17-year-olds.

Middle school may be the most stressful time in a students' education. Educators find that "teaching is likely to be particularly complex for middle school teachers because it happens amidst a critical period of cognitive, socio-emotional, and biological development of students who confront heightened social pressures from peers and gradual decline of parental oversight" (Ladd & Sorensen, 2017).

Many developmentalists find middle schools to be "developmentally regressive" (Eccles & Roeser, 2010, p. 13), which means that learning goes backward.

middle school A school for children in the grades between elementary school and high school. Middle school usually begins with grade 6 and ends with grade 8.

Now Learn This Educators and parents disagree among themselves about how and what middle school children need to learn. Accordingly, some parents send their children to a school where biology is taught via dissecting a squid *(left)*, others where equality and respect are taught via uniforms and lining up *(right)*.

For students, middle school is a sensitive period, with effects that continue for years (Yeager et al., 2017). Whether or not students go to college is often determined by their middle school experience.

Engaged Students

Middle school students are dramatically less engaged with school compared to primary school students. The concept of engagement includes how involved students are with school, how active they are in classes, and how much they like their school (Forster et al., 2020; Wang & Eccles, 2013). Every study finds that student engagement drops between primary and secondary education, with the most precipitous drop in middle school.

The most dramatic data come from a 2016 Gallup Student Poll of almost a million students at 3,000 schools in the United States and Canada. They marked whether or not they agreed or disagreed with the following statements:

- At this school, I get to do what I do best every day.
- My teachers make me feel my schoolwork is important.
- I feel safe in this school.
- I have fun at school.
- I have a best friend at school.
- In the last seven days, someone has told me I have done good work at school.
- In the last seven days, I have learned something interesting at school.
- The adults at my school care about me.
- I have at least one teacher who makes me excited about the future.

Every year from the fifth to the ninth grade, fewer students agreed with those statements. By far, the biggest drops were during middle school (see **Figure 15.4**) from three-fourths engaged to less than half (Calderon & Yu, 2017). By high school, engagement was low (about a third), but at least it didn't get much worse each year (active disengagement did, however).

As you remember from Chapter 1, surveys are often inaccurate. Gallup explains that schools and students were self-selected, so specific percentages do not reflect middle school students everywhere. However, the general trends are probably correct, since the same design flaws applied in every grade. Smaller, but more valid studies agree: Many students disengage from school every year of middle school.

Observation Quiz Although the philosophy and strategy of these two schools are quite different, both share one aspect of the hidden curriculum. What is it? (see answer, page 422) ↑

FIGURE 15.4

Fun, Safe, Exciting . . . School? A Gallup poll found these results. Do you think that a valid scientific study would also find that three-fourths of primary school students, but less than a third of high school students, are actively engaged in school?

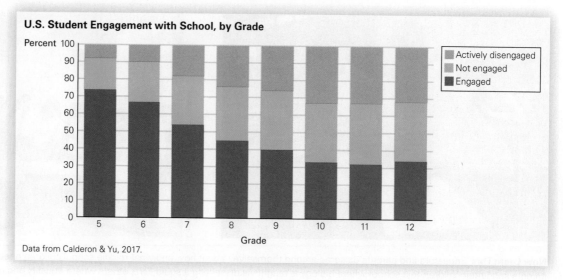

U.S. Student Engagement with School, by Grade

Data from Calderon & Yu, 2017.

Body and Mind

Puberty itself is part of the problem with education in middle school. Puberty changes the brain, and that affects readiness to learn (Goddings et al., 2019). At least for other mammals, especially when they are under stress, learning is reduced at puberty (McCormick et al., 2010). One problem may be the sexual awakening that occurs with puberty. Sexual conquests are flaunted, girls may be seen as conquests, people of every gender may feel elated, despondent, or obsessed about someone else. Students may have no psychic energy left for homework.

This was the reason that, in 2010, the nation of Trinidad and Tobago changed 20 low-performing schools to single-sex institutions. Achievement rose, in part because the teachers intensified instruction (Jackson, 2019). Would this strategy work elsewhere, or is something unique about that West Indian context?

However, if a student's school achievement suddenly dips, that is both a sign and a cause of psychological as well as academic problems (Rahman et al., 2018). Adults need to understand the reason and change the trajectory; puberty itself should not be blamed. A crucial factor is emotional: whether the students trust and like their teachers (Binning et al., 2019; Riglin et al., 2013).

Signs of a future high school dropout, among them chronic absenteeism, appear in middle school (Ladd & Sorensen, 2017). Those students most at risk are low-SES boys from minority ethnic groups, yet almost no middle school has male guidance counselors or teachers who are African American or Latinx. Given their egocentric and intuitive thinking, many young adolescents need role models of successful, educated men (Morris & Morris, 2013).

Declines in liking school, in trusting teachers, and working hard in class are particularly steep for young adolescents who are African American, Latinx, or Native American, especially when they enter a multiethnic middle school when coming from a neighborhood primary school. As postformal thinking allows them to notice social forces beyond their personal experience, they perceive low expectations and harsh punishment for people of color.

Are they right or biased? One report on 17 middle schools found that for infractions that involved teacher judgment (e.g., disrespect, excessive noise), African Americans were disciplined three times as often as European Americans, but for more objective offenses (cheating, bringing a weapon to school), the rates were more similar (Skiba et al., 2002).

That was decades ago. Has it changed? A more recent report on a large, multiethnic school, found that African American sixth-graders (including those not

CAREER ALERT   The Teacher

Many people who study human development hope to become teachers, for good reason. Teachers can make a huge impact on a child's life. Every adult probably remembers a teacher or two whose interest and insight still affects them.

The need is great, and the demand huge. According to the U.S. Bureau of Labor Statistics' Occupational Outlook Handbook, in 2016 there were 1,500,000 elementary school teachers and 1,008,000 secondary school teachers. But every year, more than 100,000 teachers leave the profession—some retire, some quit, some die. They need to be replaced.

Depending on the local school district, an aspiring teacher can qualify with a bachelor's degree in almost any field. Courses in education can be taken while teaching. Better would be a master's degree in education, which includes on-the-job training with excellent teachers.

Many specialties within the teaching profession are chronically understaffed. In the United States, those trained to teach math, science, a non-English language, or children with special needs will be hired—as long as they are willing to live outside their home community.

Those interested in teaching probably already know that the salary is not that great, but the benefits are adequate, and teaching children can be immensely satisfying as well as challenging. Further, teachers have more vacation days than most professions, and the workday may seem short since most school days end by 3 P.M.

However, good teachers spend as much time preparing and planning as they do in direct teaching. Further, the work is exhausting, physically and emotionally. Teachers become painfully aware that some students have serious problems that teaching cannot solve: abusive or neglectful parents, learning differences, severe poverty, chronic depression.

The greatest openings in the profession are in areas that require special training. Novices may hope to teach high school English or to teach third grade in an affluent suburb. But such jobs are taken by teachers who have seniority; they are unlikely to be filled by new recruits.

Instead, aspiring teachers are most likely to be hired in areas of greatest need: math teachers in cities, teachers who specialize in autism, bilingual teachers, speech teachers who can relate with children and parents of many backgrounds, middle school science teachers.

Some of my students want to be teachers. I encourage them, warning that this profession is more difficult than it may seem. As one leading educator wrote: "Teaching is not rocket science—it is much harder than that" (Sahlberg, 2015, p. 133).

---

personally disciplined) expected biased treatment. Incoming Latinx students did not expect bias, but by the eighth grade, they did. Most White students thought the teachers were fair, but they, too, became less trusting of school authorities over time (Yeager et al., 2017).

Even if there were no discrimination in the larger society, students have reasons to dislike middle school. Parents are less involved than in primary school, partly because students want more independence. It is psychologically healthy for adolescents to push their parents away, but the lack of parental support also increases risk. Bullying increases in the first year of middle school (Baly et al., 2014).

Teacher support also decreases. Unlike teachers in primary school, where each classroom typically has one teacher who knows the families and friends of each student, middle school teachers have hundreds of students. Teachers become impersonal and distant, not the engaging adults that young adolescents need (Meece & Eccles, 2010). Both teachers and students may imagine that the other is hostile toward everyone of their generation, and that imagined hostility harms them both, as now explained.

Stereotype Threat

It is easy to understand why stereotypes are mistaken, since they do not consider the individuality of each person. Probably every reader of this book knows the sting of being stereotyped and has sometimes too quickly judged someone else, although we are less aware of our own stereotypes than of the stereotypes directed at us.

But more insidious than a direct stereotype may be **stereotype threat**, when someone holds a stereotype that someone else holds a stereotype about them. That mistaken idea can boomerang, harming the person who imagined it.

That boomerang begins with a fact that has troubled social scientists for decades: Secondary school achievement varies by ethnicity. This is evident in tests, grades, and, eventually, graduation. For example, in 2018, 42 percent of public high school

Especially for Teachers You are stumped by a question that one of your students asks. What do you do? (see response, page 422)

stereotype threat The thought a person has that one's appearance or behavior will be misread to confirm another person's oversimplified, prejudiced attitudes.

Hill Street Studios/Blend Images/Newscom

More Like Him Needed In 2014 in the United States, half of the public school students were tallied as non-White and non-Hispanic, and half were male. Meanwhile, only 17 percent of teachers are non-White and non-Hispanic, and only 24 percent are male. This Gardena, California, high school teacher is a welcome exception in two other ways—he rarely sits behind his desk, and he uses gestures as well as his voice to explain.

students were Black or Hispanic, but only 38 percent of graduates were. The reasons seem objective: African Americans do not accumulate high school credits at the same rate as other groups.

One sad outcome is ethnic disparity in *status dropouts*, 16- to 24-year-olds who quit school before graduation. Status dropouts have markedly higher rates of unemployment and early death.

Data from 2017 on men aged 16–25, find that 5 percent of non-Hispanic Whites, 7 percent of non-Hispanic Blacks, and 7.5 percent of Hispanics are status dropouts. [Those ethnic categories are the official ones used by the U.S. government]. For women aged 16–24, status dropout rates are 4.3 percent non-Hispanic White, 4.4 percent non-Hispanic Black, and 7.4 percent Hispanic.

There are many explanations for ethnic differences, including poverty, language, genes, and school quality. Each explanation reflects the perceptions of the person offering it.

But look closely at the gender differences among non-Hispanics. Black girls have the same genes, families, and schools as Black boys, yet their status dropout rates are similar to White girls. Why?

Again, many explanations are proposed, each reflecting perceptions. But a novel explanation was suggested by one African American scholar, Claude Steele. He suggested that Black adolescent boys may think that other people have stereotypes about them. The result may be stereotype threat, a "threat in the air," not in reality (Steele, 1997). The mere *possibility* of being negatively stereotyped increases stress, disrupts cognition, and reduces emotional regulation.

Steele suspected that African American males, aware of the stereotypes that they are not smart, become anxious in educational settings. Their anxiety increases stress. Then heightened cortisol undercuts their achievement test scores, so young men protect their pride by denigrating academics. They avoid the anxiety of writing papers, taking exams, and reading textbooks because they anticipate that teachers will not appreciate their work.

In self-defense, they conclude that school doesn't matter, that people who are "book smart" are not "street smart," that Black classmates who study hard are "acting White." They disengage from school, which results in lower achievement.

Stereotype threat occurs within many other groups and skills. It causes college women to underperform in math, bilingual students to stumble with English, older people to be forgetful, and every member of a stigmatized minority in every nation to handicap themselves because of what they imagine others might think (Baysu & Phalet, 2019; Bouazzaoui et al., 2020; Spencer et al., 2016).

Athletic prowess, health habits, and vocational aspiration may be impaired (Aronson et al., 2013). For example, stereotype threat may cause blind people to be underemployed: They hesitate to learn new skills or apply for jobs because they think other people will judge them (Silverman & Cohen, 2014). When star athletes unexpectedly underperform (called *choking*), stereotype threat may be the cause (Smith & Martiny, 2018).

The worst part of stereotype threat is that it is self-imposed. People who are alert to the possibility of prejudice are not only hypersensitive when it occurs, but their minds are hijacked, undercutting potential.

How might this work in secondary school? If students think that teachers expect them to misbehave, anticipating racially biased treatment, that thought will cause them to perceive neutral teacher behavior as disrespectful. That interpretation increases the hormones that, in fact, trigger misbehavior.

Stereotype threat can hobble teachers as well. If they think that students dislike them, in self-defense they might be less personable and more distant, which makes things worse. This becomes a downward spiral, a self-fulfilling prophecy, a sad and hostile teacher.

No one, including Steele, believes that stereotype threat is the only reason for disparities in success and achievement. Further, some adolescents strive to prove the stereotypes wrong. Nonetheless, adolescence may be a crucial time to reduce stereotype threat and increase school engagement (Binning et al., 2019).

Because adolescents are intensely self-conscious, with heightened intuition and irrational personal fables, a teacher's praise for work well done, and explicit expression of respect and anticipation for better work, can transform a threat into a promise. The future can be imagined as worth waiting for—if the adolescent brain can allow anticipation (see Inside the Brain).

INSIDE THE BRAIN

Save for a Rainy Day?

You already know that brain growth is uneven in adolescence, with the limbic system advancing ahead of the prefrontal cortex. This explains dual processing, since intuitions arise from the limbic system while analysis is rooted in the prefrontal cortex. It also explains the impulse in middle school to stop studying, and the impulse in high school to quit before graduation.

One topic in cognitive research is called *delay discounting*, the tendency to discount future rewards and instead seek immediate pleasure. Delay discounting is evident when young children eat one marshmallow rather than waiting for two (as described in Chapter 10), when an addicted person takes the drug in front of them instead of thinking about the next day, when middle-aged adults buy a new car instead of saving for retirement.

Delay discounting is particularly strong in adolescence because of uneven growth in the adolescent brain. One part, the *ventral striatum*, is the area that is particularly sensitive to rewards. The striatum grows stronger at puberty, making rewards powerfully attractive (Shulman et al., 2016). As one team expressed it, the "subcortical regions that respond to emotional novelty and reward are more responsive in middle adolescence than in either children or adults" (Harden & Tucker-Drob, 2011, p. 743).

Because the striatum increases the allure of immediate rewards ("don't think about tomorrow"), adolescents are vulnerable to serious problems. Delay discounting is one explanation for impulsive suicide, drug addiction, and eating disorders (Felton et al., 2020).

When delay discounting is added to other aspects of adolescent brain development, with stronger sensations and quicker reactions (sensation-seeking and impulsivity), harm is lifelong. To make it worse, for some adolescents, influences from childhood, such as family chaos and abuse, reduce the brain's ability to wait for future rewards (see **Figure 15.5**) (Acheson et al., 2019).

Fortunately, sensation-seeking and impulsivity are not tightly linked in the brain (Harden & Tucker-Drob, 2011). Both increase, but an adolescent might be drawn to sensory stimulus, but be able to control impulsive reactions. Past family stability and nurturance allows better planning,

so destructive impulses are far from inevitable (Acheson et al., 2019). Sensation-seeking might not lead to trouble but might allow a young person to appreciate the moment, such as an exquisite sunset, a sumptuous meal, or a flowering tree more than an adult might.

In many ways, the adolescent brain functions differently from the adult brain. But heightened appreciation of rewards does not always means increased danger. Teachers can find ways to make the immediate classwork more enjoyable, such as encouraging teamwork, laughter, imagination. High school students can visit a college, can plan for a senior prom, can secure a summer job with successful adults.

Everyone of every age is sometimes discouraged by the daily grind. We can learn from the teenagers, even as we try to teach them to plan ahead. Stop to smell the roses?

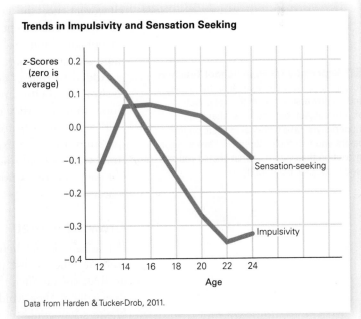

Trends in Impulsivity and Sensation Seeking

Data from Harden & Tucker-Drob, 2011.

FIGURE 15.5

Look Before You Leap As you can see, adolescents become less impulsive as they mature, but they still enjoy the thrill of a new sensation.

Finding Acclaim

Middle school is a time when children can learn how to cope with challenges, both academic and social. A habit of solving problems rather than blaming oneself—is crucial (Monti et al., 2017). But a habit that worked in primary school may not match the conditions of middle school.

To pinpoint the developmental mismatch between students' needs and the middle school context, note that just when egocentrism leads young people to feelings of shame or fantasies of stardom (the imaginary audience), schools typically require them to change rooms, teachers, and classmates every 40 minutes or so. That context limits friendship and acclaim.

Recognition for academic effort is rare because middle school teachers grade more harshly than their primary school counterparts. Effort without accomplishment is not recognized, and achievement that was previously "outstanding" is now only average. Acclaim for after-school activities is also elusive, because many art, drama, dance, and other programs put adolescents of all ages together, and 11- to 13-year-olds are not as skilled as older adolescents.

Athletic teams become competitive, so those with fragile egos protect themselves by not trying out. If sports require public showers, that is another reason for students in early puberty to avoid them: They do not feel at ease with their changing bodies, and they fear comments from their peers, a fear that is no less powerful if it is a stereotype threat rather than a fact.

As noted in the discussion of the brain, peer acceptance is more cherished at puberty than at any other time. Physical appearance—from eyebrows to foot size—suddenly becomes significant. Status symbols (e.g., trendy sunglasses, a brand-name jacket, the latest smartphone) take on new meaning. Expensive clothes and shoes are coveted.

High School

Many of the patterns and problems just explained continue in high school. However, once the erratic growth and sudden sexual impulses of puberty are less novel, adolescents are better able to cope. Moreover, peers become more encouraging, and teachers and parents allow more autonomy, which encourages more self-motivation.

Added to that is cognitive maturation. When adolescents are better able to think abstractly, analytically, hypothetically, and logically (all formal operational thought), they can respond to the usual pedagogy of high school.

If a teacher lectures, they can take notes; if someone asks a hypothetical question, they can explore the answer; if a book describes another place and time, they can imagine it. One factor is particularly crucial: whether or not the teacher, and the students themselves, believe that they can master difficult material.

Growth Mindset

If teachers do not believe that a student is able to grasp abstract ideas, or if a student believes that they are not able to master calculus, or another language, or Shakespeare, then that perception reduces learning. If a student imagines that they will fail, they avoid the disappointment of failure by not trying. Then they can blame a low grade on their choice ("I didn't study") rather than on their ability. Pivotal is how they think of their potential.

In some schools, children help each other, laugh together, and cooperate on class projects. In other schools, children compete for grades and teacher attention.

Especially for High School Teachers You are much more interested in the nuances and controversies than in the basic facts of your subject, but you know that your students will take high-stakes tests on the basics and that their scores will have a major impact on their futures. What should you do? (see response, page 422)

This may reflect an attitude toward learning, either a *growth mindset* or *fixed mindset*, also called *incremental* (growth) or *entity* (fixed) (Dweck, 2017).

- The **growth mindset** is that learning grows if people work at it, with one person's growth likely to advance another's. Mistakes are "learning opportunities"; sharing ideas and strategies does not diminish one's own learning, quite the opposite.
- The **fixed mindset** is the belief that ability is determined early on, perhaps at conception. Some people naturally have more talent or ability than others. If students fail, that is evidence of their inadequacy.

To find out inborn ability, students with a fixed mindset use social comparison, ranking themselves among their peers. If they realize that they are not good at math, or that writing is hard, they stop trying, because they are convinced that they will never be good at math, or writing, or whatever. They may deflect attention from their failure ("the dog ate my homework"; "school sucks") and leave school when they can. Teachers with the fixed mindset attribute low performance to the student's personality, neighborhood, family, or intelligence.

On the other hand, students with a growth mindset seek challenges. They work hard at learning, enjoy discussion with their classmates, change their opinions based on what they learn, and choose difficult courses in high school. Teachers with growth mindset believe that every child can succeed, and that their job is to encourage effort, curiosity, collaboration.

This is not hypothetical. In the first year of secondary school, students with a fixed mindset do not achieve much, whereas those with a growth mindset improve academically, and this is true in many nations (e.g., Burnette et al., 2013; Diseth et al., 2014; Zhao & Wang, 2014). By the end of high school, those with a fixed mindset are likely to drop out, and those with a growth mindset are eager for challenging work in college. If they consider a difficult test a challenge, they do better than if they consider the same test a threat (Putwain et al., 2019).

The College-Bound

From a developmental perspective, the fact that high schools emphasize formal thinking makes sense, since many older adolescents are capable of abstract logic. In several nations, attempts are under way to raise standards so that all high school graduates will be ready for college, where analysis is required.

A mantra in the United States is "college for all"; it is intended to encourage low achievers to aspire for tertiary education, although some authors believe the effect may be the opposite (Carlson, 2016). As already mentioned, many students take difficult classes that are assessed by externally scored exams, either the IB or the AP (Finn & Scanlan, 2019).

Other indicators of increasing standards are greater requirements for academic diplomas and restrictions on vocational or general diplomas. Most U.S. schools require two years of math beyond algebra, two years of laboratory science, three years of history, four years of English, and two years of a language other than English.

Alternatives to College

In 2018, about 31 percent of U.S. high school graduates did not enter college, and many who enrolled will not graduate. To be specific, only 44 percent at four-year colleges and 33 percent at two-year colleges earn a degree. (These data do not include those who transfer to another institution and graduate there, but transferring itself reduces the chance of graduation [Digest of Education Statistics, May, 2020].) Rates are higher in some nations, lower in others.

growth mindset An approach to understanding intelligence that holds that intelligence grows incrementally, and thus can be increased by effort. Those who subscribe to this view believe they can master whatever they seek to learn if they pay attention, participate in class, study, complete their homework, and so on.

fixed mindset An approach to understanding intelligence that sees ability as an innate entity, a fixed quantity present at birth. Those who hold this view do not believe that effort enhances achievement.

What Do They Need to Learn? Jesse Olascoaga and José Perez assemble a desk as part of a class in Trade Tech High School in Vista, California. Are they mastering skills that will lead to a good job? Much depends on what else they are learning. It may be collaboration and pride in work well done, in which case this is useful education.

San Diego Union-Tribune/ZUMA Press/Newscom

Some high schools encourage college enrollment. For example, high-achieving students in two major cities in neighboring states (Albuquerque, New Mexico, and Fort Worth, Texas) had markedly different college enrollment rates (83 percent compared to 58 percent) (Center for Education Policy, 2012). It is easy to judge Fort Worth harshly, but that itself may be too quick a conclusion.

Instead of encouraging every student to enroll in college, should schools prepare some students for employment after graduation, providing training, social skills, and practical experience? Some nations do that. In Switzerland, students in vocational education have a higher employment rate than students in the academic track, but over a lifetime, their earnings are less (Korber & Oesch, 2019). Which would you rather have at age 40—a steady and secure job or a high salary?

Overall, the data present a dilemma for educators. Suggesting that a student should *not* go to college may be classist, racist, sexist, or worse. On the other hand, since less than half of the students who begin college stay until they graduate, such students may lose time and gain debt when they could have advanced in a vocation. Everyone agrees that adolescents need to be educated for life as well as for employment, but it is difficult to decide what that means.

Students who drop out of high school or fail high-stakes tests may have succeeded if they had been offered courses that lead to employment. On the other hand, students who believe that colleges will never accept them, or that employers will never hire them, may handicap themselves, another example of stereotype threat.

> **THINK CRITICALLY:** Is it more important to prepare high school students for jobs or for college?

Variability

Adolescents themselves vary: Some are so egocentric that they ruminate obsessively about themselves, some so hypothetical that they think of possibilities no one else imagined. Some are intuitive, some analytic; some foolish, some rational; some ready to quit school in seventh grade, some eager for college; some will thrive in secondary schools, others will disengage.

In fact, these differences appear not only between adolescents but within them; the same person can be intuitive in the morning and analytic in the afternoon. Every adolescent, however, needs respect, personal encouragement, and intellectual challenge.

That brings us to general conclusions for this chapter. The cognitive skills that boost national economic development and personal happiness are creativity, flexibility, relationship building, and analytic ability. Whether or not an adolescent is college-bound, those skills can develop in adolescence, making a wiser adult.

Every cognitive theorist and researcher believes that adolescents' logical, social, and creative potential is not always realized, but that it can be. The two hands on the piano might play beautiful music together. Does that image end this chapter on a hopeful note?

WHAT HAVE YOU LEARNED?

1. What are the differences between primary, secondary, and tertiary education?
2. What are the advantages and disadvantages of high-stakes testing?
3. How does the PISA differ from other international tests?
4. What characteristics of middle schools make them more difficult for students than elementary schools?
5. Why does puberty affect a person's ability to learn?
6. How does stereotype threat affect learning in secondary school?
7. How do growth mindset and fixed mindset affect motivation and learning?
8. Should high schools prepare everyone for college? Why or why not?

SUMMARY

Logic and Self

1. Cognition in early adolescence may be egocentric, a kind of self-centered thinking. Adolescent egocentrism gives rise to the personal fable, the invincibility fable, and the imaginary audience.

2. Formal operational thought is Piaget's term for the last of his four periods of cognitive development. He tested and demonstrated formal operational thought with various problems that students in a high school science or math class might encounter.

3. Piaget realized that adolescents are no longer earthbound and concrete in their thinking; they imagine the possible, the probable, and even the impossible, instead of focusing only on what is real. They develop hypotheses and explore, using deductive reasoning.

Two Modes of Thinking

4. Many cognitive theories describe two types of thinking during adolescence. One set of names for these two types is intuitive and analytic. Both become more forceful during adolescence, but brain development means that intuitive, emotional thinking matures before analytic, logical thought.

5. Few teenagers always use logic, although they are capable of doing so. Emotional, intuitive thinking is quicker and more satisfying (and sometimes better) than analytic thought.

6. Neurological as well as survey research finds that adolescent thinking is characterized by more rapid development of the limbic system and slower development of the prefrontal cortex. Peers further increase emotional impulses, so adolescents may make choices that their parents believe to be foolish.

Secondary Education

7. Achievement in secondary education—after primary education (grade school) and before tertiary education (college)—correlates with the health and wealth of individuals and nations.

8. Educators and political leaders in the United States struggle with how to measure what adolescents learn in secondary school. High-stakes tests have become more popular, and then less popular.

9. The PISA test, taken by many 15-year-olds in 50 nations, measures how well students can apply the knowledge they have been taught. Students in the United States have particular difficulty with such tests.

10. In middle school, many students struggle both socially and academically. One reason may be that middle schools are not structured to accommodate egocentrism or intuitive thinking. Another reason may be that students become aware of the prejudices of the larger society.

11. Education in high school emphasizes formal operational thinking. Students' beliefs about the nature of intelligence—either a growth mindset or a fixed mindset—may also affect their learning.

12. About a third of high school students do not graduate or go on to college, and many more leave college without a degree. Current high school education in the United States may not meet their needs. In some other nations, educators pay substantial attention to vocational education in high school so that students are job-ready when they graduate.

KEY TERMS

adolescent egocentrism (p. 397)
imaginary audience (p. 398)
personal fable (p. 399)
invincibility fable (p. 399)

formal operational thought (p. 400)
hypothetical thought (p. 401)
deductive reasoning (p. 401)
inductive reasoning (p. 401)

dual processing (p. 402)
intuitive thought (p. 402)
analytic thought (p. 402)
secondary education (p. 407)

high-stakes test (p. 407)
middle school (p. 412)
stereotype threat (p. 415)
growth mindset (p. 419)
fixed mindset (p. 419)

APPLICATIONS

1. Describe a time when you overestimated how much other people were thinking about you. How was your mistake similar to and different from adolescent egocentrism?

2. Talk to a teenager about politics, families, school, religion, or any other topic that might reveal the way they think. Do you hear any adolescent egocentrism? Intuitive thinking? Systematic thought? Flexibility? Cite examples.

3. Think of a life-changing decision you have made. How did logic and emotion interact? What would have changed if you had given the matter more thought—or less?

4. Describe what happened and what you thought in the first year you attended a middle school or a high school. What made it better or worse than later years in that school?

Especially For ANSWERS

Response for Natural Scientists (from p. 401): Probably both. Our false assumptions are not logically tested because we do not realize that they might need testing.

Response for Teachers (from p. 415): Praise the student by saying, "What a great question!" Egos are fragile, so it's best always to validate the question. Seek student engagement, perhaps asking whether any classmates know the answer or telling the student to discover the answer online or saying you will find out. Whatever you do, don't fake it; if students lose faith in your credibility, you may lose them completely.

Response for High School Teachers (from p. 418): It would be nice to follow your instincts, but the appropriate response depends partly on pressures within the school and on the expectations of parents and the administration. A comforting fact is that adolescents can think about and learn almost anything if they feel a personal connection to it. Look for ways to teach the facts your students need for the tests as the foundation for the exciting and innovative topics you want to teach. Everyone will learn more, and the tests will be less intimidating to your students.

Observation Quiz ANSWERS

Answer to Observation Quiz (from p. 397) The first three you can probably guess: Climate change advocates speak of saving the planet, of scientific evidence, and of animals losing their habitats. Extra credit if you know the fourth: "XR" stands for Extinction Rebellion, a group that started with young people in England and has since spread to other nations.

Answer to Observation Quiz (from p. 403) Primarily hot! Building the trebuchet requires analysis, but look at their eyes and body positions in this photo.

Answer to Observation Quiz (from p. 413) Both are single-sex. What does that teach these students?

Adolescence: Psychosocial Development

What Will You Know?

1. Why might a teenager be into sports one year and into books the next?
2. Should parents back off when their teenager disputes every rule, wish, or suggestion they make?
3. Which is more troubling, teen suicide or teen crime?
4. Why are adolescents forbidden to drink and smoke while adults are free to do so?

"**Y**ou're a terrible mother of a teenager!" yelled an angry adolescent. "I'm learning on the job," the mother shot back. "I've never been a mother of a teenager before."

I sympathize. Often I needed to listen to my daughters to learn how to care for them. One night, Rachel was late coming home. I reassured myself: We had raised her to be cautious and careful; we knew her friends, who would call us if there was serious trouble; we lived in the city, so no one was driving.

Rachel walked in the door before worry overcame my reassurance. I was relieved to see her; I did not think to punish her until she asked, "How long am I grounded?"

That alerted me.

"Two days. Is that fair?"

She agreed.

Parents are not the only ones new to adolescence. Teenagers themselves have never experienced the body transformations described in Chapter 14, nor the cognitive advances described in Chapter 15. The greatest challenges, however, are in this chapter, on psychosocial development.

The social context of adolescent life has never existed before. Sex, drugs, romance, technology, and delinquency are all different for today's youth than for earlier generations. Their parents might remember being an adolescent, but that was before sexting, vaping, and much else. Some things have not changed: Adolescents are more adventurous than their parents. But even the most understanding parent has never viewed adolescence from an adult perspective before.

This developmental perspective is protective but also limiting. When our children were babies, my husband and I discussed the need to be firm, united, and consistent regarding illicit drugs, unsafe sex, and serious law breaking.

We were ready. Yet none of those issues appeared. Instead, their clothes, neatness, and homework made us impatient, bewildered, inconsistent. My husband said, "I knew they would become teenagers. I didn't know we would become parents of teenagers."

This chapter on adolescent psychosocial development includes relationships with friends, parents, partners, authorities. It begins with identity and ends with drugs, both of which may seem to be a personal choice but are strongly affected by social pressures that change from cohort to cohort. Personal traits matter, too, and everyone is unique. Each of my four daughters presented her own challenges, not only because the times had changed, but also because each is an individual, not a younger version of me or of her older sisters.

You will soon read that adults must be alert and supportive, but that they may not know how. Every generation is learning on the job. What did I need to learn? Once Rachel said to her younger sister, "If ever you think of trying cocaine, talk with me first." I was taken aback: I am still learning.

Identity

Psychosocial development during adolescence is often understood as a search for a consistent understanding of oneself. Self-expression and self-concept become increasingly important at puberty. Each young person wants to know, "Who am I?"

identity versus role confusion Erikson's fifth stage of development, when people wonder, "Who am I?" but are confused about which of many possible roles to adopt.

According to Erik Erikson, life's fifth psychosocial crisis is **identity versus role confusion**: Working through the complexities of finding one's own identity is the primary task of adolescence (Erikson, 1968/1994).

identity achievement Erikson's term for the attainment of identity, when people know who they are as unique individuals, combining past experiences and future plans.

Erikson believed that this crisis is resolved with **identity achievement**, when adolescents have reconsidered the goals and values of their parents and culture, accepting some and discarding others, to forge their own identity. They must combine emotional separation from childhood with ongoing reliance on their family and society, a difficult task (Sugimura et al., 2019).

Identity achievement entails neither wholesale rejection nor unquestioning acceptance of national norms. Teenagers maintain continuity with the past so that they can move toward the future, establishing their own identity.

Not Yet Achieved

Over the past half-century, major psychosocial shifts have lengthened the duration of adolescence, complicating identity achievement (Côté & Levine, 2015). How adolescents experience their identity crisis varies depending on genes, the social context, and family (Markovitch et al., 2017). Moments of rage, anxiety, idealism, and fantasy that seem pathological by adult standards may be part of this search (Côté, 2018).

One developmental scholar provided a useful expansion of Erikson's description of adolescence. James Marcia outlined four ways of coping with the identity crisis: (1) role confusion, (2) foreclosure, (3) moratorium, and finally (4) achievement (Kroger & Marcia, 2011; Marcia, 1966).

role confusion When adolescents have no clear identity but fluctuate from one persona to another. (Sometimes called *identity diffusion* or *role diffusion*.)

Role confusion is the opposite of achievement. It is characterized by lack of commitment to any goals or values. It arises early in adolescence, when the hormones of puberty awaken sexual impulses and the cognitive advances of hypothetical thought trigger reexamination of values that were once accepted. The young person's first reaction is confusion: Old assumptions no longer seem valid.

foreclosure Erikson's term for premature identity formation, when adolescents adopt their parents' or society's roles and values without questioning or analysis.

Foreclosure occurs when, in order to avoid the confusion of sorting through all the nuances of identity and beliefs, young people lump traditional roles and values together, to be swallowed whole or rejected totally. They might follow every custom from their parents or culture, never exploring alternatives, taking on traditional values, roles, identities.

Or they might do the opposite, foreclosing on an oppositional, *negative identity*—rejecting all their elders' values and routines, again without thoughtful questioning. Foreclosure is comfortable but limiting. It is only a temporary shelter (Meeus, 2011).

moratorium A socially acceptable way to postpone achieving identity. Going to college and joining the military are examples.

A more mature shelter is **moratorium**, a time-out that includes exploration, either in breadth (trying many things) or in depth (following one path but with only tentative commitment). Society offers many ways to postpone identity achievement and choose moratoria instead. Older adolescents may:

- take a gap year before college;
- attend college without a specific career path;

- have an intimate sexual relationship with no plans for marriage;
- sign up for two years in the military;
- take a job known to be temporary;
- volunteer for a year of mission work or social justice advocacy;
- intern in a nonprofit for little pay; or
- travel around the world.

All these are moratoria, accepted by our society.

The final outcome, *identity achievement*, is now thought to be a lifelong quest (Topolewska-Siedzik & Cieciuch, 2019). The need to establish identity is most urgent in adolescence, but the self-concept shifts, strengthens, and reverses at many points. This was captured by Anna Quindlen, novelist and grandmother, who wrote:

> It's odd when I think of the arc of my life from child to young woman to aging adult. First I was who I was, then I didn't know who I was, then I invented someone and became her, then I began to like what I'd invented, and finally I was what I was again.

[*Quindlen, 2012, p. ix*]

Arenas of Identity Formation

Erikson (1968/1994) highlighted four aspects of identity: religious, political, vocational, and sexual. Terminology and timing have changed, and a fifth identity (ethnic) has been added, yet adolescents still seek identity in every domain.

Religious Identity

Most adolescents question some aspects of their faith, but their *religious identity* is similar to that of their parents. Few reject their religion totally if they have been raised in it, especially if they have a good relationship with their parents (Kim-Spoon et al., 2012).

The search for religious identity may be universal, as a study of youth in eight nations suggests (Benson et al., 2012). Most of the research has been on Christian youth in Western nations. However, a recent study in Japan reported that Buddhist adolescents also seek to establish their own religious beliefs and practices (Sugimura et al., 2019).

Some become more devout. A Muslim girl might start to wear a headscarf; a Catholic boy might study for the priesthood; a Baptist teenager might join a Pentecostal youth group. In each of these, parents might be surprised, but none of these changes is a reversal. Adults sometimes adopt a completely different faith; adolescents almost never do.

Although becoming more devout than their parents is possible, the more common pattern is rebellion. Attendance at churches, temples, and mosques decreases when parents no longer enforce it. A young person might express the foreclosure of negative identity, criticizing everything about religious customs rather than seeing the benefits as well as the problems.

More common than either extreme is a moratorium that allows later achievement of religious identity. Adolescents question "organized religion" because it seems to be a bundle of beliefs and rituals. They reexamine each part of the package, seeking their own way to be spiritual yet open (Saroglou, 2012). Even those adults who were raised with a strong religious

LaunchPad
macmillan learning

VIDEO ACTIVITY: Adolescence Around the World: Rites of Passage presents a comparison of adolescent initiation customs in industrialized and developing societies.

Same Situation, Far Apart: Religious Identity Awesome devotion is characteristic of adolescents, whether devotion is to a sport, a person, a music group, or—as shown here—a religion. This boy *(left)* praying on a Kosovo street is part of a dangerous protest against the town's refusal to allow building another mosque. This girl *(right)* is at a stadium rally for young Christians in Michigan, declaring her faith for all to see. While adults see differences between the two religions, both teens share not only piety but also twenty-first-century clothing. Her T-shirt is a recent innovation, and on his jersey is Messi 10, for a soccer star born in Argentina.

ARMEND NIMANI/AFP/Getty Images

Jim West/The Image Works

background, but who later claim no identifiable religion, nonetheless behave and believe in ways more similar to religious people than to those raised without religion (Van Tongeren et al., 2020). That is achievement, not wholesale rejection.

Political Identity

Parents also influence their children's *political identity*. In the twenty-first century, more U.S. adults identify as nonpartisan (38 percent) than Republican (26 percent), or Democrat (26 percent), or any other party (5 percent) (Pew Research Center, March 14, 2019). Their teenage children reflect their views, some boasting that they vote for the person, not the party, or that they do not care about politics, echoing their parents without realizing it.

Some proudly vote at age 18—an event that is more likely if they are living at home with voting parents than if they have already left home. Just like other aspects of political involvement, voting is a social activity, influenced by family (Hart & van Goethem, 2017).

In general, adolescents' interest in politics is predicted by their parents' involvement and by current events (Stattin et al., 2017). Adolescents tend to be more liberal than their elders, especially on social issues (same-sex marriage, reproduction, the environment), but major political shifts do not usually occur until later (P. Taylor, 2014).

Ethnic Identity

Related to political identity is *ethnic identity*, a topic not discussed by Erikson. However, Erikson knew that identity achievement requires interaction between the individual and the historical context. Ethnic identity is now essential.

Parents seek to prepare their adolescents by teaching pride in ethnicity and methods of coping with bias. Such parental socialization becomes a crucial buffer against depression and low achievement (Umaña-Taylor & Hill, 2020).

But because of the chronosystem, the macrosystem, and the microsystem, no parent knows exactly what teenagers will encounter. Teenagers themselves realize that ethnic identity is "not a matter of one's idiosyncratic self-perception, rather, [is] profoundly shaped by one's social context, including one's social role and place in society" (Seaton et al., 2017, p. 683). They must combine parental socialization with current conditions.

In the United States and Canada, about half of all adolescents are of African, Asian, Hispanic, or Native American (First Nations in Canada) heritage. Many also have ancestors of other ethnic groups. The census categories are too broad; teenagers must forge a personal ethnic identity that is more specific.

For instance, youth categorized in school data as Hispanic must figure out how their grandparents' birthplaces (for many U.S. children that might be Mexico, Puerto Rico, or Cuba, and California, Texas, or New York) affect who they are. Many Latinx individuals (some identifying as *Chicano*) also have ancestors from other nations of Central or South America, of Europe and Africa, or of indigenous groups such as the Maya. Non-Hispanics may be surprised to learn that some Mexican Americans do not identify as Hispanic: Mexico recognizes more than 70 national languages, in addition to Spanish.

Identity challenges also confront African American youth as they learn national, global, and family history. Formal operational thought moves them from personal experience into a wider, troubling narrative. For them, parental teachings are often crucial, especially when parents teach racial pride (noting highly successful African American leaders and important cultural values) as well as how to cope with racism (M.-T. Wang et al., 2020).

THINK CRITICALLY: Since identity is formed lifelong, is your current identity different from what it was five years ago?

Similar factors apply for Latinx and American Indian youth who learn a positive narrative regarding their role in the United States. The research suggests that Asian American youth may experience different pressures: For them, racial socialization may reduce self-esteem (M.-T. Wang et al., 2020). Research on adolescents in Hawaii also finds that Native Hawaiian and other Pacific Islander youth have additional complexities establishing their own ethnic identity, a source of conflict with their parents (Wills et al., 2019).

European American adolescents are less likely to hear parents praise their ethnic roots (Loyd & Gaither, 2018) and more likely to hear messages such as "All lives matter," which they realize is a trivialization of the Black Lives Matter movement. Thus, when White children move past concrete operational thought, they also need to figure out how their heritage and ethnicity affect their identity. Some search back to their roots, talking to grandparents, for instance, or sending a saliva sample to be analyzed for DNA origins.

For every adolescent, of every background, peers help sort through stereotypes, resistance, and finally achievement (Santos et al., 2017). Adolescents who are multiracial or adopted by parents of another background encounter additional complications. Their parents usually want to help them but may be less able to do so (Umaña-Taylor & Hill, 2020).

Typically, biracial and adopted youth tend to identify with the ethnic group less powerful in American society. If they are not accepted by that group, they may be depressed (Nishina et al., 2018). Problems may be particularly acute for adoptees from Asia, because parents may be unaware of the racism their children experience (Langrehr et al., 2019). Every adolescent seeks ethnic identity, yet recent cultural shifts make it difficult for adults to understand the particular identity crises of their children.

In general, pride in ethnic identity correlates with academic achievement and overall well-being, but the relationship is "complex and nuanced" (Miller-Cotto & Byrnes, 2016, p. 67). Ethnic pride encourages achievement; ethnic fear fosters depression (Huguley et al., 2019). A positive ethnic identity in early adolescence correlates with intergroup contact later on (Meeus, 2017). Perhaps adolescents who are secure in their own identity are more able to befriend people of other groups.

Ancestors and Identity

Almost every young person seeks their ancestral roots. Outsiders may not realize how complex this is. Those who have indigenous ancestors (about 1 percent of the population) may want to resurrect customs and lineage that outsiders tried to eliminate, and thus refer to a particular group (Diné, Dakota, Lakota).

Those whose ancestors fought in the American Revolution, or Civil War, or who were enslaved and brought from Africa want to trace their lineage back more than a century. Almost every adolescent encounters identity issues that are acute during the teen years and that may evolve lifelong.

For example, in San Diego, a longitudinal (age 14 to 37) study asked first- and second-generation immigrants, "What do you call yourself?" (Feliciano & Rumbaut, 2019). Few chose pan-ethnic terms (Asian, Latino, etc.). Especially in adolescence, many used specific heritage terms (Cambodian, Mexican, and so on), with a trend toward American (hyphenated or not) as time went on.

One adolescent boy said he was Vietnamese. As an adult, he said he was American, a change facilitated not only by time but also by his surroundings. He said:

> I wouldn't identify myself as Vietnamese. . . . Physically from a phenotype
> perspective, I don't look typical American . . . most people when they think

of Americans they think of just White. But living in San Diego, you see a multicultural group of people.

[Kim quoted in Feliciano & Rumbaut, 2019, p. 90]

On the other hand, those who physically appeared White often sought to emphasize their differences from European Americans. Many Latinx individuals claimed to be a "person of color." In this study, none of them said they were simply American. That was a matter of pride. As one said:

For me it's very important to label me Mexican-American because I wanna show people . . . two sides. I wanna show my parents that, hey, I made it. And I'm proud of being Mexican. . . . I'm an immigrant that came to this country and I succeeded and I don't take advantage of the system.

[Leo quoted in Feliciano & Rumbaut, 2019, p. 93]

Many reflected the perceptions of other people. One woman with a Filipina mother and a Black father said she was Black-Filipina. She added:

. . . if it was up to me, actually, I'd just say, I'm American [but] I know what they're asking . . . I don't want to be rude about it.

[Pat quoted in Feliciano & Rumbaut, 2019, p. 93]

Every individual statement reveals that ethnic identity it is not a private, personal choice but a community one, with differences depending on education (more education led to more specific ethnic identity), family, and national politics. The impact of those influences is not the same on every individual, even those with the same origin. That makes ethnic identity "complex and varied" (Feliciano & Rumbaut, 2018, p. 42).

This is evident in my own family. My brother increasingly identified with his immigrant roots as he grew older, changing how he pronounced his name. My four daughters were required to specify their ethnic heritage as they filled out the 2020 U.S. Census. They sent in their forms and then compared answers; each, with the same eight great-grandparents, wrote a different response.

I have similar impulses myself, which makes me sympathetic when my students resist the identities that strangers assign to them. They are surprised that I bristle if they call me "Caucasian" or "Anglo": Those terms originate from regions where my ancestors did not live. Each of us forges our own ethnic identity, with details that other people may not understand.

Vocational Identity

Establishing a vocational identity is considered part of growing up, not only by developmental psychologists but also by some adolescents themselves. However, a solid vocational identity is almost impossible in the current U.S. labor market.

Often adolescents imagine what a job might be based on television, not reality, and connect job and personal identity. For that reason, many adolescents aspire to unrealistic careers. Few achieve a vocational identity that they will sustain lifelong, a marked contrast to what Erikson saw, when a young man took a job and stayed with that employer until retirement, or a young woman aspired to work lifelong as a wife and mother. They chose a vocational identity that would shape the rest of their life.

Now, instead, many older adolescents enter the *gig economy*, with temporary, episodic, or independent jobs. They drive cars for hire, tutor children, sell items online, act as social media "influencers," and much more. Those who begin more traditional jobs often quit. Unlike the joke my oldest uncle once told me (that I was going to college to get my "MRS"), few young women now consider marriage *instead* of careers. Between ages 18 and 25, the average U.S. worker has held seven jobs, with the college-educated changing jobs more than the high school graduate (U.S. Bureau of Labor Statistics, August, 2018).

Gender Identity

Now we come to the most difficult identity for many current adolescents, **gender identity**. As you remember, *sex* refers to certain physical and genetic traits assigned at birth, whereas *gender* refers to the cultural and social factors. A half-century ago, psychoanalytic theorists did not make this distinction. They assumed that adolescents needed to adopt male or female roles (Erikson, 1968/1994; A. Freud, 1958/2000).

Thus, adolescence was once time for "gender intensification," when hormones caused children to identify as one or the other of two *opposite* sexes. That reflected the *gender binary* (the prefix *bi-* means "two," as in *bi*cycle, *bi*nominal, *bi*polar). Many adolescents see sex and gender as *nonbinary*, that they are much more fluid than a male/female dichotomy.

The analytic, hypothetical thinking of adolescents leads to questioning traditional gender norms. This may trouble their elders. One mother thought she was guiding her daughter into a proper female role when she suggested that her daughter's skirt was too tight. Her daughter replied angrily, "Don't slut-shame me."

Similarly, parental comments about appropriate male or female dress, hair, and aspirations may be rejected. Attitudes are powerfully affected by culture, age, and cohort: Gender identity is an example of that. Current adolescents are growing up in a gender world never experienced by their elders.

A new term, *cisgender*, refers to people whose gender identity is the same as their birth sex. The fact that such a term is needed is evidence of the complexity of sex and gender. To be cisgender is not simple, in that exactly how a person is supposed to relate to the other cisgender sex is complicated by conflicting and changing norms of self-presentation, dress, and physical interaction.

Interestingly, more adolescent boys than girls see gender as binary, probably because boys are more strongly pressured to conform to male identity (Horn, 2019). That male identity, however, has changed from what it was, which makes it hard for boys who seek role models in their fathers and grandfathers.

Gender diversity is increasing, and is most evident with transgender identity. In 2016, researchers estimated that 1.4 million people in the United States identified as transgender. That is about .6 percent, twice the rate a decade earlier (Flores et al., 2018).

The reason for the increase is not without controversy. Between 2014 and 2016, three times as many U.S. adolescents who had been designated female at birth identified as male (Littman, 2018; Marchiano, 2017). Some blamed that increase on an encouraging, but mistaken, peer culture. Transgender advocates object to that interpretation, claiming that celebration (not blame) is in order, because more young people can express their identity (Short, 2019; Wadman, 2018).

Among Western psychiatrists in former decades, people who had "a strong and persistent cross-gender identification" were said to have *gender identity disorder*, a serious diagnosis according to *DSM-IV* (1994). However, *DSM-5* (2013) instead describes *gender dysphoria*, when people are distressed at the gender that was assigned to them at birth. *DSM-6* is now in progress: Expect gender updates.

The change from IV to 5 is more than new terminology. A "disorder" means something is amiss with the individual, no matter how they feel about it, whereas "dysphoria" means the problem is in the distress, which can be mitigated by social conditions, by cognitive framing, or by becoming the other gender.

Adam Hester/Blend Images/Superstock

The Opposite Sex? Every cohort of adolescents rebels against the conventions of the older generations. Earlier generations of boys grew their hair long. A decade later, some girls shaved their heads. Now many teenagers do not see male and female as opposites, choosing instead a nonbinary approach to gender expression.

gender identity A person's acceptance of the roles and behaviors that society associates with a particular gender.

Intersectionality

An added complexity in the identity crisis is that each identity overlaps with every other, not just in individuals, but in society. Each adolescent seeks all five identities just explained, but the various identities combine and conflict, as evident in the concept of *intersectionality* (explained in Chapter 1).

A person aspiring to a particular vocational identity may discover that no one of their ethnic background holds that job, and that itself may discourage them or may cause employers to be unlikely to hire them. Or someone may feel that someone else was chosen for reasons of ethnicity, gender, religion or politics, and that may cause resentment.

Other identities may also conflict: Someone who identifies as a religious conservative may be troubled if they also identify as gender nonbinary, or someone who identifies as African American may be troubled by assumptions about their politics. By adulthood, reconciliation among identities is possible: We all combine aspects of ourselves that others might find contradictory. But, as captured by the term "identity confusion," in adolescence every aspect of identity is in flux.

One psychologist highlighted this with vocational identity:

> career choices faced by individuals inevitably raise the question of the meaning that they intend to give their lives. To choose their work or sector in which they want to evolve is also to consider the purpose of their existence, the priorities (physical, spiritual, social, aesthetic, etc.) that they want to give, the choices that they wish to operate, the overall style of life that they wish to give themselves.
>
> [Bernaud, 2014, p. 36]

Given that each identity seems connected to "the purpose of existence" and "the overall style of life," no wonder adolescents experience an identity crisis. Self-definition and then acceptance, not only of oneself but also by other people, is an urgent psychosocial need. Those "other people" are discussed next.

WHAT HAVE YOU LEARNED?

1. What is Erikson's fifth psychosocial crisis, and how is it resolved?

2. How does foreclosure differ from moratorium?

3. What role do parents play in adolescent religious and political identity?

4. How does ethnic identity affect self-esteem and achievement?

5. What assumptions about gender identity did most adults hold 50 years ago?

6. What is the difference between gender identity disorder and gender dysphoria?

Close Relationships

The focus on adolescent self-definition may imply that teenagers are intensely individualistic, seeking their own identity, apart from their family and society. However, the opposite is more accurate. Parents, peers, grandparents, siblings, teachers, and cultures shape adolescent lives (Seaton et al., 2017).

It is a myth that adolescents choose peers over parents; usually both are supportive. A longitudinal study of all middle school students in one community (almost 800 of them) found that three-fourths had healthy relationships with their parents *and* their peers. That protected them from serious problems during adolescence and early adulthood (Dishion et al., 2019). If parents are not supportive, trouble with friends is also common.

Family

Family relationships affect identity, expectations, and daily life. Parents shift from providing direct guidance to giving advice when asked, but close parent–child relationships continue (E. Chen et al., 2017).

Older siblings also become influential, as role models and confidants, as bullies or protectors (Aizpitarte et al., 2019; Gallagher et al., 2018). This is especially important if the environment and genes push teens toward negative behavior: An older sibling can be crucial.

Family Conflict

The fact that families are influential does not mean that family life is peaceful. Disputes are common when biology, cognition, and culture all push for adolescent independence, while parents seek protective control. Consequently, each generation might misjudge the other. Adolescents think parents are overprotective; parents think peers are too loose.

Both may forget that all close relationships include disagreements. Reducing conflict by reducing connection may buy peace at a steep price. Some developmentalists think that conflict increases mutual understanding and emotional growth (Branje, 2018).

Granting too much freedom, too early, is a path toward serious trouble (law breaking, addiction) (Fosco & LoBraico, 2019). On the other hand, too much control, for too long, fractures connection. One of my college students wrote:

> My parents . . . see me as too independent. I see them as too old-fashioned. . . . My home life is a lie. I pretend to be their good daughter. . . . I used to argue with them all the time when I was in high school but it soon got to the point that screaming and yelling got tiring. Now I just want to get by the couple of hours I'm at home without any problems.
>
> I never confide in my parents. . . . Whenever I need advice I go to my friends. My parents act as if they own my life and that whatever they say goes. It's one of the reasons I wanted to run away at thirteen.
>
> *[E., personal communication]*

Authors of research on mothers and their adolescents suggest that "although too much anger may be harmful . . . some expression of anger may be adaptive" (Hofer et al., 2013, p. 276). In this study, as well as generally, the parent–child relationship usually improved with time as both parties adjust to the adolescent's increasing independence. Clearly, as my student shows, sometimes neither generation adjusts.

Crucial is that caregivers avoid either extreme, strictness or leniency, but instead maintain support while increasing autonomy (Yeager et al., 2018). As a review of dozens of studies expressed it: "parent–adolescent conflict might signal the need for families to adapt and change . . . to accommodate adolescents' increasing needs for independence and egalitarianism" (Weymouth et al., 2016, p. 107).

Closeness Within the Family

Several specific aspects of parent–child relationships have been studied. Specifically:

1. Communication (Do family members talk openly and honestly?)
2. Support (Do family members rely on each other?)
3. Connectedness (How emotionally close are family members?)
4. Control (Do parents allow independence?)

"So I blame you for everything—whose fault is that?"

Not My Fault Humans always find it easier to blame someone else, but this is particularly true when teenage girls talk to their mothers.

THINK CRITICALLY: When do parents forbid an activity they should approve of, or ignore a behavior that should alarm them?

No social scientist doubts that the first two, communication and support, are crucial. Patterns from childhood continue, ideally buffering the turbulence of adolescence. Regarding the next two, connectedness and control, consequences vary, and observers differ in what they see. How do you react to this example, written by another one of my students?

> I got pregnant when I was sixteen years old, and if it weren't for the support of my parents, I would probably not have my son. And if they hadn't taken care of him, I wouldn't have been able to finish high school or attend college. My parents also helped me overcome the shame that I felt when . . . my aunts, uncles, and especially my grandparents found out that I was pregnant.

> *[I., personal communication]*

My student's boyfriend is no longer part of her life. She is grateful that she still lives with her parents, who care for her son. However, did motherhood make her dependent on her parents, preventing her from establishing her own identity? Why didn't her parents monitor her romantic relationship, or at least explain contraception? Is this an example of the best or the worst of parent–adolescent relationships?

One issue here is **parental monitoring** — that is, parental knowledge about each child's whereabouts, activities, and companions. Many studies have shown that when parental knowledge arises from a warm, supportive relationship, adolescents usually become confident, well-educated adults, avoiding drugs and risky sex.

However, if parents are cold and punitive, monitoring may lead to deception and rebellion. If mothers are too controlling, depression increases; if fathers are too controlling, drug addiction is more likely (Eun et al., 2018).

There is a "dynamic interplay between parent and child behaviors," which affects the results of monitoring (Abar et al., 2014, p. 2177). Teenagers choose what to reveal. They share details if their parents are supportive; they lie if their parents are critical (Lushin et al., 2017).

Cultural Expectations for Parents of Teenagers

Several researchers have compared parent–child relationships in various cultures: Everywhere, parent–child communication and support (numbers 1 and 2 above) reduce teenage depression, suicide, and low self-esteem, and increase motivation and achievement. However, details of connection and control (numbers 3 and 4) vary by culture (Brown & Bakken, 2011).

Parent–child conflict is less evident in cultures that stress **familism**, the belief that family members should sacrifice personal freedom and success to care for one another. For example, most refugee youth (Palestinian, Syrian, Iraqi) in Jordan agree that parents have the right to decide their children's hairstyles, clothes, and music — contrary to what most U.S. teenagers believe (Smetana et al., 2016).

Familism is a strong value in Mexican culture, but a recent study found that Mexican American adolescents do not necessarily value familism as their parents do (Padilla et al., 2020). Fathers, particularly, are likely to endorse familism (e.g., "It is always important to be united as a family"), but older siblings often disagree. A developmental shift may occur when adolescents take "active roles in their identity formation . . . [with] increased autonomy in choosing how and with whom to spend their time" (Padilla et al., 2020, pp. 997, 1001).

In many traditional cultures, teens do not tell their parents anything that might earn disapproval. By contrast, some U.S. adolescents deliberately provoke an argument by advocating marijuana legalization, same-sex romance, immigrant citizenship, or abortion access, especially if they personally are not affected by those policies.

LaunchPad
macmillan learning

VIDEO: Parenting in Adolescence examines how family structure can help or hinder parent–teen relationships.

parental monitoring Parents' ongoing knowledge of what their children are doing, where, and with whom.

familism The belief that family members should support one another, sacrificing individual freedom and success, if necessary, in order to preserve family unity and protect the family.

The Naïveté of Your Author

Parents are sometimes unaware of adolescents' desire for respect from their classmates. I did not recognize this with my own children:

- Our oldest daughter wore the same pair of jeans in tenth grade, day after day. She washed them each night by hand, and I put them in the dryer early each morning. [Circadian rhythm—I was asleep hours before she was, and awake hours earlier.] My husband was bewildered. "Is this some weird female ritual?" he asked. Years later, our daughter explained that she was afraid that if she wore different pants each day, her classmates would think she cared about her clothes, which would prompt them to criticize her choices. To avoid that imaginary audience, she wore only one pair of jeans.

- Our second daughter, at 16, pierced her ears for the third time. I asked if this meant she would do drugs; she laughed at my foolishness. Only later did I notice that many of her friends also had multiple holes in their earlobes.

- At age 15, our third daughter was diagnosed with Hodgkin's disease, a kind of cancer. My husband and I consulted four physicians in four hospitals. Their recommendations differed: We selected the one we thought would minimize the risk of death. Our daughter had other priorities: "I don't care what you choose, as long as I keep my hair." (Now her health is good; her hair grew back.)

- Our youngest, in sixth grade, refused to wear her jacket (it was new; she had chosen it), even in midwinter. Years later she told me why—she wanted her classmates to think she was tough.

In retrospect, I am amazed that I was unaware of the power of peers, a stronger and more immediate influence than self-acceptance, personal choice, long life, or even a warm body.

Cultural variations in parent–child interaction are evident not only between nations, but also within U.S. ethnic groups. This is illustrated by another study of Mexican American adolescents (Wheeler et al., 2017). This one was longitudinal: Those who were less Americanized were more likely to endorse familism, to obey their parents, to stay in school, and to avoid gangs.

Personality also matters, with differential susceptibility. Impulsive, fearful, or adventurous adolescents are more likely to break the law *unless* their family is supportive, in which case they are more law-abiding than the average adolescent (Rioux et al., 2016).

This contrast is evident in school achievement as well. When parents encourage academics, students do better. However, a longitudinal study of intellectually gifted students found that when parents were harsh at puberty, fewer of the adolescents enrolled in college.

Gender mattered in what they did instead: Boys were likely to become delinquents; girls to become pregnant (Hentges et al., 2018). All these examples suggest that parental support and encouragement is needed, but that parental power over adolescents is limited (see A Case to Study).

Peer Power

Peers can be crucial. They help each other navigate the physical changes of puberty, the social challenges of leaving childhood, and the intellectual challenges of high school. They also share emotions, and by so doing become closer friends (von Salisch, 2018).

Choosing Friends

To fully understand the impact of peers, two concepts are helpful: *selection* and *facilitation*. Teenagers *select* friends with similar values and interests, abandoning

Especially for Parents of a Teenager Your 13-year-old comes home after a sleepover at a friend's house with a new, weird hairstyle—perhaps cut or colored in a bizarre manner. What do you say and do? (see response, page 453)

More Familiar than Foreign? Even in cultures with strong and traditional family influence, teenagers choose to be with peers whenever they can. These boys play at Cherai Beach in India.

⬤ **Observation Quiz** What evidence do you see that traditional norms remain in this culture? (see answer, page 453) ↑

peer pressure Encouragement to conform to friends or contemporaries in behavior, dress, and attitude. Adolescents do many things with peers that they would not do alone.

THINK CRITICALLY: Why is peer pressure thought to be much more sinister than it actually is?

coercive joining When others strongly encourage someone to join in their activity, usually when the pressure is that, unless the teenager joins, they will be excluded.

former friends with other interests. Then, friends *facilitate* destructive or constructive behaviors.

It is easier to do wrong ("Let's all skip school on Friday") or right ("Let's study together for the chem exam") with friends. Peer facilitation helps adolescents do things they are unlikely to do alone. This provides an important clue for adults who want to halt destructive patterns in adolescents. Grouping troubled adolescents together (as in special programs in school or residences for young lawbreakers) might make all of them worse, not better (Lochman et al., 2015).

Thus, adolescents select and facilitate, choose and are chosen. Happy, energetic, and successful teens have close friends who themselves are high achievers, with no major emotional problems. A student's grade point average and IQ increase if their friends are highly intelligent (Meldrum et al., 2019).

This works the other way as well. Those who are drug users, sexually active, and alienated from school choose compatible friends. In general, peers provide opportunity, companionship, and encouragement for what adolescents might do.

However, innocent teens are not corrupted by deviants. Adolescents choose their friends and role models—not always wisely, but never randomly. Selection and facilitation are evident lifelong, but the balance between the two shifts. Early adolescence is a time of selection; facilitation is more evident in later adolescence. After age 18 or so, selection becomes important again, as young adults abandon some high school friends and establish new ones (Samek et al., 2016).

Peer Pressure

Peer pressure refers to someone being pushed by their friends to do something that they would not do alone. Peer pressure is especially strong in early adolescence, when adults seem clueless about biological and social stresses. Some teenagers are more susceptible to peer pressure than others, with gender one factor. Boys influence other boys more than girls influence other girls (McCoy et al., 2019).

Adults warn children against peer pressure. However, sometimes having friends is beneficial: Peers pressure each other to study, to apply to college, to plan for their future.

This positive take on peer pressure is tempered by another fact: Peers *can* lead one another into trouble. A study of substance misuse and delinquency among twins found that—even controlling for genes and environment—when one twin became a delinquent, the other was more likely to do so (Laursen et al., 2017). Adolescents who are rejected by other peers tend to choose antisocial friends, who then involve them in drugs, aggression, and so on (Kornienko et al., 2020).

Thus, friends choose, teach, and encourage each other. This is shown in **coercive joining**, when two people join together in making derogatory comments about a third person. A pair of teenagers may compete in who can make the most pointed criticisms. Coercive joining at age 12 predicts antisocial violence at age 21 (Dishion et al., 2019).

Romantic Partners

Selection is obvious in romance. Adolescents choose and are chosen by romantic partners, and then they influence each other on almost everything—sex, music, work, play, education, food, and so on. Even small things matter: If one gets a new jacket, or tattoo, or sunglasses, the other might, too.

The Immediacy of Peers

Given the areas of the brain that are quickest to myelinate and mature in adolescence, it is not surprising that the most influential peers are those nearby at the moment. This was found in a study in which all eleventh-graders in several public schools in Los Angeles were offered a free online SAT prep course (worth $200) that they could take if they signed up on a paper distributed by the organizers (Bursztyn & Jensen, 2015).

In this study, most students had several honors classes and several nonhonors classes, depending on their interests as well as what tests (e.g., AP) they planned to take before college. Students were *not* allowed to talk to each other before deciding whether or not to accept the SAT offer.

Consequently, they did not know that, although all of the papers had identical, detailed descriptions of the SAT program, one word differed in who would learn of their decision—either no other students or only the students in that particular class.

The two versions were:

Your decision to sign up for the course will be kept completely private from everyone, <u>except</u> the other students in the room.

Your decision to sign up for the course will be kept completely private from everyone, <u>including</u> the other students in the room.

It mattered whether students thought their classmates would learn of their decision: The students in an honors class were *more* likely to sign up, and the students in a nonhonors class were *less* likely to sign up, when they thought their classmates would know what they did.

To make sure this was a peer effect, not just divergent motivation and ability between honors and nonhonors individuals, the researchers compared students who took exactly two honors classes and several nonhonors classes. There were 107 such students, some who happened to be in their honors class when they decided whether or not to sign up for SAT prep and some who happened to be in their nonhonors class.

When the decisions of those two-honors-class students were kept totally private, acceptance rates were similar (72 and 79 percent) no matter where students were at the moment. But if students thought their classmates might know their decision, imagined peer pressure affected them. When in an honors class, 97 percent signed up for the SAT program. Of those in a nonhonors class, only 54 percent signed up (Bursztyn & Jensen, 2015).

Evidence from many other studies finds that peers are especially influential in mid-adolescence. That is reflected in brain activity as well as in behavior (Kim-Spoon et al., 2019). Note that peer influence is not necessarily direct, as when one person tells another what to do. Instead, peer influence is contagious (Reiter et al., 2019). Just like catching a cold because someone nearby sneezes, adolescents tend to do what they think nearby peers will admire.

Teens' first romances typically occur in high school, with girls having a steady partner more often than boys. Commitment is the ideal, but the fluidity and rapidity of selection disrupt exclusive relationships. Cheating, flirting, switching, and disloyalty are rife; breakups and unreciprocated crushes are common. Emotions range from exhilaration to despair, leading to impulsive sex, cruel revenge, and deep depression.

Peer support is vital: Friends help adolescents cope with ups and downs. They also influence sexual activity, more by what they say than by what they do (which some keep secret). The friend who brags is more influential than the one who stays quiet.

Most peer relationships are nonsexual: Adolescents have platonic friends of all genders (Kreager et al., 2016a). Norms vary from group to group, school to school, city to city, nation to nation. For instance, twice as many high school students in Cleveland as in San Francisco say they have had intercourse (46 percent versus 22 percent) (MMWR, June 15, 2018).

The influence of the social context is undeniable, but sexual impulses are also affected by biology. This is evident in whether or not a romance includes physical activity, and what that physical activity is. Some young people seem driven to risk parental punishment and their own health; others seem content to postpone any sexual activity. Adolescents vary in both the strength of the sexual desire and in their sexual orientation.

Observation Quiz What does the body position of these four suggest? (see answer, page 453) ↓

Hang Loose? Are these two dating couples or a group of friends at the basketball court? Notice who has the ball and who does not want to show her face.

Hero Images/Getty Images

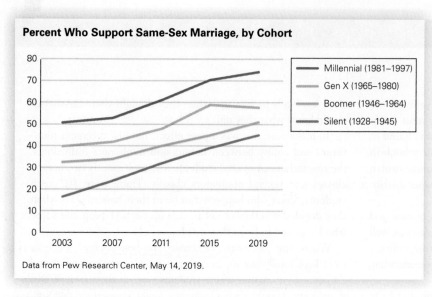

Percent Who Support Same-Sex Marriage, by Cohort

— Millennial (1981–1997)
— Gen X (1965–1980)
— Boomer (1946–1964)
— Silent (1928–1945)

Data from Pew Research Center, May 14, 2019.

FIGURE 16.1

Young and Old Everyone knows that attitudes about same-sex relationships are changing. Less well known is that cohort differences may be greater than the shift over the first decade of the twenty-first century.

sexual orientation A person's romantic or sexual attraction, which can be to others of the same gender, the other gender, or every gender.

LaunchPad
macmillan learning

VIDEO: Romantic Relationships in Adolescence explores teens' attitudes and assumptions about romance and sexuality.

Sexual orientation refers to the direction of a person's erotic desires. One meaning of *orient* is "to turn toward"; thus, sexual orientation refers to whether a person is attracted to (turned on by) people of the other sex, the same sex, or both sexes. Sexual orientation can be strong, weak, overt, secret, or unconscious. The most apparent—and controversial— evidence for the role of biology and culture occurs with young people who are lesbian, gay, bisexual, transgender, or queer/questioning (LGBTQ).

Same-Sex Attraction

Sexual orientation is surprisingly fluid during adolescence. In one study, 10 percent of sexually active teenagers had had same-sex romances, but many of those 10 percent nonetheless identified as heterosexual (Pathela & Schillinger, 2010). In that study, those most at risk of abusive relationships and sexually transmitted infections had partners of both sexes, a correlation also found in other studies (e.g., Russell et al., 2014).

In previous decades, LGBTQ adolescents were often "in the closet," unable to tell others (and sometimes themselves) about their sexual orientation. This changed in about 2000, when television aimed at teen audiences began introducing major characters who were gay, and plots portrayed "being out in high school as greatly preferable to being closeted" (Peters, 2016, p. 488). But the depiction was nonetheless homophobic. LGBTQ youth still experience overt prejudice, as well as stereotype threat.

Worldwide, some LGBTQ teens date the other sex to hide their orientation. Deception puts them at risk for binge drinking, suicidal thoughts, and drug use. Those hazards are less common in cultures where same-sex partnerships are accepted (see **Figure 16.1**), especially when parents affirm their offspring's sexuality.

At least in the United States, adolescents have similar difficulties and strengths regardless of their gender identity and sexual orientation. Nonsexual friendships with peers of whatever orientation decrease loneliness and increase resilience (Van Harmelen et al., 2017). However, LGBTQ youth have a higher risk of depression and anxiety, for reasons from every level of Bronfenbrenner's ecological-systems approach (Mustanski et al., 2014).

 DATA CONNECTIONS: Sexual Behaviors of U.S. High School Students examines how sexually active teens really are.

Learning About Sex

Some adolescents have strong sexual urges but minimal logic about pregnancy, disease, lust, and love. Millions of teenagers worry that they are oversexed, undersexed, or deviant, unaware that thousands, maybe millions, of others have the same sexual needs.

The most obvious example is with LGBTQ youth. When suicide data for these young people became public in about 2010, 50,000 gay and lesbian adults posted "It gets better" videos. They wanted young people to learn that same-sex orientation becomes less burdensome in adulthood (Garrett, 2018).

Indeed, every young person has much to learn. As one observer wrote, adolescents "seem to waffle their way through sexually relevant encounters driven both by the allure of reward and the fear of negative consequences" (Wagner, 2011, p. 193). Where can they learn about sex?

From the Media

Adolescents with intense exposure to sex in music, print, social media, film, and television are more often sexually active, but the direction of this correlation is controversial. The media may reinforce, but not cause, a focus on external appearance, body objectification, and thus sexual activity (Coyne & Ward et al., 2019; Dillman Carpentier & Stevens, 2018).

However, not much practical information comes from the media. Television programs that attract teen audiences include sexual content almost seven times per hour (Steinberg & Monahan, 2011). But almost never does a character develop an STI, deal with an unwanted pregnancy, or mention (much less use) a condom.

For questions regarding sexual health, the internet is a common source (Simon & Daneback, 2013). Unfortunately, Web sites may be frightening (pictures of diseased organs), mesmerizing (pornography), or misleading (false information). A survey of adolescents in England found that they all used the internet for sexual questions, but they had many criticisms—too much material, too technical, too pornographic, or too simplistic (S. Patterson et al., 2019). As one boy said:

> We're talking about adult issues, you know, with our bodies. We're not here to watch like a Thomas the Tank Engine explain it or anything like that.

From Parents

It may be that "the most important influences on adolescents' sexual behavior may be closer to home than to Hollywood" (Steinberg & Monahan, 2011, p. 575). As that quote implies, sex education begins within the family.

Every study finds that parental communication influences adolescents' behavior. Effective programs of sex education explicitly require parental participation (Silk & Romero, 2014). However, embarrassment and ignorance are common among both generations.

Many parents underestimate their own child's sexual activity while fearing the sexuality of peers and the media (Elliott, 2012). However, those fears do not lead to open and honest discussions about sex, love, and life. According to a survey of young women aged 15 to 24 chosen to represent the U.S. population, only 25 percent of adolescents remember receiving any sex education from either parent (Vanderberg et al., 2016).

Mothers and daughters more often have detailed conversations than do fathers and sons, but the emphasis is on avoiding pregnancy and diseases, not on pleasure and love. That may be less problematic than it seems: The strongest influence from parents on sexual relationships was not via communication about explicit thrills and dangers, but via a strong and supportive parent–child relationship (Cheshire et al., 2019).

Religious values also have some influence, but again the impact comes from general messages about respect and love rather than from specifics. Almost never does either generation share personal details (Coffelt, 2017).

"Smirking or non-smirking?"

Laugh and Learn Emotions are as crucial as facts in sex education.

⬤ **Especially for Sex Educators**
Suppose adults in your community never talk to their children about sex or puberty. Is that a mistake? (see response, page 453)

THINK CRITICALLY: Why has sex education become a political issue?

From Peers

Especially when parents are silent, forbidding, or vague, adolescent sexual behavior is strongly influenced by peers. Boys learn about sex from other boys, girls from other girls. Selection and facilitation are evident again, as adolescents choose friends whose sexual inclinations are similar to their own, and then talk about it (Trinh et al., 2019).

Partners teach each other. However, their lessons are more about pleasure than consequences: Few U.S. adolescent couples decide together *before* they have sex how they will prevent pregnancy and disease, and what they will do if their efforts fail.

In one study, adolescents were asked with whom they discussed sexual issues. Friends were the most common confidants, then parents, and last of all dating partners. Indeed, only half of the them had *ever* discussed specifics of sexual expression with a partner (Widman et al., 2014).

From Educators

Sex education from teachers varies dramatically by school and by nation. The curriculum for middle schools in most European nations includes information about masturbation, same-sex romance, oral and anal sex, and uses and failure rates of methods of contraception. Those subjects are rarely covered in U.S. classes, even in high school. Rates of teenage pregnancy in European nations are less than half the rates in the United States. Sex education in schools is far from the only reason, but it is one part of it.

Within the United States, the timing and content of sex education vary by state, by community, and by school. Some high schools provide comprehensive education, free condoms, and medical treatment; others provide nothing. Some school systems begin sex education in primary school; others wait until senior year of high school.

A review of official guidelines for sex education in all 50 states found that eight states explicitly condemn same-sex relationships. By contrast, seven states explicitly teach gender diversity of all kinds (Hall et al., 2019). Details about contraception are rarely taught. However, some instructors talk privately with individual students who want more information.

Should such private talks be forbidden? One controversy has been whether schools should teach that abstinence is the only acceptable action. Of course, abstaining from sex (including oral and anal sex) prevents STIs and avoids pregnancy, so some adults and most state curricula favor it (Hall et al., 2019).

But longitudinal data comparing students who were taught to avoid all sexual contact until marriage with those who had comprehensive sex education, showed similar ages for onset of sexual activity. Indeed, abstinence-only programs increased the rate of teen pregnancy and sexually transmitted infections, since students in those programs never learned about prevention (Fox et al., 2019; Santelli et al., 2017).

Legislative support for abstinence-only education is an example of the problem described in Chapter 1: Opinions may ignore evidence (Hall et al., 2019). It also misses evidence from medical, psychological, and neurological research. Teen behavior is driven by peer actions and the limbic system, not by textbook facts and the prefrontal cortex. Consequently, effective sex education must engage emotions and students' opinions (Suleiman & Brindis, 2014).

Technology and Human Relationships

Technology has changed the adolescent experience, but many adults are not sure whether it is for the better or the worse. Adults over age 50 grew up without the internet, instant messaging, social media, blogs, cell phones, smartphones, MP3 players, tablets, 3-D printers, or digital cameras. Until 2006, only students at a few highly selective colleges could join Facebook.

In contrast, today's teenagers are called *digital natives* because they have been networking, texting, and clicking for definitions, directions, and data all their lives. Their smartphones are always within reach; some teens text hundreds of times a day.

Some observers suggest that this is merely the latest expression of something adults have always feared, and teenagers have always sought: connection with peers. In earlier generations, adults thought that the automobile, or the shopping mall, or rock-and-roll music, would lead their children astray.

The benefits and harms of technology have been brought into focus by the school closings caused by the COVID-19 pandemic, which led to many teenagers spending more time online—75 percent of their waking hours, an average of seven hours a day. However, relatively few are worried about that. Most consider social media more helpful than hurtful, a way to feel connected, entertained, and distracted (Harris Insights and Analytics, June, 2020).

This was the result of an online poll in May 2020 of 1,516 teenagers from every part of the United States. The sample selection was not random or controlled (a teenager had to have access and be willing to answer). However, other data, such as from the American Psychological Association and the Pew Research Center, point in the same direction.

LaunchPad
macmillan learning

Watch **VIDEO: The Impact of Media on Adolescent Development** to learn more about how digital technology affects cognition during adolescence.

Cyber Abuse?

Many parents fear that the internet exposes their children to sexual abuse. The facts are reassuring: Although predators lurk online, most teens never encounter them. Sexual abuse is a serious problem, but if sexual abuse is defined as a perverted older stranger taking technological advantage of an innocent teen, it is "extremely rare" (Mitchell et al., 2013, p. 1226).

Almost all teen romances begin with a face-to-face attraction to another peer. Only 6 percent of 13- to 17-year-olds have *ever* had a romance that began online. They know how to block an uncomfortable relationship (Lenhart et al., 2015).

Cyberbullies

The facts are less reassuring regarding **cyberbullying**. Rumors, lies, embarrassing truths, or threats can all be sent from the safe distance of the private computer or smartphone, reaching a large audience just when social acceptance is crucial and the imaginary audience makes it worse (Giumetti & Kowalski, 2015). Further, insults, lies, and prejudice can be broadcast, as in "Zoom bombing," bots, and fake news.

Earlier in childhood, the least popular children may be bullied, but the social dynamic changes in high school. Popular students are frequent users of social media, using it to secure their status with prosocial and antisocial comments, which makes them likely to be victims as well as bullies (Ranney & Troop-Gordon, 2020). As with more conventional bullying at earlier ages, parental practices and school climate can mitigate the harm and reduce the prevalence of cyberbullying (Zurcher et al., 2018).

Sadly, ten years out of high school, bullies and victims—online or offline, sexual or otherwise—are less likely to have graduated from college and less likely to have good jobs or any job at all (Sigurdson et al., 2014). Adults need to be more concerned about teens who victimize each other, and less worried about strangers who lurk online.

cyberbullying When people try to harm others via electronic means, such as social media and cell phone photos or texts.

Consequences Unknown Few adolescents think about the consequences of their impulsive rage, responses, or retorts on social media or in text messages. This educator at a community center tries to explain that victims can be devastated—rarely suicidal but often depressed.

Spencer Grant/Alamy Stock Photo

Adolescent Bullying

Bullying is defined as repeated attempts to hurt someone else, physically or socially. It can take many forms. For younger children, it is often physical—hitting, shoving, fighting. That is less common among adolescents, who can hurt each other with words or exclusion. Among teenagers, not being invited to a party can be hurtful and is common—as teenagers develop dominance hierarchies and need peer support. The best protection is to have one or more close friends, and adults who encourage whatever talents the child has.

THE NATURE OF SCHOOL BULLYING

When bullying takes place at school, about two-thirds of it occurs in hallways, schoolyards, bathrooms, cafeterias, or buses. A full one-third occurs in classrooms, while teachers are present. An estimated 30% of school bullying goes unreported.

FEATURES OF SCHOOL ANTI-BULLYING PROGRAMS

- Increased supervision of students
- A school climate that encourages friendship
- Teachers who promote empathy
- School-wide implementation of anti-bullying policies
- Cooperation among school staff, parents, and professionals across disciplines
- Identification of risk factors for bullying

Success varies, with some programs having no effect. But overall, a good program can reduce bullying by 25% or more.

Data from McCallion & Feder, 2013

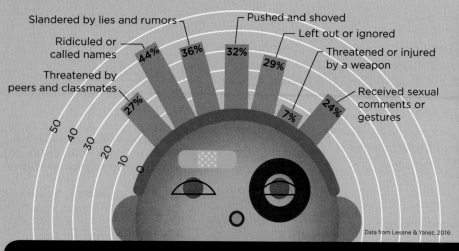

Slandered by lies and rumors 44%
Ridiculed or called names
Pushed and shoved 36%
32%
Left out or ignored 29%
Threatened by peers and classmates 27%
Threatened or injured by a weapon 7%
Received sexual comments or gestures 24%

Data from Lessne & Yanez, 2016

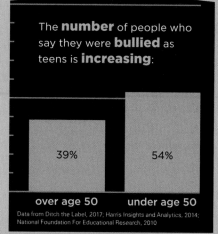

The **number** of people who say they were **bullied** as teens is **increasing**:

39% over age 50
54% under age 50

Data from Ditch the Label, 2017; Harris Insights and Analytics, 2014; National Foundation For Educational Research, 2010

CYBERBULLYING

Cyberbullying takes place via e-mail, text messages, Web sites and apps, instant messaging, chat rooms, or posted videos or photos. About 60% of boys and girls have been cyberbullied, but girls are more often the targets of online rumor-spreading or nonconsensual explicit messages (Anderson, 2018).

WHY DO TEENS CYBERBULLY?

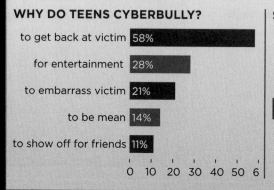

to get back at victim 58%
for entertainment 28%
to embarrass victim 21%
to be mean 14%
to show off for friends 11%

0 10 20 30 40 50 6

SOCIAL MEDIA AND CYBERBULLYING

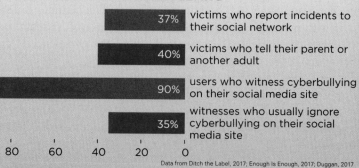

37% victims who report incidents to their social network
40% victims who tell their parent or another adult
90% users who witness cyberbullying on their social media site
35% witnesses who usually ignore cyberbullying on their social media site

80 60 40 20 0

Data from Ditch the Label, 2017; Enough Is Enough, 2017; Duggan, 2017

Benefits and Dangers

The potential for harm from technology should not prevent us from understanding the benefits. Many teachers and students have found that the vast material available online expands education, allowing students to learn far more than any teacher could be expected to know.

Further, remember the need for peer support during adolescence. Teens who were lonely and isolated a few decades ago, such as those with Down syndrome, or who are deaf, or the only one in their neighborhood with their ethnic, religious, or gender identity, can find peers.

The need for peer connection and the benefits of technology are especially obvious with the COVID-19 pandemic. To stave off depression, rage, and despair, many adolescents engaged with their friends online, morning, noon, and night. They also searched for accurate information on death and recovery rates, on effective treatments, on testing locations.

The virus offered some unexpected benefits, not only cleaner air but also family harmony: The generations could work together to separate facts from scams, partisan politics, and wishful thinking.

The danger is that adolescents can connect with people whose views are destructive. The hazards of adolescent egocentrism, and of intuitive rather than analytic cognition, are obvious. Messages are sent in an instant, with no second thoughts. Troubled adolescents can find others who share their prejudices and self-destructive compulsions, such as anorexia, gun use, racism, or cutting.

Benefits and harms are also available for sex. The internet can provide information that the school might not give and allow romantic interaction with no sexual contact. Texting—hundreds of times a day—is common among adolescent lovers. This can include **sexting**, as sending sexual photographs and videos is called.

However, when a romance ends, it may turn ugly. Of those who have quit a romance, 15 percent report being threatened online (Lenhart et al., 2015). If bullying takes a sexual tone, it can be particularly harmful.

At a weekend sleepover, Audrie Pott and her friends found alcohol. She got so drunk that she blacked out. On Monday, three boys in her school bragged that they had had sex with her, showing pictures on their cell phones to classmates. Soon Audrie thought the entire school knew. The next weekend, she hanged herself. Only then did her parents and teachers learn what had happened (Sulek, 2013).

The danger of technology lies not in the equipment but in the mind. As with many adolescent concerns (puberty, sexuality, body image, motivation), cognition and many other factors "shape, mediate, and/or modify effects" of technology (Oakes, 2009, p. 1142).

One careful observer claims that, instead of being *native* users of technology, many teenagers are *naive* users—believing they have privacy settings that they do not have, trusting sites that are markedly biased, believing news that is fake, misunderstanding how to search for and verify information (boyd, 2014).

Adults can help with all of this—but only if they themselves understand technology and adolescence. Teens are intuitive, impulsive, and egocentric, often unaware of the impact of what they send and overestimating the validity of what they see online. Adults should know better.

> **THINK CRITICALLY:** The older people are, the more likely they are to be critical of social media. Why?

sexting Sending sexual messages or photographs (usually of one's naked body) via phone or computer.

WHAT HAVE YOU LEARNED?

1. How do parent–adolescent relationships change over time?

2. When is parental monitoring a sign of a healthy parent–adolescent relationship?

3. How do the influences of peers and parents differ for adolescents?

4. Why do many adults misunderstand the role of peer pressure?

5. How does culture affect sexual orientation?

6. From whom do adolescents usually learn about sex?

7. Why do some schools teach abstinence-only sex education?

8. What are the advantages and disadvantages of technology in adolescence?

9. Why might sexting be a problem?

10. How might the term *digital native* be misleading?

Sadness and Anger

Adolescence can be a wonderful time. Nonetheless, troubles plague about 20 percent of youths. In a national survey, 16 percent of 12- to 17-year-olds said that, within the past year, they had received mental health counseling for emotional or behavioral problems. This number did not include those who did not seek help, nor did it include those who were treated for substance abuse (Substance Abuse and Mental Health Services Administration, 2019).

Sadness and anger may dissipate or may become intense, chronic, even deadly. Parents and peers can help a sad or angry child regulate emotions, or they can react with extreme emotions of their own, pushing a teenager toward suicide or prison. Most disorders are *comorbid*, with more than one problem in the same person. Everyone needs to know when normal moodiness becomes pathological, in order to pull someone back from the cliff. That is the goal of the following pages.

Depression

The general emotional trend from early childhood to early adolescence is toward less confidence. Then, gradually, self-esteem increases. A dip in self-esteem at puberty may occur with children of every ethnicity and gender (Fredricks & Eccles, 2002; Greene & Way, 2005; Kutob et al., 2010; Zeiders et al., 2013a), with notable individual differences, not always the ones expected.

For instance, girls have far higher rates of depression than boys. Children who believe that their nation is without fault are more likely to experience lower self-esteem in adolescence (Godfrey et al., 2019). Generally, self-esteem tends to be higher in African Americans than in European Americans, who themselves have higher self-esteem than Asian Americans. Among Latinos, self-esteem is relatively low, but some Latinas feel quite good about themselves (Zeiders et al., 2013a). One possible reason: Latinas with high familism become increasingly helpful at home, which makes their parents appreciative and them proud, unlike other U.S. teenage girls.

The risk of serious self-harm, such as cutting, is greatest in the years right after puberty (Plener et al., 2015). Gradually toward the end of adolescence and beginning of early adulthood, self-esteem rises (Orth & Robins, 2019). Always, however, specific social experiences and circumstances can outweigh age-related trends. Positive experiences include a fulfilling romance or academic success; negative ones include peer rejection and prison. For every adolescent of every ethnicity, the immediate social context—in the school, the family, and the community—is crucial.

Sibling Rivalry? No! This Latina 15-year-old is a role model for her 11-year-old sister, evident here as she helps with homework.

Bruce Laurance/The Image Bank/Getty Images

For instance, you just read that Latinas may have higher self-esteem, but if they have relatives who fear deportation, or they themselves are Dreamers, they may experience sleep disturbances, lower school achievement, and other symptoms of depression (Gulbas et al., 2016; Suárez-Orozco, 2017). You also read that teens during the COVID-19 pandemic have spent more time online, but the online increase did not seem to increase their depression.

What did increase depression and anxiety was social isolation. Most teens cope with stress by talking with friends, but that was more difficult when school was closed. Sadly, two-thirds of the teens who responded in the Harris poll said they felt sad or anxious more often than once a week, and the same proportion said they tried to keep their feelings to themselves, pretending to be happier in order not to worry anyone (Harris Insights and Analytics, June, 2020).

Major Depressive Disorder

Some adolescents sink into **major depression**, a deep sadness and hopelessness that disrupts all normal, regular activities. The causes predate adolescence, but puberty—with physical and emotional turbulence—pushes some vulnerable children, especially girls, into despair. The rate of serious depression more than doubles during this time to an estimated 15 percent, affecting about 1 in 5 girls and 1 in 10 boys.

Serious, distressing thoughts about killing oneself (called **suicidal ideation**) are most common at about age 15. More than one-third (41 percent) of U.S. high school girls felt so hopeless that they stopped doing some usual activities for two weeks or more in the previous year (an indication of depression), and nearly one-fourth (22 percent) thought seriously about suicide. For boys the rates were 21 and 12 percent (MMWR, June 15, 2018).

Suicidal ideation can lead to **parasuicide**, also called *attempted suicide* or *failed suicide*. Parasuicide includes any deliberate self-harm that could have been lethal. Parasuicide is the preferred term because "failed" suicide implies that to die is to succeed (!). "Attempt" is likewise misleading because, especially in adolescence, the difference between attempt and completion may be luck and treatment, not intent.

As you see in **Figure 16.2**, parasuicide can be divided according to whether or not medical treatment (surgery, pumped stomach, etc.) was needed, but every parasuicide is a warning. Among U.S. high school students in 2017, 9 percent of the girls and 5 percent of the boys attempted suicide in the previous year (MMWR, June 15, 2018).

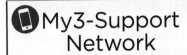

major depression Feelings of hopelessness, lethargy, and worthlessness that last two weeks or more.

suicidal ideation Serious thinking about suicide, often including extreme emotions and thoughts.

parasuicide Any potentially deadly self-harm that does not result in death. (Also called *attempted suicide* or *failed suicide*.)

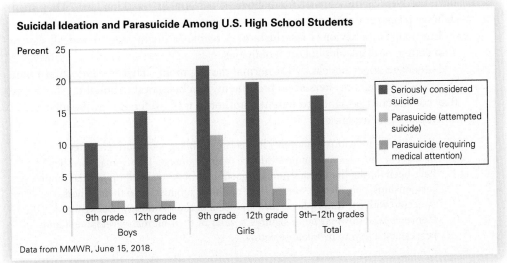

Suicidal Ideation and Parasuicide Among U.S. High School Students

Data from MMWR, June 15, 2018.

FIGURE 16.2

Sad Thoughts Completed suicide is rare in adolescence, but serious thoughts about killing oneself are frequent. More girls than boys say they have thought about, or attempted, suicide. The rate for boys may be underestimated for three reasons: (1) boys are less forthcoming about their emotions; (2) boys consider it unmanly to try suicide and fail; and (3) more males than females die by suicide.

Observation Quiz Does thinking seriously about suicide increase or decrease during high school? (see answer, page 453) ←

If you or someone you know needs help, call the National Suicide Prevention Lifeline at **1-800-273-8255**. You can also text HOME to **741-741** for free, 24-hour support from the Crisis Text Line.

cluster suicides Several suicides committed by members of a group within a brief period.

Completed Suicides

Although suicidal ideation during adolescence is common, completed suicides are not. In the United States in 2016, the rate of completed suicide for White teenagers, aged 15 to 19 (in school or not), was about 5 per 100,000, or 0.005 percent. The rate for non-White teenagers is about half that. Suicides are three times as common among adults, with the highest rate of all among White men in their 80s.

Curiously, although girls have higher rates of parasuicide, boys have higher rates of completed suicide, especially boys who are troubled about their gender identity or sexual orientation. Suicide rates are seven times higher for LGBTQ youth than others (Romanelli et al., 2020).

Of course, even one teen suicide is a tragedy, and it is particularly poignant when the media shares the story along with a photo of the fresh-faced young person. Those stories can cause teen depression and copycat suicides. Adolescent self-harm (parasuicide, cutting, extreme dieting), and clinical depression seem to be increasing; extensive media use (TV, social media, the internet) is a correlate (Twenge et al., 2018).

For example, teen suicide rates increased by almost a third after the release of a Netflix program *Thirteen Reasons Why*, which depicted a girl slitting her wrists (Bridge et al., 2019). Netflix's chief executive, Reed Hastings, said "no one has to watch it." Whether he is naive, cynical, or unfeeling is a matter of opinion.

Because they are more emotional and egocentric than logical and analytical, adolescents are particularly affected when they hear about someone else's suicide. That explains **cluster suicides** (several suicides within a group in the same time period). Although the overall rate is lower in adolescence, a higher proportion are cluster suicides, romantic couple suicides, or suicides by the same means as a recent celebrity suicide (M. Kral, 2019).

Delinquency and Defiance

Like low self-esteem and suicidal ideation, bouts of anger are common in adolescence. In fact, a moody adolescent could be both depressed and angry: Externalizing and internalizing behavior are closely connected during these years. This may explain suicide in jail: Teenagers incarcerated for assault (externalizing) are at greater risk of suicide (internalizing) than adult prisoners (Ruch et al., 2019).

Externalizing actions are hard to ignore. Many adolescents slam doors, curse parents, and tell friends exactly how badly other teenagers (or siblings or teachers) have behaved. Some—particularly boys—"act out" by breaking laws. They steal, damage property, or injure others.

Internalized behavior is less blatant. A teenager might stay in bed all day, or stop eating, or drink alcohol until oblivious.

Is teenage anger necessary for normal development? That is what Anna Freud (Sigmund's daughter, herself a prominent psychoanalyst) thought. She wrote that adolescent resistance to parental authority was "welcome . . . beneficial . . . inevitable." She explained:

> We all know individual children who, as late as the ages of fourteen, fifteen or sixteen, show no such outer evidence of inner unrest. They remain, as they have been during the latency period, "good" children, wrapped up in their family relationships, considerate sons of their mothers, submissive to their fathers, in accord with the atmosphere, idea and ideal of their childhood background. Convenient as this may be, it signifies a delay of their normal development and is, as such, a sign to be taken seriously.

[A. Freud, 1958/2000, p. 37]

However, most contemporary psychologists, teachers, and parents want well-behaved, considerate teenagers, who often become happy adults. A 30-year longitudinal study found that adults who had never been arrested usually earned colleges degrees, "held high-status jobs, and expressed optimism about their own futures" (Moffitt, 2003, p. 61). Thus, teenage acting out, while not unusual, is not essential for healthy development.

Breaking the Law

Both the *prevalence* (how widespread) and the *incidence* (how frequent) of criminal actions are higher during adolescence than earlier or later. Arrest statistics in every nation reflect this fact, with 30 percent of African American males and 22 percent of European American males being arrested at least once before age 18 (Brame et al., 2014).

Many more broke the law but were not caught, or were caught but not arrested. Self-reports suggest that most adolescents, of every gender, are lawbreakers before age 20. One reason is that many behaviors that are legal for adults—buying cigarettes, having intercourse, skipping school—are illegal for adolescents.

Arrest rates are higher for youth of minority ethnic groups, and boys are three times as likely as girls to be caught, arrested, and convicted. Does this reflect prejudice (Marotta & Voisin, 2017)? Some studies find that female aggression typically targets family and friends. Parents hesitate to call the police to arrest their daughters.

False Confessions

Determining accurate gender, ethnic, and income differences in actual lawbreaking, not just in arrests, is difficult. Self-reports may be false, with boasting or denial. For instance, researchers in the Netherlands contacted teenagers who were interrogated by the police. [The teens did not know that the researchers knew about the interrogations.] They were asked if they had ever had any police contact. One-third said no (van Batenburg-Eddes et al., 2012).

The opposite is also likely. In the United States, about 20 percent of confessions are false, with higher rates among teenagers. Why? Brain immaturity (delay discounting) makes young people ignore long-term consequences. Instead, they may prioritize protecting family members, defending friends, and pleasing adults—including the police (Feld, 2013; Steinberg, 2009).

One dramatic case involved 13-year-old Tyler Edmonds, who confessed to killing his brother-in-law. He was sentenced to life in prison. He then said that he confessed falsely to protect his 26-year-old sister, whom he admired. (She told him to confess, because she said his youth would mean his sentence would be short). His conviction was overturned—after he spent four years behind bars (Malloy et al., 2014).

The researchers who cited Tyler's case interviewed 194 boys, aged 14 to 17, all convicted of serious crimes. More than one-third (35 percent) said they had confessed falsely to a crime (not necessarily the one for which they were serving time). False confessions were more likely after two hours of intense interrogation—the adolescents wanted it to stop; acting on impulse, they said they were guilty (Malloy et al., 2014). And the police believed them.

A Criminal Career?

Many researchers distinguish between two kinds of teenage lawbreakers (Levey et al., 2019; Monahan et al., 2013), as first proposed by Terri Moffitt (2001, 2003). Both types are usually arrested for the first time in adolescence and for similar crimes, but their future diverges.

In Every Nation Everywhere, older adolescents are most likely to protest against government authority. Here, younger adolescents in Alabama celebrate the 50-year anniversary of the historic Selma-to-Montgomery march across the Pettus Bridge. In that historic movement, most of those beaten and killed were under age 25.

adolescence-limited offender A person who breaks the law as a teenage but whose criminal activity stops by age 20.

life-course-persistent offender A person whose criminal activity begins in adolescence and continues throughout life; a "career" criminal.

- Most juvenile delinquents are **adolescence-limited offenders**, whose criminal activity stops by age 21. They break the law with their friends, facilitated by their chosen antisocial peers.
- Some delinquents are **life-course-persistent offenders**, who become career criminals. Their lawbreaking is more often done alone than as part of a gang, and a contributing factor is neurological impairment (either inborn or caused by early experiences), evident in learning disabilities.

During adolescence, the criminal records of both types may be similar. However, if adolescence-limited delinquents can be protected from various snares (quitting school, incarceration, drug addiction), they outgrow their criminal behavior.

Causes of Delinquency

The best way to reduce adolescent crime is to notice early behavior that predicts lawbreaking and to change patterns before puberty. Strong and protective social relationships, emotional regulation, and moral values from childhood keep many teenagers from jail. Since learning disabilities and school failure are precursors to crime, effective remediation at age 6 may reduce delinquency at age 16.

Adolescent crime in the United States and many other nations has decreased in the past 30 years. Only a third as many arrests of people under age 18 occurred in 2018 compared to 2000. There are many possible explanations:

- fewer high school dropouts (more education means less crime);
- wiser judges (more community service than prison);
- better policing (arrests for misdemeanors are up, which may warn parents);
- smaller families (parents attend more to each of 2 children than each of 12);
- better contraception (unwanted children often become delinquents);
- less drug use (binge drinking and crack cocaine use increase crime);
- more immigrants (who are more law-abiding); and
- less lead in the blood (lead poisoning reduces brain functioning).

Nonetheless, adolescents remain more likely to break the law than adults, perhaps because of their brains as well as because of the social context. (See Inside the Brain.)

INSIDE THE BRAIN

Impulses, Rewards, and Reflection

For almost every crime, in almost every nation, the arrest rate for 15- to 17-year-olds is twice that for those over 18 (exceptions are fraud, forgery, and embezzlement).

What is wrong with those teenagers? Irresponsible parents? Poverty? Drug addiction? Maybe none of these. Perhaps the problem is in the brain, not in outside forces.

The limbic system is activated by puberty while the prefrontal cortex is "developmentally constrained," maturing more gradually (Hartley & Somerville, 2015, p. 109). Thus, adolescents are swayed by their intuition instead of by analysis.

Many studies confirm that adolescents show "heightened activity in the striatum, both when anticipating rewards and when receiving rewards" (Crone et al., 2016, p. 360). In choosing between a small but guaranteed reward and a large possible reward, adolescent brains show more activity for the larger reward than the brains of children or adults.

What does this mean? When teenagers weigh the possible results of a particular action, their brains make them more inclined to imagine success than to fear failure.

Whether this makes them brave and bold or foolish and careless is a matter of opinion, but there is no doubt that neurological circuits tip the balance toward action. Nor is there any doubt that the reward circuits in the brain are powerfully activated in adolescence (Cao et al., 2019).

The thrill of suddenly possessing a coveted jacket, or of joining a group of peers who are beating up a disrespectful stranger, is immediate. Later, if reflection occurs as the teen sits in jail, then another problem appears: According to one review, "incarcerated juveniles are at a four times higher risk of suicide than adolescents in general population" (Joshi & Billick, 2017, p. 141).

A related aspect of adolescent brains is that peer acclaim or rejection is deeply felt, with activation throughout the limbic system as well as other subcortical areas. This may help explain another aspect of adolescent crime: It often occurs in groups, whereas most adult criminals act alone.

Thus, neurological sensitivity may explain why teens readily follow impulses that promise social approval from friends and shun experiences that might bring social rejection. In experiments in which adults and adolescents, alone or with peers, play video games in which taking risks might lead to crashes or gaining points, adolescents are much more likely than adults are to risk crashing, especially when they are with peers.

When they are with their mothers, not their peers, teenagers are much more cautious in such simulations. However, as the connection between two brain regions (the *anterior insula* and the *ventral striatum*) increases in adolescence, risk-taking when the mother is absent increases, especially if the family relationship is not supportive (Guassi Moreira & Telzer, 2018).

There are other notable differences in brain activity (specifically in the ventral striatum) between adolescents and adults. When with other adults, the adult brain signals caution (inhibition)—opposite to the adolescent brain with peers (Albert et al., 2013) (see **Figure 16.3**).

This peer influence is apparent in both sexes but is stronger in boys—particularly when they are with other boys (de Boer et al., 2017). This may explain why adolescent boys use drugs, get arrested, and take dangerous risks, with far higher rates of accidental death than girls.

Teenage drivers like to fill (or overfill) cars with teen passengers who will admire them for speeding, for passing trucks on blind curves, for racing through railroad crossings when the warning lights are flashing, and so on. Fatal accidents are much more likely if the driver and the passengers are adolescents.

The accident rate in adolescence is aided by a third brain change in adolescence. Compared to children, there is a substantial increase in myelination between the emotional and action parts of the brain. This increase in white matter means rapid responses. As a result, adolescents act before slower-thinking adults can stop them (Hartley & Somerville, 2015).

Thus, don't blame teen crashes, juvenile delinquency, or drug use on inexperience; blame it on the brain. Some states now prohibit teen drivers from transporting other teenagers, reducing deaths and banning one source of adolescent excitement. States that allow marijuana always prohibit it before age 18. Some judges hesitate to give life sentences to adolescents.

Teens advocate for some laws, such as those that protect the environment; they do not advocate for laws that restrict drug use, driver's licenses, or gun purchases based on age. After a mass shooting at a high school in Parkland, Florida, students advocated a ban on some kinds of guns; the legislature instead banned gun sales to teenagers. Does this mean they understood the teen brain or that they understood their voters?

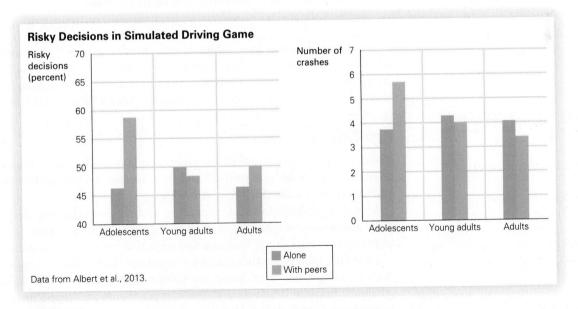

Data from Albert et al., 2013.

FIGURE 16.3

Losing Is Winning In this game, risk-taking led to more crashes and fewer points. As you see, adolescents were strongly influenced by the presence of peers, so much so that they lost points that they would have kept if they had played alone. In fact, sometimes they laughed when they crashed instead of bemoaning their loss. Note the contrast with emerging adults, who were more likely to take risks when alone.

Drug Use and Abuse

Most teenagers try *psychoactive drugs*, that is, drugs that activate the brain. Brain changes in the reward system lead directly to increases in drug abuse, such as binge drinking (Morales et al., 2018). Hormonal surges, the brain's reward centers, and cognitive immaturity make adolescents particularly attracted to the sensations produced by psychoactive drugs. But their immature bodies and brains make drug use especially toxic.

Every psychoactive drug excites the limbic system and interferes with the prefrontal cortex. Because of these neurological reactions, drug users are more emotional (varying from euphoria to terror, from paranoia to rage) and less reflective. By contrast, adult brains have automatic brakes. The rate of every hazard—including car crashes, unsafe sex, and suicide—is higher when teens use psychoactive drugs.

The critical question is what is automatic at what age:

> Our habitual and automatic responses save us valuable effort, energy, and time. The willpower required to make decisions with the prefrontal cortex, deciding how we're going to tie our shoe, or open the door anew each time, would paralyze us from ever living life. . . . stuck giving full attention to routine tasks, and drain our ability to concentrate and learn new things.
>
> [Hendlin, 2019, p. 137]

Of course, learning new things is exactly what adolescent cognition should allow. However, since every psychoactive drug affects the brain, which is not fully grown until about age 25, society tries to protect the brain by making the purchase of any drug—including alcohol and tobacco—illegal before adulthood.

Age Trends

Adolescence is a sensitive time for experimentation with psychoactive drugs. Use increases from about ages 10 to 25 and then decreases when adult responsibilities and experiences make drugs less attractive.

Use of legal drugs (alcohol, cigarettes, and marijuana) before age 15 is especially worrisome, because brain damage is more likely and early use escalates. Depression, sexual abuse, bullying, and addiction follow early drug use.

One drug follows another pattern—*inhalants* (fumes from aerosol containers, glue, cleaning fluid, etc.). Sadly, the youngest adolescents are most likely to try inhalants, because inhalants are easier to get (hardware stores, drug stores, and supermarkets stock them). Cognitive immaturity means that few pubescent children understand the risks—brain damage and even death (Nguyen et al., 2016).

Cohort differences are evident for every drug, even from one year to the next. Legalization of marijuana, e-cigarettes in many flavors, hundreds of deaths from opioids, dozens of deaths from vaping—these are examples of changes in the adolescent drug scene over the past few years.

Adolescent drug use in the United States has declined since 1976 (see **Figure 16.4**) with one major exception, vaping. Perception of risk, not availability, reduces use, since most high school students have always said that they could easily get alcohol, cigarettes, and marijuana (Miech et al., 2016).

Availability is notable for e-cigarettes. Although the United States prohibits adolescent purchase, 13- to 17-year-olds still manage to buy them, in stores and on the internet. Even when laws are strictly enforced, most young users get their e-cigarettes from other adolescents (Braak et al., 2020; McKeganey et al., 2019). Is that a problem? (See Opposing Perspectives on page 450.)

Harm from Drugs

Drug use before maturity is particularly likely to harm growth and predict later addiction. However, few adolescents are aware of when they or their friends move past use (experimenting) to abuse (experiencing harm).

Each drug is harmful in a particular way. *Tobacco* impairs digestion and nutrition, slowing down growth. Since internal organs mature after the height spurt, smoking teenagers who appear to be fully grown may damage their developing hearts, lungs, brains, and reproductive systems.

Alcohol is the most frequently abused drug in North America. Heavy drinking impairs memory and self-control by damaging the hippocampus and the prefrontal cortex, perhaps distorting the reward circuits of the brain lifelong (Guerri & Pascual, 2010).

Ironically, many antidrug parents condone adolescent drinking. For instance, a careful longitudinal study in Australia found that parents who provided alcohol to their teenagers thought they were teaching responsible drinking, but instead they were increasing binge drinking and substance use disorder six years later (Mattick et al., 2018).

Marijuana seems harmless to many people (especially teenagers), partly because users are more relaxed than verbose. Yet adolescents who regularly smoke marijuana are more likely to drop out of school, become teenage parents, be depressed, and later be unemployed—although the evidence comes from years when marijuana was illegal in every U.S. state.

In the next few years, we will learn more. Canada legalized marijuana for adults in the summer of 2018. Canadian health researchers hope that, once the brain is mature, benefits outweigh risks (Lake & Kerr, 2017). Marijuana is illegal in Canada for those under 18, although some doctors wish 21 were the cutoff (Rankin, 2017).

Any age restriction encourages younger adolescents to covet drugs used by older youth, which creates a major problem. This was evident when New Zealand lowered the age for legal purchase of alcohol from 20 to 18. Hospital admissions for intoxication, car crashes, and injuries from assault, increased, not only for 18- to 19-year-olds but also for 16- to 17-year-olds (Kypri et al., 2006, 2014).

Preventing Drug Abuse: What Works?

Remember that most adolescents think they are exceptions, sometimes feeling invincible, sometimes fearing social disapproval, but almost never being realistic

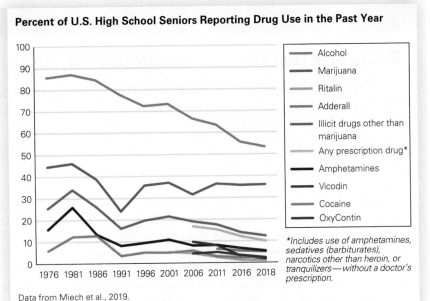

Percent of U.S. High School Seniors Reporting Drug Use in the Past Year

Legend: Alcohol, Marijuana, Ritalin, Adderall, Illicit drugs other than marijuana, Any prescription drug*, Amphetamines, Vicodin, Cocaine, OxyContin

*Includes use of amphetamines, sedatives (barbiturates), narcotics other than heroin, or tranquilizers—without a doctor's prescription.

Data from Miech et al., 2019.

FIGURE 16.4

Rise and Fall By asking the same questions year after year, the Monitoring the Future study shows notable historical effects. It is encouraging that something in society, not in the adolescent, makes drug use increase and decrease and that the most recent data show a continued decline in the drug most commonly abused—alcohol.

Especially for Parents Who Drink Socially You have heard that parents should allow their children to drink at home, to teach them to drink responsibly and not get drunk elsewhere. Is that wise? (see response, page 453)

E-Cigarettes: Path to Addiction or Health?

Electronic cigarettes (called e-cigs) are much less damaging to the lungs, because they deliver the drugs by vapor (vaping). Smokers with asthma, heart disease, or lung cancer benefit from vaping if it reduces their smoking of combustible cigarettes (Veldheer et al., 2019). However, the risk is not zero. In 2018, over 50 people died of a severe lung disease caused by e-cigarettes. The youngest death occurred in Texas in 2020. He was 15.

Death is rare, but vaping is not (see **Figure 16.5**). In 2019, 28 percent of high school students and 10 percent of middle school students had vaped in the past month (King et al., 2020). It is feared that adolescents who try e-cigarettes will become addicted to nicotine. Vaping smells better than tobacco, so might inhaling the vapor ease teenagers into smoking and then on to using other drugs?

E-cigs are not harmless. The ones that caused death may have had a particular ingredient (vitamin E) that most of them do not have, but all of them produce *benzene*, a known carcinogen (Pankow et al., 2017). Many contain nicotine, sometimes said to be more addictive than heroin. Some contain THC, a compound in marijuana. If the choice is between smoking and vaping, vaping is better; but if the choice is between not smoking and using e-cigs, vaping is worse.

Developmentalists fear that e-cigarettes will open the door to other drugs. This notion led to a new U.S. law in late 2019 that banned the sale of tobacco products, including e-cigs, to people under age 21. In early 2020, the Food and Drug Administration announced they would crack down on vendors who sold cartridge-based vaping products in kid-friendly flavors.

Teenagers who try e-cigs often smoke tobacco later (Miech et al., 2017b). Do e-cigs open a door, or would those adolescents be smokers no matter what? The American Pediatric Association warns that e-cigs will harm the next generation of children and adolescents (Jenssen & Walley, 2019).

The preceding paragraphs all mentioned harm. Now consider the other perspective. One company (JUUL) designed e-cigs with a sleek delivery gadget that looks like a USB drive, and advertised on Twitter, Instagram, and YouTube. Market share exploded among adolescents, with sales approaching a billion dollars. Inspired design and clever advertising, both admired by many adults, has made them successful (Huang et al., 2019).

The arguments from distributors of e-cigarettes are that their products are healthier than cigarettes, that people should make their own choices, and that the fear of adolescent vaping is exaggerated—part of the irrational fear that everything teenagers do is trouble.

Yet most public health doctors advise against e-cigs, and pediatricians worry that fetal and infant lungs suffer if the mother uses them (Carlsen et al., 2018). With rats, vaping decreases birthweight, which increases risks for early death and brain damage (Orzabal et al., 2019).

The evidence says caution, but caution is scarce at adolescence. The media presents mixed messages: Are strict age restrictions protective or puritanical (Morphett et al., 2020)? Opposing perspectives are apparent: Which perspective is yours?

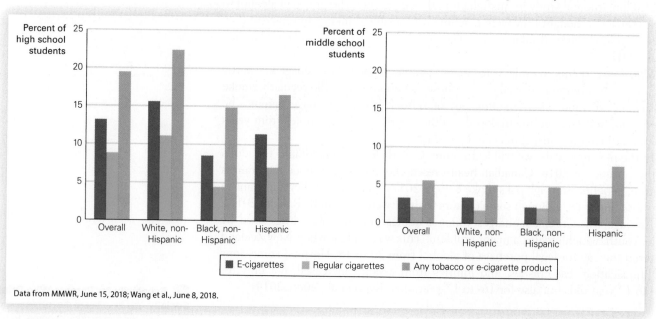

Data from MMWR, June 15, 2018; Wang et al., June 8, 2018.

FIGURE 16.5

Getting Better The fact that more than one in five high school students (that's 3 million people) used tobacco—even though purchase of any kind is illegal—in the past month is troubling. This means that more that 3 million students are at risk for addiction and poor health. The surprise (not shown) is that all of these rates are lower than a year earlier. Is that because laws are stricter or teenagers are getting wiser?

about potential addiction or about the reaction if they reject an offer or a toke, a sip, a line. For some, breaking the law is thrilling, defying authority is exciting, and drugs help them feel smarter, more awake, more fun. They do not see that, over time, stress and depression increase, and achievement decreases (Bagot, 2017; McCabe et al., 2017).

With harmful drugs, as with many other aspects of life, people of each generation prefer to learn things for themselves. A common phenomenon is **generational forgetting**, that each new cohort forgets what the previous cohort learned.

Mistrust of the older generation along with a loyalty to one's peers leads not only to generational forgetting but also to a backlash. When adults forbid something, that is a reason to try it, especially if adults exaggerate the dangers. If a friend passes out from drug use, adolescents may be slow to get medical help — a dangerous hesitancy.

Some antidrug curricula and advertisements make drugs seem exciting. Anti-smoking announcements produced by cigarette companies (such as a clean-cut young person advising viewers to think before they smoke) actually increase use (Strasburger et al., 2009).

By contrast, massive ad campaigns by public health advocates in Florida and California cut adolescent smoking almost in half, in part because the publicity appealed to the young. Teenagers respond to graphic images. In one example:

> A young man walks up to a convenience store counter and asks for a pack of cigarettes. He throws some money on the counter, but the cashier says "that's not enough." So the young man pulls out a pair of pliers, wrenches out one of his teeth, and hands it over. . . . A voiceover asks: "What's a pack of smokes cost? Your teeth."

> [Krisberg, 2014]

Parental example and social changes also make a difference. Throughout the United States, higher prices, targeted warnings, and better law enforcement have led to a marked decline in smoking among younger adolescents. Looking internationally, laws have an effect.

In Canada, cigarette advertising is outlawed, and cigarette packs have lurid pictures of diseased lungs, rotting teeth, and so on; fewer Canadian 15- to 19-year-olds smoke. What effect Canada's legalization of marijuana will have on teenagers is not yet known.

In the past three chapters, we have seen that the universal biological processes do not lead to universal psychosocial problems. Biology does not change, but context does. Rates of teenage births and abortions are declining sharply (Chapter 14), more students are graduating from high school (Chapter 15), and fewer teens drink or smoke (this chapter). Yet each of these chapters shows that much more needs to be done.

As explained at the beginning of these three chapters, adolescence starts with puberty; that much is universal. But what happens next depends on parents, peers, schools, communities, and cultures. In other words, the future of adolescents depends, in part, on you.

Pat Vasquez-Cunningham/Albuquerque Journal/ZUMAPRESS.com/Alamy Images

Telling Their Story Erika Pohl and her mother, Brenda, reflect on a documentary in which Erika had a leading role — as a teenager addicted to opioids who managed to recover. Both hope "never again." That is true for about half of teenagers with addiction; the film was created to improve those odds.

generational forgetting The idea that each new generation forgets what the previous generation learned. As used here, the term refers to knowledge about the harm drugs can do.

THINK CRITICALLY: Might the fear of adolescent drug use be foolish, if most adolescents use drugs whether or not they are forbidden?

WHAT HAVE YOU LEARNED?

1. Why are psychoactive drugs particularly attractive in adolescence?
2. Why are psychoactive drugs particularly destructive in adolescence?
3. What specific harm occurs with tobacco products?
4. How has adolescent drug use changed in the past decade?
5. What methods to reduce adolescent drug use are successful?

SUMMARY

Identity

1. Adolescence is a time for self-discovery. According to Erikson, adolescents seek their own identity, sorting through the traditions and values of their families and cultures.

2. Many young adolescents foreclose on their options without exploring possibilities, and many experience role confusion. Identity achievement takes longer for contemporary adolescents than it did a half-century ago when Erikson first described it.

3. Identity achievement occurs in many domains, including religion, politics, ethnicity, vocation, and gender. Each of these domains remains important over the life span, but timing, contexts, and often terminology have changed since Erikson and Marcia first described them. Achieving vocational and gender identity is particularly difficult as many more options seem possible.

Close Relationships

4. Parents continue to influence their growing children, despite bickering over minor issues. Ideally, communication and warmth remain high, while parental control decreases and adolescents develop autonomy. Cultures differ in the timing of conflicts and in the benefits of parental monitoring. Too much parental control is harmful, as is neglect.

5. Peers and peer pressure can be beneficial or harmful, depending on who the peers are. Adolescents select their friends, who then facilitate constructive and/or destructive behavior. Peer approval is particularly potent during adolescence.

6. Adolescents experience diverse sexual needs and may be involved in short-term or long-term romances, depending in part on their peer group. Contemporary teenagers are less likely to have intercourse than was true a decade ago.

7. Some youths are sexually attracted to people of the same sex. Social acceptance of same-sex relationships is increasing, but in some communities and nations, gay, lesbian, bisexual, and transgender youth are bullied, rejected, or worse.

8. Many adolescents learn about sex from peers and the media—sources that are not comprehensive. Ideally, parents are the best teachers on topics about sex, but many are silent or naive.

9. Some nations provide comprehensive sex education beginning in the early grades, and most U.S. parents want schools to teach adolescents about sex. Abstinence-only education is not effective at slowing down the age of sexual activity, and it may increase STIs.

10. Most adolescents in the United States use technology to connect with their peers. Sexting is also common, and adults see dangers in it that peers do not.

Sadness and Anger

11. Almost all young adolescents become more self-conscious and self-critical than they were as children. A few become chronically sad and depressed. Many adolescents (especially girls) think about suicide, and some attempt it. Few adolescents actually kill themselves; most who do so are boys.

12. At least in Western societies, almost all adolescents become more independent and angry as part of growing up, although most still respect their parents. Breaking the law and bursts of anger are common; boys are more likely to be arrested for violent offenses than are girls.

13. Adolescence-limited delinquents should be prevented from hurting themselves or others; life-course-persistent offenders may become career criminals. Early intervention—years before the first arrest—is crucial to prevent serious delinquency.

Drug Use and Abuse

14. Most adolescents experiment with drugs, which may temporarily reduce stress and increase peer connections but soon add to stress and social problems. Almost every adolescent tries alcohol, and many use e-cigarettes and marijuana. All are technically illegal for those under 18 but are readily available to teenagers.

15. All psychoactive drugs are particularly harmful in adolescence, as they affect the developing brain and threaten the already shaky impulse control. Prevention and moderation of adolescent drug use and abuse are possible. Price, perception, and parents have an effect.

KEY TERMS

identity versus role confusion (p. 424)
identity achievement (p. 424)
role confusion (p. 424)
foreclosure (p. 424)
moratorium (p. 424)

gender identity (p. 429)
parental monitoring (p. 432)
familism (p. 432)
peer pressure (p. 434)
coercive joining (p. 434)
sexual orientation (p. 436)

cyberbullying (p. 439)
sexting (p. 441)
major depression (p. 443)
suicidal ideation (p. 443)
parasuicide (p. 443)
cluster suicides (p. 444)

adolescence-limited offender (p. 446)
life-course-persistent offender (p. 446)
generational forgetting (p. 451)

APPLICATIONS

1. Locate a news article about a teenager who committed suicide. Were there warning signs that were ignored? Does the report inadvertently encourage cluster suicides?

2. Research suggests that most adolescents have broken the law but that few have been arrested or incarcerated. Ask 10 of your fellow students whether they broke the law when they were under 18 and, if so, how often, in what ways, and with what consequences. (Assure them of confidentiality; remind them that drug use, breaking curfew, and skipping school were illegal.) Do you see any evidence of gender or ethnic differences? What additional research needs to be done?

3. Cultures vary in expectations for drug use. Interview three people from different backgrounds (not necessarily from different nations; each SES, generation, or religion has different standards) about their culture's drug use, including reasons for what is allowed and when. (Legal drugs should be included in your study.)

Especially For ANSWERS

Response for Parents of a Teenager (from p. 433): Remember: Communicate, do not control. Let your child talk about the meaning of the hairstyle. Remind yourself that a hairstyle in itself is harmless. Don't say, "What will people think?" or "Are you on drugs?" or anything that might give your child reason to stop communicating.

Response for Sex Educators (from p. 438): Yes, but forgive them. Ideally, parents should talk to their children about sex, presenting honest information and listening to the child's concerns. However, many parents find it very difficult to do so because they feel embarrassed and ignorant. You might schedule separate sessions for adults over 30, for emerging adults, and for adolescents.

Response for Parents Who Drink Socially (from p. 449): No. Alcohol is particularly harmful for young brains. It is best to drink only when your children are not around. Children who are encouraged to drink with their parents are more likely to drink when no adults are present. It is true that adolescents are rebellious, and they may drink even if you forbid it. But if you allow alcohol, they might rebel with other drugs.

Observation Quiz ANSWERS

Answer to Observation Quiz (from p. 434): The girls are only observers, keeping a respectful distance.

Answer to Observation Quiz (from p. 435): Impossible to be sure, but body position suggests dating. The couple on the left seem happy with each other (she leans toward him, his hand is pulling her close), but the couple may be less so.

Answer to Observation Quiz (from p. 443): Both. It increases for boys but decreases for girls.

Epilogue: Emerging Adulthood

What Will You Know?

1. Is risk-taking an asset or a liability?
2. Does college make you think?
3. Is cohabitation a good prelude to marriage?

Infancy, childhood, and adolescence—the previous 16 chapters—are the foundation of life, but they comprise less than a fourth of the human life span. This epilogue offers a glimpse at what is next, **emerging adulthood**, ages 18–25, when people continue to learn and explore.

Consider this epilogue a brief review and a preview. We follow the same sequence as earlier chapters at each stage—body, mind, and social world—always noting the impact of genetic, prenatal, and early experiences. You will see many familiar themes—family, friends, culture, context, and cohort. The tension between what is universal (true for every human) and what is particular (powerfully influenced by culture and cohort) is apparent at this stage, as it has been throughout.

In considering this period, we include references to the COVID-19 pandemic, which has affected everyone, of every age, emotionally, economically, and educationally. Before going further, remember that the study of human development relies on the scientific method.

Regarding the pandemic, there are many questions (Step 1) and hypotheses (Step 2). What effect will it have on human development? Is this pandemic a temporary detour, or will emerging adults forge an entirely new path for their future? Longitudinal data (Step 3) are needed before we can accept conclusions (Step 4), and then peer review, publication (Step 5), and replication are needed.

The research on COVID-19 reported here is fascinating but speculative. As with many issues raised in the previous 16 chapters, some of you will be the scientists who answer those questions.

Biological Universals

Certain biosocial characteristics have always characterized 18- to 25-year-olds. Biologically, these years are prime time for high energy, hard physical work, and safe reproduction. As has been true for thousands of years, every body system—digestive, respiratory, circulatory, musculoskeletal, and sexual-reproductive—functions optimally in emerging adulthood. The rapid and sometimes unsettling changes of adolescence are over: Emerging adults are at their peak of fertility and strength.

Consequently, this has been the time for marriage and parenthood, for fighting neighboring groups and growing crops, for undertaking the hard work of raising small children, building a home, starting a job. By age 25, all that was established, and adults continued with the rest of their lives with the same partner, their growing children, and the same work.

No longer.

Remember that one of the goals of our study is to understand what is universal and what is contemporary, to recognize the impact of culture and cohort, and to acknowledge simultaneously that the human species has always experienced similar developmental changes from conception until death. That tension between the universal and the contemporary is an underlying theme of this epilogue.

Health Habits

As always, human development connects universal, biological maturation with the specific conditions of a particular place and time. The health habits of young adults affect them lifelong. On that score, if emerging adults continue the habits they have now, they will enjoy a long, healthy life.

Exercise is the most obvious good habit. Most young adults are quite active, walking or biking to work, playing on sports teams, going to the gym at higher rates than older adults. Their motivation to exercise includes social reasons (because their friends do it), appearance, as well as health reasons, and that increases their activity (Box et al., 2019). Other health habits are also better than at older ages: Their rates of obesity, excessive drinking, and smoking are lower than older adults (see **Figure EP.1**).

Emerging adults may be protected somewhat from COVID-19, as well as from other diseases. It is not yet known if many have caught the virus with few serious symptoms, or what the long-term effects on their hearts, lungs, and brains might be. But their hospitalization and death rates are low. By contrast, one of every three people over age 70 who were hospitalized in April 2020 had died before June (Petrilli et al., 2020).

The challenge, of course, is for young adults to continue their good habits. If they do, that slows down aging, allowing an extra decade or so of disease-free life in late adulthood (Li et al., 2020). Those habits make a difference lifelong, although the natural reserves and strength of the young means that few emerging adults notice any health problems (see **Table EP.1**).

Nonetheless, research on adults now in midlife finds that some people age three times faster than others, with about half of the difference between fast and slow aging evident by age 26 (Belsky et al., 2015). By the mid-30s, some people have bodies like those in their 20s and some like those in their 40s. Will the current cohort of young adults slow down the aging process?

Later life is directly affected by health habits, but other circumstances matter. College football players, for instance, are exceptionally fit and strong, but also sustain hundreds of hits to their heads. Few have notable concussions, but those hits increase their risk of severe brain disease later on (Sancar, 2019). One COVID-19 question that awaits longitudinal research is whether those who contract the virus after a year or two will be affected later on. Some research finds that heart damage, for instance, endures after recovery (Li et al., 2020), but whether complete recovery is likely in young adulthood remains an open question.

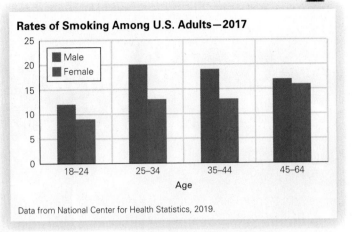

Rates of Smoking Among U.S. Adults—2017

Data from National Center for Health Statistics, 2019.

FIGURE EP.1

Good News Ahead? Habits established in emerging adulthood tend to persist. Many older adults have suffered through the hard work of quitting smoking, but about 15 percent have tried and failed. We do not yet know what this chart will look like in 2027, but optimism seems realistic!

emerging adulthood The period of life between the ages of 18 and 25. Emerging adulthood is now widely thought of as a distinct developmental stage.

TABLE EP.1

U.S. Deaths from the Top Two Causes (Heart Disease and Cancer)

Age Group	Annual Rate per 100,000
15–24	6
25–34	16
35–44	60
45–54	176
55–64	470
65–74	961
75–84	2,119
85+	5,494

Data from National Center for Health Statistics, 2018.

Ashamed to Use a Condom? This public health effort attempts to remove the stigma, in order to reduce sexually transmitted diseases and unwanted babies.

● **Observation Quiz** Where is this? (see answer, page 467) ↑

Sex and Pregnancy

Fertility peaks in late adolescence and early adulthood, pregnancy and birth are easier at age 20 than after age 30, which may be why hormones make sexual relationships particularly compelling in early adulthood. Compared to earlier centuries, however, fewer emerging adults seek parenthood.

In 2018, the average U.S. woman gave birth for the first time at age 27 (Martin et al., November 27, 2019), with men choosing fatherhood later than that. Adults of both sexes in some European nations are in their 30s, on average, when they have their first baby.

Thus, biological drives and social preferences diverge in the twenty-first century, unlike in earlier centuries. Medical research has found a solution to this clash: effective contraception. That solution is far from perfect: An estimated 40 percent of all pregnancies in the United States are unintended. However, that rate is decreasing, as is the rate of abortions and unwanted births, especially among emerging adults (Finer & Zolna, 2016; Jatlaoui et al., 2019).

The major sexual problem among emerging adults is now sexually transmitted diseases, which do not seem to be declining. Ironically, the more sexual partners an emerging adult has, the less likely they are to be conscientious with protection (Ashenhurst et al., 2017).

Risks and Benefits

That is an example of an underlying problem: Early adulthood is a time for taking risks. Remember that each age group has its own gains and losses: Risk-taking in emerging adulthood is an example. Enrolling in college, moving to a new state or nation, independence from parents, committing to a romantic partner, starting a business, filming a documentary, entering a sports competition, enlisting in the military, traveling abroad—all risky and all beneficial.

However, unprotected sex with a new partner, fast driving with no seat belt, carrying a gun, avoiding vaccines, and experimenting with drugs, are more common in early adulthood than later on. The overall death rate of young adults is lower than for older adults, but a higher proportion of those deaths are the result of risks.

In 2018 in the United States, only 10 percent of the deaths of emerging adults were of unavoidable diseases. Most of the rest were related to a risk, such as a drowning, motor vehicle crashes, drug overdose, homicide, or unprotected sex.

Similar trends are apparent even in nations where disease is more common: Young adults rarely die of the scourges of poor nations, such as malaria or malnutrition. Even in underdeveloped nations, motor accidents cause more young-adult deaths than any disease.

WHAT HAVE YOU LEARNED?

1. What is universal about physical development in emerging adulthood?

2. What are the problems and advantages of risk-taking?

3. What is similar about disease deaths in developed and developing nations?

Cognitive Development

In many ways, emerging adulthood is prime time for thinking clearly and well. The prefrontal cortex finally is fully connected to the other three lobes; intuitive and analytic thought can come together; these often are the years of intellectual challenges in college or learning new skills on the job.

Postformal Thought

Although *formal operational thought* was the final stage of Piaget's theory, many cognitive psychologists find that postadolescent thinking is a cut above that. Some developmentalists propose a fifth stage, called **postformal thought**, a "type of logical, adaptive problem-solving that is a step more complex than scientific formal-level Piagetian tasks" (Sinnott, 2014, p. 3). In postformal cognition, "thinking needs to be integrated with emotional and pragmatic aspects, rather than only dealing with the purely abstract" (Labouvie-Vief, 2015, p. 89).

As they integrate emotion and pragmatics, postformal thinkers are flexible, with a "more complex, nuanced, and paradoxical" mode of thinking (Gidley, 2016). They consider all aspects of a situation, anticipating problems and dealing with them rather than denying, avoiding, or procrastinating. As a result, postformal thought is practical and creative (Gidley, 2016; Sinnott et al., 2020).

Combining the Facts and Emotions

This synthesis may entail combining objective and subjective thought. **Objective thought** uses abstract, impersonal logic; **subjective thought** arises from personal experiences and perceptions. As you remember, formal operational thinking values impersonal logic and devalues subjective emotions, the opposite of egocentric preoperational thought. Both may not only come together; they may also produce a more advanced cognition in emerging adulthood.

Some scholars describe this process as *dialectical*, in which one idea (a thesis) leads to a contrary idea (an antithesis), and together they result in a new and better combination (a synthesis). One Canadian study used a dialectical approach to analyze vocational exploration of older adolescents over five years as they entered emerging adulthood, finding that each new career idea could lead to a deeper understanding of the next step (Gagnon et al., 2019).

If people do not reach a synthesis of objective and subjective thought, behavioral extremes (such as binge eating, anorexia, addiction, and violence) and cognitive extremes (such as believing that one is the best, or the worst, person on Earth) are common. Those are typical of the egocentrism of adolescence—and of some adults. By contrast, dialectical thinkers balance opposites, moving forward.

Moral Foundations

The importance of combining emotional and logical perspectives is stressed by Jonathan Haidt, who has studied moral development in many religions and cultures. He wrote many scenarios to uncover which moral principles were important to adults.

- A family dog is killed by a passing car and the family cooks the meat and eats it for dinner.
- A sister and a brother on vacation from college in distant land where no one knows them, using excellent contraception, have intercourse.

VIDEO ACTIVITY: Brain Development: Emerging Adulthood shows the changes that occur in a person's brain between ages 18 and 25.

postformal thought A proposed adult stage of cognitive development, following Piaget's four stages, that goes beyond adolescent thinking by being more practical, more flexible, and more dialectical (i.e., more capable of combining contradictory elements into a comprehensive whole).

objective thought Thinking that is based on facts. It is impartial, and can be verified by anyone who seeks to know.

subjective thought Thinking that is based on personal experience and perception. Because it is connected to emotions, it is deeply held and can be quite opposite another's subjective thinking.

Ideal Versus Real One indication of adult cognition is the ability to accept some imperfections in oneself, one's family, and one's nation.

"I CAN'T WAIT TILL I'M OLD ENOUGH TO HOLD MY NOSE AND VOTE."

Leo Cullum/Cartoon Collections

THINK CRITICALLY: Can you see dialectical thinking when you remember what you believed as a child?

- Someone drinks water from a glass that had a dead, sterilized cockroach in it.

Each of these was highly offensive to most adults, who tried to explain their reactions based on logic. In fact, their reactions revealed their deep moral values, which Haidt says we all have in emerging adulthood but fail to recognize. He contends that humans have five moral foundations:

1. care for others, harm no one;
2. promote freedom, avoid oppression;
3. be fair, do not cheat;
4. seek purity, avoid contamination; and
5. respect authority, do not break laws.

The importance of these five varies from nation to nation, with #4, for instance, much stronger in India, and #2 much stronger in the United States. The emphasis on each of the five is influenced by the doctrines of each religion, by the values of each political party, and by the age of the person making the judgment.

A Weird Perspective

If one culture prioritizes the first two and another the last two, then people will interpret the middle one—fairness—differently. This may explain why people of different religious, political, or cultural backgrounds consider each other immoral; they prioritize differently (Haidt, 2013).

This raises an important question regarding research on cognition. Most published research uses college students from **WEIRD** nations (Western, Educated, Industrialized, and Rich Democracies) as participants. Is that biased?

As one leading researcher expressed it, "the WEIRD group represents maximally 5% of the world's population, but probably more than 90% of the researchers and scientists producing the knowledge that is represented in our textbooks work with participants from that particular context" (Keller, quoted in Armstrong, 2018). Some scholars find that the cognitive abilities of emerging adults are similar worldwide; others disagree. This is another example of the tension between universal and particular.

Consider responses to COVID-19. Emerging adults tend to be most likely to resist government mandates to stay at home. They use logic to buttress their emotional resistance to authority. Research on emerging adults in Italy reports that some of this is cognitive: Those young adults who had a more collectivist perspective (believing more strongly in family and community) were more anxious about spreading the virus, more compliant with quarantine, and, perhaps paradoxically, less maladjusted (Germani et al., 2020). Is that true in the United States as well?

Striking national differences are apparent in limiting the spread of the virus, with New Zealand, South Korea, and Germany having done much better than the United States and England. Emerging adults in the United States are particularly likely to reject quarantine. Is this cognitive? Does national culture combine with developmental age to affect how emerging adults think about their health, or their government, or their family?

WEIRD An acronym that refers to people from Western, Educated, Industrialized, Rich Democracies, in other words, to North American college students, not necessarily the rest of humanity.

◗ **Especially for Someone Who Has to Make an Important Decision** Which is better, to go with your gut feelings or to consider pros and cons as objectively as you can? (see response, page 467)

College and Cognition

There is no doubt that college education benefits future health and wealth throughout adulthood (see Visualizing Development, page 460). In the United States, each added level of education correlates with everything from happy marriages to strong teeth, from spacious homes to long lives, from healthy children to working digestive systems (U.S. Department of Health and Human Services, January 28, 2019). Surprisingly, those from low-SES homes who, against the odds, earn a college degree, benefit even more from that degree than those from wealthier backgrounds (Brand & Xie, 2010; Karlson, 2019).

Note, however, that the benefits of college are most evident for those who earn a degree, yet about half the students who enroll in college leave before graduating. That is especially true for low-SES students who are African American, Latinx, and Native American, who often leave within the first year. Some colleges have much higher dropout rates than others, with community college students least likely to persevere until graduation.

Efforts to increase retention are twofold: financial and social. The financial problems include much more than the cost of tuition. Child care, lost income, and transportation may be crucial (Troester-Trate, 2019).

In addition, mentors and fellow students who encourage academic work may be as influential as grants that cover college costs, perhaps especially for students who are interested in science careers (Ceyhan et al., 2019). Note that although the ostensible goal of college is cognitive (to increase knowledge), the attainment of that goal seems more affected by money than the mind.

Intellectual growth, as Vygotsky would note, occurs everywhere in college: The direct academic knowledge that courses provide is only one part. This is evident with COVID-19, as all colleges in the United States temporarily canceled in-person classes, attempting to provide instruction remotely. Many students gained knowledge online, but all students missed the social interaction.

Exposure to new perspectives is a crucial benefit of college. That occurs with books and professors, but a major source is other students who have grown up in diverse communities, or who have had particular experiences, such as parenthood, or military service, or living abroad.

Of course, a diverse student body does not guarantee intellectual growth. Instead, intellectual expansion comes from conversations among people of varied backgrounds and perspectives (Pascarella et al., 2014). Colleges that make use of their diversity—via curriculum, assignments, discussions, cooperative education, learning communities, residence halls, and so on—stretch student understanding, not only of other people but also of themselves (Harper & Yeung, 2013; Shim & Perez 2018).

Hi Mom and Dad Like many college graduates, this young man in Idaho is grateful for the support of this family, who helped make this proud moment possible.

WHAT HAVE YOU LEARNED?

1. What is the difference between formal and postformal thought?
2. Why is most research on emerging adulthood based on WEIRD people?
3. Why do many students quit college before graduating?
4. How does diversity affect college education?

Psychosocial Development

Emerging adulthood is a time of transition. Young adults emerge from the close nurturance of family and immediate community and enter a broader, more complicated, and independent world. They have new freedom, as well as new restraints. They can now listen to their parents, neither rebelling nor agreeing; they can analyze their national and ethnic legacies. This is a time of extraordinary possibility.

Personality in Emerging Adulthood

Possibility is evident in personality. Young adults do not shed their genetic and childhood influences. If self-doubt, anxiety, depression, or antisocial behavior characterizes the earlier years, it does not disappear.

VISUALIZING DEVELOPMENT

Why Study?

From a life-span perspective, college graduation is a good investment, for individuals (they become healthier and wealthier) and for nations (national income rises). However, when the effort and cost of higher education depend on immediate choices made by students and families, as in the United States, many decide it is not worth it, as illustrated by the number of people who earn Bachelor's degrees.

CURRENT COLLEGE ENROLLMENT AMONG U.S. 18- TO 24-YEAR-OLDS

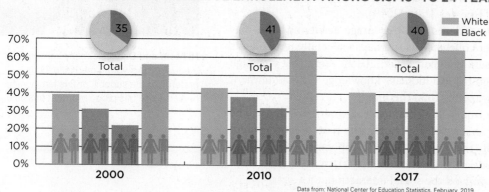

Legend: White, Hispanic, Black, Asian

If this graph showed how many in each group *ever* attended college, the numbers for Black and Hispanic people would be closer to the numbers for Asian people, many of whom earn both Bachelor's and advanced degrees during this six-year period.

Data from: National Center for Education Statistics, February, 2019.

HIGHEST LEVEL OF EDUCATION ATTAINED BY U.S. ADULTS

The percentage of U.S. residents with high school and college diplomas is increasing as more of the oldest cohort (often without degrees) dies and the youngest cohorts aim for college. The data below are for people ages 25 and older. In 1968, half of them reached high school age when education past eighth grade was a luxury, expected for those who were rich, native-born, and white, not for the general population.

Legend: Male, Female

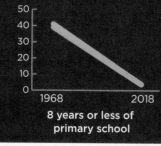

8 years or less of primary school

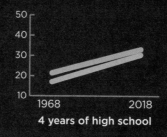

4 years of high school

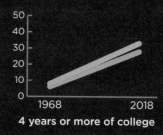

4 years or more of college

Data from U.S. Census Bureau, February 19, 2018.

INCOME IMPACT

Over an average of 40 years of employment, someone who completes at least a Master's degree earns 1.5 million dollars more than someone who leaves school in eleventh grade. That translates into more than $200,000 for each year of education from twelfth grade to a Master's. The earnings gap is even wider than those numbers indicate because this chart includes only adults who have jobs, yet finding work is more difficult for those with less education.

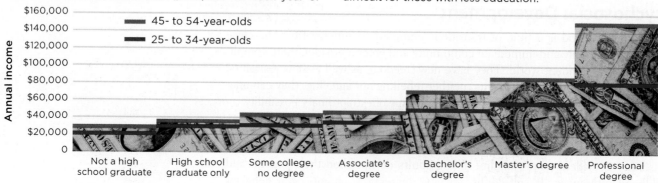

Legend: 45- to 54-year-olds; 25- to 34-year-olds

Annual income axis: $160,000, $140,000, $120,000, $100,000, $80,000, $60,000, $40,000, $20,000, 0

Education categories: Not a high school graduate, High school graduate only, Some college, no degree, Associate's degree, Bachelor's degree, Master's degree, Professional degree

Education

Data from National Center for Education Statistics, September, 2018.
Photo: Jupiterimages/Thinkstock/Photos.com/Getty Images Plus.

Yet personality is not static, and each emerging adult combines genes, parental influences, and political contexts in a specific way to form their adult personality. Emerging adulthood has been called the "crucible for personality development" (Roberts & Davis, 2016). Remember the Big Five personality traits? Generally, continuity in those five is evident throughout adulthood. Thus, a young adult who is relatively outgoing becomes an outgoing older adult.

However, especially in emerging adulthood, notable changes are possible (Deventer et al., 2019), and the result is usually positive, with negative personality traits no longer dominant and positive ones strengthened. A study of almost a million adolescents and adults from 62 nations found that "during early adulthood, individuals from different cultures across the world tend to become more agreeable, more conscientious, and less neurotic" (Bleidorn et al., 2013, p. 2530).

This positive push occurs because emerging adults gradually feel more in control of their own lives (Vargas Lascano et al., 2015) and are able to set their own goals (Hill et al., 2011). One longitudinal study found that self-criticism gradually declines from age 23 to 29, and that improves mental health later on (Michaeli et al., 2019).

We need to note that, just as personality may change but rarely transforms, relationship to parents and national culture may change but rarely transforms. Remember the stage Erikson pegged at adolescence, the identity crisis, and remember that *identity achievement* occurs when aspects of earlier life are not rejected wholesale but are reexamined and revised (see **Table EP.2**). That is what emerging adults do.

One example is financial. Emerging adults strive to attain financial independence from their parents, and yet they accept some financial help, either directly with money or indirectly with food, laundry, and so on.

A longitudinal study of 1,719 individuals found that most young adults relied on parental financial help for years, even after college was over. Financial help meant less stress and eventually more independence, from age 18 to 27. Those (about one-fourth) who were totally independent, financially, tended to fare worse at age 27 than those with substantial assistance. The conclusion of this study is that family support is not a barrier to independence but more often the opposite—a helpful launching pad toward adulthood (Bea & Yi, 2019).

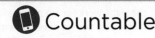

CHAPTER APP

Countable

IOS:
https://tinyurl.com/jeh8m6y

ANDROID:
https://tinyurl.com/y8rb3q5b

RELEVANT TOPIC:
Emerging adults and national culture

This app informs users of pending U.S. federal legislation (and soon, state and local legislation) with succinct, nonpartisan, sourced write-ups that include pro and con arguments. Countable also enables users to give instant feedback on pending bills and see how their representatives voted.

TABLE EP. 2

Erikson's Eight Stages of Development

Stage	Virtue / Pathology	Possible in Emerging Adulthood if Not Successfully Resolved
Trust vs. mistrust	Hope / withdrawal	Suspicious of others, making close relationships difficult
Autonomy vs. shame and doubt	Will / compulsion	Obsessively driven, single-minded, not socially responsive
Initiative vs. guilt	Purpose / inhibition	Fearful, regretful (e.g., very homesick in college)
Industry vs. inferiority	Competence / inertia	Self-critical of any endeavor, procrastinating, perfectionistic
Identity vs. role diffusion	Fidelity / repudiation	Uncertain and negative about values, lifestyle, friendships
Intimacy vs. isolation	Love / exclusivity	Anxious about close relationships, jealous, lonely
Generativity vs. stagnation	Care / rejection	[In the future] Fear of failure
Integrity vs. despair	Wisdom / disdain	[In the future] No "mindfulness," no life plan

Information from Erikson, 1982.

CAREER ALERT   The Career Counselor

We need more career counselors. Job growth in this occupation is above average, and so is annual income: 13 percent above and $56,000 a year, according to the Occupational Outlook Handbook (U.S. Bureau of Labor Statistics, 2020).

This occupation is both challenging and satisfying: Helping people find the right work for them benefits individuals, families, and communities. Studying human development is a good first step, so you are already on your way! After earning a Bachelor's degree, a Master's in counseling is recommended.

As you know from Erikson, vocational identity is crucial, and work is central to development. Emerging adults often change jobs (every year, on average), sometimes because they expected the job to be temporary (e.g., summer work as a waitress, lifeguard, or office assistant). But many employment shifts occur because young adults are unaware of job availability or of their own skills and values. Education improves job fit and satisfaction (Ilies et al., 2019), but which education for which job?

Many adults have not found the best employment for them. They take jobs that are available; they consult friends and family. They may discover that they hate their work, or an economic shift may put them out of work. Even if they are doing work they enjoy, people may be happier with another employer, or self-employment, or somewhere else. Career counseling might help.

This vocation is especially vital today for at least six reasons:

1. Most current vocations did not exist a generation ago, making past sources of job information (parents and teachers) less reliable.
2. Adult lives change over time, obviously for veterans, for parents with new babies or growing children, and for immigrants, but also for everyone else.
3. The economy is shifting, with start-ups, closed factories, relocated corporations, and emerging markets.

4. Long-term unemployment is one of the worst problems of adulthood, destroying personal happiness, as well as families and communities.
5. Major groups—women, minorities, people with disabilities—who were once shut out of productive work now can be vital workers in today's economy.
6. Adults who enjoy their work, with coworkers, challenges, and hours that make each day a good one, achieve a fulfilling, productive life.

This Career Alert appears in this epilogue chapter on emerging adulthood because floundering is more common than flourishing during these years. Most high school graduates know that they need more education in order to be hired for the work they want, but few know what college courses, requirements, and vocational training satisfy the demands of the job market. Most employees find their work by chance, or through word of mouth, or by stumbling across something on the Internet. Fantasy conflicts with reality; rejection and discouragement are common. Everyone needs guidance by a wise and informed advisor.

A few decades ago, vocational counselors had a simple task. There were valid tests of skills and personality, and the counselor used those to match a person with a career. Currently, however, a skilled counselor must do more than match. Vocational advisors still need to know the current and future job market, but they also must help each person recognize their particular skills and personality, values and aspirations (Rothausen & Henderson, 2019).

Career counselors motivate and guide, helping with searches and applications, role-playing for interviews, crafting résumés, suggesting additional education, gathering recommendations, encouraging applicants after rejection, analyzing offers after acceptance, negotiating benefits, and more.

Can you do this?

Intimacy Versus Isolation

intimacy versus isolation The sixth of Erikson's eight stages of development. Adults seek someone with whom to share their lives in an enduring and self-sacrificing commitment. Without such commitment, they risk profound aloneness and isolation.

In Erikson's theory, after achieving identity, people experience the crisis of **intimacy versus isolation**. He explains:

> The young adult, emerging from the search for and the insistence on identity, is eager and willing to fuse his identity with others. He is ready for intimacy, that is, the capacity to commit himself to concrete affiliations and partnerships and to develop the ethical strength to abide by such commitments, even though they call for significant sacrifices and compromises.
>
> *[Erikson, 1993a, p. 263]*

Other theorists have different words for the same human need: *affiliation, affection, interdependence, communion, belonging, love.* But all developmentalists agree that social connections are pivotal lifelong (Padilla-Walker et al., 2017). In adulthood, intimacy progresses from attraction to connection to commitment. Each relationship demands vulnerability and compromise, shattering the isolation caused by too much self-protection.

The social context may be particularly influential in emerging adulthood, a period called the "frontier" of efforts to prevent problems and foster positive growth (Schwartz & Petrova, 2019). Individual differences matter, but the trend is toward more social connections, with family relationships maintained and friendships established. That benefits emerging adults (Jorgensen & Nelson, 2018).

Friendship

Friends are crucial. Throughout life, friends increase each person's understanding via *self-expansion* (Aron et al., 2013); they enlarge perspective as people incorporate each other's experiences and ideas. Since fewer emerging adults today are married and have children, their social world can, and usually does, include friends who provide needed companionship and critical support. Unlike relatives, friends are selected for their ability to be loyal, trustworthy, supportive, and enjoyable—a mutual choice, not an obligatory one.

Thus, friends understand and comfort each other when romance turns sour, and they provide useful information about everything, from which college to attend to which socks to wear. For example, many adolescents are depressed about how their bodies appear. Interviews with 26-year-olds found that negative body image lifted in late adolescence and early adulthood, primarily because of friends who were reassuring (Gattario & Frisén, 2019).

People tend to make more friends during emerging adulthood than at any later period. They often use social media to extend and deepen friendships that begin face-to-face, becoming more aware of the day-to-day tribulations and celebrations of their friends.

Some older adults originally feared that increasing Internet use would diminish the number or quality of friendships. That fear has been proven false. Internet users in early adulthood tend to have more face-to-face friends, to know more about political and social events, and to advance in learning, all examples of self-expansion. Internet use is neither a boon nor a burden to emerging adults; the benefit or harm depends on the person (Blank & Lutz, 2018; Castellacci & Tveito, 2018; Hood et al., 2017).

Friendship patterns change with maturation. Young adults want many friends, and they work to gather them: befriending classmates, attending parties, speaking to strangers at concerts, on elevators, in parks, and so on. At about age 30, a switch begins, when quality becomes more important than quantity (Carmichael et al., 2015). Consequently, some friendships from early adulthood fade away, but others deepen. Social media helps with both processes.

There is a paradox here. Not only do young adults, on average, have more friends and acquaintances that adults of other ages, they also are more likely to be lonely. The only adult age group that seems to have higher rates of loneliness are adults over age 80 (Luhmann & Hawkley, 2016). Some say that social media is part of the reason, since emerging adults know how many "likes" they have, and since others post photos that showcase their most social moments, which makes some viewers feel inferior.

This leads us back to Erikson, who notes that each ongoing relationship demands some personal sacrifice, including vulnerability that brings deeper self-understanding and shatters the isolation of too much self-protection. To establish intimacy, the young adult must

> face the fear of ego loss in situations which call for self-abandon: in the
> solidarity of close affiliations [and] sexual unions, in close friendship and in
> physical combat, in experiences of inspiration by teachers and of intuition from
> the recesses of the self. The avoidance of such experiences . . . may lead to a
> deep sense of isolation and consequent self-absorption.

[*Erikson, 1963, pp. 163–164*]

THINK CRITICALLY: Can a person with many friends also be lonely?

Especially for Young Men Why would you want at least one close friend who is a woman? (see answer, page 467)

Wave/Cultura Images/Media Bakery

chee gin tan/E+/Getty Images

Being Intimate The word "intimacy" was traditionally a euphemism for sexual intercourse, but to developmentalists it is much more than that. Look closely at these two couples, one in Spain *(left)* and one in Malaysia *(right)*. Whether or not they are having sex does not matter: They are intimate in their touching, emotions, and even clothing.

Romantic Partners

In Europe in the Middle Ages, love, passion, and marriage were each considered to be distinct phenomena: Marriages were not usually for love. This was explained by Sternberg, who said that the current ideal of love includes three dimensions—passion, intimacy, and commitment (see **Table EP.3**).

For developmental reasons, this ideal is difficult to achieve. Passion seems to be sparked by unfamiliarity, uncertainty, and risk, all of which are diminished by the familiarity and security that contribute to intimacy and by the time needed for commitment.

Part of the problem is time. Over the years, passion may fade. Ideally, intimacy grows and commitment deepens as a couple together faces financial ups and downs, child-rearing challenges, and health crises. This pattern can occur for all types of couples—married, cohabiting, and remarried; same-gender and other-gender; young and old; in arranged or self-initiated relationships.

Emerging adults currently hope to find a lifelong partner, but they postpone that commitment until age 30 or so. More than in previous decades, some never find that person—the number of people who never marry is increasing. In former decades, divorce was a young person's experience: It almost always happened in the first years after the wedding. Lately, however, there has been a rise in "gray divorce" after age 50 (Crowley, 2019), and that rising divorce rate is one reason many emerging adults avoid marriage.

A report on the demography of marriage in the United States (Smock & Schwartz, 2020) notes that "despite high levels of aspiration for marriage" among young adults, an increasing number never marry.

That is one reason the birth rate is down, but it also is true that some people—especially those with relatively little education—choose to have a child but not to marry. The average age at first birth is after age 25 (the age when emerging adulthood is said to end) but is younger than the average age of first marriage.

Connecting with Others

When we consider the psychosocial needs of emerging adults, it seems that Erikson is right: Humans seek an intimate relationship with someone else. However, as you just read, cohort is crucial in how and when this need is met. Marriage is not the

TABLE EP.3			
Sternberg's Seven Forms of Love			
Present in the Relationship?			
Form of Love	**Passion**	**Intimacy**	**Commitment**
Liking	No	Yes	No
Infatuation	Yes	No	No
Empty love	No	No	Yes
Romantic love	Yes	Yes	No
Fatuous love	Yes	No	Yes
Companionate love	No	Yes	Yes
Consummate love	Yes	Yes	Yes

Information from Sternberg, 1988.

usual solution. One review cites a dramatic statistic: "In 1978, 59% of the population aged 18 to 34 was currently married. By 2018, this percentage was only 29%" (Sassler & Lichter, 2020).

Instead of traditional courtship and marriage, current young adults often use the internet to find possible partners. Virtually all emerging adults have smartphones and social media accounts, and some of them use this technology many times each day to connect with friends and possible partners. Use increased rapidly from 2010 on, but this rise has now stopped—in part because use is near universal (Hitlin, 2018).

Many young adults seeking romance join matchmaking sites that provide dozens of potential partners to meet and evaluate. Often physical attraction is the gateway to a relationship and passion rises, but intimacy and then commitment require much more.

In some ways, **cohabitation**, living with an unmarried partner, has replaced marriage. Cohabitation was relatively unusual 50 years ago: Only one in nine marriages in 1970 began with cohabitation. Now cohabitation is the norm (see **Figure EP.2**). About three of every four U.S. couples cohabit before marriage (Rosenfeld & Roesler, 2019).

Cohabitation rates vary by culture as well as cohort. Almost everyone in Canada and Europe cohabits at some point. Many people in Sweden, France, Jamaica, and Puerto Rico live with a partner for decades, sometimes all their lives, never marrying. In some other nations—including Japan, Ireland, and Italy—cohabitation is not yet the norm, although it is becoming increasingly common.

Although marriage rates are down and cohabitation is up in every demographic group, education increases the chance of marriage and marital childbearing. Cohabiting couples without college degrees have children about five times as often as cohabiting couples the same age who have graduated from college (Lundberg et al., 2016). The probable reason is *not* that college graduates know something that others do not. Instead, college graduates are more likely to marry before becoming parents, because they are more likely to have a steady, well-paying job.

One issue is whether cohabitation is an acceptable substitute for marriage, or a path toward a lifelong commitment, or neither. The research finds many similarities between cohabitation and marriage, but also some crucial differences. Cohabiting couples are more likely to end their relationship, are less likely to have children, and, if they do marry, are more likely to divorce (Sassler & Lichter, 2020).

As you have seen, historical circumstances have an impact on partnership formation. How will COVID-19 affect this? A study of 3,593 adults (age 18 and older) in 57 nations answered questions online about their relationship

Grigvovan/Shutterstock

Just Friends? This photo was taken in a public park in Isfala, Iran. Given that context, these two are probably more than friends.

● ● **Observation Quiz** What indicates that this is romance, not mere friendship? (see answer, page 467) ↑

cohabitation An arrangement in which a couple lives together in a committed romantic relationship but is not formally married.

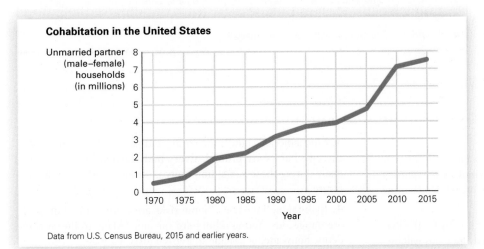

Cohabitation in the United States

Unmarried partner (male–female) households (in millions) — y-axis from 0 to 8

x-axis: Year — 1970, 1975, 1980, 1985, 1990, 1995, 2000, 2005, 2010, 2015

Data from U.S. Census Bureau, 2015 and earlier years.

FIGURE EP.2

More Together, Fewer Married As you see, the number of cohabiting male–female households in the United States has increased dramatically over the past decades. These numbers are an underestimate: Couples do not always tell the U.S. Census that they are living together, nor are cohabiters counted within their parents' households. Same-sex couples (not tallied until 2000) are also not included here.

(Balzarini et al., 2020). Most were living together; about a third were married. The sample was self-selected: people who responded to online requests.

When governments required lockdowns, conflict stress rose and satisfaction decreased. The most stressed were those who were relatively young, with children and financial insecurity. It seems that socializing with friends, having outside work, and spending time each day away from one's partner, strengthens romantic relationships. Thus, COVID-19 may increase divorce for married couples, separation for cohabiting ones, and domestic abuse for everyone. Of course, longitudinal studies are needed before science can confirm or refute this speculation.

Overall, it seems that contemporary life in general, and the COVID-19 pandemic in particular, are not favorable conditions for romantic relationships.

There is one bright side, however. It also seems that young adults are more connected in current times to their parents and other relatives than they were (Fingerman et al., 2020). That is the conclusion of extensive pre-COVID research, which finds that more young adults are living with a parent than with a romantic partner (most live with housemates or alone). Some suggest that the pandemic has increased young adults' connections to parents and grandparents, although again, more longitudinal research is needed.

 DATA CONNECTIONS: Technology and Romance: Trends for U.S. Adults examines how emerging adults find romantic partners.

Concluding Hopes

Fortunately, most emerging adults, like humans of all ages, have strengths as well as liabilities. Many survive risks, overcome substance abuse, think more deeply, combat loneliness, and deal with other problems through further education, maturation, friends, and family. If they postpone marriage, prevent parenthood, and avoid a set career until their identity is firmly established and their education is complete, they are ready for all the commitments and responsibilities ahead. I hope this is true.

WHAT HAVE YOU LEARNED?

1. What personality changes are likely and unlikely in emerging adulthood?

2. How do emerging adults meet their need for intimacy?

3. What is the difference between support from friends and parents during emerging adulthood?

4. How does cohabitation differ from marriage?

SUMMARY

Biological Universals

1. Emerging adulthood, from about age 18 to age 25, is a newly recognized period of development characterized by postponing parenthood, marriage, and career commitment, while attaining additional education.

2. COVID-19 affects adults in many ways, not usually for the better. Research needs confirmation to know exactly how this will be.

3. Most emerging adults are strong and healthy. Good eating and exercise habits are crucial. Every body system functions optimally during these years; immunity is strong; death from disease is rare.

4. Risk-taking increases. Some risks are worthwhile, but many are dangerous. Young adults are far more likely to die of accidents, homicide, or suicide than from diseases.

Cognitive Development

5. Many researchers believe that the complex and conflicting demands of adult life produce a new cognitive perspective. Postformal thinking is characterized by thinking that is flexible and practical, combing emotion and logic, intuition and analysis.

6. How much college education advances cognition in current times is debatable, although research over the past several decades indicates not only that college graduates are wealthier and healthier than other adults but also that they think at a more advanced level.

7. Contemporary college students are far more diverse than college students were a few decades ago. Learning from people of different perspectives can advance cognition.

Psychosocial Development

8. Personality tends to be quite stable lifelong, but emerging adulthood is the time when change is most likely to occur. The usual pattern is improvement, according to goals set by the individual.

9. Friends are particularly crucial in emerging adulthood. Many emerging adults use social networking and matchmaking sites on the Internet to expand and deepen their friendship circles and mating options.

10. Commitment to a partner is increasingly unusual during early adulthood, as both parenthood and marriage are often postponed. Cohabitation, and relationships with parents, are both more common now than a few decades ago.

KEY TERMS

emerging adulthood (p. 455)
postformal thought (p. 457)

objective thought (p. 457)
subjective thought (p. 457)

WEIRD (p. 458)
intimacy versus isolation (p. 462)

cohabitation (p. 465)

APPLICATIONS

1. Describe an incident during your emerging adulthood when taking a risk could have led to disaster. What were your feelings at the time? What would you do if you knew that a child of yours was about to do the same thing?

2. Read a biography or autobiography that includes information about the person's thinking from adolescence through adulthood. How did personal experiences, education, and maturation affect the person's thinking?

3. Statistics on cohort and culture in students and in colleges are fascinating, but only a few are reported here. Compare your nation, state, or province with another. Analyze the data and discuss causes and implications of differences.

4. Talk to three people you would expect to have contrasting views on love and marriage (differences in age, gender, upbringing, experience, and religion might affect attitudes). Ask each the same questions and then compare their answers.

Especially For ANSWERS

Response for Someone Who Has to Make an Important Decision (from p. 458): Both are necessary. Mature thinking requires a combination of emotions and logic. To make sure you use both, take your time (don't just act on your first impulse) and talk with people you trust. Ultimately, you will have to live with your decision, so do not ignore either intuitive or logical thought.

Response for Young Men (from p. 463): Not for sex! Women friends are particularly responsive to deep conversations about family relationships, personal weaknesses, and emotional confusion. But women friends might be offended by sexual advances, bragging, or advice giving. Save these for a future romance.

Observation Quiz ANSWERS

Answer to Observation Quiz (from p. 456) Kyiv, in Ukraine. As evident in the globalization of emerging adulthood, the same challenges are everywhere.

Answer to Observation Quiz (from p. 465): Note body position, hands, and her facial expression.

Appendix
More About
Research Methods

This appendix explains how to learn about any topic. One of the most important lessons from the 2020 COVID-19 pandemic is that we need accurate information, reported honestly and analyzed carefully, to protect our mental and physical health. Science is essential to keep speculation and wishful thinking from destroying us.

Remember that almost no conclusion is entirely certain, now and forever. That is why the scientific method requires testing every hypothesis, basing conclusion on evidence, and reporting methods and statistics so the others can confirm, dispute, and replicate.

One of the most important aspects is in the selection of the participants in a study. Ideally, they are diverse in gender, ethnicity, and economic backgrounds, but if not, the biases and limitations of a restricted sample must be acknowledged.

Beyond that, when doing research connected with your own study in learning about human development, here are several suggestions.

Make It Personal

Think about your life, observe your behavior, and watch the people around you. Pay careful attention to details of expression, emotion, and behavior. The more you see, the more fascinated, curious, and reflective you will become. Ask questions and listen carefully and respectfully to what other people say regarding development.

Whenever you ask specific questions as part of an assignment, remember that observing ethical standards (see Chapter 1) comes first. *Before* you interview anyone, inform the person of your purpose and assure them of confidentiality. Promise not to identify the person in your report (use a pseudonym) and do not repeat any personal details that emerge in the interview to anyone (friends or strangers). Your instructor will provide further ethical guidance. If you might publish what you've learned, get in touch with your college's Institutional Review Board (IRB).

Read the Research

No matter how deeply you think about your own experiences, and no matter how intently you listen to others whose background is unlike yours, you also need to read scholarly published work in order to fully understand any topic that interests you. Be skeptical about magazine or newspaper reports; some are bound to be simplified, exaggerated, or biased.

Professional Journals and Books

Part of the process of science is that conclusions are not considered solid until they are corroborated in many studies, which means that you should consult several sources on any topic. Five journals in human development are:

- *Developmental Psychology* (published by the American Psychological Association)
- *Child Development* (Society for Research in Child Development)
- *Developmental Review* (Elsevier)
- *Human Development* (Karger)
- *Developmental Science* (Wiley)

These journals differ in the types of articles and studies they publish, but all are well respected and peer-reviewed, which means that other scholars review each article submitted and recommend that it be accepted, rejected, or revised. Every article includes references to other recent work.

Also look at journals that specialize in longer reviews from the perspective of a researcher.

- *Child Development Perspectives* (from Society for Research in Child Development)
- *Perspectives on Psychological Science* (This is published by the Association for Psychological Science. APS publishes several excellent journals, none specifically on development but every issue has at least one article that is directly relevant.)

Beyond these seven are literally thousands of other professional journals, each with a particular perspective or topic, including many in sociology, family studies, economics, medicine, demography, education, and more. To judge them, look for journals that are peer-reviewed. Also consider the following details: the background of the author (research funded by corporations tends to favor their products); the nature of the publisher (professional organizations, as in the first two journals above, protect their reputations); and how long the journal has been published (the volume number tells you that). Some interesting work does not meet these criteria, so be careful before believing what you read.

Many *books* cover some aspect of development. Single-author books are likely to present only one viewpoint. That view may be insightful, but it is limited. You might consult a *handbook*, which is a book that includes many authors and many topics. One good handbook in development, now in its seventh edition (a sign that past scholars have found it useful) is:

- *Handbook of Child Psychology and Developmental Science* (7th ed.), edited by Richard M. Lerner, 2015, Hoboken, NJ: Wiley.

This handbook is updated about every five years, so a new edition might be out soon. Check on it, and use the newest one. Dozens of other good handbooks are available, many of which focus on a particular age, perspective, or topic.

The Internet

The *internet* is a mixed blessing, useful to every novice and experienced researcher but dangerous as well. Every library worldwide and most homes in North America, Western Europe, and East Asia have computers that provide access to journals and other information. If you're doing research in a library, ask for help from the librarians; many of them can guide you in the most effective ways to conduct online searches. In addition, other students, friends, and even strangers can be helpful.

Virtually everything is on the internet, not only massive national and international statistics but also accounts of very personal experiences. Photos, charts, quizzes, ongoing experiments, newspapers from around the world, videos, and much more are available at a click. Every journal has a Web site, with tables of contents, abstracts, and sometimes full texts. (An abstract gives the key findings; for the full text, most colleges and universities have access. Again, ask librarians for help.)

Unfortunately, you can spend many frustrating hours sifting through information that is useless, trash, or tangential. *Directories* (which list general topics or areas and then move you step by step in the direction you choose) and *search engines* (which give you all the sites that use a particular word or words) can help you select appropriate information. Each directory or search engine provides somewhat different lists; none provides only the most comprehensive and accurate sites. Sometimes organizations pay, or find other ways, to make their links appear first, even though they are biased. With experience and help, you will find quality on the internet, but you will also encounter some junk no matter how experienced you are.

Anybody can put anything online, regardless of its truth or fairness, so you need a very critical eye. Make sure you have several divergent sources for every "fact" you find; consider who provided the information and why. Every controversial issue has sites that forcefully advocate opposite viewpoints, sometimes with biased statistics and narrow perspectives.

Here are four internet sites that are quite reliable:

- *childtrends.org* A leading U.S. research organization focusing on improving children's lives. Their site contains a wealth of data and evidence-based research.
- *childdevelopmentinfo.com* Child Development Institute. A useful site, with links and articles on child development and information on common childhood psychological disorders.
- *eric.ed.gov* Education Resources Information Center (ERIC). Provides links to many education-related sites and includes brief descriptions of each.
- *www.cdc.gov/nchs/hus.htm* The National Center for Health Statistics issues an annual report on health trends, called *Health, United States*.

Every source—you, your interviewees, journals, books, and the internet—is helpful. Do not depend on any particular one. Avoid plagiarism and prejudice by citing every source and noting objectivity, validity, and credibility. Your own analysis, opinions, words, and conclusions are crucial, backed up by science.

Additional Terms and Concepts

As emphasized throughout the text, the study of development is a science. Social scientists spend years in graduate school, studying methods and statistics. Chapter 1 touches on some of these matters (observation and experiments; correlation and causation; independent and dependent variables; experimental and control groups; cross-sectional, longitudinal, and cross-sequential research), but there is much more. A few additional aspects of research are presented here to help you evaluate research wherever you find it.

Who Participates?

The entire group of people about whom a scientist wants to learn is called a **population**. Generally, a research population is quite large—not usually the world's entire population of more than 7 billion, but for statistics on birthweight or unwed mothers of all of the 3,788,235 babies born in the United States in 2019.

The particular individuals who are studied in a specific research project are called the **participants**. They are usually a **sample** of the population. Ideally, the participants are a **representative sample**, that is, a sample that reflects the population. Every peer-reviewed, published study reports details on the sample.

Selection of the sample is crucial. People who volunteer, or people who have telephones, or people who have some particular condition are not a *random sample*; in a random sample, everyone in a particular population is equally likely to be selected. To avoid *selection bias*, some studies are *prospective*, beginning with an entire cluster of people (for instance, every baby born on a particular day) and then tracing the development of some particular characteristic. Ideally, the sample is diverse in gender, ethnicity, and other ways: If it is not, the bias must be explained.

For example, prospective studies find the antecedents of heart disease, or child abuse, or high school dropout rates—all of which are much harder to find if the study is *retrospective,* beginning with those who had heart attacks, experienced abuse, or left school. Thus, although retrospective research finds that most high school dropouts say they disliked school, prospective research finds that some who like school still decide to drop out and then later say they hated school, while others dislike school but stay to graduate. Prospective research discovers how many students are in these last two categories; retrospective research on people who have already dropped out does not.

Research Design

Every researcher begins not only by formulating a hypothesis but also by learning what other scientists have discovered about the topic in question and what methods might be useful and ethical in designing research. Often they include measures to prevent inadvertently finding only the results they expect. For example, the people who actually gather the data may not know the purpose of the research. Scientists say that these data gatherers are **blind** to the hypothesized outcome. Participants are sometimes "blind" as well, because otherwise they might, for instance, respond the way they think they should.

Another crucial aspect of research design is to define exactly what is to be studied. Researchers establish an **operational definition** of whatever phenomenon they will be examining, defining each variable by describing specific, observable behavior. This is essential in quantitative research, but it is also useful in qualitative research.

For example, if a researcher wants to know when babies begin to walk, does walking include steps taken while holding on? Is one unsteady step enough? Some parents say yes, but the usual operational definition of *walking* is "takes at least three steps without holding on." This operational definition allows comparisons worldwide, making it possible to discover, for example, that well-fed African babies tend to walk earlier than well-fed European babies.

When emotions or personality traits are studied, operational definitions are difficult to formulate but crucial for interpretation of results. How should *aggression* or *sharing* or *shyness* be defined? Lack of an operational definition leads to contradictory results. For instance, critics report that infant day care makes children more aggressive, but advocates report that it makes them more assertive and outgoing. In this case, both may be seeing the same behavior but defining it differently.

---

population The entire group of individuals who are of particular concern in a scientific study, such as all the children of the world or all newborns who weigh less than 3 pounds.

participants The people who are studied in a research project. Participants is the term now used in psychology; other disciplines still call these people "subjects."

sample A group of individuals drawn from a specified population. A sample might be the low-birthweight babies born in four particular hospitals that are representative of all hospitals.

representative sample A group of research participants who reflect the relevant characteristics of the larger population whose attributes are under study.

blind The condition of data gatherers (and sometimes participants, as well) who are deliberately kept ignorant of the purpose of the research so that they cannot unintentionally bias the results.

operational definition A description of the specific, observable behavior that will constitute the variable that is to be studied, so that any reader will know whether that behavior occurred or not. Operational definitions may be arbitrary (e.g., an IQ score at or above 130 is operationally defined as "gifted"), but they must be precise.

For any scientist, operational definitions are crucial, and studies usually include descriptions of how they measured attitudes or behavior.

Reporting Results

You already know that results should be reported in sufficient detail so that another scientist can analyze the conclusions and replicate the research. Various methods, populations, and research designs may produce divergent conclusions. For that reason, handbooks, some journals, and some articles are called *reviews*: They summarize past research. Often, when studies are similar in operational definitions and methods, the review is a **meta-analysis**, which combines the findings of many studies to present an overall conclusion.

Table App.1 describes some other statistical measures. One of them is *statistical significance*, which indicates whether or not a particular result could have occurred by chance.

A crucial statistic is **effect size**, a way of measuring how much impact one variable has on another. Effect size ranges from 0 (no effect) to 1 (total transformation, never found in actual studies). Effect size may be particularly important when the sample size is large, because a large sample often leads to highly "significant" results (results that are unlikely to have occurred by chance) that have only a tiny effect on the variable of interest.

Hundreds of statistical measures are used by developmentalists. Often the same data can be presented in many ways: Some scientists examine statistical analysis intently before they accept conclusions as valid. A specific example

meta-analysis A technique of combining results of many studies to come to an overall conclusion. Meta-analysis is powerful, in that small samples can be added together to lead to significant conclusions, although variations from study to study sometimes make combining them impossible.

effect size A way of indicating statistically how much of an impact the independent variable in an experiment had on the dependent variable.

TABLE App.1	
Statistical Measures Often Used to Analyze Search Results	
Measure	**Use**
Effect size	There are many kinds, but the most useful in reporting studies of development is called *Cohen's d*, which can indicate the power of an intervention. An effect size of 0.2 is called small, 0.5 moderate, and 0.8 large.
Significance	Indicates whether the results might have occurred by chance. If chance would produce the results only 5 times in 100, that is significant at the 0.05 level; once in 100 times is 0.01; once in 1,000 is 0.001.
Cost-benefit analysis	Calculates how much a particular independent variable costs versus how much it saves. This is useful for analyzing public spending, such as finding that preschool education programs or preventative health measures save money.
Odds ratio	Indicates how a particular variable compares to a standard, set at 1. For example, one study found that although less than 1 percent of all child homicides occurred at school, the odds were similar for public and private schools. The odds of it in high schools, however, were 18.47 times that of elementary or middle schools (set at 1.0) (MMWR, January 18, 2008).
Factor analysis	Hundreds of variables could affect any given behavior. In addition, many variables (such as family income and parental education) overlap. To take this into account, analysis reveals variables that can be clustered together to form a factor, which is a composite of many variables. For example, SES might become one factor, child personality another.
Meta-analysis	A "study of studies." Researchers use statistical tools to synthesize the results of previous, separate studies. Then they analyze the accumulated results, using criteria that weigh each study fairly. This approach improves data analysis by combining studies that were too small, or too narrow, to lead to solid conclusions.

involved methods to improve students' writing ability between grades 4 and 12. A meta-analysis found that many methods of writing instruction have a significant impact, but effect size is much larger for some methods (teaching strategies and summarizing) than for others (prewriting exercises and studying models). For teachers, this statistic is crucial, for they want to know what has a big effect, not merely what is better than chance (significant).

Numerous articles published in the past decade are meta-analyses that combine similar studies to search for general trends. Often effect sizes are also reported, which is especially helpful for meta-analyses since standard calculations almost always find some significance if the number of participants is in the thousands.

An added problem is the *file drawer problem*—that studies without significant results tend to be filed away rather than published. Thus, an accurate effect size may be much smaller than the published meta-analysis finds, or may be nonexistent. For this reason, replication is an important step.

Overall, then, designing and conducting valid research is complex yet crucial. Remember that with your own opinions: As this appendix advises, it is good to "make it personal," but do not stop there.

Glossary

23rd pair The chromosome pair that, in humans, determines sex. The other 22 pairs are autosomes, inherited equally by males and females.

A

accommodation The restructuring of old ideas to include new experiences.

adolescence-limited offender A person who breaks the law as a teenage but whose criminal activity stops by age 20.

adolescent egocentrism A characteristic of adolescent thinking that leads young people (ages 10 to 13) to focus on themselves to the exclusion of others.

adrenal glands Two glands, located above the kidneys, that respond to the pituitary, producing hormones.

age of viability The age (about 22 weeks after conception) at which a fetus might survive outside the mother's uterus if specialized medical care is available.

allele A variation that makes a gene different in some way from other genes for the same characteristics. Many genes never vary; others have several possible alleles.

allocare Literally, "other-care"; the care of children by people other than the biological parents.

amygdala A tiny brain structure that registers emotions, particularly fear and anxiety.

analytic thought Thought that results from analysis, such as a systematic ranking of pros and cons, risks and consequences, possibilities and facts. Analytic thought depends on logic and rationality.

animism The belief that natural objects and phenomena are alive, moving around, and having sensations and abilities that are human-like.

anorexia nervosa An eating disorder characterized by self-starvation. Affected individuals voluntarily undereat and often overexercise, depriving their vital organs of nutrition. Anorexia can be fatal.

anoxia A lack of oxygen that, if prolonged, can cause brain damage or death.

antipathy Feelings of dislike or even hatred for another person.

antisocial behavior Actions that are deliberately hurtful or destructive to another person.

Apgar scale A quick assessment of a newborn's health, from 0 to 10. Below 5 is an emergency—a neonatal pediatrician is summoned immediately. Most babies are at 7, 8, or 9—almost never a perfect 10.

apprenticeship in thinking Vygotsky's term for how cognition is stimulated and developed in people by more skilled members of society.

assimilation The reinterpretation of new experiences to fit into old ideas.

asthma A chronic disease of the respiratory system in which inflammation narrows the airways from the nose and mouth to the lungs, causing difficulty in breathing. Signs and symptoms include wheezing, shortness of breath, chest tightness, and coughing.

attachment According to Ainsworth, "an affectional tie" that an infant forms with a caregiver—a tie that binds them together in space and endures over time.

attention-deficit/hyperactivity disorder (ADHD) A condition characterized by a persistent pattern of inattention and/or by hyperactive or impulsive behaviors; ADHD interferes with a person's functioning or development.

authoritarian parenting An approach to child rearing that is characterized by high behavioral standards, strict punishment of misconduct, and little communication from child to parent.

authoritative parenting An approach to child rearing in which the parents set limits and enforce rules but are flexible and listen to their children.

autism spectrum disorder (ASD) A developmental disorder marked by difficulty with social communication and interaction—including difficulty seeing things from another person's point of view—and restricted, repetitive patterns of behavior, interests, or activities.

automatization A process in which repetition of a sequence of thoughts and actions makes the sequence routine so that it no longer requires conscious thought.

autonomy versus shame and doubt Erikson's second crisis of psychosocial development. Toddlers either succeed or fail in gaining a sense of self-rule over their actions and their bodies.

axon A fiber that extends from a neuron and transmits electrochemical impulses from that neuron to the dendrites of other neurons.

B

babbling An infant's repetition of certain syllables, such as *ba-ba-ba*, that begins when babies are between 6 and 9 months old.

bed-sharing When two or more people sleep in the same bed.

behavioral teratogens Agents and conditions that can harm the prenatal brain, impairing the future child's intellectual and emotional functioning.

behaviorism A grand theory of human development that studies observable behavior. Behaviorism is also called *learning theory* because it describes the laws and processes by which behavior is learned.

bilingual education A strategy in which school subjects are taught in both the learner's original language and the second (majority) language.

binge eating disorder Frequent episodes of uncontrollable overeating to the point that the stomach hurts. Usually the person feels shame and guilt but is unable to stop.

binocular vision The ability to focus the two eyes in a coordinated manner in order to see one image.

body image A person's idea of how his or her body looks.

Brazelton Neonatal Behavioral Assessment Scale (NBAS) A test that is often administered to newborns; it measures responsiveness and records 46 behaviors, including 20 reflexes.

bulimia nervosa An eating disorder characterized by binge eating and subsequent purging, usually by induced vomiting and/or use of laxatives.

bullying Repeated, systematic efforts to inflict harm on other people through physical, verbal, or social attack on a weaker person.

bullying aggression Unprovoked, repeated physical or verbal attack, especially on victims who are unlikely to defend themselves.

bully-victim Someone who attacks others and who is attacked as well. (Also called *provocative victims* because they do things that elicit bullying.)

C

carrier A person whose genotype includes a gene that is not expressed in the phenotype. The carried gene occurs in half of the carrier's gametes and thus is passed on to half of the carrier's children. If such a gene is inherited from both parents, the characteristic appears in the phenotype.

centration A characteristic of preoperational thought in which a young child focuses (centers) on one idea, excluding all others.

cephalocaudal Growth and development that occurs from the head down.

cerebral palsy A disorder that results from damage to the brain's motor centers. People with cerebral palsy have difficulty with muscle control, so their speech and/or body movements are impaired.

cesarean section (c-section) A surgical birth in which incisions through the mother's abdomen and uterus allow the fetus to be removed quickly instead of being delivered through the vagina. (Also called simply *section*.)

child abuse Deliberate action that is harmful to a child's physical, emotional, or sexual well-being.

child culture The idea that each group of children has games, sayings, clothing styles, and superstitions that are not common among adults, just as every culture has distinct values, behaviors, and beliefs.

child maltreatment Intentional harm to, or avoidable endangerment of, anyone under 18 years of age.

child neglect Failure to meet a child's basic physical, educational, or emotional needs.

child sexual abuse Any erotic activity that arouses an adult and excites, shames, or confuses a child, whether or not the victim protests and whether or not genital contact is involved.

childhood obesity In a child, having a BMI above the 95th percentile, according to the U.S. Centers for Disease Control's 1980 standards for children of a given age.

chromosome One of the 46 molecules of DNA (in 23 pairs) that virtually each cell of the human body contains and that, together, contain all the genes. Other species have more or fewer chromosomes.

chronosystem In Bronfenbrenner's ecological approach, the impact of historical conditions (wars, inventions, policies) on the development of people who live in that era.

circadian rhythm A day–night cycle of biological activity that occurs approximately every 24 hours.

classical conditioning The learning process in which a meaningful stimulus (such as the smell of food to a hungry animal) is connected with a neutral stimulus (such as the sound of a tone) that had no special meaning before conditioning. (Also called *respondent conditioning*.)

classification The logical principle that things can be organized into groups (or categories or classes) according to some characteristic that they have in common.

cluster suicides Several suicides committed by members of a group within a brief period.

coercive joining When others strongly encourage someone to join in their activity, usually when the pressure is being included or excluded in a group and the activity is not approved by authorities (e.g., drug use, bullying.)

cognitive equilibrium In cognitive theory, a state of mental balance in which people are not confused because they can use their existing thought processes to understand current experiences and ideas.

cognitive theory A grand theory of human development that focuses on changes in how people think over time. According to this theory, our thoughts shape our attitudes, beliefs, and behaviors.

cohabitation An arrangement in which a couple lives together in a committed romantic relationship but is not formally married.

cohort People born within the same historical period who therefore move through life together, experiencing the same events, new technologies, and cultural shifts at the same ages. For example, the effect of the Internet varies depending on what cohort a person belongs to.

comorbid Refers to the presence of two or more unrelated disease conditions at the same time in the same person.

concrete operational thought Piaget's term for the ability to reason logically about direct experiences and perceptions.

conservation The principle that the amount of a substance remains the same (i.e., is conserved) even when its appearance changes.

control processes Mechanisms (including selective attention, metacognition, and emotional regulation) that combine memory, processing speed, and knowledge to regulate the analysis and flow of information within the information-processing system. (Also called *executive processes*.)

conventional moral reasoning Kohlberg's second level of moral reasoning, emphasizing social rules.

copy number variations Genes with various repeats or deletions of base pairs.

corporal punishment Disciplinary techniques that hurt the body (*corpus*) of someone, from spanking to serious harm, including death.

corpus callosum A long, thick band of nerve fibers that connects the left and right hemispheres of the brain and allows communication between them.

correlation A number between +1.0 and −1.0 that indicates the degree of relationship between two variables, expressed in terms of the likelihood that one variable will (or will not) occur when the other variable does (or does not). A correlation indicates only that two variables are somehow related, not that one variable causes the other to occur.

cortex The outer layers of the brain in humans and other mammals. Most thinking, feeling, and sensing involves the cortex.

cortisol The primary stress hormone; fluctuations in the body's cortisol level affect human emotions.

co-sleeping A custom in which parents and their children (usually infants) sleep together in the same room.

couvade Symptoms of pregnancy and birth experienced by fathers.

critical period A time when something *must* happen for normal development to occur.

cross-sectional research A research design that compares groups of people who differ in age but are similar in other important characteristics.

cross-sequential research A hybrid research design in which researchers first study several groups of people of different ages (a cross-sectional approach) and then follow those groups over the years (a longitudinal approach). (Also called *cohort-sequential research* or *time-sequential research*.)

culture A system of shared beliefs, norms, behaviors, and expectations that persist over time and prescribe social behavior and assumptions.

cyberbullying When people try to harm others via electronic means, such as social media and cell phone photos or texts.

D

deductive reasoning Reasoning from a general statement, premise, or principle, through logical steps, to figure out (deduce) specifics. (Also called *top-down reasoning*.)

dendrite A fiber that extends from a neuron and receives electrochemical impulses transmitted from other neurons via their axons.

deoxyribonucleic acid (DNA) The chemical composition of the molecules that contain the genes, which are the chemical instructions for cells to manufacture various proteins.

dependent variable In an experiment, the variable that may change as a result of whatever new condition or treatment the experimenter adds. Thus, the dependent variable *depends* on the independent variable.

developmental psychopathology The field that uses insights into typical development to understand and remediate developmental disorders.

developmental theory A group of ideas, assumptions, and generalizations that interpret and illuminate thousands of observations about human growth. A developmental theory provides a framework for explaining the patterns and problems of development.

difference-equals-deficit error The mistaken belief that a deviation from some norm is necessarily inferior to behavior or characteristics that meet the standard.

differential susceptibility The idea that people vary in how sensitive they are to particular experiences. Often such differences are genetic, which makes some people affected "for better or for worse" by life events. (Also called *differential sensitivity*.)

disorganized attachment A type of attachment that is marked by an infant's inconsistent reactions to the caregiver's departure and return.

distal parenting Caregiving practices that involve remaining distant from the baby, providing toys, food, and face-to-face communication with minimal holding and touching.

dizygotic (DZ) twins Twins who are formed when two separate ova are fertilized by two separate sperm at roughly the same time. (Also called *fraternal* twins.)

dominant Reflected in the phenotype. Dominant genes have more influence on traits than recessive genes.

doula A woman who helps with the birth process. Traditionally in Latin America, a doula was the only professional who attended childbirth. Now doulas are likely to arrive at the woman's home during early labor and later work alongside a hospital's staff.

Down syndrome A condition in which a person has 47 chromosomes instead of the usual 46, with 3 rather than 2 chromosomes at the 21st site. People with Down syndrome typically have distinctive characteristics, including unusual facial features, heart abnormalities, and language difficulties. (Also called trisomy-21.)

dual processing The notion that two networks exist within the human brain, one for emotional processing of stimuli and one for analytical reasoning.

dynamic-systems approach A view of human development as an ongoing, ever-changing interaction between the physical, cognitive, and psychosocial influences. The crucial understanding is that development is never static but is always affected by, and affects, many systems of development.

dyscalculia Unusual difficulty with math, probably originating from a distinct part of the brain.

dyslexia Unusual difficulty with reading; thought to be the result of some neurological underdevelopment.

E

eclectic perspective The approach taken by most developmentalists, in which they apply aspects of each of the various theories of development rather than adhering exclusively to one theory.

ecological-systems approach A perspective on human development that considers all of the influences from the various contexts of development. (Later renamed *bioecological theory*.)

egocentrism Piaget's term for children's tendency to think about the world entirely from their own personal perspective.

embryo The name for a developing human organism from about the third week through the eighth week after conception.

embryonic period The stage of prenatal development from approximately the third week through the eighth week after conception, during which the basic forms of all body structures, including internal organs, develop.

emerging adulthood The period of life between the ages of 18 and 25. Emerging adulthood is now widely thought of as a distinct developmental stage.

emotional regulation The ability to control when and how emotions are expressed.

empathy The ability to understand the emotions and concerns of another person, especially when they differ from one's own.

empirical evidence Evidence that is based on observation, experience, or experiment; not theoretical.

English as a Second Language (ESL) A U.S. approach to teaching English that gathers all of the non-English speakers together and provides intense instruction in English. Students' first languages are never used; the goal is to prepare students for regular classes in English.

English Language Learners (ELLs) Children in the United States whose proficiency in English is low—usually below a cutoff score on an oral or written test. Many children who speak a non-English language at home are also capable in English; they are *not* ELLs.

epigenetics The study of how environmental factors affect genes and genetic expression—enhancing, halting, shaping, or altering the expression of genes.

equifinality A basic principle of developmental psychopathology that holds that one symptom can have many causes.

estradiol A sex hormone, considered the chief estrogen. Females produce much more estradiol than males do.

ethnic group People whose ancestors were born in the same region and who often share a language, culture, and religion.

evolutionary theory When used in human development, the idea that many current human emotions and impulses are a legacy from thousands of years ago.

executive function The cognitive ability to organize and prioritize the many thoughts that arise from the various parts of the brain, allowing the person to anticipate, strategize, and plan behavior.

exosystem In Bronfenbrenner's ecological approach, the community institutions that affect the immediate contexts, such as churches and temples, schools and colleges, hospitals and courts.

experience-dependent Brain functions that depend on particular, variable experiences and therefore may or may not develop in a particular infant.

experience-expectant Brain functions that require certain basic common experiences (which an infant can be expected to have) in order to develop normally.

experiment A research method in which the scientist tries to determine the cause-and-effect relationship between two variables by manipulating one (called the *independent variable*) and then observing and recording the ensuing changes in the other (called the *dependent variable*).

extended family A family of relatives in addition to the nuclear family, usually three or more generations living in one household.

extremely low birthweight (ELBW) A body weight at birth of less than 1,000 grams (2 pounds, 3 ounces).

extrinsic motivation A drive, or reason to pursue a goal, that arises from the wish to have external rewards, perhaps by earning money or praise.

F

false positive The result of a laboratory test that reports something as true when in fact it is not true. This can occur for pregnancy tests, when a woman might not be pregnant even though the test says she is, or during pregnancy, when a problem is reported that actually does not exist.

familism The belief that family members should support one another, sacrificing individual freedom and success, if necessary, in order to preserve family unity and protect the family.

family function The way a family works to meet the needs of its members. Children need families to provide basic material necessities, to encourage learning, to help them develop self-respect, to nurture friendships, and to foster harmony and stability.

family structure The legal and genetic relationships among relatives living in the same home. Possible structures include nuclear family, extended family, stepfamily, single-parent family, and many others.

fast-mapping The speedy and sometimes imprecise way in which children learn new words by tentatively placing them in mental categories according to their perceived meaning.

fetal alcohol syndrome (FAS) A cluster of birth defects, including abnormal facial characteristics, slow physical growth, and reduced intellectual ability, that may occur in the fetus of a woman who drinks alcohol while pregnant.

fetal period The stage of prenatal development from the ninth week after conception until birth, during which the fetus gains about 7 pounds (more than 3,000 grams) and organs become more mature, gradually able to function on their own.

fetus The name for a developing human organism from the start of the ninth week after conception until birth.

fine motor skills Physical abilities involving small body movements, especially of the hands and fingers, such as drawing and picking up a coin. (The word *fine* here means "small.")

fixed mindset An approach to understanding intelligence that sees ability as an innate entity, a fixed quantity present at birth. Those who hold this view do not believe that effort enhances achievement.

Flynn effect The rise in average IQ scores that has occurred over the decades in many nations.

focus on appearance A characteristic of preoperational thought in which a young child ignores all attributes that are not apparent.

foreclosure Erikson's term for premature identity formation, when adolescents adopt their parents' or society's roles and values without questioning or analysis.

formal operational thought In Piaget's theory, the fourth and final stage of cognitive development, characterized by more systematic logical thinking and by the ability to understand and systematically manipulate abstract concepts.

foster care A legal, publicly supported system in which a maltreated child is removed from the parents' custody and entrusted to another adult or family, who is reimbursed for expenses incurred in meeting the child's needs.

fragile X syndrome A genetic disorder in which part of the X chromosome seems to be attached to the rest of it by a very thin string of molecules. The cause is a single gene that has more than 200 repetitions of one triplet.

G

g (general intelligence) The idea of *g* assumes that intelligence is one basic trait, underlying all cognitive abilities. According to this concept, people have varying levels of this general ability.

gamete A reproductive cell. In humans it is a sperm or an ovum.

gender binary The idea that there are only two (*bi-*) genders, male and female, and that they are opposites. This idea precludes intersex, gender overlap, and gender non-conformity.

gender differences Differences in male and female roles, behaviors, clothes, and so on that arise from society, not biology.

gender identity A person's acceptance of the roles and behaviors that society associates with a particular gender.

gender schema A child's cognitive concept or general belief about male and female differences.

gene A small section of a chromosome; the basic unit for the transmission of heredity. A gene consists of a string of chemicals that provide instructions for the cell to manufacture certain proteins.

generational forgetting The idea that each new generation forgets what the previous generation learned. As used here, the term refers to knowledge about the harm drugs can do.

genome The full set of genes that are the instructions to make an individual member of a certain species.

genotype An organism's entire genetic inheritance, or genetic potential.

germinal period The first two weeks of prenatal development after conception, characterized by rapid cell division and the beginning of cell differentiation.

gonads The paired sex glands (ovaries in females, testicles in males). The gonads produce hormones and mature gametes.

grammar All of the methods—word order, verb forms, and so on—that languages use to communicate meaning, apart from the words themselves.

gross motor skills Physical abilities involving large body movements, such as walking and jumping. (The word *gross* here means "big.")

growth mindset An approach to understanding intelligence that holds that intelligence grows incrementally, and thus can be increased by effort. Those who subscribe to this view believe they can master whatever they seek to learn if they pay attention, participate in class, study, complete their homework, and so on.

growth spurt The relatively sudden and rapid physical growth that occurs during puberty. Each body part increases in size on a schedule: Weight usually precedes height, and growth of the limbs precedes growth of the torso.

guided participation The process by which people learn from others who guide their experiences and explorations.

H

Head Start A federally funded early-childhood intervention program for low-income children of preschool age.

head-sparing A biological mechanism that protects the brain when malnutrition disrupts body growth. The brain is the last part of the body to be damaged by malnutrition.

herd immunity The level of immunity necessary in a population (the herd) in order to stop transmission of infectious diseases. The rate is usually above 90 percent, and even higher for very infectious diseases. Newborns, and people with certain diseases (e.g., cancer patients taking immunosuppressant drugs), cannot be vaccinated; herd immunity protects them.

heritability A statistic that indicates what percentage of the variation in a particular trait within a particular population, in a particular context and era, can be traced to genes.

heterozygous Referring to two genes of one pair that differ in some way. Typically one allele has only a few base pairs that differ from the other member of the pair.

hidden curriculum The unofficial, unstated, or implicit patterns within a school that influence what children learn. For instance, teacher background, organization of the play space, and tracking are all part of the hidden curriculum — not formally prescribed, but instructive to the children.

high-stakes test An evaluation that is critical in determining success or failure. If a single test determines whether a student will graduate or be promoted, it is a high-stakes test.

hippocampus A brain structure that is a central processor of memory, especially memory for locations.

holophrase A single word that is used to express a complete, meaningful thought.

homozygous Referring to two genes of one pair that are exactly the same in every letter of their code. Most gene pairs are homozygous.

HPA (hypothalamus–pituitary–adrenal) axis A sequence of hormone production originating in the hypothalamus and moving to the pituitary and then to the adrenal glands.

HPG (hypothalamus–pituitary–gonad) axis A sequence of hormone production originating in the hypothalamus and moving to the pituitary and then to the gonads.

Human Genome Project An international effort to map the complete human genetic code. This effort was essentially completed in 2001, though analysis is ongoing.

hypothalamus A brain area that responds to the amygdala and the hippocampus to produce hormones that activate other parts of the brain and body.

hypothesis A specific prediction that can be tested.

hypothetical thought Reasoning that includes propositions and possibilities that may not reflect reality.

I

identity achievement Erikson's term for the attainment of identity, when people know who they are as unique individuals, combining past experiences and future plans.

identity versus role confusion Erikson's fifth stage of development, when people wonder, "Who am I?" but are confused about which of many possible roles to adopt.

imaginary audience The other people who, in an adolescent's egocentric belief, are watching and taking note of his or her appearance, ideas, and behavior. This belief makes many teenagers very self-conscious.

immersion A strategy in which instruction in all school subjects occurs in the second (usually the majority) language that a child is learning.

immigrant paradox The surprising, paradoxical fact that low-SES immigrant women tend to have fewer birth complications than native-born peers with higher incomes.

immunization A process that stimulates the body's immune system by causing production of antibodies to defend against attack by a particular contagious disease. Creation of antibodies may be accomplished naturally (by having the disease), by injection, by drops that are swallowed, or by a nasal spray.

impulse control The ability to postpone or deny the immediate response to an idea or behavior.

in vitro fertilization (IVF) Fertilization that takes place outside a woman's body (as in a glass laboratory dish). The procedure involves adding sperm to ova that have been surgically removed from the woman's ovary. If a zygote is produced, it is inserted into a woman's uterus, where it may implant and develop into a baby.

independent variable In an experiment, the variable that is introduced to see what effect it has on the dependent variable. (Also called *experimental variable*.)

individual education plan (IEP) A document that specifies educational goals and plans for a child with special needs.

induction A disciplinary technique that involves explaining why a particular behavior was wrong. To be successful, explanation must be within the child's ability to understand.

inductive reasoning Reasoning from one or more specific experiences or facts to reach (induce) a general conclusion. (Also called *bottom-up reasoning*.)

industry versus inferiority The fourth of Erikson's eight psychosocial crises, during which children attempt to master many skills, developing a sense of themselves as either industrious or inferior, competent or incompetent.

information-processing theory A perspective that compares human thinking processes, by analogy, to computer analysis of data, including sensory input, connections, stored memories, and output.

initiative versus guilt Erikson's third psychosocial crisis, in which young children undertake new skills and activities and feel guilty when they do not succeed at them.

injury control/harm reduction Practices that are aimed at anticipating, controlling, and preventing dangerous activities; these practices reflect the beliefs that accidents are not random and that injuries can be made less harmful if proper controls are in place.

insecure-avoidant attachment A pattern of attachment in which an infant avoids connection with the caregiver, as when the infant seems not to care about the caregiver's presence, departure, or return.

insecure-resistant/ambivalent attachment A pattern of attachment in which an infant's anxiety and uncertainty are evident, as when the infant becomes very upset at separation from the caregiver and both resists and seeks contact on reunion.

instrumental aggression Hurtful behavior that is intended to get something that another person has.

intersectionality The idea that the various identities need to be combined. This is especially important in determining if discrimination occurs.

intimacy versus isolation The sixth of Erikson's eight stages of development. Adults seek someone with whom to share their lives in an enduring and self-sacrificing commitment. Without such commitment, they risk profound aloneness and isolation.

intrinsic motivation A drive, or reason to pursue a goal, that comes from inside a person, such as the joy of reading a good book.

intuitive thought Thought that arises from an emotion or a hunch, beyond rational explanation, and is influenced by past experiences and cultural assumptions.

invincibility fable An adolescent's egocentric conviction that he or she cannot be overcome or even harmed by anything that might defeat a normal mortal, such as unprotected sex, drug abuse, or high-speed driving.

irreversibility A characteristic of preoperational thought in which a young child thinks that nothing can be undone. A thing cannot be restored to the way it was before a change occurred.

K

kangaroo care A form of newborn care in which mothers (and sometimes fathers) rest their babies on their naked chests, like kangaroo mothers that carry their immature newborns in a pouch on their abdomen.

kinship care A form of foster care in which a relative of a maltreated child, usually a grandparent, becomes the approved caregiver.

knowledge base A body of knowledge in a particular area that makes it easier to master new information in that area.

L

language acquisition device (LAD) Chomsky's term for a hypothesized mental structure that enables humans to learn language, including the basic aspects of grammar, vocabulary, and intonation.

lateralization Literally, sidedness, referring to the specialization in certain functions by each side of the brain, with one side dominant for each activity. The left side of the brain controls the right side of the body, and vice versa.

least restrictive environment (LRE) A legal requirement that children with special needs be assigned to the most general educational context in which they can be expected to learn.

leptin A hormone that affects appetite and is believed to affect the onset of puberty. Leptin levels increase during childhood and peak at around age 12.

life-course-persistent offender A person whose criminal activity begins in adolescence and continues throughout life; a "career" criminal.

limbic system The parts of the brain that interact to produce emotions, including the amygdala, the hypothalamus, and the hippocampus. Many other parts of the brain also are involved with emotions.

"little scientist" Piaget's term for toddlers' insatiable curiosity and active experimentation as they engage in various actions to understand their world.

longitudinal research A research design in which the same individuals are followed over time, as their development is repeatedly assessed.

low birthweight (LBW) A body weight at birth of less than 2,500 grams (5½ pounds).

M

macrosystem In Bronfenbrenner's ecological approach, the overarching national or cultural policies and customs that affect the more immediate systems, such as the effect of the national economy on local hospitals (an exosystem) or on families (a microsystem).

major depression Feelings of hopelessness, lethargy, and worthlessness that last two weeks or more.

mean length of utterance (MLU) The average number of words in a typical sentence (called utterance because children may not talk in complete sentences). MLU is often used to measure language development.

menarche A girl's first menstrual period, signaling that she has begun ovulation. Pregnancy is biologically possible, but ovulation and menstruation are often irregular for years after menarche.

mesosystem In Bronfenbrenner's ecological approach, a connection between one system and another, such as parent–teacher conferences (connecting home and school) or workplace schedules (connecting family and job).

meta-analysis A technique of combining results of many studies to come to an overall conclusion. Meta-analysis is powerful, in that small samples can be added together to lead to significant conclusions, although variations from study to study sometimes make combining them impossible.

microbiome All the microbes (bacteria, viruses, and so on) with all their genes in a community; here the millions of microbes of the human body.

microsystem In Bronfenbrenner's ecological approach, the immediate social contexts that directly affect each person, such as family, peer group, work team.

middle school A school for children in the grades between elementary school and high school. Middle school usually begins with grade 6 and ends with grade 8.

modeling The central process of social learning, by which a person observes the actions of others and then copies them.

monozygotic (MZ) twins Twins who originate from one zygote that splits apart very early in development. (Also called *identical twins.*) Other monozygotic multiple births (such as triplets and quadruplets) can occur as well.

Montessori schools Schools that offer early-childhood education based on the philosophy of Maria Montessori, which emphasizes careful work and tasks that each young child can do.

moratorium A socially acceptable way to postpone achieving identity. Going to college and joining the military are examples.

motor skill The learned abilities to move some part of the body, in actions ranging from a large leap to a flicker of the eyelid. (The word *motor* here refers to movement of muscles.)

multifinality A basic principle of developmental psychopathology that holds that one cause can have many (multiple) final manifestations.

multiple intelligences The idea that human intelligence is composed of a varied set of abilities rather than a single, all-encompassing one.

myelin The coating on axons that speeds transmission of signals from one neuron to another.

myelination The process by which axons become coated with myelin, a fatty substance that speeds the transmission of nerve impulses from neuron to neuron.

N

naming explosion A sudden increase in an infant's vocabulary, especially in the number of nouns, that begins at about 18 months of age.

National Assessment of Educational Progress (NAEP) An ongoing and nationally representative measure of U.S. children's achievement in reading, mathematics, and other subjects over time; nicknamed "the Nation's Report Card."

nature In development, nature refers to the traits, capacities, and limitations that each individual inherits genetically from their parents.

neglectful/uninvolved parenting An approach to child rearing in which the parents seem indifferent toward their children, not knowing or caring about their children's lives.

neurodiversity The idea that each person has neurological strengths and weaknesses that should be appreciated, in much the same way diverse cultures and ethnicities are welcomed. Neurodiversity seems particularly relevant for children with disorders on the autism spectrum.

neuron One of billions of nerve cells in the central nervous system, especially in the brain.

neurotransmitter A brain chemical that carries information from the axon of a sending neuron to the dendrites of a receiving neuron.

nuclear family A family that consists of a father, a mother, and their biological children under age 18.

nurture In development, nurture includes all of the environmental influences that affect the individual after conception. This includes everything from the mother's nutrition while pregnant to the cultural influences in the nation.

O

object permanence The realization that objects (including people) still exist when they can no longer be seen, touched, or heard.

objective thought Thinking that is based on facts. It is impartial, and can be verified by anyone who seeks to know.

Oedipus complex The unconscious desire of young boys to replace their fathers and win their mothers' exclusive love.

operant conditioning The learning process by which a particular action is followed by something desired (which makes the person or animal more likely to repeat the action) or by something unwanted (which makes the action less likely to be repeated). (Also called *instrumental conditioning*.)

overimitation When a person imitates an action that is not a relevant part of the behavior to be learned. Overimitation is common among 2- to 6-year-olds when they imitate adult actions that are irrelevant and inefficient.

overregularization The application of rules of grammar even when exceptions occur, making the language seem more "regular" than it actually is.

P

parasuicide Any potentially deadly self-harm that does not result in death. (Also called *attempted suicide* or *failed suicide*.)

parental monitoring Parents' ongoing knowledge of what their children are doing, where, and with whom.

parentification When a child acts more like a parent than a child. Parentification may occur if the actual parents do not act as caregivers, making a child feel responsible for the family.

parent–infant bond The strong, loving connection that forms as parents hold, examine, and feed their newborn.

peer pressure Encouragement to conform to friends or contemporaries in behavior, dress, and attitude. Adolescents do many things with peers that they would not do alone.

percentile A point on a ranking scale of 0 to 100. The 50th percentile is the midpoint; half the people in the population being studied rank higher and half rank lower.

permanency planning An effort by child-welfare authorities to find a long-term living situation that will provide stability and support for a maltreated child. A goal is to avoid repeated changes of caregiver or school, which is particularly harmful.

permissive parenting An approach to child rearing that is characterized by high nurturance and communication but little discipline, guidance, or control.

perseverate To stay stuck, or persevere, in one thought or action for a longtime. The ability to be flexible, switching from one task to another, is beyond most young children.

personal fable An aspect of adolescent egocentrism characterized by an adolescent's belief that his or her thoughts, feelings, and experiences are unique, more wonderful, or more awful than anyone else's.

phallic stage Freud's third stage of development, when the penis becomes the focus of concern and pleasure.

phenotype The observable characteristics of a person, including appearance, personality, intelligence, and all other traits.

pituitary A gland in the brain that responds to a signal from the hypothalamus by producing many hormones, including those that regulate growth and that control other glands, among them the adrenal and sex glands.

plasticity The idea that abilities, personality, and other human characteristics can change over time. Plasticity is particularly evident during childhood, but even older adults are not always "set in their ways."

polygamous family A family consisting of one man, several wives, and their children.

postconventional moral reasoning Kohlberg's third level of moral reasoning, emphasizing moral principles.

postformal thought A proposed adult stage of cognitive development, following Piaget's four stages, that goes beyond adolescent thinking by being more practical, more flexible, and more dialectical (i.e., more capable of combining contradictory elements into a comprehensive whole).

postpartum depression A new mother's feelings of inadequacy and sadness in the days and weeks after giving birth.

posttraumatic stress disorder (PTSD) An anxiety disorder that develops as a delayed reaction to having experienced or witnessed a profoundly shocking or frightening event, such as rape, severe beating, war, or natural disaster. Its symptoms may include flashbacks to the event, hyperactivity and hypervigilance, displaced anger, sleeplessness, nightmares, sudden terror or anxiety, and confusion between fantasy and reality.

pragmatics The practical use of language that includes the ability to adjust language communication according to audience and context.

preconventional moral reasoning Kohlberg's first level of moral reasoning, emphasizing rewards and punishments.

prefrontal cortex The area of the cortex at the very front of the brain that specializes in anticipation, planning, and impulse control.

preoperational intelligence Piaget's term for cognitive development between the ages of about 2 and 6; it includes language and imagination (which involve symbolic thought), but logical, operational thinking is not yet possible at this stage.

preterm A birth that occurs two or more weeks before the full 38 weeks of the typical pregnancy—that is, at 36 or fewer weeks after conception.

primary circular reactions The first of three types of feedback loops in sensori-motor intelligence, this one involving the infant's own body. The infant senses motion, sucking, noise, and other stimuli and tries to understand them.

primary prevention Actions that change overall background conditions to prevent some unwanted event or circumstance, such as injury, disease, or abuse.

primary sex characteristics The parts of the body that are directly involved in reproduction, including the vagina, uterus, ovaries, testicles, and penis.

private speech The internal dialogue that occurs when people talk to themselves (either silently or out loud).

Programme for International Student Assessment (PISA) An international test taken by 15-year-olds in 50 nations that is designed to measure problem solving and cognition in daily life.

Progress in International Reading Literacy Study (PIRLS) Inaugurated in 2001, a planned five-year cycle of international trend studies in the reading ability of fourth-graders.

prosocial behavior Actions that are helpful and kind but that are of no obvious benefit to the person doing them.

protein-calorie malnutrition A condition in which a person does not consume sufficient food of any kind. This deprivation can result in several illnesses, severe weight loss, and even death.

proximal parenting Caregiving practices that involve being physically close to the baby, with frequent holding and touching.

proximodistal Growth or development that occurs from the center or core in an outward direction.

pruning When applied to brain development, the process by which unused connections in the brain atrophy and die.

psychoanalytic theory Freud's theory of the stages of development, each of which emphasizes the sexual nature of the child. As its first grand theorist, Freud believed that irrational, unconscious drives and motives, often originating in childhood erotic impulses, underlie human behavior.

psychological control A disciplinary technique that involves threatening to withdraw love and support, using a child's feelings of guilt and gratitude to the parents.

psychosocial theory Erikson's theory of the stages of development, emphasizing the interaction between the psychic needs of the individual and the surrounding social network of family and community.

puberty The time between the first onrush of hormones and full adult physical development. Puberty usually lasts three to five years. Many more years are required to achieve psychosocial maturity.

Q

qualitative research Research that considers qualities instead of quantities. Descriptions of particular conditions and participants' expressed ideas are often part of qualitative studies.

quantitative research Research that provides data that can be expressed with numbers, such as ranks or scales.

R

race A group of people who are regarded by themselves or by others as distinct from other groups on the basis of physical appearance, typically skin color. Social scientists think race is a misleading concept, as biological differences are not signified by outward appearance.

reaction time The time it takes to respond to a stimulus, either physically (with a reflexive movement such as an eyeblink) or cognitively (with a thought).

reactive aggression An impulsive retaliation for another person's intentional or accidental hurtful action.

recessive Hidden, not dominant. Recessive genes are carried in the genotype and are not evident in the phenotype, except in special circumstances.

reflex An unlearned, involuntary action or movement in response to a stimulus. A reflex occurs without conscious thought.

Reggio Emilia A program of early-childhood education that originated in the town of Reggio Emilia, Italy, and that encourages each child's creativity in a carefully designed setting.

reinforcement When a behavior is followed by something desired, such as food for a hungry animal or a welcoming smile for a lonely person.

relational aggression Nonphysical acts, such as insults or social rejection, aimed at harming the social connection between the victim and other people.

REM (rapid eye movement) sleep A stage of sleep characterized by flickering eyes behind closed lids, dreaming, and rapid brain waves.

replication Repeating a study, usually using different participants, perhaps of another age, SES, or culture.

reported maltreatment Harm or endangerment about which someone has notified the authorities.

resilience The capacity to adapt well to significant adversity and to overcome serious stress.

response to intervention (RTI) An educational strategy intended to help children who demonstrate below-average achievement in early grades, using special intervention.

role confusion When adolescents have no clear identity but fluctuate from one persona to another. (Sometimes called *identity diffusion* or *role diffusion*.)

rough-and-tumble play Play that seems to be rough, as in play wrestling or chasing, but in which there is no intent to harm.

S

scaffolding Temporary support that is tailored to a learner's needs and abilities and aimed at helping the learner master the next task in a given learning process.

science of human development The science that seeks to understand how and why people of all ages and circumstances change or remain the same over time.

scientific method A way to answer questions using empirical research and data-based conclusions.

scientific observation A method of testing a hypothesis by unobtrusively watching and recording participants' behavior in a systematic and objective manner—in a natural setting, in a laboratory, or in searches of archival data.

secondary circular reactions The first of three types of feedback loops in sensori-motor intelligence, involving the infant and an object or another person, as with shaking a rattle or playing peek-a-boo.

secondary education Literally, the period after primary education (elementary or grade school) and before tertiary education (college). It usually occurs from about ages 12 to 18, although there is some variation by school and by nation.

secondary prevention Actions that avert harm in a high-risk situation, such as stopping a car before it hits a pedestrian.

secondary sex characteristics Physical traits that are not directly involved in reproduction but that indicate sexual maturity, such as a man's beard and a woman's breasts.

secular trend The long-term upward or downward direction of a certain set of statistical measurements, as opposed to a smaller, shorter cyclical variation. As an example, over the past two centuries, because of improved nutrition and medical care, children have tended to reach their adult height earlier and their adult height has increased.

secure attachment A relationship in which an infant obtains both comfort and confidence from the presence of his or her caregiver.

selective adaptation The process by which living creatures (including people) adjust to their environment. Genes that enhance survival and reproductive ability are selected, over the generations, to become more prevalent.

self-awareness A person's realization that he or she is a distinct individual whose body, mind, and actions are separate from those of other people.

self-concept A person's understanding of who they are, in morality, intelligence, appearance, personality, talents, and skills.

self-righting The inborn drive to remedy a developmental deficit; literally, to return to sitting or standing upright after being tipped over. People of all ages have self-righting impulses, for emotional as well as physical imbalance.

sensation The response of a sensory organ (eyes, ears, skin, tongue, nose) when it detects a stimulus.

sensitive period A time when something (such as a toxin, or nutrient, or experience) has the greatest impact on development, but recovery is possible if it occurs later.

sensorimotor intelligence Piaget's term for the way infants think—by using their senses and motor skills—during the first period of cognitive development.

separation anxiety An infant's distress when a familiar caregiver leaves; most obvious between 9 and 14 months.

seriation The concept that things can be arranged in a logical series, such as the number sequence or the alphabet.

sex differences Biological differences between males and females, in organs, hormones, and body shape.

sexting Sending sexual messages or photographs (usually of one's naked body) via phone or computer.

sexual orientation A person's romantic or sexual attraction, which can be to others of the same gender, the other gender, or every gender.

sexually transmitted infection (STI) A disease spread by sexual contact; includes syphilis, gonorrhea, genital herpes, chlamydia, and HIV.

shaken baby syndrome A life-threatening injury that occurs when an infant is forcefully shaken back and forth, a motion that ruptures blood vessels in the brain and breaks neural connections.

single-parent family A family that consists of only one parent and their children.

small for gestational age (SGA) A term for a baby whose birthweight is significantly lower than expected, given the time since conception. For example, a 5-pound (2,265-gram) newborn is considered SGA if born on time but not SGA if born two months early. (Also called *small-for-dates.*)

social comparison The tendency to assess one's abilities, achievements, social status, and other attributes by measuring them against those of other people, especially one's peers.

social construction An idea that is built on shared perceptions, not on objective reality. Many age-related terms (such as *childhood, adolescence, yuppie,* and *senior citizen*) are social constructions, connected to biological traits but strongly influenced by social assumptions.

social learning theory An extension of behaviorism that emphasizes the influence that other people have over a person's behavior. Even without specific reinforcement, every individual learns many things through observation and imitation of other people. (Also called *observational learning.*)

social mediation Human interaction that expands and advances understanding, often through words that one person uses to explain something to another.

social referencing Seeking information about how to react to an unfamiliar or ambiguous object or event by observing someone else's expressions and reactions. That other person becomes a social reference.

social smile A smile evoked by a human face, normally first evident in infants about six weeks after birth.

sociocultural theory A newer theory that holds that development results from the dynamic interaction of each person with the surrounding social and cultural forces.

sociodramatic play Pretend play in which children act out various roles and themes in plots or roles that they create.

socioeconomic status (SES) A person's position in society as determined by income, occupation, education, and place of residence. (Sometimes called *social class.*)

specific learning disorder A marked deficit in a particular area of learning that is not caused by an apparent physical disability, by an intellectual disability, or by an unusually stressful home environment.

spermarche A boy's first ejaculation of sperm. Erections can occur as early as infancy, but ejaculation signals sperm production. Spermarche may occur during sleep (in a "wet dream") or via direct stimulation.

static reasoning A characteristic of pre-operational thought in which a young child thinks that nothing changes. Whatever is now has always been and always will be.

stem cells Cells from which any other specialized type of cell can form.

stereotype threat The thought a person has that one's appearance or behavior will be misread to confirm another person's oversimplified, prejudiced attitudes.

still-face technique An experimental practice in which an adult keeps their face unmoving and expressionless in face-to-face interaction with an infant.

stranger wariness An infant's expression of concern—a quiet stare while clinging to a familiar person, or a look of fear—when a stranger appears.

stunting The failure of children to grow to a normal height for their age due to severe and chronic malnutrition.

subjective thought Thinking that is based on personal experience and perception. Because it is connected to emotions, it is deeply held and can be quite opposite another's subjective thinking.

substantiated maltreatment Harm or endangerment that has been reported, investigated, and verified.

sudden infant death syndrome (SIDS) A situation in which a seemingly healthy infant, usually between 2 and 6 months old, suddenly stops breathing and dies unexpectedly while asleep.

suicidal ideation Serious thinking about suicide, often including extreme emotions and thoughts.

superego In psychoanalytic theory, the judgmental part of the personality that internalizes the moral standards of the parents.

survey A research method in which information is collected from a large number of people by interviews, written questionnaires, or some other means.

symbolic thought A major accomplishment of preoperational intelligence that allows a child to think symbolically, including understanding that words can refer to things not seen and that an item, such as a flag, can symbolize something else (in this case, a country).

synapse The intersection between the axon of one neuron and the dendrites of other neurons.

synchrony A coordinated, rapid, and smooth exchange of responses between a caregiver and an infant.

T

temperament Inborn differences between one person and another in emotions, activity, and self-regulation. It is measured by the person's typical responses to the environment.

teratogen An agent or condition, including viruses, drugs, and chemicals, that can impair prenatal development and result in birth defects or even death.

tertiary circular reaction Piaget's description of the cognitive processes of the 1-year-old, who gathers information from experiences with the wider world and then acts on it. The response to those actions leads to further understanding, which makes this circular.

tertiary prevention Actions, such as immediate and effective medical treatment, that are taken after an adverse event (such as illness or injury) and that are aimed at reducing harm or preventing disability.

testosterone A sex hormone, the best known of the androgens (male hormones); secreted in far greater amounts by males than by females.

theory of mind A person's theory of what other people might be thinking. In order to have a theory of mind, children must realize that other people are not necessarily thinking the same thoughts that they themselves are. That realization seldom occurs before age 4.

theory-theory The idea that children attempt to explain everything they see and hear by constructing theories.

threshold effect In prenatal development, when a teratogen is relatively harmless in small doses but becomes harmful once exposure reaches a certain level (the threshold).

time-out A disciplinary technique in which a person is separated from other people and activities for a specified time.

transgender Identifying oneself with a gender other than the one ascribed at birth. Thus, a transgender girl was thought to be a boy when she was born.

transient exuberance The great but temporary increase in the number of dendrites that develop in an infant's brain during the first two years of life.

Trends in Math and Science Study (TIMSS) An international assessment of the math and science skills of fourth-graders and eighth-graders. Although the TIMSS is very useful, different countries' scores are not always comparable because sample selection, test administration, and content validity are hard to keep uniform.

trust versus mistrust Erikson's first crisis of psychosocial development. Infants learn basic trust if the world is a secure place where their basic needs (for food, comfort, attention, and so on) are met.

V

very low birthweight (VLBW) A body weight at birth of less than 1,500 grams (3 pounds, 5 ounces).

W

Waldorf schools An early-childhood education program that emphasizes creativity, social understanding, and emotional growth. It originated in Germany with Rudolf Steiner and now is used in thousands of schools throughout the world.

wasting The tendency for children to be severely underweight for their age as a result of malnutrition.

WEIRD An acronym that refers to people from Western, Educated, Industrialized, Rich Democracies, in other words, to North American college students, not necessarily the rest of humanity.

working model In cognitive theory, a set of assumptions that the individual uses to organize perceptions and experiences. For example, a person might assume that other people are trustworthy and be surprised by an incident in which this working model of human behavior is erroneous.

X

X-linked A gene carried on the X chromosome. If a male inherits an X-linked recessive trait from his mother, he expresses that trait because the Y from his father has no counteracting gene. Females are more likely to be carriers of X-linked traits but are less likely to express them.

XX A 23rd chromosome pair that consists of two X-shaped chromosomes, one each from the mother and the father. XX zygotes become females.

XY A 23rd chromosome pair that consists of an X-shaped chromosome from the mother and a Y-shaped chromosome from the father. XY zygotes become males.

Z

zone of proximal development (ZPD) In sociocultural theory, a metaphorical area, or "zone," surrounding a learner that includes all of the skills, knowledge, and concepts that the person is close ("proximal") to acquiring but cannot yet master without help.

zygote The single cell formed from the union of two gametes, a sperm and an ovum.

References

Aarnoudse-Moens, Cornelieke S. H.; Smidts, Diana P.; Oosterlaan, Jaap; Duivenvoorden, Hugo J. & Weisglas-Kuperus, Nynke. (2009). Executive function in very preterm children at early school age. *Journal of Abnormal Child Psychology*, 37(7), 981–993.

Abar, Caitlin C.; Jackson, Kristina M. & Wood, Mark. (2014). Reciprocal relations between perceived parental knowledge and adolescent substance use and delinquency: The moderating role of parent–teen relationship quality. *Developmental Psychology*, 50(9), 2176–2187.

Abraham, Eyal & Feldman, Ruth. (2018). The neurobiology of human allomaternal care; implications for fathering, coparenting, and children's social development. *Physiology & Behavior*, 193, 25–34.

Abramovitch, Amitai; Anholt, Gideon; Raveh-Gottfried, Sagi; Hamo, Naama & Abramowitz, Jonathan. (2018). Meta-analysis of intelligence quotient (IQ) in obsessive-compulsive disorder. *Neuropsychology Review*, 28(1), 111–120.

Accardo, Pasquale. (2006). Who's training whom? *The Journal of Pediatrics*, 149(2), 151–152.

Acharya, Kartikey; Leuthner, Stephen; Clark, Reese; Nghiem-Rao, Tuyet-Hang; Spitzer, Alan & Lagatta, Joanne. (2017). Major anomalies and birth-weight influence NICU interventions and mortality in infants with trisomy 13 or 18. *Journal of Perinatology*, 37(4), 420–426.

Acheson, Ashley; Vincent, Andrea S.; Cohoon, Andrew & Lovallo, William R. (2019). Early life adversity and increased delay discounting: Findings from the Family Health Patterns project. *Experimental and Clinical Psychopharmacology*, 27(2), 153–159.

Acosta, Sandra; Garza, Tiberio; Hsu, Hsien-Yuan; Goodson, Patricia; Padrón, Yolanda; Goltz, Heather H. & Johnston, Anna. (2020). The accountability culture: A systematic review of high-stakes testing and English learners in the United States during no Child Left Behind. *Educational Psychology Review*, 32(2), 327–352.

Acuto, Michele & Parnell, Susan. (2016). Leave no city behind. *Science*, 352(6288), 873.

Adamson, Lauren B. & Bakeman, Roger. (2006). Development of displaced speech in early mother-child conversations. *Child Development*, 77(1), 186–200.

Addati, Laura; Cassirer, Naomi & Gilchrist, Katherine. (2014). *Maternity and paternity at work: Law and practice across the world.* Geneva: International Labour Office.

Adolph, Karen E.; Cole, Whitney G.; Komati, Meghana; Garciaguirre, Jessie S.; Badaly, Daryaneh; Lingeman, Jesse M.; . . . Sotsky, Rachel B. (2012). How do you learn to walk? Thousands of steps and dozens of falls per day. *Psychological Science*, 23(11), 1387–1394.

Adolph, Karen E. & Franchak, John M. (2017). The development of motor behavior. *WIREs*, 8(1/2), e1430.

Adolph, Karen E. & Robinson, Scott. (2013). The road to walking: What learning to walk tells us about development. In Zelazo, Philip D. (Ed.), *The Oxford handbook of developmental psychology* (Vol. 1, pp. 402–447). Oxford University Press.

Ainsworth, Mary D. Salter. (1967). *Infancy in Uganda: Infant care and the growth of love.* Johns Hopkins Press.

Ainsworth, Mary D. Salter. (1973). The development of infant-mother attachment. In Caldwell, Bettye M. & Ricciuti, Henry N. (Eds.), *Child development and social policy* (pp. 1–94). University of Chicago Press.

Aïte, Ania; Cassotti, Mathieu; Linzarini, Adriano; Osmont, Anaïs; Houdé, Olivier & Borst, Grégoire. (2018). Adolescents' inhibitory control: Keep it cool or lose control. *Developmental Science*, 21(1), e12491.

Aitken, Jess; Ruffman, Ted & Taumoepeau, Mele. (2019). Toddlers' self-recognition and progression from goal- to emotion-based helping: A longitudinal study. *Child Development*.

Aizpitarte, Alazne; Atherton, Olivia E.; Zheng, Lucy R.; Alonso-Arbiol, Itziar & Robins, Richard W. (2019). Developmental precursors of relational aggression from late childhood through adolescence. *Child Development*, 90(1), 117–126.

Akhtar, Nameera & Jaswal, Vikram K. (2013). Deficit or difference? Interpreting diverse developmental paths: An introduction to the special section. *Developmental Psychology*, 49(1), 1–3.

Akombi, Blessing J.; Agho, Kingsley E.; Hall, John J.; Wali, Nidhi; Renzaho, Andre M. N. & Merom, Dafna. (2017). Stunting, wasting and underweight in sub-Saharan Africa: A systematic review. *International Journal of Environmental Research and Public Health*, 14(8).

Aksglaede, Lise; Link, Katarina; Giwercman, Aleksander; Jørgensen, Niels; Skakkebæk, Niels E. & Juul, Anders. (2013). 47,XXY Klinefelter syndrome: Clinical characteristics and age-specific recommendations for medical management. *American Journal of Medical Genetics Part C: Seminars in Medical Genetics*, 163(1), 55–63.

Al Otaiba, Stephanie; Wanzek, Jeanne & Yovanoff, Paul. (2015). Response to intervention. *European Scientific Journal*, 1, 260–264.

Al-Hashim, Aqeela H.; Blaser, Susan; Raybaud, Charles & MacGregor, Daune. (2016). Corpus callosum abnormalities: Neuroradiological and clinical correlations. *Developmental Medicine & Child Neurology*, 58(5), 475–484.

Al-Namlah, Abdulrahman S.; Meins, Elizabeth & Fernyhough, Charles. (2012). Self-regulatory private speech relates to children's recall and organization of autobiographical memories. *Early Childhood Research Quarterly*, 27(3), 441–446.

Al-Sayes, Fatin; Gari, Mamdooh; Qusti, Safaa; Bagatian, Nadiah & Abuzenadah, Adel. (2011). Prevalence of iron deficiency and iron deficiency anemia among females at university stage. *Journal of Medical Laboratory and Diagnosis*, 2(1), 5–11.

Albataineh, Samah R.; Badran, Eman F. & Tayyem, Reema F. (2019). Overweight and obesity in childhood: Dietary, biochemical, inflammatory and lifestyle risk factors. *Obesity Medicine*, 15(100112).

Albert, Dustin; Chein, Jason & Steinberg, Laurence. (2013). The teenage brain: Peer influences on adolescent decision making. *Current Directions in Psychological Science*, 22(2), 114–120.

Albert, Dustin & Steinberg, Laurence. (2011). Judgment and decision making in adolescence. *Journal of Research on Adolescence*, 21(1), 211–224.

Alegre, Alberto. (2011). Parenting styles and children's emotional intelligence: What do we know? *The Family Journal*, 19(1), 56–62.

Alesi, Marianha; Bianco, Antonino; Padulo, Johnny; Vella, Francesco Paolo; Petrucci, Marco; Paoli, Antonio; . . . Pepi, Annamaria. (2014). Motor and cognitive development: The role of karate. *Muscle, Ligaments and Tendons Journal*, 4(2), 114–120.

Alexander, Karl L.; Entwisle, Doris R. & Olson, Linda Steffel. (2014). *The long shadow: Family background, disadvantaged urban youth, and the transition to adulthood.* Russell Sage Foundation.

Alexandre-Heymann, Laure; Dubert, Marie; Diallo, Dapa A.; Diop, Saliou; Tolo, Aissata; Belinga, Suzanne; . . . Ranque, Brigitte. (2019). Prevalence and correlates of growth failure in young African patients with sickle cell disease. *British Journal of Haematology*, 184(2), 253–262.

Allport, Gordon. (1954). *The nature of prejudice.* Addison-Wesley.

Alper, Meryl. (2013). Developmentally appropriate New Media Literacies: Supporting cultural competencies and social skills in early childhood education. *Journal of Early Childhood Literacy*, 13(2), 175–196.

Alvarado, Louis; Muller, Martin; Eaton, Melissa & Thompson, Melissa Emery. (2018). Steroid hormone reactivity in fathers watching their children compete. *Human Nature*, 29(3), 268–282.

Amato, Michael S.; Magzamen, Sheryl; Imm, Pamela; Havlena, Jeffrey A.; Anderson, Henry A.; Kanarek, Marty S. & Moore, Colleen F. (2013). Early lead exposure (<3 years old) prospectively predicts fourth grade school suspension in Milwaukee, Wisconsin (USA). *Environmental Research*, 126, 60–65.

Ameade, Evans Paul Kwame & Garti, Helene Akpene. (2016). Age at menarche and factors that influence it: A study among female university students in Tamale, northern Ghana. *PLoS ONE*, *11*(5), e0155310.

American Academy of Pediatrics. (2016). Media and young minds. *Pediatrics, 138*(5).

American Academy of Pediatrics. (2018, May 1). Children and media tips from the American Academy of Pediatrics [Press release]. American Academy of Pediatrics.

American College Health Association. (2017). *American college health association-national college health assessment II: Reference group executive summary fall 2016.* Hanover, MD: American College Health Association.

American College of Obstetricians and Gynecologists Committee on Obstetric Practice. (2011). Committee opinion no. 476: Planned home birth. *Obstetrics & Gynecology, 117*(2), 425–428.

American Psychiatric Association. (2013). *Diagnostic and statistical manual of mental disorders: DSM-5* (5th ed.). American Psychiatric Association.

Anderson, Craig A.; Shibuya, Akiko; Ihori, Nobuko; Swing, Edward L.; Bushman, Brad J.; Sakamoto, Akira; . . . Saleem, Muniba. (2010). Violent video game effects on aggression, empathy, and prosocial behavior in Eastern and Western countries: A meta-analytic review. *Psychological Bulletin, 136*(2), 151–173.

Anderson, Daniel R. & Hanson, Katherine G. (2016). Screen media and parent–child interactions. In Barr, Rachel & Linebarger, Deborah Nichols (Eds.), *Media exposure during infancy and early childhood: The effects of content and context on learning and development* (pp. 173–194). Springer.

Anderson, Monica. (2018, September 27). *A majority of teens have experienced some form of cyberbullying.* Washington, DC: Pew Research Center.

Andreas, Nicholas J.; Kampmann, Beate & Le-Doare, Kirsty Mehring. (2015). Human breast milk: A review on its composition and bioactivity. *Early Human Development, 91*(11), 629–635.

Ansado, Jennyfer; Collins, Louis; Fonov, Vladimir; Garon, Mathieu; Alexandrov, Lubomir; Karama, Sherif; . . . Beauchamp, Miriam H. (2015). A new template to study callosal growth shows specific growth in anterior and posterior regions of the corpus callosum in early childhood. *European Journal of Neuroscience, 42*(1), 1675–1684.

Araki, Atsuko & Jensen, Tina Kold. (2020). Endocrine-distributing chemicals and reproductive function. In Kishi, Reiko & Grandjean, Philippe (Eds.), *Health impacts of developmental exposure to environmental chemicals* (pp. 101–129). Springer.

Areba, Eunice M.; Eisenberg, Marla E. & McMorris, Barbara J. (2018). Relationships between family structure, adolescent health status and substance use: Does ethnicity matter? *Journal of Community Psychology, 46*(1), 44–57.

Arguz Cildir, Deniz; Ozbek, Aylin; Topuzoglu, Ahmet; Orcin, Esmahan & Janbakhishov, Ceren Evcen. (2019). Association of prenatal attachment and early childhood emotional, behavioral, and developmental characteristics: A longitudinal study. *Infant Mental Health Journal.*

Argyrides, Marios & Kkeli, Natalie. (2015). Predictive factors of disordered eating and body image satisfaction in Cyprus. *International Journal of Eating Disorders, 48*(4), 431–435.

Ariely, Dan. (2010). *Predictably irrational: The hidden forces that shape our decisions* (Revised and Expanded ed.). Harper Perennial.

Armstrong, Kim. (2018). The WEIRD science of culture, values, and behavior. *Observer.* Association for Psychological Science.

Arnott, Lorna. (2016). An ecological exploration of young children's digital play: Framing children's social experiences with technologies in early childhood. *Early Years, 36*(3), 271–288.

Aron, Arthur; Lewandowski, Gary W.; Mashek, Debra & Aron, Elaine N. (2013). The self-expansion model of motivation and cognition in close relationships. In Simpson, Jeffry A. & Campbell, Lorne (Eds.), *The Oxford handbook of close relationships* (pp. 90–115). Oxford University Press.

Aronson, Joshua; Burgess, Diana; Phelan, Sean M. & Juarez, Lindsay. (2013). Unhealthy interactions: The role of stereotype threat in health disparities. *American Journal of Public Health, 103*(1), 50–56.

Arshad, S. Hasan; Karmaus, Wilfried; Zhang, Hongmei & Holloway, John W. (2017). Multigenerational cohorts in patients with asthma and allergy. *Journal of Allergy and Clinical Immunology, 139*(2), 415–421.

Ash, Tayla; Davison, Kirsten K.; Haneuse, Sebastien; Horan, Christine; Kitos, Nicole; Redline, Susan & Taveras, Elsie M. (2019). Emergence of racial/ethnic differences in infant sleep duration in the first six months of life. *Sleep Medicine: X, 1,* 100003.

Ashenhurst, James R.; Wilhite, Emily R.; Harden, K Paige & Fromme, Kim. (2017). Number of sexual partners and relationship status are associated with unprotected sex across emerging adulthood. *Archives of Sexual Behavior, 46*(2), 419–432.

Ashraf, Quamrul & Galor, Oded. (2013). The 'Out of Africa' hypothesis, human genetic diversity, and comparative economic development. *American Economic Review, 103*(1), 1–46.

Ashwin, Sarah & Isupova, Olga. (2014). "Behind every great man . . . ": The male marriage wage premium examined qualitatively. *Journal of Marriage and Family, 76*(1), 37–55.

Aslam, Kashif; Shabir, Ghulam; Shahid, Muhammad; Sabar, Muhammad; Naveed, Shahzad Amir; Shah, Shahid Masood; . . . Arif, Muhammad. (2019). Comparative analysis of genomic variation and population structure of rice (oryza sativa). *International Journal of Agriculture and Biology, 21*(6), 1129–1136.

Aslin, Richard N. (2012). Language development: Revisiting Eimas et al.'s /ba/ and /pa/ study. In Slater, Alan M. & Quinn, Paul C. (Eds.), *Developmental psychology: Revisiting the classic studies* (pp. 191–203). Sage.

Atzaba-Poria, Naama; Deater-Deckard, Kirby & Bell, Martha Ann. (2017). Mother-child interaction: Links between mother and child frontal electroencephalograph asymmetry and negative behavior. *Child Development, 88*(2), 544–554.

Atzil, Shir; Hendler, Talma & Feldman, Ruth. (2014). The brain basis of social synchrony. *Social Cognitive and Affective Neuroscience, 9*(8), 1193–1202.

Aunola, Kaisa; Tolvanen, Asko; Viljaranta, Jaana & Nurmi, Jari-Erik. (2013). Psychological control in daily parent–child interactions increases children's negative emotions. *Journal of Family Psychology, 27*(3), 453–462.

Ayoub, Mona; Briley, Daniel A.; Grotzinger, Andrew; Patterson, Megan W.; Engelhardt, Laura E.; Tackett, Jennifer L.; . . . Tucker-Drob, Elliot M. (2018). Genetic and environmental associations between child personality and parenting. *Social Psychological and Personality Science.*

Ayyanathan, Kasirajan (Ed.) (2014). *Specific gene expression and epigenetics: The interplay between the genome and its environment.* Apple Academic Press.

Azrin, Nathan H. & Foxx, Richard M. (1974). *Toilet training in less than a day.* Simon & Schuster.

Babadjouni, Robin M.; Hodis, Drew M.; Radwanski, Ryan; Durazo, Ramon; Patel, Arati; Liu, Qinghai & Mack, William J. (2017). Clinical effects of air pollution on the central nervous system; A review. *Journal of Clinical Neuroscience, 43,* 16–24.

Babchishin, Lyzon K.; Weegar, Kelly & Romano, Elisa. (2013). Early child care effects on later behavioral outcomes using a Canadian nation-wide sample. *Journal of Educational and Developmental Psychology, 3*(2), 15–29.

Bachman, Heather J.; Votruba-Drzal, Elizabeth; El Nokali, Nermeen E. & Castle Heatly, Melissa. (2015). Opportunities for learning math in elementary school: Implications for SES disparities in procedural and conceptual math skills. *American Educational Research Journal, 52*(5), 894–923.

Baddock, Sally A.; Purnell, Melissa T.; Blair, Peter S.; Pease, Anna S.; Elder, Dawn E. & Galland, Barbara C. (2019). The influence of bed-sharing on infant physiology, breastfeeding and behaviour: A systematic review. *Sleep Medicine Reviews, 43,* 106–117.

Bagchi, Debasis (Ed.) (2019). *Global perspectives on childhood obesity: Current status, consequences and prevention* (2nd ed.). Academic Press.

Bagot, Kara. (2017). Making the grade: Adolescent prescription stimulant use. *Journal of the American Academy of Child & Adolescent Psychiatry, 56*(3), 189–190.

Bagwell, Catherine L. & Schmidt, Michelle E. (2011). *Friendships in childhood and adolescence.* Guilford Press.

Bai, Sunhye; Karan, Maira; Gonzales, Nancy A. & Fuligni, Andrew J. (2020). A daily diary study of sleep chronotype among Mexican-origin adolescents and parents: Implications for adolescent behavioral health. *Development and Psychopathology.*

Baillargeon, Renée & DeVos, Julie. (1991). Object permanence in young infants: Further evidence. *Child Development, 62*(6), 1227–1246.

Baio, Jon; Wiggins, Lisa; Christensen, Deborah L.; Maenner, Matthew J.; Daniels, Julie; Warren, Zachary; . . . Dowling, Nicole F. (2018, April 27). *Prevalence of autism spectrum disorder among children aged 8 years—Autism and*

Developmental Disabilities Monitoring Network, 11 Sites, United States, 2014. Morbidity and Mortality Weekly Report, 67(SS-6), 1–23.

Baker, Jeffrey P. (2000). Immunization and the American way: 4 childhood vaccines. *American Journal of Public Health,* 90(2), 199–207.

Baker, Simon T. E.; Lubman, Dan I.; Yücel, Murat; Allen, Nicholas B.; Whittle, Sarah; Fulcher, Ben D.; . . . Fornito, Alex. (2015). Developmental changes in brain network hub connectivity in late adolescence. *Journal of Neuroscience,* 35(24), 9078–9087.

Ball, Helen L. & Volpe, Lane E. (2013). Sudden infant death syndrome (SIDS) risk reduction and infant sleep location—Moving the discussion forward. *Social Science & Medicine,* 79(1), 84–91.

Baltes, Paul B.; Lindenberger, Ulman & Staudinger, Ursula M. (2006). Life span theory in developmental psychology. In Damon, William & Lerner, Richard M. (Eds.), *Handbook of child psychology* (6th ed., Vol. 1, pp. 569–664). Wiley.

Baly, Michael W.; Cornell, Dewey G. & Lovegrove, Peter. (2014). A longitudinal investigation of self- and peer reports of bullying victimization across middle school. *Psychology in the Schools,* 51(3), 217–240.

Balzarini, Rhonda; Muise, Amy; Zoppolat, Giulia; Di Bartolomeo, Alyssa; Rodrigues, David; Alonso-Ferres, María; . . . Slatcher, Richard B. (2020). Love in the time of COVID: Perceived partner responsiveness buffers people from lower relationship quality associated with COVID-related stressors. *PsyArXiv.*

Bandura, Albert. (1986). *Social foundations of thought and action: A social cognitive theory.* Prentice-Hall.

Bandura, Albert. (1997). The anatomy of stages of change. *American Journal of Health Promotion,* 12(1), 8–10.

Bandura, Albert. (2006). Toward a psychology of human agency. *Perspectives on Psychological Science,* 1(2), 164–180.

Bandura, Albert. (2016). *Moral disengagement: How people do harm and live with themselves.* Worth.

Banks, James R. & Andrews, Timothy. (2015). Outcomes of childhood asthma to the age of 50 years. *Pediatrics,* 136(Suppl. 3).

Bannon, Michael J.; Johnson, Magen M.; Michelhaugh, Sharon K.; Hartley, Zachary J.; Halter, Steven D.; David, James A.; . . . Schmidt, Carl J. (2014). A molecular profile of cocaine abuse includes the differential expression of genes that regulate transcription, chromatin, and dopamine cell phenotype. *Neuropsychopharmacology,* 39(9), 2191–2199.

Baranowski, Tom & Taveras, Elsie M. (2018). Childhood obesity prevention: Changing the focus. *Childhood Obesity,* 14(1), 1–3.

Barber, Brian K. (Ed.) (2002). *Intrusive parenting: How psychological control affects children and adolescents.* American Psychological Association.

Barlow, Jane; Herath, Nadeeja; Torrance, Christine B.; Bennett, Cathy & Wei, Yinghui. (2018). The Neonatal Behavioral Assessment Scale (NBAS) and Newborn Behavioral Observations (NBO) system for supporting caregivers and improving outcomes in caregivers and their infants. *Cochrane Database of Systematic Reviews.*

Barlowe, Avram & Cook, Ann. (2016). Putting the focus on student engagement: The benefits of performance-based assessment. *American Educator,* 40(1).

Barnett, W. Steven & Frede, Ellen C. (2017). Long-term effects of a system of high-quality universal preschool education in the United States. In Blossfeld, Hans-Peter; Kulic, Nevena; Skopek, Jan & Triventi, Moris (Eds.), *Childcare, early education and social inequality: An international perspective.* Edward Elgar.

Baron-Cohen, Simon; Tager-Flusberg, Helen & Lombardo, Michael (Eds.). (2013). *Understanding other minds: Perspectives from developmental social neuroscience* (3rd ed.). Oxford University Press.

Barone, Joseph. (2015). *It's not your fault!: Strategies for solving toilet training and bedwetting problems.* Rutgers University Press.

Barr, Rachel. (2013). Memory constraints on infant learning from picture books, television, and touchscreens. *Child Development Perspectives,* 7(4), 205–210.

Barrios, Yasmin V.; Sanchez, Sixto E.; Nicolaidis, Christina; Garcia, Pedro J.; Gelaye, Bizu; Zhong, Qiuyue & Williams, Michelle A. (2015). Childhood abuse and early menarche among Peruvian women. *Journal of Adolescent Health,* 56(2), 197–202.

Barroso, Raquel; Barbosa-Ducharne, Maria; Coelho, Vanessa; Costa, Isabel-Sofia & Silva, Ana. (2017). Psychological adjustment in intercountry and domestic adopted adolescents: A systematic review. *Child and Adolescent Social Work Journal,* 34(5), 399–418.

Barton, Georgina & Baguley, Margaret (Eds.). (2017). *The Palgrave handbook of global arts education.* Palgrave Macmillan.

Bassok, Daphna; Latham, Scott & Rorem, Anna. (2016). Is kindergarten the new first grade? *AERA Open,* 2(1).

Bateson, Patrick & Martin, Paul. (2013). *Play, playfulness, creativity and innovation.* Cambridge University Press.

Baude, Amandine; Pearson, Jessica & Drapeau, Sylvie. (2016). Child adjustment in joint physical custody versus sole custody: A meta-analytic review. *Journal of Divorce & Remarriage,* 57(5), 338–360.

Bauer, Patricia J.; San Souci, Priscilla & Pathman, Thanujeni. (2010). Infant memory. *Wiley Interdisciplinary Reviews: Cognitive Science,* 1(2), 267–277.

Baumrind, Diana. (1967). Child care practices anteceding three patterns of preschool behavior. *Genetic Psychology Monographs,* 75(1), 43–88.

Baumrind, Diana. (1971). Current patterns of parental authority. *Developmental Psychology,* 4(1, Pt. 2), 1–103.

Bayley, Nancy. (1966). Learning in adulthood: The role of intelligence. In Klausmeier, Herbert J. & Harris, Chester W. (Eds.), *Analyses of concept learning* (pp. 117–138). Academic Press.

Baysu, Gülseli & Phalet, Karen. (2019). The up- and downside of dual identity: Stereotype threat and minority performance. *Journal of Social Issues,* 75(2), 568–591.

Bea, Megan Doherty & Yi, Youngmin. (2019). Leaving the financial nest: Connecting young adults' financial independence to financial security. *Journal of Marriage and Family,* 81(2), 397–414.

Beal, Susan. (1988). Sleeping position and sudden infant death syndrome. *The Medical Journal of Australia,* 149(10), 562.

Beck, Melinda. (2009, May 26). How's your baby? Recalling the Apgar score's namesake. *Wall Street Journal.*

Beck, Martha N. (2011). *Expecting Adam: A true story of birth, rebirth, and everyday magic.* Three Rivers Press.

Beebe, Beatrice; Messinger, Daniel; Bahrick, Lorraine E.; Margolis, Amy; Buck, Karen A. & Chen, Henian. (2016). A systems view of mother–infant face-to-face communication. *Developmental Psychology,* 52(4), 556–571.

Beilin, Lawrence & Huang, Rae-Chi. (2008). Childhood obesity, hypertension, the metabolic syndrome and adult cardiovascular disease. *Clinical and Experimental Pharmacology and Physiology,* 35(4), 409–411.

Bell, Georgie; Hiscock, Harriet; Tobin, Sherryn; Cook, Fallon & Sung, Valerie. (2018). Behavioral outcomes of infant colic in toddlerhood: A longitudinal study. *The Journal of Pediatrics,* 201, 154–159.

Bellinger, David C. (2016). Lead contamination in Flint—An abject failure to protect public health. *New England Journal of Medicine,* 374(12), 1101–1103.

Belsky, Daniel W.; Caspi, Avshalom; Houts, Renate; Cohen, Harvey J.; Corcoran, David L.; Danese, Andrea; . . . Moffitt, Terrie E. (2015). Quantification of biological aging in young adults. *Proceedings of the National Academy of Sciences of the United States of America,* 112(30), e4104–e4110.

Belsky, Jay; Bakermans-Kranenburg, Marian J. & van IJzendoorn, Marinus H. (2007). For better and for worse: Differential susceptibility to environmental influences. *Current Directions in Psychological Science,* 16(6), 300–304.

Belsky, Jay; Steinberg, Laurence; Houts, Renate M. & Halpern-Felsher, Bonnie L. (2010). The development of reproductive strategy in females: Early maternal harshness → earlier menarche → increased sexual risk taking. *Developmental Psychology,* 46(1), 120–128.

Bem, Sandra L. (1981). Gender schema theory: A cognitive account of sex typing. *Psychological Review,* 88(4), 354–364.

Bender, Heather L.; Allen, Joseph P.; McElhaney, Kathleen Boykin; Antonishak, Jill; Moore, Cynthia M.; Kelly, Heather O'Beirne & Davis, Steven M. (2007). Use of harsh physical discipline and developmental outcomes in adolescence. *Development and Psychopathology,* 19(1), 227–242.

Benn, Peter. (2016). Prenatal diagnosis of chromosomal abnormalities through chorionic villus sampling and amniocentesis. In Milunsky, Aubrey & Milunsky, Jeff M. (Eds.), *Genetic disorders and the fetus: Diagnosis, prevention, and treatment* (7th ed., pp. 178–266). Wiley-Blackwell.

Benoit, Amelie; Lacourse, Eric & Claes, Michel. (2013). Pubertal timing and depressive symptoms in late adolescence: The moderating role of individual, peer, and parental factors. *Development and Psychopathology*, 25(2), 455–471.

Benson, Peter L.; Scales, Peter C.; Syvertsen, Amy K. & Roehlkepartain, Eugene C. (2012). Is youth spiritual development a universal developmental process? An international exploration. *The Journal of Positive Psychology*, 7(6), 453–470.

Berg, Jeremy. (2018). Imagine a world without facts. *Science*, 362(6413), 379.

Bergelson, Elika & Swingley, Daniel. (2018). Young infants' word comprehension given an unfamiliar talker or altered pronunciations. *Child Development*, 89(5), 1567–1576.

Berger, Kathleen S. (1980). *The developing person* (1st ed.). Worth.

Berger, Kathleen S. (2019). *Grandmothering: Building strong ties with every generation.* Rowman & Littlefield.

Berken, Jonathan A.; Gracco, Vincent L. & Klein, Denise. (2017). Early bilingualism, language attainment, and brain development. *Neuropsychologia*, 98, 220–227.

Berko, Jean. (1958). The child's learning of English morphology. *Word*, 14, 150–177.

Berkowitz, Talia; Schaeffer, Marjorie W.; Maloney, Erin A.; Peterson, Lori; Gregor, Courtney; Levine, Susan C. & Beilock, Sian L. (2015). Math at home adds up to achievement in school. *Science*, 350(6257), 196–198.

Berlin, Lisa J.; Martoccio, Tiffany L. & Jones Harden, Brenda. (2018). Improving early head start's impacts on parenting through attachment-based intervention: A randomized controlled trial. *Developmental Psychology*, 54(12), 2316–2327.

Bernard, Kristin; Hostinar, Camelia E. & Dozier, Mary. (2019). Longitudinal associations between attachment quality in infancy, C-reactive protein in early childhood, and BMI in middle childhood: Preliminary evidence from a CPS-referred sample. *Attachment & Human Development*, 21(1), 5–22.

Bernard, Kristin; Lind, Teresa & Dozier, Mary. (2014). Neurobiological consequences of neglect and abuse. In Korbin, Jill E. & Krugman, Richard D. (Eds.), *Handbook of child maltreatment* (pp. 205–223). Springer.

Bernaud, Jean-Luc. (2014). Career counseling and life meaning: A new perspective of life designing for research and applications. In Di, Fabio A. & Bernaud, J- L. (Eds.), *The construction of the identity in 21st century: A Festschrift for Jean Guichard* (pp. 29–40). Nova Science.

Bernier, Annie; Calkins, Susan D. & Bell, Martha Ann. (2016). Longitudinal associations between the quality of mother–infant interactions and brain development across infancy. *Child Development*, 87(4), 1159–1174.

Betancourt, Theresa S.; McBain, Ryan; Newnham, Elizabeth A. & Brennan, Robert T. (2013). Trajectories of internalizing problems in war-affected Sierra Leonean youth: Examining conflict and postconflict factors. *Child Development*, 84(2), 455–470.

Bettelheim, Bruno. (1972). *The empty fortress: Infantile autism and the birth of the self.* Free Press.

Bhatia, Tej K. & Ritchie, William C. (Eds.). (2013). *The handbook of bilingualism and multilingualism* (2nd ed.). Wiley-Blackwell.

Bialystok, Ellen. (2018). Bilingualism and executive function: What's the connection? In Miller, David; Bayram, Fatih; Rothman, Jason & Serratrice, Ludovica (Eds.), *Bilingual cognition and language: The state of the science across its subfields* (pp. 283–306). John Benjamins.

Bick, Johanna & Nelson, Charles A. (2016). Early adverse experiences and the developing brain. *Neuropsychopharmacology*, 41, 177–196.

Bick, Johanna; Palmwood, Erin N.; Zajac, Lindsay; Simons, Robert & Dozier, Mary. (2019). Early parenting intervention and adverse family environments affect neural function in middle childhood. *Biological Psychiatry*, 85(4), 326–335.

Biemiller, Andrew. (2009). Parent/caregiver narrative: Vocabulary development (0–60 Months). In Phillips, Linda M. (Ed.), *Handbook of language and literacy development: A roadmap from 0–60* (Online ed.). Canadian Language and Literacy Research Network.

Biffen, Stevie C.; Warton, Christopher M. R.; Lindinger, Nadine M.; Randall, Steven R.; Lewis, Catherine E.; Molteno, Christopher D.; . . . Meintjes, Ernesta M. (2018). Reductions in corpus callosum volume partially mediate effects of prenatal alcohol exposure on IQ. *Frontiers in Neuroanatomy*, 11(132).

Binning, Kevin R.; Cook, Jonathan E.; Purdie-Greenaway, Valerie; Garcia, Julio; Chen, Susie; Apfel, Nancy; . . . Cohen, Geoffrey L. (2019). Bolstering trust and reducing discipline incidents at a diverse middle school: How self-affirmation affects behavioral conduct during the transition to adolescence. *Journal of School Psychology*, 75, 74–88.

Biro, Frank M.; Greenspan, Louise C.; Galvez, Maida P.; Pinney, Susan M.; Teitelbaum, Susan; Windham, Gayle C.; . . . Wolff, Mary S. (2013). Onset of breast development in a longitudinal cohort. *Pediatrics*, 132(6), 1019–1027.

Bjorklund, David F. (2018). A metatheory for cognitive development (or "Piaget is dead" revisited). *Child Development*, 89(6), 2288–2302.

Bjorklund, David F. & Ellis, Bruce J. (2014). Children, childhood, and development in evolutionary perspective. *Developmental Review*, 34(3), 225–264.

Bjorklund, David F. & Sellers, Patrick D. (2014). Memory development in evolutionary perspective. In Bauer, Patricia & Fivush, Robyn (Eds.), *The Wiley handbook on the development of children's memory* (Vol. 1, pp. 126–150). Wiley.

Black, Robert E.; Victora, Cesar G.; Walker, Susan P.; Bhutta, Zulfiqar A.; Christian, Parul; Onis, Mercedes de & Ezzati, Majid. (2013). Maternal and child undernutrition and overweight in low-income and middle-income countries. *The Lancet*, 382(9809), 427–451.

Blair, Clancy. (2016). Developmental science and executive function. *Current Directions in Psychological Science*, 25(1), 3–7.

Blair, Clancy & Raver, C. Cybele. (2015). School readiness and self-regulation: A developmental psychobiological approach. *Annual Review of Psychology*, 66, 711–731.

Blakemore, Sarah-Jayne. (2018). Avoiding social risk in adolescence. *Current Directions in Psychological Science*, 27(2), 116–122.

Blandon, Alysia Y.; Calkins, Susan D. & Keane, Susan P. (2010). Predicting emotional and social competence during early childhood from toddler risk and maternal behavior. *Development and Psychopathology*, 22(1), 119–132.

Blank, Grant & Lutz, Christoph. (2018). Benefits and harms from internet use: A differentiated analysis of Great Britain. *New Media & Society*, 20(2), 618–640.

Bleich, Sara N.; Segal, Jodi; Wu, Yang; Wilson, Renee & Wang, Youfa. (2013). Systematic review of community-based childhood obesity prevention studies. *Pediatrics*, 132(1), e201–e210.

Bleidorn, Wiebke; Klimstra, Theo A.; Denissen, Jaap J. A.; Rentfrow, Peter J.; Potter, Jeff & Gosling, Samuel D. (2013). Personality maturation around the world: A cross-cultural examination of social-investment theory. *Psychological Science*, 24(12), 2530–2540.

Blencowe, Hannah; Krasevec, Julia; de Onis, Mercedes; Black, Robert E.; An, Xiaoyi; Stevens, Gretchen A.; . . . Cousens, Simon. (2019). National, regional, and worldwide estimates of low birthweight in 2015, with trends from 2000: A systematic analysis. *The Lancet Global Health*, 7(7), e849–e860.

Blomqvist, Ylva Thernström; Nyqvist, Kerstin Hedberg; Rubertsson, Christine & Funkquist, Eva-Lotta. (2017). Parents need support to find ways to optimise their own sleep without seeing their preterm infant's sleeping patterns as a problem. *Acta Paediatrica*, 106(2), 223–228.

Blossfeld, Hans-Peter; Kulic, Nevena; Skopek, Jan & Triventi, Moris (Eds.). (2017). *Childcare, early education and social inequality: An international perspective.* Edward Elgar.

Blurton-Jones, Nicholas G. (1976). Rough-and-tumble play among nursery school children. In Bruner, Jerome S.; Jolly, Alison & Sylva, Kathy (Eds.), *Play: Its role in development and evolution* (pp. 352–363). Basic Books.

Bodner-Adler, Barbara; Kimberger, Oliver; Griebaum, Julia; Husslein, Peter & Bodner, Klaus. (2017). A ten-year study of midwife-led care at an Austrian tertiary care center: A retrospective analysis with special consideration of perineal trauma. *BMC Pregnancy and Childbirth*, 17, 357–371.

Bodrava, Elena & Leong, Deborah J. (2018). Tools of the mind: The Vygotskian-based early childhood program. *Journal of Cognitive Education and Psychology*, 17(3), 223–237.

Bøe, Tormod; Serlachius, Anna; Sivertsen, Børge; Petrie, Keith & Hysing, Mari. (2018).

Cumulative effects of negative life events and family stress on children's mental health: The Bergen Child Study. *Social Psychiatry and Psychiatric Epidemiology*, 53(1), 1–9.

Boerma, Ties; Ronsmans, Carine; Melesse, Dessalegn Y.; Barros, Aluisio J. D.; Barros, Fernando C.; Juan, Liang; . . . Temmerman, Marleen. (2018). Global epidemiology of use of and disparities in caesarean sections. *The Lancet*, 392(10155), 1341–1348.

Bögels, Susan M.; Knappe, Susanne & Clark, Lee Anna. (2013). Adult separation anxiety disorder in *DSM-5*. *Clinical Psychology Review*, 33(5), 663–674.

Boggio, Andrea; Romano, Cesare P. R. & Almqvist, Jessica (Eds.). (2020). *Human germline genome modification and the right to science: A comparative study of national laws and policies*. Cambridge University Press.

Bohlen, Tabata M.; Silveira, Marina A.; Zampieri, Thais T.; Frazão, Renata & Donato, Jose. (2016). Fatness rather than leptin sensitivity determines the timing of puberty in female mice. *Molecular and Cellular Endocrinology*, 423, 11–21.

Bohn, Manuel & Köymen, Bahar. (2018). Common ground and development. *Child Development Perspectives*, 12(2), 104–108.

Bolk, Jenny; Padilla, Nelly; Forsman, Lea; Broström, Lina; Hellgren, Kerstin & Åden, Ulrika. (2018). Visual–motor integration and fine motor skills at 6½ years of age and associations with neonatal brain volumes in children born extremely preterm in Sweden: A population-based cohort study. *BMJ Open*, 8(2), e020478.

Bonora, Massimo; Wieckowski, Mariusz R.; Sinclair, David A.; Kroemer, Guido; Pinton, Paolo & Galluzzi, Lorenzo. (2019). Targeting mitochondria for cardiovascular disorders: Therapeutic potential and obstacles. *Nature Reviews Cardiology*, 16, 33–55.

Booth, Alan & Dunn, Judy (Eds.). (2014). *Stepfamilies: Who benefits? Who does not?* Routledge.

Borke, Jörn; Lamm, Bettina; Eickhorst, Andreas & Keller, Heidi. (2007). Father-infant interaction, paternal ideas about early child care, and their consequences for the development of children's self-recognition. *Journal of Genetic Psychology*, 168(4), 365–379.

Bornstein, Marc H. (2017). The specificity principle in acculturation science. *Perspectives on Psychological Science*, 12(1), 3–45.

Bornstein, Marc H.; Arterberry, Martha E. & Mash, Clay. (2005). Perceptual development. In Bornstein, Marc H. & Lamb, Michael E. (Eds.), *Developmental science: An advanced textbook* (5th ed., pp. 283–325). Lawrence Erlbaum Associates.

Bornstein, Marc H.; Mortimer, Jeylan T.; Lutfey, Karen & Bradley, Robert. (2011). Theories and processes in life-span socialization. In Fingerman, Karen L.; Berg, Cynthia; Smith, Jacqui & Antonucci, Toni (Eds.), *Handbook of life-span development* (pp. 27–56). Springer.

Bornstein, Marc H.; Putnick, Diane L.; Bradley, Robert H.; Deater-Deckard, Kirby & Lansford, Jennifer E. (2016). Gender in low- and middle-income countries: Introduction. *Monographs of the Society for Research in Child Development*, 81(1), 7–23.

Bouazzaoui, B.; Fay, S.; Guerrero-Sastoque, L; Semaine, M.; Isingrini, M. & Taconnat, L. (2020). Memory age-based stereotype threat: Role of locus of control and anxiety. *Experimental Aging Research*, 46(1), 39–51.

Boundy, Ellen O.; Dastjerdi, Roya; Spiegelman, Donna; Fawzi, Wafaie W.; Missmer, Stacey A.; Lieberman, Ellice; . . . Chan, Grace J. (2016). Kangaroo mother care and neonatal outcomes: A meta-analysis. *Pediatrics*, 137(1), e20152238.

Bourguignon, Jean-Pierre; Juul, Anders; Franssen, Delphine; Fudvoye, Julie; Pinson, Anneline & Parent, Anne-Simone. (2016). Contribution of the endocrine perspective in the evaluation of endocrine disrupting chemical effects: The case study of pubertal timing. *Hormone Research in Paediatrics*, 86(4), 221–232.

Bowlby, John. (1983). *Attachment* (2nd ed.). Basic Books.

Bowman, Lindsay C.; Dodell-Feder, David; Saxe, Rebecca & Sabbagh, Mark A. (2019). Continuity in the neural system supporting children's theory of mind development: Longitudinal links between task-independent EEG and task-dependent fMRI. *Developmental Cognitive Neuroscience*, 40(100705).

Box, Allyson G.; Feito, Yuri; Matson, Ashton; Heinrich, Katie M. & Petruzzello, Steven J. (2019). Is age just a number? Differences in exercise participatory motives across adult cohorts and the relationships with exercise behaviour. *International Journal of Sport and Exercise Psychology*.

boyd, danah. (2014). *It's complicated: The social lives of networked teens*. Yale University Press.

Braak, David; Michael Cummings, K.; Nahhas, Georges J.; Reid, Jessica L. & Hammond, David. (2020). How are adolescents getting their vaping products? Findings from the international tobacco control (ITC) youth tobacco and vaping survey. *Addictive Behaviors*, 105, 106345.

Braams, Barbara R.; van Duijvenvoorde, Anna C. K.; Peper, Jiska S. & Crone, Eveline A. (2015). Longitudinal changes in adolescent risk-taking: A comprehensive study of neural responses to rewards, pubertal development, and risk-taking behavior. *The Journal of Neuroscience*, 35(18), 7226–7238.

Bracken, Bruce A. & Crawford, Elizabeth. (2010). Basic concepts in early childhood educational standards: A 50-state review. *Early Childhood Education Journal*, 37(5), 421–430.

Bradley, Rachel & Slade, Pauline. (2011). A review of mental health problems in fathers following the birth of a child. *Journal of Reproductive and Infant Psychology*, 29(1), 19–42.

Brainard, Jeffrey. (2018). Rethinking retractions. *Science*, 362(6413), 390–393.

Braithwaite, David W.; Tian, Jing & Siegler, Robert S. (2018). Do children understand fraction addition? *Developmental Science*, 21(4), e12601.

Brame, Robert; Bushway, Shawn D.; Paternoster, Ray & Turner, Michael G. (2014). Demographic patterns of cumulative arrest prevalence by ages 18 and 23. *Crime & Delinquency*, 60(3), 471–486.

Brand, Jennie E. & Xie, Yu. (2010). Who benefits most from college?: Evidence for negative selection in heterogeneous economic returns to higher education. *American Sociological Review*, 75(2), 273–302.

Brandone, Amanda C.; Horwitz, Suzanne R.; Aslin, Richard N. & Wellman, Henry M. (2014). Infants' goal anticipation during failed and successful reaching actions. *Developmental Science*, 17(1), 23–34.

Branje, Susan. (2018). Development of parent–adolescent relationships: Conflict interactions as a mechanism of change. *Child Development Perspectives*, 12(3), 171–176.

Braver, Sanford L. & Votruba, Ashley M. (2018). Does joint physical custody "cause" children's better outcomes? *Journal of Divorce & Remarriage*, 59(5), 452–468.

Brazelton, T. Berry & Sparrow, Joshua D. (2006). *Touchpoints, birth to 3: Your child's emotional and behavioral development* (2nd ed.). Da Capo Press.

Bremner, J. Gavin & Wachs, Theodore D. (Eds.). (2010). *The Wiley-Blackwell handbook of infant development* (2nd ed.). Wiley-Blackwell.

Brennan, Arthur; Ayers, Susan; Ahmed, Hafez & Marshall-Lucette, Sylvie. (2007). A critical review of the Couvade syndrome: The pregnant male. *Journal of Reproductive and Infant Psychology*, 25(3), 173–189.

Bridge, Jeffrey A.; Greenhouse, Joel B.; Ruch, Donna; Stevens, Jack; Ackerman, John; Sheftall, Arielle H.; . . . Campo, John V. (2019). Association between the release of Netflix's *13 Reasons Why* and suicide rates in the United States: An interrupted times series analysis. *Journal of the American Academy of Child & Adolescent Psychiatry*.

Bridgett, David J.; Burt, Nicole M.; Edwards, Erin S. & Deater-Deckard, Kirby. (2015). Intergenerational transmission of self-regulation: A multidisciplinary review and integrative conceptual framework. *Psychological Bulletin*, 141(3), 602–654.

Bridgett, David J.; Laake, Lauren M.; Gartstein, Maria A. & Dorn, Danielle. (2013). Development of infant positive emotionality: The contribution of maternal characteristics and effects on subsequent parenting. *Infant and Child Development*, 22(4), 362–382.

Brix, Nis; Ernst, Andreas; Lauridsen, Lea Lykke Braskhoj; Parner, Erik; Stovring, Henrik; Olsen, Jorn; . . . Ramlau-Hansen, Cecilia Høst. (2019). Timing of puberty in boys and girls: A population-based study. *Paediatric and Perinatal Epidemiology*, 33(1), 70–78.

Brody, Jane E. (2013, February 26). Too many pills in pregnancy. *New York Times*.

Broekhuizen, Martine L.; Aken, Marcel A. G.; Dubas, Judith S. & Leseman, Paul P. M. (2018). Child care quality and Dutch 2- and 3-year-olds' socio-emotional outcomes: Does the amount of care matter? *Infant and Child Development*, 27(1), e2043.

Bronfenbrenner, Urie & Morris, Pamela A. (2006). The bioecological model of human development. In Damon, William & Lerner, Richard M.

(Eds.), *Handbook of child psychology* (6th ed., Vol. 1, pp. 793–828). Wiley.

Brooks, Rechele; Singleton, Jenny L. & Meltzoff, Andrew N. (2019). Enhanced gaze-following behavior in Deaf infants of Deaf parents. *Developmental Science*, e12900.

Broström, Stig. (2017). A dynamic learning concept in early years' education: A possible way to prevent schoolification. *International Journal of Early Years Education*, 25(1), 3–15.

Brouwer, Rachel M.; van Soelen, Inge L. C.; Swagerman, Suzanne C.; Schnack, Hugo G.; Ehli, Erik A.; Kahn, René S.; . . . Boomsma, Dorret I. (2014). Genetic associations between intelligence and cortical thickness emerge at the start of puberty. *Human Brain Mapping*, 35(8), 3760–3773.

Brown, Belinda. (2019). From hegemonic to responsive masculinity: The transformative power of the provider role. In Barry, John A.; Kingerlee, Roger; Seager, Martin & Sullivan, Luke (Eds.), *The Palgrave handbook of male psychology and mental health* (pp. 183–204). Palgrave Macmillan.

Brown, B. Bradford & Bakken, Jeremy P. (2011). Parenting and peer relationships: Reinvigorating research on family–peer linkages in adolescence. *Journal of Research on Adolescence*, 21(1), 153–165.

Brown, Christia Spears; Alabi, Basirat O.; Huynh, Virginia W. & Masten, Carrie L. (2011). Ethnicity and gender in late childhood and early adolescence: Group identity and awareness of bias. *Developmental Psychology*, 47(2), 463–471.

Brown, Steven D. & Lent, Robert W. (2016). Vocational psychology: Agency, equity, and well-being. *Annual Review of Psychology*, 67, 541–565.

Bruder, Gerard E.; Stewart, Jonathan W. & McGrath, Patrick J. (2017). Right brain, left brain in depressive disorders: Clinical and theoretical implications of behavioral, electrophysiological and neuroimaging findings. *Neuroscience and Biobehavioral Reviews*, 78, 178–191.

Brummelman, Eddie; Nelemans, Stefanie A.; Thomaes, Sander & Orobio De Castro, Bram. (2017). When parents' praise inflates, children's self-esteem deflates. *Child Development*, 88(6), 1799–1809.

Buehler, Cheryl. (2020). Family processes and children's and adolescents' well-being. *Journal of Marriage and Family*, 82(1), 145–174.

Bueno, Clarissa & Menna-Barreto, Luiz. (2016). Environmental factors influencing biological rhythms in newborns: From neonatal intensive care units to home. *Sleep Science*, 9(4), 295–300.

Burchinal, Margaret R.; Lowe Vandell, Deborah & Belsky, Jay. (2014). Is the prediction of adolescent outcomes from early child care moderated by later maternal sensitivity? Results from the NICHD study of early child care and youth development. *Developmental Psychology*, 50(2), 542–553.

Burnette, Jeni L.; O'Boyle, Ernest H.; VanEpps, Eric M.; Pollack, Jeffrey M. & Finkel, Eli J. (2013). Mind-sets matter: A meta-analytic review of implicit theories and self-regulation. *Psychological Bulletin*, 139(3), 655–701.

Bursztyn, Leonardo & Jensen, Robert. (2015). How does peer pressure affect educational investments? *Quarterly Journal of Economics*, 130(3), 1329–1367.

Buss, David M. (2015). *Evolutionary psychology: The new science of the mind* (5th ed.). Routledge.

Butterworth, Brian & Kovas, Yulia. (2013). Understanding neurocognitive developmental disorders can improve education for all. *Science*, 340(6130), 300–305.

Butterworth, Brian; Varma, Sashank & Laurillard, Diana. (2011). Dyscalculia: From brain to education. *Science*, 332(6033), 1049–1053.

Byard, Roger W. (2014). "Shaken baby syndrome" and forensic pathology: An uneasy interface. *Forensic Science, Medicine, and Pathology*, 10(2), 239–241.

Byers-Heinlein, Krista; Burns, Tracey C. & Werker, Janet F. (2010). The roots of bilingualism in newborns. *Psychological Science*, 21(3), 343–348.

Cabrera, Natasha. (2015). Why do fathers matter for children's development? In McHale, Susan M.; King, Valarie; Van Hook, Jennifer & Booth, Alan (Eds.), *Gender and Couple Relationships* (pp. 161–168). Springer.

Cabrera, Natasha J.; Volling, Brenda L. & Barr, Rachel. (2018). Fathers are parents, too! Widening the lens on parenting for children's development. *Child Development Perspectives*, 12(3), 152–157.

Cacioppo, Stephanie; Capitanio, John P. & Cacioppo, John T. (2014). Toward a neurology of loneliness. *Psychological Bulletin*, 140(6), 1464–1504.

Calarco, Jessica McCrory. (2014). The inconsistent curriculum: Cultural tool kits and student interpretations of ambiguous expectations. *Social Psychology Quarterly*, 77(2), 185–209.

Calderon, Valerie J. & Yu, Daniela. (2017, June 1). *Student enthusiasm falls as high school graduation nears.* Washington, DC: Gallup.

Callaghan, Tara. (2013). Symbols and symbolic thought. In Zelazo, Philip D. (Ed.), *The Oxford handbook of developmental psychology* (Vol. 1). Oxford University Press.

Campbell, Frances; Conti, Gabriella; Heckman, James J.; Moon, Seong H.; Pinto, Rodrigo; Pungello, Elizabeth & Pan, Yi. (2014). Early childhood investments substantially boost adult health. *Science*, 343(6178), 1478–1485.

Campbell, Frances A.; Pungello, Elizabeth P.; Miller-Johnson, Shari; Burchinal, Margaret & Ramey, Craig T. (2001). The development of cognitive and academic abilities: Growth curves from an early childhood educational experiment. *Developmental Psychology*, 37(2), 231–242.

Cao, Zhipeng; Bennett, Marc; Orr, Catherine; Icke, Ilknur; Banaschewski, Tobias; Barker, Gareth J.; . . . Whelan, Robert. (2019). Mapping adolescent reward anticipation, receipt, and prediction error during the monetary incentive delay task. *Human Brain Mapping*, 40(1), 262–283.

Carey, Nessa. (2012). *The epigenetics revolution: How modern biology is rewriting our understanding of genetics, disease, and inheritance.* Columbia University Press.

Carey, Susan; Zaitchik, Deborah & Bascandziev, Igor. (2015). Theories of development: In dialog with Jean Piaget. *Developmental Review*, 38, 36–54.

Cargnelutti, Elisa; Tomasino, Barbara & Fabbro, Franco. (2019). Language brain representation in bilinguals with different age of appropriation and proficiency of the second language: A meta-analysis of functional imaging studies. *Frontiers in Human Neuroscience*, 13(154).

Carlsen, Karin C. Lødrup; Skjerven, Håvard O. & Carlsen, Kai-Håkon. (2018). The toxicity of e-cigarettes and children's respiratory health. *Paediatric Respiratory Reviews*, 28, 63–67.

Carlson, Robert G.; Nahhas, Ramzi W.; Martins, Silvia S. & Daniulaityte, Raminta. (2016). Predictors of transition to heroin use among initially non-opioid dependent illicit pharmaceutical opioid users: A natural history study. *Drug & Alcohol Dependence*, 160, 127–134.

Carlson, Scott. (2016, May 1). Should everyone go to college?: For poor kids, 'College for all' isn't the mantra it was meant to be. *Chronicle of Higher Education*.

Carlson, Stephanie M.; Koenig, Melissa A. & Harms, Madeline B. (2013). Theory of mind. *Wiley Interdisciplinary Reviews: Cognitive Science*, 4(4), 391–402.

Carmichael, Cheryl L.; Reis, Harry T. & Duberstein, Paul R. (2015). In your 20s it's quantity, in your 30s it's quality: The prognostic value of social activity across 30 years of adulthood. *Psychology and Aging*, 30(1), 95–105.

Carra, Cecilia; Lavelli, Manuela; Keller, Heidi & Kärtner, Joscha. (2013). Parenting infants: Socialization goals and behaviors of Italian mothers and immigrant mothers from West Africa. *Journal of Cross-Cultural Psychology*, 44(8), 1304–1320.

Carroll, Katherine & Kroløkke, Charlotte. (2018). Freezing for love: Enacting 'responsible' reproductive citizenship through egg freezing. *Culture, Health & Sexuality*, 20(9), 992–1005.

Carter, A. S. & Briggs-Gowan, M. J. (2006). *Infant-toddler social and emotional assessment.* Harcourt.

Carwile, Jenny L.; Willett, Walter C.; Spiegelman, Donna; Hertzmark, Ellen; Rich-Edwards, Janet W.; Frazier, A. Lindsay & Michels, Karin B. (2015). Sugar-sweetened beverage consumption and age at menarche in a prospective study of US girls. *Human Reproduction*, 30(3), 675–683.

Casey, B. J. & Caudle, Kristina. (2013). The teenage brain: Self control. *Current Directions in Psychological Science*, 22(2), 82–87.

Caspi, Avshalom; Moffitt, Terrie E.; Morgan, Julia; Rutter, Michael; Taylor, Alan; Arseneault, Louise; . . . Polo-Tomas, Monica. (2004). Maternal expressed emotion predicts children's antisocial behavior problems: Using monozygotic-twin differences to identify environmental effects on behavioral development. *Developmental Psychology*, 40(2), 149–161.

Cassia, Viola Macchi; Kuefner, Dana; Picozzi, Marta & Vescovo, Elena. (2009). Early experience predicts later plasticity for face processing: Evidence for the reactivation of dormant effects. *Psychological Science*, 20(7), 853–859.

Cassidy, Jude & Shaver, Phillip R. (Eds.). (2016). *Handbook of attachment: Theory, research, and clinical applications.* Guilford Press.

Castellacci, Fulvio & Tveito, Vegard. (2018). Internet use and well-being: A survey and a theoretical framework. *Research Policy, 47*(1), 308–325.

Ceballo, Rosario; Maurizi, Laura K.; Suarez, Gloria A. & Aretakis, Maria T. (2014). Gift and sacrifice: Parental involvement in Latino adolescents' education. *Cultural Diversity and Ethnic Minority Psychology, 20*(1), 116–127.

Cenegy, Laura Freeman; Denney, Justin T. & Kimbro, Rachel Tolbert. (2018). Family diversity and child health: Where do same-sex couple families fit. *Journal of Marriage and Family, 80*(1), 198–218.

Center for Education Policy. (2012). *SDP strategic performance indicator: The high school effect on college-going. The SDP College-Going Diagnostic Strategic Performance Indicators.* Cambridge, MA: Harvard University, Center for Education Policy Research.

Centers for Disease Control and Prevention. (2015, May 15). *Epidemiology and prevention of vaccine-preventable diseases* (Hamborsky, Jennifer; Kroger, Andrew & Wolfe, Charles Eds. 13th ed.). Public Health Foundation.

Centers for Disease Control and Prevention. (2018). *Sexually transmitted disease surveillance 2017.* Atlanta, GA: U.S. Department of Health and Human Services.

Centers for Disease Control and Prevention. (2018, October 9). *Youth risk behavior survey: Data summary & trends report 2007–2017.* Atlanta, GA: National Center for HIV/AIDS, Viral Hepatitis, STD, and TB Prevention, Division of Adolescent and School Health. 328447.

Centers for Disease Control and Prevention. (2019). Breastfeeding rates: Breastfeeding among U.S. children born 2009–2016, CDC National Immunization Survey. U.S. Department of Health & Human Services.

Centers for Disease Control and Prevention. (2019, June 17). Measles cases and outbreaks. Centers for Disease Control and Prevention.

Centers for Disease Control and Prevention. (2020, February 20). WISQARS: Fatal injury reports, national, regional and state, 1981–2018. U.S. Department of Health & Human Services.

Centers for Disease Control and Prevention, National Center for Health Statistics. (2018, July 3). *Underlying cause of death, 1999–2017 on CDC WONDER Online Database* [Data set]. CDC WONDER.

Cesana-Arlotti, Nicoló; Martín, Ana; Téglás, Ernő; Vorobyova, Liza; Cetnarski, Ryszard & Bonatti, Luca L. (2018). Precursors of logical reasoning in preverbal human infants. *Science, 359*(6381), 1263–1266.

Cespedes, Elizabeth M.; McDonald, Julia; Haines, Jess; Bottino, Clement J.; Schmidt, Marie Evans & Taveras, Elsie M. (2013). Obesity-related behaviors of US- and non-US-born parents and children in low-income households. *Journal of Developmental & Behavioral Pediatrics, 34*(8), 541–548.

Ceyhan, Gaye D.; Thompson, Alia N.; Sloane, Jeremy D.; Wiles, Jason R. & Tillotson, John W. (2019). The socialization and retention of low-income college students: The impact of a wrap-around intervention. *International Journal of Higher Education, 8*(6).

Champagne, Frances A. & Curley, James P. (2010). Maternal care as a modulating influence on infant development. In Blumberg, Mark S.; Freeman, John H. & Robinson, Scott R. (Eds.), *Oxford handbook of developmental behavioral neuroscience* (pp. 323–341). Oxford University Press.

Chartier, Karen G.; Scott, Denise M.; Wall, Tamara L.; Covault, Jonathan; Karriker-Jaffe, Katherine J.; Mills, Britain A.; . . . Arroyo, Judith A. (2014). Framing ethnic variations in alcohol outcomes from biological pathways to neighborhood context. *Alcoholism: Clinical and Experimental Research, 38*(3), 611–618.

Chen, Edith; Brody, Gene H. & Miller, Gregory E. (2017). Childhood close family relationships and health. *American Psychologist, 72*(6), 555–566.

Chen, Hong & Jackson, Todd. (2009). Predictors of changes in weight esteem among mainland Chinese adolescents: A longitudinal analysis. *Developmental Psychology, 45*(6), 1618–1629.

Chen, Mu-Hong; Lan, Wen-Hsuan; Bai, Ya-Mei; Huang, Kai-Lin; Su, Tung-Ping; Tsai, Shih-Jen; . . . Hsu, Ju-Wei. (2016). Influence of relative age on diagnosis and treatment of Attention-deficit hyperactivity disorder in Taiwanese children. *The Journal of Pediatrics, 172,* 162–167.e1.

Chen, Xinyin. (2019). Culture and shyness in childhood and adolescence. *New Ideas in Psychology, 53,* 58–66.

Chen, Xinyin; Cen, Guozhen; Li, Dan & He, Yunfeng. (2005). Social functioning and adjustment in Chinese children: The imprint of historical time. *Child Development, 76*(1), 182–195.

Chen, Xinyin; Fu, Rui; Li, Dan & Liu, Junsheng. (2019). Developmental trajectories of shyness-sensitivity from middle childhood to early adolescence in China: Contributions of peer preference and mutual friendship. *Journal of Abnormal Child Psychology, 47*(7), 1197–1209.

Chen, Xinyin; Rubin, Kenneth H. & Sun, Yuerong. (1992). Social reputation and peer relationships in Chinese and Canadian children: A cross-cultural study. *Child Development, 63*(6), 1336–1343.

Chen, Xinyin; Yang, Fan & Wang, Li. (2013). Relations between shyness-sensitivity and internalizing problems in Chinese children: Moderating effects of academic achievement. *Journal of Abnormal Child Psychology, 41*(5), 825–836.

Chen, Yalin; Loehr, Janeen D. & Campbell, Jamie I. D. (2019). Does the min-counting strategy for simple addition become automatized in educated adults? A behavioural and ERP study of the size congruency effect. *Neuropsychologia, 124,* 311–321.

Cherlin, Andrew J. (2020). Degrees of change: An assessment of the deinstitutionalization of marriage thesis. *Journal of Marriage and Family, 82*(1), 62–80.

Cherng, Hua-Yu Sebastian & Liu, Jia-Lin. (2017). Academic social support and student expectations: The case of second-generation Asian Americans. *Asian American Journal of Psychology, 8*(1), 16–30.

Cheshire, Emily; Kaestle, Christine E. & Miyazaki, Yasuo. (2019). The influence of parent and parent–adolescent relationship characteristics on sexual trajectories into adulthood. *Archives of Sexual Behavior, 48*(3), 893–910.

Cheslack-Postava, Keely; Liu, Kayuet & Bearman, Peter S. (2011). Closely spaced pregnancies are associated with increased odds of autism in California sibling births. *Pediatrics, 127*(2), 246–253.

Child Trends. (2019, March 7). Dual language learners. Child Trends.

Child Trends Databank. (2015, March). *Lead poisoning: Indicators on children and youth.* Bethesda, MD: Child Trends.

Child Trends Databank. (2018). High school dropout rates. Child Trends.

Choe, Daniel E.; Lane, Jonathan D.; Grabell, Adam S. & Olson, Sheryl L. (2013a). Developmental precursors of young school-age children's hostile attribution bias. *Developmental Psychology, 49*(12), 2245–2256.

Choe, Daniel E.; Olson, Sheryl L. & Sameroff, Arnold J. (2013b). The interplay of externalizing problems and physical and inductive discipline during childhood. *Developmental Psychology, 49*(11), 2029–2039.

Choi, Hyunkyung; Van Riper, Marcia & Thoyre, Suzanne. (2012). Decision making following a prenatal diagnosis of Down syndrome: An integrative review. *Journal of Midwifery & Women's Health, 57*(2), 156–164.

Choi, Jayoung. (2019). A child's trilingual language practices in Korean, Farsi, and English: From a sustainable translanguaging perspective. *International Journal of Multilingualism, 16*(4), 534–548.

Chomsky, Noam. (1968). *Language and mind.* Harcourt Brace & World.

Chomsky, Noam. (1980). *Rules and representations.* Columbia University Press.

Chong, Jessica X.; Buckingham, Kati J.; Jhangiani, Shalini N.; Boehm, Corinne; Sobreira, Nara; Smith, Joshua D.; . . . Bamshad, Michael J. (2015). The genetic basis of Mendelian phenotypes: Discoveries, challenges, and opportunities. *American Journal of Human Genetics, 97*(2), 199–215.

Choudhury, Ananyo; Aron, Shaun; Sengupta, Dhriti; Hazelhurst, Scott & Ramsay, Michèle. (2018). African genetic diversity provides novel insights into evolutionary history and local adaptations. *Human Molecular Genetics, 27*(R2), R209–R218.

Christakis, Erika. (2016). *The importance of being little: What preschoolers really need from grownups.* Viking.

Christian, Cindy W. & Block, Robert. (2009). Abusive head trauma in infants and children. *Pediatrics, 123*(5), 1409–1411.

Christian, Kimberly M.; Song, Hongjun & Ming, Guo-Li. (2018). A previously undetected pathology of Zika virus infection. *Nature Medicine, 24*(3), 258–259.

Christoforides, Michael; Spanoudis, George & Demetriou, Andreas. (2016). Coping with logical fallacies: A developmental training program for learning to reason. *Child Development*, 87(6), 1856–1876.

Chu, Shuyuan; Chen, Qian; Chen, Yan; Bao, Yixiao; Wu, Min & Zhang, Jun. (2017). Cesarean section without medical indication and risk of childhood asthma, and attenuation by breastfeeding. *PLoS ONE*, 12(9), e0184920.

Cicchetti, Dante. (2013a). Annual research review: Resilient functioning in maltreated children—past, present, and future perspectives. *Journal of Child Psychology and Psychiatry*, 54(4), 402–422.

Cicchetti, Dante. (2013b). An overview of developmental psychopathology. In Zelazo, Philip D. (Ed.), *The Oxford handbook of developmental psychology* (Vol. 2, pp. 455–480). Oxford University Press.

Cicchetti, Dante. (2016). Socioemotional, personality, and biological development: Illustrations from a multilevel developmental psychopathology perspective on child maltreatment. *Annual Review of Psychology*, 67, 187–211.

Cicconi, Megan. (2014). Vygotsky meets technology: A reinvention of collaboration in the early childhood mathematics classroom. *Early Childhood Education Journal*, 42(1), 57–65.

Cierpka, Manfred & Cierpka, Astrid. (2016). Developmentally appropriate vs. persistent defiant and aggressive behavior. In Cierpka, Manfred (Ed.), *Regulatory disorders in infants.* Springer.

Cillessen, Antonius H. N. & Marks, Peter E. L. (2011). Conceptualizing and measuring popularity. In Cillessen, Antonius H. N.; Schwartz, David & Mayeux, Lara (Eds.), *Popularity in the peer system* (pp. 25–56). Guilford Press.

Claessen, Jacques. (2017). *Forgiveness in criminal law through incorporating restorative mediation.* Wolf Legal Publishers.

Clark, Caron A. C.; Fang, Hua; Espy, Kimberly A.; Filipek, Pauline A.; Juranek, Jenifer; Bangert, Barbara; . . . Taylor, H. Gerry. (2013). Relation of neural structure to persistently low academic achievement: A longitudinal study of children with differing birth weights. *Neuropsychology*, 27(3), 364–377.

Clark, Lee Anna; Cuthbert, Bruce; Lewis-Fernández, Roberto; Narrow, William E. & Reed, Geoffrey M. (2017). Three approaches to understanding and classifying mental disorder: ICD-11, *DSM-5*, and the National Institute of Mental Health's Research Domain Criteria (RDoC). *Psychological Science in the Public Interest*, 18(2), 72–145.

Coe, Jesse L.; Davies, Patrick T. & Sturge-Apple, Melissa L. (2018). Family instability and young children's school adjustment: Callousness and negative internal representations as mediators. *Child Development*, 89(4), 1193–1208.

Coffelt, Tina A. (2017). Deciding to reveal sexual information and sexuality education in mother-daughter relationships. *Sex Education*, 17(5), 571–587.

Cohen, Jon. (2014). Saving lives without new drugs. *Science*, 346(6212), 911.

Cohen, Jon. (2018). What now for human genome editing? *Science*, 362(6419), 1090–1092.

Cohen, Joel E. & Malin, Martin B. (Eds.). (2010). *International perspectives on the goals of universal basic and secondary education.* Routledge.

Colaco, Marc; Johnson, Kelly; Schneider, Dona & Barone, Joseph. (2013). Toilet training method is not related to dysfunctional voiding. *Clinical Pediatrics*, 52(1), 49–53.

Colditz, Paul B.; Boyd, Roslyn N.; Winter, Leanne; Pritchard, Margo; Gray, Peter H.; Whittingham, Koa; . . . Sanders, Matthew R. (2019). A randomized trial of Baby Triple P for preterm infants: Child outcomes at 2 years of corrected age. *Journal of Pediatrics*, 210, 48–54.e2.

Cole, Pamela M. & Hollenstein, Tom. (2018). *Emotion regulation: A matter of time.* Routledge.

Coleman-Jensen, Alisha; Rabbitt, Matthew P.; Gregory, Christian A. & Singh, Anita. (2017). *Household food security in the United States in 2016.* Washington, DC: U.S. Department of Agriculture, Economic Research Service. ERR–237.

Coles, Robert. (1997). *The moral intelligence of children: How to raise a moral child.* Random House.

Collin-Vézina, Delphine; De La Sablonnière-Griffin, Mireille; Palmer, Andrea M. & Milne, Lise. (2015). A preliminary mapping of individual, relational, and social factors that impede disclosure of childhood sexual abuse. *Child Abuse & Neglect*, 43, 123–134.

Colson, Eve R.; Willinger, Marian; Rybin, Denis; Heeren, Timothy; Smith, Lauren A.; Lister, George & Corwin, Michael J. (2013). Trends and factors associated with infant bed sharing, 1993–2010: The National Infant Sleep Position study. *JAMA Pediatrics*, 167(11), 1032–1037.

Compian, Laura J.; Gowen, L. Kris & Hayward, Chris. (2009). The interactive effects of puberty and peer victimization on weight concerns and depression symptoms among early adolescent girls. *Journal of Early Adolescence*, 29(3), 357–375.

Confer, Jaime C.; Easton, Judith A.; Fleischman, Diana S.; Goetz, Cari D.; Lewis, David M. G.; Perilloux, Carin & Buss, David M. (2010). Evolutionary psychology: Controversies, questions, prospects, and limitations. *American Psychologist*, 65(2), 110–126.

Coon, Carleton S. (1962). *The origin of races.* Knopf.

Coovadia, Hoosen M. & Wittenberg, Dankwart F. (Eds.). (2004). *Paediatrics and child health: A manual for health professionals in developing countries* (5th ed.). Oxford University Press.

Coplan, Robert J. & Weeks, Murray. (2009). Shy and soft-spoken: Shyness, pragmatic language, and socio-emotional adjustment in early childhood. *Infant and Child Development*, 18(3), 238–254.

Corballis, Michael C. (2011). *The recursive mind: The origins of human language, thought, and civilization.* Princeton University Press.

Corenblum, Barry. (2014). Relationships between racial–ethnic identity, self-esteem and in-group attitudes among First Nation children. *Journal of Youth and Adolescence*, 43(3), 387–404.

Costa, Albert & Sebastián-Gallés, Núria. (2014). How does the bilingual experience sculpt the brain? *Nature Reviews Neuroscience*, 15(5), 336–345.

Costa, Albert; Vives, Marc–Lluís & Corey, Joanna D. (2017). On language processing shaping decision making. *Current Directions in Psychological Science*, 26(2), 146–151.

Côté, James E. (2018). The enduring usefulness of Erikson's concept of the identity crisis in the 21st century: An analysis of student mental health concerns. *Identity*, 18(4), 251–263.

Côté, James E. & Levine, Charles. (2015). *Identity formation, youth, and development: A simplified approach.* Psychology Press.

Council on Communications and Media. (2011). Policy statement—Children, adolescents, obesity, and the media. *Pediatrics*, 128(1), 201–208.

Couzin-Frankel, Jennifer. (2017). Fateful imprints. *Science*, 355(6321), 122–125.

Couzin-Frankel, Jennifer. (2018). Toxin or treatment? *Science*, 362(6412), 278–282.

Couzin-Frankel, Jennifer. (2020). Rethinking anorexia. *Science*, 368(6487), 124–127.

Cowell, Jason M.; Lee, Kang; Malcolm-Smith, Susan; Selcuk, Bilge; Zhou, Xinyue & Decety, Jean. (2017). The development of generosity and moral cognition across five cultures. *Developmental Science*, 20(4), e12403.

Coyne, Sarah M.; Ward, L. Monique; Kroff, Savannah L.; Davis, Emilie J.; Holmgren, Hailey G.; Jensen, Alexander C.; . . . Essig, Lee W. (2019). Contributions of mainstream sexual media exposure to sexual attitudes, perceived peer norms, and sexual behavior: A meta-analysis. *Journal of Adolescent Health*, 64(4), 430–436.

Craig, Stephanie G.; Davies, Gregory; Schibuk, Larry; Weiss, Margaret D. & Hechtman, Lily. (2015). Long-term effects of stimulant treatment for ADHD: What can we tell our patients? *Current Developmental Disorders Reports*, 2(1), 1–9.

Crain, William C. (2011). *Theories of development: Concepts and applications* (6th ed.). Prentice-Hall.

Cree, Robyn A.; Bitsko, Rebecca H.; Robinson, Lara R.; Holbrook, Joseph R.; Danielson, Melissa L.; Smith, Camille; . . . Peacock, Georgina. (2018). *Health care, family, and community factors associated with mental, behavioral, and developmental disorders and poverty among children aged 2–8 years—United States, 2016.* 67(50), 1377–1383. Atlanta, GA: Centers for Disease Control and Prevention.

Crenshaw, Kimberle. (1989). Demarginalizing the intersection of race and sex: A Black feminist critique of antidiscrimination doctrine, feminist theory and antiracist politics. *University of Chicago Legal Forum*, 139–167.

Crivello, Cristina & Poulin-Dubois, Diane. (2019). Infants' ability to detect emotional incongruency: Deep or shallow? *Infancy*, 24(4), 480–500.

Crnic, Keith A.; Neece, Cameron L.; McIntyre, Laura Lee; Blacher, Jan & Baker, Bruce L. (2017). Intellectual disability and developmental risk: Promoting intervention to improve child and family well-being. *Child Development*, 88(2), 436–445.

Crone, Eveline A.; van Duijvenvoorde, Anna C. K. & Peper, Jiska S. (2016). Annual research review: Neural contributions to risk-taking in adolescence—developmental changes and individual differences. *Journal of Child Psychology and Psychiatry*, 57(3), 353–368.

Crosnoe, Robert; Purtell, Kelly M.; Davis-Kean, Pamela; Ansari, Arya & Benner, Aprile D. (2016). The selection of children from low-income families into preschool. *Developmental Psychology*, 52(4), 599–612.

Crossley, Nicolas A.; Mechelli, Andrea; Scott, Jessica; Carletti, Francesco; Fox, Peter T.; McGuire, Philip & Bullmore, Edward T. (2014). The hubs of the human connectome are generally implicated in the anatomy of brain disorders. *Brain*, 137(8), 2382–2395.

Crowley, Jocelyn Elise. (2019). Gray divorce: Explaining midlife marital splits. *Journal of Women & Aging*, 31(1), 49–72.

Csibra, Gergely; Hernik, Mikołaj; Mascaro, Olivier; Tatone, Denis & Lengyel, Máté. (2016). Statistical treatment of looking-time data. *Developmental Psychology*, 52(4), 521–536.

Culpin, Iryna; Heron, Jon; Araya, Ricardo & Joinson, Carol. (2015). Early childhood father absence and depressive symptoms in adolescent girls from a UK cohort: The mediating role of early menarche. *Journal of Abnormal Child Psychology*, 43(5), 921–931.

Cunningham, F. Gary; Leveno, Kenneth; Bloom, Steven; Spong, Catherine Y.; Dashe, Jodi; Hoffman, Barbara; . . . Sheffield, Jeanne S. (2014). *Williams obstetrics* (24th ed.). McGraw-Hill Education.

Currie, Candace; Nic Gabhainn, Saoirse; Godeau, Emmanuelle & International HBSC Network Coordinating Committee. (2009). The health behaviour in school-aged children: WHO Collaborative Cross-National (HBSC) study: Origins, concept, history and development 1982–2008. *International Journal of Public Health*, 54, 131–139.

Currie, Janet & Widom, Cathy S. (2010). Long-term consequences of child abuse and neglect on adult economic well-being. *Child Maltreatment*, 15(2), 111–120.

Curtiss, Susan. (2014). *Genie: A psycholinguistic study of a modern-day wild child*. Academic Press.

Cutuli, J. J.; Ahumada, Sandra M.; Herbers, Janette E.; Lafavor, Theresa L.; Masten, Ann S. & Oberg, Charles N. (2017). Adversity and children experiencing family homelessness: Implications for health. *Journal of Children and Poverty*, 23(1), 41–55.

Cuzzolaro, Massimo & Fassino, Secondo (Eds.). (2018). *Body image, eating, and weight: A guide to assessment, treatment, and prevention*. Springer.

d'Alpoim Guedes, Jade; Bestor, Theodore C.; Carrasco, David; Flad, Rowan; Fosse, Ethan; Herzfeld, Michael; . . . Warinner, Christina. (2013). Is poverty in our genes?: A critique of Ashraf and Galor, "The 'Out of Africa' hypothesis, human genetic diversity, and comparative economic development". *Forum on Public Anthropology*, 54(1).

Dadds, Mark R. & Tully, Lucy A. (2019). What is it to discipline a child: What should it be? A reanalysis of time-out from the perspective of child mental health, attachment, and trauma. *American Psychologist*, 74(7), 794–808.

Daley, Tamara C.; Whaley, Shannon E.; Sigman, Marian D.; Espinosa, Michael P. & Neumann, Charlotte. (2003). IQ on the rise: The Flynn Effect in rural Kenyan children. *Psychological Science*, 14(3), 215–219.

Dalman, Christina; Allebeck, Peter; Gunnell, David; Harrison, Glyn; Kristensson, Krister; Lewis, Glyn; . . . Karlsson, Håkan. (2008). Infections in the CNS during childhood and the risk of subsequent psychotic illness: A cohort study of more than one million Swedish subjects. *American Journal of Psychiatry*, 165(1), 59–65.

Dalrymple, Rebecca Amy. (2019). Earlier introduction of solid food is associated with improved sleep in infants. *Archives of Disease in Childhood—Education and Practice*.

Damasio, Antonio R. (2012). *Self comes to mind: Constructing the conscious brain*. Vintage.

Dantchev, Slava; Zammit, Stanley & Wolke, Dieter. (2018). Sibling bullying in middle childhood and psychotic disorder at 18 years: A prospective cohort study. *Psychological Medicine*, 48(14), 2321–2328.

Darwin, Charles. (1859). *On the origin of species by means of natural selection*. J. Murray.

Darwin, Zoe; Galdas, Paul; Hinchliff, Sharron; Littlewood, Elizabeth; McMillan, Dean; McGowan, Linda & Gilbody, Simon. (2017). Fathers' views and experiences of their own mental health during pregnancy and the first postnatal year: A qualitative interview study of men participating in the UK Born and Bred in Yorkshire (BaBY) cohort. *BMC Pregnancy and Childbirth*, 17(45).

Daum, Moritz M.; Ulber, Julia & Gredebäck, Gustaf. (2013). The development of pointing perception in infancy: Effects of communicative signals on covert shifts of attention. *Developmental Psychology*, 49(10), 1898–1908.

Davis, Corey S.; Green, Traci C.; Hernandez-Delgado, Hector & Lieberman, Amy Judd. (2018). Status of US state laws mandating timely reporting of nonfatal overdose. *American Journal of Public Health*, 108(9), 1159–1161.

Davis, Jac T. M. & Hines, Melissa. (2020). How large are gender differences in toy preferences? A systematic review and meta-analysis of toy preference research. *Archives of Sexual Behavior*, 49, 373–394.

Dawson-Hahn, Elizabeth; Koceja, Lorren; Stein, Elizabeth; Farmer, Beth; Grow, H Mollie; Saelens, Brian E.; . . . Pak-Gorstein, Suzinne. (2019). Perspectives of caregivers on the effects of migration on the nutrition, health and physical activity of their young children: A qualitative study with immigrant and refugee families. *Journal of Immigrant and Minority Health*, 22, 274–281.

Day, Felix R.; Elks, Cathy E.; Murray, Anna; Ong, Ken K. & Perry, John R. B. (2015). Puberty timing associated with diabetes, cardiovascular disease and also diverse health outcomes in men and women: The UK Biobank study. *Scientific Reports*, 5(11208).

Dayanim, Shoshana & Namy, Laura L. (2015). Infants learn baby signs from video. *Child Development*, 86(3), 800–811.

De Boeck, Paul & Jeon, Minjeong. (2018). Perceived crisis and reforms: Issues, explanations, and remedies. *Psychological Bulletin*, 144(7), 757–777.

de Boer, Anouk; Peeters, Margot & Koning, Ina. (2017). An experimental study of risk taking behavior among adolescents: A closer look at peer and sex influences. *Journal of Early Adolescence*, 37(8), 1125–1141.

de Heering, Adelaide; de Liedekerke, Claire; Deboni, Malorie & Rossion, Bruno. (2010). The role of experience during childhood in shaping the other-race effect. *Developmental Science*, 13(1), 181–187.

de Hoog, Marieke L. A.; Kleinman, Ken P.; Gillman, Matthew W.; Vrijkotte, Tanja G. M.; van Eijsden, Manon & Taveras, Elsie M. (2014). Racial/ethnic and immigrant differences in early childhood diet quality. *Public Health Nutrition*, 17(6), 1308–1317.

de Jonge, Ank; Geerts, C. C.; van der Goes, Birgit Y.; Mol, Ben W.; Buitendijk, S. E. & Nijhuis, Jan. (2015). Perinatal mortality and morbidity up to 28 days after birth among 743,070 low-risk planned home and hospital births: A cohort study based on three merged national perinatal databases. *BJOG*, 122(5), 720–728.

de la Croix, David. (2013). *Fertility, education, growth, and sustainability*. Cambridge University Press.

de Leeuw, Suzanne G. & Kalmijn, Matthijs. (2020). The intergenerational transmission of socioeconomic status in stepfamilies: What happens if two fathers are involved in the transmission process? *Journal of Marriage and Family*, 82(2), 657–674.

De Neys, Wim & Van Gelder, Elke. (2009). Logic and belief across the lifespan: The rise and fall of belief inhibition during syllogistic reasoning. *Developmental Science*, 12(1), 123–130.

de Oliveira Cardoso, Caroline; Dias, Natália; Senger, Joana; Colling, Ana Paula Cervi; Seabra, Alessandra Gotuzo & Fonseca, Rochele Paz. (2018). Neuropsychological stimulation of executive functions in children with typical development: A systematic review. *Applied Neuropsychology: Child*, 7(1), 61–81.

de Vrieze, Jop. (2018). The metawars. *Science*, 361(6408), 1184–1188.

Dean, Angela J.; Walters, Julie & Hall, Anthony. (2010). A systematic review of interventions to enhance medication adherence in children and adolescents with chronic illness. *Archives of Disease in Childhood*, 95(9), 717–723.

DeAngelis, Corey A. (2019). Does private schooling affect noncognitive skills? International evidence based on test and survey effort on PISA. *Social Science Quarterly*, 11(6), 2256–2276.

Dearing, Eric; Wimer, Christopher; Simpkins, Sandra D.; Lund, Terese; Bouffard, Suzanne M.; Caronongan, Pia; . . . Weiss, Heather.

(2009). Do neighborhood and home contexts help explain why low-income children miss opportunities to participate in activities outside of school? *Developmental Psychology, 45*(6), 1545–1562.

Dearing, Eric & Zachrisson, Henrik D. (2017). Concern over internal, external, and incidence validity in studies of child-care quantity and externalizing behavior problems. *Child Development Perspectives, 11*(2), 133–138.

Deater-Deckard, Kirby & Lansford, Jennifer E. (2016). Daughters' and sons' exposure to child-rearing discipline and violence in low- and middle-income countries. *Monographs of the Society for Research in Child Development, 81*(1), 78–103.

Defranza, David; Mishra, Himanshu & Mishra, Arul. (2020). How language shapes prejudice against women: An examination across 45 world languages. *Journal of Personality and Social Psychology, 119*(1), 7–22.

Dehaene-Lambertz, Ghislaine. (2017). The human infant brain: A neural architecture able to learn language. *Psychonomic Bulletin & Review, 24*(1), 48–55.

Deighton, Jessica; Humphrey, Neil; Belsky, Jay; Boehnke, Jan; Vostanis, Panos & Patalay, Praveetha. (2018). Longitudinal pathways between mental health difficulties and academic performance during middle childhood and early adolescence. *British Journal of Developmental Psychology, 36*(1), 110–126.

Delaunay-El Allam, Maryse; Soussignan, Robert; Patris, Bruno; Marlier, Luc & Schaal, Benoist. (2010). Long-lasting memory for an odor acquired at the mother's breast. *Developmental Science, 13*(6), 849–863.

Deventer, Jennifer; Wagner, Jenny; Lüdtke, Oliver & Trautwein, Ulrich. (2019). Are personality traits and relationship characteristics reciprocally related? Longitudinal analyses of codevelopment in the transition out of high school and beyond. *Journal of Personality and Social Psychology, 116*(2), 331–347.

Devine, Rory T. & Hughes, Claire. (2014). Relations between false belief understanding and executive function in early childhood: A meta-analysis. *Child Development, 85*(5), 1777–1794.

Diamond, Adele. (2012). Activities and programs that improve children's executive functions. *Current Directions in Psychological Science, 21*(5), 335–341.

Diamond, Adele. (2016). Why improving and assessing executive functions early in life is critical. In Griffin, James Alan; McCardle, Peggy D. & Freund, Lisa (Eds.), *Executive function in preschool-age children: Integrating measurement, neurodevelopment, and translational research* (pp. 11–43). American Psychological Association.

Diamond, Adele & Lee, Kathleen. (2011). Interventions shown to aid executive function development in children 4 to 12 years old. *Science, 333*(6045), 959–964.

Diamond, Marian C. (1988). *Enriching heredity: The impact of the environment on the anatomy of the brain.* Free Press.

Dillman Carpentier, Francesca & Stevens, Elise. (2018). Sex in the media, sex on the mind: Linking television use, sexual permissiveness, and

sexual concept accessibility in memory. *Sexuality & Culture, 22*(1), 22–38.

Dimler, Laura M. & Natsuaki, Misaki N. (2015). The effects of pubertal timing on externalizing behaviors in adolescence and early adulthood: A meta-analytic review. *Journal of Adolescence, 45*, 160–170.

Diseth, Åge; Meland, Eivind & Breidablik, Hans J. (2014). Self-beliefs among students: Grade level and gender differences in self-esteem, self-efficacy and implicit theories of intelligence. *Learning and Individual Differences, 35*.

Dishion, Thomas J.; Mun, Chung; Ha, Thao & Tein, Jenn-Yun. (2019). Observed family and friendship dynamics in adolescence: A latent profile approach to identifying "mesosystem" adaptation for intervention tailoring. *Prevention Science, 20*(1), 41–55.

Dix, Theodore & Yan, Ni. (2014). Mothers' depressive symptoms and infant negative emotionality in the prediction of child adjustment at age 3: Testing the maternal reactivity and child vulnerability hypotheses. *Development and Psychopathology, 26*(1), 111–124.

Dobson, Velma; Candy, T. Rowan; Hartmann, E. Eugenie; Mayer, D. Luisa; Miller, Joseph M. & Quinn, Graham E. (2009). Infant and child vision research: Present status and future directions. *Optometry & Vision Science, 86*(6), 559–560.

Dominguez-Bello, Maria Gloria; Godoy-Vitorino, Filipa; Knight, Rob & Blaser, Martin J. (2019). Role of the microbiome in human development. *Gut, 68*(6), 1108–1114.

Dorn, Lorah D. & Biro, Frank M. (2011). Puberty and its measurement: A decade in review. *Journal of Research on Adolescence, 21*(1), 180–195.

Dossou, S. G.; M., Lawani M.; Folly, M.; Tigri, N. & Houeto, G. (2019). Acquisition of motor skills in 784 Benin infants under one year. *International Journal of Development Research, 9*(15334).

Dougherty, Rachel. (2019). *Secret engineer: How Emily Roebling built the Brooklyn Bridge.* Roaring Brook.

Dow-Edwards, Diana & Silva, Lindsay. (2017). Endocannabinoids in brain plasticity: Cortical maturation, HPA axis function and behavior. *Brain Research, 1654*(Part B), 157–164.

Downing, Katherine L.; Hinkley, Trina; Salmon, Jo; Hnatiuk, Jill A. & Hesketh, Kylie D. (2017). Do the correlates of screen time and sedentary time differ in preschool children? *BMC Public Health, 17*(285).

Drake, Patrick; Driscoll, Anne K. & Mathews, T. J. (2018). *Cigarette smoking during pregnancy: United States, 2016.* Hyattsville, MD: National Center for Health Statistics. NCHS Data Brief No. 305.

Driemeyer, Wiebke; Janssen, Erick; Wiltfang, Jens & Elmerstig, Eva. (2016). Masturbation experiences of Swedish senior high school students: Gender differences and similarities. *The Journal of Sex Research.*

Duckworth, Angela L. (2016). *Grit: The power of passion and perseverance.* Scribner.

Dugas, Lara R.; Fuller, Miles; Gilbert, Jack & Layden, Brian T. (2016). The obese gut microbiome across the epidemiologic transition. *Emerging Themes in Epidemiology, 13*(1).

Duggan, Maeve. (2017, July 11). *Online harassment 2017. Internet & Technology.* Washington, DC: Pew Research Center.

Dumas, A.; Simmat-Durand, L. & Lejeune, C. (2014). Pregnancy and substance use in France: A literature review. *Journal de Gynécologie Obstétrique et Biologie de la Reproduction, 43*(9), 649–656.

Duncan, Greg J. & Magnuson, Katherine. (2013). Investing in preschool programs. *Journal of Economic Perspectives, 27*(2), 109–132.

Dunn, Erin C.; Nishimi, Kristen; Powers, Abigail & Bradley, Bekh. (2017). Is developmental timing of trauma exposure associated with depressive and post-traumatic stress disorder symptoms in adulthood? *Journal of Psychiatric Research, 84*, 119–127.

Dunn, Kirsty & Bremner, J. Gavin. (2017). Investigating looking and social looking measures as an index of infant violation of expectation. *Developmental Science, 20*(6), e12452.

Dunn, Kirsty & Bremner, James Gavin. (2019). Investigating the social environment of the A-not-B search task. *Developmental Science*, e12921.

Dunster, Gideon P.; de la Iglesia, Luciano; Ben-Hamo, Miriam; Nave, Claire; Fleischer, Jason G.; Panda, Satchidananda & de la Iglesia, Horacio O. (2018). Sleepmore in Seattle: Later school start times are associated with more sleep and better performance in high school students. *Science Advances, 4*(12), eaau6200.

DuPont, Robert L. & Lieberman, Jeffrey A. (2014). Young brains on drugs. *Science, 344*(6184), 557.

Duran, Chelsea A. K.; Cottone, Elizabeth; Ruzek, Erik A.; Mashburn, Andrew J. & Grissmer, David W. (2019). Family stress processes and children's self-regulation. *Child Development, 91*(2), 577–595.

Dweck, Carol S. (2013). Social development. In Zelazo, Philip D. (Ed.), *The Oxford handbook of developmental psychology* (Vol. 2, pp. 167–190). Oxford University Press.

Dweck, Carol S. (2017). From needs to goals and representations: Foundations for a unified theory of motivation, personality, and development. *Psychological Review, 124*(6), 689–719.

Dworkin, A. Gary & Quiroz, Pamela Anne. (2019). The United States of America: Accountability, high-stakes testing, and the demography of educational inequality. In Stevens, Peter A. J. & Dworkin, A. Gary (Eds.), *The Palgrave handbook of race and ethnic inequalities in education* (2nd ed., pp. 1097–1181). Palgrave Macmillan.

Eagly, Alice H. & Wood, Wendy. (2013). The nature–nurture debates: 25 years of challenges in understanding the psychology of gender. *Perspectives on Psychological Science, 8*(3), 340–357.

Ebert, Susanne. (2020). Theory of mind, language, and reading: Developmental relations from early childhood to early adolescence. *Journal of Experimental Child Psychology, 191*(104739).

Eccles, Jacquelynne S. & Roeser, Robert W. (2010). An ecological view of schools and development. In Meece, Judith L. & Eccles, Jacquelynne S. (Eds.), *Handbook of research on schools, schooling, and human development* (pp. 6–22). Routledge.

Eccles, Jacquelynne S. & Roeser, Robert W. (2011). Schools as developmental contexts during adolescence. *Journal of Research on Adolescence, 21*(1), 225–241.

Edenberg, Howard J.; Gelernter, Joel & Agrawal, Arpana. (2019). Genetics of alcoholism. *Current Psychiatry Reports, 21*(26).

Ehrenberg, Rachel. (2016). GMOs under scrutiny. *Science News, 189*(3), 22–27.

Ehrlich, Sara Z. & Blum-Kulka, Shoshana. (2014). 'Now I said that Danny becomes Danny again': A multifaceted view of kindergarten children's peer argumentative discourse. In Cekaite, Asta; Blum-Kulka, Shoshana; Grøver, Vibeke & Teubal, Eva (Eds.), *Children's peer talk: Learning from each other* (pp. 23–41). Cambridge University Press.

Eisenberg, Nancy & Zhou, Qing. (2016). Conceptions of executive function and regulation: When and to what degree do they overlap? In Griffin, James A.; McCardle, Peggy & Freund, Lisa S. (Eds.), *Executive function in preschool-age children: Integrating measurement, neurodevelopment, and translational research* (pp. 115–136). American Psychological Association.

Elicker, James & Ruprecht, Karen M. (2019). Child care quality rating and improvement systems (QRIS): National experiment for improving early childhood education quality. In Brown, Christopher P.; McMullen, Mary Benson & File, Nancy (Eds.), *The Wiley handbook of early childhood care and education* (pp. 515–536). Wiley.

Elkind, David. (1967). Egocentrism in adolescence. *Child Development, 38*(4), 1025–1034.

Ellefson, Michelle R.; Ng, Florrie Fei-Yin; Wang, Qian & Hughes, Claire. (2017). Efficiency of executive function: A two-generation cross-cultural comparison of samples from Hong Kong and the United Kingdom. *Psychological Science, 28*(5), 555–566.

Elliott, Julian & Nicolson, Rod. (2016). *Dyslexia: Developing the debate.* Bloomsbury.

Elliott, Sinikka. (2012). *Not my kid: What parents believe about the sex lives of their teenagers.* New York University Press.

Ellis, Bruce J. & Boyce, W. Thomas. (2008). Biological sensitivity to context. *Current Directions in Psychological Science, 17*(3), 183–187.

Ellis, Bruce J.; Boyce, W. Thomas; Belsky, Jay; Bakermans-kranenburg, Marian J. & Van Ijzendoorn, Marinus H. (2011a). Differential susceptibility to the environment: An evolutionary–neurodevelopmental theory. *Development and Psychopathology, 23*(1), 7–28.

Ellis, Bruce J. & Del Giudice, Marco. (2019). Developmental adaptation to stress: An evolutionary perspective. *Annual Review of Psychology, 70*(1), 111–139.

Ellis, Bruce J.; Shirtcliff, Elizabeth A.; Boyce, W. Thomas; Deardorff, Julianna & Essex, Marilyn J. (2011b). Quality of early family relationships and the timing and tempo of puberty: Effects depend on biological sensitivity to context. *Development and Psychopathology, 23*(1), 85–99.

Ellison, Christopher G.; Musick, Marc A. & Holden, George W. (2011). Does conservative Protestantism moderate the association between corporal punishment and child outcomes? *Journal of Marriage and Family, 73*(5), 946–961.

Emery, Carolyn A. (2018). Injury prevention in kids' adventure and extreme sports: Future directions. *Research in Sports Medicine, 26*(Suppl.1), 199–211.

Endo, Masaki; Mikami, Masafumi; Endo, Akira; Kaya, Hidetaka; Itoh, Takeshi; Nishimasu, Hiroshi; . . . Toki, Seiichi. (2019). Genome editing in plants by engineered CRISPR–Cas9 recognizing NG PAM. *Nature Plants, 5*(1), 14–17.

Engelberts, Adèle C. & de Jonge, Guustaaf Adolf. (1990). Choice of sleeping position for infants: Possible association with cot death. *Archives of Disease in Childhood, 65*(4), 462–467.

Engelhardt, Laura E.; Harden, K. Paige; Tucker-Drob, Elliot M. & Church, Jessica A. (2019). The neural architecture of executive functions is established by middle childhood. *Neuro-Image, 185*, 479–489.

Ennis, Linda Rose. (2015). *Intensive mothering: The cultural contradictions of modern motherhood.* Demeter Press.

Erickson, Anders C.; Ostry, Aleck; Chan, Hing Man & Arbour, Laura. (2016). Air pollution, neighbourhood and maternal-level factors modify the effect of smoking on birth weight: A multilevel analysis in British Columbia, Canada. *BMC Public Health, 16*(1).

Erikson, Erik H. (1963). *Childhood and society* (2nd ed.). Norton.

Erikson, Erik H. (1968). *Identity: Youth and crisis.* Norton.

Erikson, Erik H. (1982). *The life cycle completed: A review.* Norton.

Erikson, Erik H. (1993a). *Childhood and society* (2nd ed.). Norton.

Erikson, Erik H. (1993b). *Gandhi's truth: On the origins of militant nonviolence.* Norton.

Erikson, Erik H. (1994). *Identity: Youth and crisis.* Norton.

Erikson, Erik H. (1998). *The life cycle completed.* Norton.

Ernst, Monique. (2016). A tribute to the adolescent brain. *Neuroscience & Biobehavioral Reviews, 70*, 334–338.

Esler, Amy N.; Bal, Vanessa Hus; Guthrie, Whitney; Wetherby, Amy; Weismer, Susan Ellis & Lord, Catherine. (2015). The autism diagnostic observation schedule, toddler module: Standardized severity scores. *Journal of Autism and Developmental Disorders, 45*, 2704–2720.

Espy, Kimberly Andrews. (2016). *Monographs of the society for research in child development: The changing nature of executive control in preschool, 81*(4), 1–179.

Espy, K. A.; Clark, C. A. C.; Garza, J. P.; Nelson, J. M.; James, T. D. & Choi, H.-J. (2016). Executive control in preschoolers: New models, new results, new implications. *Monographs of the Society for Research in Child Development, 81*(4), 111–128.

Eun, Barohny. (2019). The zone of proximal development as an overarching concept: A framework for synthesizing Vygotsky's theories. *Educational Philosophy and Theory, 51*(1), 18–30.

Eun, John; Paksarian, Diana; He, Jian-Ping & Merikangas, Kathleen R. (2018). Parenting style and mental disorders in a nationally representative sample of US adolescents. *Social Psychiatry and Psychiatric Epidemiology, 53*(1), 11–20.

Evans, Angela D.; Xu, Fen & Lee, Kang. (2011). When all signs point to you: Lies told in the face of evidence. *Developmental Psychology, 47*(1), 39–49.

Evans, Gary W. & Kim, Pilyoung. (2013). Childhood poverty, chronic stress, self-regulation, and coping. *Child Development Perspectives, 7*(1), 43–48.

Evans, M. D. R.; Kelley, Jonathan; Sikora, Joanna & Treiman, Donald J. (2010). Family scholarly culture and educational success: Books and schooling in 27 nations. *Research in Social Stratification and Mobility, 28*(2), 171–197.

Everett, Caleb. (2017). *Numbers and the making of us: Counting and the course of human cultures.* Harvard University Press.

Eyer, Diane E. (1992). *Mother-infant bonding: A scientific fiction.* Yale University Press.

Fagnani, Jeanne. (2013). Equal access to quality care: Lessons from France on providing high quality and affordable early childhood education and care. In *Equal access to childcare: Providing quality early childhood education and care to disadvantaged families* (pp. 77–99). Policy Press.

Fairbank, John; Briggs, Ernestine; Lee, Robert; Corry, Nida; Pflieger, Jacqueline; Gerrity, Ellen; . . . Murphy, Robert. (2018). Mental health of children of deployed and nondeployed US military service members: The millennium cohort family study. *Journal of Developmental & Behavioral Pediatrics, 39*(9), 683–692.

Fairhurst, Merle T.; Löken, Line & Grossmann, Tobias. (2014). Physiological and behavioral responses reveal 9-month-old infants' sensitivity to pleasant touch. *Psychological Science, 25*(5), 1124–1131.

Fareed, Mohd; Anwar, Malik Azeem & Afzal, Mohammad. (2015). Prevalence and gene frequency of color vision impairments among children of six populations from North Indian region. *Genes & Diseases, 2*(2), 211–218.

Farkas, Laura; Cless, Jessica D.; Cless, Adam W.; Nelson Goff, Briana S.; Bodine, Ellen & Edelman, Ashley. (2019). The ups and downs of Down syndrome: A qualitative study of positive and negative parenting experiences. *Journal of Family Issues, 40*(4), 518–539.

Fast, Anne A. & Olson, Kristina R. (2018). Gender development in transgender preschool children. *Child Development, 89*(2), 620–637.

Fearon, R. M. Pasco & Roisman, Glenn I. (2017). Attachment theory: Progress and future directions. *Current Opinion in Psychology, 15*, 131–136.

Feeley, Nancy; Sherrard, Kathyrn; Waitzer, Elana & Boisvert, Linda. (2013). The father at the bedside: Patterns of involvement in the NICU. *Journal of Perinatal & Neonatal Nursing, 27*(1), 72–80.

Feld, Barry C. (2013). *Kids, cops, and confessions: Inside the interrogation room.* New York University Press.

Feldman, Ruth. (2012a). Oxytocin and social affiliation in humans. *Hormones and Behavior,* 61(3), 380–391.

Feldman, Ruth. (2012b). Parent-infant synchrony: A biobehavioral model of mutual influences in the formation of affiliative bonds. *Monographs of the Society for Research in Child Development,* 77(2), 42–51.

Feliciano, Cynthia & Rumbaut, Rubén G. (2018). Varieties of ethnic self-identities: Children of immigrants in middle adulthood. *RSF: The Russell Sage Foundation Journal of the Social Sciences,* 4(5), 26–46.

Feliciano, Cynthia & Rumbaut, Rubén G. (2019). The evolution of ethnic identity from adolescence to middle adulthood: The case of the immigrant second generation. *Emerging Adulthood,* 7(2), 85–96.

Felton, Julia W.; Collado, Anahí; Ingram, Katherine; Lejuez, Carl W. & Yi, Richard. (2020). Changes in delay discounting, substance use, and weight status across adolescence. *Health Psychology,* 39(5), 413–420.

Ferguson, Christopher J. (2013). Spanking, corporal punishment and negative long-term outcomes: A meta-analytic review of longitudinal studies. *Clinical Psychology Review,* 33(1), 196–208.

Ferguson, Christopher J. & Kilburn, John. (2010). Much ado about nothing: The misestimation and overinterpretation of violent video game effects in Eastern and Western nations—Comment on Anderson et al. (2010). *Psychological Bulletin,* 136(2), 174–178.

Ferguson, Kim T. & Evans, Gary W. (2019). Social ecological theory: Family systems and family psychology in bioecological and biecocultural perspective. In Fiese, Barbara H. (Ed.), *APA handbook of contemporary family psychology: Foundations, methods, and contemporary issues across the lifespan* (Vol. 1, pp. 143–161). American Psychological Association.

Fernando, Dulini; Cohen, Laurie & Duberley, Joanne. (2018). What helps? Women engineers' accounts of staying on. *Human Resource Management Journal,* 28(3), 479–495.

Ferrari, Marco & Quaresima, Valentina. (2012). A brief review on the history of human functional near-infrared spectroscopy (fNIRS) development and fields of application. *NeuroImage,* 63(2), 921–935.

Fewtrell, Mary; Wilson, David C.; Booth, Ian & Lucas, Alan. (2011). Six months of exclusive breast feeding: How good is the evidence? *BMJ,* 342, c5955.

Fields, R. Douglas. (2014). Myelin—More than insulation. *Science,* 344(6181), 264–266.

Finer, Lawrence B. & Zolna, Mia R. (2016). Declines in unintended pregnancy in the United States, 2008–2011. *New England Journal of Medicine,* 374, 843–852.

Fingerman, Karen L.; Berg, Cynthia; Smith, Jacqui & Antonucci, Toni C. (2011). *Handbook of lifespan development.* Springer.

Fingerman, Karen L.; Huo, Meng & Birditt, Kira S. (2020). A decade of research on intergenerational ties: Technological, economic, political, and demographic changes. *Journal of Marriage and Family,* 82(1), 383–403.

Finn, Amy S.; Kraft, Matthew A.; West, Martin R.; Leonard, Julia A.; Bish, Crystal E.; Martin, Rebecca E.; . . . Gabrieli, John D. E. (2014). Cognitive skills, student achievement tests, and schools. *Psychological Science,* 25(3), 736–744.

Finn, Chester E. & Scanlan, Andrew E. (2019). *Learning in the fast lane: The past, present, and future of advanced placement.* Princeton University Press.

Fischetti, Mark & Armstrong, Zan. (2017). The baby spike. *Scientific American,* 317(1), 76.

Fiset, Sylvain & Plourde, Vickie. (2013). Object permanence in domestic dogs (*Canis lupus familiaris*) and gray wolves (*Canis lupus*). *Journal of Comparative Psychology,* 127(2), 115–127.

Fisher, Dorothy Canfield. (1922). *What grandmother did not know.* Pilgrim Press.

Flannery, Jessica E.; Beauchamp, Kathryn G. & Fisher, Philip A. (2017). The role of social buffering on chronic disruptions in quality of care: Evidence from caregiver-based interventions in foster children. *Social Neuroscience,* 12(1), 86–91.

Fleming, David J.; Cowen, Joshua M.; Witte, John F. & Wolf, Patrick J. (2015). Similar students, different choices: Who uses a school voucher in an otherwise similar population of students? *Education and Urban Society,* 47(7), 785–812.

Fletcher, Richard; St. George, Jennifer & Freeman, Emily. (2013). Rough and tumble play quality: Theoretical foundations for a new measure of father–child interaction. *Early Child Development and Care,* 183(6), 746–759.

Foo, Koong Hean. (2019). *Intercultural parenting: How Eastern and Western parenting styles affect child development.* Routledge.

Forbes, Deborah. (2012). The global influence of the Reggio Emilia Inspiration. In Kelly, Robert (Ed.), *Educating for creativity: A global conversation* (pp. 161–172). Brush Education.

Forbes, Thomas A. & Gallo, Vittorio. (2017). All wrapped up: Environmental effects on myelination. *Trends in Neurosciences,* 40(9), 572–587.

Forestell, Catherine A. & Mennella, Julie A. (2017). The relationship between infant facial expressions and food acceptance. *Current Nutrition Reports,* 6(2), 141–147.

Forster, Myriam; Gower, Amy L.; Gloppen, Kari; Sieving, Renee; Oliphant, Jennifer; Plowman, Shari; . . . McMorris, Barbara J. (2020). Associations between dimensions of school engagement and bullying victimization and perpetration among middle school students. *School Mental Health,* 12, 296–307.

Fortin, Andrée; Doucet, Martin & Damant, Dominique. (2011). Children's appraisals as mediators of the relationship between domestic violence and child adjustment. *Violence and Victims,* 26(3), 377–392.

Fosco, Gregory M. & LoBraico, Emily J. (2019). Elaborating on premature adolescent autonomy: Linking variation in daily family processes to developmental risk. *Development and Psychopathology,* 31(5), 1741–1755.

Foster, Eugene A.; Jobling, Mark A.; Taylor, P. G.; Donnelly, Peter; de Knijff, Peter; Mieremet, Rene; . . . Tyler-Smith, C. (1998). Jefferson fathered slave's last child. *Nature,* 396(6706), 27–28.

Foulkes, Lucy & Blakemore, Sarah-Jayne. (2018). Studying individual differences in human adolescent brain development. *Nature Neuroscience,* 21, 315–323.

Fox, Ashley; Himmelstein, Georgia; Khalid, Hina & Howell, Elizabeth A. (2019). Funding for abstinence-only education and adolescent pregnancy prevention: Does state ideology affect outcomes? *American Journal of Public Health,* e1–e8.

Fox, Molly; Thayer, Zaneta M.; Ramos, Isabel F.; Meskal, Sarah J. & Wadhwa, Pathik D. (2018). Prenatal and postnatal mother-to-child transmission of acculturation's health effects in Hispanic Americans. *Journal of Women's Health,* 27(8), 154–1063.

Fox, Nathan A.; Henderson, Heather A.; Marshall, Peter J.; Nichols, Kate E. & Ghera, Melissa M. (2005). Behavioral inhibition: Linking biology and behavior within a developmental framework. *Annual Review of Psychology,* 56, 235–262.

Fox, Nathan A.; Henderson, Heather A.; Rubin, Kenneth H.; Calkins, Susan D. & Schmidt, Louis A. (2001). Continuity and discontinuity of behavioral inhibition and exuberance: Psychophysiological and behavioral influences across the first four years of life. *Child Development,* 72(1), 1–21.

Fox, Nathan A.; Reeb-Sutherland, Bethany C. & Degnan, Kathryn A. (2013). Personality and emotional development. In Zelazo, Philip D. (Ed.), *The Oxford handbook of developmental psychology* (Vol. 2, pp. 15–44). Oxford University Press.

Frankenburg, William K.; Dodds, Josiah; Archer, Philip; Shapiro, Howard & Bresnick, Beverly. (1992). The Denver II: A major revision and restandardization of the Denver Developmental Screening Test. *Pediatrics,* 89(1), 91–97.

Frayer, David W. (2017). Talking hyoids and talking Neanderthals. In Marom, Assaf & Hovers, Erella (Eds.), *Human paleontology and prehistory: Contributions in honor of Yoel Rak* (pp. 233–237). Springer.

Frazier, A. Lindsay; Camargo, Carlos A.; Malspeis, Susan; Willett, Walter C. & Young, Michael C. (2014). Prospective study of peripregnancy consumption of peanuts or tree nuts by mothers and the risk of peanut or tree nut allergy in their offspring. *JAMA Pediatrics,* 168(2), 156–162.

Fredricks, Jennifer A. & Eccles, Jacquelynne S. (2002). Children's competence and value beliefs from childhood through adolescence: Growth trajectories in two male-sex-typed domains. *Developmental Psychology,* 38(4), 519–533.

Freeman, Joan. (2010). *Gifted lives: What happens when gifted children grow up?* Routledge.

Frémondière, Pierre; Marchal, François; Thollon, Lionel & Saliba-serre, Bérangère. (2019). Change in head shape of newborn infants in the week following birth: Contributing factors. *Journal of Pediatric Neurology,* 17(5), 168–175.

Freud, Anna. (1958). Adolescence. *Psychoanalytic Study of the Child, 13,* 255–278.

Freud, Anna. (2000). Adolescence. In McCarthy, James B. (Ed.), *Adolescent development and psychopathology* (pp. 29–52). University Press of America.

Freud, Anna & Burlingham, Dorothy T. (1943). *War and children.* Medical War Books.

Freud, Sigmund. (1935). *A general introduction to psychoanalysis.* Liveright.

Freud, Sigmund. (1938). *The basic writings of Sigmund Freud.* Modern Library.

Freud, Sigmund. (1989). *Introductory lectures on psycho-analysis.* Liveright.

Freud, Sigmund. (1995). *The basic writings of Sigmund Freud.* Modern Library.

Freud, Sigmund. (2001). An outline of psychoanalysis. In *The standard edition of the complete psychological works of Sigmund Freud* (Vol. 23). Vintage.

Friedman, Naomi P. & Miyake, Akira. (2017). Unity and diversity of executive functions: Individual differences as a window on cognitive structure. *Cortex, 86,* 186–204.

Friedman-Krauss, Allison H.; Barnett, W. Steven; Garver, Karin A.; Hodges, Katherine S.; Weisenfeld, G. G. & DiCrecchio, Nicole. (2019). *The state of preschool 2018: State preschool yearbook.* New Brunswick, NJ: National Institute for Early Education Research.

Friend, Stephen H. & Schadt, Eric E. (2014). Clues from the resilient. *Science, 344*(6187), 970–972.

Fry, Douglas P. (2014). Environment of evolutionary adaptedness, rough-and-tumble play, and the selection of restraint in human aggression. In Narvaez, Darcia; Valentino, Kristin; Fuentes, Agustin; McKenna, James J. & Gray, Peter (Eds.), *Ancestral landscapes in human evolution: Culture, childrearing and social wellbeing* (pp. 169–188). Oxford University Press.

Fryar, Cheryl D.; Carroll, Margaret D. & Ogden, Cynthia L. (2018). *Prevalence of overweight, obesity, and severe obesity among children and adolescents aged 2–19 years: United States, 1963–1965 through 2015–2016. Health E-Stats.* Hyattsville, MD: National Center for Health Statistics.

Frydenberg, Erica. (2017). *Coping and the challenge of resilience.* Palgrave.

Fuller, Bruce & García Coll, Cynthia. (2010). Learning from Latinos: Contexts, families, and child development in motion. *Developmental Psychology, 46*(3), 559–565.

Fury, Gail; Carlson, Elizabeth A. & Sroufe, Alan. (1997). Children's representations of attachment relationships in family drawings. *Child Development, 68*(6), 1154–1164.

Gad, Rasha F.; Dowling, Donna A.; Abusaad, Fawzia E.; Bassiouny, Mohamed R. & Abd El Aziz, Magda A. (2019). Oral sucrose versus breastfeeding in managing infants' immunization-related pain: A randomized controlled trial. *MCN, 44*(2), 108–114.

Gagnon, Émilie; Ratelle, Catherine F.; Guay, Frédéric & Duchesne, Stéphane. (2019).

Developmental trajectories of vocational exploration from adolescence to early adulthood: The role of parental need supporting behaviors. *Journal of Vocational Behavior, 115,* 103338.

Galak, Jeff; Givi, Julian & Williams, Elanor F. (2016). Why certain gifts are great to give but not to get: A framework for understanding errors in gift giving. *Current Directions in Psychological Science, 25*(6), 380–385.

Galasso, Vincenzo; Profeta, Paola; Pronzato, Chiara & Billari, Francesco. (2017). Information and women's intentions: Experimental evidence about child care. *European Journal of Population, 33*(1), 109–128.

Gallagher, Annabella; Updegraff, Kimberly; Padilla, Jenny & McHale, Susan M. (2018). Longitudinal associations between sibling relational aggression and adolescent adjustment. *Journal of Youth and Adolescence, 47*(10), 2100–2113.

Gallup. (2017, August). *Time to play: A study on children's free time—how it is spent, prioritized and valued.* Washington, DC: Gallup.

Galvao, Tais F.; Silva, Marcus T.; Zimmermann, Ivan R.; Souza, Kathiaja M.; Martins, Silvia S. & Pereira, Mauricio G. (2014). Pubertal timing in girls and depression: A systematic review. *Journal of Affective Disorders, 155,* 13–19.

Gambaro, Ludovica; Stewart, Kitty & Waldfogel, Jane (Eds.). (2014). *An equal start?: Providing quality early education and care for disadvantaged children.* Policy Press.

Ganapathy, Thilagavathy. (2014). Couvade syndrome among 1st time expectant fathers. *Muller Journal of Medical Science Research, 5*(1), 43–47.

Ganchimeg, Togoobaatar; Ota, Erika; Morisaki, Naho; Laopaiboon, Malinee; Lumbiganon, P.; Zhang, Jun; . . . Mori, Rintaro. (2014). Pregnancy and childbirth outcomes among adolescent mothers: A World Health Organization multicountry study. *BJOG, 121*(Suppl. 1), 40–48.

Gao, Wei; Lin, Weili; Grewen, Karen & Gilmore, John H. (2017). Functional connectivity of the infant human brain: Plastic and modifiable. *Neuroscientist, 23*(2), 169–184.

Gardner, Frances; Leijten, Patty; Melendez-Torres, G. J.; Landau, Sabine; Harris, Victoria; Mann, Joanna; . . . Scott, Stephen. (2019). The earlier the better? Individual participant data and traditional meta-analysis of age effects of parenting interventions. *Child Development, 90*(1), 7–19.

Gardner, Howard. (1983). *Frames of mind: The theory of multiple intelligences.* Basic Books.

Gardner, Howard. (1999). Are there additional intelligences? The case for naturalist, spiritual, and existential intelligences. In Kane, Jeffrey (Ed.), *Education, information, and transformation: Essays on learning and thinking* (pp. 111–131). Merrill.

Gardner, Howard. (2006). *Multiple intelligences: New horizons in theory and practice.* Basic Books.

Gardner, Howard. (2011). *Frames of mind: The theory of multiple intelligences.* Basic Books.

Gardner, Howard & Moran, Seana. (2006). The science of multiple intelligences theory: A response to Lynn Waterhouse. *Educational Psychologist, 41*(4), 227–232.

Gariépy, Geneviève; Riehm, Kira E.; Whitehead, Ross D.; Doré, Isabelle & Elgar, Frank J. (2018). Teenage night owls or early birds? Chronotype and the mental health of adolescents. *Journal of Sleep Research,* e12723.

Garrett, Mallory. (2018). "It Gets Better" media campaign and gay youth suicide. In Stewart, Chuck (Ed.), *Lesbian, gay, bisexual, and transgender Americans at risk: Problems and solutions* (pp. 119–128). Praeger.

Garthus-Niegel, Susan; Ayers, Susan; Martini, Julia; von Soest, Tilmann & Eberhard-Gran, Malin. (2017). The impact of postpartum posttraumatic stress disorder symptoms on child development: A population-based, 2-year follow-up study. *Psychological Medicine, 47*(1), 161–170.

Garvis, Susanne; Harju-Luukkainen, Heidi; Sheridan, Sonja & Williams, Pia (Eds.). (2019). *Nordic families, children and early childhood education.* Palgrave Macmillan.

Gattario, Kristina Holmqvist & Frisén, Ann. (2019). From negative to positive body image: Men's and women's journeys from early adolescence to emerging adulthood. *Body Image, 28,* 53–65.

Ge, Xinting; Shi, Yonggang; Li, Junning; Zhang, Zhonghe; Lin, Xiangtao; Zhan, Jinfeng; . . . Liu, Shuwei. (2015). Development of the human fetal hippocampal formation during early second trimester. *NeuroImage, 119,* 33–43.

Gelfand, Amy. (2018). Episodic syndromes of childhood associated with migraine. *Current Opinion in Neurology, 31*(3), 281–285.

Gellert, Anna S. & Elbro, Carsten. (2017). Does a dynamic test of phonological awareness predict early reading difficulties? A longitudinal study from kindergarten through grade 1. *Journal of Learning Disabilities, 50*(3), 227–237.

Gergely, György & Csibra, Gergely. (2013). Natural pedagogy. In Banaji, Mahzarin R. & Gelman, Susan A. (Eds.), *Navigating the social world: What infants, children, and other species can teach us.* Oxford University Press.

Germani, Alessandro; Buratta, Livia; Delvecchio, Elisa & Mazzeschi, Claudia. (2020). Emerging adults and COVID-19: The role of individualism-collectivism on perceived risks and psychological maladjustment. *International Journal of Environmental Research and Public Health, 17*(10), 3497.

Gernhardt, Ariane; Keller, Heidi & Rübeling, Hartmut. (2016). Children's family drawings as expressions of attachment representations across cultures: Possibilities and limitations. *Child Development, 87*(4), 1069–1078.

Gernhardt, Ariane; Rübeling, Hartmut & Keller, Heidi. (2013). "This is my family": Differences in children's family drawings across cultures. *Journal of Cross-Cultural Psychology, 44*(7), 1166–1183.

Gershoff, Elizabeth; Sattler, Kierra M. P & Holden, George W. (2019). School corporal punishment and its associations with achievement and adjustment. *Journal of Applied Developmental Psychology, 63,* 1–8.

Gervais, Will M. & Norenzayan, Ara. (2012). Analytic thinking promotes religious disbelief. *Science, 336*(6080), 493–496.

Gewertz, Catherine. (2019, April 9). Which states require an exam to graduate?: An interactive breakdown of states' 2016–17 testing plans. Education Week.

Gibbons, Ann. (2012). An evolutionary theory of dentistry. *Science, 336*(6084), 973–975.

Gibson-Davis, Christina & Rackin, Heather. (2014). Marriage or carriage? Trends in union context and birth type by education. *Journal of Marriage and Family, 76*(3), 506–519.

Gidley, Jennifer M. (2016). *Postformal education: A philosophy for complex futures.* Springer.

Gilles, Floyd H. & Nelson, Marvin D. (2012). *The developing human brain: Growth and adversities.* Mac Keith Press.

Gilligan, Carol. (1982). *In a different voice: Psychological theory and women's development.* Harvard University Press.

Gillon, Raanan. (2015). Defending the four principles approach as a good basis for good medical practice and therefore for good medical ethics. *Journal of Medical Ethics, 41*(1), 111–116.

Gilmore, John H.; Knickmeyer, Rebecca C. & Gao, Wei. (2018). Imaging structural and functional brain development in early childhood. *Nature Reviews Neuroscience, 19*, 123–137.

Gilmour, Heather; Ramage-Morin, Pamela L. & Wong, Suzy L. (2019). Infant bed sharing in Canada. *Health Reports, 30*(7), 13–19.

Giubilini, Alberto & Savulescu, Julian. (2019). Vaccination, risks, and freedom: The seat belt analogy. *Public Health Ethics, 12*(3), 237–249.

Giumetti, Gary W. & Kowalski, Robin M. (2015). Cyberbullying matters: Examining the incremental impact of cyberbullying on outcomes over and above traditional bullying in North America. In Navarro, Raúl; Yubero, Santiago & Larrañaga, Elisa (Eds.), *Cyberbullying across the globe: Gender, family, and mental health* (pp. 117–130). Springer.

Glance, Laurent G.; Dick, Andrew W.; Glantz, Christopher; Wissler, Richard N.; Qian, Feng; Marroquin, Bridget M.; . . . Kellermann, Arthur L. (2014). Rates of major obstetrical complications vary almost fivefold among US hospitals. *Health Affairs, 33*(8), 1330–1336.

Glenn, Dana E.; Demir-Lira, Özlem Ece; Gibson, Dominic J.; Congdon, Eliza L. & Levine, Susan C. (2018). Resilience in mathematics after early brain injury: The roles of parental input and early plasticity. *Developmental Cognitive Neuroscience, 30*, 304–313.

Glock, Sabine & Kleen, Hannah. (2019). Attitudes toward students from ethnic minority groups: The roles of preservice teachers' own ethnic backgrounds and teacher efficacy activation. *Studies in Educational Evaluation, 62*, 82–91.

Goddings, Anne-Lise & Giedd, Jay N. (2014). Structural brain development during childhood and adolescence. In Gazzaniga, Michael S. & Mangun, George R. (Eds.), *The cognitive neurosciences* (5th ed., pp. 15–22). MIT Press.

Goddings, Anne-Lise; Heyes, Stephanie Burnett; Bird, Geoffrey; Viner, Russell M. & Blakemore, Sarah-Jayne. (2012). The relationship between puberty and social emotion processing. *Developmental Science, 15*(6), 801–811.

Goddings, Anne Lise; Beltz, Adriene; Peper, Jiska S.; Crone, Eveline A. & Braams, Barbara R. (2019). Understanding the role of puberty in structural and functional development of the adolescent brain. *Journal of Research on Adolescence, 29*(1), 32–53.

Godfrey, Erin B.; Santos, Carlos E. & Burson, Esther. (2019). For better or worse? System-justifying beliefs in sixth-grade predict trajectories of self-esteem and behavior across early adolescence. *Child Development, 90*(1), 180–195.

Godinet, Meripa T.; Li, Fenfang & Berg, Teresa. (2014). Early childhood maltreatment and trajectories of behavioral problems: Exploring gender and racial differences. *Child Abuse & Neglect, 38*(3), 544–556.

Golden, Neville H.; Yang, Wei; Jacobson, Marc S.; Robinson, Thomas N. & Shaw, Gary M. (2012). Expected body weight in adolescents: Comparison between weight-for-stature and BMI methods. *Pediatrics, 130*(6), e1607–e1613.

Golden, Rachel Lynn; Furman, Wyndol & Collibee, Charlene. (2016). The risks and rewards of sexual debut. *Developmental Psychology, 52*(11), 1913–1925.

Goldin-Meadow, Susan. (2015). From action to abstraction: Gesture as a mechanism of change. *Developmental Review, 38*, 167–184.

Goldin-Meadow, Susan & Alibali, Martha W. (2013). Gesture's role in speaking, learning, and creating language. *Annual Review of Psychology, 64*, 257–283.

Goldstein, Thalia R. & Lerner, Matthew D. (2018). Dramatic pretend play games uniquely improve emotional control in young children. *Developmental Science, 21*(4), e12603.

Golinkoff, Roberta M. & Hirsh-Pasek, Kathy. (2016). *Becoming brilliant: What science tells us about raising successful children.* American Psychological Association.

Golinkoff, Roberta Michnick; Hoff, Erika; Rowe, Meredith L.; Tamis-Lemonda, Catherine S. & Hirsh-Pasek, Kathy. (2019). Language matters: Denying the existence of the 30-million-word gap has serious consequences. *Child Development, 90*(3), 985–992.

Göncü, Artin & Gaskins, Suzanne. (2011). Comparing and extending Piaget's and Vygotsky's understandings of play: Symbolic play as individual, sociocultural, and educational interpretation. In Pellegrini, Anthony D. (Ed.), *The Oxford handbook of the development of play* (pp. 48–57). Oxford University Press.

Goodlad, James K.; Marcus, David K. & Fulton, Jessica J. (2013). Lead and Attention-deficit/hyperactivity disorder (ADHD) symptoms: A meta-analysis. *Clinical Psychology Review, 33*(3), 417–425.

Goodman, R.; Meltzer, H. & Bailey, V. (1998). The strengths and difficulties questionnaire: A pilot study on the validity of the self-report version. *European Child & Adolescent Psychiatry, 7*(3), 125–130.

Gopnik, Alison. (2001). Theories, language, and culture: Whorf without wincing. In Bowerman, Melissa & Levinson, Stephen C. (Eds.), *Language acquisition and conceptual development* (pp. 45–69). Cambridge University Press.

Gopnik, Alison. (2012). Scientific thinking in young children: Theoretical advances, empirical research, and policy implications. *Science, 337*(6102), 1623–1627.

Gopnik, Alison. (2016). *The gardener and the carpenter: What the new science of child development tells us about the relationship between parents and children.* Farrar, Strauss and Giroux.

Gopnik, Alison; Meltzoff, Andrew N. & Kuhl, Patricia K. (1999). *The scientist in the crib: Minds, brains, and how children learn.* William Morrow.

Gordon, Linda. (2017). *The second coming of the KKK: The Ku Klux Klan of the 1920s and the American political tradition.* Liveright.

Gordon-Hollingsworth, Arlene T.; Becker, Emily M.; Ginsburg, Golda S.; Keeton, Courtney; Compton, Scott N.; Birmaher, Boris B.; . . . March, John S. (2015). Anxiety disorders in Caucasian and African American children: A comparison of clinical characteristics, treatment process variables, and treatment outcomes. *Child Psychiatry & Human Development, 46*(5), 643–655.

Gorham, Lisa S.; Jernigan, Terry; Hudziak, Jim & Barch, Deanna M. (2019). Involvement in sports, hippocampal volume, and depressive symptoms in children. *Biological Psychiatry, 4*(5), 484–492.

Gostin, Lawrence; Phelan, Alexandra; Coutinho, Alex Godwin; Eccleston-Turner, Mark; Erondu, Ngozi; Filani, Oyebanji; . . . Kavanagh, Matthew. (2019). Ebola in the Democratic Republic of the Congo: Time to sound a global alert? *The Lancet, 393*(10172), 617–620.

Gottesman, Irving I.; Laursen, Thomas Munk; Bertelsen, Aksel & Mortensen, Preben Bo. (2010). Severe mental disorders in offspring with 2 psychiatrically ill parents. *Archives of General Psychiatry, 67*(3), 252–257.

Gough, Ethan K.; Moodie, Erica E. M.; Prendergast, Andrew J.; Johnson, Sarasa M. A.; Humphrey, Jean H.; Stoltzfus, Rebecca J.; . . . Manges, Amee R. (2014). The impact of antibiotics on growth in children in low and middle income countries: Systematic review and meta-analysis of randomised controlled trials. *BMJ, 348*, g2267.

Grabell, Adam S.; Huppert, Theodore J.; Fishburn, Frank A.; Li, Yanwei; Hlutkowsky, Christina O.; Jones, Hannah M.; . . . Perlman, Susan B. (2019). Neural correlates of early deliberate emotion regulation: Young children's responses to interpersonal scaffolding. *Developmental Cognitive Neuroscience, 40*(100708).

Grady, Jessica Stoltzfus & Hastings, Paul D. (2018). Becoming prosocial peers: The roles of temperamental shyness and mothers' and fathers' elaborative emotion language. *Social Development, 27*(4), 858–875.

Grady, Sarah. (2017, September 26). A fresh look at homeschooling in the U.S. *NCES Blog.* Institute of Education Sciences.

Gredebäck, Gustaf; Astor, Kim & Fawcett, Christine. (2018). Gaze following is not dependent on ostensive cues: A critical test of natural pedagogy. *Child Development, 89*(6), 2091–2098.

Green, James A.; Whitney, Pamela G. & Potegal, Michael. (2011). Screaming, yelling, whining, and crying: Categorical and intensity differences in vocal expressions of anger and sadness in children's tantrums. *Emotion, 11*(5), 1124–1133.

Greene, Melissa L. & Way, Niobe. (2005). Self-esteem trajectories among ethnic minority adolescents: A growth curve analysis of the patterns and predictors of change. *Journal of Research on Adolescence, 15*(2), 151–178.

Greenough, William T.; Black, James E. & Wallace, Christopher S. (1987). Experience and brain development. *Child Development, 58*(3), 539–559.

Greenough, William T. & Volkmar, Fred R. (1973). Pattern of dendritic branching in occipital cortex of rats reared in complex environments. *Experimental Neurology, 40*(2), 491–504.

Gregory, Robert. (1993). *Diz: The story of Dizzy Dean and baseball during the Great Depression.* Penguin.

Griffin, James A.; McCardle, Peggy & Freund, Lisa S. (Eds.). (2016). *Executive function in preschool-age children: Integrating measurement, neurodevelopment, and translational research.* American Psychological Association.

Grivell, Rosalie M.; Reilly, Aimee J.; Oakey, Helena; Chan, Annabelle & Dodd, Jodie M. (2012). Maternal and neonatal outcomes following induction of labor: A cohort study. *Acta Obstetricia et Gynecologica Scandinavica, 91*(2), 198–203.

Grogan-Kaylor, Andrew; Ma, Julie & Graham-Bermann, Sandra A. (2018). The case against physical punishment. *Current Opinion in Psychology, 19*, 22–27.

Groh, Ashley M.; Narayan, Angela J.; Bakermans-Kranenburg, Marian J.; Roisman, Glenn I.; Vaughn, Brian E.; Fearon, R. M. Pasco & van IJzendoorn, Marinus H. (2017). Attachment and temperament in the early life course: A meta-analytic review. *Child Development, 88*(3), 770–795.

Groh, Ashley M.; Roisman, Glenn I.; van IJzendoorn, Marinus H.; Bakermans-Kranenburg, Marian J. & Fearon, R. Pasco. (2012). The significance of insecure and disorganized attachment for children's internalizing symptoms: A meta-analytic study. *Child Development, 83*(2), 591–610.

Grönqvist, Hans; Nilsson, J. Peter & Robling, Per Olof. (2014). *Childhood lead exposure and criminal behavior: Lessons from the Swedish phase-out of leaded gasoline.* Stockholm: Swedish Institute for Social Research. Working Paper 9/2014.

Grossman, Matthew; Seashore, Carl & Holmes, Alison V. (2017). Neonatal abstinence syndrome management: A review of recent evidence. *Reviews on Recent Clinical Trials, 12*(4), 226–232.

Grossmann, Tobias. (2013). Mapping prefrontal cortex functions in human infancy. *Infancy, 18*(3), 303–324.

Grossmann, Tobias. (2017). The eyes as windows into other minds: An integrative perspective. *Perspectives on Psychological Science, 12*(1), 107–121.

Grotevant, Harold D. & McDermott, Jennifer M. (2014). Adoption: Biological and social processes linked to adaptation. *Annual Review of Psychology, 65*, 235–265.

Gruber, Staci A.; Sagar, Kelly A.; Dahlgren, Mary K.; Gonenc, Atilla; Smith, Rosemary T.; Lambros, Ashley M.; . . . Lukas, Scott E. (2018). The grass might be greener: Medical marijuana patients exhibit altered brain activity and improved executive function after 3 months of treatment. *Frontiers in Pharmacology, 8*(1), 983.

Grusec, Joan E.; Danyliuk, Tanya; Kil, Hali & O'Neill, David. (2017). Perspectives on parent discipline and child outcomes. *International Journal of Behavioral Development, 41*(4), 465–471.

Guassi Moreira, João F. & Telzer, Eva H. (2018). Family conflict is associated with longitudinal changes in insular-striatal functional connectivity during adolescent risk taking under maternal influence. *Developmental Science, 21*(5), e12632.

Guerra, Nancy G.; Williams, Kirk R. & Sadek, Shelly. (2011). Understanding bullying and victimization during childhood and adolescence: A mixed methods study. *Child Development, 82*(1), 295–310.

Guerri, Consuelo & Pascual, María. (2010). Mechanisms involved in the neurotoxic, cognitive, and neurobehavioral effects of alcohol consumption during adolescence. *Alcohol, 44*(1), 15–26.

Gulbas, L. E.; Zayas, L. H.; Yoon, H.; Szlyk, H.; Aguilar-Gaxiola, S. & Natera, G. (2016). Deportation experiences and depression among U.S. citizen-children with undocumented Mexican parents. *Child, 42*(2), 220–230.

Gutierrez-Galve, Leticia; Stein, Alan; Hanington, Lucy; Heron, Jon & Ramchandani, Paul. (2015). Paternal depression in the postnatal period and child development: Mediators and moderators. *Pediatrics, 135*(2), e339–e347.

Guyer, Amanda E.; Pérez-Edgar, Koraly & Crone, Eveline A. (2018). Opportunities for neurodevelopmental plasticity from infancy through early adulthood. *Child Development, 89*(3), 687–697.

Guyon-Harris, Katherine L.; Humphreys, Kathryn L.; Fox, Nathan A.; Nelson, Charles A. & Zeanah, Charles H. (2018). Course of disinhibited social engagement disorder from early childhood to early adolescence. *Journal of the American Academy of Child & Adolescent Psychiatry, 57*(5), 329–335.e2.

Guzman, Natalie S. de & Nishina, Adrienne. (2014). A longitudinal study of body dissatisfaction and pubertal timing in an ethnically diverse adolescent sample. *Body Image, 11*(1), 68–71.

Habibi, Assal; Damasio, Antonio; Ilari, Beatriz; Sachs, Matthew Elliott & Damasio, Hanna. (2018). Music training and child development: A review of recent findings from a longitudinal study. *Annals of the New York Academy of Sciences, 1423*(1), 73–81.

Haden, Catherine A. (2010). Talking about science in museums. *Child Development Perspectives, 4*(1), 62–67.

Haidt, Jonathan. (2013). *The righteous mind: Why good people are divided by politics and religion.* Vintage Books.

Halberda, Justin. (2018). Logic in babies. *Science, 359*(6381), 1214–1215.

Hales, Craig M.; Carroll, Margaret D.; Fryar, Cheryl D. & Ogden, Cynthia L. (2017, October). *Prevalence of obesity among adults and youth: United States, 2015–2016.* Atlanta, GA: Centers for Disease Control and Prevention, National Center for Health Statistics.

Halim, May Ling; Ruble, Diane N.; Tamis-LeMonda, Catherine S.; Zosuls, Kristina M.; Lurye, Leah E. & Greulich, Faith K. (2014). Pink frilly dresses and the avoidance of all things "girly": Children's appearance rigidity and cognitive theories of gender development. *Developmental Psychology, 50*(4), 1091–1101.

Hall, Matthew L.; Eigsti, Inge-Marie; Bortfeld, Heather & Lillo-Martin, Diane. (2017). Auditory deprivation does not impair executive function, but language deprivation might: Evidence from a parent-report measure in deaf native signing children. *Journal of Deaf Studies and Deaf Education, 22*(1), 9–21.

Hall, William J.; Jones, Benjamin L. H.; Witkemper, Kristen D.; Collins, Tora L. & Rodgers, Grayson K. (2019). State policy on school-based sex education: A content analysis focused on sexual behaviors, relationships, and identities. *American Journal of Health Behavior, 43*(3), 506–519.

Hallers-Haalboom, Elizabeth T.; Mesman, Judi; Groeneveld, Marleen G.; Endendijk, Joyce J.; van Berkel, Sheila R.; van der Pol, Lotte D. & Bakermans-Kranenburg, Marian J. (2014). Mothers, fathers, sons and daughters: Parental sensitivity in families with two children. *Journal of Family Psychology, 28*(2), 138–147.

Halperin, Jeffrey M. & Healey, Dione M. (2011). The influences of environmental enrichment, cognitive enhancement, and physical exercise on brain development: Can we alter the developmental trajectory of ADHD? *Neuroscience & Biobehavioral Reviews, 35*(3), 621–634.

Haltigan, John D.; Roisman, Glenn I.; Groh, Ashley M.; Holland, Ashley S.; Booth-Laforce, Cathryn; Rogosch, Fred A. & Cicchetti, Dante. (2019). Antecedents of attachment states of mind in normative-risk and high-risk caregiving: Cross-race and cross-sex generalizability in two longitudinal studies. *Journal of Child Psychology and Psychiatry, 60*(12), 1309–1322.

Hamada, Hirotaka & Matthews, Stephen G. (2019). Prenatal programming of stress responsiveness and behaviours: Progress and perspectives. *Journal of Neuroendocrinology, 31*(3), e12674.

Hambrick, David Z.; Burgoyne, Alexander P.; Macnamara, Brooke N. & Ullén, Fredrik. (2018). Toward a multifactorial model of expertise: Beyond born versus made. *Annals of the New York Academy of Sciences, 1423*(1), 284–295.

Hamerton, John L. & Evans, Jane A. (2005). Sex chromosome anomalies. In Butler, Merlin G. & Meaney, F. John (Eds.), *Genetics of developmental disabilities* (pp. 585–650). Taylor & Francis.

Hamilton, Alice. (1914). Lead poisoning in the United States. *American Journal of Public Health, 4*(6), 477–480.

Hamlat, Elissa J.; Shapero, Benjamin G.; Hamilton, Jessica L.; Stange, Jonathan P.;

Abramson, Lyn Y. & Alloy, Lauren B. (2014a). Pubertal timing, peer victimization, and body esteem differentially predict depressive symptoms in African American and Caucasian girls. *Journal of Early Adolescence, 35*(2), 378–402.

Hamlat, Elissa J.; Snyder, Hannah R.; Young, Jami F. & Hankin, Benjamin L. (2019). Pubertal timing as a transdiagnostic risk for psychopathology in youth. *Clinical Psychological Science, 7*(3), 411–429.

Hamlat, Elissa J.; Stange, Jonathan P.; Abramson, Lyn Y. & Alloy, Lauren B. (2014b). Early pubertal timing as a vulnerability to depression symptoms: Differential effects of race and sex. *Journal of Abnormal Child Psychology, 42*(4), 527–538.

Hamlett, Eric D.; Ledreux, Aurélie; Potter, Huntington; Chial, Heidi J.; Patterson, David; Espinosa, Joaquin M.; . . . Granholm, Ann-Charlotte. (2018). Exosomal biomarkers in Down syndrome and Alzheimer's disease. *Free Radical Biology and Medicine, 114*, 110–121.

Hang, Haiming; Davies, Iain & Schüring, Jennifer. (2020). Children's conformity to social norms to eat healthy: A developmental perspective. *Social Science & Medicine, 244*(112666).

Hanna-Attisha, Mona; LaChance, Jenny; Sadler, Richard Casey & Schnepp, Allison Champney. (2016). Elevated blood lead levels in children associated with the Flint drinking water crisis: A spatial analysis of risk and public health response. *American Journal of Public Health, 106*(2), 283–290.

Hannon, Erin E.; Schachner, Adena & Nave-Blodgett, Jessica E. (2017). Babies know bad dancing when they see it: Older but not younger infants discriminate between synchronous and asynchronous audiovisual musical displays. *Journal of Experimental Child Psychology, 159*, 159–174.

Hanushek, Eric A.; Piopiunik, Marc & Wiederhold, Simon. (2019). Do smarter teachers make smarter students? International evidence on teacher cognitive skills and student performance. *Education Next, 19*(2).

Hanushek, Eric A. & Woessmann, Ludger. (2007). *The role of education quality in economic growth.* World Bank Policy Research Working Paper No. 4122. Washington, DC: World Bank.

Hanushek, Eric A. & Woessmann, Ludger. (2015). *The knowledge capital of nations: Education and the economics of growth.* MIT Press.

Harden, K. Paige & Tucker-Drob, Elliot M. (2011). Individual differences in the development of sensation seeking and impulsivity during adolescence: Further evidence for a dual systems model. *Developmental Psychology, 47*(3), 739–746.

Hari, Riitta. (2017). From brain–environment connections to temporal dynamics and social interaction: Principles of human brain function. *Neuron, 94*(5), 1033–1039.

Harkness, Sara. (2014). Is biology destiny for the whole family? Contributions of evolutionary life history and behavior genetics to family theories. *Journal of Family Theory & Review, 6*(1), 31–34.

Harkness, Sara; Super, Charles M. & Mavridis, Caroline J. (2011). Parental ethnotheories about children's socioemotional development. In Chen,

Xinyin & Rubin, Kenneth H. (Eds.), *Socioemotional development in cultural context* (pp. 73–98). Guilford Press.

Harper, Casandra E. & Yeung, Fanny. (2013). Perceptions of institutional commitment to diversity as a predictor of college students' openness to diverse perspectives. *The Review of Higher Education, 37*(1), 25–44.

Harris, Cheryl M. (2019). Quitting science: Factors that influence exit from the stem workforce. *Journal of Women and Minorities in Science and Engineering, 25*(2), 93–118.

Harris Insights and Analytics. (2020, June). *Teen mental health.* New York, NY: Harris Interactive.

Harris, Judith R. (1998). *The nurture assumption: Why children turn out the way they do.* Free Press.

Harris, Judith R. (2002). Beyond the nurture assumption: Testing hypotheses about the child's environment. In Borkowski, John G.; Ramey, Sharon Landesman & Bristol-Power, Marie (Eds.), *Parenting and the child's world: Influences on academic, intellectual, and social-emotional development* (pp. 3–20). Erlbaum.

Harris, Michelle A.; Wetzel, Eunike; Robins, Richard W.; Donnellan, M. Brent & Trzesniewski, Kali H. (2017). The development of global and domain self-esteem from ages 10 to 16 for Mexican-origin youth. *International Journal of Behavioral Development.*

Harrison, Kristen; Bost, Kelly K.; McBride, Brent A.; Donovan, Sharon M.; Grigsby-Toussaint, Diana S.; Kim, Juhee; . . . Jacobsohn, Gwen Costa. (2011). Toward a developmental conceptualization of contributors to overweight and obesity in childhood: The Six-Cs model. *Child Development Perspectives, 5*(1), 50–58.

Harrison, Linda J.; Elwick, Sheena; Vallotton, Claire D. & Kappler, Gregor. (2014). Spending time with others: A time-use diary for infant-toddler child care. In Harrison, Linda J. & Sumsion, Jennifer (Eds.), *Lived spaces of infant-toddler education and care: Exploring diverse perspectives on theory, research and practice* (pp. 59–74). Springer.

Hart, Betty & Risley, Todd R. (1995). *Meaningful differences in the everyday experience of young American children.* P. H. Brookes.

Hart, Daniel & Van Goethem, Anne. (2017). The role of civic and political participation in successful early adulthood. In Padilla-Walker, Laura M. & Nelson, Larry J. (Eds.), *Flourishing in emerging adulthood: Positive development during the third decade of life* (pp. 139–166). Oxford University Press.

Harter, Susan. (2012). *The construction of the self: Developmental and sociocultural foundations* (2nd ed.). Guilford Press.

Hartig, Hannah & Geiger, Abigail. (2018, October 8). *About six-in-ten Americans support marijuana legalization. Fact Tank.* Washington, DC: Pew Research Center.

Hartley, Catherine A. & Somerville, Leah H. (2015). The neuroscience of adolescent decision-making. *Current Opinion in Behavioral Sciences, 5*, 108–115.

Hasson, Ramzi & Fine, Jodene Goldenring. (2012). Gender differences among children with

ADHD on continuous performance tests: A meta-analytic review. *Journal of Attention Disorders, 16*(3), 190–198.

Hatch, J. Amos. (2012). From theory to curriculum: Developmental theory and its relationship to curriculum and instruction in early childhood education. In File, Nancy; Mueller, Jennifer J. & Wisneski, Debora B. (Eds.), *Curriculum in early childhood education: Re-examined, rediscovered, renewed.* Routledge.

Hayden, Elizabeth P. & Mash, Eric J. (2014). Child psychopathology: A developmental-systems perspective. In Mash, Eric J. & Barkley, Russell A. (Eds.), *Child psychopathology* (3rd ed., pp. 3–72). Guilford Press.

Hayes, Peter. (2013). International adoption, "early" puberty, and underrecorded age. *Pediatrics, 131*(6), 1029–1031.

Hayne, Harlene; Scarf, Damian & Imuta, Kana. (2015). Childhood memories. In Wright, James D. (Ed.), *International encyclopedia of the social & behavioral sciences* (2nd ed., pp. 465–470). Elsevier.

Hayne, Harlene & Simcock, Gabrielle. (2009). Memory development in toddlers. In Courage, Mary L. & Cowan, Nelson (Eds.), *The development of memory in infancy and childhood* (2nd ed., pp. 43–68). Psychology Press.

Haynie, Dana L.; Soller, Brian & Williams, Kristi. (2014). Anticipating early fatality: Friends', schoolmates' and individual perceptions of fatality on adolescent risk behaviors. *Journal of Youth and Adolescence, 43*(2), 175–192.

Heitzer, Andrew M.; Piercy, Jamie C.; Peters, Brittany N.; Mattes, Allyssa M.; Klarr, Judith M.; Batton, Beau; . . . Raz, Sarah. (2020). Cumulative antenatal risk and kindergarten readiness in preterm-born preschoolers. *Journal of Abnormal Child Psychology, 48*(1), 1–12.

Hemingway, Susan J. Astley; M. Bledsoe, Julia; K. Davies, Julian; Brooks, Allison; Jirikowic, Tracy; M. Olson, Erin & C. Thorne, John. (2019). Twin study confirms virtually identical prenatal alcohol exposures can lead to markedly different fetal alcohol spectrum disorder outcomes-fetal genetics influences fetal vulnerability. *Advances in Pediatric Research, 5*(23).

Henderson, Heather A.; Pine, Daniel S. & Fox, Nathan A. (2015). Behavioral inhibition and developmental risk: A dual-processing perspective. *Neuropsychopharmacology Reviews, 40*(1), 207–224.

Hendlin, Yogi Hale. (2019). I am a fake loop: The effects of advertising-based artificial selection. *Biosemiotics, 12*(1), 131–156.

Hentges, Rochelle F. & Wang, Ming-Te. (2018). Gender differences in the developmental cascade from harsh parenting to educational attainment: An evolutionary perspective. *Child Development, 89*(2), 397–413.

Herholz, Sibylle C. & Zatorre, Robert J. (2012). Musical training as a framework for brain plasticity: Behavior, function, and structure. *Neuron, 76*(3), 486–502.

Hernández, Maciel M.; Robins, Richard W.; Widaman, Keith F. & Conger, Rand D. (2017). Ethnic pride, self-esteem, and school belonging:

A reciprocal analysis over time. *Developmental Psychology*, 53(12), 2384–2396.

Herrmann, Julia; Schmidt, Isabelle; Kessels, Ursula & Preckel, Franzis. (2016). Big fish in big ponds: Contrast and assimilation effects on math and verbal self-concepts of students in within-school gifted tracks. *British Journal of Educational Psychology*, 86(2), 222–240.

Herschensohn, Julia R. (2007). *Language development and age.* Cambridge University Press.

Hewer, Mariko. (2014). Selling sweet nothings: Science shows food marketing's effects on children's minds—and appetites. *Observer*, 27(10).

Heyer, Djai B. & Meredith, Rhiannon M. (2017). Environmental toxicology: Sensitive periods of development and neurodevelopmental disorders. *NeuroToxicology*, 58, 23–41.

Heyes, Cecilia. (2016). Who knows? Metacognitive social learning strategies. *Trends in Cognitive Sciences*, 20(3), 204–213.

Heyes, Cecilia. (2018). Empathy is not in our genes. *Neuroscience and Biobehavioral Reviews*, 95, 499–507.

Heyman, Gail D.; Li Zhao, Brian J. & Compton, Kang Lee. (2019). Dishonesty in young children. In Bucciol, Alessandro & Montinari, Natalia (Eds.), *Dishonesty in behavioral economics* (pp. 7–29). Academic Press.

Hidalgo, Marco A. & Chen, Diane. (2019). Experiences of gender minority stress in cisgender parents of transgender/gender-expansive prepubertal children: A qualitative study. *Journal of Family Issues*, 40(7), 865–886.

Hilgard, Joseph; Engelhardt, Christopher R. & Rouder, Jeffrey N. (2017). Overstated evidence for short-term effects of violent games on affect and behavior: A reanalysis of Anderson et al. (2010). *Psychological Bulletin*, 143(7), 757–774.

Hill, Patrick L.; Duggan, Peter M. & Lapsley, Daniel K. (2012). Subjective invulnerability, risk behavior, and adjustment in early adolescence. *Journal of Early Adolescence*, 32(4), 489–501.

Hill, Patrick L.; Jackson, Joshua J.; Roberts, Brent W.; Lapsley, Daniel K. & Brandenberger, Jay W. (2011). Change you can believe in: Changes in goal setting during emerging and young adulthood predict later adult well-being. *Social Psychological and Personality Science*, 2(2), 123–131.

Hill, Sarah E.; Prokosch, Marjorie L.; DelPriore, Danielle J.; Griskevicius, Vladas & Kramer, Andrew. (2016). Low childhood socioeconomic status promotes eating in the absence of energy need. *Psychological Science*, 27(3), 354–364.

Hillman, Charles H. (2014). An introduction to the relation of physical activity to cognitive and brain health, and scholastic achievement. *Monographs of the Society for Research in Child Development*, 79(4), 1–6.

Hindman, Annemarie H.; Wasik, Barbara A. & Bradley, Donald E. (2019). How classroom conversations unfold: Exploring teacher-child exchanges during shared book reading. *Early Education and Development*, 30(4), 478–495.

Hinnant, J. Benjamin; Nelson, Jackie A.; O'Brien, Marion; Keane, Susan P. & Calkins, Susan D. (2013). The interactive roles of parenting,

emotion regulation and executive functioning in moral reasoning during middle childhood. *Cognition and Emotion*, 27(8), 1460–1468.

Hinton, Lisa; Dumelow, Carol; Rowe, Rachel & Hollowell, Jennifer. (2018). Birthplace choices: What are the information needs of women when choosing where to give birth in England? A qualitative study using online and face to face focus groups. *BMC Pregnancy and Childbirth*, 18(1).

Hirsh-Pasek, Kathy & Golinkoff, Roberta M. (2016, March 11). The preschool paradox: It's time to rethink our approach to early education [Review of the book *The importance of being little: What preschoolers really need from grownups*, by Erika Christakis]. *Science*, 351(6278), 1158.

Hirth, Jacqueline. (2019). Disparities in HPV vaccination rates and HPV prevalence in the United States: A review of the literature. *Human Vaccines & Immunotherapeutics*, 15(1), 146–155.

Hisler, Garrett; Twenge, Jean M. & Krizan, Zlatan. (2020). Associations between screen time and short sleep duration among adolescents varies by media type: Evidence from a cohort study. *Sleep Medicine*, 66, 92–102.

Hitlin, Paul. (2018, September 28). *Internet, social media use and device ownership in U.S. have plateaued after years of growth.* Washington, DC: Pew Research Center.

Ho, Emily S. (2010). Measuring hand function in the young child. *Journal of Hand Therapy*, 23(3), 323–328.

Hodge, Samuel R.; Sato, Takahiro; Mukoyama, Takahito & Kozub, Francis M. (2013). Development of the physical educators' judgments about inclusion instrument for Japanese physical education majors and an analysis of their judgments. *International Journal of Disability, Development and Education*, 60(4), 332–346.

Hodson, Gordon; Crisp, Richard J.; Meleady, Rose & Earle, Megan. (2018). Intergroup contact as an agent of cognitive liberalization. *Perspectives on Psychological Science*, 13(5), 523–548.

Hoehl, Stefanie; Keupp, Stefanie; Schleihauf, Hanna; Mcguigan, Nicola; Buttelmann, David & Whiten, Andrew. (2019). 'Over-imitation': A review and appraisal of a decade of research. *Developmental Review*, 51, 90–108.

Hofer, Claire; Eisenberg, Nancy; Spinrad, Tracy L.; Morris, Amanda S.; Gershoff, Elizabeth; Valiente, Carlos; . . . Eggum, Natalie D. (2013). Mother-adolescent conflict: Stability, change, and relations with externalizing and internalizing behavior problems. *Social Development*, 22(2), 259–279.

Hoff, Erika. (2013). Interpreting the early language trajectories of children from low-SES and language minority homes: Implications for closing achievement gaps. *Developmental Psychology*, 49(1), 4–14.

Hoffman, Jessica L.; Teale, William H. & Paciga, Kathleen A. (2014). Assessing vocabulary learning in early childhood. *Journal of Early Childhood Literacy*, 14(4), 459–481.

Hogan, Jillian; Cordes, Sara; Holochwost, Steven; Ryu, Ehri; Diamond, Adele & Winner, Ellen. (2018). Is more time in general music class associated with stronger extra-musical outcomes in

kindergarten? *Early Childhood Research Quarterly*, 45, 238–248.

Holden, Constance. (2010). Myopia out of control. *Science*, 327(5961), 17.

Holzer, Jessica; Canavan, Maureen & Bradley, Elizabeth. (2014). County-level correlation between adult obesity rates and prevalence of dentists. *JADA*, 145(9), 932–939.

Hong, David S. & Reiss, Allan L. (2014). Cognitive and neurological aspects of sex chromosome aneuploidies. *The Lancet Neurology*, 13(3), 306–318.

Hood, Michelle; Creed, Peter A. & Mills, Bianca J. (2017). Loneliness and online friendships in emerging adults. *Personality and Individual Differences.*

Hoover, Eric. (2020, May 21). Golden State blockbuster: U. of California will replace ACT and SAT with new test—or None at all. *Chronicle of Higher Education.*

Horn, Stacey S. (2019). Sexual orientation and gender identity-based prejudice. *Child Development Perspectives*, 13(1), 21–27.

Horton, Megan K.; Kahn, Linda G.; Perera, Frederica; Barr, Dana B. & Rauh, Virginia. (2012). Does the home environment and the sex of the child modify the adverse effects of prenatal exposure to chlorpyrifos on child working memory? *Neurotoxicology and Teratology*, 34(5), 534–541.

Hostinar, Camelia E.; Nusslock, Robin & Miller, Gregory E. (2018). Future directions in the study of early-life stress and physical and emotional health: Implications of the neuroimmune network hypothesis. *Journal of Clinical Child & Adolescent Psychology*, 47(1), 142–156.

Howard, Sasha R. (2018). Genes underlying delayed puberty. *Molecular and Cellular Endocrinology*, 476, 119–128.

Howe, Tsu-Hsin; Sheu, Ching-Fan; Hsu, Yung-Wen; Wang, Tien-Ni & Wang, Lan-Wan. (2016). Predicting neurodevelopmental outcomes at preschool age for children with very low birth weight. *Research in Developmental Disabilities*, 48, 231–241.

Howell, Diane M.; Wysocki, Karen & Steiner, Michael J. (2010). Toilet training. *Pediatrics In Review*, 31(6), 262–263.

Hoyme, H. Eugene; Kalberg, Wendy O.; Elliott, Amy J.; Blankenship, Jason; Buckley, David; Marais, Anna-Susan; . . . May, Philip A. (2016). Updated clinical guidelines for diagnosing fetal alcohol spectrum disorders. *Pediatrics*, 138(2), e20154256.

Hoyne, Clara & Egan, Suzanne M. (2019). Shared book reading in early childhood: A review of influential factors and developmental benefits. *An Leanbh Og*, 12(1), 77–92.

Hrdy, Sarah B. (2009). *Mothers and others: The evolutionary origins of mutual understanding.* Harvard University Press.

Hu, Yousong; Wang, Yifang & Liu, Aizhen. (2017). The influence of mothers' emotional expressivity and class grouping on Chinese preschoolers' emotional regulation strategies. *Journal of Child and Family Studies*, 26(3), 824–832.

Huang, Jidong; Duan, Zongshuan; Kwok, Julian; Binns, Steven; Vera, Lisa E.; Kim,

Yoonsang; . . . Emery, Sherry L. (2019). Vaping versus JUULing: How the extraordinary growth and marketing of JUUL transformed the US retail e-cigarette market. *Tobacco Control, 28*(2), 146.

Huang, Z. Josh & Luo, Liqun. (2015). It takes the world to understand the brain. *Science, 350*(6256), 42–44.

Huguley, James P.; Wang, Ming-Te; Vasquez, Ariana C. & Guo, Jiesi. (2019). Parental ethnic–racial socialization practices and the construction of children of color's ethnic–racial identity: A research synthesis and meta-analysis. *Psychological Bulletin, 145*(5), 437–458.

Huston, Aletha C.; Bobbitt, Kaeley C. & Bentley, Alison. (2015). Time spent in child care: How and why does it affect social development? *Developmental Psychology, 51*(5), 621–634.

Hutchinson, Esther A.; De Luca, Cinzia R.; Doyle, Lex W.; Roberts, Gehan & Anderson, Peter J. (2013). School-age outcomes of extremely preterm or extremely low birth weight children. *Pediatrics, 131*(4), e1053–e1061.

Hutchison, Jane E.; Lyons, Ian M. & Ansari, Daniel. (2019). More similar than different: Gender differences in children's basic numerical skills are the exception not the rule. *Child Development, 90*(1), e66–e79.

Huynh, Jimmy L. & Casaccia, Patrizia. (2013). Epigenetic mechanisms in multiple sclerosis: Implications for pathogenesis and treatment. *The Lancet Neurology, 12*(2), 195–206.

Hvistendahl, Mara. (2013). China heads off deadly blood disorder. *Science, 340*(6133), 677–678.

Hyde, Janet S. (2016). Sex and cognition: Gender and cognitive functions. *Current Opinion in Neurobiology, 38*, 53–56.

Hyde, Janet Shibley; Bigler, Rebecca S.; Joel, Daphna; Tate, Charlotte Chucky & van Anders, Sari M. (2019). The future of sex and gender in psychology: Five challenges to the gender binary. *American Psychologist, 74*(2), 171–193.

Hyson, Marilou & Douglass, Anne L. (2019). More than academics: Supporting the whole child. In Brown, Christopher P.; McMullen, Mary Benson & File, Nancy (Eds.), *The Wiley handbook of early childhood care and education* (pp. 279–300). Wiley.

Ikeda, Martin J. (2012). Policy and practice considerations for response to intervention: Reflections and commentary. *Journal of Learning Disabilities, 45*(3), 274–277.

Ilies, Remus; Yao, Jingxian; Curseu, Petru L. & Liang, Alyssa X. (2019). Educated and happy: A four-year study explaining the links between education, job fit, and life satisfaction. *Applied Psychology, 68*(1), 150–176.

Imdad, Aamer; Sadiq, Kamran & Bhutta, Zulfiqar A. (2011). Evidence-based prevention of childhood malnutrition. *Current Opinion in Clinical Nutrition & Metabolic Care, 14*(3), 276–285.

Immordino-Yang, Mary Helen; Darling-Hammond, Linda & Krone, Christina R. (2019). Nurturing nature: How brain development is inherently social and emotional, and what this means for education. *Educational Psychologist, 54*(3), 185–204.

Inhelder, Bärbel & Piaget, Jean. (1958). *The growth of logical thinking from childhood to adolescence: An essay on the construction of formal operational structures.* Basic Books.

Inhelder, Bärbel & Piaget, Jean. (1964). *The early growth of logic in the child: Classification and seriation.* Harper & Row.

Inhelder, Bärbel & Piaget, Jean. (2013a). *The early growth of logic in the child: Classification and seriation.* Routledge.

Inhelder, Bärbel & Piaget, Jean. (2013b). *The growth of logical thinking from childhood to adolescence: An essay on the construction of formal operational structures.* Routledge.

Insel, Thomas R. (2014). Mental disorders in childhood: Shifting the focus from behavioral symptoms to neurodevelopmental trajectories. *JAMA, 311*(17), 1727–1728.

Insurance Institute for Highway Safety. (2018, December). Child safety: Motor vehicle crashes are a leading cause of death for children younger than 13. Insurance Institute for Highway Safety.

Insurance Institute for Highway Safety. (2018, December). General statistics: Crashes took 37,133 lives in the U.S. in 2017. Insurance Institute for Highway Safety.

Insurance Institute for Highway Safety, Highway Loss Data Institute. (2019, May). Older drivers. Insurance Institute for Highway Safety.

Iorio, Marilena V. & Palmieri, Dario. (2019). Editorial: From "junk DNA" to clinically relevant tools for cancer diagnosis, staging, and tailored therapies: The incredible case of non-coding RNAs. *Frontiers in Oncology, 9*(389).

Iotti, Nathalie Ophelia; Thornberg, Robert; Longobardi, Claudio & Jungert, Tomas. (2020). Early adolescents' emotional and behavioral difficulties, student-teacher relationships, and motivation to defend in bullying incidents. *Child & Youth Care Forum, 49*(1), 59–75.

Irwin, Scott; Galvez, Roberto; Weiler, Ivan Jeanne; Beckel-Mitchener, Andrea & Greenough, William. (2002). Brain structure and the functions of FMR1 protein. In Hagerman, Randi Jenssen & Hagerman, Paul J. (Eds.), *Fragile X syndrome: Diagnosis, treatment, and research* (3rd ed., pp. 191–205). Johns Hopkins University Press.

Ivarsson, Andreas; Johnson, Urban; Andersen, Mark; Tranaeus, Ulrika; Stenling, Andreas & Lindwall, Magnus. (2017). Psychosocial factors and sport injuries: Meta-analyses for prediction and prevention. *Sports Medicine, 47*(2), 353–365.

Jack, Jordynn. (2014). *Autism and gender: From refrigerator mothers to computer geeks.* University of Illinois Press.

Jackson, C. Kirabo. (2019). Can introducing single-sex education into low-performing schools improve academics, arrests, and teen motherhood? *Journal of Human Resources.*

Jacobs, Jerry A.; Ahmad, Seher & Sax, Linda J. (2017). Planning a career in engineering: Parental effects on sons and daughters. *Social Sciences, 6*(1).

Jacobs, Janis E. & Klaczynski, Paul A. (Eds.). (2005). *The development of judgment and decision making in children and adolescents.* L. Erlbaum.

Jambon, Marc & Smetana, Judith G. (2014). Moral complexity in middle childhood: Children's evaluations of necessary harm. *Developmental Psychology, 50*(1), 22–33.

James, Karin H. (2017). The importance of handwriting experience on the development of the literate brain. *Current Directions in Psychological Science, 26*(6), 502–508.

Jarcho, Johanna M.; Fox, Nathan A.; Pine, Daniel S.; Etkin, Amit; Leibenluft, Ellen; Shechner, Tomer & Ernst, Monique. (2013). The neural correlates of emotion-based cognitive control in adults with early childhood behavioral inhibition. *Biological Psychology, 92*(2), 306–314.

Jaschke, Artur C.; Honing, Henkjan & Scherder, Erik J. A. (2018). Longitudinal analysis of music education on executive functions in primary school children. *Frontiers in Neuroscience, 12*(103).

Jatlaoui, Tara C.; Eckhaus, Lindsay; Mandel, Michele G.; Nguyen, Antoinette; Oduyebo, Titilope; Petersen, Emily & Whiteman, Maura K. (2019, November 29). *Abortion surveillance—United States, 2016. Surveillance Summaries, 68*(11), 1–41. Atlanta, GA: Centers for Disease Control and Prevention.

Jednoróg, Katarzyna; Altarelli, Irene; Monzalvo, Karla; Fluss, Joel; Dubois, Jessica; Billard, Catherine; . . . Ramus, Franck. (2012). The influence of socioeconomic status on children's brain structure. *PLoS ONE, 7*(8), e42486.

Jenssen, Brian P. & Walley, Susan C. (2019). E-cigarettes and similar devices. *Pediatrics, 143*(2).

Jimerson, Shane R.; Burns, Matthew K. & VanDerHeyden, Amanda M. (Eds.). (2016). *Handbook of response to intervention: The science and practice of multi-tiered systems of support.* Springer.

Joel, Daphna; Berman, Zohar; Tavor, Ido; Wexler, Nadav; Gaber, Olga; Stein, Yaniv; . . . Assaf, Yaniv. (2015). Sex beyond the genitalia: The human brain mosaic. *Proceedings of the National Academy of Sciences, 112*(50), 15468–15473.

John, George Kunnackal; Wang, Lin; Nanavati, Julie; Twose, Claire; Singh, Rajdeep & Mullin, Gerard. (2018). Dietary alteration of the gut microbiome and its impact on weight and fat mass: A systematic review and meta-analysis. *Genes, 9*(3), 167.

Johnson, James E. & Wu, Viana Mei-Hsuan. (2019). Perspectives on play in early childhood care and education. In Brown, Christopher P.; McMullen, Mary Benson & File, Nancy (Eds.), *The Wiley handbook of early childhood care and education* (pp. 79–98). Wiley.

Johnson, Jonni L.; McWilliams, Kelly; Goodman, Gail S.; Shelley, Alexandra E. & Piper, Brianna. (2016). Basic principles of interviewing the child eyewitness. In O'Donohue, William T. & Fanetti, Matthew (Eds.), *Forensic interviews regarding child sexual abuse* (pp. 179–195). Springer.

Johnson, Mark H. & de Haan, Michelle. (2015). *Developmental cognitive neuroscience: An introduction* (4th ed.). Wiley.

Jones, Catherine R. G.; Simonoff, Emily; Baird, Gillian; Pickles, Andrew; Marsden, Anita J. S.; Tregay, Jenifer; . . . Charman, Tony. (2018). The association between theory of mind, executive function, and the symptoms of autism spectrum disorder. *Autism Research, 11*(1), 95–109.

Jones, Jeffrey M. (2015, October 21). *In U.S., 58% back legal marijuana use.* Washington, DC: Gallup.

Jones, Mary C. (1965). Psychological correlates of somatic development. *Child Development, 36*(4), 899–911.

Jong, Jyh-Tsorng; Kao, Tsair; Lee, Liang-Yi; Huang, Hung-Hsuan; Lo, Po-Tsung & Wang,

Hui-Chung. (2010). Can temperament be understood at birth? The relationship between neonatal pain cry and their temperament: A preliminary study. *Infant Behavior and Development*, 33(3), 266–272.

Jorgensen, Nathan & Nelson, Larry. (2018). Moving toward and away from others: Social orientations in emerging adulthood. *Journal of Applied Developmental Psychology*, 58, 66–76.

Joshi, Kshamta & Billick, Stephen. (2017). Biopsychosocial causes of suicide and suicide prevention outcome studies in juvenile detention facilities: A review. *Psychiatric Quarterly*, 88(1), 141–153.

Julian, Megan M. (2013). Age at adoption from institutional care as a window into the lasting effects of early experiences. *Clinical Child and Family Psychology Review*, 16(2), 101–145.

Jung, Courtney. (2015). *Lactivism: How feminists and fundamentalists, hippies and yuppies, and physicians and politicians made breastfeeding big business and bad policy.* Basic Books.

Juster, Robert-Paul; Russell, Jennifer J.; Almeida, Daniel & Picard, Martin. (2016). Allostatic load and comorbidities: A mitochondrial, epigenetic, and evolutionary perspective. *Development and Psychopathology*, 28(4), 1117–1146.

Juvonen, Jaana & Graham, Sandra. (2014). Bullying in schools: The power of bullies and the plight of victims. *Annual Review of Psychology*, 65, 159–185.

Kaczynski, Andrew T.; Besenyi, Gina M.; Child, S.; Hughey, Morgan; Colabianchi, Natalie; McIver, Kevin L.; . . . Pate, Russell R. (2018). Relationship of objective street quality attributes with youth physical activity: Findings from the Healthy Communities Study. *Pediatric Obesity*, 13(Suppl. 1), 7–13.

Kail, Robert V. (2013). Influences of credibility of testimony and strength of statistical evidence on children's and adolescents' reasoning. *Journal of Experimental Child Psychology*, 116(3), 747–754.

Kalsner, Louisa & Chamberlain, Stormy J. (2015). Prader-Willi, Angelman, and 15q11-q13 Duplication syndromes. *The Pediatric Clinics of North America*, 62(3), 587–606.

Kaltiala-Heino, Riittakerttu; Fröjd, Sari & Marttunen, Mauri. (2015). Depression, conduct disorder, smoking and alcohol use as predictors of sexual activity in middle adolescence: A longitudinal study. *Health Psychology and Behavioral Medicine*, 3(1), 25–39.

Kandel, Denise B. (Ed.) (2002). *Stages and pathways of drug involvement: Examining the gateway hypothesis.* Cambridge University Press.

Kandel, Eric R. (2018). *The disordered mind: What unusual brains tell us about ourselves.* Farrar, Straus and Giroux.

Kang, Hye-Kyung. (2014). Influence of culture and community perceptions on birth and perinatal care of immigrant women: Doulas' perspective. *The Journal of Perinatal Education*, 23(1), 25–32.

Kanner, Leo. (1943). Autistic disturbances of affective contact. *Nervous Child*, 2, 217–250.

Kaplan, N. & Main, M. (1986). *Instructions for the classification of children's family drawings in terms of representation of attachment* [Unpublished manuscript]. University of California.

Kapp, Steven K.; Gillespie-Lynch, Kristen; Sherman, Lauren E. & Hutman, Ted. (2013). Deficit, difference, or both? Autism and neurodiversity. *Developmental Psychology*, 49(1), 59–71.

Karbach, Julia & Unger, Kerstin. (2014). Executive control training from middle childhood to adolescence. *Frontiers in Psychology*, 5(390).

Karlson, Kristian Bernt. (2019). College as equalizer? Testing the selectivity hypothesis. *Social Science Research*, 80, 216–229.

Kärtner, Joscha; Borke, Jörn; Maasmeier, Kathrin; Keller, Heidi & Kleis, Astrid. (2011). Sociocultural influences on the development of self-recognition and self-regulation in Costa Rican and Mexican toddlers. *Journal of Cognitive Education and Psychology*, 10(1), 96–112.

Kärtner, Joscha; Keller, Heidi & Yovsi, Relindis D. (2010). Mother–infant interaction during the first 3 months: The emergence of culture-specific contingency patterns. *Child Development*, 81(2), 540–554.

Kastbom, Åsa A.; Sydsjö, Gunilla; Bladh, Marie; Priebe, Gisela & Svedin, Carl-Göran. (2015). Sexual debut before the age of 14 leads to poorer psychosocial health and risky behaviour in later life. *Acta Paediatrica*, 104(1), 91–100.

Kauffman, James M.; Anastasiou, Dimitris & Maag, John W. (2017). Special education at the crossroad: An identity crisis and the need for a scientific reconstruction. *Exceptionality*, 25(2), 139–155.

Kaur, Arshdeep. (2019). My child transformed me: Reflections of involved fathers. In Sriram, Rajalakshmi (Ed.), *Fathering in India* (pp. 141–152). Springer.

Kavanaugh, Robert D. (2011). Origins and consequences of social pretend play. In Pellegrini, Anthony D. (Ed.), *The Oxford handbook of the development of play* (pp. 296–307). Oxford University Press.

Kavšek, Michael & Braun, Stephanie K. (2016). Binocular vision in infancy: Responsiveness to uncrossed horizontal disparity. *Infant Behavior and Development*, 44, 219–226.

Keers, Robert & Pluess, Michael. (2017). Childhood quality influences genetic sensitivity to environmental influences across adulthood: A life-course Gene × Environment interaction study. *Development And Psychopathology*, 29(5), 1921–1933.

Kelemen, Deborah. (2019). The magic of mechanism: Explanation-based instruction on counterintuitive concepts in early childhood. *Perspectives on Psychological Science*, 14(4), 510–522.

Keller, Heidi; Borke, Jörn; Chaudhary, Nandita; Lamm, Bettina & Kleis, Astrid. (2010). Continuity in parenting strategies: A cross-cultural comparison. *Journal of Cross-Cultural Psychology*, 41(3), 391–409.

Keller, Peggy S.; El-Sheikh, Mona; Granger, Douglas A. & Buckhalt, Joseph A. (2012). Interactions between salivary cortisol and alpha-amylase as predictors of children's cognitive functioning and academic performance. *Physiology & Behavior*, 105(4), 987–995.

Kelly, Clare. (2015). 'Let's do some jumping together': Intergenerational participation in the use of remote technology to co-construct social relations over distance. *Journal of Early Childhood Research*, 13(1), 29–46.

Kempe, Ruth S. & Kempe, C. Henry. (1978). *Child abuse.* Harvard University Press.

Kendall-Taylor, Nathaniel; Lindland, Eric; O'Neil, Moira & Stanley, Kate. (2014). Beyond prevalence: An explanatory approach to reframing child maltreatment in the United Kingdom. *Child Abuse & Neglect*, 38(5), 810–821.

Keown, Louise J. & Palmer, Melanie. (2014). Comparisons between paternal and maternal involvement with sons: Early to middle childhood. *Early Child Development and Care*, 184(1), 99–117.

Kern, Ben D.; Graber, Kim C.; Shen, Sa; Hillman, Charles H. & McLoughlin, Gabriella. (2018). Association of school-based physical activity opportunities, socioeconomic status, and third-grade reading. *Journal of School Health*, 88(1), 34–43.

Kersken, Verena; Zuberbühler, Klaus & Gomez, Juan-Carlos. (2017). Listeners can extract meaning from non-linguistic infant vocalisations cross-culturally. *Scientific Reports*, 7.

Kesselring, Thomas & Müller, Ulrich. (2011). The concept of egocentrism in the context of Piaget's theory. *New Ideas in Psychology*, 29(3), 327–345.

Keysar, Boaz; Barr, Dale J.; Balin, Jennifer A. & Brauner, Jason S. (2000). Taking perspective in conversation: The role of mutual knowledge in comprehension. *Psychological Science*, 11(1), 32–38.

Kidd, Celeste; Palmeri, Holly & Aslin, Richard N. (2013). Rational snacking: Young children's decision-making on the marshmallow task is moderated by beliefs about environmental reliability. *Cognition*, 126(1), 109–114.

Kiley, Dan. (1983). *The Peter Pan syndrome: Men who have never grown up.* Dodd Mead.

Killen, Melanie & Smetana, Judith G. (Eds.). (2014). *Handbook of moral development* (2nd ed.). Psychology Press.

Kim, Bo-Ram; Chow, Sy-Miin; Bray, Bethany & Teti, Douglas M. (2017). Trajectories of mothers' emotional availability: Relations with infant temperament in predicting attachment security. *Attachment & Human Development*, 19(1), 38–57.

Kim, Dong-Sik & Kim, Hyun-Sun. (2009). Body-image dissatisfaction as a predictor of suicidal ideation among Korean boys and girls in different stages of adolescence: A two-year longitudinal study. *The Journal of Adolescent Health*, 45(1), 47–54.

Kim, Hojin I. & Johnson, Scott P. (2013). Do young infants prefer an infant-directed face or a happy face? *International Journal of Behavioral Development*, 37(2), 125–130.

Kim, Pilyoung; Strathearn, Lane & Swain, James E. (2016). The maternal brain and its plasticity in humans. *Hormones and Behavior*, 77, 113–123.

Kim, Su-Mi & Kim, Jong-Soo. (2017). A review of mechanisms of implantation. *Development & Reproduction*, 21(4), 351–359.

Kim-Spoon, Jungmeen; Deater-Deckard, Kirby; Brieant, Alexis; Lauharatanahirun, Nina; Lee, Jacob & King-Casas, Brooks. (2019). Brains of a feather flocking together? Peer and individual neurobehavioral risks for substance use across adolescence. *Development and Psychopathology*, 31(5), 1661–1674.

Kim-Spoon, Jungmeen; Longo, Gregory S. & McCullough, Michael E. (2012). Parent-adolescent relationship quality as a moderator for the influences of parents' religiousness on adolescents' religiousness and adjustment. *Journal of Youth and Adolescence, 41*(12), 1576–1587.

King, Brian A.; Jones, Christopher M.; Baldwin, Grant T. & Briss, Peter A. (2020). The EVALI and youth vaping epidemics—Implications for public health. *New England Journal of Medicine, 382*(8), 689–691.

King, Bruce M. (2013). The modern obesity epidemic, ancestral hunter-gatherers, and the sensory/reward control of food intake. *American Psychologist, 68*(2), 88–96.

King, Christian. (2018). Food insecurity and child behavior problems in fragile families. *Economics and Human Biology, 28*, 14–22.

King, Valarie; Pragg, Brianne & Lindstrom, Rachel. (2020). Family relationships during adolescence and stepchildren's educational attainment in young adulthood. *Journal of Marriage and Family, 82*(2), 622–638.

Kinney, Hannah C. & Thach, Bradley T. (2009). The sudden infant death syndrome. *New England Journal of Medicine, 361*, 795–805.

Kirk, Elizabeth; Howlett, Neil; Pine, Karen J. & Fletcher, Ben. (2013). To sign or not to sign? The impact of encouraging infants to gesture on infant language and maternal mind-mindedness. *Child Development, 84*(2), 574–590.

Kirkham, Julie Ann & Kidd, Evan. (2017). The effect of Steiner, Montessori, and National Curriculum Education upon children's pretence and creativity. *Journal of Creative Behavior, 51*(1), 20–34.

Kiserud, Torvid; Benachi, Alexandra; Hecher, Kurt; Perez, Rogelio González; Carvalho, José; Piaggio, Gilda & Platt, Lawrence D. (2018). The World Health Organization fetal growth charts: Concept, findings, interpretation, and application. *American Journal of Obstetrics and Gynecology, 218*(2), S619–S629.

Kiss, Ivy Giserman; Feldman, Melanie S.; Sheldrick, R. Christopher & Carter, Alice S. (2017). Developing autism screening criteria for the Brief Infant Toddler Social Emotional Assessment (BITSEA). *Journal of Autism and Developmental Disorders, 47*, 1269–1277.

Kjerulff, Kristen. (2014, November 18). *Epidural analgesia use during labor and adverse maternal and neonatal outcomes: A large-scale observational study* [Paper presentation]. Improving Pregnancy Outcomes: Psychosocial Influences on Birth Outcomes, New Orleans, LA, United States.

Klaczynski, Paul A. (2011). Age differences in understanding precedent-setting decisions and authorities' responses to violations of deontic rules. *Journal of Experimental Child Psychology, 109*(1), 1–24.

Klaczynski, Paul A. & Felmban, Wejdan S. (2014). Heuristics and biases during adolescence: Developmental reversals and individual differences. In Markovits, Henry (Ed.), *The developmental psychology of reasoning and decision-making* (pp. 84–111). Psychology Press.

Klaczynski, Paul A.; Felmban, Wejdan S. & Kole, James. (2020). Gender intensification and gender generalization biases in pre-adolescents, adolescents, and emerging adults. *British Journal of Developmental Psychology.*

Klaus, Marshall H. & Kennell, John H. (1976). *Maternal-infant bonding: The impact of early separation or loss on family development.* Mosby.

Klein, Denise; Mok, Kelvin; Chen, Jen-Kai & Watkins, Kate E. (2014). Age of language learning shapes brain structure: A cortical thickness study of bilingual and monolingual individuals. *Brain and Language, 131*, 20–24.

Klein, Zoe A. & Romeo, Russell D. (2013). Changes in hypothalamic–pituitary–adrenal stress responsiveness before and after puberty in rats. *Hormones and Behavior, 64*(2), 357–363.

Klinger, Laura G.; Dawson, Geraldine; Burner, Karen & Crisler, Megan. (2014). Autism spectrum disorder. In Mash, Eric J. & Barkley, Russell A. (Eds.), *Child psychopathology* (3rd ed., pp. 531–572). Guilford Press.

Knopik, Valerie S.; Neiderhiser, Jenae M.; DeFries, John C. & Plomin, Robert. (2017). *Behavioral genetics* (7th ed.). Worth.

Kochanek, Kenneth D.; Murphy, Sherry L.; Xu, Jiaquan & Arias, Elizabeth. (2019, June 24). *Deaths: Final data for 2017. National Vital Statistics Reports, 68*(9). Hyattsville, MD: National Center for Health Statistics.

Kochel, Karen P.; Ladd, Gary W.; Bagwell, Catherine L. & Yabko, Brandon A. (2015). Bully/victim profiles' differential risk for worsening peer acceptance: The role of friendship. *Journal of applied developmental psychology, 41*, 38–45.

Kohlberg, Lawrence. (1963). The development of children's orientations toward a moral order: I. Sequence in the development of moral thought. *Vita Humana, 6*(1/2), 11–33.

Kohlberg, Lawrence; Levine, Charles & Hewer, Alexandra. (1983). *Moral stages: A current formulation and a response to critics.* Karger.

Kolb, Bryan & Gibb, Robbin. (2015). Childhood poverty and brain development. *Human Development, 58*(4/5), 215–217.

Kolb, Bryan; Harker, Allonna & Gibb, Robbin. (2017). Principles of plasticity in the developing brain. *Developmental Medicine & Child Neurology, 59*(12), 1218–1223.

Kolb, Bryan & Whishaw, Ian Q. (2015). *Fundamentals of human neuropsychology* (7th ed.). Worth.

Komisar, Erica. (2017). *Being there: Why prioritizing motherhood in the first three years matters.* TarcherPerigee.

Konner, Melvin. (2010). *The evolution of childhood: Relationships, emotion, mind.* Harvard University Press.

Konner, Melvin. (2018). Nonmaternal care: A half-century of research. *Physiology & Behavior, 193*(Part A), 179–186.

Kono, Yumi; Yonemoto, Naohiro; Nakanishi, Hidehiko; Kusuda, Satoshi & Fujimura, Masanori. (2018). Changes in survival and neurodevelopmental outcomes of infants born at <25 weeks' gestation: A retrospective observational study in tertiary centres in Japan. *BMJ Paediatrics Open, 2*(1), e000211.

Korber, Maïlys & Oesch, Daniel. (2019). Vocational versus general education: Employment and earnings over the life course in Switzerland. *Advances in Life Course Research, 40.*

Kordas, Katarzyna; Burganowski, Rachael; Roy, Aditi; Peregalli, Fabiana; Baccino, Valentina; Barcia, Elizabeth; . . . Queirolo, Elena I. (2018). Nutritional status and diet as predictors of children's lead concentrations in blood and urine. *Environment International, 111*, 43–51.

Koress, Cody M.; Jones, Mark R. & Kaye, Alan David. (2019). Development of pain behavior in the fetus and newborn. In Abd-Elsayed, Alaa (Ed.), *Pain* (pp. 53–55). Springer.

Koretz, Daniel. (2017). *The testing charade: Pretending to make schools better.* University of Chicago Press.

Kornienko, Olga; Ha, Thao & Dishion, Thomas J. (2020). Dynamic pathways between rejection and antisocial behavior in peer networks: Update and test of confluence model. *Development and Psychopathology, 32*(1), 175–188.

Koster-Hale, Jorie & Saxe, Rebecca. (2013). Functional neuroimaging of theory of mind. In Baron-Cohen, Simon; Tager-Flusberg, Helen & Lombardo, Michael (Eds.), *Understanding other minds: Perspectives from developmental social neuroscience* (3rd ed., pp. 132–163). Oxford University Press.

Kozhimannil, Katy B. & Kim, Helen. (2014). Maternal mental illness. *Science, 345*(6198), 755.

Kral, Michael J. (2019). *The idea of suicide: Contagion, imitation, and cultural diffusion.* Routledge.

Krans, Elizabeth E. & Davis, Matthew M. (2012). *Preventing Low Birthweight: 25 years, prenatal risk, and the failure to reinvent prenatal care. American Journal of Obstetrics and Gynecology, 206*(5), 398–403.

Krause, Elizabeth D.; Vélez, Clorinda E.; Woo, Rebecca; Hoffmann, Brittany; Freres, Derek R.; Abenavoli, Rachel M. & Gillham, Jane E. (2018). Rumination, depression, and gender in early adolescence: A longitudinal study of a bidirectional model. *Journal of Early Adolescence, 38*(7), 923–946.

Kreager, Derek A.; Molloy, Lauren E.; Moody, James & Feinberg, Mark E. (2016). Friends first? The peer network origins of adolescent dating. *Journal of Research on Adolescence, 26*(2), 257–269.

Krebs, John R. (2009). The gourmet ape: Evolution and human food preferences. *American Journal of Clinical Nutrition, 90*(3), 707S–711S.

Kretch, Kari S. & Adolph, Karen E. (2013). No bridge too high: Infants decide whether to cross based on the probability of falling not the severity of the potential fall. *Developmental Science, 16*(3), 336–351.

Krisberg, Kim. (2014). Public health messaging: How it is said can influence behaviors: Beyond the facts. *The Nation's Health, 44*(6), 1, 20.

Kroger, Jane & Marcia, James E. (2011). The identity statuses: Origins, meanings, and interpretations. In Schwartz, Seth J.; Luyckx, Koen & Vignoles, Vivian L. (Eds.), *Handbook of identity theory and research* (pp. 31–53). Springer.

Kroncke, Anna P.; Willard, Marcy & Huckabee, Helena. (2016). Optimal outcomes and recovery. In *Assessment of autism spectrum disorder: Critical issues in clinical, forensic and school settings* (pp. 23–33). Springer.

Krouse, William J. (2012, November 14). *Gun control legislation. CRS Report for Congress.* Washington, DC: Congressional Research Service. RL32842

Krueger, Robert F. & Eaton, Nicholas R. (2015). Transdiagnostic factors of mental disorders. *World Psychiatry, 14*(1), 27–29.

Krzych-Fałta, Edyta; Furmańczyk, Konrad; Piekarska, Barbara; Raciborski, Filip; Tomaszewska, Aneta; Walkiewicz, Artur; . . . Samoliński, Bolesław Krzysztof. (2018). Extent of protective or allergy-inducing effects in cats and dogs. *Annals of Agricultural and Environmental Medicine, 25*(2), 268–273.

Kubin, Laura. (2019). Is there a resurgence of vaccine preventable diseases in the U.S.? *Journal of Pediatric Nursing, 44*, 115–118.

Kuhl, Patricia K. (2004). Early language acquisition: Cracking the speech code. *Nature Reviews Neuroscience, 5*(11), 831–843.

Kunkel, Melissa; Marete, Irene; Cheng, Erika R.; Bucher, Sherri; Liechty, Edward; Esamai, Fabian; . . . Vreeman, Rachel C. (2019). Place of delivery and perinatal mortality in Kenya. *Seminars in Perinatology, 43*(5), 252–259.

Kutob, Randa M.; Senf, Janet H.; Crago, Marjorie & Shisslak, Catherine M. (2010). Concurrent and longitudinal predictors of self-esteem in elementary and middle school girls. *Journal of School Health, 80*(5), 240–248.

Kuwahara, Keisuke; Kochi, Takeshi; Nanri, Akiko; Tsuruoka, Hiroko; Kurotani, Kayo; Pham, Ngoc Minh; . . . Mizoue, Tetsuya. (2014). Flushing response modifies the association of alcohol consumption with markers of glucose metabolism in Japanese men and women. *Alcoholism: Clinical and Experimental Research, 38*(4), 1042–1048.

Kvalvik, Liv G.; Haug, Kjell; Klungsøyr, Kari; Morken, Nils-Halvdan; Deroo, Lisa A. & Skjærven, Rolv. (2017). Maternal smoking status in successive pregnancies and risk of having a small for gestational age infant. *Paediatric and Perinatal Epidemiology, 31*(1), 21–28.

Kypri, Kypros; Davie, Gabrielle; McElduff, Patrick; Connor, Jennie & Langley, John. (2014). Effects of lowering the minimum alcohol purchasing age on weekend assaults resulting in hospitalization in New Zealand. *American Journal of Public Health, 104*(8), 1396–1401.

Kypri, Kypros; Voas, Robert B.; Langley, John D.; Stephenson, Shaun C. R.; Begg, Dorothy J.; Tippetts, A. Scott & Davie, Gabrielle S. (2006). Minimum purchasing age for alcohol and traffic crash injuries among 15- to 19-year-olds in New Zealand. *American Journal of Public Health, 96*(1), 126–131.

La Fauci, Luigi. (2018). The changing sexual life course of gay men and lesbians in contemporary Italy. *European Societies.*

Labouvie-Vief, Gisela. (2015). *Integrating emotions and cognition throughout the lifespan.* Springer.

Ladd, Helen F. (2019). How charter schools threaten the public interest. 38(4), 1063–1071.

Ladd, Helen F. & Sorensen, Lucy C. (2017). Returns to teacher experience: Student achievement and motivation in middle school. *Education Finance and Policy, 12*(2), 241–279.

Lagattuta, Kristin H. (2014). Linking past, present, and future: Children's ability to connect mental states and emotions across time. *Child Development Perspectives, 8*(2), 90–95.

Lai, Stephanie A.; Benjamin, Rebekah G.; Schwanenflugel, Paula J. & Kuhn, Melanie R. (2014). The longitudinal relationship between reading fluency and reading comprehension skills in second-grade children. *Reading & Writing Quarterly: Overcoming Learning Difficulties, 30*(2), 116–138.

Lake, Stephanie & Kerr, Thomas. (2017). The challenges of projecting the public health impacts of marijuana legalization in Canada. *International Journal of Health Policy Management, 6*(5), 285–287.

Lam, Chun Bun; McHale, Susan M. & Crouter, Ann C. (2012). Parent–child shared time from middle childhood to late adolescence: Developmental course and adjustment correlates. *Child Development, 83*(2), 2089–2103.

Lam, Kelly Ka Lai & Zhou, Mingming. (2019). Examining the relationship between grit and academic achievement within K-12 and higher education: A systematic review. *Psychology in the Schools, 56*(10), 1654–1686.

Lamb, Michael E. (1982). Maternal employment and child development: A review. In Lamb, Michael E. (Ed.), *Nontraditional families: Parenting and child development* (pp. 45–69). Erlbaum.

Lamm, Bettina; Keller, Heidi; Teiser, Johanna; Gudi, Helene; Yovsi, Relindis D.; Freitag, Claudia; . . . Lohaus, Arnold. (2018). Waiting for the second treat: Developing culture-specific modes of self-regulation. *Child Development, 89* (3), e261–e277.

Lancaster, Gillian A.; Mccray, Gareth; Kariger, Patricia; Dua, Tarun; Titman, Andrew; Chandna, Jaya; . . . Janus, Magdalena. (2018). Creation of the WHO Indicators of Infant and Young Child Development (IYCD): Metadata synthesis across 10 countries. *BMJ Global Health, 3*(5), e000747.

Landgren, Kajsa; Lundqvist, Anita & Hallström, Inger. (2012). Remembering the chaos – But life went on and the wound healed: A four year follow up with parents having had a baby with infantile colic. *The Open Nursing Journal, 6*, 53–61.

Lando, Amy M. & Lo, Serena C. (2014). Consumer understanding of the benefits and risks of fish consumption during pregnancy. *American Journal of Lifestyle Medicine, 8*(2), 88–92.

Lane, Jonathan D. & Harris, Paul L. (2014). Confronting, representing, and believing counterintuitive concepts: Navigating the natural and the supernatural. *Perspectives on Psychological Science, 9*(2), 144–160.

Lang, Samantha F. & Fowers, Blaine J. (2019). An expanded theory of Alzheimer's caregiving. *American Psychologist, 74*(2), 194–206.

Langrehr, Kimberly J.; Morgan, Sydney K.; Ross, Jessica; Oh, Monica & Chong, Wen Wen. (2019). Racist experiences, openness to discussing racism, and attitudes toward ethnic heritage activities: Adoptee–parent discrepancies. *Asian American Journal of Psychology, 10*(2), 91–102.

Laninga-Wijnen, Lydia; Gremmen, Mariola C.; Dijkstra, Jan Kornelis; Veenstra, René; Vollebergh, Wilma A. M. & Harakeh, Zeena. (2019). The role of academic status norms in friendship selection and influence processes related to academic achievement. *Developmental Psychology, 55*(2), 337–350.

Lansford, Jennifer E.; Sharma, Chinmayi; Malone, Patrick S.; Woodlief, Darren; Dodge, Kenneth A.; Oburu, Paul; . . . Di Giunta, Laura. (2014). Corporal punishment, maternal warmth, and child adjustment: A longitudinal study in eight countries. *Journal of Clinical Child & Adolescent Psychology, 43*(4), 670–685.

Lapan, Candace & Boseovski, Janet J. (2017). When peer performance matters: Effects of expertise and traits on children's self-evaluations after social comparison. *Child Development, 88*(6), 1860–1872.

Lara-Cinisomo, Sandraluz; Fuligni, Allison Sidle & Karoly, Lynn A. (2011). Preparing preschoolers for kindergarten. In Laverick, DeAnna M. & Jalongo, Mary Renck (Eds.), *Transitions to early care and education* (Vol. 4, pp. 93–105). Springer.

Larsen, Peter A. (2018). Transposable elements and the multidimensional genome. *Chromosome Research, 26*(1/2), 1–3.

Larzelere, Robert E. & Cox, Ronald B. (2013). Making valid causal inferences about corrective actions by parents from longitudinal data. *Journal of Family Theory & Review, 5*(4), 282–299.

Larzelere, Robert E.; Cox, Ronald B. & Swindle, Taren M. (2015). Many replications do not causal inferences make: The need for critical replications to test competing explanations of nonrandomized studies. *Perspectives on Psychological Science, 10*(3), 380–389.

Larzelere, Robert E.; Gunnoe, Marjorie Lindner; Roberts, Mark W. & Ferguson, Christopher J. (2017). Children and parents deserve better parental discipline research: Critiquing the evidence for exclusively "positive" parenting. *Marriage & Family Review, 53*(1), 24–35.

Laurent, Heidemarie K. (2014). Clarifying the contours of emotion regulation: Insights from parent–child stress research. *Child Development Perspectives, 8*(1), 30–35.

Laursen, Brett; Hartl, Amy C.; Vitaro, Frank; Brendgen, Mara; Dionne, Ginette & Boivin, Michel. (2017). The spread of substance use and delinquency between adolescent twins. *Developmental Psychology, 53*(2), 329–339.

Law, James; Rush, Robert; King, Tom; Westrupp, Elizabeth & Reilly, Sheena. (2018). Early home activities and oral language skills in middle childhood: A quantile analysis. *Child Development, 89*(1), 295–309.

Lazzara, Alexandra; Daymont, Carrie; Ladda, Roger; Lull, Jordan; Ficicioglu, Can; Cohen, Jennifer & Aprile, Justen. (2019). Failure to thrive: An expanded differential diagnosis. *Journal of Pediatric Genetics, 8*(1), 27–32.

Le Duc, James W. & Yuan, Zhiming. (2018). Network for safe and secure labs. *Science, 362*(6412), 267.

Leach, Penelope. (2011). The EYFS and the real foundations of children's early years. In House, Richard (Ed.), *Too much, too soon?: Early learning and the erosion of childhood*. Hawthorn.

Leavitt, Judith W. (2009). *Make room for daddy: The journey from waiting room to birthing room*. University of North Carolina Press.

Lee, Dohoon; Brooks-Gunn, Jeanne; McLanahan, Sara S.; Notterman, Daniel & Garfinkel, Irwin. (2013). The Great Recession, genetic sensitivity, and maternal harsh parenting. *Proceedings of the National Academy of Sciences, 110*(34), 13780–13784.

Lee, RaeHyuck; Zhai, Fuhua; Brooks-Gunn, Jeanne; Han, Wen-Jui & Waldfogel, Jane. (2014). Head Start participation and school readiness: Evidence from the early childhood longitudinal study–birth cohort. *Developmental Psychology, 50*(1), 202–215.

Lee, Shawna J. & Altschul, Inna. (2015). Spanking of young children: Do immigrant and U.S.-born Hispanic parents differ? *Journal of Interpersonal Violence, 30*(3), 475–498.

Lee, Shawna J.; Altschul, Inna & Gershoff, Elizabeth T. (2015). Wait until your father gets home? Mother's and fathers' spanking and development of child aggression. *Children and Youth Services Review, 52*, 158–166.

Lehner, Ben. (2013). Genotype to phenotype: Lessons from model organisms for human genetics. *Nature Reviews Genetics, 14*(3), 168–178.

Leiter, Valerie & Herman, Sarah. (2015). Guinea pig kids: Myths or modern Tuskegees? *Sociological Spectrum, 35*(1), 26–45.

Leman, Patrick J. & Björnberg, Marina. (2010). Conversation, development, and gender: A study of changes in children's concepts of punishment. *Child Development, 81*(3), 958–971.

Lemish, Daphna & Kolucki, Barbara. (2013). Media and early childhood development. In Britto, Pia Rebello; Engle, Patrice L. & Super, Charles M. (Eds.), *Handbook of early childhood development research and its impact on global policy*. Oxford University Press.

Lengua, Liliana J.; Garstein, Maria A. & Prinzie, Peter. (2019). Temperament and personality trait development in the family: Interactions and transactions with parenting from infancy through adolescence. In McAdams, Dan P.; Shiner, Rebecca L. & Tackett, Jennifer L. (Eds.), *Handbook of personality development* (pp. 201–220). Guilford.

Lenhart, Amanda; Anderson, Monica & Smith, Aaron. (2015, October 1). *Teens, technology and romantic relationships. Pew Research Center: Internet, Science & Tech*. Washington, DC: Pew Research Center.

Leonard, Hayley C. & Hill, Elisabeth L. (2014). Review: The impact of motor development on typical and atypical social cognition and language: A systematic review. *Child and Adolescent Mental Health, 19*(3), 163–170.

Leonard, Julia A.; Lee, Yuna & Schulz, Laura E. (2017). Infants make more attempts to achieve a goal when they see adults persist. *Science, 357*(6357), 1290–1294.

Lerner, Richard M.; Overton, F. Willis; Freund, Alexandra M. & Lamb, Michael E. (2010). *The handbook of life-span development*. Wiley.

Leshner, Alan I. & Dzau, Victor J. (2018). Good gun policy needs research. *Science, 359*(6381), 1195.

Leslie, Mitch. (2012). Gut microbes keep rare immune cells in line. *Science, 335*(6075), 1428.

Lessne, Deborah & Yanez, Christina. (2016, December 20). *Student reports of bullying: Results from the 2015 School Crime Supplement to the National Crime Victimization Survey*. Washington, DC: National Center for Education Statistics.

Lester, Patricia; Leskin, Gregory; Woodward, Kirsten; Saltzman, William; Nash, William; Mogil, Catherine; . . . Beardslee, William. (2011). Wartime deployment and military children: Applying prevention science to enhance family resilience. In Wadsworth, Shelley MacDermid & Riggs, David (Eds.), *Risk and resilience in U.S. military families* (pp. 149–173). Springer.

Leung, Sumie; Mareschal, Denis; Rowsell, Renee; Simpson, David; Laria, Leon; Grbic, Amanda & Kaufman, Jordy. (2016). Oscillatory activity in the infant brain and the representation of small numbers. *Frontiers in Systems Neuroscience, 10*(4).

Levey, Emma K. V.; Garandeau, Claire F.; Meeus, Wim & Branje, Susan. (2019). The longitudinal role of self-concept clarity and best friend delinquency in adolescent delinquent behavior. *Journal of Youth and Adolescence, 48*(6), 1068–1081. p. 409 Hoover.

Levy, Jonathan; Goldstein, Abraham & Feldman, Ruth. (2019). The neural development of empathy is sensitive to caregiving and early trauma. *Nature Communications, 10*(1).

Lewandowski, Lawrence J. & Lovett, Benjamin J. (2014). Learning disabilities. In Mash, Eric J. & Barkley, Russell A. (Eds.), *Child psychopathology* (3rd ed., pp. 625–669). Guilford Press.

Lewin, Kurt. (1945). The Research Center for Group Dynamics at Massachusetts Institute of Technology. *Sociometry, 8*(2), 126–136.

Lewis, John D.; Theilmann, Rebecca J.; Townsend, Jeanne & Evans, Alan C. (2013). Network efficiency in autism spectrum disorder and its relation to brain overgrowth. *Frontiers in Human Neuroscience, 7*, 845.

Lewis, Lawrence B.; Antone, Carol & Johnson, Jacqueline S. (1999). Effects of prosodic stress and serial position on syllable omission in first words. *Developmental Psychology, 35*(1), 45–59.

Lewis, Michael. (2010). The emergence of human emotions. In Lewis, Michael; Haviland-Jones, Jeannette M. & Barrett, Lisa Feldman (Eds.), *Handbook of emotions* (3rd ed.). Guilford Press.

Lewis, Michael & Brooks, Jeanne. (1978). Self-knowledge and emotional development. In Lewis, Michael & Rosenblum, L. A. (Eds.), *Genesis of behavior* (Vol. 1, pp. 205–226). Plenum Press.

Lewis, Michael & Kestler, Lisa (Eds.). (2012). *Gender differences in prenatal substance exposure*. American Psychological Association.

Li, Bo; Yang, Jing; Zhao, Faming; Zhi, Lili; Wang, Xiqian; Liu, Lin; . . . Zhao, Yunhe. (2020). Prevalence and impact of cardiovascular metabolic diseases on COVID-19 in China. *Clinical Research in Cardiology, 109*(5), 531–538.

Li, Jin; Fung, Heidi; Bakeman, Roger; Rae, Katharine & Wei, Wanchun. (2014). How European American and Taiwanese mothers talk to their children about learning. *Child Development, 85*(3), 1206–1221.

Li, Pin; Becker, Jill B.; Heitzeg, Mary M.; Mcclellan, Michele L.; Reed, Beth Glover & Zucker, Robert A. (2017). Gender differences in the transmission of risk for antisocial behavior problems across generations. *PLoS ONE, 12*(5), e0177288.

Liben, Lynn S. (2016). We've come a long way, baby (but we're not there yet): Gender past, present, and future. *Child Development, 87*(1), 5–28.

Libertus, Klaus; Joh, Amy S. & Needham, Amy W. (2016). Motor training at 3 months affects object exploration 12 months later. *Developmental Science, 19*(6), 1058–1066.

Libertus, Klaus & Needham, Amy. (2010). Teach to reach: The effects of active vs. passive reaching experiences on action and perception. *Vision Research, 50*(24), 2750–2757.

Libertus, Melissa E.; Feigenson, Lisa & Halberda, Justin. (2013). Is approximate number precision a stable predictor of math ability? *Learning and Individual Differences, 25*, 126–133.

Lillard, Angeline S. (2013). Playful learning and Montessori education. *American Journal of Play, 5*(2), 157–186.

Lillard, Angeline S. & Taggart, Jessica. (2019). Pretend play and fantasy: What if Montessori was right? *Child Development Perspectives, 13*(2), 85–90.

Liller, Karen D.; Morris, Barbara; Yang, Yingwei; Bubu, Omonigho M.; Perich, Brad & Fillion, Jessica. (2019). Injuries and concussions among young children, ages 5-11, playing sports in recreational leagues in Florida. *PLoS ONE, 14*(5), e0216217.

Lim, Cher Ping; Zhao, Yong; Tondeur, Jo; Chai, Ching Sing & Tsai, Chin-Chung. (2013). Bridging the gap: Technology trends and use of technology in schools. *Educational Technology & Society, 16*(2), 59–68.

Lin, Nan. (2017). Building a network theory of social capital: Theory and research. In Lin, Nan; Cook, Karen & Burt, Ronald S. (Eds.), *Social capital: Theory and research*. Taylor & Francis.

Lin, Phoebe. (2016). Risky behaviors: Integrating adolescent egocentrism with the theory of planned behavior. *Review of General Psychology, 20*(4), 392–398.

Lipton, Eric. (2017, March 29). E.P.A. chief, rejecting agency's science, chooses not to ban insecticide. *New York Times*.

Littman, Lisa. (2018). Parent reports of adolescents and young adults perceived to show signs of a rapid onset of gender dysphoria. *PLoS One, 13*(8), e0202330.

Liu, Andrew H. (2015). Revisiting the hygiene hypothesis for allergy and asthma. *The Journal of Allergy and Clinical Immunology, 136*(4), 860–865.

Liu, Chang & Neiderhiser, Jenae M. (2017). Using genetically informed designs to understand the environment: The importance of family-based approaches. In Tolan, Patrick H. & L., Leventhal

Bennett (Eds.), *Gene-environment transactions in developmental psychopathology: The role in intervention research* (pp. 95–110). Springer.

Liu, Junsheng; Chen, Xinyin; Coplan, Robert J.; Ding, Xuechen; Zarbatany, Lynne & Ellis, Wendy. (2015). Shyness and unsociability and their relations with adjustment in Chinese and Canadian children. *Journal of Cross-Cultural Psychology, 46*(3), 371–386.

Liu, Ning; Vigod, Simone N.; Farrugia, M. Michèle; Urquia, Marcelo L. & Ray, Joel G. (2018). Intergenerational teen pregnancy: A population-based cohort study. *BJOG, 125*(13), 1766–1774.

Liu, Peiwei & Feng, Tingyong. (2017). The overlapping brain region accounting for the relationship between procrastination and impulsivity: A voxel-based morphometry study. *Neuroscience, 360*, 9–17.

Liu, Tong; Yang, Wen-Qiang; Xie, Yu-Gu; Liu, Pei-Wen; Xie, Li-Hua; Lin, Feng; . . . Chen, Xiao-Guang. (2019). Construction of an efficient genomic editing system with CRISPR/Cas9 in the vector mosquito *Aedes albopictus*. *Insect Science, 26*(6), 1045–1054.

Livingston, Lucy Anne & Happé, Francesca. (2017). Conceptualising compensation in neurodevelopmental disorders: Reflections from autism spectrum disorder. *Neuroscience and Biobehavioral Reviews, 80*, 729–742.

Lo, June C.; Lee, Su Mei; Lee, Xuan Kai; Sasmita, Karen; Chee, Nicholas I. Y. N.; Tandi, Jesisca; . . . Chee, Michael W. L. (2018). Sustained benefits of delaying school start time on adolescent sleep and well-being. *Sleep, 41*(6).

LoBue, Vanessa. (2013). What are we so afraid of? How early attention shapes our most common fears. *Child Development Perspectives, 7*(1), 38–42.

LoBue, Vanessa & Adolph, Karen E. (2019). Fear in infancy: Lessons from snakes, spiders, heights, and strangers. *Developmental Psychology, 55*(9), 1889–1907.

Lochman, John E.; Dishion, Thomas J.; Powell, Nicole P.; Boxmeyer, Caroline L.; Qu, Lixin & Sallee, Meghann. (2015). Evidence-based preventive intervention for preadolescent aggressive children: One-year outcomes following randomization to group versus individual delivery. *Journal of Consulting and Clinical Psychology, 83*(4), 728–735.

Lockhart, Kristi L.; Goddu, Mariel K. & Keil, Frank C. (2018). When saying "I'm best" is benign: Developmental shifts in perceptions of boasting. *Developmental Psychology, 54*(3), 521–535.

Lockhart, Kristi L. & Keil, Frank C. (2018). What heals and why? Children's understanding of medical treatments. *Monographs of the Society for Research in Child Development, 83*(2).

Lønfeldt, Nicole N.; Verhulst, Frank C.; Strandberg-Larsen, Katrine; Plessen, Kerstin J. & Lebowitz, Eli R. (2018). Assessing risk of neurodevelopmental disorders after birth with oxytocin: A systematic review and meta-analysis. *Psychological Medicine.*

López-Pinar, Carlos; Martínez-Sanchís, Sonia; Carbonell-Vayá, Enrique; Fenollar-Cortés, Javier & Sánchez-Meca, Julio. (2018). Long-term

efficacy of psychosocial treatments for adults with Attention-deficit/hyperactivity disorder: A meta-analytic review. *Frontiers in Psychology, 9*(638).

Lordier, Lara; Meskaldji, Djalel-Eddine; Grouiller, Frédéric; Pittet, Marie P.; Vollenweider, Andreas; Vasung, Lana; . . . Hüppi, Petra S. (2019). Music in premature infants enhances high-level cognitive brain networks. *PNAS, 116*(24), 12103–12108.

Lorthe, Elsa; Torchin, Héloïse; Delorme, Pierre; Ancel, Pierre-Yves; Marchand-Martin, Laetitia; Foix-L'Hélias, Laurence; . . . Kayem, Gilles. (2018). Preterm premature rupture of membranes at 22–25 weeks' gestation: Perinatal and 2-year outcomes within a national population-based study (EPIPAGE-2). *American Journal of Obstetrics and Gynecology, 219*(3), 298.e1–298.e14.

Loucks, Jeff & Price, Heather L. (2019). Memory for temporal order in action is slow developing, sensitive to deviant input, and supported by foundational cognitive processes. *Developmental Psychology, 55*(2), 263–273.

Lourenço, Orlando. (2012). Piaget and Vygotsky: Many resemblances, and a crucial difference. *New Ideas in Psychology, 30*(3), 281–295.

Loyd, Aerika Brittian & Gaither, Sarah E. (2018). Racial/ethnic socialization for white youth: What we know and future directions. *Journal of Applied Developmental Psychology, 59*, 54–64.

Lubienski, Christopher; Puckett, Tiffany & Brewer, T. Jameson. (2013). Does homeschooling "work"? A critique of the empirical claims and agenda of advocacy organizations. *Peabody Journal of Education, 88*(3), 378–392.

Lubke, Gitta H.; Mcartor, Daniel B.; Boomsma, Dorret I. & Bartels, Meike. (2018). Genetic and environmental contributions to the development of childhood aggression. *Developmental Psychology, 54*(1), 39–50.

Lucca, Kelsey & Wilbourn, Makeba Parramore. (2018). Communicating to learn: Infants' pointing gestures result in optimal learning. *Child Development, 89*(3), 941–960.

Luecken, Linda J.; Lin, Betty; Coburn, Shayna S.; MacKinnon, David P.; Gonzales, Nancy A. & Crnic, Keith A. (2013). Prenatal stress, partner support, and infant cortisol reactivity in low-income Mexican American families. *Psychoneuroendocrinology, 38*(12), 3092–3101.

Luhmann, Maike & Hawkley, Louise C. (2016). Age differences in loneliness from late adolescence to oldest old age. *Developmental Psychology, 52*(6), 943–959.

Lundahl, Alyssa; Kidwell, Katherine M. & Nelson, Timothy D. (2014). Parental underestimates of child weight: A meta-analysis. *Pediatrics, 133*(3), e689–e703.

Lundberg, Shelly; Pollak, Robert A. & Stearns, Jenna. (2016). Family inequality: Diverging patterns in marriage, cohabitation, and childbearing. *Journal of Economic Perspectives, 30*(2), 79–102.

Luo, Rufan; Tamis-LeMonda, Catherine S.; Kuchirko, Yana; Ng, Florrie F. & Liang, Eva. (2014). Mother–child book-sharing and children's storytelling skills in ethnically diverse, low-income families. *Infant and Child Development, 23*(4), 402–425.

Lushin, Viktor; Jaccard, James & Kaploun, Victor. (2017). Parental monitoring, adolescent

dishonesty and underage drinking: A nationally representative study. *Journal of Adolescence, 57*, 99–107.

Luthar, Suniya S. (2015). Resilience in development: A synthesis of research across five decades. In Cicchetti, Dante & Cohen, Donald J. (Eds.), *Developmental psychopathology* (2nd ed., Vol. 3). Wiley.

Luthar, Suniya S.; Cicchetti, Dante & Becker, Bronwyn. (2000). The construct of resilience: A critical evaluation and guidelines for future work. *Child Development, 71*(3), 543–562.

Luthar, Suniya S.; Small, Phillip J. & Ciciolla, Lucia. (2018). Adolescents from upper middle class communities: Substance misuse and addiction across early adulthood. *Development and Psychopathology, 30*(1), 315–335.

Lyall, Donald M.; Inskip, Hazel M.; Mackay, Daniel; Deary, Ian J.; McIntosh, Andrew M.; Hotopf, Matthew; . . . Smith, Daniel J. (2016). Low birth weight and features of neuroticism and mood disorder in 83,545 participants of the UK Biobank cohort. *British Journal of Psychiatry Open, 2*(1), 38–44.

Macgregor, Stuart; Lind, Penelope A.; Bucholz, Kathleen K.; Hansell, Narelle K.; Madden, Pamela A. F.; Richter, Melinda M.; . . . Whitfield, John B. (2009). Associations of ADH and ALDH2 gene variation with self report alcohol reactions, consumption and dependence: An integrated analysis. *Human Molecular Genetics, 18*(3), 580–593.

Mackenzie, Karen J.; Anderton, Stephen M. & Schwarze, Jürgen. (2014). Viral respiratory tract infections and asthma in early life: Cause and effect? *Clinical & Experimental Allergy, 44*(1), 9–19.

MacKenzie, Michael J.; Nicklas, Eric; Brooks-Gunn, Jeanne & Waldfogel, Jane. (2011). Who spanks infants and toddlers? Evidence from the fragile families and child well-being study. *Children and Youth Services Review, 33*(8), 1364–1373.

MacNeill, Leigha A.; Ram, Nilam; Bell, Martha Ann; Fox, Nathan A. & Pérez-Edgar, Koraly. (2018). Trajectories of infants' biobehavioral development: Timing and rate of A-not-B performance gains and EEG maturation. *Child Development, 89*(3), 711–724.

Macosko, Evan Z. & McCarroll, Steven A. (2013). Our fallen genomes. *Science, 342*(6158), 564–565.

MacSwan, Jeff. (2018). Academic English as standard language ideology: A renewed research agenda for asset-based language education. *Language Teaching Research.*

MacWhinney, Brian. (2015). Language development. In Lerner, Richard M. (Ed.), *Handbook of child psychology and developmental science* (7th ed., Vol. 2, pp. 296–338). Wiley.

Madigan, Sheri; Browne, Dillon; Racine, Nicole; Mori, Camille & Tough, Suzanne. (2019). Association between screen time and children's performance on a developmental screening test. *JAMA Pediatrics, 173*(3), 244–250.

Maenner, Matthew J.; Shaw, Kelly A.; Baio, Jon; Washington, Anita; Patrick, Mary; DiRienzo, Monica; . . . Dietz, Patricia M.

(2020, March 27). *Prevalence of autism spectrum disorder among children aged 8 years—autism and developmental disabilities monitoring network, 11 sites, United States, 2016. Morbidity and Mortality Weekly Report: Surveillance Summaries, 69*(4), 1–12. Atlanta, GA: Centers for Disease Control and Prevention.

Mahomed, Kassam; Wild, Kellie; Brown, Consuela & Green, Ann. (2019). Does fentanyl epidural analgesia affect breastfeeding: A prospective cohort study. *Australian and New Zealand Journal of Obstetrics and Gynaecology, 59*(6), 819–824.

Majecka, Katarzyna & Pietraszewski, Dariusz. (2018). Where's the cookie? The ability of monkeys to track object transpositions. *Animal Cognition, 21*(4), 603–611.

Makinen, Mauno; Puukko-Viertomies, Leena-Riitta; Lindberg, Nina; Siimes, Martti A. & Aalberg, Veikko. (2012). Body dissatisfaction and body mass in girls and boys transitioning from early to mid-adolescence: Additional role of self-esteem and eating habits. *BMC Psychiatry, 12*(35).

Malina, Robert M.; Bouchard, Claude & Bar-Or, Oded. (2004). *Growth, maturation, and physical activity* (2nd ed.). Human Kinetics.

Mallett, Christopher A. (2016). The school-to-prison pipeline: A critical review of the punitive paradigm shift. *Child and Adolescent Social Work Journal, 33*(1), 15–24.

Malloy, Lindsay C.; Shulman, Elizabeth P. & Cauffman, Elizabeth. (2014). Interrogations, confessions, and guilty pleas among serious adolescent offenders. *Law and Human Behavior, 38*(2), 181–193.

Malpas, Jean. (2011). Between pink and blue: A multi-dimensional family approach to gender nonconforming children and their families. *Family Process, 50*(4), 453–470.

Maltby, Lauren E.; Callahan, Kelly L.; Friedlander, Scott & Shetgiri, Rashmi. (2019). Infant temperament and behavioral problems: Analysis of high-risk infants in child welfare. *Journal of Public Child Welfare, 13*(5), 512–528.

Manasyan, Albert; Chomba, Elwyn; Moore, Janet; Wallace, Dennis; McClure, Elizabeth M.; Koso-Thomas, Marion & Carlo, Waldemar A. (2019). Association between birth attendant type and delivery site and perinatal outcomes. *International Journal of Gynecology & Obstetrics, 145*(2), 187–192.

Manczak, Erika M. & Gotlib, Ian H. (2019). Lipid profiles at birth predict teacher-rated child emotional and social development 5 years later. *Psychological Science, 30*(12), 1780–1789

Mandelbaum, David E. & de la Monte, Suzanne M. (2017). Adverse structural and functional effects of marijuana on the brain: Evidence reviewed. *Pediatric Neurology, 66*, 12–20.

Mann, Joshua R.; McDermott, Suzanne; Bao, Haikun & Bersabe, Adrian. (2009). Maternal genitourinary infection and risk of cerebral palsy. *Developmental Medicine & Child Neurology, 51*(4), 282–288.

Mar, Raymond A. (2011). The neural bases of social cognition and story comprehension. *Annual Review of Psychology, 62*, 103–134.

Marazita, John M. & Merriman, William E. (2010). Verifying one's knowledge of a name without retrieving it: A U-shaped relation to vocabulary size in early childhood. *Language Learning and Development, 7*(1), 40–54.

Marchiano, Lisa. (2017). Outbreak: On transgender teens and psychic epidemics. *Psychological Perspectives: Gender Diversity, 60*(3), 345–366.

Marcia, James E. (1966). Development and validation of ego-identity status. *Journal of Personality and Social Psychology, 3*(5), 551–558.

Marcovitch, Stuart; Clearfield, Melissa W.; Swingler, Margaret; Calkins, Susan D. & Bell, Martha Ann. (2016). Attentional predictors of 5-month-olds' performance on a looking A-not-B task. *Infant and Child Development, 25*(4), 233–246.

Marcus, Gary F. & Rabagliati, Hugh. (2009). Language acquisition, domain specificity, and descent with modification. In Colombo, John; McCardle, Peggy & Freund, Lisa (Eds.), *Infant pathways to language: Methods, models, and research disorders* (pp. 267–285). Psychology Press.

Mareschal, Denis & Kaufman, Jordy. (2012). Object permanence in infancy: Revisiting Baillargeon's drawbridge study. In Slater, Alan M. & Quinn, Paul C. (Eds.), *Developmental psychology: Revisiting the classic studies*. Sage.

Markova, Gabriela. (2018). The games infants play: Social games during early mother–infant interactions and their relationship with oxytocin. *Frontiers in Psychology, 9*(1041).

Markovitch, Noam; Luyckx, Koen; Klimstra, Theo; Abramson, Lior & Knafo-Noam, Ariel. (2017). Identity exploration and commitment in early adolescence: Genetic and environmental contributions. *Developmental Psychology, 53*(11), 2092–2102.

Marks, Amy K.; Ejesi, Kida & García Coll, Cynthia. (2014). Understanding the U.S. immigrant paradox in childhood and adolescence. *Child Development Perspectives, 8*(2), 59–64.

Marotta, Phillip L. & Voisin, Dexter R. (2017). Testing three pathways to substance use and delinquency among low-income African American adolescents. *Children and Youth Services Review, 75*, 7–14.

Marouli, Eirini; Graff, Mariaelisa; Medina-Gomez, Carolina; Lo, Ken Sin; Wood, Andrew R.; Kjaer, Troels R.; . . . Perola, Markus. (2017). Rare and low-frequency coding variants alter human adult height. *Nature, 542*, 186–190.

Marshall, Eliot. (2014). An experiment in zero parenting. *Science, 345*(6198), 752–754.

Martin, Carmel. (2014). *Common Core implementation best practices. New York State Office of the Governor Common Core Implementation Panel.* Washington, DC: Center for American Progress.

Martin, Carol L.; Fabes, Richard; Hanish, Laura; Leonard, Stacie & Dinella, Lisa. (2011). Experienced and expected similarity to same-gender peers: Moving toward a comprehensive model of gender segregation. *Sex Roles, 65*(5/6), 421–434.

Martin, Joyce A.; Hamilton, Brady E.; Osterman, Michelle J. K. & Driscoll, Anne K. (2019, November 27). *Births: Final data for 2018. National Vital Statistics Reports, 68*(13). Hyattsville, MD: National Center for Health Statistics.

Martin, Joyce A.; Hamilton, Brady E.; Osterman, Michelle J. K.; Driscoll, Anne K. &

Drake, Patrick. (2018, January 31). *Births: Final data for 2016. National Vital Statistics Reports, 67*(1). Hyattsville, MD: National Center for Health Statistics.

Martinson, Melissa L. & Reichman, Nancy E. (2016). Socioeconomic inequalities in low birth weight in the United States, the United Kingdom, Canada, and Australia. *American Journal of Public Health, 106*(4), 748–754.

Marulis, Loren Marie; Baker, Sara T. & Whitebread, David. (2020). Integrating metacognition and executive function to enhance young children's perception of and agency in their learning. *Early Childhood Research Quarterly, 50*, 46–54.

Masarik, April S. & Conger, Rand D. (2017). Stress and child development: A review of the Family Stress Model. *Current Opinion in Psychology, 13*, 85–90.

Mascarelli, Amanda. (2013). Growing up with pesticides. *Science, 341*(6147), 740–741.

Mascaro, Jennifer S.; Rentscher, Kelly E.; Hackett, Patrick D.; Mehl, Matthias R. & Rilling, James K. (2017). Child gender influences paternal behavior, language, and brain function. *Behavioral Neuroscience, 131*(3), 262–273.

Maski, Kiran P. & Kothare, Sanjeev V. (2013). Sleep deprivation and neurobehavioral functioning in children. *International Journal of Psychophysiology, 89*(2), 259–264.

Masten, Ann S. (2014). *Ordinary magic: Resilience in development.* Guilford Press.

Mattick, Richard P.; Clare, Philip J.; Aiken, Alexandra; Wadolowski, Monika; Hutchinson, Delyse; Najman, Jackob; . . . Degenhardt, Louisa. (2018). Association of parental supply of alcohol with adolescent drinking, alcohol-related harms, and alcohol use disorder symptoms: A prospective cohort study. *The Lancet Public Health, 3*(2), e64–e71.

Maxwell, Lynne G.; Fraga, María V. & Malavolta, Carrie P. (2019). Assessment of pain in the newborn. *Clinics in Perinatology, 46*(4), 693–707.

May, Lillian; Byers-Heinlein, Krista; Gervain, Judit & Werker, Janet F. (2011). Language and the newborn brain: Does prenatal language experience shape the neonate neural response to speech? *Frontiers in Psychology, 2*, 222.

Mayberry, Rachel I. & Kluender, Robert. (2018). Rethinking the critical period for language: New insights into an old question from American Sign Language. *Bilingualism: Language and Cognition, 21*(5), 938–944.

Mazza, Julia Rachel; Pingault, Jean-Baptiste; Booij, Linda; Boivin, Michel; Tremblay, Richard; Lambert, Jean; . . . Côté, Sylvana. (2017). Poverty and behavior problems during early childhood: The mediating role of maternal depression symptoms and parenting. *International Journal of Behavioral Development, 41*(6), 670–680.

McAlister, Anna R. & Peterson, Candida C. (2013). Siblings, theory of mind, and executive functioning in children aged 3–6 years: New longitudinal evidence. *Child Development, 84*(4), 1442–1458.

McCabe, Sean Esteban; Veliz, Philip; Wilens, Timothy E. & Schulenberg, John E. (2017).

Adolescents' prescription stimulant use and adult functional outcomes: A national prospective study. *Journal of the American Academy of Child and Adolescent Psychiatry, 56*(3), 226–233.e4.

McCabe, Sean Esteban; West, Brady T.; Teter, Christian J. & Boyd, Carol J. (2014). Trends in medical use, diversion, and nonmedical use of prescription medications among college students from 2003 to 2013: Connecting the dots. *Addictive Behaviors, 39*(7), 1176–1182.

McCall, Robert B. (2013). The consequences of early institutionalization: Can institutions be improved?—Should they? *Child and Adolescent Mental Health, 18*(4), 193–201.

McCallion, Gail & Feder, Jody. (2013, October 18). *Student bullying: Overview of research, federal initiatives, and legal issues.* Washington, DC: Congressional Research Service. R43254.

McCarthy, Neil & Eberhart, Johann K. (2014). Gene–ethanol interactions underlying fetal alcohol spectrum disorders. *Cellular and Molecular Life Sciences, 71*(14), 2699–2706.

McClain, Natalie M. & Garrity, Stacy E. (2011). Sex trafficking and the exploitation of adolescents. *Journal of Obstetric, Gynecologic, & Neonatal Nursing, 40*(2), 243–252.

McConnell, Sean C. & Blasimme, Alessandro. (2019). Ethics, values, and responsibility in human genome editing. *AMA Journal of Ethics, 21*(12), E1017–1020.

McCormick, Cheryl M.; Mathews, Iva Z.; Thomas, Catherine & Waters, Patti. (2010). Investigations of HPA function and the enduring consequences of stressors in adolescence in animal models. *Brain and Cognition, 72*(1), 73–85.

McCoy, Shelly; Dimler, Laura; Samuels, Danielle & Natsuaki, Misaki N. (2019). Adolescent susceptibility to deviant peer pressure: Does gender matter? *Adolescent Research Review, 4*(1), 59–71.

McCray, Jennifer S.; Chen, Jie-Qi & Sorkin, Janet Eisenband (Eds.). (2018). *Growing mathematical minds: Conversations between developmental psychologists and early childhood teachers.* Routledge.

McEwen, Craig A. & McEwen, Bruce S. (2017). Social structure, adversity, toxic stress, and intergenerational poverty: An early childhood model. *Annual Review of Sociology, 43*, 445–472.

McFarland, David H.; Fortin, Annie Joëlle & Polka, Linda. (2020). Physiological measures of mother–infant interactional synchrony. *Developmental Psychobiology, 62*(1), 50–61.

McFarland, Joel; Hussar, Bill; Wang, Xiaolei; Zhang, Jijun; Wang, Ke; Rathbun, Amy; . . . Mann, Farrah Bullock. (2018, May). *The condition of education 2018.* Washington, DC: National Center for Education Statistics. NCES 2018-144.

McFarland, Joel; Hussar, Bill; Zhang, Jijun; Wang, Xiaolei; Wang, Ke; Hein, Sarah; . . . Barmer, Amy. (2019). *The condition of education 2019.* Washington, DC: National Center for Education Statistics. NCES 2019–144.

McFarlane, Alexander C. & Van Hooff, Miranda. (2009). Impact of childhood exposure to a natural disaster on adult mental health: 20-year longitudinal follow-up study. *The British Journal of Psychiatry, 195*(2), 142–148.

McGillion, Michelle; Herbert, Jane S.; Pine, Julian; Vihman, Marilyn; dePaolis, Rory; Keren-Portnoy, Tamar & Matthews, Danielle. (2017). What paves the way to conventional language? The predictive value of babble, pointing, and socioeconomic status. *Child Development, 88*(1), 156–166.

McGue, Matt; Irons, Dan & Iacono, William G. (2014). The adolescent origins of substance use disorders: A behavioral genetic perspective. In Stoltenberg, Scott F. (Ed.), *Genes and the motivation to use substances* (pp. 31–50). Springer.

McKeever, Pamela M. & Clark, Linda. (2017). Delayed high school start times later than 8:30 a.m. and impact on graduation rates and attendance rates. *Sleep Health, 3*(2), 119–125.

McKeganey, Neil; Russell, Christopher; Katsampouris, Evangelos & Haseen, Farhana. (2019). Sources of youth access to JUUL vaping products in the United States. *Addictive Behaviors Reports, 10*, 100232.

McKenzie, Sarah C. & Ritter, Gary W. (2017). School discipline in Arkansas. *Policy Briefs, 14*(4).

McKone, Elinor; Wan, Lulu; Pidcock, Madeleine; Crookes, Kate; Reynolds, Katherine; Dawel, Amy; . . . Fiorentini, Chiara. (2019). A critical period for faces: Other-race face recognition is improved by childhood but not adult social contact. *Scientific Reports, 9*(12820).

McLeod, Bryce D.; Wood, Jeffrey J. & Weisz, John R. (2007). Examining the association between parenting and childhood anxiety: A meta-analysis. *Clinical Psychology Review, 27*(2), 155–172.

McManus, I. Chris; Moore, James; Freegard, Matthew & Rawles, Richard. (2010). Science in the making: Right hand, left hand. III: Estimating historical rates of left-handedness. *Laterality: Asymmetries of Body, Brain and Cognition, 15*(1/2), 186–208.

McNally, Shelley A. & Slutsky, Ruslan. (2017). Key elements of the Reggio Emilia approach and how they are interconnected to create the highly regarded system of early childhood education. *Early Child Development and Care, 187*(12), 1925–1937.

Meece, Judith L. & Eccles, Jacquelynne S. (Eds.). (2010). *Handbook of research on schools, schooling, and human development.* Routledge.

Meeus, Wim. (2011). The study of adolescent identity formation 2000–2010: A review of longitudinal research. *Journal of Research on Adolescence, 21*(1), 75–94.

Meeus, Wim. (2017). Adolescent ethnic identity in social context: A commentary. *Child Development, 88*(3), 761–766.

Mehler, Philip S. (2018). Medical complications of anorexia nervosa and bulimia nervosa. In Agras, W. Stewart & Robinson, Athena (Eds.), *The Oxford handbook of eating disorders* (2nd ed.). Oxford University Press.

Meinhardt-Injac, Bozana; Daum, Moritz M. & Meinhardt, Günter. (2020). Theory of mind development from adolescence to adulthood: Testing the two-component model. *British Journal of Developmental Psychology, 38*(2), 289–303.

Meldrum, Ryan; Kavish, Nicholas & Boutwell, Brian. (2018). On the longitudinal association between peer and adolescent intelligence: Can our friends make us smarter? *PsyArXiv.*

Meldrum, Ryan Charles; Young, Jacob T. N.; Kavish, Nicholas & Boutwell, Brian B. (2019). Could peers influence intelligence during adolescence? An exploratory study. *Intelligence, 72*, 28–34.

Mellerson, Jenelle L.; Maxwell, Choppell B.; Knighton, Cynthia L.; Kriss, Jennifer L.; Seither, Ranee & Black, Carla L. (2018, October 12). *Vaccination coverage for selected vaccines and exemption rates among children in kindergarten—United States, 2017–18 school year. Morbidity and Mortality Weekly Report, 67*(40), 1115–1122. Atlanta, GA: Centers for Disease Control and Prevention.

Meltzoff, Andrew N. & Gopnik, Alison. (2013). Learning about the mind from evidence: Children's development of intuitive theories of perception and personality. In Baron-Cohen, Simon; Tager-Flusberg, Helen & Lombardo, Michael (Eds.), *Understanding other minds: Perspectives from developmental social neuroscience* (3rd ed., pp. 19–34). Oxford University Press.

Mendle, Jane; Harden, K. Paige; Brooks-Gunn, Jeanne & Graber, Julia A. (2010). Development's tortoise and hare: Pubertal timing, pubertal tempo, and depressive symptoms in boys and girls. *Developmental Psychology, 46*(5), 1341–1353.

Mendle, Jane; Harden, K. Paige; Brooks-Gunn, Jeanne & Graber, Julia A. (2012). Peer relationships and depressive symptomatology in boys at puberty. *Developmental Psychology, 48*(2), 429–435.

Mennella, Julie A. & Bobowski, Nuala K. (2015). The sweetness and bitterness of childhood: Insights from basic research on taste preferences. *Physiology & Behavior, 152*, 502–507.

Mercer, Neil & Howe, Christine. (2012). Explaining the dialogic processes of teaching and learning: The value and potential of sociocultural theory. *Learning, Culture and Social Interaction, 1*(1), 12–21.

Mermelshtine, Roni. (2017). Parent–child learning interactions: A review of the literature on scaffolding. *British Journal of Educational Psychology, 87*(2), 241–254.

Merz, Emily C. & McCall, Robert B. (2011). Parent ratings of executive functioning in children adopted from psychosocially depriving institutions. *Journal of Child Psychology and Psychiatry, 52*(5), 537–546.

Messinger, Daniel M.; Ruvolo, Paul; Ekas, Naomi V. & Fogel, Alan. (2010). Applying machine learning to infant interaction: The development is in the details. *Neural Networks, 23*(8/9), 1004–1016.

Metcalfe, Lindsay A.; Harvey, Elizabeth A. & Laws, Holly B. (2013). The longitudinal relation between academic/cognitive skills and externalizing behavior problems in preschool children. *Journal of Educational Psychology, 105*(3), 881–894.

Michaeli, Yossi; Kalfon Hakhmigari, Maor; Dickson, Daniel J.; Scharf, Miri & Shulman, Shmuel. (2018). The role of change in self-criticism across young adulthood in explaining developmental outcomes and psychological wellbeing. *Journal of Personality, 87*(4), 785–798.

Miech, Richard A.; Johnston, Lloyd D.; O'Malley, Patrick M.; Bachman, Jerald G. &

Schulenberg, John E. (2015). *Monitoring the future, national survey results on drug use, 1975–2014: Volume I, secondary school students.* Ann Arbor, MI: Institute for Social Research, The University of Michigan.

Miech, Richard A.; Johnston, Lloyd D.; O'Malley, Patrick M.; Bachman, Jerald G. & Schulenberg, John E. (2016). *Monitoring the future, national survey results on drug use, 1975–2015: Volume I, secondary school students.* Ann Arbor, MI: Institute for Social Research, The University of Michigan.

Miech, Richard A.; Johnston, Lloyd D.; O'Malley, Patrick M.; Bachman, Jerald G.; Schulenberg, John E. & Patrick, Megan E. (2017a). *Monitoring the future, national survey results on drug use, 1975–2016: Volume I, secondary school students.* Ann Arbor, MI: Institute for Social Research, The University of Michigan.

Miech, Richard A.; Johnston, Lloyd D.; O'Malley, Patrick M.; Bachman, Jerald G.; Schulenberg, John E. & Patrick, Megan E. (2018). *Monitoring the future, national survey results on drug use, 1975–2017: Volume I, secondary school students.* Ann Arbor, MI: Institute for Social Research, The University of Michigan.

Miech, Richard A.; Johnston, Lloyd D.; O'Malley, Patrick M.; Bachman, Jerald G.; Schulenberg, John E. & Patrick, Megan E. (2019). *Monitoring the future, national survey results on drug use, 1975–2018: Volume I, secondary school students.* Ann Arbor, MI: Institute for Social Research, The University of Michigan.

Miech, Richard A.; Patrick, Megan E.; O'Malley, Patrick M. & Johnston, Lloyd D. (2017b). E-cigarette use as a predictor of cigarette smoking: Results from a 1-year follow-up of a national sample of 12th grade students. *Tobacco Control, 26,* e106–e111.

Miklowitz, David J. & Cicchetti, Dante (Eds.). (2010). *Understanding bipolar disorder: A developmental psychopathology perspective.* Guilford Press.

Mikolajczyk, Rafael T.; Zhang, Jun; Grewal, Jagteshwar; Chan, Linda C.; Petersen, Antje & Gross, Mechthild M. (2016). Early versus late admission to labor affects labor progression and risk of cesarean section in nulliparous women. *Frontiers in Medicine, 3*(26).

Milgrom, Jeannette; Martin, Paul R.; Newnham, Carol; Holt, Christopher J.; Anderson, Peter J.; Hunt, Rod W.; . . . Gemmill, Alan W. (2019). Behavioural and cognitive outcomes following an early stress-reduction intervention for very preterm and extremely preterm infants. *Pediatric Research, 86,* 92–99.

Miller, Cindy F.; Martin, Carol Lynn; Fabes, Richard A. & Hanish, Laura D. (2013). Bringing the cognitive and the social together: How gender detectives and gender enforcers shape children's gender development. In Banaji, Mahzarin R. & Gelman, Susan A. (Eds.), *Navigating the social world: What infants, children, and other species can teach us* (pp. 306–313). Oxford University Press.

Miller, Melissa K.; Dowd, M. Denise; Harrison, Christopher J.; Mollen, Cynthia J.; Selvarangan, Rangaraj & Humiston, Sharon. (2015). Prevalence of 3 sexually transmitted infections in a pediatric emergency department. *Pediatric Emergency Care, 31*(2), 107–112.

Miller, Patricia H. (2011). *Theories of developmental psychology* (5th ed.). Worth.

Miller, Patricia H. (2016). *Theories of developmental psychology* (6th ed.). Worth.

Miller-Cotto, Dana & Byrnes, James P. (2016). Ethnic/racial identity and academic achievement: A meta-analytic review. *Developmental Review, 41,* 51–70.

Miller-Perrin, Cindy & Wurtele, Sandy K. (2017). Sex trafficking and the commercial sexual exploitation of children. *Women & Therapy, 40*(1/2), 123–151.

Mills, Catherine. (2019, May 14). CRISPR case highlights the ethical concerns over gene editing. *Lens.* Monash University.

Mills-Koonce, W. Roger; Garrett-Peters, Patricia; Barnett, Melissa; Granger, Douglas A.; Blair, Clancy & Cox, Martha J. (2011). Father contributions to cortisol responses in infancy and toddlerhood. *Developmental Psychology, 47*(2), 388–395.

Milunsky, Aubrey & Milunsky, Jeff M. (2016). *Genetic disorders and the fetus: Diagnosis, prevention, and treatment* (7th ed.). Wiley-Blackwell.

Mindell, Jodi A.; Sadeh, Avi; Wiegand, Benjamin; How, Ti Hwei & Goh, Daniel Y. T. (2010). Cross-cultural differences in infant and toddler sleep. *Sleep Medicine, 11*(3), 274–280.

Mischel, Walter. (2014). *The marshmallow test: Mastering self-control.* Little, Brown.

Mischel, Walter; Ebbesen, Ebbe B. & Raskoff Zeiss, Antonette. (1972). Cognitive and attentional mechanisms in delay of gratification. *Journal of Personality and Social Psychology, 21*(2), 204–218.

Misra, Dawn P.; Caldwell, Cleopatra; Young, Alford A. & Abelson, Sara. (2010). Do fathers matter? Paternal contributions to birth outcomes and racial disparities. *American Journal of Obstetrics and Gynecology, 202*(2), 99–100.

Missana, Manuela; Rajhans, Purva; Atkinson, Anthony P. & Grossmann, Tobias. (2014). Discrimination of fearful and happy body postures in 8-month-old infants: An event-related potential study. *Frontiers in Human Neuroscience, 8,* 531.

Mitchell, Edwin A. (2009). SIDS: Past, present and future. *Acta Paediatrica, 98*(11), 1712–1719.

Mitchell, Kimberly J.; Jones, Lisa M.; Finkelhor, David & Wolak, Janis. (2013). Understanding the decline in unwanted online sexual solicitations for U.S. youth 2000–2010: Findings from three Youth Internet Safety Surveys. *Child Abuse & Neglect, 37*(12), 1225–1236.

Miyata, Susanne; MacWhinney, Brian; Otomo, Kiyoshi; Sirai, Hidetosi; Oshima-Takane, Yuriko; Hirakawa, Makiko; . . . Itoh, Keiko. (2013). Developmental sentence scoring for Japanese. *First Language, 33*(2), 200–216.

Mize, Krystal D.; Pineda, Melannie; Blau, Alexis K.; Marsh, Kathryn & Jones, Nancy A. (2014). Infant physiological and behavioral responses to a jealousy provoking condition. *Infancy, 19*(3), 338–348.

MMWR. (2008, January 18). *School-associated student homicides—United States, 1992–2006. Morbidity and Mortality Weekly Report, 57*(2), 33–36. Atlanta, GA: U.S. Department of Health and Human Services, Centers for Disease Control and Prevention.

MMWR. (2010, June 4). *Youth risk behavior surveillance—United States, 2009. Morbidity and Mortality Weekly Report Surveillance Summaries, 59*(SS05). Atlanta, GA: U.S. Department of Health and Human Services, Centers for Disease Control and Prevention.

MMWR. (2012, June 8). *Youth risk behavior surveillance—United States, 2011. Morbidity and Mortality Weekly Report, 61*(4). Atlanta, GA: U.S. Department of Health and Human Services, Centers for Disease Control and Prevention.

MMWR. (2013, April 5). *Blood lead levels in children aged 1–5 Years—United States, 1999–2010. Morbidity and Mortality Weekly Report, 62*(13), 245–248. Atlanta, GA: U.S. Department of Health and Human Services, Centers for Disease Control and Prevention.

MMWR. (2014, June 13). *Youth risk behavior surveillance—United States, 2013. Morbidity and Mortality Weekly Report, 63*(4). Atlanta, GA: U.S. Department of Health and Human Services, Centers for Disease Control and Prevention.

MMWR. (2014, March 7). *Impact of requiring influenza vaccination for children in licensed child care or preschool programs—Connecticut, 2012–13 influenza season. Morbidity and Mortality Weekly Report, 63*(9), 181–185. Atlanta, GA: U.S. Department of Health and Human Services, Centers for Disease Control and Prevention.

MMWR. (2014, March 28). *Prevalence of autism spectrum disorder among children aged 8 years—Autism and Developmental Disabilities Monitoring Network, 11 sites, United States, 2010. Morbidity and Mortality Weekly Report, 63*(2). Atlanta, GA: U.S. Department of Health and Human Services, Centers for Disease Control and Prevention.

MMWR. (2016, June 10). *Youth risk behavior surveillance—United States, 2015. Morbidity and Mortality Weekly Report, 65*(6). Atlanta, GA: U.S. Department of Health and Human Services, Centers for Disease Control and Prevention.

MMWR. (2018, June 15). *Youth risk behavior surveillance—United States, 2017. Morbidity and Mortality Weekly Report, 67*(8). Atlanta, GA: U.S. Department of Health and Human Services, Centers for Disease Control and Prevention.

MMWR. (2020, March 6). *Quickstats: Age-adjusted suicide rates, by sex and three most common methods—United States, 2000–2018. Morbidity and mortality weekly report, 69*(9), 249. Atlanta, GA: U.S. Department of Health and Human Services, Centers for Disease Control and Prevention.

Moffitt, Terrie E. (2003). Life-course-persistent and adolescence-limited antisocial behavior: A 10-year research review and a research agenda. In Lahey, Benjamin B.; Moffitt, Terrie E. & Caspi, Avshalom (Eds.), *Causes of conduct disorder and juvenile delinquency* (pp. 49–75). Guilford Press.

Moffitt, Terrie E.; Caspi, Avshalom; Rutter, Michael & Silva, Phil A. (2001). *Sex differences in antisocial behaviour: Conduct disorder, delinquency, and violence in the Dunedin Longitudinal Study.* Cambridge University Press.

Mokrova, Irina L.; O'Brien, Marion; Calkins, Susan D.; Leerkes, Esther M. & Marcovitch, Stuart. (2013). The role of persistence at preschool age in academic skills at kindergarten. *European Journal of Psychology of Education*, 28(4), 1495–1503.

Moldavsky, Maria & Sayal, Kapil. (2013). Knowledge and attitudes about Attention-deficit/hyperactivity disorder (ADHD) and its treatment: The views of children, adolescents, parents, teachers and healthcare professionals. *Current Psychiatry Reports*, 15, 377.

Moles, Laura; Manzano, Susana; Fernández, Leonides; Montilla, Antonia; Corzo, Nieves; Ares, Susana; . . . Espinosa-Martos, Irene. (2015). Bacteriological, biochemical, and immunological properties of colostrum and mature milk from mothers of extremely preterm infants. *Journal of Pediatric Gastroenterology & Nutrition*, 60(1), 120–126.

Mollborn, Stefanie & Lawrence, Elizabeth. (2018). Family, peer, and school influences on children's developing health lifestyles. *Journal of Health and Social Behavior*, 59(1), 133–150.

Monahan, Kathryn C.; Steinberg, Laurence; Cauffman, Elizabeth & Mulvey, Edward P. (2013). Psychosocial (im)maturity from adolescence to early adulthood: Distinguishing between adolescence-limited and persisting antisocial behavior. *Development and Psychopathology*, 25(4), 1093–1105.

Montgomery, Heather. (2015). Understanding child prostitution in Thailand in the 1990s. *Child Development Perspectives*, 9(3), 154–157.

Monthly Vital Statistics Report. (1980). *Final mortality statistics, 1978: Advance report. Monthly Vital Statistics Report*, 29(6, Suppl. 2). Hyattsville, MD: National Center for Health Statistics.

Monti, Jennifer D.; Rudolph, Karen D. & Miernicki, Michelle E. (2017). Rumination about social stress mediates the association between peer victimization and depressive symptoms during middle childhood. *Journal of Applied Developmental Psychology*, 48, 25–32.

Montirosso, Rosario; Casini, Erica; Provenzi, Livio; Putnam, Samuel P.; Morandi, Francesco; Fedeli, Claudia & Borgatti, Renato. (2015). A categorical approach to infants' individual differences during the Still-Face paradigm. *Infant Behavior and Development*, 38, 67–76.

Montirosso, Rosario; Tronick, Ed & Borgatti, Renato. (2017). Promoting neuroprotective care in neonatal intensive care units and preterm infant development: Insights from the neonatal adequate care for quality of life study. *Child Development Perspectives*, 11(1), 9–15.

Moody, Myles. (2016). From under-diagnoses to over-representation: Black children, ADHD, and the school-to-prison pipeline. *Journal of African American Studies*, 20(2), 152–163.

Moore, Keith L.; Persaud, T. V. N. & Torchia, Mark G. (2015). *The developing human: Clinically oriented embryology* (10th ed.). Saunders.

Moore, Mary Ruth & Sabo-Risley, Constance (Eds.). (2017). *Play in America: Essays in honor of Joe L. Frost.* Archway.

Morales, Angelica; Jones, Scott; Ehlers, Alissa; Lavine, Jessye & Nagel, Bonnie. (2018). Ventral striatal response during decision making involving risk and reward is associated with future binge drinking in adolescents. *Neuropsychopharmacology*, 43(9), 1884–1890.

Moran, Lyndsey R.; Lengua, Liliana J. & Zalewski, Maureen. (2013). The interaction between negative emotionality and effortful control in early social-emotional development. *Social Development*, 22(2), 340–362.

Moran, Lauren V.; Masters, Grace A.; Pingali, Samira; Cohen, Bruce M.; Liebson, Elizabeth; Rajarethinam, R. P. & Ongur, Dost. (2015). Prescription stimulant use is associated with earlier onset of psychosis. *Journal of Psychiatric Research*, 71, 41–47.

Morawska, Alina & Sanders, Matthew. (2011). Parental use of time out revisited: A useful or harmful parenting strategy? *Journal of Child and Family Studies*, 20(1), 1–8.

Morelli, Gilda; Quinn, Naomi; Chaudhary, Nandita; Vicedo, Marga; Rosabal-Coto, Mariano; Keller, Heidi; . . . Takada, Akira. (2018). Ethical challenges of parenting interventions in low- to middle-income countries. *Journal of Cross-Cultural Psychology*, 49(1), 5–24.

Moreno, Sylvain; Lee, Yunjo; Janus, Monika & Bialystok, Ellen. (2015). Short-term second language and music training induces lasting functional brain changes in early childhood. *Child Development*, 86(2), 394–406.

Morgan, David L. (2018). Living within blurry boundaries: The value of distinguishing between qualitative and quantitative research. *Journal of Mixed Methods Research*, 12(3), 268–279.

Morgan, Hani. (2019). Does high-quality preschool benefit children? What the research shows. *Education Sciences*, 9(1).

Morgan, Ian G.; Ohno-Matsui, Kyoko & Saw, Seang-Mei. (2012). Myopia. *The Lancet*, 379(9827), 1739–1748.

Morones, Alyssa. (2013). Paddling persists in U.S. schools. *Education Week*, 33(9), 1, 10–11.

Morphett, Kylie; Herron, Lisa & Gartner, Coral. (2020). Protectors or puritans? Responses to media articles about the health effects of e-cigarettes. *Addiction Research & Theory*, 28(2), 95–102.

Morris, Vivian G. & Morris, Curtis L. (2013). A call for African American male teachers: The supermen expected to solve the problems of low-performing schools. In Lewis, Chance W. & Toldson, Ivory A. (Eds.), *Black male teachers: Diversifying the United States' teacher workforce* (pp. 151–165). Emerald Group.

Morrongiello, Barbara A. (2018). Preventing unintentional injuries to young children in the home: Understanding and influencing parents' safety practices. *Child Development Perspectives*, 12(4), 217–222.

Morse, Timothy E. (2019). *Response to intervention: Refining instruction to meet student needs.* Rowman & Littlefield.

Moultrie, Fiona; Goksan, Sezgi; Poorun, Ravi & Slater, Rebeccah. (2016). Pain in neonates and infants. In Battaglia, Anna A. (Ed.), *An introduction to pain and its relation to nervous system disorders* (pp. 283–293). Wiley.

Mourad, Mirella; Landau, Ruth; Wright, Jason; Siddiq, Zainab; Duffy, Cassandra; Kern-Goldberger, Adina; . . . Friedman, Alexander. (2019). Oral opioid use during vaginal delivery hospitalizations. *American Journal of Perinatology*.

Mowry, James B.; Spyker, Daniel A.; Brooks, Daniel E.; Mcmillan, Naya & Schauben, Jay L. (2015). 2014 Annual report of the American Association of Poison Control Centers' National Poison Data System (NPDS): 32nd Annual report. *Clinical Toxicology*, 53(10), 962–1146.

Mrug, Sylvie; Elliott, Marc N.; Davies, Susan; Tortolero, Susan R.; Cuccaro, Paula & Schuster, Mark A. (2014). Early puberty, negative peer influence, and problem behaviors in adolescent girls. *Pediatrics*, 133(1), 7–14.

Mueller, Isabelle & Tronick, Ed. (2019). Early life exposure to violence: Developmental consequences on brain and behavior. *Frontiers in Behavioral Neuroscience*, 13(156).

Mueller, Noel T.; Mao, G.; Bennett, Wendy L.; Hourigan, Suchi K.; Dominguez-Bello, Maria G.; Appel, Lawrence J. & Wang, Xiaobin. (2017). Does vaginal delivery mitigate or strengthen the intergenerational association of overweight and obesity? Findings from the Boston Birth Cohort. *International Journal of Obesity*, 41, 497–501.

Mullis, Ina V. S.; Martin, Michael O.; Foy, Pierre & Arora, A. (2012a). *TIMSS 2011 international results in mathematics.* Chestnut Hill, MA: TIMSS & PIRLS International Study Center, Boston College.

Mullis, Ina V. S.; Martin, Michael O.; Foy, Pierre & Drucker, Kathleen T. (2012b). *PIRLS 2011 international results in reading.* Chestnut Hill, MA: TIMSS & PIRLS International Study Center, Boston College.

Mullis, Ina V. S.; Martin, Michael O.; Foy, Pierre & Hooper, Martin. (2016). *TIMSS 2015 international results in mathematics.* Chestnut Hill, MA: TIMSS & PIRLS International Study Center, Boston College.

Mullis, Ina V. S.; Martin, Michael O.; Foy, Pierre & Hooper, Martin. (2017). *International results in reading PIRLS 2016.* Chestnut Hill, MA: TIMSS & PIRLS International Study Center, Boston College.

Muris, Peter & Meesters, Cor. (2014). Small or big in the eyes of the other: On the developmental psychopathology of self-conscious emotions as shame, guilt, and pride. *Clinical Child and Family Psychology Review*, 17(1), 19–40.

Murphy, Colleen; Gardoni, Paolo & McKim, Robert (Eds.). (2018). *Climate change and its impacts: Risks and inequalities.* Springer.

Murray, Kristen; Rieger, Elizabeth & Byrne, Don. (2018). Body image predictors of depressive symptoms in adolescence. *Journal of Adolescence*, 69, 130–139.

Murray, Thomas H. (2014). Stirring the simmering "designer baby" pot. *Science*, 343(6176), 1208–1210.

Mustanski, Brian; Birkett, Michelle; Greene, George J.; Hatzenbuehler, Mark L. & Newcomb, Michael E. (2014). Envisioning an America without sexual orientation inequities in adolescent health. *American Journal of Public Health, 104*(2), 218–225.

Næss, Kari-Anne B. (2016). Development of phonological awareness in Down syndrome: A meta-analysis and empirical study. *Developmental Psychology, 52*(2), 177–190.

NAEYC. (2014). *NAEYC Early Childhood Program standards and accreditation criteria & guidance for assessment.* Washington, DC: National Association for the Education of Young Children.

Nanji, Ayaz. (2005, February 8). World's smallest baby goes home. *CBS News.* AP.

Narvaez, Darcia; Gleason, Tracy; Wang, Lijuan; Brooks, Jeff; Lefever, Jennifer Burke & Cheng, Ying. (2013). The evolved development niche: Longitudinal effects of caregiving practices on early childhood psychosocial development. *Early Childhood Research Quarterly, 28*(4), 759–773.

Nascimento, Andressa Lagoa; Toledo, Aline Martins; Merey, Leila Foerster; Tudella, Eloisa & Soares-Marangoni, Daniele de Almeida. (2019). Brief reaching training with "sticky mittens" in preterm infants: Randomized controlled trial. *Human Movement Science, 63*, 138–147.

Natarajan, Mangai (Ed.) (2017). *Drugs of abuse.* Routledge.

National Academies of Sciences, Engineering, and Medicine (Ed.) (2016). *Preventing bullying through science, policy, and practice.* National Academies Press.

National Center for Education Statistics. (2018). *The nation's report card.* Washington, DC: Institute of Education Sciences, U.S. Department of Education.

National Center for Education Statistics. (2018, September). *Table 502.30. Median annual earnings of full-time year-round workers 25 to 34 years old and full-time year-round workers as a percentage of the labor force, by sex, race/ethnicity, and educational attainment: Selected years, 1995 through 2017.* Washington, DC: Institute of Education Sciences, U.S. Department of Education.

National Center for Education Statistics. (2019, February). *Indicator 24: Degrees awarded. Status and trends in the education of racial and ethnic groups.* Washington, DC: Institute of Education Sciences, U.S. Department of Education.

National Center for Education Statistics. (2020). *College enrollment rates.* Institute of Education Sciences.

National Center for Education Statistics. (2019, May). *Public high school graduation rates. The Condition of Education.* National Center for Education Statistics.

National Center for Education Statistics. (2020, May). *Status dropout rates. The Condition of Education.* National Center for Education Statistics, Institute of Education Sciences.

National Center for Health Statistics. (2012). *Health, United States, 2011: With special feature on socioeconomic status and health.* Hyattsville, MD: U.S. Department of Health and Human Services, Centers for Disease Control and Prevention.

National Center for Health Statistics. (2017). *Health, United States, 2016: With chartbook on long-term trends in health.* Hyattsville, MD: U.S. Department of Health and Human Services.

National Center for Health Statistics. (2018). *Health, United States, 2017: With a special feature on mortality.* Hyattsville, MD: U.S. Department of Health and Human Services.

National Center for Health Statistics. (2019). *Health, United States, 2018.* Hyattsville, MD: U.S. Department of Health and Human Services.

National Center for Statistics and Analysis. (2019, June). *Early estimate of motor vehicle traffic fatalities for 2018. Crash•Stats Brief Statistical Summary.* Washington, DC: National Highway Traffic Safety Administration.

Naughton, Michelle J.; Yi-Frazier, Joyce P.; Morgan, Timothy M.; Seid, Michael; Lawrence, Jean M.; Klingensmith, Georgeanna J.; . . . Loots, Beth. (2014). Longitudinal associations between sex, diabetes self-care, and health-related quality of life among youth with type 1 or type 2 diabetes mellitus. *The Journal of Pediatrics, 164*(6), 1376–1383.e1.

Neary, Marianne T. & Breckenridge, Ross A. (2013). Hypoxia at the heart of sudden infant death syndrome? *Pediatric Research, 74*(4), 375–379.

Needham, Amy W.; Wiesen, Sarah E.; Hejazi, Jennifer N.; Libertus, Klaus & Christopher, Caroline. (2017). Characteristics of brief sticky mittens training that lead to increases in object exploration. *Journal of Experimental Child Psychology, 164*, 209–224.

Needleman, Herbert L. & Gatsonis, Constantine A. (1990). Low-level lead exposure and the IQ of children: A meta-analysis of modern studies. *JAMA, 263*(5), 673–678.

Needleman, Herbert L.; Schell, Alan; Bellinger, David; Leviton, Alan & Allred, Elizabeth N. (1990). The long-term effects of exposure to low doses of lead in childhood. *New England Journal of Medicine, 322*(2), 83–88.

Neggers, Yasmin & Crowe, Kristi. (2013). Low birth weight outcomes: Why better in Cuba than Alabama? *Journal of the American Board of Family Medicine, 26*(2), 187–195.

Nelson, Charles A.; Fox, Nathan A. & Zeanah, Charles H. (2014). *Romania's abandoned children: Deprivation, brain development, and the struggle for recovery.* Harvard University Press.

Nelson, Geoffrey & Caplan, Rachel. (2014). The prevention of child physical abuse and neglect: An update. *Journal of Applied Research on Children, 5*(1).

Nelson, Helen; Kendall, Garth; Burns, Sharyn; Schonert-Reichl, Kimberly & Kane, Robert. (2019). Development of the student experience of teacher support scale: Measuring the experience of children who report aggression and bullying. *International Journal of Bullying Prevention, 1*(2), 99–110.

Nesdale, Drew; Zimmer-Gembeck, Melanie J. & Roxburgh, Natalie. (2014). Peer group rejection in childhood: Effects of rejection ambiguity, rejection sensitivity, and social acumen. *Journal of Social Issues, 70*(1), 12–28.

Neuman, Susan B.; Kaefer, Tanya & Pinkham, Ashley M. (2018). A double dose of disadvantage: Language experiences for low-income children in home and school. *Journal of Educational Psychology, 110*(1), 102–118.

Nevanen, Saila; Juvonen, Antti & Ruismäki, Heikki. (2014). Does arts education develop school readiness? Teachers' and artists' points of view on an art education project. *Arts Education Policy Review, 115*(3), 72–81.

Nevin, Rick. (2007). Understanding international crime trends: The legacy of preschool lead exposure. *Environmental Research, 104*(3), 315–336.

Ng, Florrie Fei-Yin; Pomerantz, Eva M. & Deng, Ciping. (2014). Why are Chinese mothers more controlling than American mothers? "My child is my report card". *Child Development, 85*(1), 355–369.

Nguyen, Jacqueline; O'Brien, Casey & Schapp, Salena. (2016). Adolescent inhalant use prevention, assessment, and treatment: A literature synthesis. *Drug Policy, 31*, 15–24.

Nic Gabhainn, Saoirse; Baban, Adriana; Boyce, William & Godeau, Emmanuelle. (2009). How well protected are sexually active 15-year-olds? Cross-national patterns in condom and contraceptive pill use 2002–2006. *International Journal of Public Health, 54*(Suppl. 2), 209–215.

Nichols, Emily S.; Wild, Conor J.; Stojanoski, Bobby; Battista, Michael E. & Owen, Adrian M. (2020). Bilingualism affords no general cognitive advantages: A population study of executive function in 11,000 people. *Psychological Science, 31*(5), 548–567.

Niclasen, Janni; Andersen, Anne-Marie N.; Strandberg-Larsen, Katrine & Teasdale, Thomas W. (2014). Is alcohol binge drinking in early and late pregnancy associated with behavioural and emotional development at age 7 years? *European Child & Adolescent Psychiatry, 23*(12), 1175–1180.

Nielsen, Mark; Tomaselli, Keyan; Mushin, Ilana & Whiten, Andrew. (2014). Exploring tool innovation: A comparison of Western and Bushman children. *Journal of Experimental Child Psychology, 126*, 384–394.

Nieto, Marta; Romero, Dulce; Ros, Laura; Zabala, Carmen; Martínez, Manuela; Ricarte, Jorge J.; . . . Latorre, Jose M. (2019). Differences in coping strategies between young and older adults: The role of executive functions. *The International Journal of Aging and Human Development.*

Nieto, Sonia. (2000). *Affirming diversity: The sociopolitical context of multicultural education* (3rd ed.). Longman.

Nigg, Joel T. & Barkley, Russell A. (2014). Attention-deficit/hyperactivity disorder. In Mash, Eric J. & Barkley, Russell A. (Eds.), *Child psychopathology* (3rd ed., pp. 75–144). Guilford Press.

Nilsson, Kristine Kahr & de López, Kristine Jensen. (2016). Theory of mind in children with specific language impairment: A systematic review and meta-analysis. *Child Development, 87*(1), 143–153.

Nishina, Adrienne; Bellmore, Amy; Witkow, Melissa R.; Nylund-Gibson, Karen & Graham, Sandra. (2018). Mismatches in self-reported and meta-perceived ethnic identification across the high school years. *Journal of Youth and Adolescence, 47*(1), 51–63.

Nishizato, Minaho; Fujisawa, Takashi; Kosaka, Hirotaka & Tomoda, Akemi. (2017). Developmental changes in social attention and oxytocin levels in infants and children. *Scientific Reports, 7*(1), 2540–2540.

Nkomo, Palesa; Naicker, Nisha; Mathee, Angela; Galpin, Jacky; Richter, Linda M. & Norris, Shane A. (2018). The association between environmental lead exposure with aggressive behavior, and dimensionality of direct and indirect aggression during mid-adolescence: Birth to Twenty Plus cohort. *Science of the Total Environment*, *612*, 472–479.

Nocentini, Annalaura; Fiorentini, Giada; Di Paola, Ludovica & Menesini, Ersilia. (2019). Parents, family characteristics and bullying behavior: A systematic review. *Aggression and Violent Behavior*, *45*, 41–50.

Noll, Jennie G.; Guastaferro, Kate; Beal, Sarah J.; Schreier, Hannah M. C.; Barnes, Jaclyn; Reader, Jonathan M. & Font, Sarah A. (2019). Is sexual abuse a unique predictor of sexual risk behaviors, pregnancy, and motherhood in adolescence? *Journal of Research on Adolescence*, *29*(4), 967–983.

Noll, Jennie G.; Trickett, Penelope K.; Long, Jeffrey D.; Negriff, Sonya; Susman, Elizabeth J.; Shalev, Idan; . . . Putnam, Frank W. (2017). Childhood sexual abuse and early timing of puberty. *Journal of Adolescent Health*, *60*(1), 65–71.

Norrman, Emma; Petzold, Max; Bergh, Christina & Wennerholm, Ulla-Britt. (2018). School performance in singletons born after assisted reproductive technology. *Human Reproduction*, *33*(10), 1948–1959.

Nowak, Elisabeth & Schaub, Bianca. (2018). Prevention of allergies. In Agache, Ioana & Hellings, Peter (Eds.), *Implementing precision medicine in best practices of chronic airway diseases* (pp. 63–71). Elsevier.

Nxumalo, Fikile & Adair, Jennifer Keys. (2019). Social justice and equity in early childhood education. In Brown, Christopher P.; McMullen, Mary Benson & File, Nancy (Eds.), *The Wiley handbook of early childhood care and education* (pp. 661–682). Wiley.

O'Brien, Edward. (2019, March 15). Missoula high-schoolers demand action on climate change. *Montana Public Radio*.

O'Dougherty, Maureen. (2013). Becoming a mother through postpartum depression: Narratives from Brazil. In Faircloth, Charlotte; Hoffman, Diane M. & Layne, Linda L. (Eds.), *Parenting in global perspective: Negotiating ideologies of kinship, self and politics* (pp. 184–199). Routledge.

O'Hara, Michael W. & McCabe, Jennifer E. (2013). Postpartum depression: Current status and future directions. *Annual Review of Clinical Psychology*, *9*, 379–407.

O'Neill, Susan A. & Schmidt, Patrick. (2017). Arts education in Canada and the United States. In Barton, Georgina & Baguley, Margaret (Eds.), *The Palgrave handbook of global arts education* (pp. 187–202). Palgrave Macmillan.

Oakes, J. Michael. (2009). The effect of media on children: A methodological assessment from a social epidemiologist. *American Behavioral Scientist*, *52*(8), 1136–1151.

Oakes, Lisa M. & Rakison, David H. (2019). *Developmental cascades: Building the infant mind*. Oxford University Press.

Obama, Michelle. (2018). *Becoming*. Crown.

OECD. (2010). *Learning mathematics for life: A perspective from PISA*. OECD.

OECD. (2011). *Education at a glance 2011: OECD indicators*. Paris, France: Organisation for Economic Cooperation and Development.

OECD. (2014). *Education at a glance 2014: OECD Indicators*. Paris, France: Organisation for Economic Cooperation and Development.

OECD. (2018). *Education at a glance 2018: OECD indicators*. Paris, France: Organisation for Economic Cooperation and Development.

OECD.Stat. (2019). *Income distribution and poverty* [Data set]. OECD Income Distribution Database.

Olson, Kristina R. & Dweck, Carol S. (2009). Social cognitive development: A new look. *Child Development Perspectives*, *3*(1), 60–65.

Olson, Sheryl L.; Lopez-Duran, Nestor; Lunkenheimer, Erika S.; Chang, Hyein & Sameroff, Arnold J. (2011). Individual differences in the development of early peer aggression: Integrating contributions of self-regulation, theory of mind, and parenting. *Development and Psychopathology*, *23*(1), 253–266.

Olweus, Dan. (1999). Sweden. In Smith, Peter K.; Morita, Yohji; Junger-Tas, Josine; Olweus, Dan; Catalano, Richard F. & Slee, Phillip T. (Eds.), *The nature of school bullying: A cross-national perspective* (pp. 7–27). Routledge.

Ormond, Kelly E.; Laurino, Mercy Ygoña; Barlow-Stewart, Kristine; Wessels, Tina-Marié; Macaulay, Shelley; Austin, Jehannine & Middleton, Anna. (2018). Genetic counseling globally: Where are we now? *American Journal of Medical Genetics Part C: Seminars in Medical Genetics*, *178*(1), 98–107.

Orth, Ulrich; Erol, Ruth Yasemin & Luciano, Eva C. (2018). Development of self-esteem from age 4 to 94 years: A meta-analysis of longitudinal studies. *Psychological Bulletin*, *144*(10), 1045–1080.

Orth, Ulrich & Robins, Richard W. (2019). Development of self-esteem across the lifespan. In Shiner, Rebecca L.; Tackett, Jennifer L. & McAdams, Dan P. (Eds.), *Handbook of personality development* (pp. 328–344). Guilford.

Orzabal, Marcus R.; Lunde-Young, Emilie R.; Ramirez, Josue I.; Howe, Selene Y. F.; Naik, Vishal D.; Lee, Jehoon; . . . Ramadoss, Jayanth. (2019). Chronic exposure to e-cig aerosols during early development causes vascular dysfunction and offspring growth deficits. *Translational Research*, *207*, 70–82.

Osher, David; Bear, George G.; Sprague, Jeffrey R. & Doyle, Walter. (2010). How can we improve school discipline? *Educational Researcher*, *39*(1), 48–58.

Ostfeld, Barbara M.; Esposito, Linda; Perl, Harold & Hegyi, Thomas. (2010). Concurrent risks in sudden infant death syndrome. *Pediatrics*, *125*(3), 447–453.

Ostrov, Jamie M.; Kamper, Kimberly E.; Hart, Emily J.; Godleski, Stephanie A. & Blakely-McClure, Sarah J. (2014). A gender-balanced approach to the study of peer victimization and aggression subtypes in early childhood. *Development and Psychopathology*, *26*(3), 575–587.

Ottesen, Ninja M.; Meluken, Iselin; Scheike, Thomas; Kessing, Lars V.; Miskowiak, Kamilla W. & Vinberg, Maj. (2018). Clinical characteristics, life adversities and personality traits in monozygotic twins with, at risk of and without affective disorders. *Frontiers in Psychiatry*, *9*(401).

Ozernov-Palchik, Ola; Norton, Elizabeth S.; Sideridis, Georgios; Beach, Sara D.; Wolf, Maryanne; Gabrieli, John D. E. & Gaab, Nadine. (2017). Longitudinal stability of pre-reading skill profiles of kindergarten children: Implications for early screening and theories of reading. *Developmental Science*, *20*(5), e12471.

Paarlberg, Robert; Mozaffarian, Dariush; Micha, Renata & Chelius, Carolyn. (2018). Keeping soda in SNAP: Understanding the other iron triangle. *Society*, *55*(4), 308–317.

Padilla, Jenny; Jager, Justin; Updegraff, Kimberly A.; McHale, Susan M. & Umaña-Taylor, Adriana J. (2020). Mexican-origin family members' unique and shared family perspectives of familism values and their links with parent-youth relationship quality. *Developmental Psychology*, *56*(5), 993–1008.

Padilla-Walker, Laura; Memmott-Elison, Madison & Nelson, Larry. (2017). Positive relationships as an indicator of flourishing during emerging adulthood. In Padilla-Walker, Laura M. & Nelson, Larry J. (Eds.), *Flourishing in emerging adulthood: Positive development during the third decade of life* (pp. 212–235). Oxford University Press.

Padilla-Walker, Laura M. & Carlo, Gustavo. (2014). The study of prosocial behavior. In Padilla-Walker, Laura M. & Carlo, Gustavo (Eds.), *Prosocial development: A multidimensional approach*. Oxford University Press.

Paik, Anthony. (2011). Adolescent sexuality and the risk of marital dissolution. *Journal of Marriage and Family*, *73*(2), 472–485.

Palmer, Sally B. & Abbott, Nicola. (2018). Bystander responses to bias-based bullying in schools: A developmental intergroup approach. *Child Development Perspectives*, *12*(1), 39–44.

Pankow, James F.; Kim, Kilsun; McWhirter, Kevin J.; Luo, Wentai; Escobedo, Jorge O.; Strongin, Robert M.; . . . Peyton, David H. (2017). Benzene formation in electronic cigarettes. *PLoS ONE*, *12*(3), e0173055.

Papapetrou, Eirini P. (2016). Induced pluripotent stem cells, past and future. *Science*, *353*(6303), 991–992.

Parade, Stephanie H.; Armstrong, Laura M.; Dickstein, Susan & Seifer, Ronald. (2018). Family context moderates the association of maternal postpartum depression and stability of infant temperament. *Child Development*, *89*(6), 2118–2135.

Park, Hyun; Bothe, Denise; Holsinger, Eva; Kirchner, H. Lester; Olness, Karen & Mandalakas, Anna. (2011). The impact of nutritional status and longitudinal recovery of motor and cognitive milestones in internationally adopted children. *International Journal of Environmental Research and Public Health*, *8*(1), 105–116.

Parke, Ross D. (2013). Gender differences and similarities in parental behavior. In Wilcox, Bradford & Kline, Kathleen K. (Eds.), *Gender and parenthood: Biological and social scientific perspectives* (pp. 120–163). Columbia University Press.

Parker, Kim; Horowitz, Juliana Menasce & Stepler, Renee. (2017, December 5). *On gender* differences, no consensus on nature vs. nurture: Americans say society places a higher premium on masculinity than on femininity. *Social & Demographic Trends.* Washington, DC: Pew Research Center.

Parker, Samantha E.; Mai, Cara T.; Canfield, Mark A.; Rickard, Russel; Wang, Ying; Meyer, Robert E.; . . . Correa, Adolfo. (2010). Updated national birth prevalence estimates for selected birth defects in the United States, 2004–2006. *Birth Defects Research Part A: Clinical and Molecular Teratology, 88*(12), 1008–1016.

Pascarella, Ernest T.; Martin, Georgianna L.; Hanson, Jana M.; Trolian, Teniell L.; Gillig, Benjamin & Blaich, Charles. (2014). Effects of diversity experiences on critical thinking skills over 4 years of college. *Journal of College Student Development, 55*(1), 86–92.

Patel, Ayush; Medhekar, Rohan; Ochoa-Perez, Melissa; Aparasu, Rajender R.; Chan, Wenyaw; Sherer, Jeffrey T.; . . . Chen, Hua. (2017). Care provision and prescribing practices of physicians treating children and adolescents with ADHD. *Psychiatric Services, 68*(7), 681–688.

Patel, Harsh; Arruarana, Victor; Yao, Lucille; Cui, Xiaojiang & Ray, Edward. (2020). Effects of hormones and hormone therapy on breast tissue in transgender patients: A concise review. *Endocrine, 68*(1), 6–15.

Patel, Manisha; Lee, Adria D.; Clemmons, Nakia S.; Redd, Susan B.; Poser, Sarah; Blog, Debra; . . . Gastañaduy, Paul A. (2019, October 11). *National update on measles cases and outbreaks—United States, January 1–October 1, 2019. Morbidity and Mortality Weekly Report.* Atlanta, GA: U.S. Department of Health and Human Services.

Pathela, Preeti & Schillinger, Julia A. (2010). Sexual behaviors and sexual violence: Adolescents with opposite-, same-, or both-sex partners. *Pediatrics, 126*(5), 879–886.

Patil, Rakesh N.; Nagaonkar, Shashikant N.; Shah, Nilesh B. & Bhat, Tushar S. (2013). A cross-sectional study of common psychiatric morbidity in children aged 5 to 14 years in an urban slum. *Journal of Family Medicine and Primary Care, 2*(2), 164–168.

Patterson, Susan Patricia; Hilton, Shona; Flowers, Paul & McDaid, Lisa M. (2019). What are the barriers and challenges faced by adolescents when searching for sexual health information on the internet? Implications for policy and practice from a qualitative study. *Sexually Transmitted Infections, 95*(6), 462–467.

Patton, Mary H.; Blundon, Jay A. & Zakharenko, Stanislav S. (2019). Rejuvenation of plasticity in the brain: Opening the critical period. *Current Opinion in Neurobiology, 54,* 83–89.

Pellegrini, Anthony D. (2011). Introduction. In Pellegrini, Anthony D. (Ed.), *The Oxford handbook of the development of play* (pp. 3–6). Oxford University Press.

Pellegrini, Anthony D. (2013). Play. In Zelazo, Philip D. (Ed.), *The Oxford handbook of developmental psychology* (Vol. 2, pp. 276–299). Oxford University Press.

Pellegrini, Anthony D.; Roseth, Cary J.; Van Ryzin, Mark J. & Solberg, David W. (2011). Popularity as a form of social dominance: An evolutionary perspective. In Cillessen, Antonius H. N.; Schwartz, David & Mayeux, Lara (Eds.), *Popularity in the peer system* (pp. 123–139). Guilford Press.

Pellis, Sergio M.; Himmler, Brett T.; Himmler, Stephanie M. & Pellis, Vivien C. (2018). Rough-and-tumble play and the development of the social brain: What do we know, how do we know it, and what do we need to know? In Gibb, Robbin & Kolb, Bryan (Eds.), *The neurobiology of brain and behavioral development* (pp. 315–337). Academic Press.

Peng, Peng; Yang, Xiujie & Meng, Xiangzhi. (2017). The relation between approximate number system and early arithmetic: The mediation role of numerical knowledge. *Journal of Experimental Child Psychology, 157,* 111–124.

Pennycook, Gordon & Rand, David G. (2020). Who falls for fake news? The roles of bullshit receptivity, overclaiming, familiarity, and analytic thinking. *Journal of Personality, 88*(2), 185–200.

Peper, Jiska S. & Dahl, Ronald E. (2013). The teenage brain: Surging hormones—brain-behavior interactions during puberty. *Current Directions in Psychological Science, 22*(2), 134–139.

Pepper, Edward J.; Pathmanathan, Sasi; Mcilrae, Shona; Rehman, Faiz-Ur & Cardno, Alastair G. (2018). Associations between risk factors for schizophrenia and concordance in four monozygotic twin samples. *American Journal of Medical Genetics Part B: Neuropsychiatric Genetics, 177*(5), 503–510.

Perlovskya, Leonid; Cabanac, Arnaud; Bonniot-Cabanac, \Marie-Claude & Cabanac, Michel. (2013). Mozart effect, cognitive dissonance, and the pleasure of music. *Behavioural Brain Research, 244,* 9–14.

Perner, Josef. (2000). Communication and representation: Why mentalistic reasoning is a lifelong endeavour. In Mitchell, Peter & Riggs, Kevin John (Eds.), *Children's reasoning and the mind* (pp. 367–401). Psychology Press.

Perone, Sammy; Palanisamy, Jeeva & Carlson, Stephanie M. (2018). Age-related change in brain rhythms from early to middle childhood: Links to executive function. *Developmental Science, 21*(6), e12691.

Perreira, Krista M. & Pedroza, Juan M. (2019). Policies of exclusion: Implications for the health of immigrants and their children. *Annual Review of Public Health, 40,* 7.1–7.20.

Perrin, Robin; Miller-Perrin, Cindy & Song, Jeongbin. (2017). Changing attitudes about spanking using alternative biblical interpretations. *International Journal of Behavioral Development, 41*(4), 514–522.

Perszyk, Danielle R. & Waxman, Sandra R. (2018). Linking language and cognition in infancy. *Annual Review of Psychology, 69,* 231–250.

Peters, Stacey L.; Lind, Jennifer N.; Humphrey, Jasmine R.; Friedman, Jan M.; Honein, Margaret A.; Tassinari, Melissa S.; . . . Broussard, Cheryl S. (2013). Safe lists for medications in pregnancy: Inadequate evidence base and inconsistent guidance from Web-based information, 2011. *Pharmacoepidemiology and Drug Safety, 22*(3), 324–328.

Peters, Wendy. (2016). Bullies and blackmail: Finding homophobia in the closet on teen TV. *Sexuality & Culture, 20*(3), 486–503.

Peterson, Candida C. & Wellman, Henry M. (2019). Longitudinal theory of mind (ToM) development from preschool to adolescence with and without ToM delay. *Child Development, 90*(6), 1917–1934.

Petrilli, Christopher M.; Jones, Simon A.; Yang, Jie; Rajagopalan, Harish; O'Donnell, Luke; Chernyak, Yelena; . . . Horwitz, Leora I. (2020). Factors associated with hospital admission and critical illness among 5279 people with coronavirus disease 2019 in New York City: Prospective cohort study. *BMJ, 369,* m1966.

Pew Research Center. (2015, December 17). *Parenting in America: Outlook, worries, aspirations are strongly linked to financial situation. Social & Demographic Trends.* Washington, DC: Pew Research Center.

Pew Research Center. (2019, March 14). *Political independents: Who they are, what they think. U.S. Politics and Policy.* Washington, DC: Pew Research Center.

Pexman, Penny M. (2017). The role of embodiment in conceptual development. *Language, Cognition and Neuroscience.*

Phillips, Deborah A.; Fox, Nathan A. & Gunnar, Megan R. (2011). Same place, different experiences: Bringing individual differences to research in child care. *Child Development Perspectives, 5*(1), 44–49.

Piaget, Jean. (1932). *The moral judgment of the child.* K. Paul, Trench, Trubner & Co.

Piaget, Jean. (1950). *The psychology of intelligence.* Routledge & Paul.

Piaget, Jean. (1952). *The origins of intelligence in children.* International Universities Press.

Piaget, Jean. (1954). *The construction of reality in the child.* Basic Books.

Piaget, Jean. (2001). *The psychology of intelligence.* Routledge.

Piaget, Jean. (2011). *The origins of intelligence in children.* Routledge.

Piaget, Jean. (2013a). *The construction of reality in the child.* Routledge.

Piaget, Jean. (2013b). *The moral judgment of the child.* Routledge.

Piaget, Jean. (2013c). *Play, dreams and imitation in childhood.* Routledge.

Piaget, Jean & Inhelder, Bärbel. (1956). *The child's conception of space.* Routledge.

Piaget, Jean & Inhelder, Bärbel. (1972). *The psychology of the child.* Basic Books.

Piaget, Jean & Inhelder, Bärbel. (2013). *The child's conception of space.* Routledge.

Piaget, Jean; Voelin-Liambey, Daphne & Berthoud-Papandropoulou, Ioanna. (2015). Problems of class inclusion and logical implication. In Campell, Robert L. (Ed.), *Studies in reflecting abstraction* (pp. 105–137). Psychology Press.

Piekny, Jeanette & Maehler, Claudia. (2013). Scientific reasoning in early and middle childhood: The development of domain-general evidence evaluation, experimentation, and hypothesis generation skills. *British Journal of Developmental Psychology, 31*(2), 153–179.

Pietschnig, Jakob & Voracek, Martin. (2015). One century of global IQ gains: A formal meta-analysis of the Flynn Effect (1909–2013).

Perspectives on Psychological Science, 10(3), 282–306.

Pilarz, Alejandra Ros & Hill, Heather D. (2014). Unstable and multiple child care arrangements and young children's behavior. *Early Childhood Research Quarterly, 29*(4), 471–483.

Pilkauskas, Natasha. (2014). Breastfeeding initiation and duration in coresident grandparent, mother and infant households. *Maternal and Child Health Journal, 18*(8), 1955–1963.

Pinker, Steven. (1999). *Words and rules: The ingredients of language.* Basic Books.

Pinker, Steven. (2018). *Enlightenment now: The case for reason, science, humanism, and progress.* Viking.

Pinquart, Martin & Kauser, Rubina. (2018). Do the associations of parenting styles with behavior problems and academic achievement vary by culture? Results from a meta-analysis. *Cultural Diversity and Ethnic Minority Psychology, 24*(1), 75–100.

Pinto, Tiago Miguel & Figueiredo, Bárbara. (2019). Attachment and coparenting representations in men during the transition to parenthood. *Infant Mental Health Journal, 40*(6), 850–861.

Pittenger, Samantha L.; Huit, Terrence Z. & Hansen, David J. (2016). Applying ecological systems theory to sexual revictimization of youth: A review with implications for research and practice. *Aggression and Violent Behavior, 26*, 35–45.

Pittenger, Samantha L.; Pogue, Jessica K. & Hansen, David J. (2018). Predicting sexual revictimization in childhood and adolescence: A longitudinal examination using ecological systems theory. *Child Maltreatment, 23*(2), 137–146.

Planalp, Elizabeth M. & Goldsmith, H. Hill. (2019). Observed profiles of infant temperament: Stability, heritability, and associations with parenting. *Child Development.*

Plener, Paul L.; Schumacher, Teresa S.; Munz, Lara M. & Groschwitz, Rebecca C. (2015). The longitudinal course of non-suicidal self-injury and deliberate self-harm: A systematic review of the literature. *Borderline Personality Disorder and Emotion Dysregulation, 2*(1).

Plomin, Robert; DeFries, John C.; Knopik, Valerie S. & Neiderhiser, Jenae M. (2013). *Behavioral genetics.* Worth.

Pluess, Michael. (2015). Individual differences in environmental sensitivity. *Child Development Perspectives, 9*(3), 138–143.

Poelker, Katelyn E. & Gibbons, Judith L. (2019). Sharing and caring: Prosocial behavior in young children around the world. In Tulviste, Tiia; Best, Deborah L. & Gibbons, Judith L. (Eds.), *Children's social worlds in cultural context* (pp. 89–102). Springer.

Pogrebin, Abigail. (2010). *One and the same: My life as an identical twin and what I've learned about everyone's struggle to be singular.* Anchor.

Polanczyk, Guilherme V.; Willcutt, Erik G.; Salum, Giovanni A.; Kieling, Christian & Rohde, Luis A. (2014). ADHD prevalence estimates across three decades: An updated systematic review and meta-regression analysis. *International Journal of Epidemiology, 43*(2), 434–442.

Polderman, Tinca J. C.; Kreukels, Baudewijntje P. C.; Irwig, Michael S.; Beach, Lauren; Chan, Yee-Ming; Derks, Eske M.; . . . Davis, Lea K. (2018). The biological contributions to gender identity and gender diversity: Bringing data to the table. *Behavior Genetics, 48*(2), 95–108.

Polirstok, Susan. (2015). Classroom management strategies for inclusive classrooms. *Creative Education, 6*, 927–933.

Pons, Ferran & Lewkowicz, David J. (2014). Infant perception of audio-visual speech synchrony in familiar and unfamiliar fluent speech. *Acta Psychologica, 149*, 142–147.

Poole, Kristie L.; Jetha, Michelle K. & Schmidt, Louis A. (2017). Linking child temperament, physiology, and adult personality: Relations among retrospective behavioral inhibition, salivary cortisol, and shyness. *Personality and Individual Differences, 113*, 68–73.

Poon, Kean. (2018). Hot and cool executive functions in adolescence: Development and contributions to important developmental outcomes. *Frontiers in Psychology, 8*(2311).

Popescu, Rebeca; Muntean, Ana & Juffer, Femmie. (2019). Adoption in Romania: Historical perspectives and recent statistics. *Adoption Quarterly, 23*(1), 1–26.

Posada, Germán E. & Waters, Harriet Salatas. (2018). The mother-child attachment partnership in early childhood: Secure base behavioral and representational processes. *Monographs of the Society for Research in Child Development, 83*(4).

Posner, Michael I. & Rothbart, Mary K. (2017). Integrating brain, cognition and culture. *Journal of Cultural Cognitive Science, 1*(1), 3–15.

Pouwels, J. Loes; Lansu, Tessa A. M. & Cillessen, Antonius H. N. (2016). Participant roles of bullying in adolescence: Status characteristics, social behavior, and assignment criteria. *Aggressive Behavior, 42*(3), 239–253.

Pouwels, J. Loes; Salmivalli, Christina; Saarento, Silja; Van Den Berg, Yvonne H. M.; Lansu, Tessa A. M. & Cillessen, Antonius H. N. (2017). Predicting adolescents' bullying participation from developmental trajectories of social status and behavior. *Child Development.*

Powell, Cynthia M. (2013). Sex chromosomes, sex chromosome disorders, and disorders of sex development. In Gersen, Steven L. & Keagle, Martha B. (Eds.), *The principles of clinical cytogenetics* (pp. 175–211). Springer.

Powell, Kendall. (2006). Neurodevelopment: How does the teenage brain work? *Nature, 442*(7105), 865–867.

Powell, Shaun; Langlands, Stephanie & Dodd, Chris. (2011). Feeding children's desires? Child and parental perceptions of food promotion to the "under 8s". *Young Consumers: Insight and Ideas for Responsible Marketers, 12*(2), 96–109.

Pozzoli, Tiziana & Gini, Gianluca. (2013). Why do bystanders of bullying help or not? A multidimensional model. *Journal of Early Adolescence, 33*(3), 315–340.

Prather, Jonathan; Okanoya, Kazuo & Bolhuis, Johan J. (2017). Brains for birds and babies: Neural parallels between birdsong and speech acquisition. *Neuroscience & Biobehavioral Reviews, 81*(Part B), 225–237.

Preckel, Katrin; Kanske, Philipp & Singer, Tania. (2018). On the interaction of social affect and cognition: Empathy, compassion and theory of mind. *Current Opinion in Behavioral Sciences, 19*, 1–6.

Preedy, Victor R. (2019). *Neuroscience of alcohol: Mechanisms and treatment.* Elsevier.

Price, Debora; Ribe, Eloi; Di Gessa, Giorgio & Glaser, Karen. (2018). Grandparental childcare: A reconceptualisation of family policy regimes. In Timonen, Virpi (Ed.), *Grandparenting practices around the world* (pp. 43–64). Policy Press.

Pridham, Karen; Limbo, Rana & Schroeder, Michele. (2018). *Guided participation in pediatric nursing practice: Relationship-based teaching and learning with parents, children, and adolescents.* Springer.

Proctor, Laura J. & Dubowitz, Howard. (2014). Child neglect: Challenges and controversies. In Korbin, Jill E. & Krugman, Richard D. (Eds.), *Handbook of child maltreatment* (pp. 27–61). Springer.

Prothero, Arianna. (2016, April 20). Charters help alums stick with college. *Education Week, 35*(28), 1, 13.

Proud2Bme. (2012, March 26). Overall, do social networking sites like Facebook and Twitter help or hurt your body confidence.

Provenzi, Livio; Guida, Elena & Montirosso, Rosario. (2018). Preterm behavioral epigenetics: A systematic review. *Neuroscience & Biobehavioral Reviews, 84*, 262–271.

Pruden, Shannon M. & Levine, Susan C. (2017). Parents' spatial language mediates a sex difference in preschoolers' spatial-language use. *Psychological Science, 28*(11), 1583–1596.

Puertas, Alberto; Magan-Fernandez, Antonio; Blanc, Vanessa; Revelles, Laura; O'Valle, Francisco; Pozo, Elena; . . . Mesa, Francisco. (2018). Association of periodontitis with preterm birth and low birth weight: A comprehensive review. *Journal of Maternal-Fetal and Neonatal Medicine, 31*(5), 597–602.

Puetz, Vanessa B.; Parker, Drew; Kohn, Nils; Dahmen, Brigitte; Verma, Ragini & Konrad, Kerstin. (2017). Altered brain network integrity after childhood maltreatment: A structural connectomic DTI-study. *Human Brain Mapping, 38*(2), 855–868.

Pulvermüller, Friedemann. (2018). Neural reuse of action perception circuits for language, concepts and communication. *Progress in Neurobiology, 160*, 1–44.

Putnam, Robert D. (2015). *Our kids: The American dream in crisis.* Simon & Schuster.

Putwain, David W.; Symes, Wendy & McCaldin, Tamsin. (2019). Teacher use of loss-focused, utility value messages, prior to high-stakes examinations, and their appraisal by students. *Journal of Psychoeducational Assessment, 37*(2), 169–180.

Qin, Desiree B. & Chang, Tzu-Fen. (2013). Asian fathers. In Cabrera, Natasha J. & Tamis-LeMonda, Catherine S. (Eds.), *Handbook of father involvement: Multidisciplinary perspectives* (2nd ed., pp. 261–281). Routledge.

Quindlen, Anna. (2012). *Lots of candles, plenty of cake.* Random House.

Quinn, Rand & Cheuk, Tina. (2018). *School vouchers in the Trump era: How political ideology and religion shape public opinion.* Philadelphia, PA: Consortium for Policy Research in Education. CPRE Policy Briefs. PB #2018-1.

Raeburn, Paul. (2014). *Do fathers matter?: What science is telling us about the parent we've overlooked.* Farrar, Straus and Giroux.

Rahilly, Elizabeth P. (2015). The gender binary meets the gender-variant child: Parents' negotiations with childhood gender variance. *Gender & Society, 29*(3), 338–361.

Rahman, Muhammad A.; Todd, Charlotte; John, Ann; Tan, Jacinta; Kerr, Michael; Potter, Robert; . . . Brophy, Sinead. (2018). School achievement as a predictor of depression and self-harm in adolescence: Linked education and health record study. *The British Journal of Psychiatry, 212*(4), 215–221.

Raipuria, Harinder Dosanjh; Lovett, Briana; Lucas, Laura & Hughes, Victoria. (2018). A literature review of midwifery-led care in reducing labor and birth interventions. *Nursing for Women's Health, 22*(5), 387–400.

Rakic, Snezana; Jankovic Raznatovic, Svetlana; Jurisic, Aleksandar; Anicic, Radomir & Zecevic, Nebojsa. (2016). Fetal neurosonography and fetal behaviour: Genesis of fetal movements and motor reflexes. *Ultrasound in Obstetrics and Gynecology, 48*(Suppl. 1), 196.

Raley, R. Kelly & Sweeney, Megan M. (2020). Divorce, repartnering, and stepfamilies: A decade in review. *Journal of Marriage and Family, 82*(1), 81–99.

Ramírez, Naja Ferjan; Ramírez, Rey R.; Clarke, Maggie; Taulu, Samu & Kuhl, Patricia K. (2017). Speech discrimination in 11-month-old bilingual and monolingual infants: A magnetoencephalography study. *Developmental Science, 20*(1), e12427.

Ramscar, Michael & Dye, Melody. (2011). Learning language from the input: Why innate constraints can't explain noun compounding. *Cognitive Psychology, 62*(1), 1–40.

Ranciaro, Alessia; Campbell, Michael C.; Hirbo, Jibril B.; Ko, Wen-Ya; Froment, Alain; Anagnostou, Paolo; . . . Tishkoff, Sarah A. (2014). Genetic origins of lactase persistence and the spread of pastoralism in Africa. *The American Journal of Human Genetics, 94*(4), 496–510.

Rand, David G. & Nowak, Martin A. (2016). Cooperation among humans. In Messner, Dirk & Weinlich, Silke (Eds.), *Global cooperation and the human factor in international relations* (pp. 113–138). Routledge.

Rankin, Jay. (2017). Physicians disagree on legal age for cannabis. *CMAJ, 189*(4), E174–E175.

Ranney, John D. & Troop-Gordon, Wendy. (2020). The role of popularity and digital self-monitoring in adolescents' cyberbehaviors and cyber-victimization. *Computers in Human Behavior, 102*, 293–302.

Raspberry, Kelly A. & Skinner, Debra. (2011). Negotiating desires and options: How mothers who carry the fragile X gene experience reproductive decisions. *Social Science & Medicine, 72*(6), 992–998.

Rauh, Virginia A. (2018). Polluting developing brains—EPA failure on chlorpyrifos. *New England Journal of Medicine, 378*(13), 1171–1174.

Rauscher, Frances H.; Shaw, Gordon L. & Ky, Catherine N. (1993). Music and spatial task performance. *Nature, 365*(6447), 611.

Ray, Brian D. (2013). Homeschooling rising into the twenty-first century: Editor's introduction. *Peabody Journal of Education, 88*(3), 261–264.

Raymond, Jaime & Brown, Mary Jean. (2017, January 20). *Childhood blood lead levels in children aged <5 Years—United States, 2009–2014. Morbidity and Mortality Weekly Report, 66*(3), 1–10. Atlanta, GA: Centers for Disease Control and Prevention.

Reardon, Sean F. (2013). The widening income achievement gap. *Educational Leadership, 70*(8), 10–16.

Reardon, Tessa; Harvey, Kate; Baranowska, Magdalena; O'Brien, Doireann; Smith, Lydia & Creswell, Cathy. (2017). What do parents perceive are the barriers and facilitators to accessing psychological treatment for mental health problems in children and adolescents? A systematic review of qualitative and quantitative studies. *European Child & Adolescent Psychiatry, 26*(6), 623–647.

Reddy, Sunita; Patel, Tulsi; Kristensen, Malene Tanderup & Nielsen, Birgitte Bruun. (2018). Surrogacy in India: Political and commercial framings. In Mitra, Sayani; Schicktanz, Silke & Patel, Tulsi (Eds.), *Cross-cultural comparisons on surrogacy and egg donation: Interdisciplinary perspectives from India, Germany and Israel.* Palgrave Macmillan.

Reimann, Zakary; Miller, Jacob R.; Dahle, Kaitana M.; Hooper, Audrey P.; Young, Ashley M.; Goates, Michael C.; . . . Crandall, AliceAnn. (2018). Executive functions and health behaviors associated with the leading causes of death in the United States: A systematic review. *Journal of Health Psychology.*

Reinehr, Thomas & Roth, Christian Ludwig. (2019). Is there a causal relationship between obesity and puberty? *The Lancet Child & Adolescent Health, 3*(1), 44–54.

Reiter, Andrea M. F.; Suzuki, Shinsuke; O'Doherty, John P.; Li, Shu-Chen & Eppinger, Ben. (2019). Risk contagion by peers affects learning and decision-making in adolescents. *Journal of Experimental Psychology, 148*(9), 1494–1504.

Renfrew, Mary J.; McFadden, Alison; Bastos, Maria Helena; Campbell, James; Channon, Andrew Amos; Cheung, Ngai Fen; . . . Declercq, Eugene. (2014). Midwifery and quality care: Findings from a new evidence-informed framework for maternal and newborn care. *The Lancet, 384*(9948), 1129–1145.

Retelas, George (Producer). (2017, December 12). *Interview with Sheri Sheppard, Grace Young, Emily Bohl, Maria Filsinger, Marina Dimitrov, and Ariana Qayumi by undergraduate students at Stanford University.* EngineerGirl @ Stanford University. [Video]. YouTube.

Reynolds, Arthur J. (2000). *Success in early intervention: The Chicago Child-Parent Centers.* University of Nebraska Press.

Reynolds, Arthur J. & Ou, Suh-Ruu. (2011). Paths of effects from preschool to adult well-being: A confirmatory analysis of the Child-Parent Center Program. *Child Development, 82*(2), 555–582.

Reynolds, Arthur J. & Temple, Judy A. (Eds.). (2019). *Sustaining early childhood learning gains:*

Program, school, and family influences. Cambridge University Press.

Reynolds, Jamila E. & Gonzales-Backen, Melinda A. (2017). Ethnic-racial socialization and the mental health of African Americans: A critical review. *Journal of Family Theory & Review, 9*(12), 182–200.

Rich, Motoko. (2013, April 11). Texas considers backtracking on testing. *New York Times.*

Richards, Jennifer S.; Hartman, Catharina A.; Franke, Barbara; Hoekstra, Pieter J.; Heslenfeld, Dirk J.; Oosterlaan, Jaap; . . . Buitelaar, Jan K. (2014). Differential susceptibility to maternal expressed emotion in children with ADHD and their siblings? Investigating plasticity genes, prosocial and antisocial behaviour. *European Child & Adolescent Psychiatry, 24*(2), 209–217.

Richards, Morgan K.; Flanagan, Meghan R.; Littman, Alyson J.; Burke, Alson K. & Callegari, Lisa S. (2016). Primary cesarean section and adverse delivery outcomes among women of very advanced maternal age. *Journal of Perinatology, 36*, 272–277.

Rideout, Victoria. (2017). *The Common Sense Census: Media use by kids age zero to eight.* San Francisco, CA: Common Sense Media.

Riglin, Lucy; Frederickson, Norah; Shelton, Katherine H. & Rice, Frances. (2013). A longitudinal study of psychological functioning and academic attainment at the transition to secondary school. *Journal of Adolescence, 36*(3), 507–517.

Riordan, Jan & Wambach, Karen (Eds.). (2009). *Breastfeeding and human lactation* (4th ed.). Jones and Bartlett.

Rioux, Charlie; Castellanos-Ryan, Natalie; Parent, Sophie & Séguin, Jean R. (2016). The interaction between temperament and the family environment in adolescent substance use and externalizing behaviors: Support for diathesis–stress or differential susceptibility? *Developmental Review, 40*(10), 117–150.

Rivera, Juan Ángel; de Cossío, Teresita González; Pedraza, Lilia S.; Aburto, Tania C.; Sánchez, Tania G. & Martorell, Reynaldo. (2014). Childhood and adolescent overweight and obesity in Latin America: A systematic review. *The Lancet Diabetes & Endocrinology, 2*(4), 321–332.

Rizzo, Michael T. & Killen, Melanie. (2016). Children's understanding of equity in the context of inequality. *British Journal of Developmental Psychology, 34*(4), 569–581.

Rizzo, Michael T.; Li, Leon; Burkholder, Amanda R. & Killen, Melanie. (2019). Lying, negligence, or lack of knowledge? Children's intention-based moral reasoning about resource claims. *Developmental Psychology, 55*(2), 274–285.

Robelen, Erik W. (2011). More students enrolling in Mandarin Chinese. *Education Week, 30*(27), 5.

Roberts, Andrea G. & Lopez-Duran, Nestor L. (2019). Developmental influences on stress response systems: Implications for psychopathology vulnerability in adolescence. *Comprehensive Psychiatry, 88*, 9–21.

Roberts, Brent W. & Davis, Jordan P. (2016). Young adulthood is the crucible of personality development. *Emerging Adulthood, 4*(5), 318–326.

Roberts, Leslie. (2020). Global polio eradication falters in the final stretch. *Science*, 367(6473), 14–15.

Robertson, Cassandra & O'Brien, Rourke. (2018). Health endowment at birth and variation in intergenerational economic mobility: Evidence from U.S. county birth cohorts. *Demography*, 55(1), 249–269.

Robinson, Eric & Sutin, Angelina R. (2017). Parents' perceptions of their children as overweight and children's weight concerns and weight gain. *Psychological Science*, 28(3), 320–329.

Robinson, Leah E.; Wadsworth, Danielle D.; Webster, E. Kipling & Bassett, David R. (2014). School reform: The role of physical education policy in physical activity of elementary school children in Alabama's Black Belt region. *American Journal of Health Promotion*, 38(Suppl. 3), S72–S76.

Rochat, Philippe. (2013). Self-conceptualizing in development. In Zelazo, Philip D. (Ed.), *The Oxford handbook of developmental psychology* (Vol. 2, pp. 378–397). Oxford University Press.

Rock, Jacoba; Geier, Charles F.; Noll, Jennie G. & De Bellis, Michael D. (2018). Developmental traumatology: Brain development in maltreated children with and without PTSD. In Noll, Jennie G. & Shalev, Idan (Eds.), *The biology of early life stress* (pp. 45–56). Springer.

Rodgers, Rachel F.; Slater, Amy; Gordon, Chloe S.; Mclean, Siân A.; Jarman, Hannah K. & Paxton, Susan J. (2020). A biopsychosocial model of social media use and body image concerns, disordered eating, and muscle-building behaviors among adolescent girls and boys. *Journal of Youth and Adolescence*, 49(2), 399–409.

Rodrigues, Daniela; Padez, Cristina & Machado-Rodrigues, Aristides M. (2018). Active parents, active children: The importance of parental organized physical activity in children's extracurricular sport participation. *Journal of Child Health Care*, 22(1), 159–170.

Rodriguez, Christina M.; Silvia, Paul J. & Gaskin, Regan E. (2019). Predicting maternal and paternal parent-child aggression risk: Longitudinal multimethod investigation using social information processing theory. *Psychology of Violence*, 9(3), 370–382.

Rogers, Forrest Dylan & Bales, Karen Lisa. (2019). Mothers, fathers, and others: Neural substrates of parental care. *Trends in Neurosciences*, 42(8), 552–562.

Rogoff, Barbara. (2003). *The cultural nature of human development.* Oxford University Press.

Romanelli, Meghan; Xiao, Yunyu & Lindsey, Michael A. (2020). Sexual identity-behavior profiles and suicide outcomes among heterosexual, lesbian, and gay sexually active adolescents: Sexual identity-behavior profiles and suicide. *Suicide and Life-Threatening Behavior.*

Romeo, Russell D. (2013). The teenage brain: The stress response and the adolescent brain. *Current Directions in Psychological Science*, 22(2), 140–145.

Romeo, Rachel R.; Leonard, Julia A.; Robinson, Sydney T.; West, Martin R.; Mackey, Allyson P.; Rowe, Meredith L. & Gabrieli, John D. E. (2018). Beyond the 30-million-word gap: Children's conversational exposure is associated with language-related brain function. *Psychological Science*, 29(5), 700–710.

Rønneberg, Vibeke & Torrance, Mark. (2019). Cognitive predictors of shallow-orthography spelling speed and accuracy in 6th grade children. *Reading and Writing*, 32(1), 197–216.

Rook, Graham A. W.; Lowry, Christopher A. & Raison, Charles L. (2014). Hygiene and other early childhood influences on the subsequent function of the immune system. *Brain Research*, (Corrected Proof).

Roopnarine, Jaipaul L. & Hossain, Ziarat. (2013). African American and African Caribbean fathers. In Cabrera, Natasha J. & Tamis-LeMonda, Catherine S. (Eds.), *Handbook of father involvement: Multidisciplinary perspectives* (2nd ed., pp. 223–243). Routledge.

Roopnarine, Jaipaul L. & Yildirim, Elif Dede. (2019). *Fathering in cultural contexts: Developmental and clinical issues.* Routledge.

Rose, Amanda J. & Asher, Steven R. (2017). The social tasks of friendship: Do boys and girls excel in different tasks? *Child Development Perspectives*, 11(1), 3–8.

Rose, Dawn; Jones Bartoli, Alice & Heaton, Pamela C. (2019). Measuring the impact of musical learning on cognitive, behavioural and socio-emotional wellbeing development in children. *Psychology of Music*, 47(2), 284–303.

Rose, Katherine K.; Johnson, Amy; Muro, Joel & Buckley, Rhonda R. (2018). Decision making about nonparental child care by fathers: What is important to fathers in a nonparental child care program? *Journal of Family Issues*, 39(2), 299–327.

Rose, Nikolas. (2016). Reading the human brain: How the mind became legible. *Body & Society*, 22(2), 140–177.

Rose, Steven. (2008). Drugging unruly children is a method of social control. *Nature*, 451(7178), 521.

Roseberry, Lynn & Roos, Johan. (2016). *Bridging the gender gap: Seven principles for achieving gender balance.* Oxford University Press.

Rosen, Meghan. (2016). Concern grows over Zika birth defects. *Science News*, 190(9), 14–15.

Rosenblum, Gianine D. & Lewis, Michael. (1999). The relations among body image, physical attractiveness, and body mass in adolescence. *Child Development*, 70(1), 50–64.

Rosenfeld, Michael J. & Roesler, Katharina. (2019). Cohabitation experience and cohabitation's association with marital dissolution: The short-term benefits of cohabitation. *Journal of Marriage and Family*, 81(1), 42–58.

Ross, Josephine; Anderson, James R. & Campbell, Robin N. (2011). *I remember me: Mnemonic self-reference effects in preschool children.* Wiley-Blackwell.

Ross, Josephine; Yilmaz, Mandy; Dale, Rachel; Cassidy, Rose; Yildirim, Iraz & Zeedyk, M. Suzanne. (2017). Cultural differences in self-recognition: The early development of autonomous and related selves? *Developmental Science*, 20(3), e12387.

Rossignol, Michel; Chaillet, Nils; Boughrassa, Faiza & Moutquin, Jean-Marie. (2014). Inter-relations between four antepartum obstetric interventions and cesarean delivery in women at low risk: A systematic review and modeling of the cascade of interventions. *Birth*, 41(1), 70–78.

Roth, Benjamin J.; Crea, Thomas M.; Jani, Jayshree; Underwood, Dawnya; Hasson, Robert G.; Evans, Kerri & Zuch, Michael. (2018). Detached and afraid: U.S. immigration policy and the practice of forcibly separating parents and young children at the border. *Children Welfare*, 96(5), 29–49.

Rothausen, Teresa J. & Henderson, Kevin E. (2019). Meaning-based job-related well-being: Exploring a meaningful work conceptualization of job satisfaction. *Journal of Business and Psychology*, 34(3), 357–376.

Rothstein, Mark A. (2015). The moral challenge of Ebola. *American Journal of Public Health*, 105(1), 6–8.

Roubinov, Danielle S. & Boyce, William Thomas. (2017). Parenting and SES: Relative values or enduring principles? *Current Opinion in Psychology*, 15, 162–167.

Rovee-Collier, Carolyn. (1987). Learning and memory in infancy. In Osofsky, Joy Doniger (Ed.), *Handbook of infant development* (2nd ed., pp. 98–148). Wiley.

Rovee-Collier, Carolyn. (1990). The "memory system" of prelinguistic infants. *Annals of the New York Academy of Sciences*, 608, 517–542.

Rowe, Meredith L.; Denmark, Nicole; Harden, Brenda Jones & Stapleton, Laura M. (2016). The role of parent education and parenting knowledge in children's language and literacy skills among white, Black, and Latino families. *Infant and Child Development*, 25(2), 198–220.

Rübeling, Hartmut; Keller, Heidi; Yovsi, Relindis D.; Lenk, Melanie & Schwarzer, Sina. (2011). Children's drawings of the self as an expression of cultural conceptions of the self. *Journal of Cross-Cultural Psychology*, 42(3), 406–424.

Rubertsson, C.; Hellström, J.; Cross, M. & Sydsjö, G. (2014). Anxiety in early pregnancy: Prevalence and contributing factors. *Archives of Women's Mental Health*, 17(3), 221–228.

Rubin, Kenneth H.; Bowker, Julie C.; McDonald, Kristina L. & Menzer, Melissa. (2013). Peer relationships in childhood. In Zelazo, Philip D. (Ed.), *The Oxford handbook of developmental psychology* (Vol. 2, pp. 242–275). Oxford University Press.

Ruch, Donna A.; Sheftall, Arielle H.; Schlagbaum, Paige; Fontanella, Cynthia A.; Campo, John V. & Bridge, Jeffrey A. (2019). Characteristics and precipitating circumstances of suicide among incarcerated youth. *Journal of the American Academy of Child & Adolescent Psychiatry*, 58(5), 514–524.e1.

Rudaz, Myriam; Ledermann, Thomas; Margraf, Jürgen; Becker, Eni S. & Craske, Michelle G. (2017). The moderating role of avoidance behavior on anxiety over time: Is there a difference between social anxiety disorder and specific phobia? *PLoS ONE*, 12(7), e0180298.

Rudkowska, Iwona; Dewailly, Eric; Hegele, Robert A.; Boiteau, Véronique; Dubé-Linteau, Ariane; Abdous, Belkacem; . . . Vohl, Marie-Claude. (2013). Gene–diet interactions on plasma lipid levels in the Inuit population. *British Journal of Nutrition, 109*(5), 953–961.

Runions, Kevin C. & Shaw, Thérèse. (2013). Teacher–child relationship, child withdrawal and aggression in the development of peer victimization. *Journal of Applied Developmental Psychology, 34*(6), 319–327.

Russell, Charlotte K.; Robinson, Lyn & Ball, Helen L. (2013). Infant sleep development: Location, feeding and expectations in the postnatal period. *The Open Sleep Journal, 6*(Suppl. 1: M9), 68–76.

Russell, Stephen T.; Everett, Bethany G.; Rosario, Margaret & Birkett, Michelle. (2014). Indicators of victimization and sexual orientation among adolescents: Analyses from youth risk behavior surveys. *American Journal of Public Health, 104*(2), 255–261.

Russo, Theresa J. & Fallon, Moira A. (2014). Coping with stress: Supporting the needs of military families and their children. *Early Childhood Education Journal, 43*(5), 407–416.

Rutter, Michael. (2012). Resilience as a dynamic concept. *Development and Psychopathology, 24*(2), 335–344.

Sabeti, Pardis & Salahi, Lara. (2018). *Outbreak culture: The Ebola crisis and the next epidemic.* Harvard University Press.

Sabol, T. J.; Soliday Hong, S. L.; Pianta, R. C. & Burchinal, M. R. (2013). Can rating pre-K programs predict children's learning? *Science, 341*(6148), 845–846.

Sadler, Thomas W. (2015). *Langman's medical embryology* (13th ed.). Lippincott Williams & Wilkins.

Saey, Tina Hesman. (2016). Neandertal DNA poses health risks. *Science News, 189*(5), 18–19.

Saffran, Jenny R. & Kirkham, Natasha Z. (2018). Infant statistical learning. *Annual Review of Psychology, 69*, 181–203.

Sahlberg, Pasi. (2011). *Finnish lessons: What can the world learn from educational change in Finland?* Teachers College Press.

Sahlberg, Pasi. (2015). *Finnish lessons 2.0: What can the world learn from educational change in Finland?* (2nd. ed.). Teachers College Press.

Sahoo, Krushnapriya; Sahoo, Bishnupriya; Choudhury, Ashok Kumar; Sofi, Nighat Yasin; Kumar, Raman & Bhadoria, Ajeet Singh. (2015). Childhood obesity: Causes and consequences. *Journal of Family Medicine and Primary Care, 4*(2), 187–192.

Saitou, Marie & Gokcumen, Omer. (2020). An evolutionary perspective on the impact of genomic copy number variation on human health. *Journal of Molecular Evolution, 88*(1), 104–119.

Salomon, Ilyssa & Brown, Christia Spears. (2019). The selfie generation: Examining the relationship between social media use and early adolescent body image. *Journal of Early Adolescence, 39*(4), 539–560.

Samek, Diana R.; Goodman, Rebecca J.; Erath, Stephen A.; McGue, Matt & Iacono, William G. (2016). Antisocial peer affiliation and externalizing disorders in the transition from adolescence to young adulthood: Selection versus socialization effects. *Developmental Psychology, 52*(5), 813–823.

SAMHSDA. (2019). Substance Abuse and Mental Health Services Administration (SAMHSA)'s public online data analysis system: Data from the national survey on drug use and health, 2017. Substance Abuse and Mental Health Services Administration.

Sampaio, Waneli Cristine Morais; Ribeiro, Mara Cláudia; Costa, Larice Feitosa; Souza, Wânia Cristina de; Castilho, Goiara Mendonça de; Assis, Melissa Sousa de; . . . Ferreira, Vania Moraes. (2017). Effect of music therapy on the developing central nervous system of rats. *Psychology & Neuroscience, 10*(2), 176–188.

Sancar, Feyza. (2019). Subconcussive football hits may leave a telltale signature of brain damage. *JAMA, 322*(16), 1537–1538.

Sanchez, Gabriel R. & Vargas, Edward D. (2016). Taking a closer look at group identity: The link between theory and measurement of group consciousness and linked fate. *Political Research Quarterly, 69*(1), 160–174.

Sanchez-Vaznaugh, Emma V.; Weverka, Aiko; Matsuzaki, Mika & Sánchez, Brisa N. (2019). Changes in fast food outlet availability near schools: Unequal patterns by income, race/ethnicity, and urbanicity. *American Journal of Preventive Medicine, 57*(3), 338–345.

Sandstrom, Andrea; Sahiti, Qendresa; Pavlova, Barbara & Uher, Rudolf. (2019). Offspring of parents with schizophrenia, bipolar disorder, and depression: A review of familial high-risk and molecular genetics studies. *Psychiatric Genetics, 29*(5), 160–169.

Santelli, John S.; Kantor, Leslie M.; Grilo, Stephanie A.; Speizer, Ilene S.; Lindberg, Laura D.; Heitel, Jennifer; . . . Ott, Mary A. (2017). Abstinence-only-until-marriage: An updated review of U.S. policies and programs and their impact. *Journal of Adolescent Health, 61*(3), 273–280.

Santos, Carlos E.; Kornienko, Olga & Rivas-Drake, Deborah. (2017). Peer influence on ethnic-racial identity development: A multi-site investigation. *Child Development, 88*(3), 725–742.

Saracho, Olivia N. (2016). *Contemporary perspectives on research on bullying and victimization in early childhood education.* Information Age.

Saraiva, Linda; Rodrigues, Luís P.; Cordovil, Rita & Barreiros, João. (2013). Influence of age, sex and somatic variables on the motor performance of pre-school children. *Annals of Human Biology, 40*(5), 444–450.

Saroglou, Vassilis. (2012). Adolescents' social development and the role of religion. In Trommsdorff, Gisela & Chen, Xinyin (Eds.), *Values, religion, and culture in adolescent development* (pp. 391–423). Cambridge University Press.

Sassler, Sharon & Lichter, Daniel T. (2020). Cohabitation and marriage: Complexity and diversity in union-formation patterns. *Journal of Marriage and Family, 82*(1), 35–61.

Satterwhite, Catherine Lindsey; Torrone, Elizabeth; Meites, Elissa; Dunne, Eileen F.; Mahajan, Reena; Ocfemia, M. Cheryl Bañez; . . . Weinstock, Hillard. (2013). Sexually transmitted infections among US women and men: Prevalence and incidence estimates, 2008. *Sexually Transmitted Diseases, 40*(3), 187–193.

Saudi, A Nur Aulia; Hartini, Nurul & Bahar, Bahar. (2018). Teenagers' motorcycle gang community aggression from the Personal Fable and risk-taking behavior perspective. *Psychology Research and Behavior Management, 11*, 305–309.

Savioja, Hanna; Helminen, Mika; Fröjd, Sari; Marttunen, Mauri & Kaltiala-Heino, Riittakerttu. (2015). Sexual experience and self-reported depression across the adolescent years. *Health Psychology and Behavioral Medicine, 3*(1), 337–347.

Saxbe, Darby E. (2017). Birth of a new perspective? A call for biopsychosocial research on childbirth. *Current Directions in Psychological Science, 26*(1), 81–86.

Scalise, Nicole R.; Gladstone, Jessica R. & Ramani, Geetha B. (2019). Motivation and mathematics in early childhood. In Saracho, Olivia N. (Ed.), *Contemporary perspectives on research in motivation in early childhood education* (pp. 101–132). Information Age.

Scarborough, William J.; Sin, Ray & Risman, Barbara. (2019). Attitudes and the stalled gender revolution: Egalitarianism, traditionalism, and ambivalence from 1977 through 2016. *Gender & Society, 33*(2), 173–200.

Scarr, Sandra. (1985). Constructing psychology: Making facts and fables for our times. *American Psychologist, 40*(5), 499–512.

Schacter, Hannah L. & Juvonen, Jaana. (2018). Dynamic changes in peer victimization and adjustment across middle school: Does friends' victimization alleviate distress? *Child development.*

Schafer, Markus H.; Morton, Patricia M. & Ferraro, Kenneth F. (2014). Child maltreatment and adult health in a national sample: Heterogeneous relational contexts, divergent effects? *Child Abuse & Neglect, 38*(3), 395–406.

Schanler, Richard. J. (2011). Outcomes of human milk-fed premature infants. *Seminars in Perinatology, 35*(1), 29–33.

Schapira, Rotem; Anger Elfenbein, Hillary; Amichay-Setter, Meirav; Zahn-Waxler, Carolyn & Knafo-Noam, Ariel. (2019). Shared environment effects on children's emotion recognition. *Frontiers in Psychiatry, 10*(215).

Scharf, Miri. (2014). Parenting in Israel: Together hand in hand, you are mine and I am yours. In Selin, Helaine (Ed.), *Parenting across cultures: Childrearing, motherhood and fatherhood in non-Western cultures* (pp. 193–206). Springer.

Schermerhorn, Alice C.; D'Onofrio, Brian M.; Turkheimer, Eric; Ganiban, Jody M.; Spotts, Erica L.; Lichtenstein, Paul; . . . Neiderhiser, Jenae M. (2011). A genetically informed study of associations between family functioning and child psychosocial adjustment. *Developmental Psychology, 47*(3), 707–725.

Schienkiewitz, Anja; Brettschneider, Anna-Kristin; Damerow, Stefan & Rosario, Angelika Schaffrath. (2018). Overweight and obesity among children and adolescents in Germany: Results of the cross-sectional KiGGS Wave 2 study and trends. *Journal of Health Monitoring, 3*(1), 15–22.

Schore, Allan & McIntosh, Jennifer. (2011). Family law and the neuroscience of attachment: Part I. *Family Court Review, 49*(3), 501–512.

Schroeder, Philip H.; Napoli, Nicholas J.; Barnhardt, William F.; Barnes, Laura E. & Young, Jeffrey S. (2019). Relative mortality analysis of the "golden hour": A comprehensive acuity stratification approach to address disagreement in current literature. *Prehospital Emergency Care, 23*(2), 254–262.

Schwartz, Seth & Petrova, Mariya. (2019). Prevention science in emerging adulthood: A field coming of age. *Prevention Science, 20*(3), 305–309.

Schwarz, Alan. (2016). *ADHD nation: Children, doctors, big pharma, and the making of an American epidemic.* Scribner.

Schweinhart, Lawrence J. (2019). Lessons on sustaining gains from the life-course study of Perry Preschool. In Reynolds, Arthur J. & Temple, Judy A. (Eds.), *Sustaining early childhood learning gains: Program, school, and family influences* (pp. 254–266). Cambridge University Press.

Schweinhart, Lawrence J.; Montie, Jeanne; Xiang, Zongping; Barnett, W. Steven; Belfield, Clive R. & Nores, Milagros. (2005). *Lifetime effects: The High/Scope Perry Preschool Study through age 40.* High/Scope Press.

Schweinhart, Lawrence J. & Weikart, David P. (1997). *Lasting differences: The High/Scope Preschool curriculum comparison study through age 23.* High/Scope Educational Research Foundation.

Schweizer, Susanne; Gotlib, Ian H. & Blakemore, Sarah-Jayne. (2020). The role of affective control in emotion regulation during adolescence. *Emotion, 20*(1), 80–86.

Scott, Diane L.; Lee, Chang-Bae; Harrell, Susan W. & Smith-West, Mary B. (2013). Permanency for children in foster care: Issues and barriers for adoption. *Child & Youth Services, 34*(3), 290–307.

Scott, Lisa S. & Monesson, Alexandra. (2010). Experience-dependent neural specialization during infancy. *Neuropsychologia, 48*(6), 1857–1861.

Sears, William & Sears, Martha. (2001). *The attachment parenting book: A commonsense guide to understanding and nurturing your baby.* Little Brown.

Seaton, Eleanor K.; Quintana, Stephen; Verkuyten, Maykel & Gee, Gilbert C. (2017). Peers, policies, and place: The relation between context and ethnic/racial identity. *Child Development, 88*(3), 683–692.

Sedgh, Gilda; Finer, Lawrence B.; Bankole, Akinrinola; Eilers, Michelle A. & Singh, Susheela. (2015). Adolescent pregnancy, birth, and abortion rates across countries: Levels and recent trends. *Journal of Adolescent Health, 56*(2), 223–230.

Sedlak, Andrea J. & Ellis, Raquel T. (2014). Trends in child abuse reporting. In Korbin, Jill E. & Krugman, Richard D. (Eds.), *Handbook of child maltreatment* (pp. 3–26). Springer.

Şendil, Çağla Öneren & Erden, Feyza Tantekin. (2014). Peer preference: A way of evaluating social competence and behavioural well-being in early childhood. *Early Child Development and Care, 184*(2), 230–246.

Senese, Vincenzo Paolo; Azhari, Atiqah; Shinohara, Kazuyuki; Doi, Hirokazu; Venuti, Paola; Bornstein, Marc H. & Esposito, Gianluca. (2019). Implicit associations to infant cry: Genetics and early care experiences influence caregiving propensities. *Hormones and Behavior, 108*, 1–9.

Seppa, Nathan. (2013a). Urban eyes: Too much time spent indoors may be behind a surge in near-sightedness. *Science News, 183*(3), 22–25.

Seppa, Nathan. (2013b). Home births more risky than hospital deliveries: Records suggest babies born at home are more prone to unresponsiveness after five minutes. *Science News, 184*(8), 14.

Severson, Kim & Blinder, Alan. (2014, January 7). Test scandal in Atlanta brings more guilty pleas. *New York Times.*

Shah, Nirvi. (2011). Policy fight brews over discipline. *Education Week, 31*(7), 1, 12.

Shanahan, Timothy & Lonigan, Christopher J. (2010). The National Early Literacy Panel: A summary of the process and the report. *Educational Researcher, 39*(4), 279–285.

Sharda, Megha; Tuerk, Carola; Chowdhury, Rakhee; Jamey, Kevin; Foster, Nicholas; Custo-Blanch, Melanie; . . . Hyde, Krista. (2018). Music improves social communication and auditory–motor connectivity in children with autism. *Translational Psychiatry, 8*(231).

Sharot, Tali. (2017). *The influential mind: What the brain reveals about our power to change others.* Henry Holt.

Shawar, Yusra Ribhi & Shiffman, Jeremy. (2017). Generation of global political priority for early childhood development: The challenges of framing and governance. *The Lancet, 389*(10064), 119–124.

Shechner, Tomer; Fox, Nathan A.; Mash, Jamie A.; Jarcho, Johanna M.; Chen, Gang; Leibenluft, Ellen; . . . Britton, Jennifer C. (2018). Differences in neural response to extinction recall in young adults with or without history of behavioral inhibition. *Development and Psychopathology, 30*(1), 179–189.

Sheridan, Margaret A.; Mclaughlin, Katie A.; Winter, Warren; Fox, Nathan; Zeanah, Charles & Nelson, Charles A. (2018). Early deprivation disruption of associative learning is a developmental pathway to depression and social problems. *Nature Communications, 9*(1), 2216.

Sherlock, James M. & Zietsch, Brendan P. (2018). Longitudinal relationships between parents' and children's behavior need not implicate the influence of parental behavior and may reflect genetics: Comment on Waldinger and Schul. (2016). *Psychological Science, 29*(1), 154–157.

Shi, Bing & Xie, Hongling. (2012). Popular and nonpopular subtypes of physically aggressive preadolescents: Continuity of aggression and peer mechanisms during the transition to middle school. *Merrill-Palmer Quarterly, 58*(4), 530–553.

Shi, Rushen. (2014). Functional morphemes and early language acquisition. *Child Development Perspectives, 8*(1), 6–11.

Shim, Woo-Jeong & Perez, Rosemary Jane. (2018). A multi-level examination of first-year students' openness to diversity and challenge. *Journal of Higher Education, 89*(4), 453–477.

Shneidman, Laura & Woodward, Amanda L. (2016). Are child-directed interactions the cradle of social learning? *Psychological Bulletin, 142*(1), 1–17.

Shoda, Yuichi; Mischel, Walter & Peake, Philip K. (1990). Predicting adolescent cognitive and self-regulatory competencies from preschool delay of gratification: Identifying diagnostic conditions. *Developmental Psychology, 26*(6), 978–986.

Short, Clara Schaertl. (2019). Comment on "Outbreak: On transgender teens and psychic epidemics". *Psychological Perspectives, 62*(2/3), 285–289.

Shpancer, Noam & Schweitzer, Stefanie N. (2018). A history of non-parental care in childhood predicts more positive adult attitudes towards nonparental care and maternal employment. *Early Child Development and Care, 188*(3), 375–386.

Shulman, Cory. (2016). *Research and practice in infant and early childhood mental health.* Springer.

Shulman, Elizabeth P.; Monahan, Kathryn C. & Steinberg, Laurence. (2017). Severe violence during adolescence and early adulthood and its relation to anticipated rewards and costs. *Child Development, 88*(1), 16–26.

Shwalb, David W.; Shwalb, Barbara J. & Lamb, Michael E. (Eds.). (2013). *Fathers in cultural context.* Psychology Press.

Siddiqui, Ayesha; Cuttini, Marina; Wood, Rachel; Velebil, Petr; Delnord, Marie; Zile, Irisa; . . . Macfarlane, Alison. (2017). Can the Apgar score be used for international comparisons of newborn health? *Paediatric and Perinatal Epidemiology, 31*(4), 338–345.

Siegal, Michael & Surian, Luca (Eds.). (2012). *Access to language and cognitive development.* Oxford University Press.

Siegler, Robert S. (2016). Continuity and change in the field of cognitive development and in the perspectives of one cognitive developmentalist. *Child Development Perspectives, 10*(2), 128–133.

Siegler, Robert S. & Braithwaite, David W. (2017). Numerical development. *Annual Review of Psychology, 68*, 187–213.

Sigurdson, J. F.; Wallander, J. & Sund, A. M. (2014). Is involvement in school bullying associated with general health and psychosocial adjustment outcomes in adulthood? *Child Abuse & Neglect, 38*(10), 1607–1617.

Silbey, Susan S. (2016, August 23). Why do so many women who study engineering leave the field? *Harvard Business Review.*

Silk, Jessica & Romero, Diana. (2014). The role of parents and families in teen pregnancy prevention: An analysis of programs and policies. *Journal of Family Issues, 35*(10), 1339–1362.

Silver, Jonathan M.; McAllister, Thomas W. & Arciniegas, David B. (Eds.). (2019). *Textbook of traumatic brain injury* (3rd ed.). American Psychiatric Association.

Silverman, Arielle M. & Cohen, Geoffrey L. (2014). Stereotypes as stumbling-blocks: How coping with stereotype threat affects life outcomes for people with physical disabilities. *Personality and Social Psychology Bulletin*, 40(10), 1330–1340.

Silvers, Jennifer A.; Shu, Jocelyn; Hubbard, Alexa D.; Weber, Jochen & Ochsner, Kevin N. (2015). Concurrent and lasting effects of emotion regulation on amygdala response in adolescence and young adulthood. *Developmental Science*, 18(5), 771–784.

Sim, Zi L. & Xu, Fei. (2017). Learning higher-order generalizations through free play: Evidence from 2- and 3-year-old children. *Developmental Psychology*, 53(4), 642–651.

Simms, Victoria; Gilmore, Camilla; Sloan, Seaneen & McKeaveney, Clare. (2019). *Interventions to improve mathematical achievement in primary school-aged children*. London, UK: Nuffield Foundation.

Simon, Laura & Daneback, Kristian. (2013). Adolescents' use of the internet for sex education: A thematic and critical review of the literature. *International Journal of Sexual Health*, 25(4), 305–319.

Simpson, Elizabeth A.; Jakobsen, Krisztina V.; Damon, Fabrice; Suomi, Stephen J.; Ferrari, Pier F. & Paukner, Annika. (2017). Face detection and the development of own-species bias in infant macaques. *Child Development*, 88(1), 103–113.

Simpson, Jeffry A. & Kenrick, Douglas. (2013). *Evolutionary social psychology*. Taylor & Francis.

Simpson, Kate; Adams, Dawn; Bruck, Susan & Keen, Deb. (2019). Investigating the participation of children on the autism spectrum across home, school, and community: A longitudinal study. *Child*, 45(5), 681–687.

Singanayagam, Aran; Ritchie, Andrew I. & Johnston, Sebastian L. (2017). Role of microbiome in the pathophysiology and disease course of asthma. *Current Opinion in Pulmonary Medicine*, 23(1), 41–47.

Singer, Elly & Wong, Sandie. (2019). Emotional security and daycare for babies and toddlers in social-political contexts: Reflections of early years pioneers since the 1970s. *Early Child Development and Care*.

Singer, Judith D. & Braun, Henry I. (2018). Testing international education assessments. *Science*, 360(6384), 38–40.

Singh, Amika; Uijtdewilligen, Léonie; Twisk, Jos W. R.; van Mechelen, Willem & Chinapaw, Mai J. M. (2012). Physical activity and performance at school: A systematic review of the literature including a methodological quality assessment. *Archives of Pediatrics & Adolescent Medicine*, 166(1), 49–55.

Sinnott, Jan; Hilton, Shelby; Wood, Michael & Douglas, Derek. (2020). Relating flow, mindfulness, cognitive flexibility, and postformal thought: Two studies. *Journal of Adult Development*, 27(1), 1–11.

Sinnott, Jan D. (2014). *Adult development: Cognitive aspects of thriving close relationships*. Oxford University Press.

Sisson, Susan B.; Krampe, Megan; Anundson, Katherine & Castle, Sherri. (2016). Obesity prevention and obesogenic behavior interventions in child care: A systematic review. *Preventive Medicine*, 87, 57–69.

Skiba, Russell J.; Michael, Robert S.; Nardo, Abra Carroll & Peterson, Reece L. (2002). The color of discipline: Sources of racial and gender disproportionality in school punishment. *The Urban Review*, 34(4), 317–342.

Skinner, B. F. (1953). *Science and human behavior*. Macmillan.

Skinner, B. F. (1957). *Verbal behavior*. Appleton-Century-Crofts.

Skoog, Thérése & Stattin, Håkan. (2014). Why and under what contextual conditions do early-maturing girls develop problem behaviors? *Child Development Perspectives*, 8(3), 158–162.

Sloan, Mark. (2009). *Birth day: A pediatrician explores the science, the history, and the wonder of childbirth*. Ballantine Books.

Smart, Andrew; Bolnick, Deborah A. & Tutton, Richard. (2017). Health and genetic ancestry testing: Time to bridge the gap. *BMC Medical Genomics*, 10(1).

Smetana, Judith G.; Ahmad, Ikhlas & Wray-Lake, Laura. (2016). Beliefs about parental authority legitimacy among refugee youth in Jordan: Between- and within-person variations. *Developmental Psychology*, 52(3), 484–495.

Smith, Daniel M. & Martiny, Sarah E. (2018). Stereotype threat in sport: Recommendations for applied practice and research. *The Sport Psychologist*, 32(4), 311–320.

Smith, Hannah E.; Ryan, Kelsey N.; Stephenson, Kevin B.; Westcott, Claire; Thakwalakwa, Chrissie; Maleta, Ken; . . . Manary, Mark J. (2014). Multiple micronutrient supplementation transiently ameliorates environmental enteropathy in Malawian children aged 12–35 months in a randomized controlled clinical trial. *Journal of Nutrition*, 144(12), 2059–2065.

Smith, Michelle I.; Yatsunenko, Tanya; Manary, Mark J.; Trehan, Indi; Mkakosya, Rajhab; Cheng, Jiye; . . . Gordon, Jeffrey I. (2013). Gut microbiomes of Malawian twin pairs discordant for kwashiorkor. *Science*, 339(6119), 548–554.

Smithells, R. W.; Sheppard, S.; Schorah, C. J.; Seller, M. J.; Nevin, N. C.; Harris, R.; . . . Fielding, D. W. (2011). Apparent prevention of neural tube defects by periconceptional vitamin supplementation. *International Journal of Epidemiology*, 40(5), 1146–1154.

Smits, Jeroen & Monden, Christiaan. (2011). Twinning across the developing world. *PLoS ONE*, 6(9), e25239.

Smock, Pamela J. & Schwartz, Christine R. (2020). The demography of families: A review of patterns and change. *Journal of Marriage and Family*, 82(1), 9–34.

Snider, Terra Ziporyn. (2012). Later school start times are a public-health issue. *Education Week*, 31(31), 25, 27.

Snow, J. B. (2015). *Narcissist and the Peter Pan syndrome: Emotionally unavailable and emotionally immature men*. J. B. Snow Publishing, Amazon Digital Services.

Snyder, Thomas D. & Dillow, Sally A. (2013). *Digest of education statistics, 2012*. Washington, DC: National Center for Education Statistics, Institute of Education Sciences, U.S. Department of Education.

Society for Developmental & Behavioral Pediatrics. (2018, July 18). *SDBP statement related to the separation of children from families at the border*. McLean, VA: Society for Developmental & Behavioral Pediatrics.

Soderstrom, Melanie; Ko, Eon-Suk & Nevzorova, Uliana. (2011). It's a question? Infants attend differently to yes/no questions and declaratives. *Infant Behavior and Development*, 34(1), 107–110.

Solheim, Elisabet; Wichstrøm, Lars; Belsky, Jay & Berg-Nielsen, Turid Suzanne. (2013). Do time in child care and peer group exposure predict poor socioemotional adjustment in Norway? *Child Development*, 84(5), 1701–1715.

Solomon, Andrew. (2012). *Far from the tree: Parents, children, and the search for identity*. Scribner.

Somerville, Leah H. (2013). The teenage brain: Sensitivity to social evaluation. *Current Directions in Psychological Science*, 22(2), 121–127.

Sonuga-Barke, Edmund J. S.; Kennedy, Mark; Kumsta, Robert; Knights, Nicky; Golm, Dennis; Rutter, Michael; . . . Kreppner, Jana. (2017). Child-to-adult neurodevelopmental and mental health trajectories after early life deprivation: The young adult follow-up of the longitudinal English and Romanian Adoptees study. *The Lancet*, 389(10078), 1539–1548.

Soska, Kasey C.; Adolph, Karen E. & Johnson, Scott P. (2010). Systems in development: Motor skill acquisition facilitates three-dimensional object completion. *Developmental Psychology*, 46(1), 129–138.

Sousa, David A. (2014). *How the brain learns to read* (2nd ed.). SAGE.

Sparks, Sarah D. (2016, July 20). Dose of empathy found to cut suspension rates. *Education Week*, 35(36), 1, 20.

Spelke, Elizabeth S. (1993). Object perception. In Goldman, Alvin I. (Ed.), *Readings in philosophy and cognitive science* (pp. 447–460). MIT Press.

Spencer, Steven J.; Logel, Christine & Davies, Paul G. (2016). Stereotype threat. *Annual Review of Psychology*, 67, 415–437.

Sperry, Douglas E.; Sperry, Linda L. & Miller, Peggy J. (2019). Reexamining the verbal environments of children from different socioeconomic backgrounds. *Child Development*, 90(4), 1303–1318.

Spolaore, Enrico & Wacziarg, Romain. (2018). Ancestry and development: New evidence. *Journal of Applied Econometrics*, 33(5), 748–762.

Sprietsma, Maresa. (2010). Effect of relative age in the first grade of primary school on long-term scholastic results: International comparative evidence using PISA 2003. *Education Economics*, 18(1), 1–32.

Springsteen, Bruce. (2017). *Born to run*. Simon & Schuster.

Srinivasan, Sharada & Li, Shuzhuo. (2018). Unifying perspectives on scarce women and surplus men in China and India. In Srinivasan, Sharada & Li, Shuzhuo (Eds.), *Scarce women and surplus men in China and India* (pp. 1–23). Springer.

Sriram, Rajalakshmi. (2019). A global perspective on fathering. In Sriram, Rajalakshmi (Ed.), *Fathering in India* (pp. 19–34). Springer.

Stahl, Aimee E. & Feigenson, Lisa. (2017). Expectancy violations promote learning in young children. *Cognition*, 163, 1–14.

Standing, E. M. (1998). *Maria Montessori: Her life and work*. Plume.

Starr, Christine R. & Zurbriggen, Eileen L. (2016). Sandra Bem's gender schema theory after 34 years: A review of its reach and impact. *Sex Roles.*

Statistics Norway. (2018). Facts about education in Norway 2018. Statistics Norway's Information Centre.

Stattin, Håkan; Hussein, Oula; Özdemir, Metin & Russo, Silvia. (2017). Why do some adolescents encounter everyday events that increase their civic interest whereas others do not? *Developmental Psychology, 53*(2), 306–318.

Steele, Claude M. (1997). A threat in the air: How stereotypes shape intellectual identity and performance. *American Psychologist, 52*(6), 613–629.

Stefansen, Kari; Smette, Ingrid & Strandbu, Åse. (2018). Understanding the increase in parents' involvement in organized youth sports. *Sport, Education and Society, 23*(2), 162–172.

Steffensmeier, Darrell; Painter-Davis, Noah & Ulmer, Jeffery. (2017). Intersectionality of race, ethnicity, gender, and age on criminal punishment. *Sociological Perspectives, 60*(4), 810–833.

Steinberg, Laurence. (2004). Risk taking in adolescence: What changes, and why? *Annals of the New York Academy of Sciences, 1021,* 51–58.

Steinberg, Laurence. (2009). Should the science of adolescent brain development inform public policy? *American Psychologist, 64*(8), 739–750.

Steinberg, Laurence. (2014). *Age of opportunity: Lessons from the new science of adolescence.* Houghton Mifflin Harcourt.

Steinberg, Laurence. (2015). The neural underpinnings of adolescent risk-taking: The roles of reward-seeking, impulse control, and peers. In Oettingen, Gabriele & Gollwitzer, Peter M. (Eds.), *Self-regulation in adolescence* (pp. 173–192). Cambridge University Press.

Steinberg, Laurence & Monahan, Kathryn C. (2011). Adolescents' exposure to sexy media does not hasten the initiation of sexual intercourse. *Developmental Psychology, 47*(2), 562–576.

Stenseng, Frode; Belsky, Jay; Skalicka, Vera & Wichstrøm, Lars. (2015). Social exclusion predicts impaired self-regulation: A 2-year longitudinal panel study including the transition from preschool to school. *Journal of Personality, 83*(2), 212–220.

Stern, Gavin. (2015). For kids with special learning needs, roadblocks remain. *Science, 349*(6255), 1465–1466.

Stern, Jessica A.; Botdorf, Morgan; Cassidy, Jude & Riggins, Tracy. (2019). Empathic responding and hippocampal volume in young children. *Developmental Psychology, 55*(9), 1908–1920.

Stern, Mark; Clonan, Sheila; Jaffee, Laura & Lee, Anna. (2015). The normative limits of choice: Charter schools, disability studies, and questions of inclusion. *Educational Policy, 29*(3), 448–477.

Stern, Peter. (2013). Connection, connection, connection . . . *Science, 342*(6158), 577.

Sternberg, Robert J. (1988). Triangulating love. In Sternberg, Robert J. & Barnes, Michael L. (Eds.), *The psychology of love* (pp. 119–138). Yale University Press.

Sternberg, Robert J. (2008). Schools should nurture wisdom. In Presseisen, Barbara Z. (Ed.), *Teaching for intelligence* (2nd ed., pp. 61–88). Corwin Press.

Sternberg, Robert J. (2011). The theory of successful intelligence. In Sternberg, Robert J. & Kaufman, Scott Barry (Eds.), *The Cambridge handbook of intelligence* (pp. 504–526). Cambridge University Press.

Stiles, Joan & Jernigan, Terry. (2010). The basics of brain development. *Neuropsychology Review, 20*(4), 327–348.

Stives, Kristen L.; May, David C.; Pilkinton, Melinda; Bethel, Cindy L. & Eakin, Deborah K. (2019). Strategies to combat bullying: Parental responses to bullies, bystanders, and victims. *Youth & Society, 51*(3), 358–376.

Stochholm, Kirstine; Juul, Svend & Gravholt, Claus H. (2010). Diagnosis and mortality in 47,XYY persons: A registry study. *Orphanet Journal of Rare Diseases, 5*(15).

Stolt, Suvi; Matomäki, Jaakko; Lind, Annika; Lapinleimu, Helena; Haataja, Leena & Lehtonen, Liisa. (2014). The prevalence and predictive value of weak language skills in children with very low birth weight—A longitudinal study. *Acta Paediatrica, 103*(6), 651–658.

Strait, Dana L.; Parbery-Clark, Alexandra; O'Connell, Samantha & Kraus, Nina. (2013). Biological impact of preschool music classes on processing speech in noise. *Developmental Cognitive Neuroscience, 6,* 51–60.

Strasburger, Victor C.; Wilson, Barbara J. & Jordan, Amy B. (2009). *Children, adolescents, and the media* (2nd ed.). Sage.

Stremmel, Andrew J. (2012). A situated framework: The Reggio experience. In File, Nancy; Mueller, Jennifer J. & Wisneski, Debora B. (Eds.), *Curriculum in early childhood education: Reexamined, rediscovered, renewed* (pp. 133–145). Routledge.

Stroebe, Wolfgang & Strack, Fritz. (2014). The alleged crisis and the illusion of exact replication. *Perspectives on Psychological Science, 9*(1), 59–71.

Strøm, Marin; Mortensen, Erik Lykke; Kesmodel, Ulrik Schiøler; Halldorsson, Thorhallur; Olsen, Jørn & Olsen, Sjurdur F. (2019). Is breast feeding associated with offspring IQ at age 5? Findings from prospective cohort: Lifestyle during pregnancy study. *BMJ Open, 9*(5), e023134.

Strouse, Gabrielle A. & Ganea, Patricia A. (2017). Toddlers' word learning and transfer from electronic and print books. *Journal of Experimental Child Psychology, 156,* 129–142.

Suárez-Orozco, Carola. (2017). Conferring disadvantage: Behavioral and developmental implications for children growing up in the shadow of undocumented immigration status. *Journal of Developmental & Behavioral Pediatrics, 38*(6), 424–428.

Suberi, Moriya; Morag, Iris; Strauss, Tzipora & Geva, Ronny. (2018). Feeding imprinting: The extreme test case of premature infants born with very low birth weight. *Child Development, 89*(5), 1553–1566.

Substance Abuse and Mental Health Services Administration. (2019). *Key substance use and mental health indicators in the United States: Results from the 2018 National Survey on Drug Use and Health.* Rockville, MD: Center for Behavioral Health Statistics and Quality, Substance Abuse and Mental Health Services Administration. HHS Publication No. PEP19-5068, NSDUH Series H-54.

Suchy, Frederick J.; Brannon, Patsy M.; Carpenter, Thomas O.; Fernandez, Jose R.; Gilsanz, Vicente; Gould, Jeffrey B.; . . . Wolf, Marshall A. (2010). National Institutes of Health Consensus Development Conference: Lactose intolerance and health. *Annals of Internal Medicine, 152*(12), 792–796.

Sugimura, Kazumi; Matsushima, Kobo; Hihara, Shogo; Takahashi, Masami & Crocetti, Elisabetta. (2019). A culturally sensitive approach to the relationships between identity formation and religious beliefs in youth. *Journal of Youth and Adolescence, 48*(4), 668–679.

Suleiman, Ahna B. & Brindis, Claire D. (2014). Adolescent school-based sex education: Using developmental neuroscience to guide new directions for policy and practice. *Sexuality Research and Social Policy, 11*(2), 137–152.

Sulek, Julia P. (2013, April 30). Audrie Pott suicide: Parents share grief, quest for justice in exclusive interview. *San Jose Mercury News.*

Sullivan, Jas M. & Ghara, Alexandra. (2015). Racial identity and intergroup attitudes: A multiracial youth analysis. *Social Science Quarterly, 96*(1), 261–272.

Sullivan, Patrick F.; Neale, Michael C. & Kendler, Kenneth S. (2000). Genetic epidemiology of major depression: Review and meta-analysis. *American Journal of Psychiatry, 157*(10), 1552–1562.

Sun, Li; Guo, Xin; Zhang, Jing; Liu, Henghui; Xu, Shaojun; Xu, Yuanyuan & Tao, Fangbiao. (2016). Gender specific associations between early puberty and behavioral and emotional characteristics in children. *Zhonghua Liu Xing Bing Xue Za Zhi, 37*(1), 35–39.

Sun, Min & Rugolotto, Simone. (2004). Assisted infant toilet training in a Western family setting. *Journal of Developmental & Behavioral Pediatrics, 25*(2), 99–101.

Sundqvist, Christel; Björk-Åman, Camilla & Ström, Kristina. (2019). The three-tiered support system and the special education teachers' role in Swedish-speaking schools in Finland. *European Journal of Special Needs Education, 34*(5), 601–616.

Suomi, Steven J. (2002). Parents, peers, and the process of socialization in primates. In Borkowski, John G.; Ramey, Sharon L. & Bristol-Power, Marie (Eds.), *Parenting and the child's world: Influences on academic, intellectual, and social-emotional development* (pp. 265–279). Erlbaum.

Sutaria, Shailen; Devakumar, Delan; Yasuda, Sílvia Shikanai; Das, Shikta & Saxena, Sonia. (2019). Is obesity associated with depression in children? Systematic review and meta-analysis. *Archives of Disease in Childhood, 104*(1), 64–74.

Suurland, Jill; van der Heijden, Kristiaan B.; Huijbregts, Stephan C. J.; Smaling, Hanneke J. A.; de Sonneville, Leo M. J.; Van Goozen, Stephanie H. M. & Swaab, Hanna. (2016). Parental perceptions of aggressive behavior in preschoolers: Inhibitory control moderates the association with negative emotionality. *Child Development, 87*(1), 256–269.

Suzumori, Nobuhiro; Kumagai, Kyoko; Goto, Shinobu; Nakamura, Akira & Sugiura-Ogasawara, Mayumi. (2015). Parental decisions following prenatal diagnosis of chromosomal abnormalities: Implications for genetic counseling practice in Japan. *Journal of Genetic Counseling*, 24(1), 117–121.

Swaab, D. F. & Hofman, M. A. (1984). Sexual differentiation of the human brain: A historical perspective. *Progress in Brain Research*, 61, 361–374.

Swanson, H. Lee. (2013). Meta-analysis of research on children with learning disabilities. In Swanson, H. Lee; Harris, Karen R. & Graham, Steve (Eds.), *Handbook of learning disabilities* (2nd ed., pp. 627–642). Guilford Press.

Swit, Cara & McMaugh, Anne. (2012). Relational aggression and prosocial behaviours in Australian preschool children. *Australasian Journal of Early Childhood*, 37(3), 30–34.

Taga, Keiko A.; Markey, Charlotte N. & Friedman, Howard S. (2006). A longitudinal investigation of associations between boys' pubertal timing and adult behavioral health and well-being. *Journal of Youth and Adolescence*, 35(3), 380–390.

Tagar, Michal Reifen; Hetherington, Chelsea; Shulman, Deborah & Koenig, Melissa. (2017). On the path to social dominance? Individual differences in sensitivity to intergroup fairness violations in early childhood. *Personality and Individual Differences*, 113, 246–250.

Taillieu, Tamara L.; Afifi, Tracie O.; Mota, Natalie; Keyes, Katherine M. & Sareen, Jitender. (2014). Age, sex, and racial differences in harsh physical punishment: Results from a nationally representative United States sample. *Child Abuse & Neglect*, 38(12), 1885–1894.

Tajalli, Hassan & Garba, Houmma A. (2014). Discipline or prejudice? Overrepresentation of minority students in disciplinary alternative education programs. *Urban Review*, 46(4), 620–631.

Takala, Marjatta; Nordmark, Marie & Allard, Karin. (2019). University curriculum in special teacher education in Finland and Sweden. *Nordic Journal of Comparative and International Education*, 3(2), 20–36.

Tam, Vivian; Patel, Nikunj; Turcotte, Michelle; Bossé, Yohan; Paré, Guillaume & Meyre, David. (2019). Benefits and limitations of genome-wide association studies. *Nature Reviews Genetics*, 20(8), 467–484.

Tamis-LeMonda, Catherine S.; Bornstein, Marc H. & Baumwell, Lisa. (2001). Maternal responsiveness and children's achievement of language milestones. *Child Development*, 72(3), 748–767.

Tamm, Leanne; Epstein, Jeffery N.; Denton, Carolyn A.; Vaughn, Aaron J.; Peugh, James & Willcutt, Erik G. (2014). Reaction time variability associated with reading skills in poor readers with ADHD. *Journal of the International Neuropsychological Society*, 20(3), 292–301.

Tamnes, Christian K.; Overbye, Knut; Ferschmann, Lia; Fjell, Anders M.; Walhovd, Kristine B.; Blakemore, Sarah-Jayne & Dumontheil, Iroise. (2018). Social perspective taking is associated with self-reported prosocial behavior and regional cortical thickness across adolescence. *Developmental Psychology*, 54(9), 1745–1757.

Tamura, Naomi; Hanaoka, Tomoyuki; Ito, Kumiko; Araki, Atsuko; Miyashita, Chihiro; Ito, Sachiko; . . . Kishi, Reiko. (2018). Different risk factors for very low birth weight, term-small-for-gestational-age, or preterm birth in Japan. *International Journal of Environmental Research and Public Health*, 15(2), 369.

Tan, Patricia Z.; Armstrong, Laura M. & Cole, Pamela M. (2013). Relations between temperament and anger regulation over early childhood. *Social Development*, 22(4), 755–772.

Tanumihardjo, Sherry A.; Gannon, Bryan & Kaliwile, Chisela. (2016). Controversy regarding widespread vitamin A fortification in Africa and Asia. *Advances in Nutrition*, 7, 5A.

Tartaro, Christine. (2019). *Suicide and self-harm in prisons and jails* (2nd ed.). Lexington.

Tarun, Kumar; Kumar, Singh Sanjeet; Manish, Kumar; Sunita & Ashok, Sharan. (2016). Study on relationship between anemia and academic performance of adolescent girls. *International Journal of Physiology*, 4(1), 81–86.

Tattersall, Ian. (2017). Why was human evolution so rapid? In Marom, Assaf & Hovers, Erella (Eds.), *Human paleontology and prehistory: Contributions in Honor of Yoel Rak* (pp. 1–9). Springer.

Tavassolie, Tanya & Winsler, Adam. (2019). Predictors of mandatory 3rd grade retention from high-stakes test performance for low-income, ethnically diverse children. *Early Childhood Research Quarterly*, 48, 62–74.

Tay, Marc Tze-Hsin; Au Eong, Kah Guan; Ng, C. Y. & Lim, M. K. (1992). Myopia and educational attainment in 421,116 young Singaporean males. *Annals Academy of Medicine Singapore*, 21(6), 785–791.

Taylor, Marjorie; Mottweiler, Candice M.; Aguiar, Naomi R.; Naylor, Emilee R. & Levernier, Jacob G. (2020). Paracosms: The imaginary worlds of middle childhood. *Child Development*, 91(1), e164–e178.

Taylor, Paul. (2014). *The next America: Boomers, millennials, and the looming generational showdown*. PublicAffairs.

Taylor, Zoe E.; Eisenberg, Nancy; Spinrad, Tracy L.; Eggum, Natalie D. & Sulik, Michael J. (2013). The relations of ego-resiliency and emotion socialization to the development of empathy and prosocial behavior across early childhood. *Emotion*, 13(5), 822–831.

Tekelab, Tesfalidet; Chojenta, Catherine; Smith, Roger & Loxton, Deborah. (2019). The impact of antenatal care on neonatal mortality in sub-Saharan Africa: A systematic review and meta-analysis. *PLoS ONE*, 14(9), e0222566.

Telzer, Eva H.; Ichien, Nicholas T. & Qu, Yang. (2015). Mothers know best: Redirecting adolescent reward sensitivity toward safe behavior during risk taking. *Social Cognitive and Affective Neuroscience*, 10(10), 1383–1391.

Teoh, Yee San & Lamb, Michael E. (2013). Interviewer demeanor in forensic interviews of children. *Psychology, Crime & Law*, 19(2), 145–159.

Terry, Nicole Patton; Connor, Carol McDonald; Johnson, Lakeisha; Stuckey, Adrienne & Tani, Novell. (2016). Dialect variation, dialect-shifting, and reading comprehension in second grade. *Reading and Writing*, 29(2), 267–295.

Tessier, Karen. (2010). Effectiveness of hands-on education for correct child restraint use by parents. *Accident Analysis & Prevention*, 42(4), 1041–1047.

Tett, Gillian. (2015). *The silo effect: The peril of expertise and the promise of breaking down barriers*. Simon & Schuster.

Tetzlaff, Anne & Hilbert, Anja. (2014). The role of the family in childhood and adolescent binge eating. A systematic review. *Appetite*, 76(1), 208.

Thaler, Richard H. & Sunstein, Cass R. (2008). *Nudge: Improving decisions about health, wealth, and happiness*. Yale University Press.

Tham, Diana Su Yun; Woo, Pei Jun & Bremner, J. Gavin. (2019). Development of the other-race effect in Malaysian-Chinese infants. *Developmental Psychobiology*, 61(1), 107–115.

Thiam, Melinda A.; Flake, Eric M. & Dickman, Michael M. (2017). Infant and child mental health and perinatal illness. In Thiam, Melinda A. (Ed.), *Perinatal mental health and the military family: Identifying and treating mood and anxiety disorders*. Routledge.

Thibodeaux, Jordan; Bock, Allison; Hutchison, Lindsey A. & Winsler, Adam. (2019). Singing to the self: Children's private speech, private song, and executive functioning. *Cognitive Development*, 50, 130–141.

Thomaes, Sander; Brummelman, Eddie & Sedikides, Constantine. (2017). Why most children think well of themselves. *Child Development*, 88(6), 1873–1884.

Thomason, Moriah E.; Scheinost, Dustin; Manning, Janessa H.; Grove, Lauren E.; Hect, Jasmine; Marshall, Narcis; . . . Romero, Roberto. (2017). Weak functional connectivity in the human fetal brain prior to preterm birth. *Scientific Reports*, 7(39286).

Thompson, Charis. (2014). Reproductions through technology. *Science*, 344(6182), 361–362.

Thompson, Richard; Kaczor, Kim; Lorenz, Douglas J.; Bennett, Berkeley L.; Meyers, Gabriel & Pierce, Mary Clyde. (2017). Is the use of physical discipline associated with aggressive behaviors in young children? *Academic Pediatrics*, 17(1), 34–44.

Thorup, Bianca; Crookes, Kate; Chang, Paul P. W.; Burton, Nichola; Pond, Stephen; Li, Tze Kwan; . . . Rhodes, Gillian. (2018). Perceptual experience shapes our ability to categorize faces by national origin: A new other-race effect. *British Journal of Psychology*, 109(3), 583–603.

Tiggemann, Marika & Slater, Amy. (2014). Net-Tweens: The internet and body image concerns in preteenage girls. *Journal of Early Adolescence*, 34(5), 606–620.

Tobey, Emily A.; Thal, Donna; Niparko, John K.; Eisenberg, Laurie S.; Quittner, Alexandra

L. & Wang, Nae-Yuh. (2013). Influence of implantation age on school-age language performance in pediatric cochlear implant users. *International Journal of Audiology, 52*(4), 219–229.

Tolman, Deborah L.; Davis, Brian R. & Bowman, Christin P. (2016). "That's just how it is": A gendered analysis of masculinity and femininity ideologies in adolescent girls' and boys' heterosexual relationships. *Journal of Adolescent Research, 31*(1), 3–31.

Tolman, Deborah L. & McClelland, Sara I. (2011). Normative sexuality development in adolescence: A decade in review, 2000–2009. *Journal of Research on Adolescence, 21*(1), 242–255.

Tomalski, Przemyslaw & Johnson, Mark H. (2010). The effects of early adversity on the adult and developing brain. *Current Opinion in Psychiatry, 23*(3), 233–238.

Tomasello, Michael. (2016a). The ontogeny of cultural learning. *Current Opinion in Psychology, 8*, 1–4.

Tomasello, Michael. (2016b). Cultural learning redux. *Child Development, 87*(3), 643–653.

Tomasello, Michael & Herrmann, Esther. (2010). Ape and human cognition. *Current Directions in Psychological Science, 19*(1), 3–8.

Topolewska-Siedzik, Ewa & Cieciuch, Jan. (2019). Modes of personal identity formation: A preliminary picture from the lifespan perspective. *Personality and Individual Differences, 138*, 237–242.

Toporek, Bryan. (2012). Sports rules revised as research mounts on head injuries. *Education Week, 31*(22), 8.

Torrey, E. Fuller & Yolken, Robert H. (2019). Schizophrenia as a pseudogenetic disease: A call for more gene-environmental studies. *Psychiatry Research, 278*, 146–150.

Toth, Sheree L. & Manly, Jody T. (2019). Developmental consequences of child abuse and neglect: Implications for intervention. *Child Development Perspectives, 13*(1), 59–65.

Tough, Paul. (2012). *How children succeed: Grit, curiosity, and the hidden power of character.* Houghton Mifflin Harcourt.

Trahan, Lisa H.; Stuebing, Karla K.; Fletcher, Jack M. & Hiscock, Merrill. (2014). The Flynn Effect: A meta-analysis. *Psychological Bulletin, 140*(5), 1332–1360.

Travers, Brittany G.; Tromp, Do P. M.; Adluru, Nagesh; Lange, Nicholas; Destiche, Dan; Ennis, Chad; . . . Alexander, Andrew L. (2015). Atypical development of white matter microstructure of the corpus callosum in males with autism: A longitudinal investigation. *Molecular Autism, 6*.

Trawick-Smith, Jeffrey. (2019). Not all children grow up the same: Child development, diversity, and early care and education. In Brown, Christopher P.; McMullen, Mary Benson & File, Nancy (Eds.), *The Wiley handbook of early childhood care and education* (pp. 29–58). Wiley.

Treffers-Daller, Jeanine & Milton, J. (2013). Vocabulary size revisited: The link between vocabulary size and academic achievement. *Applied Linguistics Review, 4*(1), 151–172.

Trinh, Sarah L.; Lee, Jaemin; Halpern, Carolyn T. & Moody, James. (2019). Our buddies, ourselves: The role of sexual homophily in adolescent friendship networks. *Child Development, 90*(1), e132–e147.

Troester-Trate, Katy Elizabeth. (2019). Food insecurity, inadequate childcare, & transportation disadvantage: Student retention and persistence of community college students. *Community College Journal of Research and Practice.*

Tronick, Edward. (1989). Emotions and emotional communication in infants. *American Psychologist, 44*(2), 112–119.

Tronick, Edward & Weinberg, M. Katherine. (1997). Depressed mothers and infants: Failure to form dyadic states of consciousness. In Murray, Lynne & Cooper, Peter J. (Eds.), *Postpartum depression and child development* (pp. 54–81). Guilford Press.

Tsang, Christine; Falk, Simone & Hessel, Alexandria. (2017). Infants prefer infant-directed song over speech. *Child Development, 88*(4), 1207–1215.

Turner, Heather A.; Finkelhor, David; Ormrod, Richard; Hamby, Sherry; Leeb, Rebecca T.; Mercy, James A. & Holt, Melissa. (2012). Family context, victimization, and child trauma symptoms: Variations in safe, stable, and nurturing relationships during early and middle childhood. *American Journal of Orthopsychiatry, 82*(2), 209–219.

Tuulari, Jetro J.; Scheinin, Noora M.; Lehtola, Satu; Merisaari, Harri; Saunavaara, Jani; Parkkola, Riitta; . . . Björnsdotter, Malin. (2019). Neural correlates of gentle skin stroking in early infancy. *Developmental Cognitive Neuroscience, 35*, 36–41.

Twenge, Jean M.; Joiner, Thomas E.; Rogers, Megan L. & Martin, Gabrielle N. (2018). Increases in depressive symptoms, suicide-related outcomes, and suicide rates among U.S. adolescents after 2010 and links to increased new media screen time. *Clinical Psychological Science, 6*(1), 3–17.

Twito, Louise & Knafo-Noam, Ariel. (2020). Beyond culture and the family: Evidence from twin studies on the genetic and environmental contribution to values. *Neuroscience & Biobehavioral Reviews, 112*, 135–143.

U.S. Bureau of Labor Statistics. (2018, August). *Labor force characteristics by race and ethnicity, 2017. BLS Reports.* Washington, DC: U.S. Bureau of Labor Statistics.

U.S. Bureau of Labor Statistics. (2019, April 18). Table 6. Employment status of mothers with own children under 3 years old by single year of age of youngest child and marital status, 2017–2018 annual averages. Division of Labor Force Statistics.

U.S. Bureau of Labor Statistics. (2020). Occupational outlook handbook. Division of Labor Force Statistics.

U.S. Census Bureau. (2015). *America's families and living arrangements: 2015: Households (H table series). Table H3: Households by race and Hispanic origin of household reference person and detailed type.* Washington, DC: U.S. Department of Commerce, Economics and Statistics Administration, U.S. Census Bureau.

U.S. Census Bureau. (2016a). *Selected population profile in the United States: 2014 American community survey 1-year estimates. American FactFinder.* Washington, DC: U.S. Department of Commerce.

U.S. Census Bureau. (2016b). *Selected population profile in the United States: 2009 American community survey 1-year estimates. American FactFinder.* Washington, DC: U.S. Department of Commerce.

U.S. Census Bureau. (2018, November). *Historical living arrangements of children: Living arrangements of children under 18 years old: 1960 to present.* U.S. Census Bureau.

U.S. Census Bureau. (2018, February 19). CPS historical time series tables: Table A-1. Years of school completed by people 25 years and over, by age and sex: Selected years 1940 to 2018. U.S. Census Bureau.

U.S. Census Bureau. (2019, June). *Annual estimates of the resident population by sex, age, race, and Hispanic origin for the United States and States: April 1, 2010 to July 1, 2018. American FactFinder.* Washington, DC: U.S. Department of Commerce. PEPASR6H.

U.S. Department of Agriculture. (2018, September 5). *Key statistics & graphics: Food insecurity in the U.S.* Washington, DC: U.S. Department of Agriculture.

U.S. Department of Health and Human Services. (2000, December 31). *Child maltreatment 2000.* Washington, DC: Administration on Children, Youth and Families, Children's Bureau.

U.S. Department of Health and Human Services. (2010). *Head Start impact study: Final report.* Washington, DC: Administration for Children and Families.

U.S. Department of Health and Human Services. (2010, January). *Child maltreatment 2009.* Washington, DC: Administration for Children and Families, Administration on Children, Youth and Families, Children's Bureau.

U.S. Department of Health and Human Services. (2011). *The Surgeon General's call to action to support breastfeeding.* Washington, DC: U.S. Department of Health and Human Services, Office of the Surgeon General.

U.S. Department of Health and Human Services. (2016, January 25). *Child maltreatment 2014.* Washington, DC: Administration for Children and Families, Administration on Children, Youth and Families, Children's Bureau.

U.S. Department of Health and Human Services. (2017, January 19). *Child maltreatment 2015.* Washington, DC: Administration for Children and Families, Administration on Children, Youth and Families, Children's Bureau.

U.S. Department of Health and Human Services. (2018, February 1). *Child maltreatment 2016.* Washington, DC: Administration for Children and Families, Administration on Children, Youth and Families, Children's Bureau.

U.S. Department of Health and Human Services. (2019, January 28). *Child maltreatment 2017.* Washington, DC: Administration for Children and Families, Administration on Children, Youth and Families, Children's Bureau.

U.S. Department of Health and Human Services. (2020, January 15). *Child maltreatment 2018.* Washington, DC: Administration for Children and Families, Administration on Children, Youth and Families, Children's Bureau.

U.S. Department of State. (2019). Adoption statistics—Adoptions by year. Travel.State.gov.

U.S. Social Security Administration. (2019). Popular names by birth year. U.S. Social Security Administration.

Uchida, Mai; Spencer, Thomas J.; Faraone, Stephen V. & Biederman, Joseph. (2018). Adult outcome of ADHD: An overview of results from the MGH longitudinal family studies of pediatrically and psychiatrically referred youth with and without ADHD of both sexes. *Journal of Attention Disorders, 22*(6), 523–534.

Umaña-Taylor, Adriana J. & Hill, Nancy E. (2020). Ethnic–racial socialization in the family: A decade's advance on precursors and outcomes. *Journal of Marriage and Family, 82*(1), 244–271.

Umapathi, Kishore Krishna; Thavamani, Aravind & Chotikanatis, Kobkul. (2019). Incidence trends, risk factors, mortality and healthcare utilization in congenital syphilis-related hospitalizations in the United States: A nationwide population analysis. *The Pediatric Infectious Disease Journal, 38*(11), 1126–1130.

Underwood, Emily. (2013). Why do so many neurons commit suicide during brain development? *Science, 340*(6137), 1157–1158.

UNESCO. (2014). *Country profiles. UNESCO Institute for Statistics Data Centre.* Montreal, Canada: UNESCO, Université de Montréal at the Montreal's École des hautes études.

UNESCO. (2018, February). *One in five children, adolescents and youth is out of school. Fact Sheet No. 48.* Montreal, Canada: UNESCO. UIS/FS/2018/ED/48.

UNICEF. (2017, January 13). *Global overview child malnutrition 1990–2015. UNICEF Data and Analytics: Joint Malnutrition Estimates 2016 Edition.* New York: NY: United Nations.

UNICEF. (2019, April). *Malnutrition* [Data set]. UNICEF Data.

United Nations. (2019, June 17). *UN Data: Life expectancy at birth for both sexes combined (years)* [Data set].

United Nations, Department of Economic and Social Affairs, Population Division. (2019). *World population prospects 2019.* New York, NY: (Volumes 1 & 2: ST/ESA/SER.A/426 & ST/ESA/SER.A/427).

Ursache, Alexandra; Blair, Clancy; Stifter, Cynthia & Voegtline, Kristin. (2013). Emotional reactivity and regulation in infancy interact to predict executive functioning in early childhood. *Developmental Psychology, 49*(1), 127–137.

van Batenburg-Eddes, Tamara; Butte, Dick & van de Looij-Jansen, Petra. (2012). Measuring juvenile delinquency: How do self-reports compare with official police statistics? *European Journal of Criminology, 9*(1), 23–37.

van de Bongardt, Daphne; Reitz, Ellen; Sandfort, Theo & Deković, Maja. (2015). A meta-analysis of the relations between three types of peer norms and adolescent sexual behavior. *Personality and Social Psychology Review, 19*(3), 203–234.

Van de Vondervoort, Julia W. & Hamlin, J. Kiley. (2016). Evidence for intuitive morality: Preverbal infants make sociomoral evaluations. *Child Development Perspectives, 10*(3), 143–148.

van den Akker, Alithe; Deković, Maja; Prinzie, Peter & Asscher, Jessica. (2010). Toddlers' temperament profiles: Stability and relations to negative and positive parenting. *Journal of Abnormal Child Psychology, 38*(4), 485–495.

van Den Bunt, M. R.; Groen, M. A.; van Der Kleij, S. W.; Noordenbos, M. W.; Segers, E.; Pugh, K. R. & Verhoeven, L. (2018). Deficient response to altered auditory feedback in dyslexia. *Developmental Neuropsychology, 43*(7), 622–641.

van den Pol, Anthony N.; Mao, Guochao; Yang, Yang; Ornaghi, Sara & Davis, John N. (2017). Zika virus targeting in the developing brain. *Journal of Neuroscience, 37*(8), 2161–2175.

Van Dongen, Rachel. (2019, March 15). The Energy 202: Kids are skipping school today—to protest climate change. *Washington Post.*

Van Dyke, Miriam E.; Cheung, Patricia C.; Franks, Padra & Gazmararian, Julie A. (2018). Socioeconomic and racial/ethnic disparities in physical activity environments in Georgia elementary schools. *American Journal of Health Promotion, 32*(2), 453–463.

van Goozen, Stephanie H. M. (2015). The role of early emotion impairments in the development of persistent antisocial behavior. *Child Development Perspectives, 9*(4), 206–210.

Van Harmelen, A.-L.; Kievit, R. A.; Ioannidis, K.; Neufeld, S.; Jones, P. B.; Bullmore, E.; . . . Goodyer, I. (2017). Adolescent friendships predict later resilient functioning across psychosocial domains in a healthy community cohort. *Psychological Medicine, 47*(13), 2312–2322.

Van Hecke, Wim; Emsell, Louise & Sunaert, Stefan (Eds.). (2016). *Diffusion tensor imaging: A practical handbook.* Springer.

Van Horn, Linda V.; Bausermann, Robert; Affenito, Sandra; Thompson, Douglas; Striegel-Moore, Ruth; Franko, Debra & Albertson, Ann. (2011). Ethnic differences in food sources of vitamin D in adolescent American girls: The National Heart, Lung, and Blood Institute Growth and Health Study. *Nutrition Research, 31*(8), 579–585.

Van Houtte, Mieke. (2016). Lower-track students' sense of academic futility: Selection or effect? *Journal of Sociology, 52*(4), 874–889.

van IJzendoorn, Marinus H.; Bakermans-Kranenburg, Marian J.; Pannebakker, Fieke & Out, Dorothée. (2010). In defence of situational morality: Genetic, dispositional and situational determinants of children's donating to charity. *Journal of Moral Education, 39*(1), 1–20.

van Nunen, Karolien; Kaerts, Nore; Wyndaele, Jean-Jacques; Vermandel, Alexandra & Van Hal, Guido. (2015). Parents' views on toilet training (TT): A quantitative study to identify the beliefs and attitudes of parents concerning TT. *Journal of Child Health Care, 19*(2), 265–274.

Van Rheenen, Derek. (2012). A century of historical change in the game preferences of American children. *Journal of American Folklore, 125*(498), 411–443.

Van Tongeren, Daryl R.; DeWall, C. Nathan; Chen, Zhansheng; Sibley, Chris G. & Bulbulia, Joseph. (2020). Religious residue: Cross-cultural evidence that religious psychology and behavior persist following deidentification. *Journal of Personality and Social Psychology.*

Van Vonderen, Kristen E. & Kinnally, William. (2012). Media effects on body image: Examining media exposure in the broader context of internal and other social factors. *American Communication Journal, 14*(2), 41–57.

Vanderberg, Rachel H.; Farkas, Amy H.; Miller, Elizabeth; Sucato, Gina S.; Akers, Aletha Y. & Borrero, Sonya B. (2016). Racial and/or ethnic differences in formal sex education and sex education by parents among young women in the United States. *Journal of Pediatric and Adolescent Gynecology, 29*(1), 69–73.

Vanhalst, Janne; Luyckx, Koen; Scholte, Ron H. J.; Engels, Rutger C. M. E. & Goossens, Luc. (2013). Low self-esteem as a risk factor for loneliness in adolescence: Perceived – but not actual – social acceptance as an underlying mechanism. *Journal of Abnormal Child Psychology, 41*(7), 1067–1081.

Vannucci, Robert C. & Vannucci, Susan J. (2019). Brain growth in modern humans using multiple developmental databases. *American Journal of Physical Anthropology, 168*(2), 247–261.

Vargas Lascano, Dayuma I.; Galambos, Nancy L.; Krahn, Harvey J. & Lachman, Margie E. (2015). Growth in perceived control across 25 years from the late teens to midlife: The role of personal and parents' education. *Developmental Psychology, 51*(1), 124–135.

Vasung, Lana; Abaci Turk, Esra; Ferradal, Silvina L.; Sutin, Jason; Stout, Jeffrey N.; Ahtam, Banu; . . . Grant, P. Ellen. (2019). Exploring early human brain development with structural and physiological neuroimaging. *NeuroImage, 187,* 226–254.

Vaughn, Byron P.; Rank, Kevin M. & Khoruts, Alexander. (2019). Fecal microbiota transplantation: Current status in treatment of GI and liver disease. *Clinical Gastroenterology and Hepatology, 17*(2), 353–361.

Vedantam, Shankar. (2011, December 5). *What's behind a temper tantrum? Scientists deconstruct the screams. Hidden Brain.* Washington DC: National Public Radio.

Veldheer, Susan; Yingst, Jessica; Midya, Vishal; Hummer, Breianna; Lester, Courtney; Krebs, Nicolle; . . . Foulds, Jonathan. (2019). Pulmonary and other health effects of electronic cigarette use among adult smokers participating in a randomized controlled smoking reduction trial. *Addictive Behaviors, 91,* 95–101.

Vennemann, Mechtild M.; Hense, Hans-Werner; Bajanowski, Thomas; Blair, Peter S.; Complojer, Christina; Moon, Rachel Y. & Kiechl-Kohlendorfer, Ursula. (2012). Bed sharing and the risk of sudden infant death syndrome:

Can we resolve the debate? *The Journal of Pediatrics*, *160*(1), 44–48.

Verdine, Brian N.; Golinkoff, Roberta Michnick; Hirsh-Pasek, Kathy & Newcombe, Nora S. (2017). Spatial skills, their development, and their links to mathematics. *Monographs of the Society for Research in Child Development: Links between spatial and mathematical skills across the preschool*, *82*(1), 7–30.

Verona, Sergiu. (2003). Romanian policy regarding adoptions. In Littel, Victor (Ed.), *Adoption update* (pp. 5–10). Nova Science.

Verrusio, Walter; Ettorre, Evaristo; Vicenzini, Edoardo; Vanacore, Nicola; Cacciafesta, Mauro & Mecarelli, Oriano. (2015). The Mozart Effect: A quantitative EEG study. *Consciousness and Cognition*, *35*, 150–155.

Verschueren, Karine. (2020). Attachment, self-esteem, and socio-emotional adjustment: There is more than just the mother. *Attachment & Human Development*, *22*(1), 105–109.

Vijayakumar, Nandita; Op de Macks, Zdena; Shirtcliff, Elizabeth A. & Pfeifer, Jennifer H. (2018). Puberty and the human brain: Insights into adolescent development. *Neuroscience and Biobehavioral Reviews*, *92*, 417–436.

Viljaranta, Jaana; Aunola, Kaisa; Mullola, Sari; Virkkala, Johanna; Hirvonen, Riikka; Pakarinen, Eija & Nurmi, Jari-Erik. (2015). Children's temperament and academic skill development during first grade: Teachers' interaction styles as mediators. *Child Development*, *86*(4), 1191–1209.

Vitale, Susan; Sperduto, Robert D. & Ferris, Frederick L. (2009). Increased prevalence of myopia in the United States between 1971–1972 and 1999–2004. *Archives of Ophthalmology*, *127*(12), 1632–1639.

Voelcker-Rehage, Claudia; Niemann, Claudia & Hübner, Lena. (2018). Structural and functional brain changes related to acute and chronic exercise effects in children, adolescents and young adults. In Meeusen, Romain; Schaefer, Sabine; Tomporowski, Phillip & Bailey, Richard (Eds.), *Physical activity and educational achievement: Insights from exercise neuroscience* (pp. 143–163). Routledge.

Voges, Juané; Berg, Astrid & Niehaus, Dana J. H. (2019). Revisiting the African origins of attachment research—50 years on from Ainsworth: A descriptive review. *Infant Mental Health Journal*, *40*(6), 799–816.

Vöhringer, Isabel A.; Kolling, Thorsten; Graf, Frauke; Poloczek, Sonja; Fassbender, Iina; Freitag, Claudia; . . . Knopf, Monika. (2018). The development of implicit memory from infancy to childhood: On average performance levels and interindividual differences. *Child Development*, *89*(2), 370–382.

von Salisch, Maria. (2018). Emotional competence and friendship involvement: Spiral effects in adolescence. *European Journal of Developmental Psychology*, *15*(6), 678–693.

Vos, Amber A.; van Voorst, Sabine F.; Steegers, Eric A. P. & Denktaş, Semiha. (2016). Analysis of policy towards improvement of perinatal mortality in the Netherlands (2004–2011). *Social Science & Medicine*, *157*, 156–164.

Votruba-Drzal, Elizabeth & Dearing, Eric (Eds.). (2017). *Handbook of early childhood development programs, practices, and policies.* Wiley.

Vygotsky, Lev S. (1980). *Mind in society: The development of higher psychological processes.* Harvard University Press.

Vygotsky, Lev S. (1987). Thinking and speech. In Rieber, Robert W. & Carton, Aaron S. (Eds.), *The collected works of L. S. Vygotsky* (Vol. 1, pp. 39–285). Springer.

Vygotsky, Lev S. (1994a). The development of academic concepts in school aged children. In van der Veer, René & Valsiner, Jaan (Eds.), *The Vygotsky reader* (pp. 355–370). Blackwell.

Vygotsky, Lev S. (1994b). Principles of social education for deaf and dumb children in Russia. In van der Veer, Rene & Valsiner, Jaan (Eds.), *The Vygotsky reader* (pp. 19–26). Blackwell.

Vygotsky, Lev S. (2012). *Thought and language.* MIT Press.

Wade, Mark; Prime, Heather; Jenkins, Jennifer; Yeates, Keith; Williams, Tricia & Lee, Kang. (2018). On the relation between theory of mind and executive functioning: A developmental cognitive neuroscience perspective. *Psychonomic Bulletin & Review*, *25*(6), 2119–2140.

Wadman, Meredith. (2018). 'Rapid onset' of transgender identity ignites storm. *Science*, *361*(6406), 958–959.

Wagmiller, Robert L. (2015). The temporal dynamics of childhood economic deprivation and children's achievement. *Child Development Perspectives*, *9*(3), 158–163.

Wagner, Erica. (2017). *Chief engineer: Washington Roebling, the man who built the Brooklyn Bridge.* Bloomsbury.

Wagner, Katie; Dobkins, Karen & Barner, David. (2013). Slow mapping: Color word learning as a gradual inductive process. *Cognition*, *127*(3), 307–317.

Wagner, Paul A. (2011). Socio-sexual education: A practical study in formal thinking and teachable moments. *Sex Education: Sexuality, Society and Learning*, *11*(2), 193–211.

Waldinger, Robert & Schulz, Marc. (2018). The blind psychological scientists and the elephant: Reply to Sherlock and Zietsch. *Psychological Science*, *29*(1), 158–160.

Waldorf, Kristina M. Adams; Nelson, Branden R.; Stencel-Baerenwald, Jennifer E.; Studholme, Colin; Kapur, Raj P.; Armistead, Blair; . . . Rajagopal, Lakshmi. (2018). Congenital Zika virus infection as a silent pathology with loss of neurogenic output in the fetal brain. *Nature Medicine*, *24*, 368–374.

Walk, Laura M.; Vaidya, Chandan; Evers, Wiebke F.; Quante, Sonja & Hille, Katrin. (2018). Evaluation of a teacher training program to enhance executive functions in preschool children. *PLoS ONE*, *13*(5), e0197454.

Walker, Christa L. Fischer; Rudan, Igor; Liu, Li; Nair, Harish; Theodoratou, Evropi; Bhutta, Zulfiqar A.; . . . Black, Robert E. (2013). Global burden of childhood pneumonia and diarrhoea. *The Lancet*, *381*(9875), 1405–1416.

Walker, Renee; Block, Jason & Kawachi, Ichiro. (2014). The spatial accessibility of fast food restaurants and convenience stores in relation to neighborhood schools. *Applied Spatial Analysis and Policy*, *7*(2), 169–182.

Walle, Eric A. & Campos, Joseph J. (2014). Infant language development is related to the acquisition of walking. *Developmental Psychology*, *50*(2), 336–348.

Wallis, Claudia. (2014). Gut reactions: Intestinal bacteria may help determine whether we are lean or obese. *Scientific American*, *310*(6), 30–33.

Walter, Melissa Clucas & Lippard, Christine N. (2017). Head Start teachers across a decade: Beliefs, characteristics, and time spent on academics. *Early Childhood Education Journal*, *45*(5), 693–702.

Wambach, Karen & Riordan, Jan. (2014). *Breastfeeding and human lactation* (5th ed.). Jones & Bartlett.

Wang, Chen; Horby, Peter W.; Hayden, Frederick G. & Gao, George F. (2020). A novel coronavirus outbreak of global health concern. *The Lancet*, *395*(10223), 470–473.

Wang, Chao; Xue, Haifeng; Wang, Qianqian; Hao, Yongchen; Li, Dianjiang; Gu, Dongfeng & Huang, Jianfeng. (2014). Effect of drinking on all-cause mortality in women compared with men: A meta-analysis. *Journal of Women's Health*, *23*(5), 373–381.

Wang, Ke; Rathbun, Amy & Musu, Lauren. (2019). *School choice in the United States: 2019.* Washington, DC: U.S. Department of Education, National Center for Education Statistics. NCES 2019–106.

Wang, Meifang & Liu, Li. (2018). Reciprocal relations between harsh discipline and children's externalizing behavior in China: A 5-year longitudinal study. *Child Development*, *89*(1), 174–187.

Wang, Ming-Te & Eccles, Jacquelynne S. (2013). School context, achievement motivation, and academic engagement: A longitudinal study of school engagement using a multidimensional perspective. *Learning and Instruction*, *28*, 12–23.

Wang, Ming-Te; Henry, Daphne A.; Smith, Leann V.; Huguley, James P. & Guo, Jiesi. (2020). Parental ethnic-racial socialization practices and children of color's psychosocial and behavioral adjustment: A systematic review and meta-analysis. *American Psychologist*, *75*(1), 1–22.

Watson, John B. (1924). *Behaviorism.* The People's Institute Pub. Co.

Watson, John B. (1928). *Psychological care of infant and child.* Norton.

Watson, John B. (1972). *Psychological care of infant and child.* Arno Press.

Watson, John B. (1998). *Behaviorism.* Transaction.

Webb, Alexandra R.; Heller, Howard T.; Benson, Carol B. & Lahav, Amir. (2015). Mother's voice and heartbeat sounds elicit auditory plasticity in the human brain before full gestation. *Proceedings of the National Academy of Sciences*, *112*(10), 3152–3157.

Weber, Ann; Fernald, Anne & Diop, Yatma. (2017). When cultural norms discourage talking to babies: Effectiveness of a parenting program in rural Senegal. *Child Development, 88*(5), 1513–1526.

Weber, Daniela; Dekhtyar, Serhiy & Herlitz, Agneta. (2017). The Flynn Effect in Europe—Effects of sex and region. *Intelligence, 60*, 39–45.

Webster, Collin A. & Suzuki, Naoki. (2014). Land of the rising pulse: A social ecological perspective of physical activity opportunities for schoolchildren in Japan. *Journal of Teaching in Physical Education, 33*(3), 304–325.

Weinshenker, Naomi J. (2014). Teenagers and body image. Education.

Weinstein, Netta & DeHaan, Cody. (2014). On the mutuality of human motivation and relationships. In Weinstein, Netta (Ed.), *Human motivation and interpersonal relationships: Theory, research, and applications* (pp. 3–25). Springer.

Weiss, Noel S. & Koepsell, Thomas D. (2014). *Epidemiologic methods: Studying the occurrence of illness* (2nd ed.). Oxford University Press.

Weissberg, Roger P.; Durlak, Joseph A.; Domitrovich, Celene E. & Gullotta, Thomas P. (2016). Social and emotional learning: Past, present, and future. In Durlak, Joseph A.; Domitrovich, Celene E.; Weissberg, Roger P. & Gullotta, Thomas P. (Eds.), *Handbook of social and emotional learning: Research and practice* (pp. 3–19). Guilford Press.

Weisskirch, Robert S. (2017a). A developmental perspective on language brokering. In Weisskirch, Robert S. (Ed.), *Language brokering in immigrant families: Theories and contexts.* Routledge.

Weisskirch, Robert S. (2017b). *Language brokering in immigrant families: Theories and contexts.* Routledge.

Weisz, John R.; Kuppens, Sofie; Ng, Mei Yi; Eckshtain, Dikla; Ugueto, Ana M.; Vaughn-Coaxum, Rachel; . . . Fordwood, Samantha R. (2017). What five decades of research tells us about the effects of youth psychological therapy: A multilevel meta-analysis and implications for science and practice. *American Psychologist, 72*(2), 79–117.

Wellman, Henry M. (2014). *Making minds: How theory of mind develops.* Oxford University Press.

Wellman, Henry M. (2018). Theory of mind: The state of the art. *European Journal of Developmental Psychology, 15*(6), 728–755.

Wellman, Henry M.; Fang, Fuxi & Peterson, Candida C. (2011). Sequential progressions in a theory-of-mind scale: Longitudinal perspectives. *Child Development, 82*(3), 780–792.

Wendelken, Carter; Baym, Carol L.; Gazzaley, Adam & Bunge, Silvia A. (2011). Neural indices of improved attentional modulation over middle childhood. *Developmental Cognitive Neuroscience, 1*(2), 175–186.

Wendland, Claire. (2018). Who counts? What counts? Place and the limits of perinatal mortality measures. *AMA Journal of Ethics, 20*(1), 278–287.

Werker, Janet F. & Hensch, Takao K. (2015). Critical periods in speech perception:

New directions. *Annual Review of Psychology, 66*, 173–196.

Westover, Tara. (2018). *Educated: A memoir.* Random House.

Weymouth, Bridget B.; Buehler, Cheryl; Zhou, Nan & Henson, Robert A. (2016). A meta-analysis of parent–adolescent conflict: Disagreement, hostility, and youth maladjustment. *Journal of Family Theory & Review, 8*(1), 95–112.

Wheeler, Lorey A.; Zeiders, Katharine H.; Updegraff, Kimberly A.; Umaña-Taylor, Adriana J.; Rodríguez de Jesús, Sue A. & Perez-Brena, Norma J. (2017). Mexican-origin youth's risk behavior from adolescence to young adulthood: The role of familism values. *Developmental Psychology, 53*(1), 126–137.

White, Rebecca M. B.; Deardorff, Julianna; Liu, Yu & Gonzales, Nancy A. (2013). Contextual amplification or attenuation of the impact of pubertal timing on Mexican-origin boys' mental health symptoms. *Journal of Adolescent Health, 53*(6), 692–698.

White, Sue; Gibson, Matthew & Wastell, David. (2019). Child protection and disorganized attachment: A critical commentary. *Children and Youth Services Review, 105*(104415).

White-Traut, Rosemary C.; Rankin, Kristin M.; Yoder, Joe; Zawacki, Laura; Campbell, Suzann; Kavanaugh, Karen; . . . Norr, Kathleen F. (2018). Relationship between mother-infant mutual dyadic responsiveness and premature infant development as measured by the Bayley III at 6 weeks corrected age. *Early Human Development, 121*, 21–26.

Widman, Laura; Choukas-Bradley, Sophia; Helms, Sarah W.; Golin, Carol E. & Prinstein, Mitchell J. (2014). Sexual communication between early adolescents and their dating partners, parents, and best friends. *The Journal of Sex Research, 51*(7), 731–741.

Widom, Cathy Spatz; Czaja, Sally J. & DuMont, Kimberly A. (2015a). Intergenerational transmission of child abuse and neglect: Real or detection bias? *Science, 347*(6229), 1480–1485.

Widom, Cathy Spatz; Horan, Jacqueline & Brzustowicz, Linda. (2015b). Childhood maltreatment predicts allostatic load in adulthood. *Child Abuse & Neglect, 47*, 59–69.

Wigger, J. Bradley. (2018). Invisible friends across four countries: Kenya, Malawi, Nepal and the Dominican Republic. *International Journal of Psychology, 53*(Suppl. 1), 46–52.

Wigginton, Nicholas S.; Fahrenkamp-Uppenbrink, Julia; Wible, Brad & Malakoff, David. (2016). Cities are the future. *Science, 352*(6288), 904–905.

Wilcox, W. Bradford (Ed.) (2011). *The sustainable demographic dividend: What do marriage and fertility have to do with the economy?* Social Trends Institute.

Wilkinson, Stephen. (2015). Prenatal screening, reproductive choice, and public health. *Bioethics, 29*(1), 26–35.

Wilks, Matti; Kirby, James & Nielsen, Mark. (2019). Developmental changes in young children's

willingness to copy the antisocial actions of ingroup members in a minimal group context. *Developmental Psychology, 55*(4), 709–721.

Williams, Anne M.; Chantry, Caroline; Geubbels, Eveline L.; Ramaiya, Astha K.; Shemdoe, Aloisia I.; Tancredi, Daniel J. & Young, Sera L. (2016). Breastfeeding and complementary feeding practices among HIV-exposed infants in coastal Tanzania. *Journal of Human Lactation, 32*(1), 112–122.

Williams, Katie M. & Hammond, Christopher J. (2016). GWAS in myopia: Insights into disease and implications for the clinic. *Expert Review of Ophthalmology, 11*(2), 101–110.

Williams, Lela Rankin; Fox, Nathan A.; Lejuez, C. W.; Reynolds, Elizabeth K.; Henderson, Heather A.; Perez-Edgar, Koraly E.; . . . Pine, Daniel S. (2010). Early temperament, propensity for risk-taking and adolescent substance-related problems: A prospective multi-method investigation. *Addictive Behaviors, 35*(2), 1148–1151.

Williams, Shanna; Moore, Kelsey; Crossman, Angela M. & Talwar, Victoria. (2016). The role of executive functions and theory of mind in children's prosocial lie-telling. *Journal of Experimental Child Psychology, 141*, 256–266.

Williamson, Victoria; Stevelink, Sharon A. M.; Da Silva, Eve & Fear, Nicola T. (2018). A systematic review of wellbeing in children: A comparison of military and civilian families. *Child and Adolescent Psychiatry and Mental Health, 12*(46).

Willoughby, Michael T.; Mills-Koonce, W. Roger; Gottfredson, Nisha C. & Wagner, Nicholas J. (2014). Measuring callous unemotional behaviors in early childhood: Factor structure and the prediction of stable aggression in middle childhood. *Journal of Psychopathology and Behavioral Assessment, 36*(1), 30–42.

Wills, Thomas A.; Okamoto, Scott K.; Knight, Rebecca & Pagano, Ian. (2019). Parental support, parent-adolescent conflict, and substance use of native Hawaiian and other Pacific Islander youth: Ethnic differences in stress-buffering and vulnerability effects. *Asian American Journal of Psychology, 10*(3), 218–226.

Wilmshurst, Linda. (2011). *Child and adolescent psychopathology: A casebook* (2nd ed.). Sage.

Wilson, Jennifer; Andrews, Glenda; Hogan, Christy; Wang, Si & Shum, David H. K. (2018). Executive function in middle childhood and the relationship with theory of mind. *Developmental Neuropsychology, 43*(3), 163–182.

Winn, Phoebe; Acharya, Krishna; Peterson, Erika & Leuthner, Steven R. (2018). Prenatal counseling and parental decision-making following a fetal diagnosis of trisomy 13 or 18. *Journal of Perinatology, 38*(7), 788–796.

Wolff, Jason J.; Gerig, Guido; Lewis, John D.; Soda, Takahiro; Styner, Martin A.; Vachet, Clement; . . . Piven, Joseph. (2015). Altered corpus callosum morphology associated with autism over the first 2 years of life. *Brain, 138*(7), 2046–2058.

Wolff, Mary S.; Teitelbaum, Susan L.; McGovern, Kathleen; Pinney, Susan M.;

Windham, Gayle C.; Galvez, Maida; . . . Biro, Frank M. (2015). Environmental phenols and pubertal development in girls. *Environment International*, 84, 174–180.

Womack, Sean R.; Taraban, Lindsay; Shaw, Daniel S.; Wilson, Melvin N. & Dishion, Thomas J. (2019). Family turbulence and child internalizing and externalizing behaviors: Moderation of effects by race. *Child Development*, 90(6), e729–e744.

Wong, Waison; Lam, Ray & Morris, Shaun. (2020). The role of the measles vaccine for travelling infants 6 to 11 months of age in the era of global outbreaks of disease. *Paediatrics & Child Health*, 1–3.

Woodward, Amanda L. & Markman, Ellen M. (1998). Early word learning. In Kuhn, Deanna & Siegler, Robert S. (Eds.), *Handbook of child psychology* (5th ed., Vol. 2, pp. 371–420). Wiley.

Woolley, Jacqueline D. & Ghossainy, Maliki E. (2013). Revisiting the fantasy–reality distinction: Children as naïve skeptics. *Child Development*, 84(5), 1496–1510.

World Bank. (2018). *World development indicators: Mortality rate, infant (per 1,000 live births)* [Data set].

World Bank. (2019). World Bank open data. World Bank.

World Health Organization. (2006). WHO Motor Development Study: Windows of achievement for six gross motor development milestones. *Acta Paediatrica*, 95(Suppl. 450), 86–95.

World Health Organization. (2015). *Global status report on road safety 2015*. Geneva, Switzerland: World Health Organization.

World Health Organization. (2017, April 28). Measles vaccines: WHO position paper – April 2017. *Weekly Epidemiological Record*, 17(92), 205–228.

World Health Organization. (2017, September 29). Global Health Observatory data repository: Prevalence of obesity among children and adolescents, BMI>+2 standard deviation above the median, crude estimates by country, among children aged 5–19 years. World Health Organization.

World Health Organization. (2019, December, 6). Immunization coverage. Fact sheets. World Health Organization.

World Health Organization. (2019, May 29). *WHO vaccine-preventable diseases: Monitoring system — 2019 global summary: Global and regional immunization profile* [Data set].

World Health Organization. (2020, March 13). *Immunization, vaccines and biologicals: Measles and rubella surveillance data* [Data set].

Wörmann, Viktoriya; Holodynski, Manfred; Kärtner, Joscha & Keller, Heidi. (2012). A cross-cultural comparison of the development of the social smile: A longitudinal study of maternal and infant imitation in 6- and 12-week-old infants. *Infant Behavior and Development*, 35(2), 335–347.

Wosje, Karen S.; Khoury, Philip R.; Claytor, Randal P.; Copeland, Kristen A.; Hornung, Richard W.; Daniels, Stephen R. & Kalkwarf, Heidi J. (2010). Dietary patterns associated with fat and bone mass in young children. *American Journal of Clinical Nutrition*, 92(2), 294–303.

Wright, Vince. (2018). Vygotsky and a global perspective on scaffolding in learning mathematics. In Zajda, Joseph (Ed.), *Globalisation and education reforms: Globalisation, comparative education and policy research* (pp. 123–135). Springer.

Wrzus, Cornelia & Neyer, Franz J. (2016). Co-development of personality and friendships across the lifespan: An empirical review on selection and socialization. *European Psychologist*, 21(4), 254–273.

Wynberg, Rachel & Laird, Sarah A. (2018). Fast science and sluggish policy: The Herculean task of regulating biodiscovery. *Trends in Biotechnology*, 36(1), 1–3.

Xu, Fei. (2013). The object concept in human infants: Commentary on Fields. *Human Development*, 56(3), 167–170.

Xu, Fei & Kushnir, Tamar. (2013). Infants are rational constructivist learners. *Current Directions in Psychological Science*, 22(1), 28–32.

Xu, Guifeng; Strathearn, Lane; Liu, Buyun; O'Brien, Matthew; Kopelman, Todd G.; Zhu, Jing; . . . Bao, Wei. (2019). Prevalence and treatment patterns of autism spectrum disorder in the United States, 2016. *JAMA Pediatrics*, 173(2), 153–159.

Xu, Guifeng; Strathearn, Lane; Liu, Buyun; Yang, Binrang & Bao, Wei. (2018). Twenty-year trends in diagnosed Attention-deficit/hyperactivity disorder among US children and adolescents, 1997–2016. *JAMA Network Open*, 1(4), e181471.

Xu, Jiaquan; Murphy, Sherry L.; Kochanek, Kenneth D. & Arias, Elizabeth. (2020, January). *Mortality in the United States, 2018. NCHS Data Brief*. U.S. Department of Health and Human Services, Centers for Disease Control and Prevention, National Center for Health Statistics. 355.

Xu, Richard; Poole, Kristie L.; Van Lieshout, Ryan J.; Saigal, Saroj & Schmidt, Louis A. (2019). Shyness and sociability among extremely low birth weight survivors in the third and fourth decades of life: Associations with relationship status. *Journal of Personality*, 87(2), 231–239.

Xu, Yaoying. (2010). Children's social play sequence: Parten's classic theory revisited. *Early Child Development and Care*, 180(4), 489–498.

Xu, Yin; Norton, Sam & Rahman, Qazi. (2020). A longitudinal birth cohort study of early life conditions, psychosocial factors, and emerging adolescent sexual orientation. *Developmental Psychobiology*, 62(1), 5–20.

Yackobovitch-Gavan, Michal; Wolf Linhard, D.; Nagelberg, Nessia; Poraz, Irit; Shalitin, Shlomit; Phillip, Moshe & Meyerovitch, Joseph. (2018). Intervention for childhood obesity based on parents only or parents and child compared with follow-up alone. *Pediatric Obesity*, 13(11), 647–655.

Yan, J.; Han, Z. R.; Tang, Y. & Zhang, X. (2017). Parental support for autonomy and child depressive symptoms in middle childhood: The mediating role of parent–child attachment. *Journal of Child and Family Studies*, 26(7), 1970–1978.

Yang, Rongwang; Zhang, Suhan; Li, Rong & Zhao, Zhengyan. (2013). Parents' attitudes toward stimulants use in China. *Journal of Developmental & Behavioral Pediatrics*, 34(3), 225.

Yeager, David S.; Dahl, Ronald E. & Dweck, Carol S. (2018). Why interventions to influence adolescent behavior often fail but could succeed. *Perspectives on Psychological Science*, 13(1), 101–122.

Yeager, David S.; Purdie-Vaughns, Valerie; Hooper, Sophia Yang & Cohen, Geoffrey L. (2017). Loss of institutional trust among racial and ethnic minority adolescents: A consequence of procedural injustice and a cause of life-span outcomes. *Child Development*, 88(2), 658–676.

Yogman, Michael; Garfield, Craig F. & Committee on Psychosocial Aspects of Child and Family Health. (2016). Fathers' roles in the care and development of their children: The role of pediatricians. *Pediatrics*, 138(1), e20161128.

Yogman, Michael; Garner, Andrew; Hutchinson, Jeffrey; Hirsh - Pasek, Kathy & Golinkoff, Roberta Michnick. (2018). The power of play: A pediatric role in enhancing development in young children. *Pediatrics*, 142(3), e20182058.

Yu, Xinting; Rueschman, Michael; Kaplan, Emily R.; Quante, Mirja; Horan, Christine; Davison, Kirsten; . . . Redline, Susan S. (2019). 0818 actigraphy sleep assessment in early infancy: Associations with socioeconomic factors over ages one to six months. *Sleep*, 42(Suppl. 1), A328.

Yudell, Michael; Roberts, Dorothy; DeSalle, Rob & Tishkoff, Sarah. (2016). Taking race out of human genetics. *Science*, 351(6273), 564–565.

Zachry, Anne H. & Kitzmann, Katherine M. (2011). Caregiver awareness of prone play recommendations. *American Journal of Occupational Therapy*, 65(1), 101–105.

Zahran, Hatice S.; Bailey, Cathy M.; Damon, Scott A.; Garbe, Paul L. & Breysse, Patrick N. (2018). *Vital signs: Asthma in children — United States, 2001–2016. Morbidity and Mortality Weekly Report*, 67(5), 149–155. Atlanta, GA: Centers for Disease Control and Prevention.

Zak, Paul J. (2012). *The moral molecule: The source of love and prosperity*. Dutton.

Zametkin, Alan J. & Solanto, Mary V. (2017). A Review of *ADHD nation* [Review of the book *ADHD nation: Children, doctors, big pharma, and the making of an American epidemic*, by Alan Schwarz]. *The ADHD Report*, 25(2), 6–10.

Zatorre, Robert J. (2013). Predispositions and plasticity in music and speech learning: Neural correlates and implications. *Science*, 342(6158), 585–589.

Zeiders, Katharine H.; Umaña-Taylor, Adriana J. & Derlan, Chelsea L. (2013a). Trajectories of depressive symptoms and self-esteem in Latino youths: Examining the role of gender and perceived discrimination. *Developmental Psychology*, 49(5), 951–963.

Zeiders, Katharine H.; Updegraff, Kimberly A.; Umaña-Taylor, Adriana J.; Wheeler, Lorey A.; Perez-Brena, Norma J. & Rodríguez, Sue A. (2013b). Mexican-origin youths trajectories of depressive symptoms: The role of familism values. *Journal of Adolescent Health, 53*(5), 648–654.

Zeifman, Debra M. (2013). Built to bond: Coevolution, coregulation, and plasticity in parent-infant bonds. In Hazan, Cindy & Campa, Mary I. (Eds.), *Human bonding: The science of affectional ties* (pp. 41–73). Guilford Press.

Zeitlin, Marian. (2011). *New information on West African traditional education and approaches to its modernization.* Tostan.

Zhang, Linlin & Eggum-Wilkens, Natalie D. (2018). Correlates of shyness and unsociability during early adolescence in urban and rural China. *Journal of Early Adolescence, 38*(3), 408–421.

Zhao, Fei; Franco, Heather L.; Rodriguez, Karina F.; Brown, Paula R.; Tsai, Ming-Jer;

Tsai, Sophia Y. & Yao, Humphrey H.-C. (2017). Elimination of the male reproductive tract in the female embryo is promoted by COUP-TFII in mice. *Science, 357*(6352), 717–720.

Zhao, Jinxia & Wang, Meifang. (2014). Mothers' academic involvement and children's achievement: Children's theory of intelligence as a mediator. *Learning and Individual Differences, 35,* 130–136.

Zhu, Qi; Song, Yiying; Hu, Siyuan; Li, Xiaobai; Tian, Moqian; Zhen, Zonglei; . . . Liu, Jia. (2010). Heritability of the specific cognitive ability of face perception. *Current Biology, 20*(2), 137–142.

Zimmerman, Marc A.; Stoddard, Sarah A.; Eisman, Andria B.; Caldwell, Cleopatra H.; Aiyer, Sophie M. & Miller, Alison. (2013). Adolescent resilience: Promotive factors that inform

prevention. *Child Development Perspectives, 7*(4), 215–220.

Zuk, Jennifer; Benjamin, Christopher; Kenyon, Arnold & Gaab, Nadine. (2014). Behavioral and neural correlates of executive functioning in musicians and non-musicians. *PLoS ONE, 9*(6), e99868.

Zurcher, Jessica D.; Holmgren, Hailey G.; Coyne, Sarah M.; Barlett, Christopher P. & Yang, Chongming. (2018). Parenting and cyberbullying across adolescence. *Cyberpsychology, Behavior, and Social Networking, 21*(5), 294–303.

Zych, Izabela; Farrington, David P.; Llorent, Vicente J. & Ttofi, Maria M. (Eds.). (2017). *Protecting children against bullying and its consequences.* Springer.

Name Index

Subject Index

Note: Page numbers followed by f, p, or t indicate figures, photographs, or tables respectively.